Molecular Biology of
THE CELL

Fifth Edition

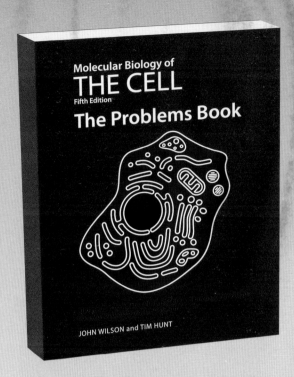

Molecular Biology of
THE CELL

Fifth Edition

Bruce Alberts

Alexander Johnson

Julian Lewis

Martin Raff

Keith Roberts

Peter Walter

With problems by

John Wilson

Tim Hunt

Garland Science
Taylor & Francis Group

Garland Science
Vice President: Denise Schanck
Assistant Editor: Sigrid Masson
Production Editor and Layout: Emma Jeffcock
Senior Publisher: Jackie Harbor
Illustrator: Nigel Orme
Designer: Matthew McClements, Blink Studio, Ltd.
Editors: Marjorie Anderson and Sherry Granum
Copy Editor: Bruce Goatly
Indexer: Merrall-Ross International, Ltd.
Permissions Coordinator: Mary Dispenza

Cell Biology Interactive
Artistic and Scientific Direction: Peter Walter
Narrated by: Julie Theriot
Production Design and Development: Michael Morales

Bruce Alberts received his Ph.D. from Harvard University and is Professor of
Biochemistry and Biophysics at the University of California, San Francisco. For
12 years, he served as President of the U.S. National Academy of Sciences (1993–2005).
Alexander Johnson received his Ph.D. from Harvard University and is Professor of
Microbiology and Immunology and Director of the Biochemistry, Cell Biology, Genetics,
and Developmental Biology Graduate Program at the University of California, San
Francisco. **Julian Lewis** received his D.Phil. from the University of Oxford and is a
Principal Scientist at the London Research Institute of Cancer Research UK.
Martin Raff received his M.D. from McGill University and is at the Medical Research
Council Laboratory for Molecular Cell Biology and the Biology Department at University
College London. **Keith Roberts** received his Ph.D. from the University of Cambridge and
is Emeritus Fellow at the John Innes Centre, Norwich. **Peter Walter** received his Ph.D.
from The Rockefeller University in New York and is Professor and Chairman of the
Department of Biochemistry and Biophysics at the University of California, San
Francisco, and an Investigator of the Howard Hughes Medical Institute.

Library of Congress Cataloging-in-Publication Data
Molecular biology of the cell / Bruce Alberts ... [et al.].-- 5th ed.
    p. cm
    ISBN 978-0-8153-4105-5 (hardcover)---ISBN 978-0-8153-4106-2 (paperback)
    1. Cytology. 2. Molecular biology. I. Alberts, Bruce.
QH581.2 .M64 2008
571.6--dc22
                                        2007005475 CIP

Published by Garland Science, Taylor & Francis Group, LLC, an informa business,
270 Madison Avenue, New York NY 10016, USA, and 2 Park Square, Milton Park,
Abingdon, OX14 4RN, UK.

Printed in the United States of America

15 14 13 12 11 10 9 8 7 6 5 4 3 2 1

Preface

In many respects, we understand the structure of the universe better than the workings of living cells. Scientists can calculate the age of the Sun and predict when it will cease to shine, but we cannot explain how it is that a human being may live for eighty years but a mouse for only two. We know the complete genome sequences of these and many other species, but we still cannot predict how a cell will behave if we mutate a previously unstudied gene. Stars may be 10^{43} times bigger, but cells are more complex, more intricately structured, and more astonishing products of the laws of physics and chemistry. Through heredity and natural selection, operating from the beginnings of life on Earth to the present day—that is, for about 20% of the age of the universe—living cells have been progressively refining and extending their molecular machinery, and recording the results of their experiments in the genetic instructions they pass on to their progeny.

With each edition of this book, we marvel at the new information that cell biologists have gathered in just a few years. But we are even more amazed and daunted at the sophistication of the mechanisms that we encounter. The deeper we probe into the cell, the more we realize how much remains to be understood. In the days of our innocence, working on the first edition, we hailed the identification of a single protein—a signal receptor, say—as a great step forward. Now we appreciate that each protein is generally part of a complex with many others, working together as a system, regulating one another's activities in subtle ways, and held in specific positions by binding to scaffold proteins that give the chemical factory a definite spatial structure. Genome sequencing has given us virtually complete molecular parts-lists for many different organisms; genetics and biochemistry have told us a great deal about what those parts are capable of individually and which ones interact with which others; but we have only the most primitive grasp of the dynamics of these biochemical systems, with all their interlocking control loops. Therefore, although there are great achievements to report, cell biologists face even greater challenges for the future.

In this edition, we have included new material on many topics, ranging from epigenetics, histone modifications, small RNAs, and comparative genomics, to genetic noise, cytoskeletal dynamics, cell-cycle control, apoptosis, stem cells, and novel cancer therapies. As in previous editions, we have tried above all to give readers a conceptual framework for the mass of information that we now have about cells. This means going beyond the recitation of facts. The goal is to learn how to put the facts to use—to reason, to predict, and to control the behavior of living systems.

To help readers on the way to an active understanding, we have for the first time incorporated end-of-chapter problems, written by John Wilson and Tim Hunt. These emphasize a quantitative approach and the art of reasoning from experiments. A companion volume, *Molecular Biology of the Cell, Fifth Edition: The Problems Book* (ISBN 978-0-8153-4110-9), by the same authors, gives complete answers to these problems and also contains more than 1700 additional problems and solutions.

A further major adjunct to the main book is the attached Media DVD-ROM disc. This provides hundreds of movies and animations, including many that are new in this edition, showing cells and cellular processes in action and bringing the text to life; the disc also now includes all the figures and tables from the main book, pre-loaded into PowerPoint® presentations. Other ancillaries available for the book include a bank of test questions and lecture outlines, available to qualified instructors, and a set of 200 full-color overhead transparencies.

Perhaps the biggest change is in the physical structure of the book. In an effort to make the standard Student Edition somewhat more portable, we are providing Chapters 21–25, covering multicellular systems, in electronic (PDF) form on the accompanying disc, while retaining in the printed volume Chapters 1–20, covering the core of the usual cell biology curriculum. But we should emphasize that the final chapters have been revised and updated as thoroughly as the rest of the book and we sincerely hope that they will be read! A Reference Edition (ISBN 978-0-8153-4111-6), containing the full set of chapters as printed pages, is also available for those who prefer it.

Full details of the conventions adopted in the book are given in the Note to the Reader that follows this Preface. As explained there, we have taken a drastic approach in confronting the different rules for the writing of gene names in different species: throughout this book, we use the same style, regardless of species, and often in defiance of the usual species-specific conventions.

As always, we are indebted to many people. Full acknowledgments for scientific help are given separately, but we must here single out some exceptionally important contributions: Julie Theriot is almost entirely responsible for Chapters 16 (Cytoskeleton) and 24 (Pathogens, Infection, and Innate Immunity), and David Morgan likewise for Chapter 17 (Cell Cycle). Wallace Marshall and Laura Attardi provided substantial help with Chapters 8 and 20, respectively, as did Maynard Olson for the genomics section of Chapter 4, Xiaodong Wang for Chapter 18, and Nicholas Harberd for the plant section of Chapter 15.

We also owe a huge debt to the staff of Garland Science and others who helped convert writers' efforts into a polished final product. Denise Schanck directed the whole enterprise and shepherded the wayward authors along the road with wisdom, skill, and kindness. Nigel Orme put the artwork into its final form and supervised the visual aspects of the book, including the back cover, with his usual flair. Matthew McClements designed the book and its front cover. Emma Jeffcock laid out its pages with extraordinary speed and unflappable efficiency, dealing impeccably with innumerable corrections. Michael Morales managed the transformation of a mass of animations, video clips, and other materials into a user-friendly DVD-ROM. Eleanor Lawrence and Sherry Granum updated and enlarged the glossary. Jackie Harbor and Sigrid Masson kept us organized. Adam Sendroff kept us aware of our readers and their needs and reactions. Marjorie Anderson, Bruce Goatly, and Sherry Granum combed the text for obscurities, infelicities, and errors. We thank them all, not only for their professional skill and dedication and for efficiency far surpassing our own, but also for their unfailing helpfulness and friendship: they have made it a pleasure to work on the book.

Lastly, and with no less gratitude, we thank our spouses, families, friends and colleagues. Without their patient, enduring support, we could not have produced any of the editions of this book.

Contents

Special Features

Detailed Contents

Chapter 16 The Cytoskeleton   965

Chapters 21–25 available on Media DVD-ROM

Chapter 21 Sexual Reproduction: Meiosis, Germ Cells, and Fertilization 1269

Chapter 22 Development of Multicellular Organisms 1305

Acknowledgments

In writing this book we have benefited greatly from the advice of many biologists and biochemists. We would like to thank the following for their suggestions in preparing this edition, as well as those who helped in preparing the first, second, third and fourth editions. (Those who helped on this edition are listed first, those who helped with the first, second, third and fourth editions follow.)

Chapter 1: W. Ford Doolittle (Dalhousie University, Canada), Jennifer Frazier (Exploratorium®, San Francisco), Douglas Kellogg (University of California, Santa Cruz), Eugene Koonin (National Institutes of Health), Mitchell Sogin (Woods Hole Institute)

Chapter 2: Michael Cox (University of Wisconsin, Madison), Christopher Mathews (Oregon State University), Donald Voet (University of Pennsylvania), John Wilson (Baylor College of Medicine)

Chapter 3: David Eisenberg (University of California, Los Angeles), Louise Johnson (University of Oxford), Steve Harrison (Harvard University), Greg Petsko (Brandeis University), Robert Stroud (University of California, San Francisco), Janet Thornton (European Bioinformatics Institute, UK)

Chapter 4: David Allis (The Rockefeller University), Adrian Bird (Wellcome Trust Centre, UK), Gary Felsenfeld (National Institutes of Health), Susan Gasser (University of Geneva, Switzerland), Eric Green (National Institutes of Health), Douglas Koshland (Carnegie Institution of Washington, Baltimore), Ulrich Laemmli (University of Geneva, Switzerland), Michael Lynch (Indiana University), Hiten Madhani (University of California, San Francisco), Elliott Margulies (National Institutes of Health), Geeta Narlikar (University of California, San Francisco), Maynard Olson (University of Washington)

Chapter 5: Elizabeth Blackburn (University of California, San Francisco), James Haber (Brandeis University), Nancy Kleckner (Harvard University), Joachim Li (University of California, San Francisco), Thomas Lindahl (Cancer Research, UK), Rodney Rothstein (Columbia University), Aziz Sancar (University of North Carolina, Chapel Hill), Bruce Stillman (Cold Spring Harbor Laboratory), Steven West (Cancer Research, UK), Rick Wood (University of Pittsburgh)

Chapter 6: Raul Andino (University of California, San Francisco), David Bartel (Massachusetts Institute of Technology), Richard Ebright (Rutgers University), Daniel Finley (Harvard University), Joseph Gall (Carnegie Institution of Washington), Michael Green (University of Massachusetts Medical School), Carol Gross (University of California, San Francisco), Christine Guthrie (University of California, San Francisco), Art Horwich (Yale University School of Medicine), Roger Kornberg (Stanford University), Reinhard Lührman (Max Planck Institute of Biophysical Chemistry, Göttingen), Quinn Mitrovich (University of California, San Francisco), (Harry Noller (University of California, Santa Cruz), Roy Parker (University of Arizona), Robert Sauer

(Massachusetts Institute of Technology), Joan Steitz (Yale University), Jack Szostak (Harvard Medical School, Howard Hughes Medical Institute), David Tollervey (University of Edinburgh, UK), Alexander Varshavsky (California Institute of Technology), Jonathan Weissman (University of California, San Francisco)

Chapter 7: Raul Andino (University of California, San Francisco), David Bartel (Massachusetts Institute of Technology), Michael Bulger (University of Rochester Medical Center), Michael Green (University of Massachusetts Medical School), Carol Gross (University of California, San Francisco), Frank Holstege (University Medical Center, The Netherlands), Roger Kornberg (Stanford University), Hiten Madhani (University of California, San Francisco), Barbara Panning (University of California, San Francisco), Mark Ptashne (Memorial Sloan-Kettering Center), Ueli Schibler (University of Geneva, Switzerland), Azim Surani (University of Cambridge)

Chapter 8: Wallace Marshall [major contribution] (University of California, San Francisco)

Chapter 9: Wolfgang Baumeister (Max Planck Institute of Biochemistry, Martinsried), Ken Sawin (The Wellcome Trust Centre for Cell Biology, UK), Peter Shaw (John Innes Centre, UK), Werner Kühlbrandt (Max Planck Institute of Biophysics, Frankfurt am Main), Ronald Vale (University of California, San Francisco), Jennifer Lippincott-Schwartz (National Institutes of Health)

Chapter 10: Ari Helenius (Swiss Federal Institute of Technology Zürich, Switzerland), Werner Kühlbrandt (Max Planck Institute of Biophysics, Frankfurt am Main), Dieter Osterhelt (Max Planck Institute of Biochemistry, Martinsried), Kai Simons (Max Planck Institute of Molecular Cell Biology and Genetics, Dresden)

Chapter 11: Wolfhard Almers (Oregon Health and Science University), Robert Edwards (University of California, San Francisco), Bertil Hille (University of Washington), Lily Jan (University of California, San Francisco), Roger Nicoll (University of California, San Francisco), Robert Stroud (University of California, San Francisco), Patrick Williamson (University of Massachusetts, Amherst)

Chapter 12: Larry Gerace (The Scripps Research Institute), Ramanujan Hegde (National Institutes of Health), Nikolaus Pfanner (University of Freiburg, Germany), Daniel Schnell (University of Massachusetts, Amherst), Karsten Weis (University of California, Berkeley), Susan Wente (Vanderbilt University

Medical Center), Pat Williamson (University of Massachusetts, Amherst)

Chapter 13: Scott Emr (University of California, San Diego), Ben Glick (University of Chicago), Ari Helenius (Swiss Federal Institute of Technology Zürich, Switzerland), Ira Mellman (Yale University), Hugh Pelham (The Medical Research Council, Cambridge), Giampietro Schiavo (London Research Institute), Graham Warren (Yale University), Marino Zerial (Max Planck Institute of Molecular Cell Biology and Genetics, Frankfurt am Main)

Chapter 14: Michael Gray (Dalhousie University), Andrew Halestrap (University of Bristol, UK), Werner Kühlbrandt (Max Planck Institute of Biophysics, Frankfurt am Main), Craig Thompson (Abramson Family Cancer Research Institute, University of Pennsylvania), Michael Yaffe (University of California, San Diego)

Chapter 15: Nicholas Harberd [substantial contribution] (John Innes Centre, UK), Henry Bourne (University of California, San Francisco), Dennis Bray (University of Cambridge), James Briscoe (National Institute for Medical Research, UK), James Ferrell (Stanford University), Matthew Freeman (Laboratory of Molecular Biology, UK), Alfred Gilman (The University of Texas Southwestern Medical Center), Sankar Ghosh (Yale University School of Medicine), Alan Hall (Memorial Sloan-Kettering Cancer Center), Carl-Henrik Heldin (Ludwig Institute for Cancer Research, Sweden), Robin Irvine (University of Cambridge), Michael Karin (University of California, San Diego), Elliott Meyerowitz (California Institute of Technology), Roel Nusse (Stanford University), Tony Pawson (Mount Sinai Hospital, Toronto), Julie Pitcher (University College London), Len Stephens (The Babraham Institute, UK)

Chapter 16: Julie Theriot [major contribution] (Stanford University), Henry Bourne (University of California, San Francisco), Larry Goldstein (University of California, San Diego), Alan Hall (MRC Laboratory for Molecular Biology and Cell Biology, UK), Joe Howard (Max Planck Institute of Molecular Cell Biology and Genetics, Dresden), Laura Machesky (The University of Birmingham, UK), Timothy Mitchison (Harvard Medical School), Ronald Vale (University of California, San Francisco)

Chapter 17: David Morgan [major contribution] (University of California, San Francisco), Arshad Desai (University of California, San Diego), Bruce Edgar (Fred Hutchinson Cancer Research Center, Seattle), Michael Glotzer (University of Chicago), Rebecca Heald (University of California, Berkeley), Eric Karsenti (European Molecular Biology Laboratory, Germany), Kim Nasmyth (University of Oxford), Jonathan Pines (Gurdon Institute, Cambridge), Charles Sherr (St. Jude Children's Hospital)

Chapter 18: Xiaodong Wang [substantial contribution] (The University of Texas Southwestern Medical School), Jerry Adams (The Walter and Eliza Hall Institute of Medical Research, Australia), Douglas Green (St. Jude Children's Hospital), Shigekazu Nagata (Kyoto University, Japan)

Chapter 19: Jeffrey Axelrod (Stanford University Medical Center), Walter Birchmeier (Max-Delbrück Center for Molecular Medicine, Germany), Keith Burridge (University of North Carolina, Chapel Hill), John Couchman (Imperial College, UK), Caroline Damsky (University of California, San Francisco), Matthias Falk (Lehigh University), David Garrod (University of Manchester, UK), Daniel Goodenough (Harvard Medical School), Martin Humphries (University of Manchester, UK), Richard Hynes (Massachusetts Institute of Technology), Ken Keegstra (Michigan State University), Morgan Sheng (Massachusetts Institute of Technology), Charles Streuli (University of Manchester, UK), Masatoshi Takeichi (RIKEN

Kobe Institute, Japan), Kenneth Yamada (National Institutes of Health)

Chapter 20: Laura Attardi [substantial contribution] (Stanford University), Anton Berns (Netherlands Cancer Institute, The Netherlands), Michael Bishop (University of California, San Francisco), Fred Bunz (Johns Hopkins), Johann De-Bono (The Institute of Cancer Research, UK), John Dick (University of Toronto, Canada), Paul Edwards (University of Cambridge), Douglas Hanahan (University of California, San Francisco), Joseph Lipsick (Stanford University School of Medicine), Scott Lowe (Cold Spring Harbor Laboratory), Bruce Ponder (University of Cambridge), Craig Thompson (University of Pennsylvania), Ian Tomlinson (Cancer Research, UK), Robert Weinberg (Massachusetts Institute of Technology)

Chapter 21: Patricia Calarco (University of California, San Francisco), John Carroll (University College London), Abby Dernburg (University of Califonia, Berkeley), Scott Hawley (Stowers Institute for Medical Research, Kansas City), Neil Hunter (University of California, Davis), Nancy Kleckner (Harvard University), Anne McLaren (Wellcome/ Cancer Research Campaign Institute, Cambridge), Diana Myles (University of California, Davis), Terry Orr-Weaver (Massachusetts Institute of Technology), Renee Reijo (University of California, San Francisco), Gerald Schatten (Pittsburgh Development Center), Azim Surani (The Gurdon Institute, UK), Paul Wassarman (Mount Sinai School of Medicine)

Chapter 22: Julie Ahringer (The Gurdon Institute, UK), Konrad Basler (University of Zürich, Switzerland), Richard Harland (University of California, Berkeley), Brigid Hogan (Duke University), Kenneth Irvine (Rutgers University), Daniel St. Johnson (The Gurdon Institute, UK), Elliott Meyerowitz (California Institute of Technology), William McGinnis (University of California, San Diego), Elizabeth Robertson (The Wellcome Trust Centre for Human Genetics, UK), Francois Schweisguth (French National Centre for Scientific Research, France), Jim Smith (The Gurdon Institute, UK), Nicolas Tapon (London Research Institute)

Chapter 23: Ralf Adams (London Research Institute), Hans Clevers (Hubrecht Institute, The Netherlands), Jeffrey Gordon (Washington University, St. Louis), Holger Gerhardt (London Research Institute), Simon Hughes (Kings College, UK), Daniel Louvard (Institut Curie, France), Bjorn Olsen (Harvard Medical School), Stuart Orkin (Harvard Medical School), Thomas Reh (University of Washington, Seattle), Austin Smith (University of Edinburgh, UK), Charles Streuli (The University of Manchester, UK), Fiona Watt (Cancer Research Institute, UK)

Chapter 24: Julie Theriot [major contribution] (Stanford University), Michael Bishop (University of California, San Francisco), Harald von Boehmer (Harvard Medical School), Lynn Enquist (Princeton University), Stan Falkow (Stanford University), Douglas Fearon (University of Cambridge), Lewis Lanier (University of California, San Francisco), Richard Locksley (University of California, San Francisco), Daniel Portnoy (University of California, Berkeley), Caetano Reis e Sousa (Cancer Research, UK), Ralph Steinman (The Rockefeller University), Gary Ward (University of Vermont)

Chapter 25: Harald von Boehmer (Harvard Medical School), Douglas Fearon (University of Cambridge), Lewis Lanier (University of California, San Francisco), Philippa Marrack (National Jewish Medical and Research Center, Denver), Michael Neuberger (University of Cambridge), Michael Nussenzweig (Rockefeller University), William Paul (National Institutes of

Health), Klaus Rajewsky (Harvard Medical School), Caetano Reis e Sousa (Cancer Research, UK), Ralph Steinman (The Rockefeller University).

Glossary Eleanor Lawrence, Sherry Granum

Readers David Kashatus (Duke University), Emmanuel Kreidl (University of Vienna, Austria), Nick Rudzik (University of Toronto, Canada), Dea Shahinas (University of Toronto, Canada)

First, second, third, and fourth editions David Agard (University of California, San Francisco), Michael Akam (University of Cambridge), Fred Alt (CBR Institute for Biomedical Research, Boston), Linda Amos (MRC Laboratory of Molecular Biology, Cambridge), Raul Andino (University of California, San Francisco), Clay Armstrong (University of Pennsylvania), Martha Arnaud (University of California, San Francisco), Spyros Artavanis-Tsakonas (Harvard Medical School), Michael Ashburner (University of Cambridge), Jonathan Ashmore (University College London), Tayna Awabdy (University of California, San Francisco), Peter Baker (deceased), David Baldwin (Stanford University), Michael Banda (University of California, San Francisco), Cornelia Bargmann (University of California, San Francisco), Ben Barres (Stanford University), David Bartel (Massachusetts Institute of Technology), Michael Bennett (Albert Einstein College of Medicine), Darwin Berg (University of California, San Diego), Merton Bernfield (Harvard Medical School), Michael Berridge (The Babraham Institute, Cambridge), David Birk (UMNDJ—Robert Wood Johnson Medical School), Michael Bishop (University of California, San Francisco), Tim Bliss (National Institute for Medical Research, London), Hans Bode (University of California, Irvine), Piet Borst (Jan Swammerdam Institute, University of Amsterdam), Henry Bourne (University of California, San Francisco), Alan Boyde (University College London), Martin Brand (University of Cambridge), Carl Branden (deceased), Andre Brandli (Swiss Federal Institute of Technology, Zurich), Mark Bretscher (MRC Laboratory of Molecular Biology, Cambridge), Marianne Bronner-Fraser (California Institute of Technology), Robert Brooks (King's College London), Barry Brown (King's College London), Michael Brown (University of Oxford), Steve Burden (New York University of Medicine), Max Burger (University of Basel), Stephen Burley (SGX Pharmaceuticals), Keith Burridge (University of North Carolina, Chapel Hill), John Cairns (Radcliffe Infirmary, Oxford), Zacheus Cande (University of California, Berkeley), Lewis Cantley (Harvard Medical School), Charles Cantor (Columbia University), Roderick Capaldi (University of Oregon), Mario Capecchi (University of Utah), Michael Carey (University of California, Los Angeles), Adelaide Carpenter (University of California, San Diego), Tom Cavalier-Smith (King's College London), Pierre Chambon (University of Strasbourg), Enrico Coen (John Innes Institute, Norwich, UK), Philip Cohen (University of Dundee, Scotland), Robert Cohen (University of California, San Francisco), Stephen Cohen (EMBL Heidelberg, Germany), Roger Cooke (University of California, San Francisco), John Cooper (Washington University School of Medicine, St. Louis), Nancy Craig (Johns Hopkins University), James Crow (University of Wisconsin, Madison), Stuart Cull-Candy (University College London), Leslie Dale (University College London), Michael Dexter (The Wellcome Trust, UK), Anthony DeFranco (University of California, San Francisco), Christopher Dobson (University of Cambridge), Russell Doolittle (University of California, San Diego), Julian Downward (Cancer Research, UK), Keith Dudley (King's College London), Graham Dunn (MRC Cell Biophysics Unit, London), Jim Dunwell (John Innes Institute, Norwich, UK), Paul Edwards (University of Cambridge), Robert Edwards (University of California, San Francisco), David Eisenberg (University of California, Los Angeles), Sarah Elgin (Washington University, St. Louis), Ruth Ellman (Institute of Cancer Research, Sutton, UK), Beverly Emerson (The Salk Institute), Charles Emerson (University of Virginia), Scott Emr (University of California, San Diego), Sharyn Endow (Duke University), Tariq Enver (Institute of Cancer Research, London), David Epel (Stanford University), Gerard Evan (University of California, Comprehensive Cancer Center), Ray Evert (University of Wisconsin, Madison), Stanley Falkow (Stanford University), Gary Felsenfeld (National Institutes of Health), Stuart Ferguson (University of Oxford), Christine Field (Harvard Medical School), Gary Firestone (University of California, Berkeley), Gerald Fischbach (Columbia University), Robert Fletterick (University of California, San Francisco), Harvey Florman (Tufts University), Judah Folkman (Harvard Medical School), Larry Fowke (University of Saskatchewan, Canada), Daniel Friend (University of California, San Francisco), Elaine Fuchs (University of Chicago), Joseph Gall (Yale University), Richard Gardner (University of Oxford), Anthony Gardner-Medwin (University College London), Peter Garland (Institute of Cancer Research, London), Walter Gehring (Biozentrum, University of Basel), Benny Geiger (Weizmann Institute of Science, Rehovot, Israel), Larry Gerace (The Scripps Research Institute), John Gerhart (University of California, Berkeley), Günther Gerisch (Max Planck Institute of Biochemistry, Martinsried), Frank Gertler (Massachusetts Institute of Technology), Sankar Ghosh (Yale University School of Medicine), Reid Gilmore (University of Massachusetts, Amherst), Bernie Gilula (deceased), Charles Gilvarg (Princeton University), Michael Glotzer (University of Vienna, Austria), Larry Goldstein (University of California, San Diego), Bastien Gomperts (University College Hospital Medical School, London), Daniel Goodenough (Harvard Medical School), Jim Goodrich (University of Colorado, Boulder), Peter Gould (Middlesex Hospital Medical School, London), Alan Grafen (University of Oxford), Walter Gratzer (King's College London), Howard Green (Harvard University), Michael Green (University of Massachusetts, Amherst), Leslie Grivell (University of Amsterdam, The Netherlands), Carol Gross (University of California, San Francisco), Frank Grosveld (Erasmus Universiteit, The Netherlands), Michael Grunstein (University of California, Los Angeles), Barry Gumbiner (Memorial Sloan-Kettering Cancer Center), Brian Gunning (Australian National University, Canberra), Christine Guthrie (University of California, San Francisco), Ernst Hafen (Universitat Zurich), David Haig (Harvard University), Alan Hall (MRC Laboratory for Molecular Biology and Cell Biology, London), Jeffrey Hall (Brandeis University), John Hall (University of Southampton, UK), Zach Hall (University of California, San Francisco), David Hanke (University of Cambridge), Nicholas Harberd (John Innes Centre, Norwich, UK), Graham Hardie (University of Dundee, Scotland), Richard Harland (University of California, Berkeley), Adrian Harris (Cancer Research, UK), John Harris (University of Otago, New Zealand), Stephen Harrison (Harvard University), Leland Hartwell (University of Washington, Seattle), Adrian Harwood (MRC Laboratory for Molecular Cell Biology and Cell Biology Unit, London), John Heath (University of Birmingham, UK), Ari Helenius (Yale University), Richard Henderson (MRC Laboratory of Molecular Biology, Cambridge), Glenn Herrick (University of Utah), Ira Herskowitz (deceased), Bertil Hille (University of Washington, Seattle), Alan Hinnebusch (National Institutes of Health, Bethesda), Nancy Hollingsworth (State University of New York, Stony Brook), Leroy Hood (Institute for Systems Biology, Seattle), John Hopfield (Princeton University), Robert Horvitz (Massachusetts Institute of Technology), David Housman (Massachusetts Institute of Technology), Jonathan Howard (University of Washington, Seattle), James Hudspeth (The Rockefeller University), Simon Hughes (King's College London), Martin Humphries (University of Manchester, UK), Tim Hunt

(Cancer Research, UK), Laurence Hurst (University of Bath, UK), Jeremy Hyams (University College London), Tony Hyman (Max Planck Institute of Molecular Cell Biology & Genetics, Dresden), Richard Hynes (Massachusetts Institute of Technology), Philip Ingham (University of Sheffield, UK), Norman Iscove (Ontario Cancer Institute, Toronto), David Ish-Horowicz (Cancer Research, UK), Lily Jan (University of California, San Francisco), Charles Janeway (deceased), Tom Jessell (Columbia University), Arthur Johnson (Texas A & M University), Andy Johnston (John Innes Institute, Norwich, UK), E.G. Jordan (Queen Elizabeth College, London), Ron Kaback (University of California, Los Angeles), Ray Keller (University of California, Berkeley), Douglas Kellogg (University of California, Santa Cruz), Regis Kelly (University of California, San Francisco), John Kendrick-Jones (MRC Laboratory of Molecular Biology, Cambridge), Cynthia Kenyon (University of California, San Francisco), Roger Keynes (University of Cambridge), Judith Kimble (University of Wisconsin, Madison), Robert Kingston (Massachusetts General Hospital), Marc Kirschner (Harvard University), Richard Klausner (National Institutes of Health), Nancy Kleckner (Harvard University), Mike Klymkowsky (University of Colorado, Boulder), Kelly Komachi (University of California, San Francisco), Eugene Koonin (National Institutes of Health), Juan Korenbrot (University of California, San Francisco), Tom Kornberg (University of California, San Francisco), Stuart Kornfeld (Washington University, St. Louis), Daniel Koshland (University of California, Berkeley), Marilyn Kozak (University of Pittsburgh), Mark Krasnow (Stanford University), Werner Kühlbrandt (Max Planck Institute for Biophysics, Frankfurt am Main), John Kuriyan (University of California, Berkeley), Robert Kypta (MRC Laboratory for Molecular Cell Biology, London), Peter Lachmann (MRC Center, Cambridge), Ulrich Laemmli (University of Geneva, Switzerland), Trevor Lamb (University of Cambridge), Hartmut Land (Cancer Research, UK), David Lane (University of Dundee, Scotland), Jane Langdale (University of Oxford), Jay Lash (University of Pennsylvania), Peter Lawrence (MRC Laboratory of Molecular Biology, Cambridge), Paul Lazarow (Mount Sinai School of Medicine), Robert J. Lefkowitz (Duke University), Michael Levine (University of California, Berkeley), Warren Levinson (University of California, San Francisco), Alex Levitzki (Hebrew University, Israel), Ottoline Leyser (University of York, UK), Joachim Li (University of California, San Francisco), Tomas Lindahl (Cancer Research, UK), Vishu Lingappa (University of California, San Francisco), Jennifer Lippincott-Schwartz (National Institutes of Health, Bethesda), Dan Littman (New York University School of Medicine), Clive Lloyd (John Innes Institute, Norwich, UK), Richard Losick (Harvard University), Robin Lovell-Badge (National Institute for Medical Research, London), Shirley Lowe (University of California, San Francisco), Laura Machesky (University of Birmingham, UK), James Maller (University of Colorado Medical School), Tom Maniatis (Harvard University), Colin Manoil (Harvard Medical School), Philippa Marrack (National Jewish Medical and Research Center, Denver), Mark Marsh (Institute of Cancer Research, London), Gail Martin (University of California, San Francisco), Paul Martin (University College London), Joan Massagué (Memorial Sloan-Kettering Cancer Center), Brian McCarthy (University of California, Irvine), Richard McCarty (Cornell University), William McGinnis (University of California, Davis), Anne McLaren (Wellcome/Cancer Research Campaign Institute, Cambridge), Frank McNally (University of California, Davis), Freiderick Meins (Freiderich Miescher Institut, Basel), Stephanie Mel (University of California, San Diego), Ira Mellman (Yale University), Barbara Meyer (University of California, Berkeley), Elliot Meyerowitz (California Institute of Technology), Chris Miller (Brandeis University), Robert Mishell (University of Birmingham, UK), Avrion Mitchison (University College London),

N.A. Mitchison (University College London), Tim Mitchison (Harvard Medical School), Peter Mombaerts (The Rockefeller University), Mark Mooseker (Yale University), David Morgan (University of California, San Francisco), Michelle Moritz (University of California, San Francisco), Montrose Moses (Duke University), Keith Mostov (University of California, San Francisco), Anne Mudge (University College London), Hans Müller-Eberhard (Scripps Clinic and Research Institute), Alan Munro (University of Cambridge), J. Murdoch Mitchison (Harvard University), Richard Myers (Stanford University), Diana Myles (University of California, Davis), Andrew Murray (Harvard University), Mark E. Nelson (University of Illinois, Urbana-Champaign), Michael Neuberger (MRC Laboratory of Molecular Biology, Cambridge), Walter Neupert (University of Munich, Germany), David Nicholls (University of Dundee, Scotland), Suzanne Noble (University of California, San Francisco), Harry Noller (University of California, Santa Cruz), Jodi Nunnari (University of California, Davis), Paul Nurse (Cancer Research, UK), Duncan O'Dell (deceased), Patrick O'Farrell (University of California, San Francisco), Maynard Olson (University of Washington, Seattle), Stuart Orkin (Children's Hospital, Boston), Terri Orr-Weaver (Massachusetts Institute of Technology), Erin O'Shea (Harvard University), William Otto (Cancer Research, UK), John Owen (University of Birmingham, UK), Dale Oxender (University of Michigan), George Palade (deceased), Barbara Panning (University of California, San Francisco), Roy Parker (University of Arizona, Tucson), William W. Parson (University of Washington, Seattle), Terence Partridge (MRC Clinical Sciences Centre, London), William E. Paul (National Institutes of Health), Tony Pawson (Mount Sinai Hospital, Toronto), Hugh Pelham (MRC Laboratory of Molecular Biology, Cambridge), Robert Perry (Institute of Cancer Research, Philadelphia), Greg Petsko (Brandeis University), Gordon Peters (Cancer Research, UK), David Phillips (The Rockefeller University), Jeremy Pickett-Heaps (The University of Melbourne, Australia), Julie Pitcher (University College London), Jeffrey Pollard (Albert Einstein College of Medicine), Tom Pollard (Yale University), Bruce Ponder (University of Cambridge), Dan Portnoy (University of California, Berkeley), James Priess (University of Washington, Seattle), Darwin Prockop (Tulane University), Dale Purves (Duke University), Efraim Racker (Cornell University), Jordan Raff (Wellcome/CRC Institute, Cambridge), Klaus Rajewsky (University of Cologne, Germany), George Ratcliffe (University of Oxford), Elio Raviola (Harvard Medical School), Martin Rechsteiner (University of Utah, Salt Lake City), David Rees (National Institute for Medical Research, London), Louis Reichardt (University of California, San Francisco), Fred Richards (Yale University), Conly Rieder (Wadsworth Center, Albany), Phillips Robbins (Massachusetts Institute of Technology), Elaine Robson (University of Reading, UK), Robert Roeder (The Rockefeller University), Joel Rosenbaum (Yale University), Janet Rossant (Mount Sinai Hospital, Toronto), Jesse Roth (National Institutes of Health), Jim Rothman (Memorial Sloan-Kettering Cancer Center), Erkki Ruoslahti (La Jolla Cancer Research Foundation), Gary Ruvkun (Massachusetts General Hospital), David Sabatini (New York University), Alan Sachs (University of California, Berkeley), Alan Sachs (University of California, Berkeley), Edward Salmon (University of North Carolina, Chapel Hill), Joshua Sanes (Harvard University), Peter Sarnow (Stanford University), Lisa Satterwhite (Duke University Medical School), Howard Schachman (University of California, Berkeley), Gottfried Schatz (Biozentrum, University of Basel), Randy Schekman (University of California, Berkeley), Richard Scheller (Stanford University), Giampietro Schiavo (Cancer Research, UK), Joseph Schlessinger (New York University Medical Center), Michael Schramm (Hebrew University), Robert Schreiber (Scripps Clinic and Research Institute), James Schwartz (Columbia

University), Ronald Schwartz (National Institutes of Health), François Schweisguth (ENS, Paris), John Scott (University of Manchester, UK), John Sedat (University of California, San Francisco), Peter Selby (Cancer Research, UK), Zvi Sellinger (Hebrew University, Israel), Gregg Semenza (Johns Hopkins University), Philippe Sengel (University of Grenoble, France), Peter Shaw (John Innes Institute, Norwich, UK), Michael Sheetz (Columbia University), David Shima (Cancer Research, UK), Samuel Silverstein (Columbia University), Kai Simons (Max Planck Institute of Molecular Cell Biology and Genetics, Dresden), Melvin I. Simon (California Institute of Technology), Jonathan Slack (Cancer Research, UK), Alison Smith (John Innes Institute, Norfolk, UK), John Maynard Smith (University of Sussex, UK), Frank Solomon (Massachusetts Institute of Technology), Michael Solursh (University of Iowa), Bruce Spiegelman (Harvard Medical School), Timothy Springer (Harvard Medical School), Mathias Sprinzl (University of Bayreuth, Germany), Scott Stachel (University of California, Berkeley), Andrew Staehelin (University of Colorado, Boulder), David Standring (University of California, San Francisco), Margaret Stanley (University of Cambridge), Martha Stark (University of California, San Francisco), Wilfred Stein (Hebrew University, Israel), Malcolm Steinberg (Princeton University), Paul Sternberg (California Institute of Technology), Chuck Stevens (The Salk Institute), Murray Stewart (MRC Laboratory of Molecular Biology, Cambridge), Monroe Strickberger (University of Missouri, St. Louis), Robert Stroud (University of California, San Francisco), Michael Stryker (University of California, San Francisco), William Sullivan (University of California, Santa Cruz), Daniel Szollosi (Institut National de la Recherche Agronomique, France), Jack Szostak (Massachusetts General Hospital), Masatoshi Takeichi (Kyoto University), Clifford Tabin (Harvard Medical School), Diethard Tautz (University of Cologne, Germany), Julie Theriot (Stanford University), Roger Thomas (University of Bristol, UK), Vernon Thornton (King's College London), Cheryll Tickle (University of Dundee, Scotland), Jim Till (Ontario Cancer Institute, Toronto), Lewis Tilney (University of Pennsylvania), Nick Tonks (Cold Spring Harbor Laboratory), Alain Townsend (Institute of Molecular Medicine, John Radcliffe Hospital, Oxford), Paul Travers (Anthony Nolan Research Institute, London), Robert Trelstad (UMDNJ, Robert Wood Johnson Medical School), Anthony Trewavas (Edinburgh University, Scotland), Nigel Unwin (MRC Laboratory of Molecular Biology, Cambridge),Victor Vacquier (University of California, San Diego), Harry van der Westen (Wageningen, The Netherlands), Tom Vanaman (University of Kentucky), Harold Varmus (Sloan-Kettering Institute), Alexander Varshavsky (California Institute of Technology), Madhu Wahi (University of California, San Francisco), Virginia Walbot (Stanford University), Frank Walsh (Glaxo-Smithkline-Beecham, UK), Trevor Wang (John Innes Institute, Norwich, UK), Yu-Lie Wang (Worcester Foundation for Biomedical Research), Anne Warner (University College London), Graham Warren (Yale University School of Medicine), Paul Wassarman (Mount Sinai School of Medicine), Fiona Watt (Cancer Research, UK), Clare Waterman-Storer (The Scripps Research Institute), Fiona Watt (Cancer Research, UK), John Watts (John Innes Institute, Norwich, UK), Klaus Weber (Max Planck Institute for Biophysical Chemistry, Göttingen), Martin Weigert (Institute of Cancer Research, Philadelphia), Harold Weintraub (deceased), Karsten Weis (University of California, Berkeley), Irving Weissman (Stanford University), Jonathan Weissman (University of California, San Francisco), Norman Wessells (Stanford University), Judy White (University of Virginia), Steven West (Cancer Research, UK), William Wickner (Dartmouth College), Michael Wilcox (deceased), Lewis T. Williams (Chiron Corporation), Keith Willison (Chester Beatty Laboratories, London), John Wilson (Baylor University), Alan Wolffe (deceased), Richard Wolfenden (University of North Carolina, Chapel Hill), Sandra Wolin (Yale University School of Medicine), Lewis Wolpert (University College London), Rick Wood (Cancer Research, UK), Abraham Worcel (University of Rochester), Nick Wright (Cancer Research, UK), John Wyke (Beatson Institute for Cancer Research, Glasgow), Keith Yamamoto (University of California, San Francisco), Charles Yocum (University of Michigan, Ann Arbor), Peter Yurchenco (UMDNJ, Robert Wood Johnson Medical School), Rosalind Zalin (University College London), Patricia Zambryski (University of California, Berkeley).

A Note to the Reader

Structure of the Book

Although the chapters of this book can be read independently of one another, they are arranged in a logical sequence of five parts. The first three chapters of **Part I** cover elementary principles and basic biochemistry. They can serve either as an introduction for those who have not studied biochemistry or as a refresher course for those who have.

Part II deals with the storage, expression and transmission of genetic information.

Part III deals with the principles of the main experimental methods for investigating cells. It is not necessary to read these two chapters in order to understand the later chapters, but a reader will find it a useful reference.

Part IV discusses the internal organization of the cell.

Part V follows the behavior of cells in multicellular systems, starting with cell–cell junctions and extracellular matrix and concluding with two chapters on the immune system. Chapters 21–25 can be found on the Media DVD-ROM which is packaged with each book, providing increased portability for students.

End-of-Chapter Problems

A selection of problems, written by John Wilson and Tim Hunt, now appears in the text at the end of each chapter. The complete solutions to these problems can be found in *Molecular Biology of the Cell, Fifth Edition: The Problems Book.*

References

A concise list of selected references is included at the end of each chapter. These are arranged in alphabetical order under the main chapter section headings. These references frequently include the original papers in which important discoveries were first reported. Chapter 8 includes several tables giving the dates of crucial developments along with the names of the scientists involved. Elsewhere in the book the policy has been to avoid naming individual scientists.

Media Codes

Media codes are integrated throughout the text to indicate when relevant videos and animations are available on the DVD-ROM. The four-letter codes are enclosed in brackets and highlighted in color, like this <ATCG>. The interface for the *Cell Biology Interactive* media player on the DVD-ROM contains a window where you enter the 4-letter code. When the code is typed into the interface, the corresponding media item will load into the media player.

Glossary Terms

Throughout the book, **boldface type** has been used to highlight key terms at the point in a chapter where the main discussion of them occurs. *Italic* is used to set off important terms with a lesser degree of emphasis. At the end of the book is the expanded **glossary**, covering technical terms that are part of the common currency of cell biology; it is intended as a first resort for a reader who encounters an unfamiliar term used without explanation.

Nomenclature for Genes and Proteins

Each species has its own conventions for naming genes; the only common feature is that they are always set in italics. In some species (such as humans), gene names are spelled out all in capital letters; in other species (such as zebrafish), all in lower case; in yet others (most mouse genes), with the first letter in upper case and rest in lower case; or (as in *Drosophila*) with different combinations of upper and lower case, according to whether the first mutant allele to be discovered gave a dominant or recessive phenotype. Conventions for naming protein products are equally varied.

This typographical chaos drives everyone crazy. It is not just tiresome and absurd; it is also unsustainable. We cannot independently define a fresh convention for each of the next few million species whose genes we may wish to study. Moreover, there are many occasions, especially in a book such as this, where we need to refer to a gene generically, without specifying the mouse version, the human version, the chick version, or the hippopotamus version, because they are all equivalent for the purposes of the discussion. What convention then should we use?

We have decided in this book to cast aside the conventions for individual species and follow a uniform rule: we write all gene names, like the names of people and places, with the first letter in upper case and the rest in lower case, but all in italics, thus: *Apc, Bazooka, Cdc2, Dishevelled, Egl1*. The corresponding protein, where it is named after the gene, will be written in the same way, but in roman rather than italic letters: Apc, Bazooka, Cdc2, Dishevelled, Egl1. When it is necessary to specify the organism, this can be done with a prefix to the gene name.

For completeness, we list a few further details of naming rules that we shall follow. In some instances an added letter in the gene name is traditionally used to distinguish between genes that are related by function or evolution; for those genes we put that letter in upper case if it is usual to do so (*LacZ, RecA, HoxA4*). We use no hyphen to separate added letters or numbers from the rest of the name. Proteins are more of a problem. Many of them have names in their own right, assigned to them before the gene was named. Such protein names take many forms, although most of them traditionally begin with a lower-case letter (actin, hemoglobin, catalase), like the names of ordinary substances (cheese, nylon), unless they are acronyms (such as GFP, for Green Fluorescent Protein, or BMP4, for Bone Morphogenetic Protein #4). To force all such protein names into a uniform style would do too much violence to established usages, and we shall simply write them in the traditional way (actin, GFP, etc.). For the corresponding gene names in all these cases, we shall nevertheless follow our standard rule: *Actin, Hemoglobin, Catalase, Bmp4, Gfp*. Occasionally in our book we need to highlight a protein name by setting it in italics for emphasis; the intention will generally be clear from the context.

For those who wish to know them, the Table below shows some of the official conventions for individual species—conventions that we shall mostly violate in this book, in the manner shown.

ORGANISM	SPECIES-SPECIFIC CONVENTION		UNIFIED CONVENTION USED IN THIS BOOK	
	GENE	PROTEIN	GENE	PROTEIN
Mouse	*Hoxa4*	Hoxa4	*HoxA4*	HoxA4
	Bmp4	BMP4	*Bmp4*	BMP4
	integrin α-1, Itgα1	integrin α1	*Integrin α1, Itgα1*	integrin α1
Human	*HOXA4*	HOXA4	*HoxA4*	HoxA4
Zebrafish	*cyclops, cyc*	Cyclops, Cyc	*Cyclops, Cyc*	Cyclops, Cyc
Caenorhabditis	*unc-6*	UNC-6	*Unc6*	Unc6
Drosophila	*seven less, sev* (named after recessive mutant phenotype)	Sevenless, SEV	*Sevenless, Sev*	Sevenless, Sev
	Deformed, Dfd (named after dominant mutant phenotype)	Deformed, DFD	*Deformed, Dfd*	Deformed, Dfd
Yeast				
Saccharomyces cerevisiae (budding yeast)	*CDC28*	Cdc28, Cdc28p	*Cdc28*	Cdc28
Schizosaccharomyces pombe (fission yeast)	*Cdc2*	Cdc2, Cdc2p	*Cdc2*	Cdc2
Arabidopsis	*GAI*	GAI	*Gai*	GAI
E. coli	*uvrA*	UvrA	*UvrA*	UvrA

Ancillaries

Molecular Biology of the Cell, Fifth Edition: The Problems Book
by John Wilson and Tim Hunt (ISBN: 978-0-8153-4110-9)

The Problems Book is designed to help students appreciate the ways in which experiments and simple calculations can lead to an understanding of how cells work. It provides problems to accompany Chapters 1–20 of *Molecular Biology of the Cell*. Each chapter of problems is divided into sections that correspond to those of the main textbook and review key terms, test for understanding basic concepts, and pose research-based problems. *Molecular Biology of the Cell, Fifth Edition: The Problems Book* should be useful for homework assignments and as a basis for class discussion. It could even provide ideas for exam questions. Solutions for all of the problems are provided on the CD-ROM which accompanies the book. Solutions for the end-of-chapter problems in the main textbook are also found in *The Problems Book*.

MBoC5 Media DVD-ROM

The DVD included with every copy of the book contains the figures, tables, and micrographs from the book, pre-loaded into PowerPoint® presentations, one for each chapter. A separate folder contains individual versions of each figure, table, and micrograph in JPEG format. The panels are available in PDF format. There are also over 125 videos, animations, molecular structure tutorials, and high-resolution micrographs on the DVD. The authors have chosen to include material that not only reinforces basic concepts but also expands the content and scope of the book. The multimedia can be accessed either as individual files or through the *Cell Biology Interactive* media player. As discussed above, the media player has been programmed to work with the Media Codes integrated throughout the book. A complete table of contents and overview of all electronic resources is contained in the *MBoC5 Media Viewing Guide*, a PDF file located on the root level of the DVD-ROM and in the Appendix of the media player. The DVD-ROM also contains Chapters 21–25 which cover multicellular systems. The chapters are in PDF format and can be easily printed or searched using Adobe® Acrobat® Reader or other PDF software.

Teaching Supplements

Upon request, teaching supplements for *Molecular Biology of the Cell* are available to qualified instructors.

MBoC5 Transparency Set

Provides 200 full-color overhead acetate transparencies of the most important figures from the book.

MBoC5 Test Questions

A selection of test questions will be available. Written by Kirsten Benjamin (Amyris Biotechnologies, Emeryville, California) and Linda Huang (University of Massachusetts, Boston), these thought questions will test students' understanding of the chapter material.

MBoC5 Lecture Outlines

Lecture outlines created from the concept heads for the text are provided.

Garland Science Classwire™

All of the teaching supplements on the DVD-ROM (these include figures in PowerPoint and JPEG format; Chapters 21–25 in PDF format; 125 videos, animations, and movies) and the test questions and lecture outlines are available to qualified instructors online at the Garland Science Classwire™ Web site. Garland Science Classwire™ offers access to other instructional resources from all of the Garland Science textbooks, and provides free online course management tools. For additional information, please visit http://www.classwire.com/garlandscience or e-mail science@garland.com. (*Classwire* is a trademark of Chalkfree, Inc.)

INTRODUCTION TO THE CELL

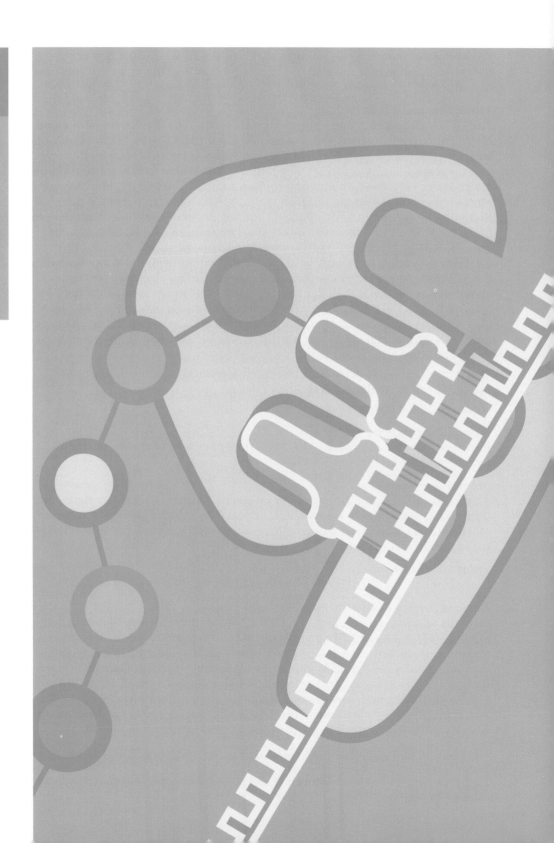

Chapter 1

Cells and Genomes

The surface of our planet is populated by living things—curious, intricately organized chemical factories that take in matter from their surroundings and use these raw materials to generate copies of themselves. The living organisms appear extraordinarily diverse. What could be more different than a tiger and a piece of seaweed, or a bacterium and a tree? Yet our ancestors, knowing nothing of cells or DNA, saw that all these things had something in common. They called that something "life," marveled at it, struggled to define it, and despaired of explaining what it was or how it worked in terms that relate to nonliving matter.

The discoveries of the past century have not diminished the marvel—quite the contrary. But they have lifted away the mystery as to the nature of life. We can now see that all living things are made of cells, and that these units of living matter all share the same machinery for their most basic functions. Living things, though infinitely varied when viewed from the outside, are fundamentally similar inside. The whole of biology is a counterpoint between the two themes: astonishing variety in individual particulars; astonishing constancy in fundamental mechanisms. In this first chapter we begin by outlining the universal features common to all life on our planet. We then survey, briefly, the diversity of cells. And we see how, thanks to the common code in which the specifications for all living organisms are written, it is possible to read, measure, and decipher these specifications to achieve a coherent understanding of all the forms of life, from the smallest to the greatest.

THE UNIVERSAL FEATURES OF CELLS ON EARTH

It is estimated that there are more than 10 million—perhaps 100 million—living species on Earth today. Each species is different, and each reproduces itself faithfully, yielding progeny that belong to the same species: the parent organism hands down information specifying, in extraordinary detail, the characteristics that the offspring shall have. This phenomenon of *heredity* is central to the definition of life: it distinguishes life from other processes, such as the growth of a crystal, or the burning of a candle, or the formation of waves on water, in which orderly structures are generated but without the same type of link between the peculiarities of parents and the peculiarities of offspring. Like the candle flame, the living organism consumes free energy to create and maintain its organization; but the free energy drives a hugely complex system of chemical processes that is specified by the hereditary information.

Most living organisms are single cells; others, such as ourselves, are vast multicellular cities in which groups of cells perform specialized functions and are linked by intricate systems of communication. But in all cases, whether we discuss the solitary bacterium or the aggregate of more than 10^{13} cells that form a human body, the whole organism has been generated by cell divisions from a single cell. The single cell, therefore, is the vehicle for the hereditary information that defines the species (**Figure 1–1**). And specified by this information, the cell includes the machinery to gather raw materials from the environment, and to construct out of them a new cell in its own image, complete with a new copy of the hereditary information. Nothing less than a cell has this capability.

Figure 1–1 The hereditary information in the fertilized egg cell determines the nature of the whole multicellular organism. (A and B) A sea urchin egg gives rise to a sea urchin. (C and D) A mouse egg gives rise to a mouse. (E and F) An egg of the seaweed *Fucus* gives rise to a *Fucus* seaweed. (A, courtesy of David McClay; B, courtesy of M. Gibbs, Oxford Scientific Films; C, courtesy of Patricia Calarco, from G. Martin, *Science* 209:768–776, 1980. With permission from AAAS; D, courtesy of O. Newman, Oxford Scientific Films; E and F, courtesy of Colin Brownlee.)

All Cells Store Their Hereditary Information in the Same Linear Chemical Code (DNA)

Computers have made us familiar with the concept of information as a measurable quantity—a million bytes (to record a few hundred pages of text or an image from a digital camera), 600 million for the music on a CD, and so on. They have also made us well aware that the same information can be recorded in many different physical forms. As the computer world has evolved, the discs and tapes that we used 10 years ago for our electronic archives have become unreadable on present-day machines. Living cells, like computers, deal in information, and it is estimated that they have been evolving and diversifying for over 3.5 billion years. It is scarcely to be expected that they should all store their information in the same form, or that the archives of one type of cell should be readable by the information-handling machinery of another. And yet it is so. All living cells on Earth, without any known exception, store their hereditary information in the form of double-stranded molecules of DNA—long unbranched paired polymer chains, formed always of the same four types of monomers. These monomers have nicknames drawn from a four-letter alphabet—A, T, C, G—and they are strung together in a long linear sequence that encodes the genetic information, just as the sequence of 1s and 0s encodes the information in a computer file. We can take a piece of DNA from a human cell and insert it into a bacterium, or a piece of bacterial DNA and insert it into a human cell, and the information will be successfully read, interpreted, and copied. Using chemical methods, scientists can read out the complete sequence of monomers in any DNA molecule—extending for millions of nucleotides—and thereby decipher the hereditary information that each organism contains.

All Cells Replicate Their Hereditary Information by Templated Polymerization

The mechanisms that make life possible depend on the structure of the double-stranded DNA molecule. Each monomer in a single DNA strand—that is, each **nucleotide**—consists of two parts: a sugar (deoxyribose) with a phosphate group attached to it, and a *base*, which may be either adenine (A), guanine (G), cytosine (C) or thymine (T) (**Figure 1–2**). Each sugar is linked to the next via the phosphate group, creating a polymer chain composed of a repetitive sugar-phosphate backbone with a series of bases protruding from it. The DNA polymer is extended by adding monomers at one end. For a single isolated strand, these can, in principle, be added in any order, because each one links to the next in the same way, through the part of the molecule that is the same for all of them. In the living cell, however, DNA is not synthesized as a free strand in isolation, but on a template formed by a preexisting DNA strand. The bases protruding from the existing strand bind to bases of the strand being synthesized, according to a strict rule defined by the complementary structures of the bases: A binds to T, and C binds to G. This base-pairing holds fresh monomers in place and thereby controls the selection of which one of the four monomers shall be added to the growing strand next. In this way, a double-stranded structure is created, consisting of two exactly complementary sequences of As, Cs, Ts, and Gs. The two strands twist around each other, forming a double helix (Figure 1–2E).

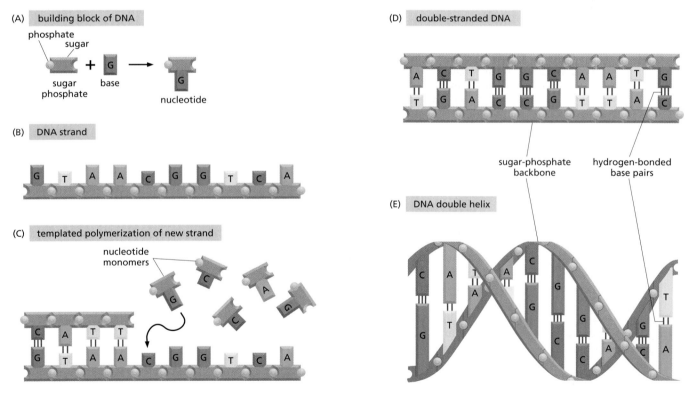

Figure 1–2 DNA and its building blocks. (A) DNA is made from simple subunits, called nucleotides, each consisting of a sugar-phosphate molecule with a nitrogen-containing sidegroup, or base, attached to it. The bases are of four types (adenine, guanine, cytosine, and thymine), corresponding to four distinct nucleotides, labeled A, G, C, and T. (B) A single strand of DNA consists of nucleotides joined together by sugar-phosphate linkages. Note that the individual sugar-phosphate units are asymmetric, giving the backbone of the strand a definite directionality, or polarity. This directionality guides the molecular processes by which the information in DNA is interpreted and copied in cells: the information is always "read" in a consistent order, just as written English text is read from left to right. (C) Through templated polymerization, the sequence of nucleotides in an existing DNA strand controls the sequence in which nucleotides are joined together in a new DNA strand; T in one strand pairs with A in the other, and G in one strand with C in the other. The new strand has a nucleotide sequence complementary to that of the old strand, and a backbone with opposite directionality: corresponding to the GTAA... of the original strand, it has ...TTAC. (D) A normal DNA molecule consists of two such complementary strands. The nucleotides within each strand are linked by strong (covalent) chemical bonds; the complementary nucleotides on opposite strands are held together more weakly, by hydrogen bonds. (E) The two strands twist around each other to form a double helix—a robust structure that can accommodate any sequence of nucleotides without altering its basic structure.

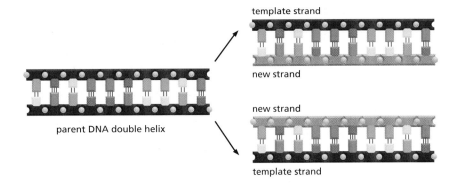

template strand

new strand

parent DNA double helix

new strand

template strand

Figure 1–3 The copying of genetic information by DNA replication. In this process, the two strands of a DNA double helix are pulled apart, and each serves as a template for synthesis of a new complementary strand.

The bonds between the base pairs are weak compared with the sugar-phosphate links, and this allows the two DNA strands to be pulled apart without breakage of their backbones. Each strand then can serve as a template, in the way just described, for the synthesis of a fresh DNA strand complementary to itself—a fresh copy, that is, of the hereditary information (**Figure 1–3**). In different types of cells, this process of **DNA replication** occurs at different rates, with different controls to start it or stop it, and different auxiliary molecules to help it along. But the basics are universal: DNA is the information store, and *templated polymerization* is the way in which this information is copied throughout the living world.

All Cells Transcribe Portions of Their Hereditary Information into the Same Intermediary Form (RNA)

To carry out its information-bearing function, DNA must do more than copy itself. It must also *express* its information, by letting it guide the synthesis of other molecules in the cell. This also occurs by a mechanism that is the same in all living organisms, leading first and foremost to the production of two other key classes of polymers: RNAs and proteins. The process (discussed in detail in Chapters 6 and 7) begins with a templated polymerization called **transcription**, in which segments of the DNA sequence are used as templates for the synthesis of shorter molecules of the closely related polymer **ribonucleic acid**, or **RNA**. Later, in the more complex process of **translation**, many of these RNA molecules direct the synthesis of polymers of a radically different chemical class—the *proteins* (**Figure 1–4**).

In RNA, the backbone is formed of a slightly different sugar from that of DNA—ribose instead of deoxyribose—and one of the four bases is slightly different—uracil (U) in place of thymine (T); but the other three bases—A, C, and G—are the same, and all four bases pair with their complementary counterparts in DNA—the A, U, C, and G of RNA with the T, A, G, and C of DNA. During transcription, RNA monomers are lined up and selected for polymerization on a template strand of DNA, just as DNA monomers are selected during replication. The outcome is a polymer molecule whose sequence of nucleotides faithfully represents a part of the cell's genetic information, even though written in a slightly different alphabet, consisting of RNA monomers instead of DNA monomers.

The same segment of DNA can be used repeatedly to guide the synthesis of many identical RNA transcripts. Thus, whereas the cell's archive of genetic information in the form of DNA is fixed and sacrosanct, the RNA transcripts are mass-produced and disposable (**Figure 1–5**). As we shall see, these transcripts function as intermediates in the transfer of genetic information: they mainly serve as **messenger RNA (mRNA)** to guide the synthesis of proteins according to the genetic instructions stored in the DNA.

RNA molecules have distinctive structures that can also give them other specialized chemical capabilities. Being single-stranded, their backbone is flexible, so that the polymer chain can bend back on itself to allow one part of the

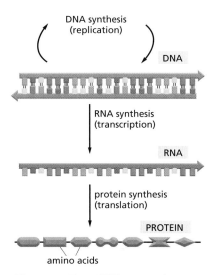

DNA synthesis (replication)

DNA

RNA synthesis (transcription)

RNA

protein synthesis (translation)

PROTEIN

amino acids

Figure 1–4 From DNA to protein. Genetic information is read out and put to use through a two-step process. First, in *transcription*, segments of the DNA sequence are used to guide the synthesis of molecules of RNA. Then, in *translation*, the RNA molecules are used to guide the synthesis of molecules of protein.

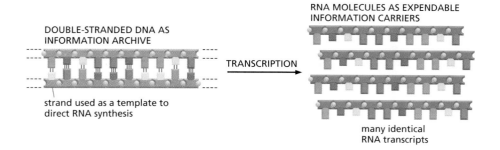

DOUBLE-STRANDED DNA AS
INFORMATION ARCHIVE

RNA MOLECULES AS EXPENDABLE
INFORMATION CARRIERS

TRANSCRIPTION

strand used as a template to
direct RNA synthesis

many identical
RNA transcripts

Figure 1–5 How genetic information is broadcast for use inside the cell. Each cell contains a fixed set of DNA molecules—its archive of genetic information. A given segment of this DNA guides the synthesis of many identical RNA transcripts, which serve as working copies of the information stored in the archive. Many different sets of RNA molecules can be made by transcribing selected parts of a long DNA sequence, allowing each cell to use its information store differently.

molecule to form weak bonds with another part of the same molecule. This occurs when segments of the sequence are locally complementary: a ...GGGG... segment, for example, will tend to associate with a ...CCCC... segment. These types of internal associations can cause an RNA chain to fold up into a specific shape that is dictated by its sequence (**Figure 1–6**). The shape of the RNA molecule, in turn, may enable it to recognize other molecules by binding to them selectively—and even, in certain cases, to catalyze chemical changes in the molecules that are bound. As we see in Chapter 6, a few chemical reactions catalyzed by RNA molecules are crucial for several of the most ancient and fundamental processes in living cells, and it has been suggested that more extensive catalysis by RNA played a central part in the early evolution of life.

All Cells Use Proteins as Catalysts

Protein molecules, like DNA and RNA molecules, are long unbranched polymer chains, formed by stringing together monomeric building blocks drawn from a standard repertoire that is the same for all living cells. Like DNA and RNA, they carry information in the form of a linear sequence of symbols, in the same way as a human message written in an alphabetic script. There are many different protein molecules in each cell, and—leaving out the water—they form most of the cell's mass.

The monomers of protein, the **amino acids**, are quite different from those of DNA and RNA, and there are 20 types, instead of 4. Each amino acid is built around the same core structure through which it can be linked in a standard way to any other amino acid in the set; attached to this core is a side group that gives each amino acid a distinctive chemical character. Each of the protein molecules, or **polypeptides**, created by joining amino acids in a particular sequence folds into a precise three-dimensional form with reactive sites on its surface (**Figure**

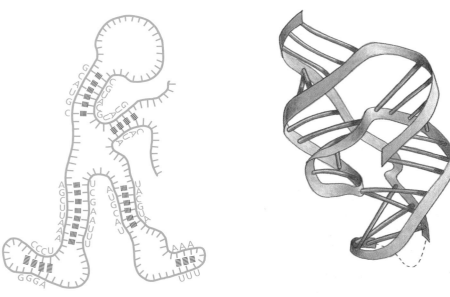

(A)

(B)

Figure 1–6 The conformation of an RNA molecule. (A) Nucleotide pairing between different regions of the same RNA polymer chain causes the molecule to adopt a distinctive shape. (B) The three-dimensional structure of an actual RNA molecule, from hepatitis delta virus, that catalyzes RNA strand cleavage. The *blue* ribbon represents the sugar-phosphate backbone; the bars represent base pairs. (B, based on A.R. Ferré D'Amaré, K. Zhou and J.A. Doudna, *Nature* 395:567–574, 1998. With permission from Macmillan Publishers Ltd.)

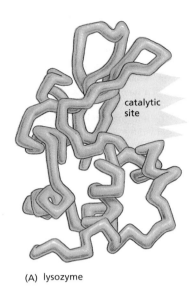

(A) lysozyme

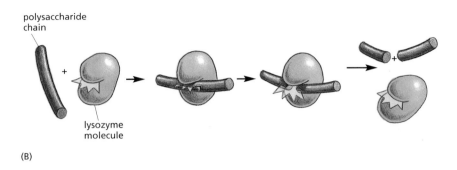

(B)

Figure 1–7 How a protein molecule acts as catalyst for a chemical reaction. (A) In a protein molecule the polymer chain folds up to into a specific shape defined by its amino acid sequence. A groove in the surface of this particular folded molecule, the enzyme lysozyme, forms a catalytic site. (B) A polysaccharide molecule *(red)*—a polymer chain of sugar monomers—binds to the catalytic site of lysozyme and is broken apart, as a result of a covalent bond-breaking reaction catalyzed by the amino acids lining the groove.

1–7A). These amino acid polymers thereby bind with high specificity to other molecules and act as **enzymes** to catalyze reactions that make or break covalent bonds. In this way they direct the vast majority of chemical processes in the cell (Figure 1–7B). Proteins have many other functions as well—maintaining structures, generating movements, sensing signals, and so on—each protein molecule performing a specific function according to its own genetically specified sequence of amino acids. Proteins, above all, are the molecules that put the cell's genetic information into action.

Thus, polynucleotides specify the amino acid sequences of proteins. Proteins, in turn, catalyze many chemical reactions, including those by which new DNA molecules are synthesized, and the genetic information in DNA is used to make both RNA and proteins. This feedback loop is the basis of the autocatalytic, self-reproducing behavior of living organisms (**Figure 1–8**).

All Cells Translate RNA into Protein in the Same Way

The translation of genetic information from the 4-letter alphabet of polynucleotides into the 20-letter alphabet of proteins is a complex process. The rules of this translation seem in some respects neat and rational, in other respects strangely arbitrary, given that they are (with minor exceptions) identical in all living things. These arbitrary features, it is thought, reflect frozen accidents in the early history of life—chance properties of the earliest organisms that were passed on by heredity and have become so deeply embedded in the constitution of all living cells that they cannot be changed without disastrous effects.

The information in the sequence of a messenger RNA molecule is read out in groups of three nucleotides at a time: each triplet of nucleotides, or *codon*, specifies (codes for) a single amino acid in a corresponding protein. Since there are 64 ($= 4 \times 4 \times 4$) possible codons, all of which occur in nature, but only 20 amino acids, there are necessarily many cases in which several codons correspond to the same amino acid. The code is read out by a special class of small RNA molecules, the **transfer RNAs** (**tRNAs**). Each type of tRNA becomes attached at one end to a specific amino acid, and displays at its other end a specific sequence of three nucleotides—an *anticodon*—that enables it to recognize, through base-pairing, a particular codon or subset of codons in mRNA (**Figure 1–9**).

For synthesis of protein, a succession of tRNA molecules charged with their appropriate amino acids have to be brought together with an mRNA molecule and matched up by base-pairing through their anticodons with each of its successive codons. The amino acids then have to be linked together to extend the growing protein chain, and the tRNAs, relieved of their burdens, have to be released. This whole complex of processes is carried out by a giant multimolecular machine, the ribosome, formed of two main chains of RNA, called **ribosomal RNAs**

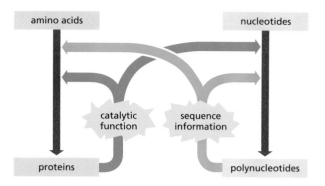

amino acids
nucleotides
catalytic function
sequence information
proteins
polynucleotides

Figure 1–8 **Life as an autocatalytic process.** Polynucleotides (nucleotide polymers) and proteins (amino acid polymers) provide the sequence information and the catalytic functions that serve—through a complex set of chemical reactions—to bring about the synthesis of more polynucleotides and proteins of the same types.

(**rRNAs**), and more than 50 different proteins. This evolutionarily ancient molecular juggernaut latches onto the end of an mRNA molecule and then trundles along it, capturing loaded tRNA molecules and stitching together the amino acids they carry to form a new protein chain (**Figure 1–10**).

The Fragment of Genetic Information Corresponding to One Protein Is One Gene

DNA molecules as a rule are very large, containing the specifications for thousands of proteins. Individual segments of the entire DNA sequence are transcribed into separate mRNA molecules, with each segment coding for a different protein. Each such DNA segment represents one **gene**. A complication is that RNA molecules transcribed from the same DNA segment can often be processed in more than one way, so as to give rise to a set of alternative versions of a protein, especially in more complex cells such as those of plants and animals. A gene therefore is defined, more generally, as the segment of DNA sequence corresponding to a single protein or set of alternative protein variants (or to a single catalytic or structural RNA molecule for those genes that produce RNA but not protein).

In all cells, the *expression* of individual genes is regulated: instead of manufacturing its full repertoire of possible proteins at full tilt all the time, the cell adjusts the rate of transcription and translation of different genes independently, according to need. Stretches of *regulatory DNA* are interspersed among the segments

Figure 1–9 **Transfer RNA.** (A) A tRNA molecule specific for the amino acid tryptophan. One end of the tRNA molecule has tryptophan attached to it, while the other end displays the triplet nucleotide sequence CCA (its anticodon), which recognizes the tryptophan codon in messenger RNA molecules. (B) The three-dimensional structure of the tryptophan tRNA molecule. Note that the codon and the anticodon in (A) are in antiparallel orientations, like the two strands in a DNA double helix (see Figure 1–2), so that the sequence of the anticodon in the tRNA is read from right to left, while that of the codon in the mRNA is read from left to right.

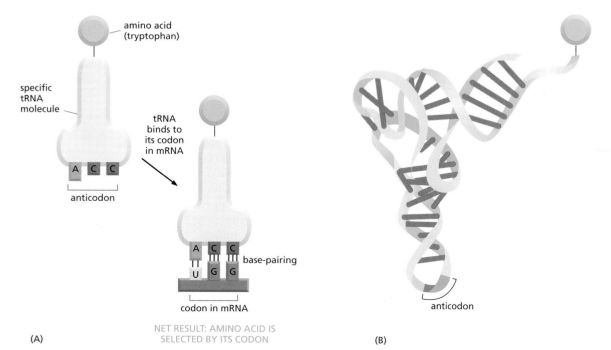

amino acid (tryptophan)

specific tRNA molecule

tRNA binds to its codon in mRNA

A C C

anticodon

A C C
base-pairing
U G G

codon in mRNA

NET RESULT: AMINO ACID IS SELECTED BY ITS CODON

anticodon

(A)

(B)

Figure 1–10 A ribosome at work. (A) The diagram shows how a ribosome moves along an mRNA molecule, capturing tRNA molecules that match the codons in the mRNA and using them to join amino acids into a protein chain. The mRNA specifies the sequence of amino acids. (B) The three-dimensional structure of a bacterial ribosome *(pale green and blue),* moving along an mRNA molecule *(orange beads),* with three tRNA molecules *(yellow, green,* and *pink)* at different stages in their process of capture and release. The ribosome is a giant assembly of more than 50 individual protein and RNA molecules. (B, courtesy of Joachim Frank, Yanhong Li and Rajendra Agarwal.)

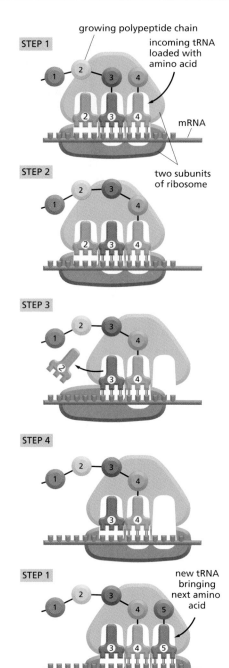

that code for protein, and these noncoding regions bind to special protein molecules that control the local rate of transcription (**Figure 1–11**). Other non-coding DNA is also present, some of it serving, for example, as punctuation, defining where the information for an individual protein begins and ends. The quantity and organization of the regulatory and other noncoding DNA vary widely from one class of organisms to another, but the basic strategy is universal. In this way, the **genome** of the cell—that is, the total of its genetic information as embodied in its complete DNA sequence—dictates not only the nature of the cell's proteins, but also when and where they are to be made.

Life Requires Free Energy

A living cell is a dynamic chemical system, operating far from chemical equilibrium. For a cell to grow or to make a new cell in its own image, it must take in free energy from the environment, as well as raw materials, to drive the necessary synthetic reactions. This consumption of free energy is fundamental to life. When it stops, a cell decays towards chemical equilibrium and soon dies.

Genetic information is also fundamental to life. Is there any connection? The answer is yes: free energy is required for the propagation of information. For example, to specify one bit of information—that is, one yes/no choice between two equally probable alternatives—costs a defined amount of free energy that can be calculated. The quantitative relationship involves some deep reasoning and depends on a precise definition of the term "free energy," discussed in Chapter 2. The basic idea, however, is not difficult to understand intuitively.

Picture the molecules in a cell as a swarm of objects endowed with thermal energy, moving around violently at random, buffeted by collisions with one another. To specify genetic information—in the form of a DNA sequence, for example—molecules from this wild crowd must be captured, arranged in a specific order defined by some preexisting template, and linked together in a fixed relationship. The bonds that hold the molecules in their proper places on the template and join them together must be strong enough to resist the disordering effect of thermal motion. The process is driven forward by consumption of free energy, which is needed to ensure that the correct bonds are made, and made robustly. In the simplest case, the molecules can be compared with spring-loaded traps, ready to snap into a more stable, lower-energy attached state when they meet their proper partners; as they snap together into the bonded arrangement, their available stored energy—their free energy—like the energy of the spring in the trap, is released and dissipated as heat. In a cell, the chemical processes underlying information transfer are more complex, but the same basic principle applies: free energy has to be spent on the creation of order.

To replicate its genetic information faithfully, and indeed to make all its complex molecules according to the correct specifications, the cell therefore requires free energy, which has to be imported somehow from the surroundings.

All Cells Function as Biochemical Factories Dealing with the Same Basic Molecular Building Blocks

Because all cells make DNA, RNA, and protein, and these macromolecules are composed of the same set of subunits in every case, all cells have to contain and

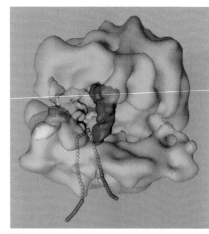

(B)

manipulate a similar collection of small molecules, including simple sugars, nucleotides, and amino acids, as well as other substances that are universally required for their synthesis. All cells, for example, require the phosphorylated nucleotide *ATP (adenosine triphosphate)* as a building block for the synthesis of DNA and RNA; and all cells also make and consume this molecule as a carrier of free energy and phosphate groups to drive many other chemical reactions.

Although all cells function as biochemical factories of a broadly similar type, many of the details of their small-molecule transactions differ, and it is not as easy as it is for the informational macromolecules to point out the features that are strictly universal. Some organisms, such as plants, require only the simplest of nutrients and harness the energy of sunlight to make from these almost all their own small organic molecules; other organisms, such as animals, feed on living things and obtain many of their organic molecules ready-made. We return to this point below.

All Cells Are Enclosed in a Plasma Membrane Across Which Nutrients and Waste Materials Must Pass

There is, however, at least one other feature of cells that is universal: each one is enclosed by a membrane—the **plasma membrane**. This container acts as a selective barrier that enables the cell to concentrate nutrients gathered from its environment and retain the products it synthesizes for its own use, while excreting its waste products. Without a plasma membrane, the cell could not maintain its integrity as a coordinated chemical system.

The molecules forming this membrane have the simple physico-chemical property of being *amphiphilic*—that is, consisting of one part that is hydrophobic (water-insoluble) and another part that is hydrophilic (water-soluble). Such molecules placed in water aggregate spontaneously, arranging their hydrophobic portions to be as much in contact with one another as possible to hide them from the water, while keeping their hydrophilic portions exposed. Amphiphilic molecules of appropriate shape, such as the phospholipid molecules that comprise most of the plasma membrane, spontaneously aggregate in water to form a *bilayer* that creates small closed vesicles (**Figure 1–12**). The phenomenon can be demonstrated in a test tube by simply mixing phospholipids and water together; under appropriate conditions, small vesicles form whose aqueous contents are isolated from the external medium.

Although the chemical details vary, the hydrophobic tails of the predominant membrane molecules in all cells are hydrocarbon polymers ($-CH_2-CH_2-CH_2-$), and their spontaneous assembly into a bilayered vesicle is but one of many examples of an important general principle: cells produce molecules whose chemical properties cause them to *self-assemble* into the structures that a cell needs.

The cell boundary cannot be totally impermeable. If a cell is to grow and reproduce, it must be able to import raw materials and export waste across its plasma membrane. All cells therefore have specialized proteins embedded in their membrane that transport specific molecules from one side to the other (**Figure 1–13**). Some of these *membrane transport proteins*, like some of the proteins that catalyze the fundamental small-molecule reactions inside the cell,

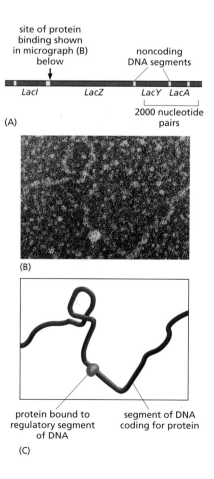

(A)

site of protein
binding shown
in micrograph (B)
below

noncoding
DNA segments

LacI   *LacZ*   *LacY* *LacA*

2000 nucleotide
pairs

(B)

(C)

protein bound to
regulatory segment
of DNA

segment of DNA
coding for protein

Figure 1–11 Gene regulation by protein binding to regulatory DNA.
(A) A diagram of a small portion of the genome of the bacterium *Escherichia coli*, containing genes (called *LacI, LacZ, LacY,* and *LacA*) coding for four different proteins. The protein-coding DNA segments *(red)* have regulatory and other noncoding DNA segments *(yellow)* between them. (B) An electron micrograph of DNA from this region, with a protein molecule (encoded by the *LacI* gene) bound to the regulatory segment; this protein controls the rate of transcription of the *LacZ, LacY,* and *LacA* genes. (C) A drawing of the structures shown in (B). (B, courtesy of Jack Griffith.)

have been so well preserved over the course of evolution that we can recognize the family resemblances between them in comparisons of even the most distantly related groups of living organisms.

The transport proteins in the membrane largely determine which molecules enter the cell, and the catalytic proteins inside the cell determine the reactions that those molecules undergo. Thus, by specifying the proteins that the cell is to manufacture, the genetic information recorded in the DNA sequence dictates the entire chemistry of the cell; and not only its chemistry, but also its form and its behavior, for these too are chiefly constructed and controlled by the cell's proteins.

A Living Cell Can Exist with Fewer Than 500 Genes

The basic principles of biological information transfer are simple enough, but how complex are real living cells? In particular, what are the minimum requirements? We can get a rough indication by considering a species that has one of the smallest known genomes—the bacterium *Mycoplasma genitalium* (**Figure 1–14**). This organism lives as a parasite in mammals, and its environment provides it with many of its small molecules ready-made. Nevertheless, it still has to make all the large molecules—DNA, RNAs, and proteins—required for the basic processes of heredity. It has only about 480 genes in its genome of 580,070 nucleotide pairs, representing 145,018 bytes of information—about as much as it takes to record the text of one chapter of this book. Cell biology may be complicated, but it is not impossibly so.

The minimum number of genes for a viable cell in today's environments is probably not less than 200–300, although there are only about 60 genes in the core set shared by all living species without any known exception.

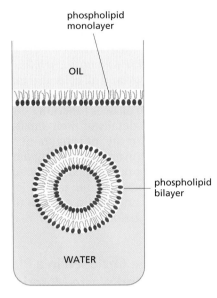

Figure 1–12 Formation of a membrane by amphiphilic phospholipid molecules. These have a hydrophilic (water-loving, phosphate) head group and a hydrophobic (water-avoiding, hydrocarbon) tail. At an interface between oil and water, they arrange themselves as a single sheet with their head groups facing the water and their tail groups facing the oil. When immersed in water, they aggregate to form bilayers enclosing aqueous compartments.

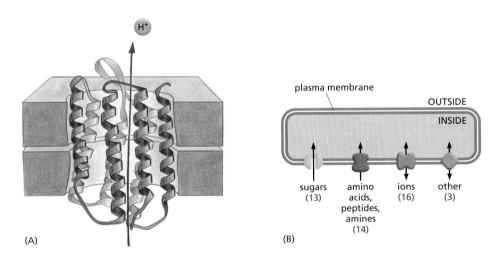

(A)

(B)

Figure 1–13 Membrane transport proteins. (A) Structure of a molecule of bacteriorhodopsin, from the archaeon (archaebacterium) *Halobacterium halobium*. This transport protein uses the energy of absorbed light to pump protons (H$^+$ ions) out of the cell. The polypeptide chain threads to and fro across the membrane; in several regions it is twisted into a helical conformation, and the helical segments are arranged to form the walls of a channel through which ions are transported. (B) Diagram of the set of transport proteins found in the membrane of the bacterium *Thermotoga maritima*. The numbers in parentheses refer to the number of different membrane transport proteins of each type. Most of the proteins within each class are evolutionarily related to one another and to their counterparts in other species.

Figure 1–14 Mycoplasma genitalium. (A) Scanning electron micrograph showing the irregular shape of this small bacterium, reflecting the lack of any rigid wall. (B) Cross section (transmission electron micrograph) of a *Mycoplasma* cell. Of the 477 genes of *Mycoplasma genitalium*, 37 code for transfer, ribosomal, and other nonmessenger RNAs. Functions are known, or can be guessed, for 297 of the genes coding for protein: of these, 153 are involved in replication, transcription, translation, and related processes involving DNA, RNA, and protein; 29 in the membrane and surface structures of the cell; 33 in the transport of nutrients and other molecules across the membrane; 71 in energy conversion and the synthesis and degradation of small molecules; and 11 in the regulation of cell division and other processes. (A, from S. Razin et al., *Infect. Immun.* 30:538–546, 1980. With permission from the American Society for Microbiology; B, courtesy of Roger Cole, in Medical Microbiology, 4th ed. [S. Baron ed.]. Galveston: University of Texas Medical Branch, 1996.)

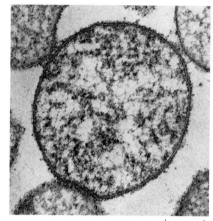

(A)

5 μm

(B)

0.2 μm

Summary

Living organisms reproduce themselves by transmitting genetic information to their progeny. The individual cell is the minimal self-reproducing unit, and is the vehicle for transmission of the genetic information in all living species. Every cell on our planet stores its genetic information in the same chemical form—as double-stranded DNA. The cell replicates its information by separating the paired DNA strands and using each as a template for polymerization to make a new DNA strand with a complementary sequence of nucleotides. The same strategy of templated polymerization is used to transcribe portions of the information from DNA into molecules of the closely related polymer, RNA. These in turn guide the synthesis of protein molecules by the more complex machinery of translation, involving a large multimolecular machine, the ribosome, which is itself composed of RNA and protein. Proteins are the principal catalysts for almost all the chemical reactions in the cell; their other functions include the selective import and export of small molecules across the plasma membrane that forms the cell's boundary. The specific function of each protein depends on its amino acid sequence, which is specified by the nucleotide sequence of a corresponding segment of the DNA—the gene that codes for that protein. In this way, the genome of the cell determines its chemistry; and the chemistry of every living cell is fundamentally similar, because it must provide for the synthesis of DNA, RNA, and protein. The simplest known cells have just under 500 genes.

THE DIVERSITY OF GENOMES AND THE TREE OF LIFE

The success of living organisms based on DNA, RNA, and protein, out of the infinitude of other chemical forms that we might conceive of, has been spectacular. They have populated the oceans, covered the land, infiltrated the Earth's crust, and molded the surface of our planet. Our oxygen-rich atmosphere, the deposits of coal and oil, the layers of iron ores, the cliffs of chalk and limestone and marble—all these are products, directly or indirectly, of past biological activity on Earth.

Living things are not confined to the familiar temperate realm of land, water, and sunlight inhabited by plants and plant-eating animals. They can be found in the darkest depths of the ocean, in hot volcanic mud, in pools beneath the frozen surface of the Antarctic, and buried kilometers deep in the Earth's crust. The creatures that live in these extreme environments are generally unfamiliar, not only because they are inaccessible, but also because they are mostly microscopic. In more homely habitats, too, most organisms are too small for us to see without special equipment: they tend to go unnoticed, unless they cause a disease or rot the timbers of our houses. Yet microorganisms make up most of the total mass of living matter on our planet. Only recently, through new methods of molecular analysis and specifically through the analysis of DNA sequences, have we begun to get a picture of life on Earth that is not grossly distorted by our biased perspective as large animals living on dry land.

In this section we consider the diversity of organisms and the relationships among them. Because the genetic information for every organism is written in the universal language of DNA sequences, and the DNA sequence of any given organism can be obtained by standard biochemical techniques, it is now possible to characterize, catalogue, and compare any set of living organisms with reference to these sequences. From such comparisons we can estimate the place of each organism in the family tree of living species—the 'tree of life'. But before describing what this approach reveals, we need first to consider the routes by which cells in different environments obtain the matter and energy they require to survive and proliferate, and the ways in which some classes of organisms depend on others for their basic chemical needs.

Cells Can Be Powered by a Variety of Free Energy Sources

Living organisms obtain their free energy in different ways. Some, such as animals, fungi, and the bacteria that live in the human gut, get it by feeding on other living things or the organic chemicals they produce; such organisms are called *organotrophic* (from the Greek word *trophe*, meaning "food"). Others derive their energy directly from the nonliving world. These fall into two classes: those that harvest the energy of sunlight, and those that capture their energy from energy-rich systems of inorganic chemicals in the environment (chemical systems that are far from chemical equilibrium). Organisms of the former class are called *phototrophic* (feeding on sunlight); those of the latter are called *lithotrophic* (feeding on rock). Organotrophic organisms could not exist without these primary energy converters, which are the most plentiful form of life.

Phototrophic organisms include many types of bacteria, as well as algae and plants, on which we—and virtually all the living things that we ordinarily see around us—depend. Phototrophic organisms have changed the whole chemistry of our environment: the oxygen in the Earth's atmosphere is a by-product of their biosynthetic activities.

Lithotrophic organisms are not such an obvious feature of our world, because they are microscopic and mostly live in habitats that humans do not frequent—deep in the ocean, buried in the Earth's crust, or in various other inhospitable environments. But they are a major part of the living world, and are especially important in any consideration of the history of life on Earth.

Some lithotrophs get energy from *aerobic* reactions, which use molecular oxygen from the environment; since atmospheric O_2 is ultimately the product of living organisms, these aerobic lithotrophs are, in a sense, feeding on the products of past life. There are, however, other lithotrophs that live anaerobically, in places where little or no molecular oxygen is present, in circumstances similar to those that must have existed in the early days of life on Earth, before oxygen had accumulated.

The most dramatic of these sites are the hot *hydrothermal vents* found deep down on the floor of the Pacific and Atlantic Oceans, in regions where the ocean floor is spreading as new portions of the Earth's crust form by a gradual upwelling of material from the Earth's interior (**Figure 1–15**). Downward-percolating seawater is heated and driven back upward as a submarine geyser, carrying with it a current of chemicals from the hot rocks below. A typical cocktail might include H_2S, H_2, CO, Mn^{2+}, Fe^{2+}, Ni^{2+}, CH_2, NH_4^+, and phosphorus-containing compounds. A dense population of microbes lives in the neighborhood of the vent, thriving on this austere diet and harvesting free energy from reactions between the available chemicals. Other organisms—clams, mussels, and giant marine worms—in turn live off the microbes at the vent, forming an entire ecosystem analogous to the system of plants and animals that we belong to, but powered by geochemical energy instead of light (**Figure 1–16**).

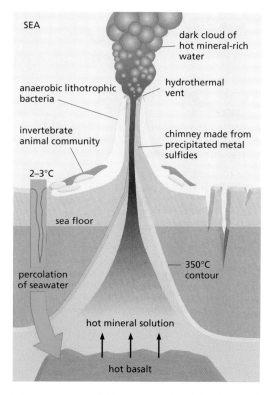

Figure 1–15 **The geology of a hot hydrothermal vent in the ocean floor.** Water percolates down toward the hot molten rock upwelling from the Earth's interior and is heated and driven back upward, carrying minerals leached from the hot rock. A temperature gradient is set up, from more than 350°C near the core of the vent, down to 2–3°C in the surrounding ocean. Minerals precipitate from the water as it cools, forming a chimney. Different classes of organisms, thriving at different temperatures, live in different neighborhoods of the chimney. A typical chimney might be a few meters tall, with a flow rate of 1–2 m/sec.

Some Cells Fix Nitrogen and Carbon Dioxide for Others

To make a living cell requires matter, as well as free energy. DNA, RNA, and protein are composed of just six elements: hydrogen, carbon, nitrogen, oxygen, sulfur, and phosphorus. These are all plentiful in the nonliving environment, in the Earth's rocks, water, and atmosphere, but not in chemical forms that allow easy incorporation into biological molecules. Atmospheric N_2 and CO_2, in particular, are extremely unreactive, and a large amount of free energy is required to drive the reactions that use these inorganic molecules to make the organic compounds needed for further biosynthesis—that is, to *fix* nitrogen and carbon dioxide, so as to make N and C available to living organisms. Many types of living cells lack the biochemical machinery to achieve this fixation, and rely on other classes of cells to do the job for them. We animals depend on plants for our supplies of

Figure 1–16 **Living organisms at a hot hydrothermal vent.** Close to the vent, at temperatures up to about 120°C, various lithotrophic species of bacteria and archaea (archaebacteria) live, directly fuelled by geochemical energy. A little further away, where the temperature is lower, various invertebrate animals live by feeding on these microorganisms. Most remarkable are the giant (2-meter) tube worms, which, rather than feed on the lithotrophic cells, live in symbiosis with them: specialized organs in the worms harbor huge numbers of symbiotic sulfur-oxidizing bacteria. These bacteria harness geochemical energy and supply nourishment to their hosts, which have no mouth, gut, or anus. The dependence of the tube worms on the bacteria for the harnessing of geothermal energy is analogous to the dependence of plants on chloroplasts for the harnessing of solar energy, discussed later in this chapter. The tube worms, however, are thought to have evolved from more conventional animals, and to have become secondarily adapted to life at hydrothermal vents. (Courtesy of Dudley Foster, Woods Hole Oceanographic Institution.)

geochemical energy and
inorganic raw materials

↓

bacteria

↓

multicellular animals e.g., tubeworms

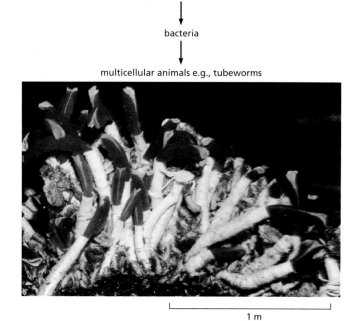

1 m

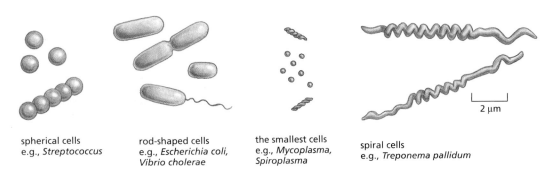

spherical cells
e.g., *Streptococcus*

rod-shaped cells
e.g., *Escherichia coli,
Vibrio cholerae*

the smallest cells
e.g., *Mycoplasma,
Spiroplasma*

spiral cells
e.g., *Treponema pallidum*

Figure 1–17 Shapes and sizes of some bacteria. Although most are small, as shown, measuring a few micrometers in linear dimension, there are also some giant species. An extreme example (not shown) is the cigar-shaped bacterium *Epulopiscium fishelsoni*, which lives in the gut of a surgeonfish and can be up to 600 µm long.

organic carbon and nitrogen compounds. Plants in turn, although they can fix carbon dioxide from the atmosphere, lack the ability to fix atmospheric nitrogen, and they depend in part on nitrogen-fixing bacteria to supply their need for nitrogen compounds. Plants of the pea family, for example, harbor symbiotic nitrogen-fixing bacteria in nodules in their roots.

Living cells therefore differ widely in some of the most basic aspects of their biochemistry. Not surprisingly, cells with complementary needs and capabilities have developed close associations. Some of these associations, as we see below, have evolved to the point where the partners have lost their separate identities altogether: they have joined forces to form a single composite cell.

The Greatest Biochemical Diversity Exists Among Procaryotic Cells

From simple microscopy, it has long been clear that living organisms can be classified on the basis of cell structure into two groups: the **eucaryotes** and the **procaryotes**. Eucaryotes keep their DNA in a distinct membrane-enclosed intracellular compartment called the nucleus. (The name is from the Greek, meaning "truly nucleated," from the words *eu*, "well" or "truly," and *karyon*, "kernel" or "nucleus".) Procaryotes have no distinct nuclear compartment to house their DNA. Plants, fungi, and animals are eucaryotes; bacteria are procaryotes, as are archaea—a separate class of procaryotic cells, discussed below.

Most procaryotic cells are small and simple in outward appearance (**Figure 1–17**), and they live mostly as independent individuals or in loosely organized communities, rather than as multicellular organisms. They are typically spherical or rod-shaped and measure a few micrometers in linear dimension. They often have a tough protective coat, called a *cell wall*, beneath which a plasma membrane encloses a single cytoplasmic compartment containing DNA, RNA, proteins, and the many small molecules needed for life. In the electron microscope, this cell interior appears as a matrix of varying texture without any discernible organized internal structure (**Figure 1–18**).

Figure 1–18 The structure of a bacterium. (A) The bacterium *Vibrio cholerae,* showing its simple internal organization. Like many other species, *Vibrio* has a helical appendage at one end—a flagellum—that rotates as a propeller to drive the cell forward. (B) An electron micrograph of a longitudinal section through the widely studied bacterium *Escherichia coli* (*E. coli*). This is related to *Vibrio* but has many flagella (not visible in this section) distributed over its surface. The cell's DNA is concentrated in the lightly stained region. (B, courtesy of E. Kellenberger.)

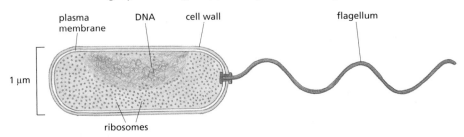

plasma membrane   DNA   cell wall   flagellum

1 µm

ribosomes

(A)

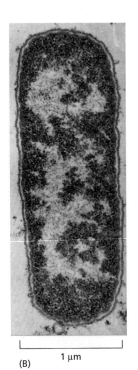

1 µm

(B)

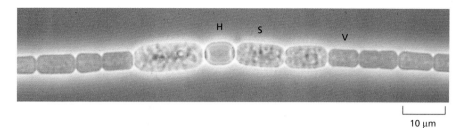

Figure 1–19 **The phototrophic bacterium *Anabaena cylindrica* viewed in the light microscope.** The cells of this species form long, multicellular filaments. Most of the cells (labeled *V)* perform photosynthesis, while others become specialized for nitrogen fixation (labeled *H),* or develop into resistant spores (labeled *S).* (Courtesy of Dave G. Adams.)

Procaryotic cells live in an enormous variety of ecological niches, and they are astonishingly varied in their biochemical capabilities—far more so than eucaryotic cells. Organotrophic species can utilize virtually any type of organic molecule as food, from sugars and amino acids to hydrocarbons and methane gas. Phototrophic species (**Figure 1–19**) harvest light energy in a variety of ways, some of them generating oxygen as a byproduct, others not. Lithotrophic species can feed on a plain diet of inorganic nutrients, getting their carbon from CO_2, and relying on H_2S to fuel their energy needs (**Figure 1–20**)—or on H_2, or Fe^{2+}, or elemental sulfur, or any of a host of other chemicals that occur in the environment.

Many parts of this world of microscopic organisms are virtually unexplored. Traditional methods of bacteriology have given us an acquaintance with those species that can be isolated and cultured in the laboratory. But DNA sequence analysis of the populations of bacteria in samples from natural habitats—such as soil or ocean water, or even the human mouth—has opened our eyes to the fact that most species cannot be cultured by standard laboratory techniques. According to one estimate, at least 99% of procaryotic species remain to be characterized.

The Tree of Life Has Three Primary Branches: Bacteria, Archaea, and Eucaryotes

The classification of living things has traditionally depended on comparisons of their outward appearances: we can see that a fish has eyes, jaws, backbone, brain, and so on, just as we do, and that a worm does not; that a rosebush is cousin to an apple tree, but less similar to a grass. As Darwin showed, we can readily interpret such close family resemblances in terms of evolution from common ancestors, and we can find the remains of many of these ancestors preserved in the fossil record. In this way, it has been possible to begin to draw a family tree of living organisms, showing the various lines of descent, as well as branch points in the history, where the ancestors of one group of species became different from those of another.

When the disparities between organisms become very great, however, these methods begin to fail. How do we decide whether a fungus is closer kin to a plant or to an animal? When it comes to procaryotes, the task becomes harder still: one microscopic rod or sphere looks much like another. Microbiologists have therefore sought to classify procaryotes in terms of their biochemistry and nutritional requirements. But this approach also has its pitfalls. Amid the bewildering variety of biochemical behaviors, it is difficult to know which differences truly reflect differences of evolutionary history.

Genome analysis has given us a simpler, more direct, and more powerful way to determine evolutionary relationships. The complete DNA sequence of an organism defines its nature with almost perfect precision and in exhaustive detail. Moreover, this specification is in a digital form—a string of letters—that can be entered straightforwardly into a computer and compared with the corresponding information for any other living thing. Because DNA is subject to random changes that accumulate over long periods of time (as we shall see shortly), the number of differences between the DNA sequences of two organisms can provide a direct, objective, quantitative indication of the evolutionary distance between them.

This approach has shown that the organisms that were traditionally classed together as "bacteria" can be as widely divergent in their evolutionary origins as

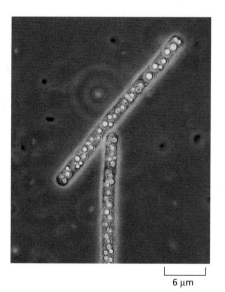

Figure 1–20 **A lithotrophic bacterium.** *Beggiatoa,* which lives in sulfurous environments, gets its energy by oxidizing H_2S and can fix carbon even in the dark. Note the yellow deposits of sulfur inside the cells. (Courtesy of Ralph W. Wolfe.)

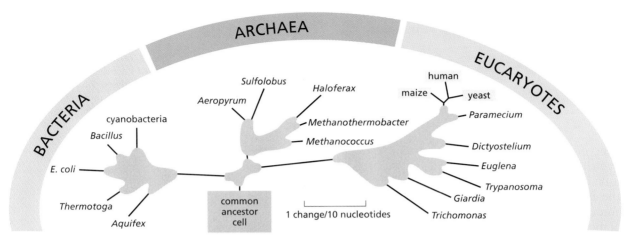

Figure 1–21 The three major divisions (domains) of the living world. Note that traditionally the word *bacteria* has been used to refer to procaryotes in general, but more recently has been redefined to refer to eubacteria specifically. The tree shown here is based on comparisons of the nucleotide sequence of a ribosomal RNA subunit in the different species, and the distances in the diagram represent estimates of the numbers of evolutionary changes that have occurred in this molecule in each lineage (see Figure 1–22). The parts of the tree shrouded in *gray cloud* represent uncertainties about details of the true pattern of species divergence in the course of evolution: comparisons of nucleotide or amino acid sequences of molecules other than rRNA, as well as other arguments, lead to somewhat different trees. There is general agreement, however, as to the early divergence of the three most basic domains—the bacteria, the archaea, and the eucaryotes.

is any procaryote from any eucaryote. It now appears that the procaryotes comprise two distinct groups that diverged early in the history of life on Earth, either before the ancestors of the eucaryotes diverged as a separate group or at about the same time. The two groups of procaryotes are called the **bacteria** (or eubacteria) and the **archaea** (or archaebacteria). The living world therefore has three major divisions or *domains*: bacteria, archaea, and eucaryotes (**Figure 1–21**).

Archaea are often found inhabiting environments that we humans avoid, such as bogs, sewage treatment plants, ocean depths, salt brines, and hot acid springs, although they are also widespread in less extreme and more homely environments, from soils and lakes to the stomachs of cattle. In outward appearance they are not easily distinguished from bacteria. At a molecular level, archaea seem to resemble eucaryotes more closely in their machinery for handling genetic information (replication, transcription, and translation), but bacteria more closely in their apparatus for metabolism and energy conversion. We discuss below how this might be explained.

Some Genes Evolve Rapidly; Others Are Highly Conserved

Both in the storage and in the copying of genetic information, random accidents and errors occur, altering the nucleotide sequence—that is, creating **mutations**. Therefore, when a cell divides, its two daughters are often not quite identical to one another or to their parent. On rare occasions, the error may represent a change for the better; more probably, it will cause no significant difference in the cell's prospects; and in many cases, the error will cause serious damage—for example, by disrupting the coding sequence for a key protein. Changes due to mistakes of the first type will tend to be perpetuated, because the altered cell has an increased likelihood of reproducing itself. Changes due to mistakes of the second type—*selectively neutral* changes—may be perpetuated or not: in the competition for limited resources, it is a matter of chance whether the altered cell or its cousins will succeed. But changes that cause serious damage lead nowhere: the cell that suffers them dies, leaving no progeny. Through endless repetition of this cycle of error and trial—of *mutation* and *natural selection*—

organisms evolve: their genetic specifications change, giving them new ways to exploit the environment more effectively, to survive in competition with others, and to reproduce successfully.

Clearly, some parts of the genome change more easily than others in the course of evolution. A segment of DNA that does not code for protein and has no significant regulatory role is free to change at a rate limited only by the frequency of random errors. In contrast, a gene that codes for a highly optimized essential protein or RNA molecule cannot alter so easily: when mistakes occur, the faulty cells are almost always eliminated. Genes of this latter sort are therefore *highly conserved*. Through 3.5 billion years or more of evolutionary history, many features of the genome have changed beyond all recognition; but the most highly conserved genes remain perfectly recognizable in all living species.

These latter genes are the ones we must examine if we wish to trace family relationships between the most distantly related organisms in the tree of life. The studies that led to the classification of the living world into the three domains of bacteria, archaea, and eucaryotes were based chiefly on analysis of one of the two main RNA components of the ribosome—the so-called small-subunit ribosomal RNA. Because translation is fundamental to all living cells, this component of the ribosome has been well conserved since early in the history of life on Earth (**Figure 1–22**).

Most Bacteria and Archaea Have 1000–6000 Genes

Natural selection has generally favored those procaryotic cells that can reproduce the fastest by taking up raw materials from their environment and replicating themselves most efficiently, at the maximal rate permitted by the available food supplies. Small size implies a large ratio of surface area to volume, thereby helping to maximize the uptake of nutrients across the plasma membrane and boosting a cell's reproductive rate.

Presumably for these reasons, most procaryotic cells carry very little superfluous baggage; their genomes are small, with genes packed closely together and minimal quantities of regulatory DNA between them. The small genome size makes it relatively easy to determine the complete DNA sequence. We now have this information for many species of bacteria and archaea, and a few species of eucaryotes. As shown in **Table 1–1**, most bacterial and archaeal genomes contain between 10^6 and 10^7 nucleotide pairs, encoding 1000–6000 genes.

A complete DNA sequence reveals both the genes an organism possesses and the genes it lacks. When we compare the three domains of the living world, we can begin to see which genes are common to all of them and must therefore have been present in the cell that was ancestral to all present-day living things, and which genes are peculiar to a single branch in the tree of life. To explain the findings, however, we need to consider a little more closely how new genes arise and genomes evolve.

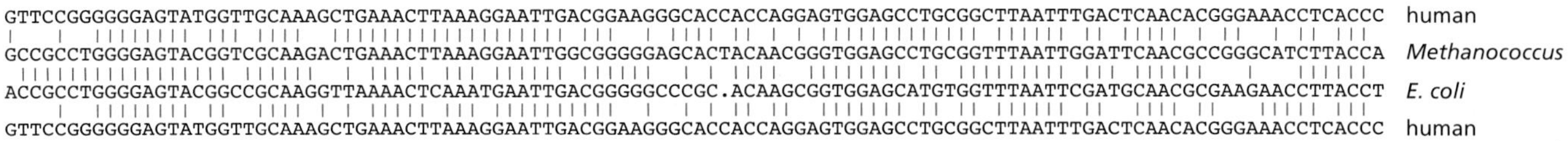

```
GTTCCGGGGGGAGTATGGTTGCAAAGCTGAAACTTAAAGGAATTGACGGAAGGGCACCACCAGGAGTGGGAGCCTGCGGCTTAATTTGACTCAACACGGGAAACCTCACCC  human
  |   |||||||| ||| |||| ||||||||||| |||||||||| ||||| ||| ||  ||| ||| |  | |||||  ||||| |||  |  || |  || |||||
GCCGCCTGGGGAGTACGGTCGCAAGACTGAAACTTAAAGGAATTGGCGGGGGGAGCACTACAACGGGTGGAGCCTGCGGTTTAATTGGATTCAACGCCGGGCATCTTACCA  Methanococcus
|||||||||||||||| ||||||| |||||| ||| ||||| ||| |||||||||||  |  |||| || |||||||||| ||| |||||||  |  |||||||
ACCGCCTGGGGAGTACGGCCGCAAGGTTAAAACTCAAATGAATTGACGGGGGCCCGC.ACAAGCGGTGGAGCATGTGGTTTAATTCGATGCAACGCGAAGAACCTTACCT  E. coli
  |  |||||||||| ||  |||| || |||| ||| ||||||||||| |  |  || || ||||||| ||  || |||||| ||  |||| || |||||| |||
GTTCCGGGGGGAGTATGGTTGCAAAGCTGAAACTTAAAGGAATTGACGGAAGGGCACCACCAGGAGTGGGAGCCTGCGGCTTAATTTGACTCAACACGGGAAACCTCACCC  human
```

Figure 1–22 Genetic information conserved since the days of the last common ancestor of all living things. A part of the gene for the smaller of the two main RNA components of the ribosome is shown. (The complete molecule is about 1500–1900 nucleotides long, depending on species.) Corresponding segments of nucleotide sequence from an archaean *(Methanococcus jannaschii)*, a bacterium *(Escherichia coli)* and a eucaryote *(Homo sapiens)* are aligned. Sites where the nucleotides are identical between species are indicated by a vertical line; the human sequence is repeated at the bottom of the alignment so that all three two-way comparisons can be seen. A dot halfway along the *E. coli* sequence denotes a site where a nucleotide has been either deleted from the bacterial lineage in the course of evolution, or inserted in the other two lineages. Note that the sequences from these three organisms, representative of the three domains of the living world, all differ from one another to a roughly similar degree, while still retaining unmistakable similarities.

Table 1–1 Some Genomes That Have Been Completely Sequenced

SPECIES	SPECIAL FEATURES	HABITAT	GENOME SIZE (1000s OF NUCLEOTIDE PAIRS PER HAPLOID GENOME)	ESTIMATED NUMBER OF GENES CODING FOR PROTEINS
BACTERIA				
Mycoplasma genitalium	has one of the smallest of all known cell genomes	human genital tract	580	468
Synechocystis sp.	photosynthetic, oxygen-generating (cyanobacterium)	lakes and streams	3573	3168
Escherichia coli	laboratory favorite	human gut	4639	4289
Helicobacter pylori	causes stomach ulcers and predisposes to stomach cancer	human stomach	1667	1590
Bacillus anthracis	causes anthrax	soil	5227	5634
Aquifex aeolicus	lithotrophic; lives at high temperatures	hydrothermal vents	1551	1544
Streptomyces coelicolor	source of antibiotics; giant genome	soil	8667	7825
Treponema pallidum	spirochete; causes syphilis	human tissues	1138	1041
Rickettsia prowazekii	bacterium most closely related to mitochondria; causes typhus	lice and humans (intracellular parasite)	1111	834
Thermotoga maritima	organotrophic; lives at very high temperatures	hydrothermal vents	1860	1877
ARCHAEA				
Methanococcus jannaschii	lithotrophic, anaerobic, methane-producing	hydrothermal vents	1664	1750
Archaeoglobus fulgidus	lithotrophic or organotrophic, anaerobic, sulfate-reducing	hydrothermal vents	2178	2493
Nanoarchaeum equitans	smallest known archaean; anaerobic; parasitic on another, larger archaean	hydrothermal and volcanic hot vents	491	552
EUCARYOTES				
Saccharomyces cerevisiae (budding yeast)	minimal model eucaryote	grape skins, beer	12,069	~6300
Arabidopsis thaliana (Thale cress)	model organism for flowering plants	soil and air	~142,000	~26,000
Caenorhabditis elegans (nematode worm)	simple animal with perfectly predictable development	soil	~97,000	~20,000
Drosophila melanogaster (fruit fly)	key to the genetics of animal development	rotting fruit	~137,000	~14,000
Homo sapiens (human)	most intensively studied mammal	houses	~3,200,000	~24,000

Genome size and gene number vary between strains of a single species, especially for bacteria and archaea. The table shows data for particular strains that have been sequenced. For eucaryotes, many genes can give rise to several alternative variant proteins, so that the total number of proteins specified by the genome is substantially greater than the number of genes.

New Genes Are Generated from Preexisting Genes

The raw material of evolution is the DNA sequence that already exists: there is no natural mechanism for making long stretches of new random sequence. In this sense, no gene is ever entirely new. Innovation can, however, occur in several ways (**Figure 1–23**):

1. *Intragenic mutation*: an existing gene can be modified by changes in its DNA sequence, through various types of error that occur mainly in the process of DNA replication.
2. *Gene duplication*: an existing gene can be duplicated so as to create a pair of initially identical genes within a single cell; these two genes may then diverge in the course of evolution.

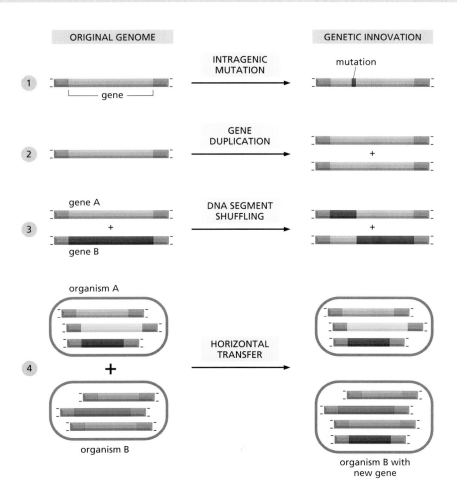

Figure 1–23 **Four modes of genetic innovation and their effects on the DNA sequence of an organism.** A special form of horizontal transfer occurs when two different types of cells enter into a permanent symbiotic association. Genes from one of the cells then may be transferred to the genome of the other, as we shall see below when we discuss mitochondria and chloroplasts.

3. *Segment shuffling*: two or more existing genes can be broken and rejoined to make a hybrid gene consisting of DNA segments that originally belonged to separate genes.

4. *Horizontal (intercellular) transfer*: a piece of DNA can be transferred from the genome of one cell to that of another—even to that of another species. This process is in contrast with the usual *vertical transfer* of genetic information from parent to progeny.

Each of these types of change leaves a characteristic trace in the DNA sequence of the organism, providing clear evidence that all four processes have occurred. In later chapters we discuss the underlying mechanisms, but for the present we focus on the consequences.

Gene Duplications Give Rise to Families of Related Genes Within a Single Cell

A cell duplicates its entire genome each time it divides into two daughter cells. However, accidents occasionally result in the inappropriate duplication of just part of the genome, with retention of original and duplicate segments in a single cell. Once a gene has been duplicated in this way, one of the two gene copies is free to mutate and become specialized to perform a different function within the same cell. Repeated rounds of this process of duplication and divergence, over many millions of years, have enabled one gene to give rise to a family of genes that may all be found within a single genome. Analysis of the DNA sequence of procaryotic genomes reveals many examples of such gene families: in *Bacillus subtilis*, for example, 47% of the genes have one or more obvious relatives (**Figure 1–24**).

When genes duplicate and diverge in this way, the individuals of one species become endowed with multiple variants of a primordial gene. This evolutionary

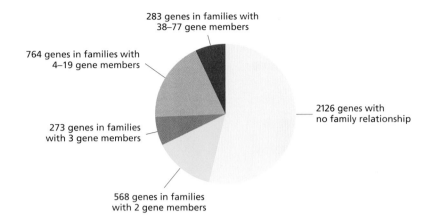

283 genes in families with
38–77 gene members

764 genes in families with
4–19 gene members

273 genes in families
with 3 gene members

2126 genes with
no family relationship

568 genes in families
with 2 gene members

Figure 1–24 **Families of evolutionarily related genes in the genome of** *Bacillus subtilis.* The biggest family consists of 77 genes coding for varieties of ABC transporters—a class of membrane transport proteins found in all three domains of the living world. (Adapted from F. Kunst et al., *Nature* 390:249–256, 1997. With permission from Macmillan Publishers Ltd.)

process has to be distinguished from the genetic divergence that occurs when one species of organism splits into two separate lines of descent at a branch point in the family tree—when the human line of descent became separate from that of chimpanzees, for example. There, the genes gradually become different in the course of evolution, but they are likely to continue to have corresponding functions in the two sister species. Genes that are related by descent in this way—that is, genes in two separate species that derive from the same ancestral gene in the last common ancestor of those two species—are called **orthologs**. Related genes that have resulted from a gene duplication event within a single genome—and are likely to have diverged in their function—are called **paralogs**. Genes that are related by descent in either way are called **homologs**, a general term used to cover both types of relationship (**Figure 1–25**).

The family relationships between genes can become quite complex (**Figure 1–26**). For example, an organism that possesses a family of paralogous genes (for example, the seven hemoglobin genes α, β, γ, δ, ε, ζ, and θ) may evolve into two separate species (such as humans and chimpanzees) each possessing the entire set of paralogs. All 14 genes are homologs, with the human hemoglobin α orthologous to the chimpanzee hemoglobin α, but paralogous to the human or chimpanzee hemoglobin β, and so on. Moreover, the vertebrate hemoglobins (the oxygen-binding proteins of blood) are homologous to the vertebrate myoglobins (the oxygen-binding proteins of muscle), as well as to more distant

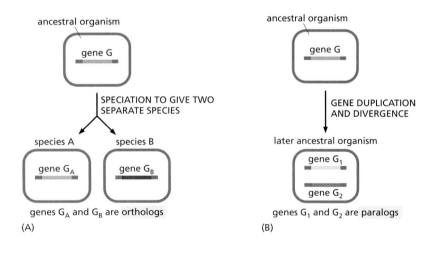

ancestral organism

gene G

SPECIATION TO GIVE TWO
SEPARATE SPECIES

species A   species B

gene G_A gene G_B

genes G_A and G_B are orthologs

(A)

ancestral organism

gene G

GENE DUPLICATION
AND DIVERGENCE

later ancestral organism

gene G_1

gene G_2

genes G_1 and G_2 are paralogs

(B)

Figure 1–25 **Paralogous genes and orthologous genes: two types of gene homology based on different evolutionary pathways.** (A) and (B) The most basic possibilities. (C) A more complex pattern of events that can occur.

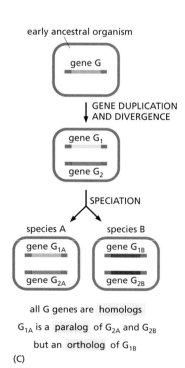

early ancestral organism

gene G

GENE DUPLICATION
AND DIVERGENCE

gene G_1

gene G_2

SPECIATION

species A   species B

gene G_{1A} gene G_{1B}

gene G_{2A} gene G_{2B}

all G genes are homologs

G_{1A} is a paralog of G_{2A} and G_{2B}

but an ortholog of G_{1B}

(C)

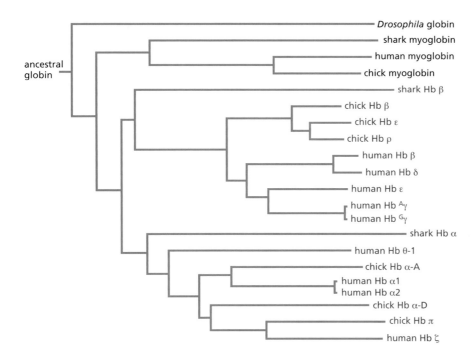

Figure 1–26 A complex family of homologous genes. This diagram shows the pedigree of the hemoglobin (Hb), myoglobin, and globin genes of human, chick, shark, and *Drosophila*. The lengths of the horizontal lines represent the amount of divergence in amino acid sequence.

genes that code for oxygen-binding proteins in invertebrates, plants, fungi, and bacteria. From the DNA sequences, it is usually easy to recognize that two genes in different species are homologous; it is much more difficult to decide, without other information, whether they stand in the precise evolutionary relationship of orthologs.

Genes Can Be Transferred Between Organisms, Both in the Laboratory and in Nature

Procaryotes also provide examples of the horizontal transfer of genes from one species of cell to another. The most obvious tell-tale signs are sequences recognizable as being derived from bacterial viruses, also called *bacteriophages* (**Figure 1–27**). **Viruses** are not themselves living cells but can act as vectors for gene transfer: they are small packets of genetic material that have evolved as parasites on the reproductive and biosynthetic machinery of host cells. They replicate in one cell, emerge from it with a protective wrapping, and then enter and infect another cell, which may be of the same or a different species. Often, the infected cell will be killed by the massive proliferation of virus particles inside it; but sometimes, the viral DNA, instead of directly generating these particles, may persist in its host for many cell generations as a relatively innocuous passenger, either as a separate intracellular fragment of DNA, known as a *plasmid*, or as a sequence inserted into the cell's regular genome. In their travels, viruses can accidentally pick up fragments of DNA from the genome of one host cell and ferry them into another cell. Such transfers of genetic material frequently occur in procaryotes, and they can also occur between eucaryotic cells of the same species.

Horizontal transfers of genes between eucaryotic cells of different species are very rare, and they do not seem to have played a significant part in eucaryote evolution (although massive transfers from bacterial to eucaryotic genomes have occurred in the evolution of mitochondria and chloroplasts, as we discuss below). In contrast, horizontal gene transfers occur much more frequently between different species of procaryotes. Many procaryotes have a remarkable capacity to take up even nonviral DNA molecules from their surroundings and thereby capture the genetic information these molecules carry. By this route, or by virus-mediated transfer, bacteria and archaea in the wild can acquire genes from neighboring cells relatively easily. Genes that confer resistance to an antibiotic or an ability to produce a toxin, for example, can be transferred from species to species and provide the recipient bacterium with a selective advantage. In this way, new and sometimes dangerous strains of bacteria have been observed to evolve in the bacterial ecosystems that inhabit hospitals or the various niches in the human body. For example, horizontal gene transfer is responsible for the spread, over the past 40 years, of penicillin-resistant strains of *Neisseria gonorrheae,* the bacterium that causes gonorrhea. On a longer time scale, the results can be even more profound; it has been estimated that at least 18% of all of the genes in the present-day genome of *E. coli* have been acquired by horizontal transfer from another species within the past 100 million years.

Sex Results in Horizontal Exchanges of Genetic Information Within a Species

Horizontal exchanges of genetic information are important in bacterial and archaeal evolution in today's world, and they may have occurred even more frequently and promiscuously in the early days of life on Earth. Such early horizontal exchanges could explain the otherwise puzzling observation that the eucaryotes seem more similar to archaea in their genes for the basic information-handling processes of DNA replication, transcription, and translation, but more similar to bacteria in their genes for metabolic processes. In any case, whether horizontal gene transfer occurred most freely in the early days of life on Earth, or has continued at a steady low rate throughout evolutionary history, it has the effect of complicating the whole concept of cell ancestry, by making each cell's genome a composite of parts derived from separate sources.

Horizontal gene transfer among procaryotes may seem a surprising process, but it has a parallel in a phenomenon familiar to us all: sex. In addition to the usual vertical transfer of genetic material from parent to offspring, sexual reproduction causes a large-scale horizontal transfer of genetic information between two initially separate cell lineages—those of the father and the mother. A key feature of sex, of course, is that the genetic exchange normally occurs only between individuals of the same species. But no matter whether they occur within a species or between species, horizontal gene transfers leave a characteristic imprint: they result in individuals who are related more closely to one set of relatives with respect to some genes, and more closely to another set of relatives with respect to others. By comparing the DNA sequences of individual human genomes, an intelligent visitor from outer space could deduce that humans reproduce sexually, even if it knew nothing about human behavior.

Sexual reproduction is widespread (although not universal), especially among eucaryotes. Even bacteria indulge from time to time in controlled sexual exchanges of DNA with other members of their own species. Natural selection has clearly favored organisms that can reproduce sexually, although evolutionary theorists dispute precisely what the selective advantage of sex is.

The Function of a Gene Can Often Be Deduced from Its Sequence

Family relationships among genes are important not just for their historical interest, but because they simplify the task of deciphering gene functions. Once the sequence of a newly discovered gene has been determined, a scientist can tap a few keys on a computer to search the entire database of known gene sequences for genes related to it. In many cases, the function of one or more of these homologs will have been already determined experimentally, and thus, since gene sequence determines gene function, one can frequently make a good guess at the function of the new gene: it is likely to be similar to that of the already-known homologs.

In this way, it is possible to decipher a great deal of the biology of an organism simply by analyzing the DNA sequence of its genome and using the information we already have about the functions of genes in other organisms that have been more intensively studied.

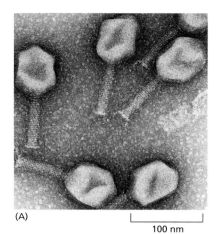

(A)

100 nm

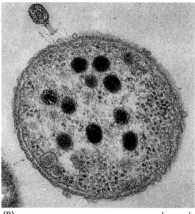

(B)

100 nm

Figure 1–27 The viral transfer of DNA from one cell to another. (A) An electron micrograph of particles of a bacterial virus, the T4 bacteriophage. The head of this virus contains the viral DNA; the tail contains the apparatus for injecting the DNA into a host bacterium. (B) A cross section of a bacterium with a T4 bacteriophage latched onto its surface. The large dark objects inside the bacterium are the heads of new T4 particles in course of assembly. When they are mature, the bacterium will burst open to release them. (A, courtesy of James Paulson; B, courtesy of Jonathan King and Erika Hartwig from G. Karp, Cell and Molecular Biology, 2nd ed. New York: John Wiley & Sons, 1999. With permission from John Wiley & Sons.)

More Than 200 Gene Families Are Common to All Three Primary Branches of the Tree of Life

Given the complete genome sequences of representative organisms from all three domains—archaea, bacteria, and eucaryotes—we can search systematically for homologies that span this enormous evolutionary divide. In this way we can begin to take stock of the common inheritance of all living things. There are considerable difficulties in this enterprise. For example, individual species have often lost some of the ancestral genes; other genes have almost certainly been acquired by horizontal transfer from another species and therefore are not truly ancestral, even though shared. In fact, genome comparisons strongly suggest that both lineage-specific gene loss and horizontal gene transfer, in some cases between evolutionarily distant species, have been major factors of evolution, at least among procaryotes. Finally, in the course of 2 or 3 billion years, some genes that were initially shared will have changed beyond recognition by current methods.

Because of all these vagaries of the evolutionary process, it seems that only a small proportion of ancestral gene families have been universally retained in a recognizable form. Thus, out of 4873 protein-coding gene families defined by comparing the genomes of 50 species of bacteria, 13 archaea, and 3 unicellular eucaryotes, only 63 are truly ubiquitous (that is, represented in all the genomes analyzed). The great majority of these universal families include components of the translation and transcription systems. This is not likely to be a realistic approximation of an ancestral gene set. A better—though still crude—idea of the latter can be obtained by tallying the gene families that have representatives in multiple, but not necessarily all, species from all three major domains. Such an analysis reveals 264 ancient conserved families. Each family can be assigned a function (at least in terms of general biochemical activity, but usually with more precision), with the largest number of shared gene families being involved in translation and in amino acid metabolism and transport (**Table 1–2**). This set of highly conserved gene families represents only a very rough sketch of the common inheritance of all modern life; a more precise reconstruction of the gene complement of the last universal common ancestor might be feasible with further genome sequencing and more careful comparative analysis.

Mutations Reveal the Functions of Genes

Without additional information, no amount of gazing at genome sequences will reveal the functions of genes. We may recognize that gene B is like gene A, but how do we discover the function of gene A in the first place? And even if we know the function of gene A, how do we test whether the function of gene B is truly the same as the sequence similarity suggests? How do we connect the world of abstract genetic information with the world of real living organisms?

The analysis of gene functions depends on two complementary approaches: genetics and biochemistry. Genetics starts with the study of mutants: we either find or make an organism in which a gene is altered, and examine the effects on the organism's structure and performance (**Figure 1–28**). Biochemistry examines the functions of molecules: we extract molecules from an organism and then study their chemical activities. By combining genetics and biochemistry and examining the chemical abnormalities in a mutant organism, it is possible to find those molecules whose production depends on a given gene. At the same time, studies of the performance of the mutant organism show us what role those molecules have in the operation of the organism as a whole. Thus, genetics and biochemistry together provide a way to relate genes, molecules, and the structure and function of the organism.

In recent years, DNA sequence information and the powerful tools of molecular biology have allowed rapid progress. From sequence comparisons, we can often identify particular subregions within a gene that have been preserved nearly unchanged over the course of evolution. These conserved subregions are likely to be the most important parts of the gene in terms of function. We can test their individual contributions to the activity of the gene product by creating in

Table 1–2 The Numbers of Gene Families, Classified by Function, That Are Common to All Three Domains of the Living World

GENE FAMILY FUNCTION	NUMBER OF "UNIVERSAL" FAMILIES
Information processing	
Translation	63
Transcription	7
Replication, recombination, and repair	13
Cellular processes and signaling	
Cell cycle control, mitosis, and meiosis	2
Defense mechanisms	3
Signal transduction mechanisms	1
Cell wall/membrane biogenesis	2
Intracellular trafficking and secretion	4
Post-translational modification, protein turnover, chaperones	8
Metabolism	
Energy production and conversion	19
Carbohydrate transport and metabolism	16
Amino acid transport and metabolism	43
Nucleotide transport and metabolism	15
Coenzyme transport and metabolism	22
Lipid transport and metabolism	9
Inorganic ion transport and metabolism	8
Secondary metabolite biosynthesis, transport, and catabolism	5
Poorly characterized	
General biochemical function predicted; specific biological role unknown	24

For the purpose of this analysis, gene families are defined as "universal" if they are represented in the genomes of at least two diverse archaea (*Archaeoglobus fulgidus* and *Aeropyrum pernix*), two evolutionarily distant bacteria (*Escherichia coli* and *Bacillus subtilis*) and one eucaryote (yeast, *Saccharomyces cerevisiae*). (Data from R.L. Tatusov, E.V. Koonin and D.J. Lipman, *Science* 278:631–637, 1997, with permission from AAAS; R.L. Tatusov et al., *BMC Bioinformatics* 4:41, 2003, with permission from BioMed Central; and the COGs database at the US National Library of Medicine.)

the laboratory mutations of specific sites within the gene, or by constructing artificial hybrid genes that combine part of one gene with part of another. Organisms can be engineered to make either the RNA or the protein specified by the gene in large quantities to facilitate biochemical analysis. Specialists in molecular structure can determine the three-dimensional conformation of the gene product, revealing the exact position of every atom in it. Biochemists can determine how each of the parts of the genetically specified molecule contributes to its chemical behavior. Cell biologists can analyze the behavior of cells that are engineered to express a mutant version of the gene.

There is, however, no one simple recipe for discovering a gene's function, and no simple standard universal format for describing it. We may discover, for example, that the product of a given gene catalyzes a certain chemical reaction, and yet have no idea how or why that reaction is important to the organism. The functional characterization of each new family of gene products, unlike the description of the gene sequences, presents a fresh challenge to the biologist's ingenuity. Moreover, we never fully understand the function of a gene until we learn its role in the life of the organism as a whole. To make ultimate sense of gene functions, therefore, we have to study whole organisms, not just molecules or cells.

Molecular Biologists Have Focused a Spotlight on *E. coli*

Because living organisms are so complex, the more we learn about any particular species, the more attractive it becomes as an object for further study. Each

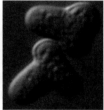

⊢————⊣
5 μm

Figure 1–28 A mutant phenotype reflecting the function of a gene. A normal yeast (of the species *Schizosaccharomyces pombe*) is compared with a mutant in which a change in a single gene has converted the cell from a cigar shape *(left)* to a T shape *(right)*. The mutant gene therefore has a function in the control of cell shape. But how, in molecular terms, does the gene product perform that function? That is a harder question, and needs biochemical analysis to answer it. (Courtesy of Kenneth Sawin and Paul Nurse.)

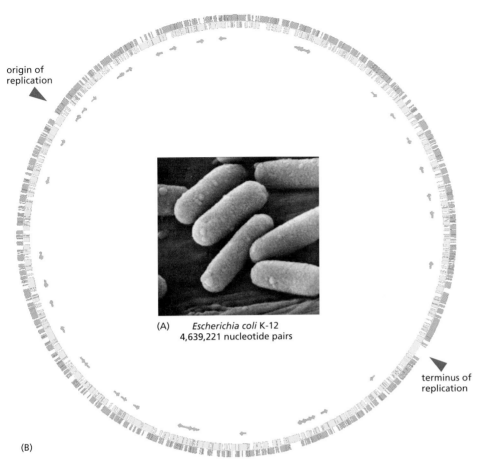

origin of
replication

terminus of
replication

(A)   *Escherichia coli* K-12
4,639,221 nucleotide pairs

(B)

Figure 1–29 **The genome of *E. coli*.**
(A) A cluster of *E. coli* cells. (B) A diagram
of the genome of *E. coli* strain K-12. The
diagram is circular because the DNA of
E. coli, like that of other procaryotes,
forms a single, closed loop. Protein-
coding genes are shown as *yellow* or
orange bars, depending on the DNA
strand from which they are transcribed;
genes encoding only RNA molecules are
indicated by *green arrows*. Some genes
are transcribed from one strand of the
DNA double helix (in a clockwise
direction in this diagram), others from the
other strand (counterclockwise).
(A, courtesy of Dr. Tony Brain and David
Parker/Photo Researchers; B, adapted
from F.R. Blattner et al., *Science*
277:1453–1462, 1997. With permission
from AAAS.)

discovery raises new questions and provides new tools with which to tackle
general questions in the context of the chosen organism. For this reason, large
communities of biologists have become dedicated to studying different aspects
of the same **model organism**.

In the enormously varied world of bacteria, the spotlight of molecular
biology has for a long time focused intensely on just one species: *Escherichia
coli*, or *E. coli* (see Figures 1–17 and 1–18). This small, rod-shaped bacterial cell
normally lives in the gut of humans and other vertebrates, but it can be grown
easily in a simple nutrient broth in a culture bottle. It adapts to variable chemical conditions and reproduces rapidly, and it can evolve by mutation and
selection at a remarkable speed. As with other bacteria, different strains of *E.
coli*, though classified as members of a single species, differ genetically to a
much greater degree than do different varieties of a sexually reproducing
organism such as a plant or animal. One *E. coli* strain may possess many hundreds of genes that are absent from another, and the two strains could have as
little as 50% of their genes in common. The standard laboratory strain *E. coli*
K-12 has a genome of approximately 4.6 million nucleotide pairs, contained in
a single circular molecule of DNA, coding for about 4300 different kinds of proteins (**Figure 1–29**).

In molecular terms, we know more about *E. coli* than about any other living
organism. Most of our understanding of the fundamental mechanisms of life—
for example, how cells replicate their DNA, or how they decode the instructions
represented in the DNA to direct the synthesis of specific proteins—has come
from studies of *E. coli*. The basic genetic mechanisms have turned out to be
highly conserved throughout evolution: these mechanisms are therefore essentially the same in our own cells as in *E. coli*.

Summary

Procaryotes (cells without a distinct nucleus) are biochemically the most diverse organisms and include species that can obtain all their energy and nutrients from inorganic chemical sources, such as the reactive mixtures of minerals released at hydrothermal vents on the ocean floor—the sort of diet that may have nourished the first living cells 3.5 billion years ago. DNA sequence comparisons reveal the family relationships of living organisms and show that the procaryotes fall into two groups that diverged early in the course of evolution: the bacteria (or eubacteria) and the archaea. Together with the eucaryotes (cells with a membrane-enclosed nucleus), these constitute the three primary branches of the tree of life. Most bacteria and archaea are small unicellular organisms with compact genomes comprising 1000–6000 genes. Many of the genes within a single organism show strong family resemblances in their DNA sequences, implying that they originated from the same ancestral gene through gene duplication and divergence. Family resemblances (homologies) are also clear when gene sequences are compared between different species, and more than 200 gene families have been so highly conserved that they can be recognized as common to most species from all three domains of the living world. Thus, given the DNA sequence of a newly discovered gene, it is often possible to deduce the gene's function from the known function of a homologous gene in an intensively studied model organism, such as the bacterium E. coli.

GENETIC INFORMATION IN EUCARYOTES

Eucaryotic cells, in general, are bigger and more elaborate than procaryotic cells, and their genomes are bigger and more elaborate, too. The greater size is accompanied by radical differences in cell structure and function. Moreover, many classes of eucaryotic cells form multicellular organisms that attain levels of complexity unmatched by any procaryote.

Because they are so complex, eucaryotes confront molecular biologists with a special set of challenges, which will concern us in the rest of this book. Increasingly, biologists meet these challenges through the analysis and manipulation of the genetic information within cells and organisms. It is therefore important at the outset to know something of the special features of the eucaryotic genome. We begin by briefly discussing how eucaryotic cells are organized, how this reflects their way of life, and how their genomes differ from those of procaryotes. This leads us to an outline of the strategy by which molecular biologists, by exploiting genetic information, are attempting to discover how eucaryotic organisms work.

Eucaryotic Cells May Have Originated as Predators

By definition, eucaryotic cells keep their DNA in an internal compartment called the nucleus. The *nuclear envelope,* a double layer of membrane, surrounds the nucleus and separates the DNA from the cytoplasm. Eucaryotes also have other features that set them apart from procaryotes (**Figure 1–30**). Their cells are, typically, 10 times bigger in linear dimension, and 1000 times larger in volume. They have a *cytoskeleton*—a system of protein filaments crisscrossing the cytoplasm and forming, together with the many proteins that attach to them, a system of girders, ropes, and motors that gives the cell mechanical strength, controls its shape, and drives and guides its movements. <GTTA> <ATGG> <TCGC> The nuclear envelope is only one part of a set of *internal membranes,* each structurally similar to the plasma membrane and enclosing different types of spaces inside the cell, many of them involved in digestion and secretion. Lacking the tough cell wall of most bacteria, animal cells and the free-living eucaryotic cells called *protozoa* can change their shape rapidly and engulf other cells and small objects by *phagocytosis* (**Figure 1–31**).

It is still a mystery how all these properties evolved, and in what sequence. One plausible view, however, is that they are all reflections of the way of life of a

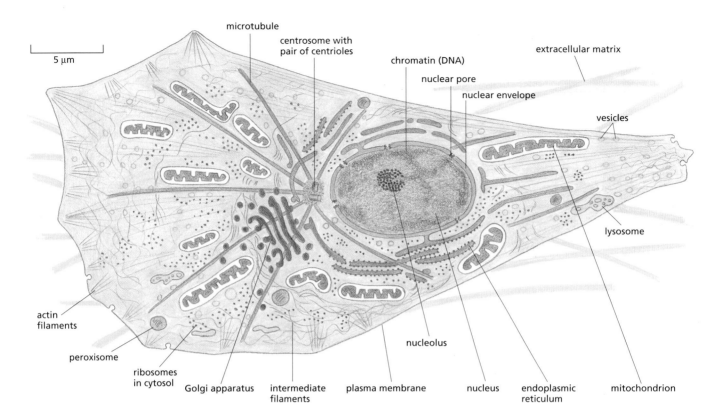

microtubule

centrosome with
pair of centrioles

chromatin (DNA)

nuclear pore

nuclear envelope

extracellular matrix

vesicles

5 µm

lysosome

actin
filaments

peroxisome

nucleolus

ribosomes
in cytosol

Golgi apparatus

intermediate
filaments

plasma membrane

nucleus

endoplasmic
reticulum

mitochondrion

primordial eucaryotic cell that was a predator, living by capturing other cells and eating them (**Figure 1–32**). Such a way of life requires a large cell with a flexible plasma membrane, as well as an elaborate cytoskeleton to support and move this membrane. It may also require that the cell's long, fragile DNA molecules be sequestered in a separate nuclear compartment, to protect the genome from damage by the movements of the cytoskeleton.

Modern Eucaryotic Cells Evolved from a Symbiosis

A predatory way of life helps to explain another feature of eucaryotic cells. Almost all such cells contain *mitochondria* (**Figure 1–33**). These small bodies in the cytoplasm, enclosed by a double layer of membrane, take up oxygen and harness energy from the oxidation of food molecules—such as sugars—to produce most of the ATP that powers the cell's activities. Mitochondria are similar in size to small bacteria, and, like bacteria, they have their own genome in the form of a circular DNA molecule, their own ribosomes that differ from those elsewhere in the eucaryotic cell, and their own transfer RNAs. It is now generally accepted that mitochondria originated from free-living oxygen-metabolizing *(aerobic)* bacteria that were engulfed by an ancestral eucaryotic cell that could otherwise make no such use of oxygen (that is, was *anaerobic*). Escaping digestion, these bacteria evolved in symbiosis with the engulfing cell and its progeny,

Figure 1–30 The major features of eucaryotic cells. The drawing depicts a typical animal cell, but almost all the same components are found in plants and fungi and in single-celled eucaryotes such as yeasts and protozoa. Plant cells contain chloroplasts in addition to the components shown here, and their plasma membrane is surrounded by a tough external wall formed of cellulose.

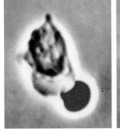

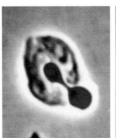
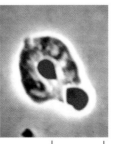

10 µm

Figure 1–31 Phagocytosis. This series of stills from a movie shows a human white blood cell (a neutrophil) engulfing a red blood cell (artificially colored *red*) that has been treated with antibody. (Courtesy of Stephen E. Malawista and Anne de Boisfleury Chevance.)

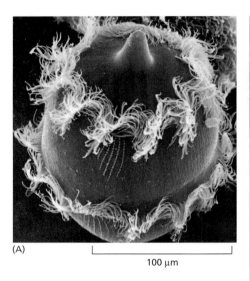

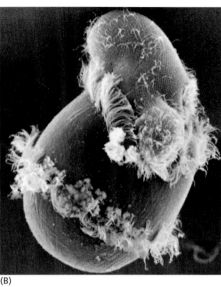

(A)

|—————————|
100 μm

(B)

Figure 1–32 A single-celled eucaryote that eats other cells. (A) *Didinium* is a carnivorous protozoan, belonging to the group known as *ciliates*. It has a globular body, about 150 μm in diameter, encircled by two fringes of cilia—sinuous, whiplike appendages that beat continually; its front end is flattened except for a single protrusion, rather like a snout. (B) *Didinium* normally swims around in the water at high speed by means of the synchronous beating of its cilia. When it encounters a suitable prey, usually another type of protozoan, it releases numerous small paralyzing darts from its snout region. Then, the *Didinium* attaches to and devours the other cell by phagocytosis, inverting like a hollow ball to engulf its victim, which is almost as large as itself. (Courtesy of D. Barlow.)

receiving shelter and nourishment in return for the power generation they performed for their hosts (**Figure 1–34**). This partnership between a primitive anaerobic eucaryotic predator cell and an aerobic bacterial cell is thought to have been established about 1.5 billion years ago, when the Earth's atmosphere first became rich in oxygen.

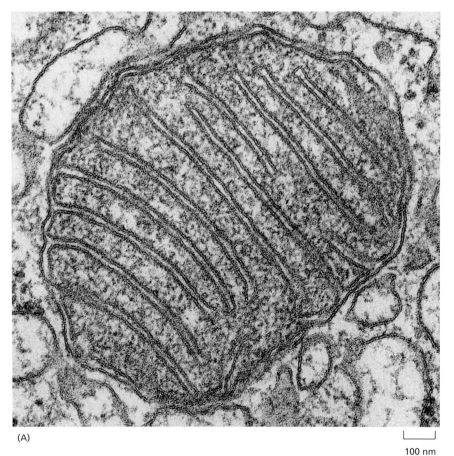

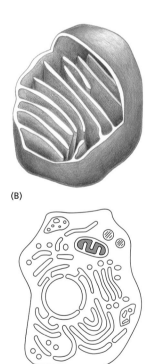

(B)

(C)

(A)

|———|
100 nm

Figure 1–33 A mitochondrion. (A) A cross section, as seen in the electron microscope. (B) A drawing of a mitochondrion with part of it cut away to show the three-dimensional structure. (C) A schematic eucaryotic cell, with the interior space of a mitochondrion, containing the mitochondrial DNA and ribosomes, colored. Note the smooth outer membrane and the convoluted inner membrane, which houses the proteins that generate ATP from the oxidation of food molecules. (A, courtesy of Daniel S. Friend.)

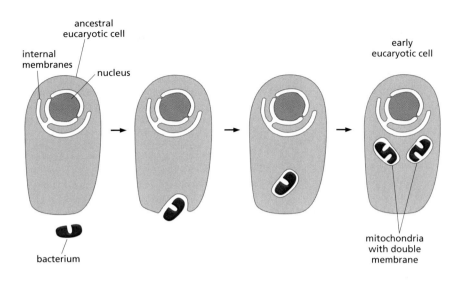

Figure 1–34 **The origin of mitochondria.** An ancestral eucaryotic cell is thought to have engulfed the bacterial ancestor of mitochondria, initiating a symbiotic relationship.

Many eucaryotic cells—specifically, those of plants and algae—also contain another class of small membrane-enclosed organelles somewhat similar to mito-chondria—the *chloroplasts* (**Figure 1–35**). Chloroplasts perform photosynthesis, using the energy of sunlight to synthesize carbohydrates from atmospheric carbon dioxide and water, and deliver the products to the host cell as food. Like mitochondria, chloroplasts have their own genome and almost certainly origi-nated as symbiotic photosynthetic bacteria, acquired by cells that already pos-sessed mitochondria (**Figure 1–36**).

A eucaryotic cell equipped with chloroplasts has no need to chase after other cells as prey; it is nourished by the captive chloroplasts it has inherited from its ancestors. Correspondingly, plant cells, although they possess the cytoskeletal equipment for movement, have lost the ability to change shape rapidly and to engulf other cells by phagocytosis. Instead, they create around themselves a tough, protective cell wall. If the ancestral eucaryote was indeed a predator on other organisms, we can view plant cells as eucaryotes that have made the transition from hunting to farming.

Fungi represent yet another eucaryotic way of life. Fungal cells, like animal cells, possess mitochondria but not chloroplasts; but in contrast with animal cells and protozoa, they have a tough outer wall that limits their ability to move

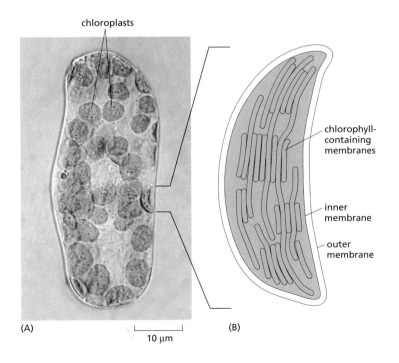

(A)

10 µm

(B)

Figure 1–35 **Chloroplasts.** These organelles capture the energy of sunlight in plant cells and some single-celled eucaryotes. (A) A single cell isolated from a leaf of a flowering plant, seen in the light microscope, showing the green chloroplasts. (B) A drawing of one of the chloroplasts, showing the highly folded system of internal membranes containing the chlorophyll molecules by which light is absorbed. (A, courtesy of Preeti Dahiya.)

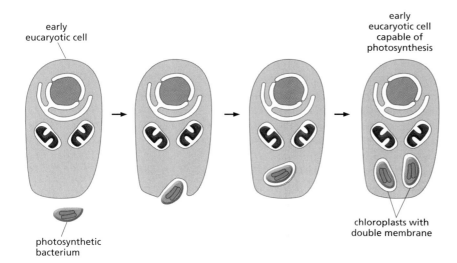

early
eucaryotic cell

early
eucaryotic cell
capable of
photosynthesis

photosynthetic
bacterium

chloroplasts with
double membrane

Figure 1–36 The origin of chloroplasts. An early eucaryotic cell, already possessing mitochondria, engulfed a photosynthetic bacterium (a cyanobacterium) and retained it in symbiosis. All present-day chloroplasts are thought to trace their ancestry back to a single species of cyanobacterium that was adopted as an internal symbiont (an endosymbiont) over a billion years ago.

rapidly or to swallow up other cells. Fungi, it seems, have turned from hunters into scavengers: other cells secrete nutrient molecules or release them upon death, and fungi feed on these leavings—performing whatever digestion is necessary extracellularly, by secreting digestive enzymes to the exterior.

Eucaryotes Have Hybrid Genomes

The genetic information of eucaryotic cells has a hybrid origin—from the ancestral anaerobic eucaryote, and from the bacteria that it adopted as symbionts. Most of this information is stored in the nucleus, but a small amount remains inside the mitochondria and, for plant and algal cells, in the chloroplasts. The mitochondrial DNA and the chloroplast DNA can be separated from the nuclear DNA and individually analyzed and sequenced. The mitochondrial and chloroplast genomes are found to be degenerate, cut-down versions of the corresponding bacterial genomes, lacking genes for many essential functions. In a human cell, for example, the mitochondrial genome consists of only 16,569 nucleotide pairs, and codes for only 13 proteins, two ribosomal RNA components, and 22 transfer RNAs.

The genes that are missing from the mitochondria and chloroplasts have not all been lost; instead, many of them have been somehow moved from the symbiont genome into the DNA of the host cell nucleus. The nuclear DNA of humans contains many genes coding for proteins that serve essential functions inside the mitochondria; in plants, the nuclear DNA also contains many genes specifying proteins required in chloroplasts.

Eucaryotic Genomes Are Big

Natural selection has evidently favored mitochondria with small genomes, just as it has favored bacteria with small genomes. By contrast, the nuclear genomes of most eucaryotes seem to have been free to enlarge. Perhaps the eucaryotic way of life has made large size an advantage: predators typically need to be bigger than their prey, and cell size generally increases in proportion to genome size. Perhaps enlargement of the genome has been driven by the accumulation of parasitic transposable elements (discussed in Chapter 5)—"selfish" segments of DNA that can insert copies of themselves at multiple sites in the genome. Whatever the explanation, the genomes of most eucaryotes are orders of magnitude larger than those of bacteria and archaea (**Figure 1–37**). And the freedom to be extravagant with DNA has had profound implications.

Eucaryotes not only have more genes than procaryotes; they also have vastly more DNA that does not code for protein or for any other functional product molecule. The human genome contains 1000 times as many nucleotide pairs as the genome of a typical bacterium, 20 times as many genes, and about 10,000

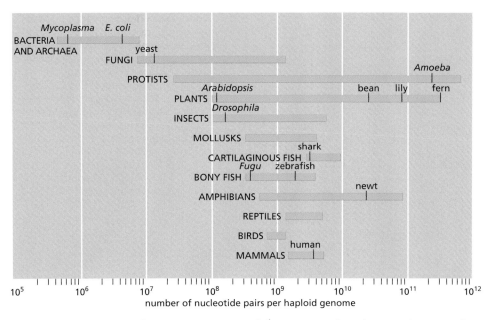

Figure 1–37 Genome sizes compared. Genome size is measured in nucleotide pairs of DNA per haploid genome, that is, per single copy of the genome. (The cells of sexually reproducing organisms such as ourselves are generally diploid: they contain two copies of the genome, one inherited from the mother, the other from the father.) Closely related organisms can vary widely in the quantity of DNA in their genomes, even though they contain similar numbers of functionally distinct genes. (Data from W.H. Li, Molecular Evolution, pp. 380–383. Sunderland, MA: Sinauer, 1997.)

times as much noncoding DNA (~98.5% of the genome for a human is noncoding, as opposed to 11% of the genome for the bacterium *E. coli*).

Eucaryotic Genomes Are Rich in Regulatory DNA

Much of our noncoding DNA is almost certainly dispensable junk, retained like a mass of old papers because, when there is little pressure to keep an archive small, it is easier to retain everything than to sort out the valuable information and discard the rest. Certain exceptional eucaryotic species, such as the puffer fish (**Figure 1–38**), bear witness to the profligacy of their relatives; they have somehow managed to rid themselves of large quantities of noncoding DNA. Yet they appear similar in structure, behavior, and fitness to related species that have vastly more such DNA.

Even in compact eucaryotic genomes such as that of puffer fish, there is more noncoding DNA than coding DNA, and at least some of the noncoding DNA certainly has important functions. In particular, it regulates the expression of adjacent genes. With this regulatory DNA, eucaryotes have evolved distinctive ways of controlling when and where a gene is brought into play. This sophisticated gene regulation is crucial for the formation of complex multicellular organisms.

The Genome Defines the Program of Multicellular Development

The cells in an individual animal or plant are extraordinarily varied. Fat cells, skin cells, bone cells, nerve cells—they seem as dissimilar as any cells could be. Yet all these cell types are the descendants of a single fertilized egg cell, and all (with minor exceptions) contain identical copies of the genome of the species.

The differences result from the way in which the cells make selective use of their genetic instructions according to the cues they get from their surroundings in the developing embryo. The DNA is not just a shopping list specifying the molecules that every cell must have, and the cell is not an assembly of all the items on the list. Rather, the cell behaves as a multipurpose machine, with sensors to receive environmental signals and with highly developed abilities to call different sets of genes into action according to the sequences of signals to which the cell has been exposed. The genome in each cell is big enough to accommodate the information that specifies an entire multicellular organism, but in any individual cell only part of that information is used.

A large fraction of the genes in the eucaryotic genome code for proteins that regulate the activities of other genes. Most of these *gene regulatory proteins* act by

Figure 1–38 The puffer fish (*Fugu rubripes*). This organism has a genome size of 400 million nucleotide pairs—about one-quarter as much as a zebrafish, for example, even though the two species of fish have similar numbers of genes. (From a woodcut by Hiroshige, courtesy of Arts and Designs of Japan.)

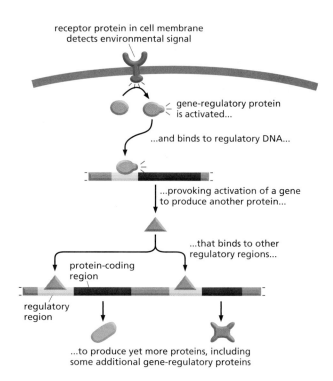

receptor protein in cell membrane
detects environmental signal

gene-regulatory protein
is activated...

...and binds to regulatory DNA...

...provoking activation of a gene
to produce another protein...

...that binds to other
regulatory regions...

protein-coding
region

regulatory
region

...to produce yet more proteins, including
some additional gene-regulatory proteins

Figure 1–39 **Controlling gene readout by environmental signals.** Regulatory DNA allows gene expression to be controlled by regulatory proteins, which are in turn the products of other genes. This diagram shows how a cell's gene expression is adjusted according to a signal from the cell's environment. The initial effect of the signal is to activate a regulatory protein already present in the cell; the signal may, for example, trigger the attachment of a phosphate group to the regulatory protein, altering its chemical properties.

binding, directly or indirectly, to the regulatory DNA adjacent to the genes that are to be controlled (**Figure 1–39**), or by interfering with the abilities of other proteins to do so. The expanded genome of eucaryotes therefore not only specifies the hardware of the cell, but also stores the software that controls how that hardware is used (**Figure 1–40**).

Cells do not just passively receive signals; rather, they actively exchange signals with their neighbors. Thus, in a developing multicellular organism, the same control system governs each cell, but with different consequences depending on the messages exchanged. The outcome, astonishingly, is a precisely patterned array of cells in different states, each displaying a character appropriate to its position in the multicellular structure.

Many Eucaryotes Live as Solitary Cells: the Protists

Many species of eucaryotic cells lead a solitary life—some as hunters (the *protozoa*), some as photosynthesizers (the unicellular *algae*), some as scavengers (the unicellular fungi, or *yeasts*). **Figure 1–41** conveys something of the variety of forms of these single-celled eucaryotes, or *protists*. The anatomy of protozoa,

Figure 1–40 **Genetic control of the program of multicellular development.** The role of a regulatory gene is demonstrated in the snapdragon *Antirrhinum*. In this example, a mutation in a single gene coding for a regulatory protein causes leafy shoots to develop in place of flowers: because a regulatory protein has been changed, the cells adopt characters that would be appropriate to a different location in the normal plant. The mutant is on the left, the normal plant on the right. (Courtesy of Enrico Coen and Rosemary Carpenter.)

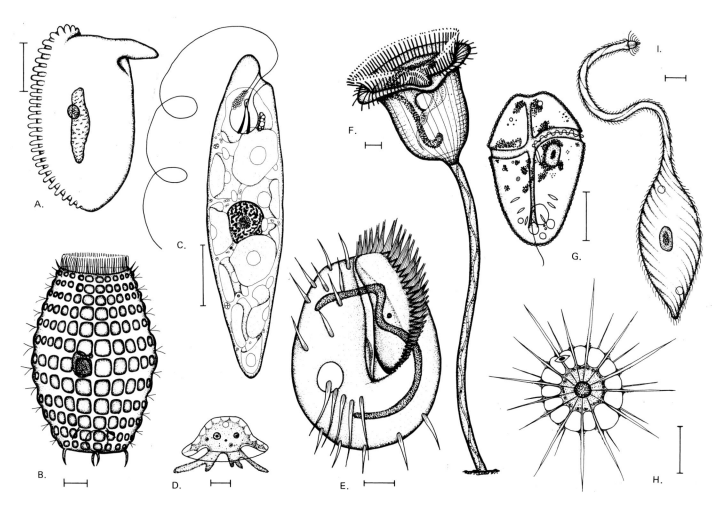

Figure 1–41 **An assortment of protists: a small sample of an extremely diverse class of organisms.** The drawings are done to different scales, but in each case the scale bar represents 10 μm. The organisms in (A), (B), (E), (F), and (I) are ciliates; (C) is a euglenoid; (D) is an amoeba; (G) is a dinoflagellate; (H) is a heliozoan. (From M.A. Sleigh, Biology of Protozoa. Cambridge, UK: Cambridge University Press, 1973.)

especially, is often elaborate and includes such structures as sensory bristles, photoreceptors, sinuously beating cilia, leglike appendages, mouth parts, stinging darts, and musclelike contractile bundles. Although they are single cells, protozoa can be as intricate, as versatile, and as complex in their behavior as many multicellular organisms (see Figure 1–32). <ATGG> <TCGC>

In terms of their ancestry and DNA sequences, protists are far more diverse than the multicellular animals, plants, and fungi, which arose as three comparatively late branches of the eucaryotic pedigree (see Figure 1–21). As with procaryotes, humans have tended to neglect the protists because they are microscopic. Only now, with the help of genome analysis, are we beginning to understand their positions in the tree of life, and to put into context the glimpses these strange creatures offer us of our distant evolutionary past.

A Yeast Serves as a Minimal Model Eucaryote

The molecular and genetic complexity of eucaryotes is daunting. Even more than for procaryotes, biologists need to concentrate their limited resources on a few selected model organisms to fathom this complexity.

To analyze the internal workings of the eucaryotic cell, without the additional problems of multicellular development, it makes sense to use a species that is unicellular and as simple as possible. The popular choice for this role of minimal model eucaryote has been the yeast *Saccharomyces cerevisiae* (**Figure 1–42**)—the same species that is used by brewers of beer and bakers of bread.

S. cerevisiae is a small, single-celled member of the kingdom of fungi and thus, according to modern views, at least as closely related to animals as it is to plants. It is robust and easy to grow in a simple nutrient medium. Like other fungi, it has a tough cell wall, is relatively immobile, and possesses mitochondria but not chloroplasts. When nutrients are plentiful, it grows and divides almost as

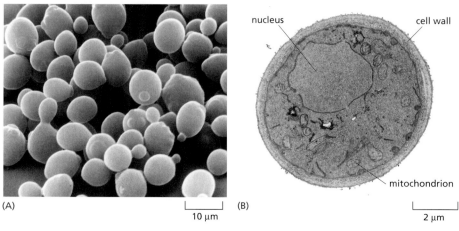

nucleus

cell wall

mitochondrion

(A)

(B)

10 μm

2 μm

Figure 1–42 **The yeast *Saccharomyces cerevisiae*.** (A) A scanning electron micrograph of a cluster of the cells. This species is also known as budding yeast; it proliferates by forming a protrusion or bud that enlarges and then separates from the rest of the original cell. Many cells with buds are visible in this micrograph. (B) A transmission electron micrograph of a cross section of a yeast cell, showing its nucleus, mitochondrion, and thick cell wall. (A, courtesy of Ira Herskowitz and Eric Schabatach.)

rapidly as a bacterium. It can reproduce either vegetatively (that is, by simple cell division), or sexually: two yeast cells that are *haploid* (possessing a single copy of the genome) can fuse to create a cell that is *diploid* (containing a double genome); and the diploid cell can undergo *meiosis* (a reduction division) to produce cells that are once again haploid (**Figure 1–43**). In contrast with higher plants and animals, the yeast can divide indefinitely in either the haploid or the diploid state, and the process leading from the one state to the other can be induced at will by changing the growth conditions.

In addition to these features, the yeast has a further property that makes it a convenient organism for genetic studies: its genome, by eucaryotic standards, is exceptionally small. Nevertheless, it suffices for all the basic tasks that every eucaryotic cell must perform. As we shall see later in this book, studies on yeasts (using both *S. cerevisiae* and other species) have provided a key to many crucial processes, including the eucaryotic cell-division cycle—the critical chain of events by which the nucleus and all the other components of a cell are duplicated and parceled out to create two daughter cells from one. The control system that governs this process has been so well conserved over the course of evolution that many of its components can function interchangeably in yeast and human cells: if a mutant yeast lacking an essential yeast cell-division-cycle gene is supplied with a copy of the homologous cell-division-cycle gene from a human, the yeast is cured of its defect and becomes able to divide normally.

The Expression Levels of All The Genes of An Organism Can Be Monitored Simultaneously

The complete genome sequence of *S. cerevisiae*, determined in 1997, consists of approximately 13,117,000 nucleotide pairs, including the small contribution (78,520 nucleotide pairs) of the mitochondrial DNA. This total is only about 2.5 times as much DNA as there is in *E. coli*, and it codes for only 1.5 times as many distinct proteins (about 6300 in all). The way of life of *S. cerevisiae* is similar in many ways to that of a bacterium, and it seems that this yeast has likewise been subject to selection pressures that have kept its genome compact.

Knowledge of the complete genome sequence of any organism—be it a yeast or a human—opens up new perspectives on the workings of the cell: things that once seemed impossibly complex now seem within our grasp. Using techniques

Figure 1–43 **The reproductive cycles of the yeast *S. cerevisiae*.** Depending on environmental conditions and on details of the genotype, cells of this species can exist in either a diploid (*2n*) state, with a double chromosome set, or a haploid (*n*) state, with a single chromosome set. The diploid form can either proliferate by ordinary cell-division cycles or undergo meiosis to produce haploid cells. The haploid form can either proliferate by ordinary cell-division cycles or undergo sexual fusion with another haploid cell to become diploid. Meiosis is triggered by starvation and gives rise to spores—haploid cells in a dormant state, resistant to harsh environmental conditions.

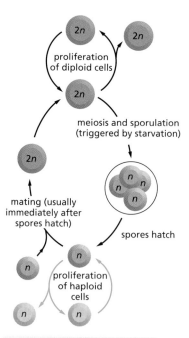

proliferation of diploid cells

meiosis and sporulation (triggered by starvation)

mating (usually immediately after spores hatch)

spores hatch

proliferation of haploid cells

BUDDING YEAST LIFE CYCLE

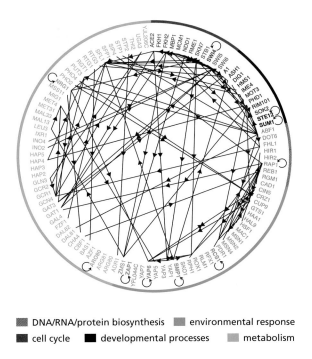

DNA/RNA/protein biosynthesis ■ environmental response
cell cycle ■ developmental processes ■ metabolism

Figure 1–44 **The network of interactions between gene regulatory proteins and the genes that code for them in a yeast cell.** Results are shown for 106 out of the total of 141 gene regulatory proteins in *Saccharomyces cerevisiae*. Each protein in the set was tested for its ability to bind to the regulatory DNA of each of the genes coding for this set of proteins. In the diagram, the genes are arranged in a circle, and an arrow pointing from gene A to gene B means that the protein encoded by A binds to the regulatory DNA of B, and therefore presumably regulates the expression of B. Small circles with arrowheads indicate genes whose products directly regulate their own expression. Genes governing different aspects of cell behavior are shown in different colors. For a multicellular plant or animal, the number of gene regulatory proteins is about 10 times greater, and the amount of regulatory DNA perhaps 100 times greater, so that the corresponding diagram would be vastly more complex. (From T.I. Lee et al., *Science* 298:799–804, 2002. With permission from AAAS.)

to be described in Chapter 8, it is now possible, for example, to monitor, simultaneously, the amount of mRNA transcript that is produced from every gene in the yeast genome under any chosen conditions, and to see how this whole pattern of gene activity changes when conditions change. The analysis can be repeated with mRNA prepared from mutants lacking a chosen gene—any gene that we care to test. In principle, this approach provides a way to reveal the entire system of control relationships that govern gene expression—not only in yeast cells, but in any organism whose genome sequence is known.

To Make Sense of Cells, We Need Mathematics, Computers, and Quantitative Information

Through methods such as these, exploiting our knowledge of complete genome sequences, we can list the genes and proteins in a cell and begin to depict the web of interactions between them (**Figure 1–44**). But how are we to turn all this information into an understanding of how cells work? Even for a single cell type belonging to a single species of organism, the current deluge of data seems overwhelming. The sort of informal reasoning on which biologists usually rely seems totally inadequate in the face of such complexity. In fact, the difficulty is more than just a matter of information overload. Biological systems are, for example, full of feedback loops, and the behavior of even the simplest of systems with feedback is remarkably difficult to predict by intuition alone (**Figure 1–45**); small

Figure 1–45 **A very simple gene regulatory circuit—a single gene regulating its own expression by the binding of its protein product to its own regulatory DNA.** Simple schematic diagrams such as this are often used to summarize what we know (as in Figure 1–44), but they leave many questions unanswered. When the protein binds, does it inhibit or stimulate transcription? How steeply does the transcription rate depend on the protein concentration? How long, on average, does a molecule of the protein remain bound to the DNA? How long does it take to make each molecule of mRNA or protein, and how quickly does each type of molecule get degraded? Mathematical modeling shows that we need quantitative answers to all these and other questions before we can predict the behavior of even this single-gene system. For different parameter values, the system may settle to a unique steady state; or it may behave as a switch, capable of existing in one or other of a set of alternative states; or it may oscillate; or it may show large random fluctuations.

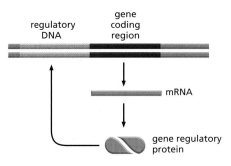

changes in parameters can cause radical changes in outcome. To go from a circuit diagram to a prediction of the behavior of the system, we need detailed
quantitative information, and to draw deductions from that information we
need mathematics and computers.

These tools for quantitative reasoning are essential, but they are not all-
powerful. You might think that, knowing how each protein influences each other
protein, and how the expression of each gene is regulated by the products of others, we should soon be able to calculate how the cell as a whole will behave, just
as an astronomer can calculate the orbits of the planets, or a chemical engineer
can calculate the flows through a chemical plant. But any attempt to perform
this feat for an entire living cell rapidly reveals the limits of our present state of
knowledge. The information we have, plentiful as it is, is full of gaps and uncertainties. Moreover, it is largely qualitative rather than quantitative. Most often,
cell biologists studying the cell's control systems sum up their knowledge in simple schematic diagrams—this book is full of them—rather than in numbers,
graphs, and differential equations. To progress from qualitative descriptions and
intuitive reasoning to quantitative descriptions and mathematical deduction is
one of the biggest challenges for contemporary cell biology. So far, the challenge
has been met only for a few very simple fragments of the machinery of living
cells—subsystems involving a handful of different proteins, or two or three
cross-regulatory genes, where theory and experiment can go closely hand in
hand. We shall discuss some of these examples later in the book.

Arabidopsis Has Been Chosen Out of 300,000 Species As a Model Plant

The large multicellular organisms that we see around us—the flowers and trees
and animals—seem fantastically varied, but they are much closer to one another
in their evolutionary origins, and more similar in their basic cell biology, than
the great host of microscopic single-celled organisms. Thus, while bacteria and
eucaryotes are separated by more than 3000 million years of divergent evolution, vertebrates and insects are separated by about 700 million years, fish and
mammals by about 450 million years, and the different species of flowering
plants by only about 150 million years.

Because of the close evolutionary relationship between all flowering plants,
we can, once again, get insight into the cell and molecular biology of this whole
class of organisms by focusing on just one or a few species for detailed analysis.
Out of the several hundred thousand species of flowering plants on Earth today,
molecular biologists have chosen to concentrate their efforts on a small weed, the
common Thale cress *Arabidopsis thaliana* (**Figure 1–46**), which can be grown
indoors in large numbers, and produces thousands of offspring per plant after
8–10 weeks. *Arabidopsis* has a genome of approximately 140 million nucleotide
pairs, about 11 times as much as yeast, and its complete sequence is known.

The World of Animal Cells Is Represented By a Worm, a Fly, a Mouse, and a Human

Multicellular animals account for the majority of all named species of living
organisms, and for the largest part of the biological research effort. Four species
have emerged as the foremost model organisms for molecular genetic studies. In
order of increasing size, they are the nematode worm *Caenorhabditis elegans*,
the fly *Drosophila melanogaster*, the mouse *Mus musculus*, and the human,
Homo sapiens. Each of these has had its genome sequenced.

Caenorhabditis elegans (**Figure 1–47**) is a small, harmless relative of the eel-
worm that attacks crops. With a life cycle of only a few days, an ability to survive
in a freezer indefinitely in a state of suspended animation, a simple body plan,
and an unusual life cycle that is well suited for genetic studies (described in
Chapter 23), it is an ideal model organism. *C. elegans* develops with clockwork
precision from a fertilized egg cell into an adult worm with exactly 959 body cells

Figure 1–46 *Arabidopsis thaliana,* **the plant chosen as the primary model for studying plant molecular genetics.** (Courtesy of Toni Hayden and the John Innes Foundation.)

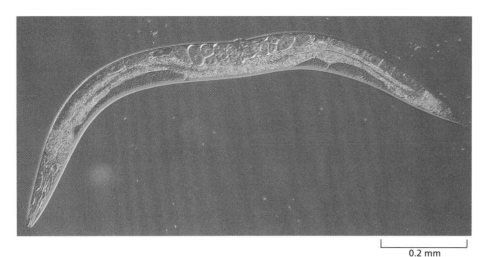

Figure 1–47 *Caenorhabditis elegans,* **the first multicellular organism to have its complete genome sequence determined.** This small nematode, about 1 mm long, lives in the soil. Most individuals are hermaphrodites, producing both eggs and sperm. The animal is viewed here using interference contrast optics, showing up the boundaries of the tissues in bright colors; the animal itself is not colored when viewed with ordinary lighting. (Courtesy of Ian Hope.)

0.2 mm

(plus a variable number of egg and sperm cells)—an unusual degree of regularity for an animal. We now have a minutely detailed description of the sequence of events by which this occurs, as the cells divide, move, and change their characters according to strict and predictable rules. The genome of 97 million nucleotide pairs codes for about 19,000 proteins, and many mutants and other tools are available for the testing of gene functions. Although the worm has a body plan very different from our own, the conservation of biological mechanisms has been sufficient for the worm to be a model for many of the developmental and cell-biological processes that occur in the human body. Studies of the worm help us to understand, for example, the programs of cell division and cell death that determine the numbers of cells in the body—a topic of great importance in developmental biology and cancer research.

Studies in *Drosophila* Provide a Key to Vertebrate Development

The fruitfly *Drosophila melanogaster* (**Figure 1–48**) has been used as a model genetic organism for longer than any other; in fact, the foundations of classical genetics were built to a large extent on studies of this insect. Over 80 years ago, it provided, for example, definitive proof that genes—the abstract units of hereditary information—are carried on chromosomes, concrete physical objects whose behavior had been closely followed in the eucaryotic cell with the light microscope, but whose function was at first unknown. The proof depended on one of the many features that make *Drosophila* peculiarly convenient for genetics—the

Figure 1–48 *Drosophila melanogaster.* Molecular genetic studies on this fly have provided the main key to understanding how all animals develop from a fertilized egg into an adult. (From E.B. Lewis, *Science* 221:cover, 1983. With permission from AAAS.)

giant chromosomes, with characteristic banded appearance, that are visible in some of its cells (**Figure 1–49**). Specific changes in the hereditary information, manifest in families of mutant flies, were found to correlate exactly with the loss or alteration of specific giant-chromosome bands.

In more recent times, *Drosophila*, more than any other organism, has shown us how to trace the chain of cause and effect from the genetic instructions encoded in the chromosomal DNA to the structure of the adult multicellular body. *Drosophila* mutants with body parts strangely misplaced or mispatterned provided the key to the identification and characterization of the genes required to make a properly structured body, with gut, limbs, eyes, and all the other parts in their correct places. Once these *Drosophila* genes were sequenced, the genomes of vertebrates could be scanned for homologs. These were found, and their functions in vertebrates were then tested by analyzing mice in which the genes had been mutated. The results, as we see later in the book, reveal an astonishing degree of similarity in the molecular mechanisms of insect and vertebrate development.

The majority of all named species of living organisms are insects. Even if *Drosophila* had nothing in common with vertebrates, but only with insects, it would still be an important model organism. But if understanding the molecular genetics of vertebrates is the goal, why not simply tackle the problem head-on? Why sidle up to it obliquely, through studies in *Drosophila*?

Drosophila requires only 9 days to progress from a fertilized egg to an adult; it is vastly easier and cheaper to breed than any vertebrate, and its genome is much smaller—about 170 million nucleotide pairs, compared with 3200 million for a human. This genome codes for about 14,000 proteins, and mutants can now be obtained for essentially any gene. But there is also another, deeper reason why genetic mechanisms that are hard to discover in a vertebrate are often readily revealed in the fly. This relates, as we now explain, to the frequency of gene duplication, which is substantially greater in vertebrate genomes than in the fly genome and has probably been crucial in making vertebrates the complex and subtle creatures that they are.

The Vertebrate Genome Is a Product of Repeated Duplication

Almost every gene in the vertebrate genome has paralogs—other genes in the same genome that are unmistakably related and must have arisen by gene duplication. In many cases, a whole cluster of genes is closely related to similar clusters present elsewhere in the genome, suggesting that genes have been duplicated in linked groups rather than as isolated individuals. According to one hypothesis, at an early stage in the evolution of the vertebrates, the entire genome underwent duplication twice in succession, giving rise to four copies of every gene. In some groups of vertebrates, such as fish of the salmon and carp families (including the zebrafish, a popular research animal), it has been suggested that there was yet another duplication, creating an eightfold multiplicity of genes.

The precise course of vertebrate genome evolution remains uncertain, because many further evolutionary changes have occurred since these ancient events. Genes that were once identical have diverged; many of the gene copies have been lost through disruptive mutations; some have undergone further rounds of local duplication; and the genome, in each branch of the vertebrate family tree, has suffered repeated rearrangements, breaking up most of the original gene orderings. Comparison of the gene order in two related organisms, such as the human and the mouse, reveals that—on the time scale of vertebrate evolution—chromosomes frequently fuse and fragment to move large blocks of DNA sequence around. Indeed, it is possible, as we shall discuss in Chapter 7, that the present state of affairs is the result of many separate duplications of fragments of the genome, rather than duplications of the genome as a whole.

There is, however, no doubt that such whole-genome duplications do occur from time to time in evolution, for we can see recent instances in which duplicated chromosome sets are still clearly identifiable as such. The frog

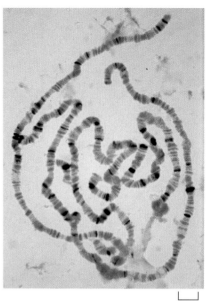

20 μm

Figure 1–49 Giant chromosomes from salivary gland cells of *Drosophila*. Because many rounds of DNA replication have occurred without an intervening cell division, each of the chromosomes in these unusual cells contains over 1000 identical DNA molecules, all aligned in register. This makes them easy to see in the light microscope, where they display a characteristic and reproducible banding pattern. Specific bands can be identified as the locations of specific genes: a mutant fly with a region of the banding pattern missing shows a phenotype reflecting loss of the genes in that region. Genes that are being transcribed at a high rate correspond to bands with a "puffed" appearance. The bands stained dark brown in the micrograph are sites where a particular regulatory protein is bound to the DNA. (Courtesy of B. Zink and R. Paro, from R. Paro, *Trends Genet.* 6:416–421, 1990. With permission from Elsevier.)

Figure 1–50 **Two species of the frog genus *Xenopus*.** *X. tropicalis*, above, has an ordinary diploid genome; *X. laevis*, below, has twice as much DNA per cell. From the banding patterns of their chromosomes and the arrangement of genes along them, as well as from comparisons of gene sequences, it is clear that the large-genome species have evolved through duplications of the whole genome. These duplications are thought to have occurred in the aftermath of matings between frogs of slightly divergent *Xenopus* species. (Courtesy of E. Amaya, M. Offield and R. Grainger, *Trends Genet.* 14:253–255, 1998. With permission from Elsevier.)

genus *Xenopus*, for example, comprises a set of closely similar species related to one another by repeated duplications or triplications of the whole genome. Among these frogs are *X. tropicalis*, with an ordinary diploid genome; the common laboratory species *X. laevis*, with a duplicated genome and twice as much DNA per cell; and *X. ruwenzoriensis*, with a sixfold reduplication of the original genome and six times as much DNA per cell (108 chromosomes, compared with 36 in *X. laevis*, for example). These species are estimated to have diverged from one another within the past 120 million years (**Figure 1–50**).

Genetic Redundancy Is a Problem for Geneticists, But It Creates Opportunities for Evolving Organisms

Whatever the details of the evolutionary history, it is clear that most genes in the vertebrate genome exist in several versions that were once identical. The related genes often remain functionally interchangeable for many purposes. This phenomenon is called **genetic redundancy**. For the scientist struggling to discover all the genes involved in some particular process, it complicates the task. If gene A is mutated and no effect is seen, it cannot be concluded that gene A is functionally irrelevant—it may simply be that this gene normally works in parallel with its relatives, and these suffice for near-normal function even when gene A is defective. In the less repetitive genome of *Drosophila*, where gene duplication is less common, the analysis is more straightforward: single gene functions are revealed directly by the consequences of single-gene mutations (the single-engined plane stops flying when the engine fails).

Genome duplication has clearly allowed the development of more complex life forms; it provides an organism with a cornucopia of spare gene copies, which are free to mutate to serve divergent purposes. While one copy becomes optimized for use in the liver, say, another can become optimized for use in the brain or adapted for a novel purpose. In this way, the additional genes allow for increased complexity and sophistication. As the genes take on divergent functions, they cease to be redundant. Often, however, while the genes acquire individually specialized roles, they also continue to perform some aspects of their original core function in parallel, redundantly. Mutation of a single gene then causes a relatively minor abnormality that reveals only a part of the gene's function (**Figure 1–51**). Families of genes with divergent but partly overlapping functions are a pervasive feature of vertebrate molecular biology, and they are encountered repeatedly in this book.

The Mouse Serves as a Model for Mammals

Mammals have typically three or four times as many genes as *Drosophila*, a genome that is 20 times larger, and millions or billions of times as many cells in their adult bodies. In terms of genome size and function, cell biology, and molecular mechanisms, mammals are nevertheless a highly uniform group of organisms. Even anatomically, the differences among mammals are chiefly a matter of size and proportions; it is hard to think of a human body part that does not have a counterpart in elephants and mice, and vice versa. Evolution plays freely with quantitative features, but it does not readily change the logic of the structure.

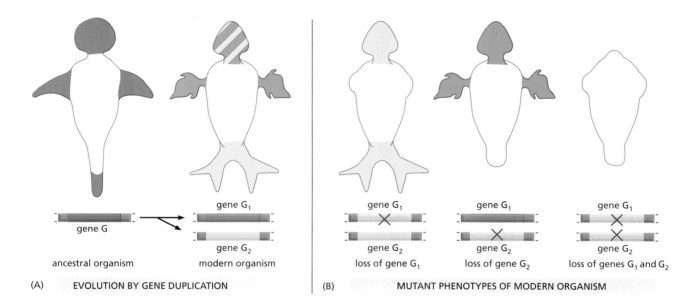

(A)   EVOLUTION BY GENE DUPLICATION

(B)   MUTANT PHENOTYPES OF MODERN ORGANISM

For a more exact measure of how closely mammalian species resemble one another genetically, we can compare the nucleotide sequences of corresponding (orthologous) genes, or the amino acid sequences of the proteins that these genes encode. The results for individual genes and proteins vary widely. But typically, if we line up the amino acid sequence of a human protein with that of the orthologous protein from, say, an elephant, about 85% of the amino acids are identical. A similar comparison between human and bird shows an amino acid identity of about 70%—twice as many differences, because the bird and the mammalian lineages have had twice as long to diverge as those of the elephant and the human (**Figure 1–52**).

The mouse, being small, hardy, and a rapid breeder, has become the foremost model organism for experimental studies of vertebrate molecular genetics. Many naturally occurring mutations are known, often mimicking the effects of corresponding mutations in humans (**Figure 1–53**). Methods have been developed, moreover, to test the function of any chosen mouse gene, or of any non-coding portion of the mouse genome, by artificially creating mutations in it, as we explain later in the book.

One made-to-order mutant mouse can provide a wealth of information for the cell biologist. It reveals the effects of the chosen mutation in a host of different contexts, simultaneously testing the action of the gene in all the different kinds of cells in the body that could in principle be affected.

Humans Report on Their Own Peculiarities

As humans, we have a special interest in the human genome. We want to know the full set of parts from which we are made, and to discover how they work. But even if you were a mouse, preoccupied with the molecular biology of mice, humans would be attractive as model genetic organisms, because of one special property: through medical examinations and self-reporting, we catalog our own genetic (and other) disorders. The human population is enormous, consisting today of some 6 billion individuals, and this self-documenting property means that a huge database of information exists on human mutations. The complete human genome sequence of more than 3 billion nucleotide pairs has now been determined, making it easier than ever before to identify at a molecular level the precise gene responsible for each human mutant characteristic.

By drawing together the insights from humans, mice, flies, worms, yeasts, plants, and bacteria—using gene sequence similarities to map out the correspondences between one model organism and another—we enrich our understanding of them all.

Figure 1–51 The consequences of gene duplication for mutational analyses of gene function. In this hypothetical example, an ancestral multicellular organism has a genome containing a single copy of gene G, which performs its function at several sites in the body, indicated in *green*. (A) Through gene duplication, a modern descendant of the ancestral organism has two copies of gene G, called G_1 and G_2. These have diverged somewhat in their patterns of expression and in their activities at the sites where they are expressed, but they still retain important similarities. At some sites, they are expressed together, and each independently performs the same old function as the ancestral gene G (alternating *green* and *yellow stripes);* at other sites, they are expressed alone and may serve new purposes. (B) Because of a functional overlap, the loss of one of the two genes by mutation *(red cross)* reveals only a part of its role; only the loss of both genes in the double mutant reveals the full range of processes for which these genes are responsible. Analogous principles apply to duplicated genes that operate in the same place (for example, in a single-celled organism) but are called into action together or individually in response to varying circumstances. Thus, gene duplications complicate genetic analyses in all organisms.

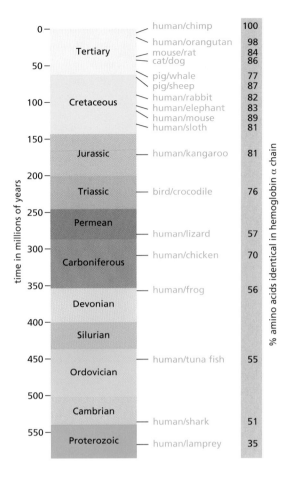

0 —	human/chimp	100
	human/orangutan	98
Tertiary	mouse/rat	84
50 —	cat/dog	86
	pig/whale	77
	pig/sheep	87
100 — Cretaceous	human/rabbit	82
	human/elephant	83
	human/mouse	89
150 —	human/sloth	81
Jurassic	human/kangaroo	81
200 —		
Triassic	bird/crocodile	76
250 — Permean		
	human/lizard	57
300 — Carboniferous	human/chicken	70
350 —	human/frog	56
Devonian		
400 —		
Silurian		
450 —	human/tuna fish	55
Ordovician		
500 —		
Cambrian		
	human/shark	51
550 — Proterozoic	human/lamprey	35

time in millions of years

% amino acids identical in hemoglobin α chain

Figure 1–52 **Times of divergence of different vertebrates.** The scale on the left shows the estimated date and geological era of the last common ancestor of each specified pair of animals. Each time estimate is based on comparisons of the amino acid sequences of orthologous proteins; the longer a pair of animals have had to evolve independently, the smaller the percentage of amino acids that remain identical. Data from many different classes of proteins have been averaged to arrive at the final estimates, and the time scale has been calibrated to match the fossil evidence that the last common ancestor of mammals and birds lived 310 million years ago. The figures on the right give data on sequence divergence for one particular protein (chosen arbitrarily)—the α chain of hemoglobin. Note that although there is a clear general trend of increasing divergence with increasing time for this protein, there are also some irregularities. These reflect the randomness within the evolutionary process and, probably, the action of natural selection driving especially rapid changes of hemoglobin sequence in some organisms that experienced special physiological demands. On average, within any particular evolutionary lineage, hemoglobins accumulate changes at a rate of about 6 altered amino acids per 100 amino acids every 100 million years. Some proteins, subject to stricter functional constraints, evolve much more slowly than this, others as much as 5 times faster. All this gives rise to substantial uncertainties in estimates of divergence times, and some experts believe that the major groups of mammals diverged from one another as much as 60 million years more recently than shown here. (Adapted from S. Kumar and S.B. Hedges, *Nature* 392:917–920, 1998. With permission from Macmillan Publishers Ltd.)

We Are All Different in Detail

What precisely do we mean when we speak of *the* human genome? Whose genome? On average, any two people taken at random differ in about one or two in every 1000 nucleotide pairs in their DNA sequence. The Human Genome Project has arbitrarily selected DNA from a small number of anonymous individuals for sequencing. The human genome—the genome of the human species—is, properly speaking, a more complex thing, embracing the entire pool of variant genes that are found in the human population and continually exchanged and reassorted in the course of sexual reproduction. Ultimately, we can hope to document this variation too. Knowledge of it will help us understand, for example, why some people are prone to one disease, others to another; why some respond well to a drug, others badly. It will also provide new clues to our history—the population movements and minglings of our ancestors, the infections they suffered, the diets they ate. All these things leave traces in the variant forms of genes that have survived in human communities.

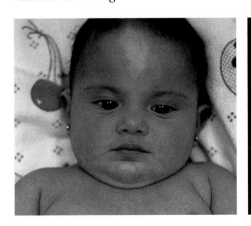

Figure 1–53 **Human and mouse: similar genes and similar development.** The human baby and the mouse shown here have similar white patches on their foreheads because both have mutations in the same gene (called *Kit*), required for the development and maintenance of pigment cells. (Courtesy of R.A. Fleischman.)

Knowledge and understanding bring the power to intervene—with humans, to avoid or prevent disease; with plants, to create better crops; with bacteria, to turn them to our own uses. All these biological enterprises are linked, because the genetic information of all living organisms is written in the same language. The new-found ability of molecular biologists to read and decipher this language has already begun to transform our relationship to the living world. The account of cell biology in the subsequent chapters will, we hope, prepare you to understand, and possibly to contribute to, the great scientific adventure of the twenty-first century.

Summary

Eucaryotic cells, by definition, keep their DNA in a separate membrane-enclosed compartment, the nucleus. They have, in addition, a cytoskeleton for support and movement, elaborate intracellular compartments for digestion and secretion, the capacity (in many species) to engulf other cells, and a metabolism that depends on the oxidation of organic molecules by mitochondria. These properties suggest that eucaryotes may have originated as predators on other cells. Mitochondria—and, in plants, chloroplasts—contain their own genetic material, and evidently evolved from bacteria that were taken up into the cytoplasm of the eucaryotic cell and survived as symbionts. Eucaryotic cells have typically 3–30 times as many genes as procaryotes, and often thousands of times more noncoding DNA. The noncoding DNA allows for complex regulation of gene expression, as required for the construction of complex multicellular organisms. Many eucaryotes are, however, unicellular—among them the yeast Saccharomyces cerevisiae, which serves as a simple model organism for eucaryotic cell biology, revealing the molecular basis of conserved fundamental processes such as the eucaryotic cell division cycle. A small number of other organisms have been chosen as primary models for multicellular plants and animals, and the sequencing of their entire genomes has opened the way to systematic and comprehensive analysis of gene functions, gene regulation, and genetic diversity. As a result of gene duplications during vertebrate evolution, vertebrate genomes contain multiple closely related homologs of most genes. This genetic redundancy has allowed diversification and specialization of genes for new purposes, but it also makes gene functions harder to decipher. There is less genetic redundancy in the nematode Caenorhabditis elegans *and the fly* Drosophila melanogaster, *which have thus played a key part in revealing universal genetic mechanisms of animal development.*

PROBLEMS

Which statements are true? Explain why or why not.

1–1   The human hemoglobin genes, which are arranged in two clusters on two chromosomes, provide a good example of an orthologous set of genes.

1–2   Horizontal gene transfer is more prevalent in single-celled organisms than in multicellular organisms.

1–3   Most of the DNA sequences in a bacterial genome code for proteins, whereas most of the sequences in the human genome do not.

Discuss the following problems.

1–4   Since it was deciphered four decades ago, some have claimed that the genetic code must be a frozen accident, while others have argued that it was shaped by natural selection. A striking feature of the genetic code is its inherent resistance to the effects of mutation. For example, a change in the third position of a codon often specifies the same amino acid or one with similar chemical properties. The natural code resists mutation more effectively (is less susceptible to error) than most other possible versions, as illustrated in **Figure Q1–1**. Only one in a million computer-generated "random" codes is more error-resistant than the natural genetic code. Does the extraordinary mutation resistance of the genetic code argue in favor of its origin as a frozen accident or as a result of natural selection? Explain your reasoning.

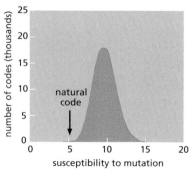

Figure Q1–1 Susceptibility of the natural code relative to millions of computer-generated codes (Problem 1–4). Susceptibility measures the average change in amino acid properties caused by random mutations. A small value indicates that mutations tend to cause minor changes. (Data courtesy of Steve Freeland.)

1–5   You have begun to characterize a sample obtained from the depths of the oceans on Europa, one of Jupiter's moons. Much to your surprise, the sample contains a life-form that grows well in a rich broth. Your preliminary analysis shows that it is cellular and contains DNA, RNA, and protein. When you show your results to a colleague, she suggests that your sample was contaminated with an organism from Earth. What approaches might you try to distinguish between contamination and a novel cellular life-form based on DNA, RNA, and protein?

1–6 It is not so difficult to imagine what it means to feed on the organic molecules that living things produce. That is, after all, what we do. But what does it mean to "feed" on sunlight, as phototrophs do? Or, even stranger, to "feed" on rocks, as lithotrophs do? Where is the "food," for example, in the mixture of chemicals (H_2S, H_2, CO, Mn^+, Fe^{2+}, Ni^{2+}, CH_4, and NH_4^+) spewed forth from a hydrothermal vent?

1–7 How many possible different trees (branching patterns) can be drawn for eubacteria, archaea, and eucaryotes, assuming that they all arose from a common ancestor?

1–8 The genes for ribosomal RNA are highly conserved (relatively few sequence changes) in all organisms on Earth; thus, they have evolved very slowly over time. Were ribosomal RNA genes "born" perfect?

1–9 Genes participating in informational processes such as replication, transcription, and translation are transferred between species much less often than are genes involved in metabolism. The basis for this inequality is unclear at present, but one suggestion is that it relates to the underlying complexity. Informational processes tend to involve large aggregates of different gene products, whereas metabolic reactions are usually catalyzed by enzymes composed of a single protein. Why would the complexity of the underlying process—informational or metabolic—have any effect on the rate of horizontal gene transfer?

1–10 The process of gene transfer from the mitochondrial to the nuclear genome can be analyzed in plants. The respiratory gene *Cox2*, which encodes subunit 2 of cytochrome oxidase, was functionally transferred to the nucleus during flowering plant evolution. Extensive analyses of plant genera have pinpointed the time of appearance of the nuclear form of the gene and identified several likely intermediates in the ultimate loss from the mitochondrial genome. A summary of *Cox2* gene distributions between mitochondria and nuclei, along with data on their transcription, is shown in a phylogenetic context in **Figure Q1–2**.
A. Assuming that transfer of the mitochondrial gene to the nucleus occurred only once (an assumption supported by the structures of the nuclear genes), indicate the point in the phylogenetic tree where the transfer occurred.
B. Are there any examples of genera in which the transferred gene and the mitochondrial gene both appear functional? Indicate them.
C. What is the minimal number of times that the mitochondrial gene has been inactivated or lost? Indicate those events on the phylogenetic tree.
D. What is the minimal number of times that the nuclear gene has been inactivated or lost? Indicate those events on the phylogenetic tree.
E. Based on this information, propose a general scheme for transfer of mitochondrial genes to the nuclear genome.

1–11 When plant hemoglobin genes were first discovered in legumes, it was so surprising to find a gene typical of animal blood that it was hypothesized that the plant gene arose

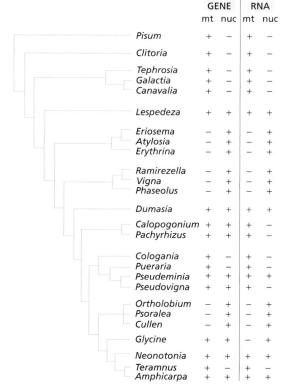

	GENE		RNA	
	mt	nuc	mt	nuc
Pisum	+	–	+	–
Clitoria	+	–	+	–
Tephrosia	+	–	+	–
Galactia	+	–	+	–
Canavalia	+	–	+	–
Lespedeza	+	+	+	+
Eriosema	–	+	–	+
Atylosia	–	+	–	+
Erythrina	–	+	–	+
Ramirezella	–	+	–	+
Vigna	–	+	–	+
Phaseolus	–	+	–	+
Dumasia	+	+	+	+
Calopogonium	+	+	+	–
Pachyrhizus	+	+	+	–
Cologania	+	–	+	–
Pueraria	+	–	+	–
Pseudeminia	+	+	+	+
Pseudovigna	+	+	+	–
Ortholobium	–	+	–	+
Psoralea	–	+	–	+
Cullen	–	+	–	+
Glycine	+	+	–	+
Neonotonia	+	+	+	+
Teramnus	+	–	+	–
Amphicarpa	+	+	+	+

Figure Q1–2 Summary of *Cox2* gene distribution and transcript data in a phylogenetic context (Problem 1–10). The presence of the intact gene or a functional transcript is indicated by (+); the absence of the intact gene or a functional transcript is indicated by (–). mt, mitochondria; nuc, nuclei.

by horizontal transfer from an animal. Many more hemoglobin genes have now been sequenced, and a phylogenetic tree based on some of these sequences is shown in **Figure Q1–3**.
A. Does this tree support or refute the hypothesis that the plant hemoglobins arose by horizontal gene transfer?
B. Supposing that the plant hemoglobin genes were originally derived from a parasitic nematode, for example, what would you expect the phylogenetic tree to look like?

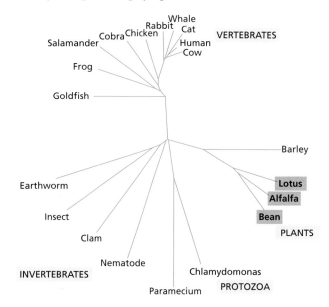

Figure Q1–3 Phylogenetic tree for hemoglobin genes from a variety of species (Problem 1–11). The legumes are highlighted in red.

1–12 Rates of evolution appear to vary in different lineages. For example, the rate of evolution in the rat lineage is significantly higher than in the human lineage. These rate differences are apparent whether one looks at changes in protein sequences that are subject to selective pressure or at changes in noncoding nucleotide sequences, which are not under obvious selection pressure. Can you offer one or more possible explanations for the slower rate of evolutionary change in the human lineage versus the rat lineage?

REFERENCES

General

Alberts B, Bray D, Hopkin K et al (2004) Essential Cell Biology, 2nd ed. New York: Garland Science.

Barton NH, Briggs DEG, Eisen JA et al (2007) Evolution. Cold Spring Harbor, NY: Cold Spring Harbor Laboratory Press.

Darwin C (1859) On the Origin of Species. London: Murray.

Graur D & Li W-H (1999) Fundamentals of Molecular Evolution, 2nd ed. Sunderland, MA: Sinauer Associates.

Madigan MT & Martinko JM (2005) Brock's Biology of Microorganisms, 11th ed. Englewood Cliffs, NJ: Prentice Hall.

Margulis L & Schwartz KV (1998) Five Kingdoms: An Illustrated Guide to the Phyla of Life on Earth, 3rd ed. New York: Freeman.

Watson JD, Baker TA, Bell SP et al (2007) Molecular Biology of the Gene, 6th ed. Menlo Park, CA: Benjamin-Cummings.

The Universal Features of Cells on Earth

Andersson SGE (2006) The bacterial world gets smaller. *Science* 314:259–260.

Brenner S, Jacob F & Meselson M (1961) An unstable intermediate carrying information from genes to ribosomes for protein synthesis. *Nature* 190:576–581.

Fraser CM, Gocayne JD, White O et al (1995) The minimal gene complement of *Mycoplasma genitalium. Science* 270:397–403.

Harris JK, Kelley ST, Spiegelman et al (2003) The genetic core of the universal ancestor. *Genome Res* 13:407–413.

Koonin EV (2005) Orthologs, paralogs, and evolutionary genomics. *Annu Rev Genet* 39:309–338.

Watson JD & Crick FHC (1953) Molecular structure of nucleic acids. A structure for deoxyribose nucleic acid. *Nature* 171:737–738.

Yusupov MM, Yusupova GZ, Baucom A et al (2001) Crystal structure of the ribosome at 5.5 Å resolution. *Science* 292:883–896.

The Diversity of Genomes and the Tree of Life

Blattner FR, Plunkett G, Bloch CA et al (1997) The complete genome sequence of *Escherichia coli* K-12. *Science* 277:1453–1474.

Boucher Y, Douady CJ, Papke RT et al (2003) Lateral gene transfer and the origins of prokaryotic groups. *Annu Rev Genet* 37:283–328.

Cole ST, Brosch R, Parkhill J et al (1998) Deciphering the biology of *Mycobacterium tuberculosis* from the complete genome sequence. *Nature* 393:537–544.

Dixon B (1994) Power Unseen: How Microbes Rule the World. Oxford: Freeman.

Kerr RA (1997) Life goes to extremes in the deep earth—and elsewhere? *Science* 276:703–704.

Lee TI, Rinaldi NJ, Robert F et al (2002) Transcriptional regulatory networks in *Saccharomyces cerevisiae. Science* 298:799–804.

Olsen GJ & Woese CR (1997) Archaeal genomics: an overview. *Cell* 89:991–994.

Pace NR (1997) A molecular view of microbial diversity and the biosphere. *Science* 276:734–740.

Woese C (1998) The universal ancestor. *Proc Natl Acad Sci USA* 95:6854–6859.

Genetic Information in Eucaryotes

Adams MD, Celniker SE, Holt RA et al (2000) The genome sequence of *Drosophila melanogaster. Science* 287:2185–2195.

Andersson SG, Zomorodipour A, Andersson JO et al (1998) The genome sequence of *Rickettsia prowazekii* and the origin of mitochondria. *Nature* 396:133–140.

The *Arabidopsis* Initiative (2000) Analysis of the genome sequence of the flowering plant *Arabidopsis thaliana. Nature* 408:796–815.

Carroll SB, Grenier JK & Weatherbee SD (2005) From DNA to Diversity: Molecular Genetics and the Evolution of Animal Design, 2nd ed. Maldon, MA: Blackwell Science.

de Duve C (2007) The origin of eukaryotes: a reappraisal. *Nature Rev Genet* 8:395–403.

Delsuc F, Brinkmann H & Philippe H (2005) Phylogenomics and the reconstruction of the tree of life. *Nature Rev Genet* 6:361–375.

DeRisi JL, Iyer VR & Brown PO (1997) Exploring the metabolic and genetic control of gene expression on a genomic scale. *Science* 278:680–686.

Gabriel SB, Schaffner SF, Nguyen H et al (2002) The structure of haplotype blocks in the human genome. *Science* 296:2225–2229.

Goffeau A, Barrell BG, Bussey H et al (1996) Life with 6000 genes. *Science* 274:546–567.

International Human Genome Sequencing Consortium (2001) Initial sequencing and analysis of the human genome. *Nature* 409:860–921.

Kellis M, Birren BW & Lander ES (2004) Proof and evolutionary analysis of ancient genome duplication in the yeast *Saccharomyces cerevisiae. Nature* 428:617–624.

Lynch M & Conery JS (2000) The evolutionary fate and consequences of duplicate genes. *Science* 290:1151–1155.

Mulley J & Holland P (2004) Comparative genomics: Small genome, big insights. *Nature* 431:916–917.

National Center for Biotechnology Information. http://www.ncbi.nlm.nih.gov/

Owens K & King MC (1999) Genomic views of human history. *Science* 286:451–453.

Palmer JD & Delwiche CF (1996) Second-hand chloroplasts and the case of the disappearing nucleus. *Proc Natl Acad Sci USA* 93:7432–7435.

Pennisi E (2004) The birth of the nucleus. *Science* 305:766–768.

Plasterk RH (1999) The year of the worm. *BioEssays* 21:105–109.

Reed FA & Tishkoff SA (2006) African human diversity, origins and migrations. *Curr Opin Genet Dev* 16:597–605.

Rubin GM, Yandell MD, Wortman JR et al (2000) Comparative genomics of the eukaryotes. *Science* 287:2204–2215.

Stillman B & Stewart D (2003) The genome of *Homo sapiens.* (Cold Spring Harbor Symp. Quant. Biol. LXVIII). Cold Spring Harbor, NY: Cold Spring Harbor Laboratory Press.

The *C. elegans* Sequencing Consortium (1998) Genome sequence of the nematode *C. elegans*: a platform for investigating biology. *Science* 282:2012–2018.

Tinsley RC & Kobel HR (eds) (1996) The Biology of *Xenopus.* Oxford: Clarendon Press.

Tyson JJ, Chen KC & Novak B (2003) Sniffers, buzzers, toggles and blinkers: dynamics of regulatory and signaling pathways in the cell. *Curr Opin Cell Biol* 15:221–231.

Venter JC, Adams MD, Myers EW et al (2001) The sequence of the human genome. *Science* 291:1304–1351.

Chapter 2

Cell Chemistry and Biosynthesis

2

It is at first sight difficult to accept the idea that each of the living creatures described in Chapter 1 is merely a chemical system. The incredible diversity of living forms, their seemingly purposeful behavior, and their ability to grow and reproduce appear to set them apart from the world of solids, liquids, and gases that chemistry normally describes. Indeed, until the nineteenth century animals were believed to contain a Vital Force—an "animus"—that was responsible for their distinctive properties.

We now know there is nothing in living organisms that disobeys chemical and physical laws. However, the chemistry of life is special. First, it is based overwhelmingly on carbon compounds, whose study is therefore known as *organic chemistry*. Second, cells are 70 percent water, and life depends largely on chemical reactions that take place in aqueous solution. Third, and most important, cell chemistry is enormously complex: even the simplest cell is vastly more complicated in its chemistry than any other chemical system known. Although cells contain a variety of small carbon-containing molecules, most of the carbon atoms in cells are incorporated into enormous *polymeric molecules*—chains of chemical subunits linked end-to-end. It is the unique properties of these macromolecules that enable cells and organisms to grow and reproduce—as well as to do all the other things that are characteristic of life.

THE CHEMICAL COMPONENTS OF A CELL

Matter is made of combinations of *elements*—substances such as hydrogen or carbon that cannot be broken down or converted into other substances by chemical means. The smallest particle of an element that still retains its distinctive chemical properties is an *atom* (**Figure 2–1**). However, the characteristics of substances other than pure elements—including the materials from which living cells are made—depend on the way their atoms are linked together in groups to form *molecules*. In order to understand how living organisms are built from inanimate matter, therefore, it is crucial to know how all of the chemical bonds that hold atoms together in molecules are formed.

Cells Are Made From a Few Types of Atoms

The **atomic weight** of an atom, or the **molecular weight** of a molecule, is its mass relative to that of a hydrogen atom. This is essentially equal to the number of protons plus neutrons that the atom or molecule contains, since the electrons are much lighter and contribute almost nothing to the total. Thus the major isotope of carbon has an atomic weight of 12 and is symbolized as ^{12}C, whereas an unstable isotope of carbon has an atomic weight of 14 and is written as ^{14}C. The mass of an atom or a molecule is often specified in *daltons*, one dalton being an atomic mass unit approximately equal to the mass of a hydrogen atom.

Atoms are so small that it is hard to imagine their size. An individual carbon atom is roughly 0.2 nm in diameter, so that it would take about 5 million of them, laid out in a straight line, to span a millimeter. One proton or neutron weighs

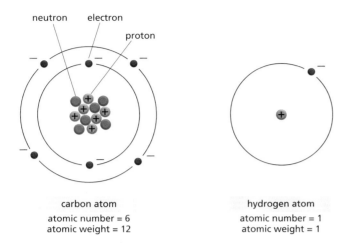

carbon atom
atomic number = 6
atomic weight = 12

hydrogen atom
atomic number = 1
atomic weight = 1

Figure 2–1 Highly schematic representations of an atom of carbon and an atom of hydrogen. The *nucleus* of every atom except hydrogen consists of both positively charged *protons* and electrically neutral *neutrons*. The number of electrons in an atom is equal to its number of protons (the *atomic number*), so that the atom has no net charge. Because it is the electrons that determine the chemical behavior of an atom, all of the atoms of a given element have the same atomic number.

Neutrons are uncharged subatomic particles of essentially the same mass as protons. They contribute to the structural stability of the nucleus—if there are too many or too few, the nucleus may disintegrate by radioactive decay—but they do not alter the chemical properties of the atom. Because of neutrons, an element can exist in several physically distinguishable but chemically identical forms, called *isotopes,* each isotope having a different number of neutrons but the same number of protons. Multiple isotopes of almost all the elements occur naturally, including some that are unstable. For example, while most carbon on Earth exists as the stable isotope carbon 12, with six protons and six neutrons, there are also small amounts of an unstable isotope, the radioactive carbon 14, whose atoms have six protons and eight neutrons. Carbon 14 undergoes radioactive decay at a slow but steady rate. This forms the basis for a technique known as carbon 14 dating, which is used in archaeology to determine the time of origin of organic materials.

The neutrons, protons, and electrons are in reality minute in relation to the atom as a whole; their size is greatly exaggerated here. In addition, the diameter of the nucleus is only about 10^{-4} that of the electron cloud. Finally, although the electrons are shown here as individual particles, in reality their behavior is governed by the laws of quantum mechanics, and there is no way of predicting exactly where an electron is at any given instant of time.

approximately $1/(6 \times 10^{23})$ gram, so one gram of hydrogen contains 6×10^{23} atoms. This huge number (6×10^{23}, called **Avogadro's number**) is the key scale factor describing the relationship between everyday quantities and quantities measured in terms of individual atoms or molecules. If a substance has a molecular weight of X, 6×10^{23} molecules of it will have a mass of X grams. This quantity is called one **mole** of the substance (**Figure 2–2**).

There are 89 naturally occurring elements, each differing from the others in the number of protons and electrons in its atoms. Living organisms, however, are made of only a small selection of these elements, four of which—carbon (C), hydrogen (H), nitrogen (N), and oxygen (O)—make up 96.5% of an organism's weight. This composition differs markedly from that of the nonliving inorganic environment (**Figure 2–3**) and is evidence of a distinctive type of chemistry.

The Outermost Electrons Determine How Atoms Interact

To understand how atoms bond together to form the molecules that make up living organisms, we focus on their electrons. Protons and neutrons are welded tightly to one another in the nucleus and change partners only under extreme conditions—during radioactive decay, for example, or in the interior of the sun or of a nuclear reactor. In living tissues, it is only the electrons of an atom that undergo rearrangements. They form the exterior of an atom and specify the rules of chemistry by which atoms combine to form molecules.

Electrons are in continuous motion around the nucleus, but motions on this submicroscopic scale obey very different laws from those familiar in everyday life. These laws dictate that electrons in an atom can exist only in certain discrete states, called orbitals, and that there is a strict limit to the number of electrons that can be accommodated in an orbital of a given type—a so-called *electron shell*. The electrons closest on average to the positive nucleus are attracted most strongly to it and occupy the innermost, most tightly bound shell. This shell holds a maximum of two electrons. The second shell is farther away from the nucleus, and its electrons are less tightly bound. This second shell holds up to eight electrons. The third shell contains electrons that are even less tightly bound; it also holds up to eight electrons. The fourth and fifth shells can hold 18 electrons each. Atoms with more than four shells are very rare in biological molecules.

The electron arrangement of an atom is most stable when all the electrons are in the most tightly bound states that are possible—that is, when they occupy the innermost shells. Therefore, with certain exceptions in the larger atoms, the electrons of an atom fill the orbitals in order—the first shell before the second, the second before the third, and so on. An atom whose outermost shell is entirely filled with electrons is especially stable and therefore chemically unreactive. Examples are helium with 2 electrons, neon with 2 + 8, and argon with 2 + 8 + 8; these are all inert gases. Hydrogen, by contrast, with only one electron and only a half-filled shell, is highly reactive. Likewise, the other atoms found in living tissues have incomplete outer electron shells and can donate, accept, or share electrons with each other to form both molecules and ions (**Figure 2–4**).

Because an unfilled electron shell is less stable than a filled one, atoms with incomplete outer shells tend to interact with other atoms in a way that causes them to either gain or lose enough electrons to achieve a completed outermost shell. This electron exchange occurs either by transferring electrons from one atom to another or by sharing electrons between two atoms. These two strategies generate two types of **chemical bonds** between atoms: an *ionic bond* is formed when electrons are donated by one atom to another, whereas a *covalent bond* is formed when two atoms share a pair of electrons (**Figure 2–5**). Often, the pair of electrons is shared unequally, with a partial transfer between two atoms that attract electrons differently—one more *electronegative* than the other: this intermediate strategy results in a *polar covalent bond*, as we shall discuss later.

An H atom, which needs only one electron to fill its shell, generally acquires it by electron sharing, forming one covalent bond with another atom; often this bond is polar—meaning that the electrons are shared unequally. The other common elements in living cells—C, N, and O, with an incomplete second shell, and P and S, with an incomplete third shell (see Figure 2–4)—generally share electrons and achieve a filled outer shell of eight electrons by forming several covalent bonds. The number of electrons that an atom must acquire or lose (either by sharing or by transfer) to fill its outer shell is known as its *valence*.

The crucial role of the outer electron shell in determining the chemical properties of an element means that, when the elements are listed in order of their atomic number, there is a periodic recurrence of elements with similar properties: an element with, say, an incomplete second shell containing one electron will behave in much the same way as an element that has filled its second shell

A **mole** is X grams of a substance, where X is its relative molecular mass (molecular weight). A mole will contain 6×10^{23} molecules of the substance.

1 mole of carbon weighs 12 g
1 mole of glucose weighs 180 g
1 mole of sodium chloride weighs 58 g

Molar solutions have a concentration of 1 mole of the substance in 1 liter of solution. A molar solution (denoted as 1 M) of glucose, for example, has 180 g/l, while a millimolar solution (1 mM) has 180 mg/l.

The standard abbreviation for gram is g; the abbreviation for liter is l.

Figure 2–2 **Moles and molar solutions.**

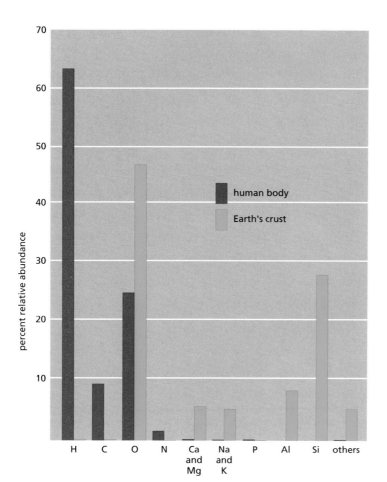

Figure 2–3 **The abundances of some chemical elements in the nonliving world (the Earth's crust) compared with their abundances in the tissues of an animal.** The abundance of each element is expressed as a percentage of the total number of atoms present including water. Thus, because of the abundance of water, more than 60% of the atoms in a living organism are hydrogen atoms. The relative abundance of elements is similar in all living things.

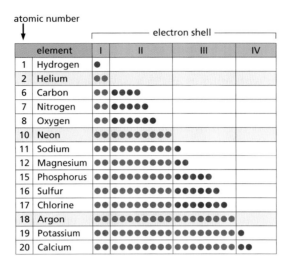

atomic number

electron shell

	element	I	II	III	IV
1	Hydrogen	●			
2	Helium	●●			
6	Carbon	●●	●●●●		
7	Nitrogen	●●	●●●●●		
8	Oxygen	●●	●●●●●●		
10	Neon	●●	●●●●●●●●		
11	Sodium	●●	●●●●●●●●	●	
12	Magnesium	●●	●●●●●●●●	●●	
15	Phosphorus	●●	●●●●●●●●	●●●●●	
16	Sulfur	●●	●●●●●●●●	●●●●●●	
17	Chlorine	●●	●●●●●●●●	●●●●●●●	
18	Argon	●●	●●●●●●●●	●●●●●●●●	
19	Potassium	●●	●●●●●●●●	●●●●●●●●	●
20	Calcium	●●	●●●●●●●●	●●●●●●●●	●●

Figure 2–4 Filled and unfilled electron shells in some common elements. All the elements commonly found in living organisms have unfilled outermost shells *(red)* and can thus participate in chemical reactions with other atoms. For comparison, some elements that have only filled shells *(yellow)* are shown; these are chemically unreactive.

and has an incomplete third shell containing one electron. The metals, for example, have incomplete outer shells with just one or a few electrons, whereas, as we have just seen, the inert gases have full outer shells. This pattern gives rise to the famous *periodic table* of the elements, presented in **Figure 2–6** with the elements found in living organisms highlighted.

Covalent Bonds Form by the Sharing of Electrons

All the characteristics of a cell depend on the molecules it contains. A **molecule** is defined as a cluster of atoms held together by **covalent bonds**; here electrons are shared between atoms to complete the outer shells, rather than being transferred between them. In the simplest possible molecule—a molecule of hydrogen (H_2)—two H atoms, each with a single electron, share two electrons, which is the number required to fill the first shell. These shared electrons form a cloud of negative charge that is densest between the two positively charged nuclei and helps to hold them together, in opposition to the mutual repulsion between like charges that would otherwise force them apart. The attractive and repulsive forces are in balance when the nuclei are separated by a characteristic distance, called the *bond length*.

Another property of any bond—covalent or noncovalent—is its *bond strength*, which is measured by the amount of energy that must be supplied to break that bond. This is often expressed in units of kilocalories per mole (kcal/mole), where a kilocalorie is the amount of energy needed to raise the temperature of one liter

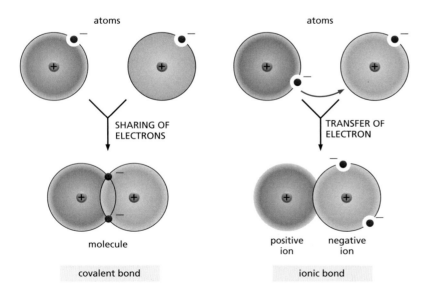

Figure 2–5 Comparison of covalent and ionic bonds. Atoms can attain a more stable arrangement of electrons in their outermost shell by interacting with one another. An ionic bond is formed when electrons are transferred from one atom to the other. A covalent bond is formed when electrons are shared between atoms. The two cases shown represent extremes; often, covalent bonds form with a partial transfer (unequal sharing of electrons), resulting in a polar covalent bond (see Figure 2–43).

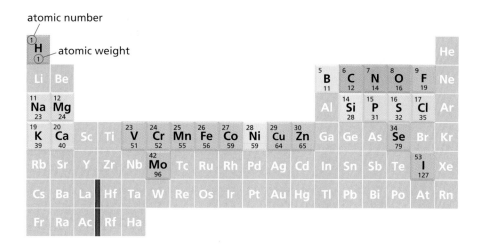

Figure 2–6 **Elements ordered by their atomic number form the periodic table.** Elements fall into groups that show similar properties based on the number of electrons each element possesses in its outer shell. For example, Mg and Ca tend to give away the two electrons in their outer shells; C, N, and O complete their second shells by sharing electrons. The four elements highlighted in *red* constitute 99% of the total number of atoms present in the human body. An additional seven elements, highlighted in *blue,* together represent about 0.9% of the total. Other elements, shown in *green,* are required in trace amounts by humans. It remains unclear whether those elements shown in *yellow* are essential in humans or not. The chemistry of life, it seems, is therefore predominantly the chemistry of lighter elements.

Atomic weights, given by the sum of the protons and neutrons in the atomic nucleus, will vary with the particular isotope of the element. The atomic weights shown here are those of the most common isotope of each element.

of water by one degree Celsius (centigrade). Thus if 1 kilocalorie must be supplied to break 6×10^{23} bonds of a specific type (that is, 1 mole of these bonds), then the strength of that bond is 1 kcal/mole. An equivalent, widely used measure of energy is the kilojoule, which is equal to 0.239 kilocalories.

To understand bond strengths, it is helpful to compare them with the average energies of the impacts that molecules are constantly experiencing from collisions with other molecules in their environment (their thermal, or heat, energy), as well as with other sources of biological energy such as light and glucose oxidation (**Figure 2–7**). Typical covalent bonds are stronger than the thermal energies by a factor of 100, so they resist being pulled apart by thermal motions and are normally broken only during specific chemical reactions with other atoms and molecules. The making and breaking of covalent bonds are violent events, and in living cells they are carefully controlled by highly specific catalysts, called *enzymes*. Noncovalent bonds as a rule are much weaker; we shall see later that they are important in the cell in the many situations where molecules have to associate and dissociate readily to carry out their functions.

Whereas an H atom can form only a single covalent bond, the other common atoms that form covalent bonds in cells—O, N, S, and P, as well as the all-important C atom—can form more than one. The outermost shell of these atoms, as we have seen, can accommodate up to eight electrons, and they form covalent bonds with as many other atoms as necessary to reach this number. Oxygen, with six electrons in its outer shell, is most stable when it acquires an extra two electrons by sharing with other atoms and therefore forms up to two covalent bonds. Nitrogen, with five outer electrons, forms a maximum of three covalent bonds, while carbon, with four outer electrons, forms up to four covalent bonds—thus sharing four pairs of electrons (see Figure 2–4).

When one atom forms covalent bonds with several others, these multiple bonds have definite arrangements in space relative to one another, reflecting the orientations of the orbits of the shared electrons. The covalent bonds of such an atom are therefore characterized by specific bond angles as well as by bond lengths and bond energies (**Figure 2–8**). The four covalent bonds that can form around a carbon atom, for example, are arranged as if pointing to the four corners of a regular tetrahedron. The precise orientation of covalent bonds forms the basis for the three-dimensional geometry of organic molecules.

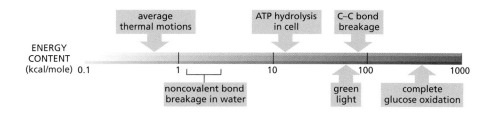

Figure 2–7 **Some energies important for cells.** Note that these energies are compared on a logarithmic scale.

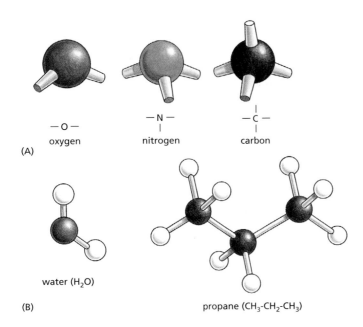

(A)

$-O-$
oxygen

$-N-$
nitrogen

$-C-$
carbon

(B)

water (H_2O)

propane (CH_3-CH_2-CH_3)

Figure 2–8 The geometry of covalent bonds. (A) The spatial arrangement of the covalent bonds that can be formed by oxygen, nitrogen, and carbon. (B) Molecules formed from these atoms have a precise three-dimensional structure, as shown here by ball-and-stick models for water and propane. A structure can be specified by the bond angles and bond lengths for each covalent linkage. The atoms are colored according to the following, generally used convention: H, *white*; C, *black*; O, *red*; N, *blue*.

There Are Different Types of Covalent Bonds

Most covalent bonds involve the sharing of two electrons, one donated by each participating atom; these are called *single bonds*. Some covalent bonds, however, involve the sharing of more than one pair of electrons. Four electrons can be shared, for example, two coming from each participating atom; such a bond is called a *double bond*. Double bonds are shorter and stronger than single bonds and have a characteristic effect on the three-dimensional geometry of molecules containing them. A single covalent bond between two atoms generally allows the rotation of one part of a molecule relative to the other around the bond axis. A double bond prevents such rotation, producing a more rigid and less flexible arrangement of atoms (**Figure 2–9** and Panel 2–1, pp. 106–107).

In some molecules, electrons are shared among three or more atoms, producing bonds that have a hybrid character intermediate between single and double bonds. The highly stable benzene molecule, for example, consists of a ring of six carbon atoms in which the bonding electrons are evenly distributed (although usually depicted as an alternating sequence of single and double bonds, as shown in Panel 2–1).

When the atoms joined by a single covalent bond belong to different elements, the two atoms usually attract the shared electrons to different degrees. Compared with a C atom, for example, O and N atoms attract electrons relatively strongly, whereas an H atom attracts electrons more weakly. By definition, a **polar** structure (in the electrical sense) is one with positive charge concentrated toward one end (the positive pole) and negative charge concentrated toward the other (the negative pole). Covalent bonds in which the electrons are shared unequally in this way are therefore known as *polar covalent bonds* (**Figure 2–10**). For example, the covalent bond between oxygen and hydrogen, –O–H, or between nitrogen and hydrogen, –N–H, is polar, whereas that between carbon and hydrogen, –C–H, has the electrons attracted much more equally by both atoms and is relatively nonpolar.

Polar covalent bonds are extremely important in biology because they create *permanent dipoles* that allow molecules to interact through electrical forces. Any large molecule with many polar groups will have a pattern of partial positive and negative charges on its surface. When such a molecule encounters a second molecule with a complementary set of charges, the two molecules will be attracted to each other by electrostatic interactions that resemble (but are weaker than) the ionic bonds discussed previously.

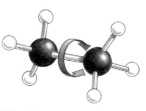

(A) ethane

(B) ethene

Figure 2–9 Carbon–carbon double bonds and single bonds compared. (A) The ethane molecule, with a single covalent bond between the two carbon atoms, illustrates the tetrahedral arrangement of single covalent bonds formed by carbon. One of the CH_3 groups joined by the covalent bond can rotate relative to the other around the bond axis. (B) The double bond between the two carbon atoms in a molecule of ethene (ethylene) alters the bond geometry of the carbon atoms and brings all the atoms into the same plane *(blue)*; the double bond prevents the rotation of one CH_2 group relative to the other.

An Atom Often Behaves as if It Has a Fixed Radius

When a covalent bond forms between two atoms, the sharing of electrons brings the nuclei of these atoms unusually close together. But most of the atoms that are rapidly jostling each other in cells are located in separate molecules. What happens when two such atoms touch?

For simplicity and clarity, atoms and molecules are usually represented schematically—either as a line drawing of the structural formula or as a ball-and-stick model. *Space-filling models,* however, give us a more accurate representation of molecular structure. In these models, a solid envelope represents the radius of the electron cloud at which strong repulsive forces prevent a closer approach of any second, non-bonded atom—the so-called *van der Waals radius* for an atom. This is possible because the amount of repulsion increases very steeply as two such atoms approach each other closely. At slightly greater distances, any two atoms will experience a weak attractive force, known as a *van der Waals attraction*. As a result, there is a distance at which repulsive and attractive forces precisely balance to produce an energy minimum in each atom's interaction with an atom of a second, non-bonded element (**Figure 2–11**).

Depending on the intended purpose, we shall represent small molecules as line drawings, ball-and-stick models, or space-filling models. For comparison, the water molecule is represented in all three ways in **Figure 2–12**. When representing very large molecules, such as proteins, we shall often need to further simplify the model used (see, for example, Panel 3–2, pp. 132–133).

Figure 2–10 Polar and nonpolar covalent bonds. The electron distributions in the polar water molecule (H_2O) and the nonpolar oxygen molecule (O_2) are compared (δ^+, partial positive charge; δ^-, partial negative charge).

Water Is the Most Abundant Substance in Cells

Water accounts for about 70% of a cell's weight, and most intracellular reactions occur in an aqueous environment. Life on Earth began in the ocean, and the conditions in that primeval environment put a permanent stamp on the chemistry of living things. Life therefore hinges on the properties of water.

In each water molecule (H_2O) the two H atoms are linked to the O atom by covalent bonds (see Figure 2–12). The two bonds are highly polar because the O is strongly attractive for electrons, whereas the H is only weakly attractive. Consequently, there is an unequal distribution of electrons in a water molecule, with a preponderance of positive charge on the two H atoms and of negative charge on the O (see Figure 2–10). When a positively charged region of one water molecule (that is, one of its H atoms) approaches a negatively charged region (that is, the O) of a second water molecule, the electrical attraction between them can result in a weak bond called a *hydrogen bond* (see Figure 2–15). These bonds are much weaker than covalent bonds and are easily broken by the random thermal motions due to the heat energy of the molecules, so each bond lasts only a short time. But the combined effect of many weak bonds can be profound. Each water molecule can form hydrogen bonds through its two H atoms to two other water molecules, producing a network in which hydrogen bonds are being continually broken and formed (Panel 2–2, pp. 108–109). It is only because of the

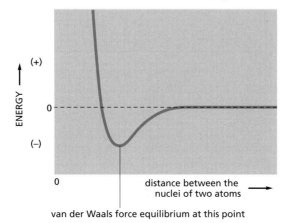

Figure 2–11 The balance of van der Waals forces between two atoms. As the nuclei of two atoms approach each other, they initially show a weak bonding interaction due to their fluctuating electric charges. However, the same atoms will strongly repel each other if they are brought too close together. The balance of these van der Waals attractive and repulsive forces occurs at the indicated energy minimum. This minimum determines the contact distance between any two noncovalently bonded atoms; this distance is the sum of their van der Waals radii. By definition, zero energy (indicated by the dotted *red* line) is the energy when the two nuclei are at infinite separation.

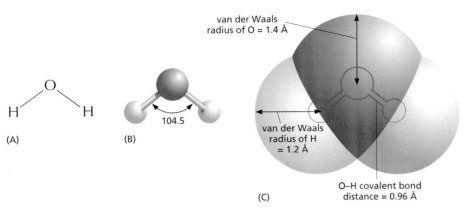

van der Waals
radius of O = 1.4 Å

104.5

(A)   (B)

van der Waals
radius of H
= 1.2 Å

O–H covalent bond
distance = 0.96 Å
(C)

Figure 2–12 Three representations of a water molecule. (A) The usual line drawing of the structural formula, in which each atom is indicated by its standard symbol, and each line represents a covalent bond joining two atoms. (B) A ball-and-stick model, in which atoms are represented by spheres of arbitrary diameter, connected by sticks representing covalent bonds. Unlike (A), bond angles are accurately represented in this type of model (see also Figure 2–8). (C) A space-filling model, in which both bond geometry and van der Waals radii are accurately represented.

hydrogen bonds that link water molecules together that water is a liquid at room temperature, with a high boiling point and high surface tension—rather than a gas.

Molecules, such as alcohols, that contain polar bonds and that can form hydrogen bonds with water dissolve readily in water. Molecules carrying plus or minus charges (ions) likewise interact favorably with water. Such molecules are termed **hydrophilic**, meaning that they are water-loving. A large proportion of the molecules in the aqueous environment of a cell necessarily fall into this category, including sugars, DNA, RNA, and most proteins. **Hydrophobic** (water-hating) molecules, by contrast, are uncharged and form few or no hydrogen bonds, and so do not dissolve in water. Hydrocarbons are an important example (see Panel 2–1, pp. 106–107). In these molecules the H atoms are covalently linked to C atoms by a largely nonpolar bond. Because the H atoms have almost no net positive charge, they cannot form effective hydrogen bonds to other molecules. This makes the hydrocarbon as a whole hydrophobic—a property that is exploited in cells, whose membranes are constructed from molecules that have long hydrocarbon tails, as we shall see in Chapter 10.

Some Polar Molecules Are Acids and Bases

One of the simplest kinds of chemical reaction, and one that has profound significance in cells, takes place when a molecule containing a highly polar covalent bond between a hydrogen and a second atom dissolves in water. The hydrogen atom in such a molecule has largely given up its electron to the companion atom and so resembles an almost naked positively charged hydrogen nucleus—in other words, a **proton (H^+)**. When water molecules surround the polar molecule, the proton is attracted to the partial negative charge on the O atom of an adjacent water molecule and can dissociate from its original partner to associate instead with the oxygen atoms of the water molecule to generate a **hydronium ion (H_3O^+)** (**Figure 2–13A**). The reverse reaction also takes place very readily, so one has to imagine an equilibrium state in which billions of protons are constantly flitting to and fro from one molecule in the solution to another.

The same type of reaction takes place in a solution of pure water itself. As illustrated in Figure 2–13B, water molecules are constantly exchanging protons with each other. As a result, pure water contains an equal, very low concentration of H_3O^+ and OH^- ions, both being present at 10^{-7} M. (The concentration of H_2O in pure water is 55.5 M.)

Substances that release protons to form H_3O^+ when they dissolve in water are termed **acids**. The higher the concentration of H_3O^+, the more acidic the solution. As H_3O^+ rises, the concentration of OH^- falls, according to the equilibrium equation for water: $[H_3O^+][OH^-] = 1.0 \times 10^{-14}$, where square brackets denote molar concentrations to be multiplied. By tradition, the H_3O^+ concentration is usually referred to as the H^+ concentration, even though nearly all H^+ in an aqueous solution is present as H_3O^+. To avoid the use of unwieldy numbers, the concentration of H^+ is expressed using a logarithmic scale called the **pH scale**, as illustrated in Panel 2–2 (pp. 108–109). Pure water has a pH of 7.0, and is neutral—that is, neither acidic (pH < 7.0) nor basic (pH > 7.0).

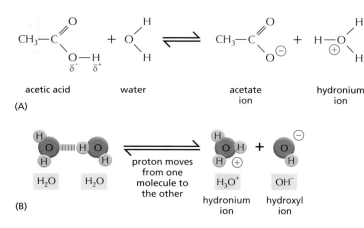

Because the proton of a hydronium ion can be passed readily to many types of molecules in cells, altering their character, the concentration of H_3O^+ inside a cell (the acidity) must be closely regulated. The interior of a cell is kept close to neutrality, and it is buffered by the presence of many chemical groups that can take up and release protons near pH 7.

The opposite of an acid is a **base**. Just as the defining property of an acid is that it donates protons to a water molecule so as to raise the concentration of H_3O^+ ions, the defining property of a base is that it accepts protons so as to lower the concentration of H_3O^+ ions, and thereby raise the concentration of hydroxyl ions (OH^-). A base can either combine with protons directly or form hydroxyl ions that immediately combine with protons to produce H_2O. Thus sodium hydroxide (NaOH) is basic (or *alkaline*) because it dissociates in aqueous solution to form Na^+ ions and OH^- ions. Other bases, especially important in living cells, contain NH_2 groups. These groups directly take up a proton from water: $-NH_2 + H_2O \rightarrow -NH_3^+ + OH^-$.

All molecules that accept protons from water will do so most readily when the concentration of H_3O^+ is high (acidic solutions). Likewise, molecules that can give up protons do so more readily if the concentration of H_3O^+ in solution is low (basic solutions), and they will tend to receive them back if this concentration is high.

Four Types of Noncovalent Attractions Help Bring Molecules Together in Cells

In aqueous solutions, covalent bonds are 10–100 times stronger than the other attractive forces between atoms, allowing their connections to define the boundaries of one molecule from another. But much of biology depends on the specific binding of different molecules to each other. This binding is mediated by a group of noncovalent attractions that are individually quite weak, but whose energies can sum to create an effective force between two separate molecules. We have previously introduced three of these attractive forces: electrostatic attractions (ionic bonds), hydrogen bonds, and van der Waals attractions. **Table 2–1** compares the strengths of these three types of *noncovalent bonds* with that of a typical covalent bond, both in the presence and in the

Table 2–1 Covalent and Noncovalent Chemical Bonds

BOND TYPE		LENGTH (nm)	STRENGTH (kcal/mole) IN VACUUM	IN WATER
Covalent		0.15	90	90
Noncovalent:	ionic*	0.25	80	3
	hydrogen	0.30	4	1
	van der Waals attraction (per atom)	0.35	0.1	0.1

*An ionic bond is an electrostatic attraction between two fully charged atoms.

absence of water. Because of their fundamental importance in all biological systems, we summarize their properties here:

- **Electrostatic attractions**. These result from the attractive forces between oppositely charged atoms. Electrostatic attractions are quite strong in the absence of water. They readily form between permanent dipoles, but are greatest when the two atoms involved are fully charged (*ionic bonds*). However, the polar water molecules cluster around both fully charged ions and polar molecules that contain permanent dipoles (**Figure 2–14**). This greatly reduces the attractiveness of these charged species for each other in most biological settings.

- **Hydrogen bonds**. The structure of a typical hydrogen bond is illustrated in **Figure 2–15**. This bond represents a special form of polar interaction in which an electropositive hydrogen atom is partially shared by two electronegative atoms. Its hydrogen can be viewed as a proton that has partially dissociated from a donor atom, allowing it to be shared by a second acceptor atom. Unlike a typical electrostatic interaction, this bond is highly directional—being strongest when a straight line can be drawn between all three of the involved atoms. As already discussed, water weakens these bonds by forming competing hydrogen-bond interactions with the involved molecules.

- **van der Waals attractions**. The electron cloud around any nonpolar atom will fluctuate, producing a flickering dipole. Such dipoles will transiently induce an oppositely polarized flickering dipole in a nearby atom. This interaction generates a very weak attraction between atoms. But since many atoms can be simultaneously in contact when two surfaces fit closely, the net result is often significant. Water does not weaken these so-called van der Waals attractions.

The fourth effect that often brings molecules together in water is not, strictly speaking, a bond at all. However, a very important **hydrophobic force** is caused by a pushing of nonpolar surfaces out of the hydrogen-bonded water network, where they would otherwise physically interfere with the highly favorable interactions between water molecules. Bringing any two nonpolar surfaces together reduces their contact with water; in this sense, the force is nonspecific. Nevertheless, we shall see in Chapter 3 that hydrophobic forces are central to the proper folding of protein molecules.

Panel 2–3 provides an overview of the four types of attractions just described. And **Figure 2–16** illustrates schematically how many such interactions can sum to hold together the matching surfaces of two macromolecules, even though each interaction by itself would be much too weak to be effective in the face of thermal motions.

A Cell Is Formed from Carbon Compounds

Having looked at the ways atoms combine into small molecules and how these molecules behave in an aqueous environment, we now examine the main classes of small molecules found in cells and their biological roles. We shall see that a few basic categories of molecules, formed from a handful of different elements, give rise to all the extraordinary richness of form and behavior shown by living things.

If we disregard water and inorganic ions such as potassium, nearly all the molecules in a cell are based on carbon. Carbon is outstanding among all the elements in its ability to form large molecules; silicon is a poor second. Because it is small and has four electrons and four vacancies in its outermost shell, a carbon atom can form four covalent bonds with other atoms. Most important, one carbon atom can join to other carbon atoms through highly stable covalent C–C

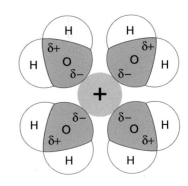

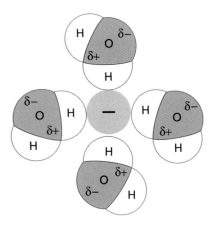

Figure 2–14 **How the dipoles on water molecules orient to reduce the affinity of oppositely charged ions or polar groups for each other.**

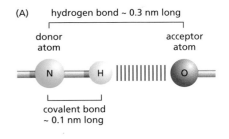

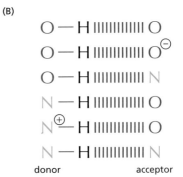

Figure 2–15 **Hydrogen bonds.** (A) Ball-and-stick model of a typical hydrogen bond. The distance between the hydrogen and the oxygen atom here is less than the sum of their van der Waals radii, indicating a partial sharing of electrons. (B) The most common hydrogen bonds in cells.

bonds to form chains and rings and hence generate large and complex molecules with no obvious upper limit to their size (see Panel 2–1, pp. 106–107). The small and large carbon compounds made by cells are called *organic molecules*.

Certain combinations of atoms, such as the methyl (–CH$_3$), hydroxyl (–OH), carboxyl (–COOH), carbonyl (–C=O), phosphate (–PO$_3^{2-}$), sulfhydryl (–SH), and amino (–NH$_2$) groups, occur repeatedly in organic molecules. Each such **chemical group** has distinct chemical and physical properties that influence the behavior of the molecule in which the group occurs. The most common chemical groups and some of their properties are summarized in Panel 2–1, pp. 106–107.

Cells Contain Four Major Families of Small Organic Molecules

The small organic molecules of the cell are carbon-based compounds that have molecular weights in the range 100–1000 and contain up to 30 or so carbon atoms. They are usually found free in solution and have many different fates. Some are used as *monomer* subunits to construct the giant polymeric *macromolecules*—the proteins, nucleic acids, and large polysaccharides—of the cell. Others act as energy sources and are broken down and transformed into other small molecules in a maze of intracellular metabolic pathways. Many small molecules have more than one role in the cell—for example, acting both as a potential subunit for a macromolecule and as an energy source. Small organic molecules are much less abundant than the organic macromolecules, accounting for only about one-tenth of the total mass of organic matter in a cell (**Table 2–2**). As a rough guess, there may be a thousand different kinds of these small molecules in a typical cell.

All organic molecules are synthesized from and are broken down into the same set of simple compounds. Both their synthesis and their breakdown occur through sequences of limited chemical changes that follow definite rules. As a consequence, the compounds in a cell are chemically related and most can be classified into a few distinct families. Broadly speaking, cells contain four major families of small organic molecules: the *sugars*, the *fatty acids*, the *amino acids*, and the *nucleotides* (**Figure 2–17**). Although many compounds present in cells do not fit into these categories, these four families of small organic molecules, together with the macromolecules made by linking them into long chains, account for a large fraction of cell mass (see Table 2–2).

Sugars Provide an Energy Source for Cells and Are the Subunits of Polysaccharides

The simplest **sugars**—the *monosaccharides*—are compounds with the general formula (CH$_2$O)$_n$, where *n* is usually 3, 4, 5, 6, 7, or 8. Sugars, and the molecules made from them, are also called *carbohydrates* because of this simple formula. Glucose, for example, has the formula C$_6$H$_{12}$O$_6$ (**Figure 2–18**). The formula, however, does not fully define the molecule: the same set of carbons, hydrogens, and

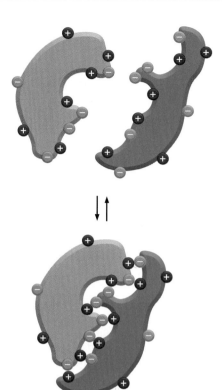

Figure 2–16 **Schematic indicating how two macromolecules with complementary surfaces can bind tightly to one another through noncovalent interactions.**

Table 2–2 **The Types of Molecules That Form a Bacterial Cell**

	PERCENT OF TOTAL CELL WEIGHT	NUMBER OF TYPES OF EACH MOLECULE
Water	70	1
Inorganic ions	1	20
Sugars and precursors	1	250
Amino acids and precursors	0.4	100
Nucleotides and precursors	0.4	100
Fatty acids and precursors	1	50
Other small molecules	0.2	~300
Macromolecules (proteins, nucleic acids, and polysaccharides)	26	~3000

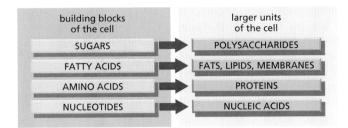

Figure 2–17 **The four main families of small organic molecules in cells.** These small molecules form the monomeric building blocks, or subunits, for most of the macromolecules and other assemblies of the cell. Some, such as the sugars and the fatty acids, are also energy sources.

oxygens can be joined together by covalent bonds in a variety of ways, creating structures with different shapes. As shown in Panel 2–4 (pp. 112–113), for example, glucose can be converted into a different sugar—mannose or galactose—simply by switching the orientations of specific OH groups relative to the rest of the molecule. Each of these sugars, moreover, can exist in either of two forms, called the D-form and the L-form, which are mirror images of each other. Sets of molecules with the same chemical formula but different structures are called *isomers*, and the subset of such molecules that are mirror-image pairs are called *optical isomers*. Isomers are widespread among organic molecules in general, and they play a major part in generating the enormous variety of sugars.

Panel 2–4 presents an outline of sugar structure and chemistry. Sugars can exist as rings or as open chains. In their open-chain form, sugars contain a number of hydroxyl groups and either one aldehyde ($_H{>}C{=}O$) or one ketone (${>}C{=}O$) group. The aldehyde or ketone group plays a special role. First, it can react with a hydroxyl group in the same molecule to convert the molecule into a ring; in the ring form the carbon of the original aldehyde or ketone group can be recognized as the only one that is bonded to two oxygens. Second, once the ring is formed, this same carbon can become further linked, via oxygen, to one of the carbons bearing a hydroxyl group on another sugar molecule. This creates a *disaccharide* such as sucrose, which is composed of a glucose and a fructose unit. Larger sugar polymers range from the *oligosaccharides* (trisaccharides, tetrasaccharides, and so on) up to giant *polysaccharides*, which can contain thousands of monosaccharide units.

The way that sugars are linked together to form polymers illustrates some common features of biochemical bond formation. A bond is formed between an –OH group on one sugar and an –OH group on another by a **condensation reaction**, in which a molecule of water is expelled as the bond is formed (**Figure 2–19**). Subunits in other biological polymers, such as nucleic acids and proteins, are also linked by condensation reactions in which water is expelled. The bonds created by all of these condensation reactions can be broken by the reverse process of **hydrolysis**, in which a molecule of water is consumed (see Figure 2–19).

Figure 2–18 **The structure of glucose, a simple sugar.** As illustrated previously for water (see Figure 2–12), any molecule can be represented in several ways. In the structural formulas shown in (A), (B) and (C), the atoms are shown as chemical symbols linked together by lines representing the covalent bonds. The *thickened lines* here are used to indicate the plane of the sugar ring, in an attempt to emphasize that the –H and –OH groups are not in the same plane as the ring. (A) The open-chain form of this sugar, which is in equilibrium with the more stable cyclic or ring form in (B). (C) The chair form is an alternative way to draw the cyclic molecule that reflects the geometry more accurately than the structural formula in (B). (D) A space-filling model, which, as well as depicting the three-dimensional arrangement of the atoms, also uses the van der Waals radii to represent the surface contours of the molecule. (E) A ball-and-stick model in which the three-dimensional arrangement of the atoms in space is shown. (H, *white*; C, *black*; O, *red*; N, *blue*.)

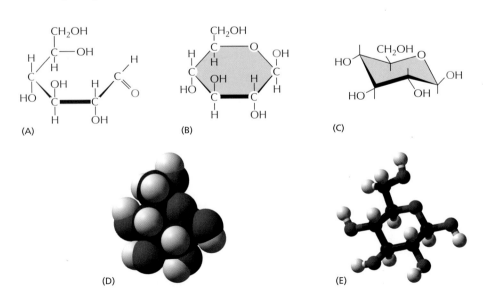

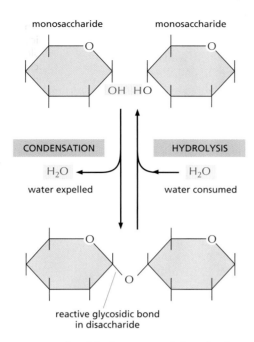

Figure 2–19 The reaction of two monosaccharides to form a disaccharide. This reaction belongs to a general category of reactions termed *condensation reactions,* in which two molecules join together as a result of the loss of a water molecule. The reverse reaction (in which water is added) is termed *hydrolysis.* Note that the reactive carbon at which the new bond is formed (on the monosaccharide on the *left* here) is the carbon joined to two oxygens as a result of sugar ring formation (see Figure 2–18). As indicated, this common type of covalent bond between two sugar molecules is known as a *glycosidic bond* (see also Figure 2–20).

Because each monosaccharide has several free hydroxyl groups that can form a link to another monosaccharide (or to some other compound), sugar polymers can be branched, and the number of possible polysaccharide structures is extremely large. Even a simple disaccharide consisting of two glucose units can exist in eleven different varieties (**Figure 2–20**), while three different hexoses ($C_6H_{12}O_6$) can join together to make several thousand trisaccharides. For this reason it is a much more complex task to determine the arrangement of sugars in a polysaccharide than to determine the nucleotide sequence of a DNA molecule, where each unit is joined to the next in exactly the same way.

The monosaccharide *glucose* is a key energy source for cells. In a series of reactions, it is broken down to smaller molecules, releasing energy that the cell can harness to do useful work, as we shall explain later. Cells use simple polysaccharides composed only of glucose units—principally *glycogen* in animals and *starch* in plants—as energy stores.

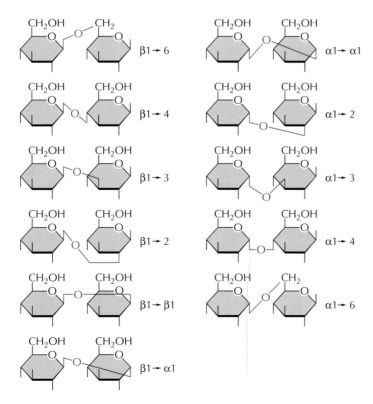

Figure 2–20 Eleven disaccharides consisting of two D-glucose units. Although these differ only in the type of linkage between the two glucose units, they are chemically distinct. Since the oligosaccharides associated with proteins and lipids may have six or more different kinds of sugar joined in both linear and branched arrangements through glycosidic bonds such as those illustrated here, the number of distinct types of oligosaccharides that can be used in cells is extremely large. For an explanation of α and β linkages, see Panel 2–4 (pp. 112–113). Short *black* lines ending "blind" indicate OH positions. (*Red* lines merely indicate disaccharide bond orientations and "corners" do not imply extra atoms.)

Sugars do not function only in the production and storage of energy. They can also be used, for example, to make mechanical supports. Thus, the most abundant organic chemical on Earth—the *cellulose* of plant cell walls—is a polysaccharide of glucose. Because the glucose–glucose linkages in cellulose differ from those in starch and glycogen, however, humans cannot digest cellulose and use its glucose. Another extraordinarily abundant organic substance, the *chitin* of insect exoskeletons and fungal cell walls, is also an indigestible polysaccharide—in this case a linear polymer of a sugar derivative called *N*-acetylglucosamine (see Panel 2–4). Other polysaccharides are the main components of slime, mucus, and gristle.

Smaller oligosaccharides can be covalently linked to proteins to form glycoproteins and to lipids to form *glycolipids*, both of which are found in cell membranes. As described in Chapter 10, most cell surfaces are clothed and decorated with glycoproteins and glycolipids in the cell membrane. The sugar side chains on these molecules are often recognized selectively by other cells. And differences between people in the details of their cell-surface sugars are the molecular basis for the different major human blood groups, termed A, B, AB, and O.

Fatty Acids Are Components of Cell Membranes, as Well as a Source of Energy

A fatty acid molecule, such as *palmitic acid*, has two chemically distinct regions (**Figure 2–21**). One is a long hydrocarbon chain, which is hydrophobic and not very reactive chemically. The other is a carboxyl (–COOH) group, which behaves as an acid (carboxylic acid): it is ionized in solution (–COO⁻), extremely hydrophilic, and chemically reactive. Almost all the fatty acid molecules in a cell are covalently linked to other molecules by their carboxylic acid group.

The hydrocarbon tail of palmitic acid is *saturated:* it has no double bonds between carbon atoms and contains the maximum possible number of hydrogens. Stearic acid, another one of the common fatty acids in animal fat, is also saturated. Some other fatty acids, such as oleic acid, have *unsaturated* tails, with one or more double bonds along their length. The double bonds create kinks in the molecules, interfering with their ability to pack together in a solid mass. It is this that accounts for the difference between hard margarine (saturated) and liquid vegetable oils (polyunsaturated). The many different fatty acids found in cells differ only in the length of their hydrocarbon chains and the number and position of the carbon–carbon double bonds (see Panel 2–5, pp. 114–115).

Fatty acids are stored in the cytoplasm of many cells in the form of droplets of *triacylglycerol* molecules, which consist of three fatty acid chains joined to a glycerol molecule (see Panel 2–5); these molecules are the animal fats found in meat, butter, and cream, and the plant oils such as corn oil and olive oil. When required to provide energy, the fatty acid chains are released from triacylglycerols and broken down into two-carbon units. These two-carbon units are identical to those derived from the breakdown of glucose and they enter the same energy-yielding reaction pathways, as will be described later in this chapter. Triglycerides serve as a concentrated food reserve in cells, because they can be broken down to produce about six times as much usable energy, weight for weight, as glucose.

Fatty acids and their derivatives such as triacylglycerols are examples of **lipids**. Lipids comprise a loosely defined collection of biological molecules that are insoluble in water, while being soluble in fat and organic solvents such as benzene. They typically contain either long hydrocarbon chains, as in the fatty acids and isoprenes, or multiple linked rings, as in the *steroids*.

The most important function of fatty acids in cells is in the construction of cell membranes. These thin sheets enclose all cells and surround their internal organelles. They are composed largely of *phospholipids*, which are small molecules that, like triacylglycerols, are constructed mainly from fatty acids and glycerol. In phospholipids the glycerol is joined to two fatty acid chains, however, rather than to three as in triacylglycerols. The "third" site on the glycerol is linked to a hydrophilic phosphate group, which is in turn attached to a small hydrophilic compound such as choline (see Panel 2–5). Each phospholipid

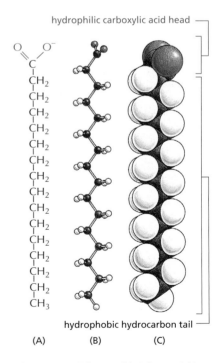

Figure 2–21 A fatty acid. A fatty acid is composed of a hydrophobic hydrocarbon chain to which is attached a hydrophilic carboxylic acid group. Palmitic acid is shown here. Different fatty acids have different hydrocarbon tails. (A) Structural formula. The carboxylic acid group is shown in its ionized form. (B) Ball-and-stick model. (C) Space-filling model.

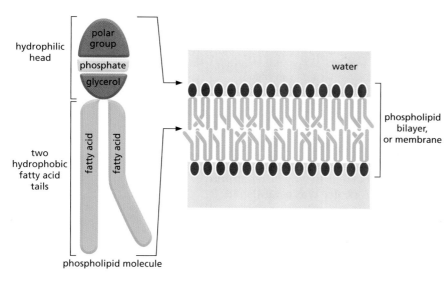

Figure 2–22 **Phospholipid structure and the orientation of phospholipids in membranes.** In an aqueous environment, the hydrophobic tails of phospholipids pack together to exclude water. Here they have formed a bilayer with the hydrophilic head of each phospholipid facing the water. Lipid bilayers are the basis for cell membranes, as discussed in detail in Chapter 10.

molecule, therefore, has a hydrophobic tail composed of the two fatty acid chains and a hydrophilic head, where the phosphate is located. This gives them different physical and chemical properties from triacylglycerols, which are predominantly hydrophobic. Molecules such as phospholipids, with both hydrophobic and hydrophilic regions, are termed *amphiphilic*.

The membrane-forming property of phospholipids results from their amphiphilic nature. Phospholipids will spread over the surface of water to form a monolayer of phospholipid molecules, with the hydrophobic tails facing the air and the hydrophilic heads in contact with the water. Two such molecular layers can readily combine tail-to-tail in water to make a phospholipid sandwich, or **lipid bilayer**. This bilayer is the structural basis of all cell membranes (**Figure 2–22**).

Amino Acids Are the Subunits of Proteins

Amino acids are a varied class of molecules with one defining property: they all possess a carboxylic acid group and an amino group, both linked to a single carbon atom called the α-carbon (**Figure 2–23**). Their chemical variety comes from the side chain that is also attached to the α-carbon. The importance of amino acids to the cell comes from their role in making **proteins**, which are polymers of amino acids joined head-to-tail in a long chain that is then folded into a three-dimensional structure unique to each type of protein. The covalent linkage between two adjacent amino acids in a protein chain forms an amide (see Panel 2–1), and it is called a **peptide bond**; the chain of amino acids is also known as a *polypeptide* (**Figure 2–24**). Regardless of the specific amino acids from which it is made, the polypeptide has an amino (NH_2) group at one end (its *N-terminus*) and a carboxyl (COOH) group at its other end (its *C-terminus*). This gives it a definite directionality—a structural (as opposed to an electrical) polarity.

Each of the 20 amino acids found commonly in proteins has a different side chain attached to the α-carbon atom (see Panel 3–1, pp. 128–129). All organisms,

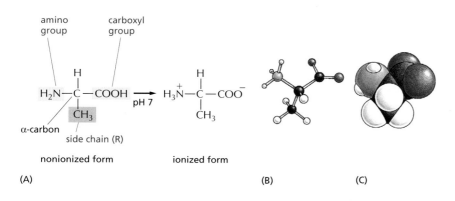

Figure 2–23 **The amino acid alanine.** (A) In the cell, where the pH is close to 7, the free amino acid exists in its ionized form; but when it is incorporated into a polypeptide chain, the charges on the amino and carboxyl groups disappear. (B) A ball-and-stick model and (C) a space-filling model of alanine (H, *white*; C, *black*; O, *red*; N, *blue*).

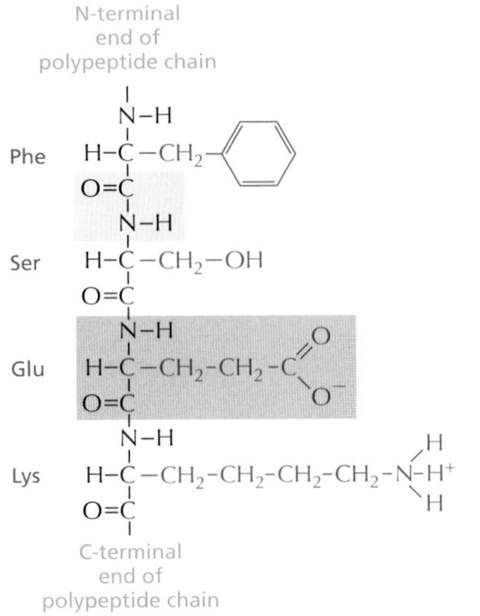

Figure 2–24 **A small part of a protein molecule.** The four amino acids shown are linked together by three peptide bonds, one of which is highlighted in *yellow*. One of the amino acids is shaded in *gray*. The amino acid side chains are shown in *red*. The two ends of a polypeptide chain are chemically distinct. One end, the N-terminus, terminates in an amino group, and the other, the C-terminus, in a carboxyl group. The sequence is always read from the N-terminal end; hence this sequence is Phe–Ser–Glu–Lys.

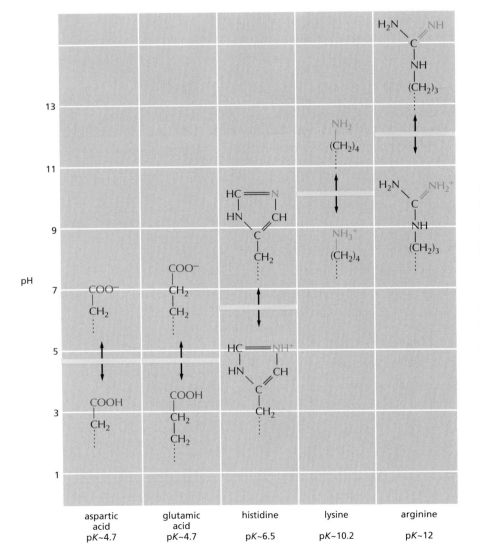

whether bacteria, archea, plants, or animals, have proteins made of the same 20 amino acids. How this precise set of 20 came to be chosen is one of the mysteries of the evolution of life; there is no obvious chemical reason why other amino acids could not have served just as well. But once the choice was established, it could not be changed; too much depended on it.

Like sugars, all amino acids, except glycine, exist as optical isomers in D- and L-forms (see Panel 3–1). But only L-forms are ever found in proteins (although D-amino acids occur as part of bacterial cell walls and in some antibiotics). The origin of this exclusive use of L-amino acids to make proteins is another evolutionary mystery.

The chemical versatility of the 20 amino acids is essential to the function of proteins. Five of the 20 amino acids have side chains that can form ions in neutral aqueous solution and thereby can carry a charge (**Figure 2–25**). The others are uncharged; some are polar and hydrophilic, and some are nonpolar and hydrophobic. As we discuss in Chapter 3, the properties of the amino acid side chains underlie the diverse and sophisticated functions of proteins.

Figure 2–25 **The charge on amino acid side chains depends on the pH.** The five different side chains that can carry a charge are shown. Carboxylic acids can readily lose H$^+$ in aqueous solution to form a negatively charged ion, which is denoted by the suffix "-ate," as in aspart*ate* or glutam*ate*. A comparable situation exists for amines, which in aqueous solution can take up H$^+$ to form a positively charged ion (which does not have a special name). These reactions are rapidly reversible, and the amounts of the two forms, charged and uncharged, depend on the pH of the solution. At a high pH, carboxylic acids tend to be charged and amines uncharged. At a low pH, the opposite is true—the carboxylic acids are uncharged and amines are charged. The pH at which exactly *half* of the carboxylic acid or amine residues are charged is known as the pK of that amino acid side chain (indicated by *yellow stripe*).

In the cell the pH is close to 7, and almost all carboxylic acids and amines are in their fully charged form.

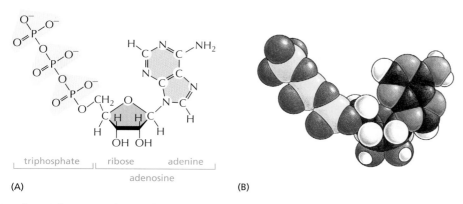

(A)

(B)

Figure 2–26 **Chemical structure of adenosine triphosphate (ATP).** (A) Structural formula. (B) Space-filling model. In (B) the colors of the atoms are C, *black;* N, *blue;* H, *white;* O, *red;* and P, *yellow.*

Nucleotides Are the Subunits of DNA and RNA

A **nucleotide** is a molecule made up of a nitrogen-containing ring compound linked to a five-carbon sugar, which in turn carries one or more phosphate groups (Panel 2–6, pp. 116–117). The five-carbon sugar can be either ribose or deoxyribose. Nucleotides containing ribose are known as ribonucleotides, and those containing deoxyribose as deoxyribonucleotides. The nitrogen-containing rings are generally referred to as *bases* for historical reasons: under acidic conditions they can each bind an H^+ (proton) and thereby increase the concentration of OH^- ions in aqueous solution. There is a strong family resemblance between the different bases. *Cytosine (C), thymine (T),* and *uracil (U)* are called pyrimidines because they all derive from a six-membered pyrimidine ring; *guanine (G)* and *adenine (A)* are *purine* compounds, and they have a second, five-membered ring fused to the six-membered ring. Each nucleotide is named for the base it contains (see Panel 2–6).

Nucleotides can act as short-term carriers of chemical energy. Above all others, the ribonucleotide **adenosine triphosphate**, or **ATP** (**Figure 2–26**), transfers energy in hundreds of different cell reactions. ATP is formed through reactions that are driven by the energy released by the oxidative breakdown of foodstuffs. Its three phosphates are linked in series by two *phosphoanhydride bonds*, whose rupture releases large amounts of useful energy. The terminal phosphate group in particular is frequently split off by hydrolysis, often transferring a phosphate to other molecules and releasing energy that drives energy-requiring biosynthetic reactions (**Figure 2–27**). Other nucleotide derivatives are carriers for the transfer of other chemical groups, as will be described later.

The most fundamental role of nucleotides in the cell, however, is in the storage and retrieval of biological information. Nucleotides serve as building blocks for the construction of *nucleic acids*—long polymers in which nucleotide subunits are covalently linked by the formation of a **phosphodiester bond** between the

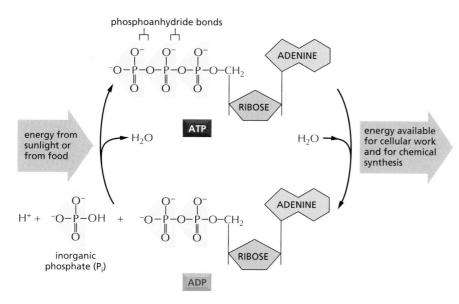

Figure 2–27 **The ATP molecule serves as an energy carrier in cells.** The energy-requiring formation of ATP from ADP and inorganic phosphate is coupled to the energy-yielding oxidation of foodstuffs (in animal cells, fungi, and some bacteria) or to the capture of light energy (in plant cells and some bacteria). The hydrolysis of this ATP back to ADP and inorganic phosphate in turn provides the energy to drive many cell reactions.

Figure 2–28 A small part of one chain of a deoxyribonucleic acid (DNA) molecule. Four nucleotides are shown. One of the phosphodiester bonds that links adjacent nucleotide residues is highlighted in *yellow,* and one of the nucleotides is shaded in *gray.* Nucleotides are linked together by a phosphodiester linkage between specific carbon atoms of the ribose, known as the 5′ and 3′ atoms. For this reason, one end of a polynucleotide chain, the *5′ end,* will have a free phosphate group and the other, the *3′ end,* a free hydroxyl group. The linear sequence of nucleotides in a polynucleotide chain is commonly abbreviated by a one-letter code, and the sequence is always read from the 5′ end. In the example illustrated the sequence is G–A–T–C.

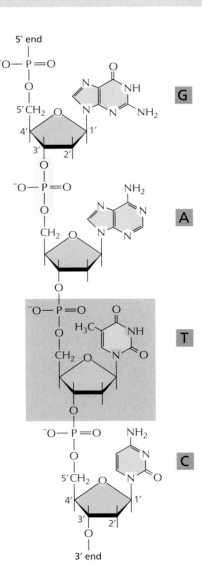

phosphate group attached to the sugar of one nucleotide and a hydroxyl group on the sugar of the next nucleotide (**Figure 2–28**). Nucleic acid chains are synthesized from energy-rich nucleoside triphosphates by a condensation reaction that releases inorganic pyrophosphate during phosphodiester bond formation.

There are two main types of nucleic acids, differing in the type of sugar in their sugar-phosphate backbone. Those based on the sugar *ribose* are known as **ribonucleic acids**, or **RNA**, and normally contain the bases A, G, C, and U. Those based on *deoxyribose* (in which the hydroxyl at the 2′ position of the ribose carbon ring is replaced by a hydrogen are known as **deoxyribonucleic acids**, or **DNA**, and contain the bases A, G, C, and T (T is chemically similar to the U in RNA, merely adding the methyl group on the pyrimidine ring; see Panel 2–6). RNA usually occurs in cells as a single polynucleotide chain, but DNA is virtually always a double-stranded molecule—a DNA double helix composed of two polynucleotide chains running antiparallel to each other and held together by hydrogen-bonding between the bases of the two chains.

The linear sequence of nucleotides in a DNA or an RNA encodes the genetic information of the cell. The ability of the bases in different nucleic acid molecules to recognize and pair with each other by hydrogen-bonding (called *base-pairing*)—G with C, and A with either T or U—underlies all of heredity and evolution, as explained in Chapter 4.

The Chemistry of Cells Is Dominated by Macromolecules with Remarkable Properties

By weight, macromolecules are the most abundant carbon-containing molecules in a living cell (**Figure 2–29** and **Table 2–3**). They are the principal building blocks from which a cell is constructed and also the components that confer the most distinctive properties of living things. The macromolecules in cells are polymers that are constructed by covalently linking small organic molecules (called *monomers)* into long chains (**Figure 2–30**). Yet they have remarkable properties that could not have been predicted from their simple constituents.

Proteins are especially abundant and versatile. They perform thousands of distinct functions in cells. Many proteins serve as *enzymes,* the catalysts that

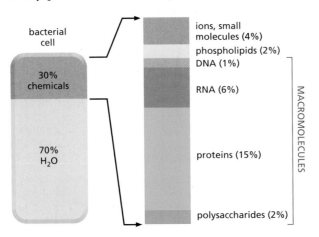

Figure 2–29 Macromolecules are abundant in cells. The approximate composition of a bacterial cell is shown by weight. The composition of an animal cell is similar (see Table 2–3).

Table 2–3 Approximate Chemical Compositions of a Typical Bacterium and a Typical Mammalian Cell

COMPONENT	PERCENT OF TOTAL CELL WEIGHT	
	E. COLI BACTERIUM	MAMMALIAN CELL
H_2O	70	70
Inorganic ions (Na$^+$, K$^+$, Mg^{2+}, Ca^{2+}, Cl$^-$, etc.)	1	1
Miscellaneous small metabolites	3	3
Proteins	15	18
RNA	6	1.1
DNA	1	0.25
Phospholipids	2	3
Other lipids	–	2
Polysaccharides	2	2
Total cell volume	2×10^{-12} cm^3	4×10^{-9} cm^3
Relative cell volume	1	2000

Proteins, polysaccharides, DNA, and RNA are macromolecules. Lipids are not generally classed as macromolecules even though they share some of their features; for example, most are synthesized as linear polymers of a smaller molecule (the acetyl group on acetyl CoA), and they self-assemble into larger structures (membranes). Note that water and protein comprise most of the mass of both mammalian and bacterial cells.

direct the many covalent bond-making and bond-breaking reactions that the cell needs. Enzymes catalyze all of the reactions whereby cells extract energy from food molecules, for example, and an enzyme called ribulose bisphosphate carboxylase helps to convert CO_2 to sugars in photosynthetic organisms, producing most of the organic matter needed for life on Earth. Other proteins are used to build structural components, such as tubulin, a protein that self-assembles to make the cell's long microtubules, or histones, proteins that compact the DNA in chromosomes. Yet other proteins act as molecular motors to produce force and movement, as in the case of myosin in muscle. Proteins perform many other functions, and we shall examine the molecular basis for many of them later in this book. Here we identify some general principles of macromolecular chemistry that make such functions possible.

Although the chemical reactions for adding subunits to each polymer are different in detail for proteins, nucleic acids, and polysaccharides, they share important features. Each polymer grows by the addition of a monomer onto the end of a growing polymer chain in a *condensation reaction*, in which a molecule of water is lost with each subunit added (see Figure 2–19). The stepwise polymerization of monomers into a long chain is a simple way to manufacture a large, complex molecule, since the subunits are added by the same reaction performed over and over again by the same set of enzymes. In a sense, the process resembles the repetitive operation of a machine in a factory—except in one crucial respect. Apart from some of the polysaccharides, most macromolecules are made from a set of monomers that are slightly different from one another—for example, the 20 different amino acids from which proteins are made. It is critical to life that the polymer chain is not assembled at random from these subunits; instead the subunits are added in a particular order, or *sequence*. The elaborate mechanisms that allow this to be accomplished by enzymes are described in detail in Chapters 5 and 6.

Noncovalent Bonds Specify Both the Precise Shape of a Macromolecule and its Binding to Other Molecules

Most of the covalent bonds in a macromolecule allow rotation of the atoms they join, giving the polymer chain great flexibility. In principle, this allows a macromolecule to adopt an almost unlimited number of shapes, or *conformations*, as

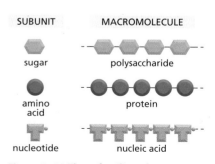

SUBUNIT　　MACROMOLECULE

sugar　　polysaccharide

amino acid　　protein

nucleotide　　nucleic acid

Figure 2–30 Three families of macromolecules. Each is a polymer formed from small molecules (called monomers) linked together by covalent bonds.

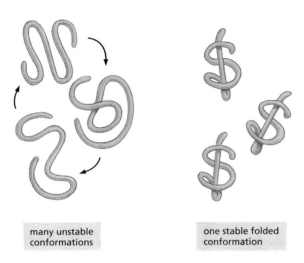

many unstable
conformations

one stable folded
conformation

Figure 2–31 Most proteins and many RNA molecules fold into only one stable conformation. If the noncovalent bonds maintaining this stable conformation are disrupted, the molecule becomes a flexible chain that usually has no biological value.

random thermal energy causes the polymer chain to writhe and rotate. However, the shapes of most biological macromolecules are highly constrained because of the many weak *noncovalent bonds* that form between different parts of the same molecule. If these noncovalent bonds are formed in sufficient numbers, the polymer chain can strongly prefer one particular conformation, determined by the linear sequence of monomers in its chain. Most protein molecules and many of the small RNA molecules found in cells fold tightly into one highly preferred conformation in this way (**Figure 2–31**).

The four types of noncovalent interactions important in biological molecules were described earlier, and they are reviewed in Panel 2–3 (pp. 110–111). Although individually very weak, these interactions cooperate to fold biological macromolecules into unique shapes. In addition, they can also add up to create a strong attraction between two different molecules when these molecules fit together very closely, like a hand in a glove. This form of molecular interaction provides for great specificity, inasmuch as the multipoint contacts required for strong binding make it possible for a macromolecule to select out—through binding—just one of the many thousands of other types of molecules present inside a cell. Moreover, because the strength of the binding depends on the number of noncovalent bonds that are formed, interactions of almost any affinity are possible—allowing rapid dissociation when necessary.

Binding of this type underlies all biological catalysis, making it possible for proteins to function as enzymes. Noncovalent interactions also allow macromolecules to be used as building blocks for the formation of larger structures. In cells, macromolecules often bind together into large complexes, thereby forming intricate machines with multiple moving parts that perform such complex tasks as DNA replication and protein synthesis (**Figure 2–32**).

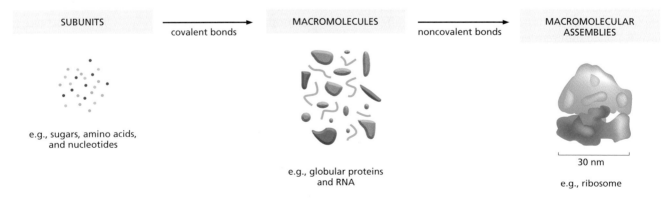

SUBUNITS	MACROMOLECULES	MACROMOLECULAR ASSEMBLIES
covalent bonds →	noncovalent bonds →	
e.g., sugars, amino acids, and nucleotides	e.g., globular proteins and RNA	30 nm
		e.g., ribosome

Figure 2–32 Small molecules, proteins, and a ribosome drawn approximately to scale. Ribosomes are a central part of the machinery that the cell uses to make proteins: each ribosome is formed as a complex of about 90 macromolecules (protein and RNA molecules).

Summary

Living organisms are autonomous, self-propagating chemical systems. They are made from a distinctive and restricted set of small carbon-based molecules that are essentially the same for every living species. Each of these molecules is composed of a small set of atoms linked to each other in a precise configuration through covalent bonds. The main categories are sugars, fatty acids, amino acids, and nucleotides. Sugars are a primary source of chemical energy for cells and can be incorporated into polysaccharides for energy storage. Fatty acids are also important for energy storage, but their most critical function is in the formation of cell membranes. Polymers consisting of amino acids constitute the remarkably diverse and versatile macromolecules known as proteins. Nucleotides play a central part in energy transfer. They are also the subunits for the informational macromolecules, RNA and DNA.

Most of the dry mass of a cell consists of macromolecules that have been produced as linear polymers of amino acids (proteins) or nucleotides (DNA and RNA), covalently linked to each other in an exact order. Most of the protein molecules and many of the RNAs fold into a unique conformation that depends on their sequence of subunits. This folding process creates unique surfaces, and it depends on a large set of weak attractions produced by noncovalent forces between atoms. These forces are of four types: electrostatic attractions, hydrogen bonds, van der Waals attractions, and an interaction between nonpolar groups caused by their hydrophobic expulsion from water. The same set of weak forces governs the specific binding of other molecules to macromolecules, making possible the myriad associations between biological molecules that produce the structure and the chemistry of a cell.

CATALYSIS AND THE USE OF ENERGY BY CELLS

One property of living things above all makes them seem almost miraculously different from nonliving matter: they create and maintain order, in a universe that is tending always to greater disorder (**Figure 2–33**). To create this order, the cells in a living organism must perform a never-ending stream of chemical reactions. In some of these reactions, small organic molecules—amino acids, sugars, nucleotides, and lipids—are being taken apart or modified to supply the many other small molecules that the cell requires. In other reactions, these small molecules are being used to construct an enormously diverse range of proteins, nucleic acids, and other macromolecules that endow living systems with all of their most distinctive properties. Each cell can be viewed as a tiny chemical factory, performing many millions of reactions every second.

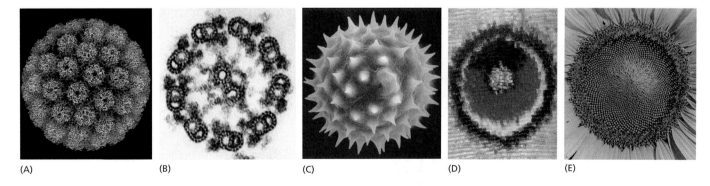

(A)   (B)   (C)   (D)   (E)

Figure 2–33 Order in biological structures. Well-defined, ornate, and beautiful spatial patterns can be found at every level of organization in living organisms. In order of increasing size: (A) protein molecules in the coat of a virus; (B) the regular array of microtubules seen in a cross section of a sperm tail; (C) surface contours of a pollen grain (a single cell); (D) close-up of the wing of a butterfly showing the pattern created by scales, each scale being the product of a single cell; (E) spiral array of seeds, made of millions of cells, in the head of a sunflower. (A, courtesy of R.A. Grant and J.M. Hogle; B, courtesy of Lewis Tilney; C, courtesy of Colin MacFarlane and Chris Jeffree; D and E, courtesy of Kjell B. Sandved.)

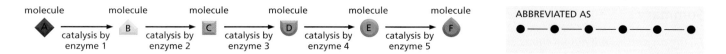

molecule A   molecule B   molecule C   molecule D   molecule E   molecule F

catalysis by enzyme 1   catalysis by enzyme 2   catalysis by enzyme 3   catalysis by enzyme 4   catalysis by enzyme 5

ABBREVIATED AS

Figure 2–34 How a set of enzyme-catalyzed reactions generates a metabolic pathway. Each enzyme catalyzes a particular chemical reaction, leaving the enzyme unchanged. In this example, a set of enzymes acting in series converts molecule A to molecule F, forming a metabolic pathway.

Cell Metabolism Is Organized by Enzymes

The chemical reactions that a cell carries out would normally occur only at much higher temperatures than those existing inside cells. For this reason, each reaction requires a specific boost in chemical reactivity. This requirement is crucial, because it allows the cell to control each reaction. The control is exerted through the specialized proteins called *enzymes*, each of which accelerates, or *catalyzes*, just one of the many possible kinds of reactions that a particular molecule might undergo. Enzyme-catalyzed reactions are usually connected in series, so that the product of one reaction becomes the starting material, or *substrate*, for the next (**Figure 2–34**). These long linear reaction pathways are in turn linked to one another, forming a maze of interconnected reactions that enable the cell to survive, grow, and reproduce (**Figure 2–35**).

Two opposing streams of chemical reactions occur in cells: (1) the *catabolic* pathways break down foodstuffs into smaller molecules, thereby generating both a useful form of energy for the cell and some of the small molecules that the cell needs as building blocks, and (2) the *anabolic*, or *biosynthetic*, pathways use the energy harnessed by catabolism to drive the synthesis of the many other molecules that form the cell. Together these two sets of reactions constitute the **metabolism** of the cell (**Figure 2–36**).

Many of the details of cell metabolism form the traditional subject of *biochemistry* and need not concern us here. But the general principles by which cells obtain energy from their environment and use it to create order are central to cell biology. We begin with a discussion of why a constant input of energy is needed to sustain living organisms.

Biological Order Is Made Possible by the Release of Heat Energy from Cells

The universal tendency of things to become disordered is a fundamental law of physics—the *second law of thermodynamics*—which states that in the universe, or in any isolated system (a collection of matter that is completely isolated from the rest of the universe), the degree of disorder only increases. This law has such profound implications for all living things that we restate it in several ways.

For example, we can present the second law in terms of probability and state that systems will change spontaneously toward those arrangements that have the greatest probability. If we consider, for example, a box of 100 coins all lying heads up, a series of accidents that disturbs the box will tend to move the arrangement toward a mixture of 50 heads and 50 tails. The reason is simple: there is a huge number of possible arrangements of the individual coins in the mixture that can achieve the 50–50 result, but only one possible arrangement that keeps all of the coins oriented heads up. Because the 50–50 mixture is therefore the most probable, we say that it is more "disordered." For the same reason,

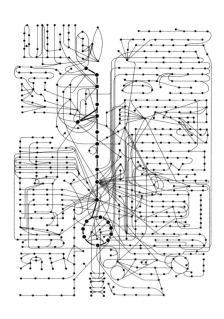

Figure 2–35 Some of the metabolic pathways and their interconnections in a typical cell. About 500 common metabolic reactions are shown diagrammatically, with each molecule in a metabolic pathway represented by a filled circle, as in the *yellow* box in Figure 2–34. The pathway that is highlighted in this diagram with larger circles and connecting lines is the central pathway of sugar metabolism, which will be discussed shortly.

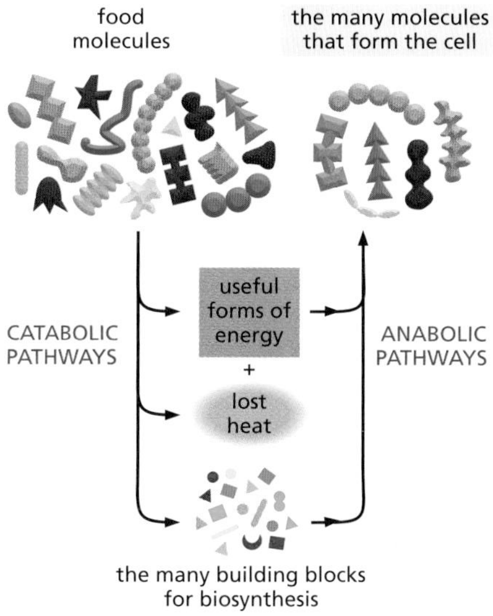

Figure 2–36 Schematic representation of the relationship between catabolic and anabolic pathways in metabolism. As suggested here, since a major portion of the energy stored in the chemical bonds of food molecules is dissipated as heat, the mass of food required by any organism that derives all of its energy from catabolism is much greater than the mass of the molecules that can be produced by anabolism.

it is a common experience that one's living space will become increasingly disordered without intentional effort: the movement toward disorder is a *spontaneous process*, requiring a periodic effort to reverse it (**Figure 2–37**).

The amount of disorder in a system can be quantified and expressed as the **entropy** of the system: the greater the disorder, the greater the entropy. Thus, another way to express the second law of thermodynamics is to say that systems will change spontaneously toward arrangements with greater entropy.

Living cells—by surviving, growing, and forming complex organisms—are generating order and thus might appear to defy the second law of thermodynamics. How is this possible? The answer is that a cell is not an isolated system: it takes in energy from its environment in the form of food, or as photons from the sun (or even, as in some chemosynthetic bacteria, from inorganic molecules alone), and it then uses this energy to generate order within itself. In the course of the chemical reactions that generate order, the cell converts part of the energy it uses into heat. The heat is discharged into the cell's environment and disorders it, so that the total entropy—that of the cell plus its surroundings—increases, as demanded by the laws of thermodynamics.

To understand the principles governing these energy conversions, think of a cell surrounded by a sea of matter representing the rest of the universe. As the cell lives and grows, it creates internal order. But it constantly releases heat energy as it synthesizes molecules and assembles them into cell structures. Heat is energy in its most disordered form—the random jostling of molecules. When the cell releases heat to the sea, it increases the intensity of molecular motions there (thermal motion)—thereby increasing the randomness, or disorder, of the sea. The second law of thermodynamics is satisfied because the increase in the amount of order inside the cell is more than compensated for by an even greater decrease in order (increase in entropy) in the surrounding sea of matter (**Figure 2–38**).

Where does the heat that the cell releases come from? Here we encounter another important law of thermodynamics. The *first law of thermodynamics* states that energy can be converted from one form to another, but that it cannot

"SPONTANEOUS" REACTION
as time elapses

ORGANIZED EFFORT REQUIRING ENERGY INPUT

Figure 2–37 An everyday illustration of the spontaneous drive toward disorder. Reversing this tendency toward disorder requires an intentional effort and an input of energy: it is not spontaneous. In fact, from the second law of thermodynamics, we can be certain that the human intervention required will release enough heat to the environment to more than compensate for the reordering of the items in this room.

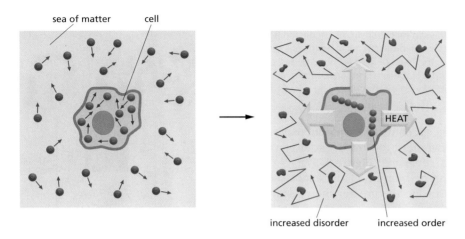

Figure 2–38 A simple thermodynamic analysis of a living cell. In the diagram on the left the molecules of both the cell and the rest of the universe (the sea of matter) are depicted in a relatively disordered state. In the diagram on the right the cell has taken in energy from food molecules and released heat by a reaction that orders the molecules the cell contains. Because the heat increases the disorder in the environment around the cell (depicted by the *jagged arrows* and distorted molecules, indicating the increased molecular motions caused by heat), the second law of thermodynamics—which states that the amount of disorder in the universe must always increase—is satisfied as the cell grows and divides. For a detailed discussion, see Panel 2–7 (pp. 118–119).

be created or destroyed. **Figure 2–39** illustrates some interconversions between different forms of energy. The amount of energy in different forms will change as a result of the chemical reactions inside the cell, but the first law tells us that the total amount of energy must always be the same. For example, an animal cell takes in foodstuffs and converts some of the energy present in the chemical bonds between the atoms of these food molecules (chemical bond energy) into the random thermal motion of molecules (heat energy). As described above, this conversion of chemical energy into heat energy is essential if the reactions that create order inside the cell are to cause the universe as a whole to become more disordered.

The cell cannot derive any benefit from the heat energy it releases unless the heat-generating reactions inside the cell are directly linked to the processes that generate molecular order. It is the tight *coupling* of heat production to an increase in order that distinguishes the metabolism of a cell from the wasteful burning of fuel in a fire. Later, we shall illustrate how this coupling occurs. For now, it is sufficient to recognize that a direct linkage of the "burning" of food molecules to the generation of biological order is required for cells to create and maintain an island of order in a universe tending toward chaos.

Photosynthetic Organisms Use Sunlight to Synthesize Organic Molecules

All animals live on energy stored in the chemical bonds of organic molecules made by other organisms, which they take in as food. The molecules in food also provide the atoms that animals need to construct new living matter. Some animals obtain their food by eating other animals. But at the bottom of the animal food chain are animals that eat plants. The plants, in turn, trap energy directly from sunlight. As a result, the sun is the ultimate source of the energy used by animal cells.

Solar energy enters the living world through **photosynthesis** in plants and photosynthetic bacteria. Photosynthesis converts the electromagnetic energy in sunlight into chemical bond energy in the cell. Plants obtain all the atoms they need from inorganic sources: carbon from atmospheric carbon dioxide, hydrogen and oxygen from water, nitrogen from ammonia and nitrates in the

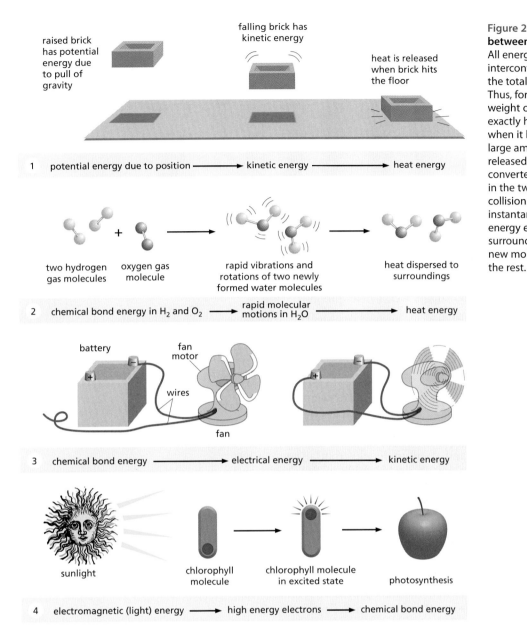

1 potential energy due to position ⟶ kinetic energy ⟶ heat energy

2 chemical bond energy in H_2 and O_2 ⟶ rapid molecular motions in H_2O ⟶ heat energy

3 chemical bond energy ⟶ electrical energy ⟶ kinetic energy

4 electromagnetic (light) energy ⟶ high energy electrons ⟶ chemical bond energy

Figure 2–39 Some interconversions between different forms of energy. All energy forms are, in principle, interconvertible. In all these processes the total amount of energy is conserved. Thus, for example, from the height and weight of the brick in (1), we can predict exactly how much heat will be released when it hits the floor. In (2), note that the large amount of chemical bond energy released when water is formed is initially converted to very rapid thermal motions in the two new water molecules; but collisions with other molecules almost instantaneously spread this kinetic energy evenly throughout the surroundings (heat transfer), making the new molecules indistinguishable from all the rest.

soil, and other elements needed in smaller amounts from inorganic salts in the soil. They use the energy they derive from sunlight to build these atoms into sugars, amino acids, nucleotides, and fatty acids. These small molecules in turn are converted into the proteins, nucleic acids, polysaccharides, and lipids that form the plant. All of these substances serve as food molecules for animals, if the plants are later eaten.

The reactions of photosynthesis take place in two stages (**Figure 2–40**). In the first stage, energy from sunlight is captured and transiently stored as chemical bond energy in specialized small molecules that act as carriers of energy and reactive chemical groups. (We discuss these "activated carrier" molecules later.) Molecular oxygen (O_2 gas) derived from the splitting of water by light is released as a waste product of this first stage.

In the second stage, the molecules that serve as energy carriers are used to help drive a *carbon fixation* process in which sugars are manufactured from carbon dioxide gas (CO_2) and water (H_2O), thereby providing a useful source of stored chemical bond energy and materials—both for the plant itself and for any animals that eat it. We describe the elegant mechanisms that underlie these two stages of photosynthesis in Chapter 14.

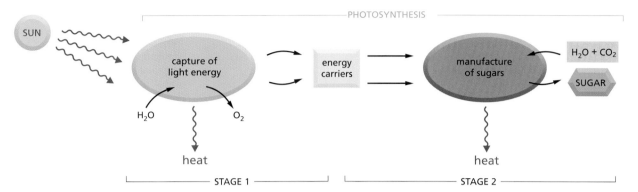

Figure 2–40 Photosynthesis. The two stages of photosynthesis. The energy carriers created in the first stage are two molecules that we discuss shortly—ATP and NADPH.

The net result of the entire process of photosynthesis, so far as the green plant is concerned, can be summarized simply in the equation

$$\text{light energy} + CO_2 + H_2O \rightarrow \text{sugars} + O_2 + \text{heat energy}$$

The sugars produced are then used both as a source of chemical bond energy and as a source of materials to make the many other small and large organic molecules that are essential to the plant cell.

Cells Obtain Energy by the Oxidation of Organic Molecules

All animal and plant cells are powered by energy stored in the chemical bonds of organic molecules, whether they are sugars that a plant has photosynthesized as food for itself or the mixture of large and small molecules that an animal has eaten. Organisms must extract this energy in usable form to live, grow, and reproduce. In both plants and animals, energy is extracted from food molecules by a process of gradual oxidation, or controlled burning.

The Earth's atmosphere contains a great deal of oxygen, and in the presence of oxygen the most energetically stable form of carbon is CO_2 and that of hydrogen is H_2O. A cell is therefore able to obtain energy from sugars or other organic molecules by allowing their carbon and hydrogen atoms to combine with oxygen to produce CO_2 and H_2O, respectively—a process called **respiration**.

Photosynthesis and respiration are complementary processes (**Figure 2–41**). This means that the transactions between plants and animals are not all one way. Plants, animals, and microorganisms have existed together on this planet for so long that many of them have become an essential part of the others' environments. The oxygen released by photosynthesis is consumed in the combustion of organic molecules by nearly all organisms. And some of the CO_2 molecules that are fixed today into organic molecules by photosynthesis in a green leaf were yesterday released into the atmosphere by the respiration of an animal—or by that of a fungus or bacterium decomposing dead organic matter. We therefore see that carbon utilization forms a huge cycle that involves the *biosphere* (all of the living organisms on Earth) as a whole, crossing boundaries

Figure 2–41 Photosynthesis and respiration as complementary processes in the living world. Photosynthesis uses the energy of sunlight to produce sugars and other organic molecules. These molecules in turn serve as food for other organisms. Many of these organisms carry out respiration, a process that uses O_2 to form CO_2 from the same carbon atoms that had been taken up as CO_2 and converted into sugars by photosynthesis. In the process, the organisms that respire obtain the chemical bond energy that they need to survive. The first cells on the Earth are thought to have been capable of neither photosynthesis nor respiration (discussed in Chapter 14). However, photosynthesis must have preceded respiration on the Earth, since there is strong evidence that billions of years of photosynthesis were required before O_2 had been released in sufficient quantity to create an atmosphere rich in this gas. (The Earth's atmosphere currently contains 20% O_2.)

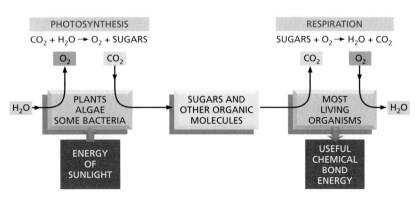

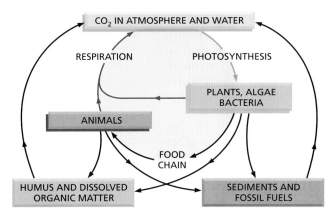

Figure 2–42 The carbon cycle. Individual carbon atoms are incorporated into organic molecules of the living world by the photosynthetic activity of bacteria and plants (including algae). They pass to animals, microorganisms, and organic material in soil and oceans in cyclic paths. CO_2 is restored to the atmosphere when organic molecules are oxidized by cells or burned by humans as fuels.

between individual organisms (**Figure 2–42**). Similarly, atoms of nitrogen, phosphorus, and sulfur move between the living and nonliving worlds in cycles that involve plants, animals, fungi, and bacteria.

Oxidation and Reduction Involve Electron Transfers

The cell does not oxidize organic molecules in one step, as occurs when organic material is burned in a fire. Through the use of enzyme catalysts, metabolism takes the molecules through a large number of reactions that only rarely involve the direct addition of oxygen. Before we consider some of these reactions and their purpose, we discuss what is meant by the process of oxidation.

Oxidation does not mean only the addition of oxygen atoms; rather, it applies more generally to any reaction in which electrons are transferred from one atom to another. Oxidation in this sense refers to the removal of electrons, and **reduction**—the converse of oxidation—means the addition of electrons. Thus, Fe^{2+} is oxidized if it loses an electron to become Fe^{3+}, and a chlorine atom is reduced if it gains an electron to become Cl^-. Since the number of electrons is conserved (no loss or gain) in a chemical reaction, oxidation and reduction always occur simultaneously: that is, if one molecule gains an electron in a reaction (reduction), a second molecule loses the electron (oxidation). When a sugar molecule is oxidized to CO_2 and H_2O, for example, the O_2 molecules involved in forming H_2O gain electrons and thus are said to have been reduced.

The terms "oxidation" and "reduction" apply even when there is only a partial shift of electrons between atoms linked by a covalent bond (**Figure 2–43**).

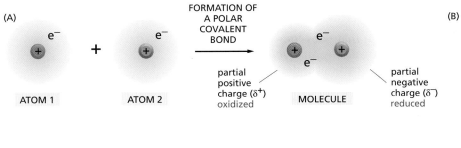

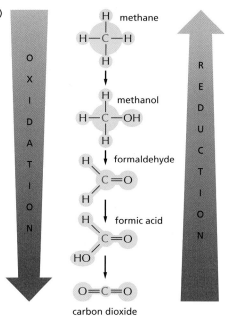

Figure 2–43 Oxidation and reduction. (A) When two atoms form a *polar* covalent bond (see p. 50), the atom ending up with a greater share of electrons is said to be reduced, while the other atom acquires a lesser share of electrons and is said to be oxidized. The reduced atom has acquired a partial negative charge (δ^-) as the positive charge on the atomic nucleus is now more than equaled by the total charge of the electrons surrounding it, and conversely, the oxidized atom has acquired a partial positive charge (δ^+). (B) The single carbon atom of methane can be converted to that of carbon dioxide by the successive replacement of its covalently bonded hydrogen atoms with oxygen atoms. With each step, electrons are shifted away from the carbon (as indicated by the *blue* shading), and the carbon atom becomes progressively more oxidized. Each of these steps is energetically favorable under the conditions present inside a cell.

When a carbon atom becomes covalently bonded to an atom with a strong affinity for electrons, such as oxygen, chlorine, or sulfur, for example, it gives up more than its equal share of electrons and forms a *polar* covalent bond: the positive charge of the carbon nucleus is now somewhat greater than the negative charge of its electrons, and the atom therefore acquires a partial positive charge and is said to be oxidized. Conversely, a carbon atom in a C–H linkage has slightly more than its share of electrons, and so it is said to be reduced (see Figure 2–43).

When a molecule in a cell picks up an electron (e^-), it often picks up a proton (H^+) at the same time (protons being freely available in water). The net effect in this case is to add a hydrogen atom to the molecule

$$A + e^- + H^+ \rightarrow AH$$

Even though a proton plus an electron is involved (instead of just an electron), such *hydrogenation* reactions are reductions, and the reverse, *dehydrogenation* reactions, are oxidations. It is especially easy to tell whether an organic molecule is being oxidized or reduced: reduction is occurring if its number of C–H bonds increases, whereas oxidation is occurring if its number of C–H bonds decreases (see Figure 2–43B).

Cells use enzymes to catalyze the oxidation of organic molecules in small steps, through a sequence of reactions that allows useful energy to be harvested. We now need to explain how enzymes work and some of the constraints under which they operate.

Enzymes Lower the Barriers That Block Chemical Reactions

Consider the reaction

$$\text{paper} + O_2 \rightarrow \text{smoke} + \text{ashes} + \text{heat} + CO_2 + H_2O$$

The paper burns readily, releasing to the atmosphere both energy as heat and water and carbon dioxide as gases, but the smoke and ashes never spontaneously retrieve these entities from the heated atmosphere and reconstitute themselves into paper. When the paper burns, its chemical energy is dissipated as heat—not lost from the universe, since energy can never be created or destroyed, but irretrievably dispersed in the chaotic random thermal motions of molecules. At the same time, the atoms and molecules of the paper become dispersed and disordered. In the language of thermodynamics, there has been a loss of *free energy*, that is, of energy that can be harnessed to do work or drive chemical reactions. This loss reflects a loss of orderliness in the way the energy and molecules were stored in the paper. We shall discuss free energy in more detail shortly, but the general principle is clear enough intuitively: chemical reactions proceed spontaneously only in the direction that leads to a loss of free energy; in other words, the spontaneous direction for any reaction is the direction that goes "downhill." A "downhill" reaction in this sense is often said to be *energetically favorable*.

Although the most energetically favorable form of carbon under ordinary conditions is CO_2, and that of hydrogen is H_2O, a living organism does not disappear in a puff of smoke, and the book in your hands does not burst into flames. This is because the molecules both in the living organism and in the book are in a relatively stable state, and they cannot be changed to a state of lower energy without an input of energy: in other words, a molecule requires **activation energy**—a kick over an energy barrier—before it can undergo a chemical reaction that leaves it in a more stable state (**Figure 2–44**). In the case of a burning book, the activation energy is provided by the heat of a lighted match. For the molecules in the watery solution inside a cell, the kick is delivered by an unusually energetic random collision with surrounding molecules—collisions that become more violent as the temperature is raised.

In a living cell, the kick over the energy barrier is greatly aided by a specialized class of proteins—the **enzymes**. Each enzyme binds tightly to one or more molecules, called **substrates**, and holds them in a way that greatly reduces the activation energy of a particular chemical reaction that the bound substrates can undergo. A substance that can lower the activation energy of a reaction is

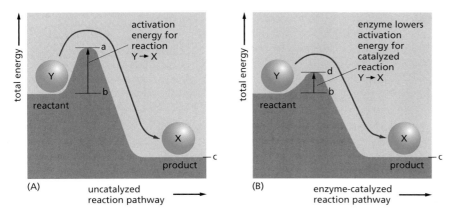

Figure 2–44 The important principle of activation energy. (A) Compound Y (a reactant) is in a relatively stable state, and energy is required to convert it to compound X (a product), even though X is at a lower overall energy level than Y. This conversion will not take place, therefore, unless compound Y can acquire enough activation energy *(energy a minus energy b)* from its surroundings to undergo the reaction that converts it into compound X. This energy may be provided by means of an unusually energetic collision with other molecules. For the reverse reaction, X → Y, the activation energy will be much larger *(energy a minus energy c);* this reaction will therefore occur much more rarely. Activation energies are always positive; note, however, that the total energy change for the energetically favorable reaction Y → X is *energy c minus energy b,* a negative number. (B) Energy barriers for specific reactions can be lowered by catalysts, as indicated by the line marked *d*. Enzymes are particularly effective catalysts because they greatly reduce the activation energy for the reactions they perform.

termed a **catalyst**; catalysts increase the rate of chemical reactions because they allow a much larger proportion of the random collisions with surrounding molecules to kick the substrates over the energy barrier, as illustrated in **Figure 2–45**. Enzymes are among the most effective catalysts known, capable of speeding up reactions by factors of 10^{14} or more. They thereby allow reactions that would not otherwise occur to proceed rapidly at normal temperatures.

Enzymes are also highly selective. Each enzyme usually catalyzes only one particular reaction: in other words, it selectively lowers the activation energy of only one of the several possible chemical reactions that its bound substrate molecules could undergo. In this way, enzymes direct each of the many different molecules in a cell along specific reaction pathways (**Figure 2–46**).

The success of living organisms is attributable to a cell's ability to make enzymes of many types, each with precisely specified properties. Each enzyme has a unique shape containing an *active site*, a pocket or groove in the enzyme into which only particular substrates will fit (**Figure 2–47**). Like all other catalysts, enzyme molecules themselves remain unchanged after participating in a reaction and therefore can function over and over again. In Chapter 3, we discuss further how enzymes work.

Figure 2–45 Lowering the activation energy greatly increases the probability of reaction. At any given instant, a population of identical substrate molecules will have a range of energies, distributed as shown on the graph. The varying energies come from collisions with surrounding molecules, which make the substrate molecules jiggle, vibrate, and spin. For a molecule to undergo a chemical reaction, the energy of the molecule must exceed the activation energy barrier for that reaction; for most biological reactions, this almost never happens without enzyme catalysis. Even with enzyme catalysis, the substrate molecules must experience a particularly energetic collision to react *(red shaded area)*. Raising the temperature can also increase the number of molecules with sufficient energy to overcome the activation energy needed for a reaction; but in contrast to enzyme catalysis, this effect is nonselective, speeding up all reactions.

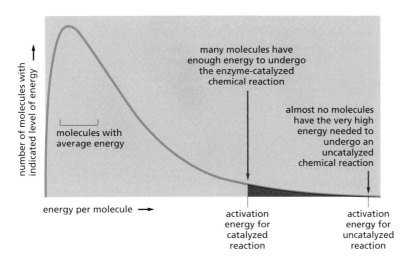

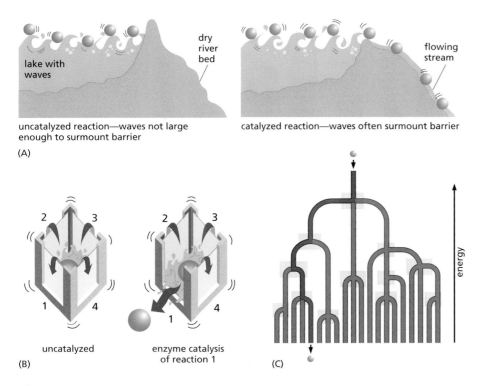

uncatalyzed reaction—waves not large enough to surmount barrier

catalyzed reaction—waves often surmount barrier

(A)

(B) uncatalyzed   enzyme catalysis of reaction 1

(C)

Figure 2–46 Floating ball analogies for enzyme catalysis. <TAAA> (A) A barrier dam is lowered to represent enzyme catalysis. The *green ball* represents a potential reactant (compound Y) that is bouncing up and down in energy level due to constant encounters with waves (an analogy for the thermal bombardment of the reactant molecule with the surrounding water molecules). When the barrier (activation energy) is lowered significantly, it allows the energetically favorable movement of the ball (the reactant) downhill. (B) The four walls of the box represent the activation energy barriers for four different chemical reactions that are all energetically favorable, in the sense that the products are at lower energy levels than the reactants. In the *left-hand* box, none of these reactions occurs because even the largest waves are not large enough to surmount any of the energy barriers. In the *right-hand* box, enzyme catalysis lowers the activation energy for reaction number 1 only; now the jostling of the waves allows passage of the reactant molecule over this energy barrier, inducing reaction 1. (C) A branching river with a set of barrier dams *(yellow boxes)* serves to illustrate how a series of enzyme-catalyzed reactions determines the exact reaction pathway followed by each molecule inside the cell.

How Enzymes Find Their Substrates: The Enormous Rapidity of Molecular Motions

An enzyme will often catalyze the reaction of thousands of substrate molecules every second. This means that it must be able to bind a new substrate molecule in a fraction of a millisecond. But both enzymes and their substrates are present in relatively small numbers in a cell. How do they find each other so fast? Rapid binding is possible because the motions caused by heat energy are enormously fast at the molecular level. These molecular motions can be classified broadly into three kinds: (1) the movement of a molecule from one place to another (*translational motion*), (2) the rapid back-and-forth movement of covalently linked atoms with respect to one another (vibrations), and (3) rotations. All of these motions help to bring the surfaces of interacting molecules together.

The rates of molecular motions can be measured by a variety of spectroscopic techniques. A large globular protein is constantly tumbling, rotating about its axis about a million times per second. Molecules are also in constant translational motion, which causes them to explore the space inside the cell very efficiently by wandering through it—a process called **diffusion**. In this way, every molecule in a cell collides with a huge number of other molecules each second. As the molecules in a liquid collide and bounce off one another, an individual molecule moves first one way and then another, its path constituting a *random walk* (**Figure 2–48**). In such a walk, the average net distance that each molecule travels (as the crow flies) from its starting point is proportional to the square root of the time involved: that is, if it takes a molecule 1 second on average to travel 1 μm, it takes 4 seconds to travel 2 μm, 100 seconds to travel 10 μm, and so on.

The inside of a cell is very crowded (**Figure 2–49**). Nevertheless, experiments in which fluorescent dyes and other labeled molecules are injected into cells

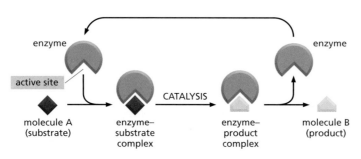

enzyme

active site

molecule A (substrate)

enzyme–substrate complex

CATALYSIS

enzyme–product complex

enzyme

molecule B (product)

Figure 2–47 How enzymes work. Each enzyme has an active site to which one or more *substrate* molecules bind, forming an enzyme–substrate complex. A reaction occurs at the active site, producing an enzyme–product complex. The *product* is then released, allowing the enzyme to bind further substrate molecules.

show that small organic molecules diffuse through the watery gel of the cytosol nearly as rapidly as they do through water. A small organic molecule, for example, takes only about one-fifth of a second on average to diffuse a distance of 10 μm. Diffusion is therefore an efficient way for small molecules to move the limited distances in the cell (a typical animal cell is 15 μm in diameter).

Since enzymes move more slowly than substrates in cells, we can think of them as sitting still. The rate of encounter of each enzyme molecule with its substrate will depend on the concentration of the substrate molecule. For example, some abundant substrates are present at a concentration of 0.5 mM. Since pure water is 55.5 M, there is only about one such substrate molecule in the cell for every 10^5 water molecules. Nevertheless, the active site on an enzyme molecule that binds this substrate will be bombarded by about 500,000 random collisions with the substrate molecule per second. (For a substrate concentration tenfold lower, the number of collisions drops to 50,000 per second, and so on.) A random encounter between the surface of an enzyme and the matching surface of its substrate molecule often leads immediately to the formation of an enzyme–substrate complex that is ready to react. A reaction in which a covalent bond is broken or formed can now occur extremely rapidly. When one appreciates how quickly molecules move and react, the observed rates of enzymatic catalysis do not seem so amazing.

Once an enzyme and substrate have collided and snuggled together properly at the active site, they form multiple weak bonds with each other that persist until random thermal motion causes the molecules to dissociate again. In general, the stronger the binding of the enzyme and substrate, the slower their rate of dissociation. However, when two colliding molecules have poorly matching surfaces, they form few noncovalent bonds and their total energy is negligible compared with that of thermal motion. In this case the two molecules dissociate as rapidly as they come together, preventing incorrect and unwanted associations between mismatched molecules, such as between an enzyme and the wrong substrate.

The Free-Energy Change for a Reaction Determines Whether It Can Occur

We must now digress briefly to introduce some fundamental chemistry. Cells are chemical systems that must obey all chemical and physical laws. Although enzymes speed up reactions, they cannot by themselves force energetically unfavorable reactions to occur. In terms of a water analogy, enzymes by themselves cannot make water run uphill. Cells, however, must do just that in order to grow and divide: they must build highly ordered and energy-rich molecules from small and simple ones. We shall see that this is done through enzymes that directly *couple* energetically favorable reactions, which release energy and produce heat, to energetically unfavorable reactions, which produce biological order.

Before examining how such coupling is achieved, we must consider more carefully the term "energetically favorable." According to the second law of thermodynamics, a chemical reaction can proceed spontaneously only if it results in a net increase in the disorder of the universe (see Figure 2–38). The criterion for an increase in disorder of the universe can be expressed most conveniently in terms of a quantity called the **free energy**, *G*, of a system. The value of *G* is of interest only when a system undergoes a *change*, and the change in *G*, denoted **Δ***G* (delta *G*), is critical. Suppose that the system being considered is a collection of molecules. As explained in Panel 2–7 (pp. 118–119), free energy has been defined such that Δ*G* directly measures the amount of disorder created in the universe when a reaction takes place that involves these molecules. *Energetically favorable reactions*, by definition, are those that decrease free energy; in other words, they have a *negative* Δ*G* and disorder the universe (**Figure 2–50**).

An example of an energetically favorable reaction on a macroscopic scale is the "reaction" by which a compressed spring relaxes to an expanded state, releasing its stored elastic energy as heat to its surroundings; an example on a microscopic scale is salt dissolving in water. Conversely, *energetically unfavorable reactions*, with a *positive* Δ*G*—such as the joining of two amino acids to

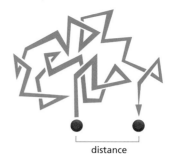

Figure 2–48 A random walk. <GGTA> Molecules in solution move in a random fashion as a result of the continual buffeting they receive in collisions with other molecules. This movement allows small molecules to diffuse rapidly from one part of the cell to another, as described in the text.

100 nm

Figure 2–49 The structure of the cytoplasm. The drawing is approximately to scale and emphasizes the crowding in the cytoplasm. Only the macromolecules are shown: RNAs are shown in *blue,* ribosomes in *green,* and proteins in *red.* Enzymes and other macromolecules diffuse relatively slowly in the cytoplasm, in part because they interact with many other macromolecules; small molecules, by contrast, diffuse nearly as rapidly as they do in water. (Adapted from D.S. Goodsell, *Trends Biochem. Sci.* 16:203–206, 1991. With permission from Elsevier.)

form a peptide bond—by themselves create order in the universe. Therefore, these reactions can take place only if they are coupled to a second reaction with a negative ΔG so large that the ΔG of the entire process is negative (**Figure 2–51**).

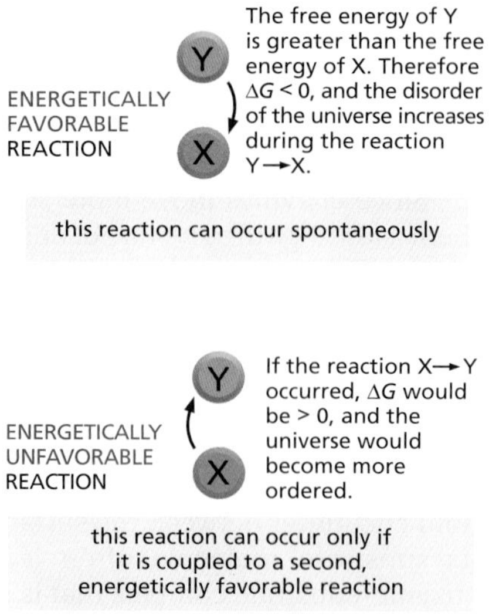

The Concentration of Reactants Influences the Free-Energy Change and a Reaction's Direction

As we have just described, a reaction $Y \rightleftharpoons X$ will go in the direction $Y \rightarrow X$ when the associated free-energy change, ΔG, is negative, just as a tensed spring left to itself will relax and lose its stored energy to its surroundings as heat. For a chemical reaction, however, ΔG depends not only on the energy stored in each individual molecule, but also on the concentrations of the molecules in the reaction mixture. Remember that ΔG reflects the degree to which a reaction creates a more disordered—in other words, a more probable—state of the universe. Recalling our coin analogy, it is very likely that a coin will flip from a head to a tail orientation if a jiggling box contains 90 heads and 10 tails, but this is a less probable event if the box has 10 heads and 90 tails.

The same is true for a chemical reaction. For a reversible reaction $Y \rightleftharpoons X$, a large excess of Y over X will tend to drive the reaction in the direction $Y \rightarrow X$; that is, there will be a tendency for there to be more molecules making the transition $Y \rightarrow X$ than there are molecules making the transition $X \rightarrow Y$. If the ratio of Y to X increases, the ΔG becomes more negative for the transition $Y \rightarrow X$ (and more positive for the transition $X \rightarrow Y$).

How much of a concentration difference is needed to compensate for a given decrease in chemical bond energy (and accompanying heat release)? The answer is not intuitively obvious, but it can be determined from a thermodynamic analysis that makes it possible to separate the concentration-dependent and the concentration-independent parts of the free-energy change. The ΔG for a given reaction can thereby be written as the sum of two parts: the first, called the *standard free-energy change*, $\Delta G°$, depends on the intrinsic characters of the reacting molecules; the second depends on their concentrations. For the simple reaction $Y \rightarrow X$ at 37°C,

$$\Delta G = \Delta G° + 0.616 \ln \frac{[X]}{[Y]} = \Delta G° + 1.42 \log \frac{[X]}{[Y]}$$

where ΔG is in kilocalories per mole, [Y] and [X] denote the concentrations of Y and X, ln is the natural logarithm, and the constant 0.616 is equal to RT: the product of the gas constant, R, and the absolute temperature, T.

Note that ΔG equals the value of $\Delta G°$ when the molar concentrations of Y and X are equal (log 1 = 0). As expected, ΔG becomes more negative as the ratio of X to Y decreases (the log of a number < 1 is negative).

Inspection of the above equation reveals that the ΔG equals the value of $\Delta G°$ when the concentrations of Y and X are equal. But as the favorable reaction $Y \rightarrow X$ proceeds, the concentration of the product X increases and the concentration of the substrate Y decreases. This change in relative concentrations will cause [X]/[Y] to become increasingly large, making the initially favorable ΔG less and less negative. Eventually, when $\Delta G = 0$, a chemical **equilibrium** will be attained; here the concentration effect just balances the push given to the reaction by $\Delta G°$, and the ratio of substrate to product reaches a constant value (**Figure 2–52**).

How far will a reaction proceed before it stops at equilibrium? To address this question, we need to introduce the **equilibrium constant**, K. The value of K is different for different reactions, and it reflects the ratio of product to substrate at equilibrium. For the reaction $Y \rightarrow X$:

$$K = \frac{[X]}{[Y]}$$

The equation that connects ΔG and the ratio [X]/[Y] allows us to connect $\Delta G°$ directly to K. Since $\Delta G = 0$ at equilibrium, the concentrations of Y and X at this point are such that:

$$\Delta G° = -1.42 \log \frac{[X]}{[Y]} \quad \text{or,} \quad \Delta G° = -1.42 \log K$$

ENERGETICALLY FAVORABLE REACTION

The free energy of Y is greater than the free energy of X. Therefore $\Delta G < 0$, and the disorder of the universe increases during the reaction $Y \rightarrow X$.

this reaction can occur spontaneously

ENERGETICALLY UNFAVORABLE REACTION

If the reaction $X \rightarrow Y$ occurred, ΔG would be > 0, and the universe would become more ordered.

this reaction can occur only if it is coupled to a second, energetically favorable reaction

Figure 2–50 The distinction between energetically favorable and energetically unfavorable reactions.

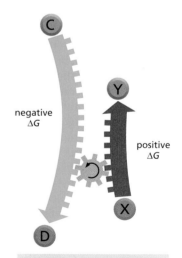

negative ΔG

positive ΔG

the energetically unfavorable reaction X→Y is driven by the energetically favorable reaction C→D, because the net free-energy change for the pair of coupled reactions is less than zero

Figure 2–51 How reaction coupling is used to drive energetically unfavorable reactions.

THE REACTION

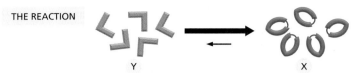

The formation of X is energetically favored in this example. In other words, the ΔG of Y → X is negative and the ΔG of X → Y is positive. But because of thermal bombardments, there will always be some X converting to Y and vice versa.

SUPPOSE WE START WITH AN EQUAL NUMBER OF Y AND X MOLECULES

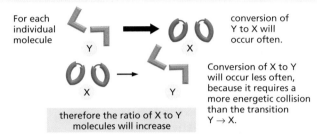

For each individual molecule

conversion of Y to X will occur often.

Conversion of X to Y will occur less often, because it requires a more energetic collision than the transition Y → X.

therefore the ratio of X to Y molecules will increase

EVENTUALLY   there will be a large enough excess of X over Y to just compensate for the slow rate of X → Y. Equilibrium will then be attained.

AT EQUILIBRIUM   the number of Y molecules being converted to X molecules each second is exactly equal to the number of X molecules being converted to Y molecules each second, so that there is no net change in the ratio of Y to X.

Figure 2–52 Chemical equilibrium. When a reaction reaches equilibrium, the forward and backward fluxes of reacting molecules are equal and opposite.

Table 2–4 Relationship Between the Standard Free-Energy Change, ΔG°, and the Equilibrium Constant

EQUILIBRIUM CONSTANT $\dfrac{[X]}{[Y]} = K$	FREE ENERGY OF X MINUS FREE ENERGY OF Y kcal/mole (kJ/mole)
10^5	−7.1 (−29.7)
10^4	−5.7 (−23.8)
10^3	−4.3 (−18.0)
10^2	−2.8 (−11.7)
10^1	−1.4 (−5.9)
1	0 (0)
10	1.4 (5.9)
10^{-2}	2.8 (11.7)
10^{-3}	4.3 (18.0)
10^{-4}	5.7 (23.8)
10^{-5}	7.1 (29.7)

Values of the equilibrium constant were calculated for the simple chemical reaction Y ⇌ X using the equation given in the text.

The ΔG° given here is in kilocalories per mole at 37°C, with kilojoules per mole in parentheses (1 kilocalorie is equal to 4.184 kilojoules). As explained in the text, ΔG° represents the free-energy difference under standard conditions (where all components are present at a concentration of 1.0 mole/liter).

From this table, we see that if there is a favorable standard free-energy change (ΔG°) of −4.3 kcal/mole (−18.0 kJ/mole) for the transition Y → X, there will be 1000 times more molecules in state X than in state Y at equilibrium (K = 1000).

Using the last equation, we can see how the equilibrium ratio of X to Y (expressed as an equilibrium constant, K) depends on the intrinsic character of the molecules, as expressed in the value of ΔG° (**Table 2–4**). Note that for every 1.4 kcal/mole (5.9 kJ/mole) difference in free energy at 37°C, the equilibrium constant changes by a factor of 10.

When an enzyme (or any catalyst) lowers the activation energy for the reaction Y → X, it also lowers the activation energy for the reaction X → Y by exactly the same amount (see Figure 2–44). The forward and backward reactions will therefore be accelerated by the same factor by an enzyme, and the equilibrium point for the reaction (and ΔG°) is unchanged (**Figure 2–53**).

For Sequential Reactions, ΔG° Values Are Additive

We can predict quantitatively the course of most reactions. A large body of thermodynamic data has been collected that makes it possible to calculate the standard change in free energy, ΔG°, for most of the important metabolic reactions of the cell. The overall free-energy change for a metabolic pathway is then simply the sum of the free-energy changes in each of its component steps. Consider, for example, two sequential reactions

$$X \rightarrow Y \quad \text{and} \quad Y \rightarrow Z$$

whose ΔG° values are +5 and –13 kcal/mole, respectively. (Recall that a mole is 6 × 10^{23} molecules of a substance.) If these two reactions occur sequentially, the ΔG° for the coupled reaction will be –8 kcal/mole. Thus, the unfavorable reaction X → Y, which will not occur spontaneously, can be driven by the favorable reaction Y → Z, provided that this second reaction follows the first.

Cells can therefore cause the energetically unfavorable transition, X → Y, to occur if an enzyme catalyzing the X → Y reaction is supplemented by a second enzyme that catalyzes the energetically *favorable* reaction, Y → Z. In effect, the reaction Y → Z will then act as a "siphon" to drive the conversion of all of molecule X to molecule Y, and thence to molecule Z (**Figure 2–54**). For example,

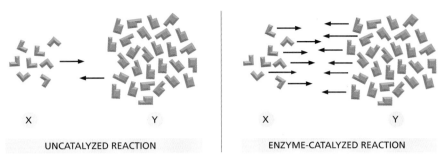

UNCATALYZED REACTION | ENZYME-CATALYZED REACTION

Figure 2–53 Enzymes cannot change the equilibrium point for reactions. Enzymes, like all catalysts, speed up the forward and backward rates of a reaction by the same factor. Therefore, for both the catalyzed and the uncatalyzed reactions shown here, the number of molecules undergoing the transition $X \rightarrow Y$ is equal to the number of molecules undergoing the transition $Y \rightarrow X$ when the ratio of Y molecules to X molecules is 3.5 to 1. In other words, the two reactions reach equilibrium at exactly the same point.

several of the reactions in the long pathway that converts sugars into CO_2 and H_2O would be energetically unfavorable if considered on their own. But the pathway nevertheless proceeds because the total $\Delta G°$ for the series of sequential reactions has a large negative value.

But forming a sequential pathway is not adequate for many purposes. Often the desired pathway is simply $X \rightarrow Y$, without further conversion of Y to some other product. Fortunately, there are other more general ways of using enzymes to couple reactions together. How these work is the topic we discuss next.

Activated Carrier Molecules Are Essential for Biosynthesis

The energy released by the oxidation of food molecules must be stored temporarily before it can be channeled into the construction of the many other molecules needed by the cell. In most cases, the energy is stored as chemical bond energy in a small set of activated "carrier molecules," which contain one or more energy-rich covalent bonds. These molecules diffuse rapidly throughout the cell and thereby carry their bond energy from sites of energy generation to the sites where energy is used for biosynthesis and other cell activities (**Figure 2–55**).

The **activated carriers** store energy in an easily exchangeable form, either as a readily transferable chemical group or as high-energy electrons, and they can serve a dual role as a source of both energy and chemical groups in biosynthetic reactions. For historical reasons, these molecules are also sometimes referred to as *coenzymes*. The most important of the activated carrier molecules are ATP and two molecules that are closely related to each other, NADH and NADPH— as we discuss in detail shortly. We shall see that cells use activated carrier molecules like money to pay for reactions that otherwise could not take place.

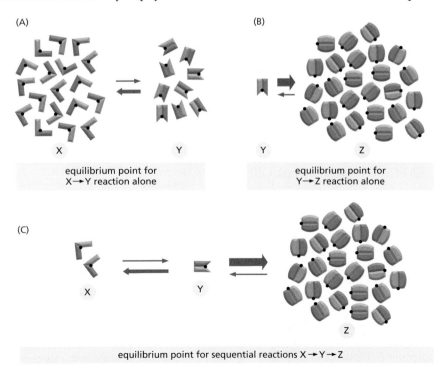

equilibrium point for sequential reactions $X \rightarrow Y \rightarrow Z$

Figure 2–54 How an energetically unfavorable reaction can be driven by a second, following reaction. (A) At equilibrium, there are twice as many X molecules as Y molecules, because X is of lower energy than Y. (B) At equilibrium, there are 25 times more Z molecules than Y molecules, because Z is of much lower energy than Y. (C) If the reactions in (A) and (B) are coupled, nearly all of the X molecules will be converted to Z molecules, as shown.

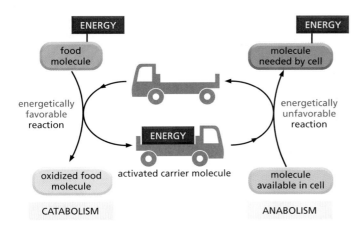

Figure 2–55 **Energy transfer and the role of activated carriers in metabolism.** By serving as energy shuttles, activated carrier molecules perform their function as go-betweens that link the breakdown of food molecules and the release of energy *(catabolism)* to the energy-requiring biosynthesis of small and large organic molecules *(anabolism)*.

The Formation of an Activated Carrier Is Coupled to an Energetically Favorable Reaction

When a fuel molecule such as glucose is oxidized in a cell, enzyme-catalyzed reactions ensure that a large part of the free energy that is released by oxidation is captured in a chemically useful form, rather than being released as heat. This is achieved by means of a **coupled reaction**, in which an energetically favorable reaction drives an energetically unfavorable one that produces an activated carrier molecule or some other useful energy store. Coupling mechanisms require enzymes and are fundamental to all the energy transactions of the cell.

The nature of a coupled reaction is illustrated by a mechanical analogy in **Figure 2–56**, in which an energetically favorable chemical reaction is represented by rocks falling from a cliff. The energy of falling rocks would normally be entirely wasted in the form of heat generated by friction when the rocks hit the ground (see the falling brick diagram in Figure 2–39). By careful design, however, part of this energy could be used instead to drive a paddle wheel that lifts a bucket of water (Figure 2–56B). Because the rocks can now reach the ground only after moving the paddle wheel, we say that the energetically favorable reaction of rock falling has been directly *coupled* to the energetically unfavorable reaction of lifting the bucket of water. Note that because part of the energy is used to do work in (B), the rocks hit the ground with less velocity than in (A), and correspondingly less energy is dissipated as heat.

Similar processes occur in cells, where enzymes play the role of the paddle wheel in our analogy. By mechanisms that will be discussed later in this chapter, they couple an energetically favorable reaction, such as the oxidation of foodstuffs, to an energetically unfavorable reaction, such as the generation of

Figure 2–56 **A mechanical model illustrating the principle of coupled chemical reactions.** The spontaneous reaction shown in (A) could serve as an analogy for the direct oxidation of glucose to CO_2 and H_2O, which produces heat only. In (B) the same reaction is coupled to a second reaction; this second reaction is analogous to the synthesis of activated carrier molecules. The energy produced in (B) is in a more useful form than in (A) and can be used to drive a variety of otherwise energetically unfavorable reactions (C).

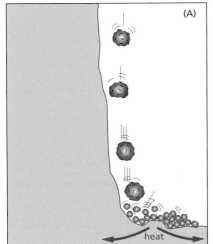

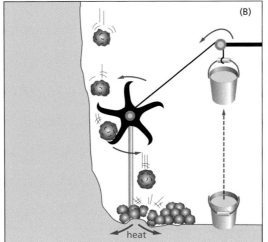

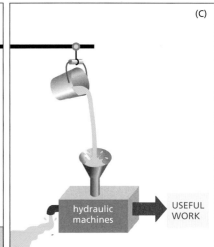

kinetic energy of falling rocks is transformed into heat energy only

part of the kinetic energy is used to lift a bucket of water, and a correspondingly smaller amount is transformed into heat

the potential kinetic energy stored in the raised bucket of water can be used to drive hydraulic machines that carry out a variety of useful tasks

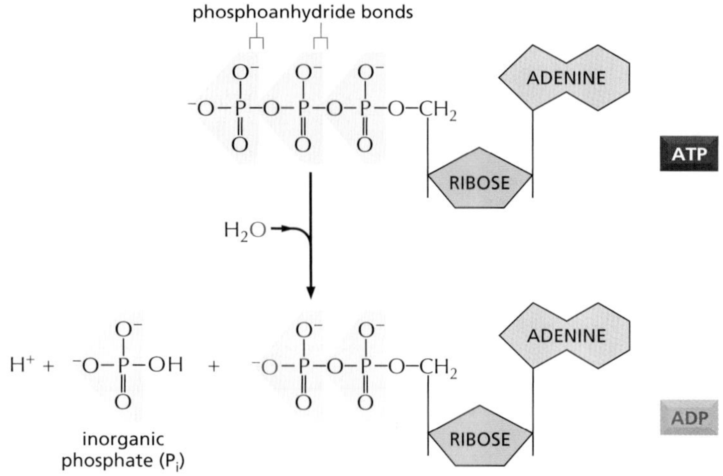

Figure 2–57 The hydrolysis of ATP to ADP and inorganic phosphate. The two outermost phosphates in ATP are held to the rest of the molecule by high-energy phosphoanhydride bonds and are readily transferred. As indicated, water can be added to ATP to form ADP and inorganic phosphate (P_i). This hydrolysis of the terminal phosphate of ATP yields between 11 and 13 kcal/mole of usable energy, depending on the intracellular conditions. The large negative ΔG of this reaction arises from several factors. Release of the terminal phosphate group removes an unfavorable repulsion between adjacent negative charges; in addition, the inorganic phosphate ion (P_i) released is stabilized by resonance and by favorable hydrogen-bond formation with water.

an activated carrier molecule. As a result, the amount of heat released by the oxidation reaction is reduced by exactly the amount of energy that is stored in the energy-rich covalent bonds of the activated carrier molecule. The activated carrier molecule in turn picks up a packet of energy of a size sufficient to power a chemical reaction elsewhere in the cell.

ATP Is the Most Widely Used Activated Carrier Molecule

The most important and versatile of the activated carriers in cells is **ATP** (adenosine triphosphate). Just as the energy stored in the raised bucket of water in Figure 2–56B can drive a wide variety of hydraulic machines, ATP is a convenient and versatile store, or currency, of energy used to drive a variety of chemical reactions in cells. ATP is synthesized in an energetically unfavorable phosphorylation reaction in which a phosphate group is added to **ADP** (adenosine diphosphate). When required, ATP gives up its energy packet through its energetically favorable hydrolysis to ADP and inorganic phosphate (**Figure 2–57**). The regenerated ADP is then available to be used for another round of the phosphorylation reaction that forms ATP.

The energetically favorable reaction of ATP hydrolysis is coupled to many otherwise unfavorable reactions through which other molecules are synthesized. We shall encounter several of these reactions later in this chapter. Many of them involve the transfer of the terminal phosphate in ATP to another molecule, as illustrated by the phosphorylation reaction in **Figure 2–58**.

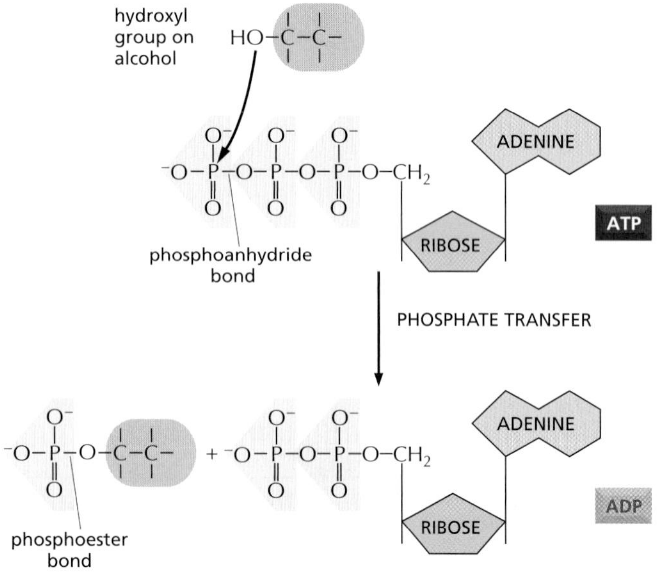

Figure 2–58 An example of a phosphate transfer reaction. Because an energy-rich phosphoanhydride bond in ATP is converted to a phosphoester bond, this reaction is energetically favorable, having a large negative ΔG. Reactions of this type are involved in the synthesis of phospholipids and in the initial steps of reactions that catabolize sugars.

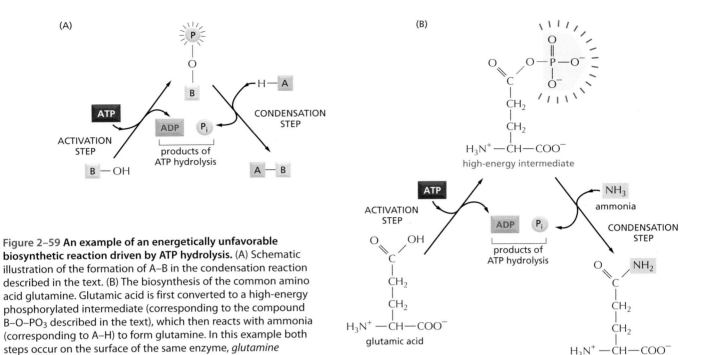

Figure 2–59 An example of an energetically unfavorable biosynthetic reaction driven by ATP hydrolysis. (A) Schematic illustration of the formation of A–B in the condensation reaction described in the text. (B) The biosynthesis of the common amino acid glutamine. Glutamic acid is first converted to a high-energy phosphorylated intermediate (corresponding to the compound B–O–PO$_3$ described in the text), which then reacts with ammonia (corresponding to A–H) to form glutamine. In this example both steps occur on the surface of the same enzyme, *glutamine synthetase*. The high energy bonds are shaded *red*.

ATP is the most abundant activated carrier in cells. As one example, it supplies energy for many of the pumps that transport substances into and out of the cell (discussed in Chapter 11). It also powers the molecular motors that enable muscle cells to contract and nerve cells to transport materials from one end of their long axons to another (discussed in Chapter 16).

Energy Stored in ATP Is Often Harnessed to Join Two Molecules Together

We have previously discussed one way in which an energetically favorable reaction can be coupled to an energetically unfavorable reaction, X → Y, so as to enable it to occur. In that scheme a second enzyme catalyzes the energetically favorable reaction Y → Z, pulling all of the X to Y in the process (see Figure 2–54). But when the required product is Y and not Z, this mechanism is not useful.

A typical biosynthetic reaction is one in which two molecules, A and B, are joined together to produce A–B in the energetically unfavorable condensation reaction

$$\text{A–H} + \text{B–OH} \rightarrow \text{A–B} + \text{H}_2\text{O}$$

There is an indirect pathway that allows A–H and B–OH to form A–B, in which a coupling to ATP hydrolysis makes the reaction go. Here energy from ATP hydrolysis is first used to convert B–OH to a higher-energy intermediate compound, which then reacts directly with A–H to give A–B. The simplest possible mechanism involves the transfer of a phosphate from ATP to B–OH to make B–OPO$_3$, in which case the reaction pathway contains only two steps:

1.   B–OH + ATP → B–O–PO$_3$ + ADP
2.   A–H + B–O–PO$_3$ → A–B + P$_i$

Net result: B–OH + ATP + A–H → A–B + ADP + P$_i$

The condensation reaction, which by itself is energetically unfavorable, is forced to occur by being directly coupled to ATP hydrolysis in an enzyme-catalyzed reaction pathway (**Figure 2–59**A).

A biosynthetic reaction of exactly this type synthesizes the amino acid glutamine (Figure 2–59B). We will see shortly that similar (but more complex) mechanisms are also used to produce nearly all of the large molecules of the cell.

NADH and NADPH Are Important Electron Carriers

Other important activated carrier molecules participate in oxidation–reduction reactions and are commonly part of coupled reactions in cells. These activated carriers are specialized to carry high-energy electrons and hydrogen atoms. The most important of these electron carriers are **NAD+** (nicotinamide adenine dinucleotide) and the closely related molecule **NADP+** (nicotinamide adenine dinucleotide phosphate). Later, we examine some of the reactions in which they participate. NAD+ and NADP+ each pick up a "packet of energy" corresponding to two high-energy electrons plus a proton (H+)—being converted to **NADH** (*reduced* nicotinamide adenine dinucleotide) and **NADPH** (*reduced* nicotinamide adenine dinucleotide phosphate), respectively. These molecules can therefore also be regarded as carriers of hydride ions (the H+ plus two electrons, or H−).

Like ATP, NADPH is an activated carrier that participates in many important biosynthetic reactions that would otherwise be energetically unfavorable. The NADPH is produced according to the general scheme shown in **Figure 2–60**A. During a special set of energy-yielding catabolic reactions, a hydrogen atom plus two electrons are removed from the substrate molecule and added to the nicotinamide ring of NADP+ to form NADPH, with a proton (H+) being released into solution. This is a typical oxidation–reduction reaction; the substrate is oxidized and NADP+ is reduced. The structures of NADP+ and NADPH are shown in Figure 2–60B.

NADPH readily gives up the hydride ion it carries in a subsequent oxidation–reduction reaction, because the nicotinamide ring can achieve a more stable arrangement of electrons without it. In this subsequent reaction, which regenerates NADP+, it is the NADPH that is oxidized and the substrate that is reduced. The NADPH is an effective donor of its hydride ion to other molecules

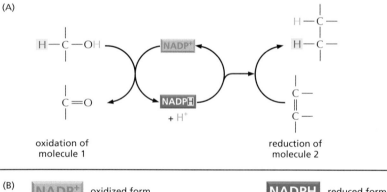

(A)

oxidation of
molecule 1

reduction of
molecule 2

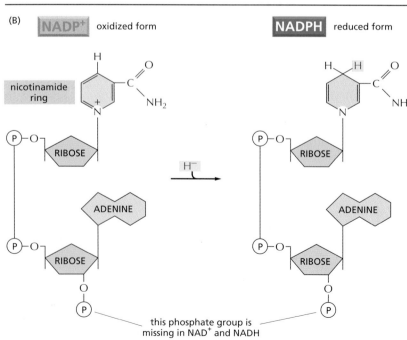

(B)   **NADP+** oxidized form   **NADPH** reduced form

nicotinamide
ring

RIBOSE

ADENINE

RIBOSE

O

P

this phosphate group is
missing in NAD+ and NADH

Figure 2–60 NADPH, an important carrier of electrons. (A) NADPH is produced in reactions of the general type shown on the left, in which two hydrogen atoms are removed from a substrate. The oxidized form of the carrier molecule, NADP+, receives one hydrogen atom plus an electron (a hydride ion); the proton (H+) from the other H atom is released into solution. Because NADPH holds its hydride ion in a high-energy linkage, the added hydride ion can easily be transferred to other molecules, as shown on the right. (B) The structures of NADP+ and NADPH. The part of the NADP+ molecule known as the nicotinamide ring accepts two electrons together with a proton (the equivalent of a hydride ion, H−), forming NADPH. The molecules NAD+ and NADH are identical in structure to NADP+ and NADPH, respectively, except that the indicated phosphate group is absent from both.

for the same reason that ATP readily transfers a phosphate: in both cases the transfer is accompanied by a large negative free-energy change. One example of the use of NADPH in biosynthesis is shown in **Figure 2–61**.

The extra phosphate group on NADPH has no effect on the electron-transfer properties of NADPH compared with NADH, being far away from the region involved in electron transfer (see Figure 2–60B). It does, however, give a molecule of NADPH a slightly different shape from that of NADH, making it possible for NADPH and NADH to bind as substrates to completely different sets of enzymes. Thus the two types of carriers are used to transfer electrons (or hydride ions) between two different sets of molecules.

Why should there be this division of labor? The answer lies in the need to regulate two sets of electron-transfer reactions independently. NADPH operates chiefly with enzymes that catalyze anabolic reactions, supplying the high-energy electrons needed to synthesize energy-rich biological molecules. NADH, by contrast, has a special role as an intermediate in the catabolic system of reactions that generate ATP through the oxidation of food molecules, as we will discuss shortly. The genesis of NADH from NAD⁺ and that of NADPH from NADP⁺ occur by different pathways and are independently regulated, so that the cell can adjust the supply of electrons for these two contrasting purposes. Inside the cell the ratio of NAD⁺ to NADH is kept high, whereas the ratio of NADP⁺ to NADPH is kept low. This provides plenty of NAD⁺ to act as an oxidizing agent and plenty of NADPH to act as a reducing agent—as required for their special roles in catabolism and anabolism, respectively.

There Are Many Other Activated Carrier Molecules in Cells

Other activated carriers also pick up and carry a chemical group in an easily transferred, high-energy linkage. For example, coenzyme A carries an acetyl group in a readily transferable linkage, and in this activated form is known as **acetyl CoA** (acetyl coenzyme A). Acetyl CoA (**Figure 2–62**) is used to add two carbon units in the biosynthesis of larger molecules.

In acetyl CoA as in other carrier molecules, the transferable group makes up only a small part of the molecule. The rest consists of a large organic portion that

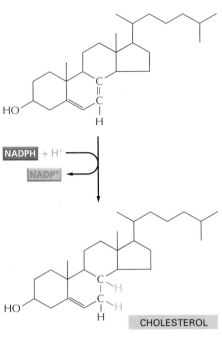

Figure 2–61 **The final stage in one of the biosynthetic routes leading to cholesterol.** As in many other biosynthetic reactions, the reduction of the C=C bond is achieved by the transfer of a hydride ion from the carrier molecule NADPH, plus a proton (H⁺) from the solution.

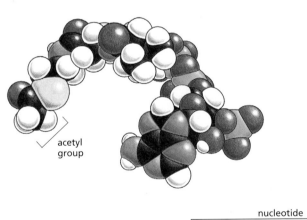

Figure 2–62 **The structure of the important activated carrier molecule acetyl CoA.** A space-filling model is shown above the structure. The sulfur atom (*yellow*) forms a thioester bond to acetate. Because this is a high-energy linkage, releasing a large amount of free energy when it is hydrolyzed, the acetate molecule can be readily transferred to other molecules.

Table 2–5 Some Activated Carrier Molecules Widely Used in Metabolism

ACTIVATED CARRIER	GROUP CARRIED IN HIGH-ENERGY LINKAGE
ATP	phosphate
NADH, NADPH, FADH$_2$	electrons and hydrogens
Acetyl CoA	acetyl group
Carboxylated biotin	carboxyl group
S-Adenosylmethionine	methyl group
Uridine diphosphate glucose	glucose

serves as a convenient "handle," facilitating the recognition of the carrier molecule by specific enzymes. As with acetyl CoA, this handle portion very often contains a nucleotide (usually adenosine), a curious fact that may be a relic from an early stage of evolution. It is currently thought that the main catalysts for early life-forms—before DNA or proteins—were RNA molecules (or their close relatives), as described in Chapter 6. It is tempting to speculate that many of the carrier molecules that we find today originated in this earlier RNA world, where their nucleotide portions could have been useful for binding them to RNA enzymes.

Figures 2–58 and 2–61 have presented examples of the type of transfer reactions powered by the activated carrier molecules ATP (transfer of phosphate) and NADPH (transfer of electrons and hydrogen). The reactions of other activated carrier molecules involve the transfer of a methyl, carboxyl, or glucose group for the purpose of biosynthesis (**Table 2–5**). These activated carriers are generated in reactions that are coupled to ATP hydrolysis, as in the example in **Figure 2–63**. Therefore, the energy that enables their groups to be used for biosynthesis ultimately comes from the catabolic reactions that generate ATP. Similar processes occur in the synthesis of the very large molecules of the cell— the nucleic acids, proteins, and polysaccharides—that we discuss next.

The Synthesis of Biological Polymers Is Driven by ATP Hydrolysis

As discussed previously, the macromolecules of the cell constitute most of its dry mass—that is, of the mass not due to water (see Figure 2–29). These

Figure 2–63 A carboxyl group transfer reaction using an activated carrier molecule. Carboxylated biotin is used by the enzyme *pyruvate carboxylase* to transfer a carboxyl group in the production of oxaloacetate, a molecule needed for the citric acid cycle. The acceptor molecule for this group transfer reaction is pyruvate. Other enzymes use biotin to transfer carboxyl groups to other acceptor molecules. Note that synthesis of carboxylated biotin requires energy that is derived from ATP—a general feature of many activated carriers.

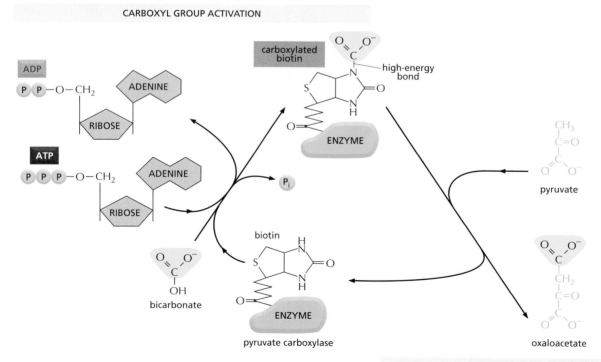

CARBOXYL GROUP ACTIVATION

CARBOXYL GROUP TRANSFER

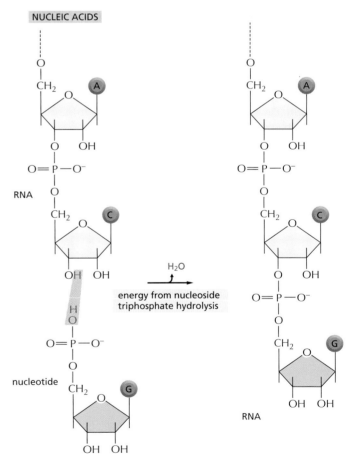

Figure 2–64 Condensation and hydrolysis as opposite reactions. The macromolecules of the cell are polymers that are formed from subunits (or monomers) by a condensation reaction and are broken down by hydrolysis. The condensation reactions are all energetically unfavorable.

molecules are made from subunits (or monomers) that are linked together in a *condensation* reaction, in which the constituents of a water molecule (OH plus H) are removed from the two reactants. Consequently, the reverse reaction—the breakdown of all three types of polymers—occurs by the enzyme-catalyzed addition of water (*hydrolysis*). This hydrolysis reaction is energetically favorable, whereas the biosynthetic reactions require an energy input (**Figure 2–64**).

The nucleic acids (DNA and RNA), proteins, and polysaccharides are all polymers that are produced by the repeated addition of a monomer onto one end of a growing chain. The synthesis reactions for these three types of macromolecules are outlined in **Figure 2–65**. As indicated, the condensation step in each case depends on energy from nucleoside triphosphate hydrolysis. And yet, except for the nucleic acids, there are no phosphate groups left in the final product molecules. How are the reactions that release the energy of ATP hydrolysis coupled to polymer synthesis?

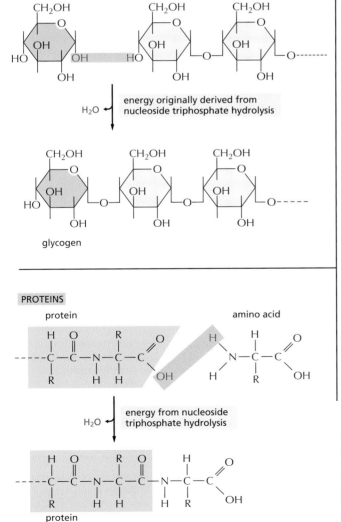

Figure 2–65 The synthesis of polysaccharides, proteins, and nucleic acids. Synthesis of each kind of biological polymer involves the loss of water in a condensation reaction. Not shown is the consumption of high-energy nucleoside triphosphates that is required to activate each monomer before its addition. In contrast, the reverse reaction—the breakdown of all three types of polymers—occurs by the simple addition of water (hydrolysis).

For each type of macromolecule, an enzyme-catalyzed pathway exists which resembles that discussed previously for the synthesis of the amino acid glutamine (see Figure 2–59). The principle is exactly the same, in that the OH group that will be removed in the condensation reaction is first activated by becoming involved in a high-energy linkage to a second molecule. However, the actual mechanisms used to link ATP hydrolysis to the synthesis of proteins and polysaccharides are more complex than that used for glutamine synthesis, since a series of high-energy intermediates is required to generate the final high-energy bond that is broken during the condensation step (discussed in Chapter 6 for protein synthesis).

Each activated carrier has limits in its ability to drive a biosynthetic reaction. The ΔG for the hydrolysis of ATP to ADP and inorganic phosphate (P_i) depends on the concentrations of all of the reactants, but under the usual conditions in a cell it is between –11 and –13 kcal/mole (between –46 and –54 kJ/mole). In principle, this hydrolysis reaction could drive an unfavorable reaction with a ΔG of, perhaps, +10 kcal/mole, provided that a suitable reaction path is available. For some biosynthetic reactions, however, even –13 kcal/mole may not be enough. In these cases the path of ATP hydrolysis can be altered so that it initially produces AMP and pyrophosphate (PP_i), which is itself then hydrolyzed in a subsequent step (**Figure 2–66**). The whole process makes available a total free-energy change of about –26 kcal/mole. An important type of biosynthetic reaction that is driven in this way is the synthesis of nucleic acids (polynucleotides) from nucleoside triphosphates, as illustrated on the right side of **Figure 2–67**.

Note that the repetitive condensation reactions that produce macromolecules can be oriented in one of two ways, giving rise to either the head polymerization or the tail polymerization of monomers. In so-called *head polymerization* the reactive bond required for the condensation reaction is carried on the end of the growing polymer, and it must therefore be regenerated each time that a monomer is added. In this case, each monomer brings with it the reactive bond that will be used in adding the *next* monomer in the series. In *tail polymerization* the reactive bond carried by each monomer is instead used immediately for its own addition (**Figure 2–68**).

We shall see in later chapters that both these types of polymerization are used. The synthesis of polynucleotides and some simple polysaccharides occurs by tail polymerization, for example, whereas the synthesis of proteins occurs by a head polymerization process.

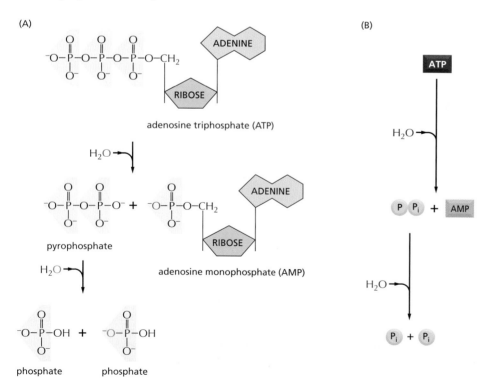

(A)

adenosine triphosphate (ATP)

pyrophosphate   adenosine monophosphate (AMP)

phosphate   phosphate

(B)

ATP

H_2O

P P_i + AMP

H_2O

P_i + P_i

Figure 2–66 **An alternative pathway of ATP hydrolysis, in which pyrophosphate is first formed and then hydrolyzed.** This route releases about twice as much free energy as the reaction shown earlier in Figure 2–57, and it forms AMP instead of ADP. (A) In the two successive hydrolysis reactions, oxygen atoms from the participating water molecules are retained in the products, as indicated, whereas the hydrogen atoms dissociate to form free hydrogen ions (H^+, not shown). (B) Diagram of overall reaction in summary form.

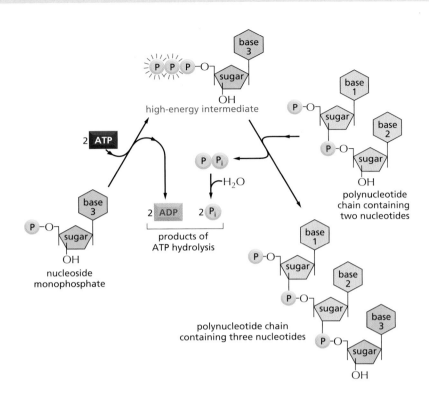

Figure 2–67 Synthesis of a polynucleotide, RNA or DNA, is a multistep process driven by ATP hydrolysis. In the first step, a nucleoside monophosphate is activated by the sequential transfer of the terminal phosphate groups from two ATP molecules. The high-energy intermediate formed—a nucleoside triphosphate—exists free in solution until it reacts with the growing end of an RNA or a DNA chain with release of pyrophosphate. Hydrolysis of the latter to inorganic phosphate is highly favorable and helps to drive the overall reaction in the direction of polynucleotide synthesis. For details, see Chapter 5.

Summary

Living cells are highly ordered and need to create order within themselves to survive and grow. This is thermodynamically possible only because of a continual input of energy, part of which must be released from the cells to their environment as heat. The energy comes ultimately from the electromagnetic radiation of the sun, which drives the formation of organic molecules in photosynthetic organisms such as green plants. Animals obtain their energy by eating these organic molecules and oxidizing them in a series of enzyme-catalyzed reactions that are coupled to the formation of ATP—a common currency of energy in all cells.

To make possible the continual generation of order in cells, the energetically favorable hydrolysis of ATP is coupled to energetically unfavorable reactions. In the biosynthesis of macromolecules, this is accomplished by the transfer of phosphate groups to form reactive phosphorylated intermediates. Because the energetically unfavorable reaction now becomes energetically favorable, ATP hydrolysis is said to drive the reaction. Polymeric molecules such as proteins, nucleic acids, and polysaccharides are assembled from small activated precursor molecules by repetitive condensation reactions that are driven in this way. Other reactive molecules, called either active carriers or coenzymes, transfer other chemical groups in the course of biosynthesis: NADPH transfers hydrogen as a proton plus two electrons (a hydride ion), for example, whereas acetyl CoA transfers an acetyl group.

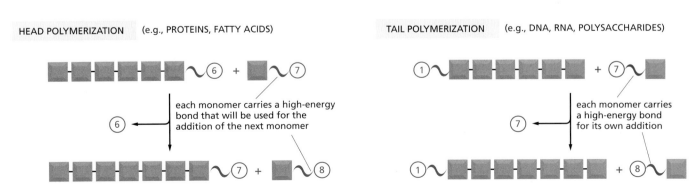

Figure 2–68 The orientation of the active intermediates in the repetitive condensation reactions that form biological polymers. The head growth of polymers is compared with its alternative, tail growth. As indicated, these two mechanisms are used to produce different types of biological macromolecules.

HOW CELLS OBTAIN ENERGY FROM FOOD

The constant supply of energy that cells need to generate and maintain the biological order that keeps them alive comes from the chemical bond energy in food molecules, which thereby serve as fuel for cells.

The proteins, lipids, and polysaccharides that make up most of the food we eat must be broken down into smaller molecules before our cells can use them—either as a source of energy or as building blocks for other molecules. Enzymatic digestion breaks down the large polymeric molecules in food into their monomer subunits—proteins into amino acids, polysaccharides into sugars, and fats into fatty acids and glycerol. After digestion, the small organic molecules derived from food enter the cytosol of cells, where their gradual oxidation begins.

Sugars are particularly important fuel molecules, and they are oxidized in small controlled steps to carbon dioxide (CO_2) and water (**Figure 2–69**). In this section we trace the major steps in the breakdown, or catabolism, of sugars and show how they produce ATP, NADH, and other activated carrier molecules in animal cells. A very similar pathway also operates in plants, fungi, and many bacteria. As we shall see, the oxidation of fatty acids is equally important for cells. Other molecules, such as proteins, can also serve as energy sources when they are funneled through appropriate enzymatic pathways.

Glycolysis Is a Central ATP-Producing Pathway

The major process for oxidizing sugars is the sequence of reactions known as **glycolysis**—from the Greek *glukus*, "sweet," and *lusis*, "rupture." Glycolysis produces ATP without the involvement of molecular oxygen (O_2 gas). It occurs in the cytosol of most cells, including many anaerobic microorganisms (those that can live without using molecular oxygen). Glycolysis probably evolved early in the history of life, before photosynthetic organisms introduced oxygen into the atmosphere. During glycolysis, a glucose molecule with six carbon atoms is converted into two molecules of *pyruvate*, each of which contains three carbon atoms. For each glucose molecule, two molecules of ATP are hydrolyzed to provide energy to drive the early steps, but four molecules of ATP are produced in the later steps. At the end of glycolysis, there is consequently a net gain of two molecules of ATP for each glucose molecule broken down.

The glycolytic pathway is outlined in **Figure 2–70** and shown in more detail in Panel 2–8 (pp. 120–121). Glycolysis involves a sequence of 10 separate reactions, each producing a different sugar intermediate and each catalyzed by a

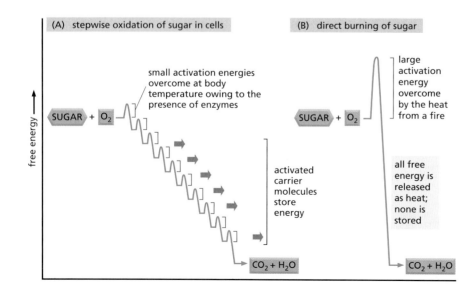

Figure 2–69 Schematic representation of the controlled stepwise oxidation of sugar in a cell, compared with ordinary burning. (A) In the cell, enzymes catalyze oxidation via a series of small steps in which free energy is transferred in conveniently sized packets to carrier molecules—most often ATP and NADH. At each step, an enzyme controls the reaction by reducing the activation energy barrier that has to be surmounted before the specific reaction can occur. The total free energy released is exactly the same in (A) and (B). But if the sugar were instead oxidized to CO_2 and H_2O in a single step, as in (B), it would release an amount of energy much larger than could be captured for useful purposes.

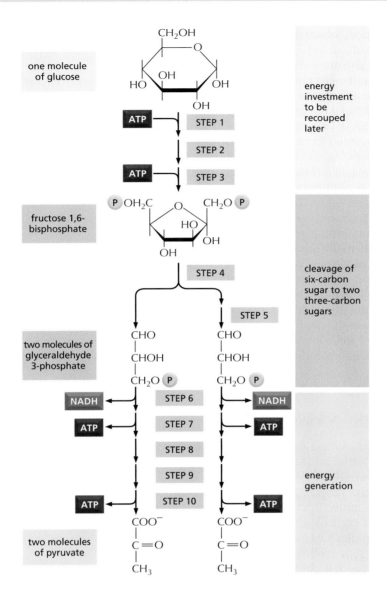

Figure 2–70 An outline of glycolysis.
<GGGC> Each of the 10 steps shown is catalyzed by a different enzyme. Note that step 4 cleaves a six-carbon sugar into two three-carbon sugars, so that the number of molecules at every stage after this doubles. As indicated, step 6 begins the energy generation phase of glycolysis. Because two molecules of ATP are hydrolyzed in the early, energy investment phase, glycolysis results in the net synthesis of 2 ATP and 2 NADH molecules per molecule of glucose (see also Panel 2–8).

different enzyme. Like most enzymes, these have names ending in *ase*—such as isomer*ase* and dehydrogen*ase*—to indicate the type of reaction they catalyze.

Although no molecular oxygen is used in glycolysis, oxidation occurs, in that electrons are removed by NAD^+ (producing NADH) from some of the carbons derived from the glucose molecule. The stepwise nature of the process releases the energy of oxidation in small packets, so that much of it can be stored in activated carrier molecules rather than all of it being released as heat (see Figure 2–69). Thus, some of the energy released by oxidation drives the direct synthesis of ATP molecules from ADP and P_i, and some remains with the electrons in the high-energy electron carrier NADH.

Two molecules of NADH are formed per molecule of glucose in the course of glycolysis. In aerobic organisms (those that require molecular oxygen to live), these NADH molecules donate their electrons to the electron-transport chain described in Chapter 14, and the NAD^+ formed from the NADH is used again for glycolysis (see step 6 in Panel 2–8, pp. 120–121).

Fermentations Produce ATP in the Absence of Oxygen

For most animal and plant cells, glycolysis is only a prelude to the final stage of the breakdown of food molecules. In these cells, the pyruvate formed by glycolysis is rapidly transported into the mitochondria, where it is converted into CO_2 plus acetyl CoA, which is then completely oxidized to CO_2 and H_2O.

In contrast, for many anaerobic organisms—which do not utilize molecular oxygen and can grow and divide without it—glycolysis is the principal source of the cell's ATP. This is also true for certain animal tissues, such as skeletal muscle, that can continue to function when molecular oxygen is limiting. In these anaerobic conditions, the pyruvate and the NADH electrons stay in the cytosol. The pyruvate is converted into products excreted from the cell—for example, into ethanol and CO_2 in the yeasts used in brewing and breadmaking, or into lactate in muscle. In this process, the NADH gives up its electrons and is converted back into NAD^+. This regeneration of NAD^+ is required to maintain the reactions of glycolysis (**Figure 2–71**).

Anaerobic energy-yielding pathways like these are called **fermentations**. Studies of the commercially important fermentations carried out by yeasts inspired much of early biochemistry. Work in the nineteenth century led in 1896 to the then startling recognition that these processes could be studied outside living organisms, in cell extracts. This revolutionary discovery eventually made it possible to dissect out and study each of the individual reactions in the fermentation process. The piecing together of the complete glycolytic pathway in the 1930s was a major triumph of biochemistry, and it was quickly followed by the recognition of the central role of ATP in cell processes. Thus, most of the fundamental concepts discussed in this chapter have been understood for many years.

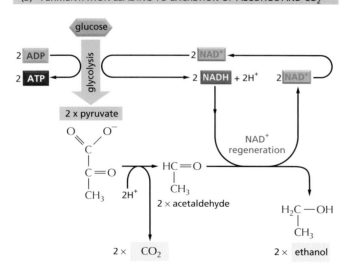

(A) FERMENTATION LEADING TO EXCRETION OF LACTATE

(B) FERMENTATION LEADING TO EXCRETION OF ALCOHOL AND CO_2

Figure 2–71 Two pathways for the anaerobic breakdown of pyruvate. (A) When there is inadequate oxygen, for example, in a muscle cell undergoing vigorous contraction, the pyruvate produced by glycolysis is converted to lactate as shown. This reaction regenerates the NAD^+ consumed in step 6 of glycolysis, but the whole pathway yields much less energy overall than complete oxidation. (B) In some organisms that can grow anaerobically, such as yeasts, pyruvate is converted via acetaldehyde into carbon dioxide and ethanol. Again, this pathway regenerates NAD^+ from NADH, as required to enable glycolysis to continue. Both (A) and (B) are examples of *fermentations*.

Glycolysis Illustrates How Enzymes Couple Oxidation to Energy Storage

Returning to the paddle-wheel analogy that we used to introduce coupled reactions (see Figure 2–56), we can now equate enzymes with the paddle wheel. Enzymes act to harvest useful energy from the oxidation of organic molecules by coupling an energetically unfavorable reaction with a favorable one. To demonstrate this coupling, we examine a step in glycolysis to see exactly how such coupled reactions occur.

Two central reactions in glycolysis (steps 6 and 7) convert the three-carbon sugar intermediate glyceraldehyde 3-phosphate (an aldehyde) into 3-phosphoglycerate (a carboxylic acid; see Panel 2–8, pp. 120–121). This entails the oxidation of an aldehyde group to a carboxylic acid group in a reaction that occurs in two steps. The overall reaction releases enough free energy to convert a molecule of ADP to ATP and to transfer two electrons from the aldehyde to NAD^+ to form NADH, while still releasing enough heat to the environment to make the overall reaction energetically favorable ($\Delta G°$ for the overall reaction is –3.0 kcal/mole).

Figure 2–72 outlines the means by which this remarkable feat of energy harvesting is accomplished. The indicated chemical reactions are precisely guided by two enzymes to which the sugar intermediates are tightly bound. In fact, as detailed in Figure 2–72, the first enzyme (glyceraldehyde 3-phosphate dehydrogenase) forms a short-lived covalent bond to the aldehyde through a reactive –SH group on the enzyme, and catalyzes its oxidation by NAD^+ in this attached state. The reactive enzyme–substrate bond is then displaced by an inorganic phosphate ion to produce a high-energy phosphate intermediate, which is released from the enzyme. This intermediate binds to the second enzyme (phosphoglycerate kinase), which catalyzes the energetically favorable transfer of the high-energy phosphate just created to ADP, forming ATP and completing the process of oxidizing an aldehyde to a carboxylic acid.

We have shown this particular oxidation process in some detail because it provides a clear example of enzyme-mediated energy storage through coupled reactions (**Figure 2–73**). Steps 6 and 7 are the only reactions in glycolysis that create a high-energy phosphate linkage directly from inorganic phosphate. As such, they account for the net yield of two ATP molecules and two NADH molecules per molecule of glucose (see Panel 2–8, pp. 120–121).

As we have just seen, ATP can be formed readily from ADP when a reaction intermediate is formed with a phosphate bond of higher-energy than the phosphate bond in ATP. Phosphate bonds can be ordered in energy by comparing the standard free-energy change $(\Delta G°)$ for the breakage of each bond by hydrolysis. **Figure 2–74** compares the high-energy phosphoanhydride bonds in ATP with the energy of some other phosphate bonds, several of which are generated during glycolysis.

Organisms Store Food Molecules in Special Reservoirs

All organisms need to maintain a high ATP/ADP ratio to maintain biological order in their cells. Yet animals have only periodic access to food, and plants need to survive overnight without sunlight, when they are unable to produce sugar from photosynthesis. For this reason, both plants and animals convert sugars and fats to special forms for storage (**Figure 2–75**).

To compensate for long periods of fasting, animals store fatty acids as fat droplets composed of water-insoluble triacylglycerols, largely in the cytoplasm of specialized fat cells, called adipocytes. For shorter-term storage, sugar is stored as glucose subunits in the large branched polysaccharide **glycogen**, which is present as small granules in the cytoplasm of many cells, including liver and muscle. The synthesis and degradation of glycogen are rapidly regulated according to need. When cells need more ATP than they can generate from the food molecules taken in from the bloodstream, they break down glycogen in a reaction that produces glucose 1-phosphate, which is rapidly converted to glucose 6-phosphate for glycolysis.

(A)

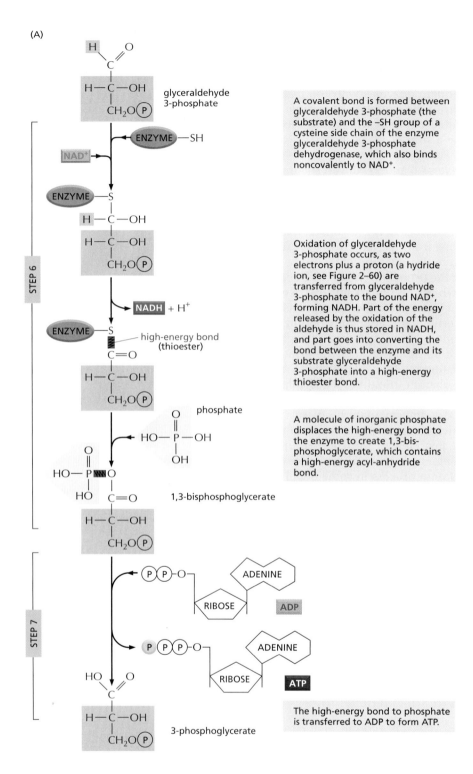

A covalent bond is formed between glyceraldehyde 3-phosphate (the substrate) and the –SH group of a cysteine side chain of the enzyme glyceraldehyde 3-phosphate dehydrogenase, which also binds noncovalently to NAD⁺.

A covalent bond is formed between glyceraldehyde 3-phosphate (the substrate) and the –SH group of a cysteine side chain of the enzyme glyceraldehyde 3-phosphate dehydrogenase, which also binds noncovalently to NAD^+.

Oxidation of glyceraldehyde 3-phosphate occurs, as two electrons plus a proton (a hydride ion, see Figure 2–60) are transferred from glyceraldehyde 3-phosphate to the bound NAD^+, forming NADH. Part of the energy released by the oxidation of the aldehyde is thus stored in NADH, and part goes into converting the bond between the enzyme and its substrate glyceraldehyde 3-phosphate into a high-energy thioester bond.

A molecule of inorganic phosphate displaces the high-energy bond to the enzyme to create 1,3-bis-phosphoglycerate, which contains a high-energy acyl-anhydride bond.

The high-energy bond to phosphate is transferred to ADP to form ATP.

(B)

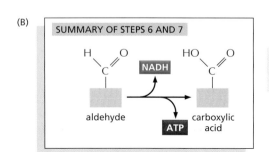

SUMMARY OF STEPS 6 AND 7

Much of the energy of oxidation has been stored in the activated carriers ATP and NADH.

Figure 2–72 Energy storage in steps 6 and 7 of glycolysis. In these steps the oxidation of an aldehyde to a carboxylic acid is coupled to the formation of ATP and NADH. (A) Step 6 begins with the formation of a covalent bond between the substrate (glyceraldehyde 3-phosphate) and an –SH group exposed on the surface of the enzyme (glyceraldehyde 3-phosphate dehydrogenase). The enzyme then catalyzes transfer of hydrogen (as a hydride ion—a proton plus two electrons) from the bound glyceraldehyde 3-phosphate to a molecule of NAD^+. Part of the energy released in this oxidation is used to form a molecule of NADH and part is used to convert the original linkage between the enzyme and its substrate to a high-energy thioester bond (shown in *red*). A molecule of inorganic phosphate then displaces this high-energy bond on the enzyme, creating a high-energy sugar-phosphate bond instead *(red)*. At this point the enzyme has not only stored energy in NADH, but also coupled the energetically favorable oxidation of an aldehyde to the energetically unfavorable formation of a high-energy phosphate bond. The second reaction has been driven by the first, thereby acting like the "paddle-wheel" coupler in Figure 2–56.

In reaction step 7, the high-energy sugar-phosphate intermediate just made, 1,3-bisphosphoglycerate, binds to a second enzyme, phosphoglycerate kinase. The reactive phosphate is transferred to ADP, forming a molecule of ATP and leaving a free carboxylic acid group on the oxidized sugar.

(B) Summary of the overall chemical change produced by reactions 6 and 7.

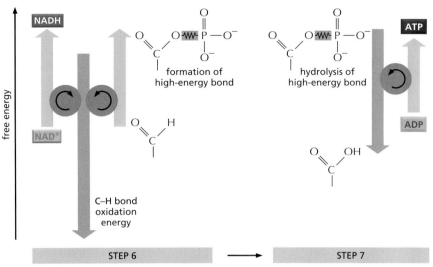

Figure 2–73 Schematic view of the coupled reactions that form NADH and ATP in steps 6 and 7 of glycolysis. The C–H bond oxidation energy drives the formation of both NADH and a high-energy phosphate bond. The breakage of the high-energy bond then drives ATP formation.

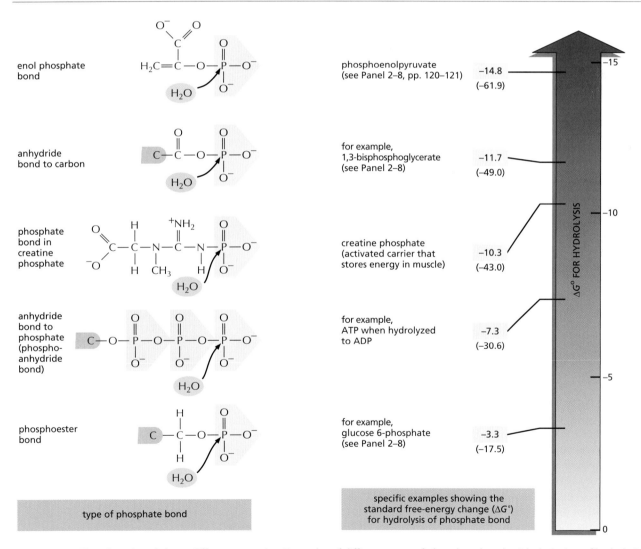

Figure 2–74 Phosphate bonds have different energies. Examples of different types of phosphate bonds with their sites of hydrolysis are shown in the molecules depicted on the left. Those starting with a *gray* carbon atom show only part of a molecule. Examples of molecules containing such bonds are given on the right, with the free-energy change for hydrolysis in kilocalories (kilojoules in parentheses). The transfer of a phosphate group from one molecule to another is energetically favorable if the standard free-energy change (ΔG°) for hydrolysis of the phosphate bond of the first molecule is more negative than that for hydrolysis of the phosphate bond in the second. Thus, a phosphate group is readily transferred from 1,3-bisphosphoglycerate to ADP to form ATP. The hydrolysis reaction can be viewed as the transfer of the phosphate group to water.

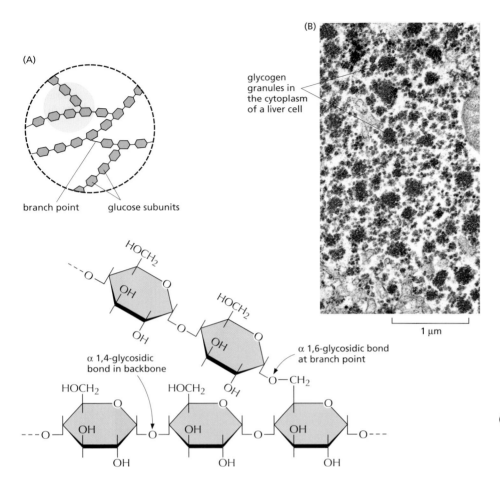

(A)

branch point   glucose subunits

α 1,4-glycosidic bond in backbone

α 1,6-glycosidic bond at branch point

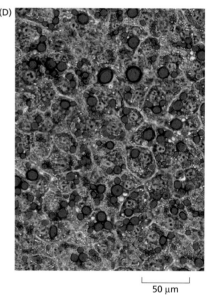

(B)

glycogen granules in the cytoplasm of a liver cell

1 μm

Figure 2–75 The storage of sugars and fats in animal and plant cells. (A) The structures of starch and glycogen, the storage form of sugars in plants and animals, respectively. Both are storage polymers of the sugar glucose and differ only in the frequency of branch points (the region in *yellow* is shown enlarged below). There are many more branches in glycogen than in starch. (B) An electron micrograph shows glycogen granules in the cytoplasm of a liver cell. (C) A thin section of a single chloroplast from a plant cell, showing the starch granules and lipid (fat droplets) that have accumulated as a result of the biosyntheses occurring there. (D) Fat droplets (stained *red*) beginning to accumulate in developing fat cells of an animal. (B, courtesy of Robert Fletterick and Daniel S. Friend; C, courtesy of K. Plaskitt; D, courtesy of Ronald M. Evans and Peter Totonoz.)

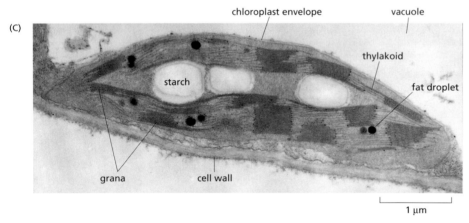

(C)

chloroplast envelope   vacuole

thylakoid

starch

fat droplet

grana   cell wall

1 μm

(D)

50 μm

Quantitatively, **fat** is far more important than glycogen as an energy store for animals, presumably because it provides for more efficient storage. The oxidation of a gram of fat releases about twice as much energy as the oxidation of a gram of glycogen. Moreover, glycogen differs from fat in binding a great deal of water, producing a sixfold difference in the actual mass of glycogen required to store the same amount of energy as fat. An average adult human stores enough glycogen for only about a day of normal activities but enough fat to last for nearly a month. If our main fuel reservoir had to be carried as glycogen instead of fat, body weight would increase by an average of about 60 pounds.

Although plants produce NADPH and ATP by photosynthesis, this important process occurs in a specialized organelle, called a chloroplast, which is isolated from the rest of the plant cell by a membrane that is impermeable to both types of activated carrier molecules. Moreover, the plant contains many other cells—such as those in the roots—that lack chloroplasts and therefore cannot produce their own sugars. Therefore, for most of its ATP production, the plant relies on an

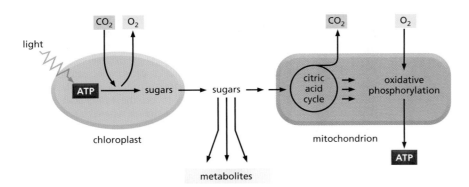

Figure 2–76 **How the ATP needed for most plant cell metabolism is made.** In plants, the chloroplasts and mitochondria collaborate to supply cells with metabolites and ATP. (For details, see Chapter 14.)

export of sugars from its chloroplasts to the mitochondria that are located in all cells of the plant. Most of the ATP needed by the plant is synthesized in these mitochondria and exported from them to the rest of the plant cell, using exactly the same pathways for the oxidative breakdown of sugars as in nonphotosynthetic organisms (**Figure 2–76**).

During periods of excess photosynthetic capacity during the day, chloroplasts convert some of the sugars that they make into fats and into **starch**, a polymer of glucose analogous to the glycogen of animals. The fats in plants are triacylglycerols, just like the fats in animals, and differ only in the types of fatty acids that predominate. Fat and starch are both stored in the chloroplast as reservoirs to be mobilized as an energy source during periods of darkness (see Figure 2–75C).

The embryos inside plant seeds must live on stored sources of energy for a prolonged period, until they germinate to produce leaves that can harvest the energy in sunlight. For this reason plant seeds often contain especially large amounts of fats and starch—which makes them a major food source for animals, including ourselves (**Figure 2–77**).

Most Animal Cells Derive Their Energy from Fatty Acids Between Meals

After a meal, most of the energy that an animal needs is derived from sugars derived from food. Excess sugars, if any, are used to replenish depleted glycogen stores, or to synthesize fats as a food store. But soon the fat stored in adipose tissue is called into play, and by the morning after an overnight fast, fatty acid oxidation generates most of the ATP we need.

Low glucose levels in the blood trigger the breakdown of fats for energy production. As illustrated in **Figure 2–78**, the triacylglycerols stored in fat droplets in adipocytes are hydrolyzed to produce fatty acids and glycerol, and the fatty acids released are transferred to cells in the body through the bloodstream. While animals readily convert sugars to fats, they cannot convert fatty acids to sugars. Instead, the fatty acids are oxidized directly.

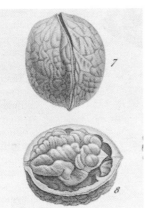
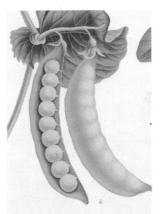

Figure 2–77 **Some plant seeds that serve as important foods for humans.** Corn, nuts, and peas all contain rich stores of starch and fat that provide the young plant embryo in the seed with energy and building blocks for biosynthesis. (Courtesy of the John Innes Foundation.)

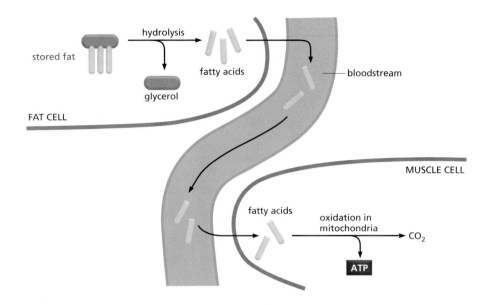

Figure 2–78 How stored fats are mobilized for energy production in animals. Low glucose levels in the blood trigger the hydrolysis of the triacylglycerol molecules in fat droplets to free fatty acids and glycerol, as illustrated. These fatty acids enter the bloodstream, where they bind to the abundant blood protein, serum albumin. Special fatty acid transporters in the plasma membrane of cells that oxidize fatty acids, such as muscle cells, then pass these fatty acids into the cytosol, from which they are moved into mitochondria for energy production (see Figure 2–80).

Sugars and Fats Are Both Degraded to Acetyl CoA in Mitochondria

In aerobic metabolism, the pyruvate that was produced by glycolysis from sugars in the cytosol is transported into the *mitochondria* of eucaryotic cells. There, it is rapidly decarboxylated by a giant complex of three enzymes, called the *pyruvate dehydrogenase complex*. The products of pyruvate decarboxylation are a molecule of CO_2 (a waste product), a molecule of NADH, and acetyl CoA (**Figure 2–79**).

The fatty acids imported from the bloodstream are moved into mitochondria, where all of their oxidation takes place (**Figure 2–80**). Each molecule of fatty acid (as the activated molecule *fatty acyl CoA*) is broken down completely by a cycle of reactions that trims two carbons at a time from its carboxyl end, generating one molecule of acetyl CoA for each turn of the cycle. A molecule of NADH and a molecule of $FADH_2$ are also produced in this process (**Figure 2–81**).

Sugars and fats are the major energy sources for most non-photosynthetic organisms, including humans. However, most of the useful energy that can be

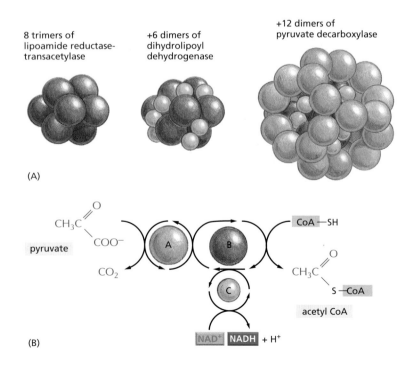

Figure 2–79 The oxidation of pyruvate to acetyl CoA and CO_2. (A) The structure of the pyruvate dehydrogenase complex, which contains 60 polypeptide chains. This is an example of a large multienzyme complex in which reaction intermediates are passed directly from one enzyme to another. In eucaryotic cells it is located in the mitochondrion. (B) The reactions carried out by the pyruvate dehydrogenase complex. The complex converts pyruvate to acetyl CoA in the mitochondrial matrix; NADH is also produced in this reaction. A, B, and C are the three enzymes *pyruvate decarboxylase, lipoamide reductase-transacetylase,* and *dihydrolipoyl dehydrogenase,* respectively. These enzymes are illustrated in (A); their activities are linked as shown.

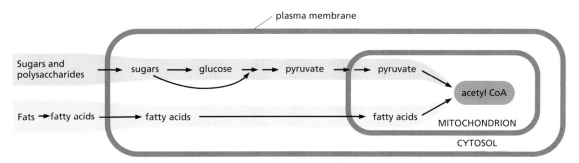

Figure 2–80 Pathways for the production of acetyl CoA from sugars and fats. The mitochondrion in eucaryotic cells is the place where acetyl CoA is produced from both types of major food molecules. It is therefore the place where most of the cell's oxidation reactions occur and where most of its ATP is made. The structure and function of mitochondria are discussed in detail in Chapter 14.

extracted from the oxidation of both types of foodstuffs remains stored in the acetyl CoA molecules that are produced by the two types of reactions just described. The citric acid cycle of reactions, in which the acetyl group in acetyl CoA is oxidized to CO_2 and H_2O, is therefore central to the energy metabolism of aerobic organisms. In eucaryotes these reactions all take place in mitochondria. We should therefore not be surprised to discover that the mitochondrion is the place where most of the ATP is produced in animal cells. In contrast, aerobic bacteria carry out all of their reactions in a single compartment, the cytosol, and it is here that the citric acid cycle takes place in these cells.

The Citric Acid Cycle Generates NADH by Oxidizing Acetyl Groups to CO_2

In the nineteenth century, biologists noticed that in the absence of air (anaerobic conditions) cells produce lactic acid (for example, in muscle) or ethanol (for example, in yeast), while in its presence (aerobic conditions) they consume O_2 and produce CO_2 and H_2O. Efforts to define the pathways of aerobic metabolism

Figure 2–81 The oxidation of fatty acids to acetyl CoA. (A) Electron micrograph of a lipid droplet in the cytoplasm *(top)*, and the structure of fats *(bottom)*. Fats are triacylglycerols. The glycerol portion, to which three fatty acids are linked through ester bonds, is shown here in *blue*. Fats are insoluble in water and form large lipid droplets in the specialized fat cells (called adipocytes) in which they are stored. (B) The fatty acid oxidation cycle. The cycle is catalyzed by a series of four enzymes in the mitochondrion. Each turn of the cycle shortens the fatty acid chain by two carbons (shown in *red*) and generates one molecule of acetyl CoA and one molecule each of NADH and $FADH_2$. The structure of $FADH_2$ is presented in Figure 2–83B. (A, courtesy of Daniel S. Friend.)

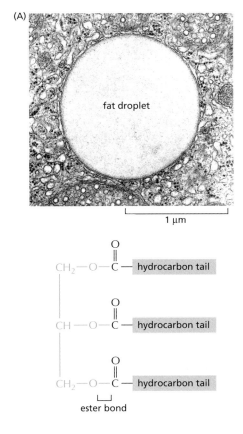

eventually focused on the oxidation of pyruvate and led in 1937 to the discovery of the **citric acid cycle**, also known as the *tricarboxylic acid cycle* or the *Krebs cycle*. The citric acid cycle accounts for about two-thirds of the total oxidation of carbon compounds in most cells, and its major end products are CO_2 and high-energy electrons in the form of NADH. The CO_2 is released as a waste product, while the high-energy electrons from NADH are passed to a membrane-bound electron-transport chain (discussed in Chapter 14), eventually combining with O_2 to produce H_2O. Although the citric acid cycle itself does not use O_2, it requires O_2 in order to proceed because there is no other efficient way for the NADH to get rid of its electrons and thus regenerate the NAD^+ that is needed to keep the cycle going.

The citric acid cycle takes place inside mitochondria in eucaryotic cells. It results in the complete oxidation of the carbon atoms of the acetyl groups in acetyl CoA, converting them into CO_2. But the acetyl group is not oxidized directly. Instead, this group is transferred from acetyl CoA to a larger, four-carbon molecule, *oxaloacetate*, to form the six-carbon tricarboxylic acid, *citric acid*, for which the subsequent cycle of reactions is named. The citric acid molecule is then gradually oxidized, allowing the energy of this oxidation to be harnessed to produce energy-rich activated carrier molecules. The chain of eight reactions forms a cycle because at the end the oxaloacetate is regenerated and enters a new turn of the cycle, as shown in outline in **Figure 2–82**.

We have thus far discussed only one of the three types of activated carrier molecules that are produced by the citric acid cycle, the NAD^+–NADH pair (see Figure 2–60). In addition to three molecules of NADH, each turn of the cycle also produces one molecule of **$FADH_2$** (reduced flavin adenine dinucleotide) from FAD and one molecule of the ribonucleotide **GTP** (guanosine triphosphate) from GDP. The structures of these two activated carrier molecules are illustrated in **Figure 2–83**. GTP is a close relative of ATP, and the transfer of its terminal phosphate group to ADP produces one ATP molecule in each cycle. Like NADH, $FADH_2$ is a carrier of high-energy electrons and hydrogen. As we discuss shortly, the energy that is stored in the readily transferred high-energy electrons of NADH and $FADH_2$ will be utilized subsequently for ATP production through the process of *oxidative phosphorylation*, the only step in the oxidative catabolism of foodstuffs that directly requires gaseous oxygen (O_2) from the atmosphere.

Panel 2–9 (pp. 122–123) presents the complete citric acid cycle. Water, rather than molecular oxygen, supplies the extra oxygen atoms required to make CO_2 from the acetyl groups entering the citric acid cycle. As illustrated in the panel,

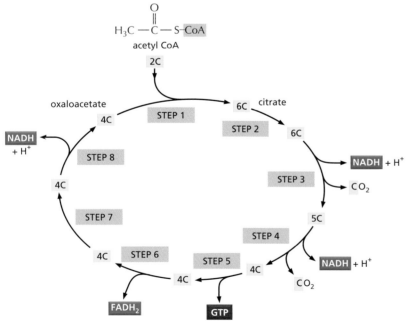

NET RESULT: ONE TURN OF THE CYCLE PRODUCES THREE NADH, ONE GTP, AND ONE $FADH_2$, AND RELEASES TWO MOLECULES OF CO_2

Figure 2–82 Simple overview of the citric acid cycle. <TAGT> The reaction of acetyl CoA with oxaloacetate starts the cycle by producing citrate (citric acid). In each turn of the cycle, two molecules of CO_2 are produced as waste products, plus three molecules of NADH, one molecule of GTP, and one molecule of $FADH_2$. The number of carbon atoms in each intermediate is shown in a *yellow box*. For details, see Panel 2–9 (pp. 122–123).

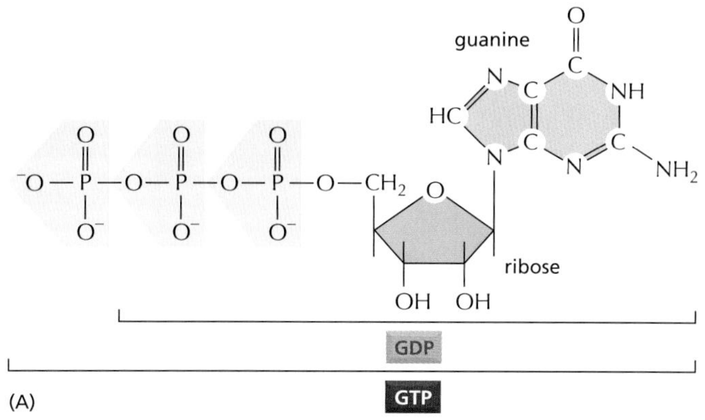

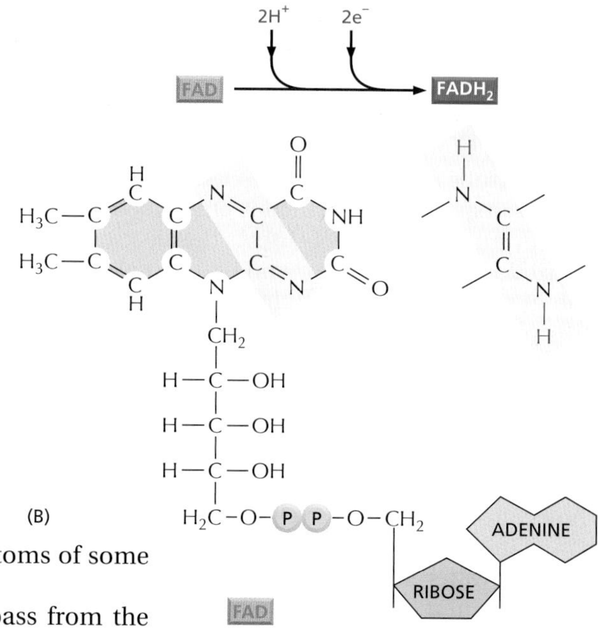

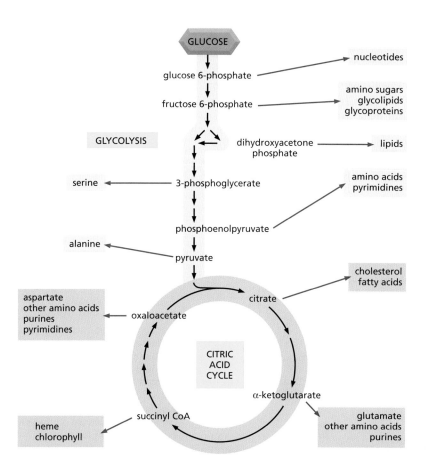

three molecules of water are split in each cycle, and the oxygen atoms of some of them are ultimately used to make CO_2.

In addition to pyruvate and fatty acids, some amino acids pass from the cytosol into mitochondria, where they are also converted into acetyl CoA or one of the other intermediates of the citric acid cycle. Thus, in the eucaryotic cell, the mitochondrion is the center toward which all energy-yielding processes lead, whether they begin with sugars, fats, or proteins.

Both the citric acid cycle and glycolysis also function as starting points for important biosynthetic reactions by producing vital carbon-containing inter-mediates, such as *oxaloacetate* and *α-ketoglutarate*. Some of these substances produced by catabolism are transferred back from the mitochondrion to the cytosol, where they serve in anabolic reactions as precursors for the synthesis of many essential molecules, such as amino acids (**Figure 2–84**).

Figure 2–83 The structures of GTP and FADH₂. (A) GTP and GDP are close relatives of ATP and ADP, respectively. (B) FADH₂ is a carrier of hydrogens and high-energy electrons, like NADH and NADPH. It is shown here in its oxidized form (FAD) with the hydrogen-carrying atoms highlighted in *yellow*.

Figure 2–84 Glycolysis and the citric acid cycle provide the precursors needed to synthesize many important biological molecules. The amino acids, nucleotides, lipids, sugars, and other molecules—shown here as products—in turn serve as the precursors for the many macromolecules of the cell. Each *black arrow* in this diagram denotes a single enzyme-catalyzed reaction; the *red arrows* generally represent pathways with many steps that are required to produce the indicated products.

Electron Transport Drives the Synthesis of the Majority of the ATP in Most Cells

Most chemical energy is released in the last step in the degradation of a food molecule. In this final process the electron carriers NADH and FADH$_2$ transfer the electrons that they have gained when oxidizing other molecules to the **electron-transport chain**, which is embedded in the inner membrane of the mitochondrion (see Figure 14–10). As the electrons pass along this long chain of specialized electron acceptor and donor molecules, they fall to successively lower energy states. The energy that the electrons release in this process pumps H$^+$ ions (protons) across the membrane—from the inner mitochondrial compartment to the outside—generating a gradient of H$^+$ ions (**Figure 2–85**). This gradient serves as a source of energy, being tapped like a battery to drive a variety of energy-requiring reactions. The most prominent of these reactions is the generation of ATP by the phosphorylation of ADP.

At the end of this series of electron transfers, the electrons are passed to molecules of oxygen gas (O$_2$) that have diffused into the mitochondrion, which simultaneously combine with protons (H$^+$) from the surrounding solution to produce water molecules. The electrons have now reached their lowest energy level, and therefore all the available energy has been extracted from the oxidized food molecule. This process, termed **oxidative phosphorylation** (**Figure 2–86**), also occurs in the plasma membrane of bacteria. As one of the most remarkable achievements of cell evolution, it is a central topic of Chapter 14.

In total, the complete oxidation of a molecule of glucose to H$_2$O and CO$_2$ is used by the cell to produce about 30 molecules of ATP. In contrast, only 2 molecules of ATP are produced per molecule of glucose by glycolysis alone.

Amino Acids and Nucleotides Are Part of the Nitrogen Cycle

So far we have concentrated mainly on carbohydrate metabolism and have not yet considered the metabolism of nitrogen or sulfur. These two elements are important constituents of biological macromolecules. Nitrogen and sulfur atoms pass from compound to compound and between organisms and their environment in a series of reversible cycles.

Although molecular nitrogen is abundant in the Earth's atmosphere, nitrogen is chemically unreactive as a gas. Only a few living species are able to incorporate it into organic molecules, a process called **nitrogen fixation**. Nitrogen fixation occurs in certain microorganisms and by some geophysical processes, such as lightning discharge. It is essential to the biosphere as a whole, for without it life could not exist on this planet. Only a small fraction of the nitrogenous compounds in today's organisms, however, is due to fresh products of nitrogen fixation from the atmosphere. Most organic nitrogen has been in circulation for

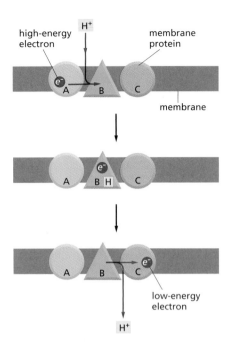

Figure 2–85 The generation of an H$^+$ gradient across a membrane by electron-transport reactions. A high-energy electron (derived, for example, from the oxidation of a metabolite) is passed sequentially by carriers A, B, and C to a lower energy state. In this diagram carrier B is arranged in the membrane in such a way that it takes up H$^+$ from one side and releases it to the other as the electron passes. The result is an H$^+$ gradient. As discussed in Chapter 14, this gradient is an important form of energy that is harnessed by other membrane proteins to drive the formation of ATP.

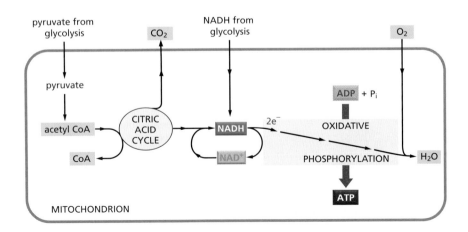

Figure 2–86 The final stages of oxidation of food molecules. Molecules of NADH and FADH$_2$ (FADH$_2$ is not shown) are produced by the citric acid cycle. These activated carriers donate high-energy electrons that are eventually used to reduce oxygen gas to water.
A major portion of the energy released during the transfer of these electrons along an electron-transfer chain in the mitochondrial inner membrane (or in the plasma membrane of bacteria) is harnessed to drive the synthesis of ATP—hence the name oxidative phosphorylation (discussed in Chapter 14).

some time, passing from one living organism to another. Thus present-day nitrogen-fixing reactions can be said to perform a "topping-up" function for the total nitrogen supply.

Vertebrates receive virtually all of their nitrogen from their dietary intake of proteins and nucleic acids. In the body these macromolecules are broken down to amino acids and the components of nucleotides, and the nitrogen they contain is used to produce new proteins and nucleic acids—or utilized to make other molecules. About half of the 20 amino acids found in proteins are essential amino acids for vertebrates (**Figure 2–87**), which means that they cannot be synthesized from other ingredients of the diet. The others can be so synthesized, using a variety of raw materials, including intermediates of the citric acid cycle as described previously. The essential amino acids are made by plants and other organisms, usually by long and energetically expensive pathways that have been lost in the course of vertebrate evolution.

The nucleotides needed to make RNA and DNA can be synthesized using specialized biosynthetic pathways. All of the nitrogens in the purine and pyrimidine bases (as well as some of the carbons) are derived from the plentiful amino acids glutamine, aspartic acid, and glycine, whereas the ribose and deoxyribose sugars are derived from glucose. There are no "essential nucleotides" that must be provided in the diet.

Amino acids not used in biosynthesis can be oxidized to generate metabolic energy. Most of their carbon and hydrogen atoms eventually form CO_2 or H_2O, whereas their nitrogen atoms are shuttled through various forms and eventually appear as urea, which is excreted. Each amino acid is processed differently, and a whole constellation of enzymatic reactions exists for their catabolism.

Sulfur is abundant on Earth in its most oxidized form, sulfate (SO_4^{2-}). To convert it to forms useful for life, sulfate must be reduced to sulfide (S^{2-}), the oxidation state of sulfur required for the synthesis of essential biological molecules. These molecules include the amino acids methionine and cysteine, coenzyme A (see Figure 2–62), and the iron-sulfur centers essential for electron transport (see Figure 14–23). The process begins in bacteria, fungi, and plants, where a special group of enzymes use ATP and reducing power to create a sulfate assimilation pathway. Humans and other animals cannot reduce sulfate and must therefore acquire the sulfur they need for their metabolism in the food that they eat.

Metabolism Is Organized and Regulated

One gets a sense of the intricacy of a cell as a chemical machine from the relation of glycolysis and the citric acid cycle to the other metabolic pathways sketched out in **Figure 2–88**. This type of chart, which was used earlier in this chapter to introduce metabolism, represents only some of the enzymatic pathways in a cell. It is obvious that our discussion of cell metabolism has dealt with only a tiny fraction of cellular chemistry.

All these reactions occur in a cell that is less than 0.1 mm in diameter, and each requires a different enzyme. As is clear from Figure 2–88, the same molecule can often be part of many different pathways. Pyruvate, for example, is a substrate for half a dozen or more different enzymes, each of which modifies it chemically in a different way. One enzyme converts pyruvate to acetyl CoA, another to oxaloacetate; a third enzyme changes pyruvate to the amino acid alanine, a fourth to lactate, and so on. All of these different pathways compete for the same pyruvate molecule, and similar competitions for thousands of other small molecules go on at the same time.

The situation is further complicated in a multicellular organism. Different cell types will in general require somewhat different sets of enzymes. And different tissues make distinct contributions to the chemistry of the organism as a whole. In addition to differences in specialized products such as hormones or antibodies, there are significant differences in the "common" metabolic pathways among various types of cells in the same organism.

Although virtually all cells contain the enzymes of glycolysis, the citric acid cycle, lipid synthesis and breakdown, and amino acid metabolism, the levels of

THE ESSENTIAL AMINO ACIDS

THREONINE

METHIONINE

LYSINE

VALINE

LEUCINE

ISOLEUCINE

HISTIDINE

PHENYLALANINE

TRYPTOPHAN

Figure 2–87 **The nine essential amino acids.** These cannot be synthesized by human cells and so must be supplied in the diet.

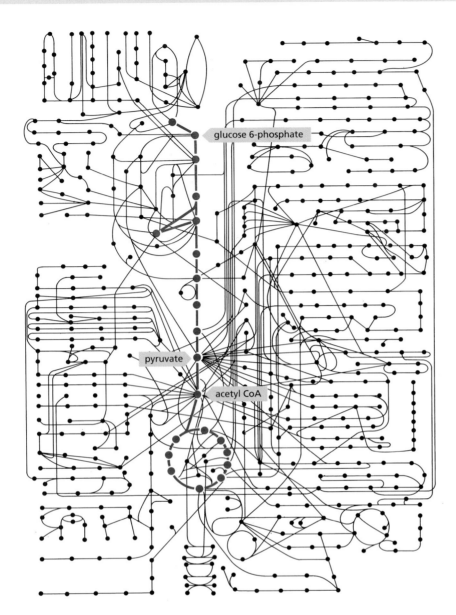

glucose 6-phosphate

pyruvate

acetyl CoA

Figure 2–88 Glycolysis and the citric acid cycle are at the center of metabolism. Some 500 metabolic reactions of a typical cell are shown schematically with the reactions of glycolysis and the citric acid cycle in *red*. Other reactions either lead into these two central pathways—delivering small molecules to be catabolized with production of energy—or they lead outward and thereby supply carbon compounds for the purpose of biosynthesis.

these processes required in different tissues are not the same. For example, nerve cells, which are probably the most fastidious cells in the body, maintain almost no reserves of glycogen or fatty acids and rely almost entirely on a constant supply of glucose from the bloodstream. In contrast, liver cells supply glucose to actively contracting muscle cells and recycle the lactic acid produced by muscle cells back into glucose. All types of cells have their distinctive metabolic traits, and they cooperate extensively in the normal state, as well as in response to stress and starvation. One might think that the whole system would need to be so finely balanced that any minor upset, such as a temporary change in dietary intake, would be disastrous.

In fact, the metabolic balance of a cell is amazingly stable. Whenever the balance is perturbed, the cell reacts so as to restore the initial state. The cell can adapt and continue to function during starvation or disease. Mutations of many kinds can damage or even eliminate particular reaction pathways, and yet—provided that certain minimum requirements are met—the cell survives. It does so because an elaborate network of *control mechanisms* regulates and coordinates the rates of all of its reactions. These controls rest, ultimately, on the remarkable abilities of proteins to change their shape and their chemistry in response to changes in their immediate environment. The principles that underlie how large molecules such as proteins are built and the chemistry behind their regulation will be our next concern.

Summary

Glucose and other food molecules are broken down by controlled stepwise oxidation to provide chemical energy in the form of ATP and NADH. There are three main sets of reactions that act in series—the products of each being the starting material for the next: glycolysis (which occurs in the cytosol), the citric acid cycle (in the mitochondrial matrix), and oxidative phosphorylation (on the inner mitochondrial membrane). The intermediate products of glycolysis and the citric acid cycle are used both as sources of metabolic energy and to produce many of the small molecules used as the raw materials for biosynthesis. Cells store sugar molecules as glycogen in animals and starch in plants; both plants and animals also use fats extensively as a food store. These storage materials in turn serve as a major source of food for humans, along with the proteins that comprise the majority of the dry mass of most of the cells in the foods we eat.

PROBLEMS

Which statements are true? Explain why or why not.

2–1 Of the original radioactivity in a sample, only about 1/1000 will remain after 10 half-lives.

2–2 A 10^{-8} M solution of HCl has a pH of 8.

2–3 Most of the interactions between macromolecules could be mediated just as well by covalent bonds as by noncovalent bonds.

2–4 Animals and plants use oxidation to extract energy from food molecules.

2–5 If an oxidation occurs in a reaction, it must be accompanied by a reduction.

2–6 Linking the energetically unfavorable reaction A → B to a second, favorable reaction B → C will shift the equilibrium constant for the first reaction.

2–7 The criterion for whether a reaction proceeds spontaneously is ΔG not $\Delta G°$, because ΔG takes into account the concentrations of the substrates and products.

2–8 Because glycolysis is only a prelude to the oxidation of glucose in mitochondria, which yields 15-fold more ATP, glycolysis is not really important for human cells.

2–9 The oxygen consumed during the oxidation of glucose in animal cells is returned as CO_2 to the atmosphere.

Discuss the following problems.

2–10 The organic chemistry of living cells is said to be special for two reasons: it occurs in an aqueous environment and it accomplishes some very complex reactions. But do you suppose it is really all that much different from the organic chemistry carried out in the top laboratories in the world? Why or why not?

2–11 The molecular weight of ethanol (CH_3CH_2OH) is 46 and its density is 0.789 g/cm^3.
A. What is the molarity of ethanol in beer that is 5% ethanol by volume? [Alcohol content of beer varies from about 4% (lite beer) to 8% (stout beer).]
B. The legal limit for a driver's blood alcohol content varies, but 80 mg of ethanol per 100 mL of blood (usually

Table Q2–1 Radioactive isotopes and some of their properties (Problem 2–12).

RADIOACTIVE ISOTOPE	EMISSION	HALF-LIFE	MAXIMUM SPECIFIC ACTIVITY (Ci/mmol)
^{14}C	β particle	5730 years	0.062
^{3}H	β particle	12.3 years	29
^{35}S	β particle	87.4 days	1490
^{32}P	β particle	14.3 days	9120

referred to as a blood alcohol level of 0.08) is typical. What is the molarity of ethanol in a person at this legal limit?
C. How many 12-oz (355-mL) bottles of 5% beer could a 70-kg person drink and remain under the legal limit? A 70-kg person contains about 40 liters of water. Ignore the metabolism of ethanol, and assume that the water content of the person remains constant.
D. Ethanol is metabolized at a constant rate of about 120 mg per hour per kg body weight, regardless of its concentration. If a 70-kg person were at twice the legal limit (160 mg/100 mL), how long would it take for their blood alcohol level to fall below the legal limit?

2–12 Specific activity refers to the amount of radioactivity per unit amount of substance, usually in biology expressed on a molar basis, for example, as Ci/mmol. [One curie (Ci) corresponds to 2.22×10^{12} disintegrations per minute (dpm).] As apparent in **Table Q2–1**, which lists properties of four isotopes commonly used in biology, there is an inverse relationship between maximum specific activity and half-life. Do you suppose this is just a coincidence or is there an underlying reason? Explain your answer.

2–13 By a convenient coincidence the ion product of water, $K_w = [H^+][OH^-]$, is a nice round number: 1.0×10^{-14} M^2.
A. Why is a solution at pH 7.0 said to be neutral?
B. What is the H^+ concentration and pH of a 1 mM solution of NaOH?
C. If the pH of a solution is 5.0, what is the concentration of OH^- ions?

2–14 Suggest a rank order for the pK values (from lowest to highest) for the carboxyl group on the aspartate side chain in the following environments in a protein. Explain your ranking.

1. An aspartate side chain on the surface of a protein with no other ionizable groups nearby.
2. An aspartate side chain buried in a hydrophobic pocket on the surface of a protein.
3. An aspartate side chain in a hydrophobic pocket adjacent to a glutamate side chain.
4. An aspartate side chain in a hydrophobic pocket adjacent to a lysine side chain.

2–15   A histidine side chain is known to play an important role in the catalytic mechanism of an enzyme; however, it is not clear whether histidine is required in its protonated (charged) or unprotonated (uncharged) state. To answer this question you measure enzyme activity over a range of pH, with the results shown in **Figure Q2–1**. Which form of histidine is required for enzyme activity?

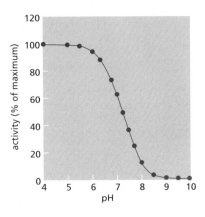

Figure Q2–1 Enzyme activity as a function of pH (Problem 2–15).

2–16   During an all-out sprint, muscles metabolize glucose anaerobically, producing a high concentration of lactic acid, which lowers the pH of the blood and of the cytosol and contributes to the fatigue sprinters experience well before their fuel reserves are exhausted. The main blood buffer against pH changes is the bicarbonate/CO_2 system.

$$CO_2 \underset{\text{(gas)}}{\rightleftharpoons} \underset{\text{(dissolved)}}{CO_2} \overset{pK_1 =}{\underset{2.3}{\rightleftharpoons}} H_2CO_3 \overset{pK_2 =}{\underset{3.8}{\rightleftharpoons}} H^+ + HCO_3^- \overset{pK_3 =}{\underset{10.3}{\rightleftharpoons}} H^+ + CO_3^{2-}$$

To improve their performance, would you advise sprinters to hold their breath or to breathe rapidly for a minute immediately before the race? Explain your answer.

2–17   The three molecules in **Figure Q2–2** contain the seven most common reactive groups in biology. Most molecules in the cell are built from these functional groups. Indicate and name the functional groups in these molecules.

2–18   "Diffusion" sounds slow—and over everyday distances it is—but on the scale of a cell it is very fast. The average instantaneous velocity of a particle in solution, that is, the velocity between collisions, is

$$v = (kT/m)^{\frac{1}{2}}$$

where $k = 1.38 \times 10^{-16}$ g cm²/K sec², T = temperature in K (37°C is 310 K), m = mass in g/molecule.

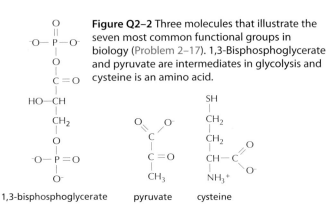

Figure Q2–2 Three molecules that illustrate the seven most common functional groups in biology (Problem 2–17). 1,3-Bisphosphoglycerate and pyruvate are intermediates in glycolysis and cysteine is an amino acid.

1,3-bisphosphoglycerate   pyruvate   cysteine

Calculate the instantaneous velocity of a water molecule (molecular mass = 18 daltons), a glucose molecule (molecular mass = 180 daltons), and a myoglobin molecule (molecular mass = 15,000 daltons) at 37°C. Just for fun, convert these numbers into kilometers/hour. Before you do any calculations, try to guess whether the molecules are moving at a slow crawl (<1 km/hr), an easy walk (5 km/hr), or a record-setting sprint (40 km/hr).

2–19   Polymerization of tubulin subunits into microtubules occurs with an increase in the orderliness of the subunits (**Figure Q2–3**). Yet tubulin polymerization occurs with an increase in entropy (decrease in order). How can that be?

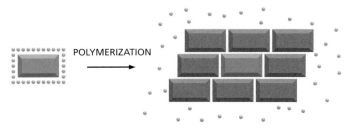

POLYMERIZATION

Figure Q2–3 Polymerization of tubulin subunits into a microtubule (Problem 2–19). The fates of one subunit (*shaded*) and its associated water molecules (*small spheres*) are shown.

2–20   A 70-kg adult human (154 lb) could meet his or her entire energy needs for one day by eating 3 moles of glucose (540 g). (We don't recommend this.) Each molecule of glucose generates 30 ATP when it is oxidized to CO_2. The concentration of ATP is maintained in cells at about 2 mM, and a 70-kg adult has about 25 liters of intracellular fluid. Given that the ATP concentration remains constant in cells, calculate how many times per day, on average, each ATP molecule in the body is hydrolyzed and resynthesized.

2–21   Assuming that there are 5×10^{13} cells in the human body and that ATP is turning over at a rate of 10^9 ATP per minute in each cell, how many watts is the human body consuming? (A watt is a joule per second, and there are 4.18 joules/calorie.) Assume that hydrolysis of ATP yields 12 kcal/mole.

2–22   Does a Snickers™ candy bar (65 g, 325 kcal) provide enough energy to climb from Zermatt (elevation 1660 m) to the top of the Matterhorn (4478 m, **Figure Q2–4**), or might

Figure Q2–4 The Matterhorn (Problem 2–22). (Courtesy of Zermatt Tourism.)

(A)

fed compound

$CH_2-CH_2-CH_2-CH_2-CH_2-CH_2-CH_2-C$

eight-carbon chain

excreted compound

CH_2-C

phenylacetate

(B)

fed compound

$CH_2-CH_2-CH_2-CH_2-CH_2-CH_2-C$

seven-carbon chain

excreted compound

C

benzoate

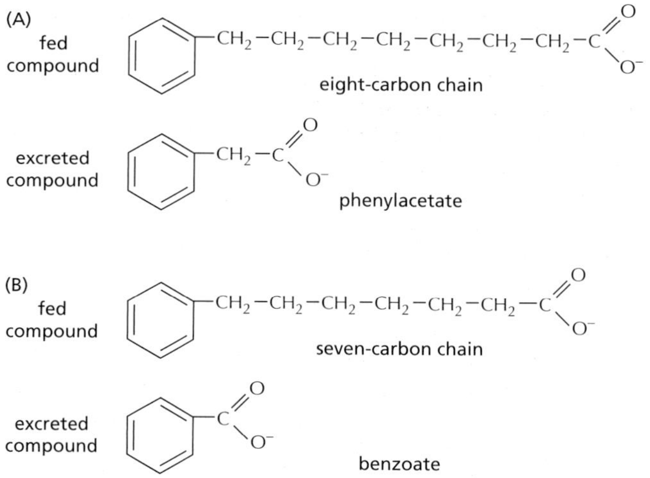

Figure Q2–5 The original labeling experiment to analyze fatty acid oxidation (Problem 2–26). (A) Fed and excreted derivatives of an even-number fatty acid chain. (B) Fed and excreted derivatives of an odd-number fatty acid chain.

you need to stop at Hörnli Hut (3260 m) to eat another one? Imagine that you and your gear have a mass of 75 kg, and that all of your work is done against gravity (that is, you are just climbing straight up). Remember from your introductory physics course that

$$\text{work (J)} = \text{mass (kg)} \times g \text{ (m/sec}^2) \times \text{height gained (m)}$$

where g is acceleration due to gravity (9.8 m/sec^2). One joule is 1 kg m^2/sec^2 and there are 4.18 kJ per kcal.

What assumptions made here will greatly underestimate how much candy you need?

2–23 At first glance, fermentation of pyruvate to lactate appears to be an optional add-on reaction to glycolysis. After all, could cells growing in the absence of oxygen not simply discard pyruvate as a waste product? In the absence of fermentation, which products derived from glycolysis would accumulate in cells under anaerobic conditions? Could the metabolism of glucose via the glycolytic pathway continue in the absence of oxygen in cells that cannot carry out fermentation? Why or why not?

2–24 In the absence of oxygen, cells consume glucose at a high, steady rate. When oxygen is added, glucose consumption drops precipitously and is then maintained at the lower rate. Why is glucose consumed at a high rate in the absence of oxygen and at a low rate in its presence?

2–25 The liver provides glucose to the rest of the body between meals. It does so by breaking down glycogen, forming glucose 6-phosphate in the penultimate step. Glucose 6-phosphate is converted to glucose by splitting off the phosphate ($\Delta G° = -3.3$ kcal/mole). Why do you suppose the liver removes the phosphate by hydrolysis, rather than reversing the reaction by which glucose 6-phosphate (G6P) is formed from glucose (glucose + ATP → G6P + ADP, $\Delta G° = -4.0$ kcal/mole)? By reversing this reaction the liver could generate both glucose and ATP.

2–26 In 1904 Franz Knoop performed what was probably the first successful labeling experiment to study metabolic pathways. He fed many different fatty acids labeled with a terminal benzene ring to dogs and analyzed their urine for excreted benzene derivatives. Whenever the fatty acid had an even number of carbon atoms, phenylacetate was excreted (**Figure Q2–5A**). Whenever the fatty acid had an odd number of carbon atoms, benzoate was excreted (Figure Q2–5B).

From these experiments Knoop deduced that oxidation of fatty acids to CO_2 and H_2O involved the removal of two-carbon fragments from the carboxylic acid end of the chain.

Can you explain the reasoning that led him to conclude that two-carbon fragments, as opposed to any other number, were removed, and that degradation was from the carboxylic acid end, as opposed to the other end?

2–27 Pathways for synthesis of amino acids in microorganisms were worked out in part by cross-feeding experiments among mutant organisms that were defective for individual steps in the pathway. Results of cross-feeding experiments for three mutants defective in the tryptophan pathway—*TrpB⁻*, *TrpD⁻*, and *TrpE⁻*—are shown in **Figure Q2–6**. The mutants were streaked on a Petri dish and allowed to grow briefly in the presence of a very small amount of tryptophan, producing three pale streaks. As shown, heavier growth was observed at points where some streaks were close to other streaks. These spots of heavier growth indicate that one mutant can cross-feed (supply an intermediate) to the other one.

From the pattern of cross-feeding shown in Figure Q2–6, deduce the order of the steps controlled by the products of the *TrpB*, *TrpD*, and *TrpE* genes. Explain your reasoning.

CROSS-FEEDING RESULT

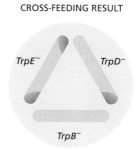

TrpE⁻   TrpD⁻

TrpB⁻

Figure Q2–6 Defining the pathway for tryptophan synthesis using cross-feeding experiments (Problem 2–27). Results of a cross-feeding experiment among mutants defective for steps in the tryptophan biosynthetic pathway. *Dark areas* on the Petri dish show regions of cell growth.

CARBON SKELETONS

Carbon has a unique role in the cell because of its ability to form strong covalent bonds with other carbon atoms. Thus carbon atoms can join to form chains.

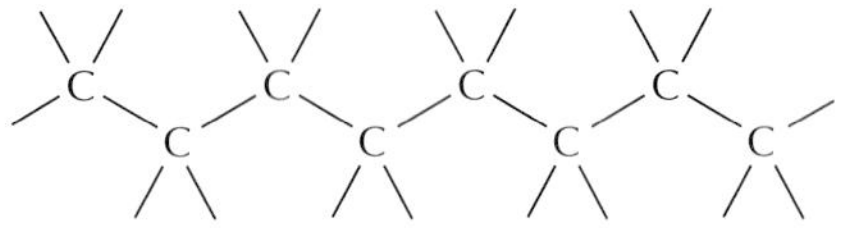

or branched trees

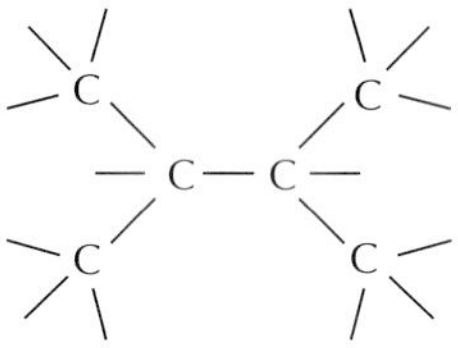

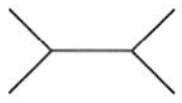

or rings

also written as

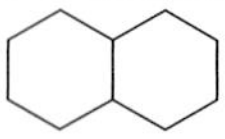

also written as

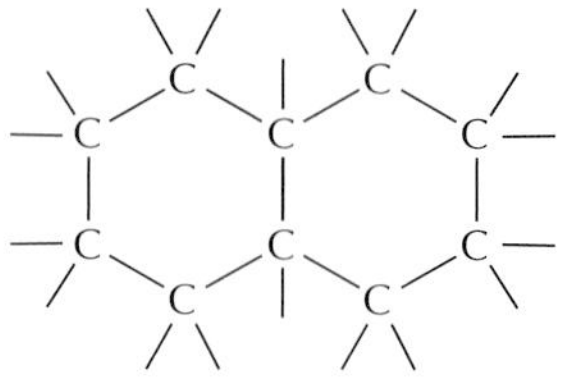

also written as

COVALENT BONDS

A covalent bond forms when two atoms come very close together and share one or more of their electrons. In a single bond one electron from each of the two atoms is shared; in a double bond a total of four electrons are shared.

Each atom forms a fixed number of covalent bonds in a defined spatial arrangement. For example, carbon forms four single bonds arranged tetrahedrally, whereas nitrogen forms three single bonds and oxygen forms two single bonds arranged as shown below.

Double bonds exist and have a different spatial arrangement.

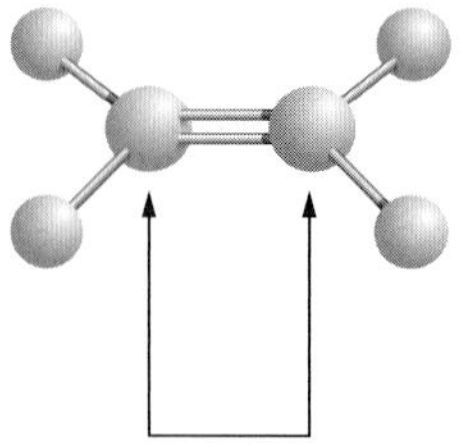

Atoms joined by two or more covalent bonds cannot rotate freely around the bond axis. This restriction is a major influence on the three-dimensional shape of many macromolecules.

HYDROCARBONS

Carbon and hydrogen combine together to make stable compounds (or chemical groups) called hydrocarbons. These are nonpolar, do not form hydrogen bonds, and are generally insoluble in water.

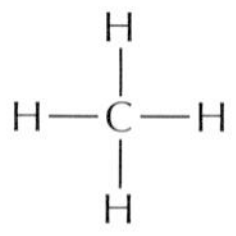

$$H—\underset{\underset{H}{|}}{\overset{\overset{H}{|}}{C}}—H$$

methane

$$H—\underset{\underset{H}{|}}{\overset{\overset{H}{|}}{C}}—$$

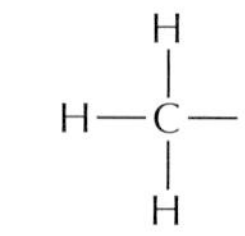

methyl **group**

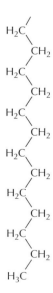

part of the hydrocarbon "tail" of a fatty acid molecule

ALTERNATING DOUBLE BONDS

The carbon chain can include double bonds. If these are on alternate carbon atoms, the bonding electrons move within the molecule, stabilizing the structure by a phenomenon called resonance.

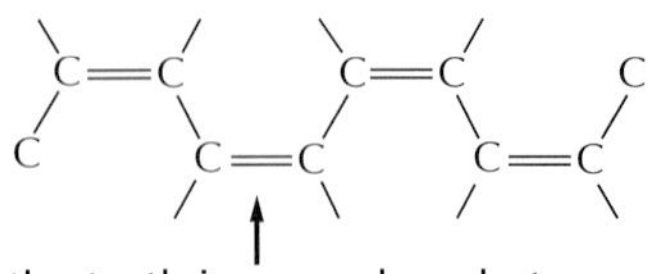

the truth is somewhere between these two structures

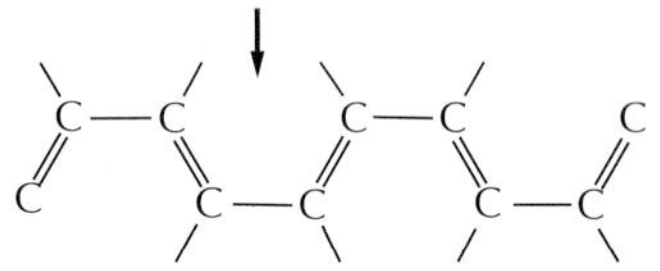

Alternating double bonds in a ring can generate a very stable structure.

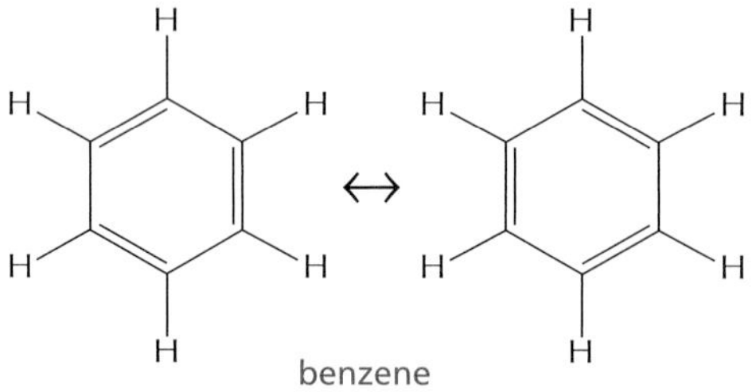

benzene

often written as

C–O CHEMICAL GROUPS

Many biological compounds contain a carbon bonded to an oxygen. For example,

alcohol

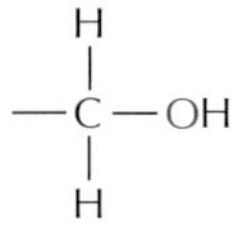

The –OH is called a hydroxyl group.

aldehyde

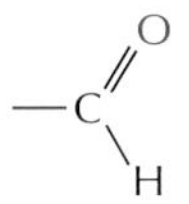

ketone

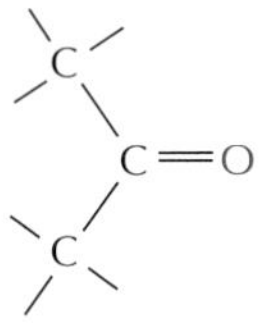

The C=O is called a carbonyl group.

carboxylic acid

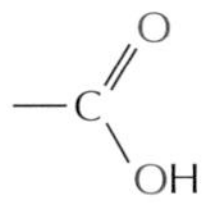

The –COOH is called a carboxyl group. In water this loses an H^+ ion to become $-COO^-$.

esters

Esters are formed by a condensation reaction between acid and an alcohol.

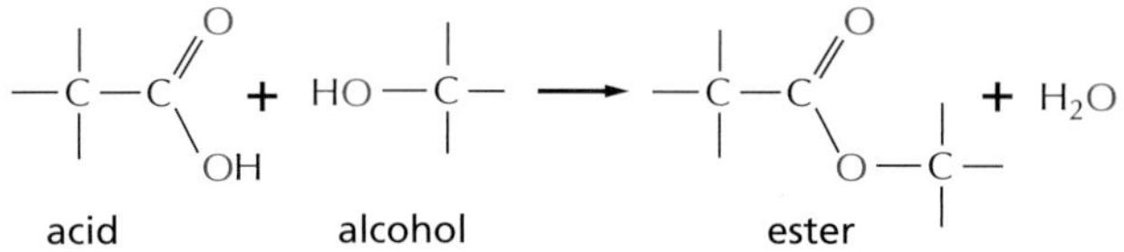

acid   alcohol   ester

C–N CHEMICAL GROUPS

Amines and amides are two important examples of compounds containing a carbon linked to a nitrogen.

Amines in water combine with an H^+ ion to become positively charged.

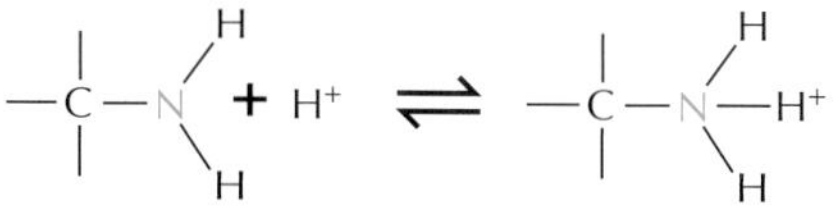

Amides are formed by combining an acid and an amine. Unlike amines, amides are uncharged in water. An example is the peptide bond that joins amino acids in a protein.

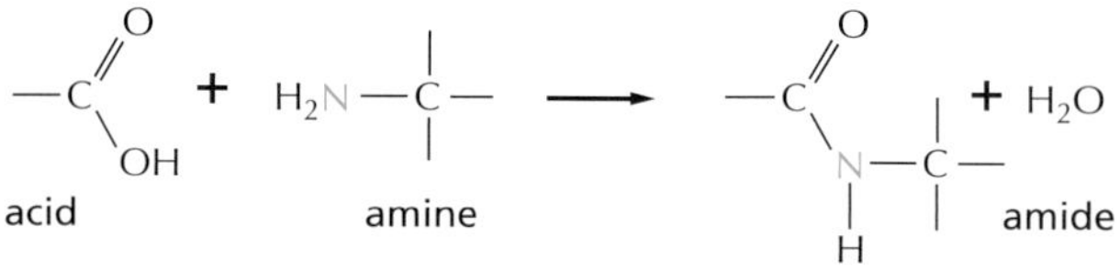

acid   amine   amide

Nitrogen also occurs in several ring compounds, including important constituents of nucleic acids: purines and pyrimidines.

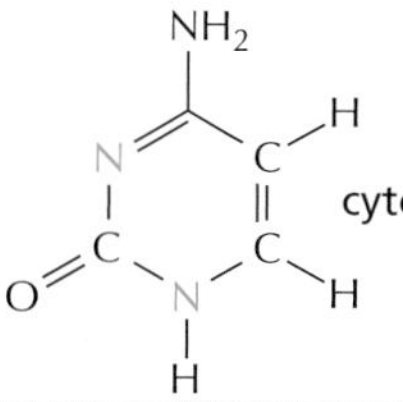

cytosine (a pyrimidine)

SULFHYDRYL GROUP

The $-\overset{|}{\underset{|}{C}}-SH$ is called a sulfhydryl group. In the amino acid cysteine the sulfhydryl group may exist in the reduced form, $-\overset{|}{\underset{|}{C}}-SH$ or more rarely in an oxidized, cross-bridging form, $-\overset{|}{\underset{|}{C}}-S-S-\overset{|}{\underset{|}{C}}-$

PHOSPHATES

Inorganic phosphate is a stable ion formed from phosphoric acid, H_3PO_4. It is often written as P_i.

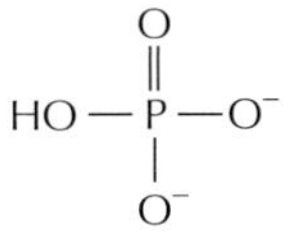

Phosphate esters can form between a phosphate and a free hydroxyl group. Phosphate groups are often attached to proteins in this way.

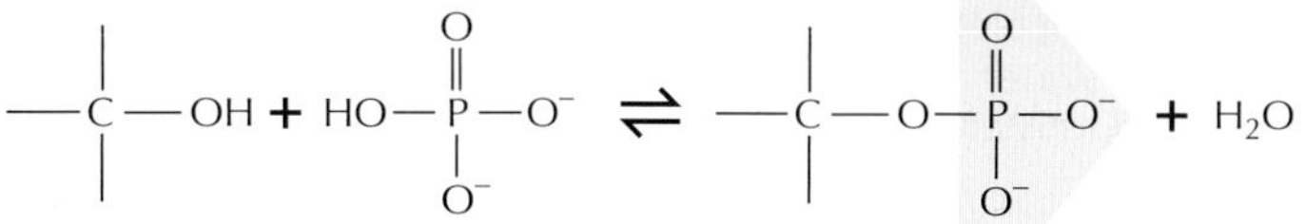

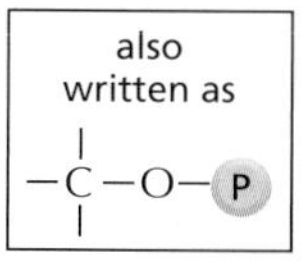

also written as

The combination of a phosphate and a carboxyl group, or two or more phosphate groups, gives an acid anhydride.

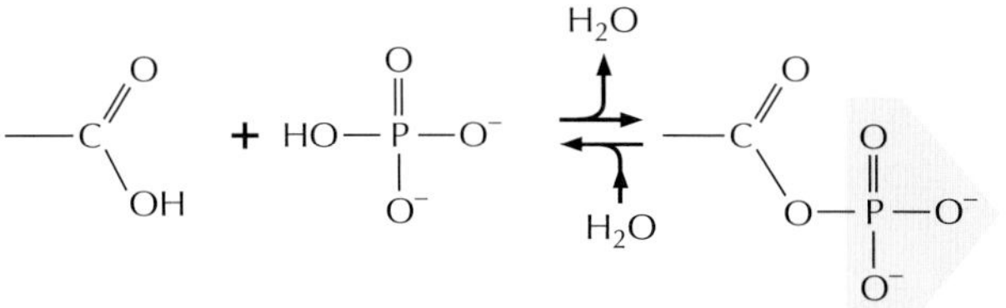

high-energy acyl phosphate bond (carboxylic–phosphoric acid anhydride) found in some metabolites

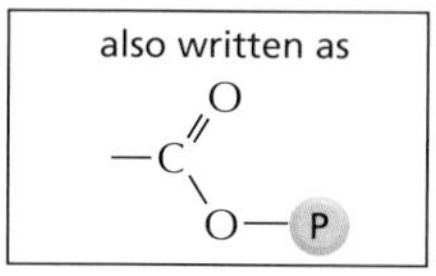

also written as

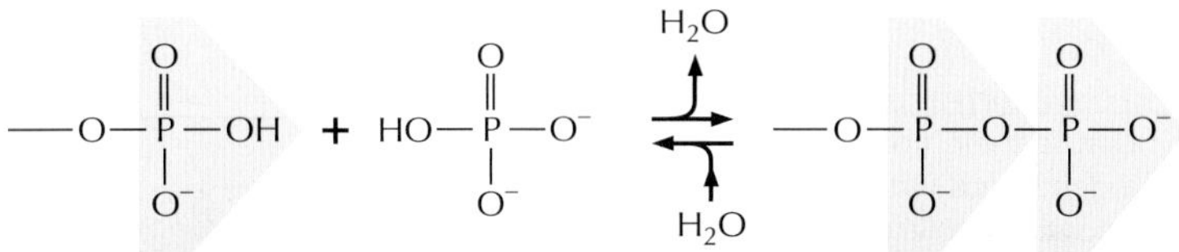

phosphoanhydride—a high-energy bond found in molecules such as ATP

also written as

WATER

Two atoms, connected by a covalent bond, may exert different attractions for the electrons of the bond. In such cases the bond is polar, with one end slightly negatively charged (δ^-) and the other slightly positively charged (δ^+).

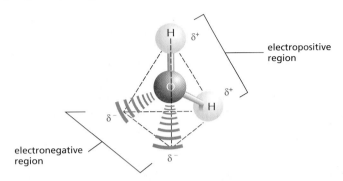

Although a water molecule has an overall neutral charge (having the same number of electrons and protons), the electrons are asymmetrically distributed, which makes the molecule polar. The oxygen nucleus draws electrons away from the hydrogen nuclei, leaving these nuclei with a small net positive charge. The excess of electron density on the oxygen atom creates weakly negative regions at the other two corners of an imaginary tetrahedron.

WATER STRUCTURE

Molecules of water join together transiently in a hydrogen-bonded lattice. Even at 37°C, 15% of the water molecules are joined to four others in a short-lived assembly known as a "flickering cluster."

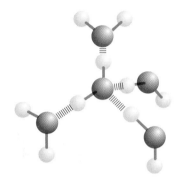

The cohesive nature of water is responsible for many of its unusual properties, such as high surface tension, specific heat, and heat of vaporization.

HYDROGEN BONDS

Because they are polarized, two adjacent H_2O molecules can form a linkage known as a hydrogen bond. Hydrogen bonds have only about 1/20 the strength of a covalent bond.

Hydrogen bonds are strongest when the three atoms lie in a straight line.

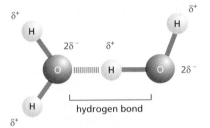

bond lengths

hydrogen bond
0.27 nm

0.10 nm
covalent bond

HYDROPHILIC MOLECULES

Substances that dissolve readily in water are termed hydrophilic. They are composed of ions or polar molecules that attract water molecules through electrical charge effects. Water molecules surround each ion or polar molecule on the surface of a solid substance and carry it into solution.

Ionic substances such as sodium chloride dissolve because water molecules are attracted to the positive (Na^+) or negative (Cl^-) charge of each ion.

Polar substances such as urea dissolve because their molecules form hydrogen bonds with the surrounding water molecules.

HYDROPHOBIC MOLECULES

Molecules that contain a preponderance of nonpolar bonds are usually insoluble in water and are termed hydrophobic. This is true, especially, of hydrocarbons, which contain many C–H bonds. Water molecules are not attracted to such molecules and so have little tendency to surround them and carry them into solution.

WATER AS A SOLVENT

Many substances, such as household sugar, dissolve in water. That is, their molecules separate from each other, each becoming surrounded by water molecules.

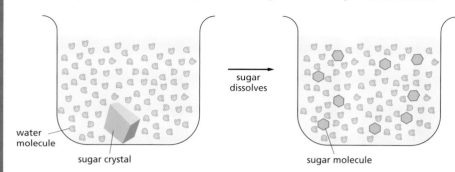

water molecule

sugar crystal

sugar dissolves

sugar molecule

When a substance dissolves in a liquid, the mixture is termed a solution. The dissolved substance (in this case sugar) is the solute, and the liquid that does the dissolving (in this case water) is the solvent. Water is an excellent solvent for many substances because of its polar bonds.

ACIDS

Substances that release hydrogen ions into solution are called acids.

$$HCl \longrightarrow H^+ + Cl^-$$

hydrochloric acid   hydrogen ion   chloride ion
(strong acid)

Many of the acids important in the cell are only partially dissociated, and they are therefore weak acids—for example, the carboxyl group (–COOH), which dissociates to give a hydrogen ion in solution

$$-\overset{\displaystyle O}{\underset{\displaystyle OH}{C}} \rightleftharpoons H^+ + -\overset{\displaystyle O}{\underset{\displaystyle O^-}{C}}$$

(weak acid)

Note that this is a reversible reaction.

HYDROGEN ION EXCHANGE

Positively charged hydrogen ions (H⁺) can spontaneously move from one water molecule to another, thereby creating two ionic species.

hydronium ion   hydroxyl ion
(water acting as   (water acting as
a weak base)   a weak acid)

often written as: $H_2O \rightleftharpoons H^+ + OH^-$

hydrogen   hydroxyl
ion   ion

Since the process is rapidly reversible, hydrogen ions are continually shuttling between water molecules. Pure water contains a steady-state concentration of hydrogen ions and hydroxyl ions (both 10^{-7} M).

pH

The acidity of a solution is defined by the concentration of H⁺ ions it possesses. For convenience we use the pH scale, where

$$pH = -\log_{10}[H^+]$$

For pure water

$$[H^+] = 10^{-7} \text{ moles/liter}$$

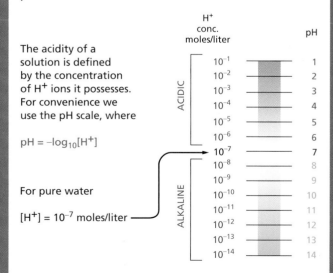

H^+ conc. moles/liter	pH
10^{-1}	1
10^{-2}	2
10^{-3}	3
10^{-4}	4
10^{-5}	5
10^{-6}	6
10^{-7}	7
10^{-8}	8
10^{-9}	9
10^{-10}	10
10^{-11}	11
10^{-12}	12
10^{-13}	13
10^{-14}	14

ACIDIC

ALKALINE

BASES

Substances that reduce the number of hydrogen ions in solution are called bases. Some bases, such as ammonia, combine directly with hydrogen ions.

$$NH_3 + H^+ \longrightarrow NH_4^+$$

ammonia   hydrogen ion   ammonium ion

Other bases, such as sodium hydroxide, reduce the number of H⁺ ions indirectly, by making OH⁻ ions that then combine directly with H⁺ ions to make H_2O.

$$NaOH \longrightarrow Na^+ + OH^-$$

sodium hydroxide   sodium   hydroxyl
(strong base)   ion   ion

Many bases found in cells are partially dissociated and are termed weak bases. This is true of compounds that contain an amino group (–NH₂), which has a weak tendency to reversibly accept an H⁺ ion from water, increasing the quantity of free OH⁻ ions.

$$-NH_2 + H^+ \rightleftharpoons -NH_3^+$$

WEAK CHEMICAL BONDS

Organic molecules can interact with other molecules through three types of short-range attractive forces known as *noncovalent bonds*: van der Waals attractions, electrostatic attractions, and hydrogen bonds. The repulsion of hydrophobic groups from water is also important for ordering biological macromolecules.

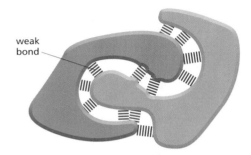

weak bond

Weak chemical bonds have less than 1/20 the strength of a strong covalent bond. They are strong enough to provide tight binding only when many of them are formed simultaneously.

HYDROGEN BONDS

As already described for water (see Panel 2–2), hydrogen bonds form when a hydrogen atom is "sandwiched" between two electron-attracting atoms (usually oxygen or nitrogen).

Hydrogen bonds are strongest when the three atoms are in a straight line:

Examples in macromolecules:

Amino acids in polypeptide chains hydrogen-bonded together.

Two bases, G and C, hydrogen-bonded in DNA or RNA.

VAN DER WAALS ATTRACTIONS

If two atoms are too close together they repel each other very strongly. For this reason, an atom can often be treated as a sphere with a fixed radius. The characteristic "size" for each atom is specified by a unique van der Waals radius. The contact distance between any two noncovalently bonded atoms is the sum of their van der Waals radii.

0.12 nm radius	0.2 nm radius	0.15 nm radius	0.14 nm radius

At very short distances any two atoms show a weak bonding interaction due to their fluctuating electrical charges. The two atoms will be attracted to each other in this way until the distance between their nuclei is approximately equal to the sum of their van der Waals radii. Although they are individually very weak, van der Waals attractions can become important when two macromolecular surfaces fit very close together, because many atoms are involved.

Note that when two atoms form a covalent bond, the centers of the two atoms (the two atomic nuclei) are much closer together than the sum of the two van der Waals radii. Thus,

0.4 nm two non-bonded carbon atoms	0.15 nm single-bonded carbons	0.13 nm double-bonded carbons

HYDROGEN BONDS IN WATER

Any molecules that can form hydrogen bonds to each other can alternatively form hydrogen bonds to water molecules. Because of this competition with water molecules, the hydrogen bonds formed between two molecules dissolved in water are relatively weak.

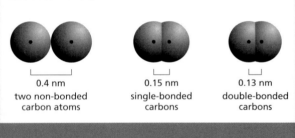

HYDROPHOBIC FORCES

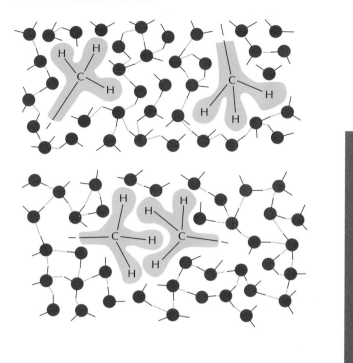

Water forces hydrophobic groups together, because doing so minimizes their disruptive effects on the hydrogen-bonded water network. Hydrophobic groups held together in this way are sometimes said to be held together by "hydrophobic bonds," even though the apparent attraction is actually caused by a repulsion from the water.

ELECTROSTATIC ATTRACTIONS IN AQUEOUS SOLUTIONS

Charged groups are shielded by their interactions with water molecules. Electrostatic attractions are therefore quite weak in water.

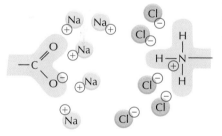

Similarly, ions in solution can cluster around charged groups and further weaken these attractions.

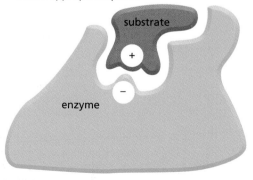

Despite being weakened by water and salt, electrostatic attractions are very important in biological systems. For example, an enzyme that binds a positively charged substrate will often have a negatively charged amino acid side chain at the appropriate place.

ELECTROSTATIC ATTRACTIONS

Attractive forces occur both between fully charged groups (ionic bond) and between the partially charged groups on polar molecules.

The force of attraction between the two charges, δ^+ and δ^-, falls off rapidly as the distance between the charges increases.

In the absence of water, electrostatic forces are very strong. They are responsible for the strength of such minerals as marble and agate, and for crystal formation in common table salt, NaCl.

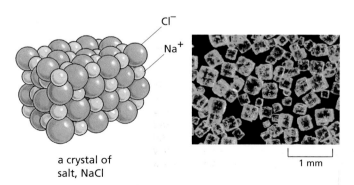

a crystal of salt, NaCl

1 mm

MONOSACCHARIDES

Monosaccharides usually have the general formula $(CH_2O)_n$, where n can be 3, 4, 5, 6, 7, or 8, and have two or more hydroxyl groups. They either contain an aldehyde group ($-C{\leq}^O_H$) and are called aldoses or a ketone group ($>C=O$) and are called ketoses.

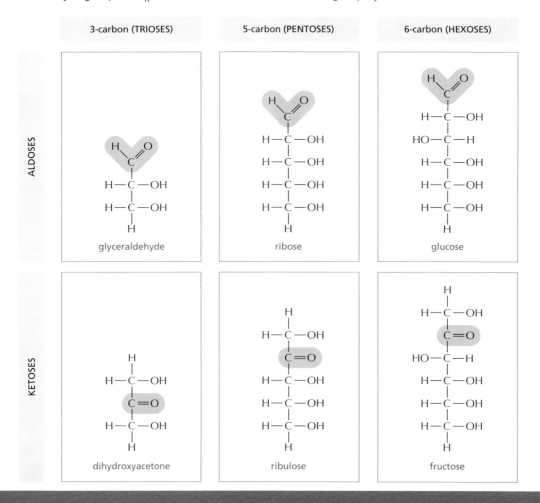

	3-carbon (TRIOSES)	5-carbon (PENTOSES)	6-carbon (HEXOSES)
ALDOSES	glyceraldehyde	ribose	glucose
KETOSES	dihydroxyacetone	ribulose	fructose

RING FORMATION

In aqueous solution, the aldehyde or ketone group of a sugar molecule tends to react with a hydroxyl group of the same molecule, thereby closing the molecule into a ring.

glucose

ribose

Note that each carbon atom has a number.

ISOMERS

Many monosaccharides differ only in the spatial arrangement of atoms—that is, they are isomers. For example, glucose, galactose, and mannose have the same formula ($C_6H_{12}O_6$) but differ in the arrangement of groups around one or two carbon atoms.

glucose

galactose

mannose

These small differences make only minor changes in the chemical properties of the sugars. But they are recognized by enzymes and other proteins and therefore can have important biological effects.

α AND β LINKS

The hydroxyl group on the carbon that carries the aldehyde or ketone can rapidly change from one position to the other. These two positions are called α and β.

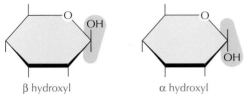

β hydroxyl   α hydroxyl

As soon as one sugar is linked to another, the α or β form is frozen.

SUGAR DERIVATIVES

The hydroxyl groups of a simple monosaccharide can be replaced by other groups. For example,

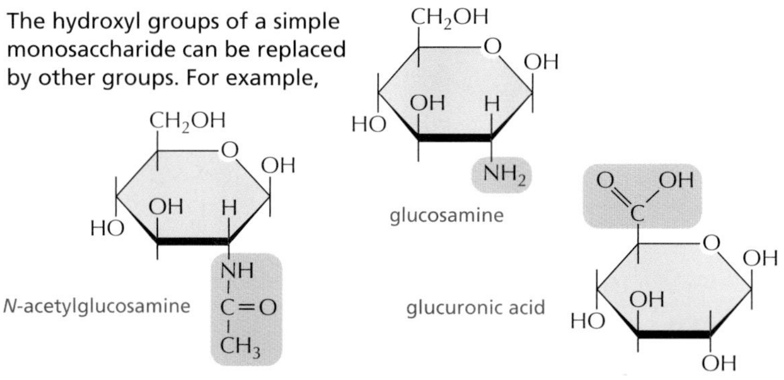

CH_2OH

OH

HO   H

NH_2

glucosamine

CH_2OH

OH

HO   H

NH

C=O

CH_3

N-acetylglucosamine

O   OH

C

O   OH

OH

HO

OH

glucuronic acid

DISACCHARIDES

The carbon that carries the aldehyde or the ketone can react with any hydroxyl group on a second sugar molecule to form a disaccharide. The linkage is called a glycosidic bond.

Three common disaccharides are

  maltose (glucose + glucose)
  lactose (galactose + glucose)
  sucrose (glucose + fructose)

The reaction forming sucrose is shown here.

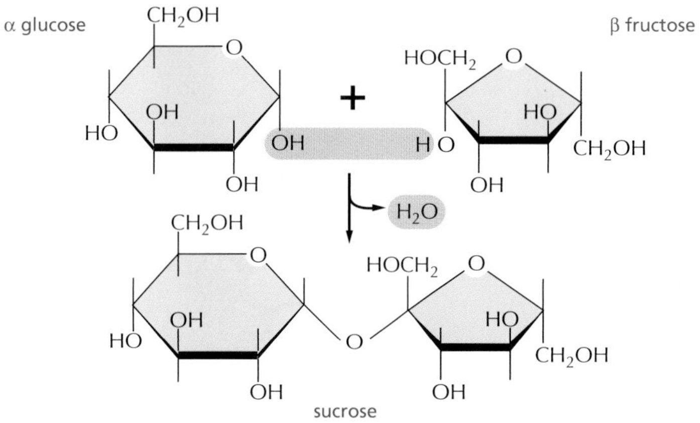

α glucose   β fructose

CH_2OH $HOCH_2$

O   O

OH   HO

HO CH_2OH

OH   +   HO

OH   OH

↓ → H_2O

CH_2OH $HOCH_2$

O   O

OH   HO

HO CH_2OH

OH   O   OH

sucrose

OLIGOSACCHARIDES AND POLYSACCHARIDES

Large linear and branched molecules can be made from simple repeating sugar subunits. Short chains are called oligosaccharides, while long chains are called polysaccharides. Glycogen, for example, is a polysaccharide made entirely of glucose units joined together.

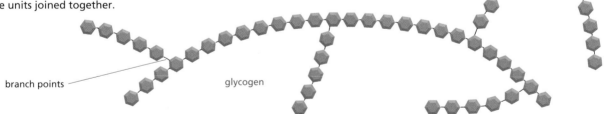

branch points   glycogen

COMPLEX OLIGOSACCHARIDES

In many cases a sugar sequence is nonrepetitive. Many different molecules are possible. Such complex oligosaccharides are usually linked to proteins or to lipids, as is this oligosaccharide, which is part of a cell-surface molecule that defines a particular blood group.

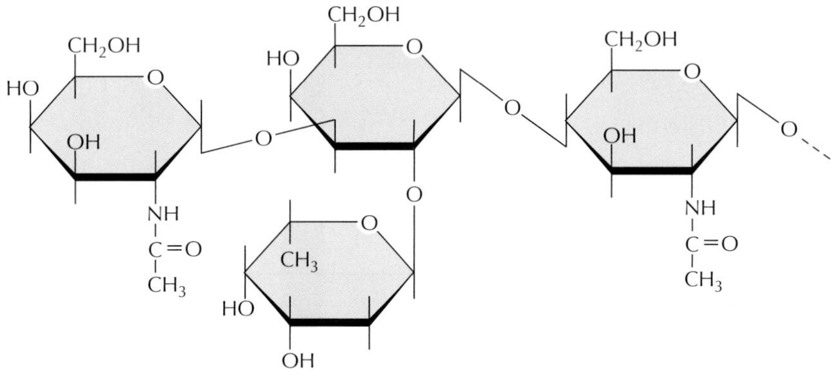

CH_2OH CH_2OH CH_2OH

HO   HO   O   O

O   OH

HO   OH   O   O

OH   OH

NH   O   O   NH

C=O CH_3 C=O

CH_3 HO CH_3

OH

COMMON FATTY ACIDS

These are carboxylic acids with long hydrocarbon tails.

COOH	COOH	COOH
CH₂	CH₂	CH₂
CH₂	CH₂	CH₂
CH₂	CH₂	CH₂
CH₂	CH₂	CH₂
CH₂	CH₂	CH₂
CH₂	CH₂	CH₂
CH₂	CH₂	CH₂
CH₂	CH₂	CH
CH₂	CH₂	CH
CH₂	CH₂	CH₂
CH₂	CH₂	CH₂
CH₂	CH₂	CH₂
CH₂	CH₂	CH₂
CH₂	CH₃	CH₂
CH₂	palmitic acid (C₁₆)	CH₂
CH₃		CH₃

stearic acid (C₁₈)   oleic acid (C₁₈)

TRIACYLGLYCEROLS

Fatty acids are stored as an energy reserve (fats and oils) through an ester linkage to glycerol to form triacylglycerols, also known as triglycerides.

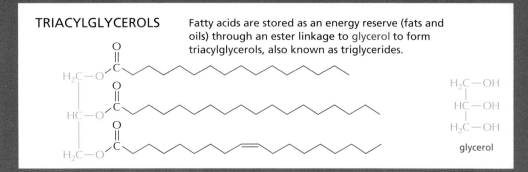

glycerol

Hundreds of different kinds of fatty acids exist. Some have one or more double bonds in their hydrocarbon tail and are said to be unsaturated. Fatty acids with no double bonds are saturated.

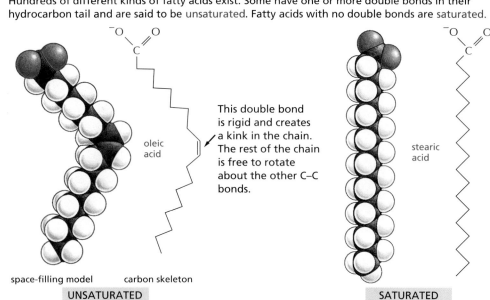

oleic acid

This double bond is rigid and creates a kink in the chain. The rest of the chain is free to rotate about the other C–C bonds.

space-filling model   carbon skeleton

UNSATURATED

stearic acid

SATURATED

CARBOXYL GROUP

If free, the carboxyl group of a fatty acid will be ionized.

But more usually it is linked to other groups to form either esters

or amides.

PHOSPHOLIPIDS

Phospholipids are the major constituents of cell membranes.

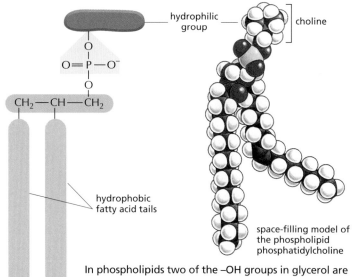

hydrophilic group

choline

hydrophobic fatty acid tails

general structure of a phospholipid

space-filling model of the phospholipid phosphatidylcholine

In phospholipids two of the –OH groups in glycerol are linked to fatty acids, while the third –OH group is linked to phosphoric acid. The phosphate is further linked to one of a variety of small polar groups (alcohols).

LIPID AGGREGATES

Fatty acids have a hydrophilic head and a hydrophobic tail.

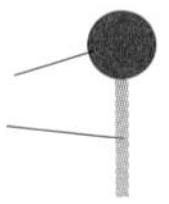

In water they can form a surface film or form small micelles.

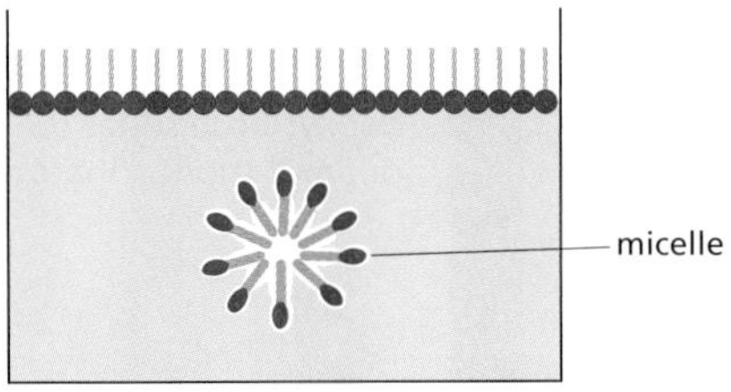

— micelle

Their derivatives can form larger aggregates held together by hydrophobic forces:

Triglycerides can form large spherical fat droplets in the cell cytoplasm.

200 nm or more

Phospholipids and glycolipids form self-sealing lipid bilayers that are the basis for all cell membranes.

|← 4 nm →|

OTHER LIPIDS

Lipids are defined as the water-insoluble molecules in cells that are soluble in organic solvents. Two other common types of lipids are steroids and polyisoprenoids. Both are made from isoprene units.

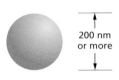

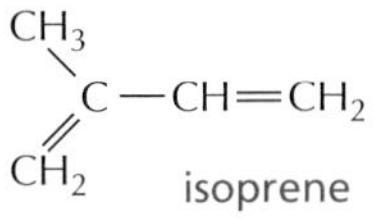

isoprene

STEROIDS

Steroids have a common multiple-ring structure.

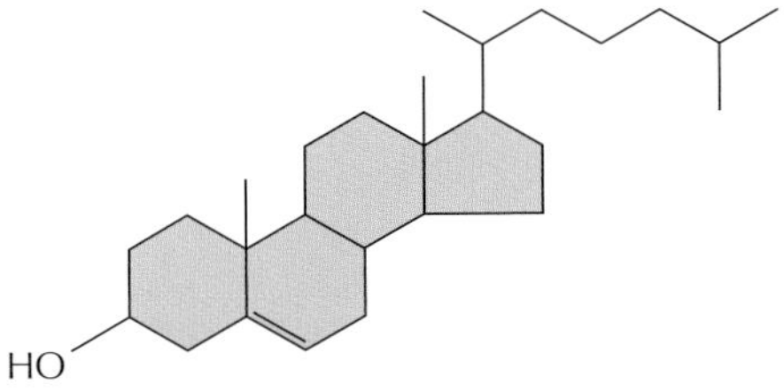

HO

cholesterol—found in many membranes

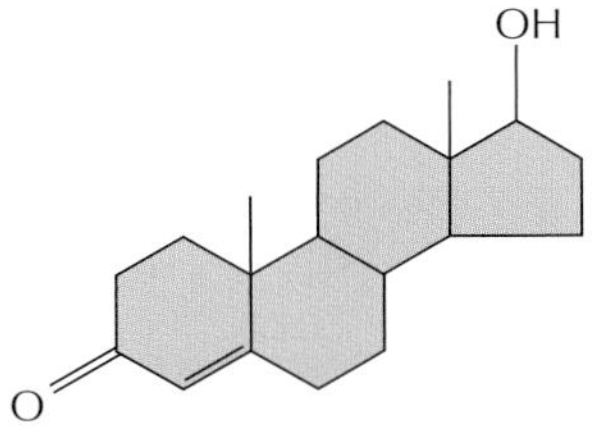

OH

O

testosterone—male steroid hormone

GLYCOLIPIDS

Like phospholipids, these compounds are composed of a hydrophobic region, containing two long hydrocarbon tails, and a polar region, which, however, contains one or more sugars and no phosphate.

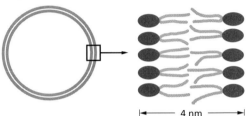

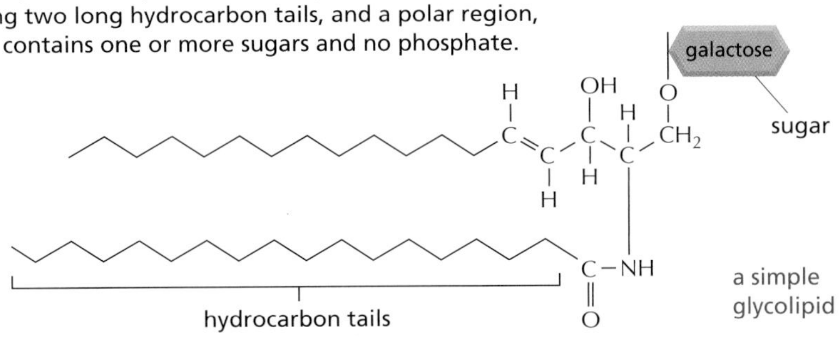

galactose

sugar

a simple glycolipid

hydrocarbon tails

POLYISOPRENOIDS

long-chain polymers of isoprene

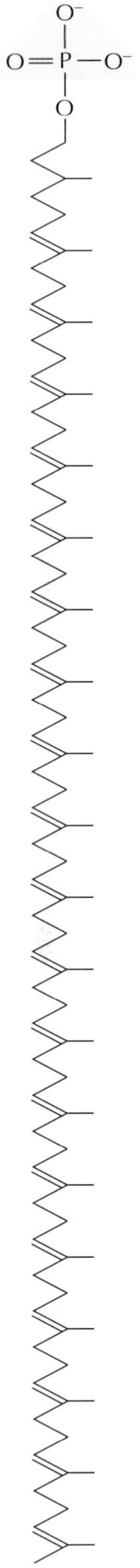

dolichol phosphate—used to carry activated sugars in the membrane-associated synthesis of glycoproteins and some polysaccharides

NUCLEOTIDES

A nucleotide consists of a nitrogen-containing base, a five-carbon sugar, and one or more phosphate groups.

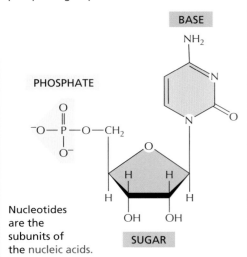

BASE

PHOSPHATE

SUGAR

Nucleotides are the subunits of the nucleic acids.

PHOSPHATES

The phosphates are normally joined to the C5 hydroxyl of the ribose or deoxyribose sugar (designated 5'). Mono-, di-, and triphosphates are common.

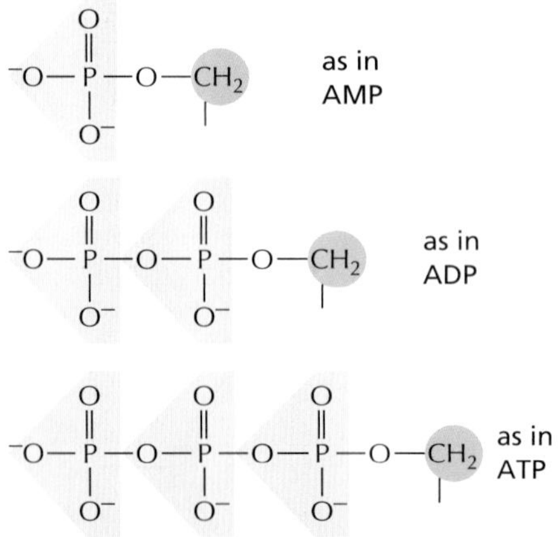

as in AMP

as in ADP

as in ATP

The phosphate makes a nucleotide negatively charged.

BASIC SUGAR LINKAGE

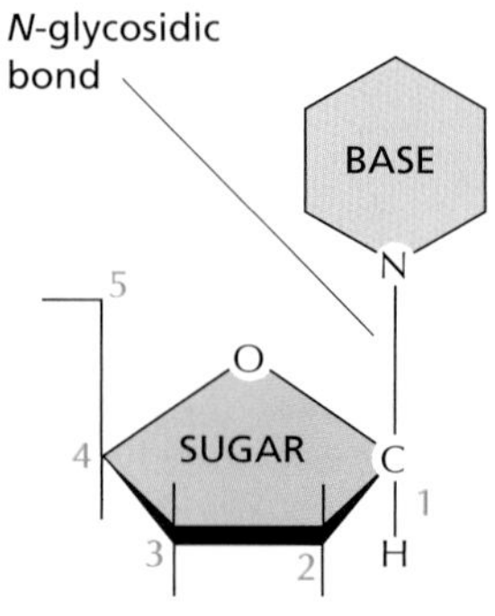

N-glycosidic bond

BASE

SUGAR

The base is linked to the same carbon (C1) used in sugar–sugar bonds.

BASES

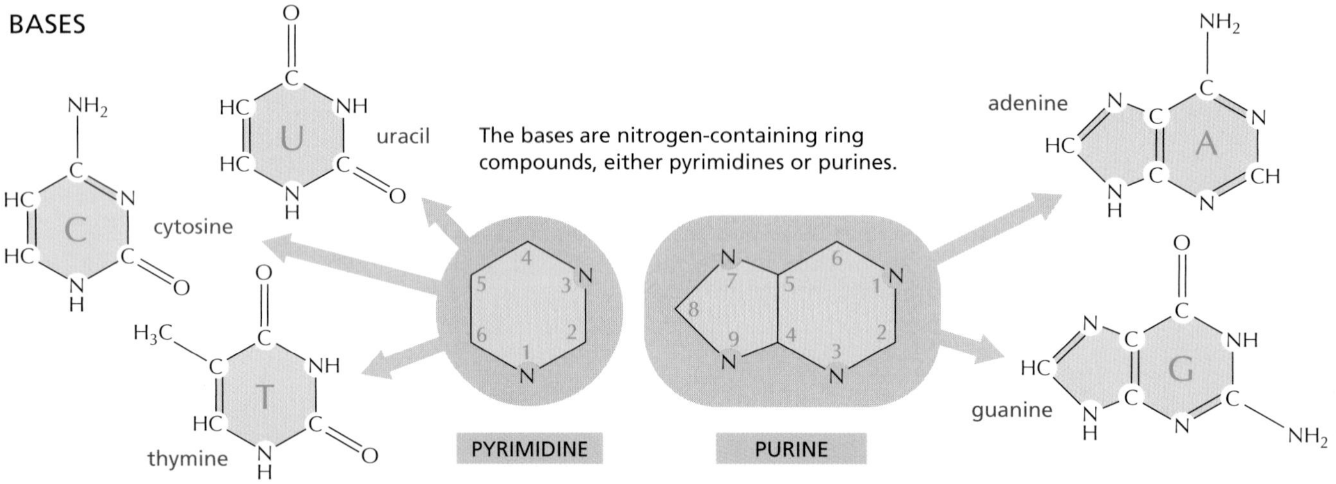

The bases are nitrogen-containing ring compounds, either pyrimidines or purines.

cytosine

uracil

thymine

PYRIMIDINE

PURINE

adenine

guanine

SUGARS

PENTOSE

a five-carbon sugar

two kinds are used

β-D-ribose
used in ribonucleic acid

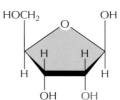

β-D-2-deoxyribose
used in deoxyribonucleic acid

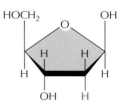

Each numbered carbon on the sugar of a nucleotide is followed by a prime mark; therefore, one speaks of the "5-prime carbon," etc.

NOMENCLATURE

A nucleoside or nucleotide is named according to its nitrogenous base.

BASE	NUCLEOSIDE	ABBR.
adenine	adenosine	A
guanine	guanosine	G
cytosine	cytidine	C
uracil	uridine	U
thymine	thymidine	T

Single letter abbreviations are used variously as shorthand for (1) the base alone, (2) the nucleoside, or (3) the whole nucleotide—the context will usually make clear which of the three entities is meant. When the context is not sufficient, we will add the terms "base", "nucleoside", "nucleotide", or—as in the examples below—use the full 3-letter nucleotide code.

AMP = adenosine monophosphate
dAMP = deoxyadenosine monophosphate
UDP = uridine diphosphate
ATP = adenosine triphosphate

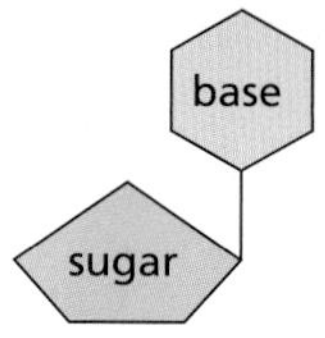

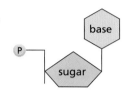

BASE + SUGAR = NUCLEOSIDE

BASE + SUGAR + PHOSPHATE = NUCLEOTIDE

NUCLEIC ACIDS

Nucleotides are joined together by a phosphodiester linkage between 5′ and 3′ carbon atoms to form nucleic acids. The linear sequence of nucleotides in a nucleic acid chain is commonly abbreviated by a one-letter code, A—G—C—T—T—A—C—A, with the 5′ end of the chain at the left.

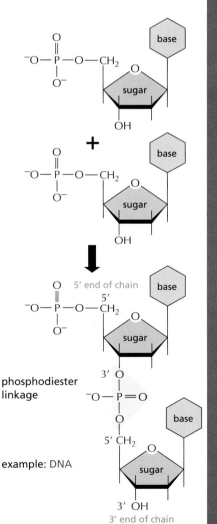

phosphodiester linkage

example: DNA

NUCLEOTIDES HAVE MANY OTHER FUNCTIONS

1 They carry chemical energy in their easily hydrolyzed phosphoanhydride bonds.

phosphoanhydride bonds

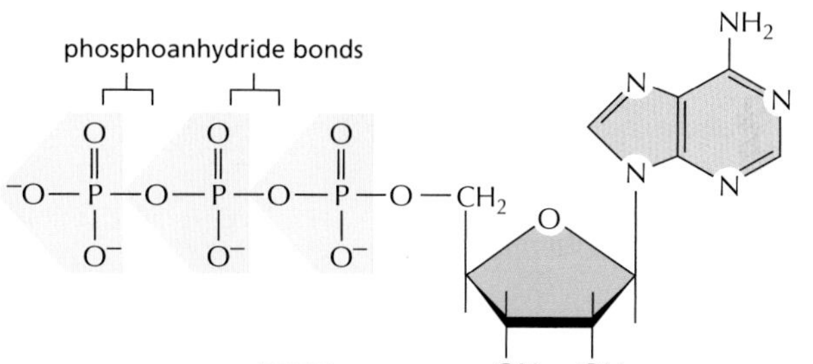

example: ATP (or ATP)

2 They combine with other groups to form coenzymes.

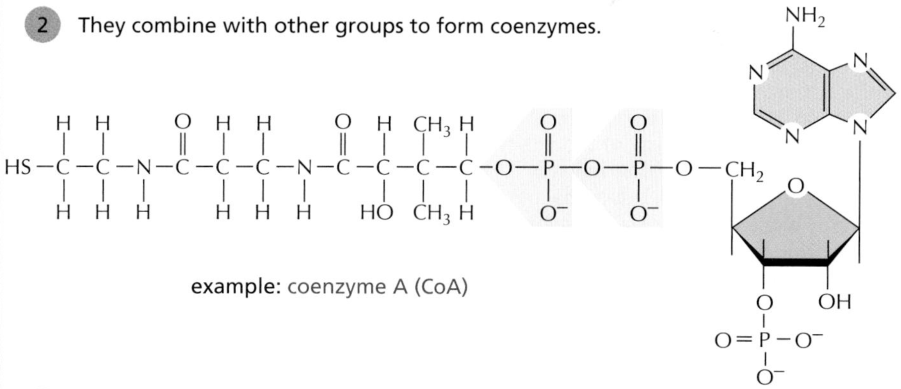

example: coenzyme A (CoA)

3 They are used as specific signaling molecules in the cell.

example: cyclic AMP (cAMP)

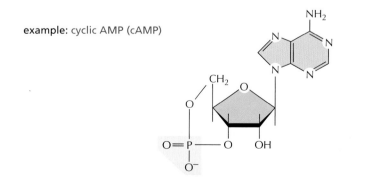

THE IMPORTANCE OF FREE ENERGY FOR CELLS

Life is possible because of the complex network of interacting chemical reactions occurring in every cell. In viewing the metabolic pathways that comprise this network, one might suspect that the cell has had the ability to evolve an enzyme to carry out any reaction that it needs. But this is not so. Although enzymes are powerful catalysts, they can speed up only those reactions that are thermodynamically possible; other reactions proceed in cells only because they are *coupled* to very favorable reactions that drive them. The question of whether a reaction can occur spontaneously, or instead needs to be coupled to another reaction, is central to cell biology. The answer is obtained by reference to a quantity called the *free energy:* the total change in free energy during a set of reactions determines whether or not the entire reaction sequence can occur. In this panel we shall explain some of the fundamental ideas—derived from a special branch of chemistry and physics called *thermodynamics*—that are required for understanding what free energy is and why it is so important to cells.

ENERGY RELEASED BY CHANGES IN CHEMICAL BONDING IS CONVERTED INTO HEAT

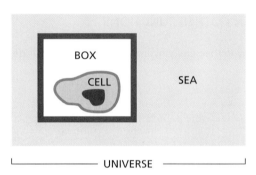

————— UNIVERSE —————

An *enclosed system* is defined as a collection of molecules that does not exchange matter with the rest of the universe (for example, the "cell in a box" shown above). Any such system will contain molecules with a total energy E. This energy will be distributed in a variety of ways: some as the translational energy of the molecules, some as their vibrational and rotational energies, but most as the bonding energies between the individual atoms that make up the molecules. Suppose that a reaction occurs in the system. The first law of thermodynamics places a constraint on what types of reactions are possible: it states that "in any process, the total energy of the universe remains constant." For example, suppose that reaction A → B occurs somewhere in the box and releases a great deal of chemical bond energy. This energy will initially increase the intensity of molecular motions (translational, vibrational, and rotational) in the system, which is equivalent to raising its temperature. However, these increased motions will soon be transferred out of the system by a series of molecular collisions that heat up first the walls of the box and then the outside world (represented by the sea in our example). In the end, the system returns to its initial temperature, by which time all the chemical bond energy released in the box has been converted into heat energy and transferred out of the box to the surroundings. According to the first law, the change in the energy in the box (ΔE_{box}, which we shall denote as ΔE) must be equal and opposite to the amount of heat energy transferred, which we shall designate as h: that is, $\Delta E = -h$. Thus, the energy in the box (E) decreases when heat leaves the system.

E also can change during a reaction as a result of work being done on the outside world. For example, suppose that there is a small increase in the volume (ΔV) of the box during a reaction. Since the walls of the box must push against the constant pressure (P) in the surroundings in order to expand, this does work on the outside world and requires energy. The energy used is $P(\Delta V)$, which according to the first law must decrease the energy in the box (E) by the same amount. In most reactions chemical bond energy is converted into both work and heat. *Enthalpy* (H) is a composite function that includes both of these ($H = E + PV$). To be rigorous, it is the change in enthalpy (ΔH) in an enclosed system, and not the change in energy, that is equal to the heat transferred to the outside world during a reaction. Reactions in which H decreases release heat to the surroundings and are said to be "exothermic," while reactions in which H increases absorb heat from the surroundings and are said to be "endothermic." Thus, $-h = \Delta H$. However, the volume change is negligible in most biological reactions, so to a good approximation

$$-h = \Delta H \cong \Delta E$$

THE SECOND LAW OF THERMODYNAMICS

Consider a container in which 1000 coins are all lying heads up. If the container is shaken vigorously, subjecting the coins to the types of random motions that all molecules experience due to their frequent collisions with other molecules, one will end up with about half the coins oriented heads down. The reason for this reorientation is that there is only a single way in which the original orderly state of the coins can be reinstated (every coin must lie heads up), whereas there are many different ways (about 10^{298}) to achieve a disorderly state in which there is an equal mixture of heads and tails; in fact, there are more ways to achieve a 50-50 state than to achieve any other state. Each state has a probability of occurrence that is proportional to the number of ways it can be realized. The second law of thermodynamics states that "systems will change spontaneously from states of lower probability to states of higher probability." Since states of lower probability are more "ordered" than states of high probability, the second law can be restated: "the universe constantly changes so as to become more disordered."

THE ENTROPY, S

The second law (but not the first law) allows one to predict the *direction* of a particular reaction. But to make it useful for this purpose, one needs a convenient measure of the probability or, equivalently, the degree of disorder of a state. The entropy (S) is such a measure. It is a logarithmic function of the probability such that the *change in entropy* (ΔS) that occurs when the reaction A $\rightarrow$ B converts one mole of A into one mole of B is

$$\Delta S = R \ln p_B / p_A$$

where p_A and p_B are the probabilities of the two states A and B, R is the gas constant (2 cal deg^{-1} $mole^{-1}$), and ΔS is measured in entropy units (eu). In our initial example of 1000 coins, the relative probability of all heads (state A) versus half heads and half tails (state B) is equal to the ratio of the number of different ways that the two results can be obtained. One can calculate that $p_A = 1$ and $p_B = 1000!(500! \times 500!) = 10^{299}$. Therefore, the entropy change for the reorientation of the coins when their container is vigorously shaken and an equal mixture of heads and tails is obtained is $R \ln (10^{298})$, or about 1370 eu per mole of such containers (6×10^{23} containers). We see that, because ΔS defined above is positive for the transition from state A to state B ($p_B / p_A > 1$), reactions with a large *increase* in S (that is, for which $\Delta S > 0$) are favored and will occur spontaneously.

As discussed in Chapter 2, heat energy causes the random commotion of molecules. Because the transfer of heat from an enclosed system to its surroundings increases the number of different arrangements that the molecules in the outside world can have, it increases their entropy. It can be shown that the release of a fixed quantity of heat energy has a greater disordering effect at low temperature than at high temperature, and that the value of ΔS for the surroundings, as defined above (ΔS_{sea}), is precisely equal to h, the amount of heat transferred to the surroundings from the system, divided by the absolute temperature (T):

$$\Delta S_{sea} = h / T$$

THE GIBBS FREE ENERGY, G

When dealing with an enclosed biological system, one would like to have a simple way of predicting whether a given reaction will or will not occur spontaneously in the system. We have seen that the crucial question is whether the entropy change for the universe is positive or negative when that reaction occurs. In our idealized system, the cell in a box, there are two separate components to the entropy change of the universe—the entropy change for the system enclosed in the box and the entropy change for the surrounding "sea"—and both must be added together before any prediction can be made. For example, it is possible for a reaction to absorb heat and thereby decrease the entropy of the sea ($\Delta S_{sea} < 0$) and at the same time to cause such a large degree of disordering inside the box ($\Delta S_{box} > 0$) that the total $\Delta S_{universe} = \Delta S_{sea} + \Delta S_{box}$ is greater than 0. In this case the reaction will occur spontaneously, even though the sea gives up heat to the box during the reaction. An example of such a reaction is the dissolving of sodium chloride in a beaker containing water (the "box"), which is a spontaneous process even though the temperature of the water drops as the salt goes into solution.

Chemists have found it useful to define a number of new "composite functions" that describe *combinations* of physical properties of a system. The properties that can be combined include the temperature (T), pressure (P), volume (V), energy (E), and entropy (S). The enthalpy (H) is one such composite function. But by far the most useful composite function for biologists is the *Gibbs free energy, G*. It serves as an accounting device that allows one to deduce the entropy change of the universe resulting from a chemical reaction in the box, while avoiding any separate consideration of the entropy change in the sea. The definition of G is

$$G = H - TS$$

where, for a box of volume V, H is the enthalpy described above ($E + PV$), T is the absolute temperature, and S is the entropy. Each of these quantities applies to the inside of the box only. The change in free energy during a reaction in the box (the G of the products minus the G of the starting materials) is denoted as ΔG and, as we shall now demonstrate, it is a direct measure of the amount of disorder that is created in the universe when the reaction occurs.

At constant temperature the change in free energy (ΔG) during a reaction equals $\Delta H - T\Delta S$. Remembering that $\Delta H = -h$, the heat absorbed from the sea, we have

$$-\Delta G = -\Delta H + T\Delta S$$
$$-\Delta G = h + T\Delta S, \text{ so } -\Delta G/T = h/T + \Delta S$$

But h/T is equal to the entropy change of the sea (ΔS_{sea}), and the ΔS in the above equation is ΔS_{box}. Therefore

$$-\Delta G/T = \Delta S_{sea} + \Delta S_{box} = \Delta S_{universe}$$

We conclude that the free-energy change is a direct measure of the entropy change of the universe. A reaction will proceed in the direction that causes the change in the free energy (ΔG) to be less than zero, because in this case there will be a positive entropy change in the universe when the reaction occurs.

For a complex set of coupled reactions involving many different molecules, the total free-energy change can be computed simply by adding up the free energies of all the different molecular species after the reaction and comparing this value with the sum of free energies before the reaction; for common substances the required free-energy values can be found from published tables. In this way one can predict the direction of a reaction and thereby readily check the feasibility of any proposed mechanism. Thus, for example, from the observed values for the magnitude of the electrochemical proton gradient across the inner mitochondrial membrane and the ΔG for ATP hydrolysis inside the mitochondrion, one can be certain that ATP synthase requires the passage of more than one proton for each molecule of ATP that it synthesizes.

The value of ΔG for a reaction is a direct measure of how far the reaction is from equilibrium. The large negative value for ATP hydrolysis in a cell merely reflects the fact that cells keep the ATP hydrolysis reaction as much as 10 orders of magnitude away from equilibrium. If a reaction reaches equilibrium, $\Delta G = 0$, the reaction then proceeds at precisely equal rates in the forward and backward direction. For ATP hydrolysis, equilibrium is reached when the vast majority of the ATP has been hydrolyzed, as occurs in a dead cell.

For each step, the part of the molecule that undergoes a change is shadowed in blue,
and the name of the enzyme that catalyzes the reaction is in a yellow box.

STEP 1   Glucose is phosphorylated by ATP to form a sugar phosphate. The negative charge of the phosphate prevents passage of the sugar phosphate through the plasma membrane, trapping glucose inside the cell.

glucose   →[hexokinase] + ATP →   glucose 6-phosphate + ADP + H⁺

STEP 2   A readily reversible rearrangement of the chemical structure (isomerization) moves the carbonyl oxygen from carbon 1 to carbon 2, forming a ketose from an aldose sugar. (See Panel 2–4.)

glucose 6-phosphate   →[phosphoglucose isomerase]   fructose 6-phosphate

STEP 3   The new hydroxyl group on carbon 1 is phosphorylated by ATP, in preparation for the formation of two three-carbon sugar phosphates. The entry of sugars into glycolysis is controlled at this step, through regulation of the enzyme *phosphofructokinase*.

fructose 6-phosphate + ATP →[phosphofructokinase] fructose 1,6-bisphosphate + ADP + H⁺

STEP 4   The six-carbon sugar is cleaved to produce two three-carbon molecules. Only the glyceraldehyde 3-phosphate can proceed immediately through glycolysis.

fructose 1,6-bisphosphate →[aldolase] dihydroxyacetone phosphate + glyceraldehyde 3-phosphate

STEP 5   The other product of step 4, dihydroxyacetone phosphate, is isomerized to form glyceraldehyde 3-phosphate.

dihydroxyacetone phosphate →[triose phosphate isomerase] glyceraldehyde 3-phosphate

STEP 6 The two molecules of glyceraldehyde 3-phosphate are oxidized. The energy generation phase of glycolysis begins, as NADH and a new high-energy anhydride linkage to phosphate are formed (see Figure 2–73).

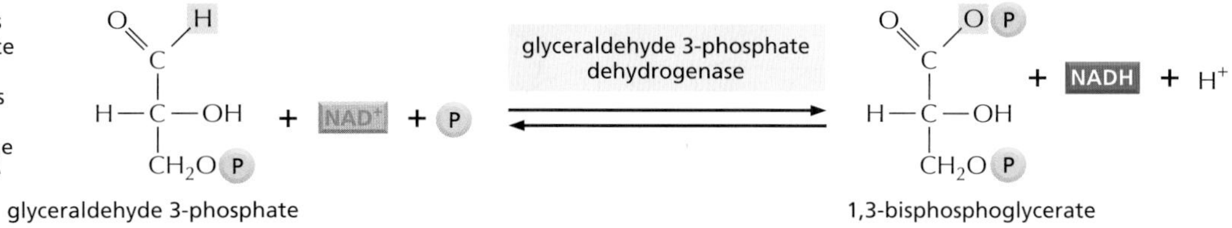

glyceraldehyde 3-phosphate + NAD$^+$ + P → (glyceraldehyde 3-phosphate dehydrogenase) → 1,3-bisphosphoglycerate + NADH + H$^+$

STEP 7 The transfer to ADP of the high-energy phosphate group that was generated in step 6 forms ATP.

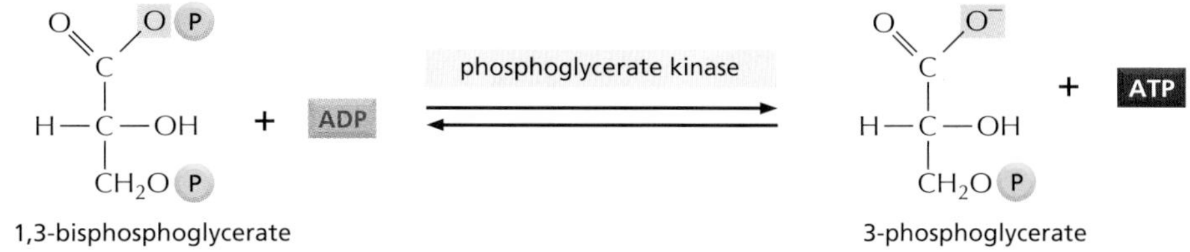

1,3-bisphosphoglycerate + ADP → (phosphoglycerate kinase) → 3-phosphoglycerate + ATP

STEP 8 The remaining phosphate ester linkage in 3-phosphoglycerate, which has a relatively low free energy of hydrolysis, is moved from carbon 3 to carbon 2 to form 2-phosphoglycerate.

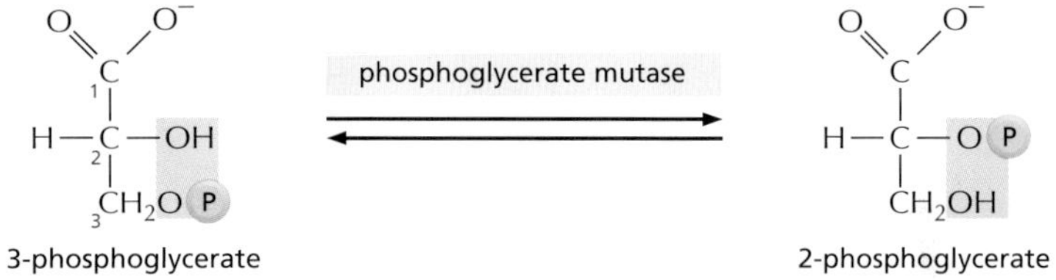

3-phosphoglycerate → (phosphoglycerate mutase) → 2-phosphoglycerate

STEP 9 The removal of water from 2-phosphoglycerate creates a high-energy enol phosphate linkage.

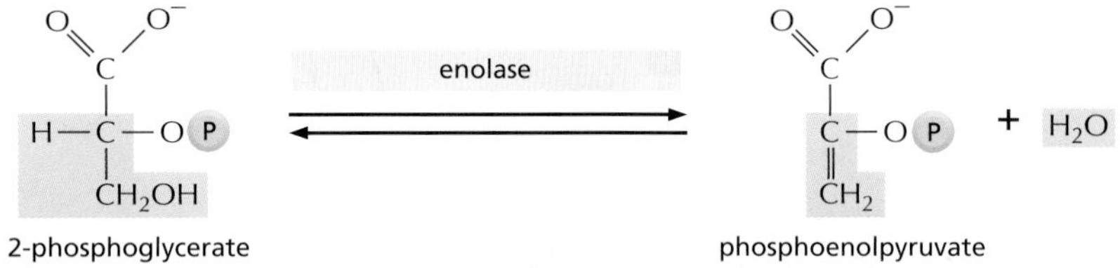

2-phosphoglycerate → (enolase) → phosphoenolpyruvate + H$_2$O

STEP 10 The transfer to ADP of the high-energy phosphate group that was generated in step 9 forms ATP, completing glycolysis.

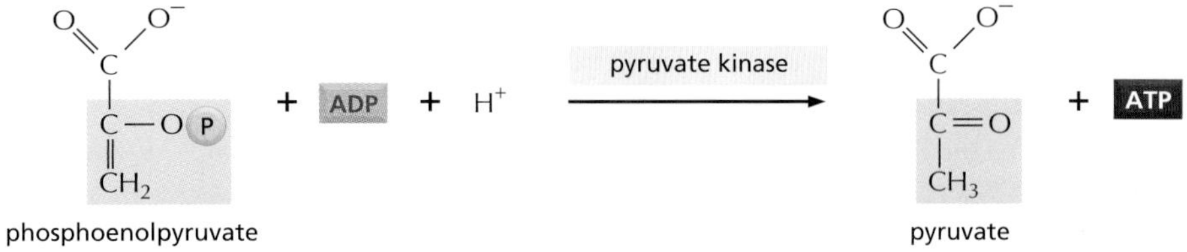

phosphoenolpyruvate + ADP + H$^+$ → (pyruvate kinase) → pyruvate + ATP

NET RESULT OF GLYCOLYSIS

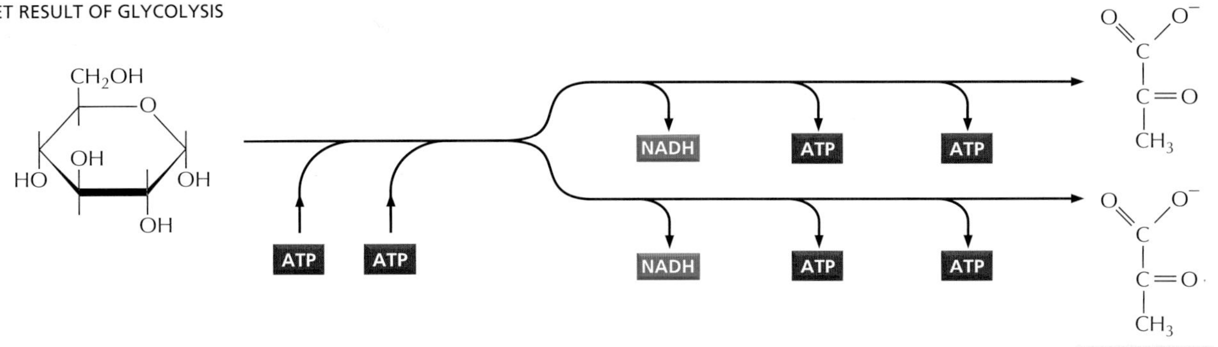

glucose → two molecules of pyruvate

In addition to the pyruvate, the net products are two molecules of ATP and two molecules of NADH

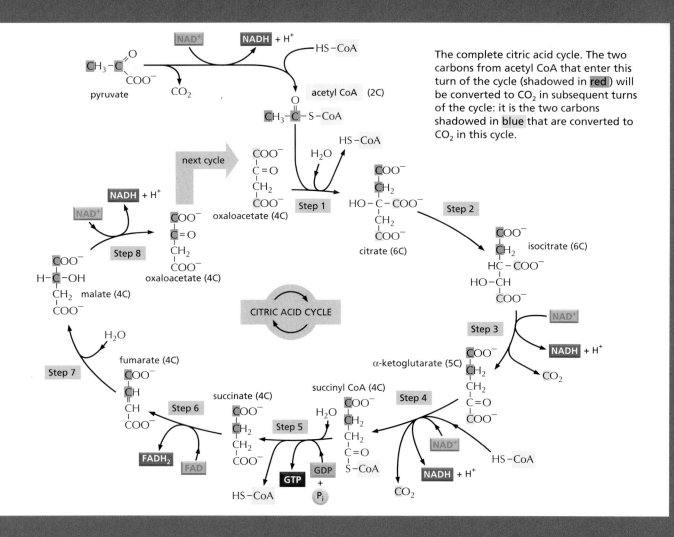

The complete citric acid cycle. The two carbons from acetyl CoA that enter this turn of the cycle (shadowed in red) will be converted to CO_2 in subsequent turns of the cycle: it is the two carbons shadowed in blue that are converted to CO_2 in this cycle.

Details of the eight steps are shown below. For each step, the part of the molecule that undergoes a change is shadowed in blue, and the name of the enzyme that catalyzes the reaction is in a yellow box.

STEP 1 After the enzyme removes a proton from the CH_3 group on acetyl CoA, the negatively charged CH_2^- forms a bond to a carbonyl carbon of oxaloacetate. The subsequent loss by hydrolysis of the coenzyme A (CoA) drives the reaction strongly forward.

acetyl CoA   oxaloacetate   S-citryl-CoA intermediate   citrate

STEP 2 An isomerization reaction, in which water is first removed and then added back, moves the hydroxyl group from one carbon atom to its neighbor.

citrate   cis-aconitate intermediate   isocitrate

STEP 3 In the first of four oxidation steps in the cycle, the carbon carrying the hydroxyl group is converted to a carbonyl group. The immediate product is unstable, losing CO_2 while still bound to the enzyme.

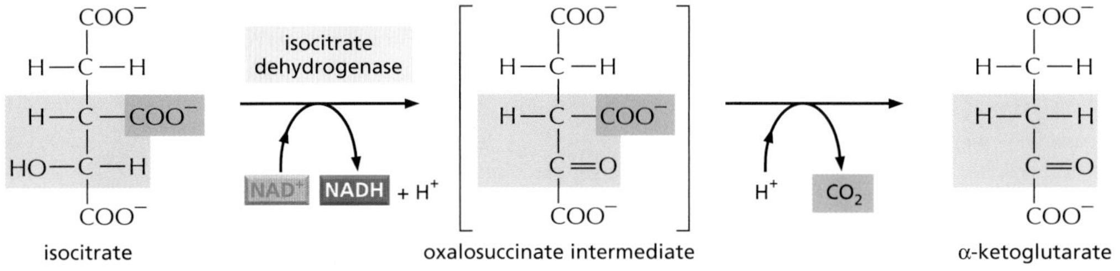

isocitrate dehydrogenase

NAD^+ $NADH$ + H^+

H^+ CO_2

isocitrate

oxalosuccinate intermediate

α-ketoglutarate

STEP 4 The α-ketoglutarate dehydrogenase complex closely resembles the large enzyme complex that converts pyruvate to acetyl CoA (pyruvate dehydrogenase). It likewise catalyzes an oxidation that produces NADH, CO_2, and a high-energy thioester bond to coenzyme A (CoA).

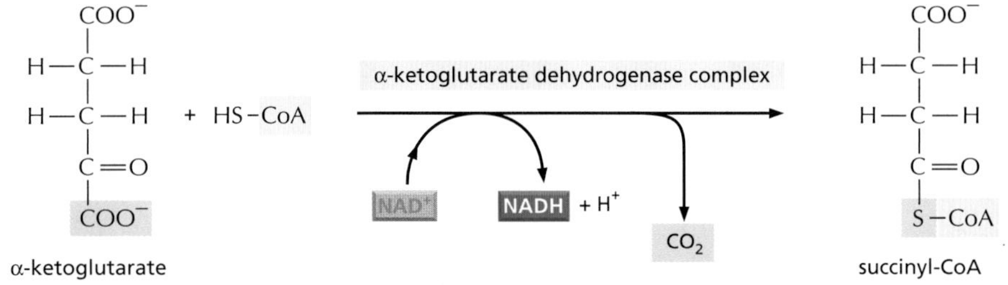

+ HS–CoA

α-ketoglutarate dehydrogenase complex

NAD^+ $NADH$ + H^+

CO_2

α-ketoglutarate

succinyl-CoA

STEP 5 A phosphate molecule from solution displaces the CoA, forming a high-energy phosphate linkage to succinate. This phosphate is then passed to GDP to form GTP. (In bacteria and plants, ATP is formed instead.)

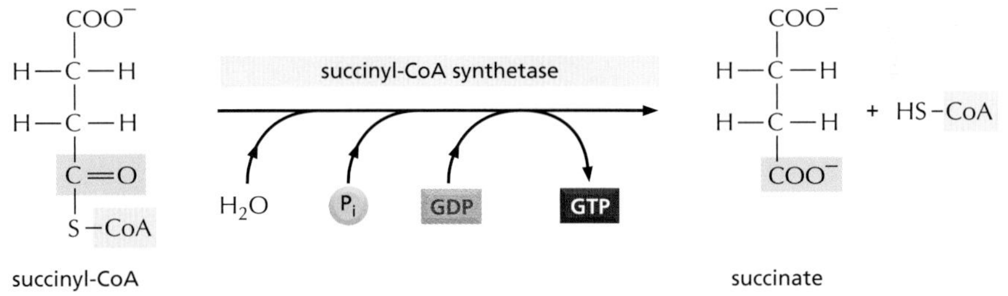

succinyl-CoA synthetase

H_2O P_i GDP GTP

succinyl-CoA

+ HS–CoA

succinate

STEP 6 In the third oxidation step in the cycle, FAD removes two hydrogen atoms from succinate.

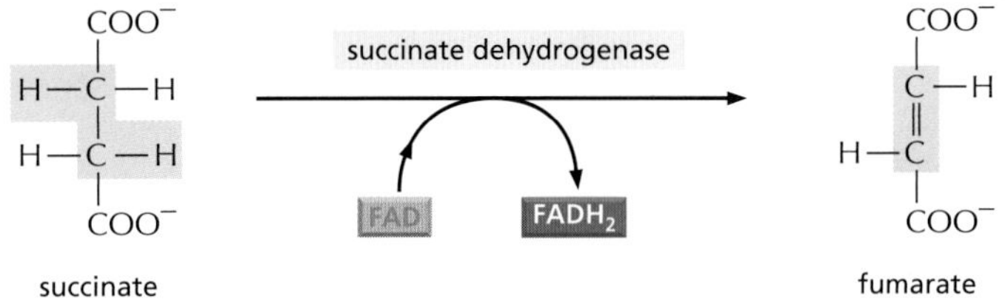

succinate dehydrogenase

FAD $FADH_2$

succinate

fumarate

STEP 7 The addition of water to fumarate places a hydroxyl group next to a carbonyl carbon.

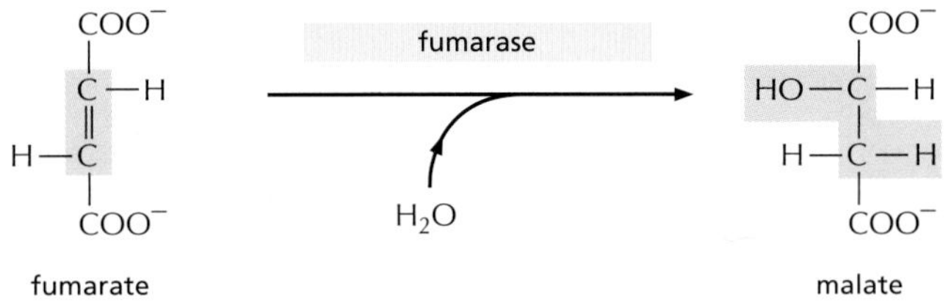

fumarase

H_2O

fumarate

malate

STEP 8 In the last of four oxidation steps in the cycle, the carbon carrying the hydroxyl group is converted to a carbonyl group, regenerating the oxaloacetate needed for step 1.

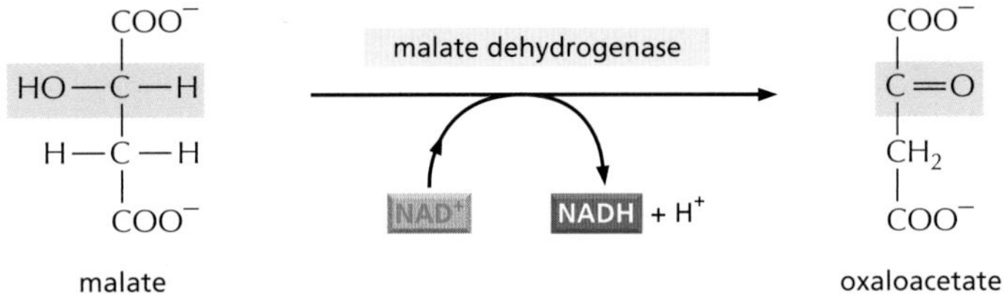

malate dehydrogenase

NAD^+ $NADH$ + H^+

malate

oxaloacetate

REFERENCES

General

Berg, JM, Tymoczko, JL & Stryer L (2006) Biochemistry, 6th ed. New York: WH Freeman.

Garrett RH & Grisham CM (2005) Biochemistry, 3rd ed. Philadelphia: Thomson Brooks/Cole.

Horton HR, Moran LA, Scrimgeour et al (2005) Principles of Biochemistry 4th ed. Upper Saddle River, NJ: Prentice Hall.

Nelson DL & Cox MM (2004) Lehninger Principles of Biochemistry, 4th ed. New York: Worth.

Nicholls DG & Ferguson SJ (2002) Bioenergetics, 3rd ed. New York: Academic Press

Mathews CK, van Holde KE & Ahern K-G (2000) Biochemistry, 3rd ed. San Francisco: Benjamin Cummings.

Moore JA (1993) Science As a Way of Knowing. Cambridge, MA: Harvard University Press.

Voet D, Voet JG & Pratt CM (2004) Fundamentals of Biochemistry, 2nd ed. New York: Wiley.

The Chemical Components of a Cell

Abeles RH, Frey PA & Jencks WP (1992) Biochemistry. Boston: Jones & Bartlett.

Atkins PW (1996) Molecules. New York: WH Freeman.

Branden C & Tooze J (1999) Introduction to Protein Structure, 2nd ed. New York: Garland Science.

Bretscher MS (1985) The molecules of the cell membrane. *Sci Am* 253:100–109.

Burley SK & Petsko GA (1988) Weakly polar interactions in proteins. *Adv Protein Chem* 39:125–189.

De Duve C (2005) Singularities: Landmarks on the Pathways of Life. Cambridge: Cambridge University Press.

Dowhan W (1997) Molecular basis for membrane phospholipid diversity: Why are there so many lipids? *Annu Rev Biochem* 66:199–232.

Eisenberg D & Kauzman W (1969) The Structure and Properties of Water. Oxford: Oxford University Press.

Fersht AR (1987) The hydrogen bond in molecular recognition. *Trends Biochem Sci* 12:301–304.

Franks F (1993) Water. Cambridge: Royal Society of Chemistry.

Henderson LJ (1927) The Fitness of the Environment, 1958 ed. Boston: Beacon.

Neidhardt FC, Ingraham JL & Schaechter M (1990) Physiology of the Bacterial Cell: A Molecular Approach. Sunderland, MA: Sinauer.

Pauling L (1960) The Nature of the Chemical Bond, 3rd ed. Ithaca, NY: Cornell University Press.

Saenger W (1984) Principles of Nucleic Acid Structure. New York: Springer.

Sharon N (1980) Carbohydrates. *Sci Am* 243:90–116.

Stillinger FH (1980) Water revisited. *Science* 209:451–457.

Tanford C (1978) The hydrophobic effect and the organization of living matter. *Science* 200:1012–1018.

Tanford C (1980) The Hydrophobic Effect. Formation of Micelles and Biological Membranes, 2nd ed. New York: John Wiley.

Catalysis and the Use of Energy by Cells

Atkins PW (1994) The Second Law: Energy, Chaos and Form. New York: Scientific American Books.

Atkins PW & De Paula JD (2006) Physical Chemistry for the Life Sciences. Oxford: Oxford University Press.

Baldwin JE & Krebs H (1981) The Evolution of Metabolic Cycles. *Nature* 291:381–382.

Berg HC (1983) Random Walks in Biology. Princeton, NJ: Princeton University Press.

Dickerson RE (1969) Molecular Thermodynamics. Menlo Park, CA: Benjamin Cummings.

Dill KA & Bromberg S (2003) Molecular Driving Forces: Statistical Thermodynamics in Chemistry and Biology. New York: Garland Science.

Dressler D & Potter H (1991) Discovering Enzymes. New York: Scientific American Library.

Einstein A (1956) Investigations on the Theory of Brownian Movement. New York: Dover.

Fruton JS (1999) Proteins, Enzymes, Genes: The Interplay of Chemistry and Biology. New Haven: Yale University Press.

Goodsell DS (1991) Inside a living cell. *Trends Biochem Sci* 16:203–206.

Karplus M & McCammon JA (1986) The dynamics of proteins. *Sci Am* 254:42–51.

Karplus M & Petsko GA (1990) Molecular dynamics simulations in biology. *Nature* 347:631–639.

Kauzmann W (1967) Thermodynamics and Statistics: with Applications to Gases. In Thermal Properties of Matter Vol 2. New York: WA Benjamin, Inc.

Kornberg A (1989) For the Love of Enzymes. Cambridge, MA: Harvard University Press.

Lavenda BH (1985) Brownian Motion. *Sci Am* 252:70–85.

Lawlor DW (2001) Photosynthesis, 3rd ed. Oxford: BIOS.

Lehninger AL (1971) The Molecular Basis of Biological Energy Transformations, 2nd ed. Menlo Park, CA: Benjamin Cummings.

Lipmann F (1941) Metabolic generation and utilization of phosphate bond energy. *Adv Enzymol* 1:99–162.

Lipmann F (1971) Wanderings of a Biochemist. New York: Wiley.

Nisbet EE & Sleep NH (2001) The habitat and nature of early life. *Nature* 409:1083–1091.

Racker E (1980) From Pasteur to Mitchell: a hundred years of bioenergetics. *Fed Proc* 39:210–215.

Schrödinger E (1944 & 1958) What is Life?: The Physical Aspect of the Living Cell and Mind and Matter, 1992 combined ed. Cambridge: Cambridge University Press.

van Holde KE, Johnson WC & Ho PS (2005) Principles of Physical Biochemistry, 2nd ed. Upper Saddle River, NJ: Prentice Hall.

Walsh C (2001) Enabling the chemistry of life. *Nature* 409:226–231.

Westheimer FH (1987) Why nature chose phosphates. *Science* 235:1173–1178.

Youvan DC & Marrs BL (1987) Molecular mechanisms of photosynthesis. *Sci Am* 256:42–49.

How Cells Obtain Energy from Food

Cramer WA & Knaff DB (1990) Energy Transduction in Biological Membranes. New York: Springer-Verlag.

Dismukes GC, Klimov VV, Baranov SV et al (2001) The origin of atmospheric oxygen on Earth: The innovation of oxygenic photosytheis. *Proc Nat Acad Sci USA* 98:2170–2175.

Fell D (1997) Understanding the Control of Metabolism. London: Portland Press.

Flatt JP (1995) Use and storage of carbohydrate and fat. *Am J Clin Nutr* 61, 952S–959S.

Friedmann HC (2004) From *Butybacterium* to *E. coli*: An essay on unity in biochemistry. *Perspect Biol Med* 47:47–66.

Fothergill-Gilmore LA (1986) The evolution of the glycolytic pathway. *Trends Biochem Sci* 11:47–51.

Heinrich R, Melendez-Hevia E, Montero F et al (1999) The structural design of glycolysis: An evolutionary approach. *Biochem Soc Trans* 27:294–298.

Huynen MA, Dandekar T & Bork P (1999) Variation and evolution of the citric-acid cycle: a genomic perspective. *Trends Microbiol* 7:281–291.

Kornberg HL (2000) Krebs and his trinity of cycles. *Nature Rev Mol Cell Biol* 1:225–228.

Krebs HA & Martin A (1981) Reminiscences and Reflections. Oxford/New York: Clarendon Press/Oxford University Press.

Krebs HA (1970) The history of the tricarboxylic acid cycle. *Perspect Biol Med* 14:154–170.

Martin BR (1987) Metabolic Regulation: A Molecular Approach. Oxford: Blackwell Scientific.

McGilvery RW (1983) Biochemistry: A Functional Approach, 3rd ed. Philadelphia: Saunders.

Morowitz HJ (1993) Beginnings of Cellular Life: Metabolism Recapitulates Biogenesis. New Haven: Yale University Press

Newsholme EA & Stark C (1973) Regulation of Metabolism. New York: Wiley.

3

Proteins

When we look at a cell through a microscope or analyze its electrical or bio-chemical activity, we are, in essence, observing proteins. Proteins constitute most of a cell's dry mass. They are not only the cell's building blocks; they also execute nearly all the cell's functions. Thus, enzymes provide the intricate molecular surfaces in a cell that promote its many chemical reactions. Proteins embedded in the plasma membrane form channels and pumps that control the passage of small molecules into and out of the cell. Other proteins carry mes-sages from one cell to another, or act as signal integrators that relay sets of sig-nals inward from the plasma membrane to the cell nucleus. Yet others serve as tiny molecular machines with moving parts: *kinesin*, for example, propels organelles through the cytoplasm; *topoisomerase* can untangle knotted DNA molecules. Other specialized proteins act as antibodies, toxins, hormones, antifreeze molecules, elastic fibers, ropes, or sources of luminescence. Before we can hope to understand how genes work, how muscles contract, how nerves conduct electricity, how embryos develop, or how our bodies function, we must attain a deep understanding of proteins.

THE SHAPE AND STRUCTURE OF PROTEINS

From a chemical point of view, proteins are by far the most structurally complex and functionally sophisticated molecules known. This is perhaps not surprising, once we realize that the structure and chemistry of each protein has been devel-oped and fine-tuned over billions of years of evolutionary history. Yet, even to experts, the remarkable versatility of proteins can seem truly amazing.

In this section, we consider how the location of each amino acid in the long string of amino acids that forms a protein determines its three-dimensional shape. Later in the chapter, we use this understanding of protein structure at the atomic level to describe how the precise shape of each protein molecule deter-mines its function in a cell.

The Shape of a Protein Is Specified by Its Amino Acid Sequence

There are 20 types of amino acids in proteins, each with different chemical prop-erties. A **protein** molecule is made from a long chain of these amino acids, each linked to its neighbor through a covalent peptide bond. Proteins are therefore also known as *polypeptides*. Each type of protein has a unique sequence of amino acids, and there are many thousands of different proteins, each with its own particular amino acid sequence.

The repeating sequence of atoms along the core of the polypeptide chain is referred to as the **polypeptide backbone**. Attached to this repetitive chain are those portions of the amino acids that are not involved in making a peptide bond and that give each amino acid its unique properties: the 20 different amino acid **side chains** (**Figure 3–1**). Some of these side chains are nonpolar and hydrophobic ("water-fearing"), others are negatively or positively charged, some readily form covalent bonds, and so on. **Panel 3–1** (pp. 128–129) shows their atomic structures and **Figure 3–2** lists their abbreviations.

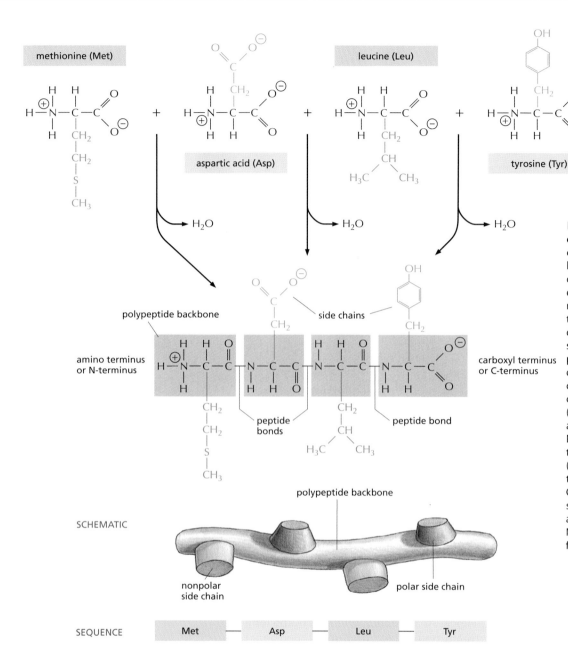

Figure 3–1 The components of a protein. A protein consists of a polypeptide backbone with attached side chains. Each type of protein differs in its sequence and number of amino acids; therefore, it is the sequence of the chemically different side chains that makes each protein distinct. The two ends of a polypeptide chain are chemically different: the end carrying the free amino group (NH_3^+, also written NH_2) is the amino terminus, or N-terminus, and that carrying the free carboxyl group (COO^-, also written COOH) is the carboxyl terminus or C-terminus. The amino acid sequence of a protein is always presented in the N-to-C direction, reading from left to right.

As discussed in Chapter 2, atoms behave almost as if they were hard spheres with a definite radius (their *van der Waals radius*). The requirement that no two atoms overlap limits greatly the possible bond angles in a polypeptide chain (**Figure 3–3**). This constraint and other steric interactions severely restrict the possible three-dimensional arrangements of atoms (or *conformations*). Nevertheless, a long flexible chain, such as a protein, can still fold in an enormous number of ways.

The folding of a protein chain is, however, further constrained by many different sets of weak *noncovalent bonds* that form between one part of the chain and another. These involve atoms in the polypeptide backbone, as well as atoms in the amino acid side chains. There are three types of weak bonds: *hydrogen bonds, electrostatic attractions*, and *van der Waals attractions*, as explained in Chapter 2 (see p. 54). Individual noncovalent bonds are 30–300 times weaker than the typical covalent bonds that create biological molecules. But many weak bonds acting in parallel can hold two regions of a polypeptide chain tightly together. In this way, the combined strength of large numbers of such noncovalent bonds determines the stability of each folded shape (**Figure 3–4**).

AMINO ACID			SIDE CHAIN
Aspartic acid	Asp	D	negative
Glutamic acid	Glu	E	negative
Arginine	Arg	R	positive
Lysine	Lys	K	positive
Histidine	His	H	positive
Asparagine	Asn	N	uncharged polar
Glutamine	Gln	Q	uncharged polar
Serine	Ser	S	uncharged polar
Threonine	Thr	T	uncharged polar
Tyrosine	Tyr	Y	uncharged polar

——————— POLAR AMINO ACIDS ———————

AMINO ACID			SIDE CHAIN
Alanine	Ala	A	nonpolar
Glycine	Gly	G	nonpolar
Valine	Val	V	nonpolar
Leucine	Leu	L	nonpolar
Isoleucine	Ile	I	nonpolar
Proline	Pro	P	nonpolar
Phenylalanine	Phe	F	nonpolar
Methionine	Met	M	nonpolar
Tryptophan	Trp	W	nonpolar
Cysteine	Cys	C	nonpolar

——————— NONPOLAR AMINO ACIDS ———————

Figure 3–2 The 20 amino acids found in proteins. Each amino acid has a three-letter and a one-letter abbreviation. There are equal numbers of polar and nonpolar side chains; however, some side chains listed here as polar are large enough to have some non-polar properties (for example, Tyr, Thr, Arg, Lys). For atomic structures, see Panel 3–1 (pp. 128–129).

A fourth weak force also has a central role in determining the shape of a protein. As described in Chapter 2, hydrophobic molecules, including the nonpolar side chains of particular amino acids, tend to be forced together in an aqueous environment in order to minimize their disruptive effect on the hydrogen-bonded network of water molecules (see p. 54 and Panel 2–2, pp. 108–109). Therefore, an important factor governing the folding of any protein is the distribution of its polar and nonpolar amino acids. The nonpolar (hydrophobic) side chains in a protein—belonging to such amino acids as phenylalanine, leucine, valine, and tryptophan—tend to cluster in the interior of the molecule (just as hydrophobic oil droplets coalesce in water to form one large droplet). This enables them to avoid contact with the water that surrounds them inside a cell. In contrast, polar groups—such as those belonging to arginine, glutamine, and histidine—tend to arrange themselves near the outside of the molecule, where they can form hydrogen bonds with water and with other polar molecules (**Figure 3–5**). Polar amino acids buried within the protein are usually hydrogen-bonded to other polar amino acids or to the polypeptide backbone.

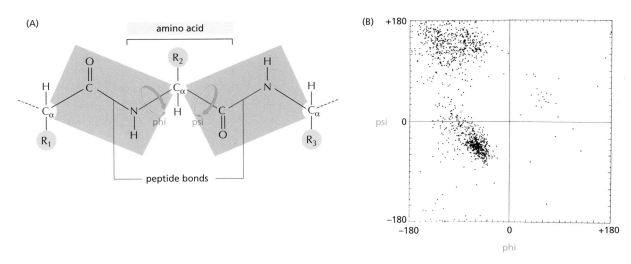

Figure 3–3 Steric limitations on the bond angles in a polypeptide chain. (A) Each amino acid contributes three bonds *(red)* to the backbone of the chain. The peptide bond is planar *(gray shading)* and does not permit rotation. By contrast, rotation can occur about the C_α–C bond, whose angle of rotation is called psi (ψ), and about the N–C_α bond, whose angle of rotation is called phi (ϕ). By convention, an R group is often used to denote an amino acid side chain *(green circles)*. (B) The conformation of the main-chain atoms in a protein is determined by one pair of ϕ and ψ angles for each amino acid; because of steric collisions between atoms within each amino acid, most pairs of ϕ and ψ angles do not occur. In this so-called Ramachandran plot, each dot represents an observed pair of angles in a protein. The cluster of dots in the bottom left quadrant represents all of the amino acids that are located in α-helix structures (see Figure 3–7A). (B, from J. Richardson, *Adv. Prot. Chem.* 34:174–175, 1981. With permission from Academic Press.)

THE AMINO ACID

The general formula of an amino acid is

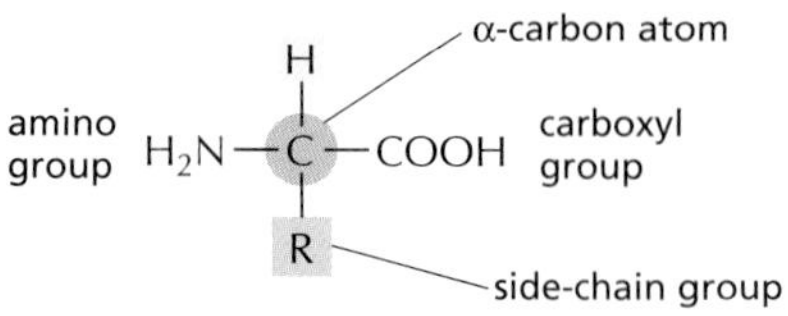

amino group H_2N — C — COOH   carboxyl group

α-carbon atom

side-chain group

R is commonly one of 20 different side chains. At pH 7 both the amino and carboxyl groups are ionized.

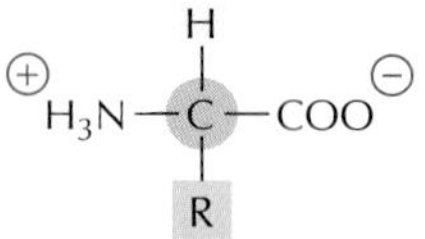

$\overset{\oplus}{H_3N}$ — C — COO$^{\ominus}$

R

OPTICAL ISOMERS

The α-carbon atom is asymmetric, which allows for two mirror image (or stereo-) isomers, L and D.

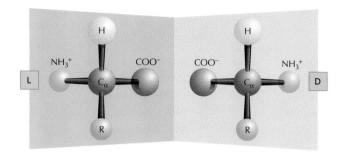

Proteins consist exclusively of L-amino acids.

FAMILIES OF AMINO ACIDS

The common amino acids are grouped according to whether their side chains are

  acidic
  basic
  uncharged polar
  nonpolar

These 20 amino acids are given both three-letter and one-letter abbreviations.

Thus: alanine = Ala = A

BASIC SIDE CHAINS

lysine
(Lys, or K)

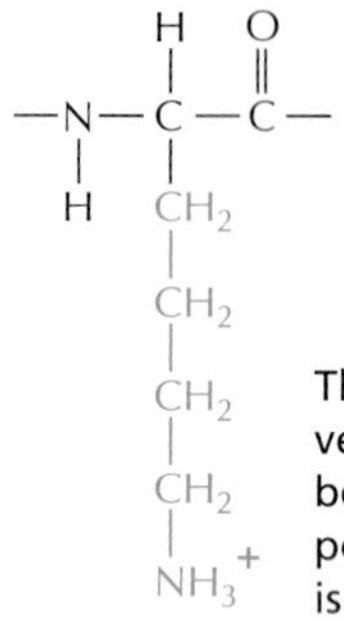

— N — C — C —
| |
H   CH$_2$
        |
       CH$_2$
        |
       CH$_2$
        |
       CH$_2$
        |
       NH$_3$$^+$

This group is very basic because its positive charge is stabilized by resonance.

arginine
(Arg, or R)

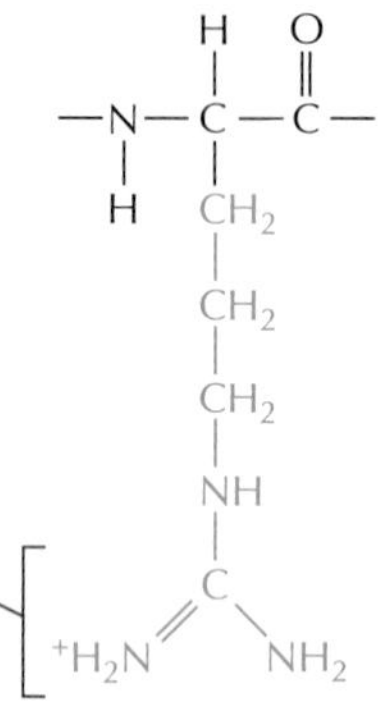

— N — C — C —
| |
H   CH$_2$
        |
       CH$_2$
        |
       CH$_2$
        |
       NH
        |
        C
      /   \
$^+$H$_2$N   NH$_2$

histidine
(His, or H)

— N — C — C —
| |
H   CH$_2$
        |
        C
      /   \\
   HN   CH
     | |
   HC = NH$^+$

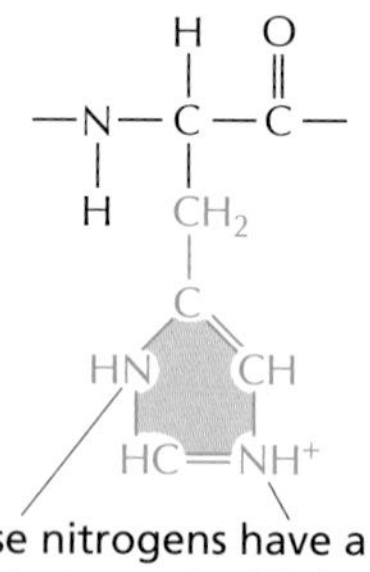

These nitrogens have a relatively weak affinity for an H$^+$ and are only partly positive at neutral pH.

PEPTIDE BONDS

Amino acids are commonly joined together by an amide linkage, called a peptide bond.

Peptide bond: The four atoms in each *gray box* form a rigid planar unit. There is no rotation around the C–N bond.

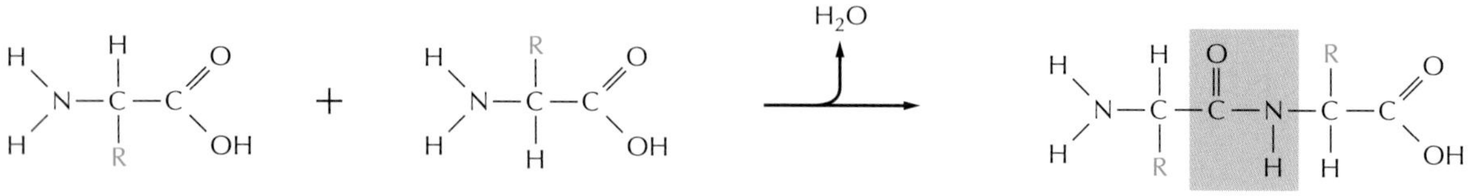

H_2O

Proteins are long polymers of amino acids linked by peptide bonds, and they are always written with the N-terminus toward the left. The sequence of this tripeptide is histidine-cysteine-valine.

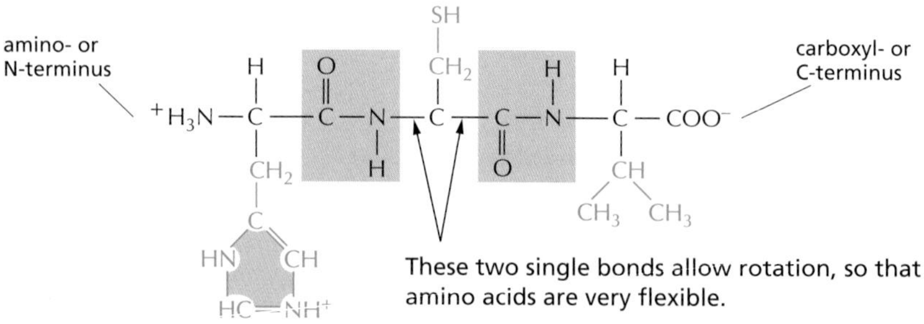

amino- or N-terminus

carboxyl- or C-terminus

These two single bonds allow rotation, so that long chains of amino acids are very flexible.

ACIDIC SIDE CHAINS

aspartic acid
(Asp, or D)

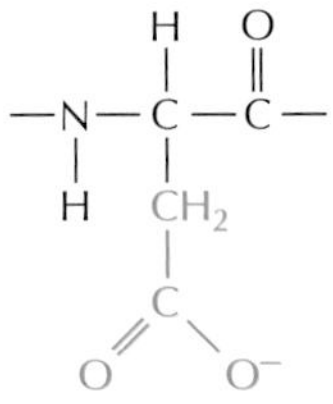

glutamic acid
(Glu, or E)

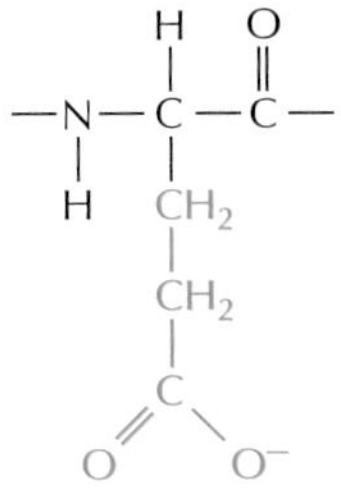

NONPOLAR SIDE CHAINS

alanine
(Ala, or A)

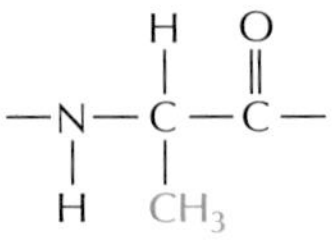

valine
(Val, or V)

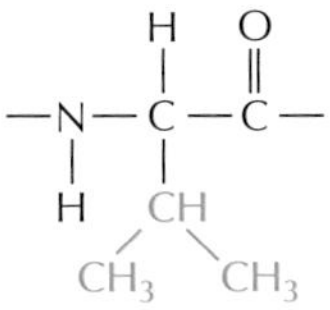

leucine
(Leu, or L)

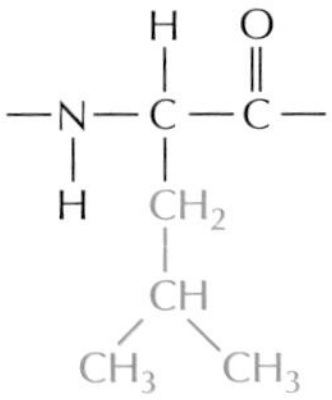

isoleucine
(Ile, or I)

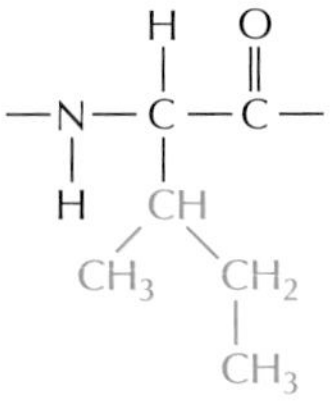

proline
(Pro, or P)

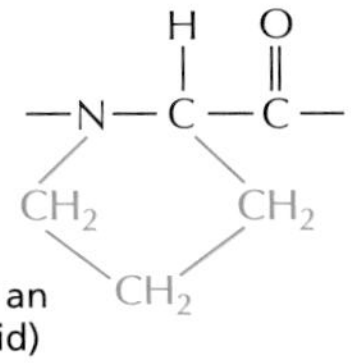

(actually an imino acid)

phenylalanine
(Phe, or F)

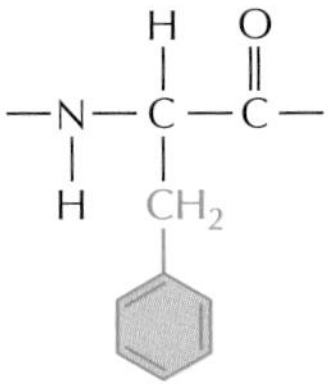

methionine
(Met, or M)

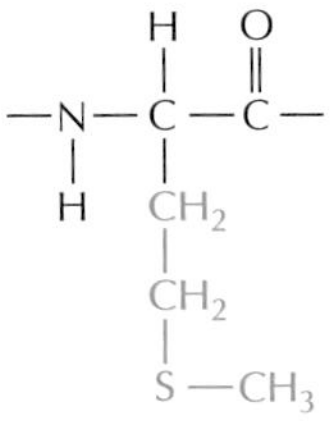

tryptophan
(Trp, or W)

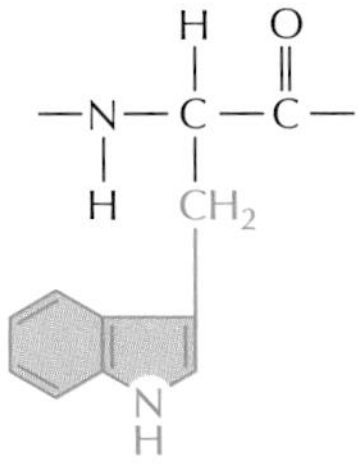

glycine
(Gly, or G)

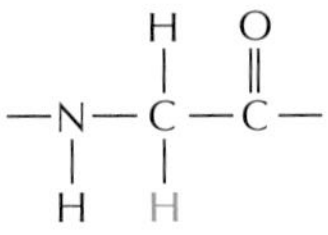

cysteine
(Cys, or C)

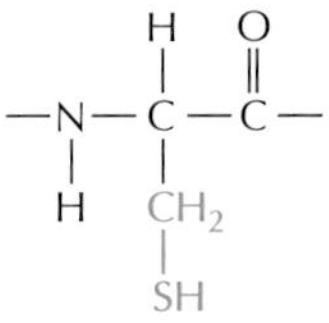

Disulfide bonds can form between two cysteine side chains in proteins.

$$- -CH_2 - S - S - CH_2 - -$$

UNCHARGED POLAR SIDE CHAINS

asparagine
(Asn, or N)

glutamine
(Gln, or Q)

Although the amide N is not charged at neutral pH, it is polar.

serine
(Ser, or S)

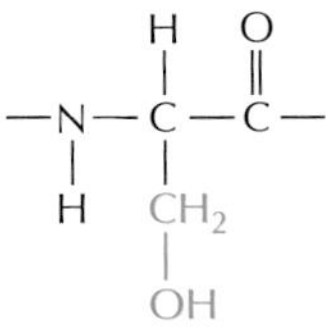

threonine
(Thr, or T)

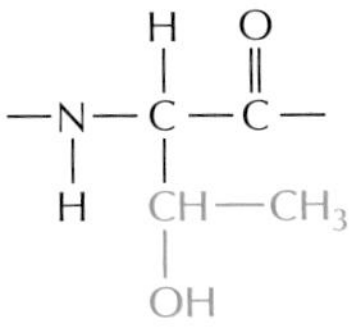

tyrosine
(Tyr, or Y)

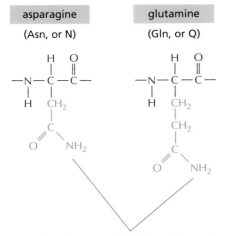

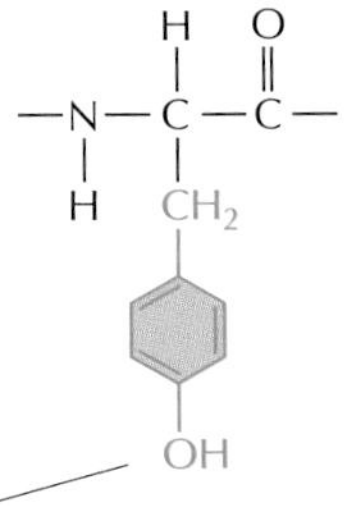

The –OH group is polar.

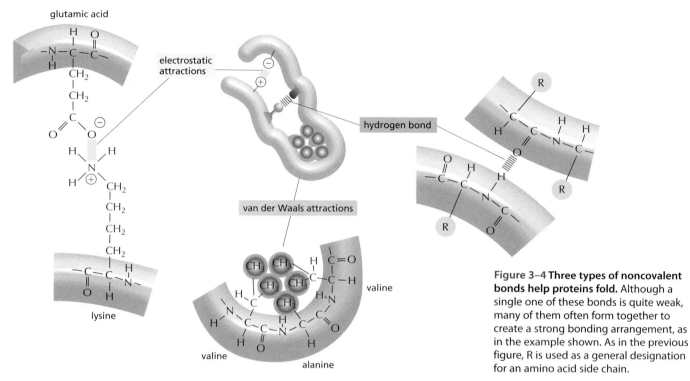

Figure 3–4 **Three types of noncovalent bonds help proteins fold.** Although a single one of these bonds is quite weak, many of them often form together to create a strong bonding arrangement, as in the example shown. As in the previous figure, R is used as a general designation for an amino acid side chain.

Proteins Fold into a Conformation of Lowest Energy

As a result of all of these interactions, most proteins have a particular three-dimensional structure, which is determined by the order of the amino acids in its chain. The final folded structure, or **conformation**, of any polypeptide chain is generally the one that minimizes its free energy. Biologists have studied protein folding in a test tube by using highly purified proteins. Treatment with certain solvents, which disrupt the noncovalent interactions holding the folded chain together, unfolds, or *denatures*, a protein. This treatment converts the protein into a flexible polypeptide chain that has lost its natural shape. When the denaturing solvent is removed, the protein often refolds spontaneously, or *renatures*, into its original conformation (**Figure 3–6**). This indicates that the amino acid sequence contains all the information needed for specifying the three-dimensional shape of a protein, which is a critical point for understanding cell function.

Each protein normally folds up into a single stable conformation. However, the conformation changes slightly when the protein interacts with other molecules in the cell. This change in shape is often crucial to the function of the protein, as we see later.

Although a protein chain can fold into its correct conformation without outside help, in a living cell special proteins called *molecular chaperones* often assist in protein folding. Molecular chaperones bind to partly folded polypeptide chains and help them progress along the most energetically favorable folding

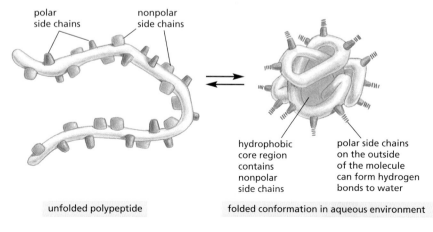

Figure 3–5 **How a protein folds into a compact conformation.** The polar amino acid side chains tend to gather on the outside of the protein, where they can interact with water; the nonpolar amino acid side chains are buried on the inside to form a tightly packed hydrophobic core of atoms that are hidden from water. In this schematic drawing, the protein contains only about 30 amino acids.

(A)

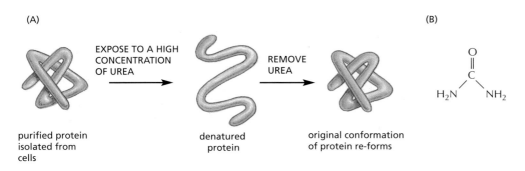

purified protein isolated from cells

EXPOSE TO A HIGH CONCENTRATION OF UREA

denatured protein

REMOVE UREA

original conformation of protein re-forms

(B)

Figure 3–6 **The refolding of a denatured protein.** (A) This type of experiment, first performed more than 40 years ago, demonstrates that a protein's conformation is determined solely by its amino acid sequence. (B) The structure of urea. Urea is very soluble in water and unfolds proteins at high concentrations, where there is about one urea molecule for every six water molecules.

pathway. In the crowded conditions of the cytoplasm, chaperones prevent the temporarily exposed hydrophobic regions in newly synthesized protein chains from associating with each other to form protein aggregates (see p. 388). However, the final three-dimensional shape of the protein is still specified by its amino acid sequence: chaperones simply make the folding process more reliable.

Proteins come in a wide variety of shapes, and they are generally between 50 and 2000 amino acids long. Large proteins usually consist of several distinct *protein domains*—structural units that fold more or less independently of each other, as we discuss below. Since the detailed structure of any protein is complicated, several different representations are used to depict the protein's structure, each emphasizing different features.

Panel 3–2 (pp. 132–133) presents four different representations of a protein domain called SH2, which has important functions in eucaryotic cells. Constructed from a string of 100 amino acids, the structure is displayed as (A) a polypeptide backbone model, (B) a ribbon model, (C) a wire model that includes the amino acid side chains, and (D) a space-filling model. Each of the three horizontal rows shows the protein in a different orientation, and the image is colored in a way that allows the polypeptide chain to be followed from its N-terminus *(purple)* to its C-terminus *(red)*. <GTGA>

Panel 3–2 shows that a protein's conformation is amazingly complex, even for a structure as small as the SH2 domain. But the description of protein structures can be simplified because they are built up from combinations of several common structural motifs, as we discuss next.

The α Helix and the β Sheet Are Common Folding Patterns

When we compare the three-dimensional structures of many different protein molecules, it becomes clear that, although the overall conformation of each protein is unique, two regular folding patterns are often found in parts of them. Both patterns were discovered more than 50 years ago from studies of hair and silk. The first folding pattern to be discovered, called the **α helix**, was found in the protein *α-keratin*, which is abundant in skin and its derivatives—such as hair, nails, and horns. Within a year of the discovery of the α helix, a second folded structure, called a **β sheet**, was found in the protein *fibroin*, the major constituent of silk. These two patterns are particularly common because they result from hydrogen-bonding between the N–H and C=O groups in the polypeptide backbone, without involving the side chains of the amino acids. Thus, many different amino acid sequences can form them. In each case, the protein chain adopts a regular, repeating conformation. **Figure 3–7** shows these two conformations, as well as the abbreviations that are used to denote them in ribbon models of proteins.

The core of many proteins contains extensive regions of β sheet. As shown in **Figure 3–8**, these β sheets can form either from neighboring polypeptide chains that run in the same orientation (parallel chains) or from a polypeptide chain that folds back and forth upon itself, with each section of the chain running in the direction opposite to that of its immediate neighbors (antiparallel chains). Both types of β sheet produce a very rigid structure, held together by hydrogen bonds that connect the peptide bonds in neighboring chains (see Figure 3–7D).

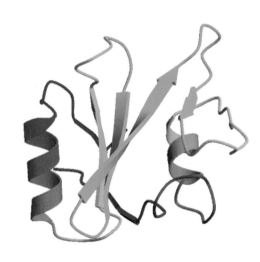

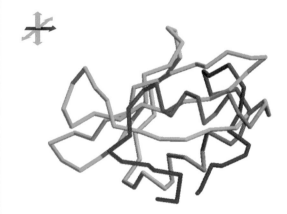

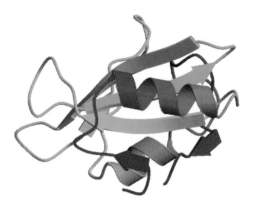

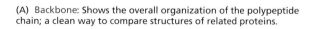

(A) Backbone: Shows the overall organization of the polypeptide chain; a clean way to compare structures of related proteins.

(B) Ribbon: Easy way to visualize secondary structures, such as α helices and β sheets.

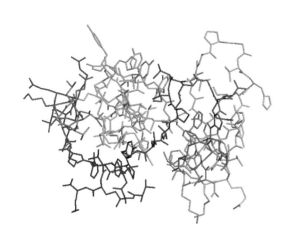

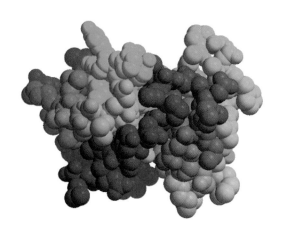

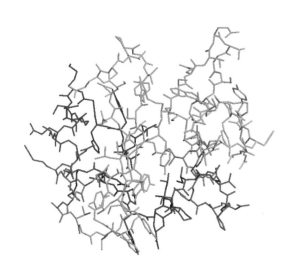

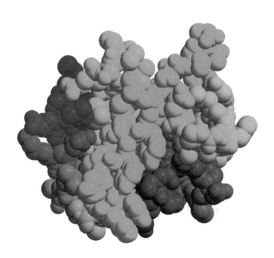

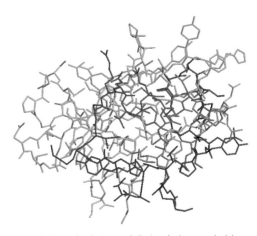

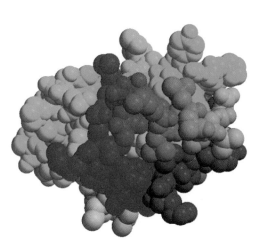

(Courtesy of David Lawson.)

(C) Wire: Highlights side chains and their relative proximities; useful for predicting which amino acids might be involved in a protein's activity, particularly if the protein is an enzyme.

(D) Space-filling: Provides contour map of the protein; gives a feel for the shape of the protein and shows which amino acid side chains are exposed on its surface. Shows how the protein might look to a small molecule, such as water, or to another protein.

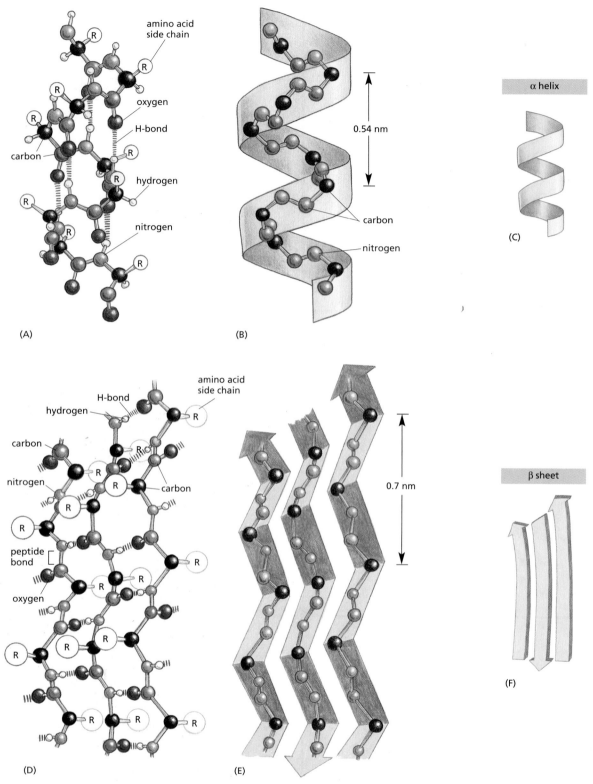

Figure 3–7 The regular conformation of the polypeptide backbone in the α helix and the β sheet. <GTAG> <TGCT>
(A, B, and C) The α helix. The N–H of every peptide bond is hydrogen-bonded to the C=O of a neighboring peptide bond located four peptide bonds away in the same chain. Note that all of the N–H groups point up in this diagram and that all of the C=O groups point down (toward the C-terminus); this gives a polarity to the helix, with the C-terminus having a partial negative and the N-terminus a partial positive charge. (D, E, and F) The β sheet. In this example, adjacent peptide chains run in opposite (antiparallel) directions. Hydrogen-bonding between peptide bonds in different strands holds the individual polypeptide chains (strands) together in a β sheet, and the amino acid side chains in each strand alternately project above and below the plane of the sheet. (A) and (D) show all the atoms in the polypeptide backbone, but the amino acid side chains are truncated and denoted by R. In contrast, (B) and (E) show the backbone atoms only, while (C) and (F) display the shorthand symbols that are used to represent the α helix and the β sheet in ribbon drawings of proteins (see Panel 3–2B).

An α helix is generated when a single polypeptide chain twists around on itself to form a rigid cylinder. A hydrogen bond forms between every fourth peptide bond, linking the C=O of one peptide bond to the N–H of another (see Figure 3–7A). This gives rise to a regular helix with a complete turn every 3.6 amino acids. Note that the protein domain illustrated in Panel 3–2 contains two α helices, as well as a three-stranded antiparallel β sheet.

Regions of α helix are especially abundant in proteins located in cell membranes, such as transport proteins and receptors. As we discuss in Chapter 10, those portions of a transmembrane protein that cross the lipid bilayer usually cross as an α helix composed largely of amino acids with nonpolar side chains. The polypeptide backbone, which is hydrophilic, is hydrogen-bonded to itself in the α helix and shielded from the hydrophobic lipid environment of the membrane by its protruding nonpolar side chains (see also Figure 3–78).

In other proteins, α helices wrap around each other to form a particularly stable structure, known as a **coiled-coil**. This structure can form when the two (or in some cases three) α helices have most of their nonpolar (hydrophobic) side chains on one side, so that they can twist around each other with these side chains facing inward (**Figure 3–9**). Long rodlike coiled-coils provide the structural framework for many elongated proteins. Examples are α-keratin, which forms the intracellular fibers that reinforce the outer layer of the skin and its appendages, and the myosin molecules responsible for muscle contraction.

Protein Domains Are Modular Units from which Larger Proteins Are Built

Even a small protein molecule is built from thousands of atoms linked together by precisely oriented covalent and noncovalent bonds, and it is extremely difficult to visualize such a complicated structure without a three-dimensional display. For

Figure 3–8 Two types of β sheet structures. (A) An antiparallel β sheet (see Figure 3–7D). (B) A parallel β sheet. Both of these structures are common in proteins.

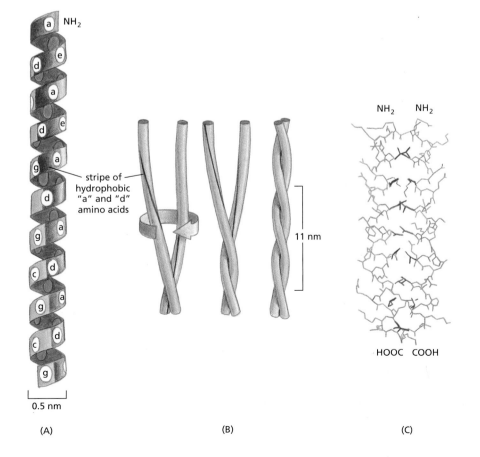

Figure 3–9 A coiled-coil. <CGGA>
(A) A single α helix, with successive amino acid side chains labeled in a sevenfold sequence, "abcdefg" (*from bottom to top*). Amino acids "a" and "d" in such a sequence lie close together on the cylinder surface, forming a "stripe" (*red*) that winds slowly around the α helix. Proteins that form coiled-coils typically have nonpolar amino acids at positions "a" and "d." Consequently, as shown in (B), the two α helices can wrap around each other with the nonpolar side chains of one α helix interacting with the nonpolar side chains of the other, while the more hydrophilic amino acid side chains are left exposed to the aqueous environment. (C) The atomic structure of a coiled-coil determined by x-ray crystallography. The *red* side chains are nonpolar.

this reason, biologists use various graphic and computer-based aids. A DVD that accompanies this book contains computer-generated images of selected proteins, displayed and rotated on the screen in a variety of formats.

Biologists distinguish four levels of organization in the structure of a protein. The amino acid sequence is known as the **primary structure**. Stretches of polypeptide chain that form α helices and β sheets constitute the protein's **secondary structure**. The full three-dimensional organization of a polypeptide chain is sometimes referred to as the **tertiary structure**, and if a particular protein molecule is formed as a complex of more than one polypeptide chain, the complete structure is designated as the **quaternary structure**.

Studies of the conformation, function, and evolution of proteins have also revealed the central importance of a unit of organization distinct from these four. This is the **protein domain**, a substructure produced by any part of a polypeptide chain that can fold independently into a compact, stable structure. A domain usually contains between 40 and 350 amino acids, and it is the modular unit from which many larger proteins are constructed.

The different domains of a protein are often associated with different functions. **Figure 3–10** shows an example—the Src protein kinase, which functions in signaling pathways inside vertebrate cells (Src is pronounced "sarc"). This protein is considered to have three domains: the SH2 and SH3 domains have regulatory roles, while the C-terminal domain is responsible for the kinase catalytic activity. Later in the chapter, we shall return to this protein, in order to explain how proteins can form molecular switches that transmit information throughout cells.

Figure 3–11 presents ribbon models of three differently organized protein domains. As these examples illustrate, the polypeptide chain tends to cross the entire domain before making a sharp turn at the surface. The central core of a domain can be constructed from α helices, from β sheets, or from various combinations of these two fundamental folding elements. <CAGT>

The smallest protein molecules contain only a single domain, whereas larger proteins can contain as many as several dozen domains, often connected to each other by short, relatively unstructured lengths of polypeptide chain.

Few of the Many Possible Polypeptide Chains Will Be Useful to Cells

Since each of the 20 amino acids is chemically distinct and each can, in principle, occur at any position in a protein chain, there are $20 \times 20 \times 20 \times 20 = 160,000$ different possible polypeptide chains four amino acids long, or 20^n different possible polypeptide chains n amino acids long. For a typical protein length of

Figure 3–10 A protein formed from multiple domains. In the Src protein shown, a C-terminal domain with two lobes (*yellow* and *orange*) forms a protein kinase enzyme, while the SH2 and SH3 domains perform regulatory functions. (A) A ribbon model, with ATP substrate in *red*. (B) A spacing-filling model, with ATP substrate in *red*. Note that the site that binds ATP is positioned at the interface of the two lobes that form the kinase. The detailed structure of the SH2 domain is illustrated in Panel 3–2 (pp. 132–133).

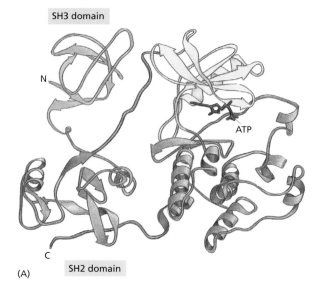

SH3 domain

N

ATP

C

SH2 domain

(A)

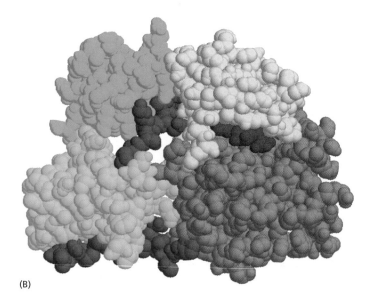

(B)

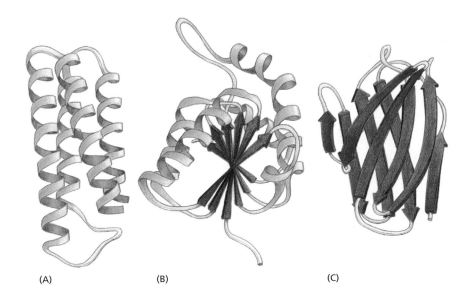

(A)   (B)   (C)

Figure 3–11 **Ribbon models of three different protein domains.** (A) Cytochrome b_{562}, a single-domain protein involved in electron transport in mitochondria. This protein is composed almost entirely of α helices. (B) The NAD-binding domain of the enzyme lactic dehydrogenase, which is composed of a mixture of α helices and parallel β sheets. (C) The variable domain of an immunoglobulin (antibody) light chain, composed of a sandwich of two antiparallel β sheets. In these examples, the α helices are shown in *green*, while strands organized as β sheets are denoted by *red arrows*.
Note how the polypeptide chain generally traverses back and forth across the entire domain, making sharp turns only at the protein surface. It is the protruding loop regions *(yellow)* that often form the binding sites for other molecules. (Adapted from drawings courtesy of Jane Richardson.)

about 300 amino acids, a cell could theoretically make more than 10^{390} (20^{300}) different polypeptide chains. This is such an enormous number that to produce just one molecule of each kind would require many more atoms than exist in the universe.

Only a very small fraction of this vast set of conceivable polypeptide chains would adopt a single, stable three-dimensional conformation—by some estimates, less than one in a billion. And yet the vast majority of proteins present in cells adopt unique and stable conformations. How is this possible? The answer lies in natural selection. A protein with an unpredictably variable structure and biochemical activity is unlikely to help the survival of a cell that contains it. Such proteins would therefore have been eliminated by natural selection through the enormously long trial-and-error process that underlies biological evolution.

Because evolution has selected for protein function in living organisms, the amino acid sequence of most present-day proteins is such that a single conformation is extremely stable. In addition, this conformation has its chemical properties finely tuned to enable the protein to perform a particular catalytic or structural function in the cell. Proteins are so precisely built that the change of even a few atoms in one amino acid can sometimes disrupt the structure of the whole molecule so severely that all function is lost.

Proteins Can Be Classified into Many Families

Once a protein had evolved that folded up into a stable conformation with useful properties, its structure could be modified during evolution to enable it to perform new functions. This process has been greatly accelerated by genetic mechanisms that occasionally duplicate genes, allowing one gene copy to evolve independently to perform a new function (discussed in Chapter 4). This type of event has occurred quite often in the past; as a result, many present-day proteins can be grouped into protein families, each family member having an amino acid sequence and a three-dimensional conformation that resemble those of the other family members.

Consider, for example, the *serine proteases*, a large family of protein-cleaving (proteolytic) enzymes that includes the digestive enzymes chymotrypsin, trypsin, and elastase, and several proteases involved in blood clotting. When the protease portions of any two of these enzymes are compared, parts of their amino acid sequences are found to match. The similarity of their three-dimensional conformations is even more striking: most of the detailed twists and turns in their polypeptide chains, which are several hundred amino acids long, are virtually identical (**Figure 3–12**). The many different serine proteases nevertheless

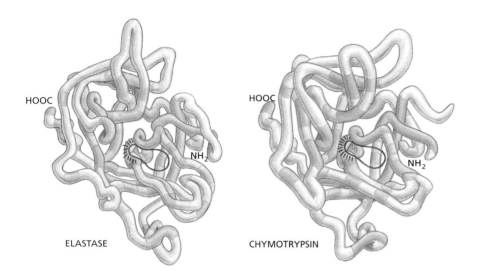

Figure 3–12 A comparison of the conformations of two serine proteases. The backbone conformations of elastase and chymotrypsin. Although only those amino acids in the polypeptide chain shaded in *green* are the same in the two proteins, the two conformations are very similar nearly everywhere. The active site of each enzyme is circled in *red;* this is where the peptide bonds of the proteins that serve as substrates are bound and cleaved by hydrolysis. The serine proteases derive their name from the amino acid serine, whose side chain is part of the active site of each enzyme and directly participates in the cleavage reaction.

have distinct enzymatic activities, each cleaving different proteins or the peptide bonds between different types of amino acids. Each therefore performs a distinct function in an organism.

The story we have told for the serine proteases could be repeated for hundreds of other protein families. In general, the structure of the different members of a protein family has been more highly conserved than has the amino acid sequence. In many cases, the amino acid sequences have diverged so far that we cannot be certain of a family relationship between two proteins without determining their three-dimensional structures. The yeast α2 protein and the *Drosophila* engrailed protein, for example, are both gene regulatory proteins in the homeodomain family (discussed in Chapter 7). Because they are identical in only 17 of their 60 amino acid residues, their relationship became certain only by comparing their three-dimensional structures (**Figure 3–13**). Many similar examples show that two proteins with more than 25% identity in their amino acid sequences usually share the same overall structure.

The various members of a large protein family often have distinct functions. Some of the amino acid changes that make family members different were no doubt selected in the course of evolution because they resulted in useful changes in biological activity, giving the individual family members the different functional properties they have today. But many other amino acid changes are effectively "neutral," having neither a beneficial nor a damaging effect on the basic structure and function of the protein. In addition, since mutation is a random process, there must also have been many deleterious changes that altered the three-dimensional structure of these proteins sufficiently to harm them.

Figure 3–13 A comparison of a class of DNA-binding domains, called homeodomains, in a pair of proteins from two organisms separated by more than a billion years of evolution. (A) A ribbon model of the structure common to both proteins. (B) A trace of the α-carbon positions. The three-dimensional structures shown were determined by x-ray crystallography for the yeast α2 protein *(green)* and the *Drosophila* engrailed protein *(red).* (C) A comparison of amino acid sequences for the region of the proteins shown in (A) and (B). *Black dots* mark sites with identical amino acids. *Orange dots* indicate the position of a three amino acid insert in the α2 protein. (Adapted from C. Wolberger et al., *Cell* 67:517–528, 1991. With permission from Elsevier.)

(A)

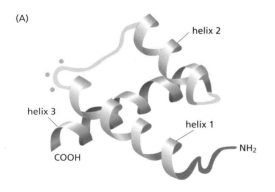

(B)

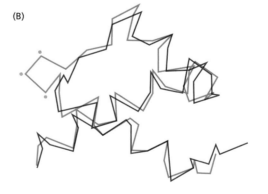

(C)

yeast

H_2N G H R F T K E N V R I L E S W F A K N I E N P Y L D T K G L E N L M K N T S L S R I Q I K N W V S N R R R K E K T I COOH
R T A F S S E O L A R L K R E F N E N - - - R Y L T E R R R Q Q L S S E L G L N E A Q I K I W F Q N K R A K I K K S

Drosophila

Such faulty proteins would have been lost whenever the individual organisms making them were at enough of a disadvantage to be eliminated by natural selection.

Protein families are readily recognized when the genome of any organism is sequenced; for example, the determination of the DNA sequence for the entire human genome has revealed that we contain about 24,000 protein-coding genes. Through sequence comparisons, we can assign the products of about 40 percent of these genes to known protein structures, belonging to more than 500 different protein families. Most of the proteins in each family have evolved to perform somewhat different functions, as for the enzymes elastase and chymotrypsin illustrated previously in Figure 3–12. These are sometimes called *paralogs* to distinguish them from the corresponding proteins in different organisms (*orthologs*, such as mouse and human elastase).

As described in Chapter 8, because of the powerful techniques of x-ray crystallography and nuclear magnetic resonance (NMR), we now know the three-dimensional shapes, or conformations, of more than 20,000 proteins. <GGCC> By carefully comparing the conformations of these proteins, structural biologists (that is, experts on the structure of biological molecules) have concluded that there are a limited number of ways in which protein domains fold up in nature—maybe as few as 2000. The structures for about 800 of these protein folds have thus far been determined. These known folds tend to be those most represented in the universe of protein structures: for example, 50 folds account for nearly three-fourths of the domain families with predicted structures. A complete catalog of the most significant protein folds that exist in living organisms would therefore seem to be within our reach.

Sequence Searches Can Identify Close Relatives

The present database of known protein sequences contains more than ten million entries, and it is growing very rapidly as more and more genomes are sequenced—revealing huge numbers of new genes that encode proteins. Powerful computer search programs are available that allow us to compare each newly discovered protein with this entire database, looking for possible relatives. Many proteins whose genes have evolved from a common ancestral gene can be identified by the discovery of statistically significant similarities in amino acid sequences.

With such a large number of proteins in the database, the search programs find many nonsignificant matches, resulting in a background noise level that makes it very difficult to pick out all but the closest relatives. Generally speaking, one requires a 30% identity in sequence to consider that two proteins match. However, we know the function of many short signature sequences ("fingerprints"), and these are widely used to find more distant relationships (**Figure 3–14**).

Protein comparisons are important because related structures often imply related functions. Many years of experimentation can be saved by discovering that a new protein has an amino acid sequence similarity with a protein of known function. Such sequence relationships, for example, first indicated that certain genes that cause mammalian cells to become cancerous are protein kinases. In the same way, many of the proteins that control pattern formation during the embryonic development of the fruit fly *Drosophila* were quickly recognized to be gene regulatory proteins.

Figure 3–14 The use of short signature sequences to find related protein domains. The two short sequences of 15 and 9 amino acids shown (*green*) can be used to search large databases for a protein domain that is found in many proteins, the SH2 domain. Here, the first 50 amino acids of the SH2 domain of 100 amino acids is compared for the human and *Drosophila* Src protein (see Figure 3–10). In the computer-generated sequence comparison (*yellow row*), exact matches between the human and *Drosophila* proteins are noted by the one-letter abbreviation for the amino acid; the positions with a similar but nonidentical amino acid are denoted by +, and nonmatches are blank. In this diagram, wherever one or both proteins contain an exact match to a position in the *green* sequences, both aligned sequences are colored *red*.

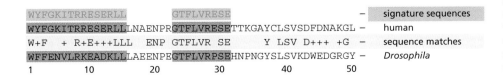

Some Protein Domains Form Parts of Many Different Proteins

As previously stated, most proteins are composed of a series of protein domains, in which different regions of the polypeptide chain have folded independently to form compact structures. Such multidomain proteins are believed to have originated from the accidental joining of the DNA sequences that encode each domain, creating a new gene. Novel binding surfaces have often been created at the juxtaposition of domains, and many of the functional sites where proteins bind to small molecules are found to be located there. In an evolutionary process called *domain shuffling*, many large proteins have evolved through the joining of preexisting domains in new combinations (**Figure 3–15**).

A subset of protein domains have been especially mobile during evolution; these seem to have particularly versatile structures and are sometimes referred to as *protein modules*. The structure of one such module, the SH2 domain, was illustrated in Panel 3–2 (pp. 132–133). Some other abundant protein domains are illustrated in **Figure 3–16**.

Each of the domains shown has a stable core structure formed from strands of β sheet, from which less-ordered loops of polypeptide chain protrude *(green)*. The loops are ideally situated to form binding sites for other molecules, as most clearly demonstrated for the immunoglobulin fold, which forms the basis for antibody molecules (see Figure 3–41). Most likely, such β-sheet-based domains have achieved their evolutionary success because they provide a convenient framework for the generation of new binding sites for ligands through small changes to their protruding loops.

A second feature of these protein domains that explains their utility is the ease with which they can be integrated into other proteins. Five of the six domains illustrated in Figure 3–16 have their N- and C-terminal ends at opposite poles of the domain. When the DNA encoding such a domain undergoes tandem duplication, which is not unusual in the evolution of genomes (discussed in Chapter 4), the duplicated modules with this "in-line" arrangement

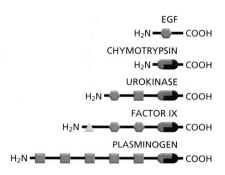

Figure 3–15 Domain shuffling. An extensive shuffling of blocks of protein sequence (protein domains) has occurred during protein evolution. Those portions of a protein denoted by the same shape and color in this diagram are evolutionarily related. Serine proteases like chymotrypsin are formed from two domains *(brown)*. In the three other proteases shown, which are highly regulated and more specialized, these two protease domains are connected to one or more domains that are similar to domains found in epidermal growth factor (EGF; *green*), to a calcium-binding protein *(yellow)*, or to a "kringle" domain *(blue)* that contains three internal disulfide bridges. Chymotrypsin is illustrated in Figure 3–12.

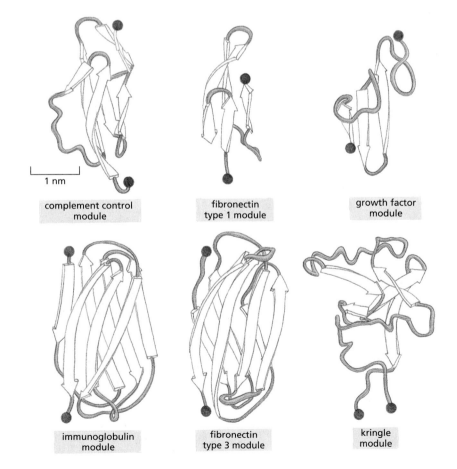

complement control module

fibronectin type 1 module

growth factor module

immunoglobulin module

fibronectin type 3 module

kringle module

Figure 3–16 The three-dimensional structures of some protein modules. In these ribbon diagrams, β-sheet strands are shown as *arrows*, and the N- and C-termini are indicated by *red spheres*. (Adapted from M. Baron, D.G. Norman and I.D. Campbell, *Trends Biochem. Sci.* 16:13–17, 1991, with permission from Elsevier, and D.J. Leahy et al., *Science* 258:987–991, 1992, with permission from AAAS.)

can be readily linked in series to form extended structures—either with themselves or with other in-line domains (**Figure 3–17**). Stiff extended structures composed of a series of domains are especially common in extracellular matrix molecules and in the extracellular portions of cell-surface receptor proteins. Other modules, including the SH2 domain and the kringle domain illustrated in Figure 3–16, are of a "plug-in" type, with their N- and C-termini close together. After genomic rearrangements, such modules are usually accommodated as an insertion into a loop region of a second protein.

A comparison of the relative frequency of domain utilization in different eucaryotes reveals that, for many common domains, such as protein kinases, this frequency is similar in organisms as diverse as yeast, plants, worms, flies, and humans (**Figure 3–18**). But there are some notable exceptions, such as the Major Histocompatibility Complex (MHC) antigen-recognition domain (see Figure 25–52) that is present in 57 copies in humans, but absent in the other four organisms just mentioned. Such domains presumably have specialized functions that are not shared with the other eucaryotes, being strongly selected for during evolution so as to give rise to the multiple copies observed. Similarly, a domain like SH2 that shows an unusual increase in its numbers in higher eucaryotes might be assumed to be especially useful for multicellularity (compare the multicellular organisms with yeast in Figure 3–18).

Certain Pairs of Domains Are Found Together in Many Proteins

We can construct a large table displaying domain usage for each organism whose genome sequence is known. For example, the human genome is estimated to contain about 1000 immunoglobulin domains, 500 protein kinase domains, 250 DNA-binding homeodomains, 300 SH3 domains, and 120 SH2 domains. Important additional information can be derived by comparing the frequencies and arrangements of domains in the more than 100 eucaryotic, bacterial, and archaeal genomes that have been completely sequenced. For example, we find that more than two-thirds of proteins consist of two or more domains, and that the same pairs of domains occur repeatedly in the same relative arrangement in a protein. Although half of all domain families are common to archaea, bacteria, and eucaryotes, only about 5 percent of the two-domain combinations are similarly shared. This pattern suggests that most proteins containing especially useful two-domain combinations arose relatively late in evolution.

The 200 most abundant two-domain combinations occur in about one-fourth of all of the proteins with recognizable domains in the complete data set. It would therefore be very useful to determine the precise three-dimensional structure for at least one protein from each common two-domain combination, so as to reveal how the domains interact in that type of protein structure.

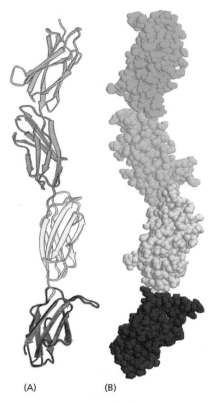

(A)   (B)

Figure 3–17 An extended structure formed from a series of in-line protein modules. Four fibronectin type 3 modules (see Figure 3–16) from the extracellular matrix molecule fibronectin are illustrated in (A) ribbon and (B) space-filling models. (Adapted from D.J. Leahy, I. Aukhil and H.P. Erickson, *Cell* 84:155–164, 1996. With permission from Elsevier.)

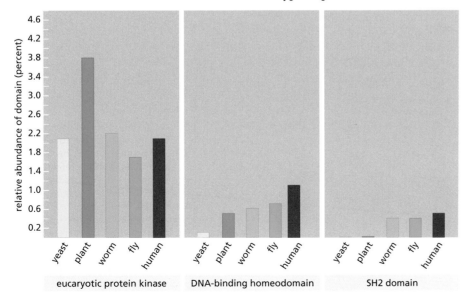

Figure 3–18 The relative frequencies of three protein domains in five eucaryotic organisms. The approximate percentages given have been determined by dividing the number of copies of each domain by the total number of distinct proteins thought to be encoded by each organism, as determined from the sequence of its genome. Thus, for SH2 domains in humans, 120/24,000 = 0.005.

The Human Genome Encodes a Complex Set of Proteins, Revealing Much That Remains Unknown

The result of sequencing the human genome has been surprising, because it reveals that our chromosomes contain only about 25,000 genes. Based on gene number alone, we would appear to be no more complex than the tiny mustard weed, *Arabidopsis*, and only about 1.3-fold more complex than a nematode worm. The genome sequences also reveal that vertebrates have inherited nearly all of their protein domains from invertebrates—with only 7 percent of identified human domains being vertebrate-specific.

Each of our proteins is on average more complicated, however (**Figure 3–19**). Domain shuffling during vertebrate evolution has given rise to many novel combinations of protein domains, with the result that there are nearly twice as many combinations of domains found in human proteins as in a worm or a fly. Thus, for example, the trypsinlike serine protease domain is linked to at least 18 other types of protein domains in human proteins, whereas it is found covalently joined to only 5 different domains in the worm. This extra variety in our proteins greatly increases the range of protein–protein interactions possible (see Figure 3–82), but how it contributes to making us human is not known.

The complexity of living organisms is staggering, and it is quite sobering to note that we currently lack even the tiniest hint of what the function might be for more than 10,000 of the proteins that have thus far been identified in the human genome. There are certainly enormous challenges ahead for the next generation of cell biologists, with no shortage of fascinating mysteries to solve.

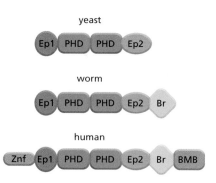

Figure 3–19 **Domain structure of a group of evolutionarily related proteins that are thought to have a similar function.** In general, there is a tendency for the proteins in more complex organisms, such as humans, to contain additional domains—as is the case for the DNA-binding protein compared here.

Larger Protein Molecules Often Contain More Than One Polypeptide Chain <GCCT>

The same weak noncovalent bonds that enable a protein chain to fold into a specific conformation also allow proteins to bind to each other to produce larger structures in the cell. Any region of a protein's surface that can interact with another molecule through sets of noncovalent bonds is called a **binding site**. A protein can contain binding sites for various large and small molecules. If a binding site recognizes the surface of a second protein, the tight binding of two folded polypeptide chains at this site creates a larger protein molecule with a precisely defined geometry. Each polypeptide chain in such a protein is called a **protein subunit**.

In the simplest case, two identical folded polypeptide chains bind to each other in a "head-to-head" arrangement, forming a symmetric complex of two protein subunits (a *dimer*) held together by interactions between two identical binding sites. The *Cro repressor protein*—a viral gene regulatory protein that binds to DNA to turn viral genes off in an infected bacterial cell—provides an example (**Figure 3–20**). Cells contain many other types of symmetric protein complexes, formed from multiple copies of a single polypeptide chain. The enzyme *neuraminidase*, for example, consists of four identical protein subunits, each bound to the next in a "head-to-tail" arrangement that forms a closed ring (**Figure 3–21**).

Many of the proteins in cells contain two or more types of polypeptide chains. *Hemoglobin*, the protein that carries oxygen in red blood cells, contains

Figure 3–20 **Two identical protein subunits binding together to form a symmetric protein dimer.** The Cro repressor protein from bacteriophage lambda binds to DNA to turn off viral genes. Its two identical subunits bind head-to-head, held together by a combination of hydrophobic forces *(blue)* and a set of hydrogen bonds *(yellow region)*. (Adapted from D.H. Ohlendorf, D.E. Tronrud and B.W. Matthews, *J. Mol. Biol.* 280:129–136, 1998. With permission from Academic Press.)

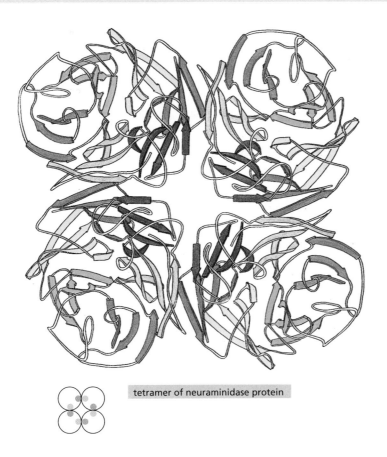

tetramer of neuraminidase protein

Figure 3–21 **A protein molecule containing multiple copies of a single protein subunit.** The enzyme neuraminidase exists as a ring of four identical polypeptide chains. Each of these chains is formed from six repeats of a four-stranded β sheet, as indicated by the colored arrows. The small diagram shows how the repeated use of the same binding interaction forms the structure.

two identical α-globin subunits and two identical β-globin subunits, symmetrically arranged (**Figure 3–22**). Such multisubunit proteins are very common in cells, and they can be very large. **Figure 3–23** shows a sample of proteins whose exact structures are known, and it compares the sizes and shapes of a few larger proteins with some of the relatively small proteins that we have thus far presented as models.

Some Proteins Form Long Helical Filaments

Some protein molecules can assemble to form filaments that may span the entire length of a cell. Most simply, a long chain of identical protein molecules can be constructed if each molecule has a binding site complementary to another region of the surface of the same molecule (**Figure 3–24**). An actin filament, for example, is a long helical structure produced from many molecules of the protein *actin* (**Figure 3–25**). Actin is very abundant in eucaryotic cells, where it constitutes one of the major filament systems of the cytoskeleton (discussed in Chapter 16).

Why is a helix such a common structure in biology? As we have seen, biological structures are often formed by linking similar subunits—such as amino acids or protein molecules—into long, repetitive chains. If all the subunits are identical, the neighboring subunits in the chain can often fit together in only one way, adjusting their relative positions to minimize the free energy of the contact between them. As a result, each subunit is positioned in exactly the same way in relation to the next, so that subunit 3 fits onto subunit 2 in the same way that subunit 2 fits onto subunit 1, and so on. Because it is very rare for subunits to join up in a straight line, this arrangement generally results in a helix—a regular structure that resembles a spiral staircase, as illustrated in **Figure 3–26**. Depending on the twist of the staircase, a helix is said to be either right-handed or left-handed (see Figure 3–26E). Handedness is not affected by turning the helix upside down, but it is reversed if the helix is reflected in the mirror.

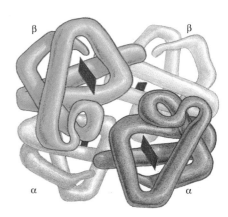

Figure 3–22 **A protein formed as a symmetric assembly of two different subunits.** Hemoglobin is an abundant protein in red blood cells that contains two copies of α-globin and two copies of β-globin. Each of these four polypeptide chains contains a heme molecule *(red),* which is the site that binds oxygen (O_2). Thus, each molecule of hemoglobin in the blood carries four molecules of oxygen.

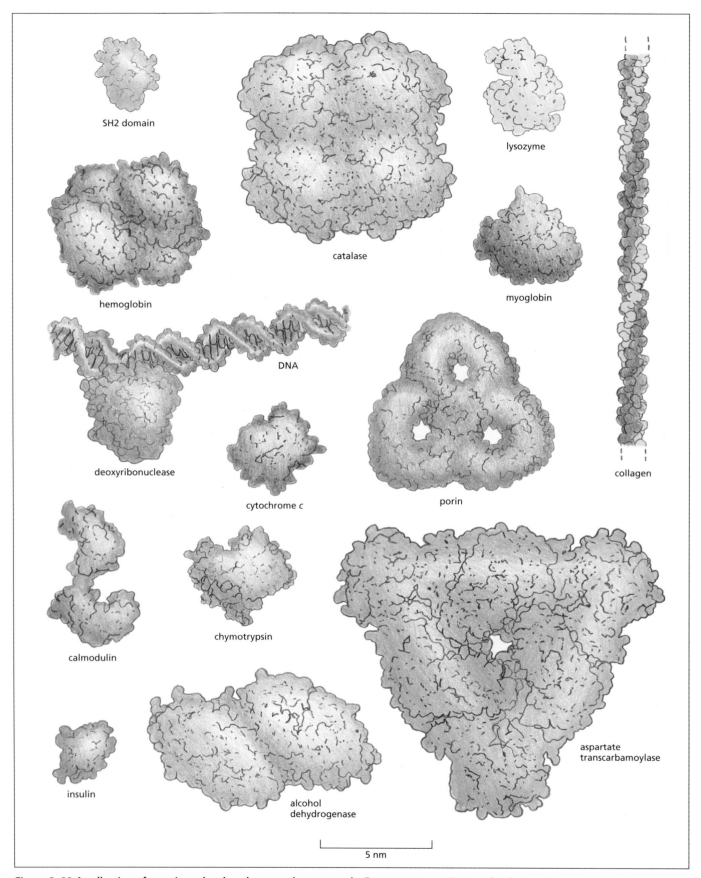

Figure 3–23 A collection of protein molecules, shown at the same scale. For comparison, a DNA molecule bound to a protein is also illustrated. These space-filling models represent a range of sizes and shapes. Hemoglobin, catalase, porin, alcohol dehydrogenase, and aspartate transcarbamoylase are formed from multiple copies of subunits. The SH2 domain *(top left)* is presented in detail in Panel 3–2 (pp. 132–133). (Adapted from David S. Goodsell, *Our Molecular Nature.* New York: Springer-Verlag, 1996. With permission from Springer Science and Business Media.)

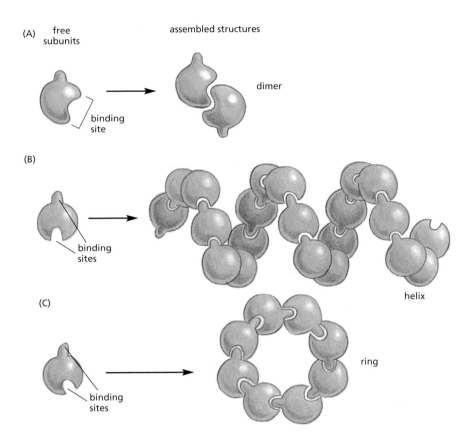

Figure 3–24 Protein assemblies.
(A) A protein with just one binding site can form a dimer with another identical protein. (B) Identical proteins with two different binding sites often form a long helical filament. (C) If the two binding sites are disposed appropriately in relation to each other, the protein subunits may form a closed ring instead of a helix. (For an example of A, see Figure 3–20; for an example of C, see Figure 3–21.)

Helices occur commonly in biological structures, whether the subunits are small molecules linked together by covalent bonds (for example, the amino acids in an α helix) or large protein molecules that are linked by noncovalent forces (for example, the actin molecules in actin filaments). This is not surprising. A helix is an unexceptional structure, and it is generated simply by placing many similar subunits next to each other, each in the same strictly repeated relationship to the one before—that is, with a fixed rotation followed by a fixed translation along the helix axis, as in a spiral staircase.

Many Protein Molecules Have Elongated, Fibrous Shapes

Most of the proteins that we have discussed so far are *globular proteins,* in which the polypeptide chain folds up into a compact shape like a ball with an irregular surface. Enzymes tend to be globular proteins: even though many are large and complicated, with multiple subunits, most have an overall rounded shape (see Figure 3–23). In contrast, other proteins have roles in the cell that require each individual protein molecule to span a large distance. These proteins generally have a relatively simple, elongated three-dimensional structure and are commonly referred to as *fibrous proteins.*

One large family of intracellular fibrous proteins consists of α-keratin, introduced when we presented the α helix, and its relatives. Keratin filaments are extremely stable and are the main component in long-lived structures such as hair, horn, and nails. An α-keratin molecule is a dimer of two identical subunits, with the long α helices of each subunit forming a coiled-coil (see Figure 3–9). The coiled-coil regions are capped at each end by globular domains containing binding sites. This enables this class of protein to assemble into ropelike *intermediate filaments*—an important component of the cytoskeleton that creates the cell's internal structural framework (see Figure 16–19).

Fibrous proteins are especially abundant outside the cell, where they are a main component of the gel-like *extracellular matrix* that helps to bind collections of cells together to form tissues. Cells secrete extracellular matrix proteins into their surroundings, where they often assemble into sheets or long fibrils.

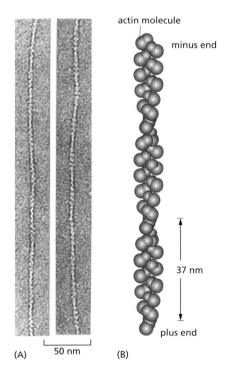

Figure 3–25 Actin filaments.
(A) Transmission electron micrographs of negatively stained actin filaments.
(B) The helical arrangement of actin molecules in an actin filament.
(A, courtesy of Roger Craig.)

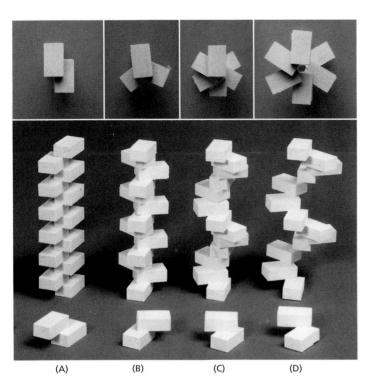

left-
handed

right-
handed

(E)

(A)   (B)   (C)   (D)

Figure 3–26 Some properties of a helix.
(A–D) A helix forms when a series of subunits bind to each other in a regular way. At the bottom, the interaction between two subunits is shown; behind them are the helices that result. These helices have two (A), three (B), and six (C and D) subunits per helical turn. The photographs at the top show the arrangement of subunits viewed from directly above the helix. Note that the helix in (D) has a wider path than that in (C), but the same number of subunits per turn. (E) A helix can be either right-handed or left-handed. As a reference, it is useful to remember that standard metal screws, which insert when turned clockwise, are right-handed. Note that a helix retains the same handedness when it is turned upside down.

Collagen is the most abundant of these proteins in animal tissues. A collagen molecule consists of three long polypeptide chains, each containing the non-polar amino acid glycine at every third position. This regular structure allows the chains to wind around one another to generate a long regular triple helix (**Figure 3–27**A). Many collagen molecules then bind to one another side-by-side and end-to-end to create long overlapping arrays—thereby generating the extremely tough collagen fibrils that give connective tissues their tensile strength, as described in Chapter 19.

Many Proteins Contain a Surprisingly Large Amount of Unstructured Polypeptide Chain

It has been well known for a long time that, in complete contrast to collagen, another abundant protein in the extracellular matrix, *elastin,* is formed as a highly disordered polypeptide. This disorder is essential for elastin's function. Its

Figure 3–27 Collagen and elastin.
(A) Collagen is a triple helix formed by three extended protein chains that wrap around one another *(bottom)*. Many rodlike collagen molecules are cross-linked together in the extracellular space to form unextendable collagen fibrils *(top)* that have the tensile strength of steel. The striping on the collagen fibril is caused by the regular repeating arrangement of the collagen molecules within the fibril. (B) Elastin polypeptide chains are cross-linked together to form rubberlike, elastic fibers. Each elastin molecule uncoils into a more extended conformation when the fiber is stretched and recoils spontaneously as soon as the stretching force is relaxed.

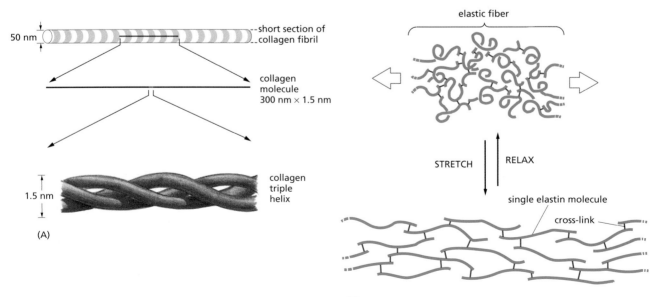

50 nm

--short section of
--collagen fibril

collagen
molecule
300 nm × 1.5 nm

1.5 nm

collagen
triple
helix

(A)

elastic fiber

STRETCH   RELAX

single elastin molecule

cross-link

(B)

relatively loose and unstructured polypeptide chains are covalently cross-linked to produce a rubberlike elastic meshwork that can be reversibly pulled from one conformation to another, as illustrated in Figure 3–27B. The elastic fibers that result enable skin and other tissues, such as arteries and lungs, to stretch and recoil without tearing.

Intrinsically unstructured regions of proteins are quite frequent in nature, having important functions in the interior of cells. As we have already seen, proteins use the short loops of polypeptide chain that generally protrude from the core region of protein domains to bind other molecules. Similarly, many proteins have much longer regions of unstructured amino acid sequences that interact with another molecule (often DNA or a protein), undergoing a structural transition to a specific folded conformation when the other molecule is bound. Other proteins appear to resemble elastin, in so far as their function requires that they remain largely unstructured. For example, the abundant nucleoporins that coat the inner surface of the nuclear pore complex form a random coil meshwork that is intimately involved in nuclear transport (see Figure 12–10). Finally, as will be discussed later in this chapter (see Figure 3–80C), unstructured regions of polypeptide chain are often used to connect the binding sites for proteins that function together to catalyze a biological reaction. Thus, for example, in facilitating cell signaling, large *scaffold proteins* use such flexible regions as "tethers" that concentrate sets of interacting proteins, often confining them to particular sites in the cell (discussed in Chapter 15).

We can recognize the unstructured regions in many proteins by their biased amino acid composition: they contain very few of the bulky hydrophobic amino acids that normally form the core of a folded protein, being composed instead of a high proportion of the amino acids Gln, Ser, Pro, Glu, and Lys. Such "natively unfolded" regions also frequently contain repeated sequences of amino acids.

Covalent Cross-Linkages Often Stabilize Extracellular Proteins

Many protein molecules are either attached to the outside of a cell's plasma membrane or secreted as part of the extracellular matrix. All such proteins are directly exposed to extracellular conditions. To help maintain their structures, the polypeptide chains in such proteins are often stabilized by covalent cross-linkages. These linkages can either tie two amino acids in the same protein together, or connect different polypeptide chains in a multisubunit protein. The most common cross-linkages in proteins are covalent sulfur–sulfur bonds. These *disulfide bonds* (also called *S–S bonds*) form as cells prepare newly synthesized proteins for export. As described in Chapter 12, their formation is catalyzed in the endoplasmic reticulum by an enzyme that links together two pairs of –SH groups of cysteine side chains that are adjacent in the folded protein (**Figure 3–28**). Disulfide bonds do not change the conformation of a protein but instead act as atomic staples to reinforce its most favored conformation. For

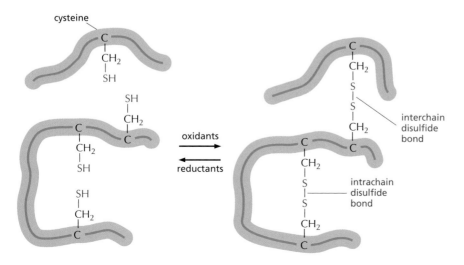

Figure 3–28 Disulfide bonds. <ATAC> This diagram illustrates how covalent disulfide bonds form between adjacent cysteine side chains. As indicated, these cross-linkages can join either two parts of the same polypeptide chain or two different polypeptide chains. Since the energy required to break one covalent bond is much larger than the energy required to break even a whole set of noncovalent bonds (see Table 2–1, p. 53), a disulfide bond can have a major stabilizing effect on a protein.

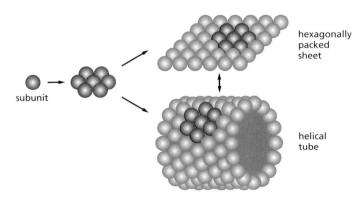

hexagonally packed sheet

helical tube

subunit

Figure 3–29 An example of the assembly of a single protein subunit requiring multiple protein–protein contacts. Hexagonally packed globular protein subunits can form either a flat sheet or a tube.

example, lysozyme—an enzyme in tears that dissolves bacterial cell walls— retains its antibacterial activity for a long time because it is stabilized by such cross-linkages.

Disulfide bonds generally fail to form in the cell cytosol, where a high concentration of reducing agents converts S–S bonds back to cysteine –SH groups. Apparently, proteins do not require this type of reinforcement in the relatively mild environment inside the cell.

Protein Molecules Often Serve as Subunits for the Assembly of Large Structures

The same principles that enable a protein molecule to associate with itself to form rings or filaments also operate to generate much larger structures in the cell—supramolecular structures such as enzyme complexes, ribosomes, protein filaments, viruses, and membranes. These large objects are not made as single, giant, covalently linked molecules. Instead they are formed by the noncovalent assembly of many separately manufactured molecules, which serve as the subunits of the final structure.

The use of smaller subunits to build larger structures has several advantages:
1. A large structure built from one or a few repeating smaller subunits requires only a small amount of genetic information.
2. Both assembly and disassembly can be readily controlled, reversible processes, because the subunits associate through multiple bonds of relatively low energy.
3. Errors in the synthesis of the structure can be more easily avoided, since correction mechanisms can operate during the course of assembly to exclude malformed subunits.

Some protein subunits assemble into flat sheets in which the subunits are arranged in hexagonal patterns. Specialized membrane proteins are sometimes arranged this way in lipid bilayers. With a slight change in the geometry of the individual subunits, a hexagonal sheet can be converted into a tube (**Figure 3–29**) or, with more changes, into a hollow sphere. Protein tubes and spheres that bind specific RNA and DNA molecules in their interior form the coats of viruses.

The formation of closed structures, such as rings, tubes, or spheres, provides additional stability because it increases the number of bonds between the protein subunits. Moreover, because such a structure is created by mutually dependent, cooperative interactions between subunits, a relatively small change that affects each subunit individually can cause the structure to assemble or disassemble. These principles are dramatically illustrated in the protein coat or *capsid* of many simple viruses, which takes the form of a hollow sphere based on an icosahedron (**Figure 3–30**). Capsids are often made of hundreds of identical protein subunits that enclose and protect the viral nucleic acid (**Figure 3–31**). The protein in such a capsid must have a particularly adaptable structure: not only must it make several different kinds of contacts to create the sphere, it must also change this arrangement to let the nucleic acid out to initiate viral replication once the virus has entered a cell.

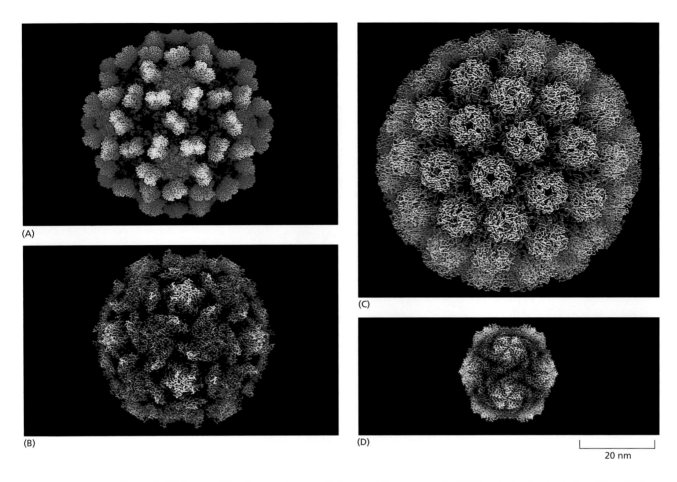

Figure 3–30 **The capsids of some viruses, all shown at the same scale.** (A) Tomato bushy stunt virus; (B) poliovirus; (C) simian virus 40 (SV40); (D) satellite tobacco necrosis virus. The structures of all of these capsids have been determined by x-ray crystallography and are known in atomic detail. (Courtesy of Robert Grant, Stephan Crainic, and James M. Hogle.)

Many Structures in Cells Are Capable of Self-Assembly

The information for forming many of the complex assemblies of macromolecules in cells must be contained in the subunits themselves, because purified subunits can spontaneously assemble into the final structure under the appropriate conditions. The first large macromolecular aggregate shown to be capable of self-assembly from its component parts was *tobacco mosaic virus (TMV)*. This virus is a long rod in which a cylinder of protein is arranged around a helical RNA core (**Figure 3–32**). If the dissociated RNA and protein subunits are mixed together in solution, they recombine to form fully active viral particles. The assembly process is unexpectedly complex and includes the formation of double rings of protein, which serve as intermediates that add to the growing viral coat.

Another complex macromolecular aggregate that can reassemble from its component parts is the bacterial ribosome. This structure is composed of about 55 different protein molecules and 3 different rRNA molecules. Incubating the individual components under appropriate conditions in a test tube causes them to spontaneously re-form the original structure. Most importantly, such reconstituted ribosomes are able to catalyze protein synthesis. As might be expected, the reassembly of ribosomes follows a specific pathway: after certain proteins have bound to the RNA, this complex is then recognized by other proteins, and so on, until the structure is complete.

It is still not clear how some of the more elaborate self-assembly processes are regulated. Many structures in the cell, for example, seem to have a precisely

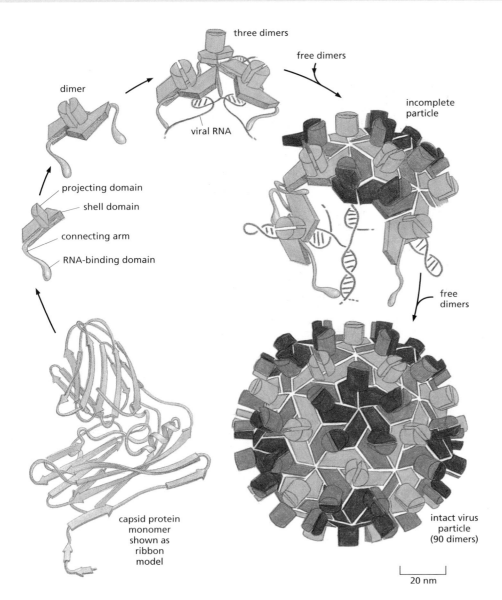

three dimers

free dimers

dimer

viral RNA

incomplete particle

projecting domain

shell domain

connecting arm

RNA-binding domain

free dimers

capsid protein monomer shown as ribbon model

intact virus particle (90 dimers)

20 nm

Figure 3–31 The structure of a spherical virus. In many viruses, identical protein subunits pack together to create a spherical shell (a capsid) that encloses the viral genome, composed of either RNA or DNA (see also Figure 3–30). For geometric reasons, no more than 60 identical subunits can pack together in a precisely symmetric way. If slight irregularities are allowed, however, more subunits can be used to produce a larger capsid that retains icosahedral symmetry. The tomato bushy stunt virus (TBSV) shown here, for example, is a spherical virus about 33 nm in diameter formed from 180 identical copies of a 386 amino acid capsid protein plus an RNA genome of 4500 nucleotides. To construct such a large capsid, the protein must be able to fit into three somewhat different environments, each of which is differently colored in the virus particle shown here. The postulated pathway of assembly is shown; the precise three-dimensional structure has been determined by x-ray diffraction. (Courtesy of Steve Harrison.)

defined length that is many times greater than that of their component macro-molecules. How such length determination is achieved is in many cases a mystery. Three possible mechanisms are illustrated in **Figure 3–33**. In the simplest

Figure 3–32 The structure of tobacco mosaic virus (TMV). (A) An electron micrograph of the viral particle, which consists of a single long RNA molecule enclosed in a cylindrical protein coat composed of identical protein subunits. (B) A model showing part of the structure of TMV. A single-stranded RNA molecule of 6395 nucleotides is packaged in a helical coat constructed from 2130 copies of a coat protein 158 amino acids long. Fully infective viral particles can self-assemble in a test tube from purified RNA and protein molecules. (A, courtesy of Robley Williams; B, courtesy of Richard J. Feldmann.)

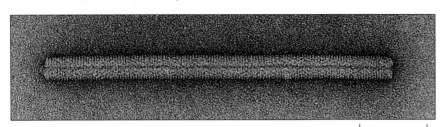

(A)

50 nm

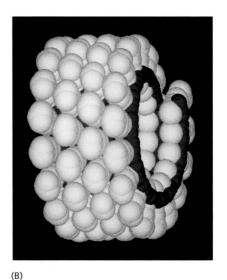

(B)

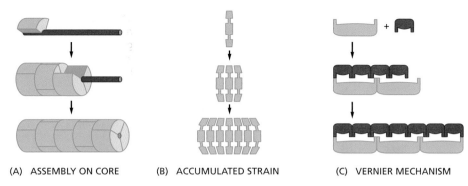

(A) ASSEMBLY ON CORE   (B) ACCUMULATED STRAIN   (C) VERNIER MECHANISM

Figure 3–33 **Three mechanisms of length determination for large protein assemblies.** (A) Coassembly along an elongated core protein or other macromolecule that acts as a measuring device. (B) Termination of assembly because of strain that accumulates in the polymeric structure as additional subunits are added, so that beyond a certain length the energy required to fit another subunit onto the chain becomes excessively large. (C) A vernier type of assembly, in which two sets of rodlike molecules differing in length form a staggered complex that grows until their ends exactly match. The name derives from a measuring device based on the same principle, used in mechanical instruments.

case, a long core protein or other macromolecule provides a scaffold that determines the extent of the final assembly. This is the mechanism that determines the length of the TMV particle, where the RNA chain provides the core. Similarly, a core protein is thought to determine the length of the thin filaments in muscle, as well as the length of the long tails of some bacterial viruses (**Figure 3–34**).

Assembly Factors Often Aid the Formation of Complex Biological Structures

Not all cellular structures held together by noncovalent bonds self-assemble. A mitochondrion, a cilium, or a myofibril of a muscle cell, for example, cannot form spontaneously from a solution of its component macromolecules. In these cases, part of the assembly information is provided by special enzymes and other proteins that perform the function of templates, guiding construction but taking no part in the final assembled structure.

Even relatively simple structures may lack some of the ingredients necessary for their own assembly. In the formation of certain bacterial viruses, for example, the head, which is composed of many copies of a single protein subunit, is assembled on a temporary scaffold composed of a second protein. Because the second protein is absent from the final viral particle, the head structure cannot spontaneously reassemble once it has been taken apart. Other examples are known in which proteolytic cleavage is an essential and irreversible step in the normal assembly process. This is even the case for some small protein assemblies, including the structural protein collagen and the hormone insulin (**Figure 3–35**). From these relatively simple examples, it seems certain that the assembly of a structure as complex as a mitochondrion or a cilium will involve temporal and spatial ordering imparted by numerous other cell components.

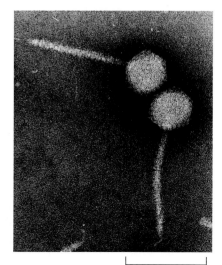

Figure 3–34 **An electron micrograph of bacteriophage lambda.** The tip of the virus tail attaches to a specific protein on the surface of a bacterial cell, after which the tightly packaged DNA in the head is injected through the tail into the cell. The tail has a precise length, determined by the mechanism shown in Figure 3–33A.

100 nm

proinsulin

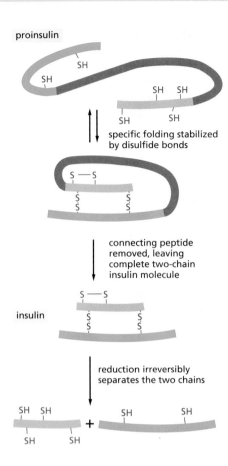

specific folding stabilized
by disulfide bonds

connecting peptide
removed, leaving
complete two-chain
insulin molecule

insulin

reduction irreversibly
separates the two chains

+

Figure 3–35 Proteolytic cleavage in insulin assembly. The polypeptide hormone insulin cannot spontaneously re-form efficiently if its disulfide bonds are disrupted. It is synthesized as a larger protein *(proinsulin)* that is cleaved by a proteolytic enzyme after the protein chain has folded into a specific shape. Excision of part of the proinsulin polypeptide chain removes some of the information needed for the protein to fold spontaneously into its normal conformation. Once insulin has been denatured and its two polypeptide chains have separated, its ability to reassemble is lost.

Summary

A protein molecule's amino acid sequence determines its three-dimensional confor-mation. Noncovalent interactions between different parts of the polypeptide chain sta-bilize its folded structure. The amino acids with hydrophobic side chains tend to clus-ter in the interior of the molecule, and local hydrogen-bond interactions between neighboring peptide bonds give rise to α helices and β sheets.

Globular regions, known as domains, are the modular units from which many proteins are constructed; such domains generally contain 40–350 amino acids. Small proteins typically consist of only a single domain, while large proteins are formed from several domains linked together by various lengths of polypeptide chain, some of which can be relatively disordered. As proteins have evolved, domains have been mod-ified and combined with other domains to construct new proteins. Thus far, about 800 different ways of folding up a domain have been observed, among more than 20,000 known protein structures.

Proteins are brought together into larger structures by the same noncovalent forces that determine protein folding. Proteins with binding sites for their own surface can assemble into dimers, closed rings, spherical shells, or helical polymers. Although mix-tures of proteins and nucleic acids can assemble spontaneously into complex struc-tures in a test tube, many biological assembly processes involve irreversible steps. Con-sequently, not all structures in the cell are capable of spontaneous reassembly after they have been dissociated into their component parts.

PROTEIN FUNCTION

We have seen that each type of protein consists of a precise sequence of amino acids that allows it to fold up into a particular three-dimensional shape, or con-formation. But proteins are not rigid lumps of material. They often have precisely engineered moving parts whose mechanical actions are coupled to chemical events. It is this coupling of chemistry and movement that gives proteins the extraordinary capabilities that underlie the dynamic processes in living cells.

In this section, we explain how proteins bind to other selected molecules and how their activity depends on such binding. We show that the ability to bind to other molecules enables proteins to act as catalysts, signal receptors, switches, motors, or tiny pumps. The examples we discuss in this chapter by no means exhaust the vast functional repertoire of proteins. You will encounter the specialized functions of many other proteins elsewhere in this book, based on similar principles.

All Proteins Bind to Other Molecules

A protein molecule's physical interaction with other molecules determines its biological properties. Thus, antibodies attach to viruses or bacteria to mark them for destruction, the enzyme hexokinase binds glucose and ATP so as to catalyze a reaction between them, actin molecules bind to each other to assemble into actin filaments, and so on. Indeed, all proteins stick, or *bind*, to other molecules. In some cases, this binding is very tight; in others it is weak and short-lived. But the binding always shows great *specificity*, in the sense that each protein molecule can usually bind just one or a few molecules out of the many thousands of different types it encounters. The substance that is bound by the protein—whether it is an ion, a small molecule, or a macromolecule such as another protein—is referred to as a **ligand** for that protein (from the Latin word *ligare*, meaning "to bind").

The ability of a protein to bind selectively and with high affinity to a ligand depends on the formation of a set of weak, noncovalent bonds—hydrogen bonds, electrostatic attractions, and van der Waals attractions—plus favorable hydrophobic interactions (see Panel 2–3, pp. 110–111). Because each individual bond is weak, effective binding occurs only when many of these bonds form simultaneously. Such binding is possible only if the surface contours of the ligand molecule fit very closely to the protein, matching it like a hand in a glove (**Figure 3–36**).

The region of a protein that associates with a ligand, known as the ligand's *binding site,* usually consists of a cavity in the protein surface formed by a particular arrangement of amino acids. These amino acids can belong to different portions of the polypeptide chain that are brought together when the protein folds (**Figure 3–37**). Separate regions of the protein surface generally provide binding sites for different ligands, allowing the protein's activity to be regulated, as we shall see later. And other parts of the protein act as a handle to position the protein in the cell—an example is the SH2 domain discussed previously, which often moves a protein containing it to particular intracellular sites in response to particular signals.

Although the atoms buried in the interior of the protein have no direct contact with the ligand, they form the framework that gives the surface its contours and its chemical and mechanical properties. Even small changes to the amino acids in the interior of a protein molecule can change its three-dimensional shape enough to destroy a binding site on the surface.

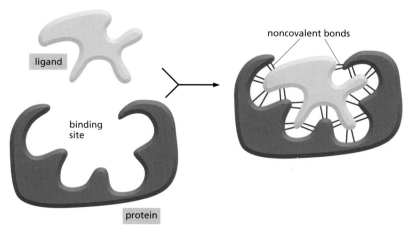

Figure 3–36 **The selective binding of a protein to another molecule.** Many weak bonds are needed to enable a protein to bind tightly to a second molecule, which is called a *ligand* for the protein. A ligand must therefore fit precisely into a protein's binding site, like a hand into a glove, so that a large number of noncovalent bonds form between the protein and the ligand.

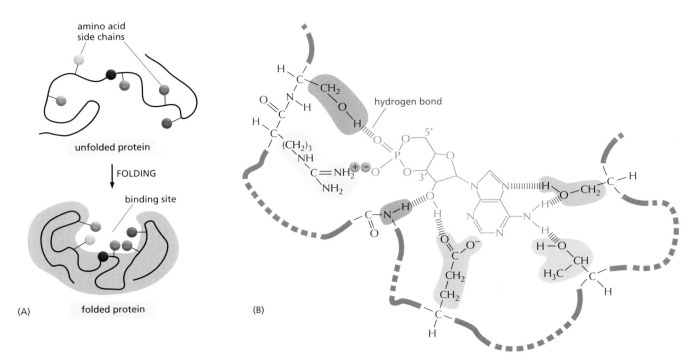

Figure 3–37 The binding site of a protein. (A) The folding of the polypeptide chain typically creates a crevice or cavity on the protein surface. This crevice contains a set of amino acid side chains disposed in such a way that they can form noncovalent bonds only with certain ligands. (B) A close-up of an actual binding site showing the hydrogen bonds and electrostatic interactions formed between a protein and its ligand. In this example, cyclic AMP is the bound ligand.

The Surface Conformation of a Protein Determines Its Chemistry

Proteins have impressive chemical capabilities because the neighboring chemical groups on their surface often interact in ways that enhance the chemical reactivity of amino acid side chains. These interactions fall into two main categories.

First, the interaction of neighboring parts of the polypeptide chain may restrict the access of water molecules to that protein's ligand-binding sites. This is important because water molecules readily form hydrogen bonds that can compete with ligands for sites on the protein surface. Proteins and their ligands form tighter hydrogen bonds (and electrostatic interactions) if the protein can exclude water molecules from its binding sites. It might be hard to imagine a mechanism that would exclude a molecule as small as water from a protein surface without affecting the access of the ligand itself. However, because of the strong tendency of water molecules to form water–water hydrogen bonds, water molecules exist in a large hydrogen-bonded network (see Panel 2–2, pp. 108–109). In effect, a protein can keep a ligand-binding site dry because it is energetically unfavorable for individual water molecules to break away from this network, as they must do to reach into a crevice on a protein's surface.

Second, the clustering of neighboring polar amino acid side chains can alter their reactivity. If protein folding forces together a number of negatively charged side chains against their mutual repulsion, for example, the affinity of the site for a positively charged ion is greatly increased. In addition, when amino acid side chains interact with one another through hydrogen bonds, normally unreactive side groups (such as the –CH$_2$OH on the serine shown in **Figure 3–38**) can become reactive, enabling them to be used to make or break selected covalent bonds.

The surface of each protein molecule therefore has a unique chemical reactivity that depends not only on which amino acid side chains are exposed, but

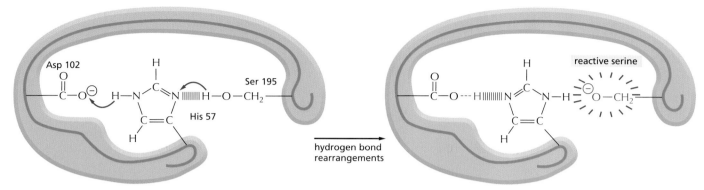

Figure 3–38 **An unusually reactive amino acid at the active site of an enzyme.** This example is the "catalytic triad" found in chymotrypsin, elastase, and other serine proteases (see Figure 3–12). The aspartic acid side chain (Asp 102) induces the histidine (His 57) to remove the proton from serine 195. This activates the serine to form a covalent bond with the enzyme substrate, hydrolyzing a peptide bond. The many convolutions of the polypeptide chain are omitted here.

Sequence Comparisons Between Protein Family Members Highlight Crucial Ligand-Binding Sites

As we have described previously, genome sequences allow us to group many of the domains in proteins into families that show clear evidence of their evolution from a common ancestor. The three-dimensional structures of the members of the same domain family are remarkably similar. For example, even when the amino acid sequence identity falls to 25%, the backbone atoms in a domain follow a common protein fold within 0.2 nanometers (2 Å).

We can therefore use a method called evolutionary tracing to identify those sites in a protein domain that are the most crucial to the domain's function. For this purpose, those amino acids that are unchanged, or nearly unchanged, in all of the known protein family members are mapped onto a model of the three-dimensional structure of one family member. When this is done, the most invariant positions often form one or more clusters on the protein surface, as illustrated in **Figure 3–39**A for the SH2 domain described previously (see Panel 3–2, pp. 132–133). These clusters generally correspond to ligand binding sites.

The SH2 domain is a module that functions in protein–protein interactions. It binds the protein containing it to a second protein that contains a phosphorylated tyrosine side chain in a specific amino acid sequence context, as shown in Figure 3–39B. The amino acids located at the binding site for the phosphorylated polypeptide have been the slowest to change during the long evolutionary process that produced the large SH2 family of peptide recognition domains. Because mutation is a random process, this result is attributed to the preferential elimination during evolution of all organisms whose SH2 domains became altered in a way that inactivated the SH2-binding site, thereby destroying the function of the SH2 domain.

Figure 3–39 **The evolutionary trace method applied to the SH2 domain.** (A) Front and back views of a space-filling model of the SH2 domain, with evolutionarily conserved amino acids on the protein surface colored *yellow*, and those more toward the protein interior colored *red*. (B) The structure of the SH2 domain with its bound polypeptide. Here, those amino acids located within 0.4 nm of the bound ligand are colored *blue*. The two key amino acids of the ligand are *yellow*, and the others are *purple*. Note the high degree of correspondance between (A) and (B). (Adapted from O. Lichtarge, H.R. Bourne and F.E. Cohen, *J. Mol. Biol.* 257:342–358, 1996. With permission from Elsevier.)

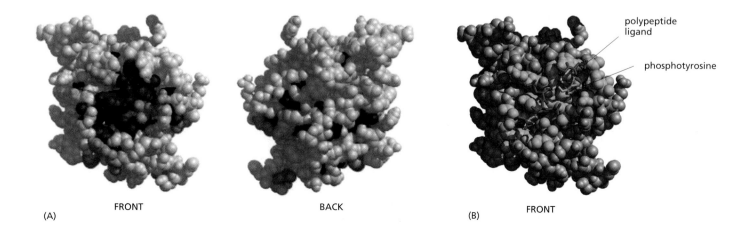

polypeptide ligand

phosphotyrosine

(A)   FRONT   BACK   (B)   FRONT

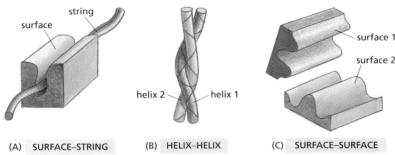

(A)   SURFACE–STRING   (B)   HELIX–HELIX   (C)   SURFACE–SURFACE

Figure 3–40 Three ways in which two proteins can bind to each other. Only the interacting parts of the two proteins are shown. (A) A rigid surface on one protein can bind to an extended loop of polypeptide chain (a "string") on a second protein. (B) Two α helices can bind together to form a coiled-coil. (C) Two complementary rigid surfaces often link two proteins together.

In this era of extensive genome sequencing, many new protein families have been discovered whose functions are unknown. Once a three-dimensional structure has been determined for one family member, evolutionary tracing allows biologists to determine binding sites for the members of that family, thereby helping to decipher protein function.

Proteins Bind to Other Proteins Through Several Types of Interfaces

Proteins can bind to other proteins in at least three ways. In many cases, a portion of the surface of one protein contacts an extended loop of polypeptide chain (a "string") on a second protein (**Figure 3–40**A). Such a surface–string interaction, for example, allows the SH2 domain to recognize a phosphorylated polypeptide loop on a second protein, as just described, and it also enables a protein kinase to recognize the proteins that it will phosphorylate (see below).

A second type of protein–protein interface forms when two α helices, one from each protein, pair together to form a coiled-coil (Figure 3–40B). This type of protein interface is found in several families of gene regulatory proteins, as discussed in Chapter 7.

The most common way for proteins to interact, however, is by the precise matching of one rigid surface with that of another (Figure 3–40C). Such interactions can be very tight, since a large number of weak bonds can form between two surfaces that match well. For the same reason, such surface–surface interactions can be extremely specific, enabling a protein to select just one partner from the many thousands of different proteins found in a cell.

Antibody Binding Sites Are Especially Versatile <GCCG>

All proteins must bind to particular ligands to carry out their various functions. The antibody family is notable for its capacity for tight selective binding (discussed in detail in Chapter 25).

Antibodies, or immunoglobulins, are proteins produced by the immune system in response to foreign molecules, such as those on the surface of an invading microorganism. Each antibody binds tightly to a particular target molecule, thereby either inactivating the target molecule directly or marking it for destruction. An antibody recognizes its target (called an **antigen**) with remarkable specificity. Because there are potentially billions of different antigens that humans might encounter, we have to be able to produce billions of different antibodies.

Antibodies are Y-shaped molecules with two identical binding sites that are complementary to a small portion of the surface of the antigen molecule. A detailed examination of the antigen-binding sites of antibodies reveals that they are formed from several loops of polypeptide chain that protrude from the ends of a pair of closely juxtaposed protein domains (**Figure 3–41**). Different antibodies generate an enormous diversity of antigen-binding sites by changing only the length and amino acid sequence of these loops, without altering the basic protein structure.

Loops of this kind are ideal for grasping other molecules. They allow a large number of chemical groups to surround a ligand so that the protein can link to

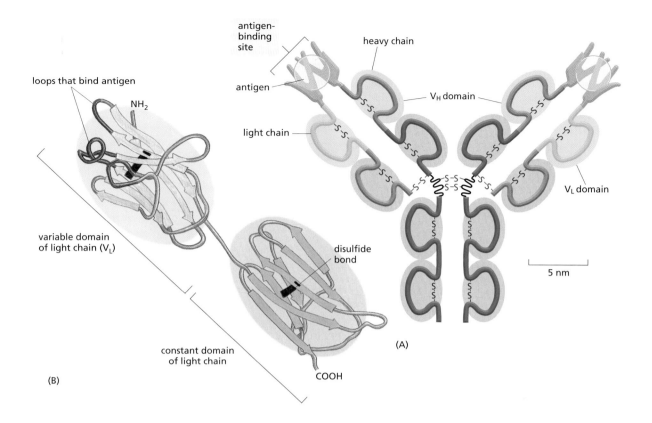

Figure 3–41 An antibody molecule.
(A) A typical antibody molecule is Y-shaped and has two identical binding sites for its antigen, one on each arm of the Y. The protein is composed of four polypeptide chains (two identical heavy chains and two identical and smaller light chains) held together by disulfide bonds. Each chain is made up of several different immunoglobulin domains, here shaded either *blue* or *gray*. The antigen-binding site is formed where a heavy-chain variable domain (V_H) and a light-chain variable domain (V_L) come close together. These are the domains that differ most in their sequence and structure in different antibodies. Each domain at the end of the two arms of the antibody molecule forms loops that bind to the antigen. In (B) we can see these fingerlike loops (*red*) contributed by the V_L domain.

it with many weak bonds. For this reason, loops often form the ligand-binding sites in proteins.

The Equilibrium Constant Measures Binding Strength

Molecules in the cell encounter each other very frequently because of their continual random thermal movements. Colliding molecules with poorly matching surfaces form few noncovalent bonds with one another, and the two molecules dissociate as rapidly as they come together. At the other extreme, when many noncovalent bonds form between two colliding molecules, the association can persist for a very long time (**Figure 3–42**). Strong interactions occur in cells whenever a biological function requires that molecules remain associated for a long time—for example, when a group of RNA and protein molecules come together to make a subcellular structure such as a ribosome.

We can measure the strength with which any two molecules bind to each other. As an example, consider a population of identical antibody molecules that suddenly encounters a population of ligands diffusing in the fluid surrounding them. At frequent intervals, one of the ligand molecules will bump into the binding site of an antibody and form an antibody–ligand complex. The population of antibody–ligand complexes will therefore increase, but not without limit: over time, a second process, in which individual complexes break apart because of thermally induced motion, will become increasingly important. Eventually, any population of antibody molecules and ligands will reach a steady state, or equilibrium, in which the number of binding (association) events per second is precisely equal to the number of "unbinding" (dissociation) events (see Figure 2–52).

From the concentrations of the ligand, antibody, and antibody–ligand complex at equilibrium, we can calculate a convenient measure—the **equilibrium constant** (*K*)—of the strength of binding (**Figure 3–43**A). The equilibrium constant for a reaction in which two molecules (A and B) bind to each other to form a complex (AB) has units of liters/mole, and half of the binding sites will be occupied by ligand when that ligand's concentration (in moles/liter) reaches a value that is equal to $1/K$. This equilibrium constant is larger the greater the binding strength, and it is a direct measure of the free-energy difference

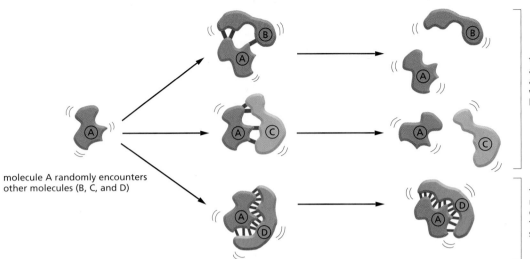

molecule A randomly encounters
other molecules (B, C, and D)

the surfaces of molecules A and B,
and A and C, are a poor match and
are capable of forming only a few
weak bonds; thermal motion rapidly
breaks them apart

the surfaces of molecules A and D
match well and therefore can form
enough weak bonds to withstand
thermal jolting; they therefore
stay bound to each other

between the bound and free states (Figure 3–43B and C). Even a change of a few noncovalent bonds can have a striking effect on a binding interaction, as shown by the example in **Figure 3–44**. (Note that the equilibrium constant, as defined here is also known as the association or affinity constant, K_a.)

We have used the case of an antibody binding to its ligand to illustrate the effect of binding strength on the equilibrium state, but the same principles apply to any molecule and its ligand. Many proteins are enzymes, which, as we now discuss, first bind to their ligands and then catalyze the breakage or formation of covalent bonds in these molecules.

Figure 3–42 How noncovalent bonds mediate interactions between macromolecules.

Enzymes Are Powerful and Highly Specific Catalysts

Many proteins can perform their function simply by binding to another molecule. An actin molecule, for example, need only associate with other actin

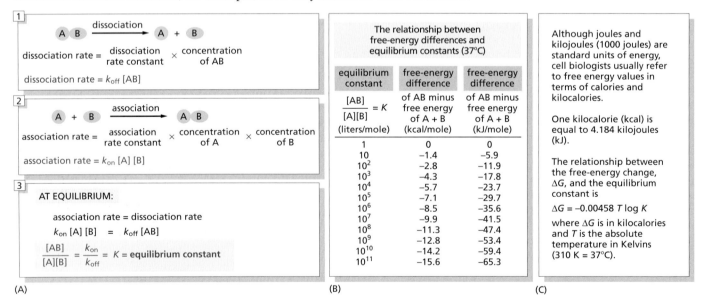

1

$$\text{A B} \xrightarrow{\text{dissociation}} \text{A} + \text{B}$$

$$\text{dissociation rate} = \frac{\text{dissociation}}{\text{rate constant}} \times \frac{\text{concentration}}{\text{of AB}}$$

$$\text{dissociation rate} = k_{\text{off}} [\text{AB}]$$

2

$$\text{A} + \text{B} \xrightarrow{\text{association}} \text{A B}$$

$$\text{association rate} = \frac{\text{association}}{\text{rate constant}} \times \frac{\text{concentration}}{\text{of A}} \times \frac{\text{concentration}}{\text{of B}}$$

$$\text{association rate} = k_{\text{on}} [\text{A}] [\text{B}]$$

3

AT EQUILIBRIUM:

association rate = dissociation rate

$$k_{\text{on}} [\text{A}] [\text{B}] = k_{\text{off}} [\text{AB}]$$

$$\frac{[\text{AB}]}{[\text{A}][\text{B}]} = \frac{k_{\text{on}}}{k_{\text{off}}} = K = \text{equilibrium constant}$$

(A)

The relationship between free-energy differences and equilibrium constants (37°C)		
equilibrium constant	free-energy difference	free-energy difference
$\dfrac{[\text{AB}]}{[\text{A}][\text{B}]} = K$	of AB minus free energy	of AB minus free energy
(liters/mole)	of A + B (kcal/mole)	of A + B (kJ/mole)
1	0	0
10	−1.4	−5.9
10^2	−2.8	−11.9
10^3	−4.3	−17.8
10^4	−5.7	−23.7
10^5	−7.1	−29.7
10^6	−8.5	−35.6
10^7	−9.9	−41.5
10^8	−11.3	−47.4
10^9	−12.8	−53.4
10^{10}	−14.2	−59.4
10^{11}	−15.6	−65.3

(B)

Although joules and kilojoules (1000 joules) are standard units of energy, cell biologists usually refer to free energy values in terms of calories and kilocalories.

One kilocalorie (kcal) is equal to 4.184 kilojoules (kJ).

The relationship between the free-energy change, ΔG, and the equilibrium constant is

$$\Delta G = -0.00458\ T \log K$$

where ΔG is in kilocalories and T is the absolute temperature in Kelvins (310 K = 37°C).

(C)

Figure 3–43 Relating binding energies to the equilibrium constant for an association reaction. (A) The equilibrium between molecules A and B and the complex AB is maintained by a balance between the two opposing reactions shown in panels 1 and 2. Molecules A and B must collide if they are to react, and the association rate is therefore proportional to the product of their individual concentrations [A] × [B]. (Square brackets indicate concentration.) As shown in panel 3, the ratio of the rate constants for the association and the dissociation reactions is equal to the equilibrium constant (K) for the reaction. (B) The equilibrium constant in panel 3 is that for the reaction A + B ⇌ AB, and the larger its value, the stronger the binding between A and B. Note that for every 1.41 kcal/mole (5.91 kJ/mole) decrease in free energy the equilibrium constant increases by a factor of 10 at 37°C.

The equilibrium constant here has units of liters/mole: for simple binding interactions it is also called the *affinity constant* or *association constant*, denoted K_a. The reciprocal of K_a is called the dissociation constant, K_d (in units of moles/liter).

molecules to form a filament. There are other proteins, however, for which ligand binding is only a necessary first step in their function. This is the case for the large and very important class of proteins called **enzymes**. As described in Chapter 2, enzymes are remarkable molecules that determine all the chemical transformations that make and break covalent bonds in cells. They bind to one or more ligands, called **substrates**, and convert them into one or more chemically modified *products*, doing this over and over again with amazing rapidity. Enzymes speed up reactions, often by a factor of a million or more, without themselves being changed—that is, they act as **catalysts** that permit cells to make or break covalent bonds in a controlled way. It is the catalysis of organized sets of chemical reactions by enzymes that creates and maintains the cell, making life possible.

We can group enzymes into functional classes that perform similar chemical reactions (**Table 3–1**). Each type of enzyme within such a class is highly specific, catalyzing only a single type of reaction. Thus, *hexokinase* adds a phosphate group to D-glucose but ignores its optical isomer L-glucose; the blood-clotting enzyme *thrombin* cuts one type of blood protein between a particular arginine and its adjacent glycine and nowhere else, and so on. As discussed in detail in Chapter 2, enzymes work in teams, with the product of one enzyme becoming the substrate for the next. The result is an elaborate network of metabolic pathways that provides the cell with energy and generates the many large and small molecules that the cell needs (see Figure 2–35).

Substrate Binding Is the First Step in Enzyme Catalysis

For a protein that catalyzes a chemical reaction (an enzyme), the binding of each substrate molecule to the protein is an essential prelude. In the simplest case, if we denote the enzyme by E, the substrate by S, and the product by P, the basic reaction path is $E + S \rightarrow ES \rightarrow EP \rightarrow E + P$. From this reaction path, we see that there is a limit to the amount of substrate that a single enzyme molecule can process in a given time. An increase in the concentration of substrate also increases the rate at which product is formed, up to a maximum value (**Figure 3–45**). At that point the enzyme molecule is saturated with substrate, and the rate of reaction (V_{max}) depends only on how rapidly the enzyme can process the substrate molecule. This maximum rate divided by the enzyme concentration is

Consider 1000 molecules of A and 1000 molecules of B in a eucaryotic cell. The concentration of both will be about 10^{-9} M.

If the equilibrium constant (K) for $A + B \rightleftharpoons AB$ is 10^{10}, then one can calculate that at equilibrium there will be

270	270	730
A molecules	B molecules	AB molecules

If the equilibrium constant is a little weaker at 10^8, which represents a loss of 2.8 kcal/mole of binding energy from the example above, or 2–3 fewer hydrogen bonds, then there will be

915	915	85
A molecules	B molecules	AB molecules

Figure 3–44 Small changes in the number of weak bonds can have drastic effects on a binding interaction. This example illustrates the dramatic effect of the presence or absence of a few weak noncovalent bonds in a biological context.

Table 3–1 Some Common Types of Enzymes

ENZYME	REACTION CATALYZED
Hydrolases	general term for enzymes that catalyze a hydrolytic cleavage reaction; *nucleases* and *proteases* are more specific names for subclasses of these enzymes.
Nucleases	break down nucleic acids by hydrolyzing bonds between nucleotides.
Proteases	break down proteins by hydrolyzing bonds between amino acids.
Synthases	synthesize molecules in anabolic reactions by condensing two smaller molecules together.
Isomerases	catalyze the rearrangement of bonds within a single molecule.
Polymerases	catalyze polymerization reactions such as the synthesis of DNA and RNA.
Kinases	catalyze the addition of phosphate groups to molecules. Protein kinases are an important group of kinases that attach phosphate groups to proteins.
Phosphatases	catalyze the hydrolytic removal of a phosphate group from a molecule.
Oxido-Reductases	general name for enzymes that catalyze reactions in which one molecule is oxidized while the other is reduced. Enzymes of this type are often more specifically named either *oxidases*, *reductases*, or *dehydrogenases*.
ATPases	hydrolyze ATP. Many proteins with a wide range of roles have an energy-harnessing ATPase activity as part of their function, for example, motor proteins such as *myosin* and membrane transport proteins such as the *sodium–potassium pump*.

Enzyme names typically end in "-ase," with the exception of some enzymes, such as pepsin, trypsin, thrombin and lysozyme that were discovered and named before the convention became generally accepted at the end of the nineteenth century. The common name of an enzyme usually indicates the substrate and the nature of the reaction catalyzed. For example, citrate synthase catalyzes the synthesis of citrate by a reaction between acetyl CoA and oxaloacetate.

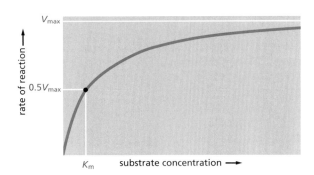

Figure 3–45 **Enzyme kinetics.** The rate of an enzyme reaction *(V)* increases as the substrate concentration increases until a maximum value *(V*~max~) is reached. At this point all substrate-binding sites on the enzyme molecules are fully occupied, and the rate of reaction is limited by the rate of the catalytic process on the enzyme surface. For most enzymes, the concentration of substrate *(K*~m~) at which the reaction rate is half-maximal (*black dot*) is a measure of how tightly the substrate is bound, with a large value of K_m corresponding to weak binding.

called the *turnover number*. The turnover number is often about 1000 substrate molecules processed per second per enzyme molecule, although turnover numbers between 1 and 10,000 are known.

The other kinetic parameter frequently used to characterize an enzyme is its K_m, the concentration of substrate that allows the reaction to proceed at one-half its maximum rate (0.5 V_{max}) (see Figure 3–45). A *low* K_m value means that the enzyme reaches its maximum catalytic rate at a *low concentration* of substrate and generally indicates that the enzyme binds to its substrate very tightly, whereas a *high* K_m value corresponds to weak binding. The methods used to characterize enzymes in this way are explained in **Panel 3–3** (pp. 162–163).

Enzymes Speed Reactions by Selectively Stabilizing Transition States

Enzymes achieve extremely high rates of chemical reaction—rates that are far higher than for any synthetic catalysts. There are several reasons for this efficiency. First, the enzyme increases the local concentration of substrate molecules at the catalytic site and holds all the appropriate atoms in the correct orientation for the reaction that is to follow. More importantly, however, some of the binding energy contributes directly to the catalysis. Substrate molecules must pass through a series of intermediate states of altered geometry and electron distribution before they form the ultimate products of the reaction. The free energy required to attain the most unstable **transition state** is called the *activation energy* for the reaction, and it is the major determinant of the reaction rate. Enzymes have a much higher affinity for the transition state of the substrate than they have for the stable form. Because this tight binding greatly lowers the energies of the transition state, the enzyme greatly accelerates a particular reaction by lowering the activation energy that is required (**Figure 3–46**).

By intentionally producing antibodies that act like enzymes, we can demonstrate that stabilizing a transition state can greatly increase a reaction rate. Consider, for example, the hydrolysis of an amide bond, which is similar to the peptide bond that joins two adjacent amino acids in a protein. In an aqueous solution, an amide bond hydrolyzes very slowly by the mechanism shown in **Figure 3–47**A. In the central intermediate, or transition state, the carbonyl carbon is bonded to four atoms arranged at the corners of a tetrahedron. By generating monoclonal antibodies that bind tightly to a stable analog of this very unstable *tetrahedral intermediate*, one can obtain an antibody that functions like an enzyme (Figure 3–47B). Because this *catalytic antibody* binds to and stabilizes the tetrahedral intermediate, it increases the spontaneous rate of amide-bond hydrolysis more than 10,000-fold.

Enzymes Can Use Simultaneous Acid and Base Catalysis

Figure 3–48 compares the spontaneous reaction rates and the corresponding enzyme-catalyzed rates for five enzymes. Rate accelerations range from 10^9 to 10^{23}. Clearly, enzymes are much better catalysts than catalytic antibodies.

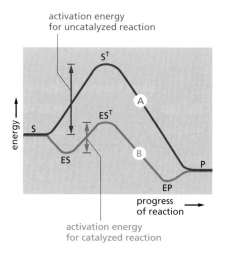

Figure 3–46 **Enzymatic acceleration of chemical reactions by decreasing the activation energy.** Often both the uncatalyzed reaction (A) and the enzyme-catalyzed reaction (B) can go through several transition states. It is the transition state with the highest energy (S^T and ES^T) that determines the activation energy and limits the rate of the reaction. (S = substrate; P = product of the reaction; ES = enzyme–substrate complex; EP = enzyme–product complex.)

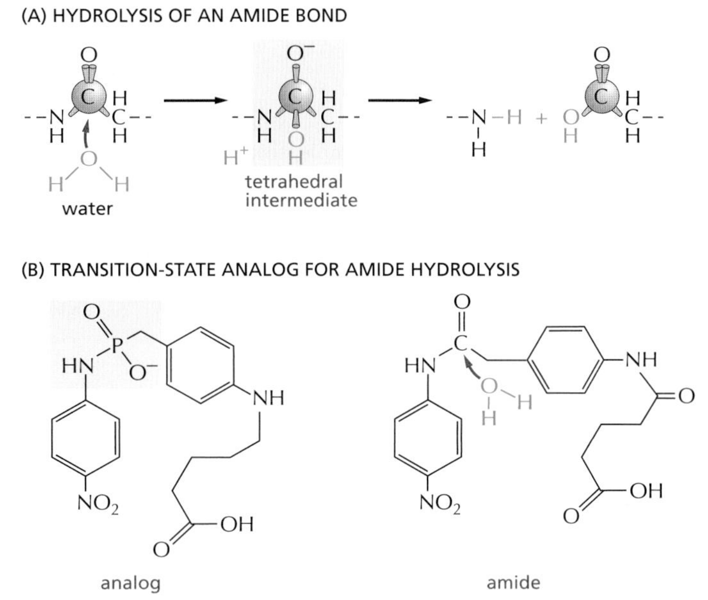

(A) HYDROLYSIS OF AN AMIDE BOND

water

tetrahedral intermediate

(B) TRANSITION-STATE ANALOG FOR AMIDE HYDROLYSIS

analog

amide

Figure 3–47 **Catalytic antibodies.** The stabilization of a transition state by an antibody creates an enzyme. (A) The reaction path for the hydrolysis of an amide bond goes through a tetrahedral intermediate, the high-energy transition state for the reaction. (B) The molecule on the left was covalently linked to a protein and used as an antigen to generate an antibody that binds tightly to the region of the molecule shown in *yellow*. Because this antibody also bound tightly to the transition state in (A), it was found to function as an enzyme that efficiently catalyzed the hydrolysis of the amide bond in the molecule on the right.

Enzymes not only bind tightly to a transition state, they also contain precisely positioned atoms that alter the electron distributions in those atoms that participate directly in the making and breaking of covalent bonds. Peptide bonds, for example, can be hydrolyzed in the absence of an enzyme by exposing a polypeptide to either a strong acid or a strong base, as illustrated in **Figure 3–49**. Enzymes are unique, however, in being able to use acid and base catalysis simultaneously, since the rigid framework of the protein binds the acidic and basic residues and prevents them from combining with each other (as they would do in solution) (Figure 3–49D).

The fit between an enzyme and its substrate needs to be precise. A small change introduced by genetic engineering in the active site of an enzyme can have a profound effect. Replacing a glutamic acid with an aspartic acid in one enzyme, for example, shifts the position of the catalytic carboxylate ion by only 1 Å (about the radius of a hydrogen atom); yet this is enough to decrease the activity of the enzyme a thousandfold.

Lysozyme Illustrates How an Enzyme Works <AGCA>

To demonstrate how enzymes catalyze chemical reactions, we examine an enzyme that acts as a natural antibiotic in egg white, saliva, tears, and other secretions. **Lysozyme** catalyzes the cutting of polysaccharide chains in the cell walls of bacteria. Because the bacterial cell is under pressure from osmotic forces, cutting even a small number of polysaccharide chains causes the cell wall to rupture and the cell to burst. Lysozyme is a relatively small and stable protein

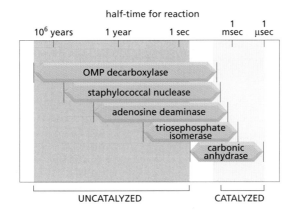

Figure 3–48 **The rate accelerations caused by five different enzymes.** (Adapted from A. Radzicka and R. Wolfenden, *Science* 267:90–93, 1995. With permission from AAAS.)

WHY ANALYZE THE KINETICS OF ENZYMES?

Enzymes are the most selective and powerful catalysts known. An understanding of their detailed mechanisms provides a critical tool for the discovery of new drugs, for the large-scale industrial synthesis of useful chemicals, and for appreciating the chemistry of cells and organisms. A detailed study of the rates of the chemical reactions that are catalyzed by a purified enzyme—more specifically how these rates change with changes in conditions such as the concentrations of substrates, products, inhibitors, and regulatory ligands—allows biochemists to figure out exactly how each enzyme works. For example, this is the way that the ATP-producing reactions of glycolysis, shown previously in Figure 2–72, were deciphered—allowing us to appreciate the rationale for this critical enzymatic pathway.

In this Panel, we introduce the important field of enzyme kinetics, which has been indispensable for deriving much of the detailed knowledge that we now have about cell chemistry.

STEADY-STATE ENZYME KINETICS

Many enzymes have only one substrate, which they bind and then process to produce products according to the scheme outlined in Figure 3–50A. In this case, the reaction is written as

$$E + S \underset{k_{-1}}{\overset{k_1}{\rightleftharpoons}} ES \overset{k_{cat}}{\longrightarrow} E + P$$

Here we have assumed that the reverse reaction, in which E + P recombine to form EP and then ES, occurs so rarely that we can ignore it. In this case, EP need not be represented, and we can express the rate of the reaction — known as its velocity, V, as

$$V = k_{cat} [ES]$$

where [ES] is the concentration of the enzyme–substrate complex, and k_{cat} is the turnover number, a rate constant that has a value equal to the number of substrate molecules processed per enzyme molecule each second.

But how does the value of [ES] relate to the concentrations that we know directly, which are the total concentration of the enzyme, $[E_o]$, and the concentration of the substrate, [S]? When enzyme and substrate are first mixed, the concentration [ES] will rise rapidly from zero to a so-called steady-state level, as illustrated below.

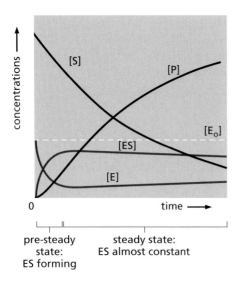

pre-steady state: ES forming   steady state: ES almost constant

At this steady state, [ES] is nearly constant, so that

rate of ES breakdown		rate of ES formation
$k_{-1} [ES] + k_{cat} [ES]$	$=$	$k_1 [E][S]$

or, since the concentration of the free enzyme, [E], is equal to $[E_o] - [ES]$,

$$[ES] = \left(\frac{k_1}{k_{-1} + k_{cat}} \right) [E][S] = \left(\frac{k_1}{k_{-1} + k_{cat}} \right) \left([E_o] - [ES] \right) [S]$$

Rearranging, and defining the constant K_m as

$$\frac{k_{-1} + k_{cat}}{k_1}$$

we get

$$[ES] = \frac{[E_o][S]}{K_m + [S]}$$

or, remembering that $V = k_{cat} [ES]$, we obtain the famous Michaelis–Menten equation

$$V = \frac{k_{cat} [E_o][S]}{K_m + [S]}$$

As [S] is increased to higher and higher levels, essentially all of the enzyme will be bound to substrate at steady state; at this point, a maximum rate of reaction, V_{max}, will be reached where $V = V_{max} = k_{cat} [E_o]$. Thus, it is convenient to rewrite the Michaelis–Menten equation as

$$V = \frac{V_{max} [S]}{K_m + [S]}$$

THE DOUBLE-RECIPROCAL PLOT

A typical plot of V versus [S] for an enzyme that follows Michaelis–Menten kinetics is shown below. From this plot, neither the value of V_{max} nor of K_m is immediately clear.

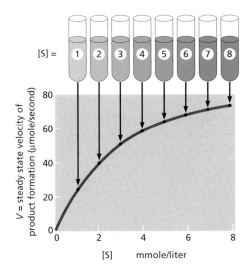

To obtain V_{max} and K_m from such data, a double-reciprocal plot is often used, in which the Michaelis–Menten equation has merely been rearranged, so that $1/V$ can be plotted versus $1/$[S].

$$1/V = \left(\frac{K_m}{V_{max}}\right)\left(\frac{1}{[S]}\right) + 1/V_{max}$$

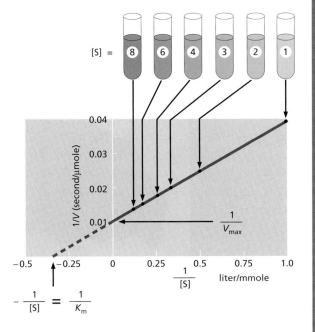

THE SIGNIFICANCE OF K_m, k_{cat}, and k_{cat}/K_m

As described in the text, K_m is an approximate measure of substrate affinity for the enzyme: it is numerically equal to the concentration of [S] at $V = 0.5\ V_{max}$. In general, a lower value of K_m means tighter substrate binding. In fact, for those cases where k_{cat} is much smaller than k_{-1}, the K_m will be equal to K_d, the dissociation constant for substrate binding to the enzyme ($K_d = 1/K_a$).

We have seen that k_{cat} is the turnover number for the enzyme. At very low substrate concentrations, where [S] $\ll K_m$, most of the enzyme is free. Thus we can think of [E] = [E$_o$], so that the Michaelis–Menten equation becomes $V = k_{cat}/K_m$ [E][S]. Thus, the ratio k_{cat}/K_m is equivalent to the rate constant for the reaction between free enzyme and free substrate.

A comparison of k_{cat}/K_m for the same enzyme with different substrates, or for two enzymes with their different substrates, is widely used as a measure of enzyme effectiveness.

For simplicity, in this Panel we have discussed enzymes that have only one substrate, such as the lysozyme enzyme described in the text (see p. 164). Most enzymes have two substrates, one of which is often an active carrier molecule—such as NADH or ATP.

A similar, but more complex, analysis is used to determine the kinetics of such enzymes—allowing the order of substrate binding and the presence of covalent intermediates along the pathway to be revealed.

SOME ENZYMES ARE DIFFUSION LIMITED

The values of k_{cat}, K_m, and k_{cat}/K_m for some selected enzymes are given below:

enzyme	substrate	k_{cat} (sec^{-1})	K_m (M)	k_{at}/K_m (sec^{-1}M^{-1})
acetylcholinesterase	acetylcholine	1.4×10^4	9×10^{-5}	1.6×10^8
catalase	H_2O_2	4×10^7	1	4×10^7
fumarase	fumarate	8×10^2	5×10^{-6}	1.6×10^8

Because an enzyme and its substrate must collide before they can react, k_{cat}/K_m has a maximum possible value that is limited by collision rates. If every collision forms an enzyme–substrate complex, one can calculate from diffusion theory that k_{cat}/K_m will be between 10^8 and 10^9 sec^{-1}M^{-1}, in the case where all subsequent steps proceed immediately. Thus, it is claimed that enzymes like acetylcholinesterase and fumarase are "perfect enzymes," each enzyme having evolved to the point where nearly every collision with its substrate converts the substrate to a product.

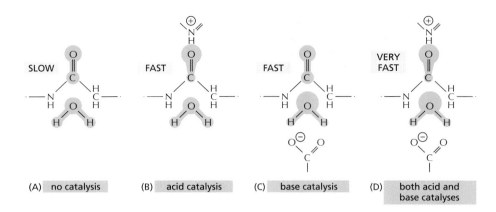

(A) no catalysis   (B) acid catalysis   (C) base catalysis   (D) both acid and base catalyses

Figure 3–49 Acid catalysis and base catalysis. (A) The start of the uncatalyzed reaction shown in Figure 3–47A, with *blue* indicating electron distribution in the water and carbonyl bonds. (B) An acid likes to donate a proton (H⁺) to other atoms. By pairing with the carbonyl oxygen, an acid causes electrons to move away from the carbonyl carbon, making this atom much more attractive to the electronegative oxygen of an attacking water molecule. (C) A base likes to take up H⁺. By pairing with a hydrogen of the attacking water molecule, a base causes electrons to move toward the water oxygen, making it a better attacking group for the carbonyl carbon. (D) By having appropriately positioned atoms on its surface, an enzyme can perform both acid catalysis and base catalysis at the same time.

that can be easily isolated in large quantities. For these reasons, it has been intensively studied, and it was the first enzyme to have its structure worked out in atomic detail by x-ray crystallography.

The reaction that lysozyme catalyzes is a hydrolysis: it adds a molecule of water to a single bond between two adjacent sugar groups in the polysaccharide chain, thereby causing the bond to break (see Figure 2–19). The reaction is energetically favorable because the free energy of the severed polysaccharide chain is lower than the free energy of the intact chain. However, the pure polysaccharide can remain for years in water without being hydrolyzed to any detectable degree. This is because there is an energy barrier to the reaction, as discussed in Chapter 2 (see Figure 2–46). A colliding water molecule can break a bond linking two sugars only if the polysaccharide molecule is distorted into a particular shape—the transition state—in which the atoms around the bond have an altered geometry and electron distribution. Because of this distortion, random collisions must supply a very large activation energy for the reaction to take place. In an aqueous solution at room temperature, the energy of collisions almost never exceeds the activation energy. Consequently, hydrolysis occurs extremely slowly, if at all.

This situation changes drastically when the polysaccharide binds to lysozyme. The active site of lysozyme, because its substrate is a polymer, is a long groove that holds six linked sugars at the same time. As soon as the polysaccharide binds to form an enzyme–substrate complex, the enzyme cuts the polysaccharide by adding a water molecule across one of its sugar–sugar bonds. The product chains are then quickly released, freeing the enzyme for further cycles of reaction (**Figure 3–50**).

The chemistry of the binding of lysozyme to its substrate is the same as that for antibody binding to its antigen—the formation of multiple noncovalent

Figure 3–50 The reaction catalyzed by lysozyme. (A) The enzyme lysozyme (E) catalyzes the cutting of a polysaccharide chain, which is its substrate (S). The enzyme first binds to the chain to form an enzyme–substrate complex (ES) and then catalyzes the cleavage of a specific covalent bond in the backbone of the polysaccharide, forming an enzyme–product complex (EP) that rapidly dissociates. Release of the severed chain (the products P) leaves the enzyme free to act on another substrate molecule. (B) A space-filling model of the lysozyme molecule bound to a short length of polysaccharide chain before cleavage. (B, courtesy of Richard J. Feldmann.)

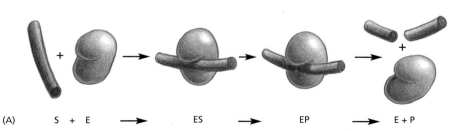

(A)   S + E   →   ES   →   EP   →   E + P

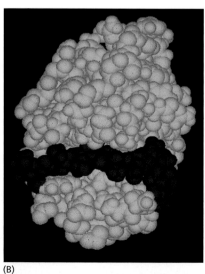

(B)

bonds. However, lysozyme holds its polysaccharide substrate in a particular way, so that it distorts one of the two sugars in the bond to be broken from its normal, most stable conformation. The bond to be broken is also held close to two amino acids with acidic side chains (a glutamic acid and an aspartic acid) within the active site.

Conditions are thereby created in the microenvironment of the lysozyme active site that greatly reduce the activation energy necessary for the hydrolysis to take place. **Figure 3–51** shows three central steps in this enzymatically catalyzed reaction.

1. The enzyme stresses its bound substrate, so that the shape of one sugar more closely resembles the shape of high-energy transition states formed during the reaction.
2. The negatively charged aspartic acid reacts with the C1 carbon atom on the distorted sugar, and the glutamic acid donates its proton to the oxygen that links this sugar to its neighbor. This breaks the sugar–sugar bond and leaves the aspartic acid side chain covalently linked to the site of bond cleavage.
3. Aided by the negatively charged glutamic acid, a water molecule reacts with the C1 carbon atom, displacing the aspartic acid side chain and completing the process of hydrolysis.

The overall chemical reaction, from the initial binding of the polysaccharide on the surface of the enzyme through the final release of the severed chains, occurs many millions of times faster than it would in the absence of enzyme.

Other enzymes use similar mechanisms to lower activation energies and speed up the reactions they catalyze. In reactions involving two or more reactants, the active site also acts like a template, or mold, that brings the substrates together in the proper orientation for a reaction to occur between them (**Figure**

Figure 3–51 Events at the active site of lysozyme. <TGGT> The top left and top right drawings show the free substrate and the free products, respectively, whereas the other three drawings show the sequential events at the enzyme active site. Note the change in the conformation of sugar D in the enzyme–substrate complex; this shape change stabilizes the oxocarbenium ion-like transition states required for formation and hydrolysis of the covalent intermediate shown in the middle panel. It is also possible that a carbonium ion intermediate forms in step 2, as the covalent intermediate shown in the middle panel has been detected only with a synthetic substrate. (See D.J. Vocadlo et al., *Nature* 412:835–838, 2001.)

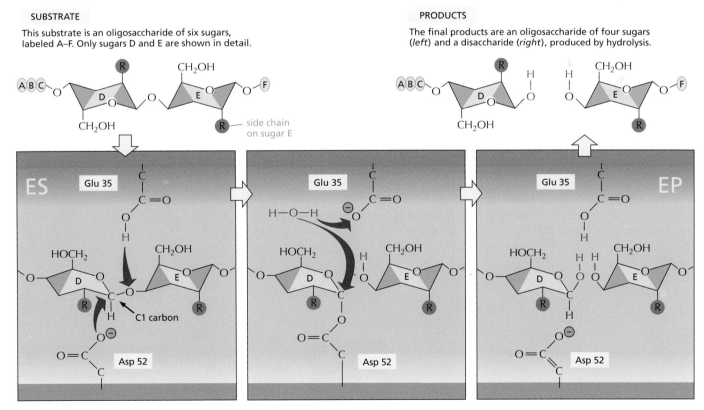

SUBSTRATE

This substrate is an oligosaccharide of six sugars, labeled A–F. Only sugars D and E are shown in detail.

PRODUCTS

The final products are an oligosaccharide of four sugars (*left*) and a disaccharide (*right*), produced by hydrolysis.

In the enzyme–substrate complex (ES), the enzyme forces sugar D into a strained conformation, with Glu 35 positioned to serve as an acid that attacks the adjacent sugar–sugar bond by donating a proton (H+) to sugar E, and Asp 52 poised to attack the C1 carbon atom.

The Asp 52 has formed a covalent bond between the enzyme and the C1 carbon atom of sugar D. The Glu 35 then polarizes a water molecule (*red*), so that its oxygen can readily attack the C1 carbon atom and displace Asp 52.

The reaction of the water molecule (*red*) completes the hydrolysis and returns the enzyme to its initial state, forming the final enzyme–product complex (EP).

(A) enzyme binds to two substrate molecules and orients them precisely to encourage a reaction to occur between them

(B) binding of substrate to enzyme rearranges electrons in the substrate, creating partial negative and positive charges that favor a reaction

(C) enzyme strains the bound substrate molecule, forcing it toward a transition state to favor a reaction

Figure 3–52 Some general strategies of enzyme catalysis. (A) Holding substrates together in a precise alignment. (B) Charge stabilization of reaction intermediates. (C) Applying forces that distort bonds in the substrate to increase the rate of a particular reaction.

3–52A). As we saw for lysozyme, the active site of an enzyme contains precisely positioned atoms that speed up a reaction by using charged groups to alter the distribution of electrons in the substrates (Figure 3–52B). In addition, when a substrate binds to an enzyme, bonds in the substrate often bend, changing the substrate shape. These changes, along with mechanical forces, drive a substrate toward a particular transition state (Figure 3–52C). Finally, like lysozyme, many enzymes participate intimately in the reaction by briefly forming a covalent bond between the substrate and a side chain of the enzyme. Subsequent steps in the reaction restore the side chain to its original state, so that the enzyme remains unchanged after the reaction (see also Figure 2–72).

Tightly Bound Small Molecules Add Extra Functions to Proteins

Although we have emphasized the versatility of proteins as chains of amino acids that perform different functions, there are many instances in which the amino acids by themselves are not enough. Just as humans employ tools to enhance and extend the capabilities of their hands, proteins often use small nonprotein molecules to perform functions that would be difficult or impossible to do with amino acids alone. Thus, the signal receptor protein *rhodopsin*, which is made by the photoreceptor cells in the retina, detects light by means of a small molecule, *retinal*, embedded in the protein (**Figure 3–53A**). Retinal changes its shape when it absorbs a photon of light, and this change causes the protein to trigger a cascade of enzymatic reactions that eventually lead to an electrical signal being carried to the brain.

Another example of a protein that contains a nonprotein portion is hemoglobin (see Figure 3–22). A molecule of hemoglobin carries four *heme* groups, ring-shaped molecules each with a single central iron atom (Figure 3–53B). Heme gives hemoglobin (and blood) its red color. By binding reversibly to oxygen gas through its iron atom, heme enables hemoglobin to pick up oxygen in the lungs and release it in the tissues.

Sometimes these small molecules are attached covalently and permanently to their protein, thereby becoming an integral part of the protein molecule itself. We shall see in Chapter 10 that proteins are often anchored to cell membranes through covalently attached lipid molecules. And membrane proteins exposed

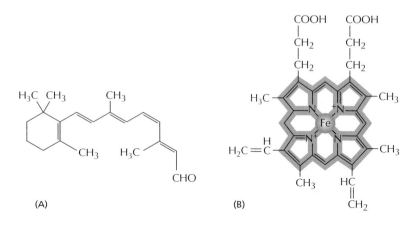

(A)   (B)

Figure 3–53 Retinal and heme. (A) The structure of retinal, the light-sensitive molecule attached to rhodopsin in the eye. (B) The structure of a heme group. The carbon-containing heme ring is *red* and the iron atom at its center is *orange*. A heme group is tightly bound to each of the four polypeptide chains in hemoglobin, the oxygen-carrying protein whose structure is shown in Figure 3–22.

Table 3–2 Many Vitamins Provide Critical Coenzymes for Human Cells

VITAMIN	COENZYME	ENZYME-CATALYZED REACTIONS REQUIRING THESE COENZYMES
Thiamine (vitamin B_1)	thiamine pyrophosphate	activation and transfer of aldehydes
Riboflavin (vitamin B_2)	FADH	oxidation–reduction
Niacin	NADH, NADPH	oxidation–reduction
Pantothenic acid	coenzyme A	acyl group activation and transfer
Pyridoxine	pyridoxal phosphate	amino acid activation; also glycogen phosphorylase
Biotin	biotin	CO_2 activation and transfer
Lipoic acid	lipoamide	acyl group activation; oxidation–reduction
Folic acid	tetrahydrofolate	activation and transfer of single carbon groups
Vitamin B_{12}	cobalamin coenzymes	isomerization and methyl group transfers

on the surface of the cell, as well as proteins secreted outside the cell, are often modified by the covalent addition of sugars and oligosaccharides.

Enzymes frequently have a small molecule or metal atom tightly associated with their active site that assists with their catalytic function. *Carboxypeptidase*, for example, an enzyme that cuts polypeptide chains, carries a tightly bound zinc ion in its active site. During the cleavage of a peptide bond by carboxypeptidase, the zinc ion forms a transient bond with one of the substrate atoms, thereby assisting the hydrolysis reaction. In other enzymes, a small organic molecule serves a similar purpose. Such organic molecules are often referred to as **coenzymes**. An example is *biotin*, which is found in enzymes that transfer a carboxylate group ($-COO^-$) from one molecule to another (see Figure 2–63). Biotin participates in these reactions by forming a transient covalent bond to the $-COO^-$ group to be transferred, being better suited to this function than any of the amino acids used to make proteins. Because it cannot be synthesized by humans, and must therefore be supplied in small quantities in our diet, biotin is a *vitamin*. Many other coenzymes are produced from vitamins (**Table 3–2**). Vitamins are also needed to make other types of small molecules that are essential components of our proteins; vitamin A, for example, is needed in the diet to make retinal, the light-sensitive part of rhodopsin.

Molecular Tunnels Channel Substrates in Enzymes with Multiple Catalytic Sites

Some of the chemical reactions catalyzed by enzymes in cells produce intermediates that are either very unstable or that could readily diffuse out of the cell through the plasma membrane if released into the cytosol. To preserve these intermediates, enzymes have evolved *molecular tunnels* that connect two or more active sites, allowing the intermediate to be rapidly processed to a final product—without ever leaving the enzyme.

Consider, for example, the enzyme carbamoyl phosphate synthetase, which uses ammonia derived from glutamine plus two molecules of ATP to convert bicarbonate (HCO_3^-) to carbamoyl phosphate—an important intermediate in several metabolic pathways (**Figure 3–54**). This enzyme contains three widely separated active sites that are connected to each other by a tunnel. The reaction starts at active site 2, located in the middle of the tunnel, where ATP is used to phosphorylate (add a phosphate group to) bicarbonate, forming carboxy phosphate. This event triggers the hydrolysis of glutamine to glutamic acid at active site 1, releasing ammonia into the tunnel. The ammonia immediately diffuses through the first half of the tunnel to active site 2, where it reacts with the carboxyphosphate to form carbamate. This unstable intermediate then diffuses through the second half of the tunnel to active site 3, where it is phosphorylated by ATP to the final product, carbamoyl phosphate.

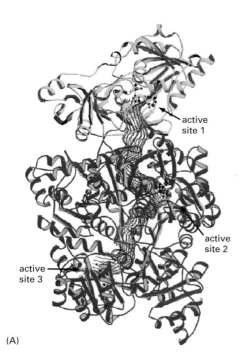

(A)

Figure 3–54 The tunneling of reaction intermediates in the enzyme carbamoyl phosphate synthetase. (A) Diagram of the structure of the enzyme, in which a *red* ribbon has been used to outline the tunnel on the inside of the protein connecting its three active sites. The small and large subunits of this dimeric enzyme are color coded *yellow* and *blue*, respectively. (B) The path of the reaction. As indicated, active site 1 produces ammonia, which diffuses through the tunnel to active site 2, where it combines with carboxy phosphate to form carbamate. This highly unstable intermediate then diffuses through the tunnel to active site 3, where it is phosphorylated by ATP to produce the final product, carbamoyl phosphate. (A, modified from F.M. Raushel, J.B. Thoden, and H.M. Holden, *Acc. Chem. Res.* 36:539–548, 2003. With permission from American Chemical Society.)

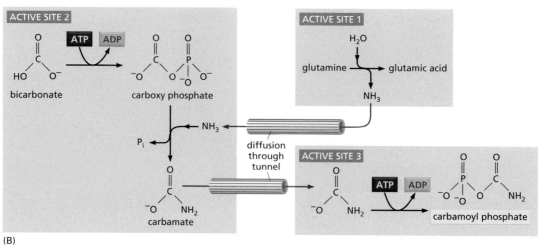

(B)

Several other well characterized enzymes contain similar molecular tunnels. Ammonia, a readily diffusable intermediate that might otherwise be lost from the cell, is the substrate most frequently channeled in the examples thus far known.

Multienzyme Complexes Help to Increase the Rate of Cell Metabolism

The efficiency of enzymes in accelerating chemical reactions is crucial to the maintenance of life. Cells, in effect, must race against the unavoidable processes of decay, which—if left unattended—cause macromolecules to run downhill toward greater and greater disorder. If the rates of desirable reactions were not greater than the rates of competing side reactions, a cell would soon die. We can get some idea of the rate at which cell metabolism proceeds by measuring the rate of ATP utilization. A typical mammalian cell "turns over" (i.e., hydrolyzes and restores by phosphorylation) its entire ATP pool once every 1 or 2 minutes. For each cell, this turnover represents the utilization of roughly 10^7 molecules of ATP per second (or, for the human body, about 1 gram of ATP every minute).

The rates of reactions in cells are rapid because enzyme catalysis is so effective. Many important enzymes have become so efficient that there is no possibility of further useful improvement. The factor that limits the reaction rate is no longer the enzyme's intrinsic speed of action; rather, it is the frequency with which the enzyme collides with its substrate. Such a reaction is said to be *diffusion-limited* (see Panel 3–3, p. 162–163).

If an enzyme-catalyzed reaction is diffusion-limited, its rate depends on the concentration of both the enzyme and its substrate. If a sequence of reactions is to occur extremely rapidly, each metabolic intermediate and enzyme involved must be present in high concentration. However, given the enormous number of different reactions performed by a cell, there are limits to the concentrations that can be achieved. In fact, most metabolites are present in micromolar (10^{-6} M) concentrations, and most enzyme concentrations are much lower. How is it possible, therefore, to maintain very fast metabolic rates?

The answer lies in the spatial organization of cell components. The cell can increase reaction rates without raising substrate concentrations by bringing the various enzymes involved in a reaction sequence together to form a large protein assembly known as a *multienzyme complex* (**Figure 3–55**). Because this allows the product of enzyme A to be passed directly to enzyme B, and so on, diffusion rates need not be limiting, even when the concentrations of the substrates in the cell as a whole are very low. It is perhaps not surprising, therefore, that such enzyme complexes are very common, and they are involved in nearly all aspects of metabolism—including the central genetic processes of DNA, RNA, and protein synthesis. In fact, few enzymes in eucaryotic cells diffuse freely in solution; instead, most seem to have evolved binding sites that concentrate them with other proteins of related function in particular regions of the cell, thereby increasing the rate and efficiency of the reactions that they catalyze.

Eucaryotic cells have yet another way of increasing the rate of metabolic reactions: using their intracellular membrane systems. These membranes can segregate particular substrates and the enzymes that act on them into the same membrane-enclosed compartment, such as the endoplasmic reticulum or the cell nucleus. If, for example, a compartment occupies a total of 10% of the volume of the cell, the concentration of reactants in that compartment may be increased by 10 times compared with a cell with the same number of enzymes and substrate molecules, but no compartmentalization. Reactions limited by the speed of diffusion can thereby be speeded up by a factor of 10.

The Cell Regulates the Catalytic Activities of its Enzymes

A living cell contains thousands of enzymes, many of which operate at the same time and in the same small volume of the cytosol. By their catalytic action, these enzymes generate a complex web of metabolic pathways, each composed of chains of chemical reactions in which the product of one enzyme becomes the substrate of the next. In this maze of pathways, there are many branch points (nodes) where different enzymes compete for the same substrate. The system is so complex (see Figure 2–88) that elaborate controls are required to regulate when and how rapidly each reaction occurs.

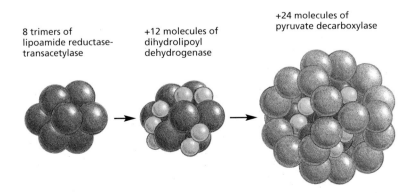

8 trimers of
lipoamide reductase-
transacetylase

+12 molecules of
dihydrolipoyl
dehydrogenase

+24 molecules of
pyruvate decarboxylase

Figure 3–55 The structure of pyruvate dehydrogenase. This enzyme complex catalyzes the conversion of pyruvate to acetyl CoA, as part of the pathway that oxidizes sugars to CO_2 and H_2O (see Figure 2–79). It is an example of a large multienzyme complex in which reaction intermediates are passed directly from one enzyme to another.

Regulation occurs at many levels. At one level, the cell controls how many molecules of each enzyme it makes by regulating the expression of the gene that encodes that enzyme (discussed in Chapter 7). The cell also controls enzymatic activities by confining sets of enzymes to particular subcellular compartments, enclosed by distinct membranes (discussed in Chapters 12 and 14). As will be discussed later in this chapter, enzymes are frequently covalently modified to control their activity. The rate of protein destruction by targeted proteolysis represents yet another important regulatory mechanism (see p. 395). But the most general process that adjusts reaction rates operates through a direct, reversible change in the activity of an enzyme in response to the specific small molecules that it encounters.

The most common type of control occurs when a molecule other than one of the substrates binds to an enzyme at a special regulatory site outside the active site, thereby altering the rate at which the enzyme converts its substrates to products. For example, in **feedback inhibition** a product produced late in a reaction pathway inhibits an enzyme that acts earlier in the pathway. Thus, whenever large quantities of the final product begin to accumulate, this product binds to the enzyme and slows down its catalytic action, thereby limiting the further entry of substrates into that reaction pathway (**Figure 3–56**). Where pathways branch or intersect, there are usually multiple points of control by different final products, each of which works to regulate its own synthesis (**Figure 3–57**). Feedback inhibition can work almost instantaneously, and it is rapidly reversed when the level of the product falls.

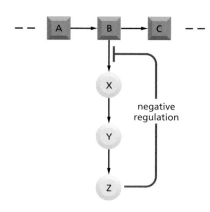

Figure 3–56 Feedback inhibition of a single biosynthetic pathway. The end-product Z inhibits the first enzyme that is unique to its synthesis and thereby controls its own level in the cell. This is an example of negative regulation.

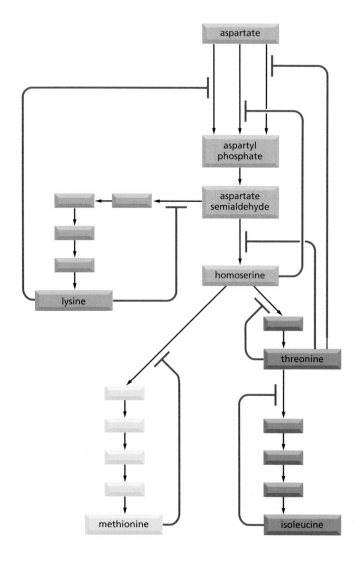

Figure 3–57 Multiple feedback inhibition. In this example, which shows the biosynthetic pathways for four different amino acids in bacteria, the *red arrows* indicate positions at which products feed back to inhibit enzymes. Each amino acid controls the first enzyme specific to its own synthesis, thereby controlling its own levels and avoiding a wasteful, or even dangerous, buildup of intermediates. The products can also separately inhibit the initial set of reactions common to all the syntheses; in this case, three different enzymes catalyze the initial reaction, each inhibited by a different product.

Feedback inhibition is *negative regulation:* it prevents an enzyme from acting. Enzymes can also be subject to *positive regulation,* in which a regulatory molecule stimulates the enzyme's activity rather than shutting the enzyme down. Positive regulation occurs when a product in one branch of the metabolic network stimulates the activity of an enzyme in another pathway. As one example, the accumulation of ADP activates several enzymes involved in the oxidation of sugar molecules, thereby stimulating the cell to convert more ADP to ATP.

Allosteric Enzymes Have Two or More Binding Sites That Interact

A striking feature of both positive and negative feedback regulation is that the regulatory molecule often has a shape totally different from the shape of the substrate of the enzyme. This is why the effect on a protein is termed *allostery* (from the Greek words *allos,* meaning "other," and *stereos,* meaning "solid" or "three-dimensional"). As biologists learned more about feedback regulation, they recognized that the enzymes involved must have at least two different binding sites on their surface—an **active site** that recognizes the substrates, and a **regulatory site** that recognizes a regulatory molecule. These two sites must somehow communicate so that the catalytic events at the active site can be influenced by the binding of the regulatory molecule at its separate site on the protein's surface.

The interaction between separated sites on a protein molecule is now known to depend on a *conformational change* in the protein: binding at one of the sites causes a shift from one folded shape to a slightly different folded shape. During feedback inhibition, for example, the binding of an inhibitor at one site on the protein causes the protein to shift to a conformation that incapacitates its active site, located elsewhere in the protein.

It is thought that most protein molecules are allosteric. They can adopt two or more slightly different conformations, and a shift from one to another caused by the binding of a ligand can alter their activity. This is true not only for enzymes but also for many other proteins, including receptors, structural proteins, and motor proteins. In all instances of allosteric regulation, each conformation of the protein has somewhat different surface contours, and the protein's binding sites for ligands are altered when the protein changes shape. Moreover as we discuss next, each ligand will stabilize the conformation that it binds to most strongly, and thus—at high enough concentrations—will tend to "switch" the protein toward the conformation that the ligand prefers.

Two Ligands Whose Binding Sites Are Coupled Must Reciprocally Affect Each Other's Binding

The effects of ligand binding on a protein follow from a fundamental chemical principle known as **linkage**. Suppose, for example, that a protein that binds glucose also binds another molecule, X, at a distant site on the protein's surface. If the binding site for X changes shape as part of the conformational change induced by glucose binding, the binding sites for X and for glucose are said to be *coupled.* Whenever two ligands prefer to bind to the *same* conformation of an allosteric protein, it follows from basic thermodynamic principles that each ligand must increase the affinity of the protein for the other. Thus, if the shift of the protein in **Figure 3–58** to the closed conformation that binds glucose best also causes the binding site for X to fit X better, then the protein will bind glucose more tightly when X is present than when X is absent.

Conversely, linkage operates in a negative way if two ligands prefer to bind to *different* conformations of the same protein. In this case, the binding of the first ligand discourages the binding of the second ligand. Thus, if a shape change caused by glucose binding decreases the affinity of a protein for molecule X, the binding of X must also decrease the protein's affinity for glucose (**Figure 3–59**). The linkage relationship is quantitatively reciprocal, so that, for example, if glucose has a very large effect on the binding of X, X has a very large effect on the binding of glucose.

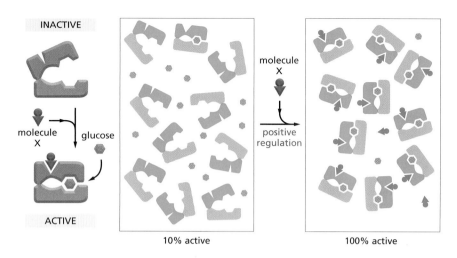

Figure 3–58 Positive regulation caused by conformational coupling between two distant binding sites. In this example, both glucose and molecule X bind best to the *closed* conformation of a protein with two domains. Because both glucose and molecule X drive the protein toward its closed conformation, each ligand helps the other to bind. Glucose and molecule X are therefore said to bind *cooperatively* to the protein.

The relationships shown in Figures 3–58 and 3–59 apply to all proteins, and they underlie all of cell biology. They seem so obvious in retrospect that we now take it for granted. But the discovery of linkage in studies of a few enzymes in the 1950s, followed by an extensive analysis of allosteric mechanisms in proteins in the early 1960s, had a revolutionary effect on our understanding of biology. Since molecule X in these examples binds at a site on the enzyme that is distinct from the site where catalysis occurs, it need have no chemical relationship to glucose or to any other ligand that binds at the active site. Moreover, as we have just seen, for enzymes that are regulated in this way, molecule X can either turn the enzyme on (positive regulation) or turn it off (negative regulation). By such a mechanism, **allosteric proteins** serve as general switches that, in principle, allow one molecule in a cell to affect the fate of any other.

Symmetric Protein Assemblies Produce Cooperative Allosteric Transitions

A single-subunit enzyme that is regulated by negative feedback can at most decrease from 90% to about 10% activity in response to a 100-fold increase in the concentration of an inhibitory ligand that it binds (**Figure 3–60**, *red line*). Responses of this type are apparently not sharp enough for optimal cell regulation, and most enzymes that are turned on or off by ligand binding consist of symmetric assemblies of identical subunits. With this arrangement, the binding of a molecule of ligand to a single site on one subunit can promote an allosteric change in the entire assembly that helps the neighboring subunits bind the same ligand. As a result, a *cooperative allosteric transition* occurs (Figure 3–60, *blue line)*, allowing a relatively small change in ligand concentration in the cell to switch the whole assembly from an almost fully active to an almost fully inactive conformation (or vice versa).

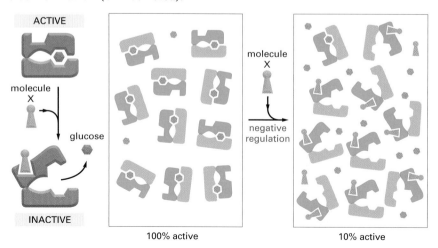

Figure 3–59 Negative regulation caused by conformational coupling between two distant binding sites. The scheme here resembles that in the previous figure, but here molecule X prefers the *open* conformation, while glucose prefers the *closed* conformation. Because glucose and molecule X drive the protein toward opposite conformations (closed and open, respectively), the presence of either ligand interferes with the binding of the other.

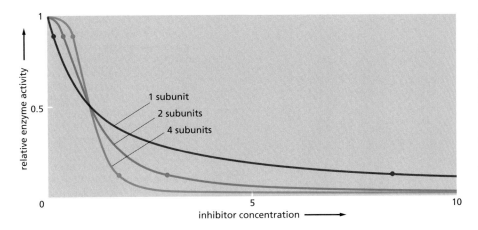

Figure 3–60 **Enzyme activity versus the concentration of inhibitory ligand for single-subunit and multisubunit allosteric enzymes.** For an enzyme with a single subunit *(red line)*, a drop from 90% enzyme activity to 10% activity (indicated by the two dots on the curve) requires a 100-fold increase in the concentration of inhibitor. The enzyme activity is calculated from the simple equilibrium relationship $K = [IP]/[I][P]$, where P is active protein, I is inhibitor, and IP is the inactive protein bound to inhibitor. An identical curve applies to any simple binding interaction between two molecules, A and B. In contrast, a multisubunit allosteric enzyme can respond in a switchlike manner to a change in ligand concentration: the steep response is caused by a cooperative binding of the ligand molecules, as explained in Figure 3–61. Here, the *green line* represents the idealized result expected for the cooperative binding of two inhibitory ligand molecules to an allosteric enzyme with two subunits, and the *blue line* shows the idealized response of an enzyme with four subunits. As indicated by the two dots on each of these curves, the more complex enzymes drop from 90% to 10% activity over a much narrower range of inhibitor concentration than does the enzyme composed of a single subunit.

The principles involved in a cooperative "all-or-none" transition are the same for all proteins, whether or not they are enzymes. But they are perhaps easiest to visualize for an enzyme that forms a symmetric dimer. In the example shown in **Figure 3–61**, the first molecule of an inhibitory ligand binds with great difficulty since its binding disrupts an energetically favorable interaction between the two identical monomers in the dimer. A second molecule of inhibitory ligand now binds more easily, however, because its binding restores the energetically favorable monomer–monomer contacts of a symmetric dimer (this also completely inactivates the enzyme).

As an alternative to this *induced fit* model for a cooperative allosteric transition, we can view such a symmetrical enzyme as having only two possible conformations, corresponding to the "enzyme on" and "enzyme off" structures in Figure 3–61. In this view, ligand binding perturbs an all-or-none equilibrium between these two states, thereby changing the proportion of active molecules. Both models represent true and useful concepts; it is the second model that we shall describe next.

The Allosteric Transition in Aspartate Transcarbamoylase Is Understood in Atomic Detail

One enzyme used in the early studies of allosteric regulation was aspartate transcarbamoylase from *E. coli*. It catalyzes the important reaction that begins the synthesis of the pyrimidine ring of C, U, and T nucleotides: carbamoyl phosphate + aspartate → *N*-carbamoylaspartate. One of the final products of this pathway, cytosine triphosphate (CTP), binds to the enzyme to turn it off whenever CTP is plentiful.

Aspartate transcarbamoylase is a large complex of six regulatory and six catalytic subunits. The catalytic subunits form two trimers, each arranged in the shape of an equilateral triangle; the two trimers face each other and are held

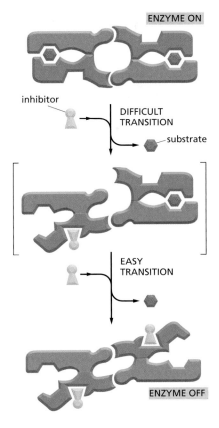

Figure 3–61 **A cooperative allosteric transition in an enzyme composed of two identical subunits.** This diagram illustrates how the conformation of one subunit can influence that of its neighbor. The binding of a single molecule of an inhibitory ligand *(yellow)* to one subunit of the enzyme occurs with difficulty because it changes the conformation of this subunit and thereby disrupts the symmetry of the enzyme. Once this conformational change has occurred, however, the energy gained by restoring the symmetric pairing interaction between the two subunits makes it especially easy for the second subunit to bind the inhibitory ligand and undergo the same conformational change. Because the binding of the first molecule of ligand increases the affinity with which the other subunit binds the same ligand, the response of the enzyme to changes in the concentration of the ligand is much steeper than the response of an enzyme with only one subunit (see Figure 3–60).

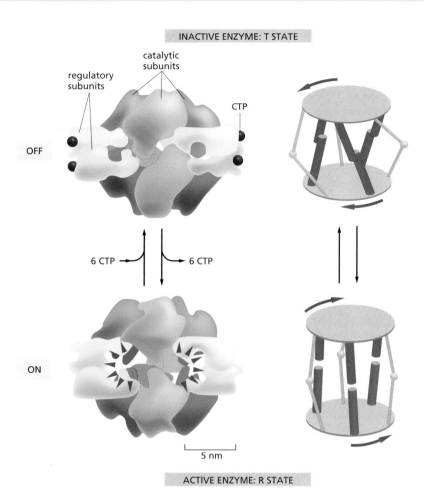

INACTIVE ENZYME: T STATE

regulatory subunits

catalytic subunits

OFF

CTP

6 CTP ⟶ ⟵ 6 CTP

ON

5 nm

ACTIVE ENZYME: R STATE

Figure 3–62 The transition between R and T states in the enzyme aspartate transcarbamoylase. <CTAA> The enzyme consists of a complex of six catalytic subunits and six regulatory subunits, and the structures of its inactive (T state) and active (R state) forms have been determined by x-ray crystallography. The enzyme is turned off by feedback inhibition when CTP concentrations rise. Each regulatory subunit can bind one molecule of CTP, which is one of the final products in the pathway. By means of this negative feedback regulation, the pathway is prevented from producing more CTP than the cell needs. (Based on K.L. Krause, K.W. Volz and W.N. Lipscomb, *Proc. Natl Acad. Sci. U.S.A.* 82:1643–1647, 1985. With permission from National Academy of Sciences.)

together by three regulatory dimers that form a bridge between them. The entire molecule is poised to undergo a concerted, all-or-none, allosteric transition between two conformations, designated as T (tense) and R (relaxed) states (**Figure 3–62**).

The binding of substrates (carbamoyl phosphate and aspartate) to the catalytic trimers drives aspartate transcarbamoylase into its catalytically active R state, from which the regulatory CTP molecules dissociate. By contrast, the binding of CTP to the regulatory dimers converts the enzyme to the inactive T state, from which the substrates dissociate. This tug-of-war between CTP and substrates is identical in principle to that described previously in Figure 3–59 for a simpler allosteric protein. But because the tug-of-war occurs in a symmetric molecule with multiple binding sites, the enzyme undergoes a cooperative allosteric transition that will turn it on suddenly as substrates accumulate (forming the R state) or shut it off rapidly when CTP accumulates (forming the T state).

A combination of biochemistry and x-ray crystallography has revealed many fascinating details of this allosteric transition. Each regulatory subunit has two domains, and the binding of CTP causes the two domains to move relative to each other, so that they function like a lever that rotates the two catalytic trimers and pulls them closer together into the T state (see Figure 3–62). When this occurs, hydrogen bonds form between opposing catalytic subunits. This helps widen the cleft that forms the active site within each catalytic subunit, thereby disrupting the binding sites for the substrates (**Figure 3–63**). Adding large amounts of substrate has the opposite effect, favoring the R state by binding in the cleft of each catalytic subunit and opposing the above conformational change. Conformations that are intermediate between R and T are unstable, so that the enzyme mostly clicks back and forth between its R and T forms, producing a mixture of these two species in proportions that depend on the relative concentrations of CTP and substrates.

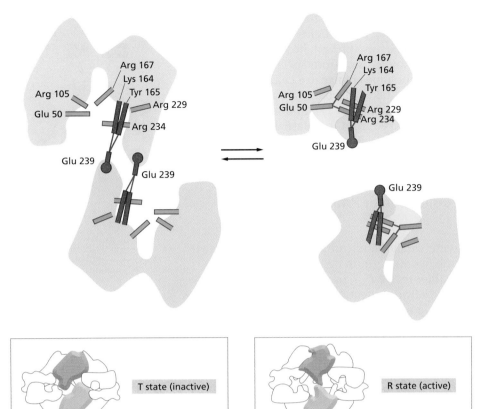

Figure 3–63 Part of the on–off switch in the catalytic subunits of aspartate transcarbamoylase. Changes in the indicated hydrogen-bonding interactions are partly responsible for switching this enzyme's active site between active *(yellow)* and inactive conformations. Hydrogen bonds are indicated by *thin red lines.* The amino acids involved in the subunit–subunit interaction in the T state are shown in *red*, while those that form the active site of the enzyme in the R state are shown in *blue.* The large drawings show the catalytic site in the interior of the enzyme; the boxed sketches show the same subunits viewed from the enzyme's external surface. (Adapted from E.R. Kantrowitz and W.N. Lipscomb, *Trends Biochem. Sci.* 15:53–59, 1990. With permission from Elsevier.)

T state (inactive)

R state (active)

Many Changes in Proteins Are Driven by Protein Phosphorylation

Proteins are regulated by more than the reversible binding of other molecules. A second method that eucaryotic cells use to regulate a protein's function is the covalent addition of a smaller molecule to one or more of its amino acid side chains. The most common such regulatory modification in higher eucaryotes is the addition of a phosphate group. We shall therefore use protein phosphorylation to illustrate some of the general principles involved in the control of protein function through the modification of amino acid side chains.

A phosphorylation event can affect the protein that is modified in two important ways. First, because each phosphate group carries two negative charges, the enzyme-catalyzed addition of a phosphate group to a protein can cause a major conformational change in the protein by, for example, attracting a cluster of positively charged amino acid side chains. This can, in turn, affect the binding of ligands elsewhere on the protein surface, dramatically changing the protein's activity. When a second enzyme removes the phosphate group, the protein returns to its original conformation and restores its initial activity.

Second, an attached phosphate group can form part of a structure that the binding sites of other proteins recognize. As previously discussed, certain protein domains, sometimes referred to as modules, appear very frequently as parts of larger proteins. One such module is the SH2 domain, described earlier, which binds to a short peptide sequence containing a phosphorylated tyrosine side chain (see Figure 3–39B). More than ten other common domains provide binding sites for attaching their protein to phosphorylated peptides in other protein molecules, each recognizing a phosphorylated amino acid side chain in a different protein context. As a result, protein phosphorylation and dephosphorylation very often drive the regulated assembly and disassembly of protein complexes (see Figure 15–22).

Reversible protein phosphorylation controls the activity, structure, and cellular localization of both enzymes and many other types of proteins in eucaryotic cells. In fact, this regulation is so extensive that more than one-third of the 10,000 or so proteins in a typical mammalian cell are thought to be phosphorylated at any given time—many with more than one phosphate. As might be expected, the addition and removal of phosphate groups from specific proteins often occur in response to signals that specify some change in a cell's state. For example, the complicated series of events that takes place as a eucaryotic cell divides is largely timed in this way (discussed in Chapter 17), and many of the signals mediating cell–cell interactions are relayed from the plasma membrane to the nucleus by a cascade of protein phosphorylation events (discussed in Chapter 15).

A Eucaryotic Cell Contains a Large Collection of Protein Kinases and Protein Phosphatases

Protein phosphorylation involves the enzyme-catalyzed transfer of the terminal phosphate group of an ATP molecule to the hydroxyl group on a serine, threonine, or tyrosine side chain of the protein (**Figure 3–64**). A **protein kinase** catalyzes this reaction, and the reaction is essentially unidirectional because of the large amount of free energy released when the phosphate–phosphate bond in ATP is broken to produce ADP (discussed in Chapter 2). A **protein phosphatase** catalyzes the reverse reaction of phosphate removal, or *dephosphorylation*. Cells contain hundreds of different protein kinases, each responsible for phosphorylating a different protein or set of proteins. There are also many different protein phosphatases; some are highly specific and remove phosphate groups from only one or a few proteins, whereas others act on a broad range of proteins and are targeted to specific substrates by regulatory subunits. The state of phosphorylation of a protein at any moment, and thus its activity, depends on the relative activities of the protein kinases and phosphatases that modify it.

The protein kinases that phosphorylate proteins in eucaryotic cells belong to a very large family of enzymes, which share a catalytic (kinase) sequence of about 290 amino acids. The various family members contain different amino acid sequences on either end of the kinase sequence (for example, see Figure 3–10), and often have short amino acid sequences inserted into loops within it (*red arrowheads* in **Figure 3–65**). Some of these additional amino acid sequences enable each kinase to recognize the specific set of proteins it phosphorylates, or to bind to structures that localize it in specific regions of the cell. Other parts of the protein regulate the activity of each kinase, so it can be turned on and off in response to different specific signals, as described below.

By comparing the number of amino acid sequence differences between the various members of a protein family, we can construct an "evolutionary tree" that is thought to reflect the pattern of gene duplication and divergence that gave rise to the family. **Figure 3–66** shows an evolutionary tree of protein kinases. Kinases with related functions are often located on nearby branches of the tree: the protein kinases involved in cell signaling that phosphorylate tyrosine side chains, for example, are all clustered in the top left corner of the tree. The other kinases shown phosphorylate either a serine or a threonine side chain, and many are organized into clusters that seem to reflect their function—in transmembrane signal transduction, intracellular signal amplification, cell-cycle control, and so on.

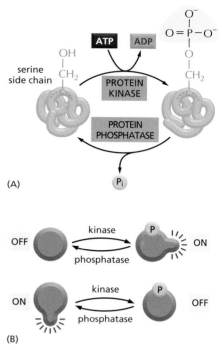

(A)

(B)

Figure 3–64 Protein phosphorylation. Many thousands of proteins in a typical eucaryotic cell are modified by the covalent addition of a phosphate group. (A) The general reaction, shown here, transfers a phosphate group from ATP to an amino acid side chain of the target protein by a protein kinase. Removal of the phosphate group is catalyzed by a second enzyme, a protein phosphatase. In this example, the phosphate is added to a serine side chain; in other cases, the phosphate is instead linked to the –OH group of a threonine or a tyrosine in the protein. (B) The phosphorylation of a protein by a protein kinase can either increase or decrease the protein's activity, depending on the site of phosphorylation and the structure of the protein.

Figure 3–65 The three-dimensional structure of a protein kinase. Superimposed on this structure are *red arrowheads* to indicate sites where insertions of 5–100 amino acids are found in some members of the protein kinase family. These insertions are located in loops on the surface of the enzyme where other ligands interact with the protein. Thus, they distinguish different kinases and confer on them distinctive interactions with other proteins. The ATP (which donates a phosphate group) and the peptide to be phosphorylated are held in the active site, which extends between the phosphate-binding loop (*yellow*) and the catalytic loop (*orange*). See also Figure 3–10. (Adapted from D.R. Knighton et al., *Science* 253:407–414, 1991. With permission from AAAS.)

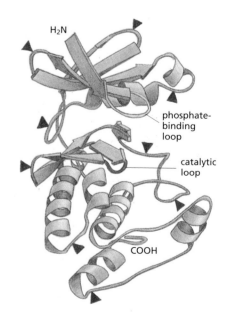

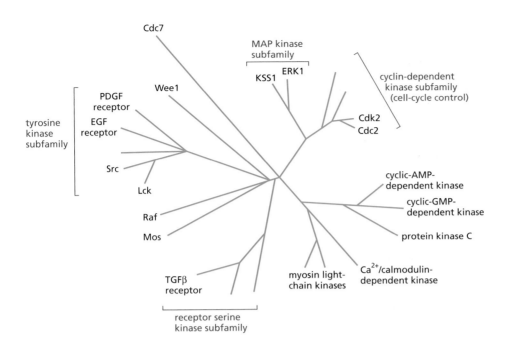

Figure 3–66 An evolutionary tree of selected protein kinases. Although a higher eucaryotic cell contains hundreds of such enzymes, and the human genome codes for more than 500, only some of those discussed in this book are shown.

As a result of the combined activities of protein kinases and protein phosphatases, the phosphate groups on proteins are continually turning over—being added and then rapidly removed. Such phosphorylation cycles may seem wasteful, but they are important in allowing the phosphorylated proteins to switch rapidly from one state to another: the more rapid the cycle, the faster a population of protein molecules can change its state of phosphorylation in response to a sudden change in the phosphorylation rate (see Figure 15–11). The energy required to drive this phosphorylation cycle is derived from the free energy of ATP hydrolysis, one molecule of which is consumed for each phosphorylation event.

The Regulation of Cdk and Src Protein Kinases Shows How a Protein Can Function as a Microchip

The hundreds of different protein kinases in a eucaryotic cell are organized into complex networks of signaling pathways that help to coordinate the cell's activities, drive the cell cycle, and relay signals into the cell from the cell's environment. Many of the extracellular signals involved need to be both integrated and amplified by the cell. Individual protein kinases (and other signaling proteins) serve as input–output devices, or "microchips," in the integration process. An important part of the input to these signal processing proteins comes from the control that is exerted by phosphates added and removed from them by protein kinases and protein phosphatases, respectively.

In general, specific sets of phosphate groups serve to activate the protein, while other sets can inactivate it. A cyclin-dependent protein kinase (Cdk) provides a good example. Kinases in this class phosphorylate serines and threonines, and they are central components of the cell-cycle control system in eucaryotic cells, as discussed in detail in Chapter 17. In a vertebrate cell, individual Cdk proteins turn on and off in succession, as a cell proceeds through the different phases of its division cycle. When a particular kinase is on, it influences various aspects of cell behavior through effects on the proteins it phosphorylates.

A Cdk protein becomes active as a serine/threonine protein kinase only when it is bound to a second protein called a *cyclin*. But, as **Figure 3–67** shows, the binding of cyclin is only one of three distinct "inputs" required to activate the Cdk. In addition to cyclin binding, a phosphate must be added to a specific threonine side chain, and a phosphate elsewhere in the protein (covalently bound to a specific tyrosine side chain) must be removed. Cdk thus monitors a specific set

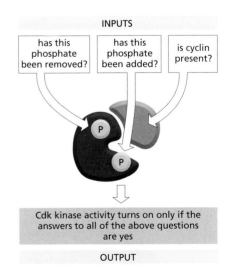

Figure 3–67 How a Cdk protein acts as an integrating device. <TAGA> The function of these central regulators of the cell cycle is discussed in Chapter 17.

Figure 3–68 The domain structure of the Src family of protein kinases, mapped along the amino acid sequence. For the three-dimensional structure of Src, see Figure 3–10.

of cell components—a cyclin, a protein kinase, and a protein phosphatase—and it acts as an input–output device that turns on if, and only if, each of these components has attained its appropriate activity state. Some cyclins rise and fall in concentration in step with the cell cycle, increasing gradually in amount until they are suddenly destroyed at a particular point in the cycle. The sudden destruction of a cyclin (by targeted proteolysis) immediately shuts off its partner Cdk enzyme, and this triggers a specific step in the cell cycle.

The Src family of protein kinases (see Figure 3–10) exhibits a similar type of microchip behavior. The *Src protein* (pronounced "sarc" and named for the type of tumor, a sarcoma, that its deregulation can cause) was the first tyrosine kinase to be discovered. It is now known to be part of a subfamily of nine very similar protein kinases, which are found only in multicellular animals. As indicated by the evolutionary tree in Figure 3–66, sequence comparisons suggest that tyrosine kinases as a group were a relatively late innovation that branched off from the serine/threonine kinases, with the Src subfamily being only one subgroup of the tyrosine kinases created in this way.

The Src protein and its relatives contain a short N-terminal region that becomes covalently linked to a strongly hydrophobic fatty acid, which holds the kinase at the cytoplasmic face of the plasma membrane. Next come two peptide-binding modules, a Src homology 3 (SH3) domain and a SH2 domain, followed by the kinase catalytic domain (**Figure 3–68**). These kinases normally exist in an inactive conformation, in which a phosphorylated tyrosine near the C-terminus is bound to the SH2 domain, and the SH3 domain is bound to an internal peptide in a way that distorts the active site of the enzyme and helps to render it inactive.

Turning the kinase on involves at least two specific inputs: removal of the C-terminal phosphate and the binding of the SH3 domain by a specific activating protein (**Figure 3–69**). Like the activation of the Cdk protein, the activation of the Src kinase signals the completion of a particular set of separate upstream events (**Figure 3–70**). Thus, both the Cdk and Src families of proteins serve as specific signal integrators, helping to generate the complex web of information-processing events that enable the cell to compute logical responses to a complex set of conditions.

Proteins That Bind and Hydrolyze GTP Are Ubiquitous Cellular Regulators

We have described how the addition or removal of phosphate groups on a protein can be used by a cell to control the protein's activity. In the examples discussed so

Figure 3–69 The activation of a Src-type protein kinase by two sequential events. (Adapted from S.C. Harrison et al., *Cell* 112:737–740, 2003. With permission from Elsevier.)

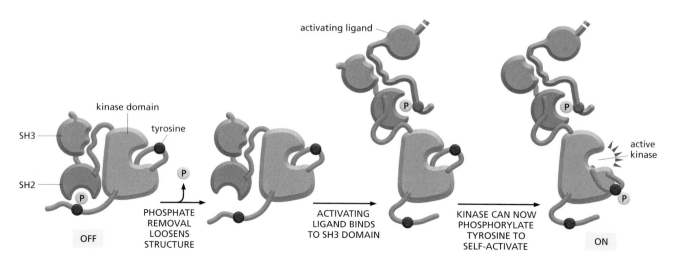

far, the phosphate is transferred from an ATP molecule to an amino acid side chain of the protein in a reaction catalyzed by a specific protein kinase. Eucaryotic cells also have another way to control protein activity by phosphate addition and removal. In this case, the phosphate is not attached directly to the protein; instead, it is a part of the guanine nucleotide GTP, which binds very tightly to the protein. In general, proteins regulated in this way are in their active conformations with GTP bound. The loss of a phosphate group occurs when the bound GTP is hydrolyzed to GDP in a reaction catalyzed by the protein itself, and in its GDP-bound state the protein is inactive. In this way, GTP-binding proteins act as on–off switches whose activity is determined by the presence or absence of an additional phosphate on a bound GDP molecule (**Figure 3–71**).

GTP-binding proteins (also called **GTPases** because of the GTP hydrolysis they catalyze) comprise a large family of proteins that all contain variations on the same GTP-binding globular domain. When the tightly bound GTP is hydrolyzed to GDP, this domain undergoes a conformational change that inactivates it. The three-dimensional structure of a prototypical member of this family, the monomeric GTPase called Ras, is shown in **Figure 3–72**.

The *Ras protein* has an important role in cell signaling (discussed in Chapter 15). In its GTP-bound form, it is active and stimulates a cascade of protein phosphorylations in the cell. Most of the time, however, the protein is in its inactive, GDP-bound form. It becomes active when it exchanges its GDP for a GTP molecule in response to extracellular signals, such as growth factors, that bind to receptors in the plasma membrane (see Figure 15–58).

Regulatory Proteins Control the Activity of GTP-Binding Proteins by Determining Whether GTP or GDP Is Bound

GTP-binding proteins are controlled by regulatory proteins that determine whether GTP or GDP is bound, just as phosphorylated proteins are turned on and off by protein kinases and protein phosphatases. Thus, Ras is inactivated by a *GTPase-activating protein (GAP)*, which binds to the Ras protein and induces it to hydrolyze its bound GTP molecule to GDP—which remains tightly bound—and inorganic phosphate (P_i), which is rapidly released. The Ras protein stays in its inactive, GDP-bound conformation until it encounters a *guanine nucleotide exchange factor (GEF)*, which binds to GDP-Ras and causes it to release its GDP. Because the empty nucleotide-binding site is immediately filled by a GTP molecule (GTP is present in large excess over GDP in cells), the GEF activates Ras by *indirectly* adding back the phosphate removed by GTP hydrolysis. Thus, in a sense, the roles of GAP and GEF are analogous to those of a protein phosphatase and a protein kinase, respectively (**Figure 3–73**).

Large Protein Movements Can Be Generated From Small Ones

The Ras protein belongs to a large superfamily of *monomeric GTPases*, each of which consists of a single GTP-binding domain of about 200 amino acids. Over the course of evolution, this domain has also become joined to larger proteins with additional domains, creating a large family of GTP-binding proteins. Family members include the receptor-associated trimeric G proteins involved in cell signaling (discussed in Chapter 15), proteins regulating the traffic of vesicles between intracellular compartments (discussed in Chapter 13), and proteins that bind to transfer RNA and are required as assembly factors for protein

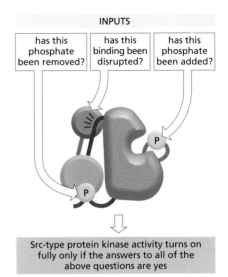

INPUTS

Src-type protein kinase activity turns on fully only if the answers to all of the above questions are yes

OUTPUT

Figure 3–70 How a Src-type protein kinase acts as an integrating device. The disruption of the SH3 domain interaction *(green)* involves replacing its binding to the indicated *red* linker region with a tighter interaction with an activating ligand, as illustrated in Figure 3–69.

Figure 3–71 GTP-binding proteins as molecular switches. The activity of a GTP-binding protein (also called a GTPase) generally requires the presence of a tightly bound GTP molecule (switch "on"). Hydrolysis of this GTP molecule produces GDP and inorganic phosphate (P_i), and it causes the protein to convert to a different, usually inactive, conformation (switch "off"). As shown here, resetting the switch requires the tightly bound GDP to dissociate, a slow step that is greatly accelerated by specific signals; once the GDP has dissociated, a molecule of GTP is quickly rebound.

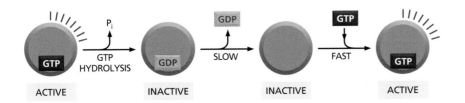

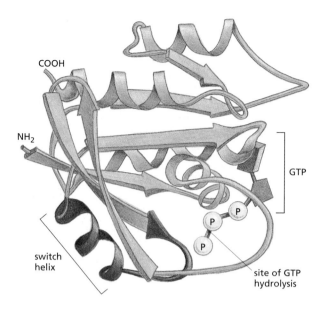

COOH

NH₂

GTP

P
P
P

switch
helix

site of GTP
hydrolysis

Figure 3–72 The structure of the Ras protein in its GTP-bound form. <GAAC> This monomeric GTPase illustrates the structure of a GTP-binding domain, which is present in a large family of GTP-binding proteins. The *red* regions change their conformation when the GTP molecule is hydrolyzed to GDP and inorganic phosphate by the protein; the GDP remains bound to the protein, while the inorganic phosphate is released. The special role of the "switch helix" in proteins related to Ras is explained next (see Figure 3–75).

synthesis on the ribosome (discussed in Chapter 6). In each case, an important biological activity is controlled by a change in the protein's conformation that is caused by GTP hydrolysis in a Ras-like domain.

The *EF-Tu protein* provides a good example of how this family of proteins works. EF-Tu is an abundant molecule that serves as an elongation factor (hence the EF) in protein synthesis, loading each aminoacyl tRNA molecule onto the ribosome. The tRNA molecule forms a tight complex with the GTP-bound form of EF-Tu (**Figure 3–74**). In this complex, the amino acid attached to the tRNA is improperly positioned for protein synthesis. The tRNA can transfer its amino acid only after the GTP bound to EF-Tu is hydrolyzed on the ribosome, allowing the EF-Tu to dissociate. Since the GTP hydrolysis is triggered by a proper fit of the tRNA to the mRNA molecule on the ribosome, the EF-Tu serves as a factor that discriminates between correct and incorrect mRNA–tRNA pairings (see Figure 6–67 for a further discussion of this function of EF-Tu).

By comparing the three-dimensional structure of EF-Tu in its GTP-bound and GDP-bound forms, we can see how the repositioning of the tRNA occurs. The dissociation of the inorganic phosphate group (P_i), which follows the reaction GTP → GDP + P_i, causes a shift of a few tenths of a nanometer at the GTP-binding site, just as it does in the Ras protein. This tiny movement, equivalent to

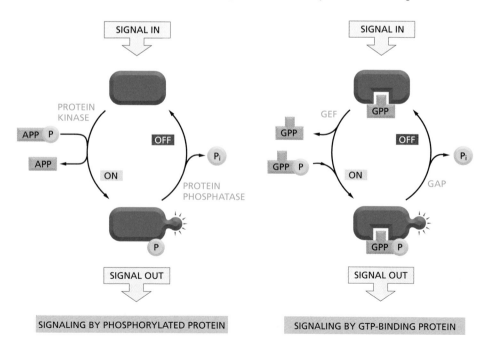

SIGNAL IN

SIGNAL IN

PROTEIN
KINASE

APP   P

APP

OFF

P_i

ON

PROTEIN
PHOSPHATASE

P

SIGNAL OUT

SIGNALING BY PHOSPHORYLATED PROTEIN

GEF

GPP

GPP

GPP   P

ON

GPP

OFF

P_i

GAP

GPP   P

SIGNAL OUT

SIGNALING BY GTP-BINDING PROTEIN

Figure 3–73 A comparison of the two major intracellular signaling mechanisms in eucaryotic cells. In both cases, a signaling protein is activated by the addition of a phosphate group and inactivated by the removal of this phosphate. To emphasize the similarities in the two pathways, ATP and GTP are drawn as APPP and GPPP, and ADP and GDP as APP and GPP, respectively. As shown in Figure 3–64, the addition of a phosphate to a protein can also be inhibitory.

Figure 3–74 **An aminoacyl tRNA molecule bound to EF-Tu.** The three domains of the EF-Tu protein are colored differently, to match Figure 3–75. This is a bacterial protein; however, a very similar protein exists in eucaryotes, where it is called EF-1. (Coordinates determined by P. Nissen et al., *Science* 270:1464–1472, 1995. With permission from AAAS.)

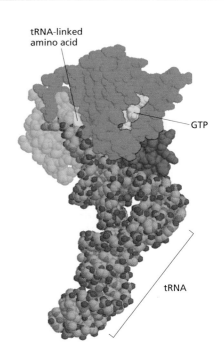

a few times the diameter of a hydrogen atom, causes a conformational change to propagate along a crucial piece of α helix, called the *switch helix*, in the Ras-like domain of the protein. The switch helix seems to serve as a latch that adheres to a specific site in another domain of the molecule, holding the protein in a "shut" conformation. The conformational change triggered by GTP hydrolysis causes the switch helix to detach, allowing separate domains of the protein to swing apart, through a distance of about 4 nm. This releases the bound tRNA molecule, allowing its attached amino acid to be used (**Figure 3–75**).

Notice in this example how cells have exploited a simple chemical change that occurs on the surface of a small protein domain to create a movement 50 times larger. Dramatic shape changes of this type also cause the very large movements that occur in motor proteins, as we discuss next.

Motor Proteins Produce Large Movements in Cells

We have seen that conformational changes in proteins have a central role in enzyme regulation and cell signaling. We now discuss proteins whose major function is to move other molecules. These **motor proteins** generate the forces responsible for muscle contraction and the crawling and swimming of cells. Motor proteins also power smaller-scale intracellular movements: they help to move chromosomes to opposite ends of the cell during mitosis (discussed in Chapter 17), to move organelles along molecular tracks within the cell (discussed

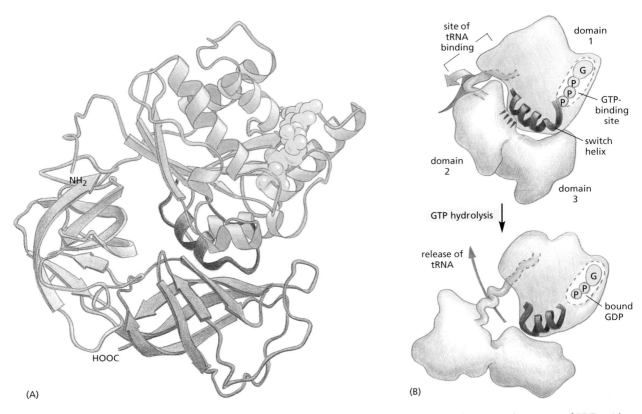

Figure 3–75 **The large conformational change in EF-Tu caused by GTP hydrolysis.** <GTAA> (A) The three-dimensional structure of EF-Tu with GTP bound. The domain at the top has a structure similar to the Ras protein, and its *red* α helix is the switch helix, which moves after GTP hydrolysis. (B) The change in the conformation of the switch helix in domain 1 causes domains 2 and 3 to rotate as a single unit by about 90° toward the viewer, which releases the tRNA that was shown bound to this structure in Figure 3–74. (A, adapted from H. Berchtold et al., *Nature* 365:126–132, 1993. With permission from Macmillan Publishers Ltd. B, courtesy of Mathias Sprinzl and Rolf Hilgenfeld.)

in Chapter 16), and to move enzymes along a DNA strand during the synthesis of a new DNA molecule (discussed in Chapter 5). All these fundamental processes depend on proteins with moving parts that operate as force-generating machines.

How do these machines work? In other words, how do cells use shape changes in proteins to generate directed movements? If, for example, a protein is required to walk along a narrow thread such as a DNA molecule, it can do this by undergoing a series of conformational changes, such as those shown in **Figure 3–76**. But with nothing to drive these changes in an orderly sequence, they are perfectly reversible, and the protein can only wander randomly back and forth along the thread. We can look at this situation in another way. Since the directional movement of a protein does work, the laws of thermodynamics (discussed in Chapter 2) demand that such movement use free energy from some other source (otherwise the protein could be used to make a perpetual motion machine). Therefore, without an input of energy, the protein molecule can only wander aimlessly.

How can the cell make such a series of conformational changes unidirectional? To force the entire cycle to proceed in one direction, it is enough to make any one of the changes in shape irreversible. Most proteins that are able to walk in one direction for long distances achieve this motion by coupling one of the conformational changes to the hydrolysis of an ATP molecule bound to the protein. The mechanism is similar to the one just discussed that drives allosteric protein shape changes by GTP hydrolysis. Because ATP (or GTP) hydrolysis releases a great deal of free energy, it is very unlikely that the nucleotide-binding protein will undergo the reverse shape change needed for moving backward—since this would require that it also reverse the ATP hydrolysis by adding a phosphate molecule to ADP to form ATP.

In the model shown in **Figure 3–77**, ATP binding shifts a motor protein from conformation 1 to conformation 2. The bound ATP is then hydrolyzed to produce ADP and inorganic phosphate (P_i), causing a change from conformation 2 to conformation 3. Finally, the release of the bound ADP and P_i drives the protein back to conformation 1. Because the energy provided by ATP hydrolysis drives the transition 2 → 3, this series of conformational changes is effectively irreversible. Thus, the entire cycle goes in only one direction, causing the protein molecule to walk continuously to the right in this example.

Many motor proteins generate directional movement in this general way, including the muscle motor protein *myosin*, which walks along actin filaments to generate muscle contraction, and the *kinesin* proteins that walk along microtubules (both discussed in Chapter 16). These movements can be rapid: some of the motor proteins involved in DNA replication (the DNA helicases) propel themselves along a DNA strand at rates as high as 1000 nucleotides per second.

Membrane-Bound Transporters Harness Energy to Pump Molecules Through Membranes

We have thus far seen how allosteric proteins can act as microchips (Cdk and Src kinases), as assembly factors (EF-Tu), and as generators of mechanical force and motion (motor proteins). Allosteric proteins can also harness energy derived from ATP hydrolysis, ion gradients, or electron transport processes to pump specific ions or small molecules across a membrane. We consider one example here; others will be discussed in Chapter 11.

The ABC transporters constitute an important class of membrane-bound pump proteins. In humans at least 48 different genes encode them. These transporters mostly function to export hydrophobic molecules from the cytoplasm,

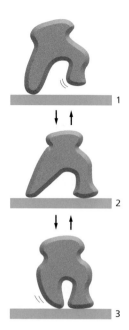

Figure 3–76 An allosteric "walking" protein. Although its three different conformations allow it to wander randomly back and forth while bound to a thread or a filament, the protein cannot move uniformly in a single direction.

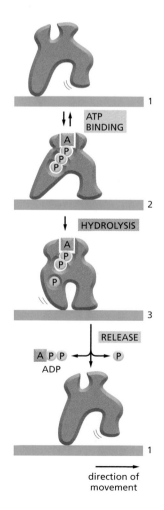

direction of movement

Figure 3–77 An allosteric motor protein. The transition between three different conformations includes a step driven by the hydrolysis of a bound ATP molecule, and this makes the entire cycle essentially irreversible. By repeated cycles, the protein therefore moves continuously to the right along the thread.

(A)

EXTRACELLULAR
SPACE

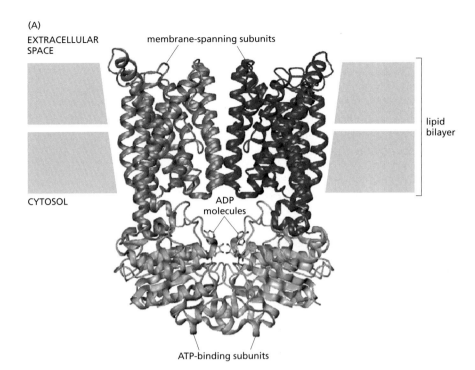

membrane-spanning subunits

lipid
bilayer

CYTOSOL

ADP
molecules

ATP-binding subunits

Figure 3–78 The ABC (ATP-binding cassette) transporter, a protein machine that pumps large hydrophobic molecules through a membrane. (A) The bacterial BtuCD protein, which imports vitamin B_{12} into *E. coli* using the energy of ATP hydrolysis. The binding of two molecules of ATP clamps together the two ATP-binding subunits. The structure is shown in its ADP-bound state, where the channel to the extracellular space can be seen to be open but the gate to the cytosol remains closed. (B) Schematic illustration of substrate pumping by ABC transporters. In bacteria, the binding of a substrate molecule to the extracellular face of the protein complex triggers ATP hydrolysis followed by ADP release, which opens the cytoplasmic gate; the pump is then reset for another cycle. In eucaryotes, an opposite process occurs, causing substrate molecules to be pumped out of the cell. (A, adapted from K.P. Locher, *Curr. Opin. Struct. Biol.* 14:426–441, 2004. With permission from Elsevier.)

(B)   A BACTERIAL ABC TRANSPORTER

substrate molecule

CYTOSOL

ATP-binding
domains

2 ATP

2 ADP + 2Pᵢ

A EUCARYOTIC ABC TRANSPORTER

CYTOSOL
substrate
molecule

ATP-binding domains

2 ATP

2 ADP + 2Pᵢ

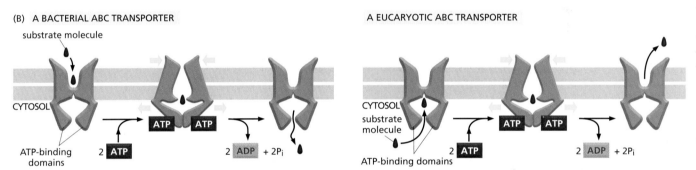

serving to remove toxic molecules at the mucosal surface of the intestinal tract, for example, or at the blood–brain barrier. The study of ABC transporters is of intense interest in clinical medicine, because the overproduction of proteins in this class contributes to the resistance of tumor cells to chemotherapeutic drugs. And in bacteria, the same type of proteins primarily function to import essential nutrients into the cell.

The ABC transporter is a tetramer, with a pair of membrane-spanning subunits linked to a pair of ATP binding subunits located just below the plasma membrane (**Figure 3–78A**). As in other examples we have discussed, the hydrolysis of the bound ATP molecules drives conformational changes in the protein, transmitting forces that cause the membrane-spanning subunits to move their bound molecules across the lipid bilayer (Figure 3–78B).

Humans have invented many different types of mechanical pumps, and it should not be surprising that cells also contain membrane-bound pumps that function in other ways. Among the most notable are the rotary pumps that couple the hydrolysis of ATP to the transport of H⁺ ions (protons). These pumps resemble miniature turbines, and they are used to acidify the interior of lysosomes and other eucaryotic organelles. Like other ion pumps that create ion gradients, they can function in reverse to catalyze the reaction ADP + P_i → ATP, if the gradient across their membrane of the ion that they transport is steep enough.

One such pump, the ATP synthase, harnesses a gradient of proton concentration produced by electron transport processes to produce most of the ATP used in the living world. This ubiquitous pump has a central role in energy conversion, and we shall discuss its three-dimensional structure and mechanism in Chapter 14.

Proteins Often Form Large Complexes That Function as Protein Machines <ACTT> <ATCG>

Large proteins formed from many domains are able to perform more elaborate functions than small, single-domain proteins. But large protein assemblies formed from many protein molecules perform the most impressive tasks. Now that it is possible to reconstruct most biological processes in cell-free systems in the laboratory, it is clear that each of the central processes in a cell—such as DNA replication, protein synthesis, vesicle budding, or transmembrane signaling—is catalyzed by a highly coordinated, linked set of 10 or more proteins. In most such *protein machines*, an energetically favorable reaction such as the hydrolysis of bound nucleoside triphosphates (ATP or GTP) drives an ordered series of conformational changes in one or more of the individual protein subunits, enabling the ensemble of proteins to move coordinately. In this way, each enzyme can be moved directly into position, as the machine catalyzes successive reactions in a series. This is what occurs, for example, in protein synthesis on a ribosome (discussed in Chapter 6)—or in DNA replication, where a large multiprotein complex moves rapidly along the DNA (discussed in Chapter 5).

Cells have evolved protein machines for the same reason that humans have invented mechanical and electronic machines. For accomplishing almost any task, manipulations that are spatially and temporally coordinated through linked processes are much more efficient than the use of individual tools.

Protein Machines with Interchangeable Parts Make Efficient Use of Genetic Information

To probe more deeply into the nature of protein machines, we shall consider a relatively simple one: the **SCF ubiquitin ligase**. This protein complex binds different "target proteins" at different times in the cell cycle, and it covalently adds multiubiquitin polypeptide chains to these proteins. Its C-shaped structure is formed from five protein subunits, the largest of which is a molecule that serves as a *scaffold protein* on which the rest of the structure is built. The structure underlies a remarkable mechanism (**Figure 3–79**). At one end of the C is an E2 ubiquitin-conjugating enzyme. At the other end is a substrate-binding arm, a subunit known as an *F-box protein*. These two subunits are separated by a gap of about 5 nm. When this protein complex is activated, the F-box protein binds to a specific site on a target protein, positioning the protein in the gap so that some of its lysine side chains contact the ubiquitin-conjugating enzyme. This enzyme can then catalyze the repeated addition of a ubiquitin polypeptide to these lysines (see Figure 3–79C), producing a polyubiquitin chain that marks the target protein for rapid destruction in a proteasome (see p. 393).

In this manner, specific proteins are targeted for rapid destruction in response to specific signals, thereby helping to drive the cell cycle (discussed in Chapter 17). The timing of the destruction often involves creating a specific pattern of phosphorylation on the target protein that is required for its recognition by the F-box subunit. It also requires the activation of an SCF ubiquitin ligase that carries the appropriate substrate-binding arm. Many of these arms (the F-box subunits) are interchangeable in the protein complex (see Figure 3–79B), and there are more than 70 human genes that encode them.

As emphasized previously, once a successful protein has evolved, its genetic information tends to be duplicated to produce a family of related proteins. Thus, for example, not only are there many F-box proteins—making possible the recognition of different sets of target proteins—but there is also a family of scaffolds (known as cullins) that give rise to a family of SCF-like ubiquitin ligases.

The pressure on organisms to minimize the number of genes (see p. 265) presumably helps to explain why RNA splicing is so prevalent in higher eucaryotes, allowing multiple related proteins to be synthesized from a single gene (discussed in Chapter 6). A protein machine like the SCF ubiquitin ligase, with its interchangeable parts, likewise makes economical use of the genetic information

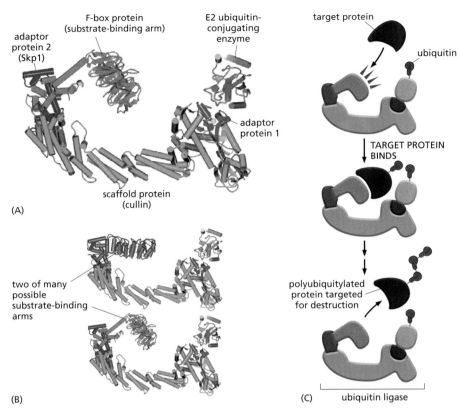

F-box protein
(substrate-binding arm)

E2 ubiquitin-
conjugating
enzyme

target protein

adaptor
protein 2
(Skp1)

ubiquitin

adaptor
protein 1

scaffold protein
(cullin)

(A)

TARGET PROTEIN
BINDS

two of many
possible
substrate-binding
arms

polyubiquitylated
protein targeted
for destruction

(B)

(C)   ubiquitin ligase

Figure 3–79 **The structure and mode of action of a SCF ubiquitin ligase.** (A) The structure of the five-protein complex that includes an E2 ubiquitin ligase. The protein denoted here as adapter protein 1 is the Rbx1/Hrt1 protein, adaptor protein 2 is the Skp1 protein, and the cullin is the Cul1 protein. (B) Comparison of the same complex with two different substrate-binding arms, the F-box proteins Skp2 *(top)* and β-trCP1 *(bottom)*, respectively. (C) The binding and ubiquitylation of a target protein by the SCF ubiquitin ligase. If, as indicated, a chain of ubiquitin molecules is added to the same lysine of the target protein, that protein is marked for rapid destruction by the proteasome. (A and B, adapted from G. Wu et al., *Mol. Cell* 11:1445–1456, 2003. With permission from Elsevier.)

in cells, inasmuch as new functions can evolve for the entire complex simply by producing an alternative version of one of its subunits.

The Activation of Protein Machines Often Involves Positioning Them at Specific Sites

As scientists have learned more of the details of cell biology, they have recognized increasing degrees of sophistication in cell chemistry. Thus, not only do we now know that protein machines play a predominant role, but it has recently become clear that most of these machines form at specific sites in the cell, being activated only where and when they are needed. Using fluorescent, GFP-tagged fusion proteins in living cells (see p. 593), cell biologists are able to follow the repositioning of individual proteins that occurs in response to specific signals. Thus, when certain extracellular signaling molecules bind to receptor proteins in the plasma membrane, they often recruit a set of other proteins to the inside surface of the plasma membrane to form protein machines that pass the signal on. As an example, **Figure 3–80**A illustrates the rapid movement of a protein kinase C (PKC) enzyme to a complex in the plasma membrane, where it associates with specific substrate proteins that it phosphorylates.

There are more than 10 distinct PKC enzymes in human cells, which differ both in their mode of regulation and in their functions. When activated, these enzymes move from the cytoplasm to different intracellular locations, forming specific complexes with other proteins that allow them to phosphorylate different protein substrates (Figure 3–80B). The SCF ubiquitin ligases can also move to specific sites of function at appropriate times. As will be explained when we discuss cell signaling in Chapter 15, the mechanisms frequently involve protein phosphorylation, as well as **scaffold proteins** that link together a set of activating, inhibiting, adaptor, and substrate proteins at a specific location in a cell.

This general phenomenon is known as *induced proximity*, and it explains the otherwise puzzling observation that slightly different forms of enzymes with the same catalytic site will often have very different biological functions. Cells change the locations of their proteins by covalently modifying them in a variety of different ways, as part of a "regulatory code" to be described next.

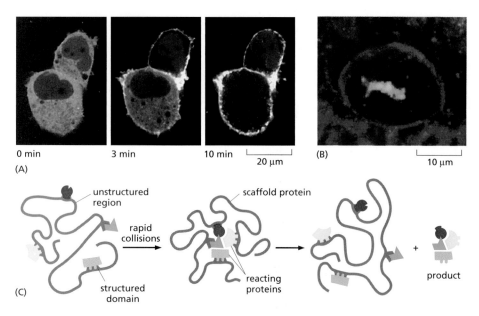

0 min   3 min   10 min
(A)   20 µm   (B)   10 µm

(C)

Figure 3–80 **The assembly of protein machines at specific sites in a cell.** (A) In response to a signal (here a phorbol ester), the gamma subspecies of protein kinase C moves rapidly from the cytosol to the plasma membrane. The protein kinase is fluorescent in these living cells because an engineered gene inside the cell encodes a fusion protein that links the kinase to green fluorescent protein (GFP). (B) The specific association of a different subspecies of protein kinase C (aPKC) with the apical tip of a differentiating neuroblast in an early *Drosophila* embryo. The kinase is stained *red*, and the cell nucleus *green*. (C) Diagram illustrating how a simple proximity created by scaffold proteins can greatly speed reactions in a cell. In this example, long unstructured regions of polypeptide chain in a large scaffold protein connect a series of structured domains that bind a set of reacting proteins. The unstructured regions serve as flexible "tethers" that greatly speed reaction rates by causing a rapid, random collision of all of the proteins that are bound to the scaffold. (For a simple example of tethering, see Figure 16–38.) (A, from N. Sakai et al, *J. Cell Biol.* 139:1465–1476, 1997. With permission from The Rockefeller University Press. B, courtesy of Andreas Wodarz, Institute of Genetics, University of Düsseldorf, Germany.)

These modifications create sites on proteins that bind them to particular scaffold proteins, thereby clustering the proteins required for particular reactions in specific regions of the cell. Most biological reactions are catalyzed by sets of 5 or more proteins, and such a clustering of proteins is often required for the reaction to occur. Scaffolds thereby allow cells to compartmentalize reactions even in the absence of membranes. Although only recently recognized as a widespread phenomenon, this type of clustering is particularly obvious in the cell nucleus (see Figure 4–69).

Many scaffolds appear to be quite different from the cullin illustrated previously in Figure 3–79: rather than holding their bound proteins in precise positions relative to each other, the interacting proteins are linked by unstructured regions of polypeptide chain. This tethers the proteins together, causing them to collide frequently with each other in random orientations—some of which will lead to a productive reaction (Figure 3–80C). In essence, this mechanism greatly speeds reactions by creating a very high local concentration of the reacting species. For this reason, the use of scaffold proteins represents an especially versatile way of controlling cell chemistry (see also Figure 15–61).

Many Proteins Are Controlled by Multisite Covalent Modification

We have thus far described only one type of posttranslational modification of proteins—that in which a phosphate is covalently attached to an amino acid side chain (see Figure 3–64). But a large number of other such modifications also occur, more than 200 distinct types being known. To give a sense of the variety, **Table 3–3** presents a subset of modifying groups with known regulatory roles. As

Table 3–3 **Some Molecules Covalently Attached to Proteins Regulate Protein Function**

MODIFYING GROUP	SOME PROMINENT FUNCTIONS
Phosphate on Ser, Thr, or Tyr	Drives the assembly of a protein into larger complexes (see Figure 15–19).
Methyl on Lys	Helps to creates histone code in chromatin through forming either mono-, di-, or tri-methyl lysine (see Figure 4–38).
Acetyl on Lys	Helps to creates histone code in chromatin (see Figure 4–38).
Palmityl group on Cys	This fatty acid addition drives protein association with membranes (see Figure 10–20).
***N*-acetylglucosamine** on Ser or Thr	Controls enzyme activity and gene expression in glucose homeostasis.
Ubiquitin on Lys	Monoubiquitin addition regulates the transport of membrane proteins in vesicles (see Figure 13–58).
	A polyubiquitin chain targets a protein for degradation (see Figure 3–79).

Ubiquitin is a 76 amino acid polypeptide; there are at least 10 other ubiquitin-related proteins, such as SUMO, that modify proteins in similar ways.

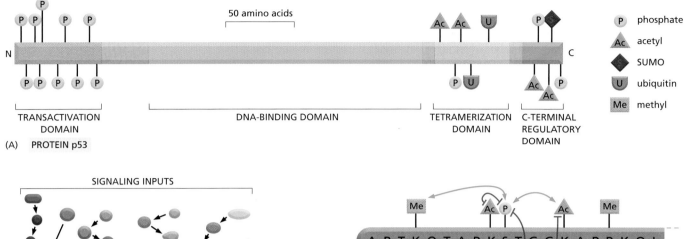

(A)   PROTEIN p53

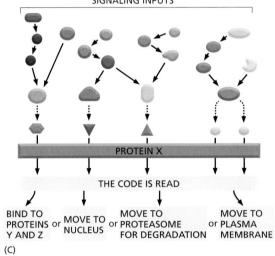

(C)

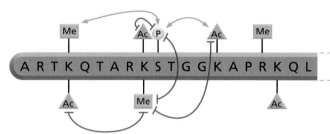

(B)   HISTONE H3

Figure 3–81 Multisite protein modification and its effects. A protein that carries a post-translational addition to more than one of its amino acid side chains can be considered to carry a combinatorial regulatory code. (A) The pattern of known covalent modifications to the protein p53; ubiquitin and SUMO are related polypeptides (see Table 3–3). (B) The possible modifications on the first 20 amino acids at the N-terminus of histone H3, showing not only their locations but also their activating *(blue)* and inhibiting *(red)* effects on the addition of neighboring covalent modifications. In addition to the effects shown, the acetylation and methylation of a lysine are mutually exclusive reactions (see Figure 4–38). (C) Diagram showing the general manner in which multisite modifications are added to (and removed from) a protein through signaling networks, and how the resulting combinatorial regulatory code on the protein is read to alter its behavior in the cell.

in phosphate addition, these groups are added and then removed from proteins according to the needs of the cell.

A large number of proteins are now known to be modified on more than one amino acid side chain, with different regulatory events producing a different pattern of such modifications. A striking example is the protein p53, which plays a central part in controlling a cell's response to adverse circumstances (see p. 1105). Through one of four different types of molecular additions, this protein can be modified at 20 different sites (**Figure 3–81**A). Because an enormous number of different combinations of these 20 modifications are possible, the protein's behavior can in principle be altered in a huge number of ways. Moreover, the pattern of modifications on a protein can determine its susceptibility to further modification, as illustrated by histone H3 in Figure 3–81B.

Cell biologists have only recently come to recognize that each protein's set of covalent modifications constitutes an important *combinatorial regulatory code.* As specific modifying groups are added to or removed from a protein, this code causes a different set of protein behaviors—changing the activity or stability of the protein, its binding partners, and its specific location within the cell (Figure 3–81C). This helps the cell respond rapidly and with great versatility to changes in its condition or environment.

A Complex Network of Protein Interactions Underlies Cell Function

There are many challenges facing cell biologists in this "post-genome" era when complete genome sequences are known. One is the need to dissect and reconstruct each one of the thousands of protein machines that exist in an organism such as ourselves. To understand these remarkable protein complexes, each must be reconstituted from its purified protein parts, so that we can study its detailed mode of operation under controlled conditions in a test tube, free from all other cell components. This alone is a massive task. But we now know that each of these subcomponents of a cell also interacts with other sets of macromolecules, creating a large network of protein–protein and protein–nucleic acid interactions throughout the cell. To understand the cell, therefore, we need to analyze most of these other interactions as well.

We can gain some idea of the complexity of intracellular protein networks from a particularly well-studied example described in Chapter 16: the many dozens of proteins that interact with the actin cytoskeleton in the yeast *Saccharomyces cerevisiae* (see Figure 16–18). The extent of such protein–protein interactions can also be estimated more generally. An enormous amount of valuable information is now freely available in protein databases on the Internet: tens of thousands of three-dimensional protein structures plus tens of millions of protein sequences derived from the nucleotide sequences of genes. Scientists have been developing new methods for mining this great resource to increase our understanding of cells. In particular, computer-based bioinformatics tools are being combined with robotics and microarray technologies (see p. 574) to allow thousands of proteins to be investigated in a single set of experiments. **Proteomics** is a term that is often used to describe such research focused on the large-scale analysis of proteins, analogous to the term *genomics* describing the large-scale analysis of DNA sequences and genes.

Biologists use two different large-scale methods to map the direct binding interactions between the many different proteins in a cell. The initial method of choice was based on genetics: through an ingenious technique known as the yeast two-hybrid screen (see Figure 8–24), tens of thousands of interactions between thousands of proteins have been mapped in yeast, a nematode, and the fruit fly *Drosophila*. More recently, a biochemical method based on affinity tagging and mass spectroscopy has gained favor (discussed in Chapter 8), because it appears to produce fewer spurious results. The results of these and other analyses that predict protein binding interactions have been tabulated and organized in Internet databases. This allows a cell biologist studying a small set of proteins to readily discover which other proteins in the same cell are thought to bind to, and thus interact with, that set of proteins. When displayed graphically as a *protein interaction map*, each protein is represented by a box or dot in a two-dimensional network, with a straight line connecting those proteins that have been found to bind to each other.

When hundreds or thousands of proteins are displayed on the same map, the network diagram becomes bewilderingly complicated, serving to illustrate how much more we have to learn before we can claim to really understand the cell. Much more useful are small subsections of these maps, centered on a few proteins of interest. Thus, **Figure 3–82** shows a network of protein–protein interactions for the five proteins that form the SCF ubiquitin ligase in a yeast cell (see Figure 3–79). Four of the subunits of this ligase are located at the bottom right of Figure 3–82. The remaining subunit, the F-box protein that serves as its substrate-binding arm, appears as a set of 15 different gene products that bind to adaptor protein 2 (the Skp1 protein). Along the top and left of the figure are sets of additional protein interactions marked with *yellow* and *green* shading: as indicated, these protein sets function at the origin of DNA replication, in cell cycle regulation, in methionine synthesis, in the kinetochore, and in vacuolar H⁺-ATPase assembly. We shall use this figure to explain how such protein interaction maps are used, and what they do and do not mean.

1. Protein interaction maps are useful for identifying the likely function of previously uncharacterized proteins. Examples are the products of the genes that have thus far only been inferred to exist from the yeast genome sequence, which are the six proteins in the figure that lack a simple three-letter abbreviation (*white letters* beginning with Y). One, the product of so-called *open reading frame* YDR196C, is located in the origin of replication group, and it is therefore likely to have a role in starting new replication forks. The remaining five in this diagram are F-box proteins that bind to Skp1; these are therefore likely to function as part of the ubiquitin ligase, serving as substrate-binding arms that recognize different target proteins.

However, as we discuss next, neither assignment can be considered certain without additional data.

2. Protein interaction networks need to be interpreted with caution because, as a result of evolution making efficient use of each organism's genetic information, the same protein can be used as part of two different protein complexes that have different types of functions. Thus, although protein A binds to protein B and protein B binds to protein C, proteins A and C need not function in the same process. For example, we know from detailed biochemical studies that the functions of Skp1 in the kinetochore and in vacuolar H⁺-ATPase assembly *(yellow shading)* are separate from its function in the SCF ubiquitin ligase. In fact, only the remaining three functions of Skp1 illustrated in the diagram—methionine synthesis, cell cycle regulation, and origin of replication *(green shading)*—involve ubiquitylation.

3. In cross-species comparisons, those proteins displaying similar patterns of interactions in the two protein interaction maps are likely to have the same function in the cell. Thus, as scientists generate more and more highly detailed maps for multiple organisms, the results will become increasingly useful for inferring protein function. These map comparisons are a particularly powerful tool for deciphering the functions of human proteins. There is a vast amount of direct information about protein function that can be obtained from genetic engineering, mutational, and

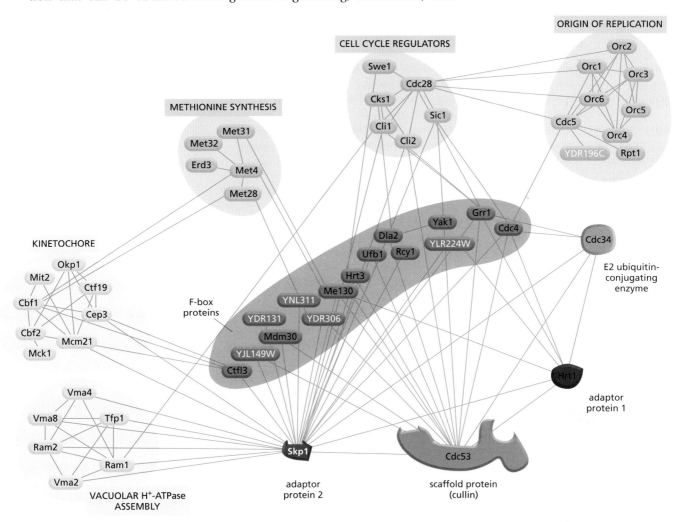

Figure 3–82 A map of some protein– protein interactions of the SCF ubiquitin ligase and other proteins in the yeast *S. cerevisiae.* The symbols and/or colors used for the 5 proteins of the ligase are those in Figure 3–79. Note that 15 different F-box proteins are shown *(purple)*; those with *white* lettering (beginning with Y) are only known from the genome sequence as open reading frames. For additional details, see text. (Courtesy of Peter Bowers and David Eisenberg, UCLA-DOE Institute for Genomics and Proteomics, UCLA.)

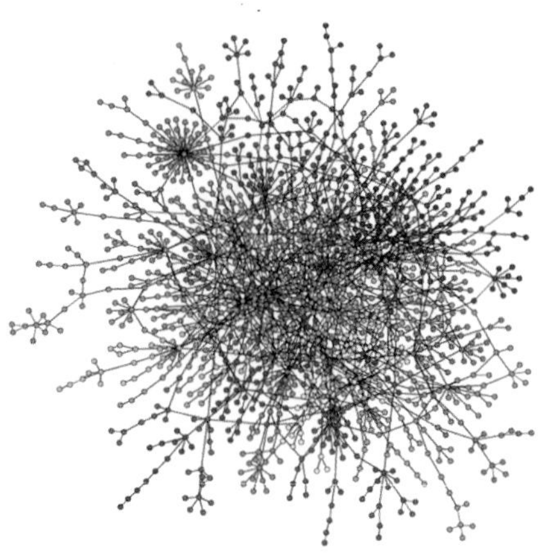

Figure 3–83 A network of protein-binding interactions in a yeast cell. Each line connecting a pair of dots (proteins) indicates a protein–protein interaction. (From A. Guimerá and M. Sales–Pardo, *Mol. Syst. Biol.* 2:42, 2006. With permission from Macmillan Publishers Ltd.)

genetic analyses in model organisms—such as yeast, worms, and flies—that is not available in humans.

The available data suggest that a typical protein in a human cell may interact with between 5 and 15 different partners. Often, each of the different domains in a multidomain protein binds to a different set of partners; in fact, we can speculate that the unusually extensive multidomain structures observed for human proteins may have evolved to facilitate these interactions. Given the enormous complexity of the interacting networks of macromolecules in cells (**Figure 3–83**), deciphering their full functional meaning may well keep scientists busy for centuries.

Summary

Proteins can form enormously sophisticated chemical devices, whose functions largely depend on the detailed chemical properties of their surfaces. Binding sites for ligands are formed as surface cavities in which precisely positioned amino acid side chains are brought together by protein folding. In this way, normally unreactive amino acid side chains can be activated to make and break covalent bonds. Enzymes are catalytic proteins that greatly speed up reaction rates by binding the high-energy transition states for a specific reaction path; they also perform acid catalysis and base catalysis simultaneously. The rates of enzyme reactions are often so fast that they are limited only by diffusion; rates can be further increased if enzymes that act sequentially on a substrate are joined into a single multienzyme complex, or if the enzymes and their substrates are confined to the same compartment of the cell.

Proteins reversibly change their shape when ligands bind to their surface. The allosteric changes in protein conformation produced by one ligand affect the binding of a second ligand, and this linkage between two ligand-binding sites provides a crucial mechanism for regulating cell processes. Metabolic pathways, for example, are controlled by feedback regulation: some small molecules inhibit and other small molecules activate enzymes early in a pathway. Enzymes controlled in this way generally form symmetric assemblies, allowing cooperative conformational changes to create a steep response to changes in the concentrations of the ligands that regulate them.

The expenditure of chemical energy can drive unidirectional changes in protein shape. By coupling allosteric shape changes to ATP hydrolysis, for example, proteins can do useful work, such as generating a mechanical force or moving for long distances in a single direction. The three-dimensional structures of proteins, determined by x-ray crystallography, have revealed how a small local change caused by nucleoside triphosphate hydrolysis is amplified to create major changes elsewhere in the protein. By such means, these proteins can serve as input–output devices that transmit information, as assembly factors, as motors, or as membrane-bound pumps. Highly efficient protein machines are formed by incorporating many different protein molecules into larger assemblies that coordinate the allosteric movements of the individual components. Such machines are now known to perform many of the most important reactions in cells.

Proteins are subjected to many reversible post-translational modifications, such as the covalent addition of a phosphate or an acetyl group to a specific amino acid side chain. The addition of these modifying groups is used to regulate the activity of a protein, changing its conformation, its binding to other proteins and its location inside the cell. A typical protein in a cell will interact with more than five different partners. Using the new technologies of proteomics, biologists can analyze thousands of proteins in one set of experiments. One important result is the production of detailed protein interaction maps, which aim at describing all of the binding interactions between the thousands of distinct proteins in a cell.

PROBLEMS

Which statements are true? Explain why or why not.

3–1 Each strand in a β sheet is a helix with two amino acids per turn.

3–2 Loops of polypeptide that protrude from the surface of a protein often form the binding sites for other molecules.

3–3 An enzyme reaches a maximum rate at high substrate concentration because it has a fixed number of active sites where substrate binds.

3–4 Higher concentrations of enzyme give rise to a higher turnover number.

3–5 Enzymes such as aspartate transcarbamoylase that undergo cooperative allosteric transitions invariably contain multiple identical subunits.

3–6 Continual addition and removal of phosphates by protein kinases and protein phosphatases is wasteful of energy—since their combined action consumes ATP—but it is a necessary consequence of effective regulation by phosphorylation.

Discuss the following problems.

3–7 Consider the following statement. "To produce one molecule of each possible kind of polypeptide chain, 300 amino acids in length, would require more atoms than exist in the universe." Given the size of the universe, do you suppose this statement could possibly be correct? Since counting atoms is a tricky business, consider the problem from the standpoint of mass. The mass of the observable universe is estimated to be about 10^{80} grams, give or take an order of magnitude or so. Assuming that the average mass of an amino acid is 110 daltons, what would be the mass of one molecule of each possible kind of polypeptide chain 300 amino acids in length? Is this greater than the mass of the universe?

3–8 A common strategy for identifying distantly related proteins is to search the database using a short signature sequence indicative of the particular protein function. Why is it better to search with a short sequence than with a long sequence? Do you not have more chances for a 'hit' in the database with a long sequence?

3–9 The so-called kelch motif consists of a four-stranded β sheet, which forms what is known as a β propeller. It is usually found to be repeated four to seven times, forming a kelch repeat domain in a multidomain protein. One such kelch repeat domain is shown in **Figure Q3–1**. Would you classify this domain as an 'in-line' or 'plug-in' type domain?

3–10 Titin, which has a molecular weight of 3×10^6 daltons, is the largest polypeptide yet described. Titin molecules extend from muscle thick filaments to the Z disc; they are thought to act as springs to keep the thick filaments centered in the sarcomere. Titin is composed of a large number of repeated immunoglobulin (Ig) sequences of 89 amino acids, each of which is folded into a domain about 4 nm in length (**Figure Q3–2A**).

You suspect that the springlike behavior of titin is caused by the sequential unfolding (and refolding) of individual Ig

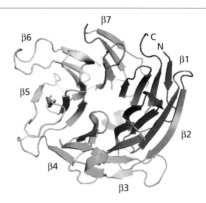

Figure Q3–1 The kelch repeat domain of galactose oxidase from *D. dendroides* (Problem 3–9). The seven individual β propellers are indicated. The N- and C-termini are indicated by N and C.

domains. You test this hypothesis using the atomic force microscope, which allows you to pick up one end of a protein molecule and pull with an accurately measured force. For a fragment of titin containing seven repeats of the Ig domain, this experiment gives the sawtooth force-versus-extension curve shown in Figure Q3–2B. When the experiment is repeated in a solution of 8 M urea (a protein denaturant), the peaks disappear and the measured extension becomes much longer for a given force. If the experiment is repeated after the protein has been cross-linked by treatment with glutaraldehyde, once again the peaks disappear but the extension becomes much smaller for a given force.

A. Are the data consistent with your hypothesis that titin's springlike behavior is due to the sequential unfolding of individual Ig domains? Explain your reasoning.

B. Is the extension for each putative domain-unfolding event the magnitude you would expect? (In an extended polypeptide chain, amino acids are spaced at intervals of 0.34 nm.)

C. Why is each successive peak in Figure Q3–2B a little higher than the one before?

D. Why does the force collapse so abruptly after each peak?

3–11 It is often said that protein complexes are made from subunits (that is, individually synthesized proteins) rather than as one long protein because the former is more likely to give a correct final structure.

A. Assuming that the protein synthesis machinery incorporates one incorrect amino acid for each 10,000 it inserts,

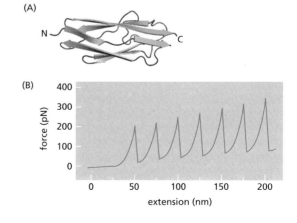

Figure Q3–2 Springlike behavior of titin (Problem 3–10). (A) The structure of an individual Ig domain. (B) Force in piconewtons versus extension in nanometers obtained by atomic force microscopy.

calculate the fraction of bacterial ribosomes that would be assembled correctly if the proteins were synthesized as one large protein versus built from individual proteins? For the sake of calculation assume that the ribosome is composed of 50 proteins, each 200 amino acids in length, and that the subunits—correct and incorrect—are assembled with equal likelihood into the complete ribosome. [The probability that a polypeptide will be made correctly, P_C, equals the fraction correct for each operation, f_C, raised to a power equal to the number of operations, n: $P_C = (f_C)^n$. For an error rate of 1/10,000, $f_C = 0.9999$.]

B.   Is the assumption that correct and incorrect subunits assemble equally well likely to be true? Why or why not? How would a change in that assumption affect the calculation in part A?

3–12   Rous sarcoma virus (RSV) carries an oncogene called *Src*, which encodes a continuously active protein tyrosine kinase that leads to unchecked cell proliferation. Normally, Src carries an attached fatty acid (myristoylate) group that allows it to bind to the cytoplasmic side of the plasma membrane. A mutant version of Src that does not allow attachment of myristoylate does not bind to the membrane. Infection of cells with RSV encoding either the normal or the mutant form of Src leads to the same high level of protein tyrosine kinase activity, but the mutant Src does not cause cell proliferation.

A.   Assuming that the normal Src is all bound to the plasma membrane and that the mutant Src is distributed throughout the cytoplasm, calculate their relative concentrations in the neighborhood of the plasma membrane. For the purposes of this calculation, assume that the cell is a sphere with a radius of 10 μm and that the mutant Src is distributed throughout, whereas the normal Src is confined to a 4-nm-thick layer immediately beneath the membrane. [For this problem, assume that the membrane has no thickness. The volume of a sphere is $(4/3)\pi r^3$.]

B.   The target (X) for phosphorylation by *Src* resides in the membrane. Explain why the mutant Src does not cause cell proliferation.

3–13   An antibody binds to another protein with an equilibrium constant, K, of 5×10^9 M^{-1}. When it binds to a second, related protein, it forms three fewer hydrogen bonds, reducing its binding affinity by 2.8 kcal/mole. What is the K for its binding to the second protein? (Free-energy change is related to the equilibrium constant by the equation $\Delta G° = -2.3\ RT \log K$, where R is 1.98×10^{-3} kcal/(mole K) and T is 310 K.)

3–14   The protein SmpB binds to a special species of tRNA, tmRNA, to eliminate the incomplete proteins made from truncated mRNAs in bacteria. If the binding of SmpB to tmRNA is plotted as fraction tmRNA bound versus SmpB concentration, one obtains a symmetrical S-shaped curve as shown in **Figure Q3–3**. This curve is a visual display of a very useful relationship between K_d and concentration, which has broad applicability. The general expression for fraction of ligand bound is derived from the equation for K_d ($K_d =$ [Pr][L]/[Pr–L]) by substituting ([L]$_{TOT}$ – [L]) for [Pr–L] and rearranging. Because the total concentration of ligand ([L]$_{TOT}$) is equal to the free ligand ([L]) plus bound ligand ([Pr–L]),

$$\text{fraction bound} = [L]/[L]_{TOT} = [Pr]/([Pr] + K_d)$$

For SmpB and tmRNA, the fraction bound = [tmRNA]/

[tmRNA]$_{TOT}$ = [SmpB]/([SmpB] + K_d). Using this relationship, calculate the fraction of tmRNA bound for SmpB concentrations equal to $10^4 K_d$, $10^3 K_d$, $10^2 K_d$, $10^1 K_d$, K_d, $10^{-1}K_d$, $10^{-2}K_d$, $10^{-3}K_d$, and $10^{-4}K_d$.

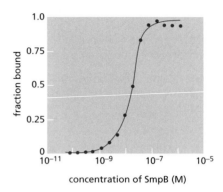

Figure Q3–3 Fraction of tmRNA bound versus SmpB concentration (Problem 3–14).

3–15   Many enzymes obey simple Michaelis–Menten kinetics, which are summarized by the equation

$$\text{rate} = V_{max}\,[S]/([S] + K_m)$$

where V_{max} = maximum velocity, [S] = concentration of substrate, and K_m = the Michaelis constant.

It is instructive to plug a few values of [S] into the equation to see how rate is affected. What are the rates for [S] equal to zero, equal to K_m, and equal to infinite concentration?

3–16   The enzyme hexokinase adds a phosphate to D-glucose but ignores its mirror image, L-glucose. Suppose that you were able to synthesize hexokinase entirely from D-amino acids, which are the mirror image of the normal L-amino acids.

A.   Assuming that the 'D' enzyme would fold to a stable conformation, what relationship would you expect it to bear to the normal 'L' enzyme?

B.   Do you suppose the 'D' enzyme would add a phosphate to L-glucose, and ignore D-glucose?

3–17   How do you suppose that a molecule of hemoglobin is able to bind oxygen efficiently in the lungs, and yet release it efficiently in the tissues?

3–18   Synthesis of the purine nucleotides AMP and GMP proceeds by a branched pathway starting with ribose 5-phosphate (R5P), as shown schematically in **Figure Q3–4**. Using the principles of feedback inhibition, propose a regulatory strategy for this pathway that ensures an adequate supply of both AMP and GMP and minimizes the buildup of the intermediates (*A–I*) when supplies of AMP and GMP are adequate.

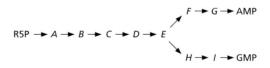

Figure Q3–4 Schematic diagram of the metabolic pathway for synthesis of AMP and GMP from R5P (Problem 3–18).

REFERENCES

General

Berg JM, Tymoczko JL & Stryer L (2006) Biochemistry, 6th ed. New York: WH Freeman.

Branden C & Tooze J (1999) Introduction to Protein Structure, 2nd ed. New York: Garland Science.

Dickerson, RE (2005) Present at the Flood: How Structural Molecular Biology Came About. Sunderland, MA: Sinauer

Kyte J (2006) Structure in Protein Chemistry. New York: Routledge.

Petsko GA & Ringe D (2004) Protein Structure and Function. London: New Science Press.

Perutz M (1992) Protein Structure: New Approaches to Disease and Therapy. New York: WH Freeman.

The Shape and Structure of Proteins

Anfinsen CB (1973) Principles that govern the folding of protein chains. *Science* 181:223–230.

Bray D (2005) Flexible peptides and cytoplasmic gels. *Genome Biol* 6:106–109.

Burkhard P, Stetefeld J & Strelkov SV (2001) Coiled coils: a highly versatile protein folding motif. *Trends Cell Biol* 11:82–88.

Caspar DLD & Klug A (1962) Physical principles in the construction of regular viruses. *Cold Spring Harb Symp Quant Biol* 27:1–24.

Doolittle RF (1995) The multiplicity of domains in proteins. *Annu Rev Biochem* 64:287–314.

Eisenberg D (2003) The discovery of the alpha-helix and beta-sheet, the principle structural features of proteins. *Proc Natl Acad Sci USA* 100:11207–11210.

Fraenkel-Conrat H & Williams RC (1955) Reconstitution of active tobacco mosaic virus from its inactive protein and nucleic acid components. *Proc Natl Acad Sci USA* 41:690–698.

Goodsell DS & Olson AJ (2000) Structural symmetry and protein function. *Annu Rev Biophys Biomol Struct* 29:105–153.

Harrison SC (1992) Viruses. *Curr Opin Struct Biol* 2:293–299.

Harrison SC (2004) Whither structural biology? *Nature Struct Mol Biol* 11:12–15.

Hudder A, Nathanson L & Deutscher MP (2003) Organization of mammalian cytoplasm. *Mol Cell Biol* 23:9318–9326.

International Human Genome Sequencing Consortium (2001) Initial sequencing and analysis of the human genome. *Nature* 409:860–921.

Meiler J & Baker D (2003) Coupled prediction of protein secondary and tertiary structure. *Proc Natl Acad Sci USA* 100:12105–12110.

Nomura M (1973) Assembly of bacterial ribosomes. *Science* 179:864–873.

Orengo CA & Thornton JM (2005) Protein families and their evolution—a structural perspective. *Annu Rev Biochem* 74:867–900.

Pauling L & Corey RB (1951) Configurations of polypeptide chains with favored orientations around single bonds: two new pleated sheets. *Proc Natl Acad Sci USA* 37:729–740.

Pauling L, Corey RB & Branson HR (1951) The structure of proteins: two hydrogen-bonded helical configurations of the polypeptide chain. *Proc Natl Acad Sci USA* 37:205–211.

Ponting CP, Schultz J, Copley RR et al. (2000) Evolution of domain families. *Adv Protein Chem* 54:185–244.

Trinick J (1992) Understanding the functions of titin and nebulin. *FEBS Lett* 307:44–48.

Vogel C, Bashton M, Kerrison ND et al (2004) Structure, function and evolution of multidomain proteins. *Curr Opin Struct Biol* 14:208–216.

Zhang C & Kim SH (2003) Overview of structural genomics: from structure to function. *Curr Opin Chem Biol* 7:28–32.

Protein Function

Alberts B (1998) The cell as a collection of protein machines: preparing the next generation of molecular biologists. *Cell* 92:291–294.

Benkovic SJ (1992) Catalytic antibodies. *Annu Rev Biochem* 61:29–54.

Berg OG & von Hippel PH (1985) Diffusion-controlled macromolecular interactions. *Annu Rev Biophys Biophys Chem* 14:131–160.

Bhattacharyya RP, Remenyi A, Yeh BJ & Lim WA (2006) Domains, motifs, and scaffolds: The role of modular interactions in the evolution and wiring of cell signaling circuits. *Annu Rev Biochem* 75:655–680.

Bourne HR (1995) GTPases: a family of molecular switches and clocks. *Philos Trans R Soc Lond B* 349:283–289.

Braden BC & Poljak RJ (1995) Structural features of the reactions between antibodies and protein antigens. *FASEB J* 9:9–16.

Dickerson RE & Geis I (1983) Hemoglobin: Structure, Function, Evolution and Pathology. Menlo Park, CA: Benjamin Cummings.

Dressler D & Potter H (1991) Discovering Enzymes. New York: Scientific American Library.

Eisenberg D, Marcotte, EM, Xenarios, I & Yeates TO (2000) Protein function in the post-genomic era. *Nature* 405:823–826.

Fersht AR (1999) Structure and Mechanisms in Protein Science: A Guide to Enzyme Catalysis. New York: WH Freeman.

Johnson, LN & Lewis RJ (2001) Structural basis for control by phosphorylation. *Chem Rev* 101:2209–2242.

Kantrowitz ER & Lipscomb WN (1988) *Escherichia coli* aspartate transcarbamoylase: the relation between structure and function. *Science* 241:669–674.

Khosla C & Harbury PB (2001) Modular enzymes. *Nature* 409:247–252.

Kim E & Sheng M (2004) PDZ domain proteins of synapses. *Nature Rev Neurosci* 5:771–781.

Koshland DE, Jr (1984) Control of enzyme activity and metabolic pathways. *Trends Biochem Sci* 9:155–159.

Kraut DA, Carroll KS & Herschlag D (2003) Challenges in enzyme mechanism and energetics. *Annu Rev Biochem* 72:517–571.

Krogan NJ, Cagney G, Yu H et al (2006) Global landscape of protein complexes in the yeast *Saccharomyces cerevisisae*. *Nature* 440:637–643.

Lichtarge O, Bourne HR & Cohen FE (1996) An evolutionary trace method defines binding surfaces common to protein families. *J Mol Biol* 257:342–358.

Marcotte EM, Pellegrini M, Ng HL et al (1999) Detecting protein function and protein–protein interactions from genome sequences. *Science* 285:751–753.

Monod J, Changeux JP & Jacob F (1963) Allosteric proteins and cellular control systems. *J Mol Biol* 6:306–329.

Pawson T & Nash P (2003) Assembly of regulatory systems through protein interaction domains. *Science* 300:445–452.

Pavletich NP (1999) Mechanisms of cyclin-dependent kinase regulation: structures of Cdks, their cyclin activators, and Cip and INK4 inhibitors. *J Mol Biol* 287:821–828.

Pellicena P & Kuriyan J (2006) Protein-protein interactions in the allosteric regulation of protein kinases. *Curr Opin Struct Biol* 16:702–709.

Perutz M (1990) Mechanisms of Cooperativity and Allosteric Regulation in Proteins. Cambridge: Cambridge University Press.

Raushel F, Thoden, JB & Holden HM (2003) Enzymes with molecular tunnels. *Acc Chem Res* 36:539–548.

Radzicka A & Wolfenden R (1995) A proficient enzyme. *Science* 267:90–93.

Sato TK, Overduin M & Emr S (2001) Location, location, location: Membrane targeting directed by PX domains. *Science* 294:1881–1885.

Schramm VL (1998) Enzymatic transition states and transition state analog design. *Annu Rev Biochem* 67:693–720.

Schultz PG & Lerner RA (1995) From molecular diversity to catalysis: lessons from the immune system. *Science* 269:1835–1842.

Vale RD & Milligan RA (2000) The way things move: looking under the hood of molecular motor proteins. *Science* 288:88–95.

Vocadlo DJ, Davies GJ, Laine R & Withers SG (2001) Catalysis by hen egg-white lysozyme proceeds via a covalent intermediate. *Nature* 412:835–838.

Walsh C (2001) Enabling the chemistry of life. *Nature* 409:226–231.

Yang XJ (2005) Multisite protein modification and intramolecular signaling. *Oncogene* 24:1653–1662.

Zhu H, Bilgin M & Snyder M (2003) Proteomics. *Annu Rev Biochem* 72:783–812.

II BASIC GENETIC MECHANISMS

Part II
Chapters

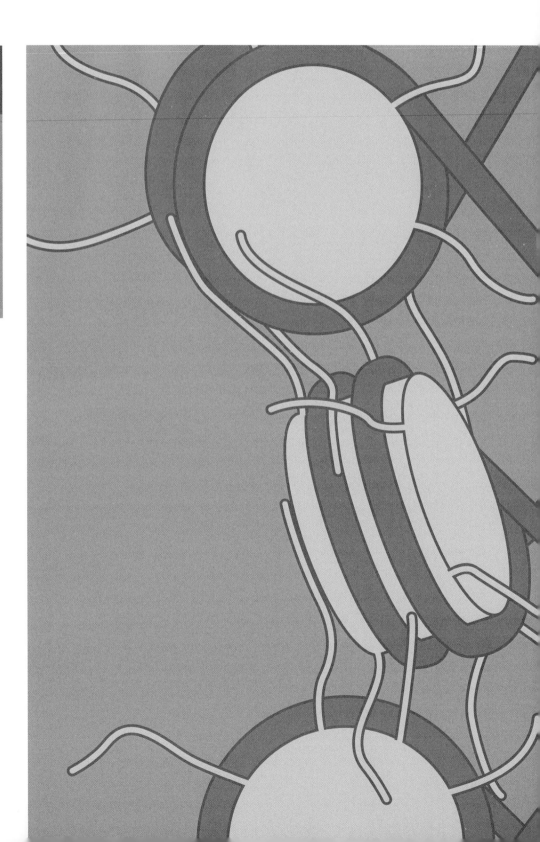

Chapter 4

DNA, Chromosomes, and Genomes

4

Life depends on the ability of cells to store, retrieve, and translate the genetic instructions required to make and maintain a living organism. This *hereditary* information is passed on from a cell to its daughter cells at cell division, and from one generation of an organism to the next through the organism's reproductive cells. These instructions are stored within every living cell as its **genes**, the information-containing elements that determine the characteristics of a species as a whole and of the individuals within it.

As soon as genetics emerged as a science at the beginning of the twentieth century, scientists became intrigued by the chemical structure of genes. The information in genes is copied and transmitted from cell to daughter cell millions of times during the life of a multicellular organism, and it survives the process essentially unchanged. What form of molecule could be capable of such accurate and almost unlimited replication and also be able to direct the development of an organism and the daily life of a cell? What kind of instructions does the genetic information contain? How can the enormous amount of information required for the development and maintenance of an organism fit within the tiny space of a cell?

The answers to several of these questions began to emerge in the 1940s. At this time, researchers discovered, from studies in simple fungi, that genetic information consists primarily of instructions for making proteins. Proteins are the macromolecules that perform most cell functions: they serve as building blocks for cell structures and form the enzymes that catalyze the cell's chemical reactions (Chapter 3), they regulate gene expression (Chapter 7), and they enable cells to communicate with each other (Chapter 15) and to move (Chapter 16). The properties and functions of a cell are determined largely by the proteins that it is able to make. With hindsight, it is hard to imagine what other type of instructions the genetic information could have contained.

Painstaking observations of cells and embryos in the late 19th century had led to the recognition that the hereditary information is carried on *chromosomes*, threadlike structures in the nucleus of a eucaryotic cell that become visible by light microscopy as the cell begins to divide (**Figure 4–1**). Later, as biochemical analysis became possible, chromosomes were found to consist of both deoxyribonucleic acid (DNA) and protein. For many decades, the DNA was thought to be merely a structural element. However, the other crucial advance made in the 1940s was the identification of DNA as the likely carrier of genetic information. This breakthrough in our understanding of cells came from studies

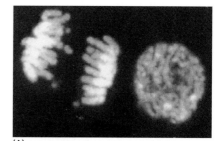

(A) dividing cell   nondividing cell

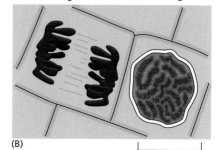

(B)

10 μm

Figure 4–1 Chromosomes in cells. (A) Two adjacent plant cells photographed through a light microscope. The DNA has been stained with a fluorescent dye (DAPI) that binds to it. The DNA is present in chromosomes, which become visible as distinct structures in the light microscope only when they become compact, sausage-shaped structures in preparation for cell division, as shown on the left. The cell on the right, which is not dividing, contains identical chromosomes, but they cannot be clearly distinguished in the light microscope at this phase in the cell's life cycle, because they are in a more extended conformation. (B) Schematic diagram of the outlines of the two cells along with their chromosomes. (A, courtesy of Peter Shaw.)

of inheritance in bacteria (**Figure 4–2**). But as the 1950s began, both how proteins could be specified by instructions in the DNA and how this information might be copied for transmission from cell to cell seemed completely mysterious. The mystery was suddenly solved in 1953, when the structure of DNA was correctly predicted by James Watson and Francis Crick. As outlined in Chapter 1, the double-helical structure of DNA immediately solved the problem of how the information in this molecule might be copied, or *replicated*. It also provided the first clues as to how a molecule of DNA might use the sequence of its subunits to encode the instructions for making proteins. Today, the fact that DNA is the genetic material is so fundamental to biological thought that it is difficult to appreciate the enormous intellectual gap that was filled.

In this chapter we begin by describing the structure of DNA. We see how, despite its chemical simplicity, the structure and chemical properties of DNA make it ideally suited as the raw material of genes. We then consider how the many proteins in chromosomes arrange and package this DNA. The packing has to be done in an orderly fashion so that the chromosomes can be replicated and apportioned correctly between the two daughter cells at each cell division. It must also allow access to chromosomal DNA for the enzymes that repair it when it is damaged and for the specialized proteins that direct the expression of its many genes. We shall also see how the packaging of DNA differs along the length of each chromosome in eucaryotes, and how it can store a valuable record of the cell's developmental history.

In the past two decades, there has been a revolution in our ability to determine the exact sequence of subunits in DNA molecules. As a result, we now know the order of the 3 billion DNA subunits that provide the information for producing a human adult from a fertilized egg, as well as the DNA sequences of thousands of other organisms. Detailed analyses of these sequences have provided exciting insights into the process of evolution, and it is with this subject that the chapter ends.

This is the first of four chapters that deal with basic genetic mechanisms—the ways in which the cell maintains, replicates, expresses, and occasionally improves the genetic information carried in its DNA. This chapter presents a broad overview of DNA and how it is packaged into chromosomes. In the following chapter (Chapter 5) we discuss the mechanisms by which the cell accurately replicates and repairs DNA; we also describe how DNA sequences can be

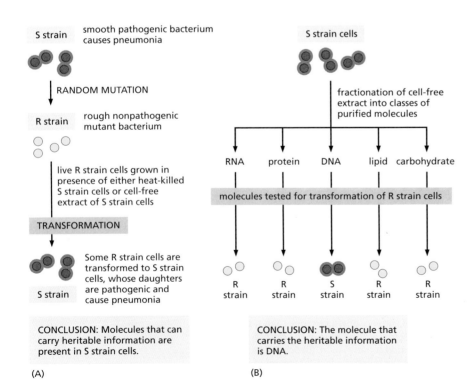

Figure 4–2 **The first experimental demonstration that DNA is the genetic material.** These experiments, carried out in the 1940s, showed that adding purified DNA to a bacterium changed its properties and that this change was faithfully passed on to subsequent generations. Two closely related strains of the bacterium *Streptococcus pneumoniae* differ from each other in both their appearance under the microscope and their pathogenicity. One strain appears smooth (S) and causes death when injected into mice, and the other appears rough (R) and is nonlethal. (A) An initial experiment shows that a substance present in the S strain can change (or transform) the R strain into the S strain and that this change is inherited by subsequent generations of bacteria. (B) This experiment, in which the R strain has been incubated with various classes of biological molecules purified from the S strain, identifies the substance as DNA.

rearranged through the process of genetic recombination. Gene expression—the process through which the information encoded in DNA is interpreted by the cell to guide the synthesis of proteins—is the main topic of Chapter 6. In Chapter 7, we describe how this gene expression is controlled by the cell to ensure that each of the many thousands of proteins and RNA molecules encrypted in its DNA are manufactured only at the proper time and place in the life of the cell.

THE STRUCTURE AND FUNCTION OF DNA

Biologists in the 1940s had difficulty in conceiving how DNA could be the genetic material because of the apparent simplicity of its chemistry. DNA was known to be a long polymer composed of only four types of subunits, which resemble one another chemically. Early in the 1950s, DNA was examined by x-ray diffraction analysis, a technique for determining the three-dimensional atomic structure of a molecule (discussed in Chapter 8). The early x-ray diffraction results indicated that DNA was composed of two strands of the polymer wound into a helix. The observation that DNA was double-stranded was of crucial significance and provided one of the major clues that led to the Watson–Crick model for DNA structure. But only when this model was proposed in 1953 did DNA's potential for replication and information encoding become apparent. In this section we examine the structure of the DNA molecule and explain in general terms how it is able to store hereditary information.

A DNA Molecule Consists of Two Complementary Chains of Nucleotides

A **deoxyribonucleic acid** (**DNA**) molecule consists of two long polynucleotide chains composed of four types of nucleotide subunits. Each of these chains is known as a *DNA chain*, or a *DNA strand. Hydrogen bonds* between the base portions of the nucleotides hold the two chains together (**Figure 4–3**). As we saw in Chapter 2 (Panel 2–6, pp. 116–117), nucleotides are composed of a five-carbon sugar to which are attached one or more phosphate groups and a nitrogen-containing base. In the case of the nucleotides in DNA, the sugar is deoxyribose attached to a single phosphate group (hence the name deoxyribonucleic acid), and the base may be either *adenine (A), cytosine (C), guanine (G),* or *thymine (T).* The nucleotides are covalently linked together in a chain through the sugars and phosphates, which thus form a "backbone" of alternating sugar–phosphate–sugar–phosphate. Because only the base differs in each of the four types of subunits, each polynucleotide chain in DNA is analogous to a necklace (the backbone) strung with four types of beads (the four bases A, C, G, and T). These same symbols (A, C, G, and T) are also commonly used to denote the four different nucleotides—that is, the bases with their attached sugar and phosphate groups.

The way in which the nucleotide subunits are linked together gives a DNA strand a chemical polarity. If we think of each sugar as a block with a protruding knob (the 5′ phosphate) on one side and a hole (the 3′ hydroxyl) on the other (see Figure 4–3), each completed chain, formed by interlocking knobs with holes, will have all of its subunits lined up in the same orientation. Moreover, the two ends of the chain will be easily distinguishable, as one has a hole (the 3′ hydroxyl) and the other a knob (the 5′ phosphate) at its terminus. This polarity in a DNA chain is indicated by referring to one end as the *3′ end* and the other as the *5′ end*.

The three-dimensional structure of DNA—the **double helix**—arises from the chemical and structural features of its two polynucleotide chains. Because these two chains are held together by hydrogen bonding between the bases on the different strands, all the bases are on the inside of the double helix, and the sugar-phosphate backbones are on the outside (see Figure 4–3). In each case, a bulkier two-ring base (a purine; see Panel 2–6, pp. 116–117) is paired with a single-ring base (a pyrimidine); A always pairs with T, and G with C (**Figure**

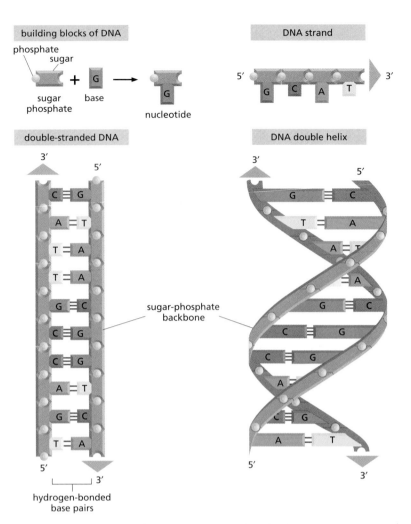

Figure 4–3 DNA and its building blocks. <CAGA> DNA is made of four types of nucleotides, which are linked covalently into a polynucleotide chain (a DNA strand) with a sugar-phosphate backbone from which the bases (A, C, G, and T) extend. A DNA molecule is composed of two DNA strands held together by hydrogen bonds between the paired bases. The *arrowheads* at the ends of the DNA strands indicate the polarities of the two strands, which run antiparallel to each other in the DNA molecule. In the diagram at the bottom left of the figure, the DNA molecule is shown straightened out; in reality, it is twisted into a double helix, as shown on the right. For details, see Figure 4–5.

4–4). This *complementary base-pairing* enables the **base pairs** to be packed in the energetically most favorable arrangement in the interior of the double helix. In this arrangement, each base pair is of similar width, thus holding the sugar-phosphate backbones an equal distance apart along the DNA molecule. To maximize the efficiency of base-pair packing, the two sugar-phosphate backbones

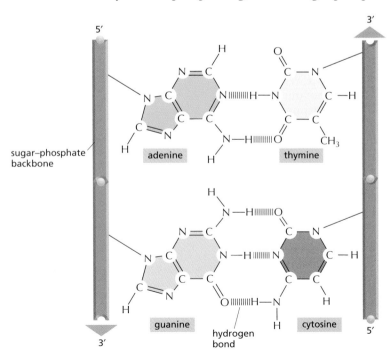

Figure 4–4 Complementary base pairs in the DNA double helix. The shapes and chemical structure of the bases allow hydrogen bonds to form efficiently only between A and T and between G and C, where atoms that are able to form hydrogen bonds (see Panel 2–3, pp. 110–111) can be brought close together without distorting the double helix. As indicated, two hydrogen bonds form between A and T, while three form between G and C. The bases can pair in this way only if the two polynucleotide chains that contain them are antiparallel to each other.

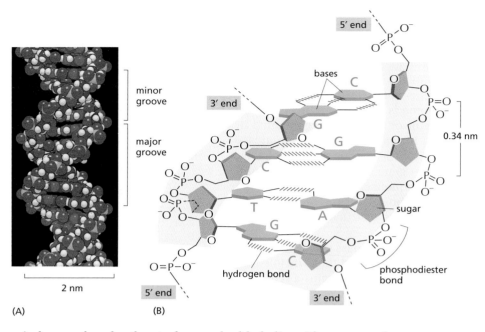

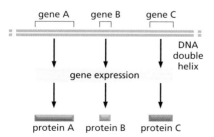

Figure 4–5 **The DNA double helix.**
(A) A space-filling model of 1.5 turns of the DNA double helix. Each turn of DNA is made up of 10.4 nucleotide pairs, and the center-to-center distance between adjacent nucleotide pairs is 3.4 nm. The coiling of the two strands around each other creates two grooves in the double helix: the wider groove is called the major groove, and the smaller the minor groove. (B) A short section of the double helix viewed from its side, showing four base pairs. The nucleotides are linked together covalently by phosphodiester bonds that join the 3′-hydroxyl (–OH) group of one sugar to the 5′-hydroxyl group of the next sugar. Thus, each polynucleotide strand has a chemical polarity; that is, its two ends are chemically different. The 5′ end of the DNA polymer is by convention often illustrated carrying a phosphate group, while the 3′-end is shown with a hydroxyl.

wind around each other to form a double helix, with one complete turn every ten base pairs (**Figure 4–5**).

The members of each base pair can fit together within the double helix only if the two strands of the helix are **antiparallel**—that is, only if the polarity of one strand is oriented opposite to that of the other strand (see Figures 4–3 and 4–4). A consequence of these base-pairing requirements is that each strand of a DNA molecule contains a sequence of nucleotides that is exactly **complementary** to the nucleotide sequence of its partner strand.

The Structure of DNA Provides a Mechanism for Heredity

Genes carry biological information that must be copied accurately for transmission to the next generation each time a cell divides to form two daughter cells. Two central biological questions arise from these requirements: how can the information for specifying an organism be carried in chemical form, and how is it accurately copied? The discovery of the structure of the DNA double helix was a landmark in twentieth-century biology because it immediately suggested answers to both questions, thereby providing a molecular explanation for the problem of heredity. We discuss these answers briefly in this section, and we shall examine them in much more detail in subsequent chapters.

DNA encodes information through the order, or sequence, of the nucleotides along each strand. Each base—A, C, T, or G—can be considered as a letter in a four-letter alphabet that spells out biological messages in the chemical structure of the DNA. As we saw in Chapter 1, organisms differ from one another because their respective DNA molecules have different nucleotide sequences and, consequently, carry different biological messages. But how is the nucleotide alphabet used to make messages, and what do they spell out?

As discussed above, it was known well before the structure of DNA was determined that genes contain the instructions for producing proteins. The DNA messages must therefore somehow encode proteins (**Figure 4–6**). This relationship immediately makes the problem easier to understand. As discussed in Chapter 3, the properties of a protein, which are responsible for its biological function, are determined by its three-dimensional structure. This structure is determined in turn by the linear sequence of the amino acids of which it is composed. The linear sequence of nucleotides in a gene must therefore somehow spell out the linear sequence of amino acids in a protein. The exact correspondence between the four-letter nucleotide alphabet of DNA and the twenty-letter amino acid alphabet of proteins—the genetic code—is not obvious from the DNA structure, and it took over a decade after the discovery of the double helix

Figure 4–6 **The relationship between genetic information carried in DNA and proteins (discussed in Chapter 1).**

before it was worked out. In Chapter 6 we will describe this code in detail in the course of elaborating the process, known as *gene expression*, through which a cell converts the nucleotide sequence of a gene first into the nucleotide sequence of an RNA molecule, and then into the amino acid sequence of a protein.

The complete set of information in an organism's DNA is called its **genome**, and it carries the information for all the proteins and RNA molecules that the organism will ever synthesize. (The term genome is also used to describe the DNA that carries this information.) The amount of information contained in genomes is staggering: for example, a typical human diploid cell contains 2 meters of DNA double helix. Written out in the four-letter nucleotide alphabet, the nucleotide sequence of a very small human gene occupies a quarter of a page of text (**Figure 4–7**), while the complete sequence of nucleotides in the human genome would fill more than a thousand books the size of this one. In addition to other critical information, it carries the instructions for roughly 24,000 distinct proteins.

At each cell division, the cell must copy its genome to pass it to both daughter cells. The discovery of the structure of DNA also revealed the principle that makes this copying possible: because each strand of DNA contains a sequence of nucleotides that is exactly complementary to the nucleotide sequence of its partner strand, each strand can act as a **template**, or mold, for the synthesis of a new complementary strand. In other words, if we designate the two DNA strands as S and S′, strand S can serve as a template for making a new strand S′, while strand S′ can serve as a template for making a new strand S (**Figure 4–8**). Thus, the genetic information in DNA can be accurately copied by the beautifully simple process in which strand S separates from strand S′, and each separated strand then serves as a template for the production of a new complementary partner strand that is identical to its former partner.

The ability of each strand of a DNA molecule to act as a template for producing a complementary strand enables a cell to copy, or *replicate*, its genome before passing it on to its descendants. In the next chapter we shall describe the elegant machinery the cell uses to perform this enormous task.

In Eucaryotes, DNA Is Enclosed in a Cell Nucleus

As described in Chapter 1, nearly all the DNA in a eucaryotic cell is sequestered in a nucleus, which in many cells occupies about 10% of the total cell volume. This compartment is delimited by a *nuclear envelope* formed by two concentric lipid bilayer membranes (**Figure 4–9**). These membranes are punctured at intervals by large nuclear pores, which transport molecules between the nucleus and the cytosol. The nuclear envelope is directly connected to the extensive membranes of the endoplasmic reticulum, which extend out from it into the cytoplasm. And it is mechanically supported by a network of intermediate filaments called the *nuclear lamina*, which forms a thin sheetlike meshwork just beneath the inner nuclear membrane (see Figure 4–9B).

The nuclear envelope allows the many proteins that act on DNA to be concentrated where they are needed in the cell, and, as we see in subsequent

Figure 4–7 The nucleotide sequence of the human β-globin gene. By convention, a nucleotide sequence is written from its 5′ end to its 3′ end, and it should be read from left to right in successive lines down the page as though it were normal English text. This gene carries the information for the amino acid sequence of one of the two types of subunits of the hemoglobin molecule, the protein that carries oxygen in the blood. A different gene, the α-globin gene, carries the information for the other type of hemoglobin subunit (a hemoglobin molecule has four subunits, two of each type). Only one of the two strands of the DNA double helix containing the β-globin gene is shown; the other strand has the exact complementary sequence. The DNA sequences highlighted in *yellow* show the three regions of the gene that specify the amino acid sequence for the β-globin protein. We shall see in Chapter 6 how the cell splices these three sequences together at the level of messenger RNA in order to synthesize a full-length β-globin protein.

```
CCCTGTGGAGCCACACCCTAGGGTTGGCCA
ATCTACTCCCAGGAGCAGGGAGGGCAGGAG
CCAGGGCTGGGCATAAAAGTCAGGGCAGAG
CCATCTATTGCTTACATTTGCTTCTGACAC
AACTGTGTTCACTAGCAACTCAAACAGACA
CCATGGTGCACCTGACTCCTGAGGAGAAGT
CTGCCGTTACTGCCCTGTGGGGCAAGGTGA
ACGTGGATGAAGTTGGTGGTGAGGCCCTGG
GCAGGTTGGTATCAAGGTTACAAGACAGGT
TTAAGGAGACCAATAGAAACTGGGCATGTG
GAGACAGAGAAGACTCTTGGGTTTCTGATA
GGCACTGACTCTCTCTGCCTATTGGTCTAT
TTTCCCACCCTTAGGCTGCTGGTGGTCTAC
CCTTGGACCCAGAGGTTCTTTGAGTCCTTT
GGGGATCTGTCCACTCCTGATGCTGTTATG
GGCAACCCTAAGGTGAAGGCTCATGGCAAG
AAAGTGCTCGGTGCCTTTAGTGATGGCCTG
GCTCACCTGGACAACCTCAAGGGCACCTTT
GCCACACTGAGTGAGCTGCACTGTGACAAG
CTGCACGTGGATCCTGAGAACTTCAGGGTG
AGTCTATGGGACCCTTGATGTTTTCTTTCC
CCTTCTTTTCTATGGTTAAGTTCATGTCAT
AGGAAGGGGAGAAGTAACAGGGTACAGTTT
AGAATGGGAAACAGACGAATGATTGCATCA
GTGTGGAAGTCTCAGGATCGTTTTAGTTTC
TTTTATTTGCTGTTCATAACAATTGTTTTC
TTTTGTTTAATTCTTGCTTTCTTTTTTTTT
CTTCTCCGCAATTTTTACTATTATACTTAA
TGCCTTAACATTGTGTATAACAAAAGGAAA
TATCTCTGAGATACATTAAGTAACTTAAAA
AAAAACTTTACACAGTCTGCCTAGTACATT
ACTATTTGGAATATATGTGTGCTTATTTGC
ATATTCATAATCTCCCTACTTTATTTTCTT
TTATTTTTAATTGATACATAATCATTATAC
ATATTTATGGGTTAAAGTGTAATGTTTTAA
TATGTGTACACATATTGACCAAATCAGGGT
AATTTTGCATTTGTAATTTTAAAAAATGCT
TTCTTCTTTTAATATACTTTTTTTGTTTATC
TTATTTCTAATACTTTCCCTAATCTCTTTC
TTTCAGGGCAATAATGATACAATGTATCAT
GCCTCTTTGCACCATTCTAAAGAATAACAG
TGATAATTTCTGGGTTAAGGCAATAGCAAT
ATTTCTGCATATAAATATTTCTGCATATAA
ATTGTAACTGATGTAAGAGGTTTCATATTG
CTAATAGCAGCTACAATCCAGCTACCATTC
TGCTTTTATTTTATGGTTGGGATAAGGCTG
GATTATTCTGAGTCCAAGCTAGGCCCTTTT
GCTAATCATGTTCATACCTCTTATCTTCCT
CCCACAGCTCCTGGGCAACGTGCTGGTCTG
TGTGCTGGCCCATCACTTTGGCAAAGAATT
CACCCCACCAGTGCAGGCTGCCTATCAGAA
AGTGGTGGCTGGTGTGGCTAATGCCCTGGC
CCACAAGTATCACTAAGCTCGCTTTCTTGC
TGTCCAATTTCTATTAAAGGTTCCTTTGTT
CCCTAAGTCCAACTACTAAACTGGGGGATA
TTATGAAGGGCCTTGAGCATCTGGATTCTG
CCTAATAAAAACATTTATTTTCATTGCAA
TGATGTATTTAAATTATTTCTGAATATTTT
ACTAAAAGGGAATGTGGGAGGTCAGTGCA
TTTAAAACATAAAGAAATGATGAGCTGTTC
AAACCTTGGGAAAATACACTATATCTTAAA
CTCCATGAAAGAAGGTGAGGCTGCAACCAG
CTAATGCACATTGGCAACAGCCCCTGATGC
CTATGCCTTATTCATCCCTCAGAAAAGGAT
TCTTGTAGAGGCTTGATTTGCAGGTTAAAG
TTTTGCTATGCTGTATTTTACATTACTTAT
TGTTTTAGCTGTCCTCATGAATGTCTTTTC
```

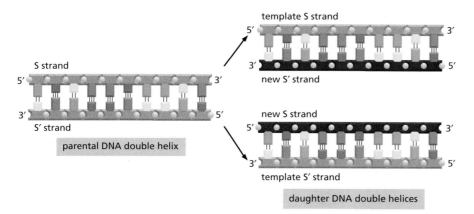

template S strand

new S' strand

new S strand

template S' strand

parental DNA double helix

daughter DNA double helices

Figure 4–8 DNA as a template for its own duplication. As the nucleotide A successfully pairs only with T, and G with C, each strand of DNA can act as a template to specify the sequence of nucleotides in its complementary strand. In this way, double-helical DNA can be copied precisely, with each parental DNA helix producing two identical daughter DNA helices.

chapters, it also keeps nuclear and cytosolic enzymes separate, a feature that is crucial for the proper functioning of eucaryotic cells. Compartmentalization, of which the nucleus is an example, is an important principle of biology; it serves to establish an environment in which biochemical reactions are facilitated by the high concentration of both substrates and the enzymes that act on them. Compartmentalization also prevents enzymes needed in one part of the cell from interfering with the orderly biochemical pathways in another.

Summary

Genetic information is carried in the linear sequence of nucleotides in DNA. Each molecule of DNA is a double helix formed from two complementary strands of nucleotides held together by hydrogen bonds between G-C and A-T base pairs. Duplication of the genetic information occurs by the use of one DNA strand as a template for the formation of a complementary strand. The genetic information stored in an organism's DNA contains the instructions for all the proteins the organism will ever synthesize and is said to comprise its genome. In eucaryotes, DNA is contained in the cell nucleus, a large membrane-bound compartment.

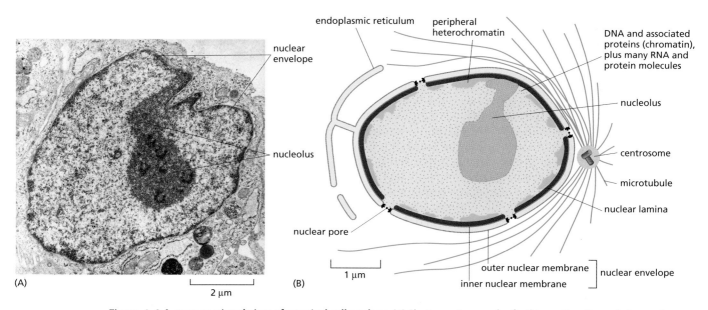

Figure 4–9 A cross-sectional view of a typical cell nucleus. (A) Electron micrograph of a thin section through the nucleus of a human fibroblast. (B) Schematic drawing, showing that the nuclear envelope consists of two membranes, the outer one being continuous with the endoplasmic reticulum membrane (see also Figure 12–8). The space inside the endoplasmic reticulum (the ER lumen) is colored *yellow*; it is continuous with the space between the two nuclear membranes. The lipid bilayers of the inner and outer nuclear membranes are connected at each nuclear pore. A sheet-like network of intermediate filaments *(brown)* inside the nucleus provides mechanical support for the nuclear envelope, forming a special supporting structure called the nuclear lamina (for details, see Chapter 12). The heterochromatin near the lamina contains specially condensed regions of DNA that will be discussed later.

CHROMOSOMAL DNA AND ITS PACKAGING IN THE CHROMATIN FIBER

The most important function of DNA is to carry genes, the information that specifies all the proteins and RNA molecules that make up an organism—including information about when, in what types of cells, and in what quantity each protein is to be made. The genomes of eucaryotes are divided up into chromosomes, and in this section we see how genes are typically arranged on each chromosome. In addition, we describe the specialized DNA sequences that are required for a chromosome to be accurately duplicated and passed on from one generation to the next.

We also confront the serious challenge of DNA packaging. If the double helices comprising all 46 chromosomes in a human cell could be laid end-to-end, they would reach approximately 2 meters; yet the nucleus, which contains the DNA, is only about 6 μm in diameter. This is geometrically equivalent to packing 40 km (24 miles) of extremely fine thread into a tennis ball! The complex task of packaging DNA is accomplished by specialized proteins that bind to and fold the DNA, generating a series of coils and loops that provide increasingly higher levels of organization, preventing the DNA from becoming an unmanageable tangle. Amazingly, although the DNA is very tightly folded, it is compacted in a way that keeps it available to the many enzymes in the cell that replicate it, repair it, and use its genes to produce RNA molecules and proteins.

Eucaryotic DNA Is Packaged into a Set of Chromosomes

In eucaryotes, the DNA in the nucleus is divided between a set of different **chromosomes**. For example, the human genome—approximately 3.2×10^9 nucleotides—is distributed over 24 different chromosomes. Each chromosome consists of a single, enormously long linear DNA molecule associated with proteins that fold and pack the fine DNA thread into a more compact structure. The complex of DNA and protein is called *chromatin* (from the Greek *chroma*, "color," because of its staining properties). In addition to the proteins involved in packaging the DNA, chromosomes are also associated with many proteins and RNA molecules required for the processes of gene expression, DNA replication, and DNA repair.

Bacteria carry their genes on a single DNA molecule, which is often circular (see Figure 1–29). This DNA is associated with proteins that package and condense the DNA, but they are different from the proteins that perform these functions in eucaryotes. Although often called the bacterial "chromosome," it does not have the same structure as eucaryotic chromosomes, and less is known about how the bacterial DNA is packaged. Therefore, our discussion of chromosome structure will focus almost entirely on eucaryotic chromosomes.

With the exception of the germ cells (discussed in Chapter 21) and a few highly specialized cell types that cannot multiply and lack DNA altogether (for example, red blood cells), each human cell contains two copies of each chromosome, one inherited from the mother and one from the father. The maternal and paternal chromosomes of a pair are called **homologous chromosomes** (**homologs**). The only nonhomologous chromosome pairs are the sex chromosomes in males, where a *Y chromosome* is inherited from the father and an *X chromosome* from the mother. Thus, each human cell contains a total of 46 chromosomes—22 pairs common to both males and females, plus two so-called sex chromosomes (X and Y in males, two Xs in females). *DNA hybridization* is a technique in which a labeled nucleic acid strand serves as a "probe" that localizes a complementary strand, as will be described in detail in Chapter 8. This technique can be used to distinguish these human chromosomes by "painting" each one a different color (**Figure 4–10**). Chromosome painting is typically done at the stage in the cell cycle called mitosis, when chromosomes are especially compacted and easy to visualize (see below).

Another more traditional way to distinguish one chromosome from another is to stain them with dyes that produce a striking and reliable pattern of bands

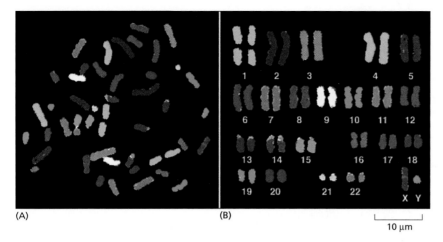

(A)   (B)

10 μm

Figure 4–10 **The complete set of human chromosomes.** These chromosomes, from a male, were isolated from a cell undergoing nuclear division (mitosis) and are therefore highly compacted. Each chromosome has been "painted" a different color to permit its unambiguous identification under the light microscope. Chromosome painting is performed by exposing the chromosomes to a collection of human DNA molecules that have been coupled to a combination of fluorescent dyes. For example, DNA molecules derived from chromosome 1 are labeled with one specific dye combination, those from chromosome 2 with another, and so on. Because the labeled DNA can form base pairs, or hybridize, only to the chromosome from which it was derived (discussed in Chapter 8), each chromosome is differently labeled. For such experiments, the chromosomes are subjected to treatments that separate the double-helical DNA into individual strands, designed to permit base-pairing with the single-stranded labeled DNA while keeping the chromosome structure relatively intact. (A) The chromosomes visualized as they originally spilled from the lysed cell. (B) The same chromosomes artificially lined up in their numerical order. This arrangement of the full chromosome set is called a karyotype. (From E. Schröck et al., *Science* 273:494–497, 1996. With permission from AAAS.)

along each mitotic chromosome (**Figure 4–11**). The structural bases for these banding patterns are not well understood. Nevertheless, the pattern of bands on each type of chromosome is unique, and it is these patterns that initially allowed each human chromosome to be identified and numbered.

The display of the 46 human chromosomes at mitosis is called the human **karyotype**. If parts of chromosomes are lost or are switched between chromosomes, these changes can be detected by changes in the banding patterns or by changes in the pattern of chromosome painting (**Figure 4–12**). Cytogeneticists use these alterations to detect chromosome abnormalities that are associated with inherited defects, as well as to characterize cancers that are associated with specific chromosome rearrangements in somatic cells (discussed in Chapter 20).

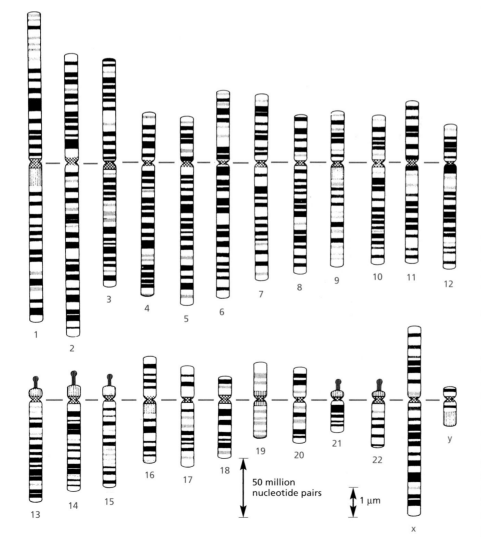

Figure 4–11 **The banding patterns of human chromosomes.** Chromosomes 1–22 are numbered in approximate order of size. A typical human somatic (non-germ-line) cell contains two of each of these chromosomes, plus two sex chromosomes—two X chromosomes in a female, one X and one Y chromosome in a male. The chromosomes used to make these maps were stained at an early stage in mitosis, when the chromosomes are incompletely compacted. The *horizontal red line* represents the position of the centromere (see Figure 4–21), which appears as a constriction on mitotic chromosomes. The red knobs on chromosomes 13, 14, 15, 21, and 22 indicate the positions of genes that code for the large ribosomal RNAs (discussed in Chapter 6). These patterns are obtained by staining chromosomes with Giemsa stain, and they can be observed under the light microscope. (For micrographs, see Figure 21–18; adapted from U. Franke, *Cytogenet. Cell Genet.* 31:24–32, 1981. With permission from Karger.)

50 million nucleotide pairs

1 μm

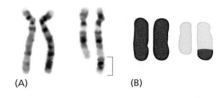

Figure 4–12 An aberrant human chromosome. (A) Two pairs of chromosomes, stained with Giemsa (see Figure 4–11), from a patient with ataxia, a disease characterized by progressive deterioration of motor skills. The patient has a normal pair of chromosome 4s *(left-hand pair)*, but one normal chromosome 12 and one aberrant chromosome 12, as seen by its greater length *(right-hand pair)*. The additional material contained on the aberrant chromosome 12 *(red bracket)* was deduced, from its pattern of bands, as a copy of part of chromosome 4 that had become attached to chromosome 12 through an abnormal recombination event, called a chromosomal translocation. (B) Drawing of the same two chromosome pairs, "painted" *red* for chromosome 4 DNA and *blue* for chromosome 12 DNA. The two techniques give rise to the same conclusion regarding the nature of the aberrant chromosome 12, but chromosome painting provides better resolution, allowing the clear identification of even short pieces of chromosomes that have become translocated. However, Giemsa staining is easier to perform. (Adapted from E. Schröck et al., *Science* 273:494–497, 1996. With permission from AAAS.)

Chromosomes Contain Long Strings of Genes

Chromosomes carry genes—the functional units of heredity. A gene is usually defined as a segment of DNA that contains the instructions for making a particular protein (or a set of closely related proteins). Although this definition holds for the majority of genes, several percent of genes produce an RNA molecule, instead of a protein, as their final product. Like proteins, these RNA molecules perform a diverse set of structural and catalytic functions in the cell, and we discuss them in detail in subsequent chapters.

As might be expected, some correlation exists between the complexity of an organism and the number of genes in its genome (see Table 1–1, p. 18). For example, some simple bacteria have only 500 genes, compared to about 25,000 for humans. Bacteria and some single-celled eucaryotes, such as yeast, have especially concise genomes; the complete nucleotide sequence of their genomes reveals that the DNA molecules that make up their chromosomes are little more than strings of closely packed genes (**Figure 4–13**). However, chromosomes from many eucaryotes (including humans) contain, in addition to genes, a large excess of interspersed DNA that does not seem to carry critical information. Sometimes called "junk DNA" to signify that its usefulness to the cell has not been demonstrated, the particular nucleotide sequence of most of this DNA may not be important. However, some of this DNA is crucial for the proper expression of certain genes, as we discuss elsewhere.

Because of differences in the amount of DNA interspersed between genes, genome sizes can vary widely (see Figure 1–37). For example, the human genome is 200 times larger than that of the yeast *S. cerevisiae*, but 30 times smaller than that of some plants and amphibians and 200 times smaller than that of a species of amoeba. Moreover, because of differences in the amount of excess DNA, the genomes of similar organisms (bony fish, for example) can vary several hundredfold in their DNA content, even though they contain roughly the same number of genes. Whatever the excess DNA may do, it seems clear that it is not a great handicap for a eucaryotic cell to carry a large amount of it.

How the genome is divided into chromosomes also differs from one eucaryotic species to the next. For example, compared with 46 for humans, somatic cells from a species of small deer contain only 6 chromosomes, while those from a species of carp contain over 100. Even closely related species with similar genome sizes can have very different numbers and sizes of chromosomes (**Figure 4–14**). Thus, there is no simple relationship between chromosome number,

Figure 4–13 The arrangement of genes in the genome of *S. cerevisiae*. *S. cerevisiae* is a budding yeast widely used for brewing and baking. The genome of this yeast cell is distributed over 16 chromosomes. A small region of one chromosome has been arbitrarily selected to show the high density of genes characteristic of this species. As indicated by the *light red shading*, some genes are transcribed from the lower strand, while others are transcribed from the upper strand. There are about 6300 genes in the complete genome, which contains somewhat more than 12 million nucleotide pairs. (For the closely packed genes of a bacterium whose genome is 4.6 million nucleotides long, see Figure 1–29).

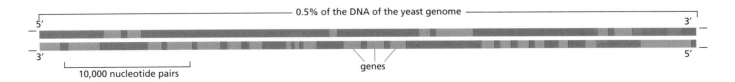

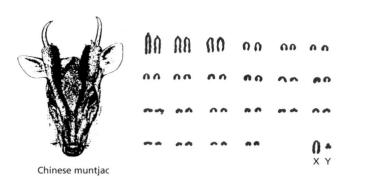

Chinese muntjac

Indian muntjac

Figure 4–14 Two closely related species of deer with very different chromosome numbers. In the evolution of the Indian muntjac, initially separate chromosomes fused, without having a major effect on the animal. These two species contain a similar number of genes. (Adapted from M.W. Strickberger, Evolution, 3rd ed. Sudbury, MA: Jones & Bartlett Publishers, 2000.)

species complexity, and total genome size. Rather, the genomes and chromosomes of modern-day species have each been shaped by a unique history of seemingly random genetic events, acted on by selection pressures over long evolutionary times.

The Nucleotide Sequence of the Human Genome Shows How Our Genes Are Arranged

In Chapter 1 we discussed, in general terms, how the information in DNA is read out and used, through RNA intermediates, to make proteins (see Figure 1–4). In 1999, it became possible for the first time to see exactly how genes are arranged along an entire vertebrate chromosome (**Figure 4–15**). Today, with the publication of the "first draft" of the entire human genome in 2001 and the "finished DNA sequence" in 2004, the genetic information in all human chromosomes is available. The sheer quantity of information provided by the Human Genome Project is staggering (**Figure 4–16** and **Table 4–1**). At its peak, the Project generated raw nucleotide sequences at a rate of 1000 nucleotides per second around the clock. It will be many decades before this information is fully analyzed, but it has already stimulated new experiments that have had major effects on the content of every chapter in this book.

The first striking feature of the human genome is how little of it (only a few percent) codes for proteins (**Figure 4–17**). Much of the remaining chromosomal

Figure 4–15 The organization of genes on a human chromosome. (A) Chromosome 22, one of the smallest human chromosomes, contains 48×10^6 nucleotide pairs and makes up approximately 1.5% of the entire human genome. Most of the left arm of chromosome 22 consists of short repeated sequences of DNA that are packaged in a particularly compact form of chromatin (heterochromatin), which is discussed later in this chapter. (B) A tenfold expansion of a portion of chromosome 22, with about 40 genes indicated. Those in *dark brown* are known genes and those in *red* are predicted genes. (C) An expanded portion of (B) shows the entire length of several genes. (D) The intron–exon arrangement of a typical gene is shown after a further tenfold expansion. Each exon *(red)* codes for a portion of the protein, while the DNA sequence of the introns *(gray)* is relatively unimportant, as discussed in detail in Chapter 6.

The human genome (3.2×10^9 nucleotide pairs) is the totality of genetic information belonging to our species. Almost all of this genome is distributed over the 22 autosomes and 2 sex chromosomes (see Figures 4–10 and 4–11) found within the nucleus. A minute fraction of the human genome (16,569 nucleotide pairs—in multiple copies per cell) is found in the mitochondria (introduced in Chapter 1, and discussed in detail in Chapter 14). The term *human genome sequence* refers to the complete nucleotide sequence of DNA in the 24 nuclear chromosomes and the mitochondria. Being diploid, a human somatic cell nucleus contains roughly twice the haploid amount of DNA, or 6.4×10^9 nucleotide pairs when not duplicating its chromosomes in preparation for division. (Adapted from International Human Genome Sequencing Consortium, *Nature* 409:860–921, 2001. With permission from Macmillan Publishers Ltd.)

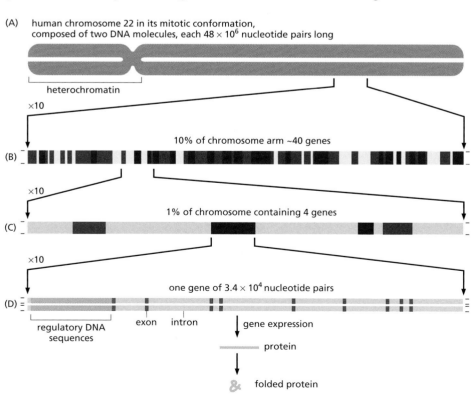

(A)   human chromosome 22 in its mitotic conformation, composed of two DNA molecules, each 48×10^6 nucleotide pairs long

heterochromatin

×10

10% of chromosome arm ~40 genes

(B)

×10

1% of chromosome containing 4 genes

(C)

×10

one gene of 3.4×10^4 nucleotide pairs

(D)

regulatory DNA sequences

exon   intron

gene expression

protein

folded protein

DNA is made up of short, mobile pieces of DNA that have gradually inserted themselves in the chromosome over evolutionary time. We discuss these *transposable elements* in detail in later chapters.

A second notable feature of the human genome is the large average gene size of 27,000 nucleotide pairs. As discussed above, a typical gene carries in its linear sequence of nucleotides the information for the linear sequence of the amino acids of a protein. Only about 1300 nucleotide pairs are required to encode a protein of average size (about 430 amino acids in humans). Most of the remaining DNA in a gene consists of long stretches of noncoding DNA that interrupt the relatively short segments of DNA that code for protein. As will be discussed in detail in Chapter 6, the coding sequences are called **exons**; the intervening (noncoding) sequences in genes are called **introns** (see Figure 4–15 and Table 4–1). The majority of human genes thus consist of a long string of alternating exons and introns, with most of the gene consisting of introns. In contrast, the majority of genes from organisms with concise genomes lack introns. This accounts for the much smaller size of their genes (about one-twentieth that of human genes), as well as for the much higher fraction of coding DNA in their chromosomes.

In addition to introns and exons, each gene is associated with *regulatory DNA sequences*, which are responsible for ensuring that the gene is turned on or off at the proper time, expressed at the appropriate level, and only in the proper type of cell. In humans, the regulatory sequences for a typical gene are spread out over tens of thousands of nucleotide pairs. As would be expected, these regulatory sequences are more compressed in organisms with concise genomes. We discuss in Chapter 7 how regulatory DNA sequences work.

Finally, the nucleotide sequence of the human genome has revealed that the critical information needed to produce a human seems to be in an alarming state of disarray. As one commentator described our genome, "In some ways it may resemble your garage/bedroom/refrigerator/life: highly individualistic, but unkempt; little evidence of organization; much accumulated clutter (referred to by the uninitiated as 'junk'); virtually nothing ever discarded; and the few patently valuable items indiscriminately, apparently carelessly, scattered throughout."

(A)

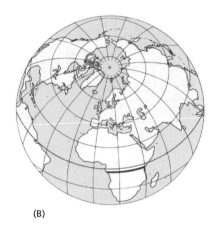

(B)

Figure 4–16 Scale of the human genome. If drawn with a 1 mm space between each nucleotide, as in (A), the human genome would extend 3200 km (approximately 2000 miles), far enough to stretch across the center of Africa, the site of our human origins (*red line* in B). At this scale, there would be, on average, a protein-coding gene every 130 m. An average gene would extend for 30 m, but the coding sequences in this gene would add up to only just over a meter.

Table 4–1 Some Vital Statistics for the Human Genome

	HUMAN GENOME
DNA length	3.2×10^9 nucleotide pairs*
Number of genes	approximately 25,000
Largest gene	2.4×10^6 nucleotide pairs
Mean gene size	27,000 nucleotide pairs
Smallest number of exons per gene	1
Largest number of exons per gene	178
Mean number of exons per gene	10.4
Largest exon size	17,106 nucleotide pairs
Mean exon size	145 nucleotide pairs
Number of pseudogenes**	more than 20,000
Percentage of DNA sequence in exons (protein coding sequences)	1.5%
Percentage of DNA in other highly conserved sequences***	3.5%
Percentage of DNA in high-copy repetitive elements	approximately 50%

* The sequence of 2.85 billion nucleotides is known precisely (error rate of only about one in 100,000 nucleotides). The remaining DNA primarily consists of short highly repeated sequences that are tandemly repeated, with repeat numbers differing from one individual to the next.

** A pseudogene is a nucleotide sequence of DNA closely resembling that of a functional gene, but containing numerous mutations that prevent its proper expression. Most pseudogenes arise from the duplication of a functional gene followed by the accumulation of damaging mutations in one copy.

*** Preserved functional regions; these include DNA encoding 5′ and 3′ UTRs (untranslated regions), structural and functional RNAs, and conserved protein-binding sites on the DNA.

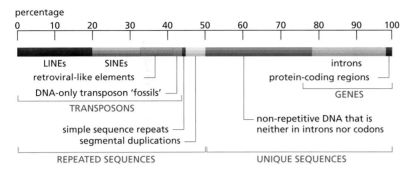

Figure 4–17 **Representation of the nucleotide sequence content of the completely sequenced human genome.** The LINEs, SINEs, retroviral-like elements, and DNA-only transposons are mobile genetic elements that have multiplied in our genome by replicating themselves and inserting the new copies in different positions. These mobile genetic elements are discussed in Chapter 5 (see Table 5–3, p. 318). Simple sequence repeats are short nucleotide sequences (less than 14 nucleotide pairs) that are repeated again and again for long stretches. Segmental duplications are large blocks of the genome (1000–200,000 nucleotide pairs) that are present at two or more locations in the genome. The most highly repeated blocks of DNA in heterochromatin have not yet been completely sequenced; therefore about 10% of human DNA sequences are not represented in this diagram. (Data courtesy of E. Margulies.)

Genome Comparisons Reveal Evolutionarily Conserved DNA Sequences

A major obstacle in interpreting the nucleotide sequences of human chromosomes is the fact that much of the sequence is probably unimportant. Moreover, the coding regions of the genome (the exons) are typically found in short segments (average size about 145 nucleotide pairs) floating in a sea of DNA whose exact nucleotide sequence is of little consequence. This arrangement makes it very difficult to identify all the exons in a stretch of DNA sequence. Even harder is the determination of where a gene begins and ends and exactly how many exons it spans.

Accurate gene identification requires approaches that extract information from the inherently low signal-to-noise ratio of the human genome. We shall describe some of them in Chapter 8. Here we discuss only one general approach, which is based on the observation that sequences that have a function are relatively conserved during evolution, whereas those without a function are free to mutate randomly. The strategy is therefore to compare the human sequence with that of the corresponding regions of a related genome, such as that of the mouse. Humans and mice are thought to have diverged from a common mammalian ancestor about 80×10^6 years ago, which is long enough for the majority of nucleotides in their genomes to have been changed by random mutational events. Consequently, the only regions that will have remained closely similar in the two genomes are those in which mutations would have impaired function and put the animals carrying them at a disadvantage, resulting in their elimination from the population by natural selection. Such closely similar regions are known as *conserved regions*. The conserved regions include both functionally important exons and regulatory DNA sequences. In contrast, *nonconserved regions* represent DNA whose sequence is unlikely to be critical for function.

The power of this method can be increased by comparing our genome with the genomes of additional animals whose genomes have been completely sequenced, including the rat, chicken, chimpanzee, and dog. By revealing in this way the results of a very long natural "experiment," lasting for hundreds of millions of years, such comparative DNA sequencing studies have highlighted the most interesting regions in these genomes. The comparisons reveal that roughly 5% of the human genome consists of "multi-species conserved sequences," as discussed in detail near the end of this chapter. Unexpectedly, only about one-third of these sequences code for proteins. Some of the conserved noncoding sequences correspond to clusters of protein-binding sites that are involved in gene regulation, while others produce RNA molecules that are not translated into protein. But the function of the majority of these sequences remains unknown. This unexpected discovery has led scientists to conclude that we understand much less about the cell biology of vertebrates than we had previously imagined. Certainly, there are enormous opportunities for new discoveries, and we should expect many surprises ahead.

Comparative studies have revealed not only that humans and other mammals share most of the same genes, but also that large blocks of our genomes contain these genes in the same order, a feature called *conserved synteny*. As a result, large blocks of our chromosomes can be recognized in other species. This allows the chromosome painting technique to be used to reconstruct the recent evolutionary history of human chromosomes (**Figure 4–18**).

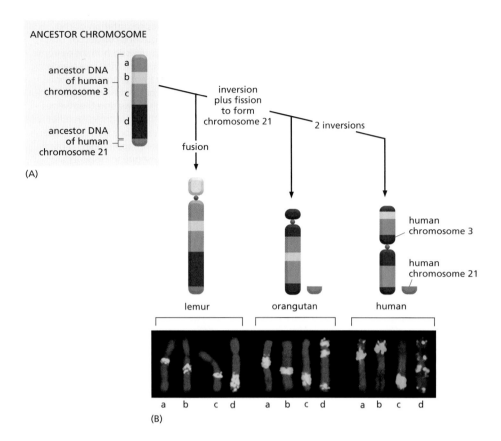

(A)

(B)

Figure 4–18 A proposed evolutionary history of human chromosome 3 and its relatives in other mammals. (A) The order of chromosome 3 segments hypothesized to be present on a chromosome of a mammalian ancestor is shown (*yellow* box). The minimum changes in this ancestral chromosome necessary to account for the appearance of each of the three modern chromosomes are indicated. (The present-day chromosomes of humans and African apes are identical at this resolution.) The *small circles* depicted in the modern chromosomes represent the positions of centromeres. A fission and inversion that leads to a change in chromosome organization is thought to occur once every $5–10 \times 10^6$ years in mammals. (B) Some of the chromosome painting experiments that led to the diagram in (A). Each image shows the chromosome most closely related to human chromosome 3, painted *green* by hybridization with different segments of DNA, lettered a, b, c, and d along the *bottom* of the figure. These letters correspond to the colored segments of the diagrams in (A), as indicated on the ancestral chromosome. (From S. Müller et al., *Proc. Natl Acad. Sci. U.S.A.* 97:206–211, 2000. With permission from National Academy of Sciences.)

Chromosomes Exist in Different States Throughout the Life of a Cell

We have seen how genes are arranged in chromosomes, but to form a functional chromosome, a DNA molecule must be able to do more than simply carry genes: it must be able to replicate, and the replicated copies must be separated and reliably partitioned into daughter cells at each cell division. This process occurs through an ordered series of stages, collectively known as the **cell cycle**, which provides for a temporal separation between the duplication of chromosomes and their segregation into two daughter cells. The cell cycle is briefly summarized in **Figure 4–19**, and it is discussed in detail in Chapter 17. Only certain

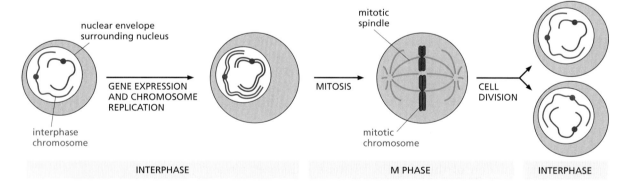

Figure 4–19 A simplified view of the eucaryotic cell cycle. During interphase, the cell is actively expressing its genes and is therefore synthesizing proteins. Also, during interphase and before cell division, the DNA is replicated and each chromosome is duplicated to produce two closely paired daughter chromosomes (a cell with only two chromosomes is illustrated here). Once DNA replication is complete, the cell can enter *M phase,* when mitosis occurs and the nucleus is divided into two daughter nuclei. During this stage, the chromosomes condense, the nuclear envelope breaks down, and the mitotic spindle forms from microtubules and other proteins. The condensed mitotic chromosomes are captured by the mitotic spindle, and one complete set of chromosomes is then pulled to each end of the cell by separating each daughter chromosome pair. A nuclear envelope re-forms around each chromosome set, and in the final step of M phase, the cell divides to produce two daughter cells. Most of the time in the cell cycle is spent in interphase; M phase is brief in comparison, occupying only about an hour in many mammalian cells.

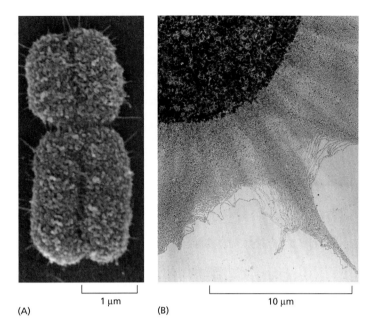

1 μm   10 μm

(A)   (B)

Figure 4–20 **A comparison of extended interphase chromatin with the chromatin in a mitotic chromosome.** (A) A scanning electron micrograph of a mitotic chromosome: a condensed duplicated chromosome in which the two new chromosomes are still linked together (see Figure 4–21). The constricted region indicates the position of the centromere, described in Figure 4–21. (B) An electron micrograph showing an enormous tangle of chromatin spilling out of a lysed interphase nucleus. Note the difference in scales. (A, courtesy of Terry D. Allen; B, courtesy of Victoria Foe.)

parts of the cycle concern us in this chapter. During *interphase* chromosomes are replicated, and during *mitosis* they become highly condensed and then are separated and distributed to the two daughter nuclei. The highly condensed chromosomes in a dividing cell are known as *mitotic chromosomes* (**Figure 4–20**A). This is the form in which chromosomes are most easily visualized; in fact, the images of chromosomes shown so far in the chapter are of chromosomes in mitosis. During cell division, this condensed state is important for the accurate separation of the duplicated chromosomes by the mitotic spindle, as discussed in Chapter 17.

During the portions of the cell cycle when the cell is not dividing, the chromosomes are extended and much of their chromatin exists as long, thin tangled threads in the nucleus so that individual chromosomes cannot be easily distinguished (Figure 4–20B). We shall refer to chromosomes in this extended state as *interphase chromosomes*. Since cells spend most of their time in interphase, and this is where their genetic information is being read out, chromosomes are of greatest interest to cell biologists when they are least visible.

Each DNA Molecule That Forms a Linear Chromosome Must Contain a Centromere, Two Telomeres, and Replication Origins

A chromosome operates as a distinct structural unit: for a copy to be passed on to each daughter cell at division, each chromosome must be able to replicate, and the newly replicated copies must subsequently be separated and partitioned correctly into the two daughter cells. These basic functions are controlled by three types of specialized nucleotide sequences in the DNA, each of which binds specific proteins that guide the machinery that replicates and segregates chromosomes (**Figure 4–21**).

Experiments in yeasts, whose chromosomes are relatively small and easy to manipulate, have identified the minimal DNA sequence elements responsible for each of these functions. One type of nucleotide sequence acts as a DNA **replication origin**, the location at which duplication of the DNA begins. Eucaryotic chromosomes contain many origins of replication to ensure that the entire chromosome can be replicated rapidly, as discussed in detail in Chapter 5.

After replication, the two daughter chromosomes remain attached to one another and, as the cell cycle proceeds, are condensed further to produce mitotic chromosomes. The presence of a second specialized DNA sequence, called a **centromere**, allows one copy of each duplicated and condensed chromosome to be pulled into each daughter cell when a cell divides. A protein

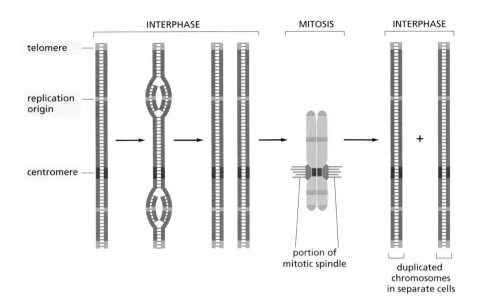

Figure 4–21 The three DNA sequences required to produce a eucaryotic chromosome that can be replicated and then segregated at mitosis. Each chromosome has multiple origins of replication, one centromere, and two telomeres. Shown here is the sequence of events that a typical chromosome follows during the cell cycle. The DNA replicates in interphase, beginning at the origins of replication and proceeding bidirectionally from the origins across the chromosome. In M phase, the centromere attaches the duplicated chromosomes to the mitotic spindle so that one copy is distributed to each daughter cell during mitosis. The centromere also helps to hold the duplicated chromosomes together until they are ready to be moved apart. The telomeres form special caps at each chromosome end.

complex called a *kinetochore* forms at the centromere and attaches the duplicated chromosomes to the mitotic spindle, allowing them to be pulled apart (discussed in Chapter 17).

The third specialized DNA sequence forms **telomeres**, the ends of a chromosome. Telomeres contain repeated nucleotide sequences that enable the ends of chromosomes to be efficiently replicated. Telomeres also perform another function: the repeated telomere DNA sequences, together with the regions adjoining them, form structures that protect the end of the chromosome from being mistaken by the cell for a broken DNA molecule in need of repair. We discuss both this type of repair and the structure and function of telomeres in Chapter 5.

In yeast cells, the three types of sequences required to propagate a chromosome are relatively short (typically less than 1000 base pairs each) and therefore use only a tiny fraction of the information-carrying capacity of a chromosome. Although telomere sequences are fairly simple and short in all eucaryotes, the DNA sequences that form centromeres and replication origins in more complex organisms are much longer than their yeast counterparts. For example, experiments suggest that human centromeres contain up to 100,000 nucleotide pairs and may not require a stretch of DNA with a defined nucleotide sequence. Instead, as we shall discuss later in this chapter, they seem to consist of a large, regularly repeating protein–nucleic acid structure that can be inherited when a chromosome replicates.

DNA Molecules Are Highly Condensed in Chromosomes

All eucaryotic organisms have special ways of packaging DNA into chromosomes. For example, if the 48 million nucleotide pairs of DNA in human chromosome 22 could be laid out as one long perfect double helix, the molecule would extend for about 1.5 cm if stretched out end to end. But chromosome 22 measures only about 2 μm in length in mitosis (see Figures 4–10 and 4–11), representing an end-to-end compaction ratio of nearly 10,000-fold. This remarkable feat of compression is performed by proteins that successively coil and fold the DNA into higher and higher levels of organization. Although much less condensed than mitotic chromosomes, the DNA of human interphase chromosomes is still tightly packed, with an overall compaction ratio of approximately 500-fold (the length of a chromosome's DNA helix divided by the end-to-end length of that chromosome).

In reading these sections it is important to keep in mind that chromosome structure is dynamic. We have seen that each chromosome condenses to an unusual degree in the M phase of the cell cycle. Much less visible, but of enormous interest and importance, specific regions of interphase chromosomes decondense as the cells gain access to specific DNA sequences for gene expression, DNA repair, and replication—and then recondense when these processes are completed. The packaging of chromosomes is therefore accomplished in a way that allows rapid localized, on-demand access to the DNA. In the next sections we discuss the specialized proteins that make this type of packaging possible.

Nucleosomes Are a Basic Unit of Eucaryotic Chromosome Structure

The proteins that bind to the DNA to form eucaryotic chromosomes are traditionally divided into two general classes: the **histones** and the *nonhistone chromosomal proteins*. The complex of both classes of protein with the nuclear DNA of eucaryotic cells is known as **chromatin**. Histones are present in such enormous quantities in the cell (about 60 million molecules of each type per human cell) that their total mass in chromatin is about equal to that of the DNA.

Histones are responsible for the first and most basic level of chromosome packing, the **nucleosome**, a protein–DNA complex discovered in 1974. When interphase nuclei are broken open very gently and their contents examined under the electron microscope, most of the chromatin is in the form of a fiber with a diameter of about 30 nm (**Figure 4–22**A). If this chromatin is subjected to treatments that cause it to unfold partially, it can be seen under the electron microscope as a series of "beads on a string" (Figure 4–22B). The string is DNA, and each bead is a "nucleosome core particle" that consists of DNA wound around a protein core formed from histones. <ACTC>

The structural organization of nucleosomes was determined after first isolating them from unfolded chromatin by digestion with particular enzymes (called nucleases) that break down DNA by cutting between the nucleosomes. After digestion for a short period, the exposed DNA between the nucleosome core particles, the *linker DNA*, is degraded. Each individual nucleosome core particle consists of a complex of eight histone proteins—two molecules each of histones H2A, H2B, H3, and H4—and double-stranded DNA that is 147 nucleotide pairs long. The *histone octamer* forms a protein core around which the double-stranded DNA is wound (**Figure 4–23**).

Each nucleosome core particle is separated from the next by a region of linker DNA, which can vary in length from a few nucleotide pairs up to about 80. (The term *nucleosome* technically refers to a nucleosome core particle plus one of its adjacent DNA linkers, but it is often used synonymously with nucleosome core particle.) On average, therefore, nucleosomes repeat at intervals of about 200 nucleotide pairs. For example, a diploid human cell with 6.4×10^9 nucleotide pairs contains approximately 30 million nucleosomes. The formation of nucleosomes converts a DNA molecule into a chromatin thread about one-third of its initial length.

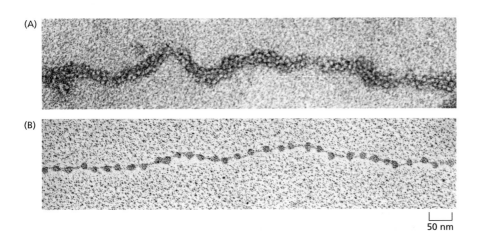

(A)

(B)

|_____|
50 nm

Figure 4–22 **Nucleosomes as seen in the electron microscope.** (A) Chromatin isolated directly from an interphase nucleus appears in the electron microscope as a thread 30 nm thick. (B) This electron micrograph shows a length of chromatin that has been experimentally unpacked, or decondensed, after isolation to show the nucleosomes. (A, courtesy of Barbara Hamkalo; B, courtesy of Victoria Foe.)

Figure 4–23 **Structural organization of the nucleosome.** A nucleosome contains a protein core made of eight histone molecules. In biochemical experiments, the nucleosome core particle can be released from isolated chromatin by digestion of the linker DNA with a nuclease, an enzyme that breaks down DNA. (The nuclease can degrade the exposed linker DNA but cannot attack the DNA wound tightly around the nucleosome core.) After dissociation of the isolated nucleosome into its protein core and DNA, the length of the DNA that was wound around the core can be determined. This length of 147 nucleotide pairs is sufficient to wrap 1.7 times around the histone core.

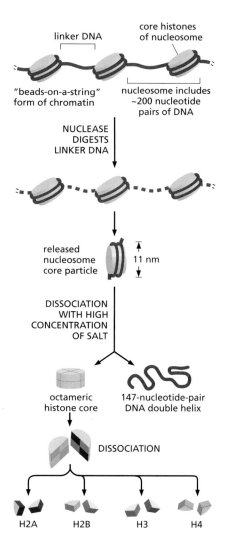

The Structure of the Nucleosome Core Particle Reveals How DNA Is Packaged

The high-resolution structure of a nucleosome core particle, solved in 1997, revealed a disc-shaped histone core around which the DNA was tightly wrapped 1.7 turns in a left-handed coil (**Figure 4–24**). All four of the histones that make up the core of the nucleosome are relatively small proteins (102–135 amino acids), and they share a structural motif, known as the *histone fold*, formed from three α helices connected by two loops (**Figure 4–25**). In assembling a nucleosome, the histone folds first bind to each other to form H3–H4 and H2A–H2B dimers, and the H3–H4 dimers combine to form tetramers. An H3–H4 tetramer then further combines with two H2A–H2B dimers to form the compact octamer core, around which the DNA is wound (**Figure 4–26**).

The interface between DNA and histone is extensive: 142 hydrogen bonds are formed between DNA and the histone core in each nucleosome. Nearly half of these bonds form between the amino acid backbone of the histones and the phosphodiester backbone of the DNA. Numerous hydrophobic interactions and salt linkages also hold DNA and protein together in the nucleosome. For example, more than one-fifth of the amino acids in each of the core histones are either lysine or arginine (two amino acids with basic side chains), and their positive

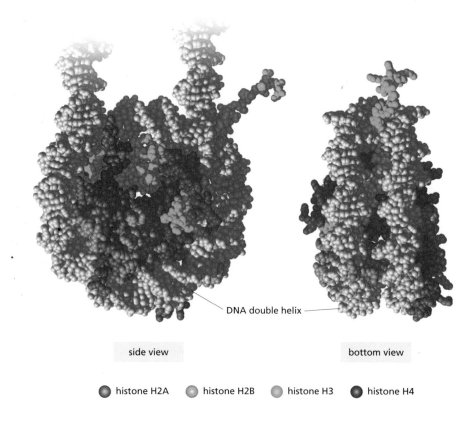

side view   bottom view

● histone H2A   ● histone H2B   ● histone H3   ● histone H4

Figure 4–24 **The structure of a nucleosome core particle, as determined by x-ray diffraction analyses of crystals.** Each histone is colored according to the scheme in Figure 4–23, with the DNA double helix in *light gray*. (From K. Luger et al., *Nature* 389:251–260, 1997. With permission from Macmillan Publishers Ltd.)

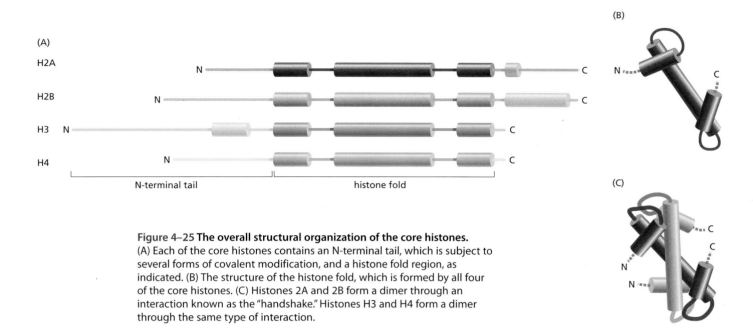

Figure 4–25 The overall structural organization of the core histones. (A) Each of the core histones contains an N-terminal tail, which is subject to several forms of covalent modification, and a histone fold region, as indicated. (B) The structure of the histone fold, which is formed by all four of the core histones. (C) Histones 2A and 2B form a dimer through an interaction known as the "handshake." Histones H3 and H4 form a dimer through the same type of interaction.

charges can effectively neutralize the negatively charged DNA backbone. These numerous interactions explain in part why DNA of virtually any sequence can be bound on a histone octamer core. The path of the DNA around the histone core is not smooth; rather, several kinks are seen in the DNA, as expected from the nonuniform surface of the core. The bending requires a substantial compression of the minor groove of the DNA helix. Certain dinucleotides in the minor groove are especially easy to compress, and some nucleotide sequences bind the nucleosome more tightly than others (**Figure 4–27**). This probably explains some striking, but unusual, cases of very precise positioning of nucleosomes along a stretch of DNA. For most of the DNA sequences found in chromosomes, however, the sequence preference of nucleosomes must be small enough to allow other factors to dominate, inasmuch as nucleosomes can occupy any one of a number of positions relative to the DNA sequence in most chromosomal regions.

In addition to its histone fold, each of the core histones has an N-terminal amino acid "tail", which extends out from the DNA–histone core (see Figure 4–26). These histone tails are subject to several different types of covalent modifications that in turn control critical aspects of chromatin structure and function, as we shall discuss shortly.

As a reflection of their fundamental role in DNA function through controlling chromatin structure, the histones are among the most highly conserved eucaryotic proteins. For example, the amino acid sequence of histone H4 from a pea and from a cow differ at only 2 of the 102 positions. This strong evolutionary conservation suggests that the functions of histones involve nearly all of their amino acids, so that a change in any position is deleterious to the cell. This suggestion has been tested directly in yeast cells, in which it is possible to mutate a given histone gene *in vitro* and introduce it into the yeast genome in place of the normal gene. As might be expected, most changes in histone sequences are lethal; the few that are not lethal cause changes in the normal pattern of gene expression, as well as other abnormalities.

Despite the high conservation of the core histones, eucaryotic organisms also produce smaller amounts of specialized variant core histones that differ in amino acid sequence from the main ones. As we shall see, these variants, combined with a surprisingly large variety of covalent modifications that can be added to the histones in nucleosomes, make possible the many different chromatin structures that are required for DNA function in higher eucaryotes.

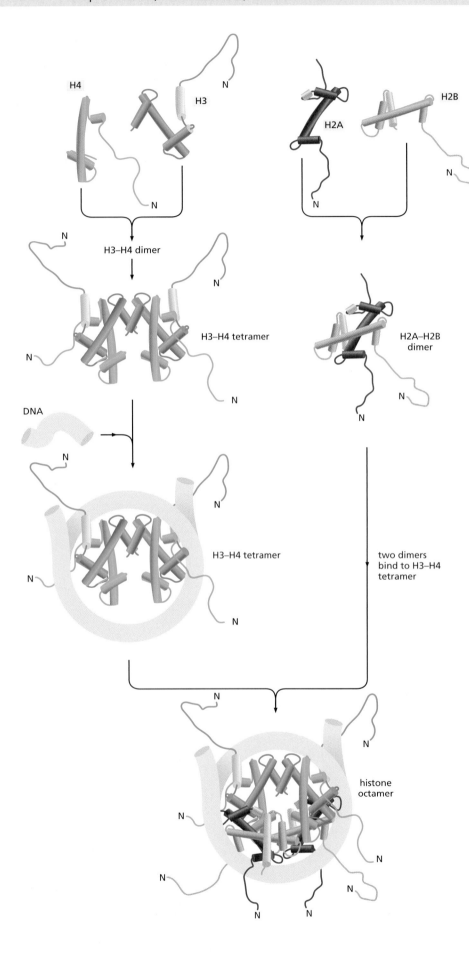

H4

H3

N

H2A

H2B

N

N

N

H3–H4 dimer

N

N

H3–H4 tetramer

H2A–H2B dimer

N

N

N

N

DNA

N

H3–H4 tetramer

two dimers bind to H3–H4 tetramer

N

N

N

histone octamer

N

N

N

N

N

N

Figure 4–26 The assembly of a histone octamer on DNA. The histone H3–H4 dimer and the H2A–H2B dimer are formed from the handshake interaction. An H3–H4 tetramer forms and binds to the DNA. Two H2A–H2B dimers are then added, to complete the nucleosome. The histones are colored as in Figures 4–24 and 4–25. Note that all eight N-terminal tails of the histones protrude from the disc-shaped core structure. Their conformations are highly flexible.

Inside the cell, the nucleosome assembly reactions shown here are mediated by *histone chaperone* proteins, some specific for H3–H4 and others specific for H2A–H2B. (Adapted from figures by J. Waterborg.)

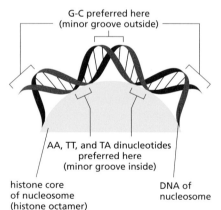

G-C preferred here (minor groove outside)

AA, TT, and TA dinucleotides preferred here (minor groove inside)

histone core of nucleosome (histone octamer)

DNA of nucleosome

Figure 4–27 The bending of DNA in a nucleosome. The DNA helix makes 1.7 tight turns around the histone octamer. This diagram illustrates how the minor groove is compressed on the inside of the turn. Owing to certain structural features of the DNA molecule, the indicated dinucleotides are preferentially accommodated in such a narrow minor groove, which helps to explain why certain DNA sequences will bind more tightly than others to the nucleosome core.

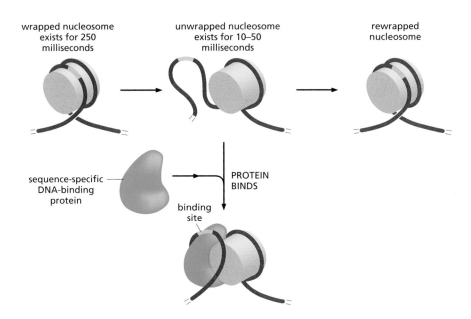

wrapped nucleosome exists for 250 milliseconds

unwrapped nucleosome exists for 10–50 milliseconds

rewrapped nucleosome

sequence-specific DNA-binding protein

PROTEIN BINDS

binding site

Figure 4–28 **Dynamic nucleosomes.** Kinetic measurements show that the DNA in an isolated nucleosome is surprisingly dynamic, rapidly uncoiling and then rewrapping around its nucleosome core. As indicated, this makes most of its bound DNA sequence accessible to other DNA-binding proteins. (Data from G. Li and J. Widom, *Nat. Struct. Mol. Biol.* 11:763–769, 2004. With permission from Macmillan Publishers Ltd.)

Nucleosomes Have a Dynamic Structure, and Are Frequently Subjected to Changes Catalyzed by ATP-Dependent Chromatin-Remodeling Complexes

For many years biologists thought that, once formed in a particular position on DNA, a nucleosome remains fixed in place because of the very tight association between its core histones and DNA. If true, this would pose problems for genetic readout mechanisms, which in principle require rapid access to many specific DNA sequences, as well as for the rapid passage of the DNA transcription and replication machinery through chromatin. But kinetic experiments show that the DNA in an isolated nucleosome unwraps from each end at rate of about 4 times per second, remaining exposed for 10 to 50 milliseconds before the partially unwrapped structure recloses. Thus, most of the DNA in an isolated nucleosome is in principle available for binding other proteins (**Figure 4–28**).

For the chromatin in a cell, a further loosening of DNA–histone contacts is clearly required, because eucaryotic cells contain a large variety of ATP-dependent *chromatin remodeling complexes*. The subunit in these complexes that hydrolyzes ATP is evolutionarily related to the DNA helicases (discussed in Chapter 5), and it binds both to the protein core of the nucleosome and to the double-stranded DNA that winds around it. By using the energy of ATP hydrolysis to move this DNA relative to the core, this subunit changes the structure of a nucleosome temporarily, making the DNA less tightly bound to the histone core. Through repeated cycles of ATP hydrolysis, the remodeling complexes can catalyze *nucleosome sliding*, and by pulling the nucleosome core along the DNA double helix in this way, they make the nucleosomal DNA available to other proteins in the cell (**Figure 4–29**). In addition, by cooperating with negatively

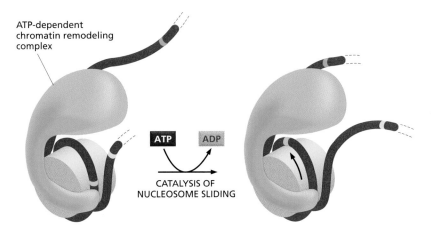

ATP-dependent chromatin remodeling complex

ATP ADP

CATALYSIS OF NUCLEOSOME SLIDING

Figure 4–29 **The nucleosome sliding catalyzed by ATP-dependent chromatin remodeling complexes.** Using the energy of ATP hydrolysis, the remodeling complex is thought to push on the DNA of its bound nucleosome and loosen its attachment to the nucleosome core. Each cycle of ATP binding, ATP hydrolysis, and release of the ADP and P_i products thereby moves the DNA with respect to the histone octamer in the direction of the arrow in this diagram. It requires many such cycles to produce the nucleosome sliding shown. (See also Figure 4–46B.)

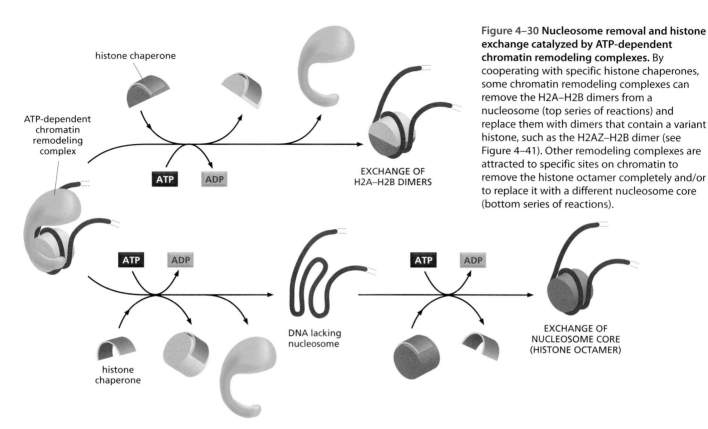

histone chaperone

ATP-dependent chromatin remodeling complex

ATP   ADP

EXCHANGE OF
H2A–H2B DIMERS

ATP   ADP

DNA lacking
nucleosome

ATP   ADP

EXCHANGE OF
NUCLEOSOME CORE
(HISTONE OCTAMER)

histone
chaperone

Figure 4–30 Nucleosome removal and histone exchange catalyzed by ATP-dependent chromatin remodeling complexes. By cooperating with specific histone chaperones, some chromatin remodeling complexes can remove the H2A–H2B dimers from a nucleosome (top series of reactions) and replace them with dimers that contain a variant histone, such as the H2AZ–H2B dimer (see Figure 4–41). Other remodeling complexes are attracted to specific sites on chromatin to remove the histone octamer completely and/or to replace it with a different nucleosome core (bottom series of reactions).

charged proteins that serve as histone chaperones, some remodeling complexes are able to remove either all or part of the nucleosome core from a nucleosome—catalyzing either an exchange of its H2A–H2B histones, or the complete removal of the octameric core from the DNA (**Figure 4–30**).

Cells contain dozens of different ATP-dependent chromatin remodeling complexes that are specialized for different roles. Most are large protein complexes that can contain 10 or more subunits. The activity of these complexes is carefully controlled by the cell. As genes are turned on and off, chromatin remodeling complexes are brought to specific regions of DNA where they act locally to influence chromatin structure (discussed in Chapter 7; see also Figure 4–46, below).

As pointed out previously, for most of the DNA sequences found in chromosomes, experiments show that a nucleosome can occupy any one of a number of positions relative to the DNA sequence. The most important influence on nucleosome positioning appears to be the presence of other tightly bound proteins on the DNA. Some bound proteins favor the formation of a nucleosome adjacent to them. Others create obstacles that force the nucleosomes to move to positions between them. The exact positions of nucleosomes along a stretch of DNA therefore depends mainly on the presence and nature of other proteins bound to the DNA. Due to the presence of ATP-dependent remodeling complexes, the arrangement of nucleosomes on DNA can be highly dynamic, changing rapidly according to the needs of the cell.

Nucleosomes Are Usually Packed Together into a Compact Chromatin Fiber

Although enormously long strings of nucleosomes form on the chromosomal DNA, chromatin in a living cell probably rarely adopts the extended "beads on a string" form. Instead, the nucleosomes are packed on top of one another, generating regular arrays in which the DNA is even more highly condensed. Thus, when nuclei are very gently lysed onto an electron microscope grid, most of the chromatin is seen to be in the form of a fiber with a diameter of about 30 nm, which is considerably wider than chromatin in the "beads on a string" form (see Figure 4–22).

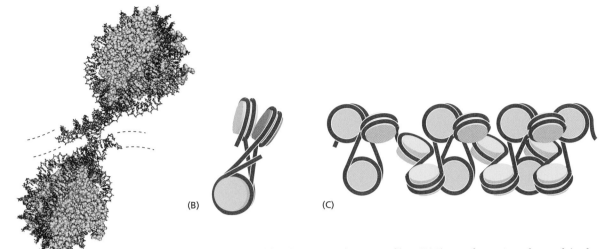

Figure 4–31 A zigzag model for the 30-nm chromatin fiber. (A) The conformation of two of the four nucleosomes in a tetranucleosome, from a structure determined by x-ray crystallography. (B) Schematic of the entire tetranucleosome; the fourth nucleosome is not visible, being stacked on the bottom nucleosome and behind it in this diagram. (C) Diagrammatic illustration of a possible zigzag structure that could account for the 30-nm chromatin fiber. (Adapted from C.L. Woodcock, *Nat. Struct. Mol. Biol.* 12:639–640, 2005. With permission from Macmillan Publishers Ltd.)

How are nucleosomes packed in the 30-nm chromatin fiber? This question has not yet been answered definitively, but important information concerning the structure has been obtained. In particular, high-resolution structural analyses have been performed on homogeneous short strings of nucleosomes, prepared from purified histones and purified DNA molecules. The structure of a tetranucleosome, obtained by X-ray crystallography, has been used to support a zigzag model for the stacking of nucleosomes in the 30-nm fiber (**Figure 4–31**). But cryoelectron microscopy of longer strings of nucleosomes supports a very different solenoidal structure with intercalated nucleosomes (**Figure 4–32**).

What causes the nucleosomes to stack so tightly on each other in a 30-nm fiber? The nucleosome to nucleosome linkages formed by histone tails, most notably the H4 tail (**Figure 4–33**) constitute one important factor. Another

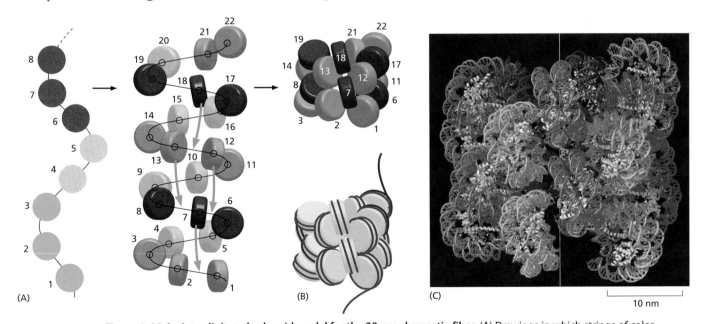

Figure 4–32 An interdigitated solenoid model for the 30-nm chromatin fiber. (A) Drawings in which strings of color-coded nucleosomes are used to illustrate how the solenoid is generated. (B) Schematic diagram of final structure in (A). (C) Structural model. The model is derived from high-resolution cryoelectron microscopy images of nucleosome arrays reconstituted from purified histones and DNA molecules of specific length and sequence. Both nucleosome octamers and a linker histone (discussed below) were used to produce regularly repeating arrays containing up to 72 nucleosomes. (Adapted from P. Robinson, L. Fairall, V. Huynh and D. Rhodes, *Proc. Natl Acad. Sci. U.S.A.* 103:6506–6511, 2006. With permission from National Academy of Sciences.)

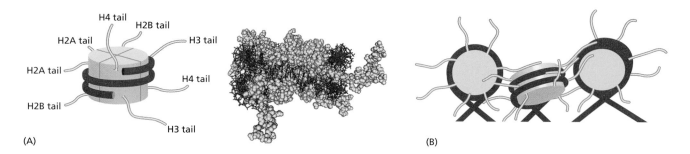

(A)

(B)

important factor is an additional histone that is often present in a 1-to-1 ratio with nucleosome cores, known as **histone H1**. This so-called linker histone is larger than the individual core histones and it has been considerably less well conserved during evolution. A single histone H1 molecule binds to each nucleosome, contacting both DNA and protein, and changing the path of the DNA as it exits from the nucleosome. Although it is not understood in detail how H1 pulls nucleosomes together into the 30-nm fiber, a change in the exit path in DNA seems crucial for compacting nucleosomal DNA so that it interlocks to form the 30-nm fiber (**Figure 4–34**). Most eucaryotic organisms make several histone H1 proteins of related but quite distinct amino acid sequences.

It is possible that the 30-nm structure found in chromosomes is a fluid mosaic of several different variations. For example, a linker histone in the H1 family was present in the nucleosomal arrays studied in Figure 4–32 but was missing from the tetranucleosome in Figure 4–31. Moreover, we saw earlier that the linker DNA that connects adjacent nucleosomes can vary in length; these differences in linker length probably introduce local perturbations into the structure. And the presence of many other DNA-binding proteins, as well as proteins that bind directly to histones, will certainly add important additional features to any array of nucleosomes.

Figure 4–33 A speculative model for the role played by histone tails in the formation of the 30-nm fiber. (A) This schematic diagram shows the approximate exit points of the eight histone tails, one from each histone protein, that extend from each nucleosome. The actual structure is shown to its right. In the high-resolution structure of the nucleosome, the tails are largely unstructured, suggesting that they are highly flexible. (B) A speculative model showing how the histone tails may help to pack nucleosomes together into the 30-nm fiber. This model is based on (1) experimental evidence that histone tails aid in the formation of the 30-nm fiber, and (2) the x-ray crystal structure of the nucleosome, in which the tails of one nucleosome contact the histone core of an adjacent nucleosome in the crystal lattice.

Summary

A gene is a nucleotide sequence in a DNA molecule that acts as a functional unit for the production of a protein, a structural RNA, or a catalytic or regulatory RNA molecule. In eucaryotes, protein-coding genes are usually composed of a string of alternating introns and exons associated with regulatory regions of DNA. A chromosome is formed from a single, enormously long DNA molecule that contains a linear array of many genes. The human genome contains 3.2×10^9 DNA nucleotide pairs, divided between 22 different autosomes and 2 sex chromosomes. Only a small percentage of this DNA codes for proteins or functional RNA molecules. A chromosomal DNA molecule also contains three other types of functionally important nucleotide sequences: replication origins and telomeres allow the DNA molecule to be efficiently replicated, while a centromere attaches the daughter DNA molecules to the mitotic spindle, ensuring their accurate segregation to daughter cells during the M phase of the cell cycle.

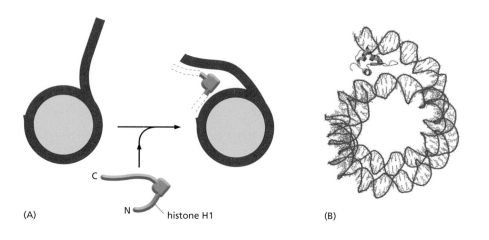

(A)

(B)

Figure 4–34 How the linker histone binds to the nucleosome. The position and structure of the globular region of histone H1 are shown. As indicated, this region constrains an additional 20 nucleotide pairs of DNA where it exits from the nucleosome core. This type of binding by H1 is thought to be important for forming the 30-nm chromatin fiber. The long C-terminal tail of histone H1 is also required for the high-affinity binding of H1 to chromatin, but neither its position or that of the N-terminal tail is known. (A) Schematic, (B) structure. (B, from D. Brown, T. Izard and T. Misteli, *Nat. Struct. Mol. Biol.* 13:250–255, 2006. With permission from Macmillan Publishers Ltd.)

The DNA in eucaryotes is tightly bound to an equal mass of histones, which form repeated arrays of DNA–protein particles called nucleosomes. The nucleosome is composed of an octameric core of histone proteins around which the DNA double helix is wrapped. Nucleosomes are spaced at intervals of about 200 nucleotide pairs, and they are usually packed together (with the aid of histone H1 molecules) into quasi-regular arrays to form a 30-nm chromatin fiber. Despite the high degree of compaction in chromatin, its structure must be highly dynamic to allow access to the DNA. There is some spontaneous DNA unwrapping and rewrapping in the nucleosome itself; however, the general strategy for reversibly changing local chromatin structure features ATP-driven chromatin remodeling complexes. Cells contain a large set of such complexes, which are targeted to specific regions of chromatin at appropriate times. The remodeling complexes collaborate with histone chaperones to allow nucleosome cores to be repositioned, reconstituted with different histones, or completely removed to expose the underlying DNA.

THE REGULATION OF CHROMATIN STRUCTURE

Having described how DNA is packaged into nucleosomes to create a chromatin fiber, we now turn to the mechanisms that create different chromatin structures in different regions of a cell's genome. We now know that mechanisms of this type are used to control many genes in eucaryotes. Most importantly, certain types of chromatin structure can be inherited; that is, the structure can be directly passed down from a cell to its decendents. Because the cell memory that results is based on an inherited protein structure rather than on a change in DNA sequence, this is a form of **epigenetic inheritance**. The prefix *epi* is Greek for "on"; this is appropriate, because epigenetics represents a form of inheritance that is superimposed on the genetic inheritance based on DNA (**Figure 4–35**).

In Chapter 7, we shall introduce the many different ways in which the expression of genes is regulated. There we discuss epigenetic inheritance in detail and present several distinct mechanisms that can produce it. Here, we are concerned with only one, that based on chromatin structure. We begin this section with an introduction to inherited chromatin structures and then describe the basis for them—the covalent modification of histones in nucleosomes. We shall see that these modifications serve as recognition sites for protein modules that bring specific protein complexes to the appropriate regions of chromatin, thereby producing specific effects on gene expression or inducing other biological functions. Through such mechanisms, chromatin structure plays a central role in the development, growth, and maintenance of eucaryotic organisms, including ourselves.

Figure 4–35 A comparison of genetic inheritance with an epigenetic inheritance based on chromatin structures. Genetic inheritance is based on the direct inheritance of DNA nucleotide sequences during DNA replication. DNA sequence changes are not only transmitted faithfully from a somatic cell to all of its descendents, but also through germ cells from one generation to the next. The field of genetics, reviewed in Chapter 8, is based on the inheritance of these changes between generations. The type of epigenetic inheritance shown here is based on other molecules bound to the DNA, and it is therefore less permanent than a change in DNA sequence; in particular, epigenetic information is usually (but not always) erased during the formation of eggs and sperm.

Only one epigenetic mechanism, that based on an inheritance of chromatin structures, is discussed in this chapter. Other epigenetic mechanisms are presented in Chapter 7, which focuses on the control of gene expression (see Figure 7–86).

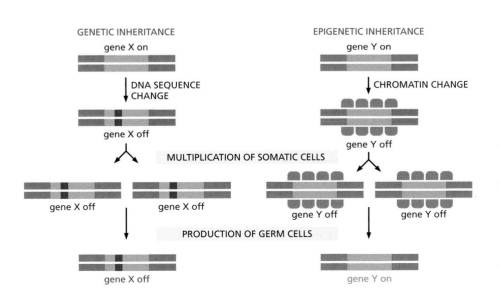

GENETIC INHERITANCE
gene X on

DNA SEQUENCE CHANGE

gene X off

EPIGENETIC INHERITANCE
gene Y on

CHROMATIN CHANGE

gene Y off

MULTIPLICATION OF SOMATIC CELLS

gene X off   gene X off

gene Y off   gene Y off

PRODUCTION OF GERM CELLS

gene X off

gene Y on

Some Early Mysteries Concerning Chromatin Structure

Thirty years ago, histones were viewed as relatively uninteresting proteins. Nucleosomes were known to cover all of the DNA in chromosomes, and they were thought to exist to allow the enormous amounts of DNA in many eucaryotic cells to be packaged into compact chromosomes. Extrapolating from what was known in bacteria, many scientists believed that gene regulation in eucaryotes would simply bypass nucleosomes, treating them as uninvolved bystanders.

But there were reasons to challenge this view. Thus, for example, biochemists had determined that mammalian chromatin consists of an approximately equal mass of histone and non-histone proteins. This would mean that, *on average*, every 200 nucleotide pairs of DNA in our cells is associated with more than 1000 amino acids of non-histone proteins (that is, a mass of protein equivalent to the total mass of the histone octamer plus histone H1). We now know that many of these proteins bind to nucleosomes, and their abundance might suggest that histones are more than just packaging proteins.

A second reason to challenge the view that histones were inconsequential to gene regulation was based on the amazingly slow rate of evolutionary change in the sequences of the four core histones. The previously mentioned fact that there are only two amino acid differences in the sequence of mammalian and pea histone H4 implies that a change in almost any one of the 102 amino acids in H4 must be deleterious to these organisms. What type of process could make the life of an organism so sensitive to the exact structure of the nucleosome core that only two amino acids had changed in more than 500 million years of random variation followed by natural selection?

Last but not least, a combination of genetics and cytology had revealed that a particular form of chromatin silences the genes that it packages without regard to nucleotide sequence—and does so in a manner that is directly inherited by both daughter cells when a cell divides. It is to this subject that we turn next.

Heterochromatin Is Highly Organized and Unusually Resistant to Gene Expression

Light-microscope studies in the 1930s distinguished two types of chromatin in the interphase nuclei of many higher eucaryotic cells: a highly condensed form, called **heterochromatin**, and all the rest, which is less condensed, called **euchromatin**. Heterochromatin represents an especially compact form of chromatin (see Figure 4–9), and we are finally beginning to understand important aspects of its molecular properties. Although present in many locations along chromosomes, it is also highly concentrated in specific regions, most notably at the centromeres and telomeres introduced previously (see Figure 4–21). In a typical mammalian cell, more than ten percent of the genome is packaged in this way.

The DNA in heterochromatin contains very few genes, and those euchromatic genes that become packaged into heterochromatin are turned off by this type of packaging. However, we know now that the term *heterochromatin* encompasses several distinct types of chromatin structures whose common feature is an especially high degree of compaction. Thus, heterochromatin should not be thought of as encapsulating "dead" DNA, but rather as creating different types of compact chromatin with distinct features that make it highly resistant to gene expression for the vast majority of genes.

When a gene that is normally expressed in euchromatin is experimentally relocated into a region of heterochromatin, it ceases to be expressed, and the gene is said to be *silenced*. These differences in gene expression are examples of **position effects**, in which the activity of a gene depends on its position relative to a nearby region of heterochromatin on a chromosome. First recognized in *Drosophila*, position effects have now been observed in many eucaryotes, including yeasts, plants, and humans.

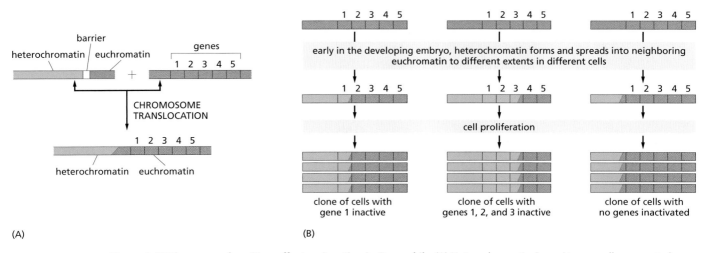

(A)

(B)

Figure 4–36 The cause of position effect variegation in *Drosophila*. (A) Heterochromatin *(green)* is normally prevented from spreading into adjacent regions of euchromatin *(red)* by special *barrier* DNA sequences, which we shall discuss shortly. In flies that inherit certain chromosomal rearrangements, however, this barrier is no longer present. (B) During the early development of such flies, heterochromatin can spread into neighboring chromosomal DNA, proceeding for different distances in different cells. This spreading soon stops, but the established pattern of heterochromatin is inherited, so that large clones of progeny cells are produced that have the same neighboring genes condensed into heterochromatin and thereby inactivated (hence the "variegated" appearance of some of these flies; see Figure 4–37). Although "spreading" is used to describe the formation of new heterochromatin close to previously existing heterochromatin, the term may not be wholly accurate. There is evidence that during expansion, heterochromatin can "skip over" some regions of chromatin, sparing the genes that lie within them from repressive effects.

The position effects associated with heterochromatin exhibit a feature called *position effect variegation*, which in retrospect provided critical clues concerning chromatin function. In *Drosophila*, chromosome breakage events that directly connect a region of heterochromatin to a region of euchromatin tend to inactivate the nearby euchromatic genes. The zone of inactivation spreads a different distance in different early cells in the fly embryo, but once the heterchromatic condition is established on a gene, it tends to be stably inherited by all of the cell's progeny (**Figure 4–36**). This remarkable phenomenon was first recognized through a detailed genetic analysis of the mottled loss of red pigment in the fly eye (**Figure 4–37**), but it shares many features with the extensive spread of heterochromatin that inactivates of one of the two X chromosomes in female mammals (see p. 473).

Extensive genetic screens have been carried out in *Drosophila*, as well as in fungi, in a search for gene products that either enhance or suppress the spread of heterochromatin and its stable inheritance—that is, for genes that when mutated serve as either enhancers or suppressors of position effect variegation. In this way, more than 50 genes have been identified that play a critical role in these processes. In recent years, the detailed characterization of the proteins produced by these genes has revealed that many are nonhistone chromosomal proteins that underlie a remarkable mechanism for eucaryotic gene control, one

Figure 4–37 The discovery of position effects on gene expression. The *White* gene in the fruit fly *Drosophila* controls eye pigment production and is named after the mutation that first identified it. Wild-type flies with a normal *White* gene *(White⁺)* have normal pigment production, which gives them *red* eyes, but if the *White* gene is mutated and inactivated, the mutant flies *(White⁻)* make no pigment and have *white* eyes. In flies in which a normal *White⁺* gene has been moved near a region of heterochromatin, the eyes are mottled, with both *red* and *white* patches. The *white* patches represent cell lineages in which the *White⁺* gene has been silenced by the effects of the heterochromatin. In contrast, the *red* patches represent cell lineages in which the *White⁺* gene is expressed. Early in development, when the heterochromatin is first formed, it spreads into neighboring euchromatin to different extents in different embryonic cells (see Figure 4–36). The presence of large patches of *red* and *white* cells reveals that the state of transcriptional activity, as determined by the packaging of this gene into chromatin in those ancestor cells, is inherited by all daughter cells.

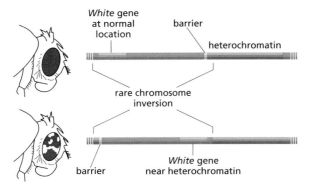

(A) LYSINE ACETYLATION AND METHYLATION ARE COMPETING REACTIONS

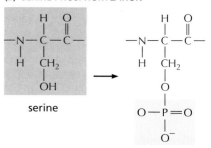

acetyl lysine   lysine   monomethyl lysine   dimethyl lysine   trimethyl lysine

Figure 4–38 Some prominent types of covalent amino acid side-chain modifications found on nucleosomal histones. (A) Three different levels of lysine methylation are shown; each can be recognized by a different binding protein and thus each can have a different significance for the cell. Note that acetylation removes the plus charge on lysine, and that, most importantly, an acetylated lysine cannot be methylated, and *vice versa*. (B) Serine phosphorylation adds a negative charge to a histone. Modifications not shown here are the mono- or di-methylation of an arginine, the phosphorylation of a threonine, the addition of ADP-ribose to a glutamic acid, and the addition of a ubiquityl, sumoyl, or biotin group to a lysine.

(B) SERINE PHOSPHORYLATION

serine

phosphoserine

that requires the precise amino acid sequences of the core histones. This mechanism of gene control therefore helps to explain the remarkably slow change in the histones over time.

The Core Histones Are Covalently Modified at Many Different Sites

The amino acid side chains of the four histones in the nucleosome core are subjected to a remarkable variety of covalent modifications, including the acetylation of lysines, the mono-, di-, and tri-methylation of lysines, and the phosphorylation of serines (**Figure 4–38**). A large number of these side-chain modifications occur on the eight relatively unstructured N-terminal "histone tails" that protrude from the nucleosome (**Figure 4–39**). However, there are also specific side-chain modifications on the nucleosome's globular core (**Figure 4–40**).

All of the above types of modifications are reversible. The modification of a particular amino acid side chain in a nucleosome is created by a specific enzyme, with most of these enzymes acting only on one or a few sites. A different enzyme is responsible for removing each side chain modification. Thus, for example, acetyl groups are added to specific lysines by a set of different histone acetyl transferases (HATs) and removed by a set of histone deacetylase complexes (HDACs). Likewise, methyl groups are added to lysine side chains by a set of different histone methyl transferases and removed by a set of histone demethylases. Each enzyme is recruited to specific sites on the chromatin at defined times in each cell's life history. For the most part, the initial recruitment of these enzymes depends on *gene regulatory proteins* that bind to specific DNA sequences along chromosomes, and these are produced at different times in the life of an organism, as described in Chapter 7. But in at least some cases, the covalent modifications on nucleosomes can persist long after the gene regulatory proteins that first induced them have disappeared, thereby carrying a memory in the cell of its developmental history. Very different patterns of covalent modifications are therefore found on different groups of nucleosomes, according to their exact position on a chromosome and the status of the cell.

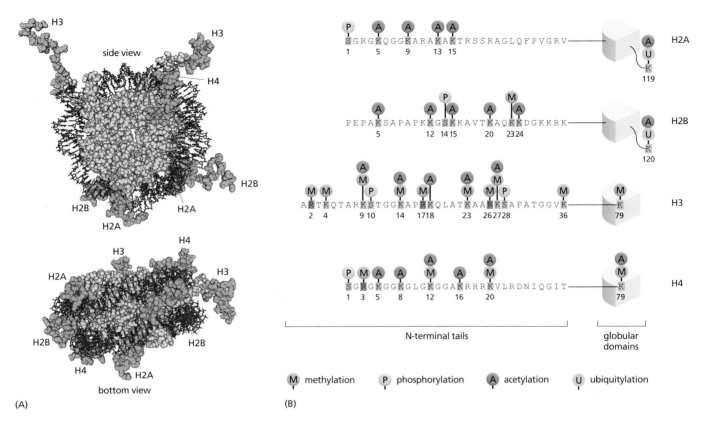

Figure 4–39 The covalent modification of core histone tails. (A) The structure of the nucleosome highlighting the location of the first 30 amino acids in each of its eight N-terminal histone tails *(green)*. (B) Well-documented modifications of the four histone core proteins are indicated. Although only a single symbol is used for methylation here (*M*), each lysine (*K*) or arginine (*R*) can be methylated in several different ways. Note also that some positions (e.g., lysine 9 of H3) can be modified either by methylation or by acetylation, but not both. Most of the modifications shown add a relatively small molecule onto the histone tails; the exception is ubiquitin, a 76 amino acid protein also used for other cell processes (see Figure 6–92). (Adapted from H. Santos-Rosa and C. Caldas, *Eur. J. Cancer* 41:2381–2402, 2005. With permission from Elsevier.)

The modifications of the histones are carefully controlled, and they have important consequences. The acetylation of lysines on the N-terminal tails tends to loosen chromatin structure, in part because adding an acetyl group to lysine removes its positive charge, thereby reducing the affinity of the tails for

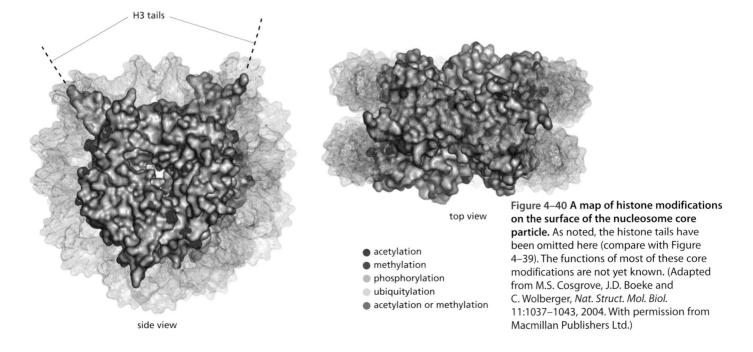

Figure 4–40 A map of histone modifications on the surface of the nucleosome core particle. As noted, the histone tails have been omitted here (compare with Figure 4–39). The functions of most of these core modifications are not yet known. (Adapted from M.S. Cosgrove, J.D. Boeke and C. Wolberger, *Nat. Struct. Mol. Biol.* 11:1037–1043, 2004. With permission from Macmillan Publishers Ltd.)

adjacent nucleosomes (see Figure 4–33). However, the most profound effect of the histone modifications is their ability to attract specific proteins to a stretch of chromatin that has been appropriately modified. These new proteins determine how and when genes will be expressed, as well as other biological functions. In this way, the precise structure of a domain of chromatin determines the expression of the genes packaged in it, and thereby the structure and function of the eucaryotic cell.

Chromatin Acquires Additional Variety Through the Site-Specific Insertion of a Small Set of Histone Variants

Despite the tight conservation of the amino acid sequences of the four core histones over hundreds of millions of years, eucaryotes also contain a few variant histones that assemble into nucleosomes. These histones are present in much smaller amounts than the major histones, and they have been less well conserved over long evolutionary times. Except for histone H4, variants exist for each of the core histones; some examples are shown in **Figure 4–41**.

The major histones are synthesized primarily during the S phase of the cell cycle (see Figure 17–4) and assembled into nucleosomes on the daughter DNA helices just behind the replication fork (see Figure 5–38). In contrast, most histone variants are synthesized throughout interphase. They are often inserted into already-formed chromatin, which requires a histone-exchange process catalyzed by the ATP-dependent chromatin remodeling complexes discussed previously. These remodeling complexes contain subunits that cause them to bind both to specific sites on chromatin and to histone chaperones that carry a particular variant. As a result, each histone variant is inserted into chromatin in a highly selective manner (see Figure 4–30).

The Covalent Modifications and the Histone Variants Act in Concert to Produce a "Histone Code" That Helps to Determine Biological Function

The number of possible distinct markings on an individual nucleosome is enormous. Even with the recognition that some of the covalent modifications are mutually exclusive (for example, it is not possible for a lysine to be both acetylated and methylated at the same time), and that other modifications are created together as a set, it is clear that thousands of combinations can exist. In addition, there is the further diversity created by nucleosomes that contain histone variants.

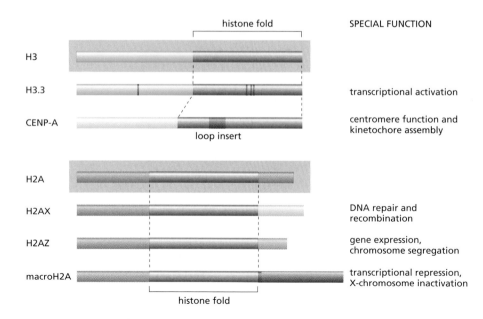

Figure 4–41 **The structure of some histone variants compared with the major histone that they replace.** These histones are inserted into nucleosomes at specific sites on chromosomes by ATP-dependent chromatin remodeling enzymes that act in concert with histone chaperones (see Figure 4–30). The CENP-A variant of histone H3 is discussed later in this chapter (see Figures 4–48 to 4–51); other variants are discussed in Chapter 7. The sequences that are colored differently in each variant are different from the corresponding sequence of the major histone. (Adapted from K. Sarma and D. Reinberg, *Nat. Rev. Mol. Cell. Biol.* 6:139–149, 2005. With permission from Macmillan Publishers Ltd.)

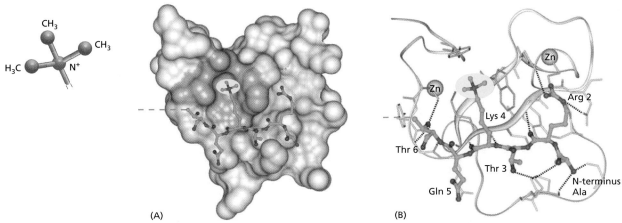

(A)   (B)

Many of the combinations appear to have a specific meaning for the cell because they determine how and when the DNA packaged in the nucleosomes is accessed, leading to the **histone code** hypothesis. For example, one type of marking signals that a stretch of chromatin has been newly replicated, another signals that the DNA in that chromatin has been damaged and needs repair, while many others signal when and how gene expression should take place. Small protein modules bind to specific marks, recognizing for example a tri-methylated lysine 4 on histone H3 (**Figure 4–42**). These modules are thought to act in concert with other modules as part of a *code-reader complex*, so as to allow particular combinations of markings on chromatin to attract additional protein complexes that execute an appropriate biological function at the right time (**Figure 4–43**).

Figure 4–42 How each mark on a nucleosome is read. The structure of a protein module that specifically recognizes histone H3 trimethylated on lysine 4 is shown. (A) Space-filling model of an ING PHD domain bound to a histone tail (*green,* with the trimethyl group highlighted in *yellow*). (B) A ribbon model showing how the N-terminal six amino acids in the H3 tail are recognized. The *dashed lines* represent hydrogen bonds. This is one of many PHD domains that recognize methylated lysines on histones; different domains bind tightly to lysines located at different positions, and they can discriminate between a mono-, di-, and tri-methylated lysine. In a similar way, other small protein modules recognize specific histone side chains that have been marked with acetyl groups, phosphate groups, and so on. (Adapted from P.V. Pena et al., *Nature* 442:100–103, 2006. With permission from Macmillan Publishers Ltd.)

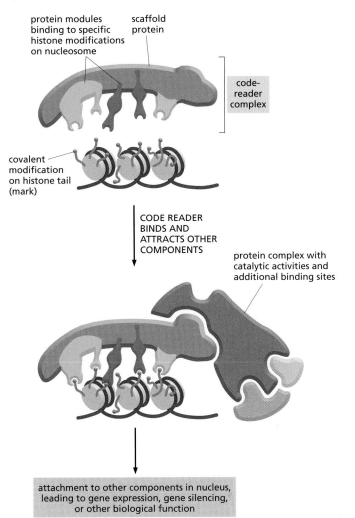

protein modules binding to specific histone modifications on nucleosome

scaffold protein

code-reader complex

covalent modification on histone tail (mark)

CODE READER BINDS AND ATTRACTS OTHER COMPONENTS

protein complex with catalytic activities and additional binding sites

attachment to other components in nucleus, leading to gene expression, gene silencing, or other biological function

Figure 4–43 Schematic diagram showing how the histone code could be read by a code-reader complex. A large protein complex that contains a series of protein modules, each of which recognizes a specific histone mark, is schematically illustrated (*green*). This "code-reader complex" will bind tightly only to a region of chromatin that contains several of the different histone marks that it recognizes. Therefore, only a specific combination of marks will cause the complex to bind to chromatin and attract additional protein complexes (*purple*) that catalyze a biological function.

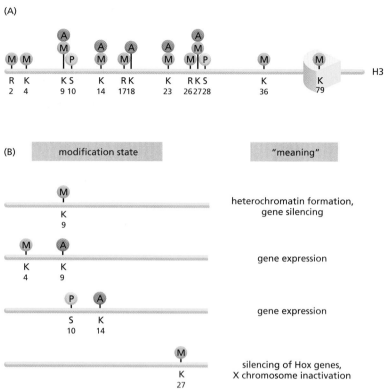

Figure 4–44 Some specific meanings of the histone code. (A) The modifications on the histone H3 N-terminal tail are shown, repeated from Figure 4–39. (B) The H3 tail can be marked by different combinations of modifications that convey a specific meaning to the stretch of chromatin where this combination occurs. Only a few of the meanings are known, including the four examples shown. To focus on just one example, the trimethylation of lysine 9 attracts the heterochromatin-specific protein HP1, which induces a spreading wave of further lysine 9 trimethylation followed by further HP1 binding, according to the general scheme that will be illustrated shortly (see Figure 4–46). Not shown is the fact that, as just implied (see Figure 4–43), reading the histone code generally involves the joint recognition of marks at other sites on the nucleosome along with the indicated H3 tail recognition. In addition, specific levels of methylation (mono-, di-, or tri-methyl groups) are required, as in Figure 4–42.

The marks on nucleosomes due to covalent additions to histones are dynamic, being constantly removed and added at rates that depend on their chromosomal locations. Because the histone tails extend outward from the nucleosome core and are likely to be accessible even when chromatin is condensed, they would seem to provide an especially suitable format for creating marks in a form that can be readily altered as a cell's needs change. Although much remains to be learned about the meaning of the many different histone code combinations, a few well-studied examples of the information that can be encoded in the histone H3 tail are listed in **Figure 4–44**.

A Complex of Code-reader and Code-writer Proteins Can Spread Specific Chromatin Modifications for Long Distances Along a Chromosome

The phenomenon of position effect variegation described previously requires that at least some modified forms of chromatin have the ability to spread for substantial distances along a chromosomal DNA molecule (see Figure 4–36). How is this possible?

The enzymes that modify (or remove modifications from) the histones in nucleosomes are part of multisubunit complexes. They can initially be brought to a particular region of chromatin by one of the sequence-specific DNA-binding proteins (gene regulatory proteins) discussed in Chapters 6 and 7 (for a specific example, see Figure 7–87). But after a modifying enzyme "writes" its mark on one or a few neighboring nucleosomes, events that resemble a chain reaction can ensue. In this case, the "code-writer" enzyme works in concert with a code-reader protein located in the same protein complex. This second protein contains a code-reader module that recognizes the mark and binds tightly to the newly modified nucleosome (see Figure 4–42), positioning its attached writer enzyme near an adjacent nucleosome. Through many such read–write cycles, the reader protein can carry the writer enzyme along the DNA—spreading the mark in a hand-over-hand manner along the chromosome (**Figure 4–45**).

In reality, the process is more complicated than the scheme just described. Both readers and writers are part of a protein complex that is likely to contain

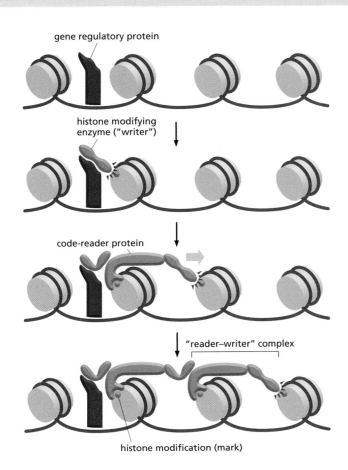

gene regulatory protein

histone modifying enzyme ("writer")

code-reader protein

"reader–writer" complex

histone modification (mark)

Figure 4–45 How the recruitment of a code-reader–writer complex can spread chromatin changes along a chromosome. The code-writer is an enzyme that creates a specific modification on one or more of the four nucleosomal histones. After its recruitment to a specific site on a chromosome by a gene regulatory protein, the writer collaborates with a code-reader protein to spread its mark from nucleosome to nucleosome by means of the indicated reader–writer complex. For this mechanism to work, the reader must recognize the same histone modification mark that the writer produces (see also Figure 4–43).

multiple readers and writers, and to require multiple marks on the nucleosome to spread. Moreover, many of these reader–writer complexes also contain an ATP-dependent chromatin remodeling protein, and the reader, writer, and remodeling proteins work in concert to either decondense or condense long stretches of chromatin as the reader moves progressively along the nucleosome-packaged DNA (**Figure 4–46**).

Some idea of the complexity of the processes just described can be derived from the results of genetic screens for mutant genes that either enhance or suppress the spreading and stability of heterochromatin in tests for position effect variegation in *Drosophila* (see Figure 4–37). As pointed out previously, more than 50 such genes are known, and most of them are likely to function as subunits in one or more reader–writer-remodeling protein complexes.

Barrier DNA Sequences Block the Spread of Reader–Writer Complexes and thereby Separate Neighboring Chromatin Domains

The above mechanism for spreading chromatin structures raises a potential problem. Inasmuch as each chromosome consists of one continuous, very long DNA molecule, what prevents a cacophony of confusing cross-talk between adjacent chromatin domains of different structure and function? Early studies of position effect variegation had suggested an answer: the existence of specific DNA sequences that separate one chromatin domain from another (see Figure 4–37). Several such *barrier* sequences have now been identified and characterized through the use of genetic engineering techniques that allow specific regions of DNA sequence to be deleted or added to chromosomes.

For example, a sequence called HS4 normally separates the active chromatin domain that contains the β-globin locus from an adjacent region of silenced, condensed chromatin in erythrocytes (see Figure 7–61). If this sequence is deleted, the β-globin locus is invaded by condensed chromatin. This chromatin silences the genes it covers, and it spreads to a different extent in different cells, causing a pattern of position effect variegation similar to that

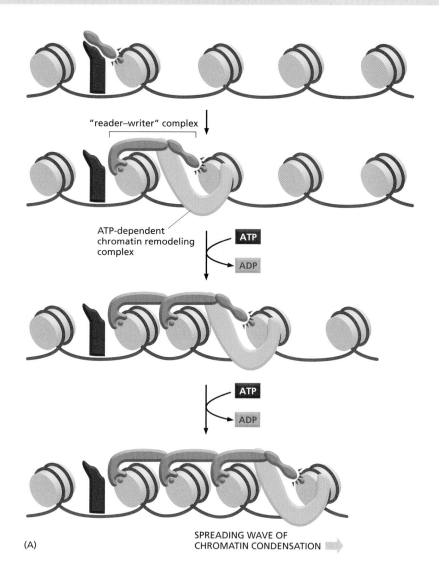

(A)

"reader–writer" complex

ATP-dependent
chromatin remodeling
complex

ATP

ADP

ATP

ADP

SPREADING WAVE OF
CHROMATIN CONDENSATION ➡

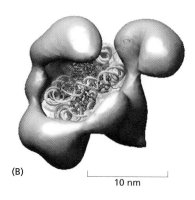

(B)

10 nm

Figure 4–46 How a complex containing reader–writer and ATP-dependent chromatin remodeling proteins can spread chromatin changes along a chromosome. (A) A spreading wave of chromatin condensation. This mechanism is identical to that in Figure 4–45, except that the reader–writer complex collaborates with an ATP-dependent chromatin remodeling protein (see Figure 4–29) to reposition nucleosomes and pack them into highly condensed arrays. This is a highly simplified view of the mechanism known to be able to spread a major form of heterochromatin for long distances along chromosomes (see Figure 4–36). The heterochromatin-specific protein HP1 plays a major role in that process. HP1 binds to trimethyl lysine 9 on histone H3, and it remains associated with the condensed chromatin as one of the readers in a reader–writer-remodeling complex that, while incompletely understood, is considerably more intricate than that shown here. (B) The actual structure of a chromatin reader-remodeling complex, showing how it is thought to interact with a nucleosome. Modeled in *gray* is the yeast RSC complex, which contains 15 subunits—including an ATP-dependent chromatin remodeling protein and at least 4 subunits with code-reader domains. (B, from A.E. Leschziner et al., *Proc. Natl Acad. Sci. U.S.A.* 104:4913–4918, 2007. With permission from National Academy of Sciences.)

observed in *Drosophila*. As described in Chapter 7, this invasion has dire consequences: the globin genes are poorly expressed, and individuals who carry such a deletion have a severe form of anemia.

The HS4 sequence is often added to both ends of a gene that is experimentally inserted into a mammalian genome, in order to protect that gene from the silencing caused by spreading heterochromatin. Analysis of this barrier sequence reveals that it contains a cluster of binding sites for histone acetylase enzymes. Since the acetylation of a lysine side chain is incompatible with the methylation of the same side chain, histone acetylases and histone deacetylases are logical candidates for the formation of barriers on the DNA that block the spread of different forms of chromatin (**Figure 4–47**). However, several other types of chromatin modifications are known that can also protect genes from silencing.

The Chromatin in Centromeres Reveals How Histone Variants Can Create Special Structures

The presence of nucleosomes carrying histone variants is thought to produce marks in chromatin that are unusually long lasting. Consider, for example, the formation and inheritance of the chromatin that forms on centromeres, the DNA region of each chromosome required for the orderly segregation of the chromosomes into daughter cells each time a cell divides (see Figure 4–21). In many complex organisms, including humans, each centromere is embedded in a stretch of special *centric heterochromatin* that persists throughout interphase, even though the centromere-mediated movement of DNA occurs only during

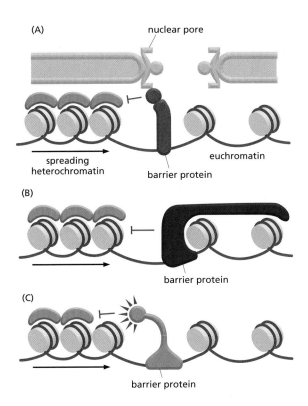

Figure 4–47 **Some mechanisms of barrier action.** These models are derived from different analyses of barrier action, and a combination of several of them may function at any one site. (A) The tethering of a region of chromatin to a large fixed site, such as the nuclear pore complex illustrated here, can form a barrier that stops the spread of heterochromatin. (B) The tight binding of barrier proteins to a group of nucleosomes can compete with heterochromatin spreading. (C) By recruiting a group of highly active histone-modifying enzymes, barriers can erase the histone marks that are required for heterochromatin to spread. For example, a potent acetylation of lysine 9 on histone H3 will compete with lysine 9 methylation, thereby preventing the HP1 protein binding needed to form some forms of heterochromatin (see Figure 4–46). (Based on A.G. West and P. Fraser, *Hum. Mol. Genet.* 14:R101–R111, 2005. With permission from Oxford University Press.)

mitosis. This chromatin contains a centromere-specific variant H3 histone, known as CENP-A (see Figure 4–41), plus additional proteins that pack the nucleosomes into particularly dense arrangements and form the kinetochore, the special structure required for attachment of the mitotic spindle.

A specific DNA sequence of approximately 125 nucleotide pairs is sufficient to serve as a centromere in the yeast *S. cerevisiae*. Despite its small size, more than a dozen different proteins assemble on this DNA sequence; the proteins include the CENP-A histone H3 variant, which, along with the three other core histones, forms a centromere-specific nucleosome. The additional proteins at the yeast centromere attach this nucleosome to a single microtubule from the yeast mitotic spindle (**Figure 4–48**).

The centromeres in more complex organisms are considerably larger than those in budding yeasts. For example, fly and human centromeres extend over hundreds of thousands of nucleotide pairs and do not seem to contain a centromere-specific DNA sequence. These centromeres largely consist of short, repeated DNA sequences, known as *alpha satellite DNA* in humans. But the same repeat sequences are also found at other (non-centromeric) positions on

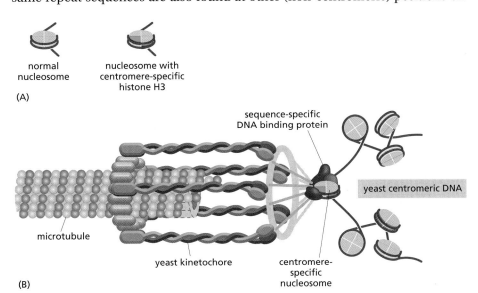

Figure 4–48 **A model for the structure of a simple centromere.** In the yeast *Saccharomyces cerevisiae,* a special centromeric DNA sequence assembles a single nucleosome in which two copies of an H3 variant histone (called CENP-A in most organisms) replaces the normal H3. Peptide sequences unique to this variant histone (see Figure 4–41) then help to assemble additional proteins, some of which form a kinetochore. This kinetochore is unusual in capturing only a single microtubule; humans have much larger centromeres and form kinetochores that can capture 20 or more microtubules (see Figure 4–50). The kinetochore is discussed in detail in Chapter 17. (Adapted from A. Joglekar et al., *Nat. Cell Biol.* 8:381–383, 2006. With permission from Macmillan Publishers Ltd.)

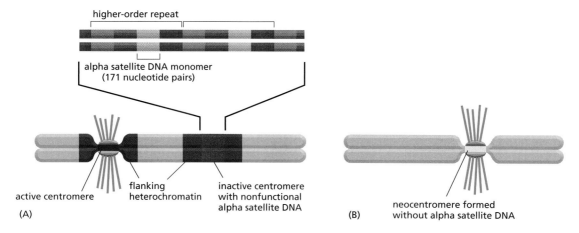

Figure 4–49 Evidence for the plasticity of human centromere formation. (A) A series of A-T-rich alpha satellite DNA sequences are repeated many thousands of times at each human centromere *(red)*, surrounded by pericentric heterochromatin *(brown)*. However, due to an ancient chromosome breakage and rejoining event, some human chromosomes contain two blocks of alpha satellite DNA, each of which presumably functioned as a centromere in its original chromosome. Usually, these dicentric chromosomes are not stably propagated because they attach improperly to the spindle and are broken apart during mitosis. In chromosomes that do survive, however, one of the centromeres has somehow inactivated, even though it contains all the necessary DNA sequences. This allows the chromosome to be stably propagated. (B) In a small fraction (1/2000) of human births, extra chromosomes are observed in cells of the offspring. Some of these extra chromosomes, which have formed from a breakage event, lack alpha satellite DNA altogether, yet new centromeres (neocentromeres) have arisen from what was originally euchromatic DNA.

chromosomes, indicating that they are not sufficient to direct centromere formation. Most strikingly, in some unusual cases, new human centromeres (called neocentromeres) have been observed to form spontaneously on fragmented chromosomes. Some of these new positions were originally euchromatic and lack alpha satellite DNA altogether (**Figure 4–49**).

It therefore seems that centromeres in complex organisms are defined by an assembly of proteins, instead of by a specific DNA sequence. When antibodies that stain specific modified nucleosomes are used to examine the stretched chromosome fibers from centromeres, one observes striking alternation of two modified forms of chromatin (**Figure 4–50**). It appears that this arrangement allows the centric heterochromatin to fold so as to position the CENP-A-containing nucleosomes on the outside of the mitotic chromosome, where they bind the set of proteins that form the kinetochore plates. These plates in turn capture a group of microtubules from the mitotic spindle in order to partition the chromosomes accurately, as described in Chapter 17.

Chromatin Structures Can Be Directly Inherited

To explain the above observations, it has been proposed that *de novo* centromere formation requires an initial seeding event, involving the formation of a specialized DNA–protein structure that contains nucleosomes formed with the CENP-A variant of histone H3. In humans, this seeding event happens more readily on arrays of alpha satellite DNA than on other DNA sequences. The H3–H4 tetramers from each nucleosome on the parental DNA helix are directly inherited by the daughter DNA helices at a replication fork (see Figure 5–38). Therefore, once a set of CENP-A-containing nucleosomes has been assembled on a stretch of DNA, it is easy to understand how a new centromere could be generated in the same place on both daughter chromosomes following each round of cell division (**Figure 4–51**).

The plasticity of centromeres may provide an important evolutionary advantage. We have seen that chromosomes evolve in part by breakage and rejoining events (see Figure 4–18). Many of these events produce chromosomes with two centromeres, or chromosome fragments with no centromeres at all. Although rare, both the inactivation of centromeres and their ability to be activated *de novo*

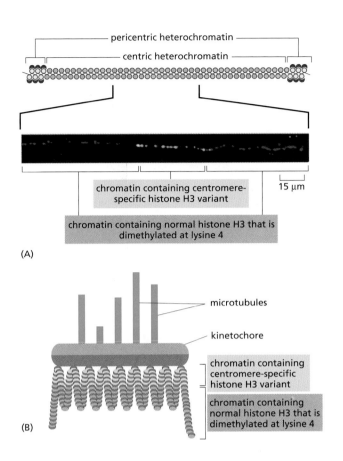

(A)

(B)

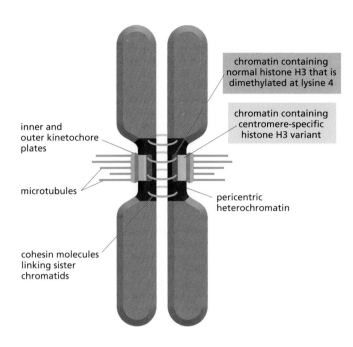

(C)

Figure 4–50 The organization and function of the chromatin that forms human centromeres. (A) By staining stretched chromatin fibers with fluorescently labeled antibodies, the alpha satellite DNA that forms centric heterochromatin in humans is seen to be packaged into alternating blocks of chromatin. One block is formed from a long string of nucleosomes containing the CENP-A H3 variant histone *(green)*; the other block contains nucleosomes that are specially marked with a dimethyl lysine 4 *(red)*. Each block is more than a thousand nucleosomes long. (B) A model for the organization of the two types of centric heterochromatin. As in yeast, the nucleosomes that contain the H3 variant histone form the kinetochore. (C) The arrangement of the centric and pericentric heterochromatin on a human metaphase chromosome, as determined by fluorescence microscopy using the same antibodies as in (A). (Adapted from B.A. Sullivan and G.H. Karpen, *Nat. Struct. Mol. Biol.* 11:1076–1083, 2004. With permission from Macmillan Publishers Ltd.)

may occasionally allow newly formed chromosomes to be maintained stably, thereby facilitating the process of chromosome evolution.

There are some striking similarities between the formation and maintenance of centromeres and the formation and maintenance of other regions of heterochromatin. In particular, the entire centromere forms as an all-or-none entity, suggesting a highly cooperative addition of proteins after a seeding event. Moreover, once formed, the structure seems to be directly inherited on the DNA as part of each round of chromosome replication.

Chromatin Structures Add Unique Features to Eucaryotic Chromosome Function

Although a great deal remains to be learned about the functions of different chromatin structures, the packaging of DNA into nucleosomes was probably crucial for the evolution of eucaryotes like ourselves. Complex multicellular organisms would appear to be possible only if the cells in different lineages can specialize by changing the accessibility and responsiveness of many hundreds of genes to genetic readout. As described in Chapter 22, each cell has a stored memory of its past developmental history in the regulatory circuits that control its many genes.

Although bacteria also require cell memory mechanisms, the complexity of the memory circuits required by higher eucaryotes is unprecedented. The packaging of selected regions of eucaryotic genomes into different forms of chromatin makes possible a type of cell memory mechanism that is not available to bacteria. The crucial feature of this uniquely eucaryotic form of gene regulation is the storage of the memory of the state of a gene on a gene-by-gene basis—in the form of local chromatin structures that can persist for various lengths of time. At one extreme are structures like centric heterochromatin that, once established, are stably inherited from one cell generation to the next (see Figure

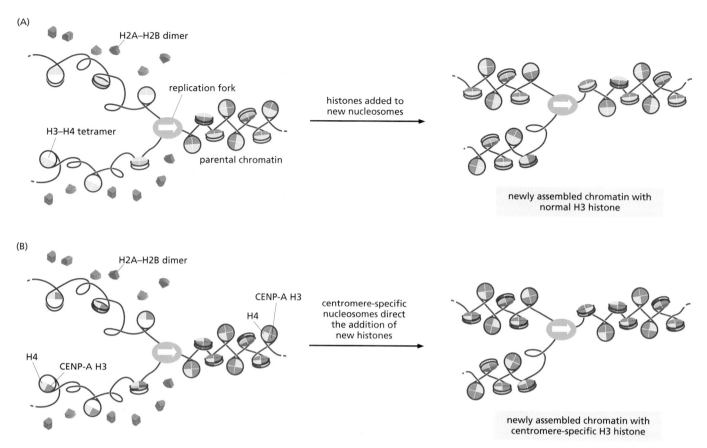

(A)

H2A–H2B dimer

replication fork

H3–H4 tetramer

parental chromatin

histones added to
new nucleosomes

newly assembled chromatin with
normal H3 histone

(B)

H2A–H2B dimer

CENP-A H3

H4

H4

CENP-A H3

centromere-specific
nucleosomes direct
the addition of
new histones

newly assembled chromatin with
centromere-specific H3 histone

Figure 4–51 A model for the direct inheritance of centromeric heterochromatin. (A) The normal assembly of chromatin on the two daughter DNA helices produced at a replication fork requires the deposition of H2A–H2B dimers onto directly inherited H3–H4 tetramers, as well as the assembly of new histone octamers (see Figure 5–38 for details). (B) At a centromere, the inheritance of H3 variant–H4 tetramers seeds the formation of new histone octamers that likewise contain the variant H3 histone. A similar seeding process could cause the adjacent blocks of centric heterochromatin (containing H3 modified at dimethyl lysine 4; see Figure 4–50) to be inherited. Although the details are not known, the seeding process is likely to involve other centromeric proteins that are inherited along with the nucleosomes (see Figure 4–52).

4–51). Closely related mechanisms that are likewise based on the direct inheritance of parental forms of chromatin by the daughter DNA helices behind the replication fork are thought to be responsible for other types of condensed chromatin (**Figure 4–52**). For example, the permanently silenced, classical type

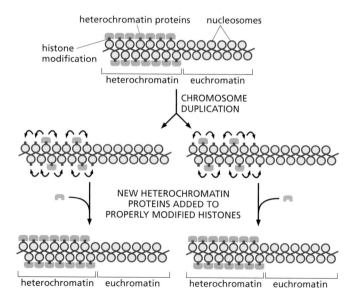

heterochromatin proteins   nucleosomes

histone
modification

heterochromatin   euchromatin

CHROMOSOME
DUPLICATION

NEW HETEROCHROMATIN
PROTEINS ADDED TO
PROPERLY MODIFIED HISTONES

heterochromatin   euchromatin

heterochromatin   euchromatin

Figure 4–52 How the packaging of DNA in chromatin can be inherited during chromosome replication. In this model, some of the specialized chromatin components are distributed to each daughter chromosome after DNA duplication, along with the specially marked nucleosomes that they bind. After DNA replication, the inherited nucleosomes that are specially modified, acting in concert with the inherited chromatin components, change the pattern of histone modification on the newly formed daughter nucleosomes nearby. This creates new binding sites for the same chromatin components, which then assemble to complete the structure. The latter process is likely to involve code reader–writer-remodeling complexes operating in a manner similar to that previously illustrated in Figure 4–46.

of heterochromatin contains the HP1 protein, whereas the condensed chromatin that coats important developmental regulatory genes is maintained by the polycomb group of proteins. The latter type of heterochromatin silences a large number of genes that encode gene regulatory proteins early in embryonic development, covering a total of about 2 percent of the human genome, and it is removed only when each individual gene is needed by the developing organism (discussed in Chapter 22). Although other types of inherited chromatin structures exist, it is not yet clear how many different types there are: the number could certainly exceed 10 (see p. 238). The fundamental importance of this mechanism for distinguishing different genes is schematically represented in (**Figure 4–53**).

Other forms of chromatin can have a shorter lifetime, much less than the division time of the cell; however, many have a built-in persistence that helps to mediate biological function.

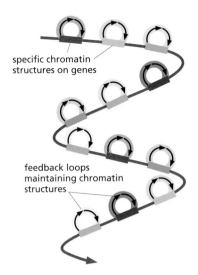

specific chromatin structures on genes

feedback loops maintaining chromatin structures

Figure 4–53 Schematic illustration of cell memory stored as chromatin-based epigenetic information in the genes of eucaryotes. Genes in eucaryotic cells can be packaged into a large variety of different chromatin structures, indicated here by different colors. At least some of these chromatin structures have a special effect on gene expression that can be directly inherited as epigenetic information when a cell divides. This allows some of the gene regulatory proteins that create different gene states to act only once, inasmuch as the state can be remembered after the regulatory protein is gone. Epigenetic information can also be stored in networks of signaling molecules that control gene expression (see Figure 7–86).

Summary

Despite the uniform assembly of chromosomal DNA into nucleosomes, a large variety of different chromatin structures are possible in eucaryotic organisms. This variety is based on a large set of reversible covalent modifications of the four histones in the nucleosome core. These modifications include the mono-, di-, and tri-methylation of many different lysine side chains, an important reaction that is incompatible with the acetylation of the same lysines. Specific combinations of the modifications mark each nucleosome with a histone code. The histone code is read when protein modules that are part of a larger protein complex bind to the modified nucleosomes in a region of chromatin. These code-reader proteins then attract additional proteins that catalyze biologically relevant functions.

Some code-reader protein complexes contain a histone-modifying enzyme, such as a histone methylase, that "writes" the same mark that the code-reader recognizes. A reader–writer-remodeling complex of this type can spread a specific form of chromatin for long distances along a chromosome. In particular, large regions of condensed heterochromatin are thought to be formed in this way. Heterochromatin is commonly found around centromeres and near telomeres, but it is also present at many other positions in chromosomes. The tight packaging of DNA into heterochromatin usually silences the genes within it.

The phenomenon of position effect variegation provides good evidence for the direct inheritance of condensed forms of chromatin by the daughter DNA helices formed at a replication fork, and a similar mechanism appears to be responsible for maintaining the specialized chromatin at centromeres. More generally, the ability to transmit specific chromatin structures from one cell generation to the next provides the basis for an epigenetic cell memory process that is likely to be critical for maintaining the complex set of different cell states required by complex multicellular organisms.

THE GLOBAL STRUCTURE OF CHROMOSOMES

Having discussed the DNA and protein molecules from which the 30-nm chromatin fiber is made, we now turn to the organization of the chromosome on a more global scale. As a 30-nm fiber, the typical human chromosome would still be 0.1 cm in length and able to span the nucleus more than 100 times. Clearly, there must be a still higher level of folding, even in interphase chromosomes. Although its molecular basis is still largely a mystery, this higher-order packaging almost certainly involves the folding of the 30-nm fiber into a series of loops and coils. This chromatin packing is fluid, frequently changing in response to the needs of the cell.

We shall begin by describing some unusual interphase chromosomes that can be easily visualized, inasmuch as certain features of these exceptional cases are thought to be representative of all interphase chromosomes. Moreover, they

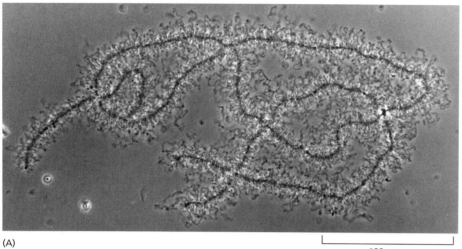

(A)

100 µm

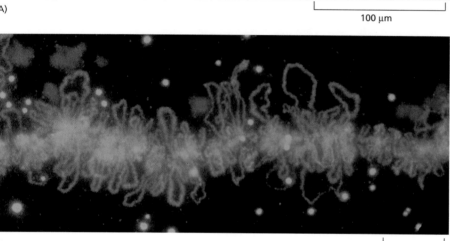

(B)

20 µm

Figure 4–54 Lampbrush chromosomes. (A) A light micrograph of lampbrush chromosomes in an amphibian oocyte. Early in oocyte differentiation, each chromosome replicates to begin meiosis, and the homologous replicated chromosomes pair to form this highly extended structure containing a total of four replicated DNA molecules, or chromatids. The lampbrush chromosome stage persists for months or years, while the oocyte builds up a supply of materials required for its ultimate development into a new individual. (B) An enlarged region of a similar chromosome, stained with a fluorescent reagent that makes the loops active in RNA synthesis clearly visible. (Courtesy of Joseph G. Gall.)

provide a unique means for investigating some fundamental aspects of chromatin structure raised in the previous section. Next we describe how a typical interphase chromosome is arranged in the cell nucleus, focusing on human cells. Finally, we conclude by discussing the additional tenfold compaction that interphase chromosomes undergo during the process of mitosis.

Chromosomes Are Folded into Large Loops of Chromatin

Insight into the structure of the chromosomes in interphase cells has been obtained from studies of the stiff and extended meiotically paired chromosomes in growing amphibian oocytes (immature eggs). These very unusual **lampbrush chromosomes** (the largest chromosomes known) are clearly visible even in the light microscope, where they are seen to be organized into a series of large chromatin loops emanating from a linear chromosomal axis (**Figure 4–54**).

The organization of a lampbrush chromosome is illustrated in **Figure 4–55**. A given loop always contains the same DNA sequence, and it remains extended in the same manner as the oocyte grows. These chromosomes are producing large amounts of RNA for the oocyte, and most of the genes present in the DNA loops are being actively expressed. The majority of the DNA, however, is not in loops but remains highly condensed in the *chromomeres* on the axis, where genes are generally not expressed.

It is thought that the interphase chromosomes of all eucaryotes are similarly arranged in loops. Although these loops are normally too small and fragile to be easily observed in a light microscope, other methods can be used to infer their presence. For example, it has become possible to assess the frequency with

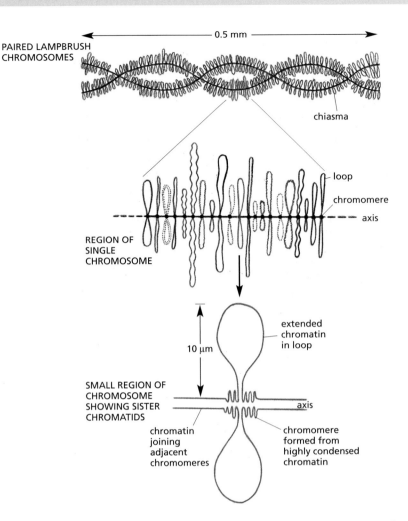

Figure 4–55 A model for the structure of a lampbrush chromosome. The set of lampbrush chromosomes in many amphibians contains a total of about 10,000 chromatin loops, although most of the DNA in each chromosome remains highly condensed in the chromomeres. Each loop corresponds to a particular DNA sequence. Four copies of each loop are present in each cell, since each of the two major units shown at the top consists of two closely apposed, newly replicated chromosomes. This four-stranded structure is characteristic of this stage of development of the oocyte, the diplotene stage of meiosis; see Figure 21–9.

which any two loci along an interphase chromosome are paired with each other, thus revealing likely candidates for the sites on chromatin that form the closely apposed bases of loop structures (**Figure 4–56**). These experiments and others suggest that the DNA in human chromosomes is organized into loops of

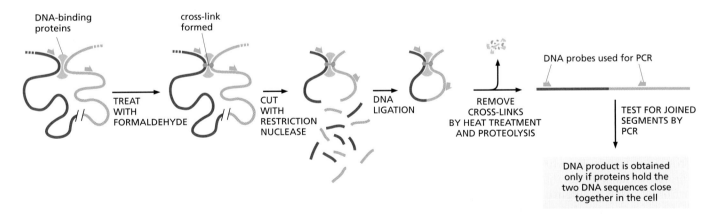

Figure 4–56 A method for determining the position of loops in interphase chromosomes. In this technique, known as the chromosome conformation capture (3C) method, cells are treated with formaldehyde to create the indicated covalent DNA–protein and DNA–DNA cross-links. The DNA is then treated with a restriction nuclease that chops the DNA into many pieces, cutting at strictly defined nucleotide sequences and forming sets of identical "cohesive ends" (see Figure 8–34). The cohesive ends can be made to join through their complementary base-pairing. Importantly, prior to the ligation step shown, the DNA is diluted so that the fragments that have been kept in close proximity to each other (through cross-linking) are the ones most likely to join. Finally, the cross-links are reversed and the newly ligated fragments of DNA are identified and quantified by PCR (the polymerase chain reaction, described in Chapter 8). By combining the frequency-of-association information generated by the 3C technique with DNA sequence information, structural models can be produced for the interphase conformation of chromosomes.

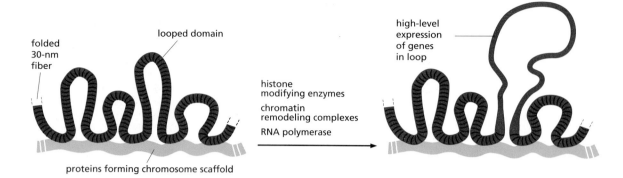

Figure 4–57 A model for the organization of an interphase chromosome. A section of an interphase chromosome is shown folded into a series of looped domains, each containing perhaps 50,000–200,000 nucleotide pairs of double-helical DNA condensed into a 30-nm fiber. The chromatin in each individual loop is further condensed through poorly understood folding processes that are reversed when the cell requires direct access to the DNA packaged in the loop. Neither the composition of the postulated chromosomal axis nor how the folded 30-nm fiber is anchored to it is clear. However, in mitotic chromosomes the bases of the chromosomal loops are enriched both in condensins and in DNA topoisomerase II enzymes, two proteins that may form much of the axis at metaphase (see Figure 4–74).

different lengths. A typical loop might contain between 50,000 and 200,000 nucleotide pairs of DNA, although loops of a million nucleotide pairs have also been suggested (**Figure 4–57**).

Polytene Chromosomes Are Uniquely Useful for Visualizing Chromatin Structures

Certain giant insect cells have grown to their enormous size through multiple cycles of DNA synthesis without cell division. Such cells with more than the normal DNA complement are said to be *polyploid* when they contain increased numbers of standard chromosomes. But in several types of secretory cells in fly larvae, all the homologous chromosome copies are held side by side, like drinking straws in a box, creating single large **polytene chromosomes**. Because polytene chromosomes can disperse to form a conventional polyploid cell in some cases, these two chromosomal states are closely related. The underlying structure of a polytene chromosome must therefore be similar to that of a normal chromosome.

Polyteny has been most studied in the larval salivary gland cells of the fruit fly *Drosophila*. When these polytene chromosomes are viewed in the light microscope, distinct alternating dark *bands* and light *interbands* are visible (**Figure 4–58**), each formed from a thousand identical DNA sequences arranged side-by-side in register. About 95% of the DNA in polytene chromosomes is in bands, and 5% is in interbands. A very thin band can contain 3000 nucleotide pairs, while a thick band may contain 200,000 nucleotide pairs in each of its chromatin strands. The chromatin in each band appears dark because the DNA is more condensed than the DNA in interbands; it may also contain a higher proportion of proteins (**Figure 4–59**).

There are approximately 3700 bands and 3700 interbands in the complete set of *Drosophila* polytene chromosomes. The bands can be recognized by their different thicknesses and spacings, and each one has been given a number to generate a chromosome "map" that has been indexed to the finished genome sequence of this fly.

The *Drosophila* polytene chromosomes provide a good starting point for examining how chromatin is organized on a large scale. In the previous section, we saw that there are many forms of chromatin, each of which contains nucleosomes with a different combination of modified histones. By reading this histone code, specific sets of non-histone proteins assemble on the nucleosomes to

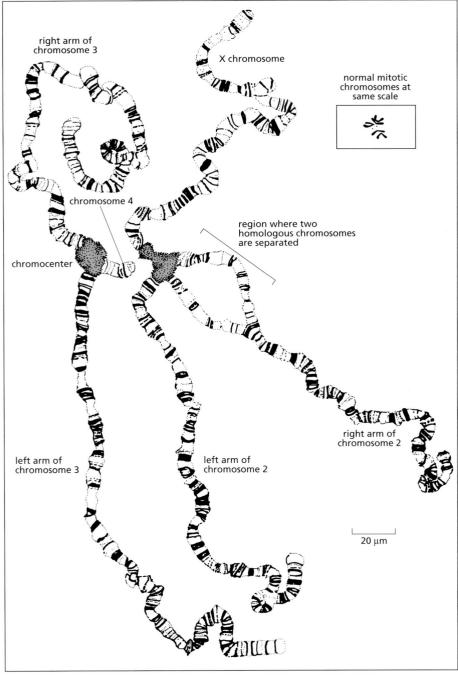

right arm of
chromosome 3

X chromosome

normal mitotic
chromosomes at
same scale

chromosome 4

region where two
homologous chromosomes
are separated

chromocenter

left arm of
chromosome 3

left arm of
chromosome 2

right arm of
chromosome 2

20 μm

Figure 4–58 The entire set of polytene chromosomes in one *Drosophila* salivary cell. In this drawing of a light micrograph, the giant chromosomes have been spread out for viewing by squashing them against a microscope slide. *Drosophila* has four chromosomes, and there are four different chromosome pairs present. But each chromosome is tightly paired with its homolog (so that each pair appears as a single structure), which is not true in most nuclei (except in meiosis). Each chromosome has undergone multiple rounds of replication, and the homologues and all their duplicates have remained in exact register with each other, resulting in huge chromatin cables many DNA strands thick.

The four polytene chromosomes are normally linked together by heterochromatic regions near their centromeres that aggregate to create a single large chromocenter *(pink region)*. In this preparation, however, the chromocenter has been split into two halves by the squashing procedure used. (Adapted from T.S. Painter, *J. Hered.* 25:465–476, 1934. With permission from Oxford University Press.)

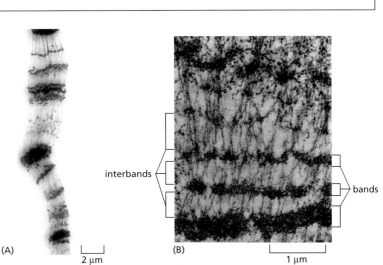

interbands

bands

(A)

2 μm

(B)

1 μm

Figure 4–59 Micrographs of polytene chromosomes from *Drosophila* salivary glands. (A) Light micrograph of a portion of a chromosome. The DNA has been stained with a fluorescent dye, but a reverse image is presented here that renders the DNA *black* rather than *white*; the bands are clearly seen to be regions of increased DNA concentration. This chromosome has been processed by a high pressure treatment so as to show its distinct pattern of bands and interbands more clearly. (B) An electron micrograph of a small section of a *Drosophila* polytene chromosome seen in thin section. Bands of very different thickness can be readily distinguished, separated by interbands, which contain less condensed chromatin. (A, adapted from D.V. Novikov, I. Kireev and A.S. Belmont, *Nat. Methods* 4:483–485, 2007. With permission from Macmillan Publishers Ltd. B, courtesy of Veikko Sorsa.)

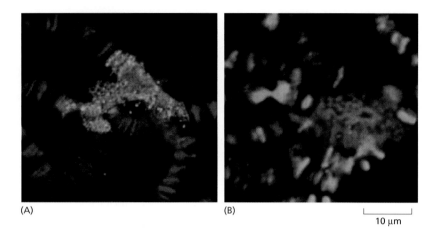

(A)   (B)

|_____|
 10 μm

Figure 4–60 **The pattern of histone modifications on *Drosophila* polytene chromosomes.** Antibodies that specifically recognize different histone modifications can reveal where each modification is found with reference to the many bands and interbands on these chromosomes. In the two preparations shown here, the positions of two different markings on histone H3 tails are compared. In both cases, the antibody labeling the modified histone is *green*, and the DNA is stained *red*. Only a small region surrounding each chromocenter is shown. (A) Dimethyl Lys 9 *(green)* is a histone modification associated with the pericentric heterochromatin. It is seen to be associated with the chromocenter. (B) Acetylated Lys 9 *(green)* is a modification that is concentrated in histones associated with active genes. It is seen to be present in numerous bands on the chromosome arms, but not in the heterochromatic chromocenter. Similar experiments can be carried out to position many other modified histones, as well as the non-histone proteins (see, for example, Figure 22–45 for chromosomes stained for Polycomb). (Adapted from A. Ebert, S. Lein, G. Schotta and G. Reuter, *Chromosome Res.* 14:377–392, 2006. With permission from Springer.)

affect biological function in different ways. Some of these non-histone proteins can spread for long distances along the DNA, imparting a similar chromatin structure to contiguous regions of the genome (see Figure 4–46). Thus, in some regions, all of the chromatin has a similar structure and is separated from neighboring domains by barrier proteins (see Figure 4–47). At low resolution, the interphase chromosome can therefore be considered as a mosaic of chromatin structures, each containing particular nucleosome modifications associated with a particular set of non-histone proteins. (At a higher level of resolution one would also emphasize the many sequence-specific DNA-binding proteins that will be described in Chapter 7).

This view of an interphase chromosome helps us to interpret the results obtained from studies of polytene chromosomes. By staining with highly specific antibodies, one can show that differently modified histones (**Figure 4–60**), as well as distinct sets of non-histone proteins, are located on different polytene chromosome bands. This suggests a powerful general strategy. By employing combinations of antibodies that bind tightly to each of the many different histone modifications that create the histone code (see Figure 4–39), it may be possible to determine which combinations of modifications specify particular types of chromatin domains. And by carrying out similar experiments with antibodies that recognize each of the hundreds of different non-histone proteins in chromatin, one can attempt to decipher the many different meanings encoded in histone modifications.

There Are Multiple Forms of Heterochromatin

Molecular studies have led to a reevaluation of our view of heterochromatin. For many decades, heterochromatin was thought to be a single entity defined by its highly condensed structure and its ability to silence genes permanently. But if we define heterochromatin as a form of compact chromatin that can silence genes, be epigenetically inherited, and spread along chromosomes to cause position effect variegation (see Figure 4–36), it is clear that different types of heterochromatin exist. In fact, we have already considered three of these types in discussing the human centromere (see Figure 4–50).

Each domain of heterochromatin is thought to be formed by the cooperative assembly of a set of non-histone proteins. For example, classical pericentromeric heterochromatin contains more than six such proteins, including heterochromatin protein 1 (HP1), whereas the so-called Polycomb form of heterochromatin contains a similar number of proteins in a non-overlapping set (PcG proteins). There are hundreds of small blocks of heterochromatin spread across the arms of *Drosophila* polytene chromosomes, as identified by their late replication (discussed in Chapter 5). Antibody staining of these regions of heterochromatin suggests that the known forms of heterochromatin can account for no more than half of the heterchromatic polytene bands (**Figure 4–61**). Thus, other types of heterochromatin must exist whose protein composition is not known. It is likely

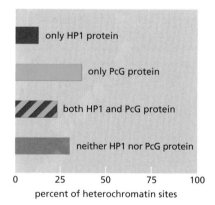

only HP1 protein

only PcG protein

both HP1 and PcG protein

neither HP1 nor PcG protein

0   25   50   75   100
percent of heterochromatin sites

Figure 4–61 **Evidence for multiple forms of heterochromatin.** In this study, 240 late-replicating sites on the *Drosophila* polytene chromosome arms were examined for the presence of two non-histone proteins. These proteins are known to help compact two different forms of heterochromatin (see text). As indicated, antibody staining suggests that roughly half of the sites are packaged into forms of heterochromatin that are different from either of these two. Experiments such as these demonstrate that we have a great deal more to learn about the packaging of DNA in eucaryotes. (Data from I.F. Zhimulev and E.S. Belyaeva, *BioEssays* 25:1040–1051, 2003. With permission from John Wiley & Sons.)

that each of these types of heterochromatin is differently regulated and has different roles in the cell.

The chromatin structure in each domain ultimately depends on the proteins that bind to specific DNA sequences, and these are known to vary depending on the cell type and its stage of development in a multicellular organism. Thus, both the pattern of chromatin domains and their individual compositions (nucleosome modifications plus non-histone proteins) can vary between tissues. These differences make different genes accessible for genetic readout, helping to explain the cell diversification that accompanies embryonic development (described in Chapter 22). Comparisons of the polytene chromosomes in two different tissues of a fly lend support to this general idea: although the patterns of bands and interbands are largely the same, there are reproducible differences.

Chromatin Loops Decondense When the Genes Within Them Are Expressed

When an insect progresses from one developmental stage to another, distinctive *chromosome puffs* arise and old puffs recede in its polytene chromosomes as new genes become expressed and old ones are turned off (**Figure 4–62**). From inspection of each puff when it is relatively small and the banding pattern is still discernible, it seems that most puffs arise from the decondensation of a single chromosome band.

The individual chromatin fibers that make up a puff can be visualized with an electron microscope. In favorable cases, loops are seen, much like those observed in the amphibian lampbrush chromosomes discussed above. When not expressed, the loop of DNA assumes a thickened structure, possibly a folded 30-nm fiber, but when gene expression is occurring, the loop becomes more extended. In electron micrographs, the chromatin located on either side of the decondensed loop appears considerably more compact, suggesting that a loop constitutes a distinct functional domain of chromatin structure.

Observations made in human cells also suggest that highly folded loops of chromatin expand to occupy an increased volume when a gene within them is expressed. For example, quiescent chromosome regions from 0.4 to 2 million nucleotide pairs in length appear as compact dots in an interphase nucleus when visualized by fluorescence microscopy using FISH or other technologies. However, the same DNA is seen to occupy a larger territory when its genes are expressed, with elongated, punctate structures replacing the original dot.

Chromatin Can Move to Specific Sites Within the Nucleus to Alter Gene Expression

New ways of visualizing individual chromosomes have shown that each of the 46 interphase chromosomes in a human cell tends to occupy its own discrete territory within the nucleus (**Figure 4–63**). However, pictures such as these present only an average view of the DNA in each chromosome. Experiments that specifically localize the heterochromatic regions of a chromosome reveal that they are often closely associated with the nuclear lamina, regardless of the chromosome examined. And DNA probes that preferentially stain gene-rich regions of human

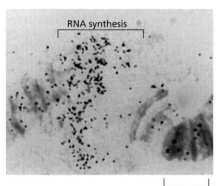

10 μm

Figure 4–62 RNA synthesis in polytene chromosome puffs. An autoradiograph of a single puff in a polytene chromosome from the salivary glands of the freshwater midge *C. tentans*. As outlined in Chapter 1 and described in detail in Chapter 6, the first step in gene expression is the synthesis of an RNA molecule using the DNA as a template. The decondensed portion of the chromosome is undergoing RNA synthesis and has become labeled with ^{3}H-uridine (see p. 603), an RNA precursor molecule that is incorporated into growing RNA chains. (Courtesy of José Bonner.)

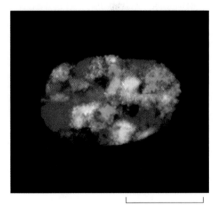

10 μm

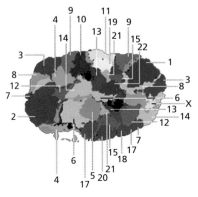

Figure 4–63 Simultaneous visualization of the chromosome territories for all of the human chromosomes in a single interphase nucleus. A FISH analysis using a different mixture of fluorochromes for marking the DNA of each chromosome, detected with seven color channels in a fluorescence microscope, allows each chromosome to be distinguished in three-dimensional reconstructions. Below the micrograph, each chromosome is identified in a schematic of the actual image. Note that the two homologous chromosomes (e.g., the two copies of chromosome 9), are not in general co-located. (From M.R. Speicher and N.P. Carter, *Nat. Rev. Genet.* 6:782–792, 2005. With permission from Macmillan Publishers Ltd.)

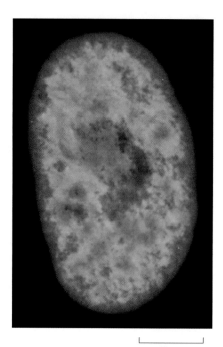

Figure 4–64 **The distribution of gene-rich regions of the human genome in an interphase nucleus.** Gene-rich regions have been visualized with a fluorescent probe that hybridizes to the Alu interspersed repeat, which is present in more than a million copies in the human genome (see Figure 5–75). For unknown reasons, these sequences cluster in chromosomal regions rich in genes. In this representation, regions enriched for the Alu sequence are *green*, regions depleted for these sequences are *red*, while the average regions are *yellow*. The gene-rich regions are seen to be depleted in the DNA near the nuclear envelope. (From A. Bolzer et al., *PLoS Biol.* 3:826–842, 2005. With permission from Public Library of Science.)

5 µm

chromosomes produce a striking picture of the interphase nucleus that presumably reflects different average positions for active and inactive genes (**Figure 4–64**).

A variety of different types of experiments have led to the conclusion that the position of a gene in the interior of the nucleus changes when it becomes highly expressed. Thus, a region that becomes very actively transcribed is often found to extend out of its chromosome territory, as if in an extended loop (**Figure 4–65**). We will see in Chapter 6 that the initiation of transcription—the first step in gene expression—requires the assembly of over 100 proteins, and it makes sense that this would occur most rapidly in regions of the nucleus particularly rich in these proteins.

More generally, it is clear that the nucleus is very heterogeneous, with functionally different regions to which portions of chromosomes can move as they are subjected to different biochemical processes—such as when their gene expression changes (**Figure 4–66**). There is evidence that some of these nuclear regions are marked with different inositol phospholipids, reminiscent of the way that the same lipids are used to distinguish different membranes in the cytoplasm (see Figure 13–11). But what these lipids are attached to in the interior of the nucleus is a mystery, as the only known lipid-rich environments are the lipid bilayers of the nuclear envelope.

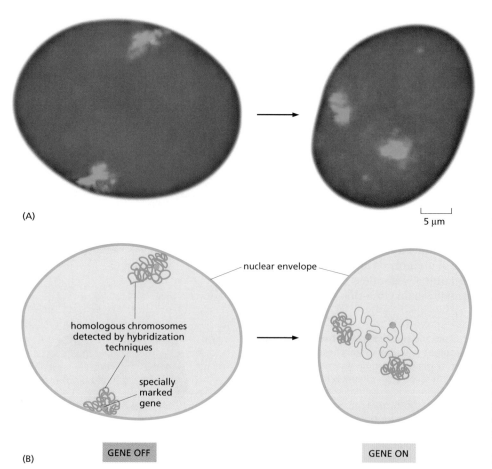

(A)

5 µm

(B)

GENE OFF   GENE ON

Figure 4–65 **An effect of high levels of gene expression on the intranuclear location of chromatin.** (A) Fluorescence micrographs of human nuclei showing how the position of a gene changes when it becomes highly transcribed. The region of the chromosome adjacent to the gene *(red)* is seen to leave its chromosomal territory *(green)* only when it is highly active. (B) Schematic representation of a large loop of chromatin that expands when the gene is on, and contracts when the gene is off. Other genes that are less actively expressed can be shown by the same methods to remain inside their chromosomal territory when transcribed. (From J.R. Chubb and W.A. Bickmore, *Cell* 112:403–406, 2003. With permission from Elsevier.)

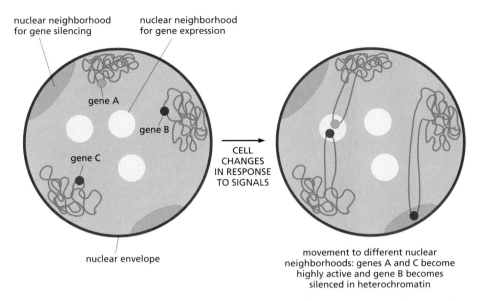

nuclear neighborhood
for gene silencing

nuclear neighborhood
for gene expression

gene A

gene B

gene C

nuclear envelope

CELL
CHANGES
IN RESPONSE
TO SIGNALS

movement to different nuclear
neighborhoods: genes A and C become
highly active and gene B becomes
silenced in heterochromatin

Figure 4–66 The movement of genes to different regions of the nucleus when their expression changes. The interior of the nucleus is very heterogeneous, and different nuclear neighborhoods are known to have distinct effects on gene expression. Movements such as those indicated here presumably reflect changes in the binding affinities that the chromatin and RNA molecules surrounding a gene have for different nuclear neighborhoods. It is thought that the movement is driven by diffusion and does not require a directed movement process, inasmuch as each region of a chromosome can be seen to undergo constant random motion when marked in a way that allows its position to be followed in a living cell.

Networks of Macromolecules Form a Set of Distinct Biochemical Environments inside the Nucleus

In Chapter 6, we describe the function of a variety of subcompartments that are present within the nucleus. The largest and most obvious of these is the nucleolus, a structure well known to microscopists even in the 19th century (see Figure 4–9). Nucleolar regions consist of networks of RNAs and proteins surrounding transcribing ribosomal RNA genes, often existing as multiple nucleoli. The nucleolus is the cell's site of ribosome assembly and maturation, as well as the place where many other specialized reactions occur.

A variety of less obvious organelles are also present inside the nucleus. For example, spherical structures called Cajal bodies and interchromatin granule clusters are present in most plant and animal cells (**Figure 4–67**). Like the nucleolus, these organelles are composed of selected protein and RNA molecules that bind together to create networks that are highly permeable to other protein and RNA molecules in the surrounding nucleoplasm (**Figure 4–68**).

Structures such as these can create distinct biochemical environments by immobilizing select groups of macromolecules, as can other networks of proteins and RNA molecules associated with nuclear pores and with the nuclear envelope. In principle, this allows the molecules that enter these spaces to be processed with great efficiency through complex reaction pathways. Highly permeable, fibrous networks of this sort can thereby impart many of the kinetic advantages of compartmentalization (see p. 186) to reactions that take place in the nucleus (**Figure 4–69**A). However, unlike the membrane-bound compartments in the cytoplasm (discussed in Chapter 12), these nuclear subcompartments—lacking a lipid bilayer membrane—can neither concentrate nor exclude specific small molecules.

The cell has a remarkable ability to construct distinct biochemical environments inside the nucleus. Those thus far mentioned facilitate various aspects of gene expression to be discussed in Chapter 6 (see Figure 6–49). Like the nucleolus, these subcompartments appear to form only as needed, and they create a high local concentration of the many different enzymes and RNA molecules needed for a particular process. In an analogous way, when DNA is damaged by irradiation, the set of enzymes needed to carry out DNA repair are observed to congregate in discrete foci inside the nucleus, creating "repair factories" (see Figure 5–60). And nuclei often contain hundreds of discrete foci representing factories for DNA or RNA synthesis.

It seems likely that all of these entities make use of the type of tethering illustrated in Figure 4–69B, where long flexible lengths of polypeptide chain (or some other polymer) are interspersed with binding sites that concentrate the multiple proteins and/or RNA molecules that are needed to catalyze a particular process. Not surprisingly, tethers are similarly used to help to speed biological processes

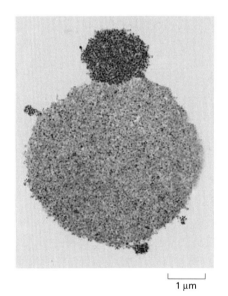

1 μm

Figure 4–67 Electron micrograph showing two very common fibrous nuclear subcompartments. The large sphere here is a Cajal body. The smaller darker sphere is an interchromatin granule cluster, also known as a spreckle (see also Figure 6–49). These "subnuclear organelles" are from the nucleus of a *Xenopus* oocyte. (From K.E. Handwerger and J.G. Gall, *Trends Cell Biol.* 16:19–26, 2006. With permission from Elsevier.)

molecular weight of fluorescent dextran in nucleus

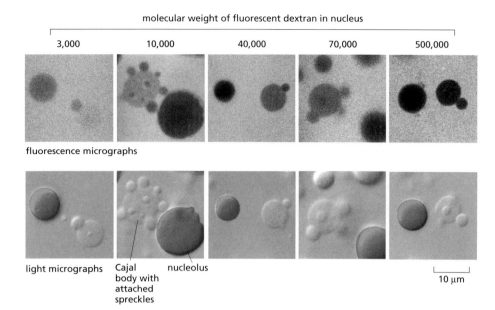

fluorescence micrographs

light micrographs   Cajal   nucleolus
                      body with
                      attached
                      spreckles

10 μm

Figure 4–68 An experiment showing that the subnuclear organelles are highly permeable to macromolecules. In these micrographs of a living oocyte nucleus, the top row compares the fluorescence of the interiors of nucleoli, Cajal bodies, and spreckles to the fluorescence of the surrounding nucleoplasm, 12 hours after fluorescent dextrans of the indicated molecular weight had been injected into the nucleoplasm. The brightness of each organelle reflects its permeability, with the most permeable organelle being the brightest. For comparison, the bottom row presents normal light micrographs of the same microscope fields, with the nucleolus in each field of view marked *brown* to distinguish it. Cajal bodies can be seen to be more permeable than nucleoli. However, quantitation shows that a great deal of dextran enters each organelle, even for the largest dextran tested. (From K.E. Handwerger, J.A. Cordero and J.G. Gall, *Mol. Biol. Cell* 16:202–211, 2005. With permission from American Society of Cell Biology.)

in the cytoplasm, increasing specific reaction rates (for example, see Figure 16–38).

Is there also is an intranuclear framework, analogous to the cytoskeleton, on which chromosomes and other components of the nucleus are organized? The *nuclear matrix*, or *scaffold*, has been defined as the insoluble material left in the nucleus after a series of biochemical extraction steps. Many of the proteins and RNA molecules that form this insoluble material are likely to be derived from the fibrous subcompartments of the nucleus just discussed, while others seem to be proteins that help to form the base of chromosomal loops or to attach chromosomes to other structures in the nucleus. Whether or not the nucleus also contains

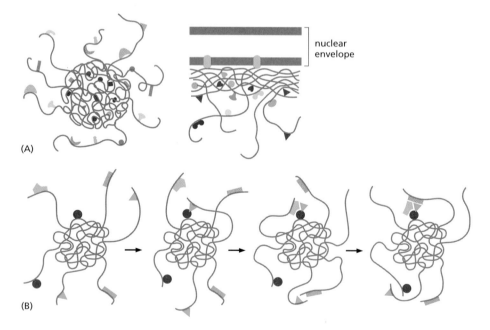

nuclear envelope

(A)

(B)

Figure 4–69 Effective compartmentalization without a bilayer membrane. (A) Schematic illustration of the organization of a spherical subnuclear organelle (*left*) and of a postulated similarly organized subcompartment just beneath the nuclear envelope (*right*). In both cases, RNAs and/or proteins (*gray*) associate to form highly porous, gel-like structures that contain binding sites for other specific proteins and RNA molecules (*colored objects*). (B) How the tethering of a selected set of proteins and RNA molecules to long flexible polymer chains, as in A, could create "staging areas" that greatly speed the rates of reactions in subcompartments of the nucleus. The reactions catalyzed will depend on the particular macromolecules that are localized by the tethering. The same type of rate accelerations are of course expected for similar subcompartments established elsewhere in the cell (see also Figure 3–80C).

Figure 4–70 **A typical mitotic chromosome at metaphase.** Each sister chromatid contains one of two identical daughter DNA molecules generated earlier in the cell cycle by DNA replication (see also Figure 17–26).

long filaments that form organized tracks on which nuclear components can move, analogous to some of the filaments in the cytoplasm, is still disputed.

Mitotic Chromosomes Are Formed from Chromatin in Its Most Condensed State

Having discussed the dynamic structure of interphase chromosomes, we now turn to mitotic chromosomes. The chromosomes from nearly all eucaryotic cells become readily visible by light microscopy during mitosis, when they coil up to form highly condensed structures. This condensation reduces the length of a typical interphase chromosome only about tenfold, but it produces a dramatic change in chromosome appearance.

Figure 4–70 depicts a typical **mitotic chromosome** at the metaphase stage of mitosis (for the stages of mitosis, see Figure 17–3). The two daughter DNA molecules produced by DNA replication during interphase of the cell-division cycle are separately folded to produce two sister chromosomes, or *sister chromatids*, held together at their centromeres (see also Figure 4–50). These chromosomes are normally covered with a variety of molecules, including large amounts of RNA–protein complexes. Once this covering has been stripped away, each chromatid can be seen in electron micrographs to be organized into loops of chromatin emanating from a central scaffolding (**Figure 4–71**). Experiments using DNA hybridization to detect specific DNA sequences demonstrate that the order of visible features along a mitotic chromosome at least roughly reflects the order of genes along the DNA molecule. Mitotic chromosome condensation can thus be thought of as the final level in the hierarchy of chromosome packaging (**Figure 4–72**).

The compaction of chromosomes during mitosis is a highly organized and dynamic process that serves at least two important purposes. First, when condensation is complete (in metaphase), sister chromatids have been disentangled from each other and lie side by side. Thus, the sister chromatids can easily separate when the mitotic apparatus begins pulling them apart. Second, the compaction of chromosomes protects the relatively fragile DNA molecules from being broken as they are pulled to separate daughter cells.

The condensation of interphase chromosomes into mitotic chromosomes begins in early M phase, and it is intimately connected with the progression of the cell cycle, as discussed in detail in Chapter 17. During M phase, gene expression shuts down, and specific modifications are made to histones that help to reorganize the chromatin as it compacts. The compaction is aided by a class of proteins called *condensins*, which use the energy of ATP hydrolysis to help coil the two DNA molecules in an interphase chromosome to produce the two chromatids of a mitotic chromosome. Condensins are large protein complexes built from SMC protein dimers: these dimers form when two stiff, elongated protein monomers join at their tails to form a hinge, leaving two globular head domains at the other end that bind DNA and hydrolyze ATP (**Figure 4–73**). When added to purified DNA, condensins can make large right-handed loops in DNA molecules in a reaction that requires ATP. Although it is not yet known how they act on chromatin, the coiling model shown in Figure 4–73C is based on the fact that condensins are a major structural component that end up at the core of metaphase chromosomes, with about one molecule of condensin for every

Figure 4–71 **A scanning electron micrograph of a region near one end of a typical mitotic chromosome.** Each knoblike projection is believed to represent the tip of a separate looped domain. Note that the two identical paired chromatids (drawn in Figure 4–70) can be clearly distinguished. (From M.P. Marsden and U.K. Laemmli, *Cell* 17:849–858, 1979. With permission from Elsevier.)

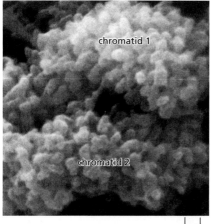

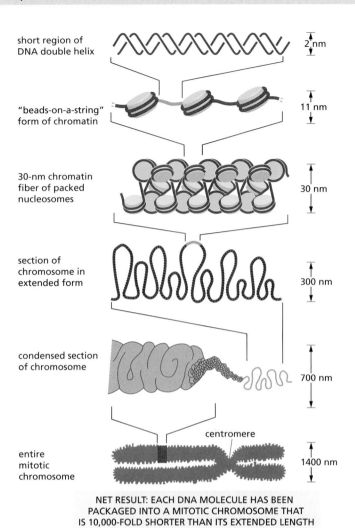

short region of DNA double helix — 2 nm

"beads-on-a-string" form of chromatin — 11 nm

30-nm chromatin fiber of packed nucleosomes — 30 nm

section of chromosome in extended form — 300 nm

condensed section of chromosome — 700 nm

entire mitotic chromosome — centromere — 1400 nm

NET RESULT: EACH DNA MOLECULE HAS BEEN PACKAGED INTO A MITOTIC CHROMOSOME THAT IS 10,000-FOLD SHORTER THAN ITS EXTENDED LENGTH

Figure 4–72 Chromatin packing. This model shows some of the many levels of chromatin packing postulated to give rise to the highly condensed mitotic chromosome.

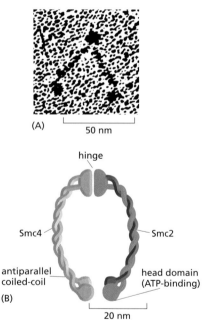

(A) 50 nm

hinge

Smc4

Smc2

antiparallel coiled-coil

head domain (ATP-binding)

(B) 20 nm

Figure 4–73 The SMC proteins in condensins. (A) Electron micrographs of a purified SMC dimer. (B) The structure of a SMC dimer. The long central region of this protein is an antiparallel coiled-coil (see Figure 3–9) with a flexible hinge in its middle. (C) A model for the way in which the SMC proteins in condensins might compact chromatin. In reality, SMC proteins are components of a much larger condensin complex. It has been proposed that, in the cell, condensins coil long strings of looped chromatin domains (see Figure 4–57). In this way, the condensins could form a structural framework that maintains the DNA in a highly organized state during metaphase of the cell cycle. (A, courtesy of H.P. Erickson; B and C, adapted from T. Hirano, *Nat. Rev. Mol. Cell Biol.* 7:311–322, 2006. With permission from Macmillan Publishers Ltd.)

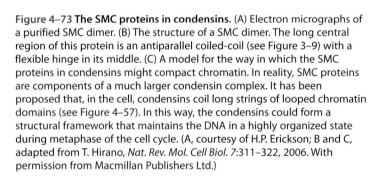

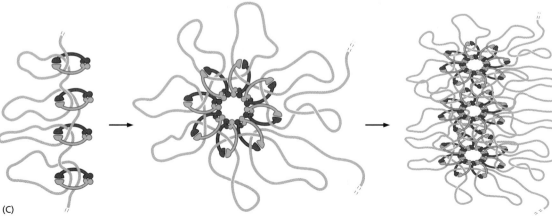

(C)

Figure 4–74 **The location of condensin in condensed mitotic chromosomes.** (A) Fluorescence micrograph of a human chromosome at mitosis, stained with an antibody that localizes condensin. In chromosomes that are this highly condensed, the condensin is seen to be concentrated in punctate structures along the chromosome axis. Similar experiments show a similar location for DNA topoisomerase II, an enzyme that makes reversible double-strand breaks in DNA that allow one DNA double helix to pass through another (see Figure 5–23). (B) Immunogold electron microscopy reveals localization of condensin (*black* dots). Here a chromatid is seen in cross section, with the chromosome axis perpendicular to the plane of the paper. (A, from K. Maeshima and U.K. Laemmli, *Dev. Cell* 4:467–480, 2003. With permission from Elsevier. B, courtesy of U.K. Laemmli, from K. Maeshima, M. Eltsov and U.K. Laemmli, *Chromosoma* 114:365–375, 2005. With permission from Springer.)

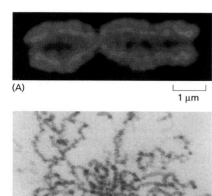

(A)
1 μm

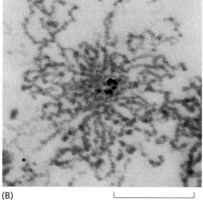

(B)
0.5 μm

10,000 nucleotides of DNA (**Figure 4–74**). When condensins are experimentally depleted from a cell, chromosome condensation still occurs, but the process is abnormal.

Summary

Chromosomes are generally decondensed during interphase, so that the details of their structure are difficult to visualize. Notable exceptions are the specialized lampbrush chromosomes of vertebrate oocytes and the polytene chromosomes in the giant secretory cells of insects. Studies of these two types of interphase chromosomes suggest that each long DNA molecule in a chromosome is divided into a large number of discrete domains organized as loops of chromatin, each loop probably consisting of a 30-nm chromatin fiber that is compacted by further folding. When genes contained in a loop are expressed, the loop unfolds and allows the cell's machinery access to the DNA.

Interphase chromosomes occupy discrete territories in the cell nucleus; that is, they are not extensively intertwined. Euchromatin makes up most of interphase chromosomes and, when not being transcribed, it probably exists as tightly folded 30-nm fibers. However, euchromatin is interrupted by stretches of heterochromatin, in which the 30-nm fibers are subjected to additional packing that usually renders it resistant to gene expression. Heterochromatin exists in several forms, some of which are found in large blocks in and around centromeres and near telomeres. But heterochromatin is also present at many other positions on chromosomes, where it can serve to regulate developmentally important genes.

The interior of the nucleus is highly dynamic, with heterochromatin often positioned near the nuclear envelope and loops of chromatin moving away from their chromosome territory when genes are very highly expressed. This reflects the existence of nuclear subcompartments, where different sets of biochemical reactions are facilitated by an increased concentration of selected proteins and RNAs. The components involved in forming a subcompartment can self-assemble into discrete organelles such as nucleoli or Cajal bodies; they can also be tethered to fixed structures such as the nuclear envelope.

During mitosis, gene expression shuts down and all chromosomes adopt a highly condensed conformation in a process that begins early in M phase to package the two DNA molecules of each replicated chromosome as two separately folded chromatids. This process is accompanied by histone modifications that facilitate chromatin packing. However, satisfactory completion of this orderly process, which reduces the end-to-end distance of each DNA molecule from its interphase length by an additional factor of ten, requires condensin proteins.

HOW GENOMES EVOLVE

In this chapter, we have discussed the structure of genes and the ways that they are packaged and arranged in chromosomes. In this final section, we provide an overview of some of the ways that genes and genomes have evolved over time to produce the vast diversity of modern-day life forms on our planet. Genome sequencing has revolutionized our view of the process of molecular evolution, uncovering an astonishing wealth of information about specific family relationships among organisms, as well as illuminating evolutionary mechanisms more generally.

It is perhaps not surprising that genes with similar functions can be found in a diverse range of living things. But the great revelation of the past 25 years has been the discovery that the actual nucleotide sequences of many genes are sufficiently well conserved that **homologous** genes—that is, genes that are similar in both their nucleotide sequence and function because of a common ancestry—can often be recognized across vast phylogenetic distances. For example, unmistakable homologs of many human genes are easy to detect in such organisms as nematode worms, fruit flies, yeasts, and even bacteria. In many cases, the resemblance is so close that the protein-coding portion of a yeast gene can be substituted with its human homolog—even though we and yeast are separated by more than a billion years of evolutionary history.

As emphasized in Chapter 3, the recognition of sequence similarity has become a major tool for inferring gene and protein function. Although finding a sequence match does not guarantee similarity in function, it has proven to be an excellent clue. Thus, it is often possible to predict the function of genes in humans for which no biochemical or genetic information is available simply by comparing their nucleotide sequences with the sequences of genes in other organisms.

In general, gene sequences are more tightly conserved than is overall genome structure. As we saw earlier, other features of genome organization such as genome size, number of chromosomes, order of genes along chromosomes, abundance and size of introns, and amount of repetitive DNA are found to differ greatly among organisms, as does the number of genes that an organism contains.

The number of genes is only very roughly correlated with the phenotypic complexity of an organism (see Table 1–1). Much of the increase in gene number observed with increasing biological complexity involves the expansion of families of closely related genes, an observation that establishes gene duplication and divergence as major evolutionary processes. Indeed, it is likely that all present-day genes are descendants—via the processes of duplication, divergence, and reassortment of gene segments—of a few ancestral genes that existed in early life forms.

Genome Alterations are Caused by Failures of the Normal Mechanisms for Copying and Maintaining DNA

Cells in the germline do not have specialized mechanisms for creating changes in the structures of their genomes: evolution depends instead on accidents and mistakes followed by nonrandom survival. Most of the genetic changes that occur result simply from failures in the normal mechanisms by which genomes are copied or repaired when damaged, although the movement of transposable DNA elements also plays an important role. As we will discuss in Chapter 5, the mechanisms that maintain DNA sequences are remarkably precise—but they are not perfect. For example, because of the elaborate DNA-replication and DNA-repair mechanisms that enable DNA sequences to be inherited with extraordinary fidelity, along a given line of descent only about one nucleotide pair in a thousand is randomly changed in the germline every million years. Even so, in a population of 10,000 diploid individuals, every possible nucleotide substitution will have been "tried out" on about 20 occasions in the course of a million years—a short span of time in relation to the evolution of species.

Errors in DNA replication, DNA recombination, or DNA repair can lead either to simple changes in DNA sequence—such as the substitution of one base pair for another—or to large-scale genome rearrangements such as deletions, duplications, inversions, and translocations of DNA from one chromosome to another. In addition to these failures of the genetic machinery, the various mobile DNA elements that will be described in Chapter 5 are an important source of genomic change (see Table 5–3, p. 318). These transposable DNA elements (*transposons*) are parasitic DNA sequences that colonize genomes and can spread within them. In the process, they often disrupt the function or alter the regulation of existing genes. On occasion, they can even create altogether novel genes through fusions between transposon sequences and segments of existing genes. Over long periods of evolutionary time, transposons have profoundly affected the structure of genomes. In fact, nearly half of the DNA in the human genome has recognizable sequence similarity with known transposon sequences, thereby indicating that these sequences are remnants of past transposition events (see Figure 4–17). Even more of our genome is no doubt derived from transposition events that occurred so long ago (>10^8 years) that the sequences can no longer be traced to transposons.

The Genome Sequences of Two Species Differ in Proportion to the Length of Time That They Have Separately Evolved

The differences between the genomes of species alive today have accumulated over more than 3 billion years. Lacking a direct record of changes over time, we can nevertheless reconstruct the process of genome evolution from detailed comparisons of the genomes of contemporary organisms.

The basic tool of comparative genomics is the phylogenetic tree. A simple example is the tree describing the divergence of humans from the great apes (**Figure 4–75**). The primary support for this tree comes from comparisons of gene or protein sequences. For example, comparisons between the sequences of human genes or proteins and those of the great apes typically reveal the fewest differences between human and chimpanzee and the most between human and orangutan.

For closely related organisms such as humans and chimpanzees, it is relatively easy to reconstruct the gene sequences of the extinct, last common ancestor of the two species (**Figure 4–76**). The close similarity between human and chimpanzee genes is mainly due to the short time that has been available for the accumulation of mutations in the two diverging lineages, rather than to functional constraints that have kept the sequences the same. Evidence for this view comes from the observation that even DNA sequences whose nucleotide order is functionally unconstrained—such as the sequences that code for the fibrinopeptides (see p. 264) or the third position of "synonymous" codons (codons specifying the same amino acid—see Figure 4–76)—are nearly identical in humans and chimpanzees.

For much less closely related organisms, such as humans and chickens (which have evolved separately for about 300 million years), the sequence conservation found in genes is largely due to **purifying selection** (that is, selection that eliminates individuals carrying mutations that interfere with important genetic functions), rather than to an inadequate time for mutations to occur. As a result, protein-coding, RNA-coding, and regulatory sequences in the DNA are often remarkably conserved. In contrast, most DNA sequences in the human and chicken genomes have diverged so far due to multiple mutations that it is often impossible to align them with one another.

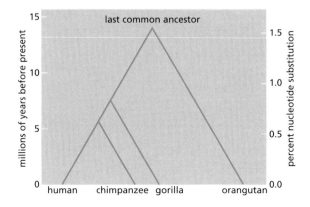

Figure 4–75 A phylogenetic tree showing the relationship between the human and the great apes based on nucleotide sequence data. As indicated, the sequences of the genomes of all four species are estimated to differ from the sequence of the genome of their last common ancestor by a little over 1.5%. Because changes occur independently on both diverging lineages, pairwise comparisons reveal twice the sequence divergence from the last common ancestor. For example, human–orangutan comparisons typically show sequence divergences of a little over 3%, while human–chimpanzee comparisons show divergences of approximately 1.2%. (Modified from F.C. Chen and W.H. Li, *Am. J. Hum. Genet.* 68:444–456, 2001. With permission from University of Chicago Press.)

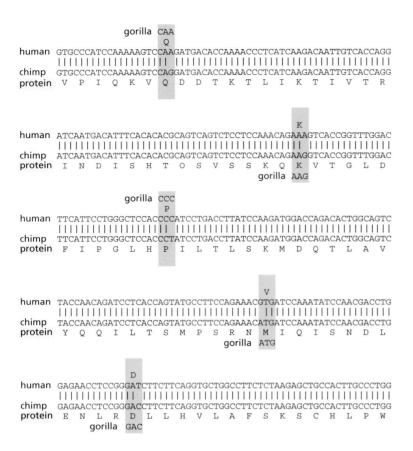

Figure 4–76 Tracing the ancestral sequence from a sequence comparison of the coding regions of human and chimpanzee leptin genes. Leptin is a hormone that regulates food intake and energy utilization in response to the adequacy of fat reserves. As indicated by the codons boxed in green, only 5 nucleotides (of 441 total) differ between these two sequences. Moreover, when the amino acids encoded by both the human and chimpanzee sequences are examined, in only one of the 5 positions does the encoded amino acid differ. For each of the 5 variant nucleotide positions, the corresponding sequence in the gorilla is also indicated. In two cases, the gorilla sequence agrees with the human sequence, while in three cases it agrees with the chimpanzee sequence.

What was the sequence of the leptin gene in the last common ancestor? An evolutionary model that seeks to minimize the number of mutations postulated to have occurred during the evolution of the human and chimpanzee genes would assume that the leptin sequence of the last common ancestor was the same as the human and chimpanzee sequences when they agree; when they disagree, it would use the gorilla sequence as a tie-breaker. For convenience, only the first 300 nucleotides of the leptin coding sequences are given. The remaining 141 are identical between humans and chimpanzees.

Phylogenetic Trees Constructed from a Comparison of DNA Sequences Trace the Relationships of All Organisms

Integration of phylogenetic trees based on molecular sequence comparisons with the fossil record has led to the best available view of the evolution of modern life forms. The fossil record remains important as a source of absolute dates based on the decay of radioisotopes in the rock formations in which fossils are found. However, precise divergence times between species are difficult to establish from the fossil record, even for species that leave good fossils with distinctive morphology.

Such integrated phylogenetic trees suggest that changes in the sequences of particular genes or proteins tend to occur at a nearly constant rate, although rates that differ from the norm by as much as twofold are observed in particular lineages. As discussed above and in Chapter 5, this "molecular clock" runs most rapidly and regularly in sequences that are not subject to purifying selection—such as intergenic regions, portions of introns that lack splicing or regulatory signals, and genes that have been irreversibly inactivated by mutation (the so-called pseudogenes). The clock runs most slowly for sequences that are subject to strong functional constraints—for example, the amino acid sequences of proteins such as actin that engage in specific interactions with large numbers of other proteins and whose structure is therefore highly constrained (see, for example, Figure 16–18).

Occasionally, rapid change is seen in a previously highly conserved sequence. As discussed later in this chapter, such episodes are especially interesting because they are thought to reflect periods of strong positive selection for mutations that conferred a selective advantage in the particular lineage where the rapid change occurred.

Molecular clocks run at rates that are determined both by mutation rates and by the degree of purifying selection on particular sequences. Therefore, a completely different calibration is required for those genes replicated and repaired by different systems within cells. Most notably, in animals, although not in plants, clocks based on functionally unconstrained mitochondrial DNA sequences run

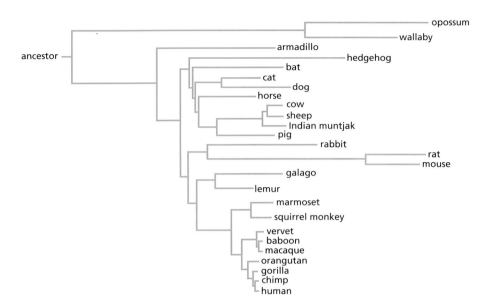

Figure 4–77 A phylogenetic tree highlighting some of the mammals whose genomes are being extensively studied. The length of each line is proportional to the number of "neutral substitutions"—representing the nucleotide changes observed in the absence of purifying selection. (Adapted from G.M. Cooper et al., *Genome Res.* 15:901–913, 2005. With permission from Cold Spring Harbor Laboratory Press.)

much faster than clocks based on functionally unconstrained nuclear sequences, due to an unusually high mutation rate in animal mitochondria.

Molecular clocks have a finer time resolution than the fossil record and are a more reliable guide to the detailed structure of phylogenetic trees than are classical methods of tree construction, which are based on comparisons of the morphology and development of different species. For example, the precise relationship among the great-ape and human lineages was not settled until sufficient molecular-sequence data accumulated in the 1980s to produce the tree that was shown in Figure 4–75. And with huge amounts of DNA sequence now determined from a variety of mammals, much better estimates of our relationship to them are being obtained (**Figure 4–77**).

A Comparison of Human and Mouse Chromosomes Shows How The Structures of Genomes Diverge

As would be expected, the human and chimpanzee genomes are much more alike than are the human and mouse genomes. Although the size of the human and mouse genomes are roughly the same and they contain nearly identical sets of genes, there has been a much longer time period over which changes have had a chance to accumulate—approximately 80 million years versus 6 million years. In addition, as indicated in Figure 4–77, rodent lineages (represented by the rat and the mouse) have unusually fast molecular clocks. Hence, these lineages have diverged from the human lineage more rapidly than otherwise expected.

As indicated by the DNA sequence comparison in **Figure 4–78**, mutation has led to extensive sequence divergence between humans and mice at all sites that are not under selection—such as most nucleotide sequences in introns. In contrast, in human–chimpanzee comparisons, nearly all sequence positions are the same simply because not enough time has elapsed since the last common ancestor for large numbers of changes to have occurred.

In contrast to the situation for humans and chimpanzees, local gene order and overall chromosome organization have diverged greatly between humans

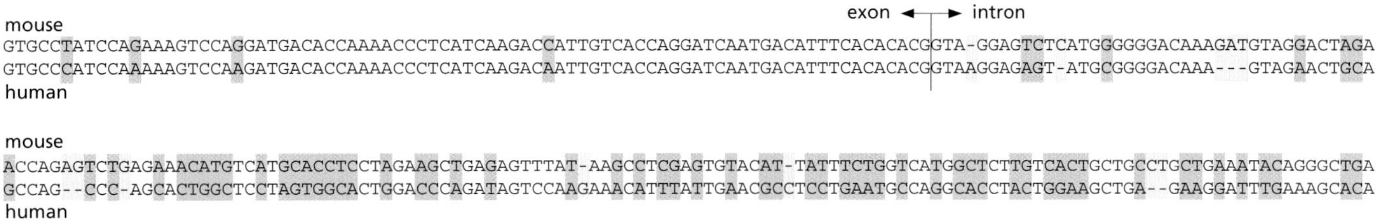

Figure 4–78 Comparison of a portion of the mouse and human leptin genes. Positions where the sequences differ by a single nucleotide substitution are boxed in *green,* and positions that differ by the addition or deletion of nucleotides are boxed in *yellow.* Note that the coding sequence of the exon is much more conserved than is the adjacent intron sequence.

exon ◄─┼─► intron

mouse
GTGCCTATCCAGAAAGTCCAGGATGACACCAAAACCCTCATCAAGACCATTGTCACCAGGATCAATGACATTTCACACACGGTA-GGAGTCTCATGGGGGGACAAAGATGTAGGACTAGA
GTGCCCATCCAAAAAGTCCAAGATGACACCAAAACCCTCATCAAGACAATTGTCACCAGGATCAATGACATTTCACACACGGTAAGGAGAGT-ATGCGGGGACAAA---GTAGAACTGCA
human

mouse
ACCAGAGTCTGAGAAACATGTCATGCACCTCCTAGAAGCTGAGAGTTTAT-AAGCCTCGAGTGTACAT-TATTTCTGGTCATGGCTCTTGTCACTGCTGCCTGCTGAAATACAGGGCTGA
GCCAG--CCC-AGCACTGGCTCCTAGTGGCACTGGACCCAGATAGTCCAAGAAACATTTATTGAACGCCTCCTGAATGCCAGGCACCTACTGGAAGCTGA--GAAGGATTTGAAAGCACA
human

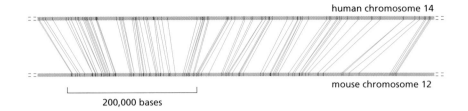

human chromosome 14

mouse chromosome 12

200,000 bases

Figure 4–79 **Comparison of a syntenic portion of mouse and human genomes.** About 90 percent of the two genomes can be aligned in this way. Note that while there is an identical order of the matched index sequences (*red marks*), there has been a net loss of DNA in the mouse lineage that is interspersed throughout the entire region. This type of net loss is typical for all such regions, and it accounts for the fact that the mouse genome contains 14 percent less DNA than does the human genome. (Adapted from Mouse Sequencing Consortium, *Nature* 420:520–573, 2002. With permission from Macmillan Publishers Ltd.)

and mice. According to rough estimates, a total of about 180 break-and-rejoin events have occurred in the human and mouse lineages since these two species last shared a common ancestor. In the process, although the number of chromosomes is similar in the two species (23 per haploid genome in the human versus 20 in the mouse), their overall structures differ greatly. Nonetheless, even after the extensive genomic shuffling, there are many large blocks of DNA in which the gene order is the same in the human and the mouse. These stretches of conserved gene order in chromosomes are referred to as regions of *synteny*.

An unexpected conclusion from a detailed comparison of the complete mouse and human genome sequences, confirmed from subsequent comparisons between the genomes of other vertebrates, is that small blocks of sequences are being deleted from and added to genomes at a surprisingly rapid rate. Thus, if we assume that our common ancestor had a genome of human size (about 3 billion nucleotide pairs), mice would have lost a total of about 45 percent of that genome from accumulated deletions during the past 80 million years, while humans would have lost about 25 percent. However, substantial sequence gains from many small chromosome duplications and from the multiplication of transposons have compensated for these deletions. As a result, our genome size is unchanged from that of the last common ancestor for humans and mice, while the mouse genome is smaller by only 0.3 billion nucleotides.

Good evidence for the loss of DNA sequences in small blocks during evolution can be obtained from a detailed comparison of most regions of synteny in the human and mouse genomes. The comparative shrinkage of the mouse genome can be clearly seen from such comparisons, with the net loss of sequences scattered throughout the long stretches of DNA that are otherwise homologous (**Figure 4–79**).

DNA is added to genomes both by the spontaneous duplication of chromosomal segments that contain tens of thousands of nucleotide pairs (as will be discussed shortly), and by active transposition (most transposition events are duplicative, because the original copy of the transposon stays where it was when a copy inserts at the new site; for example, see Figure 5–74). Comparison of the DNA sequences derived from transposons in the human and the mouse therefore readily reveals some of the sequence additions (**Figure 4–80**).

For unknown reasons, all mammals have genome sizes of about 3 billion nucleotide pairs that contain nearly identical sets of genes, even though only on the order of 150 million nucleotide pairs appear to be under sequence-specific functional constraints.

Figure 4–80 **A comparison of the β-globin gene cluster in the human and mouse genomes, showing the location of transposable elements.** This stretch of human genome contains five functional β-globin-like genes *(orange)*; the comparable region from the mouse genome has only four. The positions of the human Alu sequence are indicated by green circles, and the human L1 sequences by red circles. The mouse genome contains different but related transposable elements: the positions of B1 elements (which are related to the human Alu sequences) are indicated by *blue triangles,* and the positions of the mouse L1 elements (which are related to the human L1 sequences) are indicated by *orange triangles.* The absence of transposable elements from the globin structural genes can be attributed to purifying selection, which would have eliminated any insertion that compromised gene function. (Courtesy of Ross Hardison and Webb Miller.)

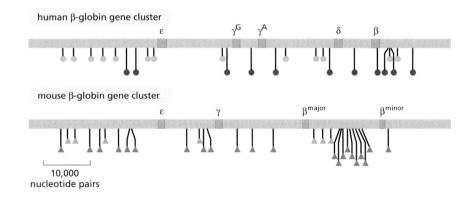

human β-globin gene cluster

ε   γ^G γ^A   δ   β

mouse β-globin gene cluster

ε   γ   β^major   β^minor

10,000 nucleotide pairs

The Size of a Vertebrate Genome Reflects the Relative Rates of DNA Addition and DNA Loss in a Lineage

Now that we know the complete sequence of a number of vertebrate genomes, we see that genome size can vary considerably, apparently without a drastic effect on the organism or its number of genes. Thus, while the mouse and dog genomes are both in the typical mammalian size range, the chicken has a genome that is only about one-third human size (one billion nucleotide pairs). A particularly notable example of an organism with a genome of anomalous size is the puffer fish, *Fugu rubripes* (**Figure 4–81**), which has a tiny genome for a vertebrate (0.4 billion nucleotide pairs compared to 1 billion or more for many other fish). The small size of the *Fugu* genome is largely due to the small size of its introns. Specifically, *Fugu* introns, as well as other noncoding segments of the *Fugu* genome, lack the repetitive DNA that makes up a large portion of the genomes of most well-studied vertebrates. Nevertheless, the positions of *Fugu* introns are nearly perfectly conserved relative to their positions in mammalian genomes (**Figure 4–82**).

While initially a mystery, we now have a simple explanation for such large differences in genome size between similar organisms: because all vertebrates experience a continuous process of DNA loss and DNA addition, the size of a genome merely depends on the balance between these opposing processes acting over millions of years. Suppose, for example, that in the lineage leading to *Fugu*, the rate of DNA addition happened to slow greatly. Over long periods of time, this would result in a major "cleansing" from this fish genome of those DNA sequences whose loss could be tolerated. In retrospect, the process of purifying selection in the *Fugu* lineage has partitioned those vertebrate DNA sequences most likely to be functional into only 400 million nucleotide pairs of DNA, providing a major resource for scientists.

We Can Reconstruct the Sequence of Some Ancient Genomes

The genomes of ancestral organisms can be inferred, but never directly observed: there are no ancient organisms alive today. Although a modern organism such as the horseshoe crab looks remarkably similar to fossil ancestors that lived 200 million years ago, there is every reason to believe that the horseshoe-crab genome has been changing during all that time at a rate similar to that occurring in other evolutionary lineages. Selection constraints must have maintained key functional properties of the horseshoe-crab genome to account for the morphological stability of the lineage. However, genome sequences reveal that the fraction of the genome subject to purifying selection is small; hence the genome of the modern horseshoe crab must differ greatly from that of its extinct ancestors, known to us only through the fossil record.

Is there any way around this problem? Can we ever hope to decipher large sections of the genome sequence of the extinct ancestors of organisms that are

Figure 4–81 **The puffer fish,** *Fugu rubripes.* (Courtesy of Byrappa Venkatesh.)

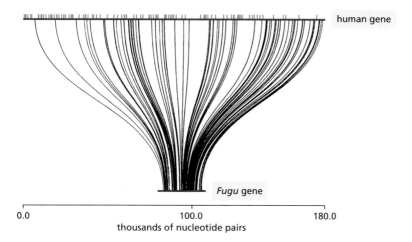

human gene

Fugu gene

0.0   100.0   180.0
thousands of nucleotide pairs

Figure 4–82 **Comparison of the genomic sequences of the human and** *Fugu* **genes encoding the protein huntingtin.** Both genes (indicated in *red*) contain 67 short exons that align in 1:1 correspondence to one another; these exons are connected by curved lines. The human gene is 7.5 times larger than the *Fugu* gene (180,000 versus 27,000 nucleotide pairs). The size difference is entirely due to larger introns in the human gene. The larger size of the human introns is due in part to the presence of retrotransposons, whose positions are represented by green vertical lines; the *Fugu* introns lack retrotransposons. In humans, mutation of the huntingtin gene causes Huntington's disease, an inherited neurodegenerative disorder. (Adapted from S. Baxendale et al., *Nat. Genet.* 10:67–76, 1995. With permission from Macmillan Publishers Ltd.)

alive today? For organisms that are as closely related as human and chimp, we saw that this may not be difficult. In that case, reference to the gorilla sequence can be be used to sort out which of the few differences between human and chimp DNA sequences was inherited from our common ancestor some 6 million years ago (see Figure 4–76). For an ancestor that has produced a large number of different organisms alive today, the DNA sequences of many species can be compared simultaneously to unscramble the ancestral sequence, allowing scientists to trace DNA sequences much farther back in time. For example, from the complete genome sequences of 20 modern mammals that will soon be obtained, it should be possible to decipher most of the genome sequence of the 100 million year-old Boreoeutherian mammal that gave rise to species as diverse as dog, mouse, rabbit, armadillo and human (see Figure 4–77).

Multispecies Sequence Comparisons Identify Important DNA Sequences of Unknown Function

The massive quantity of DNA sequence now in databases (more than a hundred billion nucleotide pairs) provides a rich resource that scientists can mine for many purposes. We have already discussed how this information can be used to unscramble the evolutionary pathways that have led to modern organisms. But sequence comparisons also provide many insights into how cells and organisms function. Perhaps the most remarkable discovery in this realm has been the observation that, although only about 1.5% of the human genome codes for proteins, about three times this amount (in total, 5% of the genome—see Table 4–1, p. 206) has been strongly conserved during mammalian evolution. This mass of conserved sequence is most clearly revealed when we align and compare DNA synteny blocks from many different species. In this way, so-called *multispecies conserved sequences* can be readily identified (**Figure 4–83**). Most of the non-coding conserved sequences discovered in this way turn out to be relatively short, containing between 50 and 200 nucleotide pairs. The strict conservation

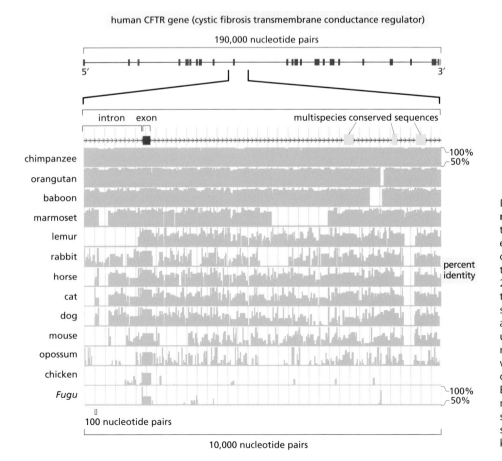

Figure 4–83 The detection of multispecies conserved sequences. In this example, genome sequences for each of the organisms shown have been compared with the indicated region of the human CFTR gene, scanning in 25 nucleotide blocks. For each organism, the percent identity across its syntenic sequences is plotted in *green*. In addition, a computational algorithm has been used to detect the sequences within this region that are most highly conserved when the sequences from all of the organisms are taken into account. Besides the exon, three other blocks of multispecies conserved sequences are shown. The function of most such sequences in the human genome is not known. (Courtesy of Eric D. Green.)

implies that they have important functions that have been maintained by purifying selection. The puzzle is to unravel what those functions are. Some of the conserved sequence that does not code for protein codes for untranslated RNA molecules that are known to have important functions, as we shall see in later chapters. Another fraction of the noncoding conserved DNA is certainly involved in regulating the transcription of adjacent genes, as discussed in Chapter 7. But we do not yet know how much of the conserved DNA can be accounted for in these ways, and the bulk of it is still a deep mystery. The solution to this mystery is bound to have profound consequences for medicine, and it reveals how much more we need to learn about the biology of vertebrate organisms.

How can cell biologists tackle this problem? The first step is to distinguish between the conserved regions that code for protein and those that do not, and then, among the latter, to focus on those that do not already have some other identified function, in coding for structural RNA molecules, for example. The next task is to discover what proteins or RNA molecules bind to these mysterious DNA sequences, how they are packaged into chromatin, and whether they ever serve as templates for RNA synthesis. Most of this task still lies before us, but a start has been made, and some remarkable insights have been obtained. One of the most intriguing concerns the evolutionary changes that have made us humans different from other animals—changes, that is, in sequences that have been conserved among our close relatives but have undergone sudden rapid change in the human sublineage.

Accelerated Changes in Previously Conserved Sequences Can Help Decipher Critical Steps in Human Evolution

As soon as both the human and the chimpanzee genome sequences became available, scientists began searching for DNA sequence changes that might account for the striking differences between us and them. With 3 billion nucleotide pairs to compare in the two species, this might seem an impossible task. But the job was made much easier by confining the search to 35,000 clearly defined multispecies conserved sequences (a total of about 5 million nucleotide pairs), representing parts of the genome that are most likely to be functionally important. Though these sequences are conserved strongly, they are not conserved perfectly, and when the version in one species is compared with that in another they are generally found to have drifted apart by a small amount corresponding simply to the time elapsed since the last common ancestor. In a small proportion of cases, however, one sees signs of a sudden evolutionary spurt. For example, some DNA sequences that have been highly conserved in other mammalian species are found to have changed exceptionally fast during the six million years of human evolution since we diverged from the chimpanzees. Such *human accelerated regions* (HARs) are thought to reflect functions that have been especially important in making us different in some useful way.

About 50 such sites were identified in one study, one-fourth of which were located near genes associated with neural development. The sequence exhibiting the most rapid change (18 changes between human and chimp, compared to only two changes between chimp and chicken) was examined further and found to encode a 118-nucleotide noncoding RNA molecule that is produced in the human cerebral cortex at a critical time during brain development (**Figure 4–84**). Although the function of this HAR1F RNA is not yet known, this exciting finding is stimulating further studies that will hopefully shed light on crucial features of the human brain.

Gene Duplication Provides an Important Source of Genetic Novelty During Evolution

Evolution depends on the creation of new genes, as well as on the modification of those that already exist. How does this occur? When we compare organisms that seem very different—a primate with a rodent, for example, or a mouse with

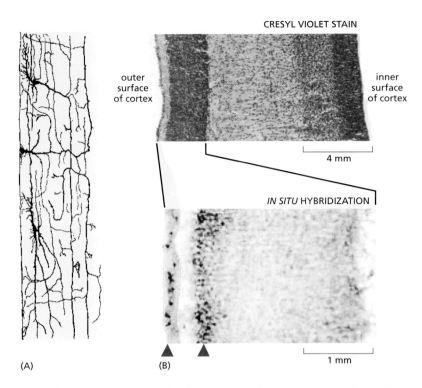

CRESYL VIOLET STAIN

outer surface of cortex

inner surface of cortex

4 mm

IN SITU HYBRIDIZATION

1 mm

(A)   (B)

Figure 4–84 Initial characterization of a new gene detected as a previously conserved DNA sequence that evolved rapidly in humans. (A) Drawing by Ramon y Cajal of the outer surface of the human neocortex, highlighting the Cajal–Retzius neurons. (B) Tissue slices from an embryonic human brain showing part of the cortex, with the region containing the Cajal–Retzius neurons highlighted in *yellow*. Upper photograph: cresyl violet stain. Lower photograph: *in situ* hybridization. The *red arrows* indicate the cells that produce HAR1F RNA as detected by *in situ* hybridization *(blue)*. HAR1F is a novel noncoding RNA that has evolved rapidly in the human lineage leading from the great apes. The Cajal–Retzius neurons make this RNA at the time when the neocortex is developing. The results are intriguing, because a large neocortex is special to humans; for the behavior of cells in forming this cortex, see Figure 22–99. (Adapted from K.S. Pollard et al., *Nature* 443:167–172, 2006. With permission from Macmillan Publishers Ltd.)

a fish—we rarely encounter genes in the one species that have no homolog in the other. Genes without homologous counterparts are relatively scarce even when we compare such divergent organisms as a mammal and a worm. On the other hand, we frequently find gene families that have different numbers of members in different species. To create such families, genes have been repeatedly duplicated, and the copies have then diverged to take on new functions that often vary from one species to another.

The genes encoding nuclear hormone receptors in humans, a nematode worm, and a fruit fly illustrate this point (**Figure 4–85**). Many of the subtypes of these nuclear receptors (also called intracellular receptors) have close homologs in all three organisms that are more similar to each other than they are to other family subtypes present in the same species. Therefore, much of the functional divergence of this large gene family must have preceded the divergence of these three evolutionary lineages. Subsequently, one major branch of the gene family underwent an enormous expansion in the worm lineage only. Similar, but smaller, lineage-specific expansions of particular subtypes are evident throughout the gene family tree.

Gene duplication occurs at high rates in all evolutionary lineages, contributing to the vigorous process of DNA addition discussed previously. In a detailed study of spontaneous duplications in yeast, duplications of 50,000 to 250,000 nucleotide pairs were commonly observed, most of which were tandemly repeated. These appeared to result from DNA replication errors that led to the inexact repair of double-strand chromosome breaks. A comparison of the human and chimpanzee genomes reveals that, since the time that these two organisms diverged, segmental duplications have added about 5 million nucleotide pairs to each genome every million years, with an average duplication size being about 50,000 nucleotide pairs (however, there are duplications five times larger, as in yeast). In fact, if one counts nucleotides, duplication events have created more differences between our two species than have single nucleotide substitutions.

Duplicated Genes Diverge

A major question in genome evolution concerns the fate of newly duplicated genes. In most cases, there is presumed to be little or no selection—at least initially—to maintain the duplicated state since either copy can provide an equivalent function. Hence, many duplication events are likely to be followed by loss-of-function mutations in one or the other gene. This cycle would functionally restore the one-gene state that preceded the duplication. Indeed, there are many examples in contemporary genomes where one copy of a duplicated gene can be seen to have become irreversibly inactivated by multiple mutations. Over time, the sequence similarity between such a **pseudogene** and the functional gene whose duplication produced it would be expected to be eroded by the accumulation of many mutations in the pseudogene—the homologous relationship eventually becoming undetectable.

An alternative fate for gene duplications is for both copies to remain functional, while diverging in their sequence and pattern of expression, thus taking on different roles. This process of "duplication and divergence" almost certainly explains the presence of large families of genes with related functions in biologically complex organisms, and it is thought to play a critical role in the evolution of increased biological complexity. An examination of many different eucaryotic genomes suggests that the probability that any particular gene will undergo a duplication event that spreads to most or all individuals in a species is approximately 1% every million years.

Whole-genome duplications offer particularly dramatic examples of the duplication–divergence cycle. A whole-genome duplication can occur quite simply: all that is required is one round of genome replication in a germline cell lineage without a corresponding cell division. Initially, the chromosome number simply doubles. Such abrupt increases in the ploidy of an organism are common, particularly in fungi and plants. After a whole-genome duplication, all genes exist as duplicate copies. However, unless the duplication event occurred so recently that there has been little time for subsequent alterations in genome structure, the results of a series of segmental duplications—occurring at different times—are very hard to distinguish from the end product of a whole-genome duplication. In mammals, for example, the role of whole-genome duplications versus a series of piecemeal duplications of DNA segments is quite uncertain. Nevertheless, it is clear that a great deal of gene duplication has ocurred in the distant past.

Analysis of the genome of the zebrafish, in which either a whole-genome duplication or a series of more local duplications occurred hundreds of millions of years ago, has cast some light on the process of gene duplication and divergence. Although many duplicates of zebrafish genes appear to have been lost by mutation, a significant fraction—perhaps as many as 30–50%—have diverged functionally while both copies have remained active. In many cases, the most obvious functional difference between the duplicated genes is that they are expressed in different tissues or at different stages of development (see Figure 22–46). One attractive theory to explain such an end result imagines that different, mildly deleterious mutations occur quickly in both copies of a duplicated gene set. For example, one copy might lose expression in a particular tissue as a result of a regulatory mutation, while the other copy loses expression in a second tissue. Following such an occurrence, both gene copies would be required to provide the full range of functions that were once supplied by a single gene; hence, both copies would now be protected from loss through inactivating mutations. Over a longer period, each copy could then undergo further changes through which it could acquire new, specialized features.

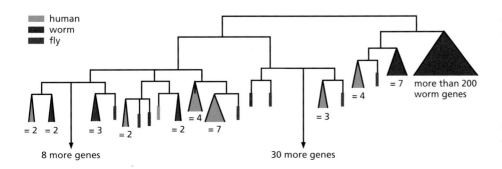

Figure 4–85 A phylogenetic tree based on the inferred protein sequences for all nuclear hormone receptors encoded in the genomes of human (*H. sapiens*), a nematode worm (*C. elegans*), and a fruit fly (*D. melanogaster*). Triangles represent protein subfamilies that have expanded within individual evolutionary lineages; the width of these triangles indicates the number of genes encoding members of these subfamilies. Colored vertical bars represent a single gene. There is no simple pattern to the historical duplications and divergences that have created the gene families encoding nuclear receptors in the three contemporary organisms. The family of nuclear hormone receptors is described in Figure 15–14. These proteins function in cell signaling and gene regulation. (Adapted from International Human Genome Sequencing Consortium, *Nature* 409:860–921, 2001. With permission from Macmillan Publishers Ltd.)

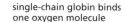

single-chain globin binds
one oxygen molecule

Figure 4–86 A comparison of the structure of one-chain and four-chain globins. The four-chain globin shown is hemoglobin, which is a complex of two α-globin and two β-globin chains. The one-chain globin in some primitive vertebrates forms a dimer that dissociates when it binds oxygen, representing an intermediate in the evolution of the four-chain globin.

oxygen-
binding site
on heme

EVOLUTION OF A
SECOND GLOBIN
CHAIN BY
GENE DUPLICATION
FOLLOWED BY
MUTATION

The Evolution of the Globin Gene Family Shows How DNA Duplications Contribute to the Evolution of Organisms

The globin gene family provides an especially good example of how DNA duplication generates new proteins, because its evolutionary history has been worked out particularly well. The unmistakable similarities in amino acid sequence and structure among the present-day globins indicate that they all must derive from a common ancestral gene, even though some are now encoded by widely separated genes in the mammalian genome.

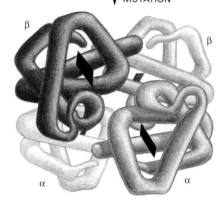

four-chain globin binds four
oxygen molecules in a
cooperative manner

We can reconstruct some of the past events that produced the various types of oxygen-carrying hemoglobin molecules by considering the different forms of the protein in organisms at different positions on the phylogenetic tree of life. A molecule like hemoglobin was necessary to allow multicellular animals to grow to a large size, since large animals could no longer rely on the simple diffusion of oxygen through the body surface to oxygenate their tissues adequately. Consequently, hemoglobin-like molecules are found in all vertebrates and in many invertebrates. The most primitive oxygen-carrying molecule in animals is a globin polypeptide chain of about 150 amino acids, which is found in many marine worms, insects, and primitive fish. The hemoglobin molecule in more complex vertebrates, however, is composed of two kinds of globin chains. It appears that about 500 million years ago, during the continuing evolution of fish, a series of gene mutations and duplications occurred. These events established two slightly different globin genes, coding for the α- and β-globin chains, in the genome of each individual. In modern vertebrates, each hemoglobin molecule is a complex of two α chains and two β chains (**Figure 4–86**). The four oxygen-binding sites in the $\alpha_2\beta_2$ molecule interact, allowing a cooperative allosteric change in the molecule as it binds and releases oxygen, which enables hemoglobin to take up and release oxygen more efficiently than the single-chain version.

Still later, during the evolution of mammals, the β-chain gene apparently underwent duplication and mutation to give rise to a second β-like chain that is synthesized specifically in the fetus. The resulting hemoglobin molecule has a higher affinity for oxygen than adult hemoglobin and thus helps in the transfer of oxygen from the mother to the fetus. The gene for the new β-like chain subsequently duplicated and mutated again to produce two new genes, ε and γ, the ε chain being produced earlier in development (to form $\alpha_2\varepsilon_2$) than the fetal γ chain, which forms $\alpha_2\gamma_2$. A duplication of the adult β-chain gene occurred still later, during primate evolution, to give rise to a δ-globin gene and thus to a minor form of hemoglobin ($\alpha_2\delta_2$) that is found only in adult primates (**Figure 4–87**).

Each of these duplicated genes has been modified by point mutations that affect the properties of the final hemoglobin molecule, as well as by changes in regulatory regions that determine the timing and level of expression of the gene. As a result, each globin is made in different amounts at different times of human development (see Figure 7–64B).

The end result of the gene duplication processes that have given rise to the diversity of globin chains is seen clearly in the human genes that arose from

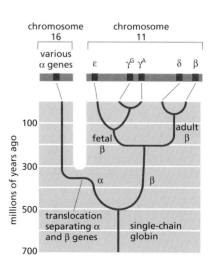

Figure 4–87 An evolutionary scheme for the globin chains that carry oxygen in the blood of animals. The scheme emphasizes the β-like globin gene family. A relatively recent gene duplication of the γ-chain gene produced γ^G and γ^A, which are fetal β-like chains of identical function. The location of the globin genes in the human genome is shown at the top of the figure (see also Figure 7–64).

the original β gene, which are arranged as a series of homologous DNA sequences located within 50,000 nucleotide pairs of one another. A similar cluster of α-globin genes is located on a separate human chromosome. Because the α- and β-globin gene clusters are on separate chromosomes in birds and mammals but are together in the frog *Xenopus*, it is believed that a chromosome translocation event separated the two gene clusters about 300 million years ago (see Figure 4–87).

There are several duplicated globin DNA sequences in the α- and β-globin gene clusters that are not functional genes but pseudogenes. These have a close sequence similarity to the functional genes but have been disabled by mutations that prevent their expression. The existence of such pseudogenes makes it clear that, as expected, not every DNA duplication leads to a new functional gene. We also know that nonfunctional DNA sequences are not rapidly discarded, as indicated by the large excess of noncoding DNA that is found in mammalian genomes.

Genes Encoding New Proteins Can Be Created by the Recombination of Exons

The role of DNA duplication in evolution is not confined to the expansion of gene families. It can also act on a smaller scale to create single genes by stringing together short duplicated segments of DNA. The proteins encoded by genes generated in this way can be recognized by the presence of repeating similar protein domains, which are covalently linked to one another in series. The immunoglobulins (**Figure 4–88**) and albumins, for example, as well as most fibrous proteins (such as collagens) are encoded by genes that have evolved by repeated duplications of a primordial DNA sequence.

In genes that have evolved in this way, as well as in many other genes, each separate exon often encodes an individual protein folding unit, or domain. It is believed that the organization of DNA coding sequences as a series of such exons separated by long introns has greatly facilitated the evolution of new proteins. The duplications necessary to form a single gene coding for a protein with repeating domains, for example, can often occur by breaking and rejoining the DNA anywhere in the long introns on either side of an exon; without introns there would be only a few sites in the original gene at which a recombinational exchange between DNA molecules could duplicate the domain. By enabling the duplication to occur by recombination at many potential sites rather than just a few, introns increase the probability of a favorable duplication event.

More generally, we know from genome sequences that the various parts of genes—both their individual exons and their regulatory elements—have served as modular elements that have been duplicated and moved about the genome to create the great diversity of living things. Thus, for example, many present-day proteins are formed as a patchwork of domains from different origins, reflecting their long evolutionary history (see Figure 3–19).

Neutral Mutations Often Spread to Become Fixed in a Population, with a Probability that Depends on Population Size

In comparisons between two species that have diverged from one another by millions of years, it makes little difference which individuals from each species are compared. For example, typical human and chimpanzee DNA sequences differ from one another by about 1%. In contrast, when the same region of the genome is sampled from two different humans, the differences are typically less than 0.1%. For more distantly related organisms, the inter-species differences overshadow intra-species variation even more dramatically. However, each "fixed difference" between the human and the chimpanzee (in other words, each difference that is now characteristic of all or nearly all individuals of each species) started out as a new mutation in a single individual. If the size of the

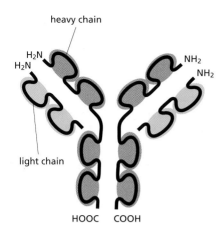

Figure 4–88 Schematic view of an antibody (immunoglobulin) molecule. This molecule is a complex of two identical heavy chains and two identical light chains. Each heavy chain contains four similar, covalently linked domains, while each light chain contains two such domains. Each domain is encoded by a separate exon, and all of the exons are thought to have evolved by the serial duplication of a single ancestral exon.

interbreeding population in which the mutation occurred is N, the initial allele frequency of a new mutation would be $1/(2N)$ for a diploid organism. How does such a rare mutation become fixed in the population, and hence become a characteristic of the species rather than of a particular individual genome?

The answer to this question depends on the functional consequences of the mutation. If the mutation has a significantly deleterious effect, it will simply be eliminated by purifying selection and will not become fixed. (In the most extreme case, the individual carrying the mutation will die without producing progeny.) Conversely, the rare mutations that confer a major reproductive advantage on individuals who inherit them can spread rapidly in the population. Because humans reproduce sexually and genetic recombination occurs each time a gamete is formed (discussed in Chapter 5), the genome of each individual who has inherited the mutation will be a unique recombinational mosaic of segments inherited from a large number of ancestors. The selected mutation along with a modest amount of neighboring sequence—ultimately inherited from the individual in which the mutation occurred—will simply be one piece of this huge mosaic.

The great majority of mutations that are not harmful are not beneficial either. These selectively neutral mutations can also spread and become fixed in a population, and they make a large contribution to the evolutionary change in genomes. Their spread is not as rapid as the spread of the rare strongly advantageous mutations. The process by which such neutral genetic variation is passed down through an idealized interbreeding population can be described mathematically by equations that are surprisingly simple. The idealized model that has proven most useful for analyzing human genetic variation assumes a constant population size and random mating, as well as selective neutrality for the mutations. While neither of the first two assumptions is a good description of human population history, they nonetheless provide a useful starting point for analyzing intra-species variation.

When a new neutral mutation occurs in a constant population of size N that is undergoing random mating, the probability that it will ultimately become fixed is approximately $1/(2N)$. For those mutations that do become fixed, the average time to fixation is approximately $4N$ generations. A detailed analysis of data on human genetic variation suggests an ancestral population size of approximately 10,000 during the period when the current pattern of genetic variation was largely established. With a population that has reached this size, the probability that a new, selectively neutral mutation would become fixed is small (5×10^{-5}), while the average time to fixation would be on the order of 800,000 years (assuming a 20-year generation time). Thus, while we know that the human population has grown enormously since the development of agriculture approximately 15,000 years ago, most of the present-day set of common human genetic variants reflects the mixture of variants that was already present long before this time, when the human population was still small enough to allow their widespread dissemination.

A Great Deal Can Be Learned from Analyses of the Variation Among Humans

Even though most of the variation among modern humans originates from variation present in a comparatively tiny group of ancestors, the number of variations encountered is very large. One important source of variation, which was missed for many years, is the presence of many duplications and deletions of large blocks of DNA. According to one estimate, when any individual human is compared with the standard reference genome in the database, one should expect to find roughly 100 differences involving long sequence blocks. Some of these "copy number variations" will be very common (**Figure 4–89**), while others will be present in only a minority of humans (**Figure 4–90**). From an initial sampling, nearly half will contain known genes. In retrospect this type of variation is not surprising, given the extensive history of DNA addition and DNA loss in vertebrate genomes (for example, see Figure 4–79).

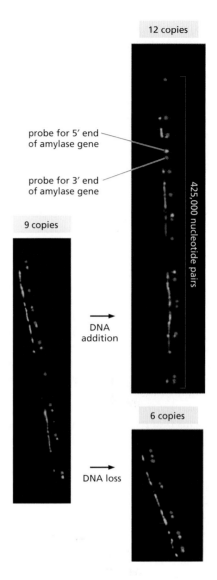

Figure 4–89 **Visualization of a frequent type of variation among humans.** About half of the humans tested had nine copies of the amylase gene (*left*), which produces an important enzyme that digests starch. In other humans, there has been either DNA loss or DNA addition to produce an altered chromosome, resulting from the deletion (loss) or the duplication (addition) of a part of this region. To obtain these images, stretched chromatin fibers have been hybridized with differently colored probes to the two ends of the amylase gene, as indicated. The *blue* lines mark the general paths of the chromatin. They have been determined by a second stain and displaced to one side for clarity. (Adapted from A.J. Iafrate et al., *Nat. Genet.* 36:949–951, 2004. With permission from Macmillan Publishers Ltd.)

The intra-species variations that have been most extensively characterized are **single-nucleotide polymorphisms** (**SNPs**). These are simply points in the genome sequence where one large fraction of the human population has one nucleotide, while another substantial fraction has another. Two human genomes sampled from the modern world population at random will differ at approximately 2.5×10^6 such sites (1 per 1300 nucleotide pairs). As will be described in the overview of genetics in Chapter 8, mapped sites in the human genome that are **polymorphic**—meaning that there is a reasonable probability (generally more than 1%) that the genomes of two individuals will differ at that site—are extremely useful for genetic analyses, in which one attempts to associate specific traits (phenotypes) with specific DNA sequences for medical or scientific purposes (see p. 560).

Against the background of ordinary SNPs inherited from our prehistoric ancestors, certain sequences with exceptionally high mutation rates stand out. A dramatic example is provided by CA repeats, which are ubiquitous in the human genome and in the genomes of other eucaryotes. Sequences with the motif $(CA)_n$ are replicated with relatively low fidelity because of a slippage that occurs between the template and the newly synthesized strands during DNA replication; hence, the precise value of n can vary over a considerable range from one genome to the next. These repeats make ideal DNA-based genetic markers, since most humans are heterozygous—carrying two values of n at any particular CA repeat, having inherited one repeat length (n) from their mother and a different repeat length from their father. While the value of n changes sufficiently rarely that most parent–child transmissions propagate CA repeats faithfully, the changes are sufficiently frequent to maintain high levels of heterozygosity in the human population. These and some other simple repeats that display exceptionally high variability therefore provide the basis for identifying individuals by DNA analysis in crime investigations, paternity suits, and other forensic applications (see Figure 8–47).

While most of the SNPs and copy number variations in the human genome sequence are thought to have no effect on phenotype, a subset of them must be

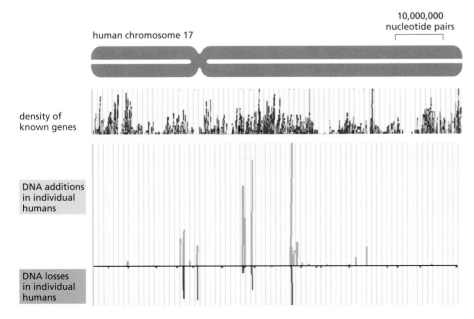

Figure 4–90 **Detection of copy number variants on human chromosome 17.** When 100 individuals were tested by a DNA microarray analysis that detects the copy number of DNA sequences throughout the entire length of this chromosome, the indicated distributions of DNA additions (*green bars*) and DNA losses (*red bars*) were observed compared with an arbitrary human reference sequence. The shortest *red* and *green bars* represent a single occurrence among the 200 chromosomes examined, whereas the longer bars indicate that the addition or loss was correspondingly more frequent. The results show preferred regions where the variations occur, and these tend to be in or near regions that already contain blocks of segmental duplications. Many of the changes include known genes. (Adapted from J.L. Freeman et al., *Genome Res.* 16:949–961, 2006. With permission from Cold Spring Harbor Laboratory Press.)

responsible for nearly all the heritable aspects of human individuality. We know that even a single nucleotide change that alters one amino acid in a protein can cause a serious disease, as for example in sickle cell anemia, which is caused by such a mutation in hemoglobin. <TTTT> We also know that gene dosage—a doubling or halving of the copy number of some genes—can have a profound effect on human development by altering the level of gene product. There is therefore every reason to suppose that some of the many differences between any two human beings will have substantial effects on human health, physiology, and behavior, whether they be SNPs or copy number variations. The major challenge in human genetics is to learn to recognize those relatively few variations that are functionally important against a large background of neutral variation in the genomes of different humans.

Summary

Comparisons of the nucleotide sequences of present-day genomes have revolutionized our understanding of gene and genome evolution. Because of the extremely high fidelity of DNA replication and DNA repair processes, random errors in maintaining the nucleotide sequences in genomes occur so rarely that only about one nucleotide in 1000 is altered every million years in any particular line of descent. Not surprisingly, therefore, a comparison of human and chimpanzee chromosomes—which are separated by about 6 million years of evolution—reveals very few changes. Not only are our genes essentially the same, but their order on each chromosome is almost identical. Although a substantial number of segmental duplications and segmental deletions have occurred in the past 6 million years, even the positions of the transposable elements that make up a major portion of our noncoding DNA are mostly unchanged.

When one compares the genomes of two more distantly related organisms—such as a human and a mouse, separated by about 80 million years—one finds many more changes. Now the effects of natural selection can be clearly seen: through purifying selection, essential nucleotide sequences—both in regulatory regions and in coding sequences (exon sequences)—have been highly conserved. In contrast, nonessential sequences (for example, much of the DNA in introns) have been altered to such an extent that an accurate alignment according to ancestry is frequently not possible.

Because of purifying selection, the comparison of the genome sequences of multiple related species is an especially powerful way to find DNA sequences with important functions. Although about 5% of the human genome has been conserved as a result of purifying selection, the function of the majority of this DNA (tens of thousands of multispecies conserved sequences) remains mysterious. Future experiments characterizing their functions should teach us a great deal about vertebrate biology.

Other sequence comparisons show that a great deal of the genetic complexity of present-day organisms is due to the expansion of ancient gene families. DNA duplication followed by sequence divergence has clearly been a major source of genetic novelty during evolution. The genomes of any two humans will differ from each other both because of nucleotide substitutions (single nucleotide polymorphisms, or SNPs) and because of inherited DNA gains and DNA losses that cause copy number variants. Understanding these differences will improve both medicine and our understanding of human biology.

PROBLEMS

Which statements are true? Explain why or why not.

4–1   Human females have 23 different chromosomes, whereas human males have 24.

4–2   In a comparison between the DNAs of related organisms such as humans and mice, identifying the conserved DNA sequences facilitates the search for functionally important regions.

4–3   The four core histones are relatively small proteins with a very high proportion of positively charged amino acids; the positive charge helps the histones bind tightly to DNA, regardless of its nucleotide sequence.

4–4   Nucleosomes bind DNA so tightly that they cannot move from the positions where they are first assembled.

4–5   Gene duplication and divergence is thought to have played a critical role in the evolution of increased biological complexity.

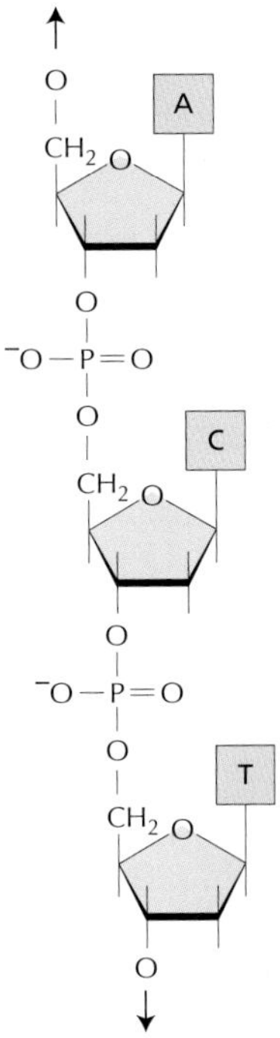

Figure Q4–1 Three nucleotides from the interior of a single strand of DNA (Problem 4–7). *Arrows* at the ends of the DNA strand indicate that the structure continues in both directions.

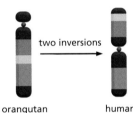

Discuss the following problems.

4–6 DNA isolated from the bacterial virus M13 contains 25% A, 33% T, 22% C, and 20% G. Do these results strike you as peculiar? Why or why not? How might you explain these values?

4–7 A segment of DNA from the interior of a single strand is shown in **Figure Q4–1**. What is the polarity of this DNA from top to bottom?

4–8 Human DNA contains 20% C on a molar basis. What are the mole percents of A, G, and T?

4–9 Chromosome 3 in orangutans differs from chromosome 3 in humans by two inversion events (**Figure Q4–2**). Draw the intermediate chromosome that resulted from the first inversion and explicitly indicate the segments included in each inversion.

Figure Q4–2 Chromosome 3 in orangutans and humans (Problem 4–9). Differently *colored blocks* indicate segments of the chromosomes that were derived by previous fusions.

4–10 Assuming that the 30-nm chromatin fiber contains about 20 nucleosomes (200 bp/nucleosome) per 50 nm of length, calculate the degree of compaction of DNA associated with this level of chromatin structure. What fraction of the 10,000-fold condensation that occurs at mitosis does this level of DNA packing represent?

4–11 In contrast to histone acetylation, which always correlates with gene activation, histone methylation can lead to either transcriptional activation or repression. How do you suppose that the same modification—methylation—can mediate different biological outcomes?

4–12 Why is a chromosome with two centromeres (a dicentric chromosome) unstable? Would a back-up centromere not be a good thing for a chromosome, giving it two chances to form a kinetochore and attach to microtubules during mitosis? Would that not help to ensure that the chromosome did not get left behind at mitosis?

4–13 HP1 proteins, a family of proteins found in heterochromatin, are implicated in gene silencing and chromatin structure. The three proteins in humans—HP1α, HP1β, and HP1γ—share a highly conserved chromodomain, which is thought to direct chromatin localization. To determine whether these proteins could bind to the histone H3 N-terminus, you have covalently attached to separate beads various versions of the H3 N-terminal peptide—unmodified,

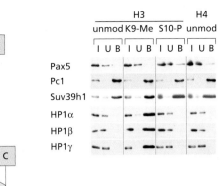

Figure Q4–3 Pull-down assays to determine binding specificity of HP1 proteins (Problem 4–13). Each protein at the left was detected by immunoblotting using a specific antibody after separation by SDS-polyacrylamide gel electrophoresis. For each histone N-terminal peptide the total input protein (I), the unbound protein (U), and the bound protein (B) are indicated. (Adapted from M. Lachner et al., *Nature* 410:116–120, 2001. With permission from Macmillan Publishers Ltd.)

Lys-9-dimethylated (K9-Me), and Ser-10-phosphorylated (S10-P)—along with an unmodified tail from histone H4. This arrangement allows you to incubate the beads with various proteins, wash away unbound proteins, and then elute bound proteins for assay by Western blotting. The results of your 'pull-down' assay for the HP1 proteins are shown in **Figure Q4–3**, along with the results from several control proteins, including Pax5, a gene regulatory protein, polycomb protein Pc1, which is known to bind to histones, and Suv39h1, a histone methyltransferase.

Based on these results, which of the proteins tested bind to the unmodified tails of histones? Do any of the HP1 proteins and control proteins selectively bind to the modified histone N-terminal peptides? What histone modification would you predict would be found in heterochromatin?

4–14 Mobile pieces of DNA—transposable elements—that insert themselves into chromosomes and accumulate during evolution make up more than 40% of the human genome. Transposable elements of four types—long interspersed elements (LINEs), short interspersed elements (SINEs), LTR retrotransposons, and DNA transposons—are inserted more or less randomly throughout the human genome. These elements are conspicuously rare at the four homeobox gene clusters, *HoxA, HoxB, HoxC,* and *HoxD,* as illustrated for *HoxD* in **Figure Q4–4**, along with an equivalent region of chromosome 22, which lacks a *Hox* cluster. Each *Hox* cluster is about 100 kb in length and contains 9 to 11 genes, whose differential expression along the anteroposterior axis of the developing embryo establishes the basic body plan for humans (and for other animals). Why do you suppose that transposable elements are so rare in the *Hox* clusters?

Figure Q4–4 Transposable elements and genes in 1 Mb regions of chromosomes 2 and 22 (Problem 4–14). Lines that project *upward* indicate exons of known genes. Lines that project *downward* indicate transposable elements; they are so numerous (constituting more than 40% of the human genome) that they merge into nearly a solid block outside the *Hox* clusters. (Adapted from E. Lander et al., *Nature* 409:860–921, 2001. With permission from Macmillan Publishers Ltd.)

REFERENCES

General

Hartwell L, Hood L, Goldberg ML et al (2006) Genetics: from Genes to Genomes, 3rd ed. Boston: McGraw Hill.

Olson MV (2002) The Human Genome Project: a player's perspective. *J Mol Biol* 319:931–942.

Strachan T & Read AP (2004) Human Molecular Genetics. New York: Garland Science.

Wolffe A (1999) Chromatin: Structure and Function, 3rd ed. New York: Academic Press.

The Structure and Function of DNA

Avery OT, MacLeod CM & McCarty M (1944) Studies on the chemical nature of the substance inducing transformation of *pneumococcal* types. *J Exp Med* 79:137–158.

Meselson M & Stahl FW (1958) The replication of DNA in *E. coli*. *Proc Natl Acad Sci USA* 44:671–682.

Watson JD & Crick FHC (1953) Molecular structure of nucleic acids. A structure for deoxyribose nucleic acids. *Nature* 171:737–738.

Chromosomal DNA and Its Packaging in the Chromatin Fiber

Jin J, Cai Y, Li B et al (2005) In and out: histone variant exchange in chromatin. *Trends Biochem Sci* 30:680–687.

Kornberg RD & Lorch Y (1999) Twenty-five years of the nucleosome, fundamental particle of the eukaryote chromosome. *Cell* 98:285–294.

Li G, Levitus M, Bustamante C & Widom J (2005) Rapid spontaneous accessibility of nucleosomal DNA. *Nature Struct Mol Biol* 12:46–53.

Lorch Y, Maier-Davis B & Kornberg RD (2006) Chromatin remodeling by nucleosome disassembly *in vitro*. *Proc Natl Acad Sci USA* 103:3090–3093.

Luger K, Mader AW, Richmond RK et al (1997) Crystal structure of the nucleosome core particle at 2.8 Å resolution. *Nature* 389:251–260.

Luger K & Richmond TJ (1998) The histone tails of the nucleosome. *Curr Opin Genet Dev* 8:140–146.

Malik H S & Henikoff S (2003) Phylogenomics of the nucleosome *Nature Struct Biol* 10:882-891.

Ried T, Schrock E, Ning Y & Wienberg J (1998) Chromosome painting: a useful art. *Hum Mol Genet* 7:1619–1626.

Robinson PJ & Rhodes R (2006) Structure of the 30 nm chromatin fibre: A key role for the linker histone. *Curr Opin Struct Biol* 16:1–8.

Saha A, Wittmeyer J & Cairns BR (2006) Chromatin remodeling: the industrial revolution of DNA around histones. *Nature Rev Mol Cell Biol* 7:437–446.

Woodcock CL (2006) Chromatin architecture. *Curr Opin Struct Biol* 16:213–220.

The Regulation of Chromatin Structure

Egger G, Liang, G, Aparicio A & Jones PA (2004) Epigenetics in human disease and prospects for epigenetic therapy. *Nature* 429:457–463.

Henikoff S (1990) Position-effect variegation after 60 years. *Trends Genet* 6:422–426.

Henikoff S & Ahmad K (2005) Assembly of variant histones into chromatiin. *Annu Rev Cell Dev Biol* 21:133–153.

Gaszner M & Felsenfeld G (2006) Insulators: exploiting transcriptional and epigenetic mechanisms. *Nature Rev Genet* 7:703–713.

Hake SB & Allis CD (2006) Histone H3 variants and their potential role in indexing mammalian genomes: the "H3 barcode hypothesis." *Proc Natl Acad Sci USA* 103:6428–6435.

Jenuwein T (2006) The epigenetic magic of histone lysine methylation. *FEBS J* 273:3121–3135.

Martin C & Zhang Y (2005) The diverse functions of histone lysine methylation. *Nature Rev Mol Cell Biol* 6:838–849.

Mellone B, Erhardt S & Karpen GH (2006) The ABCs of centromeres. *Nature Cell Biol* 8:427–429.

Peterson CL & Laniel MA (2004) Histones and histone modifications. *Curr Biol* 14:R546–R551.

Ruthenburg AJ, Allis CD & Wysocka J (2007) Methylation of lysine 4 on histone H3: intricacy of writing and reading a single epigenetic mark. *Mol Cell* 25:15–30.

Shahbazian MD & Grunstein M (2007) Functions of site-specific histone acetylation and deacetylation. *Annu Rev Biochem* 76:75–100.

The Global Structure of Chromosomes

Akhtar A & Gasser SM (2007) The nuclear envelope and transcriptional control. *Nature Rev Genet* 8:507–517.

Callan HG (1982) Lampbrush chromosomes. *Proc Roy Soc Lond Ser B* 21:417–448.

Chakalova L, Debrand E, Mitchel JA et al (2005) Replication and transcription: shaping the landscape of the genome. *Nature Rev Genet* 6:669–678.

Cremer T, Cremer M, Dietzel S et al (2006) Chromosome territories—a functional nuclear landscape. *Curr Opin Cell Biol* 18:307–316.

Ebert A, Lein S, Schotta G & Reuter G (2006) Histone modification and the control of heterochromatic gene silencing in *Drosophila*. *Chromosome Res* 14:377–392.

Fraser P & Bickmore W (2007) Nuclear organization of the genome and the potential for gene regulation. *Nature* 447:413–417.

Handwerger KE & Gall JG (2006) Subnuclear organelles: new insights into form and function. *Trends Cell Biol* 16:19–26.

Hirano T (2006) At the heart of the chromosome: SMC proteins in action. *Nature Rev Mol Cell Biol* 7:311–322.

Lamond AI & Spector DL (2003) Nuclear speckles: a model for nuclear organelles. *Nature Rev Mol Cell Biol* 4:605–612.

Maeshima K & Laemmli UK (2003) A two-step scaffolding model for mitotic chromosome assembly. *Dev Cell* 4:467–480.

Sims JK, Houston SI, Magazinnik T & Rice JC (2006) A trans-tail histone code defined by monomethylated H4 Lys-20 and H3 Lys-9 demarcates distinct regions of silent chromatin. *J Biol Chem* 281:12760–12766.

Speicher MR & Carter NP (2005) The new cytogenetics: blurring the boundaries with molecular biology. *Nature Rev Genet* 6:782–792.

Zhimulev IF (1998) Morphology and structure of polytene chromosomes. *Adv Genet* 37:1–566.

How Genomes Evolve

Batzer MA & Deininger PL (2002) ALU repeats and human genomic diversity. *Nature Rev Genet* 3:370–379.

Blanchette M, Green ED, Miller W, & Haussler D (2004) Reconstructing large regions of an ancestral mammalian genome in silico. *Genome Res* 14:2412–2423.

Cheng Z, Ventura M, She X et al (2005) A genome-wide comparison of recent chimpanzee and human segmental duplications. *Nature* 437:88–93.

Feuk L, Carson AR & Scherer S (2006) Structural variation in the human genome. *Nature Rev Genet* 7:85–97.

International Human Genome Sequencing Consortium (2001) Initial sequencing and analysis of the human genome. *Nature* 409:860–921.

International Human Genome Consortium (2004) Finishing the euchromatic sequence of the human genome. *Nature* 431:931–945.

Koszul R, Caburet S, Dujon B & Fischer G (2004) Eucaryotic genome evolution through the spontaneous duplication of large chromosomal segments. *EMBO J* 23:234–243.

Margulies EH, NISC Comparative Sequencing Program & Green ED (2003) Detecting highly conserved regions of the human genome by multispecies sequence comparisons. *Cold Spring Harbor Symp Quant Biol* 68:255–263.

Mouse Genome Sequencing Consortium (2002) Initial sequencing and comparative analysis of the mouse genome. *Nature* 420:520–562.

Pollard KS, Salama SR, Lambert N et al (2006) An RNA gene expressed during cortical development evolved rapidly in humans. *Nature* 443:167–172.

Sharp AJ, Cheng Z & Eichler EE (2007) Structural variation of the human genome. *Annu Rev Genomics Hum Genet* 7:407–442.

Siepel A, Bejerano G, Pedersen JS et al (2005) Evolutionarily conserved elements in vertebrate, insect, worm, and yeast genomes. *Genome Res* 15:1034–1050.

The International HapMap Consortium (2005) A haplotype map of the human genome. *Nature* 437:1299–1320.

The ENCODE Project Consortium (2007) Identification and analysis of functional elements in 1% of the human genome by the ENCODE pilot project. *Nature* 447:799–816.

Chapter 5

DNA Replication, Repair, and Recombination

The ability of cells to maintain a high degree of order in a chaotic universe depends upon the accurate duplication of vast quantities of genetic information carried in chemical form as DNA. This process, called *DNA replication*, must occur before a cell can produce two genetically identical daughter cells. Maintaining order also requires the continued surveillance and repair of this genetic information because DNA inside cells is repeatedly damaged by chemicals and radiation from the environment, as well as by thermal accidents and reactive molecules generated inside the cell. In this chapter, we describe the protein machines that replicate and repair the cell's DNA. These machines catalyze some of the most rapid and accurate processes that take place within cells, and their mechanisms clearly demonstrate the elegance and efficiency of cell chemistry.

While the short-term survival of a cell can depend on preventing changes in its DNA, the long-term survival of a species requires that DNA sequences be changeable over many generations. Despite the great efforts that cells make to protect their DNA, occasional changes in DNA sequences do occur. Over time, these changes provide the genetic variation upon which selection pressures act during the evolution of organisms.

We begin this chapter with a brief discussion of the changes that occur in DNA as it is passed down from generation to generation. Next, we discuss the cell mechanisms—DNA replication and DNA repair—that are responsible for minimizing these changes. Finally, we consider some of the most intriguing pathways that alter DNA sequences—those of DNA recombination, including the movement of special DNA sequences in chromosomes called transposable elements.

THE MAINTENANCE OF DNA SEQUENCES

Although, as just pointed out, occasional genetic changes enhance the long-term survival of a species, the survival of the individual demands a high degree of genetic stability. Only rarely do the cell's DNA-maintenance processes fail, resulting in permanent change in the DNA. Such a change is called a **mutation**, and it can destroy an organism if it occurs in a vital position in the DNA sequence.

Mutation Rates Are Extremely Low

The **mutation rate**, the rate at which observable changes occur in DNA sequences, can be determined directly from experiments carried out with a bacterium such as *Escherichia coli*—a resident of our intestinal tract and a commonly used laboratory organism (discussed in Chapter 1). Under laboratory conditions, *E. coli* divides about once every 40 minutes, and a single cell can generate a very large population—several billion—in less than a day. In such a population, it is possible to detect the small fraction of bacteria that have suffered a damaging mutation in a particular gene, if that gene is not required for the bacterium's survival. For example, the mutation rate of a gene specifically required for cells to use the sugar lactose as an energy source can be determined when the cells are grown in the presence of a different sugar, such as glucose. The fraction of damaged genes underestimates the actual mutation rate because many mutations are *silent* (for example, those that change a codon but not the amino acid it specifies, or those that change an amino acid without affecting the activity of the protein coded for by the gene). After correcting for these silent mutations, one finds that a single gene that encodes an average-sized protein ($\sim 10^3$ coding nucleotide pairs) accumulates a mutation (not necessarily one that would inactivate the protein) about once in about 10^6 bacterial cell generations. Stated differently, bacteria display a mutation rate of about 1 nucleotide change per 10^9 nucleotides per cell generation.

Recently it has become possible to measure the germ line mutation rate directly in more complex, sexually reproducing organisms such as the nemotode *C. elegans*. These worms, whose generation time is 4 days, were grown for many generations using their self-fertilization mode of reproduction (discussed in Chapter 22). The DNA sequence of a large region of the genome was then determined for many different descendent worms and compared with that of the progenitor worm. This analysis showed, on average, two new mutations (mostly short insertions and deletions) arise in the haploid genome each generation. When the number of cell divisions needed to produce sperm and eggs is taken into account, the mutation rate is roughly 1 mutation per 10^9 nucleotides per cell division, a rate remarkably similar to that in the asexually reproducing *E. coli* described above.

Direct measurement of the germ-line mutation rate in mammals is more difficult, but indirect estimates can be obtained. One way is to compare the amino acid sequences of the same protein in several species. The fraction of the amino acids that differ between any two species can then be compared with the estimated number of years since that pair of species diverged from a common ancestor, as determined from the fossil record. Using this method, one can calculate the number of years that elapse, on average, before an inherited change in the amino acid sequence of a protein becomes fixed in an organism. Because each such change usually reflects the alteration of a single nucleotide in the DNA sequence of the gene encoding that protein, we can use this value to estimate the average number of years required to produce a single, stable mutation in the gene.

These calculations will nearly always substantially underestimate the actual mutation rate, because many mutations will spoil the function of the protein and vanish from the population because of natural selection—that is, by the preferential death of the organisms that contain them. But the sequence of one family of protein fragments does not seem to matter, allowing the genes that encode them to accumulate mutations without being selected against. These are the *fibrinopeptides*, fragments twenty amino acids long that are discarded when the protein *fibrinogen* is activated to form *fibrin* during blood clotting. Since the function of fibrinopeptides apparently does not depend on their amino acid sequence, fibrinopeptides can tolerate almost any amino acid change. Sequence comparisons of the fibrinopeptides can therefore be used to estimate the mutation rate in the germ line. As determined from these studies, a typical protein of 400 amino acids will suffer an amino acid alteration roughly once every 200,000 years.

Another way to estimate mutation rates in humans is to use DNA sequencing to compare corresponding nucleotide sequences directly from closely related species in regions of the genome that do not appear to carry critical information. As expected, such comparisons produce estimates of the mutation rate that agree with those obtained from the fibrinopeptide studies.

E. coli, worms and humans differ greatly in their modes of reproduction and in their generation times. Yet, when the mutation rates of each are normalized to a single round of DNA replication, they are found to be similar: roughly 1 nucleotide change per 10^9 nucleotides each time that DNA is replicated.

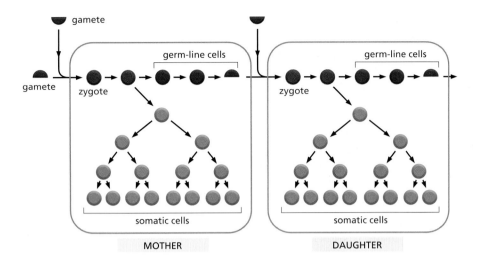

Figure 5–1 Germ-line cells and somatic cells carry out fundamentally different functions. In sexually reproducing organisms, the germ-line cells *(red)* propagate genetic information into the next generation. Somatic cells *(blue)*, which form the body of the organism, are necessary for the survival of germ-line cells but do not themselves leave any progeny.

Low Mutation Rates Are Necessary for Life as We Know It

Since many mutations are deleterious, no species can afford to allow them to accumulate at a high rate in its germ cells. Although the observed mutation frequency is low, it is nevertheless thought to limit the number of essential proteins that any organism can encode to perhaps 50,000. By an extension of the same argument, a mutation frequency tenfold higher would limit an organism to about 5000 essential genes. In this case, evolution would have been limited to organisms considerably less complex than a fruit fly.

The cells of a sexually reproducing organism are of two types: **germ cells** and **somatic cells**. The germ cells transmit genetic information from parent to offspring; the somatic cells form the body of the organism (**Figure 5–1**). We have seen that germ cells must be protected against high rates of mutation to maintain the species. However, the somatic cells of multicellular organisms must also be protected from genetic change to safeguard each individual. Nucleotide changes in somatic cells can give rise to variant cells, some of which, through natural selection, proliferate rapidly at the expense of the rest of the organism. In an extreme case, the result is an uncontrolled cell proliferation known as cancer, a disease that causes more than 20% of the deaths each year in Europe and North America. These deaths are due largely to an accumulation of changes in the DNA sequences of somatic cells (discussed in Chapter 23). A significant increase in the mutation frequency would presumably cause a disastrous increase in the incidence of cancer by accelerating the rate at which somatic cell variants arise. Thus, both for the perpetuation of a species with a large number of genes (germ cell stability) and for the prevention of cancer resulting from mutations in somatic cells (somatic cell stability), multicellular organisms like ourselves depend on the remarkably high fidelity with which their DNA sequences are replicated and maintained.

Summary

In all cells, DNA sequences are maintained and replicated with high fidelity. The mutation rate, approximately 1 nucleotide change per 10^9 nucleotides each time the DNA is replicated, is roughly the same for organisms as different as bacteria and humans. Because of this remarkable accuracy, the sequence of the human genome (approximately 3×10^9 nucleotide pairs) is changed by only about 3 nucleotides each time a cell divides. This allows most humans to pass accurate genetic instructions from one generation to the next, and also to avoid the changes in somatic cells that lead to cancer.

DNA REPLICATION MECHANISMS

All organisms must duplicate their DNA with extraordinary accuracy before each cell division. In this section, we explore how an elaborate "replication machine" achieves this accuracy, while duplicating DNA at rates as high as 1000 nucleotides per second.

Base-Pairing Underlies DNA Replication and DNA Repair

As introduced in Chapter 1, *DNA templating* is the mechanism the cell uses to copy the nucleotide sequence of one DNA strand into a complementary DNA sequence (**Figure 5–2**). This process entails the recognition of each nucleotide in the DNA *template strand* by a free (unpolymerized) complementary nucleotide, and it requires the separation of the two strands of the DNA helix. This separation exposes the hydrogen-bond donor and acceptor groups on each DNA base for base-pairing with the appropriate incoming free nucleotide, aligning it for its enzyme-catalyzed polymerization into a new DNA chain.

The first nucleotide-polymerizing enzyme, **DNA polymerase**, was discovered in 1957. The free nucleotides that serve as substrates for this enzyme were found to be deoxyribonucleoside triphosphates, and their polymerization into DNA required a single-stranded DNA template. **Figure 5–3** and **Figure 5–4** illustrate the stepwise mechanism of this reaction.

The DNA Replication Fork Is Asymmetrical

During DNA replication inside a cell, each of the two original DNA strands serves as a template for the formation of an entire new strand. Because each of the two daughters of a dividing cell inherits a new DNA double helix containing one original and one new strand (**Figure 5–5**), the DNA double helix is said to be replicated "semiconservatively" by DNA polymerase. How is this feat accomplished?

Analyses carried out in the early 1960s on whole replicating chromosomes revealed a localized region of replication that moves progressively along the parental DNA double helix. Because of its Y-shaped structure, this active region is called a **replication fork** (**Figure 5–6**). At the replication fork, a multienzyme complex that contains the DNA polymerase synthesizes the DNA of both new daughter strands.

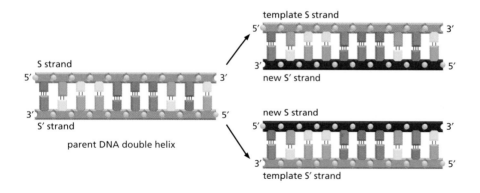

Figure 5–2 The DNA double helix acts as a template for its own duplication. Because the nucleotide A will pair successfully only with T, and G only with C, each strand of DNA can serve as a template to specify the sequence of nucleotides in its complementary strand by DNA base-pairing. In this way, a double-helical DNA molecule can be copied precisely.

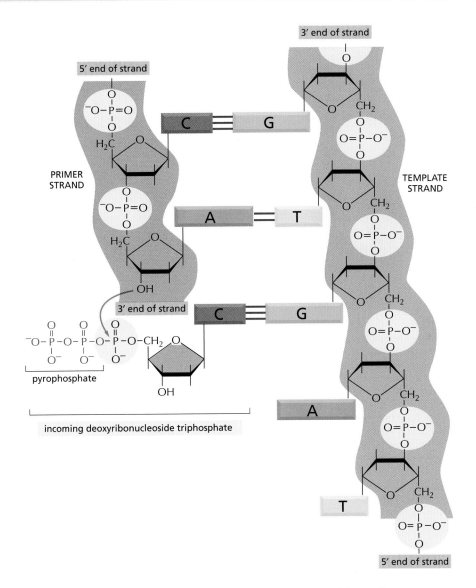

Figure 5–3 **The chemistry of DNA synthesis.** The addition of a deoxyribonucleotide to the 3′ end of a polynucleotide chain (the *primer strand*) is the fundamental reaction by which DNA is synthesized. As shown, base-pairing between an incoming deoxyribonucleoside triphosphate and an existing strand of DNA (the *template strand*) guides the formation of the new strand of DNA and causes it to have a complementary nucleotide sequence.

Initially, the simplest mechanism of DNA replication seemed to be the continuous growth of both new strands, nucleotide by nucleotide, at the replication fork as it moves from one end of a DNA molecule to the other. But because of the antiparallel orientation of the two DNA strands in the DNA double helix (see Figure 5–2), this mechanism would require one daughter strand to polymerize in the 5′-to-3′ direction and the other in the 3′-to-5′ direction. Such a replication fork would require two distinct types of DNA polymerase enzymes. However, all of the many DNA polymerases that have been discovered can synthesize only in the 5′-to-3′ direction.

How, then, can a DNA strand grow in the 3′-to-5′ direction? The answer was first suggested by the results of an experiment performed in the late 1960s. Researchers added highly radioactive ^{3}H-thymidine to dividing bacteria for a few seconds, so that only the most recently replicated DNA—that just behind the replication fork—became radiolabeled. This experiment revealed the transient existence of pieces of DNA that were 1000–2000 nucleotides long, now

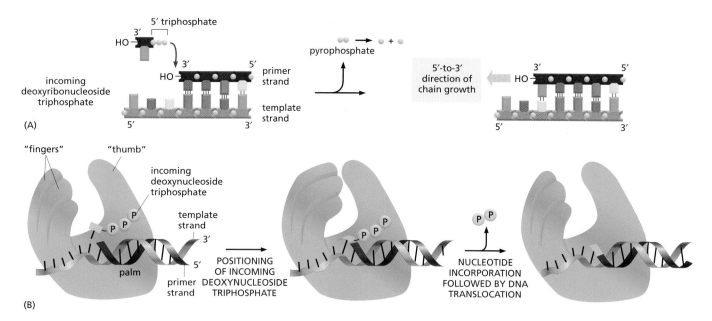

(A)

"fingers"   "thumb"

incoming
deoxynucleoside
triphosphate

template
strand

palm

primer
strand

POSITIONING
OF INCOMING
DEOXYNUCLEOSIDE
TRIPHOSPHATE

NUCLEOTIDE
INCORPORATION
FOLLOWED BY DNA
TRANSLOCATION

(B)

Figure 5–4 DNA synthesis catalyzed by DNA polymerase. (A) As indicated, DNA polymerase catalyzes the stepwise addition of a deoxyribonucleotide to the 3′-OH end of a polynucleotide chain, the *primer strand* that is paired to a second *template strand*. The newly synthesized DNA strand therefore polymerizes in the 5′-to-3′ direction as shown in the previous figure. Because each incoming deoxyribonucleoside triphosphate must pair with the template strand to be recognized by the DNA polymerase, this strand determines which of the four possible deoxyribonucleotides (A, C, G, or T) will be added. The reaction is driven by a large, favorable free-energy change, caused by the release of pyrophosphate and its subsequent hydrolysis to two molecules of inorganic phosphate. (B) The shape of a DNA polymerase molecule, as determined by x-ray crystallography. Roughly speaking DNA polymerases resemble a right hand in which the palm, fingers, and thumb grasp the DNA and form the active site. In the sequence shown, the correct positioning of an incoming deoxynucleoside triphosphate causes the fingers of the polymerase to tighten, thereby initiating the nucleotide addition reaction. Dissociation of pyrophosphate causes release of the fingers and translocation of the DNA by one nucleotide so the active site of the polymerase is ready to receive the next deoxynucleoside triphosphate.

commonly known as *Okazaki fragments,* at the growing replication fork. (Similar replication intermediates were later found in eucaryotes, where they are only 100–200 nucleotides long.) The Okazaki fragments were shown to be polymerized only in the 5′-to-3′ chain direction and to be joined together after their synthesis to create long DNA chains.

A replication fork therefore has an asymmetric structure (**Figure 5–7**). The DNA daughter strand that is synthesized continuously is known as the **leading strand**. Its synthesis slightly precedes the synthesis of the daughter strand that is synthesized discontinuously, known as the **lagging strand**. For the lagging strand, the direction of nucleotide polymerization is opposite to the overall direction of DNA chain growth. The synthesis of this strand by a discontinuous "backstitching" mechanism means that DNA replication requires only the 5′-to-3′ type of DNA polymerase.

The High Fidelity of DNA Replication Requires Several Proofreading Mechanisms

As discussed above, the fidelity of copying DNA during replication is such that only about 1 mistake occurs for every 10^9 nucleotides copied. This fidelity is much higher than one would expect from the accuracy of complementary base-

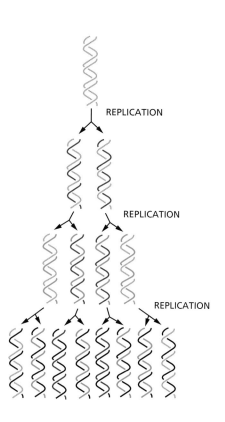

REPLICATION

REPLICATION

REPLICATION

Figure 5–5 The semiconservative nature of DNA replication. In a round of replication, each of the two strands of DNA is used as a template for the formation of a complementary DNA strand. The original strands therefore remain intact through many cell generations.

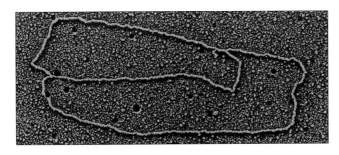

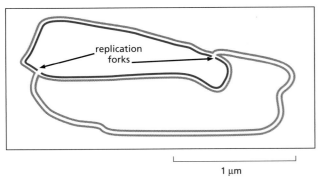

1 µm

Figure 5–6 **Two replication forks moving in opposite directions on a circular chromosome.** An active zone of DNA replication moves progressively along a replicating DNA molecule, creating a Y-shaped DNA structure known as a replication fork: the two arms of each Y are the two daughter DNA molecules, and the stem of the Y is the parental DNA helix. In this diagram, parental strands are *orange;* newly synthesized strands are *red*. (Micrograph courtesy of Jerome Vinograd.)

pairing. The standard complementary base pairs (see Figure 4–4) are not the only ones possible. For example, with small changes in helix geometry, two hydrogen bonds can form between G and T in DNA. In addition, rare tautomeric forms of the four DNA bases occur transiently in ratios of 1 part to 10^4 or 10^5. These forms mispair without a change in helix geometry: the rare tautomeric form of C pairs with A instead of G, for example.

If the DNA polymerase did nothing special when a mispairing occurred between an incoming deoxyribonucleoside triphosphate and the DNA template, the wrong nucleotide would often be incorporated into the new DNA chain, producing frequent mutations. The high fidelity of DNA replication, however, depends not only on the initial base-pairing but also on several "proofreading" mechanisms that act sequentially to correct any initial mispairings that might have occurred.

DNA polymerase performs the first proofreading step just before a new nucleotide is added to the growing chain. Our knowledge of this mechanism comes from studies of several different DNA polymerases, including one produced by a bacterial virus, T7, that replicates inside *E. coli*. The correct nucleotide has a higher affinity for the moving polymerase than does the incorrect nucleotide, because the correct pairing is more energetically favorable. Moreover, after nucleotide binding, but before the nucleotide is covalently added to the growing chain, the enzyme must undergo a conformational change in which its "fingers" tighten around the active site (see Figure 5–4). Because this change occurs more readily with correct than incorrect base-pairing, it allows the polymerase to "double-check" the exact base-pair geometry before it catalyzes the addition of the nucleotide.

The next error-correcting reaction, known as *exonucleolytic proofreading*, takes place immediately after those rare instances in which an incorrect nucleotide is covalently added to the growing chain. DNA polymerase enzymes

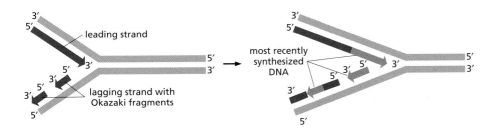

Figure 5–7 **The structure of a DNA replication fork.** Because both daughter DNA strands are polymerized in the 5′-to-3′ direction, the DNA synthesized on the lagging strand must be made initially as a series of short DNA molecules, called *Okazaki fragments*. On the lagging strand, the Okazaki fragments are synthesized sequentially, with those nearest the fork being the most recently made.

Figure 5–8 Exonucleolytic proofreading by DNA polymerase during DNA replication. In this example, the mismatch is due to the incorporation of a rare, transient tautomeric form of C, indicated by an asterisk. But the same proofreading mechanism applies to any misincorporation at the growing 3′-OH end. The part of DNA polymerase that removes the misincorporated nucleotide is a specialized member of a large class of enzymes, known as *exonucleases,* that cleave nucleotides one at a time from the ends of polynucleotides.

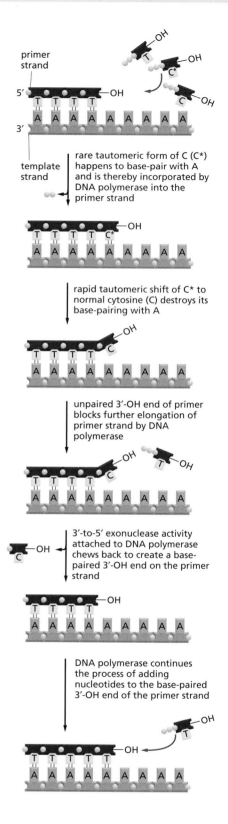

rare tautomeric form of C (C*) happens to base-pair with A and is thereby incorporated by DNA polymerase into the primer strand

rapid tautomeric shift of C* to normal cytosine (C) destroys its base-pairing with A

unpaired 3′-OH end of primer blocks further elongation of primer strand by DNA polymerase

3′-to-5′ exonuclease activity attached to DNA polymerase chews back to create a base-paired 3′-OH end on the primer strand

DNA polymerase continues the process of adding nucleotides to the base-paired 3′-OH end of the primer strand

are highly discriminating in the types of DNA chains they will elongate: they absolutely require a previously formed base-paired 3′-OH end of a *primer strand* (see Figure 5–4). Those DNA molecules with a mismatched (improperly base-paired) nucleotide at the 3′-OH end of the primer strand are not effective as templates because the polymerase cannot extend such a strand. DNA polymerase molecules correct such a mismatched primer strand by means of a separate catalytic site (either in a separate subunit or in a separate domain of the polymerase molecule, depending on the polymerase). This *3′-to-5′ proofreading exonuclease* clips off any unpaired residues at the primer terminus, continuing until enough nucleotides have been removed to regenerate a correctly base-paired 3′-OH terminus that can prime DNA synthesis. In this way, DNA polymerase functions as a "self-correcting" enzyme that removes its own polymerization errors as it moves along the DNA (**Figure 5–8** and **Figure 5–9**).

The self-correcting properties of the DNA polymerase depend on its requirement for a perfectly base-paired primer terminus, and it is apparently not possible for such an enzyme to start synthesis *de novo*. By contrast, the RNA polymerase enzymes involved in gene transcription do not need such an efficient exonucleolytic proofreading mechanism: errors in making RNA are not passed on to the next generation, and the occasional defective RNA molecule that is produced has no long-term significance. RNA polymerases are thus able to start new polynucleotide chains without a primer.

There is an error frequency of about 1 mistake for every 10^4 polymerization events both in RNA synthesis and in the separate process of translating mRNA sequences into protein sequences. This error rate is 100,000 times greater than that in DNA replication, where a series of proofreading processes makes the process unusually accurate (**Table 5–1**).

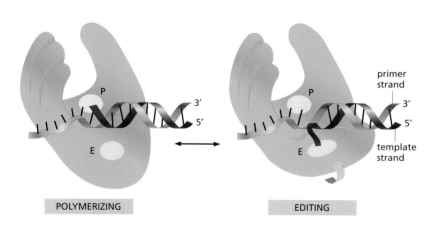

POLYMERIZING   EDITING

Figure 5–9 Editing by DNA polymerase. Outline of the structures of DNA polymerase complexed with the DNA template in the polymerizing mode *(left)* and the editing mode *(right)*. The catalytic site for the exonucleolytic (E) and the polymerization (P) reactions are indicated. In the editing mode, the newly synthesized DNA transiently unpairs from the template and the polymerase undergoes a conformational change, moving the editing catalytic site into place to remove the most recently added nucleotide. <GATT>

Table 5–1 The Three Steps That Give Rise to High-Fidelity DNA Synthesis

REPLICATION STEP	ERRORS PER NUCLEOTIDE
$5' \rightarrow 3'$ polymerization	1 in 10^5
$3' \rightarrow 5'$ exonucleolytic proofreading	1 in 10^2
Strand-directed mismatch repair	1 in 10^2
Combined	1 in 10^9

The third step, strand-directed mismatch repair, is described later in this chapter.

Only DNA Replication in the 5′-to-3′ Direction Allows Efficient Error Correction

The need for accuracy probably explains why DNA replication occurs only in the 5′-to-3′ direction. If there were a DNA polymerase that added deoxyribonucleoside triphosphates in the 3′-to-5′ direction, the growing 5′-chain end, rather than the incoming mononucleotide, would provide the activating triphosphate needed for the covalent linkage. In this case, the mistakes in polymerization could not be simply hydrolyzed away, because the bare 5′-chain end thus created would immediately terminate DNA synthesis (**Figure 5–10**). It is therefore possible to correct a mismatched base only if it has been added to the 3′ end of a DNA chain. Although the backstitching mechanism for DNA replication (see

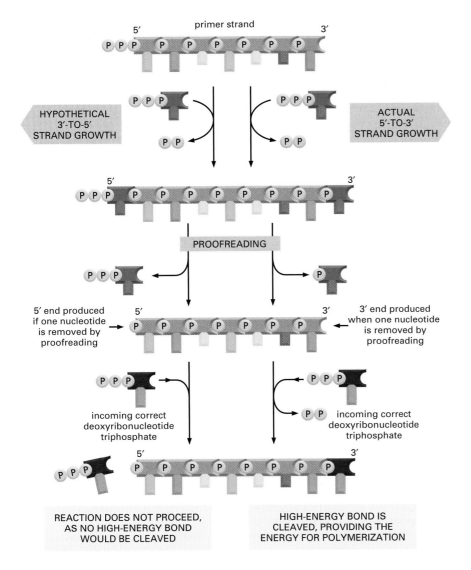

Figure 5–10 An explanation for the 5′-to-3′ direction of DNA chain growth. Growth in the 5′-to-3′ direction, shown on the right, allows the chain to continue to be elongated when a mistake in polymerization has been removed by exonucleolytic proofreading (see Figure 5–8). In contrast, exonucleolytic proofreading in the hypothetical 3′-to-5′ polymerization scheme, shown on the left, would block further chain elongation. For convenience, only the primer strand of the DNA double helix is shown.

Figure 5–7) seems complex, it preserves the 5′-to-3′ direction of polymerization that is required for exonucleolytic proofreading.

Despite these safeguards against DNA replication errors, DNA polymerases occasionally make mistakes. However, as we shall see later, cells have yet another chance to correct these errors by a process called *strand-directed mismatch repair*. Before discussing this mechanism, however, we describe the other types of proteins that function at the replication fork.

A Special Nucleotide-Polymerizing Enzyme Synthesizes Short RNA Primer Molecules on the Lagging Strand

For the leading strand, a special primer is needed only at the start of replication: once a replication fork is established, the DNA polymerase is continuously presented with a base-paired chain end on which to add new nucleotides. On the lagging side of the fork, however, every time the DNA polymerase completes a short DNA Okazaki fragment (which takes a few seconds), it must start synthesizing a completely new fragment at a site further along the template strand (see Figure 5–7). A special mechanism produces the base-paired primer strand required by this DNA polymerase molecule. The mechanism involves an enzyme called **DNA primase**, which uses ribonucleoside triphosphates to synthesize short **RNA primers** on the lagging strand (**Figure 5–11**). In eucaryotes, these primers are about 10 nucleotides long and are made at intervals of 100–200 nucleotides on the lagging strand.

The chemical structure of RNA was introduced in Chapter 1 and is described in detail in Chapter 6. Here, we note only that RNA is very similar in structure to DNA. A strand of RNA can form base pairs with a strand of DNA, generating a DNA/RNA hybrid double helix if the two nucleotide sequences are complementary. Thus the same templating principle used for DNA synthesis guides the synthesis of RNA primers. Because an RNA primer contains a properly base-paired nucleotide with a 3′-OH group at one end, it can be elongated by the DNA polymerase at this end to begin an Okazaki fragment. The synthesis of each Okazaki fragment ends when this DNA polymerase runs into the RNA primer attached to the 5′ end of the previous fragment. To produce a continuous DNA chain from the many DNA fragments made on the lagging strand, a special DNA repair system acts quickly to erase the old RNA primer and replace it with DNA. An enzyme called **DNA ligase** then joins the 3′ end of the new DNA fragment to the 5′ end of the previous one to complete the process (**Figures 5–12 and 5–13**).

Why might an erasable RNA primer be preferred to a DNA primer that would not need to be erased? The argument that a self-correcting polymerase cannot start chains *de novo* also implies its converse: an enzyme that starts chains anew cannot be efficient at self-correction. Thus, any enzyme that primes the synthesis of Okazaki fragments will of necessity make a relatively inaccurate copy (at least 1 error in 10^5). Even if the copies retained in the final product constituted as little as 5% of the total genome (for example, 10 nucleotides per 200-nucleotide DNA fragment), the resulting increase in the overall mutation rate would be enormous. It therefore seems likely that the use of RNA rather than DNA for priming brings a powerful advantage to the cell: the ribonucleotides in the primer automatically mark these sequences as "suspect copy" to be efficiently removed and replaced.

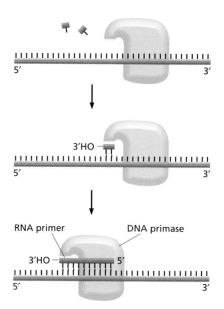

Figure 5–11 RNA primer synthesis. A schematic view of the reaction catalyzed by *DNA primase,* the enzyme that synthesizes the short RNA primers made on the lagging strand using DNA as a template. Unlike DNA polymerase, this enzyme can start a new polynucleotide chain by joining two nucleoside triphosphates together. The primase synthesizes a short polynucleotide in the 5′-to-3′ direction and then stops, making the 3′ end of this primer available for the DNA polymerase.

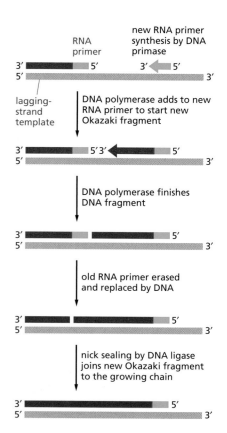

Figure 5–12 The synthesis of one of many DNA fragments on the lagging strand. In eucaryotes, RNA primers are made at intervals spaced by about 200 nucleotides on the lagging strand, and each RNA primer is approximately 10 nucleotides long. This primer is erased by a special DNA repair enzyme (an RNAse H) that recognizes an RNA strand in an RNA/DNA helix and fragments it; this leaves gaps that are filled in by DNA polymerase and DNA ligase.

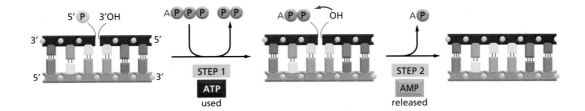

Special Proteins Help to Open Up the DNA Double Helix in Front of the Replication Fork

For DNA synthesis to proceed, the DNA double helix must be opened up ahead of the replication fork so that the incoming deoxyribonucleoside triphosphates can form base pairs with the template strand. However, the DNA double helix is very stable under physiological conditions; the base pairs are locked in place so strongly that it requires temperatures approaching that of boiling water to separate the two strands in a test tube. For this reason, two additional types of replication proteins—DNA helicases and single-strand DNA-binding proteins—are needed to open the double helix and thus provide the appropriate single-stranded DNA template for the DNA polymerase to copy.

DNA helicases were first isolated as proteins that hydrolyze ATP when they are bound to single strands of DNA. As described in Chapter 3, the hydrolysis of ATP can change the shape of a protein molecule in a cyclical manner that allows the protein to perform mechanical work. DNA helicases use this principle to propel themselves rapidly along a DNA single strand. When they encounter a region of double helix, they continue to move along their strand, thereby prying apart the helix at rates of up to 1000 nucleotide pairs per second (**Figures 5–14 and 5–15**).

The two strands of DNA have opposite polarities, and, in principle, a helicase could unwind the DNA double helix by moving in the 5′ to 3′ direction along one strand or in the 3′ to 5′ direction along the other. In fact, both types of DNA helicase exist. In the best understood replication systems in bacteria, a helicase moving 5′ to 3′ along the lagging-strand template appears to have the predominant role, for reasons that will become clear shortly.

Single-strand DNA-binding (SSB) proteins, also called *helix-destabilizing proteins*, bind tightly and cooperatively to exposed single-stranded DNA without covering the bases, which therefore remain available for templating. These proteins are unable to open a long DNA helix directly, but they aid helicases by stabilizing the unwound, single-stranded conformation. In addition, their cooperative binding coats and straightens out the regions of single-stranded DNA on the lagging-strand template, thereby preventing the formation of the short hairpin helices that readily form in single-strand DNA (**Figures 5–16 and 5–17**). These hairpin helices can impede the DNA synthesis catalyzed by DNA polymerase.

A Sliding Ring Holds a Moving DNA Polymerase onto the DNA

On their own, most DNA polymerase molecules will synthesize only a short string of nucleotides before falling off the DNA template. The tendency to dissociate

Figure 5–13 The reaction catalyzed by DNA ligase. This enzyme seals a broken phosphodiester bond. As shown, DNA ligase uses a molecule of ATP to activate the 5′ end at the nick (step 1) before forming the new bond (step 2). In this way, the energetically unfavorable nick-sealing reaction is driven by being coupled to the energetically favorable process of ATP hydrolysis.

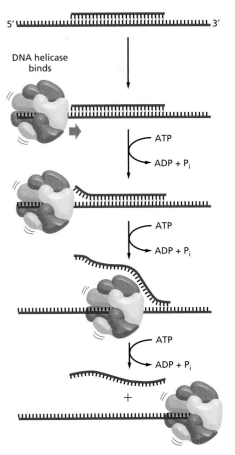

Figure 5–14 An assay used to test for DNA helicase enzymes. A short DNA fragment is annealed to a long DNA single strand to form a region of DNA double helix. The double helix is melted as the helicase runs along the DNA single strand, releasing the short DNA fragment in a reaction that requires the presence of both the helicase protein and ATP. The rapid stepwise movement of the helicase is powered by its ATP hydrolysis (see Figure 3–77). As indicated, many DNA helicases are composed of six subunits.

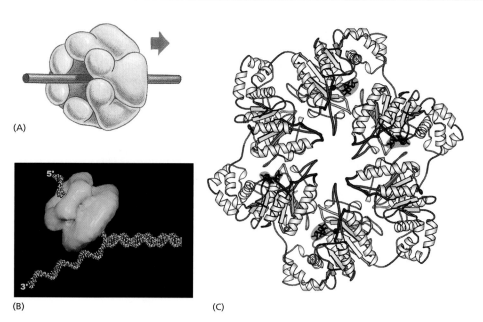

(A)

(B)

(C)

Figure 5–15 The structure of a DNA helicase. <TGCC> (A) A schematic diagram of the protein as a hexameric ring. (B) Schematic diagram showing a DNA replication fork and helicase to scale. (C) Detailed structure of the bacteriophage T7 replicative helicase, as determined by x-ray diffraction. Six identical subunits bind and hydrolyze ATP in an ordered fashion to propel this molecule like a rotary engine along a DNA single strand that passes through the central hole. *Red* indicates bound ATP molecules in the structure. (B, courtesy of Edward H. Egelman; C, from M.R. Singleton et al., *Cell* 101:589–600, 2000. With permission from Elsevier.)

quickly from a DNA molecule allows a DNA polymerase molecule that has just finished synthesizing one Okazaki fragment on the lagging strand to be recycled quickly, so as to begin the synthesis of the next Okazaki fragment on the same strand. This rapid dissociation, however, would make it difficult for the polymerase to synthesize the long DNA strands produced at a replication fork were it not for an accessory protein that functions as a regulated **sliding clamp**. This clamp keeps the polymerase firmly on the DNA when it is moving, but releases it as soon as the polymerase runs into a double-stranded region of DNA.

How can a sliding clamp prevent the polymerase from dissociating without at the same time impeding the polymerase's rapid movement along the DNA molecule? The three-dimensional structure of the clamp protein, determined by x-ray diffraction, reveals that it forms a large ring around the DNA double helix. One side of the ring binds to the back of the DNA polymerase, and the whole ring slides freely along the DNA as the polymerase moves. The assembly of the clamp around the DNA requires ATP hydrolysis by a special protein complex, the **clamp loader**, which hydrolyzes ATP as it loads the clamp on to a primer–template junction (**Figure 5–18**).

On the leading-strand template, the moving DNA polymerase is tightly bound to the clamp, and the two remain associated for a very long time. The DNA polymerase on the lagging-strand template also makes use of the clamp, but each time the polymerase reaches the 5′ end of the preceding Okazaki fragment, the polymerase releases itself from the clamp and dissociates from the

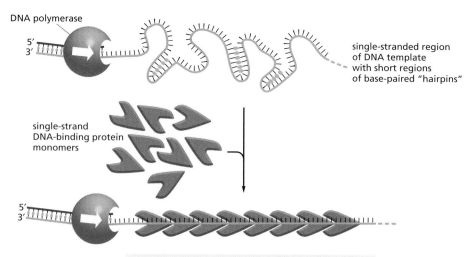

DNA polymerase

5′
3′

single-stranded region
of DNA template
with short regions
of base-paired "hairpins"

single-strand
DNA-binding protein
monomers

5′
3′

cooperative protein binding straightens region of chain

Figure 5–16 The effect of single-strand DNA-binding proteins (SSB proteins) on the structure of single-stranded DNA. Because each protein molecule prefers to bind next to a previously bound molecule, long rows of this protein form on a DNA single strand. This *cooperative binding* straightens out the DNA template and facilitates the DNA polymerization process. The "hairpin helices" shown in the bare, single-stranded DNA result from a chance matching of short regions of complementary nucleotide sequence; they are similar to the short helices that typically form in RNA molecules (see Figure 1–6).

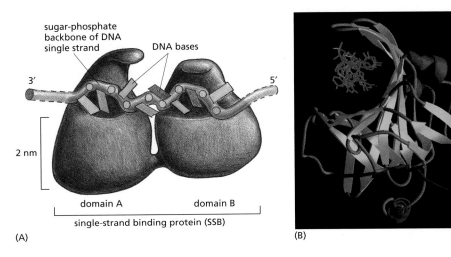

(A)

(B)

Figure 5–17 **The structure of the human single-strand binding protein bound to DNA.** (A) A front view of the two DNA-binding domains of RPA protein, which cover a total of eight nucleotides. Note that the DNA bases remain exposed in this protein–DNA complex. (B) A diagram showing the three-dimensional structure, with the DNA strand *(red)* viewed end-on. (B, from A. Bochkarev et al., *Nature* 385:176–181, 1997. With permission from Macmillan Publishers Ltd.)

template. This polymerase molecule then associates with a new clamp that is assembled on the RNA primer of the next Okazaki fragment.

The Proteins at a Replication Fork Cooperate to Form a Replication Machine

Although we have discussed DNA replication as though it were performed by a mix of proteins all acting independently, in reality most of the proteins are held together in a large and orderly multienzyme complex that rapidly synthesizes

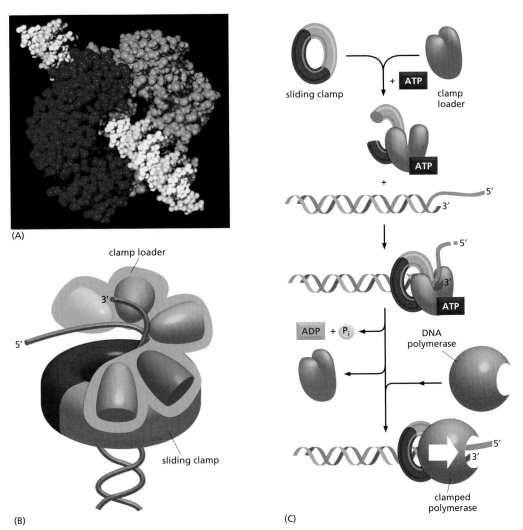

(A)

(B)

(C)

Figure 5–18 **The regulated sliding clamp that holds DNA polymerase on the DNA.** (A) The structure of the clamp protein from *E. coli*, as determined by X-ray crystallography, with a DNA helix added to indicate how the protein fits around DNA. (B) The structure of the five-subunit clamp loader resembles a screw nut, with its threads matching the grooves of DNA. It appears to tighten around the primer junction until its further progress is blocked by the 3′ end of the primer, at which point the loader hydrolyzes ATP and releases the clamp. (C) Schematic illustration showing how the clamp is assembled to hold a moving DNA polymerase molecule on the DNA. In the simplified reaction shown here, the clamp loader dissociates into solution once the clamp has been assembled. At a true replication fork, the clamp loader remains close to the lagging-strand polymerase, ready to assemble a new clamp at the start of each new Okazaki fragment (see Figure 5–19). (A, from X.P. Kong et al., *Cell* 69:425–437, 1992. With permission from Elsevier; C, from G.D. Bowman, M. O'Donnell and J. Kuriyan, *Nature* 429:708–709, 2004. With permission from Macmillan Publishers Ltd.)

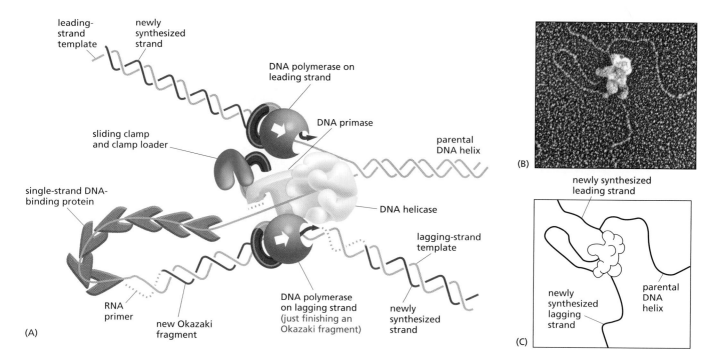

(A)

(B)

(C)

Figure 5–19 An active replication fork.
(A) This schematic diagram shows a current view of the arrangement of replication proteins at a replication fork when DNA is being synthesized. The lagging-strand DNA has been folded to bring the lagging-strand DNA polymerase molecule into a complex with the leading-strand DNA polymerase molecule. This folding also brings the 3′ end of each completed Okazaki fragment close to the start site for the next Okazaki fragment. Because the lagging-strand DNA polymerase molecule remains bound to the rest of the replication proteins, it can be reused to synthesize successive Okazaki fragments. In this diagram, it is about to let go of its completed DNA fragment and move to the RNA primer that will be synthesized nearby, as required to start the next DNA fragment. Additional proteins (not shown) help to hold the different protein components of the fork together, enabling them to function as a well-coordinated protein machine. <AATA> <CCCG> (B) An electron micrograph showing the replication machine from the bacteriophage T4 as it moves along a template synthesizing DNA behind it. (C) An interpretation of the micrograph is given in the sketch: note especially the DNA loop on the lagging strand. Apparently, the replication proteins became partly detached from the very front of the replication fork during the preparation of this sample for electron microscopy. (B, courtesy of Jack Griffith; see P.D. Chastain et al., *J. Biol. Chem.* 278:21276–21285, 2003.)

DNA. This complex can be likened to a tiny sewing machine composed of protein parts and powered by nucleoside triphosphate hydrolyses. Like a sewing machine, the replication complex probably remains stationary with respect to its immediate surroundings; the DNA can be thought of as a long strip of cloth being rapidly threaded through it. Although the replication complex has been most intensively studied in *E. coli* and several of its viruses, a very similar complex also operates in eucaryotes, as we see below.

Figure 5–19A summarizes the functions of the subunits of the replication machine. At the front of the replication fork, DNA helicase opens the DNA helix. Two DNA polymerase molecules work at the fork, one on the leading strand and one on the lagging strand. Whereas the DNA polymerase molecule on the leading strand can operate in a continuous fashion, the DNA polymerase molecule on the lagging strand must restart at short intervals, using a short RNA primer made by a DNA primase molecule. The close association of all these protein components increases the efficiency of replication and is made possible by a folding back of the lagging strand as shown in Figure 5–19A. This arrangement also facilitates the loading of the polymerase clamp each time that an Okazaki fragment is synthesized: the clamp loader and the lagging-strand DNA polymerase molecule are kept in place as a part of the protein machine even when they detach from their DNA template. The replication proteins are thus linked together into a single large unit (total molecular weight >10^6 daltons), enabling DNA to be synthesized on both sides of the replication fork in a coordinated and efficient manner.

On the lagging strand, the DNA replication machine leaves behind a series of unsealed Okazaki fragments, which still contain the RNA that primed their synthesis at their 5′ ends. This RNA is removed and the resulting gap is filled in by DNA repair enzymes that operate behind the replication fork (see Figure 5–12).

A Strand-Directed Mismatch Repair System Removes Replication Errors That Escape from the Replication Machine

As stated previously, bacteria such as *E. coli* are capable of dividing once every 40 minutes, making it relatively easy to screen large populations to find a rare mutant cell that is altered in a specific process. One interesting class of mutants contains alterations in so-called *mutator genes,* which greatly increase the rate of spontaneous mutation. Not surprisingly, one such mutant makes a defective form of the 3′-to-5′ proofreading exonuclease that is a part of the DNA polymerase enzyme (see Figures 5–8 and 5–9). The mutant DNA polymerase no longer proofreads effectively, and many replication errors that would otherwise have been removed accumulate in the DNA.

The study of other *E. coli* mutants exhibiting abnormally high mutation rates has uncovered a proofreading system that removes replication errors made by the polymerase that have been missed by the proofreading exonuclease. This **strand-directed mismatch repair** system detects the potential for distortion in the DNA helix from the misfit between noncomplementary base pairs.

If the proofreading system simply recognized a mismatch in newly replicated DNA and randomly corrected one of the two mismatched nucleotides, it would mistakenly "correct" the original template strand to match the error exactly half the time, thereby failing to lower the overall error rate. To be effective, such a proofreading system must be able to distinguish and remove the mismatched nucleotide only on the newly synthesized strand, where the replication error occurred.

The strand-distinction mechanism used by the mismatch proofreading system in *E. coli* depends on the methylation of selected A residues in the DNA. Methyl groups are added to all A residues in the sequence GATC, but not until some time after the A has been incorporated into a newly synthesized DNA chain. As a result, the only GATC sequences that have not yet been methylated are in the new strands just behind a replication fork. The recognition of these unmethylated GATCs allows the new DNA strands to be transiently distinguished from old ones, as required if their mismatches are to be selectively removed. The three-step process involves recognition of a mismatch, excision of the segment of DNA containing the mismatch from the newly synthesized strand, and resynthesis of the excised segment using the old strand as a template. This strand-directed mismatch repair system reduces the number of errors made during DNA replication by an additional factor of 100 (see Table 5–1, p. 271).

A similar mismatch proofreading system functions in human cells (**Figure 5–20**). The importance of this system in humans is seen in individuals who inherit one defective copy of a mismatch repair gene (along with a functional gene on the other copy of the chromosome). These people have a marked predisposition for certain types of cancers. For example, in a type of colon cancer called *hereditary nonpolyposis colon cancer (HNPCC)*, spontaneous mutation of the remaining functional gene produces a clone of somatic cells that, because they are deficient in mismatch proofreading, accumulate mutations unusually rapidly. Most cancers arise in cells that have accumulated multiple mutations

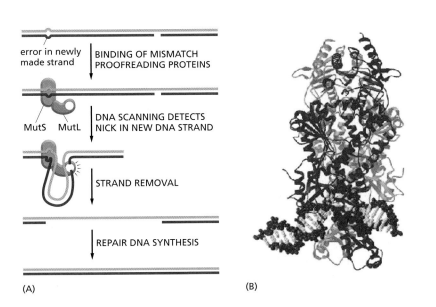

(A) (B)

Figure 5–20 A model for strand-directed mismatch repair in eucaryotes. (A) The two proteins shown are present in both bacteria and eucaryotic cells: MutS binds specifically to a mismatched base pair, while MutL scans the nearby DNA for a nick. Once MutL finds a nick, it triggers the degradation of the nicked strand all the way back through the mismatch. Because nicks are largely confined to newly replicated strands in eucaryotes, replication errors are selectively removed. In bacteria, the mechanism is the same, except that an additional protein in the complex (MutH) nicks unmethylated (and therefore newly replicated) GATC sequences, thereby beginning the process illustrated here. (B) The structure of the MutS protein bound to a DNA mismatch. This protein is a dimer, which grips the DNA double helix as shown, kinking the DNA at the mismatched base pair. It seems that the MutS protein scans the DNA for mismatches by testing for sites that can be readily kinked, which are those without a normal complementary base pair. (B, from G. Obmolova et al., *Nature* 407:703–710, 2000. With permission from Macmillan Publishers Ltd.)

(see Figure 20–11), and cells deficient in mismatch proofreading therefore have a greatly enhanced chance of becoming cancerous. Fortunately, most of us inherit two good copies of each gene that encodes a mismatch proofreading protein; this protects us, because it is highly unlikely for both copies to become mutated in the same cell.

In eucaryotes, the mechanism for distinguishing the newly synthesized strand from the parental template strand at the site of a mismatch does not depend on DNA methylation. Indeed, some eucaryotes—including yeasts and *Drosophila*—do not methylate their DNA. Newly synthesized lagging-strand DNA transiently contains *nicks* (before they are sealed by DNA ligase) and biochemical experiments reveal that such nicks (also called *single-strand breaks*) provide the signal that directs the mismatch proofreading system to the appropriate strand (see Figure 5–20). This idea also requires that the newly synthesized DNA on the leading strand be transiently nicked; how this occurs is uncertain.

DNA Topoisomerases Prevent DNA Tangling During Replication

As a replication fork moves along double-stranded DNA, it creates what has been called the "winding problem." Every 10 base pairs replicated at the fork corresponds to one complete turn about the axis of the parental double helix. Therefore, for a replication fork to move, the entire chromosome ahead of the fork would normally have to rotate rapidly (**Figure 5–21**). This would require large amounts of energy for long chromosomes, and an alternative strategy is used instead: a swivel is formed in the DNA helix by proteins known as **DNA topoisomerases**.

A DNA topoisomerase can be viewed as a reversible nuclease that adds itself covalently to a DNA backbone phosphate, thereby breaking a phosphodiester bond in a DNA strand. This reaction is reversible, and the phosphodiester bond re-forms as the protein leaves.

One type of topoisomerase, called *topoisomerase I*, produces a transient single-strand break (or nick); this break in the phosphodiester backbone allows the two sections of DNA helix on either side of the nick to rotate freely relative to each other, using the phosphodiester bond in the strand opposite the nick as a swivel point (**Figure 5–22**). Any tension in the DNA helix will drive this rotation in the direction that relieves the tension. As a result, DNA replication can occur with the rotation of only a short length of helix—the part just ahead of the fork. Because the covalent linkage that joins the DNA topoisomerase protein to a DNA phosphate retains the energy of the cleaved phosphodiester bond, resealing is rapid and does not require additional energy input. In this respect, the rejoining mechanism differs from that catalyzed by the enzyme DNA ligase, discussed previously (see Figure 5–13).

A second type of DNA topoisomerase, *topoisomerase II*, forms a covalent linkage to both strands of the DNA helix at the same time, making a transient *double-strand break* in the helix. These enzymes are activated by sites on chromosomes where two double helices cross over each other. Once a topoisomerase II molecule binds to such a crossing site, the protein uses ATP hydrolysis to perform the following set of reactions efficiently: (1) it breaks one double helix reversibly to create a DNA "gate;" (2) it causes the second, nearby double helix to pass through this break; and (3) it then reseals the break and dissociates from the DNA (**Figure 5–23**). In this way, type II DNA topoisomerases can efficiently separate two interlocked DNA circles (**Figure 5–24**).

The same reaction also prevents the severe DNA tangling problems that would otherwise arise during DNA replication. This role is nicely illustrated by

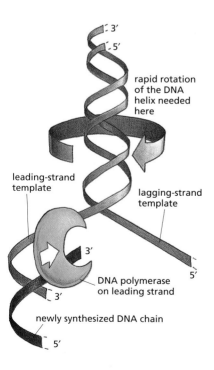

rapid rotation of the DNA helix needed here

leading-strand template

lagging-strand template

DNA polymerase on leading strand

newly synthesized DNA chain

Figure 5–21 The "winding problem" that arises during DNA replication. For a bacterial replication fork moving at 500 nucleotides per second, the parental DNA helix ahead of the fork must rotate at 50 revolutions per second.

mutant yeast cells that produce, in place of the normal topoisomerase II, a version that is inactive above 37°C. When the mutant cells are warmed to this temperature, their daughter chromosomes remain intertwined after DNA replication and are unable to separate. The enormous usefulness of topoisomerase II

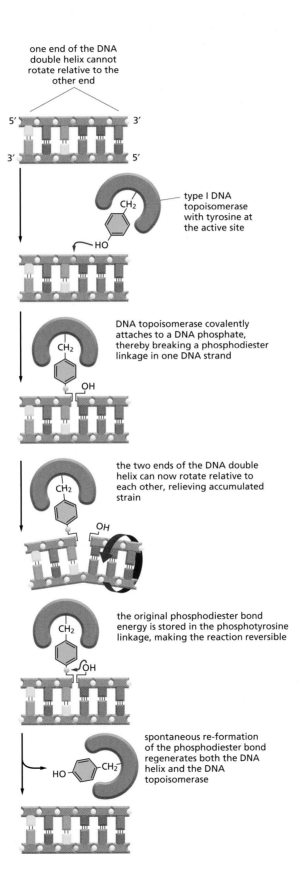

one end of the DNA double helix cannot rotate relative to the other end

type I DNA topoisomerase with tyrosine at the active site

DNA topoisomerase covalently attaches to a DNA phosphate, thereby breaking a phosphodiester linkage in one DNA strand

the two ends of the DNA double helix can now rotate relative to each other, relieving accumulated strain

the original phosphodiester bond energy is stored in the phosphotyrosine linkage, making the reaction reversible

spontaneous re-formation of the phosphodiester bond regenerates both the DNA helix and the DNA topoisomerase

Figure 5–22 **The reversible DNA nicking reaction catalyzed by a eucaryotic DNA topoisomerase I enzyme.** As indicated, these enzymes transiently form a single covalent bond with DNA; this allows free rotation of the DNA around the covalent backbone bonds linked to the *blue* phosphate.

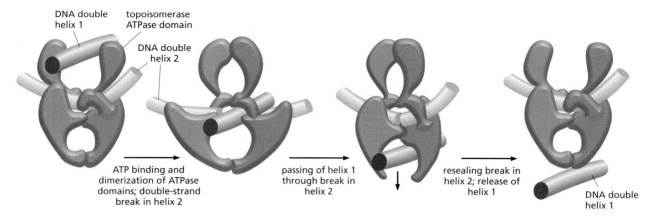

Figure 5–23 A model for topoisomerase II action. As indicated, ATP binding to the two ATPase domains causes them to dimerize and drives the reactions shown. Because a single cycle of this reaction can occur in the presence of a non-hydrolyzable ATP analog, ATP hydrolysis is thought to be needed only to reset the enzyme for each new reaction cycle. This model is based on the structure of enzyme in combination with biochemical experiments. (Modified from J.M. Berger, *Curr. Opin. Struct. Biol.* 8:26–32, 1998. With permission from Elsevier.)

for untangling chromosomes can readily be appreciated by anyone who has struggled to remove a tangle from a fishing line without the aid of scissors.

DNA Replication Is Fundamentally Similar in Eucaryotes and Bacteria

Much of what we know about DNA replication was first derived from studies of purified bacterial and bacteriophage multienzyme systems capable of DNA replication *in vitro*. The development of these systems in the 1970s was greatly facilitated by the prior isolation of mutants in a variety of replication genes; these mutants were exploited to identify and purify the corresponding replication proteins. The first mammalian replication system that accurately replicated DNA *in vitro* was described in the mid-1980s, and mutations in genes encoding nearly all of the replication components have now been isolated and analyzed in the yeast *Saccharomyces cerevisiae*. As a result, much is known about the detailed enzymology of DNA replication in eucaryotes, and it is clear that the fundamental features of DNA replication—including replication fork geometry and the use of a multiprotein replication machine—have been conserved during the long evolutionary process that separated bacteria from eucaryotes.

There are more protein components in eucaryotic replication machines than there are in the bacterial analogs, even though the basic functions are the same. Thus, for example, the eucaryotic single-strand binding (SSB) protein is formed from three subunits, whereas only a single subunit is found in bacteria. Similarly, the eucaryotic DNA primase is incorporated into a multisubunit enzyme that also contains a DNA polymerase called DNA polymerase α-primase. This protein complex begins each Okazaki fragment on the lagging strand with RNA and then extends the RNA primer with a short length of DNA. At this point, the two main eucaryotic replicative polymerases, δ and ε, come into play and complete each Okazaki fragment while simultaneously extending the leading strand. Exactly how the tasks of leading and lagging strand synthesis are distributed between these two DNA polymerases is not yet understood.

As we see in the next section, the eucaryotic replication machinery has the added complication of having to replicate through nucleosomes, the repeating structural unit of chromosomes discussed in Chapter 4. Nucleosomes are spaced at intervals of about 200 nucleotide pairs along the DNA, which may explain why new Okazaki fragments are synthesized on the lagging strand at intervals of 100–200 nucleotides in eucaryotes, instead of 1000–2000 nucleotides as in bacteria. Nucleosomes may also act as barriers that slow down the movement of DNA polymerase molecules, which may be why eucaryotic replication forks move only about one-tenth as fast as bacterial replication forks.

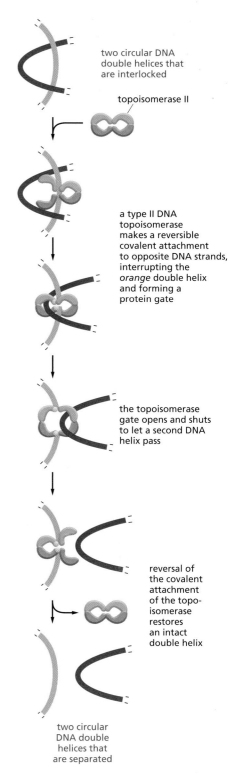

Figure 5–24 The DNA-helix-passing reaction catalyzed by DNA topoisomerase II. Identical reactions are used to untangle DNA inside the cell. Unlike type I topoisomerases, type II enzymes use ATP hydrolysis and some of the bacterial versions can introduce superhelical tension into DNA. Type II topoisomerases are largely confined to proliferating cells in eucaryotes; partly for that reason, they have been popular targets for anticancer drugs.

two circular DNA double helices that are interlocked

topoisomerase II

a type II DNA topoisomerase makes a reversible covalent attachment to opposite DNA strands, interrupting the *orange* double helix and forming a protein gate

the topoisomerase gate opens and shuts to let a second DNA helix pass

reversal of the covalent attachment of the topo-isomerase restores an intact double helix

two circular DNA double helices that are separated

Summary

DNA replication takes place at a Y-shaped structure called a replication fork. A self-correcting DNA polymerase enzyme catalyzes nucleotide polymerization in a 5′-to-3′ direction, copying a DNA template strand with remarkable fidelity. Since the two strands of a DNA double helix are antiparallel, this 5′-to-3′ DNA synthesis can take place continuously on only one of the strands at a replication fork (the leading strand). On the lagging strand, short DNA fragments must be made by a "backstitching" process. Because the self-correcting DNA polymerase cannot start a new chain, these lagging-strand DNA fragments are primed by short RNA primer molecules that are subsequently erased and replaced with DNA.

DNA replication requires the cooperation of many proteins. These include (1) DNA polymerase and DNA primase to catalyze nucleoside triphosphate polymerization; (2) DNA helicases and single-strand DNA-binding (SSB) proteins to help in opening up the DNA helix so that it can be copied; (3) DNA ligase and an enzyme that degrades RNA primers to seal together the discontinuously synthesized lagging-strand DNA fragments; and (4) DNA topoisomerases to help to relieve helical winding and DNA tangling problems. Many of these proteins associate with each other at a replication fork to form a highly efficient "replication machine," through which the activities and spatial movements of the individual components are coordinated.

THE INITIATION AND COMPLETION OF DNA REPLICATION IN CHROMOSOMES

We have seen how a set of replication proteins rapidly and accurately generates two daughter DNA double helices behind a replication fork. But how is this replication machinery assembled in the first place, and how are replication forks created on a double-stranded DNA molecule? In this section, we discuss how cells initiate DNA replication and how they carefully regulate this process to ensure that it takes place not only at the proper positions on the chromosome but also at the appropriate time in the life of the cell. We also discuss a few of the special problems that the replication machinery in eucaryotic cells must overcome. These include the need to replicate the enormously long DNA molecules found in eucaryotic chromosomes, as well as the difficulty of copying DNA molecules that are tightly complexed with histones in nucleosomes.

DNA Synthesis Begins at Replication Origins

As discussed previously, the DNA double helix is normally very stable: the two DNA strands are locked together firmly by many hydrogen bonds formed between the bases on each strand. To be used as a template, the double helix must be opened up and the two strands separated to expose unpaired bases. As we shall see, the process of DNA replication is begun by special *initiator proteins* that bind to double-stranded DNA and pry the two strands apart, breaking the hydrogen bonds between the bases.

The positions at which the DNA helix is first opened are called **replication origins** (**Figure 5–25**). In simple cells like those of bacteria or yeast, origins are specified by DNA sequences several hundred nucleotide pairs in length. This DNA contains both short sequences that attract initiator proteins and stretches of DNA that are especially easy to open. We saw in Figure 4–4 that an A-T base pair is held together by fewer hydrogen bonds than a G-C base pair. Therefore, DNA rich in A-T base pairs is relatively easy to pull apart, and regions of DNA enriched in A-T pairs are typically found at replication origins.

Although the basic process of replication fork initiation depicted in Figure 5–25 is fundamentally the same for bacteria and eucaryotes, the detailed way in which this process is performed and regulated differs between these two groups of organisms. We first consider the simpler and better-understood case in bacteria and then turn to the more complex situation found in yeasts, mammals, and other eucaryotes.

Bacterial Chromosomes Typically Have a Single Origin of DNA Replication

The genome of *E. coli* is contained in a single circular DNA molecule of 4.6×10^6 nucleotide pairs. DNA replication begins at a single origin of replication, and the two replication forks assembled there proceed (at approximately 500–1000 nucleotides per second) in opposite directions until they meet up roughly halfway around the chromosome (**Figure 5–26**). The only point at which *E. coli* can control DNA replication is initiation: once the forks have been assembled at the origin, they synthesize DNA at relatively constant speed until replication is finished. Therefore, it is not surprising that the initiation of DNA replication is highly regulated. The process begins when initiator proteins bind in multiple copies to specific sites in the replication origin, wrapping the DNA around the proteins to form a large protein–DNA complex. This complex then attracts a DNA helicase bound to a helix loader, and the helicase is placed around an adjacent DNA single strand whose bases have been exposed by the assembly of the initiator protein–DNA complex. The helicase loader is analogous to the clamp loader we encountered above; it has the additional job of keeping the helicase in an inactive form until it is properly loaded onto a nascent replication fork. Once the helicase is loaded, it begins to unwind DNA, exposing enough single-stranded DNA for primase to synthesize the RNA primer that begins the leading strand (**Figure 5–27**). This quickly leads to the assembly of the remaining proteins to create two replication forks, with protein complexes that move, with respect to the replication origin, in opposite directions. These protein machines continue to synthesize DNA until all of the DNA template downstream of each fork has been replicated.

In *E. coli*, the interaction of the initiator protein with the replication origin is carefully regulated, with initiation occurring only when sufficient nutrients are available for the bacterium to complete an entire round of replication. Not only is the activity of the initiator protein controlled, but an origin of replication that has just been used experiences a "refractory period," caused by a delay in the methylation of newly synthesized A nucleotides. Further initiation of replication is blocked until these As are methylated (**Figure 5–28**).

Eucaryotic Chromosomes Contain Multiple Origins of Replication

We have seen how two replication forks begin at a single replication origin in bacteria and proceed in opposite directions, moving away from the origin until all of the DNA in the single circular chromosome is replicated. The bacterial genome is sufficiently small for these two replication forks to duplicate the genome in about 40 minutes. Because of the much greater size of most eucaryotic chromosomes, a different strategy is required to allow their replication in a timely manner.

A method for determining the general pattern of eucaryotic chromosome replication was developed in the early 1960s. Human cells growing in culture are labeled for a short time with ³H-thymidine so that the DNA synthesized during this period becomes highly radioactive. The cells are then gently lysed, and the DNA is streaked on the surface of a glass slide coated with a photographic emulsion. Development of the emulsion reveals the pattern of labeled DNA through

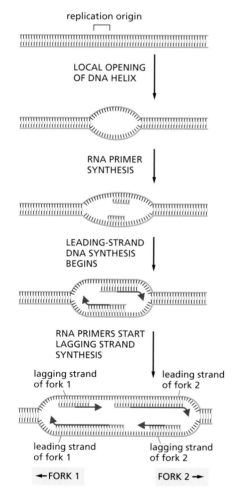

Figure 5–25 A replication bubble formed by replication fork initiation. This diagram outlines the major steps in the initiation of replication forks at replication origins. The structure formed at the last step, in which both strands of the parental DNA helix have been separated from each other and serve as templates for DNA synthesis, is called a *replication bubble*.

Figure 5–26 DNA replication of a bacterial genome. It takes *E. coli* about 40 minutes to duplicate its genome of 4.6×10^6 nucleotide pairs. For simplicity, no Okazaki fragments are shown on the lagging strand. What happens as the two replication forks approach each other and collide at the end of the replication cycle is not well understood, although the replication machines are disassembled as part of the process.

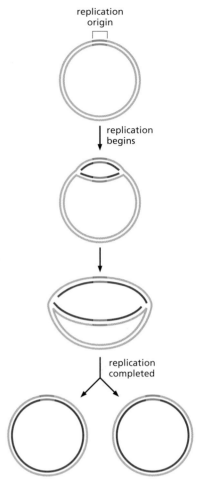

a technique known as *autoradiography*. The time allotted for radioactive labeling is chosen to allow each replication fork to move several micrometers along the DNA, so that the replicated DNA can be detected in the light microscope as lines of silver grains, even though the DNA molecule itself is too thin to be visible. In this way, both the rate and the direction of replication-fork movement can be determined (**Figure 5–29**). From the rate at which tracks of replicated DNA increase in length with increasing labeling time, the replication forks are estimated to travel at about 50 nucleotides per second. This is approximately one-tenth of the rate at which bacterial replication forks move, possibly reflecting the increased difficulty of replicating DNA that is packaged tightly in chromatin.

An average-size human chromosome contains a single linear DNA molecule of about 150 million nucleotide pairs. It would take 0.02 seconds/nucleotide × 150×10^6 nucleotides = 3.0×10^6 seconds (about 800 hours) to replicate such a DNA molecule from end to end with a single replication fork moving at a rate of 50 nucleotides per second. As expected, therefore, the autoradiographic experiments just described reveal that many forks are moving simultaneously on each eucaryotic chromosome.

Further experiments of this type have shown the following: (1) Replication origins tend to be activated in clusters, called *replication units*, of perhaps 20–80

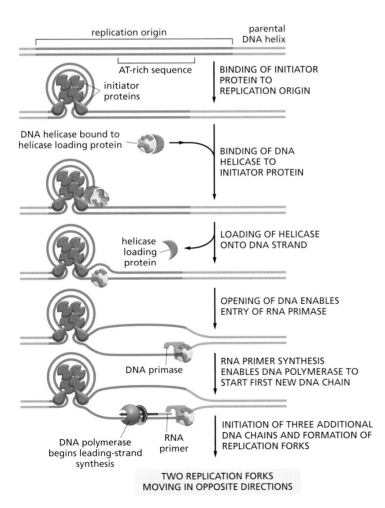

Figure 5–27 The proteins that initiate DNA replication in bacteria. The mechanism shown was established by studies *in vitro* with mixtures of highly purified proteins. For *E. coli* DNA replication, the major initiator protein, the helicase, and the primase are the DnaA, DnaB, and DnaG proteins, respectively. In the first step, several molecules of the initiator protein bind to specific DNA sequences in the replication origin and form a compact structure in which the DNA is wrapped around the protein. Next, the helicase is brought in by a helicase loading protein (the DnaC protein), which inhibits the helicase until it is properly loaded at the replication origin. The helicase loading protein thereby prevents the helicase from inappropriately entering other single-stranded stretches of DNA in the bacterial genome. Aided by single-strand binding protein (not shown), the loaded helicase opens up the DNA thereby enabling primase to enter and synthesize the primer for the first DNA chain. Subsequent steps (not shown) result in the initiation of three additional DNA chains and the final assembly of two complete replication forks.

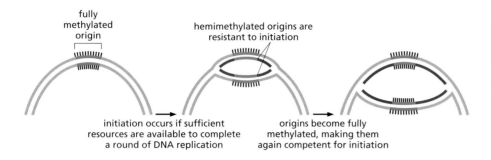

Figure 5–28 Methylation of the *E. coli* replication origin creates a refractory period for DNA initiation. DNA methylation occurs at GATC sequences, 11 of which are found in the origin of replication (spanning approximately 250 nucleotide pairs). In its hemimethylated state, the origin of replication is bound by an inhibitor protein (Seq A, not shown), which blocks access of the origin to initiator proteins. Eventually (about 20 minutes after replication is initiated), the hemimethylated origins become fully methylated by a DNA methylase enzyme; Seq A then dissociates.

A single enzyme, the *Dam* methylase, is responsible for methylating all *E. coli* GATC sequences. A lag in methylation after the replication of GATC sequences is also used by the *E. coli* mismatch proofreading system to distinguish the newly synthesized DNA strand from the parental DNA strand; in that case, the relevant GATC sequences are scattered throughout the chromosome, and they are not bound by Seq A.

origins. (2) New replication units seem to be activated at different times during the cell cycle until all of the DNA is replicated, a point that we return to below. (3) Within a replication unit, individual origins are spaced at intervals of 30,000–250,000 nucleotide pairs from one another. (4) As in bacteria, replication forks are formed in pairs and create a replication bubble as they move in opposite directions away from a common point of origin, stopping only when they collide head-on with a replication fork moving in the opposite direction (or when they reach a chromosome end). In this way, many replication forks operate independently on each chromosome and yet form two complete daughter DNA helices.

In Eucaryotes DNA Replication Takes Place During Only One Part of the Cell Cycle

When growing rapidly, bacteria replicate their DNA continually, and they can begin a new round before the previous one is complete. In contrast, DNA replication in most eucaryotic cells occurs only during a specific part of the cell division cycle, called the *DNA synthesis phase* or **S phase** (**Figure 5–30**). In a mammalian cell, the S phase typically lasts for about 8 hours; in simpler eucaryotic cells such as yeasts, the S phase can be as short as 40 minutes. By its end, each chromosome has been replicated to produce two complete copies, which remain joined together at their centromeres until the *M phase* (M for *mitosis*), which soon follows. In Chapter 17, we describe the control system that runs the cell cycle and explain why entry into each phase of the cycle requires the cell to have successfully completed the previous phase.

In the following sections, we explore how chromosome replication is coordinated within the S phase of the cell cycle.

Figure 5–29 The experiments that demonstrated the pattern in which replication forks are formed and move on eucaryotic chromosomes. The new DNA made in human cells in culture was labeled briefly with a pulse of highly radioactive thymidine (^{3}H-thymidine). (A) In this experiment, the cells were lysed, and the DNA was stretched out on a glass slide that was subsequently covered with a photographic emulsion. After several months the emulsion was developed, revealing a line of silver grains over the radioactive DNA. The *brown* DNA in this figure is shown only to help with the interpretation of the autoradiograph; the unlabeled DNA is invisible in such experiments. (B) This experiment was the same except that a further incubation in unlabeled medium allowed additional DNA, with a lower level of radioactivity, to be replicated. The pairs of dark tracks in (B) were found to have silver grains tapering off in opposite directions, demonstrating bidirectional fork movement from a central replication origin where a replication bubble forms (see Figure 5–25). A replication fork is thought to stop only when it encounters a replication fork moving in the opposite direction or when it reaches the end of the chromosome; in this way, all the DNA is eventually replicated.

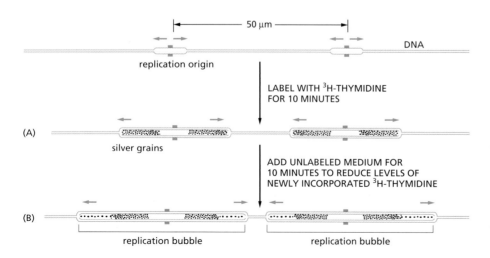

Figure 5–30 **The four successive phases of a standard eucaryotic cell cycle.** During the G_1, S, and G_2 phases, the cell grows continuously. During M phase growth stops, the nucleus divides, and the cell divides in two. DNA replication is confined to the part of the cell cycle known as S phase. G_1 is the gap between M phase and S phase; G_2 is the gap between S phase and M phase.

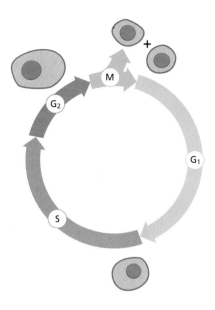

Different Regions on the Same Chromosome Replicate at Distinct Times in S Phase

In mammalian cells, the replication of DNA in the region between one replication origin and the next should normally require only about an hour to complete, given the rate at which a replication fork moves and the largest distances measured between replication origins. Yet S phase usually lasts for about 8 hours in a mammalian cell. This implies that the replication origins are not all activated simultaneously and that the DNA in each replication unit (which, as we noted above, contains a cluster of about 20–80 replication origins) is replicated during only a small part of the total S-phase interval.

Are different replication units activated at random, or are different regions of the genome replicated in a specified order? One way to answer this question is to use the thymidine analog bromodeoxyuridine (BrdU) to label the newly synthesized DNA in synchronized cell populations, adding it for different short periods throughout S phase. Later, during M phase, those regions of the mitotic chromosomes that have incorporated BrdU into their DNA can be recognized by their altered staining properties or by means of anti-BrdU antibodies. The results show that different regions of each chromosome are replicated in a reproducible order during S phase (**Figure 5–31**). Moreover, as one would expect from the clusters of replication forks seen in DNA autoradiographs (see Figure 5–29), the timing of replication is coordinated over large regions of the chromosome.

Much more sophisticated methods now exist for monitoring DNA replication initiation and tracking the movement of DNA replication forks in cells. These approaches use DNA microarrays—grids the size of a postage stamp studded with tens of thousands of fragments of known DNA sequence. As we will see in detail in Chapter 8, each different DNA fragment is placed at a unique position on the microarray, and whole genomes can thereby be represented in an orderly manner. If a DNA sample from a group of cells in S phase is broken up and hybridized to a microarray representing that organism's genome, the amount of each DNA sequence can be determined. Because a segment of a genome that has been replicated will contain twice as much DNA as an unreplicated segment, replication fork initiation and fork movement can be accurately monitored in this way (**Figure 5–32**). Although this method provides much greater precision, it leads to many of the same conclusions reached from the earlier studies.

Highly Condensed Chromatin Replicates Late, While Genes in Less Condensed Chromatin Tend to Replicate Early

It seems that the order in which replication origins are activated depends, in part, on the chromatin structure in which the origins reside. We saw in Chapter 4 that heterochromatin is a particularly condensed state of chromatin, while euchromatin chromatin has a less condensed conformation that is apparently required to allow transcription. Heterochromatin tends to be replicated very late in S phase, suggesting that the timing of replication is related to the packing of the DNA in chromatin. This suggestion is supported by an examination of the two X chromosomes in a female mammalian cell. While these two chromosomes contain essentially the same DNA sequences, one is active for DNA transcription and the other is not (discussed in Chapter 7). Nearly all of the inactive X chromosome is condensed into heterochromatin, and its DNA replicates late in S phase. Its active homolog is less condensed and replicates throughout S phase.

early S 0–2 hours	middle S 3–5 hours	late S 6–8 hours

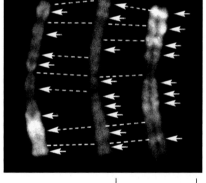

5 μm

Figure 5–31 **Different regions of a chromosome are replicated at different times in S phase.** These light micrographs show stained mitotic chromosomes in which the replicating DNA has been differentially labeled during different defined intervals of the preceding S phase. In these experiments, cells were first grown in the presence of BrdU (a thymidine analog) and absence of thymidine to label the DNA uniformly. The cells were then briefly pulsed with thymidine in the absence of BrdU during early, middle, or late S phase. Because the DNA made during the thymidine pulse is a double helix with thymidine on one strand and BrdU on the other, it stains more darkly than the remaining DNA (which has BrdU on both strands) and shows up as a bright band *(arrows)* on these negatives. Broken lines connect corresponding positions on the three identical copies of the chromosome shown. (Courtesy of Elton Stubblefield.)

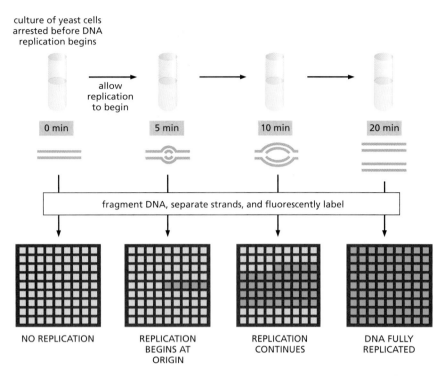

culture of yeast cells
arrested before DNA
replication begins

allow
replication
to begin

0 min   5 min   10 min   20 min

fragment DNA, separate strands, and fluorescently label

NO REPLICATION | REPLICATION BEGINS AT ORIGIN | REPLICATION CONTINUES | DNA FULLY REPLICATED

Figure 5–32 Use of DNA microarrays to monitor the formation and progress of replication forks in the budding yeast genome. For this experiment, a population of cells is sychronized so that they all begin replication at the same time. DNA is collected and hybridized to the microarray; DNA that has been replicated once gives a hybridization signal *(dark green squares)* twice as high as that of unreplicated DNA *(light green squares)*. The spots on these microarrays represent consecutive sequences along a segment of a yeast chromosome arranged left to right, top to bottom. Only 81 spots are shown here, but the actual arrays contain tens of thousands of sequences that span the entire yeast genome. As can be seen, replication begins at an origin and proceeds bidirectionally. For simplicity only one origin is shown here. In yeast cells, replication begins at hundreds of origins located throughout the genome.

These findings suggest that regions of the genome whose chromatin is least condensed are replicated first. Replication forks seem to move at comparable rates throughout S phase, so the extent of chromosome condensation seems to influence the time at which replication forks are initiated, rather than their speed once formed.

Well-defined DNA Sequences Serve as Replication Origins in a Simple Eucaryote, the Budding Yeast

Having seen that a eucaryotic chromosome is replicated using many origins of replication, each of which "fires" at a characteristic time in S phase of the cell cycle, we turn to the nature of these origins of replication. We saw earlier in this chapter that replication origins have been precisely defined in bacteria as specific DNA sequences that attract initiator proteins, which then assemble the DNA replication machinery. By analogy, one would expect the replication origins in eucaryotic chromosomes to be specific DNA sequences too.

The search for eucaryotic DNA sequences that carry all the information necessary to specify a replication origin has been most productive in the budding yeast *S. cerevisiae*. Powerful selection methods to find them have been devised that make use of mutant yeast cells defective for an essential gene. These cells can survive in a selective medium only if they are provided with DNA that carries a functional copy of the missing gene. If a circular bacterial plasmid containing this gene is introduced into the mutant yeast cells directly, the plasmid will not be able to replicate because it lacks a functional origin. If random pieces of yeast DNA are inserted into this plasmid, however, only the rare plasmid DNA molecules that happen to contain a yeast replication origin can replicate. The yeast cells that carry such plasmids are able to proliferate because they have been provided with the essential gene in a form that can be replicated and passed on to progeny cells (**Figure 5–33**). A DNA sequence identified by its presence in a plasmid isolated from these surviving yeast cells is called an *autonomously replicating sequence (ARS)*. Most ARSs have been shown to be authentic chromosomal origins of replication, thereby validating the strategy used to obtain them.

For budding yeast, the location of every origin of replication on each chromosome has been determined. The particular chromosome shown in **Figure 5–34**—chromosome III from the yeast *S. cerevisiae*—is one of the smallest

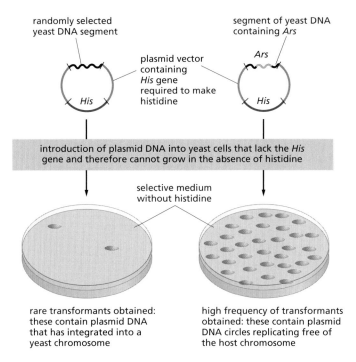

Figure 5–33 A strategy used to identify DNA sequences that are sufficient for initiating DNA replication. Each of the yeast DNA sequences identified in this way was called an autonomously replicating sequence *(Ars)*, since it enables a plasmid that contains it to replicate in the host cell without having to be incorporated into a host cell chromosome.

chromosomes known, with a length less than 1/100 that of a typical human chromosome. Its major origins are spaced an average of 30,000 nucleotide pairs apart; this density of origins should permit this chromosome to be replicated in about 10 minutes if all the origins fire at once. As we have previously discussed, mammalian origins are spaced further apart, typically every 100,000 to 250,000 nucleotide pairs.

Genetic experiments in *S. cerevisiae* have tested the effect of deleting various replication origins on chromosome III. Removing a few origins has little effect, because replication forks that begin at neighboring origins of replication can continue into the regions that lack their own origins. The deletion of more replication origins, however, results in the loss of the chromosome as the cells divide, because it is replicated too slowly. Many eucaryotes carry an excess of origins of replication, presumably to ensure that the complete genome can still be replicated in a timely manner if some of the origins fail to function.

A Large Multisubunit Complex Binds to Eucaryotic Origins of Replication

The minimal DNA sequence required for directing DNA replication initiation in the yeast *S. cerevisiae* has been determined by testing smaller and smaller DNA fragments in the experiment shown in Figure 5–33. Each DNA sequence that can serve as an origin of replication is found to contain (1) a binding site for a large, multisubunit initiator protein called **ORC**, for **origin recognition complex**, (2) a stretch of DNA that is rich in As and Ts and therefore easy to unwind, and (3) at least one binding site for proteins that help attract ORC to the origin DNA (**Figure 5–35**). In bacteria, once the initiator protein is properly bound to the single origin of replication the assembly of the replication forks follows more or less automatically. In eucaryotes, the situation is significantly different because of a profound problem eucaryotes have in replicating chromosomes with so many origins of replication (an estimated 400 in yeast and 10,000 in

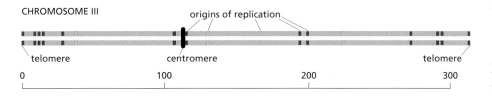

Figure 5–34 The origins of DNA replication on chromosome III of the yeast *S. cerevisiae*. This chromosome, one of the smallest eucaryotic chromosomes known, carries a total of 180 genes. As indicated, it contains 19 replication origins, although they are used with different efficiencies. Those in *red* are typically used less than 10% of the time, while those in *green* are used in 90% of S phases.

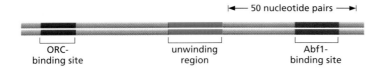

Figure 5–35 An origin of replication in yeast. Comprising about 150 nucleotide pairs, this yeast origin (identified by the procedure shown in Figure 5–33) has a binding site for ORC, and one for Abf1, an auxiliary protein that facilitates ORC binding. All origins contain binding sites for ORC but the auxiliary proteins are different from one origin to the next. Most origins, like the one depicted, also contain a stretch of DNA that is especially easy to unwind.

humans, for example). With so many places to begin replication, how is the process regulated to ensure that all the DNA is copied once and only once?

The answer lies in the way that the ORC complex, once bound to an origin of replication, is sequentially activated and deactivated. This matter is discussed in detail in Chapter 17, when we consider the cellular machinery that underlies the cell division cycle. The ORC–origin interaction persists throughout the entire cell cycle, dissociating only briefly immediately following replication of the origin DNA, and other proteins that bind to it regulate origin activity. These include the DNA helicase and two helicase loading proteins, Cdc6 and Cdt1, which are assembled onto an ORC–DNA complex to form a *prereplicative complex* at each origin during G_1 phase (**Figure 5–36**). The passage of a cell from G_1 to S phase is triggered by the activation of protein kinases (Cdks) that lead to dissociation of the helicase loading proteins, activation of the helicase, unwinding of the origin DNA, and loading of the remaining replication proteins including DNA polymerases (see Figure 5–36).

The protein kinases that trigger DNA replication simultaneously prevent all assembly of new prereplicative complexes until the next M phase resets the entire cycle (for details, see pp. 1067–1069). This strategy provides a single window of opportunity for prereplicative complexes to form (G_1 phase, when Cdk activity is low) and a second window for them to be activated and subsequently disassembled (S phase, when Cdk activity is high). Because these two phases of the cell cycle are mutually exclusive and occur in a prescribed order, each origin of replication can fire once and only once a cell cycle.

The Mammalian DNA Sequences That Specify the Initiation of Replication Have Been Difficult to Identify

Compared with the situation in budding yeasts, DNA sequences that specify replication origins in other eucaryotes have been more difficult to define. Recently, however, it has been possible to identify specific human DNA sequences, each several thousand nucleotide pairs in length, that are sufficient to serve as replication origins. These origins continue to function when moved to a different chromosomal region by recombinant DNA methods, as long as they are placed in a region where the chromatin is relatively uncondensed. One of these origins is from the β-globin gene cluster. At its normal position in the genome, the function of this origin depends critically upon distant DNA sequences (**Figure 5–37**). As discussed in Chapter 7, this distant DNA is needed for expression of all genes in the β-globin cluster, and its effects on both transcription and origin function may reflect its long-range decondensation of chromatin structure.

A human ORC homologous to that in yeast cells is required for replication initiation. Many of the other proteins that function in the initiation process in yeast likewise have central roles in humans. It therefore seems likely that the yeast and human initiation mechanisms will turn out to be similar in outline. However, the binding sites for the ORC protein seem to be less specific in humans than they are in yeast, which may explain why the replication origins of humans are less sharply defined. In fact, chromatin structure, rather than DNA sequences, may have the central role in defining mammalian origins of replication. Thus, as is true in many other areas of cell biology, the mechanism of DNA replication initiation in yeast may vividly highlight the core process, while the situation in humans represents an elaborate variation on the theme.

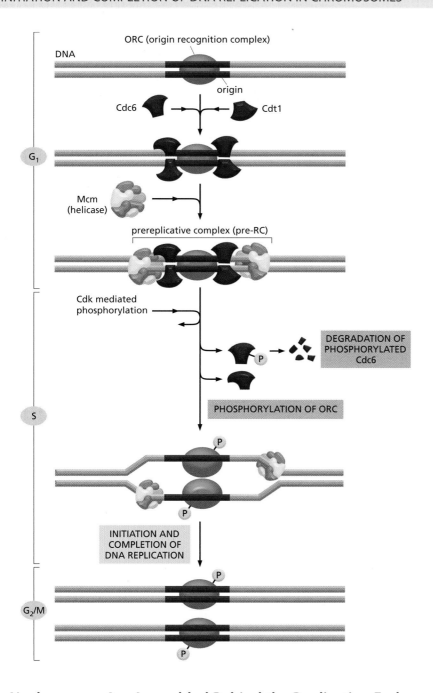

Figure 5–36 The mechanism of DNA replication initiation in eucaryotes. This mechanism ensures that each origin of replication is activated only once per cell cycle. An origin of replication can be used only if a prereplicative complex forms there in G_1 phase. At the beginning of S phase, cyclin-dependent kinases (Cdks) phosphorylate various replication proteins, causing both disassembly of the prereplicative complex and initiation of DNA replication. A new prereplicative complex cannot form at the origin until the cell progresses to the next G_1 phase.

New Nucleosomes Are Assembled Behind the Replication Fork

There are several additional aspects of DNA replication that are specific to eucaryotes. As discussed in Chapter 4, eukaryotic chromosomes are composed of roughly equal mixtures of DNA and protein. Chromosome duplication therefore requires not only the replication of DNA, but also the synthesis and assembly of new chromosomal proteins onto the DNA behind each replication fork. Although we are far from understanding this process in detail, we are beginning to learn how the fundamental unit of chromatin packaging, the nucleosome, is duplicated. The cell requires a large amount of new histone protein, approximately equal in mass to the newly synthesized DNA, to make the new nucleosomes in each cell cycle. For this reason, most eucaryotic organisms possess multiple copies of the gene for each histone. Vertebrate cells, for example, have about 20 repeated gene sets, most containing the genes that encode all five histones (H1, H2A, H2B, H3, and H4).

Unlike most proteins, which are made continuously throughout interphase, histones are synthesized mainly in S phase, when the level of histone mRNA increases about fiftyfold as a result of both increased transcription and

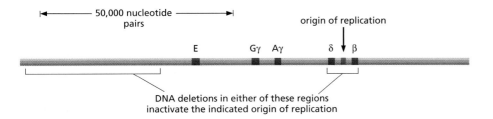

DNA deletions in either of these regions
inactivate the indicated origin of replication

Figure 5–37 Deletions that inactivate an origin of replication in humans. These two deletions are found separately in two individuals who suffer from *thalassemia,* a disorder caused by the failure to express one or more of the genes in the β-globin gene cluster shown. In both of these deletion mutants, the DNA in this region is replicated by forks that begin at replication origins outside the β-globin gene cluster.

decreased mRNA degradation. The major histone mRNAs are degraded within minutes when DNA synthesis stops at the end of S phase. The mechanism depends on special properties of the 3′ ends of these mRNAs, as discussed in Chapter 7. In contrast, the histone proteins themselves are remarkably stable and may survive for the entire life of a cell. The tight linkage between DNA synthesis and histone synthesis appears to reflect a feedback mechanism that monitors the level of free histone to ensure that the amount of histone made exactly matches the amount of new DNA synthesized.

As a replication fork advances, it must somehow pass through the parental nucleosomes. *In vitro* studies show that the replication apparatus has a poorly understood intrinsic ability to pass through parental nucleosomes without displacing them from the DNA. However, to replicate chromosomes efficiently in the cell, chromatin-remodeling proteins (discussed in Chapter 4), which destabilize the DNA–histone interface, are required. Aided by such complexes, replication forks can transit even highly condensed heterochromatin efficiently.

As the replication fork passes through chromatin, most of the old histones remain DNA-bound and are distributed to the daughter DNA helices behind a replication fork. But since the amount of DNA has doubled, an equal amount of new histones is also needed to complete the packaging of DNA into chromatin. The old and the new histones are combined in an intriguing way. When a nucleosome is transited by a replication fork, the histone octamer appears to be broken into an H3-H4 tetramer and two H2A-H2B dimers (see Figure 4–26). The H3-H4 tetramer remains associated with DNA and is distributed at random to one or the other daughter duplexes, but the H2A-H2B dimers are released from DNA. Freshly made H3-H4 tetramers are added to the newly synthesized DNA to fill in the "spaces," and H2AB dimers—half of which are old and half new—are then added at random to complete the nucleosomes (**Figure 5–38**).

The orderly and rapid addition of new H3-H4 tetramers and H2A-H2B dimers behind a replication fork requires **histone chaperones** (also called *chromatin assembly factors*). These multisubunit complexes bind the highly basic histones and release them for assembly only in the appropriate context. These histone chaperones, along with their cargoes, are directed to newly replicated DNA through a specific interaction with the eucaryotic sliding clamp, called *PCNA* (see Figure 5–38B). These clamps are left behind moving replication forks and remain on the DNA long enough for the histone chaperones to complete their tasks.

The Mechanisms of Eucaryotic Chromosome Duplication Ensure That Patterns of Histone Modification Can Be Inherited

We saw in Chapter 4 that histones are subject to many types of covalent modifications and that the patterns of these modifications can carry important information regarding the fate of underlying DNA. It makes little intuitive sense for these patterns to be erased each time a cell divides, but since this information is encoded in the histone proteins rather than the DNA, special mechansims are needed to preserve and duplicate it. We have seen that histone H3-H4 tetramers are distributed randomly to the two daughter chromosomes that emerge behind a moving replication fork. The tails as well as other regions of H3 and H4 can be extensively modified (see Figure 4–39), and thus each daughter chromosome is seeded with the memory of the parental pattern of H3 and H4 modification.

(A)

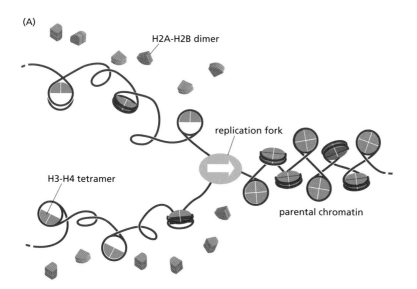

(B)

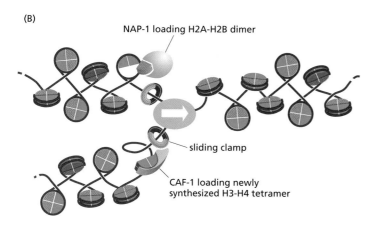

Figure 5–38 Distribution of parental and newly synthesized histones behind a eucaryotic replication fork. (A) The distribution of parental H3-H4 tetramers to the daughter DNA molecules is apparently random, with roughly equal numbers inherited by each daughter. In contrast, H2A-H2B dimers are released from the DNA as the replication fork passes. (B) Histone chaperones (NAP1 and CAF1) restore the full complement of histones to daughter molecules. Although some daughter nucleosomes contain only parental histones or only newly synthesized histones, most are hybrids of old and new. (Adapted from J.D. Watson et al., Molecular Biology of the Gene, 5th ed. Cold Spring Harbor: Cold Spring Harbor Laboratory Press, 2004.)

Once the nucleosome assembly behind a replication fork has been completed, the parental patterns of H3-H4 modification can be reinforced through histone modification enzymes in reader–writer complexes that recognize the same type of modification they create (**Figure 5–39**).

The faithful duplication of patterns of histone modification may be responsible for many examples of *epigenetic inheritance*, in which a heritable change in a cell's phenotype occurs without a change in the nucleotide sequence of DNA. We shall revisit the topic of epigenetics in Chapter 7 when we consider how decisions made by a cell are "remembered" by its progeny cells many generations later.

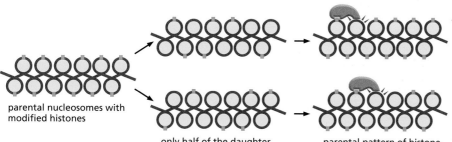

Figure 5–39 Strategy through which parental patterns of histone H3 and H4 modification can be inherited by daughter chromosomes. Although it is unlikely that this mechanism applies to all histone modifications, it does pertain to some (see Figure 4–51). For example, a number of histone methylase complexes specifically recognize N-terminal histone tails that have been previously methylated at the same site that the methylase modifies.

Telomerase Replicates the Ends of Chromosomes

We saw earlier that synthesis of the lagging strand at a replication fork must occur discontinuously through a backstitching mechanism that produces short DNA fragments. This mechanism encounters a special problem when the replication fork reaches an end of a linear chromosome: there is no place to produce the RNA primer needed to start the last Okazaki fragment at the very tip of a linear DNA molecule.

Bacteria solve this "end-replication" problem by having circular DNA molecules as chromosomes (see Figure 5–27). Eucaryotes solve it in an ingenious way: they have specialized nucleotide sequences at the ends of their chromosomes that are incorporated into structures called telomeres (see Chapter 4). Telomeres contain many tandem repeats of a short sequence that is similar in organisms as diverse as protozoa, fungi, plants, and mammals. In humans, the sequence of the repeat unit is GGGTTA, and it is repeated roughly a thousand times at each telomere.

Telomere DNA sequences are recognized by sequence-specific DNA-binding proteins that attract an enzyme, called **telomerase**, that replenishes these sequences each time a cell divides. Telomerase recognizes the tip of an existing telomere DNA repeat sequence and elongates it in the 5′-to-3′ direction, using an RNA template that is a component of the enzyme itself to synthesize new copies of the repeat (**Figure 5–40**). The enzymatic portion of telomerase resembles other *reverse transcriptases*, enzymes that synthesize DNA using an RNA template (see Figure 5–72). After extension of the parental DNA strand by telomerase, replication of the lagging strand at the chromosome end can be completed by the conventional DNA polymerases, using these extensions as a template to synthesize the complementary strand (**Figure 5–41**).

The mechanism just described, aided by a nuclease that chews back the 5′ end, ensures that the 3′ DNA end at each telomere is always longer than the 5′ end with which it is paired, leaving a protruding single-stranded end (see Figure 5–41). This protruding end has been shown to loop back and tuck its single-stranded terminus into the duplex DNA of the telomeric repeat sequence to form a *t-loop* (**Figure 5–42**). In broad outline this reaction resembles strand invasion during homologous recombination, discussed below, and it may have evolved from these ancient recombination systems. T-loops provide the normal ends of chromosomes with a unique structure, which protects them from degradative enzymes and clearly distinguishes them from the ends of the broken DNA molecules that the cell rapidly repairs (see Figure 5–51).

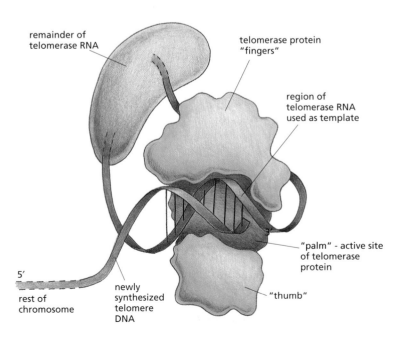

remainder of telomerase RNA

telomerase protein "fingers"

region of telomerase RNA used as template

"palm" - active site of telomerase protein

"thumb"

5′

rest of chromosome

newly synthesized telomere DNA

Figure 5–40 The structure of a portion of telomerase. Telomerase is a large protein–RNA complex. The RNA *(blue)* contains a templating sequence for synthesizing new DNA telomere repeats. The synthesis reaction itself is carried out by the reverse transcriptase domain of the protein, shown in *green*. A reverse transcriptase is a special form of polymerase enzyme that uses an RNA template to make a DNA strand; telomerase is unique in carrying its own RNA template with it at all times. Telomerase also has several additional protein domains (not shown) that are needed to assemble the enzyme at the ends of chromosomes properly. (Modified from J. Lingner and T.R. Cech, *Curr. Opin. Genet. Dev.* 8:226–232, 1998. With permission from Elsevier.)

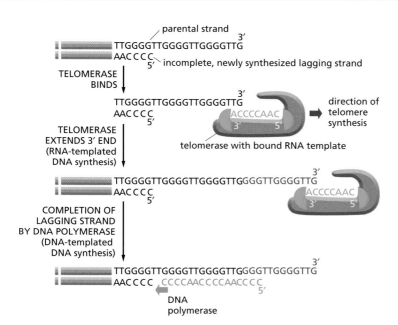

parental strand

TELOMERASE BINDS

TELOMERASE EXTENDS 3′ END (RNA-templated DNA synthesis)

telomerase with bound RNA template

direction of telomere synthesis

COMPLETION OF LAGGING STRAND BY DNA POLYMERASE (DNA-templated DNA synthesis)

DNA polymerase

Figure 5–41 Telomere replication. Shown here are the reactions that synthesize the repeating G-rich sequences that form the ends of the chromosomes (telomeres) of diverse eucaryotic organisms. The 3′ end of the parental DNA strand is extended by RNA-templated DNA synthesis; this allows the incomplete daughter DNA strand that is paired with it to be extended in its 5′ direction. This incomplete, lagging strand is presumed to be completed by DNA polymerase α, which carries a DNA primase as one of its subunits. The telomere sequence illustrated is that of the ciliate *Tetrahymena,* in which these reactions were first discovered.

Telomere Length Is Regulated by Cells and Organisms

Because the processes that grow and shrink each telomere sequence are only approximately balanced, a chromosome end contains a variable number of telomeric repeats. Not surprisingly, experiments show that cells that proliferate indefinitely (such as yeast cells) have homeostatic mechanisms that maintain the number of these repeats within a limited range (**Figure 5–43**).

In the somatic cells of humans, the telomere repeats have been proposed to provide each cell with a counting mechanism that helps prevent the unlimited proliferation of wayward cells in adult tissues. The simplest form of this idea holds that our somatic cells are born with a full complement of telomeric repeats. Some stem cells, notably those in tissues that must be replenished throughout life—such as those of bone marrow or skin—retain full telomerase activity. However, in many other types of cells, the level of telomerase is turned down so that the enzyme cannot quite keep up with chromosome duplication. Such cells lose 100–200 nucleotides from each telomere every time they divide. After many cell generations, the descendent cells will inherit defective chromosomes (because their tips cannot be replicated completely) and consequently will withdraw permanently from the cell cycle and cease dividing—a process called *replicative cell senescence* (discussed in Chapter 17). In theory, such a mechanism could provide a safeguard against the uncontrolled cell proliferation of abnormal cells in somatic tissues, thereby helping to protect us from cancer.

The idea that telomere length acts as a "measuring stick" to count cell divisions and thereby regulate the cell's lifetime has been tested in several ways. For certain types of human cells grown in tissue culture, the experimental results support such a theory. Human fibroblasts normally proliferate for about 60 cell divisions in culture before undergoing replicative senescence. Like most other somatic cells in humans, fibroblasts produce only low levels of telomerase, and their telomeres gradually shorten each time they divide. When telomerase is provided to the fibroblasts by inserting an active telomerase gene, telomere length is maintained and many of the cells now continue to proliferate indefinitely. It therefore seems clear that telomere shortening can count cell divisions and trigger replicative senescence in some human cells.

It has been proposed that this type of control on cell proliferation is important for the maintenance of tissue architecture and that it is also somehow responsible for the aging of animals like ourselves. These ideas have been tested by producing transgenic mice that lack telomerase entirely. The telomeres in mouse chromosomes are about five times longer than human telomeres, and the mice must therefore be bred through three or more generations before their telomeres have shrunk to the normal human length. It is therefore perhaps not

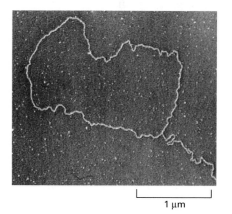

Figure 5–42 A t-loop at the end of a mammalian chromosome. Electron micrograph of the DNA at the end of an interphase human chromosome. The chromosome was fixed, deproteinated, and artificially thickened before viewing. The loop seen here is approximately 15,000 nucleotide pairs in length. The insertion of the single-stranded 3′ end into the duplex repeats to form a t-loop is thought to be carried out and maintained by specialized proteins. (From J.D. Griffith et al., *Cell* 97:503–514, 1999. With permission from Elsevier.)

1 μm

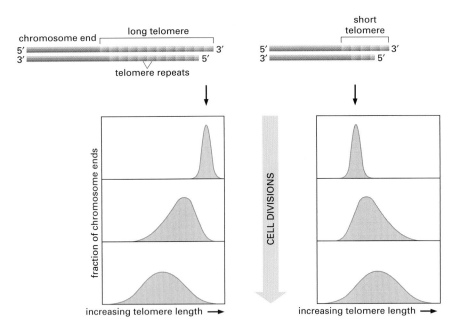

Figure 5–43 A demonstration that yeast cells control the length of their telomeres. In this experiment, the telomere at one end of a particular chromosome is artificially made either longer (left) or shorter (right) than average. After many cell divisions, the chromosome recovers, showing an average telomere length and a length distribution that is typical of the other chromosomes in the yeast cell. A similar feedback mechanism for controlling telomere length has been proposed for the cells in the germ-line cells of animals.

surprising that the first generations of mice develop normally. However, the mice in later generations develop progressively more defects in some of their highly proliferative tissues. In addition, these mice show signs of premature aging and have a pronounced tendency to develop tumors. In these and other respects these mice resemble humans with the genetic disease *dyskeratosis congenita*. Individuals afflicted with this disease carry one functional and one nonfunctional copy of the telomerase RNA gene; they have prematurely shortened telomeres and typically die of progressive bone marrow failure. They also develop lung scarring and liver cirrhosis and show abnormalities in various epidermal structures including skin, hair follicles, and nails.

The above observations demonstrate that controlling cell proliferation by telomere shortening poses a risk to an organism, because not all of the cells that begin losing the ends of their chromosomes will stop dividing. Some apparently become genetically unstable, but continue to divide giving rise to variant cells that can lead to cancer. Thus, one can question whether the observed down-regulation of telomerase in most human somatic cells provides an evolutionary advantage, as suggested by those who postulate that telomere shortening protects us from cancer and other proliferative diseases.

Summary

The proteins that initiate DNA replication bind to DNA sequences at a replication origin to catalyze the formation of a replication bubble with two outward-moving replication forks. The process begins when an initiator protein–DNA complex is formed that subsequently loads a DNA helicase onto the DNA template. Other proteins are then added to form the multienzyme "replication machine" that catalyzes DNA synthesis at each replication fork.

In bacteria and some simple eucaryotes, replication origins are specified by specific DNA sequences that are only several hundred nucleotide pairs long. In other eucaryotes, such as humans, the sequences needed to specify an origin of DNA replication seem to be less well defined, and the origin can span several thousand nucleotide pairs.

Bacteria typically have a single origin of replication in a circular chromosome. With fork speeds of up to 1000 nucleotides per second, they can replicate their genome in less than an hour. Eucaryotic DNA replication takes place in only one part of the cell cycle, the S phase. The replication fork in eucaryotes moves about 10 times more slowly than the bacterial replication fork, and the much longer eucaryotic chromosomes each require many replication origins to complete their replication in an S phase, which typically lasts for 8 hours in human cells. The different replication origins in these eucaryotic chromosomes are activated in a sequence, determined in part by the structure of the chromatin, with the most condensed regions of chromatin typically beginning their replication last. After the replication fork has passed, chromatin structure is re-formed by the addition of new histones to the old histones that are directly inherited by each daughter DNA molecule. The mechanism of chromosome duplication allows the parental patterns of histone modification to be passed on to daughter chromosomes, thus providing a means of epigenetic inheritance.

Eucaryotes solve the problem of replicating the ends of their linear chromosomes with a specialized end structure, the telomere, maintained by a special nucleotide polymerizing enzyme called telomerase. Telomerase extends one of the DNA strands at the end of a chromosome by using an RNA template that is an integral part of the enzyme itself, producing a highly repeated DNA sequence that typically extends for thousands of nucleotide pairs at each chromosome end.

DNA REPAIR

Maintaining the genetic stability that an organism needs for its survival requires not only an extremely accurate mechanism for replicating DNA, but also mechanisms for repairing the many accidental lesions that occur continually in DNA. Most such spontaneous changes in DNA are temporary because they are immediately corrected by a set of processes that are collectively called **DNA repair**. Of the thousands of random changes created every day in the DNA of a human cell by heat, metabolic accidents, radiation of various sorts, and exposure to substances in the environment, only a few accumulate as mutations in the DNA sequence. For example, we now know that fewer than one in 1000 accidental base changes in DNA results in a permanent mutation; the rest are eliminated with remarkable efficiency by DNA repair.

The importance of DNA repair is evident from the large investment that cells make in DNA repair enzymes. For example, analysis of the genomes of bacteria and yeasts has revealed that several percent of the coding capacity of these organisms is devoted solely to DNA repair functions. The importance of DNA repair is also demonstrated by the increased rate of mutation that follows the inactivation of a DNA repair gene. Many DNA repair proteins and the genes that encode them—which we now know operate in a wide range of organisms, including humans—were originally identified in bacteria by the isolation and characterization of mutants that displayed an increased mutation rate or an increased sensitivity to DNA-damaging agents.

Recent studies of the consequences of a diminished capacity for DNA repair in humans have linked many human diseases with decreased repair (**Table 5–2**).

Table 5–2 **Some Inherited Syndromes with Defects in DNA Repair**

NAME	PHENOTYPE	ENZYME OR PROCESS AFFECTED
MSH2, 3, 6, MLH1, PMS2	colon cancer	mismatch repair
Xeroderma pigmentosum (XP) groups A–G	skin cancer, UV sensitivity, neurological abnormalities	nucleotide excision–repair
XP variant	UV sensitivity, skin cancer	translesion synthesis by DNA polymerase η
Ataxia telangiectasia (AT)	leukemia, lymphoma, γ-ray sensitivity, genome instability	ATM protein, a protein kinase activated by double-strand breaks
BRCA2	breast, ovarian, and prostate cancer	repair by homologous recombination
Werner syndrome	premature aging, cancer at several sites, genome instability	accessory 3'-exonuclease and DNA helicase
Bloom syndrome	cancer at several sites, stunted growth, genome instability	accessory DNA helicase for replication
Fanconi anemia groups A–G	congenital abnormalities, leukemia, genome instability	DNA interstrand cross-link repair
46 BR patient	hypersensitivity to DNA-damaging agents, genome instability	DNA ligase I

Thus, we saw previously that defects in a human gene that normally functions to repair the mismatched base pairs resulting from DNA replication errors can lead to an inherited predisposition to certain cancers, reflecting an increased mutation rate. In another human disease, *xeroderma pigmentosum* (XP), the afflicted individuals have an extreme sensitivity to ultraviolet radiation because they are unable to repair certain DNA photoproducts. This repair defect results in an increased mutation rate that leads to serious skin lesions and an increased susceptibility to certain cancers.

Without DNA Repair, Spontaneous DNA Damage Would Rapidly Change DNA Sequences

Although DNA is a highly stable material, as required for the storage of genetic information, it is a complex organic molecule that is susceptible, even under normal cell conditions, to spontaneous changes that would lead to mutations if left unrepaired (**Figure 5–44**). For example, the DNA of each human cell loses about 5000 purine bases (adenine and guanine) every day because their N-glycosyl linkages to deoxyribose hydrolyze, a spontaneous reaction called *depurination*. Similarly, a spontaneous *deamination* of cytosine to uracil in DNA occurs at a rate of about 100 bases per cell per day (**Figure 5–45**). DNA bases are also occasionally damaged by an encounter with reactive metabolites produced in the cell (including reactive forms of oxygen) or by exposure to chemicals in the environment. Likewise, ultraviolet radiation from the sun can produce a covalent linkage between two adjacent pyrimidine bases in DNA to form, for example, thymine dimers (**Figure 5–46**). If left uncorrected when the DNA is replicated, most of these changes would be expected to lead either to the deletion of one or more base pairs or to a base-pair substitution in the daughter DNA chain (**Figure 5–47**). The mutations would then be propagated throughout subsequent cell generations. Such a high rate of random changes in the DNA sequence would have disastrous consequences for an organism.

The DNA Double Helix Is Readily Repaired

The double-helical structure of DNA is ideally suited for repair because it carries two separate copies of all the genetic information—one in each of its two strands. Thus, when one strand is damaged, the complementary strand retains an intact copy of the same information, and this copy is generally used to restore the correct nucleotide sequences to the damaged strand.

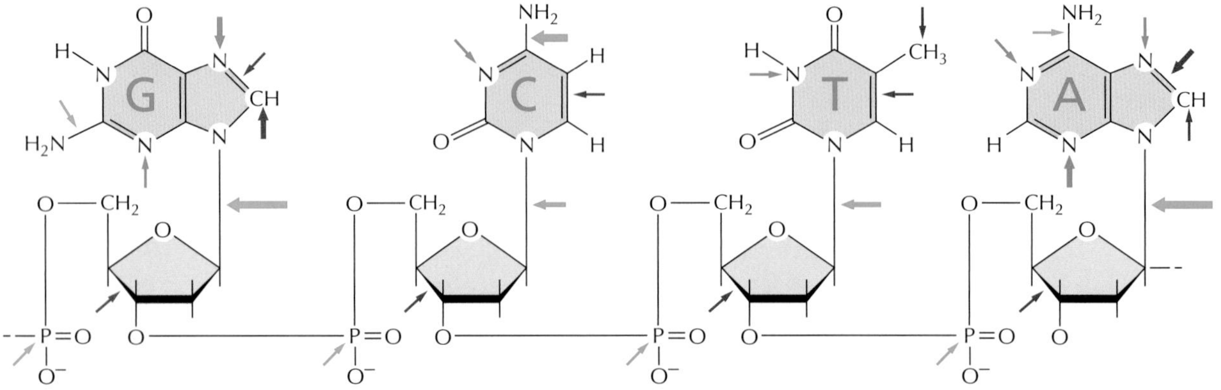

Figure 5–44 A summary of spontaneous alterations likely to require DNA repair. The sites on each nucleotide that are known to be modified by spontaneous oxidative damage (*red arrows*), hydrolytic attack (*blue arrows*), and uncontrolled methylation by the methyl group donor *S*-adenosylmethionine (*green arrows*) are shown, with the width of each arrow indicating the relative frequency of each event. (After T. Lindahl, *Nature* 362:709–715, 1993. With permission from Macmillan Publishers Ltd.)

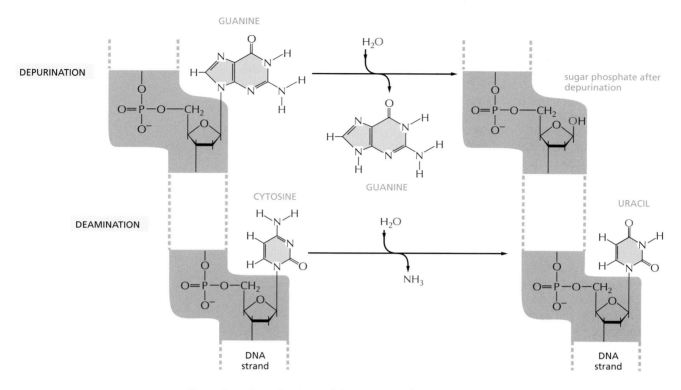

Figure 5–45 Depurination and deamination. These two reactions are the most frequent spontaneous chemical reactions known to create serious DNA damage in cells. Depurination can release guanine (shown here), as well as adenine, from DNA. The major type of deamination reaction converts cytosine to an altered DNA base, uracil (shown here), but deamination occurs on other bases as well. These reactions normally take place in double-helical DNA; for convenience, only one strand is shown.

An indication of the importance of a double-stranded helix to the safe storage of genetic information is that all cells use it; only a few small viruses use single-stranded DNA or RNA as their genetic material. The types of repair processes described in this section cannot operate on such nucleic acids, and once damaged, the chance of a permanent nucleotide change occurring in these single-stranded genomes of viruses is thus very high. It seems that only organisms with tiny genomes (and therefore tiny targets for DNA damage) can afford to encode their genetic information in any molecule other than a DNA double helix.

DNA Damage Can Be Removed by More Than One Pathway

Cells have multiple pathways to repair their DNA using different enzymes that act upon different kinds of lesions. **Figure 5–48** shows two of the most common pathways. In both, the damage is excised, the original DNA sequence is restored by a DNA polymerase that uses the undamaged strand as its template, and a remaining break in the double helix is sealed by DNA ligase (see Figure 5–13).

The two pathways differ in the way in which they remove the damage from DNA. The first pathway, called **base excision repair**, involves a battery of enzymes called *DNA glycosylases*, each of which can recognize a specific type of altered base in DNA and catalyze its hydrolytic removal. There are at least six types of these enzymes, including those that remove deaminated Cs, deaminated As, different types of alkylated or oxidized bases, bases with opened rings, and bases in which a carbon–carbon double bond has been accidentally converted to a carbon–carbon single bond. How is an altered base detected within the context of the double helix? A key step is an enzyme-mediated "flipping-out"

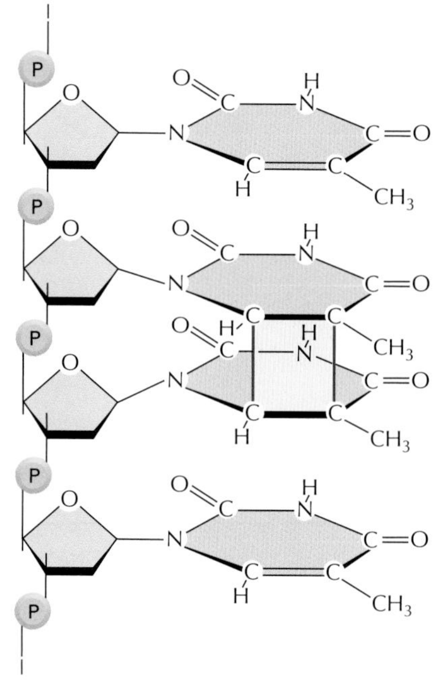

Figure 5–46 The most common type of thymine dimer. This type of damage occurs in the DNA of cells exposed to ultraviolet irradiation (as in sunlight). A similar dimer will form between any two neighboring pyrimidine bases (C or T residues) in DNA.

of the altered nucleotide from the helix, which allows the DNA glycosylase to probe all faces of the base for damage (**Figure 5–49**). It is thought that these enzymes travel along DNA using base-flipping to evaluate the status of each base. Once an enzyme finds the damaged base that it recognizes, it removes the base from its sugar.

The "missing tooth" created by DNA glycosylase action is recognized by an enzyme called *AP endonuclease* (AP for *apurinic* or *apyrimidinic, endo* to signify that the nuclease cleaves within the polynucleotide chain), which cuts the phosphodiester backbone, after which the damage is removed and the resulting gap repaired (see Figure 5–48A). Depurination, which is by far the most frequent type of damage suffered by DNA, also leaves a deoxyribose sugar with a missing base. Depurinations are directly repaired beginning with AP endonuclease, following the bottom half of the pathway in Figure 5–48A.

The second major repair pathway is called **nucleotide excision repair**. This mechanism can repair the damage caused by almost any large change in the structure of the DNA double helix. Such "bulky lesions" include those created by the covalent reaction of DNA bases with large hydrocarbons (such as the carcinogen benzopyrene), as well as the various pyrimidine dimers (T-T, T-C, and C-C) caused by sunlight. In this pathway, a large multienzyme complex scans the DNA for a distortion in the double helix, rather than for a specific base change. Once it finds a bulky lesion, it cleaves the phosphodiester backbone of the abnormal strand on both sides of the distortion, and a DNA helicase peels away a single-strand oligonucleotide containing the lesion. The large gap produced in the DNA helix is then repaired by DNA polymerase and DNA ligase (Figure 5–48B).

An alternative to base and nucleotide excision-repair processes is direct chemical reversal of DNA damage, and this strategy is employed for the rapid removal of certain highly mutagenic or cytotoxic lesions. For example, the alkylation lesion O^6-methylguanine has its methyl group removed by direct transfer to a cysteine residue in the repair protein itself, which is destroyed in the reaction. In another example, methyl groups in the alkylation lesions 1-methyladenine and 3-methylcytosine are "burnt off" by an iron-dependent demethylase, with release of formaldehyde from the methylated DNA and regeneration of the native base.

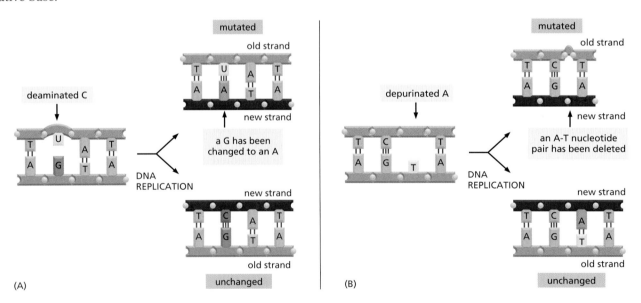

(A)   (B)

Figure 5–47 How chemical modifications of nucleotides produce mutations. (A) Deamination of cytosine, if uncorrected, results in the substitution of one base for another when the DNA is replicated. As shown in Figure 5–45, deamination of cytosine produces uracil. Uracil differs from cytosine in its base-pairing properties and preferentially base pairs with adenine. The DNA replication machinery therefore adds an adenine when it encounters a uracil on the template strand. (B) Depurination can lead to the loss of a nucleotide pair. When the replication machinery encounters a missing purine on the template strand, it may skip to the next complete nucleotide as illustrated here, thus producing a nucleotide deletion in the newly synthesized strand. Many other types of DNA damage (see Figure 5–44), if left uncorrected, also produce mutations when the DNA is replicated.

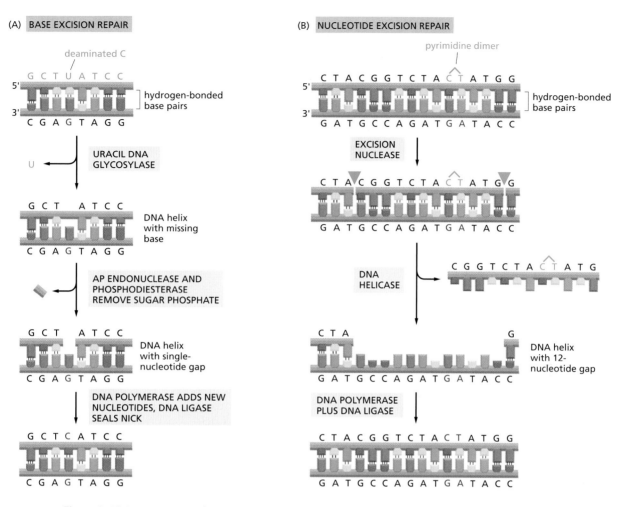

Figure 5–48 A comparison of two major DNA repair pathways. (A) *Base excision repair.* This pathway starts with a DNA glycosylase. Here the enzyme uracil DNA glycosylase removes an accidentally deaminated cytosine in DNA. After the action of this glycosylase (or another DNA glycosylase that recognizes a different kind of damage), the sugar phosphate with the missing base is cut out by the sequential action of AP endonuclease and a phosphodiesterase. (These same enzymes begin the repair of depurinated sites directly.) The gap of a single nucleotide is then filled by DNA polymerase and DNA ligase. The net result is that the U that was created by accidental deamination is restored to a C. AP endonuclease is so-named because it recognizes any site in the DNA helix that contains a deoxyribose sugar with a missing base; such sites can arise either by the loss of a purine *(ap*urinic sites) or by the loss of a pyrimidine *(ap*yrimidinic sites). (B) *Nucleotide excision repair.* In bacteria, after a multienzyme complex has recognized a lesion such as a pyrimidine dimer (see Figure 5–46), one cut is made on each side of the lesion, and an associated DNA helicase then removes the entire portion of the damaged strand. The excision repair machinery in bacteria leaves the gap of 12 nucleotides shown. In humans, once the damaged DNA is recognized, a helicase is recruited to unwind the DNA duplex locally. Next, the excision nuclease enters and cleaves on either side of the damage, leaving a gap of about 30 nucleotides. The nucleotide excision repair machinery in both bacteria and humans can recognize and repair many different types of DNA damage.

Coupling DNA Repair to Transcription Ensures That the Cell's Most Important DNA Is Efficiently Repaired

All of a cell's DNA is under constant surveillance for damage, and the repair mechanisms we have described act on all parts of the genome. However, cells have a way of directing DNA repair to the DNA sequences that are most urgently needed. They do this by linking RNA polymerase, the enzyme that transcribes DNA into RNA as the first step in gene expression, to the repair of DNA damage. RNA polymerase stalls at DNA lesions and, through the use of coupling proteins, directs the repair machinery to these sites. In bacteria, where genes are relatively short, the stalled RNA polymerase can be dissociated from the DNA, the DNA is repaired, and the gene is transcribed again from the beginning. In eucaryotes, where genes can be enormously long, a more complex reaction is used to "back up" the RNA polymerase, repair the damage, and then restart the polymerase.

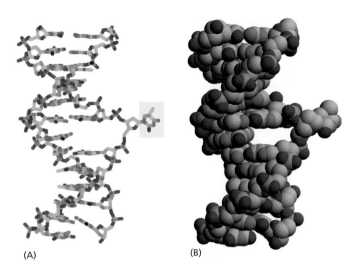

(A)   (B)

Figure 5–49 The recognition of an unusual nucleotide in DNA by base-flipping. The DNA glycosylase family of enzymes recognizes specific bases in the conformation shown. Each of these enzymes cleaves the glycosyl bond that connects a particular recognized base *(yellow)* to the backbone sugar, removing it from the DNA. (A) Stick model; (B) space-filling model.

Transcription-coupled repair works with base excision, nucleotide excision, and other repair machinery to direct repair immediately to the cell's most important DNA sequences, namely those being expressed when the damage occurs. Remarkably, this type of repair is specific for the template strand of transcribed DNA; the other strand is repaired with the same speed and efficiency as DNA that is not being transcribed at all. Transcription-coupled repair is particularly advantageous in humans, because only a small fraction of our genome is transcribed at any given time. Its importance is seen in individuals with Cockayne's syndrome, which is caused by a defect in transcription-coupled repair. These individuals suffer from growth retardation, skeletal abnormalities, progressive neural retardation, and severe sensitivity to sunlight. Most of these problems are thought to arise from RNA polymerase molecules that become permanently stalled at sites of DNA damage that lie in important genes.

The Chemistry of the DNA Bases Facilitates Damage Detection

The DNA double helix seems to be optimally constructed for repair. As noted above, it contains a backup copy of all genetic information. Equally importantly, the nature of the four bases in DNA makes the distinction between undamaged and damaged bases very clear. For example, every possible deamination event in DNA yields an "unnatural" base, which can be directly recognized and removed by a specific DNA glycosylase. Hypoxanthine, for example, is the simplest purine base capable of pairing specifically with C, but hypoxanthine is the direct deamination product of A (**Figure 5–50**A). The addition of a second amino group to hypoxanthine produces G, which cannot be formed from A by spontaneous deamination, and whose deamination product (xanthine) is likewise unique.

As discussed in Chapter 6, RNA is thought, on an evolutionary time-scale, to have served as the genetic material before DNA, and it seems likely that the genetic code was initially carried in the four nucleotides A, C, G, and U. This raises the question of why the U in RNA was replaced in DNA by T (which is 5-methyl U). We have seen that the spontaneous deamination of C converts it to U, but that this event is rendered relatively harmless by uracil DNA glycosylase. However, if DNA contained U as a natural base, the repair system would find it difficult to distinguish a deaminated C from a naturally occurring U.

A special situation occurs in vertebrate DNA, in which selected C nucleotides are methylated at specific C-G sequences that are associated with inactive genes (discussed in Chapter 7). The accidental deamination of these methylated C nucleotides produces the natural nucleotide T (Figure 5–50B) in a mismatched base pair with a G on the opposite DNA strand. To help in repairing deaminated methylated C nucleotides, a special DNA glycosylase recognizes a mismatched base pair involving T in the sequence T-G and removes the T. This DNA repair mechanism must be relatively ineffective, however, because methylated C nucleotides are common sites for mutations in vertebrate DNA. It is striking that, even though only about 3% of the C nucleotides in human DNA are methylated, mutations in these methylated nucleotides account for about one-third of the single-base mutations that have been observed in inherited human diseases.

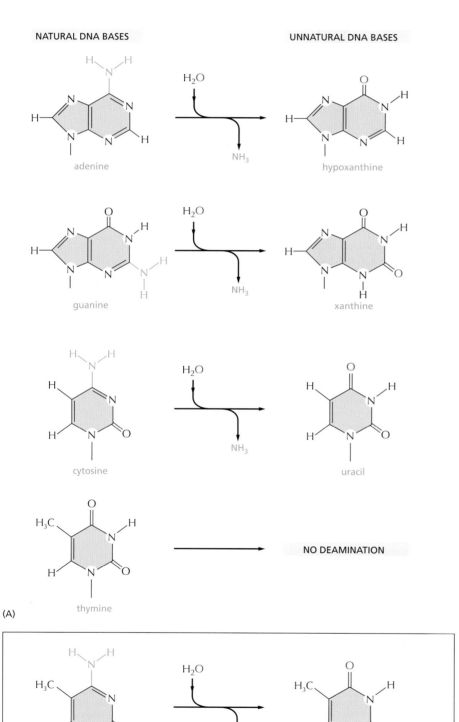

NATURAL DNA BASES   UNNATURAL DNA BASES

adenine → hypoxanthine

guanine → xanthine

cytosine → uracil

thymine → NO DEAMINATION

(A)

5-methyl cytosine → thymine

(B)

Figure 5–50 The deamination of DNA nucleotides. In each case, the oxygen atom that is added in this reaction with water is colored *red*. (A) The spontaneous deamination products of A and G are recognizable as unnatural when they occur in DNA and thus are readily recognized and repaired. The deamination of C to U was also illustrated in Figure 5–45; T has no amino group to remove. (B) About 3% of the C nucleotides in vertebrate DNAs are methylated to help in controlling gene expression (discussed in Chapter 7). When these 5-methyl C nucleotides are accidentally deaminated, they form the natural nucleotide T. However, this T will be paired with a G on the opposite strand, forming a mismatched base pair.

Special DNA Polymerases Are Used in Emergencies to Repair DNA

If a cell's DNA is heavily damaged, the repair mechanisms that we have discussed are often insufficient to cope with it. In these cases, a different strategy is called into play, one that entails some risk to the cell. The highly accurate replicative DNA polymerases stall when they encounter damaged DNA, and in emergencies cells employ versatile but less accurate back-up polymerases to replicate through the DNA damage.

Human cells contain more than 10 such DNA polymerases, some of which can recognize a specific type of DNA damage and specifically add the nucleotide required to restore the initial sequence. Others make only "good guesses," especially when the template base has been extensively damaged. These enzymes are not as accurate as the normal replicative polymerases when they copy a normal DNA sequence. For one thing, the backup polymerases lack exonucleolytic proofreading activity; in addition, many are much less discriminating than the replicative polymerase in choosing which nucleotide to initially incorporate. Presumably, for this reason, each such polymerase molecule is given a chance to add only one or a few nucleotides. Although the details of these fascinating reactions are still being worked out, they provide elegant testimony to the care with which organisms maintain the integrity of their DNA.

Double-Strand Breaks Are Efficiently Repaired

An especially dangerous type of DNA damage occurs when both strands of the double helix are broken, leaving no intact template strand to enable accurate repair. Ionizing radiation, replication errors, oxidizing agents, and other metabolites produced in the cell cause breaks of this type. If these lesions were left unrepaired, they would quickly lead to the breakdown of chromosomes into smaller fragments and to the loss of genes when the cell divides. However, two distinct mechanisms have evolved to ameliorate this type of damage (**Figure 5–51**). The simplest to understand is **nonhomologous end-joining**, in which the broken ends are simply brought together and rejoined by DNA ligation, generally with the loss of one or more nucleotides at the site of joining (**Figure 5–52**). This end-joining mechanism, which might be seen as a "quick and dirty" solution to the repair of double-strand breaks, is common in mammalian somatic cells. Although a change in the DNA sequence (a mutation) results at the site of breakage, so little of the mammalian genome codes for proteins that this mechanism is apparently an acceptable solution to the problem of rejoining broken

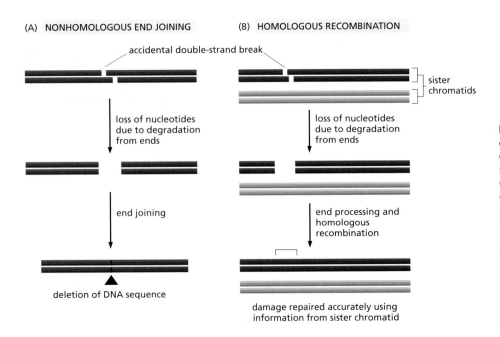

(A) NONHOMOLOGOUS END JOINING

(B) HOMOLOGOUS RECOMBINATION

accidental double-strand break

sister chromatids

loss of nucleotides due to degradation from ends

loss of nucleotides due to degradation from ends

end joining

end processing and homologous recombination

deletion of DNA sequence

damage repaired accurately using information from sister chromatid

Figure 5–51 Two different ways to repair double-strand breaks. (A) Nonhomologous end-joining alters the original DNA sequence when repairing a broken chromosome. These alterations can be either deletions (as shown) or short insertions. (B) Repairing double-strand breaks by homologous recombination is more difficult to accomplish, but this type of repair restores the original DNA sequence. It typically takes place after the DNA has been duplicated but before the cell has divided. Details of the homologous recombination pathway will be presented later (see Figure 5–61).

chromosomes. By the time a human reaches the age of 70, the typical somatic cell contains over 2000 such "scars", distributed throughout its genome, representing places where DNA has been inaccurately repaired by nonhomologous end-joining. As previously discussed, the specialized structure of telomeres prevents the natural ends of chromosomes from being mistaken for broken DNA and repaired.

A much more accurate type of double-strand break repair occurs in newly replicated DNA (Figure 5–51B). Here, the DNA is repaired using the sister chromatid as a template. This reaction is an example of *homologous recombination,* and we consider its mechanism later in this chapter. Most organisms employ both nonhomologous end-joining and homologous recombination to repair double-strand breaks in DNA. Nonhomologous end-joining predominates in humans; homologous recombination is used only during and shortly after DNA replication (in S and G_2 phases), when sister chromatids are available to serve as templates.

DNA Damage Delays Progression of the Cell Cycle

We have just seen that cells contain multiple enzyme systems that can recognize and repair many types of DNA damage. Because of the importance of maintaining intact, undamaged DNA from generation to generation, eukaryotic cells have an additional mechanism that maximizes the effectiveness of their DNA repair enzymes: they delay progression of the cell cycle until DNA repair is complete. As discussed in detail in Chapter 17, the orderly progression of the cell cycle is maintained through the use of *checkpoints* that ensure the completion of one step before the next step can begin. At several of these cell-cycle checkpoints, the cycle stops if damaged DNA is detected. Thus, in mammalian cells, the presence of DNA damage can block entry from G_1 into S phase, it can slow S phase once it has begun, and it can block the transition from S phase to M phase. These delays facilitate DNA repair by providing the time needed for the repair to reach completion.

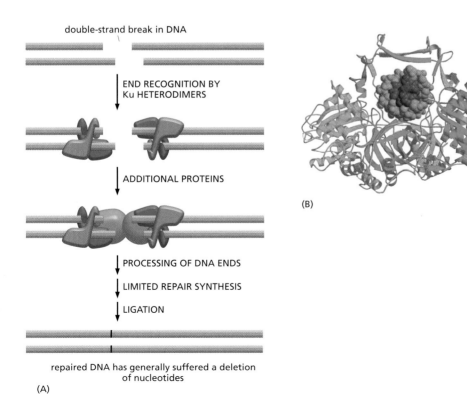

double-strand break in DNA

END RECOGNITION BY Ku HETERODIMERS

ADDITIONAL PROTEINS

(B)

PROCESSING OF DNA ENDS

LIMITED REPAIR SYNTHESIS

LIGATION

repaired DNA has generally suffered a deletion of nucleotides

(A)

Figure 5–52 **Nonhomologous end-joining.** (A) A central role is played by the Ku protein, a heterodimer that grasps the broken chromosome ends. The additional proteins shown are needed to hold the broken ends together while they are processed and eventually joined covalently. (B) Three-dimensional structure of a Ku heterodimer bound to the end of a duplex DNA fragment. The Ku protein is also essential for V(D)J joining, a specific recombination process through which antibody and T cell receptor diversity is generated in developing B and T cells (discussed in Chapter 25). V(D)J joining and nonhomologous end-joining show many similarities in mechanism but the former relies on specific double-strand breaks produced deliberately by the cell. (B, from J.R. Walker, R.A. Corpina and J. Goldberg, *Nature* 412:607–614, 2001. With permission from Macmillan Publishers Ltd.)

DNA damage also results in an increased synthesis of some DNA repair enzymes. The importance of special signaling mechanisms that respond to DNA damage is indicated by the phenotype of humans who are born with defects in the gene that encodes the *ATM protein*. These individuals have the disease *ataxia telangiectasia (AT)*, the symptoms of which include neurodegeneration, a predisposition to cancer, and genome instability. The ATM protein is a large kinase needed to generate the intracellular signals that produce a response to many types of spontaneous DNA damage, and individuals with defects in this protein therefore suffer from the effects of unrepaired DNA lesions.

Summary

Genetic information can be stored stably in DNA sequences only because a large set of DNA repair enzymes continuously scan the DNA and replace any damaged nucleotides. Most types of DNA repair depend on the presence of a separate copy of the genetic information in each of the two strands of the DNA double helix. An accidental lesion on one strand can therefore be cut out by a repair enzyme and a corrected strand resynthesized by reference to the information in the undamaged strand.

Most of the damage to DNA bases is excised by one of two major DNA repair pathways. In base excision repair, the altered base is removed by a DNA glycosylase enzyme, followed by excision of the resulting sugar phosphate. In nucleotide excision repair, a small section of the DNA strand surrounding the damage is removed from the DNA double helix as an oligonucleotide. In both cases, the gap left in the DNA helix is filled in by the sequential action of DNA polymerase and DNA ligase, using the undamaged DNA strand as the template. Some types of DNA damage can be repaired by a different strategy—the direct chemical reversal of the damage—which is carried out by specialized repair proteins.

Other critical repair systems—based on either nonhomologous end-joining or homologous recombination—reseal the accidental double-strand breaks that occur in the DNA helix. In most cells, an elevated level of DNA damage causes a delay in the cell cycle via checkpoint mechanisms, which ensure that DNA damage is repaired before a cell divides.

HOMOLOGOUS RECOMBINATION

In the two preceding sections, we discussed the mechanisms that allow the DNA sequences in cells to be maintained from generation to generation with very little change. In this section, we further explore one of these mechanisms, *homologous recombination*. Although crucial for accurately repairing double-strand breaks (see Figure 5–51B) and other types of DNA damage, homologous recombination, as we shall see, can also rearrange DNA sequences. These rearrangements often alter the particular versions of genes present in an individual genome, as well as the timing and the level of their expression. In a population, the type of genetic variation produced by this and other types of genetic recombination is crucial for facilitating the evolution of organisms in response to a changing environment.

Homologous Recombination Has Many Uses in the Cell

In **homologous recombination** (also known as *general recombination*), genetic exchange takes place between a pair of homologous DNA sequences, that is, DNA sequences similar or identical in nucleotide sequence. Homologous recombination has many uses in the cell, but three are of paramount importance. The most widespread use is in accurately repairing double-strand breaks, as introduced in the previous section (Figure 5–51B). Although double-strand breaks can result from radiation and reactive chemicals, many arise from DNA replication forks that become stalled or broken. This application of homologous

Figure 5–53 Repair of a broken replication fork by homologous recombination. When a moving replication fork encounters a single-strand break, it will collapse but can be repaired by homologous recombination. As shown, the initial strand invasion requires a free 3′ end generated by a nuclease that degrades the 5′ end of the complementary strand. Recombination then begins with strand invasion, as described in detail in subsequent figures. Arrowheads represent 3′ ends of DNA strands. Green strands represent the new DNA synthesis that takes place after the replication fork has broken. Note that, in this mechanism, the fork moves past the site that was nicked on the original template by using an undamaged copy of the site as its template. (Adapted from M.M. Cox, *Proc. Natl Acad. Sci. U.S.A.* 98:8173–8180, 2001. With permission from National Academy of Sciences.)

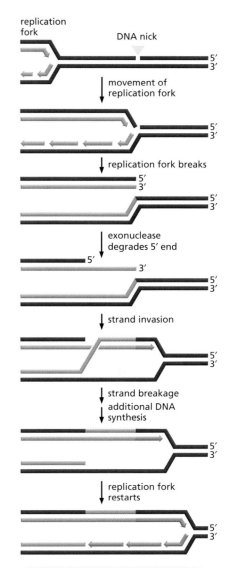

BLOCK TO REPLICATION OVERCOME

recombination is essential for every proliferating cell, because accidents occur during nearly every round of DNA replication.

Many types of events can cause the replication fork to break during the replication process. Consider just one example: a single-strand nick or gap in the parental DNA helix just ahead of a replication fork. When the fork reaches this lesion, it falls apart—resulting in one broken and one intact daughter chromosome. However, a series of recombination reactions, which can begin with a *strand invasion* process that triggers DNA synthesis by DNA polymerase, can flawlessly repair the broken chromosome (**Figure 5–53**).

In addition, homologous recombination is used to exchange bits of genetic information between two different chromosomes to create new combinations of DNA sequences in each chromosome. The potential evolutionary benefit of this type of gene mixing is that it creates an array of new, perhaps beneficial, combinations of genes. During meiosis in fungi, plants, and animals, homologous recombination also plays an important mechanical role in assuring accurate chromosome segregation. In this section, we consider only the universal roles of homologous recombination, those of repairing DNA damage and of mediating genetic exchange. Its more specialized mechanical role in chromosome segregation during meiosis will be discussed in Chapter 21.

Homologous Recombination Has Common Features in All Cells

The current view of homologous recombination as a critical DNA repair mechanism in all cells evolved slowly from its initial discovery as a key component in the specialized process of meiosis in plants and animals. The subsequent recognition that homologous recombination also occurs in less complex unicellular organisms made it much more amenable to molecular analyses. Thus, most of what we know about the biochemistry of genetic recombination was originally derived from studies of bacteria, especially of *E. coli* and its viruses, as well as from experiments with simple eucaryotes such as yeasts. For these organisms with short generation times and relatively small genomes, it was possible to isolate a large set of mutants with defects in their recombination processes. The protein altered in each mutant was then identified and, ultimately, studied biochemically. More recently, close relatives of these proteins have been discovered and extensively characterized in *Drosophila*, mice, and humans. These studies reveal that the fundamental processes that catalyze homologous recombination are common to all cells, as we shall now discuss.

DNA Base-Pairing Guides Homologous Recombination

The hallmark of homologous recombination is that it takes place only between DNA duplexes that have extensive regions of sequence similarity (homology). Not surprisingly, base-pairing underlies this requirement, and two DNA duplexes that are undergoing homologous recombination "sample" each other's DNA sequence by engaging in extensive base-pairing between a single strand from one DNA duplex and the complementary single strand from the other. The match need not be perfect, but it must be very close for homologous recombination to succeed.

In its simplest form, this type of base-pairing interaction can be mimicked in a test tube by allowing a DNA double helix to re-form from its separated single strands. This process, called *DNA renaturation* or **hybridization**, occurs when a rare random collision juxtaposes complementary nucleotide sequences on two matching DNA single strands, allowing the formation of a short stretch of double helix between them. This relatively slow helix nucleation step is followed by a very rapid "zippering" step, as the region of double helix is extended to maximize the number of base-pairing interactions (**Figure 5–54**).

Formation of a new double helix in this way requires that the annealing strands be in an open, unfolded conformation. For this reason, *in vitro* hybridization reactions are performed either at high temperature or in the presence of an organic solvent such as formamide; these conditions "melt out" the short hairpin helices that result from the base-pairing interactions that occur within a single strand that folds back on itself. Most cells cannot survive such harsh conditions and instead use single-strand DNA-binding proteins (see p. 273) to melt out the hairpin helices. Single-strand DNA-binding proteins are essential for DNA replication (as described earlier) as well as for homologous recombination; they bind tightly and cooperatively to the sugar-phosphate backbone of all single-stranded DNA regions of DNA, holding them in an extended conformation with the bases exposed (see Figures 5–16 and 5–17). In this extended conformation, a DNA single strand can base-pair efficiently either with a nucleoside triphosphate molecule (in DNA replication) or with a complementary section of another DNA single strand (as part of a genetic recombination process).

DNA hybridization creates a region of DNA helix formed from strands that originate from two different DNA molecules. The formation of such a region, known as a *heteroduplex*, is an essential step in any homologous recombination process. Because the vast majority of DNA inside the cell is double-stranded, the "test tube" model of DNA hybridization cannot fully explain how this process occurs in the cell. Indeed, special mechanisms are required to start homologous recombination between two double-stranded DNA molecules of similar nucleotide sequence. Crucial to these mechanisms are proteins that allow DNA hybridization to occur in the cell through *strand invasion*—the pairing of a region of single-stranded DNA with a complementary strand in a different DNA double helix—as we describe next.

Figure 5–54 DNA hybridization. DNA double helices re-form from their separated strands in a reaction that depends on the random collision of two complementary DNA strands. The vast majority of such collisions are not productive, as shown on the left, but a few result in a short region where complementary base pairs have formed (helix nucleation). A rapid zippering then leads to the formation of a complete double helix. Through this trial-and-error process, a DNA strand will find its complementary partner even in the midst of millions of nonmatching DNA strands.

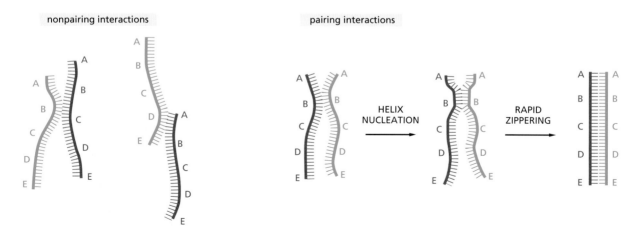

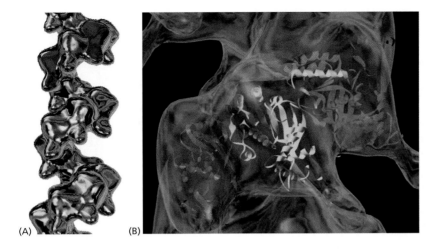

(A)   (B)

Figure 5–55 **The structure of the RecA and Rad51 protein–DNA filaments.**
(A) The Rad51 protein, a human homolog of the bacterial RecA protein, is bound to a DNA single strand. Three successive protein monomers in this helical filament are colored. (B) A short section of the RecA filament, with the three-dimensional structure of the protein fitted to the image of the filament determined by electron microscopy. There are about six RecA monomers per turn of the helix, holding 18 nucleotides of single-stranded DNA that is stretched out by the protein. The exact path of the DNA in this structure is not known. (A, courtesy of Edward Egelman; B, from X. Yu et al., *J. Mol. Biol.* 283:985–992, 1998. With permission from Academic Press.)

The RecA Protein and its Homologs Enable a DNA Single Strand to Pair with a Homologous Region of DNA Double Helix

Because extensive base-pair interactions cannot occur between two intact DNA double helices, the DNA hybridization that is critical for homologous recombination can begin only after a DNA strand from one DNA helix is freed from pairing with its complementary strand, thereby making its nucleotide available for pairing with a second DNA helix. In the example that was previously illustrated in Figure 5–53, this free single strand is formed when a replication fork encounters a DNA nick, falls apart (creating a new double-stranded end), and an exonuclease degrades the 5′ end at the break, leaving an unpaired single strand at its 3′ end. In other applications of homologous recombination, single stranded regions are generated in similar ways, as we discuss later.

The single-strand at the 3′ DNA end is acted upon by several specialized proteins that direct it to invade a homologous DNA duplex. Of central importance is the **RecA protein**, its name in *E.coli*, and its homolog **Rad51**, its name in virtually all eucaryotic organisms (**Figure 5–55**). Like a single-strand DNA-binding protein, the RecA type of protein binds tightly and in long cooperative clusters to single-stranded DNA forming a nucleoprotein filament. Because each RecA monomer has more than one DNA-binding site, a RecA filament can hold a single strand and a double helix together (**Figure 5–56**). This arrangement allows RecA to catalyze a multistep *DNA synapsis* reaction that occurs between a DNA double helix and a homologous region of single-stranded DNA. In the first step, the RecA protein intertwines the DNA single strand and the DNA duplex in a sequence-independent manner. Next, the DNA single strand "searches" the duplex for homologous sequences. Exactly how this searching and eventual recognition occurs is not understood, but it may involve transient base pairs formed between the single strand and bases that flip out from the duplex DNA (see Figure 5–49). Once a homologous sequence has been located, a strand invasion occurs: the single strand displaces one strand of the duplex as

Figure 5–56 **DNA synapsis catalyzed by the RecA protein.** *In vitro* experiments show that several types of complex are formed between a DNA single strand *(red)* covered with RecA protein *(blue)* and a DNA double helix *(green)*. First a non-base-paired complex is formed, which is converted into a "joint molecule" as a homologous sequence is found. This complex is dynamic and spins out a DNA heteroduplex (one strand *green* and the other strand *red*) plus a displaced single strand from the original helix *(green)*. Thus, the structure shown in this diagram migrates to the left, reeling in the "input DNAs" while producing the "output DNAs." (Adapted from S.C. West, *Annu. Rev. Biochem.* 61:603–640, 1992. With permission from Annual Reviews.)

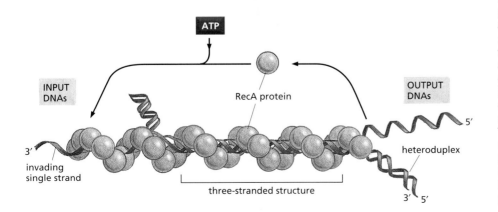

ATP

INPUT DNAs   RecA protein   OUTPUT DNAs

3′   heteroduplex

invading single strand   three-stranded structure   3′ 5′   5′

it forms conventional base pairs with the other strand. The result is a **heterodu-plex**—a region of DNA double helix formed by the pairing of two DNA strands that were initially part of two different DNA molecules (see Figure 5–56).

The search for homology and the invasion of a single strand into a DNA duplex are the critical reactions that initiate homologous recombination. They require, in addition to RecA-like proteins and single-strand binding proteins, several proteins with specialized functions. For example, Rad52 protein displaces the single-strand binding proteins allowing the binding of Rad51 molecules, and in addition, promotes the annealing of complementary single-strands (**Figure 5–57**).

The short heteroduplex region formed, where the invading single strand has paired with its complementary strand in the DNA duplex, is often greatly enlarged by a process called branch migration, as we now discuss.

Branch Migration Can either Enlarge Heteroduplex Regions or Release Newly Synthesized DNA as a Single Strand

Once a strand invasion reaction has occurred, the point of strand exchange (the "branch point") can move through a process called *branch migration* (**Figure 5–58**). In this reaction, an unpaired region of one of the single strands displaces a paired region of the other single strand, moving the branch point without changing the total number of DNA base pairs. Although spontaneous branch migration can occur, it proceeds equally in both directions, so it makes little net progress (Figure 5–58A). Specialized DNA helicases, however, can catalyze unidirectional branch migration, readily producing a region of heteroduplex DNA that can be thousands of base pairs long (Figure 5–58B).

In a related reaction, DNA synthesis catalyzed by DNA polymerase can drive a unidirectional branch migration process through which the newly synthesized DNA is displaced as a single strand, mimicking the way that a newly synthesized RNA chain is released by RNA polymerase. This form of DNA synthesis appears to be used in several homologous recombination processes, including the double-strand break repair processes to be described next.

Homologous Recombination Can Flawlessly Repair Double-Stranded Breaks in DNA

Earlier in this chapter, we discussed the dire problems created by double-strand breaks in DNA, and we saw that cells can repair these breaks in two ways. Non-homologous end-joining (see Figure 5–51) occurs without a template and creates

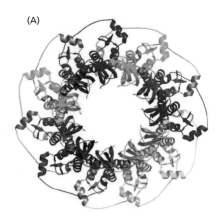

(A)

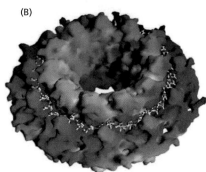

(B)

Figure 5–57 Structure of a portion of the Rad52 protein. (A) This doughnut-shaped structure is composed of 11 subunits. (B) Single-stranded DNA has been modeled into the deep groove running along the protein surface. The bases of the DNA are exposed, a configuration that is proposed to mediate the annealing of two complementary single strands. (From M.R. Singleton et al., *Proc. Natl Acad. Sci. U.S.A.* 99:13492–13497, 2002. With permission from the National Academy of Sciences.)

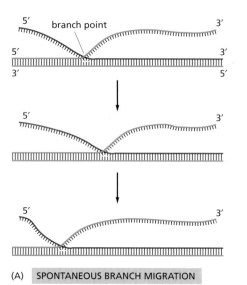

(A)   SPONTANEOUS BRANCH MIGRATION

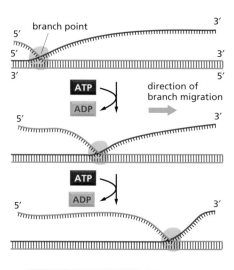

(B)   PROTEIN-DIRECTED BRANCH MIGRATION

Figure 5–58 Two types of DNA branch migration observed in experiments *in vitro*. (A) Spontaneous branch migration is a back-and-forth, random-walk process, and it therefore makes little progress over long distances. (B) Protein-directed branch migration requires energy and it moves the branch point at a uniform rate in one direction.

a mutation at the site at which two DNA duplexes are joined. It can also inadvertently join together segments from two different chromosomes creating chromosome translocations, many of which have serious consequences for the cell. In contrast to nonhomologous end-joining, homologous recombination can repair double-stranded breaks accurately, without any loss or alteration of nucleotides at the site of repair (**Figure 5–59**). In most cells, recombination-mediated double-strand break repair occurs only after the cell has replicated its DNA, when one nearby daughter DNA duplex can serve as the template for repair of the other.

Homologous recombination can also be used to repair many other types of DNA damage, making it perhaps the most versatile DNA repair mechanism available to the cell; the "all-purpose" nature of recombinational repair probably explains why its mechanism and the proteins that carry it out have been conserved in virtually all cells on Earth.

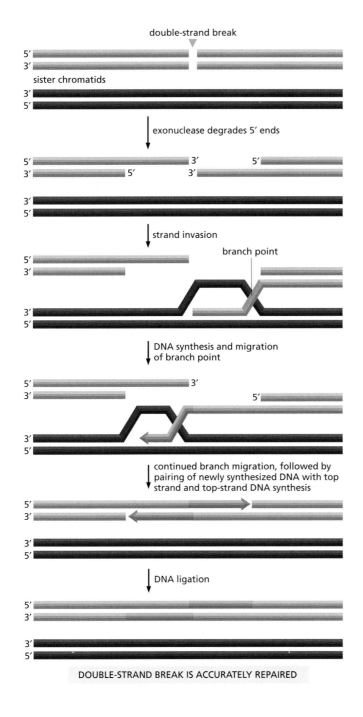

DOUBLE-STRAND BREAK IS ACCURATELY REPAIRED

Figure 5–59 Mechanism of double-strand break repair by homologous recombination. This is the preferred method for repairing DNA double-strand breaks that arise shortly after the DNA has been replicated and the two sister chromatids are still held together. This reaction uses many of the same proteins as that of Figure 5–53 and proceeds through the same basic intermediate steps. In general, homologous recombination can be regarded as a flexible series of reactions, with the exact pathway differing from one case to the next. For example, the length of the repair "patch" can vary considerably depending on the extent of 5′ processing, branch migration, and new DNA synthesis. (See M. McVey, J. LaRocque, M.D. Adams, and J. Sekelsky, *Proc. Natl Acad. Sci. U.S.A.* 101:15694–15699, 2004.)

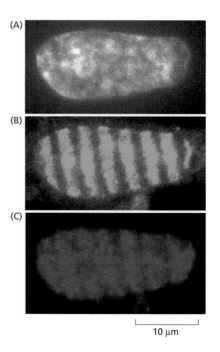

Figure 5–60 Experiment demonstrating the rapid localization of repair proteins to DNA double-strand breaks. Human fibroblasts were x-irradiated to produce DNA double-strand breaks. Before the x-rays struck the cells, they were passed through a microscopic grid with x-ray-absorbing "bars" spaced 1 μm apart. This produced a striped pattern of DNA damage, allowing a comparison of damaged and undamaged DNA in the same nucleus. (A) Total DNA in a fibroblast nucleus stained with the dye DAPI. (B) Sites of new DNA synthesis indicated by incorporation of BudR (a thymine analog) and subsequent staining with FITC-coupled antibodies to BudR. (C) Localization of the Mre11 complex to damaged DNA as visualized by Texas-red complexed to antibodies against the Mre11 subunit. It has been proposed that the Mre11 complex initially recognizes double-strand breaks in the cell and then mobilizes additional proteins to repair the breaks through homologous recombination (see Figure 5–59). (A), (B) and (C) were processed 30 minutes after x-irradiation. (From B.E. Nelms et al., *Science* 280:590–592, 1998. With permission from AAAS.)

Cells Carefully Regulate the Use of Homologous Recombination in DNA Repair

Although homologous recombination neatly solves the problem of accurately repairing double-strand breaks and other types of DNA damage, it does present some dangers to the cell and therefore must be tightly regulated. For example, the DNA sequence in one chromosomal homolog can be rendered non-functional by "repairing" it using the other chromosomal homolog as the template. A *loss of heterozygosity* of this type is frequently a critical step in the formation of cancers (as discussed in Chapter 20), and cells have poorly understood mechanisms to minimize it. Although relatively rare in normal cells, loss of heterozygosity can be viewed as an unfortunate side effect of the versatility of homologous recombination.

Another type of control over recombination repair, which is found in nearly all eucaryotic cells, prevents recombination-based "repair" in the absence of DNA damage. The enzymes that catalyze recombination repair are normally made at relatively high levels in eucaryotes and are dispersed throughout the nucleus. In response to DNA damage, they rapidly converge on the sites of damage and eventually form "repair factories" where many DNA lesions are apparently brought together and repaired (**Figure 5–60**). This rapid mobilization of repair proteins to DNA damage is tightly controlled by the cell and requires a series of additional proteins. Two of these, the *Brca1* and *Brca2* proteins, were first discovered because mutations in their genes lead to a greatly increased frequency of breast cancer. Whereas the removal of a protein essential for homologous recombination (such as the human Rad51 protein) will kill a cell, an alteration in an accessory protein can lead to inefficient repair. The subsequent accumulation of DNA damage can, in a small proportion of cells, give rise to a cancer. Brca2 binds to the Rad51 protein preventing its polymerization on DNA and thereby maintaining it in an inactive form. It is thought that Brca2 helps to bring Rad51 protein rapidly to sites of damage and, once in place, to release it in its active form.

In Chapter 20, we shall see that both too much and too little homologous recombination can lead to cancer in humans, the former through an enhanced loss of heterozygosity and the latter through an increased mutation rate caused by inefficient DNA repair. Clearly, a delicate balance has evolved that keeps this process in check on undamaged DNA, while still allowing it to act efficiently and rapidly on DNA lesions as soon as they arise.

Bacteria also carefully regulate their recombination and other DNA repair enzymes, but this occurs largely by controlling their intracellular levels. In response to severe DNA damage, *E. coli* increases the transcription of many DNA repair enzymes as part of the so-called *SOS response*. These include nucleotide excision repair enzymes, error-prone DNA polymerases that can use damaged DNA as a template, and proteins that mediate homologous recombination.

Studies of mutant bacteria deficient in different parts of the SOS response demonstrate that the newly synthesized proteins have two effects. First, as would be expected, the induction of these additional DNA repair enzymes increases cell survival after DNA damage. Second, several of the induced proteins transiently increase the mutation rate by increasing the number of errors made in copying DNA sequences. The errors are caused by the production of low-fidelity DNA polymerases that can efficiently use damaged DNA as a template for DNA synthesis. While this "error-prone" DNA repair can be harmful to individual bacterial cells, it is presumed to be advantageous in the long term because it produces a burst of genetic variability in the bacterial population that increases the likelihood of a mutant cell arising that is better able to survive in the altered environment.

Holliday Junctions Are Often Formed During Homologous Recombination Events

Homologous recombination can be viewed as a group of related reactions that use single-strand invasion, branch migration, and limited DNA synthesis to exchange DNA between two double helices of similar nucleotide sequence. Having discussed its role in accurately repairing damaged DNA, we now turn to homologous recombination as a means to generate DNA molecules of novel sequence. During this process a special DNA intermediate often forms that contains four DNA strands shared between two DNA helices. In this key intermediate, known as a **Holliday junction**, or *cross-strand exchange,* two DNA strands switch partners between two double helices. The Holliday junction can adopt multiple conformations, and a special set of recombination proteins binds to, and thereby stabilizes, the open, symmetric isomer (**Figure 5–61**). By using the energy of ATP hydrolysis to coordinate two branch migration reactions, these proteins can move the point at which the two DNA helices are joined rapidly along the two helices (**Figure 5–62**).

The four-stranded DNA structures produced by homologous recombination are only transiently present in cells. Thus, to regenerate two separate DNA helices, and thus end the recombination process, the strands connecting the two helices in a Holliday junction must eventually be cut, a process referred to as *resolution.* In bacteria, where we understand this process the best, a specialized endonuclease (called RuvC) cleaves the Holliday junctions leaving nicks in the DNA that DNA ligase can seal easily. However, during the meiotic processes that produce germ cells in eucaryotes (sperm and egg in animals), the resolution mechanisms appear to be much more complicated.

As we discuss in Chapter 21, extensive homologous recombination occurs as an integral part of the process whereby chromosomes are parceled out to germ cells during meiosis. Both chromosome *crossing over* and *gene conversion* result from these recombination events, producing hybrid chromosomes that contain genetic information from both the maternal and paternal homologs (**Figure 5–63A**).

As we shall see next, in meiosis the crossing over and gene conversion are both generated by homologous recombination mechanisms that, at their core, resemble those used to repair double-strand breaks.

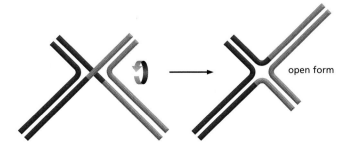

open form

Figure 5–61 **A Holliday junction.** The initially formed structure is usually drawn with two crossing (inside) strands and two noncrossing (outside) strands. An isomerization of the Holliday junction produces an open, symmetrical structure <CTAG>. This is the form that is bound by the RuvA and RuvB proteins (see Figure 5–62). The Holliday junction is named for the scientist who first proposed its formation.

MOVES IN

MOVES
OUT

MOVES
OUT

MOVES IN

Figure 5–62 **Enzyme-catalyzed double branch migration at a Holliday junction.** In *E. coli*, a tetramer of the RuvA protein *(green)* and two hexamers of the RuvB protein *(pale gray)* bind to the open form of the junction. The RuvB protein, which resembles the hexameric helicases used in DNA replication (Figure 5–15), uses the energy of ATP hydrolysis to move the crossover point rapidly along the paired DNA helices, extending the heteroduplex region as shown. (Image courtesy of P. Artymiuk; modified from S.C. West, *Cell* 94:699–701, 1998. With permission from Elsevier.)

Meiotic Recombination Begins with a Programmed Double Strand Break

Homologous recombination in meiosis starts with a bold stroke: a specialized protein (called Spo11 in budding yeast) breaks both strands of the DNA double helix in one of the recombining chromosomes. Like a topoisomerase, the reaction of Spo11 with DNA leaves the protein covalently bound to the broken DNA (see Figure 5–22). A specialized nuclease then rapidly processes the ends bound by Spo11, removing the protein and leaving protruding 3′ single-strand ends. At this point, a series of strand invasions and branch migrations take place that frequently produce an intermediate consisting of two closely spaced Holliday junctions, often called a double Holliday junction (**Figure 5–64**).

Although some of the same proteins that function in double-strand break repair are used in meiosis, these proteins are directed by several meiosis-specific proteins to perform their tasks somewhat differently, resulting in the different DNA intermediates formed (compare Figure 5–59 with Figure 5–64). Another important difference is that, in meiosis, recombination occurs preferentially between maternal and paternal chromosomal homologs rather than between the newly replicated, identical DNA duplexes that pair in double-strand break repair.

There are two different ways to resolve the double Holliday intermediate shown in Figure 5–64. In the conceptually simplest resolution ("noncrossover"), the original pairs of crossing strands are cut at both Holliday junctions in the same way, which causes the two original helices to separate from one another in a form unaltered except for the region between the two junctions (see Figure 5–64, left; in that region, each helix contains a short region of

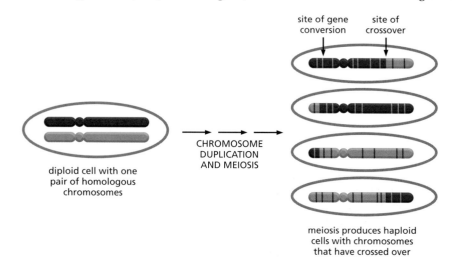

site of gene
conversion

site of
crossover

diploid cell with one
pair of homologous
chromosomes

CHROMOSOME
DUPLICATION
AND MEIOSIS

meiosis produces haploid
cells with chromosomes
that have crossed over

Figure 5–63 **Chromosome crossing over occurs in meiosis.** Meiosis is the process by which a diploid cell gives rise to four haploid germ cells, as described in detail in Chapter 21. Meiosis produces germ cells in which the paternal and maternal genetic information (*red* and *blue*) has been reassorted through chromosome crossovers. In addition, many short regions of gene conversion occur, as indicated.

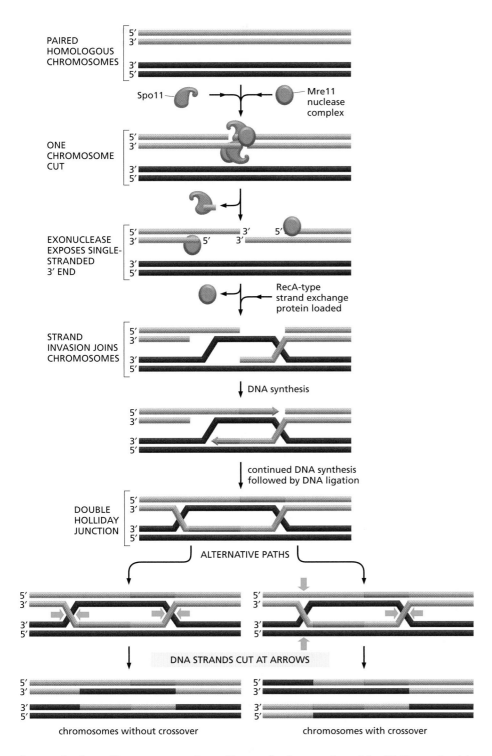

Figure 5–64 Homologous recombination in meiosis can generate crossovers. Once the meiosis-specific protein Spo11 and the Mre11 complex break the duplex DNA and process the ends, homologous recombination proceeds through a double Holliday junction. Many of the steps that produce chromosome crossovers in meiosis resemble those used to repair double-strand breaks (Figure 5–59). However, in meiosis, the process is tightly coupled to other meiotic events and it is directed by proteins, such as Spo11, that are only produced in meiotic cells.

heteroduplex adjacent to a region of homoduplex produced by DNA synthesis). If, on the other hand, the two Holliday junctions are resolved oppositely (one cleaved on the original pair of crossing strands and the other on the non-crossing strands), the outcome is much more profound. In this type of resolution ("crossover"), the portions of each chromosome upstream and downstream from the two Holliday junctions are swapped, creating two chromosomes that have crossed over (see Figure 5–64, right).

Relatively few of the Spo11-mediated double-strand breaks become crossovers; the majority (90% in humans, for example) are resolved as non-crossovers. It is not understood how this choice is made, but it apparently happens early in the recombination process, before the Holliday junctions are formed. The relatively few crossovers that do form are distributed along chromosomes such that the presence of a crossover in one position somehow inhibits crossing over in

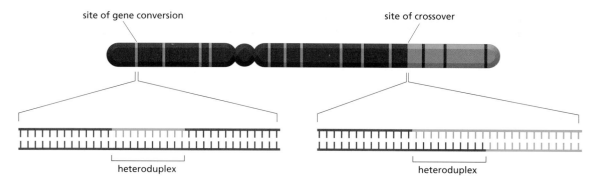

site of gene conversion   site of crossover

heteroduplex   heteroduplex

Figure 5–65 Heteroduplexes formed during meiosis. Heteroduplex DNA is present at sites of recombination that are resolved either as crossovers or non-crossovers.

neighboring regions. Termed *crossover control*, this fascinating but poorly understood regulatory mechanism presumably ensures the roughly even distribution of crossover points along chromosomes. For many organisms, roughly two crossovers per chromosome occur during each meiosis, one on each arm. As discussed in detail in Chapter 21, these crossovers play an important mechanical role in the proper segregation of chromosomes during meiosis.

Whether a meiotic recombination event is resolved as a crossover or a non-crossover, the recombination machinery leaves behind a *heteroduplex region* where a strand from the parental homolog is base-paired with a strand from the maternal homolog (**Figure 5–65**). These heteroduplex regions can tolerate a small percentage of mismatched base pairs, and they often extend for thousands of nucleotide pairs. Because of the many non-crossover events in meiosis, they produce scattered sites in the germ cells where short DNA sequences from one homolog have been pasted into the other homolog. And in all cases, they mark sites of potential *gene conversion*—that is, sites where the four haploid chromosomes produced by meiosis contain three copies of a short DNA sequence from one homolog and only one copy of this sequence from the other homolog (see Figure 5–63), as will now be explained.

Homologous Recombination Often Results in Gene Conversion

In sexually reproducing organisms, it is a fundamental law of genetics that each parent makes an equal genetic contribution to an offspring, which inherits one complete set of nuclear genes from the father and one complete set from the mother. Underlying this law is the highly accurate parceling out of chromosomes to the germ cells (eggs and sperm) that takes place during meiosis. Thus, when a diploid cell undergoes meiosis to produce four haploid germ cells (discussed in Chapter 21), exactly half of the genes distributed among these four cells should be maternal (genes that the diploid cell inherited from its mother) and the other half paternal (genes that the diploid cell inherited from its father) In some organisms (fungi, for example), it is possible to recover and analyze all four of the haploid gametes produced from a single cell by meiosis. Studies in such organisms have revealed rare cases in which the parceling out of genes violates the standard rules of genetics. Occasionally, for example, meiosis yields three copies of the maternal version of a gene and only one copy of the paternal allele (see Figure 5–63). Alternative versions of the same gene are called **alleles**, and the divergence from their expected distribution during meiosis is known as **gene conversion**. Genetic studies show that only small sections of DNA typically undergo gene conversion, and in many cases only a part of a gene is changed.

Several pathways in the cell can lead to gene conversion. First, the DNA synthesis that accompanies the early steps of homologous recombination will produce regions of the double Holliday junction where three copies of the sequence on one homolog are present (see *green* strands at the bottom of Figure 5–64); these will produce sites of gene conversion once the Holliday junction is resolved. In addition, if the two strands that make up a heteroduplex region do not have identical nucleotide sequences, mismatched pairs will result. These can be repaired by the cell's mismatch repair system, described earlier (see Figure 5–20). When used during recombination, however, the mismatch repair system makes

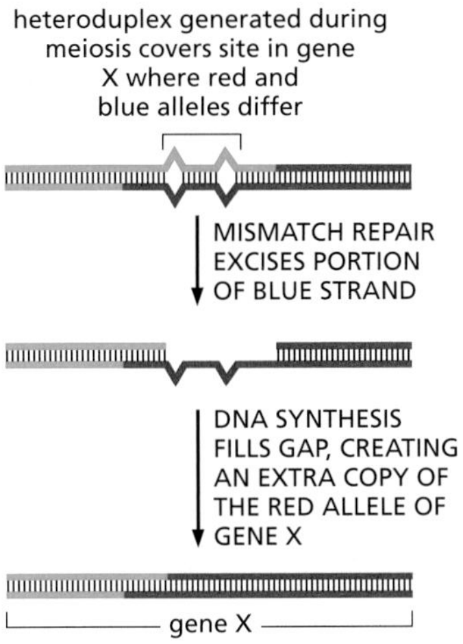

Figure 5–66 Gene conversion caused by mismatch correction. In this process, heteroduplex DNA is formed at the sites of homologous recombination between maternal and paternal chromosomes. If the maternal and paternal DNA sequences are slightly different, the heteroduplex region will include some mismatched base pairs, which may then be corrected by the DNA mismatch repair machinery (see Figure 5–20). Such repair can "erase" nucleotides on either the paternal or the maternal strand. The consequence of this mismatch repair is gene conversion, detected as a deviation from the segregation of equal copies of maternal and paternal alleles that normally occurs in meiosis.

heteroduplex generated during meiosis covers site in gene X where red and blue alleles differ

MISMATCH REPAIR EXCISES PORTION OF BLUE STRAND

DNA SYNTHESIS FILLS GAP, CREATING AN EXTRA COPY OF THE RED ALLELE OF GENE X

gene X

no distinction between the paternal and maternal strand and will randomly choose which strand to repair. As a consequence of this repair, one allele will be "lost" and the other duplicated (**Figure 5–66**), resulting in net "conversion" of one allele to the other. Thus, gene conversion, originally regarded as a mysterious deviation from the rules of genetics, can be seen as a straightforward consequence of the mechanisms of homologous recombination and DNA repair.

Mismatch Proofreading Prevents Promiscuous Recombination Between Two Poorly Matched DNA Sequences

We have seen that homologous recombination relies on the pairing of complementary (or nearly complementary) DNA strands that initially come from separate DNA duplexes. But what controls how precise the matching must be? This is particularly crucial for recombination events that lead to crossovers. For example, the human genome contains many sets of closely related DNA sequences, and if crossing over were permitted between all of them, it would create havoc in the cell.

Although we do not completely understand how cells prevent inappropriate crossovers, we do know that components of the same mismatch proofreading system that removes replication errors (see Figure 5–20) and is responsible for some types of gene conversion (see Figure 5–66) have the additional role of interrupting genetic recombination between poorly matched DNA sequences. It is thought that the mismatch proofreading system normally recognizes the mispaired bases in an initial strand exchange, and—if there are significant mismatches—it prevents the subsequent steps (particularly branch migration) required to form a crossover. This type of recombinational proofreading is thought to prevent promiscuous recombination events that would otherwise scramble the human genome (**Figure 5–67**). Although controversial, it has also

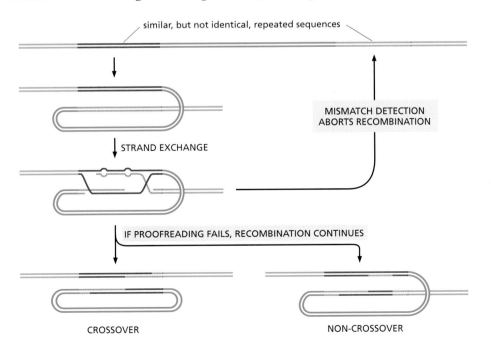

similar, but not identical, repeated sequences

MISMATCH DETECTION ABORTS RECOMBINATION

STRAND EXCHANGE

IF PROOFREADING FAILS, RECOMBINATION CONTINUES

CROSSOVER

NON-CROSSOVER

Figure 5–67 The mechanism that prevents general recombination from destabilizing a genome that contains repeated sequences. Components of the mismatch proofreading system, diagrammed in Figure 5–20, have the additional role of recognizing mismatches and preventing inappropriate recombination. If allowed to proceed, such recombination would produce deletions (*left*) or gene conversions (*right*) where information from one of the original repeated sequences has been lost.

been proposed that recombinational proofreading helps to preserve speciation, particularly among bacteria, by blocking genetic exchange between closely related species. For example, the genomes of *E. coli* and *Salmonella typhimurium* are 80% identical in nucleotide sequences, yet this proofreading step blocks recombination between their genomes.

Summary

Homologous recombination (also called general recombination) results in the transfer of genetic information between two DNA duplex segments of similar nucleotide sequence. This process is essential for the error-free repair of chromosome damage in all cells, and it is also responsible for the crossing-over of chromosomes that occurs during meiosis. The recombination event is guided by a specialized set of proteins. Although it can occur anywhere on a DNA molecule, it always requires extensive base-pairing interactions between complementary strands of the two interacting DNA duplexes.

In meiosis, homologous recombination is initiated by double-strand breaks that are intentionally produced along each chromosome. These breaks are processed into single-stranded 3′ ends which, in a reaction catalyzed by RecA-family proteins, invade a homologous partner DNA duplex. Branch migration accompanied by limited DNA synthesis then leads to the formation of four-stranded structures known as Holliday junctions. Each recombination reaction ends when DNA cutting resolves these recombination intermediates. The result can either be two chromosomes that have crossed-over (that is, chromosomes in which the DNA on either side of the site of DNA pairing originates from two different homologs), or two non-crossover chromosomes. In the latter case, the two chromosomes that result are identical to the original two homologs, except for relatively minor DNA sequence changes at the site of recombination. Unlike the situation in meiosis, the homologous recombination reactions that flawlessly repair double-strand breaks rarely produce crossover products.

TRANSPOSITION AND CONSERVATIVE SITE-SPECIFIC RECOMBINATION

We have seen that, through homologous recombination, rearrangements occur between DNA segments that can result in the exchange of DNA sequences between chromosomes. However, the order of genes on the interacting chromosome typically remains the same following homologous recombination, inasmuch as the recombining sequences must be very similar for the process to occur. In this section, we describe two very different types of recombination— **transposition** (also called *transpositional recombination*) and **conservative site-specific recombination**—that do not require substantial regions of DNA homology. These two types of recombination events can alter gene order along a chromosome, and cause unusual types of mutations that add new information to genomes.

Transposition and conservative site-specific recombination are largely dedicated to moving a wide variety of specialized segments of DNA, collectively termed *mobile genetic elements*, from one position in a genome to another. We will see that mobile genetic elements can range in size from a few hundred to tens of thousands of nucleotide pairs, and each typically carries a unique set of genes. Often, one of these genes encodes a specialized enzyme that catalyzes the movement of only that element, thereby making this type of recombination possible.

Virtually all cells contain mobile genetic elements (known informally as "jumping genes"). As explained in Chapter 4, over evolutionary timescales, they have had a profound effect on the shaping of modern genomes. For example, nearly half of the human genome can be traced to these elements (see Figure 4–17). Over time, random mutation has altered their nucleotide sequences, and, as a result, only a few of the many copies of these elements in our DNA are still active and capable of movement. The remainder are molecular fossils whose existence provides striking clues about our own evolutionary history.

Mobile genetic elements are often considered to be molecular parasites (they are also termed "selfish DNA") that persist because cells cannot get rid of them; they certainly have come close to overrunning our own genome. However, mobile DNA elements can provide benefits to the cell. For example, the genes they carry are sometimes advantageous, as in the case of antibiotic resistance in bacterial cells discussed below. The movement of mobile genetic elements also produces many of the genetic variants upon which evolution depends, because, in addition to moving themselves, mobile genetic elements occasionally rearrange neighboring sequences of the host genome. Thus, spontaneous mutations observed in *Drosophila*, humans, and other organisms are often due to the movement of mobile genetic elements. While the vast majority of these mutations will be deleterious to the organism, some will result in increased fitness and tend to spread throughout the population. It is almost certain that much of the variety of life we see around us originally arose from the movement of mobile genetic elements.

In this section, we introduce mobile genetic elements and describe the mechanisms that enable them to move around a genome. We shall see that some of these elements move through transposition mechanisms and others through conservative site-specific recombination. We begin with transposition, as there are many more examples of this type of movement known.

Through Transposition, Mobile Genetic Elements Can Insert Into Any DNA Sequence

Mobile elements that move by way of transposition are called **transposons**, or **transposable elements**. In transposition, a specific enzyme, usually encoded by the transposon itself and typically called a *transposase*, acts on a specific DNA sequence at each end of the transposon, causing it to insert into a new target DNA site. Most transposons are only modestly selective in choosing their target site, and they can therefore insert themselves into many different locations in the genome. In particular, there is no general requirement for homology between the ends of the element and the target sequence. Most transposons move only rarely. In bacteria, where it is possible to measure the frequency accurately, transposons typically move once every 10^5 cell divisions. In most cases, transposition appears to be a rare stochastic process, albeit one often linked to the passage of a replication fork.

On the basis of their structure and transposition mechanism, transposons can be grouped into three large classes: *DNA-only transposons, retroviral-like retrotransposons,* and *nonretroviral retrotransposons.* Each class will be discussed in detail below. For reference purposes, the differences between them are briefly outlined in **Table 5–3**.

DNA-Only Transposons Move by Both Cut-and-Paste and Replicative Mechanisms

DNA-only transposons predominate in bacteria, and they are largely responsible for the spread of antibiotic resistance in bacterial strains. When antibiotics like penicillin and streptomycin first became widely available in the 1950s, most bacteria that caused human disease were susceptible to them. Fifty years later, the situation has changed dramatically—antibiotics such as penicillin (and its modern derivatives) are no longer effective against many modern bacterial strains, including those causing gonorrhea and bacterial pneumonia. The spread of antibiotic resistance is due largely to genes that encode antibiotic-inactivating enzymes that are carried on transposons (**Figure 5–68**). Although these mobile elements can transpose only within cells that already carry them, they can be moved from one cell to another through other mechanisms known collectively as horizontal gene transfer (Figure 1–23). Once introduced into a new cell, a transposon can insert itself into the genome and be faithfully passed on to all its progeny cells through the normal processes of DNA replication and cell division.

Table 5–3 Three Major Classes of Transposable Elements

CLASS DESCRIPTION AND STRUCTURE	SPECIALIZED ENZYMES REQUIRED FOR MOVEMENT	MODE OF MOVEMENT	EXAMPLES
DNA-only transposons			
short inverted repeats at each end	transposase	moves as DNA, either by cut-and-paste or replicative pathways	P element (*Drosophila*) Ac-Ds (maize) Tn3 and Tn10 (*E. coli*) Tam3 (snapdragon)
Retroviral-like retrotransposons			
directly repeated long terminal repeats (LTRs) at each end	reverse transcriptase and integrase	moves via an RNA intermediate produced by a promoter in the LTR	Copia (*Drosophila*) Ty1 (yeast) THE1 (human) Bs1 (maize)
Nonretroviral retrotransposons			
Poly A at 3′ end of RNA transcript; 5′ end is often truncated	reverse transcriptase and endonuclease	moves via an RNA intermediate that is often produced from a neighboring promoter	F element (*Drosophila*) L1 (human) Cin4 (maize)

These elements range in length from 1000 to about 12,000 nucleotide pairs. Each family contains many members, only a few of which are listed here. In addition to transposable elements, some viruses can move in and out of host cell chromosomes by transpositional mechanisms. These viruses are related to the first two classes of transposons.

DNA-only transposons, so named because they exist only as DNA during their movement, can relocate from a donor site to a target site either by *cut-and-paste transposition* or by *replicative transposition*. Because it is conceptually simpler, we discuss the cut-and-paste mechanism first. The process begins when each of the special short DNA sequences that mark the two ends of the element binds a molecule of transposase. The two transposase molecules come together to form a multimeric "transpososome" that produces a DNA loop juxtaposing the two ends of the element (**Figure 5–69**). The transposase then introduces cuts at the base of the loop and removes the element completely from its original chromosome, forming the central intermediate in the transposition process (**Figure 5–70**). To complete the DNA movement, the transposase catalyzes a direct attack of the element's two DNA ends on a target DNA molecule, breaking two phosphodiester bonds in the target molecule and creating two new ones as it joins the element and target DNAs together. Because this DNA joining reaction begins and ends with the same number of phosphodiester bonds, it can occur without the input of additional energy. We will see in the next chapter that this same type of phosphodiester bond rearrangement (called *transesterification*) underlies another fundamental process in molecular biology, RNA splicing.

Because the breaks made in the two target DNA strands are staggered (*red arrowheads* in Figure 5–69), the product DNA molecule initially contains two

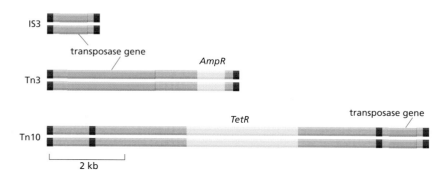

Figure 5–68 Three of the many DNA-only transposons found in bacteria. Each of these mobile DNA elements contains a gene that encodes a *transposase*, an enzyme that conducts at least some of the DNA breakage and joining reactions needed for the element to move. Each transposon also carries short DNA sequences (indicated in *red*) that are recognized only by the transposase encoded by that element and are necessary for movement of the element. In addition, two of the three mobile elements shown carry genes that encode enzymes that inactivate the antibiotics ampicillin (*AmpR*) and tetracycline (*TetR*). The transposable element Tn10, shown in the bottom diagram, is thought to have evolved from the chance landing of two much shorter mobile elements on either side of a tetracycline-resistance gene: the wide use of tetracycline as an antibiotic has selected for the spread of this transposon through bacterial populations.

short, single-stranded gaps, one at each end of the inserted transposon. A host-cell DNA polymerase and DNA ligase fill in and seal these gaps to complete the recombination process. This produces a short duplication of the target DNA sequence at the insertion site; these flanking direct repeat sequences, whose length is different for different transposons, serve as convenient records of prior transposition events.

When a cut-and-paste DNA-only transposon is excised from its original location, it leaves behind a "hole" in the chromosome. This lesion can be perfectly healed by recombinational double-strand break repair (see Figure 5–59), provided that the chromosome has just been replicated and an identical copy of the damaged host sequence is available. In this case, the repair process will restore the transposon to its original position. Alternatively, in diploid organisms, the damaged chromosome can be recombinationally repaired using the chromosomal homolog, in which case the transposon will not be restored but a loss of heterozygosity could occur at the site of repair. As the third possibility, a nonhomologous end-joining reaction can reseal the break; in this case, the DNA sequence that originally flanked the transposon is altered, producing a mutation at the chromosomal site from which the transposon was excised (see Figure 5–52).

Remarkably, the same mechanism used to excise cut-and-paste transposons from DNA has been found to operate in developing immune systems of vertebrates, catalyzing the DNA rearrangments that produce antibody and T cell receptor diversity. Known as *V(D)J recombination*, this process will be discussed in Chapter 25. Found only in vertebrates, V(D)J recombination is a relatively recent evolutionary novelty, but it is believed to have evolved from the much more ancient cut-and-paste transposons.

Some DNA-only transposons move by a mechanism called *replicative transposition*. In this case, the transposon DNA is replicated during transposition: one copy remains at the original site while the other is inserted at a new chromosomal location. Although the mechanism used is more complex, it is closely related to the cut-and-paste mechanism just described; indeed, some transposons can move by both pathways.

Some Viruses Use a Transposition Mechanism to Move Themselves into Host Cell Chromosomes

Certain viruses are considered mobile genetic elements because they use transposition mechanisms to integrate their genomes into that of their host cell. However, unlike transposons, these viruses encode proteins that package their genetic information into virus particles that can infect other cells. Many of the

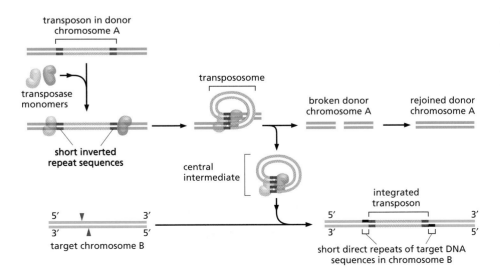

Figure 5–69 Cut-and-paste transposition. DNA-only transposons can be recognized in chromosomes by the "inverted repeat DNA sequences" (*red*) present at their ends. These sequences, which can be as short as 20 nucleotides, are all that is necessary for the DNA between them to be transposed by the particular transposase enzyme associated with the element. The cut-and-paste movement of a DNA-only transposable element from one chromosomal site to another begins when the transposase brings the two inverted DNA sequences together, forming a DNA loop. Insertion into the target chromosome, catalyzed by the transposase, occurs at a random site through the creation of staggered breaks in the target chromosome (*red arrowheads*). Following the transposition reaction, the single-strand gaps created by the staggered break are repaired by DNA polymerase and ligase (*purple*). As a result, the insertion site is marked by a short direct repeat of the target DNA sequence, as shown. Although the break in the donor chromosome (*green*) is repaired, this process often alters the DNA sequence, causing a mutation at the original site of the excised transposable element (not shown).

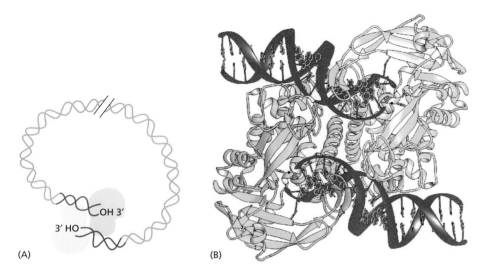

(A)   (B)

Figure 5–70 The structure of the central intermediate formed by a cut-and-paste transposase. (A) Schematic view of the overall structure. (B) The detailed structure of a transposase holding the two DNA ends, whose 3′-OH groups are poised to attack a target chromosome. One domain of the transposase recognizes the DNA sequence at the transposon end while a different domain carries out the DNA breakage–joining chemistry (B, from D.R. Davies et al., *Science* 289:77–85, 2000. With permission from AAAS.)

viruses that insert themselves into a host chromosome do so by employing one of the top two mechanisms listed in Table 5–3. Indeed, much of our knowledge of these mechanisms has come from studies of particular viruses that employ them.

A virus that infects a bacterium is known as a **bacteriophage**. The *bacteriophage Mu* not only uses DNA-based transposition to integrate its genome into its host cell chromosome, it also uses replicative transposition to replicate its genome. Transposition also has a key role in the life cycle of many other viruses. Most notable are the **retroviruses**, which include the human AIDS virus, HIV. Outside the cell, a retrovirus exists as a single-stranded RNA genome packed into a protein capsid along with a virus-encoded **reverse transcriptase** enzyme. During the infection process, the viral RNA enters a cell and is converted to a double-stranded DNA molecule by the action of this crucial enzyme, which is able to polymerize DNA on either an RNA or a DNA template (**Figure 5–71** and **Figure 5–72**). The term *retrovirus* refers to the virus's ability to reverse the usual flow of genetic information, which normally is from DNA to RNA (see Figure 1–5).

Once the reverse transcriptase has produced a double-stranded DNA molecule, specific sequences near its two ends are held together by a virus-encoded transposase called *integrase*. Integrase creates activated 3′-OH viral DNA ends that can directly attack a target DNA molecule through a mechanism similar to that used by the cut-and-paste DNA-only transposons (**Figure 5–73**). In fact, detailed analyses of the three-dimensional structures of bacterial transposases and HIV integrase have revealed remarkable similarities in these enzymes, even though their amino acid sequences have diverged considerably.

Retroviral-like Retrotransposons Resemble Retroviruses, but Lack a Protein Coat

A large family of transposons called **retroviral-like retrotransposons** (see Table 5–3) move themselves in and out of chromosomes by a mechanism that is identical to that used by retroviruses. These elements are present in organisms as diverse as yeasts, flies, and mammals; unlike viruses, they have no intrinsic ability to leave their resident cell but are passed along to all descendants of that cell through the normal process of DNA replication and cell division. The first step in their transposition is the transcription of the entire transposon, producing an RNA copy of the element that is typically several thousand nucleotides long. This transcript, which is translated as a messenger RNA by the host cell, encodes a reverse transcriptase enzyme. This enzyme makes a double-stranded DNA copy of the RNA molecule via an RNA/DNA hybrid intermediate, precisely mirroring the early stages of infection by a retrovirus (see Figure 5–71). Like retroviruses, the linear double-stranded DNA molecule then integrates into a site on the chromosome by using an integrase enzyme that is also encoded by the element (see Figure 5–73).

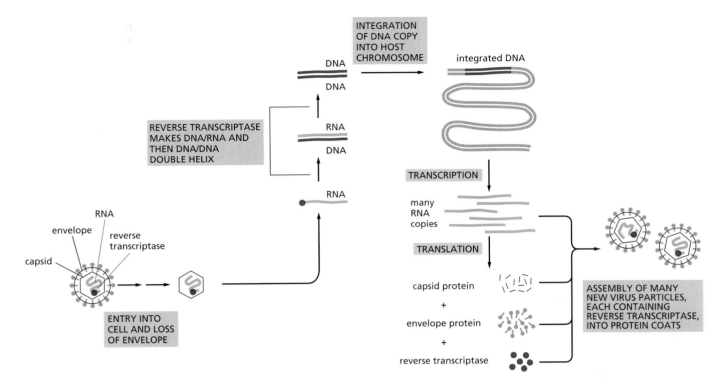

Figure 5–71 The life cycle of a retrovirus. The retrovirus genome consists of an RNA molecule of about 8500 nucleotides; two such molecules are typically packaged into each viral particle. The enzyme reverse transcriptase first makes a DNA copy of the viral RNA molecule and then a second DNA strand, generating a double-stranded DNA copy of the RNA genome. The integration of this DNA double helix into the host chromosome is then catalyzed by a virus-encoded integrase enzyme (see Figure 5–73). This integration is required for the synthesis of new viral RNA molecules by the host cell RNA polymerase, the enzyme that transcribes DNA into RNA (discussed in Chapter 6).

A Large Fraction of the Human Genome Is Composed of Nonretroviral Retrotransposons

A significant fraction of many vertebrate chromosomes is made up of repeated DNA sequences. In human chromosomes, these repeats are mostly mutated and truncated versions of **nonretroviral retrotransposons**, the third major type of transposon (see Table 5–3). Although most of these transposons are immobile, a few retain the ability to move. Relatively recent movements of the *L1 element* (sometimes referred to as a LINE or long interspersed nuclear element) have been identified, some of which result in human disease; for example, a particular type of hemophilia results from an *L1* insertion into the gene encoding the blood clotting protein Factor VIII (see Figure 6–25).

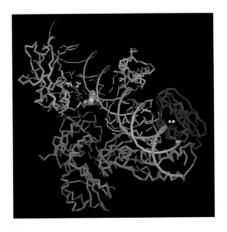

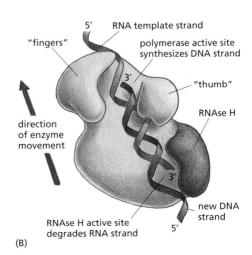

Figure 5–72 Reverse transcriptase.
(A) The three-dimensional structure of the enzyme from HIV (the human AIDS virus) determined by X-ray crystallography. (B) A model showing the enzyme's activity on an RNA template. Note that the polymerase domain *(yellow in B)* has a covalently attached RNAse H (H for "hybrid") domain *(red)* that degrades the RNA strand in an RNA/DNA helix. This activity helps the polymerase to convert the initial hybrid helix into a DNA double helix (A, courtesy of Tom Steitz; B, adapted from L.A. Kohlstaedt et al., *Science* 256:1783–1790, 1992. With permission from AAAS.)

(A)

(B)

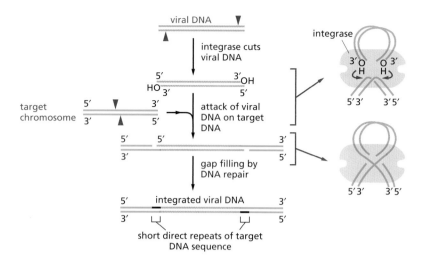

Figure 5–73 Transposition by either a retrovirus (such as HIV) or a retroviral-like retrotransposon. The process begins with a double-stranded DNA molecule *(orange)* produced by reverse transcriptase (see Figure 5–71). In an initial step, the integrase enzyme forms a DNA loop and cuts one strand at each end of the viral DNA sequence, exposing new 3'-OH groups. Each of these 3'-OH ends then directly attack a phosphodiester bond on opposite strands of a randomly selected site on a target chromosome *(red arrowheads* on *blue* DNA). This inserts the viral DNA sequence into the target chromosome, leaving short gaps on each side that are filled in by DNA repair processes. Because of the gap filling, this mechanism (like that of cut-and-paste transposition) leaves short repeats of target DNA sequence *(black)* on each side of the integrated DNA segment.

Nonretroviral retrotransposons are found in many organisms and move via a distinct mechanism that requires a complex of an endonuclease and a reverse transcriptase. As illustrated in **Figure 5–74**, the RNA and reverse transcriptase have a much more direct role in the recombination event than they do in the retroviral-like retrotransposons described above.

Inspection of the human genome sequence has revealed that the bulk of nonretroviral retrotransposons—for example, the many copies of the *Alu* element, a member of the SINE (short interspersed nuclear element) family—do not carry their own endonuclease or reverse transcriptase genes. Nonetheless they have successfully amplified themselves to become major constituents of our genome, presumably by pirating enzymes encoded by other transposons.

The *L1* and *Alu* elements seem to have multiplied in the human genome relatively recently (**Figure 5–75**). Thus, for example, the mouse contains sequences closely related to *L1* and *Alu*, but their placement in mouse chromosomes differs from that in human chromosomes (see Figure 4–80). Together the LINEs and SINEs make up about 40% of the human genome (see Figure 4–17).

Different Transposable Elements Predominate in Different Organisms

We have described several types of transposable elements: (1) DNA-only transposons, the movement of which is based on DNA breaking and joining reactions; (2) retroviral-like retrotransposons, which also move via DNA breakage and joining, but where RNA has a key role as a template to generate the DNA recombination substrate; and (3) nonretroviral retrotransposons, in which an RNA copy of the element is central to the incorporation of the element into the target DNA, acting as a direct template for a DNA target-primed reverse transcription event.

Intriguingly, different types of transposons predominate in different organisms. For example, the vast majority of bacterial transposons are DNA-only types, with a few related to the nonretroviral retrotransposons also present. In yeasts, the main mobile elements that have been observed are retroviral-like retrotransposons. In *Drosophila*, DNA-based, retroviral, and nonretroviral

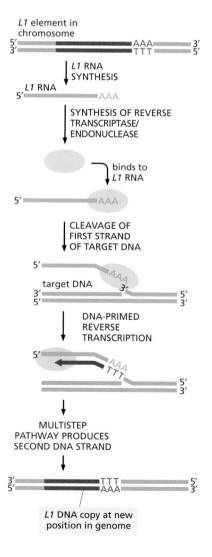

Figure 5–74 Transposition by a nonretroviral retrotransposon. Transposition by the *L1* element *(red)* begins when an endonuclease attached to the *L1* reverse transcriptase *(green)* and the *L1* RNA *(blue)* nicks the target DNA at the point at which insertion will occur. This cleavage releases a 3'-OH DNA end in the target DNA, which is then used as a primer for the reverse transcription step shown. This generates a single-stranded DNA copy of the element that is directly linked to the target DNA. In subsequent reactions, further processing of the single-stranded DNA copy results in the generation of a new double-stranded DNA copy of the *L1* element that is inserted at the site of the initial nick.

transposons are all found. Finally, the human genome contains all three types of transposon, but as discussed below, their evolutionary histories are strikingly different.

Genome Sequences Reveal the Approximate Times that Transposable Elements Have Moved

The nucleotide sequence of the human genome provides a rich "fossil record" of the activity of transposons over evolutionary time spans. By carefully comparing the nucleotide sequences of the approximately 3 million transposable element remnants in the human genome, it has been possible to broadly reconstruct the movements of transposons in our ancestors' genomes over the past several hundred million years. For example, the DNA-only transposons appear to have been very active well before the divergence of humans and Old World monkeys (25–35 million years ago); but, because they gradually accumulated inactivating mutations, they have been dormant in the human lineage since that time. Likewise, although our genome is littered with relics of retroviral-like transposons, none appear to be active today. Only a single family of retroviral-like retrotransposons is believed to have transposed in the human genome since the divergence of human and chimpanzee approximately 6 million years ago. The nonretroviral retrotransposons are also ancient, but in contrast to other types, some are still moving in our genome, as mentioned previously. For example, it is estimated that *de novo* movement of an *Alu* element is seen once every 100–200 human births. The movement of nonretroviral retrotransposons is responsible for a small but significant fraction of new human mutations—perhaps two mutations out of every thousand.

The situation in mice is significantly different. Although the mouse and human genomes contain roughly the same density of the three types of transposons, both types of retrotransposons are still actively transposing in the mouse genome, being responsible for approximately 10% of new mutations.

Although we are only beginning to understand how the movements of transposons has shaped the genomes of present-day mammals, it has been proposed that bursts in transposition activity could have been responsible for critical speciation events during the radiation of the mammalian lineages from a common ancestor, a process that began approximately 170 million years ago. At this point, we can only wonder how many of our uniquely human qualities arose from the past activity of the many mobile genetic elements whose remnants are found today scattered throughout our chromosomes.

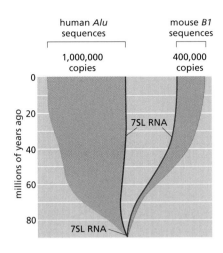

Figure 5–75 The expansion of the abundant *Alu* and *B1* sequences found in the human and mouse genomes, respectively. Both of these transposable DNA sequences are thought to have evolved from the 7SL RNA gene, which encodes the SRP RNA (see Figure 12–39). On the basis of their positions in the two genomes and the sequence similarity of these highly repeated elements, the major expansions in copy number seem to have occurred independently in mice and humans (see Figure 4–80). (Adapted from P.L. Deininger and G.R. Daniels, *Trends Genet.* 2:76–80, 1986, with permission from Elsevier, and International Human Genome Sequencing Consortium, *Nature* 409:860–921, 2001, with permission from Macmillan Publishers Ltd.)

Conservative Site-specific Recombination Can Reversibly Rearrange DNA

A different kind recombination mechanism, known as *conservative site-specific recombination,* mediates the rearrangements of other types of mobile DNA elements. In this pathway, breakage and joining occur at two special sites, one on each participating DNA molecule. Depending on the positions and relative orientations of the two recombination sites, DNA integration, DNA excision, or DNA inversion can occur (**Figure 5–76**). Conservative site-specific recombination is carried out by specialized enzymes that break and rejoin two DNA double helices at specific sequences on each DNA molecule. The same enzyme system that joins two DNA molecules can often take them apart again, precisely restoring the sequence of the two original DNA molecules (see Figure 5–76A).

Several features distinguish conservative site-specific recombination from transposition. First, conservative site-specific recombination requires specialized DNA sequences on both the donor and recipient DNA (hence the term site-specific). These sequences contain recognition sites for the particular recombinase that will catalyze the rearrangement. In contrast, transposition requires only that the transposon have a specialized sequence; for most transposons, the recipient DNA can be of any sequence. Second, the reaction mechanisms are

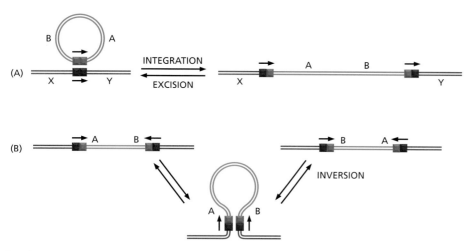

Figure 5–76 Two types of DNA rearrangement produced by conservative site-specific recombination. The only difference between the reactions in (A) and (B) is the relative orientation of the two short DNA sites (indicated by *arrows*) at which a site-specific recombination event occurs. (A) Through an integration reaction, a circular DNA molecule can become incorporated into a second DNA molecule; by the reverse reaction (excision), it can exit to reform the original DNA circle. Many bacterial viruses move in and out of their host chromosomes in this way (see Figure 5–77). (B) Conservative site-specific recombination can also invert a specific segment of DNA in a chromosome. A well-studied example of DNA inversion through site-specific recombination occurs in the bacterium *Salmonella typhimurium*, an organism that is a major cause of food poisoning in humans; the inversion of a DNA segment changes the type of flagellum that is produced by the bacterium (see Figure 7–64).

fundamentally different. The recombinases that catalyze conservative site-specific recombination resemble topoisomerases in the sense that they form transient high-energy covalent bonds with the DNA and use this energy to complete the DNA rearrangements. Thus all the phosphate bonds that are broken during a recombination event are restored upon its completion (hence the term conservative). Transposition, in contrast, uses a transesterification reaction that does not proceed through a covalently joined protein–DNA intermediate. This process leaves gaps in the DNA that must be resealed by DNA polymerase and ligase, both of which require the input of energy in the form nucleotide hydrolysis.

Conservative Site-Specific Recombination Was Discovered in Bacteriophage λ

A bacterial virus, *bacteriophage lambda*, was the first mobile DNA element of any type to be understood in biochemical detail. When this virus enters a cell, it directs the synthesis of a virus-encoded recombinase enzyme called *lambda integrase*. This enzyme mediates the covalent joining of the viral DNA to the bacterial chromosome, causing the virus to become part of this chromosome so that it is replicated automatically—as part of the host's DNA. The recombination process begins when several molecules of the integrase protein bind tightly to a specific DNA sequence on the circular bacteriophage chromosome, along with several host proteins. This DNA–protein complex can now bind to an attachment-site DNA sequence on the bacterial chromosome, bringing the bacterial and bacteriophage chromosomes together. The integrase then catalyzes the required cutting and resealing reactions that result in recombination. Because of a short region of sequence homology in the two joined sequences, a tiny heteroduplex joint is formed at this point of exchange (**Figure 5–77**).

   The same type of site-specific recombination mechanism enables bacteriophage lambda DNA to exit from its integration site in the *E. coli* chromosome in response to specific signals and multiply rapidly within the bacterial cell (**Figure 5–78**). Excision is catalyzed by a complex of integrase enzyme and host factors with a second bacteriophage protein, excisionase, which is produced by the virus only when its host cell is stressed—in which case, it is in the bacteriophage's interest to abandon the host cell and multiply again as a virus particle.

Conservative Site-Specific Recombination Can Be Used to Turn Genes On or Off

When the special sites recognized by a conservative site-specific recombination enzyme are inverted in their orientation, the DNA sequence between them is inverted rather than excised (see Figure 5–76). Many bacteria use such an inversion of a DNA sequence to control the gene expression of particular genes—for example, by assembling an active gene from separated coding segments. This

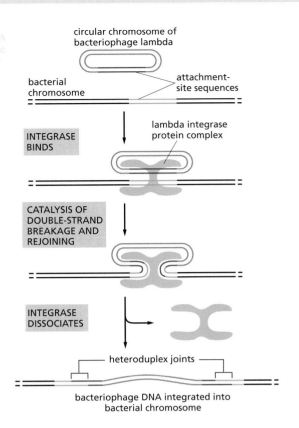

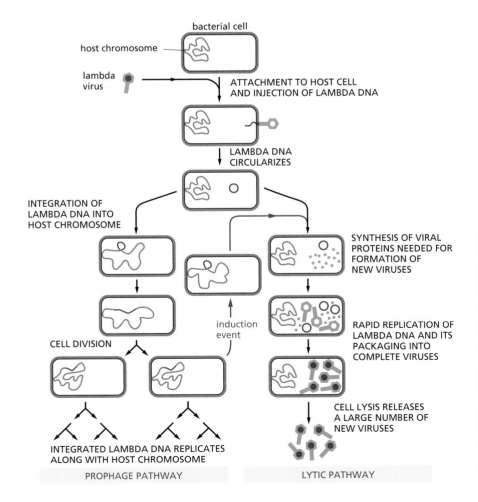

Figure 5–77 The insertion of a circular bacteriophage lambda DNA chromosome into the bacterial chromosome. In this example of conservative site-specific recombination, the lambda integrase enzyme binds to a specific attachment-site DNA sequence on each chromosome, where it makes cuts that bracket a short homologous DNA sequence. The integrase then switches the partner strands and reseals them to form a heteroduplex joint that is seven nucleotide pairs long. A total of four strand-breaking and strand-joining reactions are required; for each of them, the energy of the cleaved phosphodiester bond is stored in a transient covalent linkage between the DNA and the enzyme, so that DNA strand resealing occurs without a requirement for ATP or DNA ligase.

Figure 5–78 The life cycle of bacteriophage lambda. The double-stranded DNA lambda genome contains 50,000 nucleotide pairs and encodes 50–60 different proteins. When the lambda DNA enters the cell, its two ends join to form a circular DNA molecule. This bacteriophage can multiply in *E. coli* by a lytic pathway, which destroys the cell, or it can enter a latent prophage state. Damage to a cell carrying a lambda prophage induces the prophage to exit from the host chromosome and shift to lytic growth *(red arrows)*. Both the entrance of the lambda DNA into, and its exit from, the bacterial chromosome are accomplished by conservative site-specific recombination catalyzed by the lambda integrase enzyme (see Figure 5–77).

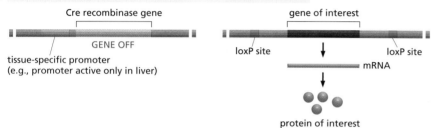

(A)   IN SPECIFIC TISSUE (e.g., LIVER)

Cre recombinase gene

GENE ON

→ mRNA

↓

Cre recombinase made
only in liver cells

gene of interest

loxP site   loxP site

↓

+

gene of interest deleted from chromosome
and lost as liver cells divide

(B)   IN OTHER TISSUES, THE GENE OF INTEREST IS EXPRESSED NORMALLY

Cre recombinase gene

GENE OFF

tissue-specific promoter
(e.g., promoter active only in liver)

gene of interest

loxP site   loxP site

↓ mRNA

↓

protein of interest

Figure 5–79 How a conservative site-specific recombination enzyme from bacteria can be used to delete specific genes from particular mouse tissues. This approach requires the insertion of two specially engineered DNA molecules into the animal's germ line. The first contains the gene for a recombinase (in this case the Cre recombinase from the bacteriophage P1) under the control of a tissue-specific promoter, which ensures that the recombinase is expressed only in that tissue. The second DNA molecule contains the gene of interest flanked by recognition sites (in this case loxP sites) for the recombinase. The mouse is engineered so that this is the only copy of this gene. Therefore, if the recombinase is expressed only in the liver, the gene of interest will be deleted there, and only there.

As described in Chapter 7, many tissue-specific promoters are known; moreover, many of these promoters are active only at specific times in development. Thus, it is possible to study the effect of deleting specific genes at many different times during the development of each tissue.

type of gene control has the advantage of being directly inheritable, since the new DNA arrangement is transferred to daughter chromosomes automatically when a cell divides. We will encounter a specific example of this use of conservative site-specific recombination in Chapter 7 (see Figure 7–64).

Bacterial conservative site-specific recombinases have also become powerful tools for cell and developmental biologists. To decipher the roles of specific genes and proteins in complex multicellular organisms, genetic engineering techniques are used to produce mice carrying a gene encoding a site-specific recombination enzyme plus a carefully designed target DNA with the DNA sites that this enzyme recognizes. At an appropriate time, the gene encoding the enzyme can be activated to rearrange the target DNA sequence. Such a rearrangement is widely used to delete a specific gene in a particular tissue of the mouse (**Figure 5–79**). It is particularly useful when the gene of interest plays a key role in the early development of many tissues, and a complete deletion of the gene from the germ line would cause death very early in embryogenesis. The same strategy can also be used to inappropriately express any specific gene in a tissue of interest; here, the triggered deletion joins a strong transcriptional promoter to the gene of interest. With this tool one can in principle determine the influence of any protein in any desired tissue of an intact animal.

Summary

The genomes of nearly all organisms contain mobile genetic elements that can move from one position in the genome to another by either transpositional or conservative site-specific recombination processes. In most cases, this movement is random and happens at a very low frequency. Mobile genetic elements include transposons, which move within a single cell (and its descendants), plus those viruses whose genomes can integrate into the genomes of their host cells.

There are three classes of transposons: the DNA-only transposons, the retroviral-like retrotransposons, and the nonretroviral retrotransposons. All but the last have close relatives among the viruses. Although viruses and transposable elements can be viewed as parasites, many of the new arrangements of DNA sequences that their site-specific recombination events produce have created the genetic variation crucial for the evolution of cells and organisms.

PROBLEMS

Which statements are true? Explain why or why not.

5–1 No two cells in your body have the identical nucleotide sequence.

5–2 In *E. coli*, where the replication fork travels at 500 nucleotide pairs per second, the DNA ahead of the fork must rotate at nearly 3000 revolutions per minute.

5–3 When bidirectional replication forks from adjacent origins meet, a leading strand always runs into a lagging strand.

5–4 DNA repair mechanisms all depend on the existence of two copies of the genetic information, one in each of the two homologous chromosomes.

Discuss the following problems.

5–5 To determine the reproducibility of mutation frequency measurements, you do the following experiment. You inoculate each of 10 cultures with a single *E. coli* bacterium, allow the cultures to grow until each contains 10^6 cells, and then measure the number of cells in each culture that carry a mutation in your gene of interest. You were so surprised by the initial results that you repeated the experiment to confirm them. Both sets of results display the same extreme variability, as shown in **Table Q5–1**. Assuming that the rate of mutation is constant, why do you suppose there is so much variation in the frequencies of mutant cells in different cultures?

Table Q5–1 Frequencies of mutant cells in multiple cultures (Problem 5–5).

	CULTURE (mutant cells/10^6 cells)									
EXPERIMENT	1	2	3	4	5	6	7	8	9	10
1		4	0	257	1	2	32	0	0	2	1
2		128	0		1	4	0	0	66	5	0	2

5–6 DNA repair enzymes preferentially repair mismatched bases on the newly synthesized DNA strand, using the old DNA strand as a template. If mismatches were repaired instead without regard for which strand served as template, would mismatch repair reduce replication errors? Would such an indiscriminate mismatch repair result in fewer mutations, more mutations, or the same number of mutations as there would have been without any repair at all? Explain your answers.

5–7 If DNA polymerase requires a perfectly paired primer in order to add the next nucleotide, how is it that any mismatched nucleotides "escape" the polymerase and become substrates for mismatch repair enzymes?

5–8 The laboratory you joined is studying the life cycle of an animal virus that uses a circular, double-stranded DNA as its genome. Your project is to define the location of the origin(s) of replication and to determine whether replication proceeds in one or both directions away from an origin (unidirectional or bidirectional replication). To accomplish your goal, you isolated replicating molecules, cleaved them with a restriction nuclease that cuts the viral genome at one site to produce a linear molecule from the circle, and examined the resulting molecules in the electron microscope. Some of the molecules you observed are illustrated schematically in **Figure Q5–1**. (Note that it is impossible to distinguish the orientation of one DNA molecule from another in the electron microscope.)

You must present your conclusions to the rest of the lab tomorrow. How will you answer the two questions your advisor posed for you? Is there a single, unique origin of replication or several origins? Is replication unidirectional or bidirectional?

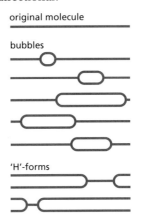

Figure Q5–1 Parental and replicating forms of an animal virus (Problem 5–8).

original molecule

bubbles

'H'-forms

5–9 If you compare the frequency of the sixteen possible dinucleotide sequences in *E. coli* and human cells, there are no striking differences except for one dinucleotide, 5'-CG-3'. The frequency of CG dinucleotides in the human genome is significantly lower than in *E. coli* and significantly lower than expected by chance. Why do you suppose that CG dinucleotides are underrepresented in the human genome?

5–10 With age, somatic cells are thought to accumulate genomic "scars" as a result of the inaccurate repair of double-strand breaks by nonhomologous end-joining (NHEJ). Estimates based on the frequency of breaks in primary human fibroblasts suggest that by age 70 each human somatic cell may carry some 2000 NHEJ-induced mutations due to inaccurate repair. If these mutations were distributed randomly around the genome, how many genes would you expect to be affected? Would you expect cell function to be compromised? Why or why not? (Assume that 2% of the genome—1.5% coding and 0.5% regulatory—is crucial information.)

5–11 Draw the structure of the double Holliday junction that would result from strand invasion by both ends of the broken duplex into the intact homologous duplex shown in **Figure Q5–2**. Label the left end of each strand in the Holliday junction 5' or 3' so that the relationship to the parental and recombinant duplexes is clear. Indicate how DNA synthesis would be used to fill in any single-strand gaps in your double Holliday junction.

Figure Q5–2 A broken duplex with single-strand tails ready to invade an intact homologous duplex (Problem 5–11).

5'   3'

5'   3'

5–12 Why is it that recombination between similar, but nonidentical, repeated sequences poses a problem for human cells? How does the mismatch-repair system protect against such recombination events?

REFERENCES

General

Biological Response to DNA Damage (2000) *Cold Spring Harb Symp Quant Biol* 65.

Brown TA (2002) Genomes 2, 2nd ed. New York: Wiley-Liss.

Friedberg EC, Walker GC, Siede W et al (2005) DNA Repair and Mutagenesis. Washington, DC: ASM Press.

Hartwell L, Hood L, Goldberg ML et al (2006) Genetics: from Genes to Genomes. Boston: McGraw Hill.

Stent GS (1971) Molecular Genetics: An Introductory Narrative. San Francisco: WH Freeman.

Watson J, Baker T, Bell S et al (2004) Molecular Biology of the Gene, 5th ed. Plainview, New York: Cold Spring Harbor Laboratory Press.

The Maintenance of DNA Sequences

Cooper GM, Brudno M, Stone ES et al (2004) Characterization of evolutionary rates and constraints in three mammalian genomes. *Genome Res* 14:539–548.

Crow JF (2000) The origins, patterns and implications of human spontaneous mutation. *Nature Rev Genet* 1:40–47.

Hedges SB (2002) The origin and evolution of model organisms. *Nature Rev Genet* 3:838–849.

King MC, Wilson AC (1965) Evolution at two levels in humans and chimpanzees. *Science* 188:107–16.

DNA Replication Mechanisms

Alberts B (1998) The cell as a collection of protein machines: preparing the next generation of molecular biologists. *Cell* 92:291–294.

Dillingham MS (2006) Replicative helicases: a staircase with a twist. *Curr Biol* 16:R844–R847.

Indiani C & O'Donnell M (2006) The replication clamp-loading machine at work in the three domains of life. *Nature Rev Mol Cell Biol* 7:751–761.

Kornberg A (1960) Biological synthesis of DNA. *Science* 131:1503–1508.

Li JJ & Kelly TJ (1984) SV40 DNA replication *in vitro*. *Proc Natl Acad Sci USA* 81:6973.

Meselson M & Stahl FW (1958) The replication of DNA in *E. coli*. *Proc Natl Acad Sci USA* 44:671–682.

Modrich P & Lahue R (1996) Mismatch repair in replication fidelity, genetic recombination, and cancer biology. *Annu Rev Biochem* 65:101–133.

Mott ML & Berger JM (2007) DNA replication initiation: mechanisms and regulation in bacteria. *Nature Rev Microbiol* 5:343–354.

O'Donnell M (2006) Replisome architecture and dynamics in *E. coli*. *J Biol Chem* 281:10653–10656.

Okazaki R, Okazaki T, Sakabe K et al. (1968) Mechanism of DNA chain growth. I. Possible discontinuity and unusual secondary structure of newly synthesized chains. *Proc Natl Acad Sci USA* 59:598–605.

Raghuraman MK, Winzeler EA, Collingwood D et al (2001) Replication dynamics of the yeast genome. *Science* 294:115–121.

Rao PN & Johnson RT (1970) Mammalian cell fusion: studies on the regulation of DNA synthesis and mitosis. *Nature* 225:159.

Wang JC (2002) Cellular roles of DNA topoisomerases: a molecular perspective. *Nature Rev Mol Cell Biol* 3:430–440.

The Initiation and Completion of DNA Replication in Chromosomes

Chan SR & Blackburn EH (2004) Telomeres and telomerase. *Philos Trans R Soc Lond B Bio Sci* 359:109–121.

Costa S & Blow JJ (2007) The elusive determinants of replication origins. *EMBO Rep* 8:332–334.

Groth A, Rocha W & Almouzni G (2007) Chromatin challenges during DNA replication and repair. *Cell* 128:721–733.

Machida YJ, Hamlin JL & Dutta A (2005) Right place, right time, and only once: replication initiation in metazoans. *Cell* 123:13–24.

O'Donnell M & Kuriyan J (2005) Clamp loaders and replication initiation. *Curr Opin Struct Biol* 16:35–41.

Robinson NP & Bell SD (2005) Origins of DNA replication in the three domains of life. *FEBS J* 272:3757–3766.

Smogorzewska A & de Lange T (2004) Regulation of telomerase by telomeric proteins. *Annu Rev Biochem* 73:177–208.

DNA Repair

Barnes DE, & Lindahl T (2004) Repair and genetic consequences of endogenous DNA base damage in mammalian cells. *Annu Rev Genet* 38:445–476.

Harrison JC & Haber JE (2006) Surviving the breakup: the DNA damage checkpoint. *Annu Rev Genet* 40:209–235.

Heller RC & Marians KJ (2006) Replisome assembly and the direct restart of stalled replication. *Nature Rev Mol Cell Biol* 7:932–43.

Lieber M, Ma Y et al (2003) Mechanism and regulation of human non-homologous DNA end-joining. *Nature Rev Mol Cell Biol* 4:712–720.

Lindahl T (1993) Instability and decay of the primary structure of DNA. *Nature* 362:709–715.

Prakash S & Prakash L (2002) Translesion DNA synthesis in eukaryotes: a one- or two-polymerase affair. *Genes Dev* 16:1872–1883.

Sancar A, Lindsey-Boltz LA et al. (2004) Molecular mechanisms of mammalian DNA repair and the DNA damage checkpoints. *Annu Rev Biochem* 73:39–85.

Svejstrup JQ (2002) Mechanisms of transcription-coupled DNA repair. *Nature Rev Mol Cell Biol* 3:21–29.

Wyman C & Kanaar R (2006) DNA double-strand break repair: all's well that ends well. *Annu Rev Genet* 40:363–383.

Homologous Recombination

Adams MD, McVey M & Sekelsky JJ (2003) *Drosophila* BLM in double-strand break repair by synthesis-dependent strand annealing. *Science* 299:265–267.

Cox MM (2001) Historical overview: searching for replication help in all of the rec places. *Proc Natl Acad Sci USA* 98:8173–8180.

Holliday R (1990) The history of the DNA heteroduplex. *BioEssays* 12:133–142.

Lisby M, Bartow JH et al (2004) Choreography of the DNA damage response: spatiotemporal relationships among checkpoint and repair proteins. *Cell* 118:699–713.

McEachern MJ & Haber JE (2006) Break-induced replication and recombinational telomere elongation in yeast. *Annu Rev Biochem* 75:111–135.

Michel B, Gromponee G et al (2004) Multiple pathways process stalled replication forks. *Proc Natl Acad Sci USA* 101:12783–12788.

Szostak JW, Orr-Weaver TK, Rothstein RJ et al (1983) The double-strand break repair model for recombination. *Cell* 33:25–35.

West SC (2003) Molecular views of recombination proteins and their control. *Nature Rev Mol Cell Biol* 4:435–445.

Transposition and Site-Specific Recombination

Campbell AM (1993) Thirty years ago in genetics: prophage insertion into bacterial chromosomes. *Genetics* 133:433–438.

Comfort NC (2001) From controlling elements to transposons: Barbara McClintock and the Nobel Prize. *Trends Biochem Sci* 26:454–457.

Craig NL (1996) Transposition, in *Escherichia coli* and *Salmonella*, pp 2339–2362. Washington, DC: ASM Press.

Gottesman M (1999) Bacteriophage lambda: the untold story. *J Mol Biol* 293:177–180.

Grindley ND, Whiteson KL & Rice PA (2006) Mechanisms of site-specific recombination. *Annu Rev Biochem* 75:567–605.

Varmus H (1988) Retroviruses. *Science* 240:1427–1435.

Zickler D & Kleckner N (1999) Meiotic chromosomes: integrating structure and function. *Annu Rev Genet* 33:603–754.

How Cells Read the Genome: From DNA to Protein

Only when the structure of DNA was discovered in the early 1950s did it become clear how the hereditary information in cells is encoded in DNA's sequence of nucleotides. The progress since then has been astounding. Within fifty years we knew the complete genome sequences for many organisms, including humans. We therefore know the maximum amount of information that is required to produce a complex organism like ourselves. The limits on the hereditary information needed for life constrain the biochemical and structural features of cells and make it clear that biology is not infinitely complex.

In this chapter, we explain how cells decode and use the information in their genomes. Much has been learned about how the genetic instructions written in an alphabet of just four "letters"—the four different nucleotides in DNA—direct the formation of a bacterium, a fruit fly, or a human. Nevertheless, we still have a great deal to discover about how the information stored in an organism's genome produces even the simplest unicellular bacterium with 500 genes, let alone how it directs the development of a human with approximately 25,000 genes. An enormous amount of ignorance remains; many fascinating challenges therefore await the next generation of cell biologists.

The problems that cells face in decoding genomes can be appreciated by considering a small portion of the genome of the fruit fly *Drosophila melanogaster* (**Figure 6–1**). Much of the DNA-encoded information present in this and other genomes specifies the linear order—the sequence—of amino acids for every protein the organism makes. As described in Chapter 3, the amino acid sequence in turn dictates how each protein folds to give a molecule with a distinctive shape and chemistry. When a cell makes a particular protein, it must decode accurately the corresponding region of the genome. Additional information encoded in the DNA of the genome specifies exactly when in the life of an organism and in which cell types each gene is to be expressed into protein. Since proteins are the main constituents of cells, the decoding of the genome determines not only the size, shape, biochemical properties, and behavior of cells, but also the distinctive features of each species on Earth.

One might have predicted that the information present in genomes would be arranged in an orderly fashion, resembling a dictionary or a telephone directory. Although the genomes of some bacteria seem fairly well organized, the genomes of most multicellular organisms, such as our *Drosophila* example, are surprisingly disorderly. Small bits of coding DNA (that is, DNA that codes for protein) are interspersed with large blocks of seemingly meaningless DNA. Some sections of the genome contain many genes and others lack genes altogether. Proteins that work closely with one another in the cell often have their genes located on different chromosomes, and adjacent genes typically encode proteins that have little to do with each other in the cell. Decoding genomes is therefore no simple matter. Even with the aid of powerful computers, it is still difficult for researchers to locate definitively the beginning and end of genes in the DNA sequences of complex genomes, much less to predict when each gene is expressed in the life of the organism. Although the DNA sequence of the human genome is known, it will probably take at least a decade to identify every gene and determine the precise amino acid sequence of the protein it produces. Yet the cells in our body do this thousands of times a second.

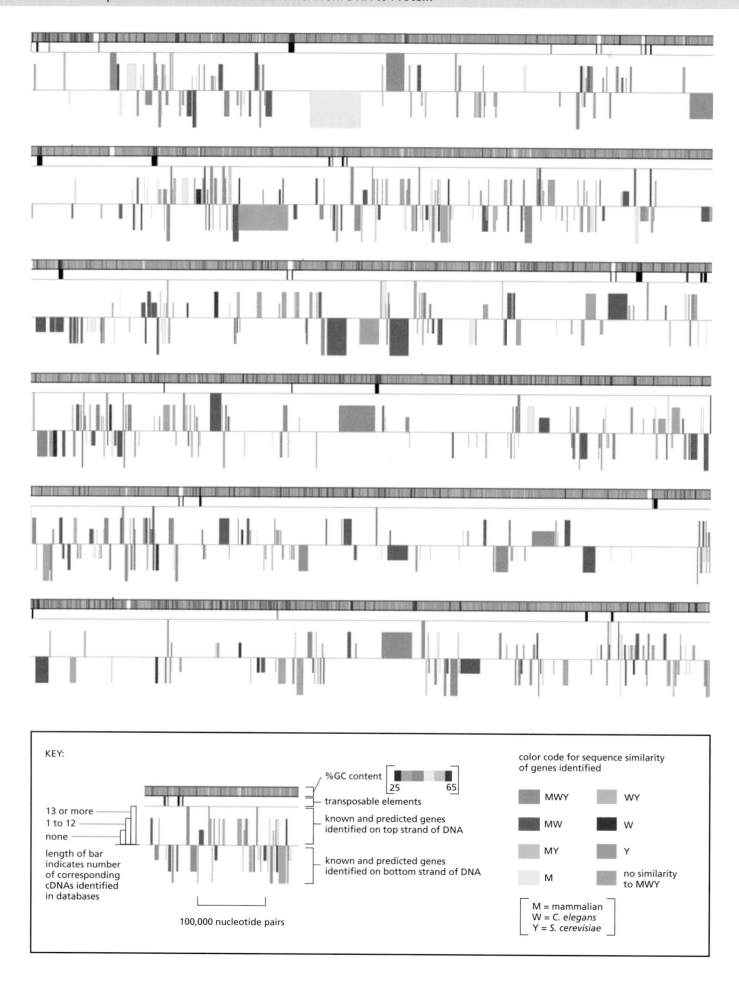

KEY:

%GC content
25   65

transposable elements

13 or more
1 to 12
none

known and predicted genes
identified on top strand of DNA

length of bar
indicates number
of corresponding
cDNAs identified
in databases

known and predicted genes
identified on bottom strand of DNA

100,000 nucleotide pairs

color code for sequence similarity
of genes identified

MWY   WY

MW   W

MY   Y

M   no similarity
             to MWY

M = mammalian
W = C. elegans
Y = S. cerevisiae

Figure 6–1 *(opposite page)* **Schematic depiction of a portion of chromosome 2 from the genome of the fruit fly** *Drosophila melanogaster.* This figure represents approximately 3% of the total *Drosophila* genome, arranged as six contiguous segments. As summarized in the key, the symbolic representations are: *black vertical lines* of various thicknesses: locations of transposable elements, with thicker bars indicating clusters of elements; *colored boxes:* genes (both known and predicted) coded on one strand of DNA (boxes *above* the midline) and genes coded on the other strand (boxes *below* the midline). The length of each gene box includes both its exons (protein-coding DNA) and its introns (noncoding DNA) (see Figure 4–15); its height is proportional to the number of known cDNAs that match the gene. (As described in Chapter 8, cDNAs are DNA copies of mRNA molecules, and large collections of the nucleotide sequences of cDNAs have been deposited in a variety of databases, the more matches, the higher the confidence that the predicted gene is transcribed into RNA and is thus a genuine gene.) The color of each gene box indicates whether a closely related gene is known to occur in other organisms. For example, MWY means the gene has close relatives in mammals, in the nematode worm *Caenorhabditis elegans*, and in the yeast *Saccharomyces cerevisiae.* MW indicates the gene has close relatives in mammals and the worm but not in yeast. The *rainbow-colored bar* indicates percent G–C base pairs; across many different genomes, this percentage shows a striking regional variation, whose origin and significance are uncertain. (From M.D. Adams et al., *Science* 287:2185–2195, 2000. With permission from AAAS.)

The DNA in genomes does not direct protein synthesis itself, but instead uses RNA as an intermediary. When the cell needs a particular protein, the nucleotide sequence of the appropriate portion of the immensely long DNA molecule in a chromosome is first copied into RNA (a process called *transcription*). It is these RNA copies of segments of the DNA that are used directly as templates to direct the synthesis of the protein (a process called *translation*). The flow of genetic information in cells is therefore from DNA to RNA to protein (**Figure 6–2**). All cells, from bacteria to humans, express their genetic information in this way—a principle so fundamental that it is termed the *central dogma* of molecular biology.

Despite the universality of the central dogma, there are important variations in the way in which information flows from DNA to protein. Principal among these is that RNA transcripts in eucaryotic cells are subject to a series of processing steps in the nucleus, including *RNA splicing*, before they are permitted to exit from the nucleus and be translated into protein. These processing steps can critically change the "meaning" of an RNA molecule and are therefore crucial for understanding how eucaryotic cells read their genomes. Finally, although we focus on the production of the proteins encoded by the genome in this chapter, we see that for many genes RNA is the final product. Like proteins, many of these RNAs fold into precise three-dimensional structures that have structural, catalytic, and regulatory roles in the cell.

We begin this chapter with the first step in decoding a genome: the process of transcription by which an RNA molecule is produced from the DNA of a gene. We then follow the fate of this RNA molecule through the cell, finishing when a correctly folded protein molecule has been formed. At the end of the chapter, we consider how the present quite complex scheme of information storage, transcription, and translation might have arisen from simpler systems in the earliest stages of cell evolution.

FROM DNA TO RNA

Transcription and translation are the means by which cells read out, or express, the genetic instructions in their genes. Because many identical RNA copies can be made from the same gene, and each RNA molecule can direct the synthesis of many identical protein molecules, cells can synthesize a large amount of protein rapidly when necessary. But each gene can also be transcribed and translated with a different efficiency, allowing the cell to make vast quantities of some proteins and tiny quantities of others (**Figure 6–3**). Moreover, as we see in the next chapter, a cell can change (or regulate) the expression of each of its genes according to the needs of the moment—most commonly by controlling the production of its RNA.

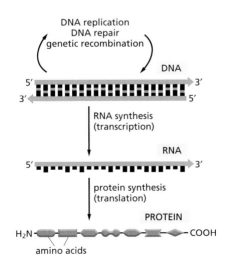

Figure 6–2 The pathway from DNA to protein. The flow of genetic information from DNA to RNA (transcription) and from RNA to protein (translation) occurs in all living cells.

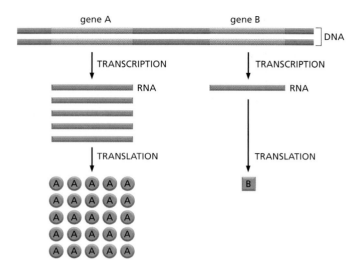

Figure 6–3 Genes can be expressed with different efficiencies. In this example, gene A is transcribed and translated much more efficiently than gene B. This allows the amount of protein A in the cell to be much greater than that of protein B.

Portions of DNA Sequence Are Transcribed into RNA

The first step a cell takes in reading out a needed part of its genetic instructions is to copy a particular portion of its DNA nucleotide sequence—a gene—into an RNA nucleotide sequence. The information in RNA, although copied into another chemical form, is still written in essentially the same language as it is in DNA—the language of a nucleotide sequence. Hence the name **transcription**.

Like DNA, RNA is a linear polymer made of four different types of nucleotide subunits linked together by phosphodiester bonds (**Figure 6–4**). It differs from DNA chemically in two respects: (1) the nucleotides in RNA are *ribonucleotides*—that is, they contain the sugar ribose (hence the name *ribo*nucleic acid) rather than deoxyribose; (2) although, like DNA, RNA contains the bases adenine (A), guanine (G), and cytosine (C), it contains the base uracil (U) instead of the thymine (T) in DNA. Since U, like T, can base-pair by hydrogen-bonding with A (**Figure 6–5**), the complementary base-pairing properties described for DNA in Chapters 4 and 5 apply also to RNA (in RNA, G pairs with C, and A pairs with U). We also find other types of base pairs in RNA: for example, G occasionally pairs with U.

Figure 6–4 The chemical structure of RNA. (A) RNA contains the sugar ribose, which differs from deoxyribose, the sugar used in DNA, by the presence of an additional –OH group. (B) RNA contains the base uracil, which differs from thymine, the equivalent base in DNA, by the absence of a –CH$_3$ group. (C) A short length of RNA. The phosphodiester chemical linkage between nucleotides in RNA is the same as that in DNA.

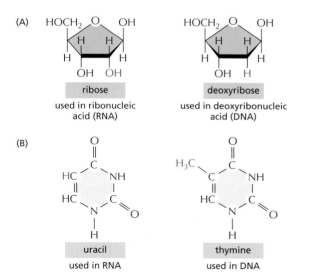

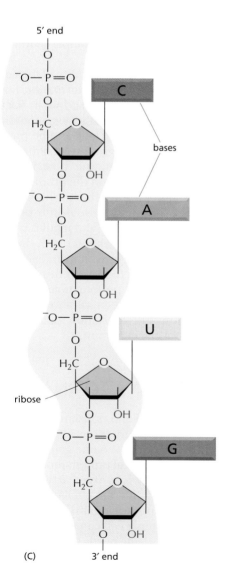

Although these chemical differences are slight, DNA and RNA differ quite dramatically in overall structure. Whereas DNA always occurs in cells as a double-stranded helix, RNA is single-stranded. An RNA chain can therefore fold up into a particular shape, just as a polypeptide chain folds up to form the final shape of a protein (**Figure 6–6**). As we see later in this chapter, the ability to fold into complex three-dimensional shapes allows some RNA molecules to have precise structural and catalytic functions.

Transcription Produces RNA Complementary to One Strand of DNA

The RNA in a cell is made by DNA transcription, a process that has certain similarities to the process of DNA replication discussed in Chapter 5. Transcription begins with the opening and unwinding of a small portion of the DNA double helix to expose the bases on each DNA strand. One of the two strands of the DNA double helix then acts as a template for the synthesis of an RNA molecule. As in DNA replication, the nucleotide sequence of the RNA chain is determined by the complementary base-pairing between incoming nucleotides and the DNA template. When a good match is made, the incoming ribonucleotide is covalently linked to the growing RNA chain in an enzymatically catalyzed reaction. The RNA chain produced by transcription—the *transcript*—is therefore elongated one nucleotide at a time, and it has a nucleotide sequence that is exactly complementary to the strand of DNA used as the template (**Figure 6–7**).

Transcription, however, differs from DNA replication in several crucial ways. Unlike a newly formed DNA strand, the RNA strand does not remain hydrogen-bonded to the DNA template strand. Instead, just behind the region where the ribonucleotides are being added, the RNA chain is displaced and the DNA helix re-forms. Thus, the RNA molecules produced by transcription are released from the DNA template as single strands. In addition, because they are copied from only a limited region of the DNA, RNA molecules are much shorter than DNA molecules. A DNA molecule in a human chromosome can be up to 250 million nucleotide-pairs long; in contrast, most RNAs are no more than a few thousand nucleotides long, and many are considerably shorter.

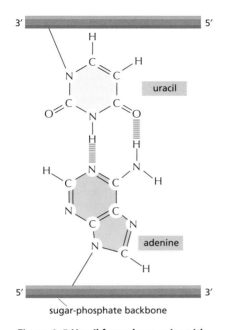

Figure 6–5 Uracil forms base pairs with adenine. The absence of a methyl group in U has no effect on base-pairing; thus, U–A base pairs closely resemble T–A base pairs (see Figure 4–4).

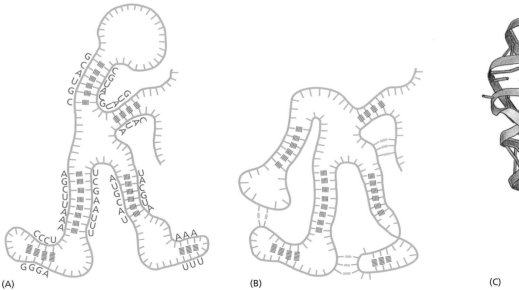

(A)   (B)   (C)

Figure 6–6 RNA can fold into specific structures. RNA is largely single-stranded, but it often contains short stretches of nucleotides that can form conventional base pairs with complementary sequences found elsewhere on the same molecule. These interactions, along with additional "nonconventional" base-pair interactions, allow an RNA molecule to fold into a three-dimensional structure that is determined by its sequence of nucleotides. <AATC> (A) Diagram of a folded RNA structure showing only conventional base-pair interactions. (B) Structure with both conventional *(red)* and nonconventional *(green)* base-pair interactions. (C) Structure of an actual RNA, a portion of a group I intron (see Figure 6–36). Each conventional base-pair interaction is indicated by a "rung" in the double helix. Bases in other configurations are indicated by broken rungs.

The enzymes that perform transcription are called **RNA polymerases**. Like the DNA polymerase that catalyzes DNA replication (discussed in Chapter 5), RNA polymerases catalyze the formation of the phosphodiester bonds that link the nucleotides together to form a linear chain. The RNA polymerase moves stepwise along the DNA, unwinding the DNA helix just ahead of the active site for polymerization to expose a new region of the template strand for complementary base-pairing. In this way, the growing RNA chain is extended by one nucleotide at a time in the 5′-to-3′ direction (**Figure 6–8**). The substrates are nucleoside triphosphates (ATP, CTP, UTP, and GTP); as in DNA replication, the hydrolysis of high-energy bonds provides the energy needed to drive the reaction forward (see Figure 5–4).

The almost immediate release of the RNA strand from the DNA as it is synthesized means that many RNA copies can be made from the same gene in a relatively short time, with the synthesis of additional RNA molecules being started before the first RNA is completed (**Figure 6–9**). When RNA polymerase molecules follow hard on each other's heels in this way, each moving at about 20 nucleotides per second (the speed in eucaryotes), over a thousand transcripts can be synthesized in an hour from a single gene.

Although RNA polymerase catalyzes essentially the same chemical reaction as DNA polymerase, there are some important differences between the activities of the two enzymes. First, and most obviously, RNA polymerase catalyzes the linkage of ribonucleotides, not deoxyribonucleotides. Second, unlike the DNA polymerases involved in DNA replication, RNA polymerases can start an RNA chain without a primer. This difference may exist because transcription need not be as accurate as DNA replication (see Table 5–1, p. 271). Unlike DNA, RNA does not permanently store genetic information in cells. RNA polymerases make about one mistake for every 10^4 nucleotides copied into RNA (compared with an error rate for direct copying by DNA polymerase of about one in 10^7 nucleotides), and the consequences of an error in RNA transcription are much less significant than that in DNA replication.

Although RNA polymerases are not nearly as accurate as the DNA polymerases that replicate DNA, they nonetheless have a modest proofreading mechanism. If an incorrect ribonucleotide is added to the growing RNA chain, the polymerase can back up, and the active site of the enzyme can perform an excision reaction that resembles the reverse of the polymerization reaction,

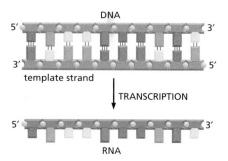

Figure 6–7 DNA transcription produces a single-stranded RNA molecule that is complementary to one strand of DNA.

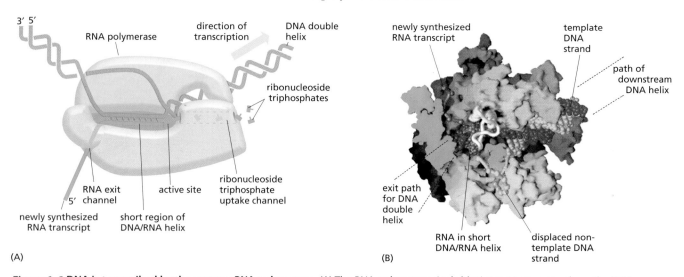

(A)

(B)

Figure 6–8 DNA is transcribed by the enzyme RNA polymerase. (A) The RNA polymerase *(pale blue)* moves stepwise along the DNA, unwinding the DNA helix at its active site. As it progresses, the polymerase adds nucleotides (represented as *small "T" shapes)* one by one to the RNA chain at the polymerization site, using an exposed DNA strand as a template. The RNA transcript is thus a complementary copy of one of the two DNA strands. A short region of DNA/RNA helix (approximately nine nucleotide pairs in length) is therefore formed only transiently, and a "window" of DNA/RNA helix therefore moves along the DNA with the polymerase. The incoming nucleotides are in the form of ribonucleoside triphosphates (ATP, UTP, CTP, and GTP), and the energy stored in their phosphate–phosphate bonds provides the driving force for the polymerization reaction (see Figure 5–4). (B) The structure of a bacterial RNA polymerase, as determined by x-ray crystallography. Four different subunits, indicated by different colors, comprise this RNA polymerase. The DNA strand used as a template is *red*, and the nontemplate strand is *yellow*. (A, adapted from a figure courtesy of Robert Landick; B, courtesy of Seth Darst.)

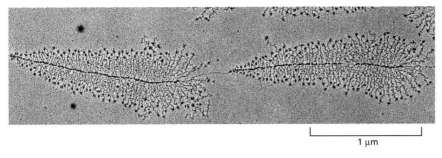

Figure 6–9 Transcription of two genes as observed under the electron microscope. The micrograph shows many molecules of RNA polymerase simultaneously transcribing each of two adjacent genes. Molecules of RNA polymerase are visible as a series of dots along the DNA with the newly synthesized transcripts (fine threads) attached to them. The RNA molecules (ribosomal RNAs) shown in this example are not translated into protein but are instead used directly as components of ribosomes, the machines on which translation takes place. The particles at the 5′ end (the free end) of each rRNA transcript are believed to reflect the beginnings of ribosome assembly. From the lengths of the newly synthesized transcripts, it can be deduced that the RNA polymerase molecules are transcribing from left to right. (Courtesy of Ulrich Scheer.)

1 μm

except that water instead of pyrophosphate is used and a nucleoside monophosphate is released.

Given that DNA and RNA polymerases both carry out template-dependent nucleotide polymerization, it might be expected that the two types of enzymes would be structurally related. However, x-ray crystallographic studies of both types of enzymes reveal that, other than containing a critical Mg²⁺ ion at the catalytic site, they are virtually unrelated to each other; indeed template-dependent nucleotide polymerizing enzymes seem to have arisen independently twice during the early evolution of cells. One lineage led to the modern DNA polymerases and reverse transcriptases discussed in Chapter 5, as well as to a few single-subunit RNA polymerases from viruses. The other lineage formed all of the modern cellular RNA polymerases (**Figure 6–10**), which we discuss in this chapter.

Cells Produce Several Types of RNA

The majority of genes carried in a cell's DNA specify the amino acid sequence of proteins; the RNA molecules that are copied from these genes (which ultimately direct the synthesis of proteins) are called **messenger RNA** (**mRNA**) molecules. The final product of a minority of genes, however, is the RNA itself. Careful analysis of the complete DNA sequence of the genome of the yeast *S. cerevisiae* has uncovered well over 750 genes (somewhat more than 10% of the total number of yeast genes) that produce RNA as their final product. These RNAs, like proteins, serve as enzymatic and structural components for a wide variety of processes in

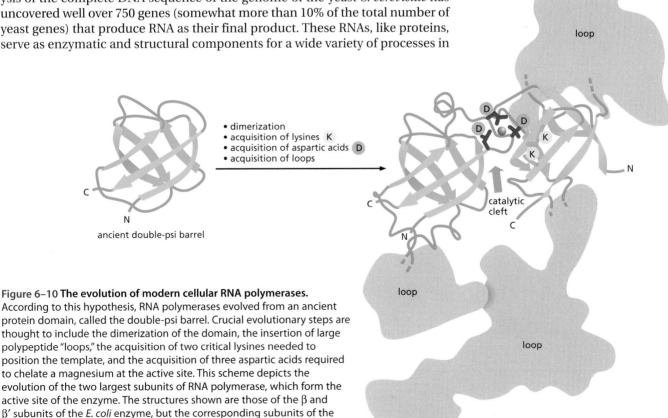

- dimerization
- acquisition of lysines K
- acquisition of aspartic acids D
- acquisition of loops

ancient double-psi barrel

catalytic cleft

Figure 6–10 The evolution of modern cellular RNA polymerases. According to this hypothesis, RNA polymerases evolved from an ancient protein domain, called the double-psi barrel. Crucial evolutionary steps are thought to include the dimerization of the domain, the insertion of large polypeptide "loops," the acquisition of two critical lysines needed to position the template, and the acquisition of three aspartic acids required to chelate a magnesium at the active site. This scheme depicts the evolution of the two largest subunits of RNA polymerase, which form the active site of the enzyme. The structures shown are those of the β and β′ subunits of the *E. coli* enzyme, but the corresponding subunits of the eucaryotic enzyme are closely related. (Adapted from L.M. Iyer, E.V. Koonin and L. Aravind, *BMC Struct. Biol.* 3:1, 2003.)

the cell. In Chapter 5 we encountered one of those RNAs, the template carried by the enzyme telomerase. Although many of these noncoding RNAs are still mysterious, we shall see in this chapter that *small nuclear RNA (snRNA)* molecules direct the splicing of pre-mRNA to form mRNA, that *ribosomal RNA (rRNA)* molecules form the core of ribosomes, and that *transfer RNA (tRNA)* molecules form the adaptors that select amino acids and hold them in place on a ribosome for incorporation into protein. Finally, we shall see in Chapter 7 that *microRNA (miRNA)* molecules and *small interfering RNA (siRNA)* molecules serve as key regulators of eucaryotic gene expression (**Table 6–1**).

Each transcribed segment of DNA is called a *transcription unit.* In eucaryotes, a transcription unit typically carries the information of just one gene, and therefore codes for either a single RNA molecule or a single protein (or group of related proteins if the initial RNA transcript is spliced in more than one way to produce different mRNAs). In bacteria, a set of adjacent genes is often transcribed as a unit; the resulting mRNA molecule therefore carries the information for several distinct proteins.

Overall, RNA makes up a few percent of a cell's dry weight. Most of the RNA in cells is rRNA; mRNA comprises only 3–5% of the total RNA in a typical mammalian cell. The mRNA population is made up of tens of thousands of different species, and there are on average only 10–15 molecules of each species of mRNA present in each cell.

Signals Encoded in DNA Tell RNA Polymerase Where to Start and Stop

To transcribe a gene accurately, RNA polymerase must recognize where on the genome to start and where to finish. The way in which RNA polymerases perform these tasks differs somewhat between bacteria and eucaryotes. Because the processes in bacteria are simpler, we discuss them first.

The initiation of transcription is an especially important step in gene expression because it is the main point at which the cell regulates which proteins are to be produced and at what rate. The bacterial RNA polymerase core enzyme is a multisubunit complex that sythesizes RNA using a DNA template as a guide. A detachable subunit called *sigma (σ) factor* associates with the core enzyme and assists it in reading the signals in the DNA that tell it where to begin transcribing (**Figure 6–11**). Together, σ factor and core enzyme are known as the **RNA polymerase holoenzyme**; this complex adheres only weakly to bacterial DNA when

Table 6–1 Principal Types of RNAs Produced in Cells

TYPE OF RNA	FUNCTION
mRNAs	messenger RNAs, code for proteins
rRNAs	ribosomal RNAs, form the basic structure of the ribosome and catalyze protein synthesis
tRNAs	transfer RNAs, central to protein synthesis as adaptors between mRNA and amino acids
snRNAs	small nuclear RNAs, function in a variety of nuclear processes, including the splicing of pre-mRNA
snoRNAs	small nucleolar RNAs, used to process and chemically modify rRNAs
scaRNAs	small cajal RNAs, used to modify snoRNAs and snRNAs
miRNAs	microRNAs, regulate gene expression typically by blocking translation of selective mRNAs
siRNAs	small interfering RNAs, turn off gene expression by directing degradation of selective mRNAs and the establishment of compact chromatin structures
Other noncoding RNAs	function in diverse cell processes, including telomere synthesis, X-chromosome inactivation, and the transport of proteins into the ER

the two collide, and a holoenzyme typically slides rapidly along the long DNA molecule until it dissociates again. However, when the polymerase holoenzyme slides into a region on the DNA double helix called a **promoter**, a special sequence of nucleotides indicating the starting point for RNA synthesis, the polymerase binds tightly to this DNA. The polymerase holoenzyme, through its σ factor, recognizes the promoter DNA sequence by making specific contacts with the portions of the bases that are exposed on the outside of the helix (step 1 in Figure 6–11).

After the RNA polymerase holoenzyme binds tightly to the promoter DNA in this way, it opens up the double helix to expose a short stretch of nucleotides on each strand (step 2 in Figure 6–11). Unlike a DNA helicase reaction (see Figure 5–14), this limited opening of the helix does not require the energy of ATP hydrolysis. Instead, the polymerase and DNA both undergo reversible structural changes that result in a state more energetically favorable than that of the initial binding. With the DNA unwound, one of the two exposed DNA strands acts as a template for complementary base-pairing with incoming ribonucleotides, two of which are joined together by the polymerase to begin an RNA chain (step 3 in Figure 6–11). After the first ten or so nucleotides of RNA have been synthesized (a relatively inefficient process during which polymerase synthesizes and discards short RNA oligomers), the core enzyme breaks its interactions with the promoter DNA, weakens its interactions with σ factor, and begins to move down the DNA, synthesizing RNA (steps 4 and 5 in Figure 6–11). Chain elongation continues (at a speed of approximately 50 nucleotides/sec for bacterial RNA polymerases) until the enzyme encounters a second signal in the DNA, the **terminator** (described below), where the polymerase halts and releases both the newly made RNA chain and the DNA template (step 7 in Figure 6–11). After the polymerase core enzyme has been released at a terminator, it reassociates with a free σ factor to form a holoenzyme that can begin the process of transcription again.

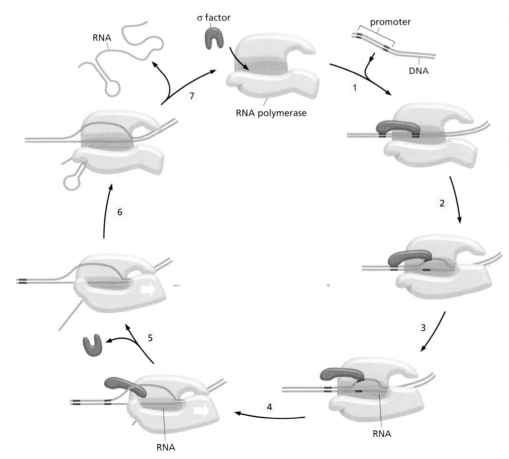

σ factor

RNA

RNA polymerase

promoter

DNA

RNA

RNA

Figure 6–11 The transcription cycle of bacterial RNA polymerase. In step 1, the RNA polymerase holoenzyme (polymerase core enzyme plus σ factor) assembles and then locates a promoter (see Figure 6–12). The polymerase unwinds the DNA at the position at which transcription is to begin (step 2) and begins transcribing (step 3). This initial RNA synthesis (sometimes called "abortive initiation") is relatively inefficient. However, once RNA polymerase has managed to synthesize about 10 nucleotides of RNA, it breaks its interactions with the promoter DNA and weakens, and eventually breaks, its interaction with σ. The polymerase now shifts to the elongation mode of RNA synthesis (step 4), moving rightward along the DNA in this diagram. During the elongation mode (step 5), transcription is highly processive, with the polymerase leaving the DNA template and releasing the newly transcribed RNA only when it encounters a termination signal (steps 6 and 7). Termination signals are typically encoded in DNA, and many function by forming an RNA structure that destabilizes the polymerase's hold on the RNA (step 7). In bacteria, all RNA molecules are synthesized by a single type of RNA polymerase and the cycle depicted in the figure therefore applies to the production of mRNAs as well as structural and catalytic RNAs. (Adapted from a figure courtesy of Robert Landick.)

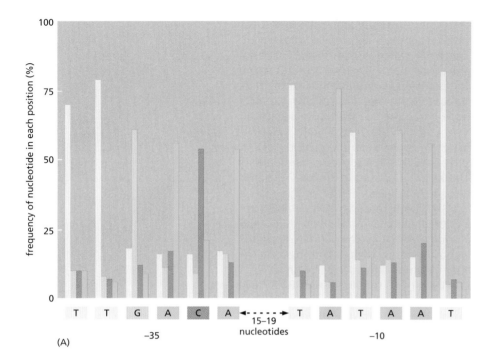

(A)

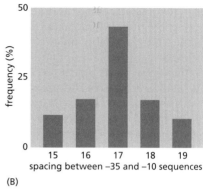

(B)

Figure 6–12 Consensus sequence for the major class of *E. coli* promoters. (A) The promoters are characterized by two hexameric DNA sequences, the –35 sequence and the –10 sequence named for their approximate location relative to the start point of transcription (designated +1). For convenience, the nucleotide sequence of a single strand of DNA is shown; in reality the RNA polymerase recognizes the promoter as double-stranded DNA. On the basis of a comparison of 300 promoters, the frequencies of the four nucleotides at each position in the –35 and –10 hexamers are given. The consensus sequence, shown *below* the graph, reflects the most common nucleotide found at each position in the collection of promoters. The sequence of nucleotides between the –35 and –10 hexamers shows no significant similarities among promoters. (B) The distribution of spacing between the –35 and –10 hexamers found in *E. coli* promoters.

The information displayed in these two graphs applies to *E. coli* promoters that are recognized by RNA polymerase and the major σ factor (designated σ70). As we shall see in the next chapter, bacteria also contain minor σ factors, each of which recognizes a different promoter sequence. Some particularly strong promoters recognized by RNA polymerase and σ70 have an additional sequence, located upstream (to the *left*, in the figure) of the –35 hexamer, which is recognized by another subunit of RNA polymerase.

The process of transcription initiation is complex and requires that the RNA polymerase holoenzyme and the DNA undergo a series of conformational changes. We can view these changes as opening up and positioning the DNA in the active site followed by a successive tightening of the enzyme around the DNA and RNA to ensure that it does not dissociate before it has finished transcribing a gene. If an RNA polymerase does dissociate prematurely, it cannot resume synthesis but must start over again at the promoter.

How do the termination signals in the DNA stop the elongating polymerase? For most bacterial genes a termination signal consists of a string of A–T nucleotide pairs preceded by a two-fold symmetric DNA sequence, which, when transcribed into RNA, folds into a "hairpin" structure through Watson–Crick base-pairing (see Figure 6–11). As the polymerase transcribes across a terminator, the formation of the hairpin may help to "pull" the RNA transcript from the active site. The DNA–RNA hybrid in the active site, which is held together at terminators predominantly by U–A base pairs (which are less stable than G–C base pairs because they form two rather than three hydrogen bonds per base pair), is not strong enough to hold the RNA in place, and it dissociates causing the release of the polymerase from the DNA (step 7 in Figure 6–11). Thus, in some respects, transcription termination seems to involve a reversal of the structural transitions that happen during initiation. The process of termination also is an example of a common theme in this chapter: the folding of RNA into specific structures affects many steps in decoding the genome.

Transcription Start and Stop Signals Are Heterogeneous in Nucleotide Sequence

As we have just seen, the processes of transcription initiation and termination involve a complicated series of structural transitions in protein, DNA, and RNA molecules. The signals encoded in DNA that specify these transitions are often difficult for researchers to recognize. Indeed, a comparison of many different bacterial promoters reveals a surprising degree of variation. Nevertheless, they all contain related sequences, reflecting in part aspects of the DNA that are recognized directly by the σ factor. These common features are often summarized in the form of a *consensus sequence* (**Figure 6–12**). A **consensus nucleotide sequence** is derived by comparing many sequences with the same basic function and tallying up the most common nucleotide found at each position. It therefore serves as a summary or "average" of a large number of individual nucleotide sequences.

The DNA sequences of individual bacterial promoters differ in ways that determine their strength (the number of initiation events per unit time of the promoter). Evolutionary processes have fine-tuned each to initiate as often as necessary and have thereby created a wide spectrum of promoters. Promoters for genes that code for abundant proteins are much stronger than those associated with genes that encode rare proteins, and their nucleotide sequences are responsible for these differences.

Like bacterial promoters, transcription terminators also have a wide range of sequences, with the potential to form a simple hairpin RNA structure being the most important common feature. Since an almost unlimited number of nucleotide sequences have this potential, terminator sequences are even more heterogeneous than promoter sequences.

We have discussed bacterial promoters and terminators in some detail to illustrate an important point regarding the analysis of genome sequences. Although we know a great deal about bacterial promoters and terminators and can construct consensus sequences that summarize their most salient features, their variation in nucleotide sequence makes it difficult to definitively locate them simply by analysis of the nucleotide sequence of a genome. It is even more difficult to locate analogous sequences in eucaryotic genomes, due in part to the excess DNA carried in them. Often, we need additional information, some of it from direct experimentation, to locate and accurately interpret the short DNA signals contained in genomes.

Since DNA is double-stranded, two different RNA molecules could in principle be transcribed from any gene, using each of the two DNA strands as a template. However, a gene typically has only a single promoter, and because the promoter's nucleotide sequence is asymmetric (see Figure 6–12), the polymerase can bind in only one orientation. The polymerase synthesizes RNA in the 5′-to-3′ direction, and it can therefore only transcribe one strand per gene (**Figure 6–13**). Genome sequences reveal that the DNA strand used as the template for RNA synthesis varies from gene to gene depending on the location and orientation of the promoter (**Figure 6–14**).

Having considered transcription in bacteria, we now turn to the situation in eucaryotes, where the synthesis of RNA molecules is a much more elaborate affair.

Transcription Initiation in Eucaryotes Requires Many Proteins

In contrast to bacteria, which contain a single type of RNA polymerase, eucaryotic nuclei have three: *RNA polymerase I, RNA polymerase II,* and *RNA polymerase III.* The three polymerases are structurally similar to one another (and to the bacterial enzyme) and share some common subunits, but they transcribe different types of genes (**Table 6–2**). RNA polymerases I and III transcribe the genes encoding transfer RNA, ribosomal RNA, and various small RNAs. RNA polymerase II transcribes most genes, including all those that encode proteins, and our subsequent discussion therefore focuses on this enzyme.

Although eucaryotic RNA polymerase II has many structural similarities to bacterial RNA polymerase (**Figure 6–15**), there are several important differences in the way in which the bacterial and eucaryotic enzymes function, two of which concern us immediately.

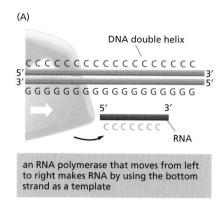

(A)

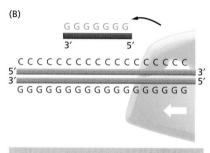

(B)

Figure 6–13 The importance of RNA polymerase orientation. The DNA strand serving as template must be traversed in a 3′-to-5′ direction. Thus, the direction of RNA polymerase movement determines which of the two DNA strands is to serve as a template for the synthesis of RNA, as shown in (A) and (B). Polymerase direction is, in turn, determined by the orientation of the promoter sequence, the site at which the RNA polymerase begins transcription.

Figure 6–14 Directions of transcription along a short portion of a bacterial chromosome. Some genes are transcribed using one DNA strand as a template, while others are transcribed using the other DNA strand. The direction of transcription is determined by the promoter at the beginning of each gene *(green arrowheads)*. This diagram shows approximately 0.2% (9000 base pairs) of the *E. coli* chromosome. The genes transcribed from *left* to *right* use the bottom DNA strand as the template; those transcribed from *right* to *left* use the top strand as the template.

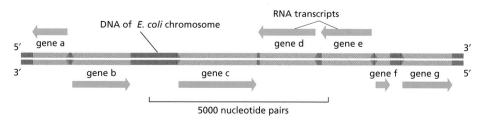

Table 6–2 The Three RNA Polymerases in Eucaryotic Cells

TYPE OF POLYMERASE	GENES TRANSCRIBED
RNA polymerase I	5.8S, 18S, and 28S rRNA genes
RNA polymerase II	all protein-coding genes, plus snoRNA genes, miRNA genes, siRNA genes, and most snRNA genes
RNA polymerase III	tRNA genes, 5S rRNA genes, some snRNA genes and genes for other small RNAs

The rRNAs are named according to their "S" values, which refer to their rate of sedimentation in an ultracentrifuge. The larger the S value, the larger the rRNA.

1. While bacterial RNA polymerase requires only a single additional protein (σ factor) for transcription initiation to occur *in vitro*, eucaryotic RNA polymerases require many additional proteins, collectively called the *general transcription factors*.
2. Eucaryotic transcription initiation must deal with the packing of DNA into nucleosomes and higher-order forms of chromatin structure, features absent from bacterial chromosomes.

RNA Polymerase II Requires General Transcription Factors

The **general transcription factors** help to position eucaryotic RNA polymerase correctly at the promoter, aid in pulling apart the two strands of DNA to allow transcription to begin, and release RNA polymerase from the promoter into the elongation mode once transcription has begun. <CTAT> The proteins are "general" because they are needed at nearly all promoters used by RNA polymerase II; consisting of a set of interacting proteins, they are designated as *TFII* (for transcription factor for polymerase II), and are denoted arbitrarily as TFIIB, TFIID, and so on. In a broad sense, the eucaryotic general transcription factors carry out functions equivalent to those of the σ factor in bacteria; indeed, portions of TFIIF have the same three-dimensional structure as the equivalent portions of σ.

Figure 6–16 illustrates how the general transcription factors assemble at promoters used by RNA polymerase II, and **Table 6–3** summarizes their activities. The assembly process begins when the general transcription factor TFIID binds to a short double-helical DNA sequence primarily composed of T and A nucleotides. For this reason, this sequence is known as the TATA sequence, or **TATA box**, and the subunit of TFIID that recognizes it is called TBP (for TATA-binding protein). The TATA box is typically located 25 nucleotides upstream from the transcription start site. It is not the only DNA sequence that signals the start of transcription (**Figure 6–17**), but for most polymerase II promoters it is the most important. The binding of TFIID causes a large distortion in the DNA

Figure 6–15 Structural similarity between a bacterial RNA polymerase and a eucaryotic RNA polymerase II. Regions of the two RNA polymerases that have similar structures are indicated in *green*. The eucaryotic polymerase is larger than the bacterial enzyme (12 subunits instead of 5), and some of the additional regions are shown in *gray*. The *blue* spheres represent Zn atoms that serve as structural components of the polymerases, and the *red* sphere represents the Mg atom present at the active site, where polymerization takes place. The RNA polymerases in all modern-day cells (bacteria, archaea, and eucaryotes) are closely related, indicating that the basic features of the enzyme were in place before the divergence of the three major branches of life. (Courtesy of P. Cramer and R. Kornberg.)

Figure 6–16 Initiation of transcription of a eucaryotic gene by RNA polymerase II. To begin transcription, RNA polymerase requires several general transcription factors. (A) The promoter contains a DNA sequence called the TATA box, which is located 25 nucleotides away from the site at which transcription is initiated. (B) Through its subunit TBP, TFIID recognizes and binds the TATA box, which then enables the adjacent binding of TFIIB (C). For simplicity the DNA distortion produced by the binding of TFIID (see Figure 6–18) is not shown. (D) The rest of the general transcription factors, as well as the RNA polymerase itself, assemble at the promoter. (E) TFIIH then uses ATP to pry apart the DNA double helix at the transcription start point, locally exposing the template strand. TFIIH also phosphorylates RNA polymerase II, changing its conformation so that the polymerase is released from the general factors and can begin the elongation phase of transcription. As shown, the site of phosphorylation is a long C-terminal polypeptide tail, also called the C-terminal domain (CTD), that extends from the polymerase molecule. The assembly scheme shown in the figure was deduced from experiments performed *in vitro*, and the exact order in which the general transcription factors assemble on promoters may vary from gene to gene *in vivo*. The general transcription factors have been highly conserved in evolution; some of those from human cells can be replaced in biochemical experiments by the corresponding factors from simple yeasts.

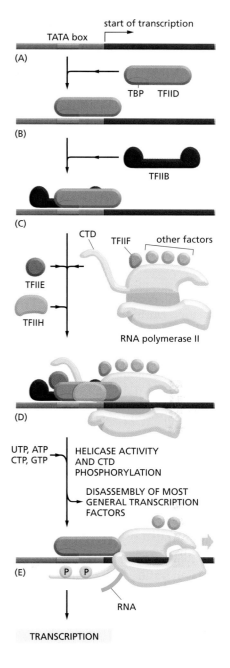

of the TATA box (**Figure 6–18**). This distortion is thought to serve as a physical landmark for the location of an active promoter in the midst of a very large genome, and it brings DNA sequences on both sides of the distortion together to allow for subsequent protein assembly steps. Other factors then assemble, along with RNA polymerase II, to form a complete *transcription initiation complex* (see Figure 6–16). The most complicated of the general transcription factors is TFIIH. Consisting of 9 subunits, it is nearly as large as RNA polymerase II itself and, as we shall see shortly, performs several enzymatic steps needed for the initiation of transcription.

After forming a transcription initiation complex on the promoter DNA, RNA polymerase II must gain access to the template strand at the transcription start point. TFIIH, which contains a DNA helicase as one of its subunits, makes this step possible by hydrolyzing ATP and unwinding the DNA, thereby exposing the template strand. Next, RNA polymerase II, like the bacterial polymerase, remains at the promoter synthesizing short lengths of RNA until it undergoes a series of conformational changes that allow it to move away from the promoter and enter the elongation phase of transcription. A key step in this transition is the addition of phosphate groups to the "tail" of the RNA polymerase (known as the CTD or C-terminal domain). In humans, the CTD consists of 52 tandem repeats of a seven-amino-acid sequence, which extend from the RNA polymerase core structure. During transcription initiation, the serine located at the

Table 6–3 The General Transcription Factors Needed for Transcription Initiation by Eucaryotic RNA Polymerase II

NAME	NUMBER OF SUBUNITS	ROLES IN TRANSITION INITIATION
TFIID		
TBP subunit	1	recognizes TATA box
TAF subunits	~11	recognizes other DNA sequences near the transcription start point; regulates DNA-binding by TBP
TFIIB	1	recognizes BRE element in promoters; accurately positions RNA polymerase at the start site of transcription
TFIIF	3	stabilizes RNA polymerase interaction with TBP and TFIIB; helps attract TFIIE and TFIIH
TFIIE	2	attracts and regulates TFIIH
TFIIH	9	unwinds DNA at the transcription start point, phosphorylates Ser5 of the RNA polymerase CTD; releases RNA polymerase from the promoter

TFIID is composed of TBP and ~11 additional subunits called TAFs (TBP-associated factors); CTD, C-terminal domain.

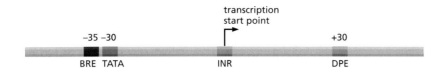

element	consensus sequence	general transcription factor
BRE	G/C G/C G/A C G C C	TFIIB
TATA	T A T A A/T A A/T	TBP
INR	C/T C/T A N T/A C/T C/T	TFIID
DPE	A/G G A/T C G T G	TFIID

Figure 6–17 Consensus sequences found in the vicinity of eucaryotic RNA polymerase II start points. The name given to each consensus sequence *(first column)* and the general transcription factor that recognizes it *(last column)* are indicated. N indicates any nucleotide, and two nucleotides separated by a slash indicate an equal probability of either nucleotide at the indicated position. In reality, each consensus sequence is a shorthand representation of a histogram similar to that of Figure 6–12.

For most RNA polymerase II transcription start points, only two or three of the four sequences are present. For example, many polymerase II promoters have a TATA box sequence, but those that do not typically have a "strong" INR sequence. Although most of the DNA sequences that influence transcription initiation are located upstream of the transcription start point, a few, such as the DPE shown in the figure, are located in the transcribed region.

fifth position in the repeat sequence (Ser5) is phosphorylated by TFIIH, which contains a protein kinase in another of its subunits (see Figure 6–16D and E). The polymerase can then disengage from the cluster of general transcription factors. During this process, it undergoes a series of conformational changes that tighten its interaction with DNA, and it acquires new proteins that allow it to transcribe for long distances, and in some cases for many hours, without dissociating from DNA.

Once the polymerase II has begun elongating the RNA transcript, most of the general transcription factors are released from the DNA so that they are available to initiate another round of transcription with a new RNA polymerase molecule. As we see shortly, the phosphorylation of the tail of RNA polymerase II also causes components of the RNA-processing machinery to load onto the polymerase and thus be positioned to modify the newly transcribed RNA as it emerges from the polymerase.

Polymerase II Also Requires Activator, Mediator, and Chromatin-Modifying Proteins

Studies of the behavior of RNA polymerase II and its general transcription factors on purified DNA templates *in vitro* established the model for transcription initiation just described. However, as discussed in Chapter 4, DNA in eucaryotic cells is packaged into nucleosomes, which are further arranged in higher-order

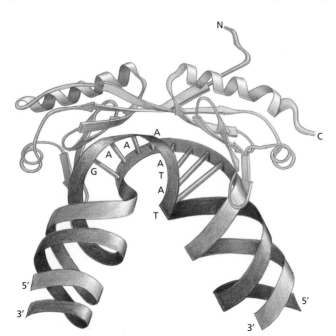

Figure 6–18 Three-dimensional structure of TBP (TATA-binding protein) bound to DNA. The TBP is the subunit of the general transcription factor TFIID that is responsible for recognizing and binding to the TATA box sequence in the DNA *(red)*. The unique DNA bending caused by TBP—two kinks in the double helix separated by partly unwound DNA—may serve as a landmark that helps to attract the other general transcription factors. TBP is a single polypeptide chain that is folded into two very similar domains *(blue and green)*. (Adapted from J.L. Kim et al., *Nature* 365:520–527, 1993. With permission from Macmillan Publishers Ltd.)

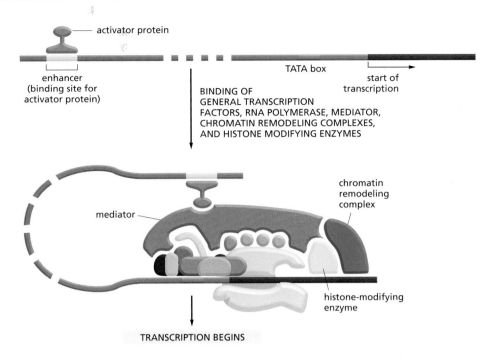

enhancer
(binding site for
activator protein)

activator protein

TATA box

start of
transcription

BINDING OF
GENERAL TRANSCRIPTION
FACTORS, RNA POLYMERASE, MEDIATOR,
CHROMATIN REMODELING COMPLEXES,
AND HISTONE MODIFYING ENZYMES

mediator

chromatin
remodeling
complex

histone-modifying
enzyme

TRANSCRIPTION BEGINS

Figure 6–19 Transcription initiation by RNA polymerase II in a eucaryotic cell. Transcription initiation *in vivo* requires the presence of transcriptional activator proteins. As described in Chapter 7, these proteins bind to specific short sequences in DNA. Although only one is shown here, a typical eucaryotic gene has many activator proteins, which together determine its rate and pattern of transcription. Sometimes acting from a distance of several thousand nucleotide pairs (indicated by the dashed DNA molecule), these gene regulatory proteins help RNA polymerase, the general transcription factors, and the mediator all to assemble at the promoter. In addition, activators attract ATP-dependent chromatin remodeling complexes and histone acetylases.

As discussed in Chapter 4, the "default" state of chromatin is probably the 30-nm filament (see Figure 4–22), and this is likely to be a form of DNA upon which transcription is initiated. For simplicity, it is not shown in the figure.

chromatin structures. As a result, transcription initiation in a eucaryotic cell is more complex and requires even more proteins than it does on purified DNA. First, gene regulatory proteins known as *transcriptional activators* must bind to specific sequences in DNA and help to attract RNA polymerase II to the start point of transcription (**Figure 6–19**). We discuss the role of activators in Chapter 7, because they are one of the main ways in which cells regulate expression of their genes. Here we simply note that their presence on DNA is required for transcription initiation in a eucaryotic cell. Second, eucaryotic transcription initiation *in vivo* requires the presence of a protein complex known as *Mediator*, which allows the activator proteins to communicate properly with the polymerase II and with the general transcription factors. Finally, transcription initiation in a eucaryotic cell typically requires the local recruitment of chromatin-modifying enzymes, including chromatin remodeling complexes and histone-modifying enzymes. As discussed in Chapter 4, both types of enzymes can allow greater access to the DNA present in chromatin, and by doing so, they facilitate the assembly of the transcription initiation machinery onto DNA. We will revisit the role of these enzymes in transcription initiation in Chapter 7.

As illustrated in Figure 6–19, many proteins (well over 100 individual subunits) must assemble at the start point of transcription to initiate transcription in a eucaryotic cell. The order of assembly of these proteins does not seem to follow a prescribed pathway; rather, the order differs from gene to gene. Indeed, some of these different protein complexes may interact with each other away from the DNA and be brought to DNA as preformed subassemblies. To begin transcribing, RNA polymerase II must be released from this large complex of proteins, and, in addition to the steps described in Figure 6–16, this often requires the *in situ* proteolysis of the activator protein. We return to some of these issues in Chapter 7, where we discuss how eucaryotic cells can regulate the process of transcription initiation.

Transcription Elongation Produces Superhelical Tension in DNA

Once it has initiated transcription, RNA polymerase does not proceed smoothly along a DNA molecule; rather, it moves jerkily, pausing at some sequences and rapidly transcribing through others. Elongating RNA polymerases, both bacterial and eucaryotic, are associated with a series of *elongation factors*, proteins that decrease the likelihood that RNA polymerase will dissociate before it reaches the end of a gene. These factors typically associate with RNA polymerase

(A)
DNA with free end

(B)
DNA with fixed ends

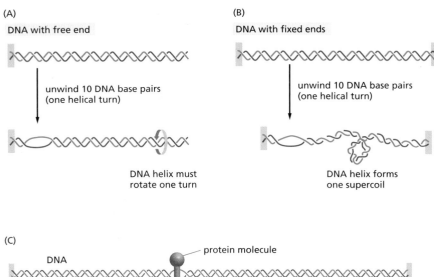

unwind 10 DNA base pairs
(one helical turn)

DNA helix must
rotate one turn

unwind 10 DNA base pairs
(one helical turn)

DNA helix forms
one supercoil

(C)

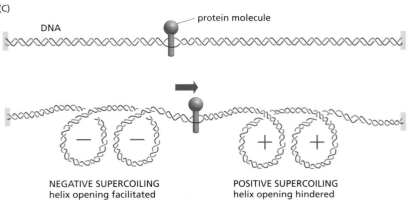

DNA

protein molecule

NEGATIVE SUPERCOILING
helix opening facilitated

POSITIVE SUPERCOILING
helix opening hindered

Figure 6–20 Superhelical tension in DNA causes DNA supercoiling. (A) For a DNA molecule with one free end (or a nick in one strand that serves as a swivel), the DNA double helix rotates by one turn for every 10 nucleotide pairs opened. (B) If rotation is prevented, superhelical tension is introduced into the DNA by helix opening. One way of accommodating this tension would be to increase the helical twist from 10 to 11 nucleotide pairs per turn in the double helix that remains; the DNA helix, however, resists such a deformation in a springlike fashion, preferring to relieve the superhelical tension by bending into supercoiled loops. As a result, one DNA supercoil forms in the DNA double helix for every 10 nucleotide pairs opened. The supercoil formed in this case is a positive supercoil. (C) Supercoiling of DNA is induced by a protein tracking through the DNA double helix. The two ends of the DNA shown here are unable to rotate freely relative to each other, and the protein molecule is assumed also to be prevented from rotating freely as it moves. Under these conditions, the movement of the protein causes an excess of helical turns to accumulate in the DNA helix ahead of the protein and a deficit of helical turns to arise in the DNA behind the protein, as shown.

shortly after initiation and help polymerases to move through the wide variety of different DNA sequences that are found in genes. Eucaryotic RNA polymerases must also contend with chromatin structure as they move along a DNA template, and they are typically aided by ATP-dependent chromatin remodeling complexes (see pp. 215–216). These complexes may move with the polymerase or may simply seek out and rescue the occasional stalled polymerase. In addition, some elongation factors associated with eucaryotic RNA polymerase facilitate transcription through nucleosomes without requiring additional energy. It is not yet understood in detail how this is accomplished, but these proteins can transiently dislodge H2A–H2B dimers from the nucleosome core, replacing them as the polymerase moves through the nucleosome.

There is yet another barrier to elongating polymerases, both bacterial and eucaryotic. To discuss this issue, we need first to consider a subtle property inherent in the DNA double helix called **DNA supercoiling**. DNA supercoiling represents a conformation that DNA adopts in response to superhelical tension; conversely, creating various loops or coils in the helix can create such tension. **Figure 6–20** illustrates the topological constraints that cause DNA supercoiling. There are approximately 10 nucleotide pairs for every helical turn in a DNA double helix. Imagine a helix whose two ends are fixed with respect to each other (as they are in a DNA circle, such as a bacterial chromosome, or in a tightly clamped loop, as is thought to exist in eucaryotic chromosomes). In this case, one large DNA supercoil will form to compensate for each 10 nucleotide pairs that are opened (unwound). The formation of this supercoil is energetically favorable because it restores a normal helical twist to the base-paired regions that remain, which would otherwise need to be overwound because of the fixed ends.

RNA polymerase also creates superhelical tension as it moves along a stretch of DNA that is anchored at its ends (see Figure 6–20C). As long as the polymerase is not free to rotate rapidly (and such rotation is unlikely given the size of RNA polymerases and their attached transcripts), a moving polymerase generates

(A) EUCARYOTES

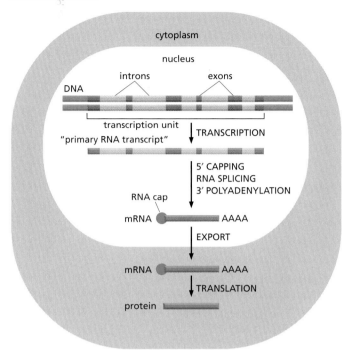

(B) PROCARYOTES

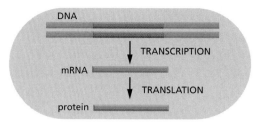

Figure 6–21 **Summary of the steps leading from gene to protein in eucaryotes and bacteria.** The final level of a protein in the cell depends on the efficiency of each step and on the rates of degradation of the RNA and protein molecules. (A) In eucaryotic cells the RNA molecule resulting from transcription contains both coding (exon) and noncoding (intron) sequences. Before it can be translated into protein, the two ends of the RNA are modified, the introns are removed by an enzymatically catalyzed RNA splicing reaction, and the resulting mRNA is transported from the nucleus to the cytoplasm. Although the steps in this figure are depicted as occurring one at a time, in a sequence, in reality they can occur concurrently. For example, the RNA cap is added and splicing typically begins before transcription has been completed. Because of the coupling between transcription and RNA processing, primary transcripts—the RNAs that would, in theory, be produced if no processing had occurred—are found only rarely. (B) In procaryotes the production of mRNA is much simpler. The 5′ end of an mRNA molecule is produced by the initiation of transcription, and the 3′ end is produced by the termination of transcription. Since procaryotic cells lack a nucleus, transcription and translation take place in a common compartment. In fact, the translation of a bacterial mRNA often begins before its synthesis has been completed.

positive superhelical tension in the DNA in front of it and negative helical tension behind it. For eucaryotes, this situation is thought to provide a bonus: the positive superhelical tension ahead of the polymerase makes the DNA helix more difficult to open, but this tension should facilitate the unwrapping of DNA in nucleosomes, as the release of DNA from the histone core helps to relax positive superhelical tension.

Any protein that propels itself alone along a DNA strand of a double helix tends to generate superhelical tension. In eucaryotes, DNA topoisomerase enzymes rapidly remove this superhelical tension (see p. 278). But in bacteria a specialized topoisomerase called *DNA gyrase* uses the energy of ATP hydrolysis to pump supercoils continuously into the DNA, thereby maintaining the DNA under constant tension. These are *negative supercoils*, having the opposite handedness from the *positive supercoils* that form when a region of DNA helix opens (see Figure 6–20B). Whenever a region of helix opens, it removes these negative supercoils from bacterial DNA, reducing the superhelical tension. DNA gyrase therefore makes the opening of the DNA helix in bacteria energetically favorable compared with helix opening in DNA that is not supercoiled. For this reason, it usually facilitates those genetic processes in bacteria, including the initiation of transcription by bacterial RNA polymerase, that require helix opening (see Figure 6–11).

Transcription Elongation in Eucaryotes Is Tightly Coupled to RNA Processing

We have seen that bacterial mRNAs are synthesized solely by the RNA polymerase starting and stopping at specific spots on the genome. The situation in eucaryotes is substantially different. In particular, transcription is only the first of several steps needed to produce an mRNA. Other critical steps are the covalent modification of the ends of the RNA and the removal of *intron sequences* that are discarded from the middle of the RNA transcript by the process of *RNA splicing* (**Figure 6–21**).

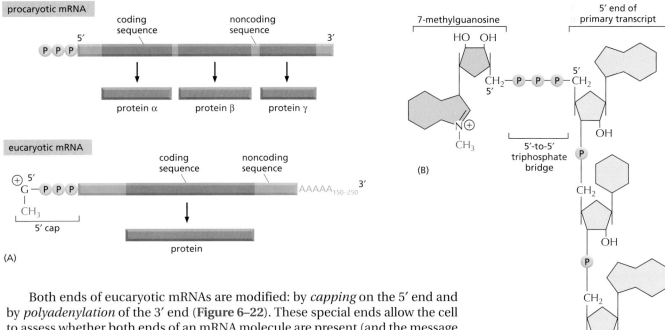

Figure 6–22 **A comparison of the structures of procaryotic and eucaryotic mRNA molecules.** (A) The 5′ and 3′ ends of a bacterial mRNA are the unmodified ends of the chain synthesized by the RNA polymerase, which initiates and terminates transcription at those points, respectively. The corresponding ends of a eucaryotic mRNA are formed by adding a 5′ cap and by cleavage of the pre-mRNA transcript and the addition of a poly-A tail, respectively. The figure also illustrates another difference between the procaryotic and eucaryotic mRNAs: bacterial mRNAs can contain the instructions for several different proteins, whereas eucaryotic mRNAs nearly always contain the information for only a single protein. (B) The structure of the cap at the 5′ end of eucaryotic mRNA molecules. Note the unusual 5′-to-5′ linkage of the 7-methyl G to the remainder of the RNA. Many eucaryotic mRNAs carry an additional modification: the 2′-hydroxyl group on the second ribose sugar in the mRNA is methylated (not shown).

Both ends of eucaryotic mRNAs are modified: by *capping* on the 5′ end and by *polyadenylation* of the 3′ end (**Figure 6–22**). These special ends allow the cell to assess whether both ends of an mRNA molecule are present (and the message is therefore intact) before it exports the RNA sequence from the nucleus and translates it into protein. RNA splicing joins together the different portions of a protein coding sequence, and it provides higher eucaryotes with the ability to synthesize several different proteins from the same gene.

An ingenious mechanism couples all of the above RNA processing steps to transcription elongation. As discussed previously, a key step in transcription initiation by RNA polymerase II is the phosphorylation of the RNA polymerase II tail, called the CTD (C-terminal domain). This phosphorylation proceeds gradually as the RNA polymerase initiates transcription and moves along the DNA. It not only helps dissociate the RNA polymerase II from other proteins present at the start point of transcription, but also allows a new set of proteins to associate with the RNA polymerase tail that function in transcription elongation and RNA processing. As discussed next, some of these processing proteins seem to "hop" from the polymerase tail onto the nascent RNA molecule to begin processing it as it emerges from the RNA polymerase. Thus, we can view RNA polymerase II in its elongation mode as an RNA factory that both transcribes DNA into RNA and processes the RNA it produces (**Figure 6–23**). Fully extended, the CTD is nearly 10 times longer than the remainder of RNA polymerase and, in effect, it serves as a tether, holding a variety of proteins close by until they are needed. This strategy, which speeds up the rate of subsequent reactions, is one commonly observed in the cell (see Figures 4–69 and 16–38).

RNA Capping Is the First Modification of Eucaryotic Pre-mRNAs

As soon as RNA polymerase II has produced about 25 nucleotides of RNA, the 5′ end of the new RNA molecule is modified by addition of a cap that consists of a modified guanine nucleotide (see Figure 6–22B). Three enzymes, acting in succession, perform the capping reaction: one (a phosphatase) removes a phosphate from the 5′ end of the nascent RNA, another (a guanyl transferase) adds a GMP in a reverse linkage (5′ to 5′ instead of 5′ to 3′), and a third (a methyl transferase) adds a methyl group to the guanosine (**Figure 6–24**). Because all three enzymes bind to the RNA polymerase tail phosphorylated at serine-5 position, the modification added by TFIIH during transcription initiation, they are poised to modify the 5′ end of the nascent transcript as soon as it emerges from the polymerase.

The 5′-methyl cap signifies the 5′ end of eucaryotic mRNAs, and this landmark helps the cell to distinguish mRNAs from the other types of RNA molecules present in the cell. For example, RNA polymerases I and III produce uncapped

Figure 6–23 Eucaryotic RNA polymerase II as an "RNA factory." As the polymerase transcribes DNA into RNA, it carries pre-mRNA-processing proteins on its tail that are transferred to the nascent RNA at the appropriate time. The tail, known as the CTD, contains 52 tandem repeats of a seven amino acid sequence, and there are two serines in each repeat. The capping proteins first bind to the RNA polymerase tail when it is phosphorylated on Ser5 of the heptad repeat late in the process of transcription initiation (see Figure 6–16). This strategy ensures that the RNA molecule is efficiently capped as soon as its 5′ end emerges from the RNA polymerase. As the polymerase continues transcribing, its tail is extensively phosphorylated on the Ser2 positions by a kinase associated with the elongating polymerase and is eventually dephosphorylated at Ser5 positions. These further modifications attract splicing and 3′-end processing proteins to the moving polymerase, positioning them to act on the newly synthesized RNA as it emerges from the RNA polymerase. There are many RNA-processing enzymes, and not all travel with the polymerase. For RNA splicing, for example, the tail carries only a few critical components; once transferred to an RNA molecule, they serve as a nucleation site for the remaining components.

When RNA polymerase II finishes transcribing a gene, it is released from DNA, soluble phosphatases remove the phosphates on its tail, and it can reinitiate transcription. Only the dephosphorylated form of RNA polymerase II is competent to begin RNA synthesis at a promoter.

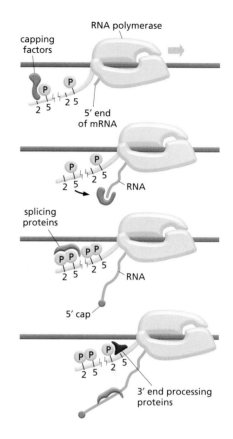

RNAs during transcription, in part because these polymerases lack a CTD. In the nucleus, the cap binds a protein complex called CBC (cap-binding complex), which, as we discuss in subsequent sections, helps the RNA to be properly processed and exported. The 5′-methyl cap also has an important role in the translation of mRNAs in the cytosol, as we discuss later in the chapter.

RNA Splicing Removes Intron Sequences from Newly Transcribed Pre-mRNAs

As discussed in Chapter 4, the protein coding sequences of eucaryotic genes are typically interrupted by noncoding intervening sequences (introns). Discovered in 1977, this feature of eucaryotic genes came as a surprise to scientists, who had been, until that time, familiar only with bacterial genes, which typically consist of a continuous stretch of coding DNA that is directly transcribed into mRNA. In marked contrast, eucaryotic genes were found to be broken up into small pieces of coding sequence (*expressed sequences* or **exons**) interspersed with much longer *intervening sequences* or **introns**; thus, the coding portion of a eucaryotic gene is often only a small fraction of the length of the gene (**Figure 6–25**).

Both intron and exon sequences are transcribed into RNA. The intron sequences are removed from the newly synthesized RNA through the process of **RNA splicing**. The vast majority of RNA splicing that takes place in cells functions in the production of mRNA, and our discussion of splicing focuses on this so-called precursor-mRNA (or pre-mRNA) splicing. Only after 5′ and 3′ end processing and splicing have taken place is such RNA termed mRNA.

Each splicing event removes one intron, proceeding through two sequential phosphoryl-transfer reactions known as transesterifications; these join two exons while removing the intron as a "lariat" (**Figure 6–26**). Since the number of high-energy phosphate bonds remains the same, these reactions could in

Figure 6–24 The reactions that cap the 5′ end of each RNA molecule synthesized by RNA polymerase II. The final cap contains a novel 5′-to-5′ linkage between the positively charged 7-methyl G residue and the 5′ end of the RNA transcript (see Figure 6–22B). The letter N represents any one of the four ribonucleotides, although the nucleotide that starts an RNA chain is usually a purine (an A or a G). (After A.J. Shatkin, *BioEssays* 7:275–277, 1987. With permission from ICSU Press.)

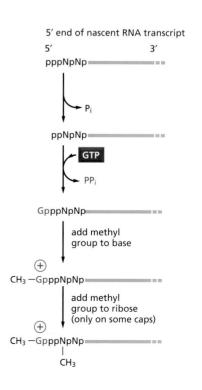

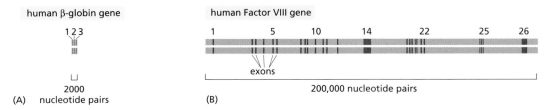

Figure 6–25 Structure of two human genes showing the arrangement of exons and introns. (A) The relatively small β-globin gene, which encodes one of the subunits of the oxygen-carrying protein hemoglobin, contains 3 exons (see also Figure 4–7). (B) The much larger Factor VIII gene contains 26 exons; it codes for a protein (Factor VIII) that functions in the blood-clotting pathway. The most prevalent form of hemophilia results from mutations in this gene.

principle take place without nucleoside triphosphate hydrolysis. However, the machinery that catalyzes pre-mRNA splicing is complex, consisting of 5 additional RNA molecules and as many as 200 proteins, and it hydrolyzes many ATP molecules per splicing event. This additional complexity ensures that splicing is accurate, while at the same time being flexible enough to deal with the enormous variety of introns found in a typical eucaryotic cell.

It may seem wasteful to remove large numbers of introns by RNA splicing. In attempting to explain why it occurs, scientists have pointed out that the exon–intron arrangement would seem to facilitate the emergence of new and useful proteins over evolutionary time scales. Thus, the presence of numerous introns in DNA allows genetic recombination to readily combine the exons of different genes (see p. 140), enabling genes for new proteins to evolve more easily by the combination of parts of preexisting genes. The observation, described in Chapter 3, that many proteins in present-day cells resemble patchworks composed from a common set of protein *domains*, supports this idea.

RNA splicing also has a present-day advantage. The transcripts of many eucaryotic genes (estimated at 75% of genes in humans) are spliced in more than one way, thereby allowing the same gene to produce a corresponding set of different proteins (**Figure 6–27**). Rather than being the wasteful process it may have seemed at first sight, RNA splicing enables eucaryotes to increase the already enormous coding potential of their genomes. We shall return to this idea again in this chapter and the next, but we first need to describe the cellular machinery that performs this remarkable task.

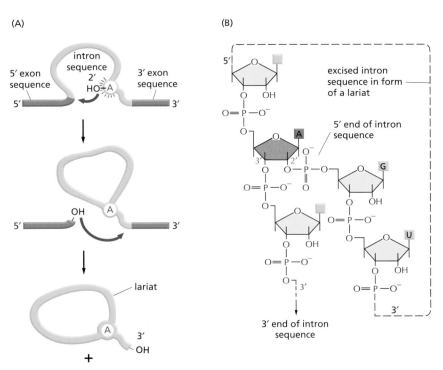

Figure 6–26 The pre-mRNA splicing reaction. (A) In the first step, a specific adenine nucleotide in the intron sequence (indicated in *red*) attacks the 5′ splice site and cuts the sugar-phosphate backbone of the RNA at this point. The cut 5′ end of the intron becomes covalently linked to the adenine nucleotide, as shown in detail in (B), thereby creating a loop in the RNA molecule. The released free 3′-OH end of the exon sequence then reacts with the start of the next exon sequence, joining the two exons together and releasing the intron sequence in the shape of a *lariat*. The two exon sequences thereby become joined into a continuous coding sequence; the released intron sequence is eventually degraded.

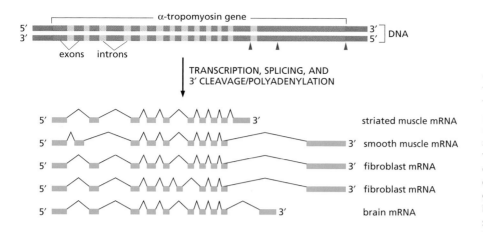

Figure 6–27 Alternative splicing of the α-tropomyosin gene from rat. α-Tropomyosin is a coiled-coil protein (see Figure 3–9) that regulates contraction in muscle cells. The primary transcript can be spliced in different ways, as indicated in the figure, to produce distinct mRNAs, which then give rise to variant proteins. Some of the splicing patterns are specific for certain types of cells. For example, the α-tropomyosin made in striated muscle is different from that made from the same gene in smooth muscle. The arrowheads in the top part of the figure mark the sites where cleavage and poly-A addition form the 3′ ends of the mature mRNAs.

Nucleotide Sequences Signal Where Splicing Occurs

The mechanism of pre-mRNA splicing shown in Figure 6–26 implies that the splicing machinery must recognize three portions of the precursor RNA molecule: the 5′ splice site, the 3′ splice site, and the branch point in the intron sequence that forms the base of the excised lariat. Not surprisingly, each site has a consensus nucleotide sequence that is similar from intron to intron and provides the cell with cues for where splicing is to take place (**Figure 6–28**). However, these consensus sequences are relatively short and can accommodate a high degree of sequence variability; as we shall see shortly, the cell incorporates additional types of information to ultimately choose exactly where, on each RNA molecule, splicing is to take place.

The high variability of the splicing consensus sequences presents a special challenge for scientists attempting to decipher genome sequences. Introns range in size from about 10 nucleotides to over 100,000 nucleotides, and choosing the precise borders of each intron is a difficult task even with the aid of powerful computers. The possibility of alternative splicing compounds the problem of predicting protein sequences solely from a genome sequence. This difficulty is one of the main barriers to identifying all of the genes in a complete genome sequence, and it is one of the primary reasons why we know only the approximate number of genes in the human genome.

RNA Splicing Is Performed by the Spliceosome

Unlike the other steps of mRNA production we have discussed, key steps in RNA splicing are performed by RNA molecules rather than proteins. Specialized RNA molecules recognize the nucleotide sequences that specify where splicing is to occur and also participate in the chemistry of splicing. These RNA molecules are relatively short (less than 200 nucleotides each), and there are five of them (U1, U2, U4, U5, and U6) involved in the major form of pre-mRNA splicing. Known as

Figure 6–28 The consensus nucleotide sequences in an RNA molecule that signal the beginning and the end of most introns in humans. Only the three blocks of nucleotide sequences shown are required to remove an intron sequence; the rest of the intron can be occupied by any nucleotides. Here A, G, U, and C are the standard RNA nucleotides; R stands for purines (A or G); and Y stands for pyrimidines (C or U). The A highlighted in *red* forms the branch point of the lariat produced by splicing. Only the GU at the start of the intron and the AG at its end are invariant nucleotides in the splicing consensus sequences. Several different nucleotides can occupy the remaining positions (even the branch point A), although the indicated nucleotides are preferred. The distances along the RNA between the three splicing consensus sequences are highly variable; however, the distance between the branch point and 3′ splice junction is typically much shorter than that between the 5′ splice junction and the branch point.

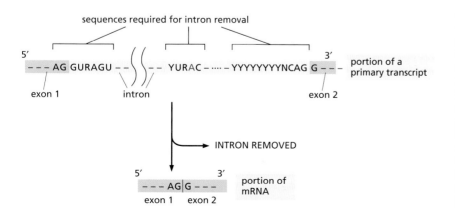

snRNAs (**small nuclear RNAs**), each is complexed with at least seven protein subunits to form a snRNP (small nuclear ribonucleoprotein). These snRNPs form the core of the **spliceosome**, the large assembly of RNA and protein molecules that performs pre-mRNA splicing in the cell.

The spliceosome is a complex and dynamic machine. When studied *in vitro*, a few components of the spliceosome assemble on pre-mRNA and, as the splicing reaction proceeds, new components enter as those that have already performed their tasks are jettisoned (**Figure 6–29**). However, many scientists believe that, inside the cell, the spliceosome is a preexisting, loose assembly of all the components—capturing, splicing and releasing RNA as a coordinated unit, and undergoing extensive rearrangements each time a splice is made. During the splicing reaction, recognition of the 5′ splice junction, the branch-point site, and the 3′ splice junction is performed largely through base-pairing between the snRNAs and the consensus RNA sequences in the pre-mRNA substrate (**Figure**

Figure 6–29 The pre-mRNA splicing mechanism. RNA splicing is catalyzed by an assembly of snRNPs (shown as *colored circles*) plus other proteins (most of which are not shown), which together constitute the spliceosome. The spliceosome recognizes the splicing signals on a pre-mRNA molecule, brings the two ends of the intron together, and provides the enzymatic activity for the two reaction steps (see Figure 6–26).

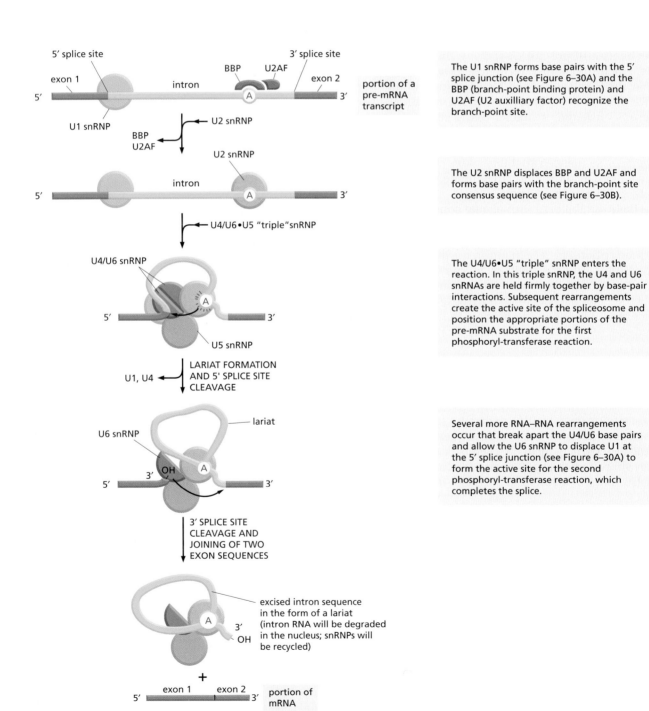

The U1 snRNP forms base pairs with the 5′ splice junction (see Figure 6–30A) and the BBP (branch-point binding protein) and U2AF (U2 auxilliary factor) recognize the branch-point site.

The U2 snRNP displaces BBP and U2AF and forms base pairs with the branch-point site consensus sequence (see Figure 6–30B).

The U4/U6•U5 "triple" snRNP enters the reaction. In this triple snRNP, the U4 and U6 snRNAs are held firmly together by base-pair interactions. Subsequent rearrangements create the active site of the spliceosome and position the appropriate portions of the pre-mRNA substrate for the first phosphoryl-transferase reaction.

Several more RNA–RNA rearrangements occur that break apart the U4/U6 base pairs and allow the U6 snRNP to displace U1 at the 5′ splice junction (see Figure 6–30A) to form the active site for the second phosphoryl-transferase reaction, which completes the splice.

6–30). In the course of splicing, the spliceosome undergoes several shifts in which one set of base-pair interactions is broken and another is formed in its place. For example, U1 is replaced by U6 at the 5′ splice junction (see Figure 6–30A). This type of RNA–RNA rearrangement (in which the formation of one RNA–RNA interaction requires the disruption of another) occurs several times during the splicing reaction. It permits the checking and rechecking of RNA sequences before the chemical reaction is allowed to proceed, thereby increasing the accuracy of splicing.

The Spliceosome Uses ATP Hydrolysis to Produce a Complex Series of RNA–RNA Rearrangements

Although ATP hydrolysis is not required for the chemistry of RNA splicing *per se*, it is required for the assembly and rearrangements of the spliceosome. Some of the additional proteins that make up the spliceosome use the energy of ATP hydrolysis to break existing RNA–RNA interactions to allow the formation of new ones. In fact, all the steps shown previously in Figure 6–29—except the association of BBP with the branch-point site and U1 snRNP with the 5′ splice site—require ATP hydrolysis and additional proteins. Each successful splice requires as many as 200 proteins, if we include those that form the snRNPs.

The ATP-requiring RNA–RNA rearrangements that take place in the spliceosome occur within the snRNPs themselves and between the snRNPs and the pre-mRNA substrate. One of the most important functions of these rearrangements is the creation of the active catalytic site of the spliceosome. The strategy of creating an active site only after the assembly and rearrangement of splicing components on a pre-mRNA substrate is a particularly effective way to prevent wayward splicing.

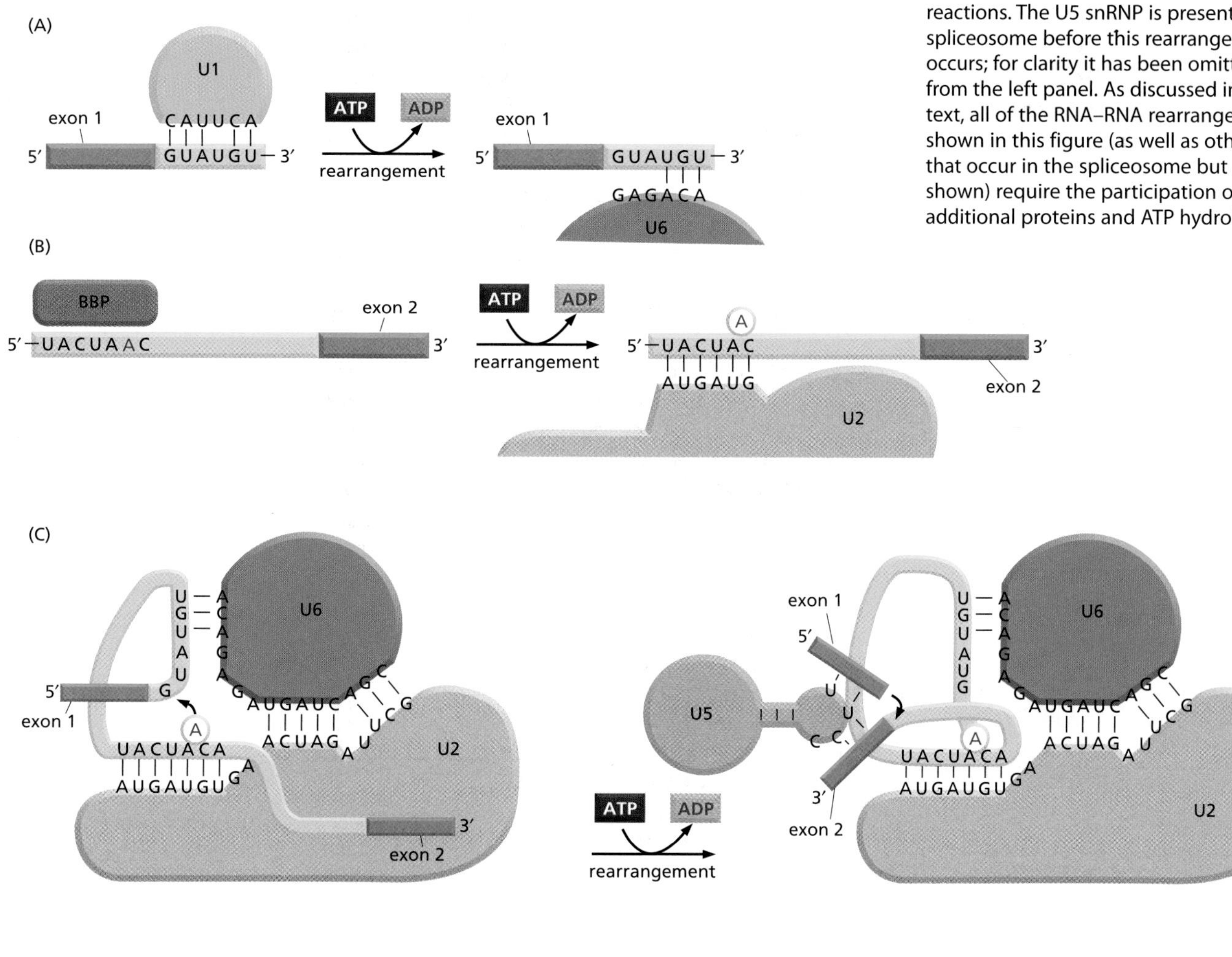

Figure 6–30 **Several of the rearrangements that take place in the spliceosome during pre-mRNA splicing.** Shown here are the details for the yeast *Saccharomyces cerevisiae*, in which the nucleotide sequences involved are slightly different from those in human cells. (A) The exchange of U1 snRNP for U6 snRNP occurs before the first phosphoryl-transfer reaction (see Figure 6–29). This exchange requires the 5′ splice site to be read by two different snRNPs, thereby increasing the accuracy of 5′ splice site selection by the spliceosome. (B) The branch-point site is first recognized by BBP and subsequently by U2 snRNP; as in (A), this "check and recheck" strategy provides increased accuracy of site selection. The binding of U2 to the branch point forces the appropriate adenine (in *red*) to be unpaired and thereby activates it for the attack on the 5′ splice site (see Figure 6–29). This, in combination with recognition by BBP, is the way in which the spliceosome accurately chooses the adenine that is ultimately to form the branch point. (C) After the first phosphoryl-transfer reaction *(left)* has occurred, a series of rearrangements brings the two exons into close proximity for the second phosphoryl-transfer reaction *(right)*. The snRNAs both position the reactants and provide (either all or in part) the catalytic sites for the two reactions. The U5 snRNP is present in the spliceosome before this rearrangement occurs; for clarity it has been omitted from the left panel. As discussed in the text, all of the RNA–RNA rearrangements shown in this figure (as well as others that occur in the spliceosome but are not shown) require the participation of additional proteins and ATP hydrolysis.

Perhaps the most surprising feature of the spliceosome is the nature of the catalytic site itself: it is largely (if not exclusively) formed by RNA molecules instead of proteins. In the last section of this chapter we discuss in general terms the structural and chemical properties of RNA that allow it to perform catalysis; here we need only consider that the U2 and U6 snRNAs in the spliceosome form a precise three-dimensional RNA structure that juxtaposes the 5′ splice site of the pre-mRNA with the branch-point site and probably performs the first transesterification reaction (see Figure 6–30C). In a similar way, the 5′ and 3′ splice junctions are brought together (an event requiring the U5 snRNA) to facilitate the second transesterification.

Once the splicing chemistry is completed, the snRNPs remain bound to the lariat. The disassembly of these snRNPs from the lariat (and from each other) requires another series of RNA–RNA rearrangements that require ATP hydrolysis, thereby returning the snRNAs to their original configuration so that they can be used again in a new reaction. At the completion of a splice, the spliceosome directs a set of proteins to bind to the mRNA near the position formerly occupied by the intron. Called the *exon junction complex (EJC)*, these proteins mark the site of a successful splicing event and, as we shall see later in this chapter, influence the subsequent fate of the mRNA.

Other Properties of Pre-mRNA and Its Synthesis Help to Explain the Choice of Proper Splice Sites

As we have seen, intron sequences vary enormously in size, with some being in excess of 100,000 nucleotides. If splice-site selection were determined solely by the snRNPs acting on a preformed, protein-free RNA molecule, we would expect splicing mistakes—such as exon skipping and the use of "cryptic" splice sites—to be very common (**Figure 6–31**). The fidelity mechanisms built into the spliceosome, however, are supplemented by two additional strategies that increase the accuracy of splicing. The first is simply a consequence of the early stages of splicing occurring while the pre-mRNA molecules are being synthesized by RNA polymerase II. As transcription proceeds, the phosphorylated tail of RNA polymerase carries several components of the spliceosome (see Figure 6–23), and these components are transferred directly from the polymerase to the RNA as RNA is synthesized. This strategy helps the cell keep track of introns and exons: for example, the snRNPs that assemble at a 5′ splice site are initially presented with only a single 3′ splice site since the sites further downstream have not yet been synthesized. The coordination of transcription with splicing is especially important in preventing inappropriate exon skipping.

A strategy called "exon definition" is another way cells choose the appropriate splice sites. Exon size tends to be much more uniform than intron size, averaging about 150 nucleotide pairs across a wide variety of eucaryotic organisms (**Figure 6–32**). According to the exon definition idea, the splicing machinery initially seeks out the relatively homogenously sized exon sequences. As RNA synthesis proceeds, a group of additional components (most notably SR proteins, so-named because they contain a domain rich in serines and arginines) assemble on exon sequences and help to mark off each 3′ and 5′ splice site starting at the 5′ end of the RNA (**Figure 6–33**). These proteins, in turn, recruit U1 snRNA, which

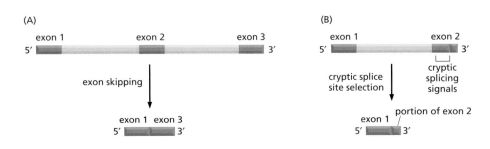

Figure 6–31 **Two types of splicing errors.** (A) Exon skipping. (B) Cryptic splice-site selection. Cryptic splicing signals are nucleotide sequences of RNA that closely resemble true splicing signals.

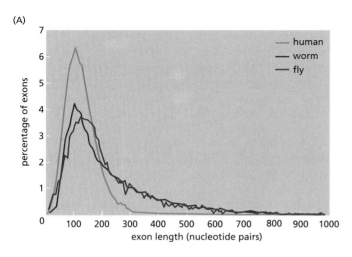

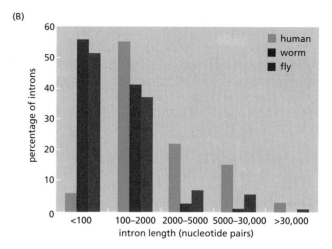

marks the downstream exon boundary, and U2AF, which specifies the upstream one. By specifically marking the exons in this way and thereby taking advantage of the relatively uniform size of exons, the cell increases the accuracy with which it deposits the initial splicing components on the nascent RNA and thereby helps to avoid cryptic splice sites. How the SR proteins discriminate exon sequences from intron sequences is not understood in detail; however, it is known that some of the SR proteins bind preferentially to specific RNA sequences in exons, termed *splicing enhancers*. In principle, since any one of several different codons can be used to code for a given amino acid, there is freedom to adjust the exon nucleotide sequence so as to form a binding site for an SR protein, without necessarily affecting the amino acid sequence that the exon specifies.

Both the marking of exon and intron boundaries and the assembly of the spliceosome begin on an RNA molecule while it is still being elongated by RNA polymerase at its 3′ end. However, the actual chemistry of splicing can take place much later. This delay means that intron sequences are not necessarily removed from a pre-mRNA molecule in the order in which they occur along the RNA chain. It also means that, although spliceosome assembly is co-transcriptional, the splicing reactions sometimes occur posttranscriptionally—that is, after a complete pre-mRNA molecule has been made.

Figure 6–32 Variation in intron and exon lengths in the human, worm, and fly genomes. (A) Size distribution of exons. (B) Size distribution of introns. Note that exon length is much more uniform than intron length. (Adapted from International Human Genome Sequencing Consortium, *Nature* 409:860–921, 2001. With permission from Macmillan Publishers Ltd.)

A Second Set of snRNPs Splice a Small Fraction of Intron Sequences in Animals and Plants

Simple eucaryotes such as yeasts have only one set of snRNPs that perform all pre-mRNA splicing. However, more complex eucaryotes such as flies, mammals, and plants have a second set of snRNPs that direct the splicing of a small fraction of their intron sequences. This minor form of spliceosome recognizes a different set of RNA sequences at the 5′ and 3′ splice junctions and at the branch point; it is called the *U12-type spliceosome* because of the involvement of the

Figure 6–33 The exon definition idea. According to one proposal, SR proteins bind to each exon sequence in the pre-mRNA and thereby help to guide the snRNPs to the proper intron/exon boundaries. This demarcation of exons by the SR proteins occurs co-transcriptionally, beginning at the CBC (cap-binding complex) at the 5′ end. As indicated, the intron sequences in the pre-mRNA, which can be extremely long, are packaged into hnRNP (heterogeneous nuclear ribonucleoprotein) complexes that compact them into more manageable structures and perhaps mask cryptic splice sites. It has been proposed that hnRNP proteins may preferentially associate with intron sequences and that this preference may also help the spliceosome distinguish introns from exons. However, as shown, at least some hnRNP proteins also bind to exon sequences. (Adapted from R. Reed, *Curr. Opin. Cell Biol.* 12:340–345, 2000. With permission from Elsevier.)

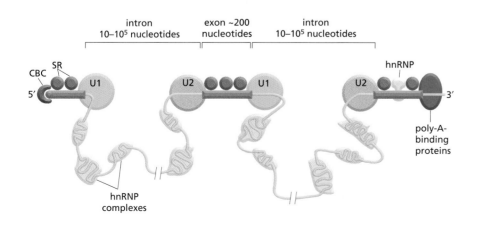

(A)

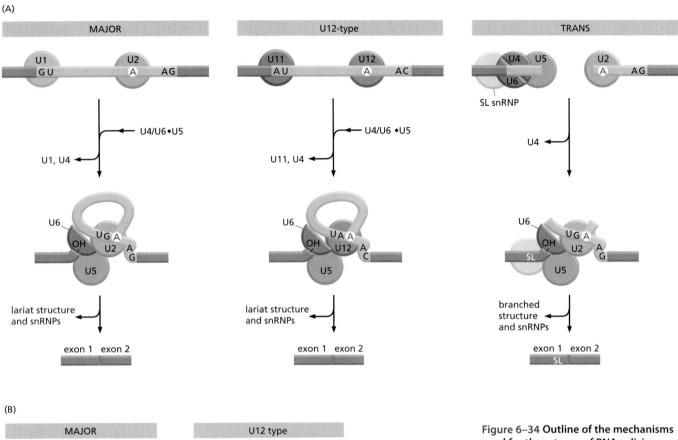

(B)

Figure 6–34 **Outline of the mechanisms used for three types of RNA splicing.**
(A) Three types of spliceosomes. The major spliceosome *(left)*, the U12-type spliceosome *(middle)*, and the trans-spliceosome *(right)* are each shown at two stages of assembly. Introns removed by the U12-type spliceosome have a different set of consensus nucleotide sequences from those removed by the major spliceosome. In humans, it is estimated that 0.1% of introns are removed by the U12-type spliceosome. In trans-splicing, which does not occur in humans, the SL snRNP is consumed in the reaction because a portion of the SL snRNA becomes the first exon of the mature mRNA. (B) The major U6 snRNP and the U6 snRNP specific to the U12-type spliceosome both recognize the 5′ splice junction, but they do so through a different set of base-pair interactions. The sequences shown are from humans. (Adapted from Y.T. Yu et al., The RNA World, pp. 487–524. Cold Spring Harbor, New York: Cold Spring Harbor Laboratory Press, 1999.)

U12 SnRNP (**Figure 6–34**A). Despite recognizing different nucleotide sequences, the snRNPs in this spliceosome make the same types of RNA–RNA interactions with the pre-mRNA and with each other as do the major snRNPs (Figure 6–34B). Although, as we have seen, components of the major spliceosomes travel with RNA polymerase II as it transcribes genes, this may not be the case for the U12 spliceosome. It is possible that U12-mediated splicing is thereby delayed, and this presents the cell with a way to co-regulate splicing of the several hundred genes whose expression requires this spliceosome. A number of mammalian mRNAs contain a mixture of introns, some removed by the major spliceosome and others by the minor spliceosome, and it has been proposed that this arrangement permits particularly complex patterns of alternative splicing to occur.

A few eucaryotic organisms exhibit a particular variation on splicing, called **trans-splicing**. These organisms include the single-celled trypanosomes—protozoans that cause African sleeping sickness in humans—and the model multicellular organism, the nematode worm. In trans-splicing, exons from two separate RNA transcripts are spliced together to form a mature mRNA molecule (see Figure 6–34A). Trypanosomes produce all of their mRNAs in this way, whereas trans-splicing accounts for only about 1% of nematode mRNAs. In both cases, a single exon is spliced onto the 5′ end of many different RNA transcripts produced by the cell; in this way, all of the products of trans-splicing have the same 5′ exon and different 3′ exons. Many of the same snRNPs that function in conventional splicing are used in this reaction, although trans-splicing uses a unique snRNP (called the SL RNP) that brings in the common exon (see Figure 6–34).

Figure 6–35 Abnormal processing of the β-globin primary RNA transcript in humans with the disease β thalassemia. In the examples shown, the disease is caused by splice-site mutations (*black arrowheads*) found in the genomes of affected patients. The *dark blue boxes* represent the three normal exon sequences; the *red lines* indicate the 5′ and 3′ splice sites. The *light blue boxes* depict new nucleotide sequences included in the final mRNA molecule as a result of the mutation. Note that when a mutation leaves a normal splice site without a partner, an exon is skipped or one or more abnormal cryptic splice sites nearby is used as the partner site, as in (C) and (D). (Adapted in part from S.H. Orkin, in The Molecular Basis of Blood Diseases [G. Stamatoyannopoulos et al., eds.], pp. 106–126. Philadelphia: Saunders, 1987.)

(A) NORMAL ADULT β-GLOBIN PRIMARY RNA TRANSCRIPT

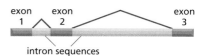

normal mRNA is formed from three exons

(B) SOME SINGLE-NUCLEOTIDE CHANGES THAT DESTROY A NORMAL SPLICE SITE CAUSE EXON SKIPPING

mRNA with exon 2 missing

(C) SOME SINGLE-NUCLEOTIDE CHANGES THAT DESTROY NORMAL SPLICE SITES ACTIVATE CRYPTIC SPLICE SITES

mRNA with extended exon 3

(D) SOME SINGLE-NUCLEOTIDE CHANGES THAT CREATE NEW SPLICE SITES CAUSE NEW EXONS TO BE INCORPORATED

mRNA with extra exon inserted between exon 2 and exon 3

We do not know why even a few organisms use trans-splicing; however, it is thought that the common 5′ exon may aid in the translation of the mRNA. Thus, the mRNAs produced by trans-splicing in nematodes seem to be translated with especially high efficiency.

RNA Splicing Shows Remarkable Plasticity

We have seen that the choice of splice sites depends on such features of the pre-mRNA transcript as the affinity of the three signals on the RNA (the 5′ and 3′ splice junctions and the branch point) for the splicing machinery, the co-transcriptional assembly of the spliceosome, and the "bookkeeping" that underlies exon definition. We do not know how accurate splicing normally is because, as we see later, there are several quality control systems that rapidly destroy mRNAs whose splicing goes awry. However, we do know that, compared with other steps in gene expression, splicing is unusually flexible. For example, a mutation in a nucleotide sequence critical for splicing of a particular intron does not necessarily prevent splicing of that intron altogether. Instead, the mutation typically creates a new pattern of splicing (**Figure 6–35**). Most commonly, an exon is simply skipped (Figure 6–35B). In other cases, the mutation causes a cryptic splice junction to be efficiently used (Figure 6–35C). Apparently, the splicing machinery has evolved to pick out the best possible pattern of splice junctions, and if the optimal one is damaged by mutation, it will seek out the next best pattern, and so on. This flexibility in the process of RNA splicing suggests that changes in splicing patterns caused by random mutations have been an important pathway in the evolution of genes and organisms.

The plasticity of RNA splicing also means that the cell can regulate the pattern of RNA splicing. Earlier in this section we saw that alternative splicing can give rise to different proteins from the same gene. Some examples of alternative splicing are constitutive; that is, the alternatively spliced mRNAs are produced continuously by cells of an organism. However, in many cases, the cell regulates the splicing patterns so that different forms of the protein are produced at different times and in different tissues (see Figure 6–27). In Chapter 7 we return to this issue to discuss some specific examples of regulated RNA splicing.

Spliceosome-Catalyzed RNA Splicing Probably Evolved from Self-splicing Mechanisms

When the spliceosome was first discovered, it puzzled molecular biologists. Why do RNA molecules instead of proteins perform important roles in splice site recognition and in the chemistry of splicing? Why is a lariat intermediate used rather than the apparently simpler alternative of bringing the 5′ and 3′ splice sites together in a single step, followed by their direct cleavage and rejoining? The answers to these questions reflect the way in which the spliceosome is believed to have evolved.

As discussed briefly in Chapter 1 (and in more detail in the final section of this chapter), it is likely that early cells used RNA molecules rather than proteins as their major catalysts and that they stored their genetic information in RNA rather than in DNA sequences. RNA-catalyzed splicing reactions presumably had important roles in these early cells. As evidence, some *self-splicing RNA introns* (that is, intron sequences in RNA whose splicing out can occur in the absence of proteins or any other RNA molecules) remain today—for example, in the nuclear rRNA genes of the ciliate *Tetrahymena*, in a few bacteriophage T4 genes, and in some mitochondrial and chloroplast genes.

A self-splicing intron sequence can be identified in a test tube by incubating a pure RNA molecule that contains the intron sequence and observing the splicing reaction. Two major classes of self-splicing intron sequences can be distinguished in this way. *Group I intron sequences* begin the splicing reaction by binding a G nucleotide to the intron sequence; this G is thereby activated to form the attacking group that will break the first of the phosphodiester bonds cleaved during splicing (the bond at the 5′ splice site). In *group II intron sequences,* an especially reactive A residue in the intron sequence is the attacking group, and a lariat intermediate is generated. Otherwise the reaction pathways for the two types of self-splicing intron sequences are the same. Both are presumed to represent vestiges of very ancient mechanisms (**Figure 6–36**).

For both types of self-splicing reactions, the nucleotide sequence of the intron is critical; the intron RNA folds into a specific three-dimensional structure, which brings the 5′ and 3′ splice junctions together and provides precisely positioned reactive groups to perform the chemistry (see Figure 6–6C). Because the chemistries of their splicing reactions are so similar, it has been proposed that the pre-mRNA splicing mechanism of the spliceosome evolved from group

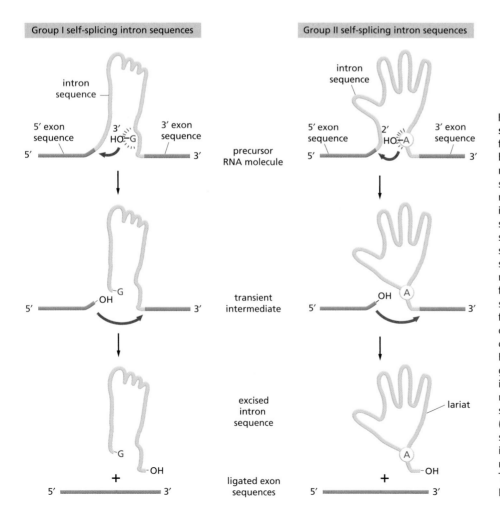

Figure 6–36 The two known classes of self-splicing intron sequences. The figure emphasizes the similarities between the two mechanisms. Both are normally aided by proteins in the cell that speed up the reaction, but the catalysis is nevertheless mediated by the RNA in the intron sequence. The group I intron sequences bind a free G nucleotide to a specific site on the RNA to initiate splicing, while the group II intron sequences use an especially reactive A nucleotide in the intron sequence itself for the same purpose. Both types of self-splicing reactions require the intron to be folded into a highly specific three-dimensional structure that provides the catalytic activity for the reaction (see Figure 6–6). The mechanism used by group II intron sequences releases the intron as a lariat structure and closely resembles the pathway of pre-mRNA splicing catalyzed by the spliceosome (compare with Figure 6–29). The spliceosome performs most RNA splicing in eucaryotic cells, and self-splicing RNAs represent unusual cases. (Adapted from T.R. Cech, *Cell* 44:207–210, 1986. With permission from Elsevier.)

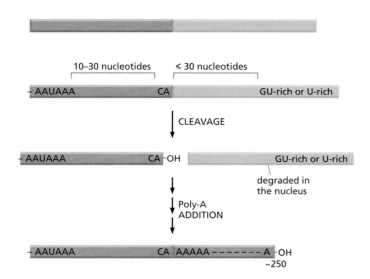

Figure 6–37 Consensus nucleotide sequences that direct cleavage and polyadenylation to form the 3′ end of a eucaryotic mRNA. These sequences are encoded in the genome; specific proteins recognize them after they are transcribed into RNA. The hexamer AAUAAA is bound by CPSF, the GU-rich element beyond the cleavage site is bound by CstF (see Figure 6–38), and the CA sequence is bound by a third factor required for the cleavage step. Like other consensus nucleotide sequences discussed in this chapter (see Figure 6–12), the sequences shown in the figure represent a variety of individual cleavage and polyadenylation signals.

II self-splicing. According to this idea, when the spliceosomal snRNPs took over the structural and chemical roles of the group II introns, the strict sequence constraints on intron sequences would have disappeared, thereby permitting a vast expansion in the number of different RNAs that could be spliced.

RNA-Processing Enzymes Generate the 3′ End of Eucaryotic mRNAs

As previously explained, the 5′ end of the pre-mRNA produced by RNA polymerase II is capped almost as soon as it emerges from the RNA polymerase. Then, as the polymerase continues its movement along a gene, the spliceosome assembles on the RNA and delineates the intron and exon boundaries. The long C-terminal tail of the RNA polymerase coordinates these processes by transferring capping and splicing components directly to the RNA as it emerges from the enzyme. We see in this section that, as RNA polymerase II reaches the end of a gene, a similar mechanism ensures that the 3′ end of the pre-mRNA is appropriately processed.

As might be expected, the position of the 3′ end of each mRNA molecule is ultimately specified by a signal encoded in the genome (**Figure 6–37**). These signals are transcribed into RNA as the RNA polymerase II moves through them, and they are then recognized (as RNA) by a series of RNA-binding proteins and RNA-processing enzymes (**Figure 6–38**). Two multisubunit proteins, called CstF (cleavage stimulation factor) and CPSF (cleavage and polyadenylation specificity factor), are of special importance. Both of these proteins travel with the RNA polymerase tail and are transferred to the 3′-end processing sequence on an RNA molecule as it emerges from the RNA polymerase.

Once CstF and CPSF bind to specific nucleotide sequences on the emerging RNA molecule, additional proteins assemble with them to create the 3′ end of the mRNA. First, the RNA is cleaved (see Figure 6–38). Next an enzyme called poly-A polymerase (PAP) adds, one at a time, approximately 200 A nucleotides to the 3′ end produced by the cleavage. The nucleotide precursor for these additions is ATP, and the same type of 5′-to-3′ bonds are formed as in conventional RNA synthesis (see Figure 6–4). Unlike the usual RNA polymerases, poly-A polymerase does not require a template; hence the poly-A tail of eucaryotic mRNAs

Figure 6–38 Some of the major steps in generating the 3′ end of a eucaryotic mRNA. This process is much more complicated than the analogous process in bacteria, where the RNA polymerase simply stops at a termination signal and releases both the 3′ end of its transcript and the DNA template (see Figure 6–11).

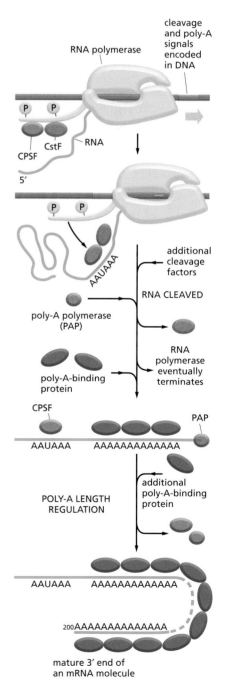

is not directly encoded in the genome. As the poly-A tail is synthesized, proteins called poly-A-binding proteins assemble onto it and, by a poorly understood mechanism, determine the final length of the tail. Some poly-A-binding proteins remain bound to the poly-A tail as the mRNA travels from the nucleus to the cytosol and they help to direct the synthesis of a protein on the ribosome, as we see later in this chapter.

After the 3′ end of a eucaryotic pre-mRNA molecule has been cleaved, the RNA polymerase II continues to transcribe, in some cases for hundreds of nucleotides. But the polymerase soon releases its grip on the template and transcription terminates. After 3′-end cleavage has occurred, the newly synthesized RNA that emerges from the polymerases lacks a 5′ cap; this unprotected RNA is rapidly degraded by a 5′ → 3′ exonuclease, which is carried along on the polymerase tail. Apparently, it is this RNA degradation that eventually causes the RNA polymerase to dissociate from the DNA.

Mature Eucaryotic mRNAs Are Selectively Exported from the Nucleus

We have seen how eucaryotic pre-mRNA synthesis and processing take place in an orderly fashion within the cell nucleus. However, these events create a special problem for eucaryotic cells, especially those of complex organisms where the introns are vastly longer than the exons. Of the pre-mRNA that is synthesized, only a small fraction—the mature mRNA—is of further use to the cell. The rest—excised introns, broken RNAs, and aberrantly processed pre-mRNAs—is not only useless but potentially dangerous. How, then, does the cell distinguish between the relatively rare mature mRNA molecules it wishes to keep and the overwhelming amount of debris from RNA processing?

The answer is that, as an RNA molecule is processed, it loses certain proteins and acquires others, thereby signifying the successful completion of each of the different steps. For example, we have seen that acquisition of the cap-binding complexes, the exon junction complexes, and the poly-A-binding proteins mark the completion of capping, splicing, and poly-A addition, respectively. A properly completed mRNA molecule is also distinguished by the proteins it lacks. For example, the presence of a snRNP would signify incomplete or aberrant splicing. Only when the proteins present on an mRNA molecule collectively signify that processing was successfully completed is the mRNA exported from the nucleus into the cytosol, where it can be translated into protein. Improperly processed mRNAs, and other RNA debris are retained in the nucleus, where they are eventually degraded by the nuclear **exosome**, a large protein complex whose interior is rich in 3′-to-5′ RNA exonucleases. Eucaryotic cells thus export only useful RNA molecules to the cytoplasm, while debris is disposed of in the nucleus.

Of all the proteins that assemble on pre-mRNA molecules as they emerge from transcribing RNA polymerases, the most abundant are the hnRNPs (heterogeneous nuclear ribonuclear proteins) (see Figure 6–33). Some of these proteins (there are approximately 30 of them in humans) unwind the hairpin helices from the RNA so that splicing and other signals on the RNA can be read more easily. Others preferentially package the RNA contained in the very long intron sequences typically found in genes of complex organisms. They may therefore play an important role in distinguishing mature mRNA from the debris left over from RNA processing.

Successfully processed mRNAs are guided through the **nuclear pore complexes** (NPCs)—aqueous channels in the nuclear membrane that directly connect the nucleoplasm and cytosol (**Figure 6–39**). Small molecules (less than 50,000 daltons) can diffuse freely through these channels. However, most of the macromolecules in cells, including mRNAs complexed with proteins, are far too large to pass through the channels without a special process. The cell uses energy to actively transport such macromolecules in both directions through the nuclear pore complexes.

As explained in detail in Chapter 12, macromolecules are moved through nuclear pore complexes by *nuclear transport receptors,* which, depending on the

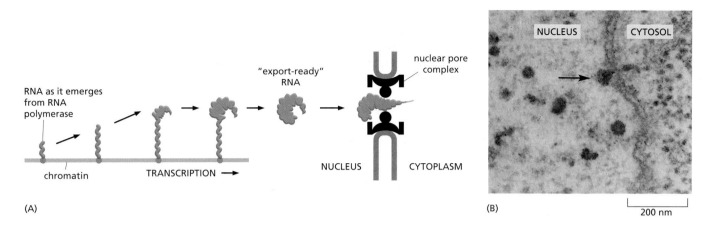

(A)

(B)

Figure 6–39 **Transport of a large mRNA molecule through the nuclear pore complex.** (A) The maturation of an mRNA molecule as it is synthesized by RNA polymerase and packaged by a variety of nuclear proteins. This drawing of an unusually abundant RNA, called the Balbiani Ring mRNA, is based on EM micrographs such as that shown in (B). Balbiani Rings are found in the cells of certain insects. (A, adapted from B. Daneholt, *Cell* 88:585–588, 1997. With permission from Elsevier; B, from B.J. Stevens and H. Swift, *J. Cell Biol.* 31:55–77, 1966. With permission from The Rockefeller University Press.)

identity of the macromolecule, escort it from the nucleus to the cytoplasm or vice versa. For mRNA export to occur, a specific nuclear transport receptor must be loaded onto the mRNA, a step that, at least in some organisms, takes place in concert with 3′ cleavage and polyadenylation. Once it helps to move an RNA molecule through the nuclear pore complex, the transport receptor dissociates from the mRNA, re-enters the nucleus, and exports a new mRNA molecule (**Figure 6–40**).

The export of mRNA–protein complexes from the nucleus can be observed with the electron microscope for the unusually abundant mRNA of the insect Balbiani Ring genes. As these genes are transcribed, the newly formed RNA is seen to be packaged by proteins, including hnRNPs, SR proteins, and components of the spliceosome. This protein–RNA complex undergoes a series of structural transitions, probably reflecting RNA processing events, culminating in a curved fiber (see Figure 6–39). This curved fiber moves through the nucleoplasm and enters the nuclear pore complex (with its 5′ cap proceeding first), and it then undergoes another series of structural transitions as it moves through the

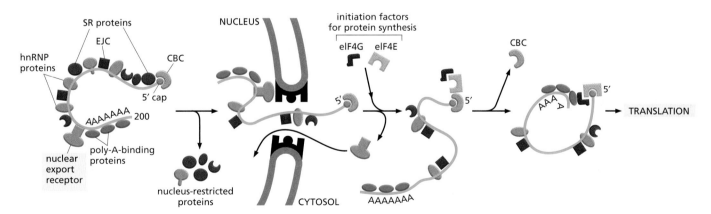

Figure 6–40 **Schematic illustration of an "export-ready" mRNA molecule and its transport through the nuclear pore.** As indicated, some proteins travel with the mRNA as it moves through the pore, whereas others remain in the nucleus. The nuclear export receptor for mRNAs is a complex of proteins that is deposited when the mRNA has been correctly spliced and polyadenylated. When the mRNA is exported to the cytosol, the export receptor dissociates from the mRNA and is re-imported into the nucleus, where it can be used again. Just after it leaves the nucleus, and before it loses the cap-binding complex (CBC) the mRNA is subjected to a final check, called *nonsense-mediated decay*, which is described later in the chapter. Once it passes this test the mRNA continues to shed previously bound proteins and acquire new ones before it is efficiently translated into protein. EJC, exon junction complex.

pore. These and other observations reveal that the pre-mRNA–protein and mRNA–protein complexes are dynamic structures that gain and lose numerous specific proteins during RNA synthesis, processing, and export (see Figure 6–40).

As we have seen, some of these proteins mark the different stages of mRNA maturation; other proteins deposited on the mRNA while it is still in the nucleus can affect the fate of the RNA after it is transported to the cytosol. Thus, the stability of an mRNA in the cytosol, the efficiency with which it is translated into protein, and its ultimate destination in the cytosol can all be determined by proteins acquired in the nucleus that remain bound to the RNA after it leaves the nucleus. We will discuss these issues in Chapter 7 when we turn to the post-transcriptional control of gene expression.

We have seen that RNA synthesis and processing are closely coupled in the cell, and it might be expected that export from the nucleus is somehow integrated with these two processes. Although the Balbiani Ring RNAs can be seen to move through the nucleoplasm and out through the nuclear pores, other mRNAs appear to be synthesized and processed in close proximity to nuclear pore complexes. In these cases, which may represent the majority of eucaryotic genes, mRNA synthesis, processing, and transport all appear to be tightly coupled; the mRNA can thus be viewed as emerging from the nuclear pore as a newly manufactured car might emerge from an assembly line. Later in this chapter, we will see that the cell performs an additional quality-control check on each mRNA before it is allowed to be efficiently translated into protein.

Before discussing what happens to mRNAs after they leave the nucleus, we briefly consider how the synthesis and processing of noncoding RNA molecules occurs. Although there are many other examples, our discussion focuses on the rRNAs that are critically important for the translation of mRNAs into protein.

Many Noncoding RNAs Are Also Synthesized and Processed in the Nucleus

A few percent of the dry weight of a mammalian cell is RNA; of that, only about 3–5% is mRNA. A fraction of the remainder represents intron sequences before they have been degraded, but the bulk of the RNA in cells performs structural and catalytic functions (see Table 6–1, p. 336). The most abundant RNAs in cells are the ribosomal RNAs (rRNAs), constituting approximately 80% of the RNA in rapidly dividing cells. As discussed later in this chapter, these RNAs form the core of the ribosome. Unlike bacteria—in which a single RNA polymerase synthesizes all RNAs in the cell—eucaryotes have a separate, specialized polymerase, RNA polymerase I, that is dedicated to producing rRNAs. RNA polymerase I is similar structurally to the RNA polymerase II discussed previously; however, the absence of a C-terminal tail in polymerase I helps to explain why its transcripts are neither capped nor polyadenylated. As mentioned earlier, this difference helps the cell distinguish between noncoding RNAs and mRNAs.

Because multiple rounds of translation of each mRNA molecule can provide an enormous amplification in the production of protein molecules, many of the proteins that are very abundant in a cell can be synthesized from genes that are present in a single copy per haploid genome. In contrast, the RNA components of the ribosome are final gene products, and a growing mammalian cell must synthesize approximately 10 million copies of each type of ribosomal RNA in each cell generation to construct its 10 million ribosomes. The cell can produce adequate quantities of ribosomal RNAs only because it contains multiple copies of the **rRNA genes** that code for ribosomal RNAs (**rRNAs**). Even *E. coli* needs seven copies of its rRNA genes to meet the cell's need for ribosomes. Human cells contain about 200 rRNA gene copies per haploid genome, spread out in small clusters on five different chromosomes (see Figure 4–11), while cells of the frog *Xenopus* contain about 600 rRNA gene copies per haploid genome in a single cluster on one chromosome (**Figure 6–41**).

There are four types of eucaryotic rRNAs, each present in one copy per ribosome. Three of the four rRNAs (18S, 5.8S, and 28S) are made by chemically modifying and cleaving a single large precursor rRNA (**Figure 6–42**); the fourth (5S

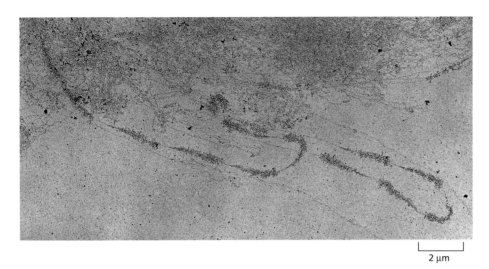

Figure 6–41 Transcription from tandemly arranged rRNA genes, as seen in the electron microscope. The pattern of alternating transcribed gene and nontranscribed spacer is readily seen. A higher-magnification view of rRNA genes is shown in Figure 6–9. (From V.E. Foe, *Cold Spring Harbor Symp. Quant. Biol.* 42:723–740, 1978. With permission from Cold Spring Harbor Laboratory Press.)

2 μm

RNA) is synthesized from a separate cluster of genes by a different polymerase, RNA polymerase III, and does not require chemical modification.

Extensive chemical modifications occur in the 13,000-nucleotide-long precursor rRNA before the rRNAs are cleaved out of it and assembled into ribosomes. These include about 100 methylations of the 2′-OH positions on nucleotide sugars and 100 isomerizations of uridine nucleotides to pseudouridine (**Figure 6–43**A). The functions of these modifications are not understood in detail, but many probably aid in the folding and assembly of the final rRNAs and some may subtly alter the function of ribosomes. Each modification is made at a specific position in the precursor rRNA. These positions are specified by about 150 "guide RNAs," which position themselves through base-pairing to the precursor rRNA and thereby bring an RNA-modifying enzyme to the appropriate position (Figure 6–43B). Other guide RNAs promote cleavage of the precursor rRNAs into the mature rRNAs, probably by causing conformational changes in the precursor rRNA that expose these sites to nucleases. All of these guide RNAs are members of a large class of RNAs called **small nucleolar RNAs** (or **snoRNAs**), so named because these RNAs perform their functions in a subcompartment of the nucleus called the nucleolus. Many snoRNAs are encoded in the introns of other genes, especially those encoding ribosomal proteins. They are therefore synthesized by RNA polymerase II and processed from excised intron sequences.

Recently several snoRNA-like RNAs have been identified that are synthesized only in cells of the brain. These are believed to direct the modification of mRNAs, instead of rRNAs, and are likely to represent a new, but poorly understood, type of gene regulatory mechanism.

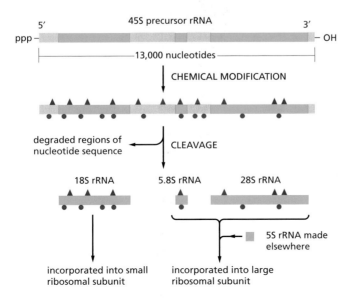

Figure 6–42 The chemical modification and nucleolytic processing of a eucaryotic 45S precursor rRNA molecule into three separate ribosomal RNAs. Two types of chemical modifications (color-coded as indicated in Figure 6–43) are made to the precursor rRNA before it is cleaved. Nearly half of the nucleotide sequences in this precursor rRNA are discarded and degraded in the nucleus. The rRNAs are named according to their "S" values, which refer to their rate of sedimentation in an ultracentrifuge. The larger the S value, the larger the rRNA.

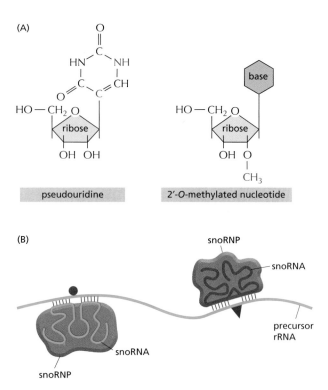

Figure 6–43 Modifications of the precursor rRNA by guide RNAs. (A) Two prominent covalent modifications occur after rRNA synthesis; the differences from the initially incorporated nucleotide are indicated by *red* atoms. Pseudouridine is an isomer of uridine; the base has been "rotated" relative to the sugar. (B) As indicated, snoRNAs determine the sites of modification by base-pairing to complementary sequences on the precursor rRNA. The snoRNAs are bound to proteins, and the complexes are called snoRNPs. snoRNPs contain both the guide sequences and the enzymes that modify the rRNA.

The Nucleolus Is a Ribosome-Producing Factory

The nucleolus is the most obvious structure seen in the nucleus of a eucaryotic cell when viewed in the light microscope. Consequently, it was so closely scrutinized by early cytologists that an 1898 review could list some 700 references. We now know that the nucleolus is the site for the processing of rRNAs and their assembly into ribosome subunits. Unlike many of the major organelles in the cell, the nucleolus is not bound by a membrane (**Figure 6–44**); instead, it is a large aggregate of macromolecules, including the rRNA genes themselves, precursor rRNAs, mature rRNAs, rRNA-processing enzymes, snoRNPs, ribosomal proteins and partly assembled ribosomes. The close association of all these components presumably allows the assembly of ribosomes to occur rapidly and smoothly.

Figure 6–44 Electron micrograph of a thin section of a nucleolus in a human fibroblast, showing its three distinct zones. (A) View of entire nucleus. (B) High-power view of the nucleolus. It is believed that transcription of the rRNA genes takes place between the fibrillar center and the dense fibrillar component and that processing of the rRNAs and their assembly into the two subunits of the ribosome proceeds outward from the dense fibrillar component to the surrounding granular components. (Courtesy of E.G. Jordan and J. McGovern.)

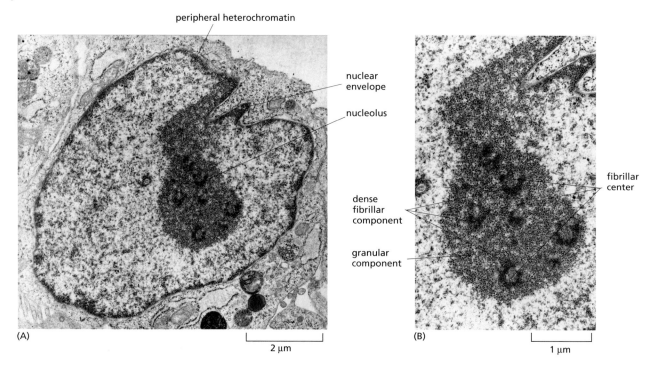

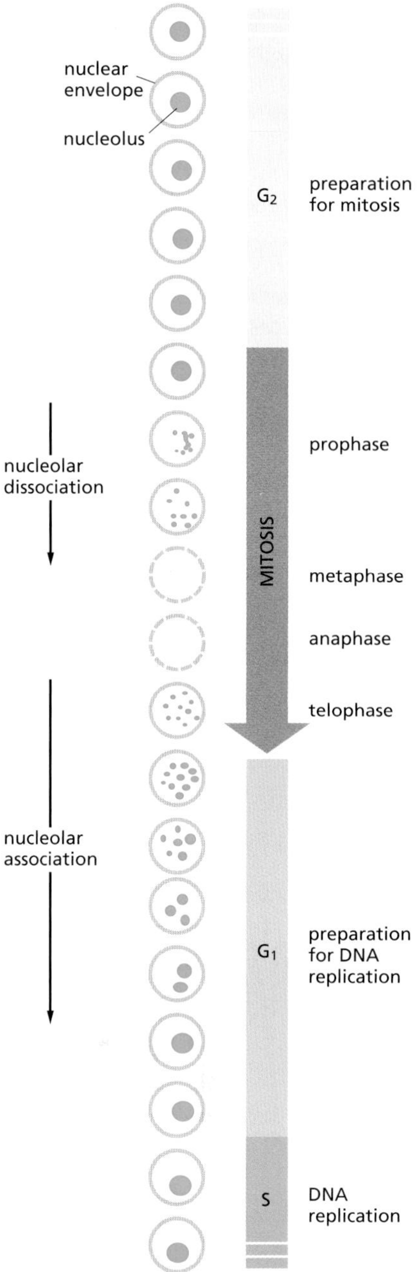

Figure 6–45 Changes in the appearance of the nucleolus in a human cell during the cell cycle. Only the cell nucleus is represented in this diagram. In most eucaryotic cells the nuclear envelope breaks down during mitosis, as indicated by the dashed circles.

Various types of RNA molecules play a central part in the chemistry and structure of the nucleolus, suggesting that it may have evolved from an ancient structure present in cells dominated by RNA catalysis. In present-day cells, the rRNA genes also have an important role in forming the nucleolus. In a diploid human cell, the rRNA genes are distributed into 10 clusters, located near the tips of five different chromosome pairs (see Figure 4–11). During interphase these 10 chromosomes contribute DNA loops (containing the rRNA genes) to the nucleolus; in M-phase, when the chromosomes condense, the nucleolus disappears. Finally, in the telophase part of mitosis, as chromosomes return to their semi-dispersed state, the tips of the 10 chromosomes coalesce and the nucleolus reforms (**Figure 6–45** and **Figure 6–46**). The transcription of the rRNA genes by RNA polymerase I is necessary for this process. As might be expected, the size of the nucleolus reflects the number of ribosomes that the cell is producing. Its size therefore varies greatly in different cells and can change in a single cell, occupying 25% of the total nuclear volume in cells that are making unusually large amounts of protein.

Ribosome assembly is a complex process, the most important features of which are outlined in **Figure 6–47**. In addition to its important role in ribosome biogenesis, the nucleolus is also the site where other RNAs are produced and other RNA–protein complexes are assembled. For example, the U6 snRNP, which functions in pre-mRNA splicing (see Figure 6–29), is composed of one RNA molecule and at least seven proteins. The U6 snRNA is chemically modified by snoRNAs in the nucleolus before its final assembly there into the U6 snRNP. Other important RNA–protein complexes, including telomerase (encountered in Chapter 5) and the signal recognition particle (which we discuss in Chapter 12), are also believed to be assembled at the nucleolus. Finally, the tRNAs (transfer RNAs) that carry the amino acids for protein synthesis are processed there as well; like the rRNA genes, those encoding tRNAs are clustered in the nucleolus. Thus, the nucleolus can be thought of as a large factory at which many different noncoding RNAs are transcribed, processed, and assembled with proteins to form a large variety of ribonucleoprotein complexes.

The Nucleus Contains a Variety of Subnuclear Structures

Although the nucleolus is the most prominent structure in the nucleus, several other nuclear bodies have been observed and studied (**Figure 6–48**). These include Cajal bodies (named for the scientist who first described them in 1906), GEMS (Gemini of Cajal bodies), and interchromatin granule clusters (also called

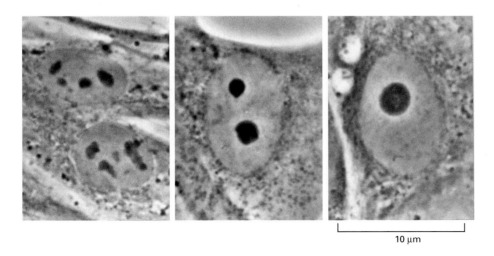

10 μm

Figure 6–46 Nucleolar fusion. These light micrographs of human fibroblasts grown in culture show various stages of nucleolar fusion. After mitosis, each of the 10 human chromosomes that carry a cluster of rRNA genes begins to form a tiny nucleolus, but these rapidly coalesce as they grow to form the single large nucleolus typical of many interphase cells. (Courtesy of E.G. Jordan and J. McGovern.)

nuclear envelope

nucleolus

G_2 preparation for mitosis

nucleolar dissociation

prophase

MITOSIS

metaphase

anaphase

telophase

nucleolar association

G_1 preparation for DNA replication

S DNA replication

"speckles"). Like the nucleolus, these other nuclear structures lack membranes and are highly dynamic; their appearance is probably the result of the tight association of protein and RNA components involved in the synthesis, assembly, and storage of macromolecules involved in gene expression. Cajal bodies and GEMS resemble one another and are frequently paired in the nucleus; it is not clear whether they truly represent distinct structures. These are likely to be the locations in which snoRNAs and snRNAs undergo covalent modifications and final assembly with proteins. A group of guide RNAs, termed *small Cajal RNAs (scaRNAs)*, selects the sites of these modifications through base pairing. Cajal bodies/GEMS may also be sites where the snRNPs are recycled and their RNAs are "reset" after the rearrangements that occur during splicing (see p. 352). In contrast, the interchromatin granule clusters have been proposed to be stockpiles of fully mature snRNPs and other RNA processing components that are ready to be used in the production of mRNA (**Figure 6–49**).

Scientists have had difficulties in working out the function of these small subnuclear structures, in part because their appearances differ between organisms and can change dramatically as cells traverse the cell cycle or respond to changes in their environment. Much of the progress now being made depends on genetic tools—examination of the effects of designed mutations in model

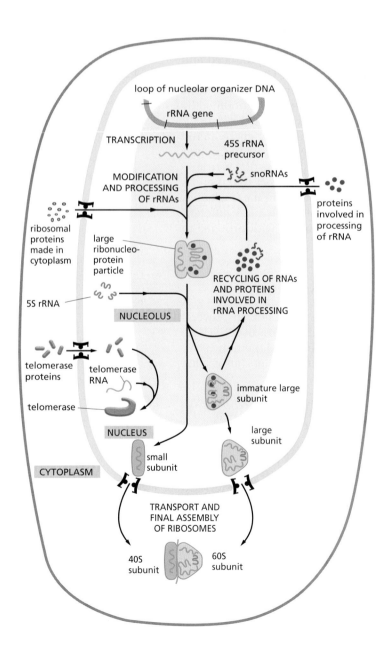

Figure 6–47 The function of the nucleolus in ribosome and other ribonucleoprotein synthesis. The 45S precursor rRNA is packaged in a large ribonucleoprotein particle containing many ribosomal proteins imported from the cytoplasm. While this particle remains at the nucleolus, selected pieces are added and others discarded as it is processed into immature large and small ribosomal subunits. The two ribosomal subunits are thought to attain their final functional form only as each is individually transported through the nuclear pores into the cytoplasm. Other ribonucleoprotein complexes, including telomerase shown here, are also assembled in the nucleolus.

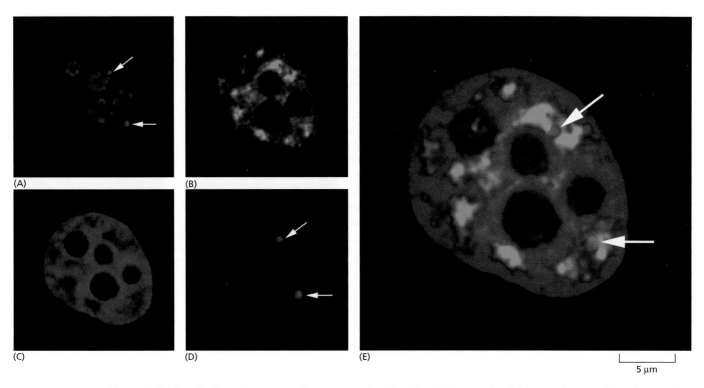

5 μm

Figure 6–48 Visualization of some prominent nuclear bodies. (A)–(D) Micrographs of the same human cell nucleus, each processed to show a particular set of nuclear structures. (E) All four images enlarged and superimposed. (A) shows the location of the protein fibrillarin (a component of several snoRNPs), which is present at both nucleoli and Cajal bodies, the latter indicated by arrows. (B) shows interchromatin granule clusters or "speckles" detected by using antibodies against a protein involved in pre-mRNA splicing. (C) is stained to show bulk chromatin. (D) shows the location of the protein coilin, which is present at Cajal bodies (*arrows*; see also Figure 4–67). (From J.R. Swedlow and A.I. Lamond, *Gen. Biol.* 2:1–7, 2001. With permission from BioMed Central. Micrographs courtesy of Judith Sleeman.)

organisms or of spontaneous mutations in humans. As one example, GEMS contain the SMN (survival of motor neurons) protein. Certain mutations of the gene encoding this protein are the cause of inherited spinal muscular atrophy, a human disease characterized by a wasting away of the muscles. The disease seems to be caused by a defect in snRNP production. A more complete loss of snRNPs would be expected to be lethal.

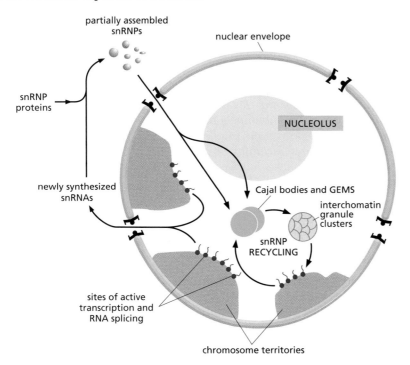

Figure 6–49 Schematic view of subnuclear structures. A typical vertebrate nucleus has several Cajal bodies, which are proposed to be the sites where snRNPs and snoRNPs undergo their final modifications. Interchromatin granule clusters are proposed to be storage sites for fully mature snRNPs. A typical vertebrate nucleus has 20–50 interchromatin granule clusters.

After their initial synthesis, snRNAs are exported from the nucleus to undergo 5′ and 3′ end-processing and assemble with the seven common snRNP proteins (called Sm proteins). These complexes are reimported into the nucleus and the snRNPs undergo their final modification by scaRNAs at Cajal bodies. In addition, snoRNAs chemically modify the U6 snRNP at the nucleolus. The sites of active transcription and splicing (approximately 2000–3000 sites per vertebrate nucleus) correspond to the "perichromatin fibers" seen under the electron microscope. (Adapted from J.D. Lewis and D. Tollervey, *Science* 288:1385–1389, 2000. With permission from AAAS.)

Given the importance of nuclear subdomains in RNA processing, it might have been expected that pre-mRNA splicing would occur in a particular location in the nucleus, as it requires numerous RNA and protein components. However, the assembly of splicing components on pre-mRNA is co-transcriptional; thus, splicing must occur at many locations along chromosomes. Although a typical mammalian cell may be expressing on the order of 15,000 genes, transcription and RNA splicing may be localized to only several thousand sites in the nucleus. These sites themselves are highly dynamic and probably result from the association of transcription and splicing components to create small "assembly lines" with a high local concentration of these components. Interchromatin granule clusters—which contain stockpiles of RNA-processing components—are often observed next to sites of transcription, as though poised to replenish supplies. Thus, the nucleus seems to be highly organized into subdomains, with snRNPs, snoRNPs, and other nuclear components moving between them in an orderly fashion according to the needs of the cell (see Figure 6–48; also see Figure 4–69).

Summary

Before the synthesis of a particular protein can begin, the corresponding mRNA molecule must be produced by transcription. Bacteria contain a single type of RNA polymerase (the enzyme that carries out the transcription of DNA into RNA). An mRNA molecule is produced when this enzyme initiates transcription at a promoter, synthesizes the RNA by chain elongation, stops transcription at a terminator, and releases both the DNA template and the completed mRNA molecule. In eucaryotic cells, the process of transcription is much more complex, and there are three RNA polymerases— polymerase I, II, and III—that are related evolutionarily to one another and to the bacterial polymerase.

RNA polymerase II synthesizes eucaryotic mRNA. This enzyme requires a series of additional proteins, the general transcription factors, to initiate transcription on a purified DNA template, and still more proteins (including chromatin-remodeling complexes and histone-modifying enzymes) to initiate transcription on its chromatin templates inside the cell.

During the elongation phase of transcription, the nascent RNA undergoes three types of processing events: a special nucleotide is added to its 5′ end (capping), intron sequences are removed from the middle of the RNA molecule (splicing), and the 3′ end of the RNA is generated (cleavage and polyadenylation). Each of these processes is initiated by proteins that travel along with RNA polymerase II by binding to sites on its long, extended C-terminal tail. Splicing is unusual in that many of its key steps are carried out by specialized RNA molecules rather than proteins. Properly processed mRNAs are passed through nuclear pore complexes into the cytosol, where they are translated into protein.

For some genes, RNA is the final product. In eucaryotes, these genes are usually transcribed by either RNA polymerase I or RNA polymerase III. RNA polymerase I makes the ribosomal RNAs. After their synthesis as a large precursor, the rRNAs are chemically modified, cleaved, and assembled into the two ribosomal subunits in the nucleolus—a distinct subnuclear structure that also helps to process some smaller RNA–protein complexes in the cell. Additional subnuclear structures (including Cajal bodies and interchromatin granule clusters) are sites where components involved in RNA processing are assembled, stored, and recycled.

FROM RNA TO PROTEIN

In the preceding section we have seen that the final product of some genes is an RNA molecule itself, such as those present in the snRNPs and in ribosomes. However, most genes in a cell produce mRNA molecules that serve as intermediaries on the pathway to proteins. In this section we examine how the cell converts the information carried in an mRNA molecule into a protein molecule. This feat of translation was a focus of attention of biologists in the late 1950s, when it was posed as the "coding problem": how is the information in a linear sequence of nucleotides in RNA translated into the linear sequence of a chemically quite different set of units—the amino acids in proteins? This fascinating question stimulated great excitement among scientists at the time. Here was a cryptogram set up by nature that, after more than 3 billion years of evolution, could finally be solved by one of the products of evolution—human beings. And indeed, not only has the code been cracked step by step, but in the year 2000 the structure of the elaborate machinery by which cells read this code—the ribosome—was finally revealed in atomic detail.

An mRNA Sequence Is Decoded in Sets of Three Nucleotides

Once an mRNA has been produced by transcription and processing, the information present in its nucleotide sequence is used to synthesize a protein. Transcription is simple to understand as a means of information transfer: since DNA and RNA are chemically and structurally similar, the DNA can act as a direct template for the synthesis of RNA by complementary base-pairing. As the term *transcription* signifies, it is as if a message written out by hand is being converted, say, into a typewritten text. The language itself and the form of the message do not change, and the symbols used are closely related.

In contrast, the conversion of the information in RNA into protein represents a **translation** of the information into another language that uses quite different symbols. Moreover, since there are only 4 different nucleotides in mRNA and 20 different types of amino acids in a protein, this translation cannot be accounted for by a direct one-to-one correspondence between a nucleotide in RNA and an amino acid in protein. The nucleotide sequence of a gene, through the intermediary of mRNA, is translated into the amino acid sequence of a protein by rules that are known as the **genetic code**. This code was deciphered in the early 1960s.

The sequence of nucleotides in the mRNA molecule is read in consecutive groups of three. RNA is a linear polymer of four different nucleotides, so there are $4 \times 4 \times 4 = 64$ possible combinations of three nucleotides: the triplets AAA, AUA, AUG, and so on. However, only 20 different amino acids are commonly found in proteins. Either some nucleotide triplets are never used, or the code is redundant and some amino acids are specified by more than one triplet. The second possibility is, in fact, the correct one, as shown by the completely deciphered genetic code in **Figure 6–50**. Each group of three consecutive nucleotides in RNA is called a **codon**, and each codon specifies either one amino acid or a stop to the translation process.

This genetic code is used universally in all present-day organisms. Although a few slight differences in the code have been found, these are chiefly in the DNA of mitochondria. Mitochondria have their own transcription and protein synthesis systems that operate quite independently from those of the rest of the cell, and it is understandable that their small genomes have been able to accommodate minor changes to the code (discussed in Chapter 14).

	AGA						GGA			UUA					CCA	AGC				
	AGG						GGC			UUG					CCC	AGU				
GCA	CGA						GGG		AUA	CUA					CCG	UCA	ACA			GUA	
GCC	CGC						GGU		AUC	CUC				UUC	CCU	UCC	ACC			GUC	UAA
GCG	CGG	GAC	AAC	UGC	GAA	CAA		CAC	AUC	CUG	AAA		UUU		UCG	ACG		UAC	GUG	UAG
GCU	CGU	GAU	AAU	UGU	GAG	CAG		CAU	AUU	CUU	AAG	AUG			UCU	ACU	UGG	UAU	GUU	UGA
Ala	Arg	Asp	Asn	Cys	Glu	Gln	Gly	His	Ile	Leu	Lys	Met	Phe	Pro	Ser	Thr	Trp	Tyr	Val	stop
A	R	D	N	C	E	Q	G	H	I	L	K	M	F	P	S	T	W	Y	V	

Figure 6–50 **The genetic code.** The standard one-letter abbreviation for each amino acid is presented below its three-letter abbreviation (see Panel 3–1, pp. 128–129, for the full name of each amino acid and its structure). By convention, codons are always written with the 5′-terminal nucleotide to the left. Note that most amino acids are represented by more than one codon, and that there are some regularities in the set of codons that specifies each amino acid. Codons for the same amino acid tend to contain the same nucleotides at the first and second positions, and vary at the third position. Three codons do not specify any amino acid but act as termination sites (stop codons), signaling the end of the protein-coding sequence. One codon—AUG—acts both as an initiation codon, signaling the start of a protein-coding message, and also as the codon that specifies methionine.

In principle, an RNA sequence can be translated in any one of three different **reading frames**, depending on where the decoding process begins (**Figure 6–51**). However, only one of the three possible reading frames in an mRNA encodes the required protein. We see later how a special punctuation signal at the beginning of each RNA message sets the correct reading frame at the start of protein synthesis.

tRNA Molecules Match Amino Acids to Codons in mRNA

The codons in an mRNA molecule do not directly recognize the amino acids they specify: the group of three nucleotides does not, for example, bind directly to the amino acid. Rather, the translation of mRNA into protein depends on adaptor molecules that can recognize and bind both to the codon and, at another site on their surface, to the amino acid. These adaptors consist of a set of small RNA molecules known as **transfer RNAs** (**tRNAs**), each about 80 nucleotides in length.

We saw earlier in this chapter that RNA molecules can fold into precise three-dimensional structures, and the tRNA molecules provide a striking example. Four short segments of the folded tRNA are double-helical, producing a molecule that looks like a cloverleaf when drawn schematically (**Figure 6–52**). For example, a 5′-GCUC-3′ sequence in one part of a polynucleotide chain can form a relatively strong association with a 5′-GAGC-3′ sequence in another region of the same molecule. The cloverleaf undergoes further folding to form a compact L-shaped structure that is held together by additional hydrogen bonds between different regions of the molecule.

Two regions of unpaired nucleotides situated at either end of the L-shaped molecule are crucial to the function of tRNA in protein synthesis. One of these

Figure 6–51 The three possible reading frames in protein synthesis. In the process of translating a nucleotide sequence *(blue)* into an amino acid sequence *(red)*, the sequence of nucleotides in an mRNA molecule is read from the 5′ end to the 3′ end in consecutive sets of three nucleotides. In principle, therefore, the same RNA sequence can specify three completely different amino acid sequences, depending on the reading frame. In reality, however, only one of these reading frames contains the actual message.

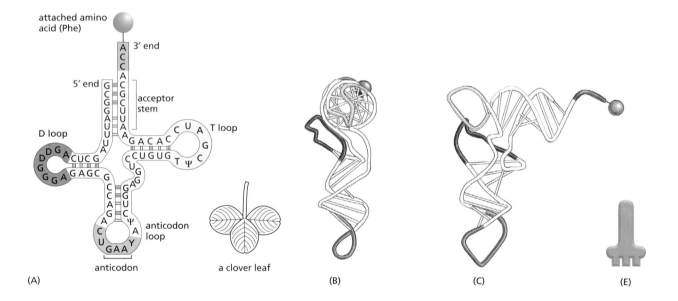

(A)   (B)   (C)   (E)

5′ GCGGAUUUAGCUCAGDDGGGAGAGCGCCAGACUGAAYAΨCUGGAGGUCCUGUGTΨCGAUCCACAGAAUUCGCACCA 3′

(D)   anticodon

Figure 6–52 A tRNA molecule. A tRNA specific for the amino acid phenylalanine (Phe) is depicted in various ways. (A) The cloverleaf structure showing the complementary base-pairing *(red lines)* that creates the double-helical regions of the molecule. The anticodon is the sequence of three nucleotides that base-pairs with a codon in mRNA. The amino acid matching the codon/anticodon pair is attached at the 3′ end of the tRNA. tRNAs contain some unusual bases, which are produced by chemical modification after the tRNA has been synthesized. For example, the bases denoted Ψ (pseudouridine—see Figure 6–43) and D (dihydrouridine—see Figure 6–55) are derived from uracil. (B and C) Views of the L-shaped molecule, based on x-ray diffraction analysis. Although this diagram shows the tRNA for the amino acid phenylalanine, all other tRNAs have similar structures. <CGCA> (D) The linear nucleotide sequence of the molecule, color-coded to match (A), (B), and (C). (E) The tRNA icon we use in this book.

regions forms the **anticodon**, a set of three consecutive nucleotides that pairs with the complementary codon in an mRNA molecule. The other is a short single-stranded region at the 3′ end of the molecule; this is the site where the amino acid that matches the codon is attached to the tRNA.

We have seen in the previous section that the genetic code is redundant; that is, several different codons can specify a single amino acid (see Figure 6–50). This redundancy implies either that there is more than one tRNA for many of the amino acids or that some tRNA molecules can base-pair with more than one codon. In fact, both situations occur. Some amino acids have more than one tRNA and some tRNAs are constructed so that they require accurate base-pairing only at the first two positions of the codon and can tolerate a mismatch (or *wobble*) at the third position (**Figure 6–53**). This wobble base-pairing explains why so many of the alternative codons for an amino acid differ only in their third nucleotide (see Figure 6–50). In bacteria, wobble base-pairings make it possible to fit the 20 amino acids to their 61 codons with as few as 31 kinds of tRNA molecules. The exact number of different kinds of tRNAs, however, differs from one species to the next. For example, humans have nearly 500 tRNA genes but, among them, only 48 different anticodons are represented.

tRNAs Are Covalently Modified Before They Exit from the Nucleus

Like most other eucaryotic RNAs, tRNAs are covalently modified before they are allowed to exit from the nucleus. Eucaryotic tRNAs are synthesized by RNA polymerase III. Both bacterial and eucaryotic tRNAs are typically synthesized as larger precursor tRNAs, which are then trimmed to produce the mature tRNA. In addition, some tRNA precursors (from both bacteria and eucaryotes) contain introns that must be spliced out. This splicing reaction differs chemically from pre-mRNA splicing; rather than generating a lariat intermediate, tRNA splicing uses a cut-and-paste mechanism that is catalyzed by proteins (**Figure 6–54**). Trimming and splicing both require the precursor tRNA to be correctly folded in its cloverleaf configuration. Because misfolded tRNA precursors will not be processed properly, the trimming and splicing reactions are thought to act as quality-control steps in the generation of tRNAs.

All tRNAs are modified chemically—nearly 1 in 10 nucleotides in each mature tRNA molecule is an altered version of a standard G, U, C, or A ribonucleotide. Over 50 different types of tRNA modifications are known; a few are shown in **Figure 6–55**. Some of the modified nucleotides—most notably inosine, produced by the deamination of adenosine—affect the conformation and base-pairing of the anticodon and thereby facilitate the recognition of the appropriate mRNA codon by the tRNA molecule (see Figure 6–53). Others affect the accuracy with which the tRNA is attached to the correct amino acid.

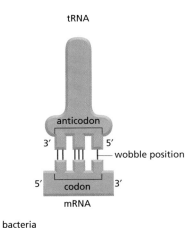

Figure 6–53 Wobble base-pairing between codons and anticodons. If the nucleotide listed in the first column is present at the third, or wobble, position of the codon, it can base-pair with any of the nucleotides listed in the second column. Thus, for example, when inosine (I) is present in the wobble position of the tRNA anticodon, the tRNA can recognize any one of three different codons in bacteria and either of two codons in eucaryotes. The inosine in tRNAs is formed from the deamination of guanine (see Figure 6–55), a chemical modification that takes place after the tRNA has been synthesized. The nonstandard base pairs, including those made with inosine, are generally weaker than conventional base pairs. Note that codon–anticodon base pairing is more stringent at positions 1 and 2 of the codon: here only conventional base pairs are permitted. The differences in wobble base-pairing interactions between bacteria and eucaryotes presumably result from subtle structural differences between bacterial and eucaryotic ribosomes, the molecular machines that perform protein synthesis. (Adapted from C. Guthrie and J. Abelson, in The Molecular Biology of the Yeast *Saccharomyces*: Metabolism and Gene Expression, pp. 487–528. Cold Spring Harbor, New York: Cold Spring Harbor Laboratory Press, 1982.)

bacteria

wobble codon base	possible anticodon bases
U	A, G, or I
C	G or I
A	U or I
G	C or U

eucaryotes

wobble codon base	possible anticodon bases
U	A, G, or I
C	G or I
A	U
G	C

Figure 6–54 Structure of a tRNA-splicing endonuclease docked to a precursor tRNA. The endonuclease (a four-subunit enzyme) removes the tRNA intron *(blue)*. A second enzyme, a multifunctional tRNA ligase (not shown), then joins the two tRNA halves together. (Courtesy of Hong Li, Christopher Trotta and John Abelson.)

Specific Enzymes Couple Each Amino Acid to Its Appropriate tRNA Molecule

We have seen that, to read the genetic code in DNA, cells make a series of different tRNAs. We now consider how each tRNA molecule becomes linked to the one amino acid in 20 that is its appropriate partner. Recognition and attachment of the correct amino acid depends on enzymes called **aminoacyl-tRNA synthetases**, which covalently couple each amino acid to its appropriate set of tRNA molecules (**Figure 6–56** and **Figure 6–57**). Most cells have a different synthetase enzyme for each amino acid (that is, 20 synthetases in all); one attaches glycine to all tRNAs that recognize codons for glycine, another attaches alanine to all tRNAs that recognize codons for alanine, and so on. Many bacteria, however, have fewer than 20 synthetases, and the same synthetase enzyme is responsible for coupling more than one amino acid to the appropriate tRNAs. In these cases, a single synthetase places the identical amino acid on two different types of tRNAs, only one of which has an anticodon that matches the amino acid. A second enzyme then chemically modifies each "incorrectly" attached amino acid so that it now corresponds to the anticodon displayed by its covalently linked tRNA.

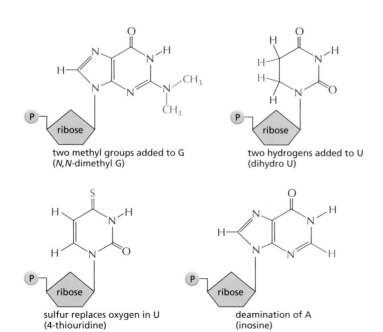

two methyl groups added to G
(*N,N*-dimethyl G)

two hydrogens added to U
(dihydro U)

sulfur replaces oxygen in U
(4-thiouridine)

deamination of A
(inosine)

Figure 6–55 A few of the unusual nucleotides found in tRNA molecules. These nucleotides are produced by covalent modification of a normal nucleotide after it has been incorporated into an RNA chain. Two other types of modified nucleotides are shown in Figure 6–43. In most tRNA molecules about 10% of the nucleotides are modified (see Figure 6–52).

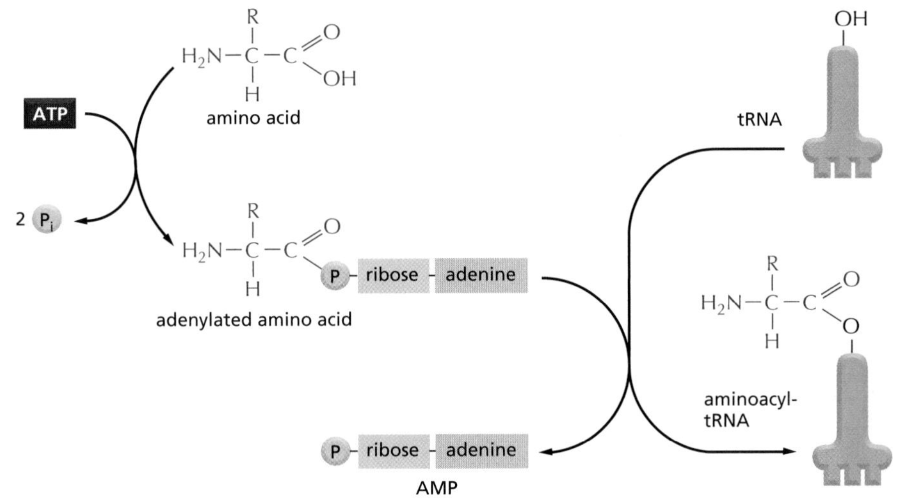

Figure 6–56 Amino acid activation. An amino acid is activated for protein synthesis by an aminoacyl-tRNA synthetase enzyme in two steps. As indicated, the energy of ATP hydrolysis is used to attach each amino acid to its tRNA molecule in a high-energy linkage. The amino acid is first activated through the linkage of its carboxyl group directly to an AMP moiety, forming an *adenylated amino acid*; the linkage of the AMP, normally an unfavorable reaction, is driven by the hydrolysis of the ATP molecule that donates the AMP. Without leaving the synthetase enzyme, the AMP-linked carboxyl group on the amino acid is then transferred to a hydroxyl group on the sugar at the 3′ end of the tRNA molecule. This transfer joins the amino acid by an activated ester linkage to the tRNA and forms the final aminoacyl-tRNA molecule. The synthetase enzyme is not shown in this diagram.

The synthetase-catalyzed reaction that attaches the amino acid to the 3′ end of the tRNA is one of many reactions coupled to the energy-releasing hydrolysis of ATP (see pp. 79–81), and it produces a high-energy bond between the tRNA and the amino acid. The energy of this bond is used at a later stage in protein synthesis to link the amino acid covalently to the growing polypeptide chain.

The aminoacyl-tRNA synthetase enzymes and the tRNAs are equally important in the decoding process (**Figure 6–58**). This was established by an experiment in which one amino acid (cysteine) was chemically converted into a different amino acid (alanine) after it already had been attached to its specific tRNA. When such "hybrid" aminoacyl-tRNA molecules were used for protein synthesis in a cell-free system, the wrong amino acid was inserted at every point in the protein chain where that tRNA was used. Although, as we shall see, cells have several quality control mechanisms to avoid this type of mishap, the experiment establishes that the genetic code is translated by two sets of adaptors that act sequentially. Each matches one molecular surface to another with great specificity, and it is their combined action that associates each sequence of three nucleotides in the mRNA molecule—that is, each codon—with its particular amino acid.

Editing by tRNA Synthetases Ensures Accuracy

Several mechanisms working together ensure that the tRNA synthetase links the correct amino acid to each tRNA. The synthetase must first select the correct amino acid, and most synthetases do so by a two-step mechanism. First, the correct amino acid has the highest affinity for the active-site pocket of its synthetase

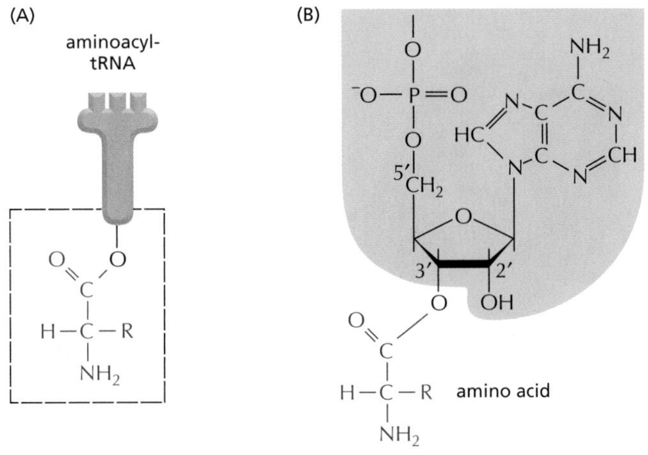

Figure 6–57 The structure of the aminoacyl-tRNA linkage. The carboxyl end of the amino acid forms an ester bond to ribose. Because the hydrolysis of this ester bond is associated with a large favorable change in free energy, an amino acid held in this way is said to be activated. (A) Schematic drawing of the structure. The amino acid is linked to the nucleotide at the 3′ end of the tRNA (see Figure 6–52). (B) Actual structure corresponding to the boxed region in (A). There are two major classes of synthetase enzymes: one links the amino acid directly to the 3′-OH group of the ribose, and the other links it initially to the 2′-OH group. In the latter case, a subsequent transesterification reaction shifts the amino acid to the 3′ position. As in Figure 6–56, the "R group" indicates the side chain of the amino acid.

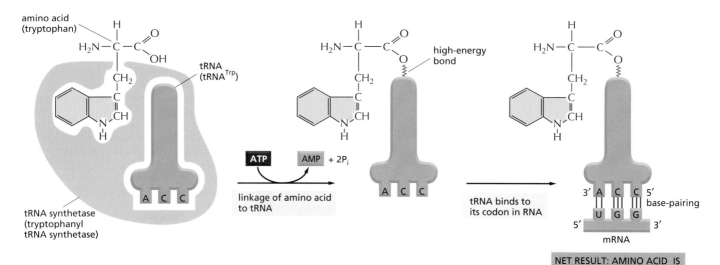

NET RESULT: AMINO ACID IS SELECTED BY ITS CODON

and is therefore favored over the other 19. In particular, amino acids larger than the correct one are effectively excluded from the active site. However, accurate discrimination between two similar amino acids, such as isoleucine and valine (which differ by only a methyl group), is very difficult to achieve by a one-step recognition mechanism. A second discrimination step occurs after the amino acid has been covalently linked to AMP (see Figure 6–56). When tRNA binds the synthetase, it tries to force the amino acid into a second pocket in the synthetase, the precise dimensions of which exclude the correct amino acid but allow access by closely related amino acids. Once an amino acid enters this editing pocket, it is hydrolyzed from the AMP (or from the tRNA itself if the aminoacyl-tRNA bond has already formed), and is released from the enzyme. This hydrolytic editing, which is analogous to the exonucleolytic proofreading by DNA polymerases (**Figure 6–59**), raises the overall accuracy of tRNA charging to approximately one mistake in 40,000 couplings.

Figure 6–58 The genetic code is translated by means of two adaptors that act one after another. The first adaptor is the aminoacyl-tRNA synthetase, which couples a particular amino acid to its corresponding tRNA; the second adaptor is the tRNA molecule itself, whose *anticodon* forms base pairs with the appropriate *codon* on the mRNA. An error in either step would cause the wrong amino acid to be incorporated into a protein chain. In the sequence of events shown, the amino acid tryptophan (Trp) is selected by the codon UGG on the mRNA.

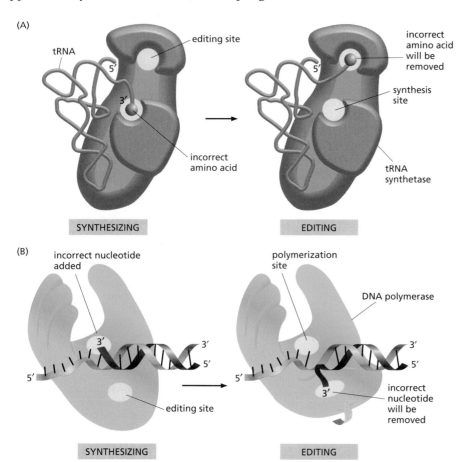

Figure 6–59 Hydrolytic editing. (A) tRNA synthetases remove their own coupling errors through hydrolytic editing of incorrectly attached amino acids. As described in the text, the correct amino acid is rejected by the editing site. (B) The error-correction process performed by DNA polymerase shows some similarities; however, it differs in so far as the removal process depends strongly on a mispairing with the template (see Figure 5–8).

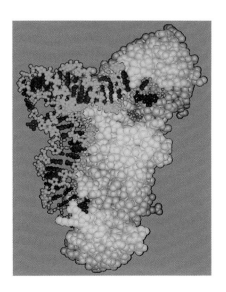

Figure 6–60 **The recognition of a tRNA molecule by its aminoacyl-tRNA synthetase.** For this tRNA (tRNA^{Gln}), specific nucleotides in both the anticodon *(bottom)* and the amino acid-accepting arm allow the correct tRNA to be recognized by the synthetase enzyme *(blue)*. A bound ATP molecule is *yellow*. (Courtesy of Tom Steitz.)

The tRNA synthetase must also recognize the correct set of tRNAs, and extensive structural and chemical complementarity between the synthetase and the tRNA allows the synthetase to probe various features of the tRNA (**Figure 6–60**). Most tRNA synthetases directly recognize the matching tRNA anticodon; these synthetases contain three adjacent nucleotide-binding pockets, each of which is complementary in shape and charge to a nucleotide in the anticodon. For other synthetases, the nucleotide sequence of the acceptor stem is the key recognition determinant. In most cases, however, the synthetase "reads" the nucleotides at several different positions on the tRNA.

Amino Acids Are Added to the C-terminal End of a Growing Polypeptide Chain

Having seen that amino acids are first coupled to tRNA molecules, we now turn to the mechanism that joins amino acids together to form proteins. The fundamental reaction of protein synthesis is the formation of a peptide bond between the carboxyl group at the end of a growing polypeptide chain and a free amino group on an incoming amino acid. Consequently, a protein is synthesized stepwise from its N-terminal end to its C-terminal end. Throughout the entire process the growing carboxyl end of the polypeptide chain remains activated by its covalent attachment to a tRNA molecule (forming a peptidyl-tRNA). Each addition disrupts this high-energy covalent linkage, but immediately replaces it with an identical linkage on the most recently added amino acid (**Figure 6–61**). In this way, each amino acid added carries with it the activation energy for the addition of the next amino acid rather than the energy for its own addition—an example of the "head growth" type of polymerization described in Figure 2–68.

The RNA Message Is Decoded in Ribosomes

The synthesis of proteins is guided by information carried by mRNA molecules. To maintain the correct reading frame and to ensure accuracy (about 1 mistake every 10,000 amino acids), protein synthesis is performed in the **ribosome**, a complex catalytic machine made from more than 50 different proteins (the *ribosomal proteins*) and several RNA molecules, the **ribosomal RNAs** (**rRNAs**). <CGCC> A typical eucaryotic cell contains millions of ribosomes in its cytoplasm (**Figure 6–62**). Eucaryotic ribosome subunits are assembled at the nucleolus, when newly transcribed and modified rRNAs associate with ribosomal

Figure 6–61 **The incorporation of an amino acid into a protein.** A polypeptide chain grows by the stepwise addition of amino acids to its C-terminal end. The formation of each peptide bond is energetically favorable because the growing C-terminus has been activated by the covalent attachment of a tRNA molecule. The peptidyl-tRNA linkage that activates the growing end is regenerated during each addition. The amino acid side chains have been abbreviated as R_1, R_2, R_3, and R_4; as a reference point, all of the atoms in the second amino acid in the polypeptide chain are shaded *gray*. The figure shows the addition of the fourth amino acid *(red)* to the growing chain.

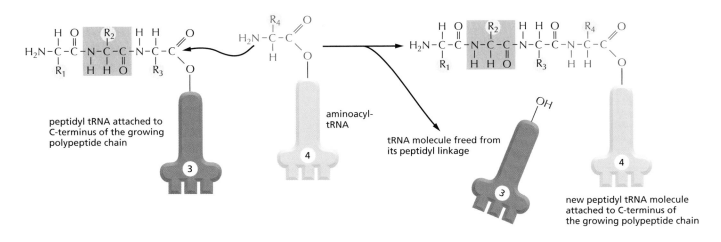

peptidyl tRNA attached to C-terminus of the growing polypeptide chain

aminoacyl-tRNA

tRNA molecule freed from its peptidyl linkage

new peptidyl tRNA molecule attached to C-terminus of the growing polypeptide chain

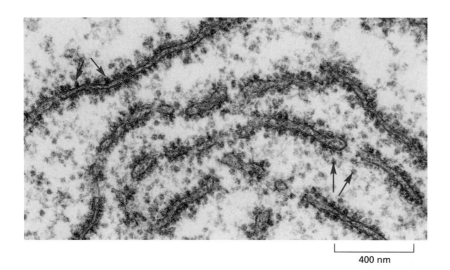

Figure 6–62 **Ribosomes in the cytoplasm of a eucaryotic cell.** This electron micrograph shows a thin section of a small region of cytoplasm. The ribosomes appear as black dots *(red arrows)*. Some are free in the cytosol; others are attached to membranes of the endoplasmic reticulum. (Courtesy of Daniel S. Friend.)

400 nm

proteins, which have been transported into the nucleus after their synthesis in the cytoplasm. The two ribosomal subunits are then exported to the cytoplasm, where they join together to synthesize proteins.

Eucaryotic and procaryotic ribosomes have similar designs and functions. Both are composed of one large and one small subunit that fit together to form a complete ribosome with a mass of several million daltons (**Figure 6–63**). The small subunit provides the framework on which the tRNAs can be accurately

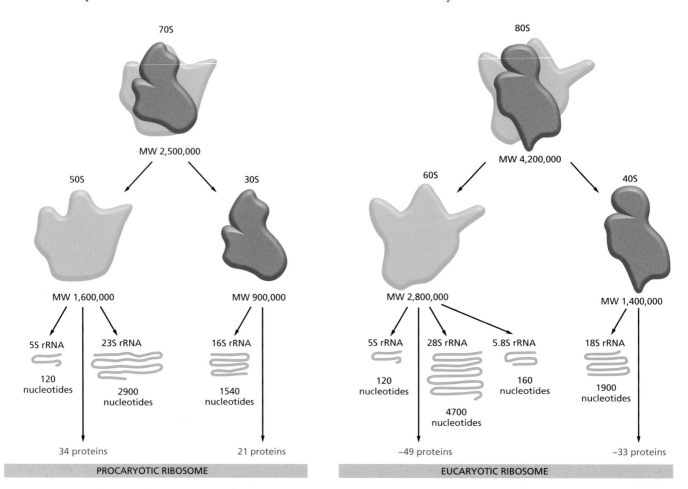

Figure 6–63 **A comparison of procaryotic and eucaryotic ribosomes.** Despite differences in the number and size of their rRNA and protein components, both procaryotic and eucaryotic ribosomes have nearly the same structure and they function similarly. Although the 18S and 28S rRNAs of the eucaryotic ribosome contain many nucleotides not present in their bacterial counterparts, these nucleotides are present as multiple insertions that form extra domains and leave the basic structure of each rRNA largely unchanged.

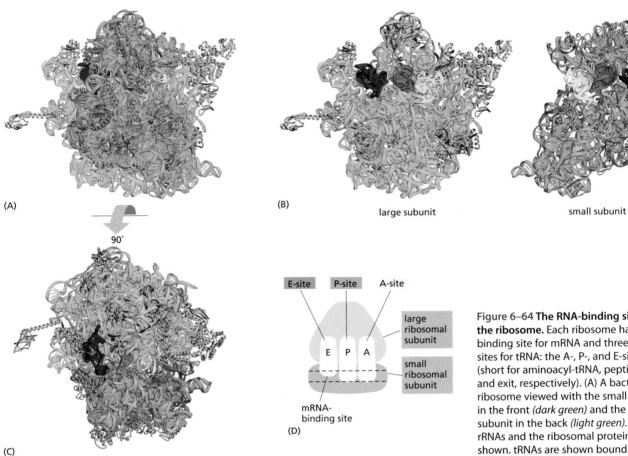

(A)

90°

(B)

large subunit

small subunit

(C)

E-site   P-site   A-site

large
ribosomal
subunit

E   P   A

small
ribosomal
subunit

mRNA-
binding site

(D)

Figure 6–64 The RNA-binding sites in the ribosome. Each ribosome has one binding site for mRNA and three binding sites for tRNA: the A-, P-, and E-sites (short for aminoacyl-tRNA, peptidyl-tRNA, and exit, respectively). (A) A bacterial ribosome viewed with the small subunit in the front *(dark green)* and the large subunit in the back *(light green)*. Both the rRNAs and the ribosomal proteins are shown. tRNAs are shown bound in the E-site *(red)*, the P-site *(orange)* and the A-site *(yellow)*. Although all three tRNA sites are shown occupied here, during the process of protein synthesis not more than two of these sites are thought to contain tRNA molecules at any one time (see Figure 6–66). (B) Large and small ribosomal subunits arranged as though the ribosome in (A) were opened like a book. (C) The ribosome in (A) rotated through 90° and viewed with the large subunit on top and small subunit on the bottom. (D) Schematic representation of a ribosome (in the same orientation as C), which will be used in subsequent figures. (A, B, and C, adapted from M.M. Yusupov et al., *Science* 292:883–896, 2001. With permission from AAAS; courtesy of Albion Baucom and Harry Noller.)

matched to the codons of the mRNA (see Figure 6–58), while the large subunit catalyzes the formation of the peptide bonds that link the amino acids together into a polypeptide chain (see Figure 6–61).

When not actively synthesizing proteins, the two subunits of the ribosome are separate. They join together on an mRNA molecule, usually near its 5′ end, to initiate the synthesis of a protein. The mRNA is then pulled through the ribosome; as its codons enter the core of the ribosome, the mRNA nucleotide sequence is translated into an amino acid sequence using the tRNAs as adaptors to add each amino acid in the correct sequence to the end of the growing polypeptide chain. When a stop codon is encountered, the ribosome releases the finished protein, and its two subunits separate again. These subunits can then be used to start the synthesis of another protein on another mRNA molecule.

Ribosomes operate with remarkable efficiency: in one second, a single ribosome of a eucaryotic cell adds about 2 amino acids to a polypeptide chain; the ribosomes of bacterial cells operate even faster, at a rate of about 20 amino acids per second. How does the ribosome choreograph the many coordinated movements required for efficient translation? A ribosome contains four binding sites for RNA molecules: one is for the mRNA and three (called the A-site, the P-site, and the E-site) are for tRNAs (**Figure 6–64**). A tRNA molecule is held tightly at the A- and P-sites only if its anticodon forms base pairs with a complementary codon (allowing for wobble) on the mRNA molecule that is threaded through the ribosome (**Figure 6–65**). The A- and P-sites are close enough together for their two tRNA molecules to be forced to form base pairs with adjacent codons on the mRNA molecule. This feature of the ribosome maintains the correct reading frame on the mRNA.

Once protein synthesis has been initiated, each new amino acid is added to the elongating chain in a cycle of reactions containing four major steps: tRNA

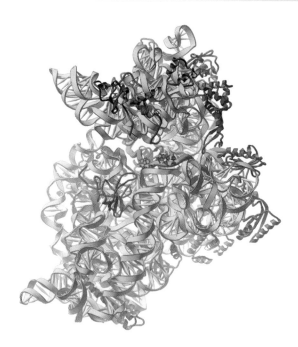

Figure 6–65 The path of mRNA *(blue)* **through the small ribosomal subunit.** The orientation is the same as that in the right-hand panel of Figure 6–64B. (Courtesy of Harry F. Noller, based on data in G.Z. Yusopova et al., *Cell* 106:233–241, 2001. With permission from Elsevier.)

binding, peptide bond formation, large subunit and small subunit translocation. As a result of the two translocation steps, the entire ribosome moves three nucleotides along the mRNA and is positioned to start the next cycle. (**Figure 6–66**). Our description of the chain elongation process begins at a point at which some amino acids have already been linked together and there is a tRNA molecule in the P-site on the ribosome, covalently joined to the end of the growing polypeptide. In step 1, a tRNA carrying the next amino acid in the chain binds to the ribosomal A-site by forming base pairs with the mRNA codon positioned there, so that the P-site and the A-site contain adjacent bound tRNAs. In step 2, the carboxyl end of the polypeptide chain is released from the tRNA at the P-site (by breakage of the high-energy bond between the tRNA and its amino acid) and joined to the free amino group of the amino acid linked to the tRNA at the A-site, forming a new peptide bond. This central reaction of protein synthesis is catalyzed by a *peptidyl transferase* contained in the large ribosomal subunit. In step 3, the large subunit moves relative to the mRNA held by the small subunit, thereby shifting the acceptor stems of the two tRNAs to the E- and P-sites of the large subunit. In step 4, another series of conformational changes moves the small subunit and its bound mRNA exactly three nucleotides, resetting the ribosome so it is ready to receive the next aminoacyl-tRNA. Step 1 is then repeated with a new incoming aminoacyl-tRNA, and so on. <CGTT>

This four-step cycle is repeated each time an amino acid is added to the polypeptide chain, as the chain grows from its amino to its carboxyl end.

Figure 6–66 Translating an mRNA molecule. Each amino acid added to the growing end of a polypeptide chain is selected by complementary base-pairing between the anticodon on its attached tRNA molecule and the next codon on the mRNA chain. Because only one of the many types of tRNA molecules in a cell can base-pair with each codon, the codon determines the specific amino acid to be added to the growing polypeptide chain. The four-step cycle shown is repeated over and over during the synthesis of a protein. In step 1, an aminoacyl-tRNA molecule binds to a vacant A-site on the ribosome and a spent tRNA molecule dissociates from the E-site. In step 2, a new peptide bond is formed. In step 3, the large subunit translocates relative to the small subunit, leaving the two tRNAs in hybrid sites: P on the large subunit and A on the small, for one; E on the large subunit and P on the small, for the other. In step 4, the small subunit translocates carrying its mRNA a distance of three nucleotides through the ribosome. This "resets" the ribosome with a fully empty A-site, ready for the next aminoacyl-tRNA molecule to bind. As indicated, the mRNA is translated in the 5′-to-3′ direction, and the N-terminal end of a protein is made first, with each cycle adding one amino acid to the C-terminus of the polypeptide chain.

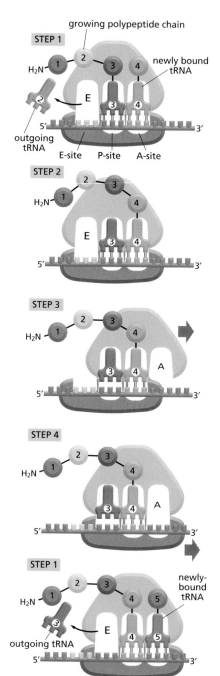

Elongation Factors Drive Translation Forward and Improve Its Accuracy

The basic cycle of polypeptide elongation shown in outline in Figure 6–66 has an additional feature that makes translation especially efficient and accurate. Two *elongation factors* enter and leave the ribosome during each cycle, each hydrolyzing GTP to GDP and undergoing conformational changes in the process. These factors are called EF-Tu and EF-G in bacteria, and EF1 and EF2 in eucaryotes. Under some conditions *in vitro*, ribosomes can be forced to synthesize proteins without the aid of these elongation factors and GTP hydrolysis, but this synthesis is very slow, inefficient, and inaccurate. Coupling the GTP hydrolysis-driven changes in the elongation factors to transitions between different states of the ribosome speeds up protein synthesis enormously. Although these ribosomal states are not yet understood in detail, they almost certainly involve RNA structure rearrangements in the ribosome core. The cycles of elongation factor association, GTP hydrolysis, and dissociation ensure that all such changes occur in the "forward" direction so that translation can proceed efficiently (**Figure 6–67**).

As shown previously, EF-Tu simultaneously binds GTP and aminoacyl-tRNAs (see Figure 3–74). In addition to helping move translation forward, EF-Tu (EF1 in eucaryotes) increases the accuracy of translation in several ways. First, as it escorts an incoming aminoacyl-tRNA to the ribosome, EF-Tu checks whether the tRNA–amino acid match is corrrect. Exactly how this is accomplished is not well understood. According to one idea, correct tRNA–amino acid matches have a narrowly defined affinity for EF-Tu, which allows EF-Tu to discriminate, albeit crudely, among many different amino acid–tRNA combinations, selectively bringing the correct ones with it into the ribosome. Second, EF-Tu monitors the initial interaction between the anticodon of an incoming aminoacyl-tRNA and the codon of the mRNA in the A-site. Aminoacyl-tRNAs are "bent" when bound to the GTP-form of EF-Tu; this bent conformation allows codon pairing but prevents incorporation of the amino acid into the growing polypeptide chain. However, if the codon–anticodon match is correct, the ribosome rapidly triggers the hydrolysis of the GTP molecule, whereupon EF-Tu releases its grip on the tRNA and dissociates from the ribosome, allowing the tRNA to donate its amino acid for protein synthesis. But how is the "correctness" of the codon–anticodon match assessed? This feat is carried out by the ribosome itself through an RNA-based mechanism. The rRNA in the small subunit of the ribosome forms a series of hydrogen bonds with the codon–anticodon pair that allows determination of its correctness (**Figure 6–68**). In essence, the rRNA folds around the codon–anticodon pair, and its final closure—which occurs only when the correct anticodon is in place—triggers GTP hydrolysis. Remarkably, this induced fit mechanism can distinguish correct from incorrect codon–anticodon interactions despite the rules for wobble base-pairing summarized in Figure 6–53. From this example, as for RNA splicing, one gets a sense of the highly sophisticated forms of molecular recognition that can be achieved solely by RNA.

The interactions of EF-Tu, tRNA, and the ribosome just described introduce critical proofreading steps into protein synthesis at the initial tRNA selection stage. But after GTP is hydrolyzed and EF-Tu dissociates from the ribosome, there is an additional opportunity for the ribosome to prevent an incorrect amino acid from being added to the growing chain. Following GTP hydrolysis, there is a short time delay as the amino acid carried by the tRNA moves into position on the ribosome. This time delay is shorter for correct than incorrect codon–anticodon pairs. Moreover, incorrectly matched tRNAs dissociate more

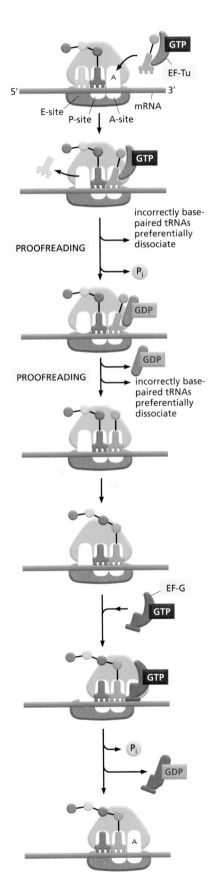

Figure 6–67 Detailed view of the translation cycle. The outline of translation presented in Figure 6–66 has been expanded to show the roles of two elongation factors EF-Tu and EF-G, which drive translation in the forward direction. As explained in the text, EF-Tu also provides two opportunities for proofreading of the codon–anticodon match. In this way, incorrectly paired tRNAs are selectively rejected, and the accuracy of translation is improved.

rapidly than those correctly bound because their interaction with the codon is weaker. Thus, most incorrectly bound tRNA molecules (as well as a significant number of correctly bound molecules) will leave the ribosome without being used for protein synthesis. All of these proofreading steps, taken together, are largely responsible for the 99.99% accuracy of the ribosome in translating RNA into protein.

The Ribosome Is a Ribozyme

The ribosome is a large complex composed of two-thirds RNA and one-third protein. The determination, in 2000, of the entire three-dimensional conformation of its large and small subunits is a major triumph of modern structural biology. The findings confirm earlier evidence that rRNAs—and not proteins—are responsible for the ribosome's overall structure, its ability to position tRNAs on the mRNA, and its catalytic activity in forming covalent peptide bonds. The ribosomal RNAs are folded into highly compact, precise three-dimensional structures that form the compact core of the ribosome and determine its overall shape (**Figure 6–69**).

In marked contrast to the central positions of the rRNAs, the ribosomal proteins are generally located on the surface and fill in the gaps and crevices of the folded RNA (**Figure 6–70**). Some of these proteins send out extended regions of polypeptide chain that penetrate short distances into holes in the RNA core (**Figure 6–71**). The main role of the ribosomal proteins seems to be to stabilize the RNA core, while permitting the changes in rRNA conformation that are necessary for this RNA to catalyze efficient protein synthesis. The proteins probably also aid in the initial assembly of the rRNAs that make up the core of the ribosome.

Not only are the A-, P-, and E-binding sites for tRNAs formed primarily by ribosomal RNAs, but the catalytic site for peptide bond formation is also formed by RNA, as the nearest amino acid is located more than 1.8 nm away.

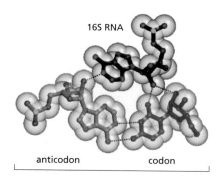

Figure 6–68 Recognition of correct codon–anticodon matches by the small subunit rRNA of the ribosome. Shown is the interaction between a nucleotide of the small subunit rRNA and the first nucleotide pair of a correctly paired codon–anticodon; similar interactions occur between other nucleotides of the rRNA and the second and third positions of the codon–anticodon pair. The small-subunit rRNA can form this network of hydrogen bonds only with correctly matched codon–anticodon pairs. As explained in the text, this codon–anticodon monitoring by the small-subunit rRNA increases the accuracy of protein synthesis. (From J.M. Ogle et al., *Science* 292:897–902, 2001. With permission from AAAS.)

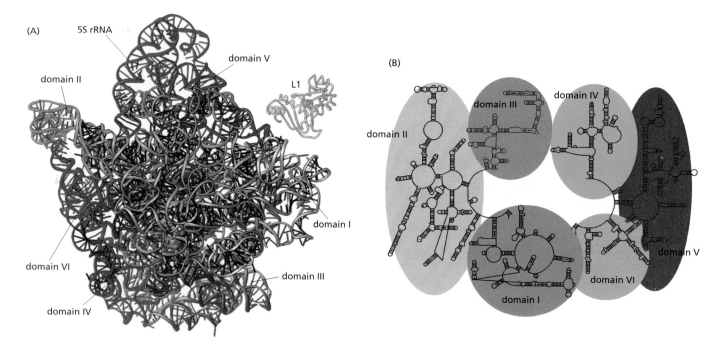

Figure 6–69 Structure of the rRNAs in the large subunit of a bacterial ribosome, as determined by x-ray crystallography. (A) Three-dimensional conformations of the large-subunit rRNAs (5S and 23S) as they appear in the ribosome. One of the protein subunits of the ribosome (L1) is also shown as a reference point, since it forms a characteristic protrusion on the ribosome. (B) Schematic diagram of the secondary structure of the 23S rRNA, showing the extensive network of base-pairing. The structure has been divided into six "domains" whose colors correspond to those in (A). The secondary-structure diagram is highly schematized to represent as much of the structure as possible in two dimensions. To do this, several discontinuities in the RNA chain have been introduced, although in reality the 23S RNA is a single RNA molecule. For example, the base of Domain III is continuous with the base of Domain IV even though a gap appears in the diagram. (Adapted from N. Ban et al., *Science* 289:905–920, 2000. With permission from AAAS.)

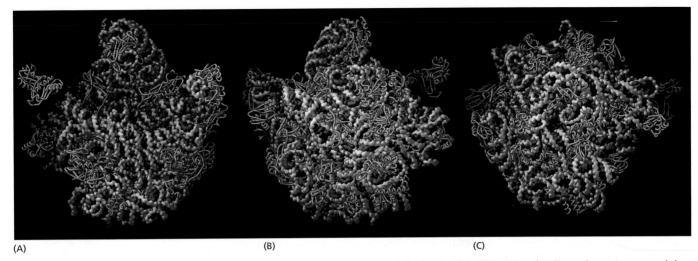

(A)   (B)   (C)

Figure 6–70 Location of the protein components of the bacterial large ribosomal subunit. The rRNAs (5S and 23S) are shown in *gray* and the large-subunit proteins (27 of the 31 total) in *gold*. For convenience, the protein structures depict only the polypeptide backbones. (A) Interface with the small subunit, the same view shown in Figure 6–64B. (B) Side opposite to that shown in (A), obtained by rotating (A) by 180° around a vertical axis. (C) Further slight rotation of (B) through a diagonal axis, allowing a view into the peptide exit channel in the center of the structure. (From N. Ban et al., *Science* 289:905–920, 2000. With permission from AAAS.)

This discovery came as a surprise to biologists because, unlike proteins, RNA does not contain easily ionizable functional groups that can be used to catalyze sophisticated reactions like peptide bond formation. Moreover, metal ions, which are often used by RNA molecules to catalyze chemical reactions (as discussed later in the chapter), were not observed at the active site of the ribosome. Instead, it is believed that the 23S rRNA forms a highly structured pocket that, through a network of hydrogen bonds, precisely orients the two reactants (the growing peptide chain and an aminoacyl-tRNA) and thereby greatly accelerates their covalent joining. In addition, the tRNA in the P site contributes to the active site, perhaps supplying a functional OH group that participates directly in the catalysis. This mechanism may ensure that catalysis occurs only when the tRNA is properly positioned in the ribosome.

RNA molecules that possess catalytic activity are known as **ribozymes**. We saw earlier in this chapter how other ribozymes function in self-splicing reactions (for example, see Figure 6–36). In the final section of this chapter, we consider what the ability of RNA molecules to function as catalysts for a wide variety of different reactions might mean for the early evolution of living cells. For now, we merely note that there is good reason to suspect that RNA rather than protein molecules served as the first catalysts for living cells. If so, the ribosome, with its RNA core, may be a relic of an earlier time in life's history—when protein synthesis evolved in cells that were run almost entirely by ribozymes.

Nucleotide Sequences in mRNA Signal Where to Start Protein Synthesis

The initiation and termination of translation share features of the translation elongation cycle described above. The site at which protein synthesis begins on the mRNA is especially crucial, since it sets the reading frame for the whole length of the message. An error of one nucleotide either way at this stage would cause every subsequent codon in the message to be misread, resulting in a nonfunctional protein with a garbled sequence of amino acids. The initiation step is also important because for most genes it is the last point at which the cell can decide whether the mRNA is to be translated and the protein synthesized; the rate of initiation is thus one determinant of the rate at which any protein is synthesized. We shall see in Chapter 7 that cells use several mechanisms to regulate translation initiation.

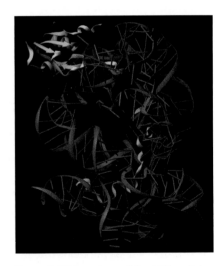

Figure 6–71 Structure of the L15 protein in the large subunit of the bacterial ribosome. The globular domain of the protein lies on the surface of the ribosome and an extended region penetrates deeply into the RNA core of the ribosome. The L15 protein is shown in *yellow* and a portion of the ribosomal RNA core is shown in *red*. (From D. Klein, P.B. Moore and T.A. Steitz, *J. Mol. Biol.* 340:141–147, 2004. With permission from Academic Press.)

The translation of an mRNA begins with the codon AUG, and a special tRNA is required to start translation. This **initiator tRNA** always carries the amino acid methionine (in bacteria, a modified form of methionine—formylmethionine—is used), with the result that all newly made proteins have methionine as the first amino acid at their N-terminus, the end of a protein that is synthesized first. This methionine is usually removed later by a specific protease. The initiator tRNA can be specially recognized by initiation factors because it has a nucleotide sequence distinct from that of the tRNA that normally carries methionine.

In eucaryotes, the initiator tRNA–methionine complex (Met–tRNAi) is first loaded into the small ribosomal subunit along with additional proteins called **eucaryotic initiation factors**, or **eIFs** (**Figure 6–72**). Of all the aminoacyl-tRNAs in the cell, only the methionine-charged initiator tRNA is capable of tightly binding the small ribosome subunit without the complete ribosome being present and it binds directly to the P-site. Next, the small ribosomal subunit binds to the 5′ end of an mRNA molecule, which is recognized by virtue of its 5′ cap and its two bound initiation factors, eIF4E (which directly binds the cap) and eIF4G (see Figure 6–40). The small ribosomal subunit then moves forward (5′ to 3′) along the mRNA, searching for the first AUG. Additional initiation factors that act as ATP-powered helicases facilitate the ribosome's movement through RNA secondary structure. In 90% of mRNAs, translation begins at the first AUG encountered by the small subunit. At this point, the initiation factors dissociate, allowing the large ribosomal subunit to assemble with the complex and complete the ribosome. The initiator tRNA is still bound to the P-site, leaving the A-site vacant. Protein synthesis is therefore ready to begin (see Figure 6–72).

The nucleotides immediately surrounding the start site in eucaryotic mRNAs influence the efficiency of AUG recognition during the above scanning process. If this recognition site differs substantially from the consensus recognition sequence (5′-ACCAUGG-3′), scanning ribosomal subunits will sometimes ignore the first AUG codon in the mRNA and skip to the second or third AUG codon instead. Cells frequently use this phenomenon, known as "leaky scanning," to produce two or more proteins, differing in their N-termini, from the same mRNA molecule. It allows some genes to produce the same protein with and without a signal sequence attached at its N-terminus, for example, so that the protein is directed to two different compartments in the cell.

The mechanism for selecting a start codon in bacteria is different. Bacterial mRNAs have no 5′ caps to signal the ribosome where to begin searching for the start of translation. Instead, each bacterial mRNA contains a specific ribosome-binding site (called the Shine–Dalgarno sequence, named after its discoverers) that is located a few nucleotides upstream of the AUG at which translation is to begin. This nucleotide sequence, with the consensus 5′-AGGAGGU-3′, forms base pairs with the 16S rRNA of the small ribosomal subunit to position the initiating AUG codon in the ribosome. A set of translation initiation factors orchestrates this interaction, as well as the subsequent assembly of the large ribosomal subunit to complete the ribosome.

Unlike a eucaryotic ribosome, a bacterial ribosome can therefore readily assemble directly on a start codon that lies in the interior of an mRNA molecule, so long as a ribosome-binding site precedes it by several nucleotides. As a result, bacterial mRNAs are often *polycistronic*—that is, they encode several different proteins, each of which is translated from the same mRNA molecule (**Figure 6–73**). In contrast, a eucaryotic mRNA generally encodes only a single protein.

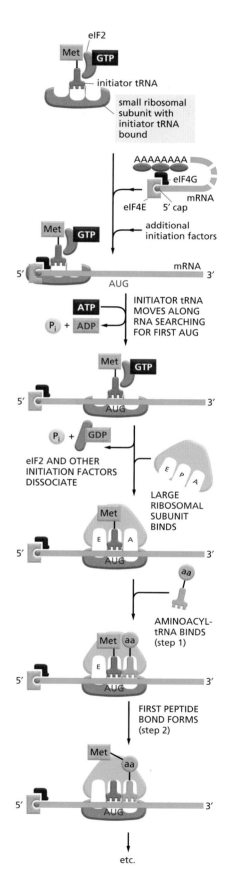

Figure 6–72 The initiation of protein synthesis in eucaryotes. Only three of the many translation initiation factors required for this process are shown. Efficient translation initiation also requires the poly-A tail of the mRNA bound by poly-A-binding proteins which, in turn, interact with eIF4G. In this way, the translation apparatus ascertains that both ends of the mRNA are intact before initiating protein synthesis (see Figure 6–40). Although only one GTP hydrolysis event is shown in the figure, a second is known to occur just before the large and small ribosomal subunits join.

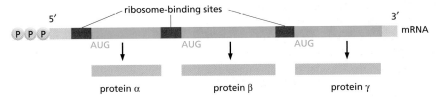

Figure 6–73 **Structure of a typical bacterial mRNA molecule.** Unlike eucaryotic ribosomes, which typically require a capped 5′ end, procaryotic ribosomes initiate transcription at ribosome-binding sites (Shine–Dalgarno sequences), which can be located anywhere along an mRNA molecule. This property of ribosomes permits bacteria to synthesize more than one type of protein from a single mRNA molecule.

Stop Codons Mark the End of Translation

The end of the protein-coding message is signaled by the presence of one of three *stop codons* (UAA, UAG, or UGA) (see Figure 6–50). These are not recognized by a tRNA and do not specify an amino acid, but instead signal to the ribosome to stop translation. Proteins known as *release factors* bind to any ribosome with a stop codon positioned in the A site, forcing the peptidyl transferase in the ribosome to catalyze the addition of a water molecule instead of an amino acid to the peptidyl-tRNA (**Figure 6–74**). This reaction frees the carboxyl end of the growing polypeptide chain from its attachment to a tRNA molecule, and since only this attachment normally holds the growing polypeptide to the ribosome, the completed protein chain is immediately released into the cytoplasm. The ribosome then releases the mRNA and separates into the large and small subunits, which can assemble on this or another mRNA molecule to begin a new round of protein synthesis.

Release factors are an example of *molecular mimicry*, whereby one type of macromolecule resembles the shape of a chemically unrelated molecule. In this case, the three-dimensional structure of release factors (made entirely of protein) resembles the shape and charge distribution of a tRNA molecule (**Figure 6–75**). This shape and charge mimicry helps them enter the A-site on the ribosome and cause translation termination.

During translation, the nascent polypeptide moves through a large, water-filled tunnel (approximately 10 nm × 1.5 nm) in the large subunit of the ribosome (see Figure 6–70C). The walls of this tunnel, made primarily of 23S rRNA, are a patchwork of tiny hydrophobic surfaces embedded in a more extensive hydrophilic surface. This structure is not complementary to any peptide, and thus provides a "Teflon" coating through which a polypeptide chain can easily slide. The dimensions of the tunnel suggest that nascent proteins are largely unstructured as they pass through the ribosome, although some α-helical regions of the protein can form before leaving the ribosome tunnel. As it leaves the ribosome, a newly synthesized protein must fold into its proper three-dimensional conformation to be useful to the cell, and later in this chapter we discuss how this folding occurs. First, however, we describe several additional aspects of the translation process itself.

Proteins Are Made on Polyribosomes

The synthesis of most protein molecules takes between 20 seconds and several minutes. During this very short period, however, it is usual for multiple initiations to take place on each mRNA molecule being translated. As soon as the preceding ribosome has translated enough of the nucleotide sequence to move out

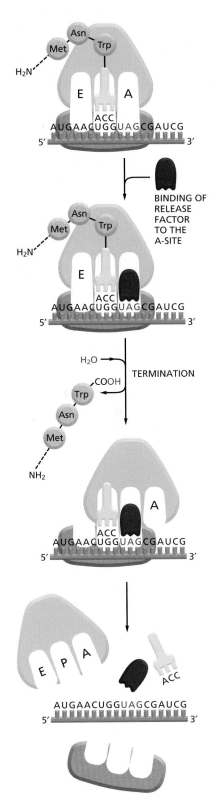

Figure 6–74 **The final phase of protein synthesis.** The binding of a release factor to an A-site bearing a stop codon terminates translation. The completed polypeptide is released and, in a series of reactions that requires additional proteins and GTP hydrolysis (not shown), the ribosome dissociates into its two separate subunits.

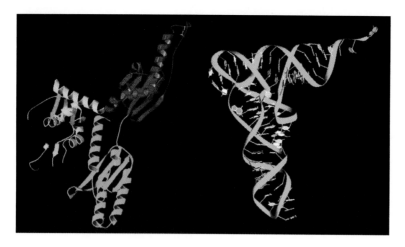

Figure 6–75 **The structure of a human translation release factor (eRF1) and its resemblance to a tRNA molecule.** The protein is on the *left* and the tRNA on the *right*. (From H. Song et al., *Cell* 100:311–321, 2000. With permission from Elsevier.)

of the way, the 5′ end of the mRNA is threaded into a new ribosome. The mRNA molecules being translated are therefore usually found in the form of *polyribosomes* (or *polysomes*): large cytoplasmic assemblies made up of several ribosomes spaced as close as 80 nucleotides apart along a single mRNA molecule (**Figure 6–76**). These multiple initiations allow the cell to make many more protein molecules in a given time than would be possible if each had to be completed before the next could start. <GAAG>

Both bacteria and eucaryotes use polysomes, and both employ additional strategies to speed up the overall rate of protein synthesis even further. Because bacterial mRNA does not need to be processed and is accessible to ribosomes while it is being made, ribosomes can attach to the free end of a bacterial mRNA molecule and start translating it even before the transcription of that RNA is complete, following closely behind the RNA polymerase as it moves along DNA. In eucaryotes, as we have seen, the 5′ and 3′ ends of the mRNA interact (see Figures 6–40 and 6–76A); therefore, as soon as a ribosome dissociates, its two subunits are in an optimal position to reinitiate translation on the same mRNA molecule.

There Are Minor Variations in the Standard Genetic Code

As discussed in Chapter 1, the genetic code (shown in Figure 6–50) applies to all three major branches of life, providing important evidence for the common

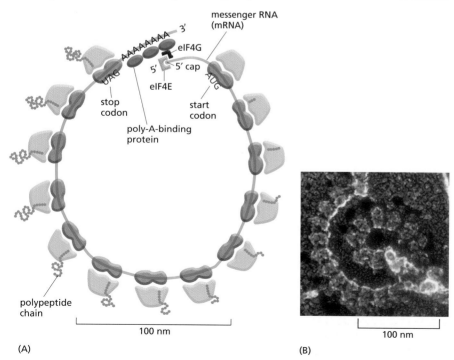

(A)

(B)

Figure 6–76 **A polyribosome.** (A) Schematic drawing showing how a series of ribosomes can simultaneously translate the same eucaryotic mRNA molecule. (B) Electron micrograph of a polyribosome from a eucaryotic cell. (B, courtesy of John Heuser.)

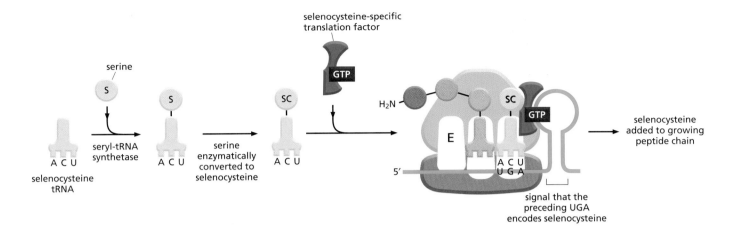

Figure 6–77 Incorporation of
selenocysteine into a growing
polypeptide chain. A specialized tRNA is
charged with serine by the normal seryl-
tRNA synthetase, and the serine is
subsequently converted enzymatically to
selenocysteine. A specific RNA structure
in the mRNA (a stem and loop structure
with a particular nucleotide sequence)
signals that selenocysteine is to be
inserted at the neighboring UGA codon.
As indicated, this event requires the
participation of a selenocysteine-specific
translation factor.

ancestry of all life on Earth. Although rare, there are exceptions to this code. For example, *Candida albicans,* the most prevalent human fungal pathogen, translates the codon CUG as serine, whereas nearly all other organisms translate it as leucine. Mitochondria (which have their own genomes and encode much of their translational apparatus) often deviate from the standard code. For example, in mammalian mitochondria AUA is translated as methionine, whereas in the cytosol of the cell it is translated as isoleucine (see Table 14–3, p. 862). This type of deviation in the genetic code is "hardwired" into the organisms or the organelles in which it occurs.

A different type of variation, sometimes called *translation recoding,* occurs in many cells. In this case, other nucleotide sequence information present in an mRNA can change the meaning of the genetic code at a particular site in the mRNA molecule. The standard code allows cells to manufacture proteins using only 20 amino acids. However, bacteria, archaea, and eucaryotes have available to them a twenty-first amino acid that can be incorporated directly into a growing polypeptide chain through translation recoding. Selenocysteine, which is essential for the efficient function of a variety of enzymes, contains a selenium atom in place of the sulfur atom of cysteine. Selenocysteine is enzymatically produced from a serine attached to a special tRNA molecule that base-pairs with the UGA codon, a codon normally used to signal a translation stop. The mRNAs for proteins in which selenocysteine is to be inserted at a UGA codon carry an additional nucleotide sequence in the mRNA nearby that causes this recoding event (**Figure 6–77**).

Another form of recoding, *translational frameshifting,* allows more than one protein to be synthesized from a single mRNA. Retroviruses, members of a large group of eucaryotic-infecting pathogens, commonly use translational frameshifting to make both the capsid proteins *(Gag proteins)* and the viral reverse transcriptase and integrase *(Pol proteins)* from the same RNA transcript (see Figure 5–73). The virus needs many more copies of the Gag proteins than it does of the Pol proteins. This quantitative adjustment is achieved by encoding the *Pol* genes just after the *Gag* genes but in a different reading frame. Small amounts of the Pol gene products are made because, on occasion, an upstream translational frameshift allows the Gag protein stop codon to be bypassed. This frameshift occurs at a particular codon in the mRNA and requires a specific *recoding signal,* which seems to be a structural feature of the RNA sequence downstream of this site (**Figure 6–78**).

Inhibitors of Procaryotic Protein Synthesis Are Useful as Antibiotics

Many of the most effective antibiotics used in modern medicine are compounds made by fungi that inhibit bacterial protein synthesis. Fungi and bacteria compete for many of the same environmental niches, and millions of years of coevolution has resulted in fungi producing potent bacterial inhibitors. Some of these

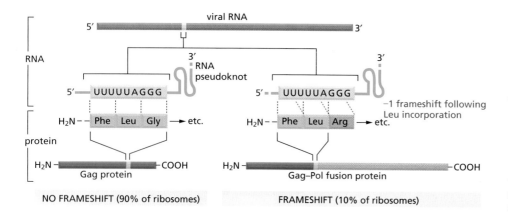

Figure 6–78 The translational frameshifting that produces the reverse transcriptase and integrase of a retrovirus. The viral reverse transcriptase and integrase are produced by proteolytic processing of a large protein (the Gag–Pol fusion protein) consisting of both the Gag and Pol amino acid sequences. Proteolytic processing of the more abundant Gag protein produces the viral capsid proteins. Both the Gag and the Gag–Pol fusion proteins start with identical mRNA, but whereas the Gag protein terminates at a stop codon downstream of the sequence shown, translation of the Gag–Pol fusion protein bypasses this stop codon, allowing the synthesis of the longer Gag–Pol fusion protein. The stop-codon-bypass is made possible by a controlled translational frameshift, as illustrated. Features in the local RNA structure (including the RNA loop shown) cause the tRNA^{Leu} attached to the C-terminus of the growing polypeptide chain occasionally to slip backward by one nucleotide on the ribosome, so that it pairs with a UUU codon instead of the UUA codon that had initially specified its incorporation; the next codon (AGG) in the new reading frame specifies an arginine rather than a glycine. This controlled slippage is due in part to a *pseudoknot* that forms in the viral mRNA (see Figure 6–102). The sequence shown is from the human AIDS virus, HIV. (Adapted from T. Jacks et al., *Nature* 331:280–283, 1988. With permission from Macmillan Publishers Ltd.)

drugs exploit the structural and functional differences between bacterial and eucaryotic ribosomes so as to interfere preferentially with the function of bacterial ribosomes. Thus humans can take high dosages of some of these compounds without undue toxicity. Many antibiotics lodge in pockets in the ribosomal RNAs and simply interfere with the smooth operation of the ribosome (**Figure 6–79**). **Table 6–4** lists some of the more common antibiotics of this kind along with several other inhibitors of protein synthesis, some of which act on eucaryotic cells and therefore cannot be used as antibiotics.

Because they block specific steps in the processes that lead from DNA to protein, many of the compounds listed in Table 6–4 are useful for cell biological studies. Among the most commonly used drugs in such investigations are *chloramphenicol*, *cycloheximide*, and *puromycin*, all of which specifically inhibit protein synthesis. In a eucaryotic cell, for example, chloramphenicol inhibits protein synthesis on ribosomes only in mitochondria (and in chloroplasts in plants), presumably reflecting the procaryotic origins of these organelles (discussed in Chapter 14). Cycloheximide, in contrast, affects only ribosomes in the cytosol. Puromycin is especially interesting because it is a structural analog of a tRNA molecule linked to an amino acid and is therefore another example of molecular mimicry; the ribosome mistakes it for an authentic amino acid and covalently incorporates it at the C-terminus of the growing peptide chain, thereby causing the premature termination and release of the polypeptide. As might be expected, puromycin inhibits protein synthesis in both procaryotes and eucaryotes.

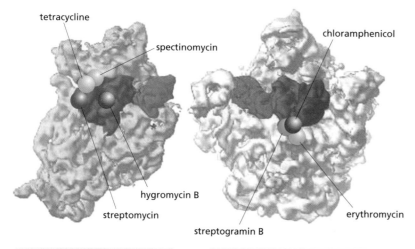

Figure 6–79 Binding sites for antibiotics on the bacterial ribosome. The small *(left)* and large *(right)* subunits of the ribosome are arranged as though the ribosome has been opened like a book; the bound tRNA molecules are shown in *purple* (see Figure 6–64). Most of the antibiotics shown bind directly to pockets formed by the ribosomal RNA molecules. Hygromycin B induces errors in translation, spectinomycin blocks the translocation of the peptidyl-tRNA from the A-site to the P-site, and streptogramin B prevents elongation of nascent peptides. Table 6–4 lists the inhibitory mechanisms of the other antibiotics shown in the figure. (Adapted from J. Poehlsgaard and S. Douthwaite, *Nat. Rev. Microbiol.* 3:870–881, 2005. With permission from Macmillan Publishers Ltd.)

Table 6–4 Inhibitors of Protein or RNA Synthesis

INHIBITOR	SPECIFIC EFFECT
Acting only on bacteria	
Tetracycline	blocks binding of aminoacyl-tRNA to A-site of ribosome
Streptomycin	prevents the transition from translation initiation to chain elongation and also causes miscoding
Chloramphenicol	blocks the peptidyl transferase reaction on ribosomes (step 2 in Figure 6–66)
Erythromycin	binds in the exit channel of the ribosome and thereby inhibits elongation of the peptide chain
Rifamycin	blocks initiation of RNA chains by binding to RNA polymerase (prevents RNA synthesis)
Acting on bacteria and eucaryotes	
Puromycin	causes the premature release of nascent polypeptide chains by its addition to the growing chain end
Actinomycin D	binds to DNA and blocks the movement of RNA polymerase (prevents RNA synthesis)
Acting on eucaryotes but not bacteria	
Cycloheximide	blocks the translocation reaction on ribosomes (step 3 in Figure 6–66)
Anisomycin	blocks the peptidyl transferase reaction on ribosomes (step 2 in Figure 6–66)
α-Amanitin	blocks mRNA synthesis by binding preferentially to RNA polymerase II

The ribosomes of eucaryotic mitochondria (and chloroplasts) often resemble those of bacteria in their sensitivity to inhibitors. Therefore, some of these antibiotics can have a deleterious effect on human mitochondria.

Accuracy in Translation Requires the Expenditure of Free Energy

Translation by the ribosome is a compromise between the opposing constraints of accuracy and speed. We have seen, for example, that the accuracy of translation (1 mistake per 10^4 amino acids joined) requires time delays each time a new amino acid is added to a growing polypeptide chain, producing an overall speed of translation of 20 amino acids incorporated per second in bacteria. Mutant bacteria with a specific alteration in the small ribosomal subunit have longer delays and translate mRNA into protein with an accuracy considerably higher than this; however, protein synthesis is so slow in these mutants that the bacteria are barely able to survive.

We have also seen that attaining the observed accuracy of protein synthesis requires the expenditure of a great deal of free energy; this is expected, since, as discussed in Chapter 2, there is a price to be paid for any increase in order in the cell. In most cells, protein synthesis consumes more energy than any other biosynthetic process. At least four high-energy phosphate bonds are split to make each new peptide bond: two are consumed in charging a tRNA molecule with an amino acid (see Figure 6–56), and two more drive steps in the cycle of reactions occurring on the ribosome during synthesis itself (see Figure 6–67). In addition, extra energy is consumed each time that an incorrect amino acid linkage is hydrolyzed by a tRNA synthetase (see Figure 6–59) and each time that an incorrect tRNA enters the ribosome, triggers GTP hydrolysis, and is rejected (see Figure 6–67). To be effective, these proofreading mechanisms must also allow an appreciable fraction of correct interactions to be removed; for this reason, proofreading is even more costly in energy than it might seem.

Quality Control Mechanisms Act to Prevent Translation of Damaged mRNAs

In eucaryotes, mRNA production involves both transcription and a series of elaborate RNA-processing steps; these take place in the nucleus, segregated from ribosomes, and only when the processing is complete are the mRNAs transported to the cytoplasm to be translated (see Figure 6–40). However, this scheme is not foolproof, and some incorrectly processed mRNAs are inadvertently sent to the cytoplasm. In addition, mRNAs that were flawless when they left the nucleus can become broken or otherwise damaged in the cytosol. The danger of translating damaged or incompletely processed mRNAs (which would produce truncated or otherwise aberrant proteins) is apparently so great that the cell has several backup measures to prevent this from happening.

To avoid translating broken mRNAs, the 5′ cap and the poly-A tail are both recognized by the translation-initiation machinery before translation begins (see Figure 6–72). To help ensure that mRNAs are properly spliced before they are translated, the exon junction complex (EJC), which is deposited on the mRNA following splicing (see Figure 6–40), stimulates the subsequent translation of the mRNA.

But the most powerful mRNA surveillance system, called **nonsense-mediated mRNA decay**, eliminates defective mRNAs before they can be efficiently translated into protein. This mechanism is brought into play when the cell determines that an mRNA molecule has a nonsense (stop) codon (UAA, UAG, or UGA) in the "wrong" place—a situation likely to arise in an mRNA molecule that has been improperly spliced. Aberrant splicing will usually result in the random introduction of a nonsense codon into the reading frame of the mRNA, especially in organisms, such as humans, that have a large average intron size (see Figure 6–32B).

This surveillance mechanism begins as an mRNA molecule is being transported from the nucleus to the cytosol. As its 5′ end emerges from the nuclear pore, the mRNA is met by a ribosome, which begins to translate it. As translation proceeds, the exon junction complexes (EJC) bound to the mRNA at each splice-site are apparently displaced by the moving ribosome. The normal stop codon will be within the last exon, so by the time the ribosome reaches it and stalls, no more EJCs should be bound to the mRNA. If this is the case, the mRNA "passes inspection" and is released to the cytosol where it can be translated in earnest (**Figure 6–80**). However, if the ribosome reaches a premature stop codon and stalls, it senses that EJCs remain and the bound mRNA molecule is rapidly degraded. In this way, the first round of translation allows the cell to test the fitness of each mRNA molecule as it exits the nucleus.

Nonsense-mediated decay may have been especially important in evolution, allowing eucaryotic cells to more easily explore new genes formed by DNA rearrangements, mutations, or alternative patterns of splicing—by selecting only those mRNAs for translation that can produce a full-length protein. Nonsense-mediated decay is also important in cells of the developing immune system, where the extensive DNA rearrangements that occur (see Figure 25–36) often generate premature termination codons. The surveillance system degrades the mRNAs produced from such rearranged genes, thereby avoiding the potential toxic effects of truncated proteins.

Figure 6–80 Nonsense-mediated mRNA decay. As shown on the right, the failure to correctly splice a pre-mRNA often introduces a premature stop codon into the reading frame for the protein. The introduction of such an "in-frame" stop codon is particularly likely to occur in mammals, where the introns tend to be very long. When translated, these abnormal mRNAs produce aberrant proteins, which could damage the cell. However, as shown at the bottom right of the figure, these abnormal RNAs are destroyed by the nonsense-mediated decay mechanism. According to one model, an mRNA molecule, bearing exon junction complexes (EJCs) to mark successfully completed splices, is first met by a ribosome that performs a "test" round of translation. As the mRNA passes through the tight channel of the ribosome, the EJCs are stripped off, and successful mRNAs are released to undergo multiple rounds of translation *(left side)*. However, if an in-frame stop codon is encountered before the final exon junction complex is reached *(right side)*, the mRNA undergoes nonsense-mediated decay, which is triggered by the Upf proteins *(green)* that bind to each EJC. Note that, to trigger nonsense-mediated decay, the premature stop codon must be in the same reading frame as that of the normal protein. (Adapted from J. Lykke-Andersen et al., *Cell* 103:1121–1131, 2000. With permission from Elsevier.)

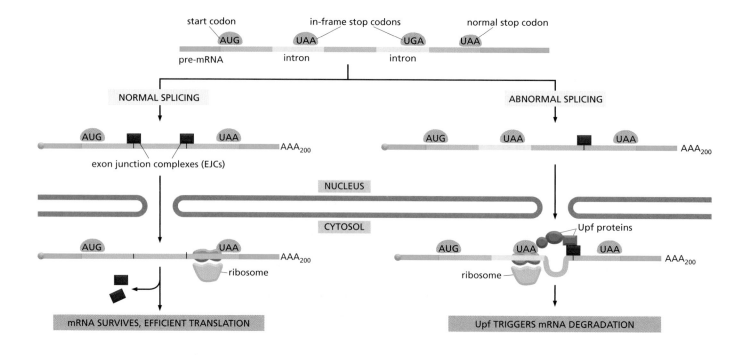

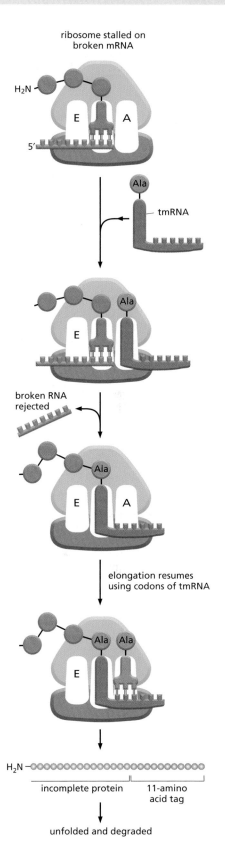

Figure 6–81 The rescue of a bacterial ribosome stalled on an incomplete mRNA molecule. The tmRNA shown is a 363-nucleotide RNA with both tRNA and mRNA functions, hence its name. It carries an alanine and can enter the vacant A-site of a stalled ribosome to add this alanine to a polypeptide chain, mimicking a tRNA although no codon is present to guide it. The ribosome then translates 10 codons from the tmRNA, completing an 11 amino acid tag on the protein. Proteases recognize this tag and degrade the entire protein. Although the example shown in the figure is from bacteria, eucaryotes can employ a similar strategy.

Finally, the nonsense-mediated surveillance pathway plays an important role in mitigating the symptoms of many inherited human diseases. As we have seen, inherited diseases are usually caused by mutations that spoil the function of a key protein, such as hemoglobin or one of the blood clotting factors. Approximately one-third of all genetic disorders in humans result from non-sense mutations or mutations (such as frameshift mutations or splice-site muta-tions) that place nonsense mutations into the gene's reading frame. In individu-als that carry one mutant and one functional gene, nonsense-mediated decay eliminates the aberrant mRNA and thereby prevents a potentially toxic protein from being made. Without this safeguard, individuals with one functional and one mutant "disease gene" would likely suffer much more severe symptoms.

We saw earlier in this chapter that bacteria lack the elaborate mRNA pro-cessing found in eucaryotes and that translation often begins before the synthe-sis of the RNA molecule is completed. Yet bacteria also have quality control mechanisms to deal with incompletely synthesized and broken mRNAs. When the bacterial ribosome translates to the end of an incomplete RNA it stalls and does not release the RNA. Rescue comes in the form of a special RNA (called tmRNA), which enters the A-site of the ribosome and is itself translated, releas-ing the ribosome. The special 11 amino acid tag thus added to the C-terminus of the truncated protein signals to proteases that the entire protein is to be degraded (**Figure 6–81**).

Some Proteins Begin to Fold While Still Being Synthesized

The process of gene expression is not over when the genetic code has been used to create the sequence of amino acids that constitutes a protein. To be useful to the cell, this new polypeptide chain must fold up into its unique three-dimen-sional conformation, bind any small-molecule cofactors required for its activity, be appropriately modified by protein kinases or other protein-modifying enzymes, and assemble correctly with the other protein subunits with which it functions (**Figure 6–82**).

The information needed for all of the steps listed above is ultimately con-tained in the sequence of linked amino acids that the ribosome produces when it translates an mRNA molecule into a polypeptide chain. As discussed in Chap-ter 3, when a protein folds into a compact structure, it buries most of its hydrophobic residues in an interior core. In addition, large numbers of nonco-valent interactions form between various parts of the molecule. It is the sum of all of these energetically favorable arrangements that determines the final fold-ing pattern of the polypeptide chain—as the conformation of lowest free energy (see p. 130).

Through many millions of years of evolution, the amino acid sequence of each protein has been selected not only for the conformation that it adopts but also for an ability to fold rapidly. For some proteins, this folding begins immedi-ately, as the protein spins out of the ribosome, starting from the N-terminal end. In these cases, as each protein domain emerges from the ribosome, within a few seconds it forms a compact structure that contains most of the final secondary features (α helices and β sheets) aligned in roughly the right conformation (**Fig-ure 6–83**). For many protein domains, this unusually dynamic and flexible state called a *molten globule*, is the starting point for a relatively slow process in which many side-chain adjustments occur that eventually form the correct tertiary

Figure 6–82 Steps in the creation of a functional protein. As indicated, translation of an mRNA sequence into an amino acid sequence on the ribosome is not the end of the process of forming a protein. To function, the completed polypeptide chain must fold correctly into its three-dimensional conformation, bind any cofactors required, and assemble with its partner protein chains (if any). Noncovalent bond formation drives these changes. As indicated, many proteins also require covalent modifications of selected amino acids. Although the most frequent modifications are protein glycosylation and protein phosphorylation, more than 100 different types of covalent modifications are known (see, for example, Figure 3–81).

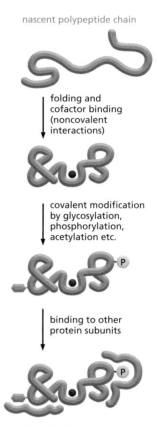

nascent polypeptide chain

folding and cofactor binding (noncovalent interactions)

covalent modification by glycosylation, phosphorylation, acetylation etc.

binding to other protein subunits

mature functional protein

structure. It takes several minutes to synthesize a protein of average size, and for some proteins much of the folding process is complete by the time the ribosome releases the C-terminal end of a protein (**Figure 6–84**).

Molecular Chaperones Help Guide the Folding of Most Proteins

Most proteins probably do not begin to fold during their synthesis. Instead, they are met at the ribosome by a special class of proteins called **molecular chaperones**. Molecular chaperones are useful for cells because there are many different paths that can be taken to convert an unfolded or partially folded protein to its final compact conformation. For many proteins, some of the intermediates formed along the way would aggregate and be left as off-pathway dead ends without the intervention of a chaperone (**Figure 6–85**).

Many molecular chaperones are called *heat-shock proteins* (designated *Hsp*), because they are synthesized in dramatically increased amounts after a brief exposure of cells to an elevated temperature (for example, 42°C for cells that normally live at 37°C). This reflects the operation of a feedback system that responds to an increase in misfolded proteins (such as those produced by elevated temperatures) by boosting the synthesis of the chaperones that help these proteins refold.

There are several major families of eucaryotic molecular chaperones, including the Hsp60 and Hsp70 proteins. Different family members function in different organelles. Thus, as discussed in Chapter 12, mitochondria contain their own Hsp60 and Hsp70 molecules that are distinct from those that function in the cytosol; and a special Hsp70 (called *BIP*) helps to fold proteins in the endoplasmic reticulum.

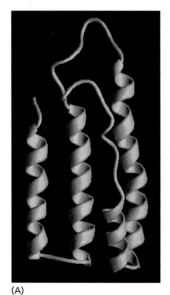

(A)   (B)

Figure 6–83 The structure of a molten globule. (A) A molten globule form of cytochrome b_{562} is more open and less highly ordered than the final folded form of the protein, shown in (B). Note that the molten globule contains most of the secondary structure of the final form, although the ends of the α helices are unravelled and one of the helices is only partly formed. (Courtesy of Joshua Wand, from Y. Feng et al., *Nat. Struct. Biol.* 1:30–35, 1994. With permission from Macmillan Publishers Ltd.)

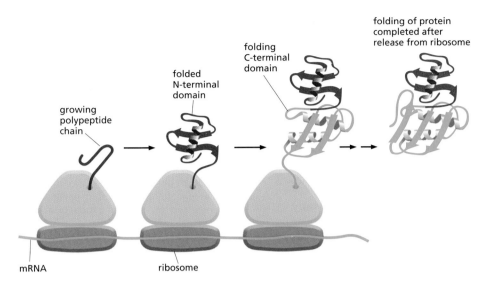

Figure 6–84 Co-translational protein folding. A growing polypeptide chain is shown acquiring its secondary and tertiary structure as it emerges from a ribosome. The N-terminal domain folds first, while the C-terminal domain is still being synthesized. This protein has not achieved its final conformation at the time it is released from the ribosome. (Modified from A.N. Federov and T.O. Baldwin, *J. Biol. Chem.* 272:32715–32718, 1997.)

The Hsp60 and Hsp70 proteins each work with their own small set of associated proteins when they help other proteins to fold. Hsps share an affinity for the exposed hydrophobic patches on incompletely folded proteins, and they hydrolyze ATP, often binding and releasing their protein substrate with each cycle of ATP hydrolysis. In other respects, the two types of Hsp proteins function differently. The Hsp70 machinery acts early in the life of many proteins, binding to a string of about seven hydrophobic amino acids before the protein leaves the ribosome (**Figure 6–86**). In contrast, Hsp60-like proteins form a large barrel-shaped structure that acts after a protein has been fully synthesized. This type of chaperone, sometimes called a *chaperonin*, forms an "isolation chamber" into which misfolded proteins are fed, preventing their aggregation and providing them with a favorable environment in which to attempt to refold (**Figure 6–87**).

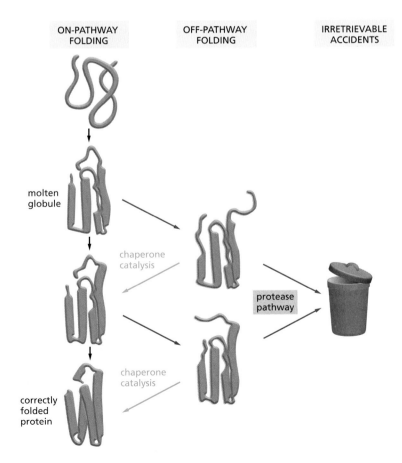

Figure 6–85 A current view of protein folding. Each domain of a newly synthesized protein rapidly attains a "molten globule" state. Subsequent folding occurs more slowly and by multiple pathways, often involving the help of a molecular chaperone. Some molecules may still fail to fold correctly; as explained in the text, specific proteases recognize and degrade these molecules.

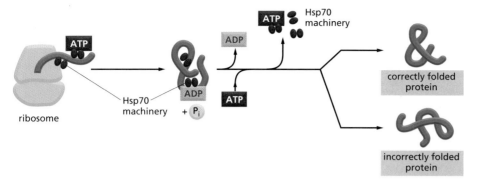

Figure 6–86 The Hsp70 family of molecular chaperones. These proteins act early, recognizing a small stretch of hydrophobic amino acids on a protein's surface. Aided by a set of smaller Hsp40 proteins (not shown), ATP-bound Hsp70 molecules grasp their target protein and then hydrolyze ATP to ADP, undergoing conformational changes that cause the Hsp70 molecules to associate even more tightly with the target. After the Hsp40 dissociates, the rapid rebinding of ATP induces the dissociation of the Hsp70 protein after ADP release. In reality, repeated cycles of Hsp protein binding and release help the target protein to refold, as schematically illustrated in Figure 6–85.

The chaperones shown in Figures 6–86 and 6–87 often use many cycles of ATP hydrolysis to fold a single polypeptide chain correctly. Although some of this energy expenditure is used to perform mechanical work, probably much more is expended to ensure that protein folding is accurate. Just as we saw for transcription, splicing, and translation, the expenditure of free energy can be used by cells to improve the accuracy of a biological process. In the case of protein folding, ATP hydrolysis allows chaperones to recognize a wide variety of misfolded structures, to halt any further misfolding and to recommence folding of a protein in an orderly way.

Although our discussion focuses on only two types of chaperones, the cell has a variety of others. The enormous diversity of proteins in cells presumably requires a wide range of chaperones with versatile surveillance and correction capabilities.

Exposed Hydrophobic Regions Provide Critical Signals for Protein Quality Control

If radioactive amino acids are added to cells for a brief period, the newly synthesized proteins can be followed as they mature into their final functional form.

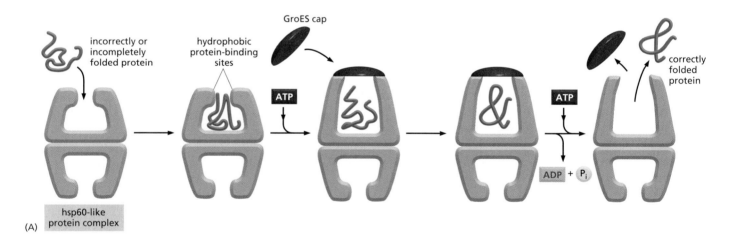

(A)

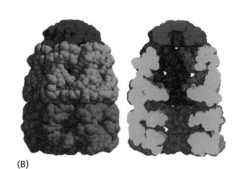

(B)

Figure 6–87 The structure and function of the Hsp60 family of molecular chaperones. (A) The catalysis of protein refolding. A misfolded protein is initially captured by hydrophobic interactions along one rim of the barrel. The subsequent binding of ATP plus a protein cap increases the diameter of the barrel rim, which may transiently stretch (partly unfold) the client protein. This also confines the protein in an enclosed space, where it has a new opportunity to fold. After about 15 seconds, ATP hydrolysis occurs, weakening the complex. Subsequent binding of another ATP molecule ejects the protein, whether folded or not, and the cycle repeats. This type of molecular chaperone is also known as a chaperonin; it is designated as Hsp60 in mitochondria, TCP1 in the cytosol of vertebrate cells, and GroEL in bacteria. As indicated, only half of the symmetrical barrel operates on a client protein at any one time. (B) The structure of GroEL bound to its GroES cap, as determined by X-ray crystallography. On the *left* is shown the outside of the barrel-like structure and on the *right* a cross section through its center. (B, adapted from B. Bukau and A.L. Horwich, *Cell* 92:351–366, 1998. With permission from Elsevier.)

This type of experiment demonstrates that the Hsp70 proteins act first, beginning when a protein is still being synthesized on a ribosome, and the Hsp60-like proteins act only later to help fold completed proteins. But how does the cell distinguish misfolded proteins, which require additional rounds of ATP-catalyzed refolding, from those with correct structures?

Before answering, we need to pause to consider the post-translational fate of proteins more broadly. Usually, if a protein has a sizable exposed patch of hydrophobic amino acids on its surface, it is abnormal: it has either failed to fold correctly after leaving the ribosome, suffered an accident that partly unfolded it at a later time, or failed to find its normal partner subunit in a larger protein complex. Such a protein is not merely useless to the cell, it can be dangerous. Many proteins with an abnormally exposed hydrophobic region can form large aggregates in the cell. We shall see that, in rare cases, such aggregates do form and cause severe human diseases. Normally, however, powerful protein quality control mechanisms prevent such disasters.

Given this background, it is not surprising that cells have evolved elaborate mechanisms that recognize the hydrophobic patches on proteins and minimize the damage they cause. Two of these mechanisms depend on the molecular chaperones just discussed, which bind to the patch and attempt to repair the defective protein by giving it another chance to fold. At the same time, by covering the hydrophobic patches, these chaperones transiently prevent protein aggregation. Proteins that very rapidly fold correctly on their own do not display such patches and the chaperones bypass them.

Figure 6–88 outlines all of the quality control choices that a cell makes for a difficult-to-fold, newly synthesized protein. As indicated, when attempts to refold a protein fail, a third mechanism is called into play that completely destroys the protein by proteolysis. The proteolytic pathway begins with the recognition of an abnormal hydrophobic patch on a protein's surface, and it ends with the delivery of the entire protein to a protein destruction machine, a complex protease known as the *proteasome*. As described next, this process depends on an elaborate protein-marking system that also carries out other central functions in the cell by destroying selected normal proteins.

The Proteasome Is a Compartmentalized Protease with Sequestered Active Sites

The proteolytic machinery and the chaperones compete with one another to reorganize a misfolded protein. If a newly synthesized protein folds rapidly, at most only a small fraction of it is degraded. In contrast, a slowly folding protein is vulnerable to the proteolytic machinery for a longer time, and many more of its molecules are destroyed before the remainder attain the proper folded state. Due to mutations or to errors in transcription, RNA splicing, and translation, some proteins never fold properly. It is particularly important that the cell destroy these potentially harmful proteins.

The apparatus that deliberately destroys aberrant proteins is the **proteasome**, an abundant ATP-dependent protease that constitutes nearly 1% of cell protein. Present in many copies dispersed throughout the cytosol and the nucleus, the proteasome also destroys aberrant proteins of the endoplasmic

Figure 6–88 The processes that monitor protein quality following protein synthesis. A newly synthesized protein sometimes folds correctly and assembles on its own with its partner proteins, in which case the quality control mechanisms leave it alone. Incompletely folded proteins are helped to refold by molecular chaperones: first by a family of Hsp70 proteins, and then in some cases, by Hsp60-like proteins. For both types of chaperones, the client proteins are recognized by an abnormally exposed patch of hydrophobic amino acids on their surface. These "protein-rescue" processes compete with another mechanism that, upon recognizing an abnormally exposed patch, marks the protein for destruction by the proteasome. The combined activity of all of these processes is needed to prevent massive protein aggregation in a cell, which can occur when many hydrophobic regions on proteins clump together nonspecifically.

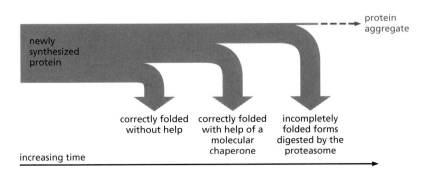

(A)   (B)

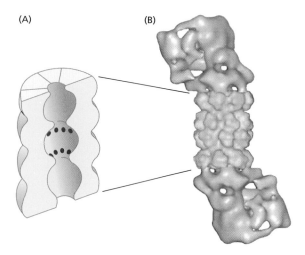

Figure 6–89 The proteasome. (A) A cut-away view of the structure of the central 20S cylinder, as determined by x-ray crystallography, with the active sites of the proteases indicated by *red dots*. (B) The entire proteasome, in which the central cylinder *(yellow)* is supplemented by a 19S cap *(blue)* at each end. The cap structure has been determined by computer processing of electron microscope images. The complex cap (also called the regulatory particle) selectively binds proteins that have been marked by ubiquitin for destruction; it then uses ATP hydrolysis to unfold their polypeptide chains and feed them through a narrow channel (see Figure 6–91) into the inner chamber of the 20S cylinder for digestion to short peptides. (B, from W. Baumeister et al., *Cell* 92:367–380, 1998. With permission from Elsevier.)

reticulum (ER). An ER-based surveillance system detects proteins that fail either to fold or to be assembled properly after they enter the ER, and *retrotranslocates* them back to the cytosol for degradation (discussed in Chapter 12).

Each proteasome consists of a central hollow cylinder (the 20S core proteasome) formed from multiple protein subunits that assemble as a quasi-cylindrical stack of four heptameric rings (**Figure 6–89**). Some of the subunits are distinct proteases whose active sites face the cylinder's inner chamber. The design prevents these highly efficient proteases from running rampant through the cell. Each end of the cylinder is normally associated with a large protein complex (the 19S cap), which contains a six-subunit protein ring, through which target proteins are threaded into the proteasome core where they are degraded (**Figure 6–90**). The threading reaction, driven by ATP hydrolysis, unfolds the target proteins as they move through the cap, exposing them to the proteases lining the proteasome core (**Figure 6–91**). The proteins that make up the ring structure in the proteasome cap belong to a large class of protein "unfoldases" known as *AAA proteins*. Many of them function as hexamers, and it is possible that they share mechanistic features with the ATP-dependent unwinding of DNA by DNA helicases (see Figure 5–15).

A crucial property of the proteasome, and one reason for the complexity of its design, is the *processivity* of its mechanism: in contrast to a "simple" protease that cleaves a substrate's polypeptide chain just once before dissociating, the proteasome keeps the entire substrate bound until all of it is converted into short peptides.

The 19S caps also act as regulated "gates" at the entrances to the inner proteolytic chamber, and they are responsible for binding a targeted protein substrate to the proteasome. With a few exceptions, the proteasomes act on proteins that have been specifically marked for destruction by the covalent attachment of a recognition tag formed from a small protein called *ubiquitin* (**Figure 6–92**A). Ubiquitin exists in cells either free or covalently linked to

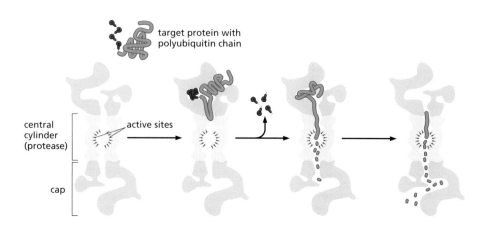

target protein with
polyubiquitin chain

central
cylinder
(protease)   active sites

cap

Figure 6–90 Processive protein digestion by the proteasome. The proteasome cap recognizes a substrate protein, in this case marked by a polyubiquitin chain (see Figure 6–92), and subsequently translocates it into the proteasome core, where it is digested. At an early stage, the ubiquitin is cleaved from the substrate protein and is recycled. Translocation into the core of the proteasome is mediated by a ring of ATP-dependent proteins that unfold the substrate protein as it is threaded through the ring and into the proteasome core (see Figure 6–91). (From S. Prakash and A. Matouschek, *Trends Biochem. Sci.* 29:593–600, 2004. With permission from Elsevier.)

(A)

(B)

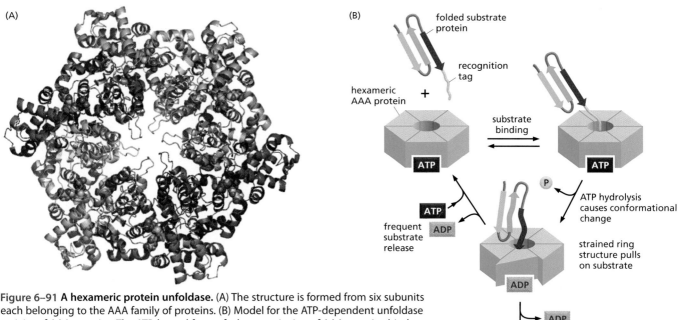

Figure 6–91 A hexameric protein unfoldase. (A) The structure is formed from six subunits each belonging to the AAA family of proteins. (B) Model for the ATP-dependent unfoldase activity of AAA proteins. The ATP-bound form of a hexameric ring of AAA proteins binds a folded substrate protein that has been marked for unfolding (and eventual destruction) by a recognition tag such as a polyubiquitin chain (see below) or the peptide added to mark incompletely synthesized proteins (see Figure 6–81). A conformational change, made irreversible by ATP hydrolysis, pulls the substrate into the central core and strains the ring structure. At this point, the substrate protein, which is being tugged upon, can partially unfold and enter further into the pore or it can maintain its structure and dissociate. Very stable protein substrates may require hundreds of cycles of ATP hydrolysis and dissociation before they are successfully pulled into the AAA ring. Once unfolded, the substrate protein moves relatively quickly through the pore by successive rounds of ATP hydrolysis. (A, from X. Zhang et al., *Mol. Cell* 6:1473–1484, 2000, and A.N. Lupas and J. Martin, *Curr. Opin. Struct. Biol.* 12:746–753, 2002; B, from R.T. Sauer et al., *Cell* 119:9–18, 2004. All with permission from Elsevier.)

many different intracellular proteins. For many proteins, tagging by ubiquitin results in their destruction by the proteasome. However, in other cases, ubiquitin tagging has an entirely different meaning. Ultimately, it is the number of ubiquitin molecules added and the way in which they are linked together that determines how the cell interprets the ubiquitin message (**Figure 6–93**). In the following sections, we emphasize the role of ubiquitylation in signifying protein degradation.

An Elaborate Ubiquitin-Conjugating System Marks Proteins for Destruction

Ubiquitin is prepared for conjugation to other proteins by the ATP-dependent *ubiquitin-activating enzyme* (E1), which creates an activated, E1-bound ubiquitin that is subsequently transferred to one of a set of *ubiquitin-conjugating* (E2) enzymes (Figure 6–92B). The E2 enzymes act in conjunction with accessory (E3) proteins. In the E2–E3 complex, called *ubiquitin ligase*, the E3 component binds to specific degradation signals, called degrons, in protein substrates, helping E2 to form a *polyubiquitin* chain linked to a lysine of the substrate protein. In this chain, the C-terminal residue of each ubiquitin is linked to a specific lysine of the preceding ubiquitin molecule (see Figure 6–93), producing a linear series of ubiquitin–ubiquitin conjugates (Figure 6–92C). It is this polyubiquitin chain on a target protein that is recognized by a specific receptor in the proteasome.

There are roughly 30 structurally similar but distinct E2 enzymes in mammals, and hundreds of different E3 proteins that form complexes with specific E2 enzymes. The ubiquitin–proteasome system thus consists of many distinct but similarly organized proteolytic pathways, which have in common both the

E1 enzyme at the "top" and the proteasome at the "bottom," and differ by the compositions of their E2–E3 ubiquitin ligases and accessory factors. Distinct ubiquitin ligases recognize different degradation signals, and therefore target distinct subsets of intracellular proteins for destruction.

Denatured or otherwise misfolded proteins, as well as proteins containing oxidized or other abnormal amino acids, are recognized and destroyed because abnormal proteins tend to present on their surface amino acid sequences or conformational motifs that are recognized as degradation signals by a set of E3 molecules in the ubiquitin–proteasome system; these sequences must of course be buried and therefore inaccessible in the normal counterparts of these proteins. However, a proteolytic pathway that recognizes and destroys abnormal proteins must be able to distinguish between *completed* proteins that have "wrong" conformations and the many growing polypeptides on ribosomes (as well as polypeptides just released from ribosomes) that have not yet achieved their normal folded conformation. This is not a trivial problem; the ubiquitin–proteasome system is thought to destroy many of the nascent and newly formed protein molecules not because these proteins are abnormal as such, but because they transiently expose degradation signals that are buried in their mature (folded) state.

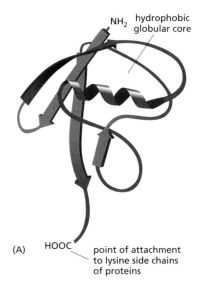

Figure 6–92 Ubiquitin and the marking of proteins with polyubiquitin chains. (A) The three-dimensional structure of ubiquitin; this relatively small protein contains 76 amino acids. (B) The C-terminus of ubiquitin is initially activated through its high-energy thioester linkage to a cysteine side chain on the E1 protein. This reaction requires ATP, and it proceeds via a covalent AMP-ubiquitin intermediate. The activated ubiquitin on E1, also known as the ubiquitin-activating enzyme, is then transferred to the cysteines on a set of E2 molecules. These E2s exist as complexes with an even larger family of E3 molecules. (C) The addition of a polyubiquitin chain to a target protein. In a mammalian cell there are several hundred distinct E2–E3 complexes, many of which recognize a specific degradation signal on target proteins by means of the E3 component. The E2s are called ubiquitin-conjugating enzymes. The E3s have been referred to traditionally as ubiquitin ligases, but it is more accurate to reserve this name for the functional E2–E3 complex. The detailed structure of such a complex is presented in Figure 3–79.

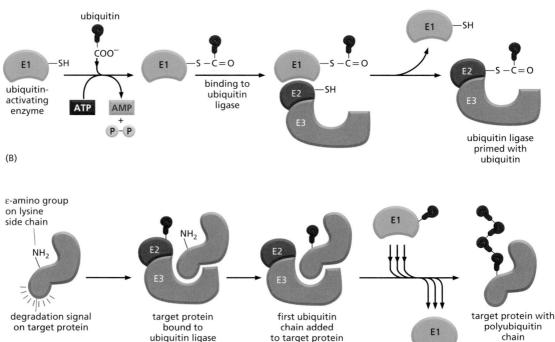

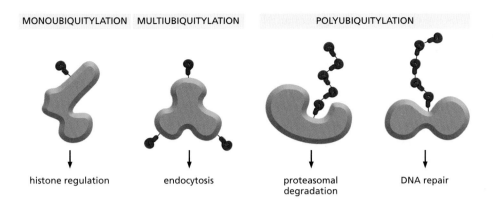

MONOUBIQUITYLATION   MULTIUBIQUITYLATION   POLYUBIQUITYLATION

histone regulation   endocytosis   proteasomal   DNA repair
                                             degradation

Figure 6–93 **The marking of proteins by ubiquitin.** Each modification pattern shown can have a specific meaning to the cell. The two types of polyubiquitylation differ in the way the ubiquitin molecules are linked together. Linkage through Lys48 signifies degradation by the proteasome whereas that through Lys63 has other meanings. Ubiquitin markings are "read" by proteins that specifically recognize each type of modification.

Many Proteins Are Controlled by Regulated Destruction

One function of intracellular proteolytic mechanisms is to recognize and eliminate misfolded or otherwise abnormal proteins, as just described. Yet another function of these proteolytic pathways is to confer short lifetimes on specific normal proteins whose concentrations must change promptly with alterations in the state of a cell. Some of these short-lived proteins are degraded rapidly at all times, while many others are *conditionally* short-lived, that is, they are metabolically stable under some conditions but become unstable upon a change in the cell's state. For example, mitotic cyclins are long-lived throughout the cell cycle until their sudden degradation at the end of mitosis, as explained in Chapter 17.

How is such a regulated destruction of a protein controlled? Several mechanisms are illustrated through specific examples that appear later in this book. In one general class of mechanism (**Figure 6–94**A), the activity of a ubiquitin ligase is turned on either by E3 phosphorylation or by an allosteric transition in an E3 protein caused by its binding to a specific small or large molecule. For example, the anaphase-promoting complex (APC) is a multisubunit ubiquitin ligase that is activated by a cell-cycle-timed subunit addition at mitosis. The activated APC then causes the degradation of mitotic cyclins and several other regulators of the metaphase–anaphase transition (see Figure 17–44).

Alternatively, in response either to intracellular signals or to signals from the environment, a degradation signal can be created in a protein, causing its rapid ubiquitylation and destruction by the proteasome. One common way to create such a signal is to phosphorylate a specific site on a protein that unmasks a normally hidden degradation signal. Another way to unmask such a signal is by the regulated dissociation of a protein subunit. Finally, powerful degradation signals can be created by cleaving a single peptide bond, provided that this cleavage creates a new N-terminus that is recognized by a specific E3 as a "destabilizing" N-terminal residue (Figure 6–94B).

The N-terminal type of degradation signal arises because of the "N-end rule," which relates the lifetime of a protein *in vivo* to the identity of its N-terminal residue. There are 12 destabilizing residues in the N-end rule of the yeast *S. cerevisiae* (Arg, Lys, His, Phe, Leu, Tyr, Trp, Ile, Asp, Glu, Asn, and Gln), out of the 20 standard amino acids. The destabilizing N-terminal residues are recognized by a special ubiquitin ligase that is conserved from yeast to humans.

As we have seen, all proteins are initially synthesized bearing methionine (or formylmethionine in bacteria), as their N-terminal residue, which is a stabilizing residue in the N-end rule. Special proteases, called methionine aminopeptidases, will often remove the first methionine of a nascent protein, but they will do so only if the second residue is also stabilizing according to N-end rule. Therefore, it was initially unclear how N-end rule substrates form *in vivo*. However, it is now understood that these substrates are formed by site-specific proteases. For example, a subunit of cohesin, a protein complex that holds sister chromatids together, is cleaved by a highly specific protease during the metaphase–anaphase transition. This cell-cycle-regulated cleavage allows separation of the sister chromatids and leads to the completion of mitosis (see

Figure 17–44). The C-terminal fragment of the cleaved subunit bears an N-terminal arginine, a destabilizing residue in the N-end rule. Mutant cells lacking the N-end rule pathway exhibit a greatly increased frequency of chromosome loss, presumably because a failure to degrade this fragment of the cohesin subunit interferes with the formation of new chromatid-associated cohesin complexes in the next cell cycle.

Abnormally Folded Proteins Can Aggregate to Cause Destructive Human Diseases

Many inherited human diseases (for example, sickle-cell anemia (see p. 1495) and α-1-antitrypsin deficiency, a condition that often leads to liver disease and emphysema) result from mutant proteins that escape the cell's quality controls, fold abnormally, and form aggregates. By absorbing critical macromolecules, these aggregates can severely damage cells and even cause cell death. Often, the inheritance of a single mutant allele of a gene can cause disease, since the normal copy of the gene cannot protect the cell from the destructive properties of the aggregate.

In normal humans, the gradual decline of the cell's protein quality controls can also cause disease by permitting normal proteins to form aggregates (**Figure 6–95**). In some cases, the protein aggregates are released from dead cells and accumulate in the extracellular matrix that surrounds the cells in a tissue, and in extreme cases they can also damage tissues. Because the brain is composed of a highly organized collection of nerve cells, it is especially vulnerable.

(A)   ACTIVATION OF A UBIQUITIN LIGASE

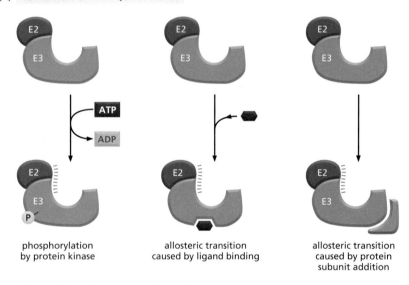

phosphorylation
by protein kinase

allosteric transition
caused by ligand binding

allosteric transition
caused by protein
subunit addition

(B)   ACTIVATION OF A DEGRADATION SIGNAL

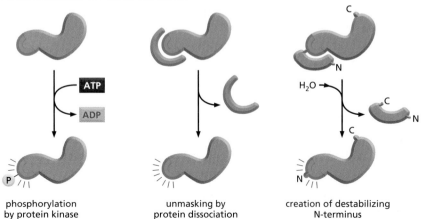

phosphorylation
by protein kinase

unmasking by
protein dissociation

creation of destabilizing
N-terminus

Figure 6–94 Two general ways of inducing the degradation of a specific protein. (A) Activation of a specific E3 molecule creates a new ubiquitin ligase. (B) Creation of an exposed degradation signal in the protein to be degraded. This signal binds a ubiquitin ligase, causing the addition of a polyubiquitin chain to a nearby lysine on the target protein. All six pathways shown are known to be used by cells to induce the movement of selected proteins into the proteasome.

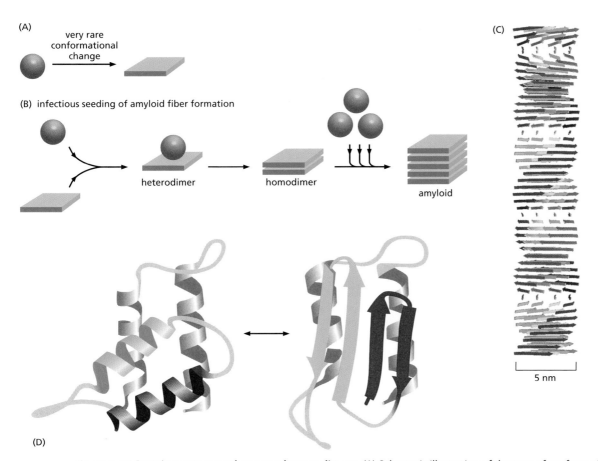

Figure 6–95 Protein aggregates that cause human disease. (A) Schematic illustration of the type of conformational change in a protein that produces material for a cross-beta filament. (B) Diagram illustrating the self-infectious nature of the protein aggregation that is central to prion diseases. PrP (prion protein) is highly unusual because the misfolded version of the protein, called PrP*, induces the normal PrP protein it contacts to change its conformation, as shown. Most of the human diseases caused by protein aggregation are caused by the overproduction of a variant protein that is especially prone to aggregation, but the protein aggregate cannot spread from one animal to another. (C) Drawing of a cross-beta filament, a common type of protease-resistant protein aggregate found in many human neurological diseases. Because the hydrogen-bond interactions in a β sheet form between polypeptide backbone atoms (see Figure 3–9), a number of different abnormally folded proteins can produce this structure. (D) One of several possible models for the conversion of PrP to PrP*, showing the likely change of two α helices into four β strands. Although the structure of the normal protein has been determined accurately, the structure of the infectious form is not yet known with certainty because the aggregation has prevented the use of standard structural techniques. (C, courtesy of Louise Serpell, adapted from M. Sunde et al., *J. Mol. Biol.* 273:729–739, 1997. With permission from Academic Press; D, adapted from S.B. Prusiner, *Trends Biochem. Sci.* 21:482–487, 1996. With permission from Elsevier.)

Not surprisingly, therefore, protein aggregates primarily cause neurodegenerative diseases. Prominent among these are Huntington's disease and Alzheimer's disease—the latter causing age-related dementia in more than 20 million people in today's world.

For a particular type of protein aggregate to survive, grow, and damage an organism, it must be highly resistant to proteolysis both inside and outside the cell. Many of the protein aggregates that cause problems form fibrils built from a series of polypeptide chains that are layered one over the other as a continuous stack of β sheets. This so-called *cross-beta filament* (Figure 6–95C), a structure particularly resistant to proteolysis, is observed in many of the neurological disorders caused by protein aggregates, where it produces distinctly staining deposits known as *amyloids*.

One particular variety of these pathologies has attained special notoriety. These are the **prion diseases**. Unlike Huntington's or Alzheimer's, prion diseases can spread from one organism to another, providing that the second organism eats a tissue containing the protein aggregate. A set of diseases—called scrapie in sheep, Creutzfeldt–Jacob disease (CJD) in humans, and bovine spongiform encephalopathy (BSE) in cattle—are caused by a misfolded, aggregated form of a protein called PrP (for prion protein). The PrP is normally located on the outer surface of the plasma membrane, most prominently in neurons. Its normal

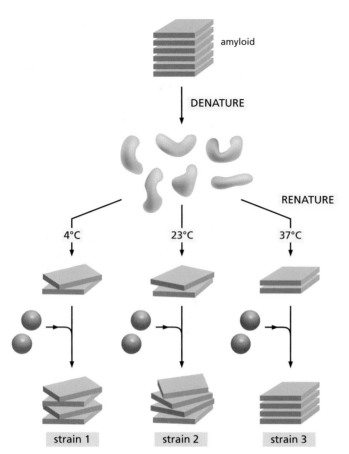

amyloid

DENATURE

RENATURE

4°C   23°C   37°C

strain 1   strain 2   strain 3

Figure 6–96 Creation of different prion strains *in vitro*. In this experiment, amyloid fibers were denatured and the components renatured at different temperatures. This treatment produced three distinctive types of amyloids, each of which could self-propagate when new subunits are added.

function is not known. However, PrP has the unfortunate property of being convertible to a very special abnormal conformation (see Figure 6–95A). This conformation not only forms protease-resistant, cross-beta filaments; it is also "infectious" because it converts normally folded molecules of PrP to the same pathological form. This property creates a positive feedback loop that propagates the abnormal form of PrP, called PrP* (see Figure 6–95B) and thereby allows the pathological conformation to spread rapidly from cell to cell in the brain, eventually causing death in both animals and humans. It can be dangerous to eat the tissues of animals that contain PrP*, as witnessed by the spread of BSE (commonly referred to as "mad cow disease") from cattle to humans in Great Britain. Fortunately, in the absence of PrP*, PrP is extraordinarily difficult to convert to its abnormal form.

Although very few proteins have the potential to misfold into an infectious conformation, another example causes an otherwise mysterious "protein-only inheritance" observed in yeast cells. The ability to study infectious proteins in yeast has clarified another remarkable feature of prions. These protein molecules can form several distinctively different types of aggregates from the same polypeptide chain. Moreover, each type of aggregate can be infectious, forcing normal protein molecules to adopt the same type of abnormal structure. Thus, several different "strains" of infectious particles can arise from the same polypeptide chain (**Figure 6–96**). How a single polypeptide sequence can adopt multiple aggregate forms is not fully understood; it is possible that all prion aggregates resemble cross-beta filaments (see Figure 6–95C) where the structure is held together predominantly with main peptide chain interactions. This would leave the amino acid side chains free to adopt different conformations and, if the structures are self-propagating, the existence of different strains could be explained.

Finally, although prions were discovered because they cause disease, they also appear to have some positive roles in the cell. For example, some species of fungi use prion transformations to establish different types of cells. Although the idea is controversial, it has even been proposed that prions have a role in consolidating memories in complex, multicellular organisms like ourselves.

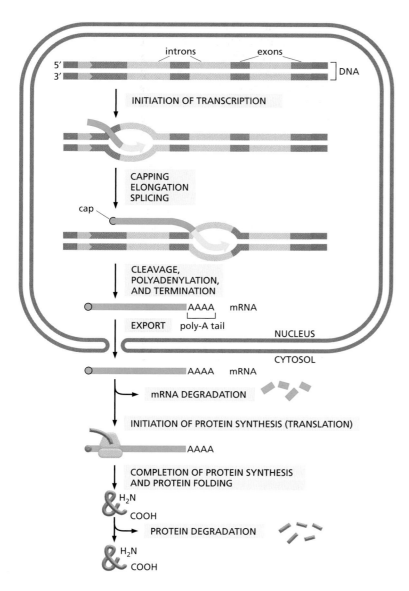

Figure 6–97 The production of a protein by a eucaryotic cell. The final level of each protein in a eucaryotic cell depends upon the efficiency of each step depicted.

There Are Many Steps From DNA to Protein

We have seen so far in this chapter that many different types of chemical reactions are required to produce a properly folded protein from the information contained in a gene (**Figure 6–97**). The final level of a properly folded protein in a cell therefore depends upon the efficiency with which each of the many steps is performed.

In the following chapter, we shall see that cells have the ability to change the levels of their proteins according to their needs. In principle, any or all of the steps in Figure 6–97 could be regulated for each individual protein. As we shall see in Chapter 7, there are examples of regulation at each step from gene to protein. However, the initiation of transcription is the most common point for a cell to regulate the expression of each of its genes. This makes sense, inasmuch as the most efficient way to keep a gene from being expressed is to block the very first step—the transcription of its DNA sequence into an RNA molecule.

Summary

The translation of the nucleotide sequence of an mRNA molecule into protein takes place in the cytoplasm on a large ribonucleoprotein assembly called a ribosome. The amino acids used for protein synthesis are first attached to a family of tRNA molecules, each of which recognizes, by complementary base-pair interactions, particular sets of three nucleotides in the mRNA (codons). The sequence of nucleotides in the mRNA is then read from one end to the other in sets of three according to the genetic code.

To initiate translation, a small ribosomal subunit binds to the mRNA molecule at a start codon (AUG) that is recognized by a unique initiator tRNA molecule. A large ribosomal subunit binds to complete the ribosome and begin protein synthesis. During this phase, aminoacyl-tRNAs—each bearing a specific amino acid—bind sequentially to the appropriate codons in mRNA through complementary base pairing between tRNA anticodons and mRNA codons. Each amino acid is added to the C-terminal end of the growing polypeptide in four sequential steps: aminoacyl-tRNA binding, followed by peptide bond formation, followed by two ribosome translocation steps. Elongation factors use GTP hydrolysis to drive these reactions forward and to improve the accuracy of amino acid selection. The mRNA molecule progresses codon by codon through the ribosome in the 5′-to-3′ direction until it reaches one of three stop codons. A release factor then binds to the ribosome, terminating translation and releasing the completed polypeptide.

Eucaryotic and bacterial ribosomes are closely related, despite differences in the number and size of their rRNA and protein components. The rRNA has the dominant role in translation, determining the overall structure of the ribosome, forming the binding sites for the tRNAs, matching the tRNAs to codons in the mRNA, and creating the active site of the peptidyl transferase enzyme that links amino acids together during translation.

In the final steps of protein synthesis, two distinct types of molecular chaperones guide the folding of polypeptide chains. These chaperones, known as Hsp60 and Hsp70, recognize exposed hydrophobic patches on proteins and serve to prevent the protein aggregation that would otherwise compete with the folding of newly synthesized proteins into their correct three-dimensional conformations. This protein folding process must also compete with an elaborate quality control mechanism that destroys proteins with abnormally exposed hydrophobic patches. In this case, ubiquitin is covalently added to a misfolded protein by a ubiquitin ligase, and the resulting polyubiquitin chain is recognized by the cap on a proteasome that moves the entire protein to the interior of the proteasome for proteolytic degradation. A closely related proteolytic mechanism, based on special degradation signals recognized by ubiquitin ligases, is used to determine the lifetimes of many normally folded proteins. By this method, selected normal proteins are removed from the cell in response to specific signals.

THE RNA WORLD AND THE ORIGINS OF LIFE

We have seen that the expression of hereditary information requires extraordinarily complex machinery and proceeds from DNA to protein through an RNA intermediate. This machinery presents a central paradox: if nucleic acids are required to synthesize proteins and proteins are required, in turn, to synthesize nucleic acids, how did such a system of interdependent components ever arise? One view is that an *RNA world* existed on Earth before modern cells arose (**Figure 6–98**). According to this hypothesis, RNA both stored genetic information and catalyzed the chemical reactions in primitive cells. Only later in evolutionary time did DNA take over as the genetic material and proteins become the major catalyst and structural component of cells. If this idea is correct, then the transition out of the RNA world was never complete; as we have seen in this chapter, RNA still catalyzes several fundamental reactions in modern-day cells, which can be viewed as molecular fossils of an earlier world.

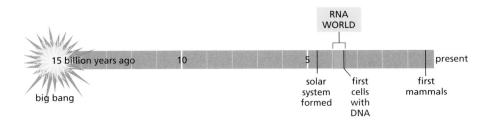

Figure 6–98 Time line for the universe, suggesting the early existence of an RNA world of living systems.

In this section we present some of the arguments in support of the RNA world hypothesis. We will see that several of the more surprising features of modern-day cells, such as the ribosome and the pre-mRNA splicing machinery, are most easily explained by viewing them as descendants of a complex network of RNA-mediated interactions that dominated cell metabolism in the RNA world. We also discuss how DNA may have taken over as the genetic material, how the genetic code may have arisen, and how proteins may have eclipsed RNA to perform the bulk of biochemical catalysis in modern-day cells.

Life Requires Stored Information

It has been proposed that the first "biological" molecules on Earth were formed by metal-based catalysis on the crystalline surfaces of minerals. In principle, an elaborate system of molecular synthesis and breakdown (metabolism) could have existed on these surfaces long before the first cells arose. Although controversial, many scientists believe that an extensive phase of "chemical evolution" took place on the prebiotic Earth, during which small molecules that could catalyze their own synthesis competed with each other for raw materials.

But life requires much more than this. As described in Chapter 1, *heredity* is perhaps the central feature of life. Not only must a cell use raw materials to create a network of catalyzed reactions, it must do so according to an elaborate set of instructions encoded in the hereditary information. The replication of this information ensures that the complex metabolism of cells can accurately reproduce itself. Another crucial feature of life is the genetic variability that results from changes in the hereditary information. This variability, acted upon by selective pressures, is responsible for the great diversity of life on our planet.

Thus, the emergence of life requires a way to store information, a way to duplicate it, a way to change it, and a way to convert the information through catalysis into favorable chemical reactions. But how could such a system begin to be formed? In present-day cells the most versatile catalysts are polypeptides, composed of many different amino acids with chemically diverse side chains and, consequently, able to adopt diverse three-dimensional forms that bristle with reactive chemical groups. Polypeptides also carry information, in the order of their amino acid subunits. But there is no known way in which a polypeptide can reproduce itself by directly specifying the formation of another of precisely the same sequence.

Polynucleotides Can Both Store Information and Catalyze Chemical Reactions

Polynucleotides have one property that contrasts with those of polypeptides: they can directly guide the formation of copies of their own sequence. This capacity depends on complementary base pairing of nucleotide subunits, which enables one polynucleotide to act as a template for the formation of another. As we have seen in this and the preceding chapter, such complementary templating mechanisms lie at the heart of DNA replication and transcription in modern-day cells.

But the efficient synthesis of polynucleotides by such complementary templating mechanisms requires catalysts to promote the polymerization reaction: without catalysts, polymer formation is slow, error-prone, and inefficient. Today, template-based nucleotide polymerization is rapidly catalyzed by protein enzymes—such as the DNA and RNA polymerases. How could such polymerization be catalyzed before proteins with the appropriate enzymatic specificity existed? The beginnings of an answer to this question came from the discovery in 1982 that RNA molecules themselves can act as catalysts. We have seen in this chapter, for example, that a molecule of RNA catalyzes one of the central reactions in the cell, the covalent joining of amino acids to form proteins. The unique potential of RNA molecules to act both as information carrier and as catalyst forms the basis of the RNA world hypothesis.

Figure 6–99 An RNA molecule that can catalyze its own synthesis. This hypothetical process would require catalysis of the production of both a second RNA strand of complementary nucleotide sequence and the use of this second RNA molecule as a template to form many molecules of RNA with the original sequence. The *red* rays represent the active site of this hypothetical RNA enzyme.

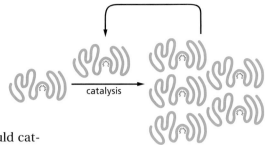

RNA therefore has all the properties required of a molecule that could catalyze a variety of chemical reactions, including those that lead to its own synthesis (**Figure 6–99**). Although self-replicating systems of RNA molecules have not been found in nature, scientists are confident that they can be constructed in the laboratory. While this demonstration would not prove that self-replicating RNA molecules were essential in the origin of life on Earth, it would certainly indicate that such a scenario is possible.

A Pre-RNA World May Predate the RNA World

Although RNA seems well suited to form the basis for a self-replicating set of biochemical catalysts, it is not clear that RNA was the first kind of molecule to do so. From a purely chemical standpoint, it is difficult to imagine how long RNA molecules could be formed initially by purely nonenzymatic means. For one thing, the precursors of RNA, the ribonucleotides, are difficult to form nonenzymatically. Moreover, the formation of RNA requires that a long series of 3′-to-5′ phosphodiester linkages assemble in the face of a set of competing reactions, including hydrolysis, 2′-to-5′ linkages, and 5′-to-5′ linkages. Given these problems, it has been suggested that the first molecules to possess both catalytic activity and information storage capabilities may have been polymers that resemble RNA but are chemically simpler (**Figure 6–100**). We do not have any remnants of these compounds in present-day cells, nor do such compounds leave fossil records. Nonetheless, the relative simplicity of these "RNA-like polymers" suggests that one of them, rather than RNA itself, may have been the first biopolymer on Earth capable of both information storage and catalytic activity.

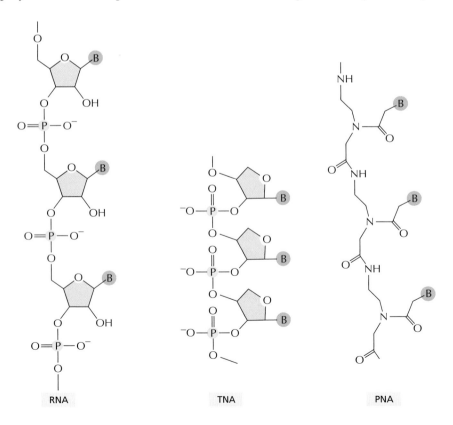

Figure 6–100 Structures of RNA and two related information-carrying polymers. In each case, B indicates a purine or pyrimidine base. The polymer TNA (threose nucleic acid) has a 4-carbon sugar unit in contrast to the 5-carbon ribose in RNA. In PNA (peptide nucleic acid), the ribose phosphate backbone of RNA has been replaced by the peptide backbone found in proteins. Like RNA, TNA and PNA can form double helices through complementary base-pairing, and each could therefore in principle serve as a template for its own synthesis.

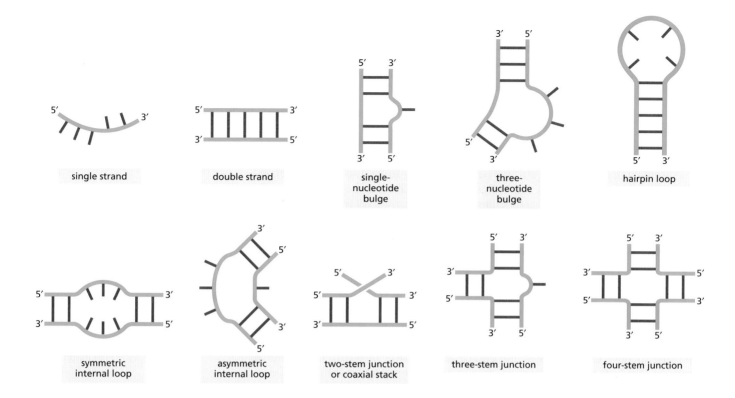

Figure 6–101 **Common elements of RNA secondary structure.** Conventional, complementary base-pairing interactions are indicated by *red* "rungs" in double-helical portions of the RNA.

If the pre-RNA world hypothesis is correct, then a transition to the RNA world must have occurred, presumably through the synthesis of RNA using one of these simpler polymers as both template and catalyst. While the details of the pre-RNA and RNA worlds will likely remain unknown, we know for certain that RNA molecules can catalyze a wide variety of chemical reactions, and we now turn to the properties of RNA that make this possible.

Single-Stranded RNA Molecules Can Fold into Highly Elaborate Structures

We have seen that complementary base-pairing and other types of hydrogen bonds can occur between nucleotides in the same chain, causing an RNA molecule to fold up in a unique way determined by its nucleotide sequence (see, for example, Figures 6–6, 6–52, and 6–69). Comparisons of many RNA structures have revealed conserved motifs, short structural elements that are used over and over again as parts of larger structures. **Figure 6–101** shows some of these RNA secondary structural motifs, and **Figure 6–102** shows a few common examples of more complex and often longer-range interactions, known as RNA tertiary interactions.

Figure 6–102 **Examples of RNA tertiary interactions.** Some of these interactions can join distant parts of the same RNA molecule or bring two separate RNA molecules together.

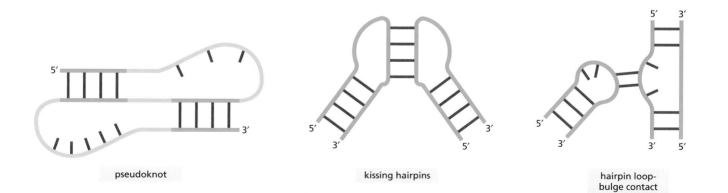

Figure 6–103 A ribozyme. This simple RNA molecule catalyzes the cleavage of a second RNA at a specific site. This ribozyme is found embedded in larger RNA genomes—called viroids—which infect plants. The cleavage, which occurs in nature at a distant location on the same RNA molecule that contains the ribozyme, is a step in the replication of the viroid genome. Although not shown in the figure, the reaction requires a Mg molecule positioned at the active site. (Adapted from T.R. Cech and O.C. Uhlenbeck, *Nature* 372:39–40, 1994. With permission from Macmillan Publishers Ltd.)

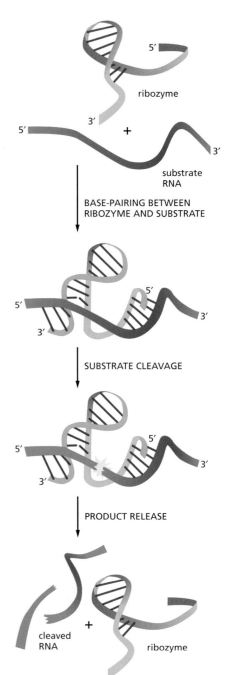

Protein catalysts require a surface with unique contours and chemical properties on which a given set of substrates can react (discussed in Chapter 3). In exactly the same way, an RNA molecule with an appropriately folded shape can serve as an enzyme (**Figure 6–103**). Like some proteins, many of these ribozymes work by positioning metal ions at their active sites. This feature gives them a wider range of catalytic activities than the limited chemical groups of the polynucleotide chain.

Relatively few catalytic RNAs are known to exist in modern-day cells, however, and much of our inference about the RNA world has come from experiments in which large pools of RNA molecules of random nucleotide sequences are generated in the laboratory. Those rare RNA molecules with a property specified by the experimenter are then selected out and studied (**Figure 6–104**). Such experiments have created RNAs that can catalyze a wide variety of biochemical reactions (**Table 6–5**), with reaction rate enhancements approaching those of proteins. Given these findings, it is not clear why protein catalysts greatly outnumber ribozymes in modern cells. Experiments have shown that RNA molecules may have more difficulty than proteins in binding to flexible, hydrophobic substrates; moreover, the availability of 20 types of amino acids over four types of bases may provide proteins with a greater number of catalytic strategies.

Like proteins, RNAs can undergo conformational changes, either in response to small molecules or to other RNAs. We saw several examples of this in the ribosome and the spliceosome, and we will see others in Chapter 7 when we discuss *riboswitches*. One of the most dramatic RNA conformational changes has been observed with an artificial ribozyme which can exist in two entirely different conformations, each with a different catalytic activity (**Figure 6–105**). Since the discovery of catalysis by RNA, it has become clear that RNA is an enormously versatile molecule, and it is therefore not unreasonable to contemplate the past existence of an RNA world with a very high level of biochemical sophistication.

Self-Replicating Molecules Undergo Natural Selection

The three-dimensional folded structure of a polynucleotide affects its stability, its actions on other molecules, and its ability to replicate. Therefore, certain polynucleotides will be especially successful in any self-replicating mixture. Because errors inevitably occur in any copying process, new variant sequences of these polynucleotides will be generated over time.

Certain catalytic activities would have had a cardinal importance in the early evolution of life. Consider in particular an RNA molecule that helps to catalyze the process of templated polymerization, taking any given RNA molecule as a template (**Figure 6–106**). Such a molecule, by acting on copies of itself, can replicate. At the same time, it can promote the replication of other types of RNA molecules in its neighborhood (**Figure 6–107**). If some of these neighboring RNAs have catalytic actions that help the survival of RNA in other ways (catalyzing ribonucleotide production, for example), a set of different types of RNA molecules, each specialized for a different activity, could evolve into a cooperative system that replicates with unusually great efficiency.

But for any of these cooperative systems to evolve, they must be present together in a compartment. For example, a set of mutually beneficial RNAs (such

Figure 6–104 *In vitro* selection of a synthetic ribozyme. Beginning with a large pool of nucleic acid molecules synthesized in the laboratory, those rare RNA molecules that possess a specified catalytic activity can be isolated and studied. Although a specific example (that of an autophosphorylating ribozyme) is shown, variations of this procedure have been used to generate many of the ribozymes listed in Table 6–5. During the autophosphorylation step, the RNA molecules are kept sufficiently dilute to prevent the "cross"-phosphorylation of additional RNA molecules. In reality, several repetitions of this procedure are necessary to select the very rare RNA molecules with this catalytic activity. Thus, the material initially eluted from the column is converted back into DNA, amplified many fold (using reverse transcriptase and PCR as explained in Chapter 8), transcribed back into RNA, and subjected to repeated rounds of selection. (Adapted from J.R. Lorsch and J.W. Szostak, *Nature* 371:31–36, 1994. With permission from Macmillan Publishers Ltd.)

large pool of double-stranded DNA molecules, each with a different, randomly generated nucleotide sequence

TRANSCRIPTION BY RNA POLYMERASE AND FOLDING OF RNA MOLECULES

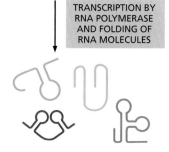

large pool of single-stranded RNA molecules, each with a different, randomly generated nucleotide sequence

ATP γS — ADDITION OF ATP DERIVATIVE CONTAINING A SULFUR IN PLACE OF AN OXYGEN

ADP

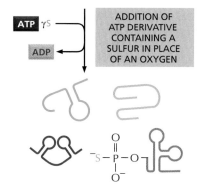

only the rare RNA molecules able to phosphorylate themselves incorporate sulfur

CAPTURE OF PHOSPHORYLATED MATERIAL ON COLUMN MATERIAL THAT BINDS TIGHTLY TO THE SULFUR GROUP

discard RNA molecules that fail to bind to the column

ELUTION OF BOUND MOLECULES

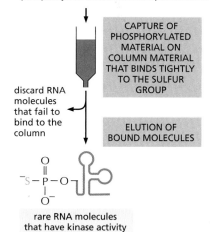

rare RNA molecules that have kinase activity

as those of Figure 6–107) could replicate themselves only if all the RNAs remained in the neighborhood of the RNA that is specialized for templated polymerization. Moreover, compartmentalization would bar parasitic RNA molecules from entering the system. Selection of a set of RNA molecules according to the quality of the self-replicating systems they generated could not therefore occur efficiently until some form of compartment evolved to contain them.

An early, crude form of compartmentalization may have been simple adsorption on surfaces or particles. The need for more sophisticated types of containment is easily fulfilled by a class of small molecules that has the simple physicochemical property of being *amphiphilic*, that is, consisting of one part that is hydrophobic (water insoluble) and another part that is hydrophilic (water soluble). When such molecules are placed in water, they aggregate, arranging their hydrophobic portions as much in contact with one another as possible and their hydrophilic portions in contact with the water. Amphiphilic molecules of appropriate shape aggregate spontaneously to form *bilayers*, creating small closed vesicles whose aqueous contents are isolated from the external medium (**Figure 6–108**). The phenomenon can be demonstrated in a test tube by simply mixing phospholipids and water together: under appropriate conditions, small vesicles will form. All present-day cells are surrounded by a *plasma membrane* consisting of amphiphilic molecules—mainly phospholipids—in this configuration; we discuss these molecules in detail in Chapter 10.

The spontaneous assembly of a set of amphiphilic molecules, enclosing a self-replicating mixture of RNAs (or pre-RNAs) and other molecules (**Figure**

Table 6–5 Some Biochemical Reactions That Can Be Catalyzed by Ribozymes

ACTIVITY	RIBOZYMES
Peptide bond formation in protein synthesis	ribosomal RNA
RNA cleavage, RNA ligation	self-splicing RNAs; RNase P; also *in vitro* selected RNA
DNA cleavage	self-splicing RNAs
RNA splicing	self-splicing RNAs, perhaps RNAs of the spliceosome
RNA polymerizaton	*in vitro* selected RNA
RNA and DNA phosphorylation	*in vitro* selected RNA
RNA aminoacylation	*in vitro* selected RNA
RNA alkylation	*in vitro* selected RNA
Amide bond formation	*in vitro* selected RNA
Glycosidic bond formation	*in vitro* selected RNA
Oxidation/reduction reactions	*in vitro* selected RNA
Carbon–carbon bond formation	*in vitro* selected RNA
Phosphoamide bond formation	*in vitro* selected RNA
Disulfide exchange	*in vitro* selected RNA

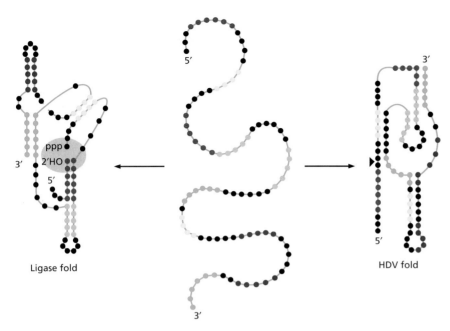

Ligase fold

HDV fold

Figure 6–105 An RNA molecule that folds into two different ribozymes. This 88-nucleotide RNA, created in the laboratory, can fold into a ribozyme that carries out a self-ligation reaction *(left)* or a self-cleavage reaction *(right)*. The ligation reaction forms a 2′,5′ phosphodiester linkage with the release of pyrophosphate. This reaction seals the gap *(gray shading)*, which was experimentally introduced into the RNA molecule. In the reaction carried out by the HDV fold, the RNA is cleaved at this same position, indicated by the *arrowhead*. This cleavage resembles that used in the life cycle of HDV, a hepatitis B satellite virus, hence the name of the fold. Each nucleotide is represented by a *colored dot*, with the colors used simply to clarify the two different folding patterns. The folded structures illustrate the secondary structures of the two ribozymes with regions of base-pairing indicated by close oppositions of the *colored dots*. Note that the two ribozyme folds have no secondary structure in common. (Adapted from E.A. Schultes and D.P. Bartel, *Science* 289:448–452, 2000. With permission from AAAS.)

6–109), presumably formed the first membrane-bounded cells. Although it is not clear at what point in the evolution of biological catalysts this might have occurred, once RNA molecules were sealed within a closed membrane they could begin to evolve in earnest as carriers of genetic instructions: new variants could be selected not merely on the basis of their own structure, but also according to their effect on the other molecules in the same compartment. The nucleotide sequences of the RNA molecules could now be expressed in the character of a unitary living cell.

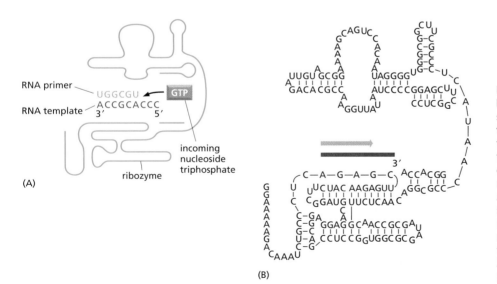

(A)

(B)

Figure 6–106 A ribozyme created in the laboratory that can catalyze templated synthesis of RNA from nucleoside triphosphates. (A) Schematic diagram of the ribozyme showing one step of the templated polymerization reaction it catalyzes. (B) Nucleotide sequence of the ribozyme with base pairings indicated. Although relatively inefficient (it can only synthesize short lengths of RNA), this ribozyme adds the correct base, as specified by the template, over 95% of the time. (From W.K. Johnston et al., *Science* 292:1319–1325, 2001. With permission from AAAS.)

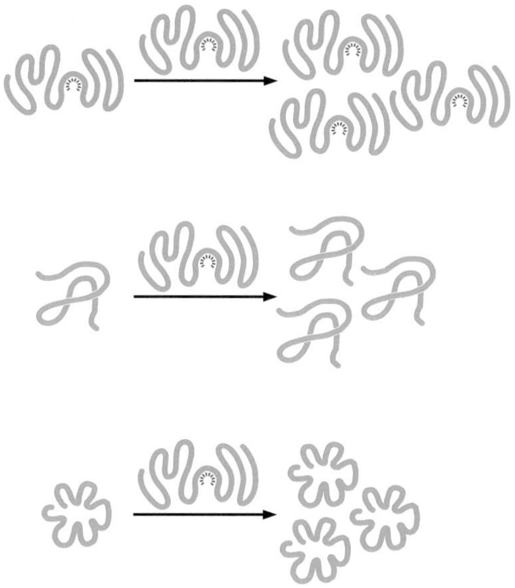

Figure 6–107 **A family of mutually supportive RNA molecules.** One molecule is a ribozyme that replicates itself as well as the other RNA molecules. The other molecules would catalyze secondary tasks needed for the survival of the cooperative system, for example, by synthesizing ribonucleotides for RNA synthesis or phospholipids for compartmentalization.

How Did Protein Synthesis Evolve?

The molecular processes underlying protein synthesis in present-day cells seem inextricably complex. Although we understand most of them, they do not make conceptual sense in the way that DNA transcription, DNA repair, and DNA replication do. It is especially difficult to imagine how protein synthesis evolved because it is now performed by a complex interlocking system of protein and RNA molecules; obviously the proteins could not have existed until an early version of the translation apparatus was already in place. The RNA world hypothesis is especially appealing because the use of RNA in both information and catalysis seems both economic and conceptually simple. As attractive as this idea is for envisioning early life, it does not explain how the modern-day system of protein synthesis arose. Although we can only speculate on the origins of modern protein synthesis and the genetic code, several experimental observations have provided plausible scenarios.

In modern cells, some short peptides (such as antibiotics) are synthesized without the ribosome; peptide synthetase enzymes assemble these peptides, with their proper sequence of amino acids, without mRNAs to guide their synthesis. It is plausible that this non-coded, primitive version of protein synthesis first developed during the RNA world where it would have been catalyzed by RNA molecules. This idea presents no conceptual difficulties because, as we have seen, rRNA catalyzes peptide bond formation in present-day cells. We also know that ribozymes created in the laboratory can perform specific aminoacylation reactions; that is, they can match specific amino acids to specific tRNAs. It is therefore possible that tRNA-like adapters, each matched to a specific amino acid, could have arisen in the RNA world, marking the beginnings of a genetic code.

In principle, other RNAs (the precursors to mRNAs) could have served as crude templates to direct the nonrandom polymerization of a few different amino acids. Any RNA that helped guide the synthesis of a useful polypeptide would have a great advantage in the evolutionary struggle for survival. We can envision a relatively nonspecific peptidyl transferase ribozyme, which, over time, grew larger and acquired the ability to position charged tRNAs accurately on RNA templates—leading eventually to the modern ribosome. Once protein

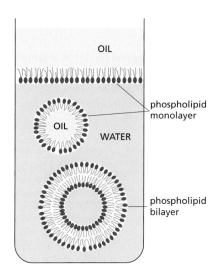

Figure 6–108 **Formation of membrane by phospholipids.** Because these molecules have hydrophilic heads and lipophilic tails, they align themselves at an oil/water interface with their heads in the water and their tails in the oil. In the water they associate to form closed bilayer vesicles in which the lipophilic tails are in contact with one another and the hydrophilic heads are exposed to the water.

Figure 6–109 Encapsulation of RNA by simple amphiphilic molecules. For these experiments, the clay mineral montmorillonite was used to bring together RNA and fatty acids. (A) A montmorillonite particle, coated by RNA *(red)* has become trapped inside a fatty acid vesicle *(green)*. (B) RNA *(red)* in solution has been encapsulated by fatty acids *(green)*. These experiments show that montmorillonite can greatly accelerate the spontaneous generation of vesicles from amphiphilic molecules and trap RNA inside them. It has been hypothesized that conceptually similar actions may have led to the first primitive cells on Earth. (From M.M. Hanczyc et al., *Science* 302:618–622, 2003. With permission from AAAS.)

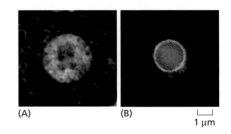

(A)   (B)   1 μm

synthesis evolved, the transition to a protein-dominated world could proceed, with proteins eventually taking over the majority of catalytic and structural tasks because of their greater versatility, with 20 rather than 4 different subunits. Although the scenarios just discussed are highly speculative, the known properties of RNA molecules are consistent with these ideas.

All Present-Day Cells Use DNA as Their Hereditary Material

If the evolutionary speculations embodied in the RNA world hypothesis are correct, early cells would have differed fundamentally from the cells we know today in having their hereditary information stored in RNA rather than in DNA (**Figure 6–110**). Evidence that RNA arose before DNA in evolution can be found in the chemical differences between them. Ribose, like glucose and other simple carbohydrates, can be formed from formaldehyde (HCHO), a simple chemical which is readily produced in laboratory experiments that attempt to simulate conditions on the primitive Earth. The sugar deoxyribose is harder to make, and in present-day cells it is produced from ribose in a reaction catalyzed by a protein enzyme, suggesting that ribose predates deoxyribose in cells. Presumably, DNA appeared on the scene later, but then proved more suitable than RNA as a permanent repository of genetic information. In particular, the deoxyribose in its sugar-phosphate backbone makes chains of DNA chemically more stable than chains of RNA, so that much greater lengths of DNA can be maintained without breakage.

The other differences between RNA and DNA—the double-helical structure of DNA and the use of thymine rather than uracil—further enhance DNA stability by making the many unavoidable accidents that occur to the molecule much easier to repair, as discussed in detail in Chapter 5 (see pp. 296–297 and 300–301).

Summary

From our knowledge of present-day organisms and the molecules they contain, it seems likely that the development of the directly autocatalytic mechanisms fundamental to living systems began with the evolution of families of molecules that could catalyze their own replication. With time, a family of cooperating RNA catalysts probably developed the ability to direct the synthesis of polypeptides. DNA is likely to have been a late addition: as the accumulation of additional protein catalysts allowed more efficient and complex cells to evolve, the DNA double helix replaced RNA as a more stable molecule for storing the increased amounts of genetic information required by such cells.

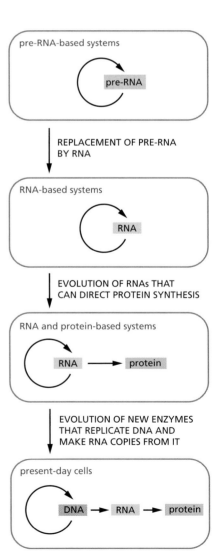

Figure 6–110 The hypothesis that RNA preceded DNA and proteins in evolution. In the earliest cells, pre-RNA molecules would have had combined genetic, structural, and catalytic functions and RNA would have gradually taken over these functions. In present-day cells, DNA is the repository of genetic information, and proteins perform the vast majority of catalytic functions in cells. RNA primarily functions today as a go-between in protein synthesis, although it remains a catalyst for a small number of crucial reactions.

PROBLEMS

Which statements are true? Explain why or why not.

6–1 The consequences of errors in transcription are less than those of errors in RNA replication.

6–2 Since introns are largely genetic "junk," they do not have to be removed precisely from the primary transcript during RNA splicing.

6–3 Wobble pairing occurs between the first position in the codon and the third position in the anticodon.

6–4 Protein enzymes are thought to greatly outnumber ribozymes in modern cells because they catalyze a much greater variety of reactions at much faster rates than ribozymes.

Discuss the following problems.

6–5 In which direction along the template must the RNA polymerase in **Figure Q6–1** be moving to have generated the supercoiled structures that are shown? Would you expect supercoils to be generated if the RNA polymerase were free to rotate about the axis of the DNA as it progressed along the template?

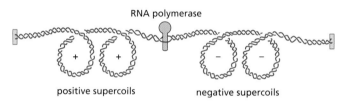

Figure Q6–1 Supercoils around a moving RNA polymerase (Problem 6–5).

6–6 Phosphates are attached to the CTD (C-terminal domain) of RNA polymerasse II during transcription. What are the various roles of RNA polymerase II CTD phosphorylation?

6–7 The human α-tropomyosin gene is alternatively spliced to produce several forms of α-tropomyosin mRNA in various cell types (**Figure Q6–2**). For all forms of the mRNA, the encoded protein sequence is the same for exons 1 and

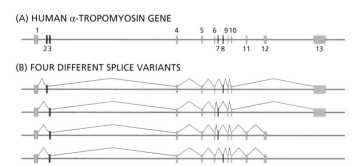

Figure Q6–2 Alternatively spliced mRNAs from the human α-tropomyosin gene (Problem 6–7). (A) Exons in the human α-tropomyosin gene. The locations and relative sizes of exons are shown by the *blue* and *red rectangles*. (B) Splicing patterns for four α-tropomyosin mRNAs. Splicing is indicated by *lines* connecting the exons that are included in the mRNA.

10. Exons 2 and 3 are alternative exons used in different mRNAs, as are exons 7 and 8. Which of the following statements about exons 2 and 3 is the most accurate? Is that statement also the most accurate one for exons 7 and 8? Explain your answers.

A. Exons 2 and 3 must have the same number of nucleotides.

B. Exons 2 and 3 must each contain an integral number of codons (that is, the number of nucleotides divided by 3 must be an integer).

C. Exons 2 and 3 must each contain a number of nucleotides that when divided by 3 leaves the same remainder (that is, 0, 1, or 2).

6–8 After treating cells with a chemical mutagen, you isolate two mutants. One carries alanine and the other carries methionine at a site in the protein that normally contains valine (**Figure Q6–3**). After treating these two mutants again with the mutagen, you isolate mutants from each that now carry threonine at the site of the original valine (Figure Q6–3). Assuming that all mutations involve single nucleotide changes, deduce the codons that are used for valine, methionine, threonine, and alanine at the affected site. Would you expect to be able to isolate valine-to-threonine mutants in one step?

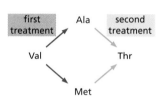

Figure Q6–3 Two rounds of mutagenesis and the altered amino acids at a single position in a protein (Problem 6–8).

6–9 The elongation factor EF-Tu introduces two short delays between codon–anticodon base-pairing and formation of the peptide bond. These delays increase the accuracy of protein synthesis. Describe these delays and explain how they improve the fidelity of translation.

6–10 Both Hsp60-like and Hsp70 molecular chaperones share an affinity for exposed hydrophobic patches on proteins, using them as indicators of incomplete folding. Why do you suppose hydrophobic patches serve as critical signals for the folding status of a protein?

6–11 Most proteins require molecular chaperones to assist in their correct folding. How do you suppose the chaperones themselves manage to fold correctly?

6–12 What is so special about RNA that makes it such an attractive evolutionary precursor to DNA and protein? What is it about DNA that makes it a better material than RNA for storage of genetic information?

6–13 If an RNA molecule could form a hairpin with a symmetric internal loop, as shown in **Figure Q6–4**, could the complement of this RNA form a similar structure? If so, would there be any regions of the two structures that are identical? Which ones?

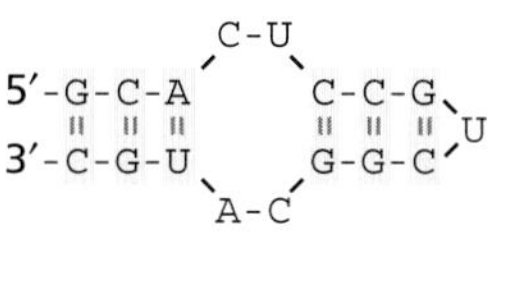

```
            C–U
5'-G-C-A        C-C-G
   ║ ║ ║        ║ ║ ║   U
3'-C-G-U        G-G-C
            A–C
```

Figure Q6–4 An RNA hairpin with a symmetric internal loop (Problem 6–13).

REFERENCES

General

Berg JM, Tymoczko JL & Stryer L (2006) Biochemistry, 6th ed. New York: WH Freeman.

Brown TA (2002) Genomes 2, 2nd ed. New York: Wiley-Liss.

Gesteland RF, Cech TR & Atkins JF (eds) (2006) The RNA World, 3rd ed. Cold Spring Harbor, NY: Cold Spring Harbor Laboratory Press.

Hartwell L, Hood L, Goldberg ML et al (2006) Genetics: from Genes to Genomes, 3rd ed. Boston: McGraw Hill.

Lodish H, Berk A, Kaiser C et al (2007) Molecular Cell Biology, 6th ed. New York: WH Freeman.

Stent GS (1971) Molecular Genetics: An Introductory Narrative. San Francisco: WH Freeman.

The Genetic Code (1966) *Cold Spring Harb. Symp Quant Biol* 31.

The Ribosome (2001) *Cold Spring Harb Symp Quant Biol* 66.

Watson JD, Baker TA, Bell SP et al (2003) Molecular Biology of the Gene, 5th ed. Menlo Park, CA: Benjamin Cummings.

From DNA to RNA

Bentley DL (2005) Rules of engagement: co-transcriptional recruitment of pre-mRNA processing factors. *Curr Opin Cell Biol* 17:251–256.

Berget SM, Moore C & Sharp PA (1977) Spliced segments at the 5′ terminus of adenovirus 2 late mRNA. *Proc Natl Acad Sci USA* 74:3171–3175.

Black DL (2003). Mechanisms of alternative pre-messenger RNA splicing. *Annu Rev Biochem* 72:291–336.

Brenner S, Jacob F & Meselson M (1961) An unstable intermediate carrying information from genes to ribosomes for protein synthesis. *Nature* 190:576–581.

Cate JH, Gooding AR, Podell E et al. (1996) Crystal structure of a group I ribozyme domain: principles of RNA packing. *Science* 273:1678–1685.

Chow LT, Gelinas RE, Broker TR et al (1977) An amazing sequence arrangement at the 5′ ends of adenovirus 2 messenger RNA. *Cell* 12:1–8.

Cramer P (2002) Multisubunit RNA polymerases. *Curr Opin Struct Biol* 12:89–97.

Daneholt B (1997) A look at messenger RNP moving through the nuclear pore. *Cell* 88:585–588.

Dreyfuss G, Kim VN, & Kataoka N (2002) Messenger-RNA-binding proteins and the messages they carry. *Nature Rev Mol Cell Biol* 3:195–205.

Houseley J, LaCava J & Tollervey D (2006) RNA-quality control by the exosome. *Nature Rev Mol Cell Biol* 7:529–539.

Izquierdo JM, & Valcárcel J (2006) A simple principle to explain the evolution of pre-mRNA splicing. *Genes Dev* 20:1679–1684.

Kornberg RD (2005) Mediator and the mechanism of transcriptional activation. *Trends Biochem Sci* 30:235–239.

Malik S & Roeder RG (2005) Dynamic regulation of pol II transcription by the mammalian Mediator complex. *Trends Biochem Sci* 30:256–263.

Matsui T, Segall J, Weil PA & Roeder RG (1980) Multiple factors required for accurate initiation of transcription by purified RNA polymerase II. *J Biol Chem* 255:11992–11996.

Patel AA & Steitz JA (2003) Splicing double: insights from the second spliceosome. *Nature Rev Mol Cell Biol* 4|:960–970.

Phatnani HP & Greenleaf AL (2006) Phosphorylation and functions of the RNA polymerase II CTD. *Genes Dev* 20:2922–2936.

Query CC & Konarska MM (2006) Splicing fidelity revisited. *Nature Struct Mol Biol* 13:472–474.

Ruskin B, Krainer AR, Maniatis T et al (1984) Excision of an intact intron as a novel lariat structure during pre-mRNA splicing *in vitro*. *Cell* 38:317–331.

Spector DL (2003) The dynamics of chromosome organization and gene regulation. *Annu Rev Biochem* 72:573–608.

Staley JP & Guthrie C (1998) Mechanical devices of the spliceosome: motors, clocks, springs, and things. *Cell* 92:315–326.

Thomas MC & Chiang CM (2006) The general transcription machinery and general cofactors. *Critical Rev Biochem Mol Biol* 41:105–178.

Wang D, Bushnell DA, Westover KD et al (2006) Structural basis of transcription: role of the trigger loop in substrate specificity and catalysis. *Cell* 127:941–954.

From RNA to Protein

Allen GS & Frank J (2007) Structural insights on the translation initiation complex: ghosts of a universal initiation complex. *Mol Microbiol* 63:941–950.

Anfinsen CB (1973) Principles that govern the folding of protein chains. *Science* 181:223–230.

Brunelle JL, Youngman EM, Sharma D et al (2006) The interaction between C75 of tRNA and the A loop of the ribosome stimulates peptidyl transferase activity. *RNA* 12:33–39.

Chien P, Weissman JS, & DePace AH (2004). Emerging principles of conformation-based prion inheritance. *Annu Rev Biochem* 73:617–656.

Crick FHC (1966) The genetic code: III. *Sci Am* 215:55–62.

Hershko A, Ciechanover A & Varshavsky A (2000) The ubiquitin system. *Nature Med* 6:1073–1081.

Ibba M & Soll D (2000) Aminoacyl-tRNA synthesis. *Annu Rev Biochem* 69:617–650.

Kozak M (1992) Regulation of translation in eukaryotic systems. *Annu Rev Cell Biol* 8:197–225.

Kuzmiak HA, & Maquat LE (2006) Applying nonsense-mediated mRNA decay research to the clinic: progress and challenges. *Trends Mol Med* 12:306–316.

Moore PB & Steitz TA (2005) The ribosome revealed. *Trends Biochem Sci* 30:281–283.

Noller HF (2005) RNA structure: reading the ribosome. *Science* 309:1508–1514.

Ogle JM, Carter AP & Ramakrishnan V (2003) Insights into the decoding mechanism from recent ribosome structures. *Trends Biochem Sci* 28:259–266.

Prusiner SB (1998) Nobel lecture. Prions. *Proc Natl Acad Sci USA* 95:13363–13383.

Rehwinkel J, Raes J & Izaurralde E (2006) Nonsense-mediated mRNA decay: Target genes and functional diversification of effectors. *Trends Biochem Sci* 31:639–646.

Sauer RT, Bolon DN, Burton BM et al (2004) Sculpting the proteome with AAA(+) proteases and disassembly machines. *Cell* 119:9–18.

Shorter J & Lindquist S (2005) Prions as adaptive conduits of memory and inheritance. *Nature Rev Genet* 6:435–450.

Varshavsky A (2005) Regulated protein degradation. *Trends in Biochem Sci* 30:283–286.

Voges D, Zwickl P & Baumeister W (1999) The 26S proteasome: a molecular machine designed for controlled proteolysis. *Annu Rev Biochem* 68:1015–1068.

Weissmann C (2005) Birth of a prion: spontaneous generation revisited. *Cell* 122:165–168.

Young JC, Agashe VR, Siegers K et al (2004) Pathways of chaperone-mediated protein folding in the cytosol. *Nature Rev Mol Cell Biol* 5:781–791.

The RNA World and the Origins of Life

Joyce GF (1992) Directed molecular evolution. *Sci Am* 267:90–97.

Orgel L (2000) Origin of life. A simpler nucleic acid. *Science* 290:1306–1307.

Kruger K, Grabowski P, Zaug P et al (1982) Self-splicing RNA: Autoexcision and autocyclization of the ribosomal RNA intervening seuence of Tetrahymena. *Cell* 31:147–157.

Silverman SK (2003) Rube Goldberg goes (ribo)nuclear? Molecular switches and sensors made from RNA. *RNA* 9:377–383.

Szostak JW, Bartel DP & Luisi PL (2001) Synthesizing life. *Nature* 409:387–390.

Chapter 7

Control of Gene Expression

7

An organism's DNA encodes all of the RNA and protein molecules required to construct its cells. Yet a complete description of the DNA sequence of an organism—be it the few million nucleotides of a bacterium or the few billion nucleotides of a human—no more enables us to reconstruct the organism than a list of English words enables us to reconstruct a play by Shakespeare. In both cases, the problem is to know how the elements in the DNA sequence or the words on the list are used. Under what conditions is each gene product made, and, once made, what does it do?

In this chapter we discuss the first half of this problem—the rules and mechanisms by which a subset of the genes is selectively expressed in each cell. The mechanisms that control the expression of genes operate at many levels, and we discuss the different levels in turn. We begin with an overview of some basic principles of gene control in multicellular organisms.

AN OVERVIEW OF GENE CONTROL

The different cell types in a multicellular organism differ dramatically in both structure and function. If we compare a mammalian neuron with a lymphocyte, for example, the differences are so extreme that it is difficult to imagine that the two cells contain the same genome (**Figure 7–1**). For this reason, and because cell differentiation is often irreversible, biologists originally suspected that genes might be selectively lost when a cell differentiates. We now know, however, that cell differentiation generally depends on changes in gene expression rather than on any changes in the nucleotide sequence of the cell's genome.

The Different Cell Types of a Multicellular Organism Contain the Same DNA

The cell types in a multicellular organism become different from one another because they synthesize and accumulate different sets of RNA and protein molecules. Evidence that they generally do this without altering the sequence of their DNA comes from a classic set of experiments in frogs. When the nucleus of a fully differentiated frog cell is injected into a frog egg whose nucleus has been removed, the injected donor nucleus is capable of directing the recipient egg to produce a normal tadpole (**Figure 7–2**A). Because the tadpole contains a full range of differentiated cells that derived their DNA sequences from the nucleus of the original donor cell, it follows that the differentiated donor cell cannot have lost any important DNA sequences. A similar conclusion has been reached in experiments performed with various plants. Here differentiated pieces of plant tissue are placed in culture and then dissociated into single cells. Often, one of these individual cells can regenerate an entire adult plant (Figure 7–2B). Finally, this same principle has been demonstrated in mammals, including sheep, cattle, pigs, goats, dogs, and mice by introducing nuclei from somatic cells into enucleated eggs; when placed into surrogate mothers, some of these eggs (called reconstructed zygotes) develop into healthy animals (Figure 7–2C).

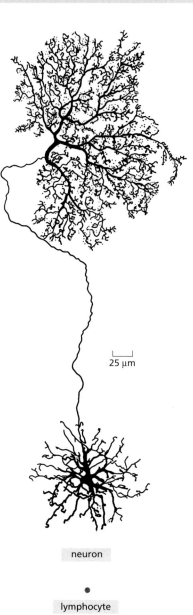

Figure 7–1 **A mammalian neuron and a lymphocyte.** The long branches of this neuron from the retina enable it to receive electrical signals from many cells and carry those signals to many neighboring cells. The lymphocyte is a white blood cell involved in the immune response to infection and moves freely through the body. Both of these cells contain the same genome, but they express different RNAs and proteins. (From B.B. Boycott, Essays on the Nervous System [R. Bellairs and E.G. Gray, eds.]. Oxford, UK: Clarendon Press, 1974.)

25 μm

neuron

lymphocyte

Further evidence that large blocks of DNA are not lost or rearranged during vertebrate development comes from comparing the detailed banding patterns detectable in condensed chromosomes at mitosis (see Figure 4–11). By this criterion the chromosome sets of differentiated cells in the human body appear to be identical. Moreover, comparisons of the genomes of different cells based on recombinant DNA technology have confirmed, as a general rule, that the changes in gene expression that underlie the development of multicellular organisms do not rely on changes in the DNA sequences of the corresponding genes. There are, however, a few cases where DNA rearrangements of the genome take place during the development of an organism—most notably, in generating the diversity of the immune system of mammals, which we discuss in Chapter 25.

Different Cell Types Synthesize Different Sets of Proteins

As a first step in understanding cell differentiation, we would like to know how many differences there are between any one cell type and another. Although we still do not have a detailed answer to this fundamental question, we can make certain general statements.

1. Many processes are common to all cells, and any two cells in a single organism therefore have many proteins in common. These include the structural proteins of chromosomes, RNA polymerases, DNA repair enzymes, ribosomal proteins, enzymes involved in the central reactions of metabolism, and many of the proteins that form the cytoskeleton.

2. Some proteins are abundant in the specialized cells in which they function and cannot be detected elsewhere, even by sensitive tests. Hemoglobin, for example, can be detected only in red blood cells.

3. Studies of the number of different mRNAs suggest that, at any one time, a typical human cell expresses 30–60% of its approximately 25,000 genes. When the patterns of mRNAs in a series of different human cell lines are compared, it is found that the level of expression of almost every active gene varies from one cell type to another. A few of these differences are striking, like that of hemoglobin noted above, but most are much more subtle. Even genes that are expressed in all cell types vary in their level of expression from one cell type to the next. The patterns of mRNA abundance (determined using DNA microarrays, discussed in Chapter 8) are so characteristic of cell type that they can be used to type human cancer cells of uncertain tissue origin (**Figure 7–3**).

4. Although the differences in mRNAs among specialized cell types are striking, they nonetheless underestimate the full range of differences in the pattern of protein production. As we shall see in this chapter, there are many steps after transcription at which gene expression can be regulated. For example, alternative splicing can produce a whole family of proteins from a single gene. Finally, proteins can be covalently modified after they are synthesized. Therefore a better way of appreciating the radical differences in gene expression between cell types is through methods that directly display the levels of proteins and their post-translational modifications (**Figure 7–4**).

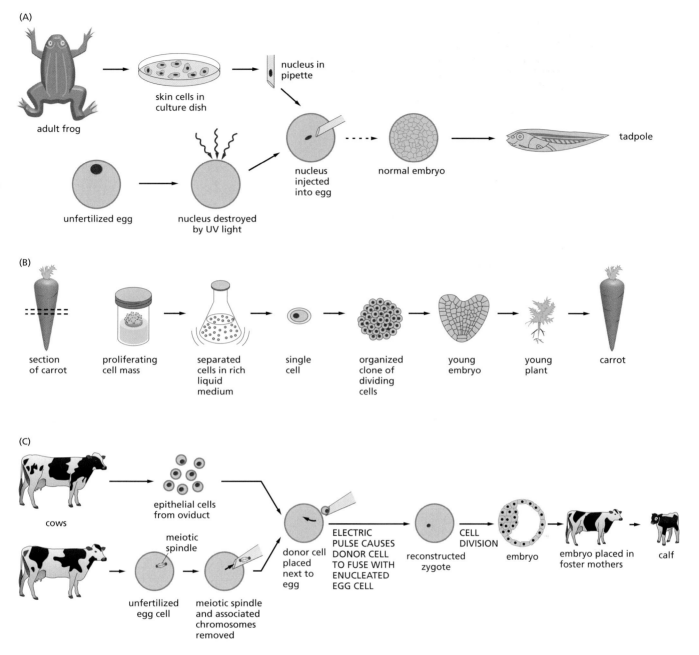

Figure 7–2 Evidence that a differentiated cell contains all the genetic instructions necessary to direct the formation of a complete organism. (A) The nucleus of a skin cell from an adult frog transplanted into an enucleated egg can give rise to an entire tadpole. The *broken arrow* indicates that, to give the transplanted genome time to adjust to an embryonic environment, a further transfer step is required in which one of the nuclei is taken from the early embryo that begins to develop and is put back into a second enucleated egg. (B) In many types of plants, differentiated cells retain the ability to "dedifferentiate," so that a single cell can form a clone of progeny cells that later give rise to an entire plant. (C) A differentiated cell nucleus from an adult cow introduced into an enucleated egg from a different cow can give rise to a calf. Different calves produced from the same differentiated cell donor are genetically identical and are therefore clones of one another. (A, modified from J.B. Gurdon, *Sci. Am.* 219:24–35, 1968. With permission from Scientific American.)

External Signals Can Cause a Cell to Change the Expression of Its Genes

Most of the specialized cells in a multicellular organism are capable of altering their patterns of gene expression in response to extracellular cues. If a liver cell is exposed to a glucocorticoid hormone, for example, the production of several specific proteins is dramatically increased. Glucocorticoids are released in the body during periods of starvation or intense exercise and signal the liver to increase the

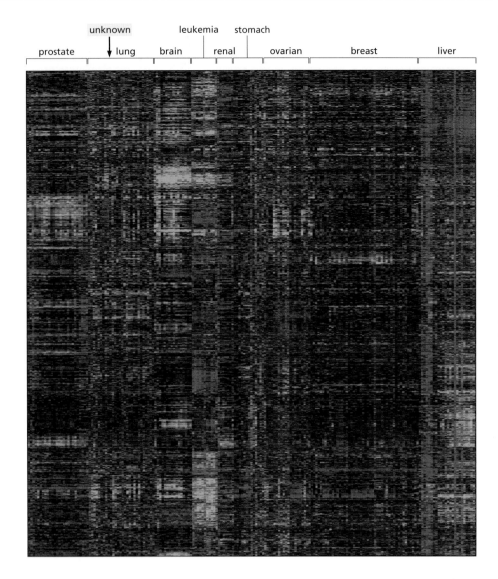

unknown   leukemia   stomach

prostate   lung   brain   renal   ovarian   breast   liver

Figure 7–3 Differences in mRNA expression patterns among different types of human cancer cells. This figure summarizes a very large set of measurements in which the mRNA levels of 1800 selected genes (arranged *top* to *bottom*) were determined for 142 different human tumor cell lines (arranged *left* to *right*), each from a different patient. Each small *red* bar indicates that the given gene in the given tumor is transcribed at a level significantly higher than the average across all the cell lines. Each small *green* bar indicates a less-than-average expression level, and each *black* bar denotes an expression level that is close to average across the different tumors. The procedure used to generate these data—mRNA isolation followed by hybridization to DNA microarrays—is described in Chapter 8 (pp. 574–575). The figure shows that the relative expression levels of each of the 1800 genes analyzed vary among the different tumors (seen by following a given gene from *left* to *right* across the figure). This analysis also shows that each type of tumor has a characteristic gene expression pattern. This information can be used to "type" cancer cells of unknown tissue origin by matching the gene expression profiles to those of known tumors. For example, the unknown sample in the figure has been identified as a lung cancer. (Courtesy of Patrick O. Brown, David Botstein, and the Stanford Expression Collaboration.)

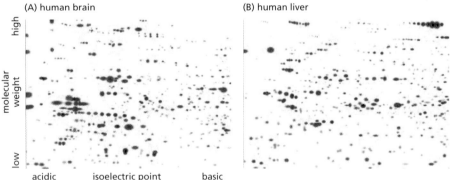

(A) human brain   (B) human liver

high

molecular weight

low

acidic   isoelectric point   basic

Figure 7–4 Differences in the proteins expressed by two human tissues. In each panel, the proteins are displayed using two-dimensional polyacrylamide-gel electrophoresis (see pp. 521–522). The proteins have been separated by molecular weight (*top* to *bottom*) and isoelectric point, the pH at which the protein has no net charge (*right* to *left*). The protein spots artificially colored *red* are common to both samples; those in *blue* are specific to one of the two tissues. The differences between the two tissue samples vastly outweigh their similarities: even for proteins that are shared between the two tissues, their relative abundances are usually different. Note that this technique separates proteins by both size and charge; therefore a protein that has, for example, several different phosphorylation states will appear as a series of *horizontal spots* (see *upper right-hand* portion of *right* panel). Only a small portion of the complete protein spectrum is shown for each sample. Although two-dimensional gel electrophoresis provides a simple way to visualize the differences between two protein samples, methods based on mass spectrometry (see pp. 519–521) provide much more detailed information and are therefore more commonly used. (Courtesy of Tim Myers and Leigh Anderson, Large Scale Biology Corporation.)

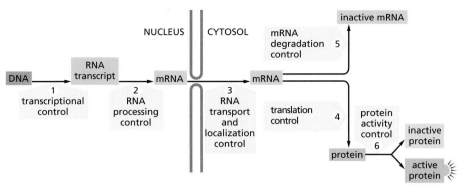

Figure 7–5 Six steps at which eucaryotic gene expression can be controlled. Controls that operate at steps 1 through 5 are discussed in this chapter. Step 6, the regulation of protein activity, occurs largely through covalent post-translational modifications including phosphorylation, acetylation, and ubiquitylation (see Table 3–3, p. 186) and is discussed in many chapters throughout the book.

production of glucose from amino acids and other small molecules; the set of proteins whose production is induced includes enzymes such as tyrosine amino-transferase, which helps to convert tyrosine to glucose. When the hormone is no longer present, the production of these proteins drops to its normal level.

Other cell types respond to glucocorticoids differently. Fat cells, for example, reduce the production of tyrosine aminotransferase, while some other cell types do not respond to glucocorticoids at all. These examples illustrate a general feature of cell specialization: different cell types often respond differently to the same extracellular signal. Underlying such adjustments that occur in response to extracellular signals, there are features of the gene expression pattern that do not change and give each cell type its permanently distinctive character.

Gene Expression Can Be Regulated at Many of the Steps in the Pathway from DNA to RNA to Protein

If differences among the various cell types of an organism depend on the particular genes that the cells express, at what level is the control of gene expression exercised? As we saw in the previous chapter, there are many steps in the pathway leading from DNA to protein. We now know that all of them can in principle be regulated. Thus a cell can control the proteins it makes by (1) controlling when and how often a given gene is transcribed (**transcriptional control**), (2) controlling the splicing and processing of RNA transcripts (**RNA processing control**), (3) selecting which completed mRNAs are exported from the nucleus to the cytosol and determining where in the cytosol they are localized (**RNA transport and localization control**), (4) selecting which mRNAs in the cytoplasm are translated by ribosomes (**translational control**), (5) selectively destabilizing certain mRNA molecules in the cytoplasm (**mRNA degradation control**), or (6) selectively activating, inactivating, degrading, or locating specific protein molecules after they have been made (**protein activity control**) (**Figure 7–5**).

For most genes transcriptional controls are paramount. This makes sense because, of all the possible control points illustrated in Figure 7–5, only transcriptional control ensures that the cell will not synthesize superfluous intermediates. In the following sections we discuss the DNA and protein components that perform this function by regulating the initiation of gene transcription. We shall return at the end of the chapter to the many additional ways of regulating gene expression.

Summary

The genome of a cell contains in its DNA sequence the information to make many thousands of different protein and RNA molecules. A cell typically expresses only a fraction of its genes, and the different types of cells in multicellular organisms arise because different sets of genes are expressed. Moreover, cells can change the pattern of genes they express in response to changes in their environment, such as signals from other cells. Although all of the steps involved in expressing a gene can in principle be regulated, for most genes the initiation of RNA transcription is the most important point of control.

DNA-BINDING MOTIFS IN GENE REGULATORY PROTEINS

How does a cell determine which of its thousands of genes to transcribe? As outlined in Chapter 6, the transcription of each gene is controlled by a regulatory region of DNA relatively near the site where transcription begins. Some regulatory regions are simple and act as switches thrown by a single signal. Many others are complex and resemble tiny microprocessors, responding to a variety of signals that they interpret and integrate in order to switch their neighboring gene on or off. Whether complex or simple, these switching devices are found in all cells and are composed of two types of fundamental components: (1) short stretches of DNA of defined sequence and (2) *gene regulatory proteins* that recognize and bind to this DNA.

We begin our discussion of gene regulatory proteins by describing how they were discovered.

Gene Regulatory Proteins Were Discovered Using Bacterial Genetics

Genetic analyses in bacteria carried out in the 1950s provided the first evidence for the existence of **gene regulatory proteins** (often loosely called "transcription factors") that turn specific sets of genes on or off. One of these regulators, the *lambda repressor*, is encoded by a bacterial virus, *bacteriophage lambda*. The repressor shuts off the viral genes that code for the protein components of new virus particles and thereby enables the viral genome to remain a silent passenger in the bacterial chromosome, multiplying with the bacterium when conditions are favorable for bacterial growth (see Figure 5–78). The lambda repressor was among the first gene regulatory proteins to be characterized, and it remains one of the best understood, as we discuss later. Other bacterial regulators respond to nutritional conditions by shutting off genes encoding specific sets of metabolic enzymes when they are not needed. The *Lac repressor*, the first of these bacterial proteins to be recognized, turns off the production of the proteins responsible for lactose metabolism when this sugar is absent from the medium.

The first step toward understanding gene regulation was the isolation of mutant strains of bacteria and bacteriophage lambda that were unable to shut off specific sets of genes. It was proposed at the time, and later proven, that most of these mutants were deficient in proteins acting as specific repressors for these sets of genes. Because these proteins, like most gene regulatory proteins, are present in small quantities, it was difficult and time-consuming to isolate them. They were eventually purified by fractionating cell extracts. Once isolated, the proteins were shown to bind to specific DNA sequences close to the genes that they regulate. The precise DNA sequences that they recognized were then determined by a combination of classical genetics and methods for studying protein–DNA interactions discussed later in this chapter.

The Outside of the DNA Helix Can Be Read by Proteins

As discussed in Chapter 4, the DNA in a chromosome consists of a very long double helix (**Figure 7–6**). Gene regulatory proteins must recognize specific nucleotide sequences embedded within this structure. It was originally thought that these proteins might require direct access to the hydrogen bonds between base pairs in the interior of the double helix to distinguish between one DNA

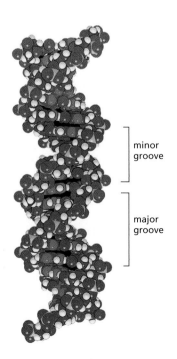

minor groove

major groove

Figure 7–6 Double-helical structure of DNA. A space-filling model of DNA showing the major and minor grooves on the outside of the double helix. The atoms are colored as follows: carbon, *dark blue*; nitrogen, *light blue*; hydrogen, *white*; oxygen, *red*; phosphorus, *yellow*.

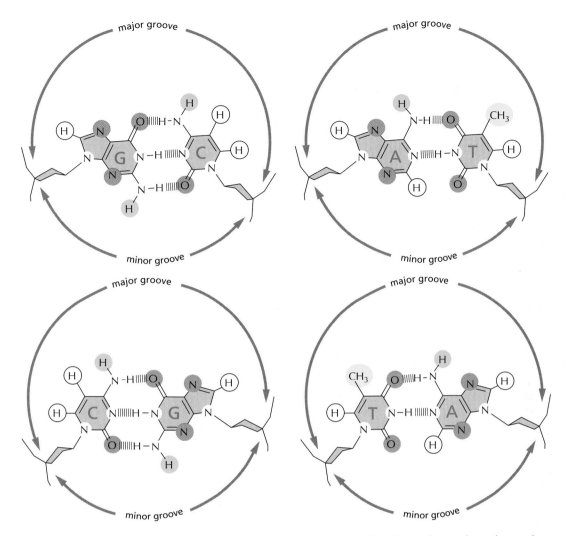

Figure 7–7 How the different base pairs in DNA can be recognized from their edges without the need to open the double helix. The four possible configurations of base pairs are shown, with potential hydrogen bond donors indicated in *blue,* potential hydrogen bond acceptors in *red,* and hydrogen bonds of the base pairs themselves as a series of short parallel *red* lines. Methyl groups, which form hydrophobic protuberances, are shown in *yellow,* and hydrogen atoms that are attached to carbons, and are therefore unavailable for hydrogen bonding, are *white.* (From C. Branden and J. Tooze, Introduction to Protein Structure, 2nd ed. New York: Garland Publishing, 1999.)

sequence and another. It is now clear, however, that the outside of the double helix is studded with DNA sequence information that gene regulatory proteins can recognize without having to open the double helix. The edge of each base pair is exposed at the surface of the double helix, presenting a distinctive pattern of hydrogen bond donors, hydrogen bond acceptors, and hydrophobic patches for proteins to recognize in both the major and minor groove (**Figure 7–7**). But only in the major groove are the patterns markedly different for each of the four base-pair arrangements (**Figure 7–8**). For this reason, gene regulatory proteins generally make specific contacts with the major groove—as we shall see.

Figure 7–8 A DNA recognition code. The edge of each base pair, seen here looking directly at the major or minor groove, contains a distinctive pattern of hydrogen bond donors, hydrogen bond acceptors, and methyl groups. From the major groove, each of the four base-pair configurations projects a unique pattern of features. From the minor groove, however, the patterns are similar for G–C and C–G as well as for A–T and T–A. The color code is the same as that in Figure 7–7. (From C. Branden and J. Tooze, Introduction to Protein Structure, 2nd ed. New York: Garland Publishing, 1999.)

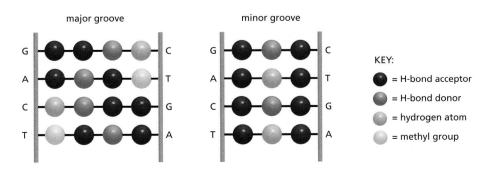

KEY:
● = H-bond acceptor
● = H-bond donor
○ = hydrogen atom
○ = methyl group

Short DNA Sequences Are Fundamental Components of Genetic Switches

A specific nucleotide sequence can be "read" as a pattern of molecular features on the surface of the DNA double helix. Particular nucleotide sequences, each typically less than 20 nucleotide pairs in length, function as fundamental components of genetic switches by serving as recognition sites for the binding of specific gene regulatory proteins. Thousands of such DNA sequences have been identified, each recognized by a different gene regulatory protein (or by a set of related gene regulatory proteins). Some of the gene regulatory proteins that are discussed in the course of this chapter are listed in **Table 7–1**, along with the DNA sequences that they recognize.

We now turn to the gene regulatory proteins themselves, the second fundamental component of genetic switches. We begin with the structural features that allow these proteins to recognize short, specific DNA sequences contained in a much longer double helix.

Gene Regulatory Proteins Contain Structural Motifs That Can Read DNA Sequences

Molecular recognition in biology generally relies on an exact fit between the surfaces of two molecules, and the study of gene regulatory proteins has provided some of the clearest examples of this principle. A gene regulatory protein recognizes a specific DNA sequence because the surface of the protein is extensively

Table 7–1 Some Gene Regulatory Proteins and the DNA Sequences That They Recognize

	NAME	DNA SEQUENCE RECOGNIZED*
Bacteria	Lac repressor	5′ AATTGTGAGCGGATAACAATT 3′ TTAACACTCGCCTATTGTTAA
	CAP	TGTGAGTTAGCTCACT ACACTCAATCGAGTGA
	Lambda repressor	TATCACCGCCAGAGGT ATAGTGGCGGTCTCCAT
Yeast	Gal4	CGGAGGACTGTCCTCCG GCCTCCTGACAGGAGGC
	Matα2	CATGTAATT GTACATTAA
	Gcn4	ATGACTCAT TACTGAGTA
Drosophila	Kruppel	AACGGGTTAA TTGCCCAATT
	Bicoid	GGGATTAGA CCCTAATCT
Mammals	Sp1	GGGCGG CCCGCC
	Oct1 Pou domain	ATGCAAAT TACGTTTA
	GATA1	TGATAG ACTATC
	MyoD	CAAATG GTTTAC
	p53	GGGCAAGTCT CCCGTTCAGA

*For convenience, only one recognition sequence, rather than a consensus sequence (see Figure 6–12), is given for each protein.

complementary to the special surface features of the double helix in that region. In most cases the protein makes a series of contacts with the DNA, involving hydrogen bonds, ionic bonds, and hydrophobic interactions. Although each individual contact is weak, the 20 or so that are typically formed at the protein–DNA interface add together to ensure that the interaction is both highly specific and very strong (**Figure 7–9**). In fact, DNA–protein interactions include some of the tightest and most specific molecular interactions known in biology.

Although each example of protein–DNA recognition is unique in detail, x-ray crystallographic and NMR spectroscopic studies of several hundred gene regulatory proteins have revealed that many of them contain one or another of a small set of DNA-binding structural motifs. These motifs generally use either α helices or β sheets to bind to the major groove of DNA; this groove, as we have seen, contains sufficient information to distinguish one DNA sequence from any other. The fit is so good that it has been suggested that the dimensions of the basic structural units of nucleic acids and proteins evolved together to permit these molecules to interlock.

The Helix–Turn–Helix Motif Is One of the Simplest and Most Common DNA-Binding Motifs

The first DNA-binding protein motif to be recognized was the **helix–turn–helix**. Originally identified in bacterial proteins, this motif has since been found in many hundreds of DNA-binding proteins from both eucaryotes and procaryotes. It is constructed from two α helices connected by a short extended chain of amino acids, which constitutes the "turn" (**Figure 7–10**). The two helices are held at a fixed angle, primarily through interactions between the two helices. The more C-terminal helix is called the *recognition helix* because it fits into the major groove of DNA; its amino acid side chains, which differ from protein to protein, play an important part in recognizing the specific DNA sequence to which the protein binds.

Outside the helix–turn–helix region, the structure of the various proteins that contain this motif can vary enormously (**Figure 7–11**). Thus each protein "presents" its helix–turn–helix motif to the DNA in a unique way, a feature thought to enhance the versatility of the helix–turn–helix motif by increasing the number of DNA sequences that the motif can be used to recognize. Moreover, in most of these proteins, parts of the polypeptide chain outside the helix–turn–helix domain also make important contacts with the DNA, helping to fine-tune the interaction.

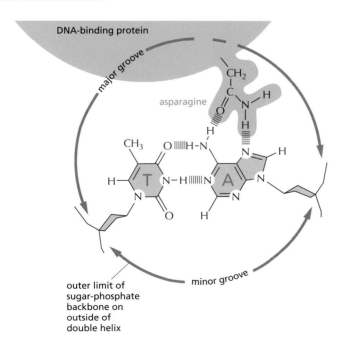

Figure 7–9 **The binding of a gene regulatory protein to the major groove of DNA.** Only a single contact is shown. Typically, the protein–DNA interface would consist of 10–20 such contacts, involving different amino acids, each contributing to the strength of the protein–DNA interaction.

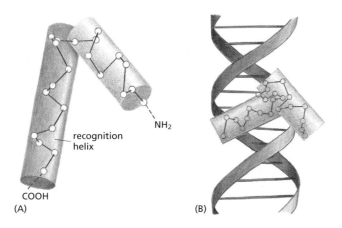

(A)

recognition helix

NH₂

COOH

(B)

Figure 7–10 The DNA-binding helix–turn–helix motif. The motif is shown in (A), where each *white* circle denotes the central carbon of an amino acid. The C-terminal α helix *(red)* is called the recognition helix because it participates in sequence-specific recognition of DNA. As shown in (B), this helix fits into the major groove of DNA, where it contacts the edges of the base pairs (see also Figure 7–7). The N-terminal α-helix *(blue)* functions primarily as a structural component that helps to position the recognition helix.

The group of helix–turn–helix proteins shown in Figure 7–11 demonstrates a common feature of many sequence-specific DNA-binding proteins. They bind as symmetric dimers to DNA sequences that are composed of two very similar "half-sites," which are also arranged symmetrically (**Figure 7–12**). This arrangement allows each protein monomer to make a nearly identical set of contacts and enormously increases the binding affinity: as a first approximation, doubling the number of contacts doubles the free energy of the interaction and thereby *squares* the affinity constant.

Homeodomain Proteins Constitute a Special Class of Helix–Turn–Helix Proteins

Not long after the first gene regulatory proteins were discovered in bacteria, genetic analyses in the fruit fly *Drosophila* led to the characterization of an important class of genes, the *homeotic selector genes*, that play a critical part in orchestrating fly development. As discussed in Chapter 22, they have since proved to have a fundamental role in the development of higher animals as well. Mutations in these genes can cause one body part in the fly to be converted into another, showing that the proteins they encode control critical developmental decisions.

When the nucleotide sequences of several homeotic selector genes were determined in the early 1980s, each proved to code for an almost identical stretch of 60 amino acids that defines this class of proteins and is termed the **homeodomain**. When the three-dimensional structure of the homeodomain was determined, it was seen to contain a helix–turn–helix motif related to that of

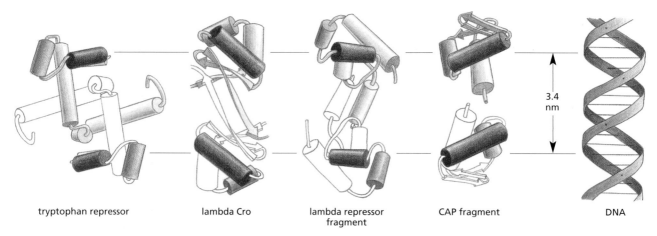

tryptophan repressor   lambda Cro   lambda repressor fragment   CAP fragment   DNA

3.4 nm

Figure 7–11 Some helix–turn–helix DNA-binding proteins. All of the proteins bind DNA as dimers in which the two copies of the recognition helix *(red cylinder)* are separated by exactly one turn of the DNA helix (3.4 nm). The other helix of the helix–turn–helix motif is colored *blue*, as in Figure 7–10. The lambda repressor and Cro proteins control bacteriophage lambda gene expression, and the tryptophan repressor and the catabolite activator protein (CAP) control the expression of sets of *E. coli* genes.

Figure 7–12 A specific DNA sequence recognized by the bacteriophage lambda Cro protein. The nucleotides labeled in *green* in this sequence are arranged symmetrically, allowing each half of the DNA site to be recognized in the same way by each protein monomer, also shown in *green*. See Figure 7–11 for the actual structure of the protein.

```
5′ T A A C A C C G T G C G T G T T G 3′
   | | | | | | | | | | | | | | | | | |
3′ A T T G T G G C A C G C A C A A C 5′
```

the bacterial gene regulatory proteins, providing one of the first indications that the principles of gene regulation established in bacteria are relevant to higher organisms as well. More than 60 homeodomain proteins have now been discovered in *Drosophila* alone, and homeodomain proteins have been identified in virtually all eucaryotic organisms that have been studied, from yeasts to plants to humans.

The structure of a homeodomain bound to its specific DNA sequence is shown in **Figure 7–13**. Whereas the helix–turn–helix motif of bacterial gene regulatory proteins is often embedded in different structural contexts, the helix–turn–helix motif of homeodomains is always surrounded by the same structure (which forms the rest of the homeodomain), suggesting that the motif is always presented to DNA in the same way. Indeed, structural studies have shown that a yeast homeodomain protein and a *Drosophila* homeodomain protein have very similar conformations and recognize DNA in almost exactly the same manner, although they are identical at only 17 of 60 amino acid positions (see Figure 3–13).

There Are Several Types of DNA-Binding Zinc Finger Motifs

The helix–turn–helix motif is composed solely of amino acids. A second important group of DNA-binding motifs includes one or more zinc atoms as structural components. Although all such zinc-coordinated DNA-binding motifs are called **zinc fingers**, this description refers only to their appearance in schematic drawings dating from their initial discovery (**Figure 7–14**A). Subsequent structural studies have shown that they fall into several distinct structural groups, two of which we consider here. The first type was initially discovered in the protein that activates the transcription of a eucaryotic ribosomal RNA gene. It has a simple structure, in which the zinc holds an α helix and a β sheet together (Figure 7–14B). This type of zinc finger is often found in tandem clusters so that the α helix of each can contact the major groove of the DNA, forming a nearly continuous stretch of α helices along the groove. In this way, a strong and specific DNA–protein interaction is built up through a repeating basic structural unit (**Figure 7–15**).

Another type of zinc finger is found in the large family of intracellular receptor proteins (discussed in detail in Chapter 15). It forms a different type of

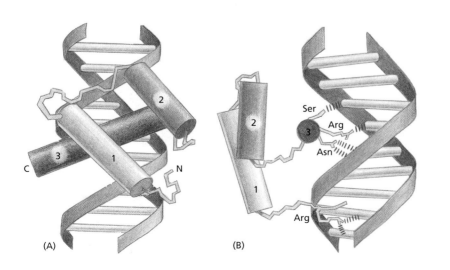

(A)   (B)

Figure 7–13 A homeodomain bound to its specific DNA sequence. Two different views of the same structure are shown. (A) The homeodomain is folded into three α helices, which are packed tightly together by hydrophobic interactions. The part containing helices 2 and 3 closely resembles the helix–turn–helix motif. (B) The recognition helix (helix 3, *red*) forms important contacts with the major groove of DNA. The asparagine (Asn) of helix 3, for example, contacts an adenine, as shown in Figure 7–9. A flexible arm attached to helix 1 forms contacts with nucleotide pairs in the minor groove. The homeodomain shown here is from a yeast gene regulatory protein, but it closely resembles homeodomains from many eucaryotic organisms. <ACGT> (Adapted from C. Wolberger et al., *Cell* 67:517–528, 1991. With permission from Elsevier.)

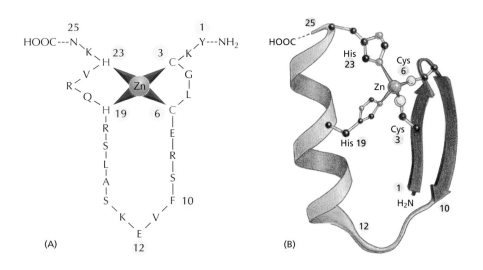

(A)

(B)

Figure 7–14 **One type of zinc finger protein.** This protein belongs to the Cys–Cys–His–His family of zinc finger proteins, named after the amino acids that grasp the zinc. (A) Schematic drawing of the amino acid sequence of a zinc finger from a frog protein of this class. (B) The three-dimensional structure of this same type of zinc finger is constructed from an antiparallel β sheet (amino acids 1 to 10) followed by an α helix (amino acids 12 to 24). The four amino acids that bind the zinc (Cys 3, Cys 6, His 19, and His 23) hold one end of the α helix firmly to one end of the β sheet. (Adapted from M.S. Lee et al., *Science* 245:635–637, 1989. With permission from AAAS.)

structure (similar in some respects to the helix–turn–helix motif) in which two α helices are packed together with zinc atoms (**Figure 7–16**). Like the helix–turn–helix proteins, these proteins usually form dimers that allow one of the two α helices of each subunit to interact with the major groove of the DNA. Although the two types of zinc finger structures discussed in this section are structurally distinct, they share two important features: both use zinc as a structural element, and both use an α helix to recognize the major groove of the DNA.

β sheets Can Also Recognize DNA

In the DNA-binding motifs discussed so far, α helices are the primary mechanism used to recognize specific DNA sequences. One large group of gene regulatory proteins, however, has evolved an entirely different recognition strategy. In this case, a two-stranded β sheet, with amino acid side chains extending from the sheet toward the DNA, reads the information on the surface of the major groove (**Figure 7–17**). As in the case of a recognition α helix, this β-sheet motif can be used to recognize many different DNA sequences; the exact DNA sequence recognized depends on the sequence of amino acids that make up the β sheet.

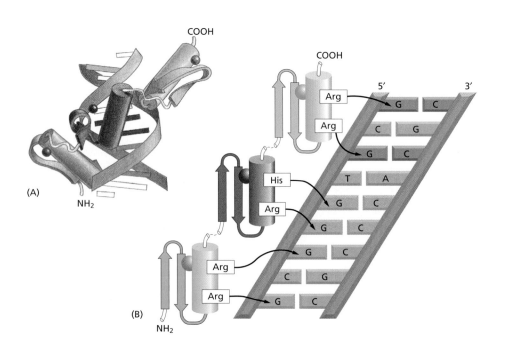

(A)

(B)

Figure 7–15 **DNA binding by a zinc finger protein.** (A) The structure of a fragment of a mouse gene regulatory protein bound to a specific DNA site. This protein recognizes DNA by using three zinc fingers of the Cys–Cys–His–His type (see Figure 7–14) arranged as direct repeats. <ATCT> (B) The three fingers have similar amino acid sequences and contact the DNA in similar ways. In both (A) and (B) the zinc atom in each finger is represented by a small sphere. (Adapted from N. Pavletich and C. Pabo, *Science* 252:810–817, 1991. With permission from AAAS.)

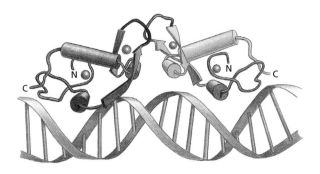

Figure 7–16 **A dimer of the zinc finger domain of the intracellular receptor family bound to its specific DNA sequence.** Each zinc finger domain contains two atoms of Zn (indicated by the small *gray spheres*); one stabilizes the DNA recognition helix (shown in *brown* in one subunit and *red* in the other), and one stabilizes a loop (shown in *purple*) involved in dimer formation. Each Zn atom is coordinated by four appropriately spaced cysteine residues. Like the helix–turn–helix proteins shown in Figure 7–11, the two recognition helices of the dimer are held apart by a distance corresponding to one turn of the DNA double helix. The specific example shown is a fragment of the glucocorticoid receptor. This is the protein through which cells detect and respond transcriptionally to the glucocorticoid hormones produced in the adrenal gland in response to stress. (Adapted from B.F. Luisi et al., *Nature* 352:497–505, 1991. With permission from Macmillan Publishers Ltd.)

Some Proteins Use Loops That Enter the Major and Minor Grooves to Recognize DNA

A few DNA-binding proteins use protruding peptide loops to read nucleotide sequences, rather than α helices and β sheets. For example, p53, a critical *tumor suppressor* in humans, recognizes nucleotide pairs from both the major and minor grooves using such loops (**Figure 7–18**). The normal role of the p53 protein is to tightly regulate cell growth and proliferation. Its importance can be appreciated by the fact that nearly half of all human cancers have acquired somatic mutations in the gene for p53; this step is key to the progression of many tumors, as we shall see in Chapter 20. Many of the p53 mutations observed in cancer cells destroy or alter its DNA-binding properties; indeed, Arg 248, which contacts the minor groove of DNA (see Figure 7–18) is the most frequently mutated p53 residue in human cancers.

The Leucine Zipper Motif Mediates Both DNA Binding and Protein Dimerization

Many gene regulatory proteins recognize DNA as homodimers, probably because, as we have seen, this is a simple way of achieving strong specific binding (see Figure 7–12). Usually, the portion of the protein responsible for dimerization is distinct from the portion that is responsible for DNA binding. One motif, however, combines these two functions elegantly and economically. It is called the **leucine zipper motif**, so named because of the way the two α helices, one from each monomer, are joined together to form a short coiled-coil (see Figure 3–9). The helices are held together by interactions between hydrophobic amino acid side chains (often on leucines) that extend from one side of each helix. Just beyond the dimerization interface the two α helices separate from each other to form a Y-shaped structure, which allows their side chains to contact the major groove of DNA. The dimer thus grips the double helix like a clothespin on a clothesline (**Figure 7–19**).

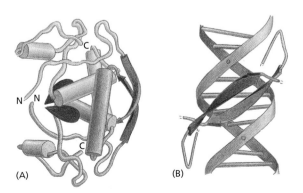

(A)   (B)

Figure 7–17 **The bacterial Met repressor protein.** The bacterial Met repressor regulates the genes encoding the enzymes that catalyze methionine synthesis. When this amino acid is abundant, it binds to the repressor, causing a change in the structure of the protein that enables it to bind to DNA tightly, shutting off the synthesis of the enzyme. (A) In order to bind to DNA tightly, the Met repressor must be complexed with *S*-adenosyl methionine, outlined in *red*. One subunit of the dimeric protein is shown in *green*, while the other is shown in *blue*. The two-stranded β sheet that binds to DNA is formed by one strand from each subunit and is shown in *dark green* and *dark blue*. (B) Simplified diagram of the Met repressor bound to DNA, showing how the two-stranded β sheet of the repressor binds to the major groove of DNA. For clarity, the other regions of the repressor have been omitted. (A, adapted from S. Phillips, *Curr. Opin. Struct. Biol.* 1:89–98, 1991, with permission from Elsevier; B, adapted from W. Somers and S. Phillips, *Nature* 359:387–393, 1992, with permission from Macmillan Publishers Ltd.)

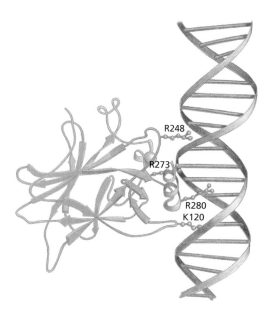

Figure 7–18 **DNA recognition by the p53 protein.** The most important DNA contacts are made by arginine 248 and lysine 120, which extend from the protruding loops entering the minor and major grooves. The folding of the p53 protein requires a zinc atom (shown as a sphere), but the way in which the zinc is grasped by the protein is completely different from that of the zinc finger proteins, described previously.

Heterodimerization Expands the Repertoire of DNA Sequences That Gene Regulatory Proteins Can Recognize

Many of the gene regulatory proteins we have seen thus far bind DNA as homodimers, that is, dimers made up of two identical subunits. However, many gene regulatory proteins can also associate with nonidentical partners to form heterodimers composed of two different subunits. Because heterodimers typically form from two proteins with distinct DNA-binding specificities, the mixing and matching of gene regulatory proteins in this way greatly expands the repertoire of DNA-binding specificities that these proteins can display. As illustrated in **Figure 7–20**, three distinct DNA-binding specificities could, in principle, be generated from two types of leucine zipper monomers, while six could be created from three types of monomers, and so on.

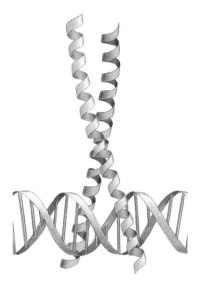

Figure 7–19 **A leucine zipper dimer bound to DNA.** Two α-helical DNA-binding domains *(bottom)* dimerize through their α-helical leucine zipper region *(top)* to form an inverted Y-shaped structure. Each arm of the Y is formed by a single α helix, one from each monomer, that mediates binding to a specific DNA sequence in the major groove of DNA. <TGTT> Each α helix binds to one-half of a symmetric DNA structure. The structure shown is of the yeast Gcn4 protein, which regulates transcription in response to the availability of amino acids in the environment. (Adapted from T.E. Ellenberger et al., *Cell* 71:1223–1237, 1992. With permission from Elsevier.)

DNA

Figure 7–20 **Heterodimerization of leucine zipper proteins can alter their DNA-binding specificity.** Leucine zipper homodimers bind to symmetric DNA sequences, as shown in the left-hand and center drawings. These two proteins recognize different DNA sequences, as indicated by the *red* and *blue* regions in the DNA. The two different monomers can combine to form a heterodimer, which now recognizes a hybrid DNA sequence, composed from one *red* and one *blue* region.

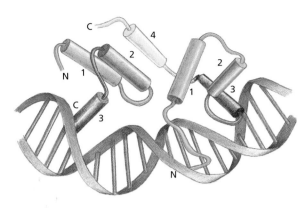

Figure 7–21 **A heterodimer composed of two homeodomain proteins bound to its DNA recognition site.** The *yellow* helix 4 of the protein on the right (Matα2) is unstructured in the absence of the protein on the left (Mata1), forming a helix only upon heterodimerization. The DNA sequence is recognized jointly by both proteins; some of the protein–DNA contacts made by Matα2 were shown in Figure 7–13. These two proteins are from budding yeast, where the heterodimer specifies a particular cell type (see Figure 7–65). The helices are numbered in accordance with Figure 7–13. (Adapted from T. Li et al., *Science* 270:262–269, 1995. With permission from AAAS.)

There are, however, limits to this promiscuity: for example, if all the many types of leucine zipper proteins in a typical eucaryotic cell formed heterodimers, the amount of "cross-talk" between the gene regulatory circuits of a cell would presumably be so great as to cause chaos. Whether or not a particular heterodimer can form depends on how well the hydrophobic surfaces of the two leucine zipper α helices mesh with each other, which in turn depends on the exact amino acid sequences of the two zipper regions. Thus, each leucine zipper protein in the cell can form dimers with only a small set of other leucine zipper proteins.

Heterodimerization is an example of **combinatorial control**, in which combinations of different proteins, rather than individual proteins, control a cell process. Heterodimerization as a mechanism for combinatorial control of gene expression occurs in many different types of gene regulatory proteins (**Figure 7–21**). Combinatorial control is a major theme that we shall encounter repeatedly in this chapter, and the formation of heterodimeric gene regulatory complexes is only one of many ways in which proteins work in combinations to control gene expression.

Certain combinations of gene regulatory proteins have become "hardwired" in the cell; for example, two distinct DNA-binding domains can, through gene rearrangements occurring over evolutionary time scales, become joined into a single polypeptide chain that displays a novel DNA-binding specificity (**Figure 7–22**).

The Helix–Loop–Helix Motif Also Mediates Dimerization and DNA Binding

Another important DNA-binding motif, related to the leucine zipper, is the **helix–loop–helix (HLH) motif**, which differs from the helix–turn–helix motif discussed earlier. An HLH motif consists of a short α helix connected by a loop to a second, longer α helix. The flexibility of the loop allows one helix to fold back

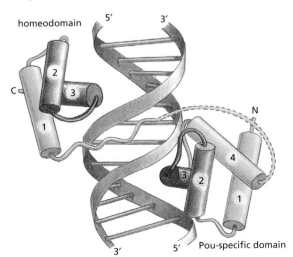

Figure 7–22 **Two DNA-binding domains covalently joined by a flexible polypeptide.** The structure shown (called a Pou-domain) consists of both a homeodomain and a helix–turn–helix structure joined by a flexible polypeptide "leash," indicated by the broken lines. A single gene encodes the entire protein, which is synthesized as a continuous polypeptide chain. The covalent joining of two structures in this way results in a large increase in the affinity of the protein for its specific DNA sequence compared with the DNA affinity of either separate structure. The group of mammalian gene regulatory proteins exemplified by this structure regulate the production of growth factors, immunoglobulins, and other molecules involved in development. The particular example shown is from the Oct1 protein. (Adapted from J.D. Klemm et al., *Cell* 77:21–32, 1994. With permission from Elsevier.)

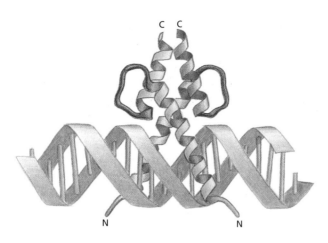

Figure 7–23 **A helix–loop–helix (HLH) dimer bound to DNA.** The two monomers are held together in a four-helix bundle: each monomer contributes two α helices connected by a flexible loop of protein *(red)*. A specific DNA sequence is bound by the two α helices that project from the four-helix bundle. (Adapted from A.R. Ferre-D'Amare et al., *Nature* 363:38–45, 1993. With permission from Macmillan Publishers Ltd.)

and pack against the other. As shown in **Figure 7–23**, this two-helix structure binds both to DNA and to the HLH motif of a second HLH protein. The second HLH protein can be the same (creating a homodimer) or different (creating a heterodimer). In either case, two α helices that extend from the dimerization interface make specific contacts with the DNA.

Several HLH proteins lack the α-helical extension responsible for binding to DNA. These truncated proteins can form heterodimers with full-length HLH proteins, but the heterodimers are unable to bind DNA tightly because they form only half of the necessary contacts. Thus, in addition to creating active dimers, heterodimerization provides cells with a widely used way to hold specific gene regulatory proteins in check (**Figure 7–24**).

It Is Not Yet Possible to Predict the DNA Sequences Recognized by All Gene Regulatory Proteins

The various DNA-binding motifs that we have discussed provide structural frameworks from which specific amino acid side chains extend to contact specific base pairs in the DNA. It is reasonable to ask, therefore, whether there is a simple amino acid–base pair recognition code: is a G–C base pair, for example, always contacted by a particular amino acid side chain? The answer is no, although certain types of amino acid-base interactions appear much more frequently than others (**Figure 7–25**). As we saw in Chapter 3, protein surfaces of virtually any shape and chemistry can be made from just 20 different amino acids, and a gene regulatory protein uses different patterns of these to create a surface that is precisely complementary to a particular DNA sequence. We know that the same base pair can thereby be recognized in many ways depending on its context (**Figure 7–26**). Nevertheless, molecular biologists are beginning to understand the principles of protein–DNA recognition well enough to design new proteins that will recognize a given DNA sequence.

Having outlined the general features of gene regulatory proteins, we turn to some of the methods that are now used to study them.

active HLH homodimer   inactive HLH heterodimer

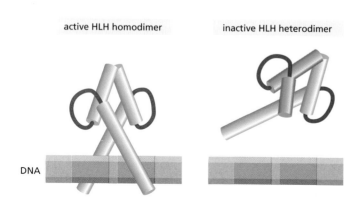

DNA

Figure 7–24 **Inhibitory regulation by truncated HLH proteins.** The HLH motif is responsible for both dimerization and DNA binding. On the *left,* an HLH homodimer recognizes a symmetric DNA sequence. On the *right,* the binding of a full-length HLH protein *(blue)* to a truncated HLH protein *(green)* that lacks the DNA-binding α helix generates a heterodimer that is unable to bind DNA tightly. If present in excess, the truncated protein molecule blocks the homodimerization of the full-length HLH protein and thereby prevents it from binding to DNA.

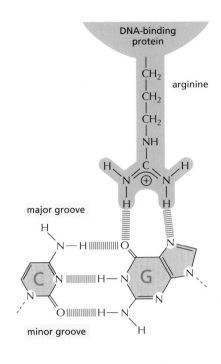

Figure 7–25 One of the most common protein–DNA interactions. Because of its specific geometry of hydrogen-bond acceptors (see Figure 7–7), the side chain of arginine unambiguously recognizes guanine. Figure 7–9 shows another common protein–DNA interaction.

A Gel-Mobility Shift Assay Readily Detects Sequence-Specific DNA-Binding Proteins

Genetic analyses, which provided a route to the gene regulatory proteins of bacteria, yeast, and *Drosophila*, are much more difficult in vertebrates. Therefore, the isolation of vertebrate gene regulatory proteins had to await the development of different approaches. Many of these approaches rely on the detection in a cell extract of a DNA-binding protein that specifically recognizes a DNA sequence known to control the expression of a particular gene. One of the most common ways to detect and study sequence-specific DNA-binding proteins is based on the effect of a bound protein on the migration of DNA molecules in an electric field.

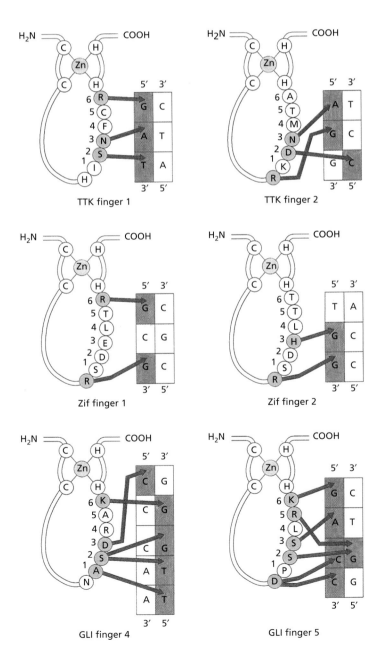

Figure 7–26 Summary of sequence-specific interactions between six different zinc fingers and their DNA recognition sequences. Even though all six Zn fingers have the same overall structure (see Figure 7–14), each binds to a different DNA sequence. The numbered amino acids form the α helix that recognizes DNA (Figures 7–14 and 7–15), and those that make sequence-specific DNA contacts are *green*. Bases contacted by protein are *orange*. Although arginine–guanine contacts are common (see Figure 7–25), guanine can also be recognized by serine, histidine, and lysine, as shown. Moreover, the same amino acid (serine, in this example) can recognize more than one base. Two of the Zn fingers depicted are from the TTK protein (a *Drosophila* protein that functions in development); two are from the mouse protein (Zif268) that was shown in Figure 7–15; and two are from a human protein (GLI) whose aberrant forms can cause certain types of cancers. (Adapted from C. Branden and J. Tooze, Introduction to Protein Structure, 2nd ed. New York: Garland Publishing, 1999.)

A DNA molecule is highly negatively charged and will therefore move rapidly toward a positive electrode when it is subjected to an electric field. When analyzed by polyacrylamide-gel electrophoresis (see p. 534), DNA molecules are separated according to their size because smaller molecules are able to penetrate the fine gel meshwork more easily than large ones. Protein molecules bound to a DNA molecule will cause it to move more slowly through the gel; in general, the larger the bound protein, the greater the retardation of the DNA molecule. This phenomenon provides the basis for the **gel-mobility shift assay**, which allows even trace amounts of a sequence-specific DNA-binding protein to be readily detected. In this assay, a short DNA fragment of specific length and sequence (produced either by DNA cloning or by chemical synthesis, as discussed in Chapter 8) is radioactively labeled and mixed with a cell extract; the mixture is then loaded onto a polyacrylamide gel and subjected to electrophoresis. If the DNA fragment corresponds to a chromosomal region where, for example, several sequence-specific proteins bind, autoradiography (see pp. 602–603) will reveal a series of DNA bands, each retarded to a different extent and representing a distinct DNA–protein complex. The proteins responsible for each band on the gel can then be separated from one another by subsequent fractionations of the cell extract (**Figure 7–27**). Once a sequence-specific DNA protein has been purified, the gel-mobility shift assay can be used to study the strength and specificity of its interactions with different DNA sequences, the lifetime of DNA–protein complexes, and other properties critical to the functioning of the protein in the cell.

DNA Affinity Chromatography Facilitates the Purification of Sequence-Specific DNA-Binding Proteins

A particularly powerful protein-purification method called **DNA affinity chromatography** can be used once the DNA sequence that a gene regulatory protein recognizes has been determined. A double-stranded oligonucleotide of the correct sequence is synthesized by chemical methods and linked to an insoluble porous matrix such as agarose; the matrix with the oligonucleotide attached is

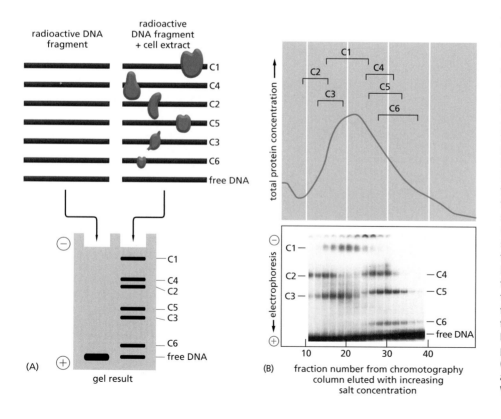

(A) gel result

(B) fraction number from chromotography column eluted with increasing salt concentration

Figure 7–27 A gel-mobility shift assay. The principle of the assay is shown schematically in (A). In this example an extract of an antibody-producing cell line is mixed with a radioactive DNA fragment containing about 160 nucleotides of a regulatory DNA sequence from a gene encoding the light chain of the antibody made by the cell line. The effect of the proteins in the extract on the mobility of the DNA fragment is analyzed by polyacrylamide-gel electrophoresis followed by autoradiography. The free DNA fragments migrate rapidly to the bottom of the gel, while those fragments bound to proteins are retarded; the finding of six retarded bands suggests that the extract contains six different sequence-specific DNA-binding proteins (indicated as C1–C6) that bind to this DNA sequence. (For simplicity, any DNA fragments with more than one protein bound have been omitted from the figure.) In (B) a standard chromatographic technique (see pp. 512–513) was used to fractionate the extract *(top),* and each fraction was mixed with the radioactive DNA fragment, applied to one lane of a polyacrylamide gel, and analyzed as in (A). (B, modified from C. Scheidereit, A. Heguy and R.G. Roeder, *Cell* 51:783–793, 1987. With permission from Elsevier.)

then used to construct a column that selectively binds proteins that recognize the particular DNA sequence (**Figure 7–28**). Purifications as great as 10,000-fold can be achieved by this means with relatively little effort.

Although most gene regulatory proteins are present at very low levels in the cell, enough pure protein can usually be isolated by affinity chromatography to obtain a partial amino acid sequence by mass spectrometry or other means (discussed in Chapter 8). If the complete genome sequence of the organism is known, the partial amino acid sequence can be used to identify the gene. The gene not only provides the complete amino acid sequence of the protein; it also provides the means to produce the protein in unlimited amounts through genetic engineering techniques, also discussed in Chapter 8.

The DNA Sequence Recognized by a Gene Regulatory Protein Can Be Determined Experimentally

Gene regulatory proteins can be discovered before the DNA sequence they recognize is known. For example, many of the *Drosophila* homeodomain proteins were discovered through the isolation of mutations that altered fly development. This allowed the genes encoding the proteins to be identified, and the proteins could then be overexpressed in cultured cells and easily purified. *DNA footprinting* is one method of determining the DNA sequences recognized by a gene regulatory protein once it has been purified. This strategy also requires a purified fragment of duplex DNA that contains somewhere within it a recognition site for the protein. Short recognition sequences can occur by chance on any long DNA fragment, although it is often necessary to use DNA corresponding to a regulatory region for a gene known to be controlled by the protein of interest. DNA footprinting is based on nucleases or chemicals that randomly cleave DNA at every phosphodiester bond. A bound gene regulatory protein blocks the phosphodiester bonds from attack, thereby revealing the protein's precise recognition site as a protected zone, or footprint (**Figure 7–29**).

A second way of determining the DNA sequences recognized by a gene regulatory protein requires no prior knowledge of what genes the protein might

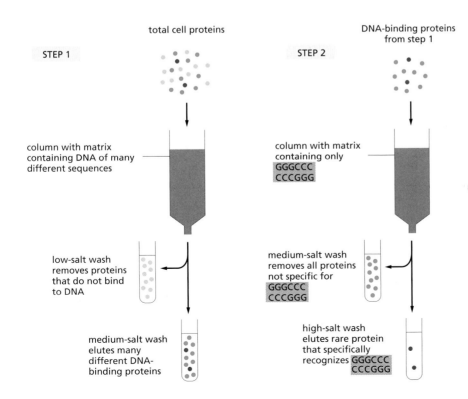

Figure 7–28 **DNA affinity chromatography.** In the first step, all the proteins that can bind DNA are separated from the remainder of the cell proteins on a column containing a huge number of different DNA sequences. Most sequence-specific DNA-binding proteins have a weak (nonspecific) affinity for bulk DNA and are therefore retained on the column. This affinity is due largely to ionic attractions, and the proteins can be washed off the DNA by a solution that contains a moderate concentration of salt. In the second step, the mixture of DNA-binding proteins is passed through a column that contains only DNA of a particular sequence. Typically, all the DNA-binding proteins will stick to the column, the great majority by nonspecific interactions. These are again eluted by solutions of moderate salt concentration, leaving on the column only those proteins (typically one or only a few) that bind specifically and therefore very tightly to the particular DNA sequence. These remaining proteins can be eluted from the column by solutions containing a very high concentration of salt.

(A)

region of DNA protected
by DNA-binding protein

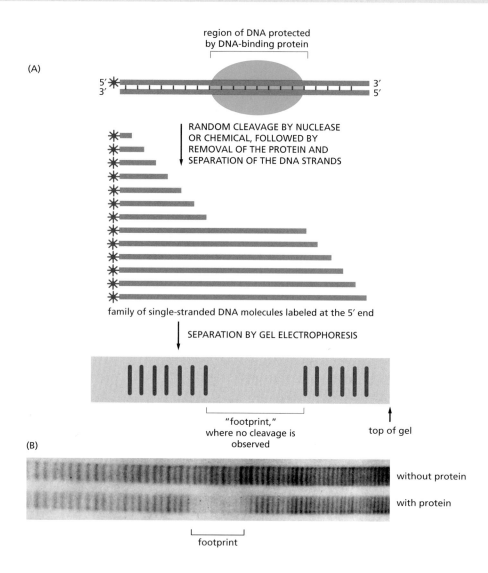

family of single-stranded DNA molecules labeled at the 5′ end

RANDOM CLEAVAGE BY NUCLEASE
OR CHEMICAL, FOLLOWED BY
REMOVAL OF THE PROTEIN AND
SEPARATION OF THE DNA STRANDS

SEPARATION BY GEL ELECTROPHORESIS

"footprint,"
where no cleavage is
observed

top of gel

(B)

without protein

with protein

footprint

Figure 7–29 DNA footprinting.
(A) Schematic of the method. A DNA fragment is labeled at one end with ^{32}P, a procedure described in Figure 8–34; next, the DNA is cleaved with a nuclease or chemical that makes random, single-stranded cuts. After the DNA molecule is denatured to separate its two strands, the resultant fragments from the labeled strand are separated on a gel and detected by autoradiography (see Figure 8–33). The pattern of bands from DNA cut in the presence of a DNA-binding protein is compared with that from DNA cut in its absence. When protein is present, it covers the nucleotides at its binding site and protects their phosphodiester bonds from cleavage. As a result, those labeled fragments that would otherwise terminate in the binding site are missing, leaving a gap in the gel pattern called a "footprint." In the example shown, the DNA-binding protein protects seven phosphodiester bonds from the DNA cleaving agent. (B) An actual footprint used to determine the binding site for a gene regulatory protein from humans. The cleaving agent was a small, iron-containing organic molecule that normally cuts at every phosphodiester bond with nearly equal frequency. (B, courtesy of Michele Sawadogo and Robert Roeder.)

regulate. Here, the purified protein is used to select, from a large, randomly generated pool of different short DNA fragments, only those that bind tightly to it. After several rounds of such selection, the nucleotide sequences of the tightly bound DNAs are determined, and a consensus DNA recognition sequence for the gene regulatory protein can be formulated (**Figure 7–30**). Once the DNA sequence recognized by a gene regulatory protein is known, computerized genome searches can identify candidate genes whose transcription the gene

Figure 7–30 A method for determining the DNA sequence recognized by a gene regulatory protein. A purified gene regulatory protein is mixed with millions of different short DNA fragments, each with a different sequence of nucleotides. A collection of such DNA fragments can be produced by programming a DNA synthesizer, a machine that chemically synthesizes DNA of any desired sequence (discussed in Chapter 8). For example, there are 4^{11}, or approximately 4.2 million, possible sequences for a DNA fragment of 11 nucleotides. The double-stranded DNA fragments that bind tightly to the gene regulatory protein are then separated from the DNA fragments that fail to bind. One method for accomplishing this separation is through gel-mobility shifts, as illustrated in Figure 7–27. After separation of the DNA–protein complexes from the free DNA, the DNA fragments are removed from the protein and typically used for several additional rounds of the same selection process (not shown). The nucleotide sequences of those DNA fragments that remain through multiple rounds of binding and release can be determined, and a consensus DNA recognition sequence can thus be generated.

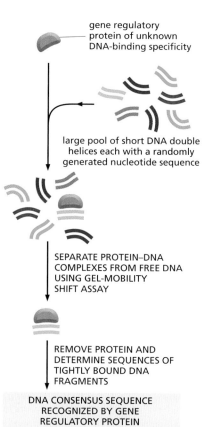

gene regulatory
protein of unknown
DNA-binding specificity

large pool of short DNA double
helices each with a randomly
generated nucleotide sequence

SEPARATE PROTEIN–DNA
COMPLEXES FROM FREE DNA
USING GEL-MOBILITY
SHIFT ASSAY

REMOVE PROTEIN AND
DETERMINE SEQUENCES OF
TIGHTLY BOUND DNA
FRAGMENTS

DNA CONSENSUS SEQUENCE
RECOGNIZED BY GENE
REGULATORY PROTEIN

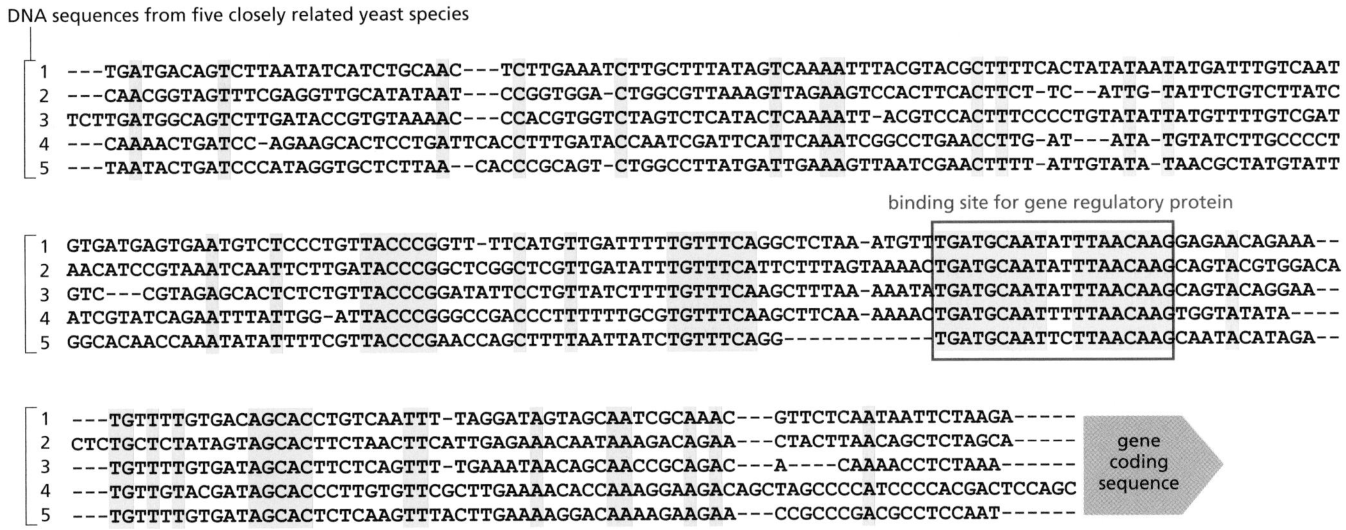

DNA sequences from five closely related yeast species

```
1  ---TGATGACAGTCTTAATATCATCTGCAAC---TCTTGAAATCTTGCTTTATAGTCAAAATTTACGTACGCTTTTCACTATATAATATGATTTGTCAAT
2  ---CAACGGTAGTTTCGAGGTTGCATATAAT---CCGGTGGA-CTGGCGTTAAAGTTAGAAGTCCACTTCACTTCT-TC--ATTG-TATTCTGTCTTATC
3  TCTTGATGGCAGTCTTGATACCGTGTAAAAC---CCACGTGGTCTAGTCTCATACTCAAAATT-ACGTCCACTTTCCCCTGTATATTATGTTTTGTCGAT
4  ---CAAAACTGATCC-AGAAGCACTCCTGATTCACCTTTGATACCAATCGATTCATTCAAATCGGCCTGAACCTTG-AT---ATA-TGTATCTTGCCCCT
5  ---TAATACTGATCCCATAGGTGCTCTTAA--CACCCGCAGT-CTGGCCTTATGATTGAAAGTTAATCGAACTTTT-ATTGTATA-TAACGCTATGTATT
```

binding site for gene regulatory protein

```
1  GTGATGAGTGAATGTCTCCCTGTTACCCGGTT-TTCATGTTGATTTTTGTTTCAGGCTCTAA-ATGTTTGATGCAATATTTAACAAGGAGAACAGAAA--
2  AACATCCGTAAATCAATTCTTGATACCCGGCTCGGCTCGTTGATATTTGTTTCATTCTTTAGTAAAACTGATGCAATATTTAACAAGCAGTACGTGGACA
3  GTC---CGTAGAGCACTCTCTGTTACCCGGATATTCCTGTTATCTTTTGTTTCAAGCTTTAA-AAATATGATGCAATATTTAACAAGCAGTACAGGAA--
4  ATCGTATCAGAATTTATTGG-ATTACCCGGGCCGACCCTTTTTTGCGTGTTTCAAGCTTCAA-AAAACTGATGCAATTTTTAACAAGTGGTATATA----
5  GGCACAACCAAATATATTTTCGTTACCCGAACCAGCTTTTAATTATCTGTTTCAGG-----------TGATGCAATTCTTAACAAGCAATACATAGA--
```

```
1  ---TGTTTTGTGACAGCACCTGTCAATTT-TAGGATAGTAGCAATCGCAAAC---GTTCTCAATAATTCTAAGA-----
2  CTCTGCTCTATAGTAGCACTTCTAACTTCATTGAGAAACAATAAAGACAGAA---CTACTTAACAGCTCTAGCA-----
3  ---TGTTTTGTGATAGCACTTCTCAGTTT-TGAAATAACAGCAACCGCAGAC---A----CAAAACCTCTAAA------
4  ---TGTTGTACGATAGCACCCTTGTGTTCGCTTGAAAACACCAAAGGAAGACAGCTAGCCCCATCCCCACGACTCCAGC
5  ---TGTTTTGTGATAGCACTCTCAAGTTTACTTGAAAAGGACAAAAGAAGAA---CCGCCCGACGCCTCCAAT------
```

gene
coding
sequence

Figure 7–31 Phylogenetic footprinting. This example compares DNA sequences upstream of the same gene from five closely related yeasts; identical nucleotides are highlighted in *yellow*. Phylogenetic footprinting reveals DNA recognition sites for regulatory proteins, as they are typically more conserved than surrounding sequences. Only the region upstream of a particular gene is shown in this example, but the approach is typically used to analyze entire genomes. The gene regulatory proteins that bind to the site outlined in *red* are shown in Figure 7–21. Some of the shorter phylogenetic footprints in this example represent binding sites for additional gene regulatory proteins, not all of which have been identified. (From M. Kellis et al., *Nature* 423:241–254, 2003, with permission from Macmillan Publishers Ltd., and D.J. Galgoczy et al., *Proc. Natl Acad. Sci. U.S.A.* 101:18069–18074, 2004, with permission from National Academy of Sciences.)

regulatory protein of interest might control. However, this strategy is not foolproof. For example, many organisms produce a set of closely related gene regulatory proteins that recognize very similar DNA sequences, and this approach cannot resolve them. In most cases, predictions of the sites of action of gene regulatory proteins obtained from searching genome sequences must, in the end, be tested experimentally.

Phylogenetic Footprinting Identifies DNA Regulatory Sequences Through Comparative Genomics

The widespread availability of complete genome sequences provides a surprisingly simple method for identifying important regulatory sites on DNA, even when the gene regulatory protein that binds them is unknown. In this approach, genomes from several closely related species are compared. If the species are chosen properly, the protein-coding portions of the genomes will be very similar, but the regions between sequences that encode protein or RNA molecules will have diverged considerably, as most of this sequence is functionally irrelevant and therefore not constrained in evolution. Among the exceptions are the regulatory sequences that control gene transcription. These stand out as conserved islands in a sea of nonconserved nucleotides (**Figure 7–31**). Although the identity of the gene regulatory proteins that recognize the conserved DNA sequences must be determined by other means, phylogenetic footprinting is a powerful method for identifying many of the DNA sequences that control gene expression.

Chromatin Immunoprecipitation Identifies Many of the Sites That Gene Regulatory Proteins Occupy in Living Cells

A gene regulatory protein will not occupy all of its potential DNA-binding sites in the genome at a particular time. Under some conditions, the protein may not be synthesized, and so will be absent from the cell; it may be present but lacking a heterodimer partner; or it may be excluded from the nucleus until an appropriate signal is received from the cell's environment. Even if the gene regulatory

Figure 7–32 **Chromatin immunoprecipitation.** This method allows the identification of all the sites in a genome that a gene regulatory protein occupies *in vivo*. For the amplification of DNA by a polymerase chain reaction (PCR), see Figure 8–45. The identities of the precipitated, amplified DNA fragments can be determined by hybridizing the mixture of fragments to DNA microarrays, as described in Chapter 8.

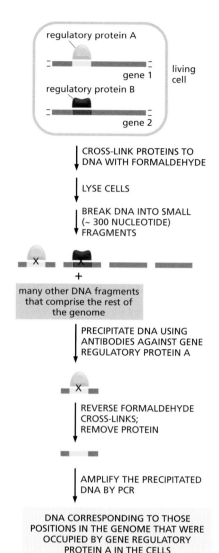

protein is present in the nucleus and is competent to bind DNA, components of chromatin or other gene regulatory proteins that can bind to the same or overlapping DNA sequences may occlude many of its potential binding sites on DNA.

Chromatin immunoprecipitation provides one way of empirically determining the sites on DNA that a given gene regulatory protein occupies under a particular set of conditions (**Figure 7–32**). In this approach, proteins are covalently cross-linked to DNA in living cells, the cells are broken open, and the DNA is mechanically sheared into small fragments. Antibodies directed against a given gene regulatory protein are then used to purify DNA that became covalently cross-linked to that protein in the cell. If this DNA is hybridized to microarrays that contain the entire genome displayed as a series of discrete DNA fragments (see Figure 8–73), the precise genomic location of each precipitated DNA fragment can be determined. In this way, all the sites occupied by the gene regulatory protein in the original cells can be mapped on the cell's genome (**Figure 7–33**).

Chromatin immunoprecipitation is also routinely used to identify the positions along a genome that are packaged by the various types of modified histones (discussed in Chapter 4). In this case, antibodies specific to the particular histone modification of interest are employed.

Summary

Gene regulatory proteins recognize short stretches of double-helical DNA of defined sequence and thereby determine which of the thousands of genes in a cell will be transcribed. Thousands of gene regulatory proteins have been identified in a wide variety of organisms. Although each of these proteins has unique features, most bind to DNA as homodimers or heterodimers and recognize DNA through one of a small number of structural motifs. The common motifs include the helix–turn–helix, the homeodomain, the leucine zipper, the helix–loop–helix, and zinc fingers of several types. The precise amino acid sequence that is folded into a motif determines the particular DNA sequence that a gene regulatory protein recognizes. Heterodimerization increases the range of DNA sequences that can be recognized. Powerful techniques are now available for identifying and isolating these proteins, the genes that encode them, and the DNA sequences they recognize, and for mapping all of the genes that they regulate on a genome.

HOW GENETIC SWITCHES WORK

In the previous section, we described the basic components of genetic switches: gene regulatory proteins and the specific DNA sequences that these proteins recognize. We shall now discuss how these components operate to turn genes on and off in response to a variety of signals.

In the mid-twentieth century, the idea that genes could be switched on and off was revolutionary. This concept was a major advance, and it came originally from the study of how *E. coli* bacteria adapt to changes in the composition of their growth medium. Parallel studies of the lambda bacteriophage led to many of the same conclusions and helped to establish the underlying mechanism. Many of the same principles apply to eukaryotic cells. However, the enormous complexity of gene regulation in higher organisms, combined with the packaging

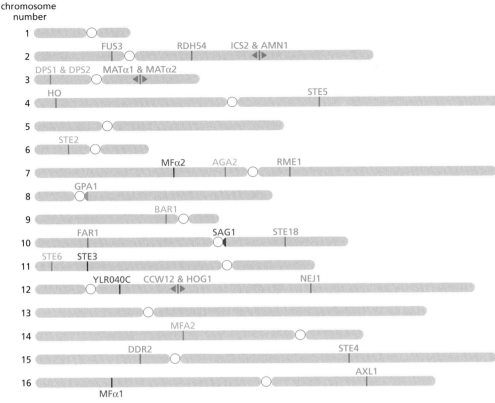

Figure 7–33 **A gene regulatory circuit: the complete set of genes controlled by three key regulatory proteins in budding yeast, as deduced from the DNA sites where the regulatory proteins bind.** The regulatory proteins—called Mata1, Matα1, and Matα2—specify the two different haploid mating types (analogous to male and female gametes) of this unicellular organism. The 16 chromosomes in the yeast genome are shown *(gray)*, with colored bars indicating sites where various combinations of the three regulatory proteins bind. Above each binding site is the name of the protein product of the regulated target gene. Matα1, acting in a complex with another protein, Mcm1, activates expression of the genes marked in *red;* Matα2, acting in a complex with Mcm1, represses the genes marked in *blue;* and Mata1 in a complex with Matα2 represses the genes marked in *green* (see Figures 7–21 and 7–65). *Double arrowheads* represent divergently transcribed genes, which are controlled by the indicated gene regulatory proteins. This complete map of bound regulatory proteins was determined using a combination of genome-wide chromatin immunoprecipitation (see Figure 7–32) and phylogenetic footprinting (see Figure 7–29). Such determinations of complete transcriptional circuits show that transcriptional networks are not infinitely complex, although they may appear that way initially. This type of study also helps to reveal the overall logic of the transcriptional circuits used by modern cells. (From D.J. Galgoczy et al., *Proc. Natl Acad. Sci. U.S.A.* 101:18069–18074, 2004. With permission from National Academy of Sciences.)

of their DNA into chromatin, creates special challenges and some novel opportunities for control—as we shall see. We begin with the simplest example—an on–off switch in bacteria that responds to a single signal.

The Tryptophan Repressor Is a Simple Switch That Turns Genes On and Off in Bacteria

The chromosome of the bacterium *E. coli*, a single-celled organism, is a single circular DNA molecule of about 4.6×10^6 nucleotide pairs. This DNA encodes approximately 4300 proteins, although the cell makes only a fraction of these at any one time. The expression of many genes is regulated according to the available food in the environment. This is illustrated by the five *E. coli* genes that code for enzymes that manufacture the amino acid tryptophan. These genes are arranged as a single **operon**; that is, they are adjacent to one another on the chromosome and are transcribed from a single *promoter* as one long mRNA molecule (**Figure 7–34**). But when tryptophan is present in the growth medium and enters the cell (when the bacterium is in the gut of a mammal that has just eaten a meal of protein, for example), the cell no longer needs these enzymes and shuts off their production.

The molecular basis for this switch is understood in considerable detail. As described in Chapter 6, a promoter is a specific DNA sequence that directs RNA polymerase to bind to DNA, to open the DNA double helix, and to begin

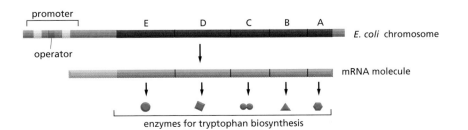

Figure 7–34 **The clustered genes in *E. coli* that code for enzymes that manufacture the amino acid tryptophan.** These five genes of the *Trp* operon—denoted as *TrpA, B, C, D,* and *E*—are transcribed as a single mRNA molecule, which allows their expression to be controlled coordinately. Clusters of genes transcribed as a single mRNA molecule are common in bacteria. Each such cluster is called an operon.

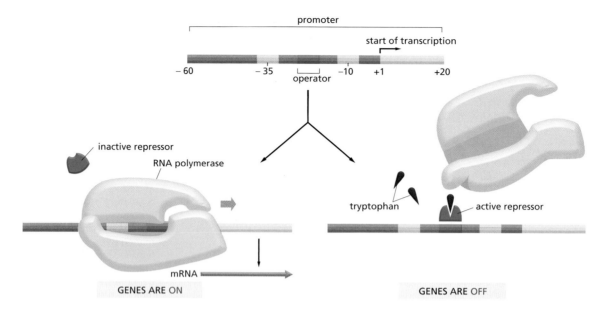

synthesizing an RNA molecule. Within the promoter that directs transcription of the tryptophan biosynthetic genes lies a regulator element called an **operator** (see Figure 7–34). This is simply a short region of regulatory DNA of defined nucleotide sequence that is recognized by a repressor protein, in this case the **tryptophan repressor**, a member of the helix–turn–helix family (see Figure 7–11). The promoter and operator are arranged so that when the tryptophan repressor occupies the operator, it blocks access to the promoter by RNA polymerase, thereby preventing expression of the tryptophan-producing enzymes (**Figure 7–35**).

The block to gene expression is regulated in an ingenious way: to bind to its operator DNA, the repressor protein has to have two molecules of the amino acid tryptophan bound to it. As shown in **Figure 7–36**, tryptophan binding tilts the helix–turn–helix motif of the repressor so that it is presented properly to the DNA major groove; without tryptophan, the motif swings inward and the protein is unable to bind to the operator. Thus, the tryptophan repressor and operator form a simple device that switches production of the tryptophan biosynthetic enzymes on and off according to the availability of free tryptophan.

Figure 7–35 Switching the tryptophan genes on and off. If the level of tryptophan inside the cell is low, RNA polymerase binds to the promoter and transcribes the five genes of the tryptophan (*Trp*) operon. If the level of tryptophan is high, however, the tryptophan repressor is activated to bind to the operator, where it blocks the binding of RNA polymerase to the promoter. Whenever the level of intracellular tryptophan drops, the repressor releases its tryptophan and becomes inactive, allowing the polymerase to begin transcribing these genes. The promoter includes two key blocks of DNA sequence information, the –35 and –10 regions highlighted in *yellow* (see Figure 6–12).

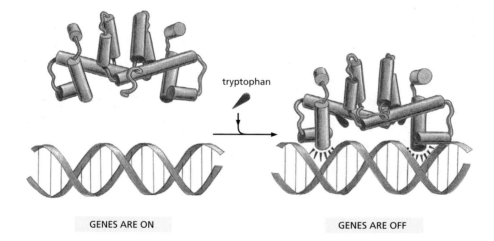

Figure 7–36 The binding of tryptophan to the tryptophan repressor protein changes its conformation. This structural change enables this gene regulatory protein to bind tightly to a specific DNA sequence (the operator), thereby blocking transcription of the genes encoding the enzymes required to produce tryptophan (the *Trp* operon). The three-dimensional structure of this bacterial helix–turn–helix protein, as determined by x-ray diffraction with and without tryptophan bound, is illustrated. Tryptophan binding increases the distance between the two recognition helices in the homodimer, allowing the repressor to fit snugly on the operator. (Adapted from R. Zhang et al., *Nature* 327:591–597, 1987. With permission from Macmillan Publishers Ltd.)

Because the active, DNA-binding form of the protein serves to turn genes off, this mode of gene regulation is called **negative control**, and the gene regulatory proteins that function in this way are called *transcriptional repressors* or *gene repressor proteins*.

Transcriptional Activators Turn Genes On

We saw in Chapter 6 that purified *E. coli* RNA polymerase (including its σ subunit) can bind to a promoter and initiate DNA transcription. Many bacterial promoters, however, are only marginally functional on their own, either because they are recognized poorly by RNA polymerase or because the polymerase has difficulty opening the DNA helix and beginning transcription. In either case these poorly functioning promoters can be rescued by gene regulatory proteins that bind to a nearby site on the DNA and contact the RNA polymerase in a way that dramatically increases the probability that a transcript will be initiated. Because the active, DNA-binding form of such a protein turns genes on, this mode of gene regulation is called **positive control**, and the gene regulatory proteins that function in this manner are known as *transcriptional activators* or *gene activator proteins*. In some cases, bacterial gene activator proteins aid RNA polymerase in binding to the promoter by providing an additional contact surface for the polymerase. In other cases, they contact RNA polymerase and facilitate its transition from the initial DNA-bound conformation of polymerase to the actively transcribing form by stabilizing a transition state of the enzyme. Like repressors, gene activator proteins must be bound to DNA to exert their effects. In this way, each regulatory protein acts selectively, controlling only those genes that bear a DNA sequence recognized by it.

DNA-bound activator proteins can increase the rate of transcription initiation up to 1000-fold, a value consistent with a relatively weak and nonspecific interaction between the activator and RNA polymerase. For example, a 1000-fold change in the affinity of RNA polymerase for its promoter corresponds to a change in ΔG of ~4 kcal/mole, which could be accounted for by just a few weak, noncovalent bonds. Thus gene activator proteins can work simply by providing a few favorable interactions that help to attract RNA polymerase to the promoter.

As in negative control by a transcriptional repressor, a transcriptional activator can operate as part of a simple on–off genetic switch. The bacterial activator protein *CAP (catabolite activator protein)*, for example, activates genes that enable *E. coli* to use alternative carbon sources when glucose, its preferred carbon source, is unavailable. Falling levels of glucose cause an increase in the intracellular signaling molecule cyclic AMP, which binds to the CAP protein, enabling it to bind to its specific DNA sequence near target promoters and thereby turn on the appropriate genes. In this way the expression of a target gene is switched on or off, depending on whether cyclic AMP levels in the cell are high or low, respectively. **Figure 7–37** summarizes the different ways that positive and negative control can be used to regulate genes.

Transcriptional activators and transcriptional repressors are similar in design. The tryptophan repressor and the transcriptional activator CAP, for example, both use a helix–turn–helix motif (see Figure 7–11) and both require a small cofactor in order to bind DNA. In fact, some bacterial proteins (including CAP and the bacteriophage lambda repressor) can act as either activators or repressors, depending on the exact placement of the DNA sequence they recognize in relation to the promoter: if the binding site for the protein overlaps the promoter, the polymerase cannot bind and the protein acts as a repressor (**Figure 7–38**).

A Transcriptional Activator and a Transcriptional Repressor Control the *Lac* Operon

More complicated types of genetic switches combine positive and negative controls. The *Lac operon* in *E. coli*, for example, unlike the *Trp operon*, is under both

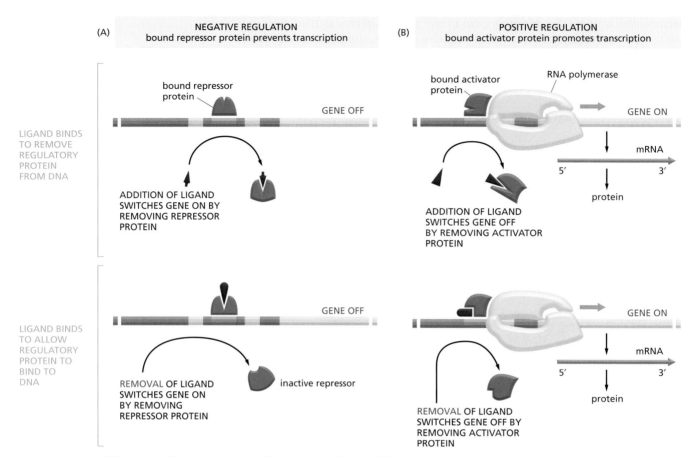

(A) NEGATIVE REGULATION
bound repressor protein prevents transcription

(B) POSITIVE REGULATION
bound activator protein promotes transcription

LIGAND BINDS TO REMOVE REGULATORY PROTEIN FROM DNA

bound repressor protein

GENE OFF

ADDITION OF LIGAND SWITCHES GENE ON BY REMOVING REPRESSOR PROTEIN

bound activator protein

RNA polymerase

GENE ON

mRNA

5′ 3′

protein

ADDITION OF LIGAND SWITCHES GENE OFF BY REMOVING ACTIVATOR PROTEIN

LIGAND BINDS TO ALLOW REGULATORY PROTEIN TO BIND TO DNA

GENE OFF

REMOVAL OF LIGAND SWITCHES GENE ON BY REMOVING REPRESSOR PROTEIN

inactive repressor

GENE ON

mRNA

5′ 3′

protein

REMOVAL OF LIGAND SWITCHES GENE OFF BY REMOVING ACTIVATOR PROTEIN

Figure 7–37 Summary of the mechanisms by which specific gene regulatory proteins control gene transcription in procaryotes. (A) Negative regulation; (B) positive regulation. Note that the addition of an "inducing" ligand can turn on a gene either by removing a gene repressor protein from the DNA *(upper left panel)* or by causing a gene activator protein to bind *(lower right panel)*. Likewise, the addition of an "inhibitory" ligand can turn off a gene either by removing a gene activator protein from the DNA *(upper right panel)* or by causing a gene repressor protein to bind *(lower left panel).*

negative and positive transcriptional controls by the Lac repressor protein and CAP, respectively. The *Lac* operon codes for proteins required to transport the disaccharide lactose into the cell and to break it down. CAP, as we have seen, enables bacteria to use alternative carbon sources such as lactose in the absence of glucose. It would be wasteful, however, for CAP to induce expression of the *Lac* operon if lactose is not present, and the Lac repressor ensures that the *Lac* operon is shut off in the absence of lactose. This arrangement enables the control region of the *Lac* operon to respond to and integrate two different signals, so that the operon is highly expressed only when two conditions are met: lactose must be present and glucose must be absent. In any of the other three possible signal combinations, the cluster of genes is held in the off state (**Figure 7–39**).

The simple logic of this genetic switch first attracted the attention of biologists over 50 years ago. As explained above, the molecular basis of the switch was uncovered by a combination of genetics and biochemistry, providing the first glimpse into how gene expression is controlled.

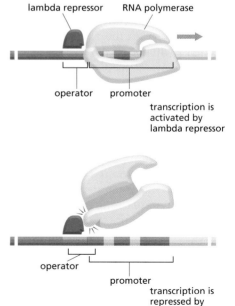

lambda repressor   RNA polymerase

operator   promoter

transcription is activated by lambda repressor

operator

promoter

transcription is repressed by lambda repressor

Figure 7–38 Some bacterial gene regulatory proteins can act as either a transcriptional activator or a repressor, depending on the precise placement of their DNA-binding sites. An example is the bacteriophage lambda repressor. For some genes, the protein acts as a transcriptional activator by providing a favorable contact for RNA polymerase *(top).* At other genes *(bottom),* the operator is located one base pair closer to the promoter, and, instead of helping polymerase, the repressor now competes with it for binding to the DNA. The lambda repressor recognizes its operator by a helix–turn–helix motif, as shown in Figure 7–11.

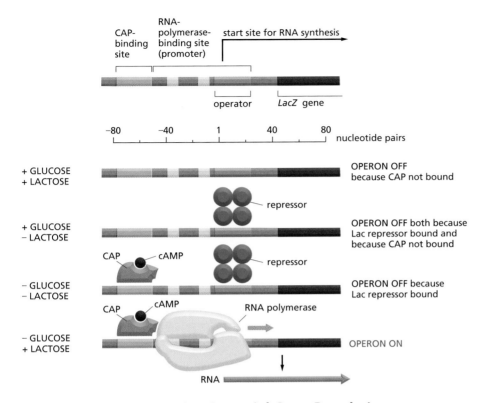

Figure 7–39 Dual control of the *Lac* operon. Glucose and lactose levels control the initiation of transcription of the *Lac* operon through their effects on CAP and the *Lac* repressor protein, respectively. *LacZ*, the first gene of the *Lac* operon, encodes the enzyme β-galactosidase, which breaks down the disaccharide lactose to galactose and glucose. Lactose addition increases the concentration of allolactose, an isomer of lactose, which binds to the repressor protein and removes it from the DNA. Glucose addition decreases the concentration of cyclic AMP; because cyclic AMP no longer binds to CAP, this gene activator protein dissociates from the DNA, turning off the operon.

This figure summarizes the essential features of the *Lac* operon, but in reality the situation is more complex. There are several Lac repressor binding sites located at different positions along the DNA. Although the one illustrated exerts the greatest effect, the others are required for full repression (see Figure 7–40). In addition, expression of the *Lac* operon never completely shuts down. A small amount of the enzyme β-galactosidase is required to convert lactose to allolactose, thereby permitting the Lac repressor to be inactivated when lactose is added to the growth medium.

DNA Looping Occurs During Bacterial Gene Regulation

The control of the Lac operon as shown in Figure 7–39 is simple and economical, but the continued study of this and other examples of bacterial gene regulation revealed a new feature of gene regulation, known as *DNA looping*. The *Lac* operon was originally thought to contain a single operator, but subsequent work revealed additional, secondary operators located nearby. A single tetrameric molecule of the Lac repressor can bind two operators simultaneously, looping out the intervening DNA. The ability to bind simultaneously to two operators strengthens the overall interaction of the Lac repressor with DNA and thereby leads to greater levels of repression in the cell (**Figure 7–40**).

DNA looping also allows two different proteins bound along a DNA double helix to contact one another readily. The DNA can be thought of as a tether, helping one DNA-bound protein interact with another even though thousands of nucleotide pairs may separate the binding sites for the two proteins (**Figure 7–41**). We shall see below that DNA looping is especially important in eucaryotic

Figure 7–40 DNA looping can stabilize protein–DNA interactions. The Lac repressor, a tetramer, can simultaneously bind to two operators. The *Lac* operon has a total of three operators, but for simplicity, only two are shown here, the main operator (O_m) and an auxiliary operator (O_a). The figure shows all the possible states of the Lac repressor bound to these two operators. At the concentrations of Lac repressor in the cell, and in the absence of lactose, the state in the lower right is the most stable, and to dissociate completely from the DNA, the Lac repressor must first pass through an intermediate where it is bound to only a single operator. In these states, the local concentration of the repressor is very high in relation to the free operator, and the reaction to the double-bound form is favored over the dissociation reaction. In this way, even a low-affinity site (O_a) can increase the occupancy of a high-affinity site (O_m) and give higher levels of gene repression in the cell. (Adapted from J.M.G. Vilar and S. Leibler, *J. Mol. Biol.* 331:981–989, 2003. With permission from Academic Press.)

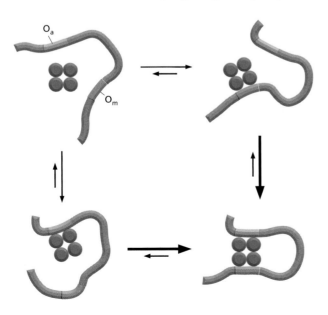

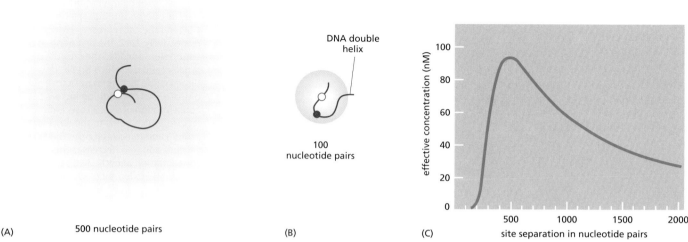

(A)   500 nucleotide pairs

(B)

(C)

Figure 7–41 Binding of two proteins to separate sites on the DNA double helix can greatly increase their probability of interacting. (A) The tethering of one protein to the other via an intervening DNA loop of 500 nucleotide pairs increases their frequency of collision. The intensity of the *blue* coloring at each point in space indicates the probability that the *red* protein will be located at that distance from the *white* protein. (B) The flexibility of DNA is such that an average sequence makes a smoothly graded 90° bend (a curved turn) about once every 200 nucleotide pairs. Thus, when only 100 nucleotide pairs tethers two proteins, the contact between those proteins is relatively restricted. In such cases the protein interaction is facilitated when the two protein-binding sites are separated by a multiple of about 10 nucleotide pairs, which places both proteins on the same side of the DNA helix (which has about 10 nucleotides per turn) and thus on the inside of the DNA loop, where they can best reach each other. (C) The theoretical effective concentration of the *red* protein at the site where the *white* protein is bound, as a function of their separation. Experiments suggest that the actual effective concentrations at short distances are greater than those predicted here. (C, courtesy of Gregory Bellomy, modified from M.C. Mossing and M.T. Record, *Science* 233:889–892, 1986. With permission from AAAS.)

gene regulation. However, it also plays crucial roles in many examples of bacterial gene regulation in addition to that of the *Lac* operon. For example, DNA looping readily allows the bacterial gene activator protein NtrC to contact RNA polymerase directly even though the two proteins are bound several hundred nucleotide pairs apart (**Figure 7–42**).

Bacteria Use Interchangeable RNA Polymerase Subunits to Help Regulate Gene Transcription

We have seen the importance of gene regulatory proteins that bind to sequences of DNA and signal to RNA polymerase whether or not to start the synthesis of an

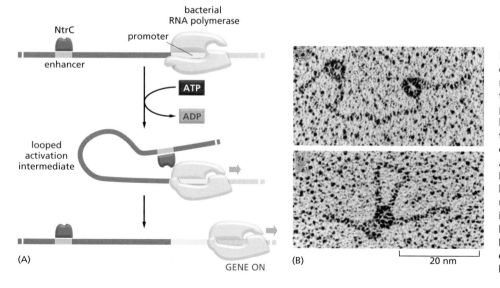

Figure 7–42 Gene activation at a distance. (A) NtrC is a bacterial gene regulatory protein that activates transcription by directly contacting RNA polymerase and causing a transition between the initial DNA-bound form of the polymerase and the transcriptionally competent form (discussed in Chapter 6). As indicated, the transition stimulated by NtrC requires the energy from ATP hydrolysis, although this requirement is unusual for bacterial transcription initiation. (B) The interaction of NtrC and RNA polymerase, with the intervening DNA looped out, can be seen in the electron microscope. (B, courtesy of Harrison Echols and Sydney Kustu.)

Table 7–2 Sigma Factors of *E. coli*

SIGMA FACTOR	PROMOTERS RECOGNIZED
σ^{70}	most genes
σ^{32}	genes induced by heat shock
σ^{28}	genes for stationary phase and stress response
σ^{28}	genes involved in motility and chemotaxis
σ^{54}	genes for nitrogen metabolism
σ^{24}	genes dealing with misfolded proteins in the periplasm

The sigma factor designations refer to their approximate molecular weights, in kilodaltons.

RNA chain. Although this is one of the main ways in which both eucaryotes and procaryotes control transcription initiation, some bacteria and their viruses use an additional strategy based on interchangeable subunits of RNA polymerase. As described in Chapter 6, a sigma (σ) subunit is required for the bacterial RNA polymerase to recognize a promoter. Most bacteria produce a whole range of sigma subunits, each of which can interact with the RNA polymerase core and direct it to a different set of promoters (**Table 7–2**). This scheme permits one large set of genes to be turned off and a new set to be turned on simply by replacing one sigma subunit with another; the strategy is efficient because it bypasses the need to deal with genes one by one. Indeed, some bacteria code for nearly one hundred different sigma subunits and therefore rely heavily on this form of gene regulation. Bacterial viruses often use it subversively to take over the host polymerase and activate several sets of viral genes rapidly and sequentially (**Figure 7–43**).

Complex Switches Have Evolved to Control Gene Transcription in Eucaryotes

Bacteria and eucaryotes share many principles of gene regulation, including the key role played by gene regulatory proteins that bind tightly to short stretches of DNA, the importance of weak protein–protein actions in gene activation, and the versatility afforded by DNA looping. However, by comparison, gene regulation in eucaryotes involves many more proteins, much longer stretches of DNA, and often seems bewilderingly complex. This increased complexity provides the eucaryotic cell with an important advantage. Genetic switches in bacteria, as we have seen, typically respond to one or a few signals. But in eucaryotes it is common for dozens of signals to converge on a single promoter, with the transcription machinery integrating all these different signals to produce the appropriate level of mRNA. We begin our description of eucaryotic gene regulation by outlining the main features that distinguish it from gene regulation in bacteria.

- As discussed in Chapter 6, eucaryotic RNA polymerase II, which transcribes all the protein-coding genes, requires five general transcription factors (27 subunits *in toto*, see Table 6–3, p. 341), whereas bacterial RNA polymerase needs only a single general transcription factor, the σ subunit. As we have seen, the stepwise assembly of the general transcription factors at a eucaryotic promoter provides, in principle, multiple steps at which the

Figure 7–43 Interchangeable RNA polymerase subunits as a strategy to control gene expression in a bacterial virus. The bacterial virus SPO1, which infects the bacterium *B. subtilis,* uses the bacterial polymerase to transcribe its early genes immediately after the viral DNA enters the cell. One of the early genes, called 28, encodes a sigmalike factor that binds to RNA polymerase and displaces the bacterial sigma factor. This new form of polymerase specifically initiates transcription of the SPO1 "middle" genes. One of the middle genes encodes a second sigmalike factor, 34, that displaces the 28 product and directs RNA polymerase to transcribe the "late" genes. This last set of genes produces the proteins that package the virus chromosome into a virus coat and lyse the cell. By this strategy, sets of virus genes are expressed in the order in which they are needed; this ensures a rapid and efficient viral replication.

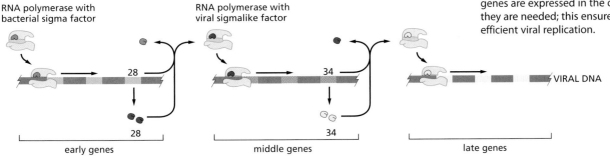

RNA polymerase with bacterial sigma factor

RNA polymerase with viral sigmalike factor

28

34

VIRAL DNA

28

34

early genes

middle genes

late genes

cell can speed up or slow down the rate of transcription initiation in response to gene regulatory proteins.

- Eucaryotic cells lack operons—sets of related genes transcribed as a unit—and therefore must regulate each gene individually.
- Each bacterial gene is typically controlled by one or only a few gene regulatory proteins, but it is common in eucaryotes for genes to be controlled by many (sometimes hundreds) of different regulatory proteins. This complexity is possible because, as we shall see, many eucaryotic gene regulatory proteins can act over very large distances (tens of thousands of nucleotide pairs) along DNA, allowing an almost unlimited number of them to influence the expression of a single gene.
- A central component of gene regulation in eucaryotes is *Mediator*, a 24-subunit complex, which serves as an intermediary between gene regulatory proteins and RNA polymerase (see Figure 6–19). Mediator provides an extended contact area for gene regulatory proteins compared to that provided by RNA polymerase alone, as in bacteria.
- The packaging of eucaryotic DNA into chromatin provides many opportunities for transcriptional regulation not available to bacteria.

Having discussed the general transcription factors for RNA polymerase II in Chapter 6 (see pp. 340–343), we focus here on the last four of these features and how they are used to control eucaryotic gene expression.

A Eucaryotic Gene Control Region Consists of a Promoter Plus Regulatory DNA Sequences

Because the typical eucaryotic gene regulatory protein controls transcription when bound to DNA far away from the promoter, the DNA sequences that control the expression of a gene are often spread over long stretches of DNA. We use the term **gene control region** to describe the whole expanse of DNA involved in regulating and initiating transcription of a gene, including the **promoter**, where the general transcription factors and the polymerase assemble, and all of the **regulatory sequences** to which gene regulatory proteins bind to control the rate of the assembly processes at the promoter (**Figure 7–44**). In animals and plants, it is not unusual to find the regulatory sequences of a gene dotted over distances as great as 50,000 nucleotide pairs. Much of this DNA serves as "spacer" sequences that gene regulatory proteins do not directly recognize, but this DNA may provide the flexibility needed for efficient DNA looping. In this context, it is important to remember that, like other regions of eucaryotic chromosomes, most of the DNA in gene control regions is packaged into nucleosomes and higher-order forms of chromatin, thereby compacting its length and altering its properties.

In this chapter, we shall loosely use the term **gene** to refer only to a segment of DNA that is transcribed into RNA (see Figure 7–44). However, the classical view of a gene includes the gene control region as well, making most eucaryotic genes considerably larger. The discovery of alternative RNA splicing has further complicated the definition of a gene—a point we discussed briefly in Chapter 6 and will return to later in this chapter.

It is the gene regulatory proteins that allow the genes of an organism to be turned on or off individually. In contrast to the small number of general transcription factors, which are abundant proteins that assemble on the promoters of all genes transcribed by RNA polymerase II, there are thousands of different gene regulatory proteins. For example, of the roughly 25,000 human genes, an estimated 8% (~2000 genes) encode gene regulatory proteins. Most of these recognize DNA sequences using one of the DNA-binding motifs described previously. Not surprisingly, the eucaryotic cell regulates each of its many genes in a unique way. Given the sheer number of genes in eucaryotes and the complexity of their regulation, it has been difficult to formulate simple rules for gene regulation that apply in every case. We can, however, make some generalizations about how gene regulatory proteins, once bound to gene control regions on DNA, set in motion the train of events that lead to gene activation or repression.

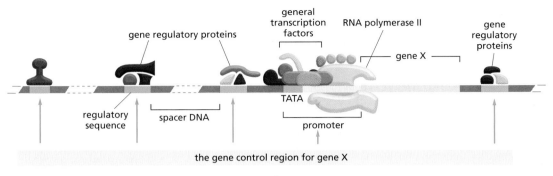

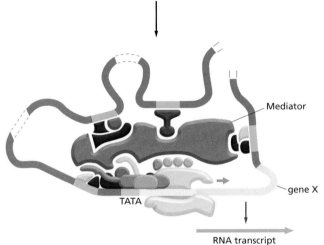

Figure 7–44 **The gene control region for a typical eucaryotic gene.** The *promoter* is the DNA sequence where the general transcription factors and the polymerase assemble (see Figure 6–16). The *regulatory sequences* serve as binding sites for gene regulatory proteins, whose presence on the DNA affects the rate of transcription initiation. These sequences can be located adjacent to the promoter, far upstream of it, or even within introns or downstream of the gene. As shown in the lower panel, DNA looping allows gene regulatory proteins bound at any of these positions to interact with the proteins that assemble at the promoter. Many gene regulatory proteins act through Mediator, while others influence the general transcription factors and RNA polymerase directly. Although not shown here, many gene regulatory proteins also influence the chromatin structure of the DNA control region thereby affecting transcription initiation indirectly (see Figure 4–45). As noted in the text, for simplicity, "gene X" refers here to the coding sequence within the gene.

Whereas Mediator and the general transcription factors are the same for all polymerase II transcribed genes, the gene regulatory proteins and the locations of their binding sites relative to the promoter differ for each gene.

Eucaryotic Gene Activator Proteins Promote the Assembly of RNA Polymerase and the General Transcription Factors at the Startpoint of Transcription

The DNA sites to which eucaryotic gene activator proteins bind were originally called *enhancers* because their presence "enhanced" the rate of transcription initiation. It came as a surprise when it was first discovered that these activator proteins could be bound tens of thousands of nucleotide pairs away from the promoter, but, as we have seen, DNA looping provides at least one explanation for this initially puzzling observation.

The simplest gene activator proteins have a modular design consisting of two distinct domains. One domain usually contains one of the structural motifs discussed previously that recognizes a specific DNA sequence. The second domain—sometimes called an *activation domain*—accelerates the rate of transcription initiation. This type of modular design was first revealed by experiments in which genetic engineering techniques were used to create a chimeric protein containing the activation domain of one protein fused to the DNA-binding domain of a different protein (**Figure 7–45**).

Once bound to DNA, how do eucaryotic gene activator proteins increase the rate of transcription initiation? As we will see shortly, there are several mechanisms by which this can occur, and, in many cases, these different mechanisms work in concert at a single promoter. But, regardless of the precise biochemical pathway, the ultimate function of activators is to attract, position, and modify the general transcription factors, Mediator, and RNA polymerase II at the promoter so that transcription can begin. They do this both by acting directly on these components and, indirectly, by changing the chromatin structure around the promoter.

Some activator proteins bind directly to one or more of the general transcription factors, accelerating their assembly on a promoter that is linked through DNA to that activator. Others interact with Mediator and attract it to DNA where it can then facilitate assembly of RNA polymerase and the general transcription factors at the promoter (see Figure 7–44). In this sense, eucaryotic

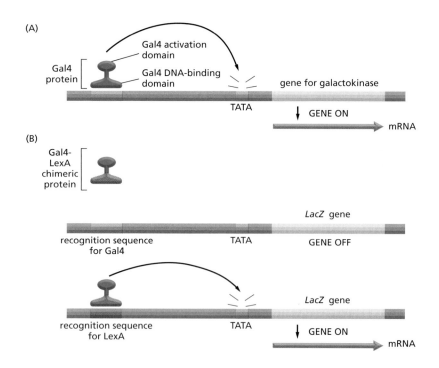

(A)

Gal4 activation domain

Gal4 protein

Gal4 DNA-binding domain

gene for galactokinase

TATA

GENE ON

mRNA

(B)

Gal4-LexA chimeric protein

LacZ gene

recognition sequence for Gal4

TATA

GENE OFF

LacZ gene

recognition sequence for LexA

TATA

GENE ON

mRNA

Figure 7–45 The modular structure of a gene activator protein. Outline of an experiment that reveals the presence of independent DNA-binding and transcription-activating domains in the yeast gene activator protein Gal4. A functional activator can be reconstituted from the C-terminal portion of the yeast Gal4 protein if it is attached to the DNA-binding domain of a bacterial gene regulatory protein (the LexA protein) by genetic engineering techniques. When the resulting bacterial–yeast hybrid protein is produced in yeast cells, it will activate transcription from yeast genes provided that the specific DNA-binding site for the bacterial protein has been inserted next to them. (A) Gal4 is normally responsible for activating the transcription of yeast genes that code for the enzymes that convert galactose to glucose. (B) A chimeric gene regulatory protein, produced by genetic engineering techniques, requires a LexA recognition sequence to activate transcription. In the experiment shown, the control region for one of the genes controlled by LexA was fused to the *E. coli LacZ* gene, which codes for the enzyme β-galactosidase (see Figure 7–39). β-Galactosidase is very simple to detect biochemically and thus provides a convenient way to monitor the expression level specified by a gene control region. As used here, *LacZ* is said to serve as a *reporter gene,* since it "reports" the activity of a gene control region.

activators resemble those of bacteria in recruiting RNA polymerase to specific sites on DNA so it can begin transcribing.

Eucaryotic Gene Activator Proteins Also Modify Local Chromatin Structure

The general transcription factors, Mediator, and RNA polymerase seem unable on their own to assemble on a promoter that is packaged in standard nucleosomes. Indeed, it has been proposed such packaging may have evolved to prevent "leaky" transcription. In addition to their direct actions in assembling the transcription machinery at the promoter, gene activator proteins also promote transcription initiation by changing the chromatin structure of the regulatory sequences and promoters of genes.

As we saw in Chapter 4, four of the most important ways of locally altering chromatin are through covalent histone modifications, nucleosome remodeling, nucleosome removal, and nucleosome replacement. Gene activator proteins use all four of these mechanisms by attracting histone modification enzymes, ATP-dependent chromatin remodeling complexes, and histone chaperones to alter the chromatin structure of promoters they control (**Figure 7–46**). In general terms, these local alterations in chromatin structure are believed to make the underlying DNA more accessible, thereby facilitating the assembly of the general transcription factors, Mediator, and RNA polymerase at the promoter. Local chromatin modification also allows additional gene regulatory proteins to bind to the control region of the gene. However, the most important role of covalent histone modifications in transcription is probably not in directly changing chromatin structure; rather, as discussed in Chapter 4, these modifications provide favorable interactions for the binding of a large set of proteins that read a "histone code." For transcription initiation, these proteins include other histone-modifying enzymes (reader–writer complexes), chromatin remodeling complexes, and at least one of the general transcription factors (**Figure 7–47**).

The alterations of chromatin structure that occur during transcription initiation can persist for variable lengths of time. In some cases, as soon as the gene regulatory protein dissociates from DNA, the chromatin modifications are rapidly reversed, restoring the gene to its pre-activated state. This rapid reversal is especially important for genes that the cell must quickly switch on and off in

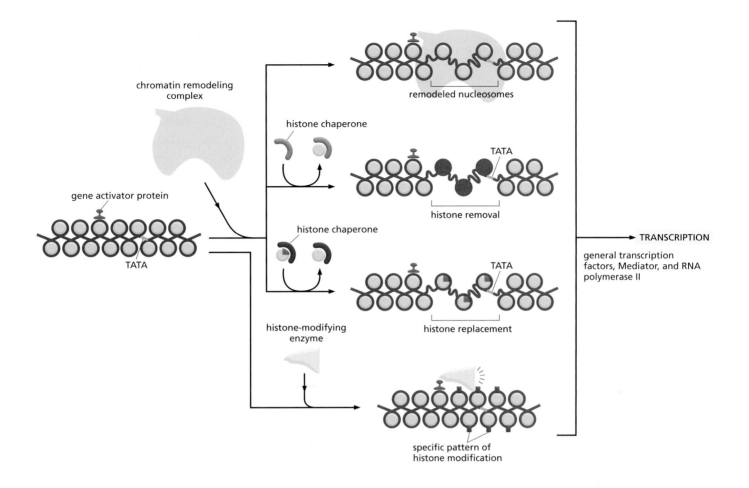

Figure 7–46 Four ways eucaryotic activator proteins can direct local alterations in chromatin structure to stimulate transcription initiation. Although shown as separate pathways, these mechanisms often work together during the activation of a gene. For example, prior acetylation of histones makes it easier for histone chaperones to remove them from nucleosomes. A few patterns of histone modification that promote transcription initiation are listed in Figure 4–44, and a specific example is given in Figure 7–47. Nucleosome remodeling and histone removal favor transcription initiation by increasing the accessibility of DNA and thereby facilitating the binding of Mediator, RNA polymerase, and the general transcription factors as well as additional activator proteins. Transcription initiation and the formation of a compact chromatin structure can be regarded as competing biochemical assembly reactions, and enzymes that increase— even transiently—the accessibility of DNA in chromatin will tend to favor transcription initiation.

response to external signals, such as the glucocorticoid hormone discussed earlier in this chapter. However, in other cases, the altered chromatin structure seems to persist, even after the gene regulatory protein that directed its establishment has dissociated from DNA. In principle, this memory can extend into the next cell generation because, as discussed in Chapter 4, chromatin structure can be self-renewing (see Figure 4–52). It is interesting to consider the possibility that different histone modifications persist for different times in order to provide the cell with a mechanism for long-, medium-, and short-term memory of gene expression patterns.

A special type of chromatin modification occurs as RNA polymerase II transcribes through a gene. In most cases, the nucleosomes just ahead of the polymerase are acetylated by writer complexes carried by the polymerase, removed by histone chaperones, and deposited behind the moving polymerase. They are then rapidly deacetylated and methylated, also by reader–writer complexes that are carried by the polymerase, leaving behind nucleosomes that are especially resistant to transcription. Although this remarkable process may seem counterintuitive, it likely evolved to prevent spurious transcription re-initiation behind a moving polymerase, which is, in essence, clearing a path through chromatin. Later in this chapter, when we discuss *RNA interference*, the potential dangers to the cell of such inappropriate transcription will become especially obvious.

We have just seen that gene activator proteins can profoundly influence chromatin structure. However, even before these activator proteins are brought into play, many genes are "poised" to become rapidly activated. For example, the regulatory regions for many genes are "marked" by a short, nucleosome-free region flanked by nucleosomes that contain the histone variant H2AZ. This arrangement, which is specified by DNA sequence, allows free access of gene regulatory proteins to the nucleosome-free region; in addition, the H2AZ-containing nucleosomes are thought to be easily disassembled, thus further facilitating transcription initiation.

Figure 7–47 **Writing and reading the histone code during transcription initiation.** In this example, taken from the human interferon gene promoter, a gene activator protein binds to DNA packaged into chromatin and first attracts a histone acetyl transferase to acetylate lysine 9 of histone H3 and lysine 8 of histone H4. Next, a histone kinase, attracted by the gene activator protein, phosphorylates serine 10 of histone H3, but can only do so after lysine 9 has been acetylated. The serine modification then signals the histone acetyl transferase to acetylate position K14 of histone H3. At this point the histone code for transcription initiation, set into motion by the binding of the gene activator protein, has been written. Note that the writing is sequential, with each histone modification depending on a prior modification.

The final reading of the code occurs when the general transcription factor TFIID and the chromatin remodeling complex SWI/SNF bind, both of which strongly promote the subsequent steps of transcription initiation. TFIID and SWI/SNF both recognize acetylated histone tails through a *bromodomain*, a protein domain specialized to read this particular mark on histones; a bromodomain is carried in a subunit of each protein complex. (Adapted from T. Agalioti, G. Chen and D. Thanos, *Cell* 111:381–392, 2002. With permission from Elsevier.)

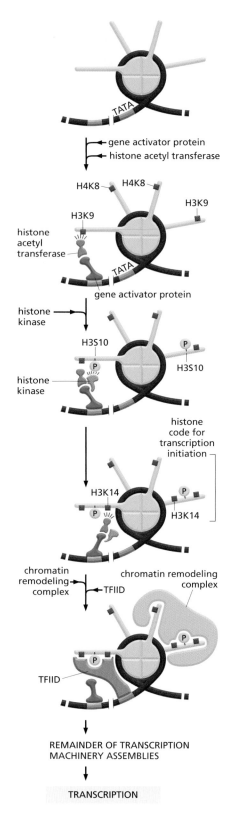

Gene Activator Proteins Work Synergistically

We have seen that eucaryotic gene activator proteins can influence different steps in transcription initiation. In general, where several factors work together to enhance a reaction rate, the joint effect is not merely the sum of the enhancements that each factor alone contributes, but the product. If, for example, factor A lowers the free-energy barrier for a reaction by a certain amount and thereby speeds up the reaction 100-fold, and factor B, by acting on another aspect of the reaction, does likewise, then A and B acting in parallel will lower the barrier by a double amount and speed up the reaction 10,000-fold. Even if A and B work simply by attracting the same protein, the affinity of that protein for the reaction site increases multiplicatively. Thus, gene activator proteins often exhibit *transcriptional synergy*, where several activator proteins working together produce a transcription rate that is much higher than that of the sum of the activators working alone (**Figure 7–48**). It is not difficult to see how multiple gene regulatory proteins, each binding to a different regulatory DNA sequence, work together to control the final rate of transcription of a eucaryotic gene.

Since gene activator proteins can influence many different steps on the pathway to transcriptional activation, it is worth considering whether these steps always occur in a prescribed order. For example, does histone modification always precede chromatin remodeling, as in the example of Figure 7–47? Does Mediator enter before or after RNA polymerase? The answers to these questions appear to be different for different genes—and even for the same gene under the influence of different gene regulatory proteins (**Figure 7–49**).

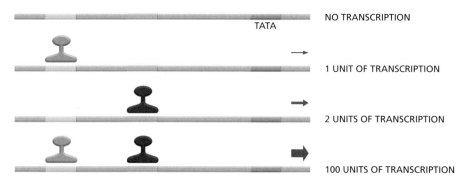

Figure 7–48 **Transcriptional synergy.** This experiment compares the rate of transcription produced by three experimentally constructed regulatory regions in a eucaryotic cell and reveals transcriptional synergy, the greater than additive effect of multiple activators. Transcriptional synergy is typically observed between different gene activator proteins from the same organism and even between activator proteins from different eucaryotic species when they are experimentally introduced into the same cell. This last observation reflects the high degree of conservation of the elaborate machinery responsible for eucaryotic transcription initiation.

Figure 7–49 An order of events leading to transcription initiation of a specific gene. In this well-studied example from the budding yeast *S. cerevisiae,* the steps toward transcription initiation occur in a particular order; however, this order differs from one gene to the next. For example, at another gene, histone modification occurs first, followed by RNA polymerase recruitment, followed by chromatin remodeling complex recruitment. Figure 7–47 illustrates yet another possible order of events.

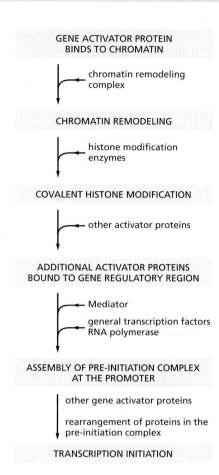

Whatever the precise mechanisms and the order in which they are carried out, a gene regulatory protein must be bound to DNA either directly or indirectly to influence transcription of its target gene, and the rate of transcription of a gene ultimately depends upon the spectrum of regulatory proteins bound upstream and downstream of its transcription start site.

Eucaryotic Gene Repressor Proteins Can Inhibit Transcription in Various Ways

Like bacteria, eucaryotes use **gene repressor proteins** in addition to activator proteins to regulate transcription of their genes. However, because of differences in the way that eucaryotes and bacteria initiate transcription, eucaryotic repressors have many more possible mechanisms of action. We saw in Chapter 4 that large regions of the genome can be shut down by the packaging of DNA into heterochromatin. However, eucaryotic genes are rarely organized along the genome according to function, so this strategy is not generally useful for most examples of gene regulation. Instead, most eucaryotic repressors must work on a gene-by-gene basis. Unlike bacterial repressors, most eucaryotic repressors do not directly compete with the RNA polymerase for access to the DNA; rather they use a variety of other mechanisms, some of which are illustrated in **Figure 7–50**. Like gene activator proteins, many eucaryotic repressor proteins act through more than one mechanism at a given target gene, thereby ensuring robust and efficient repression.

Gene repression is especially important to animals and plants whose growth depends on elaborate and complex developmental programs. Misexpression of a single gene at a critical time can have disastrous consequences for the individual. For this reason, many of the genes encoding the most important developmental regulatory proteins are kept tightly repressed when they are not needed.

Eucaryotic Gene Regulatory Proteins Often Bind DNA Cooperatively

So far we have seen that when eucaryotic activator and repressor proteins bind to specific DNA sequences, they set in motion a complex series of events that culminate in transcription initiation or its opposite, repression. However, these proteins rarely recognize DNA as individual polypeptides. In reality, efficient DNA binding in the eucaryotic cell typically requires several sequence-specific DNA proteins acting together. For example, two gene regulatory proteins with a weak affinity for each other might cooperate to bind to a DNA sequence, neither protein having a sufficient affinity for DNA to bind to the DNA site on its own. In one well-studied case, the DNA-bound protein dimer creates a distinct surface that is recognized by a third protein that carries an activator domain that stimulates transcription. This example illustrates an important general point: protein–protein interactions that are too weak to form complexes in solution can do so on DNA, with the DNA sequence acting as a "crystallization" site or seed for the assembly of a protein complex.

As shown in **Figure 7–51**, an individual gene regulatory protein can often participate in more than one type of regulatory complex. A protein might function, for example, in one case as part of a complex that activates transcription and in another case as part of a complex that represses transcription. Thus, individual eucaryotic gene regulatory proteins are not necessarily dedicated activators or repressors; instead, they function as regulatory parts that are used to build complexes whose function depends on the final assembly of all of the individual components. This final assembly, in turn, depends both on the arrangement of control region DNA sequences and on the particular gene regulatory proteins present in active form in the cell. Each eucaryotic gene is therefore regulated by a "committee" of proteins, all of which must be present to express the gene at its proper level.

In some cases, the precise DNA sequence to which a regulatory protein binds directly can affect the conformation of this protein and thereby influence its subsequent transcriptional activity. When bound to one type of DNA sequence, for example, a steroid hormone receptor protein interacts with a co-repressor and ultimately turns off transcription. When bound to a slightly different DNA sequence, it assumes a different conformation and interacts with a coactivator, thereby stimulating transcription.

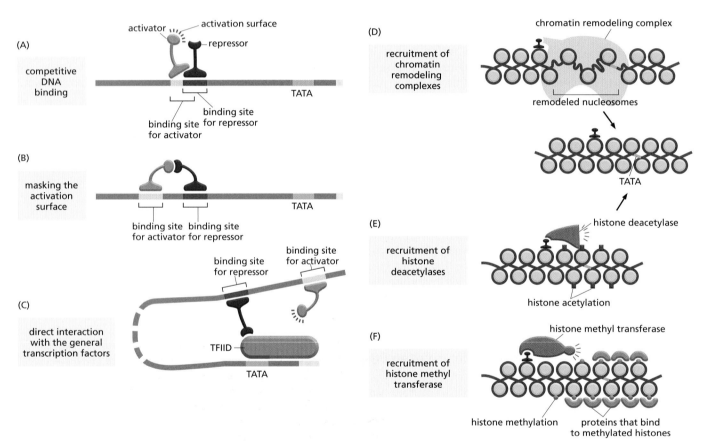

Figure 7–50 Six ways in which eucaryotic gene repressor proteins can operate. (A) Gene activator proteins and gene repressor proteins compete for binding to the same regulatory DNA sequence. (B) Both proteins can bind DNA, but the repressor binds to the activation domain of the activator protein, thereby preventing it from carrying out its activation functions. In a variation of this strategy, the repressor binds tightly to the activator without having to be bound to DNA directly. (C) The repressor blocks assembly of the general transcription factors. Some repressors also act at late stages in transcription initiation, for example, by preventing the release of the RNA polymerase from the general transcription factors. (D) The repressor recruits a chromatin remodeling complex which returns the nucleosomal state of the promoter region to its pre-transcriptional form. (E) The repressor attracts a histone deacetylase to the promoter. As we have seen, histone acetylation can stimulate transcription initiation (Figure 7–47), and the repressor simply reverses this modification. (F) The repressor attracts a histone methyl transferase which modifies certain positions on histones which, in turn, are bound by proteins that maintain the chromatin in a transcriptionally silent form. For example, in *Drosophila*, the histone methyl transferase Suv39 methylates the K9 position of histone H3, a modification that is bound by the HP1 protein. In another example, E(z) methylates the K27 position of H3, and this modification is bound by the Polycomb protein. HP1 and Polycomb recognize methylated lysines through a *chromodomain*. They can act locally to turn off specific genes, as shown here, or can occupy a whole region of a chromosome to repress a cluster of genes. A seventh mechanism of negative control—inactivation of a transcriptional activator by heterodimerization—is illustrated in Figure 7–24. For simplicity, nucleosomes have been omitted from (A)–(C), and the scales of (D)–(F) have been reduced relative to (A)–(C).

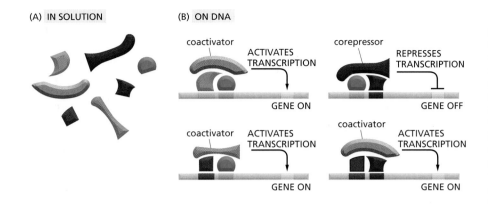

Figure 7–51 Eucaryotic gene regulatory proteins often assemble into complexes on DNA. Seven gene regulatory proteins are shown in (A). The nature and function of the complex they form depends on the specific DNA sequence that seeds their assembly. In (B), some assembled complexes activate gene transcription, while another represses transcription. Note that both the *red* and the *green* proteins are shared by both activating and repressing complexes. Proteins that do not themselves bind DNA but assemble on other DNA-bound gene regulatory proteins are often termed coactivators or co-repressors. However, these terms are somewhat confusing because they encompass an enormous variety of proteins including histone readers and writers, chromatin remodeling complexes, and many other classes of proteins. Some have no intrinsic activity themselves but simply serve as a "scaffolding" to attract those that do.

Typically, a few relatively short stretches of nucleotide sequence guide the assembly of a group of regulatory proteins on DNA (see Figure 7–51). However, in some extreme cases of regulation by committee a more elaborate protein–DNA structure is formed (**Figure 7–52**). Since the final assembly requires the presence of many gene regulatory proteins that bind DNA, it provides a simple way to ensure that a gene is expressed only when the cell contains the correct combination of these proteins. We saw earlier how the formation of heterodimers in solution provides a mechanism for the combinatorial control of gene expression. The assembly of complexes of gene regulatory proteins on DNA provides a second important mechanism for combinatorial control, one that offers far richer opportunities.

Complex Genetic Switches That Regulate *Drosophila* Development Are Built Up from Smaller Modules

Given that gene regulatory proteins can be positioned at multiple sites along long stretches of DNA, that these proteins can assemble into complexes at each site, and that the complexes influence the chromatin structure as well as the recruitment and assembly of the general transcription machinery at the promoter, there would seem to be almost limitless possibilities for the elaboration of control devices to regulate eucaryotic gene transcription.

A particularly striking example of a complex, multicomponent genetic switch is that controlling the transcription of the *Drosophila Even-skipped* (*Eve*) gene, whose expression plays an important part in the development of the *Drosophila* embryo. If this gene is inactivated by mutation, many parts of the embryo fail to form, and the embryo dies early in development. As discussed in Chapter 22, at the stage of development when *Eve* begins to be expressed, the embryo is a single giant cell containing multiple nuclei in a common cytoplasm. This cytoplasm is not uniform, however: it contains a mixture of gene regulatory proteins that are distributed unevenly along the length of the embryo, thus providing positional information that distinguishes one part of the embryo from another (**Figure 7–53**). (The way these differences are initially set up is discussed in Chapter 22.) Although the nuclei are initially identical, they rapidly begin to express different genes because they are exposed to different gene regulatory proteins. The nuclei near the anterior end of the developing embryo, for example, are exposed to a set of gene regulatory proteins that is distinct from the set that influences nuclei at the posterior end of the embryo.

The regulatory DNA sequences controlling the *Eve* gene are designed to read the concentrations of gene regulatory proteins at each position along the length of the embryo and to interpret this information in such a way that the *Eve* gene is expressed in seven stripes, each initially five to six nuclei wide and positioned precisely along the anterior–posterior axis of the embryo (**Figure 7–54**). How is this remarkable feat of information processing carried out? Although not all of the molecular details are understood, several general principles have emerged from studies of *Eve* and other *Drosophila* genes that are similarly regulated.

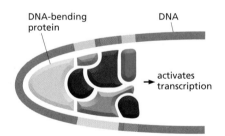

Figure 7–52 Schematic depiction of a committee of gene regulatory proteins bound to an enhancer. The protein shown in *yellow* is called an architectural protein since its main role is to bend the DNA to allow the cooperative assembly of the other components. The structure depicted here is based on that found in the control region of the gene that codes for a subunit of the T cell receptor (discussed in Chapter 25), and it activates transcription at a nearby promoter. Only certain cells of the developing immune system, which eventually give rise to mature T cells, have the complete set of proteins needed to form this structure.

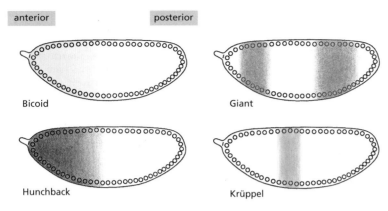

anterior · posterior

Bicoid

Giant

Hunchback

Krüppel

Figure 7–53 The nonuniform distribution of four gene regulatory proteins in an early *Drosophila* embryo. At this stage the embryo is a syncytium, with multiple nuclei in a common cytoplasm. Although the detail is not shown in these drawings, all of these proteins are concentrated in the nuclei.

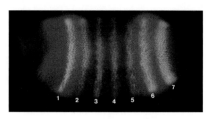

Figure 7–54 The seven stripes of the protein encoded by the *Even-skipped* (*Eve*) gene in a developing *Drosophila* embryo. Two and one-half hours after fertilization, the egg was fixed and stained with antibodies that recognize the Eve protein (*green*) and antibodies that recognize the Giant protein (*red*). Where Eve and Giant proteins are both present, the staining appears *yellow*. At this stage in development, the egg contains approximately 4000 nuclei. The Eve and Giant proteins are both located in the nuclei, and the Eve stripes are about four nuclei wide. The staining pattern of the Giant protein is also shown in Figure m7–52/7–53. (Courtesy of Michael Levine.)

The regulatory region of the *Eve* gene is very large (approximately 20,000 nucleotide pairs). It is formed from a series of relatively simple regulatory modules, each of which contains multiple regulatory sequences and is responsible for specifying a particular stripe of *Eve* expression along the embryo. This modular organization of the *Eve* gene control region is revealed by experiments in which a particular regulatory module (say, that specifying stripe 2) is removed from its normal setting upstream of the *Eve* gene, placed in front of a reporter gene (see Figure 7–45), and reintroduced into the *Drosophila* genome. When developing embryos derived from flies carrying this genetic construct are examined, the reporter gene is found to be expressed in precisely the position of stripe 2 (**Figure 7–55**). Similar experiments reveal the existence of other regulatory modules, each of which specifies either one of the other six stripes or some other part of the *Eve* expression pattern normally displayed at later stages of development (see Figure 22–39).

The *Drosophila Eve* Gene Is Regulated by Combinatorial Controls

A detailed study of the stripe 2 regulatory module has provided insights into how it reads and interprets positional information. It contains recognition sequences for two gene regulatory proteins (Bicoid and Hunchback) that activate *Eve* transcription and two (Krüppel and Giant) that repress it (**Figure 7–56**). (The gene regulatory proteins of *Drosophila* often have colorful names reflecting the phenotype that results if the gene encoding the protein is inactivated by mutation.) The relative concentrations of these four proteins determine whether the protein complexes that form at the stripe 2 module activate transcription of the *Eve* gene. **Figure 7–57** shows the distributions of the four gene regulatory proteins across the region of a *Drosophila* embryo where stripe 2 forms. It is thought that either of the two repressor proteins, when bound to the DNA, will turn off the

Figure 7–55 Experiment demonstrating the modular construction of the *Eve* gene regulatory region.
(A) A 480-nucleotide-pair piece of the Eve regulatory region was removed and inserted upstream of a test promoter that directs the synthesis of the enzyme β-galactosidase (the product of the *E. coli LacZ* gene). (B) When this artificial construct was reintroduced into the genome of *Drosophila* embryos, the embryos expressed β-galactosidase (detectable by histochemical staining) precisely in the position of the second of the seven Eve stripes (C). (B and C, courtesy of Stephen Small and Michael Levine.)

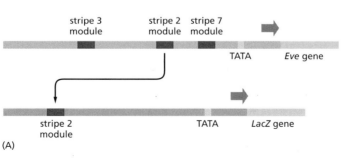

(A)

stripe 3 module · stripe 2 module · stripe 7 module · TATA · *Eve* gene

stripe 2 module · TATA · *LacZ* gene

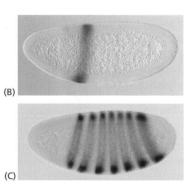

(B)

(C)

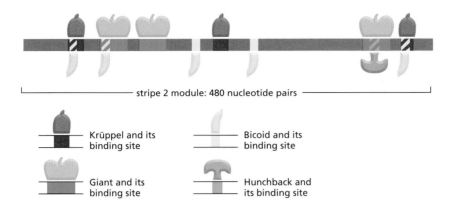

stripe 2 module: 480 nucleotide pairs

Krüppel and its
binding site

Giant and its
binding site

Bicoid and its
binding site

Hunchback and
its binding site

stripe 2 module, whereas both Bicoid and Hunchback must bind for its maximal activation. This simple regulatory unit thereby combines these four positional signals so as to turn on the stripe 2 module (and therefore the expression of the *Eve* gene) only in those nuclei that are located where the levels of both Bicoid and Hunchback are high and both Krüppel and Giant are absent. This combination of activators and repressors occurs in only one region of the early embryo; everywhere else, therefore, the stripe 2 module is silent.

We have thus far discussed two mechanisms of combinatorial control of gene expression—heterodimerization of gene regulatory proteins in solution (see Figure 7–20) and the assembly of combinations of gene regulatory proteins into small complexes on DNA (see Figure 7–51). It is likely that both mechanisms participate in the complex regulation of *Eve* expression. In addition, the regulation of stripe 2 just described illustrates a third type of combinatorial control.

Figure 7–56 Close-up view of the *Eve* stripe 2 unit. The segment of the *Eve* gene control region identified in the previous figure contains regulatory sequences, each of which binds one or another of four gene regulatory proteins. It is known from genetic experiments that these four regulatory proteins are responsible for the proper expression of *Eve* in stripe 2. Flies that are deficient in the two gene activators Bicoid and Hunchback, for example, fail to express efficiently *Eve* in stripe 2. In flies deficient in either of the two gene repressors, Giant and Krüppel, stripe 2 expands and covers an abnormally broad region of the embryo. The DNA binding sites for these gene regulatory proteins were determined by cloning the genes encoding the proteins, overexpressing the proteins in *E. coli*, purifying them, and performing DNA-footprinting experiments (see Figure 7–29). The *top* diagram indicates that, in some cases, the binding sites for the gene regulatory proteins overlap and the proteins can compete for binding to the DNA. For example, binding of Krüppel and binding of Bicoid to the site at the far right are thought to be mutually exclusive.

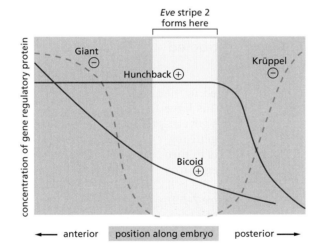

Eve stripe 2
forms here

Giant ⊖

Hunchback ⊕

Krüppel ⊖

Bicoid ⊕

concentration of gene regulatory protein

← anterior   position along embryo   posterior →

Figure 7–57 Distribution of the gene regulatory proteins responsible for ensuring that *Eve* is expressed in stripe 2. The distributions of these proteins were visualized by staining a developing *Drosophila* embryo with antibodies directed against each of the four proteins (see Figures 7–53 and 7–54). The expression of *Eve* in stripe 2 occurs only at the position where the two activators (Bicoid and Hunchback) are present and the two repressors (Giant and Krüppel) are absent. In fly embryos that lack Krüppel, for example, stripe 2 expands posteriorly. Likewise, stripe 2 expands posteriorly if the DNA-binding sites for Krüppel in the stripe 2 module (see Figure 7–56) are inactivated by mutation.

The *Eve* gene itself encodes a gene regulatory protein, which, after its pattern of expression is set up in seven stripes, regulates the expression of other *Drosophila* genes. As development proceeds, the embryo is thus subdivided into finer and finer regions that eventually give rise to the different body parts of the adult fly, as discussed in Chapter 22.

This example from *Drosophila* embryos is unusual in that the nuclei are exposed directly to positional cues in the form of concentrations of gene regulatory proteins. In embryos of most other organisms, individual nuclei are in separate cells, and extracellular positional information must either pass across the plasma membrane or, more usually, generate signals in the cytosol in order to influence the genome.

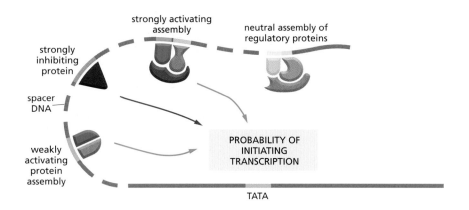

Figure 7–58 **The integration of multiple inputs at a promoter.** Multiple sets of gene regulatory proteins can work together to influence transcription initiation at a promoter, as they do in the *Eve* stripe 2 module illustrated in Figure 7–56. It is not yet understood in detail how the cell achieves integration of multiple inputs, but it is likely that the final transcriptional activity of the gene results from a competition between activators and repressors that act by the mechanisms summarized in Figures 7–46 and 7–50.

Because the individual regulatory sequences in the *Eve* stripe 2 module are strung out along the DNA, many sets of gene regulatory proteins can be bound simultaneously at separate sites and influence the promoter of a gene. The promoter integrates the transcriptional cues provided by all of the bound proteins (**Figure 7–58**).

The regulation of *Eve* expression is an impressive example of combinatorial control. Seven combinations of gene regulatory proteins—one combination for each stripe—activate *Eve* expression, while many other combinations (all those found in the interstripe regions of the embryo) keep the stripe elements silent. The other stripe regulatory modules are thought to be constructed similarly to those described for stripe 2, being designed to read positional information provided by other combinations of gene regulatory proteins. The entire gene control region, strung out over 20,000 nucleotide pairs of DNA, binds more than 20 different regulatory proteins. A large and complex control region is thereby built from a series of smaller modules, each of which consists of a unique arrangement of short DNA sequences recognized by specific gene regulatory proteins.

Complex Mammalian Gene Control Regions Are Also Constructed from Simple Regulatory Modules

Perhaps 8% of the coding capacity of a mammalian genome is devoted to the synthesis of proteins that serve as regulators of gene transcription. This large number of genes reflects the exceedingly complex network of controls governing expression of mammalian genes. Each gene is regulated by a set of gene regulatory proteins; each of those proteins is the product of a gene that is in turn regulated by a whole set of other proteins, and so on. Moreover, the regulatory protein molecules are themselves influenced by signals from outside the cell, which can make them active or inactive in a whole variety of ways (**Figure 7–59**). Thus, we can view the pattern of gene expression in a cell as the result of a complicated molecular computation that the intracellular gene control network performs in response to information from the cell's surroundings. We shall discuss these issues further in Chapters 15 and 22, which deal with cell signaling and development, but the complexity is remarkable even at the level of an individual genetic switch regulating the activity of a single gene. It is not unusual, for example, to find a mammalian gene with a control region that is 100,000 nucleotide pairs in length, in which many modules, each containing a number of regulatory sequences that bind gene regulatory proteins, are interspersed with long stretches of other noncoding DNA.

One of the best-understood examples of a complex mammalian regulatory region is found in the human β-globin gene, which is expressed exclusively in red blood cells. A complex array of gene regulatory proteins controls the expression of the gene, some acting as activators and others as repressors (**Figure 7–60**). The concentrations (or activities) of many of these gene regulatory proteins change during development, and only a particular combination of all the proteins triggers transcription of the gene. The human β-globin gene is part of a cluster of globin genes (**Figure 7–61A**) that are all transcribed exclusively in erythroid cells,

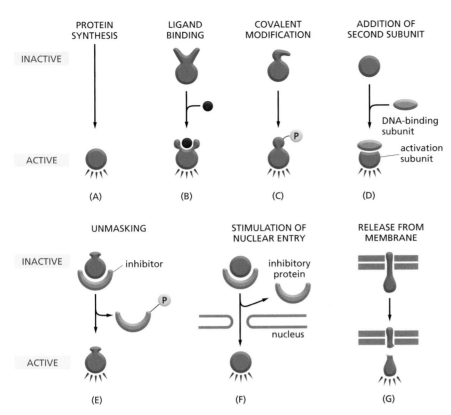

Figure 7–59 Some ways in which the activity of gene regulatory proteins is regulated in eucaryotic cells. (A) The protein is synthesized only when needed and is rapidly degraded by proteolysis so that it does not accumulate. **(B)** Activation by ligand binding. **(C)** Activation by covalent modification. Phosphorylation is shown here, but many other modifications are possible (see Table 3–3, p. 186). **(D)** Formation of a complex between a DNA-binding protein and a separate protein with a transcription-activating domain. **(E)** Unmasking of an activation domain by the phosphorylation of an inhibitor protein. **(F)** Stimulation of nuclear entry by removal of an inhibitory protein that otherwise keeps the regulatory protein from entering the nucleus. **(G)** Release of a gene regulatory protein from a membrane bilayer by regulated proteolysis.

that is, cells of the red blood cell lineage, but at different stages of mammalian development (see Figure 7–61B). The ε-globin gene is expressed in the early embryo, γ in the later embryo and fetus, and δ and β primarily in the adult. The gene products differ slightly in their oxygen-binding properties, suiting them for the different oxygenation conditions in the embryo, fetus, and adult. Each of the globin genes has its own set of regulatory proteins that are necessary to turn the gene on at the appropriate time.

The globin genes are unusual in that, at the appropriate time and place, they are transcribed at extremely high rates: indeed, red blood cells are little more

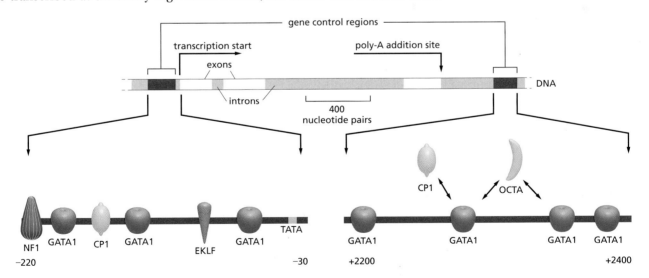

Figure 7–60 Model for the control of the human β-globin gene. The diagram shows some of the gene regulatory proteins that control expression of the gene during red blood cell development (see Figure 7–61). Some of the gene regulatory proteins shown, such as CP1, are found in many types of cells, while others, such as GATA1, are present in only a few types of cells—including red blood cells—and therefore are thought to contribute to the cell-type specificity of β-globin gene expression. As indicated by the *double-headed arrows,* several of the binding sites for GATA1 overlap those of other gene regulatory proteins; it is thought that by binding to these sites, GATA1 excludes binding of other proteins. Once bound to DNA, the gene regulatory proteins recruit chromatin remodeling complexes, histone modifying enzymes, the general transcription factors, Mediator, and RNA polymerase to the promoter. (Adapted from B. Emerson, in Gene Expression: General and Cell-Type Specific [M. Karin, ed.], pp. 116–161. Boston: Birkhauser, 1993.)

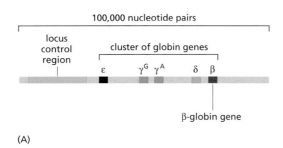

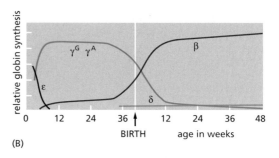

(A)   (B)

than bags of hemoglobin that was synthesized by precursor cells. To achieve this extraordinarily high level of transcription, the globin genes, in addition to their individual regulatory sequences, share a control region called the *locus control region* (LCR), which lies far upstream from the gene cluster and is needed for the proper expression of each gene in the cluster (see Figure 7–61A). The importance of the LCR can be seen in patients with a certain type of thalassemia, a severe inherited form of anemia. In these patients, the β-globin locus has suffered a deletion that removes all or part of the LCR. Although the β-globin and its nearby regulatory region are intact, the gene remains transcriptionally silent, even in erythroid cells.

The way in which the LCR functions is not understood in detail, but it is known that the gene regulatory proteins that bind the LCR interact, through DNA looping, with proteins bound to the control regions of the globin genes they regulate. In this way, the proteins bound at the LCR help attract chromatin remodeling complexes, histone-modifying enzymes, and components of the transcription machinery that act in conjunction with the specific regulatory regions of each individual globin gene. In addition, the LCR includes a *barrier sequence* (see Figure 4–47) that prevents the spread of neighboring heterochromatin into the β-globin locus, as discussed in Chapter 4. This dual feature distinguishes the globin LCR from many other types of regulatory sequences in the human genome; however, the globin genes are not alone in having an LCR, as LCRs are also present upstream of other highly transcribed, cell-type-specific genes. We should probably think of LCRs, not as unique DNA elements with specialized properties, but rather as especially powerful combinations of more fundamental types of regulatory sequences.

Insulators Are DNA Sequences That Prevent Eucaryotic Gene Regulatory Proteins from Influencing Distant Genes

All genes have control regions, which dictate at which times, under what conditions, and in what tissues the gene will be expressed. We have also seen that eucaryotic gene regulatory proteins can act across very long stretches of DNA. How, then, are control regions of different genes kept from interfering with one another? In other words, what keeps a gene regulatory protein bound on the control region of one gene from inappropriately influencing the transcription of adjacent genes?

To avoid such cross-talk, several types of DNA elements function to compartmentalize the genome into discrete regulatory domains. In Chapter 4 we discussed barrier sequences that prevent the spread of heterochromatin into genes that need to be expressed. A second type of DNA element, called an insulator, prevents enhancers from running amok and activating inappropriate genes (**Figure 7–62**). An insulator can apparently block the communication

Figure 7–61 The cluster of β-like globin genes in humans. (A) The large chromosomal region shown spans 100,000 nucleotide pairs and contains the five globin genes and a locus control region (LCR). (B) Changes in the expression of the β-like globin genes at various stages of human development. Each of the globin chains encoded by these genes combines with an α-globin chain to form the hemoglobin in red blood cells (see Figure 4–86). (A, after F. Grosveld, G.B. van Assendelft, D.R. Greaves and G. Kollias, *Cell* 51:975–985, 1987. With permission from Elsevier.)

Figure 7–62 Schematic diagram summarizing the properties of insulators and barrier sequences. Insulators directionally block the action of enhancers *(left-hand side)*, and barrier sequences prevent the spread of heterochromatin *(right-hand side)*. Thus gene B is properly regulated and gene B's enhancer is prevented from influencing the transcription of gene A. How barrier sequences are likely to function is depicted in Figure 4–47. It is not yet understood how insulators exert their effects; one possibility is that they serve as "decoys," tying up the transcriptional machinery and preventing it from interacting with an authentic enhancer. Another is that they anchor DNA to the nuclear envelope, thereby interfering with DNA looping between an enhancer and an inappropriate promoter.

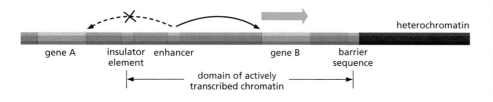

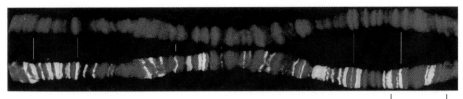

10 μm

Figure 7–63 Localization of a *Drosophila* insulator-binding protein on polytene chromosomes. A polytene chromosome (discussed in Chapter 4) was stained with propidium iodide *(red)* to show its banding patterns—with bands appearing *bright red* and interbands as dark gaps in the pattern *(top)*. The positions on this polytene chromosome that are bound by a particular insulator protein (called BEAF) are stained *bright green* using antibodies directed against the protein *(bottom)*. BEAF is preferentially localized to interband regions, reflecting its role in organizing chromosomes into structural, as well as functional, domains. For convenience, these two micrographs of the same polytene chromosome are arranged as mirror images. (Courtesy of Uli Laemmli, from K. Zhao et al., *Cell* 81:879–889, 1995. With permission from Elsevier.)

between an enhancer and a promoter, but, to do so, it must be located between the two. Although proteins that bind to insulators have been identified, how they directionally neutralize enhancers is still a mystery.

Even though their mechanisms are not understood in detail, the distribution of insulators and barrier sequences in a genome is thought to divide it into independent domains of gene regulation and chromatin structure (**Figure 7–63**). Aspects of this organization can be visualized by staining whole chromosomes for the specialized proteins that bind these DNA elements.

Although chromosomes are organized into orderly domains that discourage enhancers from acting indiscriminately, there are special circumstances where an enhancer located one chromosome has been found to activate a gene located on a second chromosome. A remarkable example occurs in the regulation of the mammalian olfactory receptors. These are the proteins expressed by sensory neurons that allow mammals to discriminate accurately among many thousands of distinct smells (see p. 917). Humans, for example, have 350 olfactory receptor genes, and they are carefully regulated so that only one of these genes is expressed in each sensory neuron. The olfactory receptor genes are dispersed among many different chromosomes, but there is only a single enhancer for all of them. Once this enhancer activates a receptor gene by associating with its regulatory region, it remains stably associated thereby precluding activation of any of the other receptor genes. Although there is much we do not understand about this mechanism, it does indicate the extreme versatility of transcriptional regulation strategies.

Gene Switches Rapidly Evolve

We have seen that the control regions of eucaryotic genes are often spread out over long stretches of DNA, whereas those of procaryotic genes are typically clustered around the start point of transcription. It seems likely that the close-packed arrangement of bacterial genetic switches developed from more extended forms of switches in response to the evolutionary pressure on bacteria to maintain a small genome size. This compression comes at a price, however, as it restricts the complexity and adaptability of the control device. In contrast, the extended form of eucaryotic control regions—with discrete regulatory modules separated by long stretches of spacer DNA—facilitates the reshuffling of regulatory modules during evolution, both to create new regulatory circuits and to modify old ones. As we saw in Chapters 1 and 4, and we shall see again in Chapter 22, changes in gene regulation—rather than the acquisition of new genes—underlie much of the wide variety of life on Earth. Unraveling the history of how modern gene control regions have evolved presents a fascinating challenge to biologists, with many clues available in present-day genomes.

Summary

Gene regulatory proteins switch the transcription of individual genes on and off in cells. In procaryotes these proteins usually bind to specific DNA sequences close to the RNA polymerase start site and, depending on the nature of the regulatory protein and the precise location of its binding site relative to the start site, either activate or repress transcription of the gene. The flexibility of the DNA helix, however, also allows proteins bound at distant sites to affect the RNA polymerase at the promoter by the looping out of the intervening DNA. The regulation of higher eucaryotic genes is much more complex, commensurate with a larger genome size and the large variety of cell types that are formed. A single eucaryotic gene is typically controlled by many gene regulatory proteins bound to sequences that can be thousands of nucleotide pairs from the promoter that directs transcription of the gene. Eucaryotic activators and repressors act by a wide variety of mechanisms—generally altering chromatin structure and controlling the assembly of the general transcription factors, Mediator, and RNA polymerase at the promoter. The time and place that each gene is transcribed, as well as its rates of transcription under different conditions, are determined by the spectrum of gene regulatory proteins that bind to the regulatory region of the gene.

THE MOLECULAR GENETIC MECHANISMS THAT CREATE SPECIALIZED CELL TYPES

Although all cells must be able to switch genes on and off in response to changes in their environments, the cells of multicellular organisms have evolved this capacity to an extreme degree and in highly specialized ways to form an organized array of differentiated cell types. In particular, once a cell in a multicellular organism becomes committed to differentiate into a specific cell type, the cell maintains this choice through many subsequent cell generations, which means that it remembers the changes in gene expression involved in the choice. This phenomenon of *cell memory* is a prerequisite for the creation of organized tissues and for the maintenance of stably differentiated cell types. In contrast, other changes in gene expression in eucaryotes, as well as most in bacteria, are only transient. The tryptophan repressor, for example, switches off the tryptophan genes in bacteria only in the presence of tryptophan; as soon as tryptophan is removed from the medium, the genes are switched back on, and the descendants of the cell will have no memory that their ancestors had been exposed to tryptophan. Even in bacteria, however, a few types of changes in gene expression can be inherited stably.

In this section we shall examine not only cell memory mechanisms, but also how gene regulatory devices can be combined to create "logic circuits" through which cells integrate signals, keep time, remember events in their past, and adjust the levels of gene expression over entire chromosomes. We begin by considering some of the best-understood genetic mechanisms of cell differentiation, which operate in bacterial and yeast cells.

DNA Rearrangements Mediate Phase Variation in Bacteria

We have seen that cell differentiation in higher eucaryotes usually occurs without detectable changes in DNA sequence. In some procaryotes, in contrast, a stably inherited pattern of gene regulation is achieved by DNA rearrangements that activate or inactivate specific genes. Since a change in DNA sequence will be copied faithfully during all subsequent DNA replication cycles, an altered state of gene activity will be inherited by all the progeny of the cell in which the rearrangement occurred. Some of these DNA rearrangements are, however, reversible so that occasional individuals can switch back to the original DNA configuration. The result is an alternating pattern of gene activity that can be detected by observations over long time periods and many generations.

A well-studied example of this differentiation mechanism occurs in *Salmonella* bacteria and is known as **phase variation**. Although this mode of differentiation has no known counterpart in higher eucaryotes, it can have considerable impact on animals because disease-causing bacteria use it to evade detection by the immune system. The switch in *Salmonella* gene expression is brought about by the occasional inversion of a specific 1000-nucleotide-pair piece of DNA. This change alters the expression of the cell-surface protein flagellin, for which the bacterium has two different genes (**Figure 7–64**). A site-specific recombination enzyme catalyzes the inversion and thereby changes the orientation of a promoter that is located within the inverted DNA segment. With the promoter in one orientation, the bacteria synthesize one type of flagellin;

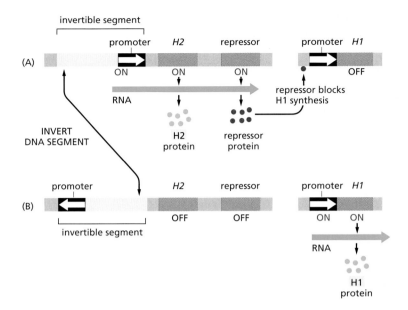

Figure 7–64 Switching gene expression by DNA inversion in bacteria. Alternating transcription of two flagellin genes in a *Salmonella* bacterium is caused by a simple site-specific recombination event that inverts a small DNA segment containing a promoter. (A) In one orientation, the promoter activates transcription of the *H2* flagellin gene as well as that of a repressor protein that blocks the expression of the *H1* flagellin gene. (B) When the promoter is inverted, it no longer turns on *H2* or the repressor, and the *H1* gene, which is thereby released from repression, is expressed instead. The recombination mechanism is activated only rarely (about once in every 10^5 cell divisions). Therefore, the production of one or other flagellin tends to be faithfully inherited in each clone of cells.

with the promoter in the other orientation, they synthesize the other type. Because inversions occur only rarely, entire clones of bacteria will have one type of flagellin or the other.

Phase variation almost certainly evolved because it protects the bacterial population against the immune response of its vertebrate host. If the host makes antibodies against one type of flagellin, a few bacteria whose flagellin has been altered by gene inversion will still be able to survive and multiply.

Bacteria isolated from the wild very often exhibit phase variation for one or more phenotypic traits. Standard laboratory strains of bacteria lose these "instabilities" over time, and underlying mechanisms have been studied in only a few cases. Not all involve DNA inversion. A bacterium that causes a common sexually transmitted human disease *(Neisseria gonorrhoeae)*, for example, avoids immune attack by means of a heritable change in its surface properties that arises from gene conversion (discussed in Chapter 5) rather than by inversion. This mechanism transfers DNA sequences from a library of silent "gene cassettes" to a site in the genome where the genes are expressed; it has the advantage of creating many variants of the major bacterial surface protein.

A Set of Gene Regulatory Proteins Determines Cell Type in a Budding Yeast

Because they are so easy to grow and to manipulate genetically, yeasts have served as model organisms for studying the mechanisms of gene control in eucaryotic cells. The common baker's yeast, *Saccharomyces cerevisiae*, has attracted special interest because of its ability to differentiate into three distinct cell types. *S. cerevisiae* is a single-celled eucaryote that exists in either a haploid or a diploid state. Diploid cells form by a process known as **mating**, in which two haploid cells fuse. In order for two haploid cells to mate, they must differ in *mating type* (sex). In *S. cerevisiae* there are two mating types, α and **a**, which are specialized for mating with each other. Each produces a specific diffusible signaling molecule (mating factor) and a specific cell-surface receptor protein. These jointly enable a cell to recognize and be recognized by its opposite cell type, with which it then fuses. The resulting diploid cells, called **a**/α, are distinct from either parent: they are unable to mate but can form spores (sporulate) when they run out of food, giving rise to haploid cells by the process of meiosis (discussed in Chapter 21).

The mechanisms by which these three cell types are established and maintained illustrate several of the strategies we have discussed for changing the pattern of gene expression. The mating type of the haploid cell is determined by a single locus, the **mating-type (Mat) locus**, which in an **a**-type cell encodes a single gene regulatory protein, Mata1, and in an α cell encodes two gene regulatory proteins, Matα1 and Matα2. The Mata1 protein has no effect in the **a**-type haploid cell that produces it, but becomes important later in the diploid cell that results from mating. In contrast, the Matα2 protein acts in the α cell as a transcriptional repressor that turns off the **a**-specific genes, while the Matα1 protein acts as a transcriptional activator that turns on the α-specific genes. Once cells of the two mating types have fused, the combination of the Mata1 and Matα2 regulatory proteins generates a completely new pattern of gene expression, unlike that of either parent cell. **Figure 7–65** illustrates the mechanism by which the mating-type-specific genes are expressed in different patterns in the three cell types. This was among the first examples of combinatorial gene control to be identified, and it remains one of the best understood at the molecular level.

Although in most laboratory strains of *S. cerevisiae*, the **a** and α cell types are stably maintained through many cell divisions, some strains isolated from the wild can switch repeatedly between the **a** and α cell types by a mechanism of gene rearrangement whose effects are reminiscent of the DNA rearrangements in *N. gonorrhoeae*, although the exact mechanism seems to be peculiar to yeast. On either side of the *Mat* locus in the yeast chromosome, there is a silent locus encoding the mating-type gene regulatory proteins: the silent locus on one side encodes Matα1 and Matα2; the silent locus on the other side encodes Mata1. In approximately every other cell division, the active gene in the *Mat* locus is excised and replaced by a newly synthesized copy of the silent locus determining the opposite mating type. Because the change removes one gene from the active "slot" and replaces it by another, this mechanism is called the *cassette mechanism*. The change is reversible because, although the original gene at the *Mat* locus is discarded, a silent copy remains in the genome. New DNA copies made from the silent genes function as disposable cassettes that will be inserted in alternation into the *Mat* locus, which serves as the "playing head" (**Figure 7–66**).

The silent cassettes are packaged into a specialized form of chromatin and maintained in a transcriptionally inactive form. The study of these cassettes—

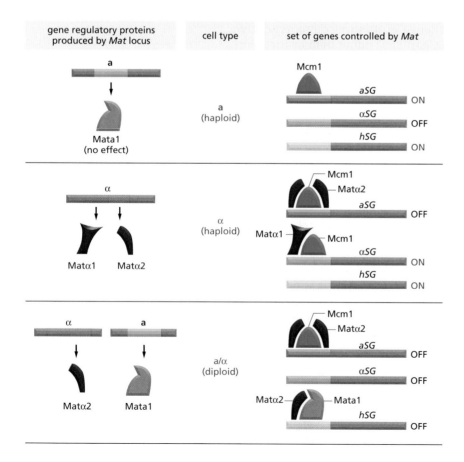

Figure 7–65 Control of cell type in yeasts. Three gene regulatory proteins (Matα1, Matα2, and Mata1) produced by the *Mat* locus determine yeast cell type. Different sets of genes are transcribed in haploid cells of type a, in haploid cells of type α, and in diploid cells (type a/α). The haploid cells express a set of haploid-specific genes *(hSG)* and either a set of α-specific genes *(αSG)* or a set of a-specific genes *(aSG)*. The diploid cells express none of these genes. The Mat regulatory proteins control many target genes in each type of cell by binding, in various combinations, to specific regulatory sequences upstream of these genes. Note that the Matα1 protein is a gene activator protein, whereas the Matα2 protein is a gene repressor protein. Both work in combination with a gene regulatory protein called Mcm1 that is present in all three cell types. In the diploid cell type, Matα2 and Mata1 form a heterodimer (shown in detail in Figure 7–21) that turns off a set of genes (including the gene encoding the Matα1 activator protein) different from that turned off by the Matα2 and Mcm1 proteins. This relatively simple system of gene regulatory proteins is an example of combinatorial control of gene expression.

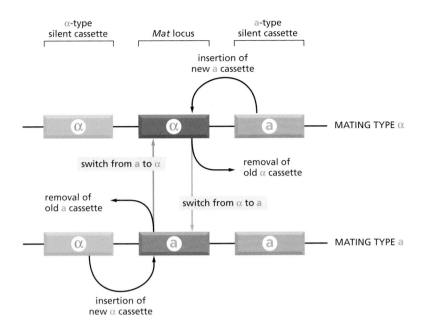

Figure 7–66 Cassette model of yeast mating-type switching. Cassette switching occurs by a gene-conversion process that involves a specialized enzyme (the HO endonuclease) that makes a double-stranded cut at a specific DNA sequence in the *Mat* locus. The DNA near the cut is then excised and replaced by a copy of the silent cassette of opposite mating type. The mechanism of this specialized form of gene conversion is similar to the repair of double-stranded breaks discussed in Chapter 5 (pp. 308–309).

which has been ongoing for nearly 40 years—has provided many of the key insights into the role of chromatin structure in gene regulation.

Two Proteins That Repress Each Other's Synthesis Determine the Heritable State of Bacteriophage Lambda

As we saw at the beginning of the present chapter, the nucleus of a single differentiated cell contains all the genetic information needed to construct a whole vertebrate or plant. This observation eliminates the possibility that an irreversible change in DNA sequence is a major mechanism of cell differentiation in these higher eucaryotes although such changes do occur in lymphocyte differentiation (discussed in Chapter 25). *Reversible* DNA sequence changes, resembling those just described for *Salmonella* and yeasts, in principle could still be responsible for some of the inherited changes in gene expression observed in higher organisms, but there is currently no evidence that such mechanisms are widely used.

Other mechanisms that we have already touched upon in this chapter, however, are also capable of producing patterns of gene regulation that can be inherited by subsequent cell generations. One of the simplest examples is found in the bacterial virus (bacteriophage) lambda where a switch causes the virus to flip-flop between two stable self-maintaining states. This type of switch can be viewed as a prototype for similar, but more complex, switches that operate in the development of higher eucaryotes.

We mentioned earlier that bacteriophage lambda can in favorable conditions become integrated into the *E. coli* cell DNA, to be replicated automatically each time the bacterium divides. Alternatively, the virus can multiply in the cytoplasm, killing its host (see Figure 5–78). Proteins encoded by the bacteriophage genome mediate the switch between these two states. The genome contains a total of about 50 genes, which are transcribed in very different patterns in the two states. A virus destined to integrate, for example, must produce the lambda *integrase* protein, which is needed to insert the lambda DNA into the bacterial chromosome, but must repress the production of the viral proteins responsible for virus multiplication. Once one transcriptional pattern or the other has been established, it is stably maintained.

At the heart of this complex gene regulatory switching mechanism are two gene regulatory proteins synthesized by the virus: the **lambda repressor protein** (cI protein), which we have already encountered, and the **Cro protein**. These proteins repress each other's synthesis, an arrangement that gives rise to just two stable states (**Figure 7–67**). In state 1 (the *prophage state)* the lambda

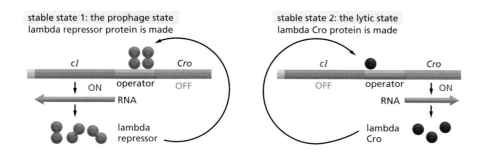

Figure 7–67 A simplified version of the regulatory system that determines the mode of growth of bacteriophage lambda in the _E. coli_ host cell. In stable state 1 (the prophage state) the bacteriophage synthesizes a repressor protein, which activates its own synthesis and turns off the synthesis of several other bacteriophage proteins, including the Cro protein. In state 2 (the lytic state) the bacteriophage synthesizes the Cro protein, which turns off the synthesis of the repressor protein, so that many bacteriophage proteins are made and the viral DNA replicates freely in the _E. coli_ cell, eventually producing many new bacteriophage particles and killing the cell. This example shows how two gene regulatory proteins can be combined in a circuit to produce two heritable states. Both the lambda repressor and the Cro protein recognize the operator through a helix–turn–helix motif (see Figure 7–11).

repressor occupies the operator, blocking the synthesis of Cro and also activating its own synthesis. In state 2 (the _lytic state)_ the Cro protein occupies a different site in the operator, blocking the synthesis of repressor but allowing its own synthesis. In the prophage state most of the DNA of the stably integrated bacteriophage is not transcribed; in the lytic state, this DNA is extensively transcribed, replicated, packaged into new bacteriophage, and released by host cell lysis.

When the host bacteria are growing well, an infecting virus tends to adopt state 1, allowing the DNA of the virus to multiply along with the host chromosome. When the host cell is damaged, an integrated virus converts from state 1 to state 2 in order to multiply in the cell cytoplasm and make a quick exit. This conversion is triggered by the host response to DNA damage, which inactivates the repressor protein. In the absence of such interference, however, the lambda repressor both turns off production of the Cro protein and turns on its own synthesis, and this _positive feedback loop_ helps to maintain the prophage state.

Simple Gene Regulatory Circuits Can Be Used to Make Memory Devices

Positive feedback loops provide a simple general strategy for cell memory—that is, for the establishment and maintenance of heritable patterns of gene transcription. **Figure 7–68** shows the basic principle, stripped to its barest essentials. Eucaryotic cells use many variations of this simple strategy. Several gene regulatory proteins that are involved in establishing the _Drosophila_ body plan (discussed in Chapter 22), for example, stimulate their own transcription, thereby creating a positive feedback loop that promotes their continued synthesis; at the same time many of these proteins repress the transcription of genes encoding other important gene regulatory proteins. In this way, a few gene regulatory proteins that reciprocally affect one another's synthesis and activities can specify a sophisticated pattern of inherited behavior.

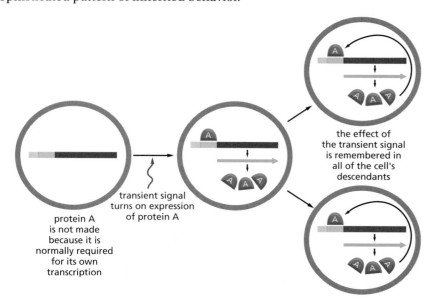

Figure 7–68 Schematic diagram showing how a positive feedback loop can create cell memory. Protein A is a gene regulatory protein that activates its own transcription. All of the descendants of the original cell will therefore "remember" that the progenitor cell had experienced a transient signal that initiated the production of the protein.

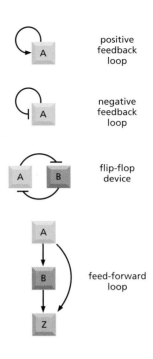

Figure 7–69 Common types of network motifs in transcriptional circuits. A and B represent gene regulatory proteins, arrows indicate positive transcriptional control, and lines with bars depict negative transcriptional control. More detailed descriptions of positive feedback loops and flip-flop devices are given in Figures 7–70 and 7–71, respectively. In feed-forward loops, A and B represent regulatory proteins that both activate the transcription of a target gene, Z.

Transcription Circuits Allow the Cell to Carry Out Logic Operations

Simple gene regulatory switches can be combined to create all sorts of control devices, just as simple electronic switching elements in a computer can be linked to perform many types of operations. The analysis of gene regulatory circuits has revealed that certain simple types of arrangements are found over and over again in cells from widely different species. For example, positive and *negative feedback loops* are especially common in all cells (**Figure 7–69**). As we have seen, the former provides a simple memory device; the latter is often used to keep the expression of a gene close to a standard level regardless of variations in biochemical conditions inside a cell. Suppose, for example, that a transcriptional repressor protein binds to the regulatory region of its own gene and exerts a strong negative feedback, such that transcription occurs at a very low rate when the concentration of repressor protein is above some critical value (determined by its affinity for its DNA binding site), and at a very high rate when it is below the critical value. The concentration of the protein will then be held close to the critical value, since any circumstance that causes a fall below that value will lead to a steep increase in synthesis, and any rise above that value will cause synthesis to switch off. Such adjustments will, however, take time, so that an abrupt change of conditions will cause a disturbance of gene expression that is strong but transient. As we discuss in Chapter 15, the negative feedback system can thus function as a detector of sudden change. Alternatively, if there is a delay in the feedback loop, the result may be spontaneous oscillations in the expression of the gene (see Figure 15–28). The quantitative details of the negative feedback loop determine which of these possible behaviors will occur.

With two or more genes, the possible range of control circuits and circuit behaviors rapidly becomes more complex. Bacteriophage lambda, as we have seen, exemplifies a common type of two-gene circuit that can flip-flop between expression of one gene and expression of the other. Another common circuit arrangement is called a *feed-forward* loop (see Figure 7–69); among other things, this can serve as a filter, responding to input signals that are prolonged but disregarding those that are brief (**Figure 7–70**). A cell can use these various network motifs as miniature logic devices to process information in surprisingly sophisticated ways.

The simple types of devices just illustrated are combined in a typical eucaryotic cell to create exceedingly complex circuits (**Figure 7–71**). Each cell in a developing multicellular organism is equipped with this control machinery, and must, in effect, use the intricate system of interlocking transcriptional switches to compute how it should behave at each time point in response to the many different past and present inputs received. We are only beginning to understand

Figure 7–70 How a feed-forward loop can measure the duration of a signal. (A) In this theoretical example the gene activator proteins A and B are both required for transcription of Z, and A becomes active only when an input signal is present. (B) If the input signal to A is brief, A does not stay active long enough for B to accumulate, and the Z gene is not transcribed. (C) If the signal to A persists, B accumulates, A remains active, and Z is transcribed. This arrangement allows the cell to ignore rapid fluctuations of the input signal and respond only to persistent levels. This strategy could be used, for example, to distinguish between random noise and a true signal.

The behavior shown here was computed for one particular set of parameter values describing the quantitative properties of A, B, and Z and their syntheses. With different values of these parameters, feed-forward loops can in principle perform other types of "calculations." Many feed-forward loops have been discovered in cells, and theoretical analysis helps researchers to appreciate and subsequently test the different ways in which they may function. (Adapted from S.S. Shen-Orr, R. Milo, S. Mangan and U. Alon, *Nat. Genet.* 31:64–68, 2002. With permission from Macmillan Publishers Ltd.)

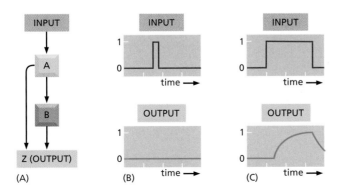

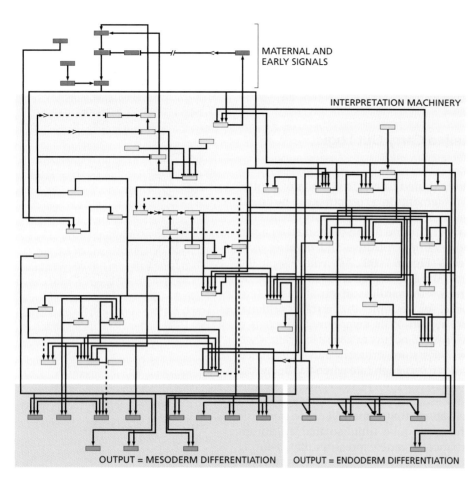

MATERNAL AND EARLY SIGNALS

INTERPRETATION MACHINERY

OUTPUT = MESODERM DIFFERENTIATION   OUTPUT = ENDODERM DIFFERENTIATION

Figure 7–71 The exceedingly complex gene circuit that specifies a portion of the developing sea urchin embryo. Each *colored small box* represents a different gene. Those in *yellow* code for gene regulatory proteins and those in *green* and *blue* code for proteins that give cells of the mesoderm and endoderm, respectively, their specialized characteristics. Genes depicted in *gray* are largely active in the mother and provide the egg with cues needed for proper development. *Arrows* depict instances in which a gene regulatory protein activates the transcription of another gene. Lines ending in *bars* indicate examples of gene repression.

how to study such complex intracellular control networks. Indeed, without quantitative information far more precise and complete than we yet have, it is impossible to predict the behavior of a system such as that shown in Figure 7–71: the circuit diagram by itself is not enough.

Synthetic Biology Creates New Devices from Existing Biological Parts

Our discussion has focused on naturally occurring transcriptional circuits, but it is also instructive to design and construct artificial circuits in the laboratory and introduce them into cells to examine their behavior. **Figure 7–72** shows, for example, how an engineered bacterial cell can switch between three states in a prescribed order, thus functioning as an oscillator or simple clock. The construction of such new devices from existing parts is often termed *synthetic biology*. Scientists use this approach to test whether they truly understand the properties of the component parts; if so, they should be able to combine these parts in novel ways and accurately predict the characteristics of the new device. The fact that these predictions usually fail illustrates how far we are from truly understanding the detailed workings of the cell. There are many large gaps in our knowledge that will require skillful application of the quantitative approaches of the physical sciences to complex biological systems.

Circadian Clocks Are Based on Feedback Loops in Gene Regulation

Life on Earth evolved in the presence of a daily cycle of day and night, and many present-day organisms (ranging from archaea to plants to humans) have come to possess an internal rhythm that dictates different behaviors at different times of day. These behaviors range from the cyclical change in metabolic enzyme

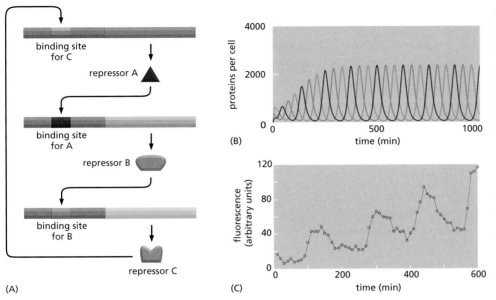

Figure 7–72 **A simple gene oscillator or "clock" designed in the laboratory.** (A) Recombinant DNA techniques were used to make three artificial genes, each coding for a different bacterial repressor protein, and each controlled by the product of another gene in the set, so as to create a regulatory circuit as shown. These repressors (denoted A, B, and C in the figure) are the Lac repressor (see Figure 7–39), the Tet repressor, which regulates genes in response to tetracycline, and the Lambda repressor (see Figure 7–67). When introduced into a bacterial cell, the three genes form a delayed negative feedback circuit: the product of gene A, for example, acts via genes B and C to indirectly inhibit its own expression. The delayed negative feedback gives rise to oscillations. (B) Computer prediction of the oscillatory behavior. The cell cycles repetitively through a series of states, expressing A, then B, then C, then A again, and so on, as each gene product in turn escapes from inhibition by the previous one and represses the next. (C) Actual oscillations observed in a cell containing the three artificial genes in (A), demonstrated with a fluorescent reporter of the expression of one of these genes. The increasing amplitude of the fluorescence signal reflects the growth of the bacterial cell. (Adapted from M.B. Elowitz and S. Leibler, *Nature* 403:335–338, 2000. With permission from Macmillan Publishers Ltd.)

activities of a fungus to the elaborate sleep–wake cycles of humans. The internal oscillators that control such diurnal rhythms are called circadian clocks.

By carrying its own circadian clock, an organism can anticipate the regular daily changes in its environment and take appropriate action in advance. Of course, the internal clock cannot be perfectly accurate, and so it must be capable of being reset by external cues such as the light of day. Thus, circadian clocks keep running even when the environmental cues (changes in light and dark) are removed, but the period of this free-running rhythm is generally a little less or a little more than 24 hours. External signals indicating the time of day cause small adjustments in the running of the clock, so as to keep the organism in synchrony with its environment. Following more drastic shifts, circadian cycles become gradually reset (entrained) by the new cycle of light and dark, as anyone who has experienced jet lag can attest.

We might expect that the circadian clock in a creature such as a human would itself be a complex multicellular device, with different groups of cells responsible for different parts of the oscillation mechanism. Remarkably, however, it turns out that in almost all organisms, including humans, the timekeepers are individual cells. Thus, a clock that operates in each member of a specialized group of brain cells (the SCN cells in the suprachiasmatic nucleus of the hypothalamus) controls our diurnal cycles of sleeping and waking, body temperature, and hormone release. Even if these cells are removed from the brain and dispersed in a culture dish, they will continue to oscillate individually, showing a cyclic pattern of gene expression with a period of approximately 24 hours. In the intact body, the SCN cells receive neural cues from the retina, entraining them to the daily cycle of light and dark, and they send information about the time of day to another brain area, the pineal gland, which relays the time signal to the rest of the body by releasing the hormone melatonin in time with the clock.

Although the SCN cells have a central role as timekeepers in mammals, they are not the only cells in the mammalian body that have an internal circadian rhythm or an ability to reset it in response to light. Similarly, in *Drosophila*, many different types of cells, including those of the thorax, abdomen, antenna, leg, wing, and testis all continue a circadian cycle when they have been dissected away from the rest of the fly. The clocks in these isolated tissues, like those in the SCN cells, can be reset by externally imposed light and dark cycles.

The working of circadian clocks, therefore, is a fundamental problem in cell biology. Although we do not yet understand all the details, studies in a wide variety of organisms have revealed many of the basic principles and molecular components. For animals, much of what we know has come from searches in *Drosophila* for mutations that make the fly's circadian clock run fast, or slow, or not at all; and this work has led to the discovery that many of the same components are involved in the circadian clock of mammals.

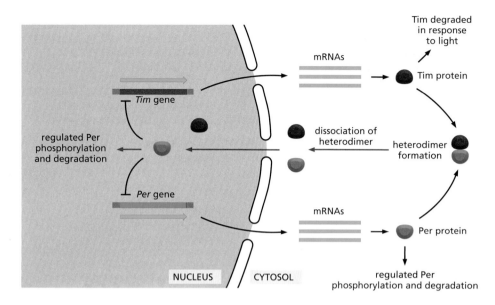

Figure 7–73 Simplified outline of the mechanism of the circadian clock in *Drosophila* cells. A central feature of the clock is the periodic accumulation and decay of two gene regulatory proteins, Tim (short for timeless, based on the phenotype of a gene mutation) and Per (short for period). The mRNAs encoding these proteins are translated in the cytosol, and, when each protein has accumulated to critical levels, they form a heterodimer. After a time delay, the heterodimer dissociates and Tim and Per are transported into the nucleus, where they regulate a number of gene products that mediate effects of the clock. Once in the nucleus, Per also represses the *Tim* and *Per* genes, creating a feedback system that causes the levels of Tim and Per to fall. In addition to this transcriptional feedback, the clock depends on a set of other proteins. For example, the controlled degradation of Per indicated in the diagram imposes delays in the periodic accumulation of Tim and Per, which are crucial to the functioning of the clock. Steps at which specific delays are imposed are shown in *red*.

Entrainment (or resetting) of the clock occurs in response to new light–dark cycles. Although most *Drosophila* cells do not have true photoreceptors, light is sensed by intracellular flavoproteins, also called cryptochromes. In the presence of light, these proteins associate with the Tim protein and cause its degradation, thereby resetting the clock. (Adapted from J.C. Dunlap, *Science* 311:184–186, 2006. With permission from AAAS.)

The mechanism of the clock in *Drosophila* is briefly outlined in **Figure 7–73**. At the heart of the oscillator is a transcriptional feedback loop that has a time delay built into it: accumulation of certain key gene products switches off the transcription of their genes, but with a delay, so that—crudely speaking—the cell oscillates between a state in which the products are present and transcription is switched off, and one in which the products are absent and transcription is switched on.

Despite the relative simplicity of the basic principle behind circadian clocks, the details are complex. One reason for this complexity is that clocks must be buffered against changes in temperature, which typically speed up or slow down macromolecular association. They must also run accurately but be capable of being reset. Although it is not yet understood how biological clocks run at a constant speed despite changes in temperature, the mechanism for resetting the *Drosophila* clock is the light-induced destruction of one of the key gene regulatory proteins (see Figure 7–73).

A Single Gene Regulatory Protein Can Coordinate the Expression of a Set of Genes

Cells need to be able to switch genes on and off individually but they also need to coordinate the expression of large groups of different genes. For example, when a quiescent eucaryotic cell receives a signal to divide, many hitherto unexpressed genes are turned on together to set in motion the events that lead eventually to cell division (discussed in Chapter 17). One way bacteria coordinate the expression of a set of genes is to cluster them together in an *operon* under control of a single promoter (see Figure 7–34). In eucaryotes, however, each gene is transcribed from a separate promoter.

How, then, do eucaryotes coordinate gene expression? This is an especially important question because, as we have seen, most eucaryotic gene regulatory proteins act as part of a regulatory protein committee, all of whose members are necessary to express the gene in the right cell, at the right time, in response to the proper signals, and to the proper level. How, then, can a eucaryotic cell rapidly and decisively switch whole groups of genes on or off?

The answer is that even though control of gene expression is combinatorial, the effect of a single gene regulatory protein can still be decisive in switching any particular gene on or off, simply by completing the combination needed to maximally activate or repress that gene. This situation is analogous to dialing in the final number of a combination lock: the lock will spring open with only this simple addition if all of the other numbers have been previously entered. Moreover, the same number can complete the combination for many different locks. Analogously, the addition of a particular protein can turn on many different genes.

An example in humans is the control of gene expression by the *glucocorticoid receptor protein*. To bind to regulatory sites in DNA, this gene regulatory protein must first form a complex with a molecule of a glucocorticoid steroid hormone, such as cortisol (see Figure 15–13). The body releases this hormone during times of starvation and intense physical activity, and among its other activities, it stimulates liver cells to increase the production of glucose from amino acids and other small molecules. To respond in this way, liver cells increase the expression of many different genes that code for metabolic enzymes and other products. Although these genes all have different and complex control regions, their maximal expression depends on the binding of the hormone–glucocorticoid receptor complex to a regulatory site in the DNA of each gene. When the body has recovered and the hormone is no longer present, the expression of each of these genes drops to its normal level in the liver. In this way a single gene regulatory protein can control the expression of many different genes (**Figure 7–74**).

The effects of the glucocorticoid receptor are not confined to cells of the liver. In other cell types, activation of this gene regulatory protein by hormone also causes changes in the expression levels of many genes; the genes affected, however, are often different from those affected in liver cells. As we have seen, each cell type has an individualized set of gene regulatory proteins, and because of combinatorial control, these critically influence the action of the glucocorticoid receptor. Because the receptor is able to assemble with many different sets of cell-type-specific gene regulatory proteins, switching it on with hormone produces a different spectrum of effects in each cell type.

Expression of a Critical Gene Regulatory Protein Can Trigger the Expression of a Whole Battery of Downstream Genes

The ability to switch many genes on or off coordinately is important not only in the day-to-day regulation of cell function. It is also the means by which eucaryotic cells differentiate into specialized cell types during embryonic development. The development of muscle cells provides a striking example.

Figure 7–74 A single gene regulatory protein can coordinate the expression of several different genes. The action of the glucocorticoid receptor is illustrated schematically. On the *left* is a series of genes, each of which has various gene activator proteins bound to its regulatory region. However, these bound proteins are not sufficient on their own to fully activate transcription. On the *right* is shown the effect of adding an additional gene regulatory protein—the glucocorticoid receptor in a complex with glucocorticoid hormone—that can bind to the regulatory region of each gene. The glucocorticoid receptor completes the combination of gene regulatory proteins required for maximal initiation of transcription, and the genes are now switched on as a set. In the absence of the hormone, the glucocorticoid receptor is unavailable to bind to DNA.

In addition to activating gene expression, the hormone-bound form of the glucocorticoid receptor represses transcription of certain genes, depending on the gene regulatory proteins already present on their control regions. The effect of the glucocorticoid receptor on any given gene therefore depends upon the type of cell, the gene regulatory proteins contained within it, and the regulatory region of the gene. The structure of the DNA-binding portion of the glucocorticoid receptor is shown in Figure 7–16.

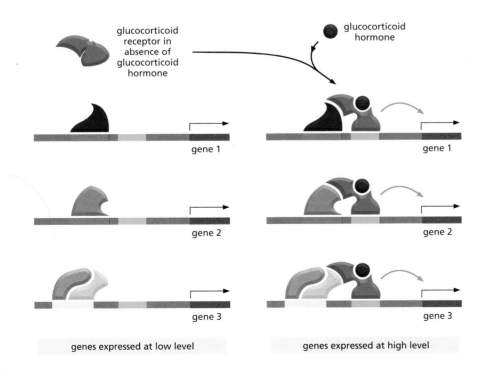

glucocorticoid receptor in absence of glucocorticoid hormone

glucocorticoid hormone

gene 1

gene 1

gene 2

gene 2

gene 3

gene 3

genes expressed at low level

genes expressed at high level

As described in Chapter 16, a mammalian skeletal muscle cell is a highly distinctive giant cell, formed by the fusion of many muscle precursor cells called *myoblasts*, and therefore containing many nuclei. The mature muscle cell synthesizes a large number of characteristic proteins, including specific types of actin, myosin, tropomyosin, and troponin (all part of the contractile apparatus), creatine phosphokinase (for the specialized metabolism of muscle cells), and acetylcholine receptors (to make the membrane sensitive to nerve stimulation). In proliferating myoblasts, these muscle-specific proteins and their mRNAs are absent or are present in very low concentrations. As myoblasts begin to fuse with one another, the corresponding genes are all switched on coordinately as part of a general transformation of the pattern of gene expression.

This entire program of muscle differentiation can be triggered in cultured skin fibroblasts and certain other cell types by introducing any one of a family of helix–loop–helix proteins—the so-called myogenic proteins (MyoD, Myf5, MyoG, and Mrf4)—that are normally expressed only in muscle cells (**Figure 7–75**A). Binding sites for these regulatory proteins are present in the regulatory DNA sequences adjacent to many muscle-specific genes, and the myogenic proteins thereby directly activate the transcription of these genes. In addition, the myogenic proteins stimulate their own transcription as well as that of various other gene regulatory proteins involved in muscle development, creating an elaborate series of positive feedback and feed-forward loops that amplify and maintain the muscle developmental program, even after the initiating signal has disappeared (Figure 7–75B; see also Chapter 22).

It is probable that those cell types that are converted to muscle cells by the addition of myogenic proteins have already accumulated a number of gene regulatory proteins that can cooperate with the myogenic proteins to switch on muscle-specific genes. Other cell types fail to be converted to muscle by myogenin or its relatives; these cells presumably have not accumulated the other gene regulatory proteins required.

The conversion of one cell type (fibroblast) to another (skeletal muscle) by a single gene regulatory protein reemphasizes one of the most important principles discussed in this chapter: differences in gene expression can produce dramatic differences between cell types—in size, shape, chemistry, and function.

Combinatorial Gene Control Creates Many Different Cell Types in Eucaryotes

We have already discussed how multiple gene regulatory proteins can act in combination to regulate the expression of an individual gene. But, as the example of the myogenic proteins shows, combinatorial gene control means more than this: not only does each gene respond to many gene regulatory proteins that control it, but each regulatory protein contributes to the control of many genes. Moreover, although some gene regulatory proteins are specific to a single cell type, most are switched on in a variety of cell types, at several sites in the body, and at several times in development. This point is illustrated schematically in **Figure 7–76**, which shows how combinatorial gene control makes it possible to generate a great deal of biological complexity even with relatively few gene regulatory proteins.

With combinatorial control, a given gene regulatory protein does not necessarily have a single, simply definable function as commander of a particular battery of genes or specifier of a particular cell type. Rather, gene regulatory proteins can be likened to the words of a language: they are used with different meanings in a variety of contexts and rarely alone; it is the well-chosen combination that conveys the information that specifies a gene regulatory event.

One requirement of combinatorial control is that many gene regulatory proteins must be able to work together to influence the final rate of transcription. Experiments demonstrate that even unrelated gene regulatory proteins from widely different eucaryotic species can cooperate when introduced into the same cell. This situation reflects the high degree of conservation of the transcription

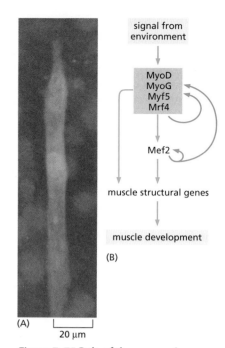

(A) 20 μm

(B)

Figure 7–75 Role of the myogenic regulatory proteins in muscle development. (A) The effect of expressing the MyoD protein in fibroblasts. As shown in this immunofluorescence micrograph, fibroblasts from the skin of a chick embryo have been converted to muscle cells by the experimentally induced expression of the *MyoD* gene. The fibroblasts that have been induced to express the *MyoD* gene have fused to form elongated multinucleate muscle-like cells, which are stained *green* with an antibody that detects a muscle-specific protein. Fibroblasts that do not express the *MyoD* gene are barely visible in the background. (B) Simplified scheme showing some of the gene regulatory proteins involved in skeletal muscle development. External signals result in the synthesis of the four closely related myogenic gene regulatory proteins, MyoD, Myf5, MyoG, and Mrf4. These gene regulatory proteins activate their own as well as each other's synthesis in a complex series of feedback loops, only some of which are shown in the figure. These proteins in turn directly activate transcription of muscle structural genes as well as the *Mef2* gene, which encodes an additional gene regulatory protein. Mef2 acts in combination with the myogenic proteins in a feed-forward loop to further activate the transcription of muscle structural genes, as well as forming an additional positive feedback loop that helps to maintain transcription of the myogenic genes. (A, courtesy of Stephen Tapscott and Harold Weintraub; B, adapted from J.D. Molkentin and E.N. Olson, *Proc. Natl Acad. Sci. U.S.A.* 93:9366–9373, 1996. With permission from National Academy of Sciences.)

machinery. It seems that the multifunctional, combinatorial mode of action of gene regulatory proteins has put a tight constraint on their evolution: they must interlock with other gene regulatory proteins, the general transcription factors, Mediator, RNA polymerase, and the chromatin-modifying enzymes.

An important consequence of combinatorial gene control is that the effect of adding a new gene regulatory protein to a cell will depend on the cell's past history, since this history will determine which gene regulatory proteins are already present. Thus, during development a cell can accumulate a series of gene regulatory proteins that need not initially alter gene expression. The addition of the final members of the requisite combination of gene regulatory proteins completes the regulatory message, and can lead to large changes in gene expression. Such a scheme, as we have seen, helps to explain how the addition of a single regulatory protein to a fibroblast can produce the dramatic transformation of the fibroblast into a muscle cell. It also can account for the important difference, discussed in Chapter 22, between the process of *cell determination*—where a cell becomes committed to a particular developmental fate—and the process of *cell differentiation*, in which a committed cell expresses its specialized character.

A Single Gene Regulatory Protein Can Trigger the Formation of an Entire Organ

We have seen that even though combinatorial control is the norm for eucaryotic genes, a single gene regulatory protein, if it completes the appropriate combination, can be decisive in switching a whole set of genes on or off, and we have

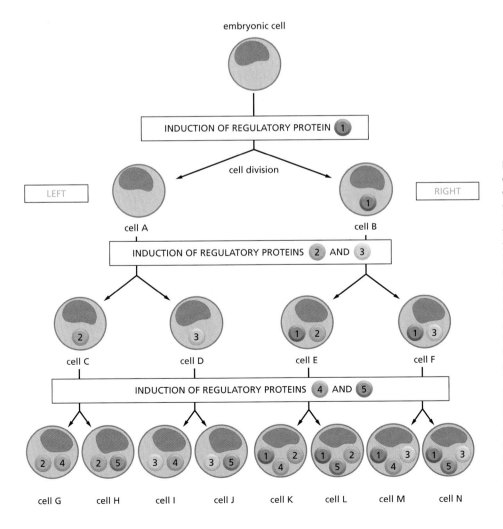

Figure 7–76 The importance of combinatorial gene control for development. Combinations of a few gene regulatory proteins can generate many cell types during development. In this simple, idealized scheme a "decision" to make one of a pair of different gene regulatory proteins (shown as numbered *circles*) is made after each cell division. Sensing its relative position in the embryo, the daughter cell toward the *left side* of the embryo is always induced to synthesize the even-numbered protein of each pair, while the daughter cell toward the *right side* of the embryo is induced to synthesize the odd-numbered protein. The production of each gene regulatory protein is assumed to be self-perpetuating once it has become initiated (see Figure 7–68). In this way, through cell memory, the final combinatorial specification is built up step by step. In this purely hypothetical example, five different gene regulatory proteins have created eight final cell types (G–N).

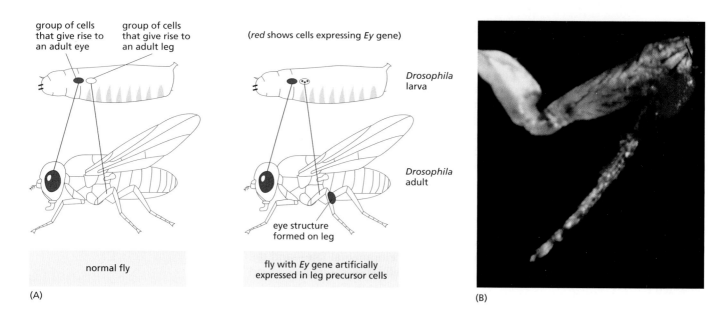

group of cells that give rise to an adult eye

group of cells that give rise to an adult leg

(*red* shows cells expressing *Ey* gene)

Drosophila larva

Drosophila adult

eye structure formed on leg

normal fly

fly with *Ey* gene artificially expressed in leg precursor cells

(A)

(B)

seen how this can convert one cell type into another. A dramatic extension of the principle comes from studies of eye development in *Drosophila*, mice, and humans. Here, a gene regulatory protein, called Ey (short for Eyeless) in flies and Pax6 in vertebrates, is crucial. When expressed in the proper context, Ey can trigger the formation of not just a single cell type but a whole organ (an eye), composed of different types of cells, all properly organized in three-dimensional space.

The most striking evidence for the role of Ey comes from experiments in fruit flies in which the *Ey* gene is artificially expressed early in development in groups of cells that normally will go on to form leg parts. This abnormal gene expression causes eyes to develop in the legs (**Figure 7–77**).

The *Drosophila* eye is composed of thousands of cells, and the question of how a regulatory protein coordinates the construction of a whole organ is a central topic in *developmental biology*. As discussed in Chapter 22, it involves cell–cell interactions as well as intracellular gene regulatory proteins. Here, we note that Ey directly controls the expression of many other genes by binding to their regulatory regions. Some of the genes controlled by Ey themselves code for gene regulatory proteins that, in turn, control the expression of other genes. Moreover, some of these regulatory gene products act back on *Ey* itself to create a positive feedback loop that ensures the continued synthesis of the Ey protein as the cells divide and further differentiate (**Figure 7–78**). In this way, the action of just one regulatory protein can permanently turn on a cascade of gene regulatory proteins and cell–cell interaction mechanisms, whose actions result in an organized group of many different types of cells. One can begin to imagine how, by repeated applications of this principle, a complex organism is assembled piece by piece.

Figure 7–77 Expression of the *Drosophila* Ey gene in precursor cells of the leg triggers the development of an eye on the leg. (A) Simplified diagrams showing the result when a fruit fly larva contains either the normally expressed *Ey* gene *(left)* or an *Ey* gene that is additionally expressed artificially in cells that normally give rise to leg tissue *(right)*. (B) Photograph of an abnormal leg that contains a misplaced eye (see also Figure 22–2). (B, courtesy of Walter Gehring.)

Figure 7–78 Gene regulatory proteins that specify eye development in *Drosophila*. Toy *(Twin of eyeless)* and Ey *(Eyeless)* encode similar gene regulatory proteins, Toy and Ey, either of which, when ectopically expressed, can trigger eye development. In normal eye development, expression of *Ey* requires the *Toy* gene. Once its transcription is activated by Toy, Ey activates the transcription of *So (Sine oculis)* and *Eya (Eyes absent)*, which act together to switch on the *Dac (Dachshund)* gene. As indicated by the *green arrows,* some of the gene regulatory proteins form a series of interlocking positive feedback loops that reinforce the initial commitment to eye development. The Ey protein is known to bind directly to numerous target genes for eye development, including those encoding lens crystallins, rhodopsins, and other photoreceptor proteins. (Adapted from T. Czerny et al., *Mol. Cell* 3:297–307, 1999. With permission from Elsevier.)

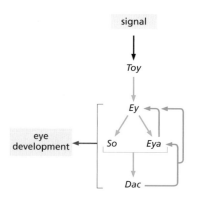

signal

Toy

Ey

eye development

So   Eya

Dac

The Pattern of DNA Methylation Can Be Inherited When Vertebrate Cells Divide

Thus far, we have emphasized the regulation of gene transcription by proteins that associate with DNA. However, DNA itself can be covalently modified, and in the following sections we shall see that this, too, provides opportunities for the regulation of gene expression. In vertebrate cells, the methylation of cytosine provides a powerful mechanism through which gene expression patterns are passed on to progeny cells. The methylated form of cytosine, 5-methylcytosine (5-methyl C), has the same relation to cytosine that thymine has to uracil, and the modification likewise has no effect on base-pairing (**Figure 7–79**). **DNA methylation** in vertebrate DNA is restricted to cytosine (C) nucleotides in the sequence CG, which is base-paired to exactly the same sequence (in opposite orientation) on the other strand of the DNA helix. Consequently, a simple mechanism permits the existing pattern of DNA methylation to be inherited directly by the daughter DNA strands. An enzyme called *maintenance methyltransferase* acts preferentially on those CG sequences that are base-paired with a CG sequence that is already methylated. As a result, the pattern of DNA methylation on the parental DNA strand serves as a template for the methylation of the daughter DNA strand, causing this pattern to be inherited directly following DNA replication (**Figure 7–80**).

The stable inheritance of DNA methylation patterns can be explained by maintenance DNA methyltransferases. DNA methylation patterns, however, are dynamic during vertebrate development. Shortly after fertilization there is a genome-wide wave of demethylation, when the vast majority of methyl groups are lost from the DNA. This demethylation may occur either by suppression of maintenance DNA methyltransferase activity, resulting in the passive loss of methyl groups during each round of DNA replication, or by a specific demethyl-ating enzyme. Later in development, new methylation patterns are established by several *de novo DNA methyltransferases* that are directed to DNA by sequence-specific DNA-binding proteins where they modify adjacent unmethy-lated CG nucleotides. Once the new patterns of methylation are established, they can be propagated through rounds of DNA replication by the maintenance methyl transferases.

DNA methylation has several uses in the vertebrate cell. Perhaps its most important role is to work in conjunction with other gene expression control mechanisms to establish a particularly efficient form of gene repression that can be faithfully passed on to progeny cells (**Figure 7–81**). This combination of mechanisms ensures that unneeded eucaryotic genes can be repressed to very high degrees. For example, the rate at which a vertebrate gene is transcribed can vary 10^6-fold between one tissue and another. The unexpressed vertebrate genes are much less "leaky" in terms of transcription than bacterial genes, in which the largest known differences in transcription rates between expressed and unex-pressed gene states are about 1000-fold.

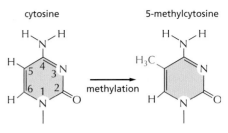

Figure 7–79 Formation of 5-methyl-cytosine occurs by methylation of a cytosine base in the DNA double helix. In vertebrates this event is confined to selected cytosine (C) nucleotides located in the sequence CG.

Figure 7–80 How DNA methylation patterns are faithfully inherited. In vertebrate DNAs a large fraction of the cytosine nucleotides in the sequence CG are methylated (see Figure 7–79). Because of the existence of a methyl-directed methylating enzyme (the maintenance methyltransferase), once a pattern of DNA methylation is established, that pattern of methylation is inherited in the progeny DNA, as shown.

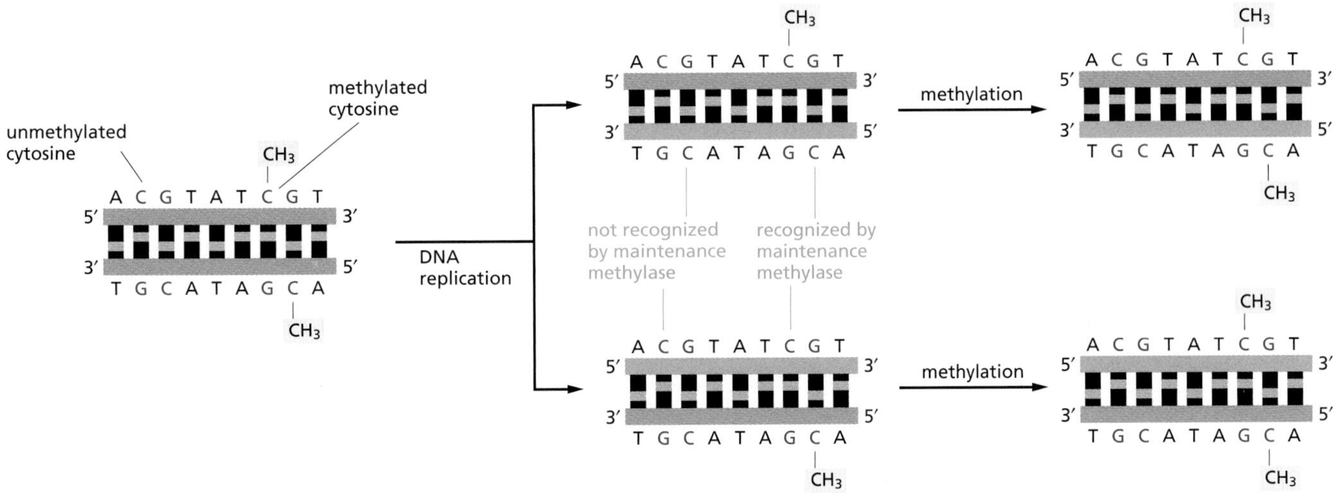

How DNA methylation helps to repress gene expression is not understood in detail, but two general mechanisms have emerged. DNA methylation of the promoter region of a gene or of its regulatory sequences can interfere directly with the binding of proteins required for transcription initiation. In addition, the cell has a repertoire of proteins that specifically bind to methylated DNA (see Figure 7–81), thereby blocking access of other proteins. One reflection of the importance of DNA methylation to humans is the widespread involvement of errors in this mechanism in cancer progression (see Chapter 20).

We shall return to the topic of gene silencing by DNA methylation later in this chapter, when we discuss X-chromosome inactivation and other examples of large-scale gene silencing. First, however, we describe some of the other ways in which DNA methylation affects our genomes.

Genomic Imprinting Is Based on DNA Methylation

Mammalian cells are diploid, containing one set of genes inherited from the father and one set from the mother. The expression of a small minority of genes depends on whether they have been inherited from the mother or the father: while the paternally inherited gene copy is active, the maternally inherited gene copy is silent, or vice-versa. This phenomenon is called **genomic imprinting**. The gene for *insulin-like growth factor-2* (*Igf2*) is a well-studied example of an

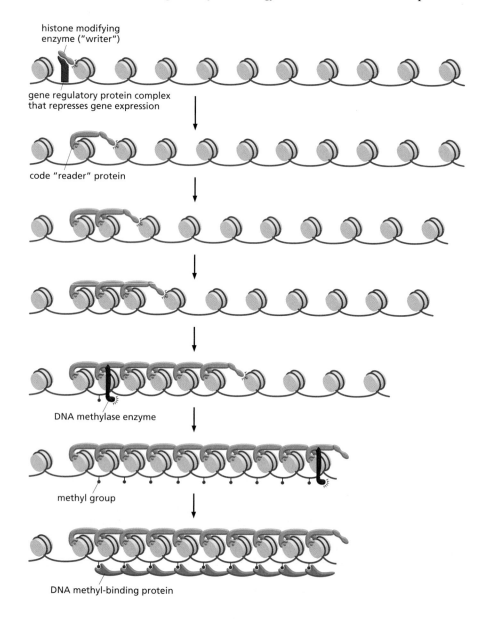

histone modifying enzyme ("writer")

gene regulatory protein complex that represses gene expression

code "reader" protein

DNA methylase enzyme

methyl group

DNA methyl-binding protein

Figure 7–81 Multiple mechanisms contribute to stable gene repression. In this schematic example, histone reader and writer proteins, under the direction of gene regulatory proteins, establish a repressive form of chromatin. A *de novo* DNA methylase is attracted by the histone reader and methylates nearby cytosines in DNA, which are, in turn, bound by DNA methyl-binding proteins. During DNA replication, some of the modified (*blue* dot) histones will be inherited by one daughter chromosome, some by the other, and in each daughter they can induce reconstruction of the same pattern of chromatin modifications (see Figure 5–39). At the same time, the mechanism shown in Figure 7–80 will cause both daughter chromosomes to inherit the same methylation pattern. The two inheritance mechanisms will be mutually reinforcing, if DNA methylation stimulates the activity of the histone writer. This scheme can account for the inheritance by daughter cells of both the histone and the DNA modifications. It can also explain the tendency of some chromatin modifications to spread along a chromosome (see Figure 4–45).

imprinted gene. *Igf2* is required for prenatal growth, and mice that do not express *Igf2* at all are born half the size of normal mice. However, only the paternal copy of *Igf2* is transcribed, and only this gene copy matters for the phenotype. As a result, mice with a mutated paternally derived *Igf2* gene are stunted, while mice with a mutated maternally derived *Igf2* gene are normal.

In the early embryo, genes subject to imprinting are marked by methylation according to whether they were derived from a sperm or an egg chromosome. In this way, DNA methylation is used as a mark to distinguish two copies of a gene that may be otherwise identical (**Figure 7–82**). Because imprinted genes are somehow protected from the wave of demethylation that takes place shortly after fertilization (see p. 467), this mark enables somatic cells to "remember" the

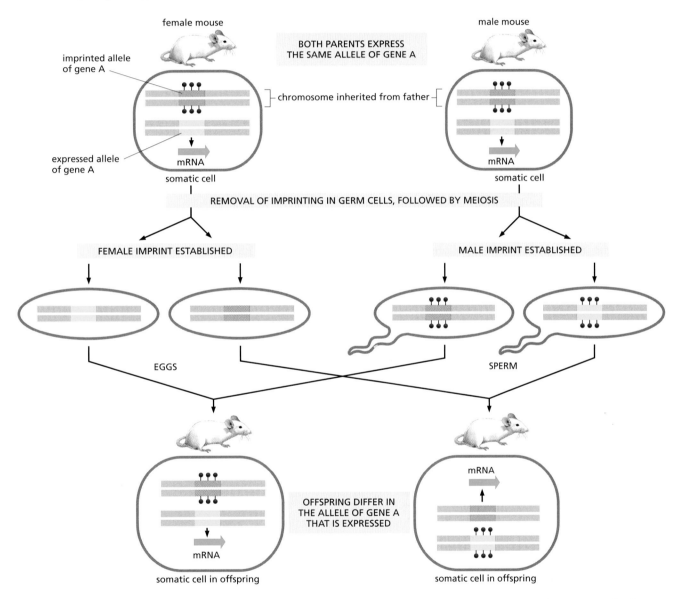

Figure 7–82 Imprinting in the mouse. The *top* portion of the figure shows a pair of homologous chromosomes in the somatic cells of two adult mice, one male and one female. In this example, both mice have inherited the top homolog from their father and the bottom homolog from their mother, and the paternal copy of a gene subject to imprinting (indicated in *orange*) is methylated, preventing its expression. The maternally derived copy of the same gene *(yellow)* is expressed. The remainder of the figure shows the outcome of a cross between these two mice. During germ cell formation, but before meiosis, the imprints are erased and then, much later in germ cell development, they are reimposed in a sex-specific pattern (*middle* portion of figure). In eggs produced from the female, neither allele of the A gene is methylated. In sperm from the male, both alleles of gene A are methylated. Shown at the *bottom* of the figure are two of the possible imprinting patterns inherited by the progeny mice; the mouse on the *left* has the same imprinting pattern as each of the parents, whereas the mouse on the *right* has the opposite pattern. If the two alleles of A gene are distinct, these different imprinting patterns can cause phenotypic differences in the progeny mice, even though they carry exactly the same DNA sequences of the two A gene alleles. Imprinting provides an important exception to classical genetic behavior, and several hundred mouse genes are thought to be affected in this way. However, the majority of mouse genes are not imprinted, and therefore the rules of Mendelian inheritance apply to most of the mouse genome.

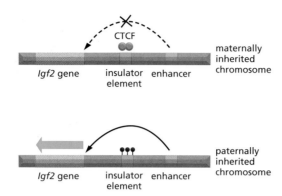

Figure 7–83 Mechanism of imprinting of the mouse *Igf2* gene. On chromosomes inherited from the female, a protein called CTCF binds to an insulator (see Figure 7–62), blocking communication between the enhancer *(green)* and the *Igf2* gene *(orange)*. IGF2 is therefore not expressed from the maternally inherited chromosome. Because of imprinting, the insulator on the male-derived chromosome is methylated; this inactivates the insulator, by blocking the binding of the CTCF protein, and allows the enhancer to activate transcription of the *Igf2* gene. In other examples of imprinting, methylation blocks gene expression by interfering with the binding of proteins required for a gene's transcription.

parental origin of each of the two copies of the gene and to regulate their expression accordingly. In most cases, the methyl imprint silences nearby gene expression. In some cases, however, the methyl imprint can activate expression of a gene. In the case of *Igf2*, for example, methylation of an insulator element (see Figure 7–62) on the paternally derived chromosome blocks its function and allows a distant enhancer to activate transcription of the *Igf2* gene. On the maternally derived chromosome, the insulator is not methylated and the *Igf2* gene is therefore not transcribed (**Figure 7–83**).

Why imprinting should exist at all is a mystery. In vertebrates, it is restricted to placental mammals, and many of the imprinted genes are involved in fetal development. One idea is that imprinting reflects a middle ground in the evolutionary struggle between males to produce larger offspring and females to limit offspring size. Whatever its purpose might be, imprinting provides startling evidence that features of DNA other than its sequence of nucleotides can be inherited.

CG-Rich Islands Are Associated with Many Genes in Mammals

Because of the way in which DNA repair enzymes work, methylated C nucleotides in the genome tend to be eliminated in the course of evolution. Accidental deamination of an unmethylated C gives rise to U (see Figure 5–45), which is not normally present in DNA and thus is recognized easily by the DNA repair enzyme uracil DNA glycosylase, excised, and then replaced with a C (as discussed in Chapter 5). But accidental deamination of a 5-methyl C cannot be repaired in this way, for the deamination product is a T and so is indistinguishable from the other, nonmutant T nucleotides in the DNA. Although a special repair system exists to remove these mutant T nucleotides, many of the deaminations escape detection, so that those C nucleotides in the genome that are methylated tend to mutate to T over evolutionary time.

During the course of evolution, more than three out of every four CGs have been lost in this way, leaving vertebrates with a remarkable deficiency of this dinucleotide. The CG sequences that remain are very unevenly distributed in the genome; they are present at 10–20 times their average density in selected regions, called **CG islands**, which are 1000–2000 nucleotide pairs long. These islands, with some important exceptions, seem to remain unmethylated in all cell types. They often surround the promoters of the so-called *housekeeping genes*—those genes that code for the many proteins that are essential for cell viability and are therefore expressed in most cells (**Figure 7–84**).

The distribution of CG islands (also called CpG islands to distinguish the CG dinucleotides from the CG base pair) can be explained if we assume that CG methylation was adopted in vertebrates primarily as a way of maintaining DNA in a transcriptionally inactive state (see Figure 7–81). In vertebrates, new methyl-C to T mutations can be transmitted to the next generation only if they occur in the germ line, the cell lineage that gives rise to sperm or eggs. Most of the DNA in vertebrate germ cells is inactive and highly methylated. Over long

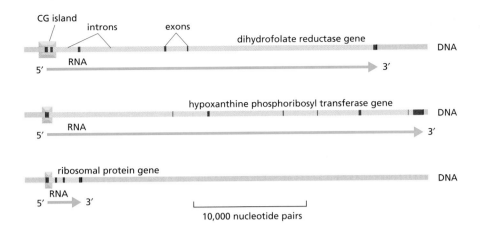

Figure 7–84 **The CG islands surrounding the promoter in three mammalian housekeeping genes.** The *yellow boxes* show the extent of each island. As for most genes in mammals (see Figure 6–25), the exons *(dark red)* are very short relative to the introns *(light red)*. (Adapted from A.P. Bird, *Trends Genet.* 3:342–347, 1987. With permission from Elsevier.)

periods of evolutionary time, the methylated CG sequences in these inactive regions have presumably been lost through spontaneous deamination events that were not properly repaired. However, promoters of genes that remain active in the germ cell lineages (including most housekeeping genes) are kept unmethylated, and therefore spontaneous deaminations of Cs that occur within them can be accurately repaired. Such regions are preserved in modern-day vertebrate cells as CG islands (**Figure 7–85**). In addition, any mutation of a CG sequence in the genome that destroyed the function or regulation of a gene in the adult would be selected against, and some CG islands are presumably the result of a higher than normal density of critical CG sequences for these genes.

The mammalian genome contains an estimated 20,000 CG islands. Most of the islands mark the 5′ ends of transcription units and thus, presumably, of genes. The presence of CG islands thereby provides a convenient way of identifying genes in the DNA sequences of vertebrate genomes.

Epigenetic Mechanisms Ensure That Stable Patterns of Gene Expression Can Be Transmitted to Daughter Cells

As we have seen, once a cell in an organism differentiates into a particular cell type, it generally remains specialized in that way; if it divides, its daughters inherit the same specialized character. For example, liver cells, pigment cells, and endothelial cells (discussed in Chapter 23) divide many times in the life of an individual, each of them faithfully producing daughter cells of the same type. Such differentiated cells must remember their specific pattern of gene expression and pass it on to their progeny through all subsequent cell divisions.

We have already described several ways of enabling daughter cells to "remember" what kind of cells they are supposed to be. One of the simplest is through a positive feedback loop in which a key gene regulatory protein activates, either directly or indirectly, the transcription of its own gene (see Figures 7–68 and 7–69). Interlocking positive feedback loops provide even more stability by buffering the circuit against fluctuations in the level of any one gene regulatory protein (Figures 7–75B and 7–78). We also saw above that DNA methylation can serve as a means for propagating gene expression patterns to descendants (see Figure 7–80).

Figure 7–85 **A mechanism to explain both the marked overall deficiency of CG sequences and their clustering into CG islands in vertebrate genomes.** A *black line* marks the location of a CG dinucleotide in the DNA sequence, while a red *"lollipop"* indicates the presence of a methyl group on the CG dinucleotide. CG sequences that lie in regulatory sequences of genes that are transcribed in germ cells are unmethylated and therefore tend to be retained in evolution. Methylated CG sequences, on the other hand, tend to be lost through deamination of 5-methyl C to T, unless the CG sequence is critical for survival.

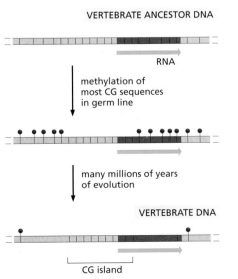

Positive feedback loops and DNA methylation are common to both bacteria and eucaryotes; but eucaryotes also have available to them another means of maintaining a differentiated state through many cell generations. As we saw in Chapter 4, chromatin structure itself can be faithfully propagated from parent to daughter cell. There are several mechanisms to bring this about, but the simplest is based on the covalent modifications of histones. As we have seen, these modifications form a "histone code," with different patterns of modification serving as binding sites for different reader proteins. If these proteins, in turn, serve as (or attract) writer enzymes that replicate the very modification patterns that attracted them in the first place, then the distribution of active and silent regions of chromatin can be faithfully propagated (see Figure 5–39). In a sense, self-sustaining modification of histones is a form of positive feedback loop that is tied to the DNA but does not require the participation of the underlying DNA sequences.

The ability of a daughter cell to retain a memory of the gene expression patterns that were present in the parent cell is an example of **epigenetic inheritance**. This term has subtly different meanings in different branches of biology, but we will use it in its broadest sense to cover any heritable difference in the phenotype of a cell or an organism that does not result from changes in the nucleotide sequence of DNA (see Figure 4–35). We have just discussed three of the most important mechanisms underlying epigenetic changes, but others also exist (**Figure 7–86**). Cells often combine these mechanisms to ensure that patterns of gene expression are maintained and inherited accurately and reliably— over a period of up to a hundred years or more, in our own case.

For more than half a century, biologists have been preoccupied with DNA as the carrier of heritable information—and rightly so. However, it has become clear that human chromosomes also carry a great deal of information that is epigenetic, and not contained in the sequence of the DNA itself. Imprinting is one

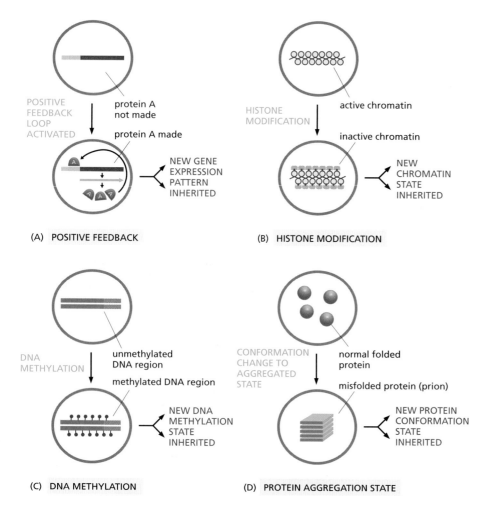

(A) POSITIVE FEEDBACK

(B) HISTONE MODIFICATION

(C) DNA METHYLATION

(D) PROTEIN AGGREGATION STATE

Figure 7–86 **Four distinct mechanisms that can produce an epigenetic form of inheritance in an organism.** (For the inheritance of histone modifications, see Figure 4–52; for the inheritance of protein conformations, see Figure 6–95.)

example. Another is seen in the phenomenon of *mono-allelic expression*, in which only one of the two copies of certain human genes is expressed. For many such genes, the decision of which allele to express and which to silence is random, but once made, it is passed on to progeny cells. Below, we will see an extreme example of this phenomenon in X-chromosome inactivation.

The net effect of random and environmentally triggered epigenetic changes in humans can be seen by comparing identical twins: their genomes have the same sequence of nucleotides, but when their histone modification and DNA methylation patterns are compared, many differences are observed. Because these differences are roughly correlated not only with age but also with the time that the twins have spent apart from each other, it is believed that some of these changes are the result of environmental factors (**Figure 7–87**). Although studies of the *epigenome* are in early stages, the idea that environmental events can be permanently registered by our cells is a fascinating one that presents an important challenge to the next generation of biological scientists.

Figure 7–87 Identical twins raised apart from one another. (Courtesy of Nancy L. Segal.)

Chromosome-Wide Alterations in Chromatin Structure Can Be Inherited

We have seen that chromatin states and DNA methylation can be heritable, serving to establish and preserve patterns of gene expression for many cell generations. Perhaps the most striking example of this effect occurs in mammals, in which an alteration in the chromatin structure of an entire chromosome can modulate the levels of expression of all genes on that chromosome.

Males and females differ in their *sex chromosomes*. Females have two X chromosomes, whereas males have one X and one Y chromosome. As a result, female cells contain twice as many copies of X-chromosome genes as do male cells. In mammals, the X and Y sex chromosomes differ radically in gene content: the X chromosome is large and contains more than a thousand genes, whereas the Y chromosome is small and contains less than 100 genes. Mammals have evolved a *dosage compensation* mechanism to equalize the dosage of X-chromosome gene products between males and females. Mutations that interfere with dosage compensation are lethal: the correct ratio of X chromosome to *autosome* (non-sex chromosome) gene products is critical for survival.

Mammals achieve dosage compensation by the transcriptional inactivation of one of the two X chromosomes in female somatic cells, a process known as **X-inactivation**. Early in the development of a female embryo, when it consists of a few thousand cells, one of the two X chromosomes in each cell becomes highly condensed into a type of heterochromatin. The condensed X chromosome can be easily seen under the light microscope in interphase cells; it was originally called a *Barr body* and is located near the nuclear membrane (**Figure 7–88**). As a result of X-inactivation, two X chromosomes can coexist within the same nucleus, exposed to the same diffusible gene regulatory proteins, yet differ entirely in their expression.

The initial choice of which X chromosome to inactivate, the maternally inherited one (X_m) or the paternally inherited one (X_p), is random. Once either X_p or X_m has been inactivated, it remains silent throughout all subsequent cell divisions of that cell and its progeny, indicating that the inactive state is faithfully maintained through many cycles of DNA replication and mitosis. Because X-inactivation is random and takes place after several thousand cells have already formed in the embryo, every female is a mosaic of clonal groups of cells

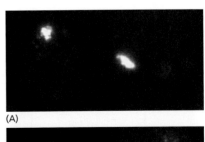

(A)

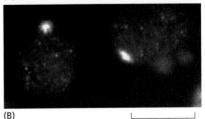

(B)

10 μm

Figure 7–88 X-chromosome inactivation in female cells. (A) Only the inactive X chromosome is coated with XIST RNA, visualized here by *in situ* hybridization to fluorescently labeled RNAs of complementary nucleotide sequence. The panel shows the nuclei of two adjacent cells. (B) The same sample, stained with antibodies against a component of the Polycomb group complex, which coats the X chromosome and helps to silence expression of its genes. (From B. Panning, *Methods Enzymol.* 376:419–428, 2004. With permission from Academic Press.)

in which either X_p or X_m is silenced (**Figure 7–89**). These clonal groups are distributed in small clusters in the adult animal because sister cells tend to remain close together during later stages of development. For example, X-chromosome inactivation causes the red and black "tortoise-shell" coat coloration of some female cats. In these cats, one X chromosome carries a gene that produces red hair color, and the other X chromosome carries an allele of the same gene that results in black hair color; it is the random X-inactivation that produces patches of cells of two distinctive colors. In contrast to the females, male cats of this genetic stock are either solid red or solid black, depending on which X chromosome they inherit from their mothers.

Although X-chromosome inactivation is maintained over thousands of cell divisions, it is not always permanent. In particular, it is reversed during germ-cell formation, so that all haploid oocytes contain an active X chromosome and can express X-linked gene products.

How is an entire chromosome transcriptionally inactivated? X-chromosome inactivation is initiated and spreads from a single site in the middle of the X chromosome, the **X-inactivation center** (**XIC**). Encoded within the XIC is an unusual RNA molecule, *XIST RNA*, which is expressed solely from the inactive X chromosome and whose expression is necessary for X-inactivation. The XIST RNA is not translated into protein and remains in the nucleus, where it eventually coats the entire inactive X chromosome. The spread of XIST RNA from the XIC over the entire chromosome correlates with the spread of gene silencing, indicating that XIST RNA drives the formation and spread of heterochromatin (**Figure 7–90**). Curiously, about 10% of the genes on the X-chromosome escape this silencing and remain active.

In addition to containing XIST RNA, the X-chromosome heterochromatin is characterized by a specific variant of histone 2A, by hypoacetylation of histones

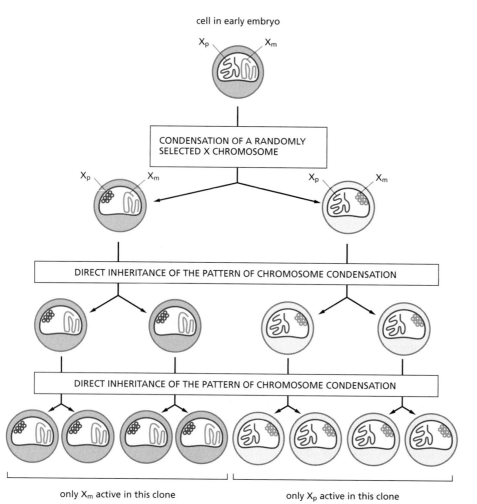

only X_m active in this clone   only X_p active in this clone

Figure 7–89 **X-inactivation.** The clonal inheritance of a condensed inactive X chromosome that occurs in female mammals.

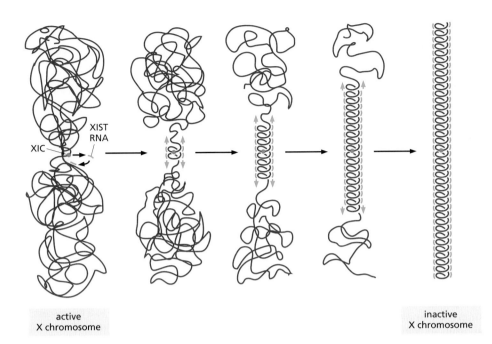

active
X chromosome

inactive
X chromosome

Figure 7–90 Mammalian X-chromosome inactivation. X-chromosome inactivation begins with the synthesis of XIST (X-inactivation specific transcript) RNA from the XIC (X-inactivation center) locus. The association of XIST RNA with one of a female's two X chromosomes is correlated with the condensation of that chromosome. Early in embryogenesis, both XIST association and chromosome condensation gradually move from the XIC locus outward to the chromosome ends. The details of how this occurs remain to be deciphered.

H3 and H4, by ubiquitylation of histone 2A, by methylation of a specific position on histone H3, and by specific patterns of methylation on the underlying DNA (for a suggestion of how these features may be causally linked, see Figure 7–81). The combination of such modifications presumably makes most of the inactive X chromosome unusually resistant to transcription. Because these modifications are, at least in principle, self-propagating, it is easy to see how, once formed, an inactive X chromosome can be stably maintained through many cell divisions.

Many features of mammalian X-chromosome inactivation remain to be discovered. How is the initial decision made as to which X chromosome to inactivate? What mechanism prevents the other X chromosome from also being inactivated? How does XIST RNA coordinate the formation of heterochromatin? How do some genes on the X chromosome escape inactivation? We are just beginning to understand this mechanism of gene regulation that is crucial for the survival of our own species.

X-chromosome inactivation in females is only one of the ways in which sexually reproducing organisms achieve dosage compensation. In *Drosophila*, most of the genes on the single X chromosome present in male cells are transcribed at approximately twofold higher levels than their counterparts in female cells. This male-specific "up-regulation" of transcription results from an alteration in chromatin structure over the entire male X chromosome. A dosage-compensation complex, containing several histone-modifying enzymes as well as two noncoding RNAs transcribed from the X chromosome, assembles at hundreds of positions along the X chromosome and produces patterns of histone modification that are thought to upregulate transcription—through effects on either initiation or elongation—at most genes on the male X chromosome.

The nematode worm uses a third strategy for dosage compensation. Here, the two sexes are male (with one X chromosome) and hermaphrodite (with two X chromosomes), and dosage compensation occurs by an approximately twofold "down-regulation" of transcription from each of the two X chromosomes in cells of the hermaphrodite. This is brought about through chromosome-wide structural changes in the X chromosomes of hermaphrodites (**Figure 7–91**). A dosage-compensation complex, which is completely different from that of *Drosophila* and resembles instead the *condensin* complex that compacts chromosomes during mitosis and meiosis (see Figure 17–27), assembles along each X chromosome of hermaphrodites and, by an unknown mechanism, superimposes a global twofold repression on the normal expression level of each gene.

Although the strategy and components used to cause dosage compensation differ between mammals, flies, and worms, they all involve structural alterations over the entire X chromosome. It seems likely that features of chromosome structure that are quite general were independently adapted and harnessed during evolution to overcome a highly specific problem in gene regulation encountered by sexually reproducing animals.

The Control of Gene Expression is Intrinsically Noisy

So far in this chapter we have discussed gene expression as though it were a strictly deterministic process, so that, if only one knew the concentrations of all the relevant regulatory proteins and other control molecules, the level of gene expression would be precisely predictable. In reality, there is a large amount of random variation in the behavior of cells. In part, this is because there are random fluctuations in the environment, and these disturb the concentrations of regulatory molecules inside the cell in unpredictable ways. Another possible cause, in some cases, may be chaotic behavior of the intracellular control system: mathematical analysis shows that even quite simple control systems may be acutely sensitive to the control parameters, in such a way that, for example, a tiny difference of initial conditions may lead to a radically different long-term outcome. But in addition to these causes of unpredictability, there is a further, more fundamental reason why all cell behavior is inescapably random to some degree.

Cells are chemical systems consisting of relatively small numbers of molecules, and chemical reactions at the level of individual molecules occur in an essentially random, or *stochastic*, manner. A given molecule has a certain probability per unit time of undergoing a chemical reaction, but whether it will actually do so at any given moment is unpredictable, depending on random thermal collisions and the probabilistic rules of quantum mechanics. The smaller the number of molecules governing a process inside the cell, the more severely it will be affected by the randomness of chemical events at the single-molecule level. Thus there is some degree of randomness in every aspect of cell behavior, but certain processes are liable to be random in the extreme.

The control of transcription, in particular, depends on the precise chemical condition of the gene. Consider a simple idealized case, in which a gene is transcribed so long as it has a transcriptional activator protein bound to its regulatory region, and transcriptionally silent when this protein is not bound. The association/dissociation reaction between the regulatory DNA and the protein is stochastic: if the bound state has a half-life $t_{1/2}$ of an hour, the gene may remain activated sometimes for 30 minutes or less, sometimes for a couple of hours or more at a stretch, before the activator protein dissociates. In this way, transcription will flicker on and off in an essentially random way. The average rate of flickering, and the ratio of the average time spent in the "on" state to the average time spent in the "off" state, will be determined by the k_{off} and k_{on} values for the binding reaction and by the concentration of the activator protein in the cell. The quantity of gene transcripts accumulated in the cell will fluctuate accordingly; if the lifetime of the transcripts is long compared with $t_{1/2}$, the fluctuations will be smoothed out; if it is short, they will be severe.

One way to demonstrate such random fluctuations in the expression of individual gene copies is to genetically engineer cells in which one copy of a gene control region is linked to a sequence coding for a green fluorescent reporter protein, while another copy is linked similarly to a sequence coding for a red fluorescent reporter. Although both these gene constructs are in the same cell and experiencing the same environment, they fluctuate independently in their level of expression. As a result, in a population of cells that all carry the same pair of constructs, some cells appear green, others red, and still others a mixture of the two colors, and thus in varying shades of yellow (see Figure 8–75). More generally, cell fate decisions are often made in a stochastic manner, presumably as a result of such random fluctuations; we shall encounter an example in Chapter 23, where we discuss the genesis of the different types of white blood cells.

In some types of cells, and for some aspects of cell behavior, randomness in the control of gene transcription, such as we have just described, seems to be the major source of random variability; in other cell types, other sources of random variation predominate. Where noise in a control system would be

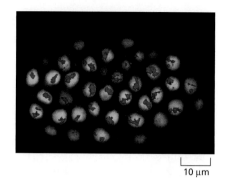

10 µm

Figure 7–91 Localization of dosage compensation proteins to the X chromosomes of *C. elegans* hermaphrodite (XX) nuclei. This image shows many nuclei from a developing embryo. Total DNA is stained *blue* with the DNA-intercalating dye DAPI, and the Sdc2 protein is stained *red* using anti-Sdc2 antibodies coupled to a fluorescent dye. This experiment shows that the Sdc2 protein associates with only a limited set of chromosomes, identified by other experiments to be the two X chromosomes. Sdc2 is bound along the entire length of the X chromosome and recruits the dosage-compensation complex. (From H.E. Dawes et al., *Science* 284:1800–1804, 1999. With permission from AAAS.)

harmful, special control mechanisms have evolved to minimize its effects; the feed-forward loop discussed earlier is an example of such a device, serving to filter out the effects of rapid fluctuations in a control signal. But in all cells, some degree of randomness is inevitable. It is a fundamental feature of cell behavior.

Summary

The many types of cells in animals and plants are created largely through mechanisms that cause different sets of genes to be transcribed in different cells. Since specialized animal cells can maintain their unique character through many cell division cycles and even when grown in culture, the gene regulatory mechanisms involved in creating them must be stable once established and heritable when the cell divides. These features reflect the cell's memory of its developmental history. Bacteria and yeasts also exhibit cell memory and provide unusually accessible model systems in which to study gene regulatory mechanisms.

Direct or indirect positive feedback loops, which enable gene regulatory proteins to perpetuate their own synthesis, provide the simplest mechanism for cell memory. Transcription circuits also provide the cell with the means to carry out logic operations and measure time. Simple transcription circuits combined into large regulatory networks drive highly sophisticated programs of embryonic development.

In eucaryotes, the transcription of any particular gene is generally controlled by a combination of gene regulatory proteins. It is thought that each type of cell in a higher eucaryotic organism contains a specific set of gene regulatory proteins that ensures the expression of only those genes appropriate to that type of cell. A given gene regulatory protein may be active in a variety of circumstances and is typically involved in the regulation of many different genes.

Unlike bacteria, eucaryotic cells use inherited states of chromatin condensation as an additional mechanism to regulate gene expression and to create cell memory. An especially dramatic case is the inactivation of an entire X chromosome in female mammals. DNA methylation can also silence genes in a heritable manner in eucaryotes. In addition, it underlies the phenomenon of genomic imprinting in mammals, in which the expression of a gene depends on whether it was inherited from the mother or the father.

POST-TRANSCRIPTIONAL CONTROLS

In principle, every step required for the process of gene expression can be controlled. Indeed, one can find examples of each type of regulation, and many genes are regulated by multiple mechanisms. As we have seen, controls on the initiation of gene transcription are a critical form of regulation for all genes. But other controls can act later in the pathway from DNA to protein to modulate the amount of gene product that is made—and in some cases, to determine the exact amino acid sequence of the protein product. These **post-transcriptional controls**, which operate after RNA polymerase has bound to the gene's promoter and begun RNA synthesis, are crucial for the regulation of many genes.

In the following sections, we consider the varieties of post-transcriptional regulation in temporal order, according to the sequence of events that an RNA molecule might experience after its transcription has begun (**Figure 7–92**).

Transcription Attenuation Causes the Premature Termination of Some RNA Molecules

It has long been known that the expression of certain genes in bacteria is inhibited by premature termination of transcription, a phenomenon called **transcription attenuation**. In some of these cases the nascent RNA chain adopts a structure that causes it to interact with the RNA polymerase in such a way as to abort its transcription. When the gene product is required, regulatory proteins bind to the nascent RNA chain and interfere with attenuation, allowing the transcription of a complete RNA molecule.

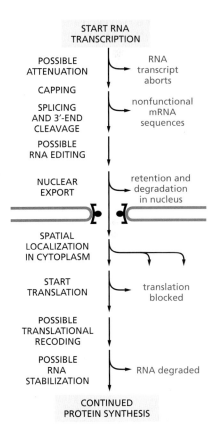

Figure 7–92 Post-transcriptional controls on gene expression. The final synthesis rate of a protein can, in principle, be controlled at any of the steps shown. RNA splicing, RNA editing, and translation recoding (described in Chapter 6) can also alter the sequence of amino acids in a protein, making it possible for the cell to produce more than one protein variant from the same gene. Only a few of the steps depicted here are likely to be critical for the regulation of any one particular protein.

Transcription attenuation also operates in eucaryotes. A well-studied example occurs during the life cycle of HIV, the human immunodeficiency virus, causative agent of acquired immune deficiency syndrome, or AIDS. Once it has been integrated into the host genome, the viral DNA is transcribed by the cell's RNA polymerase II (see Figure 5–71). However, the host polymerase usually terminates transcription after synthesizing transcripts of several hundred nucleotides and therefore does not efficiently transcribe the entire viral genome. When conditions for viral growth are optimal, a virus-encoded protein called Tat, which binds to a specific stem-loop structure in the nascent RNA that contains a "bulged base," prevents this premature termination. Once bound to this specific RNA structure (called Tar), Tat assembles several cell proteins that allow the RNA polymerase to continue transcribing. The normal role of at least some of these proteins is to prevent pausing and premature termination by RNA polymerase when it transcribes normal cell genes. Eucaryotic genes often contain long introns; to transcribe a gene efficiently, RNA polymerase II cannot afford to linger at nucleotide sequences that happen to promote pausing. Thus, a normal cell mechanism has apparently been adapted by HIV to permit efficient transcription of its genome to be controlled by a single viral protein.

Riboswitches Might Represent Ancient Forms of Gene Control

In Chapter 6, we discussed the idea that, before modern cells arose on Earth, RNA both stored hereditary information and catalyzed chemical reactions. The recent discovery of *riboswitches* shows that RNA can also form control devices that regulate gene expression. Riboswitches are short sequences of RNA that change their conformation on binding small molecules, such as metabolites. Each riboswitch recognizes a specific small molecule and the resulting conformational change is used to regulate gene expression. Riboswitches are often located near the 5′ end of mRNAs, and they fold while the mRNA is being synthesized blocking or permitting progress of the RNA polymerase according to whether the regulatory small molecule is bound (**Figure 7–93**).

Riboswitches are particularly common in bacteria, in which they sense key small metabolites in the cell and adjust gene expression accordingly. Perhaps their most remarkable feature is the high specificity and affinity with which each recognizes only the appropriate small molecule; in many cases, every chemical feature of the small molecule is read by the RNA (Figure 7–93C). Moreover, the

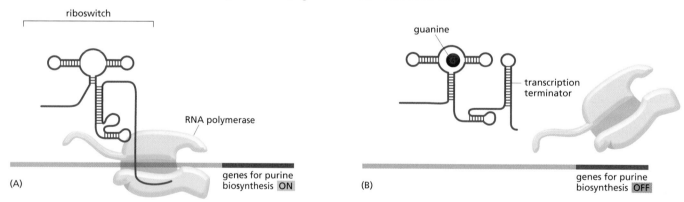

Figure 7–93 A riboswitch that responds to guanine. (A) In this example from bacteria, the riboswitch controls expression of the purine biosynthetic genes. When guanine levels in cells are low, an elongating RNA polymerase transcribes the purine biosynthetic genes, and the enzymes needed for guanine synthesis are therefore expressed. (B) When guanine is abundant, it binds the riboswitch, causing it to undergo a conformational change that forces the RNA polymerase to terminate transcription (see Figure 6–11). (C) Guanine *(red)* bound to the riboswitch. Only those nucleotides that form the guanine-binding pocket are shown. Many other riboswitches exist, including those that recognize *S*-adenosyl methione, coenzyme B₁₂, flavin mononucleotide, adenine, lysine, and glycine. (Adapted from M. Mandal and R.R. Breaker, *Nat. Rev. Mol. Cell Biol.* 5:451–63, 2004, with permission from Macmillan Publishers Ltd., and C.K. Vanderpool and S. Gottesman, *Mol. Microbiol.* 54:1076–1089, 2004, with permission from Blackwell Publishing.)

binding affinities observed are as tight as those typically observed between small molecules and proteins.

Riboswitches are perhaps the most economical examples of gene control devices, inasmuch as they bypass the need for regulatory proteins altogether. In the example shown in Figure 7–93, the riboswitch controls transcription elongation, but they also regulate other steps in gene expression, as we shall see later in this chapter. Clearly, highly sophisticated gene control devices can be made from short sequences of RNA.

Alternative RNA Splicing Can Produce Different Forms of a Protein from the Same Gene

As discussed in Chapter 6, RNA splicing shortens the transcripts of many eucaryotic genes by removing the intron sequences from the mRNA precursor. We also saw that a cell can splice an RNA transcript differently and thereby make different polypeptide chains from the same gene—a process called **alternative RNA splicing** (see Figure 6–27 and **Figure 7–94**). A substantial proportion of animal genes (estimated at 40% in flies and 75% in humans) produce multiple proteins in this way.

When different splicing possibilities exist at several positions in the transcript, a single gene can produce dozens of different proteins. In one extreme case, a *Drosophila* gene may produce as many as 38,000 different proteins from a single gene through alternative splicing (**Figure 7–95**), although only a fraction of these forms have thus far been experimentally observed. Considering that the *Drosophila* genome has approximately 14,000 identified genes, it is clear that the protein complexity of an organism can greatly exceed the number of its genes. This example also illustrates the perils in equating gene number with an organism's complexity. For example, alternative splicing is relatively rare in single-celled budding yeasts but very common in flies. Budding yeast has ~6200 genes, only about 300 of which are subject to splicing, and nearly all of these have only a single intron. To say that flies have only 2–3 times as many genes as yeasts greatly underestimates the difference in complexity of these two genomes.

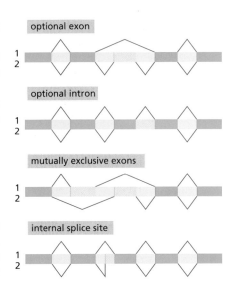

Figure 7–94 **Four patterns of alternative RNA splicing.** In each case a single type of RNA transcript is spliced in two alternative ways to produce two distinct mRNAs (1 and 2). The *dark blue boxes* mark exon sequences that are retained in both mRNAs. The *light blue boxes* mark possible exon sequences that are included in only one of the mRNAs. The boxes are joined by *red lines* to indicate where intron sequences (*yellow*) are removed. (Adapted with permission from A. Andreadis, M.E. Gallego and B. Nadal-Ginard, *Annu. Rev. Cell Biol.* 3:207–242, 1987. With permission from Annual Reviews.)

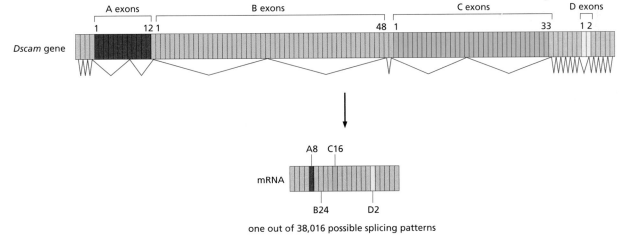

one out of 38,016 possible splicing patterns

Figure 7–95 **Alternative splicing of RNA transcripts of the *Drosophila Dscam* gene.** DSCAM proteins are axon guidance receptors that help to direct growth cones to their appropriate targets in the developing nervous system. The final mRNA contains 24 exons, four of which (denoted A, B, C, and D) are present in the *Dscam* gene as arrays of alternative exons. Each RNA contains 1 of 12 alternatives for exon A (*red*), 1 of 48 alternatives for exon B (*green*), 1 of 33 alternatives for exon C (*blue*), and 1 of 2 alternatives for exon D (*yellow*). If all possible splicing combinations are used, 38,016 different proteins could in principle be produced from the *Dscam* gene. This figure shows only one of the many possible splicing patterns (indicated by the red line and by the mature mRNA below it). Each variant Dscam protein would fold into roughly the same structure [predominantly a series of extracellular immunoglobulin-like domains linked to a membrane-spanning region (see Figure 25–74)], but the amino acid sequence of the domains would vary according to the splicing pattern. It is suspected that this receptor diversity contributes to the formation of complex neural circuits, but the precise properties and functions of the many Dscam variants are not yet understood. (Adapted from D.L. Black, *Cell* 103:367–370, 2000. With permission from Elsevier.)

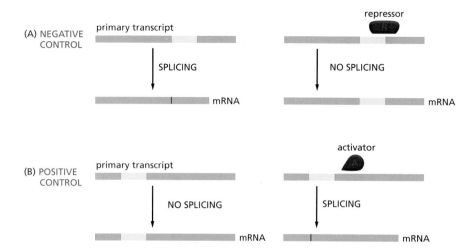

Figure 7–96 Negative and positive control of alternative RNA splicing. (A) In negative control, a repressor protein binds to the pre-mRNA transcript and blocks access of the splicing machinery to a splice junction. This often results in the use of a cryptic splice site, thereby producing an altered pattern of splicing (not shown). (B) In positive control, the splicing machinery is unable to remove a particular intron sequence efficiently without assistance from an activator protein. Because the nucleotide sequences that bind these activators can be located many nucleotide pairs from the splice junctions they control, they are often called *splicing enhancers*.

In some cases alternative RNA splicing occurs because there is an *intron sequence ambiguity:* the standard spliceosome mechanism for removing intron sequences (discussed in Chapter 6) is unable to distinguish cleanly between two or more alternative pairings of 5′ and 3′ splice sites, so that different choices are made by chance on different transcripts. Where such constitutive alternative splicing occurs, several versions of the protein encoded by the gene are made in all cells in which the gene is expressed.

In many cases, however, alternative RNA splicing is regulated rather than constitutive. In the simplest examples, regulated splicing is used to switch from the production of a nonfunctional protein to the production of a functional one. The transposase that catalyzes the transposition of the *Drosophila* P element, for example, is produced in a functional form in germ cells and a nonfunctional form in somatic cells of the fly, allowing the P element to spread throughout the genome of the fly without causing damage in somatic cells (see Figure 5–69). The difference in transposon activity has been traced to the presence of an intron sequence in the transposase RNA that is removed only in germ cells.

In addition to switching from the production of a functional protein to the production of a nonfunctional one, the regulation of RNA splicing can generate different versions of a protein in different cell types, according to the needs of the cell. Tropomyosin, for example, is produced in specialized forms in different types of cells (see Figure 6–27). Cell-type-specific forms of many other proteins are produced in the same way.

RNA splicing can be regulated either negatively, by a regulatory molecule that prevents the splicing machinery from gaining access to a particular splice site on the RNA, or positively, by a regulatory molecule that helps direct the splicing machinery to an otherwise overlooked splice site (**Figure 7–96**).

Because of the plasticity of RNA splicing, the blocking of a "strong" splicing site will often expose a "weak" site and result in a different pattern of splicing. Likewise, activating a suboptimal splice site can result in alternative splicing by suppressing a competing splice site. Thus the splicing of a pre-mRNA molecule can be thought of as a delicate balance between competing splice sites—a balance that can easily be tipped by regulatory proteins.

The Definition of a Gene Has Had to Be Modified Since the Discovery of Alternative RNA Splicing

The discovery that eucaryotic genes usually contain introns and that their coding sequences can be assembled in more than one way raised new questions about the definition of a gene. A gene was first clearly defined in molecular terms in the early 1940s from work on the biochemical genetics of the fungus *Neurospora*. Until then, a gene had been defined operationally as a region of the genome that segregates as a single unit during meiosis and gives rise to a definable phenotypic trait, such as a red or a white eye in *Drosophila* or a round or wrinkled seed in peas. The work on *Neurospora* showed that most genes correspond to a region of the genome that directs the synthesis of a single enzyme. This led to the hypothesis that one gene encodes one polypeptide chain. The hypothesis proved fruitful for subsequent research; as more was learned about the mechanism of gene expression in the 1960s, a gene became identified as that stretch of DNA that was transcribed into the RNA coding for a single polypeptide chain (or a single structural RNA such as a tRNA or an rRNA molecule). The discovery of split genes and introns in the late 1970s could be readily accommodated by the original definition of a gene, provided that a single polypeptide chain was specified by the RNA transcribed from any one DNA sequence. But it is now clear that many DNA sequences in higher eucaryotic cells can produce a set of distinct (but related) proteins by means of alternative RNA splicing. How, then, is a gene to be defined?

In those relatively rare cases in which a single transcription unit produces two very different eucaryotic proteins, the two proteins are considered to be produced by distinct genes that overlap on the chromosome. It seems unnecessarily complex, however, to consider most of the protein variants produced by alternative RNA splicing as being derived from overlapping genes. A more sensible alternative is to modify the original definition to count as a gene any DNA sequence that is transcribed as a single unit and encodes one set of closely related polypeptide chains (protein isoforms). This definition of a gene also accommodates those DNA sequences that encode protein variants produced by post-transcriptional processes other than RNA splicing, such as translational frameshifting (see Figure 6–78), regulated poly-A addition, and RNA editing (to be discussed below).

Sex Determination in *Drosophila* Depends on a Regulated Series of RNA Splicing Events

We now turn to one of the best-understood examples of regulated RNA splicing. In *Drosophila* the primary signal for determining whether the fly develops as a male or female is the ratio of the number of X chromosomes (X) to the number of autosomal sets (A). Individuals with an X/A ratio of 1 (normally two X chromosomes and two sets of autosomes) develop as females, whereas those with an X/A a ratio of 0.5 (normally one X chromosome and two sets of autosomes) develop as males. This ratio is assessed early in development and is remembered thereafter by each cell. Three crucial gene products transmit information about this ratio to the many other genes that specify male and female characteristics (**Figure 7–97**). As explained in **Figure 7–98**, sex determination in *Drosophila* depends on a cascade of regulated RNA splicing events that involves these three gene products.

Although *Drosophila* sex determination provides one of the best-understood examples of a regulatory cascade based on RNA splicing, it is not clear why the fly should use this strategy. Other organisms (the nematode, for example) use an entirely different scheme for sex determination—one based on transcriptional and translational controls. Moreover, the *Drosophila* male-determination pathway requires that a number of nonfunctional RNA molecules be continually produced, which seems unnecessarily wasteful. One speculation is that this RNA-splicing cascade, like the ribsoswitches discussed above, represents an ancient control strategy, left over from the early stage of evolution in which RNA

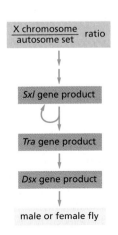

Figure 7–97 Sex determination in *Drosophila*. The gene products shown act in a sequential cascade to determine the sex of the fly according to the X-chromosome/autosome set ratio (X/A). The genes are called *Sex-lethal (Sxl)*, *Transformer (Tra)*, and *Doublesex (Dsx)* because of the phenotypes that result when the gene is inactivated by mutation. The function of these gene products is to transmit the information about the X/A ratio to the many other genes that create the sex-related phenotypes. These other genes function as two alternative sets: those that specify female features and those that specify male features (see Figure 7–98).

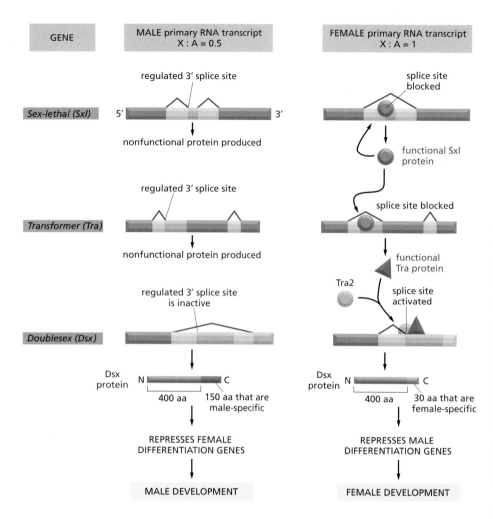

Figure 7–98 The cascade of changes in gene expression that determines the sex of a fly through alternative RNA splicing. An X-chromosome/autosome set ratio of 0.5 results in male development. Male is the "default" pathway in which the *Sxl* and *Tra* genes are both transcribed, but the RNAs are spliced constitutively to produce only nonfunctional RNA molecules, and the *Dsx* transcript is spliced to produce a protein that turns off the genes that specify female characteristics. An X/A ratio of 1 triggers the female differentiation pathway in the embryo by transiently activating a promoter within the *Sxl* gene that causes the synthesis of a special class of *Sxl* transcripts that are constitutively spliced to give functional Sxl protein. Sxl is a splicing regulatory protein with two sites of action: (1) it binds to the constitutively produced *Sxl* RNA transcript, causing a female-specific splice that continues the production of a functional Sxl protein, and (2) it binds to the constitutively produced *Tra* RNA and causes an alternative splice of this transcript, which now produces an active Tra regulatory protein. The Tra protein acts with the constitutively produced Tra2 protein to produce the female-specific spliced form of the *Dsx* transcript; this encodes the female form of the Dsx protein, which turns off the genes that specify male features.

The components in this pathway were all initially identified through the study of *Drosophila* mutants that are altered in their sexual development. The *Dsx* gene, for example, derives its name *(Doublesex)* from the observation that a fly lacking this gene product expresses both male-specific and female-specific features. Note that this pathway includes both negative and positive control of splicing (see Figure 7–96). Sxl binds to the pyrimidine-rich stretch of nucleotides that is part of the standard splicing consensus sequence and blocks access by the normal splicing factor, U2AF (see Figure 6–29). Tra binds to specific RNA sequences in an exon and with Tra2 activates a normally suboptimal splicing signal by the binding of U2AF.

was the predominant biological molecule, and controls of gene expression would have had to be based almost entirely on RNA–RNA interactions.

A Change in the Site of RNA Transcript Cleavage and Poly-A Addition Can Change the C-terminus of a Protein

We saw in Chapter 6 that the 3′ end of a eucaryotic mRNA molecule is not formed by the termination of RNA synthesis by the RNA polymerase. Instead, it results from an RNA cleavage reaction that is catalyzed by additional factors while the transcript is elongating (see Figure 6–37). A cell can control the site of this cleavage so as to change the C-terminus of the resultant protein.

A well-studied example is the switch from the synthesis of membrane-bound to secreted antibody molecules that occurs during the development of B lymphocytes (see Figure 25–17). Early in the life history of a B lymphocyte, the antibody it produces is anchored in the plasma membrane, where it serves as a receptor for antigen. Antigen stimulation causes B lymphocytes to multiply and to begin secreting their antibody. The secreted form of the antibody is identical to the membrane-bound form except at the extreme C-terminus. In this part of the protein, the membrane-bound form has a long string of hydrophobic amino acids that traverses the lipid bilayer of the membrane, whereas the secreted form has a much shorter string of hydrophilic amino acids. The switch from membrane-bound to secreted antibody therefore requires a different nucleotide sequence at the 3′ end of the mRNA; this difference is generated through a change in the length of the primary RNA transcript, caused by a change in the site of RNA cleavage, as shown in **Figure 7–99**. This change is caused by an increase in the concentration of a subunit of CstF, the protein that binds to the G/U-rich sequences of RNA cleavage and poly-A addition sites and promotes

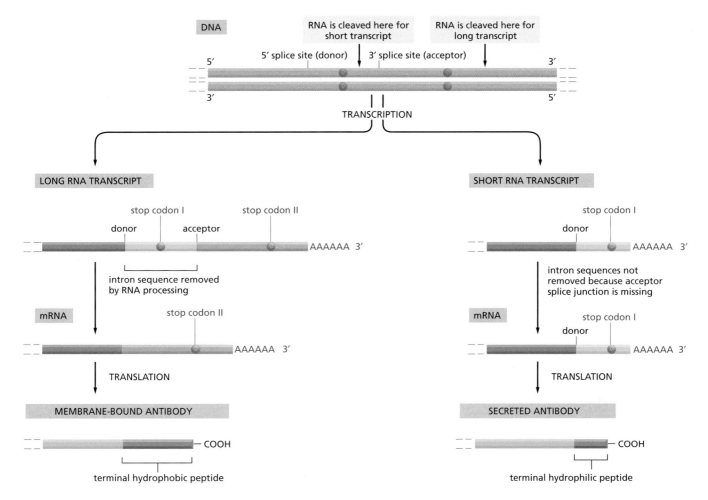

Figure 7–99 Regulation of the site of RNA cleavage and poly-A addition determines whether an antibody molecule is secreted or remains membrane-bound. In unstimulated B lymphocytes *(left)*, a long RNA transcript is produced, and the intron sequence near its 3′ end is removed by RNA splicing to give rise to an mRNA molecule that codes for a membrane-bound antibody molecule. In contrast, after antigen stimulation *(right)*, the primary RNA transcript is cleaved upstream from the splice site in front of the last exon sequence. As a result, some of the intron sequence that is removed from the long transcript remains as a coding sequence in the short transcript. These are the nucleotide sequences that encode the hydrophilic C-terminal portion of the secreted antibody molecule.

RNA cleavage (see Figures 6–37 and 6–38). The first cleavage-poly-A addition site that an RNA polymerase transcribing the antibody gene encounters is suboptimal and is usually skipped in unstimulated B lymphocytes, leading to production of the longer RNA transcript. When activated to produce antibodies, the B lymphocyte increases its CstF concentration; as a result, cleavage now occurs at the suboptimal site, and the shorter transcript is produced. In this way, a change in concentration of a general RNA processing factor can have a dramatic effect on the expression of a particular gene.

RNA Editing Can Change the Meaning of the RNA Message

The molecular mechanisms used by cells are a continual source of surprises. An example is the process of **RNA editing**, which alters the nucleotide sequences of RNA transcripts once they are synthesized and thereby changes the coded message they carry. The most dramatic form of RNA editing was discovered in RNA transcripts that code for proteins in the mitochondria of trypanosomes. Here, one or more U nucleotides are inserted (or, less frequently, removed) from selected regions of a transcript, altering both the original reading frame and the sequence and thereby changing the meaning of the message. For some genes the editing is so extensive that over half of the nucleotides in the mature mRNA are U nucleotides that were inserted during the editing process. A set of 40- to 80-nucleotide-long RNA molecules that are transcribed separately contains the information that specifies exactly how the initial RNA transcript is to be altered. These so-called *guide RNAs* have a 5′ end that is complementary in sequence to one end of the region of the transcript to be edited, followed by a sequence that specifies the set of nucleotides to be inserted into the transcript (**Figure 7–100**). The editing mechanism is remarkably complex: at each edited position, the RNA is broken, U nucleotides are added to the broken 3′ end, and the RNA is ligated.

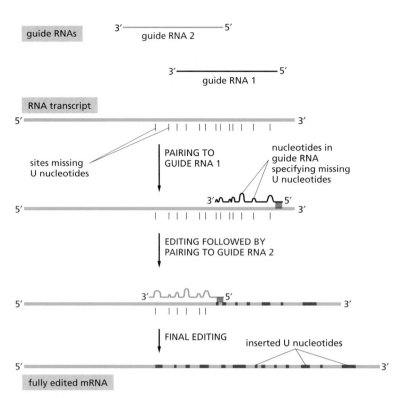

Figure 7–100 RNA editing in the mitochondria of trypanosomes. Editing generally starts near the 3′ end and progresses toward the 5′ end of the RNA transcript, as shown, because the "anchor sequence" at the 5′ end of most guide RNAs can pair only with edited sequences. The U nucleotides are added by a specialized enzyme called a uridylyl transferase.

RNA editing of a more refined type occurs in mammals. Here, two principal types of editing occur, the deamination of adenine to produce inosine (A-to-I editing) and the deamination of cytosine to produce uracil (C-to-U editing; see Figure 5–50). Because these chemical modifications alter the pairing properties of the bases (I pairs with C, and U pairs with A), they can have profound effects on the meaning of the RNA. If the edit occurs in a coding region, it can change the amino acid sequence of the protein or produce a truncated protein. Edits that occur outside coding sequences can affect the pattern of pre-mRNA splicing, the transport of mRNA from the nucleus to the cytosol, or the efficiency with which the RNA is translated.

The process of A-to-I editing is particularly prevalent in humans, where it is estimated to affect over 1000 genes. Enzymes called *ADARs* (adenosine deaminases acting on RNA) perform this type of editing; these enzymes recognize a double-stranded RNA structure that is formed through base pairing between the site to be edited and a complementary sequence located elsewhere on the same RNA molecule, typically in a 3′ intron (**Figure 7–101**). These complementary sequences specify whether the mRNA is to be edited, and if so, exactly where the edit should be made. An especially important example of A-to-I editing takes place in the mRNA that codes for a transmitter-gated ion channel in the brain. A single edit changes a glutamine to an arginine; the affected amino acid lies on the inner wall of the channel, and the editing change alters the Ca^{2+} permeability of the channel. The importance of this edit in mice has been demonstrated by deleting the relevant ADAR gene. The mutant mice are prone to epileptic seizures and die during or shortly after weaning. If the gene for the gated ion channel is mutated to produce the edited form of the protein directly, mice lacking the ADAR develop normally, showing that editing of the ion channel RNA is normally crucial for proper brain development.

C-to-U editing, which is carried out by a different set of enzymes, is also crucial in mammals. For example, in certain cells of the gut, the mRNA for apolipoprotein B undergoes a C-to-U edit that creates a premature stop codon and therefore produces a shorter form of the protein. In cells of the liver, the editing enzyme is not expressed, and the full-length apolipoprotein B is produced. The two protein isoforms have different properties, and each plays a specialized role in lipid metabolism that is specific to the organ that produces it.

Why RNA editing exists at all is a mystery. One idea is that it arose in evolution to correct "mistakes" in the genome. Another is that it arose as a somewhat

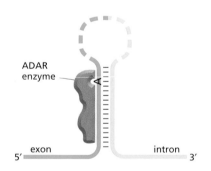

Figure 7–101 Mechanism of A-to-I RNA editing in mammals. RNA sequences carried on the same RNA molecule signal the position of an edit. Typically, a sequence complementary to the position of the edit is present in an intron, and the resulting double-stranded RNA structure attracts the A-to-I editing enzyme ADAR. This type of editing takes place in the nucleus, before the pre-mRNA has been fully processed. Mice and humans have three ADAR enzymes: ADR1 is required in the liver for proper red blood cell development, ADR2 is required for proper brain development (as described in the text), and the role of ADR3 is uncertain.

slapdash way for the cell to produce subtly different proteins from the same gene. A third possibility is that RNA editing originally evolved as a defense mechanism against retroviruses and retrotransposons and was later adapted by the cell to change the meanings of certain mRNAs. Indeed, RNA editing still plays important roles in cell defense. Some retroviruses, including HIV (see Figure 5–71), are extensively edited after they infect cells. This hyperediting creates many harmful mutations in the viral RNA genome and also causes viral mRNAs to be retained in the nucleus, where they are eventually degraded. Although some modern retroviruses protect themselves against this defense mechanism, it presumably helps to hold many viruses in check.

Primates have much higher levels of A-to-I editing than do other mammals, and most of this takes place on RNAs that are transcribed from the highly abundant Alu elements. It has been proposed that A-to-I editing has stopped these mobile elements from completely overtaking our genomes by inactivating the RNA transcripts they require to proliferate (see Figure 5–74). If this idea is correct, RNA editing may have had a profound impact on the shaping of the modern human genome.

RNA Transport from the Nucleus Can Be Regulated

It has been estimated that in mammals only about one-twentieth of the total mass of RNA synthesized ever leaves the nucleus. We saw in Chapter 6 that most mammalian RNA molecules undergo extensive processing and that the "leftover" RNA fragments (excised introns and RNA sequences 3′ to the cleavage/poly-A site) are degraded in the nucleus. Incompletely processed and otherwise damaged RNAs are also eventually degraded as part of the quality control system of RNA production.

As described in Chapter 6, the export of RNA molecules from the nucleus is delayed until processing has been completed. However, mechanisms that deliberately override this control point can be used to regulate gene expression. This strategy forms the basis for one of the best-understood examples of **regulated nuclear transport** of mRNA, which occurs in the human AIDS virus, HIV.

As we saw in Chapter 5, HIV, once inside the cell, directs the formation of a double-stranded DNA copy of its genome, which is then inserted into the genome of the host (see Figure 5–71). Once inserted, the viral DNA is transcribed as one long RNA molecule by the host cell's RNA polymerase II. This transcript is then spliced in many different ways to produce over 30 different species of mRNA, which in turn are translated into a variety of different proteins (**Figure 7–102**). In order to make progeny virus, entire, unspliced viral transcripts must be exported from the nucleus to the cytosol, where they are packaged into viral capsids and serve as the viral genome (see Figure 5–71). This large transcript, as well as alternatively spliced HIV mRNAs that the virus needs to move to the cytoplasm for protein synthesis, still carries complete introns. The host cell's normal block to the nuclear export of unspliced RNAs therefore presents a special problem for HIV.

The block is overcome in an ingenious way. The virus encodes a protein (called Rev) that binds to a specific RNA sequence (called the Rev responsive element, RRE) located within a viral intron. The Rev protein interacts with a nuclear export receptor (exportin 1), which directs the movement of viral RNAs through nuclear pores into the cytosol despite the presence of intron sequences. We discuss in detail the way in which export receptors function in Chapter 12.

Figure 7–102 The compact genome of HIV, the human AIDS virus. The positions of the nine HIV genes are shown in *green*. The *red double line* indicates a DNA copy of the viral genome that has become integrated into the host DNA *(gray)*. Note that the coding regions of many genes overlap, and that those of *Tat* and *Rev* are split by introns. The *blue line* at the bottom of the figure represents the pre-mRNA transcript of the viral DNA and shows the locations of all the possible splice sites *(arrows)*. There are many alternative ways of splicing the viral transcript; for example the *Env* mRNAs retain the intron that has been spliced out of the *Tat* and *Rev* mRNAs. The Rev response element (RRE) is indicated by a blue ball and stick. It is a 234-nucleotide long stretch of RNA that folds into a defined structure; Rev recognizes a particular hairpin within this larger structure.

The *Gag* gene codes for a protein that is cleaved into several smaller proteins that form the viral capsid. The *Pol* gene codes for a protein that is cleaved to produce reverse transcriptase (which transcribes RNA into DNA), as well as the integrase involved in integrating the viral genome (as double-stranded DNA) into the host genome. Pol is produced by ribosomal frameshifting of translation that begins at *Gag* (see Figure 6–78). The *Env* gene codes for the envelope proteins (see Figure 5–71). Tat, Rev, Vif, Vpr, Vpu, and Nef are small proteins with a variety of functions. For example, Rev regulates nuclear export (see Figure 7–103) and Tat regulates the elongation of transcription across the integrated viral genome (see p. 478).

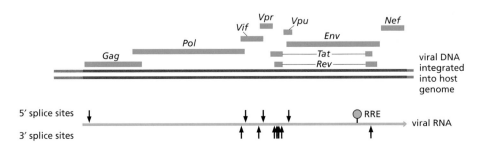

(A) early HIV synthesis

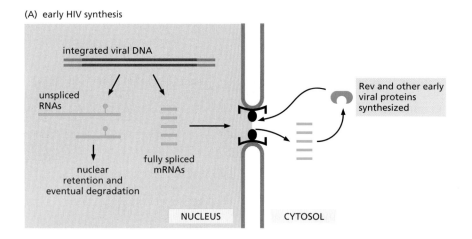

(B) late HIV synthesis

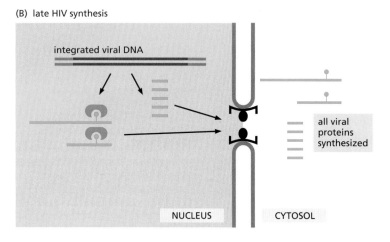

Figure 7–103 Regulation of nuclear export by the HIV Rev protein. Early in HIV infection (A), only the fully spliced RNAs (which contain the coding sequences for Rev, Tat, and Nef) are exported from the nucleus and translated. Once sufficient Rev protein has accumulated and been transported into the nucleus (B), unspliced viral RNAs can be exported from the nucleus. Many of these RNAs are translated into protein, and the full-length transcripts are packaged into new viral particles.

The regulation of nuclear export by Rev has several important consequences for HIV growth and pathogenesis. In addition to ensuring the nuclear export of specific unspliced RNAs, it divides the viral infection into an early phase (in which Rev is translated from a fully spliced RNA and all of intron-containing viral RNAs are retained in the nucleus and degraded) and a late phase (in which unspliced RNAs are exported due to Rev function). This timing helps the virus replicate by providing the gene products in roughly the order in which they are needed (**Figure 7–103**). Regulation by Rev may also help the HIV virus to achieve latency, a condition in which the HIV genome has become integrated into the host cell genome but the production of viral proteins has temporarily ceased. If, after its initial entry into a host cell, conditions became unfavorable for viral transcription and replication, Rev is made at levels too low to promote export of unspliced RNA. This situation stalls the viral growth cycle until conditions improve, Rev levels increase, and the virus enters the replication cycle.

Some mRNAs Are Localized to Specific Regions of the Cytoplasm

Once a newly made eucaryotic mRNA molecule has passed through a nuclear pore and entered the cytosol, it is typically met by ribosomes, which translate it into a polypeptide chain (see Figure 6–40). Once the first round of translation "passes" the nonsense-mediated decay test (see Figure 6–80), the mRNA is usually translated in earnest. If the mRNA encodes a protein that is destined to be secreted or expressed on the cell surface, a signal sequence at the protein's amino terminus will direct it to the endoplasmic reticulum (ER); components of the cell's protein-sorting apparatus recognize the signal sequence as soon as it emerges from the ribosome and direct the entire complex of ribosome, mRNA, and nascent protein to the membrane of the ER, where the remainder of the polypeptide chain is synthesized, as discussed in Chapter 12. In other cases free

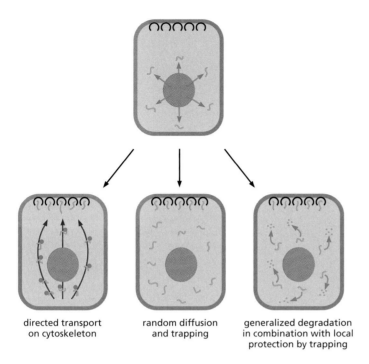

directed transport
on cytoskeleton

random diffusion
and trapping

generalized degradation
in combination with local
protection by trapping

Figure 7–104 Three mechanisms for the localization of mRNAs. The mRNA to be localized leaves the nucleus through nuclear pores *(top)*. Some localized mRNAs *(left diagram)* travel to their destination by associating with cytoskeletal motors. As described in Chapter 16, these motors use the energy of ATP hydrolysis to move unidirectionally along filaments in the cytoskeleton *(red)*. At their destination, anchor proteins *(black)* hold the mRNAs in place. Other mRNAs randomly diffuse through the cytosol and are simply trapped and therefore concentrated at their sites of localization *(center diagram)*. Some of these mRNAs *(right diagram)* are degraded in the cytosol unless they have bound, through random diffusion, a localized protein complex that anchors and protects the mRNA from degradation *(black)*. Each of these mechanisms requires specific signals on the mRNA, which are typically located in the 3′ UTR (see Figure 7–105). In many cases of mRNA localization, additional mechanisms block the translation of the mRNA until it is properly localized. (Adapted from H.D. Lipshitz and C.A. Smibert, *Curr. Opin. Genet. Dev.* 10:476–488, 2000. With permission from Elsevier.)

ribosomes in the cytosol synthesize the entire protein, and signals in the completed polypeptide chain may then direct the protein to other sites in the cell.

Some mRNAs are themselves directed to specific intracellular locations before their efficient translation begins, allowing the cell to position its mRNAs close to the sites where the encoded protein is needed. This strategy provides the cell with many advantages. For example, it allows the establishment of asymmetries in the cytosol of the cell, a key step in many stages of development. Localized mRNA, coupled with translation control, also allows the cell to regulate gene expression independently in its different parts. This feature is particularly important in large, highly polarized cells such as neurons, where growth cones must respond to signals without waiting to involve the distant nucleus. RNA localization has been observed in many organisms, including unicellular fungi, plants, and animals, and it is likely to be a common mechanism that cells use to concentrate high-level production of proteins at specific sites.

Several distinct mechanisms for mRNA localization have been discovered (**Figure 7–104**), but all of them require specific signals in the mRNA itself. These signals are usually concentrated in the 3′ *untranslated region (UTR)*, the region of RNA that extends from the stop codon that terminates protein synthesis to the start of the poly-A tail (**Figure 7–105**). This mRNA localization is usually coupled with translational controls to ensure that the mRNA remains quiescent until it has been moved into place.

The *Drosophila* egg exhibits an especially striking example of mRNA localization. The mRNA encoding the bicoid gene regulatory protein is localized by attachment to the cytoskeleton at the anterior tip of the developing egg. When fertilization triggers the translation of this mRNA, it generates a gradient of the bicoid protein that plays a crucial part in directing the development of the anterior part of the embryo (shown in Figure 7–53 and discussed in more detail in Chapter 22). Many mRNAs in somatic cells are also localized in a similar way. The mRNA that encodes actin, for example, is localized to the actin-filament-rich cell cortex in mammalian fibroblasts by means of a 3′ UTR signal.

We saw in Chapter 6 that mRNA molecules exit from the nucleus bearing numerous markings in the form of RNA modifications (the 5′ cap and the 3′ poly-A tail) and bound proteins (exon-junction complexes, for example) that signify the successful completion of the different pre-mRNA processing steps. As just described, the 3′ UTR of an mRNA can be thought of as a "zip code" that directs mRNAs to different places in the cell. Below we will also see that mRNAs carry information specifying their average lifetime in the cytosol and the efficiency with which they are translated into protein. In a broad sense, the untranslated regions of eucaryotic mRNAs resemble the transcriptional control regions of genes: their nucleotide sequences contain information specifying the way the RNA is to be used, and proteins interpret this information by binding specifically to these sequences. Thus, over and above the specification of the amino acid sequences of proteins, mRNA molecules are rich with many additional types of information.

The 5′ and 3′ Untranslated Regions of mRNAs Control Their Translation

Once an mRNA has been synthesized, one of the most common ways of regulating the levels of its protein product is to control the step that initiates translation. Even though the details of translation initiation differ between eucaryotes and bacteria (as we saw in Chapter 6), they each use some of the same basic regulatory strategies.

In bacterial mRNAs, a conserved stretch of six nucleotides, the *Shine-Dalgarno sequence*, is always found a few nucleotides upstream of the initiating AUG codon. This sequence forms base pairs with the 16S RNA in the small ribosomal subunit, correctly positioning the initiating AUG codon in the ribosome. Because this interaction strongly contributes to the efficiency of initiation, it provides the bacterial cell with a simple way to regulate protein synthesis through **translational control** mechanisms. These mechanisms, carried out by proteins or by RNA molecules, generally involve either exposing or blocking the Shine–Dalgarno sequence (**Figure 7–106**).

Eucaryotic mRNAs do not contain a Shine–Dalgarno sequence. Instead, as discussed in Chapter 6, the selection of an AUG codon as a translation start site is largely determined by its proximity to the cap at the 5′ end of the mRNA molecule, which is the site at which the small ribosomal subunit binds to the mRNA and begins scanning for an initiating AUG codon. Despite the differences in translation initiation, eucaryotes use similar strategies to regulate translation. For example, translational repressors bind to the 5′ end of the mRNA and thereby inhibit translation initiation. Other repressors recognize nucleotide sequences in the 3′ UTR of specific mRNAs and decrease translation initiation by interfering with the communication between the 5′ cap and 3′ poly-A tail, a step required for efficient translation (see Figure 6–72). A particularly important type of translational control in eucaryotes relies on small RNAs (termed *microRNAs* or *miRNAs*) that bind to mRNAs and reduce protein output. The miRNAs are synthesized and processed in a specialized way, and we shall return to them later in the chapter.

The Phosphorylation of an Initiation Factor Regulates Protein Synthesis Globally

Eucaryotic cells decrease their overall rate of protein synthesis in response to a variety of stressful situations, including deprivation of growth factors or nutrients, infection by viruses, and sudden increases in temperature. Much of this decrease is caused by the phosphorylation of the translation initiation factor eIF2 by specific protein kinases that respond to the changes in conditions.

The normal function of eIF2 was outlined in Chapter 6. It forms a complex with GTP and mediates the binding of the methionyl initiator tRNA to the small ribosomal subunit, which then binds to the 5′ end of the mRNA and begins scanning along the mRNA. When an AUG codon is recognized, the eIF2 protein hydrolyzes the bound GTP to GDP, causing a conformational change in the protein and releasing it from the small ribosomal subunit. The large ribosomal subunit then joins the small one to form a complete ribosome that begins protein synthesis (see Figure 6–71).

Because eIF2 binds very tightly to GDP, a guanine nucleotide exchange factor (see Figure 3–73), designated eIF2B, is required to cause GDP release so that a new GTP molecule can bind and eIF2 can be reused (**Figure 7–107**A). The

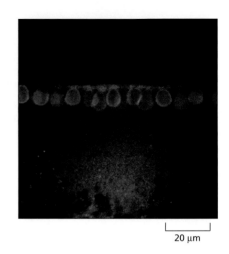

20 μm

Figure 7–105 An experiment demonstrating the importance of the 3′ UTR in localizing mRNAs to specific regions of the cytoplasm. For this experiment, two different fluorescently labeled RNAs were prepared by transcribing DNA *in vitro* in the presence of fluorescently labeled derivatives of UTP. One RNA (labeled with a *red* fluorochrome) contains the coding region for the *Drosophila* hairy protein and includes the adjacent 3′ UTR (see Figure 6–22). The other RNA (labeled *green*) contains the hairy coding region with the 3′ UTR deleted. The two RNAs were mixed and injected into a *Drosophila* embryo at a stage of development when multiple nuclei reside in a common cytoplasm (see Figure 7–53). When the fluorescent RNAs were visualized 10 minutes later, the full-length hairy RNA *(red)* was localized to the apical side of nuclei *(blue)* but the transcript missing the 3′ UTR *(green)* failed to localize. Hairy is one of many gene regulatory proteins that specifies positional information in the developing *Drosophila* embryo (discussed in Chapter 22). The localization of its mRNA (shown in this experiment to depend on its 3′ UTR) is thought to be critical for proper fly development. (Courtesy of Simon Bullock and David Ish-Horowicz.)

reuse of eIF2 is inhibited when it is phosphorylated—the phosphorylated eIF2 binds to eIF2B unusually tightly, inactivating eIF2B. There is more eIF2 than eIF2B in cells, and even a fraction of phosphorylated eIF2 can trap nearly all of the eIF2B. This prevents the reuse of the nonphosphorylated eIF2 and greatly slows protein synthesis (Figure 7–107B).

Regulation of the level of active eIF2 is especially important in mammalian cells; eIF2 is part of the mechanism that allows cells to enter a nonproliferating, resting state (called G_0)—in which the rate of total protein synthesis is reduced to about one-fifth the rate in proliferating cells (discussed in Chapter 17).

Initiation at AUG Codons Upstream of the Translation Start Can Regulate Eucaryotic Translation Initiation

We saw in Chapter 6 that eucaryotic translation typically begins at the first AUG downstream of the 5′ end of the mRNA, which is the first AUG encountered by a scanning small ribosomal subunit. But the nucleotides immediately surrounding the AUG also influence the efficiency of translation initiation. If the recognition site is poor enough, scanning ribosomal subunits will sometimes ignore the first

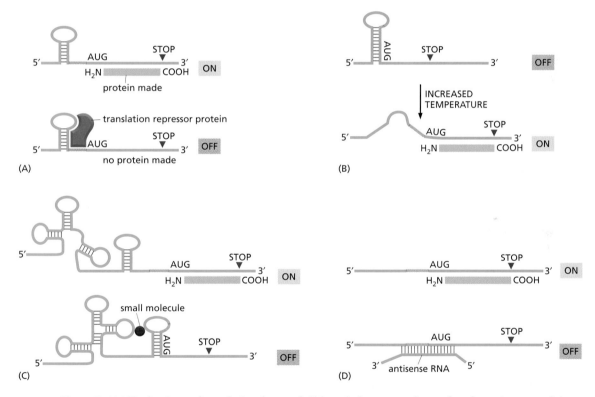

Figure 7–106 **Mechanisms of translational control.** Although these examples are from bacteria, many of the same principles operate in eucaryotes. (A) Sequence-specific RNA binding proteins repress translation of specific mRNAs by blocking access of the ribosome to the Shine–Dalgarno sequence (*orange*). For example, some ribosomal proteins repress translation of their own RNA. This mechanism comes into play only when the ribosomal proteins are produced in excess over ribosomal RNA and are therefore not incorporated into ribosomes, and it allows the cell to maintain correctly balanced quantities of the various components needed to form ribosomes. In these cases, the regulatory RNA sequence present on the mRNA often matches the RNA sequence that the protein recognizes during ribosome assembly. (B) An RNA "thermosensor" permits efficient translation initiation only at elevated temperatures in which the stem-loop structure has been melted. An example occurs in the human pathogen *Listeria monocytogenes,* in which the translation of its virulence genes increases at 37°C, the temperature of the host. (C) Binding of a small molecule to a riboswitch causes a structural rearrangement of the RNA, sequestering the Shine–Dalgarno sequence *(orange)* and blocking translation initiation. In many bacteria, S-adenosyl methionine acts in this manner to block production of the enzymes that synthesize it. (D) An "antisense" RNA produced elsewhere from the genome base-pairs with a specific mRNA, and blocks its translation. Many bacteria regulate expression of iron-storage proteins in this way. When iron is abundant, an antisense transcript is down-regulated, thereby allowing efficient translation of genes encoding the storage proteins. Antisense RNAs are used extensively by eucaryotic cells to regulate gene expression. The mechanism is somewhat different from that shown here and is discussed in detail later in this chapter.

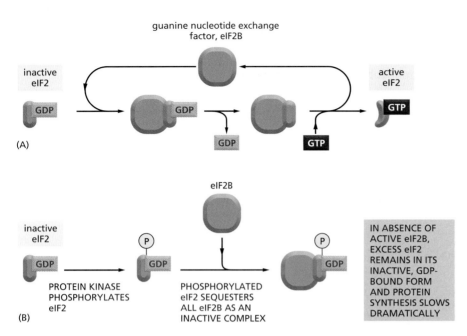

AUG codon in the mRNA and skip to the second or third AUG codon instead. This phenomenon, known as "leaky scanning," is a strategy frequently used to produce two or more closely related proteins, differing only in their amino termini, from the same mRNA. Very importantly it allows some genes to produce the same protein with and without a signal sequence attached at its amino terminus so that the protein is directed to two different locations in the cell (for example, to both mitochondria and the cytosol). In some cases, the cell can regulate the relative abundance of the protein isoforms produced by leaky scanning; for example, a cell-type-specific increase in the abundance of the initiation factor eIF4F favors the use of the AUG closest to the 5′ end of the mRNA.

Another type of control found in eukaryotes uses one or more short open reading frames (nucleotide sequences free from stop codons) that lie between the 5′ end of the mRNA and the beginning of the gene. Open reading frames (ORFs) will be discussed more fully in Chapter 8: for present purposes an ORF can be considered a stretch of DNA that begins with a start codon (ATG) and ends with a stop codon, with no stop codons in between, and thus could in principle encode a polypeptide. Often, the amino acid sequences coded by these upstream open reading frames (uORFs) are not important; rather the uORFs serve a purely regulatory function. An uORF present on an mRNA molecule will generally decrease translation of the downstream gene by trapping a scanning ribosome initiation complex and causing the ribosome to translate the uORF and dissociate from the mRNA before it reaches the protein-coding sequences.

When the activity of a general translation factor (such as the eIF2 discussed above) is reduced, one might expect that the translation of all mRNAs would be reduced equally. Contrary to this expectation, however, the phosphorylation of eIF2 can have selective effects, even enhancing the translation of specific mRNAs that contain uORFs. This can enable yeast cells, for example, to adapt to starvation for specific nutrients by shutting down the synthesis of all proteins except those that are required for synthesis of the nutrients that are missing. The details of this mechanism have been worked out for a specific yeast mRNA that encodes a protein called Gcn4, a gene regulatory protein that is required for the activation of many genes that encode proteins that are important for amino acid synthesis.

The *Gcn4* mRNA contains four short uORFs, and these are responsible for selectively increasing the translation of *Gcn4* in response to the eIF2 phosphorylation provoked by amino acid starvation. The mechanism by which *Gcn4* translation is increased is complex. In outline, the small subunit of the ribosome moves along the mRNA, encountering each of the uORFs but directing translation of only a subset of them; if the last uORF is translated, as is the case in normal unstarved cells, the ribosomes dissociate at the end of the uORF, and translation of *Gcn4* is inefficient. The global decrease in eIF2 activity brought about by nutrient starvation (see Figure 7–108) makes it more likely that a scanning small ribosomal subunit will move through the fourth uORF before it acquires a molecule of eIF2 (see Figure 6–72). Such a ribosomal subunit is free to initiate translation on the actual *Gcn4* sequences, and the increased amount of this gene regulatory protein that results leads to the production of a set of proteins that increase amino acid synthesis inside the cell.

Internal Ribosome Entry Sites Provide Opportunities for Translation Control

Although approximately 90% of eucaryotic mRNAs are translated beginning with the first AUG downstream from the 5′ cap, certain AUGs, as we saw in the previous section, can be skipped over during the scanning process. In this section, we discuss yet another way that cells can initiate translation at positions distant from the 5′ end of the mRNA, using a specialized type of RNA sequence called an **internal ribosome entry site** (**IRES**). An IRES can occur in many different places in an mRNA and, in some unusual cases, two distinct protein-coding sequences are carried in tandem on the same eucaryotic mRNA; translation of the first occurs by the usual scanning mechanism, and translation of the second occurs through an IRES. IRESs are typically several hundred nucleotides in length and fold into specific structures that bind many, but not all, of the same proteins that are used to initiate normal 5′ cap-dependent translation (**Figure 7–108**). In fact, different IRESs require different subsets of initiation factors. However, all of them bypass the need for a 5′ cap structure and the translation initiation factor that recognizes it, eIF4E.

Some viruses use IRESs as part of a strategy to get their own mRNA molecules translated while blocking normal 5′-cap-dependent translation of host mRNAs. On infection, these viruses produce a protease (encoded in the viral genome) that cleaves the host cell translation factor eIF4G, rendering it unable to bind to eIF4E, the cap-binding complex. This shuts down most of the host cell's translation and effectively diverts the translation machinery to the IRES sequences, which are present on many viral mRNAs. The truncated eIF4G remains competent to initiate translation at these internal sites and may even stimulate the translation of certain IRES-containing viral mRNAs. A selective activation of IRES-mediated translation can also occur on host cell mRNAs. For example, when mammalian cells enter the programmed cell death pathway (discussed in Chapter 18), eIF4G is cleaved, and a general decrease in translation ensues. However, some proteins critical for the control of cell death seem to be translated from IRES-containing mRNAs, allowing their continued synthesis. In this way, the IRES mechanism allows translation of selected mRNAs at a high rate despite a general decrease in the cell's overall capacity to initiate protein synthesis.

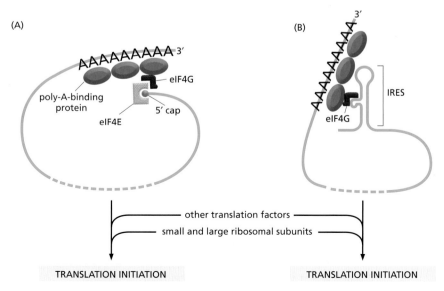

Figure 7–108 **Two mechanisms of translation initiation.** (A) The normal, cap-dependent mechanism requires a set of initiation factors whose assembly on the mRNA is stimulated by the presence of a 5′ cap and a poly-A tail (see also Figure 6–72). (B) The IRES-dependent mechanism seen mainly in viruses, requires only a subset of the normal translation initiating factors, and these assemble directly on the folded IRES. (Adapted from A. Sachs, *Cell* 101:243–245, 2000. With permission from Elsevier.)

Changes in mRNA Stability Can Regulate Gene Expression

Most mRNAs in a bacterial cell are very unstable, having half-lives of less than a couple of minutes. Exonucleases, which degrade in the 3'-to-5' direction, are usually responsible for the rapid destruction of these mRNAs. Because its mRNAs are both rapidly synthesized and rapidly degraded, a bacterium can adapt quickly to environmental changes.

As a general rule, the mRNAs in eucaryotic cells are more stable. Some, such as that encoding β-globin, have half-lives of more than 10 hours, but most have considerably shorter half-lives, typically less than 30 minutes. The mRNAs that code for proteins such as growth factors and gene regulatory proteins, whose production rates need to change rapidly in cells, have especially short half-lives.

Two general mechanisms exist for destroying eucaryotic mRNAs. Both begin with the gradual shortening of the poly-A tail by an exonuclease, a process that starts as soon as the mRNA reaches the cytoplasm. In a broad sense, this poly-A shortening acts as a timer that counts down the lifetime of each mRNA. Once the poly-A tail is reduced to a critical length (about 25 nucleotides in humans), the two pathways diverge. In one, the 5' cap is removed (a process called decapping) and the "exposed" mRNA is rapidly degraded from its 5' end. In the other, the mRNA continues to be degraded from the 3' end, through the poly-A tail into the coding sequences (**Figure 7–109**). Most eucaryotic mRNAs are degraded by both mechanisms.

Nearly all mRNAs are subject to these two types of decay, and the specific sequences of each mRNA determine how fast each step occurs and therefore how long each mRNA will persist in the cell and be able to produce protein. The 3' UTR sequences are especially important in controlling mRNA lifetimes, and they often carry binding sites for specific proteins that increase or decrease the rate of poly-A shortening, decapping, or 3'-to-5' degradation. The half-life of an mRNA is also affected by how efficiently it is translated. Poly-A shortening and decapping compete directly with the machinery that translates the mRNA; therefore any factors that affect the translation efficiency of an mRNA will tend to have the opposite effect on its degradation (**Figure 7–110**).

Although poly-A shortening controls the half-life of most eucaryotic mRNAs, some can be degraded by a specialized mechanism that bypasses this step altogether. In these cases, specific nucleases cleave the mRNA internally, effectively decapping one end and removing the poly-A tail from the other so that both halves are rapidly degraded. The mRNAs that are destroyed in this way carry specific nucleotide sequences, often in the 3' UTRs, that serve as recognition sequences for these endonucleases. This strategy makes it especially simple to tightly regulate the stability of these mRNAs by blocking the endonuclease site in response to extracellular signals. For example, the addition of iron to cells decreases the stability of the mRNA that encodes the receptor protein that binds the iron-transporting protein transferrin, causing less of this receptor to be made. This effect is mediated by the iron-sensitive RNA-binding protein aconitase (which also controls ferritin mRNA translation). Aconitase can bind to the 3' UTR of the transferrin receptor mRNA and increase receptor production by

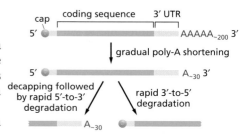

Figure 7–109 Two mechanisms of eucaryotic mRNA decay. A critical threshold of poly-A tail length induces 3'-to-5' degradation, which may be triggered by the loss of the poly-A binding proteins (see Figure 6–40). As shown in Figure 7–110, the deadenylase associates with both the 3' poly-A tail and the 5' cap, and this arrangement may signal decapping after poly-A shortening. Although 5' to 3' and 3' to 5' degradation are shown here on separate RNA molecules, these two processes can occur together on the same molecule. (Adapted from C.A. Beelman and R. Parker, *Cell* 81:179–183, 1995. With permission from Elsevier.)

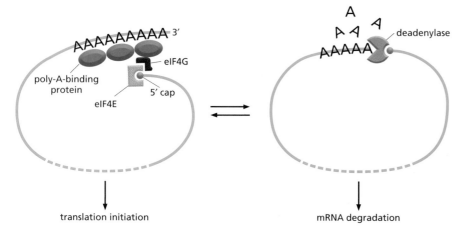

Figure 7–110 The competition between mRNA translation and mRNA decay. The same two features of an mRNA molecule, its 5' cap and the 3' poly-A tail, are used in both translation initiation and deadenylation-dependent mRNA decay (see Figure 7–109). The deadenylase that shortens the poly-A tail in the 3'-to-5' direction associates with the 5' cap. As described in Chapter 6 (see Figure 6–72), the translation initiation machinery also associates with both the 5' cap and the poly-A tail. (Adapted from M. Gao et al., *Mol. Cell* 5:479–488, 2000. With permission from Elsevier.)

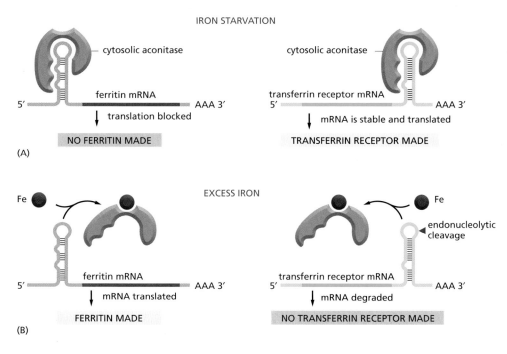

IRON STARVATION

EXCESS IRON

(A)

(B)

Figure 7–111 **Two post-translational controls mediated by iron.** (A) During iron starvation, the binding of aconitase to the 5′ UTR of the ferritin mRNA blocks translation initiation; its binding to the 3′ UTR of the transferrin receptor mRNA blocks an endonuclease cleavage site and thereby stabilizes the mRNA. (B) In response to an increase in iron concentration in the cytosol, a cell increases its synthesis of ferritin in order to bind the extra iron and decreases its synthesis of transferrin receptors in order to import less iron across the plasma membrane. Both responses are mediated by the same iron-responsive regulatory protein, aconitase, which recognizes common features in a stem-and-loop structure in the mRNAs encoding ferritin and transferrin receptor. Aconitase dissociates from the mRNA when it binds iron. But because the transferrin receptor and ferritin are regulated by different types of mechanisms, their levels respond oppositely to iron concentrations even though they are regulated by the same iron-responsive regulatory protein. (Adapted from M.W. Hentze et al., *Science* 238:1570–1573, 1987 and J.L. Casey et al., *Science* 240:924–928, 1988. With permission from AAAS.)

blocking endonucleolytic cleavage of the mRNA. On the addition of iron, aconitase is released from the mRNA, exposing the cleavage site and thereby decreasing the stability of the mRNA (**Figure 7–111**).

Cytoplasmic Poly-A Addition Can Regulate Translation

The initial polyadenylation of an RNA molecule (discussed in Chapter 6) occurs in the nucleus, apparently automatically for nearly all eucaryotic mRNA precursors. As we have just seen, the poly-A tails on most mRNAs gradually shorten in the cytosol, and the RNAs are eventually degraded. In some cases, however, the poly-A tails of specific mRNAs are lengthened in the cytosol, and this mechanism provides an additional form of translation regulation.

Maturing oocytes and eggs provide the most striking example. Many of the normal mRNA degradation pathways seem to be disabled in these giant cells, so that the cells can build up large stores of mRNAs in preparation for fertilization. Many mRNAs are stored in the cytoplasm with only 10 to 30 As at their 3′ end, and in this form they are not translated. At specific times during oocyte maturation and just after fertilization, when the cell requires the proteins encoded by these mRNAs, poly-A is added to selected mRNAs by a cytosolic poly-A polymerase, greatly stimulating their translation.

Small Noncoding RNA Transcripts Regulate Many Animal and Plant Genes

In the previous chapter, we introduced the central dogma, according to which the flow of genetic information proceeds from DNA through RNA to protein (Figure 6–2). But we have seen that RNA molecules perform many critical tasks in the cell besides serving as intermediate carriers of genetic information. A series of recent, striking discoveries has revealed that noncoding RNAs are far more prevalent than previously imagined and play previously unanticipated, but widespread, roles in regulating gene expression.

Of special importance to animals and plants is a type of short noncoding RNA called **microRNA** (miRNA). Humans, for example, express more than 400 different miRNAs, and these appear to regulate at least one-third of all human protein-coding genes. Once made, miRNAs base-pair with specific mRNAs and regulate their stability and their translation. The miRNA precursors are synthesized by RNA

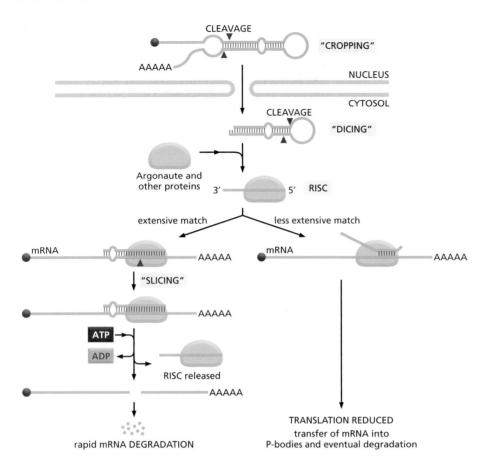

Figure 7–112 miRNA processing and mechanism of action. The precursor miRNA, through complementarity between one part of its sequence and another, forms a double-stranded structure. This is cropped while still in the nucleus, and then exported to the cytosol, where it is further cleaved by the Dicer enzyme to form the miRNA proper. Argonaute, in conjunction with other components of RISC, initially associates with both strands of the miRNA and cleaves and discards one of them. The other strand guides RISC to specific mRNAs through base pairing. If the RNA:RNA match is extensive, as is commonly seen in plants, Argonaute cleaves the target mRNA, causing its rapid degradation. In animals, the miRNA-mRNA match often does not extend beyond a short 7-nucleotide "seed" region near the 5′ end of the miRNA. This less extensive base pairing leads to inhibition of translation, mRNA destabilization, and transfer of the mRNA to P-bodies, where it is eventually degraded.

polymerase II and are capped and polyadenylated. They then undergo a special type of processing, after which the miRNA is assembled with a set of proteins to form an *RNA-induced silencing complex* or *RISC*. Once formed, the RISC seeks out its target mRNAs by searching for complementary nucleotide sequences (**Figure 7–112**). This search is greatly facilitated by the Argonaute protein, a component of RISC, which displays the 5′ region of the miRNA so that it is optimally positioned for base-pairing to another RNA molecule (**Figure 7–113**). In animals, the extent of base-pairing is typically seven nucleotide pairs, and it usually takes place in the 3′ UTR of the target mRNA.

Once an mRNA has been bound by an miRNA, several outcomes are possible. If the base-pairing is extensive, the mRNA is cleaved by the Argonaute protein, effectively removing its poly-A tail and exposing it to exonucleases (see Figure 7–109). Following cleavage of the mRNA, RISC (with its associated miRNA) is released, and it can seek out additional mRNAs. Thus, a single miRNA can act catalytically to destroy many complementary mRNAs. The miRNAs can be thought of as guide sequences that bring destructive nucleases into contact with specific mRNAs.

If the base-pairing between the miRNA and the mRNA is less extensive, Argonaute does not slice the mRNA; rather, translation of the mRNA is repressed and the mRNA is destabilized. This effect is associated with shortening of the poly-A tail and the movement of the mRNA to cytosolic structures called *processing bodies (P-bodies)*. Here the mRNAs are sequestered from ribosomes and eventually decapped and degraded. P-bodies are dynamic structures composed of large assemblies of mRNAs and RNA-degrading enzymes, and they are believed to be the sites in the cell where the final destruction of most mRNAs, even those not controlled by miRNAs, takes place (**Figure 7–114**).

Several features make miRNAs especially useful regulators of gene expression. First, a single miRNA can regulate a whole set of different mRNAs so long as the mRNAs carry a common sequence in their UTRs. This situation is common in humans, where some miRNAs control hundreds of different mRNAs. Second, regulation by miRNAs can be combinatorial. When the base-pairing between the miRNA and mRNA fails to trigger cleavage, additional miRNAs binding to the

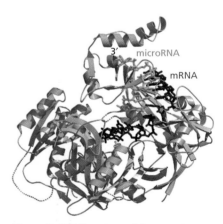

Figure 7–113 Structure of Argonaute protein bound to a perfectly base-paired miRNA and mRNA. (Adapted from N.H. Tolia and L. Joshua-Tor, *Nat. Chem. Biol.* 3:36–43, 2007. With permission from Macmillan Publishers Ltd.)

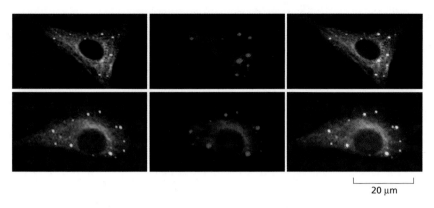

Figure 7–114 Visualization of P-bodies. Human cells were stained with antibodies to a component of the mRNA decapping enzyme Dcp1a *(left panels)* and to the Argonaute protein *(middle panels).* The merged image *(right panels)* shows that the two proteins co-localize to foci in the cytoplasm called P-bodies. (Adapted from J. Liu et al., *Nat. Cell Biol.* 7:643-644, 2005. With permission from Macmillan Publishers Ltd.)

20 μm

same mRNA lead to further reductions in its translation. As discussed earlier in this chapter for gene regulatory proteins, combinatorial control greatly expands the possibilities available to the cell by linking gene expression to a combination of different regulators rather than a single regulator. Third, an miRNA occupies relatively little space in the genome when compared with a protein. Indeed, their small size is one reason that miRNAs were discovered only recently. Although we are only beginning to understand the full impact of miRNAs, it is clear that they represent a very important part of the cell's equipment for regulating the expression of its genes.

RNA Interference Is a Cell Defense Mechanism

Many of the proteins that participate in the miRNA regulatory mechanisms just described also serve a second function as a defense mechanism: they orchestrate degradation of foreign RNA molecules, specifically those that occur in double-stranded form. Termed **RNA interference** (**RNAi**), this mechanism is found in a wide variety of organisms, including single-celled fungi, plants, and worms, suggesting that it is evolutionarily ancient. Many transposable elements and viruses produce double-stranded RNA, at least transiently, in their life cycles, and RNAi helps to keep these potentially dangerous invaders in check. As we shall see, RNAi has also provided scientists with a powerful experimental technique to turn off the expression of individual genes.

The presence of double-stranded RNA in the cell triggers RNAi by attracting a protein complex containing *Dicer*, the same nuclease that processes miRNAs (see Figure 7–112). This protein complex cleaves the double-stranded RNA into small (approximately 23- nucleotide-pair) fragments called **small interfering RNAs** (**siRNA**). These double-stranded siRNAs are then bound by Argonaute and other components of the RISC, as we saw above for miRNAs, and one strand of the duplex RNA is cleaved by Argonaute and discarded. The single-stranded siRNA molecule that remains directs RISC back to complementary RNA molecules produced by the virus or transposable element; because the match is exact, Argonaute cleaves these molecules, leading to their rapid destruction (**Figure 7–115**).

Each time RISC cleaves a new RNA molecule, it is released; thus as we saw for miRNAs, a single RNA molecule can act catalytically to destroy many complementary RNAs. Some organisms employ an additional mechanism that amplifies the RNAi response even further. In these organisms, RNA-dependent RNA polymerases can convert the products of siRNA-mediated cleavage into more double-stranded RNA. This amplification ensures that, once initiated, RNA interference can continue even after all the initiating double-stranded RNA has been degraded or diluted out. For example, it permits progeny cells to continue carrying out RNA interference that was provoked in the parent cells.

In some organisms, the RNA interference activity can be spread by the transfer of RNA fragments from cell to cell. This is particularly important in plants (whose cells are linked by fine connecting channels, as discussed in Chapter 19), because it allows an entire plant to become resistant to an RNA virus after only a few of its cells have been infected. In a broad sense, the RNAi

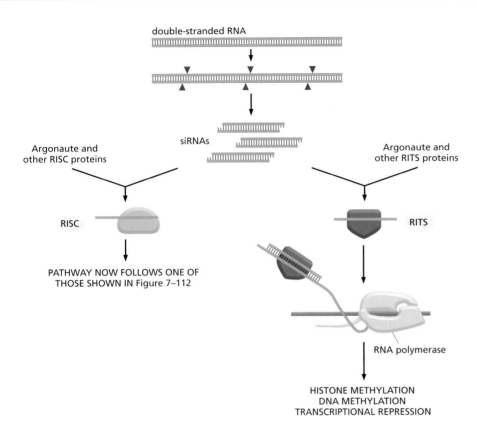

double-stranded RNA

siRNAs

Argonaute and
other RISC proteins

RISC

PATHWAY NOW FOLLOWS ONE OF
THOSE SHOWN IN Figure 7–112

Argonaute and
other RITS proteins

RITS

RNA polymerase

HISTONE METHYLATION
DNA METHYLATION
TRANSCRIPTIONAL REPRESSION

Figure 7–115 siRNA-mediated heterochromatin formation. In many organisms, double-stranded RNA can trigger both the destruction of complementary mRNAs *(left)* and transcriptional silencing *(right)*. The change in chromatin structure induced by the bound RITS (RNA-induced transcriptional silencing) complex resembles that in Figure 7–81.

response resembles certain aspects of animal immune systems; in both, an invading organism elicits a customized response, and—through amplification of the "attack" molecules—the host becomes systemically protected.

RNA Interference Can Direct Heterochromatin Formation

The RNA interference pathway just described does not necessarily stop with the destruction of target RNA molecules. In some cases, the RNA interference machinery can selectively shut off synthesis of the target RNAs. For this remarkable mechanism to occur, the short siRNAs produced by the Dicer protein are assembled with a group of proteins (including Argonaute) to form the RITS (RNA-induced transcriptional silencing) complex. Using single-stranded siRNA as a guide sequence, this complex binds complementary RNA transcripts as they emerge from a transcribing RNA polymerase II (see Figure 7–115). Positioned on the genome in this manner, the RITS complex attracts proteins that covalently modify nearby histones and eventually directs the formation and spread of heterochromatin to prevent further transcription initiation. In some cases, the RITS complex also induces the methylation of DNA, which, as we have seen, can repress gene expression even further. Because heterochromatin and DNA methylation can be self-propagating, an initial RNA interference signal can continue to silence gene expression long after all the siRNA molecules have dissipated.

RNAi-directed heterochromatin formation is an important cell defense mechanism that limits the accumulation of transposable elements in the genome by maintaining them in a transcriptionally silent form. However, this same mechanism is also used in many normal processes in the cell. For example, in many organisms, the RNA interference machinery maintains the heterochromatin formed around centromeres. Centromeric DNA sequences are transcribed in both directions, producing complementary RNA transcripts that can base-pair to form double-stranded RNA. This double-stranded RNA triggers the RNA interference pathway and stimulates heterochromatin formation at centromeres. This heterochromatin, in turn, is necessary for the centromeres to segregate chromosomes accurately during mitosis (see Figure 4–50).

RNA Interference Has Become a Powerful Experimental Tool

Although it probably arose initially as a defense mechanism, RNA interference has become thoroughly integrated into many aspects of normal cell biology, ranging from the control of gene expression to the structure of chromosomes. It has also been developed by scientists into a powerful experimental tool that allows almost any gene to be inactivated by evoking an RNAi response to it. This technique, carried out in cultured cells, and in some cases, whole animals and plants, has revolutionized genetic approaches in cell and molecular biology. We shall discuss it in more detail in the following chapter (see pp. 571–572). RNAi also has great potential in treating human disease. Since many human disorders result from the misexpression of genes, the ability to turn these genes off by experimentally introducing complementary siRNA molecules holds great medical promise. Remarkably, the mechanism of RNA interference was discovered only recently, and we are still being surprised by its mechanistic details and by its broad biological implications.

Summary

Many steps in the pathway from RNA to protein are regulated by cells in order to control gene expression. Most genes are regulated at multiple levels, in addition to being controlled at the initiation stage of transcription. The regulatory mechanisms include (1) attenuation of the RNA transcript by its premature termination, (2) alternative RNA splice-site selection, (3) control of 3′-end formation by cleavage and poly-A addition, (4) RNA editing, (5) control of transport from the nucleus to the cytosol, (6) localization of mRNAs to particular parts of the cell, (7) control of translation initiation, and (8) regulated mRNA degradation. Most of these control processes require the recognition of specific sequences or structures in the RNA molecule being regulated, a task performed by either regulatory proteins or regulatory RNA molecules. A particularly widespread form of post-transcriptional control is RNA interference, where guide RNAs base-pair with mRNAs. RNA interference can cause mRNAs to be either destroyed or translationally repressed. It can also cause specific genes to be packaged into heterochromatin.

PROBLEMS

Which statements are true? Explain why or why not.

7–1 In terms of its biochemical function, the helix-loop-helix motif is more closely related to the leucine zipper motif than it is to the helix-turn-helix motif.

7–2 Reversible genetic rearrangements are a common way of regulating gene expression in procaryotes and mammalian cells.

7–3 CG islands are thought to have arisen during evolution because they were associated with portions of the genome that remained active, hence unmethylated, in the germline.

Discuss the following problems.

7–4 A small portion of a two-dimensional display of proteins from human brain is shown in **Figure Q7–1**. These proteins were separated on the basis of size in one dimension and electrical charge (isoelectric point) in the other. Not all protein spots on such displays are products of different genes; some represent modified forms of a protein that migrate to different positions. Pick out a couple of sets of spots that could represent proteins that differ by the number of phosphates they carry. Explain the basis for your selection.

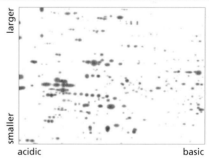

Figure Q7–1 Two-dimensional separation of proteins from the human brain (Problem 7–4). The proteins were displayed using two-dimensional gel electrophoresis. Only a small portion of the protein spectrum is shown. (Courtesy of Tim Myers and Leigh Anderson, Large Scale Biology Corporation.)

7–5 DNA microarray analysis of the patterns of mRNA abundance in different human cell types shows that the level of expression of almost every active gene is different. The patterns of mRNA abundance are so characteristic of cell type that they can be used to determine the tissue of origin of cancer cells, even though the cells may have metastasized to different parts of the body. By definition, however, cancer cells are different from their noncancerous precursor cells. How do you suppose then that patterns of mRNA expression might be used to determine the tissue source of a human cancer?

7–6 The nucleus of a eucaryotic cell is much larger than a bacterium, and it contains much more DNA. As a consequence, a DNA-binding protein in a eucaryotic cell must be able to select its specific binding site from among many more unrelated sequences than does a DNA-binding protein in a bacterium. Does this present any special problems for eucaryotic gene regulation?

Consider the following situation. Assume that the eucaryotic nucleus and the bacterial cell each have a single copy of the same DNA-binding site. In addition, assume that the nucleus is 500 times the volume of the bacterium, and has 500 times as much DNA. If concentration of the gene regulatory protein that binds the site were the same in the nucleus and in the bacterium, would the regulatory protein find its binding site equally as well in the eucaryotic nucleus as it does in the bacterium? Explain your answer.

7–7 DNA-binding proteins often find their specific sites much faster than would be anticipated by simple three-dimensional diffusion. The Lac repressor, for example, associates with the *Lac* operator—its DNA-binding site—more than 100 times faster than expected from this model. Clearly, the repressor must find the operator by mechanisms that reduce the dimensionality or volume of the search in order to hasten target acquisition.

Several techniques have been used to investigate this problem. One of the most elegant used strongly fluorescent RNA polymerase molecules that could be followed individually. An array of DNA molecules was aligned in parallel and anchored to a glass slide. Fluorescent RNA polymerase molecules were then allowed to flow across them at an oblique angle (**Figure Q7–2**A). Traces of individual RNA polymerases showed that about half flowed in the same direction as the bulk and about half deviated from the bulk flow in a characteristic manner (Figure Q7–2B). If the RNA polymerase molecules were first incubated with short DNA fragments containing a strong promoter, all the traces followed the bulk flow.

A. Offer an explanation for why some RNA polymerase molecules deviated from the bulk flow as shown in Figure Q7–2B. Why did incubation with short DNA fragments containing a strong promoter eliminate traces that deviated from the bulk flow?

B. Do these results suggest an explanation for how site-specific DNA-binding molecules manage to find their sites faster than expected by diffusion?

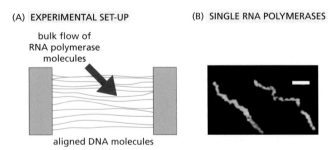

(A) EXPERIMENTAL SET-UP (B) SINGLE RNA POLYMERASES

bulk flow of RNA polymerase molecules

aligned DNA molecules

Figure Q7–2 Interactions of individual RNA polymerase molecules with DNA (Problem 7–7). (A) Experimental set-up. DNA molecules are aligned and anchored to a glass slide, and highly fluorescent RNA polymerase molecules are allowed to flow across them. (B) Traces of two individual RNA polymerase molecules. The one on the *left* has traveled with the bulk flow, and the one on the *right* has deviated from it. The scale bar is 10 μm. (B, reprinted from H. Kabata et al., *Science* 262:1561–1563, 1993. With permission from AAAS.

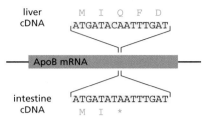

liver cDNA

```
    M   I   Q   F   D
  ATGATACAATTTGAT
```

ApoB mRNA

intestine cDNA

```
  ATGATATAATTTGAT
    M   I   *
```

Figure Q7–3 Location of the sequence differences in cDNA clones from ApoB RNA isolated from liver and intestine (Problem 7–9). The encoded amino acid sequences, in the one-letter code, are shown aligned with the cDNA sequences.

C. Based on your explanation, would you expect a site-specific DNA-binding molecule to find its target site faster in a population of short DNA molecules or in a population of long DNA molecules? Assume that the concentration of target sites is identical and that there is one target site per DNA molecule.

7–8 Most people who are completely blind have circadian rhythms that are 'free-running;' that is, their rhythms are not synchronized to environmental time cues and they oscillate on a cycle of about 24.5 hours. Why do you suppose the circadian clocks of blind people are not entrained to the same 24-hour clock as the majority of the population? Can you guess what symptoms might be associated with a free-running circadian clock? Do you suppose that blind people have trouble sleeping?

7–9 In humans, two closely related forms of apolipoprotein B (ApoB) are found in blood as constituents of the plasma lipoproteins. ApoB48 (molecular mass, 48 kilodaltons) is synthesized by the intestine and is a key component of chylomicrons, the large lipoprotein particles responsible for delivery of dietary triglycerides to adipose tissue for storage. ApoB100 (molecular mass, 100,000 kilodaltons) is synthesized in the liver for formation of the much smaller, very low-density lipoprotein particles used in the distribution of triglycerides to meet energy needs. A classic set of studies defined the surprising relationship between these two proteins.

Sequences of cloned cDNA copies of the mRNAs from these two tissues revealed a single difference: cDNAs from intestinal cells had a T, as part of a stop codon, at a point where the cDNAs from liver cells had a C, as part of a glutamine codon (**Figure Q7–3**). To verify the differences in the mRNAs and to search for corresponding differences in the genome, RNA and DNA were isolated from intestinal and liver cells and then subjected to PCR amplification, using oligonucleotides that flanked the region of interest. The amplified DNA segments from the four samples were tested for the presence of the T or C by hybridization to oligonucleotides containing either the liver cDNA sequence (oligo-Q) or the intestinal cDNA sequence (oligo-STOP). The results are shown in **Table Q7–1**.

Are the two forms of ApoB produced by transcriptional control from two different genes, by a processing control of the RNA transcript from a single gene, or by differential cleavage of the protein product from a single gene? Explain your reasoning.

Table Q7–1 Hybridization of specific oligonucleotides to the amplified segments from liver and intestine RNA and DNA (Problem 7–9).

	RNA		DNA	
	LIVER	**INTESTINE**	**LIVER**	**INTESTINE**
Oligo-Q	+	–	+	+
Oligo-STOP	–	+	–	–

Hybridization is indicated by +; its absence by –.

REFERENCES

General

Brown TA (2002) Genomes 2, 2nd ed. New York: Wiley-Liss.

Epigenetics (2004) *Cold Spring Harb Symp Quant Biol* 69.

Hartwell L, Hood L, Goldberg ML et al (2006) Genetics: from Genes to Genomes, 3rd ed. Boston: McGraw Hill.

Lodish H, Berk A, Kaiser CL et al (2007) Molecular Cell Biology, 6th ed. New York: WH Freeman.

McKnight SL & Yamamoto KR (eds) (1993) Transcriptional Regulation. Cold Spring Harbor, NY: Cold Spring Harbor Laboratory Press.

Mechanisms of Transcription (1998) *Cold Spring Harb Symp Quant Biol* 63.

Ptashne M & Gann A (2002) Genes and Signals. Cold Spring Harbor Laboratory Press, Cold Spring Harbor.

Regulatory RNAs (2006) *Cold Spring Harb Symp Quant Biol* 71.

Watson JD, Baker TA, Bell SP et al (2003) Molecular Biology of the Gene, 5th ed. Menlo Park, CA: Benjamin Cummings.

An Overview of Gene Control

Campbell KH, McWhir J, Ritchie WA & Wilmut I (1996) Sheep cloned by nuclear transfer from a cultured cell line. *Nature* 380:64–66.

Davidson EH (2006) The Regulatory Genome: Gene Regulatory Networks in Development and Evolution. Burlington, MA: Elsevier.

Gurdon JB (1992) The generation of diversity and pattern in animal development. *Cell* 68:185–199.

Levine M & Tjian R (2003) Transcription regulation and animal diversity. *Nature* 424:147–151.

Ross DT, Scherf U, Eisen MB et al (2000) Systematic variation in gene expression patterns in human cancer cell lines. *Nature Genet* 24:227–235.

DNA-binding Motifs in Gene Regulatory Proteins

Gehring WJ, Affolter M & Burglin T (1994) Homeodomain proteins. *Annu Rev Biochem* 63:487–526.

Harbison CT, Gordon DB, Lee TI et al (2004) Transcriptional regulatory code of a eukaryotic genome. *Nature* 431:99–104.

Luscombe NM, Austin SE, Berman HM, et al (2000) An overview of the structures of protein–DNA complexes. *Gen Biol* 1:reviews 001.1–001.37.

McKnight SL (1991) Molecular zippers in gene regulation. *Sci Am* 264:54–64.

Pabo CO & Sauer RT (1992) Transcription factors: structural families and principles of DNA recognition. *Annu Rev Biochem* 61:1053–1095.

Rhodes D & Klug A (1993) Zinc fingers. *Sci Am* 268:56–65.

Seeman NC, Rosenberg JM & Rich A (1976) Sequence-specific recognition of double helical nucleic acids by proteins. *Proc Natl Acad Sci USA* 73:804–808.

How Genetic Switches Work

Becker PB & Hörz W (2002) ATP-dependent nucleosome remodeling. *Annu Rev Biochem* 71:247–273.

Beckwith J (1987) The operon: an historical account. In *Escherichia coli* and *Salmonella typhimurium*: Cellular and Molecular Biology (Neidhart FC, Ingraham JL, Low KB et al eds), vol 2, pp 1439–1443. Washington, DC: ASM Press.

Gaszner M & Felsenfeld G (2006) Insulators: exploiting transcriptional and epigenetic mechanisms. *Nature Rev Genet* 7:703–713.

Gilbert W and Müller-Hill B (1967) The lac operator is DNA. *Proc Natl Acad Sci USA* 58:2415.

Green MR (2005) Eukaryotic transcription activation: right on target. *Mol Cell* 18:399–402.

Jacob F & Monod J (1961) Genetic regulatory mechanisms in the synthesis of proteins. *J Mol Biol* 3:318–356.

Lawson CL, Swigon D, Murakami KS et al (2004) Catabolite activator protein: DNA binding and transcription activation. *Curr Opin Struct Biol* 14:10–20.

Millar CB, & Grunstein M (2006) Genome-wide patterns of histone modifications in yeast. *Nature Rev Mol Cell Biol* 7:657–666.

Narlikar GJ, Fan HY & Kingston RE (2002) Cooperation between complexes that regulate chromatin structure and transcription. *Cell* 108:475–487.

Oehler S, Eismann ER, Krämer H et al (1990) The three operators of the lac operon cooperate in repression. *EMBO J* 9:973–979.

Ptashne M (2004) A Genetic Switch: Phage and Lambda Revisited, 3rd ed. Cold Spring Harbor, NY: Cold Spring Harbor Laboratory Press.

Ptashne M (1967) Specific binding of the lambda phage repressor to lambda DNA. *Nature* 214:232–234.

St Johnston D & Nusslein-Volhard C (1992) The origin of pattern and polarity in the *Drosophila* embryo. *Cell* 68:201–219.

Strahl BD & Allis CD (2000) The language of covalent histone modifications. *Nature* 403:41–45.

The Molecular Genetic Mechanisms that Create Specialized Cell Types

Alon, U (2006) An Introduction to Systems Biology: Design Principles of Biological Circuits (Chapman & Hall/Crc Mathematical and Computational Biology Series) TF Chapman.

Bell-Pedersen D, Cassone VM, Earnest DJ et al (2005) Circadian rhythms from multiple oscillators: lessons from diverse organisms. *Nature Rev Genet* 6:544–556.

Bernstein BE, Meissner A & Lander ES (2007) The mammalian epigenome. *Cell* 128:669–681.

Herskowitz I (1989) A regulatory hierarchy for cell specialization in yeast. *Nature* 342:749–757.

Klose RJ & Bird AP (2006) Genomic DNA methylation: the mark and its mediators. *Trends Biochem Sci* 31:89–97.

Meyer BJ (2000) Sex in the worm: counting and compensating X-chromosome dose. *Trends Genet* 16:247–253.

Surani MA (2001) Reprogramming of genome function through epigenetic inheritance. *Nature* 414:122–128.

Tapscott SJ (2005) The circuitry of a master switch: MyoD and the regulation of skeletal muscle gene transcription. *Development* 132:2685–2695.

Post-transcriptional Controls

Bass BL (2002) RNA editing by adenosine deaminases that act on RNA. *Annu Rev Biochem* 71:817–846.

Blencowe BJ (2006) Alternative splicing: new insights from global analyses. *Cell* 126:37–47.

Brennecke J, Stark A, Russell RB et al (2005) Principles of microRNA-target recognition. *PLoS Biology* 3.

Fire A, Xu S, Montgomery MK et al (1998) Potent and specific genetic interference by double-stranded RNA in *Caenorhabditis elegans*. *Nature* 391:806–811.

Frankel AD & Young JAT (1998) HIV-1: fifteen proteins and an RNA. *Annu Rev Biochem* 67:1–25.

Gottesman S (2004) The small RNA regulators of *Escherichia coli*: roles and mechanisms. *Annu Rev Microbiol* 58:303–328.

Mello CC & Conte D (2004) Revealing the world of RNA interference. *Nature* 431:338–342.

Parker R & Sheth U (2007) P bodies and the control of mRNA translation and degradation. *Mol Cell* 25:635–646.

Stuart KD, Schnaufer A, Ernst NL et al (2005) Complex management: RNA editing in trypanosomes. *Trends Biochem Sci* 30:97–105.

Tolia NH & Joshua-Tor L (2007) Slicer and the argonautes. *Nature Chem Biol* 3:36–43.

Tomari Y & Zamore PD (2005) Perspective: machines for RNAi. *Genes Dev* 19:517–529.

Valencia-Sanchez MA, Liu J, Hannon GJ et al (2006) Control of translation and mRNA degradation by miRNAs and siRNAs. *Genes Dev* 20:515–524.

Verdel A & Moazed D (2005) RNAi-directed assembly of heterochromatin in fission yeast. *FEBS Letters* 579:5872–5878.

Wilhelm JE & Smibert, CA (2005) Mechanisms of translational regulation in *Drosophila*. *Biol Cell* 97:235–252.

Winkler WC & Breaker RR (2005) Regulation of bacterial gene expression by riboswitches. *Annu Rev Microbiol* 59:487–517.

METHODS

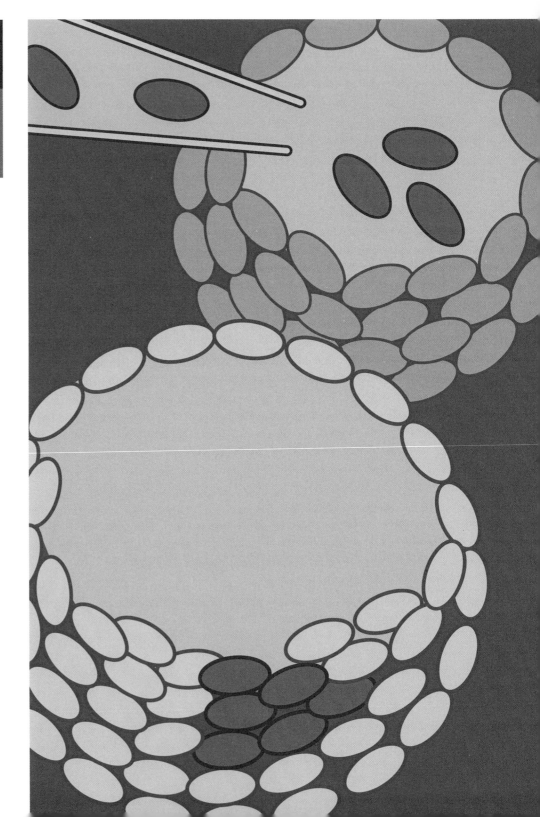

Chapter 8

Manipulating Proteins, DNA, and RNA

8

Progress in science is often driven by advances in technology. The entire field of cell biology, for example, came into being when optical craftsmen learned to grind small lenses of sufficiently high quality to observe cells and their sub-structures. Innovations in lens grinding, rather than any conceptual or philosophical advance, allowed Hooke and van Leeuwenhoek to discover a previously unseen cellular world, where tiny creatures tumble and twirl in a small droplet of water (**Figure 8–1**).

The 21st century promises to be a particularly exciting time for biology. New methods for analyzing proteins, DNA, and RNA are fueling an information explosion and allowing scientists to study cells and their macromolecules in previously unimagined ways. We now have access to the sequences of many billions of nucleotides, providing the complete molecular blueprints for hundreds of organisms—from microbes and mustard weeds to worms, flies, mice, dogs, chimpanzees, and humans. And powerful new techniques are helping us to decipher that information, allowing us not only to compile huge, detailed catalogs of genes and proteins but also to begin to unravel how these components work together to form functional cells and organisms. The long-range goal is nothing short of obtaining a complete understanding of what takes place inside a cell as it responds to its environment and interacts with its neighbors. We want to know which genes are switched on, which mRNA transcripts are present, and which proteins are active—where they are located, with what other proteins and other molecules they associate, and to which pathways or networks they belong. We also want to understand how the cell successfully manages this staggering number of variables and how it chooses among an almost unlimited number of possibilities in performing its diverse biological tasks. Such information will permit us to begin to build a framework for delineating, and eventually predicting, how genes and proteins operate to lay the foundations for life.

In this chapter, we present some of the principal methods used to study the molecular components of cells, particularly proteins, DNA, and RNA. We consider how to separate cells of different types from tissues, how to grow cells outside the body, and how to disrupt cells and isolate their organelles and constituent macromolecules in pure form. We also present the latest techniques used to determine protein structure, function, and interactions, and we discuss the breakthroughs in DNA technology that continue to revolutionize our understanding of cell function.

The techniques and methods described in this chapter have made possible the discoveries that are presented throughout this book, and they are currently being used by tens of thousands of scientists each day.

ISOLATING CELLS AND GROWING THEM IN CULTURE

Although the organelles and large molecules in a cell can be visualized with microscopes, understanding how these components function requires a detailed biochemical analysis. Most biochemical procedures require that large

In This Chapter

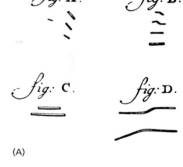

(A)

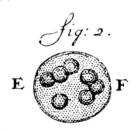

(B)

Figure 8–1 Microscopic life. A sample of "diverse animalcules" seen by van Leeuwenhoek using his simple microscope. (A) Bacteria seen in material he excavated from between his teeth. Those in fig. B he described as "swimming first forward and then backwards" (1692). (B) The eucaryotic green alga *Volvox* (1700). (Courtesy of the John Innes Foundation.)

numbers of cells be physically disrupted to gain access to their components. If the sample is a piece of tissue, composed of different types of cells, heterogeneous cell populations will be mixed together. To obtain as much information as possible about the cells in a tissue, biologists have developed ways of dissociating cells from tissues and separating them according to type. These manipulations result in a relatively homogeneous population of cells that can then be analyzed—either directly or after their number has been greatly increased by allowing the cells to proliferate in culture.

Cells Can Be Isolated from Intact Tissues

Intact tissues provide the most realistic source of material, as they represent the actual cells found within the body. The first step in isolating individual cells is to disrupt the extracellular matrix and cell–cell junctions that hold the cells together. For this purpose, a tissue sample is typically treated with proteolytic enzymes (such as trypsin and collagenase) to digest proteins in the extracellular matrix and with agents (such as ethylenediaminetetraacetic acid, or EDTA) that bind, or chelate, the Ca^{2+} on which cell–cell adhesion depends. The tissue can then be teased apart into single cells by gentle agitation.

For some biochemical preparations, the protein of interest can be obtained in sufficient quantity without having to separate the tissue or organ into cell types. Examples include the preparation of histones from calf thymus, actin from rabbit muscle, or tubulin from cow brain. In other cases, obtaining the desired protein requires enrichment for a specific cell type of interest. Several approaches are used to separate the different cell types from a mixed cell suspension. The most general cell-separation technique uses an antibody coupled to a fluorescent dye to label specific cells. An antibody is chosen that specifically binds to the surface of only one cell type in the tissue. The labeled cells can then be separated from the unlabeled ones in an electronic *fluorescence-activated cell sorter*. In this remarkable machine, individual cells traveling single file in a fine stream pass through a laser beam, and the fluorescence of each cell is rapidly measured. A vibrating nozzle generates tiny droplets, most containing either one cell or no cells. The droplets containing a single cell are automatically given a positive or a negative charge at the moment of formation, depending on whether the cell they contain is fluorescent; they are then deflected by a strong electric field into an appropriate container. Occasional clumps of cells, detected by their increased light scattering, are left uncharged and are discarded into a waste container. Such machines can accurately select 1 fluorescent cell from a pool of 1000 unlabeled cells and sort several thousand cells each second (**Figure 8–2**).

Selected cells can also be obtained by carefully dissecting them from thin tissue slices that have been prepared for microscopic examination (discussed in Chapter 9). In one approach, a tissue section is coated with a thin plastic film and a region containing the cells of interest is irradiated with a focused pulse from an infrared laser. This light pulse melts a small circle of the film, binding the cells underneath. These captured cells are then removed for further analysis. The technique, called *laser capture microdissection*, can be used to separate and analyze cells from different areas of a tumor, allowing their properties or molecular composition to be compared with neighboring normal cells. A related method uses a laser beam to directly cut out a group of cells and catapult them into an appropriate container for future analysis (**Figure 8–3**).

A uniform population of cells obtained by any of these or other separation methods can be used directly for biochemical analysis. After breaking open the cells by mechanical disruption, detergents, and other methods, cytoplasm or individual organelles can be extracted and then specific molecules purified.

Cells Can Be Grown in Culture

Although molecules can be extracted from whole tissues, this is often not the most convenient source of material, requiring, for example, early-morning trips

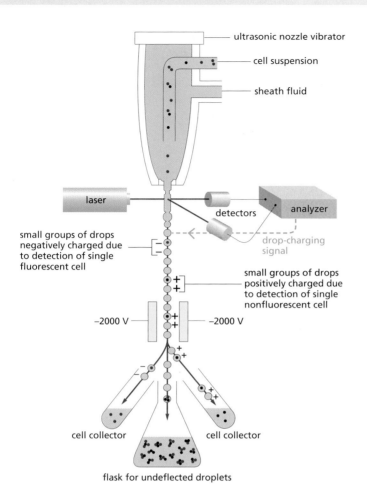

ultrasonic nozzle vibrator

cell suspension

sheath fluid

laser

detectors

analyzer

small groups of drops
negatively charged due
to detection of single
fluorescent cell

drop-charging
signal

small groups of drops
positively charged due
to detection of single
nonfluorescent cell

–2000 V

–2000 V

cell collector

cell collector

flask for undeflected droplets

Figure 8–2 A fluorescence-activated cell sorter. A cell passing through the laser beam is monitored for fluorescence. Droplets containing single cells are given a negative or positive charge, depending on whether the cell is fluorescent or not. The droplets are then deflected by an electric field into collection tubes according to their charge. Note that the cell concentration must be adjusted so that most droplets contain no cells and flow to a waste container together with any cell clumps.

to a slaughterhouse. The problem is not only a question of convenience. The livestock commonly used as organ sources are not amenable to genetic manipulation. Moreover, the complexity of intact tissues and organs is an inherent disadvantage when trying to purify particular molecules. Cells grown in culture provide a more homogeneous population of cells from which to extract material, and they are also much more convenient to work with in the laboratory. Given appropriate surroundings, most plant and animal cells can live, multiply, and even express differentiated properties in a tissue-culture dish. The cells can be watched continuously under the microscope or analyzed biochemically, and the effects of adding or removing specific molecules, such as hormones or growth factors, can be systematically explored. In addition, by mixing two cell types, the interactions between one cell type and another can be studied.

Experiments performed on cultured cells are sometimes said to be carried out *in vitro* (literally, "in glass") to contrast them with experiments using intact organisms, which are said to be carried out *in vivo* (literally, "in the living organism"). These terms can be confusing, however, because they are often used in a very different sense by biochemists. In the biochemistry lab, *in vitro* refers to reactions carried out in a test tube in the absence of living cells, whereas *in vivo*

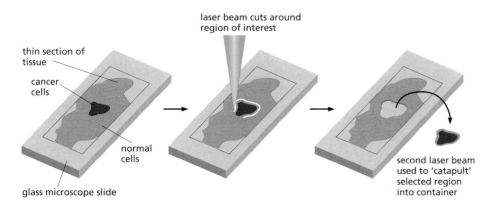

laser beam cuts around
region of interest

thin section of
tissue

cancer
cells

normal
cells

glass microscope slide

second laser beam
used to 'catapult'
selected region
into container

Figure 8–3 Microdissection techniques to select cells from tissue slices. This method uses a laser beam to excise a region of interest and eject it into a container, and it permits the isolation of even a single cell from a tissue sample.

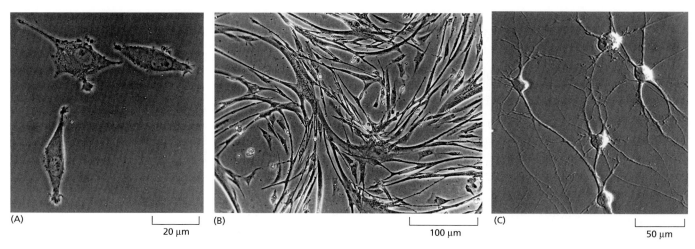

(A) 20 µm (B) 100 µm (C) 50 µm

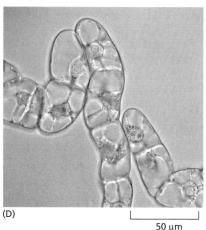

(D) 50 µm

Figure 8–4 **Light micrographs of cells in culture.** (A) Mouse fibroblasts. (B) Chick myoblasts fusing to form multinucleate muscle cells. (C) Purified rat retinal ganglion nerve cells. (D) Tobacco cells in liquid culture. (A, courtesy of Daniel Zicha; B, courtesy of Rosalind Zalin; C, from A. Meyer-Franke et al., *Neuron* 15:805–819, 1995. With permission from Elsevier; D, courtesy of Gethin Roberts.)

refers to any reaction taking place inside a living cell, even if that cell is growing in culture.

Tissue culture began in 1907 with an experiment designed to settle a controversy in neurobiology. The hypothesis under examination was known as the neuronal doctrine, which states that each nerve fiber is the outgrowth of a single nerve cell and not the product of the fusion of many cells. To test this contention, small pieces of spinal cord were placed on clotted tissue fluid in a warm, moist chamber and observed at regular intervals under the microscope. After a day or so, individual nerve cells could be seen extending long, thin filaments (axons) into the clot. Thus, the neuronal doctrine received strong support, and the foundation was laid for the cell-culture revolution.

These original experiments on nerve fibers used cultures of small tissue fragments called explants. Today, cultures are more commonly made from suspensions of cells dissociated from tissues using the methods described earlier. Unlike bacteria, most tissue cells are not adapted to living suspended in fluid and require a solid surface on which to grow and divide. For cell cultures, this support is usually provided by the surface of a plastic tissue-culture dish. Cells vary in their requirements, however, and many do not proliferate or differentiate unless the culture dish is coated with materials that cells like to adhere to, such as polylysine or extracellular matrix components.

Cultures prepared directly from the tissues of an organism are called *primary cultures*. These can be made with or without an initial fractionation step to separate different cell types. In most cases, cells in primary cultures can be removed from the culture dish and recultured repeatedly in so-called secondary cultures; in this way, they can be repeatedly subcultured *(passaged)* for weeks or months. Such cells often display many of the differentiated properties appropriate to their origin (**Figure 8–4**): fibroblasts continue to secrete collagen; cells derived from embryonic skeletal muscle fuse to form muscle fibers that contract spontaneously in the culture dish; nerve cells extend axons that are electrically excitable and make synapses with other nerve cells; and epithelial cells form extensive sheets with many of the properties of an intact epithelium. Because these properties are maintained in culture, they are accessible to study in ways that are often not possible in intact tissues.

Cell culture is not limited to animal cells. When a piece of plant tissue is cultured in a sterile medium containing nutrients and appropriate growth regulators, many of the cells are stimulated to proliferate indefinitely in a disorganized manner, producing a mass of relatively undifferentiated cells called a *callus*. If the nutrients and growth regulators are carefully manipulated, one can induce the formation of a shoot and then root apical meristems within the callus, and, in many species, regenerate a whole new plant. Similar to animal cells, callus cultures can be mechanically dissociated into single cells, which will grow and divide as a suspension culture (see Figure 8–4D).

Eucaryotic Cell Lines Are a Widely Used Source of Homogeneous Cells

The cell cultures obtained by disrupting tissues tend to suffer from a problem—eventually the cells die. Most vertebrate cells stop dividing after a finite number of cell divisions in culture, a process called *replicative cell senescence* (discussed in Chapter 17). Normal human fibroblasts, for example, typically divide only 25–40 times in culture before they stop. In these cells, the limited proliferation capacity reflects a progressive shortening and uncapping of the cell's telomeres, the repetitive DNA sequences and associated proteins that cap the ends of each chomosome (discussed in Chapter 5). Human somatic cells in the body have turned off production of the enzyme, called *telomerase,* that normally maintains the telomeres, which is why their telomeres shorten with each cell division. Human fibroblasts can often be coaxed to proliferate indefinitely by providing them with the gene that encodes the catalytic subunit of telomerase; in this case, they can be propagated as an "immortalized" cell line.

Some human cells, however, cannot be immortalized by this trick. Although their telomeres remain long, they still stop dividing after a limited number of divisions because the culture conditions eventually activate cell-cycle *checkpoint mechanisms* (discussed in Chapter 17) that arrest the cell cycle—a process sometimes called "culture shock." In order to immortalize these cells, one has to do more than introduce telomerase. One must also inactivate the checkpoint mechanisms. This can be done by introducing certain cancer-promoting oncogenes, such as those derived from tumor viruses (discussed in Chapter 20). Unlike human cells, most rodent cells do not turn off production of telomerase and therefore their telomeres do not shorten with each cell division. Therefore, if culture shock can be avoided, some rodent cell types will divide indefinitely in culture. In addition, rodent cells often undergo genetic changes in culture that inactivate their checkpoint mechanisms, thereby spontaneously producing immortalized cell lines.

Cell lines can often be most easily generated from cancer cells, but these cultures differ from those prepared from normal cells in several ways, and are referred to as *transformed cell lines.* Transformed cell lines often grow without attaching to a surface, for example, and they can proliferate to a much higher density in a culture dish. Similar properties can be induced experimentally in normal cells by transforming them with a tumor-inducing virus or chemical. The resulting transformed cell lines can usually cause tumors if injected into a susceptible animal (although it is usually only a small subpopulation, called cancer stem cells, that can do so—discussed in Chapter 20).

Both transformed and nontransformed cell lines are extremely useful in cell research as sources of very large numbers of cells of a uniform type, especially since they can be stored in liquid nitrogen at –196°C for an indefinite period and retain their viability when thawed. It is important to keep in mind, however, that the cells in both types of cell lines nearly always differ in important ways from their normal progenitors in the tissues from which they were derived.

Some widely used cell lines are listed in **Table 8–1**. Different lines have different advantages; for example, the PtK epithelial cell lines derived from the rat kangaroo, unlike many other cell lines which round up during mitosis, remain flat during mitosis, allowing the mitotic apparatus to be readily observed in action.

Embryonic Stem Cells Could Revolutionize Medicine

Among the most promising cell lines to be developed—from a medical point of view—are embryonic stem (ES) cells. These remarkable cells, first harvested from the inner cell mass of the early mouse embryo, can proliferate indefinitely

Table 8–1 Some Commonly Used Cell Lines

CELL LINE*	CELL TYPE AND ORIGIN
3T3	fibroblast (mouse)
BHK21	fibroblast (Syrian hamster)
MDCK	epithelial cell (dog)
HeLa	epithelial cell (human)
PtK1	epithelial cell (rat kangaroo)
L6	myoblast (rat)
PC12	chromaffin cell (rat)
SP2	plasma cell (mouse)
COS	kidney (monkey)
293	kidney (human); transformed with adenovirus
CHO	ovary (Chinese hamster)
DT40	lymphoma cell for efficient targeted recombination (chick)
R1	embryonic stem cell (mouse)
E14.1	embryonic stem cell (mouse)
H1, H9	embryonic stem cell (human)
S2	macrophage-like cell (*Drosophila*)
BY2	undifferentiated meristematic cell (tobacco)

*Many of these cell lines were derived from tumors. All of them are capable of indefinite replication in culture and express at least some of the special characteristics of their cell's of origin.

in culture and yet retain an unrestricted developmental potential. If the cells from the culture dish are put back into an early embryonic environment, they can give rise to all the cell types in the body, including germ cells (**Figure 8–5**). Their descendants in the embryo are able to integrate perfectly into whatever site they come to occupy, adopting the character and behavior that normal cells would show at that site.

Cells with properties similar to those of mouse ES cells can now be derived from early human embryos, creating a potentially inexhaustible supply of cells that might be used to replace and repair damaged mature human tissue. Experiments in mice suggest that it may be possible, in the future, to use ES cells to produce specialized cells for therapy—to replace the skeletal muscle fibers that degenerate in victims of muscular dystrophy, the nerve cells that die in patients with Parkinson's disease, the insulin-secreting cells that are destroyed in type I

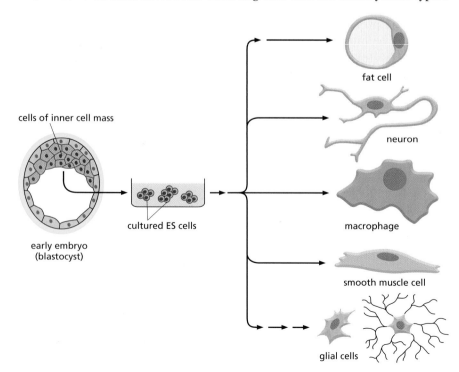

Figure 8–5 Embryonic stem (ES) cells derived from an embryo. These cultured cells can give rise to all of the cell types of the body. ES cells are harvested from the inner cell mass of an early embryo and can be maintained indefinitely as stem cells (discussed in Chapter 23) in culture. If they are put back into an embryo, they will integrate perfectly and differentiate to suit whatever environment they find themselves. The cells can also be kept in culture as an immortal cell line; they can then be supplied with different hormones or growth factors to encourage them to differentiate into specific cell types. (Based on E. Fuchs and J.A. Segré, *Cell* 100:143–155, 2000. With permission from Elsevier.)

diabetics, and the cardiac muscle cells that die during a heart attack. Perhaps one day it may even become possible to grow entire organs from ES cells by a recapitulation of embryonic development. It is important not to transplant ES cells by themselves into adults, as they can produce tumors called teratomas.

There is another major problem associated with the use of ES-cell-derived cells for tissue repair. If the transplanted cells differ genetically from the cells of the patient into whom they are grafted, the patient's immune system will reject and destroy those cells. This problem can be avoided, of course, if the cells used for repair are derived from the patient's own body. As discussed in Chapter 23, many adult tissues contain stem cells dedicated to continuous production of just one or a few specialized cell types, and a great deal of stem-cell research aims to manipulate the behavior of these adult stem cells for use in tissue repair.

ES cell technology, however, in theory at least, also offers another way around the problem of immune rejection, using a strategy known as "therapeutic cloning," as we explain next.

Somatic Cell Nuclear Transplantation May Provide a Way to Generate Personalized Stem Cells

The term "cloning" has been used in confusing ways as a shorthand term for several quite distinct types of procedures. It is important to understand the distinctions, particularly in the context of public debates about the ethics of stem cell research.

As biologists define the term, a *clone* is simply a set of individuals that are genetically identical because they have descended from a single ancestor. The simplest type of cloning is the cloning of cells. Thus, one can take a single epidermal stem cell from the skin and let it grow and divide in culture to obtain a large clone of genetically identical epidermal cells, which can, for example, be used to help reconstruct the skin of a badly burned patient. This kind of cloning is no more than an extension by artificial means of the processes of cell proliferation and repair that occur in a normal human body.

The cloning of entire multicellular animals, called *reproductive cloning,* is a very different enterprise, involving a far more radical departure from the ordinary course of nature. Normally, each individual animal has both a mother and a father, and is not genetically identical to either of them. In reproductive cloning, the need for two parents and sexual union is bypassed. For mammals, this difficult feat has been achieved in sheep and some other domestic animals by *somatic cell nuclear transplantation*. The procedure begins with an unfertilized egg cell. The nucleus of this haploid cell is sucked out and replaced by a nucleus from a regular diploid somatic cell. The diploid donor cell is typically taken from a tissue of an adult individual. The hybrid cell, consisting of a diploid donor nucleus in a host egg cytoplasm, is allowed to develop for a short while in culture. In a small proportion of cases, this procedure can give rise to an early embryo, which is then put into the uterus of a foster mother (**Figure 8–6**). If the experimenter is lucky, development continues like that of a normal embryo,

Figure 8–6 Reproductive and therapeutic cloning. Cells from adult tissue can be used for reproductive cloning or for generating personalized ES cells (so-called therapeutic cloning).

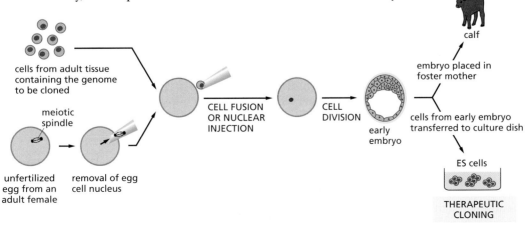

giving rise, eventually, to a whole new animal. An individual produced in this way, by reproductive cloning, should be genetically identical to the adult individual that donated the diploid cell (except for the small amount of genetic information in mitochondria, which is inherited solely from the egg cytoplasm).

Therapeutic cloning, which is very different from reproductive cloning, employs the technique of somatic cell nuclear transplantation to produce personalized ES cells (see Figure 8–6). In this case, the very early embryo generated by nuclear transplantation, consisting of about 200 cells, is not transferred to the uterus of a foster mother. Instead, it is used as a source from which ES cells are derived in culture, with the aim of generating various cell types that can be used for tissue repair. Cells obtained by this route are genetically nearly identical to the donor of the original nucleus, so they can be grafted back into the donor, without fear of immunological rejection.

Somatic cell nuclear transfer has an additional potential benefit—for studying inherited human diseases. ES cells that have received a somatic nucleus from an individual with an inherited disorder can be used to directly study the way in which the disease develops as the ES cells are induced to differentiate into distinct cell types. "Disease-specific" ES cells and their differentiated progeny can also be used to study the progression of such diseases and to test and develop new drugs to treat the disorders. These strategies are still in their infancy, and some countries outlaw certain aspects of the research. It remains to be seen whether human ES cells can be produced by nuclear transfer and whether human ES cells will fulfill the great hopes that medical scientists have for them.

Hybridoma Cell Lines Are Factories That Produce Monoclonal Antibodies

As we see throughout this book, antibodies are particularly useful tools for cell biology. Their great specificity allows precise visualization of selected proteins among the many thousands that each cell typically produces. Antibodies are often produced by inoculating animals with the protein of interest and subsequently isolating the antibodies specific to that protein from the serum of the animal. However, only limited quantities of antibodies can be obtained from a single inoculated animal, and the antibodies produced will be a heterogeneous mixture of antibodies that recognize a variety of different antigenic sites on a macromolecule that differs from animal to animal. Moreover, antibodies specific for the antigen will constitute only a fraction of the antibodies found in the serum. An alternative technology, which allows the production of an infinite quantity of identical antibodies and greatly increases the specificity and convenience of antibody-based methods, is the production of monoclonal antibodies by hybridoma cell lines.

This technology, developed in 1975, has revolutionized the production of antibodies for use as tools in cell biology, as well as for the diagnosis and treatment of certain diseases, including rheumatoid arthritis and cancer. The procedure requires hybrid cell technology (**Figure 8–7**), and it involves propagating a clone of cells from a single antibody-secreting B lymphocyte to obtain a homogeneous preparation of antibodies in large quantities. B lymphocytes normally have a limited life-span in culture, but individual antibody-producing B lymphocytes from an immunized mouse or rat, when fused with cells derived from a transformed B lymphocyte cell line, can give rise to hybrids that have both the ability to make a particular antibody and the ability to multiply indefinitely in culture. These **hybridomas** are propagated as individual clones, each of which provides a permanent and stable source of a single type of **monoclonal antibody** (**Figure 8–8**). Each type of monoclonal antibody recognizes a single type of antigenic site—for example, a particular cluster of five or six amino acid side chains on the surface of a protein. Their uniform specificity makes monoclonal antibodies much more useful than conventional antisera for most purposes.

An important advantage of the hybridoma technique is that monoclonal antibodies can be made against molecules that constitute only a minor component of a complex mixture. In an ordinary antiserum made against such a mix-

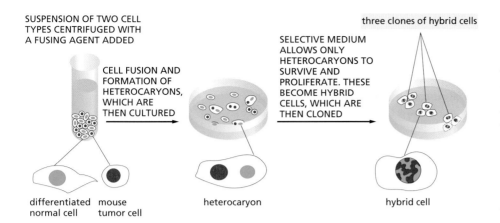

SUSPENSION OF TWO CELL TYPES CENTRIFUGED WITH A FUSING AGENT ADDED

CELL FUSION AND FORMATION OF HETEROCARYONS, WHICH ARE THEN CULTURED

three clones of hybrid cells

SELECTIVE MEDIUM ALLOWS ONLY HETEROCARYONS TO SURVIVE AND PROLIFERATE. THESE BECOME HYBRID CELLS, WHICH ARE THEN CLONED

differentiated normal cell   mouse tumor cell

heterocaryon

hybrid cell

Figure 8–7 The production of hybrid cells. It is possible to fuse one cell with another to form a *heterocaryon,* a combined cell with two separate nuclei. Typically, a suspension of cells is treated with certain inactivated viruses or with polyethylene glycol, each of which alters the plasma membranes of cells in a way that induces them to fuse. Eventually, a heterocaryon proceeds to mitosis and produces a hybrid cell in which the two separate nuclear envelopes have been disassembled, allowing all the chromosomes to be brought together in a single large nucleus. Such hybrid cells can give rise to immortal hybrid cell lines. If one of the parent cells was from a tumor cell line, the hybrid cell is called a hybridoma.

ture, the proportion of antibody molecules that recognize the minor component would be too small to be useful. But if the B lymphocytes that produce the various components of this antiserum are made into hybridomas, it becomes possible to screen individual hybridoma clones from the large mixture to select one that produces the desired type of monoclonal antibody and to propagate the selected hybridoma indefinitely so as to produce that antibody in unlimited quantities. In principle, therefore, a monoclonal antibody can be made against any protein in a biological sample. Once an antibody has been made, it can be used to localize the protein in cells and tissues, to follow its movement, and to purify the protein to study its structure and function.

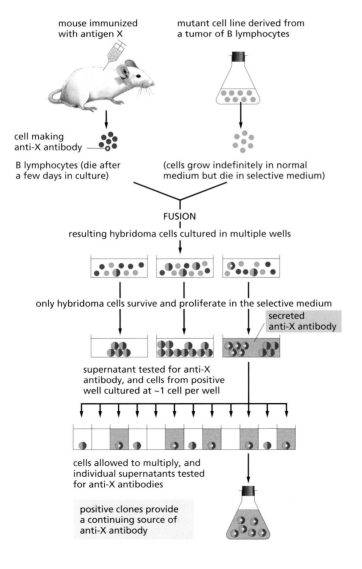

mouse immunized with antigen X

mutant cell line derived from a tumor of B lymphocytes

cell making anti-X antibody

B lymphocytes (die after a few days in culture)

(cells grow indefinitely in normal medium but die in selective medium)

FUSION

resulting hybridoma cells cultured in multiple wells

only hybridoma cells survive and proliferate in the selective medium

secreted anti-X antibody

supernatant tested for anti-X antibody, and cells from positive well cultured at ~1 cell per well

cells allowed to multiply, and individual supernatants tested for anti-X antibodies

positive clones provide a continuing source of anti-X antibody

Figure 8–8 Preparation of hybridomas that secrete monoclonal antibodies against a particular antigen. Here, the antigen of interest is designated as "antigen X." The selective growth medium used after the cell fusion step contains an inhibitor (aminopterin) that blocks the normal biosynthetic pathways by which nucleotides are made. The cells must therefore use a bypass pathway to synthesize their nucleic acids. This pathway is defective in the mutant cell line derived from the B cell tumor, but it is intact in the normal cells obtained from the immunized mouse. Because neither cell type used for the initial fusion can survive and proliferate on its own, only the hybridoma cells do so.

Summary

Tissues can be dissociated into their component cells, from which individual cell types can be purified and used for biochemical analysis or for the establishment of cell cultures. Many animal and plant cells survive and proliferate in a culture dish if they are provided with a suitable culture medium containing nutrients and appropriate signal molecules. Although many animal cells stop dividing after a finite number of cell divisions, cells that have been immortalized through spontaneous mutations or genetic manipulation can be maintained indefinitely as cell lines. Embryonic stem cells can proliferate indefinitely in a culture dish, while retaining the ability to differentiate into all the different cell types of the body. They therefore hold great medical promise. Hybridoma cells are widely employed to produce unlimited quantities of uniform monoclonal antibodies, which are used to detect and purify cell proteins, as well as to diagnose and treat diseases.

PURIFYING PROTEINS

The challenge of isolating a single type of protein from the thousands of other proteins present in a cell is a formidable one, but must be overcome in order to study protein function *in vitro*. As we shall see later in this chapter, *recombinant DNA technology* can enormously simplify this task by "tricking" cells into producing large quantities of a given protein, thereby making its purification much easier. Whether the source of the protein is an engineered cell or a natural tissue, a purification procedure usually starts with subcellular fractionation to reduce the complexity of the material, and is then followed by purification steps of increasing specificity.

Cells Can Be Separated into Their Component Fractions

In order to purify a protein, it must first be extracted from inside the cell. Cells can be broken up in various ways: they can be subjected to osmotic shock or ultrasonic vibration, forced through a small orifice, or ground up in a blender. These procedures break many of the membranes of the cell (including the plasma membrane and endoplasmic reticulum) into fragments that immediately reseal to form small closed vesicles. If carefully carried out, however, the disruption procedures leave organelles such as nuclei, mitochondria, the Golgi apparatus, lysosomes, and peroxisomes largely intact. The suspension of cells is thereby reduced to a thick slurry (called a *homogenate* or *extract*) that contains a variety of membrane-enclosed organelles, each with a distinctive size, charge, and density. Provided that the homogenization medium has been carefully chosen (by trial and error for each organelle), the various components—including the vesicles derived from the endoplasmic reticulum, called microsomes—retain most of their original biochemical properties.

The different components of the homogenate must then be separated. Such cell fractionations became possible only after the commercial development in the early 1940s of an instrument known as the *preparative ultracentrifuge*, which rotates extracts of broken cells at high speeds (**Figure 8–9**). This treatment separates cell components by size and density: in general, the largest units experience the largest centrifugal force and move the most rapidly. At relatively low speed, large components such as nuclei sediment to form a pellet at the bottom of the centrifuge tube; at slightly higher speed, a pellet of mitochondria is deposited; and at even higher speeds and with longer periods of centrifugation, first the small closed vesicles and then the ribosomes can be collected (**Figure 8–10**). All of these fractions are impure, but many of the contaminants can be removed by resuspending the pellet and repeating the centrifugation procedure several times.

Centrifugation is the first step in most fractionations, but it separates only components that differ greatly in size. A finer degree of separation can be achieved by layering the homogenate in a thin band on top of a dilute salt solution that fills

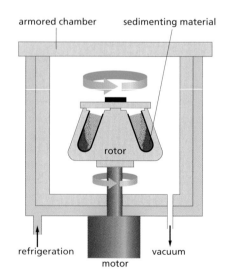

Figure 8–9 **The preparative ultracentrifuge.** The sample is contained in tubes that are inserted into a ring of cylindrical holes in a metal *rotor*. Rapid rotation of the rotor generates enormous centrifugal forces, which cause particles in the sample to sediment. The vacuum reduces friction, preventing heating of the rotor and allowing the refrigeration system to maintain the sample at 4°C.

Figure 8–10 Cell fractionation by centrifugation. Repeated centrifugation at progressively higher speeds will fractionate homogenates of cells into their components. In general, the smaller the subcellular component, the greater is the centrifugal force required to sediment it. Typical values for the various centrifugation steps referred to in the figure are:
low speed: 1000 times gravity for 10 minutes
medium speed: 20,000 times gravity for 20 minutes
high speed: 80,000 times gravity for 1 hour
very high speed: 150,000 times gravity for 3 hours

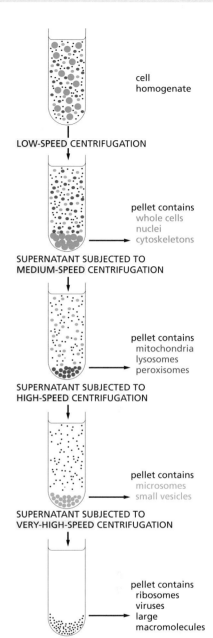

cell homogenate

LOW-SPEED CENTRIFUGATION

pellet contains
whole cells
nuclei
cytoskeletons

SUPERNATANT SUBJECTED TO
MEDIUM-SPEED CENTRIFUGATION

pellet contains
mitochondria
lysosomes
peroxisomes

SUPERNATANT SUBJECTED TO
HIGH-SPEED CENTRIFUGATION

pellet contains
microsomes
small vesicles

SUPERNATANT SUBJECTED TO
VERY-HIGH-SPEED CENTRIFUGATION

pellet contains
ribosomes
viruses
large
macromolecules

a centrifuge tube. When centrifuged, the various components in the mixture move as a series of distinct bands through the salt solution, each at a different rate, in a process called *velocity sedimentation* (**Figure 8–11**A). For the procedure to work effectively, the bands must be protected from convective mixing, which would normally occur whenever a denser solution (for example, one containing organelles) finds itself on top of a lighter one (the salt solution). This is achieved by augmenting the solution in the tube with a shallow gradient of sucrose prepared by a special mixing device. The resulting density gradient—with the dense end at the bottom of the tube—keeps each region of the salt solution denser than any solution above it, and it thereby prevents convective mixing from distorting the separation.

When sedimented through such dilute sucrose gradients, different cell components separate into distinct bands that can be collected individually. The relative rate at which each component sediments depends primarily on its size and shape—normally being described in terms of its sedimentation coefficient, or S value. Present-day ultracentrifuges rotate at speeds of up to 80,000 rpm and produce forces as high as 500,000 times gravity. These enormous forces drive even small macromolecules, such as tRNA molecules and simple enzymes, to sediment at an appreciable rate and allow them to be separated from one another by size.

The ultracentrifuge is also used to separate cell components on the basis of their buoyant density, independently of their size and shape. In this case the sample is sedimented through a steep density gradient that contains a very high concentration of sucrose or cesium chloride. Each cell component begins to move down the gradient as in Figure 8–11A, but it eventually reaches a position where the density of the solution is equal to its own density. At this point the component floats and can move no farther. A series of distinct bands is thereby produced in the centrifuge tube, with the bands closest to the bottom of the tube containing the components of highest buoyant density (Figure 8–11B). This method, called *equilibrium sedimentation*, is so sensitive that it can separate macromolecules that have incorporated heavy isotopes, such as ^{13}C or ^{15}N, from the same macromolecules that contain the lighter, common isotopes (^{12}C or ^{14}N). In fact, the cesium-chloride method was developed in 1957 to separate the labeled from the unlabeled DNA produced after exposure of a growing population of bacteria to nucleotide precursors containing ^{15}N; this classic experiment provided direct evidence for the semiconservative replication of DNA (see Figure 5–5).

Cell Extracts Provide Accessible Systems to Study Cell Functions

Studies of organelles and other large subcellular components isolated in the ultracentrifuge have contributed enormously to our understanding of the functions of different cell components. Experiments on mitochondria and chloroplasts purified by centrifugation, for example, demonstrated the central function of these organelles in converting energy into forms that the cell can use. Similarly, resealed vesicles formed from fragments of rough and smooth endoplasmic reticulum (microsomes) have been separated from each other and analyzed as functional models of these compartments of the intact cell.

Similarly, highly concentrated cell extracts, especially extracts of *Xenopus laevis* (African clawed frog) oocytes, have played a critical role in the study of

(A)

VELOCITY SEDIMENTATION

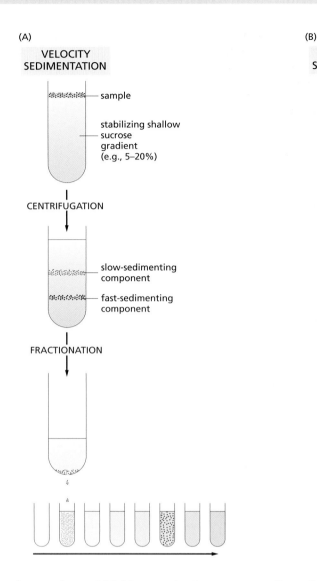

sample

stabilizing shallow sucrose gradient (e.g., 5–20%)

CENTRIFUGATION

slow-sedimenting component

fast-sedimenting component

FRACTIONATION

(B)

EQUILIBRIUM SEDIMENTATION

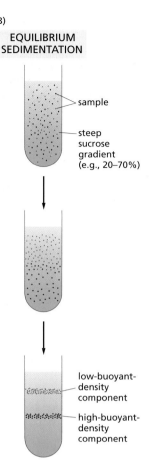

sample

steep sucrose gradient (e.g., 20–70%)

low-buoyant-density component

high-buoyant-density component

Figure 8–11 Comparison of velocity sedimentation and equilibrium sedimentation. (A) In velocity sedimentation, subcellular components sediment at different speeds according to their size and shape when layered over a dilute solution containing sucrose. To stabilize the sedimenting bands against convective mixing caused by small differences in temperature or solute concentration, the tube contains a continuous shallow gradient of sucrose, which increases in concentration toward the bottom of the tube (typically from 5% to 20% sucrose). After centrifugation, the different components can be collected individually, most simply by puncturing the plastic centrifuge tube and collecting drops from the bottom, as illustrated here. (B) In equilibrium sedimentation, subcellular components move up or down when centrifuged in a gradient until they reach a position where their density matches their surroundings. Although a sucrose gradient is shown here, denser gradients, which are especially useful for protein and nucleic acid separation, can be formed from cesium chloride. The final bands, at equilibrium, can be collected as in (A).

such complex and highly organized processes as the cell-division cycle, the separation of chromosomes on the mitotic spindle, and the vesicular-transport steps involved in the movement of proteins from the endoplasmic reticulum through the Golgi apparatus to the plasma membrane.

Cell extracts also provide, in principle, the starting material for the complete separation of all of the individual macromolecular components of the cell. We now consider how this separation is achieved, focusing on proteins.

Proteins Can Be Separated by Chromatography

Proteins are most often fractionated by **column chromatography**, in which a mixture of proteins in solution is passed through a column containing a porous solid matrix. The different proteins are retarded to different extents by their interaction with the matrix, and they can be collected separately as they flow out of the bottom of the column (**Figure 8–12**). Depending on the choice of matrix, proteins can be separated according to their charge *(ion-exchange chromatography)*, their hydrophobicity *(hydrophobic chromatography)*, their size *(gel-filtration chromatography)*, or their ability to bind to particular small molecules or to other macromolecules *(affinity chromatography)*.

Many types of matrices are commercially available (**Figure 8–13**). Ion-exchange columns are packed with small beads that carry either a positive or a negative charge, so that proteins are fractionated according to the arrangement of charges on their surface. Hydrophobic columns are packed with beads from which hydrophobic side chains protrude, selectively retarding proteins with exposed hydrophobic regions. Gel-filtration columns, which separate proteins according to their size, are packed with tiny porous beads: molecules that are small enough to enter the pores linger inside successive beads as they pass down the column, while larger molecules remain in the solution flowing between the beads and therefore move more rapidly, emerging from the column first. Besides providing a means of separating molecules, gel-filtration chromatography is a convenient way to determine their size.

Inhomogeneities in the matrices (such as cellulose), which cause an uneven flow of solvent through the column, limit the resolution of conventional column chromatography. Special chromatography resins (usually silica-based) composed of tiny spheres (3–10 μm in diameter) can be packed with a special apparatus to form a uniform column bed. Such **high-performance liquid chromatography** (**HPLC**) columns attain a high degree of resolution. In HPLC, the solutes equilibrate very rapidly with the interior of the tiny spheres, and so solutes with different affinities for the matrix are efficiently separated from one another even at very fast flow rates. HPLC is therefore the method of choice for separating many proteins and small molecules.

Affinity Chromatography Exploits Specific Binding Sites on Proteins

If one starts with a complex mixture of proteins, the types of column chromatography just discussed do not produce very highly purified fractions: a single passage through the column generally increases the proportion of a given protein in the mixture no more than twentyfold. Because most individual proteins represent less than 1/1000 of the total cell protein, it is usually necessary to use several different types of columns in succession to attain sufficient purity (**Figure 8–14**). A more efficient procedure, known as **affinity chromatography**, takes advantage of the biologically important binding interactions that occur on protein surfaces. If a substrate molecule is covalently coupled to an inert matrix such as a polysaccharide bead, the enzyme that operates on that substrate will often be specifically retained by the matrix and can then be eluted (washed out) in nearly pure form. Likewise, short DNA oligonucleotides of a specifically

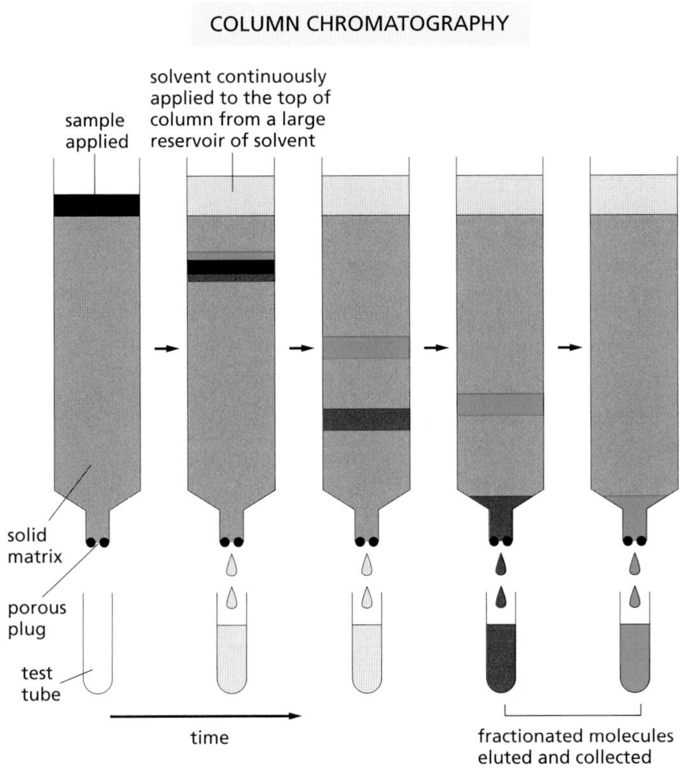

Figure 8–12 **The separation of molecules by column chromatography.** The sample, a solution containing a mixture of different molecules, is applied to the top of a cylindrical glass or plastic column filled with a permeable solid matrix, such as cellulose. A large amount of solvent is then pumped slowly through the column and collected in separate tubes as it emerges from the bottom. Because various components of the sample travel at different rates through the column, they are fractionated into different tubes.

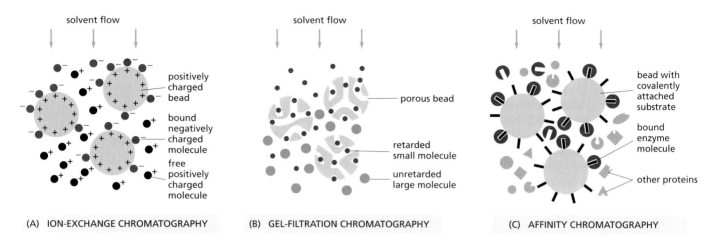

Figure 8–13 Three types of matrices used for chromatography. (A) In ion-exchange chromatography, the insoluble matrix carries ionic charges that retard the movement of molecules of opposite charge. Matrices used for separating proteins include diethylaminoethylcellulose (DEAE-cellulose), which is positively charged, and carboxymethylcellulose (CM-cellulose) and phosphocellulose, which are negatively charged. Analogous matrices based on agarose or other polymers are also frequently used. The strength of the association between the dissolved molecules and the ion-exchange matrix depends on both the ionic strength and the pH of the solution that is passing down the column, which may therefore be varied systematically (as in Figure 8–14) to achieve an effective separation. (B) In gel-filtration chromatography, the matrix is inert but porous. Molecules that are small enough to penetrate into the matrix are thereby delayed and travel more slowly through the column than larger molecules that cannot penetrate. Beads of cross-linked polysaccharide (dextran, agarose, or acrylamide) are available commercially in a wide range of pore sizes, making them suitable for the fractionation of molecules of various molecular weights, from less than 500 daltons to more than 5×10^6 daltons. (C) Affinity chromatography uses an insoluble matrix that is covalently linked to a specific ligand, such as an antibody molecule or an enzyme substrate, that will bind a specific protein. Enzyme molecules that bind to immobilized substrates on such columns can be eluted with a concentrated solution of the free form of the substrate molecule, while molecules that bind to immobilized antibodies can be eluted by dissociating the antibody–antigen complex with concentrated salt solutions or solutions of high or low pH. High degrees of purification can be achieved in a single pass through an affinity column.

designed sequence can be immobilized in this way and used to purify DNA-binding proteins that normally recognize this sequence of nucleotides in chromosomes (see Figure 7–28). Alternatively, specific antibodies can be coupled to a matrix to purify protein molecules recognized by the antibodies. Because of the great specificity of all such affinity columns, 1000- to 10,000-fold purifications can sometimes be achieved in a single pass.

Genetically-Engineered Tags Provide an Easy Way to Purify Proteins

Using the recombinant DNA methods discussed in subsequent sections, any gene can be modified to produce its protein with a special recognition tag attached to it, so as to make subsequent purification of the protein by affinity chromatography simple and rapid. Often the recognition tag is itself an antigenic determinant, or *epitope*, which can be recognized by a highly specific antibody. The antibody, can then be used both to localize the protein in cells and to purify it (**Figure 8–15**). Other types of tags are specifically designed for protein purification. For example, the amino acid histidine binds to certain metal ions, including nickel and copper. If genetic engineering techniques are used to attach a short string of histidines to one end of a protein, the slightly modified protein can be retained selectively on an affinity column containing immobilized nickel ions. Metal affinity chromatography can thereby be used to purify the modified protein from a complex molecular mixture.

In other cases, an entire protein is used as the recognition tag. When cells are engineered to synthesize the small enzyme glutathione S-transferase (GST) attached to a protein of interest, the resulting **fusion protein** can be purified from the other contents of the cell with an affinity column containing glutathione, a

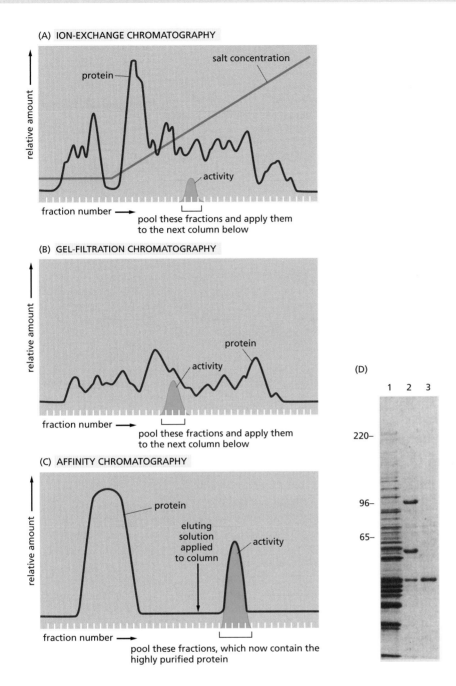

Figure 8–14 Protein purification by chromatography. Typical results obtained when three different chromatographic steps are used in succession to purify a protein. In this example, a homogenate of cells was first fractionated by allowing it to percolate through an ion-exchange resin packed into a column (A). The column was washed to remove all unbound contaminants, and the bound proteins were then eluted by passing a solution containing a gradually increasing concentration of salt onto the top of the column. Proteins with the lowest affinity for the ion-exchange resin passed directly through the column and were collected in the earliest fractions eluted from the bottom of the column. The remaining proteins were eluted in sequence according to their affinity for the resin—those proteins binding most tightly to the resin requiring the highest concentration of salt to remove them. The protein of interest was eluted in several fractions and was detected by its enzymatic activity. The fractions with activity were pooled and then applied to a second, gel-filtration column (B). The elution position of the still-impure protein was again determined by its enzymatic activity, and the active fractions were pooled and purified to homogeneity on an affinity column (C) that contained an immobilized substrate of the enzyme. (D) Affinity purification of cyclin-binding proteins from *S. cerevisiae*, as analyzed by SDS polyacrylamide-gel electrophoresis, which is described below in Figure 8–18. Lane 1 is a total cell extract; lane 2 shows the proteins eluted from an affinity column containing cyclin B2; lane 3 shows one major protein eluted from a cyclin B3 affinity column. Proteins in lanes 2 and 3 were eluted from the affinity columns with salt, and the gel was stained with Coomassie blue. The scale at the left shows the molecular weights of marker proteins, in kilodaltons. (D, from D. Kellogg et al., *J. Cell Biol.* 130:675–685, 1995. With permisison from The Rockefeller University Press.)

substrate molecule that binds specifically and tightly to GST. If the purification is carried out under conditions that do not disrupt protein–protein interactions, the fusion protein can be isolated in association with the proteins it interacts with inside the cell (**Figure 8–16**).

As a further refinement of purification methods using recognition tags, an amino acid sequence that forms a cleavage site for a highly specific proteolytic enzyme can be engineered between the protein of choice and the recognition tag. Because the amino acid sequences at the cleavage site are very rarely found by chance in proteins, the tag can later be cleaved off without destroying the purified protein.

This type of specific cleavage is used in an especially powerful purification methodology known as *tandem affinity purification tagging (tap-tagging)*. Here, one end of a protein is engineered to contain two recognition tags that are separated by a protease cleavage site. The tag on the very end of the construct is chosen to bind irreversibly to an affinity column, allowing the column to be washed extensively to remove all contaminating proteins. Protease cleavage then releases the protein, which is then further purified using the second tag.

Figure 8–15 **Epitope tagging for the localization or purification of proteins.** Using standard genetic engineering techniques, a short peptide tag can be added to a protein of interest. If the tag is itself an antigenic determinant, or *epitope*, it can be targeted by an appropriate commercially available antibody. The antibody, suitably labeled, can be used to determine the location of the protein in cells or to purify it by immunoprecipitation or affinity chromatography. In immunoprecipitation, antibodies directed against the epitope tag are added to a solution containing the tagged protein; the antibodies specifically cross-link the tagged protein molecules and precipitate them out of solution as antibody–protein complexes.

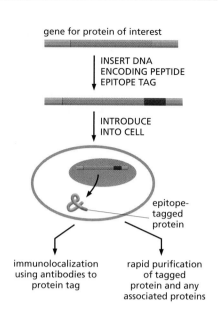

Because this two-step strategy provides an especially high degree of protein purification with relatively little effort, it is used extensively in cell biology. Thus, for example, a set of approximately 6000 yeast strains, each with a different gene fused to DNA that encodes a tap-tag, has been constructed to allow any yeast protein to be rapidly purified.

Purified Cell-free Systems Are Required for the Precise Dissection of Molecular Functions

It is important to study biological processes free from all of the complex side reactions that occur in a living cell by using **purified cell-free systems**. To make this possible, cell homogenates are fractionated with the aim of purifying each of the individual macromolecules that are needed to catalyze a biological process of interest. For example, the experiments to decipher the mechanisms of protein synthesis began with a cell homogenate that could translate RNA molecules to produce proteins. Fractionation of this homogenate, step by step, produced in turn the ribosomes, tRNAs, and various enzymes that together constitute the protein-synthetic machinery. Once individual pure components were available, each could be added or withheld separately to define its exact role in the overall process.

A major goal for cell biologists is the reconstitution of every biological process in a purified cell-free system. Only in this way can one define all of the components needed for the process and control their concentrations, as required to work out their precise mechanism of action. Although much remains to be done, a great deal of what we know today about the molecular biology of the cell has been discovered by studies in such cell-free systems. They have been used, for example, to decipher the molecular details of DNA replication and DNA transcription, RNA splicing, protein translation, muscle contraction, and particle transport along microtubules, and many other processes that occur in cells.

Summary

Populations of cells can be analyzed biochemically by disrupting them and fractionating their contents, allowing functional cell-free systems to be developed. Highly purified cell-free systems are needed for determining the molecular details of complex cell processes, requiring extensive purification of all the proteins and other components involved. The proteins in soluble cell extracts can be purified by column

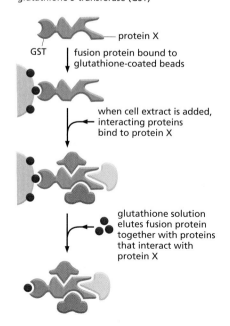

Figure 8–16 **Purification of protein complexes by using a GST-tagged fusion protein.** GST fusion proteins, produced in cells with recombinant DNA techniques, can be captured on an affinity column containing beads coated with glutathione. Proteins not bound to the beads are washed away. The fusion protein, along with other proteins in the cell that are bound tightly to it, can then be eluted with glutathione. The identities of these additional proteins can be determined by mass spectrometry (see Figure 8–21). Affinity columns can also be made to contain antibodies against GST or another convenient small protein or epitope tag (see Figure 8–15).

chromatography; depending on the type of column matrix, biologically active proteins can be separated on the basis of their molecular weight, hydrophobicity, charge characteristics, or affinity for other molecules. In a typical purification, the sample is passed through several different columns in turn—the enriched fractions obtained from one column are applied to the next. Recombinant DNA techniques, to be described later, allow special recognition tags to be attached to proteins, thereby greatly simplifying their purification.

ANALYZING PROTEINS

Proteins perform most processes in cells: they catalyze metabolic reactions, use nucleotide hydrolysis to do mechanical work, and serve as the major structural elements of the cell. The great variety of protein structures and functions has stimulated the development of a multitude of techniques to study them.

Proteins Can Be Separated by SDS Polyacrylamide-Gel Electrophoresis

Proteins usually possess a net positive or negative charge, depending on the mixture of charged amino acids they contain. An electric field applied to a solution containing a protein molecule causes the protein to migrate at a rate that depends on its net charge and on its size and shape. The most popular application of this property is **SDS polyacrylamide-gel electrophoresis (SDS-PAGE)**. It uses a highly cross-linked gel of polyacrylamide as the inert matrix through which the proteins migrate. The gel is prepared by polymerization of monomers; the pore size of the gel can be adjusted so that it is small enough to retard the migration of the protein molecules of interest. The proteins themselves are not in a simple aqueous solution but in one that includes a powerful negatively charged detergent, sodium dodecyl sulfate, or SDS (**Figure 8–17**). Because this detergent binds to hydrophobic regions of the protein molecules, causing them to unfold into extended polypeptide chains, the individual protein molecules are released from their associations with other proteins or lipid molecules and rendered freely soluble in the detergent solution. In addition, a reducing agent such as β-mercaptoethanol (see Figure 8–17) is usually added to break any S–S linkages in the proteins, so that all of the constituent polypeptides in multisubunit proteins can be analyzed separately.

What happens when a mixture of SDS-solubilized proteins is run through a slab of polyacrylamide gel? Each protein molecule binds large numbers of the negatively charged detergent molecules, which mask the protein's intrinsic charge and cause it to migrate toward the positive electrode when a voltage is applied. Proteins of the same size tend to move through the gel with similar speeds because (1) their native structure is completely unfolded by the SDS, so that their shapes are the same, and (2) they bind the same amount of SDS and therefore have the same amount of negative charge. Larger proteins, with more charge, are subjected to larger electrical forces and also to a larger drag. In free solution, the two effects would cancel out, but, in the mesh of the polyacrylamide gel, which acts as a molecular sieve, large proteins are retarded much more than small ones. As a result, a complex mixture of proteins is fractionated into a series of discrete protein bands arranged in order of molecular weight (**Figure 8–18**). The major proteins are readily detected by staining the proteins in the gel with a dye such as Coomassie blue. Even minor proteins are seen in gels treated with a silver or gold stain, so that as little as 10 ng of protein can be detected in a band.

SDS-PAGE is widely used because it can separate all types of proteins, including those that are normally insoluble in water—such as the many proteins in membranes. And because the method separates polypeptides by size, it provides information about the molecular weight and the subunit composition of proteins. **Figure 8–19** presents a photograph of a gel that has been used to analyze each of the successive stages in the purification of a protein.

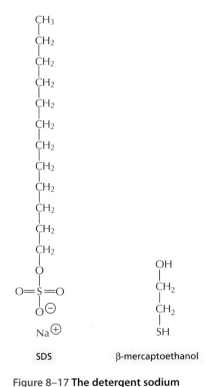

Figure 8–17 The detergent sodium dodecyl sulfate (SDS) and the reducing agent β-mercaptoethanol. These two chemicals are used to solubilize proteins for SDS polyacrylamide-gel electrophoresis. The SDS is shown here in its ionized form.

(A)

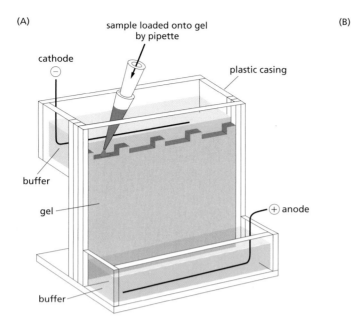

(B)

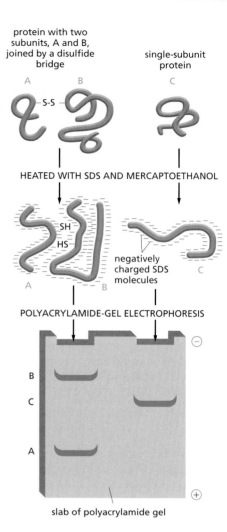

Figure 8–18 SDS polyacrylamide-gel electrophoresis (SDS-PAGE). (A) An electrophoresis apparatus. (B) Individual polypeptide chains form a complex with negatively charged molecules of sodium dodecyl sulfate (SDS) and therefore migrate as a negatively charged SDS–protein complex through a porous gel of polyacrylamide. Because the speed of migration under these conditions is greater the smaller the polypeptide, this technique can be used to determine the approximate molecular weight of a polypeptide chain as well as the subunit composition of a protein. If the protein contains a large amount of carbohydrate, however, it will move anomalously on the gel and its apparent molecular weight estimated by SDS-PAGE will be misleading.

Specific Proteins Can Be Detected by Blotting with Antibodies

A specific protein can be identified after its fractionation on a polyacrylamide gel by exposing all the proteins present on the gel to a specific antibody that has been coupled to a radioactive isotope, to an easily detectable enzyme, or to a fluorescent dye. For convenience, this procedure is normally carried out after transferring (by "blotting") all of the separated proteins present in the gel onto a sheet of nitrocellulose paper or nylon membrane. Placing the membrane over the gel and driving the proteins out of the gel with a strong electric field transfers the protein onto the membrane. The membrane is then soaked in a solution of labeled antibody to reveal the protein of interest. This method of detecting proteins is called **Western blotting**, or **immunoblotting** (**Figure 8–20**).

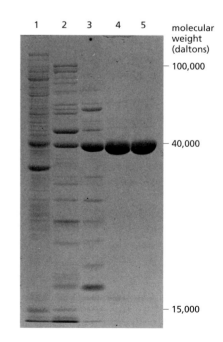

Figure 8–19 Analysis of protein samples by SDS polyacrylamide-gel electrophoresis. The photograph shows a Coomassie-stained gel that has been used to detect the proteins present at successive stages in the purification of an enzyme. The leftmost lane (lane 1) contains the complex mixture of proteins in the starting cell extract, and each succeeding lane analyzes the proteins obtained after a chromatographic fractionation of the protein sample analyzed in the previous lane (see Figure 8–14). The same total amount of protein (10 µg) was loaded onto the gel at the top of each lane. Individual proteins normally appear as sharp, dye-stained bands; a band broadens, however, when it contains too much protein. (From T. Formosa and B.M. Alberts, *J. Biol. Chem.* 261:6107–6118, 1986.)

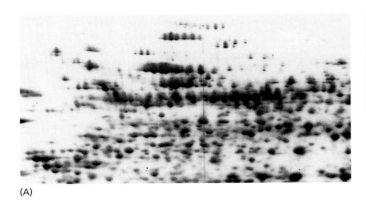

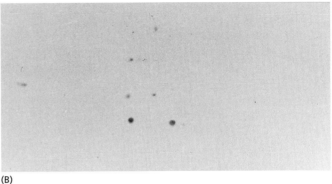

(A)   (B)

Figure 8–20 **Western blotting.** All the proteins from dividing tobacco cells in culture are first separated by two-dimensional polyacrylamide-gel electrophoresis (described in Figure 8–23). In (A), the positions of the proteins are revealed by a sensitive protein stain. In (B), the separated proteins on an identical gel were then transferred to a sheet of nitrocellulose and exposed to an antibody that recognizes only those proteins that are phosphorylated on threonine residues during mitosis. The positions of the dozen or so proteins that are recognized by this antibody are revealed by an enzyme-linked second antibody. This technique is also known as immunoblotting (or Western blotting). (From J.A. Traas et al., *Plant J.* 2:723–732, 1992. With permission from Blackwell Publishing.)

Mass Spectrometry Provides a Highly Sensitive Method for Identifying Unknown Proteins

A frequent problem in cell biology and biochemistry is the identification of a protein or collection of proteins that has been obtained by one of the purification procedures discussed in the preceding pages (see, for example, Figure 8–16). Because the genome sequences of most common experimental organisms are now known, catalogues of all the proteins produced in those organisms are available. The task of identifying an unknown protein (or collection of unknown proteins) thus reduces to matching some of the amino acid sequences present in the unknown sample with known catalogued genes. This task is now performed almost exclusively by using mass spectrometry in conjunction with computer searches of databases.

Charged particles have very precise dynamics when subjected to electrical and magnetic fields in a vacuum. Mass spectrometry exploits this principle to separate ions according to their mass-to-charge ratio. It is an enormously sensitive technique. It requires very little material and is capable of determining the precise mass of intact proteins and of peptides derived from them by enzymatic or chemical cleavage. Masses can be obtained with great accuracy, often with an error of less than one part in a million. The most commonly used form of the technique is called *matrix-assisted laser desorption ionization–time-of-flight spectrometry (MALDI-TOF)*. In this approach, the proteins in the sample are first broken into short peptides. These peptides are mixed with an organic acid and then dried onto a metal or ceramic slide. A laser then blasts the sample, ejecting the peptides from the slide in the form of an ionized gas, in which each molecule carries one or more positive charges. The ionized peptides are accelerated in an electric field and fly toward a detector. Their mass and charge determines the time it takes them to reach the detector: large peptides move more slowly, and more highly charged molecules move more quickly. By analyzing those ionized peptides that bear a single charge, the precise masses of peptides present in the original sample can be determined. MALDI-TOF can also be used to accurately measure the mass of intact proteins as large as 200,000 daltons. This information is then used to search genomic databases, in which the masses of all proteins and of all their predicted peptide fragments have been tabulated from the genomic sequences of the organism (**Figure 8–21**A). An unambiguous match to a particular open reading frame can often be made by knowing the mass of only a few peptides derived from a given protein.

MALDI-TOF provides accurate molecular weight measurements for proteins and peptides. Moreover, by employing two mass spectrometers in tandem (an arrangement known as MS/MS), it is possible to directly determine the

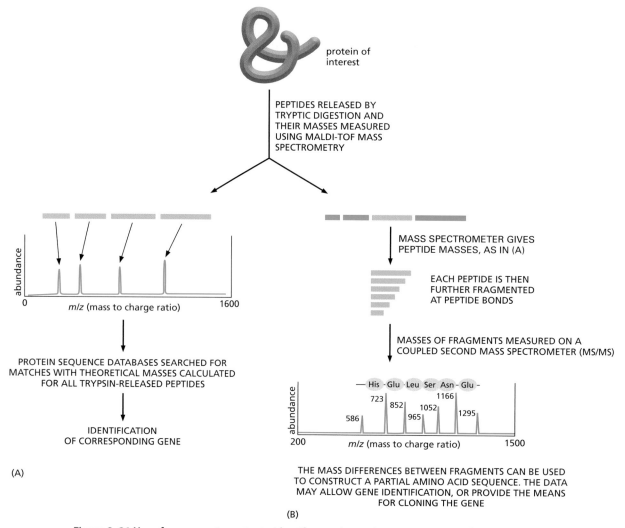

Figure 8–21 Use of mass spectrometry to identify proteins and to sequence peptides. An isolated protein is digested with trypsin and the peptide fragments are then loaded into the mass spectrometer. Two different approaches can then be used to identify the protein. (A) In the first method, peptide masses are measured precisely using MALDI-TOF mass spectrometry. Sequence databases are then searched to find the gene that encodes a protein whose calculated tryptic digest profile matches these values. (B) Mass spectrometry can also be used to determine directly the amino acid sequence of peptide fragments. In this example, tryptic peptides are first separated based on mass within a mass spectrometer. Each peptide is then further fragmented, primarily by cleaving its peptide bonds. This treatment generates a nested set of peptides, each differing in size by one amino acid. These fragments are fed into a second coupled mass spectrometer, and their masses are determined. The difference in masses between two closely related peptides can be used to deduce the "missing" amino acid. By repeated applications of this procedure, a partial amino acid sequence of the original protein can be determined. For simplicity, the analysis shown begins with a single species of purified protein. In reality, mass spectrometry is usually carried out on mixtures of proteins, such as those obtained for affinity chromatography experiments (see Figure 8–16), and can identify all the proteins present in the mixtures. As explained in the text, mass spectrometry can also detect post-translational modifications of proteins.

amino acid sequences of individual peptides in a complex mixture. As described above, the protein sample is first broken down into smaller peptides, which are separated from each other by mass spectrometry. Each peptide is then further fragmented through collisions with high-energy gas atoms. This method of fragmentation preferentially cleaves the peptide bonds, generating a ladder of fragments, each differing by a single amino acid. The second mass spectrometer then separates these fragments and displays their masses. The amino acid sequence of a peptide can then be deduced from these differences in mass (Figure 8–21B).

MS/MS is particularly useful for detecting and precisely mapping post-translational modifications of proteins, such as phosphorylations or acetylations. Because these modifications impart a characteristic mass increase to an amino acid, they are easily detected by mass spectrometry. As described in

Chapter 3, proteomics, a general term that encompasses many different experimental techniques, is the characterization of all proteins in the cell, including all protein–protein interactions and all post-translational modifications. In combination with the rapid purification techniques discussed in the last section, mass spectrometry has emerged as the most powerful method for mapping both the post-translational modifications of a given protein and the proteins that remain associated with it during purification.

Two-Dimensional Separation Methods are Especially Powerful

Because different proteins can have similar sizes, shapes, masses, and overall charges, most separation techniques such as SDS polyacrylamide-gel electrophoresis or ion-exchange chromatography cannot typically display all the proteins in a cell or even in an organelle. In contrast, **two-dimensional gel electrophoresis**, which combines two different separation procedures, can resolve up to 2000 proteins—the total number of different proteins in a simple bacterium—in the form of a two-dimensional protein map.

In the first step, the proteins are separated by their intrinsic charges. The sample is dissolved in a small volume of a solution containing a nonionic (uncharged) detergent, together with β-mercaptoethanol and the denaturing reagent urea. This solution solubilizes, denatures, and dissociates all the polypeptide chains but leaves their intrinsic charge unchanged. The polypeptide chains are then separated in a pH gradient by a procedure called *isoelectric focusing*, which takes advantage of the variation in the net charge on a protein molecule with the pH of its surrounding solution. Every protein has a characteristic isoelectric point, the pH at which the protein has no net charge and therefore does not migrate in an electric field. In isoelectric focusing, proteins are separated electrophoretically in a narrow tube of polyacrylamide gel in which a gradient of pH is established by a mixture of special buffers. Each protein moves to a position in the gradient that corresponds to its isoelectric point and remains there (**Figure 8–22**). This is the first dimension of two-dimensional polyacrylamide-gel electrophoresis.

In the second step, the narrow gel containing the separated proteins is again subjected to electrophoresis but in a direction that is at a right angle to the direction used in the first step. This time SDS is added, and the proteins separate according to their size, as in one-dimensional SDS-PAGE: the original narrow gel is soaked in SDS and then placed on one edge of an SDS polyacrylamide-gel slab, through which each polypeptide chain migrates to form a discrete spot. This is the second dimension of two-dimensional polyacrylamide-gel electrophoresis. The only proteins left unresolved are those that have both identical sizes and identical isoelectric points, a relatively rare situation. Even trace amounts of each polypeptide chain can be detected on the gel by various staining procedures—or by autoradiography if the protein sample was initially labeled with a radioisotope (**Figure 8–23**). The technique has such great resolving power that it can distinguish between two proteins that differ in only a single charged amino acid.

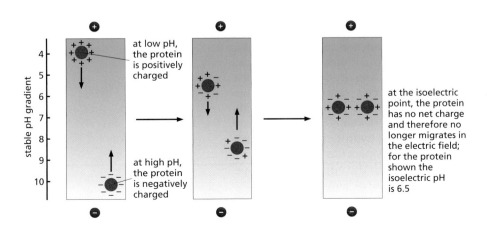

Figure 8–22 Separation of protein molecules by isoelectric focusing. At low pH (high H^+ concentration), the carboxylic acid groups of proteins tend to be uncharged (–COOH) and their nitrogen-containing basic groups fully charged (for example, $-NH_3^+$), giving most proteins a net positive charge. At high pH, the carboxylic acid groups are negatively charged (–COO$^-$) and the basic groups tend to be uncharged (for example, $-NH_2$), giving most proteins a net negative charge. At its *isoelectric pH*, a protein has no net charge since the positive and negative charges balance. Thus, when a tube containing a fixed pH gradient is subjected to a strong electric field in the appropriate direction, each protein species present migrates until it forms a sharp band at its isoelectric pH, as shown.

basic ← stable pH gradient → acidic

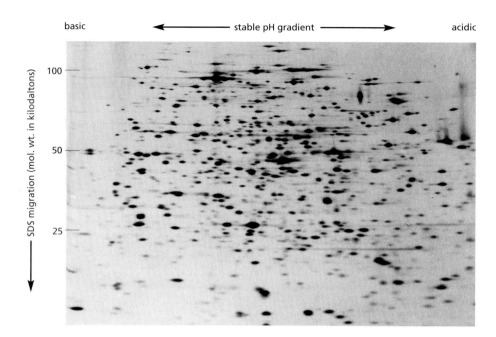

SDS migration (mol. wt. in kilodaltons)

100

50

25

Figure 8–23 Two-dimensional polyacrylamide-gel electrophoresis. All the proteins in an *E. coli* bacterial cell are separated in this gel, in which each spot corresponds to a different polypeptide chain. The proteins were first separated on the basis of their isoelectric points by isoelectric focusing from left to right. They were then further fractionated according to their molecular weights by electrophoresis from top to bottom in the presence of SDS. Note that different proteins are present in very different amounts. The bacteria were fed with a mixture of radioisotope-labeled amino acids so that all of their proteins were radioactive and could be detected by autoradiography (see pp. 602–603). (Courtesy of Patrick O'Farrell.)

A different, even more powerful, "two-dimensional" technique is now available when the aim is to determine all of the proteins present in an organelle or another complex mixture of proteins. Because the technique relies on mass spectroscopy, it requires that the proteins be from an organism with a completely sequenced genome. First, the mixture of proteins present is digested with trypsin to produce short peptides. Next, these peptides are separated by a series of automated liquid chromatography steps. As the second dimension, each separated peptide is fed directly into a tandem mass spectrometer (MS/MS) that allows its amino acid sequence, as well as any post-translational modifications, to be determined. This arrangement, in which a tandem mass spectrometer (MS/MS) is attached to the output of an automated liquid chromatography (LC) system, is referred to as LC-MS/MS. It is now becoming routine to subject an entire organelle preparation to LC-MS/MS analysis and to identify hundreds of proteins and their modifications. Of course, no organelle isolation procedure is perfect, and some of the proteins identified will be contaminating proteins. These can often be excluded by analyzing neighboring fractions from the organelle purification and "subtracting" them out from the peak organelle fractions.

Hydrodynamic Measurements Reveal the Size and Shape of a Protein Complex

Most proteins in a cell act as part of larger complexes, and knowledge of the size and shape of these complexes often leads to insights regarding their function. This information can be obtained in several important ways. Sometimes, a complex can be directly visualized using electron microscopy, as described in Chapter 9. A complementary approach relies on the hydrodynamic properties of a complex, that is, its behavior as it moves through a liquid medium. Usually, two separate measurements are made. One measure is the velocity of a complex as it moves under the influence of a centrifugal field produced by an ultracentrifuge (see Figure 8–11A). The sedimentation constant (or S-value) obtained depends on both the size and the shape of the complex and does not, by itself, convey especially useful information. However, once a second hydrodynamic measurement is performed—by charting the migration of a complex through a gel-filtration chromatography column (see Figure 8–13B)—both the approximate shape of a complex and its molecular weight can be calculated.

Molecular weight can also be determined more directly by using an analytical ultracentrifuge, a complex device that allows protein absorbance measurements to be made on a sample while it is subjected to centrifugal forces. In this approach, the sample is centrifuged until it reaches equilibrium, where the centrifugal force on a protein complex exactly balances its tendency to diffuse away. Because this balancing point is dependent on a complex's molecular weight but not on its particular shape, the molecular weight can be directly calculated, as needed to determine the stoichiometry of each protein in a protein complex.

Sets of Interacting Proteins Can Be Identified by Biochemical Methods

Because most proteins in the cell function as part of complexes with other proteins, an important way to begin to characterize the biological role of an unknown protein is to identify all of the other proteins to which it specifically binds.

One method for identifying proteins that bind to one another tightly is *co-immunoprecipitation*. In this case, an antibody recognizes a specific target protein; reagents that bind to the antibody and are coupled to a solid matrix then drag the complex out of solution to the bottom of a test tube. If the original target protein is associated tightly enough with another protein when it is captured by the antibody, the partner precipitates as well. This method is useful for identifying proteins that are part of a complex inside cells, including those that interact only transiently—for example, when extracellular signal molecules stimulate cells (discussed in Chapter 15). Another method frequently used to identify a protein's binding partners is protein affinity chromatography (see Figure 8–13C). To employ this technique to capture interacting proteins, a target protein is attached to polymer beads that are packed into a column. When the proteins in a cell extract are washed through this column, those proteins that interact with the target protein are retained by the affinity matrix. These proteins can then be eluted and their identity determined by mass spectrometry.

In addition to capturing protein complexes on columns or in test tubes, researchers are developing high-density protein arrays to investigate protein interactions. These arrays, which contain thousands of different proteins or antibodies spotted onto glass slides or immobilized in tiny wells, allow one to examine the biochemical activities and binding profiles of a large number of proteins at once. For example, if one incubates a fluorescently labeled protein with arrays containing thousands of immobilized proteins, the spots that remain fluorescent after extensive washing each contain a protein to which the labeled protein specifically binds.

Protein–Protein Interactions Can Also Be Identified by a Two-Hybrid Technique in Yeast

Thus far, we have emphasized biochemical approaches to the study of protein–protein interactions. However, a particularly powerful strategy, called the **two-hybrid system**, relies on exploiting the cell's own mechanisms to reveal protein–protein interactions.

The technique takes advantage of the modular nature of gene activator proteins (see Figure 7–45). These proteins both bind to specific DNA sequences and activate gene transcription, and these activities are often performed by two separate protein domains. Using recombinant DNA techniques, two such protein domains are used to create separate "bait" and "prey" fusion proteins. To create the "bait" fusion protein, the DNA sequence that codes for a target protein is fused with DNA that encodes the DNA-binding domain of a gene activator protein. When this construct is introduced into yeast, the cells produce the fusion protein, with the target protein attached to this DNA-binding domain (**Figure 8–24**). This fusion protein binds to the regulatory region of a reporter gene, where it serves as "bait" to fish for proteins that interact with the target protein. To search for potential binding partners (potential prey for the bait), the candidate proteins also have to be constructed as fusion proteins: DNA encoding the activation domain of a gene activator protein is fused to a large

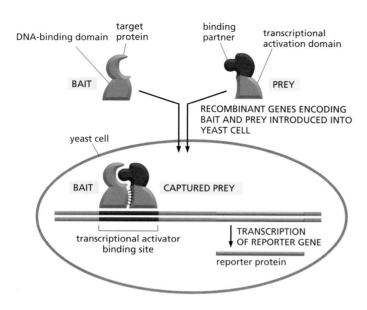

Figure 8–24 **The yeast two-hybrid system for detecting protein–protein interactions.** The target protein is fused to a DNA-binding domain that directs the fusion protein to the regulatory region of a reporter gene as "bait." When this target protein binds to another specially designed protein in the cell nucleus ("prey"), their interaction brings together two halves of a transcriptional activator, which then switches on the expression of the reporter gene.

number of different genes. Members of this collection of genes—encoding potential "prey"—are introduced individually into yeast cells containing the bait. If the yeast cell receives a DNA clone that expresses a prey partner for the bait protein, the two halves of a transcriptional activator are united, switching on the reporter gene (see Figure 8–24).

This ingenious technique sounds complex, but the two-hybrid system is relatively simple to use in the laboratory. Although the protein–protein interactions occur in the yeast cell nucleus, proteins from every part of the cell and from any organism can be studied in this way. The two-hybrid system has been scaled up to map the interactions that occur among all of the proteins an organism produces. In this case, a set of bait and prey fusions is produced for every cell protein, and every bait/prey combination can be monitored. In this way protein interaction maps have been generated for most of the proteins in yeast, *C. elegans,* and *Drosophila.*

Combining Data Derived from Different Techniques Produces Reliable Protein-Interaction Maps

As previously discussed in Chapter 3, extensive protein-interaction maps can be very useful for identifying the functions of proteins (see Figure 3–82). For this reason, both the two-hybrid method and the biochemical technique discussed earlier known as tap-tagging (see pp. 515–516) have been automated to determine the interactions between thousands of proteins. Unfortunately, different results are found in different experiments, and many of the interactions detected in one laboratory are not detected in another. Therefore, the most useful protein-interaction maps are those that combine data from many experiments, requiring that each interaction in the map be confirmed by more than one technique.

Optical Methods Can Monitor Protein Interactions in Real Time

Once two proteins—or a protein and a small molecule—are known to associate, it becomes important to characterize their interaction in more detail. Proteins can associate with each other more or less permanently (like the subunits of RNA polymerase or the proteosome), or engage in transient encounters that may last only a few milliseconds (like a protein kinase and its substrate).

To understand how a protein functions inside a cell, we need to determine how tightly it binds to other proteins, how rapidly it dissociates from them, and how covalent modifications, small molecules, or other proteins influence these interactions. Such studies of protein dynamics often employ optical methods.

Certain amino acids (for example, tryptophan) exhibit weak fluorescence that can be detected with sensitive fluorimeters. In many cases, the fluorescence intensity, or the emission spectrum of fluorescent amino acids located in a protein–protein interface, will change when the proteins associate. When this change can be detected by fluorimetry, it provides a sensitive and quantitative measure of protein binding.

A particularly useful method for monitoring the dynamics of a protein's binding to other molecules is called **surface plasmon resonance (SPR)**. The SPR method has been used to characterize a wide variety of molecular interactions, including antibody-antigen binding, ligand-receptor coupling, and the binding of proteins to DNA, carbohydrates, small molecules, and other proteins.

SPR detects binding interactions by monitoring the reflection of a beam of light off the interface between an aqueous solution of potential binding molecules and a biosensor surface carrying an immobilized bait protein. The bait protein is attached to a very thin layer of metal that coats one side of a glass prism (**Figure 8–25**). A light beam is passed through the prism; at a certain angle, called the *resonance angle*, some of the energy from the light interacts with the cloud of electrons in the metal film, generating a plasmon—an oscillation of the electrons at right angles to the plane of the film, bouncing up and down between its upper and lower surfaces like a weight on a spring. The plasmon, in turn, generates an electrical field that extends a short distance—about the wavelength of the light—above and below the metal surface. Any change in the composition of

(A)

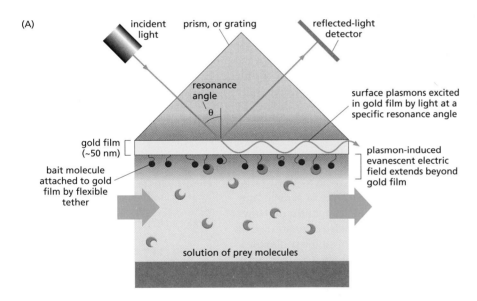

(B) The binding of prey molecules to bait molecules increases the refractive index of the surface layer. This alters the resonance angle for plasmon induction, which can be measured by a detector.

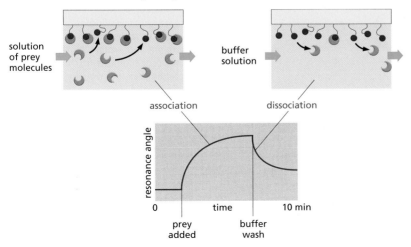

Figure 8–25 Surface plasmon resonance.
(A) SPR can detect binding interactions by monitoring the reflection of a beam of light off the interface between an aqueous solution of potential binding molecules *(green)* and a biosensor surface coated with an immobilized bait protein *(red)*.
(B) A solution of prey proteins is allowed to flow past the immobilized bait protein. Binding of prey molecules to the bait protein produces a measurable change in the resonance angle, as does their dissociation when a buffer solution washes them off. These changes, monitored in real time, reflect the association and dissociation of the molecular complexes.

the environment within the range of the electrical field will cause a measurable change in the resonance angle.

To measure binding, a solution containing proteins (or other molecules) that might interact with the immobilized bait protein is allowed to flow past the biosensor surface. Proteins binding to the bait change the composition of the molecular complexes on the metal surface, causing a change in the resonance angle (see Figure 8–25). The changes in the resonance angle are monitored in real time and reflect the kinetics of the association—or dissociation—of molecules with the bait protein. The association rate (k_{on}) is measured as the molecules interact, and the dissociation rate (k_{off}) is determined as buffer washes the bound molecules from the sensor surface. A binding constant (K) is calculated by dividing k_{off} by k_{on}. In addition to determining the kinetics, SPR can be used to determine the number of molecules that are bound in each complex: the magnitude of the SPR signal change is proportional to the mass of the immobilized complex.

The SPR method is particularly useful because it requires only small amounts of the protein, the protein does not have to be labeled in any way, and the interactions of the protein with other molecules can be monitored in real time.

A third optical method for probing protein interactions uses *green fluorescent protein* (discussed in detail below) and its derivatives of different colors. In this application, two proteins of interest are each labeled with a different fluorochrome, such that the emission spectrum of one fluorochrome overlaps the absorption spectrum of the second fluorochrome. If the two proteins—and their attached fluorochromes—come very close to each other (within about 1–10 nm), the energy of the absorbed light is transferred from one fluorochrome to the other. The energy transfer, called **fluorescence resonance energy transfer** (**FRET**), is determined by illuminating the first fluorochrome and measuring emission from the second (**Figure 8–26**). This technique is especially powerful because, when combined with fluorescence microscopy, it can be used to characterize protein-protein interactions at specific locations inside living cells.

Some Techniques Can Monitor Single Molecules

The biochemical methods described so far in this chapter are used to study large populations of molecules, a limitation that reflects the small size of typical biological molecules relative to the sensitivity of the methods to detect them. However, the recent development of highly sensitive and precise measurement methods has created a new branch of biophysics—the study of single molecules. Single-molecule studies are particularly important in cell biology because many processes rely on the activities of only a few critical molecules in the cell.

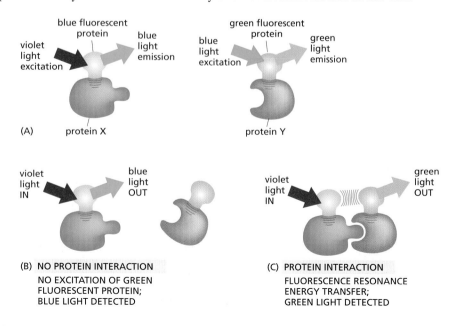

(A) protein X — blue fluorescent protein; violet light excitation; blue light emission — protein Y; green fluorescent protein; blue light excitation; green light emission

(B) NO PROTEIN INTERACTION
violet light IN; blue light OUT
NO EXCITATION OF GREEN FLUORESCENT PROTEIN; BLUE LIGHT DETECTED

(C) PROTEIN INTERACTION
violet light IN; green light OUT
FLUORESCENCE RESONANCE ENERGY TRANSFER; GREEN LIGHT DETECTED

Figure 8–26 Fluorescence resonance energy transfer (FRET). To determine whether (and when) two proteins interact inside a cell, the proteins are first produced as fusion proteins attached to different color variants of green fluorescent protein (GFP). (A) In this example, protein X is coupled to a blue fluorescent protein, which is excited by violet light (370–440 nm) and emits blue light (440–480 nm); protein Y is coupled to a green fluorescent protein, which is excited by blue light and emits green light (510 nm). (B) If protein X and Y do not interact, illuminating the sample with violet light yields fluorescence from the blue fluorescent protein only. (C) When protein X and protein Y interact, FRET can now occur. Illuminating the sample with violet light excites the blue fluorescent protein, whose emission in turn excites the green fluorescent protein, resulting in an emission of green light. The fluorochromes must be quite close together—within about 1–10 nm of one another—for FRET to occur. Because not every molecule of protein X and protein Y is bound at all times, some blue light may still be detected. But as the two proteins begin to interact, emission from the donor GFP falls as the emission from the acceptor GFP rises.

The first example of a technique for studying the function of single protein molecules was the use of a patch electrode to measure current flow through single ion channels (see Figure 11–33). Another approach is to attach the protein to a larger structure, such as a polystyrene bead, which can then be observed by conventional microscopy. This strategy has been particularly useful in measuring the movements of motor proteins. For example, molecules of the motor protein kinesin (discussed in Chapter 16) can be attached to a bead, and by observing the kinesin-attached bead moving along a microtubule, the step size of the motor (that is, the distance moved for each ATP molecule hydrolyzed) can be measured. As we will see in Chapter 9, optical microscopes have a limited resolution due to the diffraction of light, but computational and optical methods can be used to determine the position of a bead to a much finer precision than the resolution limit of the microscope. Using such techniques, extremely small movements—on the order of nanometers—can easily be detected and quantified.

Another advantage of attaching molecules to large beads is that these beads can serve as "handles" by which the molecules can be manipulated. This allows forces to be applied to the molecules, and their response observed. For example, the speed or step size of a motor can be measured as a function of the force it is pulling against. As discussed in the next chapter, a focused laser beam can be used as "optical tweezers" to generate a mechanical force on a bead, allowing motor proteins to be studied under an applied force (see Figure 9–35). <CGCG> Beads can also be manipulated using magnetic fields, a technology known as "magnetic tweezers." If multiple beads are present in a magnetic field, they will all experience the same force, potentially allowing large numbers of beads to be manipulated in parallel in a single experiment.

While beads can be used as markers to track protein movements, it is clearly preferable to be able to visualize the proteins themselves. In the next chapter, we shall see that recent refinements in microscopy have now made this possible.

Protein Function Can Be Selectively Disrupted With Small Molecules

Chemical inhibitors have contributed to the development of cell biology. For example, the microtubule inhibitor colchicine is routinely used to test whether microtubules are required for a given biological process; it also led to the first purification of tubulin several decades ago. In the past, these small molecules were usually natural products; that is, they were synthesized by living creatures. Although, as a whole, natural products have been extraordinarily useful in science and medicine (see, for example, Table 6–4, p. 385), they acted on a limited number of biological processes. However, the recent development of methods to synthesize hundreds of thousands of small molecules and to carry out large-scale automated screens holds the promise of identifying chemical inhibitors for virtually any biological process. In such approaches, large collections of small chemical compounds are simultaneously tested, either on living cells or in cell-free assays. Once an inhibitor is identified, it can be used as a probe to identify, through affinity chromatography (see Figure 8–13C) or other means, the protein to which the inhibitor binds. This general strategy, often called **chemical biology**, has successfully identified inhibitors of many proteins that carry out key processes in cell biology. The kinesin protein that functions in mitosis, for example, was identified by this method (**Figure 8–27**). Chemical inhibitors give the cell biologist great control over the timing of inhibition, as drugs can be rapidly added to or removed from cells, allowing protein function to be switched on or off quickly.

Protein Structure Can Be Determined Using X-Ray Diffraction

The main technique that has been used to discover the three-dimensional structure of molecules, including proteins, at atomic resolution is **x-ray crystallography**. X-rays, like light, are a form of electromagnetic radiation, but they have a much shorter wavelength, typically around 0.1 nm (the diameter of a hydrogen

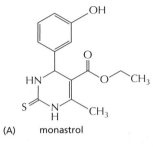

(A)　　monastrol

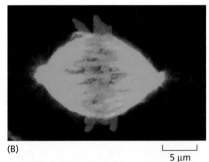

(B) 　　　　　　　　　　　　 5 μm

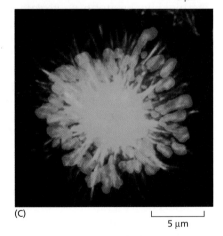

(C) 　　　　　　　　　　　　 5 μm

Figure 8–27 Small-molecule inhibitors for manipulating living cells. (A) Chemical structure of monastrol, a kinesin inhibitor identified in a large-scale screen for small molecules that disrupt mitosis. (B) Normal mitotic spindle seen in an untreated cell. The microtubules are stained *green* and chromosomes *blue*. (C) Monopolar spindle that forms in cells treated with monastrol. (B and C, from T.U. Mayer et al., *Science* 286:971–974, 1999. With permission from AAAS.)

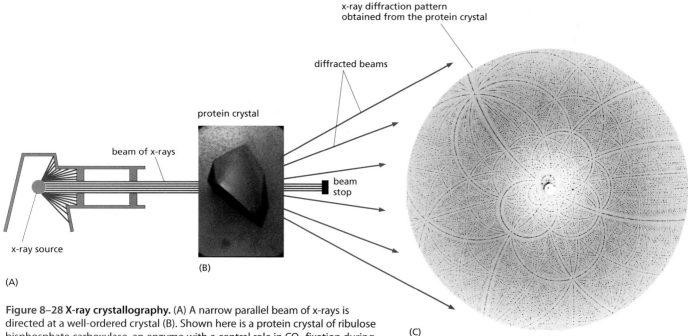

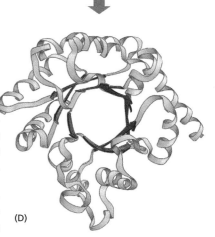

Figure 8–28 X-ray crystallography. (A) A narrow parallel beam of x-rays is directed at a well-ordered crystal (B). Shown here is a protein crystal of ribulose bisphosphate carboxylase, an enzyme with a central role in CO_2 fixation during photosynthesis. The atoms in the crystal scatter some of the beam, and the scattered waves reinforce one another at certain points and appear as a pattern of diffraction spots (C). This diffraction pattern, together with the amino acid sequence of the protein, can be used to produce an atomic model (D). The complete atomic model is hard to interpret, but this simplified version, derived from the x-ray diffraction data, shows the protein's structural features clearly (α helices, *green*; β strands, *red*). The components pictured in A to D are not shown to scale. (B, courtesy of C. Branden; C, courtesy of J. Hajdu and I. Andersson; D, adapted from original provided by B. Furugren.)

atom). If a narrow parallel beam of x-rays is directed at a sample of a pure protein, most of the x-rays pass straight through it. A small fraction, however, are scattered by the atoms in the sample. If the sample is a well-ordered crystal, the scattered waves reinforce one another at certain points and appear as diffraction spots when recorded by a suitable detector (**Figure 8–28**).

The position and intensity of each spot in the x-ray diffraction pattern contain information about the locations of the atoms in the crystal that gave rise to it. Deducing the three-dimensional structure of a large molecule from the diffraction pattern of its crystal is a complex task and was not achieved for a protein molecule until 1960. But in recent years x-ray diffraction analysis has become increasingly automated, and now the slowest step is likely to be the generation of suitable protein crystals. This step requires large amounts of very pure protein and often involves years of trial and error to discover the proper crystallization conditions; the pace has greatly accelerated with the use of recombinant DNA techniques to produce pure proteins and robotic techniques to test large numbers of crystallization conditions.

Analysis of the resulting diffraction pattern produces a complex three-dimensional electron-density map. Interpreting this map—translating its contours into a three-dimensional structure—is a complicated procedure that requires knowledge of the amino acid sequence of the protein. Largely by trial and error, the sequence and the electron-density map are correlated by computer to give the best possible fit. The reliability of the final atomic model depends on the resolution of the original crystallographic data: 0.5 nm resolution might produce a low-resolution map of the polypeptide backbone, whereas a resolution of 0.15 nm allows all of the non-hydrogen atoms in the molecule to be reliably positioned.

A complete atomic model is often too complex to appreciate directly, but simplified versions that show a protein's essential structural features can be readily derived from it (see Panel 3–2, pp. 132–133). The three-dimensional structures of about 20,000 different proteins have now been determined by x-ray crystallography or by NMR spectroscopy (see below)—enough to begin to see families of common structures emerging. These structures or protein folds often seem to be more conserved in evolution than are the amino acid sequences that form them (see Figure 3–13).

X-ray crystallographic techniques can also be applied to the study of macromolecular complexes. In a recent triumph, the method was used to determine the structure of the ribosome, a large and complex machine made of several RNAs and more than 50 proteins (see Figure 6–64). The determination required the use of a synchrotron, a radiation source that generates x-rays with the intensity needed to analyze the crystals of such large macromolecular complexes. <GGCC>

NMR Can Be Used to Determine Protein Structure in Solution

Nuclear magnetic resonance (NMR) spectroscopy has been widely used for many years to analyze the structure of small molecules. This technique is now also increasingly applied to the study of small proteins or protein domains. Unlike x-ray crystallography, NMR does not depend on having a crystalline sample. It simply requires a small volume of concentrated protein solution that is placed in a strong magnetic field; indeed, it is the main technique that yields detailed evidence about the three-dimensional structure of molecules in solution.

Certain atomic nuclei, particularly hydrogen nuclei, have a magnetic moment or spin: that is, they have an intrinsic magnetization, like a bar magnet. The spin aligns along the strong magnetic field, but it can be changed to a misaligned, excited state in response to applied radiofrequency (RF) pulses of electromagnetic radiation. When the excited hydrogen nuclei return to their aligned state, they emit RF radiation, which can be measured and displayed as a spectrum. The nature of the emitted radiation depends on the environment of each hydrogen nucleus, and if one nucleus is excited, it influences the absorption and emission of radiation by other nuclei that lie close to it. It is consequently possible, by an ingenious elaboration of the basic NMR technique known as two-dimensional NMR, to distinguish the signals from hydrogen nuclei in different amino acid residues, and to identify and measure the small shifts in these signals that occur when these hydrogen nuclei lie close enough together to interact. Because the size of such a shift reveals the distance between the interacting pair of hydrogen atoms, NMR can provide information about the distances between the parts of the protein molecule. By combining this information with a knowledge of the amino acid sequence, it is possible in principle to compute the three-dimensional structure of the protein (**Figure 8–29**).

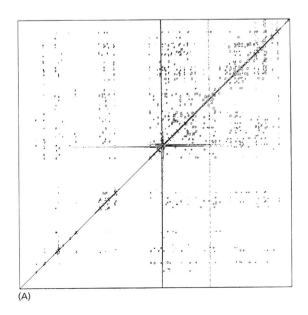

(A)

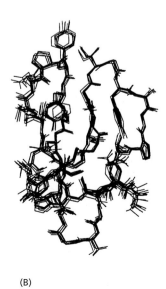

(B)

Figure 8–29 **NMR spectroscopy.** (A) An example of the data from an NMR machine. This two-dimensional NMR spectrum is derived from the C-terminal domain of the enzyme cellulase. The spots represent interactions between hydrogen atoms that are near neighbors in the protein and hence reflect the distance that separates them. Complex computing methods, in conjunction with the known amino acid sequence, enable possible compatible structures to be derived. (B) Ten structures of the enzyme, which all satisfy the distance constraints equally well, are shown superimposed on one another, giving a good indication of the probable three-dimensional structure. (Courtesy of P. Kraulis.)

For technical reasons the structure of small proteins of about 20,000 daltons or less can be most readily determined by NMR spectroscopy. Resolution decreases as the size of a macromolecule increases. But recent technical advances have now pushed the limit to about 100,000 daltons, thereby making the majority of proteins accessible for structural analysis by NMR.

Because NMR studies are performed in solution, this method also offers a convenient means of monitoring changes in protein structure, for example during protein folding or when the protein binds to another molecule. NMR is also used widely to investigate molecules other than proteins and is valuable, for example, as a method to determine the three-dimensional structures of RNA molecules and the complex carbohydrate side chains of glycoproteins.

Some landmarks in the development of x-ray crystallography and NMR are listed in **Table 8–2**.

Protein Sequence and Structure Provide Clues About Protein Function

Having discussed methods for purifying and analyzing proteins, we now turn to a common situation in cell and molecular biology: an investigator has identified a gene important for a biological process but has no direct knowledge of the biochemical properties of its protein product.

Thanks to the proliferation of protein and nucleic acid sequences that are catalogued in genome databases, the function of a gene—and its encoded protein—can often be predicted by simply comparing its sequence with those of previously characterized genes (see Figure 3–14). Because amino acid sequence

Table 8–2 Landmarks in the Development of X-ray Crystallography and NMR and Their Application to Biological Molecules

1864	**Hoppe-Seyler** crystallizes, and names, the protein hemoglobin.
1895	**Röntgen** observes that a new form of penetrating radiation, which he names x-rays, is produced when cathode rays (electrons) hit a metal target.
1912	**Von Laue** obtains the first x-ray diffraction patterns by passing x-rays through a crystal of zinc sulfide. **W.L. Bragg** proposes a simple relationship between an x-ray diffraction pattern and the arrangement of atoms in a crystal that produce the pattern.
1926	**Summer** obtains crystals of the enzyme urease from extracts of jack beans and demonstrates that proteins possess catalytic activity.
1931	**Pauling** publishes his first essays on 'The Nature of the Chemical Bond,' detailing the rules of covalent bonding.
1934	**Bernal and Crowfoot** present the first detailed x-ray diffraction patterns of a protein obtained from crystals of the enzyme pepsin.
1935	**Patterson** develops an analytical method for determining interatomic spacings from x-ray data.
1941	**Astbury** obtains the first x-ray diffraction pattern of DNA.
1946	**Block** and **Purcell** describe NMR.
1951	**Pauling and Corey** propose the structure of a helical conformation of a chain of L-amino acids—the α helix—and the structure of the β sheet, both of which were later found in many proteins.
1953	**Watson and Crick** propose the double-helix model of DNA, based on x-ray diffraction patterns obtained by **Franklin and Wilkins**.
1954	**Perutz** and colleagues develop heavy-atom methods to solve the phase problem in protein crystallography.
1960	**Kendrew** describes the first detailed structure of a protein (sperm whale myoglobin) to a resolution of 0.2 nm, and **Perutz** presents a lower-resolution structure of the larger protein hemoglobin.
1966	**Phillips** describes the structure of lysozyme, the first enzyme to have its structure analyzed in detail.
1971	**Jeener** proposes the use of two-dimensional NMR, and **Wuthrich** and colleagues first use the method to solve a protein structure in the early 1980s.
1976	**Kim and Rich** and **Klug** and colleagues describe the detailed three-dimensional structure of tRNA determined by x-ray diffraction.
1977–1978	**Holmes and Klug** determine the structure of tobacco mosaic virus (TMV), and **Harrison** and **Rossman** determine the structure of two small spherical viruses.
1985	**Michel, Deisenhofer** and colleagues determine the first structure of a transmembrane protein (a bacterial reaction center) by x-ray crystallography. **Henderson** and colleagues obtain the structure of bacteriorhodopsin, a transmembrane protein, by high-resolution electron-microscopy methods between 1975 and 1990.

```
Score =  399 bits (1025), Expect = e-111
Identities = 198/290 (68%), Positives = 241/290 (82%), Gaps = 1/290

Query:  57  MENFQKVEKIGEGTYGVVYKARNKLTGEVVALKKIRLDTETEGVPSTAIREISLLKELNH  116
            ME ++KVEKIGEGTYGVVYKA +K T E +ALKKIRL+ E EGVPSTAIREISLLKE+NH
Sbjct:   1  MEQYEKVEKIGEGTYGVVYKALDKATNETIALKKIRLEQEDEGVPSTAIREISLLKEMNH  60

Query: 117  PNIVKLLDVIHTENKLYLVFEFLHQDLKKFMDASALTGIPLPLIKSYLFQLLQGLAFCHS  176
            NIV+L DV+H+E ++YLVFE+L  DLKKFMD+         LIKSYL+Q+L G+A+CHS
Sbjct:  61· GNIVRLHDVVHSEKRIYLVFEYLDLDLKKFMDSCPEFAKNPTLIKSYLYQILHGVAYCHS  120

Query: 177  HRVLHRDLKPQNLLINTE-GAIKLADFGLARAFGVPVRTYTHEVVTLWYRAPEILLGCKY  235
            HRVLHRDLKPQNLLI+      A+KLADFGLARAFG+PVRT+THEVVTLWYRAPEILLG +
Sbjct: 121  HRVLHRDLKPQNLLIDRRTNALKLADFGLARAFGIPVRTFTHEVVTLWYRAPEILLGARQ  180

Query: 236  YSTAVDIWSLGCIFAEMVTRRALFPGDSEIDQLFRIFRTLGTPDEVVWPGVTSMPDYKPS  295
            YST VD+WS+GCIFAEMV ++ LFPGDSEID+LF+IFR LGTP+E  WPGV+ +PD+K +
Sbjct: 181  YSTPVDVWSVGCIFAEMVNQKPLFPGDSEIDELFKIFRILGTPNEQSWPGVSCLPDFKTA  240

Query: 296  FPKWARQDFSKVVPPLDEDGRSLLSQMLHYDPNKRISAKAALAHPFFQDV  345
            FP+W  QD + VVP LD  G  LLS+ML Y+P+KRI+A+ AL H +F+D+
Sbjct: 241  FPRWQAQDLATVVPNLDPAGLDLLSKMLRYEPSKRITARQALEHEYFKDL  290
```

Figure 8–30 Results of a BLAST search. Sequence databases can be searched to find similar amino acid or nucleic acid sequences. Here, a search for proteins similar to the human cell-cycle regulatory protein Cdc2 *(Query)* locates maize Cdc2 *(Sbjct)*, which is 68% identical (and 82% similar) to human Cdc2 in its amino acid sequence. The alignment begins at residue 57 of the Query protein, suggesting that the human protein has an N-terminal region that is absent from the maize protein. The *green* blocks indicate differences in sequence, and the *yellow* bar summarizes the similarities: when the two amino acid sequences are identical, the residue is shown; conservative amino acid substitutions are indicated by a plus sign (+). Only one small gap has been introduced—indicated by the *red arrow* at position 194 in the Query sequence—to align the two sequences maximally. The alignment score *(Score),* which is expressed in two different types of units, takes into account penalties for substitutions and gaps; the higher the alignment score, the better the match. The significance of the alignment is reflected in the *Expectation* (E) value, which specifies how often a match this good would be expected to occur by chance. The lower the E value, the more significant the match; the extremely low value here (e^{-111}) indicates certain significance. E values much higher than 0.1 are unlikely to reflect true relatedness. For example, an E value of 0.1 means there is a 1 in 10 likelihood that such a match would arise solely by chance.

determines protein structure, and structure dictates biochemical function, proteins that share a similar amino acid sequence usually have the same structure and usually perform similar biochemical functions, even when they are found in distantly related organisms. In modern cell biology, the study of a newly discovered protein usually begins with a search for previously characterized proteins that are similar in their amino acid sequences.

Searching a collection of known sequences for homologous genes or proteins is typically done over the World Wide Web, and it simply involves selecting a database and entering the desired sequence. A sequence alignment program—the most popular are BLAST and FASTA—scans the database for similar sequences by sliding the submitted sequence along the archived sequences until a cluster of residues falls into full or partial alignment (**Figure 8–30**). The results of even a complex search—which can be performed on either a nucleotide or an amino acid sequence—are returned within minutes. Such comparisons can predict the functions of individual proteins, families of proteins, or even most of the protein complement of a newly sequenced organism.

As was explained in Chapter 3, many proteins that adopt the same conformation and have related functions are too distantly related to be identified as clearly homologous from a comparison of their amino acid sequences alone (see Figure 3–13). Thus, an ability to reliably predict the three dimensional structure of a protein from its amino acid sequence would improve our ability to infer protein function from the sequence information in genomic databases. In recent years, major progress has been made in predicting the precise structure of a protein. These predictions are based, in part, on our knowledge of tens of thousands of protein structures that have already been determined by x-ray crystallography and NMR spectroscopy and, in part, on computations using our knowledge of the physical forces acting on the atoms. However, it remains a substantial and important challenge to predict the structures of proteins that are large or have multiple domains, or to predict structures at the very high levels of resolution needed to assist in computer-based drug discovery.

While finding homologous sequences and structures for a new protein will provide many clues about its function, it is usually necessary to test these insights through direct experimentation. However, the clues generated from sequence comparisons typically point the investigator in the correct experimental direction, and their use has therefore become one of the most important strategies in modern cell biology.

Summary

Most proteins function in concert with other proteins, and many methods exist for identifying and studying protein–protein interactions. Small-molecule inhibitors allow the functions of proteins they act upon to be studied in living cells. Because proteins with similar structures often have similar functions, the biochemical activity of a protein can often be predicted by searching databases for previously characterized proteins that are similar in their amino acid sequences.

ANALYZING AND MANIPULATING DNA

Until the early 1970s, DNA was the most difficult biological molecule for the biochemist to analyze. Enormously long and chemically monotonous, the string of nucleotides that forms the genetic material of an organism could be examined only indirectly, by protein or RNA sequencing or by genetic analysis. Today, the situation has changed entirely. From being the most difficult macromolecule of the cell to analyze, DNA has become the easiest. It is now possible to isolate a specific region of almost any genome, to produce a virtually unlimited number of copies of it, and to determine the sequence of its nucleotides in a few hours. At the height of the Human Genome Project, large facilities with automated machines were generating DNA sequences at the rate of 1000 nucleotides per second, around the clock. By related techniques, an isolated gene can be altered (engineered) at will and transferred back into the germ line of an animal or plant, so as to become a functional and heritable part of the organism's genome.

These technical breakthroughs in **genetic engineering**—the ability to manipulate DNA with precision in a test tube or an organism—have had a dramatic impact on all aspects of cell biology by facilitating the study of cells and their macromolecules in previously unimagined ways. **Recombinant DNA technology** comprises a mixture of techniques, some newly developed and some borrowed from other fields such as microbial genetics (**Table 8–3**). Central to the technology are the following key techniques:

1. Cleavage of DNA at specific sites by restriction nucleases, which greatly facilitates the isolation and manipulation of individual genes.
2. DNA ligation, which makes it possible to design and construct DNA molecules that are not found in nature.
3. DNA cloning through the use of either cloning vectors or the polymerase chain reaction, in which a portion of DNA is repeatedly copied to generate many billions of identical molecules.
4. Nucleic acid hybridization, which makes it possible to find a specific sequence of DNA or RNA with great accuracy and sensitivity on the basis of its ability to selectively bind a complementary nucleic acid sequence.
5. Rapid determination of the sequence of nucleotides of any DNA (even entire genomes), making it possible to identify genes and to deduce the amino acid sequence of the proteins they encode.
6. Simultaneous monitoring of the level of mRNA produced by every gene in a cell, using nucleic acid microarrays, in which tens of thousands of hybridization reactions take place simultaneously.

In this section, we describe each of these basic techniques, which together have revolutionized the study of cell biology.

Restriction Nucleases Cut Large DNA Molecules into Fragments

Unlike a protein, a gene does not exist as a discrete entity in cells, but rather as a small region of a much longer DNA molecule. Although the DNA molecules in a cell can be randomly broken into small pieces by mechanical force, a fragment containing a single gene in a mammalian genome would still be only one among a hundred thousand or more DNA fragments, indistinguishable in their average size. How could such a gene be purified? Because all DNA molecules consist of an approximately equal mixture of the same four nucleotides, they cannot be readily separated, as proteins can, on the basis of their different charges and binding properties.

The solution to all of these problems began to emerge with the discovery of **restriction nucleases**. These enzymes, which can be purified from bacteria, cut the DNA double helix at specific sites defined by the local nucleotide sequence, thereby cleaving a long double-stranded DNA molecule into fragments of

Table 8–3 Some Major Steps in the Development of Recombinant DNA and Transgenic Technology

1869	**Miescher** first isolates DNA from white blood cells harvested from pus-soaked bandages obtained from a nearby hospital.
1944	**Avery** provides evidence that DNA, rather than protein, carries the genetic information during bacterial transformation.
1953	**Watson and Crick** propose the double-helix model for DNA structure based on x-ray results of **Franklin and Wilkins**.
1955	**Kornberg** discovers DNA polymerase, the enzyme now used to produce labeled DNA probes.
1961	**Marmur and Doty** discover DNA renaturation, establishing the specificity and feasibility of nucleic acid hydridization reactions.
1962	**Arber** provides the first evidence for the existence of DNA restriction nucleases, leading to their purification and use in DNA sequence characterization by **Nathans and H. Smith**.
1966	**Nirenberg**, **Ochoa**, and **Khorana** elucidate the genetic code.
1967	**Gellert** discovers DNA ligase, the enzyme used to join DNA fragments together.
1972–1973	DNA cloning techniques are developed by the laboratories of **Boyer**, **Cohen**, **Berg**, and their colleagues at Stanford University and the University of California at San Francisco.
1975	**Southern** develops gel-transfer hybridization for the detection of specific DNA sequences.
1975–1977	**Sanger and Barrell** and **Maxam and Gilbert** develop rapid DNA-sequencing methods.
1981–1982	**Palmiter and Brinster** produce transgenic mice; **Spradling and Rubin** produce transgenic fruit flies.
1982	**GenBank**, NIH's public genetic sequence database, is established at Los Alamos National Laboratory.
1985	**Mullis** and co-workers invent the polymerase chain reaction (PCR).
1987	**Capecchi** and **Smithies** introduce methods for performing targeted gene replacement in mouse embryonic stem cells.
1989	**Fields and Song** develop the yeast two-hybrid system for identifying and studying protein interactions.
1989	**Olson** and colleagues describe sequence-tagged sites, unique stretches of DNA that are used to make physical maps of human chromosomes.
1990	**Lipman** and colleagues release BLAST, an algorithm used to search for homology between DNA and protein sequences.
1990	**Simon** and colleagues study how to efficiently use bacterial artificial chromosomes, BACs, to carry large pieces of cloned human DNA for sequencing.
1991	**Hood and Hunkapillar** introduce new automated DNA sequence technology.
1995	**Venter** and colleagues sequence the first complete genome, that of the bacterium *Haemophilus influenzae*.
1996	**Goffeau** and an international consortium of researchers announce the completion of the first genome sequence of a eucaryote, the yeast *Saccharomyces cerevisiae*.
1996–1997	**Lockhart** and colleagues and **Brown and DeRisi** produce DNA microarrays, which allow the simultaneous monitoring of thousands of genes.
1998	**Sulston and Waterston** and colleagues produce the first complete sequence of a multicellular organism, the nematode worm *Caenorhabditis elegans*.
2001	Consortia of researchers announce the completion of the draft human genome sequence.
2004	Publication of the "finished" human genome sequence.

strictly defined sizes. Different restriction nucleases have different sequence specificities, and it is relatively simple to find an enzyme that can create a DNA fragment that includes a particular gene. The size of the DNA fragment can then be used as a basis for partial purification of the gene from a mixture.

Different species of bacteria make different restriction nucleases, which protect them from viruses by degrading incoming viral DNA. Each bacterial nuclease recognizes a specific sequence of four to eight nucleotides in DNA. These sequences, where they occur in the genome of the bacterium itself, are protected from cleavage by methylation at an A or a C nucleotide; the sequences in foreign DNA are generally not methylated and so are cleaved by the restriction nucleases. Large numbers of restriction nucleases have been purified from various species of bacteria; several hundred, most of which recognize different nucleotide sequences, are now available commercially.

Some restriction nucleases produce staggered cuts, which leave short single-stranded tails at the two ends of each fragment (**Figure 8–31**). Ends of this type are known as *cohesive ends*, as each tail can form complementary base pairs with the tail at any other end produced by the same enzyme (**Figure 8–32**). The cohesive ends generated by restriction enzymes allow any two DNA fragments to be easily joined together, as long as the fragments were generated with the same restriction nuclease (or with another nuclease that produces the same cohesive ends). DNA molecules produced by splicing together two or more DNA fragments are called **recombinant DNA** molecules.

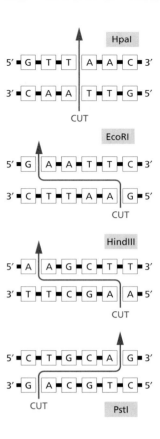

Figure 8–31 The DNA nucleotide sequences recognized by four widely used restriction nucleases. As in the examples shown, such sequences are often six base pairs long and "palindromic" (that is, the nucleotide sequence is the same if the helix is turned by 180 degrees around the center of the short region of helix that is recognized). The enzymes cut the two strands of DNA at or near the recognition sequence. For the genes encoding some enzymes, such as HpaI, the cleavage leaves blunt ends; for others, such as EcoRI, HindIII, and PstI, the cleavage is staggered and creates cohesive ends. Restriction nucleases are obtained from various species of bacteria: HpaI is from *Haemophilus parainfluenzae*, EcoRI is from *Escherichia coli*, HindIII is from *Haemophilus influenzae*, and PstI is from *Providencia stuartii*.

Gel Electrophoresis Separates DNA Molecules of Different Sizes

The same types of gel electrophoresis methods that have proved so useful in the analysis of proteins can determine the length and purity of DNA molecules. The procedure is actually simpler than for proteins: because each nucleotide in a nucleic acid molecule already carries a single negative charge (on the phosphate group), there is no need to add the negatively charged detergent SDS that is required to make protein molecules move uniformly toward the positive electrode. For DNA fragments less than 500 nucleotides long, specially designed polyacrylamide gels allow the separation of molecules that differ in length by as little as a single nucleotide (**Figure 8–33**A). The pores in polyacrylamide gels, however, are too small to permit very large DNA molecules to pass; to separate these by size, the much more porous gels formed by dilute solutions of agarose (a polysaccharide isolated from seaweed) are used (Figure 8–33B). These DNA separation methods are widely used for both analytical and preparative purposes.

A variation of agarose-gel electrophoresis, called *pulsed-field gel electrophoresis,* makes it possible to separate even extremely long DNA molecules. Ordinary gel electrophoresis fails to separate such molecules because the steady electric field stretches them out so that they travel end-first through the gel in snakelike configurations at a rate that is independent of their length. In pulsed-field gel electrophoresis, by contrast, the direction of the electric field changes periodically, which forces the molecules to reorient before continuing to move snakelike through the gel. This reorientation takes much more time for larger molecules, so that longer molecules move more slowly than shorter ones. As a consequence, even entire bacterial or yeast chromosomes separate into discrete bands in pulsed-field gels and so can be sorted and identified on the basis of their size (Figure 8–33C). Although a typical mammalian chromosome of 10^8 base pairs is too large to be sorted even in this way, large segments of these chromosomes are readily separated and identified if the chromosomal DNA is first cut with a restriction nuclease selected to recognize sequences that occur only rarely (once every 10,000 or more nucleotide pairs).

The DNA bands on agarose or polyacrylamide gels are invisible unless the DNA is labeled or stained in some way. One sensitive method of staining DNA is to expose it to the dye *ethidium bromide,* which fluoresces under ultraviolet light when it is bound to DNA (see Figure 8–33B,C). An even more sensitive detection method incorporates a radioisotope into the DNA molecules before electrophoresis; ^{32}P is often used as it can be incorporated into DNA phosphates and emits an energetic β particle that is easily detected by autoradiography, as in Figure 8–33. (For a discussion of radioisotopes, see p. 601).

Figure 8–32 The use of restriction nucleases to produce DNA fragments that can be easily joined together. Fragments with the same cohesive ends can readily join by complementary base-pairing between their cohesive ends, as illustrated. The two DNA fragments that join in this example were both produced by the EcoRI restriction nuclease, whereas the three other fragments were produced by different restriction nucleases that generated different cohesive ends (see Figure 8–31). Blunt-ended fragments, like those generated by HpaI (see Figure 8–31), can be spliced together with more difficulty.

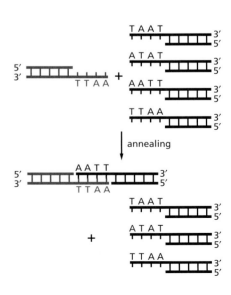

Figure 8–33 Gel electrophoresis techniques for separating DNA molecules by size. In the three examples shown, electrophoresis is from top to bottom, so that the largest—and thus slowest-moving—DNA molecules are near the top of the gel. (A) A polyacrylamide gel with small pores was used to fractionate single-stranded DNA. In the size range 10 to 500 nucleotides, DNA molecules that differ in size by only a single nucleotide can be separated from each other. In the example, the four lanes represent sets of DNA molecules synthesized in the course of a DNA-sequencing procedure. The DNA to be sequenced has been artificially replicated from a fixed start site up to a variable stopping point, producing a set of partial replicas of differing lengths. (Figure 8–50 explains how such sets of partial replicas are synthesized.) Lane 1 shows all the partial replicas that terminate in a G, lane 2 all those that terminate in an A, lane 3 all those that terminate in a T, and lane 4 all those that terminate in a C. Since the DNA molecules used in these reactions were radiolabeled, their positions can be determined by autoradiography, as shown.

(B) An agarose gel with medium-sized pores was used to separate double-stranded DNA molecules. This method is most useful in the size range 300 to 10,000 nucleotide pairs. These DNA molecules are fragments produced by cleaving the genome of a bacterial virus with a restriction nuclease, and they have been detected by their fluorescence when stained with the dye ethidium bromide. (C) The technique of pulsed-field agarose gel electrophoresis was used to separate 16 different yeast (*Saccharomyces cerevisiae*) chromosomes, which range in size from 220,000 to 2.5 million nucleotide pairs. The DNA was stained as in (B). DNA molecules as large as 10^7 nucleotide pairs can be separated in this way. (A, courtesy of Leander Lauffer and Peter Walter; B, courtesy of Ken Kreuzer; C, from D. Vollrath and R.W. Davis, *Nucleic Acids Res.* 15:7865–7876, 1987. With permission from Oxford University Press.)

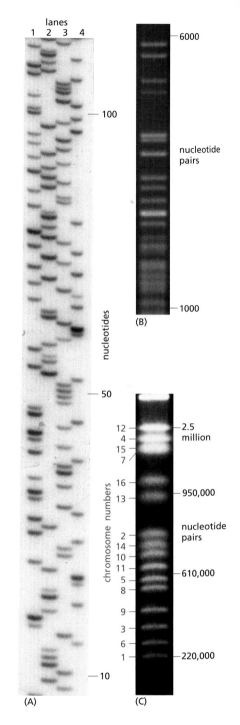

Purified DNA Molecules Can Be Specifically Labeled with Radioisotopes or Chemical Markers *in vitro*

Two procedures are widely used to label isolated DNA molecules. In the first method, a DNA polymerase copies the DNA in the presence of nucleotides that are either radioactive (usually labeled with ^{32}P) or chemically tagged (**Figure 8–34**A). In this way, "DNA probes" containing many labeled nucleotides can be produced for nucleic acid hybridization reactions (discussed below). The second procedure uses the bacteriophage enzyme polynucleotide kinase to transfer a single ^{32}P-labeled phosphate from ATP to the 5′ end of each DNA chain (Figure 8–34B). Because only one ^{32}P atom is incorporated by the kinase into each DNA strand, the DNA molecules labeled in this way are often not radioactive enough to be used as DNA probes; because they are labeled at only one end, however, they have been invaluable for other applications, including DNA footprinting, as discussed in Chapter 7.

Radioactive labeling methods are being replaced by labeling with molecules that can be detected chemically or through fluorescence. To produce such nonradioactive DNA molecules, specially modified nucleotide precursors are used (Figure 8–34C). A DNA molecule made in this way is allowed to bind to its complementary DNA sequence by hybridization, as discussed in the next section, and is then detected with an antibody (or other ligand) that specifically recognizes its modified side chain (**Figure 8–35**).

Nucleic Acid Hybridization Reactions Provide a Sensitive Way to Detect Specific Nucleotide Sequences

When an aqueous solution of DNA is heated at 100°C or exposed to a very high pH (pH ≥ 13), the complementary base pairs that normally hold the two strands of the double helix together are disrupted and the double helix rapidly dissociates into two single strands. This process, called *DNA denaturation*, was for many years thought to be irreversible. In 1961, however, it was discovered that complementary single strands of DNA readily re-form double helices by a process called **hybridization** (also called *DNA renaturation*) if they are kept for a

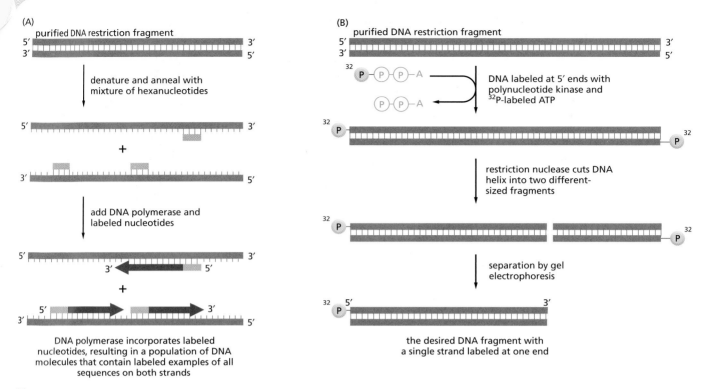

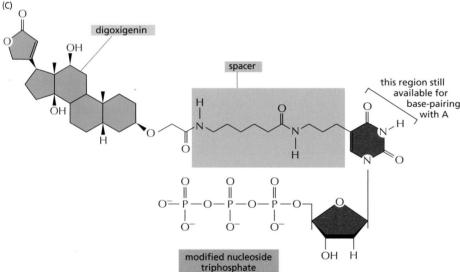

Figure 8–34 Methods for labeling DNA molecules *in vitro*. (A) A purified DNA polymerase enzyme labels all the nucleotides in a DNA molecule and can thereby produce highly radioactive DNA probes. (B) Polynucleotide kinase labels only the 5′ ends of DNA strands; therefore, when labeling is followed by restriction nuclease cleavage, as shown, DNA molecules containing a single 5′-end-labeled strand can be readily obtained. (C) The method in (A) is also used to produce nonradioactive DNA molecules that carry a specific chemical marker that can be detected with an appropriate antibody. The modified nucleotide shown can be incorporated into DNA by DNA polymerase, allowing the DNA molecule to serve as a probe that can be readily detected. The base on the nucleoside triphosphate shown is an analog of thymine, in which the methyl group on T has been replaced by a spacer arm linked to the plant steroid digoxigenin. An anti-digoxygenin antibody coupled to a visible marker such as a fluorescent dye is used to visualize the probe. Other chemical labels such as biotin can be attached to nucleotides and used in essentially the same way.

prolonged period at 65°C. Similar hybridization reactions can occur between any two single-stranded nucleic acid chains (DNA/DNA, RNA/RNA, or RNA/DNA), provided that they have complementary nucleotide sequences. These specific hybridization reactions are widely used to detect and characterize specific nucleotide sequences in both RNA and DNA molecules.

Single-stranded DNA molecules used to detect complementary sequences are known as **probes**; these molecules, which carry radioactive or chemical markers to facilitate their detection, can range from fifteen to thousands of nucleotides long. Hybridization reactions using DNA probes are so sensitive and selective that they can detect complementary sequences present at a concentration as low as one molecule per cell. It is thus possible to determine how many copies of any DNA sequence are present in a particular DNA sample. The same technique can be used to search for related but nonidentical genes. To find a gene of interest in an organism whose genome has not yet been sequenced, for example, a portion of a known gene can be used as a probe (**Figure 8–36**).

Figure 8–35 *In situ* hybridization to locate specific genes on chromosomes. Here, six different DNA probes have been used to mark the locations of their respective nucleotide sequences on human chromosome 5 at metaphase. The probes have been chemically labeled and detected with fluorescent antibodies. Both copies of chromosome 5 are shown, aligned side by side. Each probe produces two dots on each chromosome, since a metaphase chromosome has replicated its DNA and therefore contains two identical DNA helices. (Courtesy of David C. Ward.)

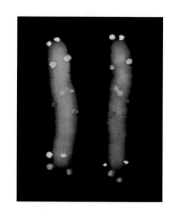

Alternatively, DNA probes can be used in hybridization reactions with RNA rather than DNA to find out whether a cell is expressing a given gene. In this case a DNA probe that contains part of the gene's sequence is hybridized with RNA purified from the cell in question to see whether the RNA includes nucleotide sequences matching the probe DNA and, if so, in what quantities. In somewhat more elaborate procedures, the DNA probe is treated with specific nucleases after the hybridization is complete, to determine the exact regions of the DNA probe that have paired with the RNA molecules. One can thereby determine the start and stop sites for RNA transcription, as well as the precise boundaries of the intron and exon sequences in a gene (**Figure 8–37**).

Today, the positions of intron/exon boundaries are usually determined by sequencing the *complementary DNA (cDNA)* sequences that represent the mRNAs expressed in a cell and comparing them with the nucleotide sequence of the genome. We describe later how cDNAs are prepared from mRNAs.

The hybridization of DNA probes to RNAs allows one to determine whether or not a particular gene is being transcribed; moreover, when the expression of a gene changes, one can determine whether the change is due to transcriptional or post-transcriptional controls (see Figure 7–92). These tests of gene expression were initially performed with one DNA probe at a time. *DNA microarrays* now allow the simultaneous monitoring of hundreds or thousands of genes at a time, as we discuss later. Hybridization methods are in such wide use in cell biology today that it is difficult to imagine how we could study gene structure and expression without them.

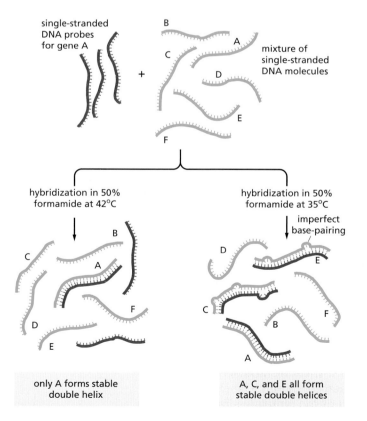

Figure 8–36 Stringent versus nonstringent hybridization conditions. To use a DNA probe to find an identical match, stringent hybridization conditions are used; the reaction temperature is kept just a few degrees below that at which a perfect DNA helix denatures in the solvent used (its *melting temperature*), so that all imperfect helices formed are unstable. When a DNA probe is being used to find DNAs with related, as well as identical, sequences, less stringent conditions are used; hybridization is performed at a lower temperature, which allows even imperfectly paired double helices to form. Only the lower-temperature hybridization conditions can be used to search for genes that are nonidentical but related to gene A (C and E in this example).

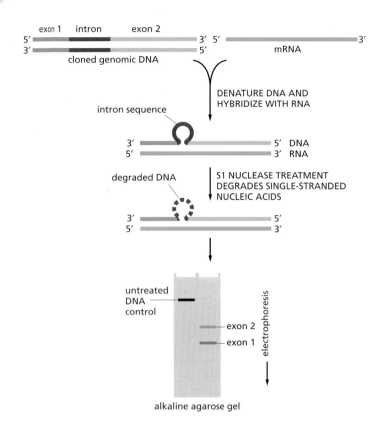

exon 1   intron   exon 2

cloned genomic DNA

mRNA

DENATURE DNA AND
HYBRIDIZE WITH RNA

intron sequence

degraded DNA

S1 NUCLEASE TREATMENT
DEGRADES SINGLE-STRANDED
NUCLEIC ACIDS

untreated
DNA
control

exon 2
exon 1

electrophoresis

alkaline agarose gel

Figure 8–37 **The use of nucleic acid hybridization to determine the region of a cloned DNA fragment that is present in an mRNA molecule.** The method shown requires a nuclease that cuts the DNA chain only where it is not base-paired to a complementary RNA chain. The positions of the introns in eucaryotic genes are mapped by the method shown. For this type of analysis, the DNA is electrophoresed through a denaturing agarose gel, which causes it to migrate as single-stranded molecules. The location of each end of an RNA molecule can be determined using similar methods.

Northern and Southern Blotting Facilitate Hybridization with Electrophoretically Separated Nucleic Acid Molecules

In a complex mixture of nucleic acids, DNA probes are often used to detect only those molecules with sequences that are complementary to all or part of the probe. Gel electrophoresis can be used to fractionate the many different RNA or DNA molecules in a crude mixture according to their size before the hybridization reaction is performed; if the probe binds to molecules of only one or a few sizes, one can be certain that the hybridization was indeed specific. Moreover, the size information obtained can be invaluable in itself. An example illustrates this point.

Suppose that one wishes to determine the nature of the defect in a mutant mouse that produces abnormally low amounts of albumin, a protein that liver cells normally secrete into the blood in large amounts. First, one collects identical samples of liver tissue from mutant and normal mice (the latter serving as controls) and disrupts the cells in a strong detergent to inactivate nucleases that might otherwise degrade the nucleic acids. Next, one separates the RNA and DNA from all of the other cell components: the proteins present are completely denatured and removed by repeated extractions with phenol—a potent organic solvent that is partly miscible with water; the nucleic acids, which remain in the aqueous phase, are then precipitated with alcohol to separate them from the small molecules of the cell. Then, one separates the DNA from the RNA by their different solubilities in alcohols and degrades any contaminating nucleic acid of the unwanted type by treatment with a highly specific enzyme—either an RNase or a DNase. The mRNAs are typically separated from bulk RNA by retention on a chromatography column that specifically binds the poly-A tails of mRNAs.

To analyze the albumin-encoding mRNAs, a technique called **Northern blotting** is used. First, the intact mRNA molecules purified from mutant and control liver cells are fractionated on the basis of their sizes into a series of bands by gel electrophoresis. Then, to make the RNA molecules accessible to DNA probes, a replica of the pattern of RNA bands on the gel is made by transferring ("blotting") the fractionated RNA molecules onto a sheet of nitrocellulose or nylon paper. The paper is then incubated in a solution containing a labeled DNA probe, the sequence of which corresponds to part of the template strand that

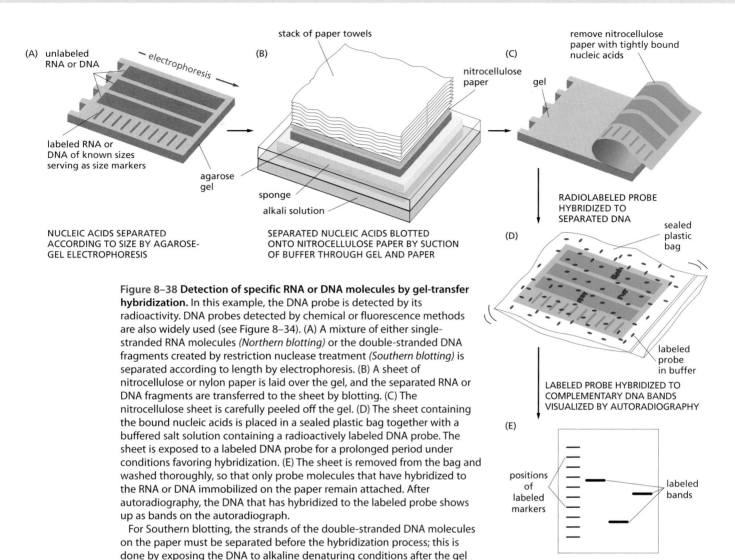

(A) unlabeled RNA or DNA
electrophoresis

labeled RNA or DNA of known sizes serving as size markers

agarose gel

NUCLEIC ACIDS SEPARATED ACCORDING TO SIZE BY AGAROSE-GEL ELECTROPHORESIS

(B) stack of paper towels

nitrocellulose paper

sponge
alkali solution

SEPARATED NUCLEIC ACIDS BLOTTED ONTO NITROCELLULOSE PAPER BY SUCTION OF BUFFER THROUGH GEL AND PAPER

(C) remove nitrocellulose paper with tightly bound nucleic acids

gel

RADIOLABELED PROBE HYBRIDIZED TO SEPARATED DNA

(D) sealed plastic bag

labeled probe in buffer

LABELED PROBE HYBRIDIZED TO COMPLEMENTARY DNA BANDS VISUALIZED BY AUTORADIOGRAPHY

(E) positions of labeled markers

labeled bands

Figure 8–38 Detection of specific RNA or DNA molecules by gel-transfer hybridization. In this example, the DNA probe is detected by its radioactivity. DNA probes detected by chemical or fluorescence methods are also widely used (see Figure 8–34). (A) A mixture of either single-stranded RNA molecules *(Northern blotting)* or the double-stranded DNA fragments created by restriction nuclease treatment *(Southern blotting)* is separated according to length by electrophoresis. (B) A sheet of nitrocellulose or nylon paper is laid over the gel, and the separated RNA or DNA fragments are transferred to the sheet by blotting. (C) The nitrocellulose sheet is carefully peeled off the gel. (D) The sheet containing the bound nucleic acids is placed in a sealed plastic bag together with a buffered salt solution containing a radioactively labeled DNA probe. The sheet is exposed to a labeled DNA probe for a prolonged period under conditions favoring hybridization. (E) The sheet is removed from the bag and washed thoroughly, so that only probe molecules that have hybridized to the RNA or DNA immobilized on the paper remain attached. After autoradiography, the DNA that has hybridized to the labeled probe shows up as bands on the autoradiograph.

For Southern blotting, the strands of the double-stranded DNA molecules on the paper must be separated before the hybridization process; this is done by exposing the DNA to alkaline denaturing conditions after the gel has been run (not shown).

produces albumin mRNA. The RNA molecules that hybridize to the labeled DNA probe on the paper (because they are complementary to part of the normal albumin gene sequence) are then located by detecting the bound probe by autoradiography or by chemical means (**Figure 8–38**). The sizes of the hybridized RNA molecules can be determined by reference to RNA standards of known sizes that are electrophoresed side by side with the experimental sample. In this way, one might discover that liver cells from the mutant mice make albumin mRNA in normal amounts and of normal size; alternatively, you might find that they make it in normal size but in greatly reduced amounts. Another possibility is that the mutant albumin mRNA molecules are abnormally short; in this case the gel blot could be retested with a series of shorter DNA probes, each corresponding to small portions of the gene, to reveal which part of the normal RNA is missing.

The original gel-transfer hybridization method, called **Southern blotting**, analyzes DNA rather than RNA. (It was named after its inventor, and the Northern and Western blotting techniques were named with reference to it.) Here, isolated DNA is first cut into readily separable fragments with restriction nucleases. The double-stranded fragments are then separated on the basis of size by gel electrophoresis, and those complementary to a DNA probe are identified by blotting and hybridization, as just described for RNA (see Figure 8–38). To characterize the structure of the albumin gene in the mutant mice, an albumin-specific DNA probe would be used to construct a detailed *restriction map* of the genome in the region of the albumin gene (such a map consists of the pattern of DNA fragments produced by various restriction nucleases). From this map one could determine if the albumin gene has been rearranged in the defective animals—for example, by the deletion or the insertion of a short DNA sequence; most single-base changes, however, could not be detected in this way.

Genes Can Be Cloned Using DNA Libraries

Any DNA fragment can be cloned. In molecular biology, the term **DNA cloning** is used in two senses. In one sense, it literally refers to the act of making many identical copies of a DNA molecule—the amplification of a particular DNA sequence. However, the term also describes the isolation of a particular stretch of DNA (often a particular gene) from the rest of a cell's DNA, because this isolation is greatly facilitated by making many identical copies of the DNA of interest. As discussed earlier in this chapter, cloning, particularly when used in the context of developmental biology, can also refer to the generation of many genetically identical cells starting from a single cell or even to the generation of genetically identical organisms. In all cases, cloning refers to the act of making many genetically identical copies; in this section, we will use the term cloning (or DNA cloning or gene cloning) to refer to methods designed to generate many identical copies of a segment of nucleic acid.

DNA cloning in its most general sense can be accomplished in several ways. The simplest involves inserting a particular fragment of DNA into the purified DNA genome of a self-replicating genetic element—generally a virus or a plasmid. A DNA fragment containing a human gene, for example, can be joined in a test tube to the chromosome of a bacterial virus, and the new recombinant DNA molecule can then be introduced into a bacterial cell, where the inserted DNA fragment will be replicated along with the DNA of the virus. Starting with only one such recombinant DNA molecule that infects a single cell, the normal replication mechanisms of the virus can produce more than 10^{12} identical virus DNA molecules in less than a day, thereby amplifying the amount of the inserted human DNA fragment by the same factor. A virus or plasmid used in this way is known as a *cloning vector*, and the DNA propagated by insertion into it is said to have been *cloned*.

To isolate a specific gene, one often begins by constructing a *DNA library*—a comprehensive collection of cloned DNA fragments from a cell, tissue, or organism. This library includes (one hopes) at least one fragment that contains the gene of interest. Libraries can be constructed with either a virus or a plasmid vector and are generally housed in a population of bacterial cells. The principles underlying the methods used for cloning genes are the same for either type of cloning vector, although the details may differ. Today, most cloning is performed with plasmid vectors.

The **plasmid vectors** most widely used for gene cloning are small circular molecules of double-stranded DNA derived from larger plasmids that occur naturally in bacterial cells. They generally account for only a minor fraction of the total host bacterial cell DNA, but they can easily be separated owing to their small size from chromosomal DNA molecules, which are large and precipitate as a pellet upon centrifugation. For use as cloning vectors, the purified plasmid DNA circles are first cut with a restriction nuclease to create linear DNA molecules. The genomic DNA to be used in constructing the library is cut with the same restriction nuclease, and the resulting restriction fragments (including those containing the gene to be cloned) are then added to the cut plasmids and annealed via their cohesive ends to form recombinant DNA circles. These recombinant molecules containing foreign DNA inserts are then covalently sealed with the enzyme DNA ligase (**Figure 8–39**).

In the next step in preparing the library, the recombinant DNA circles are introduced into bacterial cells that have been made transiently permeable to DNA. These bacterial cells are now said to be *transfected* with the plasmids. As the cells grow and divide, doubling in number every 30 minutes, the recombinant plasmids also replicate to produce an enormous number of copies of DNA circles containing the foreign DNA (**Figure 8–40**). Many bacterial plasmids carry genes for antibiotic resistance (discussed in Chapter 24), a property that can be

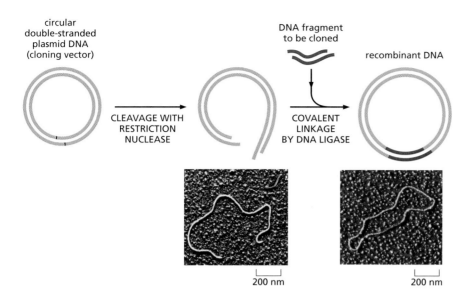

Figure 8–39 **The insertion of a DNA fragment into a bacterial plasmid with the enzyme DNA ligase.** The plasmid is cut open with a restriction nuclease (in this case one that produces cohesive ends) and is mixed with the DNA fragment to be cloned (which has been prepared with the same restriction nuclease). DNA ligase and ATP are added. The cohesive ends base-pair, and DNA ligase seals the nicks in the DNA backbone, producing a complete recombinant DNA molecule. (Micrographs courtesy of Huntington Potter and David Dressler.)

exploited to select those cells that have been successfully transfected; if the bacteria are grown in the presence of the antibiotic, only cells containing plasmids will survive. Each original bacterial cell that was initially transfected contains, in general, a different foreign DNA insert; this insert is inherited by all of the progeny cells of that bacterium, which together form a small colony in a culture dish.

For many years, plasmids were used to clone fragments of DNA of 1000 to 30,000 nucleotide pairs. Larger DNA fragments are more difficult to handle and were harder to clone. Then researchers began to use *yeast artificial chromosomes* (*YACs*), which could accommodate very large pieces of DNA (**Figure 8–41**). Today, new plasmid vectors based on the naturally occurring F plasmid of *E. coli* are used to clone DNA fragments of 300,000 to 1 million nucleotide pairs. Unlike smaller bacterial plasmids, the F plasmid—and its derivative, the **bacterial artificial chromosome** (**BAC**)—is present in only one or two copies per *E. coli* cell. The fact that BACs are kept in such low numbers in bacterial cells may contribute to their ability to maintain large cloned DNA sequences stably: with only a few BACs present, it is less likely that the cloned DNA fragments will become scrambled by recombination with sequences carried on other copies of the plasmid. Because of their stability, ability to accept large DNA inserts, and ease of handling, BACs are now the preferred vector for building DNA libraries of complex organisms—including those representing the human and mouse genomes.

Two Types of DNA Libraries Serve Different Purposes

Cleaving the entire genome of a cell with a specific restriction nuclease and cloning each fragment as just described produces a very large number of DNA fragments—on the order of a million for a mammalian genome. The fragments are distributed among millions of different colonies of transfected bacterial cells.

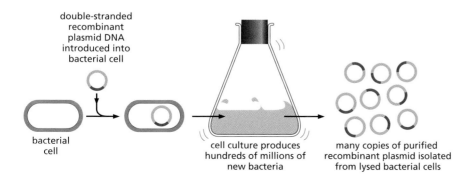

Figure 8–40 **The amplification of the DNA fragments inserted into a plasmid.** To produce large amounts of the DNA of interest, the recombinant plasmid DNA in Figure 8–39 is introduced into a bacterium by transfection, where it will replicate many millions of times as the bacterium multiplies.

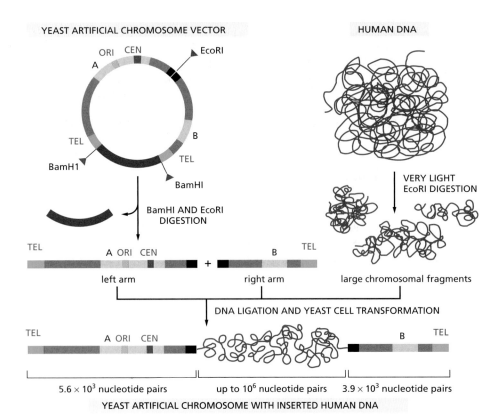

YEAST ARTIFICIAL CHROMOSOME VECTOR

HUMAN DNA

BamHI AND EcoRI DIGESTION

VERY LIGHT EcoRI DIGESTION

TEL   A ORI CEN   B   TEL

left arm   right arm   large chromosomal fragments

DNA LIGATION AND YEAST CELL TRANSFORMATION

TEL   A ORI CEN   B   TEL

5.6×10^3 nucleotide pairs   up to 10^6 nucleotide pairs 3.9×10^3 nucleotide pairs

YEAST ARTIFICIAL CHROMOSOME WITH INSERTED HUMAN DNA

Figure 8–41 The making of a yeast artificial chromosome (YAC). A YAC vector allows the cloning of very large DNA molecules. TEL, CEN, and ORI are the telomere, centromere, and origin of replication sequences, respectively, for the yeast *Saccharomyces cerevisiae*; all of these are required to propagate the YAC. BamHI and EcoRI are sites where the corresponding restriction nucleases cut the DNA double helix. The sequences denoted A and B encode enzymes that serve as selectable markers to allow the easy isolation of yeast cells that have taken up the artificial chromosome. Because bacteria divide more rapidly than yeasts, most large-scale cloning projects now use *E. coli* as the means for amplifying DNA. (Adapted from D.T. Burke, G.F. Carle and M.V. Olson, *Science* 236:806–812, 1987. With permission from AAAS.)

When working with BACs rather than typical plasmids, larger fragments can be inserted, and so fewer transfected bacterial cells are required to cover the genome. In either case, each of the colonies is composed of a clone of cells derived from a single ancestor cell, and therefore harbors many copies of a particular stretch of the fragmented genome (**Figure 8–42**). Such a plasmid is said to contain a **genomic DNA clone**, and the entire collection of plasmids is called a **genomic DNA library**. But because the genomic DNA is cut into fragments at random, only some fragments contain genes. Many of the genomic DNA clones obtained from the DNA of a higher eucaryotic cell contain only noncoding DNA, which, as we discussed in Chapter 4, makes up most of the DNA in such genomes.

An alternative strategy is to begin the cloning process by selecting only those DNA sequences that are transcribed into mRNA and thus are presumed to correspond to protein-encoding genes. This is done by extracting the mRNA from cells and then making a DNA copy of each mRNA molecule present—a so-called *complementary DNA*, or *cDNA*. The copying reaction is catalyzed by the reverse transcriptase enzyme of retroviruses, which synthesizes a complementary DNA chain on an RNA template. The single-stranded cDNA molecules synthesized by the reverse transcriptase are converted into double-stranded cDNA molecules by DNA polymerase, and these molecules are inserted into a plasmid or virus vector and cloned (**Figure 8–43**). Each clone obtained in this way is called a **cDNA clone**, and the entire collection of clones derived from one mRNA preparation constitutes a **cDNA library**.

Figure 8–44 illustrates some important differences between genomic DNA clones and cDNA clones. Genomic clones represent a random sample of all of the DNA sequences in an organism and, with very rare exceptions, are the same regardless of the cell type used to prepare them. By contrast, cDNA clones contain only those regions of the genome that have been transcribed into mRNA. Because the cells of different tissues produce distinct sets of mRNA molecules, a distinct cDNA library is obtained for each type of cell used to prepare the library.

Figure 8–42 Construction of a human genomic DNA library. A genomic library is usually stored as a set of bacteria, each bacterium carrying a different fragment of human DNA. For simplicity, cloning of just a few representative fragments (*colored*) is shown. In reality, all of the *gray* DNA fragments would also be cloned.

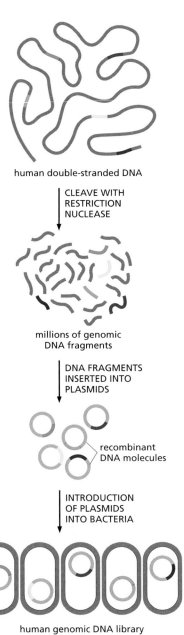

human double-stranded DNA

CLEAVE WITH RESTRICTION NUCLEASE

millions of genomic DNA fragments

DNA FRAGMENTS INSERTED INTO PLASMIDS

recombinant DNA molecules

INTRODUCTION OF PLASMIDS INTO BACTERIA

human genomic DNA library

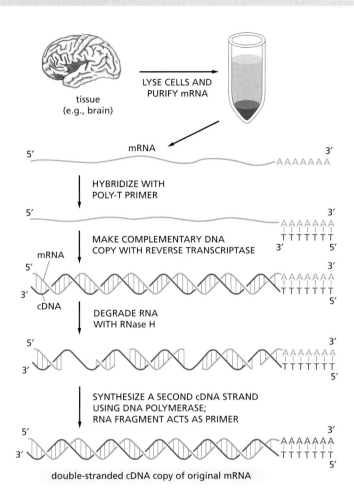

Figure 8–43 The synthesis of cDNA. Total mRNA is extracted from a particular tissue, and the enzyme reverse transcriptase produces DNA copies (cDNA) of the mRNA molecules (see p. 320). For simplicity, the copying of just one of these mRNAs into cDNA is illustrated. A short oligonucleotide complementary to the poly-A tail at the 3′ end of the mRNA (discussed in Chapter 6) is first hybridized to the RNA to act as a primer for the reverse transcriptase, which then copies the RNA into a complementary DNA chain, thereby forming a DNA/RNA hybrid helix. Treating the DNA/RNA hybrid with RNase H (see Figure 5–12) creates nicks and gaps in the RNA strand. The enzyme DNA polymerase then copies the remaining single-stranded cDNA into double-stranded cDNA. The fragment of the original mRNA is the primer for this synthesis reaction, as shown. Because the DNA polymerase used to synthesize the second DNA strand can synthesize through the bound RNA molecules, the RNA fragment that is base-paired to the 3′ end of the first DNA strand usually acts as the primer for the final product of the second strand synthesis. This RNA is eventually degraded during subsequent cloning steps. As a result, the nucleotide sequences at the extreme 5′ ends of the original mRNA molecules are often absent from cDNA libraries.

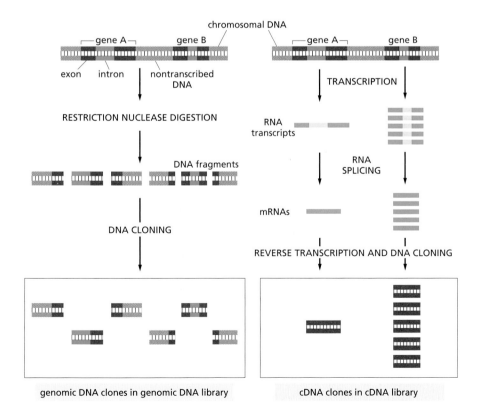

Figure 8–44 The differences between cDNA clones and genomic DNA clones derived from the same region of DNA. In this example, gene A is infrequently transcribed, whereas gene B is frequently transcribed, and both genes contain introns (green). In the genomic DNA library, both the introns and the nontranscribed DNA (pink) are included in the clones, and most clones contain, at most, only part of the coding sequence of a gene (red). In the cDNA clones, the intron sequences (yellow) have been removed by RNA splicing during the formation of the mRNA (blue), and a continuous coding sequence is therefore present in each clone. Because gene B is transcribed more frequently than gene A in the cells from which the cDNA library was made, it is represented much more frequently than A in the cDNA library. In contrast, A and B are in principle represented equally in the genomic DNA library.

cDNA Clones Contain Uninterrupted Coding Sequences

There are several advantages in using a cDNA library for gene cloning. First, specialized cells produce large quantities of some proteins. In this case, the mRNA encoding the protein is likely to be produced in such large quantities that a cDNA library prepared from the cells is highly enriched for the cDNA molecules encoding the protein, greatly reducing the problem of identifying the desired clone in the library (see Figure 8–44). Hemoglobin, for example, is made in large amounts by developing erythrocytes (red blood cells); for this reason the globin genes were among the first to be cloned.

By far the most important advantage of cDNA clones is that they contain the uninterrupted coding sequence of a gene. As we have seen, eucaryotic genes usually consist of short coding sequences of DNA (exons) separated by much longer noncoding sequences (introns); the production of mRNA entails the removal of the noncoding sequences from the initial RNA transcript and the splicing together of the coding sequences. Neither bacterial nor yeast cells will make these modifications to the RNA produced from a gene of a higher eucaryotic cell. Thus, when the aim of the cloning is either to deduce the amino acid sequence of the protein from the DNA sequence or to produce the protein in bulk by expressing the cloned gene in a bacterial or yeast cell, it is much preferable to start with cDNA. cDNA libraries have an additional use: as described in Chapter 7, many mRNAs from humans and other complex organisms are alternatively spliced, and a cDNA library often represents many, if not all, of the alternatively spliced mRNAs produced from a given cell line or tissue.

Genomic and cDNA libraries are inexhaustible resources, which are widely shared among investigators. Today, many such libraries are also available from commercial sources.

Genes Can Be Selectively Amplified by PCR

Now that so many genome sequences are available, genes can be cloned directly without the need to first construct DNA libraries. A technique called the **polymerase chain reaction** (**PCR**) makes this rapid cloning possible. Starting with an entire genome, PCR allows the DNA from a selected region to be amplified several billionfold, effectively "purifying" this DNA away from the remainder of the genome.

To begin, a pair of DNA oligonucleotides, chosen to flank the desired nucleotide sequence of the gene, are synthesized by chemical methods. These oligonucleotides are then used to prime DNA synthesis on single strands generated by heating the DNA from the entire genome. The newly synthesized DNA is produced in a reaction catalyzed *in vitro* by a purified DNA polymerase, and the primers remain at the 5′ ends of the final DNA fragments that are made (**Figure 8–45**A).

Nothing special is produced in the first cycle of DNA synthesis; the power of the PCR method is revealed only after repeated rounds of DNA synthesis. Every cycle doubles the amount of DNA synthesized in the previous cycle. Because each cycle requires a brief heat treatment to separate the two strands of the template DNA double helix, the technique requires the use of a special DNA polymerase, isolated from a thermophilic bacterium, that is stable at much higher temperatures than normal so that it is not denatured by the repeated heat treatments. With each round of DNA synthesis, the newly generated fragments serve as templates in their turn, and within a few cycles the predominant product is a single species of DNA fragment whose length corresponds to the distance between the two original primers (see Figure 8–45B).

In practice, effective DNA amplification requires 20–30 reaction cycles, with the products of each cycle serving as the DNA templates for the next—hence the term polymerase "chain reaction." A single cycle requires only about 5 minutes, and the entire procedure can be easily automated. PCR thereby makes possible the "cell-free molecular cloning" of a DNA fragment in a few hours, compared with the several days required for standard cloning procedures. This technique is now used routinely to clone DNA from genes of interest directly—starting either from genomic DNA or from mRNA isolated from cells (**Figure 8–46**).

The PCR method is extremely sensitive; it can detect a single DNA molecule in a sample. Trace amounts of RNA can be analyzed in the same way by first transcribing them into DNA with reverse transcriptase. The PCR cloning technique has largely replaced Southern blotting for the diagnosis of genetic diseases and for the detection of low levels of viral infection. It also has great promise in forensic medicine as a means of analyzing minute traces of blood or other tissues—

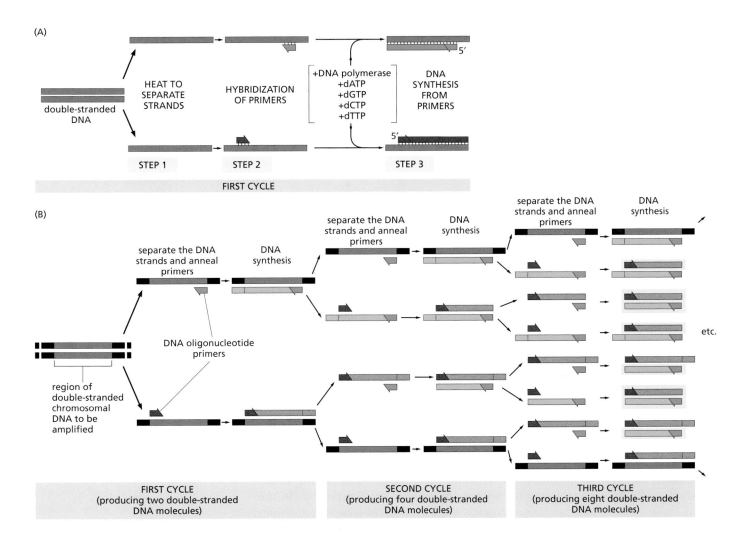

Figure 8–45 Amplification of DNA by the PCR technique. <TACG> Knowledge of the DNA sequence to be amplified is used to design two synthetic, primer DNA oligonucleotides. One primer is complementary to the sequence on one strand of the DNA double helix, and one is complementary to the sequence on the other strand, but at the opposite end of the region to be amplified. These oligonucleotides serve as primers for *in vitro* DNA synthesis, which is performed by a DNA polymerase, and they determine the segment of the DNA to be amplified. (A) PCR starts with a double-stranded DNA, and each cycle of the reaction begins with a brief heat treatment to separate the two strands (step 1). After strand separation, cooling of the DNA in the presence of a large excess of the two primer DNA oligonucleotides allows these primers to hybridize to complementary sequences in the two DNA strands (step 2). This mixture is then incubated with DNA polymerase and the four deoxyribonucleoside triphosphates to synthesize DNA, starting from the two primers (step 3). The entire cycle is then begun again by a heat treatment to separate the newly synthesized DNA strands. (B) As the procedure is performed over and over again, the newly synthesized fragments serve as templates in their turn, and within a few cycles the predominant DNA is identical to the sequence bracketed by and including the two primers in the original template. Of the DNA put into the original reaction, only the sequence bracketed by the two primers is amplified because there are no primers attached anywhere else. In the example illustrated in (B), three cycles of reaction produce 16 DNA chains, 8 of which *(boxed in yellow)* are the same length as and correspond exactly to one or the other strand of the original bracketed sequence shown at the far left; the other strands contain extra DNA downstream of the original sequence, which is replicated in the first few cycles. After four more cycles, 240 of the 256 DNA chains correspond exactly to the original bracketed sequence, and after several more cycles, essentially all of the DNA strands have this unique length.

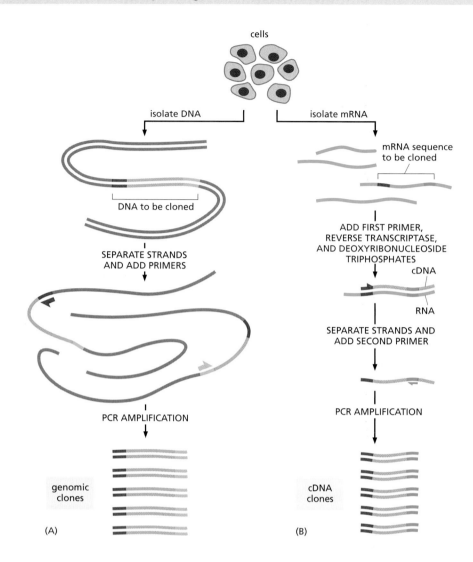

Figure 8–46 Use of PCR to obtain a genomic or cDNA clone. (A) To obtain a genomic clone using PCR, chromosomal DNA is first purified from cells. PCR primers that flank the stretch of DNA to be cloned are added, and many cycles of the reaction are completed (see Figure 8–45). Since only the DNA between (and including) the primers is amplified, PCR provides a way to obtain a short stretch of chromosomal DNA selectively in a virtually pure form. (B) To use PCR to obtain a cDNA clone of a gene, mRNA is first purified from cells. The first primer is then added to the population of mRNAs, and reverse transcriptase is used to make a complementary DNA strand. The second primer is then added, and the single-stranded cDNA molecule is amplified through many cycles of PCR, as shown in Figure 8–45. For both types of cloning, the nucleotide sequence of at least part of the region to be cloned must be known beforehand.

even as little as a single cell—and identifying the person from whom the sample came by his or her genetic "fingerprint" (**Figure 8–47**).

Cells Can Be Used As Factories to Produce Specific Proteins

The vast majority of the thousands of different proteins in a cell, including many with crucially important functions, are present in very small amounts. In the past, for most of them, it has been extremely difficult, if not impossible, to obtain more than a few micrograms of pure material. One of the most important contributions of DNA cloning and genetic engineering to cell biology is that they have made it possible to produce any of the cell's proteins in nearly unlimited amounts.

Large amounts of a desired protein are produced in living cells by using **expression vectors** (**Figure 8–48**). These are generally plasmids that have been designed to produce a large amount of a stable mRNA that can be efficiently translated into protein in the transfected bacterial, yeast, insect, or mammalian cell. To prevent the high level of the foreign protein from interfering with the transfected cell's growth, the expression vector is often designed to delay the synthesis of the foreign mRNA and protein until shortly before the cells are harvested and lysed (**Figure 8–49**).

Because the desired protein made from an expression vector is produced inside a cell, it must be purified away from the host-cell proteins by chromatography after cell lysis; but because it is such a plentiful species in the cell lysate (often 1–10% of the total cell protein), the purification is usually easy to accomplish in only a few steps. As we saw above, many expression vectors have been

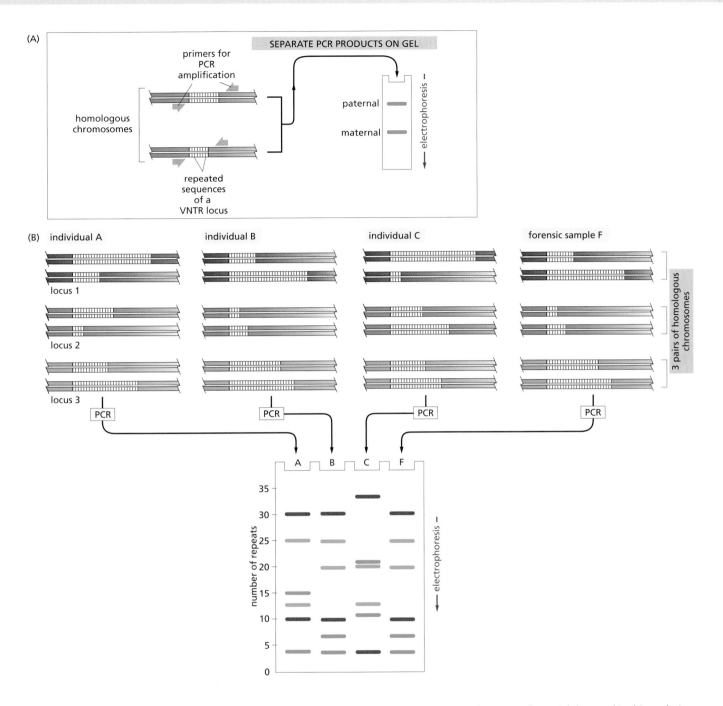

Figure 8–47 How PCR is used in forensic science. (A) The DNA sequences that create the variability used in this analysis contain runs of short, repeated sequences, such as CACACA . . ., which are found in various positions (loci) in the human genome. The number of repeats in each run can be highly variable in the population, ranging from 4 to 40 in different individuals. A run of repeated nucleotides of this type is commonly referred to as a *hypervariable microsatellite* sequence— also known as a VNTR *(variable number of tandem repeat)* sequence. Because of the variability in these sequences at each locus, individuals usually inherit a different variant from their mother and from their father; two unrelated individuals therefore do not usually contain the same pair of sequences. A PCR analysis using primers that bracket the locus produces a pair of bands of amplified DNA from each individual, one band representing the maternal variant and the other representing the paternal variant. The length of the amplified DNA, and thus the position of the band it produces after electrophoresis, depends on the exact number of repeats at the locus. (B) In the schematic example shown here, the same three VNTR loci are analyzed (requiring three different pairs of specially selected oligonucleotide primers) from three suspects (individuals A, B, and C), producing six DNA bands for each person after polyacrylamide-gel electrophoresis. Although some individuals have several bands in common, the overall pattern is quite distinctive for each. The band pattern can therefore serve as a "fingerprint" to identify an individual nearly uniquely. The fourth lane (F) contains the products of the same reactions carried out on a forensic sample. The starting material for such a PCR can be a single hair or a tiny sample of blood that was left at the crime scene. When examining the variability at 5–10 different VNTR loci, the odds that two random individuals would share the same genetic pattern by chance can be approximately 1 in 10 billion. In the case shown here, individuals A and C can be eliminated from further enquiries, whereas individual B remains a clear suspect for committing the crime. A similar approach is now routinely used for paternity testing.

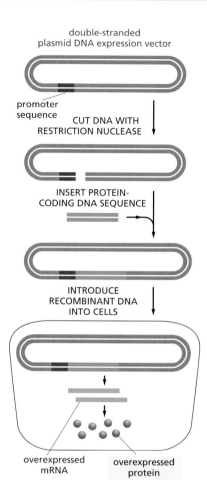

double-stranded
plasmid DNA expression vector

promoter
sequence   CUT DNA WITH
RESTRICTION NUCLEASE

INSERT PROTEIN-
CODING DNA SEQUENCE

INTRODUCE
RECOMBINANT DNA
INTO CELLS

overexpressed
mRNA   overexpressed
protein

Figure 8–48 Production of large amounts of a protein from a protein-coding DNA sequence cloned into an expression vector and introduced into cells. A plasmid vector has been engineered to contain a highly active promoter, which causes unusually large amounts of mRNA to be produced from an adjacent protein-coding gene inserted into the plasmid vector. Depending on the characteristics of the cloning vector, the plasmid is introduced into bacterial, yeast, insect, or mammalian cells, where the inserted gene is efficiently transcribed and translated into protein.

designed to add a molecular tag—a cluster of histidine residues or a small marker protein—to the expressed protein to allow easy purification by affinity chromatography (see Figure 8–16). A variety of expression vectors are available, each engineered to function in the type of cell in which the protein is to be made. In this way, cells can be induced to make vast quantities of medically useful proteins—such as human insulin and growth hormone, interferon, and viral antigens for vaccines. More generally, these methods make it possible to produce every protein—even those that may be present in only a few copies per cell—in large enough amounts to be used in the kinds of detailed structural and functional studies that we discussed earlier.

DNA technology also can produce large amounts of any RNA molecule whose gene has been isolated. Studies of RNA splicing, protein synthesis, and RNA-based enzymes, for example, are greatly facilitated by the availability of pure RNA molecules. Most RNAs are present in only tiny quantities in cells, and they are very difficult to purify away from other cell components—especially from the many thousands of other RNAs present in the cell. But any RNA of interest can be synthesized efficiently *in vitro* by transcription of its DNA sequence (produced by one of the methods just described) with a highly efficient viral RNA polymerase. The single species of RNA produced is then easily purified away from the DNA template and the RNA polymerase.

Proteins and Nucleic Acids Can Be Synthesized Directly by Chemical Reactions

Chemical reactions have been devised to synthesize directly specific sequences of amino acids or nucleic acids. These methodologies provide direct sources of biological molecules and do not rely on any cells or enzymes. Chemical synthesis is the method of choice for obtaining nucleic acids in the range of 100 nucleotides or fewer, which are particularly useful in the PCR-based approaches discussed above. Chemical synthesis is also routinely used to produce specific peptides that, when chemically coupled to other proteins, are used to generate antibodies against the peptide.

DNA Can Be Rapidly Sequenced

Methods that allow the nucleotide sequence of any DNA fragment to be determined simply and quickly have made it possible to determine the DNA sequences of tens of thousands of genes, and many complete genomes (see Table 1–1, p. 18). The volume of DNA sequence information is now so large (many tens of billions of nucleotides) that powerful computers must be used to store and analyze it.

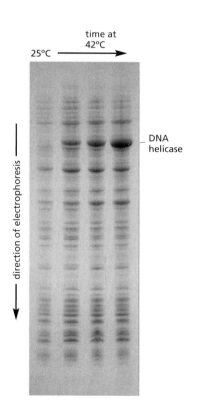

time at
42°C
25°C ⟶

direction of electrophoresis

DNA
helicase

Figure 8–49 Production of large amounts of a protein by using a plasmid expression vector. In this example, bacterial cells have been transfected with the coding sequence for an enzyme, DNA helicase; transcription from this coding sequence is under the control of a viral promoter that becomes active only at temperatures of 37°C or higher. The total cell protein has been analyzed by SDS polyacrylamide-gel electrophoresis, either from bacteria grown at 25°C (no helicase protein made) or after a shift of the same bacteria to 42°C for up to 2 hours (helicase protein has become the most abundant protein species in the lysate). (Courtesy of Jack Barry.)

Large-volume DNA sequencing was made possible through the development in the mid-1970s of the **dideoxy method** for sequencing DNA, which is based on *in vitro* DNA synthesis performed in the presence of chain-terminating dideoxyribonucleoside triphosphates (**Figure 8–50**).

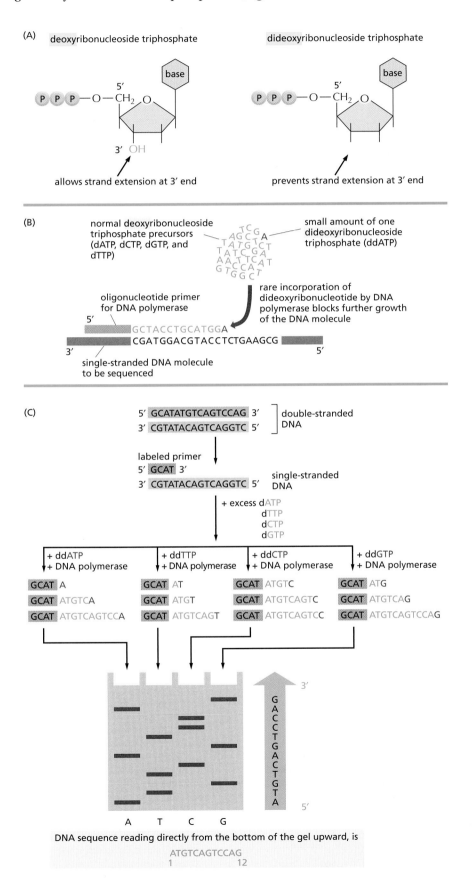

Figure 8–50 The enzymatic—or dideoxy—method of sequencing DNA. (A) This method relies on the use of dideoxyribonucleoside triphosphates, derivatives of the normal deoxyribonucleoside triphosphates that lack the 3′ hydroxyl group. (B) Purified DNA is synthesized *in vitro* in a mixture that contains single-stranded molecules of the DNA to be sequenced *(gray)*, the enzyme DNA polymerase, a short primer DNA *(orange)* to enable the polymerase to start DNA synthesis, and the four deoxyribonucleoside triphosphates (dATP, dCTP, dGTP, dTTP: *blue* A, C, G, and T). If a dideoxyribonucleotide analog *(red)* of one of these nucleotides is also present in the nucleotide mixture, it can become incorporated into a growing DNA chain. Because this chain now lacks a 3′ OH group, the addition of the next nucleotide is blocked, and the DNA chain terminates at that point. In the example illustrated, a small amount of dideoxyATP (ddATP, symbolized here as a *red* A) has been included in the nucleotide mixture. It competes with an excess of the normal deoxyATP (dATP, *blue* A), so that ddATP is occasionally incorporated, at random, into a growing DNA strand. This reaction mixture will eventually produce a set of DNAs of different lengths complementary to the template DNA that is being sequenced and terminating at each of the different As. The exact lengths of the DNA synthesis products can then be used to determine the position of each A in the growing chain. (C) To determine the complete sequence of a DNA fragment, the double-stranded DNA is first separated into its single strands and one of the strands is used as the template for sequencing. Four different chain-terminating dideoxyribonucleoside triphosphates (ddATP, ddCTP, ddGTP, ddTTP, again shown in *red*) are used in four separate DNA synthesis reactions on copies of the same single-stranded DNA template *(gray)*. Each reaction produces a set of DNA copies that terminate at different points in the sequence. The products of these four reactions are separated by electrophoresis in four parallel lanes of a polyacrylamide gel (labeled here as A, T, C, and G). The newly synthesized fragments are detected by a label (either radioactive or fluorescent) that has been incorporated either into the primer or into one of the deoxyribonucleoside triphosphates used to extend the DNA chain. In each lane, the bands represent fragments that have terminated at a given nucleotide (e.g., A in the leftmost lane) but at different positions in the DNA. By reading off the bands in order, starting at the bottom of the gel and working across all lanes, the DNA sequence of the newly synthesized strand can be determined. The sequence is given in the *green arrow* to the right of the gel. This sequence is complementary to the template strand *(gray)* from the original double-stranded DNA molecule, and identical to a portion of the *green* 5′-to-3′ strand.

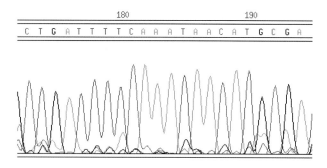

Although the same basic method is still used today, many improvements have been made. DNA sequencing is now completely automated: robotic devices mix the reagents and then load, run, and read the order of the nucleotide bases from the gel. Chain-terminating nucleotides that are each labeled with a different colored fluorescent dye facilitate these tasks; in this case, all four synthesis reactions can be performed in the same tube, and the products can be separated in a single lane of a gel. A detector positioned near the bottom of the gel reads and records the color of the fluorescent label on each band as it passes through a laser beam (**Figure 8–51**). A computer then reads and stores this nucleotide sequence. Some modern systems dispense with the traditional gel entirely, separating nucleic acids by capillary electrophoresis, a method that facilitates rapid automation.

Nucleotide Sequences Are Used to Predict the Amino Acid Sequences of Proteins

Now that DNA sequencing is so rapid and reliable, it has become the preferred method for determining, indirectly, the amino acid sequences of most proteins. Given a nucleotide sequence that encodes a protein, the procedure is quite straightforward. Although in principle there are six different reading frames in which a DNA sequence can be translated into protein (three on each strand), the correct one is generally recognizable as the only one lacking frequent stop codons (**Figure 8–52**). As we saw when we discussed the genetic code in Chap-

Figure 8–51 Automated DNA sequencing. Shown at the bottom is a tiny part of the raw data from an automated DNA-sequencing run as it appears on the computer screen. Each prominant colored peak represents a nucleotide in the DNA sequence—a clear stretch of nucleotide sequence can be read here between positions 173 and 194 from the start of the sequence. The small peaks along the baseline represent background "noise" and, as long as they are much lower than the "signal" peaks, they are ignored. This particular example is taken from the international project that determined the complete nucleotide sequence of the genome of the plant *Arabidopsis*. (Courtesy of George Murphy.)

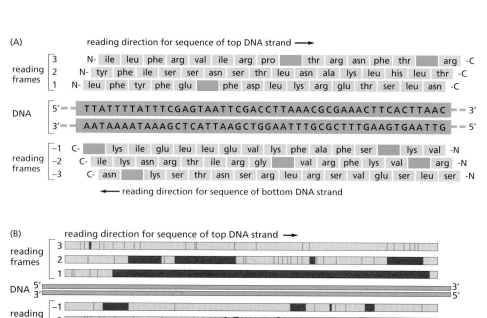

Figure 8–52 Finding the regions in a DNA sequence that encode a protein. (A) Any region of the DNA sequence can, in principle, code for six different amino acid sequences, because any one of three different reading frames can be used to interpret the nucleotide sequence on each strand. Note that a nucleotide sequence is always read in the 5′-to-3′ direction and encodes a polypeptide from the N-terminus to the C-terminus. For a random nucleotide sequence read in a particular frame, a stop signal for protein synthesis is encountered, on average, about once every 20 amino acids. In this sample sequence of 48 base pairs, each such signal (stop codon) is colored *blue*, and only reading frame 2 lacks a stop signal. (B) Search of a 1700 base-pair DNA sequence for a possible protein-encoding sequence. The information is displayed as in (A), with each stop signal for protein synthesis denoted by a *blue line*. In addition, all of the regions between possible start and stop signals for protein synthesis (see p. 381) are displayed as *red bars*. Only reading frame 1 actually encodes a protein, which is 475 amino acid residues long.

ter 6, a random sequence of nucleotides, read in frame, will encode a stop signal for protein synthesis about once every 20 amino acids. Nucleotide sequences that encode a stretch of amino acids much longer than this are candidates for presumptive exons, and they can be translated (by computer) into amino acid sequences and checked against databases for similarities to known proteins from other organisms. If necessary, a limited amount of amino acid sequence can then be determined from the purified protein to confirm the sequence predicted from the DNA.

The problem comes, however, in determining which nucleotide sequences—within a whole genome—represent genes that encode proteins. Identifying genes is easiest when the DNA sequence is from a bacterial or archaeal chromosome, which lacks introns, or from a cDNA clone. The location of genes in these nucleotide sequences can be predicted by examining the DNA for certain distinctive features (discussed in Chapter 6). Briefly, these genes that encode proteins are identified by searching the nucleotide sequence for *open reading frames (ORFs)* that begin with an initiation codon, usually ATG, and end with a termination codon, TAA, TAG, or TGA. To minimize errors, computers used to search for ORFs are often directed to count as genes only those sequences that are longer than, say, 100 codons in length.

For more complex genomes, such as those of animals and plants, the presence of large introns embedded within the coding portion of genes complicates the process. In many multicellular organisms, including humans, the average exon is only 150 nucleotides long. Thus one must also search for other features that signal the presence of a gene, for example, sequences that signal an intron/exon boundary or distinctive upstream regulatory regions. Recent efforts to solve the exon prediction problem have turned to artificial intelligence algorithms, in which the computer learns, based on known examples, what sets of features are most indicative of an exon boundary.

A second major approach to identifying the coding regions in chromosomes is through the characterization of the nucleotide sequences of the detectable mRNAs (using the corresponding cDNAs). The mRNAs (and the cDNAs produced from them) lack introns, regulatory DNA sequences, and the nonessential "spacer" DNA that lies between genes. It is therefore useful to sequence large numbers of cDNAs to produce a very large database of the coding sequences of an organism. These sequences are then readily used to distinguish the exons from the introns in the long chromosomal DNA sequences that correspond to genes.

The Genomes of Many Organisms Have Been Fully Sequenced

Owing in large part to the automation of DNA sequencing, the genomes of many organisms have been fully sequenced; these include plant chloroplasts and animal mitochondria, large numbers of bacteria, and archaea, and many of the model organisms that are studied routinely in the laboratory, including many yeasts, a nematode worm, the fruit fly *Drosophila*, the model plant *Arabidopsis*, the mouse, dog, chimpanzee, and, last but not least, humans. Researchers have also deduced the complete DNA sequences for a wide variety of human pathogens. These include the bacteria that cause cholera, tuberculosis, syphilis, gonorrhea, Lyme disease, and stomach ulcers, as well as hundreds of viruses—including smallpox virus and Epstein–Barr virus (which causes infectious mononucleosis). Examination of the genomes of these pathogens provides clues about what makes them virulent and will also point the way to new and more effective treatments.

Haemophilus influenzae (a bacterium that can cause ear infections and meningitis in children) was the first organism to have its complete genome sequence—all 1.8 million nucleotide pairs—determined by the *shotgun sequencing method*, the most common strategy used today. In the shotgun method, long sequences of DNA are broken apart randomly into many shorter fragments. Each fragment is then sequenced and a computer is used to order these pieces into a whole chromosome or genome, using sequence overlap to guide the assembly. The shotgun method is the technique of choice for sequencing small genomes. Although larger, more repetitive genome sequences are more challenging to assemble, the shotgun method—in combination with the analysis of large DNA fragments cloned in BACs—has played a key role in their sequencing as well.

With new sequences appearing at a steadily accelerating pace in the scientific literature, comparison of the complete genome sequences of different organisms allows us to trace the evolutionary relationships among genes and organisms, and to discover genes and predict their functions (discussed in Chapters 3 and 4). Assigning functions to genes often involves comparing their sequences with related sequences from model organisms that have been well characterized in the laboratory, such as the bacterium *E. coli*, the yeasts *S. cerevisiae* and *S. pombe*, the nematode worm *C. elegans*, and the fruit fly *Drosophila* (discussed in Chapter 1).

Although the organisms whose genomes have been sequenced share many biochemical pathways and possess many proteins that are homologous in their amino acid sequence or structure, the functions of a very large number of newly identified proteins remain unknown. Depending on the organism, some 15–40% of the proteins encoded by a sequenced genome do not resemble any protein that has been studied biochemically. This observation underscores a limitation of the emerging field of genomics: although comparative analysis of genomes reveals a great deal of information about the relationships between genes and organisms, it often does not provide immediate information about how these genes function, or what roles they have in the physiology of an organism. Comparison of the full gene complement of several thermophilic bacteria, for example, does not reveal why these bacteria thrive at temperatures exceeding 70°C. And examination of the genome of the incredibly radioresistant bacterium *Deinococcus radiodurans* does not explain how this organism can survive a blast of radiation that can shatter glass. Further biochemical and genetic studies, like those described in the other sections of this chapter, are required to determine how genes, and the proteins they produce, function in the context of living organisms.

Summary

DNA cloning allows a copy of any specific part of a DNA or RNA sequence to be selected from the millions of other sequences in a cell and produced in unlimited amounts in pure form. DNA sequences can be amplified after cutting chromosomal DNA with a restriction nuclease and inserting the resulting DNA fragments into the chromosome of a self-replicating genetic element such as a virus or a plasmid. Plasmid vectors are generally used, and the resulting "genomic DNA library" is housed in millions of bacterial cells, each carrying a different cloned DNA fragment. Individual cells from this library that are allowed to proliferate produce large amounts of a single cloned DNA fragment. The polymerase chain reaction (PCR) allows DNA cloning to be performed directly with a thermostable DNA polymerase—provided that the DNA sequence of interest is already known.

The procedures used to obtain DNA clones that correspond in sequence to mRNA molecules are the same except that a DNA copy of the mRNA sequence, called cDNA, is first made. Unlike genomic DNA clones, cDNA clones lack intron sequences, making them the clones of choice for analyzing the protein product of a gene.

Nucleic acid hybridization reactions provide a sensitive means of detecting a gene or any other nucleotide sequence of interest. Under stringent hybridization conditions (a combination of solvent and temperature at which even a perfect double helix is barely stable), two strands can pair to form a "hybrid" helix only if their nucleotide sequences are almost perfectly complementary. The enormous specificity of this hybridization reaction allows any single-stranded sequence of nucleotides to be labeled with a radioisotope or chemical and used as a probe to find a complementary partner strand, even in a cell or cell extract that contains millions of different DNA and RNA sequences. Probes of this type are widely used to detect the nucleic acids corresponding to specific genes, both to facilitate their purification and characterization, and to localize them in cells, tissues, and organisms.

The nucleotide sequence of DNA can be determined rapidly and simply by using highly automated techniques based on the dideoxy method for sequencing DNA. This technique has made it possible to determine the complete DNA sequences of the genomes of many organisms. Comparison of the genome sequences of different organisms allows us to trace the evolutionary relationships among genes and organisms, and it has proved valuable for discovering new genes and predicting their functions.

Taken together, these techniques for analyzing and manipulating DNA have made it possible to identify, isolate, and sequence genes from any organism of interest. Related technologies allow scientists to produce the protein products of these genes in the large quantities needed for detailed analyses of their structure and function, as well as for medical purposes.

STUDYING GENE EXPRESSION AND FUNCTION

Ultimately, one wishes to determine how genes—and the proteins they encode—function in the intact organism. Although it may seem counterintuitive, one of the most direct ways to find out what a gene does is to see what happens to the organism when that gene is missing. Studying mutant organisms that have acquired changes or deletions in their nucleotide sequences is a time-honored practice in biology and forms the basis of the important field of **genetics**. Because mutations can disrupt cell processes, mutants often hold the key to understanding gene function. In the classical genetic approach, one begins by isolating mutants that have an interesting or unusual appearance: fruit flies with white eyes or curly wings, for example. Working backward from the **phenotype**—the appearance or behavior of the individual—one then determines the organism's **genotype**, the form of the gene responsible for that characteristic (**Panel 8–1**).

Today, with numerous genome sequences available, the exploration of gene function often begins with a DNA sequence. Here, the challenge is to translate sequence into function. One approach, discussed earlier in the chapter, is to search databases for well-characterized proteins that have similar amino acid sequences to the protein encoded by a new gene, and from there employ some of the methods described in the previous section to explore the gene's function further. But to determine directly a gene's function in a cell or organism, the most effective approach involves studying mutants that either lack the gene or express an altered version of it. Determining which cell processes have been disrupted or compromised in such mutants will usually shed light on a gene's biological role.

In this section, we describe several approaches to determining a gene's function, starting from a DNA sequence or an organism with an interesting phenotype. We begin with the classical genetic approach, which starts with a *genetic screen* for isolating mutants of interest and then proceeds toward identification of the gene or genes responsible for the observed phenotype. We then describe the set of techniques that are collectively called *reverse genetics*, in which one begins with a gene or gene sequence and attempts to determine its function. This approach often involves some intelligent guesswork—searching for homologous sequences and determining when and where a gene is expressed—as well as generating mutant organisms and characterizing their phenotype.

Classical Genetics Begins by Disrupting a Cell Process by Random Mutagenesis

Before the advent of gene cloning technology, most genes were identified by the abnormalities produced when the gene was mutated. This classical genetic approach—identifying the genes responsible for mutant phenotypes—is most easily performed in organisms that reproduce rapidly and are amenable to genetic manipulation, such as bacteria, yeasts, nematode worms, and fruit flies. Although spontaneous mutants can sometimes be found by examining extremely large populations—thousands or tens of thousands of individual

GENES AND PHENOTYPES

Gene: a functional unit of inheritance, usually corresponding to the segment of DNA coding for a single protein.

Genome: all of an organism's DNA sequences.

locus: the site of the gene in the genome

alleles: alternative forms of a gene

Wild-type: the normal, naturally occurring type

Mutant: differing from the wild-type because of a genetic change (a mutation)

GENOTYPE: the specific set of alleles forming the genome of an individual

PHENOTYPE: the visible character of the individual

homozygous A/A   heterozygous a/A   homozygous a/a

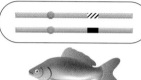

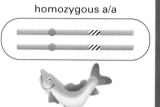

allele A is dominant (relative to a); allele a is recessive (relative to A)

In the example above, the phenotype of the heterozygote is the same as that of one of the homozygotes; in cases where it is different from both, the two alleles are said to be co-dominant.

CHROMOSOMES

a chromosome at the beginning of the cell cycle, in G$_1$ phase; the single long bar represents one long double helix of DNA

centromere

short "p" arm   long "q" arm

a chromosome near the end of the cell cycle, in metaphase; it is duplicated and condensed, consisting of two identical sister chromatids (each containing one DNA double helix) joined at the centromere.

short "p" arm   long "q" arm

pair of autosomes

maternal 1
paternal 1
maternal 3
paternal 3
paternal 2
maternal 2
Y
X

sex chromosomes

A normal diploid chromosome set, as seen in a metaphase spread, prepared by bursting open a cell at metaphase and staining the scattered chromosomes. In the example shown schematically here, there are three pairs of autosomes (chromosomes inherited symmetrically from both parents, regardless of sex) and two sex chromosomes—an X from the mother and a Y from the father. The numbers and types of sex chromosomes and their role in sex determination are variable from one class of organisms to another, as is the number of pairs of autosomes.

THE HAPLOID–DIPLOID CYCLE OF SEXUAL REPRODUCTION

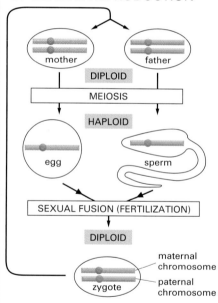

mother   father

DIPLOID

MEIOSIS

HAPLOID

egg   sperm

SEXUAL FUSION (FERTILIZATION)

DIPLOID

maternal chromosome

zygote

paternal chromosome

For simplicity, the cycle is shown for only one chromosome/chromosome pair.

MEIOSIS AND GENETIC RECOMBINATION

maternal chromosome

A   B

paternal chromosome

a   b

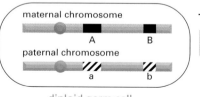

diploid germ cell

genotype $\dfrac{AB}{ab}$

MEIOSIS AND RECOMBINATION

genotype Ab

A   b

site of crossing-over

genotype aB

a   B

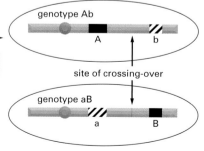

haploid gametes (eggs or sperm)

The greater the distance between two loci on a single chromosome, the greater is the chance that they will be separated by crossing over occurring at a site between them. If two genes are thus reassorted in x% of gametes, they are said to be separated on a chromosome by a genetic map distance of x map units (or x centimorgans).

TYPES OF MUTATIONS

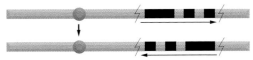

POINT MUTATION: maps to a single site in the genome, corresponding to a single nucleotide pair or a very small part of a single gene

INVERSION: inverts a segment of a chromosome

lethal mutation: causes the developing organism to die prematurely.

conditional mutation: produces its phenotypic effect only under certain conditions, called the *restrictive* conditions. Under other conditions—the *permissive* conditions—the effect is not seen. For a *temperature-sensitive* mutation, the restrictive condition typically is high temperature, while the permissive condition is low temperature.

loss-of-function mutation: either reduces or abolishes the activity of the gene. These are the most common class of mutations. Loss-of-function mutations are usually *recessive*—the organism can usually function normally as long as it retains at least one normal copy of the affected gene.

null mutation: a loss-of-function mutation that completely abolishes the activity of the gene.

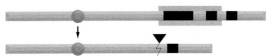

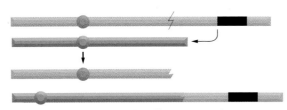

DELETION: deletes a segment of a chromosome

TRANSLOCATION: breaks off a segment from one chromosome and attaches it to another

gain-of-function mutation: increases the activity of the gene or makes it active in inappropriate circumstances; these mutations are usually *dominant.*

dominant-negative mutation: dominant-acting mutation that blocks gene activity, causing a loss-of-function phenotype even in the presence of a normal copy of the gene. This phenomenon occurs when the mutant gene product interferes with the function of the normal gene product.

suppressor mutation: suppresses the phenotypic effect of another mutation, so that the double mutant seems normal. An *intragenic* suppressor mutation lies within the gene affected by the first mutation; an *extragenic* suppressor mutation lies in a second gene—often one whose product interacts directly with the product of the first.

TWO GENES OR ONE?

Given two mutations that produce the same phenotype, how can we tell whether they are mutations in the same gene? If the mutations are recessive (as they most often are), the answer can be found by a complementation test.

In the simplest type of complementation test, an individual who is homozygous for one mutation is mated with an individual who is homozygous for the other. The phenotype of the offspring gives the answer to the question.

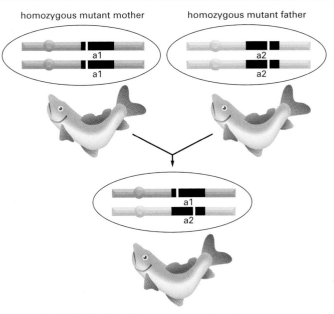

| COMPLEMENTATION: MUTATIONS IN TWO DIFFERENT GENES | NONCOMPLEMENTATION: TWO INDEPENDENT MUTATIONS IN THE SAME GENE |

homozygous mutant mother   homozygous mutant father

homozygous mutant mother   homozygous mutant father

hybrid offspring shows normal phenotype: one normal copy of each gene is present

hybrid offspring shows mutant phenotype: no normal copies of the mutated gene are present

organisms—isolating mutant individuals is much more efficient if one generates mutations with chemicals or radiation that damage DNA. By treating organisms with such mutagens, very large numbers of mutant individuals can be created quickly and then screened for a particular defect of interest, as we discuss shortly.

An alternative approach to chemical or radiation mutagenesis is called *insertional mutagenesis*. This method relies on the fact that exogenous DNA inserted randomly into the genome can produce mutations if the inserted fragment interrupts a gene or its regulatory sequences. The inserted DNA, whose sequence is known, then serves as a molecular tag that aids in the subsequent identification and cloning of the disrupted gene (**Figure 8–53**). In *Drosophila*, the use of the transposable P element to inactivate genes has revolutionized the study of gene function in the fly. Transposable elements (see Table 5–3, p. 318) have also been used to generate mutations in bacteria, yeast, mice, and the flowering plant *Arabidopsis*.

Such classical genetic studies are well suited for dissecting biological processes in experimental organisms, but how can we study gene function in humans? Unlike the genetically accessible organisms we have been discussing, humans do not reproduce rapidly, and they cannot be intentionally treated with mutagens. Moreover, any human with a serious defect in an essential process, such as DNA replication, would die long before birth.

There are two main ways that we can study human genes. First, because genes and gene functions have been so highly conserved throughout evolution, the study of less complex model organisms reveals critical information about similar genes and processes in humans. The corresponding human genes can then be studied further in cultured human cells. Second, many mutations that are not lethal—tissue-specific defects in lysosomes or cell-surface receptors, for example—have arisen spontaneously in the human population. Analyses of the phenotypes of the affected individuals, together with studies of their cultured cells, have provided many unique insights into important human cell functions. Although such mutations are rare, they are very efficiently discovered because of a unique human property: the mutant individuals call attention to themselves by seeking special medical care.

Figure 8–53 Insertional mutant of the snapdragon, Antirrhinum. A mutation in a single gene coding for a regulatory protein causes leafy shoots to develop in place of flowers. The mutation allows cells to adopt a character that would be appropriate to a different part of the normal plant. The mutant plant is on the *left,* the normal plant on the *right.* (Courtesy of Enrico Coen and Rosemary Carpenter.)

Genetic Screens Identify Mutants with Specific Abnormalities

Once a collection of mutants in a model organism such as yeast or fly has been produced, one generally must examine thousands of individuals to find the altered phenotype of interest. Such a search is called a **genetic screen**, and the larger the genome, the less likely it is that any particular gene will be mutated. Therefore, the larger the genome of an organism, the bigger the screening task becomes. The phenotype being screened for can be simple or complex. Simple phenotypes are easiest to detect: one can screen many organisms rapidly, for example, for mutations that make it impossible for the organism to survive in the absence of a particular amino acid or nutrient.

More complex phenotypes, such as defects in learning or behavior, may require more elaborate screens (**Figure 8–54**). But even genetic screens that are used to dissect complex physiological systems should be as simple as possible in design, and, if possible, should permit the simultaneous examination of large numbers of mutants. As an example, one particularly elegant screen was designed to search for genes involved in visual processing in zebrafish. The basis of this screen, which monitors the fishes' response to motion, is a change in behavior. Wild-type fish tend to swim in the direction of a perceived motion, whereas mutants with defects in their visual processing systems swim in random directions—a behavior that is easily detected. One mutant discovered in this screen is called *lakritz*, which is missing 80% of the retinal ganglion cells that help to relay visual signals from the eye to the brain. As the cellular organization of the zebrafish retina is similar to that of all vertebrates, the study of such mutants should also provide insights into visual processing in humans.

Because defects in genes that are required for fundamental cell processes—RNA synthesis and processing or cell-cycle control, for example—are usually

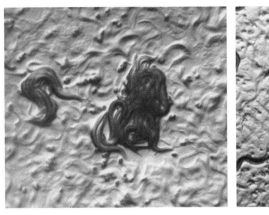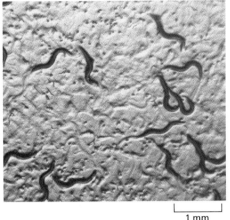

1 mm

Figure 8–54 **A behavioral phenotype detected in a genetic screen.** (A) Wild-type *C. elegans* engage in social feeding. The worms migrate around until they encounter their neighbors and commence feeding on bacteria. (B) Mutant animals feed by themselves. (Courtesy of Cornelia Bargmann, *Cell* 94: cover, 1998. With permission from Elsevier.)

lethal, the functions of these genes are often studied in individuals with **conditional mutations**. The mutant individuals function normally as long as "permissive" conditions prevail, but demonstrate abnormal gene function when subjected to "nonpermissive" (restrictive) conditions. In organisms with *temperature-sensitive mutations,* for example, the abnormality can be switched on and off experimentally simply by changing the temperature; thus, a cell containing a temperature-sensitive mutation in a gene essential for survival will die at a nonpermissive temperature but proliferate normally at the permissive temperature (**Figure 8–55**). The temperature-sensitive gene in such a mutant usually contains a point mutation that causes a subtle change in its protein product.

Many temperature-sensitive mutations were found in the bacterial genes that encode the proteins required for DNA replication. The mutants were identified by screening populations of mutagen-treated bacteria for cells that stop making DNA when they are warmed from 30°C to 42°C. These mutants were later used to identify and characterize the corresponding DNA replication proteins (discussed in Chapter 5). Similarly, screens for temperature-sensitive mutations led to the identification of many proteins involved in regulating the cell cycle, as well as many proteins involved in moving proteins through the secretory pathway in yeast (see Panel 13–1). Related screening approaches demonstrated the function of enzymes involved in the principal metabolic pathways of bacteria and yeast (discussed in Chapter 2) and identified many of the gene products responsible for the orderly development of the *Drosophila* embryo (discussed in Chapter 22).

Mutations Can Cause Loss or Gain of Protein Function

Gene mutations are generally classed as "loss of function" or "gain of function." A loss of function mutation results in a gene product that either does not work or works too little; thus, it reveals the normal function of the gene. A gain of function mutation results in a gene product that works too much, works at the wrong time or place, or works in a new way (**Figure 8–56**).

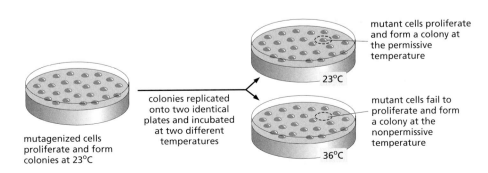

mutant cells proliferate and form a colony at the permissive temperature

23°C

mutant cells fail to proliferate and form a colony at the nonpermissive temperature

36°C

colonies replicated onto two identical plates and incubated at two different temperatures

mutagenized cells proliferate and form colonies at 23°C

Figure 8–55 **Screening for temperature-sensitive bacterial or yeast mutants.** Mutagenized cells are plated out at the permissive temperature. They divide and form colonies, which are transferred to two identical Petri dishes by replica plating. One of these plates is incubated at the permissive temperature, the other at the nonpermissive temperature. Cells containing a temperature-sensitive mutation in a gene essential for proliferation can divide at the normal, permissive temperature but fail to divide at the elevated, nonpermissive temperature.

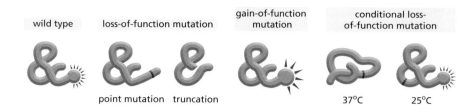

Figure 8–56 Gene mutations that affect their protein product in different ways. In this example, the wild-type protein has a specific cell function denoted by the *red rays*. Mutations that eliminate this function, increase the function, or render the function sensitive to higher temperatures are shown. The temperature-sensitive conditional mutant protein carries an amino acid substitution *(red)* that prevents its proper folding at 37°C, but allows the protein to fold and function normally at 25°C. Such conditional mutations are especially useful for studying essential genes; the organism can be grown under the permissive condition and then moved to the nonpermissive condition to study the function of the gene.

An important early step in the genetic analysis of any mutant cell or organism is to determine whether the mutation causes a loss or a gain of function. A standard test is to determine whether the mutation is *dominant* or *recessive*. A dominant mutation is one that still causes the mutant phenotype, in the presence of a single copy of the wild-type gene. A recessive mutation is one that is no longer able to cause the mutant phenotype in the presence of a single wild-type copy of the gene. Although cases have been described in which a loss-of-function mutation is dominant or a gain-of-function mutation is recessive, in the vast majority of cases, recessive mutations are loss of function, and dominant mutations are gain of function. It is easy to determine if a mutation is dominant or recessive. One simply mates a mutant with a wild-type to obtain diploid cells or organisms. The progeny from the mating will be heterozygous for the mutation. If the mutant phenotype is no longer observed, one can conclude that the mutation is recessive and is very likely to be a loss-of-function mutation.

Complementation Tests Reveal Whether Two Mutations Are in the Same Gene or Different Genes

A large-scale genetic screen can turn up many different mutations that show the same phenotype. These defects might lie in different genes that function in the same process, or they might represent different mutations in the same gene. Alternative forms of a gene are known as **alleles**. The most common difference between alleles is a substitution of a single nucleotide pair, but different alleles can also bear deletions, substitutions, and duplications. How can we tell, then, whether two mutations that produce the same phenotype occur in the same gene or in different genes? If the mutations are recessive—if, for example, they represent a loss of function of a particular gene—a **complementation test** can be used to ascertain whether the mutations fall in the same or in different genes. To test complementation in a diploid organism, an individual that is homozygous for one mutation—that is, it possesses two identical alleles of the mutant gene in question—is mated with an individual that is homozygous for the other mutation. If the two mutations are in the same gene, the offspring show the mutant phenotype, because they still will have no normal copies of the gene in question (see Panel 8–1). If, in contrast, the mutations fall in different genes, the resulting offspring show a normal phenotype, because they retain one normal copy (and one mutant copy) of each gene; the mutations thereby complement one another and restore a normal phenotype. Complementation testing of mutants identified during genetic screens has revealed, for example, that 5 genes are required for yeast to digest the sugar galactose, 20 genes are needed for *E. coli* to build a functional flagellum, 48 genes are involved in assembling bacteriophage T4 viral particles, and hundreds of genes are involved in the development of an adult nematode worm from a fertilized egg.

Genes Can Be Ordered in Pathways by Epistasis Analysis

Once a set of genes involved in a particular biological process has been identified, the next step is often to determine in which order the genes function. Gene order is perhaps easiest to explain for metabolic pathways, where, for example, enzyme A is necessary to produce the substrate for enzyme B. In this case, we would say that the gene encoding enzyme A acts before (upstream of) the gene encoding enzyme B in the pathway. Similarly, where one protein regulates the activity of another protein, we would say that the former gene acts before the latter. Gene order can, in many cases, be determined purely by genetic analysis without any knowledge of the mechanism of action of the gene products involved.

Suppose we have a biosynthetic process consisting of a sequence of steps, such that performance of a step B is conditional on completion of the preceding step A; and suppose *gene A* is required for step A, and *gene B* is required for step B. Then a null mutation (a mutation that abolishes function) in gene *A* will arrest the process at step A, regardless of whether gene *B* is functional or not, whereas a null mutation in gene *B* will cause arrest at step B only if gene *A* is still active. In such a case, gene *A* is said to be *epistatic* to gene *B*. By comparing the phenotypes of the different combinations of mutations, we can therefore discover the order in which the genes act. This type of analysis is called **epistasis analysis**. As an example, the pathway of protein secretion in yeast has been analyzed in this way. Different mutations in this pathway cause proteins to accumulate aberrantly in the endoplasmic reticulum (ER) or in the Golgi apparatus. When a yeast cell is engineered to carry both a mutation that blocks protein processing in the ER *and* a mutation that blocks processing in the Golgi apparatus, proteins accumulate in the ER. This indicates that proteins must pass through the ER before being sent to the Golgi before secretion (**Figure 8–57**). Strictly speaking, an epistasis analysis can only provide information about gene order in a pathway when both mutations are null alleles. When the mutations retain partial function, their epistasis interactions can be difficult to interpret.

Sometimes, a double mutant will show a new or more severe phenotype than either single mutant alone. This type of genetic interaction is called a *synthetic* phenotype, and if the phenotype is death of the organism, it is called *synthetic lethality*. In most cases, a synthetic phenotype indicates that the two genes act in two different parallel pathways, either of which is capable of mediating the same cell process. Thus, when both pathways are disrupted in the double mutant, the process fails altogether, and the synthetic phenotype is observed.

Genes Identified by Mutations Can Be Cloned

Once the mutant organisms are produced in a genetic screen, the next task is identifying the gene or genes responsible for the altered phenotype. If the phenotype has been produced by insertional mutagenesis, locating the disrupted gene is fairly simple. DNA fragments containing the insertion (a transposon or a retrovirus, for example) are collected and amplified by PCR, and the nucleotide sequence of the flanking DNA is determined. Genome databases can then be searched for open reading frames containing this flanking sequence.

If a DNA-damaging chemical was used to generate the mutations, identifying the inactivated gene is often more laborious, but it can be accomplished by several different approaches. In one, the first step is to experimentally determine the gene's location in the genome. To map a newly discovered gene, its rough chromosomal location is first determined by assessing how far the gene lies from other known genes in the genome. Estimating the distance between genetic loci is usually done by *linkage analysis*, a technique that relies on the tendency for genes that lie near one another on a chromosome to be inherited together. Even closely linked genes, however, can be separated by

Figure 8–57 **Using genetics to determine the order of function of genes.** In normal cells, secretory proteins are loaded into vesicles, which fuse with the plasma membrane to secrete their contents into the extracellular medium. Two mutants, A and B, fail to secrete proteins. In mutant A, secretory proteins accumulate in the ER. In mutant B, secretory proteins accumulate in the Golgi. In the double mutant AB, proteins accumulate in the ER; this indicates that the gene defective in mutant A acts before the gene defective in mutant B in the secretory pathway.

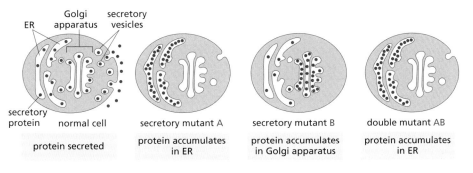

secretory protein | normal cell
protein secreted

secretory mutant A
protein accumulates in ER

secretory mutant B
protein accumulates in Golgi apparatus

double mutant AB
protein accumulates in ER

recombination during meiosis. The larger the distance between two genetic loci, the greater the chance that they will be separated by a crossover (see Panel 8–1). By calculating the recombination frequency between two genes, the approximate distance between them can be determined. If the position of one gene in the genome is known, that of the second gene can thereby be estimated.

Because genes are not always located close enough to one another to allow a precise pinpointing of their position, linkage analyses often rely on physical markers along the genome for estimating the location of an unknown gene. These markers are generally short stretches of nucleotides, with a known sequence and genome location, that can exist in at least two allelic forms. The simplest markers are *single-nucleotide polymorphisms (SNPs)*, short sequences that differ by one nucleotide pair among individuals in a population. SNPs can be detected by hybridization techniques. Many such physical markers, distributed all along the length of chromosomes, have been collected for a variety of organisms. If the distribution of these markers is sufficiently dense, one can, through a linkage analysis that tests for the tight co-inheritance of one or more SNPs with the mutant phenotype, narrow the potential location of a gene to a chromosomal region that may contain only a few gene sequences. These are then considered candidate genes, and their structure and function can be tested directly to determine which gene is responsible for the original mutant phenotype.

Human Genetics Presents Special Problems and Opportunities

Although genetic experimentation on humans is considered unethical and is legally banned, humans do suffer from a large variety of genetic disorders. The linkage analysis described above can be used to identify the genes responsible for these heritable conditions. Such studies require DNA samples from a large number of families affected by the disease. These samples are examined for the presence of physical markers such as SNPs that seem to be closely linked to the disease gene, in that they are always inherited by individuals who have the disease and not by their unaffected relatives. The disease gene is then located as described above (**Figure 8–58**). The genes for cystic fibrosis and Huntington's disease, for example, were discovered in this way.

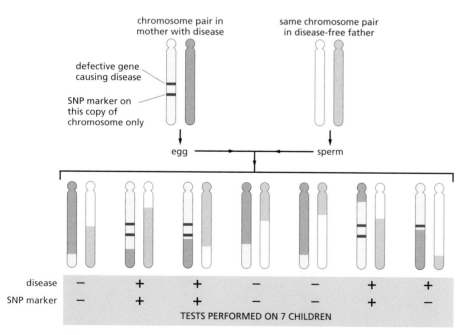

disease   −   +   +   −   −   +   +

SNP marker   −   +   +   −   −   +   −

TESTS PERFORMED ON 7 CHILDREN

CONCLUSION: gene causing disease is co-inherited with SNP marker from diseased mother in 75% of the diseased progeny. If this same correlation is observed in other families that have been examined, the gene causing disease is mapped to this chromosome close to the SNP. Note that an SNP that is either far away from the gene on the same chromosome or located on a different chromosome from the gene of interest will be co-inherited only 50% of the time.

Figure 8–58 Genetic linkage analysis using physical markers on DNA to find a human gene. In this example, the co-inheritance of a specific human phenotype (here a genetic disease) with an SNP marker. If individuals who inherit the disease nearly always inherit a particular SNP marker, then the gene causing the disease and the SNP are likely to be close together on the chromosome, as shown here. To prove that an observed linkage is statistically significant, hundreds of individuals may need to be examined. Note that the linkage will not be absolute unless the SNP marker is located in the gene itself. Thus, occasionally the SNP will be separated from the disease gene by crossing over during meiosis in the formation of the egg or sperm: this has happened in the case of the chromosome pair on the far right. When working with a sequenced genome, this procedure would be repeated with SNPs located on either side of the initial SNP, until a 100% co-inheritance is found.

Note that the egg and sperm will each contribute only one chromosome of each pair from the parent to the child.

Human Genes Are Inherited in Haplotype Blocks, Which Can Aid in the Search for Mutations That Cause Disease

With the complete human genome sequence in hand, we can now study human genetics in a way that was impossible only a few years ago. For example, we can begin to identify those DNA differences that distinguish one individual from another. No two humans (with the exception of identical twins) have the same genome. Each of us carries a set of polymorphisms—differences in nucleotide sequence—that make us unique. These polymorphisms can be used as markers for building genetic maps and performing genetic analyses to link particular polymorphisms with specific diseases or predispositions to disease.

The problem is that any two humans typically differ by about 0.1% in their nucleotide sequences (approximately one nucleotide difference every 1000 nucleotides). This translates to about 3 million differences between one person and another. Theoretically, one would need to search through all 3 million of those polymorphisms to identify the one or two that are responsible for a particular heritable disease or disease predisposition. To reduce the number of polymorphisms we need to examine, researchers are taking advantage of the recent discovery that human genes tend to be inherited in blocks.

The human species is relatively young, and it is thought that we are descended from a relatively small population of individuals who lived in Africa about 10,000 years ago. Because only a few hundred generations separate us from this ancestral population, large segments of human chromosomes have passed from parent to child unaltered by the recombination events that occur in meiosis. In fact, we observe that certain sets of alleles (including SNPs) are inherited in large blocks within chromosomes. These ancestral chromosome segments—sets of alleles that have been inherited in clusters with little genetic rearrangement across the generations—are called **haplotype blocks**. Like genes, SNPs, and other genetic markers—which exist in different alleleic forms—haplotype blocks also come in a limited number of "flavors" that are common in the human population, each of which represents an allele combination passed down from a shared ancestor long ago.

Researchers are now constructing a human genome map based on these haplotype blocks—called a **haplotype map** (**hapmap**). Geneticists hope that the human haplotype map will make the search for disease-causing and disease-susceptibility genes a much more manageable task. Instead of searching through each of the many millions of SNPs in the human population, one need only search through a considerably smaller set of selected SNPs to identify the haplotype block that appears to be inherited by individuals with the disease. (These searches still involve DNA samples from large numbers of people, and SNPs are now typically scored using robotic technologies.) If a specific haplotype block is more common among people with the disease than in unaffected individuals, the mutation linked to that disease will likely be located in that same segment of DNA (**Figure 8–59**). Researchers can then zero in on the specific region within the block to search for the specific gene associated with the disease. This approach should, in principle, allow one to analyze the genetics of those common diseases in which multiple genes confer susceptibility.

A detailed examination of haplotype blocks can even tell us whether a particular allele has been favored by natural selection. As a rule, when a new allele of a gene arises that does not confer a selective advantage on the individual, it will take a long time for that allele to become common in the population. The more common—and therefore older—such an allele is, the smaller should be the haplotype block that surrounds it, because it will have had many chances of being separated from its neighboring variations by the recombination events that occur in meiosis generation after generation.

A new allele may quickly spread in a population, however, if it confers some dramatic advantage on the organism. For example, mutations or variations that make an organism more resistant to an infection might be selected for because organisms with this variation would be more likely to survive and pass the

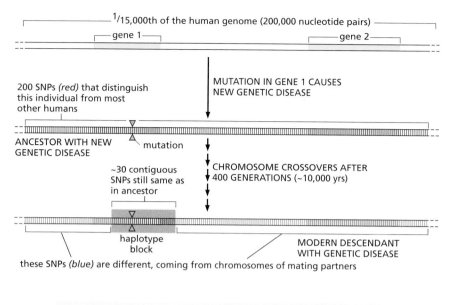

CONCLUSION: A MUTATION IN GENE 1 CAUSES THE DISEASE

Figure 8–59 Tracing the inheritance of SNPs within haplotype blocks to reveal the location of a disease-causing gene. An ancestor who acquires a disease-causing mutation in gene 1 will pass that mutation along to his or her descendants. Part of this gene is embedded within a haplotype block *(red shading)*—a cluster of variations (about 30 SNPs) that have been passed along from the ancestor in a continuous chunk. In the 400 generations that separate the ancestor from modern descendants with the disease, SNPs located over most of the ancestral 200,000-nucleotide-pair region shown have been shuffled by meiotic recombination in the descendant genome *(blue)*. (Note that the overlap of yellow and red is seen as *orange,* and the overlap of yellow and blue is seen as *green.*) The 30 SNPs within the haplotype block, however, have been inherited as a group, as no crossover events have yet separated them. To locate a gene that causes the inherited disease, the SNP patterns in a number of people who have the disease need to be analyzed. An individual with the disease will retain the ancestral pattern of SNPs located within the haplotype block shown, revealing that the disease-causing mutation is likely to lie within that haplotype block—thus in gene 1. The beauty of using haplotype maps for this type of linkage analysis is that only a fraction of the total SNPs need to be examined: one should be able to locate genes after searching through only about 10% of the 3 million useful SNPs present in the human genome.

mutation on to their offspring. Working with haplotype maps of individual genes, researchers have detected such positive selection for two human genes that confer resistance to malaria. The alleles that confer resistance are widespread in the population, but they are embedded in unusually large haplotype blocks, suggesting that they rose to prominence recently in the human gene pool (**Figure 8–60**).

In revealing the paths along which humans evolved, the human haplotype map provides a new window into our past; in helping us discover the genes that make us susceptible or resistant to disease, the map may also provide a rough guide to our individual futures.

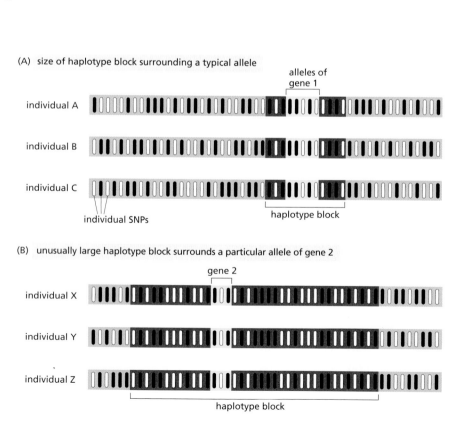

Figure 8–60 Identification of alleles that have been selected for in fairly recent human history by the unusually large haplotype blocks in which they are embedded. The SNPs are indicated in this diagram by vertical bars, which are shown as *white* or *black* according to their DNA sequence. Haplotype blocks are shaded in *red,* genes in *yellow,* and the rest of the chromosome in *blue.* These data suggest that this particular allele of gene 2 arose relatively recently in human history.

Complex Traits Are Influenced by Multiple Genes

A concert pianist might have an aunt who plays the violin. In another family, the parents and the children might all be fat. In a third family, the grandmother might be an alcoholic, and her grandson might abuse drugs. To what extent are such characteristics—musical ability, obesity, and addiction—inherited genetically? This is a very difficult question to answer. Some traits or diseases "run in families" but appear in only a few relatives or with no easily discernible pattern.

Characteristics that do not follow simple (sometimes called Mendelian) patterns of inheritance but have a genetically inherited component are termed **complex traits**. These traits are often **polygenic**; that is, they are influenced by multiple genes, each of which makes a small contribution to the phenotype in question. The effects of these genes are additive, which means that, together, they produce a continuum of varying features within the population. Individually, the genes that contribute to a polygenic trait are distributed to offspring in simple patterns, but because they all influence the phenotype, the pattern of traits inherited by offspring is often highly complex.

A simple example of a polygenic trait is eye color, which is determined by enzymes that control the distribution and production of the pigment melanin: the more melanin produced, the darker the eye color. Because numerous genes contribute to the formation of melanin, eye color in humans shows enormous variation, from the palest gray to a dark chocolate brown.

Although diseases based on mutations in single genes (for example, sickle-cell anemia and hemophilia) were some of the earliest recognized human inherited phenotypes, only a small fraction of human traits are dictated by single genes. The most obvious human phenotypes—from height, weight, eye color, and hair color to intelligence, temperament, sociability, and humor—arise from the interaction of many genes. Multiple genes also almost certainly underlie a propensity for the most common human diseases: diabetes, heart disease, high blood pressure, allergies, asthma, and various mental illnesses, including major depression and schizophrenia. Researchers are exploring new strategies—including the use of the haplotype maps discussed earlier—to understand the complex interplay between genes that act together to determine many of our most "human" traits.

Reverse Genetics Begins with a Known Gene and Determines Which Cell Processes Require Its Function

As we have seen, classical genetics starts with a mutant phenotype (or, in the case of humans, a range of characteristics) and identifies the mutations (and consequently the genes) responsible for it. Recombinant DNA technology, in combination with genome sequencing, has made possible a different type of genetic approach. Instead of beginning with a mutant organism and using it to identify a gene and its protein, an investigator can start with a particular gene and proceed to make mutations in it, creating mutant cells or organisms so as to analyze the gene's function. Because this approach reverses the traditional direction of genetic discovery—proceeding from genes to mutations, rather than vice versa—it is commonly referred to as **reverse genetics**.

Reverse genetics begins with a cloned gene, a protein with interesting properties that has been isolated from a cell, or simply a genome sequence. If the starting point is a protein, the gene encoding it is first identified and, if necessary, its nucleotide sequence is determined. The gene sequence can then be altered *in vitro* to create a mutant version. This engineered mutant gene, together with an appropriate regulatory region, is transferred into a cell where it can integrate into a chromosome, becoming a permanent part of the cell's genome. All of the descendants of the modified cell will now contain the mutant gene.

If the original cell used for the gene transfer is a fertilized egg, whole multi-cellular organisms can be obtained that contain the mutant gene, provided that the mutation does not cause lethality. In some of these animals, the altered gene will be incorporated into the germ cells—a *germ-line mutation*—allowing the mutant gene to be passed on to their progeny.

Genes Can Be Engineered in Several Ways

We have seen that mutant organisms lacking a particular gene may quickly reveal the function of the protein it encodes. For this reason, a gene "knockout"—in which both copies of the gene in a diploid organism have been inactivated or deleted—is a particularly useful type of mutation. However, there are many more types of genetic alterations available to the experimenter. For example, by altering the regulatory region of a gene before it is reintegrated into the genome, one can create mutant organisms in which the gene product is expressed at abnormally high levels, in the wrong tissue, or at the wrong time in development (**Figure 8–61**). By placing the gene under the control of an *inducible promoter*, the gene can be switched on or off at any time, and the effects observed. Inducible promoters that function in only a specific tissue can be used to monitor the effects of shutting the gene off (or turning it on) in that particular tissue. Finally, *dominant-negative* mutations are often employed particularly in those organisms in which it is simpler to add an altered gene to the genome than to replace the endogenous genes with it. The dominant-negative strategy exploits the fact that most proteins function as parts of larger protein complexes. The inclusion of just one nonfunctional component can often inactivate such complexes. Therefore, by designing a gene that produces large quantities of a mutant protein that is inactive but still able to assemble into the complex, it is often possible to produce a cell in which all the complexes are inactivated despite the presence of the normal protein (**Figure 8–62**).

As noted in the earlier discussion of classical genetics, if a protein is required for the survival of the cell (or the organism), a dominant-negative mutant will be inviable, making it impossible to test the function of the protein. To avoid this problem in reverse genetics, one can couple the mutant gene to an inducible promoter in order to produce the faulty gene product only on command—for example, in response to an increase in temperature or to the presence of a specific signal molecule.

In studying the action of a gene and the protein it encodes, one does not always wish to make drastic changes—flooding cells with huge quantities of the protein or eliminating a gene product entirely. It is sometimes useful to make slight changes in a protein's structure so that one can begin to dissect which portions of a protein are important for its function. The activity of an enzyme, for example, can be studied by changing a single amino acid in its active site. Special techniques are required to alter genes (and thus their protein products) in such subtle ways. The first step is often the chemical synthesis of a short DNA molecule containing the desired altered portion of the gene's nucleotide sequence. This synthetic DNA oligonucleotide is hybridized with single-stranded plasmid DNA that contains the DNA sequence to be altered, using conditions that allow imperfectly matched DNA strands to pair. The synthetic oligonucleotide will now serve as a primer for DNA synthesis by DNA polymerase, thereby generating a DNA double helix that incorporates the altered

Figure 8–61 Ectopic misexpression of Wnt, a signaling protein that affects development of the body axis in the early *Xenopus* embryo. In this experiment, mRNA coding for Wnt was injected into the ventral vegetal blastomere, inducing a second body axis (discussed in Chapter 22). (From S. Sokol et al., *Cell* 67:741–752, 1991. With permission from Elsevier.)

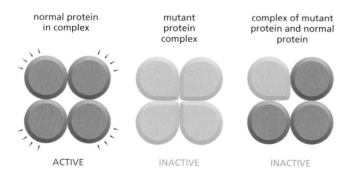

normal protein in complex	mutant protein complex	complex of mutant protein and normal protein
ACTIVE	INACTIVE	INACTIVE

Figure 8–62 A dominant-negative effect of a protein. Here, a gene is engineered to produce a mutant protein that prevents the normal copies of the same protein from performing their function. In this simple example, the normal protein must form a multisubunit complex to be active, and the mutant protein blocks function by forming a mixed complex that is inactive. In this way, a single copy of a mutant gene located anywhere in the genome can inactivate the normal products produced by other gene copies.

sequence into one of its two strands. After transfection, plasmids that carry the fully modified gene sequence are obtained. The modified DNA is then inserted into an expression vector so that the redesigned protein can be produced in the appropriate type of cells for detailed studies of its function. By changing selected amino acids in a protein in this way—a technique called **site-directed mutagenesis**—one can determine exactly which parts of the polypeptide chain are important for such processes as protein folding, interactions with other proteins, and enzymatic catalysis (**Figure 8–63**).

Engineered Genes Can Be Inserted into the Germ Line of Many Organisms

Altered genes can be introduced into cells in a variety of ways. DNA can be microinjected into mammalian cells with a glass micropipette or introduced by a virus that has been engineered to carry foreign genes. In plant cells, genes are frequently introduced by a technique called particle bombardment: DNA samples are painted onto tiny gold beads and then literally shot through the cell wall with a specially modified gun. *Electroporation* is the method of choice for introducing DNA into bacteria and some other cells. In this technique, a brief electric shock renders the cell membrane temporarily permeable, allowing foreign DNA to enter the cytoplasm.

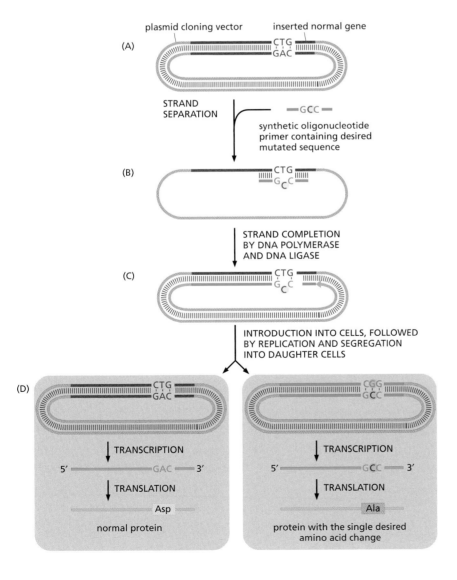

Figure 8–63 The use of a synthetic oligonucleotide to modify the protein-coding region of a gene by site-directed mutagenesis. (A) A recombinant plasmid containing a gene insert is separated into its two DNA strands. A synthetic oligonucleotide primer corresponding to part of the gene sequence but containing a single altered nucleotide at a predetermined point is added to the single-stranded DNA under conditions that permit imperfect DNA hybridization (see Figure 8–36). (B) The primer hybridizes to the DNA, forming a single mismatched nucleotide pair. (C) The recombinant plasmid is made double-stranded by *in vitro* DNA synthesis (starting from the primer) followed by sealing by DNA ligase. (D) The double-stranded DNA is introduced into a cell, where it is replicated. Replication using one strand of the template produces a normal DNA molecule, but replication using the other strand (the one that contains the primer) produces a DNA molecule carrying the desired mutation. Only half of the progeny cells will end up with a plasmid that contains the desired mutant gene. However, a progeny cell that contains the mutated gene can be identified, separated from other cells, and cultured to produce a pure population of cells, all of which carry the mutated gene. Only one of the many changes that can be engineered in this way is shown here. With an oligonucleotide of the appropriate sequence, more than one amino acid substitution can be made at a time, or one or more amino acids can be inserted or deleted. Although not shown in this figure, it is also possible to create a site-directed mutation by using the appropriate oligonucleotides and PCR (instead of plasmid replication) to amplify the mutated gene.

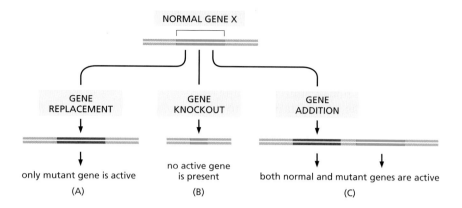

NORMAL GENE X

GENE REPLACEMENT

GENE KNOCKOUT

GENE ADDITION

only mutant gene is active

(A)

no active gene is present

(B)

both normal and mutant genes are active

(C)

Figure 8–64 Gene replacement, gene knockout, and gene addition. A normal gene can be altered in several ways to produce a transgenic organism. (A) The normal gene *(green)* can be completely replaced by a mutant copy of the gene *(red)*. This provides information on the activity of the mutant gene without interference from the normal gene, and thus the effects of small and subtle mutations can be determined. (B) The normal gene can be inactivated completely, for example, by making a large deletion in it. (C) A mutant gene can simply be added to the genome. In some organisms this is the easiest type of genetic engineering to perform. This approach can provide useful information when the introduced mutant gene overrides the function of the normal gene, as with a dominant-negative mutation (see Figure 8–62).

Unlike higher eucaryotes (which are multicellular and diploid), bacteria, yeasts, and the cellular slime mold *Dictyostelium* generally exist as haploid single cells. In these organisms, an artificially introduced DNA molecule carrying a mutant gene can, with a relatively high frequency, replace the single copy of the normal gene by homologous recombination; it is therefore easy to produce cells in which the mutant gene has replaced the normal gene (**Figure 8–64**A). In this way, cells can be made in order to miss a particular protein or produce an altered form of it. The ability to perform direct gene replacements in lower eucaryotes, combined with the power of standard genetic analyses in these haploid organisms, explains in large part why studies in these types of cells have been so important for working out the details of those cell processes that are shared by all eucaryotes.

Animals Can Be Genetically Altered

Gene additions and replacements are also possible, but more difficult to perform, in animals and plants. Animals and plants that have been genetically engineered by either gene insertion, gene deletion, or gene replacement are called **transgenic organisms**, and any foreign or modified genes that are added are called **transgenes**. We concentrate our discussion on transgenic mice, as enormous progress is being made in this area. If a DNA molecule carrying a mutated mouse gene is transferred into a mouse cell, it usually inserts into the chromosomes at random, but about once in a thousand times, it replaces one of the two copies of the normal gene by homologous recombination. By exploiting these rare "gene targeting" events, any specific gene can be altered or inactivated in a mouse cell by a direct gene replacement. In the special case in which both copies of the gene of interest is completely inactivated or deleted, the resulting animal is called a **"knockout" mouse**.

The technique works as follows. In the first step, a DNA fragment containing a desired mutant gene (or a DNA fragment designed to interrupt a target gene) is inserted into a vector and then introduced into cultured embryonic stem (ES) cells (see Figure 8–5), which are capable of producing cells of many different types. After a period of cell proliferation, the rare colonies of cells in which a homologous recombination event is likely to have caused a gene replacement to occur are isolated. The correct colonies among these are identified by PCR or by Southern blotting: they contain recombinant DNA sequences in which the inserted fragment has replaced all or part of one copy of the normal gene. In the second step, individual ES cells from the identified colony are taken up into a fine micropipette and injected into an early mouse embryo. The transfected ES cells collaborate with the cells of the host embryo to produce a normal-looking mouse; large parts of this chimeric animal, including—in favorable cases—cells of the germ line, often derive from the transfected ES cells (**Figure 8–65**).

The mice with the transgene in their germ line are then bred to produce both a male and a female animal, each heterozygous for the gene replacement (that is, they have one normal and one mutant copy of the gene). When these two mice are mated, one-fourth of their progeny will be homozygous for the altered gene. Studies of these homozygotes allow the function of the altered gene—or the effects of eliminating the gene's activity—to be examined in the absence of the corresponding normal gene.

The ability to prepare transgenic mice lacking a known normal gene has been a major advance, and the technique is now being used to determine the functions of all mouse genes (**Figure 8–66**). A special technique is used to produce conditional mutants, in which a selected gene becomes disrupted in a specific tissue at a certain time in development. The strategy takes advantage of a site-specific recombination system to excise—and thus disable—the target gene in a particular place or at a particular time. The most common of these recombination systems, called **Cre/lox**, is widely used to engineer gene replacements in mice and in plants (see Figure 5–79). In this case, the target gene in ES cells is replaced by a fully functional version of the gene that is flanked by a pair of the

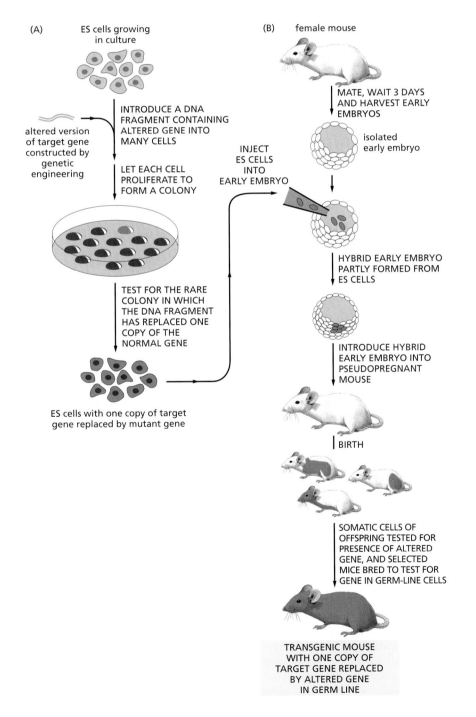

Figure 8–65 Summary of the procedures used for making gene replacements in mice. In the first step (A), an altered version of the gene is introduced into cultured ES (embryonic stem) cells. Only a few rare ES cells will have their corresponding normal genes replaced by the altered gene through a homologous recombination event. Although the procedure is often laborious, these rare cells can be identified and cultured to produce many descendants, each of which carries an altered gene in place of one of its two normal corresponding genes. In the next step of the procedure (B), these altered ES cells are injected into a very early mouse embryo; the cells are incorporated into the growing embryo, and a mouse produced by such an embryo will contain some somatic cells (indicated by *orange*) that carry the altered gene. Some of these mice will also contain germ-line cells that contain the altered gene; when bred with a normal mouse, some of the progeny of these mice will contain one copy of the altered gene in all of their cells. If two such mice are bred (not shown), some of the progeny will contain two altered genes (one on each chromosome) in all of their cells.

If the original gene alteration completely inactivates the function of the gene, these homozygous mice are known as knockout mice. When such mice are missing genes that function during development, they often die with specific defects long before they reach adulthood. These lethal defects are carefully analyzed to help determine the normal function of the missing gene.

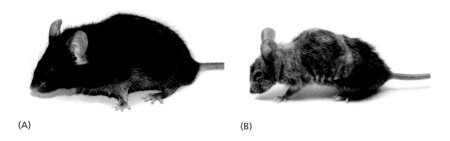

(A)   (B)

Figure 8–66 Transgenic mice engineered to express a mutant DNA helicase show premature aging. The helicase, encoded by the *Xpd* gene, is involved in both transcription and DNA repair. Compared with a wild-type mouse of the same age (A), a transgenic mouse that expresses a defective version of *Xpd* (B) exhibits many of the symptoms of premature aging, including osteoporosis, emaciation, early graying, infertility, and reduced life-span. The mutation in *Xpd* used here impairs the activity of the helicase and mimics a mutation that in humans causes trichothiodystrophy, a disorder characterized by brittle hair, skeletal abnormalities, and a very reduced life expectancy. These results indicate that an accumulation of DNA damage can contribute to the aging process in both humans and mice. (From J. de Boer et al., *Science* 296:1276–1279, 2002. With permission from AAAS.)

short DNA sequences, called *lox sites*, that are recognized by the *Cre recombinase* protein. The transgenic mice that result are phenotypically normal. They are then mated with transgenic mice that express the Cre recombinase gene under the control of an inducible promoter. In the specific cells or tissues in which Cre is switched on, it catalyzes recombination between the lox sequences—excising a target gene and eliminating its activity. Similar recombination systems are used to generate conditional mutants in *Drosophila* (see Figure 22–49).

Transgenic Plants Are Important for Both Cell Biology and Agriculture

A damaged plant can often repair itself by a process in which mature differentiated cells "dedifferentiate," proliferate, and then redifferentiate into other cell types. In some circumstances, the dedifferentiated cells can even form an apical meristem, which can then give rise to an entire new plant, including gametes. This remarkable developmental plasticity of plant cells can be exploited to generate transgenic plants from cells growing in culture.

When a piece of plant tissue is cultured in a sterile medium containing nutrients and appropriate growth regulators, many of the cells are stimulated to proliferate indefinitely in a disorganized manner, producing a mass of relatively undifferentiated cells called a *callus*. By carefully manipulating the nutrients and growth regulators, one can induce the formation of shoot and then root apical meristems within the callus, and, in many species, regenerate a whole new plant.

Callus cultures can also be mechanically dissociated into single cells, which will grow and divide as a suspension culture. In several plants—including tobacco, petunia, carrot, potato, and *Arabidopsis*—a single cell from such a suspension culture can be grown into a small clump (a clone) from which a whole plant can be regenerated. Such a cell, which has the ability to give rise to all parts of the organism, is considered **totipotent**. Just as mutant mice can be derived by the genetic manipulation of ES cells in culture, so transgenic plants can be created from single totipotent plant cells that have been transfected with DNA in culture (**Figure 8–67**).

The ability to produce transgenic plants has greatly accelerated progress in many areas of plant cell biology. It has had an important role, for example, in isolating receptors for growth regulators and in analyzing the mechanisms of morphogenesis and of gene expression in plants. It has also opened up many new possibilities in agriculture that could benefit both farmer and consumer. It has made it possible, for example, to modify the lipid, starch, and protein stored in seeds, to impart pest and virus resistance to plants, and to create modified plants that tolerate extreme habitats such as salt marshes or water-stressed soil.

Many of the major advances in understanding animal development have come from studies on the fruit fly *Drosophila* and the nematode worm *C. elegans*, which are amenable to classical genetic analysis, as well as to experimental manipulation. Progress in plant developmental biology has, in the past, been relatively slow by comparison. Many of the plants that have proved most amenable to genetic analysis—such as maize and tomato—have long life cycles and very large genomes, making both classical and molecular genetic analysis very time-consuming. Increasing attention is consequently being paid to a fast-growing small weed, the common wall cress *(Arabidopsis thaliana)*, which has

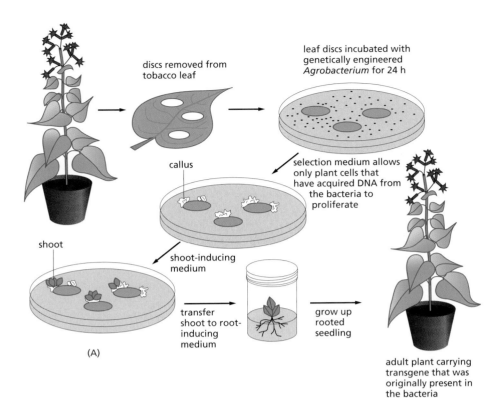

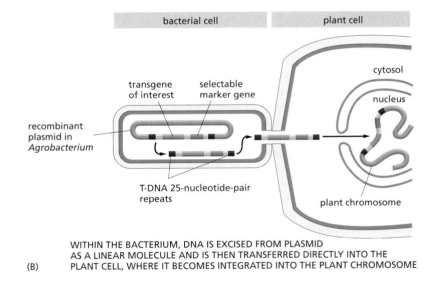

Figure 8–67 **A procedure used to make a transgenic plant.** (A) Outline of the process. A disc is cut out of a leaf and incubated in culture with *Agrobacterium* cells that carry a recombinant plasmid that contains both a selectable marker gene and a desired transgene. The wounded cells at the edge of the disc release substances that attract the *Agrobacterium* cells and cause them to inject DNA into these cells. Only those plant cells that take up the appropriate DNA and express the selectable marker gene survive to proliferate and form a callus. The manipulation of growth regulators and nutrients supplied to the callus induces it to form shoots, which subsequently root and grow into adult plants carrying the transgene. (B) The preparation of the recombinant plasmid and its transfer to plant cells. An *Agrobacterium* plasmid that normally carries the T-DNA sequence is modified by substituting a selectable marker gene (such as the kanamycin-resistance gene) and a desired transgene between the 25-nucleotide-pair T-DNA repeats. When the *Agrobacterium* recognizes a plant cell, it efficiently passes a DNA strand that carries these sequences into the plant cell, using the special machinery that normally transfers the plasmid's T-DNA sequence.

several major advantages as a "model plant" (see Figures 1–46 and 22–112). The relatively small *Arabidopsis* genome was the first plant genome to be completely sequenced, and the pace of research on this organism now rivals that of the model animals.

Large Collections of Tagged Knockouts Provide a Tool for Examining the Function of Every Gene in an Organism

Extensive collaborative efforts are underway to assemble comprehensive libraries of mutations in a variety of model organisms, including *S. cerevisiae, C. elegans, Drosophila, Arabidopsis,* and the mouse. The ultimate aim in each case is to produce a collection of mutant strains in which every gene in the organism has been systematically deleted or altered in such a way that it can be conditionally disrupted. Collections of this type will provide an invaluable resource for investi-

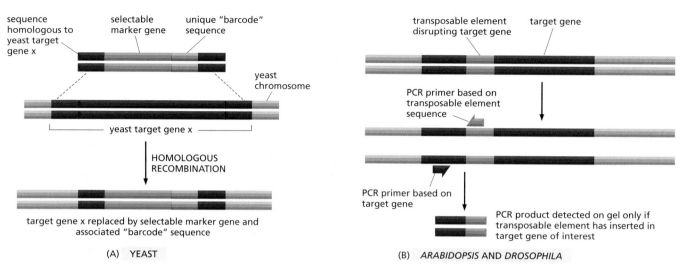

Figure 8–68 Making collections of mutant organisms. (A) A deletion cassette for use in yeast contains DNA sequences *(red)* homologous to each end of a target gene x, a selectable marker gene *(blue)*, and a unique "barcode" sequence approximately 20 nucleotide pairs in length *(green)*. This DNA is introduced into yeast cells, where it readily replaces the target gene by homologous recombination. By using a collection of such cassettes, each specific for one gene, a library of yeast mutants can be constructed containing a mutant for every gene. (B) A similar approach can be taken to prepare tagged knockout mutants in *Arabidopsis* and *Drosophila*. In this case, mutations are generated by the accidental insertion of a transposable element into a target gene. The total DNA from the resulting organism can be collected and quickly screened for disruption of a gene of interest by using PCR primers that bind to the transposable element and to the target gene. A PCR product is detected on the gel only if the transposable element has inserted into the target gene (see Figure 8–45).

gating gene function on a genomic scale. In some cases, each of the individual mutations within the collection will express a distinct molecular tag—in the form of a unique DNA sequence—designed to make identification of the altered gene rapid and routine.

In *S. cerevisiae*, the task of generating a complete set of 6000 mutants, each missing only one gene, is made simpler by yeast's propensity for homologous recombination. For each gene, a "deletion cassette" is prepared. The cassette consists of a special DNA molecule that contains 50 nucleotides identical in sequence to each end of the targeted gene, surrounding a selectable marker. In addition, a special "barcode" sequence tag is embedded in this DNA molecule to facilitate the later rapid identification of each resulting mutant strain (**Figure 8–68**). A large mixture of such gene knockout mutants can then be grown under various selective test conditions—such as nutritional deprivation, a temperature shift, or the presence of various drugs—and the cells that survive can be rapidly identified by their unique sequence tags. By assessing how well each mutant in the mixture fares, one can begin to assess which genes are essential, useful, or irrelevant for growth under the various conditions.

The challenge in deriving information from the study of such yeast mutants lies in deducing a gene's activity or biological role based on a mutant phenotype. Some defects—an inability to live without histidine, for example—point directly to the function of the wild-type gene. Other connections may not be so obvious. What might a sudden sensitivity to cold indicate about the role of a particular gene in the yeast cell? Such problems are even greater in organisms that are more complex than yeast. The loss of function of a single gene in the mouse, for example, may affect many different tissue types at different stages of development—whereas the loss of other genes may have no obvious effect. Adequately characterizing mutant phenotypes in mice often requires a thorough examination, along with extensive knowledge of mouse anatomy, histology, pathology, physiology, and complex behavior.

The insights generated by examination of mutant libraries, however, will be great. For example, studies of an extensive collection of mutants in *Mycoplasma genitalium*—the organism with the smallest known genome—have identified the minimum complement of genes essential for cellular life. Analysis of the mutant pool suggests that growth under laboratory conditions requires about three-quarters of the 480 protein-coding genes in *M. genitalium*. Approximately

100 of these essential genes are of unknown function, which suggests that a surprising number of the basic molecular mechanisms that underlie life have yet to be discovered.

RNA Interference Is a Simple and Rapid Way to Test Gene Function

Although knocking out a gene in an organism and studying the consequences is perhaps the most powerful approach for understanding the functions of the gene, a much easier way to inactivate genes has been recently discovered. Called **RNA interference** (**RNAi**, for short), this method exploits a natural mechanism used in many plants, animals, fungi, and protozoa to protect themselves against certain viruses and transposable elements (see Figure 7–115). The technique introduces into a cell or organism a double-stranded RNA molecule whose nucleotide sequence matches that of part of the gene to be inactivated. After the RNA is processed, it hybridizes with the mRNA produced by the target gene and directs its degradation. The cell subsequently uses small fragments of this degraded RNA to produce more double-stranded RNA, which directs the continued elimination of the target mRNA. Because these short RNA fragments can be passed on to progeny cells, RNAi can cause heritable changes in gene expression. But, as we saw in Chapter 7, there is a second mechanism through which RNAi can stably inactivate genes. RNA fragments produced by degradation in the cytosol can enter the nucleus and interact with the target gene itself, directing its packaging into a transcriptionally repressed form of chromatin. This dual mode of controlling gene expression makes RNAi an especially effective tool for shutting down genes, one at a time.

RNAi is frequently used to inactivate genes in *Drosophila* and mammalian cell culture lines. Indeed, sets of 15,000 *Drosophila* RNAi molecules (one for every gene) allow researchers, in several months, to test the role of every fly gene in any process that can be monitored using cultured cells. Soon, it will be possible to carry out the same type of analysis with the 25,000 mouse and human genes. RNAi has also been widely used to study gene function in the nematode, *C. elegans*. When working with worms, introducing the double-stranded RNA is quite simple: the RNA can be injected directly into the intestine of the animal, or the worm can be fed with *E. coli* engineered to produce the RNA (**Figure 8–69**). The RNA is distributed throughout the body of the worm, where it inhibits expression of the target gene in different tissue types. Because the entire genome of *C. elegans* has been sequenced, RNAi is being used to help in assigning functions to the entire complement of worm genes.

More recently, a related technique has also been widely applied to mice. In this case, the RNAi is not injected or fed to the mouse; rather, recombinant DNA techniques are used to make transgenic animals that express the RNAi under the control of an inducible promoter. Often this is a specially designed RNA that can fold back on itself and, through base pairing, produce a double-stranded region

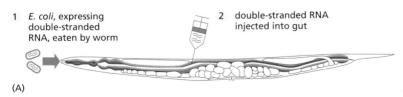

1 *E. coli*, expressing double-stranded RNA, eaten by worm

2 double-stranded RNA injected into gut

(A)

(B)

(C)

20 μm

Figure 8–69 Dominant-negative mutation created by RNA interference. (A) Double-stranded RNA (dsRNA) can be introduced into *C. elegans* (1) by feeding the worms with *E. coli* expressing the dsRNA or (2) by injecting dsRNA directly into the gut. (B) Wild-type worm embryo shortly after the egg has been fertilized. The egg and sperm pronuclei (*red* arrowheads) have migrated and come together in the posterior half of the embryo. (C) Worm embryo at the same stage in which a gene involved in cell division has been inactivated by RNAi. The two pronuclei have failed to migrate. (B and C, from P. Gönczy et al., *Nature* 408:331–336, 2000. With permission from Macmillan Publishers Ltd.)

that is recognized by the RNAi machinery. The process inactivates only the genes that exactly match the RNAi sequence. Depending on the inducible promoter used, the RNAi can be produced only in a specified tissue or only at a particular time in development, allowing the functions of the target genes to be analyzed in elaborate detail.

RNAi has made reverse genetics simple and efficient in many organisms, but it has several potential limitations compared with true genetic knockouts. For unknown reasons, RNAi does not efficiently inactivate all genes. Moreover, within whole organisms, certain tissues may be resistant to the action of RNAi (for example, neurons in nematodes). Another problem arises because many organisms contain large gene families, the members of which exhibit sequence similarity. RNAi therefore sometimes produces "off-target" effects, inactivating related genes in addition to the targeted gene. One strategy to avoid such problems is to use multiple small RNA molecules matched to different regions of the same gene. Ultimately, the results of any RNAi experiment must be viewed as a strong clue to, but not necessarily a proof of, normal gene function.

Reporter Genes and *In Situ* Hybridization Reveal When and Where a Gene Is Expressed

Important insights into gene function can often be obtained by examining when and where a gene is expressed in the cell or in the whole organism. Determining the pattern and timing of gene expression can be accomplished by replacing the coding portion of the gene under study with a reporter gene. In most cases, the expression of the reporter gene is then monitored by tracking the fluorescence or enzymatic activity of its protein product (see Figures 9–26 and 9–27).

As discussed in detail in Chapter 7, regulatory DNA sequences, located upstream or downstream of the coding region, control gene expression. These regulatory sequences, which determine exactly when and where the gene is expressed, can be easily studied by placing a reporter gene under their control and introducing these recombinant DNA molecules into cells (**Figure 8–70**).

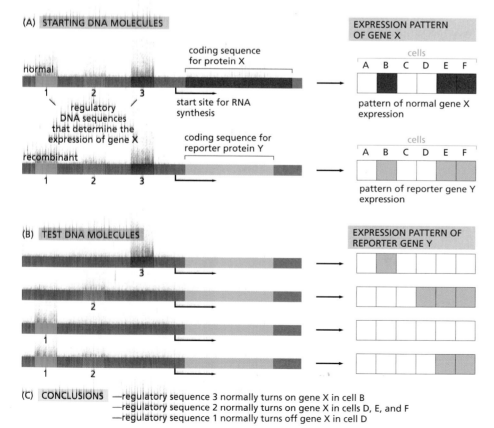

(A) STARTING DNA MOLECULES

coding sequence for protein X

normal

1 2 3

regulatory DNA sequences that determine the expression of gene X

start site for RNA synthesis

coding sequence for reporter protein Y

recombinant

1 2 3

EXPRESSION PATTERN OF GENE X

cells
A B C D E F

pattern of normal gene X expression

cells
A B C D E F

pattern of reporter gene Y expression

(B) TEST DNA MOLECULES

3

2

1

1 2

EXPRESSION PATTERN OF REPORTER GENE Y

(C) CONCLUSIONS —regulatory sequence 3 normally turns on gene X in cell B
—regulatory sequence 2 normally turns on gene X in cells D, E, and F
—regulatory sequence 1 normally turns off gene X in cell D

Figure 8–70 Using a reporter protein to determine the pattern of a gene's expression. (A) In this example, the coding sequence for protein X is replaced by the coding sequence for reporter protein Y. The expression pattern for X and Y are the same. (B) Various fragments of DNA containing candidate regulatory sequences are added in combinations to produce test DNA molecules encoding reporter gene Y. These recombinant DNA molecules are then tested for expression after their transfection into a variety of different types of mammalian cells. The results are summarized in (C).

For experiments in eucaryotic cells, two commonly used reporter proteins are the enzyme β-galactosidase *(β-gal)* (see Figure 7–55B) and green fluorescent protein or GFP (see Figure 9–26). Figure 7–55B shows an example in which the *β-gal* gene is used to monitor the activity of the *Eve* gene regulatory sequence in a *Drosophila* embryo.

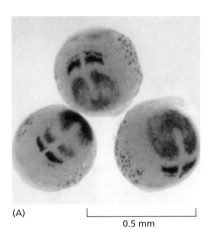
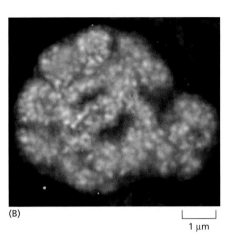

(A) 0.5 mm   (B) 1 μm

Figure 8–71 *In situ* hybridization for RNA localization. (A) Expression pattern of *DeltaC* mRNA in the early zebrafish embryo. This gene codes for a ligand in the Notch signaling pathway (discussed in Chapter 15), and the pattern shown here reflects its role in the development of somites—the future segments of the vertebrate trunk and tail. (B) High-resolution RNA *in situ* localization reveals the sites within the nucleolus of a pea cell where ribosomal RNA is synthesized. The sausage-like structures, 0.5–1 μm in diameter, correspond to the loops of chromosomal DNA that contain the genes encoding rRNA. Each small white spot represents transcription of a single rRNA gene. (A, courtesy of Yun-Jin Jiang; B, courtesy of Peter Shaw.)

It is also possible to directly observe the time and place that the mRNA product of a gene is expressed. Although this strategy often provides the same general information as the reporter gene approaches discussed above, there are instances where it provides additional information; for example, when the gene is transcribed but the mRNA is not immediately translated, or when the gene's final product is RNA rather than protein. This procedure, called ***in situ hybridization***, relies on the principles of nucleic acid hybridization described earlier. Typically, tissues are gently fixed so that their RNA is retained in an exposed form that can hybridize with a labeled complementary DNA or RNA probe. In this way, the patterns of differential gene expression can be observed in tissues, and the location of specific RNAs in cells can be determined (**Figure 8–71**). In the *Drosophila* embryo, for example, such patterns have provided new insights into the mechanisms that create distinctions between cells in different positions during development (described in Chapter 22).

Using similar approaches, it is also possible to visualize specific DNA sequences in cells. In this case, tissue, cell, or even chromosome preparations are briefly exposed to high pH to disrupt their nucleotide pairs, and nucleic acid probes are added, allowed to hybridize with the cells' DNA, and then visualized (see Figure 8–35).

Expression of Individual Genes Can Be Measured Using Quantitative RT-PCR

Although reporter genes and *in situ* hybridization reveal patterns of gene expression, it is often desirable to quantitate gene expression by directly measuring mRNA levels in cells. Although Northern blots (see Figure 8–38) can be adapted to this purpose, a more accurate method is based on the principles of PCR (**Figure 8–72**). This method, called **quantitative RT-PCR** (reverse transcription-polymerase chain reaction), begins with the total population of mRNA molecules purified from a tissue or a cell culture. It is important that no DNA be present in the preparation; it must be purified away or enzymatically degraded. Two DNA primers that specifically match the gene of interest are added, along with reverse transcriptase, DNA polymerase, and the four deoxynucleoside triphosphates needed for DNA synthesis. The first round of synthesis is the reverse transcription of the mRNA into DNA using one of the primers. Next, a series of heating and cooling cycles allows the amplification of that DNA strand by conventional PCR (see Figure 8–45). The quantitative part of this method relies on a direct relationship between the rate at which the PCR product is generated and the original concentration of the mRNA species of interest. By adding chemical dyes to the PCR reaction that fluoresce only when bound to double-stranded DNA, a simple fluorescence measurement can be used to track the progress of the reaction and thereby accurately deduce the starting concentration of the mRNA that is amplified (see Figure 8–72). Although it seems complicated, this quantitative RT-PCR technique (sometimes called *real time PCR*) is relatively fast and simple

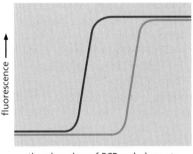

time (number of PCR cycles) →

Figure 8–72 **RNA levels can be measured by quantitative RT-PCR.** The fluorescence measured is generated by a dye that fluoresces only when bound to the double-stranded DNA products of the RT-PCR reaction (see Figure 8–46B). The red sample has a higher concentration of the mRNA being measured than does the blue sample, since it requires fewer PCR cycles to reach the same half-maximal concentration of double-stranded DNA. Based on this difference, the relative amounts of the mRNA in the two samples can be precisely determined.

Figure 8–73 Using DNA microarrays to monitor the expression of thousands of genes simultaneously. To prepare the microarray, DNA fragments—each corresponding to a gene—are spotted onto a slide by a robot. Prepared arrays are also widely available commercially. In this example, mRNA is collected from two different cell samples for a direct comparison of their relative levels of gene expression; the two samples, for example, could be from cells treated with a hormone and untreated cells of the same type. These samples are converted to cDNA and labeled, one with a red fluorochrome, the other with a green fluorochrome. The labeled samples are mixed and then allowed to hybridize to the microarray. After incubation, the array is washed and the fluorescence scanned. In the portion of a microarray shown, which represents the expression of 110 yeast genes, *red* spots indicate that the gene in sample 1 is expressed at a higher level than the corresponding gene in sample 2; *green* spots indicate that expression of the gene is higher in sample 2 than in sample 1. *Yellow* spots reveal genes that are expressed at equal levels in both cell samples. Dark spots indicate little or no expression in either sample of the gene whose fragment is located at that position in the array. (Microarray courtesy of J.L. DeRisi et al., *Science* 278:680–686, 1997. With permission from AAAS.)

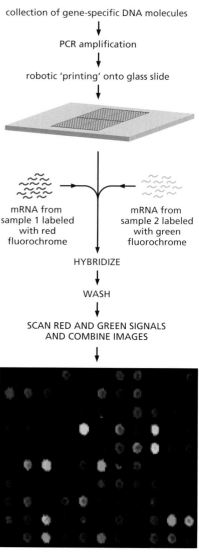

small region of microarray representing expression of 110 genes from yeast

to perform in the laboratory; it has displaced Northern blotting as the method of choice for quantifying mRNA levels from any given gene.

Microarrays Monitor the Expression of Thousands of Genes at Once

So far we have discussed techniques that can be used to monitor the expression of only a single gene (or relatively few genes) at a time. Developed in the 1990s, **DNA microarrays** have revolutionized the analysis of gene expression by monitoring the RNA products of thousands of genes at once. By examining the expression of so many genes simultaneously, we can now begin to identify and study the gene expression patterns that underlie cell physiology: we can see which genes are switched on (or off) as cells grow, divide, differentiate, or respond to hormones or to toxins.

DNA microarrays are little more than glass microscope slides studded with a large number of DNA fragments, each containing a nucleotide sequence that serves as a probe for a specific gene. The most dense arrays may contain tens of thousands of these fragments in an area smaller than a postage stamp, allowing thousands of hybridization reactions to be performed in parallel (**Figure 8–73**). Some microarrays are prepared from large DNA fragments that have been generated by PCR and then spotted onto the slides by a robot. Others contain short oligonucleotides that are synthesized on the surface of the glass wafer with techniques similar to those that are used to etch circuits onto computer chips. In either case, the exact sequence—and position—of every probe on the chip is known. Thus, any nucleotide fragment that hybridizes to a probe on the array can be identified as the product of a specific gene simply by detecting the position at which it is bound.

To use a DNA microarray to monitor gene expression, mRNA from the cells being studied is first extracted and converted to cDNA (see Figure 8–43). The cDNA is then labeled with a fluorescent probe. The microarray is incubated with this labeled cDNA sample and hybridization is allowed to occur (see Figure 8–73). The array is then washed to remove cDNA that is not tightly bound, and the positions in the microarray to which labeled DNA fragments have bound are identified by an automated scanning-laser microscope. The array positions are then matched to the particular gene whose sample of DNA was spotted in this location.

Typically the fluorescent DNA from the experimental samples (labeled, for example, with a red fluorescent dye) are mixed with a reference sample of cDNA fragments labeled with a differently colored fluorescent dye (green, for example). Thus, if the amount of RNA expressed from a particular gene in the cells of interest is increased relative to that of the reference sample, the resulting spot is red. Conversely, if the gene's expression is decreased relative to the reference sample, the spot is green. If there is no change compared to the reference sample, the spot

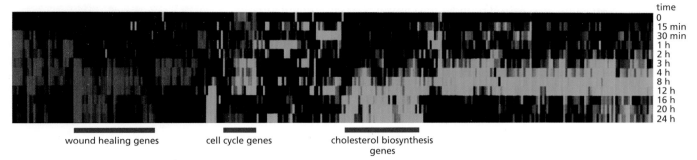

Figure 8–74 Using cluster analysis to identify sets of genes that are coordinately regulated. Genes that belong to the same cluster may be involved in common pathways or processes. To perform a cluster analysis, microarray data are obtained from cell samples exposed to a variety of different conditions, and genes that show coordinate changes in their expression pattern are grouped together. In this experiment, human fibroblasts were deprived of serum for 48 hours; serum was then added back to the cultures at time 0 and the cells were harvested for microarray analysis at different time points. Of the 8600 genes analyzed on the DNA microarray, just over 300 showed threefold or greater variation in their expression patterns in response to serum re-introduction. Here, *red* indicates an increase in expression; *green* is a decrease in expression. On the basis of the results of many microarray experiments, the 8600 genes have been grouped in clusters based on similar patterns of expression. The results of this analysis show that genes involved in wound healing are turned on in response to serum, while genes involved in regulating cell cycle progression and cholesterol biosynthesis are shut down. (From M.B. Eisen et al., *Proc. Natl Acad. Sci. U.S.A.* 95:14863–14868, 1998. With permission from National Academy of Sciences.)

is yellow. Using such an internal reference, gene expression profiles can be tabulated with great precision.

So far, DNA microarrays have been used to examine everything from the changes in gene expression that make strawberries ripen to the gene expression "signatures" of different types of human cancer cells (see Figure 7–3); or from changes that occur as cells progress through the cell cycle to those made in response to sudden shifts in temperature. Indeed, because microarrays allow the simultaneous monitoring of large numbers of genes, they can detect subtle changes in a cell, changes that might not be manifested in its outward appearance or behavior.

Comprehensive studies of gene expression also provide an additional layer of information that is useful for predicting gene function. Earlier, we discussed how identifying a protein's interaction partners can yield clues about that protein's function. A similar principle holds true for genes: information about a gene's function can be deduced by identifying genes that share its expression pattern. Using a technique called *cluster analysis*, one can identify sets of genes that are coordinately regulated. Genes that are turned on or turned off together under different circumstances are likely to work in concert in the cell: they may encode proteins that are part of the same multiprotein machine, or proteins that are involved in a complex coordinated activity, such as DNA replication or RNA splicing. Characterizing a gene whose function is unknown by grouping it with known genes that share its transcriptional behavior is sometimes called "guilt by association." Cluster analyses have been used to analyze the gene expression profiles that underlie many interesting biological processes, including wound healing in humans (**Figure 8–74**).

In addition to monitoring the level of mRNA corresponding to every gene in a genome, DNA microarrays have many other uses. For example, they can be used to monitor the progression of DNA replication in a cell (see Figure 5–32) and, when combined with immunoprecipitation, can pinpoint every position in the genome occupied by a given gene regulatory protein (see Figure 7–32). Microarrays can also be used to quickly identify disease-causing microbes by hybridizing DNA from infected tissues to an array containing genomic DNA sequences from large collections of pathogens.

Single-Cell Gene Expression Analysis Reveals Biological "Noise"

The methods for monitoring mRNAs just described give average expression levels for each mRNA across a large population of cells. By using a fluorescent reporter protein whose expression is under the control of a promoter of interest, it is also possible to accurately measure expression levels in individual cells. These new approaches have revealed a startling amount of variability, often called *biological noise,* between the individual cells in a homogeneous population of cells. These studies have also revealed the presence of distinct subpopulations of cells whose existence would be masked if only the average across a whole population were considered. For example, a bimodal distribution of expression levels would indicate that the cells can exist in two distinct states (**Figure 8–75**), with the average expression level of the population being somewhere between them. The behavior of individual cells has important implications for understanding biology, for example, by revealing that some cells constantly and rapidly switch back and forth between two states.

Currently, there are two approaches for monitoring gene expression in individual cells. In the imaging approach, live cells are mounted on a slide and viewed through a fluorescence microscope. This method has the advantage that a given cell can be followed over time, allowing temporal changes in expression to be measured. The second approach, flow cytometry, works by streaming a dilute suspension of cells past an illuminator and measuring the fluorescence of individual cells as they flow past the detector (see Figure 8–2). Although it has the advantage that the expression levels of very large numbers of cells can be measured with precision, flow cytometry does not allow a given cell to be tracked over time; hence, it is complementary to the imaging methods.

Figure 8–75 Different levels of gene expression in individual cells within a population of *E. coli* bacteria. For this experiment, two different reporter proteins (one fluorescing *green*, the other *red*) controlled by a copy of the same promoter, have been introduced into all of the bacteria. When illuminated, some cells express only one gene copy, and so appear either *red* or *green*, while others express both gene copies, and so appear *yellow*. This experiment also reveals variable levels of fluorescence, indicating variable levels of gene expression within an apparently uniform population of cells. (From M.B. Elowitz, A.J. Levine, E.O. Siggia and P.S. Swain, *Science* 297:1183–1186, 2002. With permission from AAAS.)

Summary

Genetics and genetic engineering provide powerful tools for the study of gene function in both cells and organisms. In the classical genetic approach, random mutagenesis is coupled with screening to identify mutants that are deficient in a particular biological process. These mutants are then used to locate and study the genes responsible for that process.

Gene function can also be ascertained by reverse genetic techniques. DNA engineering methods can be used to alter genes and to re-insert them into a cell's chromosomes so that they become a permanent part of the genome. If the cell used for this gene transfer is a fertilized egg (for an animal) or a totipotent plant cell in culture, transgenic organisms can be produced that express the mutant gene and pass it on to their progeny. Especially important for cell biology is the ability to alter cells and organisms in highly specific ways—allowing one to discern the effect on the cell or the organism of a designed change in a single protein or RNA molecule.

Many of these methods are being expanded to investigate gene function on a genome-wide scale. The generation of mutant libraries in which every gene in an organism has been systematically deleted or disrupted provides invaluable tools for exploring the role of each gene in the elaborate molecular collaboration that gives rise to life. Technologies such as DNA microarrays can monitor the expression of thousands of genes simultaneously, providing detailed, comprehensive snapshots of the dynamic patterns of gene expression that underlie complex cell processes.

PROBLEMS

Which statements are true? Explain why or why not.

8–1   Because a monoclonal antibody recognizes a specific antigenic site (epitope), it binds only to the specific protein against which it was made.

8–2   Given the inexorable progress of technology, it seems inevitable that the sensitivity of detection of molecules will ultimately be pushed beyond the yoctomole level (10^{-24} mole).

8–3   Surface plasmon resonance (SPR) measures association (k_{on}) and dissociation (k_{off}) rates between molecules in real time, using small amounts of unlabeled molecules, but it does not give the information needed to determine the binding constant (K).

8–4   If each cycle of PCR doubles the amount of DNA synthesized in the previous cycle, then 10 cycles will give a 10^3-fold amplification, 20 cycles will give a 10^6-fold amplification, and 30 cycles will give a 10^9-fold amplification.

Discuss the following problems.

8–5   A common step in the isolation of cells from a sample of animal tissue is to treat it with trypsin, collagenase, and EDTA. Why is such a treatment necessary, and what does each component accomplish? And why does this treatment not kill the cells?

8–6 Do you suppose it would be possible to raise an antibody against another antibody? Explain your answer.

8–7 Distinguish between velocity sedimentation and equilibrium sedimentation. For what general purpose is each technique used? Which do you suppose might be best suited for separating two proteins of different size?

8–8 Tropomyosin, at 93 kd, sediments at 2.6 S, whereas the 65-kd protein, hemoglobin, sediments at 4.3 S. (The sedimentation coefficient S is a linear measure of the rate of sedimentation: both increase or decrease in parallel.) These two proteins are shown as α-carbon backbone models in **Figure Q8–1**. How is it that the bigger protein sediments more slowly than the smaller one? Can you think of an analogy from everyday experience that might help you with this problem?

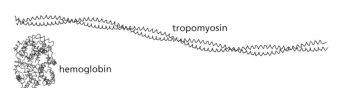

Figure Q8–1 Backbone models of tropomyosin and hemoglobin (Problem 8–8).

8–9 In the classic paper that demonstrated the semi-conservative replication of DNA, Meselson and Stahl began by showing that DNA itself will form a band when subjected to equilibrium sedimentation. They mixed randomly fragmented *E. coli* DNA with a solution of CsCl so that the final solution had a density of 1.71 g/mL. As shown in **Figure Q8–2**, with increasing length of centrifugation at 70,000 times gravity, the DNA, which was initially dispersed throughout the centrifuge tube, became concentrated over time into a discrete band in the middle.

A. Describe what is happening with time and explain why the DNA forms a discrete band.

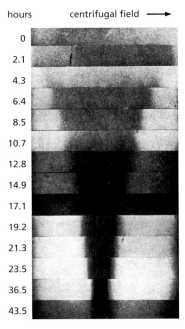

Figure Q8–2 Ultraviolet absorption photographs showing successive stages in the banding of *E. coli* DNA (Problem 8–9). DNA, which absorbs UV light, shows up as *dark regions* in the photographs. The bottom of the centrifuge tube is on the *right*. (From M. Meselson and F.W. Stahl, *Proc. Natl Acad. Sci. U.S.A.* 44:671–682, 1958. With permission from National Academy of Sciences.)

B. What is the buoyant density of the DNA? (The density of the solution at which DNA "floats" at equilibrium defines the "buoyant density" of the DNA.)

C. Even if the DNA were centrifuged for twice as long—or even longer—the width of the band remains about what is shown at the bottom of Figure Q8–2. Why does the band not become even more compressed? Suggest some possible reasons to explain the thickness of the DNA band at equilibrium.

8–10 Hybridoma technology allows one to generate monoclonal antibodies to virtually any protein. Why is it then that tagging proteins with epitopes is such a commonly used technique, especially since an epitope tag has the potential to interfere with the function of the protein?

8–11 How many copies of a protein need to be present in a cell in order for it to be visible as a band on a gel? Assume that you can load 100 μg of cell extract onto a gel and that you can detect 10 ng in a single band by silver staining. The concentration of protein in cells is about 200 mg/mL, and a typical mammalian cell has a volume of about 1000 μm^3 and a typical bacterium a volume of about 1 μm^3. Given these parameters, calculate the number of copies of a 120-kd protein that would need to be present in a mammalian cell and in a bacterium in order to give a detectable band on a gel. You might try an order-of-magnitude guess before you make the calculations.

8–12 You want to amplify the DNA between the two stretches of sequence shown in **Figure Q8–3**. Of the listed primers choose the pair that will allow you to amplify the DNA by PCR.

DNA to be amplified

5'-GACCTGTGGAAGC ———————— CATACGGGATTGA-3'
3'-CTGGACACCTTCG ———————— GTATGCCCTAACT-5'

primers

(1) 5'-GACCTGTCCAAGC-3'   (5) 5'-CATACGGGATTGA-3'
(2) 5'-CTGGACACCTTCG-3'   (6) 5'-GTATGCCCTAACT-3'
(3) 5'-CGAAGGTGTCCAG-3'   (7) 5'-TGTTAGGGCATAC-3'
(4) 5'-GCTTCCACAGGTC-3'   (8) 5'-TCAATCCCGTATG-3'

Figure Q8–3 DNA to be amplified and potential PCR primers (Problem 8–12).

8–13 In the very first round of PCR using genomic DNA, the DNA primers prime synthesis that terminates only when the cycle ends (or when a random end of DNA is encountered). Yet, by the end of 20 to 30 cycles—a typical amplification—the only visible product is defined precisely by the ends of the DNA primers. In what cycle is a double-stranded fragment of the correct size first generated?

8–14 Explain the difference between a gain-of-function mutation and a dominant-negative mutation. Why are both these types of mutation usually dominant?

8–15 Discuss the following statement: "We would have no idea today of the importance of insulin as a regulatory hormone if its absence were not associated with the devastating human disease diabetes. It is the dramatic consequences of its absence that focused early efforts on the identification of insulin and the study of its normal role in physiology."

REFERENCES

General

Ausubel FM, Brent R, Kingston RE et al (eds) (2002) Short Protocols in Molecular Biology, 5th ed. New York: Wiley.

Brown TA (2002) Genomes 2, 2nd ed. New York: Wiley-Liss.

Spector DL, Goldman RD & Leinwand LA (eds) (1998) *Cells: A Laboratory Manual.* Cold Spring Harbor, NY: Cold Spring Harbor Laboratory Press.

Watson JD, Caudy AA, Myers RM & Witkowski JA (2007) Recombinant DNA: Genes and Genomes—A Short Course, 3rd ed. New York: WH Freeman.

Isolating Cells and Growing Them in Culture

Emmert-Buck MR, Bonner RF, Smith PD et al (1996) Laser capture microdissection. *Science* 274:998–1001.

Ham RG (1965) Clonal growth of mammalian cells in a chemically defined, synthetic medium. *Proc Natl Acad Sci USA* 53:288–293.

Harlow E & Lane D (1999) *Using Antibodies: A Laboratory Manual.* Cold Spring Harbor, NY: Cold Spring Harbor Laboratory Press.

Herzenberg LA, Sweet RG & Herzenberg LA (1976) Fluorescence-activated cell sorting. *Sci Am* 234:108–116.

Levi-Montalcini R (1987) The nerve growth factor thirty-five years later. *Science* 237:1154–1162.

Lerou PH & Daley GQ (2005) Therapeutic potential of embryonic stem cells. *Blood Rev* 19:321–31.

Milstein C (1980) Monoclonal antibodies. *Sci Am* 243:66–74.

Purifying Proteins

de Duve C & Beaufay H (1981) A short history of tissue fractionation. *J Cell Biol* 91:293s–299s.

Krogan NJ, Cagney G, Yu H et al (2006) Global landscape of protein complexes in the yeast *Saccharomyces cerevisiae. Nature* 440:637–43.

Laemmli UK (1970) Cleavage of structural proteins during the assembly of the head of bacteriophage T4. *Nature* 227:680–685.

Nirenberg MW & Matthaei JH (1961) The dependence of cell-free protein synthesis in *E. coli* on naturally occurring or synthetic polyribonucleotides. *Proc Natl Acad Sci. USA* 47:1588–1602.

O'Farrell PH (1975) High-resolution two-dimensional electrophoresis of proteins. *J Biol Chem* 250:4007–4021.

Palade G (1975) Intracellular aspects of the process of protein synthesis. *Science* 189:347–358.

Scopes RK & Cantor CR (1994) Protein Purification: Principles and Practice, 3rd ed. New York: Springer-Verlag.

Analyzing Proteins

Branden C & Tooze J (1999) Introduction to Protein Structure, 2nd ed. New York: Garland Science.

Fields S & Song O (1989) A novel genetic system to detect protein–protein interactions. *Nature* 340:245–246.

Giepmans BN, Adams SR et al (2006) The fluorescent toolbox for assessing protein location and function. *Science* 312:217–24.

Kendrew JC (1961) The three-dimensional structure of a protein molecule. *Sci Am* 205:96–111.

Knight ZA & Shokat KM (2007) Chemical genetics: Where genetics and pharmacology meet. *Cell* 128:425–30.

Rigaut G, Shevchenko A, Rutz B et al (1999) A generic protein purification method for protein complex characterization and proteome exploration. *Nature Biotechnol* 17:1030–1032.

Washburn MP, Wolters D and Yates JR (2001) Large-scale analysis of the yeast proteome by multidimensional protein identification technology. *Nature Biotechnol* 19:242–7.

Wuthrich K (1989) Protein structure determination in solution by nuclear magnetic resonance spectroscopy. *Science* 243:45–50.

Isolating, Cloning, and Sequencing DNA

Adams MD, Celniker SE, Holt RA et al (2000) The genome sequence of *Drosophila melanogaster. Science* 287:2185–2195.

Alwine JC, Kemp DJ & Stark GR (1977) Method for detection of specific RNAs in agarose gels by transfer to diabenzyloxymethyl-paper and hybridization with DNA probes. *Proc Natl Acad Sci USA* 74:5350–5354.

Blattner FR, Plunkett G, Bloch CA et al (1997) The complete genome sequence of *Escherichia coli* K-12. *Science* 277:1453–1474.

Cohen S, Chang A, Boyer H & Helling R (1973) Construction of biologically functional bacterial plasmids *in vitro. Proc Natl Acad Sci USA* 70:3240–3244.

International Human Genome Sequencing Consortium (2000) Initial sequencing and analysis of the human genome. *Nature* 409:860–921.

International Human Genome Sequencing Consortium (2006) The DNA sequence, annotation and analysis of human chromosome 3. *Nature* 440:1194–1198.

Jackson D, Symons R & Berg P (1972) Biochemical method for inserting new genetic information into DNA of simian virus 40: circular SV40 DNA molecules containing lambda phage genes and the galactose operon of *Escherichia coli. Proc Natl Acad Sci USA* 69:2904–2909.

Maniatis T et al (1978) The isolation of structural genes from libraries of eukaryotic DNA. *Cell* 15:687–701.

Mullis KB (1990). The unusual origin of the polymerase chain reaction. *Sci Am* 262:56–61.

Nathans D & Smith HO (1975) Restriction endonucleases in the analysis and restructuring of dna molecules. *Annu Rev Biochem* 44:273–93.

Saiki RK, Gelfand DH, Stoffel S et al (1988) Primer-directed enzymatic amplification of DNA with a thermostable DNA polymerase. *Science* 239:487–491.

Sambrook J, Russell D (2001) Molecular Cloning: A Laboratory Manual, 3rd ed. Cold Spring Harbor, NY: Cold Spring Harbor Laboratory Press.

Sanger F, Nicklen S & Coulson AR (1977) DNA sequencing with chain-terminating inhibitors. *Proc Natl Acad Sci USA* 74:5463–5467.

Smith M (1994) Nobel lecture. Synthetic DNA and biology. *Biosci Rep* 14:51–66.

Southern EM (1975) Detection of specific sequences among DNA fragments separated by gel electrophoresis. *J Mol Biol* 98:503–517.

The *Arabidopsis* Genome Initiative (2000) Analysis of the genome sequence of the flowering plant *Arabidopsis thaliana. Nature* 408:796–815.

The *C. elegans* Sequencing Consortium (1998) Genome sequence of the nematode *C. elegans*: a platform for investigating biology. *Science* 282:2012–2018.

Venter JC, Adams MA, Myers EW et al (2000) The sequence of the human genome. *Science* 291:1304–1351.

Studying Gene Expression and Function

Boone C, Bussey H & Andrews BJ (2007) Exploring genetic interactions and networks with yeast. *Nature Rev Genet* 8:437–449.

Botstein D, White RL, Skolnick M & Davis RW (1980) Construction of a genetic linkage map in man using restriction fragment length polymorphisms. *Am J Hum Genet* 32:314–331.

DeRisi JL, Iyer VR & Brown PO (1997) Exploring the metabolic and genetic control of gene expression on a genomic scale. *Science* 278:680–686.

International HapMap Consortium (2005) A haplotype map of the human genome. *Nature* 437:1299–320.

Lockhart DJ & Winzeler EA (2000) Genomics, gene expression and DNA arrays. *Nature* 405:827–836.

Mello CC & Conte D (2004) Revealing the world of RNA interference. *Nature* 431:338–342.

Nusslein-Volhard C & Weischaus E (1980) Mutations affecting segment number and polarity in *Drosophila. Nature* 287:795–801.

Palmiter RD & Brinster RL (1985) Transgenic mice. *Cell* 41:343–345.

Rubin GM & Sprading AC (1982) Genetic transformation of *Drosophila* with transposable element vectors. *Science* 218:348–353.

Sabeti PC, Schaffner SF, Fry B et al (2006) Positive natural selection in the human lineage. *Science* 312:1614–1620.

Weigel D & Glazebrook J (2001) *Arabidopsis:* A Laboratory Manual. Cold Spring Harbor, NY: Cold Spring Harbor Laboratory Press.

Chapter 9

Visualizing Cells

9

Because cells are small and complex, it is hard to see their structure, hard to discover their molecular composition, and harder still to find out how their various components function. The tools at our disposal determine what we can learn about cells, and the introduction of new techniques has frequently resulted in major advances in cell biology. To understand contemporary cell biology, therefore, it is necessary to know something of its methods.

In this chapter, we briefly describe some of the principal microscopy methods used to study cells. Understanding the structural organization of cells is an essential prerequisite for learning how cells function. Optical microscopy will be our starting point because cell biology began with the light microscope, and it is still an essential tool. In recent years optical microscopy has become ever more important, largely owing to the development of methods for the specific labeling and imaging of individual cellular constituents and the reconstruction of their three-dimensional architecture. An important advantage of optical microscopy is that light is relatively nondestructive. By tagging specific cell components with fluorescent probes, such as intrinsically fluorescent proteins, we can thus watch their movement, dynamics, and interactions in living cells. Optical microscopy is limited in resolution by the wavelength of visible light. By using a beam of electrons instead, electron microscopy can image the macromolecular complexes within cells at almost atomic resolution, and in three dimensions.

Although optical microscopy and electron microscopy are important methods, it is what they have enabled scientists to discover about the structural architecture of the cell that makes them interesting. Use this chapter as a reference and read it in conjunction with the later chapters of the book rather than viewing it as an introduction to them.

LOOKING AT CELLS IN THE LIGHT MICROSCOPE

A typical animal cell is 10–20 μm in diameter, which is about one-fifth the size of the smallest particle visible to the naked eye. Only after good light microscopes became available in the early part of the nineteenth century did Schleiden and Schwann propose that all plant and animal tissues were aggregates of individual cells. Their discovery in 1838, known as the **cell doctrine**, marks the formal birth of cell biology.

Animal cells are not only tiny, but they are also colorless and translucent. Consequently, the discovery of their main internal features depended on the development, in the latter part of the nineteenth century, of a variety of stains that provided sufficient contrast to make those features visible. Similarly, the far more powerful electron microscope introduced in the early 1940s required the development of new techniques for preserving and staining cells before the full complexities of their internal fine structure could begin to emerge. To this day, microscopy relies as much on techniques for preparing the specimen as on the performance of the microscope itself. In the following discussions, we therefore consider both instruments and specimen preparation, beginning with the light microscope.

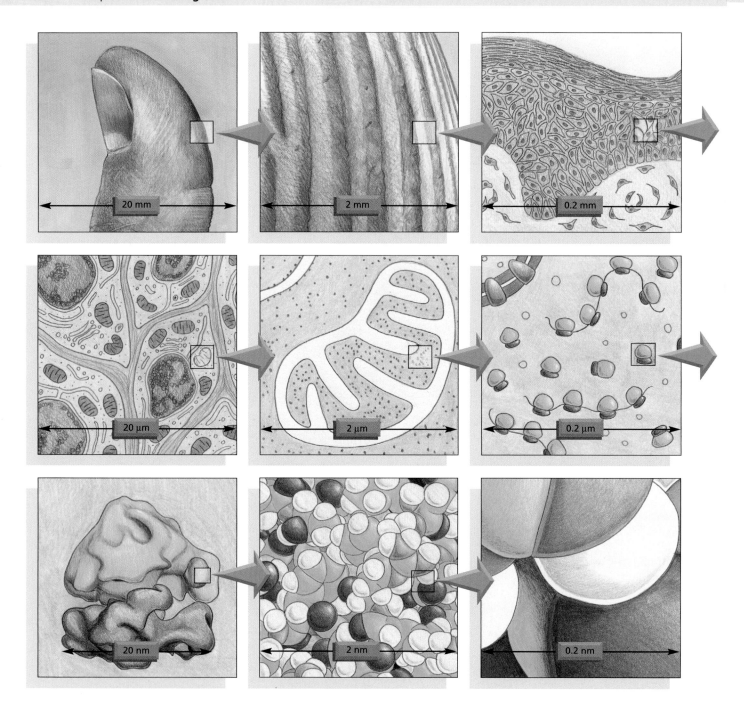

The series of images in **Figure 9–1** illustrate an imaginary progression from a thumb to a cluster of atoms. Each successive image represents a tenfold increase in magnification. The naked eye could see features in the first two panels, the resolution of the light microscope would extend to about the fourth panel, and the electron microscope to between about the seventh and eighth panel. **Figure 9–2** shows the sizes of various cellular and subcellular structures and the ranges of size that different types of microscopes can visualize.

Figure 9–1 A sense of scale between living cells and atoms. Each diagram shows an image magnified by a factor of ten in an imaginary progression from a thumb, through skin cells, to a ribosome, to a cluster of atoms forming part of one of the many protein molecules in our body. Atomic details of macromolecules, as shown in the last two panels, are usually beyond the power of the electron microscope.

The Light Microscope Can Resolve Details 0.2 μm Apart

A fundamental limitation of all microscopes is that a given type of radiation cannot be used to probe structural details much smaller than its own wavelength. The ultimate limit to the resolution of a light microscope is therefore set by the wavelength of visible light, which ranges from about 0.4 μm (for violet) to 0.7 μm (for deep red). In practical terms, bacteria and mitochondria, which are about 500 nm (0.5 μm) wide, are generally the smallest objects whose shape we can clearly discern in the **light microscope**; smaller details than this are obscured by effects resulting from the wavelike nature of light. To understand why this occurs, we must follow the path of a beam of light waves as it passes through the lenses of a microscope (**Figure 9–3**).

Because of its wave nature, light does not follow exactly the idealized straight ray paths that geometrical optics predict. Instead, light waves travel through an optical system by several slightly different routes, so that they interfere with one another and cause *optical diffraction* effects. If two trains of waves reaching the same point by different paths are precisely *in phase*, with crest matching crest and trough matching trough, they will reinforce each other so as to increase brightness. In contrast, if the trains of waves are *out of phase*, they will interfere with each other in such a way as to cancel each other partly or entirely (**Figure 9–4**). The interaction of light with an object changes the phase relationships of the light waves in a way that produces complex interference effects. At high magnification, for example, the shadow of an edge that is evenly illuminated with light of uniform wavelength appears as a set of parallel lines (**Figure 9–5**), whereas that of a circular spot appears as a set of concentric rings. For the same reason, a single point seen through a microscope appears as a blurred disc, and two point objects close together give overlapping images and may merge into one. No amount of refinement of the lenses can overcome this limitation imposed by the wavelike nature of light.

The limiting separation at which two objects appear distinct—the so-called **limit of resolution**—depends on both the wavelength of the light and the *numerical aperture* of the lens system used. The numerical aperture is a measure of the width of the entry pupil of the microscope, scaled according to its distance from the object; the wider the microscope opens its eye, so to speak, the more sharply it can see (**Figure 9–6**). Under the best conditions, with violet light

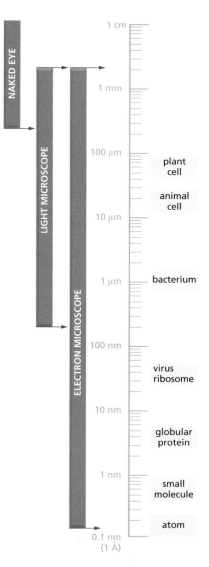

Figure 9–2 Resolving power. Sizes of cells and their components are drawn on a logarithmic scale, indicating the range of objects that can be readily resolved by the naked eye and in the light and electron microscopes. The following units of length are commonly employed in microscopy:

μm (micrometer) = 10^{-6} m
nm (nanometer) = 10^{-9} m
Å (Ångström unit) = 10^{-10} m

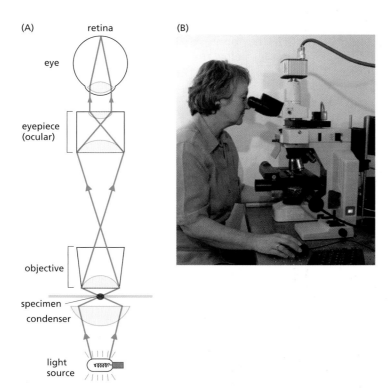

Figure 9–3 A light microscope. (A) Diagram showing the light path in a compound microscope. Light is focused on the specimen by lenses in the condenser. A combination of objective lenses and eyepiece lenses are arranged to focus an image of the illuminated specimen in the eye.
(B) A modern research light microscope. (B, courtesy of Andrew Davies.)

TWO WAVES IN PHASE   TWO WAVES OUT OF PHASE

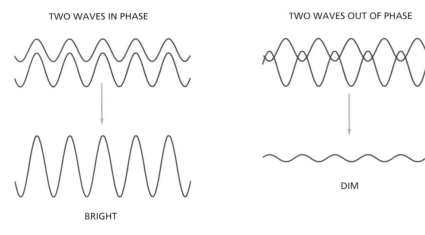

BRIGHT   DIM

Figure 9–4 Interference between light waves. When two light waves combine in phase, the amplitude of the resultant wave is larger and the brightness is increased. Two light waves that are out of phase cancel each other partly and produce a wave whose amplitude, and therefore brightness, is decreased.

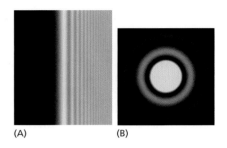

(A)   (B)

Figure 9–5 Images of an edge and of a point of light. (A) The interference effects, or fringes, seen at high magnification when light of a specific wavelength passes the edge of a solid object placed between the light source and the observer. (B) The image of a point source of light. Diffraction spreads this out into a complex, circular pattern, whose width depends on the numerical aperture of the optical system: the smaller the aperture the bigger (more blurred) the diffracted image. Two point sources can be just resolved when the center of the image of one lies on the first dark ring in the image of the other: this defines the limit of resolution.

(wavelength = 0.4 μm) and a numerical aperture of 1.4, the light microscope can theoretically achieve a limit of resolution of just under 0.2 μm. Microscope makers at the end of the nineteenth century achieved this resolution and it is only rarely matched in contemporary, factory-produced microscopes. Although it is possible to *enlarge* an image as much as we want—for example, by projecting it onto a screen—it is never possible to resolve two objects in the light microscope that are separated by less than about 0.2 μm; they will appear as a single object. Notice the difference between *resolution*, discussed above, and *detection*. If a small object, below the resolution limit, itself emits light, then we may still be able to see or detect it. Thus, we can see a single fluorescently labeled microtubule even though it is about ten times thinner than the resolution limit of the light microscope. Diffraction effects, however, will cause it to appear blurred and at least 0.2 μm thick (see Figure 9–17). Because of the bright light they emit we can detect or see the stars in the night sky, even though they are far below the angular resolution of our unaided eyes. They all appear as similar points of light,

LENSES

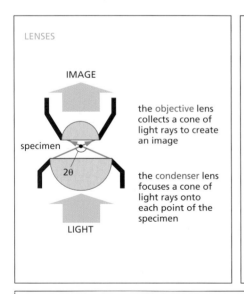

IMAGE

the objective lens collects a cone of light rays to create an image

specimen

2θ

the condenser lens focuses a cone of light rays onto each point of the specimen

LIGHT

RESOLUTION: the resolving power of the microscope depends on the width of the cone of illumination and therefore on both the condenser and the objective lens. It is calculated using the formula

$$\text{resolution} = \frac{0.61\,\lambda}{n\sin\theta}$$

where:

θ = half the angular width of the cone of rays collected by the objective lens from a typical point in the specimen (since the maximum width is 180°, sin θ has a maximum value of 1)

n = the refractive index of the medium (usually air or oil) separating the specimen from the objective and condenser lenses

λ = the wavelength of light used (for white light a figure of 0.53 μm is commonly assumed)

NUMERICAL APERTURE: n sin θ in the equation above is called the numerical aperture of the lens (NA) and is a function of its light-collecting ability. For dry lenses this cannot be more than 1, but for oil-immersion lenses it can be as high as 1.4. The higher the numerical aperture, the greater the resolution and the brighter the image (brightness is important in fluorescence microscopy). However, this advantage is obtained at the expense of very short working distances and a very small depth of field.

Figure 9–6 Numerical aperture. The path of light rays passing through a transparent specimen in a microscope illustrates the concept of numerical aperture and its relation to the limit of resolution.

differing only in their color or brightness. Using sensitive detection methods, we can detect and follow the behavior of even a single fluorescent protein molecule with a light microscope.

We see next how we can exploit interference and diffraction to study unstained cells in the living state.

Living Cells Are Seen Clearly in a Phase-Contrast or a Differential-Interference-Contrast Microscope

Microscopists have always been challenged by the possibility that some components of the cell may be lost or distorted during specimen preparation. The only certain way to avoid the problem is to examine cells while they are alive, without fixing or freezing. For this purpose, light microscopes with special optical systems are especially useful.

When light passes through a living cell, the phase of the light wave is changed according to the cell's refractive index: a relatively thick or dense part of the cell, such as a nucleus, retards light passing through it. The phase of the light, consequently, is shifted relative to light that has passed through an adjacent thinner region of the cytoplasm. The **phase-contrast microscope** and, in a more complex way, the **differential-interference-contrast microscope** exploit the interference effects produced when these two sets of waves recombine, thereby creating an image of the cell's structure (**Figure 9–7**). Both types of light microscopy are widely used to visualize living cells. <TCAA>

A simpler way to see some of the features of a living cell is to observe the light that is scattered by its various components. In the **dark-field microscope**, the illuminating rays of light are directed from the side so that only scattered light enters the microscope lenses. Consequently, the cell appears as a bright object against a dark background. With a normal **bright-field microscope**, light passing through a cell in culture forms the image directly. **Figure 9–8** compares images of the same cell obtained by four kinds of light microscopy.

Phase-contrast, differential-interference-contrast, and dark-field microscopy make it possible to watch the movements involved in such processes as mitosis and cell migration. Since many cellular motions are too slow to be seen in real time, it is often helpful to make time-lapse movies. Here, the camera records successive frames separated by a short time delay, so that when the resulting picture series is played at normal speed, events appear greatly speeded up.

Images Can Be Enhanced and Analyzed by Digital Techniques

In recent years electronic, or digital, imaging systems, and the associated technology of **image processing**, have had a major impact on light microscopy. Certain practical limitations of microscopes, relating to imperfections in the optical

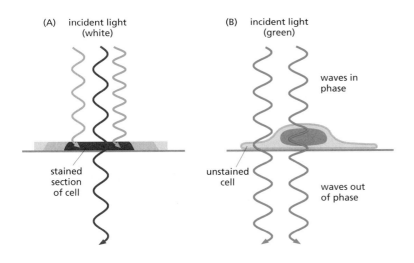

Figure 9–7 Two ways to obtain contrast in light microscopy. (A) The stained portion of the cell will absorb light of some wavelengths, which depend on the stain, but will allow other wavelengths to pass through it. A colored image of the cell is thereby obtained that is visible in the normal bright-field light microscope. (B) Light passing through the unstained, living cell experiences very little change in amplitude, and the structural details cannot be seen even if the image is highly magnified. The phase of the light, however, is altered by its passage through either thicker or denser parts of the cell, and small phase differences can be made visible by exploiting interference effects using a phase-contrast or a differential-interference-contrast microscope.

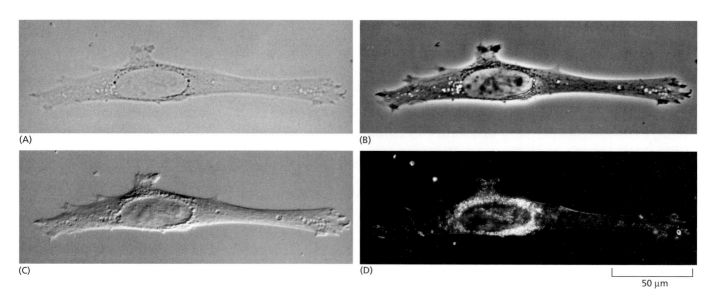

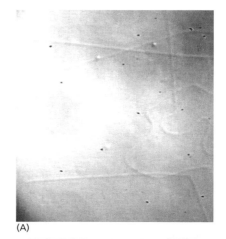

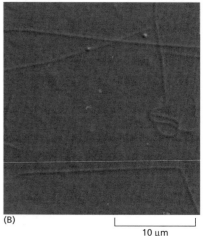

Figure 9–8 Four types of light microscopy. Four images are shown of the same fibroblast cell in culture. All images can be obtained with most modern microscopes by interchanging optical components. (A) Bright-field microscopy. (B) Phase-contrast microscopy. (C) Nomarski differential-interference-contrast microscopy. (D) Dark-field microscopy.

system have been largely overcome. Electronic imaging systems have also circumvented two fundamental limitations of the human eye: the eye cannot see well in extremely dim light, and it cannot perceive small differences in light intensity against a bright background. To increase our ability to observe cells in low light conditions, we can attach a sensitive digital camera to a microscope. These cameras contain a charge-coupled device (CCD), similar to those found in consumer digital cameras. Such CCD cameras are often cooled to reduce image noise. It is then possible to observe cells for long periods at very low light levels, thereby avoiding the damaging effects of prolonged bright light (and heat). Such low-light cameras are especially important for viewing fluorescent molecules in living cells, as explained below.

Because images produced by CCD cameras are in electronic form, they can be readily digitized, fed to a computer, and processed in various ways to extract latent information. Such image processing makes it possible to compensate for various optical faults in microscopes to attain the theoretical limit of resolution. Moreover, by digital image processing, contrast can be greatly enhanced to overcome the eye's limitations in detecting small differences in light intensity. Although this processing also enhances the effects of random background irregularities in the optical system, digitally subtracting an image of a blank area of the field removes such defects. This procedure reveals small transparent objects that were previously impossible to distinguish from the background.

The high contrast attainable by computer-assisted differential-interference-contrast microscopy makes it possible to see even very small objects such as single microtubules (**Figure 9–9**), which have a diameter of 0.025 μm, less than one-tenth the wavelength of light. Individual microtubules can also be seen in a fluorescence microscope if they are fluorescently labeled (see Figure 9–15). In both cases, however, the unavoidable diffraction effects badly blur the image so that the microtubules appear at least 0.2 μm wide, making it impossible to distinguish a single microtubule from a bundle of several microtubules.

Figure 9–9 Image processing. (A) Unstained microtubules are shown here in an unprocessed digital image, captured using differential-interference-contrast microscopy. (B) The image has now been processed, first by digitally subtracting the unevenly illuminated background, and second by digitally enhancing the contrast. The result of this image processing is a picture that is easier to interpret. Note that the microtubules are dynamic and some have changed length or position between the before-and-after images. (Courtesy of Viki Allan.)

Intact Tissues Are Usually Fixed and Sectioned before Microscopy

Because most tissue samples are too thick for their individual cells to be examined directly at high resolution, they must be cut into very thin transparent slices, or *sections*. To first immobilize, kill, and preserve the cells within the tissue they must be treated with a *fixative*. Common fixatives include formaldehyde and glutaraldehyde, which form covalent bonds with the free amino groups of proteins, cross-linking them so they are stabilized and locked into position.

Because tissues are generally soft and fragile, even after fixation, they need to be embedded in a supporting medium before sectioning. The usual embedding media are waxes or resins. In liquid form these media both permeate and surround the fixed tissue; they can then be hardened (by cooling or by polymerization) to form a solid block, which is readily sectioned with a microtome. This is a machine with a sharp blade that operates like a meat slicer (**Figure 9–10**). The sections (typically 1–10 μm thick) are then laid flat on the surface of a glass microscope slide.

There is little in the contents of most cells (which are 70% water by weight) to impede the passage of light rays. Thus, most cells in their natural state, even if fixed and sectioned, are almost invisible in an ordinary light microscope. There are three main approaches to working with thin tissue sections that reveal the cells themselves or specific components within them.

First, and traditionally, sections can be stained with organic dyes that have some specific affinity for particular subcellular components. The dye *hematoxylin,* for example, has an affinity for negatively charged molecules and therefore reveals the distribution of DNA, RNA, and acidic proteins in a cell (**Figure 9–11**). The chemical basis for the specificity of many dyes, however, is not known.

Second, sectioned tissues can be used to visualize specific patterns of differential gene expression. *In situ* hybridization, discussed earlier (p. 573), reveals the cellular distribution and abundance of specific expressed RNA molecules in sectioned material or in whole mounts of small organisms or organs (**Figure 9–12**). A third and very sensitive approach, generally and widely applicable for localizing proteins of interest, depends on using fluorescent probes and markers, as we explain next.

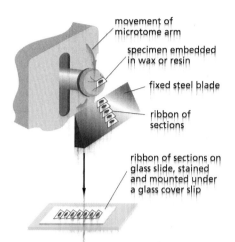

movement of microtome arm

specimen embedded in wax or resin

fixed steel blade

ribbon of sections

ribbon of sections on glass slide, stained and mounted under a glass cover slip

Figure 9–10 **Making tissue sections.** This illustration shows how an embedded tissue is sectioned with a microtome in preparation for examination in the light microscope.

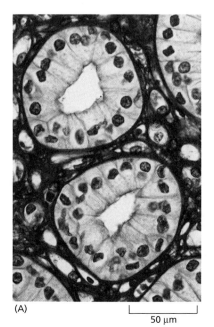

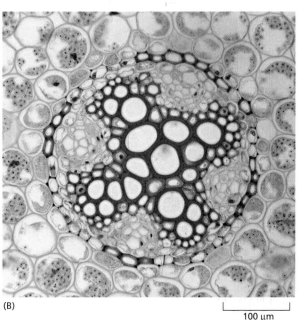

(A)   50 μm   (B)   100 μm

Figure 9–11 **Staining of cellular components.** (A) This section of cells in the urine-collecting ducts of the kidney was stained with a combination of dyes, hematoxylin and eosin, commonly used in histology. Each duct is made of closely packed cells (with nuclei stained *red*) that form a ring. The ring is surrounded by extracellular matrix, stained *purple*. (B) This section of a young plant root is stained with two dyes, safranin and fast green. The fast green stains the cellulosic cell walls while the safranin stains the lignified xylem cell walls bright red. (A, from P.R. Wheater et al., Functional Histology, 2nd ed. London: Churchill Livingstone, 1987; B, courtesy of Stephen Grace.)

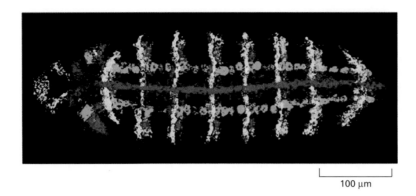

100 μm

Figure 9–12 RNA *in situ* hybridization.
As described in chapter 8 (see Figure
8–71), it is possible to visualize the
distribution of different RNAs in tissues
using *in situ* hybridization. Here, the
transcription pattern of five different
genes involved in patterning the early fly
embryo is revealed in a single embryo.
Each RNA probe has been fluorescently
labeled in a different way, some directly
and some indirectly, and the resulting
images false-colored and combined to
see each individual transcript most
clearly. The genes whose expression
pattern is revealed here are *wingless
(yellow), engrailed (blue), short
gastrulation (red), intermediate
neuroblasts defective (green),* and *muscle
specific homeobox (purple).* (From
D. Kosman et al., *Science* 305:846, 2004.
With permission from AAAS.)

Specific Molecules Can Be Located in Cells by Fluorescence Microscopy

Fluorescent molecules absorb light at one wavelength and emit it at another, longer wavelength. If we illuminate such a compound at its absorbing wavelength and then view it through a filter that allows only light of the emitted wavelength to pass, it will glow against a dark background. Because the background is dark, even a minute amount of the glowing fluorescent dye can be detected. The same number of molecules of an ordinary stain viewed conventionally would be practically invisible because the molecules would give only the faintest tinge of color to the light transmitted through this stained part of the specimen.

The fluorescent dyes used for staining cells are visualized with a **fluorescence microscope**. This microscope is similar to an ordinary light microscope except that the illuminating light, from a very powerful source, is passed through two sets of filters—one to filter the light before it reaches the specimen and one to filter the light obtained from the specimen. The first filter passes only the wavelengths that excite the particular fluorescent dye, while the second filter blocks out this light and passes only those wavelengths emitted when the dye fluoresces (**Figure 9–13**).

Fluorescence microscopy is most often used to detect specific proteins or other molecules in cells and tissues. A very powerful and widely used technique is to couple fluorescent dyes to antibody molecules, which then serve as highly specific and versatile staining reagents that bind selectively to the particular macromolecules they recognize in cells or in the extracellular matrix. Two fluorescent dyes that have been commonly used for this purpose are *fluorescein*, which emits an intense green fluorescence when excited with blue light, and

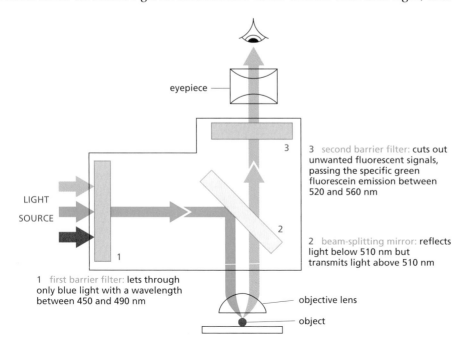

eyepiece

3 second barrier filter: cuts out
unwanted fluorescent signals,
passing the specific green
fluorescein emission between
520 and 560 nm

LIGHT
SOURCE

2 beam-splitting mirror: reflects
light below 510 nm but
transmits light above 510 nm

1 first barrier filter: lets through
only blue light with a wavelength
between 450 and 490 nm

objective lens

object

Figure 9–13 The optical system of a fluorescence microscope. A filter set consists of two barrier filters (1 and 3) and a dichroic (beam-splitting) mirror (2). This example shows the filter set for detection of the fluorescent molecule fluorescein. High-numerical-aperture objective lenses are especially important in this type of microscopy because, for a given magnification, the brightness of the fluorescent image is proportional to the fourth power of the numerical aperture (see also Figure 9–6).

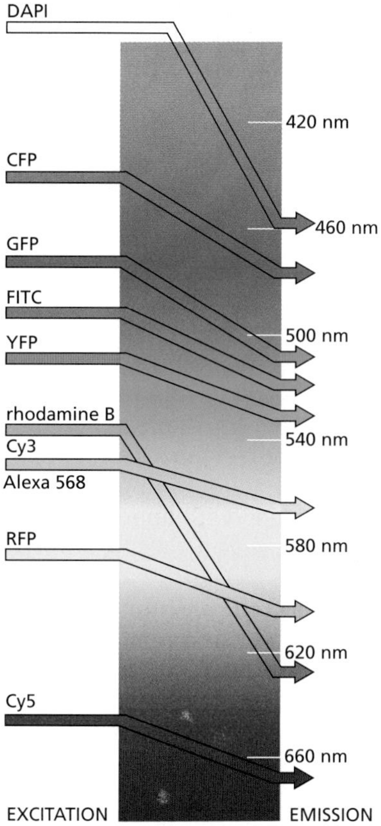

Figure 9–14 Fluorescent probes. The maximum excitation and emission wavelengths of several commonly used fluorescent probes are shown in relation to the corresponding colors of the spectrum. The photon emitted by a fluorescent molecule is necessarily of lower energy (longer wavelength) than the photon absorbed and this accounts for the difference between the excitation and emission peaks. CFP, GFP, YFP and RFP are cyan, green, yellow and red fluorescent proteins respectively. These are not dyes, and are discussed in detail later in the chapter. DAPI is widely used as a general fluorescent DNA probe, which absorbs UV light and fluoresces bright blue. FITC is an abbreviation for fluorescence isothiocyanate, a widely used derivative of fluorescein, which fluoresces bright green. The other probes are all commonly used to fluorescently label antibodies and other proteins.

rhodamine, which emits a deep red fluorescence when excited with green–yellow light (**Figure 9–14**). By coupling one antibody to fluorescein and another to rhodamine, the distributions of different molecules can be compared in the same cell; the two molecules are visualized separately in the microscope by switching back and forth between two sets of filters, each specific for one dye. As shown in **Figure 9–15**, three fluorescent dyes can be used in the same way to distinguish between three types of molecules in the same cell. Many newer fluorescent dyes, such as Cy3, Cy5, and the Alexa dyes, have been specifically developed for fluorescence microscopy (see Figure 9–14). These organic fluorochromes have some disadvantages. They are excited only by light of precise, but different, wavelengths, and additionally they fade fairly rapidly when continuously illuminated. More stable inorganic fluorochromes have recently been developed, however. Tiny crystals of semiconductor material, called nanoparticles, or *quantum dots*, can all be excited to fluoresce by a broad spectrum of blue light. Their emitted light has a color that depends on the exact size of the nanocrystal, between 2 and 10 nm in diameter, and additionally the fluorescence fades only slowly with time (**Figure 9–16**). These nanoparticles, when coupled to other probes such as antibodies, are therefore ideal for tracking molecules over time. If introduced into a living cell, in an embryo for example, the progeny of that cell can be followed many days later by their fluorescence, allowing cell lineages to be tracked.

Fluorescence microscopy methods, discussed later in the chapter, can be used to monitor changes in the concentration and location of specific molecules inside *living* cells (see p. 592).

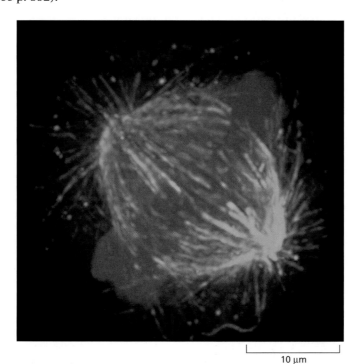

10 μm

Figure 9–15 Multiple-fluorescent-probe microscopy. In this composite micrograph of a cell in mitosis, three different fluorescent probes have been used to stain three different cellular components. <GTCT> The spindle microtubules are revealed with a *green* fluorescent antibody, centromeres with a *red* fluorescent antibody and the DNA of the condensed chromosomes with the *blue* fluorescent dye DAPI. (Courtesy of Kevin F. Sullivan.)

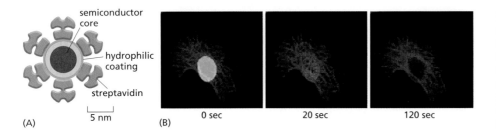

(A)   5 nm

(B)   0 sec   20 sec   120 sec

Figure 9–16 Fluorescent nanoparticles or quantum dots. Quantum dots are tiny nanoparticles of cadmium selenide, a semiconductor, with a coating to make them water-soluble (A). They can be coupled to protein probes such as antibodies or streptavidin and, when introduced into a cell, will bind to a protein of interest. Different-sized quantum dots emit light of different colors—the larger the dot the longer the wavelength—but they are all excited by the same blue light. (B) Quantum dots can keep shining for weeks, unlike most fluorescent organic dyes. In this cell, a nuclear protein is labeled (green) with an organic fluorescent dye (Alexa 488), while microtubules are stained (red) with quantum dots bound to streptavidin. On continuous exposure to blue light the fluorescent dye fades quickly while the quantum dots continue to fluoresce. (B, from X. Wu et al., *Nat. Biotechnol.* 21:41–46, 2003. With permission from Macmillan Publishers Ltd.)

Antibodies Can Be Used to Detect Specific Molecules

Antibodies are proteins produced by the vertebrate immune system as a defense against infection (discussed in Chapter 24). They are unique among proteins because they are made in billions of different forms, each with a different binding site that recognizes a specific target molecule (or *antigen*). The precise antigen specificity of antibodies makes them powerful tools for the cell biologist. When labeled with fluorescent dyes, antibodies are invaluable for locating specific molecules in cells by fluorescence microscopy (**Figure 9–17**); labeled with electron-dense particles such as colloidal gold spheres, they are used for similar purposes in the electron microscope (discussed below).

When we use antibodies as probes to detect and assay specific molecules in cells we frequently amplify the fluorescent signal they produce by chemical methods. For example, although a marker molecule such as a fluorescent dye can be linked directly to an antibody used for specific recognition—the *primary antibody*—a stronger signal is achieved by using an unlabeled primary antibody and then detecting it with a group of labeled *secondary antibodies* that bind to it (**Figure 9–18**). This process is called *indirect immunocytochemistry*.

The most sensitive amplification methods use an enzyme as a marker molecule attached to the secondary antibody. The enzyme alkaline phosphatase, for example, in the presence of appropriate chemicals, produces inorganic phosphate that in turn leads to the local formation of a colored precipitate. This reveals the location of the secondary antibody and hence the location of the antibody–antigen complex. Since each enzyme molecule acts catalytically to generate many thousands of molecules of product, even tiny amounts of antigen can be detected. An enzyme-linked immunosorbent assay (ELISA) based on this principle is frequently used in medicine as a sensitive test—for pregnancy or for various types of infections, for example. Although the enzyme amplification makes enzyme-linked methods very sensitive, diffusion of the colored precipitate away from the enzyme limits the spatial resolution of this method for

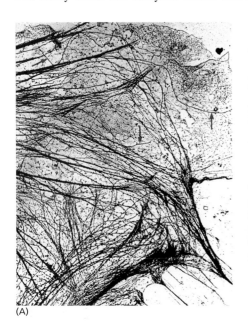

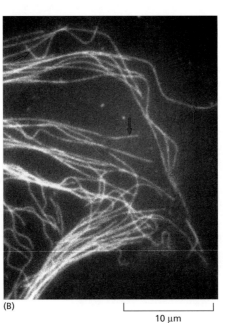

(A)   (B)

10 μm

Figure 9–17 Immunofluorescence. (A) A transmission electron micrograph of the periphery of a cultured epithelial cell showing the distribution of microtubules and other filaments. (B) The same area stained with fluorescent antibodies against tubulin, the protein that assembles to form microtubules, using the technique of indirect immunocytochemistry (see Figure 9–18). *Red arrows* indicate individual microtubules that are readily recognizable in both images. Note that, because of diffraction effects, the microtubules in the light microscope appear 0.2 μm wide rather than their true width of 0.025 μm. (From M. Osborn, R. Webster and K. Weber, *J. Cell Biol.* 77:R27–R34, 1978. With permission from The Rockefeller University Press.)

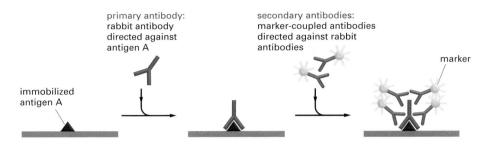

primary antibody:
rabbit antibody
directed against
antigen A

secondary antibodies:
marker-coupled antibodies
directed against rabbit
antibodies

marker

immobilized
antigen A

Figure 9–18 **Indirect immunocyto-chemistry.** This detection method is very sensitive because many molecules of the secondary antibody recognize each primary antibody. The secondary antibody is covalently coupled to a marker molecule that makes it readily detectable. Commonly used marker molecules include fluorescent dyes (for fluorescence microscopy), the enzyme horseradish peroxidase (for either conventional light microscopy or electron microscopy), colloidal gold spheres (for electron microscopy), and the enzymes alkaline phosphatase or peroxidase (for biochemical detection).

microscopy, and fluorescent labels are usually used for the most precise optical localization.

Antibodies are made most simply by injecting a sample of the antigen several times into an animal such as a rabbit or a goat and then collecting the antibody-rich serum. This *antiserum* contains a heterogeneous mixture of antibodies, each produced by a different antibody-secreting cell (a B lymphocyte). The different antibodies recognize various parts of the antigen molecule (called an antigenic determinant, or epitope), as well as impurities in the antigen preparation. Removing the unwanted antibody molecules that bind to other molecules sharpens the specificity of an antiserum for a particular antigen; an antiserum produced against protein X, for example, when passed through an affinity column of antigens X, will bind to these antigens, allowing other antibodies to pass through the column. Purified anti-X antibody can subsequently be eluted from the column. Even so, the heterogeneity of such antisera sometimes limits their usefulness. The use of monoclonal antibodies largely overcomes this problem (see Figure 8–8). However, monoclonal antibodies can also have problems. Since they are single-antibody protein species, they show almost perfect specificity for a single site or epitope on the antigen, but the accessibility of the epitope, and thus the usefulness of the antibody, may depend on the specimen preparation. For example, some monoclonal antibodies will react only with unfixed antigens, others only after the use of particular fixatives, and still others only with proteins denatured on SDS polyacrylamide gels, and not with the proteins in their native conformation.

Imaging of Complex Three-Dimensional Objects Is Possible with the Optical Microscope

For ordinary light microscopy, as we have seen, a tissue has to be sliced into thin sections to be examined; the thinner the section, the crisper the image. The process of sectioning loses information about the third dimension. How, then, can we get a picture of the three-dimensional architecture of a cell or tissue, and how can we view the microscopic structure of a specimen that, for one reason or another, cannot first be sliced into sections? Although an optical microscope is focused on a particular focal plane within complex three-dimensional specimens, all the other parts of the specimen, above and below the plane of focus, are also illuminated and the light originating from these regions contributes to the image as "out-of-focus" blur. This can make it very hard to interpret the image in detail and can lead to fine image structure being obscured by the out-of-focus light.

Two distinct but complementary approaches solve this problem: one is computational, the other is optical. These three-dimensional microscopic imaging methods make it possible to focus on a chosen plane in a thick specimen while rejecting the light that comes from out-of-focus regions above and below that plane. Thus one sees a crisp, thin *optical section*. From a series of such optical sections taken at different depths and stored in a computer, it is easy to reconstruct a three-dimensional image. The methods do for the microscopist what the CT scanner does (by different means) for the radiologist investigating a human body: both machines give detailed sectional views of the interior of an intact structure.

(A)   (B)

5 µm

Figure 9–19 Image deconvolution.
(A) A light micrograph of the large polytene chromosomes from *Drosophila*, stained with a fluorescent DNA-binding dye. (B) The same field of view after image deconvolution clearly reveals the banding pattern on the chromosomes. Each band is about 0.25 µm thick, approaching the resolution limit of the light microscope. (Courtesy of the John Sedat Laboratory.)

The computational approach is often called *image deconvolution*. To understand how it works, remember that the wavelike nature of light means that the microscope lens system produces a small blurred disc as the image of a point light source (see Figure 9–5), with increased blurring if the point source lies above or below the focal plane. This blurred image of a point source is called the *point spread function*. An image of a complex object can then be thought of as being built up by replacing each point of the specimen by a corresponding blurred disc, resulting in an image that is blurred overall. For deconvolution, we first obtain a series of (blurred) images, usually with a cooled CCD camera, focusing the microscope in turn on a series of focal planes—in effect, a (blurred) three-dimensional image. The stack of digital images is then processed by computer to remove as much of the blur as possible. Essentially the computer program uses the microscope's point spread function to determine what the effect of the blurring would have been on the image, and then applies an equivalent "deblurring" (deconvolution), turning the blurred three-dimensional image into a series of clean optical sections. The computation required is quite complex, and used to be a serious limitation. However, with faster and cheaper computers, the image deconvolution method is gaining in power and popularity. **Figure 9–19** shows an example.

The Confocal Microscope Produces Optical Sections by Excluding Out-of-Focus Light

The confocal microscope achieves a result similar to that of deconvolution, but does so by manipulating the light before it is measured; thus it is an analog technique rather than a digital one. The optical details of the **confocal microscope** are complex, but the basic idea is simple, as illustrated in **Figure 9–20**, and the results are far superior to those obtained by conventional light microscopy (**Figure 9–21**).

The microscope is generally used with fluorescence optics (see Figure 9–13), but instead of illuminating the whole specimen at once, in the usual way, the optical system at any instant focuses a spot of light onto a single point at a specific depth in the specimen. It requires a very bright source of pinpoint illumination that is usually supplied by a laser whose light has been passed through a pinhole. The fluorescence emitted from the illuminated material is collected and brought to an image at a suitable light detector. A pinhole aperture is placed in front of the detector, at a position that is *confocal* with the illuminating pinhole—that is, precisely where the rays emitted from the illuminated point in the specimen come to a focus. Thus, the light from this point in the specimen converges on this aperture and enters the detector.

By contrast, the light from regions out of the plane of focus of the spotlight is also out of focus at the pinhole aperture and is therefore largely excluded from the detector (see Figure 9–20). To build up a two-dimensional image, data from

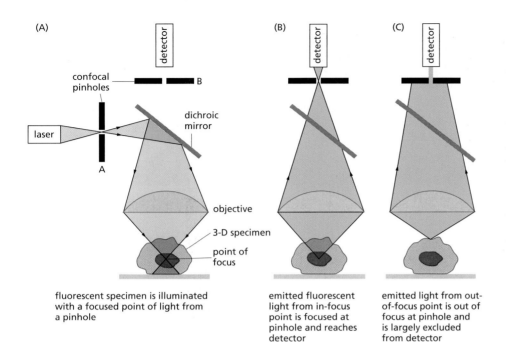

(A) fluorescent specimen is illuminated with a focused point of light from a pinhole

(B) emitted fluorescent light from in-focus point is focused at pinhole and reaches detector

(C) emitted light from out-of-focus point is out of focus at pinhole and is largely excluded from detector

Figure 9–20 **The confocal fluorescence microscope.** This simplified diagram shows that the basic arrangement of optical components is similar to that of the standard fluorescence microscope shown in Figure 9–13, except that a laser is used to illuminate a small pinhole whose image is focused at a single point in the specimen (A). Emitted fluorescence from this focal point in the specimen is focused at a second (confocal) pinhole (B). Emitted light from elsewhere in the specimen is not focused at the pinhole and therefore does not contribute to the final image (C). By scanning the beam of light across the specimen, a very sharp two-dimensional image of the exact plane of focus is built up that is not significantly degraded by light from other regions of the specimen.

each point in the plane of focus are collected sequentially by scanning across the field in a raster pattern (as on a television screen) and are displayed on a video screen. Although not shown in Figure 9–20, the scanning is usually done by deflecting the beam with an oscillating mirror placed between the dichroic mirror and the objective lens in such a way that the illuminating spotlight and the confocal pinhole at the detector remain strictly in register.

The confocal microscope has been used to resolve the structure of numerous complex three-dimensional objects (**Figure 9–22**), including the networks of cytoskeletal fibers in the cytoplasm and the arrangements of chromosomes and genes in the nucleus.

The relative merits of deconvolution methods and confocal microscopy for three-dimensional optical microscopy are still the subject of debate. Confocal microscopes are generally easier to use than deconvolution systems and the final optical sections can be seen quickly. In contrast, the cooled CCD (charge-coupled device) cameras used for deconvolution systems are extremely efficient at collecting small amounts of light, and they can be used to make detailed three-dimensional images from specimens that are too weakly stained or too easily damaged by the bright light used for confocal microscopy.

Both methods, however, have another drawback; neither is good at coping with thick specimens. Deconvolution methods quickly become ineffective any deeper than about 40 μm into a specimen, while confocal microscopes can only

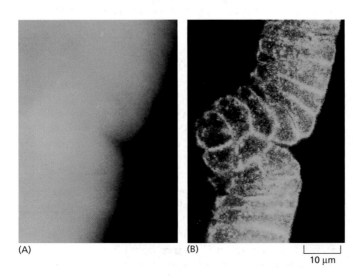

(A)   **(B)**   10 μm

Figure 9–21 **Conventional and confocal fluorescence microscopy compared.** These two micrographs are of the same intact gastrula-stage *Drosophila* embryo that has been stained with a fluorescent probe for actin filaments. (A) The conventional, unprocessed image is blurred by the presence of fluorescent structures above and below the plane of focus. (B) In the confocal image, this out-of-focus information is removed, resulting in a crisp optical section of the cells in the embryo. (Courtesy of Richard Warn and Peter Shaw.)

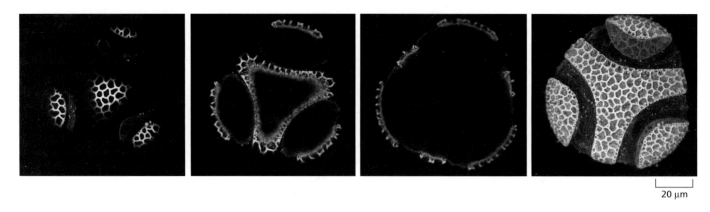

20 μm

obtain images up to a depth of about 150 μm. Special confocal microscopes can now take advantage of the way in which fluorescent molecules are excited, to probe even deeper into a specimen. Fluorescent molecules are usually excited by a single high-energy photon, of shorter wavelength than the emitted light, but they can in addition be excited by the absorption of two (or more) photons of lower energy, as long as they both arrive within a femtosecond or so of each other. The use of this longer-wavelength excitation has some important advantages. In addition to reducing background noise, red or near infrared light can penetrate deeper within a specimen. Multiphoton confocal microscopes, constructed to take advantage of this "**two-photon**" effect, can typically obtain sharp images even at a depth of 0.5 mm within a specimen. This is particularly valuable for studies of living tissues, notably in imaging the dynamic activity of synapses and neurons just below the surface of living brains (**Figure 9–23**).

Fluorescent Proteins Can Be Used to Tag Individual Proteins in Living Cells and Organisms

Even the most stable cellular structures must be assembled, disassembled, and reorganized during the cell's life cycle. Other structures, often enormous on the molecular scale, rapidly change, move, and reorganize themselves as the cell conducts its internal affairs and responds to its environment. Complex, highly organized pieces of molecular machinery move components around the cell, controlling traffic into and out of the nucleus, from one organelle to another, and into and out of the cell itself.

Various techniques have been developed to make specific components of living cells visible in the microscope. Most of these methods use fluorescent proteins, and they require a trade-off between structural preservation and efficient labeling. All of the fluorescent molecules discussed so far are made outside the cell and then artificially introduced into it. Now new opportunities have been opened up by the discovery of genes coding for protein molecules that are themselves inherently fluorescent. Genetic engineering then enables the creation of lines of cells or organisms that make their own visible tags and labels, without the introduction of foreign molecules. These cellular exhibitionists display their inner workings in glowing fluorescent color.

Foremost among the fluorescent proteins used for these purposes by cell biologists is the **green fluorescent protein** (**GFP**), isolated from the jellyfish *Aequoria victoria*. This protein is encoded in the normal way by a single gene that can be cloned and introduced into cells of other species. The freshly translated protein is not fluorescent, but within an hour or so (less for some alleles of

Figure 9–22 Three-dimensional reconstruction from confocal microscope images. Pollen grains, in this case from a passion flower, have a complex sculptured cell wall that contains fluorescent compounds. Images obtained at different depths through the grain, using a confocal microscope, can be recombined to give a three-dimensional view of the whole grain, shown on the right. Three selected individual optical sections from the full set of 30, each of which shows little contribution from its neighbors, are shown on the left. (Courtesy of Brad Amos.)

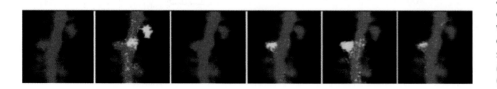

Figure 9–23 Multi-photon imaging. Infrared laser light causes less damage to living cells and can also penetrate further, allowing microscopists to peer deeper into living tissues. The two-photon effect, in which a fluorochrome can be excited by two coincident infrared photons instead of a single high-energy photon, allows us to see nearly 0.5 mm inside the cortex of a live mouse brain. A dye, whose fluorescence changes with the calcium concentration, reveals active synapses *(yellow)* on the dendritic spines *(red)* that change as a function of time. (Courtesy of Karel Svoboda.)

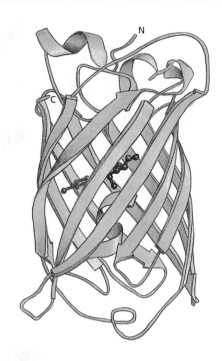

Figure 9-24 **Green fluorescent protein (GFP).** The structure of GFP, shown here schematically, highlights the eleven β strands that form the staves of a barrel. Buried within the barrel is the active chromophore *(dark green)* that is formed post-translationally from the protruding side chains of three amino acid residues. (Adapted from M. Ormö et al., *Science* 273:1392–1395, 1996. With permission from AAAS.)

the gene, more for others) it undergoes a self-catalyzed post-translational modification to generate an efficient and bright fluorescent center, shielded within the interior of a barrel-like protein (**Figure 9–24**). Extensive site-directed mutagenesis performed on the original gene sequence has resulted in useful fluorescence in organisms ranging from animals and plants to fungi and microbes. The fluorescence efficiency has also been improved, and variants have been generated with altered absorption and emission spectra in the blue–green–yellow range. Recently a family of related fluorescent proteins discovered in corals, has extended the range into the red region of the spectrum (see Figure 9–14).

One of the simplest uses of GFP is as a reporter molecule, a fluorescent probe to monitor gene expression. A transgenic organism can be made with the GFP-coding sequence placed under the transcriptional control of the promoter belonging to a gene of interest, giving a directly visible readout of the gene's expression pattern in the living organism (**Figure 9–25**). In another application, a peptide location signal can be added to the GFP to direct it to a particular cellular compartment, such as the endoplasmic reticulum or a mitochondrion, lighting up these organelles so they can be observed in the living state (see Figure 12–35B).

The GFP DNA-coding sequence can also be inserted at the beginning or end of the gene for another protein, yielding a chimeric product consisting of that protein with a GFP domain attached. In many cases, this GFP-fusion protein behaves in the same way as the original protein, directly revealing its location and activities by means of its genetically encoded contrast (**Figure 9–26**). <TAAT> It is often possible to prove that the GFP-fusion protein is functionally equivalent to the untagged protein, for example by using it to rescue a mutant lacking that protein. GFP tagging is the clearest and most unequivocal way of showing the distribution and dynamics of a protein in a living organism (**Figure 9–27**). <TTCT>

Protein Dynamics Can Be Followed in Living Cells

Fluorescent proteins are now exploited, not just to see where in a cell a particular protein is located, but also to uncover its kinetic properties and to find out whether it might interact with other proteins. We now describe three techniques in which GFP and its relatives are used in this way.

The first is the monitoring of interactions between one protein and another by **fluorescence resonance energy transfer (FRET)**. In this technique, whose principles have been described earlier (see Figure 8–26), the two molecules of interest are each labeled with a different fluorochrome, chosen so that the emission spectrum of one fluorochrome overlaps with the absorption spectrum of the other. If the two proteins bind so as to bring their fluorochromes into very close proximity (closer than about 5 nm), one fluorochrome transfers the energy of the absorbed light directly to the other. Thus, when the complex is illuminated at the excitation wavelength of the first fluorochrome, fluorescent light is pro-

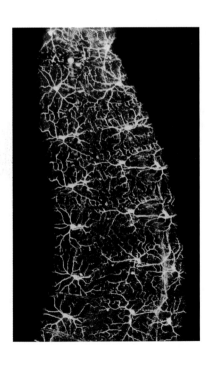

Figure 9-25 **Green fluorescent protein (GFP) as a reporter.** For this experiment, carried out in the fruit fly, the GFP gene was joined (using recombinant DNA techniques) to a fly promoter that is active only in a specialized set of neurons. This image of a live fly embryo was captured by a fluorescence microscope and shows approximately 20 neurons, each with long projections (axons and dendrites) that communicate with other (nonfluorescent) cells. These neurons are located just under the surface of the animal and allow it to sense its immediate environment. (From W.B. Grueber et al., *Curr. Biol.* 13:618–626, 2003. With permission from Elsevier.)

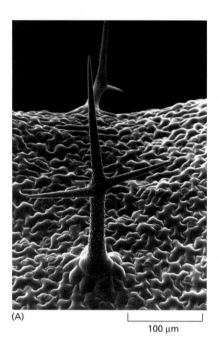

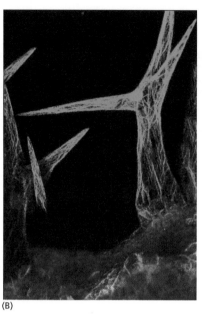

(A)

(B)

100 μm

Figure 9–26 **GFP-tagged proteins.**
(A) The upper surface of the leaves of *Arabidopsis* plants are covered with huge branched single-cell hairs that rise up from the surface of the epidermis. These hairs, or trichomes, can be imaged in the scanning electron microscope. (B) If an *Arabidopsis* plant is transformed with a DNA sequence coding for talin (an actin-binding protein), fused to a DNA sequence coding for GFP, the fluorescent talin protein produced binds to actin filaments in all the living cells of the transgenic plant. Confocal microscopy can reveal the dynamics of the entire actin cytoskeleton of the trichome (*green*). The red fluorescence arises from chlorophyll in cells within the leaf below the epidermis. (A, courtesy of Paul Linstead; B, courtesy of Jaideep Mathur.)

duced at the emission wavelength of the second. This method can be used with two different spectral variants of GFP as fluorochromes to monitor processes such as the interaction of signaling molecules with their receptors, or proteins in macromolecular complexes (**Figure 9–28**).

The complexity and rapidity of many intracellular processes, such as the actions of signaling molecules or the movements of cytoskeletal proteins, make them difficult to study at a single-cell level. Ideally, we would like to be able to introduce any molecule of interest into a living cell at a precise time and location and follow its subsequent behavior, as well as the response of the cell to that molecule. Microinjection is limited by the difficulty of controlling the place and time of delivery. A more powerful approach involves synthesizing an inactive form of the fluorescent molecule of interest, introducing it into the cell, and then activating it suddenly at a chosen site in the cell by focusing a spot of light on it. This process is referred to as **photoactivation**. Inactive photosensitive precursors of this type, often called *caged molecules*, have been made for many fluorescent molecules. A microscope can be used to focus a strong pulse of light from a laser on any tiny region of the cell, so that the experimenter can control exactly where and when the fluorescent molecule is photoactivated.

One class of caged fluorescent proteins is made by attaching a photoactivatable fluorescent tag to a purified protein. It is important that the modified protein remain biologically active: labeling with a caged fluorescent dye adds a bulky group to the surface of a protein, which can easily change the protein's properties. A satisfactory labeling protocol is usually found by trial and error. Once a biologically active labeled protein has been produced, it needs to be introduced into the living cell (see Figure 9–34), where its behavior can be followed. Tubulin, labeled with caged fluorescein for example, when injected into a dividing cell, can be incorporated into microtubules of the mitotic spindle.

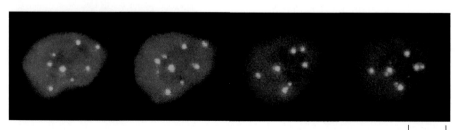

5 μm

Figure 9–27 **Dynamics of GFP tagging.** This sequence of micrographs shows a set of three-dimensional images of a living nucleus taken over the course of an hour. Tobacco cells have been stably transformed with GFP fused to a spliceosomal protein that is concentrated in small nuclear bodies called Cajal bodies (see Figure 6–48). The fluorescent Cajal bodies, easily visible in a living cell with confocal microscopy, are dynamic structures that move around within the nucleus. (Courtesy of Kurt Boudonck, Liam Dolan, and Peter Shaw.)

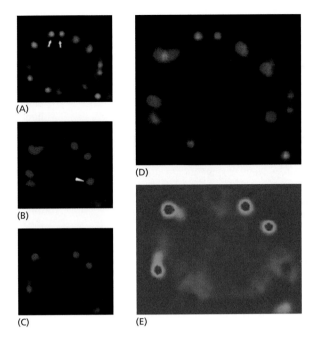

(A)

(B)

(C)

(D)

(E)

Figure 9–28 Fluorescence resonance energy transfer (FRET) imaging. This experiment shows that a protein called Sla1p can interact tightly with another protein, called Abp1p, which is involved in cortical actin attachment at the surface of a budding yeast cell. Sla1p is expressed in the yeast cell (A) as a fusion protein with a yellow variant of GFP (YFP), while Abp1p is expressed as a fusion protein (B) with a cyan variant of GFP (CFP). The FRET signal (see also Figure 8–26), displayed here in *red* (C), is obtained by exciting the CFP but recording only the fluorescence emitted from the YFP, which will occur only when the two molecules are tightly associated (within 0.5 nm). The spots at the cortex (D), seen when (A), (B), and (C) are superimposed, are of three sorts, those where Sla1p is found alone *(arrows* in A), those where Abp1p is found alone *(arrowhead* in B), and those where they are closely associated and generate a FRET signal, shown in the false-colored and corrected image (E). Since Sla1p was already known to form part of the endocytic machinery, this experiment physically connects that process with the process of actin attachment to the cell cortex. (From D.T. Warren et al., *J. Cell Sci.* 115:1703–1715, 2002. With permission from The Company of Biologists.)

When a small region of the spindle is illuminated with a laser, the labeled tubulin becomes fluorescent, so that its movement along the spindle microtubules can be readily followed (**Figure 9–29**).

A more recent development in photoactivation is the discovery that the genes encoding GFP and related fluorescent proteins can be mutated to produce protein variants, usually with a single amino acid change, that fluoresce only weakly under normal excitation conditions, but can be induced to fluoresce strongly by activating them with a strong pulse of light at a different wavelength. In principle the microscopist can then follow the local *in vivo* behavior of any protein that can be expressed as a fusion with one of these GFP variants. These genetically encoded, photoactivateable fluorescent proteins thus avoid the need to introduce the probe into the cell, and allow the lifetime and behaviour of any protein to be studied independently of other newly synthesized proteins (**Figure 9–30**).

A third way to exploit GFP fused to a protein of interest is to use a strong focussed beam of light from a laser to extinguish the GFP fluorescence in a specified region of the cell. By analyzing the way in which the remaining fluorescent protein molecules move into the bleached area as a function of time, we can obtain information about the protein's kinetic parameters. This technique, usually carried out with a confocal microscope, is known as **fluorescence recovery after photobleaching (FRAP)** and, like photoactivation, can deliver valuable quantitative data about the protein of interest, such as diffusion coefficients <ATGT>, active transport rates, or binding and dissociation rates from other proteins (**Figure 9–31**).

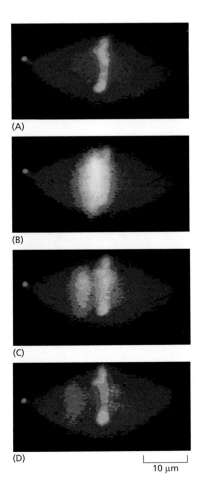

(A)

(B)

(C)

(D)

$\vdash\!\!-\!\!-\!\!\dashv$
10 µm

Figure 9–29 Determining microtubule flux in the mitotic spindle with caged fluorescein linked to tubulin. (A) A metaphase spindle formed *in vitro* from an extract of *Xenopus* eggs has incorporated three fluorescent markers: rhodamine-labeled tubulin *(red)* to mark all the microtubules, a *blue* DNA-binding dye that labels the chromosomes, and caged-fluorescein-labeled tubulin, which is also incorporated into all the microtubules but is invisible because it is nonfluorescent until activated by ultraviolet light. (B) A beam of UV light uncages the caged-fluorescein-labeled tubulin locally, mainly just to the left side of the metaphase plate. Over the next few minutes (after 1.5 minutes in C, after 2.5 minutes in D), the uncaged-fluorescein–tubulin signal moves toward the left spindle pole, indicating that tubulin is continuously moving poleward even though the spindle (visualized by the *red* rhodamine-labeled tubulin fluorescence) remains largely unchanged. (From K.E. Sawin and T.J. Mitchison, *J. Cell Biol.* 112:941–954, 1991. With permission from The Rockefeller University Press.)

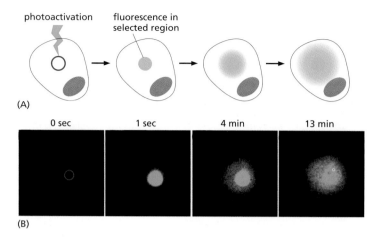

(A)

0 sec   1 sec   4 min   13 min

(B)

Light-Emitting Indicators Can Measure Rapidly Changing Intracellular Ion Concentrations

One way to study the chemistry of a single living cell is to insert the tip of a fine, glass, ion-sensitive **microelectrode** directly into the cell interior through the plasma membrane. This technique is used to measure the intracellular concentrations of common inorganic ions, such as H^+, Na^+, K^+, Cl^-, and Ca^{2+}. However, ion-sensitive microelectrodes reveal the ion concentration only at one point in a cell, and for an ion present at a very low concentration, such as Ca^{2+}, their responses are slow and somewhat erratic. Thus, these microelectrodes are not ideally suited to record the rapid and transient changes in the concentration of cytosolic Ca^{2+} that have an important role in allowing cells to respond to extracellular signals. Such changes can be analyzed with **ion-sensitive indicators**, whose light emission reflects the local concentration of the ion. Some of these indicators are luminescent (emitting light spontaneously), while others are fluorescent (emitting light on exposure to light).

Aequorin is a luminescent protein isolated from a marine jellyfish; it emits light in the presence of Ca^{2+} and responds to changes in Ca^{2+} concentration in the range of 0.5–10 µM. If microinjected into an egg, for example, aequorin emits a flash of light in response to the sudden localized release of free Ca^{2+} into the cytoplasm that occurs when the egg is fertilized (**Figure 9–32**). Aequorin has also been expressed transgenically in plants and other organisms to provide a method of monitoring Ca^{2+} in all their cells without the need for microinjection, which can be a difficult procedure.

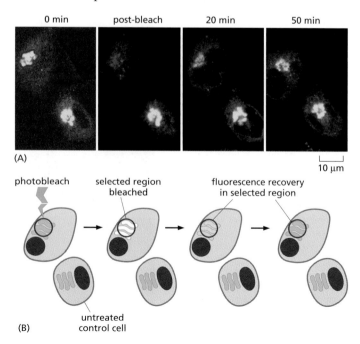

0 min   post-bleach   20 min   50 min

(A)

|_____|
10 µm

photobleach   selected region   fluorescence recovery
                     bleached   in selected region

untreated
control cell
(B)

Figure 9–30 Photoactivation. Photoactivation is the light-induced activation of an inert molecule to an active state. In this experiment a photoactivatable variant of GFP is expressed in a cultured animal cell. Before activation (time 0), little or no GFP fluorescence is detected in the selected region *(red circle)* when excited by blue light at 488 nm. After activation of the GFP however, using a UV laser pulse at 413 nm, it rapidly fluoresces brightly in the selected region *(green)*. The movement of GFP, as it diffuses out of this region, can be measured. Since only the photoactivated proteins are fluorescent within the cell, the trafficking, turnover and degradative pathways of proteins can be monitored. (B, from J. Lippincott-Schwartz and G.H. Patterson, *Science* 300:87–91, 2003. With permission from AAAS.)

Figure 9–31 Fluorescence recovery after photobleaching (FRAP). A strong focused pulse of laser light will extinguish, or bleach, the fluorescence of GFP. By selectively photobleaching a set of fluorescently tagged protein molecules within a defined region of a cell, the microscopist can monitor recovery over time, as the remaining fluorescent molecules move into the bleached region. The experiment shown in (A) uses monkey cells in culture that express galactosyltransferase, an enzyme that constantly recycles between the Golgi apparatus and the endoplasmic reticulum. The Golgi apparatus in one of the two cells is selectively photobleached, while the production of new fluorescent protein is blocked by treating the cells with cycloheximide. The recovery, resulting from fluorescent enzyme molecules moving from the ER to the Golgi, can then be followed over a period of time. (B) Schematic diagram of the experiment shown in (A). (A, from J. Lippincott-Schwartz et al., *Histochem. Cell Biol.* 116:97–107, 2001. With permission from Springer-Verlag.)

Figure 9–32 **Aequorin, a luminescent protein.** The luminescent protein aequorin emits light in the presence of free Ca^{2+}. Here, an egg of the medaka fish has been injected with aequorin, which has diffused throughout the cytosol, and the egg has then been fertilized with a sperm and examined with the help of a very sensitive camera. The four photographs were taken looking down on the site of sperm entry at intervals of 10 seconds and reveal a wave of release of free Ca^{2+} into the cytosol from internal stores just beneath the plasma membrane. This wave sweeps across the egg starting from the site of sperm entry, as indicated in the diagrams on the *left*. (Photographs reproduced from J.C. Gilkey, L.F. Jaffe, E.B. Ridgway and G.T. Reynolds, *J. Cell Biol.* 76:448–466, 1978. With permission from The Rockefeller University Press.)

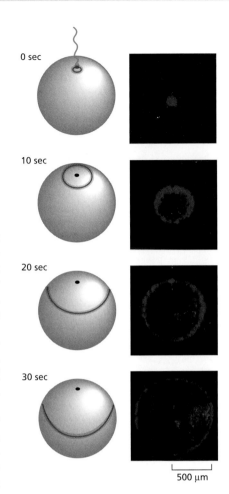

0 sec

10 sec

20 sec

30 sec

500 μm

Bioluminescent molecules like aequorin emit tiny amounts of light—at best, a few photons per indicator molecule—that are difficult to measure. Fluorescent indicators produce orders of magnitude more photons per molecule; they are therefore easier to measure and can give better spatial resolution. Fluorescent Ca^{2+} indicators have been synthesized that bind Ca^{2+} tightly and are excited by or emit light at slightly different wavelengths when they are free of Ca^{2+} than when they are in their Ca^{2+}-bound form. By measuring the ratio of fluorescence intensity at two excitation or emission wavelengths, we can determine the concentration ratio of the Ca^{2+}-bound indicator to the Ca^{2+}-free indicator, thereby providing an accurate measurement of the free Ca^{2+} concentration. <CGTC> Indicators of this type are widely used for second-by-second monitoring of changes in intracellular Ca^{2+} concentrations in the different parts of a cell viewed in a fluorescence microscope (**Figure 9–33**). <AGGA>

Similar fluorescent indicators measure other ions; some detect H^+, for example, and hence measure intracellular pH. Some of these indicators can enter cells by diffusion and thus need not be microinjected; this makes it possible to monitor large numbers of individual cells simultaneously in a fluorescence microscope. New types of indicators, used in conjunction with modern image-processing methods, are leading to similarly rapid and precise methods for analyzing changes in the concentrations of many types of small molecules in cells.

Several Srategies Are Available by Which Membrane-Impermeant Substances Can Be Introduced into Cells

It is often useful to introduce membrane-impermeant molecules into a living cell, whether they are antibodies that recognize intracellular proteins, normal cell proteins tagged with a fluorescent label, or molecules that influence cell behavior. One approach is to microinject the molecules into the cell through a glass micropipette.

When microinjected into a cell, antibodies can block the function of the molecule that they recognize. Anti-myosin-II antibodies injected into a fertilized sea urchin egg, for example, prevent the egg cell from dividing in two, even though nuclear division occurs normally. This observation demonstrates that this myosin has an essential role in the contractile process that divides the cytoplasm during cell division, but that it is not required for nuclear division.

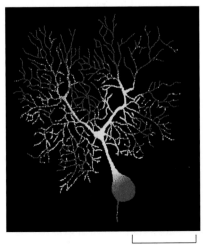

Figure 9–33 **Visualizing intracellular Ca^{2+} concentrations by using a fluorescent indicator.** The branching tree of dendrites of a Purkinje cell in the cerebellum receives more than 100,000 synapses from other neurons. The output from the cell is conveyed along the single axon seen leaving the cell body at the bottom of the picture. This image of the intracellular Ca^{2+} concentration in a single Purkinje cell (from the brain of a guinea pig) was taken with a low-light camera and the Ca^{2+}-sensitive fluorescent indictor fura-2. The concentration of free Ca^{2+} is represented by different colors, *red* being the highest and *blue* the lowest. The highest Ca^{2+} levels are present in the thousands of dendritic branches. (Courtesy of D.W. Tank, J.A. Connor, M. Sugimori and R.R. Llinas.)

100 μm

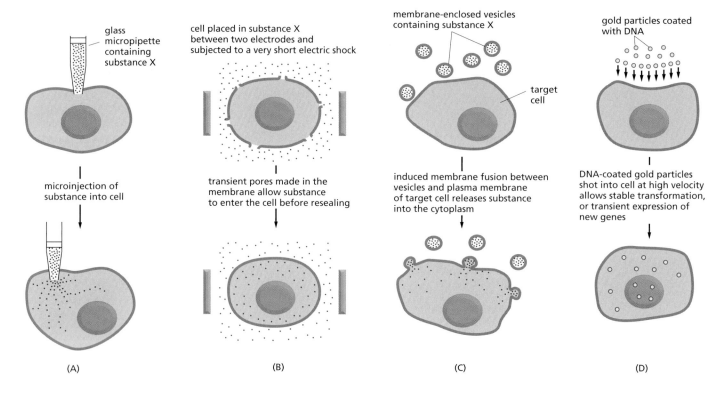

Figure 9–34 Methods of introducing a membrane-impermeant substance into a cell. (A) The substance is injected through a micropipette, either by applying pressure or, if the substance is electrically charged, by applying a voltage that drives the substance into the cell as an ionic current (a technique called *iontophoresis*). (B) The cell membrane is made transiently permeable to the substance by disrupting the membrane structure with a brief but intense electric shock (2000 V/cm for 200 μsec, for example).(C) Membrane-enclosed vesicles are loaded with the desired substance and then induced to fuse with the target cells. (D) Gold particles coated with DNA are used to introduce a novel gene into the nucleus.

Microinjection, although widely used, demands that each cell be injected individually; therefore, it is possible to study at most only a few hundred cells at a time. Other approaches allow large populations of cells to be permeabilized simultaneously. Partly disrupting the structure of the cell plasma membrane, for example, makes it more permeable; this is usually accomplished by using a powerful electric shock or a chemical such as a low concentration of detergent. The electrical technique has the advantage of creating large pores in the plasma membrane without damaging intracellular membranes. Depending on the cell type and the size of the electric shock, the pores allow even macromolecules to enter (and leave) the cytosol rapidly. This process of *electroporation* is valuable also in molecular genetics, as a way of introducing DNA molecules into cells. With a limited treatment, a large fraction of the cells repair their plasma membrane and survive.

A third method for introducing large molecules into cells is to cause membrane-enclosed vesicles that contain these molecules to fuse with the cell's plasma membrane thus delivering their cargo. Thus method is used routinely to deliver nucleic acids into mammalian cells, either DNA for transfection studies or RNA for RNAi experiments (discussed in Chapter 8). In the medical field it is also being explored as a method for the targeted delivering of new pharmaceuticals.

Finally, DNA and RNA can also be physically introduced into cells by simply blasting them in at high velocity, coated onto tiny gold particles. Living cells, shot with these nucleic-acid-coated gold particles (typically less than 1 μm in diameter) can successfully incorporate the introduced RNA (used for transient expression studies or RNAi, for example) or DNA (for stable transfection). All four of these methods, illustrated in **Figure 9–34**, are used widely in cell biology.

Light Can Be Used to Manipulate Microscopic Objects As Well As to Image Them

Photons carry a small amount of momentum. This means that an object that absorbs or deflects a beam of light experiences a small force. With ordinary light sources, this radiation pressure is too small to be significant. But it is important on a cosmic scale (helping prevent gravitational collapse inside stars), and, more

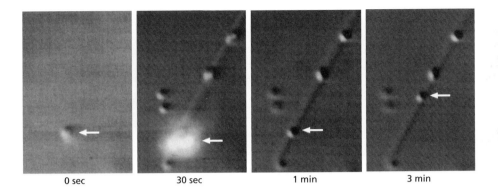

0 sec   30 sec   1 min   3 min

Figure 9–35 Optical tweezers. A focused laser beam can be used to trap microscopic particles and move them about at will. <CGCG> In this experiment, such optical tweezers are used to pick up a small silica bead (0.2 nm, *arrow*), coated with few kinesin molecules (0 sec), and place it on an isolated ciliary axoneme that is built from microtubules (30 sec). The bright halo seen here is the reflection of the laser at the interface between the water and the coverslip. The kinesin on the released bead (1 min) couples ATP hydrolysis to movement along the microtubules of the axoneme, and powers the transport of the bead along it (3 min). (From S.M. Block et al., *Nature* 348:348–352, 1990. With permission from Macmillan Publishers Ltd.)

modestly, in the cell biology lab, where an intense focused laser beam can exert large enough forces to push small objects around inside a cell. If the laser beam is focused on an object having a higher refractive index than its surroundings, the beam is refracted, causing very large numbers of photons to change direction. The pattern of photon deflection holds the object at the focus of the beam; if it begins to drift away from this position, radiation pressure pushes it back by acting more strongly on one side than the other. Thus, by steering a focused laser beam, usually an infrared laser, which is minimally absorbed by the cellular constituents, one can create "**optical tweezers**" to move subcellular objects like organelles and chromosomes around. This method, sometimes referred to as laser tweezers <CGCG> <CACA>, has been used to measure the forces exerted by single actin–myosin molecules, by single microtubule motors, and by RNA polymerase (**Figure 9–35**).

Intense focused laser beams that are more strongly absorbed by biological material can also be used more straightforwardly as optical knives—to kill individual cells, to cut or burn holes in them, or to detach one intracellular component from another. In these ways, optical devices can provide a basic toolkit for cellular microsurgery.

Single Molecules Can Be Visualized by Using Total Internal Reflection Fluorescence Microscopy

While beads can be used as markers to track protein movements, it is clearly preferable to be able to visualize the proteins themselves. In principle this can be accomplished by labeling the protein with a fluorescent molecule, either by chemically attaching a small fluorescent molecule to isolated protein molecules or by expressing fluorescent protein fusion constructs (see p. 593). In ordinary microscopes, however, single fluorescent molecules cannot be reliably detected. The limitation has nothing to do with the resolution limit, but instead arises from the interference of light emitted by out-of-focus molecules that tends to blot out the fluorescence from the particular molecule of interest. This problem can be solved by the use of a specialized optical technique called total internal reflectance fluorescence (TIRF) microscopy. In a TIRF microscope, laser light shines onto the coverslip surface at the precise critical angle at which total internal reflection occurs (**Figure 9–36**A). Because of total internal reflection, the light does not enter the sample, and the majority of fluorescent molecules are not, therefore, illuminated. However, electromagnetic energy does extend, as an evanescent field, for a very short distance beyond the surface of the coverslip and into the specimen, allowing just those molecules in the layer closest to the surface to become excited. When these molecules fluoresce, their emitted light is no longer competing with out-of-focus light from the overlying molecules, and can now be detected. TIRF has allowed several dramatic experiments, for instance imaging of single motor proteins moving along microtubules or single actin filaments forming and branching, although at present the technique is restricted to a thin layer within only 100–200 nm of the cell surface (Figure 9–36B and C).

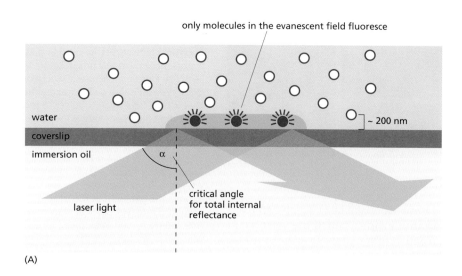

(A)

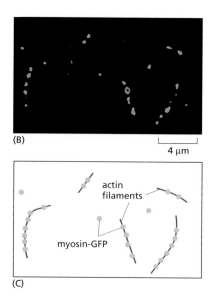

(B)

⊢————⊣ 4 μm

(C)

only molecules in the evanescent field fluoresce

water

coverslip

immersion oil

critical angle for total internal reflectance

laser light

α

~ 200 nm

actin filaments

myosin-GFP

Figure 9–36 **TIRF microscopy allows the detection of single fluorescent molecules.** (A) TIRF microscopy uses excitatory laser light to illuminate the coverslip surface at the critical angle at which all the light is reflected by the glass–water interface. Some electromagnetic energy extends a short distance across the interface as an evanescent wave that excites just those molecules that are very close to the surface. (B) TIRF microscopy is used here to image individual myosin-GFP molecules *(green dots)* attached to non-fluorescent actin filaments (C), which are invisible but stuck to the surface of the coverslip. (Courtesy of Dmitry Cherny and Clive R. Bagshaw.)

Individual Molecules Can Be Touched and Moved Using Atomic Force Microscopy

While TIRF allows single molecules to be visualized, it is strictly a passive observation method. In order to probe molecular function, it is ultimately useful to be able to manipulate individual molecules themselves, and atomic force microscopy (AFM) provides a method to do just that. In an AFM device, an extremely small and sharply pointed tip, of silicon or silicon nitride, is made using nanofabrication methods similar to those used in the semiconductor industry. The tip of the AFM is attached to a springy cantilever arm mounted on a highly precise positioning system that allows it to be moved over very small distances. In addition to this precise movement capability, the AFM is able to measure the mechanical force felt by its tip as it moves over the surface (**Figure 9–37**A). When AFM was first developed, it was intended as an imaging technology to measure molecular-scale features on a surface. When used in this mode, the probe is scanned over the surface, moving up and down as necessary to maintain a constant interaction force with the surface, thus revealing any objects such as proteins that might be present on the otherwise flat surface (see Figures 10–14 and 10–32). AFM is not limited to simply imaging surfaces, however, and can also be used to pick up and move single molecules, in a molecular-scale version of the optical tweezers described above. Using this technology, the mechanical properties of individual protein molecules can be measured in detail. For example, AFM has been used to unfold a single protein molecule in order to measure the energetics of domain folding (Figure 9–37B). The full potential to probe proteins mechanically, as well as to assemble individual proteins into defined arrangements using AFM, is only now starting to be explored, but it seems likely that this tool will become increasingly important in the future.

Molecules Can Be Labeled with Radioisotopes

As we have just seen, in cell biology it is often important to determine the quantities of specific molecules and to know where they are in the cell and how their level or location changes in response to extracellular signals. The molecules of interest range from small inorganic ions, such as Ca^{2+} or H^+, to large macromolecules, such as specific proteins, RNAs, or DNA sequences. We have so far described how sensitive fluorescence methods can be used for assaying these types of molecules, as well as for following the dynamic behavior of many of them in living cells. In ending this section, we describe how radioisotopes are used to trace the path of specific molecules through the cell.

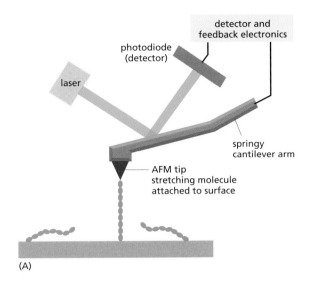

(A)

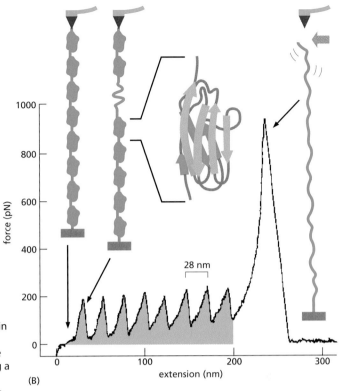

(B)

Figure 9–37 Single protein molecules can be manipulated by atomic force microscopy. (A) Schematic diagram of the key components of an atomic force microscope (AFM), showing the force-sensing tip attached to one end of a single protein molecule in the experiment described in (B). (B) Titin is an enormous protein molecule that provides muscle with its passive elasticity (see Figure 16–76). The extensibility of this protein can be tested directly, using a short artificially produced protein that contains eight repeated Ig-domains from one region of the titin protein. In this experiment the tip of the AFM is used to pick up, and progressively stretch, a single molecule until it eventually ruptures. As force is applied, each Ig-domain suddenly begins to unfold, and the force needed in each case (about 200 pN) can be recorded. The region of the force–extension curve shown in *green* records the sequential unfolding event for each of the eight protein domains. (Adapted from W.A. Linke et al., *J. Struct. Biol.* 137:194–205, 2002. With permission from Elsevier.)

Most naturally occurring elements are a mixture of slightly different isotopes. These differ from one another in the mass of their atomic nuclei, but because they have the same number of protons and electrons, they have the same chemical properties. In radioactive isotopes, or radioisotopes, the nucleus is unstable and undergoes random disintegration to produce a different atom. In the course of these disintegrations, either energetic subatomic particles, such as electrons, or radiations, such as gamma-rays, are given off. By using chemical synthesis to incorporate one or more radioactive atoms into a small molecule of interest, such as a sugar or an amino acid, the fate of that molecule (and of specific atoms in it) can be traced during any biological reaction.

Although naturally occurring radioisotopes are rare (because of their instability), they can be produced in large amounts in nuclear reactors, where stable atoms are bombarded with high-energy particles. As a result, radioisotopes of many biologically important elements are readily available (**Table 9–1**). The radiation they emit is detected in various ways. Electrons (β particles) can be detected in a Geiger counter by the ionization they produce in a gas, or they can be measured in a scintillation counter by the small flashes of light they induce in a scintillation fluid. These methods make it possible to measure accurately the quantity of a particular radioisotope present in a biological specimen. Using either light or electron microscopy, it is also possible to determine the location of a radioisotope in a specimen by autoradiography, as we describe below. All of these methods of detection are extremely sensitive: in favorable circumstances, nearly every disintegration—and therefore every radioactive atom that decays—can be detected.

Table 9–1 Some Radioisotopes in Common Use in Biological Research

ISOTOPE	HALF-LIFE
^{32}P	14 days
^{131}I	8.1 days
^{35}S	87 days
^{14}C	5570 years
^{45}Ca	164 days
^{3}H	12.3 years

The isotopes are arranged in decreasing order of the energy of the β radiation (electrons) they emit. ^{131}I also emits γ radiation. The half-life is the time required for 50% of the atoms of an isotope to disintegrate.

Radioisotopes Are Used to Trace Molecules in Cells and Organisms

One of the earliest uses of radioactivity in biology was to trace the chemical pathway of carbon during photosynthesis. Unicellular green algae were maintained in an atmosphere containing radioactively labeled CO_2 ($^{14}CO_2$), and at various times after they had been exposed to sunlight, their soluble contents were separated by paper chromatography. Small molecules containing ^{14}C atoms derived from CO_2 were detected by a sheet of photographic film placed over the dried paper chromatogram. In this way most of the principal components in the photosynthetic pathway from CO_2 to sugar were identified.

Radioactive molecules can be used to follow the course of almost any process in cells. In a typical experiment the cells are supplied with a precursor molecule in radioactive form. The radioactive molecules mix with the preexisting unlabeled ones; both are treated identically by the cell as they differ only in the weight of their atomic nuclei. Changes in the location or chemical form of the radioactive molecules can be followed as a function of time. The resolution of such experiments is often sharpened by using a pulse-chase labeling protocol, in which the radioactive material (the pulse) is added for only a very brief period and then washed away and replaced by nonradioactive molecules (the chase). Samples are taken at regular intervals, and the chemical form or location of the radioactivity is identified for each sample (**Figure 9–38**). Pulse-chase experiments, combined with autoradiography, have been important, for example, in elucidating the pathway taken by secreted proteins from the ER to the cell exterior.

Radioisotopic labeling is a uniquely valuable way of distinguishing between molecules that are chemically identical but have different histories—for example, those that differ in their time of synthesis. In this way, for example, it was shown that almost all of the molecules in a living cell are continually being degraded and replaced, even when the cell is not growing and is apparently in a steady state. This "turnover," which sometimes takes place very slowly, would be almost impossible to detect without radioisotopes.

Today, nearly all common small molecules are available in radioactive form from commercial sources, and virtually any biological molecule, no matter how complicated, can be radioactively labeled. Compounds can be made with radioactive atoms incorporated at particular positions in their structure, enabling the separate fates of different parts of the same molecule to be followed during biological reactions (**Figure 9–39**).

As mentioned previously, one of the important uses of radioactivity in cell biology is to localize a radioactive compound in sections of whole cells or tissues by autoradiography. In this procedure, living cells are briefly exposed to a pulse of a specific radioactive compound and then incubated for a variable period—to allow them time to incorporate the compound—before being fixed and processed for light or electron microscopy. Each preparation is then overlaid with a thin film of photographic emulsion and left in the dark for several days, during which the radioisotope decays. The emulsion is then developed, and the position of the radioactivity in each cell is indicated by the position of the developed silver grains (see Figure 5–29). If cells are exposed to ^3H-thymidine, a radioactive precursor of DNA, for example, it can be shown that DNA is made in the nucleus

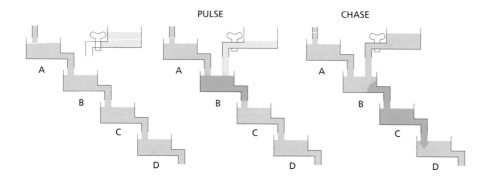

Figure 9–38 **The logic of a typical pulse-chase experiment using radioisotopes.** The chambers labeled A, B, C, and D represent either different compartments in the cell (detected by autoradiography or by cell-fractionation experiments) or different chemical compounds (detected by chromatography or other chemical methods).

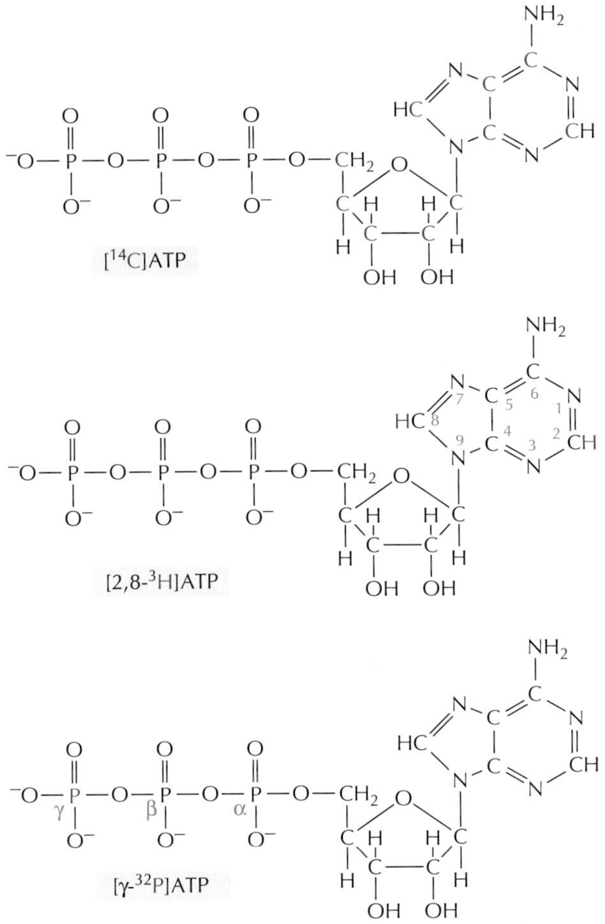

$[^{14}C]ATP$

$[2,8-^3H]ATP$

$[\gamma-^{32}P]ATP$

Figure 9–39 Radioisotopically labeled molecules. Three commercially available radioactive forms of ATP, with the radioactive atoms shown in *red*. The nomenclature used to identify the position and type of the radioactive atoms is also shown.

and remains there (**Figure 9–40**). By contrast, if cells are exposed to ^3H-uridine, a radioactive precursor of RNA, it is found that RNA is initially made in the nucleus (see Figure 4–62) and then moves rapidly into the cytoplasm. Radiolabeled molecules can also be detected by autoradiography after they are separated from other molecules by gel electrophoresis: the positions of both proteins (see Figure 8–23) and nucleic acids (see Figure 8–33A) are commonly detected on gels in this way.

Summary

Many light-microscope techniques are available for observing cells. Cells that have been fixed and stained can be studied in a conventional light microscope, whereas antibodies coupled to fluorescent dyes can be used to locate specific molecules in cells in a fluorescence microscope. Living cells can be seen with phase-contrast, differential-interference-contrast, dark-field, or bright-field microscopes. All forms of light microscopy are facilitated by digital image-processing techniques, which enhance sensitivity and refine the image. Confocal microscopy and image deconvolution both provide thin optical sections and can be used to reconstruct three-dimensional images.

Techniques are now available for detecting, measuring, and following almost any desired molecule in a living cell. Fluorescent indicator dyes can be introduced to measure the concentrations of specific ions in individual cells or in different parts of a cell. Fluorescent proteins are especially versatile probes that can be attached to other proteins by genetic manipulation. Virtually any protein of interest can be genetically engineered as a fluorescent-fusion protein, and then imaged in living cells by fluorescence microscopy. The dynamic behavior and interactions of many molecules can now be followed in living cells by variations on the use of fluorescent protein tags, in some cases at the level of single molecules. Radioactive isotopes of various elements can also be used to follow the fate of specific molecules both biochemically and microscopically.

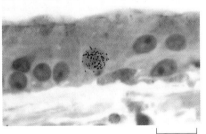

20 μm

Figure 9–40 Autoradiography. This tissue has been exposed for a short period to ^3H-thymidine. Cells that are replicating their DNA incorporate this radioactively labeled DNA precursor into their nuclei and can subsequently be visualized by autoradiography. The silver grains, seen here as *black dots* in the photographic emulsion over the section, reveal which cell was making new DNA. The labeled nucleus shown here is in the sensory epithelium from the inner ear of a chicken. (Courtesy of Mark Warchol and Jeffrey Corwin.)

LOOKING AT CELLS AND MOLECULES IN THE ELECTRON MICROSCOPE

Light microscopy is limited in the fineness of detail that it can reveal. Microscopes using other types of radiation—in particular, electron microscopes—can resolve much smaller structures than is possible with visible light. This higher resolution comes at a cost: specimen preparation for electron microscopy is much more complex and it is harder to be sure that what we see in the image corresponds precisely to the actual structure being examined. It is now possible, however, to use very rapid freezing to preserve structures faithfully for electron microscopy. Digital image analysis can be used to reconstruct three-dimensional objects by combining information either from many individual particles or from multiple tilted views of a single object. Together these approaches are extending the resolution and scope of electron microscopy to the point at which we can begin to faithfully image the structures of individual macromolecules and the complexes they form.

Figure 9–41 The limit of resolution of the electron microscope. This transmission electron micrograph of a thin layer of gold shows the individual files of atoms in the crystal as bright spots. The distance between adjacent files of gold atoms is about 0.2 nm (2 Å). (Courtesy of Graham Hills.)

The Electron Microscope Resolves the Fine Structure of the Cell

The relationship between the limit of resolution and the wavelength of the illuminating radiation (see Figure 9–6) holds true for any form of radiation, whether it is a beam of light or a beam of electrons. With electrons, however, the limit of resolution can be made very small. The wavelength of an electron decreases as its velocity increases. In an **electron microscope** with an accelerating voltage of 100,000 V, the wavelength of an electron is 0.004 nm. In theory the resolution of such a microscope should be about 0.002 nm, which is 100,000 times that of the light microscope. Because the aberrations of an electron lens are considerably harder to correct than those of a glass lens, however, the practical resolving power of most modern electron microscopes is, at best, 0.1 nm (1 Å) (**Figure 9–41**). This is because only the very center of the electron lenses can be used, and the effective numerical aperture is tiny. Furthermore, problems of specimen preparation, contrast, and radiation damage have generally limited the normal effective resolution for biological objects to 1 nm (10 Å). This is nonetheless about 200 times better than the resolution of the light microscope. Moreover, in recent years, the performance of electron microscopes has been improved by the development of electron illumination sources called field emission guns. These very bright and coherent sources can substantially improve the resolution achieved.

In overall design the transmission electron microscope (TEM) is similar to a light microscope, although it is much larger and "upside down" (**Figure 9–42**).

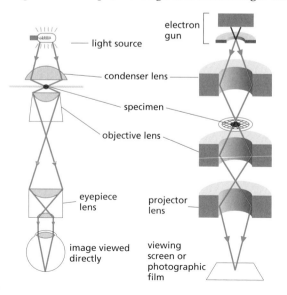

Figure 9–42 The principal features of a light microscope and a transmission electron microscope. These drawings emphasize the similarities of overall design. Whereas the lenses in the light microscope are made of glass, those in the electron microscope are magnetic coils. The electron microscope requires that the specimen be placed in a vacuum. The inset shows a transmission electron microscope in use. (Photograph courtesy of FEI Company Ltd.)

The source of illumination is a filament or cathode that emits electrons at the top of a cylindrical column about 2 m high. Since electrons are scattered by collisions with air molecules, air must first be pumped out of the column to create a vacuum. The electrons are then accelerated from the filament by a nearby anode and allowed to pass through a tiny hole to form an electron beam that travels down the column. Magnetic coils placed at intervals along the column focus the electron beam, just as glass lenses focus the light in a light microscope. The specimen is put into the vacuum, through an airlock, into the path of the electron beam. As in light microscopy, the specimen is usually stained—in this case, with *electron-dense* material, as we see in the next section. Some of the electrons passing through the specimen are scattered by structures stained with the electron-dense material; the remainder are focused to form an image, in a manner analogous to the way an image is formed in a light microscope. The image can be observed on a phosphorescent screen or recorded, either on a photographic plate or with a high-resolution digital camera. Because the scattered electrons are lost from the beam, the dense regions of the specimen show up in the image as areas of reduced electron flux, which look dark.

Biological Specimens Require Special Preparation for the Electron Microscope

In the early days of its application to biological materials, the electron microscope revealed many previously unimagined structures in cells. But before these discoveries could be made, electron microscopists had to develop new procedures for embedding, cutting, and staining tissues.

Since the specimen is exposed to a very high vacuum in the electron microscope, living tissue is usually killed and preserved by fixation—first with *glutaraldehyde*, which covalently cross-links protein molecules to their neighbors, and then with *osmium tetroxide*, which binds to and stabilizes lipid bilayers as well as proteins (**Figure 9–43**). Because electrons have very limited penetrating power, the fixed tissues normally have to be cut into extremely thin sections (50–100 nm thick, about 1/200 the thickness of a single cell) before they are viewed. This is achieved by dehydrating the specimen and permeating it with a monomeric resin that polymerizes to form a solid block of plastic; the block is then cut with a fine glass or diamond knife on a special microtome. These *thin sections*, free of water and other volatile solvents, are placed on a small circular metal grid for viewing in the microscope (**Figure 9–44**). <AACA>

The steps required to prepare biological material for viewing in the electron microscope have challenged electron microscopists from the beginning. How can we be sure that the image of the fixed, dehydrated, resin-embedded specimen finally seen bears any relation to the delicate aqueous biological system that was originally present in the living cell? The best current approaches to this problem depend on rapid freezing. If an aqueous system is cooled fast enough to a low enough temperature, the water and other components in it do not have time to rearrange themselves or crystallize into ice. Instead, the water is supercooled into a rigid but noncrystalline state—a "glass"—called vitreous ice. This state can be achieved by slamming the specimen onto a polished copper block cooled by liquid helium, by plunging it into or spraying it with a jet of a coolant such as liquid propane, or by cooling it at high pressure.

Some frozen specimens can be examined directly in the electron microscope using a special, cooled specimen holder. In other cases the frozen block can be fractured to reveal interior surfaces, or the surrounding ice can be sublimed away to expose external surfaces. However, we often want to examine thin sections, and stain them to yield adequate contrast in the electron microscope image (discussed further below). A compromise is therefore to rapid-freeze the tissue, then replace the water, maintained in the vitreous (glassy) state, by organic solvents, and finally embed the tissue in plastic resin, cut sections, and stain. Although technically still difficult, this approach stabilizes and preserves the tissue in a condition very close to its original living state (**Figure 9–45**).

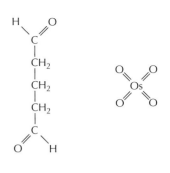

glutaraldehyde   osmium tetroxide

Figure 9–43 Two common chemical fixatives used for electron microscopy. The two reactive aldehyde groups of glutaraldehyde enable it to cross-link various types of molecules, forming covalent bonds between them. Osmium tetroxide forms cross-linked complexes with many organic compounds, and in the process becomes reduced. This reaction is especially useful for fixing cell membranes, since the C=C double bonds present in many fatty acids react with osmium tetroxide.

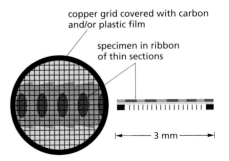

copper grid covered with carbon and/or plastic film

specimen in ribbon of thin sections

3 mm

Figure 9–44 The copper grid that supports the thin sections of a specimen in a TEM.

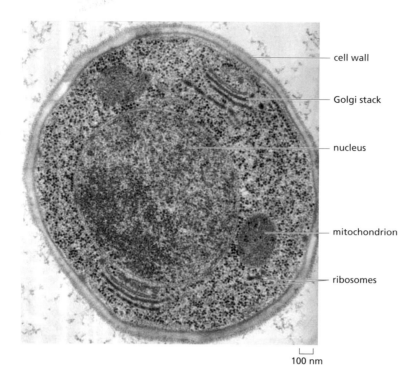

cell wall

Golgi stack

nucleus

mitochondrion

ribosomes

100 nm

Figure 9–45 Thin section of a cell. This thin section is of a yeast cell that has been very rapidly frozen and the vitreous ice replaced by organic solvents and then by plastic resin. The nucleus, mitochondria, cell wall, Golgi stacks, and ribosomes can all be readily seen in a state that is presumed to be as life-like as possible. (Courtesy of Andrew Staehelin.)

Contrast in the electron microscope depends on the atomic number of the atoms in the specimen: the higher the atomic number, the more electrons are scattered and the greater the contrast. Biological tissues are composed of atoms of very low atomic number (mainly carbon, oxygen, nitrogen, and hydrogen). To make them visible, they are usually impregnated (before or after sectioning) with the salts of heavy metals such as uranium and lead. The degree of impregnation, or "staining," with these salts reveals different cellular constituents with various degrees of contrast. Lipids, for example, tend to stain darkly after osmium fixation, revealing the location of cell membranes.

Specific Macromolecules Can Be Localized by Immunogold Electron Microscopy

We have seen how antibodies can be used in conjunction with fluorescence microscopy to localize specific macromolecules. An analogous method—**immunogold electron microscopy**—can be used in the electron microscope. The usual procedure is to incubate a thin section with a specific primary antibody, and then with a secondary antibody to which a colloidal gold particle has been attached. The gold particle is electron-dense and can be seen as a black dot in the electron microscope (**Figure 9–46**).

Thin sections often fail to convey the three-dimensional arrangement of cellular components in the TEM and can be very misleading: a linear structure such as a microtubule may appear in section as a pointlike object, for example, and a section through protruding parts of a single irregularly shaped solid body may give the appearance of two or more separate objects. The third dimension can be reconstructed from serial sections (**Figure 9–47**), but this is still a lengthy and tedious process.

Even thin sections, however, have a significant depth compared with the resolution of the electron microscope, so they can also be misleading in an opposite way. The optical design of the electron microscope—the very small aperture used—produces a large depth of field, so the image seen corresponds to a superimposition (a projection) of the structures at different depths. A further complication for immunogold labeling is that the antibodies and colloidal gold particles do not penetrate into the resin used for embedding; therefore, they detect antigens only at the surface of the section. This means that first, the sensitivity of detection is low, since antigen molecules present in the deeper parts of the

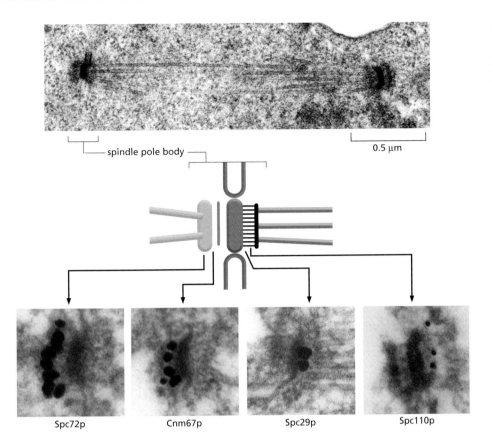

spindle pole body

0.5 µm

Spc72p   Cnm67p   Spc29p   Spc110p

Figure 9–46 Localizing proteins in the electron microscope. Immunogold electron microscopy is used here to localize four different protein components to particular locations within the spindle pole body of yeast. At the top is a thin section of a yeast mitotic spindle showing the spindle microtubules that cross the nucleus, and connect at each end to spindle pole bodies embedded in the nuclear envelope. A diagram of the components of a single spindle pole body is shown below. Antibodies against four different proteins of the spindle pole body are used, together with colloidal gold particles *(black dots)*, to reveal where within the complex structure each protein is located. (Courtesy of John Kilmartin.)

section are not detected, and second, we may get a false impression of which structures contain the antigen and which do not. A solution to this problem is to label the specimen before embedding it in plastic, when cells and tissues are still fully accessible to labeling reagents. Extremely small gold particles, about 1 nm in diameter, work best for this procedure. Such small gold particles are usually not directly visible in the final sections, so additional silver or gold is nucleated around the tiny 1 nm gold particles in a chemical process very much like photographic development.

Images of Surfaces Can Be Obtained by Scanning Electron Microscopy

A **scanning electron microscope (SEM)** directly produces an image of the three-dimensional structure of the surface of a specimen. The SEM is usually a smaller, simpler, and cheaper device than a transmission electron microscope. Whereas the TEM uses the electrons that have passed through the specimen to form an

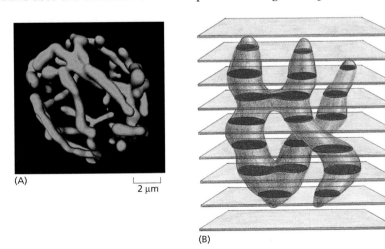

(A)

2 µm

(B)

Figure 9–47 A three-dimensional reconstruction from serial sections. (A) A three-dimensional reconstruction of the mitochondrial compartment of a live yeast cell, assembled from a stack of optical sections, shows its complex branching structure. Single thin sections of such a structure in the electron microscope sometimes give misleading impressions. In this example (B), most sections through a cell containing a branched mitochondrion seem to contain two or three separate mitochondria (compare Figure 9–45). Sections 4 and 7, moreover, might be interpreted as showing a mitochondrion in the process of dividing. The true three-dimensional shape, however, can be reconstructed from serial sections. (A, courtesy of Stefan Hell.)

image, the SEM uses electrons that are scattered or emitted from the specimen's surface. The specimen to be examined is fixed, dried, and coated with a thin layer of heavy metal. Alternatively, it can be rapidly frozen, and then transferred to a cooled specimen stage for direct examination in the microscope. Often an entire plant part or small animal can be put into the microscope with very little preparation (**Figure 9–48**). The specimen, prepared in any of these ways, is then scanned with a very narrow beam of electrons. The quantity of electrons scattered or emitted as this primary beam bombards each successive point of the metallic surface is measured and used to control the intensity of a second beam, which moves in synchrony with the primary beam and forms an image on a television screen. In this way, a highly enlarged image of the surface as a whole is built up (**Figure 9–49**).

The SEM technique provides great depth of field; moreover, since the amount of electron scattering depends on the angle of the surface relative to the beam, the image has highlights and shadows that give it a three-dimensional appearance (see Figure 9–48 and **Figure 9–50**). Only surface features can be examined, however, and in most forms of SEM, the resolution attainable is not very high (about 10 nm, with an effective magnification of up to 20,000 times). As a result, the technique is usually used to study whole cells and tissues rather than subcellular organelles. <AACG> Very high-resolution SEMs have, however, been developed with a bright coherent-field emission gun as the electron source. This type of SEM can produce images that rival TEM images in resolution (**Figure 9–51**).

Metal Shadowing Allows Surface Features to Be Examined at High Resolution by Transmission Electron Microscopy

The TEM can also be used to study the surface of a specimen—and generally at a higher resolution than in the SEM—to reveal the shape of individual macromolecules for example. As in scanning electron microscopy, a thin film of a heavy metal such as platinum is evaporated onto the dried specimen. In this case, however, the metal is sprayed from an oblique angle so as to deposit a coat-

1 mm

Figure 9–48 A developing wheat flower, or spike. This delicate flower spike was rapidly frozen, coated with a thin metal film, and examined in the frozen state in a SEM. This micrograph, which is at a low magnification, demonstrates the large depth of focus of the SEM. (Courtesy of Kim Findlay.)

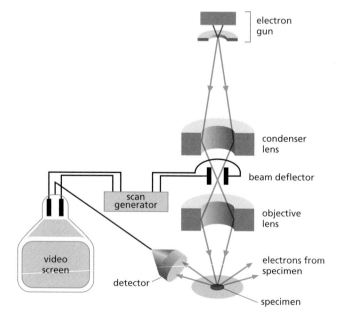

Figure 9–49 The scanning electron microscope. In a SEM, the specimen is scanned by a beam of electrons brought to a focus on the specimen by the electromagnetic coils that act as lenses. The detector measures the quantity of electrons scattered or emitted as the beam bombards each successive point on the surface of the specimen and controls the intensity of successive points in an image built up on a video screen. The SEM creates striking images of three-dimensional objects with great depth of focus and a resolution between 3 nm and 20 nm depending on the instrument. (Photograph courtesy of Andrew Davies.)

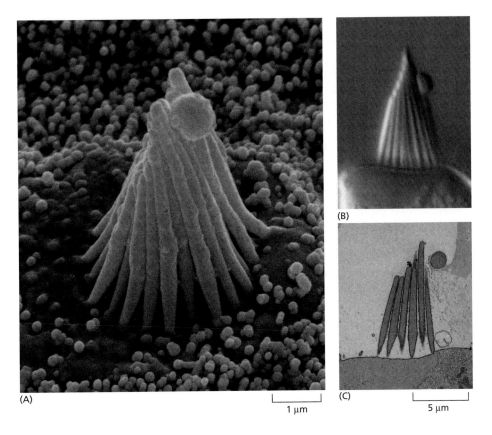

(B)

(A)   (C)

1 μm   5 μm

Figure 9–50 **Scanning electron microscopy.** (A) A scanning electron micrograph of the stereocilia projecting from a hair cell in the inner ear of a bullfrog. <CATA> For comparison, the same structure is shown by (B) differential-interference-contrast light microscopy and (C) thin-section transmission electron microscopy. (Courtesy of Richard Jacobs and James Hudspeth.)

ing that is thicker in some places than others—a process known as *metal shadowing* because a shadow effect is created that gives the image a three-dimensional appearance.

Some specimens coated in this way are thin enough or small enough for the electron beam to penetrate them directly. This is the case for individual molecules, macromolecular complexes, and viruses—all of which can be dried down, before shadowing, onto a flat supporting film made of a material that is relatively transparent to electrons, such as carbon or plastic. The internal structure of cells can also be imaged using metal shadowing. In this case samples are very rapidly frozen (as described above) and then cracked open with a knife blade. The ice level at the fractured surface is lowered by the sublimation of ice in a vacuum as the temperature is raised—in a process called freeze-drying. The parts of the cell exposed by this *etching* process are then shadowed as before to make a metal replica. The organic material of the cell remains must be dissolved away after shadowing to leave only the thin metal *replica* of the surface of the specimen. The replica is then reinforced with a film of carbon so it can be placed on a grid and examined in the transmission electron microscope in the ordinary way (**Figure 9–52**). This technique exposes structures in the interior of the cell and can reveal their three-dimensional organization with exceptional clarity (**Figure 9–53**).

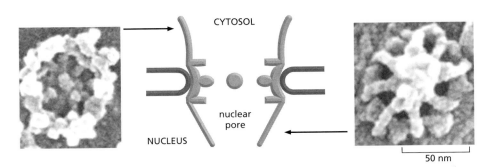

CYTOSOL

nuclear pore

NUCLEUS

50 nm

Figure 9–51 **The nuclear pore.** Rapidly frozen nuclear envelopes were imaged in a high-resolution SEM, equipped with a field emission gun as the source of electrons. These views of each side of a nuclear pore represent the limit of resolution of the SEM, and should be compared with Figure 12–9. (Courtesy of Martin Goldberg and Terry Allen.)

Negative Staining and Cryoelectron Microscopy Both Allow Macromolecules to Be Viewed at High Resolution

Although isolated macromolecules, such as DNA or large proteins, can be visualized readily in the electron microscope if they are shadowed with a heavy metal to provide contrast, finer detail can be seen by using **negative staining**. In this technique, the molecules, supported on a thin film of carbon, are mixed with a solution of a heavy-metal salt such as uranyl acetate. After the sample has dried, a very thin film of metal salt covers the carbon film everywhere except where it has been excluded by the presence of an adsorbed macromolecule. Because the macromolecule allows electrons to pass through it much more readily than does the surrounding heavy-metal stain, a reversed or negative image of the molecule is created. Negative staining is especially useful for viewing large macromolecular aggregates such as viruses or ribosomes, and for seeing the subunit structure of protein filaments (**Figure 9–54**).

Shadowing and negative staining can provide high-contrast surface views of small macromolecular assemblies, but the size of the smallest metal particles in the shadow or stain used limits the resolution of both techniques. Recent methods provide an alternative that has allowed us to visualize directly at high resolution even the interior features of three-dimensional structures such as viruses and organelles. In this technique, called **cryoelectron microscopy**, rapid freezing to form vitreous ice is again the key. A very thin (about 100 nm) film of an aqueous suspension of virus or purified macromolecular complex is prepared on a microscope grid. The specimen is then rapidly frozen by plunging it into a coolant. A special sample holder is used to keep this hydrated specimen at –160°C in the vacuum of the microscope, where it can be viewed directly without fixation, staining, or drying. Unlike negative staining, in which what we see is the envelope of stain exclusion around the particle, hydrated cryoelectron microscopy produces an image from the macromolecular structure itself. However, to extract the maximum amount of structural information, special image-processing techniques must be used, as we describe next.

Multiple Images Can Be Combined to Increase Resolution

Any image, whether produced by an electron microscope or by an optical microscope, is made by particles—electrons or photons—striking a detector of some sort. But these particles are governed by quantum mechanics, so the numbers reaching the detector are predictable only in a statistical sense. In the limit of very large numbers of particles, the distribution at the detector is accurately determined by the imaged specimen. However, with smaller numbers of particles, this underlying structure in the image is obscured by the statistical fluctuations in the numbers of particles detected in each region. The term *noise* describes the random variability that confuses the underlying image of the spec-

Figure 9–52 The preparation of a metal-shadowed replica of the surface of a specimen. Note that the thickness of the metal reflects the surface contours of the original specimen.

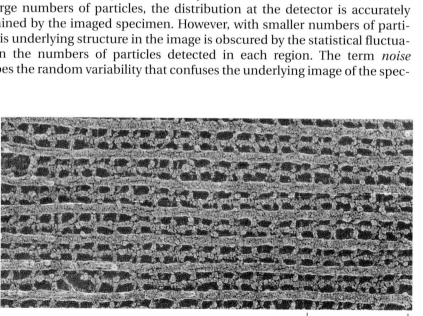

0.1 μm

Figure 9–53 A regular array of protein filaments in an insect muscle. To obtain this image, the muscle cells were rapidly frozen to liquid helium temperature, fractured through the cytoplasm, and subjected to deep etching. A metal-shadowed replica was then prepared and examined at high magnification. (Courtesy of Roger Cooke and John Heuser.)

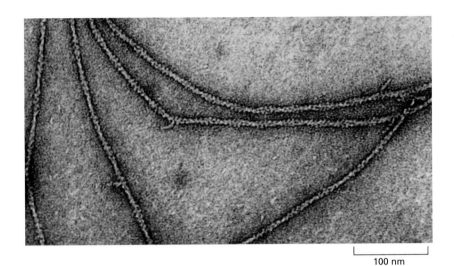

Figure 9–54 Negatively stained actin filaments. In this transmission electron micrograph, each filament is about 8 nm in diameter and is seen, on close inspection, to be composed of a helical chain of globular actin molecules. (Courtesy of Roger Craig.)

100 nm

imen itself. Noise is important in light microscopy at low light levels, but it is a particularly severe problem for electron microscopy of unstained macromolecules. A protein molecule can tolerate a dose of only a few tens of electrons per square nanometer without damage, and this dose is orders of magnitude below what is needed to define an image at atomic resolution.

The solution is to obtain images of many identical molecules—perhaps tens of thousands of individual images—and combine them to produce an averaged image, revealing structural details that were hidden by the noise in the original images. This procedure is called **single-particle reconstruction**. Before combining all the individual images, however, they must be aligned with each other. Sometimes it is possible to induce proteins and complexes to form crystalline arrays, in which each molecule is held in the same orientation in a regular lattice. In this case, the alignment problem is easily solved, and several protein structures have been determined at atomic resolution by this type of electron crystallography. In principle, however, crystalline arrays are not absolutely required. With the help of a computer, the digital images of randomly distributed and unaligned molecules can be processed and combined to yield high-resolution reconstructions. <TATT> Although structures that have some intrinsic symmetry make the task of alignment easier and more accurate, this technique has has also been used for objects, like ribosomes, with no symmetry. **Figure 9–55** shows the structure of an icosahedral virus that has been determined at high resolution by the combination of many particles and multiple views.

With well-ordered crystalline arrays, a resolution of 0.3 nm has been achieved by electron microscopy—enough to begin to see the internal atomic

Figure 9–55 Single-particle reconstruction. Spherical protein shells of the hepatitis B virus are preserved in a thin film of ice (A) and imaged in the transmission electron microscope. Thousands of individual particles were combined by single-particle reconstruction to produce the three-dimensional map of the icosahedral particle shown in (B). The two views of a single protein dimer (C), forming the spikes on the surface of the shell, show that the resolution of the reconstruction (0.74 nm) is sufficient to resolve the complete fold of the polypeptide chain. (A, courtesy of B. Böttcher, S.A. Wynne, and R.A. Crowther; B and C, from B. Böttcher, S.A. Wynne, and R.A. Crowther, *Nature* 386:88–91, 1997. With permission from Macmillan Publishers Ltd.)

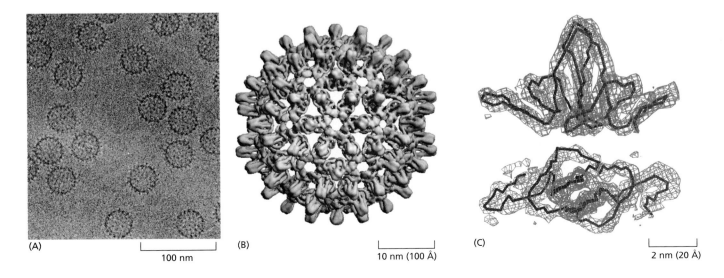

(A) 100 nm

(B) 10 nm (100 Å)

(C) 2 nm (20 Å)

arrangements in a protein and to rival x-ray crystallography in resolution. With single-particle reconstruction, the present limit is about 0.5 nm, enough to identify protein subunits and domains, and limited protein secondary structure. Although electron microscopy is unlikely to supersede x-ray crystallography (discussed in Chapter 8) as a method for macromolecular structure determination, it has some very clear advantages. First, it does not absolutely require crystalline specimens. Second, it can deal with extremely large complexes—structures that may be too large or too variable to crystallize satisfactorily.

The analysis of large and complex macromolecular structures is helped considerably if the atomic structure of one or more of the subunits is known, for example from x-ray crystallography. Molecular models can then be mathematically "fitted" into the envelope of the structure determined at lower resolution using the electron microscope. **Figure 9–56** shows the structure of a ribosome with the location of a bound release factor displayed in this way (see also Figures 6–74 and 6–75).

Different Views of a Single Object Can Be Combined to Give a Three-dimensional Reconstruction

The detectors used to record images from electron microscopes produce two-dimensional pictures. Because of the large depth of field of the microscope, all the parts of the three-dimensional specimen are in focus, and the resulting image is a projection of the structure along the viewing direction. The lost information in the third dimension can be recovered if we have views of the same specimen from many different directions. The computational methods for this technique were worked out in the 1960s, and they are widely used in medical computed tomography (CT) scans. In a CT scan, the imaging equipment is moved around the patient to generate the different views. In **electron-microscope (EM) tomography**, the specimen holder is tilted in the microscope, which achieves the same result. In this way, we can arrive at a three-dimensional reconstruction, in a chosen standard orientation, by combining a set of different views of a single object in the microscope's field of view. Each individual view will be very noisy, but by combining them in three dimensions and taking an average, the noise can be largely eliminated, yielding a clear view of the molecular structure. Starting with thick plastic sections of embedded material, three-dimensional reconstructions, or *tomograms*, <ATCC> <CGAT> are used extensively to describe the detailed anatomy of small regions of the cell, such as the Golgi apparatus (**Figure 9–57**) or the cytoskeleton. Increasingly, however, microscopists are applying EM tomography to unstained frozen hydrated sections, and even to rapidly frozen whole cells or organelles (**Figure 9–58**). Electron microscopy now provides a robust bridge between the scale of the single molecule and that of the whole cell.

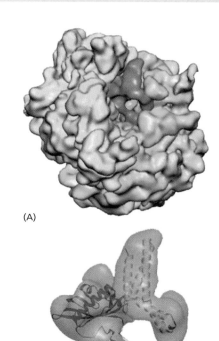

(A)

(B)

Figure 9–56 Single-particle reconstruction and molecular model fitting. Bacterial ribosomes, with and without the release factor required for peptide release from the ribosome, were used here to derive high-resolution three-dimensional cryo-EM maps at a resolution of better than 1 nm. Images of nearly 20,000 separate ribosomes, preserved in ice, were used to produce single particle reconstructions. In (A) the 30S ribosomal subunit (*yellow*) and the 50S subunit (*blue*) can be distinguished from the additional electron density that can be attributed to the release factor RF2 (*pink*). The known molecular structure of RF2 has then been modeled into this electron density (B). (From U.B.S. Rawat et al., *Nature* 421:87–90, 2003. With permission from Macmillan Publishers Ltd.)

Summary

Determining the detailed structure of the membranes and organelles in cells requires the higher resolution attainable in a transmission electron microscope. Specific macromolecules can be localized with colloidal gold linked to antibodies. Three-dimensional views of the surfaces of cells and tissues are obtained by scanning electron microscopy. The shapes of isolated macromolecules that have been shadowed with a heavy metal or outlined by negative staining can also be readily determined by electron microscopy. Using computational methods, either multiple images or views from different directions can be combined to produce detailed reconstructions of macromolecules and molecular complexes through the techniques of electron tomography and single-particle reconstruction, often applied to cryo-preserved specimens. The resolution obtained with these methods means that atomic structures of individual macromolecules can often be "fitted" to the images derived by electron microscopy, and that the TEM is increasingly able to completely bridge the gap between structures determined by x-ray crystallography and those determined in the light microscope.

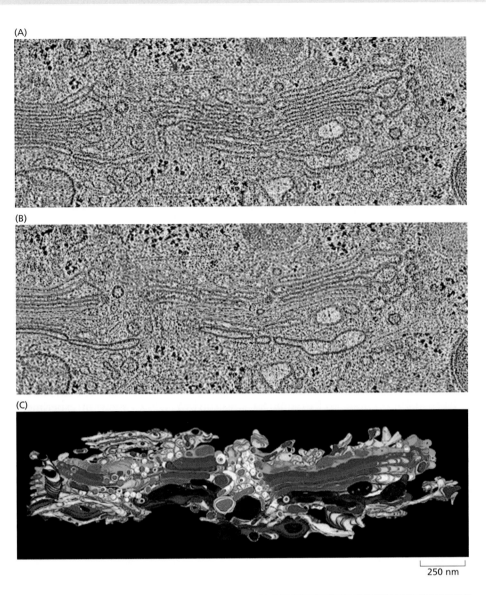

(A)

(B)

(C)

250 nm

Figure 9–57 Electron microscope (EM) tomography. Samples that have been rapidly frozen, and then freeze-substituted and embedded in plastic, preserve their structure in a condition that is very close to their original living state. This experiment shows an analysis of the three-dimensional structure of the Golgi apparatus from a rat kidney cell prepared in this way. Several thick sections (250 nm) of the cell have been tilted in a high-voltage electron microscope, along two different axes, and about 160 different views recorded. The digital data were combined using EM tomography methods to produce a final three-dimensional reconstruction at a resolution of about 7 nm. The computer can then present very thin slices of the complete three-dimensional data set, or tomogram, and two serial slices, each only 4 nm thick, are shown here (A) and (B). Very little changes from one slice to the next, but using the full data set, and by manually color coding the membranes (B), a full three-dimensional picture of the complete Golgi complex and its associated vesicles can be presented (C). (From M.S. Ladinsky et al., *J.Cell Biol.* 144:1135–1149, 1999. With permission from The Rockefeller University Press.)

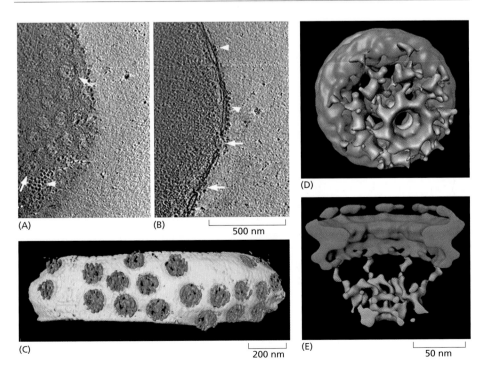

(A)   (B)

500 nm

(C)

200 nm

(D)

(E)

50 nm

Figure 9–58 Combining cryo-EM tomography and single-particle reconstruction. In addition to sections, the technique of EM tomography may also be applied to small unfixed specimens that are rapidly frozen and examined, while still frozen, using a tilting stage in the microscope. In this experiment the small nuclei of *Dictyostelium* are gently isolated and then very rapidly frozen before a series of tilted views of them is recorded. These different digital views are combined by EM tomography methods to produce a three-dimensional tomogram. Two thin digital slices (10 nm) through this tomogram are shown, in which top views (A) and side views (B) of individual nuclear pores can be seen. In the three-dimensional model (C), a surface rendering of the pores *(blue)* can be seen embedded in the nuclear envelope *(yellow)*. From a series of tomograms it was possible to extract data sets for nearly 300 separate nuclear pores, whose structures could then be averaged using the techniques of single particle reconstruction. The surface-rendered view of one of these reconstructed pores is shown from the nuclear face in (D) and in section in (E) and should be compared with Figure 12–10. The pore complex is colored *blue* and the nuclear basket *brown.* (From M. Beck et al., *Science* 306:1387–1390, 2004. With permission from AAAS.)

PROBLEMS

Which statements are true? Explain why or why not.

9–1 Because the DNA double helix is only 10 nm wide—well below the resolution of the light microscope—it is impossible to see chromosomes in living cells without special stains.

9–2 A fluorescent molecule, having absorbed a single photon of light at one wavelength, always emits it at a longer wavelength.

9–3 Caged molecules can be introduced into a cell and then activated by a strong pulse of laser light at the precise time and cellular location chosen by the experimenter.

Discuss the following problems.

9–4 The diagrams in **Figure Q9–1** show the paths of light rays passing through a specimen with a dry lens and with an oil-immersion lens. Offer an explanation for why oil-immersion lenses should give better resolution. Air, glass, and oil have refractive indices of 1.00, 1.51, and 1.51, respectively.

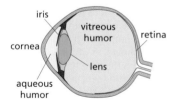

Figure Q9–1 Paths of light rays through dry and oil-immersion lenses (Problem 9–4). The *white circle* at the origin of the light rays is the specimen.

9–5 **Figure Q9–2** shows a diagram of the human eye. The refractive indices of the components in the light path are: cornea 1.38, aqueous humor 1.33, crystalline lens 1.41, and vitreous humor 1.38. Where does the main refraction—the main focusing—occur? What role do you suppose the lens plays?

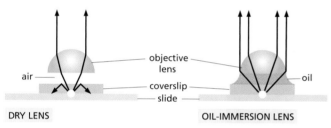

Figure Q9–2 Diagram of the human eye (Problem 9–5).

9–6 Why do humans see so poorly underwater? And why do goggles help?

9–7 Explain the difference between resolution and magnification.

9–8 Antibodies that bind to specific proteins are important tools for defining the locations of molecules in cells. The sensitivity of the primary antibody—the antibody that reacts with the target molecule—is often enhanced by using labeled secondary antibodies that bind to it. What are the advantages and disadvantages of using secondary antibodies that carry fluorescent tags versus those that carry bound enzymes?

9–9 **Figure Q9–3** shows a series of modified GFPs that emit light in a range of colors. How do you suppose the exact same chromophore can fluoresce at so many different wavelengths?

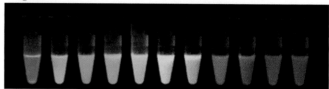

Figure Q9–3 A rainbow of colors produced by modified GFPs (Problem 9–9). (From R.F. Service, *Science* 306:1457, 2004. With permission from AAAS.)

9–10 Consider a fluorescent detector designed to report the cellular location of active protein tyrosine kinases. A blue (cyan) fluorescent protein (CFP) and a yellow fluorescent protein (YFP) were fused to either end of a hybrid protein domain. The hybrid protein segment consisted of a substrate peptide recognized by the Abl protein tyrosine kinase and a phosphotyrosine binding domain (**Figure Q9–4**A). Stimulation of the CFP domain does not cause emission by the YFP domain when the domains are separated. When the CFP and YFP domains are brought close together, however, fluorescence resonance energy transfer (FRET) allows excitation of CFP to stimulate emission by YFP. FRET shows up experimentally as an increase in the ratio of emission at 526 nm versus 476 nm (YFP/CFP) when CFP is excited by 434-nm light.

Incubation of the reporter protein with Abl protein tyrosine kinase in the presence of ATP gave an increase in YFP/CFP emission (Figure Q9–4B). In the absence of ATP or the Abl protein, no FRET occurred. FRET was also eliminated by addition of a tyrosine phosphatase (Figure Q9–4B). Describe as best you can how the reporter protein detects active Abl protein tyrosine kinase.

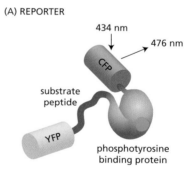

(A) REPORTER

(B) FRET

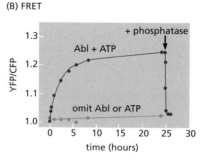

Figure Q9–4 Fluorescent reporter protein designed to detect tyrosine phosphorylation (Problem 9–10). (A) Domain structure of reporter protein. Four domains are indicated: CFP, YFP, tyrosine kinase substrate peptide, and a phosphotyrosine-binding domain. (B) FRET assay. YFP/CFP is normalized to 1.0 at time zero. The reporter was incubated in the presence (or absence) of Abl and ATP for the indicated times. *Arrow* indicates time of addition of a tyrosine phosphatase. (From A.Y. Ting, K.H. Klain, R.L. Klemke and R.Y. Tsien, *Proc. Natl Acad. Sci. U.S.A.* 98:15003–15008, 2001. With permission from National Academy of Sciences.)

9–11 The practical resolving power of modern electron microscopes is around 0.1 nm. The major reason for this constraint is the small numerical aperture ($n \sin \theta$), which is limited by θ (half the angular width of rays collected at the objective lens). Assuming that the wavelength (λ) of the electron is 0.004 nm and that the refractive index (n) is 1.0, calculate the value for θ. How does that value compare with a θ of 60°, which is typical for light microscopes?

$$resolution = \frac{0.61\,\lambda}{n\,\sin\,\theta}$$

9–12 It is difficult to tell bumps from pits just by looking at the pattern of shadows. Consider **Figure Q9–5**, which shows a set of shaded circles. In Figure Q9–5A the circles appear to be bumps; however, when the picture is simply turned upside down (Figure Q9–5B), the circles seem to be pits. This is a classic illusion. The same illusion is present in metal shadowing, as shown in the two electron micrographs in Figure Q9–5. In one the membrane appears to be covered in bumps, while in the other the membrane looks heavily pitted. Is it possible for an electron microscopist to be sure that one view is correct, or is it all arbitrary? Explain your reasoning.

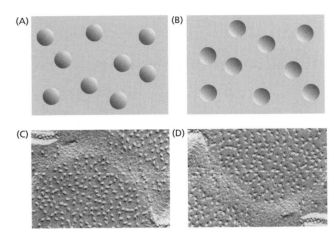

Figure Q9–5 Bumps and pits (Problem 9–12). (A) Shaded circles that look like bumps. (B) Shaded circles that look like pits. (C) An electron micrograph oriented so that it appears to be covered with bumps. (D) An electron micrograph oriented so that it appears to be covered with pits. (C and D, courtesy of Andrew Staehelin.)

REFERENCES

General

Celis JE, Carter N, Simons K et al (eds) (2005) Cell Biology: A Laboratory Handbook, 3rd ed. San Diego: Academic Press. (Volume 3 of this four volume set covers the practicalities of most of the current light and electron imaging methods that are used in cell biology, while volume 4 covers the transfer of molecules into cells.)

Pawley BP (ed) (2006) Handbook of Biological Confocal Microscopy, 3rd ed. New York: Springer Science.

Looking at Cells in the Light Microscope

Adams MC, Salmon WC, Gupton SL et al (2003) A high-speed multispectral spinning-disk confocal microscope system for fluorescent speckle microscopy of living cells. *Methods* 29:29–41.

Agard DA, Hiraoka Y, Shaw P & Sedat JW (1989) Fluorescence microscopy in three dimensions. In Methods in Cell Biology, vol 30: Fluorescence Microscopy of Living Cells in Culture, part B (DL Taylor, Y-L Wang eds). San Diego: Academic Press.

Centonze VE (2002) Introduction to multiphoton excitation imaging for the biological sciences. *Methods Cell Biol* 70:129–48.

Chalfie M, Tu Y, Euskirchen G et al (1994) Green fluorescent protein as a marker for gene expression. *Science* 263:802–805.

Giepmans BN, Adams SR, Ellisman MH & Tsien RY (2006) The fluorescent toolbox for assessing protein location and function. *Science* 312:217–224.

Harlow E & Lane D (1988) Antibodies. A Laboratory Manual. Cold Spring Harbor, NY: Cold Spring Harbor Laboratory Press.

Haugland RP (ed) (1996) Handbook of Fluorescent Probes and Research Chemicals, 8th ed. Eugene, OR: Molecular Probes, Inc. (Available online at http://www.probes.com/)

Jaiswai JK & Simon SM (2004) Potentials and pitfalls of fluorescent quantum dots for biological imaging. *Trends Cell Biol* 14:497–504.

Jares-Erijman EA & Jovin TM (2003) FRET imaging. *Nature Biotech* 21:1387–1395.

Lippincott-Schwartz J, Altan-Bonnet N & Patterson G (2003) Photobleaching and photoactivation: following protein dynamics in living cells. *Nature Cell Biol* 5:S7–S14.

Minsky M (1988) Memoir on inventing the confocal scanning microscope. *Scanning* 10:128–138.

Miyawaki A, Sawano A & Kogure T (2003) Lighting up cells: labeling proteins with fluorophores. *Nature Cell Biol* 5:S1–S7.

Parton RM & Read ND (1999) Calcium and pH imaging in living cells. In Light Microscopy in Biology. A Practical Approach, 2nd ed. (Lacey AJ ed) Oxford: Oxford University Press.

Sako Y & Yanagida T (2003) Single-molecule visualization in cell biology. *Nature Rev Mol Cell Biol* 4:SS1–SS5.

Sekar RB & Periasamy A (2003) Fluorescence resonance energy transfer (FRET) microscopy imaging of live cell protein localizations. *J Cell Biol* 160:629–633.

Shaner NC, Steinbach PA & Tsien RY (2005) A guide to choosing fluorescent proteins. *Nature Methods* 2:905–909.

Sheetz MP (ed) (1997) Laser Tweezers in Cell Biology. *Methods Cell Biol* 55.

Sluder G & Wolf DE (2007) Video Microscopy 3rd ed. *Methods Cell Biol* 81.

Stevens DJ & Allan (2003) Light Microscopy Techniques for Live Cell Imaging. *Science* 300:82–86.

Tsien RY (2003) Imagining imaging's future. *Nature Rev Mol Cell Rev* 4:SS16–SS21.

Weiss DG, Maile W, Wick RA & Steffen W (1999) Video microscopy. In Light Microscopy in Biology: A Practical Approach, 2nd ed. (AJ Lacey ed) Oxford: Oxford University Press.

White JG, Amos WB & Fordham M (1987) An evaluation of confocal versus conventional imaging of biological structures by fluorescence light microscopy. *J Cell Biol* 105:41–48.

Zernike F (1955) How I discovered phase contrast. *Science* 121:345–349.

Looking at Cells and Molecules in the Electron Microscope

Allen TD & Goldberg MW (1993) High resolution SEM in cell biology. *Trends Cell Biol* 3:203–208.

Baumeister W (2002) Electron tomography: towards visualizing the molecular organization of the cytoplasm. *Curr Opin Struct Biol* 12:679–684.

Böttcher B, Wynne SA & Crowther RA (1997) Determination of the fold of the core protein of hepatits B virus by electron cryomicroscopy. *Nature* 386:88–91.

Dubochet J, Adrian M, Chang J-J et al (1988) Cryoelectron microscopy of vitrified specimens. *Q Rev Biophys* 21:129–228.

Frank J (2003) Electron microscopy of functional ribosome complexes. *Biopolymers* 68:223–233.

Hayat MA (2000) Principles and Techniques of Electron Microscopy, 4th ed. Cambridge: Cambridge University Press.

Heuser J (1981) Quick-freeze, deep-etch preparation of samples for 3D electron microscopy. *Trends Biochem Sci* 6:64–68.

Lippincott-Schwartz J & Patterson GH (2003) Development and use of fluorescent protein markers in living cells. *Science* 300:87–91.

McIntosh R, Nicastro D & Mastronarde D (2005) New views of cells in 3D: an introduction to electron tomography. *Trends Cell Biol* 15:43–51.

McDonald KL & Auer M (2006) High pressure freezing, cellular tomography, and structural cell biology. *Biotechniques* 41:137–139.

Pease DC & Porter KR (1981) Electron microscopy and ultramicrotomy. *J Cell Biol* 91:287s–292s.

Unwin PNT & Henderson R (1975) Molecular structure determination by electron microscopy of unstained crystal specimens. *J Mol Biol* 94:425–440.

IV

INTERNAL ORGANIZATION OF THE CELL

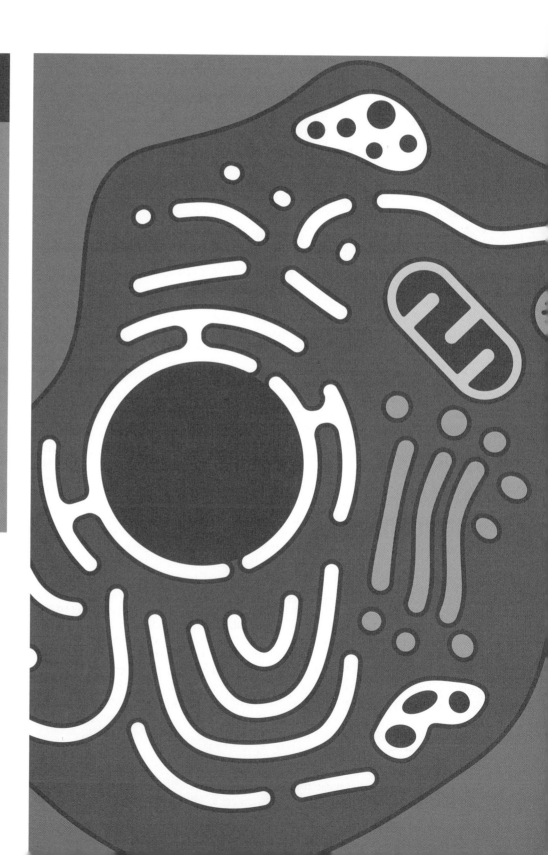

Chapter 10

Membrane Structure

Cell membranes are crucial to the life of the cell. The **plasma membrane** encloses the cell, defines its boundaries, and maintains the essential differences between the cytosol and the extracellular environment. Inside eucaryotic cells, the membranes of the endoplasmic reticulum, Golgi apparatus, mitochondria, and other membrane-enclosed organelles maintain the characteristic differences between the contents of each organelle and the cytosol. Ion gradients across membranes, established by the activities of specialized membrane proteins, can be used to synthesize ATP, to drive the transmembrane movement of selected solutes, or, as in nerve and muscle cells, to produce and transmit electrical signals. In all cells, the plasma membrane also contains proteins that act as sensors of external signals, allowing the cell to change its behavior in response to environmental cues, including signals from other cells; these protein sensors, or *receptors*, transfer information—rather than molecules—across the membrane.

Despite their differing functions, all biological membranes have a common general structure: each is a very thin film of lipid (fatty) and protein molecules, held together mainly by noncovalent interactions (**Figure 10–1**). Cell membranes are dynamic, fluid structures, and most of their molecules move about in the plane of the membrane. The lipid molecules are arranged as a continuous double layer about 5 nm thick. This *lipid bilayer* provides the basic fluid structure of the membrane and serves as a relatively impermeable barrier to the passage of most water-soluble molecules. Protein molecules that span the lipid bilayer (*transmembrane proteins*; see Figure 10–1) mediate nearly all of the other functions of the membrane, transporting specific molecules across it, for example, or catalyzing membrane-associated reactions such as ATP synthesis. In the plasma membrane, some transmembrane proteins serve as structural links that connect the cytoskeleton through the lipid bilayer to either the extracellular matrix or an adjacent cell, while others serve as receptors to detect and transduce chemical signals in the cell's environment. As would be expected, it takes many different membrane proteins to enable a cell to function and interact with its environment, and it is estimated that about 30% of the proteins encoded in an animal cell's genome are membrane proteins.

In this chapter, we consider the structure and organization of the two main constituents of biological membranes—the lipids and the proteins. Although we focus mainly on the plasma membrane, most concepts discussed apply to the various internal membranes in cells as well. The functions of cell membranes are considered in later chapters: their role in ATP synthesis, for example, is discussed in Chapter 14; their role in the transmembrane transport of small molecules in Chapter 11; and their roles in cell signaling and cell adhesion in Chapters 15 and 19, respectively. In Chapters 12 and 13, we discuss the internal membranes of the cell and the protein traffic through and between them.

THE LIPID BILAYER

The **lipid bilayer** provides the basic structure for all cell membranes. It is easily seen by electron microscopy, and its structure is attributable exclusively to the

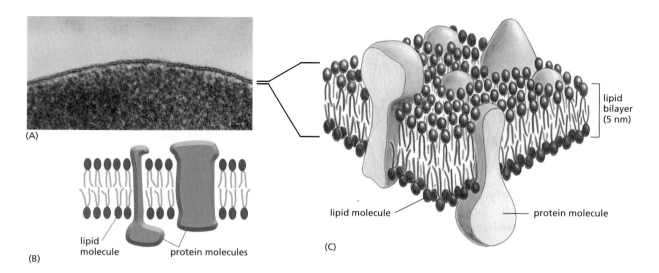

(A)

(B)

lipid molecule

protein molecules

(C)

lipid molecule

protein molecule

lipid bilayer (5 nm)

special properties of the lipid molecules, which assemble spontaneously into bilayers even under simple artificial conditions.

Phosphoglycerides, Sphingolipids, and Sterols Are the Major Lipids in Cell Membranes

Lipid molecules constitute about 50% of the mass of most animal cell membranes, nearly all of the remainder being protein. There are approximately 5×10^6 lipid molecules in a 1 µm × 1 µm area of lipid bilayer, or about 10^9 lipid molecules in the plasma membrane of a small animal cell. All of the lipid molecules in cell membranes are **amphiphilic**—that is, they have a **hydrophilic** ("water-loving") or *polar* end and a **hydrophobic** ("water-fearing") or *nonpolar* end.

The most abundant membrane lipids are the **phospholipids**. These have a polar head group and two hydrophobic *hydrocarbon tails*. In animal, plant, and bacterial cells, the tails are usually fatty acids, and they can differ in length (they normally contain between 14 and 24 carbon atoms). One tail typically has one or more *cis*-double bonds (that is, it is *unsaturated)*, while the other tail does not (that is, it is *saturated)*. As shown in **Figure 10–2**, each *cis*-double bond creates a small kink in the tail. Differences in the length and saturation of the fatty acid tails influence how phospholipid molecules pack against one another, thereby affecting the fluidity of the membrane, as we discuss later.

The main phospholipids in most animal cell membranes are the **phosphoglycerides**, which have a three-carbon *glycerol* backbone (see Figure 10–2). Two long-chain fatty acids are linked through ester bonds to adjacent carbon atoms of the glycerol, and the third carbon atom is attached to a phosphate group, which in turn is linked to one of several different types of head group. By combining several different fatty acids and head groups, cells make many different phosphoglycerides. *Phosphatidylethanolamine, phosphatidylserine,* and *phosphatidylcholine* are the main ones in mammalian cell membranes (**Figure 10–3A–C)**

Another important phospholipid, called *sphingomyelin,* is built from *sphingosine* rather than glycerol (Figure 10–3D–E). Sphingosine is a long acyl chain with an amino group (NH₂) and two hydroxyl groups (OH) at one end of the molecule. In sphingomyelin, a fatty acid tail is attached to the amino group, and a phosphocholine group is attached to the terminal hydroxyl group, leaving one hydroxyl group free. The free hydroxyl group contributes to the polar properties of the adjacent head group, as it can form hydrogen bonds with the head group of a neighboring lipid, with a water molecule, or with a membrane protein. Together, the phospholipids phosphatidylcholine, phosphatidylethanolamine, phosphatidylserine, and sphingomyelin constitute more than half the mass of lipid in most mammalian cell membranes (see Table 10–1).

Figure 10–1 Three views of a cell membrane. <TAGC> (A) An electron micrograph of a plasma membrane (of a human red blood cell) seen in cross section. (B and C) These drawings show two-dimensional and three-dimensional views of a cell membrane and the general disposition of its lipid and protein constituents. (A, courtesy of Daniel S. Friend.)

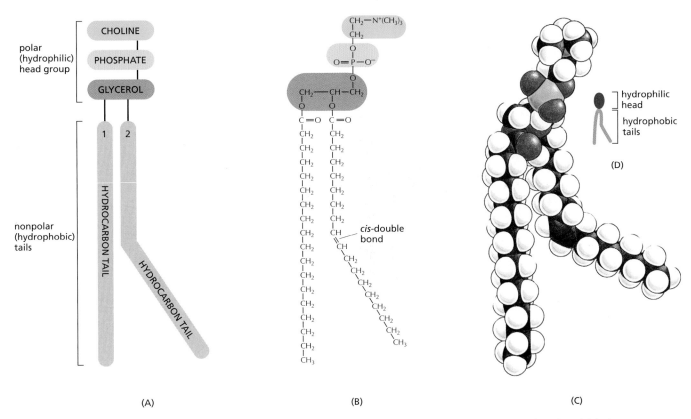

Figure 10–2 **The parts of a phosphoglyceride molecule.** This example is a phosphatidylcholine, represented (A) schematically, (B) by a formula, (C) as a space-filling model, and (D) as a symbol. The kink resulting from the *cis*-double bond is exaggerated for emphasis.

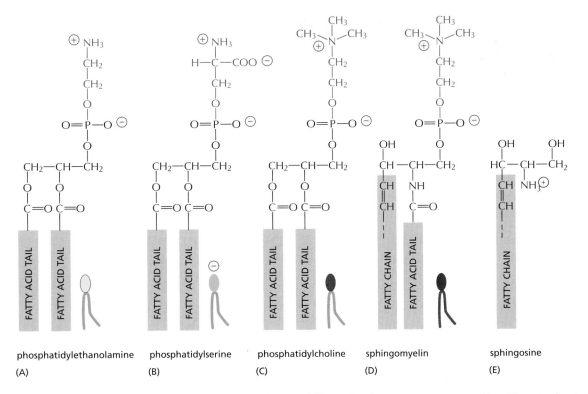

Figure 10–3 **Four major phospholipids in mammalian plasma membranes.** Different head groups are represented by different colors in the symbols. The lipid molecules shown in (A–C) are phosphoglycerides, which are derived from glycerol. The molecule in (D) is sphingomyelin, which is derived from sphingosine (E) and is therefore a sphingolipid. Note that only phosphatidylserine carries a net negative charge, the importance of which we discuss later; the other three are electrically neutral at physiological pH, carrying one positive and one negative charge.

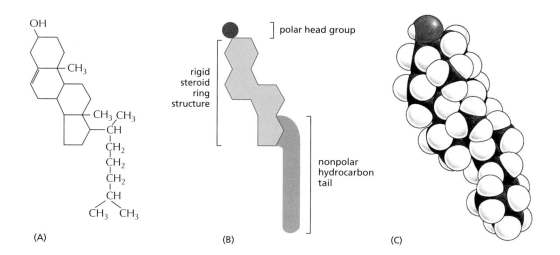

Figure 10–4 The structure of cholesterol.
Cholesterol is represented (A) by a
formula, (B) by a schematic drawing, and
(C) as a space-filling model.

In addition to phospholipids, the lipid bilayers in many cell membranes
contain *cholesterol* and *glycolipids*. Eucaryotic plasma membranes contain
especially large amounts of **cholesterol** (**Figure 10–4**)—up to one molecule for
every phospholipid molecule. Cholesterol is a sterol. It contains a rigid ring
structure, to which is attached a single polar hydroxyl group and a short nonpolar hydrocarbon chain. The cholesterol molecules orient themselves in the
bilayer with their hydroxyl group close to the polar head groups of adjacent
phospholipid molecules (**Figure 10–5**).

Phospholipids Spontaneously Form Bilayers

The shape and amphiphilic nature of the phospholipid molecules cause them to
form bilayers spontaneously in aqueous environments. As discussed in Chapter
2, hydrophilic molecules dissolve readily in water because they contain charged
groups or uncharged polar groups that can form either favorable electrostatic
interactions or hydrogen bonds with water molecules. Hydrophobic molecules,
by contrast, are insoluble in water because all, or almost all, of their atoms are
uncharged and nonpolar and therefore cannot form energetically favorable
interactions with water molecules. If dispersed in water, they force the adjacent
water molecules to reorganize into ice-like cages that surround the hydrophobic
molecule (**Figure 10–6**). Because these cage structures are more ordered than
the surrounding water, their formation increases the free energy. This free
energy cost is minimized, however, if the hydrophobic molecules (or the
hydrophobic portions of amphiphilic molecules) cluster together so that the
smallest number of water molecules is affected.

The hydrophobic and hydrophilic regions of lipid molecules behave in the
same way. Thus, lipid molecules spontaneously aggregate to bury their
hydrophobic hydrocarbon tails in the interior and expose their hydrophilic
heads to water. Depending on their shape, they can do this in either of two ways:
they can form spherical *micelles*, with the tails inward, or they can form double-
layered sheets, or *bilayers*, with the hydrophobic tails sandwiched between the
hydrophilic head groups (**Figure 10–7**).

Being cylindrical, phospholipid molecules spontaneously form bilayers in
aqueous environments. In this energetically most favorable arrangement, the
hydrophilic heads face the water at each surface of the bilayer, and the
hydrophobic tails are shielded from the water in the interior. The same forces
that drive phospholipids to form bilayers also provide a self-healing property. A
small tear in the bilayer creates a free edge with water; because this is energetically unfavorable, the lipids tend to rearrange spontaneously to eliminate the free
edge. (In eucaryotic plasma membranes, the fusion of intracellular vesicles
repairs larger tears.) The prohibition of free edges has a profound consequence:
the only way for a bilayer to avoid having edges is by closing in on itself and forming a sealed compartment (**Figure 10–8**). This remarkable behavior, fundamental

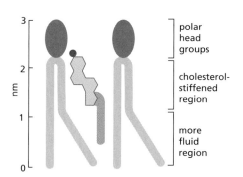

**Figure 10–5 Cholesterol in a lipid
bilayer.** Schematic drawing of a
cholesterol molecule interacting with two
phospholipid molecules in one
monolayer of a lipid bilayer.

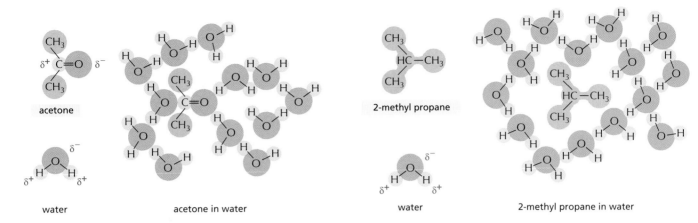

Figure 10–6 How hydrophilic and hydrophobic molecules interact differently with water. (A) Because acetone is polar, it can form favorable electrostatic interactions with water molecules, which are also polar. Thus, acetone readily dissolves in water. (B) By contrast, 2-methyl propane is entirely hydrophobic. Because it cannot form favorable interactions with water, it would force adjacent water molecules to reorganize into icelike cage structures, which increases the free energy. This compound is therefore virtually insoluble in water. The symbol δ^- indicates a partial negative charge, and δ^+ indicates a partial positive charge. Polar atoms are shown in color and nonpolar groups are shown in *gray*.

to the creation of a living cell, follows directly from the shape and amphiphilic nature of the phospholipid molecule.

A lipid bilayer also has other characteristics that make it an ideal structure for cell membranes. One of the most important of these is its fluidity, which is crucial to many membrane functions.

The Lipid Bilayer Is a Two-dimensional Fluid

Around 1970, researchers first recognized that individual lipid molecules are able to diffuse freely within lipid bilayers. The initial demonstration came from studies of synthetic lipid bilayers. Two types of preparations have been very useful in such studies: (1) bilayers made in the form of spherical vesicles, called **liposomes**, which can vary in size from about 25 nm to 1 µm in diameter

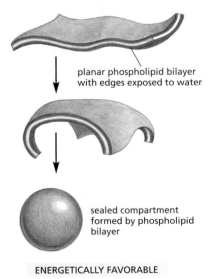

ENERGETICALLY UNFAVORABLE

planar phospholipid bilayer with edges exposed to water

sealed compartment formed by phospholipid bilayer

ENERGETICALLY FAVORABLE

Figure 10–8 The spontaneous closure of a phospholipid bilayer to form a sealed compartment. The closed structure is stable because it avoids the exposure of the hydrophobic hydrocarbon tails to water, which would be energetically unfavorable.

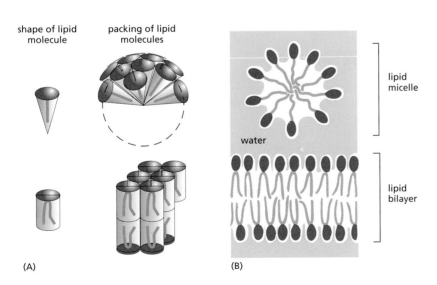

shape of lipid molecule

packing of lipid molecules

water

lipid micelle

lipid bilayer

(A)

(B)

Figure 10–7 Packing arrangements of lipid molecules in an aqueous environment. (A) Cone-shaped lipid molecules *(above)* form micelles, whereas cylinder-shaped phospholipid molecules *(below)* form bilayers. (B) A lipid micelle and a lipid bilayer seen in cross section. Lipid molecules spontaneously form one or the other structure in water, depending on their shape.

depending on how they are produced (**Figure 10–9**); and (2) planar bilayers, called **black membranes**, formed across a hole in a partition between two aqueous compartments (**Figure 10–10**).

Various techniques have been used to measure the motion of individual lipid molecules and their components. One can construct a lipid molecule, for example, with a fluorescent dye or a small gold particle attached to its polar head group and follow the diffusion of even individual molecules in a membrane. Alternatively, one can modify a lipid head group to carry a "spin label," such as a nitroxyl group ($\supset$N–O); this contains an unpaired electron whose spin creates a paramagnetic signal that can be detected by electron spin resonance (ESR) spectroscopy. (The principles of this technique are similar to those of nuclear magnetic resonance, discussed in Chapter 8.) The motion and orientation of a spin-labeled lipid in a bilayer can be deduced from the ESR spectrum. Such studies show that phospholipid molecules in synthetic bilayers very rarely migrate from the monolayer (also called a *leaflet*) on one side to that on the other. This process, known as "flip-flop," occurs less than once a month for any individual molecule, although cholesterol is an exception to this rule and can flip-flop rapidly. In contrast, lipid molecules readily exchange places with their neighbors *within* a monolayer (~10^7 times per second). This gives rise to a rapid lateral diffusion, with a diffusion coefficient *(D)* of about 10^{-8} cm²/sec, which means that an average lipid molecule diffuses the length of a large bacterial cell (~2 μm) in about 1 second. These studies have also shown that individual lipid molecules rotate very rapidly about their long axis and have flexible hydrocarbon chains. Computer simulations show that lipid molecules in membranes are very disordered, presenting an irregular surface of variously spaced and oriented head groups to the water phase on either side of the bilayer (**Figure 10–11**).

Similar mobility studies on labeled lipid molecules in isolated biological membranes and in living cells give results similar to those in synthetic bilayers. They demonstrate that the lipid component of a biological membrane is a two-dimensional liquid in which the constituent molecules are free to move laterally. As in synthetic bilayers, individual phospholipid molecules are normally confined to their own monolayer. This confinement creates a problem for their synthesis. Phospholipid molecules are manufactured in only one monolayer of a membrane, mainly in the cytosolic monolayer of the endoplasmic reticulum membrane. If none of these newly made molecules could migrate reasonably promptly to the noncytosolic monolayer, new lipid bilayer could not be made. The problem is solved by a special class of transmembrane enzymes called *phospholipid translocators*, which catalyze the rapid flip-flop of phospholipids from one monolayer to the other, as discussed in Chapter 12.

The Fluidity of a Lipid Bilayer Depends on Its Composition

The fluidity of cell membranes has to be precisely regulated. Certain membrane transport processes and enzyme activities, for example, cease when the bilayer viscosity is experimentally increased beyond a threshold level.

The fluidity of a lipid bilayer depends on both its composition and its temperature, as is readily demonstrated in studies of synthetic bilayers. A synthetic bilayer made from a single type of phospholipid changes from a liquid state to a

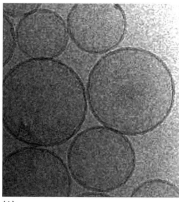

(A)

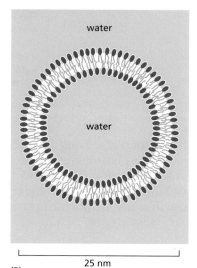

25 nm

(B)

Figure 10–9 Liposomes. (A) An electron micrograph of unfixed, unstained phospholipid vesicles—liposomes—in water rapidly frozen to liquid nitrogen temperature. (B) A drawing of a small spherical liposome seen in cross section. Liposomes are commonly used as model membranes in experimental studies. (A, from P. Frederik and W. Hubert, *Meth. Enzymol.* 391:431, 2005. With permission from Elsevier.)

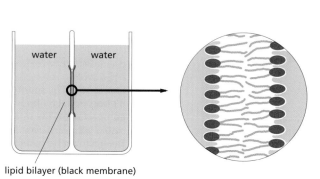

lipid bilayer (black membrane)

Figure 10–10 A cross-sectional view of a black membrane, a synthetic lipid bilayer. This planar bilayer appears black when it forms across a small hole in a partition separating two aqueous compartments. Black membranes are used to measure the permeability properties of synthetic membranes.

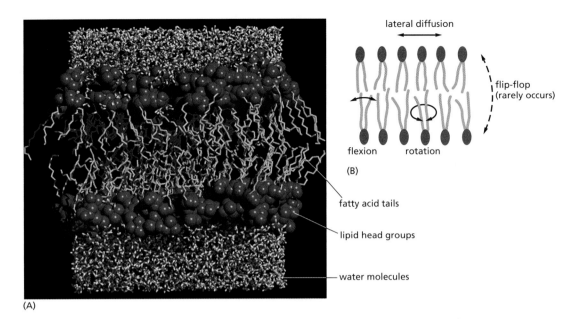

Figure 10–11 **The mobility of phospholipid molecules in an artificial lipid bilayer.** <CACA> Starting with a model of 100 phosphatidylcholine molecules arranged in a regular bilayer, a computer calculated the position of every atom after 300 picoseconds of simulated time. From these theoretical calculations (taking weeks of processor time in 1995), a model of the lipid bilayer emerges that accounts for almost all of the measurable properties of a synthetic lipid bilayer, such as its thickness, number of lipid molecules per membrane area, depth of water penetration, and unevenness of the two surfaces. Note that the tails in one monolayer can interact with those in the other monolayer, if the tails are long enough. (B) The different motions of a lipid molecule in a bilayer. (A, based on S.W. Chiu et al., *Biophys. J.* 69:1230–1245, 1995. With permission from the Biophysical Society.)

two-dimensional rigid crystalline (or gel) state at a characteristic freezing point. This change of state is called a *phase transition*, and the temperature at which it occurs is lower (that is, the membrane becomes more difficult to freeze) if the hydrocarbon chains are short or have double bonds. A shorter chain length reduces the tendency of the hydrocarbon tails to interact with one another, in both the same and opposite monolayer, and *cis*-double bonds produce kinks in the hydrocarbon chains that make them more difficult to pack together, so that the membrane remains fluid at lower temperatures (**Figure 10–12**). Bacteria, yeasts, and other organisms whose temperature fluctuates with that of their environment adjust the fatty acid composition of their membrane lipids to maintain a relatively constant fluidity. As the temperature falls, for instance, the cells of those organisms synthesize fatty acids with more *cis*-double bonds, and they avoid the decrease in bilayer fluidity that would otherwise result from the temperature drop.

Cholesterol modulates the properties of lipid bilayers. When mixed with phospholipids, it enhances the permeability-barrier properties of the lipid bilayer. It inserts into the bilayer with its hydroxyl group close to the polar head groups of the phospholipids, so that its rigid, platelike steroid rings interact with—and partly immobilize—those regions of the hydrocarbon chains closest to the polar head groups (see Figure 10–5). By decreasing the mobility of the first

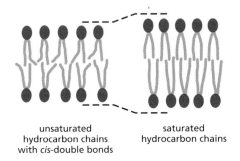

unsaturated
hydrocarbon chains
with *cis*-double bonds

saturated
hydrocarbon chains

Figure 10–12 **The influence of *cis*-double bonds in hydrocarbon chains.** The double bonds make it more difficult to pack the chains together, thereby making the lipid bilayer more difficult to freeze. In addition, because the hydrocarbon chains of unsaturated lipids are more spread apart, lipid bilayers containing them are thinner than bilayers formed exclusively from saturated lipids.

Table 10–1 Approximate Lipid Compositions of Different Cell Membranes

	PERCENTAGE OF TOTAL LIPID BY WEIGHT					
LIPID	LIVER CELL PLASMA MEMBRANE	RED BLOOD CELL PLASMA MEMBRANE	MYELIN	MITOCHONDRION (INNER AND OUTER MEMBRANES)	ENDOPLASMIC RETICULUM	E. COLI BACTERIUM
Cholesterol	17	23	22	3	6	0
Phosphatidylethanolamine	7	18	15	28	17	70
Phosphatidylserine	4	7	9	2	5	trace
Phosphatidylcholine	24	17	10	44	40	0
Sphingomyelin	19	18	8	0	5	0
Glycolipids	7	3	28	trace	trace	0
Others	22	14	8	23	27	30

few CH_2 groups of the hydrocarbon chains of the phospholipid molecules, cholesterol makes the lipid bilayer less deformable in this region and thereby decreases the permeability of the bilayer to small water-soluble molecules. Although cholesterol tightens the packing of the lipids in a bilayer, it does not make membranes any less fluid. At the high concentrations found in most eucaryotic plasma membranes, cholesterol also prevents the hydrocarbon chains from coming together and crystallizing.

Table 10–1 compares the lipid compositions of several biological membranes. Note that bacterial plasma membranes are often composed of one main type of phospholipid and contain no cholesterol; their mechanical stability is enhanced by an overlying cell wall (see Figure 11–18). In archaea, lipids usually contain 20–25-carbon-long prenyl chains instead of fatty acids, prenyl and fatty acid chains are similarly hydrophobic and flexible (see Figure 10–20F). Thus, lipid bilayers can be built from molecules with similar features but different molecular designs. The plasma membranes of most eucaryotic cells are more varied than those of procaryotes and archaea, not only in containing large amounts of cholesterol but also in containing a mixture of different phospholipids.

Analysis of membrane lipids by mass spectrometry has revealed that the lipid composition of a typical cell membrane is much more complex than originally thought. According to these studies, membranes are composed of a bewildering variety of 500–1000 different lipid species. While some of this complexity reflects the combinatorial variation in head groups, hydrocarbon chain lengths, and desaturation of the major phospholipid classes, membranes also contain many structurally distinct minor lipids, at least some of which have important functions. The *inositol phospholipids*, for example, are present in small quantities but have crucial functions in guiding membrane traffic and in cell signaling (discussed in Chapters 13 and 15, respectively). Their local synthesis and destruction are regulated by a large number of enzymes, which create both small intracellular signaling molecules and lipid docking sites on membranes that recruit specific proteins from the cytosol, as we discuss later.

Despite Their Fluidity, Lipid Bilayers Can Form Domains of Different Compositions

Because a lipid bilayer is a two-dimensional fluid, we might expect most types of lipid molecules in it to be randomly distributed in their own monolayer. The van der Waals attractive forces between neighboring hydrocarbon tails are not selective enough to hold groups of phospholipid molecules together. With certain lipid mixtures, however, different lipids can come together transiently, creating a dynamic patchwork of different domains. In synthetic lipid bilayers composed of phosphatidylcholine, sphingomyelin, and cholesterol, van der Waals forces between the long and saturated hydrocarbon chains of the sphingomyelin molecules can be just strong enough to hold the adjacent molecules together transiently (**Figure 10–13**).

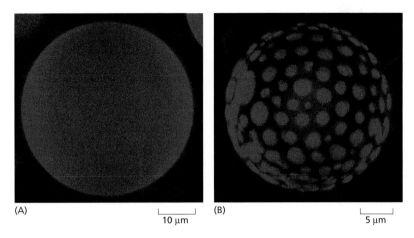

(A)

|— — —|
10 μm

(B)

|— — —|
5 μm

Figure 10–13 **Lateral phase separation in artifical lipid bilayers.** (A) Giant liposomes produced from a 1:1 mixture of phosphatidylcholine and spingomyelin form uniform bilayers, whereas (B) liposomes produced from a 1:1:1 mixture of phosphatidylcholine, spingomyelin, and cholesterol form bilayers with two immiscible phases. The liposomes are stained with trace concentrations of a fluorescent dye that preferentially partitions into one of the phases. The average size of the domains formed in these giant artificial liposomes is much larger than that expected in biological membranes, where rafts may be as small as a few nanometers in diameter. (A, from N. Kahya et al., *J. Struct. Biol.* 147:77–89, 2004. With permission from Elsevier; B, courtesy of Petra Schwille.)

There has been a long debate among scientists whether the lipid molecules in the plasma membrane of animal cells can transiently assemble into specialized domains, called **lipid rafts**. Certain specialized regions of the plasma membrane, such as the *caveolae* involved in endocytosis (discussed in Chapter 13), are enriched in sphingolipids and cholesterol, and it is thought that the specific proteins that assemble there help stabilize these rafts. Because the hydrocarbon chains of sphingolipids are longer and straighter than those of other membrane lipids, raft domains are thicker than other parts of the bilayer (see Figure 10–12) and better accommodate certain membrane proteins (**Figure 10–14**). Thus, the lateral segregation of proteins and of lipids into raft domains would, in principle, be a mutually stabilizing process. In this way, lipid rafts could help organize membrane proteins—concentrating them either for transport in membrane vesicles (discussed in Chapter 13) or for working together in protein assemblies, as when they convert extracellular signals into intracellular ones (discussed in Chapter 15).

Lipid Droplets Are Surrounded by a Phospholipid Monolayer

Most cells store an excess of lipids in **lipid droplets**, from where they can be retrieved as building blocks for membrane synthesis or as a food source. Fat cells, also called adipocytes, are specialized for lipid storage (see Figure 14–34). They contain vast numbers of large lipid droplets, from which fatty acids can be liberated on demand and exported to other cells through the bloodstream. Lipid droplets store neutral lipids, such as triacylglycerides and cholesterol esters, which are synthesized from fatty acids and cholesterol by enzymes in the

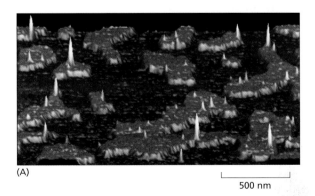

(A)

|— — —|
500 nm

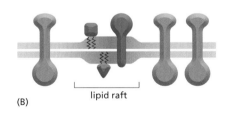

(B)

lipid raft

Figure 10–14 **The effects of lipid rafts in artificial lipid bilayers.** (A) The surface contours of a synthetic bilayer containing lipid rafts, analyzed by atomic force microscopy. Note that the raft areas, shown in *orange*, are thicker than the rest of the bilayer; as in Figure 10–13, the rafts primarily contain sphingomyelin and cholesterol. The sharp, yellow spikes are incorporated protein molecules, which are attached to the bilayer by a glycosylphosphatidyl-inositol (GPI) anchor (illustrated in Figure 10–19, example 6), and preferentially partition into the raft domains. (B) Because of both their increased thickness and lipid composition, rafts are thought to concentrate specific membrane proteins (*dark green*). (A, from D.E. Saslowsky et al., *J. Biol. Chem.* 277:26966–26970, 2002. With permission from the American Society for Biochemistry and Molecular Biology.)

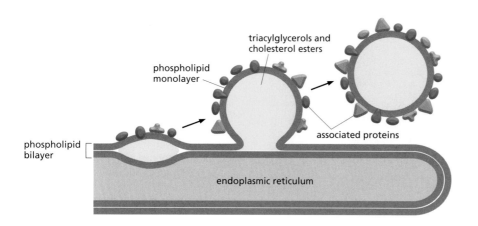

triacylglycerols and
cholesterol esters

phospholipid
monolayer

associated proteins

phospholipid
bilayer

endoplasmic reticulum

Figure 10–15 **A model for the formation of lipid droplets.** Neutral lipids are deposited between the two monolayers of the endoplasmic reticulum membrane. There, they aggregate into a three-dimensional droplet, which buds and pinches off from the endoplasmic reticulum membrane as a unique organelle, surrounded by a single monolayer of phospholipids and associated proteins. (Adapted from S. Martin and R.G. Parton, *Nat. Rev. Mol. Cell Biol.* 7:373–378, 2006. With permission from Macmillan Publishers Ltd.)

endoplasmic reticulum membrane. Because these lipids do not contain hydrophilic head groups, they are exclusively hydrophobic molecules, which aggregate into three-dimensional droplets rather than into bilayers.

Lipid droplets are unique organelles because they are surrounded by a single monolayer of phospholipids, which contains a large variety of proteins. Some of the proteins are enzymes involved in lipid metabolism, but the functions of most are unknown. Lipid droplets form rapidly when cells are exposed to high concentrations of fatty acids. They form from discrete regions of the endoplasmic reticulum membrane where many enzymes of lipid metabolism are concentrated. **Figure 10–15** shows one model of how lipid droplets may form and acquire their surrounding monolayer of phospholipids and proteins.

The Asymmetry of the Lipid Bilayer Is Functionally Important

The lipid compositions of the two monolayers of the lipid bilayer in many membranes are strikingly different. In the human red blood cell membrane, for example, almost all of the phospholipid molecules that have choline—$(CH_3)_3N^+CH_2CH_2OH$—in their head group (phosphatidylcholine and sphingomyelin) are in the outer monolayer, whereas almost all that contain a terminal primary amino group (phosphatidylethanolamine and phosphatidylserine) are in the inner monolayer (**Figure 10–16**). Because the negatively charged phosphatidylserine is located in the inner monolayer, there is a significant difference in charge between the two halves of the bilayer. We discuss in Chapter 12 how membrane-bound phospholipid translocators generate and maintain lipid asymmetry.

Lipid asymmetry is functionally important, especially in converting extracellular signals into intracellular ones (discussed in Chapter 15). Many cytosolic proteins bind to specific lipid head groups found in the cytosolic monolayer of the lipid bilayer. The enzyme *protein kinase C (PKC)*, for example, is activated in response to various extracellular signals. It binds to the cytosolic face of the plasma membrane, where phosphatidylserine is concentrated, and requires this negatively charged phospholipid for its activity.

EXTRACELLULAR SPACE

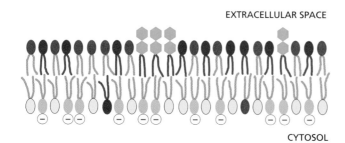

CYTOSOL

Figure 10–16 **The asymmetrical distribution of phospholipids and glycolipids in the lipid bilayer of human red blood cells.** The colors used for the phospholipid head groups are those introduced in Figure 10–3. In addition, glycolipids are drawn with hexagonal polar head groups *(blue)*. Cholesterol (not shown) is thought to be distributed roughly equally in both monolayers.

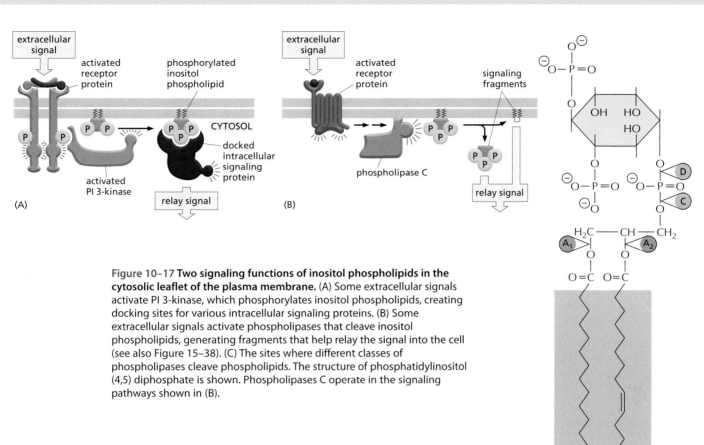

Figure 10–17 Two signaling functions of inositol phospholipids in the cytosolic leaflet of the plasma membrane. (A) Some extracellular signals activate PI 3-kinase, which phosphorylates inositol phospholipids, creating docking sites for various intracellular signaling proteins. (B) Some extracellular signals activate phospholipases that cleave inositol phospholipids, generating fragments that help relay the signal into the cell (see also Figure 15–38). (C) The sites where different classes of phospholipases cleave phospholipids. The structure of phosphatidylinositol (4,5) diphosphate is shown. Phospholipases C operate in the signaling pathways shown in (B).

In other cases, specific lipid head groups must first be modified to create protein-binding sites at a particular time and place. *Phosphatidylinositol*, for instance, is one of the minor phospholipids that are concentrated in the cytosolic monolayer of cell membranes. Various lipid kinases can add phosphate groups at distinct positions in the inositol ring, creating binding sites that recruit specific proteins from the cytosol to the membrane. An important example of such a lipid kinase is *phosphoinositide 3-kinase (PI 3-kinase)*, which is activated in response to extracellular signals and helps to recruit specific intracellular signaling proteins to the cytosolic face of the plasma membrane (**Figure 10–17A**). Similar lipid kinases phosphorylate inositol phospholipids in intracellular membranes and thereby help to recruit proteins that guide membrane transport.

Phospholipids in the plasma membrane are used in yet another way to convert extracellular signals into intracellular ones. The plasma membrane contains various *phospholipases* that are activated by extracellular signals to cleave specific phospholipid molecules, generating fragments of these molecules that act as short-lived intracellular mediators. *Phospholipase C*, for example, cleaves an inositol phospholipid in the cytosolic monolayer of the plasma membrane to generate two fragments, one of which remains in the membrane and helps activate protein kinase C, while the other is released into the cytosol and stimulates the release of Ca^{2+} from the endoplasmic reticulum (Figure 10–17B–C).

Animals exploit the phospholipid asymmetry of their plasma membranes to distinguish between live and dead cells. When animal cells undergo apoptosis (a form of programmed cell death, discussed in Chapter 18), phosphatidylserine, which is normally confined to the cytosolic monolayer of the plasma membrane lipid bilayer, rapidly translocates to the extracellular monolayer. The phosphatidylserine exposed on the cell surface signals neighboring cells, such as macrophages, to phagocytose the dead cell and digest it. The translocation of the phosphatidylserine in apoptotic cells is thought to occur by two mechanisms:

1. The phospholipid translocator that normally transports this lipid from the noncytosolic monolayer to the cytosolic monolayer is inactivated.
2. A "scramblase" that transfers phospholipids nonspecifically in both directions between the two monolayers is activated.

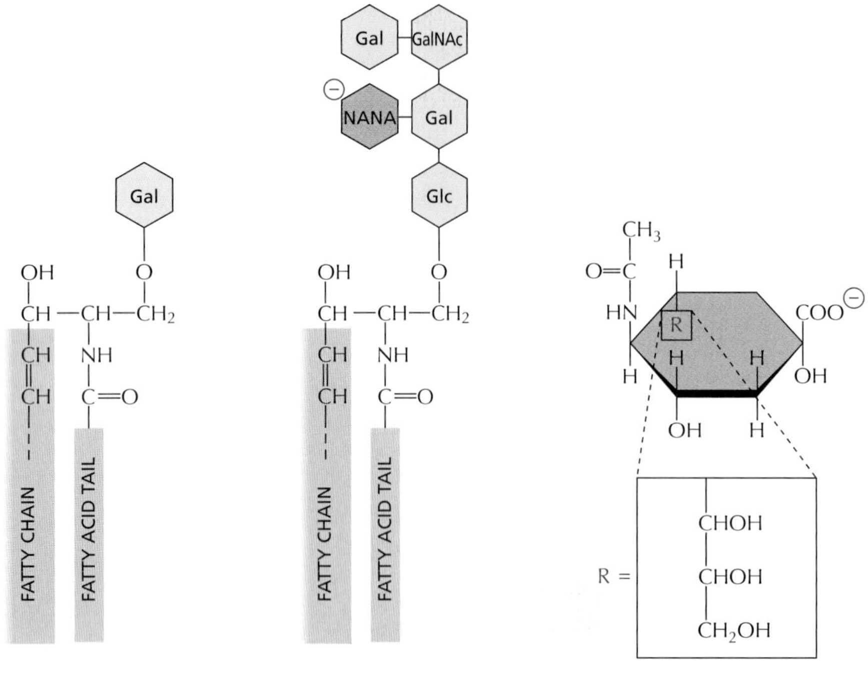

Figure 10–18 Glycolipid molecules.
(A) Galactocerebroside is called a *neutral glycolipid* because the sugar that forms its head group is uncharged. (B) A ganglioside always contains one or more negatively charged sialic acid residues (also called *N*-acetylneuraminic acid, or NANA), whose structure is shown in (C). Whereas in bacteria and plants almost all glycolipids are derived from glycerol, as are most phospholipids, in animal cells almost all glycolipids are based on sphingosine, as is the case for sphingomyelin (see Figure 10–3). Gal = galactose; Glc = glucose, GalNAc = *N*-acetylgalactosamine; these three sugars are uncharged.

(A) galactocerebroside   (B) G$_{M1}$ ganglioside   (C) sialic acid (NANA)

Glycolipids Are Found on the Surface of All Plasma Membranes

Sugar-containing lipid molecules called **glycolipids**, found exclusively in the noncytosolic monolayer of the lipid bilayer, have the most extreme asymmetry in their membrane distribution. In animal cells they are made from sphingosine, just like sphingomyelin. These intriguing molecules tend to self-associate, partly through hydrogen bonds between their sugars and partly through van der Waals forces between their long and straight hydrocarbon chains, and they may preferentially partition into lipid rafts. The asymmetric distribution of glycolipids in the bilayer results from the addition of sugar groups to the lipid molecules in the lumen of the Golgi apparatus. Thus, the compartment in which they are manufactured is topologically equivalent to the exterior of the cell (discussed in Chapter 12). As they are delivered to the plasma membrane, the sugar groups are exposed at the cell surface (see Figure 10–16), where they have important roles in interactions of the cell with its surroundings.

Glycolipids probably occur in all animal cell plasma membranes, where they generally constitute about 5% of the lipid molecules in the outer monolayer. They are also found in some intracellular membranes. The most complex of the glycolipids, the **gangliosides**, contain oligosaccharides with one or more sialic acid residues, which give gangliosides a net negative charge (**Figure 10–18**). The most abundant of the more than 40 different gangliosides that have been identified are in the plasma membrane of nerve cells, where gangliosides constitute 5–10% of the total lipid mass; they are also found in much smaller quantities in other cell types.

Hints as to the functions of glycolipids come from their localization. In the plasma membrane of epithelial cells, for example, glycolipids are confined to the exposed apical surface, where they may help to protect the membrane against the harsh conditions frequently found there (such as low pH and high concentrations of degradative enzymes). Charged glycolipids, such as gangliosides, may be important because of their electrical effects: their presence alters the electrical field across the membrane and the concentrations of ions—especially Ca^{2+}—at the membrane surface. Glycolipids are also thought to function in cell-recognition processes, in which membrane-bound carbohydrate-binding proteins *(lectins)* bind to the sugar groups on both glycolipids and glycoproteins in the process of cell–cell adhesion (discussed in Chapter 19). Surprisingly, however, mutant mice that are deficient in all of their complex gangliosides show no obvious abnormalities, although the males cannot transport testosterone normally in the testes and are consequently sterile.

Whatever their normal function, some glycolipids provide entry points for certain bacterial toxins. The ganglioside G_{M1} (see Figure 10–18), for example, acts as a cell-surface receptor for the bacterial toxin that causes the debilitating diarrhea of cholera. Cholera toxin binds to and enters only those cells that have G_{M1} on their surface, including intestinal epithelial cells. Its entry into a cell leads to a prolonged increase in the concentration of intracellular cyclic AMP (discussed in Chapter 15), which in turn causes a large efflux of Na^+ and water into the intestine.

Summary

Biological membranes consist of a continuous double layer of lipid molecules in which membrane proteins are embedded. This lipid bilayer is fluid, with individual lipid molecules able to diffuse rapidly within their own monolayer. The membrane lipid molecules are amphiphilic. When placed in water they assemble spontaneously into bilayers, which form sealed compartments.

Cells contain 500–1000 different lipid species. There are three major classes of membrane lipids—phospholipids, cholesterol, and glycolipids—and hundreds of minor classes. The lipid compositions of the inner and outer monolayers are different, reflecting the different functions of the two faces of a cell membrane. Different mixtures of lipids are found in the membranes of cells of different types, as well as in the various membranes of a single eucaryotic cell. Inositol phospholids are a minor class of phospholipids, which in the cytosolic leaflet of the plasma membrane lipid bilayer play an important part in cell signaling: in response to extracellular signals, specific lipid kinases phosphorylate the head groups of these lipids to form docking sites for cytosolic signaling proteins, whereas specific phospholipases cleave certain inositol phospholipids to generate small intracellular signaling molecules.

MEMBRANE PROTEINS

Although the lipid bilayer provides the basic structure of biological membranes, the membrane proteins perform most of the membrane's specific tasks and therefore give each type of cell membrane its characteristic functional properties. Accordingly, the amounts and types of proteins in a membrane are highly variable. In the myelin membrane, which serves mainly as electrical insulation for nerve cell axons, less than 25% of the membrane mass is protein. By contrast, in the membranes involved in ATP production (such as the internal membranes of mitochondria and chloroplasts), approximately 75% is protein. A typical plasma membrane is somewhere in between, with protein accounting for about half of its mass. Because lipid molecules are small compared with protein molecules, there are always many more lipid molecules than protein molecules in cell membranes—about 50 lipid molecules for each protein molecule in cell membranes that are 50% protein by mass. Membrane proteins vary widely in structure and in the way they associate with the lipid bilayer, which reflects their diverse functions.

Membrane Proteins Can Be Associated with the Lipid Bilayer in Various Ways

Figure 10–19 shows the different ways in which membrane proteins can associate with the membrane. Many extend through the lipid bilayer, with part of their mass on either side (Figure 10–19, examples 1, 2, and 3). Like their lipid neighbors, these **transmembrane proteins** are amphiphilic, having hydrophobic and hydrophilic regions. Their hydrophobic regions pass through the membrane and interact with the hydrophobic tails of the lipid molecules in the interior of the bilayer, where they are sequestered away from water. Their hydrophilic regions are exposed to water on either side of the membrane. The covalent attachment of a fatty acid chain that inserts into the cytosolic monolayer of the

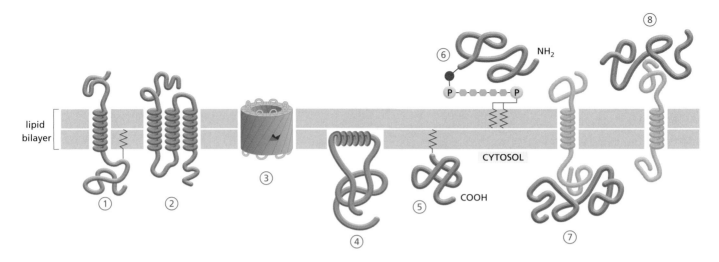

lipid bilayer

CYTOSOL

COOH

lipid bilayer increases the hydrophobicity of some of these transmembrane proteins (see Figure 10–19, example 1).

Other membrane proteins are located entirely in the cytosol and are associated with the cytosolic monolayer of the lipid bilayer, either by an amphiphilic α helix exposed on the surface of the protein (Figure 10–19, example 4) or by one or more covalently attached lipid chains (Figure 10–19, example 5). Yet other membrane proteins are entirely exposed at the external cell surface, being attached to the lipid bilayer only by a covalent linkage (via a specific oligosaccharide) to phosphatidylinositol in the outer lipid monolayer of the plasma membrane (Figure 10–19, example 6).

The lipid-linked proteins in example 5 in Figure 10–19 are made as soluble proteins in the cytosol and are subsequently anchored to the membrane by the covalent attachment of a lipid group. The proteins in example 6, however, are made as single-pass transmembrane proteins in the endoplasmic reticulum (ER). While still in the ER, the transmembrane segment of the protein is cleaved off and a **glycosylphosphatidylinositol (GPI) anchor** is added, leaving the protein bound to the noncytosolic surface of the membrane solely by this anchor (discussed in Chapter 12). Transport vesicles eventually deliver the protein to the plasma membrane (discussed in Chapter 13). Proteins bound to the plasma membrane by a GPI anchor can be readily distinguished by the use of an enzyme called phosphatidylinositol-specific phospholipase C. This enzyme cuts these proteins free from their anchors, thereby releasing them from the membrane.

Some membrane proteins do not extend into the hydrophobic interior of the lipid bilayer at all; they are instead bound to either face of the membrane by noncovalent interactions with other membrane proteins (Figure 10–19, examples 7 and 8). Many of the proteins of this type can be released from the membrane by relatively gentle extraction procedures, such as exposure to solutions of very high or low ionic strength or of extreme pH, which interfere with protein–protein interactions but leave the lipid bilayer intact; these proteins are referred to as **peripheral membrane proteins**. Transmembrane proteins and many proteins held in the bilayer by lipid groups or hydrophobic polypeptide regions that insert into the hydrophobic core of the lipid bilayer cannot be released in these ways. These proteins are called **integral membrane proteins**.

Lipid Anchors Control the Membrane Localization of Some Signaling Proteins

How a membrane protein is associated with the lipid bilayer reflects the function of the protein. Only transmembrane proteins can function on both sides of the bilayer or transport molecules across it. Cell-surface receptors, for example, are transmembrane proteins that bind signal molecules in the extracellular space and generate different intracellular signals on the opposite side of the plasma membrane. To transfer small hydrophilic molecules across a membrane,

Figure 10–19 Various ways in which membrane proteins associate with the lipid bilayer. Most transmembrane proteins are thought to extend across the bilayer as (1) a single α helix, (2) as multiple α helices, or (3) as a rolled-up β sheet (a β barrel). Some of these "single-pass" and "multipass" proteins have a covalently attached fatty acid chain inserted in the cytosolic lipid monolayer (1). Other membrane proteins are exposed at only one side of the membrane. (4) Some of these are anchored to the cytosolic surface by an amphiphilic α helix that partitions into the cytosolic monolayer of the lipid bilayer through the hydrophobic face of the helix. (5) Others are attached to the bilayer solely by a covalently attached lipid chain—either a fatty acid chain or a prenyl group (see Figure 10–20)—in the cytosolic monolayer or, (6) via an oligosaccharide linker, to phosphatidylinositol in the noncytosolic monolayer—called a GPI anchor. (7, 8) Finally, many proteins are attached to the membrane only by noncovalent interactions with other membrane proteins. The way in which the structure in (5) is formed is illustrated in Figure 10–20, while the way in which the GPI anchor shown in (6) is formed is illustrated in Figure 12–56. The details of how membrane proteins become associated with the lipid bilayer are discussed in Chapter 12.

a membrane transport protein must provide a path for the molecules to cross the hydrophobic permeability barrier of the lipid bilayer; the molecular architecture of multipass membrane proteins is ideally suited for this task, as we discuss in Chapter 11.

Proteins that function on only one side of the lipid bilayer, by contrast, are often associated exclusively with either the lipid monolayer or a protein domain on that side. Some intracellular signaling proteins, for example, that are involved in converting extracellular signals into intracellular ones are bound to the cytosolic half of the plasma membrane by one or more covalently attached lipid groups, which can be fatty acid chains or *prenyl groups* (**Figure 10–20**). In some cases, myristic acid, a saturated 14-carbon fatty acid, is added to the N-terminal amino group of the protein during its synthesis on the ribosome. All members of the *Src family* of cytoplasmic protein tyrosine kinases (discussed in Chapter 15) are myristoylated in this way. Membrane attachment through a single lipid anchor is not very strong, however, and a second lipid group is often added to anchor proteins more firmly to a membrane. For most Src kinases, the second lipid modification is the attachment of palmitic acid, a saturated 16-carbon fatty acid, to a cysteine side chain of the protein. This modification occurs in response to an extracellular signal and helps recruit the kinases to the plasma membrane. When the signaling pathway is turned off, the palmitic acid is removed, allowing the kinase to return to the cytosol. Other intracellular signaling proteins, such as the Ras family small GTPases (discussed in Chapter 15), use a combination of prenyl group and palmitic acid attachment to recruit the proteins to the plasma membrane.

In Most Transmembrane Proteins the Polypeptide Chain Crosses the Lipid Bilayer in an α-Helical Conformation

A transmembrane protein always has a unique orientation in the membrane. This reflects both the asymmetric manner in which it is inserted into the lipid

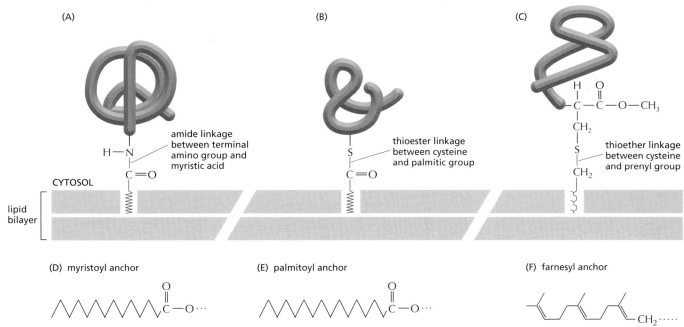

Figure 10–20 Membrane protein attachment by a fatty acid chain or a prenyl group. The covalent attachment of either type of lipid can help localize a water-soluble protein to a membrane after its synthesis in the cytosol. (A) A fatty acid chain (myristic acid) is attached via an amide linkage to an N-terminal glycine. (B) A fatty acid chain (palmitic acid) is attached via a thioester linkage to a cysteine. (C) A prenyl group (either farnesyl or a longer geranylgeranyl group) is attached via a thioether linkage to a cysteine residue that is initially located four residues from the protein's C-terminus. After prenylation, the terminal three amino acids are cleaved off, and the new C-terminus is methylated before insertion of the anchor into the membrane (not shown). The structures of the lipid anchors are shown below: (D) a myristoyl anchor (a 14-carbon saturated fatty acid chain), (E) a palmitoyl anchor (a 16-carbon saturated fatty acid chain), and (F) a farnesyl anchor (a 15-carbon unsaturated hydrocarbon chain).

Figure 10–21 A segment of a transmembrane polypeptide chain crossing the lipid bilayer as an α helix. <GTAG> Only the α-carbon backbone of the polypeptide chain is shown, with the hydrophobic amino acids in *green* and *yellow*. The polypeptide segment shown is part of the bacterial photosynthetic reaction center illustrated in Figure 10–34, the structure of which was determined by x-ray diffraction. (Based on data from J. Deisenhofer et al., *Nature* 318:618–624, 1985, and H. Michel et al., *EMBO J.* 5:1149–1158, 1986. All with permission from Macmillan Publishers Ltd.)

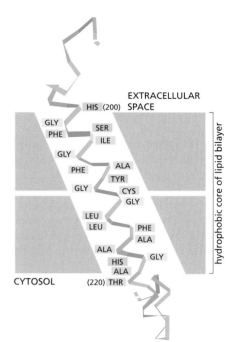

bilayer in the ER during its biosynthesis (discussed in Chapter 12) and the different functions of its cytosolic and noncytosolic domains. These domains are separated by the membrane-spanning segments of the polypeptide chain, which contact the hydrophobic environment of the lipid bilayer and are composed largely of amino acids with nonpolar side chains. Because the peptide bonds themselves are polar and because water is absent, all peptide bonds in the bilayer are driven to form hydrogen bonds with one another. The hydrogen-bonding between peptide bonds is maximized if the polypeptide chain forms a regular α helix as it crosses the bilayer, and this is how most membrane-spanning segments of polypeptide chains traverse the bilayer (**Figure 10–21**).

In **single-pass transmembrane proteins**, the polypeptide chain crosses only once (see Figure 10–19, example 1), whereas in **multipass transmembrane proteins**, the polypeptide chain crosses multiple times (see Figure 10–19, example 2). An alternative way for the peptide bonds in the lipid bilayer to satisfy their hydrogen-bonding requirements is for multiple transmembrane strands of a polypeptide chain to be arranged as a β sheet that is rolled up into a closed barrel (a so-called *β barrel*; see Figure 10–19, example 3). This form of multipass transmembrane structure is seen in the *porin proteins* that we discuss later.

Rapid progress in the x-ray crystallography of membrane proteins has enabled us to determine the three-dimensional structure of many of them. The structures confirm that it is often possible to predict from the protein's amino acid sequence which parts of the polypeptide chain extend across the lipid bilayer. Segments containing about 20–30 amino acids with a high degree of hydrophobicity are long enough to span a lipid bilayer as an α helix, and they can often be identified in *hydropathy plots* (**Figure 10–22**). From such plots, it is estimated that about 20% of the kind of an organism's proteins are transmembrane proteins, emphasizing their importance. Hydropathy plots cannot identify the membrane-spanning segments of a β barrel, as 10 amino acids or fewer are sufficient to traverse a lipid bilayer as an extended β strand and only every other amino acid side chain is hydrophobic.

The strong drive to maximize hydrogen-bonding in the absence of water means that a polypeptide chain that enters the bilayer is likely to pass entirely through it before changing direction, since chain bending requires a loss of regular hydrogen-bonding interactions. But multipass membrane proteins can also contain regions that fold into the membrane from either side, squeezing into spaces between transmembrane α helices without contacting the hydrophobic core of the lipid bilayer. Because such regions of the polypeptide chain interact only with other polypeptide regions, they do not need to maximize hydrogen-bonding; they can therefore have a variety of secondary structures, including helices that extend only part way across the lipid bilayer (**Figure 10–23**). Such regions are important for the function of some membrane proteins, including the K+ and water channels; the regions contribute to the walls of the pores traversing the membrane and confer substrate specificity on the channels, as we discuss in Chapter 11. These regions cannot be identified in hydropathy plots and are only revealed by x-ray crystallography, electron diffraction (a technique similar to x-ray diffraction but performed on two-dimensional arrays of proteins), or NMR studies of the protein's three-dimensional structure.

Transmembrane α Helices Often Interact with One Another

The transmembrane α helices of many single-pass membrane proteins do not contribute to the folding of the protein domains on either side of the membrane.

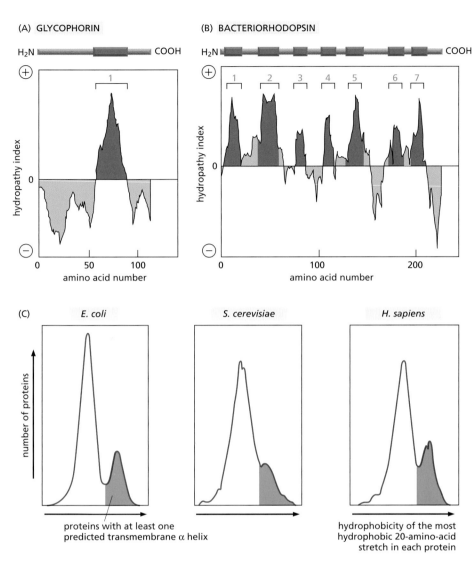

(A) GLYCOPHORIN

H₂N — COOH

hydropathy index

amino acid number

(B) BACTERIORHODOPSIN

H₂N — COOH

hydropathy index

amino acid number

(C)

E. coli

S. cerevisiae

H. sapiens

number of proteins

proteins with at least one predicted transmembrane α helix

hydrophobicity of the most hydrophobic 20-amino-acid stretch in each protein

Figure 10–22 Using hydropathy plots to localize potential α-helical membrane-spanning segments in a polypeptide chain. The free energy needed to transfer successive segments of a polypeptide chain from a nonpolar solvent to water is calculated from the amino acid composition of each segment using data obtained from model compounds. This calculation is made for segments of a fixed size (usually around 10–20 amino acids), beginning with each successive amino acid in the chain. The "hydropathy index" of the segment is plotted on the Y axis as a function of its location in the chain. A positive value indicates that free energy is required for transfer to water (i.e., the segment is hydrophobic), and the value assigned is an index of the amount of energy needed. Peaks in the hydropathy index appear at the positions of hydrophobic segments in the amino acid sequence. (A and B) Two examples of membrane proteins discussed later in this chapter are shown. Glycophorin (A) has a single membrane-spanning α helix and one corresponding peak in the hydropathy plot. Bacteriorhodopsin (B) has seven membrane-spanning α helices and seven corresponding peaks in the hydropathy plot. (C) The proportion of predicted membrane proteins encoded by the genomes of E. coli, S. cerevisiae, and human. The area shaded in *green* indicates the fraction of proteins that contain at least one predicted transmembrane helix. The data for E. coli and S. cerevisiae represent the whole genome; the data for human represent only part of the genome; in each case, the area under the curve is proportional to the number of genes analyzed. (A, adapted from D. Eisenberg, *Annu. Rev. Biochem.* 53:595–624, 1984. With permission from Annual Reviews; C, adapted from D. Boyd et al., *Protein Sci.* 7:201–205, 1998. With permission from The Protein Society.)

As a consequence, it is often possible to engineer cells to produce the cytosolic or extracellular domains of these proteins as water-soluble protein. This approach has been invaluable to study the structure and function of these domains, especially of those in transmembrane receptor proteins (discussed in Chapter 15). A transmembrane α helix, even in a single-pass membrane protein, however, often does more than just anchor the protein to the lipid bilayer. Many single-pass membrane proteins form homodimers, which are held together by strong and highly specific interactions between the two transmembrane α helices; the sequence of the hydrophobic amino acids of these helices contains the information that directs the protein–protein interaction.

Similarly, the transmembrane α helices in multipass membrane proteins occupy specific positions in the folded protein structure that are determined by interactions between the neighboring helices. These interactions are crucial for the structure and function of the many channels and transporters that move molecules across lipid bilayers. In many cases, one can use proteases to cut the loops of the polypeptide chain that link the transmembrane α helices on either side of the bilayer and the helices stay together and function normally. In some

Figure 10–23 Two α helices in the aquaporin water channel, each of which spans only halfway through the lipid bilayer. In the membrane, the protein forms a tetramer of four such two-helix segments, such that the colored surface shown here is buried at an interface formed by protein–protein interactions. The mechanism by which the channel allows the passage of water molecules across the lipid bilayer is discussed in more detail in Chapter 11.

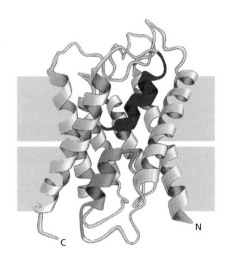

N

C

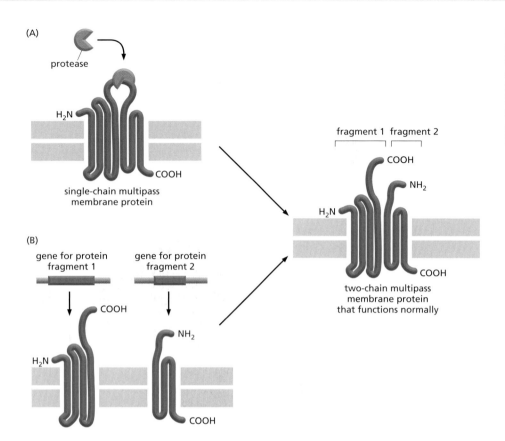

Figure 10–24 Converting a single-chain multipass protein into a two-chain multipass protein. (A) Proteolytic cleavage of one loop to create two fragments that stay together and function normally. (B) Expression of the same two fragments from separate genes gives rise to a similar protein that functions normally.

cases, one can even express engineered genes encoding separate pieces of a multipass protein in living cells, and one finds that the separate pieces assemble properly to form a functional transmembrane protein (**Figure 10–24**), emphasizing the exquisite specificity with which transmembrane α helices can interact.

In multipass membrane proteins, neighboring transmembrane helices in the folded structure of the protein shield many of the transmembrane helices from the membrane lipids. Why, then, are these shielded helices nevertheless composed primarily of hydrophobic amino acids? The answer lies in the way in which multipass proteins are integrated into the membrane during their biosynthesis. As we discuss in Chapter 12, transmembrane α helices are inserted into the lipid bilayer sequentially by a protein translocator. After leaving the translocator, each helix is transiently surrounded by lipids in the bilayer, which requires that the helix be hydrophobic. It is only as the protein folds up into its final structure that contacts are made between adjacent helices and protein–protein contacts replace some of the protein–lipid contacts (**Figure 10–25**).

Some β Barrels Form Large Transmembrane Channels

Multipass transmembrane proteins that have their transmembrane segments arranged as a *β barrel* rather than as an α helix are comparatively rigid and tend to crystallize readily. Thus, some of them were among the first multipass

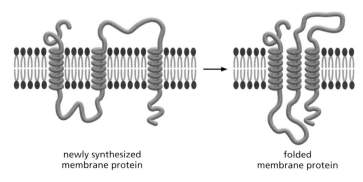

newly synthesized
membrane protein

folded
membrane protein

Figure 10–25 Steps in the folding of a multipass transmembrane protein. When the newly synthesized transmembrane α helices are released into the lipid bilayer, they are initially surrounded by lipid molecules. As the protein folds, contacts between the helices displace some of the lipid molecules surrounding the helices.

membrane protein structures to be determined by x-ray crystallography. The number of β strands in a β barrel varies widely, from as few as 8 strands to as many as 22 (**Figure 10–26**).

β barrel proteins are abundant in the outer membrane of mitochondria, chloroplasts, and many bacteria. Some are pore-forming proteins, which create water-filled channels that allow selected small hydrophilic molecules to cross the lipid bilayer of the bacterial outer membrane. The porins are well-studied examples (example 3 in Figure 10–26). The porin barrel is formed from a 16-strand, antiparallel β sheet, which is sufficiently large to roll up into a cylindrical structure. Polar amino acid side chains line the aqueous channel on the inside, while nonpolar side chains project from the outside of the barrel to interact with the hydrophobic core of the lipid bilayer. Loops of the polypeptide chain often protrude into the lumen of the channel, narrowing it so that only certain solutes can pass. Some porins are therefore highly selective: *maltoporin*, for example, preferentially allows maltose and maltose oligomers to cross the outer membrane of *E. coli*.

The *FepA protein* is a more complex example of a β barrel transport protein (example 4 in Figure 10–26). It transports iron ions across the bacterial outer membrane. It is constructed from 22 β strands, and a large globular domain completely fills the inside of the barrel. Iron ions bind to this domain, which is thought to undergo a large conformational change to transfer the iron across the membrane.

Not all β barrel proteins are transport proteins. Some form smaller barrels that are completely filled by amino acid side chains that project into the center of the barrel. These proteins function as receptors or enzymes (examples 1 and 2 in Figure 10–26), and the barrel serves as a rigid anchor, which holds the protein in the membrane and orients the cytosolic loops that form binding sites for specific intracellular molecules.

Although β barrel proteins have various functions, they are largely restricted to bacterial, mitochondrial, and chloroplast outer membranes. Most multipass transmembrane proteins in eucaryotic cells and in the bacterial plasma membrane are constructed from transmembrane α helices. The helices can slide against each other, allowing conformational changes in the protein that can open and shut ion channels, transport solutes, or transduce extracellular signals into intracellular ones. In β barrel proteins, by contrast, hydrogen bonds bind each β strand rigidly to its neighbors, making conformational changes within the wall of the barrel unlikely.

Figure 10–26 β barrels formed from different numbers of β strands. <TGCT> (1) The *E. coli* OmpA protein serves as a receptor for a bacterial virus. (2) The *E. coli* OMPLA protein is an enzyme (a lipase) that hydrolyzes lipid molecules. The amino acids that catalyze the enzymatic reaction (shown in *red*) protrude from the outside surface of the barrel. (3) A porin from the bacterium *Rhodobacter capsulatus* forms a water-filled pore across the outer membrane. The diameter of the channel is restricted by loops (shown in *blue*) that protrude into the channel. (4) The *E. coli* FepA protein transports iron ions. The inside of the barrel is completely filled by a globular protein domain (shown in *blue*) that contains an iron-binding site (not shown). This domain is thought to change its conformation to transport the bound iron, but the molecular details of the changes are not known.

Many Membrane Proteins Are Glycosylated

Most transmembrane proteins in animal cells are glycosylated. As in glycolipids, the sugar residues are added in the lumen of the ER and the Golgi apparatus

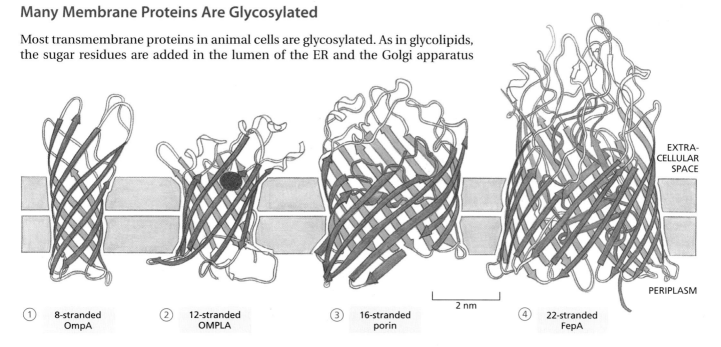

EXTRA-CELLULAR SPACE

2 nm

PERIPLASM

① 8-stranded OmpA   ② 12-stranded OMPLA   ③ 16-stranded porin   ④ 22-stranded FepA

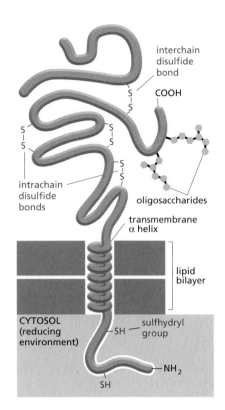

Figure 10–27 **A single-pass transmembrane protein.** Note that the polypeptide chain traverses the lipid bilayer as a right-handed α helix and that the oligosaccharide chains and disulfide bonds are all on the noncytosolic surface of the membrane. The sulfhydryl groups in the cytosolic domain of the protein do not normally form disulfide bonds because the reducing environment in the cytosol maintains these groups in their reduced (–SH) form.

(discussed in Chapters 12 and 13). For this reason, the oligosaccharide chains are always present on the noncytosolic side of the membrane. Another important difference between proteins (or parts of proteins) on the two sides of the membrane results from the reducing environment of the cytosol. This environment decreases the likelihood that intrachain or interchain disulfide (S–S) bonds will form between cysteines on the cytosolic side of membranes. These bonds form on the noncytosolic side, where they can help stabilize either the folded structure of the polypeptide chain or its association with other polypeptide chains (**Figure 10–27**).

Because most plasma membrane proteins are glycosylated, carbohydrates extensively coat the surface of all eucaryotic cells. These carbohydrates occur as oligosaccharide chains covalently bound to membrane proteins (glycoproteins) and lipids (glycolipids). They also occur as the polysaccharide chains of integral membrane *proteoglycan* molecules. Proteoglycans, which consist of long polysaccharide chains linked covalently to a protein core, are found mainly outside the cell, as part of the extracellular matrix (discussed in Chapter 19). But, for some proteoglycans, the protein core either extends across the lipid bilayer or is attached to the bilayer by a glycosylphosphatidylinositol (GPI) anchor.

The terms cell coat or glycocalyx are sometimes used to describe the carbohydrate-rich zone on the cell surface. This **carbohydrate layer** can be visualized by various stains, such as ruthenium red (**Figure 10–28**A), as well as by its affinity for carbohydrate-binding proteins called **lectins**, which can be labeled with a fluorescent dye or some other visible marker. Although most of the sugar groups are attached to intrinsic plasma membrane molecules, the carbohydrate layer also contains both glycoproteins and proteoglycans that have been secreted into the extracellular space and then adsorbed onto the cell surface (Figure 10–28B). Many of these adsorbed macromolecules are components of the extracellular matrix, so that the boundary between the plasma membrane and the extracellular matrix is often not sharply defined. One of the many functions of the carbohydrate layer is to protect cells against mechanical and chemical damage; it also keeps various other cells at a distance, preventing unwanted protein–protein interactions.

The oligosaccharide side chains of glycoproteins and glycolipids are enormously diverse in their arrangement of sugars. Although they usually contain fewer than 15 sugars, they are often branched, and the sugars can be bonded together by various covalent linkages—unlike the amino acids in a polypeptide chain, which are all linked by identical peptide bonds. Even three sugars can be put together to form hundreds of different trisaccharides. Both the diversity and the exposed position of the oligosaccharides on the cell surface make them especially well suited to function in specific cell-recognition processes. As we discuss in Chapter 19, plasma membrane-bound lectins that recognize specific oligosaccharides on cell-surface glycolipids and glycoproteins mediate a variety of transient cell–cell adhesion processes, including those occurring in sperm–egg interactions, blood clotting, lymphocyte recirculation, and inflammatory responses.

Membrane Proteins Can Be Solubilized and Purified in Detergents

In general, only agents that disrupt hydrophobic associations and destroy the lipid bilayer can solubilize transmembrane proteins (and some other tightly bound membrane proteins). The most useful of these for the membrane biochemist are **detergents**, which are small amphiphilic molecules of variable

(A)

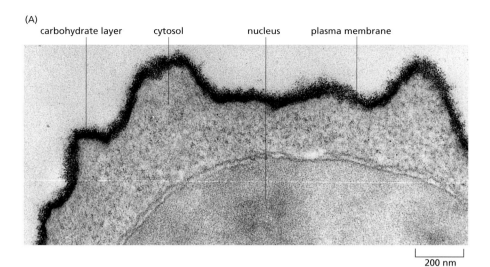

carbohydrate layer   cytosol   nucleus   plasma membrane

200 nm

(B)

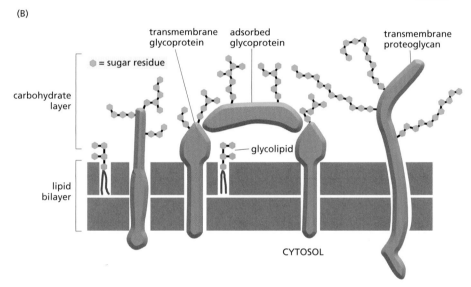

Figure 10–28 The carbohydrate layer on the cell surface. This electron micrograph of the surface of a lymphocyte stained with ruthenium red emphasizes the thick carbohydrate-rich layer surrounding the cell. (B) The carbohydrate layer is made up of the oligosaccharide side chains of glycolipids and integral membrane glycoproteins and the polysaccharide chains on integral membrane proteoglycans. In addition, adsorbed glycoproteins, and adsorbed proteoglycans (not shown) contribute to the carbohydrate layer in many cells. Note that all of the carbohydrate is on the noncytosolic surface of the membrane. (A, courtesy of Audrey M. Glauert and G.M.W. Cook.)

structure. Detergents are much more soluble in water than lipids. Their polar (hydrophilic) ends can be either charged (ionic), as in *sodium dodecyl sulfate (SDS)*, or uncharged (nonionic), as in *octylglucoside* and Triton (**Figure 10–29**A). At low concentration, detergents are monomeric in solution, but when their concentration is increased above a threshold, called the *critical micelle concentration* or *CMC*, they aggregate to form micelles (Figure 10–29B–C). Detergent molecules rapidly diffuse in and out of micelles, keeping the concentration of monomer in the solution constant, no matter how many micelles are present. Both the CMC and the average number of detergent molecules in a micelle are characteristic properties of each detergent, but they also depend on the temperature, pH, and salt concentration. Detergent solutions are therefore complex systems and are difficult to study.

When mixed with membranes, the hydrophobic ends of detergents bind to the hydrophobic regions of the membrane proteins, where they displace lipid molecules with a collar of detergent molecules. Since the other end of the detergent molecule is polar, this binding tends to bring the membrane proteins into solution as detergent–protein complexes (**Figure 10–30**). Usually, some lipid molecules also remain attached to the protein.

Strong ionic detergents, such as SDS, can solubilize even the most hydrophobic membrane proteins. This allows the proteins to be analyzed by *SDS polyacrylamide-gel electrophoresis* (discussed in Chapter 8), a procedure that has revolutionized the study of membrane proteins. Such strong detergents unfold (denature) proteins by binding to their internal "hydrophobic cores,"

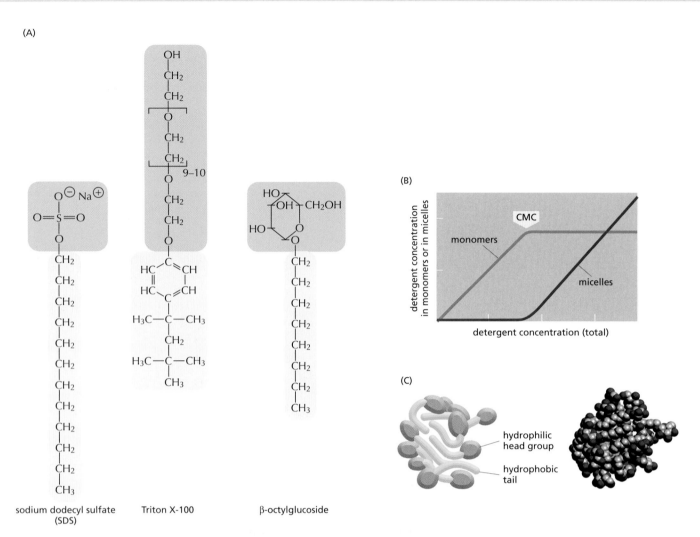

Figure 10–29 The structure and function of detergent micelles. <GCGA> (A) Three commonly used detergents are sodium dodecyl sulfate (SDS), an anionic detergent, and Triton X-100 and β-octylglucoside, two nonionic detergents. Triton X-100 is a mixture of compounds in which the region in brackets is repeated between 9 and 10 times. The hydrophobic portion of each detergent is shown in *yellow,* and the hydrophilic portion is shown in *orange.* (B) At low concentration, detergent molecules are monomeric in solution. As their concentration is increased beyond the critical micelle concentration (CMC), some of the detergent molecules form micelles. Note that the concentration of detergent monomer stays constant above the CMC. (C) Because they have both polar and nonpolar ends, detergent molecules are amphiphilic; and because they are cone-shaped, they form micelles rather than bilayers (see Figure 10–7). Detergent micelles have irregular shapes, and, due to packing constraints, the hydrophobic tails are partially exposed to water. The space-filling model shows the structure of a micelle composed of 20 β-octylglucoside molecules, predicted by molecular dynamics calculations (B, adapted from G. Gunnarsson, B. Jönsson and H. Wennerström, *J. Phys. Chem.* 84:3114–3121, 1980; C, from S. Bogusz, R.M. Venable and R.W. Pastor, *J. Phys. Chem. B.* 104:5462–5470, 2000. With permission from the American Chemical Society.)

thereby rendering the proteins inactive and unusable for functional studies. Nonetheless, proteins can be readily separated and purified in their SDS-denatured form. In some cases, removal of the detergent allows the purified protein to renature, with recovery of functional activity.

Many hydrophobic membrane proteins can be solubilized and then purified in an active form by the use of mild detergents. These detergents cover the hydrophobic regions on membrane-spanning segments that become exposed after lipid removal but do not unfold the protein. If the detergent concentration of a solution of solubilized membrane proteins is reduced (by dilution, for example), membrane proteins do not remain soluble. In the presence of an excess of phospholipid molecules in such a solution, membrane proteins incorporate into small liposomes that form spontaneously. In this way, functionally active membrane protein systems can be reconstituted from purified components, providing a powerful means of analyzing the activities of membrane transporters, ion channels, signaling receptors, and so on (**Figure 10–31**). Such

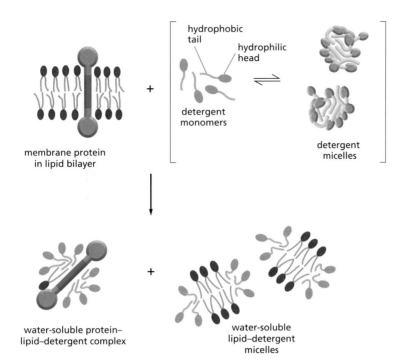

Figure 10–30 Solubilizing membrane proteins with a mild nonionic detergent. The detergent disrupts the lipid bilayer and brings the proteins into solution as protein–lipid– detergent complexes. The phospholipids in the membrane are also solubilized by the detergent.

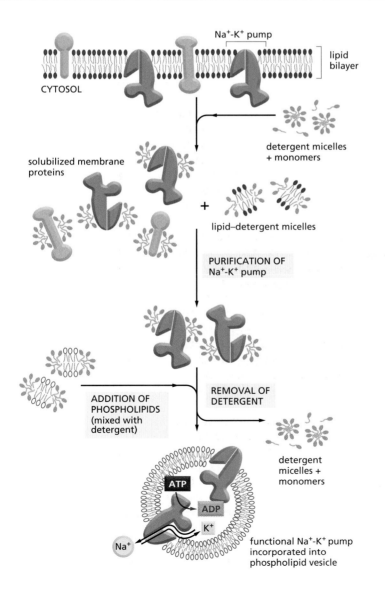

Figure 10–31 The use of mild nonionic detergents for solubilizing, purifying, and reconstituting functional membrane protein systems. In this example, functional Na⁺-K⁺ pump molecules are purified and incorporated into phospholipid vesicles. The Na⁺-K⁺ pump is an ion pump that is present in the plasma membrane of most animal cells; it uses the energy of ATP hydrolysis to pump Na⁺ out of the cell and K⁺ in, as discussed in Chapter 11.

functional reconstitution, for example, provided proof for the hypothesis that the transmembrane ATPases use H^+ gradients in mitochondrial, chloroplast, and bacterial membranes to synthesize ATP.

Detergents have also played a crucial part in the purification and crystallization of membrane proteins. The development of new detergents and new expression systems producing large quantities of membrane proteins from cDNA clones has led to a rapid increase in the number of structures of membrane proteins and protein complexes that are known.

Bacteriorhodopsin Is a Light-Driven Proton Pump That Traverses the Lipid Bilayer as Seven α Helices

In Chapter 11, we consider how multipass transmembrane proteins mediate the selective transport of small hydrophilic molecules across cell membranes. But a detailed understanding of how a membrane transport protein actually works requires precise information about its three-dimensional structure in the bilayer. *Bacteriorhodopsin* was the first membrane transport protein whose structure was determined. It has remained the prototype of many multipass membrane proteins with a similar structure, and it merits a brief digression here.

The "purple membrane" of the archaean *Halobacterium salinarum* is a specialized patch in the plasma membrane that contains a single species of protein molecule, **bacteriorhodopsin** (**Figure 10–32**). Each bacteriorhodopsin molecule contains a single light-absorbing group, or chromophore (called *retinal*), which gives the protein its purple color. Retinal is vitamin A in its aldehyde form and is identical to the chromophore found in *rhodopsin* of the photoreceptor cells of the vertebrate eye (discussed in Chapter 15). Retinal is covalently linked to a lysine side chain of the bacteriorhodopsin protein. When activated by a single photon of light, the excited chromophore changes its shape and causes a series of small conformational changes in the protein, resulting in the transfer of one H^+ from the inside to the outside of the cell (**Figure 10–33**). In bright light, each bacteriorhodopsin molecule can pump several hundred protons per second. The light-driven proton transfer establishes an H^+ gradient across the plasma membrane, which in turn drives the production of ATP by a second protein in the cell's plasma membrane. The energy stored in the H^+ gradient also drives other energy-requiring processes in the cell. Thus, bacteriorhodopsin converts solar energy into a proton gradient, which provides energy to the archaeal cell.

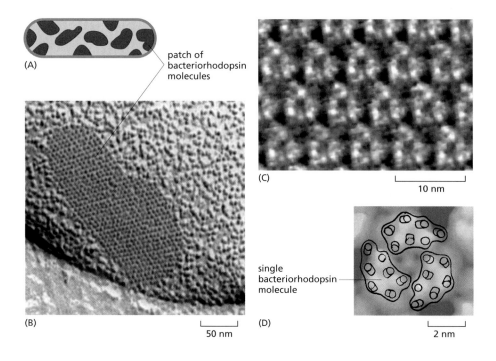

(A)

patch of bacteriorhodopsin molecules

(B)

50 nm

(C)

10 nm

single bacteriorhodopsin molecule

(D)

2 nm

Figure 10–32 Patches of purple membrane, which contain bacteriorhodopsin in the archaean *Halobacterium salinarum*. (A) These archaea live in saltwater pools, where they are exposed to sunlight. They have evolved a variety of light-activated proteins, including bacteriorhodopsin, which is a light-activated proton pump in the plasma membrane. (B) The bacteriorhodopsin molecules in the purple membrane patches are tightly packed into two-dimensional crystalline arrays. (C) Details of the molecular surface visualized by atomic force microscopy. With this technique individual bacteriorhodopsin molecules can be seen. (D) Outline of the approximate locations of three bacteriorhodopsin monomers and their individual α helices in the images shown in (B). (B–D, courtesy of Dieter Oesterhelt.)

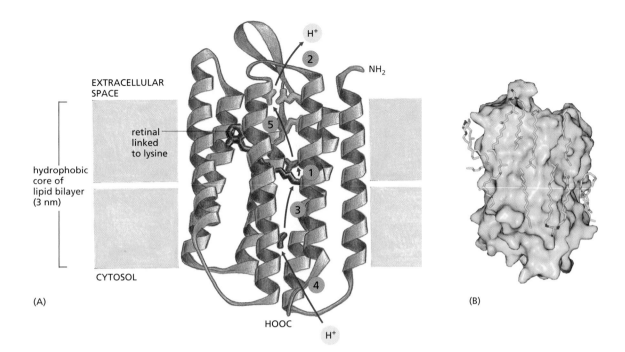

EXTRACELLULAR
SPACE

retinal
linked
to lysine

hydrophobic
core of
lipid bilayer
(3 nm)

CYTOSOL

(A)

(B)

The numerous bacteriorhodopsin molecules in the purple membrane are arranged as a planar two-dimensional crystal. The regular packing has made it possible to determine the three-dimensional structure and orientation of bacteriorhodopsin in the membrane to moderate resolution (3 Å) by an approach that uses a combination of electron microscopy and electron diffraction analysis. This procedure, known as **electron crystallography**, is analogous to the study of three-dimensional crystals of soluble proteins by x-ray diffraction analysis. It has provided the first structural views of many membrane proteins that were found to be difficult to crystallize from detergent solutions. For bacteriorhodopsin, the structure obtained by electron crystallography was later confirmed and extended to very high resolution by x-ray crystallography. Each bacteriorhodopsin molecule is folded into seven closely packed α helices (each containing about 25 amino acids), which pass through the lipid bilayer at slightly different angles. By obtaining very well ordered protein crystals and freezing them at very low temperatures, it has also been possible to determine the structures of some of the protein's intermediate conformations during its H+ pumping cycle.

Bacteriorhodopsin is a member of a large superfamily of membrane proteins with similar structures but different functions. For example, rhodopsin in rod cells of the vertebrate retina and many cell-surface receptor proteins that bind extracellular signal molecules are also built from seven transmembrane α helices. These proteins function as signal transducers rather than as transporters: each responds to an extracellular signal by activating a GTP-binding protein (G protein) inside the cell and are therefore called *G-protein-coupled receptors* (*GPCRs*), as we discuss in Chapter 15. Although the structures of bacteriorhodopsins and GPCRs are strikingly similar, they show no sequence similarity and thus probably belong to two evolutionarily distant branches of an ancient protein family.

The high-resolution crystal structure of bacteriorhodopsin reveals many lipid molecules that are bound in specific places on the protein surface (Figure 10–33B). Interactions with specific lipids are thought to help stabilize many membrane proteins, which work best and crystallize more readily if some of the lipids remain bound during detergent extraction, or if specific lipids are added back to the proteins in detergent solutions. The specificity of these lipid–protein interactions helps explain why eucaryotic membranes contain such a variety of lipids, with head groups that differ in size, shape, and charge. We can think of the membrane lipids as constituting a two-dimensional solvent for the proteins in the membrane, just as water constitutes a three-dimensional solvent for proteins

Figure 10–33 The three-dimensional structure of a bacteriorhodopsin molecule. <TTAA> (A) The polypeptide chain crosses the lipid bilayer seven times as α helices. The location of the retinal chromophore (*purple*) and the probable pathway taken by protons during the light-activated pumping cycle are shown. The first and key step is the passing of a H+ from the chromophore to the side chain of aspartic acid 85 (*red*, located next to the chromophore) that occurs upon absorption of a photon by the chromophore. Subsequently, other H+ transfers—in the numerical order indicated and utilizing the hydrophilic amino acid side chains that line a path through the membrane—complete the pumping cycle and return the enzyme to its starting state. Color code: glutamic acid (*orange*), aspartic acid (*red*), arginine (*blue*). (B) The high-resolution crystal structure of bacteriorhodopsin shows many lipid molecules (*yellow* with *red* head groups) that are tightly bound to specific places on the surface of the protein. (A, adapted from H. Luecke et al., *Science* 286:255–260, 1999. With permission from AAAS; B, from H. Luecke et al., *J. Mol. Biol.* 291:899–911, 1999. With permission from Academic Press.)

in an aqueous solution. Some membrane proteins can function only in the presence of specific lipid head groups, just as many enzymes in aqueous solution require a particular ion for activity.

Membrane Proteins Often Function as Large Complexes

Many membrane proteins function as part of multicomponent complexes, several of which have been studied by x-ray crystallography. One is a bacterial *photosynthetic reaction center*, which was the first transmembrane protein complex to be crystallized and analyzed by x-ray diffraction. The results of this analysis were of general importance to membrane biology because they showed for the first time how multiple polypeptides associate in a membrane to form a complex protein machine (**Figure 10–34**). In Chapter 14, we discuss how such photosynthetic complexes function to capture light energy and use it to pump protons across the membrane. Many of the membrane protein complexes involved in photosynthesis, proton pumping, and electron transport are even larger than the photosynthetic reaction center. The enormous photosystem II complex from cyanobacteria, for example, contains 19 protein subunits and well over 60 transmembrane helices. Membrane proteins are often arranged in large complexes, not only for harvesting various forms of energy, but also for transducing extracellular signals into intracellular ones (discussed in Chapter 15).

Many Membrane Proteins Diffuse in the Plane of the Membrane

Like most membrane lipids, membrane proteins do not tumble *(flip-flop)* across the lipid bilayer, but they do rotate about an axis perpendicular to the plane of the bilayer *(rotational diffusion)*. In addition, many membrane proteins are able

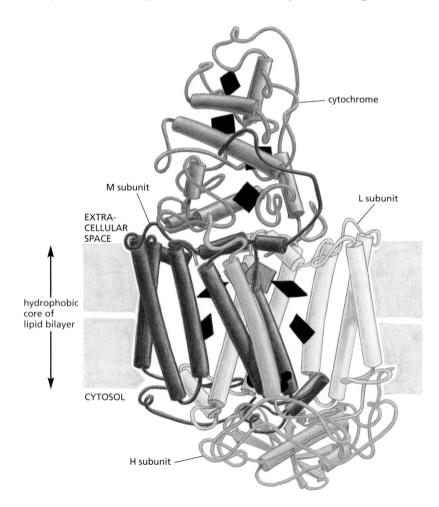

Figure 10–34 **The three-dimensional structure of the photosynthetic reaction center of the bacterium *Rhodopseudomonas viridis*.** <ATCA> The structure was determined by x-ray diffraction analysis of crystals of this transmembrane protein complex. The complex consists of four subunits L, M, H, and a cytochrome. The L and M subunits form the core of the reaction center, and each contains five α helices that span the lipid bilayer. The locations of the various electron carrier coenzymes are shown in *black*. Note that the coenzymes are arranged in the spaces between the helices. The *special pair* of chlorophyll molecules (discussed in Chapter 14) is shown in *turquoise*. (Adapted from a drawing by J. Richardson based on data from J. Deisenhofer et al., *Nature* 318:618–624, 1985. With permission from Macmillan Publishers Ltd.)

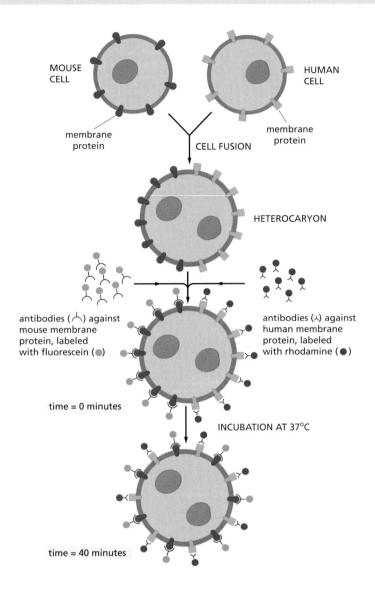

Figure 10–35 An experiment demonstrating the diffusion of proteins in the plasma membrane of mouse–human hybrid cells. The mouse and human proteins are initially confined to their own halves of the newly formed heterocaryon plasma membrane, but they intermix over time. The two antibodies used to visualize the proteins can be distinguished in a fluorescence microscope because fluorescein is green and rhodamine is red. (Based on L.D. Frye and M. Edidin, *J. Cell Sci.* 7:319–335, 1970. With permission from The Company of Biologists.)

to move laterally within the membrane *(lateral diffusion)*. An experiment in which mouse cells were artificially fused with human cells to produce hybrid cells *(heterocaryons)* provided the first direct evidence that some plasma membrane proteins are mobile in the plane of the membrane. Two differently labeled antibodies were used to distinguish selected mouse and human plasma membrane proteins. Although at first the mouse and human proteins were confined to their own halves of the newly formed heterocaryon, the two sets of proteins diffused and mixed over the entire cell surface in about half an hour (**Figure 10–35**).

The lateral diffusion rates of membrane proteins can be measured by using the technique of *fluorescence recovery after photobleaching (FRAP)*. The method usually involves marking the membrane protein of interest with a specific fluorescent group. This can be done either with a fluorescent ligand such as a fluorophore-labeled antibody that binds to the protein or with recombinant DNA technology to express the protein fused to green fluorescent protein (GFP) (discussed in Chapter 9). The fluorescent group is then bleached in a small area of membrane by a laser beam, and the time taken for adjacent membrane proteins carrying unbleached ligand or GFP to diffuse into the bleached area is measured (**Figure 10–36**A). A complementary technique is *fluorescence loss in photobleaching (FLIP)*. Here, a laser beam continuously irradiates a small area of membrane to bleach all the fluorescent molecules that diffuse into it, thereby gradually depleting the surrounding membrane of fluorescently labeled molecules (Figure 10–36B). From such FRAP and FLIP measurements, we can calculate the diffusion coefficient for the marked cell-surface protein. The values

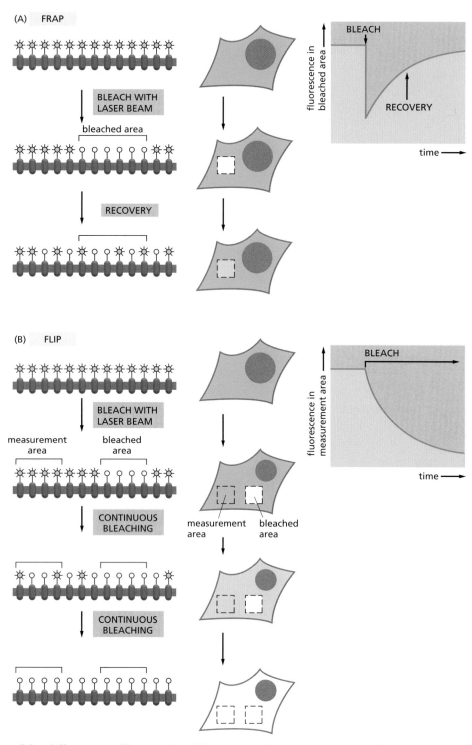

(A) FRAP

BLEACH WITH
LASER BEAM

bleached area

RECOVERY

(B) FLIP

BLEACH WITH
LASER BEAM

measurement
area

bleached
area

CONTINUOUS
BLEACHING

measurement
area

bleached
area

CONTINUOUS
BLEACHING

BLEACH

RECOVERY

fluorescence in bleached area

time

BLEACH

fluorescence in measurement area

time

Figure 10–36 Measuring the rate of lateral diffusion of a membrane protein by photobleaching techniques. <ATGT> A specific protein of interest can be expressed as a fusion protein with green fluorescent protein (GFP), which is intrinsically fluorescent. (A) In the FRAP technique, fluorescent molecules are bleached in a small area using a laser beam. The fluorescence intensity recovers as the bleached molecules diffuse away and unbleached molecules diffuse into the irradiated area (shown here in side and top views). The diffusion coefficient is calculated from a graph of the rate of recovery: the greater the diffusion coefficient of the membrane protein, the faster the recovery. (B) In the FLIP technique, a small area of membrane is irradiated continuously, and fluorescence is measured in a separate area. Fluorescence in the second area progressively decreases as fluorescent proteins diffuse out and bleached molecules diffuse in; eventually, all of the fluorescent protein molecules are bleached, as long as they are mobile and not permanently anchored to the cytoskeleton or extracellular matrix.

of the diffusion coefficients for different membrane proteins in different cells are highly variable, because interactions with other proteins impede the diffusion of the proteins to varying degrees. Measurements of proteins that are minimally impeded in this way indicate that cell membranes have a viscosity comparable to that of olive oil.

One drawback to the FRAP and FLIP techniques is that they monitor the movement of large populations of molecules in a relatively large area of membrane; one cannot follow individual protein molecules. If a protein fails to migrate into a bleached area, for example, one cannot tell whether the molecule is truly immobile or just restricted in its movement to a very small region of membrane—perhaps by cytoskeletal proteins. *Single-particle tracking* techniques overcome this problem by labeling individual membrane molecules with antibodies coupled to fluorescent dyes or tiny gold particles and tracking their

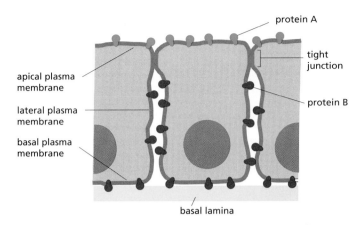

apical plasma membrane

lateral plasma membrane

basal plasma membrane

protein A

tight junction

protein B

basal lamina

Figure 10–37 How membrane molecules can be restricted to a particular membrane domain. In this drawing of an epithelial cell, protein A (in the apical membrane) and protein B (in the basal and lateral membranes) can diffuse laterally in their own domains but are prevented from entering the other domain, at least partly by the specialized cell junction called a tight junction. Lipid molecules in the outer (noncytosolic) monolayer of the plasma membrane are likewise unable to diffuse between the two domains; lipids in the inner (cytosolic) monolayer, however, are able to do so (not shown). The basal lamina is a thin mat of extracellular matrix that separates epithelial sheets from other tissues (discussed in Chapter 19).

movement by video microscopy. Using single-particle tracking, one can record the diffusion path of a single membrane protein molecule over time. Results from all of these techniques indicate that plasma membrane proteins differ widely in their diffusion characteristics, as we now discuss.

Cells Can Confine Proteins and Lipids to Specific Domains Within a Membrane

The recognition that biological membranes are two-dimensional fluids was a major advance in understanding membrane structure and function. It has become clear, however, that the picture of a membrane as a lipid sea in which all proteins float freely is greatly oversimplified. Many cells confine membrane proteins to specific regions in a continuous lipid bilayer. We have already discussed how bacteriorhodopsin molecules in the purple membrane of *Halobacterium* assemble into large two-dimensional crystals, in which individual protein molecules are relatively fixed in relationship to one another (see Figure 10–32); large aggregates of this kind diffuse very slowly.

In epithelial cells, such as those that line the gut or the tubules of the kidney, certain plasma membrane enzymes and transport proteins are confined to the apical surface of the cells, whereas others are confined to the basal and lateral surfaces (**Figure 10–37**). This asymmetric distribution of membrane proteins is often essential for the function of the epithelium, as we discuss in Chapters 11 and 19. The lipid compositions of these two membrane domains are also different, demonstrating that epithelial cells can prevent the diffusion of lipid as well as protein molecules between the domains. Experiments with labeled lipids, however, suggest that only lipid molecules in the outer monolayer of the membrane are confined in this way. The barriers set up by a specific type of intercellular junction (called a *tight junction*, discussed in Chapter 19) maintain the separation of both protein and lipid molecules. Clearly, the membrane proteins that form these intercellular junctions cannot be allowed to diffuse laterally in the interacting membranes.

A cell can also create membrane domains without using intercellular junctions. The mammalian spermatozoon, for instance, is a single cell that consists of several structurally and functionally distinct parts covered by a continuous plasma membrane. When a sperm cell is examined by immunofluorescence microscopy with a variety of antibodies, each of which react with a specific cell-surface molecule, the plasma membrane is found to consist of at least three distinct domains (**Figure 10–38**). Some of the membrane molecules are able to diffuse freely within the confines of their own domain. The molecular nature of the "fence" that prevents the molecules from leaving their domain is not known. Many other cells have similar membrane fences that confine membrane protein diffusion to certain membrane domains. The plasma membrane of nerve cells, for example, contains a domain enclosing the cell body and dendrites, and another enclosing the axon. In this case, it is thought that a belt of actin filaments tightly associated with the plasma membrane at the cell-body–axon junction forms part of the barrier.

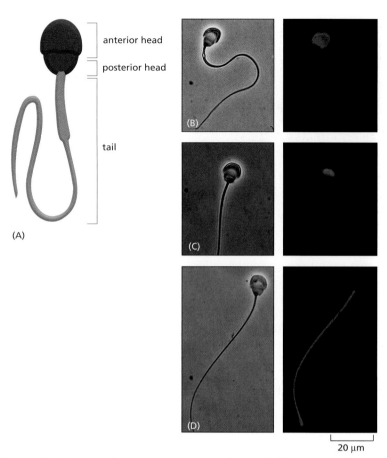

anterior head

posterior head

tail

(A)

(B)

(C)

(D)

20 μm

Figure 10–38 Three domains in the plasma membrane of a guinea pig sperm. (A) A drawing of a guinea pig sperm. In the three pairs of micrographs, phase-contrast micrographs are on the *left*, and the same cell is shown with cell-surface immunofluorescence staining on the *right*. Different monoclonal antibodies selectively label cell-surface molecules on (B) the anterior head, (C) the posterior head, and (D) the tail. (Micrographs courtesy of Selena Carroll and Diana Myles.)

Figure 10–39 shows four common ways of immobilizing specific membrane proteins through protein–protein interactions.

The Cortical Cytoskeleton Gives Membranes Mechanical Strength and Restricts Membrane Protein Diffusion

As shown in Figure 10–39B and C, a common way in which a cell restricts the lateral mobility of specific membrane proteins is to tether them to macromolecular assemblies on either side of the membrane. The characteristic biconcave shape of a red blood cell (**Figure 10–40**), for example, results from interactions of its plasma membrane proteins with an underlying *cytoskeleton*, which consists mainly of a meshwork of the filamentous protein **spectrin**. Spectrin is a long, thin, flexible rod about 100 nm in length. Being the principal component of the red cell cytoskeleton, it maintains the structural integrity and shape of the plasma membrane, which is the red cell's only membrane, as the cell has no nucleus or other organelles. The spectrin cytoskeleton is riveted to the membrane through various membrane proteins. The final result is a deformable, net-like meshwork that covers the entire cytosolic surface of the red cell membrane (**Figure 10–41**). This spectrin-based cytoskeleton enables the red cell to withstand the stress on its membrane as it is forced through narrow capillaries. Mice and humans with genetic abnormalities in spectrin are anemic and have red cells that are spherical (instead of concave) and fragile; the severity of the anemia increases with the degree of spectrin deficiency.

An analogous but much more elaborate and complicated cytoskeletal network exists beneath the plasma membrane of most other cells in our body. This network, which constitutes the cortical region (or **cortex**) of the cytoplasm, is rich in actin filaments, which are attached to the plasma membrane in numerous ways. The cortex of nucleated cells contains proteins that are structurally homologous to spectrin and the other components of the red cell cytoskeleton. We discuss the cortical cytoskeleton in nucleated cells and its interactions with the plasma membrane in Chapter 16.

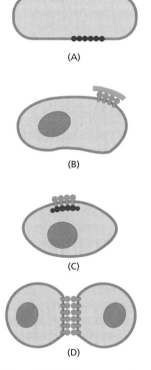

(A)

(B)

(C)

(D)

Figure 10–39 Four ways of restricting the lateral mobility of specific plasma membrane proteins. (A) The proteins can self-assemble into large aggregates (as seen for bacteriorhodopsin in the purple membrane of *Halobacterium*); they can be tethered by interactions with assemblies of macromolecules (B) outside or (C) inside the cell; or they can interact with proteins on the surface of another cell (D).

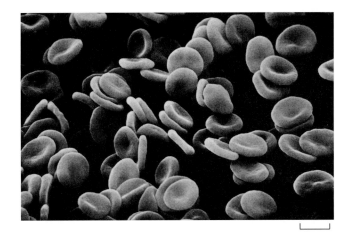

Figure 10–40 A scanning electron micrograph of human red blood cells. <GTAC> The cells have a biconcave shape and lack a nucleus and other organelles. (Courtesy of Bernadette Chailley.)

5 μm

The cortical cytoskeletal network underlying the plasma membrane restricts diffusion of not only the proteins that are directly anchored to it. Because the *cytoskeletal filaments* are often closely apposed to the cytosolic membrane surface, they can form mechanical barriers that obstruct the free diffusion of membrane proteins. These barriers partition the membrane into small domains, or *corrals* (**Figure 10–42**), which can be either permanent, as in the sperm (see Figure 10–38), or transient. The barriers can be detected when the diffusion of individual membrane proteins is followed by high-speed, single-particle tracking. The proteins diffuse rapidly but are confined within an individual corral; occasionally,

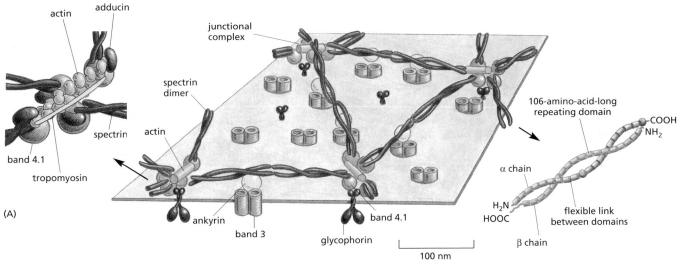

100 nm

Figure 10–41 The spectrin-based cytoskeleton on the cytosolic side of the human red blood cell plasma membrane. (A) The arrangement shown in the drawing has been deduced mainly from studies on the interactions of purified proteins *in vitro*. Spectrin dimers (enlarged in the box on the *right*) are linked together into a netlike meshwork by "junctional complexes" (enlarged in the box on the *left*). Each spectrin heterodimer consists of two antiparallel, loosely intertwined, flexible polypeptide chains called α and β. The two chains are attached noncovalently to each other at multiple points, including at both ends. Both the α and β chains are composed largely of repeating domains. The junctional complexes are composed of short actin filaments (containing 13 actin monomers), *band 4.1*, *adducin*, and a *tropomyosin* molecule that probably determines the length of the actin filaments. The cytoskeleton is linked to the membrane through two transmembrane proteins—a multipass protein called *band 3* and a single-pass protein called *glycophorin*. The spectrin tetramers bind to some band 3 proteins via *ankyrin* molecules, and to glycophorin and band 3 (not shown) via band 4.1 proteins. (B) The electron micrograph shows the cytoskeleton on the cytosolic side of a red blood cell membrane after fixation and negative staining. The spectrin meshwork has been purposely stretched out to allow the details of its structure to be seen. In a normal cell, the meshwork shown would be much more crowded and occupy only about one-tenth of this area. (B, courtesy of T. Byers and D. Branton, *Proc. Natl Acad. Sci. U.S.A.* 82:6153–6157, 1985. With permission from National Academy of Sciences.)

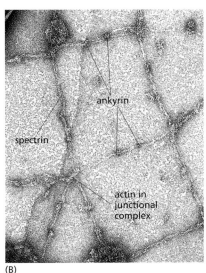

(B)

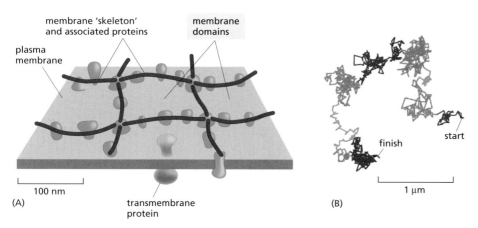

<image-description>membrane 'skeleton' and associated proteins; membrane domains; plasma membrane; transmembrane protein; 100 nm; (A); finish; start; 1 μm; (B)</image-description>

Figure 10–42 Corraling of membrane proteins by cortical cytoskeletal filaments. (A) How cytoskeletal filaments are thought to provide diffusion barriers that divide the membrane into small domains, or corrals. (B) High-speed, single-particle tracking was used to follow the paths of a fluorescently-labeled membrane protein over time. The trace shows that membrane proteins diffuse within a tightly delimited membrane domain (shown by different colors of the trace) and only infrequently escape into a neighboring domain. (Adapted from A. Kusumi et al., *Annu. Rev. Biophys. Biomol. Struct.* 34:351–378, 2005. With permission from Annual Reviews.)

however, thermal motions cause a few cortical filaments to detach transiently from the membrane, allowing the protein to escape into an adjacent corral.

The extent to which a transmembrane protein is confined within a corral depends on its association with other proteins and the size of its cytoplasmic domain; proteins with a large cytosolic domain will have a harder time passing through barriers. When a cell-surface receptor binds its extracellular signal molecules, for example, large protein complexes build up on the cytosolic domain of the receptor, making it more difficult for the receptor to escape from its corral. It is thought that corraling helps concentrate activated signaling complexes, increasing the speed and efficiency of the signaling process (discussed in Chapter 15).

Summary

Whereas the lipid bilayer determines the basic structure of biological membranes, proteins are responsible for most membrane functions, serving as specific receptors, enzymes, transport proteins, and so on. Many membrane proteins extend across the lipid bilayer. Some of these transmembrane proteins are single-pass proteins, in which the polypeptide chain crosses the bilayer as a single α helix. Others are multipass proteins, in which the polypeptide chain crosses the bilayer multiple times—either as a series of α helices or as a β sheet in the form of a closed barrel. All proteins responsible for the transmembrane transport of ions and other small water-soluble molecules are multipass proteins. Some membrane-associated proteins do not span the bilayer but instead are attached to either side of the membrane. Many of these are bound by noncovalent interactions with transmembrane proteins, but others are bound via covalently attached lipid groups. In the plasma membrane of all eucaryotic cells, most of the proteins exposed on the cell surface and some of the lipid molecules in the outer lipid monolayer have oligosaccharide chains covalently attached to them. Like the lipid molecules in the bilayer, many membrane proteins are able to diffuse rapidly in the plane of the membrane. However, cells have ways of immobilizing specific membrane proteins, as well as ways of confining both membrane protein and lipid molecules to particular domains in a continuous lipid bilayer.

PROBLEMS

Which statements are true? Explain why or why not.

10–1 Although lipid molecules are free to diffuse in the plane of the bilayer, they cannot flip-flop across the bilayer unless enzyme catalysts called phospholipid translocators are present in the membrane.

10–2 Whereas all the carbohydrate in the plasma membrane faces outward on the external surface of the cell, all the carbohydrate on internal membranes faces toward the cytosol.

10–3 Although membrane domains with different protein compositions are well known, there are at present no examples of membrane domains that differ in lipid composition.

Discuss the following problems.

10–4 When a lipid bilayer is torn, why does it not seal itself by forming a "hemi-micelle" cap at the edges, as shown in **Figure Q10–1**?

10–5 Margarine is made from vegetable oil by a chemical process. Do you suppose this process converts saturated fatty acids to unsaturated ones, or vice versa? Explain your answer.

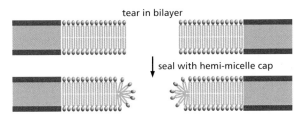

Figure Q10–1 A torn lipid bilayer sealed with a hypothetical "hemi-micelle" cap (Problem 10–4).

10–6 If a lipid raft is typically 70 nm in diameter and each lipid molecule has a diameter of 0.5 nm, about how many lipid molecules would there be in a lipid raft composed entirely of lipid? At a ratio of 50 lipid molecules per protein molecule (50% protein by mass) how many proteins would be in a typical raft? (Neglect the loss of lipid from the raft that would be required to accommodate the protein.)

10–7 A classic paper studied the behavior of lipids in the two monolayers of a membrane by labeling individual molecules with nitroxide groups, which are stable free radicals (**Figure Q10–2**). These spin-labeled lipids can be detected by electron spin-resonance (ESR) spectroscopy, a technique that does not harm living cells. Spin-labeled lipids are introduced into small lipid vesicles, which are then fused with cells, thereby transferring the labeled lipids into the plasma membrane.

The two spin-labeled phospholipids shown in Figure Q10–2 were incorporated into intact human red cell membranes in this way. To determine whether they were introduced equally into the two monolayers of the bilayer, ascorbic acid (vitamin C), which is a water-soluble reducing agent that does not cross membranes, was added to the medium to destroy any nitroxide radicals exposed on the outside of the cell. The ESR signal was followed as a function of time in the presence and absence of ascorbic acid as indicated in **Figure Q10–3**A and B.

A. Ignoring for the moment the difference in extent of loss of ESR signal, offer an explanation for why phospholipid 1 (Figure Q10–3A) reacts faster with ascorbate than does phospholipid 2 (Figure Q10–3B). Note that phospholipid 1 reaches a plateau in about 15 minutes, whereas phospholipid 2 takes almost an hour.

B. To investigate the difference in extent of loss of ESR signal with the two phospholipids, the experiments were repeated using red cell ghosts that had been resealed to make them impermeable to ascorbate (Figure Q10–3C and D). Resealed red cell ghosts are missing all of their cytoplasm but have an intact plasma membrane. In these experiments the loss of ESR signal for both phospholipids was negligible in the absence of ascorbate and reached a plateau at 50% in the presence of ascorbate. What

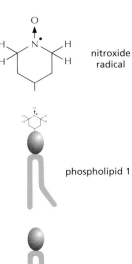

Figure Q10–2 Structures of two nitroxide-labeled lipids (Problem 10–7). The nitroxide radical is shown at the *top*, and its position of attachment to the phospholipids is shown *below*.

do you suppose might account for the difference in extent of loss of ESR signal in experiments with red cell ghosts (Figure Q10–3C and D) versus those with normal red cells (Figure Q10–3A and B).

C. Were the spin-labeled phospholipids introduced equally into the two monolayers of the red cell membrane?

10–8 Monomeric single-pass transmembrane proteins span a membrane with a single α helix that has characteristic chemical properties in the region of the bilayer. Which of the three 20-amino acid sequences listed below is the most likely candidate for such a transmembrane segment? Explain the reasons for your choice. (See back of book for one-letter amino acid code; FAMILY VW is a convenient mnemonic for hydrophobic amino acids.)

A. I T L I Y F G V M A G V I G T I L L I S
B. I T P I Y F G P M A G V I G T P L L I S
C. I T E I Y F G R M A G V I G T D L L I S

10–9 You are studying the binding of proteins to the cytoplasmic face of cultured neuroblastoma cells and have found a method that gives a good yield of inside-out vesicles from the plasma membrane. Unfortunately, your preparations are contaminated with variable amounts of right-side-out vesicles. Nothing you have tried avoids this problem. A friend suggests that you pass your vesicles over an affinity column made of lectin coupled to solid beads. What is the point of your friend's suggestion?

10–10 Glycophorin, a protein in the plasma membrane of the red blood cell, normally exists as a homodimer that is held together entirely by interactions between its transmembrane domains. Since transmembrane domains are hydrophobic, how is it that they can associate with one another so specifically?

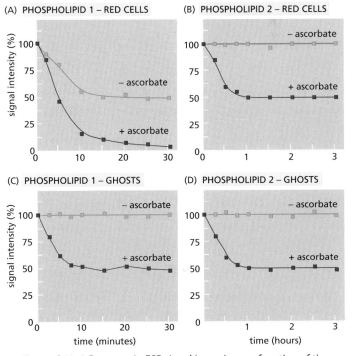

Figure Q10–3 Decrease in ESR signal intensity as a function of time in intact red cells and red cell ghosts in the presence and absence of ascorbate (Problem 10–7). (A and B) Phospholipid 1 and phospholipid 2 in intact red cells. (C and D) Phospholipid 1 and phospholipid 2 in red cell ghosts.

REFERENCES

General

Bretscher MS (1973) Membrane structure:some general principles. *Science* 181:622–629.

Edidin M (2003) Lipids on the frontier:a century of cell-membrane bilayers. *Nat Rev Mol Cell Biol* 4:414–418.

Jacobson K et al (1995) Revisiting the fluid mosaic model of membranes. *Science* 268:1441–1442.

Lipowsky R & Sackmann E (eds) (1995) The structure and dynamics of membranes. Amsterdam: Elsevier.

Singer SJ & Nicolson GL (1972) The fluid mosaic model of the structure of cell membranes. *Science* 175:720–731.

The Lipid Bilayer

Bevers EM, Comfurius P & Zwaal RF (1999) Lipid translocation across the plasma membrane of mammalian cells. *Biochim Biophys Acta* 1439:317–330.

Devaux PF (1993) Lipid transmembrane asymmetry and flip-flop in biological membranes and in lipid bilayers. *Curr Opin Struct Biol* 3:489–494.

Dowhan W (1997) Molecular basis for membrane phospholipid diversity:why are there so many lipids? *Annu Rev Biochem* 66:199–232.

Hakomori Si SI (2002) Inaugural Article:The glycosynapse. *Proc Natl Acad Sci USA* 99:225–232.

Harder T & Simons K (1997) Caveolae, DIGs, and the dynamics of sphingolipid-cholesterol microdomains. *Curr Opin Cell Biol* 9:534–542.

Hazel JR (1995) Thermal adaptation in biological membranes:is homeoviscous adaptation the explanation? *Annu Rev Physiol* 57:19–42.

Ichikawa S & Hirabayashi Y (1998) Glucosylceramide synthase and glycosphingolipid synthesis. *Trends Cell Biol* 8:198–202.

Kornberg RD & McConnell HM (1971) Lateral diffusion of phospholipids in a vesicle membrane. *Proc Natl Acad Sci USA* 68:2564–2568.

Mansilla MC, Cybulski LE & de Mendoza D (2004) Control of membrane lipid fluidity by molecular thermosensors. *J Bacteriol* 186:6681–6688.

McConnell HM & Radhakrishnan A (2003) Condensed complexes of cholesterol and phospholipids. *Biochim Biophys Acta* 1610:159–73.

Pomorski T & Menon AK (2006) Lipid flippases and their biological functions. *Cell Mol Life Sci* 63:2908–2921.

Rothman JE & Lenard J (1977) Membrane asymmetry. *Science* 195:743–53.

Simons K & Vaz WL (2004) Model systems, lipid rafts, and cell membranes. *Annu Rev Biophys Biomol Struct* 33:269–95.

Tanford C (1980) The Hydrophobic Effect: Formation of Micelles and Biological Membranes. New York: Wiley.

van Meer G (2005) Cellular lipidomics. *EMBO J* 24:3159–3165.

Membrane Proteins

Bennett V & Baines AJ (2001) Spectrin and ankyrin-based pathways: metazoan inventions for integrating cells into tissues. *Physiol Rev* 81:1353–1392.

Bijlmakers MJ & Marsh M (2003) The on-off story of protein palmitoylation. *Trends Cell Biol* 13:32–42.

Branden C & Tooze J (1999) Introduction to Protein Structure, 2nd ed. New York: Garland Science.

Bretscher MS & Raff MC (1975) Mammalian plasma membranes. *Nature* 258:43–49.

Buchanan SK (1999) Beta-barrel proteins from bacterial outer membranes:structure, function and refolding. *Curr Opin Struct Biol* 9:455–461.

Chen Y, Lagerholm BC & Jacobson K (2006) Methods to measure the lateral diffusion of membrane lipids and proteins. *Methods* 39:147–153.

Curranb AR & Engelman DM (2003) Sequence motifs, polar interactions and conformational changes in helical membrane proteins. *Curr Opin Struct Biol* 13:412

Deisenhofer J & Michel H (1991) Structures of bacterial photosynthetic reaction centers. *Annu Rev Cell Biol* 7:1–23.

Drickamer K & Taylor ME (1993) Biology of animal lectins. *Annu Rev Cell Biol* 9:237–64.

Drickamer K & Taylor ME (1998) Evolving views of protein glycosylation. *Trends Biochem Sci* 23:321–324.

Frye LD & Edidin M (1970) The rapid intermixing of cell surface antigens after formation of mouse-human heterokaryons. *J Cell Sci* 7:319–335.

Helenius A and Simons K (1975) Solubilization of membranes by detergents. *Biochim Biophys Acta* 415:29–79.

Henderson R & Unwin PN (1975) Three-dimensional model of purple membrane obtained by electron microscopy. *Nature* 257:28–32.

Kyte J & Doolittle RF (1982) A simple method for displaying the hydropathic character of a protein. *J Mol Biol* 157:105–132.

le Maire M, Champeil P et al (2000) Interaction of membrane proteins and lipids with solubilizing detergents. *Biochim Biophys Acta* 1508:86–111.

Lee AG (2003) Lipid-protein interactions in biological membranes:a structural perspective. *Biochim Biophys Acta* 1612:1–40.

Marchesi VT, Furthmayr H et al (1976) The red cell membrane. *Annu Rev Biochem* 45:667–698.

Nakada C, Ritchie K & Kusumi A (2003) Accumulation of anchored proteins forms membrane diffusion barriers during neuronal polarization. *Nat Cell Biol* 5:626–632.

Oesterhelt D (1998) The structure and mechanism of the family of retinal proteins from halophilic archaea. *Curr Opin Struct Biol* 8:489–500.

Reig N & van der Goot FG (2006) About lipids and toxins. *FEBS Lett* 580:5572–5579.

Reithmeier RAF (1993) The erythrocyte anion transporter (band 3) *Curr Opin Cell Biol* 7:707–714.

Rodgers W & Glaser M (1993) Distributions of proteins and lipids in the erythrocyte membrane. *Biochemistry* 32:12591–12598.

Sharon N & Lis H (2004) History of lectins:from hemagglutinins to biological recognition molecules. *Glycobiology* 14:53R–62R.

Sheetz MP (2001) Cell control by membrane-cytoskeleton adhesion. *Nat Rev Mol Cell Biol* 2:392–396.

Silvius JR (1992) Solubilization and functional reconstitution of biomembrane components. *Annu Rev Biophys Biomol Struct* 21:323–348.

Steck TL (1974) The organization of proteins in the human red blood cell membrane. A review. *J Cell Biol* 62:1–19.

Subramaniam S (1999) The structure of bacteriorhodopsin:an emerging consensus. *Curr Opin Struct Biol* 9:462–468.

Viel A & Branton D (1996) Spectrin:on the path from structure to function. *Curr Opin Cell Biol* 8:49–55.

Wallin E & von Heijne G (1998) Genome-wide analysis of integral membrane proteins from eubacterial, archaean, and eukaryotic organisms. *Protein Sci* 7:1029–1038.

White SH & Wimley WC (1999) Membrane protein folding and stability:physical principles. *Annu Rev Biophys Biomol Struct* 28:319–365.

Membrane Transport of Small Molecules and the Electrical Properties of Membranes

11

Because of its hydrophobic interior, the lipid bilayer of cell membranes prevents the passage of most polar molecules. This barrier function allows the cell to maintain concentrations of solutes in its cytosol that differ from those in the extracellular fluid and in each of the intracellular membrane-enclosed compartments. To benefit from this barrier, however, cells have had to evolve ways of transferring specific water-soluble molecules and ions across their membranes in order to ingest essential nutrients, excrete metabolic waste products, and regulate intracellular ion concentrations. Cells use specialized transmembrane proteins to transport inorganic ions and small water-soluble organic molecules across the lipid bilayer. Cells can also transfer macromolecules and even large particles across their membranes, but the mechanisms involved in most of these cases differ from those used for transferring small molecules, and they are discussed in Chapters 12 and 13. The importance of membrane transport is reflected in the large number of genes in all organisms that code for transport proteins, which make up 15–30% of the membrane proteins in all cells. Some specialized mammalian cells devote up to two-thirds of their total metabolic energy consumption to membrane transport processes.

We begin this chapter by describing some general principles of how small water-soluble molecules traverse cell membranes. We then consider, in turn, the two main classes of membrane proteins that mediate this traffic of molecules back and forth across lipid bilayers: *transporters*, which have moving parts to transport specific molecules across membranes, and *channels*, which form a narrow hydrophilic pore, allowing passive transmembrane movement, primarily of small inorganic ions. Transporters can be coupled to a source of energy to catalyze active transport, and a combination of selective passive permeability and active transport creates large differences in the composition of the cytosol compared with that of either the extracellular fluid (**Table 11–1**) or the fluid within membrane-enclosed organelles. By generating ionic concentration differences across the lipid bilayer, cell membranes can store potential energy in the form of electrochemical gradients, which drive various transport processes, convey electrical signals in electrically excitable cells, and (in mitochondria, chloroplasts, and bacteria) make most of the cell's ATP. We focus our discussion mainly on transport across the plasma membrane, but similar mechanisms operate across the other membranes of the eucaryotic cell, as discussed in later chapters.

In the last part of the chapter, we concentrate mainly on the functions of ion channels in neurons (nerve cells). In these cells, channel proteins perform at their highest level of sophistication, enabling networks of neurons to carry out all the human brain's astonishing feats.

PRINCIPLES OF MEMBRANE TRANSPORT

We begin this section by describing the permeability properties of protein-free, synthetic lipid bilayers. We then introduce some of the terms used to describe the various forms of membrane transport and some strategies for characterizing the proteins and processes involved.

Table 11–1 A Comparison of Ion Concentrations Inside and Outside a Typical Mammalian Cell

COMPONENT	INTRACELLULAR CONCENTRATION (mM)	EXTRACELLULAR CONCENTRATION (mM)
Cations		
Na^+	5–15	145
K^+	140	5
Mg^{2+}	0.5	1–2
Ca^{2+}	10^{-4}	1–2
H^+	7×10^{-5} ($10^{-7.2}$ M or pH 7.2)	4×10^{-5} ($10^{-7.4}$ M or pH 7.4)
Anions*		
Cl^-	5–15	110

*The cell must contain equal quantities of positive and negative charges (that is, it must be electrically neutral). Thus, in addition to Cl^-, the cell contains many other anions not listed in this table; in fact, most cell constituents are negatively charged (HCO_3^-, PO_4^{3-}, proteins, nucleic acids, metabolites carrying phosphate and carboxyl groups, etc.). The concentrations of Ca^{2+} and Mg^{2+} given are for the free ions. There is a total of about 20 mM Mg^{2+} and 1–2 mM Ca^{2+} in cells, but both are mostly bound to proteins and other substances and, for Ca^{2+}, stored within various organelles.

Protein-Free Lipid Bilayers Are Highly Impermeable to Ions

Given enough time, virtually any molecule will diffuse across a protein-free lipid bilayer down its concentration gradient. The rate of diffusion, however, varies enormously, depending partly on the size of the molecule but mostly on its relative solubility in oil. In general, the smaller the molecule and the more soluble it is in oil (the more hydrophobic, or nonpolar, it is), the more rapidly it will diffuse across a lipid bilayer. Small nonpolar molecules, such as O_2 and CO_2, readily dissolve in lipid bilayers and therefore diffuse rapidly across them. Small uncharged polar molecules, such as water or urea, also diffuse across a bilayer, albeit much more slowly (**Figure 11–1**). By contrast, lipid bilayers are highly impermeable to charged molecules (ions), no matter how small: the charge and high degree of hydration of such molecules prevents them from entering the hydrocarbon phase of the bilayer. Thus, synthetic lipid bilayers are 10^9 times more permeable to water than to even such small ions as Na^+ or K^+ (**Figure 11–2**).

There Are Two Main Classes of Membrane Transport Proteins: Transporters and Channels

Like synthetic lipid bilayers, cell membranes allow water and nonpolar molecules to permeate by simple diffusion. Cell membranes, however, also have to allow the passage of various polar molecules, such as ions, sugars, amino acids, nucleotides, and many cell metabolites that cross synthetic lipid bilayers only very slowly. Special **membrane transport proteins** transfer such solutes across cell membranes. These proteins occur in many forms and in all types of biological membranes. Each protein transports a particular class of molecule (such as ions, sugars, or amino acids) and often only certain molecular species of the class. Studies in the 1950s found that bacteria with a single-gene mutation were unable to transport sugars across their plasma membrane, thereby demonstrating the specificity of membrane transport proteins. We now know that humans with similar mutations suffer from various inherited diseases that hinder the transport of a specific solute in the kidney, intestine, or other cell type. Individuals with the inherited disease *cystinuria*, for example, cannot transport certain amino acids (including cystine, the disulfide-linked dimer of cysteine) from either the urine or the intestine into the blood; the resulting accumulation of cystine in the urine leads to the formation of cystine stones in the kidneys.

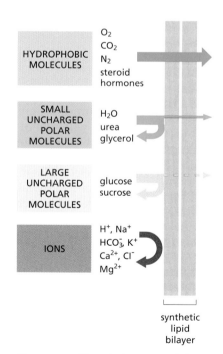

Figure 11–1 The relative permeability of a synthetic lipid bilayer to different classes of molecules. The smaller the molecule and, more importantly, the less strongly it associates with water, the more rapidly the molecule diffuses across the bilayer.

Figure 11–2 **Permeability coefficients for the passage of various molecules through synthetic lipid bilayers.** The rate of flow of a solute across the bilayer is directly proportional to the difference in its concentration on the two sides of the membrane. Multiplying this concentration difference (in mol/cm³) by the permeability coefficient (in cm/sec) gives the flow of solute in moles per second per square centimeter of bilayer. A concentration difference of tryptophan of 10^{-4} mol/cm³ ($10^{-4}/10^{-3}$ L = 0.1 M), for example, would cause a flow of 10^{-4} mol/cm³ $\times 10^{-7}$ cm/sec = 10^{-11} mol/sec through 1 cm² of bilayer, or 6×10^4 molecules/sec through 1 µm² of bilayer.

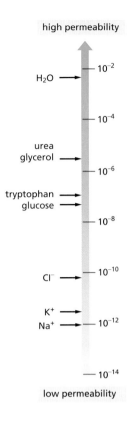

All membrane transport proteins that have been studied in detail have been found to be multipass transmembrane proteins—that is, their polypeptide chains traverse the lipid bilayer multiple times. By forming a continuous protein pathway across the membrane, these proteins enable specific hydrophilic solutes to cross the membrane without coming into direct contact with the hydrophobic interior of the lipid bilayer.

Transporters and channels are the two major classes of membrane transport proteins (**Figure 11–3**). **Transporters** (also called *carriers*, or *permeases)* bind the specific solute to be transported and undergo a series of conformational changes to transfer the bound solute across the membrane. **Channels**, in contrast, interact with the solute to be transported much more weakly. They form aqueous pores that extend across the lipid bilayer; when open, these pores allow specific solutes (usually inorganic ions of appropriate size and charge) to pass through them and thereby cross the membrane. Not surprisingly, transport through channels occurs at a much faster rate than transport mediated by transporters. Although water can diffuse across synthetic lipid bilayers, all cells contain specific channel proteins (called *water channels*, or *aquaporins*) that greatly increase the permeability of these membranes to water, as we discuss later.

Active Transport Is Mediated by Transporters Coupled to an Energy Source

All channels and many transporters allow solutes to cross the membrane only passively ("downhill"), a process called **passive transport**, or **facilitated diffusion**. In the case of transport of a single uncharged molecule, the difference in the concentration on the two sides of the membrane—its *concentration gradient*—drives passive transport and determines its direction (**Figure 11–4A**).

If the solute carries a net charge, however, both its concentration gradient and the electrical potential difference across the membrane, the *membrane potential*, influence its transport. The concentration gradient and the electrical gradient combine to form a net driving force, the **electrochemical gradient**, for each charged solute (Figure 11–4B). We discuss electrochemical gradients in more detail in Chapter 14. In fact, almost all plasma membranes have an electrical potential difference (voltage gradient) across them, with the inside usually negative with respect to the outside. This potential difference favors the entry of positively charged ions into the cell but opposes the entry of negatively charged ions.

Figure 11–3 **Transporters and channel proteins.** (A) A transporter alternates between two conformations, so that the solute-binding site is sequentially accessible on one side of the bilayer and then on the other. (B) In contrast, a channel protein forms a water-filled pore across the bilayer through which specific solutes can diffuse.

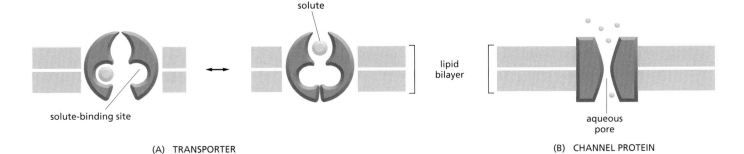

(A) TRANSPORTER

(B) CHANNEL PROTEIN

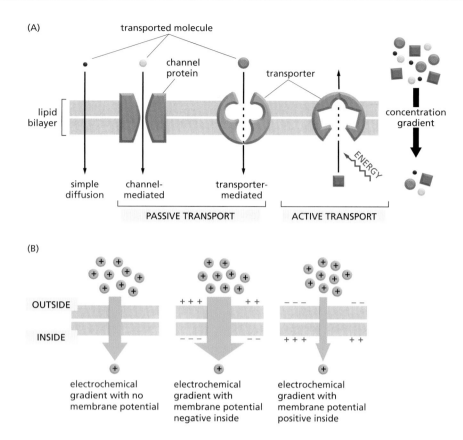

(A)

transported molecule

channel protein

transporter

lipid bilayer

concentration gradient

ENERGY

simple diffusion

channel-mediated

transporter-mediated

PASSIVE TRANSPORT

ACTIVE TRANSPORT

(B)

OUTSIDE

INSIDE

electrochemical gradient with no membrane potential

electrochemical gradient with membrane potential negative inside

electrochemical gradient with membrane potential positive inside

Figure 11–4 Passive and active transport compared. (A) Passive transport down an electrochemical gradient occurs spontaneously, either by simple diffusion through the lipid bilayer or by facilitated diffusion through channels and passive transporters. By contrast, active transport requires an input of metabolic energy and is always mediated by transporters that harvest metabolic energy to pump the solute against its electrochemical gradient. (B) An electrochemical gradient combines the membrane potential and the concentration gradient; they can work additively to increase the driving force on an ion across the membrane *(middle)* or can work against each other *(right)*.

Cells also require transport proteins that will actively pump certain solutes across the membrane against their electrochemical gradients ("uphill"); this process, known as **active transport**, is mediated by transporters, which are also called *pumps*. In active transport, the pumping activity of the transporter is directional because it is tightly coupled to a source of metabolic energy, such as ATP hydrolysis or an ion gradient, as discussed later. Thus, transmembrane movement of small molecules mediated by transporters can be either active or passive, whereas that mediated by channels is always passive.

Summary

Lipid bilayers are highly impermeable to most polar molecules. To transport small water-soluble molecules into or out of cells or intracellular membrane-enclosed compartments, cell membranes contain various membrane transport proteins, each of which is responsible for transferring a particular solute or class of solutes across the membrane. There are two classes of membrane transport proteins—transporters and channels. Both form continuous protein pathways across the lipid bilayer. Whereas transmembrane movement mediated by transporters can be either active or passive, solute flow through channel proteins is always passive.

TRANSPORTERS AND ACTIVE MEMBRANE TRANSPORT

The process by which a transporter transfers a solute molecule across the lipid bilayer resembles an enzyme–substrate reaction, and in many ways transporters behave like enzymes. In contrast to ordinary enzyme–substrate reactions, however, the transporter does not modify the transported solute but instead delivers it unchanged to the other side of the membrane.

Each type of transporter has one or more specific binding sites for its solute (substrate). It transfers the solute across the lipid bilayer by undergoing

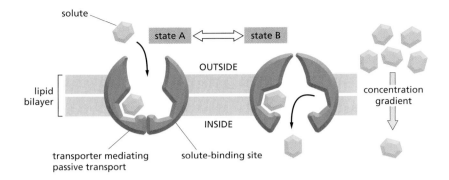

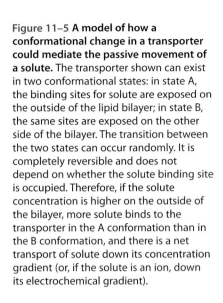

Figure 11–5 **A model of how a conformational change in a transporter could mediate the passive movement of a solute.** The transporter shown can exist in two conformational states: in state A, the binding sites for solute are exposed on the outside of the lipid bilayer; in state B, the same sites are exposed on the other side of the bilayer. The transition between the two states can occur randomly. It is completely reversible and does not depend on whether the solute binding site is occupied. Therefore, if the solute concentration is higher on the outside of the bilayer, more solute binds to the transporter in the A conformation than in the B conformation, and there is a net transport of solute down its concentration gradient (or, if the solute is an ion, down its electrochemical gradient).

reversible conformational changes that alternately expose the solute-binding site first on one side of the membrane and then on the other. **Figure 11–5** shows a schematic model of how a transporter operates. When the transporter is saturated (that is, when all solute-binding sites are occupied), the rate of transport is maximal. This rate, referred to as V_{max} (V for velocity), is characteristic of the specific carrier. V_{max} measures the rate with which the carrier can flip between its two conformational states. In addition, each transporter has a characteristic affinity for its solute, reflected in the K_m of the reaction, which is equal to the concentration of solute when the transport rate is half its maximum value (**Figure 11–6**). As with enzymes, the binding of solute can be blocked specifically by either competitive inhibitors (which compete for the same binding site and may or may not be transported) or noncompetitive inhibitors (which bind elsewhere and specifically alter the structure of the transporter).

As we discuss below, it requires only a relatively minor modification of the model shown in Figure 11–5 to link a transporter to a source of energy in order to pump a solute uphill against its electrochemical gradient. Cells carry out such active transport in three main ways (**Figure 11–7**):

1. *Coupled transporters* couple the uphill transport of one solute across the membrane to the downhill transport of another.
2. *ATP-driven pumps* couple uphill transport to the hydrolysis of ATP.
3. *Light-driven pumps*, which are found mainly in bacteria and archaea, couple uphill transport to an input of energy from light, as with bacteriorhodopsin (discussed in Chapter 10).

Amino acid sequence comparisons suggest that, in many cases, there are strong similarities in molecular design between transporters that mediate active transport and those that mediate passive transport. Some bacterial transporters, for example that use the energy stored in the H⁺ gradient across the plasma membrane to drive the active uptake of various sugars are structurally similar to the transporters that mediate passive glucose transport into most animal cells. This suggests an evolutionary relationship between various transporters. Given the importance of small metabolites and sugars as energy sources, it is not surprising that the superfamily of transporters is an ancient one.

We begin our discussion of active transport by considering transporters that are driven by ion gradients. These proteins have a crucial role in the transport of small metabolites across membranes in all cells. We then discuss ATP-driven pumps, including the Na⁺ pump that is found in the plasma membrane of almost all cells.

Figure 11–6 **The kinetics of simple diffusion and transporter-mediated diffusion.** Whereas the rate of simple diffusion is always proportional to the solute concentration, the rate of transporter-mediated diffusion reaches a maximum (V_{max}) when the transporter is saturated. The solute concentration when transport is at half its maximal value approximates the binding constant (K_m) of the transporter for the solute and is analogous to the K_m of an enzyme for its substrate. The graph applies to a transporter moving a single solute; the kinetics of coupled transport of two or more solutes is more complex.

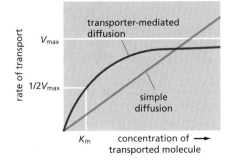

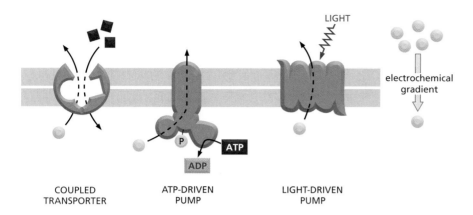

Figure 11–7 Three ways of driving active transport. The actively transported molecule is shown in *yellow,* and the energy source is shown in *red.*

Active Transport Can Be Driven by Ion Gradients

Some transporters simply mediate the movement of a single solute from one side of the membrane to the other at a rate determined by their V_{max} and K_m; they are called **uniporters**. Others function as *coupled transporters,* in which the transfer of one solute strictly depends on the transport of a second. Coupled transport involves either the simultaneous transfer of a second solute in the same direction, performed by **symporters** (also called *co-transporters*), or the transfer of a second solute in the opposite direction, performed by **antiporters** (also called *exchangers*) (**Figure 11–8**).

The tight coupling between the transfer of two solutes allows these coupled transporters to harvest the energy stored in the electrochemical gradient of one solute, typically an ion, to transport the other. In this way, the free energy released during the movement of an inorganic ion down an electrochemical gradient is used as the driving force to pump other solutes uphill, against their electrochemical gradient. This principle can work in either direction; some coupled transporters function as symporters, others as antiporters. In the plasma membrane of animal cells, Na^+ is the usual co-transported ion, the electrochemical gradient of which provides a large driving force for the active transport of a second molecule. The Na^+ that enters the cell during transport is subsequently pumped out by an ATP-driven Na^+ pump in the plasma membrane (as we discuss later), which, by maintaining the Na^+ gradient, indirectly drives the transport. (For this reason ion-driven carriers are said to mediate *secondary active transport,* whereas ATP-driven carriers are said to mediate *primary active transport.*)

Intestinal and kidney epithelial cells, for example, contain a variety of symporters that are driven by the Na^+ gradient across the plasma membrane. Each Na^+-driven symporter is specific for importing a small group of related sugars or amino acids into the cell, and the solute and Na^+ bind to different sites on the transporter. Because the Na^+ tends to move into the cell down its electrochemical gradient, the sugar or amino acid is, in a sense, "dragged" into the cell with it. The greater the electrochemical gradient for Na^+, the greater the rate of solute

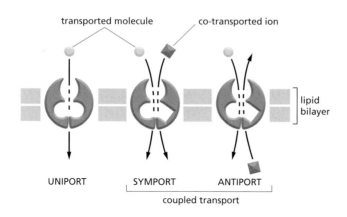

Figure 11–8 Three types of transporter-mediated movement. <ACCC> This schematic diagram shows transporters functioning as uniporters, symporters, and antiporters.

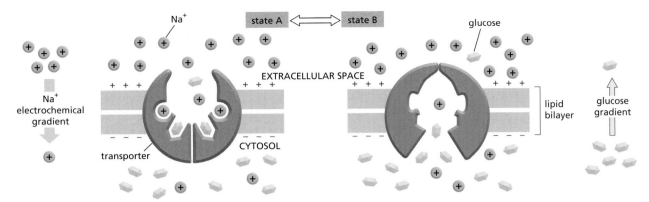

Figure 11–9 One way in which a glucose transporter can be driven by a Na$^+$ gradient. As in the model shown in Figure 11–5, the transporter oscillates between two alternate states, A and B. In the A state, the protein is open to the extracellular space; in the B state, it is open to the cytosol. Binding of Na$^+$ and glucose is cooperative—that is, the binding of either ligand induces a conformational change that increases the protein's affinity for the other ligand. Since the Na$^+$ concentration is much higher in the extracellular space than in the cytosol, glucose is more likely to bind to the transporter in the A state. Therefore, both Na$^+$ and glucose enter the cell (via an A → B transition) much more often than they leave it (via a B → A transition). The overall result is the net transport of both Na$^+$ and glucose into the cell. Note that, because the binding is cooperative, if one of the two solutes is missing, the other fails to bind to the transporter. Thus, the transporter undergoes a conformational switch between the two states only if both solutes or neither are bound.

entry; conversely, if the Na$^+$ concentration in the extracellular fluid is reduced, solute transport decreases (**Figure 11–9**).

In bacteria and yeasts, as well as in many membrane-enclosed organelles of animal cells, most active transport systems driven by ion gradients depend on H$^+$ rather than Na$^+$ gradients, reflecting the predominance of H$^+$ pumps and the virtual absence of Na$^+$ pumps in these membranes. The electrochemical H$^+$ gradient drives the active transport of many sugars and amino acids across the plasma membrane and into bacterial cells. One well-studied H$^+$-driven symporter is **lactose permease**, which transports lactose across the plasma membrane of *E. coli*. Structural and biophysical studies of the permease, as well as extensive analyses of mutant forms of the protein, have led to a detailed model of how the symporter works. The permease consists of 12 loosely packed transmembrane α helices. During the transport cycle, some of the helices undergo sliding motions that cause them to tilt. These motions alternately open and close a crevice between the helices, exposing the binding sites for lactose and H$^+$, first on one side of the membrane and then on the other (**Figure 11–10**).

Transporters in the Plasma Membrane Regulate Cytosolic pH

Most proteins operate optimally at a particular pH. Lysosomal enzymes, for example, function best at the low pH (~5) found in lysosomes, whereas cytosolic enzymes function best at the close to neutral pH (~7.2) found in the cytosol. It is therefore crucial that cells control the pH of their intracellular compartments.

Most cells have one or more types of Na$^+$-driven antiporters in their plasma membrane that help to maintain the cytosolic pH at about 7.2. These transporters use the energy stored in the Na$^+$ gradient to pump out excess H$^+$, which either leaks in or is produced in the cell by acid-forming reactions. Two mechanisms are used: either H$^+$ is directly transported out of the cell or HCO$_3^-$ is brought into the cell to neutralize H$^+$ in the cytosol (according to the reaction HCO$_3^-$ + H$^+$ → H$_2$O + CO$_2$). One of the antiporters that uses the first mechanism is a *Na$^+$–H$^+$ exchanger*, which couples an influx of Na$^+$ to an efflux of H$^+$. Another, which uses a combination of the two mechanisms, is a *Na$^+$-driven Cl$^-$–HCO$_3^-$ exchanger* that couples an influx of Na$^+$ and HCO$_3^-$ to an efflux of Cl$^-$ and H$^+$ (so

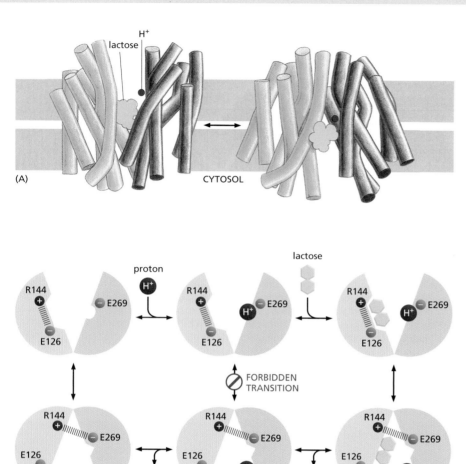

Figure 11–10 The molecular mechanism of the bacterial lactose permease suggested from its crystal structure. (A) The 12 transmembrane helices of the permease are clustered into two lobes, shown in two shades of *green*. The loops that connect the helices on either side of the membrane are omitted for clarity. During transport, the helices slide and tilt in the membrane, exposing binding sites for the disaccharide lactose *(yellow)* and H+ to either side of the membrane. (B) In one conformational state, the H+- and lactose-binding sites are accessible to the extracellular space *(top row)*; in the other, they are exposed to the cytosol *(bottom row)*. Loading the solutes on the extracellular side is favored because arginine (R) 144 forms a bond with glutamic acid (E) 126, leaving E269 free to accept H+. Unloading the solutes on the cytosolic side is favored because R144 forms a bond with E269, which destabilizes the bound H+. In addition, the lactose-binding site is partially disrupted due to the rearrangement of the helices. Because the transition between the two protonated states *(middle)* is forbidden, H+ can only be transported when a lactose is also transported. In this way, the electrochemical H+ gradient drives lactose import. (Adapted from J. Abramson et al., *Science* 301: 610–615, 2003. With permission from AAAS.)

that $NaHCO_3$ comes in and HCl goes out). The Na^+-driven Cl^-–HCO_3^- exchanger is twice as effective as the Na^+–H^+ exchanger: it pumps out one H^+ and neutralizes another for each Na^+ that enters the cell. If HCO_3^- is available, as is usually the case, this antiporter is the most important transporter regulating the cytosolic pH. The pH inside the cell regulates both exchangers; when the pH in the cytosol falls, both exchangers increase their activity.

A *Na^+-independent Cl^-–HCO_3^- exchanger* adjusts the cytosolic pH in the reverse direction. Like the Na^+-dependent transporters, pH regulates the Na^+-independent Cl^-–HCO_3^- exchanger, but the exchanger's activity increases as the cytosol becomes too alkaline. The movement of HCO_3^- in this case is normally out of the cell, down its electrochemical gradient, which decreases the pH of the cytosol. A Na^+-independent Cl^-–HCO_3^- exchanger in the membrane of red blood cells (called band 3 protein—see Figure 10–41) facilitates the quick discharge of CO_2 (as HCO_3^-) as the cells pass through capillaries in the lung.

The intracellular pH is not entirely regulated by these coupled transporters: ATP-driven H^+ pumps are also used to control the pH of many intracellular compartments. As discussed in Chapter 13, H^+ pumps maintain the low pH in lysosomes, as well as in endosomes and secretory vesicles. These H^+ pumps use the energy of ATP hydrolysis to pump H^+ into these organelles from the cytosol.

An Asymmetric Distribution of Transporters in Epithelial Cells Underlies the Transcellular Transport of Solutes

In epithelial cells, such as those that absorb nutrients from the gut, transporters are distributed nonuniformly in the plasma membrane and thereby contribute to the **transcellular transport** of absorbed solutes. By the action of

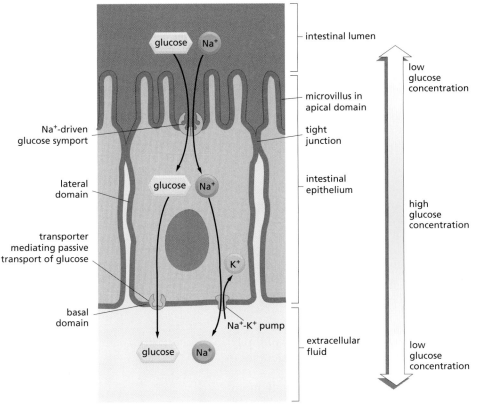

Figure 11–11 Transcellular transport. <GGAT> The transcellular transport of glucose across an intestinal epithelial cell depends on the nonuniform distribution of transporters in the cell's plasma membrane. The process shown here results in the transport of glucose from the intestinal lumen to the extracellular fluid (from where it passes into the blood). Glucose is pumped into the cell through the apical domain of the membrane by a Na^+-powered glucose symporter. Glucose passes out of the cell (down its concentration gradient) by passive movement through a different glucose transporter in the basal and lateral membrane domains. The Na^+ gradient driving the glucose symport is maintained by a Na^+ pump in the basal and lateral plasma membrane domains, which keeps the internal concentration of Na^+ low. Adjacent cells are connected by impermeable tight junctions, which have a dual function in the transport process illustrated: they prevent solutes from crossing the epithelium between cells, allowing a concentration gradient of glucose to be maintained across the cell sheet, and they also serve as diffusion barriers within the plasma membrane, which help confine the various transporters to their respective membrane domains (see Figure 10–37).

the transporters in these cells, solutes are moved across the epithelial cell layer into the extracellular fluid from where they pass into the blood. As shown in **Figure 11–11**, Na^+-linked symporters located in the apical (absorptive) domain of the plasma membrane actively transport nutrients into the cell, building up substantial concentration gradients for these solutes across the plasma membrane. Na^+-independent transport proteins in the basal and lateral (basolateral) domain allow the nutrients to leave the cell passively down these concentration gradients.

In many of these epithelial cells, the plasma membrane area is greatly increased by the formation of thousands of microvilli, which extend as thin, fingerlike projections from the apical surface of each cell. Such microvilli can increase the total absorptive area of a cell as much as 25-fold, thereby enhancing its transport capabilities.

As we have seen, ion gradients have a crucial role in driving many essential transport processes in cells. Ion pumps that use the energy of ATP hydrolysis establish and maintain these gradients, as we discuss next.

There Are Three Classes of ATP-Driven Pumps

ATP-driven pumps are often called *transport ATPases* because they hydrolyze ATP to ADP and phosphate and use the energy released to pump ions or other solutes across a membrane. There are three principal classes of ATP-driven pumps (**Figure 11–12**), and representatives of each are found in all procaryotic and eucaryotic cells.

1. **P-type pumps** are structurally and functionally related multipass transmembrane proteins. They are called "P-type" because they phosphorylate themselves during the pumping cycle. This class includes many of the ion pumps that are responsible for setting up and maintaining gradients of Na^+, K^+, H^+, and Ca^{2+} across cell membranes.

2. **F-type pumps** are turbine-like proteins, constructed from multiple different subunits. They differ structurally from P-type ATPases and are found in the plasma membrane of bacteria, the inner membrane of mitochondria,

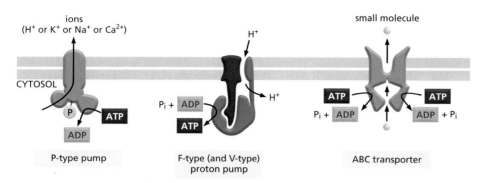

Figure 11–12 Three types of ATP-driven pumps. The different molecular designs of the pumps are cartooned here. Like any enzyme, pumps can work in reverse: when the electrochemical gradients of the solutes are reversed and the ATP/ADP ratio is low, they can synthesize ATP from ADP, as shown for the F-type ATPase, which normally works in this mode.

and the thylakoid membrane of chloroplasts. They are often called *ATP synthases* because they normally work in reverse: instead of using ATP hydrolysis to drive H$^+$ transport, they use the H$^+$ gradient across the membrane to drive the synthesis of ATP from ADP and phosphate. The H$^+$ gradient is generated either during the electron-transport steps of oxidative phosphorylation (in aerobic bacteria and mitochondria), during photosynthesis (in chloroplasts), or by the light-activated H$^+$ pump (bacteriorhodopsin) in *Halobacterium*. We discuss these proteins in detail in Chapter 14.

Structurally related to the F-type ATPases is a distinct family of *V-type ATPases* that normally pump H$^+$ rather than synthesize ATP. They pump H$^+$ into organelles, such as lysosomes, synaptic vesicles, and plant vacuoles to acidify the interior of these organelles (see Figure 13–36).

3. **ABC transporters** primarily pump small molecules across cell membranes, in contrast to P-type and the F- or V-type ATPases, which exclusively transport ions.

For the remainder of this section, we focus on P-type pumps and ABC transporters.

The Ca^{2+} Pump Is the Best-Understood P-type ATPase

Eucaryotic cells maintain very low concentrations of free Ca^{2+} in their cytosol (~10^{-7} M) in the face of a very much higher extracellular Ca^{2+} concentration (~10^{-3} M). Even a small influx of Ca^{2+} significantly increases the concentration of free Ca^{2+} in the cytosol, and the flow of Ca^{2+} down its steep concentration gradient in response to extracellular signals is one means of transmitting these signals rapidly across the plasma membrane (discussed in Chapter 15). It is important, therefore, that the cell maintain a steep Ca^{2+} gradient across its plasma membrane. Ca^{2+} transporters that actively pump Ca^{2+} out of the cell help maintain the gradient. One of these is a P-type Ca^{2+} ATPase; the other is an antiporter (called a *Na$^+$–Ca^{2+} exchanger*) that is driven by the Na$^+$ electrochemical gradient across the membrane (see Figure 15–41).

The best-understood P-type transport ATPase is the **Ca^{2+} pump**, or **Ca^{2+} ATPase**, in the *sarcoplasmic reticulum* (SR) membrane of skeletal muscle cells. The SR is a specialized type of endoplasmic reticulum that forms a network of tubular sacs in the muscle cell cytoplasm and serves as an intracellular store of Ca^{2+}. (When an action potential depolarizes the muscle cell plasma membrane, Ca^{2+} is released into the cytosol from the SR through *Ca^{2+}-release channels*, stimulating the muscle to contract, as discussed in Chapter 16.) The Ca^{2+} pump, which accounts for about 90% of the membrane protein of the SR, moves Ca^{2+} from the cytosol back into the SR. The endoplasmic reticulum of nonmuscle cells contains a similar Ca^{2+} pump, but in smaller quantities.

The three-dimensional structure of the SR Ca^{2+} pump has been determined by x-ray crystallography. This structure and the analysis of a related fungal H$^+$ pump have provided the first views of P-type transport ATPases, which are all thought to have similar structures. They contain 10 transmembrane α helices, three of which line a central channel that spans the lipid bilayer. In the unphosphorylated state, two helices are disrupted and form a cavity that binds two Ca^{2+}

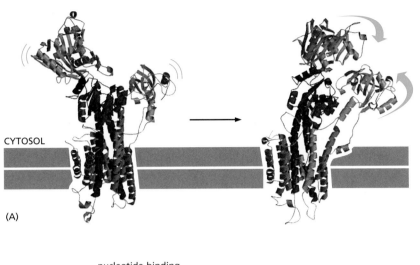

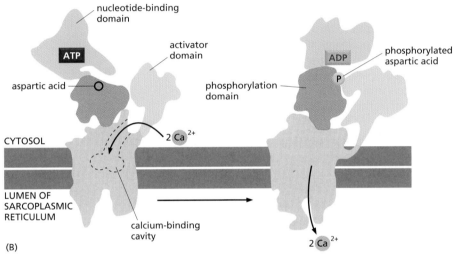

(A)

(B)

nucleotide-binding domain

activator domain

ATP

aspartic acid

CYTOSOL

phosphorylation domain

ADP

phosphorylated aspartic acid

P

2 Ca²⁺

LUMEN OF SARCOPLASMIC RETICULUM

calcium-binding cavity

2 Ca²⁺

Figure 11–13 **A model of how the sarcoplasmic reticulum Ca²⁺ pump moves Ca²⁺.** (A) The structures of the unphosphorylated, Ca²⁺-bound state *(left)* and the phosphorylated, Ca²⁺-free state *(right)* were determined by x-ray crystallography. (B) The model shows how ATP binding and hydrolysis cause drastic conformational changes, bringing the nucleotide-binding and phosphorylation domains into close proximity. This change is thought to cause a 90° rotation of the activator domain, which leads to a rearrangement of the transmembrane helices. The rearrangement of the helices disrupts the Ca²⁺-binding cavity and releases the Ca²⁺ into the lumen of the sarcoplasmic reticulum. (Adapted from C. Toyoshima et al., *Nature* 405:647–655, 2000. With permission from Macmillan Publishers Ltd.)

ions and is accessible from the cytosolic side of the membrane. The binding of ATP to a binding site on the same side of the membrane and the subsequent transfer of the terminal phosphate group of the ATP to an aspartic acid of an adjacent domain lead to a drastic rearrangement of the transmembrane helices. The rearrangement disrupts the Ca²⁺-binding site and releases the Ca²⁺ ions on the other side of the membrane, into the lumen of the SR (**Figure 11–13**). An essential characteristic of all P-type pumps is that the pump transiently phosphorylates itself during the pumping cycle.

The Plasma Membrane P-type Na⁺-K⁺ Pump Establishes the Na⁺ Gradient Across the Plasma Membrane

The concentration of K⁺ is typically 10–30 times higher inside cells than outside, whereas the reverse is true of Na⁺ (see Table 11–1, p. 652). A **Na⁺-K⁺ pump**, or **Na⁺ pump**, found in the plasma membrane of virtually all animal cells, maintains these concentration differences. The pump operates as an ATP-driven antiporter, actively pumping Na⁺ out of the cell against its steep electrochemical gradient and pumping K⁺ in (**Figure 11–14**). Because the pump hydrolyzes ATP to pump Na⁺ out and K⁺ in, it is also known as a **Na⁺-K⁺ ATPase**. The pump belongs to the family of P-type ATPases and functions very similarly to the Ca²⁺ pump (**Figure 11–15**).

We mentioned earlier that the Na⁺ gradient produced by the Na⁺-K⁺ pump drives the transport of most nutrients into animal cells and also has a crucial role in regulating cytosolic pH. A typical animal cell devotes almost one-third of its energy to fueling this pump, and the pump consumes even more energy in

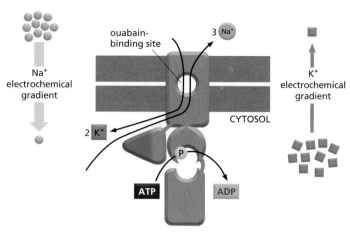

Figure 11–14 The Na⁺-K⁺ pump. <GAGT> This transporter actively pumps Na⁺ out of and K⁺ into a cell against their electrochemical gradients. For every molecule of ATP hydrolyzed inside the cell, three Na⁺ are pumped out and two K⁺ are pumped in. The specific inhibitor ouabain and K⁺ compete for the same site on the extracellular side of the pump.

electrically active nerve cells, which, as we shall see, repeatedly gain small amounts of Na⁺ and lose small amounts of K⁺ during the propagation of nerve impulses.

Like any enzyme, the Na⁺-K⁺ pump can be driven in reverse, in this case to produce ATP. When the Na⁺ and K⁺ gradients are experimentally increased to such an extent that the energy stored in their electrochemical gradients is greater than the chemical energy of ATP hydrolysis, these ions move down their electrochemical gradients and ATP is synthesized from ADP and phosphate by the Na⁺-K⁺ pump. Thus, the phosphorylated form of the pump (step 2 in Figure 11–15) can relax by either donating its phosphate to ADP (step 2 to step 1) or changing its conformation (step 2 to step 3). Whether the overall change in free energy is used to synthesize ATP or to pump Na⁺ out of the cell depends on the relative concentrations of ATP, ADP, and phosphate, as well as on the electro-chemical gradients for Na⁺ and K⁺.

Since the Na⁺-K⁺ pump drives three positively charged ions out of the cell for every two it pumps in, it is *electrogenic*. It drives a net current across the

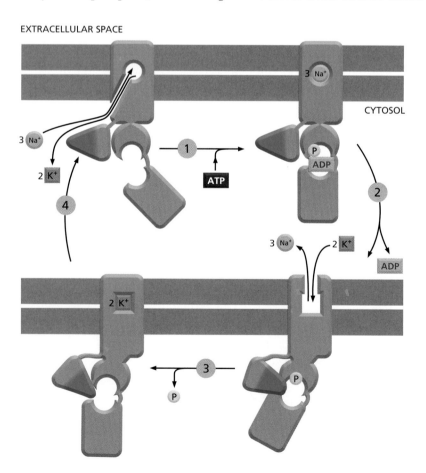

Figure 11–15 A model of the pumping cycle of the Na⁺-K⁺ pump. (1) The binding of intracellular Na⁺ and the subsequent phosphorylation by ATP of the cytoplasmic face of the pump induce a conformational change in the protein that (2) transfers the Na⁺ across the membrane and releases it on the outside of the cell. (3) Then, the binding of K⁺ on the extracellular surface and the subsequent dephosphorylation of the pump return the protein to its original conformation, which (4) transfers the K⁺ across the membrane and releases it into the cytosol. These changes in conformation are analogous to the A ↔ B transitions shown in Figure 11–5, except that here the Na⁺-dependent phosphorylation and the K⁺-dependent dephosphorylation of the protein cause the conformational transitions to occur in an orderly manner, enabling the protein to do useful work. Although for simplicity the diagram shows only one Na⁺- and one K⁺-binding site, in the real pump there are three Na⁺- and two K⁺-binding sites.

membrane, tending to create an electrical potential, with the cell's inside being negative relative to the outside. This electrogenic effect of the pump, however, seldom contributes more than 10% to the membrane potential. The remaining 90%, as we discuss later, depends only indirectly on the Na^+-K^+ pump.

On the other hand, the Na^+-K^+ pump does have a direct role in controlling the solute concentration inside the cell and thereby helps regulate **osmolarity** (or *tonicity*) of the cytosol. All cells contain specialized water channel proteins called *aquaporins* (discussed in detail on p. 673) in their plasma membrane to facilitate water flow across this membrane. Thus, water moves into or out of cells down its concentration gradient, a process called *osmosis*. As explained in **Panel 11–1**, cells contain a high concentration of solutes, including numerous negatively charged organic molecules that are confined inside the cell (the so-called *fixed anions*) and their accompanying cations that are required for charge balance. This tends to create a large osmotic gradient that tends to "pull" water into the cell. Animal cells counteract this effect by an opposite osmotic gradient due to a high concentration of inorganic ions—chiefly Na^+ and Cl^-—in the extracellular fluid. The Na^+-K^+ pump helps maintain osmotic balance by pumping out the Na^+ that leaks in down its steep electrochemical gradient. The Cl^- is kept out by the membrane potential.

In the special case of human red blood cells, which lack a nucleus and other organelles and have a plasma membrane that has an unusually high permeability to water, osmotic water movements can greatly influence cell volume, and the Na^+-K^+ pump plays an important part in maintaining red cell volume. If these cells are placed in a *hypotonic solution* (that is, a solution having a low solute concentration and therefore a high water concentration), there is net movement of water into the cells, causing them to swell and burst (lyse); conversely, if the cells are placed in a *hypertonic solution*, they shrink (**Figure 11–16**). The role of the Na^+-K^+ pump in controlling red cell volume is indicated by the observation that the cells swell, and may eventually burst, if they are treated with *ouabain*, which inhibits the Na^+-K^+ pump. For most animal cells, however, osmosis and the Na^+-K^+ pump have only minor roles in regulating cell volume. This is because most of the cytoplasm is in a gel-like state and resists large changes in its volume in response to changes in osmolarity.

Nonanimal cells cope with their osmotic problems in various ways. Plant cells and many bacteria are prevented from bursting by the semirigid cell wall that surrounds their plasma membrane. In amoebae, the excess water that flows in osmotically is collected in contractile vacuoles, which periodically discharge their contents to the exterior (see Panel 11–1). Bacteria have also evolved strategies that allow them to lose ions, and even macromolecules, quickly when subjected to an osmotic shock.

ABC Transporters Constitute the Largest Family of Membrane Transport Proteins

The last type of carrier protein that we discuss is the family of the **ABC transporters**, so named because each member contains two highly conserved ATPase domains or ATP-binding "cassettes" (**Figure 11–17**). ATP binding leads to dimerization of the two ATP-binding domains, and ATP hydrolysis leads to

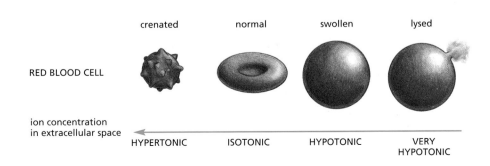

Figure 11–16 **Response of a human red blood cell to changes in osmolarity of the extracellular fluid.** <GTAC> The cell swells or shrinks as water moves into or out of the cell down its concentration gradient.

SOURCES OF INTRACELLULAR OSMOLARITY

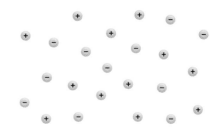

1 Macromolecules themselves contribute very little to the osmolarity of the cell interior since, despite their large size, each one counts only as a single molecule and there are relatively few of them compared to the number of small molecules in the cell. However, most biological macromolecules are highly charged, and they attract many inorganic ions of opposite charge. Because of their large numbers, these counterions make a major contribution to intracellular osmolarity.

2 As the result of active transport and metabolic processes, the cell contains a high concentration of small organic molecules, such as sugars, amino acids, and nucleotides, to which its plasma membrane is impermeable. Because most of these metabolites are charged, they also attract counterions. Both the small metabolites and their counterions make a further major contribution to intracellular osmolarity.

3 The osmolarity of the extracellular fluid is usually due mainly to small inorganic ions. These leak slowly across the plasma membrane into the cell. If they were not pumped out, and if there were no other molecules inside the cell that interacted with them so as to influence their distribution, they would eventually come to equilibrium with equal concentrations inside and outside the cell. However, the presence of charged macromolecules and metabolites in the cell that attract these ions gives rise to the Donnan effect: it causes the total concentration of inorganic ions (and therefore their contribution to the osmolarity) to be greater inside than outside the cell at equilibrium.

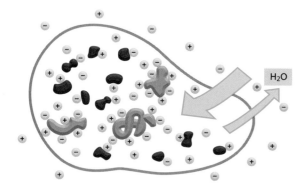

THE PROBLEM

Because of the above factors, a cell that does nothing to control its osmolarity will have a higher concentration of solutes inside than outside. As a result, water will be higher in concentration outside the cell than inside. This difference in water concentration across the plasma membrane will cause water to move continuously into the cell by osmosis.

THE SOLUTION

Animal cells and bacteria control their intracellular osmolarity by actively pumping out inorganic ions, such as Na⁺, so that their cytoplasm contains a lower total concentration of inorganic ions than the extracellular fluid, thereby compensating for their excess of organic solutes.

Plant cells are prevented from swelling by their rigid walls and so can tolerate an osmotic difference across their plasma membranes: an internal turgor pressure is built up, which at equilibrium forces out as much water as enters.

Many protozoa avoid becoming swollen with water, despite an osmotic difference across the plasma membrane, by periodically extruding water from special contractile vacuoles.

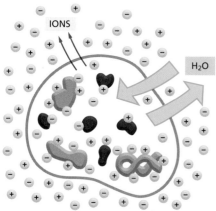

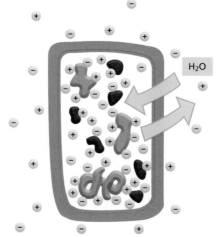

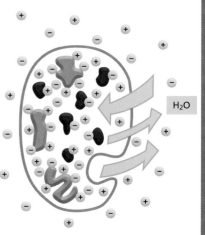

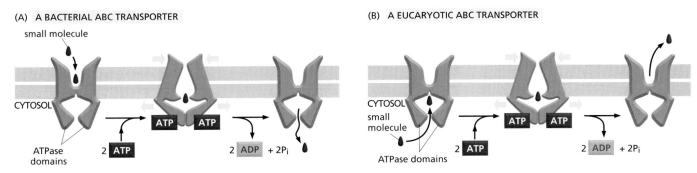

Figure 11–17 **Typical ABC transporters in procaryotes (A) and eucaryotes (B).** Transporters consist of multiple domains: typically, two hydrophobic domains, each built of six membrane-spanning segments that form the translocation pathway and provide substrate specificity, and two ATPase domains (also called ATP-binding cassettes) protruding into the cytosol. In some cases, the two halves of the transporter are formed by a single polypeptide, whereas in other cases they are formed by two or more separate polypeptides that assemble into a similar structure (see Figure 10–24). Without ATP bound, the transporter exposes a substrate-binding site to either the extracellular space (in procaryotes) or the intracellular space (in eucaryotes or procaryotes). ATP binding induces a conformational change that exposes the substrate-binding pocket to the opposite face; ATP hydrolysis followed by ADP dissociation returns the transporter to its original conformation. Most individual ABC transporters are unidirectional. Both importing and exporting ABC transporters are found in bacteria, but in eucaryotes almost all ABC transporters export substances from the cytosol—either to the extracellular space or to a membrane-bound intracellular compartment such as the ER or the mitochondria.

their dissociation. These structural changes in the cytosolic domains are thought to be transmitted to the transmembrane segments, driving cycles of conformational changes that alternately expose substrate-binding sites on one side of the membrane and then on the other. In this way, ABC transporters use ATP binding and hydrolysis to transport small molecules across the bilayer.

ABC transporters constitute the largest family of membrane transport proteins and are of great clinical importance. The first of these proteins to be characterized was found in bacteria. We have already mentioned that the plasma membranes of all bacteria contain transporters that use the H⁺ gradient across the membrane to pump a variety of nutrients into the cell. Bacteria also have transport ATPases that use the energy of ATP hydrolysis to import certain small molecules. In bacteria such as *E. coli*, which have double membranes (**Figure 11–18**), the transport ATPases are located in the inner membrane, and an auxiliary mechanism operates to capture the nutrients and deliver them to the transporters (**Figure 11–19**).

In *E. coli*, 78 genes (an amazing 5% of the bacterium's genes) encode ABC transporters, and animal genomes encode more. Although each transporter is thought to be specific for a particular molecule or class of molecules, the variety of substrates transported by this superfamily is great and includes inorganic ions, amino acids, mono- and polysaccharides, peptides, and even proteins. Whereas bacterial ABC transporters are used for both import and export, those identified in eucaryotes seem mostly specialized for export.

Indeed, the first eucaryotic ABC transporters identified were discovered because of their ability to pump hydrophobic drugs out of the cytosol. One of these transporters is the **multidrug resistance (MDR) protein**, the overexpression

Figure 11–18 **A small section of the double membrane of an *E. coli* bacterium.** The inner membrane is the cell's plasma membrane. Between the inner and outer lipid bilayer membranes is a highly porous, rigid peptidoglycan layer, composed of protein and polysaccharide, that constitutes the bacterial cell wall. It is attached to lipoprotein molecules in the outer membrane and fills the *periplasmic space* (only a little of the peptidoglycan layer is shown). This space also contains a variety of soluble protein molecules. The dashed threads (shown in *green*) at the top represent the polysaccharide chains of the special lipopolysaccharide molecules that form the external monolayer of the outer membrane; for clarity, only a few of these chains are shown. Bacteria with double membranes are called *Gram-negative* because they do not retain the dark blue dye used in Gram staining. Bacteria with single membranes (but thicker cell walls), such as staphylococci and streptococci, retain the blue dye and are therefore called *Gram-positive*; their single membrane is analogous to the inner (plasma) membrane of Gram-negative bacteria.

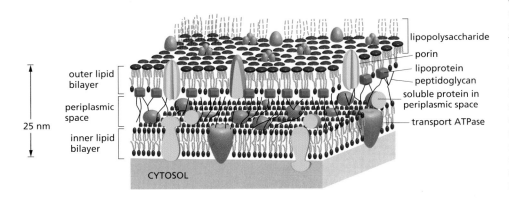

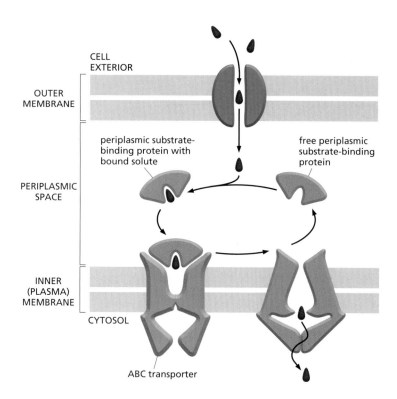

CELL EXTERIOR

OUTER MEMBRANE

periplasmic substrate-binding protein with bound solute

free periplasmic substrate-binding protein

PERIPLASMIC SPACE

INNER (PLASMA) MEMBRANE

CYTOSOL

ABC transporter

Figure 11–19 The auxiliary transport system associated with transport ATPases in bacteria with double membranes. The solute diffuses through channel-forming proteins (porins) in the outer membrane and binds to a *periplasmic substrate-binding protein,* which undergoes a conformational change that enables it to bind to an ABC transporter in the plasma membrane. The ABC transporter then picks up the solute and actively transfers it across the plasma membrane in a reaction driven by ATP hydrolysis. The peptidoglycan is omitted for simplicity; its porous structure allows the substrate-binding proteins and water-soluble solutes to move through it by simple diffusion.

of which in human cancer cells can make the cells simultaneously resistant to a variety of chemically unrelated cytotoxic drugs that are widely used in cancer chemotherapy. Treatment with any one of these drugs can result in the selective survival and overgrowth of those cancer cells that express more of the MDR transporter. These cells can pump drugs out of the cell very efficiently and are therefore relatively resistant to the toxic effects of the anticancer drugs. Selection for cancer cells with resistance to one drug can thereby lead to resistance to a wide variety of anti-cancer drugs. Some studies indicate that up to 40% of human cancers develop multidrug resistance, making it a major hurdle in the battle against cancer.

A related and equally sinister phenomenon occurs in the protist *Plasmodium falciparum*, which causes malaria. More than 200 million people are infected worldwide with this parasite, which remains a major cause of human death, killing more than a million people every year. The development of resistance to the antimalarial drug *chloroquine* hampers the control of malaria. The resistant *P. falciparum* have amplified a gene encoding an ABC transporter that pumps out the chloroquine.

In most vertebrate cells, an ABC transporter in the endoplasmic reticulum (ER) membrane actively transports a wide variety of peptides from the cytosol into the ER lumen. These peptides are produced by protein degradation in proteasomes (discussed in Chapter 6). They are carried from the ER to the cell surface, where they are displayed for scrutiny by cytotoxic T lymphocytes, which will kill the cell if the peptides are derived from a virus or other microorganisms lurking in the cytosol of an infected cell (discussed in Chapter 25).

Yet another member of the ABC transporter family is the *cystic fibrosis transmembrane conductance regulator* protein (CFTR), which was discovered through studies of the common genetic disease *cystic fibrosis*. This disease is caused by a mutation in the gene encoding CFTR, which functions as a Cl^- channel in the plasma membrane of epithelial cells. CFTR regulates ion concentrations in the extracellular fluid, especially in the lung. One in 27 Caucasians carries a gene encoding a mutant form of this protein; in 1 in 2900, both copies of the gene are mutated, causing the disease. In contrast to other ABC transporters, ATP binding and hydrolysis do not drive the transport process. Instead, they control the opening and closing of the Cl^- channel, which provides a passive conduit for Cl^- to move down its electrochemical gradient. Thus, ABC proteins can apparently function as either transporters or channels.

Summary

Transporters bind specific solutes and transfer them across the lipid bilayer by undergoing conformational changes that expose the solute-binding site sequentially on one side of the membrane and then on the other. Some transporters simply move a single solute "downhill," whereas others can act as pumps to move a solute "uphill" against its electrochemical gradient, using energy provided by ATP hydrolysis, by a downhill flow of another solute (such as Na^+ or H^+), or by light to drive the requisite series of conformational changes in an orderly manner. Transporters belong to a small number of protein families. Each family contains proteins of similar amino acid sequences that are thought to have evolved from a common ancestral protein and to operate by a similar mechanism. The family of P-type transport ATPases, which includes Ca^{2+} and Na^+-K^+ pumps, is an important example; each of these ATPases sequentially phosphorylates and dephosphorylates itself during the pumping cycle. The superfamily of ABC transporters is the largest family of membrane transport proteins and is especially important clinically. It includes proteins that are responsible for cystic fibrosis, as well as for drug resistance in both cancer cells and malaria-causing parasites.

ION CHANNELS AND THE ELECTRICAL PROPERTIES OF MEMBRANES

Unlike carrier proteins, channel proteins form hydrophilic pores across membranes. One class of channel proteins found in virtually all animals forms *gap junctions* between two adjacent cells; each plasma membrane contributes equally to the formation of the channel, which connects the cytoplasm of the two cells. These channels are discussed in Chapter 19 and will not be considered further here. Both gap junctions and *porins*, the channel-forming proteins of the outer membranes of bacteria, mitochondria, and chloroplasts (discussed in Chapter 10), have relatively large and permissive pores, which would be disastrous if they directly connected the inside of a cell to an extracellular space. Indeed, many bacterial toxins do exactly that to kill other cells (discussed in Chapter 24).

In contrast, most channel proteins in the plasma membrane of animal and plant cells that connect the cytosol to the cell exterior necessarily have narrow, highly selective pores that can open and close rapidly. Because these proteins are concerned specifically with inorganic ion transport, they are referred to as **ion channels**. For transport efficiency, ion channels have an advantage over carriers in that up to 100 million ions can pass through one open channel each second—a rate 10^5 times greater than the fastest rate of transport mediated by any known carrier protein. However, channels cannot be coupled to an energy source to perform active transport, so the transport that they mediate is always passive (downhill). Thus, the function of ion channels is to allow specific inorganic ions—primarily Na^+, K^+, Ca^{2+}, or Cl^-—to diffuse rapidly down their electrochemical gradients across the lipid bilayer. As we shall see, the ability to control ion fluxes through these channels is essential for many cell functions. Nerve cells (neurons), in particular, have made a specialty of using ion channels, and we shall consider how they use many different ion channels to receive, conduct, and transmit signals.

Ion Channels Are Ion-Selective and Fluctuate Between Open and Closed States

Two important properties distinguish ion channels from simple aqueous pores. First, they show *ion selectivity*, permitting some inorganic ions to pass, but not others. This suggests that their pores must be narrow enough in places to force permeating ions into intimate contact with the walls of the channel so that only ions of appropriate size and charge can pass. The permeating ions have to shed most or all of their associated water molecules to pass, often in single file,

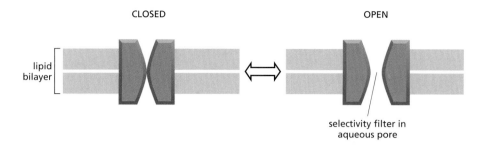

CLOSED   OPEN

lipid bilayer

selectivity filter in aqueous pore

Figure 11–20 A typical ion channel, which fluctuates between closed and open conformations. The channel protein shown here in cross section forms a hydrophilic pore across the lipid bilayer only in the "open" conformational state. Polar groups are thought to line the wall of the pore, while hydrophobic amino acid side chains interact with the lipid bilayer (not shown). The pore narrows to atomic dimensions in one region (the selectivity filter), where the ion selectivity of the channel is largely determined.

through the narrowest part of the channel, which is called the *selectivity filter;* this limits their rate of passage (**Figure 11–20**). Thus, as the ion concentration increases, the flux of the ion through a channel increases proportionally but then levels off (saturates) at a maximum rate.

The second important distinction between ion channels and simple aqueous pores is that ion channels are not continuously open. Instead, they are *gated,* which allows them to open briefly and then close again (**Figure 11–21**). Moreover, with prolonged (chemical or electrical) stimulation, most channels go into a closed "desensitized" or "inactivated" state, in which they are refractory to further opening until the stimulus has been removed, as we discuss later. In most cases, the gate opens in response to a specific stimulus. The main types of stimuli that are known to cause ion channels to open are a change in the voltage across the membrane *(voltage-gated channels)*, a mechanical stress *(mechanically gated channels)*, or the binding of a ligand *(ligand-gated channels)*. The ligand can be either an extracellular mediator—specifically, a neurotransmitter *(transmitter-gated channels)*—or an intracellular mediator such as an ion *(ion-gated channels)* or a nucleotide *(nucleotide-gated channels)*. In addition, protein phosphorylation and dephosphorylation regulates the activity of many ion channels; this type of channel regulation is discussed, together with nucleotide-gated ion channels, in Chapter 15.

More than 100 types of ion channels have been described thus far, and new ones are still being discovered, each characterized by the ions it conducts, the mechanism by which it is gated, and its abundance and localization in the cell. Ion channels are responsible for the electrical excitability of muscle cells, and they mediate most forms of electrical signaling in the nervous system. A single neuron might typically contain 10 or more kinds of ion channels, located in different domains of its plasma membrane. But ion channels are not restricted to electrically excitable cells. They are present in all animal cells and are found in plant cells and microorganisms: they propagate the leaf-closing response of the mimosa plant, for example, and allow the single-celled *Paramecium* to reverse direction after a collision.

Perhaps the most common ion channels are those that are permeable mainly to K⁺. These channels are found in the plasma membrane of almost all

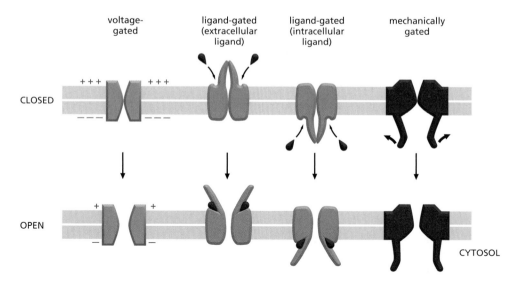

voltage-gated   ligand-gated (extracellular ligand)   ligand-gated (intracellular ligand)   mechanically gated

CLOSED

OPEN

CYTOSOL

Figure 11–21 The gating of ion channels. This drawing shows different kinds of stimuli that open ion channels. Mechanically gated channels often have cytoplasmic extensions that link the channel to the cytoskeleton (not shown).

animal cells. An important subset of K^+ channels opens even in an unstimulated or "resting" cell, and hence these channels are sometimes called **K^+ leak channels**. Although this term applies to many different K^+ channels, depending on the cell type, they serve a common purpose. By making the plasma membrane much more permeable to K^+ than to other ions, they have a crucial role in maintaining the membrane potential across all plasma membranes.

The Membrane Potential in Animal Cells Depends Mainly on K^+ Leak Channels and the K^+ Gradient Across the Plasma Membrane

A **membrane potential** arises when there is a difference in the electrical charge on the two sides of a membrane, due to a slight excess of positive ions over negative ones on one side and a slight deficit on the other. Such charge differences can result both from active electrogenic pumping (see p. 662) and from passive ion diffusion. As we discuss in Chapter 14, electrogenic H^+ pumps in the mitochondrial inner membrane generate most of the membrane potential across this membrane. Electrogenic pumps also generate most of the electrical potential across the plasma membrane in plants and fungi. In typical animal cells, however, passive ion movements make the largest contribution to the electrical potential across the plasma membrane.

As explained earlier, the Na^+-K^+ pump helps maintain an osmotic balance across the animal cell membrane by keeping the intracellular concentration of Na^+ low. Because there is little Na^+ inside the cell, other cations have to be plentiful there to balance the charge carried by the cell's fixed anions—the negatively charged organic molecules that are confined inside the cell. This balancing role is performed largely by K^+, which is actively pumped into the cell by the Na^+-K^+ pump and can also move freely in or out through the *K^+ leak channels* in the plasma membrane. Because of the presence of these channels, K^+ comes almost to equilibrium, where an electrical force exerted by an excess of negative charges attracting K^+ into the cell balances the tendency of K^+ to leak out down its concentration gradient. The membrane potential is the manifestation of this electrical force, and we can calculate its equilibrium value from the steepness of the K^+ concentration gradient. The following argument may help to make this clear.

Suppose that initially there is no voltage gradient across the plasma membrane (the membrane potential is zero) but the concentration of K^+ is high inside the cell and low outside. K^+ will tend to leave the cell through the K^+ leak channels, driven by its concentration gradient. As K^+ begins to move out, each ion leaves behind an unbalanced negative charge, thereby creating an electrical field, or membrane potential, which will tend to oppose the further efflux of K^+. The net efflux of K^+ halts when the membrane potential reaches a value at which this electrical driving force on K^+ exactly balances the effect of its concentration gradient—that is, when the electrochemical gradient for K^+ is zero. Although Cl^- ions also equilibrate across the membrane, the membrane potential keeps most of these ions out of the cell because their charge is negative.

The equilibrium condition, in which there is no net flow of ions across the plasma membrane, defines the **resting membrane potential** for this idealized cell. A simple but very important formula, the **Nernst equation**, quantifies the equilibrium condition and, as explained in **Panel 11–2**, makes it possible to calculate the theoretical resting membrane potential if we know the ratio of internal and external ion concentrations. As the plasma membrane of a real cell is not exclusively permeable to K^+ and Cl^-, however, the actual resting membrane potential is usually not exactly equal to that predicted by the Nernst equation for K^+ or Cl^-.

The Resting Potential Decays Only Slowly When the Na^+-K^+ Pump Is Stopped

Only a minute number of ions must move across the plasma membrane to set up the membrane potential. Thus, we can think of the membrane potential as arising from movements of charge that leave ion *concentrations* practically unaffected

THE NERNST EQUATION AND ION FLOW

The flow of any ion through a membrane channel protein is driven by the electrochemical gradient for that ion. This gradient represents the combination of two influences: the voltage gradient and the concentration gradient of the ion across the membrane. When these two influences just balance each other the electrochemical gradient for the ion is zero and there is no *net* flow of the ion through the channel. The voltage gradient (membrane potential) at which this equilibrium is reached is called the equilibrium potential for the ion. It can be calculated from an equation that will be derived below, called the Nernst equation.

The Nernst equation is

$$V = \frac{RT}{zF} \ln \frac{C_o}{C_i}$$

where

V = the equilibrium potential in volts (internal potential minus external potential)

C_o and C_i = outside and inside concentrations of the ion, respectively

R = the gas constant (2 cal mol^{-1} K^{-1})

T = the absolute temperature (K)

F = Faraday's constant (2.3×10^4 cal V^{-1} mol^{-1})

z = the valence (charge) of the ion

$\ln$ = logarithm to the base e

The Nernst equation is derived as follows:

A molecule in solution (a solute) tends to move from a region of high concentration to a region of low concentration simply due to the random movement of molecules, which results in their equilibrium. Consequently, movement down a concentration gradient is accompanied by a favorable free-energy change ($\Delta G < 0$), whereas movement up a concentration gradient is accompanied by an unfavorable free-energy change ($\Delta G > 0$). (Free energy is introduced in Chapter 2, and discussed in the context of redox reactions in Panel 14–1, p. 830.)

The free-energy change per mole of solute moved across the plasma membrane (ΔG_{conc}) is equal to $-RT \ln C_o / C_i$.

If the solute is an ion, moving it into a cell across a membrane whose inside is at a voltage V relative to the outside will cause an additional free-energy change (per mole of solute moved) of $\Delta G_{volt} = zFV$.

At the point where the concentration and voltage gradients just balance,

$$\Delta G_{conc} + \Delta G_{volt} = 0$$

and the ion distribution is at equilibrium across the membrane.

Thus,

$$zFV - RT \ln \frac{C_o}{C_i} = 0$$

and, therefore,

$$V = \frac{RT}{zF} \ln \frac{C_o}{C_i}$$

or, using the constant that converts natural logarithms to base 10,

$$V = 2.3 \frac{RT}{zF} \log_{10} \frac{C_o}{C_i}$$

For a univalent ion,

$$2.3 \frac{RT}{F} = 58 \text{ mV at } 20°C \quad \text{and} \quad 61.5 \text{ mV at } 37°C$$

Thus, for such an ion at 37°C,

$$V = +61.5 \text{ mV for } C_o / C_i = 10,$$

whereas

$$V = 0 \text{ for } C_o / C_i = 1.$$

The K$^+$ equilibrium potential (V_K), for example, is

$$61.5 \log_{10}([K^+]_o / [K^+]_i) \text{ millivolts}$$
$$(-89 \text{ mV for a typical cell where } [K^+]_o = 5 \text{ mM}$$
$$\text{and } [K^+]_i = 140 \text{ mM}).$$

At V_K, there is no net flow of K$^+$ across the membrane.

Similarly, when the membrane potential has a value of

$$61.5 \log_{10}([Na^+]_o / [Na^+]_i),$$
the Na$^+$ equilibrium potential (V_{Na}),

there is no net flow of Na$^+$.

For any particular membrane potential, V_M, the net force tending to drive a particular type of ion out of the cell, is proportional to the difference between V_M and the equilibrium potential for the ion: hence,

$$\text{for K}^+ \text{ it is } V_M - V_K$$
$$\text{and for Na}^+ \text{ it is } V_M - V_{Na}.$$

The number of ions that go to form the layer of charge adjacent to the membrane is minute compared with the total number inside the cell. For example, the movement of 6000 Na$^+$ ions across 1 μm^2 of membrane will carry sufficient charge to shift the membrane potential by about 100 mV.

Because there are about 3×10^7 Na$^+$ ions in a typical cell (1 μm^3 of bulk cytoplasm), such a movement of charge will generally have a negligible effect on the ion concentration gradients across the membrane.

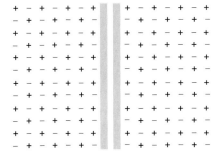

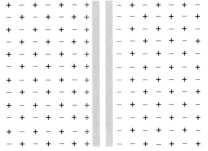

exact balance of charges on each side of the membrane; membrane potential = 0

a few of the positive ions *(red)* cross the membrane from right to left, leaving their negative counterions *(red)* behind; this sets up a nonzero membrane potential

Figure 11–22 The ionic basis of a membrane potential. A small flow of ions carries sufficient charge to cause a large change in the membrane potential. The ions that give rise to the membrane potential lie in a thin (< 1 nm) surface layer close to the membrane, held there by their electrical attraction to their oppositely charged counterparts (counterions) on the other side of the membrane. For a typical cell, 1 microcoulomb of charge (6×10^{12} monovalent ions) per square centimeter of membrane, transferred from one side of the membrane to the other, changes the membrane potential by roughly 1 V. This means, for example, that in a spherical cell of diameter 10 μm, the number of K^+ ions that have to flow out to alter the membrane potential by 100 mV is only about 1/100,000 of the total number of K^+ ions in the cytosol.

and result in only a very slight discrepancy in the number of positive and negative ions on the two sides of the membrane (**Figure 11–22**). Moreover, these movements of charge are generally rapid, taking only a few milliseconds or less.

Consider the change in the membrane potential in a real cell after the sudden inactivation of the Na^+-K^+ pump. A slight drop in the membrane potential occurs immediately. This is because the pump is electrogenic and, when active, makes a small direct contribution to the membrane potential by pumping out three Na^+ for every two K^+ that it pumps in. However, switching off the pump does not abolish the major component of the resting potential, which is generated by the K^+ equilibrium mechanism described above. This component of the membrane potential persists as long as the Na^+ concentration inside the cell stays low and the K^+ ion concentration high—typically for many minutes. But the plasma membrane is somewhat permeable to all small ions, including Na^+. Therefore, without the Na^+-K^+ pump, the ion gradients set up by the pump will eventually run down, and the membrane potential established by diffusion through the K^+ leak channels will fall as well. As Na^+ enters, the osmotic balance is upset, and water seeps into the cell (see Panel 11–1, p. 664), and the cell eventually comes to a new resting state where Na^+, K^+, and Cl^- are all at equilibrium across the membrane. The membrane potential in this state is much less than it was in the normal cell with an active Na^+-K^+ pump.

The resting potential of an animal cell varies between –20 mV and –120 mV, depending on the organism and cell type. Although the K^+ gradient always has a major influence on this potential, the gradients of other ions (and the disequilibrating effects of ion pumps) also have a significant effect: the more permeable the membrane for a given ion, the more strongly the membrane potential tends to be driven toward the equilibrium value for that ion. Consequently, changes in a membrane's permeability to ions can cause significant changes in the membrane potential. This is one of the key principles relating the electrical excitability of cells to the activities of ion channels.

To understand how ion channels select their ions and how they open and close, we need to know their atomic structure. The first ion channel to be crystallized and studied by x-ray diffraction was a bacterial K^+ channel. The details of its structure revolutionized our understanding of ion channels.

The Three-Dimensional Structure of a Bacterial K^+ Channel Shows How an Ion Channel Can Work

Scientists were puzzled by the remarkable ability of ion channels to combine exquisite ion selectivity with a high conductance. K^+ leak channels, for example, conduct K^+ 10,000-fold better than Na^+, yet the two ions are both featureless spheres and have similar diameters (0.133 nm and 0.095 nm, respectively). A single amino acid substitution in the pore of an animal cell K^+ channel can result in a loss of ion selectivity and cell death. We cannot explain the normal K^+ selectivity by pore size, because Na^+ is smaller than K^+. Moreover, the high conductance rate is incompatible with the channel's having selective, high-affinity K^+-binding sites, as the binding of K^+ ions to such sites would greatly slow their passage.

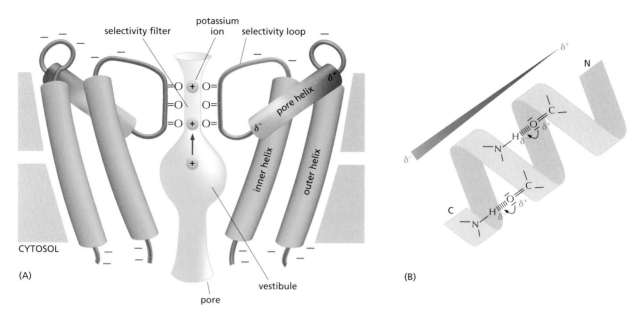

Figure 11–23 The structure of a bacterial K+ channel. <ATTA> (A) Two transmembrane α helices from only two of the four identical subunits are shown. From the cytosolic side, the pore opens up into a vestibule in the middle of the membrane. The vestibule facilitates transport by allowing the K+ ions to remain hydrated even though they are halfway across the membrane. The narrow selectivity filter links the vestibule to the outside of the cell. Carbonyl oxygens line the walls of the selectivity filter and form transient binding sites for dehydrated K+ ions. The positions of the K+ ions in the pore were determined by soaking crystals of the channel protein in a solution containing rubidium ions, which are more electron-dense but only slightly larger than K+ ions; from the differences in the diffraction patterns obtained with K+ ions and with rubidium ions in the channel, the positions of the ions could be calculated. Two K+ ions occupy sites in the selectivity filter, while a third K+ ion is located in the center of the vestibule, where it is stabilized by electrical interactions with the more negatively charged ends of the pore helices. The ends of the four pore helices (only two of which are shown) point precisely toward the center of the vestibule, thereby guiding K+ ions into the selectivity filter. Negatively charged amino acids (indicated by *red* minus signs) are concentrated near the channel entrance and exit. (B) Because of the polarity of the hydrogen bonds (*red*) that link adjacent turns of an α helix, every α helix has an electric dipole along its axis, with a more negatively charged C-terminal end (δ^-) and a more positively charged N-terminal end (δ^+). (A, adapted from D.A. Doyle et al., *Science* 280:69–77, 1998. With permission from AAAS.)

The puzzle was solved when the structure of a *bacterial K+ channel* was determined by x-ray crystallography. The channel is made from four identical transmembrane subunits, which together form a central pore through the membrane (**Figure 11–23**). Negatively charged amino acids concentrated at the cytosolic entrance to the pore are thought to attract cations and repel anions, making the channel cation-selective. Each subunit contributes two transmembrane α helices, which are tilted outward in the membrane and together form a cone, with its wide end facing the outside of the cell where K+ ions exit from the channel. The polypeptide chain that connects the two transmembrane helices forms a short α helix (the *pore helix)* and a crucial loop that protrudes into the wide section of the cone to form the **selectivity filter**. The selectivity loops from the four subunits form a short, rigid, narrow pore, which is lined by the carbonyl oxygen atoms of their polypeptide backbones. Because the selectivity loops of all known K+ channels have similar amino acid sequences, it is likely that they form a closely similar structure. The crystal structure shows two K+ ions in single file within the selectivity filter, separated by about 0.8 nm. Mutual repulsion between the two ions is thought to help move them through the pore into the extracellular fluid.

The structure of the selectivity filter explains the ion selectivity of the channel. A K+ ion must lose almost all of its bound water molecules to enter the filter, where it interacts instead with the carbonyl oxygens lining the filter; the oxygens are rigidly spaced at the exact distance to accommodate a K+ ion. A Na+ ion, in contrast, cannot enter the filter because the carbonyl oxygens are too far away from the smaller Na+ ion to compensate for the energy expense associated with the loss of water molecules required for entry (**Figure 11–24**).

Structural studies of K+ channels and other channels have also indicated some general principles of how channels may open and close. This gating seems

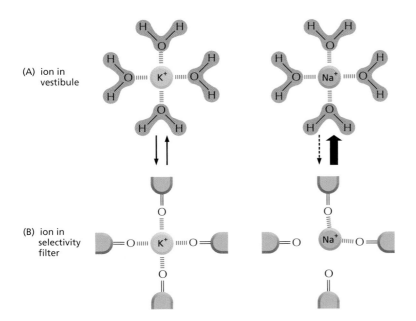

Figure 11–24 **K$^+$ specificity of the selectivity filter in a K$^+$ channel.** The drawing shows K$^+$ and Na$^+$ ions (A) in the vestibule and (B) in the selectivity filter of the pore, viewed in cross section. In the vestibule, the ions are hydrated. In the selectivity filter, they have lost their water, and the carbonyl oxygens are placed precisely to accommodate a dehydrated K$^+$ ion. The dehydration of the K$^+$ ion requires energy, which is precisely balanced by the energy regained by the interaction of the ion with the carbonyl oxygens that serve as surrogate water molecules. Because the Na$^+$ ion is too small to interact with the oxygens, it can enter the selectivity filter only at a great energetic expense. The filter therefore selects K$^+$ ions with high specificity. (Adapted from D.A. Doyle et al., *Science* 280:69–77, 1998. With permission from AAAS.)

to involve movement of the helices in the membrane so that they either obstruct (in the closed state) or free (in the open state) the path for ion movement. Depending on the particular type of channel, helices are thought to tilt, rotate, or bend during gating. The structure of a closed K$^+$ channel shows that by tilting the inner helices, the pore constricts like a diaphragm at its cytosolic end (**Figure 11–25**). Bulky hydrophobic amino acid side chains block the small opening that remains, preventing the entry of ions.

Most ion channels are constructed from multiple identical subunits, each of which contributes to a common central pore. A recently determined crystal structure of a Cl$^-$ channel, however, has revealed that some ion channels are built very differently. Although the protein is a dimer formed by two identical subunits, each of the subunits contains its own pore through which Cl$^-$ ions move. In the center of the membrane, amino acid side chains form a selectivity filter, which is conceptually similar to that in K$^+$ channels. But, unlike the filter in K$^+$ channels, different regions of the protein contribute the side chains, and they are not symmetrically arranged (**Figure 11–26**).

Aquaporins Are Permeable to Water But Impermeable to Ions

We discussed earlier that procaryotic and eucaryotic cells have **water channels**, or **aquaporins**, embedded in their plasma membrane to allow water to move readily across this membrane. Aquaporins are especially abundant in cells that must transport water at particularly high rates, such as the epithelial cells of the kidney.

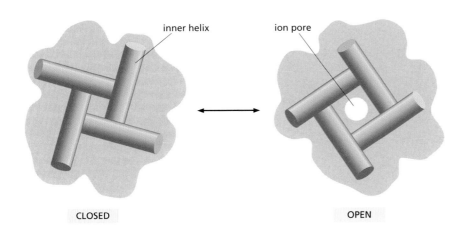

inner helix   ion pore

CLOSED   OPEN

Figure 11–25 **A model for the gating of a bacterial K$^+$ channel.** The channel is viewed in cross section. To adopt the closed conformation, the four inner transmembrane helices that line the pore on the cytosolic side of the selectivity filter (see Figure 11–23) rearrange to close the cytosolic entrance to the channel. (Adapted from E. Perozo et al., *Science* 285:73–78, 1999. With permission from AAAS.)

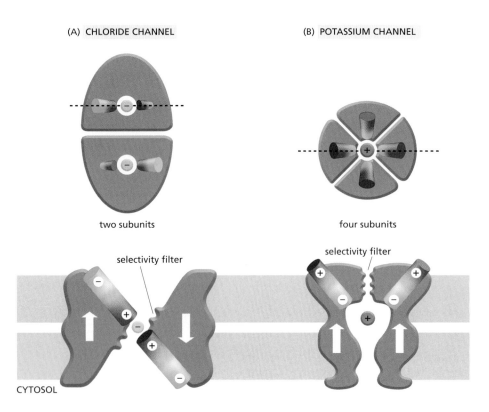

(A) CHLORIDE CHANNEL

(B) POTASSIUM CHANNEL

two subunits

four subunits

selectivity filter

selectivity filter

CYTOSOL

Figure 11–26 Comparison of Cl⁻ and K⁺ channel architectures. (A) The Cl⁻ channel is a "double-channel" dimer built of two identical subunits, each of which contains its own ion-conducting pore. The upper cartoon is a schematic view of the extracellular face of the channel, showing the two identical ion-conducting pores. The lower cartoon shows a cross section through one subunit, viewed from within the membrane. (*Dotted black line* in top cartoon indicates the plane of section.) The subunit is a single polypeptide chain consisting of two portions, which, though similar, span the membrane with opposite orientations *(white arrows)*: each portion contributes one pore helix oriented such that its positively charged end points towards a centrally positioned selectivity filter. Both elements, the selectivity filter and the helix dipoles, contribute to the selectivity of the channel for negatively charged Cl⁻ ions. (B) By contrast, the K⁺ channel is a tetramer built of four identical subunits, each of which contributes to a centrally located pore. All four subunits have the same orientation in the membrane *(white arrows)*. Four pore helices, one contributed by each subunit, point their negatively charged ends towards a vestibule, stabilizing a positively charged K⁺ ion there (also see Figure 11–23). (Lower cartoons in A and B, adapted from R. Dutzler et al., *Nature* 415:287–294, 2002. With permission from Macmillan Publishers Ltd.)

Aquaporins must solve a problem that is opposite to that facing ion channels. To avoid disrupting ion gradients across membranes, they have to allow the rapid passage of water molecules while completely blocking the passage of ions. The crystal structure of an aquaporin reveals how it achieves this remarkable selectivity. The channels have a narrow pore that allows water molecules to traverse the membrane in single file, following the path of carbonyl oxygens that line one side of the pore (**Figure 11–27**A and B). Hydrophobic amino acids line the other side of the pore. The pore is too narrow for any hydrated ion to enter, and the energy cost of dehydrating an ion would be enormous because the hydrophobic wall of the pore cannot interact with a dehydrated ion to compensate for the loss of water. This design readily explains why the aquaporins cannot conduct K⁺, Na⁺, Ca²⁺, or Cl⁻ ions. To understand why these channels are also impermeable to H⁺, recall that most protons are present in cells as H_3O^+,

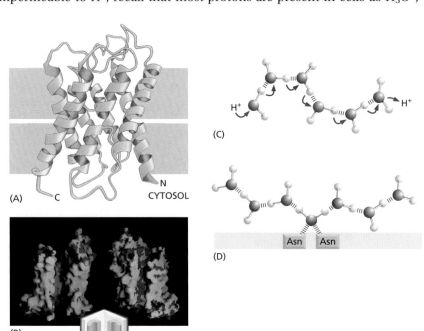

(A) CYTOSOL
N
C

(B)

(C) H⁺ ... H⁺

(D) Asn Asn

Figure 11–27 The structure of aquaporins. (A) A ribbon diagram of an aquaporin monomer. In the membrane, aquaporins form tetramers, with each monomer containing a pore in its center (not shown). (B) A space-filling model of an aquaporin monomer, which has been cut and opened like a book, so that the inside of the pore is visible. Hydrophilic amino acids lining the pores are colored *red* and *blue,* whereas hydrophobic amino acids lining the pore are colored *yellow.* The amino acids not involved in forming the pore are shown in *green.* Note that one face of the pore is lined with hydrophilic amino acids, which provide transient hydrogen bonds to water molecules; these bonds help line up the transiting water molecules in a single row and orient them as they traverse the membrane. By contrast, the other side of the pore is devoid of such amino acids, providing a *hydrophobic slide* that does not allow hydrogen bonds to form. (C and D) A model explaining why aquaporins are impermeable to H⁺. (C) In water, H⁺ diffuses extremely rapidly by being relayed from one water molecule to the next. (D) Two strategically placed asparagines in the center of each aquaporin pore help tether a central water molecule such that both valencies on its oxygen are occupied, thereby preventing an H⁺ relay. (A and B, adapted from R.M. Stroud et al., *Curr. Opin. Struct. Biol.* 13:424–431, 2003. With permission from Elsevier.)

which diffuses through water extremely rapidly, using a molecular relay mechanism that requires the making and breaking of hydrogen bonds between adjacent water molecules (Figure 11–27C). Aquaporins contain two strategically placed asparagines, which bind to the oxygen atom of the central water molecule in the line of water molecules traversing the pore. Because both valences of this oxygen are unavailable for hydrogen bonding, the central water molecule cannot participate in an H^+ relay, and the pore is therefore impermeable to protons (Figure 11–27C and D).

Some bacterial water channels similar to aquaporins also conduct glycerol and small sugars, which interact with similarly positioned carbonyl oxygens lining the pore. Such transient contacts that solutes make with the pore walls ensure that the transport is highly specific, without significantly impeding the speed with which the solute passes. Each individual aquaporin channel passes about 10^9 water molecules per second.

The cells that make most sophisticated use of channels are neurons. Before discussing how they do so, we digress briefly to describe how a typical neuron is organized.

The Function of a Neuron Depends on Its Elongated Structure

The fundamental task of a **neuron**, or **nerve cell**, is to receive, conduct, and transmit signals. To perform these functions, neurons are often extremely elongated. A single neuron in a human, extending, for example, from the spinal cord to a muscle in the foot, may be as long as 1 meter. Every neuron consists of a cell body (containing the nucleus) with a number of thin processes radiating outward from it. Usually one long **axon** conducts signals away from the cell body toward distant targets, and several shorter branching **dendrites** extend from the cell body like antennae, providing an enlarged surface area to receive signals from the axons of other neurons (**Figure 11–28**), although the cell body itself also receives signals. A typical axon divides at its far end into many branches, passing on its message to many target cells simultaneously. Likewise, the extent of branching of the dendrites can be very great—in some cases sufficient to receive as many as 100,000 inputs on a single neuron.

Despite the varied significance of the signals carried by different classes of neurons, the form of the signal is always the same, consisting of changes in the electrical potential across the neuron's plasma membrane. The signal spreads because an electrical disturbance produced in one part of the cell spreads to other parts, although the disturbance becomes weaker with increasing distance from its source, unless the neuron expends energy to amplify it as it travels. Over short distances this attenuation is unimportant; in fact, many small neurons conduct their signals passively, without amplification. For long-distance communication, however, such passive spread is inadequate. Thus, larger neurons employ an active signaling mechanism, which is one of their most striking features. An electrical stimulus that exceeds a certain threshold strength triggers an explosion of electrical activity that propagates rapidly along the neuron's plasma

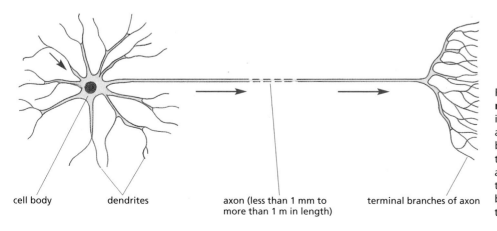

cell body   dendrites   axon (less than 1 mm to more than 1 m in length)   terminal branches of axon

Figure 11–28 **A typical vertebrate neuron.** The arrows indicate the direction in which signals are conveyed. The single axon conducts signals away from the cell body, while the multiple dendrites (and the cell body) receive signals from the axons of other neurons. The nerve terminals end on the dendrites or cell body of other neurons or on other cell types, such as muscle or gland cells.

membrane and is sustained by automatic amplification all along the way. This traveling wave of electrical excitation, known as an **action potential**, or *nerve impulse*, can carry a message without attenuation from one end of a neuron to the other at speeds of 100 meters per second or more. Action potentials are the direct consequence of the properties of voltage-gated cation channels, as we now discuss.

Voltage-Gated Cation Channels Generate Action Potentials in Electrically Excitable Cells

The plasma membrane of all electrically excitable cells—not only neurons, but also muscle, endocrine, and egg cells—contains **voltage-gated cation channels**, which are responsible for generating the action potentials. An action potential is triggered by a *depolarization* of the plasma membrane—that is, by a shift in the membrane potential to a less negative value inside. (We shall see later how the action of a neurotransmitter causes depolarization.) In nerve and skeletal muscle cells, a stimulus that causes sufficient depolarization promptly opens the **voltage-gated Na⁺ channels**, allowing a small amount of Na⁺ to enter the cell down its electrochemical gradient. The influx of positive charge depolarizes the membrane further, thereby opening more Na⁺ channels, which admit more Na⁺ ions, causing still further depolarization. This self-amplification process (an example of *positive feedback*, discussed in Chapter 15), continues until, within a fraction of a millisecond, the electrical potential in the local region of membrane has shifted from its resting value of about –70 mV to almost as far as the Na⁺ equilibrium potential of about +50 mV (see Panel 11–2, p. 670). At this point, when the net electrochemical driving force for the flow of Na⁺ is almost zero, the cell would come to a new resting state, with all of its Na⁺ channels permanently open, if the open conformation of the channel were stable. Two mechanisms that act in concert to save the cell from such a permanent electrical spasm: the Na⁺ channels inactivate and voltage-gated K⁺ channels open.

The Na⁺ channels have an automatic inactivating mechanism, which causes the channels to reclose rapidly even though the membrane is still depolarized. The Na⁺ channels remain in this *inactivated* state, unable to reopen, until after the membrane potential has returned to its initial negative value. The whole cycle from initial stimulus to the return to the original resting state takes a few milliseconds or less. The Na⁺ channel can therefore exist in three distinct states—closed, open, and inactivated. **Figure 11–29** shows how the changes in state contribute to the rise and fall of the action potential.

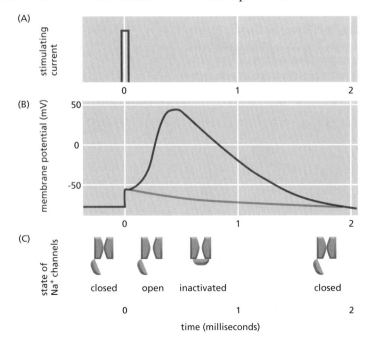

Figure 11–29 An action potential. <CGAG> (A) An action potential is triggered by a brief pulse of current, which (B) partially depolarizes the membrane, as shown in the plot of membrane potential versus time. The *green curve* shows how the membrane potential would have simply relaxed back to the resting value after the initial depolarizing stimulus if there had been no voltage-gated Na⁺ channels in the membrane; this relatively slow return of the membrane potential to its initial value of –70 mV in the absence of open Na⁺ channels occurs because of the efflux of K⁺ through voltage-gated K⁺ channels, which open in response to membrane depolarization and drive the membrane back toward the K⁺ equilibrium potential. The *red curve* shows the course of the action potential that is caused by the opening and subsequent inactivation of voltage-gated Na⁺ channels, whose state is shown in (C). The membrane cannot fire a second action potential until the Na⁺ channels have returned to the closed conformation; until then, the membrane is refractory to stimulation.

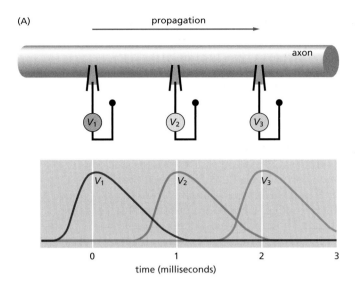

Figure 11–30 **The propagation of an action potential along an axon.** (A) The voltages that would be recorded from a set of intracellular electrodes placed at intervals along the axon. (B) The changes in the Na⁺ channels and the current flows *(orange arrows)* that give rise to the traveling disturbance of the membrane potential. The region of the axon with a depolarized membrane is shaded in *blue*. Note that an action potential can only travel away from the site of depolarization, because Na⁺-channel inactivation prevents the depolarization from spreading backward.

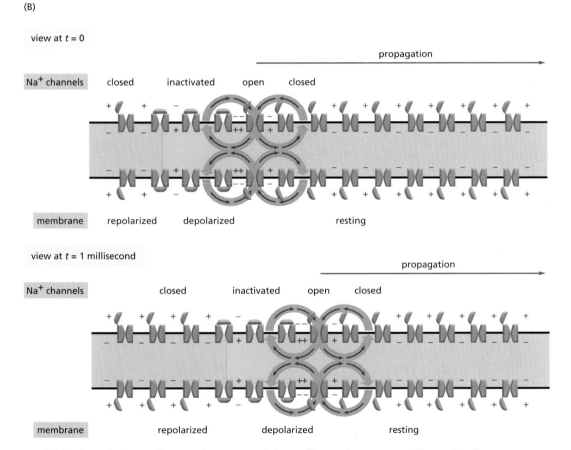

This description of an action potential applies only to a small patch of plasma membrane. The self-amplifying depolarization of the patch, however, is sufficient to depolarize neighboring regions of membrane, which then go through the same cycle. In this way, the action potential sweeps like a wave from the initial site of depolarization over the entire plasma membrane, as shown in **Figure 11–30**.

Voltage-gated K⁺ channels provide a second mechanism in most nerve cells to help bring the activated plasma membrane more rapidly back toward its original negative potential, ready to transmit a second impulse. These channels open in response to membrane depolarization in much the same way that the Na⁺ channels do, but with slightly slower kinetics; for this reason they are sometimes called *delayed K⁺ channels*. Once the K⁺ channels open, the efflux of K⁺ rapidly overwhelms the transient influx of Na⁺ and quickly drives the membrane back toward the K⁺ equilibrium potential, even before the inactivation of the Na⁺ channels is complete.

Like the Na⁺ channel, the voltage-gated K⁺ channels automatically inactivate. Studies of mutant voltage-gated K⁺ channels show that the N-terminal 20 amino acids of the channel protein are required for rapid inactivation: altering this region changes the kinetics of channel inactivation, and removal of the region abolishes inactivation. Amazingly, in the latter case, exposing the cytoplasmic face of the plasma membrane to a small synthetic peptide corresponding to the missing N-terminus restores inactivation. These findings suggest that the N-terminus of each K⁺ channel subunit acts like a tethered ball that occludes the cytoplasmic end of the pore soon after it opens, thereby inactivating the channel (**Figure 11–31**). A similar mechanism is thought to operate in the rapid inactivation of voltage-gated Na⁺ channels (which we discuss later), although a different segment of the protein seems to be involved.

The electrochemical mechanism of the action potential was first established by a famous series of experiments carried out in the 1940s and 1950s. Because the techniques for studying electrical events in small cells had not yet been developed, the experiments exploited the giant neurons in the squid. Despite the many technical advances made since then, the logic of the original analysis continues to serve as a model for present-day work. **Panel 11–3** summarizes some of the key original experiments.

Myelination Increases the Speed and Efficiency of Action Potential Propagation in Nerve Cells

The axons of many vertebrate neurons are insulated by a **myelin sheath**, which greatly increases the rate at which an axon can conduct an action potential. The importance of myelination is dramatically demonstrated by the demyelinating disease *multiple sclerosis*, in which the immune system destroys myelin sheaths in some regions of the central nervous system; in the affected regions, the propagation of nerve impulses is greatly slowed, often with devastating neurological consequences.

Myelin is formed by specialized supporting cells, called **glial cells. Schwann cells** myelinate axons in peripheral nerves, and **oligodendrocytes** do so in the central nervous system. These glial cells wrap layer upon layer of their own plasma membrane in a tight spiral around the axon (**Figure 11–32**A and B), thereby insulating the axonal membrane so that little current can leak across it.

Figure 11–31 The "ball-and-chain" model of rapid inactivation of a voltage-gated K⁺ channel. When the membrane is depolarized, the channel opens and begins to conduct ions. If the depolarization is maintained, the open channel adopts an inactive conformation, in which the pore is occluded by the N-terminal 20 amino acid "ball," which is linked to the channel proper by a segment of unfolded polypeptide chain that serves as the "chain." For simplicity, only two balls are shown; in fact, there are four, one from each subunit. A similar mechanism, using a different segment of the polypeptide chain, is thought to operate in Na⁺ channel inactivation. Internal forces stabilize each state against small disturbances, but a sufficiently violent collision with other molecules can cause the channel to flip from one of these states to another. The state of lowest energy depends on the membrane potential because the different conformations have different charge distributions. When the membrane is at rest (highly polarized), the closed conformation has the lowest free energy and is therefore most stable; when the membrane is depolarized, the energy of the *open* conformation is lower, so the channel has a high probability of opening. But the free energy of the *inactivated* conformation is lower still; therefore, after a randomly variable period spent in the open state, the channel becomes inactivated. Thus, the open conformation corresponds to a metastable state that can exist only transiently. The *red arrows* indicate the sequence that follows a sudden depolarization; the *black arrow* indicates the return to the original conformation as the lowest energy state after the membrane is repolarized.

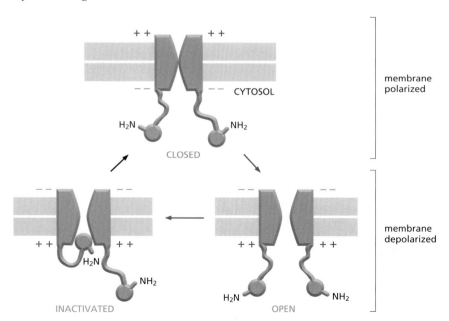

1. Action potentials are recorded with an intracellular electrode

The squid giant axon is about 0.5–1 mm in diameter and several centi-meters long. An electrode in the form of a glass capillary tube containing a conducting solution can be thrust down the axis of the axon so that its tip lies deep in the cytoplasm. With its help, one can measure the voltage difference between the inside and the outside of the axon—that is, the membrane potential—as an action potential sweeps past the electrode. The action potential is triggered by a brief electrical stimulus to one end of the axon. It does not matter which end, because the excitation can travel in either direction; and it does not matter how big the stimulus is, as long as it exceeds a certain threshold: the action potential is all or none.

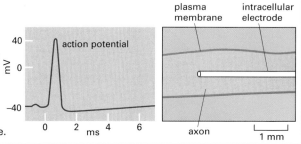

2. Action potentials depend only on the neuronal plasma membrane and on gradients of Na⁺ and K⁺ across it

The three most plentiful ions, both inside and outside the axon, are Na^+, K^+, and Cl^-. As in other cells, the Na^+-K^+ pump maintains a concentration gradient: the concentration of Na^+ is about 9 times lower inside the axon than outside, while the concentration of K^+ is about 20 times higher inside than outside. Which ions are important for the action potential?

The squid giant axon is so large and robust that it is possible to extrude the gel-like cytoplasm from it, like toothpaste from a tube,

and then to perfuse it internally with pure artificial solutions of Na^+, K^+, and Cl^- or SO_4^{2-}. Remarkably, if (and only if) the concentrations of Na^+ and K^+ inside and outside approximate those found naturally, the axon will still propagate action potentials of the normal form. The important part of the cell for electrical signaling, therefore, must be the plasma membrane; the important ions are Na^+ and K^+; and a sufficient source of free energy to power the action potential must be provided by the concentration gradients of these ions across the membrane, because all other sources of metabolic energy have presumably been removed by the perfusion.

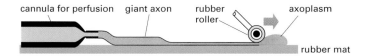

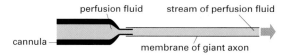

3. At rest, the membrane is chiefly permeable to K⁺; during the action potential, it becomes transiently permeable to Na⁺

At rest the membrane potential is close to the equilibrium potential for K^+. When the external concentration of K^+ is changed, the resting potential changes roughly in accordance with the Nernst equation for K^+ (see Panel 11–2). At rest, therefore, the membrane is chiefly permeable to K^+: K^+ leak channels provide the main ion pathway through the membrane.

If the external concentration of Na^+ is varied, there is no effect on the resting potential. However, the height of the peak of the action potential varies roughly in accordance with the Nernst equation for Na^+. During the action potential, therefore, the membrane appears to be chiefly permeable to Na^+: Na^+ channels have opened. In the aftermath of the action potential, the

membrane potential reverts to a negative value that depends on the external concentration of K^+ and is even closer to the K^+ equilibrium potential than the resting potential is: the membrane has lost most of its permeability to Na^+ and has become even more permeable to K^+ than before—that is, Na^+ channels have closed, and additional K^+ channels have opened.

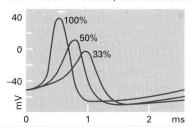

The form of the action potential when the external medium contains 100%, 50%, or 33% of the normal concentration of Na^+.

4. Voltage clamping reveals how the membrane potential controls opening and closing of ion channels

The membrane potential can be held constant ("voltage clamped") throughout the axon by passing a suitable current through a bare metal wire inserted along the axis of the axon while monitoring the membrane potential with another intracellular electrode. When the membrane is abruptly shifted from the resting potential and held in a depolarized state (A), Na^+ channels rapidly open until the Na^+ permeability of the membrane is much greater than the K^+ permeability; they then close again spontaneously, even though the membrane potential is clamped and unchanging. K^+ channels also open but with a delay, so that the K^+ permeability increases as the Na^+ permeability falls (B). If the experiment is now very promptly repeated, by returning the membrane briefly to the resting potential and then quickly depolarizing it again, the response is different: prolonged depolarization has caused the Na^+ channels to enter an inactivated state, so that the second depolarization fails to cause a rise and fall similar to the first. Recovery from this state requires a

relatively long time—about 10 milliseconds—spent at the repolarized (resting) membrane potential.

In a normal unclamped axon, an inrush of Na^+ through the opened Na^+ channels produces the spike of the action potential; inactivation of Na^+ channels and opening of K^+ channels bring the membrane rapidly back down to the resting potential.

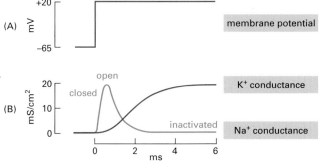

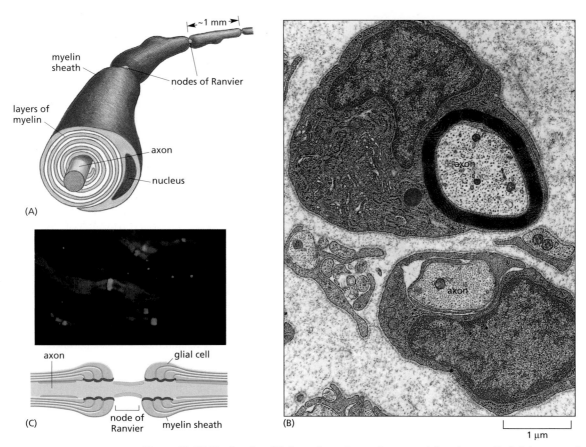

Figure 11–32 Myelination. (A) A myelinated axon from a peripheral nerve. Each Schwann cell wraps its plasma membrane concentrically around the axon to form a segment of myelin sheath about 1 mm long. For clarity, the membrane layers of the myelin in this drawing are shown less compacted than they are in reality (see part B). (B) An electron micrograph of a section from a nerve in the leg of a young rat. Two Schwann cells can be seen: one near the bottom is just beginning to myelinate its axon; the one above it has formed an almost mature myelin sheath. (C) Fluorescence micrograph and diagram of individual myelinated axons teased apart in a nerve. Three different proteins are detected by staining with antibodies. Voltage-gated Na^+ channels (stained in *green*) are concentrated in the axonal membrane at the nodes of Ranvier. An extracellular protein (called *Caspr*, stained in *red*) marks the end of each myelin sheath. Caspr assembles at the junctions where the glial cell plasma membrane tightly abuts the axon to provide the electrical seal. Voltage-gated K^+ channels (stained in *blue*) localize to regions in the axon plasma membrane that are close to the nodes. (B, from Cedric S. Raine, in Myelin [P. Morell, ed.]. New York: Plenum, 1976; C, from M.N. Rasband and P. Shrager, *J. Physiol.* 525:63–73, 2000. With permission from Blackwell Publishing.)

The myelin sheath is interrupted at regularly spaced *nodes of Ranvier,* where almost all the Na^+ channels in the axon are concentrated (see Figure 11–32C). Because the ensheathed portions of the axonal membrane have excellent cable properties (in other words, they behave electrically much like well-designed underwater telegraph cables), a depolarization of the membrane at one node almost immediately spreads passively to the next node. Thus, an action potential propagates along a myelinated axon by jumping from node to node, a process called *saltatory conduction.* This type of conduction has two main advantages: action potentials travel faster, and metabolic energy is conserved because the active excitation is confined to the small regions of axonal plasma membrane at nodes of Ranvier.

Patch-Clamp Recording Indicates That Individual Gated Channels Open in an All-or-Nothing Fashion

Neuron and skeletal muscle cell plasma membranes contain many thousands of voltage-gated Na^+ channels, and the current crossing the membrane is the sum of the currents flowing through all of these. An intracellular microelectrode can

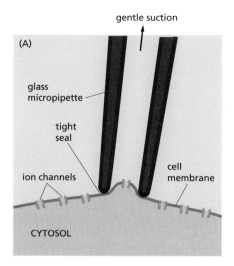

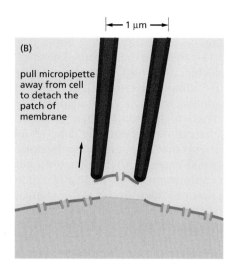

gentle suction

(A)

glass micropipette

tight seal

ion channels

cell membrane

CYTOSOL

|← 1 μm →|

(B)

pull micropipette away from cell to detach the patch of membrane

Figure 11–33 The technique of patch-clamp recording. Because of the extremely tight seal between the micropipette and the membrane, current can enter or leave the micropipette only by passing through the channels in the *patch* of membrane covering its tip. The term *clamp* is used because an electronic device is employed to maintain, or "clamp," the membrane potential at a set value while recording the ionic current through individual channels. The current through these channels can be recorded with the patch still attached to the rest of the cell, as in (A), or detached, as in (B). The advantage of the detached patch is that it is easy to alter the composition of the solution on either side of the membrane to test the effect of various solutes on channel behavior. A detached patch can also be produced with the opposite orientation, so that the cytoplasmic surface of the membrane faces the inside of the pipette.

record this aggregate current, as shown in Figure 11–30. Remarkably, however, it is also possible to record current flowing through individual channels. **Patch-clamp recording**, developed in the 1970s and 80s has revolutionized the study of ion channels, making it possible to examine transport through a single molecule of channel protein in a small patch of membrane covering the mouth of a micropipette (**Figure 11–33**). With this simple but powerful technique, one can study the detailed properties of ion channels in all sorts of cell types. This work has led to the discovery that even cells that are not electrically excitable usually have a variety of gated ion channels in their plasma membrane. Many of these cells, such as yeasts, are too small to be investigated by the traditional electro-physiologist's method of impalement with an intracellular microelectrode.

Patch-clamp recording indicates that individual voltage-gated Na+ channels open in an all-or-nothing fashion. A channel opens and closes at random, but when open, the channel always has the same large conductance, allowing more than 1000 ions to pass per millisecond. Therefore, the aggregate current crossing the membrane of an entire cell does not indicate the *degree* to which a typical individual channel is open but rather the *total number* of channels in its membrane that are open at any one time (**Figure 11–34**).

Some simple physical principles allow us to understand voltage-gating. The interior of the resting neuron or muscle cell is at an electrical potential about 50–100 mV more negative than the external medium. Although this potential difference seems small, it exists across a plasma membrane only about 5 nm thick, so that the resulting voltage gradient is about 100,000 V/cm. Proteins in the

Figure 11–34 Patch-clamp measurements for a single voltage-gated Na+ channel. A tiny patch of plasma membrane was detached from an embryonic rat muscle cell, as in Figure 11–33. (A) The membrane was depolarized by an abrupt shift of potential. (B) Three current records from three experiments performed on the same patch of membrane. Each major current step in (B) represents the opening and closing of a single channel. A comparison of the three records shows that, whereas the durations of channel opening and closing vary greatly, the rate at which current flows through an open channel is practically constant. The minor fluctuations in the current records arise largely from electrical noise in the recording apparatus. Current is measured in picoamperes (pA). By convention, the electrical potential on the outside of the cell is defined as zero. (C) The sum of the currents measured in 144 repetitions of the same experiment. This aggregate current is equivalent to the usual Na+ current that would be observed flowing through a relatively large region of membrane containing 144 channels. A comparison of (B) and (C) reveals that the time course of the aggregate current reflects the probability that any individual channel will be in the open state; this probability decreases with time as the channels in the depolarized membrane adopt their inactivated conformation. (Data from J. Patlak and R. Horn, *J. Gen. Physiol.* 79:333–351, 1982. With permission from The Rockefeller University Press.)

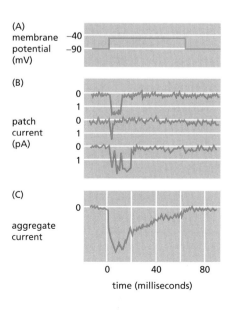

(A) membrane potential (mV)   −40 −90

(B)

patch current (pA)   0 1   0 1   0 1

(C)

aggregate current   0

0   40   80

time (milliseconds)

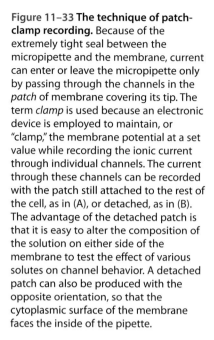

membrane are thus subjected to a very large electrical field that can profoundly affect their conformation. These proteins, like all others, have many charged groups, as well as polarized bonds between their various atoms. The electrical field therefore exerts forces on the molecular structure. For many membrane proteins the effects of changes in the membrane electrical field are probably insignificant, but voltage-gated ion channels can adopt alternative conformations whose stabilities depend on the strength of the field. Voltage-gated Na^+, K^+, and Ca^{2+} channels, for example, have characteristic positively charged amino acids in one of their transmembrane segments that respond to depolarization by moving outward, triggering conformational changes that open the channel. Each conformation can "flip" to another conformation if given a sufficient jolt by the random thermal movements of the surroundings, and it is the relative stability of the closed, open, and inactivated conformations against flipping that is altered by changes in the membrane potential (see legend to Figure 11–31).

Voltage-Gated Cation Channels Are Evolutionarily and Structurally Related

Na^+ channels are not the only kind of voltage-gated cation channel that can generate an action potential. The action potentials in some muscle, egg, and endocrine cells, for example, depend on *voltage-gated Ca^{2+} channels* rather than on Na^+ channels.

There is a surprising amount of structural and functional diversity within each of these three classes, generated both by multiple genes and by the alternative splicing of RNA transcripts produced from the same gene. Nonetheless, the amino acid sequences of the known voltage-gated Na^+, K^+, and Ca^{2+} channels show striking similarities, demonstrating that they all belong to a large superfamily of evolutionarily and structurally related proteins and share many of the design principles. Whereas the single-celled yeast *S. cerevisiae* contains a single gene that codes for a voltage-gated K^+ channel, the genome of the worm *C. elegans* contains 68 genes that encode different but related K^+ channels. This complexity indicates that even a simple nervous system made up of only 302 neurons uses a large number of different ion channels to compute its responses.

Humans who inherit mutant genes encoding ion channel proteins can suffer from a variety of nerve, muscle, brain, or heart diseases, depending in which cells the channel encoded by the mutant gene normally functions. Mutations in genes that encode voltage-gated Na^+ channels in skeletal muscle cells, for example, can cause *myotonia*, a condition in which there is a delay in muscle relaxation after voluntary contraction, causing painful muscle spasms. In some cases, this occurs because the abnormal channels fail to inactivate normally; as a result, Na^+ entry persists after an action potential finishes and repeatedly reinitiates membrane depolarization and muscle contraction. Similarly, mutations that affect Na^+ or K^+ channels in the brain can cause *epilepsy*, in which excessive synchronized firing of large groups of neurons cause epileptic seizures (convulsions, or fits).

Transmitter-Gated Ion Channels Convert Chemical Signals into Electrical Ones at Chemical Synapses

Neuronal signals are transmitted from cell to cell at specialized sites of contact known as **synapses**. The usual mechanism of transmission is indirect. The cells are electrically isolated from one another, the *presynaptic cell* being separated from the *postsynaptic cell* by a narrow *synaptic cleft*. A change of electrical potential in the presynaptic cell triggers it to release small signal molecules known as **neurotransmitters**, which are stored in membrane-enclosed synaptic vesicles and released by exocytosis. The neurotransmitter diffuses rapidly across the synaptic cleft and provokes an electrical change in the postsynaptic cell by binding to *transmitter-gated ion channels* (**Figure 11–35**) and opening them. After the neurotransmitter has been secreted, it is rapidly removed: it is either

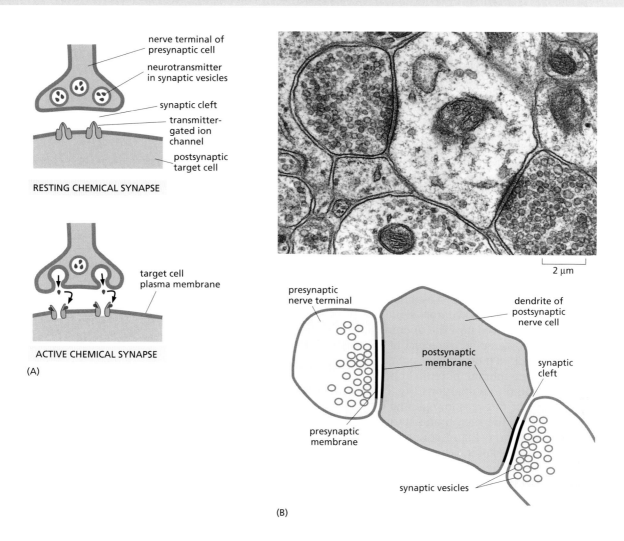

RESTING CHEMICAL SYNAPSE

- nerve terminal of presynaptic cell
- neurotransmitter in synaptic vesicles
- synaptic cleft
- transmitter-gated ion channel
- postsynaptic target cell

ACTIVE CHEMICAL SYNAPSE

- target cell plasma membrane

(A)

2 µm

- presynaptic nerve terminal
- dendrite of postsynaptic nerve cell
- postsynaptic membrane
- synaptic cleft
- presynaptic membrane
- synaptic vesicles

(B)

destroyed by specific enzymes in the synaptic cleft or taken up by the nerve terminal that released it or by surrounding glial cells. Reuptake is mediated by a variety of Na^+-dependent neurotransmitter transporters; in this way, neurotransmitters are recycled, allowing cells to keep up with high rates of release. Rapid removal ensures both spatial and temporal precision of signaling at a synapse. It decreases the chances that the neurotransmitter will influence neighboring cells, and it clears the synaptic cleft before the next pulse of neurotransmitter is released, so that the timing of repeated, rapid signaling events can be accurately communicated to the postsynaptic cell. As we shall see, signaling via such *chemical synapses* is far more versatile and adaptable than direct electrical coupling via gap junctions at *electrical synapses* (discussed in Chapter 19), which are also used by neurons but to a much smaller extent.

Transmitter-gated ion channels are specialized for rapidly converting extracellular chemical signals into electrical signals at chemical synapses. The channels are concentrated in the plasma membrane of the postsynaptic cell in the region of the synapse and open transiently in response to the binding of neurotransmitter molecules, thereby producing a brief permeability change in the membrane (see Figure 11–35A). Unlike the voltage-gated channels responsible for action potentials, transmitter-gated channels are relatively insensitive to the membrane potential and therefore cannot by themselves produce a self-amplifying excitation. Instead, they produce local permeability changes, and hence changes of membrane potential, that are graded according to the amount of neurotransmitter released at the synapse and how long it persists there. An action potential can be triggered from this site only if the local membrane potential depolarizes enough to open a sufficient number of nearby voltage-gated cation channels that are present in the same target cell membrane.

Figure 11–35 A chemical synapse. <CTGA> (A) When an action potential reaches the nerve terminal in a presynaptic cell, it stimulates the terminal to release its neurotransmitter. The neurotransmitter molecules are contained in synaptic vesicles and are released to the cell exterior when the vesicles fuse with the plasma membrane of the nerve terminal. The released neurotransmitter binds to and opens the transmitter-gated ion channels concentrated in the plasma membrane of the postsynaptic target cell at the synapse. The resulting ion flows alter the membrane potential of the target cell, thereby transmitting a signal from the excited nerve. (B) A thin section electron micrograph of two nerve terminal synapses on a dendrite of a postsynaptic cell. (B, courtesy of Cedric Raine.)

Chemical Synapses Can Be Excitatory or Inhibitory

Transmitter-gated ion channels differ from one another in several important ways. First, as receptors, they have a highly selective binding site for the neurotransmitter that is released from the presynaptic nerve terminal. Second, as channels, they are selective in the type of ions that they let pass across the plasma membrane; this determines the nature of the postsynaptic response. **Excitatory neurotransmitters** open cation channels, causing an influx of Na^+ that depolarizes the postsynaptic membrane toward the threshold potential for firing an action potential. **Inhibitory neurotransmitters**, by contrast, open either Cl^- channels or K^+ channels, and this suppresses firing by making it harder for excitatory influences to depolarize the postsynaptic membrane. Many transmitters can be either excitatory or inhibitory, depending on where they are released, what receptors they bind to, and the ionic conditions that they encounter. *Acetylcholine*, for example, can either excite or inhibit, depending on the type of acetylcholine receptors it binds to. Usually, however, acetylcholine, *glutamate*, and *serotonin* are used as excitatory transmitters, and *γ-aminobutyric acid (GABA)* and *glycine* are used as inhibitory transmitters. Glutamate, for instance, mediates most of the excitatory signaling in the vertebrate brain.

We have already discussed how the opening of cation channels depolarizes a membrane. We can understand the effect of opening Cl^- channels as follows. The concentration of Cl^- is much higher outside the cell than inside (see Table 11–1, p. 652), but the membrane potential opposes its influx. In fact, for many neurons, the equilibrium potential for Cl^- is close to the resting potential—or even more negative. For this reason, opening Cl^- channels tends to buffer the membrane potential; as the membrane starts to depolarize, more negatively charged Cl^- ions enter the cell and counteract the depolarizatrion. Thus, the opening of Cl^- channels makes it more difficult to depolarize the membrane and hence to excite the cell. The opening of K^+ channels has a similar effect. The effects of toxins that block their action demonstrate the importance of inhibitory neurotransmitters: strychnine, for example, by binding to glycine receptors and blocking the inhibitory action of glycine, causes muscle spasms, convulsions, and death.

However, not all chemical signaling in the nervous system operates through ligand-gated ion channels. Many of the signaling molecules that are secreted by nerve terminals, including a large variety of neuropeptides, bind to receptors that regulate ion channels only indirectly. We discuss these so-called *G-protein-coupled receptors* and *enzyme-coupled receptors* in detail in Chapter 15. Whereas signaling mediated by excitatory and inhibitory neurotransmitters binding to transmitter-gated ion channels is generally immediate, simple, and brief, signaling mediated by ligands binding to G-protein-coupled receptors and enzyme-coupled receptors tends to be far slower and more complex, and longer lasting in its consequences.

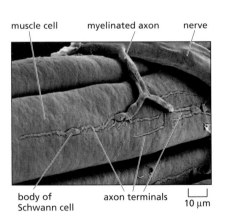

muscle cell   myelinated axon   nerve

body of   axon terminals   10 μm
Schwann cell

Figure 11–36 A low-magnification scanning electron micrograph of a neuromuscular junction in a frog. The termination of a single axon on a skeletal muscle cell is shown. (From J. Desaki and Y. Uehara, *J. Neurocytol.* 10:101–110, 1981. With permission from Kluwer Academic Publishers.)

The Acetylcholine Receptors at the Neuromuscular Junction Are Transmitter-Gated Cation Channels

The best-studied example of a transmitter-gated ion channel is the *acetylcholine receptor* of skeletal muscle cells. This channel is opened transiently by acetylcholine released from the nerve terminal at a **neuromuscular junction**—the specialized chemical synapse between a motor neuron and a skeletal muscle cell (**Figure 11–36**). This synapse has been intensively investigated because it is readily accessible to electrophysiological study, unlike most of the synapses in the central nervous system.

The **acetylcholine receptor** has a special place in the history of ion channels. It was the first ion channel to be purified, the first to have its complete amino acid sequence determined, the first to be functionally reconstituted in synthetic lipid bilayers, and the first for which the electrical signal of a single open channel was recorded. Its gene was also the first ion channel gene to be cloned and sequenced, and its three-dimensional structure has been determined, albeit at only moderate resolution. There were at least two reasons for the rapid progress in purifying and characterizing this receptor. First, an unusually rich source of the acetylcholine receptors exists in the electric organs of electric fish and rays (these organs are modified muscles designed to deliver a large electric shock to prey). Second, certain neurotoxins (such as *α-bungarotoxin*) in the venom of certain snakes bind with high affinity ($K_a = 10^9$ liters/mole) and specificity to the receptor and can therefore be used to purify it by affinity chromatography. Fluorescent or radiolabeled α-bungarotoxin can also be used to localize and count acetylcholine receptors. In this way, researchers have shown that the receptors are densely packed in the muscle cell plasma membrane at a neuromuscular junction (about 20,000 such receptors per μm^2), with relatively few receptors elsewhere in the same membrane.

The acetylcholine receptor of skeletal muscle is composed of five transmembrane polypeptides, two of one kind and three others, encoded by four separate genes. The four genes are strikingly similar in sequence, implying that they evolved from a single ancestral gene. The two identical polypeptides in the pentamer each contribute to one of two binding sites for acetylcholine that are nestled between adjoining subunits. When two acetylcholine molecules bind to the pentameric complex, they induce a conformational change: the helices that line the pore rotate to disrupt a ring of hydrophobic amino acids that blocks ion flow in the closed state. With ligand bound, the channel still flickers between open and closed states, but now it has a 90% probability of being open. This state continues until hydrolysis by a specific enzyme *(acetylcholinesterase)* located at the neuromuscular junction lowers the concentration of acetylcholine sufficiently. Once freed of its bound neurotransmitter, the acetylcholine receptor reverts to its initial resting state. If the presence of acetylcholine persists for a prolonged time as a result of excessive nerve stimulation, the channel inactivates (**Figure 11–37**).

The general shape of the acetylcholine receptor and the likely arrangement of its subunits have been determined by electron microscopy (**Figure 11–38**). The five subunits are arranged in a ring, forming a water-filled transmembrane channel that consists of a narrow pore through the lipid bilayer, which widens into vestibules at both ends. Clusters of negatively charged amino acids at either end of the pore help to exclude negative ions and encourage any positive ion of diameter less than 0.65 nm to pass through. The normal traffic consists chiefly of Na^+ and K^+, together with some Ca^{2+}. Thus, unlike voltage-gated cation channels, such as the K^+ channel discussed earlier, there is little selectivity among cations, and the relative contributions of the different cations to the current through the channel depend chiefly on their concentrations and on the electrochemical driving forces. When the muscle cell membrane is at its resting potential, the net driving force for K^+ is near zero, since the voltage gradient nearly balances the K^+ concentration gradient across the membrane (see Panel 11–2, p. 670). For Na^+, in contrast, the voltage gradient and the concentration gradient both act in the same direction to drive the ion into the cell. (The same is true for Ca^{2+}, but the extracellular concentration of Ca^{2+} is so much lower than that of Na^+ that Ca^{2+} makes only a small contribution to the total inward current.)

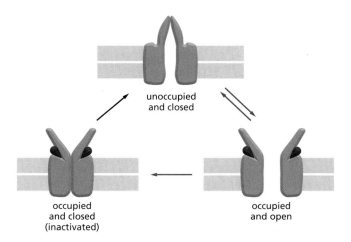

unoccupied
and closed

occupied
and closed
(inactivated)

occupied
and open

Figure 11–37 Three conformations of the acetylcholine receptor. The binding of two acetylcholine molecules opens this transmitter-gated ion channel. It then maintains a high probability of being open until the acetylcholine has been hydrolyzed. In the persistent presence of acetylcholine, however, the channel inactivates (desensitizes). Normally, the acetylcholine is rapidly hydrolyzed and the channel closes within about 1 millisecond, well before significant desensitization occurs. Desensitization would occur after about 20 milliseconds in the continued presence of acetylcholine.

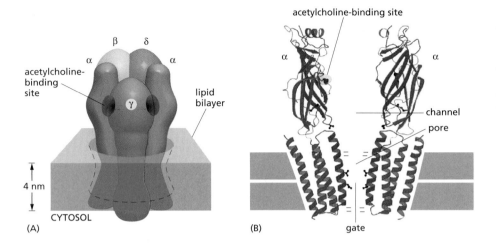

Figure 11–38 A model for the structure of the acetylcholine receptor. (A) Five homologous subunits (α, α, β, γ, δ) combine to form a transmembrane aqueous pore. The pore is lined by a ring of five transmembrane α helices, one contributed by each subunit. In its closed conformation, the pore is thought to be occluded by the hydrophobic side chains of five leucines, one from each α helix, which form a gate near the middle of the lipid bilayer. The negatively charged side chains at either end of the pore ensure that only positively charged ions pass through the channel. (B) Both of the α subunits contribute to an acetylcholine-binding site nestled between adjoining subunits; when acetylcholine binds to both sites, the channel undergoes a conformational change that opens the gate, possibly by rotating the helices containing the occluding leucines to move outward. In the structural drawing *(right)*, the parts of the channel that move in response to AChR binding to open the pore are colored in *blue*. (Adapted from N. Unwin, *Cell* 72[Suppl.]:31–41, 1993. With permission from Elsevier.)

Therefore, the opening of the acetylcholine receptor channels leads to a large net influx of Na^+ (a peak rate of about 30,000 ions per channel each millisecond). This influx causes a membrane depolarization that signals the muscle to contract, as discussed below.

Transmitter-Gated Ion Channels Are Major Targets for Psychoactive Drugs

The ion channels that open directly in response to the neurotransmitters acetylcholine, serotonin, GABA, and glycine contain subunits that are structurally similar and probably form transmembrane pores in the same way, even though they have distinct neurotransmitter-binding specificities and ion selectivities. These channels are all built from homologous polypeptide subunits, which probably assemble as a pentamer resembling the acetylcholine receptor. Glutamate-gated ion channels are constructed from a distinct family of subunits and are thought to form tetramers resembling the K^+ channels discussed earlier.

For each class of transmitter-gated ion channel, there are alternative forms of each type of subunit, either encoded by distinct genes or generated by alternative RNA splicing of the same gene product. The subunits assemble in different combinations to form an extremely diverse set of distinct channel subtypes, with different ligand affinities, different channel conductances, different rates of opening and closing, and different sensitivities to drugs and toxins. Vertebrate neurons, for example, have acetylcholine-gated ion channels that differ from those of muscle cells in that they are usually formed from two subunits of one type and three of another; but there are at least nine genes coding for different versions of the first type of subunit and at least three coding for different versions of the second, with further diversity due to alternative RNA splicing. Subsets of acetylcholine-sensitive neurons performing different functions in the brain express different combinations of these subunits. This, in principle, and already to some extent in practice, makes it possible to design drugs targeted against narrowly defined groups of neurons or synapses, thereby specifically influencing particular brain functions.

Indeed, transmitter-gated ion channels have for a long time been important targets for drugs. A surgeon, for example, can relax muscles for the duration of an operation by blocking the acetylcholine receptors on skeletal muscle cells with *curare*, a drug from a plant that was originally used by South American Indians to make poison arrows. Most drugs used to treat insomnia, anxiety, depression, and schizophrenia exert their effects at chemical synapses, and many of these act by binding to transmitter-gated channels. Both barbiturates and tranquilizers, such as Valium and Librium, for example, bind to GABA receptors, potentiating the inhibitory action of GABA by allowing lower concentrations of this neurotransmitter to open Cl^- channels. The new molecular biology of ion channels, by revealing both their diversity and the details of their structure, holds out the hope of designing a new generation of psychoactive drugs that will act still more selectively to alleviate the miseries of mental illness.

In addition to ion channels, many other components of the synaptic signaling machinery are potential targets for psychoactive drugs. As mentioned earlier, after release into the synaptic cleft, many neurotransmitters are cleared by reuptake mechanisms mediated by Na^+-driven transporters. The inhibition of such a transporter prolongs the effect of the transmitter and thereby strengthens synaptic transmission. Many antidepressant drugs, including Prozac, for example, inhibit the uptake of serotonin; others inhibit the uptake of both serotonin and norepinephrine.

Ion channels are the basic molecular components from which neuronal devices for signaling and computation are built. To provide a glimpse of how sophisticated the functions of these devices can be, we consider several examples that demonstrate how groups of ion channels work together in synaptic communication between electrically excitable cells.

Neuromuscular Transmission Involves the Sequential Activation of Five Different Sets of Ion Channels

The following process, in which a nerve impulse stimulates a muscle cell to contract, illustrates the importance of ion channels to electrically excitable cells. This apparently simple response requires the sequential activation of at least five different sets of ion channels, all within a few milliseconds (**Figure 11–39**).

1. The process is initiated when the nerve impulse reaches the nerve terminal and depolarizes the plasma membrane of the terminal. The depolarization transiently opens voltage-gated Ca^{2+} channels in this membrane. As the Ca^{2+} concentration outside cells is more than 1000 times greater than the free Ca^{2+} concentration inside, Ca^{2+} flows into the nerve terminal. The increase in Ca^{2+} concentration in the cytosol of the nerve terminal triggers the local release of acetylcholine into the synaptic cleft.

2. The released acetylcholine binds to acetylcholine receptors in the muscle cell plasma membrane, transiently opening the cation channels associated with them. The resulting influx of Na^+ causes a local membrane depolarization.

3. The local depolarization of the muscle cell plasma membrane opens voltage-gated Na^+ channels in this membrane, allowing more Na^+ to enter, which further depolarizes the membrane. This, in turn, opens neighboring voltage-gated Na^+ channels and results in a self-propagating depolarization (an action potential) that spreads to involve the entire plasma membrane (see Figure 11–30).

Figure 11–39 The system of ion channels at a neuromuscular junction. These gated ion channels are essential for the stimulation of muscle contraction by a nerve impulse. The various channels are numbered in the sequence in which they are activated, as described in the text.

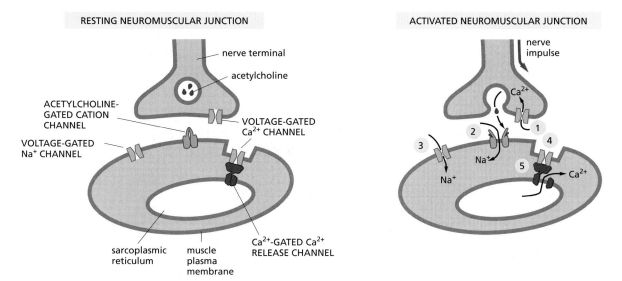

RESTING NEUROMUSCULAR JUNCTION

nerve terminal

acetylcholine

ACETYLCHOLINE-GATED CATION CHANNEL

VOLTAGE-GATED Ca^{2+} CHANNEL

VOLTAGE-GATED Na^+ CHANNEL

sarcoplasmic reticulum

muscle plasma membrane

Ca^{2+}-GATED Ca^{2+} RELEASE CHANNEL

ACTIVATED NEUROMUSCULAR JUNCTION

nerve impulse

Ca^{2+}

Na^+

Na^+

Ca^{2+}

4. The generalized depolarization of the muscle cell plasma membrane activates voltage-gated Ca^{2+} channels in specialized regions (the transverse [T] tubules—discussed in Chapter 16) of this membrane.

5. This, in turn, causes Ca^{2+}-gated *Ca^{2+} release channels* in an adjacent region of the sarcoplasmic reticulum (SR) membrane to open transiently and release the Ca^{2+} stored in the SR into the cytosol. The T-tubule and SR membranes are closely apposed with the two types of channels joined together in a specialized structure (see Figure 16–77). It is the sudden increase in the cytosolic Ca^{2+} concentration that causes the myofibrils in the muscle cell to contract.

Whereas the activation of muscle contraction by a motor neuron is complex, an even more sophisticated interplay of ion channels is required for a neuron to integrate a large number of input signals at synapses and compute an appropriate output, as we now discuss.

Single Neurons Are Complex Computation Devices

In the central nervous system, a single neuron can receive inputs from thousands of other neurons, and can in turn form synapses with many thousands of other cells. Several thousand nerve terminals, for example, make synapses on an average motor neuron in the spinal cord; its cell body and dendrites are almost completely covered with them (**Figure 11–40**). Some of these synapses transmit signals from the brain or spinal cord; others bring sensory information from muscles or from the skin. The motor neuron must combine the information received from all these sources and react either by firing action potentials along its axon or by remaining quiet.

Of the many synapses on a neuron, some tend to excite it, others to inhibit it. Neurotransmitter released at an excitatory synapse causes a small depolarization in the postsynaptic membrane called an *excitatory postsynaptic potential (excitatory PSP)*, while neurotransmitter released at an inhibitory synapse generally causes a small hyperpolarization called an *inhibitory PSP*. The membrane of the dendrites and cell body of most neurons contains a relatively low density of voltage-gated Na^+ channels, and an individual excitatory PSP is generally too small to

Figure 11–40 A motor neuron cell body in the spinal cord. (A) Many thousands of nerve terminals synapse on the cell body and dendrites. These deliver signals from other parts of the organism to control the firing of action potentials along the single axon of this large cell. (B) Micrograph showing a nerve cell body and its dendrites stained with a fluorescent antibody that recognizes a cytoskeletal protein *(green)*. Thousands of axon terminals *(red)* from other nerve cells (not visible) make synapses on the cell body and dendrites; they are stained with a fluorescent antibody that recognizes a protein in synaptic vesicles. (B, courtesy of Olaf Mundigl and Pietro de Camilli.)

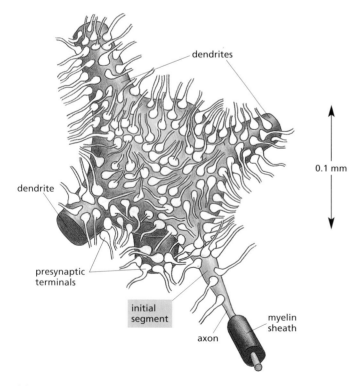

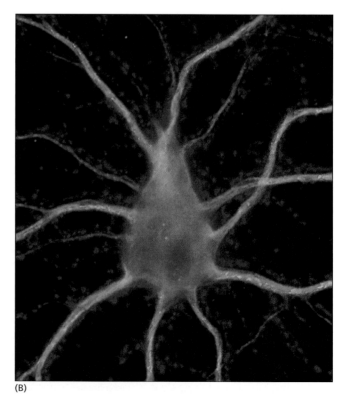

dendrites

0.1 mm

dendrite

presynaptic terminals

initial segment

myelin sheath

axon

(A)

(B)

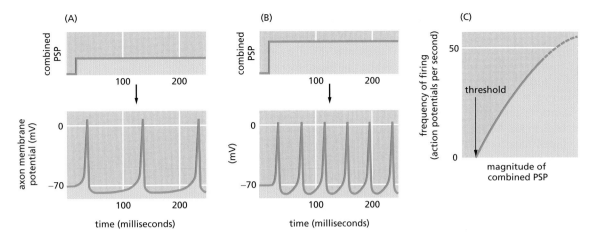

Figure 11–41 The magnitude of the combined postsynaptic potential (PSP) is reflected in the frequency of firing of action potentials. When successive action potentials arrive at the same synapse, each PSP produced adds to the preceding one to produce a larger combined PSP. A comparison of (A) and (B) shows how the firing frequency of an axon increases with an increase in the combined PSP, while (C) summarizes the general relationship.

trigger an action potential. Instead, each incoming signal is reflected in a local PSP of graded magnitude, which decreases with distance from the site of the synapse. If signals arrive simultaneously at several synapses in the same region of the dendritic tree, the total PSP in that neighborhood will be roughly the sum of the individual PSPs, with inhibitory PSPs making a negative contribution to the total. The PSPs from each neighborhood spread passively and converge on the cell body. For long-distance transmission, the combined magnitude of the PSP is then translated, or *encoded*, into the *frequency* of firing of action potentials (**Figure 11–41**). This encoding is achieved by a special set of gated ion channels that are present at high density at the base of the axon, adjacent to the cell body, in a region known as the *initial segment*, or *axon hillock* (see Figure 11–40).

Neuronal Computation Requires a Combination of at Least Three Kinds of K⁺ Channels

We have seen that the intensity of stimulation a neuron receives is encoded for long-distance transmission by the frequency of action potentials that the neuron fires: the stronger the stimulation, the higher the frequency of action potentials. Action potentials are initiated at the **initial segment**, a unique region of each neuron with plentiful voltage-gated Na⁺ channels. But to perform its special function of encoding, the membrane of the initial segment also contains at least four other classes of ion channels—three selective for K⁺ and one selective for Ca²⁺. The three varieties of K⁺ channels have different properties; we shall refer to them as *delayed, rapidly inactivating,* and *Ca²⁺-activated K⁺ channels.*

To understand the need for multiple types of channels, consider first what would happen if the only voltage-gated ion channels present in the nerve cell were the Na⁺ channels. Below a certain threshold level of synaptic stimulation, the depolarization of the initial segment membrane would be insufficient to trigger an action potential. With gradually increasing stimulation, the threshold would be crossed, the Na⁺ channels would open, and an action potential would fire. The action potential would be terminated in the usual way by inactivation of the Na⁺ channels. Before another action potential could fire, these channels would have to recover from their inactivation. But that would require a return of the membrane voltage to a very negative value, which would not occur as long as the strong depolarizing stimulus (from PSPs) was maintained. An additional channel type is needed, therefore, to repolarize the membrane after each action potential to prepare the cell to fire again.

The **delayed K⁺ channels** perform this task, as discussed previously in relation to the propagation of the action potential (see p. 677). They are voltage-gated, but because of their slower kinetics they open only during the falling phase of the action potential, when the Na^+ channels are inactive. Their opening permits an efflux of K^+ that drives the membrane back toward the K^+ equilibrium potential, which is so negative that the Na^+ channels rapidly recover from their inactivated state. Repolarization of the membrane also closes the delayed K^+ channels. The initial segment is now reset so that the depolarizing stimulus from synaptic inputs can fire another action potential. In this way, sustained stimulation of the dendrites and cell body leads to repetitive firing of the axon.

Repetitive firing in itself, however, is not enough. The frequency of the firing has to reflect the intensity of the stimulation, and a simple system of Na^+ channels and delayed K^+ channels is inadequate for this purpose. Below a certain threshold level of steady stimulation, the cell will not fire at all; above that threshold level, it will abruptly begin to fire at a relatively rapid rate. The **rapidly inactivating K^+ channels** solve the problem. These, too, are voltage-gated and open when the membrane is depolarized, but their specific voltage sensitivity and kinetics of inactivation are such that they act to reduce the rate of firing at levels of stimulation that are only just above the threshold required for firing. Thus, they remove the discontinuity in the relationship between the firing rate and the intensity of stimulation. The result is a firing rate that is proportional to the strength of the depolarizing stimulus over a very broad range (see Figure 11–41C).

The process of encoding is usually further modulated by the two other types of ion channels in the initial segment that were mentioned at the outset, namely voltage-gated Ca^{2+} channels and Ca^{2+}-activated K^+ channels. They act together to decrease the response of the cell to an unchanging, prolonged stimulation—a process called **adaptation**. These Ca^{2+} channels are similar to the Ca^{2+} channels that mediate the release of neurotransmitter from presynaptic axon terminals; they open when an action potential fires, transiently allowing Ca^{2+} into the initial segment.

The **Ca^{2+}-activated K^+ channel** is both structurally and functionally different from any of the channel types described earlier. It opens in response to a raised concentration of Ca^{2+} at the cytoplasmic face of the nerve cell membrane. Suppose we apply a strong depolarizing stimulus for a long time, triggering a long train of action potentials. Each action potential permits a brief influx of Ca^{2+} through the voltage-gated Ca^{2+} channels, so that the intracellular Ca^{2+} concentration gradually builds up to a level high enough to open the Ca^{2+}-activated K^+ channels. Because the resulting increased permeability of the membrane to K^+ makes the membrane harder to depolarize, it increases the delay between one action potential and the next. In this way, a neuron that is stimulated continuously for a prolonged period becomes gradually less responsive to the constant stimulus.

Such adaptation, which can also occur by other mechanisms, allows a neuron—indeed, the nervous system generally—to react sensitively to *change*, even against a high background level of steady stimulation. It is one of the strategies that help us, for example, to feel a light touch on the shoulder and yet ignore the constant pressure of our clothing. We discuss adaptation as a general feature in cell signaling processes in more detail in Chapter 15.

Other neurons do different computations, reacting to their synaptic inputs in myriad ways, reflecting the different assortments of members of the various ion channel families that reside in their membranes. There are several hundred genes that code for ion channels in the human genome, with over 150 encoding voltage-gated channels alone. Further complexity is introduced by alternative splicing of RNAs and assembling channels from different combinations of diverse subunits. The multiplicity of ion channels evidently allows for many different types of neurons, the electrical behavior of which is specifically tuned to the particular tasks that they must perform.

One of the crucial properties of the nervous system is its ability to learn and remember, which seems to depend largely on long-term changes in specific synapses. We end this chapter by considering a remarkable type of ion channel that is thought to have a special role in some forms of learning and memory. It is located at many synapses in the central nervous system, where it is gated by both voltage and the excitatory neurotransmitter glutamate. It is also the site of action of the psychoactive drug phencyclidine, or angel dust.

Long-Term Potentiation (LTP) in the Mammalian Hippocampus Depends on Ca^{2+} Entry Through NMDA-Receptor Channels

Practically all animals can learn, but mammals seem to learn exceptionally well (or so we like to think). In a mammal's brain, the region called the *hippocampus* has a special role in learning. When it is destroyed on both sides of the brain, the ability to form new memories is largely lost, although previous long-established memories remain. Correspondingly, some synapses in the hippocampus show marked functional alterations with repeated use: whereas occasional single action potentials in the presynaptic cells leave no lasting trace, a short burst of repetitive firing causes **long-term potentiation (LTP)**, such that subsequent single action potentials in the presynaptic cells evoke a greatly enhanced response in the postsynaptic cells. The effect lasts hours, days, or weeks, according to the number and intensity of the bursts of repetitive firing. Only the synapses that were activated exhibit LTP; synapses that have remained quiet on the same postsynaptic cell are not affected. However, while the cell is receiving a burst of repetitive stimulation via one set of synapses, if a single action potential is delivered at *another* synapse on its surface, that latter synapse also will undergo LTP, even though a single action potential delivered there at another time would leave no such lasting trace.

The underlying rule in such synapses seems to be that *LTP occurs on any occasion when a presynaptic cell fires (once or more) at a time when the post-synaptic membrane is strongly depolarized* (either through recent repetitive firing of the same presynaptic cell or by other means). This rule reflects the behavior of a particular class of ion channels in the postsynaptic membrane. Glutamate is the main excitatory neurotransmitter in the mammalian central nervous system, and glutamate-gated ion channels are the most common of all transmitter-gated channels in the brain. In the hippocampus, as elsewhere, most of the depolarizing current responsible for excitatory PSPs is carried by glutamate-gated ion channels, called **AMPA receptors**, that operate in the standard way. But the current has, in addition, a second and more intriguing component, which is mediated by a separate subclass of glutamate-gated ion channels known as **NMDA receptors**, so named because they are selectively activated by the artificial glutamate analog N-methyl-D-aspartate. The NMDA-receptor channels are doubly gated, opening only when two conditions are satisfied simultaneously: glutamate must be bound to the receptor, and the membrane must be strongly depolarized. The second condition is required for releasing the Mg^{2+} that normally blocks the resting channel. This means that NMDA receptors are normally activated only when AMPA receptors are activated as well and depolarize the membrane. The NMDA receptors are critical for LTP. When they are selectively blocked with a specific inhibitor, or in transgenic animals in which the gene has been knocked out, LTP does not occur, even though ordinary synaptic transmission continues. Such animals exhibit specific deficits in their learning abilities but behave almost normally otherwise.

How do NMDA receptors mediate such a remarkable effect? The answer is that these channels, when open, are highly permeable to Ca^{2+}, which acts as an intracellular mediator in the postsynaptic cell, triggering a cascade of changes that are responsible for LTP. Thus, LTP is prevented when Ca^{2+} levels are held artificially low in the postsynaptic cell by injecting the Ca^{2+} chelator EGTA into it, and LTP can be induced by artificially raising intracellular Ca^{2+} levels. Among the long-term changes that increase the sensitivity of the postsynaptic cell to glutamate is the insertion of new AMPA receptors into the plasma membrane (**Figure 11–42**). Evidence also indicates that changes can occur in the presynaptic cell as well, so that it releases more glutamate than normal when it is activated subsequently.

If synapses expressed only LTP they would quickly become saturated and, thus, be of limited value as an information storage device. In fact, synapses also exhibit **long-term depression (LTD)**, which surprisingly also requires NMDA receptor activation and a rise in Ca^{2+}. How does Ca^{2+} trigger opposite effects at the same synapse? It turns out that this bidirectional control of synaptic strength depends on the magnitude of the rise in Ca^{2+}: high Ca^{2+} levels activate protein kinases and LTP, whereas modest Ca^{2+} levels activate protein phosphatases and LTD.

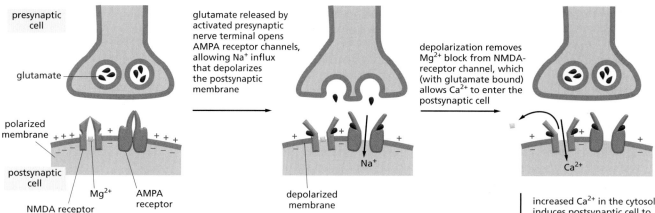

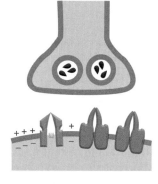

There is evidence that NMDA receptors have an important role in learning and related phenomena in other parts of the brain, as well as in the hippocampus. In Chapter 21 we see, moreover, that NMDA receptors have a crucial role in adjusting the anatomical pattern of synaptic connections in the light of experience during the development of the nervous system.

Thus, neurotransmitters released at synapses, besides relaying transient electrical signals, can also alter concentrations of intracellular mediators that bring about lasting changes in the efficacy of synaptic transmission. However, it is still uncertain how these changes endure for weeks, months, or a lifetime in the face of the normal turnover of cell constituents.

Some of the ion channels that we have discussed are summarized in **Table 11–2**.

Figure 11–42 The signaling events in long-term potentiation. Although not shown, evidence suggests that changes can also occur in the presynaptic nerve terminals in LTP, which may be stimulated by retrograde signals from the postsynaptic cell.

Summary

Ion channels form aqueous pores across the lipid bilayer and allow inorganic ions of appropriate size and charge to cross the membrane down their electrochemical gradients at rates about 1000 times greater than those achieved by any known transporter. The channels are "gated" and usually open transiently in response to a specific perturbation in the membrane, such as a change in membrane potential (voltage-gated channels) or the binding of a neurotransmitter (transmitter-gated channels).

K⁺-selective leak channels have an important role in determining the resting membrane potential across the plasma membrane in most animal cells. Voltage-gated cation channels are responsible for the generation of self-amplifying action potentials in electrically excitable cells, such as neurons and skeletal muscle cells. Transmitter-gated ion channels convert chemical signals to electrical signals at chemical synapses. Excitatory neurotransmitters, such as acetylcholine and glutamate, open transmitter-gated cation channels and thereby depolarize the postsynaptic membrane toward the threshold level for firing an action potential. Inhibitory neurotransmitters, such as GABA and glycine, open transmitter-gated Cl⁻ or K⁺ channels and thereby suppress firing by keeping the postsynaptic membrane polarized. A subclass of glutamate-gated ion channels, called NMDA-receptor channels, is highly permeable to Ca²⁺, which can trigger the long-term changes in synapses such as LTP and LTD that are thought to be involved in some forms of learning and memory.

Table 11–2 Some Ion Channel Families

CHANNEL TYPE	REPRESENTATIVE EXAMPLE	
Voltage-gated cation channels	voltage-gated Na⁺ channels voltage-gated K⁺ channels (including delayed and early) voltage-gated Ca²⁺ channels	
Transmitter-gated ion channels	acetylcholine-gated cation channels glutamate-gated Ca²⁺ channels serotonin-gated cation channels	excitatory
	GABA-gated Cl⁻ channels glycine-gated Cl⁻ channels	inhibitory

Ion channels work together in complex ways to control the behavior of electrically excitable cells. A typical neuron, for example, receives thousands of excitatory and inhibitory inputs, which combine by spatial and temporal summation to produce a postsynaptic potential (PSP) in the cell body. The magnitude of the PSP is translated into the rate of firing of action potentials by a mixture of cation channels in the membrane of the initial segment.

PROBLEMS

Which statements are true? Explain why or why not.

11–1 Transport by transporters can be either active or passive, whereas transport by channels is always passive.

11–2 Transporters saturate at high concentrations of the transported molecule when all their binding sites are occupied; channels, on the other hand, do not bind the ions they transport and thus the flux of ions through a channel does not saturate.

11–3 The membrane potential arises from movements of charge that leave ion concentrations practically unaffected, causing only a very slight discrepancy in the number of positive and negative ions on the two sides of the membrane.

Discuss the following problems.

11–4 Order Ca^{2+}, CO_2, ethanol, glucose, RNA, and H_2O according to their ability to diffuse through a lipid bilayer, beginning with the one that crosses the bilayer most readily. Explain your order.

11–5 How is it possible for some molecules to be at equilibrium across a biological membrane and yet not be at the same concentration on both sides?

11–6 Ion transporters are "linked" together—not physically, but as a consequence of their actions. For example, cells can raise their intracellular pH, when it becomes too acidic, by exchanging external Na^+ for internal H^+, using a Na^+–H^+ antiporter. The change in internal Na^+ is then redressed using the Na^+-K^+ pump.
A. Can these two transporters, operating together, normalize both the H^+ and the Na^+ concentrations inside the cell?
B. Does the linked action of these two pumps cause imbalances in either the K^+ concentration or the membrane potential? Why or why not?

11–7 Microvilli increase the surface area of intestinal cells, providing more efficient absorption of nutrients. Microvilli are shown in profile and cross section in **Figure Q11–1**. From the dimensions given in the figure, estimate the increase in surface area that microvilli provide (for the portion of the plasma membrane in contact with the lumen of the gut) relative to the corresponding surface of a cell with a "flat" plasma membrane.

11–8 According to Newton's laws of motion, an ion exposed to an electric field in a vacuum would experience a constant acceleration from the electric driving force, just as a falling body in a vacuum constantly accelerates due to gravity. In water, however, an ion moves at constant velocity in an electric field. Why do you suppose that is?

11–9 The "ball-and-chain" model for the rapid inactivation of voltage-gated K^+ channels has been elegantly confirmed for the *shaker* K^+ channel from *Drosophila melanogaster*. (The *shaker* K^+ channel in *Drosophila* is named after a mutant form that causes excitable behavior—even anesthetized flies keep twitching.) Deletion of the N-terminal amino acids from the normal *shaker* channel gives rise to a channel that opens in response to membrane depolarization, but stays open instead of rapidly closing as the normal channel does. A peptide (MAAVAGLYGLGEDRQHRKKQ) that corresponds to the deleted N-terminus can inactivate the open channel at 100 μM.

Is the concentration of free peptide (100 μM) that is required to inactivate the defective K^+ channel anywhere near the normal local concentration of the tethered ball on a normal channel? Assume that the tethered ball can explore a hemisphere [volume = $(2/3)\pi r^3$] with a radius of 21.4 nm, the length of the polypeptide "chain" (**Figure Q11–2**). Calculate the concentration for one ball in this hemisphere. How does that value compare with the concentration of free peptide needed to inactivate the channel?

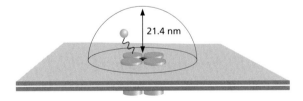

Figure Q11–2 A "ball" tethered by a "chain" to a voltage-gated K^+ channel (Problem 11–9).

11–10 The squid giant axon occupies a unique position in the history of our understanding of cell membrane potentials and nerve action. When an electrode is stuck into an intact giant axon, the membrane potential registers –70 mV. When the axon, suspended in a bath of seawater, is stimulated to conduct a nerve impulse, the membrane potential changes transiently from –70 mV to +40 mV.

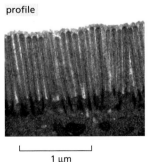
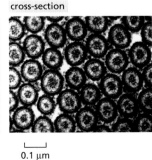

profile   cross-section

1 μm   0.1 μm

Figure Q11–1 Microvilli of intestinal epithelial cells in profile and cross section (Problem 11–7). (Left panel, from Rippel Electron Microscope Facility, Dartmouth College; Right panel, from David Burgess.)

For univalent ions and at 20°C (293 K), the Nernst equation reduces to

$$V = 58 \text{ mV} \times \log (C_0/C_i)$$

where C_0 and C_i are the concentrations outside and inside, respectively.

Using this equation, calculate the potential across the resting membrane (1) assuming that it is due solely to K^+ and (2) assuming that it is due solely to Na^+. (The Na^+ and K^+ concentrations in axon cytoplasm and in seawater are given in **Table Q11–1**.) Which calculation is closer to the measured resting potential? Which calculation is closer to the measured action potential? Explain why these assumptions approximate the measured resting and action potentials.

Table Q11–1 Ionic composition of seawater and of cytoplasm from the squid giant axon (Problem 11–10).

ION	CYTOPLASM	SEAWATER
Na^+	65 mM	430 mM
K^+	344 mM	9 mM

REFERENCES

General

Martonosi AN (ed) (1985) The enzymes of Biological membranes, vol 3: Membrane transport. 2nd edn. New York: Penum Press.
Stein WD (1990) Channels, carriers and pumps: An introduction to membrane transport. San Diego: Academic Press

Principles of Membrane Transport

Al-Awqati Q (1999) One hundred years of membrane permeability: does Overton still rule? *Nature Cell Biol* 1: E201–E202.
Forrest LR & Sansom MS (2000) Membrane simulations: bigger and better? *Curr Opin Struct Biol* 10:174–181.
Gouaux E and MacKinnon R (2005) Principles of selective ion transport in channels and pumps. *Science* 310:1461–1465.
Mitchell P (1977) Vectorial chemiosmotic processes. *Annu Rev Biochem* 46:996–1005.
Tanford C (1983) Mechanism of free energy coupling in active transport. *Annu Rev Biochem* 52:379–409.

Carrier Proteins and Active Membrane Transport

Almers W & Stirling C (1984) Distribution of transport proteins over animal cell membranes. *J Membr Biol* 77:169–186.
Baldwin SA & Henderson PJ (1989) Homologies between sugar transporters from eukaryotes and prokaryotes. *Annu Rev Physiol* 51:459–471.
Borst P & Elferink RO (2002) Mammalian ABC transporters in health and disease. *Annu Rev Biochem* 71:537–592.
Carafoli E & Brini M (2000) Calcium pumps: structural basis for and mechanism of calcium transmembrane transport. *Curr Opin Chem Biol* 4:152–161.
Dean M, Rzhetsky A et al (2001) The human ATP-binding cassette (ABC) transporter superfamily. *Genome Res* 11:1156–1166.
Doige CA & Ames GF (1993) ATP-dependent transport systems in bacteria and humans: relevance to cystic fibrosis and multidrug resistance. *Annu Rev Microbiol* 47:291–319.
Gadsby DC, Vergani P & Csanady L (2006) The ABC protein turned chloride channel whose failure causes cystic fibrosis. *Nature* 440:477–83.
Higgins CF (2007) Multiple molecular mechanisms for multidrug resistance transporters. *Nature* 446:749–57.
Kaback HR, Sahin-Toth M et al (2001) The kamikaze approach to membrane transport. *Nature Rev Mol Cell Biol* 2:610–620.
Kühlbrandt W (2004) Biology, structure and mechanism of P-type ATPases. *Nature Rev Mol Cell Biol* 5:282–295.
Lodish HF (1986) Anion-exchange and glucose transport proteins: structure, function, and distribution. *Harvey Lect* 82:19–46.
Pedersen PL & Carafoli E (1987) Ion motive ATPases. 1. Ubiquity, properties, and significance to cell function. *Trends Biochem Sci* 12:146–150.
Romero MF & Boron WF (1999) Electrogenic Na^+/HCO_3^- cotransporters: cloning and physiology. *Annu Rev Physiol* 61:699–723.
Saier MH, Jr (2000) Vectorial metabolism and the evolution of transport systems. *J Bacteriol* 182:5029–5035.
Scarborough GA (2003) Rethinking the P-type ATPase problem. *Trends Biochem Sci* 28:581–584.

Stein WD (2002) Cell volume homeostasis: ionic and nonionic mechanisms. The sodium pump in the emergence of animal cells. *Int Rev Cytol* 215:231–258.

Ion Channels and the Electrical Properties of Membranes

Armstrong C (1998) The vision of the pore. *Science* 280:56–57.
Choe S (2002) Potassium channel structures. *Nature Rev Neurosci* 3:115–21.
Choe S, Kreusch A & Pfaffinger PJ (1999) Towards the three-dimensional structure of voltage-gated potassium channels. *Trends Biochem Sci* 24:345–349.
Franks NP & Lieb WR (1994) Molecular and cellular mechanisms of general anaesthesia. *Nature* 367:607–614.
Greengard P (2001) The neurobiology of slow synaptic transmission. *Science* 294:1024–30.
Hille B (2001) Ionic Channels of Excitable Membranes, 3rd ed. Sunderland, MA: Sinauer.
Hucho F, Tsetlin VI & Machold J (1996) The emerging three-dimensional structure of a receptor. The nicotinic acetylcholine receptor. *Eur J Biochem* 239:539–557.
Hodgkin AL & Huxley AF (1952) A quantitative description of membrane current and its application to conduction and excitation in nerve. *J Physiol* 117:500–544.
Hodgkin AL & Huxley AF (1952) Currents carried by sodium and potassium ions through the membrane of the giant axon of *Loligo*. *J Physiol* 116:449–472.
Jessell TM & Kandel ER (1993) Synaptic transmission: a bidirectional and self-modifiable form of cell–cell communication. *Cell* 72[Suppl]:1–30.
Kandel ER, Schwartz JH & Jessell TM (2000) Principles of Neural Science, 4th ed. New York: McGraw-Hill.
Karlin A (2002) Emerging structure of the nicotinic acetylcholine receptors. *Nature Rev Neurosci* 3:102–114.
Katz B (1966) Nerve, Muscle and Synapse. New York: McGraw-Hill.
King LS, Kozono D & Agre P (2004) From structure to disease: the evolving tale of aquaporin biology. *Nature Rev Mol Cell Biol* 5:687–698.
MacKinnon R (2003) Potassium channels. *FEBS Lett* 555:62–65.
Malenka RC & Nicoll RA (1999) Long-term potentiation—a decade of progress? *Science* 285:1870–1874.
Moss SJ & Smart TG (2001) Constructing inhibitory synapses. *Nature Rev Neurosci* 2:240–250.
Neher E and Sakmann B (1992) The patch clamp technique. *Sci Am* 266:44–51.
Nicholls JG, Fuchs PA, Martin AR & Wallace BG (2000) From Neuron to Brain, 4th ed. Sunderland, MA: Sinauer.
Numa S (1987) A molecular view of neurotransmitter receptors and ionic channels. *Harvey Lect* 83:121–165.
Scannevin RH & Huganir RL (2000) Postsynaptic organization and regulation of excitatory synapses. *Nature Rev Neurosci* 1:133–141.
Seeburg PH (1993) The molecular biology of mammalian glutamate receptor channels. *Trends Neurosci* 16:359–365.
Snyder SH (1996) Drugs and the Brain. New York: WH Freeman/ Scientific American Books.
Stevens CF (2004) Presynaptic function. *Curr Opin Neurobiol* 14:341–345.
Tsien RW, Lipscombe D, Madison DV et al (1988) Multiple types of neuronal calcium channels and their selective modulation. *Trends Neurosci* 11:431–438.
Unwin N (2003) Structure and action of the nicotinic acetylcholine receptor explored by electron microscopy. *FEBS Lett* 555:91–95.

Chapter 12

Intracellular Compartments and Protein Sorting

12

Unlike a bacterium, which generally consists of a single intracellular compartment surrounded by a plasma membrane, a eucaryotic cell is elaborately subdivided into functionally distinct, membrane-enclosed compartments. Each compartment, or **organelle**, contains its own characteristic set of enzymes and other specialized molecules, and complex distribution systems transport specific products from one compartment to another. To understand the eucaryotic cell, it is essential to know how the cell creates and maintains these compartments, what occurs in each of them, and how molecules move between them.

Proteins confer upon each compartment its characteristic structural and functional properties. They catalyze the reactions that occur in each organelle and selectively transport small molecules into and out of its interior, or **lumen**. Proteins also serve as organelle-specific surface markers that direct new deliveries of proteins and lipids to the appropriate organelle.

An animal cell contains about 10 billion (10^{10}) protein molecules of perhaps 10,000 kinds, and the synthesis of almost all of them begins in the cytosol. Each newly synthesized protein is then delivered specifically to the cell compartment that requires it. The intracellular transport of proteins is the central theme of both this chapter and the next. By tracing the protein traffic from one compartment to another, one can begin to make sense of the otherwise bewildering maze of intracellular membranes.

THE COMPARTMENTALIZATION OF CELLS

In this brief overview of the compartments of the cell and the relationships between them, we organize the organelles conceptually into a small number of discrete families, discuss how proteins are directed to specific organelles, and explain how proteins cross organelle membranes.

All Eucaryotic Cells Have the Same Basic Set of Membrane-Enclosed Organelles

Many vital biochemical processes take place in or on membrane surfaces. Membrane-bound enzymes, for example, catalyze lipid metabolism, and oxidative phosphorylation and photosynthesis both require a membrane to couple the transport of H^+ to the synthesis of ATP. In addition to providing increased membrane area to host biochemical reactions, intracellular membrane systems form enclosed compartments that are separate from the cytosol, thus creating functionally specialized aqueous spaces within the cell. Because the lipid bilayer of organelle membranes is impermeable to most hydrophilic molecules, the membrane of each organelle must contain membrane transport proteins to import and export specific metabolites. Each organelle membrane must also have a mechanism for importing, and incorporating into the organelle, the specific proteins that make the organelle unique.

Figure 12–1 illustrates the major intracellular compartments common to eucaryotic cells. The *nucleus* contains the genome (aside from mitochondrial

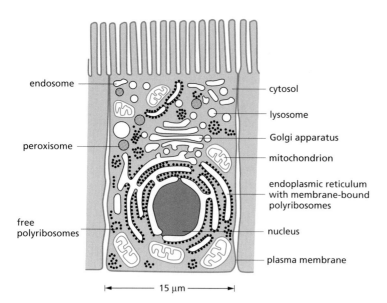

Figure 12–1 The major intracellular compartments of an animal cell. <ATCC> The cytosol *(gray),* endoplasmic reticulum, Golgi apparatus, nucleus, mitochondrion, endosome, lysosome, and peroxisome are distinct compartments isolated from the rest of the cell by at least one selectively permeable membrane.

and chloroplast DNA) and it is the principal site of DNA and RNA synthesis. The surrounding **cytoplasm** consists of the cytosol and the cytoplasmic organelles suspended in it. The **cytosol** constitutes a little more than half the total volume of the cell, and it is the site of protein synthesis and degradation. It also performs most of the cell's intermediary metabolism—that is, the many reactions that degrade some small molecules and synthesize others to provide the building blocks for macromolecules (discussed in Chapter 2).

About half the total area of membrane in a eucaryotic cell encloses the labyrinthine spaces of the *endoplasmic reticulum (ER)*. The *rough ER* has many ribosomes bound to its cytosolic surface; these synthesize both soluble and integral membrane proteins, most of which are destined either for secretion to the cell exterior or for other organelles. We shall see that, whereas proteins are transported into other organelles only after their synthesis is complete, they are transported into the ER as they are synthesized. This explains why the ER membrane is unique in having ribosomes tethered to it. The ER also produces most of the lipid for the rest of the cell and functions as a store for Ca^{2+} ions. Regions of the ER that lack bound ribosomes are called *smooth ER*. The ER sends many of its proteins and lipids to the *Golgi apparatus*, which consists of organized stacks of disclike compartments called *Golgi cisternae*. The Golgi apparatus receives lipids and proteins from the ER and dispatches them to various destinations, usually covalently modifying them *en route.*

Mitochondria and (in plants) *chloroplasts* generate most of the ATP that cells use to drive reactions requiring an input of free energy; chloroplasts are a specialized version of *plastids,* which can also have other functions in plant cells, such as the storage of food or pigment molecules. *Lysosomes* contain digestive enzymes that degrade defunct intracellular organelles, as well as macromolecules and particles taken in from outside the cell by endocytosis. On their way to lysosomes, endocytosed material must first pass through a series of organelles called *endosomes*. Finally, *peroxisomes* are small vesicular compartments that contain enzymes used in various oxidation reactions.

In general, each membrane-enclosed organelle performs the same set of basic functions in all cell types. But to serve the specialized functions of cells, these organelles vary in abundance and can have additional properties that differ from cell type to cell type.

On average, the membrane-enclosed compartments together occupy nearly half the volume of a cell (**Table 12–1**), and a large amount of intracellular membrane is required to make them all. In liver and pancreatic cells, for example, the endoplasmic reticulum has a total membrane surface area that is, respectively, 25 times and 12 times that of the plasma membrane (**Table 12–2**). In terms of its area and mass, the plasma membrane is only a minor membrane in most eucaryotic cells, and organelles are packed tightly in the cytosol (**Figure 12–2**).

Table 12–1 Relative Volumes Occupied by the Major Intracellular Compartments in a Liver Cell (Hepatocyte)

INTRACELLULAR COMPARTMENT	PERCENTAGE OF TOTAL CELL VOLUME
Cytosol	54
Mitochondria	22
Rough ER cisternae	9
Smooth ER cisternae plus Golgi cisternae	6
Nucleus	6
Peroxisomes	1
Lysosomes	1
Endosomes	1

Membrane-enclosed organelles often have characteristic positions in the cytosol. In most cells, for example, the Golgi apparatus is located close to the nucleus, whereas the network of ER tubules extends from the nucleus throughout the entire cytosol. These characteristic distributions depend on interactions of the organelles with the cytoskeleton. The localization of both the ER and the Golgi apparatus, for instance, depends on an intact microtubule array; if the microtubules are experimentally depolymerized with a drug, the Golgi apparatus fragments and disperses throughout the cell, and the ER network collapses toward the cell center (discussed in Chapter 16).

Evolutionary Origins Explain the Topological Relationships of Organelles

To understand the relationships between the compartments of the cell, it is helpful to consider how they might have evolved. The precursors of the first eucaryotic cells are thought to have been simple organisms that resembled bacteria, which generally have a plasma membrane but no internal membranes. The plasma membrane in such cells therefore provides all membrane-dependent functions, including the pumping of ions, ATP synthesis, protein secretion, and lipid synthesis. Typical present-day eucaryotic cells are 10–30 times larger in linear dimension and 1000–10,000 times greater in volume than a typical bacterium

Table 12–2 Relative Amounts of Membrane Types in Two Kinds of Eucaryotic Cells

MEMBRANE TYPE	PERCENTAGE OF TOTAL CELL MEMBRANE	
	LIVER HEPATOCYTE*	PANCREATIC EXOCRINE CELL*
Plasma membrane	2	5
Rough ER membrane	35	60
Smooth ER membrane	16	<1
Golgi apparatus membrane	7	10
Mitochondria		
Outer membrane	7	4
Inner membrane	32	17
Nucleus		
Inner membrane	0.2	0.7
Secretory vesicle membrane	not determined	3
Lysosome membrane	0.4	not determined
Peroxisome membrane	0.4	not determined
Endosome membrane	0.4	not determined

*These two cells are of very different sizes: the average hepatocyte has a volume of about 5000 μm^3 compared with 1000 μm^3 for the pancreatic exocrine cell. Total cell membrane areas are estimated at about 110,000 μm^2 and 13,000 μm^2, respectively.

rough endoplasmic reticulum nucleus lysosomes

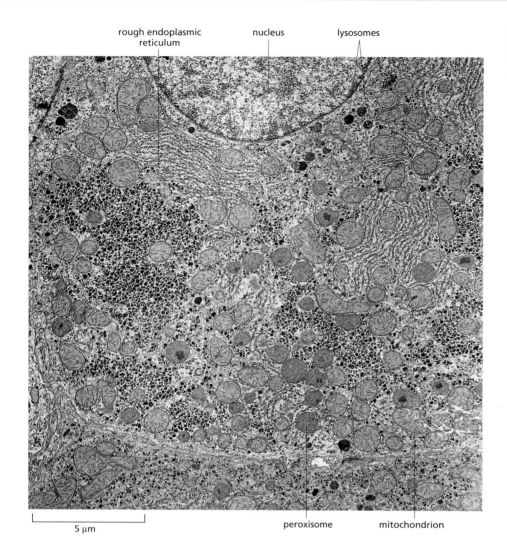

5 μm   peroxisome mitochondrion

Figure 12–2 An electron micrograph of part of a liver cell seen in cross section. Examples of most of the major intracellular compartments are indicated. (Courtesy of Daniel S. Friend.)

such as *E. coli*. The profusion of internal membranes can be seen, in part, as an adaptation to this increase in size: the eucaryotic cell has a much smaller ratio of surface area to volume, and its plasma membrane therefore presumably has too small an area to sustain the many vital functions that membranes perform. The extensive internal membrane systems of a eucaryotic cell alleviate this problem.

The evolution of internal membranes evidently accompanied the specialization of membrane function. Consider, for example, the generation of *thylakoid vesicles* in chloroplasts. These vesicles form when chloroplasts in green leaf cells develop from proplastids, small precursor organelles that are present in all immature plant cells. These organelles are surrounded by a double membrane and develop according to the needs of the differentiated cells: they turn into chloroplasts in leaf cells, for example, but into organelles that store starch, fat, or pigments in other cell types (**Figure 12–3**A). When they convert into chloroplasts, specialized membrane patches form, invaginate, and pinch off from the inner membrane of the proplastid. The vesicles that result form a new specialized compartment, the *thylakoid,* which harbors all of the chloroplast's photosynthetic machinery (Figure 12–3B).

Other compartments in eucaryotic cells may have originated in a conceptually similar way (**Figure 12–4**). The invagination and pinching off of specialized intracellular membrane structures from the plasma membrane creates organelles with an interior that is topologically equivalent to the exterior of the cell. We shall see that this topological relationship holds for all of the organelles involved in the secretory and endocytic pathways, including the ER, Golgi apparatus, endosomes, and lysosomes. We can therefore think of all of these organelles as members of the same family. As we discuss in detail in the next chapter, their interiors communicate extensively with one another and with the outside of the cell via *transport vesicles*, which bud off from one organelle and fuse with another (**Figure 12–5**).

As described in Chapter 14, mitochondria and plastids differ from the other membrane-enclosed organelles because they contain their own genomes. The nature of these genomes, and the close resemblance of the proteins in these organelles to those in some present-day bacteria, strongly suggest that mitochondria and plastids evolved from bacteria that were engulfed by other cells with which they initially lived in symbiosis (discussed in Chapters 1 and 14). According to the hypothetical scheme shown in Figure 12–4B, the inner membrane of mitochondria and plastids corresponds to the original plasma membrane of the bacterium, while the lumen of these organelles evolved from the bacterial cytosol. As we might expect from such an endocytic origin, these two organelles are surrounded by a double membrane, and they remain isolated from the extensive vesicular traffic that connects the interiors of most of the other membrane-enclosed organelles to each other and to the outside of the cell.

The evolutionary scheme described above groups the intracellular compartments in eucaryotic cells into four distinct families: (1) the nucleus and the cytosol, which communicate with each other through *nuclear pore complexes* and are thus topologically continuous (although functionally distinct); (2) all organelles that function in the secretory and endocytic pathways—including the ER, Golgi apparatus, endosomes, and lysosomes, the numerous classes of transport intermediates such as transport vesicles that move between them, and possibly peroxisomes; (3) the mitochondria; and (4) the plastids (in plants only).

Proteins Can Move Between Compartments in Different Ways

The synthesis of all proteins begins on ribosomes in the cytosol, except for the few that are synthesized on the ribosomes of mitochondria and plastids. Their subsequent fate depends on their amino acid sequence, which can contain **sorting signals** that direct their delivery to locations outside the cytosol. Most proteins do not have a sorting signal and consequently remain in the cytosol as permanent residents. Many others, however, have specific sorting signals that direct their transport from the cytosol into the nucleus, the ER, mitochondria, plastids, or peroxisomes; sorting signals can also direct the transport of proteins from the ER to other destinations in the cell.

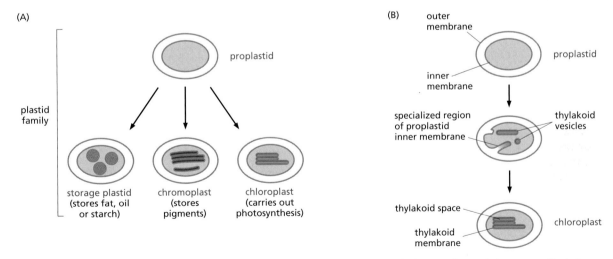

Figure 12–3 Development of plastids. (A) Proplastids are inherited with the cytoplasm of plant egg cells. As immature plant cells differentiate, the proplastids develop according to the needs of the specialized cell: they can become chloroplasts (in green leaf cells), storage plastids that accumulate starch (e.g., in potato tubers) or oil and lipid droplets (e.g., in fatty seeds), or chromoplasts that harbor pigments (e.g., in flower petals). (B) Development of the thylakoid. As chloroplasts develop, patches of specialized membrane in the proplastid inner membrane invaginate and pinch off to form thylakoid vesicles, which then develop into the mature thylakoid. The thylakoid membrane forms a separate compartment, the thylakoid space, which is structurally and functionally distinct from the rest of the chloroplast. Thylakoids can grow and divide autonomously as chloroplasts proliferate.

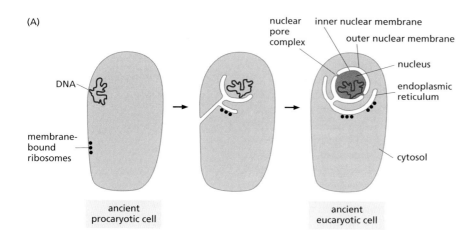

(A)

(B)

Figure 12–4 Hypothetical schemes for the evolutionary origins of some organelles. The origins of mitochondria, chloroplasts, ER, and the cell nucleus might explain the topological relationships of these compartments in eucaryotic cells.

(A) A possible pathway for the evolution of the cell nucleus and the ER. In some bacteria the single DNA molecule that comprises the bacterial chromosome is attached to an invagination of the plasma membrane. Such an invagination in a very ancient procaryotic cell could have rearranged to form an envelope around the DNA, while still allowing the DNA access to the cell cytosol (as is required for DNA to direct protein synthesis). This envelope is presumed to have eventually pinched off completely from the plasma membrane, producing a nuclear compartment surrounded by a double membrane.

As illustrated, gated passageways called nuclear pore complexes (NPCs) penetrate the nuclear envelope. Because it is surrounded by two membranes that are in continuity where they are penetrated by NPCs, the nuclear compartment is topologically equivalent to the cytosol; in fact, during mitosis the nuclear contents mix with the cytosol. The lumen of the ER is continuous with the space between the inner and outer nuclear membranes, and it is topologically equivalent to the extracellular space (see Figure 12–5).

(B) Mitochondria (and plastids) are thought to have originated when a bacterium was engulfed by a larger pre-eucaryotic cell. This could explain why they contain their own genomes and why the lumens of these organelles remain isolated from the membrane traffic that interconnects the lumens of many other intracellular compartments.

To understand the general principles by which sorting signals operate, it is important to distinguish three fundamentally different ways by which proteins move from one compartment to another. These three mechanisms are described below, and their sites of action in the cell are outlined in **Figure 12–6**. We discuss the first two mechanisms in this chapter, and the third (*green arrows* in Figure 12–6) in Chapter 13.

1. In **gated transport**, proteins move between the cytosol and the nucleus (which are topologically equivalent) through nuclear pore complexes in the nuclear envelope. The nuclear pore complexes function as selective gates that actively transport specific macromolecules and macromolecular assemblies, although they also allow free diffusion of smaller molecules.

2. In **transmembrane transport**, transmembrane *protein translocators* directly transport specific proteins across a membrane from the cytosol into a space that is topologically distinct. The transported protein molecule

Figure 12–5 Topological relationships between compartments of the secretory and endocytic pathways in a eucaryotic cell. Topologically equivalent spaces are shown in *red*. In principle, cycles of membrane budding and fusion permit the lumen of any of these organelles to communicate with any other and with the cell exterior by means of transport vesicles. *Blue arrows* indicate the extensive outbound and inbound vesicular traffic (discussed in Chapter 13). Some organelles, most notably mitochondria and (in plant cells) plastids, do not take part in this communication and are isolated from the traffic between organelles shown here.

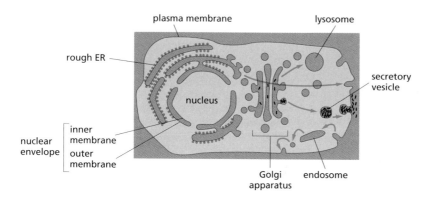

Figure 12–6 **A simplified "roadmap" of protein traffic.** Proteins can move from one compartment to another by gated transport *(red)*, transmembrane transport *(blue)*, or vesicular transport *(green)*. The sorting signals that direct a given protein's movement through the system, and thereby determine its eventual location in the cell, are contained in each protein's amino acid sequence. The journey begins with the synthesis of a protein on a ribosome in the cytosol and terminates when the protein reaches its final destination. At each intermediate station *(boxes)*, a decision is made as to whether the protein is to be retained in that compartment or transported further. In principle, a sorting signal could be required for either retention in or exit from a compartment.

We shall refer to this figure often as a guide in this chapter and the next, highlighting in color the particular pathway being discussed.

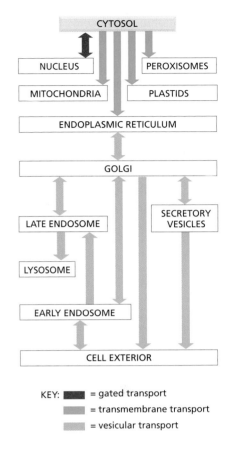

KEY: ■■ = gated transport

       ▐ = transmembrane transport

       ▐ = vesicular transport

usually must unfold to snake through the translocator. The initial transport of selected proteins from the cytosol into the ER lumen or mitochondria, for example, occurs in this way.

3. In **vesicular transport**, membrane-enclosed transport intermediates—which may be small, spherical transport vesicles or larger, irregularly shaped organelle fragments—ferry proteins from one compartment to another. The transport vesicles and fragments become loaded with a cargo of molecules derived from the lumen of one compartment as they bud and pinch off from its membrane; they discharge their cargo into a second compartment by fusing with the membrane enclosing that compartment (**Figure 12–7**). The transfer of soluble proteins from the ER to the Golgi apparatus, for example, occurs in this way. Because the transported proteins do not cross a membrane, vesicular transport can move proteins only between compartments that are topologically equivalent (see Figure 12–5).

Each mode of protein transfer is usually guided by sorting signals in the transported protein, which are recognized by complementary *sorting receptors*. If a large protein is to be imported into the nucleus, for example, it must possess a sorting signal that receptor proteins recognize to guide it through the nuclear pore complex. If a protein is to be transferred directly across a membrane, it must possess a sorting signal that the membrane translocator recognizes. Likewise, if a protein is to be loaded into a certain type of vesicle or retained in certain organelles, a complementary receptor in the appropriate membrane must recognize its sorting signal.

Signal Sequences Direct Proteins to the Correct Cell Address

Most protein sorting signals reside in a stretch of amino acid sequence, typically 15–60 residues long. These **signal sequences** are often found at the N-terminus; specialized **signal peptidases** remove the signal sequence from the finished protein once the sorting process is complete. Signal sequences can also be internal stretches of amino acids, which remain part of the protein. In some cases, sorting signals are composed of multiple internal amino acid sequences that form a specific three-dimensional arrangement of atoms on the protein's surface, called a **signal patch**.

Each signal sequence specifies a particular destination in the cell. Proteins destined for initial transfer to the ER usually have a signal sequence at their N-terminus, which characteristically includes a sequence composed of about 5–10

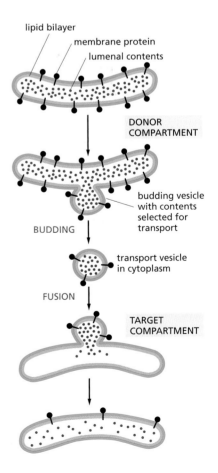

Figure 12–7 **Vesicle budding and fusion during vesicular transport.** Transport vesicles bud from one compartment (donor) and fuse with another (target) compartment. In the process, soluble components *(red dots)* are transferred from lumen to lumen. Note that membrane is also transferred and that the original orientation of both proteins and lipids in the donor-compartment membrane is preserved in the target-compartment membrane. Thus, membrane proteins retain their asymmetric orientation, with the same domains always facing the cytosol.

hydrophobic amino acids. Many of these proteins will in turn pass from the ER to the Golgi apparatus, but those with a specific signal sequence of four amino acids at their C-terminus are recognized as ER residents and are returned to the ER. Proteins destined for mitochondria have signal sequences of yet another type, in which positively charged amino acids alternate with hydrophobic ones. Finally, many proteins destined for peroxisomes have a signal sequence of three characteristic amino acids at their C-terminus.

Table 12–3 presents some specific signal sequences. Experiments in which the peptide is transferred from one protein to another by genetic engineering techniques have demonstrated the importance of each of these signal sequences for protein targeting. Placing the N-terminal ER signal sequence at the beginning of a cytosolic protein, for example, redirects the protein to the ER. Signal sequences are therefore both necessary and sufficient for protein targeting. Even though their amino acid sequences can vary greatly, the signal sequences of all proteins having the same destination are functionally interchangeable; physical properties, such as hydrophobicity, often seem to be more important in the signal-recognition process than the exact amino acid sequence.

Signal sequences are recognized by complementary sorting receptors that guide proteins to their appropriate destination, where the receptors unload their cargo. The receptors function catalytically: after completing one round of targeting, they return to their point of origin to be reused. Most sorting receptors recognize classes of proteins rather than an individual protein species. They can therefore be viewed as public transportation systems, dedicated to delivering different components to their correct location in the cell.

Panel 12–1 describes the main ways of studying how proteins are directed from the cytosol to a specific compartment and the mechanism of their translocation across membranes.

Most Organelles Cannot Be Constructed *De Novo:* They Require Information in the Organelle Itself

When a cell reproduces by division, it has to duplicate its organelles. In general, cells do this by incorporating new molecules into the existing organelles, thereby enlarging them; the enlarged organelles then divide and are distributed to the two daughter cells. Thus, each daughter cell inherits a complete set of specialized cell membranes from its mother. This inheritance is essential because a cell could not make such membranes from scratch. If the ER were completely removed from a cell, for example, how could the cell reconstruct it? As we discuss later, the membrane proteins that define the ER and perform many of its functions are themselves products of the ER. A new ER could not be made without an existing ER or, at least, a membrane that specifically contains the protein

Table 12–3 Some Typical Signal Sequences

FUNCTION OF SIGNAL SEQUENCE	EXAMPLE OF SIGNAL SEQUENCE
Import into nucleus	-Pro-Pro-Lys-Lys-Lys-Arg-Lys-Val-
Export from nucleus	-Leu-Ala-Leu-Lys-Leu-Ala-Gly-Leu-Asp-Ile-
Import into mitochondria	^+H_3N-Met-Leu-Ser-Leu-Arg-Gln-Ser-Ile-Arg-Phe-Phe-Lys-Pro-Ala-Thr-Arg-Thr-Leu-Cys-Ser-Ser-Arg-Tyr-Leu-Leu-
Import into plastid	^+H_3N-Met-Val-Ala-Met-Ala-Met-Ala-Ser-Leu-Gln-Ser-Ser-Met-Ser-Ser-Leu-Ser-Leu-Ser-Ser-Asn-Ser-Phe-Leu-Gly-Gln-Pro-Leu-Ser-Pro-Ile-Thr-Leu-Ser-Pro-Phe-Leu-Gln-Gly-
Import into peroxisomes	-Ser-Lys-Leu-COO$^-$
Import into ER	^+H_3N-Met-Met-Ser-Phe-Val-Ser-Leu-Leu-Leu-Val-Gly-Ile-Leu-Phe-Trp-Ala-Thr-Glu-Ala-Glu-Gln-Leu-Thr-Lys-Cys-Glu-Val-Phe-Gln-
Return to ER	-Lys-Asp-Glu-Leu-COO$^-$

Some characteristic features of the different classes of signal sequences are highlighted in color. Where they are known to be important for the function of the signal sequence, positively charged amino acids are shown in *red* and negatively charged amino acids are shown in *green*. Similarly, important hydrophobic amino acids are shown in *white* and important hydroxylated amino acids are shown in *blue*. ^+H_3N indicates the N-terminus of a protein; COO$^-$ indicates the C-terminus.

A TRANSFECTION APPROACH FOR DEFINING SIGNAL SEQUENCES

One way to show that a signal sequence is required and sufficient to target a protein to a specific intracellular compartment is to create a fusion protein in which the signal sequence is attached by genetic engineering techniques to a protein that normally resides in the cytosol. After the cDNA encoding this protein is transfected into cells, the location of the fusion protein is determined by immunostaining or by cell fractionation.

signal sequence a or b   gene encoding cytosolic protein

plasmid used to transfect cells

A   B

A   B

signal sequence a (▦) directs fusion protein to organelle A

signal sequence b (▬) directs fusion protein to organelle B

By altering the signal sequence using site-directed mutagenesis, we can determine which structural features are important for its function.

A BIOCHEMICAL APPROACH FOR STUDYING THE MECHANISM OF PROTEIN TRANSLOCATION

In this approach, a labeled protein containing a specific signal sequence is transported into isolated organelles *in vitro*. The labeled protein is usually produced by cell-free translation of a purified mRNA encoding the protein; radioactive amino acids are used to label the newly synthesized protein so that it can be distinguished from the many other proteins that are present in the *in vitro* translation system.

Three methods are commonly used to test if the labeled protein has been translocated into the organelle:

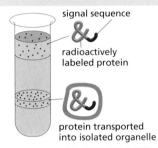

1. The labeled protein co-fractionates with the organelle during centrifugation.

signal sequence

radioactively labeled protein

protein transported into isolated organelle

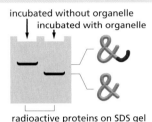

2. The signal sequence is removed by a specific protease that is present inside the organelle.

incubated without organelle
incubated with organelle

radioactive proteins on SDS gel

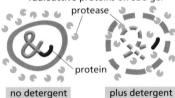

3. The protein is protected from digestion when proteases are added to the incubation medium, but is susceptible if a detergent is first added to disrupt the organelle membrane.

protease

protein

no detergent   plus detergent

By exploiting such *in vitro* assays, one can determine what components (proteins, ATP, GTP, etc.) are required for the translocation process.

GENETIC APPROACHES FOR STUDYING THE MECHANISM OF PROTEIN TRANSLOCATION

wild-type yeast cell

histidinol   histidine

→ enzyme in cytosol: cell lives without histidine as nutrient

translocation apparatus

ER

engineered yeast cell

histidinol

→ enzyme targeted to ER: cell dies without histidine as nutrient

mutant engineered cell

histidinol   histidine

→ not all enzyme taken up into ER: cell lives without histidine as nutrient

mutant translocation apparatus

Yeast cells with mutations in genes that encode components of the translocation machinery have been useful for studying protein translocation. Because mutant cells that cannot translocate any proteins across their membranes will die, the challenge is to find mutations that cause only a partial defect in protein translocation.

One experimental strategy uses genetic engineering to design special yeast cells. The enzyme histidinol dehydrogenase, for example, normally resides in the cytosol, where it is required to produce the essential amino acid histidine from its precursor histidinol. A yeast strain is constructed in which the histidinol dehydrogenase gene is replaced by a re-engineered gene encoding a fusion protein with an added signal sequence that misdirects the enzyme into the endoplasmic reticulum (ER). When such cells are grown without histidine, they die because all of the histidinol dehydrogenase is sequestered in the ER, where it is of no use. Cells with a mutation that only partially inactivates the mechanism for translocating proteins from the cytosol to the ER, however, will survive because the cytosol retains enough of the dehydrogenase to produce histidine.

Often one obtains a cell in which the mutant protein in the translocation machinery still functions partially at normal temperature but is completely inactive at higher temperature. A cell carrying such a temperature-sensitive mutation dies at higher temperature, whether or not histidine is present, as it cannot transport any protein into the ER. The normal gene that was disabled by the temperature-sensitive mutation can be identified by transfecting the mutant cells with a yeast plasmid vector into which random yeast genomic DNA fragments have been cloned: the specific DNA fragment that rescues the mutant cells when they are grown at high temperature should encode the wild-type version of the mutant gene.

translocators required to import selected proteins into the ER from the cytosol (including the ER-specific translocators themselves). The same is true for mitochondria and plastids (see Figure 12–6).

Thus, it seems that the information required to construct an organelle does not reside exclusively in the DNA that specifies the organelle's proteins. Information in the form of at least one distinct protein that preexists in the organelle membrane is also required, and this information is passed from parent cell to progeny cell in the form of the organelle itself. Presumably, such information is essential for the propagation of the cell's compartmental organization, just as the information in DNA is essential for the propagation of the cell's nucleotide and amino acid sequences.

As we discuss in more detail in Chapter 13, however, the ER sheds a constant stream of membrane vesicles that incorporate only a subset of ER proteins and therefore have a different composition from the ER itself. Similarly, the plasma membrane constantly produces various types of specialized endocytic vesicles. Thus, some organelles can form from other organelles and do not have to be inherited at cell division.

Summary

Eucaryotic cells contain intracellular membranes that enclose nearly half the cell's total volume in separate intracellular compartments called organelles. The main types of organelles present in all eucaryotic cells are the endoplasmic reticulum, Golgi apparatus, nucleus, mitochondria, lysosomes, endosomes, and peroxisomes; plant cells also contain plastids, such as chloroplasts. Each organelle contains a distinct set of proteins, which mediate its unique functions.

Each newly synthesized organelle protein must find its way from a ribosome in the cytosol, where the protein is made, to the organelle where it functions. It does so by following a specific pathway, guided by sorting signals in its amino acid sequence that function as signal sequences or signal patches. Sorting signals are recognized by complementary sorting receptors, which deliver the protein to the appropriate target organelle. Proteins that function in the cytosol do not contain sorting signals and therefore remain there after they are synthesized.

During cell division, organelles such as the ER and mitochondria are distributed intact to each daughter cell. These organelles contain information that is required for their construction, and so they cannot be made de novo.

THE TRANSPORT OF MOLECULES BETWEEN THE NUCLEUS AND THE CYTOSOL

The **nuclear envelope** encloses the DNA and defines the *nuclear compartment*. This envelope consists of two concentric membranes, which are penetrated by nuclear pore complexes (**Figure 12–8**). Although the inner and outer nuclear membranes are continuous, they maintain distinct protein compositions. The **inner nuclear membrane** contains specific proteins that act as anchoring sites for chromatin and for the *nuclear lamina*, a protein meshwork that provides structural support for the nuclear envelope. The inner membrane is surrounded by the **outer nuclear membrane**, which is continuous with the membrane of the ER. Like the membrane of the ER (discussed later), the outer nuclear membrane is studded with ribosomes engaged in protein synthesis. The proteins made on these ribosomes are transported into the space between the inner and outer nuclear membranes (the *perinuclear space*), which is continuous with the ER lumen.

Bidirectional traffic occurs continuously between the cytosol and the nucleus. The many proteins that function in the nucleus—including histones, DNA and RNA polymerases, gene regulatory proteins, and RNA-processing proteins—are selectively imported into the nuclear compartment from the cytosol, where they are made. At the same time, tRNAs and mRNAs are synthesized in the

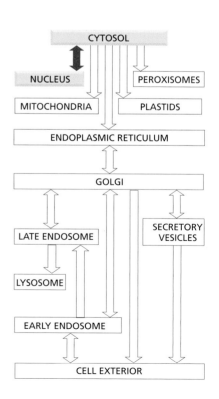

nuclear compartment and then exported to the cytosol. Like the import process, the export process is selective; mRNAs, for example, are exported only after they have been properly modified by RNA-processing reactions in the nucleus. In some cases, the transport process is complex. Ribosomal proteins, for instance, are made in the cytosol and imported into the nucleus, where they assemble with newly made ribosomal RNA into particles. The particles are then exported to the cytosol, where they assemble into ribosomes. Each of these steps requires selective transport across the nuclear envelope.

Nuclear Pore Complexes Perforate the Nuclear Envelope

Large, elaborate structures known as **nuclear pore complexes (NPCs)** perforate the nuclear envelope of all eucaryotes. In animal cells, each NPC has an estimated molecular mass of about 125 million daltons and is composed of about 30 different **NPC proteins**, or *nucleoporins*, which are present in multiple copies and are arranged with a striking octagonal symmetry (**Figure 12–9**).

The nuclear envelope of a typical mammalian cell contains 3000–4000 NPCs, and the total traffic that passes through each NPC is enormous: each NPC can transport up to 500 macromolecules per second and can transport in both directions at the same time. How it coordinates the bidirectional flow of macromolecules to avoid congestion and head-on collisions is not known.

Each NPC contains one or more aqueous passages, through which small water-soluble molecules can diffuse passively. Researchers have determined the effective size of these passages by injecting labeled water-soluble molecules of different sizes into the cytosol and then measuring their rate of diffusion into the nucleus. Small molecules (5000 daltons or less) diffuse in so fast that we can consider the nuclear envelope freely permeable to them. Large proteins, however, traverse the NPC much more slowly; the larger a protein, the more slowly it passes through the NPC. Proteins larger than 60,000 daltons can barely enter by passive diffusion. This size cut-off to free diffusion is thought to result from the NPC structure (**Figure 12–10**). Many NPC proteins that line the central pore contain extensive unstructured regions, which are thought to form a disordered tangle (much like a kelp bed in the ocean), blocking the central opening in the NPC to the passage of large macromolecules, but leaving small openings to allow the diffusion of smaller molecules.

Because many cell proteins are too large to diffuse passively through the NPCs, the nuclear compartment and the cytosol can maintain different complements of proteins. Mature cytosolic ribosomes, for example, are about 30 nm in diameter and thus cannot diffuse through the NPC, confining protein synthesis to the cytosol. But how does the nucleus export newly made ribosomal subunits or import large molecules, such as DNA and RNA polymerases, which have subunit molecular weights of 100,000–200,000 daltons? As we discuss next, these and many other protein and RNA molecules bind to specific receptor proteins that ferry large molecules actively through NPCs.

Nuclear Localization Signals Direct Nuclear Proteins to the Nucleus

When proteins are experimentally extracted from the nucleus and reintroduced into the cytosol, even the very large ones reaccumulate efficiently in the nucleus. Sorting signals called **nuclear localization signals** are responsible for the selectivity of this active nuclear import process. The signals have been precisely defined by using recombinant DNA technology for numerous nuclear proteins, as well as for proteins that enter the nucleus only transiently (**Figure 12–11**). In many proteins, signals consist of one or two short sequences that are rich in the positively charged amino acids lysine and arginine (see Table 12–3, p. 702), with the precise sequence varying for different nuclear proteins. Other nuclear proteins contain different signals, some of which are not yet characterized.

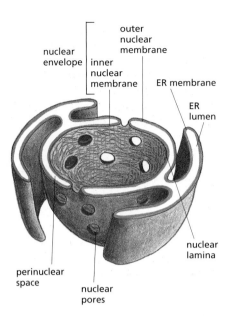

Figure 12–8 The nuclear envelope. The double-membrane envelope is penetrated by pores in which NPCs are positioned and is continuous with the endoplasmic reticulum. The ribosomes that are normally bound to the cytosolic surface of the ER membrane and outer nuclear membrane are not shown. The nuclear lamina is a fibrous meshwork underlying the inner membrane.

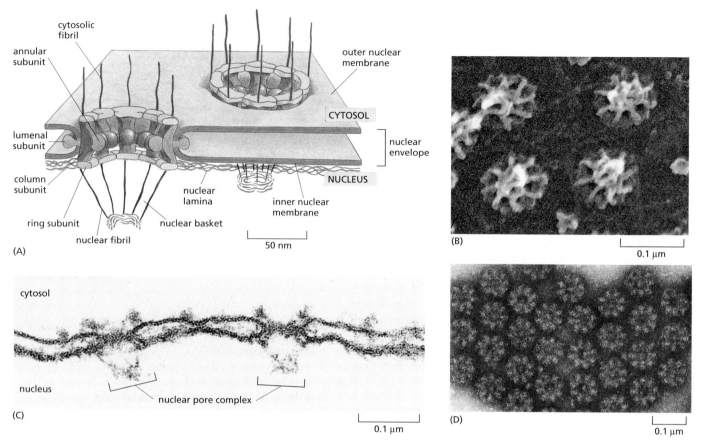

Figure 12–9 The arrangement of NPCs in the nuclear envelope. (A) A small region of the nuclear envelope. In cross section, an NPC seems to have four structural building blocks: column subunits, which form the bulk of the pore wall; annular subunits, which are centrally located; lumenal subunits, which contain transmembrane proteins that anchor the complex to the nuclear membrane; and ring subunits, which form the cytosolic and nuclear faces of the complex. In addition, fibrils protrude from both the cytosolic and the nuclear sides of the NPC. On the nuclear side, the fibrils converge to form basketlike structures. Immunoelectron microscopic studies show that the proteins that make up the core of the NPC are oriented symmetrically across the nuclear envelope, so that the nuclear and cytosolic sides of the core look identical. In contrast, the proteins that make up the fibrils are different on the cytosolic and nuclear sides of the NPC. The eight-fold rotational and two-fold transverse symmetry of the core NPC explains how such a huge structure can be formed from only about 30 different proteins: many of these proteins are present in 16 copies (or multiples of 16). Disordered domains of the core proteins (not shown) are thought to extend toward the center of the NPC, blocking the passive diffusion of large macromolecules. (B) A scanning electron micrograph of the nuclear side of the nuclear envelope of an oocyte (see also Figure 9–55). (C) An electron micrograph showing a side view of two NPCs *(brackets);* note that the inner and outer nuclear membranes are continuous at the edges of the pore. (D) An electron micrograph showing face-on views of negatively stained NPCs. The membrane has been removed by detergent extraction. Note that some of the NPCs contain material in their center, which is thought to be macromolecules in transit through these NPCs. (B, from M.W. Goldberg and T.D. Allen, *J. Cell Biol.* 119:1429–1440, 1992. With permission from The Rockefeller University Press; C, courtesy of Werner Franke and Ulrich Scheer; D, courtesy of Ron Milligan.)

Nuclear localization signals can be located almost anywhere in the amino acid sequence and are thought to form loops or patches on the protein surface. Many function even when linked as short peptides to lysine side chains on the surface of a cytosolic protein, suggesting that the precise location of the signal within the amino acid sequence of a nuclear protein is not important. Moreover, as long as one of the protein subunits of a multicomponent complex displays a nuclear localization signal, the complex can be imported into the nucleus.

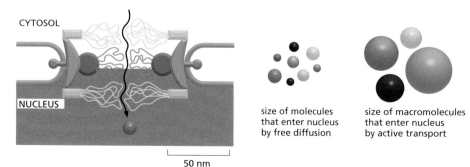

Figure 12–10 A model for the gated diffusion barrier of the NPC. The drawing shows a side view of an NPC. Unstructured regions of the proteins lining the central pore form a tangled meshwork, which blocks the passive diffusion of large macromolecules. During active transport, however, even particles up to 39 nm in diameter can pass through NPCs.

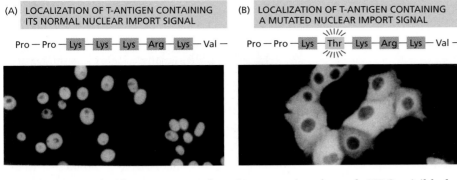

(A) LOCALIZATION OF T-ANTIGEN CONTAINING ITS NORMAL NUCLEAR IMPORT SIGNAL

Pro — Pro — Lys — Lys — Lys — Arg — Lys — Val —

(B) LOCALIZATION OF T-ANTIGEN CONTAINING A MUTATED NUCLEAR IMPORT SIGNAL

Pro — Pro — Lys — Thr — Lys — Arg — Lys — Val —

Figure 12–11 The function of a nuclear localization signal. Immunofluorescence micrographs showing the cell location of SV40 virus T-antigen containing or lacking a short sequence that serves as a nuclear localization signal. (A) The normal T-antigen protein contains the lysine-rich sequence indicated and is imported to its site of action in the nucleus, as indicated by immunofluorescence staining with antibodies against the T-antigen. (B) T-antigen with an altered nuclear localization signal (a threonine replacing a lysine) remains in the cytosol. (From D. Kalderon, B. Roberts, W. Richardson and A. Smith, *Cell* 39:499–509, 1984. With permission from Elsevier.)

One can make the transport of nuclear proteins through NPCs visible by coating gold particles with a nuclear localization signal, injecting the particles into the cytosol, and then following their fate by electron microscopy (**Figure 12–12**). Transport begins when the particles bind to tentaclelike fibrils that extend from the rim of the NPC into the cytoplasm, and then proceed through the center of the NPC. Presumably, the unstructured regions of the NPC proteins that form a diffusion barrier for large molecules (mentioned earlier) are pushed away to allow the coated gold particles to squeeze through.

Macromolecular transport across NPCs differs fundamentally from the transport of proteins across the membranes of other organelles, in that it occurs through a large aqueous pore rather than through a protein transporter spanning one or more lipid bilayers. For this reason, fully folded nuclear proteins can be transported into the nucleus through an NPC, and newly formed ribosomal subunits are transported out of the nucleus as an assembled particle. By contrast, proteins have to be extensively unfolded to be transported into most other organelles, as we discuss later. In the electron microscope, however, very large particles traversing the NPC seem to become compressed as they squeeze through the pore, indicating that they undergo restructuring during transport. The export of some very large mRNAs has been extensively studied, as discussed in Chapter 6 (see Figure 6–39).

Nuclear Import Receptors Bind to Both Nuclear Localization Signals and NPC proteins

To initiate nuclear import, most nuclear localization signals must be recognized by **nuclear import receptors**, which are encoded by a family of related genes. Each family member encodes a receptor protein that is specialized for the transport of a subset of cargo proteins—those that contain nuclear localization signals to which the receptor can bind (**Figure 12–13A**).

The import receptors are soluble cytosolic proteins that bind both to the nuclear localization signal on the protein to be transported and to NPC proteins, some of which form the fibrils of the NPC that extend into the cytosol. These fibrils, as well as many of the NPC proteins that line the center of the NPC and contribute to the diffusion barrier, include a large number of short amino-acid repeats that contain phenylalanine and glycine and are therefore called *FG-repeats* (named after the one-letter code for these amino acids, discussed in Chapter 3). FG-repeats serve as binding sites for the import receptors. They are thought to line the path through the NPCs taken by the import receptors and their bound cargo proteins. The receptor–cargo complexes move along the transport path by repeatedly binding, dissociating, and then re-binding to adjacent FG-repeat sequences. In this way, the complexes hop from one NPC protein to another to traverse the tangled interior of the NPC. Once inside the nucleus, the import receptors dissociate from their cargo and return to the cytosol.

Nuclear import receptors do not always bind to nuclear proteins directly. Additional adaptor proteins sometimes form a bridge between the import receptors and the nuclear localization signals on the proteins to be transported (Figure 12–13B). Some adaptor proteins are structurally related to nuclear import receptors, suggesting a common evolutionary origin. By using a variety of different

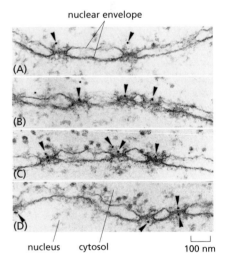

nuclear envelope

(A)

(B)

(C)

(D)

nucleus   cytosol   100 nm

Figure 12–12 Visualizing active import through NPCs. This series of electron micrographs shows colloidal gold spheres *(arrowheads)* coated with peptides containing nuclear localization signals entering the nucleus through NPCs. Gold particles were injected into living cells, which then were fixed and prepared for electron microscopy at various times after injection. (A) Gold particles are first seen in proximity to the cytosolic fibrils of the NPCs. (B, C) They migrate to the center of the NPCs, where they are first seen exclusively on the cytosolic face. (D) They then appear on the nuclear face. These gold particles have much larger diameters than the diffusion channel in the NPC and are imported by active transport. (From N. Panté and U. Aebi, *Science* 273:1729–1732, 1996. With permission from AAAS.)

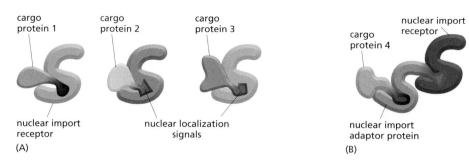

(A)

(B)

Figure 12–13 **Nuclear import receptors.** (A) Many nuclear import receptors bind both to NPC proteins and to a nuclear localization signal on the cargo proteins they transport. Cargo proteins 1, 2, and 3 in this example contain different nuclear localization signals, and therefore each binds to different nuclear import receptors. (B) Cargo protein 4 requires an adaptor protein to bind to its nuclear import receptor. The adaptors are structurally related to nuclear import receptors and recognize nuclear localization signals on cargo proteins. They also contain a nuclear localization signal that binds them to an import receptor.

import receptors and adaptors, cells are able to recognize the broad repertoire of nuclear localization signals that are displayed on nuclear proteins.

Nuclear Export Works Like Nuclear Import, But in Reverse

The nuclear export of large molecules, such as new ribosomal subunits and RNA molecules, occurs through NPCs and also depends on a selective transport system. The transport system relies on **nuclear export signals** on the macromolecules to be exported, as well as on complementary **nuclear export receptors**. These receptors bind to both the export signal and NPC proteins to guide their cargo through the NPC to the cytosol.

Many nuclear export receptors are structurally related to nuclear import receptors, and they are encoded by the same gene family of **nuclear transport receptors**, or *karyopherins*. In yeast, there are 14 genes encoding members of this family; in animal cells, the number is significantly larger. It is often not possible to tell from their amino acid sequence alone whether a particular family member works as a nuclear import or nuclear export receptor. As might be expected, therefore, the import and export transport systems work in similar ways but in opposite directions: the import receptors bind their cargo molecules in the cytosol, release them in the nucleus, and are then exported to the cytosol for reuse, while the export receptors function in the opposite fashion.

The Ran GTPase Imposes Directionality on Transport Through NPCs

The import of nuclear proteins through the NPC concentrates specific proteins in the nucleus and thereby increases order in the cell. The cell obtains the energy it needs for this process through the hydrolysis of GTP by the monomeric GTPase **Ran**. Ran is found in both the cytosol and the nucleus, and it is required for both nuclear import and export.

Like other GTPases, Ran is a molecular switch that can exist in two conformational states, depending on whether GDP or GTP is bound (discussed in Chapter 3). Two Ran-specific regulatory proteins trigger the conversion between the two states: a cytosolic *GTPase-activating protein (GAP)* triggers GTP hydrolysis and thus converts Ran-GTP to Ran-GDP, and a nuclear *guanine exchange factor (GEF)* promotes the exchange of GDP for GTP and thus converts Ran-GDP to Ran-GTP. Because *Ran-GAP* is located in the cytosol and *Ran-GEF* is located in the nucleus, the cytosol contains mainly Ran-GDP, and the nucleus contains mainly Ran-GTP (**Figure 12–14**).

This gradient of the two conformational forms of Ran drives nuclear transport in the appropriate direction (**Figure 12–15**). Docking of nuclear import receptors to FG-repeats on the cytosolic side of the NPC, for example, occurs whether or not these receptors are loaded with an appropriate cargo. Import receptors then hop from FG-repeat to FG-repeat. If they reach the nuclear side of the pore complex, Ran-GTP binds to them, and, if they arrive loaded with cargo molecules, the Ran-GTP binding causes the import receptors to release their cargo (**Figure 12–16**). Because the Ran-GDP in the cytosol does not bind to the cargo receptors, unloading occurs only on the nuclear side of the NPC. In this way, the nuclear localization of Ran-GTP creates the directionality.

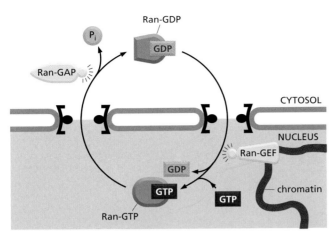

Figure 12–14 **The compartmentalization of Ran-GDP and Ran-GTP.** Localization of Ran-GDP in the cytosol and Ran-GTP in the nucleus results from the localization of two Ran regulatory proteins: Ran GTPase-activating protein (Ran-GAP) is located in the cytosol and Ran guanine nucleotide exchange factor (Ran-GEF) binds to chromatin and is therefore located in the nucleus.

Having discharged its cargo in the nucleus, the empty import receptor with Ran-GTP bound is transported back through the pore complex to the cytosol. There, Ran-GAP triggers Ran-GTP to hydrolyze its bound GTP, thereby converting it to Ran-GDP. The import receptor is then ready for another cycle of nuclear import.

Nuclear export occurs by a similar mechanism, except that Ran-GTP in the nucleus promotes cargo binding to the export receptor, rather than dissociating it. Once the export receptor moves through the pore to the cytosol, its Ran-GTP encounters Ran-GAP and hydrolyzes GTP. As a result, the export receptor releases both its cargo and Ran-GDP in the cytosol. Free export receptors are then returned to the nucleus to complete the cycle (see Figure 12–15).

Transport Through NPCs Can Be Regulated by Controlling Access to the Transport Machinery

Some proteins, such as those that bind newly made mRNAs in the nucleus, contain both nuclear localization signals and nuclear export signals. These proteins continually shuttle back and forth between the nucleus and the cytosol. The relative rates of their import and export determine the steady-state localization of such *shuttling proteins*. If the rate of import exceeds the rate of export, a protein will be located mainly in the nucleus. Conversely, if the rate of export exceeds the rate of import, a protein will be located mainly in the cytosol. Thus, changing the rate of import, export, or both, can change the location of a protein.

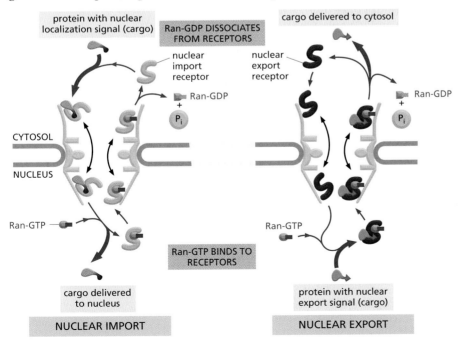

Figure 12–15 **A model explaining how GTP hydrolysis by Ran in the cytosol provides directionality to nuclear transport.** Movement through the NPC of loaded nuclear transport receptors occurs along the FG-repeats displayed by certain NPC proteins. The differential localization of Ran-GTP in the nucleus and Ran-GDP in the cytosol provides directionality *(red arrows)* to both nuclear import *(left)* and nuclear export *(right)*. Ran-GAP stimulates the hydrolysis of GTP to produce Ran-GDP on the cytosolic side of the NPC (see Figure 12–14).

Ran-GDP is imported into the nucleus by its own import receptor, which is specific for the GDP-bound conformation of Ran. The Ran-GDP receptor is structurally unrelated to the main family of nuclear transport receptors. However, it also binds to FG-repeats in NPC proteins and hops across the NPC.

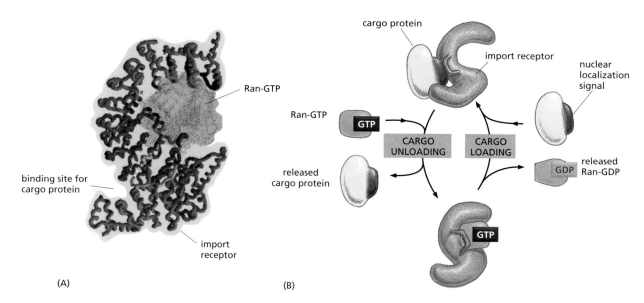

(A)

(B)

Some shuttling proteins move continuously into and out of the nucleus. In other cases, however, the transport is stringently controlled. As discussed in Chapter 7, cells control the activity of some gene regulatory proteins by keeping them out of the nucleus until they are needed there (**Figure 12–17**). In many cases, cells control transport by regulating nuclear localization and export signals; these can be turned on or off, often by phosphorylation of amino acids close to the signal sequences (**Figure 12–18**).

Other gene regulatory proteins are bound to inhibitory cytosolic proteins that either anchor them in the cytosol (through interactions with the cytoskeleton or specific organelles) or mask their nuclear localization signals so that they cannot interact with nuclear import receptors. An appropriate stimulus releases the gene regulatory protein from its cytosolic anchor or mask, and it is then transported into the nucleus. One important example is the latent gene regulatory protein that controls the expression of proteins involved in cholesterol metabolism. The protein is made and stored in an inactive form as a transmembrane protein in the ER. When deprived of cholesterol, the protein exits the ER to the Golgi apparatus where it meets specific proteases that cleave it, releasing its cytosolic domain into the cytosol. This domain is then imported into the nucleus, where it activates the transcription of genes required for both cholesterol import and synthesis.

As we discussed in detail in Chapter 6, cells control the export of mRNA from the nucleus in a similar way. Proteins that guide the mRNA out of the nucleus load onto the RNA as transcription and splicing proceed. Upon entry into the cytosol, the proteins are stripped off and rapidly returned to the nucleus. Other RNAs, such as snRNAs and tRNAs, are exported by different nuclear export receptors.

During Mitosis the Nuclear Envelope Disassembles

The **nuclear lamina**, located on the nuclear side of the inner nuclear membrane, is a meshwork of interconnected protein subunits called **nuclear lamins**. The lamins are a special class of intermediate filament proteins (discussed in Chapter 16) that polymerize into a two-dimensional lattice (**Figure 12–19**). The nuclear lamina gives shape and stability to the nuclear envelope, to which it is anchored by attachment to both the NPCs and integral membrane proteins of the inner nuclear membrane. The lamina also interacts directly with chromatin, which

Figure 12–16 How the binding of Ran-GTP can cause nuclear import receptors to release their cargo. (A) The structure of a nuclear transport receptor with bound Ran-GTP. The receptor contains repeated α-helical motifs that stack on top of each other, forming a flexible springlike structure, which can adopt multiple conformations in response to the binding of cargo proteins and Ran-GTP. Note that cargo proteins and Ran-GTP bind to different regions of the coiled spring. The Ran-GTP partly covers a conserved loop *(red)* of the receptor, which, in the Ran-free state, is thought to be important for cargo binding. (B) The cycle of loading in the cytosol and unloading in the nucleus of a nuclear import receptor. (A, adapted from Y.M. Chook and G. Blobel, *Nature* 399:230–237, 1999. With permission from Macmillan Publishers Ltd.)

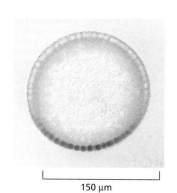

Figure 12–17 The control of nuclear transport in the fly embryo. The gene regulatory protein *dorsal* is expressed uniformly throughout this early *Drosophila* embryo, which is shown in cross section. The protein has been stained with an enzyme-coupled antibody that yields a brown product. It is active only in cells at the ventral side *(bottom)* of the embryo, where it is found in nuclei. (Courtesy of Siegfried Roth.)

150 µm

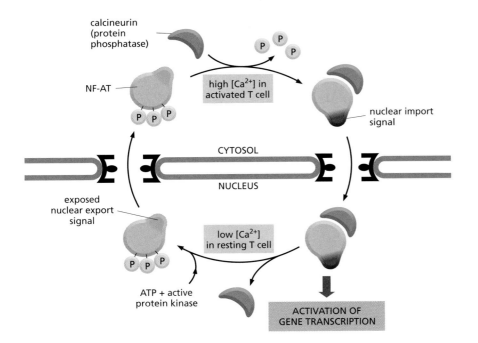

Figure 12–18 **The control of nuclear import during T cell activation.** <AGTT> The nuclear factor of activated T cells (NF-AT) is a gene regulatory protein that, in the resting T cell, is found in the cytosol in a phosphorylated state. When T cells are activated by foreign antigen (discussed in Chapter 25), the intracellular Ca^{2+} concentration increases. In high Ca^{2+}, the protein phosphatase calcineurin binds to NF-AT and dephosphorylates it. The dephosphorylation exposes nuclear import signals and blocks a nuclear export signal. The complex of NF-AT and calcineurin is then imported into the nucleus, where NF-AT activates the transcription of numerous genes required for T cell activation.

The response shuts off when Ca^{2+} levels decrease, releasing NF-AT from calcineurin. Rephosphorylation of NF-AT inactivates the nuclear import signals and re-exposes the nuclear export signal, causing NF-AT to relocate to the cytosol. Some of the most potent immunosuppressive drugs, including cyclosporin A and FK506, inhibit the ability of calcineurin to dephosphorylate NF-AT and thereby block the nuclear accumulation of NF-AT and T cell activation.

itself interacts with integral membrane proteins of the inner nuclear membrane. Together with the lamina, these inner membrane proteins provide structural links between the DNA and the nuclear envelope.

When a nucleus disassembles during mitosis, the nuclear lamina depolymerizes. The disassembly is at least partly a consequence of direct phosphorylation of the nuclear lamins by the cyclin-dependent protein kinase Cdk that is activated at the onset of mitosis (discussed in Chapter 16). At the same time, proteins of the inner nuclear membrane are phosphorylated, and the NPCs disassemble and disperse in the cytosol. During this process, some NPC proteins become bound to nuclear import receptors, which play an important part in the reassembly of NPCs at the end of mitosis. Nuclear envelope membrane proteins—no longer tethered to the pore complexes, lamina, or chromatin—disperse throughout the ER membrane. The dynein motor protein, which moves along microtubules (discussed in Chapter 17), actively participates in tearing the nuclear envelope off the chromatin. Together, these processes break down the barriers that normally separate the nucleus and cytosol, and the nuclear proteins that are not bound to membranes or chromosomes intermix completely with those of the cytosol (**Figure 12–20**).

Later in mitosis, the nuclear envelope reassembles on the surface of the chromosomes. In addition to its crucial role in nuclear transport, Ran-GTPase also acts as a positional marker for chromatin during cell division, when the nuclear and cytosolic components intermix. Because Ran-GEF remains bound to chromatin when the nuclear envelope breaks down, Ran molecules close to chromatin are mainly in their GTP-bound conformation. By contrast, Ran molecules further away have a high likelihood of encountering Ran-GAP, which is distributed throughout the cytosol; these Ran molecules are therefore mainly in their GDP-bound conformation.

Chromatin in mitotic cells is therefore surrounded by a cloud of Ran-GTP. This cloud locally displaces nuclear import receptors from NPC proteins, which starts the assembly process of NPCs attached to the chromosome surface. At the same time, inner nuclear membrane proteins and dephosphorylated lamins bind again to chromatin. ER membranes wrap around groups of chromosomes and continue fusing until they form a sealed nuclear envelope. During this process, the NPCs start actively reimporting proteins that contain nuclear localization signals. Because the nuclear envelope is initially closely applied to the surface of the chromosomes, the newly formed nucleus excludes all proteins except those initially bound to the mitotic chromosomes and those that are selectively imported through NPCs. In this way, all other large proteins are kept out of the newly assembled nucleus.

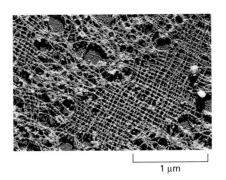

Figure 12–19 **The nuclear lamina.** An electron micrograph of a portion of the nuclear lamina in a *Xenopus* oocyte prepared by freeze-drying and metal shadowing. The lamina is formed by a regular lattice of specialized intermediate filaments. (Courtesy of Ueli Aebi.)

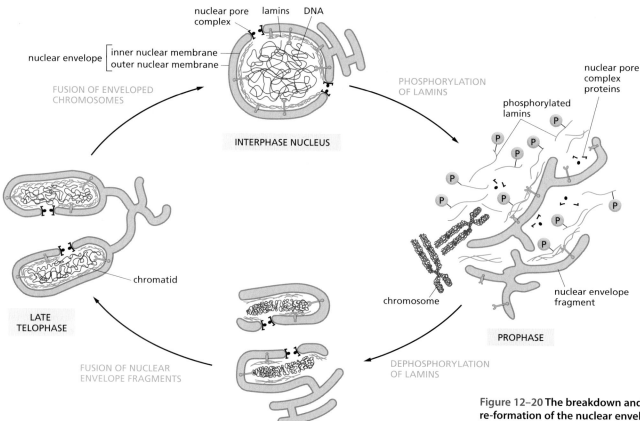

Figure 12–20 The breakdown and re-formation of the nuclear envelope during mitosis. Phosphorylation of the lamins is thought to trigger the disassembly of the nuclear lamina, which helps the nuclear envelope to break up. Dephosphorylation of the lamins is thought to help reverse the process. An analogous phosphorylation and dephosphorylation cycle occurs for some nucleoporins and proteins of the inner nuclear membrane, and some of these dephosphorylations are also involved in the reassembly process shown.

As indicated, the nuclear envelope initially reforms around individual decondensing chromatids. Eventually as decondensation progresses, these structures fuse to form a single complete nucleus.

Nuclear localization signals are not cleaved off after transport into the nucleus, presumably because nuclear proteins need to be imported repeatedly, once after every cell division. In contrast, once a protein molecule has been imported into any of the other membrane-enclosed organelles, it is passed on from generation to generation within that compartment and does not need to be translocated ever again; the signal sequence on these molecules is often removed after protein translocation.

As we discuss in Chapter 17, the cloud of Ran-GTP surrounding chromatin is also important in assembling the mitotic spindle in a dividing cell.

Summary

The nuclear envelope consists of an inner and an outer nuclear membrane. The outer membrane is continuous with the ER membrane, and the space between it and the inner membrane is continuous with the ER lumen. RNA molecules, which are made in the nucleus, and ribosomal subunits, which are assembled there, are exported to the cytosol; in contrast, all the proteins that function in the nucleus are synthesized in the cytosol and are then imported. The extensive traffic of materials between the nucleus and cytosol occurs through nuclear pore complexes (NPCs), which provide a direct passageway across the nuclear envelope. Small molecules diffuse passively through the NPCs, but large macromolecules have to be actively transported.

Proteins containing nuclear localization signals are actively transported inward through NPCs, while RNA molecules and newly made ribosomal subunits contain nuclear export signals, which direct their outward active transport through NPCs. Some proteins, including the nuclear import and export receptors, continually shuttle between the cytosol and nucleus. Ran-GTPase provides both the free energy and the directionality for nuclear transport. Cells regulate the transport of nuclear proteins and RNA molecules through the NPCs by controlling the access of these molecules to the transport machinery. Because nuclear localization signals are not removed, nuclear proteins can be imported repeatedly, as is required each time that the nucleus reassembles after mitosis.

THE TRANSPORT OF PROTEINS INTO MITOCHONDRIA AND CHLOROPLASTS

Mitochondria and chloroplasts are double-membrane-enclosed organelles (discussed in Chapter 14). They specialize in ATP synthesis, using energy derived from electron transport and oxidative phosphorylation in mitochondria and from photosynthesis in chloroplasts. Although both organelles contain their own DNA, ribosomes, and other components required for protein synthesis, most of their proteins are encoded in the cell nucleus and imported from the cytosol. Each imported protein must reach the particular organelle subcompartment in which it functions.

There are two subcompartments in mitochondria: the internal **matrix space** and the **intermembrane space**. These compartments are formed by the two concentric mitochondrial membranes: the **inner membrane**, which encloses the matrix space and forms extensive invaginations called *cristae*, and the **outer membrane**, which is in contact with the cytosol (**Figure 12–21A**). Chloroplasts have the same two subcompartments plus an additional subcompartment, the *thylakoid space*, which is surrounded by the *thylakoid membrane* (Figure 12–21B). Each of the subcompartments in mitochondria and chloroplasts contains a distinct set of proteins.

New mitochondria and chloroplasts are produced by the growth of preexisting organelles, followed by fission (discussed in Chapter 14). The growth depends mainly on the import of proteins from the cytosol. The imported proteins must be transported across a number of membranes in succession and end up in the appropriate place. The process of protein movement across membranes is often called *protein translocation*. This section explains how it occurs.

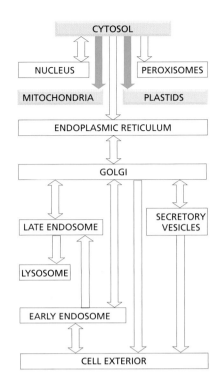

Translocation into Mitochondria Depends on Signal Sequences and Protein Translocators

Proteins imported into **mitochondria** are usually taken up from the cytosol within seconds or minutes of their release from ribosomes. Thus, in contrast to the protein translocation into the ER, described later, mitochondrial proteins are first fully synthesized as **mitochondrial precursor proteins** in the cytosol and then translocated into mitochondria by a *post-translational* mechanism. One or more signal sequences direct all mitochondrial precursor proteins to their appropriate mitochondrial subcompartment. Many proteins entering the matrix space contain a signal sequence at their N-terminus that a signal peptidase rapidly removes after import. Others, including all outer membrane and many inner membrane and intermembrane space proteins, have an internal signal sequence that is not removed. The signal sequences are both necessary and sufficient for the import and correct localization of the proteins: when genetic engineering techniques are used to link these signals to a cytosolic protein, the signals direct the protein to the correct mitochondrial subcompartment.

(A) MITOCHONDRION

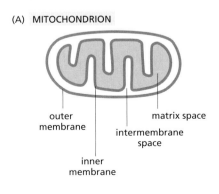

outer membrane
matrix space
intermembrane space
inner membrane

(B) CHLOROPLAST

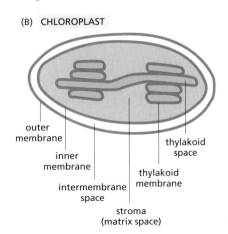

outer membrane
inner membrane
intermembrane space
thylakoid space
thylakoid membrane
stroma (matrix space)

Figure 12–21 The subcompartments of mitochondria and chloroplasts. In contrast to the cristae of mitochondria (A), the thylakoids of chloroplasts (B) are not connected to the inner membrane and therefore form a compartment with a separate internal space (see Figure 12–3).

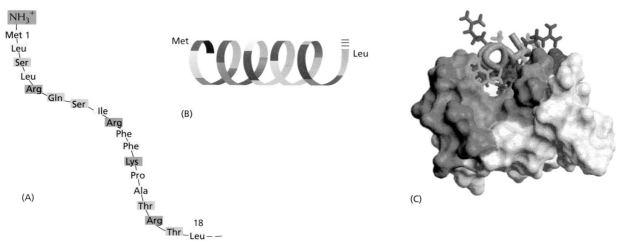

Figure 12–22 A signal sequence for mitochondrial protein import. Cytochrome oxidase is a large multiprotein complex located in the inner mitochondrial membrane, where it functions as the terminal enzyme in the electron-transport chain (discussed in Chapter 14). (A) The first 18 amino acids of the precursor to subunit IV of this enzyme serve as a signal sequence for import of the subunit into the mitochondrion. (B) When the signal sequence is folded as an α helix, the positively charged residues *(red)* are clustered on one face of the helix, while the nonpolar residues *(yellow)* are clustered primarily on the opposite face. Amino acids with uncharged polar side chains are shaded *blue*. Signal sequences that direct proteins into the matrix space always have the potential to form such an amphiphilic α helix, which is recognized by specific receptor proteins on the mitochondrial surface. (C) The structure of a signal sequence of alcohol dehydrogenase, another mitochondrial matrix enzyme, bound to an import receptor was determined by NMR. The amphiphilic α helix binds with its hydrophobic face to a hydrophobic groove in the receptor. (C, adapted from Y. Abe et al., *Cell* 100:551–560, 2000. With permission from Elsevier.)

The signal sequences that direct precursor proteins into the mitochondrial matrix space are best understood. They all form an amphiphilic α helix, in which positively charged residues cluster on one side of the helix, while uncharged hydrophobic residues cluster on the opposite side. Specific receptor proteins that initiate protein translocation recognize this configuration rather than the precise amino acid sequence of the signal sequence (**Figure 12–22**).

Multisubunit protein complexes that function as **protein translocators** mediate protein translocation across mitochondrial membranes. The **TOM complex** transfers proteins across the outer membrane, and two **TIM complexes** (TIM23 and TIM22) transfer proteins across the inner membrane (**Figure 12–23**). (TOM and TIM stand for translocase of the outer and inner mitochondrial membranes, respectively.) These complexes contain some components that act as receptors for mitochondrial precursor proteins, and other components that form the translocation channels.

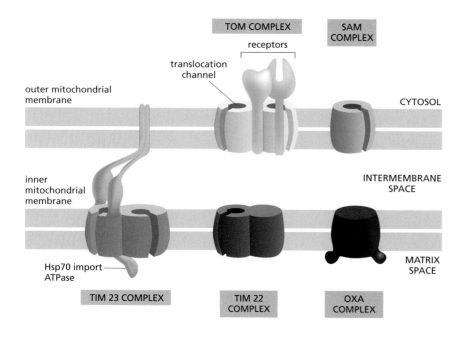

Figure 12–23 The protein translocators in the mitochondrial membranes. The TOM, TIM, SAM, and OXA complexes are multimeric membrane protein assemblies that catalyze protein transport across mitochondrial membranes. The protein components of the TIM22 and TIM23 complexes that line the import channel are structurally related, suggesting a common evolutionary origin of both TIM complexes. As indicated, one of the core components of the TIM23 complex contains a hydrophobic α-helical extension that is inserted into the outer mitochondrial membrane; the complex is therefore unusual in that it simultaneously spans two membranes.

On the matrix side, the TIM23 complex is bound to a protein complex containing mitochondrial Hsp70, which acts as an import ATPase using ATP hydrolysis to pull proteins through the pore.

The TOM complex is required for the import of all nucleus-encoded mitochondrial proteins. It initially transports their signal sequences into the intermembrane space and helps to insert transmembrane proteins into the outer membrane. β-Barrel proteins, which are particularly abundant in the outer membrane, are then passed on to an additional translocator, the **SAM complex**, which helps them to fold properly in the outer membrane. The TIM23 complex transports some soluble proteins into the matrix space and helps to insert transmembrane proteins into the inner membrane. The TIM22 complex mediates the insertion of a subclass of inner membrane proteins, including the transporter that moves ADP, ATP, and phosphate in and out of mitochondria. Yet another protein translocator in the inner mitochondrial membrane, the **OXA complex**, mediates the insertion of those inner membrane proteins that are synthesized within mitochondria. It also helps to insert some imported inner membrane proteins that are initially transported into the matrix space by the other complexes.

Mitochondrial Precursor Proteins Are Imported as Unfolded Polypeptide Chains

We have learned almost everything we know about the molecular mechanism of protein import into mitochondria from analyses of cell-free, reconstituted transport systems, in which purified mitochondria in a test tube import radiolabeled mitochondrial precursor proteins. By changing the conditions in the test tube, it is possible to establish the biochemical requirements for the import.

Mitochondrial precursor proteins do not fold into their native structures after they are synthesized; instead, they remain unfolded in the cytosol through interactions with other proteins. Some of these interacting proteins are general *chaperone proteins* of the *Hsp70 family* (discussed in Chapter 6), whereas others are dedicated to mitochondrial precursor proteins and bind directly to their signal sequences. All the interacting proteins help to prevent the precursor proteins from aggregating or folding up spontaneously before they engage with the TOM complex in the outer mitochondrial membrane. As a first step in the import process, the import receptors of the TOM complex bind the signal sequence of the mitochondrial precursor protein. The interacting proteins are then stripped off, and the unfolded polypeptide chain is fed—signal sequence first—into the translocation channel.

In principle, a protein could reach the mitochondrial matrix space by either crossing the two membranes all at once or crossing one at a time. One can distinguish between these possibilities by cooling a cell-free mitochondrial import system to arrest the proteins at an intermediate step in the translocation process. The result is that the arrested proteins no longer contain their N-terminal signal sequence, indicating that the N-terminus must be in the matrix space where the signal peptidase is located, but the bulk of the protein can still be attacked from outside the mitochondria by externally added proteolytic enzymes (**Figure 12–24**). Clearly, the precursor proteins can pass through both mitochondrial membranes at once to enter the matrix space (**Figure 12–25**). It is

Figure 12–24 Proteins transiently span the inner and outer mitochondrial membranes during their translocation into the matrix space. When isolated mitochondria are incubated with a precursor protein at 5°C, the precursor is only partly translocated across mitochondrial membranes. The N-terminal signal sequence (*red*) is cleaved off in the matrix space; but most of the polypeptide chain remains outside the mitochondrion, where it is accessible to externally added proteolytic enzymes. Upon warming to 25°C, the translocation is completed. Once inside the mitochondrion, the polypeptide chain is protected from added proteolytic enzymes. As a control, when detergents are added to disrupt the mitochondrial membranes, the imported proteins can be readily digested by the same protease treatment.

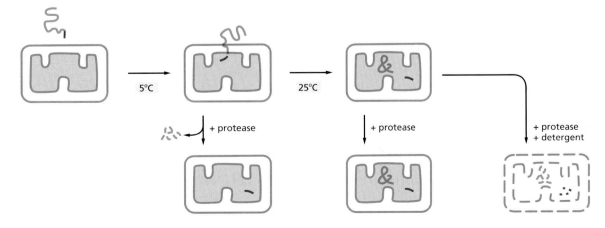

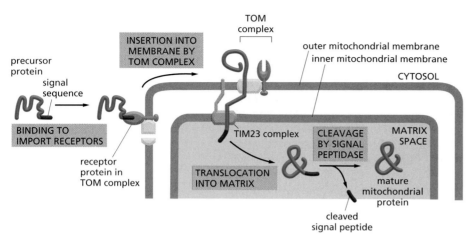

Figure 12–25 Protein import by mitochondria. <ACGG> The N-terminal signal sequence of the mitochondrial precursor protein is recognized by receptors of the TOM complex. The protein is then translocated through the TIM23 complex so that it transiently spans both mitochondrial membranes. The signal sequence is cleaved off by a signal peptidase in the matrix space to form the mature protein. The free signal sequence is then rapidly degraded (not shown).

thought that the TOM complex first transports the signal sequence across the outer membrane to the intermembrane space, where it binds to a TIM complex, opening the channel in the complex. The polypeptide chain then either enters the matrix space or inserts into the inner membrane.

Although the TOM and TIM complexes usually work together to transport precursor proteins across both membranes at the same time, they can work independently. In isolated outer membranes, for example, the TOM complex can translocate the signal sequence of precursor proteins across the membrane. Similarly, if the outer membrane is experimentally disrupted in isolated mitochondria, the exposed TIM23 complex can efficiently import precursor proteins into the matrix space.

ATP Hydrolysis and a Membrane Potential Drive Protein Import Into the Matrix Space

Directional transport requires energy, which in most biological systems is supplied by ATP hydrolysis. ATP hydrolysis fuels mitochondrial protein import at two discrete sites, one outside the mitochondria and one in the matrix space (**Figure 12–26**). In addition, protein import requires another energy source, which is the membrane potential across the inner mitochondrial membrane.

The first requirement for energy occurs at the initial stage of the translocation process, when the unfolded precursor protein, associated with chaperone proteins, interacts with the import receptors of the TOM complex. As discussed in Chapter 6, the binding and release of newly synthesized polypeptides from the Hsp70 family of chaperone proteins requires ATP hydrolysis. The requirement for Hsp70 and ATP in the cytosol can be bypassed if the precursor protein is artificially unfolded prior to adding it to purified mitochondria.

Once the signal sequence has passed through the TOM complex and is bound to a TIM complex, further translocation through the TIM translocation channel requires the membrane potential, which is the electrical component of the electrochemical H$^+$ gradient across the inner membrane (see Figure 11–4). Pumping of H$^+$ from the matrix space to the intermembrane space, driven by

Figure 12–26 The role of energy in protein import into the mitochondrial matrix space. (1) Bound *cytosolic Hsp70* is released from the protein in a step that depends on ATP hydrolysis. After initial insertion of the signal sequence and of adjacent portions of the polypeptide chain into the TOM complex, the signal sequence interacts with a TIM complex. (2) The signal sequence is then translocated into the matrix space in a process that requires a membrane potential across the inner membrane. (3) *Mitochondrial Hsp70*, which is part of an import ATPase complex, binds to regions of the polypeptide chain as they become exposed in the matrix space, pulling the protein through the translocation channel.

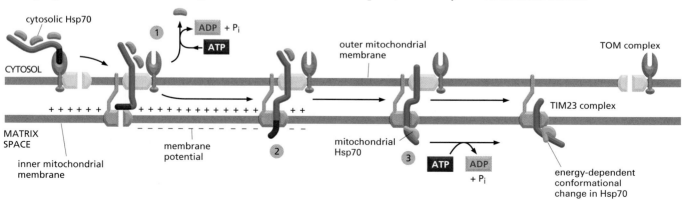

electron transport processes in the inner membrane (discussed in Chapter 14), maintains the electrochemical gradient. The energy in the electrochemical H^+ gradient across the inner membrane not only helps drive most of the cell's ATP synthesis, but it also drives the translocation of the positively charged signal sequences through the TIM complexes by electrophoresis.

Mitochondrial Hsp70 also plays a crucial part in the import process. Mitochondria containing mutant forms of the protein fail to import precursor proteins. The Hsp70 is part of a multisubunit protein assembly that is bound to the matrix side of the TIM23 complex and acts as a motor to pull the precursor protein into the matrix space. Like its cytosolic cousin, mitochondrial Hsp70 has a high affinity for unfolded polypeptide chains, and it binds tightly to an imported protein as soon as the protein emerges from the TIM translocator in the matrix space. The Hsp70 then releases the protein in an ATP-dependent step. This energy-driven cycle of binding and subsequent release is thought to provide the final driving force needed to complete protein import after a protein has initially inserted into the TIM23 complex (see Figure 12–26).

After the initial interaction with mitochondrial Hsp70, many imported matrix proteins are passed on to another chaperone protein, *mitochondrial Hsp60*. As discussed in Chapter 6, Hsp60 helps the unfolded polypeptide chain to fold by binding and releasing it through cycles of ATP hydrolysis.

Bacteria and Mitochondria Use Similar Mechanisms to Insert Porins into their Outer Membrane

The outer mitochondrial membrane, like the outer membrane of Gram-negative bacteria (see Figure 11–18), contains abundant pore-forming proteins called porins and is thus freely permeable to inorganic ions and metabolites (but not to most proteins). Porins are β-barrel proteins (see Figure 10–26) and are first imported through the TOM complex (**Figure 12–27**). In contrast to other outer membrane proteins, which are anchored in the membrane through α-helical regions, the TOM complex cannot integrate porins into the lipid bilayer. Instead, porins are first transported into the intermembrane space, where they transiently bind specialized chaperone proteins, which keep the porins from aggregating. They then bind to the SAM complex in the outer membrane, which both inserts them into the outer membrane and helps them fold properly.

One of the central subunits of the SAM complex is homologous to a bacterial outer membrane protein that helps insert β-barrel proteins into the bacterial outer membrane from the periplasmic space (the topological equivalent of the intermembrane space in mitochondria). This conserved pathway for inserting β-barrel proteins is further evidence for the endosymbiotic origin of mitochondria.

Transport Into the Inner Mitochondrial Membrane and Intermembrane Space Occurs Via Several Routes

The same mechanism that transports proteins into the matrix space, using the TOM and TIM23 translocators (see Figure 12–25), also mediates the initial translocation of many proteins that are destined for the inner mitochondrial

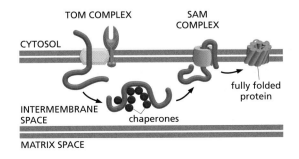

Figure 12–27 **Integration of porins into the outer mitochondrial membrane.** After translocation through the TOM complex, β-barrel proteins bind to chaperones in the intermembrane space. The SAM complex then inserts the unfolded polypeptide chain into the outer membrane.

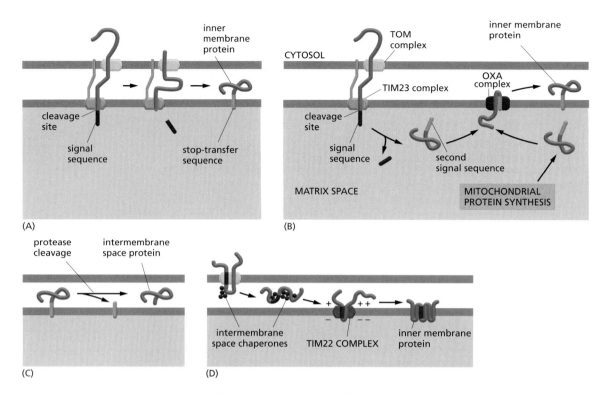

Figure 12–28 Protein import from the cytosol into the inner mitochondrial membrane and intermembrane space. (A) The N-terminal signal sequence (*red*) initiates import into the matrix space (see Figure 12–25). A hydrophobic sequence (*orange*) that follows the matrix-targeting signal binds to the TIM23 translocator in the inner membrane and stops translocation. The remainder of the protein is then pulled into the intermembrane space through the TOM translocator in the outer membrane, and the hydrophobic sequence is released into the inner membrane. (B) A second route for protein integration into the inner membrane first delivers the protein completely into the matrix space. Cleavage of the signal sequence *(red)* used for the initial translocation unmasks an adjacent hydrophobic signal sequence *(orange)* at the new N-terminus. This signal then directs the protein into the inner membrane, using the same OXA-dependent pathway that inserts proteins that are encoded by the mitochondrial genome and translated in the matrix space. (C) Some soluble proteins of the intermembrane space also use the pathways shown in (A) and (B) before they are released into the intermembrane space by a second signal peptidase, which has its active site in the intermembrane space and removes the hydrophobic signal sequence. (D) Metabolite transporters contain internal signal sequences and snake through the TOM complex as a loop. They then bind to the chaperones in the intermembrane space, which guide the proteins to the TIM22 complex. The TIM22 complex is specialized for the insertion of multipass inner membrane proteins.

membrane or the intermembrane space. In the most common translocation route taken, only the N-terminal signal sequence of the transported protein actually enters the matrix space (**Figure 12–28**A). A hydrophobic amino acid sequence, strategically placed after the N-terminal signal sequence, acts as a *stop-transfer sequence,* preventing further translocation across the inner membrane. The TOM complex pulls the remainder of the protein through the outer membrane into the intermembrane space; the signal sequence is cleaved off in the matrix; and the hydrophobic sequence, released from TIM23, remains anchored in the inner membrane.

In another transport route to the inner membrane or intermembrane space, the TIM23 complex initially translocates the entire protein into the matrix space (Figure 12–28B). A matrix signal peptidase then removes the N-terminal signal sequence, exposing a hydrophobic sequence at the new N-terminus. This signal sequence guides the protein to the OXA complex, which inserts the protein into the inner membrane (see Figure 12–23). As mentioned earlier, the OXA complex is primarily used to insert proteins that are encoded and translated in the mitochondrion into the inner membrane, and only a few imported proteins use this pathway. Translocators that are closely related to the OXA complex are found in the plasma membrane of bacteria and in the thylakoid membrane of chloroplasts, where they are thought to help insert membrane proteins by a similar mechanism.

Many proteins that use these pathways to the inner membrane remain anchored there through their hydrophobic signal sequence (see Figure 12–28A and B). Others, however, are released into the intermembrane space by a protease that removes the membrane anchor (Figure 12–28C). Many of these cleaved proteins remain attached to the outer surface of the inner membrane as peripheral subunits of protein complexes that also contain transmembrane proteins.

Mitochondria are the principal sites of ATP synthesis in the cell, but they also contain many metabolic enzymes, such as those of the citric acid cycle. Thus, in addition to proteins, mitochondria must also transport small metabolites across their membranes. While the outer membrane contains porins, which make the membrane freely permeable to such small molecules, the inner membrane does not. Instead, a family of metabolite-specific transporters transfers a vast number of small molecules across the inner membrane. In yeast cells, these transporters comprise a family of 35 different proteins, the most abundant of which transport ATP, ADP, and phosphate. These are multipass transmembrane proteins, which do not have cleavable signal sequences at their N-termini but instead contain internal signal sequences. They cross the TOM complex in the outer membrane, and intermembrane space chaperones guide them to the TIM22 complex, which inserts them into the inner membrane by a process that requires the membrane potential, but not mitochondrial Hsp70 or ATP (Figure 12–28D). An energetically favorable partitioning of the hydrophobic transmembrane regions into the inner membrane is also likely to help drive this process.

Two Signal Sequences Direct Proteins to the Thylakoid Membrane in Chloroplasts

Protein transport into **chloroplasts** resembles transport into mitochondria. Both processes occur post-translationally, use separate translocation complexes in each membrane, require energy, and use amphiphilic N-terminal signal sequences that are removed after use. With the exception of some of the chaperone molecules, however, the protein components that form the translocation complexes differ. Moreover, whereas mitochondria harness the electrochemical H^+ gradient across their inner membrane to drive transport, chloroplasts, which have an electrochemical H^+ gradient across their thylakoid membrane but not their inner membrane, use GTP and ATP hydrolysis to power import across their double membrane. The functional similarities may thus result from convergent evolution, reflecting the common requirements for translocation across a double membrane.

Although the signal sequences for import into chloroplasts superficially resemble those for import into mitochondria, the same plant cells have both mitochondria and chloroplasts, so proteins must partition appropriately between them. In plants, for example, a bacterial enzyme can be directed specifically to mitochondria if it is experimentally joined to an N-terminal signal sequence of a mitochondrial protein; the same enzyme joined to an N-terminal signal sequence of a chloroplast protein ends up in chloroplasts. Thus, the import receptors on each organelle distinguish between the different signal sequences.

Chloroplasts have an extra membrane-enclosed compartment, the **thylakoid**. Many chloroplast proteins, including the protein subunits of the photosynthetic system and of the ATP synthase (discussed in Chapter 14) are located in the thylakoid membrane. Like the precursors of some mitochondrial proteins, the transport of these precursor proteins from the cytosol to their final destination occurs in two steps. First, they pass across the double membrane at special contact sites into the matrix space (called the **stroma** in chloroplasts) and then they translocate either into the thylakoid membrane or into the thylakoid space (**Figure 12–29**A). The precursors of these proteins have a hydrophobic thylakoid signal sequence following the N-terminal chloroplast signal sequence. After the N-terminal signal sequence has been used to import the protein into the stroma,

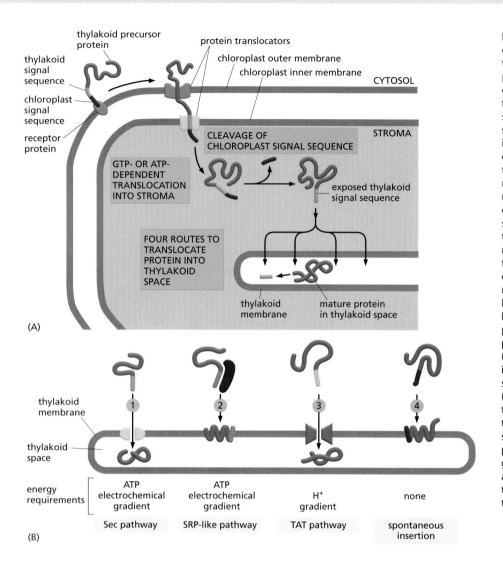

(A)

(B)

Figure 12–29 Translocation of chloroplast precursor proteins into the thylakoid space. (A) The precursor protein contains an N-terminal chloroplast signal sequence *(red)*, followed immediately by a thylakoid signal sequence *(orange)*. The chloroplast signal sequence initiates translocation into the stroma through a membrane contact site by a mechanism similar to that used for the translocation of mitochondrial precursor proteins into the matrix space. The signal sequence is then cleaved off, unmasking the thylakoid signal sequence, which initiates translocation across the thylakoid membrane. (B) Translocation into the thylakoid space or thylakoid membrane can occur by any one of at least four routes: (1) a *Sec pathway*, so called because it uses components that are homologs of Sec proteins, which mediate protein translocation across the bacterial plasma membrane (discussed later), (2) an *SRP-like pathway*, so called because it uses a chloroplast homolog of the signal recognition particle, or SRP (discussed later); (3) a *TAT* (twin arginine translocation) *pathway*, so called because two arginines are critical in the signal sequences that direct proteins into this pathway, which depends on the H^+ gradient across the thylakoid membrane; and (4) a *spontaneous insertion pathway* that seems not to require any protein translocator.

a stromal signal peptidase removes it, unmasking the thylakoid signal sequence that initiates transport across the thylakoid membrane. There are at least four routes by which proteins cross or become integrated into the thylakoid membrane, distinguished by their need for different stromal chaperones and energy sources (Figure 12–29B).

Summary

Although mitochondria and chloroplasts have their own genetic systems, they produce only a small proportion of their own proteins. Instead, the two organelles import most of their proteins from the cytosol, using similar mechanisms. In both cases, proteins are transported in an unfolded state across both outer and inner membranes simultaneously into the matrix space or stroma. Both ATP hydrolysis and a membrane potential across the inner membrane drive translocation into mitochondria, whereas GTP and ATP hydrolysis drive translocation into chloroplasts. Chaperone proteins of the cytosolic Hsp70 family maintain the precursor proteins in an unfolded state, and a second set of Hsp70 proteins in the matrix space or stroma pull the polypeptide chain into the organelle. Only proteins that contain a specific signal sequence are translocated. The signal sequence can either be located at the N-terminus and cleaved off after import or be internal and retained. Transport into the inner membrane sometimes uses a second, hydrophobic signal sequence that is unmasked when the first signal sequence is removed. In chloroplasts, import from the stroma into the thylakoid can occur by several routes, distinguished by the chaperones and energy source used.

PEROXISOMES

Peroxisomes differ from mitochondria and chloroplasts in many ways. Most notably, they are surrounded by only a single membrane, and they do not contain DNA or ribosomes. Thus, lacking a genome, all of their proteins are encoded in the nucleus. Peroxisomes acquire most of these proteins by selective import from the cytosol, although some of them enter the peroxisome membrane via the ER.

Because we do not discuss peroxisomes elsewhere, we shall digress to consider some of the functions of this diverse family of organelles, before discussing their biosynthesis. All eucaryotic cells have peroxisomes. They contain oxidative enzymes, such as *catalase* and *urate oxidase*, at such high concentrations that, in some cells, the peroxisomes stand out in electron micrographs because of the presence of a crystalloid core (**Figure 12–30**).

Like mitochondria, peroxisomes are major sites of oxygen utilization. One hypothesis is that peroxisomes are a vestige of an ancient organelle that performed all the oxygen metabolism in the primitive ancestors of eucaryotic cells. When the oxygen produced by photosynthetic bacteria first accumulated in the atmosphere, it would have been highly toxic to most cells. Peroxisomes might have lowered the intracellular concentration of oxygen, while also exploiting its chemical reactivity to perform useful oxidation reactions. According to this view, the later development of mitochondria rendered peroxisomes largely obsolete because many of the same biochemical reactions—which had formerly been carried out in peroxisomes without producing energy—were now coupled to ATP formation by means of oxidative phosphorylation. The oxidation reactions performed by peroxisomes in present-day cells would therefore be those whose functions were not taken over by mitochondria.

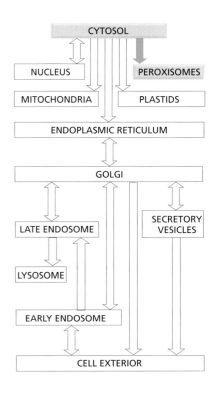

Peroxisomes Use Molecular Oxygen and Hydrogen Peroxide to Perform Oxidation Reactions

Peroxisomes are so named because they usually contain one or more enzymes that use molecular oxygen to remove hydrogen atoms from specific organic substrates (designated here as R) in an oxidation reaction that produces *hydrogen peroxide (H_2O_2)*:

$$RH_2 + O_2 \rightarrow R + H_2O_2$$

Catalase uses the H_2O_2 generated by other enzymes in the organelle to oxidize a variety of other substrates—including phenols, formic acid, formaldehyde, and alcohol—by the "peroxidation" reaction: $H_2O_2 + R'H_2 \rightarrow R' + 2H_2O$. This type of oxidation reaction is particularly important in liver and kidney cells, where the peroxisomes detoxify various toxic molecules that enter the bloodstream. About 25% of the ethanol we drink is oxidized to acetaldehyde in this way. In addition, when excess H_2O_2 accumulates in the cell, catalase converts it to H_2O through the reaction

$$2H_2O_2 \rightarrow 2H_2O + O_2$$

A major function of the oxidation reactions performed in peroxisomes is the breakdown of fatty acid molecules. The process called *β oxidation* shortens the alkyl chains of fatty acids sequentially in blocks of two carbon atoms at a time, thereby converting the fatty acids to acetyl CoA. The peroxisomes then export the acetyl CoA to the cytosol for reuse in biosynthetic reactions. In mammalian cells, β oxidation occurs in both mitochondria and peroxisomes; in yeast and plant cells, however, this essential reaction occurs exclusively in peroxisomes.

An essential biosynthetic function of animal peroxisomes is to catalyze the first reactions in the formation of *plasmalogens*, which are the most abundant class of phospholipids in myelin (**Figure 12–31**). Plasmalogen deficiencies cause profound abnormalities in the myelination of nerve cell axons, which is why many peroxisomal disorders lead to neurological disease.

Peroxisomes are unusually diverse organelles, and even in the various cell types of a single organism they may contain different sets of enzymes. They also adapt remarkably to changing conditions. Yeasts grown on sugar, for example,

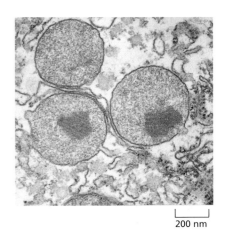

Figure 12–30 An electron micrograph of three peroxisomes in a rat liver cell. The paracrystalline, electron-dense inclusions are composed of the enzyme urate oxidase. (Courtesy of Daniel S. Friend.)

Figure 12–31 The structure of a plasmalogen. Plasmalogens are very abundant in the myelin sheaths that insulate the axons of nerve cells. They make up some 80–90% of the myelin membrane phospholipids. In addition to an ethanolamine head group and a long-chain fatty acid attached to the same glycerol phosphate backbone used for phospholipids, plasmalogens contain an unusual fatty alcohol that is attached through an ether linkage *(bottom left)*.

have small peroxisomes. But when some yeasts are grown on methanol they develop large peroxisomes that oxidize methanol; and when grown on fatty acids they develop large peroxisomes that break down fatty acids to acetyl CoA by β oxidation.

Peroxisomes are also important in plants. Two types of plant peroxisomes have been studied extensively. One is present in leaves, where it participates in *photorespiration* (discussed in Chapter 14) (**Figure 12–32**A). The other type of peroxisome is present in germinating seeds, where it converts the fatty acids stored in seed lipids into the sugars needed for the growth of the young plant. Because this conversion of fats to sugars is accomplished by a series of reactions known as the *glyoxylate cycle*, these peroxisomes are also called *glyoxysomes* (Figure 12–32B). In the glyoxylate cycle, two molecules of acetyl CoA produced by fatty acid breakdown in the peroxisome are used to make succinic acid, which then leaves the peroxisome and is converted into glucose in the cytosol. The glyoxylate cycle does not occur in animal cells, and animals are therefore unable to convert the fatty acids in fats into carbohydrates.

A Short Signal Sequence Directs the Import of Proteins into Peroxisomes

A specific sequence of three amino acids (Ser–Lys–Leu) located at the C-terminus of many peroxisomal proteins functions as an import signal (see Table 12–3, p. 702). Other peroxisomal proteins contain a signal sequence near the N-terminus. If either sequence is attached to a cytosolic protein, the protein is imported into peroxisomes. The import process is still poorly understood, although it is known to involve both soluble receptor proteins in the cytosol, which recognize the targeting signals, and docking proteins on the cytosolic surface of the peroxisome. At least 23 distinct proteins, called **peroxins**, participate in the import process, which is driven by ATP hydrolysis. A complex of at least six different peroxins forms a membrane translocator. Because even oligomeric proteins do not have to unfold to be imported into peroxisomes, the mechanism differs from that used by mitochondria and chloroplasts. At least one soluble import receptor, the peroxin Pex5, accompanies its cargo all the way into peroxisomes and, after cargo release, cycles back to the cytosol. These aspects of peroxisomal protein import resemble protein transport into the nucleus.

Figure 12–32 Electron micrographs of two types of peroxisomes found in plant cells. (A) A peroxisome with a paracrystalline core in a tobacco leaf mesophyll cell. Its close association with chloroplasts is thought to facilitate the exchange of materials between these organelles during photorespiration. (B) Peroxisomes in a fat-storing cotyledon cell of a tomato seed 4 days after germination. Here the peroxisomes (glyoxysomes) are associated with the lipid droplets that store fat, reflecting their central role in fat mobilization and gluconeogenesis during seed germination. (A, from S.E. Frederick and E.H. Newcomb, *J. Cell Biol.* 43:343–353, 1969. With permission from The Rockefeller Press; B, from W.P. Wergin, P.J. Gruber and E.H. Newcomb, *J. Ultrastruct. Res.* 30:533–557, 1970. With permission from Academic Press.)

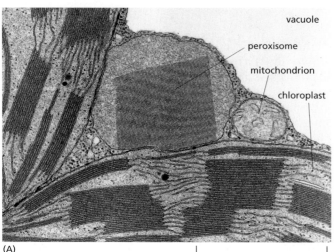

(A)

1 μm

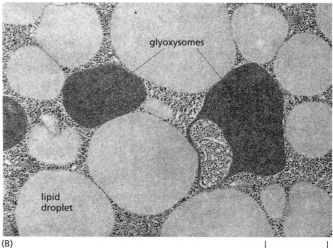

(B)

1 μm

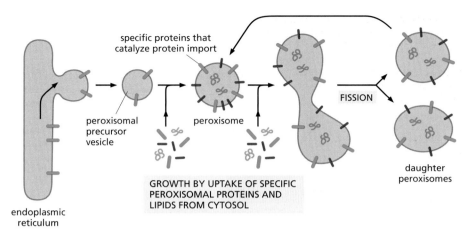

specific proteins that
catalyze protein import

peroxisomal
precursor
vesicle

peroxisome

FISSION

daughter
peroxisomes

GROWTH BY UPTAKE OF SPECIFIC
PEROXISOMAL PROTEINS AND
LIPIDS FROM CYTOSOL

endoplasmic
reticulum

Figure 12–33 A model that explains how peroxisomes proliferate and how new peroxisomes may arise. Peroxisome precursor vesicles are thought to bud from the ER. The machinery that drives the budding reaction and that selects only peroxisomal proteins for packaging into these vesicles—and not ER proteins or proteins destined for other locations in the cell—is not known. Peroxisome precursor vesicles may then fuse with one another or with preexisting peroxisomes. The peroxisome membrane contains import receptor proteins. Cytosolic ribosomes synthesize peroxisomal proteins, including new copies of the import receptor, and then import them into the organelle. Presumably, the lipids required for growth are also imported, although some may derive directly from the ER in the membrane of peroxisome precursor vesicles. (We discuss later the transport of lipids made in the ER through the cytosol to other organelles.)

The importance of this import process and of peroxisomes is demonstrated by the inherited human disease *Zellweger syndrome*, in which a defect in importing proteins into peroxisomes leads to a profound peroxisomal deficiency. These individuals, whose cells contain "empty" peroxisomes, have severe abnormalities in their brain, liver, and kidneys, and they die soon after birth. A mutation in the gene encoding peroxin Pex2, a peroxisomal integral membrane protein involved in protein import, causes one form of the disease. A defective receptor for the N-terminal import signal causes a milder inherited peroxisomal disease.

It has long been debated whether new peroxisomes arise from preexisting ones by organelle growth and fission—and therefore replicate in an autonomous way as mentioned earlier for mitochondria and plastids—or whether they derive as a specialized compartment from the endoplasmic reticulum (ER). Aspects of both views may be true (**Figure 12–33**). Most peroxisomal membrane proteins are made in the cytosol and insert into the membrane of preexisting ones, yet others are first integrated into the ER membrane from where they may bud in specialized peroxisomal precursor vesicles. New precursor vesicles may then fuse with one another and begin importing additional peroxisomal proteins using their own protein import machinery to grow into mature peroxisomes, which can enter into a cycle of growth and fission.

Summary

Peroxisomes are specialized for carrying out oxidation reactions using molecular oxygen. They generate hydrogen peroxide, which they employ for oxidative purposes—and contain catalase to destroy the excess. Like mitochondria and plastids, peroxisomes are self-replicating organelles. Because they do not contain DNA or ribosomes, however, all of their proteins are encoded in the cell nucleus. Some of these proteins are conveyed to peroxisomes via the ER, but most are synthesized in the cytosol. A specific sequence of three amino acids near the C-terminus of many of the cytosolic proteins functions as a peroxisomal import signal. The mechanism of protein import differs from that of mitochondria and chloroplasts, in that even oligomeric proteins are imported from the cytosol without unfolding.

THE ENDOPLASMIC RETICULUM

All eucaryotic cells have an **endoplasmic reticulum (ER)**. Its membrane typically constitutes more than half of the total membrane of an average animal cell (see Table 12–2, p. 697). The ER is organized into a netlike labyrinth of branching tubules and flattened sacs that extends throughout the cytosol (**Figure 12–34**). The tubules and sacs interconnect, and their membrane is continuous with the outer nuclear membrane. Thus, the ER and nuclear membranes form a continuous sheet enclosing a single internal space, called the **ER lumen** or the *ER cisternal space*, which often occupies more than 10% of the total cell volume (see Table 12–1, p. 697).

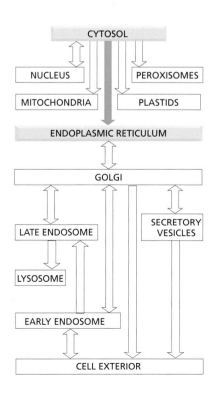

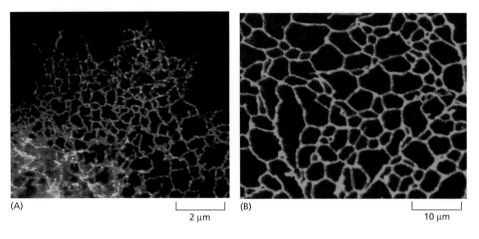

(A)

2 µm

(B)

10 µm

Figure 12–34 Fluorescent micrographs of the endoplasmic reticulum. <TCGT> (A) Part of the ER network in a cultured mammalian cell, stained with an antibody that binds to a protein retained in the ER. The ER extends as a network throughout the entire cytosol, so that all regions of the cytosol are close to some portion of the ER membrane. (B) Part of an ER network in a living plant cell that was genetically engineered to express a fluorescent protein in the ER. (A, courtesy of Hugh Pelham; B, courtesy of Petra Boevink and Chris Hawes.)

The ER has a central role in both lipid and protein biosynthesis, and it also serves as an intracellular Ca^{2+} store that is used in many cell signaling responses (discussed in Chapter 15). The ER membrane is the site of production of all the transmembrane proteins and lipids for most of the cell's organelles, including the ER itself, the Golgi apparatus, lysosomes, endosomes, secretory vesicles, and the plasma membrane. The ER membrane also makes most of the lipids for mitochondrial and peroxisomal membranes. In addition, almost all of the proteins that will be secreted to the cell exterior—plus those destined for the lumen of the ER, Golgi apparatus, or lysosomes—are initially delivered to the ER lumen.

The ER Is Structurally and Functionally Diverse

While the various functions of the ER are essential to every cell, their relative importance varies greatly between individual cell types. To meet different functional demands, distinct regions of the ER become highly specialized. We observe such functional specialization as dramatic changes in ER structure, and different cell types can therefore possess characteristically different types of ER membrane. One of the most remarkable ER specializations is the *rough ER*.

Mammalian cells begin to import most proteins into the ER before complete synthesis of the polypeptide chain—that is, import is a **co-translational** process (**Figure 12–35A**). In contrast, the import of proteins into mitochondria, chloroplasts, nuclei, and peroxisomes is a **post-translational** process (Figure 12–35B). In co-translational transport, the ribosome that is synthesizing the protein is attached directly to the ER membrane, enabling one end of the protein to be translocated into the ER while the rest of the polypeptide chain is being assembled. These membrane-bound ribosomes coat the surface of the ER, creating regions termed **rough endoplasmic reticulum**, or **rough ER** (**Figure 12–36A**).

Regions of ER that lack bound ribosomes are called **smooth endoplasmic reticulum**, or **smooth ER**. Most cells have scanty regions of smooth ER, and the

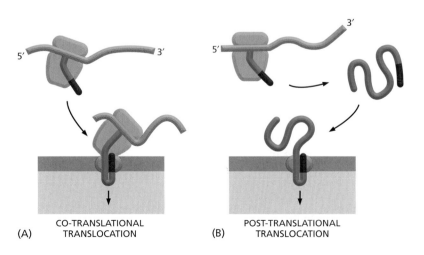

(A)   CO-TRANSLATIONAL TRANSLOCATION

(B)   POST-TRANSLATIONAL TRANSLOCATION

Figure 12–35 Co-translational and post-translational protein translocation. (A) Ribosomes bind to the ER membrane during co-translational translocation. (B) By contrast, ribosomes complete the synthesis of a protein and release it prior to post-translational translocation.

inner nuclear membrane

nucleus | outer nuclear membrane | ER membrane

rough ER | smooth ER

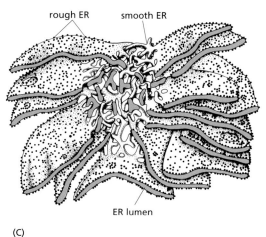

ER lumen

(C)

ER membrane

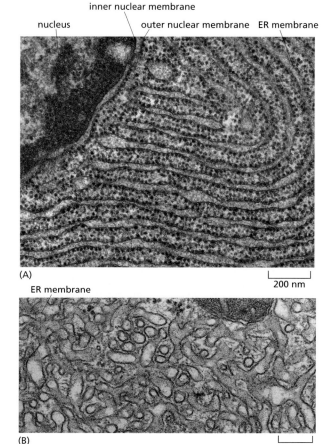

(A)

200 nm

(B)

200 nm

Figure 12–36 The rough and smooth ER. (A) An electron micrograph of the rough ER in a pancreatic exocrine cell that makes and secretes large amounts of digestive enzymes every day. The cytosol is filled with closely packed sheets of ER membrane studded with ribosomes. At the top left is a portion of the nucleus and its nuclear envelope; note that the outer nuclear membrane, which is continuous with the ER, is also studded with ribosomes. (B) Abundant smooth ER in a steroid-hormone-secreting cell. This electron micrograph is of a testosterone-secreting Leydig cell in the human testis. (C) A three-dimensional reconstruction of a region of smooth ER and rough ER in a liver cell. The rough ER forms oriented stacks of flattened cisternae, each having a lumenal space 20–30 nm wide. The smooth ER membrane is connected to these cisternae and forms a fine network of tubules 30–60 nm in diameter. (A, courtesy of Lelio Orci; B, courtesy of Daniel S. Friend; C, after R.V. Krstić, Ultrastructure of the Mammalian Cell. New York: Springer-Verlag, 1979.)

ER is often partly smooth and partly rough. Areas of smooth ER from which transport vesicles carrying newly synthesized proteins and lipids bud off for transport to the Golgi apparatus are called *transitional ER*. In certain specialized cells, the smooth ER is abundant and has additional functions. It is prominent, for example, in cells that specialize in lipid metabolism, such as cells that synthesize steroid hormones from cholesterol; the expanded smooth ER accommodates the enzymes that make cholesterol and modify it to form the hormones (Figure 12–36B).

The main cell type in the liver, the *hepatocyte*, also has abundant smooth ER. It is the principal site of production of *lipoprotein particles*, which carry lipids via the bloodstream to other parts of the body. The enzymes that synthesize the lipid components of the particles are located in the membrane of the smooth ER, which also contains enzymes that catalyze a series of reactions to detoxify both lipid-soluble drugs and various harmful compounds produced by metabolism. The most extensively studied of these *detoxification reactions* are carried out by the *cytochrome P450* family of enzymes, which catalyze a series of reactions in which water-insoluble drugs or metabolites that would otherwise accumulate to toxic levels in cell membranes are rendered sufficiently water-soluble to leave the cell and be excreted in the urine. Because the rough ER alone cannot house enough of these and other necessary enzymes, a major portion of the membrane in a hepatocyte normally consists of smooth ER (Figure 12–36C; see Table 12–2).

Another crucially important function of the ER in most eucaryotic cells is to sequester Ca^{2+} from the cytosol. The release of Ca^{2+} into the cytosol from the ER, and its subsequent reuptake, occurs in many rapid responses to extracellular signals, as discussed in Chapter 15. A Ca^{2+} pump transports Ca^{2+} from the cytosol into the ER lumen. A high concentration of Ca^{2+}-binding proteins in the ER facilitates Ca^{2+} storage. In some cell types, and perhaps in most, specific regions of the ER are specialized for Ca^{2+} storage. Muscle cells have an abundant, modified smooth ER, called the *sarcoplasmic reticulum*. The release and

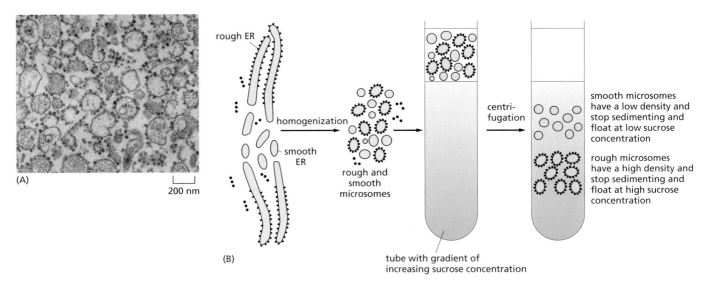

Figure 12–37 The isolation of purified rough and smooth microsomes from the ER. (A) A thin section electron micrograph of the purified rough ER fraction shows an abundance of ribosome-studded vesicles. (B) When sedimented to equilibrium through a gradient of sucrose, the two types of microsomes separate from each other on the basis of their different densities. (A, courtesy of George Palade.)

reuptake of Ca^{2+} by the sarcoplasmic reticulum trigger myofibril contraction and relaxation, respectively, during each round of muscle contraction (discussed in Chapter 16).

To study the functions and biochemistry of the ER, it is necessary to isolate it. This may seem to be a hopeless task because the ER is intricately interleaved with other components of the cytoplasm. Fortunately, when tissues or cells are disrupted by homogenization, the ER breaks into fragments, which reseal to form small (~100–200 nm in diameter) closed vesicles called **microsomes**. Microsomes are relatively easy to purify. To the biochemist, microsomes represent small authentic versions of the ER, still capable of protein translocation, protein glycosylation, Ca^{2+} uptake and release, and lipid synthesis. Microsomes derived from rough ER are studded with ribosomes and are called *rough microsomes*. The ribosomes are always found on the outside surface, so the interior of the microsome is biochemically equivalent to the lumenal space of the ER (**Figure 12–37A**).

Many vesicles similar in size to rough microsomes, but lacking attached ribosomes, are also found in cell homogenates. Such *smooth microsomes* are derived in part from smooth portions of the ER and in part from vesiculated fragments of the plasma membrane, Golgi apparatus, endosomes, and mitochondria (the ratio depending on the tissue). Thus, whereas rough microsomes are clearly derived from rough portions of ER, it is not easy to determine the origins of "smooth microsomes" prepared from disrupted cells. The smooth microsomes prepared from liver or muscle cells are an exception. Because of the unusually large quantities of smooth ER or sarcoplasmic reticulum, respectively, most of the smooth microsomes in homogenates of these tissues are derived from the smooth ER. The ribosomes attached to rough microsomes make them more dense than smooth microsomes. As a result, we can use equilibrium centrifugation to separate the rough and smooth microsomes (Figure 12–37B). Microsomes have been invaluable in elucidating the molecular aspects of ER function, as we discuss next.

Signal Sequences Were First Discovered in Proteins Imported into the Rough ER

The ER captures selected proteins from the cytosol as they are being synthesized. These proteins are of two types: *transmembrane proteins,* which are only partly translocated across the ER membrane and become embedded in it, and *water-soluble proteins,* which are fully translocated across the ER membrane and are released into the ER lumen. Some of the transmembrane proteins function in the ER, but many are destined to reside in the plasma membrane or the membrane of another organelle. The water-soluble proteins are destined either for secretion or for residence in the lumen of an organelle. All of these proteins,

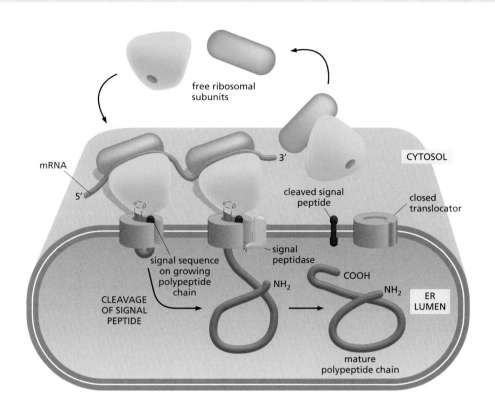

Figure 12–38 The signal hypothesis. A simplified view of protein translocation across the ER membrane, as originally proposed. When the ER signal sequence emerges from the ribosome, it directs the ribosome to a translocator on the ER membrane that forms a pore in the membrane through which the polypeptide is translocated. A signal peptidase is closely associated with the translocator and clips off the signal sequence during translation, and the mature protein is released into the lumen of the ER immediately after synthesis. The translocator is closed until the ribosome has bound, so that the permeability barrier of the ER membrane is maintained at all times.

regardless of their subsequent fate, are directed to the ER membrane by an **ER signal sequence**, which initiates their translocation by a common mechanism.

Signal sequences (and the signal sequence strategy of protein sorting) were first discovered in the early 1970s in secreted proteins that are translocated across the ER membrane as a first step toward their eventual discharge from the cell. In the key experiment, the mRNA encoding a secreted protein was translated by ribosomes *in vitro*. When microsomes were omitted from this cell-free system, the protein synthesized was slightly larger than the normal secreted protein, the extra length being the N-terminal *leader peptide*. In the presence of microsomes derived from the rough ER, however, a protein of the correct size was produced. According to the *signal hypothesis*, the leader is a signal sequence that directs the secreted protein to the ER membrane and is then cleaved off by a *signal peptidase* in the ER membrane before the polypeptide chain has been completed (**Figure 12–38**). Cell-free systems in which proteins are imported into microsomes have provided powerful procedures for identifying, purifying, and studying the various components of the molecular machinery responsible for the ER import process.

A Signal-Recognition Particle (SRP) Directs ER Signal Sequences to a Specific Receptor in the Rough ER Membrane

The ER signal sequence is guided to the ER membrane by at least two components: a **signal-recognition particle** (**SRP**), which cycles between the ER membrane and the cytosol and binds to the signal sequence, and an **SRP receptor** in the ER membrane. The SRP is a complex particle, consisting of six different polypeptide chains bound to a single small RNA molecule (**Figure 12–39**). SRP and its receptor are found in all cells, indicating that this protein-targeting mechanism arose early in evolution and has been conserved.

ER signal sequences vary greatly in amino acid sequence, but each has eight or more nonpolar amino acids at its center (see Table 12–3, p. 702). How can the SRP bind specifically to so many different sequences? The answer has come from the crystal structure of the SRP protein, which shows that the signal-sequence-binding site is a large hydrophobic pocket lined by methionines. Because methionines have unbranched, flexible side chains, the pocket

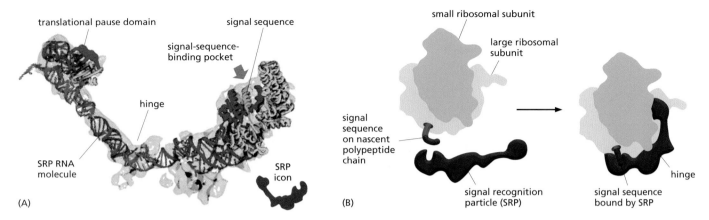

Figure 12–39 The signal-recognition particle (SRP). (A) A mammalian SRP is a rodlike complex containing six protein subunits and one RNA molecule (shown in *red*). The SRP RNA forms the backbone that links the domain of the SRP containing the signal sequence binding pocket to the domain responsible for pausing translation. The three-dimensional outline of the SRP (shown in *gray*) was determined by cryo-electron microscopy. Where known, crystal structures of individual SRP pieces are fitted into the envelope and shown as ribbon diagrams. A bound signal sequence is shown as a *green* helix. (B) SRP bound to the ribsosome visualized by cryo-electron microscopy. SRP binds to the large ribosomal subunit so that its signal sequence binding pocket is positioned near the nascent chain exit site and its translational pause domain is positioned at the interface between the ribosomal subunits, where it interferes with elongation factor binding. (Adapted from M. Halic et al., *Nature* 427:808–814, 2004. With permission from Macmillan Publishers Ltd.)

is sufficiently plastic to accommodate hydrophobic signal sequences of different sequences, sizes, and shapes.

The SRP is a rodlike structure that wraps around the large ribosomal subunit, with one end binding to the ER signal sequence as it emerges as part of the newly made polypeptide chain from the ribosome; the other end blocks the elongation factor binding site at the interface between the large and small ribosomal subunits (Figure 12–39). This block halts protein synthesis as soon as the signal peptide has emerged from the ribosome. The transient pause presumably gives the ribosome enough time to bind to the ER membrane before completion of the polypeptide chain, thereby ensuring that the protein is not released into the cytosol. This safety device may be especially important for secreted and lysosomal hydrolases that could wreak havoc in the cytosol; cells that secrete large amounts of hydrolases, however, take the added precaution of having high concentrations of hydrolase inhibitors in their cytosol. The pause also ensures that large portions of a protein that could fold into a compact structure are not made before reaching the translocator in the ER membrane. Thus, in contrast to the post-translational import of proteins into mitochondria and chloroplasts, chaperone proteins are not required to keep the protein unfolded.

Once formed, the SRP–ribosome complex binds to the SRP receptor, which is an integral membrane protein complex embedded in the rough ER membrane. This interaction brings the SRP–ribosome complex to a protein translocator. The SRP and SRP receptor are then released, and the translocator transfers the growing polypeptide chain across the membrane (**Figure 12–40**).

This co-translational transfer process creates two spatially separate populations of ribosomes in the cytosol. **Membrane-bound ribosomes**, attached to the cytosolic side of the ER membrane, are engaged in the synthesis of proteins that are being concurrently translocated into the ER. **Free ribosomes**, unattached to any membrane, synthesize all other proteins encoded by the nuclear genome. Membrane-bound and free ribosomes are structurally and functionally identical. They differ only in the proteins they are making at any given time.

Since many ribosomes can bind to a single mRNA molecule, a *polyribosome* is usually formed, which becomes attached to the ER membrane, directed there by the signal sequences on multiple growing polypeptide chains (**Figure 12–41A**). The individual ribosomes associated with such an mRNA molecule can return to the cytosol when they finish translation and intermix with the pool of

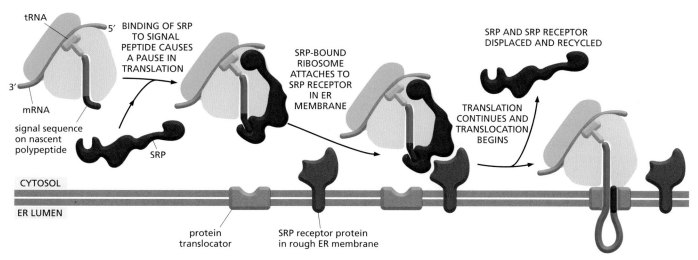

Figure 12–40 How ER signal sequences and SRP direct ribosomes to the ER membrane. <TTCC> The SRP and its receptor are thought to act in concert. The SRP binds to both the exposed ER signal sequence and the ribosome, thereby inducing a pause in translation. The SRP receptor in the ER membrane, which is composed of two different polypeptide chains, binds the SRP–ribosome complex and directs it to the translocator. In a poorly understood reaction, the SRP and SRP receptor are then released, leaving the ribosome bound to the translocator in the ER membrane. The translocator then inserts the polypeptide chain into the membrane and transfers it across the lipid bilayer. Because one of the SRP proteins and both chains of the SRP receptor contain GTP-binding domains, it is thought that conformational changes that occur during cycles of GTP binding and hydrolysis (discussed in Chapter 15) ensure that SRP release occurs only after the ribosome has become properly engaged with the translocator in the ER membrane. The translocator is closed until the ribosome has bound, so that the permeability barrier of the ER membrane is maintained at all times.

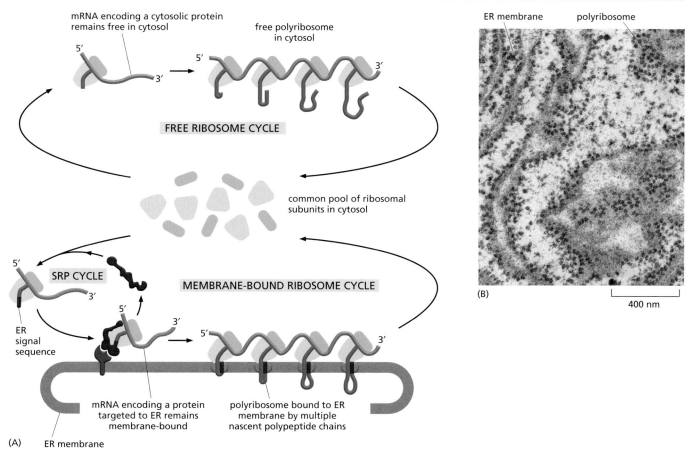

Figure 12–41 Free and membrane-bound ribosomes. (A) A common pool of ribosomes synthesizes the proteins that stay in the cytosol and those that are transported into the ER. The ER signal sequence on a newly formed polypeptide chain binds to SRP, which directs the translating ribosome to the ER membrane. The mRNA molecule remains permanently bound to the ER as part of a polyribosome, while the ribosomes that move along it are recycled; at the end of each round of protein synthesis the ribosomal subunits are released and rejoin the common pool in the cytosol. (B) A thin-section electron micrograph of polyribosomes attached to the ER membrane. The plane of section in some places cuts through the ER roughly parallel to the membrane, giving a face-on view of the rosettelike pattern of the polyribosomes. (B, courtesy of George Palade.)

free ribosomes. The mRNA itself, however, remains attached to the ER membrane by a changing population of ribosomes, each transiently held at the membrane by the translocator (Figure 12–41B).

The Polypeptide Chain Passes Through an Aqueous Pore in the Translocator

It has long been debated whether polypeptide chains are transferred across the ER membrane in direct contact with the lipid bilayer or through a pore in a protein translocator. The debate ended with the identification of the translocator, which was shown to form a water-filled pore in the membrane through which the polypeptide chain passes. The core of the translocator, called the **Sec61 complex**, is built from three subunits that are highly conserved from bacteria to eucaryotic cells. Recently, the structure of the Sec61 complex was determined by x-ray crystallography. The structure suggests that α helices contributed by the largest subsunit surround a central pore through which the polypeptide chain may traverse the membrane (**Figure 12–42**). The pore is gated by a short helix that is thought to keep the translocator closed when it is idle and to move aside when it is engaged in passing a polypeptide chain. According to this view, the pore is a dynamic gated structure that opens only transiently when a polypeptide chain traverses the membrane. In an idle translocator it is important to keep the pore closed, so that the membrane remains impermeable to ions, such as Ca^{2+}, which otherwise would leak out of the ER.

The structure of the Sec61 complex suggests that the pore can also open along a seam on its side. This opening allows a translocating peptide chain lateral access into the hydrophobic core of the membrane, a process that is important both for the release of a cleaved signal peptide into the membrane (see Figure 12–38) and for the integration of membrane proteins into the bilayer, as we discuss later.

In eucaryotic cells, four Sec61 complexes form a large translocator assembly that can be visualized on the ribosomes after detergent solubilization of ER membranes (**Figure 12–43**). It is likely that not all four Sec61 complexes in the eucaryotic translocator participate in protein translocation directly. Some of them may be inactive, providing binding sites for the ribosome and for accessory proteins that help polypeptide chains fold as they enter the ER. According to one view, the bound ribosome forms a tight seal with the translocator, with the space inside the ribosome continuous with the lumen of the ER so that no molecules can escape from the ER. Alternatively, the structure of the Sec61 complex suggests that the pore in the translocator could form a tight-fitting diaphragm around the translocating chain that prevents the escape of other molecules.

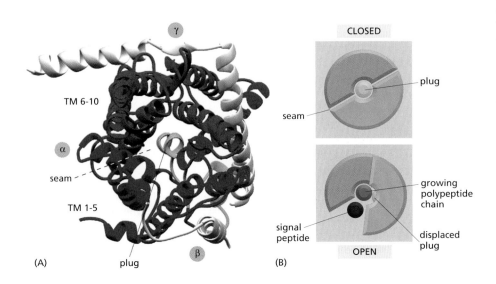

(A)   (B)

Figure 12–42 Structure of the Sec61 complex. (A) A top view of the structure of the Sec61 complex of the archae *Methanococcus jannaschii* seen from the cytosol side of the membrane. The Sec61α subunit is shown in *blue* and *red;* the two smaller β- and γ-subunits are shown in *gray*. The *yellow* short helix is thought to form a plug that seals the pore when the translocator is closed. To open, the complex rearranges itself to move the plug helix out of the way. It is thought that the pore of the Sec61 complex can also open sideways at a seam. (B) Model of the closed and open states of the translocator are shown, illustrating how a signal sequence could be released into the membrane after opening of the seam. (A, from B. van der Berg et al., *Nature* 427:36–44, 2004. With permission from Macmillan Publishers Ltd.)

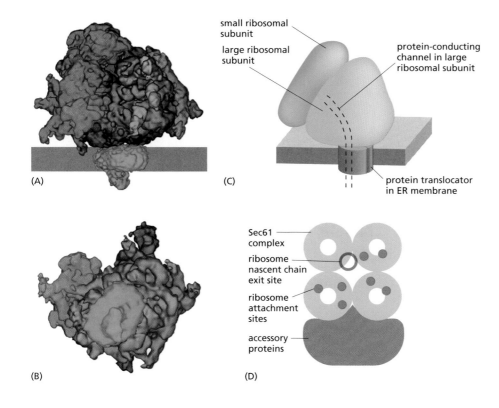

small ribosomal subunit

large ribosomal subunit

protein-conducting channel in large ribosomal subunit

(A)

(C)

protein translocator in ER membrane

Sec61 complex

ribosome nascent chain exit site

ribosome attachment sites

accessory proteins

(B)

(D)

Figure 12–43 A ribosome bound to the eucaryotic protein translocator.
(A) A reconstruction of the complex from electron microscopic images viewed from the side. (B) A view of the translocator seen from the ER lumen. The translocator is thought to contain four copies of the Sec61 complex. (C) A schematic drawing of a membrane-bound ribosome attached to the translocator with the tunnel in the large ribosomal subunit, through which the growing polypeptide chain exits from the ribosome (see Figure 6–70). (D) A schematic model of how four Sec61 complexes and various accessory proteins may be arranged in the translocator. The *red dots* indicate ribosome attachment sites that are seen in the electron microscopic reconstructions and the *red circle* shows the position of the end of the ribosome exit tunnel. It is not known which of the four Sec61 complexes are actively participating in the protein translocation event. Domains of the accessory proteins extend across the membrane and form the lumenal bulge seen in the side view in (A). (A, B, and C, adapted from J.F. Ménétret et al., *J. Mol. Biol.* 348:445–457, 2005. With permission from Academic Press.)

Translocation Across the ER Membrane Does Not Always Require Ongoing Polypeptide Chain Elongation

As we have seen, translocation of proteins into mitochondria, chloroplasts, and peroxisomes occurs post-translationally, after the protein has been made and released into the cytosol, whereas translocation across the ER membrane usually occurs during translation (co-translationally). This explains why ribosomes are bound to the ER but usually not to other organelles.

Some completely synthesized proteins, however, are imported into the ER, demonstrating that translocation does not always require ongoing translation. Post-translational protein translocation is especially common across the yeast ER membrane and the bacterial plasma membrane (which is thought to be evolutionarily related to the ER; see Figure 12–4). To function in post-translational translocation, the translocator needs accessory proteins that feed the polypeptide chain into the pore and drive translocation (**Figure 12–44**). In bacteria, a translocation motor protein, the *SecA ATPase*, attaches to the cytosolic side of the translocator, where it undergoes cyclic conformational changes driven by ATP hydrolysis. Each time an ATP is hydrolyzed, a portion of the SecA protein inserts into the pore of the translocator, pushing a short segment of the passenger protein with it. As a result of this ratchet mechanism, the SecA protein progressively pushes the polypeptide chain of the transported protein across the membrane.

Eucaryotic cells use a different set of accessory proteins that associate with the Sec61 complex. These proteins span the ER membrane and use a small domain on the lumenal side of the ER membrane to deposit a Hsp70-like chaperone protein (called *BiP*, for *b*inding *p*rotein) onto the polypeptide chain as it emerges from the pore into the ER lumen. Cycles of BiP binding and release drive unidirectional translocation, as described earlier for the mitochondrial Hsp70 proteins that pull proteins across mitochondrial membranes.

Proteins that are transported into the ER by a post-translational mechanism are first released into the cytosol, where they bind to chaperone proteins to prevent folding, as discussed earlier for proteins destined for mitochondria and chloroplasts.

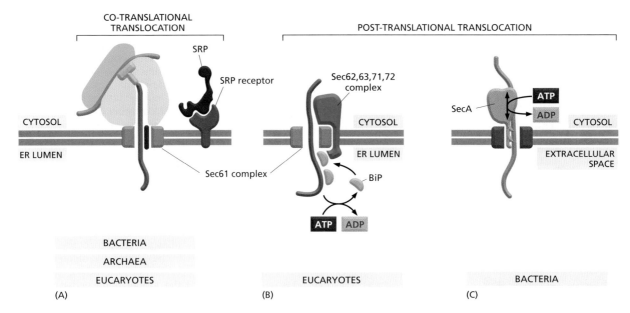

Figure 12–44 Three ways in which protein translocation can be driven through structurally similar translocators.
(A) Co-translational translocation. The ribosome is brought to the membrane by the SRP and SRP receptor and engages with the Sec61 protein translocator. The growing polypeptide chain is threaded across the membrane as it is made. No additional energy is needed, as the only path available to the growing chain is to cross the membrane. (B) Post-translational translocation in eucaryotic cells. An additional complex composed of the Sec62, Sec63, Sec71, and Sec72 proteins is attached to the Sec61 translocator and deposits BiP molecules onto the translocating chain as it emerges into the ER lumen. ATP-driven cycles of BiP binding and release pull the protein into the lumen, a mechanism that closely resembles the mechanism of mitochondrial import in Figure 12–26. (C) Post-translational translocation in bacteria. The completed polypeptide chain is fed from the cytosolic side into a translocator in the plasma membrane by the SecA ATPase. ATP-hydrolysis-driven conformational changes drive a pistonlike motion in SecA, each cycle pushing about 20 amino acids of the protein chain through the pore of the translocator. The Sec pathway used for protein translocation across the thylakoid membrane in chloroplasts uses a similar mechanism (see Figure 12–29B).

Whereas the Sec61 translocator, SRP, and SRP receptor are found in all organisms, SecA is found exclusively in bacteria, and the Sec62, Sec63, Sec71, and Sec72 proteins are found exclusively in eucaryotic cells. (Adapted from P. Walter and A.E. Johnson, *Annu. Rev. Cell Biol.* 10:87–119, 1994. With permission from Annual Reviews.)

In Single-Pass Transmembrane Proteins, a Single Internal ER Signal Sequence Remains in the Lipid Bilayer as a Membrane-Spanning α Helix

The signal sequence in the growing polypeptide chain is thought to trigger the opening of the pore in the protein translocator: after the signal sequence is released from the SRP and the growing chain has reached a sufficient length, the signal sequence binds to a specific site inside the pore itself, thereby opening the pore. An ER signal sequence is therefore recognized twice: first by an SRP in the cytosol and then by a binding site in the pore of the protein translocator, where it serves as a **start-transfer signal** (or start-transfer peptide) that opens the pore (illustrated for an ER soluble protein in **Figure 12–45**). Dual recognition may help to ensure that only appropriate proteins enter the lumen of the ER.

While bound in the translocation pore, a signal sequence is in contact not only with the Sec61 complex, which forms the walls of the pore, but also with the hydrophobic lipid core of the membrane. This was shown in chemical cross-linking experiments in which signal sequences and the hydrocarbon chains of lipids could be covalently linked together. When the nascent polypeptide chain grows long enough, an ER signal peptidase cleaves off signal sequences and releases them from the pore into the membrane, where they are rapidly degraded to amino acids by other proteases in the ER membrane. To release the signal sequence into the membrane, the translocator has to open laterally. The translocator is therefore gated in two directions: it can open to form a pore across the membrane to let the hydrophilic portions of proteins cross the lipid bilayer, and it can open laterally within the membrane to let hydrophobic portions of

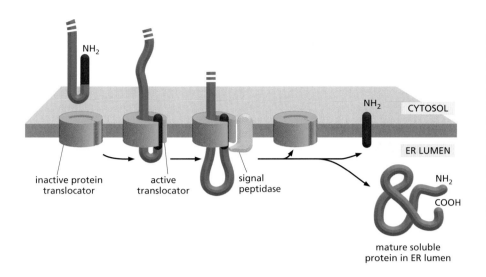

Figure 12–45 **A model to explain how a soluble protein is translocated across the ER membrane.** On binding an ER signal sequence (which acts as a start-transfer signal), the translocator opens its pore, allowing the transfer of the polypeptide chain across the lipid bilayer as a loop. After the protein has been completely translocated, the pore closes, but the translocator now opens laterally within the lipid bilayer, allowing the hydrophobic signal sequence to diffuse into the bilayer, where it is rapidly degraded. (In this figure and the three figures that follow, the ribosomes have been omitted for clarity.)

proteins partition into the lipid bilayer. Lateral gating of the pore is an essential step during the integration of membrane proteins.

The integration of membrane proteins requires that some parts of the polypeptide chain be translocated across the lipid bilayer whereas others are not. Despite this additional complexity, all modes of insertion of membrane proteins are variants of the sequence of events just described for transferring a soluble protein into the lumen of the ER. We begin by describing the three ways in which **single-pass transmembrane proteins** (see Figure 10–19) become inserted into the ER.

In the simplest case, an N-terminal signal sequence initiates translocation, just as for a soluble protein, but an additional hydrophobic segment in the polypeptide chain stops the transfer process before the entire polypeptide chain is translocated. This **stop-transfer signal** anchors the protein in the membrane after the ER signal sequence (the start-transfer signal) has been released from the translocator and has been cleaved off (**Figure 12–46**). The lateral gating mechanism transfers the stop-transfer sequence into the bilayer, and it remains there as a single α-helical membrane-spanning segment, with the N-terminus of the protein on the lumenal side of the membrane and the C-terminus on the cytosolic side.

In the other two cases, the signal sequence is internal, rather than at the N-terminal end of the protein. Like the N-terminal ER signal sequences, the SRP also

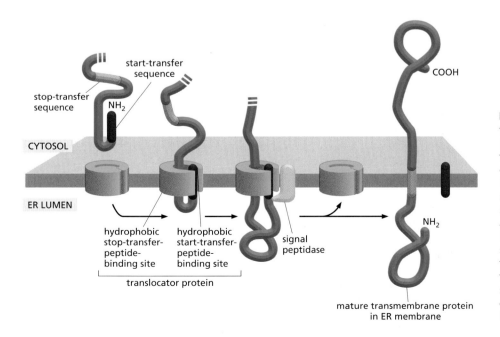

Figure 12–46 **How a single-pass transmembrane protein with a cleaved ER signal sequence is integrated into the ER membrane.** In this protein the co-translational translocation process is initiated by an N-terminal ER signal sequence *(red)* that functions as a start-transfer signal, as in Figure 12–45. In addition to this start-transfer sequence, however, the protein also contains a stop-transfer sequence *(orange)*. When the stop-transfer sequence enters the translocator and interacts with a binding site, the translocator changes its conformation and discharges the protein laterally into the lipid bilayer.

binds to an internal signal sequence. The SRP brings the ribosome making the protein to the ER membrane and serves as a start-transfer signal that initiates the protein's translocation. After release from the translocator, the internal start-transfer sequence remains in the lipid bilayer as a single membrane-spanning α helix.

Internal start-transfer sequences can bind to the translocation apparatus in either of two orientations; this in turn determines which protein segment (the one preceding or the one following the start-transfer sequence) is moved across the membrane into the ER lumen. In one case, the resulting membrane protein has its C-terminus on the lumenal side (Pathway A in **Figure 12–47**), while in the other, it has its N-terminus on the lumenal side (Pathway B in Figure 12–47). The orientation of the start-transfer sequence depends on the distribution of nearby charged amino acids, as described in the figure legend.

Combinations of Start-Transfer and Stop-Transfer Signals Determine the Topology of Multipass Transmembrane Proteins

In **multipass transmembrane proteins**, the polypeptide chain passes back and forth repeatedly across the lipid bilayer (see Figure 10–19). It is thought that an internal signal sequence serves as a start-transfer signal in these proteins to initiate translocation, which continues until the translocator encounters a stop-transfer sequence. In double-pass transmembrane proteins, for example, the polypeptide can then be released into the bilayer (**Figure 12–48**).

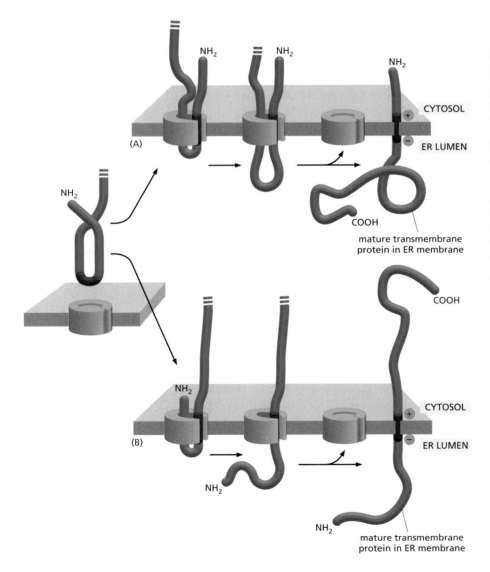

Figure 12–47 Integration of a single-pass transmembrane protein with an internal signal sequence into the ER membrane. An internal ER signal sequence that functions as a start-transfer signal can bind to the translocator in one of two different ways, leading to a membrane protein that has either its C-terminus (pathway A) or its N-terminus (pathway B) in the ER lumen. Proteins are directed into either pathway by features in the polypeptide chain flanking the internal start transfer sequence: if there are more positively charged amino acids immediately *preceding* the hydrophobic core of the start-transfer sequence than there are following it, the membrane protein is inserted into the translocator in the orientation shown in pathway A, whereas if there are more positively charged amino acids immediately *following* the hydrophobic core of the start-transfer sequence than there are preceding it, the membrane protein is inserted into the translocator in the orientation shown in pathway B. Because translocation cannot start before a start-transfer sequence appears outside the ribosome, translocation of the N-terminal portion of the protein shown in (B) can occur only after this portion has been fully synthesized.

Note that there are two ways to insert a single-pass membrane-spanning protein whose N-terminus is located in the ER lumen: that shown in Figure 12–46 and that shown in (B) here.

mature transmembrane protein in ER membrane

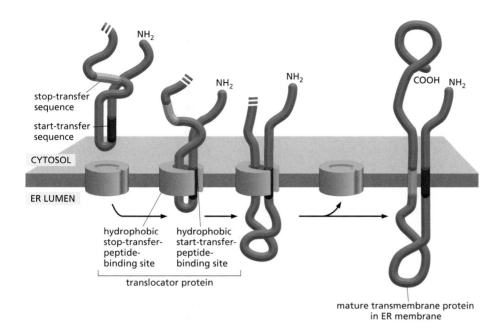

Figure 12–48 Integration of a double-pass transmembrane protein with an internal signal sequence into the ER membrane. In this protein, an internal ER signal sequence acts as a start-transfer signal (as in Figure 12–47) and initiates the transfer of the C-terminal part of the protein. At some point after a stop-transfer sequence has entered the translocator, the translocator discharges the sequence laterally into the membrane.

In more complex multipass proteins, in which many hydrophobic α helices span the bilayer, a second start-transfer sequence reinitiates translocation further down the polypeptide chain until the next stop-transfer sequence causes polypeptide release, and so on for subsequent start-transfer and stop-transfer sequences (**Figure 12–49**).

Whether a given hydrophobic signal sequence functions as a start-transfer or stop-transfer sequence must depend on its location in a polypeptide chain, since its function can be switched by changing its location in the protein by using recombinant DNA techniques. Thus, the distinction between start-transfer and stop-transfer sequences results mostly from their relative order in the growing polypeptide chain. It seems that the SRP begins scanning an unfolded polypeptide chain for hydrophobic segments at its N-terminus and proceeds toward the C-terminus, in the direction that the protein is synthesized. By recognizing the first appropriate hydrophobic segment to emerge from the ribosome, the SRP sets the "reading frame" for membrane integration: after the SRP initiates translocation, the translocator recognizes the next appropriate hydrophobic segment in the direction of transfer as a stop-transfer sequence, causing the region of the polypeptide chain in between to be threaded across the membrane. A similar scanning process continues until all of the hydrophobic regions in the protein have been inserted into the membrane.

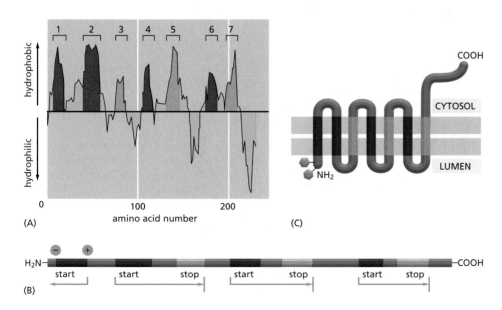

Figure 12–49 The insertion of the multipass membrane protein rhodopsin into the ER membrane. Rhodopsin is the light-sensitive protein in rod photoreceptor cells in the mammalian retina (discussed in Chapter 15). (A) A hydrophobicity plot identifies seven short hydrophobic regions in rhodopsin. (B) The hydrophobic region nearest the N-terminus serves as a start-transfer sequence that causes the preceding N-terminal portion of the protein to pass across the ER membrane. Subsequent hydrophobic sequences function in alternation as start-transfer and stop-transfer sequences. (C) The final integrated rhodopsin has its N-terminus located in the ER lumen and its C-terminus located in the cytosol. The *blue hexagons* represent covalently attached oligosaccharides. Arrows indicate the paired start and stop signals inserted into the translocator.

Because membrane proteins are always inserted from the cytosolic side of the ER in this programmed manner, all copies of the same polypeptide chain will have the same orientation in the lipid bilayer. This generates an asymmetrical ER membrane in which the protein domains exposed on one side are different from those exposed on the other side. This asymmetry is maintained during the many membrane budding and fusion events that transport the proteins made in the ER to other cell membranes (discussed in Chapter 13). Thus, the way in which a newly synthesized protein is inserted into the ER membrane determines the orientation of the protein in all of the other membranes as well.

When proteins dissociate from a membrane and are then reconstituted into artificial lipid vesicles, a random mixture of right-side-out and inside-out protein orientations usually results. Thus, the protein asymmetry observed in cell membranes seems not to be an inherent property of the protein, but instead results solely from the process by which proteins are inserted into the ER membrane from the cytosol.

Translocated Polypeptide Chains Fold and Assemble in the Lumen of the Rough ER

Many of the proteins in the lumen of the ER are in transit, *en route* to other destinations; others, however, normally reside there and are present at high concentrations. These **ER resident proteins** contain an **ER retention signal** of four amino acids at their C-terminus that is responsible for retaining the protein in the ER (see Table 12–3; discussed in Chapter 13). Some of these proteins function as catalysts that help the many proteins that are translocated into the ER to fold and assemble correctly.

One important ER resident protein is *protein disulfide isomerase (PDI)*, which catalyzes the oxidation of free sulfhydryl (SH) groups on cysteines to form disulfide (S–S) bonds. Almost all cysteines in protein domains exposed to either the extracellular space or the lumen of organelles in the secretory and endocytic pathways are disulfide-bonded. By contrast, disulfide bonds form only very rarely in domains exposed to the cytosol because of the reducing environment there.

Another ER resident protein is the chaperone protein **BiP**. We have already discussed how BiP pulls proteins post-translationally into the ER through the ER translocator. Like other chaperones, BiP recognizes incorrectly folded proteins, as well as protein subunits that have not yet assembled into their final oligomeric complexes. It does so by binding to exposed amino acid sequences that would normally be buried in the interior of correctly folded or assembled polypeptide chains. An example of a BiP-binding site is a stretch of alternating hydrophobic and hydrophilic amino acids that would normally be buried in a β sheet. The bound BiP both prevents the protein from aggregating and helps to keep it in the ER (and thus out of the Golgi apparatus and later parts of the secretory pathway). Like some other members of the Hsp70 family of proteins, which bind unfolded proteins and facilitate their import into mitochondria and chloroplasts, BiP hydrolyzes ATP to provide the energy needed to help proteins translocate post-translationally into the ER. It also helps these and other proteins fold.

Most Proteins Synthesized in the Rough ER Are Glycosylated by the Addition of a Common *N*-Linked Oligosaccharide

The covalent addition of sugars to proteins is one of the major biosynthetic functions of the ER. About half of all eucaryotic proteins are glycosylated. Most of the soluble and membrane-bound proteins that are made in the ER—including those destined for transport to the Golgi apparatus, lysosomes, plasma membrane, or extracellular space—are **glycoproteins**. In contrast, very few proteins in the cytosol are glycosylated, and those that are carry a much simpler sugar modification, in which a single *N*-acetylglucosamine group is added to a serine or threonine residue of the protein.

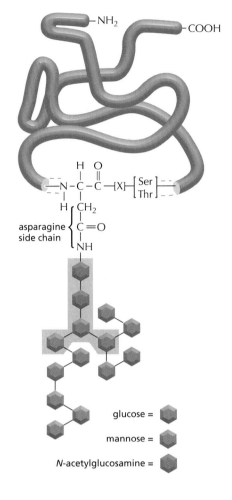

Figure 12–50 The asparagine-linked (N-linked) precursor oligosaccharide that is added to most proteins in the rough ER membrane. The five sugars in the *gray box* form the "core region" of this oligosaccharide. For many glycoproteins, only the core sugars survive the extensive oligosaccharide trimming process that takes place in the Golgi apparatus. Only asparagines in the sequences Asn-X-Ser and Asn-X-Thr (where X is any amino acid except proline) become glycosylated. These two sequences occur much less frequently in glycoproteins than in nonglycosylated cytosolic proteins; evidently there has been selective pressure against these sequences during protein evolution, presumably because glycosylation at too many sites would interfere with protein folding.

An important advance in understanding the process of **protein glycosylation** was the discovery that a preformed *precursor oligosaccharide* (composed of *N*-acetylglucosamine, mannose, and glucose and containing a total of 14 sugars) is transferred *en bloc* to proteins in the ER. Because this oligosaccharide is transferred to the side-chain NH₂ group of an asparagine amino acid in the protein, it is said to be *N-linked* or *asparagine-linked* (**Figure 12–50**). The transfer is catalyzed by a membrane-bound enzyme complex, an *oligosaccharyl transferase,* which has its active site exposed on the lumenal side of the ER membrane; this explains why cytosolic proteins are not glycosylated in this way. A special lipid molecule called **dolichol** holds the precursor oligosaccharide in the ER membrane. It transfers its oligosaccharide chain to the target asparagine in a single enzymatic step immediately after that amino acid has reached the ER lumen during protein translocation (**Figure 12–51**). One copy of oligosaccharyl transferase is associated with each protein translocator, allowing it to scan and glycosylate the incoming polypeptide chains efficiently.

The precursor oligosaccharide is linked to the dolichol lipid by a high-energy pyrophosphate bond, which provides the activation energy that drives the glycosylation reaction illustrated in Figure 12–51. The entire precursor oligosaccharide is built up sugar by sugar on this membrane-bound lipid molecule and is then transferred to a protein. The sugars are first activated in the cytosol by the formation of *nucleotide-sugar intermediates,* which then donate their sugar (directly or indirectly) to the lipid in an orderly sequence. Partway

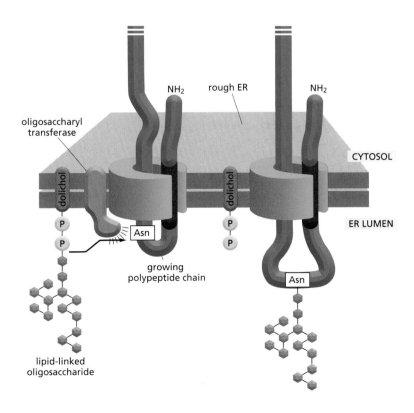

Figure 12–51 Protein glycosylation in the rough ER. Almost as soon as a polypeptide chain enters the ER lumen, it is glycosylated on target asparagine amino acids. The precursor oligosaccharide shown in Figure 12–50 is transferred to the asparagine as an intact unit in a reaction catalyzed by a membrane-bound *oligosaccharyl transferase* enzyme. As with signal peptidase, one copy of this enzyme is associated with each protein translocator in the ER membrane. (The ribosome is not shown.)

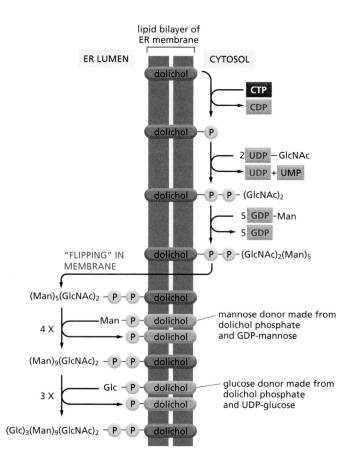

Figure 12–52 Synthesis of the lipid-linked precursor oligosaccharide in the rough ER membrane. The oligosaccharide is assembled sugar by sugar onto the carrier lipid dolichol (a polyisoprenoid; see Panel 2–5, pp. 114–115). Dolichol is long and very hydrophobic: its 22 five-carbon units can span the thickness of a lipid bilayer more than three times, so that the attached oligosaccharide is firmly anchored in the membrane. The first sugar is linked to dolichol by a pyrophosphate bridge. This high-energy bond activates the oligosaccharide for its eventual transfer from the lipid to an asparagine side chain of a nascent polypeptide on the lumenal side of the rough ER. As indicated, the synthesis of the oligosaccharide starts on the cytosolic side of the ER membrane and continues on the lumenal face after the $(Man)_5(GlcNAc)_2$ lipid intermediate is flipped across the bilayer by a transporter. All the subsequent glycosyl transfer reactions on the lumenal side of the ER involve transfers from dolichol-P-glucose and dolichol-P-mannose; these activated, lipid-linked monosaccharides are synthesized from dolichol phosphate and UDP-glucose or GDP-mannose (as appropriate) on the cytosolic side of the ER and are then thought to be flipped across the ER membrane. GlcNAc = N-acetylglucosamine; Man = mannose; Glc = glucose.

through this process, the lipid-linked oligosaccharide is flipped, with the help of a transporter, from the cytosolic to the lumenal side of the ER membrane (**Figure 12–52**).

All of the diversity of the *N*-linked oligosaccharide structures on mature glycoproteins results from the later modification of the original precursor oligosaccharide. While still in the ER, three glucoses (see Figure 12–50) and one mannose are quickly removed from the oligosaccharides of most glycoproteins. We shall return to the importance of glucose trimming shortly. This oligosaccharide "trimming" or "processing" continues in the Golgi apparatus and is discussed in Chapter 13.

The *N*-linked oligosaccharides are by far the most common oligosaccharides, being found on 90% of all glycoproteins. Less frequently, oligosaccharides are linked to the hydroxyl group on the side chain of a serine, threonine, or hydroxylysine amino acid. These *O-linked oligosaccharides* are formed in the Golgi apparatus.

Oligosaccharides Are Used as Tags to Mark the State of Protein Folding

It has long been debated why glycosylation is such a common modification of proteins that enter the ER. One particularly puzzling observation has been that some proteins require *N*-linked glycosylation for proper folding in the ER, yet the precise location of the oligosaccharides attached to the protein's surface does not seem to matter. A clue to the role of glycosylation in protein folding came from studies of two ER chaperone proteins, which are called **calnexin** and **calreticulin** because they require Ca^{2+} for their activities. These chaperones are carbohydrate-binding proteins, or *lectins,* which bind to oligosaccharides on incompletely folded proteins and retain them in the ER. Like other chaperones, they prevent incompletely folded proteins from becoming irreversibly aggregated. Both calnexin and calreticulin also promote the association of incompletely folded proteins with another ER chaperone, which binds to cysteines that have not yet formed disulfide bonds.

Calnexin and calreticulin recognize *N*-linked oligosaccharides that contain a single terminal glucose, and therefore bind proteins only after two of the three glucoses on the precursor oligosaccharide have been removed by ER glucosidases. When the third glucose has been removed, the protein dissociates from its chaperone and can leave the ER.

How, then, do calnexin and calreticulin distinguish properly folded from incompletely folded proteins? The answer lies in yet another ER enzyme, a glucosyl transferase that keeps adding a glucose to those oligosaccharides that have lost their last glucose. It adds the glucose, however, only to oligosaccharides that are attached to unfolded proteins. Thus, an unfolded protein undergoes continuous cycles of glucose trimming (by glucosidase) and glucose addition (by glycosyl transferase), maintaining an affinity for calnexin and calreticulin until it has achieved its fully folded state (**Figure 12–53**).

Improperly Folded Proteins Are Exported from the ER and Degraded in the Cytosol

Despite all the help from chaperones, many protein molecules (more than 80% for some proteins) translocated into the ER fail to achieve their properly folded or oligomeric state. Such proteins are exported from the ER back into the cytosol, where they are degraded. The mechanism of retrotranslocation, also called *dislocation*, is still unknown but is likely to be similar to other post-translational modes of translocation. For example, like translocation into mitochondria or chloroplasts, chaperone proteins are probably necessary to keep the polypeptide chain in an unfolded state prior to and during transport. Similarly, a source of energy is required to provide directionality to the transport and to pull the protein into the cytosol. Finally, a translocator, possibly composed of some of the same components that are used for forward transport into the ER (such as Sec61), is presumably necessary.

Selecting proteins from the ER for degradation is a challenging process. Misfolded proteins or unassembled protein subunits should be degraded, but folding

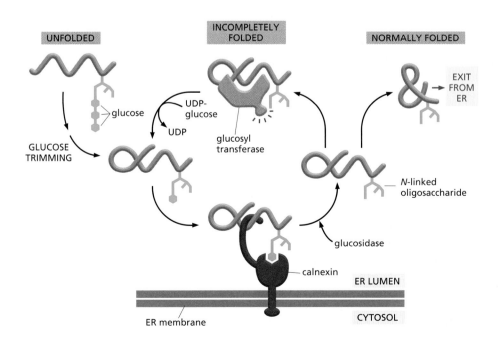

Figure 12–53 **The role of *N*-linked glycosylation in ER protein folding.** The ER-membrane-bound chaperone protein calnexin binds to incompletely folded proteins containing one terminal glucose on *N*-linked oligosaccharides, trapping the protein in the ER. Removal of the terminal glucose by a glucosidase releases the protein from calnexin. A glucosyl transferase is the crucial enzyme that determines whether the protein is folded properly or not: if the protein is still incompletely folded, the enzyme transfers a new glucose from UDP-glucose to the *N*-linked oligosaccharide, renewing the protein's affinity for calnexin and retaining it in the ER. The cycle repeats until the protein has folded completely. Calreticulin functions similarly, except that it is a soluble ER resident protein. Another ER chaperone, ERp57 (not shown), collaborates with calnexin and calreticulin in retaining an incompletely folded protein in the ER.

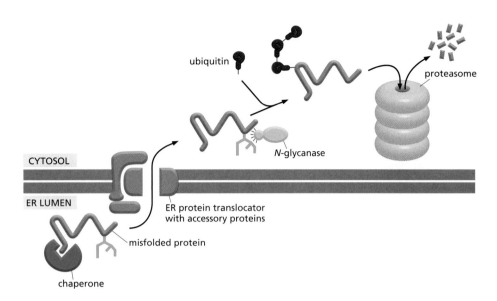

Figure 12–54 The export and degradation of misfolded ER proteins. Misfolded soluble proteins in the ER lumen are translocated back into the cytosol, where they are deglycosylated, ubiquitylated, and degraded in proteasomes. Misfolded membrane proteins follow a similar pathway.

intermediates of newly made proteins should not. Help in making this distinction comes from the *N*-linked oligosaccharides, which serve as timers that measure how long a protein has spent in the ER. The slow trimming of a particular mannose on the core-oligosaccharide tree by an enzyme (a mannosidase) in the ER is thought to create a new oligosaccharide structure that the retrotranslocation apparatus recognizes. Proteins that fold and exit from the ER faster than the action of the mannosidase would therefore escape degradation.

Once a misfolded protein has been retrotranslocated into the cytosol, an *N*-glycanase removes its oligosaccharide chains *en bloc*. The deglycosylated polypeptide is rapidly ubiquitylated by ER-bound ubiquitin-conjugating enzymes and is then fed into proteasomes (discussed in Chapter 6), where it is degraded (**Figure 12–54**).

Misfolded Proteins in the ER Activate an Unfolded Protein Response

Cells carefully monitor the amount of misfolded protein in various compartments. An accumulation of misfolded proteins in the cytosol, for example, triggers a *heat-shock response* (discussed in Chapter 6), which stimulates the transcription of genes encoding cytosolic chaperones that help to refold the proteins. Similarly, an accumulation of misfolded proteins in the ER triggers an **unfolded protein response**, which includes an increased transcription of genes encoding ER chaperones, proteins involved in retrotranslocation and protein degradation in the cytosol, and many other proteins that help to increase the protein folding capacity of the ER.

How do misfolded proteins in the ER signal to the nucleus? There are three parallel pathways that execute the unfolded protein response (**Figure 12–55**A). The first pathway, which was initially discovered in yeast cells, is particularly remarkable. Misfolded proteins in the ER activate a transmembrane protein kinase in the ER, which causes the kinase to oligomerize and phosphorylate itself. (Some cell-surface receptors in the plasma membrane are activated in a similar way, as discussed in Chapter 15). The oligomerization and autophosphorylation activates an endoribonuclease domain in the cytosolic portion of the same molecule, which cleaves a specific, cytosolic RNA molecule at two positions, excising an intron. The separated exons are then joined by an RNA ligase, generating a spliced mRNA, which is translated to produce an active gene regulatory protein. This protein activates the transcription of the genes encoding the proteins that mediate the unfolded protein response (Figure 12–55B).

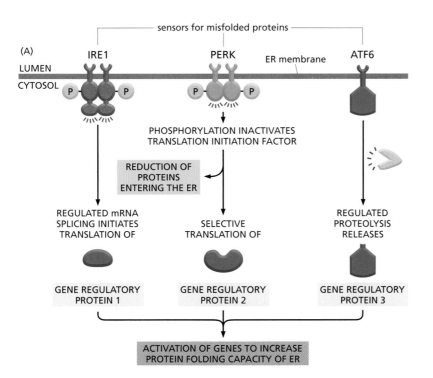

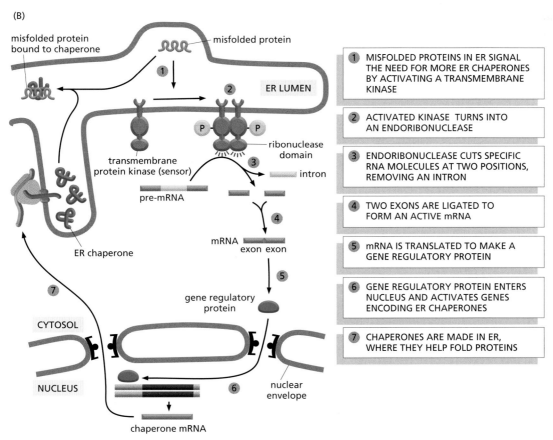

Figure 12–55 **The unfolded protein response.** (A) By three parallel intracellular signaling pathways, the accumulation of misfolded proteins in the ER lumen signals to the nucleus to activate the transcription of genes that encode proteins that help the cell to cope with the abundance of misfolded proteins in the ER. (B) Regulated mRNA splicing is a key regulatory switch in Pathway 1 of the unfolded protein response.

Misfolded proteins also activate a second transmembrane kinase in the ER, which inhibits a translation initiation factor by phosphorylating it and thereby reduces the production of new proteins throughout the cell. One consequence of the reduction in protein translation is to reduce the flux of proteins into the ER, thereby limiting the load of proteins that need to be folded there. Some proteins, however, are preferentially translated when translation initiation factors are scarce (see p. 490), and one of these is a gene regulatory protein that helps activate the transcription of the genes encoding proteins active in the unfolded protein response.

Finally, a third gene regulatory protein is initially synthesized as an integral ER membrane protein. Because it is covalently tethered to the membrane, it cannot activate the transcription of genes in the nucleus. When misfolded proteins accumulate in the ER, the transmembrane protein is transported to the Golgi apparatus, where it encounters proteases that cleave off its cytosolic domain, which can now migrate to the nucleus and help activate the transcription of the genes encoding proteins involved in the unfolded protein response. The relative importance of each of these three pathways differs in different cell types, enabling each cell type to tailor the unfolded protein response to its particular needs.

Some Membrane Proteins Acquire a Covalently Attached Glycosylphosphatidylinositol (GPI) Anchor

As discussed in Chapter 10, several cytosolic enzymes catalyze the covalent addition of a single fatty acid chain or prenyl group to selected proteins. The attached lipids help to direct these proteins to cell membranes. A related process is catalyzed by ER enzymes, which covalently attach a **glycosylphosphatidyl-inositol (GPI) anchor** to the C-terminus of some membrane proteins destined for the plasma membrane. This linkage forms in the lumen of the ER, where, at the same time, the transmembrane segment of the protein is cleaved off (**Figure 12–56**). A large number of plasma membrane proteins are modified in this way. Since they are attached to the exterior of the plasma membrane only by their GPI anchors, they can in principle be released from cells in soluble form in response to signals that activate a specific phospholipase in the plasma membrane. Trypanosome

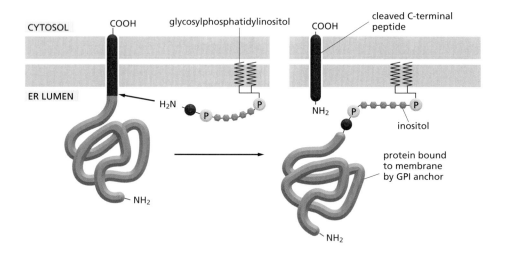

Figure 12–56 The attachment of a GPI anchor to a protein in the ER. Immediately after the completion of protein synthesis, the precursor protein remains anchored in the ER membrane by a hydrophobic C-terminal sequence of 15–20 amino acids; the rest of the protein is in the ER lumen. Within less than a minute, an enzyme in the ER cuts the protein free from its membrane-bound C-terminus and simultaneously attaches the new C-terminus to an amino group on a preassembled GPI intermediate. The sugar chain contains an inositol attached to the lipid from which the GPI anchor derives its name. It is followed by a glucosamine and three mannoses. The terminal mannose links to a phosphoethanolamine that provides the amino group to attach the protein. The signal that specifies this modification is contained within the hydrophobic C-terminal sequence and a few amino acids adjacent to it on the lumenal side of the ER membrane; if this signal is added to other proteins, they too become modified in this way. Because of the covalently linked lipid anchor, the protein remains membrane-bound, with all of its amino acids exposed initially on the lumenal side of the ER and eventually on the cell exterior.

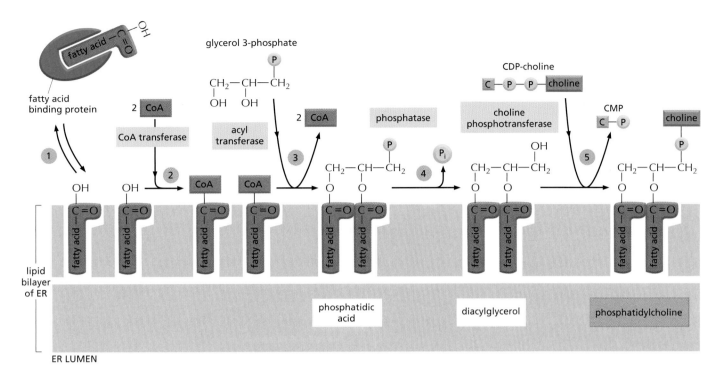

Figure 12–57 The synthesis of phosphatidylcholine. As illustrated, this phospholipid is synthesized from glycerol 3-phosphate, cytidine-diphosphocholine (CDP-choline), and fatty acids delivered to the ER by fatty acid binding protein.

parasites, for example, use this mechanism to shed their coat of GPI-anchored surface proteins when attacked by the immune system. GPI anchors may also be used to direct plasma membrane proteins into *lipid rafts* and thus segregate the proteins from other membrane proteins, as we discuss in Chapter 13.

The ER Assembles Most Lipid Bilayers

The ER membrane synthesizes nearly all of the major classes of lipids, including both phospholipids and cholesterol, required for the production of new cell membranes. The major phospholipid made is *phosphatidylcholine* (also called *lecithin*), which can be formed in three steps from choline, two fatty acids, and glycerol phosphate (**Figure 12–57**). Each step is catalyzed by enzymes in the ER membrane that have their active sites facing the cytosol, where all of the required metabolites are found. Thus, phospholipid synthesis occurs exclusively in the cytosolic leaflet of the ER membrane. Because fatty acids are not soluble in water, they are shepherded from their sites of synthesis to the ER by a fatty acid binding protein in the cytosol. After arrival in the ER membrane and activation with CoA, acyl transferases successively add two fatty acids to glycerol phosphate to produce phosphatidic acid. Phosphatidic acid is sufficiently water-insoluble to remain in the lipid bilayer, and it cannot be extracted from the bilayer by the fatty acid binding proteins. It is therefore this first step that enlarges the ER lipid bilayer. The later steps determine the head group of a newly formed lipid molecule and therefore the chemical nature of the bilayer, but they do not result in net membrane growth. The two other major membrane phospholipids—phosphatidyl ethanolamine and phosphatidylserine—as well as the minor phospholipid phosphatidylinositol (PI), are all synthesized in this way.

Because phospholipid synthesis takes place in the cytosolic half of the ER lipid bilayer, there needs to be a mechanism that transfers some of the newly formed phospholipid molecules to the lumenal leaflet of the bilayer. In synthetic lipid bilayers, lipids do not "flip-flop" in this way. In the ER, however, phospholipids equilibrate across the membrane within minutes, which is almost 100,000

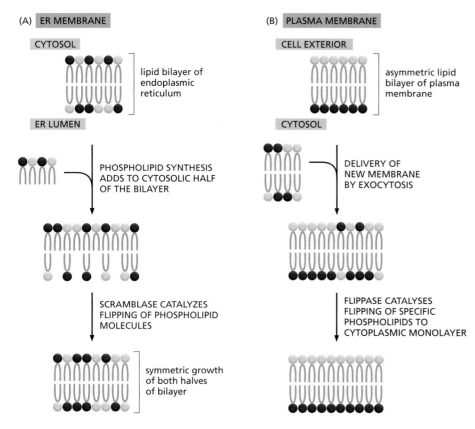

Figure 12–58 The role of phospholipid translocators in lipid bilayer synthesis. (A) Because new lipid molecules are added only to the cytosolic half of the bilayer and lipid molecules do not flip spontaneously from one monolayer to the other, a membrane-bound phospholipid translocator (called a scramblase) is required to transfer lipid molecules from the cytosolic half to the lumenal half so that the membrane grows as a bilayer. The scramblase is not specific for particular phospholipid head groups and therefore equilibrates the different phospholipids between the two monolayers. (B) Fueled by ATP hydrolysis, a head-group-specific flippase in the plasma membrane actively flips phosphatidylserine and phosphatidyl-ethanolamine directionally from the extracellular to the cytosolic leaflet, creating the characteristically asymmetric lipid bilayer of the plasma membrane of animal cells (see Figure 10–16).

times faster than can be accounted for by spontaneous "flip-flop." This rapid trans-bilayer movement is mediated by a poorly characterized phospholipid translocator called a *scramblase*, which equilibrates phospholipids between the two leaflets of the lipid bilayer (**Figure 12–58**). Thus, the different types of phospholipids are thought to be equally distributed between the two leaflets of the ER membrane.

The plasma membrane contains a different type of phospholipid translocator that belongs to the family of P-type pumps (discussed in Chapter 11). These *flippases* specifically remove phospholipids containing in their head groups free amino groups (phosphatidylserine and phosphatidylethanolamine—see Figure 10–3) from the extracellular leaflet and use the energy of ATP hydrolysis to flip them directionally into the leaflet facing the cytosol. The plasma membrane therefore has a highly asymmetric phospholipid composition, which is actively maintained by the flippases (see Figure 10–16). The plasma membrane also contains a scramblase but, in contrast to the ER scramblase, which is always active, the plasma membrane enzyme is regulated and only activated in some situations, such as in apoptosis and in activated platelets, where it acts to abolish lipid asymmetry; the resulting exposure of phosphotidylserine on the surface of apoptotic cells serves as a signal for phagocytic cells to ingest and degrade the dead cell.

The ER also produces cholesterol and ceramide (**Figure 12–59**). *Ceramide* is made by condensing the amino acid serine with a fatty acid to form the amino alcohol *sphingosine* (see Figure 10–3); a second fatty acid is then added to form

$$
\begin{array}{cc}
\text{OH} & \text{OH} \\
| & | \\
\text{CH} \rule{0.5cm}{0.4pt} \text{CH} \rule{0.5cm}{0.4pt} \text{CH}_2 \\
| & | \\
\text{CH} & \text{NH} \\
\| & | \\
\text{CH} & \text{C} = \text{O} \\
| & | \\
(\text{CH}_2)_{12} & (\text{CH}_2)_{16} \\
| & | \\
\text{CH}_3 & \text{CH}_3 \\
\end{array}
$$

CERAMIDE

Figure 12–59 The structure of ceramide.

ceramide. The ceramide is exported to the Golgi apparatus, where it serves as a precursor for the synthesis of two types of lipids: oligosaccharide chains are added to form *glycosphingolipids* (glycolipids; see Figure 10–18), and phosphocholine head groups are transferred from phosphatidylcholine to other ceramide molecules to form *sphingomyelin* (discussed in Chapter 10). Thus, both glycolipids and sphingomyelin are produced relatively late in the process of membrane synthesis. Because they are produced by enzymes exposed to the Golgi lumen, they are found exclusively in the noncytosolic leaflet of the lipid bilayers that contain them.

As discussed in Chapter 13, the plasma membrane and the membranes of the Golgi apparatus, lysosomes, and endosomes all form part of a membrane system that communicates with the ER by means of transport vesicles, which transfer both proteins and lipids. Mitochondria and plastids, however, do not belong to this system, and they therefore require different mechanisms to import proteins and lipids for growth. We have already seen that they import most of their proteins from the cytosol. Although mitochondria modify some of the lipids they import, they do not synthesize lipids *de novo*; instead, their lipids have to be imported from the ER, either directly, or indirectly by way of other cell membranes. In either case, special mechanisms are required for the transfer.

The details of how lipid distribution between different membranes is catalyzed and regulated are not known. Water-soluble carrier proteins called *phospholipid exchange proteins* (or *phospholipid transfer proteins*) are thought to transfer individual phospholipid molecules between membranes, functioning much like fatty acid binding proteins that shepherd fatty acids through the cytosol. In addition, mitochondria are often seen in close juxtaposition to ER membranes in electron micrographs, and there may be specific lipid transfer mechanisms that operate between adjacent membranes.

Summary

The extensive ER network serves as a factory for the production of almost all of the cell's lipids. In addition, a major portion of the cell's protein synthesis occurs on the cytosolic surface of the ER: all proteins destined for secretion and all proteins destined for the ER itself, the Golgi apparatus, the lysosomes, the endosomes, and the plasma membrane are first imported into the ER from the cytosol. In the ER lumen, the proteins fold and oligomerize, disulfide bonds are formed, and N-linked oligosaccharides are added. The pattern of N-linked glycosylation is used to indicate the extent of protein folding, so that proteins leave the ER only when they are properly folded. Proteins that do not fold or oligomerize correctly are translocated back into the cytosol, where they are deglycosylated, ubiquitylated, and degraded in proteasomes. If misfolded proteins accumulate in excess in the ER, they trigger an unfolded protein response, which activates appropriate genes in the nucleus to help the ER to cope.

Only proteins that carry a special ER signal sequence are imported into the ER. The signal sequence is recognized by a signal recognition particle (SRP), which binds both the growing polypeptide chain and a ribosome and directs them to a receptor protein on the cytosolic surface of the rough ER membrane. This binding to the ER membrane initiates the translocation process by threading a loop of polypeptide chain across the ER membrane through the hydrophilic pore in a transmembrane protein translocator.

Soluble proteins—destined for the ER lumen, for secretion, or for transfer to the lumen of other organelles—pass completely into the ER lumen. Transmembrane proteins destined for the ER or for other cell membranes are translocated partway across the ER membrane and remain anchored there by one or more membrane-spanning α-helical regions in their polypeptide chains. These hydrophobic portions of the protein can act either as start-transfer or stop-transfer signals during the translocation process. When a polypeptide contains multiple, alternating start-transfer and stop-transfer signals, it will pass back and forth across the bilayer multiple times as a multipass transmembrane protein.

The asymmetry of protein insertion and glycosylation in the ER establishes the sidedness of the membranes of all the other organelles that the ER supplies with membrane proteins.

PROBLEMS

Which statements are true? Explain why or why not.

12–1 Like the lumen of the ER, the interior of the nucleus is topologically equivalent to the outside of the cell.

12–2 Membrane-bound and free ribosomes, which are structurally and functionally identical, differ only in the proteins they happen to be making at a particular time.

12–3 To avoid the inevitable congestion that would occur if two-way traffic through a single pore were allowed, nuclear pore complexes are specialized so that some mediate import while others mediate export.

12–4 Peroxisomes are found in only a few specialized types of eucaryotic cell.

12–5 In multipass transmembrane proteins the odd-numbered transmembrane segments (counting from the N-terminus) act as start-transfer signals and the even-numbered segments act as stop-transfer signals.

Discuss the following problems.

12–6 What is the fate of a protein with no sorting signal?

12–7 Imagine that you have engineered a set of genes, each encoding a protein with a pair of conflicting signal sequences that specify different compartments. If the genes were expressed in a cell, predict which signal would win out for the following combinations. Explain your reasoning.
A. Signals for import into nucleus and import into ER.
B. Signals for import into peroxisomes and import into ER.
C. Signals for import into mitochondria and retention in ER.
D. Signals for import into nucleus and export from nucleus.

12–8 The rough ER is the site of synthesis of many different classes of membrane protein. Some of these proteins remain in the ER, whereas others are sorted to compartments such as the Golgi apparatus, lysosomes, and the plasma membrane. One measure of the difficulty of the sorting problem is the degree of "purification" that must be achieved during transport from the ER. Are proteins bound for the plasma membrane common or rare among all ER membrane proteins?

A few simple considerations allow one to answer this question. In a typical growing cell that is dividing once every 24 hours, the equivalent of one new plasma membrane must transit the ER every day. If the ER membrane is 20 times the area of a plasma membrane, what is the ratio of plasma membrane proteins to other membrane proteins in the ER? (Assume that all proteins on their way to the plasma membrane remain in the ER for 30 minutes on average before exiting, and that the ratio of proteins to lipids in the ER and plasma membranes is the same.)

12–9 Before nuclear pore complexes were well understood, it was unclear whether nuclear proteins diffused passively into the nucleus and accumulated there by binding to residents of the nucleus such as chromosomes, or were actively transported and accumulated regardless of their affinity for nuclear components.

A classic experiment that addressed this problem used several forms of radioactive nucleoplasmin, which is a large

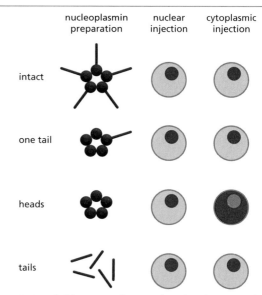

Figure Q12–1 Cell location of injected nucleoplasmin and nucleoplasmin components (Problem 12–9). Schematic diagrams of autoradiographs show the cytoplasm and nucleus with the location of nucleoplasmin indicated by the *red* areas.

pentameric protein involved in chromatin assembly. In this experiment, either the intact protein or the nucleoplasmin heads, tails, or heads with a single tail were injected into the cytoplasm of a frog oocyte or into the nucleus (**Figure Q12–1**). All forms of nucleoplasmin, except heads, accumulated in the nucleus when injected into the cytoplasm, and all forms were retained in the nucleus when injected there.
A. What portion of the nucleoplasmin molecule is responsible for localization in the nucleus?
B. How do these experiments distinguish between active transport, in which a nuclear localization signal triggers transport by the nuclear pore complex, and passive diffusion, in which a binding site for a nuclear component allows accumulation in the nucleus?

12–10 Assuming that 32 million histone octamers are required to package the human genome, how many histone molecules must be transported per second per pore complex in cells whose nuclei contain 3000 nuclear pores and are dividing once per day?

12–11 The structures of Ran-GDP and Ran-GTP (actually Ran-GppNp, a stable GTP analog) are strikingly different, as shown in **Figure Q12–2**. Not surprisingly, Ran-GDP binds to a different set of proteins than does Ran-GTP.

To look at the uptake of Ran itself into the nuclei of permeabilized cells, you attach a red fluorescent tag to a cysteine side chain in Ran to make it visible. This modified Ran supports normal nuclear uptake. Fluorescent Ran-GDP is taken up by nuclei only if cytoplasm is added, whereas a

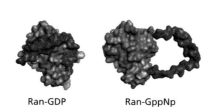

Figure Q12–2 Structures of Ran-GDP and Ran-GppNp (Problem 12–11). The *red portions* of the structures show the segment of Ran that differs most dramatically when GDP or the GTP analog is bound.

Ran-GDP   Ran-GppNp

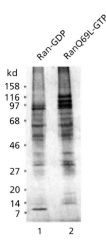

Figure Q12–3 Proteins eluted from Ran-GDP (lane 1) and RanQ69L-GTP (lane 2) affinity columns (Problem 12–11). The molecular weights of marker proteins are shown on the *left*. (From K. Ribbeck et al., *EMBO J.* 17:6587–6598, 1998. With permission from Macmillan Publishers Ltd.)

mutant form, RanQ69L-GTP, which is unable to hydrolyze GTP, is not taken up in the presence or absence of cytoplasm. To identify the cytoplasmic protein that is crucial for Ran-GDP uptake, you construct affinity columns with bound Ran-GDP or bound RanQ69L-GTP and pass cytoplasm through them. Cytoplasm passed over a Ran-GDP column no longer supports nuclear uptake of Ran-GDP, whereas cytoplasm passed over a column of RanQ69L-GTP retains this activity. You elute the bound proteins from each column and analyze them on an SDS polyacrylamide gel, looking for differences that might identify the factor that is required for nuclear uptake of Ran (**Figure Q12–3**).

A. Why did you use RanQ69L-GTP instead of Ran-GTP in these experiments? Would Ran-GppNp in place of RanQ69L-GTP have achieved the same purpose?

B. Which of the many proteins eluted from the two different affinity columns is a likely candidate for the factor that promotes nuclear import of Ran-GDP?

C. What other protein or proteins would you predict that the Ran-GDP import factor would be likely to bind in order to carry out its function?

D. How might you confirm that the factor you have identified is necessary for promoting the nuclear uptake of Ran?

12–12 Components of the TIM complex, the multisubunit protein translocator in the mitochondrial inner membrane, are much less abundant than those of the TOM complex. They were initially identified by using a genetic trick. The yeast *Ura3* gene, whose product is normally located in the cytosol where it is essential for synthesis of uracil, was modified to carry an import signal for the mitochondrial matrix. A population of cells carrying the modified *Ura3* gene was then grown in the absence of uracil. Most cells died, but the rare cells that grew were shown to be defective for mitochondrial import. Explain how this selection identifies cells with defects in components required for mitochondrial import. Why do normal cells with the modified *Ura3* gene not grow in the absence of uracil? Why do cells that are defective for mitochondrial import grow in the absence of uracil?

12–13 If the enzyme dihydrofolate reductase (DHFR), which is normally located in the cytosol, is engineered to carry a mitochondrial targeting sequence at its N-terminus, it is efficiently imported into mitochondria. If the modified DHFR is first incubated with methotrexate, which binds tightly to the active site, the enzyme remains in the cytosol. How do you suppose that the binding of methotrexate interferes with mitochondrial import?

12–14 Why do mitochondria need a fancy translocator to import proteins across the outer membrane? Their outer membranes already have large pores formed by porins.

12–15 Catalase, an enzyme normally found in peroxisomes, is present in normal amounts in cells that do not have visible peroxisomes. It is possible to determine the location of catalase in such cells using immunofluorescence microscopy with antibodies specific for catalase. Fluorescence micrographs of normal cells and peroxisome-deficient cells are shown in **Figure Q12–4**. Where is catalase located in cells without peroxisomes (Figure Q12–4B)? Why does catalase show up as small dots of fluorescence in normal cells (Figure Q12–4A)?

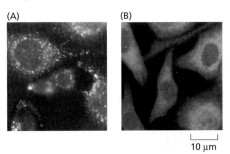

Figure Q12–4 Location of catalase in cells as determined by immunofluorescence microscopy (Problem 12–15). (A) Normal cells. (B) Peroxisome-deficient cells. Cells were reacted with antibodies specific for catalase, washed, and then stained with a fluorescein-labeled second antibody that is specific for the catalase-specific antibody. The two panels are at the same magnification. (From N. Kinoshita et al., *J. Biol. Chem.* 273:24122–24130, 1998. With permission from the American Society for Biochemistry and Molecular Biology.)

12–16 Examine the multipass transmembrane protein shown in **Figure Q12–5**. What would you predict would be the effect of converting the first hydrophobic transmembrane segment to a hydrophilic segment? Sketch the arrangement of the modified protein in the ER membrane.

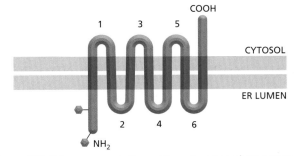

Figure Q12–5 Arrangement of a multipass transmembrane protein in the ER membrane (Problem 12–16). *Hexagons* represent covalently attached oligosaccharides.

12–17 All new phospholipids are added to the cytoplasmic leaflet of the ER membrane, yet the ER membrane has a symmetrical distribution of different phospholipids in its two leaflets. By contrast, the plasma membrane, which receives all its membrane components ultimately from the ER, has a very asymmetrical distribution of phospholipids in the two leaflets of its lipid bilayer. How is symmetry generated in the ER membrane, and how is asymmetry generated in the plasma membrane?

REFERENCES

General

Palade G (1975) Intracellular aspects of the process of protein synthesis. *Science* 189:347–358.

The Compartmentalization of Cells

Blobel G (1980) Intracellular protein topogenesis. *Proc Natl Acad Sci USA* 77:1496–1500.

De Duve C (2007) The origin of eukaryotes: a reappraisal. *Nature Rev Genet* 8:395–403.

Schatz G & Dobberstein B (1996) Common principles of protein translocation across membranes. *Science* 271:1519–1526.

Warren G & Wickner W (1996) Organelle inheritance. *Cell* 84:395–400.

The Transport of Molecules Between the Nucleus and the Cytosol

Adam SA & Gerace L (1991) Cytosolic proteins that specifically bind nuclear location signals are receptors for nuclear import. *Cell* 66:837–847.

Bednenko J, Cingolani G & Gerace L (2003) Nucleocytoplasmic transport: navigating the channel. *Traffic* 4:127–135.

Chook YM & Blobel G (2001) Karyopherins and nuclear import. *Curr Opin Struct Biol* 11:703–715.

Cole CN & Scarcelli JJ (2006) Transport of messenger RNA from the nucleus to the cytoplasm. *Curr Opin Cell Biol* 18:299–306.

Fahrenkrog B, Koser J & Aebi U (2004) The nuclear pore complex: a jack of all trades? *Trends Biochem Sci* 29:175–182.

Görlich D & Kutay U (1999) Transport between the cell nucleus and the cytoplasm. *Annu Rev Cell Dev Biol* 15:607–660.

Hetzer MW, Walther TC & Mattaj IW (2005) Pushing the envelope: structure, function, and dynamics of the nuclear periphery. *Annu Rev Cell Dev Biol* 21:347–380.

Komeili A & O'Shea EK (2000) Nuclear transport and transcription. *Curr Opin Cell Biol* 12:355–360.

Kuersten S, Ohno M & Mattaj IW (2001) Nucleocytoplasmic transport: Ran, beta and beyond. *Trends Cell Biol* 11:497–503.

Tran EJ & Wente SR (2006) Dynamic nuclear pore complexes: life on the edge. *Cell* 125:1041–1053.

Weis K (2002) Nucleocytoplasmic transport: cargo trafficking across the border. *Curr Opin Cell Biol* 14:328–335.

The Transport of Proteins Into Mitochondria and Chloroplasts

Jarvis P & Robinson C (2004) Mechanisms of protein import and routing in chloroplasts. *Curr Biol* 14:R1064–R1077.

Kessler F & Schnell DJ (2004) Chloroplast protein import: solve the GTPase riddle for entry. *Trends Cell Biol* 14:334–338.

Koehler CM, Merchant S & Schatz G (1999) How membrane proteins travel across the mitochondrial intermembrane space. *Trends Biochem Sci* 24:428–432.

Mokranjac D & Neupert W (2005) Protein import into mitochondria. *Biochem Soc Trans* 33:1019–1023.

Prakash S & Matouschek A (2004) Protein unfolding in the cell. *Trends Biochem Sci* 29:593–600.

Soll J & Schleiff E (2004) Protein import into chloroplasts. *Nature Rev Mol Cell Biol* 5:198–208.

Truscott KN, Brandner K & Pfanner N (2003) Mechanisms of protein import into mitochondria. *Curr Biol* 13:R326–R337.

Peroxisomes

Fujiki Y (2000) Peroxisome biogenesis and peroxisome biogenesis disorders. *FEBS Lett* 476:42–46.

Lazarow PB (2003) Peroxisome biogenesis: advances and conundrums. *Curr Opin Cell Biol* 15:489–497.

van der Zand A, Braakman I & Tabak HF (2006) The return of the peroxisome. *J Cell Sci* 119:989–994.

The Endoplasmic Reticulum

Adelman MR, Sabatini DD et al (1973) Ribosome-membrane interaction. Nondestructive disassembly of rat liver rough microsomes into ribosomal and membranous components. *J Cell Biol* 56:206–229.

Bernales S, Papa FR & Walter P (2006) Intracellular signaling by the unfolded protein response. *Annu Rev Cell Dev Biol* 22:487–508.

Bishop WR & Bell RM (1988) Assembly of phospholipids into cellular membranes: biosynthesis, transmembrane movement and intracellular translocation. *Annu Rev Cell Biol* 4:579–610.

Blobel G & Dobberstein B (1975) Transfer of proteins across membranes. I. Presence of proteolytically processed and unprocessed nascent immunoglobulin light chains on membrane-bound ribosomes of murine myeloma. *J Cell Biol* 67:835–851.

Borgese N, Mok W & Sabatini DD (1974) Ribosomal-membrane interaction: in vitro binding of ribosomes to microsomal membranes. *J Mol Biol* 88:559–580.

Daleke DL (2003) Regulation of transbilayer plasma membrane phospholipid asymmetry. *J Lipid Res* 44:233–242.

Deshaies RJ, Sanders SL & Schekman R (1991) Assembly of yeast Sec proteins involved in translocation into the endoplasmic reticulum into a membrane-bound multisubunit complex. *Nature* 349:806–808.

Ellgaard L & Helenius A (2003) Quality control in the endoplasmic reticulum. *Nature Rev Mol Cell Biol* 4:181–191.

Ferguson MA (1999) The structure, biosynthesis and functions of glycosylphosphatidylinositol anchors, and the contributions of trypanosome research. *J Cell Sci* 112:2799–2809.

Gething MJ (1999) Role and regulation of the ER chaperone BiP. *Semin Cell Dev Biol* 10:465–472.

Görlich D, Prehn S et al (1992) A mammalian homolog of SEC61p and SECYp is associated with ribosomes and nascent polypeptides during translocation. *Cell* 71:489–503.

Helenius J & Aebi M (2002) Transmembrane movement of dolichol linked carbohydrates during N-glycoprotein biosynthesis in the endoplasmic reticulum. *Semin Cell Dev Biol* 13:171–178.

Johnson AE & van Waes MA (1999) The translocon: a dynamic gateway at the ER membrane. *Annu Rev Cell Dev Biol* 15:799–842.

Keenan RJ, Freymann DM & Walter P (2001) The signal recognition particle. *Annu Rev Biochem* 70:755–775.

Kostova Z & Wolf DH (2003) For whom the bell tolls: protein quality control of the endoplasmic reticulum and the ubiquitin-proteasome connection. *EMBO J* 22:2309–2317.

Levine T & Loewen C (2006) Inter-organelle membrane contact sites: through a glass, darkly. *Curr Opin Cell Biol* 18:371–378.

Marciniak SJ & Ron D (2006) Endoplasmic reticulum stress signaling in disease. *Physiol Rev* 86:1133–1149.

Milstein C, Brownlee GG et al (1972) A possible precursor of immunoglobulin light chains. *Nature New Biol* 239:117–120.

Römisch K (2005) Endoplasmic reticulum-associated degradation. *Annu Rev Cell Dev Biol* 21:435–456.

Simon SM & Blobel G (1991) A protein-conducting channel in the endoplasmic reticulum. *Cell* 65:371–380.

Staehelin LA (1997) The plant ER: a dynamic organelle composed of a large number of discrete functional domains. *Plant J* 11:1151–1165.

Trombetta ES & Parodi AJ (2003) Quality control and protein folding in the secretory pathway. *Annu Rev Cell Dev Biol* 19:649–676.

Tsai B, Ye Y & Rapoport TA (2002) Retro-translocation of proteins from the endoplasmic reticulum into the cytosol. *Nature Rev Mol Cell Biol* 3:246–255.

Shibata Y, Voeltz GK & Rapoport TA (2006) Rough sheets and smooth tubules. *Cell* 126:435–439

White SH & von Heijne G (2004) The machinery of membrane protein assembly. *Curr Opin Struct Biol* 14:397–404.

Yan A & Lennarz WJ (2005) Unraveling the mechanism of protein N-glycosylation. *J Biol Chem* 280:3121–3124.

Chapter 13

Intracellular Vesicular Traffic

13

Every cell must eat, communicate with the world around it, and quickly respond to changes in its environment. To help accomplish these tasks, cells continually adjust the composition of their plasma membrane in rapid response to need. They use an elaborate internal membrane system to add and remove cell-surface proteins embedded in the membrane, such as receptors, ion channels, and transporters. Through the process of *exocytosis,* the *biosynthetic–secretory pathway* delivers newly synthesized proteins, carbohydrates, and lipids to either the plasma membrane or the extracellular space. By the converse process of *endocytosis* (**Figure 13–1**) cells remove plasma membrane components and deliver them to internal compartments called *endosomes,* from where they can be recycled to the same or different regions of the plasma membrane or can be delivered to lysosomes for degradation. Cells also use endocytosis to capture important nutrients, such as vitamins, lipids, cholesterol, and iron; these are taken up together with the macromolecules to which they bind and are then released in endosomes or lysosomes and transported into the cytosol, where they are used in various biosynthetic processes.

The interior space, or *lumen,* of each membrane-enclosed compartment along the biosynthetic–secretory and endocytic pathways is topologically equivalent to the lumen of most other membrane-enclosed compartments and to the cell exterior. Proteins can travel in this space without having to cross a membrane, being passed from one compartment to another by means of numerous membrane-enclosed transport containers. Some of these containers are small spherical *vesicles,* while others are larger irregular vesicles or tubules formed from the donor compartment. We shall use the term **transport vesicle** to apply to all forms of these containers.

Within a eucaryotic cell, transport vesicles continually bud off from one membrane and fuse with another, carrying membrane components and soluble molecules, which are referred to as **cargo** (**Figure 13–2**). This membrane traffic flows along highly organized, directional routes, which allow the cell to secrete, eat, and remodel its plasma membrane. The biosynthetic–secretory pathway leads outward from the endoplasmic reticulum (ER) toward the Golgi apparatus and cell surface, with a side route leading to lysosomes, while the endocytic pathway leads inward from the plasma membrane. In each case, the flow of membrane between compartments is balanced, with retrieval pathways balancing the flow in the opposite direction, bringing membrane and selected proteins back to the compartment of origin (**Figure 13–3**).

To perform its function, each transport vesicle that buds from a compartment must be selective. It must take up only the appropriate molecules and must fuse only with the appropriate target membrane. A vesicle carrying cargo from the Golgi apparatus to the plasma membrane, for example, must exclude proteins that are to stay in the Golgi apparatus, and it must fuse only with the plasma membrane and not with any other organelle.

We begin this chapter by considering the molecular mechanisms of budding and fusion that underlie all vesicular transport. We then discuss the fundamental problem of how, in the face of this transport, the cell maintains the differences between the compartments. Finally, we consider the function of the Golgi apparatus, lysosomes, secretory vesicles, and endosomes, as we trace the pathways that connect these organelles.

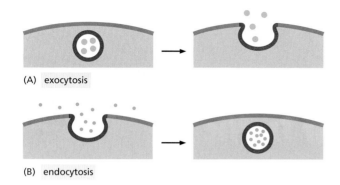

(A)   exocytosis

(B)   endocytosis

Figure 13–1 Exocytosis and endocytosis. (A) In exocytosis, a transport vesicle fuses with the plasma membrane. Its content is released into the extracellular space, while the vesicle membrane (*red*) becomes continuous with the plasma membrane. (B) In endocytosis, a plasma membrane patch (*red*) is internalized forming a transport vesicle. Its content derives from the extracellular space.

THE MOLECULAR MECHANISMS OF MEMBRANE TRANSPORT AND THE MAINTENANCE OF COMPARTMENTAL DIVERSITY

Vesicular transport mediates a continuous exchange of components between the ten or more chemically distinct, membrane-enclosed compartments that collectively comprise the biosynthetic-secretory and endocytic pathways. With this massive exchange, how can each compartment maintain its special identity? To answer this question, we must first consider what defines the character of a compartment. Above all, it is the composition of the enclosing membrane: molecular markers displayed on the cytosolic surface of the membrane serve as guidance cues for incoming traffic to ensure that transport vesicles fuse only with the correct compartment. Many of these membrane markers, however, are found on more than one compartment, and it is the specific combination of marker molecules that gives each compartment its unique molecular address.

How are these membrane markers kept at high concentration on one compartment and at low concentration on another? To answer this question, we need to consider how patches of membrane, enriched or depleted in specific membrane components, bud off from one compartment and transfer to another. **Panel 13–1** outlines some of the genetic and biochemical strategies that have been used to study the molecular machinery involved in vesicular transport.

We begin by discussing how cells segregate proteins into separate membrane domains by assembling a special protein coat on the membrane's cytosolic face. We consider how coats form, what they are made of, and how they are

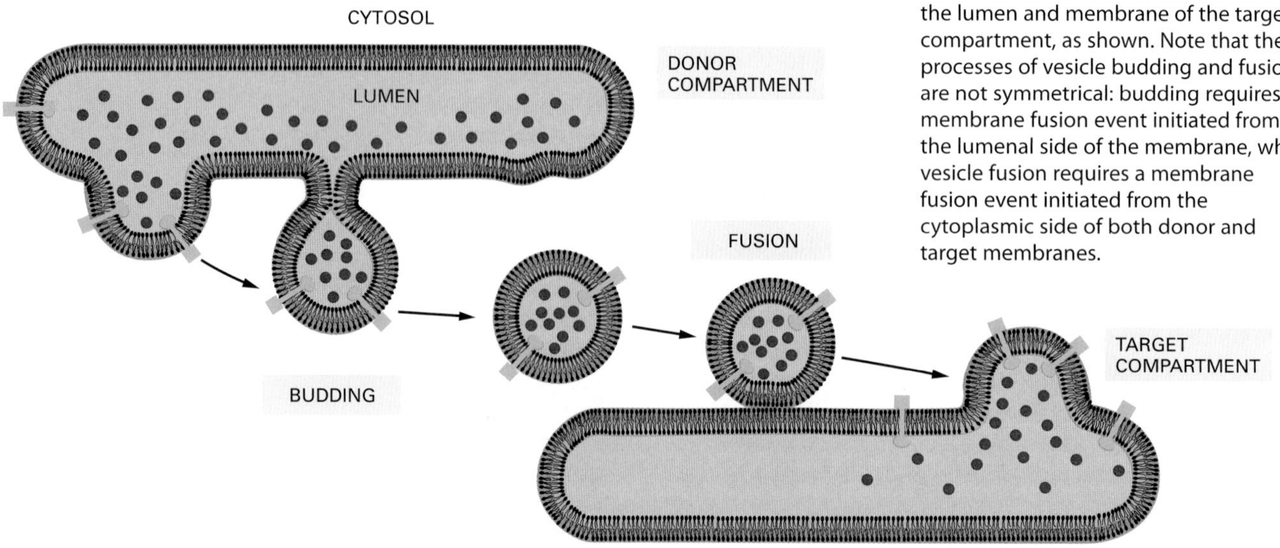

Figure 13–2 Vesicular transport. Transport vesicles bud off from one compartment and fuse with another. As they do so, they carry material as cargo from the *lumen* (the space within a membrane-enclosed compartment) and membrane of the donor compartment to the lumen and membrane of the target compartment, as shown. Note that the processes of vesicle budding and fusion are not symmetrical: budding requires a membrane fusion event initiated from the lumenal side of the membrane, while vesicle fusion requires a membrane fusion event initiated from the cytoplasmic side of both donor and target membranes.

CYTOSOL

LUMEN

DONOR COMPARTMENT

BUDDING

FUSION

TARGET COMPARTMENT

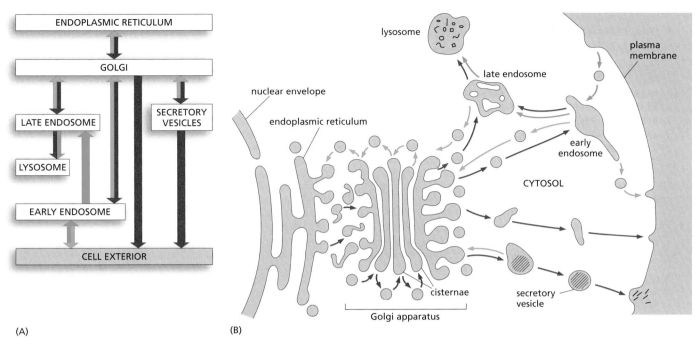

Figure 13–3 A "road-map" of the biosynthetic–secretory and endocytic pathways. (A) In this diagram, which was introduced in Chapter 12, the endocytic and biosynthetic–secretory pathways are illustrated with *green* and *red* arrows, respectively. In addition, *blue* arrows denote retrieval pathways for the backflow of selected components. (B) The compartments of the eucaryotic cell involved in vesicular transport. The lumen of each membrane-enclosed compartment is topologically equivalent to the outside of the cell. All compartments shown communicate with one another and the outside of the cell by means of transport vesicles. In the biosynthetic–secretory pathway *(red arrows)*, protein molecules are transported from the ER to the plasma membrane or (via endosomes) to lysosomes. In the endocytic pathway *(green arrows)*, molecules are ingested in vesicles derived from the plasma membrane and delivered to early endosomes and then (via late endosomes) to lysosomes. Many endocytosed molecules are retrieved from early endosomes and returned to the cell surface for reuse; similarly, some molecules are retrieved from the early and late endosomes and returned to the Golgi apparatus, and some are retrieved from the Golgi apparatus and returned to the ER. All of these retrieval pathways are shown with *blue arrows,* as in part (A).

used to extract specific components from a membrane for delivery to another compartment. Finally, we discuss how transport vesicles dock at the appropriate target membrane and fuse with it to deliver their cargo.

There Are Various Types of Coated Vesicles

Most transport vesicles form from specialized, coated regions of membranes. They bud off as **coated vesicles**, which have a distinctive cage of proteins covering their cytosolic surface. Before the vesicles fuse with a target membrane, they discard their coat, as is required for the two cytosolic membrane surfaces to interact directly and fuse.

The coat performs two main functions. First, it concentrates specific membrane proteins in a specialized patch, which then gives rise to the vesicle membrane. In this way, it selects the appropriate molecules for transport. Second, the coat molds the forming vesicle. Coat proteins assemble into a curved, basketlike lattice that deforms the membrane patch and thereby shapes the vesicle. This may explain why vesicles with the same type of coat often have a relatively uniform size and shape.

There are three well-characterized types of coated vesicles, distinguished by their coat proteins: *clathrin-coated, COPI-coated,* and *COPII-coated* (**Figure 13–4**). Each type is used for different transport steps. Clathrin-coated vesicles, for example, mediate transport from the Golgi apparatus and from the plasma membrane, whereas COPI- and COPII-coated vesicles most commonly mediate transport from the ER and from the Golgi cisternae (**Figure 13–5**). There is, however, much more variety in coated vesicles and their functions than this short list suggests. As we discuss below, there are several types of clathrin-coated vesicles,

CELL-FREE SYSTEMS FOR STUDYING THE COMPONENTS AND MECHANISM OF VESICULAR TRANSPORT

Vesicular transport can be reconstituted in cell-free systems. This was first achieved for the Golgi stack. When Golgi stacks are isolated from cells and incubated with cytosol and with ATP as a source of energy, transport vesicles bud from their rims and appear to transport proteins between cisternae. By following the progressive processing of the oligosaccharides on a glycoprotein as it moves from one Golgi compartment to the next, it is possible to follow the process of vesicular transport.

To follow the transport, two distinct populations of Golgi stacks are incubated together. The "donor" population is isolated from mutant cells that lack the enzyme N-acetylglucosamine (GlcNAc) transferase I and that have been infected with a virus; because of the mutation, the major viral glycoprotein fails to be modified with GlcNAc in

the Golgi apparatus of the mutant cells. The "acceptor" Golgi stacks are isolated from uninfected wild-type cells and thus contain a good copy of GlcNAc transferase I, but lack the viral glycoprotein. In the mixture of Golgi stacks the viral glycoprotein acquires GlcNAc, indicating that it must have been transported between the Golgi stacks—presumably by vesicles that bud from the *cis* compartment of the donor Golgi and fuse with the medial compartment of the acceptor Golgi or by homotypic fusion between the cisternae of two Golgi stacks. This transport-dependent glycosylation is monitored by measuring the transfer of ^{3}H-GlcNAc from UDP-^{3}H-GlcNAc to the viral glycoprotein. Transport occurs only when ATP and cytosol are added. By fractionating the cytosol, a number of specific cytosolic proteins have been identified that are required for the budding and fusion of transport vesicles.

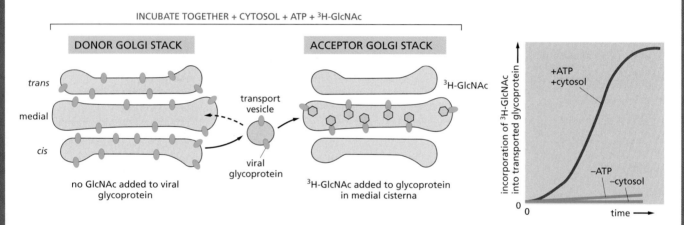

INCUBATE TOGETHER + CYTOSOL + ATP + ^{3}H-GlcNAc

DONOR GOLGI STACK

trans

medial

cis

no GlcNAc added to viral glycoprotein

transport vesicle

viral glycoprotein

ACCEPTOR GOLGI STACK

^{3}H-GlcNAc

^{3}H-GlcNAc added to glycoprotein in medial cisterna

incorporation of ^{3}H-GlcNAc into transported glycoprotein

+ATP +cytosol

−ATP −cytosol

0

time ⟶

Similar cell-free systems have been used to study transport from the medial to the *trans* Golgi network, from the *trans* Golgi network to the plasma membrane, from endosomes to lysosomes, from the *trans* Golgi network to late endosomes, and to study homotypic fusion between like compartments—such as endosomes and immature secretory vesicles.

GENETIC APPROACHES FOR STUDYING VESICULAR TRANSPORT

Genetic studies of mutant yeast cells defective for secretion have identified more than 25 genes that are involved in the secretory pathway. Many of the mutant genes encode *temperature-sensitive* proteins. These function normally at 25°C, but when the mutant cells (A–I) are shifted to an elevated temperature, such as 35°C, they fail to transport proteins from the ER to the Golgi apparatus, others from one Golgi cisterna to another, and still others from the Golgi apparatus to the vacuole (the yeast lysosome) or to the plasma membrane.

Once a protein required for secretion has been identified in this way, a phenomenon called *multicopy suppression* can be used to identify genes that encode other proteins that interact with it. At high temperatures, a temperature-sensitive mutant protein often has too low an affinity for its normal interaction partners. If the interacting proteins are produced at much higher concentration than normal, however, sufficient binding occurs to cure the defect.

To create an experimental paradigm in which such high concentrations of ligand are present, mutant yeast cells (with a temperature-sensitive mutation in a gene involved in vesicular transport) are transfected with a yeast plasmid vector into which random normal yeast genomic DNA fragments have been cloned. Because these plasmids are maintained in cells at high copy number, any cells that happen to carry plasmids with intact genes will overproduce the normal gene product, allowing rare cells to survive at the high temperature. The relevant DNA fragments, which presumably encode proteins that interact with the original mutant protein, can then be isolated from the surviving cell clones.

The genetic and biochemical approaches complement each other, and many of the proteins involved in vesicular transport have been identified independently by biochemical studies of mammalian cell-free systems and by genetic studies in yeast.

GFP FUSION PROTEINS HAVE REVOLUTIONIZED THE STUDY OF INTRACELLULAR TRANSPORT

One way to follow the whereabouts of a protein in living cells is to construct fusion proteins, in which green fluorescent protein (GFP) is attached by genetic engineering techniques to the protein of interest. When a cDNA encoding such a fusion protein is expressed in a cell, the protein is readily visible in a fluorescent microscope, so that it can be followed in living cells in real time. Fortunately, for most proteins studied the addition of GFP does not perturb the protein's function.

GFP fusion proteins are widely used to study the location and movement of proteins in cells. GFP fused to proteins that shuttle in and out of the nucleus, for example, facilitates studies of nuclear transport and its regulation. GFP fused to mitochondrial or Golgi proteins is used to study the behavior of these organelles. GFP fused to plasma membrane proteins allows measurement of the kinetics of their movement from the ER through the secretory pathway. Dramatic examples of such experiments can be seen as movies on the DVD that accompanies this book.

The study of GFP fusion proteins is often combined with FRAP and FLIP techniques (discussed in Chapter 10), in which the GFP in selected regions of the cell is bleached by strong laser light. The rate of diffusion of unbleached GFP fusion proteins into that area can then be determined to provide measurement of the protein's diffusion or transport in the cell. In this way, for example, it was determined that many Golgi enzymes recycle between the Golgi apparatus and the ER. <GCTG>

(A–D, courtesy of Jennifer Lippincott-Schwartz Lab.)

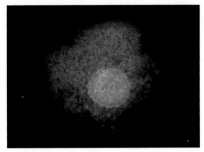

(A) In this experiment, cultured cells express a GFP fusion protein consisting of GFP attached to a viral coat protein—called vesicular stomatitis virus coat protein. The viral protein is an integral membrane protein that normally moves through the secretory pathway from the ER to the cell surface, where the virus would be assembled if cells also expressed the other viral components. The viral protein contains a mutation that allows export from the ER only at a low temperature. Thus, at the high temperature shown, the fusion protein labels the ER.

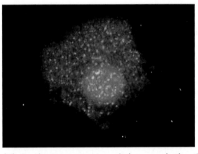

(B) As the temperature is lowered, the GFP fusion protein rapidly accumulates at ER exit sites.

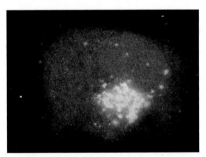

(C) The fusion protein then moves to the Golgi apparatus.

(D) Finally, the fusion protein is delivered to the plasma membrane where the delivered protein diffuses into the plasma membrane (the arrows bracket a fusion event). From such studies the kinetics of each step in the pathway can be determined.

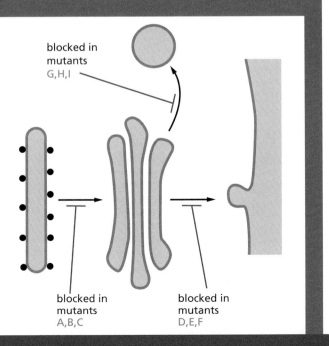

blocked in mutants G,H,I

blocked in mutants A,B,C

blocked in mutants D,E,F

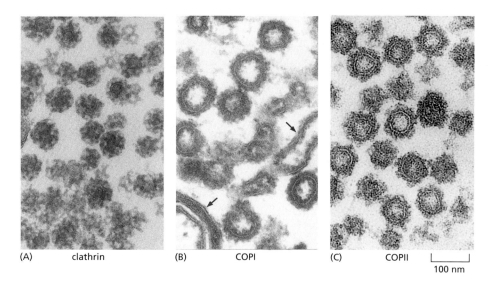

Figure 13–4 **Electron micrograph of clathrin-coated, COPI-coated, and COPII-coated vesicles.** All are shown in electron micrographs at the same scale. (A) Clathrin-coated vesicles. (B) Golgi cisternae (*arrows*) from a cell-free system in which COPI-coated vesicles bud in the test tube. (C) COPII-coated vesicles. (A and B, from L. Orci, B. Glick and J. Rothman, *Cell* 46:171–184, 1986. With permission from Elsevier; C, courtesy of Charles Barlowe and Lelio Orci.)

(A) clathrin (B) COPI (C) COPII 100 nm

each specialized for a different transport step, and the COPI- and COPII-coated vesicles may be similarly diverse.

The Assembly of a Clathrin Coat Drives Vesicle Formation

Clathrin-coated vesicles, the first coated vesicles to be identified, transport material from the plasma membrane and between endosomal and Golgi compartments. **COPI-coated vesicles** and **COPII-coated vesicles** transport material early in the secretory pathway: COPII-coated vesicles bud from the ER, and COPI-coated vesicles bud from Golgi compartments (see Figure 13–5). We discuss clathrin-coated vesicles first, as they provide a good example of how vesicles form.

The major protein component of clathrin-coated vesicles is **clathrin** itself. Each clathrin subunit consists of three large and three small polypeptide chains that together form a three-legged structure called a *triskelion*. Clathrin triskelions assemble into a basketlike convex framework of hexagons and pentagons to form coated pits on the cytosolic surface of membranes (**Figure 13–6**). Under appropriate conditions, isolated triskelions spontaneously self-assemble into typical polyhedral cages in a test tube, even in the absence of the membrane vesicles that these baskets normally enclose (**Figure 13–7**). Thus, the clathrin triskelions determine the geometry of the clathrin cage.

Adaptor proteins, another major coat component in clathrin-coated vesicles, form a discrete second layer of the coat, positioned between the clathrin cage and the membrane. They bind the clathrin coat to the membrane and trap

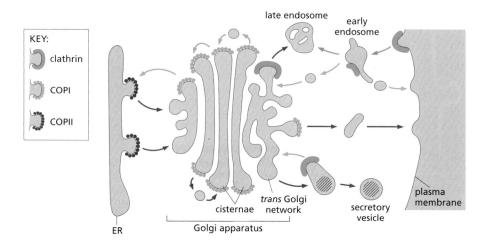

KEY:
- clathrin
- COPI
- COPII

late endosome
early endosome
trans Golgi network
cisternae
Golgi apparatus
ER
secretory vesicle
plasma membrane

Figure 13–5 **Use of different coats in vesicular traffic.** Different coat proteins select different cargo and shape the transport vesicles that mediate the various steps in the biosynthetic– secretory and endocytic pathways. When the same coats function in different places in the cell, they usually incorporate different coat protein subunits that modify their properties (not shown). Many differentiated cells have additional pathways beside those shown here, including a sorting pathway from the *trans* Golgi network to the apical surface of epithelial cells and a specialized recycling pathway for proteins of synaptic vesicles in the synapses of neurons.

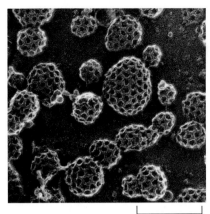

Figure 13–6 **Clathrin-coated pits and vesicles.** This rapid-freeze, deep-etch electron micrograph shows numerous clathrin-coated pits and vesicles on the inner surface of the plasma membrane of cultured fibroblasts. The cells were rapidly frozen in liquid helium, fractured, and deep-etched to expose the cytoplasmic surface of the plasma membrane. (From J. Heuser, *J. Cell Biol.* 84:560–583, 1980. With permission from The Rockefeller University Press.)

0.2 µm

various transmembrane proteins, including transmembrane receptors that capture soluble cargo molecules inside the vesicle—so-called *cargo receptors*. In this way, a selected set of transmembrane proteins, together with the soluble proteins that interact with them, are packaged into each newly formed clathrin-coated transport vesicle (**Figure 13–8**).

There are several types of adaptor proteins. The best characterized have four different protein subunits; others are single-chain proteins. Each type of adaptor protein is specific for a different set of cargo receptors and its use leads to the formation of distinct clathrin-coated vesicles. Clathrin-coated vesicles budding from different membranes use different adaptor proteins and thus package different receptors and cargo molecules.

The sequential assembly of adaptor complexes and the clathrin coat on the cytosolic surface of the membrane generate forces that result in the formation of a clathrin-coated vesicle. Lateral interactions between adaptor complexes and between clathrin molecules aid in forming the vesicle.

Not All Coats Form Basketlike Structures

Not all coats are as regular and universal as the examples of clathrin or COP coats suggest. Some coats may be better described as specialized protein assemblies that form patches dedicated to specific cargo proteins. An example is a coat called **retromer**, which assembles on endosomes and forms vesicles that return *acid hydrolase receptors*, such as the *mannose-6-phosphate receptor*, to the Golgi apparatus (**Figure 13–9**). We discuss later the important role of these receptors in delivering enzymes into new lysosomes.

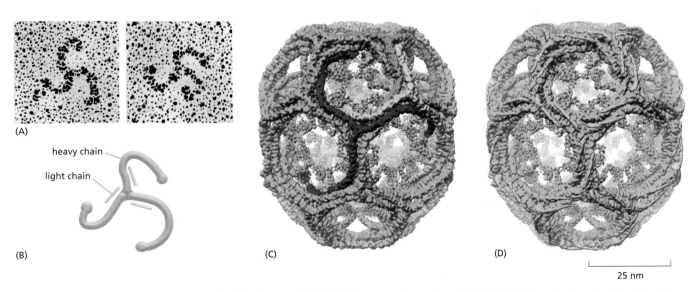

(A)

heavy chain

light chain

(B)

(C)

(D)

25 nm

Figure 13–7 **The structure of a clathrin coat.** <TATT> (A) Electron micrographs of clathrin triskelions shadowed with platinum. Each triskelion is composed of three clathrin heavy chains and three clathrin light chains as shown in (B). (C and D) A cryoelectron micrograph taken of a clathrin coat composed of 36 triskelions organized in a network of 12 pentagons and 6 hexagons, with heavy chains (C) and light chains (D) highlighted. The interwoven legs of the clathrin triskelions form an outer shell into which the N-terminal domains of the triskelions protrude to form an inner layer visible through the openings. It is this inner layer that contacts the adaptor proteins shown in Figure 13–8. Although the coat shown is too small to enclose a membrane vesicle, the clathrin coats on vesicles are constructed in a similar way, from 12 pentagons and a larger number of hexagons, resembling the architecture of a soccer ball. (A, from E. Ungewickell and D. Branton, *Nature* 289:420–422, 1981; C and D, from A. Fotin et al., *Nature* 432:573–579, 2004. All with permission from Macmillan Publishers Ltd.)

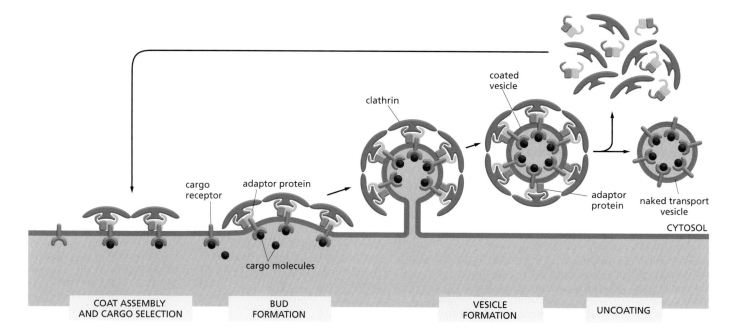

| COAT ASSEMBLY AND CARGO SELECTION | BUD FORMATION | VESICLE FORMATION | UNCOATING |

Retromer is a multiprotein complex that assembles into a coat on endosomal membranes only when:

1. it can bind to the cytoplasmic tails of the cargo receptors,
2. it can interact directly with a curved phospholipid bilayer, and
3. it can bind to a specific phosphorylated phosphatidylinositol lipid (a *phosphoinositide*), which acts as an endosomal marker, as we discuss next.

Because these three requirements must be met simultaneously, retromer is thought to act as a *coincidence detector* and only assemble at the right time and place. Upon binding as a dimer, it stabilizes the membrane curvature, which makes the binding of additional retromers in its proximity more likely. The cooperative assembly of retromer then leads to the formation and budding of a transport vesicle, which delivers its cargo to the Golgi apparatus.

The adaptor proteins found in clathrin coats also bind to phosphoinositides, which not only have a major role in directing when and where coats assemble in the cell, but are used much more widely as molecular markers of compartment identity. This helps to control membrane trafficking events, as we discuss next.

Figure 13–8 The assembly and disassembly of a clathrin coat. The assembly of the coat introduces curvature into the membrane, which leads in turn to the formation of uniformly sized coated buds. The adaptor proteins bind clathrin triskelions and membrane-bound cargo receptors, thereby mediating the selective recruitment of both membrane and cargo molecules into the vesicle. The clathrin coat is rapidly lost shortly after the vesicle forms.

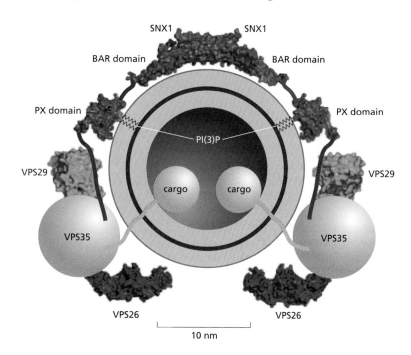

Figure 13–9 A model for retromer assembly on endosomal membranes. The four retromer subunits, SNX1, VPS29, VPS35, and VPS26, form coated domains on endosome membranes that capture cargo molecules, including transmembrane proteins such as acid hydrolase receptors, into vesicles that return to the *trans* Golgi network. VPS35 binds to the cytoplasmic tails of the transmembrane cargo proteins. SNX1 protein contains different protein modules: a *PX domain* that binds to the phosphorylated phosphatidylinositol PI(3)P, and a *BAR domain* that mediates dimerization and attachment to highly curved membranes. Both PX and BAR domains are protein modules that are found in many proteins, where they carry out similar functions. With the exception of the PI(3)P, which is enlarged for visibility, the membrane and other components are drawn approximately to scale. (Adapted from J.S. Bonifacino and R. Rojas, *Nat. Rev. Mol. Cell Biol.* 7:568–579, 2006. With permission from Macmillan Publishers Ltd.)

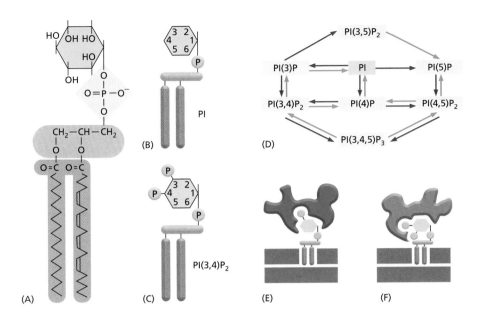

Figure 13–10 Phosphatidylinositol (PI) and phosphoinositides (PIPs). (A, B) The structure of PI shows the free hydroxyl groups in the inositol sugar that can in principle be modified. (C) Phosphorylation of one, two, or three of the hydroxyl groups on PI by PI and PIP kinases produces a variety of PIP species. They are named according to the ring position (in parentheses) and the number of phosphate groups (subscript) added to PI. PI(3,4)P$_2$ is shown. (D) Animal cells have several PI and PIP kinases and a similar number of PIP phosphatases, which are localized to different organelles, where they are regulated to catalyze the production of particular PIPs. The red and green arrows show the kinase and phosphatase reactions, respectively. (E, F) Phosphoinositide head groups are recognized by protein domains that discriminate between the different forms. In this way, selected groups of proteins containing such domains are recruited to regions of membrane in which these phosphoinositides are present. PI(3)P and PI(4,5)P$_2$ are shown. (D, modified from M.A. de Matteis and A. Godi, Nat. Cell Biol. 6:487–492, 2004. With permission from Macmillan Publishers Ltd.)

Phosphoinositides Mark Organelles and Membrane Domains

Although inositol phospholipids typically comprise less than 10% of the total phospholipids in a membrane, they have important regulatory functions. They can undergo rapid cycles of phosphorylation and dephosphorylation at the 3′, 4′, and 5′ positions of their inositol sugar head groups to produce various types of **phosphoinositides (PIPs)**. The interconversion of phosphatidylinositol (PI) and PIPs is highly compartmentalized: different organelles in the endocytic and biosynthetic–secretory pathways have distinct sets of PI and PIP kinases and PIP phosphatases (**Figure 13–10**). The distribution, regulation, and local balance of these enzymes determine the steady-state distribution of each PIP species. As a consequence, the distribution of PIPs varies from organelle to organelle, and often within a continuous membrane from one region to another, thereby defining specialized membrane domains.

Many proteins involved at different steps in vesicular transport contain domains that bind with high specificity to the head groups of particular PIPs, distinguishing one phosphorylated form from another. Local control of the PI and PIP kinases and PIP phosphatases can therefore be used to rapidly control the binding of proteins to a membrane or membrane domain. The production of a particular type of PIP recruits proteins containing matching PIP-binding domains. The PIP-binding proteins then help regulate vesicle formation and other steps in membrane transport (**Figure 13–11**). The same strategy is widely used to recruit specific intracellular signaling proteins to the plasma membrane in response to extracellular signals (discussed in Chapter 15).

Cytoplasmic Proteins Regulate the Pinching-Off and Uncoating of Coated Vesicles

As a clathrin-coated bud grows, soluble cytoplasmic proteins, including **dynamin**, assemble as a ring around the neck of each bud (**Figure 13–12**). Dynamin contains a PI(4,5)P$_2$-binding domain, which tethers the protein to the membrane, and a GTPase domain, which regulates the rate at which vesicles pinch off from the membrane. The pinching-off process brings the two noncytosolic leaflets of the membrane into close proximity and fuses them, sealing off the forming vesicle. To perform this task, dynamin recruits other proteins to the neck of the budding vesicle, which, together with dynamin, help to bend the patch of membrane—by directly distorting the bilayer structure, by changing its lipid composition through the recruitment of lipid-modifying enzymes, or by both mechanisms.

Once released from the membrane, the vesicle rapidly loses its clathrin coat. A PIP phosphatase that is co-packaged into clathrin-coated vesicles depletes

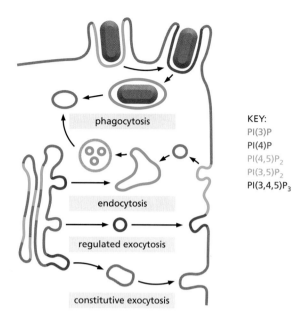

KEY:
PI(3)P
PI(4)P
PI(4,5)P$_2$
PI(3,5)P$_2$
PI(3,4,5)P$_3$

Figure 13–11 The intracellular location of phosphoinositides. Different types of PIPs are located in different membranes and membrane domains, where they are often associated with specific vesicular transport events. The membrane of secretory vesicles, for example, contains PI(4)P. When the vesicles fuse with the plasma membrane, a PI 5-kinase that is localized there converts the PI(4)P into PI(4,5)P$_2$. The PI(4,5)P$_2$, in turn, helps recruit adaptor proteins, which initiate the formation of a clathrin-coated pit, as the first step in clathrin-mediated endocyosis. Once the clathrin-coated vesicle buds off from the plasma membrane, a PI(5)P phosphatase hydrolyzes PI(4,5)P$_2$, which weakens the binding of the adaptor proteins, promoting vesicle uncoating. We discuss phagocytosis and the distinction between regulated and constitutive exocytosis later in the chapter. (Modified from M.A. de Matteis and A. Godi, *Nat. Cell Biol.* 6:487–492, 2004. With permission from Macmillan Publishers Ltd.)

PI(4,5)P$_2$ from the membrane, which weakens the binding of the adaptor proteins. In addition, an Hsp70 chaperone protein functions as an uncoating ATPase, using the energy of ATP hydrolysis to peel off the clathrin coat. *Auxillin*, another vesicle protein, is believed to activate the ATPase. Because the coated bud persists much longer than the coat on the vesicle, additional control mechanisms must somehow prevent the clathrin from being removed before it has formed a complete vesicle (discussed below).

Monomeric GTPases Control Coat Assembly

To balance the vesicular traffic to and from a compartment, coat proteins must assemble only when and where they are needed. While local production of PIPs plays a major part in regulating the assembly of clathrin coats on the plasma membrane and Golgi apparatus, cells have additional ways of regulating coat formation. *Coat-recruitment GTPases*, for example, control the assembly of clathrin coats on endosomes and the COPI and COPII coats on Golgi and ER membranes.

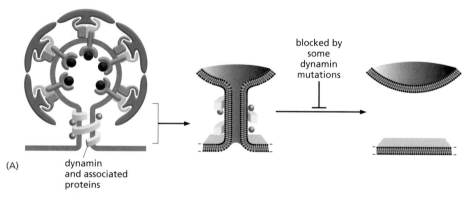

blocked by some dynamin mutations

(A) dynamin and associated proteins

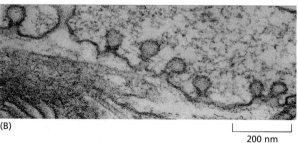

(B)

|— 200 nm —|

Figure 13–12 The role of dynamin in pinching off clathrin-coated vesicles. (A) The dynamin assembles into a ring around the neck of the forming bud. The dynamin ring is thought to recruit other proteins to the vesicle neck, which, together with dynamin, destabilize the interacting lipid bilayers so that the noncytoplasmic leaflets flow together. The newly formed vesicle then pinches off from the membrane. Specific mutations in dynamin can either enhance or block the pinching-off process. (B) Dynamin was discovered as the protein defective in the *shibire* mutant of *Drosophila*. These mutant flies become paralyzed because clathrin-mediated endocytosis stops, and the synaptic vesicle membrane fails to recycle, blocking neurotransmitter release. Deeply invaginated clathrin-coated pits form in the fly's nerve cells, with a ring of mutant dynamin assembled around the neck, as shown in this thin-section electron micrograph. The pinching-off process fails because membrane fusion does not take place. (B, from J.H. Koenig and K. Ikeda, *J. Neurosci.* 9:3844–3860, 1989. With permission from the Society of Neuroscience.)

Many steps in vesicular transport depend on a variety of GTP-binding proteins that control both the spatial and temporal aspects of membrane exchange. As discussed in Chapter 3, GTP-binding proteins regulate most processes in eucaryotic cells. They act as molecular switches, which flip between an active state with GTP bound and an inactive state with GDP bound. Two classes of proteins regulate the flipping: *guanine-nucleotide-exchange factors (GEFs)* activate the proteins by catalyzing the exchange of GDP for GTP, and *GTPase-activating proteins (GAPs)* inactivate the proteins by triggering the hydrolysis of the bound GTP to GDP (see Figure 3–71) Although both monomeric GTP-binding proteins (monomeric GTPases) and trimeric GTP-binding proteins (G proteins) have important roles in vesicular transport, the roles of the monomeric GTPases are better understood, and we focus on them here.

Coat-recruitment GTPases are members of a family of monomeric GTPases. They include the **Arf proteins**, which are responsible for both COPI coat assembly and clathrin coat assembly at Golgi membranes, and the **Sar1 protein**, which is responsible for COPII coat assembly at the ER membrane. Coat-recruitment GTPases are usually found in high concentration in the cytosol in an inactive, GDP-bound state. When a COPII-coated vesicle is to bud from the ER membrane, a specific Sar1-GEF embedded in the ER membrane binds to cytosolic Sar1, causing the Sar1 to release its GDP and bind GTP in its place. (Recall that GTP is present in much higher concentration in the cytosol than GDP and therefore will spontaneously bind after GDP is released.) In its GTP-bound state, the Sar1 protein exposes an amphiphilic helix, which inserts into the cytoplasmic leaflet of the lipid bilayer of the ER membrane. The tightly bound Sar1 now recruits coat protein subunits to the ER membrane to initiate budding (**Figure 13–13**). Other GEFs and coat-recruitment GTPases operate in a similar way on other membranes.

Some coat protein subunits also interact, albeit more weakly, with the head groups of certain lipid molecules, in particular phosphatidic acid and phosphoinositides, as well as with the cytoplasmic tail of some of the transmembrane

Figure 13–13 Formation of a COPII-coated vesicle. (A) The Sar1 protein is a coat-recruitment GTPase. Inactive, soluble Sar1-GDP binds to a Sar1-GEF in the ER membrane, causing the Sar1 to release its GDP and bind GTP. The Sar1-GEF (also called Sec12) was discovered as one of the temperature-sensitive mutants that block ER to Golgi transport (see Panel 13–1). A GTP-triggered conformational change in Sar1 exposes an amphipathic helix, which inserts into the cytoplasmic leaflet of the ER membrane, initiating a process of membrane curvature. (B) GTP-bound Sar1 binds to a complex of two COPII coat proteins, called Sec23/24. The crystal structure of the Sec23/24 proteins together with Sar1 predicts how the amphiphilic helix of Sar1 anchors the complex to the membrane. Sec24 has several different binding sites for cytoplasmic tails of cargo receptors. The entire surface of the complex that attaches to the membrane is gently curved to match the diameter of COPII-coated vesicles. (C) A complex of two additional COPII coat proteins, called Sec13/31, forms the outer shell of the coat. Like clathrin, Sec13/31 alone can assemble into symmetrical cages with appropriate dimensions to enclose a COPII-coated vesicle. (D) Membrane-bound, active Sar1-GTP recruits COPII subunits to the membrane. This causes the membrane to form a bud, which includes selected transmembrane proteins. A subsequent membrane-fusion event pinches off the coated vesicle.

Other coated vesicles are thought to form in a similar way. The coat-recruitment GTPase Arf also contains a regulated amphiphilic helix but, in contrast to Sar1, the helix also has a covalently attached fatty acid chain that contributes to its hydrophobicity. As in Sar1, the regulated amphiphilic helix is retracted in the GDP-bound state and exposed in the GTP-bound state. As we discuss later, *Rab GTPases* regulate their membrane attachment in a similar manner (see Figure 13–14). (C, modified from S.M. Stagg et al., *Nature* 439:234–238, 2006. With permission from Macmillan Publishers Ltd.)

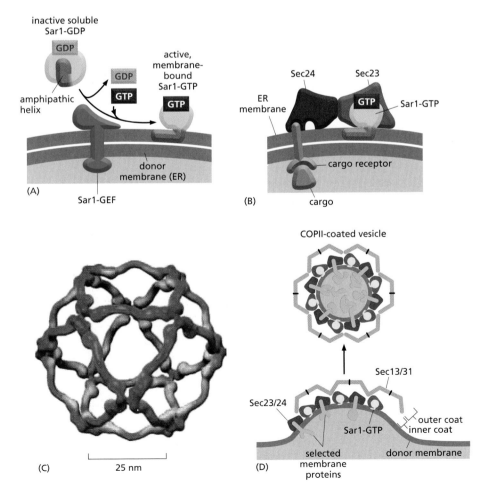

(A) inactive soluble Sar1-GDP — active, membrane-bound Sar1-GTP — amphipathic helix — donor membrane (ER) — Sar1-GEF

(B) Sec24 — Sec23 — ER membrane — Sar1-GTP — cargo receptor — cargo

(C) 25 nm

(D) COPII-coated vesicle — Sec13/31 — Sec23/24 — Sar1-GTP — outer coat inner coat — selected membrane proteins — donor membrane proteins they recruit into the bud. Together, these protein–lipid and protein–protein and interactions tightly bind the coat to the membrane, causing the membrane to deform into a bud, which then pinches off as a coated vesicle.

The coat-recruitment GTPases also have a role in coat disassembly. The hydrolysis of bound GTP to GDP causes the GTPase to change its conformation so that its hydrophobic tail pops out of the membrane, causing the vesicle's coat to disassemble. Although it is not known what triggers the GTP hydrolysis process, it has been proposed that the GTPases work like timers, which hydrolyze GTP at a slow but predictable rate. COPII coats, for example, accelerate GTP hydrolysis by Sar1, thereby triggering coat disassembly at a certain time after coat assembly has begun. Thus, a fully formed vesicle will be produced only when bud formation occurs faster than the timed disassembly process; otherwise, disassembly will be triggered before a vesicle pinches off, and the process will have to start again at a more appropriate time and place.

Not All Transport Vesicles Are Spherical

Although vesicle budding at various locations in the cell has many similarities, each cell membrane poses its own special challenges. The plasma membrane, for example, is comparatively flat and stiff, owing to its cholesterol-rich lipid composition and underlying cortical cytoskeleton. Thus, clathrin coats have to produce considerable force to introduce curvature, especially at the neck of the bud where dynamin and its associated proteins facilitate the sharp bends required for the pinching-off process. In contrast, vesicle budding from many intracellular membranes occurs preferentially at regions where the membranes are already curved, such as the rims of the Golgi cisternae or ends of membrane tubules. In these places, the primary function of the coats is to capture the appropriate cargo proteins rather than to deform the membrane.

Transport vesicles occur in various sizes and shapes. When living cells are genetically engineered to express fluorescent membrane components, the endosomes and *trans* Golgi network are seen in a fluorescence microscope to continually send out long tubules. Coat proteins assemble onto the tubules and help recruit specific cargo. The tubules then either regress or pinch off with the help of dynamin-like proteins to form transport vesicles. Depending on the relative efficiencies of tubule formation and the pinching-off process, vesicles of different sizes and shapes are produced. Thus, vesicular transport does not necessarily occur only through uniformly sized spherical vesicles, but can involve larger portions of a donor organelle.

Tubules have a much higher surface-to-volume ratio than the organelles from which they form. They are therefore relatively enriched in membrane proteins compared with soluble cargo proteins. As we discuss later, this property of tubules is used for sorting proteins in endosomes.

Rab Proteins Guide Vesicle Targeting

To ensure an orderly flow of vesicular traffic, transport vesicles must be highly selective in recognizing the correct target membrane with which to fuse. Because of the diversity and crowding of membrane systems in the cytoplasm, a vesicle is likely to encounter many potential target membranes before it finds the correct one. Specificity in targeting is ensured because all transport vesicles display surface markers that identify them according to their origin and type of cargo, and target membranes display complementary receptors that recognize the appropriate markers. This crucial process depends on two types of proteins: *Rab proteins* direct the vesicle to specific spots on the correct target membrane, and then *SNARE* proteins mediate the fusion of the lipid bilayers.

Rab proteins play a central part in the specificity of vesicular transport. Like the coat-recruitment GTPases discussed earlier (see Figure 13–13), they also are monomeric GTPases. With over 60 known members, they are the largest subfamily of such GTPases. Each Rab protein is associated with one or more membrane-

Table 13–1 Subcellular Locations of Some Rab Proteins

PROTEIN	ORGANELLE
Rab1	ER and Golgi complex
Rab2	*cis* Golgi network
Rab3A	synaptic vesicles, secretory granules
Rab4/Rab11	recycling endosomes
Rab5A	plasma membrane, clathrin-coated vesicles, early endosomes
Rab5C	early endosomes
Rab6	medial and *trans* Golgi cisternae
Rab7	late endosomes
Rab8	early endosomes
Rab9	late endosomes, *trans* Golgi network

enclosed organelles of the biosynthetic–secretory or endocytic pathways, and each of these organelles has at least one Rab protein on its cytosolic surface (**Table 13–1**). Their highly selective distribution on these membrane systems makes Rab proteins ideal molecular markers for identifying each membrane type and guiding vesicular traffic between them. Rab proteins can function on transport vesicles, on target membranes, or both.

Like the coat-recruitment GTPases, Rab proteins cycle between a membrane and the cytosol and regulate the reversible assembly of protein complexes on the membrane. In their GDP-bound state they are inactive and bound to another protein (*Rab-GDP dissociation inhibitor* or *GDI*) that keeps them soluble in the cytosol, while in their GTP-bound state they are active and tightly associated with the membrane of an organelle or transport vesicle. Membrane bound Rab-GEFs activate Rab proteins on both transport vesicle and target membranes, as activate Rab molecules are usually required on both sides. Once in its GTP-bound state and membrane-bound through a hydrophobic lipid anchor, Rab proteins bind to other proteins, called **Rab effectors**, which facilitate vesicle transport, membrane tethering, and fusion (**Figure 13–14**). The GTP hydrolysis sets the concentration of active Rab and, consequently, the concentration of its effectors on the membrane.

In contrast to the highly conserved structure of Rab proteins, the structures of Rab effectors vary greatly. Some Rab effectors, for example, are *motor proteins* that propel vesicles along actin filaments or microtubules to their target membrane. Others are *tethering proteins*, some of which have long threadlike domains that serve as "fishing lines" that can extend to link two membranes more than 200 nm apart; other tethering proteins are large protein complexes that link two membranes that are closer together. Rab effectors can also interact with SNAREs, thus coupling membrane tethering to fusion.

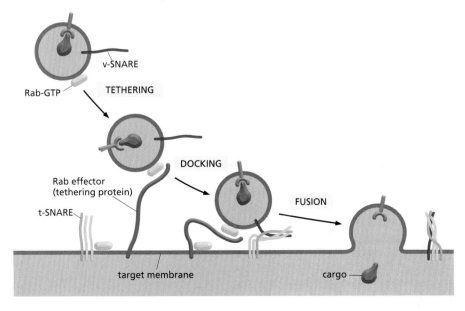

Figure 13–14 Tethering of a vesicle to a target membrane. Rab effector proteins interact via active Rab proteins (Rab-GTPs, *yellow*) located on the target membrane, vesicle membrane, or both, to establish the first connection between the two membranes that are going to fuse. In the example shown here, the Rab effector is a filamentous tethering protein (*green*). Next, SNARE proteins on the two membranes (*red* and *blue*) pair to dock the vesicle to the target membrane and catalyze the fusion of the two apposed lipid bilayers.

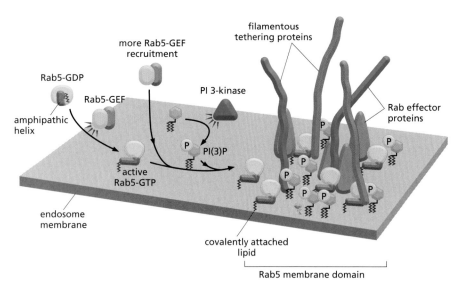

Figure 13–15 The formation of a Rab5 domain on the endosome membrane. A Rab5-specific GEF in the endosome membrane binds a Rab5 protein and induces it to exchange GDP for GTP. GTP binding alters the conformation of the Rab protein, exposing an amphiphilic helix and a covalently attached lipid group, which together anchor the Rab5-GTP to the membrane. Active Rab5 activates PI 3-kinase, which converts PI into PI(3)P. PI(3)P and active Rab5 together bind a variety of Rab effector proteins that contain PI(3)P-binding sites, including filamentous tethering proteins that catch incoming clathrin-coated vesicles from the plasma membrane. Active Rab5 also recruits more Rab5-GEF, further enhancing the assembly of the Rab5 domain on the membrane.

Controlled cycles of GTP hydrolysis and GDP–GTP exchange are thought to regulate the size and activity of such Rab domains dynamically. Unlike SNAREs, which are integral membrane proteins, the GDP/GTP cycle coupled to the membrane/cytosol translocation cycle endows the Rab machinery with the ability to undergo assembly and disassembly on the membrane. For example, during transport from early to late endosomes, Rab5 can be moved and replaced by Rab7, thus committing cargo to degradation. (Adapted from M. Zerial and H. McBride, *Nat. Rev. Mol. Cell Biol.* 2:107–117, 2001. With permission from Macmillan Publishers Ltd.)

The same Rab proteins can bind to multiple effectors. The assembly of Rab proteins and their effectors on a membrane is cooperative and results in the formation of large, specialized membrane patches. Rab5, for example, assembles on endosomal membranes and mediates the capture of clathrin-coated vesicles arriving from the plasma membrane. It recruits tethering proteins to catch the incoming vesicles. Rab5-GEF initially recruits Rab5 to the endosome and converts it to its active GTP-bound form, which becomes anchored to the membrane (**Figure 13–15**). Active Rab5 recruits more Rab5-GEF to the endosome, thereby stimulating the recruitment of more Rab5 to the same site. In addition, active Rab5 activates a PI 3-kinase, which locally converts PI to PI(3)P, which in turn binds some of the Rab effectors. This type of positive feedback greatly enhances the assembly process and helps to establish functionally distinct membrane domains within a continuous membrane.

The endosomal membrane provides a striking example of how different Rab proteins and their effectors help to create multiple specialized membrane domains, each fulfilling a particular set of functions. Thus, while the Rab5 domain receives incoming vesicles from the plasma membrane, distinct Rab11 and Rab4 domains in the same membrane are thought to organize the budding of recycling vesicles that return proteins from the endosome to the plasma membrane. Once assembled, these domains coexist in the same membrane over extended periods.

SNAREs Mediate Membrane Fusion

Once a transport vesicle has been tethered to its target membrane, it unloads its cargo by membrane fusion. Membrane fusion requires bringing the lipid bilayers of two membranes to within 1.5 nm of each other so that they can join. When the membranes are in such close apposition, lipids can flow from one bilayer to the other. For this close approach, water must be displaced from the hydrophilic surface of the membrane—a process that is highly energetically unfavorable. It seems likely that specialized *fusion proteins* that overcome this energy barrier catalyze all membrane fusions in cells. We have already discussed the role of dynamin in a related task during the pinching-off of clathrin-coated vesicles (see Figure 13–12).

The **SNARE proteins** (also called **SNAREs**, for short) catalyze the membrane fusion reactions in vesicular transport. They also provide an additional layer of specificity in the transport process by helping to ensure that only correctly targeted vesicles fuse. There are at least 35 different SNAREs in an animal cell, each associated with a particular organelle in the biosynthetic–secretory or endocytic pathway. These transmembrane proteins exist as complementary sets—with **v-SNAREs** usually found on vesicle membranes and **t-SNAREs** usually found on target membranes (see Figure 13–14). A v-SNARE is a single polypeptide chain,

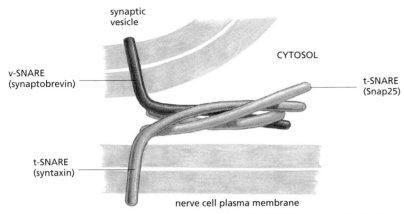

Figure 13–16 The structure of a trans-SNARE complex. The SNAREs responsible for docking synaptic vesicles at the plasma membrane of nerve terminals consist of three proteins. The v-SNARE *synaptobrevin* and the t-SNARE *syntaxin* are both transmembrane proteins, and each contributes one α helix to the complex. The t-SNARE *Snap25* is a peripheral membrane protein that contributes two α helices to the four-helix bundle. They are connected by a loop (omitted from the figure) that lies parallel to the membrane and has fatty acyl chains attached to anchor it there. In many other t-SNARES, each α helix is a separate protein anchored in the membrane by a transmembrane helix. Trans-SNARE complexes always consist of four tightly intertwined α helices, three contributed by a t-SNARE and one by a v-SNARE. The t-SNAREs are composed of multiple chains, one of which is always a transmembrane protein and contributes one helix, and one or two additional light chains that may or may not be transmembrane proteins and that contribute the remaining two helices to the four-helix bundle of the trans-SNARE complex. The crystal structure of a stable complex of the four intertwining α helices contributed by these proteins is modeled here in the context of the whole proteins. The α helices are shown as rods for simplicity. (Adapted from R.B. Sutton et al., *Nature* 395:347–353, 1998. With permission from Macmillan Publishers Ltd.)

whereas a t-SNARE is composed of two or three proteins (**Figure 13–16**). The v-SNAREs and t-SNAREs have characteristic helical domains, and when a v-SNARE interacts with a t-SNARE, the helical domains of one wrap around the helical domains of the other to form a stable four-helix bundle. The resulting *trans-SNARE complexes* lock the two membranes together.

SNAREs have been best characterized in neurons, where they mediate the docking and fusion of synaptic vesicles at the nerve terminal's plasma membrane in the process of neurotransmitter release. The bacteria that cause tetanus and botulism secrete powerful proteolytic neurotoxins that enter specific neurons and cleave SNARE proteins in the nerve terminals. In this way, the toxins block synaptic transmissions, often leading to death.

It is thought that the trans-SNARE complexes catalyze membrane fusion by using the energy that is freed when the interacting helices wrap around each other to pull the membrane faces together, simultaneously squeezing out water molecules from the interface (**Figure 13–17**). When liposomes containing purified v-SNAREs are mixed with liposomes containing complementary t-SNAREs, their membranes fuse, albeit slowly. In the cell, other proteins recruited to the fusion site, presumably Rab effectors, cooperate with SNAREs to accelerate fusion.

Fusion does not always follow immediately after v-SNAREs and t-SNAREs pair. As we discuss later, in the process of regulated exocytosis, fusion is delayed until secretion is triggered by a specific extracellular signal. In this case, a localized influx of Ca^{2+} triggers fusion, presumably by releasing inhibitory proteins that prevent the complete zipping-up of the trans-SNARE complexes.

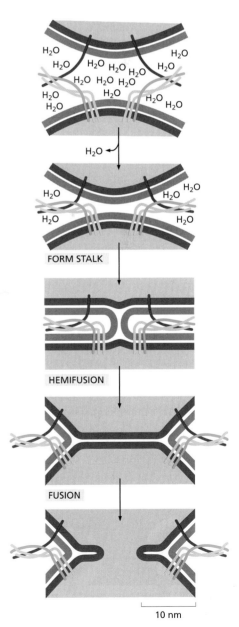

Figure 13–17 A model for how SNARE proteins may catalyze membrane fusion. Bilayer fusion occurs in multiple steps. A tight pairing between v- and t-SNAREs forces lipid bilayers into close apposition and expels water molecules from the interface. Lipid molecules in the two interacting leaflets of the bilayers then flow between the membranes to form a connecting stalk. Lipids of the other two leaflets then contact each other, forming a new bilayer, which widens the fusion zone (*hemifusion*, or half-fusion). Rupture of the new bilayer completes the fusion reaction.

Rab proteins, which can regulate the availability of SNARE proteins, exert an additional layer of control. t-SNAREs in target membranes are often associated with inhibitory proteins that must be released before the t-SNARE can function. Rab proteins and their effectors trigger the release of such SNARE inhibitory proteins. In this way, SNARE proteins are concentrated and activated in the correct location on the membrane, where tethering proteins capture incoming vesicles. Rab proteins thus speed up the process by which appropriate SNARE proteins in two membranes find each other.

For vesicular transport to operate normally, transport vesicles must incorporate the appropriate SNARE and Rab proteins. Not surprisingly, therefore, many transport vesicles will form only if they incorporate the appropriate complement of SNARE and Rab proteins in their membrane. How this crucial control process operates during vesicle budding remains a mystery.

Interacting SNAREs Need to Be Pried Apart Before They Can Function Again

Most SNARE proteins in cells have already participated in multiple rounds of vesicular transport and are sometimes present in a membrane as stable complexes with partner SNAREs. The complexes have to disassemble before the SNAREs can mediate new rounds of transport. A crucial protein called **NSF** cycles between membranes and the cytosol and catalyzes the disassembly process. It is an ATPase that uses the energy of ATP hydrolysis to unravel the intimate interactions between the helical domains of paired SNARE proteins (**Figure 13–18**). The requirement for NSF-mediated reactivation of SNAREs by SNARE complex disassembly helps prevent membranes from fusing indiscriminately: if the t-SNAREs in a target membrane were always active, any membrane containing an appropriate v-SNARE might fuse whenever the two membranes made contact. It is not known how the activity of NSF is controlled so that the SNARE machinery is activated at the right time and place, but Rab effectors are likely candidates to play a part in this process.

Viral Fusion Proteins and SNAREs May Use Similar Fusion Mechanisms

Membrane fusion is important in other processes beside vesicular transport. The plasma membranes of a sperm and an egg fuse during fertilization (discussed in Chapter 21), and myoblasts fuse with one another during the development of multinucleate muscle fibers (discussed in Chapter 22). Likewise, mitochondria fuse and fragment in a dynamic way (discussed in Chapter 14). All cell membrane fusions require special proteins and are tightly regulated to ensure that only appropriate membranes fuse. The controls are crucial for maintaining both the identity of cells and the individuality of each type of intracellular compartment.

The membrane fusions catalyzed by viral fusion proteins are the best understood. These proteins have a crucial role in permitting the entry of enveloped viruses (which have a lipid-bilayer-based membrane coat) into the cells that they infect (discussed in Chapters 5 and 24). For example, viruses such as the human immunodeficiency virus (HIV), which causes AIDS, bind to cell-surface

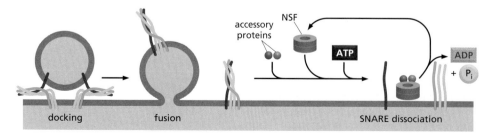

Figure 13–18 Dissociation of SNARE pairs by NSF after a membrane fusion cycle. After a v-SNARE and t-SNARE have mediated the fusion of a transport vesicle on a target membrane, the NSF binds to the SNARE complex and, with the help of two accessory proteins, hydrolyzes ATP to pry the SNAREs apart.

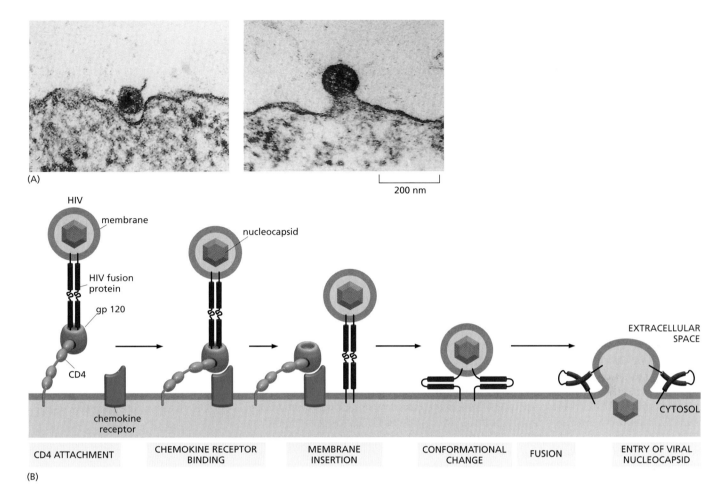

Figure 13–19 **The entry of enveloped viruses into cells.** (A) Electron micrographs showing how HIV enters a cell by fusing its membrane with the plasma membrane of the cell. (B) A model for the fusion process. HIV binds first to the CD4 protein on the surface of the target cell. The viral gp120 protein, which is bound to the HIV fusion protein, mediates this interaction. A second cell-surface protein on the target cell, which normally serves as a receptor for chemokines (discussed in Chapters 24 and 25), now interacts with gp120. This interaction releases the HIV fusion protein from gp120, allowing the previously buried hydrophobic fusion peptide to insert into the plasma membrane of the target cell. The fusion protein, which is a trimer (not shown), thus becomes transiently anchored as an integral membrane protein in the two opposing membranes. The fusion protein then spontaneously rearranges, collapsing into a tightly packed six-helix bundle. The energy released by this rearrangement in multiple copies of the fusion protein pulls the two membranes together, overcoming the high activation energy barrier that normally prevents membrane fusion. Thus, like a mouse trap, the HIV fusion protein contains a reservoir of potential energy, which is released and harnessed to do mechanical work. (A, from B.S. Stein et al., *Cell* 49:659–668, 1987. With permission from Elsevier; B, adapted from a drawing by Wayne Hendrickson.)

receptors and then fuse with the plasma membrane of the target cell (**Figure 13–19**). This fusion event allows the viral nucleic acid inside the nucleocapsid to enter the cytosol, where it replicates. Other viruses, such as the influenza virus, first enter the cell by receptor-mediated endocytosis (discussed later) and are delivered to endosomes; the low pH in endosomes activates a fusion protein in the viral envelope that catalyzes the fusion of the viral and endosomal membranes, releasing the viral nucleic acid into the cytosol.

The three-dimensional structures of the fusion proteins of HIV and influenza virus provide valuable insights into the molecular mechanism of the membrane fusion catalyzed by these proteins. Exposure of the HIV fusion protein to receptors on the target cell membrane, or exposure of the influenza fusion protein to low pH, uncovers previously buried hydrophobic regions. These regions, called *fusion peptides*, then insert directly into the lipid bilayer of the target cell membrane. In this way, the fusion proteins transiently become integral membrane proteins in two separate lipid bilayers. Structural rearrangements in the fusion proteins then bring the two lipid bilayers into very close apposition and destabilize them so that the bilayers fuse (see Figure 13–19). Thus, viral fusion proteins and SNAREs promote lipid bilayer fusion in similar ways.

Summary

Directed and selective transport of particular membrane components from one membrane-enclosed compartment of a eucaryotic cell to another maintains the differences between those compartments. Transport vesicles, which can be spherical, tubular, or irregularly shaped, bud from specialized coated regions of the donor membrane. The assembly of the coat helps to collect specific membrane and soluble cargo molecules for transport and to drive the formation of the vesicle.

There are various types of coated vesicles. The best characterized are clathrin-coated vesicles, which mediate transport from the plasma membrane and the trans *Golgi network, and COPI- and COPII-coated vesicles, which mediate transport between Golgi cisternae and between the ER and the Golgi apparatus, respectively. In clathrin-coated vesicles, adaptor proteins link the clathrin to the vesicle membrane and also trap specific cargo molecules for packaging into the vesicle. The coat is shed rapidly after budding, enabling the vesicle to fuse with its appropriate target membrane.*

Local synthesis of phosphoinositides creates binding sites that trigger coat assembly and vesicle budding. In addition, monomeric GTPases help regulate various steps in vesicular transport, including both vesicle budding and docking. The coat-recruitment GTPases, including Sar1 and the Arf proteins, regulate coat assembly and disassembly. A large family of Rab proteins functions as vesicle targeting GTPases. Rab proteins are recruited to transport vesicles and target membranes. The assembly and disassembly of Rab proteins and their effectors in specialized membrane domains are dynamically controlled by GTP binding and hydrolysis. Active Rab proteins recruit Rab effectors, such as motor proteins, which transport vesicles on actin filaments or microtubules, and filamentous tethering proteins, which help ensure that the vesicles deliver their contents only to the appropriate membrane-enclosed compartment. Complementary v-SNARE proteins on transport vesicles and t-SNARE proteins on the target membrane form stable trans-SNARE complexes, which force the two membranes into close apposition so that their lipid bilayers can fuse.

TRANSPORT FROM THE ER THROUGH THE GOLGI APPARATUS

As discussed in Chapter 12, newly synthesized proteins cross the ER membrane from the cytosol to enter the biosynthetic–secretory pathway. During their subsequent transport, from the ER to the Golgi apparatus and from the Golgi apparatus to the cell surface and elsewhere, these proteins are successively modified as they pass through a series of compartments. Transfer from one compartment to the next involves a delicate balance between forward and backward (retrieval) transport pathways. Some transport vesicles select cargo molecules and move them to the next compartment in the pathway, while others retrieve escaped proteins and return them to a previous compartment where they normally function. Thus, the pathway from the ER to the cell surface consists of many sorting steps, which continuously select membrane and soluble lumenal proteins for packaging and transport—in vesicles or organelle fragments that bud from the ER and Golgi apparatus.

In this section we focus mainly on the **Golgi apparatus** (also called the **Golgi complex**). It is a major site of carbohydrate synthesis, as well as a sorting and dispatching station for products of the ER. The cell makes many polysaccharides in the Golgi apparatus, including the pectin and hemicellulose of the cell wall in plants and most of the glycosaminoglycans of the extracellular matrix in animals (discussed in Chapter 19). The Golgi apparatus also lies on the exit route from the ER, and a large proportion of the carbohydrates that it makes are attached as oligosaccharide side chains to the many proteins and lipids that the ER sends to it. A subset of these oligosaccharide groups serve as tags to direct specific proteins into vesicles that then transport them to lysosomes. But most proteins and lipids, once they have acquired their appropriate oligosaccharides in the Golgi apparatus, are recognized in other ways for targeting into the transport vesicles going to other destinations.

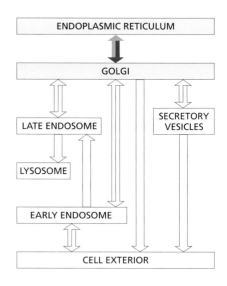

Proteins Leave the ER in COPII-Coated Transport Vesicles

To initiate their journey along the biosynthetic–secretory pathway, proteins that have entered the ER and are destined for the Golgi apparatus or beyond are first packaged into small COPII-coated transport vesicles. These vesicles bud from specialized regions of the ER called *ER exit sites*, whose membrane lacks bound ribosomes. Most animal cells have ER exit sites dispersed throughout the ER network.

Originally, it was thought that all proteins that are not tethered in the ER enter transport vesicles by default. It is now clear, however, that entry into vesicles that leave the ER is usually a selective process. Many membrane proteins are actively recruited into such vesicles, where they become concentrated. It is thought that these cargo proteins display exit (transport) signals on their cytosolic surface that components of the COPII coat recognize (**Figure 13–20**); these coat components act as cargo receptors and are recycled back to the ER after they have delivered their cargo to the Golgi apparatus. Soluble cargo proteins in the ER lumen, by contrast, have exit signals that attach them to transmembrane cargo receptors, which in turn bind through exit signals in their cytoplasmic tails to components of the COPII coat. At a lower rate, proteins without exit signals can also enter transport vesicles, so that even proteins that normally function in the ER (so-called *ER resident proteins*) slowly leak out of the ER and are delivered to the Golgi apparatus. Similarly, secretory proteins that are made in high concentrations may leave the ER without the help of exit signals or cargo receptors.

The exit signals that direct soluble proteins out of the ER for transport to the Golgi apparatus and beyond are not well understood. Some transmembrane proteins that serve as cargo receptors for packaging some secretory proteins into COPII-coated vesicles are lectins that bind to oligosaccharides. The ERGIC53 lectin, for example, binds to mannose and is thought to recognize this sugar on two secreted blood-clotting factors (Factor V and Factor VIII), thereby packaging the proteins into transport vesicles in the ER. ERGIC53's role in protein transport was identified because humans who lack it owing to an inherited mutation have lowered serum levels of Factors V and VIII, and they therefore bleed excessively.

Only Proteins That Are Properly Folded and Assembled Can Leave the ER

To exit from the ER, proteins must be properly folded and, if they are subunits of multimeric protein complexes, they may need to be completely assembled. Those that are misfolded or incompletely assembled remain in the ER, where they are bound to chaperone proteins (discussed in Chapter 6), such as *BiP* or *calnexin*. The chaperones may cover up the exit signals or somehow anchor the proteins in

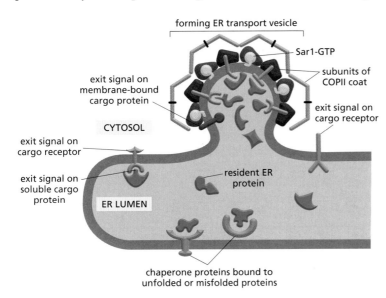

Figure 13–20 The recruitment of cargo molecules into ER transport vesicles. By binding directly or indirectly to the COPII coat, membrane and soluble cargo proteins, respectively, become concentrated in the transport vesicles as they leave the ER. Membrane proteins are packaged into budding transport vesicles through interactions of exit signals on their cytosolic tails with the COPII coat. Some of the membrane proteins that the coat traps function as cargo receptors, binding soluble proteins in the lumen and helping to package them into vesicles. A typical 50-nm transport vesicle contains about 200 membrane proteins, which can be of many different types. As indicated, unfolded or incompletely assembled proteins are bound to chaperones and retained in the ER compartment.

forming ER transport vesicle

Sar1-GTP

subunits of COPII coat

exit signal on membrane-bound cargo protein

exit signal on cargo receptor

CYTOSOL

exit signal on cargo receptor

resident ER protein

exit signal on soluble cargo protein

ER LUMEN

chaperone proteins bound to unfolded or misfolded proteins

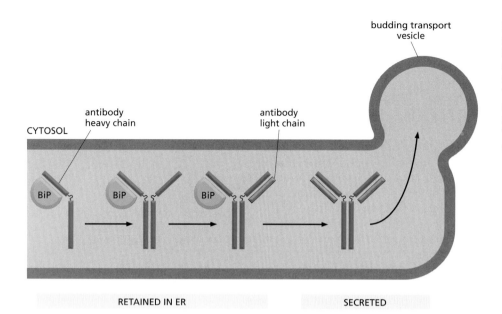

Figure 13–21 Retention of incompletely assembled antibody molecules in the ER. Antibodies are made up of two heavy and two light chains (discussed in Chapter 25), which assemble in the ER. The chaperone BiP is thought to bind to all incompletely assembled antibody molecules and to cover up an exit signal. Thus, only completely assembled antibodies leave the ER and are secreted.

the ER (**Figure 13–21**). Such failed proteins are eventually transported back into the cytosol, where they are degraded by proteasomes (discussed in Chapters 6 and 12). This quality-control step prevents the onward transport of misfolded or mis-assembled proteins that could potentially interfere with the functions of normal proteins. There is a surprising amount of corrective action. More than 90% of the newly synthesized subunits of the T cell receptor (discussed in Chapter 25) and of the acetylcholine receptor (discussed in Chapter 11), for example, are normally degraded without ever reaching the cell surface where they function. Thus, cells must make a large excess of many protein molecules to produce a select few that fold, assemble, and function properly.

The process of continual degradation of a portion of ER proteins also pro-vides an early warning system to alert the immune system when a virus infects cells. Using specialized ABC-type transporters, the ER imports peptide frag-ments of viral proteins produced by proteases in the proteasome. The foreign peptides are loaded onto class I MHC proteins in the ER lumen and then trans-ported to the cell surface. T lymphocytes then recognize the peptides as non-self antigens and kill the infected cells (discussed in Chapter 25).

Sometimes, however, there are drawbacks to the stringent quality-control mechanism. The predominant mutations that cause cystic fibrosis, a common inherited disease, result in the production of a slightly misfolded form of a plasma membrane protein important for Cl⁻ transport. Although the mutant protein would function perfectly normally if it reached the plasma membrane, it remains in the ER. This devastating disease thus results not because the muta-tion inactivates the protein but because the active protein is discarded before it reaches the plasma membrane.

Vesicular Tubular Clusters Mediate Transport from the ER to the Golgi Apparatus

After transport vesicles have budded from ER exit sites and have shed their coat, they begin to fuse with one another. This fusion of membranes from the same compartment is called *homotypic fusion,* to distinguish it from *heterotypic fusion,* in which a membrane from one compartment fuses with the membrane of a different compartment. As with heterotypic fusion, homotypic fusion requires a set of matching SNAREs. In this case, however, the interaction is sym-metrical, with both membranes contributing v-SNAREs and t-SNAREs (**Figure 13–22**).

The structures formed when ER-derived vesicles fuse with one another are called *vesicular tubular clusters,* because they have a convoluted appearance in

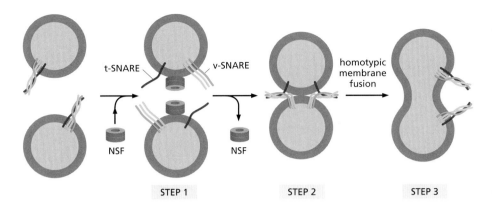

Figure 13–22 **Homotypic membrane fusion.** <AAAA> In step 1, NSF pries apart identical pairs of v-SNAREs and t-SNAREs in both membranes (see Figure 13–18). In steps 2 and 3, the separated matching SNAREs on adjacent identical membranes interact, which leads to membrane fusion and the formation of one continuous compartment called a vesicular tubular cluster. Subsequently, the compartment grows by further homotypic fusion with vesicles from the same kind of membrane, displaying matching SNAREs. Homotypic fusion is not restricted to the formation of vesicular tubular clusters; in a similar process, endosomes fuse to generate larger endosomes. Rab proteins help regulate the extent of homotypic fusion and hence the size of the compartments in a cell (not shown).

the electron microscope (**Figure 13–23**A). These clusters constitute a new compartment that is separate from the ER and lacks many of the proteins that function in the ER. They are generated continually and function as transport containers that bring material from the ER to the Golgi apparatus. The clusters are relatively short-lived because they move quickly along microtubules to the Golgi apparatus, with which they fuse to deliver their contents (Figure 13–23B).

As soon as vesicular tubular clusters form, they begin to bud off transport vesicles of their own. Unlike the COPII-coated vesicles that bud from the ER, these vesicles are COPI-coated. They carry back to the ER resident proteins that have escaped, as well as proteins such as cargo receptors that participated in the ER budding reaction and are being returned. This retrieval process demonstrates the exquisite control mechanisms that regulate coat assembly reactions. The COPI coat assembly begins only seconds after the COPII coats have been shed. It remains a mystery how this switch in coat assembly is controlled.

The *retrieval* (or *retrograde*) *transport* continues as the vesicular tubular clusters move towards the Golgi apparatus. Thus, the clusters continuously mature, gradually changing their composition as selected proteins are returned to the ER. A similar retrieval process continues from the Golgi apparatus, after the vesicular tubular clusters have delivered their cargo.

The Retrieval Pathway to the ER Uses Sorting Signals

The retrieval pathway for returning escaped proteins back to the ER depends on *ER retrieval signals*. Resident ER membrane proteins, for example, contain signals that bind directly to COPI coats and are thus packaged into COPI-coated

Figure 13–23 **Vesicular tubular clusters.** (A) An electron micrograph section of vesicular tubular clusters forming from the ER membrane. Many of the vesicle-like structures seen in the micrograph are cross sections of tubules that extend above and below the plane of this thin section and are interconnected. (B) Vesicular tubular clusters move along microtubules to carry proteins from the ER to the Golgi apparatus. COPI coats mediate the budding of vesicles that return to the ER from these clusters. As indicated, the coats quickly disassemble after the vesicles have formed. (A, courtesy of William Balch.)

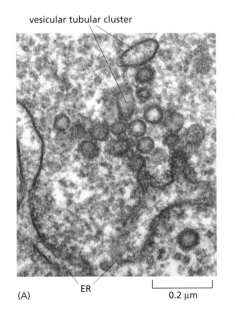

(A)

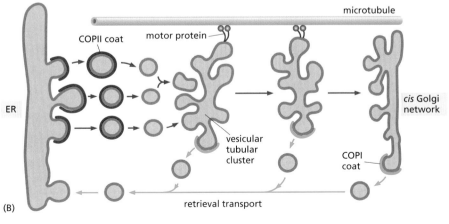

(B)

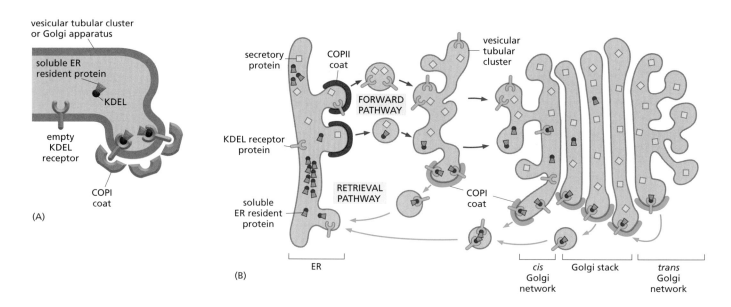

(A)

(B)

Figure 13–24 A model for the retrieval of soluble ER resident proteins. ER resident proteins that escape from the ER are returned by vesicular transport. (A) The KDEL receptor present in vesicular tubular clusters and the Golgi apparatus captures the soluble ER resident proteins and carries them in COPI-coated transport vesicles back to the ER. Upon binding its ligands in this environment, the KDEL receptor may change conformation, so as to facilitate its recruitment into budding COPI-coated vesicles. (B) The retrieval of ER proteins begins in vesicular tubular clusters and continues from all parts of the Golgi apparatus. In the environment of the ER, the ER resident proteins dissociate from the KDEL receptor, which is then returned to the Golgi apparatus for reuse.

transport vesicles for retrograde delivery to the ER. The best-characterized retrieval signal of this type consists of two lysines, followed by any two other amino acids, at the extreme C-terminal end of the ER membrane protein. It is called a *KKXX sequence*, based on the single-letter amino acid code.

Soluble ER resident proteins, such as BiP, also contain a short retrieval signal at their C-terminal end, but it is different: it consists of a Lys-Asp-Glu-Leu or a similar sequence. If this signal (called the *KDEL sequence*) is removed from BiP by genetic engineering, the protein is slowly secreted from the cell. If the signal is transferred to a protein that is normally secreted, the protein is now efficiently returned to the ER, where it accumulates.

Unlike the retrieval signals on ER membrane proteins, which can interact directly with the COPI coat, soluble ER resident proteins must bind to specialized receptor proteins such as the *KDEL receptor*—a multipass transmembrane protein that binds to the KDEL sequence and packages any protein displaying it into COPI-coated retrograde transport vesicles (**Figure 13–24**). To accomplish this task, the KDEL receptor itself must cycle between the ER and the Golgi apparatus, and its affinity for the KDEL sequence must differ in these two compartments. The receptor must have a high affinity for the KDEL sequence in vesicular tubular clusters and the Golgi apparatus, so as to capture escaped, soluble ER resident proteins that are present there at low concentration. It must have a low affinity for the KDEL sequence in the ER, however, to unload its cargo in spite of the very high concentration of KDEL-containing resident proteins in the ER.

How does the affinity of the KDEL receptor change depending on the compartment in which it resides? The answer is not known, but it may be related to the different ionic conditions and pH in the different compartments, which are regulated by ion transporters in the compartment membrane. As we discuss later, pH-sensitive protein–protein interactions form the basis for many of the sorting steps in the cell.

Most membrane proteins that function at the interface between the ER and Golgi apparatus, including v- and t-SNAREs and some cargo receptors, enter the retrieval pathway back to the ER. Whereas the recycling of some of these proteins is mediated by signals, as just described, for others no specific signal seems to be required. Thus, while retrieval signals increase the efficiency of the retrieval process, some proteins randomly enter budding vesicles destined for the ER and are returned to the ER at a slower rate. Many Golgi enzymes cycle constantly between the ER and the Golgi, but their rate of return to the ER is slow enough for most of the protein to be found in the Golgi apparatus.

Many Proteins Are Selectively Retained in the Compartments in Which They Function

The KDEL retrieval pathway only partly explains how ER resident proteins are maintained in the ER. As expected, cells that express genetically modified ER resident proteins, from which the KDEL sequence has been experimentally removed, secrete these proteins. But the rate of secretion is much slower than for a normal secretory protein. It seems that a mechanism that is independent of their KDEL signal anchors ER resident proteins and that only those proteins that escape this retention mechanism are captured and returned via the KDEL receptor. A suggested retention mechanism is that ER resident proteins bind to one another, thus forming complexes that are too big to enter transport vesicles efficiently. Because ER resident proteins are present in the ER at very high concentrations (estimated to be millimolar), relatively low-affinity interactions would suffice to tie up most of the proteins in such complexes.

Aggregation of proteins that function in the same compartment—called *kin recognition*—is a general mechanism that compartments use to organize and retain their resident proteins. Golgi enzymes that function together, for example, also bind to each other and are thereby restrained from entering transport vesicles leaving the Golgi apparatus.

The Golgi Apparatus Consists of an Ordered Series of Compartments

Because of its large and regular structure, the Golgi apparatus was one of the first organelles described by early light microscopists. It consists of a collection of flattened, membrane-enclosed compartments, called *cisternae,* that somewhat resemble a stack of pita breads. Each Golgi stack usually consists of four to six cisternae (**Figure 13–25**), although some unicellular flagellates can have up to

Figure 13–25 The Golgi apparatus.
(A) Three-dimensional reconstruction from electron micrographs of the Golgi apparatus in a secretory animal cell. The *cis* face of the Golgi stack is that closest to the ER. (B) A thin-section electron micrograph emphasizing the transitional zone between the ER and the Golgi apparatus in an animal cell. (C) An electron micrograph of a Golgi apparatus in a plant cell (the green alga *Chlamydomonas*) seen in cross section. In plant cells, the Golgi apparatus is generally more distinct and more clearly separated from other intracellular membranes than in animal cells. (A, redrawn from A. Rambourg and Y. Clermont, *Eur. J. Cell Biol.* 51:189–200, 1990. With permission from Wissenschaftliche Verlagsgesellschaft; B, courtesy of Brij J. Gupta; C, courtesy of George Palade.)

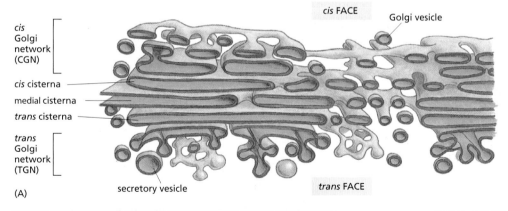

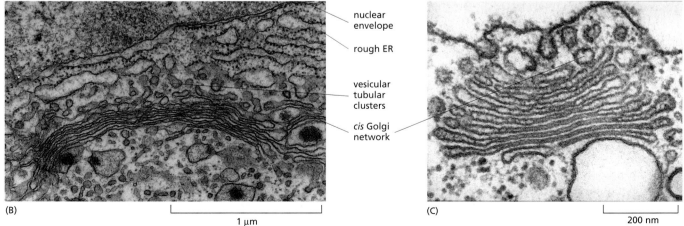

(A)

cis FACE

Golgi vesicle

cis Golgi network (CGN)

cis cisterna

medial cisterna

trans cisterna

trans Golgi network (TGN)

secretory vesicle

trans FACE

(B)

1 μm

nuclear envelope

rough ER

vesicular tubular clusters

cis Golgi network

(C)

200 nm

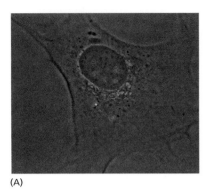 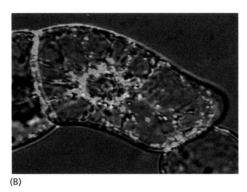

(A)   (B)

Figure 13–26 Localization of the Golgi apparatus in animal and plant cells. (A) The Golgi apparatus in a cultured fibroblast stained with a fluorescent antibody that recognizes a Golgi resident protein (*red*). The Golgi apparatus is polarized, facing the direction in which the cell was crawling before fixation. (B) The Golgi apparatus in a plant cell that is expressing a fusion protein consisting of a resident Golgi enzyme fused to green fluorescent protein. The bright *orange* spots (false color) are Golgi stacks. (A, courtesy of John Henley and Mark McNiven; B, courtesy of Chris Hawes.)

60. In animal cells, tubular connections between corresponding cisternae link many stacks, thus forming a single complex, which is usually located near the cell nucleus and close to the centrosome (**Figure 13–26**A). This localization depends on microtubules. If microtubules are experimentally depolymerized, the Golgi apparatus reorganizes into individual stacks that are found throughout the cytoplasm, adjacent to ER exit sites. Some cells, including most plant cells, have hundreds of individual Golgi stacks dispersed throughout the cytoplasm (Figure 13–26B).

During their passage through the Golgi apparatus, transported molecules undergo an ordered series of covalent modifications. Each Golgi stack has two distinct faces: a *cis* **face** (or entry face) and a *trans* **face** (or exit face). Both *cis* and *trans* faces are closely associated with special compartments, each composed of a network of interconnected tubular and cisternal structures: the *cis* **Golgi network** (**CGN**) and the *trans* **Golgi network** (**TGN**), respectively. The CGN is a collection of fused vesicular tubular clusters arriving from the ER. Proteins and lipids enter the *cis* Golgi network and exit from the *trans* Golgi network bound for the cell surface or another compartment. Both networks are important for protein sorting: proteins entering the CGN can either move onward in the Golgi apparatus or be returned to the ER. Similarly, proteins exiting from the TGN move onward and are sorted according to their next destination: lysosomes, secretory vesicles, or the cell surface. They also can be returned to an earlier compartment.

As described in Chapter 12, a single species of *N-linked oligosaccharide* is attached *en bloc* to many proteins in the ER and then trimmed while the protein is still in the ER. The oligosaccharide intermediates created by the trimming reactions serve to help proteins fold and to help transport misfolded proteins to the cytosol for degradation. Thus, they play an important role in controlling the quality of proteins exiting from the ER. Once these ER functions have been fulfilled, the cell is free to redesign the oligosaccharides for new functions. This happens in the Golgi apparatus, which generates the heterogeneous oligosaccharide structures seen in mature proteins. Upon arrival in the CGN, proteins pass through the *cis* Golgi network, before entering the first of the Golgi processing compartments (the *cis* Golgi cisternae). They then move to the next compartment (the medial cisternae) and finally to the *trans* cisternae, where glycosylation is completed. The lumen of the *trans* cisternae is thought to be continuous with the TGN, the place where proteins are segregated into different transport packages and dispatched to their final destinations.

The oligosaccharide processing steps occur in an organized sequence in the Golgi stack, with each cisterna containing a characteristic abundance of processing enzymes. Proteins are modified in successive stages as they move from cisterna to cisterna across the stack, so that the stack forms a multistage processing unit. This compartmentalization might seem unnecessary, since each oligosaccharide processing enzyme can accept a glycoprotein as a substrate only after it has been properly processed by the preceding enzyme. Nonetheless, it is clear that processing occurs in a spatial as well as a biochemical sequence: enzymes catalyzing early processing steps are concentrated in the cisternae toward the *cis* face of the Golgi stack, whereas enzymes catalyzing later processing steps are concentrated in the cisternae toward the *trans* face.

Figure 13–27 Molecular compartmentalization of the Golgi apparatus. A series of electron micrographs shows the Golgi apparatus (A) unstained, (B) stained with osmium, which is preferentially reduced by the cisternae of the *cis* compartment, and (C and D) stained to reveal the location of specific enzymes. Nucleoside diphosphatase is found in the *trans* Golgi cisternae (C), while acid phosphatase is found in the *trans* Golgi network (D). Note that usually more than one cisterna is stained. The enzymes are therefore thought to be highly enriched rather than precisely localized to a specific cisterna. (Courtesy of Daniel S. Friend.)

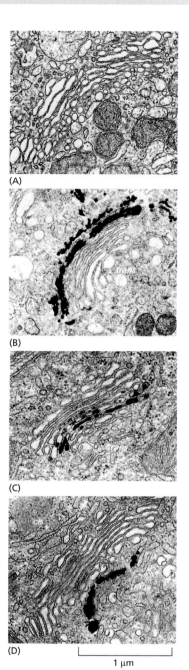

(A)

(B)

(C)

(D)

1 μm

Investigators discovered the functional differences between the *cis*, medial, and *trans* subdivisions of the Golgi apparatus by localizing the enzymes involved in processing *N*-linked oligosaccharides in distinct regions of the organelle, both by physical fractionation of the organelle and by labeling the enzymes in electron microscope sections with antibodies. The removal of mannose residues and the addition of *N*-acetylglucosamine, for example, were shown to occur in the medial compartment, while the addition of galactose and sialic acid was found to occur in the *trans* compartment and the *trans* Golgi network (**Figure 13–27**). **Figure 13–28** summarizes the functional compartmentalization of the Golgi apparatus.

The Golgi apparatus is especially prominent in cells that are specialized for secretion of glycoproteins, such as the goblet cells of the intestinal epithelium, which secrete large amounts of polysaccharide-rich mucus into the gut (**Figure 13–29**). In such cells, unusually large vesicles are found on the *trans* side of the Golgi apparatus, which faces the plasma membrane domain where secretion occurs.

Oligosaccharide Chains Are Processed in the Golgi Apparatus

Whereas the ER lumen is full of soluble lumenal resident proteins and enzymes, the resident proteins in the Golgi apparatus are all membrane bound. The enzymatic reactions in the Golgi apparatus seem to be carried out entirely on its membrane surface. All of the Golgi glycosidases and glycosyl transferases are single-pass transmembrane proteins, many of which are organized in multienzyme complexes. Thus, the two synthetic organelles in the biosynthetic–secretory pathway, the ER and the Golgi apparatus, are organized in fundamentally different ways.

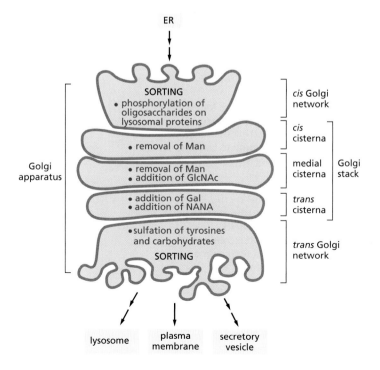

Figure 13–28 Oligosaccharide processing in Golgi compartments. The localization of each processing step shown was determined by a combination of techniques, including biochemical subfractionation of the Golgi apparatus membranes and electron microscopy after staining with antibodies specific for some of the processing enzymes. Processing enzymes are not restricted to a particular cisterna; instead, their distribution is graded across the stack—such that early-acting enzymes are present mostly in the *cis* Golgi cisternae and later-acting enzymes are mostly in the *trans* Golgi cisternae.

Figure 13–29 A goblet cell of the small intestine. This cell is specialized for secreting mucus, a mixture of glycoproteins and proteoglycans synthesized in the ER and Golgi apparatus. Like all epithelial cells, goblet cells are highly polarized, with the apical domain of their plasma membrane facing the lumen of the gut and the basolateral domain facing the basal lamina. The Golgi apparatus is also highly polarized, which facilitates the discharge of mucus by exocytosis at the apical domain of the plasma membrane. (After R.V. Krstic, Illustrated Encyclopedia of Human Histology. New York: Springer-Verlag, 1984. With permission from Springer-Verlag.)

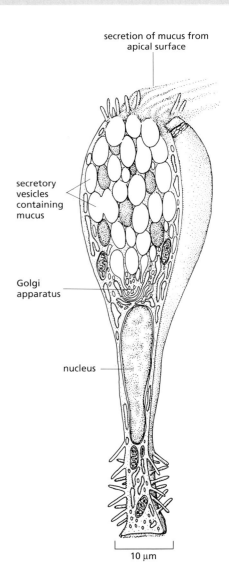

Two broad classes of *N*-linked oligosaccharides, the **complex oligosaccharides** and the **high-mannose oligosaccharides**, are attached to mammalian glycoproteins (**Figure 13–30**). Sometimes, both types are attached (in different places) to the same polypeptide chain.

Complex oligosaccharides are generated when the original *N*-linked oligosaccharide added in the ER is trimmed and further sugars are added. By contrast, high-mannose oligosaccharides are trimmed but have no new sugars added to them in the Golgi apparatus. They contain just two *N*-acetylglucosamines and many mannose residues, often approaching the number originally present in the lipid-linked oligosaccharide precursor added in the ER. Complex oligosaccharides can contain more than the original two *N*-acetylglucosamines, as well as a variable number of galactose and sialic acids and, in some cases, fucose. Sialic acid is of special importance because it is the only sugar in glycoproteins that bears a negative charge. Whether a given oligosaccharide remains high-mannose or is processed depends largely on its position in the protein. If the oligosaccharide is

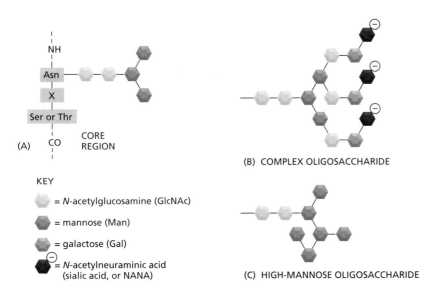

Figure 13–30 The two main classes of asparagine-linked (*N*-linked) oligosaccharides found in mature mammalian glycoproteins. (A) Both complex oligosaccharides and high-mannose oligosaccharides share a common *core region* derived from the original *N*-linked oligosaccharide added in the ER (see Figure 12–50) and typically containing two *N*-acetylglucosamines (GlcNAc) and three mannoses (Man). (B) Each complex oligosaccharide consists of a *core region*, together with a *terminal region* that contains a variable number of copies of a special trisaccharide unit (*N-acetylglucosamine–galactose–sialic acid*) linked to the core mannoses. Frequently, the terminal region is truncated and contains only GlcNAc and galactose (Gal) or just GlcNAc. In addition, a fucose residue may be added, usually to the core GlcNAc attached to the asparagine (Asn). Thus, although the steps of processing and subsequent sugar addition are rigidly ordered, complex oligosaccharides can be heterogeneous. Moreover, although the complex oligosaccharide shown has three terminal branches, two and four branches are also common, depending on the glycoprotein and the cell in which it is made. (C) High-mannose oligosaccharides are not trimmed back all the way to the core region and contain additional mannose residues. Hybrid oligosaccharides with one Man branch and one GlcNAc and Gal branch are also found (not shown).

The three amino acids indicated in (A) constitute the sequence recognized by the oligosaccharyl transferase enzyme that adds the initial oligosaccharide to the protein. Ser = serine; Thr = threonine; X = any amino acid, except proline.

accessible to the processing enzymes in the Golgi apparatus, it is likely to be converted to a complex form; if it is inaccessible because its sugars are tightly held to the protein's surface, it is likely to remain in a high-mannose form. The processing that generates complex oligosaccharide chains follows the highly ordered pathway shown in **Figure 13–31**.

Beyond these commonalities in oligosaccharide processing that are shared among most cells, the products of the carbohydrate modifications carried out in the Golgi apparatus are highly complex and have given rise to a new field called glycobiology. The human genome, for example, encodes hundreds of different Golgi glycosyl transferases, which are expressed differently from one cell type to another, resulting in a variety of glycosylated forms of a given protein or lipid in different cell types and at varying stages of differentiation, depending on the spectrum of enzymes expressed by the cell. The complexity of modifications is not limited to *N*-linked oligosaccharides but also occurs on *O-linked sugars*, as we discuss next.

Proteoglycans Are Assembled in the Golgi Apparatus

In addition to the *N*-linked oligosaccharide alterations made to proteins as they pass through the Golgi cisternae *en route* from the ER to their final destinations, many proteins are also modified in other ways. Some proteins have sugars added to the hydroxyl groups of selected serine or threonine side chains. This **O-linked glycosylation** (**Figure 13–32**), like the extension of *N*-linked oligosaccharide chains, is catalyzed by a series of glycosyl transferase enzymes that use the sugar nucleotides in the lumen of the Golgi apparatus to

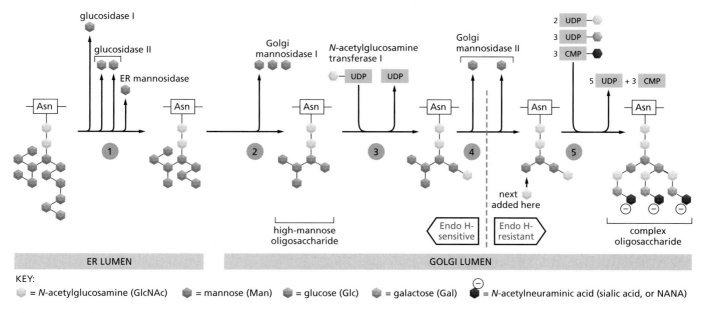

Figure 13–31 Oligosaccharide processing in the ER and the Golgi apparatus. The processing pathway is highly ordered, so that each step shown depends on the previous one. Step 1: Processing begins in the ER with the removal of the glucoses from the oligosaccharide initially transferred to the protein. Then a mannosidase in the ER membrane removes a specific mannose. Step 2: The remaining steps occur in the Golgi stack, where Golgi mannosidase I first removes three more mannoses. Step 3: *N*-acetylglucosamine transferase I then adds an *N*-acetylglucosamine. Step 4: Mannosidase II then removes two additional mannoses. This yields the final core of three mannoses that is present in a complex oligosaccharide. At this stage, the bond between the two *N*-acetylglucosamines in the core becomes resistant to attack by a highly specific endoglycosidase *(Endo H)*. Since all later structures in the pathway are also Endo H-resistant, treatment with this enzyme is widely used to distinguish complex from high-mannose oligosaccharides. Step 5: Finally, as shown in Figure 13–30, additional *N*-acetylglucosamines, galactoses, and sialic acids are added. These final steps in the synthesis of a complex oligosaccharide occur in the cisternal compartments of the Golgi apparatus. Three types of glycosyl transferase enzymes act sequentially, using sugar substrates that have been activated by linkage to the indicated nucleotide. The membranes of the Golgi cisternae contain specific carrier proteins that allow each sugar nucleotide to enter in exchange for the nucleoside phosphates that are released after the sugar is attached to the protein on the lumenal face.

Note that as a biosynthetic organelle, the Golgi apparatus differs fundamentally from the ER: all sugars are assembled inside the lumen from sugar nucleotides. By contrast, in the ER, the *N*-linked precursor oligosaccharide is assembled partly in the cytosol and partly in the lumen, and all lumenal reactions use dolichol-linked sugars as their substrates (see Figure 12–52).

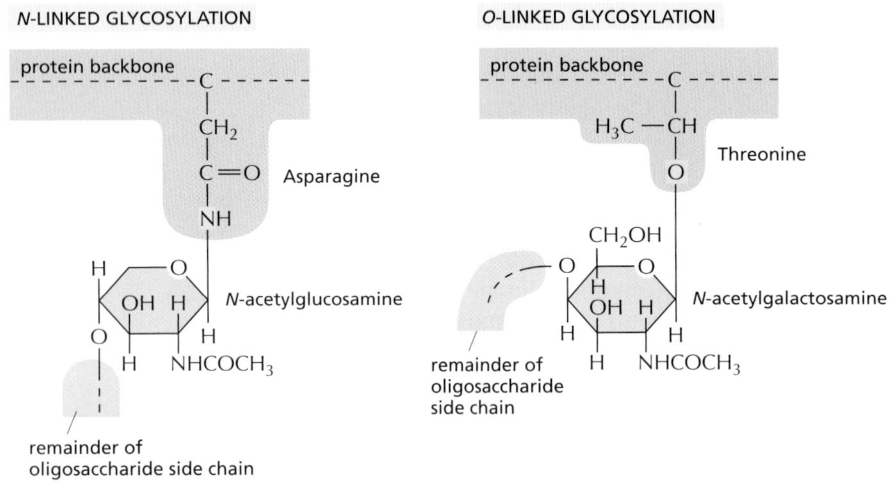

Figure 13–32 N- and O-linked glycosylation. In each case, only the single sugar group that is directly attached to the protein chain is shown.

add sugar residues to a protein one at a time. Usually, *N*-acetylgalactosamine is added first, followed by a variable number of additional sugar residues, ranging from just a few to 10 or more.

The Golgi apparatus confers the heaviest *O*-linked glycosylation of all on *mucins*, the glycoproteins in mucus secretions (see Figure 13–29), and on *proteoglycan core proteins*, which it modifies to produce **proteoglycans.** As discussed in Chapter 19, this process involves the polymerization of one or more *glycosaminoglycan chains* (long unbranched polymers composed of repeating disaccharide units; see Figure 19–58) via a xylose link onto serines on a core protein. Many proteoglycans are secreted and become components of the extracellular matrix, while others remain anchored to the extracellular face of the plasma membrane. Still others form a major component of slimy materials, such as the mucus that is secreted to form a protective coating on the surface of many epithelia.

The sugars incorporated into glycosaminoglycans are heavily sulfated in the Golgi apparatus immediately after these polymers are made, thus adding a significant portion of their characteristically large negative charge. Some tyrosines in proteins also become sulfated shortly before they exit from the Golgi apparatus. In both cases, the sulfation depends on the sulfate donor 3′-phosphoadenosine-5′-phosphosulfate (PAPS) (**Figure 13–33**), which is transported from the cytosol into the lumen of the *trans* Golgi network.

What Is the Purpose of Glycosylation?

There is an important difference between the construction of an oligosaccharide and the synthesis of other macromolecules such as DNA, RNA, and protein. Whereas nucleic acids and proteins are copied from a template in a repeated series of identical steps using the same enzyme or set of enzymes, complex carbohydrates require a different enzyme at each step, each product being recognized as the exclusive substrate for the next enzyme in the series. The vast abundance of glycoproteins and the complicated pathways that have evolved to synthesize them suggest that the oligosaccharides on glycoproteins and glycosphingolipids have very important functions.

N-linked glycosylation, for example, is prevalent in all eucaryotes, including yeasts. *N*-linked oligosaccharides also occur in a very similar form in archaeal cell wall proteins, suggesting that the whole machinery required for their synthesis is evolutionarily ancient. *N*-linked glycosylation promotes protein folding in two ways. First, it has a direct role in making folding intermediates more soluble, thereby preventing their aggregation. Second, the sequential modifications of the *N*-linked oligosaccharide establish a "glyco-code" that marks the progression of protein folding and mediates the binding of the protein to chaperones (discussed in Chapter 12) and lectins—for example, in guiding ER-to-Golgi transport. As we discuss later, lectins also participate in protein sorting in the *trans* Golgi network.

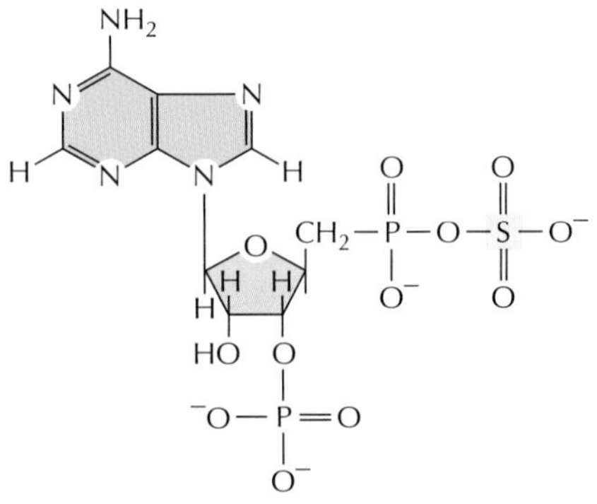

3′-phosphoadenosine-5′-phosphosulfate (PAPS)

Figure 13–33 The structure of PAPS.

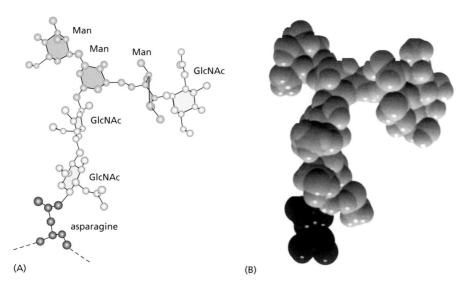

Man

Man   Man

GlcNAc

GlcNAc

GlcNAc

asparagine

(A)   (B)

Figure 13–34 **The three-dimensional structure of a small *N*-linked oligosaccharide.** The structure was determined by x-ray crystallographic analysis of a glycoprotein. This oligosaccharide contains only 6 sugars, whereas there are 14 sugars in the *N*-linked oligosaccharide that is initially transferred to proteins in the ER (see Figures 12–50 and 12–51). (A) A backbone model showing all atoms except hydrogens. (B) A space-filling model, with the asparagine indicated by dark atoms. (B, courtesy of Richard Feldmann.)

Because chains of sugars have limited flexibility, even a small *N*-linked oligosaccharide protruding from the surface of a glycoprotein (**Figure 13–34**) can limit the approach of other macromolecules to the protein surface. In this way, for example, the presence of oligosaccharides tends to make a glycoprotein more resistant to digestion by proteolytic enzymes. It may be that the oligosaccharides on cell-surface proteins originally provided an ancestral cell with a protective coat. Compared to the rigid bacterial cell wall, a mucus coat has the advantage that it leaves the cell with the freedom to change shape and move.

The sugar chains have since become modified to serve other purposes as well. The mucus coat of lung and intestinal cells, for example, protects against many pathogens. The recognition of sugar chains by *lectins* in the extracellular space is important in many developmental processes and in cell–cell recognition: *selectins*, for example, are lectins that function in cell–cell adhesion during lymphocyte migration, as discussed in Chapter 19. The presence of oligosaccharides may modify a protein's antigenic properties, making glycosylation an important factor in the production of proteins for pharmaceutical purposes.

Glycosylation can also have important regulatory roles. Signaling through the cell-surface signaling receptor Notch, for example, determines the cell's fate in development. Notch is a transmembrane protein that is *O*-glycosylated by addition of a single fucose to some serines, threonines, and hydroxylysines. Some cell types express an additional glycosyl transferase that adds an *N*-acetylglucosamine to each of these fucoses in the Golgi apparatus. This addition changes the specificity of the Notch receptor for the cell-surface signal proteins that activate the receptor.

Transport Through the Golgi Apparatus May Occur by Vesicular Transport or by Cisternal Maturation

It is still uncertain how the Golgi apparatus achieves and maintains its polarized structure and how molecules move from one cisterna to another. Functional evidence from *in vitro* transport assays, and the finding of abundant transport vesicles in the vicinity of Golgi cisternae, initially led to the view that these vesicles transport proteins between the cisternae, budding from one cisterna and fusing with the next. According to this **vesicular transport model**, the Golgi apparatus is a relatively static structure, with its enzymes held in place, while the molecules in transit move through the cisternae in sequence, carried by transport vesicles (**Figure 13–35**A). A retrograde flow of vesicles retrieves escaped ER and Golgi proteins and returns them to preceding compartments. Directional flow could be achieved because forward-moving cargo molecules are selectively allowed access to forward-moving vesicles. Although both forward-moving and retrograde types of vesicles are likely to be COPI-coated, the coats may contain different adaptor proteins that confer selectivity on the packaging of cargo

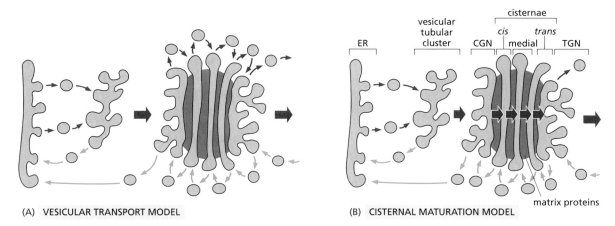

(A) VESICULAR TRANSPORT MODEL   (B) CISTERNAL MATURATION MODEL

molecules. An alternative possibility is that the transport vesicles shuttling between Golgi cisternae are not directional at all, instead transporting cargo material randomly back and forth; directional flow would then occur because of the continual input at the *cis* cisterna and output at the *trans* cisterna.

A different hypothesis, called the **cisternal maturation model**, views the Golgi as a dynamic structure in which the cisternae themselves move. The vesicular tubular clusters that arrive from the ER fuse with one another to become a *cis* Golgi network. According to this model, this network then progressively matures to become a *cis* cisterna, then a medial cisterna, and so on. Thus, at the *cis* face of a Golgi stack, new *cis* cisternae would continually form and then migrate through the stack as they mature (Figure 13–35B). This model is supported by microscopic observations demonstrating that large structures such as collagen rods in fibroblasts and scales in certain algae—which are much too large to fit into classical transport vesicles—move progressively through the Golgi stack.

In the cisternal maturation model, retrograde flow explains the characteristic distribution of Golgi enzymes. Everything moves continuously forward with the maturing cisterna, including the processing enzymes that belong in the early Golgi apparatus. But budding COPI-coated vesicles continually collect the appropriate enzymes, almost all of which are membrane proteins, and carry them back to the earlier cisterna where they function. A newly formed *cis* cisterna would therefore receive its normal complement of resident enzymes primarily from the cisterna just ahead of it and would later pass those enzymes back to the next *cis* cisterna that forms.

As we discuss later, when a cisterna finally moves forward to become part of the *trans* Golgi network, various types of coated vesicles bud off it until this network disappears, to be replaced by a maturing cisterna just behind. At the same time, other transport vesicles are continually retrieving membrane from post-Golgi compartments and returning this membrane to the *trans* Golgi network.

The vesicular transport and the cisternal maturation models are not mutually exclusive. Indeed, evidence suggests that transport may occur by a combination of the two mechanisms, in which some cargo is moved forward rapidly in transport vesicles, whereas other cargo is moved forward more slowly as the Golgi apparatus constantly renews itself through cisternal maturation.

Golgi Matrix Proteins Help Organize the Stack

The unique architecture of the Golgi apparatus depends on both the microtubule cytoskeleton, as already mentioned, and cytoplasmic Golgi matrix proteins, which form a scaffold between adjacent cisternae and give the Golgi stack its structural integrity. Some of the matrix proteins form long, filamentous tethers, which are thought to help retain Golgi transport vesicles close to the organelle. When the cell prepares to divide, mitotic protein kinases phosphorylate the Golgi matrix proteins, causing the Golgi apparatus to fragment and disperse throughout the cytosol. The Golgi fragments are then distributed evenly to

Figure 13–35 Two possible models explaining the organization of the Golgi apparatus and the transport of proteins from one cisterna to the next. It is likely that the transport through the Golgi apparatus in the forward direction *(red arrows)* involves elements of both models. (A) In the vesicular transport model, Golgi cisternae are static organelles, which contain a characteristic complement of resident enzymes. The passing of molecules from *cis* to *trans* through the Golgi is accomplished by forward-moving transport vesicles, which bud from one cisterna and fuse with the next in a *cis*-to-*trans* direction. (B) According to the alternative cisternal maturation model, each Golgi cisterna matures as it migrates outward through the stack. At each stage, the Golgi resident proteins that are carried forward in a cisterna are moved backward to an earlier compartment in COPI-coated vesicles. When a newly formed cisterna moves to a medial position, for example, "left-over" *cis* Golgi enzymes would be extracted and transported retrogradely to a new *cis* cisterna behind. Likewise, the medial enzymes would be received by retrograde transport from the cisternae just ahead. In this way, a *cis* cisterna would mature to a medial cisterna as it moves outward.

the two daughter cells, where the matrix proteins are dephosphorylated, leading to the reassembly of the Golgi apparatus.

Remarkably, the Golgi matrix proteins can assemble into appropriately localized stacks near the centrosome even when Golgi membrane proteins are experimentally prevented from leaving the ER. This observation suggests that the matrix proteins are largely responsible for both the structure and location of the Golgi apparatus.

Summary

Correctly folded and assembled proteins in the ER are packaged into COPII-coated transport vesicles that pinch off from the ER membrane. Shortly thereafter, the vesicles shed their coat and fuse with one another to form vesicular tubular clusters. The clusters then move on microtubule tracks to the Golgi apparatus, where they fuse with one another to form the cis-Golgi network. Any resident ER proteins that escape from the ER are returned there from the vesicular tubular clusters and Golgi apparatus by retrograde transport in COPI-coated vesicles.

The Golgi apparatus, unlike the ER, contains many sugar nucleotides, which glycosyl transferase enzymes use to perform glycosylation reactions on lipid and protein molecules as they pass through the Golgi apparatus. The mannoses on the N-linked oligosaccharides that are added to proteins in the ER are often initially removed, and further sugars are added. Moreover, the Golgi apparatus is the site where O-linked glycosylation occurs and where glycosaminoglycan chains are added to core proteins to form proteoglycans. Sulfation of the sugars in proteoglycans and of selected tyrosines on proteins also occurs in a late Golgi compartment.

The Golgi apparatus modifies the many proteins and lipids that it receives from the ER and then distributes them to the plasma membrane, lysosomes, and secretory vesicles. The Golgi apparatus is a polarized organelle, consisting of one or more stacks of disc-shaped cisternae, each stack organized as a series of at least three functionally distinct compartments, termed cis, medial, *and* trans *cisternae. The* cis *and* trans *cisternae are both connected to special sorting stations, called the* cis *Golgi network and the* trans *Golgi network, respectively. Proteins and lipids move through the Golgi stack in the* cis-to-trans *direction. This movement may occur by vesicular transport, by progressive maturation of the* cis *cisternae as they migrate continuously through the stack, or, most likely, by a combination of these two mechanisms. Continual retrograde vesicular transport from more distal cisternae is thought to keep the enzymes concentrated in the cisternae where they are needed. The finished new proteins end up in the* trans *Golgi network, which packages them in transport vesicles and dispatches them to their specific destinations in the cell.*

TRANSPORT FROM THE *TRANS* GOLGI NETWORK TO LYSOSOMES

The *trans* Golgi network sorts all of the proteins that pass through the Golgi apparatus (except those that are retained there as permanent residents) according to their final destination. The sorting mechanism is especially well understood for those proteins destined for the lumen of lysosomes, and in this section we consider this selective transport process. We begin with a brief account of lysosome structure and function.

Lysosomes Are the Principal Sites of Intracellular Digestion

Lysosomes are membrane-enclosed compartments filled with soluble hydrolytic enzymes that control intracellular digestion of macromolecules. Lysosomes contain about 40 types of hydrolytic enzymes, including proteases, nucleases, glycosidases, lipases, phospholipases, phosphatases, and sulfatases. All are **acid hydrolases**. For optimal activity, they need to be activated by

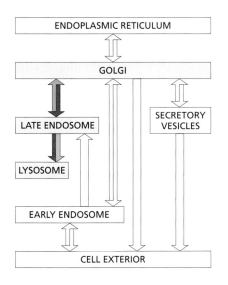

proteolytic cleavage and require an acid environment, which the lysosome provides by maintaining a pH of about 4.5–5.0 in its interior. By this arrangement, the contents of the cytosol are doubly protected against attack by the cell's own digestive system: the membrane of the lysosome keeps the digestive enzymes out of the cytosol, but even if they leak out, they can do little damage at the cytosolic pH of about 7.2.

Like all other intracellular organelles, the lysosome not only contains a unique collection of enzymes, but also has a unique surrounding membrane. Most of the lysosomal membrane proteins, for example, are unusually highly glycosylated, which helps to protect them from the lysosomal proteases in the lumen. Transport proteins in the lysosomal membrane carry the final products of the digestion of macromolecules—such as amino acids, sugars, and nucleotides—to the cytosol, where the cell can either reuse or excrete them.

A *vacuolar H^+ ATPase* in the lysosomal membrane uses the energy of ATP hydrolysis to pump H^+ into the lysosome, thereby maintaining the lumen at its acidic pH (**Figure 13–36**). The lysosomal H^+ pump belongs to the family of *V-type ATPases* and has a similar architecture to the mitochondrial and chloroplast ATP synthases (F-type ATPases), which convert the energy stored in H^+ gradients into ATP (see Figure 11–12). By contrast to these enzymes, however, the vacuolar H^+ ATPase exclusively works in reverse, pumping H^+ into the organelle. Similar or identical V-type ATPases acidify all endocytic and exocytic organelles, including lysosomes, endosomes, selected compartments of the Golgi apparatus, and many transport and secretory vesicles. In addition to providing a low-pH environment that is suitable for reactions occurring in the organelle lumen, the H^+ gradient provides a source of energy that drives the transport of small metabolites across the organelle membrane.

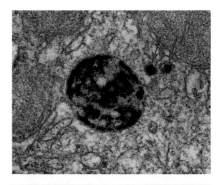

Figure 13–36 Lysosomes. The acid hydrolases are hydrolytic enzymes that are active under acidic conditions. A V-type ATPase in the membrane pumps H^+ into the lysosome, maintaining its lumen at an acidic pH.

Lysosomes Are Heterogeneous

Lysosomes were initially discovered by the biochemical fractionation of cell extracts; only later were they seen clearly in the electron microscope. Although extraordinarily diverse in shape and size, staining them with specific antibodies shows they are members of a single family of organelles. They can also be identified by histochemistry, using the precipitate produced by the action of an acid hydrolase on its substrate to indicate which organelles contain the hydrolase (**Figure 13–37**). By this criterion, lysosomes are found in all eucaryotic cells.

The heterogeneity of lysosomal morphology contrasts with the relatively uniform structures of most other cell organelles. The diversity reflects the wide variety of digestive functions that acid hydrolases mediate, including the breakdown of intra- and extracellular debris, the destruction of phagocytosed microorganisms, and the production of nutrients for the cell. The diversity of lysosomal morphology, however, also reflects the way lysosomes form: late endosomes contain material received from both the plasma membrane by endocytosis and newly synthesized lysosomal hydrolases, and they therefore already bear a resemblance to lysosomes. Late endosomes fuse with preexisting lysosomes to form structures that are sometimes referred to as *endolysosomes,* which then fuse with one another (**Figure 13–38**). When the majority of the endocytosed material within an endolysosome has been digested so that only resistant or slowly digestable residues remain, these organelles become "classical" lysosomes. These are relatively dense, round, and small, but they can enter

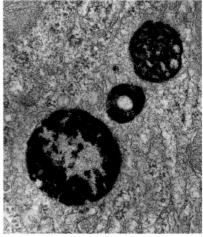

Figure 13–37 Histochemical visualization of lysosomes. These electron micrographs show two sections of a cell stained to reveal the location of acid phosphatase, a marker enzyme for lysosomes. The larger membrane-enclosed organelles, containing dense precipitates of lead phosphate, are lysosomes. Their diverse morphology reflects variations in the amount and nature of the material they are digesting. The precipitates are produced when tissue fixed with glutaraldehyde (to fix the enzyme in place) is incubated with a phosphatase substrate in the presence of lead ions. *Red arrows* in the top panel indicate two small vesicles thought to be carrying acid hydrolases from the Golgi apparatus. (Courtesy of Daniel S. Friend.)

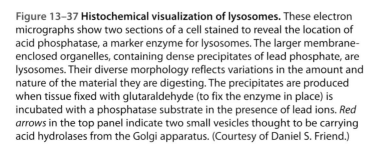

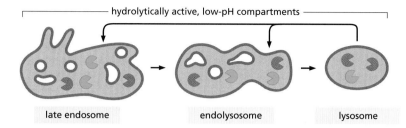

hydrolytically active, low-pH compartments

late endosome   endolysosome   lysosome

Figure 13–38 A model for lysosome maturation. The heterogeneity of lysosomal morphology reflects, in part, the different nature of the materials delivered to the organelle, as well as the different stages in the maturation cycle shown here.

the cycle again by fusing with late endosomes or endolysosomes. Thus, there is no real distinction between late endosomes and lysosomes: they are the same except that they are in different stages of a maturation cycle. For this reason, lysosomes are sometimes viewed as a heterogeneous collection of distinct organelles, the common feature of which is a high content of hydrolytic enzymes. It is especially hard to apply a narrower definition than this in plant cells, as we discuss next.

Plant and Fungal Vacuoles Are Remarkably Versatile Lysosomes

Most plant and fungal cells (including yeasts) contain one or several very large, fluid-filled vesicles called **vacuoles**. They typically occupy more than 30% of the cell volume, and as much as 90% in some cell types (**Figure 13–39**). Vacuoles are related to animal cell lysosomes and contain a variety of hydrolytic enzymes, but their functions are remarkably diverse. The plant vacuole can act as a storage organelle for both nutrients and waste products, as a degradative compartment, as an economical way of increasing cell size (**Figure 13–40**), and as a controller of *turgor pressure* (the osmotic pressure that pushes outward on the cell wall and keeps the plant from wilting). The same cell may have different vacuoles with distinct functions, such as digestion and storage.

The vacuole is important as a homeostatic device, enabling plant cells to withstand wide variations in their environment. When the pH in the environment drops, for example, the flux of H^+ into the cytosol is balanced, at least in part, by an increased transport of H^+ into the vacuole, which tends to keep the pH in the cytosol constant. Similarly, many plant cells maintain an almost constant turgor pressure despite large changes in the tonicity of the fluid in their

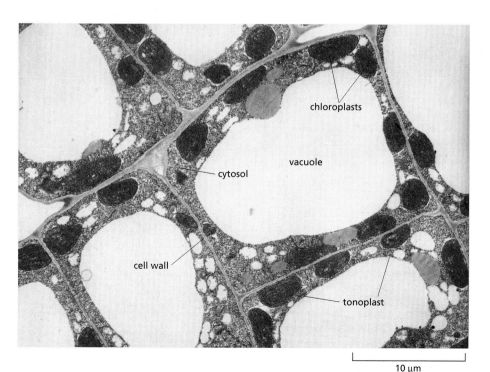

chloroplasts

vacuole

cytosol

cell wall

tonoplast

10 μm

Figure 13–39 The plant cell vacuole. This electron micrograph of cells in a young tobacco leaf shows the cytosol as a thin layer, containing chloroplasts, pressed against the cell wall by the enormous vacuole. The vacuole membrane is called the tonoplast. (Courtesy of J. Burgess.)

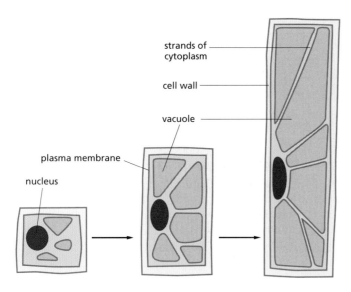

strands of cytoplasm

cell wall

vacuole

plasma membrane

nucleus

Figure 13–40 The role of the vacuole in controlling the size of plant cells. A plant cell can achieve a large increase in volume without increasing the volume of the cytosol. Localized weakening of the cell wall orients a turgor-driven cell enlargement that accompanies the uptake of water into an expanding vacuole. The cytosol is eventually confined to a thin peripheral layer, which is connected to the nuclear region by strands of cytoplasm stabilized by bundles of actin filaments (not shown).

immediate environment. They do so by changing the osmotic pressure of the cytosol and vacuole—in part by the controlled breakdown and resynthesis of polymers such as polyphosphate in the vacuole, and in part by altering the transport rates of sugars, amino acids, and other metabolites across the plasma membrane and the vacuolar membrane. The turgor pressure regulates the activities of distinct transporters in each membrane to control these fluxes.

Humans often harvest substances stored in plant vacuoles. In different species, these range from rubber to opium to the flavoring of garlic. Many stored products have a metabolic function. Proteins, for example, can be preserved for years in the vacuoles of the storage cells of many seeds, such as those of peas and beans. When the seeds germinate, these proteins are hydrolyzed and the resulting amino acids provide a food supply for the developing embryo. Anthocyanin pigments stored in vacuoles color the petals of many flowers so as to attract pollinating insects, while noxious molecules released from vacuoles when a plant is eaten or damaged provide a defense against predators.

Multiple Pathways Deliver Materials to Lysosomes

Lysosomes are usually meeting places where several streams of intracellular traffic converge. A route that leads outwards from the ER via the Golgi apparatus delivers most digestive enzymes, while at least three paths from different sources feed substances into lysosomes for digestion.

The best studied of these degradation paths in lysosomes is the one followed by macromolecules taken up from extracellular fluid by *endocytosis*. As discussed in detail later, endocytosed molecules are initially delivered in vesicles to small, irregularly shaped intracellular organelles called *early endosomes*. Here endocytosed materials first meet the lysosomal hydrolases, which are delivered to the endosome from the Golgi apparatus. Some of the ingested molecules are selectively retrieved and recycled to the plasma membrane, while others pass on into *late endosomes*. The interior of the late endosomes is mildly acidic (pH ~6), and it is the site where the hydrolytic digestion of the endocytosed molecules begins. As discussed above, mature lysosomes form by a maturation process from late endosomes, accompanied by a further decrease in internal pH. As lysosomes mature, endosomal membrane proteins are selectively retrieved from the developing lysosome by transport vesicles that deliver these proteins back to endosomes or the TGN.

All cell types use a second degradation pathway in lysosomes to dispose of obsolete parts of the cell itself—a process called **autophagy**. In a liver cell, for example, an average mitochondrion has a lifetime of about 10 days, and electron microscopic images of normal cells reveal lysosomes containing (and presumably digesting) mitochondria, as well as other organelles. The process seems to begin with the enclosure of an organelle by a double membrane of unknown origin, creating an *autophagosome*, which then fuses with a lysosome (or a late endosome). The process is highly regulated, and selected cell components can somehow be marked for lysosomal destruction during cell remodeling. For example, the smooth ER that proliferates in a liver cell in a detoxification response to lipid-soluble drugs such as phenobarbital (discussed in Chapter 12) is selectively removed by autophagy after the drug is withdrawn.

Similarly, other obsolete organelles, including senescent peroxisomes or mitochondria, can be selectively targeted for degradation by autophagy. Under starvation conditions, large portions of the cytosol are nonselectively captured into autophagosomes. Metabolites derived from the digestion of the captured material help the cell survive when external nutrients are limiting.

In addition to maintaining basic cell functions in balance and helping to dispose of obsolete parts, autophagy also has a role in development and health. It helps restructure differentiating cells by disposing of no longer needed parts and helps defend against invading viruses and bacteria. Autophagy is uniquely suited as a mechanism that can remove whole organelles or large protein aggregates, which other mechanisms such as proteasomal degradation cannot handle.

We still know very little about the events that lead to the formation of autophagosomes, or how the process of autophagy is controlled and targeted at specific organelles. More than 25 different proteins have been identified in yeast and animal cells that participate in the process. Autophagy can be divided into four general steps: (1) nucleation and extension of a delimiting membrane into a crescent-shaped structure that engulfs a portion of the cytoplasm, (2) closure of the autophagosome into a sealed double-membrane-bounded compartment, (3) fusion of the new compartment with lysosomes, and (4) digestion of the inner membrane of the autophagosome and its contents (**Figure 13–41**). Many mysteries remain to be solved, including identifying the membrane system from which the vesicles that form the autophagosomal envelope derive, and how some target organelles can be enclosed so selectively.

As we discuss later, the third pathway that brings materials to lysosomes for degradation is found mainly in cells specialized for the *phagocytosis* of large particles and microorganisms. Such professional phagocytes (macrophages and neutrophils in vertebrates) engulf objects to form a *phagosome*, which is then converted to a lysosome in the manner described for the autophagosome. **Figure 13–42** summarizes the three pathways.

A Mannose 6-Phosphate Receptor Recognizes Lysosomal Proteins in the *Trans* Golgi Network

We now consider the pathway that delivers lysosomal hydrolases and membrane proteins to lysosomes. Both classes of proteins are co-translationally transported into the rough ER and then transported through the Golgi apparatus to the TGN. The transport vesicles that deliver these proteins to endosomes (from where the proteins are moved on to lysosomes) bud from the TGN. The vesicles incorporate the lysosomal proteins and exclude the many other proteins being packaged into different transport vesicles for delivery elsewhere.

Figure 13–41 A model of autophagy. After a nucleation event in the cytoplasm, a crescent of autophagosomal membrane grows by fusion of vesicles of unknown origin that extend its edges. Eventually, a membrane fusion event closes the autophagosome, sequestering a portion of the cytoplasm of the cell in a double membrane. The autophagosome then fuses with lysosomes containing acid hydrolases that digest its content.

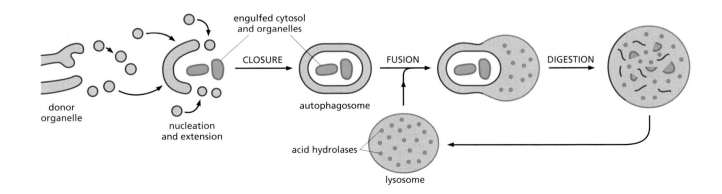

(A)

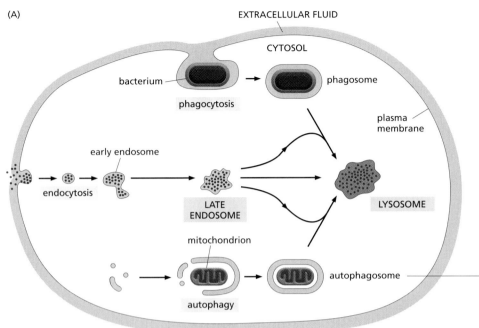

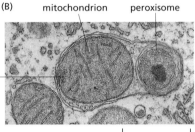

Figure 13–42 **Three pathways to degradation in lysosomes.** (A) Materials in each pathway are derived from a different source. Note that the autophagosome has a double membrane. (B) An electron micrograph of an autophagosome containing a mitochondrion and a peroxisome. (B, courtesy of Daniel S. Friend, from D.W. Fawcett, A Textbook of Histology, 12th ed. New York: Chapman and Hall, 1994. With permission from Kluwer.)

How are lysosomal proteins recognized and selected in the TGN with the required accuracy? We know the answer for the lysosomal hydrolases. They carry a unique marker in the form of *mannose 6-phosphate (M6P)* groups, which are added exclusively to the *N*-linked oligosaccharides of these soluble lysosomal enzymes as they pass through the lumen of the *cis* Golgi network (**Figure 13–43**). Transmembrane **M6P receptor proteins**, which are present in the TGN, recognize the M6P groups. The receptor proteins bind to lysosomal hydrolases on the lumenal side of the membrane and to adaptor proteins in assembling clathrin coats on the cytosolic side. In this way, they help package the hydrolases into clathrin-coated vesicles that bud from the TGN. The vesicles shed their coat and deliver their contents to early endosomes.

The M6P Receptor Shuttles Between Specific Membranes

The M6P receptor protein binds its specific oligosaccharide at pH 6.5–6.7 in the TGN and releases it at pH 6, which is the pH in the interior of late endosomes. Thus, as the pH drops during endosomal maturation, the lysosomal hydrolases dissociate from the M6P receptor and eventually begin to digest the material delivered by endocytosis. An acid phosphatase removes the phosphate group from the mannose, thereby destroying the sorting signal and contributing to the release of the lysosomal hydrolases from the M6P receptor. Having released their bound enzymes, the M6P receptors are retrieved into retromer-coated transport vesicles that bud from endosomes; the receptors are then returned to the membrane of the TGN for reuse (**Figure 13–44**). Transport in either direction requires signals in the cytoplasmic tail of the M6P receptor that direct this protein to the endosome or back to the Golgi apparatus. These signals are recognized by the retromer complex (see Figure 13–9) that recruits M6P receptors into vesicles in endosomes. The recycling of the M6P receptor resembles the recycling of the KDEL receptor discussed earlier, although it differs in the type of coated vesicles that mediate the transport.

Not all the hydrolase molecules that are tagged with M6P get to lysosomes. Some escape the normal packaging process in the *trans* Golgi network and are transported "by default" to the cell surface, where they are secreted into the extracellular fluid. Some M6P receptors, however, also take a detour to the plasma membrane, where they recapture the escaped lysosomal hydrolases and return them by *receptor-mediated endocytosis* to lysosomes via early and late endosomes. As lysosomal hydrolases require an acidic milieu to work, they can do little harm in the extracellular fluid, which usually has a neutral pH of 7.4.

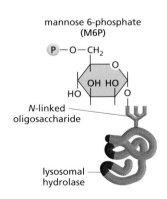

Figure 13–43 **The structure of mannose 6-phosphate on a lysosomal hydrolase.**

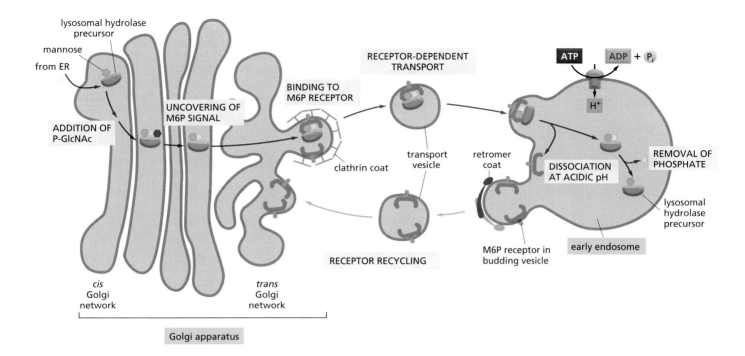

Figure 13–44 The transport of newly
synthesized lysosomal hydrolases to
lysosomes. The sequential action of two
enzymes in the *cis* and *trans* Golgi
network adds mannose 6-phosphate
(M6P) groups to the precursors of
lysosomal enzymes (see Figure 13–45).
They then segregate from all other types
of proteins in the TGN because
monomeric adaptor proteins in the
clathrin coat bind the M6P receptors,
which, in turn, bind the modified
lysosomal hydrolases. The clathrin-coated
vesicles bud off from the TGN, shed their
coat, and fuse with early endosomes. At
the lower pH of the endosome, the
hydrolases dissociate from the M6P
receptors, and the empty receptors are
recycled in retromer-coated vesicles to
the Golgi apparatus for further rounds of
transport. In the endosomes, the
phosphate is removed from the mannose
sugars attached to the hydrolases, further
ensuring that the hydrolases do not
return to the Golgi apparatus with the
receptor.

A Signal Patch in the Hydrolase Polypeptide Chain Provides the Cue for M6P Addition

The sorting system that segregates lysosomal hydrolases and dispatches them to endosomes works because M6P groups are added only to the appropriate glycoproteins in the Golgi apparatus. This requires specific recognition of the hydrolases by the Golgi enzymes responsible for adding M6P. Since all glycoproteins leave the ER with identical *N*-linked oligosaccharide chains, the signal for adding the M6P units to oligosaccharides must reside somewhere in the polypeptide chain of each hydrolase. Genetic engineering experiments have revealed that the recognition signal is a cluster of neighboring amino acids on each protein's surface, known as a *signal patch*.

Two enzymes act sequentially to catalyze the addition of M6P groups to lysosomal hydrolases. The first is a GlcNAc phosphotransferase in the *cis* Golgi that specifically binds the hydrolase and adds GlcNAc-phosphate to one or two of the mannose residues on each oligosaccharide chain (**Figure 13–45**). A second enzyme in the *trans* Golgi then cleaves off the GlcNAc residue, leaving behind a newly created M6P marker. Since most lysosomal hydrolases contain multiple oligosaccharides, they acquire many M6P residues, providing a high-affinity signal for the M6P receptor.

Defects in the GlcNAc Phosphotransferase Cause a Lysosomal Storage Disease in Humans

Genetic defects that affect one or more of the lysosomal hydrolases cause a number of human **lysosomal storage diseases**. The defects result in an accumulation of undigested substrates in lysosomes, with severe pathological consequences, most often in the nervous system. In most cases, there is a mutation in a structural gene that codes for an individual lysosomal hydrolase. This occurs in *Hurler's disease*, for example, in which the enzyme required for the breakdown of certain types of glycosaminoglycan chains is defective or missing. The most severe form of lysosomal storage disease, however, is a very rare disorder called *inclusion-cell disease* (*I-cell disease*). In this condition, almost all of the hydrolytic enzymes are missing from the lysosomes of fibroblasts, and their undigested substrates accumulate in lysosomes, which consequently form large *inclusions* in the patients' cells.

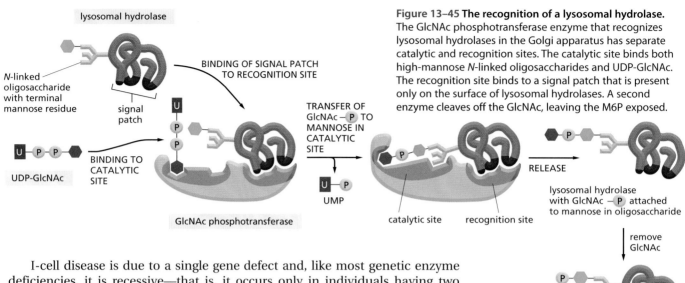

Figure 13–45 The recognition of a lysosomal hydrolase. The GlcNAc phosphotransferase enzyme that recognizes lysosomal hydrolases in the Golgi apparatus has separate catalytic and recognition sites. The catalytic site binds both high-mannose N-linked oligosaccharides and UDP-GlcNAc. The recognition site binds to a signal patch that is present only on the surface of lysosomal hydrolases. A second enzyme cleaves off the GlcNAc, leaving the M6P exposed.

I-cell disease is due to a single gene defect and, like most genetic enzyme deficiencies, it is recessive—that is, it occurs only in individuals having two copies of the defective gene. In patients with I-cell disease, all the hydrolases missing from lysosomes are found in the blood. Because they fail to sort properly in the Golgi apparatus, the hydrolases are secreted rather than transported to lysosomes. The missorting has been traced to a defective or missing GlcNAc-phosphotransferase. Because lysosomal enzymes are not phosphorylated in the *cis* Golgi network, the M6P receptors do not segregate them into the appropriate transport vesicles in the TGN. Instead, the lysosomal hydrolases are carried to the cell surface and secreted by a default pathway.

In I-cell disease, the lysosomes in some cell types, such as hepatocytes, contain a normal complement of lysosomal enzymes, implying that there is another pathway for directing hydrolases to lysosomes that is used by some cell types but not others. The nature of this M6P-independent pathway is unknown. Similarly, an M6P-independent pathway in all cells sorts the membrane proteins of lysosomes from the TGN for transport to late endosomes, and those proteins are therefore normal in I-cell disease. These membrane proteins exit from the TGN in clathrin-coated vesicles that are distinct from those that transport the M6P-tagged hydrolases and use different adaptor proteins.

It is unclear why cells need more than one sorting pathway to construct lysosomes, although it is perhaps not surprising that different mechanisms operate for soluble and membrane-bound lysosomal proteins, especially since—unlike M6P receptors—those membrane proteins are lysosomal residents and need not be returned to the TGN.

Some Lysosomes Undergo Exocytosis

Targeting of material to lysosomes is not necessarily the end of the pathway. *Lysosomal secretion* of undigested content enables all cells to eliminate undigestible debris. For most cells, this seems to be a minor pathway, used only when the cells are stressed. Some cell types, however, contain specialized lysosomes that have acquired the necessary machinery for fusion with the plasma membrane. *Melanocytes* in the skin, for example, produce and store pigments in their lysosomes. These pigment-containing *melanosomes* release their pigment into the extracellular space of the epidermis by exocytosis. The pigment is then taken up by keratinocytes, leading to normal skin pigmentation. In some genetic disorders, defects in melanosome exocytosis block this transfer process, leading to forms of hypopigmentation (albinism).

Summary

Lysosomes are specialized for the intracellular digestion of macromolecules. They contain unique membrane proteins and a wide variety of soluble hydrolytic enzymes that operate best at pH 5, which is the internal pH of lysosomes. An ATP-driven H⁺ pump in the lysosomal membrane maintains this low pH. Newly synthesized lysosomal proteins are transferred into the lumen of the ER, transported through the Golgi apparatus, and then carried from the trans *Golgi network to late endosomes by means of clathrin-coated transport vesicles.*

The lysosomal hydrolases contain N-linked oligosaccharides that are covalently modified in a unique way in the cis *Golgi network so that their mannose residues are phosphorylated. These mannose 6-phosphate (M6P) groups are recognized by an M6P receptor protein in the* trans *Golgi network that segregates the hydrolases and helps package them into budding transport vesicles that deliver their contents to endosomes. The M6P receptors shuttle back and forth between the* trans *Golgi network and these endosomes. The low pH in endosomes and removal of the phosphate from the M6P group cause the lysosomal hydrolases to dissociate from these receptors, making the transport of the hydrolases unidirectional. A separate transport system uses clathrin-coated vesicles to deliver resident lysosomal membrane proteins from the* trans *Golgi network.*

TRANSPORT INTO THE CELL FROM THE PLASMA MEMBRANE: ENDOCYTOSIS

The routes that lead inward from the cell surface to lysosomes start with the process of **endocytosis**, by which cells take up macromolecules, particulate substances, and, in specialized cases, even other cells. In this process, the material to be ingested is progressively enclosed by a small portion of the plasma membrane, which first invaginates and then pinches off to form an *endocytic vesicle* containing the ingested substance or particle. Two main types of endocytosis differ according to the size of the endocytic vesicles formed. In *phagocytosis* ("cell eating"), large particles are ingested via large vesicles called *phagosomes* (generally >250 nm in diameter). In *pinocytosis* ("cell drinking"), fluid and solutes are ingested via small *pinocytic vesicles* (about 100 nm in diameter). Most eucaryotic cells are continually ingesting fluid and solutes by pinocytosis; large particles are ingested most efficiently by specialized phagocytic cells.

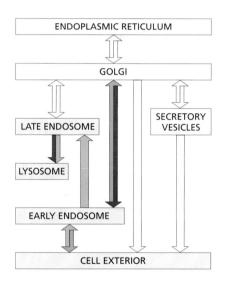

Specialized Phagocytic Cells Can Ingest Large Particles

Phagocytosis is a special form of endocytosis in which a cell uses large endocytic vesicles called **phagosomes** to ingest large particles such as microorganisms and dead cells. <TCAT> In protozoa, phagocytosis is a form of feeding: large particles taken up into phagosomes end up in lysosomes, and the products of the subsequent digestive processes pass into the cytosol to be used as food. However, few cells in multicellular organisms are able to ingest such large particles efficiently. In the gut of animals, for example, extracellular processes break down food particles, and cells import the small hydrolysis products.

Phagocytosis is important in most animals for purposes other than nutrition, and it is carried out mainly by specialized cells—so-called *professional phagocytes*. In mammals, two classes of white blood cells act as professional phagocytes—**macrophages** and **neutrophils**. These cells develop from hemopoietic stem cells (discussed in Chapter 23), and they ingest invading microorganisms to defend us against infection. Macrophages also have an important role in scavenging senescent cells and cells that have died by apoptosis (discussed in Chapter 18). In quantitative terms, the clearance of senescent and dead cells is by far the most important: our macrophages phagocytose more than 10¹¹ senescent red blood cells in each of us every day, for example.

Whereas the endocytic vesicles involved in pinocytosis are small and uniform, phagosomes have diameters that are determined by the size of the ingested particle, and they can be almost as large as the phagocytic cell itself (**Figure 13–46**). The phagosomes fuse with lysosomes inside the cell, and the ingested material is then degraded. Any indigestible substances will remain in lysosomes, forming *residual bodies*, which can be excreted from cells by exocytosis, as we have previously discussed. Some of the internalized plasma membrane components never reach the lysosome, because they are retrieved from the phagosome in transport vesicles and returned to the plasma membrane.

To be phagocytosed, particles must first bind to the surface of the phagocyte. Not all particles that bind are ingested, however. Phagocytes have a variety of specialized surface receptors that are functionally linked to the phagocytic machinery of the cell. Phagocytosis is a triggered process. That is, it requires the activation of receptors that transmit signals to the cell interior and initiate the response. By contrast, pinocytosis is a constitutive process. It occurs continuously, regardless of the needs of the cell. The best-characterized triggers of phagocytosis are antibodies, which protect us by binding to the surface of infectious microorganisms to form a coat that exposes the tail region on the exterior of each antibody molecule. This tail region is called the Fc region (discussed in Chapter 25). The antibody coat is recognized by specific *Fc receptors* on the surface of macrophages and neutrophils, whose binding induces the phagocytic cell to extend pseudopods that engulf the particle and fuse at their tips to form a phagosome (**Figure 13–47**A). Localized actin polymerization, initiated by Rho-family GTPases and their activating Rho-GEFs (discussed in Chapters 15 and 16), shapes the pseudopods. An active Rho GTPase switches on the kinase activity of local PI kinases, and initial actin polymerization occurs in response to an accumulation of $PI(4,5)P_2$ in the membrane (see Figure 13–11). To seal off the phagosome and complete its engulfment, actin is depolymerized at its base as $PI(4,5)P_2$ is subjected to a PI 3-kinase, which converts it to $PI(3,4,5)P_3$. $PI(3,4,5)P_3$ is required for closure of the phagosome and may also contribute to reshaping the actin network to help drive the invagination of the forming phagosome (Figure 13–47B). In this way, the ordered generation and consumption of specific phosphoinositides guides sequential steps in phagosome formation.

Several other classes of receptors that promote phagocytosis have been characterized. Some recognize *complement* components, which collaborate with antibodies in targeting microbes for destruction (discussed in Chapter 24). Others directly recognize oligosaccharides on the surface of certain microorganisms. Still others recognize cells that have died by apoptosis. Apoptotic cells lose the asymmetric distribution of phospholipids in their plasma membrane. As a consequence, negatively charged phosphatidylserine, which is normally confined to the cytosolic leaflet of the lipid bilayer, is now exposed on the outside of the cell, where it helps to trigger the phagocytosis of the dead cell.

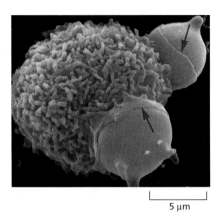

⊢——————⊣
5 μm

Figure 13–46 Phagocytosis by a macrophage. A scanning electron micrograph of a mouse macrophage phagocytosing two chemically altered red blood cells. The *red arrows* point to edges of thin processes (pseudopods) of the macrophage that are extending as collars to engulf the red cells. (Courtesy of Jean Paul Revel.)

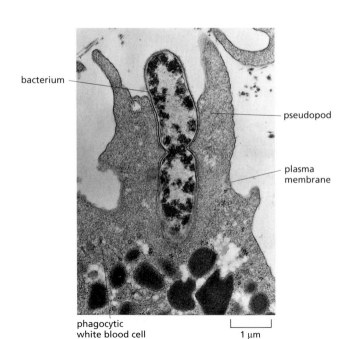

phagocytic
white blood cell
(A)

⊢————⊣
1 μm

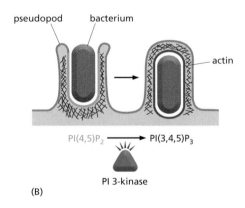

(B)

Figure 13–47 A neutrophil reshaping a plasma membrane during phagocytosis. (A) An electron micrograph of a neutrophil phagocytosing a bacterium, which is in the process of dividing. (B) Pseudopod extension and phagosome formation are driven by actin polymerization and reorganization, which respond to the accumulation of specific phosphoinositides in the membrane of the forming phagosome. (A, courtesy of Dorothy F. Bainton, Phagocytic Mechanisms in Health and Disease. New York: Intercontinental Medical Book Corporation, 1971.)

Remarkably, macrophages will also phagocytose a variety of inanimate particles—such as glass or latex beads and asbestos fibers—yet they do not phagocytose live animal cells. Living animal cells seem to display "don't-eat-me" signals in the form of cell-surface proteins that bind to inhibiting receptors on the surface of macrophages. The inhibitory receptors recruit tyrosine phosphatases that antagonize the intracellular signaling events required to initiate phagocytosis, thereby locally inhibiting the phagocytic process. Thus phagocytosis, like many other cell processes, depends on a balance between positive signals that activate the process and negative signals that inhibit it. Apoptotic cells are thought both to gain "eat-me" signals (such as extracellularly exposed phosphatidylserine) and to lose their "don't-eat-me" signals, causing them to be very rapidly phagocytosed by macrophages.

Pinocytic Vesicles Form from Coated Pits in the Plasma Membrane

Virtually all eucaryotic cells continually ingest bits of their plasma membrane in the form of small pinocytic (endocytic) vesicles, which are later returned to the cell surface. The rate at which plasma membrane is internalized in this process of **pinocytosis** varies between cell types, but it is usually surprisingly large. A macrophage, for example, ingests 25% of its own volume of fluid each hour. This means that it must ingest 3% of its plasma membrane each minute, or 100% in about half an hour. Fibroblasts endocytose at a somewhat lower rate (1% of their plasma membrane per minute), whereas some amoebae ingest their plasma membrane even more rapidly. Since a cell's surface area and volume remain unchanged during this process, it is clear that the same amount of membrane being removed by endocytosis is being added to the cell surface by the converse process of *exocytosis*. In this sense, endocytosis and exocytosis are linked processes that can be considered to constitute an *endocytic–exocytic cycle*. The coupling between exocytosis and endocytosis is particularly strict in specialized structures characterized by high membrane turnover, such as the neuronal synapse.

The endocytic part of the cycle often begins at **clathrin-coated pits**. These specialized regions typically occupy about 2% of the total plasma membrane area. The lifetime of a clathrin-coated pit is short: within a minute or so of being formed, it invaginates into the cell and pinches off to form a clathrin-coated vesicle (**Figure 13–48**). It has been estimated that about 2500 clathrin-coated vesicles leave the plasma membrane of a cultured fibroblast every minute. The coated vesicles are even more transient than the coated pits: within seconds of being formed, they shed their coat and are able to fuse with early endosomes. Since extracellular fluid is trapped in clathrin-coated pits as they invaginate to

Figure 13–48 The formation of clathrin-coated vesicles from the plasma membrane. These electron micrographs illustrate the probable sequence of events in the formation of a clathrin-coated vesicle from a clathrin-coated pit. The clathrin-coated pits and vesicles shown are larger than those seen in normal-sized cells. They take up lipoprotein particles into a very large hen oocyte to form yolk. The lipoprotein particles bound to their membrane-bound receptors appear as a dense, fuzzy layer on the extracellular surface of the plasma membrane—which is the inside surface of the vesicle. (Courtesy of M.M. Perry and A.B. Gilbert, *J. Cell Sci.* 39:257–272, 1979. With permission from The Company of Biologists.)

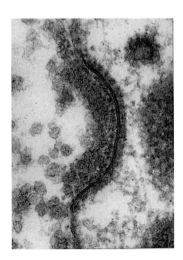
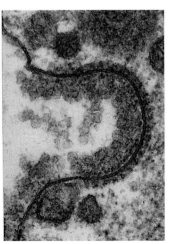
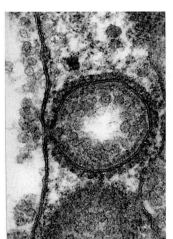
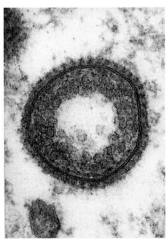

0.1 μm

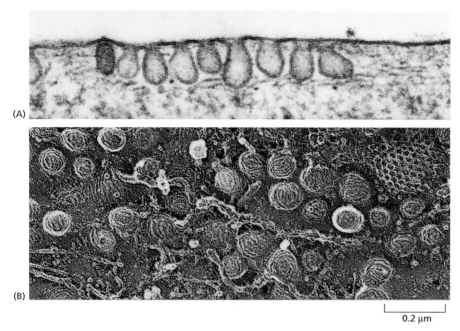

Figure 13–49 Caveolae in the plasma membrane of a fibroblast. (A) This electron micrograph shows a plasma membrane with a very high density of caveolae. Note that no cytosolic coat is visible. (B) This rapid-freeze deep-etch image demonstrates the characteristic "cauliflower" texture of the cytosolic face of the caveolae membrane. The regular texture is thought to result from aggregates of caveolin in the membrane. A clathrin-coated pit is also seen at the upper right. (Courtesy of R.G.W. Anderson, from K.G. Rothberg et al., *Cell* 68:673–682, 1992. With permission from Elsevier.)

0.2 µm

form coated vesicles, any substance dissolved in the extracellular fluid is internalized—a process called *fluid-phase endocytosis.*

Not All Pinocytic Vesicles Are Clathrin-Coated

In addition to clathrin-coated pits and vesicles, there are other, less well-understood mechanisms by which cells can form pinocytic vesicles. One of these pathways initiates at **caveolae** (from the Latin for "little cavities"), originally recognized by their ability to transport molecules across endothelial cells, which form the inner lining of blood vessels. Caveolae are present in the plasma membrane of most cell types, and in some of these they are seen in the electron microscope as deeply invaginated flasks (**Figure 13–49**). They are thought to form from membrane microdomains, or *lipid rafts,* which are patches of the plasma membrane that are especially rich in cholesterol, glycosphingolipids, and GPI-anchored membrane proteins (see Figure 10–14). The major structural proteins in caveolae are **caveolins**, which are a family of unusual integral membrane proteins that each insert a hydrophobic loop into the membrane from the cytosolic side but do not extend across the membrane.

In contrast to clathrin-coated and COPI- or COPII-coated vesicles, caveolae are thought to invaginate and collect cargo proteins by virtue of the lipid composition of the calveolar membrane, rather than by the assembly of a cytosolic protein coat. Caveolins may stabilize these raft domains, into which certain plasma membrane proteins partition. Caveolae pinch off from the plasma membrane using dynamin, and they deliver their contents either to an endosome-like compartment (called a *caveosome*) or to the plasma membrane on the opposite side of a polarized cell (in a process called *transcytosis,* which we discuss later). Because caveolins are integral membrane proteins, they do not dissociate from the vesicles after endocytosis; instead they are delivered to the target compartments, where they are maintained as discrete membrane domains. Some animal viruses such as SV40 and papilloma virus (which causes warts) enter cells in vesicles derived from caveolae. The viruses are first delivered to caveosomes, and they move from there in specialized transport vesicles to the ER. The viral genome exits from the ER across the ER membrane into the cytosol, from where it is imported into the nucleus to start the infection cycle.

Endocytic vesicles can also bud from caveolin-free raft domains on the plasma membrane and deliver their cargo to caveosomes. Molecules that enter the cell through caveosomes avoid endosomes and lysosomes and are therefore shielded from exposure to low pH and lysosomal hydrolases; it is unknown how they move from caveosomes to other destinations in the cell.

Cells Use Receptor-Mediated Endocytosis to Import Selected Extracellular Macromolecules

In most animal cells, clathrin-coated pits and vesicles provide an efficient pathway for taking up specific macromolecules from the extracellular fluid. In this process, called **receptor-mediated endocytosis**, the macromolecules bind to complementary transmembrane receptor proteins, accumulate in coated pits, and then enter the cell as receptor–macromolecule complexes in clathrin-coated vesicles (see Figure 13–48). Because ligands are selectively captured by receptors, receptor-mediated endocytosis provides a selective concentrating mechanism that increases the efficiency of internalization of particular ligands more than a hundredfold. In this way, even minor components of the extracellular fluid can be specifically taken up in large amounts without taking in a large volume of extracellular fluid. A particularly well-understood and physiologically important example is the process that mammalian cells use to take up cholesterol.

Many animals cells take up cholesterol through receptor-mediated endocytosis and, in this way, acquire most of the cholesterol they require to make new membrane. If the uptake is blocked, cholesterol accumulates in the blood and can contribute to the formation in blood vessel (artery) walls of *atherosclerotic plaques*, deposits of lipid and fibrous tissue that can cause strokes and heart attacks by blocking arterial blood flow. In fact, it was a study of humans with a strong genetic predisposition for *atherosclerosis* that first revealed the mechanism of receptor-mediated endocytosis.

Most cholesterol is transported in the blood as cholesteryl esters in the form of lipid–protein particles known as **low-density lipoproteins** (**LDLs**) (**Figure 13–50**). When a cell needs cholesterol for membrane synthesis, it makes transmembrane receptor proteins for LDL and inserts them into its plasma membrane. Once in the plasma membrane, the *LDL receptors* diffuse until they associate with clathrin-coated pits that are in the process of forming (**Figure 13–51A**). Since coated pits constantly pinch off to form coated vesicles, any LDL particles bound to LDL receptors in the coated pits are rapidly internalized in coated vesicles. After shedding their clathrin coats, the vesicles deliver their contents to early endosomes, which are located near the cell periphery. Once the LDL and LDL receptors encounter the low pH in the endosomes, LDL is released from its receptor and is delivered via late endosomes to lysosomes. There, the cholesteryl esters in the LDL particles are hydrolyzed to free cholesterol, which is now available to the cell for new membrane synthesis. If too much free cholesterol accumulates in a cell, the cell shuts off both its own cholesterol synthesis and the synthesis of LDL receptors, so that it ceases either to make or to take up cholesterol.

This regulated pathway for cholesterol uptake is disrupted in individuals who inherit defective genes encoding LDL receptors. The resulting high levels of blood cholesterol predispose these individuals to develop atherosclerosis

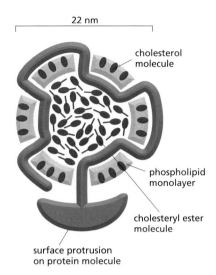

Figure 13–50 A low-density lipoprotein (LDL) particle. Each spherical particle has a mass of 3×10^6 daltons. It contains a core of about 1500 cholesterol molecules esterified to long-chain fatty acids. A lipid monolayer composed of about 800 phospholipid and 500 unesterified cholesterol molecules surrounds the core of cholesterol esters. A single molecule of a 500,000-dalton protein organizes the particle and mediates the specific binding of LDL to cell-surface LDL receptors.

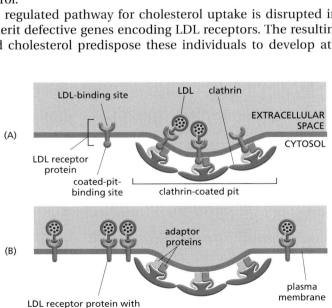

Figure 13–51 Normal and mutant LDL receptors. (A) LDL receptors binding to a coated pit in the plasma membrane of a normal cell. The human LDL receptor is a single-pass transmembrane glycoprotein composed of about 840 amino acids, only 50 of which are on the cytoplasmic side of the membrane. (B) A mutant cell in which the LDL receptors are abnormal and lack the site in the cytoplasmic domain that enables them to bind to adaptor proteins in the clathrin-coated pits. Such cells bind LDL but cannot ingest it. In most human populations, 1 in 500 individuals inherits one defective LDL receptor gene and, as a result, has an increased risk of a heart attack caused by atherosclerosis.

prematurely, and many would die at an early age of heart attacks resulting from coronary artery disease if they were not treated with drugs that lower the level of blood cholesterol. In some cases, the receptor is lacking altogether. In others, the receptors are defective—in either the extracellular binding site for LDL or the intracellular binding site that attaches the receptor to the coat of a clathrin-coated pit (see Figure 13–51B). In the latter case, normal numbers of LDL receptors are present, but they fail to become localized in clathrin-coated pits. Although LDL binds to the surface of these mutant cells, it is not internalized, directly demonstrating the importance of clathrin-coated pits for the receptor-mediated endocytosis of cholesterol.

More than 25 distinct receptors are known to participate in receptor-mediated endocytosis of different types of molecules. They all apparently use clathrin-dependent internalization routes and are guided into clathrin-coated pits by signals in their cytoplasmic tails that bind to adaptor proteins in the clathrin coat. Many of these receptors, like the LDL receptor, enter coated pits irrespective of whether they have bound their specific ligands. Others enter preferentially when bound to a specific ligand, suggesting that a ligand-induced conformational change is required for them to activate the signal sequence that guides them into the pits. Since most plasma membrane proteins fail to become concentrated in clathrin-coated pits, the pits serve as molecular filters, preferentially collecting certain plasma membrane proteins (receptors) over others.

Electron-microscope studies of cultured cells exposed simultaneously to different labeled ligands demonstrate that many kinds of receptors can cluster in the same coated pit, whereas some other receptors cluster in different clathrin-coated pits. The plasma membrane of one clathrin-coated pit can probably accommodate up to 1000 receptors of assorted varieties. Although all of the receptor–ligand complexes that use this endocytic pathway are apparently delivered to the same endosomal compartment, the subsequent fates of the endocytosed molecules vary, as we discuss next.

Endocytosed Materials That Are Not Retrieved from Endosomes End Up in Lysosomes

The endosomal compartments of a cell can be complex. They can be made visible in the electron microscope by adding a readily detectable tracer molecule, such as the enzyme peroxidase, to the extracellular medium and leaving the cells for various lengths of time to take it up by endocytosis. The distribution of the molecule after its uptake reveals the endosomal compartments as a set of heterogeneous, membrane-enclosed tubes extending from the periphery of the cell to the perinuclear region, where it is often close to the Golgi apparatus. Two sequential sets of endosomes can be readily distinguished in such labeling experiments. The tracer molecule appears within a minute or so in **early endosomes**, just beneath the plasma membrane. After 5–15 minutes, it has moved to **late endosomes**, close to the Golgi apparatus and near the nucleus. Early and late endosomes differ in their protein compositions. The transition from early to late endosomes is accompanied by the release of Rab5 and the binding of Rab7, for example.

As mentioned earlier, a vacuolar H^+ ATPase in the endosomal membrane, which pumps H^+ into endosomes from the cytosol, keeps the lumen of the endosomal compartments acidic (pH ~6). In general, later endosomes are more acidic than early endosomes. This gradient of acidic environments has a crucial role in the function of these organelles.

We have already seen how endocytosed materials mix in early endosomes with newly synthesized acid hydrolases and eventually end up being degraded in lysosomes. Many molecules, however, are specifically diverted from this journey to destruction. They are instead recycled from the early endosomes back to the plasma membrane via transport vesicles. Only molecules that are not retrieved from endosomes in this way are delivered to lysosomes for degradation.

Although mild digestion may start in early endosomes, many hydrolases are synthesized and delivered there as proenzymes, called *zymogens*, which contain extra inhibitory domains at their N-terminus that keep the hydrolase inactive until these domains are proteolytically removed. The hydrolases are activated when late endosomes become endolysosomes as the result of fusion with pre-existing lysosomes, which contain a full complement of active hydrolases that digest off the inhibitory domains from the newly synthesized enzymes. Moreover, the pH in early endosomes is not low enough to activate lysosomal hydrolases optimally. By these means, cells can retrieve most membrane proteins from early endosomes and recycle them back to the plasma membrane.

Specific Proteins Are Retrieved from Early Endosomes and Returned to the Plasma Membrane

Early endosomes form a compartment that acts as the main sorting station in the endocytic pathway, just as the *cis* and *trans* Golgi networks serve this function in the biosynthetic–secretory pathway. In the mildly acidic environment of the early endosome, many internalized receptor proteins change their conformation and release their ligand, as already discussed for the M6P receptors. Those endocytosed ligands that dissociate from their receptors in the early endosome are usually doomed to destruction in lysosomes, along with the other soluble contents of the endosome. Some other endocytosed ligands, however, remain bound to their receptors, and thereby share the fate of the receptors.

The fates of receptors—and of any ligands remaining bound to them—vary according to the specific type of receptor. (1) Most receptors are recycled and return to the same plasma membrane domain from which they came; (2) some proceed to a different domain of the plasma membrane, thereby mediating *transcytosis*; and (3) some progress to lysosomes, where they are degraded (**Figure 13–52**).

The LDL receptor follows the first pathway. It dissociates from its ligand, LDL, in the early endosome and is recycled back to the plasma membrane for reuse, leaving the discharged LDL to be carried to lysosomes (**Figure 13–53**). The recycling transport vesicles bud from long, narrow tubules that extend from the

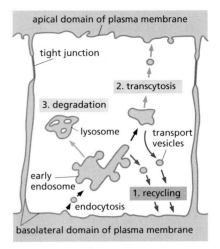

Figure 13–52 Possible fates for transmembrane receptor proteins that have been endocytosed. Three pathways from the endosomal compartment in an epithelial cell are shown. Retrieved receptors are returned (1) to the same plasma membrane domain from which they came *(recycling)* or (2) to a different domain of the plasma membrane *(transcytosis)*. (3) Receptors that are not specifically retrieved from endosomes follow the pathway from the endosomal compartment to lysosomes, where they are degraded *(degradation)*. The formation of oligomeric aggregates in the endosomal membrane may be one of the signals that guide receptors into the degradative pathway. If the ligand that is endocytosed with its receptor stays bound to the receptor in the acidic environment of the endosome, it follows the same pathway as the receptor; otherwise it is delivered to lysosomes.

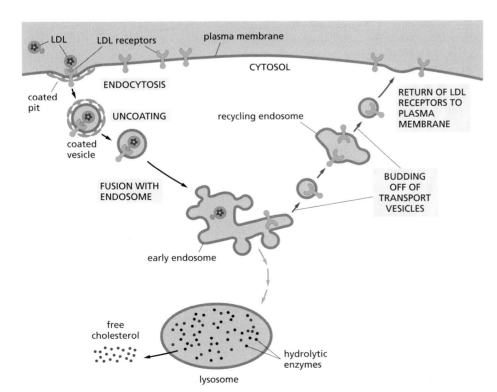

Figure 13–53 The receptor-mediated endocytosis of LDL. <GCTA> Note that the LDL dissociates from its receptors in the acidic environment of the early endosome. After a number of steps (shown in Figure 13–55), the LDL ends up in lysosomes, where it is degraded to release free cholesterol. In contrast, the LDL receptors are returned to the plasma membrane via clathrin-coated transport vesicles that bud off from the tubular region of the early endosome, as shown. For simplicity, only one LDL receptor is shown entering the cell and returning to the plasma membrane. Whether it is occupied or not, an LDL receptor typically makes one round trip into the cell and back to the plasma membrane every 10 minutes, making a total of several hundred trips in its 20-hour lifespan.

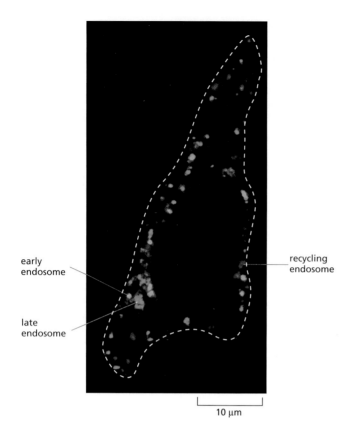

early
endosome

late
endosome

recycling
endosome

10 μm

Figure 13–54 Sorting of membrane proteins in the endocytic pathway. Transferrin receptors mediate iron uptake and constitutively cycle between endosomes and the plasma membrane. By contrast, activated opioid receptors are down-regulated by endocytosis followed by degradation in lysosomes; they are activated by opiates such as morphine and heroin, as well as by endogenous peptides called enkephalins and endorphins. Endocytosis of both types of receptors starts in clathrin-coated pits. The receptors are then delivered to early endosomes, where their pathways part: transferrin receptors are sorted to the recycling endosomes, whereas opioid receptors are sorted to late endosomes. The micrograph shows both receptors—labeled with different fluorescent dyes—30 min after endocytosis (transferrin receptors are labeled in *red* and opioid receptors in *green*). At this time, some early endosomes still contain both receptors and are seen as *yellow*, due to the overlap of red and green light emitted from the fluorescent dyes. By contrast, recycling endosomes and late endosomes are selectively enriched in either transferrin or opioid receptors, respectively, and therefore appear as distinct red and green structures. (Courtesy of Mark von Zastrow.)

early endosomes. It is likely that the geometry of these tubules helps the sorting process: because tubules have a large membrane area enclosing a small volume, membrane proteins become enriched over soluble proteins. Transport vesicles that return material to the plasma membrane begin budding from the tubules, but tubular portions of the early endosome also pinch off and fuse with one another to form *recycling endosomes,* which serve as way-stations for the traffic between early endosomes and the plasma membrane. This recycling pathway operates continuously, compensating for the continuous endocytosis occurring at the plasma membrane.

The **transferrin receptor** follows a similar recycling pathway as the LDL receptor, but unlike the LDL receptor it also recycles its ligand. Transferrin is a soluble protein that carries iron in the blood. Cell-surface transferrin receptors deliver transferrin with its bound iron to early endosomes by receptor-mediated endocytosis. The low pH in the endosome induces transferrin to release its bound iron, but the iron-free transferrin itself (called apotransferrin) remains bound to its receptor. The receptor–apotransferrin complex enters the tubular extensions of the early endosome and from there is recycled back to the plasma membrane (**Figure 13–54**). When the apotransferrin returns to the neutral pH of the extracellular fluid, it dissociates from the receptor and is thereby freed to pick up more iron and begin the cycle again. Thus, transferrin shuttles back and forth between the extracellular fluid and the endosomal compartment, avoiding lysosomes and delivering iron to the cell interior, as needed for cells to grow and proliferate.

The second pathway that endocytosed receptors can follow from endosomes is taken by many signaling receptors, including opioid receptors (see Figure 13–54) and the receptor that binds *epidermal growth factor (EGF)*. EGF is a small, extracellular signal protein that stimulates epidermal and various other cells to divide. Unlike LDL receptors, EGF receptors accumulate in clathrin-coated pits only after binding EGF, and most of them do not recycle but are degraded in lysosomes, along with the ingested EGF. EGF binding therefore first activates intracellular signaling pathways and then leads to a decrease in the concentration of EGF receptors on the cell surface, a process called *receptor down-regulation* that reduces the cell's subsequent sensitivity to EGF (see Figure 15–29).

Clathrin-dependent receptor-mediated endocytosis is highly regulated. The receptors are first covalently modified with the small protein ubiquitin. But, unlike *polyubiquitylation*, which adds a chain of ubiquitins that typically targets a protein for degradation in proteasomes (discussed in Chapter 6), ubiquitin tagging for sorting into the clathrin-dependent endocytic pathway adds one or more single ubiquitin molecules to the protein—a process called *monoubiquitylation* or *multiubiquitylation*, respectively. Ubiquitin-binding proteins recognize the attached ubiquitin and help direct the modified receptors into clathrin-coated pits. After delivery to the endosome, other ubiquitin-binding proteins recognize the ubiquitin and help mediate sorting steps.

Multivesicular Bodies Form on the Pathway to Late Endosomes

As previously stated, many of the endocytosed molecules move from the early to the late endosomal compartment. In this process, early endosomes migrate slowly along microtubules toward the cell interior, while shedding membrane tubules and vesicles that recycle material to the plasma membrane and TGN. At the same time, the membrane enclosing the migrating endosomes forms invaginating buds that pinch off and form internal vesicles; they are then called **multivesicular bodies** (**Figure 13–55**). Multivesicular bodies eventually fuse with a late endosomal compartment or with each other to become late endosomes. At the end of this pathway, the late endosomes convert to endolysosomes and lysosomes as a result of both their fusion with preexisting lysosomes and progressive acidification (**Figure 13–56**).

The multivesicular bodies carry those endocytosed membrane proteins that are to be degraded. As part of the protein-sorting process, receptors destined for degradation, such as occupied EGF receptors described previously, selectively partition into the invaginating membrane of the multivesicular bodies. In this way, both the receptors and any signaling proteins strongly bound to them are made fully accessible to the digestive enzymes that will degrade them (**Figure 13–57**). In addition to endocytosed membrane proteins, multivesicular bodies also contain the soluble content of early endosomes destined for late endosomes and digestion in lysosomes.

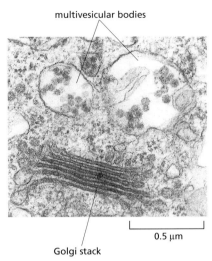

Figure 13–55 Electron micrograph of a multivesicular body in a plant cell. The large amount of internal membrane will be delivered to the vacuole, the plant equivalent of the lysosome, for digestion.

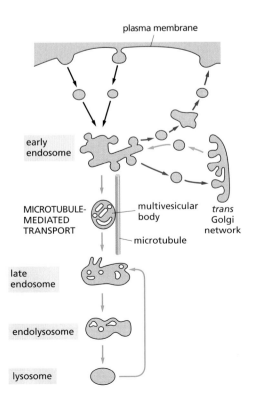

Figure 13–56 Details of the endocytic pathway from the plasma membrane to lysosomes. Maturation of early endosomes to late endosomes occurs through the formation of multivesicular bodies, which contain large amounts of invaginated membrane and internal vesicles (hence their name). Multivesicular bodies move inward along microtubules, continually shedding transport vesicles that recycle components to the plasma membrane. They gradually convert into late endosomes, either by fusing with each other or by fusing with preexisting late endosomes. The late endosomes no longer send vesicles to the plasma membrane.

Figure 13–57 The sequestration of endocytosed proteins into internal membranes of multivesicular bodies. Eventually, proteases and lipases in lysosomes digest all of the internal membranes within multivesicular bodies produced by the invaginations. The invagination processes are essential to achieve complete digestion of endocytosed membrane proteins: because the outer membrane of the multivesicular body becomes continuous with the lysosomal membrane, for example, lysosomal hydrolases could not digest the cytosolic domains of endocytosed transmembrane proteins such as the EGF receptor shown here, if the protein were not localized in internal vesicles.

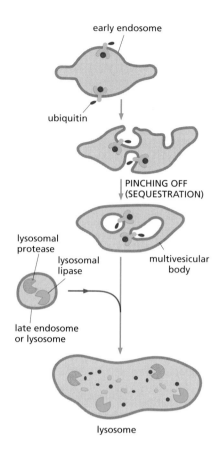

Sorting into the internal vesicles of a multivesicular body requires one or multiple ubiquitin tags, which are added to the cytosolic domains of membrane proteins. These tags initially help guide the proteins into clathrin-coated vesicles. Once delivered to the endosomal membrane, the ubiquitin tags are recognized again, this time by a series of cytosolic protein complexes, called *ESCRT-0, -I, -II,* and *-III,* which bind sequentially, handing the ubiquitylated cargo from one complex to the next, and ultimately mediate the sorting process into the internal vesicles of multivesicular bodies (**Figure 13–58**). Membrane invagination into multivesicular bodies also depends on a lipid kinase that phosphorylates phosphatidylinositol to produce PI(3)P, which serves as an additional docking site for the ESCRT complexes; these complexes require both PI(3)P and the presence of ubiquitylated cargo proteins to bind to the endosomal membrane. A second PI kinase adds another phosphate group to PI(3)P, producing PI(3,5)P$_2$, which is required for ESCRT-III to form large multimeric assemblies on the membrane. It is not known how the assembly of ESCRT complexes ultimately

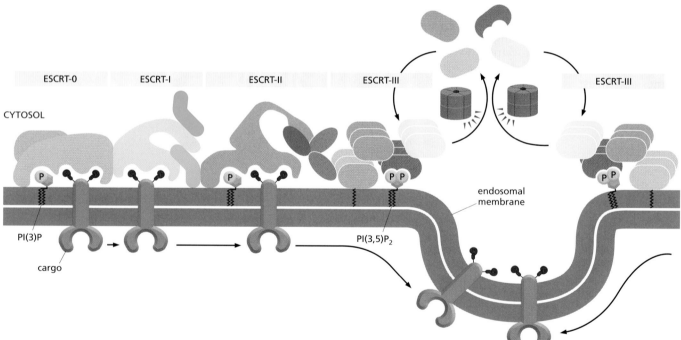

Figure 13–58 Sorting of endocytosed membrane proteins into the internal vesicles of a multivesicular body. A series of complex binding events passes the ubiquitylated cargo proteins sequentially from one ESCRT complex to the next, enventually concentrating them in membrane areas that bud away from the cytosol into the lumen of the endosome to form the internal membrane vesicles of the multivesicular body. ESCRT complexes are soluble in the cytosol and are recruited to the membrane as needed. First, ESCRT-0 binds both the ubiquitin attached to the cargo protein and to PI(3)P head groups. ESCRT-0 dissociates from the membrane, handing the ubiquitylated cargo protein over to the ESCRT-I complex; next ESCRT-I dissociates, handing the cargo protein over to ESCRT-II complex; and finally ESCRT-II dissociates and ESCRT-III complexes assemble on the membrane. By contrast to ESCRT-0, -I, and -II, ESCRT-III does not bind to the ubiquitylated cargo directly. Instead its assembly into expansive multimeric structures is thought to confine the cargo molecules into specialized membrane areas that then invaginate, leaving the ESCRT components on the endosome surface. An AAA-ATPase (*red* cylinders) then disassembles the ESCRT-III complexes so that they can be reused.

drives the invagination and pinching-off processes required to form the internal vesicles but the ESCRT complexes themselves are not part of the invaginating membranes.

Mutant cells compromised in ESCRT function display signaling defects. In such cells, activated receptors cannot be down-regulated by endocytosis and packaged into multivesicular bodies and therefore mediate prolonged signaling, which can lead to uncontrolled cell proliferation and cancer.

The same ESCRT machinery that drives the internal budding from the endosomal membrane to form multivesicular bodies is also used by HIV, ebola, and other enveloped viruses to bud from the plasma membrane into the extracellular space. The two processes are topologically equivalent, as they both involve budding away from the cytosolic surface of the membrane (**Figure 13–59**).

Transcytosis Transfers Macromolecules Across Epithelial Cell Sheets

Some receptors on the surface of polarized epithelial cells transfer specific macromolecules from one extracellular space to another by **transcytosis** (**Figure 13–60**). These receptors are endocytosed and then follow a pathway from endosomes to a different plasma membrane domain (see Figure 13–52). A newborn rat, for example, obtains antibodies from its mother's milk (which help protect it against infection) by transporting them across the epithelium of its gut. The lumen of the gut is acidic, and, at this low pH, the antibodies in the milk bind to specific receptors on the apical (absorptive) surface of the gut epithelial cells. The receptor–antibody complexes are internalized via clathrin-coated pits and vesicles and are delivered to early endosomes. The complexes remain intact and are retrieved in transport vesicles that bud from the early

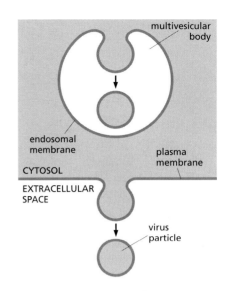

Figure 13–59 ESCRT complexes in multivesicular body formation and virus budding. In the two topologically equivalent processes indicated by the arrows, ESCRT complexes shape membranes into buds that bulge away from the cytosol.

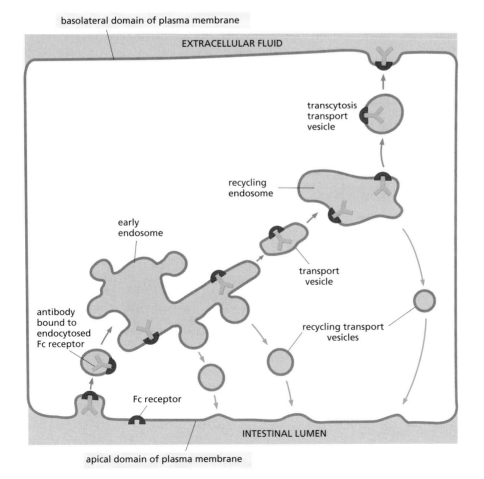

Figure 13–60 The role of recycling endosomes in transcytosis. Recycling endosomes form a way-station on the transcytotic pathway. In the example shown here, an antibody receptor on a gut epithelial cell binds antibody and is endocytosed, eventually carrying the antibody to the basolateral plasma membrane. The receptor is called an Fc receptor because it binds the Fc part of the antibody (discussed in Chapter 25).

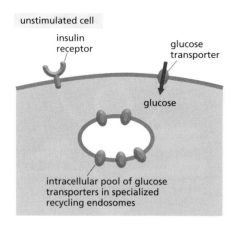

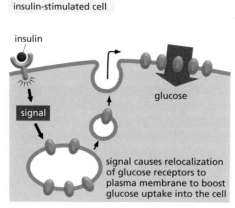

Figure 13–61 **Storage of plasma membrane proteins in recycling endosomes.** Recycling endosomes can serve as an intracellular pool of specialized plasma membrane proteins that can be mobilized when needed. In the example shown, insulin binding to the insulin receptor triggers an intracellular signaling pathway that causes the rapid insertion of glucose transporters into the plasma membrane of a fat or muscle cell, greatly increasing glucose intake.

endosome and subsequently fuse with the basolateral domain of the plasma membrane. On exposure to the neutral pH of the extracellular fluid that bathes the basolateral surface of the cells, the antibodies dissociate from their receptors and eventually enter the newborn's bloodstream.

The transcytotic pathway from the early endosome to the plasma membrane is not direct. The receptors first move from the early endosome to an intermediate endosomal compartment, the **recycling endosome** described previously (see Figure 13–60). The variety of pathways that different receptors follow from endosomes implies that, in addition to binding sites for their ligands and binding sites for coated pits, many receptors also possess sorting signals that guide them into the appropriate type of transport vesicle leaving the endosome and moving to the appropriate target membrane in the cell.

A unique property of recycling endosomes is that cells can regulate the exit of membrane proteins from the compartment. Thus, cells can adjust the flux of proteins through the transcytotic pathway according to need. Although the mechanism is uncertain, this regulation allows recycling endosomes to play an important part in adjusting the concentration of specific plasma membrane proteins. Fat cells and muscle cells, for example, contain large intracellular pools of the glucose transporters that are responsible for the uptake of glucose across the plasma membrane. These membrane transport proteins are stored in specialized recycling endosomes until the hormone *insulin* stimulates the cell to increase its rate of glucose uptake. In response to the insulin signal, transport vesicles rapidly bud from the recycling endosome and deliver large numbers of glucose transporters to the plasma membrane, thereby greatly increasing the rate of glucose uptake into the cell (**Figure 13–61**).

Epithelial Cells Have Two Distinct Early Endosomal Compartments but a Common Late Endosomal Compartment

In polarized epithelial cells, endocytosis occurs from both the *basolateral domain* and the *apical domain* of the plasma membrane. Material endocytosed from either domain first enters an early endosomal compartment that is unique to that domain. This arrangement allows endocytosed receptors to be recycled back to their original membrane domain, unless they contain sorting signals that mark them for transcytosis to the other domain. Molecules endocytosed from either plasma membrane domain that are not retrieved from the early endosomes end up in a common late endosomal compartment near the cell center and are eventually degraded in lysosomes (**Figure 13–62**).

Whether cells contain a few connected or many unconnected endosomal compartments seems to depend on the cell type and the physiological state of the cell. Like many other membrane-enclosed organelles, endosomes of the same type can readily fuse with one another (an example of homotypic fusion, discussed earlier) to create large continuous endosomes.

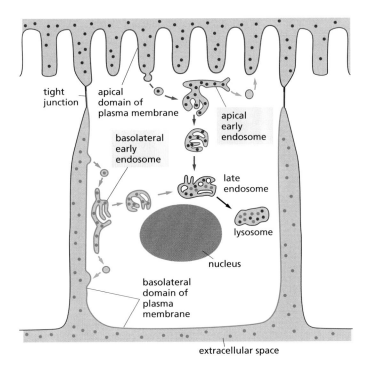

Figure 13–62 The two distinct early endosomal compartments in an epithelial cell. The basolateral and the apical domains of the plasma membrane communicate with separate early endosomal compartments. However, endocytosed molecules from both domains that do not contain signals for recycling or transcytosis meet in a common late endosomal compartment before being digested in lysosomes.

Summary

Cells ingest fluid, molecules, and particles by endocytosis, in which localized regions of the plasma membrane invaginate and pinch off to form endocytic vesicles. Many of the endocytosed molecules and particles eventually end up in lysosomes, where they are degraded. Endocytosis occurs both constitutively and as a triggered response to extra-cellular signals. Endocytosis is so extensive in many cells that a large fraction of the plasma membrane is internalized every hour. The cells remain the same size because most of the plasma membrane components (proteins and lipid) that are endocytosed are continually returned to the cell surface by exocytosis. This large-scale endocytic–exocytic cycle is mediated largely by clathrin-coated pits and vesicles.

Many cell-surface receptors that bind specific extracellular macromolecules become tagged with ubiquitin, which guides them into clathrin-coated pits. As a result, these receptors and their ligands are efficiently internalized in clathrin-coated vesicles, a process called receptor-mediated endocytosis. The coated vesicles rapidly shed their clathrin coats and fuse with early endosomes.

Most of the ligands dissociate from their receptors in the acidic environment of the endosome and eventually end up in lysosomes, while most of the receptors are recycled via transport vesicles back to the cell surface for reuse. But receptor–ligand complexes can follow other pathways from the endosomal compartment. In some cases, both the receptor and the ligand end up being degraded in lysosomes, resulting in receptor down-regulation; in these cases, the ubiquitin-tagged receptors recruit various ESCRT complexes, which drive the invagination and pinching-off of endosomal membrane vesicles to form multivesicular bodies. In other cases, both receptor and ligand are transferred to a different plasma membrane domain, causing the ligand to be released at a surface of the cell that differs from the membrane where it originated, a process called transcytosis. The transcytosis pathway involves recycling endosomes, where endocytosed plasma membrane proteins can be stored until they are needed.

TRANSPORT FROM THE *TRANS* GOLGI NETWORK TO THE CELL EXTERIOR: EXOCYTOSIS

Having considered the cell's internal digestive system and the various types of incoming membrane traffic that converge on lysosomes, we now return to the Golgi apparatus and examine the secretory pathways that lead out to the cell exterior. Transport vesicles destined for the plasma membrane normally leave

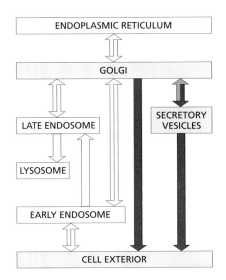

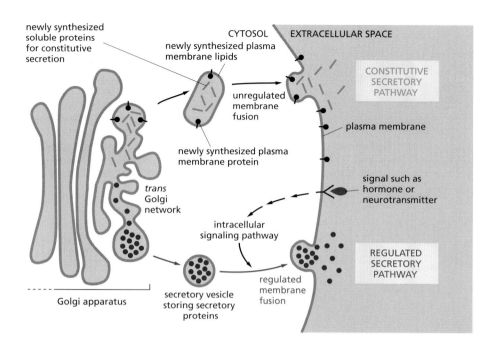

Figure 13–63 The constitutive and regulated secretory pathways. <ACAG> The two pathways diverge in the *trans* Golgi network. The constitutive secretory pathway operates in all cells. Many soluble proteins are continually secreted from the cell by this pathway, which also supplies the plasma membrane with newly synthesized lipids and proteins. Specialized secretory cells also have a regulated secretory pathway, by which selected proteins in the *trans* Golgi network are diverted into secretory vesicles, where the proteins are concentrated and stored until an extracellular signal stimulates their secretion. The regulated secretion of small molecules, such as histamine and neurotransmitters, occurs by a similar pathway; these molecules are actively transported from the cytosol into preformed secretory vesicles. There they are often complexed to specific macromolecules (proteoglycans, for histamine), so that they can be stored at high concentration without generating an excessively high osmotic pressure.

the TGN in a steady stream as irregularly shaped tubules. The membrane proteins and the lipids in these vesicles provide new components for the cell's plasma membrane, while the soluble proteins inside the vesicles are secreted to the extracellular space. The fusion of the vesicles with the plasma membrane is called **exocytosis**. In this way, for example, cells produce and secrete most of the proteoglycans and glycoproteins of the *extracellular matrix*, which is discussed in Chapter 19.

All cells require this **constitutive secretory pathway**, which operates continuously. Specialized secretory cells, however, have a second secretory pathway in which soluble proteins and other substances are initially stored in *secretory vesicles* for later release. This is the **regulated secretory pathway**, found mainly in cells specialized for secreting products rapidly on demand—such as hormones, neurotransmitters, or digestive enzymes (**Figure 13–63**). In this section, we consider the role of the Golgi apparatus in both of these secretory pathways and compare the two mechanisms of secretion.

Many Proteins and Lipids Seem to Be Carried Automatically from the Golgi Apparatus to the Cell Surface

A cell capable of regulated secretion must separate at least three classes of proteins before they leave the *trans* Golgi network—those destined for lysosomes (via endosomes), those destined for secretory vesicles, and those destined for immediate delivery to the cell surface (**Figure 13–64**). We have already noted that proteins destined for lysosomes are tagged for packaging into specific departing vesicles (with mannose 6-phosphate in the case of lysosomal hydrolases), and analogous signals are thought to direct *secretory proteins* into secretory vesicles. The non-selective constitutive secretory pathway transports most other proteins directly to the cell surface. Because entry into this pathway does not require a particular signal, it is also called the **default pathway**. Thus, in an unpolarized cell such as a white blood cell or a fibroblast, it seems that any protein in the lumen of the Golgi apparatus is automatically carried by the constitutive pathway to the cell surface unless it is specifically returned to the ER, retained as a resident protein in the Golgi apparatus itself, or selected for the pathways that lead to regulated secretion or to lysosomes. In polarized cells, where different products have to be delivered to different domains of the cell surface, we shall see that the options are more complex.

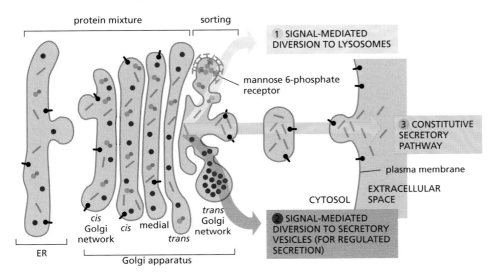

Figure 13–64 The three best-understood pathways of protein sorting in the *trans* Golgi network. (1) Proteins with the mannose 6-phosphate (M6P) marker are diverted to lysosomes (via endosomes) in clathrin-coated transport vesicles (see Figure 13–44). (2) Proteins with signals directing them to secretory vesicles are concentrated in such vesicles as part of a regulated secretory pathway that is present only in specialized secretory cells. (3) In unpolarized cells, a constitutive secretory pathway delivers proteins with no special features to the cell surface. In polarized cells, such as epithelial cells, however, secreted and plasma membrane proteins are selectively directed to either the apical or the basolateral plasma membrane domain, so a specific signal must mediate at least one of these two pathways, as we discuss later.

Secretory Vesicles Bud from the *Trans* Golgi Network

Cells that are specialized for secreting some of their products rapidly on demand concentrate and store these products in **secretory vesicles** (often called *secretory granules* or *dense-core vesicles* because they have dense cores when viewed in the electron microscope). Secretory vesicles form from the *trans* Golgi network, and they release their contents to the cell exterior by exocytosis in response to specific signals. The secreted product can be either a small molecule (such as histamine) or a protein (such as a hormone or digestive enzyme).

Proteins destined for secretory vesicles (called *secretory proteins)* are packaged into appropriate vesicles in the *trans* Golgi network by a mechanism that is thought to involve the selective aggregation of the secretory proteins. Clumps of aggregated, electron-dense material can be detected by electron microscopy in the lumen of the *trans* Golgi network. The signal that directs secretory proteins into such aggregates is not known, but it is thought to be composed of signal patches that are common to proteins of this class. When a gene encoding a secretory protein is artificially expressed in a secretory cell that normally does not make the protein, the foreign protein is appropriately packaged into secretory vesicles. This observation shows that, although the proteins that an individual cell expresses and packages in secretory vesicles differ, they all contain common sorting signals, which function properly even when the proteins are expressed in cells that do not normally make them.

It is unclear how the aggregates of secretory proteins are segregated into secretory vesicles. Secretory vesicles have unique proteins in their membrane, some of which might serve as receptors for aggregated protein in the *trans* Golgi network. The aggregates are much too big, however, for each molecule of the secreted protein to be bound by its own cargo receptor, as proposed for transport of the lysosomal enzymes. The uptake of the aggregates into secretory vesicles may therefore more closely resemble the uptake of particles by phagocytosis at the cell surface, where the plasma membrane zippers up around large structures.

Initially, most of the membrane of the secretory vesicles that leave the *trans* Golgi network is only loosely wrapped around the clusters of aggregated secretory proteins. Morphologically, these **immature secretory vesicles** resemble dilated *trans* Golgi cisternae that have pinched off from the Golgi stack. As the vesicles mature, they can fuse with one another and their contents become concentrated (**Figure 13–65**A), probably as the result of both the continuous retrieval of membrane that is recycled back to late endosomes and the TGN and the progressive acidification of the vesicle lumen that results from the increasing concentration of ATP-driven H^+ pumps in the vesicle membrane. Recall that V-type ATPases acidify all endocytic and exocytic organelles (see Figure 13–36). The degree of concentration of proteins during the formation and maturation

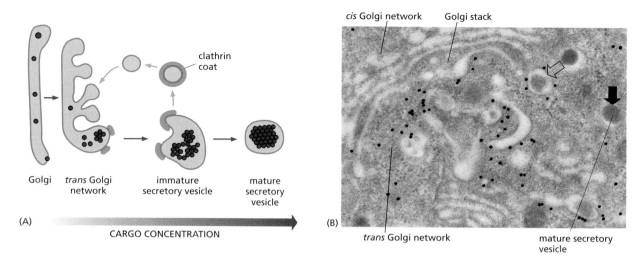

Figure 13–65 The formation of secretory vesicles. (A) Secretory proteins become segregated and highly concentrated in secretory vesicles by two mechanisms. First, they aggregate in the ionic environment of the *trans* Golgi network; often the aggregates become more condensed as secretory vesicles mature and their lumen becomes more acidic. Second, clathrin-coated vesicles retrieve excess membrane and lumenal content present in immature secretory vesicles as the secretory vesicles mature. (B) This electron micrograph shows secretory vesicles forming from the *trans* Golgi network in an insulin-secreting β cell of the pancreas. An antibody conjugated to gold spheres *(black dots)* has been used to locate clathrin molecules. The immature secretory vesicles *(open arrow)*, which contain insulin precursor protein (proinsulin), contain clathrin patches. Clathrin coats are no longer seen on the mature secretory vesicle, which has a highly condensed core *(solid arrow)*. (Courtesy of Lelio Orci.)

of secretory vesicles is only a small part of the total 200–400-fold concentration of these proteins that occurs after they leave the ER. Secretory and membrane proteins become concentrated as they move from the ER through the Golgi apparatus because of an extensive retrograde retrieval process mediated by COPI-coated transport vesicles that exclude them (see Figure 13–24).

Membrane recycling is important for returning Golgi components to the Golgi apparatus, as well as for concentrating the contents of secretory vesicles. The vesicles that mediate this retrieval originate as clathrin-coated buds on the surface of immature secretory vesicles, often being seen even on budding secretory vesicles that have not yet been severed from the Golgi stack (see Figure 13–65B).

Because the final mature secretory vesicles are so densely filled with contents, the secretory cell can disgorge large amounts of material promptly by exocytosis when triggered to do so (**Figure 13–66**).

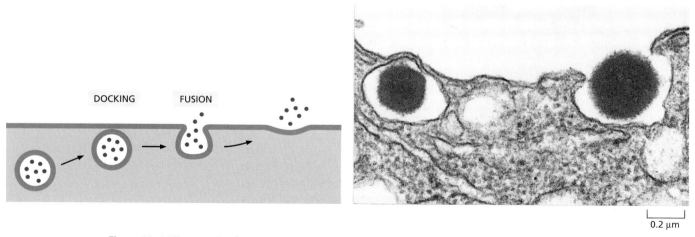

Figure 13–66 Exocytosis of secretory vesicles. The electron micrograph shows the release of insulin from a secretory vesicle of a pancreatic β cell. (Courtesy of Lelio Orci, from L. Orci, J.-D. Vassali and A. Perrelet, *Sci. Am.* 256:85–94, 1988. With permission from Scientific American.)

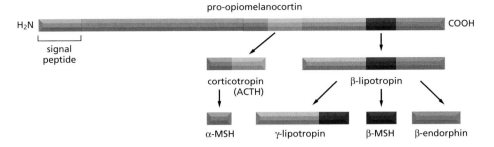

Figure 13–67 Alternative processing pathways for the prohormone pro-opiomelanocortin. The initial cleavages are made by proteases that cut next to pairs of positively charged amino acids (Lys-Arg, Lys-Lys, Arg-Lys, or Arg-Arg pairs). Trimming reactions then produce the final secreted products. Different cell types produce different concentrations of individual processing enzymes, so that the same prohormone precursor is cleaved to produce different peptide hormones. In the anterior lobe of the pituitary gland, for example, only corticotropin (ACTH) and β-lipotropin are produced from proopiomelanocortin, whereas in the intermediate lobe of the pituitary gland mainly α-melanocyte stimulating hormone (α-MSH), γ-lipotropin, β-MSH, and β-endorphin are produced.

Proteins Are Often Proteolytically Processed During the Formation of Secretory Vesicles

Concentration is not the only process to which secretory proteins are subject as the secretory vesicles mature. Many polypeptide hormones and neuropeptides, as well as many secreted hydrolytic enzymes, are synthesized as inactive protein precursors. Proteolysis is necessary to liberate the active molecules from these precursors. The cleavages begin in the *trans* Golgi network, and they continue in the secretory vesicles and sometimes in the extracellular fluid after secretion has occurred. Many secreted polypeptides have, for example, an N-terminal *pro-peptide* that is cleaved off to yield the mature protein. These proteins are thus synthesized as *pre-pro-proteins*, the *pre-peptide* consisting of the ER signal peptide that is cleaved off earlier in the rough ER (see Figure 12–38). In other cases, peptide-signaling molecules are made as *polyproteins* that contain multiple copies of the same amino acid sequence. In still more complex cases, a variety of peptide-signaling molecules are synthesized as parts of a single polyprotein that acts as a precursor for multiple end products, which are individually cleaved from the initial polypeptide chain. The same polyprotein may be processed in various ways to produce different peptides in different cell types (**Figure 13–67**).

Why is proteolytic processing so common in the secretory pathway? Some of the peptides produced in this way, such as the *enkephalins* (five-amino-acid neuropeptides with morphine-like activity), are undoubtedly too short in their mature forms to be co-translationally transported into the ER lumen or to include the necessary signal for packaging into secretory vesicles. In addition, for secreted hydrolytic enzymes—or any other protein whose activity could be harmful inside the cell that makes it—delaying activation of the protein until it reaches a secretory vesicle or until after it has been secreted has a clear advantage: it prevents it from acting prematurely inside the cell in which it is synthesized.

Secretory Vesicles Wait Near the Plasma Membrane Until Signaled to Release Their Contents

Once loaded, a secretory vesicle has to reach the site of secretion, which in some cells is far away from the Golgi apparatus. Nerve cells are the most extreme example. Secretory proteins, such as peptide neurotransmitters (neuropeptides) that are to be released from nerve terminals at the end of the axon, are made and packaged into vesicles in the cell body, where the ribosomes, ER, and Golgi apparatus are located. They must then travel along the axon to the nerve terminals, which can be a meter or more away. As discussed in Chapter 16, motor proteins propel the vesicles along axonal microtubules, whose uniform orientation guides the vesicles in the proper direction. Microtubules also guide vesicles to the cell surface for constitutive exocytosis.

Whereas vesicles containing materials for constitutive release fuse with the plasma membrane once they arrive there, secretory vesicles in the regulated pathway wait at the membrane until the cell receives a signal to secrete, and they then fuse. The signal is often a chemical messenger, such as a hormone, that binds to receptors on the cell surface. The resulting activation of the receptors

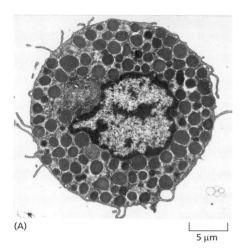 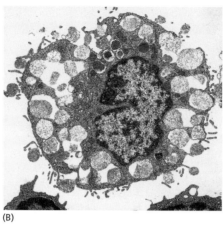

(A)   (B)

5 µm

Figure 13–68 Electron micrographs of exocytosis in rat mast cells. (A) An unstimulated mast cell. (B) This cell has been activated to secrete its stored histamine by a soluble extracellular stimulant. Histamine-containing secretory vesicles are dark, while those that have released their histamine are light. The material remaining in the spent vesicles consists of a network of proteoglycans to which the stored histamine was bound. Once a secretory vesicle has fused with the plasma membrane, the secretory vesicle membrane often serves as a target to which other secretory vesicles fuse. Thus, the cell in (B) contains several large cavities lined by the fused membranes of many spent secretory vesicles, which are now in continuity with the plasma membrane. This continuity is not always apparent in one plane of section through the cell. (From D. Lawson, C. Fewtrell, B. Gomperts and M. Raff, *J. Exp. Med.* 142:391–402, 1975. With permission from The Rockefeller University Press.)

generates intracellular signals, often including a transient increase in the concentration of free Ca^{2+} in the cytosol. In nerve terminals, the initial signal for exocytosis is usually an electrical excitation (an action potential) triggered by a chemical transmitter binding to receptors elsewhere on the same cell surface. When the action potential reaches the nerve terminals, it causes an influx of Ca^{2+} through voltage-gated Ca^{2+} channels. The binding of Ca^{2+} ions to specific sensors then triggers the secretory vesicles (called synaptic vesicles) to fuse with the plasma membrane and release their contents to the extracellular space (see Figure 11–35).

The speed of transmitter release (taking only milliseconds) indicates that the proteins mediating the fusion reaction do not undergo complex, multistep rearrangements. After vesicles have been docked to the presynaptic plasma membrane, they undergo a priming step, which prepares them for rapid fusion. The SNAREs may be partly paired, but their helices are not fully wound into the final four-helix bundle required for fusion (see Figure 13–18). Other proteins are thought to keep the SNAREs from completing the fusion reaction until the Ca^{2+} influx releases this brake. At a typical synapse, only few of the docked vesicles seem to be primed and ready for exocytosis. The use of only a few vesicles at a time allows each synapse to fire over and over again in quick succession. With each firing, new synaptic vesicles become primed to replace those that have fused and released their contents.

Regulated Exocytosis Can Be a Localized Response of the Plasma Membrane and Its Underlying Cytoplasm

Histamine is a small molecule secreted by *mast cells*. It is released by the regulated pathway in response to specific ligands that bind to receptors on the mast cell surface (see Figure 25–27). Histamine causes many of the unpleasant symptoms that accompany allergic reactions, such as itching and sneezing. When mast cells are incubated in fluid containing a soluble stimulant, massive exocytosis occurs all over the cell surface (**Figure 13–68**). But if the stimulating ligand is artificially attached to a solid bead so that it can interact only with a localized region of the mast cell surface, exocytosis is now restricted to the region where the cell contacts the bead (**Figure 13–69**).

This experiment shows that individual segments of the plasma membrane can function independently in regulated exocytosis. As a result, the mast cell, unlike a nerve cell, does not respond as a whole when it is triggered; the activation of receptors, the resulting intracellular signals, and the subsequent exocytosis are all localized in the particular region of the cell that has been excited. Similarly, localized exocytosis enables a killer lymphocyte to deliver the proteins that induce the death of a single infected target cell precisely, without endangering normal neighboring cells (see Figure 25–46).

nucleus

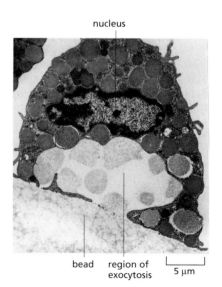

bead   region of exocytosis   5 µm

Figure 13–69 Exocytosis as a localized response. This electron micrograph shows a mast cell that has been activated to secrete histamine by a stimulant coupled to a large solid bead. Exocytosis has occurred only in the region of the cell that is in contact with the bead. (From D. Lawson, C. Fewtrell and M. Raff, *J. Cell Biol.* 79:394–400, 1978. With permission from The Rockefeller University Press.)

Secretory Vesicle Membrane Components Are Quickly Removed from the Plasma Membrane

When a secretory vesicle fuses with the plasma membrane, its contents are discharged from the cell by exocytosis, and its membrane becomes part of the plasma membrane. Although this should greatly increase the surface area of the plasma membrane, it does so only transiently, because membrane components are removed from the surface by endocytosis almost as fast as they are added by exocytosis, a process reminiscent of the exocytosis–endocytosis cycle discussed earlier. After their removal from the plasma membrane, the proteins of the secretory vesicle membrane are either recycled or shuttled to lysosomes for degradation. The amount of secretory vesicle membrane that is temporarily added to the plasma membrane can be enormous: in a pancreatic acinar cell discharging digestive enzymes for delivery to the gut lumen, about 900 μm^2 of vesicle membrane is inserted into the apical plasma membrane (whose area is only 30 μm^2) when the cell is stimulated to secrete.

Control of membrane traffic thus has a major role in maintaining the composition of the various membranes of the cell. To maintain each membrane-enclosed compartment in the secretory and endocytotic pathways at a constant size, the balance between the outward and inward flows of membrane needs to be precisely regulated. For cells to grow, the forward flow needs to be greater than the retrograde flow, so that the membrane can increase in area. For cells to maintain a constant size, the forward and retrograde flows must be equal. We still know very little about the mechanisms that coordinate these flows.

Some Regulated Exocytosis Events Serve to Enlarge the Plasma Membrane

An important task of regulated exocytosis is to deliver more membrane to enlarge the surface area of a cell's plasma membrane when such a need arises. A spectacular example is the plasma membrane expansion that occurs during the cellularization process of a fly embryo, which initially is a single cell containing about 6000 nuclei surrounded by a single plasma membrane. Within tens of minutes, the embryo is converted into the same number of cells. This process of cellularization requires a vast amount of new plasma membrane, which is added by a carefully orchestrated fusion of vesicles, eventually forming the plasma membranes that enclose the separate cells. Similar vesicle fusion events are required to enlarge the plasma membrane at the cleavage furrows of other animal and plant cells during *cytokinesis*, the process by which the two daughter cells separate after mitosis (discussed in Chapter 17).

Many cells, especially those subjected to mechanical stresses, frequently experience small ruptures to their plasma membranes during their life span. In a remarkable process thought to involve both homotypic vesicle–vesicle fusion and exocytosis, a temporary cell-surface patch is quickly fashioned from locally available internal-membrane sources, such as lysosomes. In addition to providing an emergency barrier against leaks, the patch also helps to reduce membrane tension over the wounded area allowing the "normal" bilayer to flow back together to restore continuity and seal the puncture. In wound repair, the fusion and exocytosis of vesicles is triggered by the sudden increase of Ca^{2+}, which is abundant in the extracellular space and rushes into the cells as soon as the plasma membrane is punctured. **Figure 13–70** shows three examples in which regulated exocytosis leads to plasma membrane expansion.

Polarized Cells Direct Proteins from the *Trans* Golgi Network to the Appropriate Domain of the Plasma Membrane

Most cells in tissues are *polarized* and have two (and sometimes more) distinct plasma membrane domains that are the targets of different types of vesicles.

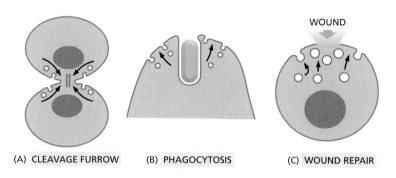

(A) CLEAVAGE FURROW   (B) PHAGOCYTOSIS   (C) WOUND REPAIR

Figure 13–70 Three examples of regulated exocytosis leading to plasma membrane enlargement. (A, B) The vesicles fusing with the plasma membrane during cytokinesis and phagocytosis are thought to be derived from endosomes, whereas (C) those involved in wound repair may be derived from lysosomes. Plasma membrane wound repair is very important in cells subjected to mechanical stresses, such as muscle cells.

This raises the general problem of how the delivery of membrane from the Golgi apparatus is organized so as to maintain the differences between one cell-surface domain and another. A typical epithelial cell has an *apical domain,* which faces an internal cavity, or the outside world, and often has specialized features such as cilia or a brush border of microvilli; it also has a *basolateral domain,* which covers the rest of the cell. The two domains are separated by a ring of *tight junctions* (see Figure 19–24), which prevent proteins and lipids (in the outer leaflet of the lipid bilayer) from diffusing between the two domains, so that the compositions of the two domains are different.

A nerve cell is another example of a polarized cell. The plasma membrane of its axon and nerve terminals is specialized for signaling to other cells, whereas the plasma membrane of its cell body and dendrites is specialized for receiving signals from other nerve cells. The two domains have distinct protein compositions. Studies of protein traffic in nerve cells in culture suggest that, with regard to vesicular transport from the *trans* Golgi network to the cell surface, the plasma membrane of the nerve cell body and dendrites resembles the basolateral membrane of a polarized epithelial cell, while the plasma membrane of the axon and its nerve terminals resembles the apical membrane of such a cell (**Figure 13–71**). Thus, some proteins that are targeted to a specific domain in the epithelial cell are also targeted to the corresponding domain in the nerve cell.

Different Strategies Guide Membrane Proteins and Lipids Selectively to the Correct Plasma Membrane Domains

In principle, differences between plasma membrane domains need not depend on the targeted delivery of the appropriate membrane components. Instead, membrane components could be delivered to all regions of the cell surface indiscriminately but then be selectively stabilized in some locations and selectively eliminated in others. Although this strategy of random delivery followed by selective retention or removal seems to be used in certain cases, deliveries are often specifically directed to the appropriate membrane domain. Epithelial cells, for example, frequently secrete one set of products—such as digestive enzymes or mucus in cells lining the gut—at their apical surface and another set of products—such as components of the basal lamina—at their basolateral sur-

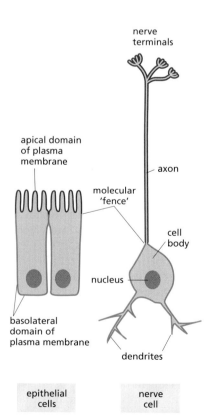

Figure 13–71 A comparison of two types of polarized cells. In terms of the mechanisms used to direct proteins to them, the plasma membrane of the nerve cell body and dendrites resembles the basolateral plasma membrane domain of a polarized epithelial cell, whereas the plasma membrane of the axon and its nerve terminals resembles the apical domain of an epithelial cell. The different membrane domains of both the epithelial cell and the nerve cell are separated by a molecular fence, consisting of a meshwork of membrane proteins tightly associated with the underlying actin cytoskeleton; this barrier—called a tight junction in the epithelial cell and an axonal hillock in neurons—keeps membrane proteins from diffusing between the two distinct domains.

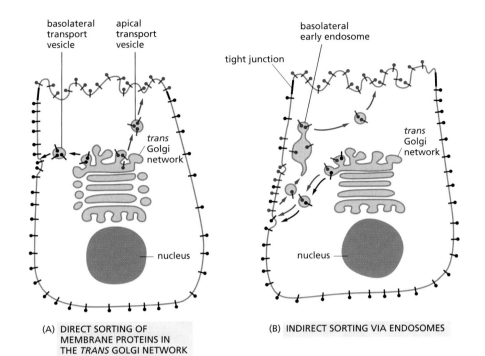

(A) DIRECT SORTING OF MEMBRANE PROTEINS IN THE *TRANS* GOLGI NETWORK

(B) INDIRECT SORTING VIA ENDOSOMES

Figure 13–72 **Two ways of sorting plasma membrane proteins in a polarized epithelial cell.** Newly synthesized proteins can reach their proper plasma membrane domain by either (A) a direct specific pathway or (B) an indirect pathway. In the indirect pathway, a protein is retrieved from the inappropriate plasma membrane domain by endocytosis and then transported to the correct domain via early endosomes—that is, by transcytosis. The indirect pathway is used in liver hepatocytes to deliver proteins to the apical domain that lines bile ducts. However, in other cases, the direct pathway is used, as in the case of the lipid-raft-dependent mechanism in epithelial cells described in the text.

face. Thus, cells must have ways of directing vesicles carrying different cargoes to different plasma membrane domains.

By examining polarized epithelial cells in culture, it has been found that proteins from the ER destined for different domains travel together until they reach the TGN. Here they are separated and dispatched in secretory or transport vesicles to the appropriate plasma membrane domain (**Figure 13–72**).

The apical plasma membrane of most epithelial cells is greatly enriched in glycosphingolipids, which help protect this exposed surface from damage—for example, from the digestive enzymes and low pH in sites such as the gut or stomach, respectively. Similarly, plasma membrane proteins that are linked to the lipid bilayer by a glycosylphosphatidylinositol (GPI) anchor (discussed in Chapter 12) are found predominantly in the apical plasma membrane. If recombinant DNA techniques are used to attach a GPI anchor to a protein that would normally be delivered to the basolateral surface, the protein is usually delivered to the apical surface instead. GPI-anchored proteins are thought to be directed to the apical membrane because they associate with glycosphingolipids in lipid rafts that form in the membrane of the TGN. As discussed in Chapter 10, lipid rafts form in the TGN and plasma membrane when glycosphingolipids and cholesterol self-associate into microaggregates (see Figure 10–14). Having selected a unique set of cargo molecules, the rafts then bud from the *trans* Golgi network into transport vesicles destined for the apical plasma membrane. Thus, similar to the selective partitioning of some membrane proteins into the specialized lipid domains in caveolae at the plasma membrane, lipid domains may also participate in protein sorting in the TGN.

Membrane proteins destined for delivery to the basolateral membrane contain sorting signals in their cytosolic tail. When present in an appropriate structural context, these signals are recognized by coat proteins that package them into appropriate transport vesicles in the TGN. The same basolateral signals that are recognized in the TGN also function in endosomes to redirect the proteins back to the basolateral plasma membrane after they have been endocytosed.

Synaptic Vesicles Can Form Directly from Endocytic Vesicles

Nerve cells (and some endocrine cells) contain two types of secretory vesicles. As for all secretory cells, these cells package proteins and peptides in dense-cored

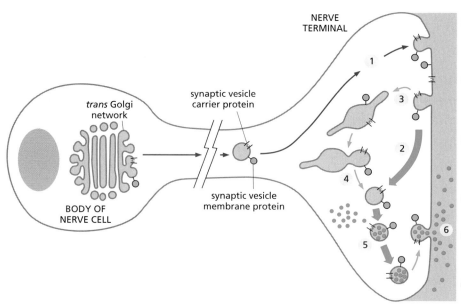

Figure 13–73 The formation of synaptic vesicles. These tiny uniform vesicles are found only in nerve cells and in some endocrine cells, where they store and secrete small-molecule neurotransmitters. The import of neurotransmitter directly into the small endocytic vesicles that form from the plasma membrane is mediated by membrane carrier proteins that function as antiports, being driven by a H^+ gradient maintained by proton pumps in the vesicle membrane.

secretory vesicles in the standard way for release by the regulated secretory pathway. In addition, however, they use another specialized class of tiny (~50-nm diameter) secretory vesicles called **synaptic vesicles**, which are generated in a different way. In nerve cells, these vesicles store small neurotransmitter molecules, such as acetylcholine, glutamate, glycine, and γ-aminobutyric acid (GABA), which mediate rapid signaling from cell to cell at chemical synapses. As discussed earlier, when an action potential arrives at a nerve terminal, it triggers the vesicles to release their contents within a fraction of a millisecond. Some neurons fire more than 1000 times per second, releasing neurotransmitters each time. This rapid release is possible because some of the vesicles are docked and primed for fusion, which will occur only when an action potential causes an influx of Ca^{2+} into the terminal.

Only a small proportion of the synaptic vesicles in the nerve terminal fuse with the plasma membrane in response to each action potential. But for the nerve terminal to respond rapidly and repeatedly, the vesicles need to be replenished very quickly after they discharge. Thus, most synaptic vesicles are generated not from the Golgi membrane in the nerve cell body but by local recycling from the plasma membrane in the nerve terminals. It is thought that the membrane components of the synaptic vesicles are initially delivered to the plasma membrane by the constitutive secretory pathway and then retrieved by endocytosis. But instead of fusing with endosomes, most of the endocytic vesicles immediately fill with transmitter to become synaptic vesicles.

The membrane components of a synaptic vesicle include transporters specialized for the uptake of neurotransmitter from the cytosol, where the small-molecule neurotransmitters that mediate fast synaptic signaling are synthesized. Once filled with neurotransmitter, the vesicles return to the plasma membrane, where they wait until the cell is stimulated. After they have released their contents, their membrane components are retrieved in the same way and used again (**Figure 13–73**).

Because synaptic vesicles are abundant and relatively uniform in size, they can be purified in large numbers and, consequently, are the best-characterized organelle of the cell. Careful quantitative proteomic analyses have identified all the components of a synaptic vesicle (**Figure 13–74**).

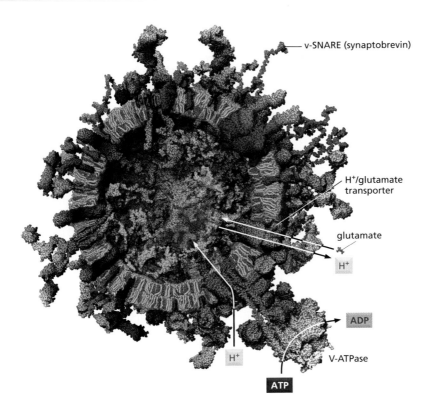

Figure 13–74 A scale model of a synaptic vesicle. The illustration shows a section through a synaptic vesicle in which proteins and lipids are drawn to scale based on their known stoichiometry and either their known or approximated structures. Only 70% of the membrane proteins estimated to be present in the membrane are shown; a complete model would therefore show a membrane that is even more crowded than this picture suggests. Each synaptic vesicle membrane contains 7000 phospholipid molecules and 5700 cholesterol molecules. Each also contains close to 50 different integral membrane protein molecules, which vary widely in their relative abundance and together contribute about 600 transmembrane α-helices. The v-SNARE synaptobrevin is the most abundant protein in the vesicle (~70 copies/vesicle). Because of its filamentous structure, it extends from the dense forest of the cytosolic protein domains, which almost completely cover the vesicle surface. By contrast, the V-ATPase, which uses ATP hydrolysis to pump H^+ into the vesicle lumen, is present in 1–2 copies per vesicle. The H^+ gradient provides the energy for neurotransmitter import by an H^+/neurotransmitter antiport, which loads each vesicle with 1800 neurotransmitter molecules, such as glutamate, one of which is shown to scale. (Adapted from S. Takamori et al., *Cell*, 127:831–846, 2006. With permission from Elsevier.) <ACCA>

Summary

Cells can secrete molecules by exocytosis in either a constitutive or a regulated fashion. Whereas the regulated pathways operate only in specialized secretory cells, a constitutive secretory pathway operates in all eucaryotic cells, characterized by continual vesicular transport from the TGN to the plasma membrane. In the regulated pathways, the molecules are stored either in secretory vesicles or in synaptic vesicles, which do not fuse with the plasma membrane to release their contents until they receive an appropriate signal. Secretory vesicles containing proteins for secretion bud from the TGN. The secretory proteins become concentrated during the formation and maturation of the secretory vesicles. Synaptic vesicles, which are confined to nerve cells and some endocrine cells, form from both endocytic vesicles and from endosomes, and they mediate the regulated secretion of small-molecule neurotransmitters.

Proteins are delivered from the TGN to the plasma membrane by the constitutive pathway unless they are diverted into other pathways or retained in the Golgi apparatus. In polarized cells, the transport pathways from the TGN to the plasma membrane operate selectively to ensure that different sets of membrane proteins, secreted proteins, and lipids are delivered to the different domains of the plasma membrane.

PROBLEMS

Which statements are true? Explain why or why not.

13–1   In all events involving fusion of a vesicle to a target membrane, the cytosolic leaflets of the vesicle and target bilayers always fuse together, as do the leaflets that are not in contact with the cytosol.

13–2   There is one strict requirement for the exit of a protein from the ER: it must be correctly folded.

13–3   All of the glycoproteins and glycolipids in intracellular membranes have their oligosaccharide chains facing the lumenal side, and all those in the plasma membrane have their oligosaccharide chains facing the outside of the cell.

13–4   During transcytosis, vesicles that form from coated pits on the apical surface fuse with the plasma membrane on the basolateral surface, and in that way transport molecules across the epithelium.

Discuss the following problems.

13–5   In a nondividing cell such as a liver cell, why must the flow of membrane between compartments be balanced, so that the retrieval pathways match the outward flow? Would you expect the same balanced flow in a gut epithelial cell, which is actively dividing?

13–6   Yeast, and many other organisms, make a single type of clathrin heavy chain and a single type of clathrin light chain; thus, they make a single kind of clathrin coat. How is it, then, that a single clathrin coat can be used for three different transport pathways—Golgi to late endosomes, plasma membrane to early endosomes, and immature secretory vesicles to Golgi—that each involves different specialized cargo proteins?

13–7   How can it possibly be true that complementary pairs of specific SNAREs uniquely mark vesicles and their target membranes? After vesicle fusion, the target membrane will contain a mixture of t-SNAREs and v-SNAREs. Initially, these SNAREs will be tightly bound to one another, but NSF can pry them apart, reactivating them. What do you suppose prevents target membranes from accumulating a population of v-SNAREs equal to or greater than their population of t-SNAREs?

13–8   Viruses are the ultimate scavengers—a necessary consequence of their small genomes. Wherever possible, they make use of the cell's machinery to accomplish the steps involved in their own reproduction. Many different viruses have membrane coverings. These so-called enveloped viruses gain access to the cytosol by fusing with a cell membrane. Why do you suppose that each of these viruses encodes its own special fusion protein, rather than making use of a cell's SNAREs?

13–9   For fusion of a vesicle with its target membrane to occur, the membranes have to be brought to within 1.5 nm so that the two bilayers can join (**Figure Q13–1**). Assuming that the relevant portions of the two membranes at the fusion site are circular regions 1.5 nm in diameter, calculate the number of water molecules that would remain between the membranes. (Water is 55.5 M and the volume of a cylinder is $\pi r^2 h$.) Given that an average phospholipid occupies a

membrane surface area of 0.2 nm², how many phospholipids would be present in each of the opposing monolayers at the fusion site? Are there sufficient water molecules to bind to the hydrophilic head groups of this number of phospholipids? (It is estimated that 10–12 water molecules are normally associated with each phospholipid head group at the exposed surface of a membrane.)

13–10   SNAREs exist as complementary partners that carry out membrane fusions between appropriate vesicles and their target membranes. In this way, a vesicle with a particular variety of v-SNARE will fuse only with a membrane that carries the complementary t-SNARE. In some instances, however, fusions of identical membranes (homotypic fusions) are known to occur. For example, when a yeast cell forms a bud, vesicles derived from the mother cell's vacuole move into the bud where they fuse with one another to form a new vacuole. These vesicles carry both v-SNAREs and t-SNAREs. Are both types of SNAREs essential for this homotypic fusion event?

To test this point, you have developed an ingenious assay for fusion of vacuolar vesicles. You prepare vesicles from two different mutant strains of yeast: strain B has a defective gene for vacuolar alkaline phosphatase (Pase); strain A is defective for the protease that converts the precursor of alkaline phosphatase (pro-Pase) into its active form (Pase) (**Figure Q13–2**A). Neither strain has active alkaline phosphatase, but when extracts of the strains are mixed, vesicle fusion generates active alkaline phosphatase, which can be easily measured (Figure Q13–2).

Now you delete the genes for the vacuolar v-SNARE, t-SNARE, or both in each of the two yeast strains. You prepare vacuolar vesicles from each and test them for their ability to fuse, as measured by the alkaline phosphatase assay (Figure Q13–2B).

What do these data say about the requirements for v-SNAREs and t-SNAREs in the fusion of vacuolar vesicles? Does it matter which kind of SNARE is on which vesicle?

13–11   If you were to remove the ER retrieval signal from protein disulfide isomerase (PDI), which is normally a soluble resident of the ER lumen, where would you expect the modified PDI to be located?

13–12   The KDEL receptor must shuttle back and forth between the ER and the Golgi apparatus to accomplish its task of ensuring that soluble ER proteins are retained in the ER lumen. In which compartment does the KDEL receptor bind its ligands more tightly? In which compartment does it bind its ligands more weakly? What is thought to be the basis for its different binding affinities in the two compartments? If you were designing the system, in which compartment would you have the highest concentration of KDEL receptor? Would you predict that the KDEL receptor,

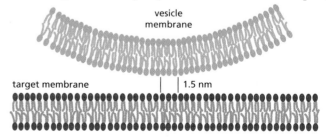

Figure Q13–1 Close approach of a vesicle and its target membrane in preparation for fusion (Problem 13–9).

Figure Q13–2 SNARE requirements for vesicle fusion (Problem 13–10). (A) Scheme for measuring the fusion of vacuolar vesicles. (B) Results of fusions of vesicles with different combinations of v-SNAREs and t-SNAREs. The SNAREs present on the vesicles of the two strains are indicated as v (v-SNARE) and t (t-SNARE).

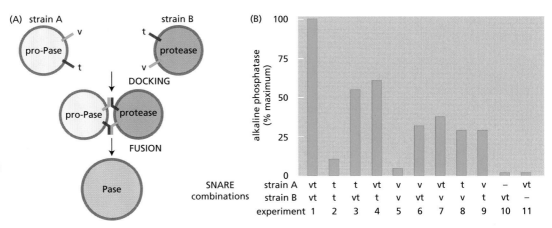

which is a transmembrane protein, would itself possess an ER retrieval signal?

13–13 How does the low pH of lysosomes protect the rest of the cell from lysosomal enzymes in case the lysosome breaks?

13–14 Melanosomes are specialized lysosomes that store pigments for eventual release by exocytosis. Various cells such as skin and hair cells then take up the pigment, which accounts for their characteristic pigmentation. Mouse mutants that have defective melanosomes often have pale or unusual coat colors. One such light-colored mouse, the *Mocha* mouse (**Figure Q13–3**), has a defect in the gene for one of the subunits of the adaptor protein complex AP3, which is associated with coated vesicles budding from the *trans* Golgi network. How might the loss of AP3 cause a defect in melanosomes?

normal mouse   *Mocha* mouse

Figure Q13–3 A normal mouse and the *Mocha* mouse (Problem 13–14). In addition to its light coat color, the *Mocha* mouse has a poor sense of balance. (Courtesy of Margit Burmeister.)

13–15 Patients with Hunter's syndrome or with Hurler's syndrome rarely live beyond their teens. These patients accumulate glycosaminoglycans in lysosomes due to the lack of specific lysosomal enzymes necessary for their degradation. When cells from patients with the two syndromes are fused, glycosaminoglycans are degraded properly, indicating that the cells are missing different degradative enzymes. Even if the cells are just cultured together, they still correct each other's defects. Most surprising of all, the medium from a culture of Hurler's cells corrects the defect in Hunter's cells (and vice versa). The corrective factors in the media are inactivated by treatment with proteases, by treatment with periodate, which destroys carbohydrate, and by treatment with alkaline phosphatase, which removes phosphates.

A. What do you suppose the corrective factors are? Beginning with the donor patient's cells, describe the route by which the factors reach the medium and subsequently enter the recipient cells to correct the lysosomal defects.

B. Why do you suppose the treatments with protease, periodate, and alkaline phosphatase inactivate the corrective factors?

C. Would you expect a similar sort of correction scheme to work for mutant cytosolic enzymes?

13–16 A macrophage ingests the equivalent of 100% of its plasma membrane each half hour by endocytosis. What is the rate at which membrane is returned by exocytosis?

13–17 Cells take up extracellular molecules by receptor-mediated endocytosis and by fluid-phase endocytosis. A classic paper compared the efficiencies of these two pathways by incubating human cells for various periods of time in a range of concentrations of either ^{125}I-labeled epidermal growth factor (EGF), to measure receptor-mediated endocytosis, or horseradish peroxidase (HRP), to measure fluid-phase endocytosis. Both EGF and HRP were found to be present in small vesicles with an internal radius of 20 nm. The uptake of HRP was linear (**Figure Q13–4**A), while that of EGF was initially linear but reached a plateau at higher concentrations (Figure Q13–4B).

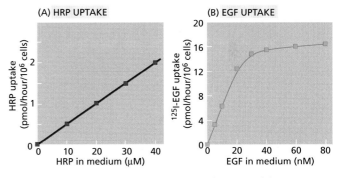

Figure Q13–4 Uptake of HRP and EGF as a function of their concentration in the medium (Problem 13–17).

A. Explain why the shapes of the curves in Figure Q13–4 are different for HRP and EGF.

B. From the curves in Figure Q13–4, estimate the difference in the uptake rates for HRP and EGF when both are present at 40 nM. What would the difference be if both were present at 40 μM?

C. Calculate the average number of HRP molecules that get taken up by each endocytic vesicle (radius 20 nm) when the medium contains 40 μM HRP. [The volume of a sphere is $(4/3)\pi r^3$.]

D. The scientists who did these experiments said at the time, "These calculations clearly illustrate how cells can internalize EGF by endocytosis while excluding all but insignificant quantities of extracellular fluid." What do you think they meant.

REFERENCES

General

Bonifacino JS & Glick BS (2004) The mechanisms of vesicle budding and fusion. *Cell* 116:153–66.

Mellman I & Warren G (2000) The road taken: past and future foundations of membrane traffic. *Cell* 100:99–112.

Pelham HR (1999) The Croonian Lecture 1999. Intracellular membrane traffic: getting proteins sorted. *Philos Trans R Soc Lond B Biol Sci* 354:1471–1478.

Rothman JE & Wieland FT (1996) Protein sorting by transport vesicles. *Science* 272:227–234.

Schekman RW (1994) Regulation of membrane traffic in the secretory pathway. *Harvey Lect* 90:41–57.

The Molecular Mechanisms of Membrane Transport and the Maintencance of Compartment Diversity

Di Paolo G & De Camilli P (2006) Phosphoinositides in cell regulation and membrane dynamics. *Nature* 443:651–657.

Grosshans BL, Ortiz D & Novick P (2006) Rabs and their effectors: achieving specificity in membrane traffic. *Proc Natl Acad Sci USA* 103:11821–11827.

Gurkan C, Stagg SM & Balch WE (2006) The COPII cage: unifying principles of vesicle coat assembly. *Nature Rev Mol Cell Biol* 7:727–738.

Jahn R & Scheller RH (2006) SNAREs—engines for membrane fusion. *Nature Rev Mol Cell Biol* 7:631–643.

Kirchhausen T (2000) Three ways to make a vesicle. *Nature Rev Mol Cell Biol* 1:187–198.

McNew JA, Parlati F, Sollner T & Rothman JE (2000) Compartmental specificity of cellular membrane fusion encoded in SNARE proteins. *Nature* 407:153–159.

Pfeffer SR (1999) Transport-vesicle targeting: tethers before SNAREs. *Nature Cell Biol* 1:E17–E22.

Robinson MS & Bonifacino JS (2001) Adaptor-related proteins. *Curr Opin Cell Biol* 13:444–453.

Seaman MN (2005) Recycle your receptors with retromer. *Trends Cell Biol* 15:68–75.

Springer S, Spang A & Schekman RW (1999) A primer on vesicle budding. *Cell* 97:145–148.

Transport from the ER Through the Golgi Apparatus

Bannykh SI, Nishimura N & Balch WE (1998) Getting into the Golgi. *Trends Cell Biol* 8:21–25.

Ellgaard L & Helenius A (2003) Quality control in the endoplasmic reticulum. *Nature Rev Mol Cell Biol* 4:181–191.

Farquhar MG & Palade GE (1998) The Golgi apparatus: 100 years of progress and controversy. *Trends Cell Biol* 8:2–10.

Glick BS (2000) Organization of the Golgi apparatus. *Curr Opin Cell Biol* 12:450–456.

Ladinsky MS, Mastronarde DN et al (1999) Golgi structure in three dimensions: functional insights from the normal rat kidney cell. *J Cell Biol* 144:1135–1149.

Lee MC, Miller EA, Goldberg J et al (2004) Bi-directional protein transport between the ER and Golgi. *Annu Rev Cell Dev Biol* 20:87–123.

Pelham HR & Rothman JE (2000) The debate about transport in the Golgi—two sides of the same coin? *Cell* 102:713–719.

Warren G & Malhotra V (1998) The organisation of the Golgi apparatus. *Curr Opin Cell Biol* 10:493–498.

Wells L & Hart GW (2003) O-GlcNAc turns twenty: functional implications for post-translational modification of nuclear and cytosolic proteins with a sugar. *FEBS Lett* 546:154–158.

Zeuschner D, Geerts WJ & Klumperman J (2006) Immuno-electron tomography of ER exit sites reveals the existence of free COPII-coated transport carriers. *Nature Cell Biol* 8:377–383.

Transport from the *Trans* Golgi Network to Lysosomes

Andrews NW (2000) Regulated secretion of conventional lysosomes. *Trends Cell Biol* 10:316–321.

de Duve C (2005) The lysosome turns fifty. *Nature Cell Biol* 7:847–849.

Kornfeld S & Mellman I (1989) The biogenesis of lysosomes. *Annu Rev Cell Biol* 5:483–525.

Levine B & Klionsky DJ (2004) Development by self-digestion: molecular mechanisms and biological functions of autophagy. *Dev Cell* 6:463–477.

Noda T, Suzuki K & Ohsumi Y (2002) Yeast autophagosomes: *de novo* formation of a membrane structure. *Trends Cell Biol* 12:231–235.

Futerman AH & van Meer G (2004) The cell biology of lysosomal storage disorders. *Nature Rev Mol Cell Biol* 5:554–565.

Peters C & von Figura K (1994) Biogenesis of lysosomal membranes. *FEBS Lett* 346:108–114.

Rouille Y, Rohn W & Hoflack B (2000) Targeting of lysosomal proteins. *Semin Cell Dev Biol* 11:165–171.

Transport into the Cell from the Plasma Membrane: Endocytosis

Anderson RG (1998) The caveolae membrane system. *Annu Rev Biochem* 67:199–225.

Bonifacino JS & Traub LM (2003) Signals for sorting of transmembrane proteins to endosomes and lysosomes. *Annu Rev Biochem* 72:395–447.

Brown MS & Goldstein JL (1986) A receptor-mediated pathway for cholesterol homeostasis. *Science* 232:34–47.

Conner SD & Schmid SL (2003) Regulated portals of entry into the cell. *Nature* 422:37–44.

Gruenberg J & Stenmark H (2004) The biogenesis of multivesicular endosomes. *Nature Rev Mol Cell Biol* 5:317–323.

Hicke L (2001) A new ticket for entry into budding vesicles-ubiquitin. *Cell* 106:527–530.

Katzmann DJ, Odorizzi G & Emr S (2002) Receptor downregulation and multivesicular-body sorting. *Nature Rev Mol Cell Biol* 3:893–905.

Maxfield FR & McGraw TE (2004) Endocytic recycling. *Nature Rev Mol Cell Biol* 5:121–132.

Mellman I (1996) Endocytosis and molecular sorting. *Annu Rev Cell Dev Biol* 12:575–625.

Pelkmans L & Helenius A (2003) Insider information:what viruses tell us about endocytosis. *Curr Opin Cell Biol* 15:414–422.

Tjelle TE, Lovdal T et al (2000) Phagosome dynamics and function. *BioEssays* 22:255–263.

Yeung T, Ozdamar B & Grinstein S (2006) Lipid metabolism and dynamics during phagocytosis. *Curr Opin Cell Biol* 18:429–437.

Transport from the *Trans* Golgi Network to the Cell Exterior: Exocytosis

Burgess TL & Kelly RB (1987) Constitutive and regulated secretion of proteins. *Annu Rev Cell Biol* 3:243–293.

Dietrich C, Volovyk ZN & Jacobson K (2001) Partitioning of Thy-1, GM1, and cross-linked phospholipid analogs into lipid rafts reconstituted in supported model membrane monolayers. *Proc Natl Acad Sci USA* 98:10642–10647.

Martin TF (1997) Stages of regulated exocytosis. *Trends Cell Biol* 7:271–276.

Mostov K, Su T et al (2003) Polarized epithelial membrane traffic: conservation and plasticity. *Nature Cell Biol* 5:287–293.

Murthy VN & De Camilli P (2003) Cell biology of the presynaptic terminal. *Annu Rev Neurosci* 26:701–728.

Schuck S & Simons K (2004) Polarized sorting in epithelial cells: raft clustering and the biogenesis of the apical membrane. *J Cell Sci* 117(Pt 25):5955–5964.

Simons K & Ikonen E (1997) Functional rafts in cell membranes. *Nature* 387:569–572.

Sudhof TC (2004) The synaptic vesicle cycle. *Annu Rev Neurosci* 27:509–547.

Tooze SA (1998) Biogenesis of secretory granules in the trans-Golgi network of neuroendocrine and endocrine cells. *Biochim Biophys Acta* 1404:231–244.

Traub LM & Kornfeld S (1997) The trans-Golgi network: a late secretory sorting station. *Curr Opin Cell Biol* 9:527–533.

Chapter 14

Energy Conversion: Mitochondria and Chloroplasts

14

Through a set of reactions that occur in the cytosol, energy derived from the partial oxidation of energy-rich carbohydrate molecules is used to form ATP, the chemical energy currency of cells (discussed in Chapter 2). But a much more efficient method of energy generation appeared very early in the history of life. This process requires a membrane, and it enables cells to acquire energy from a wide variety of sources. For example, it is central to the conversion of light energy into chemical bond energy in photosynthesis, as well as to the aerobic respiration that enables us to use oxygen to produce large amounts of ATP from food molecules.

Procaryotes use their plasma membrane to produce ATP. But the plasma membrane in eucaryotic cells is reserved for the transport processes described in Chapter 11. Eucaryotes instead use the specialized membranes inside *energy-converting organelles* to produce most of their ATP. The membrane-enclosed organelles are **mitochondria**, which are present in the cells of virtually all eucaryotic organisms (including fungi, animals, plants, algae, and protozoa), and **plastids**—most notably **chloroplasts**—which occur only in plants and algae. In electron micrographs the most striking morphological feature of both mitochondria and chloroplasts is the large amount of internal membrane they contain. This internal membrane provides the framework for an elaborate set of electron-transport processes that produce most of the cell's ATP.

The common pathway used by mitochondria, chloroplasts, and procaryotes to harness energy for biological purposes operates by a process known as **chemiosmotic coupling**—reflecting a link between the chemical bond-forming reactions that generate ATP ("chemi") and membrane-transport processes ("osmotic"). The coupling process occurs in two linked stages, both of which are performed by protein complexes embedded in a membrane:

Stage 1. High-energy electrons (derived from the oxidation of food molecules, from the action of sunlight, or from other sources discussed later) are transferred along a series of electron carriers embedded in the membrane. These electron transfers release energy that is used to pump protons (H^+, derived from the water that is plentiful in cells) across the membrane and thus generate an *electrochemical proton gradient*. As discussed in Chapter 11, an ion gradient across a membrane is a form of stored energy, which can be harnessed to do useful work when the ions are allowed to flow back across the membrane down their electrochemical gradient.

Stage 2. H^+ flows back down its electrochemical gradient through a protein machine called *ATP synthase*, which catalyzes the energy-requiring synthesis of ATP from ADP and inorganic phosphate (P_i). This ubiquitous enzyme plays the role of a turbine, permitting the proton gradient to drive the production of ATP (**Figure 14–1**).

The electrochemical proton gradient also drives other membrane-embedded protein machines (**Figure 14–2**). In eucaryotes, special proteins couple the "downhill" H^+ flow to the transport of specific metabolites into and out of the organelles. In bacteria, the electrochemical proton gradient drives more than ATP synthesis and transport processes; as a store of directly usable energy, it also drives the rapid rotation of the bacterial flagellum, which enables the bacterium to swim.

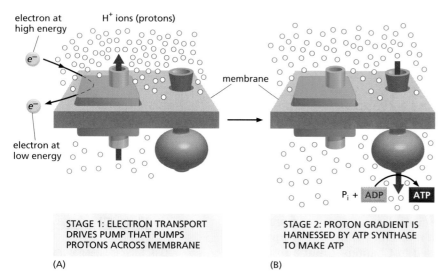

(A)   (B)

Figure 14–1 Harnessing energy for life. (A) The essential requirements for chemiosmosis are a membrane—in which are embedded a pump protein and an ATP synthase, plus a source of high-energy electrons (e^-). The protons (H^+) shown are freely available from water molecules. The pump harnesses the energy of electron transfer (details not shown here) to pump protons, creating an electrochemical proton gradient across the membrane. (B) This proton gradient serves as an energy store that can be used to drive ATP synthesis by the ATP synthase enzyme. The *red arrow* shows the direction of proton movement at each stage.

The entire set of proteins in the membrane, together with the small molecules involved in the orderly sequence of electron transfers, is called an **electron-transport chain**. The mechanism of electron transport is analogous to an electric cell driving a current through a set of electric pumps. However, in biological systems, electrons are carried between one site and another not by conducting wires, but by diffusible molecules that can pick up electrons at one location and deliver them to another. For mitochondria, the first of these electron carriers is NAD^+, which takes up two electrons (plus an H^+) to become NADH, a water-soluble small molecule that ferries electrons from the sites where food molecules are degraded to the inner mitochondrial membrane.

Figure 14–3 compares the electron-transport processes in mitochondria, which convert energy from chemical fuels, with those in chloroplasts, which convert energy from sunlight. In the mitochondrion, electrons—which have been released from a carbohydrate food molecule in the course of its degradation to CO_2—are transferred through the membrane by a chain of electron carriers, finally reducing oxygen gas (O_2) to form water. The free energy released as the electrons flow down this path from a high-energy state to a low-energy state drives a series of three H^+ pumps in the inner mitochondrial membrane, and it is the third H^+ pump in the series that catalyzes the transfer of the electrons to O_2 (see Figure 14–3A).

Although the chloroplast can be described in similar terms, and several of its main components are similar to those of the mitochondrion, the chloroplast membrane contains some crucial components not found in the mitochondrial membrane. Foremost among these are the *photosystems*, where the green pigment chlorophyll captures light energy and harnesses it to drive the transfer of electrons, much as photocells in solar panels absorb light energy and use it to drive an electric current. The electron-motive force generated by the chloroplast photosystems drives electron transfer in the direction opposite to that in mitochondria: electrons are taken from water to produce O_2, and they are donated to CO_2 (via NADPH, a compound closely related to NADH) to synthesize carbohydrate. Thus, the chloroplast generates O_2 and carbohydrate, whereas the mitochondrion consumes them (see Figure 14–3B).

It is thought that the energy-converting organelles of eukaryotes evolved from procaryotes that were engulfed by primitive eucaryotic cells and developed a symbiotic relationship with them (discussed in Chapter 12). This hypothesis explains why mitochondria and chloroplasts contain their own DNA, which codes for some of their proteins. Since their initial uptake by a host cell, these

Figure 14–2 Chemiosmotic coupling. Energy from sunlight or the oxidation of foodstuffs is first used to create an electrochemical proton gradient across a membrane. This gradient serves as a versatile energy store that drives energy-requiring reactions in mitochondria, chloroplasts, and bacteria.

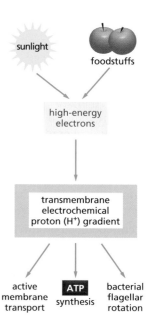

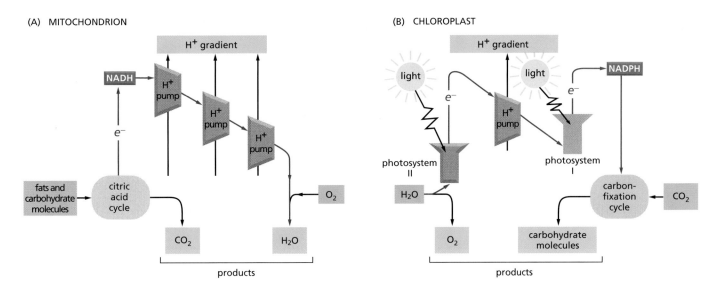

Figure 14–3 Electron-transport processes. (A) The mitochondrion converts energy from chemical fuels. **(B)** The chloroplast converts energy from sunlight. Inputs are *light green,* products are *blue,* and the path of electron flow is indicated by *red arrows.* Each of the protein complexes *(orange)* is embedded in a membrane. Note that the electron-motive force generated by the two chloroplast photosystems enables the chloroplast to drive electron transfer from H_2O to carbohydrate, and that this is *opposite* to the energetically favorable direction of electron transfer in a mitochondrion. Thus, whereas carbohydrate molecules and O_2 are inputs for the mitochondrion, they are products of the chloroplast.

organelles have lost much of their own genomes and become heavily dependent on proteins that are encoded by genes in the nucleus, synthesized in the cytosol, and then imported into the organelle. Conversely, the host cells have become dependent on these organelles for much of the ATP they need for biosyntheses, ion pumping, and movement; they have also become dependent on selected biosynthetic reactions that occur inside these organelles.

THE MITOCHONDRION

Mitochondria occupy a substantial portion of the cytoplasmic volume of eucaryotic cells, and they have been essential for the evolution of complex animals. Without mitochondria, present-day animal cells would have to depend on anaerobic glycolysis for all of their ATP. When glycolysis converts glucose to pyruvate, it releases only a small fraction of the total free energy that is potentially available from glucose oxidation. In mitochondria, the metabolism of sugars is completed: the pyruvate is imported into the mitochondrion and oxidized by O_2 to CO_2 and H_2O. This allows 15 times more ATP to be made than is produced by glycolysis alone.

Mitochondria are usually depicted as stiff, elongated cylinders with a diameter of 0.5–1 μm, resembling bacteria. Time-lapse microcinematography of living cells, however, shows that mitochondria are remarkably mobile and plastic organelles, constantly changing their shape (**Figure 14–4**) and even fusing with one another and then separating again. As they move about in the cytoplasm, they often seem to be associated with microtubules (**Figure 14–5**), which can determine the unique orientation and distribution of mitochondria in different types of cells. Thus, the mitochondria in some cells form long moving filaments or chains. In others they remain fixed in one position where they provide ATP directly to a site of unusually high ATP consumption—packed between adjacent myofibrils in a cardiac muscle cell, for example, or wrapped tightly around the flagellum in a sperm (**Figure 14–6**).

Mitochondria are large enough to be seen in the light microscope, and they were first identified during the nineteenth century. Real progress in understanding their function, however, depended on procedures developed in 1948 for isolating intact mitochondria. For technical reasons, many of these biochemical

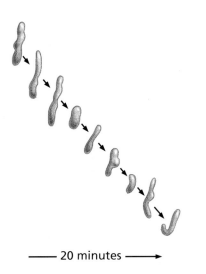

← 20 minutes →

Figure 14–4 Mitochondrial plasticity. Rapid changes of shape are often observed when an individual mitochondrion is followed in a living cell.

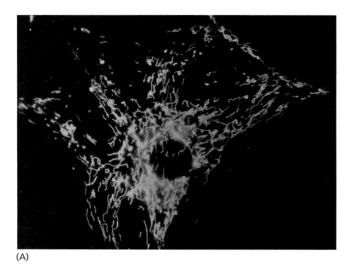

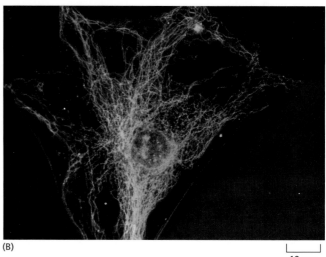

(A)

(B)

10 μm

studies were performed with mitochondria purified from liver; each liver cell contains 1000–2000 mitochondria, which in total occupy about one-fifth of the cell volume.

The Mitochondrion Contains an Outer Membrane, an Inner Membrane, and Two Internal Compartments

Each mitochondrion is enclosed by two highly specialized membranes, which have very different functions. Together they create two separate mitochondrial compartments: the internal **matrix** and a much narrower **intermembrane space**. If purified mitochondria are gently disrupted and then fractionated into separate components (**Figure 14–7**), the biochemical composition of each of the two membranes and of the spaces enclosed by them can be determined. Each contains a unique collection of proteins. Most of these 1000 or so different mitochondrial proteins are encoded in the nucleus and imported into the mitochondrion from the cytoplasm by specialized protein translocases of the outer (TOM, translocase of the outer membrane) and inner (TIM) mitochondrial membrane (discussed in Chapter 12).

The **outer membrane** contains many *porin* molecules, a type of transport protein that forms large aqueous channels through the lipid bilayer (discussed in Chapter 11). This membrane thus resembles a sieve that is permeable to all molecules of 5000 daltons or less, including small proteins. Such molecules can enter the intermembrane space, but most of them cannot pass the impermeable inner membrane. Thus, whereas the intermembrane space is chemically equivalent to the cytosol with respect to the small molecules it contains, the matrix contains a highly selected set of these molecules.

Figure 14–5 The relationship between mitochondria and microtubules. (A) A light micrograph of chains of elongated mitochondria in a living mammalian cell in culture. The cell was stained with a fluorescent dye (rhodamine 123) that specifically labels mitochondria in living cells. (B) An immunofluorescence micrograph of the same cell stained (after fixation) with fluorescent antibodies that bind to microtubules. Note that the mitochondria tend to be aligned along microtubules. (Courtesy of Lan Bo Chen.)

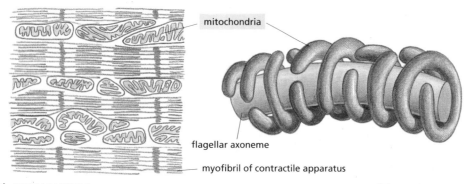

mitochondria

flagellar axoneme

myofibril of contractile apparatus

(A) CARDIAC MUSCLE

(B) SPERM TAIL

Figure 14–6 Localization of mitochondria near sites of high ATP utilization in cardiac muscle and a sperm tail. Cardiac muscle (A) in the wall of the heart is the most heavily used muscle in the body, and its continual contractions require a reliable energy supply. It has limited built in energy stores and has to depend on a steady supply of ATP from the copious mitochondria aligned close to the contractile myofibrils (see p. 1031). During the development of the flagellum of the sperm tail (B), microtubules wind helically around the axoneme, where they are thought to help localize the mitochondria in the tail; these microtubules then disappear, and the mitochondria fuse with one another to create the structure shown.

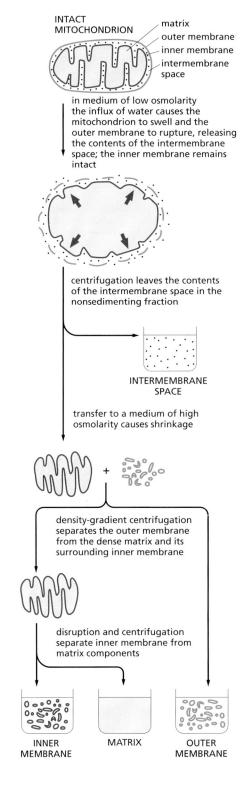

Figure 14–7 Biochemical fractionation of purified mitochondria into separate components. These techniques have made it possible to study the different proteins in each mitochondrial compartment. The method shown allows the processing of large numbers of mitochondria at the same time. It takes advantage of the fact that, in a solution of low osmotic strength, water flows into mitochondria and greatly expands the matrix space *(yellow)*. While the cristae of the inner membrane unfold to accommodate the expansion, the outer membrane—which has no folds—breaks, releasing a structure composed of only the inner membrane and the matrix.

As we explain in detail later, the major working part of the mitochondrion is the matrix and the **inner membrane** that surrounds it. The inner membrane is highly specialized. Its lipid bilayer contains a high proportion of the "double" phospholipid *cardiolipin,* which has four fatty acids rather than two and may help to make the membrane especially impermeable to ions (see Figure 14–65). This membrane also contains a variety of transport proteins that make it selectively permeable to those small molecules that are metabolized or required by the many mitochondrial enzymes concentrated in the matrix. The matrix enzymes include those that metabolize pyruvate and fatty acids to produce acetyl CoA and those that oxidize acetyl CoA in the *citric acid cycle.* The principal end products of this oxidation are CO_2, which is released from the cell as waste, and NADH, which is the main source of electrons for transport along the **respiratory chain**—the name given to the electron-transport chain in mitochondria. The enzymes of the respiratory chain are embedded in the inner mitochondrial membrane, and they are essential to the process of *oxidative phosphorylation,* which generates most of the animal cell's ATP.

As illustrated in **Figure 14–8**, the inner membrane is usually highly convoluted, forming a series of infoldings, known as **cristae**, that project into the matrix. These convolutions greatly increase the area of the inner membrane, so that in a liver cell, for example, it constitutes about one-third of the total cell membrane. The number of cristae is three times greater in the mitochondrion of a cardiac muscle cell than in the mitochondrion of a liver cell, presumably because of the greater demand for ATP in heart cells. There are also substantial differences in the mitochondrial enzymes of different cell types. In this chapter, we largely ignore these differences and focus instead on the enzymes and properties that are common to all mitochondria.

The Citric Acid Cycle Generates High-Energy Electrons

Mitochondria can use both pyruvate and fatty acids as fuel. Pyruvate comes from glucose and other sugars, whereas fatty acids come from fats. Both of these fuel molecules are transported across the inner mitochondrial membrane and are then converted to the crucial metabolic intermediate *acetyl CoA* by enzymes located in the mitochondrial matrix. The acetyl groups in acetyl CoA are then oxidized in the matrix via the **citric acid cycle**, described in Chapter 2. The cycle converts the carbon atoms in acetyl CoA to CO_2, which the cell releases as a waste product. Most importantly, this oxidation generates high-energy electrons, carried by the activated carrier molecules NADH and $FADH_2$ (**Figure 14–9**). These high-energy electrons are then transferred to the inner mitochondrial membrane, where they enter the electron-transport chain; the loss of electrons from NADH and $FADH_2$ also regenerates the NAD^+ and FAD that is needed for continued oxidative metabolism. **Figure 14–10** presents the entire sequence of reactions schematically.

A Chemiosmotic Process Converts Oxidation Energy into ATP

Although the citric acid cycle is considered to be part of aerobic metabolism, it does not itself use the oxygen. Only in the final catabolic reactions that take place on the inner mitochondrial membrane is molecular oxygen (O_2) directly

Matrix. This large internal space contains a highly concentrated mixture of hundreds of enzymes, including those required for the oxidation of pyruvate and fatty acids and for the citric acid cycle. The matrix also contains several identical copies of the mitochondrial DNA genome, special mitochondrial ribosomes, tRNAs, and various enzymes required for expression of the mitochondrial genes.

Inner membrane. The inner membrane is folded into numerous cristae, greatly increasing its total surface area. It contains proteins with three types of functions: (1) those that carry out the oxidation reactions of the electron-transport chain, (2) the ATP synthase that makes ATP in the matrix, and (3) transport proteins that allow the passage of metabolites into and out of the matrix. An electrochemical gradient of H^+, which drives the ATP synthase, is established across this membrane, so the membrane must be impermeable to ions and most small charged molecules.

Outer membrane. Because it contains a large channel-forming protein (a porin, VDAC), the outer membrane is permeable to all molecules of 5000 daltons or less. Other proteins in this membrane include enzymes involved in mitochondrial lipid synthesis and enzymes that convert lipid substrates into forms that are subsequently metabolized in the matrix, import receptors for mitochondrial proteins, and enzymatic machinery for division and fusion of the organelle.

Intermembrane space. This space contains several enzymes that use the ATP passing out of the matrix to phosphorylate other nucleotides.

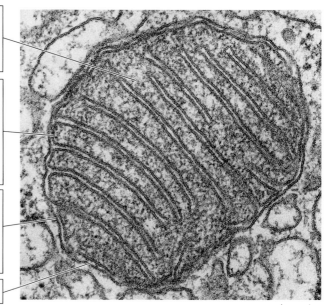

100 nm

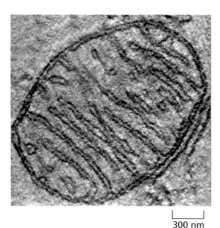

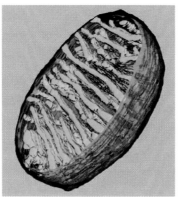

300 nm

Figure 14–8 The structure of a mitochondrion. <CGAT> In the liver, an estimated 67% of the total mitochondrial protein is located in the matrix, 21% is located in the inner membrane, 6% in the outer membrane, and 6% in the intermembrane space. As indicated below, each of these four regions contains a special set of proteins that mediate distinct functions. (Large micrograph courtesy of Daniel S. Friend; small micrograph and three-dimensional reconstruction from T.G. Frey, C.W. Renken and G.A. Perkins, *Biochim. Biophys. Acta* 1555:196–203, 2002. With permission from Elsevier.)

consumed. Nearly all the energy available from burning carbohydrates, fats, and other foodstuffs in the earlier stages of their oxidation is initially saved in the form of high-energy electrons removed from substrates by NAD^+ and FAD. These electrons, carried by NADH and $FADH_2$, then combine with O_2 by means of the

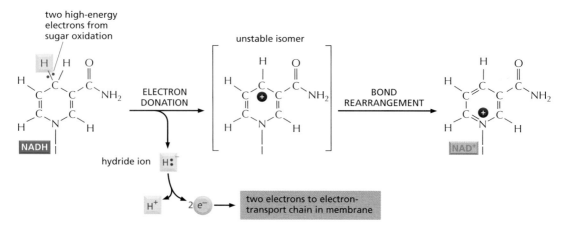

Figure 14–9 How NADH donates electrons. In this diagram, the high-energy electrons are shown as two *red dots* on a *yellow* hydrogen atom. A hydride ion (H^-, a hydrogen atom with an extra electron) is removed from NADH and is converted into a proton and two high-energy electrons: $H^- \rightarrow H^+ + 2e^-$. Only the ring that carries the electrons in a high-energy linkage is shown; for the complete structure and the conversion of NAD^+ back to NADH, see the structure of the closely related NADPH in Figure 2–60. Electrons are also carried in a similar way by $FADH_2$, whose structure is shown in Figure 2–83.

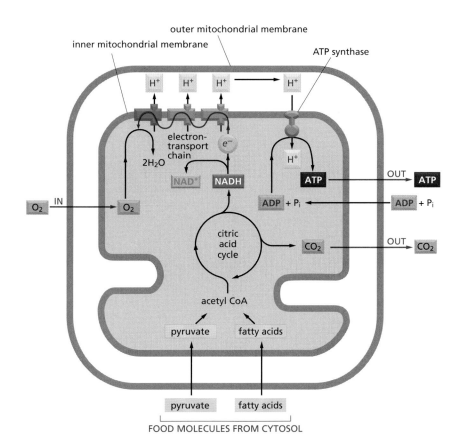

outer mitochondrial membrane

inner mitochondrial membrane

ATP synthase

FOOD MOLECULES FROM CYTOSOL

Figure 14–10 A summary of energy-generating metabolism in mitochondria. Pyruvate and fatty acids enter the mitochondrion *(bottom)* and are broken down to acetyl CoA. The acetyl CoA is then metabolized by the citric acid cycle, which reduces NAD$^+$ to NADH (and FAD to FADH$_2$, not shown). In the process of oxidative phosphorylation, high-energy electrons from NADH (and FADH$_2$) are then passed along the electron-transport chain in the inner membrane to oxygen (O$_2$). This electron transport generates a proton gradient across the inner membrane, which drives the production of ATP by ATP synthase (see Figure 14–1). The NADH generated by glycolysis in the cytosol also passes electrons to the respiratory chain (not shown). Since NADH cannot pass across the inner mitochondrial membrane, the electron transfer from cytosolic NADH must be accomplished indirectly by means of one of several "shuttle" systems that transport another reduced compound into the mitochondrion; after being oxidized, this compound is returned to the cytosol, where it is reduced by NADH again (see also Figure 14–32).

respiratory chain embedded in the inner mitochondrial membrane. The inner membrane harnesses the large amount of energy released to drive the conversion of ADP + P$_i$ to ATP. For this reason, the term **oxidative phosphorylation** is used to describe this last series of reactions (**Figure 14–11**).

As previously mentioned, the generation of ATP by oxidative phosphorylation via the respiratory chain depends on a chemiosmotic process. When it was first proposed in 1961, this mechanism explained a long-standing puzzle in cell biology. Nonetheless, the idea was so novel that it was some years before enough supporting evidence accumulated to make it generally accepted.

In the remainder of this section we shall briefly describe the type of reactions that make oxidative phosphorylation possible, saving the details of the respiratory chain for later.

NADH Transfers its Electrons to Oxygen Through Three Large Respiratory Enzyme Complexes

Although the respiratory chain harvests energy by a different mechanism than that used in other catabolic reactions, the principle is the same. The energetically favorable reaction H$_2$ + ½ O$_2$ → H$_2$O is made to occur in many small steps, so that most of the energy released can be stored instead of being lost to the environment as heat. The hydrogen atoms are first separated into protons and electrons. The electrons pass through a series of electron carriers in the inner mitochondrial membrane. At several steps along the way, protons and electrons are transiently recombined. But only at the end of the electron-transport chain are the protons returned permanently, when they are used to neutralize the negative charges created by the final addition of the electrons to the oxygen molecule (**Figure 14–12**).

The process of electron transport begins when the hydride ion is removed from NADH (to regenerate NAD$^+$) and is converted into a proton and two electrons (H$^-$ → H$^+$ + 2e^-). The two electrons are passed to the first of the more than 15 different electron carriers in the respiratory chain. The electrons start with very high energy and gradually lose it as they pass along the chain. For the most

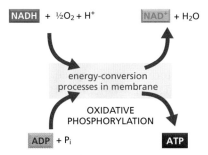

NADH + ½O$_2$ + H$^+$ NAD$^+$ + H$_2$O

energy-conversion processes in membrane

OXIDATIVE PHOSPHORYLATION

ADP + P$_i$ ATP

Figure 14–11 The major net energy conversion catalyzed by the mitochondrion. In this process of oxidative phosphorylation, the inner mitochondrial membrane serves as a device that changes one form of chemical bond energy to another, converting a major part of the energy of NADH (and FADH$_2$) oxidation into phosphate-bond energy in ATP.

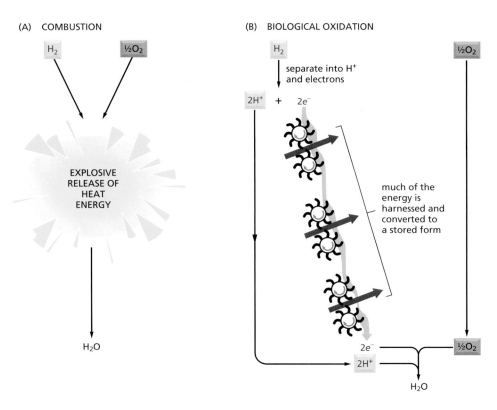

(A) COMBUSTION

(B) BIOLOGICAL OXIDATION

Figure 14–12 A comparison of biological oxidations with combustion. (A) Most of the energy would be released as heat if hydrogen were simply burned. (B) In biological oxidation by contrast, most of the released energy is stored in a form useful to the cell by means of the electron-transport chain in the inner mitochondrial membrane (the respiratory chain). The mitochondrion releases the rest of the oxidation energy as heat. In reality, the protons and electrons shown as being derived from H_2 are removed from hydrogen atoms that are covalently linked to NADH or $FADH_2$ molecules.

part, the electrons pass from one metal ion to another, each of these ions being tightly bound to a protein molecule that alters the electron affinity of the metal ion (discussed in detail later). Most of the proteins involved are grouped into three large *respiratory enzyme complexes,* each containing transmembrane proteins that hold the complex firmly in the inner mitochondrial membrane. Each complex in the chain has a greater affinity for electrons than its predecessor, and electrons pass sequentially from one complex to another until they are finally transferred to oxygen, which has the greatest affinity of all for electrons.

As Electrons Move Along the Respiratory Chain, Energy Is Stored as an Electrochemical Proton Gradient Across the Inner Membrane

The close association of the electron carriers with protein molecules makes oxidative phosphorylation possible. The proteins guide the electrons along the respiratory chain so that the electrons move sequentially from one enzyme complex to another. The transfer of electrons is coupled to oriented H^+ uptake and release, as well as to allosteric changes in energy-converting protein pumps. The net result is the pumping of H^+ across the inner membrane—from the matrix to the intermembrane space, driven by the energetically favorable flow of electrons. This movement of H^+ has two major consequences:

1. It generates a pH gradient across the inner mitochondrial membrane, with the pH higher in the matrix than in the cytosol, where the pH is generally close to 7. (Since small molecules equilibrate freely across the outer membrane of the mitochondrion, the pH in the intermembrane space is the same as in the cytosol.)

2. It generates a voltage gradient *(membrane potential)* across the inner mitochondrial membrane, with the inside negative and the outside positive (as a result of the net outflow of positive ions).

The pH gradient (ΔpH) drives H^+ back into the matrix, thereby reinforcing the effect of the membrane potential (ΔV), which acts to attract any positive ion into the matrix and to push any negative ion out. Together, the ΔpH and the ΔV are said to constitute an **electrochemical proton gradient** (**Figure 14–13**). The

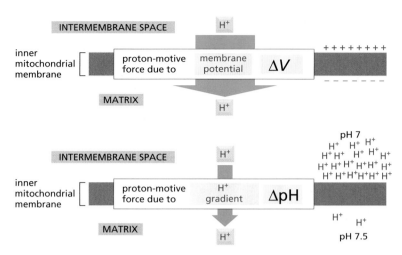

Figure 14–13 **The two components of the electrochemical proton gradient.** The total proton-motive force across the inner mitochondrial membrane consists of a large force due to the membrane potential (traditionally designated $\Delta\psi$ by experts, but designated ΔV in this text) and a smaller force due to the H^+ gradient (ΔpH). Both forces act to drive H^+ into the matrix.

electrochemical proton gradient exerts a **proton-motive force**, which can be measured in units of millivolts (mV). In a typical cell, the proton-motive force across the inner membrane of a respiring mitochondrion is about 180 to 190 mV (inside negative), and it is made up of a membrane potential of about 160 to 170 mV and a pH gradient of about 0.3 to 0.5 pH units (each ΔpH of 1 pH unit has an effect equivalent to a membrane potential of about 60 mV).

The Proton Gradient Drives ATP Synthesis

The electrochemical proton gradient across the inner mitochondrial membrane drives ATP synthesis in the critical process of oxidative phosphorylation (**Figure 14–14**). This is made possible by the membrane-bound enzyme **ATP synthase**, mentioned previously. This enzyme creates a hydrophilic pathway across the inner mitochondrial membrane that allows protons to flow down their electrochemical gradient. As these ions thread their way through the ATP synthase, they are used to drive the energetically unfavorable reaction between ADP and P_i that makes ATP (see Figure 2–27). The ATP synthase is of ancient origin; the same enzyme occurs in the mitochondria of animal cells, the chloroplasts of plants and algae, and in the plasma membrane of bacteria and archaea.

Figure 14–15 shows the structure of ATP synthase. Also called the F_0F_1 ATPase, it is a multisubunit protein with a mass of more than 500,000 daltons that works by rotary catalysis. A large enzymatic portion, shaped like a lollipop head and composed of a ring of 6 subunits, projects on the matrix side of the inner mitochondrial membrane. An elongated arm holds this head in place by tying it to a group of transmembrane proteins that form a "stator" in the membrane. This stator is in contact with a "rotor" composed of a ring of 10 to 14 identical transmembrane protein subunits. As protons pass through a narrow channel formed at the stator–rotor contact, their movement causes the rotor ring to spin. This spinning also turns a stalk attached to the rotor (*blue* in Figure 14–15B), which is thereby made to turn rapidly inside the lollipop head. As a result, the energy of proton flow down a gradient has been converted into the

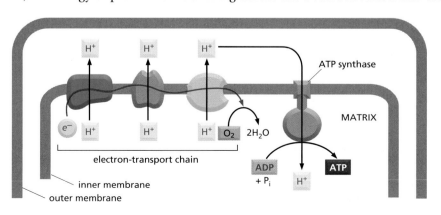

Figure 14–14 **The general mechanism of oxidative phosphorylation.** As a high-energy electron is passed along the electron-transport chain, some of the energy released drives the three respiratory enzyme complexes that pump H^+ out of the matrix. The resulting electrochemical proton gradient across the inner membrane drives H^+ back through the ATP synthase, a transmembrane protein complex that uses the energy of the H^+ flow to synthesize ATP from ADP and P_i in the matrix.

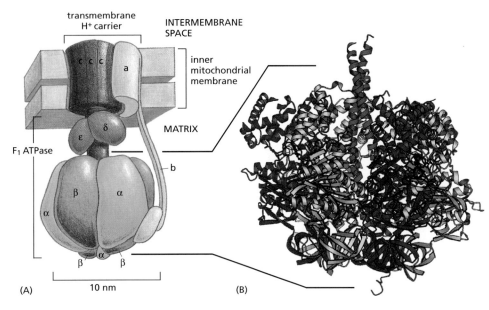

(A)   10 nm   (B)

Figure 14–15 ATP synthase. <ATCG> <GAGA> (A) The enzyme is composed of a head portion, called the F_1 ATPase, and a transmembrane H^+ carrier, called F_0. Both F_1 and F_0 are formed from multiple subunits, as indicated. A rotating stalk is fixed to a rotor *(red)* formed by a ring of 10–14 c subunits in the membrane. The stator *(green)* is formed from transmembrane a subunits, tied to other subunits that create an elongated arm. This arm fixes the stator to a ring of 3α and 3β subunits that forms the head, which likewise cannot rotate. (B) The three-dimensional structure of the F_1 ATPase, determined by x-ray crystallography. This part of the ATP synthase derives its name from its ability to carry out the reverse of the ATP synthesis reaction—namely, the hydrolysis of ATP to ADP and P_i, when detached from the transmembrane portion. (B, courtesy of John Walker, from J.P. Abrahams et al., *Nature* 370:621–628, 1994. With permission from Macmillan Publishers Ltd.)

mechanical energy of two sets of proteins rubbing against each other: rotating stalk proteins pushing against a stationary ring of head proteins.

Three of the six subunits in the head contain binding sites for ADP and inorganic phosphate. These are driven to form ATP as mechanical energy is converted into chemical bond energy through the repeated changes in protein conformation that the rotating stalk creates. In this way, the ATP synthase is able to produce more than 100 molecules of ATP per second, generating 3 molecules of ATP per revolution. The number of proton-translocating subunits in the rotor is different in different ATP synthases, and it is this number that determines the number of protons that need to pass through this marvelous device to make each molecule of ATP (its "gear ratio", which is generally a non-integral number between 3 and 5).

The Proton Gradient Drives Coupled Transport Across the Inner Membrane

The electrochemical proton gradient drives other processes besides ATP synthesis. In mitochondria, many charged small molecules, such as pyruvate, ADP, and P_i, are pumped into the matrix from the cytosol, while others, such as ATP, must be moved in the opposite direction. Transporters that bind these molecules can couple their transport to the energetically favorable flow of H^+ into the mitochondrial matrix. Thus, for example, pyruvate and inorganic phosphate (P_i) are co-transported inward with H^+ as the H^+ moves into the matrix.

ADP and ATP are co-transported in opposite directions by a single transporter protein. Since an ATP molecule has one more negative charge than ADP, each nucleotide exchange results in a total of one negative charge being moved out of the mitochondrion. Thus, the voltage difference across the membrane drives this ADP–ATP co-transporter (**Figure 14–16**).

We have seen how, in eucaryotic cells, the electrochemical proton gradient across the inner mitochondrial membrane is used to drive both the formation of ATP and the transport of metabolites across the membrane. In bacteria, a similar gradient across the bacterial plasma membrane is harnessed to drive these two types of processes. In motile bacteria, this gradient also drives the rapid rotation of the bacterial flagellum, which propels the bacterium along (**Figure 14–17**).

Proton Gradients Produce Most of the Cell's ATP

As stated previously, glycolysis alone produces a net yield of 2 molecules of ATP for every molecule of glucose that is metabolized, and this is the total energy

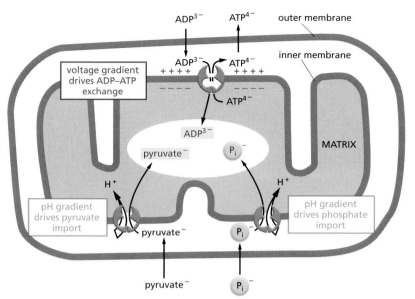

Figure 14–16 Some of the active transport processes driven by the electrochemical proton gradient across the inner mitochondrial membrane. Pyruvate, inorganic phosphate (P$_i$), and ADP are moved into the matrix, while ATP is pumped out. The charge on each of the transported molecules is indicated for comparison with the membrane potential, which is negative inside, as shown. The outer membrane is freely permeable to all of these compounds. The active transport of molecules across membranes by transporter proteins is discussed in Chapter 11.

yield for the fermentation processes that occur in the absence of O$_2$ (discussed in Chapter 2). During oxidative phosphorylation, each pair of electrons donated by the NADH produced in mitochondria is thought to provide energy for the formation of about 2.5 molecules of ATP, after subtracting the energy needed for transporting this ATP to the cytosol. Oxidative phosphorylation also produces 1.5 ATP molecules per electron pair from FADH$_2$, or from the NADH molecules produced by glycolysis in the cytosol. From the product yields of glycolysis and the citric acid cycle summarized in **Table 14–1A**, we can calculate that the complete oxidation of one molecule of glucose—starting with glycolysis and ending with oxidative phosphorylation—gives a net yield of about 30 ATPs.

In conclusion, the vast majority of the ATP produced from the oxidation of glucose in an animal cell is produced by chemiosmotic mechanisms in the mitochondrial membrane. Oxidative phosphorylation in the mitochondrion also produces a large amount of ATP from the NADH and the FADH$_2$ that is derived from the oxidation of fats (Table 14–1B; see also Figure 2–81).

Mitochondria Maintain a High ATP:ADP Ratio in Cells

Because of the carrier protein in the inner mitochondrial membrane that exchanges ATP for ADP, the ADP molecules produced by ATP hydrolysis in the

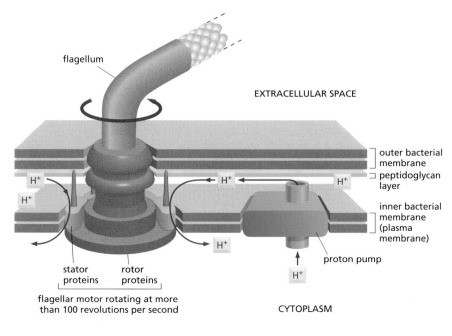

Figure 14–17 The rotation of the bacterial flagellum driven by H$^+$ flow. The flagellum is attached to a series of protein rings *(orange)*, which are embedded in the outer and inner membranes and rotate with the flagellum. The rotation is driven by a flow of protons through an outer ring of proteins (the stator) by mechanisms that may resemble those used by the ATP synthase.

Table 14–1 Product Yields from the Oxidation of Sugars and Fats

A. NET PRODUCTS FROM OXIDATION OF ONE MOLECULE OF GLUCOSE
In cytosol (glycolysis) 1 glucose → 2 pyruvate + 2 NADH + 2 ATP In mitochondrion (pyruvate dehydrogenase and citric acid cycle) 2 pyruvate → 2 acetyl CoA + 2 NADH 2 acetyl CoA → 6 NADH + 2 FADH$_2$ + 2 GTP
Net result in mitochondrion 2 pyruvate → 8 NADH + 2 FADH$_2$ + 2 GTP
B. NET PRODUCTS FROM OXIDATION OF ONE MOLECULE OF PALMITOYL COA (ACTIVATED FORM OF PALMITATE, A FATTY ACID)
In mitochondrion (fatty acid oxidation and citric acid cycle) 1 palmitoyl CoA → 8 acetyl CoA + 7 NADH + 7 FADH$_2$ 8 acetyl CoA → 24 NADH + 8 FADH$_2$ + 8 GTP
Net result in mitochondrion 1 palmitoyl CoA → 31 NADH + 15 FADH$_2$ + 8 GTP

cytosol rapidly enter mitochondria for recharging, while the ATP molecules formed in the mitochondrial matrix by oxidative phosphorylation are rapidly pumped into the cytosol where they are needed. A typical ATP molecule in the human body shuttles out of a mitochondrion and back into it (as ADP) for recharging more than once per minute, and the cell maintains a concentration of ATP that is about 10 times higher than the concentration of ADP.

As discussed in Chapter 2, biosynthetic enzymes often drive energetically unfavorable reactions by coupling them to the energetically favorable hydrolysis of ATP (see Figure 2–59). The ATP pool is therefore used to drive cellular processes in much the same way that a battery can be used to drive electric engines. If the activity of the mitochondria is blocked, ATP levels fall and the cell's battery runs down; eventually, energetically unfavorable reactions are no longer driven, and the cell dies. The poison cyanide, which blocks electron transport in the inner mitochondrial membrane, causes death in exactly this way.

It might seem that cellular processes would stop only when the concentration of ATP reaches zero; but, in fact, life is more demanding: it depends on cells maintaining a concentration of ATP that is high compared with the concentrations of ADP and P$_i$. To explain why, we must consider some elementary principles of thermodynamics.

A Large Negative Value of ΔG for ATP Hydrolysis Makes ATP Useful to the Cell

In Chapter 2, we introduced the concept of free energy *(G)*. The free-energy change for a reaction, ΔG, determines whether this reaction will occur in a cell. We showed on p. 76 that the ΔG for a given reaction can be written as the sum of two parts: the first, called the *standard free-energy change*, $\Delta G°$, depends on the intrinsic characters of the reacting molecules; the second depends on their concentrations. For the simple reaction A → B,

$$\Delta G = \Delta G° + RT \ln \frac{[B]}{[A]}$$

where [A] and [B] denote the concentrations of A and B, and ln is the natural logarithm. $\Delta G°$ is therefore only a reference value that equals the value of ΔG when the molar concentrations of A and B are equal (ln 1 = 0).

In Chapter 2, ATP was described as the major "activated energy carrier molecule" in cells. The large, favorable free-energy change (large negative ΔG) for its hydrolysis is used, through *coupled reactions*, to drive other chemical reactions that would otherwise not occur (see pp. 79–87). The ATP hydrolysis

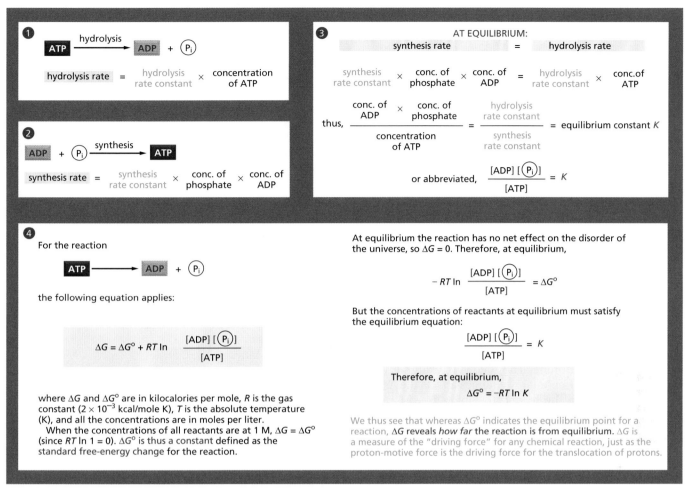

Figure 14–18 The basic relationship between free-energy changes and equilibrium in the ATP hydrolysis reaction. The rate constants in boxes 1 and 2 are determined from experiments in which product accumulation is measured as a function of time. The equilibrium constant shown here, *K*, is in units of moles per liter. (See Panel 2–7, pp. 118–119, for a discussion of free energy and see Figure 3–43 for a discussion of the equilibrium constant.)

reaction produces two products, ADP and inorganic phosphate (P_i); it is therefore of the type A → B + C, where, as described in **Figure 14–18**,

$$\Delta G = \Delta G° + RT\ln\frac{[B][C]}{[A]}$$

When ATP is hydrolyzed to ADP and P_i under the conditions that normally exist in a cell, the free-energy change is roughly –11 to –13 kcal/mole (–46 to –54 kJ/mole). This extremely favorable ΔG depends on having a high concentration of ATP in the cell compared with the concentration of ADP and P_i. When ATP, ADP, and P_i are all present at the same concentration of 1 mole/liter (so-called standard conditions), the ΔG for ATP hydrolysis is the standard free-energy change ($\Delta G°$), which is only –7.3 kcal/mole (–30.5 kJ/mol). At much lower concentrations of ATP relative to ADP and P_i, ΔG becomes zero. At this point, the rate at which ADP and P_i will join to form ATP will be equal to the rate at which ATP hydrolyzes to form ADP and P_i. In other words, when $\Delta G = 0$, the reaction is at *equilibrium* (see Figure 14–18).

It is ΔG, not $\Delta G°$, that indicates how far a reaction is from equilibrium and determines whether it can be used to drive other reactions. Because the efficient conversion of ADP to ATP in mitochondria maintains such a high concentration of ATP relative to ADP and P_i, the ATP-hydrolysis reaction in cells is kept very far from equilibrium and ΔG is correspondingly very negative. Without this large disequilibrium, ATP hydrolysis could not be used to direct the reactions of the cell; for example, many biosynthetic reactions would run backward rather than forward at low ATP concentrations.

ATP Synthase Can Function in Reverse to Hydrolyze ATP and Pump H⁺

In addition to harnessing the flow of H⁺ down an electrochemical proton gradient to make ATP, the ATP synthase can work in reverse: it can use the energy of ATP hydrolysis to pump H⁺ across the inner mitochondrial membrane (**Figure 14–19**). It thus acts as a *reversible coupling device,* interconverting electrochemical proton gradient and chemical bond energies. The direction of action at any instant depends on the balance between the steepness of the electrochemical proton gradient and the local ΔG for ATP hydrolysis, as we now explain.

The exact number of protons needed to make each ATP molecule depends on the number of subunits in the ring of transmembrane proteins that forms the base of the rotor (see Figure 14–15). However, to illustrate the principles involved, let us assume that one molecule of ATP is made by the ATP synthase for every 3 protons driven through it. Whether the ATP synthase works in its ATP-synthesizing or its ATP-hydrolyzing direction at any instant depends, in this case, on the exact balance between the favorable free-energy change for moving the three protons across the membrane into the matrix, ΔG_{3H^+} (which is less than zero), and the unfavorable free-energy change for ATP *synthesis* in the matrix, $\Delta G_{ATP\ synthesis}$ (which is greater than zero). As just discussed, the value of $\Delta G_{ATP\ synthesis}$ depends on the exact concentrations of the three reactants ATP, ADP, and P_i in the mitochondrial matrix (see Figure 14–18). The value of ΔG_{3H^+}, in contrast, is directly proportional to the value of the proton-motive force across the inner mitochondrial membrane. The following example will show how the balance between these two free-energy changes affects the ATP synthase.

As explained in the legend to Figure 14–19, a single H⁺ moving into the matrix down an electrochemical gradient of 200 mV liberates 4.6 kcal/mole (19.2 kJ/mol) of free energy, while the movement of three protons liberates three times this much free energy (ΔG_{3H^+} = –13.8 kcal/mole; 57.7 kJ/mol). Thus, if the proton-motive force remains constant at 200 mV, the ATP synthase synthesizes ATP until a ratio of ATP to ADP and P_i is reached where $\Delta G_{ATP\ synthesis}$ is just equal to +13.8 kcal/mole (57.7 kJ/mol; here $\Delta G_{ATP\ synthesis} + \Delta G_{3H^+} = 0$). At this point there is no further net ATP synthesis or hydrolysis by the ATP synthase.

Suppose the energy-requiring reactions in the cytosol suddenly hydrolyze a large amount of ATP, causing the ATP:ADP ratio in the matrix to fall. Now the value of $\Delta G_{ATP\ synthesis}$ will decrease (see Figure 14–18), and ATP synthase will begin to synthesize ATP again to restore the original ATP:ADP ratio. Alternatively, if the proton-motive force drops suddenly and is then maintained at a constant 160 mV, ΔG_{3H^+} will change to –11.0 kcal/mole (–46 kJ/mol). As a result, ATP synthase will start hydrolyzing some of the ATP in the matrix until a new balance of ATP to ADP and P_i is reached (where $\Delta G_{ATP\ synthesis}$ = +11.0 kcal/mole, or +46 kJ/mol), and so on.

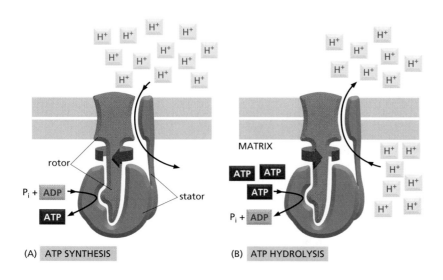

(A) ATP SYNTHESIS   (B) ATP HYDROLYSIS

Figure 14–19 The ATP synthase is a reversible coupling device that can convert the energy of the electrochemical proton gradient into chemical-bond energy, or vice versa. <ATCG> The ATP synthase can either (A) synthesize ATP by harnessing the proton-motive force or (B) pump protons against their electrochemical gradient by hydrolyzing ATP. The direction of operation at any given instant depends on the net free-energy change (ΔG) for the coupled processes of H⁺ translocation across the membrane and the synthesis of ATP from ADP and P_i. Measurement of the torque that the ATP synthase can produce when hydrolyzing ATP reveals that the synthase can pump 60 times more strongly than a diesel-engine of equal weight.

The free-energy change (ΔG) for ATP hydrolysis depends on the concentrations of the three reactants ATP, ADP, and P_i (see Figure 14–18); the ΔG for ATP synthesis is the negative of this value. The ΔG for proton translocation across the membrane is proportional to the proton-motive force. The conversion factor between them is the faraday. Thus, ΔG_{H^+} = –0.023 (proton-motive force), where ΔG_{H^+} is in kcal/mole and the proton-motive force is in mV. For an electrochemical proton gradient (proton-motive force) of 200 mV, ΔG_{H^+} = –4.6 kcal/mole (–19.2kJ/mole).

In many bacteria, the ATP synthase is routinely reversed in a transition between aerobic and anaerobic metabolism, as we shall see later. And the V-type ATPases that acidify organelles, which are architecturally similar to the ATP synthase, normally function in reverse (see Figure 13–36). Other membrane transport proteins that couple the transmembrane movement of an ion to ATP synthesis or hydrolysis share the same type of reversibility. Both the Na^+-K^+ pump and the Ca^{2+} pump described in Chapter 11, for example, normally hydrolyze ATP and use the energy released to move their specific ions across a membrane. If either of these pumps is exposed to an abnormally steep gradient of the ions it transports, however, it will act in reverse—synthesizing ATP from ADP and P_i instead of hydrolyzing it. Thus, the ATP synthase is by no means unique in its ability to convert the electrochemical energy stored in a transmembrane ion gradient directly into phosphate-bond energy in ATP.

Summary

The mitochondrion performs most cellular oxidations and produces the bulk of the animal cell's ATP. A mitochondrion is enclosed by two concentric membranes, and its major working part is the inner-most space (the matrix) and the inner membrane that surrounds it. The matrix contains a large variety of enzymes—including those that convert pyruvate and fatty acids to acetyl CoA and those that oxidize this acetyl CoA to CO_2 through the citric acid cycle. These oxidation reactions produce large amounts of NADH (and $FADH_2$). The electron-transport chain (respiratory chain) located in the inner mitochondrial membrane then harnesses the energy available from combining molecular oxygen with the reactive electrons carried by the NADH and $FADH_2$.

The respiratory chain uses energy derived from electron transport to pump H^+ out of the matrix to create a transmembrane electrochemical proton (H^+) gradient, which includes contributions from both a membrane potential and a pH difference. The large amount of free energy released when H^+ flows back into the matrix (across the inner membrane) provides the basis for ATP production in the matrix by a remarkable protein machine—the ATP synthase, a reversible coupling device between proton flows and ATP synthesis or hydrolysis. The transmembrane electrochemical gradient also drives the active transport of selected metabolites across the mitochondrial inner membrane, including an efficient ATP–ADP exchange between the mitochondrion and the cytosol that keeps the cell's ATP pool highly charged. The resulting high ratio of ATP to its hydrolysis products makes the free-energy change for ATP hydrolysis extremely favorable, allowing this hydrolysis reaction to drive a large number of energy-requiring processes throughout the cell.

ELECTRON-TRANSPORT CHAINS AND THEIR PROTON PUMPS

Having considered in general terms how a mitochondrion uses electron transport to create an electrochemical proton gradient, we now turn to the mechanisms that underlie this membrane-based energy-conversion process. In doing so, we also accomplish a larger purpose. As emphasized at the beginning of this chapter, mitochondria, chloroplasts, archaea, and bacteria use very similar chemiosmotic mechanisms. In fact, these mechanisms underlie the function of nearly all living organisms—including anaerobes that derive all their energy from electron transfers between two inorganic molecules.

Our goal in this section is to explain how electron-transport process can pump protons across a membrane. We start with some of the basic principles on which this process depends.

Protons Are Unusually Easy to Move

Although protons resemble other positive ions such as Na^+ and K^+ in normally requiring proteins to move them across membranes, in some respects they are

Figure 14–20 How protons behave in water. (A) Protons move very rapidly along a chain of hydrogen-bonded water molecules. In this diagram, proton jumps are indicated by *blue arrows,* and hydronium ions are indicated by *green shading.* As discussed in Chapter 2, naked protons rarely exist as such; they are instead associated with a water molecule in the form of a hydronium ion, H_3O^+. At a neutral pH (pH 7.0), the hydronium ions are present at a concentration of 10^{-7} M. However, for simplicity, we usually refer to this as an H^+ concentration of 10^{-7} M (see Panel 2–2, pp. 108–109). (B) Electron transfer can result in the transfer of entire hydrogen atoms, because protons are readily accepted from or donated to water inside cells. In this example, molecule A picks up an electron plus a proton when it is reduced, and B loses an electron plus a proton when it is oxidized.

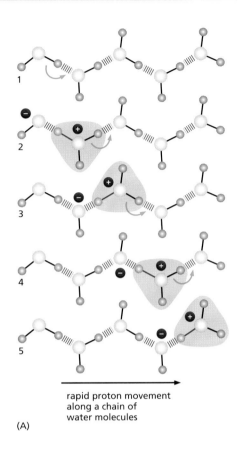

rapid proton movement along a chain of water molecules

(A)

unique. Hydrogen atoms are by far the most abundant type of atom in living organisms; they are plentiful not only in all carbon-containing biological molecules, but also in the water molecules that surround them. The protons in water are highly mobile, flickering through the hydrogen-bonded network of water molecules by rapidly dissociating from one water molecule to associate with its neighbor, as illustrated in **Figure 14–20A**. Protons are thought to move across a protein pump embedded in a lipid bilayer in a similar way: they transfer from one amino acid side chain to another, following a special channel through the protein.

Protons are also special with respect to electron transport. Whenever a molecule is reduced by acquiring an electron, the electron (e^-) brings with it a negative charge. In many cases, the addition of a proton (H^+) from water rapidly neutralizes this charge, so that the net effect of the reduction is to transfer an entire hydrogen atom, $H^+ + e^-$ (Figure 14–20B). Similarly, when a molecule is oxidized, a hydrogen atom removed from it can be readily dissociated into its constituent electron and proton—allowing the electron to be transferred separately to a molecule that accepts electrons, while the proton is passed to the water. Therefore, in a membrane in which electrons are being passed along an electron-transport chain, pumping protons from one side of the membrane to another can be relatively simple. The electron carrier merely needs to be arranged in the membrane in a way that causes it to pick up a proton from one side of the membrane when it accepts an electron, and to release the proton on the other side of the membrane as the electron is passed to the next carrier molecule in the chain (**Figure 14–21**).

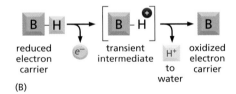

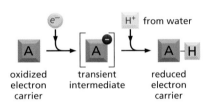

(B)

The Redox Potential Is a Measure of Electron Affinities

In biochemical reactions, any electrons removed from one molecule are always passed to another, so that whenever one molecule is oxidized, another is reduced. Like any other chemical reaction, the tendency of such oxidation–reduction reactions, or **redox reactions**, to proceed spontaneously depends on the free-energy change (ΔG) for the electron transfer, which in turn depends on the relative affinities of the two molecules for electrons.

Because electron transfers provide most of the energy for living things, it is worth taking the time to understand them. As discussed in Chapter 2, acids and bases donate and accept protons (see Panel 2–2, pp. 108–109). Acids and bases exist in conjugate acid–base pairs, in which the acid is readily converted into the base by the loss of a proton. For example, acetic acid (CH_3COOH) is converted into its conjugate base (CH_3COO^-) in the reaction:

$$CH_3COOH \rightleftharpoons CH_3COO^- + H^+$$

In exactly the same way, pairs of compounds such as NADH and NAD^+ are called **redox pairs**, since NADH is converted to NAD^+ by the loss of electrons in the reaction:

$$NADH \rightleftharpoons NAD^+ + H^+ + 2e^-$$

NADH is a strong electron donor: because its electrons are held in a high-energy linkage, the free-energy change for passing its electrons to many other molecules is favorable (see Figure m14–9/14–9). It is difficult to form a high-energy linkage. Therefore its redox partner, NAD^+, is of necessity a weak electron acceptor.

We can measure the tendency to transfer electrons from any redox pair experimentally. All that is required is the formation of an electrical circuit linking a 1:1 (equimolar) mixture of the redox pair to a second redox pair that has been arbitrarily selected as a reference standard, so that we can measure the voltage difference between them (**Panel 14–1**, p. 830). This voltage difference is defined as the **redox potential**; as defined, electrons move spontaneously from a redox pair like NADH/NAD$^+$ with a low redox potential (a low affinity for electrons) to a redox pair like O_2/H_2O with a high redox potential (a high affinity for electrons). Thus, NADH is a good molecule for donating electrons to the respiratory chain, while O_2 is well suited to act as the "sink" for electrons at the end of the pathway. As explained in Panel 14–1, the difference in redox potential, $\Delta E_0'$, is a direct measure of the standard free-energy change ($\Delta G°$) for the transfer of an electron from one molecule to another.

Electron Transfers Release Large Amounts of Energy

As just discussed, those pairs of compounds that have the most negative redox potentials have the weakest affinity for electrons and therefore contain carriers with the strongest tendency to donate electrons. Conversely, those pairs that have the most positive redox potentials have the strongest affinity for electrons and therefore contain carriers with the strongest tendency to accept electrons. A 1:1 mixture of NADH and NAD$^+$ has a redox potential of –320 mV, indicating that NADH has a strong tendency to donate electrons; a 1:1 mixture of H_2O and $\frac{1}{2}O_2$ has a redox potential of +820 mV, indicating that O_2 has a strong tendency to accept electrons. The difference in redox potential is 1.14 volts (1140 mV), which means that the transfer of each electron from NADH to O_2 under these standard conditions is enormously favorable, where $\Delta G° = -26.2$ kcal/mole (110 kJ/mol), or twice this amount for the two electrons transferred per NADH molecule (see Panel 14–1). If we compare this free-energy change with that for the formation of the phosphoanhydride bonds in ATP, where $\Delta G° = -7.3$ kcal/mole (–30.5 kJ/mol; see Figure 2–75), we see that the oxidization of one NADH molecule releases more than enough energy to synthesize several molecules of ATP from ADP and P_i.

Living systems could certainly have evolved enzymes that would allow NADH to donate electrons directly to O_2 to make water in the reaction:

$$2H^+ + 2e^- + \tfrac{1}{2}O_2 \rightarrow H_2O$$

But because of the huge free-energy drop, this reaction would proceed with almost explosive force and nearly all of the energy would be released as heat. Cells perform this reaction much more gradually by passing the high-energy electrons from NADH to O_2 via the many electron carriers in the electron-transport chain. Each successive carrier in the chain holds its electrons more tightly, so that the highly energetically favorable reaction $2H^+ + 2e^- + \frac{1}{2}O_2 \rightarrow H_2O$ occurs in many small steps. This stepwise process allows the cell to store nearly half of the energy that is released.

Spectroscopic Methods Identified Many Electron Carriers in the Respiratory Chain

Many of the electron carriers in the respiratory chain absorb visible light and change color when they are oxidized or reduced. In general, each has an absorption spectrum and reactivity that are distinct enough to allow its behavior to be traced spectroscopically, even in crude mixtures. It was therefore possible to purify these components long before their exact functions were known. Thus, the **cytochromes** were discovered in 1925 as compounds that undergo rapid oxidation and reduction in living organisms as disparate as bacteria, yeasts, and insects. By observing cells and tissues with a spectroscope, researchers identified three types of cytochromes by their distinctive absorption spectra and designated them as cytochromes *a*, *b*, and *c*. This nomenclature has survived, even though cells are now known to contain several cytochromes of each type and the classification into types is not functionally important.

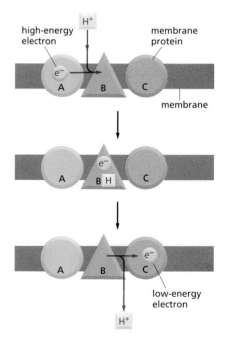

Figure 14–21 How protons can be pumped across membranes. As an electron passes along an electron-transport chain embedded in a lipid-bilayer membrane, it can bind and release a proton at each step. In this diagram, electron carrier B picks up a proton (H$^+$) from one side of the membrane when it accepts an electron (e$^-$) from carrier A; it releases the proton to the other side of the membrane when it donates its electron to carrier C.

HOW REDOX POTENTIALS ARE MEASURED

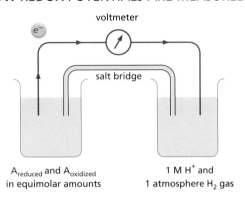

voltmeter

e^-

salt bridge

$A_{reduced}$ and $A_{oxidized}$
in equimolar amounts

1 M H^+ and
1 atmosphere H_2 gas

One beaker (*left*) contains substance A, with an equimolar mixture of the reduced ($A_{reduced}$) and oxidized ($A_{oxidized}$) members of its redox pair. The other beaker contains the hydrogen reference standard ($2H^+ + 2e^- \rightleftharpoons H_2$), whose redox potential is arbitrarily assigned as zero by international agreement. (A salt bridge formed from a concentrated KCl solution allows the ions K^+ and Cl^- to move between the two beakers, as required to neutralize the charges in each beaker when electrons flow between them.) The metal wire (*red*) provides a resistance-free path for electrons, and a voltmeter then measures the redox potential of substance A. If electrons flow from $A_{reduced}$ to H^+, as indicated here, the redox pair formed by substance A is said to have a negative redox potential. If they instead flow from H_2 to $A_{oxidized}$, the redox pair is said to have a positive redox potential.

SOME STANDARD REDOX POTENTIALS AT pH7

By convention, the redox potential for a redox pair is designated E. For the standard state, with all reactants at a concentration of 1 M, including H^+, one can determine a standard redox potential, designated E_0. Since biological reactions occur at pH 7, biologists use a different standard state in which $A_{reduced} = A_{oxidized}$ and $H^+ = 10^{-7}$ M. This standard redox potential is designated E'_0. A few examples of special relevance to oxidative phosphorylation are given here.

redox reactions	redox potential E'_0
NADH $\rightleftharpoons$ NAD$^+$ + H$^+$ + 2e$^-$	–320 mV
reduced ubiquinone $\rightleftharpoons$ oxidized ubiquinone + 2H$^+$ + 2e$^-$	+30 mV
reduced cytochrome *c* $\rightleftharpoons$ oxidized cytochrome *c* + e$^-$	+230 mV
H$_2$O $\rightleftharpoons$ ½O$_2$ + 2H$^+$ + 2e$^-$	+820 mV

CALCULATION OF $\Delta G°$ FROM REDOX POTENTIALS

$\Delta E'_0 = E'_0$ (acceptor) $- E'_0$ (donor) $= +350$ mV

e^-

1:1 mixture of
NADH and NAD$^+$

1:1 mixture of oxidized
and reduced ubiquinone

$\Delta G° = -8$ kcal/mole

$\Delta G° = -n(0.023)\ \Delta E'_0$, where n is the number of electrons transferred across a redox potential change of $\Delta E'_0$ millivolts (mV)

Example: The transfer of one electron from NADH to ubiquinone has a favorable $\Delta G°$ of –8.0 kcal/mole (–1.9 kJ/mole), whereas the transfer of one electron from ubiquinone to oxygen has an even more favorable $\Delta G°$ of –18.2 kcal/mole (–4.35 kJ/mole). The $\Delta G°$ value for the transfer of one electron from NADH to oxygen is the sum of these two values, –26.2 kcal/mole.

THE EFFECT OF CONCENTRATION CHANGES

The actual free-energy change for a reaction, ΔG, depends on the concentration of the reactants and generally is different from the standard free-energy change, $\Delta G°$. The standard redox potentials are for a 1:1 mixture of the redox pair. For example, the standard redox potential of –320 mV is for a 1:1 mixture of NADH and NAD$^+$. But when there is an excess of NADH over NAD$^+$, electron transfer from NADH to an electron acceptor becomes more favorable. This is reflected by a more negative redox potential and a more negative ΔG for electron transfer.

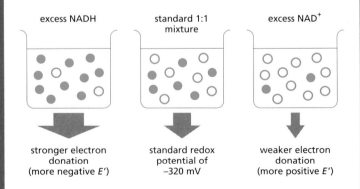

excess NADH

standard 1:1
mixture

excess NAD$^+$

stronger electron
donation
(more negative E')

standard redox
potential of
–320 mV

weaker electron
donation
(more positive E')

The cytochromes constitute a family of colored proteins that are related by the presence of a bound *heme group*, whose iron atom changes from the ferric oxidation state (Fe^{3+}) to the ferrous oxidation state (Fe^{2+}) whenever it accepts an electron. The heme group consists of a *porphyrin ring* with a tightly bound iron atom held by four nitrogen atoms at the corners of a square (**Figure 14–22**). A similar porphyrin ring is responsible for the red color of blood and for the green color of leaves, being bound to iron in hemoglobin and to magnesium in chlorophyll, respectively.

In *iron–sulfur proteins*, a second major family of electron carriers, either two or four iron atoms are bound to an equal number of sulfur atoms and to cysteine side chains, forming an **iron–sulfur center** on the protein (**Figure 14–23**). There are more iron–sulfur centers than cytochromes in the respiratory chain. But their spectroscopic detection requires electron paramagnetic resonance (EPR) spectroscopy, and they are less completely characterized. Like the cytochromes, these centers carry one electron at a time.

The simplest of the electron carriers in the respiratory chain—and the only one that is not part of a protein—is a quinone (called *ubiquinone*, or *coenzyme Q*). A **quinone (Q)** is a small hydrophobic molecule that is freely mobile in the lipid bilayer and can pick up or donate either one or two electrons; upon reduction, it picks up a proton from the medium along with each electron it carries (**Figure 14–24**).

In addition to six different hemes linked to cytochromes, more than seven iron–sulfur centers, and ubiquinone, there are also two copper atoms and a flavin serving as electron carriers tightly bound to respiratory-chain proteins in the pathway from NADH to oxygen. This pathway involves more than 60 different proteins in all.

As we would expect, these electron carriers have higher and higher affinities for electrons (greater redox potentials) as the electrons move along the respiratory chain. The redox potentials have been fine-tuned during evolution by the binding of each electron carrier in a particular protein context, which can alter its normal affinity for electrons. However, because iron–sulfur centers have a relatively low affinity for electrons, they predominate in the early part of the respiratory chain; in contrast, the cytochromes predominate further down the chain, where a higher affinity for electrons is required.

The order of the individual electron carriers in the chain was determined by sophisticated spectroscopic measurements (**Figure 14–25**), and many of the proteins were initially isolated and characterized as individual polypeptides. A major advance in understanding the respiratory chain, however, was the later realization that most of the proteins are organized into three large enzyme complexes.

The Respiratory Chain Includes Three Large Enzyme Complexes Embedded in the Inner Membrane

Membrane proteins are difficult to purify as intact complexes because they are insoluble in aqueous solutions, and some of the detergents required to solubilize them can destroy normal protein–protein interactions. In the early 1960s, however, researchers discovered that relatively mild ionic detergents, such as

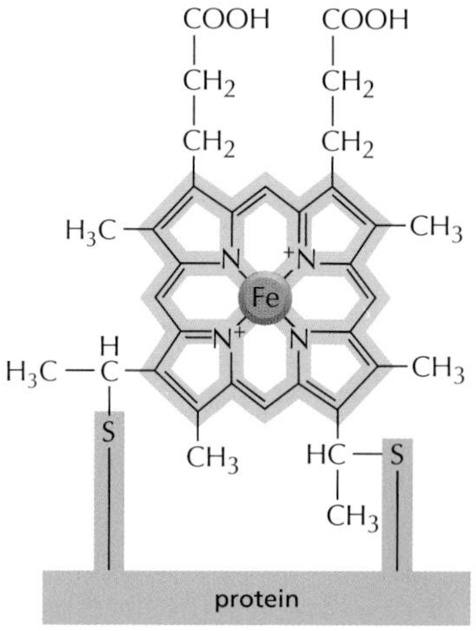

Figure 14–22 The structure of the heme group attached covalently to cytochrome c. The porphyrin ring is shown in *blue*. There are five different cytochromes in the respiratory chain. Because the hemes in different cytochromes have slightly different structures and are held by their respective proteins in different ways, each of the cytochromes has a different affinity for an electron.

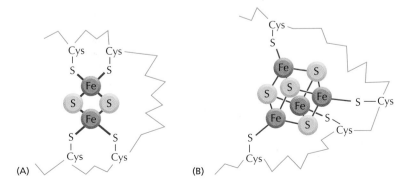

(A)　　　　　　(B)

Figure 14–23 The structures of two types of iron–sulfur centers. (A) A center of the 2Fe2S type. (B) A center of the 4Fe4S type. Although they contain multiple iron atoms, each iron–sulfur center can carry only one electron at a time. There are more than seven different iron–sulfur centers in the respiratory chain.

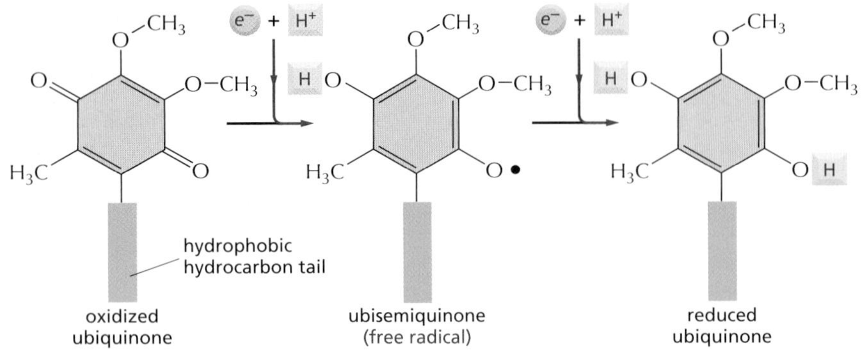

oxidized   ubisemiquinone   reduced
ubiquinone   (free radical)   ubiquinone

Figure 14–24 Quinone electron carriers. Ubiquinone in the respiratory chain picks up one H^+ from the aqueous environment for every electron it accepts, and it can carry either one or two electrons as part of a hydrogen atom *(yellow)*. When reduced ubiquinone donates its electrons to the next carrier in the chain, these protons are released. A long hydrophobic tail confines ubiquinone to the membrane and consists of 6–10 five-carbon isoprene units, the number depending on the organism. The corresponding electron carrier in the photosynthetic membranes of chloroplasts is plastoquinone, which is almost identical in structure. For simplicity, we refer to both ubiquinone and plastoquinone in this chapter as quinone (abbreviated as Q).

deoxycholate, can solubilize selected components of the inner mitochondrial membrane in their native form. This permitted the identification and purification of the three major membrane-bound **respiratory enzyme complexes** in the pathway from NADH to oxygen. Each of these purified complexes can be inserted into lipid bilayer vesicles and shown to pump protons across the bilayer as electrons pass through it. In the mitochondrion, the three complexes are asymmetrically oriented in the inner membrane, and they are linked in series as electron-transport-driven H^+ pumps that pump protons out of the matrix (**Figure 14–26**):

1. The **NADH dehydrogenase complex** (generally known as complex I) is the largest of the respiratory enzyme complexes, containing more than 40 polypeptide chains. It accepts electrons from NADH and passes them through a flavin and at least seven iron–sulfur centers to ubiquinone. Ubiquinone then transfers its electrons to a second respiratory enzyme complex, the cytochrome b-c_1 complex.

2. The **cytochrome b-c_1 complex** contains at least 11 different polypeptide chains and functions as a dimer. Each monomer contains three hemes bound to cytochromes and an iron–sulfur protein. The complex accepts electrons from ubiquinone and passes them on to cytochrome c, which carries its electron to the cytochrome oxidase complex.

3. The **cytochrome oxidase complex** also functions as a dimer; each monomer contains 13 different polypeptide chains, including two cytochromes and two copper atoms. The complex accepts one electron at a time from cytochrome c and passes them four at a time to oxygen.

The cytochromes, iron–sulfur centers, and copper atoms can carry only one electron at a time. Yet each NADH donates two electrons, and each O_2 molecule must receive four electrons to produce water. There are several electron-collecting and electron-dispersing points along the electron-transport chain that coordinate these changes in electron number. The most obvious of these is cytochrome oxidase.

An Iron–Copper Center in Cytochrome Oxidase Catalyzes Efficient O_2 Reduction

Because oxygen has a high affinity for electrons, it releases a large amount of free energy when it is reduced to form water. Thus, the evolution of cellular

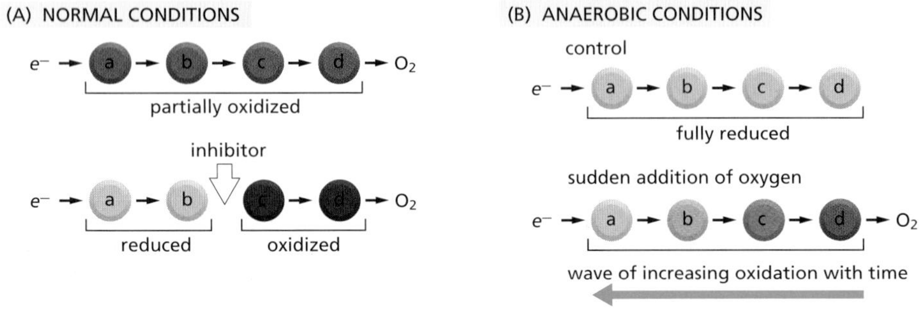

(A) NORMAL CONDITIONS

(B) ANAEROBIC CONDITIONS

Figure 14–25 The general methods used to determine the path of electrons along an electron-transport chain. The extent of oxidation of electron carriers a, b, c, and d is continuously monitored by following their distinct spectra, which differ in their oxidized and reduced states. In this diagram, darker red indicates an increased degree of oxidation. (A) Under normal conditions, where oxygen is abundant, all carriers are in a partly oxidized state. The addition of a specific inhibitor causes the downstream carriers to become more oxidized *(red)* and the upstream carriers to become more reduced. (B) In the absence of oxygen, all carriers are in their fully reduced state *(gray)*. The sudden addition of oxygen converts each carrier to its partly oxidized form with a delay that is greatest for the most upstream carriers.

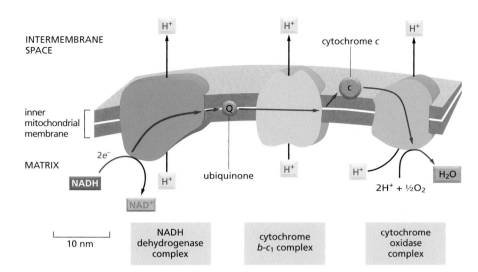

Figure 14–26 The path of electrons through the three respiratory enzyme complexes. The approximate size and shape of each complex are shown. During the transfer of electrons from NADH to oxygen *(red lines)*, ubiquinone and cytochrome *c* serve as mobile carriers that ferry electrons from one complex to the next. As indicated, protons are pumped across the membrane by each of the respiratory enzyme complexes.

respiration, in which O_2 is converted to water, enabled organisms to harness much more energy than can be derived from anaerobic metabolism. This is presumably why all higher organisms respire. The ability of biological systems to use O_2 in this way, however, requires a very sophisticated chemistry. We can tolerate O_2 in the air we breathe because it has trouble picking up its first electron; this fact has allowed cells to control its initial reaction through enzymatic catalysis. But once a molecule of O_2 has picked up one electron to form a superoxide radical (O_2^-), it becomes dangerously reactive and rapidly takes up an additional three electrons wherever it can find them. The cell can use O_2 for respiration only because cytochrome oxidase holds onto oxygen at a special bimetallic center, where it remains clamped between a heme-linked iron atom and a copper atom until it has picked up a total of four electrons. Only then can the two oxygen atoms of the oxygen molecule be safely released as two molecules of water (**Figure 14–27**).

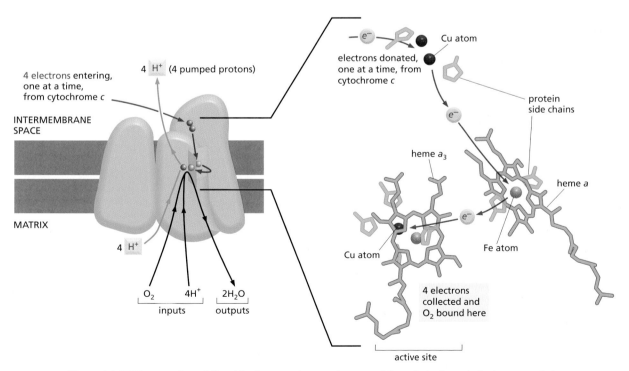

Figure 14–27 The reaction of O_2 with electrons in cytochrome oxidase. As indicated, the iron atom in heme *a* serves as an electron queuing point; this heme feeds four electrons into an O_2 molecule held at the bimetallic center active site, which is formed by the other heme-linked iron (in heme a_3) and a closely opposed copper atom. Note that four protons are pumped out of the matrix for each O_2 molecule that undergoes the reaction $4e^- + 4H^+ + O_2 \rightarrow 2H_2O$.

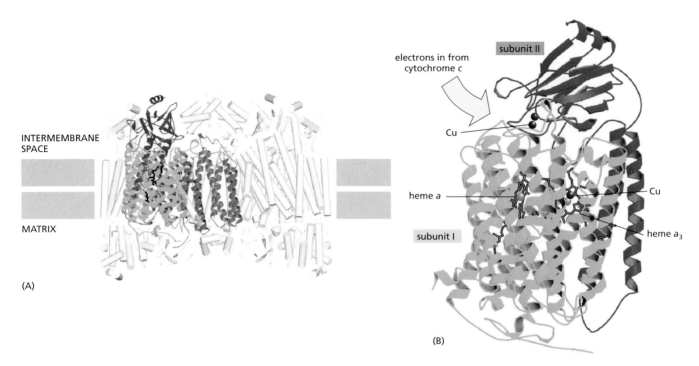

(A)

(B)

The cytochrome oxidase reaction accounts for about 90% of the total oxygen uptake in most cells. This protein complex is therefore crucial for all aerobic life. Cyanide and azide are extremely toxic because they bind tightly to the cell's cytochrome oxidase complexes to stop electron transport, thereby greatly reducing ATP production.

Although the cytochrome oxidase in mammals contains 13 different protein subunits, most of these seem to have a subsidiary role, helping to regulate either the activity or the assembly of the three subunits that form the core of the enzyme. The complete structure of this large enzyme complex has been determined by x-ray crystallography, as illustrated in **Figure 14–28**. The atomic resolution structures, combined with mechanistic studies of the effect of precisely tailored mutations introduced into the enzyme by genetic engineering of the yeast and bacterial proteins, are revealing the detailed mechanisms of this finely tuned protein machine.

Figure 14–28 The molecular structure of cytochrome oxidase. This protein is a dimer formed from a monomer with 13 different protein subunits (monomer mass of 204,000 daltons). The three colored subunits are encoded by the mitochondrial genome, and they form the functional core of the enzyme. As electrons pass through this protein on the way to its bound O_2 molecule, they cause the protein to pump protons across the membrane (see Figure 14–27). (A) The entire protein is shown, positioned in the inner mitochondrial membrane. (B) The electron carriers are located in subunits I and II, as indicated.

Electron Transfers in the Inner Mitochondrial Membrane Are Mediated by Electron Tunneling during Random Collisions

The two components that carry electrons between the three major enzyme complexes of the respiratory chain—ubiquinone and cytochrome c—diffuse rapidly in the plane of the inner mitochondrial membrane. The expected rate of random collisions between these mobile carriers and the more slowly diffusing enzyme complexes can account for the observed rates of electron transfer (each complex donates and receives an electron about once every 5–20 milliseconds).

The ordered transfer of electrons along the respiratory chain is due entirely to the specificity of the functional interactions between the components of the chain: each electron carrier is able to interact only with the carrier adjacent to it in the sequence shown in Figure 14–26, with no short circuits.

Electrons move between the molecules that carry them in biological systems not only by moving along covalent bonds within a molecule, but also by jumping across a gap as large as 2 nm. The jumps occur by electron "tunneling," a quantum-mechanical property that is critical for the processes we are discussing. Insulation prevents short circuits that would otherwise occur when an electron carrier with a low redox potential collides with a carrier with a high redox potential. This insulation seems to be provided by carrying an electron deep enough inside a protein to prevent its tunneling interactions with an inappropriate partner.

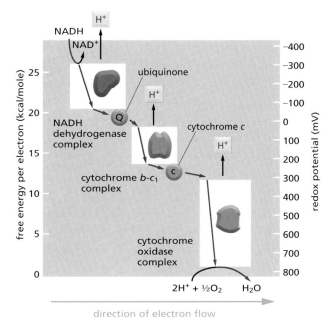

Figure 14–29 Redox potential changes along the mitochondrial electron-transport chain. The redox potential (designated E'_0) increases as electrons flow down the respiratory chain to oxygen. The standard free-energy change, $\Delta G°$, for the transfer of each of the two electrons donated by an NADH molecule can be obtained from the left-hand ordinate ($\Delta G = -n(0.023)\,\Delta E'_0$, where n is the number of electrons transferred across a redox potential change of $\Delta E'_0$ mV). Electrons flow through a respiratory enzyme complex by passing in sequence through the multiple electron carriers in each complex. As indicated, part of the favorable free-energy change is harnessed by each enzyme complex to pump H^+ across the inner mitochondrial membrane. The NADH dehydrogenase and cytochrome b-c_1 complexes each pump two H^+ per electron, whereas the cytochrome oxidase complex pumps one.
  Note that NADH is not the only source of electrons for the respiratory chain. The flavin $FADH_2$ is also generated by fatty acid oxidation (see Figure 2–81) and by the citric acid cycle (see Figure 2–82). Its two electrons are passed directly to ubiquinone, bypassing NADH dehydrogenase; they therefore cause less H^+ pumping than the two electrons transported from NADH.

We discuss next how the changes in redox potential from one electron carrier to the next are harnessed to pump protons out of the mitochondrial matrix.

A Large Drop in Redox Potential Across Each of the Three Respiratory Enzyme Complexes Provides the Energy for H^+ Pumping

We have previously discussed how the redox potential reflects electron affinities (see p. 76). **Figure 14–29** presents an outline of the redox potentials measured along the respiratory chain. These potentials drop in three large steps, one across each major respiratory complex. The change in redox potential between any two electron carriers is directly proportional to the free energy released when an electron transfers between them. Each enzyme complex acts as an energy-conversion device by harnessing some of this free-energy change to pump H^+ across the inner membrane, thereby creating an electrochemical proton gradient as electrons pass through that complex. This conversion can be demonstrated by purifying each respiratory enzyme complex and incorporating it separately into liposomes: when an appropriate electron donor and acceptor are added so that electrons can pass through the complex, protons are translocated across the liposome membrane.

The H^+ Pumping Occurs by Distinct Mechanisms in the Three Major Enzyme Complexes

Some respiratory enzyme complexes pump one H^+ per electron across the inner mitochondrial membrane, whereas others pump two. The detailed mechanism by which electron transport is coupled to H^+ pumping is different for the three different enzyme complexes. In the cytochrome b-c_1 complex, the quinones clearly have a role. As mentioned previously, a quinone picks up a H^+ from the aqueous medium along with each electron it carries and liberates it when it releases the electron (see Figure 14–24). Since ubiquinone is freely mobile in the lipid bilayer, it could accept electrons near the inside surface of the membrane and donate them to the cytochrome b-c_1 complex near the outside surface, thereby transferring one H^+ across the bilayer for every electron transported. Two protons are pumped per electron in the cytochrome b-c_1 complex, however. The complicated series of electron transfers that make this possible are still being worked out at the atomic level, aided by the complete structure of the cytochrome b-c_1 complex determined by x-ray crystallography (**Figure 14–30**).

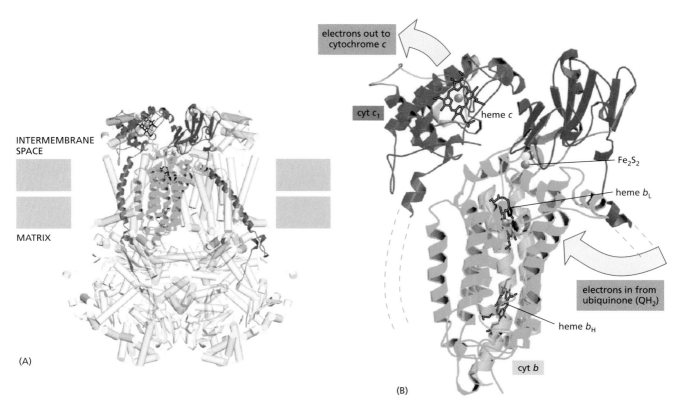

Figure 14–30 The atomic structure of cytochrome b-c₁. This protein is a dimer. The 240,000-dalton monomer is composed of 11 different protein molecules in mammals. The three colored proteins form the functional core of the enzyme: cytochrome b *(green),* cytochrome c₁ *(blue),* and the Rieske protein containing an iron–sulfur center *(purple).* (A) The interaction of these three proteins across the two monomers. (B) Their electron carriers, along with the entrance and exit sites for electrons. The electrons initially donated by ubiquinone follow a complex path of electron and proton transfer reactions through the protein complex that enhances the storage of redox energy. This process, in which some of the electrons are recycled back into the quinone pool, is known as the *Q cycle.*

Electron transport causes allosteric changes in protein conformations that can also pump H⁺, just as H⁺ is pumped when ATP is hydrolyzed by the ATP synthase running in reverse. For both the NADH dehydrogenase complex and the cytochrome oxidase complex, it seems likely that electron transport drives sequential allosteric changes in protein conformation by altering the redox state of the components. These conformational changes in turn cause the protein to pump H⁺ across the mitochondrial inner membrane. This type of H⁺ pumping requires at least three distinct conformations for the pump protein; a general mechanism is presented in **Figure 14–31.**

Now that we have discussed the mechanistic basis for electron transport and proton pumping, we are ready to consider how the respiratory chain is regulated to make it optimally useful to the cell.

H⁺ Ionophores Uncouple Electron Transport from ATP Synthesis

Since the 1940s, several substances—such as 2,4-dinitrophenol—have been known to act as *uncoupling agents,* uncoupling electron transport from ATP synthesis. The addition of these low-molecular-weight organic compounds to cells stops ATP synthesis by mitochondria without blocking their uptake of oxygen. In the presence of an uncoupling agent, electron transport and H⁺ pumping continue at a rapid rate, but no H⁺ gradient is generated. The explanation for this effect is both simple and elegant: uncoupling agents are lipid-soluble weak acids that act as diffusible H⁺ carriers in the lipid bilayer (H⁺ ionophores), and they provide a pathway for the flow of H⁺ across the inner mitochondrial membrane that bypasses the ATP synthase. As a result of this short-circuiting, the proton-motive force is dissipated completely, and ATP can no longer be made.

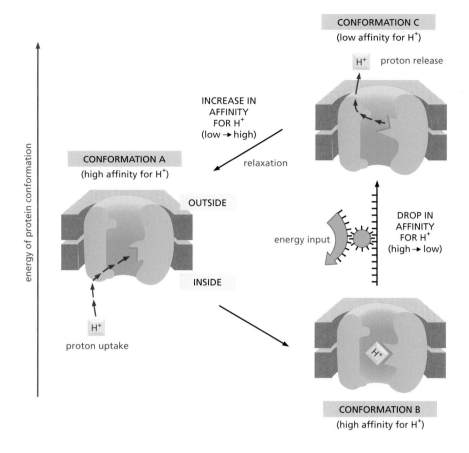

Figure 14–31 A general model for H⁺ pumping. This model for H⁺ pumping by a transmembrane protein is based on mechanisms that are thought to be used by NADH dehydrogenase and cytochrome oxidase, as well as by the light-driven procaryotic proton pump, bacteriorhodopsin. The protein is driven through a cycle of three conformations: A, B, and C. As indicated by their vertical spacing, these protein conformations have different energies. In conformation A, the protein has a high affinity for H⁺, causing it to pick up a H⁺ on the inside of the membrane. In conformation C, the protein has a low affinity for H⁺, causing it to release a H⁺ on the outside of the membrane. The transition from conformation B to conformation C that releases the H⁺ is energetically unfavorable, and it occurs only because it is driven by being allosterically coupled to an energetically favorable reaction occurring elsewhere on the protein *(blue arrow)*. The other two conformational changes, A → B and C → A, lead to states of lower energy, and they proceed spontaneously. Because the overall cycle A → B → C → A releases free energy, H⁺ is pumped from the inside (the matrix in mitochondria) to the outside (the intermembrane space in mitochondria). For NADH dehydrogenase and cytochrome oxidase, the energy required for the transition B → C is provided by electron transport, whereas for bacteriorhodopsin this energy is provided by light (see Figure 10–33). For yet other proton pumps, the energy is derived from ATP hydrolysis. In all cases, at least three distinct conformations are required to create a vectorial pumping process, for the same reason that at least three conformations are required to create a protein that can walk in a single direction along a filament (see Figure 3–77).

Respiratory Control Normally Restrains Electron Flow Through the Chain

The addition of an uncoupler such as dinitrophenol to cells causes mitochondria to increase their rate of electron transport substantially, resulting in an increase in oxygen uptake that reflects the existence of **respiratory control**. The control is thought to act via a direct inhibitory influence of the electrochemical proton gradient on the rate of electron transport. When an uncoupler collapses the gradient, electron transport is free to run unchecked at the maximal rate. As the gradient increases, electron transport becomes more difficult, and the process slows. Moreover, if an artificially large electrochemical proton gradient is experimentally created across the inner membrane, normal electron transport stops completely, and a *reverse electron flow* can be detected in some sections of the respiratory chain. This observation suggests that respiratory control reflects a simple balance between the free-energy change for electron-transport-linked proton pumping and the free-energy change for electron transport—that is, the magnitude of the electrochemical proton gradient affects both the rate and the direction of electron transport, just as it affects the directionality of the ATP synthase (see Figure 14–19).

Respiratory control is just one part of an elaborate interlocking system of feedback controls that coordinate the rates of glycolysis, fatty acid breakdown, the citric acid cycle, and electron transport. The rates of all of these processes are adjusted to the ATP:ADP ratio, increasing whenever an increased utilization of ATP causes the ratio to fall. The ATP synthase in the inner mitochondrial membrane, for example, works faster as the concentrations of its substrates ADP and P_i increase. As it speeds up, the enzyme lets more H⁺ flow into the matrix and thereby dissipates the electrochemical proton gradient more rapidly. The falling gradient, in turn, enhances the rate of electron transport.

Similar controls, including feedback inhibition of several key enzymes by ATP, act to adjust the rates of NADH production to the rate of NADH utilization by the respiratory chain, and so on. As a result of these multiple control mechanisms, the body oxidizes fats and sugars 5–10 times more rapidly during a period of strenuous exercise than during a period of rest.

Natural Uncouplers Convert the Mitochondria in Brown Fat into Heat-Generating Machines

In some specialized fat cells, mitochondrial respiration is normally uncoupled from ATP synthesis. In these cells, known as brown fat cells, most of the energy of oxidation is dissipated as heat rather than being converted into ATP. The inner membranes of the large mitochondria in these cells contain a special transport protein, known as an uncoupling protein, that allows protons to move down their electrochemical gradient without passing through ATP synthase. This uncoupling protein is switched on when heat generation is required, causing the cells to oxidize their fat stores at a rapid rate and produce more heat than ATP. Tissues containing brown fat serve as "heating pads," helping to revive hibernating animals and to protect sensitive areas of newborn human babies from the cold.

The Mitochondrion Has Many Critical Roles in Cell Metabolism

Cells are largely composed of macromolecules, which are constantly in need of repair or replacement as the cell ages. Even for cells and organisms that are not growing, those molecules that decay must be replaced by biosyntheses. Throughout this chapter, we emphasize the critical role of mitochondria in producing the ATP that cells need to maintain themselves as highly organized entities in a universe that is always driving toward increasing disorder (discussed in Chapter 2). In addition to ATP, however, the biosynthesis in the cytosol requires a constant supply of reducing power in the form of NADPH and of carbon skeletons. Most descriptions of this type of biosynthesis state that the needed carbon skeletons come directly from the breakdown of sugars, whereas the NADPH is produced in the cytosol by a side pathway for the breakdown of sugars (the pentose phosphate pathway, an alternative to glycolysis). But under conditions where foodstuffs are abundant, and plenty of ATP is available, mitochondria also generate both carbon skeletons and NADPH needed for cell growth. For this purpose, excess citrate produced in the mitochondrial matrix by the citric acid cycle is transported down its electrochemical gradient to the cytosol, where it is metabolized to produce both NADPH and carbon skeletons for biosyntheses. Thus, for example, as part of a cell's response to growth signals, large amounts of acetyl CoA are produced in the cytosol from citrate exported from mitochondria, accelerating the production of the fatty acids and sterols that build new membranes.

Mitochondria are also critical for buffering the redox potential in the cytosol. Cells need a constant supply of the electron acceptor NAD^+ for the central reaction in glycolysis that converts glyceraldehyde 3-phosphate to 1,3-bisphosphoglycerate (see Figure 2–72). This NAD^+ is converted to NADH in the process, and the NAD^+ needs to be regenerated by transferring the high-energy NADH electrons somewhere.

The NADH electrons will eventually be used to help drive oxidative phosphorylation inside the mitochondrion. But the mitochondrial inner membrane is impermeable to NADH. The electrons are therefore passed from the NADH to smaller molecules in the cytosol that can move through the inner mitochondrial membrane. Once in the matrix, these smaller molecules transfer their electrons to NAD^+ to form mitochondrial NADH, after which they are returned to the cytosol for recharging. This so-called shuttle system is bypassed in some specialized cells, such as insect flight muscle, that produce especially large amounts of ATP by aerobic glycolysis. Here the high-energy electrons derived from glyceraldehyde 3-phosphate are passed directly to the outer surface of the mitochondrial inner membrane, enabling them to enter the electron transport chain more rapidly and directly—but with loss of some of the usable energy.

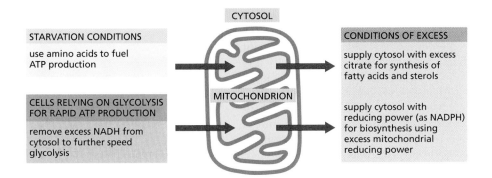

Figure 14–32 **Critical roles of mitochondria in cell metabolism besides ATP production.** The many essential metabolic reactions carried out by mitochondria, such as those illustrated here, emphasize the inadequacy of our emphasis on mitochondria as the cell furnace that oxidizes pyruvate and fatty acids to feed oxidative phosphorylation.

Under conditions of starvation, proteins in our bodies are broken down to amino acids, and the amino acids are imported into mitochondria and oxidized to produce NADH for ATP production. Thus, by carrying out different reactions under different conditions, the mitochondrion has many critical functions in maintaining cellular metabolism (**Figure 14–32**).

Bacteria Also Exploit Chemiosmotic Mechanisms to Harness Energy

Bacteria use enormously diverse energy sources. Some, like animal cells, are aerobic; they synthesize ATP from sugars they oxidize to CO_2 and H_2O by glycolysis, the citric acid cycle, and a respiratory chain in their plasma membrane that is similar to the one in the inner mitochondrial membrane. Others are strict anaerobes, deriving their energy either from glycolysis alone (by fermentation) or from an electron-transport chain that employs a molecule other than oxygen as the final electron acceptor. The alternative electron acceptor can be a nitrogen compound (nitrate or nitrite), a sulfur compound (sulfate or sulfite), or a carbon compound (fumarate or carbonate), for example. A series of electron carriers in the plasma membrane that are comparable to those in mitochondrial respiratory chains transfer the electrons to these acceptors.

Despite this diversity, the plasma membrane of the vast majority of bacteria contains an ATP synthase that is very similar to the one in mitochondria. In bacteria that use an electron-transport chain to harvest energy, the electron-transport chain pumps H^+ out of the cell and thereby establishes a proton-motive force across the plasma membrane that drives the ATP synthase to make ATP. In other bacteria, the ATP synthase works in reverse, using the ATP produced by glycolysis to pump H^+ and establish a proton gradient across the plasma membrane. The ATP used for this process is generated by fermentation processes (discussed in Chapter 2).

Thus, most bacteria, including the strict anaerobes, maintain a proton gradient across their plasma membrane. It can be harnessed to drive a flagellar motor, and it is used to pump Na^+ out of the bacterium via a Na^+-H^+ antiporter that takes the place of the Na^+-K^+ pump of eucaryotic cells. This gradient is also used for the active inward transport of nutrients, such as most amino acids and many sugars: each nutrient is dragged into the cell along with one or more protons through a specific symporter (**Figure 14–33**). In animal cells, by contrast,

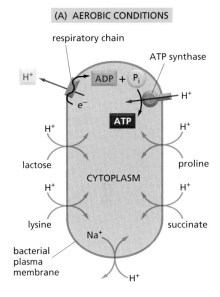

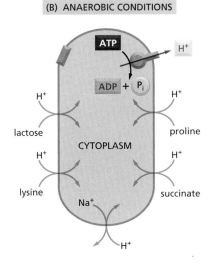

Figure 14–33 **The importance of H^+-driven transport in bacteria.** A proton-motive force generated across the plasma membrane pumps nutrients into the cell and expels Na^+. (A) In an aerobic bacterium, a respiratory chain produces an electrochemical proton gradient across the plasma membrane; this gradient is then used both to transport some nutrients into the cell and to make ATP. (B) The same bacterium growing under anaerobic conditions derives its ATP from glycolysis. The ATP synthase then hydrolyzes some of this ATP to establish an electrochemical proton gradient that drives those transport processes that depend on the respiratory chain in (A).

most inward transport across the plasma membrane is driven by the Na^+ gradient (high Na^+ outside, low Na^+ inside) that is established by the Na^+-K^+ pump.

Some unusual bacteria have adapted to live in a very alkaline environment and yet must maintain their cytoplasm at a physiological pH. For these cells, any attempt to generate an electrochemical H^+ gradient would be opposed by a large H^+ concentration gradient in the wrong direction (H^+ higher inside than outside). Presumably for this reason, some of these bacteria substitute Na^+ for H^+ in all of their chemiosmotic mechanisms. The respiratory chain pumps Na^+ out of the cell, the transport systems and flagellar motor are driven by an inward flux of Na^+, and a Na^+-driven ATP synthase synthesizes ATP. The existence of such bacteria demonstrates that the principle of chemiosmosis is more fundamental than the proton-motive force on which it is normally based.

Summary

The respiratory chain embedded in the inner mitochondrial membrane contains three respiratory enzyme complexes through which electrons pass on their way from NADH to O_2. Each complex can be purified, inserted into synthetic lipid vesicles, and then shown to pump H^+ when electrons are transported through it. In these complexes, electrons are transferred along a series of protein-bound electron carriers, including hemes and iron-sulfur centers. Energy released as the electrons move to lower and lower energy levels is used to drive allosteric changes in each respiratory enzyme complex that help to pump protons. Electrons are carried between enzyme complexes by the mobile electron carriers ubiquinone and cytochrome c to complete the electron-transport chain. The path of electron flow is NADH → NADH dehydrogenase complex → ubiquinone → cytochrome b-c_1 complex → cytochrome c → cytochrome oxidase complex → molecular oxygen (O_2).

The coupling of the energetically favorable transport of electrons to the pumping of H^+ out of the matrix creates an electrochemical proton gradient. This gradient is harnessed to make ATP by the ATP synthase, through which H^+ flows back into the matrix. The universal presence of ATP synthase in mitochondria, chloroplasts, and procaryotes testifies to the central importance of chemiosmotic mechanisms in cells.

CHLOROPLASTS AND PHOTOSYNTHESIS

All animals and most microorganisms rely on the continual uptake of large amounts of organic compounds from their environment. These compounds provide both the carbon skeletons for biosynthesis and the metabolic energy that drives cellular processes. It is likely that the first organisms on the primitive Earth had access to an abundance of the organic compounds produced by geochemical processes, but it is clear that most of these original compounds were used up billions of years ago. Since that time, nearly all of the organic materials required by living cells have been produced by *photosynthetic organisms,* including many types of photosynthetic bacteria.

The most advanced photosynthetic bacteria are the cyanobacteria, which have minimal nutrient requirements. They use electrons from water and the energy of sunlight to convert atmospheric CO_2 into organic compounds—a process called *carbon fixation*. In the course of splitting water [in the overall reaction $nH_2O + nCO_2 \xrightarrow{light} (CH_2O)_n + nO_2$], they also liberate into the atmosphere the oxygen required for oxidative phosphorylation. As we see in this section, it is thought that the evolution of cyanobacteria from more primitive photosynthetic bacteria eventually made possible the development of abundant aerobic life forms.

In plants and algae, which developed much later, photosynthesis occurs in a specialized intracellular organelle—the **chloroplast**. Chloroplasts perform photosynthesis during the daylight hours. Photosynthetic cells use the immediate products of photosynthesis, NADPH and ATP, to produce many organic molecules. In plants, the products include a low-molecular-weight sugar (usually sucrose) that these cells export to meet the metabolic needs of the many non-photosynthetic cells of the organism.

Biochemical and genetic evidence strongly suggests that chloroplasts are descended from oxygen-producing photosynthetic bacteria that were endocytosed and lived in symbiosis with primitive eucaryotic cells. Mitochondria are also generally believed to be descended from an endocytosed bacterium. The many differences between chloroplasts and mitochondria presumably reflect their different bacterial ancestors, as well as their subsequent evolutionary divergence. Nevertheless, the fundamental mechanisms involved in light-driven ATP synthesis in chloroplasts are similar to those that we have already discussed for respiration-driven ATP synthesis in mitochondria.

The Chloroplast Is One Member of the Plastid Family of Organelles

Chloroplasts are the most prominent members of the **plastid** family of organelles. Plastids are present in all living plant cells, each cell type having its own characteristic complement. All plastids share certain features. Most notably, all plastids in a particular plant species contain multiple copies of the same relatively small genome. In addition, an envelope composed of two concentric membranes encloses each plastid.

As discussed in Chapter 12 (see Figure 12–3), all plastids develop from *proplastids,* small organelles in the immature cells of plant meristems (**Figure 14–34**A). Proplastids develop according to the requirements of each differentiated cell, and the type that is present is determined in large part by the nuclear genome. If a leaf is grown in darkness, its proplastids enlarge and develop into *etioplasts,* which have a semicrystalline array of internal membranes containing a yellow chlorophyll precursor instead of chlorophyll. When exposed to light, the etioplasts rapidly develop into chloroplasts by converting this precursor to chlorophyll and by synthesizing new membrane pigments, photosynthetic enzymes, and components of the electron-transport chain.

Leucoplasts are plastids present in many epidermal and internal tissues that do not become green and photosynthetic. They are little more than enlarged proplastids. A common form of leucoplast is the *amyloplast* (Figure 14–34B), which accumulates the polysaccharide starch in storage tissues—a source of sugar for future use. In some plants, such as potatoes, the amyloplasts can grow to be as large as an average animal cell.

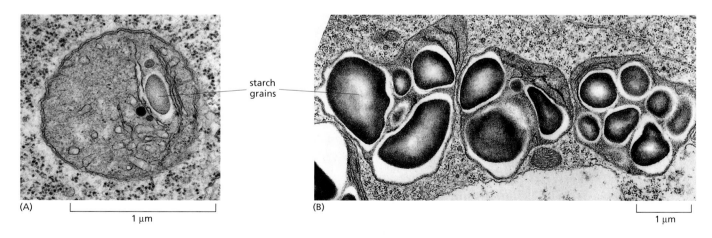

(A) 1 µm

starch grains

(B) 1 µm

Figure 14–34 **Plastid diversity.** (A) A proplastid from a root tip cell of a bean plant. Note the double membrane; the inner membrane has also generated the relatively sparse internal membranes present. (B) Three amyloplasts (a form of leucoplast), or starch-storing plastids, in a root tip cell of soybean. (From B. Gunning and M. Steer, Plant Cell Biology: Structure and Function. Sudbury, MA: Jones & Bartlett, 1996.)

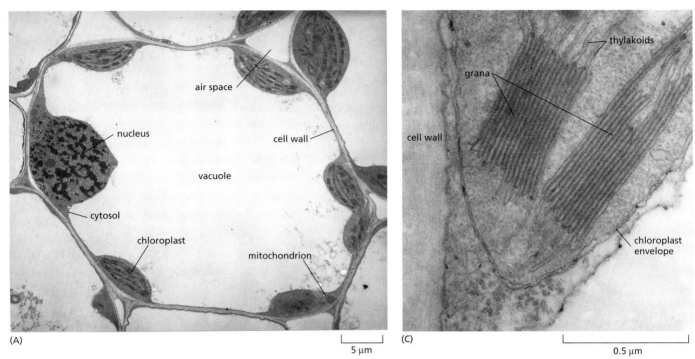

(A)

5 μm

(C)

0.5 μm

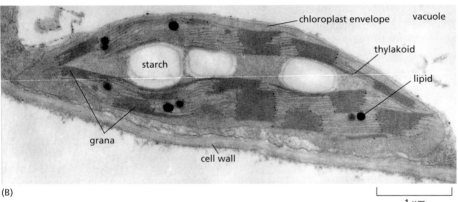

(B)

1 μm

Figure 14–35 **Electron micrographs of chloroplasts.** (A) In a wheat leaf cell, a thin rim of cytoplasm—containing chloroplasts, the nucleus, and mitochondria—surrounds a large vacuole. (B) A thin section of a single chloroplast, showing the chloroplast envelope, starch granules, and lipid (fat) droplets that have accumulated in the stroma as a result of the biosyntheses occurring there. (C) A high-magnification view of two grana. A granum is a stack of thylakoids. (Courtesy of K. Plaskitt.)

Plastids are not just sites for photosynthesis and the deposition of storage materials. Plants have also used their plastids to compartmentalize their intermediary metabolism. Purine and pyrimidine synthesis, most amino acid synthesis, and all of the fatty acid synthesis of plants takes place in the plastids, whereas in animal cells these compounds are produced in the cytosol.

Chloroplasts Resemble Mitochondria But Have an Extra Compartment

Chloroplasts use chemiosmotic mechanisms to carry out their energy interconversions in much the same way that mitochondria do. Although much larger (**Figure 14–35**A), they are organized on the same principles. They have a highly permeable outer membrane; a much less permeable inner membrane, in which membrane transport proteins are embedded; and a narrow intermembrane space in between. Together, these membranes form the chloroplast envelope (Figure 14–35B,C). The inner membrane surrounds a large space called the **stroma**, which is analogous to the mitochondrial matrix and contains many metabolic enzymes. Like the mitochondrion, the chloroplast has its own genome and genetic system. The stroma therefore also contains a special set of ribosomes, RNAs, and the chloroplast DNA.

There is, however, an important difference between the organization of mitochondria and that of chloroplasts. The inner membrane of the chloroplast is not folded into cristae and does not contain electron-transport chains.

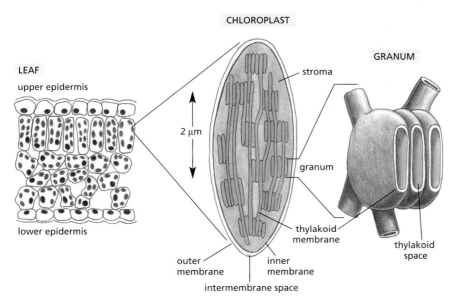

LEAF

upper epidermis

lower epidermis

CHLOROPLAST

stroma

granum

thylakoid membrane

outer membrane

inner membrane

intermembrane space

2 µm

GRANUM

thylakoid space

Figure 14–36 The chloroplast. This photosynthetic organelle contains three distinct membranes (the outer membrane, the inner membrane, and the thylakoid membrane) that define three separate internal compartments (the intermembrane space, the stroma, and the thylakoid space). The thylakoid membrane contains all the energy-generating systems of the chloroplast, including its chlorophyll. In electron micrographs, this membrane seems to be broken up into separate units that enclose individual flattened vesicles (see Figure 14–35), but these are probably joined into a single, highly folded membrane in each chloroplast. As indicated, the individual thylakoids are interconnected, and they tend to stack to form grana.

Instead, the electron-transport chains, photosynthetic light-capturing systems, and ATP synthase are all contained in the *thylakoid membrane*, a third distinct membrane that forms a set of flattened disclike sacs, the *thylakoids* (**Figure 14–36**). The lumen of each thylakoid is thought to be connected with the lumen of other thylakoids, thereby defining a third internal compartment called the *thylakoid space*, which is separated by the thylakoid membrane from the stroma that surrounds it. Thylakoid membranes interact with each other to form numerous local stacks called *grana*.

Figure 14–37 highlights the structural similarities and differences between mitochondria and chloroplasts. An important difference is that the head of the ATP synthase, where the ATP is made, protrudes into the stroma from the thylakoid membrane in a chloroplast, whereas it protrudes into the matrix from the inner mitochondrial membrane in a mitochondrion.

Chloroplasts Capture Energy from Sunlight and Use It to Fix Carbon

We can group the many reactions that occur during photosynthesis in plants into two broad categories:

1. In the **photosynthetic electron-transfer** reactions (also called the "light reactions"), energy derived from sunlight energizes an electron in the green organic pigment *chlorophyll*, enabling the electron to move along an electron-transport chain in the thylakoid membrane in much the same way that an electron moves along the respiratory chain in mitochondria.

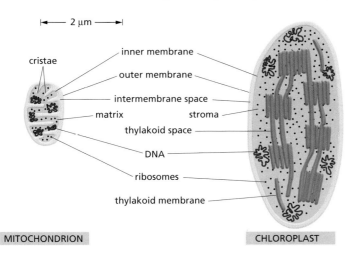

2 µm

cristae

inner membrane

outer membrane

intermembrane space

matrix

stroma

thylakoid space

DNA

ribosomes

thylakoid membrane

MITOCHONDRION

CHLOROPLAST

Figure 14–37 A mitochondrion and chloroplast compared. A chloroplast is generally much larger than a mitochondrion and contains, in addition to an outer and inner membrane, a thylakoid membrane enclosing a thylakoid space. Unlike the chloroplast inner membrane, the inner mitochondrial membrane is folded into cristae to increase its surface area.

The chlorophyll obtains its electrons from water (H_2O), producing O_2 as a by-product. During the electron-transport process, H^+ is pumped across the thylakoid membrane, and the resulting electrochemical proton gradient drives the synthesis of ATP in the stroma. As the final step in this series of reactions, high-energy electrons are loaded (together with H^+) onto $NADP^+$, converting it to NADPH. All of these reactions are confined to the chloroplast.

2. In the **carbon-fixation reactions** (also called the "dark reactions"), the ATP and the NADPH produced by the photosynthetic electron-transfer reactions serve as the source of energy and reducing power, respectively, to drive the conversion of CO_2 to carbohydrate. The carbon-fixation reactions, which begin in the chloroplast stroma and continue in the cytosol, produce sucrose and many other organic molecules in the leaves of the plant. The sucrose is exported to other tissues as a source of both organic molecules and energy for growth.

Thus, the formation of ATP, NADPH, and O_2 (which requires light energy directly) and the conversion of CO_2 to carbohydrate (which requires light energy only indirectly) are separate processes (**Figure 14–38**), although elaborate feedback mechanisms interconnect the two. Several of the chloroplast enzymes required for carbon fixation, for example, are inactivated in the dark and reactivated by light-stimulated electron-transport processes.

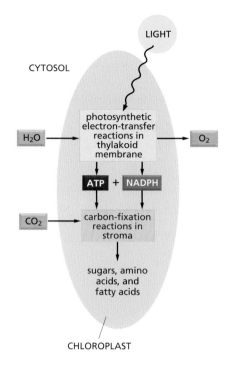

Figure 14–38 The reactions of photosynthesis in a chloroplast. Water is oxidized and oxygen is released in the photosynthetic electron-transfer reactions, while carbon dioxide is assimilated (fixed) to produce sugars and a variety of other organic molecules in the carbon-fixation reactions.

Carbon Fixation Is Catalyzed by Ribulose Bisphosphate Carboxylase

We have seen earlier in this chapter how cells produce ATP by using the large amount of free energy released when carbohydrates are oxidized to CO_2 and H_2O. Clearly, therefore, the reverse reaction, in which CO_2 and H_2O combine to make carbohydrate, must be a very unfavorable one that can occur only if it is coupled to other, very favorable reactions that drive it.

Figure 14–39 illustrates the central reaction of **carbon fixation**, in which an atom of inorganic carbon is converted to organic carbon: CO_2 from the atmosphere combines with the five-carbon compound ribulose 1,5-bisphosphate plus water to yield two molecules of the three-carbon compound 3-phosphoglycerate. This "carbon-fixing" reaction, which was discovered in 1948, is catalyzed in the chloroplast stroma by a large enzyme called *ribulose bisphosphate carboxylase*. Since each molecule of the complex works sluggishly (processing only about 3 molecules of substrate per second compared to 1000 molecules per second for a typical enzyme), the reaction requires an unusually large number of enzyme molecules. Ribulose bisphosphate carboxylase often constitutes more than 50% of the total chloroplast protein, and it is thought to be the most abundant protein on Earth.

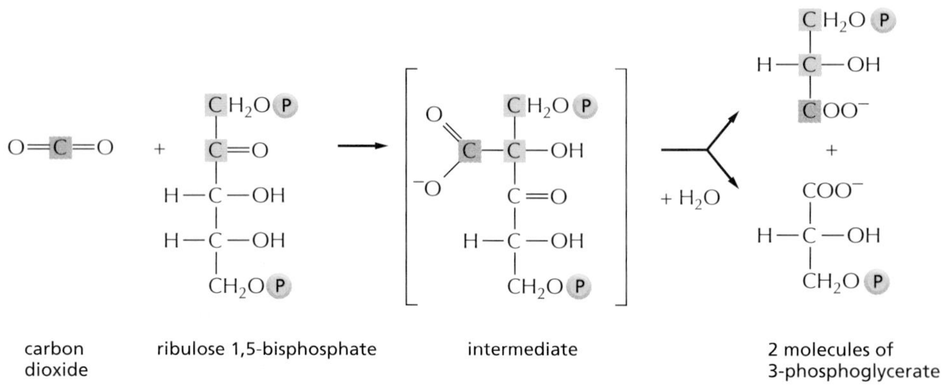

Figure 14–39 The initial reaction in carbon fixation. This reaction, in which carbon dioxide is converted into organic carbon, is catalyzed in the chloroplast stroma by the abundant enzyme ribulose bisphosphate carboxylase. The product is 3-phosphoglycerate, which is also an intermediate in glycolysis. The two carbon atoms shaded in *blue* are used to produce phosphoglycolate when the same enzyme adds oxygen instead of CO_2 (see text).

carbon dioxide   ribulose 1,5-bisphosphate   intermediate   2 molecules of 3-phosphoglycerate

Each CO$_2$ Molecule That Is Fixed Consumes Three Molecules of ATP and Two Molecules of NADPH

The actual reaction in which CO$_2$ is fixed is energetically favorable because of the reactivity of the energy-rich compound *ribulose 1,5-bisphosphate*, to which each molecule of CO$_2$ is added (see Figure 14–39). The elaborate metabolic pathway that produces ribulose 1,5-bisphosphate requires both NADPH and ATP; it was worked out in one of the first successful applications of radioisotopes as tracers in biochemistry. This **carbon-fixation cycle** (also called the **Calvin cycle**) is outlined in **Figure 14–40**. It starts when 3 molecules of CO$_2$ are fixed by ribulose bisphosphate carboxylase to produce 6 molecules of 3-phosphoglycerate (containing $6 \times 3 = 18$ carbon atoms in all: 3 from the CO$_2$ and 15 from ribulose 1,5-bisphosphate). The 18 carbon atoms then undergo a cycle of reactions that regenerates the 3 molecules of ribulose 1,5-bisphosphate used in the initial carbon-fixation step (containing $3 \times 5 = 15$ carbon atoms). This leaves 1 molecule of *glyceraldehyde 3-phosphate* (3 carbon atoms) as the net gain.

Each CO$_2$ molecule converted into carbohydrate consumes a total of 3 molecules of ATP and 2 molecules of NADPH. The net equation is:

$$3CO_2 + 9ATP + 6NADPH + water \rightarrow$$
$$glyceraldehyde\ 3\text{-}phosphate + 8P_i + 9ADP + 6NADP^+$$

Thus, the formation of organic molecules from CO$_2$ and H$_2$O requires both *phosphate-bond energy* (as ATP) and *reducing power* (as NADPH). We return to this important point later.

The glyceraldehyde 3-phosphate produced in chloroplasts by the carbon-fixation cycle is a three-carbon sugar that also serves as a central intermediate in glycolysis. Much of it is exported to the cytosol, where it can be converted into

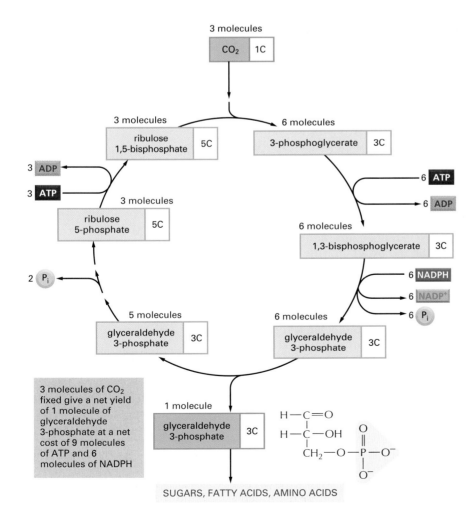

3 molecules of CO$_2$ fixed give a net yield of 1 molecule of glyceraldehyde 3-phosphate at a net cost of 9 molecules of ATP and 6 molecules of NADPH

SUGARS, FATTY ACIDS, AMINO ACIDS

Figure 14–40 The carbon-fixation cycle, which forms organic molecules from CO$_2$ and H$_2$O. The number of carbon atoms in each type of molecule is indicated in the *white box*. There are many intermediates between glyceraldehyde 3-phosphate and ribulose 5-phosphate, but they have been omitted here for clarity. The entry of water into the cycle is also not shown.

fructose 6-phosphate and glucose 1-phosphate by the reversal of several reactions in glycolysis (see Panel 2–8, pp. 120–121). The glucose 1-phosphate is then converted to the sugar nucleotide UDP-glucose, and this combines with the fructose 6-phosphate to form sucrose phosphate, the immediate precursor of the disaccharide sucrose. **Sucrose** is the major form in which sugar is transported between plant cells: just as glucose is transported in the blood of animals, sucrose is exported from the leaves via vascular bundles, providing the carbohydrate required by the rest of the plant.

Most of the glyceraldehyde 3-phosphate that remains in the chloroplast is converted to *starch* in the stroma. Like glycogen in animal cells, **starch** is a large polymer of glucose that serves as a carbohydrate reserve (see Figure 14–34B). The production of starch is regulated so that it is produced and stored as large grains in the chloroplast stroma during periods of excess photosynthetic capacity. This occurs through reactions in the stroma that are the reverse of those in glycolysis: they convert glyceraldehyde 3-phosphate to glucose 1-phosphate, which is then used to produce the sugar nucleotide ADP-glucose, the immediate precursor of starch. At night the plant breaks down the starch to help support the metabolic needs of the plant. Starch provides an important part of the diet of all animals that eat plants.

Carbon Fixation in Some Plants Is Compartmentalized to Facilitate Growth at Low CO_2 Concentrations

Although ribulose bisphosphate carboxylase preferentially adds CO_2 to ribulose 1,5-bisphosphate, it can use O_2 as a substrate in place of CO_2, and if the concentration of CO_2 is low, it will add O_2 to ribulose 1,5-bisphosphate instead (see Figure 14–39). This is the first step in a pathway called **photorespiration**, whose ultimate effect is to use up O_2 and liberate CO_2 without the production of useful energy stores. In many plants, about one-third of the CO_2 fixed is lost again as CO_2 because of photorespiration.

Photorespiration can be a serious liability for plants in hot, dry conditions, which cause them to close their stomata (the gas exchange pores in their leaves, each of which is called a stoma) to avoid excessive water loss. This in turn causes the CO_2 levels in the leaf to fall precipitously, thereby favoring photorespiration. A special adaptation, however, occurs in the leaves of many plants, such as corn and sugar cane, that grow in hot, dry environments. In these plants, the carbon-fixation cycle occurs only in the chloroplasts of specialized *bundle-sheath cells*, which contain all of the plant's ribulose bisphosphate carboxylase. These cells are protected from the air and are surrounded by a specialized layer of *mesophyll cells* that use the energy harvested by their chloroplasts to "pump" CO_2 into the bundle-sheath cells. This supplies the ribulose bisphosphate carboxylase with a high concentration of CO_2, thereby greatly reducing photorespiration.

The CO_2 pump is produced by a reaction cycle that begins in the cytosol of the mesophyll cells. A CO_2-fixation step is catalyzed by an enzyme that binds carbon dioxide (as bicarbonate) and combines it with an activated three-carbon molecule (phosphoenol-pyruvate) to produce a four-carbon molecule. The four-carbon molecule diffuses into the bundle-sheath cells, where it is broken down to release the CO_2 and generate a molecule with three carbons. The pumping cycle is completed when this three-carbon molecule is returned to the mesophyll cells and converted back to its original activated form. Because the CO_2 is initially captured by converting it into a compound containing four carbons, the CO_2-pumping plants are called *C_4 plants*. All other plants are called *C_3 plants* because they capture CO_2 into the three-carbon compound 3-phosphoglycerate (**Figure 14–41**).

As with any vectorial transport process, pumping CO_2 into the bundle-sheath cells in C_4 plants costs energy (ATP is hydrolyzed; see Figure 14–41B). In hot, dry environments, however, this cost can be much less than the energy lost by photorespiration in C_3 plants, so C_4 plants have a potential advantage. Moreover, because C_4 plants can perform photosynthesis at a lower concentration of CO_2 inside the leaf, they need to open their stomata less often and therefore can

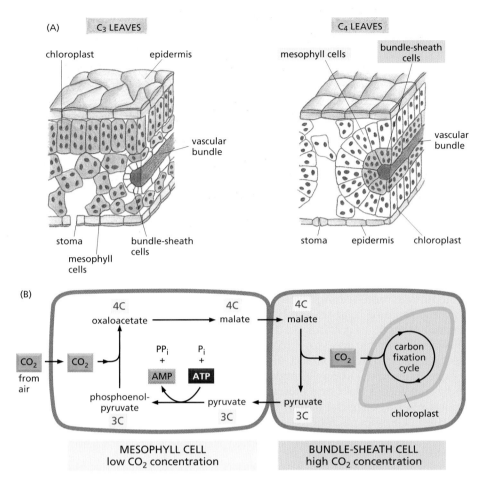

Figure 14–41 The CO_2 pumping in C_4 plants. (A) Comparative leaf anatomy in a C_3 plant and a C_4 plant. The cells with *green* cytosol in the leaf interior contain chloroplasts that perform the normal carbon-fixation cycle. In C_4 plants, the mesophyll cells are specialized for CO_2 pumping rather than for carbon fixation, and they thereby create a high ratio of CO_2 to O_2 in the bundle-sheath cells, which are the only cells in these plants where the carbon-fixation cycle occurs. The vascular bundles carry the sucrose made in the leaf to other tissues. (B) How carbon dioxide is concentrated in bundle-sheath cells by the harnessing of ATP energy in mesophyll cells.

fix about twice as much net carbon as C_3 plants per unit of water lost. This type of carbon fixation has evolved independently in several different plant lineages. Although the vast majority of plant species are C_3 plants, C_4 plants such as corn and sugar cane are much more effective at converting sunlight energy into biomass than C_3 plants such as cereal grains. They are therefore of special importance in world agriculture.

Photosynthesis Depends on the Photochemistry of Chlorophyll Molecules

Having discussed the carbon-fixation reactions, we now return to the question of how the photosynthetic electron-transfer reactions in the chloroplast generate the ATP and the NADPH needed to drive the production of carbohydrates from CO_2 and H_2O. Sunlight absorbed by **chlorophyll** molecules supplies the required energy (**Figure 14–42**). The process of energy conversion begins when a quantum of light (a photon) excites a chlorophyll molecule, causing an electron in the chlorophyll to move from one molecular orbital to another of higher energy. Such an excited molecule is unstable and tends to return quickly to its original, unexcited state. This can happen in one of three ways:

1. By converting the extra energy into heat (molecular motions) or to some combination of heat and light of a longer wavelength (fluorescence); this is what happens when an isolated chlorophyll molecule in solution absorbs light energy.
2. By transferring the energy—but not the electron—directly to a neighboring chlorophyll molecule by a process called *resonance energy transfer*.
3. By transferring the negatively charged high-energy electron to another nearby molecule, an *electron acceptor*, after which the positively charged chlorophyll returns to its original state by taking up a low-energy electron from some other molecule, an *electron donor*.

In the process of photosynthesis, the last two mechanisms are greatly facilitated by two different protein complexes: resonance energy transfer by an *antenna complex* and high-energy electron transfer by a *photochemical reaction center*. These two types of protein complexes, acting in concert, make most of the life on Earth possible. We shall now describe how they work.

A Photochemical Reaction Center Plus an Antenna Complex Form a Photosystem

Large multiprotein complexes called **photosystems** catalyze the conversion of the light energy captured in excited chlorophyll molecules to useful forms. A photosystem consists of two closely linked components: an antenna complex, consisting of proteins bound to a large set of pigment molecules that capture light energy and feed it to the reaction center; and a photochemical reaction center, consisting of a complex of proteins and chlorophyll molecules that enable light energy to be converted into chemical energy.

The **antenna complex** is important for capturing light energy. In chloroplasts it consists of a number of distinct membrane protein complexes (known as light-harvesting complexes); together, these proteins bind several hundred chlorophyll molecules per reaction center, orienting them precisely in the thylakoid membrane. The antenna complex also contains accessory pigments called *carotenoids*, which protect the chlorophylls from oxidation and can help collect light of other wavelengths. When light excites a chlorophyll molecule in the antenna complex, the energy is rapidly transferred from one molecule to another by resonance energy transfer until it reaches a special pair of chlorophyll molecules in the photochemical reaction center. Each antenna complex thereby acts as a funnel, collecting light energy and directing it to a specific site where it can be used effectively (**Figure 14–43**).

The **photochemical reaction center** is a transmembrane protein–pigment complex that lies at the heart of photosynthesis. It is thought to have evolved more than 3 billion years ago in primitive photosynthetic bacteria. The special pair of chlorophyll molecules in the reaction center acts as an irreversible trap for excitation quanta because the excited electron is immediately passed to a neighboring chain of electron acceptors in the protein complex (**Figure 14–44**). By moving the high-energy electron rapidly away from the chlorophylls, the photochemical reaction center transfers it to an environment where it is much more stable. The electron is thereby suitably positioned for subsequent reactions. These require more time to complete, and they result in the production of light-generated high-energy electrons that are fed into electron transport chains.

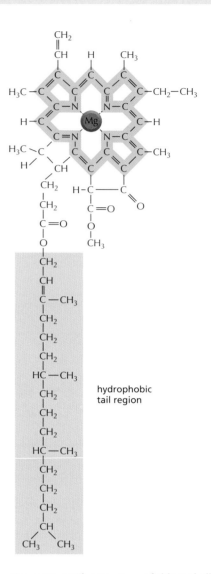

Figure 14–42 The structure of chlorophyll. A magnesium atom is held in a porphyrin ring, which is related to the porphyrin ring that binds iron in heme (see Figure 14–22). Electrons are delocalized over the bonds shown in *blue*.

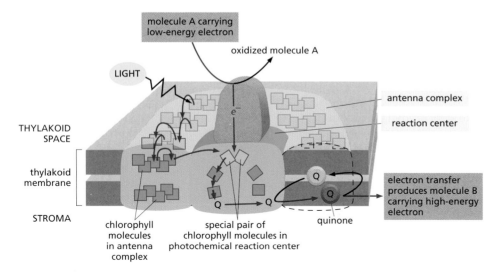

Figure 14–43 The antenna complex and photochemical reaction center in a photosystem. <GCAC> The antenna complex is a collector of light energy in the form of excited electrons. The energy of the excited electrons is funneled, through a series of resonance energy transfers, to a special pair of chlorophyll molecules in the photochemical reaction center. The reaction center then produces a high-energy electron that can be passed rapidly to the electron-transport chain in the thylakoid membrane, via a quinone.

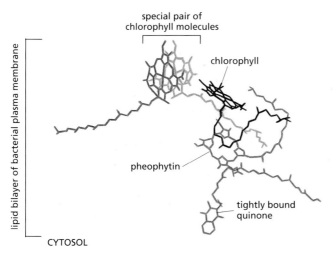

Figure 14–44 The arrangement of the electron carriers in the photochemical reaction center of a purple bacterium. <ATCA> The pigment molecules shown are held in the interior of a transmembrane protein and are surrounded by the lipid bilayer of the bacterial plasma membrane. An electron in the special pair of chlorophyll molecules is excited by resonance from an antenna complex chlorophyll, and the excited electron is then transferred stepwise from the special pair to the quinone (see also Figure 14–45). A similar arrangement of electron carriers is present in the reaction centers of plants (see Figure 14–47).

In a Reaction Center, Light Energy Captured by Chlorophyll Creates a Strong Electron Donor from a Weak One

The electron transfers involved in the photochemical reactions just described have been analyzed extensively by rapid spectroscopic methods. **Figure 14–45** illustrates, in a general way, how light provides the energy needed to transfer an electron from a weak electron donor (a molecule with a strong affinity for electrons) to a molecule that is a strong electron donor in its reduced form (a

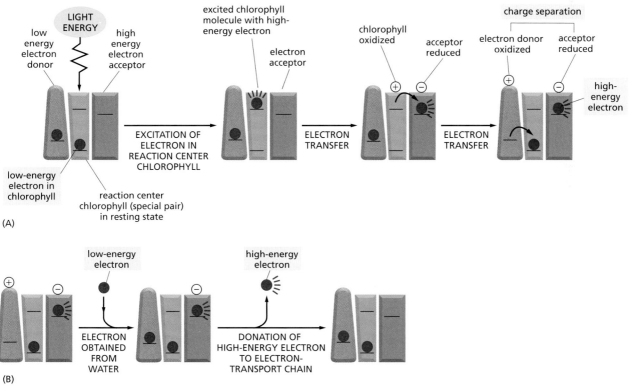

Figure 14–45 How light energy is harvested by a reaction center chlorophyll molecule. (A) The initial events in a reaction center create a charge separation. A pigment–protein complex holds a chlorophyll molecule of the special pair *(blue)* precisely positioned so that both a potential low-energy electron donor *(orange)* and a potential high-energy electron acceptor *(green)* are immediately available. When light energizes an electron in the chlorophyll molecule *(red electron)*, the excited electron is immediately passed to the electron acceptor and is thereby partially stabilized. The positively charged chlorophyll molecule then quickly attracts the low-energy electron from the electron donor and returns to its resting state, creating a larger charge separation that further stabilizes the high-energy electron. These reactions require less than 10^{-6} second to complete. (B) In the final stage of this process, which follows the steps in (A), the photosynthetic reaction center is restored to its original resting state by acquiring a new low-energy electron and then transferring the high-energy electron derived from chlorophyll to an electron transport chain in the membrane. As will be discussed subsequently, the ultimate source of low-energy electrons for photosystem II in the chloroplast is water; as a result, light produces high-energy electrons in the thylakoid membrane from low-energy electrons in water.

molecule with a weak affinity for electrons). The special pair of chlorophyll molecules in the reaction center is poised to pass each excited electron to a precisely positioned neighboring molecule in the same protein complex (an electron acceptor). The chlorophyll molecule that loses an electron becomes positively charged, but it rapidly regains an electron from an adjacent electron donor to return to its unexcited, uncharged state (Figure 14–45A, *orange* electron). Then, in slower reactions, the electron donor has its missing electron replaced, and the high-energy electron that was generated by the excited chlorophyll is transferred to the electron-transport chain (Figure 14–45B). The excitation energy in chlorophyll that would normally be released as fluorescence or heat is thereby used instead to create a strong electron donor (a molecule carrying a high-energy electron) where none had been before.

The photosystem of purple bacteria is somewhat simpler than the evolutionarily related photosystems in chloroplasts, and it has served as a good model for working out reaction details. The reaction center in this photosystem is a large protein–pigment complex that can be solubilized with detergent and purified in active form. In a major triumph of structure analysis, its complete three-dimensional structure was determined by x-ray crystallography (see Figure 10–34). This structure, combined with kinetic data, provides the best picture we have of the initial electron-transfer reactions that occur during photosynthesis. **Figure 14–46** shows the actual sequence of electron transfers that take place, for comparison with Figure 14–45A.

In the purple bacterium, the electron used to fill the electron-deficient hole created by the light-induced charge separation comes from a cyclic flow of electrons transferred through a cytochrome (see *orange box* in Figure 14–45); the strong electron donor produced is a quinone. One of the two photosytems in the chloroplasts of higher plants likewise produces a quinone carrying high-energy electrons. However, as we discuss next, because water provides the electrons for this photosystem, photosynthesis in plants—unlike that in purple bacteria—releases large quantities of oxygen gas.

Noncyclic Photophosphorylation Produces Both NADPH and ATP

Photosynthesis in plants and cyanobacteria produces both ATP and NADPH directly by a two-step process called **noncyclic photophosphorylation**. Because two photosystems—called photosystems I and II—work in series to energize an electron to a high-enough energy state, the electron can be transferred all the

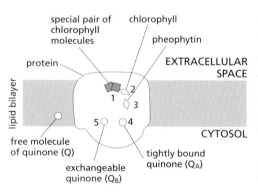

Figure 14–46 The electron transfers that occur in the photochemical reaction center of a purple bacterium. <ATCA> A similar set of reactions occurs in the evolutionarily related photosystem II in plants. At the top left is an orientating diagram showing the molecules that carry electrons, which are those in Figure 14–45, plus an exchangeable quinone (Q_B) and a freely mobile quinone (Q) dissolved in the lipid bilayer. Electron carriers 1–5 are each bound in a specific position on a 596-amino-acid transmembrane protein formed from two separate subunits (see Figure 10–34). After excitation by a photon of light, a high-energy electron passes from pigment molecule to pigment molecule, very rapidly creating a stable *charge separation*, as shown in the sequence of steps A–C, in which the pigment molecule carrying a high-energy electron is colored *red*. Steps D and E then occur progressively. After a second photon has repeated this sequence with a second electron, the exchangeable quinone is released into the bilayer carrying two high-energy electrons. This quinone quickly loses its charge by picking up two protons (see Figure 14–24).

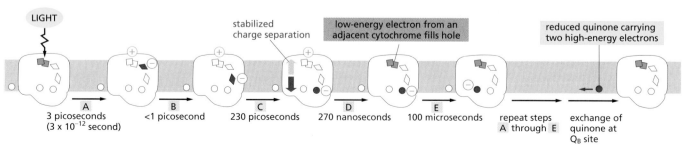

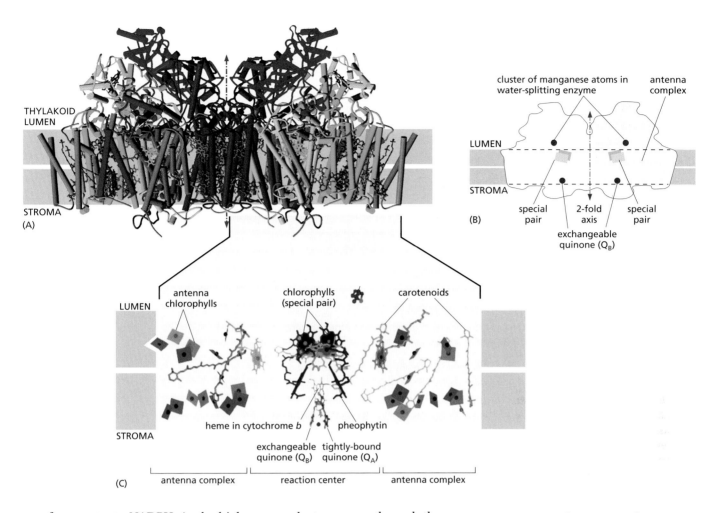

THYLAKOID LUMEN

STROMA

(A)

cluster of manganese atoms in water-splitting enzyme

antenna complex

LUMEN

STROMA

(B)

special pair

2-fold axis

special pair

exchangeable quinone (Q_B)

LUMEN

antenna chlorophylls

chlorophylls (special pair)

carotenoids

heme in cytochrome b

pheophytin

STROMA

exchangeable quinone (Q_B)

tightly-bound quinone (Q_A)

(C)

antenna complex

reaction center

antenna complex

way from water to NADPH. As the high-energy electrons pass through the coupled photosystems to generate NADPH, some of their energy is siphoned off for ATP synthesis.

The first of the two photosystems—paradoxically called *photosystem II* for historical reasons—has the unique ability to withdraw electrons from water. The oxygens of two water molecules bind to a cluster of four manganese atoms on the lumenal surface of the photosystem II reaction center complex (**Figure 14–47**). This cluster enables electrons to be removed one at a time from the water, as required to fill the electron-deficient holes created by light in chlorophyll molecules in the reaction center. As soon as four electrons have been removed from the two water molecules (requiring four quanta of light), O_2 is released. Photosystem II thus catalyzes the reaction $2H_2O + 4$ photons $\rightarrow 4H^+ + 4e^- + O_2$. As we discussed for the electron-transport chain in mitochondria, which uses O_2 and produces water, the mechanism ensures that no partly oxidized water molecules are released as dangerous, highly reactive oxygen radicals. Essentially all the oxygen in the Earth's atmosphere has been produced in this way.

The core of the reaction center in photosystem II is homologous to the bacterial reaction center just described, and it likewise produces strong electron donors in the form of reduced quinone molecules dissolved in the lipid bilayer of the membrane. The quinones pass their electrons to a H^+ pump called the *cytochrome b_6-f complex,* which resembles the cytochrome b-c_1 complex in the respiratory chain of mitochondria. The cytochrome b_6-f complex pumps H^+ into the thylakoid space across the thylakoid membrane (or out of the cytosol across the plasma membrane in cyanobacteria), and the resulting electrochemical gradient drives the synthesis of ATP by an ATP synthase (**Figure 14–48**).

The final electron acceptor in this electron-transport chain is the second photosystem, *photosystem I,* which accepts an electron into the electron-deficient hole created by light in the chlorophyll molecule in its reaction center. Each electron that enters photosystem I is finally boosted to a very high-energy

Figure 14–47 The structure of photosystem II in plants and cyanobacteria. The structure shown is a dimer, organized around a two fold axis (*red dotted* arrows). Each monomer is composed of 16 integral membrane protein subunits plus three subunits in the lumen, with a total of 36 bound chlorophylls, 7 carotenoids, two pheophytins, two hemes, two plastoquinones, and one manganese cluster in an oxygen-evolving water-splitting center. (A) The complete three-dimensional structure of the dimer. (B) Schematic of the dimer with a few central features indicated. (C) A monomer drawn to show only the non-protein molecules in the structure, thereby highlighting the protein-bound pigments and electron carriers; *green* structures are chlorophylls. (Adapted from K.N. Ferreira et al., *Science* 303:1831–1838, 2004. With permission from AAAS.)

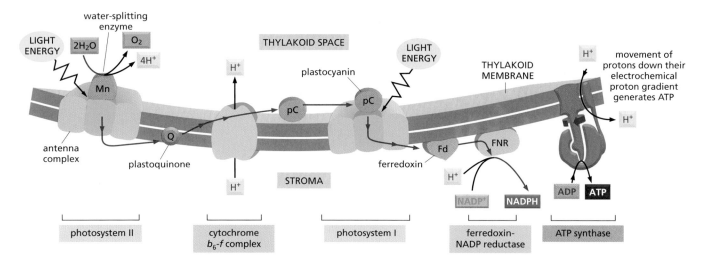

Figure 14–48 Electron flow during photosynthesis in the thylakoid membrane. The mobile electron carriers in the chain are plastoquinone (which closely resembles the ubiquinone of mitochondria), plastocyanin (a small copper-containing protein), and ferredoxin (a small protein containing an iron–sulfur center). The cytochrome b_6-f complex resembles the b-c_1 complex of mitochondria and the b-c complex of bacteria (see Figure 14–73): all three complexes accept electrons from quinones and pump H^+ across the membrane. The H^+ released by water oxidation into the thylakoid space, and the H^+ consumed during NADPH formation in the stroma, also contribute to the generation of the electrochemical H^+ gradient. As illustrated, this gradient drives ATP synthesis by an ATP synthase present in this same membrane.

level that allows it to be passed to the iron–sulfur center in ferredoxin and then to $NADP^+$ to generate NADPH.

The scheme for photosynthesis just discussed is known as the *Z scheme*. By means of its two electron-energizing steps, one catalyzed by each photosystem, an electron is passed from water, which normally holds on to its electrons very tightly (redox potential = +820 mV), to NADPH, which normally holds on to its electrons loosely (redox potential = –320 mV) (**Figure 14–49**).

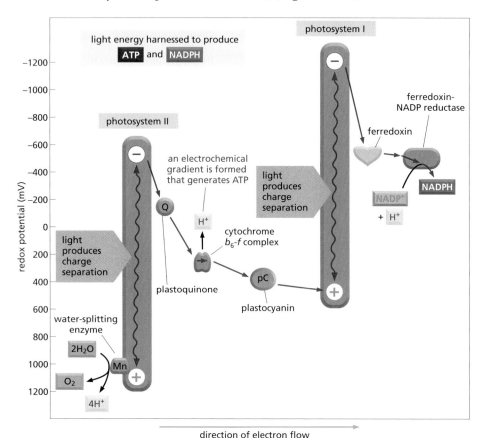

Figure 14–49 Changes in redox potential during photosynthesis. The redox potential for each molecule is indicated by its position along the vertical axis. Note that photosystem II passes electrons derived from water to photosystem I. The net electron flow through the two photosystems in series is from water to $NADP^+$, and it produces NADPH as well as ATP.

The ATP is synthesized by an ATP synthase that harnesses the electrochemical proton gradient produced by the three sites of H^+ activity that are highlighted in Figure 14–48. This Z scheme for ATP production is called noncyclic photophosphorylation, to distinguish it from a cyclic scheme that utilizes only photosystem I (see the text).

A single quantum of visible light does not have enough energy to move an electron all the way from the bottom of photosystem II to the top of photosystem I, which is presumably the energy change required to pass an electron efficiently from water to $NADP^+$. The use of two separate photosystems in series makes the energy from two quanta of light available for this purpose. In addition, there is enough energy left over to enable the electron-transport chain that links the two photosystems to pump H^+ across the thylakoid membrane (or the plasma membrane of cyanobacteria), so that the ATP synthase can harness some of the light-derived energy for ATP production.

Chloroplasts Can Make ATP by Cyclic Photophosphorylation Without Making NADPH

In the noncyclic photophosphorylation scheme just discussed, high-energy electrons leaving photosystem II are harnessed to generate ATP and are passed on to photosystem I to drive the production of NADPH. This produces slightly more than 1 molecule of ATP for every pair of electrons that passes from H_2O to $NADP^+$ to generate a molecule of NADPH. But carbon fixation requires 1.5 molecules of ATP per NADPH molecule (see Figure 14–40). To produce extra ATP, chloroplasts can switch photosystem I into a cyclic mode so that it produces ATP instead of NADPH. In this process, called *cyclic photophosphorylation,* the high-energy electrons from photosystem I are transferred to the cytochrome b_6-f complex rather than being passed on to $NADP^+$. From the b_6-f complex, the electrons are passed back to photosystem I at a low energy. The only net result, besides the conversion of some light energy to heat, is that H^+ is pumped across the thylakoid membrane by the b_6-f complex as electrons pass through it, thereby increasing the electrochemical proton gradient that drives the ATP synthase. (This is analogous to the cyclic process that occurs in purple nonsulfur bacteria in Figure 14–73, below.)

To summarize, cyclic photophosphorylation involves only photosystem I, and it produces ATP without the formation of either NADPH or O_2. The relative activities of cyclic and noncyclic electron flows are regulated by the cell to determine how much light energy is converted into reducing power (NADPH) and how much into high-energy phosphate bonds (ATP).

Photosystems I and II Have Related Structures, and Also Resemble Bacterial Photosystems

The mechanisms of fundamental cell processes such as DNA replication or respiration generally turn out to be the same in eucaryotic cells and in bacteria, even though the number of protein components involved is considerably greater in eucaryotes. Eucaryotes evolved from procaryotes, and the additional proteins were presumably selected for during evolution because they provided an extra degree of efficiency, regulation that was useful to the cell, or both.

Photosystems provide a clear example of this type of evolution. The atomic structures of both eucaryotic photosystems have been revealed by a combination of electron and x-ray crystallography, and the close relationship of photosystem I, photosystem II, and the photochemical reaction center of purple bacteria has been clearly demonstrated from such analyses (**Figure 14–50**).

The Proton-Motive Force Is the Same in Mitochondria and Chloroplasts

The presence of the thylakoid space separates a chloroplast into three rather than the two internal compartments of a mitochondrion. The net effect of H^+ translocation in the two organelles is, however, similar. As illustrated in **Figure 14–51**, in chloroplasts exposed to light, H^+ is pumped out of the stroma (pH 7.5) into the thylakoid space (pH ~4.5), creating a gradient of 3 pH units. This represents a proton-motive force of about 180 mV across the thylakoid membrane, and it drives ATP synthesis by the ATP synthase embedded in this membrane. The force is nearly the same as that across the inner mitochondrial membrane, but nearly all of it is contributed by the pH gradient rather than by a membrane potential, unlike the case in mitochondria.

For both mitochondria and chloroplasts, the catalytic site of the ATP synthase is at a pH of about 7.5 and is located in a large organelle compartment (matrix or stroma) that is packed full of soluble enzymes. It is here that all of the organelle's ATP is made (see Figure 14–51).

Carrier Proteins in the Chloroplast Inner Membrane Control Metabolite Exchange with the Cytosol

If chloroplasts are isolated in a way that leaves their inner membrane intact, this membrane can be shown to have a selective permeability, reflecting the presence of specific transporters. Most notably, much of the glyceraldehyde 3-phosphate produced by CO_2 fixation in the chloroplast stroma is transported out of the chloroplast by an efficient antiport system that exchanges three-carbon sugar phosphates for an inward flux of inorganic phosphate.

Glyceraldehyde 3-phosphate normally provides the cytosol with an abundant source of carbohydrate, which the cell uses as the starting point for many other biosyntheses—including the production of sucrose for export. But this is not all that this molecule provides. Once the glyceraldehyde 3-phosphate reaches the cytosol, it is readily converted (by part of the glycolytic pathway) to 1,3-phosphoglycerate and then 3-phosphoglycerate (see p. 92), generating one molecule of ATP and one of NADH. (A similar two-step reaction, but working in reverse, forms glyceraldehyde 3-phosphate in the carbon-fixation cycle; see Figure 14–40.) As a result, the export of glyceraldehyde 3-phosphate from the chloroplast provides

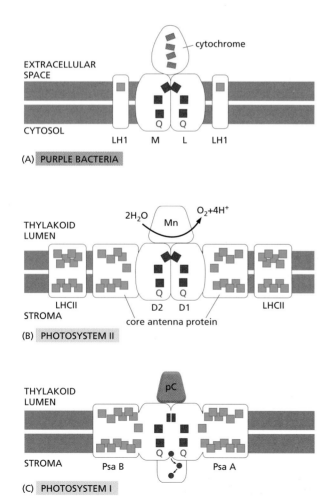

(A) **PURPLE BACTERIA**

(B) **PHOTOSYSTEM II**

(C) **PHOTOSYSTEM I**

Figure 14–50 Three types of photosynthetic reaction centers compared. Pigments involved in light harvesting are colored *green;* those involved in the central photochemical events are colored *red*. (A) The photochemical reaction center of purple bacteria, whose detailed structure is illustrated in Figure 10–34, contains two related protein subunits, L and M, that bind the pigments involved in the central process illustrated in Figure 14–46. Low-energy electrons are fed into the excited chlorophylls by a cytochrome. LH1 is a protein–pigment complex involved in light harvesting. (B) Photosystem II contains the D_1 and D_2 proteins, which are homologous to the L and M subunits in (A). Low-energy electrons from water are fed into the excited chlorophylls by a manganese cluster. LHCII is a light-harvesting complex that feeds energy into the core antenna proteins (see Figure 14–47). (C) Photosystem I contains the Psa A and Psa B proteins, each of which is equivalent to a fusion of the D_1 or D_2 protein to a core antenna protein of photosystem II. Low-energy electrons are fed into the excited chlorophylls by loosely bound plastocyanin (pC). As indicated, in photosystem I, high-energy electrons are passed from a nonmobile quinone (Q) through a series of three iron-sulfur centers *(red circles)*. (Modified from K. Rhee, E. Morris, J. Barber and W. Kühlbrandt, *Nature* 396:283–286, 1998; W. Kühlbrandt, *Nature* 411:896–899, 2001. With permission from Macmillan Publishers Ltd.)

Figure 14–51 A comparison of the flow of H⁺ and the orientation of the ATP synthase in mitochondria and chloroplasts. Those compartments with similar pH values have been colored the same. The proton-motive force across the thylakoid membrane consists almost entirely of the pH gradient; a high permeability of this membrane to Mg^{2+} and Cl^- ions allows the flow of these ions to dissipate most of the membrane potential. Mitochondria presumably need a large membrane potential because they could not tolerate having their matrix at pH 10, as would be required to generate their proton-motive force without one.

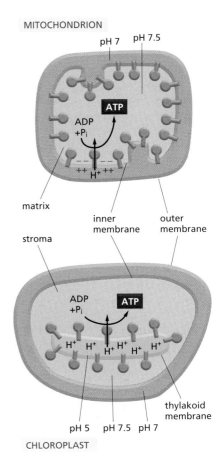

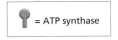

= ATP synthase

not only the main source of fixed carbon to the rest of the cell, but also the reducing power and ATP needed for metabolism outside the chloroplast.

Chloroplasts Also Perform Other Crucial Biosyntheses

The chloroplast performs many biosyntheses in addition to photosynthesis. Enzymes in the chloroplast stroma make all of the cell's fatty acids and a set of amino acids, for example. Similarly, the reducing power of light-activated electrons drives the reduction of nitrite (NO_2^-) to ammonia (NH_3) in the chloroplast; this ammonia provides the plant with nitrogen for the synthesis of amino acids and nucleotides. The metabolic importance of the chloroplast for plants and algae therefore extends far beyond its role in photosynthesis.

Summary

Chloroplasts and photosynthetic bacteria obtain high-energy electrons by means of photosystems that capture the electrons that are excited when sunlight is absorbed by chlorophyll molecules. Photosystems are composed of an antenna complex that funnels energy to a photochemical reaction center, where a precisely ordered complex of proteins and pigments allows electron carriers to capture the energy of an excited chlorophyll electron. The most completely understood photochemical reaction center is that of purple photosynthetic bacteria, which contain only a single photosystem. In contrast, there are two distinct photosystems in chloroplasts and cyanobacteria. The two photosystems are normally linked in series, and they transfer electrons from water to NADP⁺ to form NADPH, with the concomitant production of a transmembrane electrochemical proton gradient. In these linked photosystems, molecular oxygen (O_2) is generated as a by-product of removing four low-energy electrons from two specifically positioned water molecules. The detailed three-dimensional structures of photosystems I and II show a striking degree of homology to the structure of the photosystems of purple photosynthetic bacteria, demonstrating a remarkable degree of conservation over billions of years of evolution.

Compared with mitochondria, chloroplasts have an additional internal membrane (the thylakoid membrane) and a third internal space (the thylakoid space). All electron-transport processes occur in the thylakoid membrane: to make ATP, H⁺ is pumped into the thylakoid space, and a backflow of H⁺ through an ATP synthase then produces the ATP in the chloroplast stroma. This ATP is used in conjunction with the NADPH made by photosynthesis to drive a large number of biosynthetic reactions in the chloroplast stroma, including the all-important carbon-fixation cycle, which creates carbohydrate from CO_2. Along with some other important chloroplast products, this carbohydrate is exported to the cell cytosol, where—as glyceraldehyde 3-phosphate—it provides organic carbon, ATP, and reducing power to the rest of the cell.

THE GENETIC SYSTEMS OF MITOCHONDRIA AND PLASTIDS

The view that mitochondria and plastids evolved from bacteria that were engulfed by ancestral cells can explain why both types of organelles contain their own genomes, as well as their own biosynthetic machinery for making RNA and organelle proteins. Mitochondria and plastids are never made from scratch, but instead arise by the growth and division of an existing mitochondrion or plastid. On average, each organelle must double in mass in each cell generation and then be distributed into each daughter cell. Even nondividing cells must replenish organelles that are degraded as part of the continual process of organelle turnover, or produce additional organelles as the need arises. Organelle growth and proliferation is a complicated process because mitochondrial and plastid proteins are encoded in two places: the nuclear genome and the separate genomes harbored in the organelles themselves (**Figure 14–52**). In Chapter 12, we discuss how selected proteins and lipids are imported into mitochondria and chloroplasts from the cytosol. Here we describe how the organelle genomes are maintained and the contributions they make to organelle biogenesis.

Mitochondria and Chloroplasts Contain Complete Genetic Systems

The biosynthesis of mitochondria and plastids requires contributions from two separate genetic systems. Special genes in nuclear DNA encode most of the proteins in mitochondria and chloroplasts. The organelle imports these proteins from the cytosol after they have been synthesized on cytosolic ribosomes. The organelle DNA encodes other organelle proteins that are synthesized on ribosomes within the organelle, using organelle-produced mRNA to specify their amino acid sequence (**Figure 14–53**). The protein traffic between the cytosol and these organelles seems to be unidirectional, as proteins are normally not exported from mitochondria or chloroplasts to the cytosol. An exception occurs under special conditions when a cell is about to undergo apoptosis. As will be discussed in detail in Chapter 19, the mitochondrion releases intermembrane space proteins (including cytochrome *c*) through its outer mitochondrial membrane as part of an elaborate signaling pathway that is triggered to cause cells to undergo programmed cell death.

The processes of organelle DNA transcription, protein synthesis, and DNA replication (**Figure 14–54**) take place where the genome is located: in the matrix of mitochondria and the stroma of chloroplasts. Although the proteins that

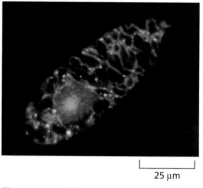

Figure 14–52 Mitochondrial and nuclear DNA stained with a fluorescent dye. This micrograph shows the distribution of the nuclear genome *(red)* and the multiple small mitochondrial genomes (bright *yellow* spots) in a *Euglena gracilis* cell. The DNA is stained with ethidium bromide, a fluorescent dye that emits red light. In addition, the mitochondrial matrix space is stained with a green fluorescent dye that reveals the mitochondria as a branched network extending throughout the cytosol. The superposition of the *green* matrix and the *red* DNA gives the mitochondrial genomes their *yellow* color. (Courtesy of Y. Hayashi and K. Ueda, *J. Cell Sci.* 93:565–570, 1989. With permission from The Company of Biologists.)

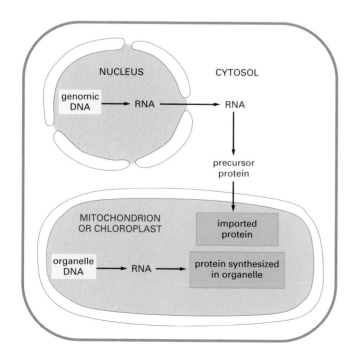

Figure 14–53 The production of mitochondrial and chloroplast proteins by two separate genetic systems. Most of the proteins in these organelles are encoded by the nucleus and must be imported from the cytosol.

mediate these genetic processes are unique to the organelle, the nuclear genome encodes most of them. This is all the more surprising because the protein-synthesis machinery of the organelles resembles that of bacteria rather than that of eucaryotes. The resemblance is particularly close in chloroplasts. For example, chloroplast ribosomes are very similar to *E. coli* ribosomes, both in their structure and in their sensitivity to various antibiotics (such as chloramphenicol, streptomycin, erythromycin, and tetracycline). In addition, protein synthesis in chloroplasts starts with *N*-formyl methionine, as in bacteria, and not with the methionine used for this purpose in the cytosol of eucaryotic cells. Although mitochondrial genetic systems are much less similar to those of present-day bacteria than are the genetic systems of chloroplasts, their ribosomes are also sensitive to antibacterial antibiotics, and protein synthesis in mitochondria also starts with *N*-formyl methionine.

Organelle Growth and Division Determine the Number of Mitochondria and Plastids in a Cell

In mammalian cells, mitochondrial DNA makes up less than 1% of the total cellular DNA. In other cells, however, such as the leaves of higher plants or the very large egg cells of amphibians, a much larger fraction of the cellular DNA may be present in mitochondria or chloroplasts (**Table 14–2**), and a large fraction of RNA and protein synthesis takes place there.

Mitochondria and plastids are large enough to be observed by light microscopy in living cells. For example, mitochondria can be visualized by expressing a genetically engineered fusion of a mitochondrial protein linked to the green fluorescent protein (GFP) in cells, or cells can be incubated with a fluorescent dye that is specifically taken up by mitochondria because of the electrochemical gradient across their membranes. Such images demonstrate that the mitochondria in living cells are dynamic—frequently dividing, fusing, and changing shape (**Figure 14–55**), as mentioned previously. Division (fission) and fusion of these organelles are topologically complex processes, because the organelles are enclosed by a double membrane and the mitochondrion must maintain the integrity of its separate mitochondrial compartments during these processes (**Figure 14–56**).

The number and shape of mitochondria vary dramatically in different cell types and can change in the same cell type under different physiological conditions, ranging from multiple spherical or cylindrically shaped organelles to a single organelle with a branched structure (a reticulum). The arrangement is controlled by the relative rates of mitochondrial division and fusion, which are regulated by dedicated GTPases that reside on mitochondrial membranes. In

1 µm

Figure 14–54 An electron micrograph of an animal mitochondrial DNA molecule caught during the process of DNA replication. The circular DNA genome has replicated only between the two points marked by *red* arrows. The newly synthesized DNA is colored *yellow*. (Courtesy of David A. Clayton.)

Table 14–2 Relative Amounts of Organelle DNA in Some Cells and Tissues

ORGANISM	TISSUE OR CELL TYPE	DNA MOLECULES PER ORGANELLE	ORGANELLES PER CELL	ORGANELLE DNA AS PERCENTAGE OF TOTAL CELLULAR DNA
MITOCHONDRIAL DNA				
Rat	liver	5–10	1000	1
Yeast*	vegetative	2–50	1–50	15
Frog	egg	5–10	10^7	99
CHLOROPLAST DNA				
Chlamydomonas	vegetative	80	1	7
Maize	leaves	0–300**	20–40	0–15**

*The large variation in the number and size of mitochondria per cell in yeasts is due to mitochondrial fusion and fission.
**In maize, the amount of chloroplast DNA drops precipitously in mature leaves, after cell division ceases: the chloroplast DNA is degraded and stable mRNAs persist to provide for protein synthesis.

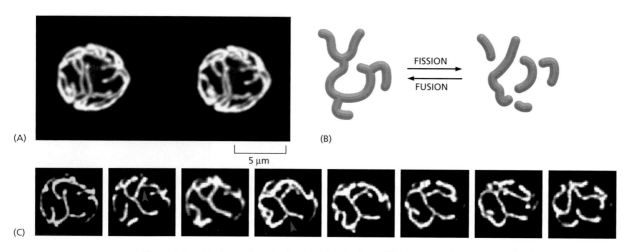

5 µm

(A)   (B)

(C)

Figure 14–55 A dynamic mitochondrial reticulum. (A) In yeast cells, mitochondria form a continuous reticulum underlying the plasma membrane (stereo pair). (B) A balance between fission and fusion determines the arrangement of the mitochondria in different cells. (C) Time-lapse fluorescent microscopy shows the dynamic behavior of the mitochondrial network in a yeast cell. In addition to shape changes, fission and fusion constantly remodel the network *(red arrows)*. The pictures were taken at 3-minute intervals. (A and C, from J. Nunnari et al., *Mol. Biol. Cell* 8:1233–1242, 1997. With permission by the American Society for Cell Biology.)

addition, the total organelle mass per cell can be regulated according to need. For example, a large increase in mitochondria (as much as 5–10-fold) occurs when a resting skeletal muscle is repeatedly stimulated to contract for a prolonged period.

There can be many copies of the mitochondrial and plastid genomes in the space enclosed by each organelle's inner membrane. The degree of organelle fragmentation determines the number of these genomes present in a single organelle; generally, a single compartment houses many genomes (see Table 14–2). In most cells, the replication of the organelle DNA is not limited to the S phase of the cell cycle, when the nuclear DNA replicates, but occurs throughout the cell cycle—out of phase with cell division. Individual organelle DNA molecules seem to be selected at random for replication, so that in a given cell cycle, some may replicate more than once and others not at all. Nonetheless, under constant conditions, the process is regulated to ensure that the total number of organelle DNA molecules doubles in every cell cycle, as required if each cell type is to maintain a constant amount of organelle DNA.

In special circumstances, the cell can precisely control organelle division. In some algae that contain only one or a few chloroplasts, for example, the organelle divides just before the cell does, in a plane that is identical to the future plane of cell division.

Figure 14–56 Mitochondrial fission and fusion. These processes involve both outer and inner mitochondrial membranes. (A) During fusion and fission, the mitochondrion maintains both its matrix and its intermembrane space compartments. Different membrane fusion machines are thought to operate at the outer and inner membranes. The fission process resembles that of bacterial cell division (discussed in Chapter 16). The pathway shown has been postulated from static views such as that shown in (B). (B) An electron micrograph of a dividing mitochondrion in a liver cell. (B, courtesy of Daniel S. Friend.)

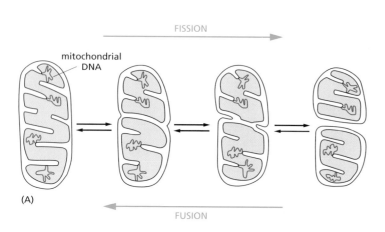

FISSION

mitochondrial DNA

(A)

FUSION

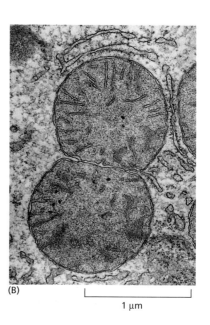

(B)

1 µm

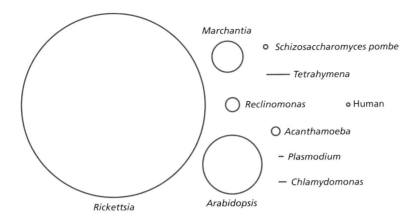

Figure 14–57 Various sizes of mitochondrial genomes. The complete DNA sequences for more than 500 mitochondrial genomes have been determined. The lengths of a few of these mitochondrial DNAs are shown to scale as circles for those genomes thought to be circular and lines for linear genomes. The largest circle represents the genome of *Rickettsia prowazekii*, a small pathogenic bacterium whose genome most closely resembles that of mitochondria. The size of mitochondrial genomes does not correlate well with the number of proteins encoded in them: while human mitochondrial DNA encodes 13 proteins, the 22-fold larger mitochondrial DNA of *Arabidopsis* encodes only 32 proteins—that is, about 2.5-fold as many as human mitochondrial DNA. The extra DNA that is found in *Arabidopsis*, *Marchantia*, and other plant mitochondria may be "junk DNA". The mitochondrial DNA of the protozoan *Reclinomonas americana* has 98 genes, more than the mitochondrion of any other organism analyzed so far. (Adapted from M.W. Gray et al., *Science* 283:1476–1481, 1999. With permission from AAAS.)

Mitochondria and Chloroplasts Have Diverse Genomes

The multiple copies of mitochondrial and chloroplast DNA contained within the matrix or stroma of these organelles are usually distributed in several clusters, called *nucleoids*. Nucleoids are thought to be attached to the inner mitochondrial membrane. The DNA structure in nucleoids is likely to resemble that in bacteria rather than that in eucaryotic chromatin; as in bacteria, for example, there are no histones.

The size range of organelle DNAs is similar to that of viral DNAs. Mitochondrial DNA molecules range in size from less than 6000 nucleotide pairs in *Plasmodium falciparum* (the human malaria parasite) to more than 300,000 nucleotide pairs in some land plants (**Figure 14–57**). The chloroplast genome of land plants ranges in size from 70,000 to 200,000 nucleotide pairs. Chloroplast chromosomes had long been thought to be circular DNA molecules that contain a single copy of the genome, but it is now recognized that such molecules are outnumbered in most chloroplasts by linear DNA molecules formed from multiple copies of the genome, joined end to end. The mixture of structures observed for chloroplast chromosomes is best explained as originating from a DNA replication process that is closely linked to DNA recombination events, analogous to the replication processes observed for viruses such as Herpes. Similar types of mixed structures may also exist for many mitochondrial genomes. However, in mammals, the mitochondrial genome is a simple DNA circle of about 16,500 base pairs (less than 0.001% of the size of the nuclear genome), and it is nearly the same size in animals as diverse as *Drosophila* and sea urchins.

Mitochondria and Chloroplasts Probably Both Evolved from Endosymbiotic Bacteria

The procaryotic character of the organelle genetic systems, especially striking in chloroplasts, suggests that mitochondria and chloroplasts evolved from bacteria that were endocytosed more than 1 billion years ago. According to one version of this *endosymbiont hypothesis*, eucaryotic cells started out as anaerobic organisms without mitochondria or chloroplasts and then established a stable endosymbiotic relation with a bacterium, whose oxidative phosphorylation system they subverted for their own use (**Figure 14–58**). According to this hypothesis, the endocytic event that led to the development of mitochondria occurred when oxygen entered the atmosphere in substantial amounts, more than 1.5×10^9 years ago, before animals and plants separated (see Figure 14–71). The chloroplast seems to have been similarly derived later, from endocytosis of an oxygen-producing photosynthetic organism that resembled a cyanobacterium.

Most of the genes encoding present-day mitochondrial and chloroplast proteins are in the cell nucleus. Thus, an extensive transfer of genes from organelle to nuclear DNA must have occurred during eucaryote evolution. A successful

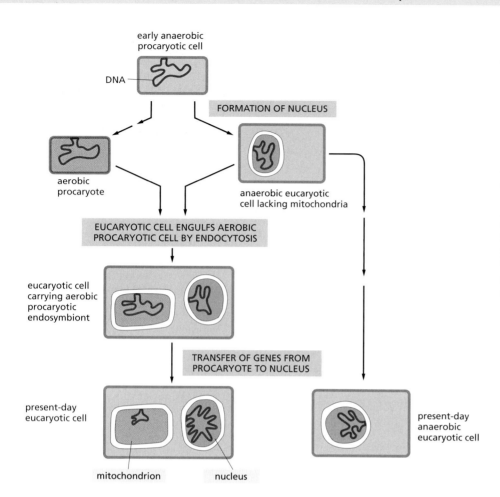

Figure 14–58 One possible evolutionary pathway for the origin of mitochondria. There are anaerobic single-celled eucaryotes present on the Earth today that lack a mitochondrial genome, including the protozoan *Giardia*. However, all those examined appear to contain some type of membrane-enclosed organelle remnant. In these remnants, mitochondrial-like reactions, such as the production of iron-sulfur centers, occur. Thus far, we know of no living example of the anaerobic cell postulated in this figure to have first engulfed the precursors of mitochondria. For this and other reasons, alternative pathways have been postulated in which the combination of a bacterium and an archaea led to both the first eukaryotic cells and the first mitochondria.

transfer of this type is expected to be rare, because a gene moved from organelle DNA needs to change to become a functional nuclear gene: it must adapt to the nuclear and cytoplasmic transcription and translation requirements, and also acquire a signal sequence so that the encoded protein can be delivered to the organelle after its synthesis in the cytosol. Nevertheless, there is evidence that such gene transfers to the nucleus continue to occur in some organisms today.

Gene transfer explains why many of the nuclear genes encoding mitochondrial and chloroplast proteins resemble bacterial genes. The amino acid sequence of the chicken mitochondrial enzyme *superoxide dismutase,* for example, resembles the corresponding bacterial enzyme much more than it resembles the superoxide dismutase found in the cytosol of the same eucaryotic cell.

Gene transfer seems to have been a gradual process. When mitochondrial genomes encoding different numbers of proteins are compared, a pattern of sequential reduction of encoded mitochondrial functions emerges (**Figure 14–59**). The smallest and presumably most highly evolved mitochondrial genomes, for example, encode only a few inner-membrane proteins involved in electron-transport reactions, plus ribosomal RNAs and some tRNAs. Mitochondrial genomes that have remained more complex tend to contain this same subset of genes, plus others. The most complex genomes are characterized by the presence of many extra genes compared with animal and yeast mitochondrial genomes. Many of these genes encode components of the mitochondrial genetic system, such as RNA polymerase subunits and ribosomal proteins; these genes are instead found in the cell nucleus in organisms that have reduced their mitochondrial DNA content.

What type of bacterium gave rise to the mitochondrion? From sequence comparisons, it seems that mitochondria have descended from a particular type of purple photosynthetic bacterium that had previously lost its ability to perform photosynthesis and was left with only a respiratory chain.

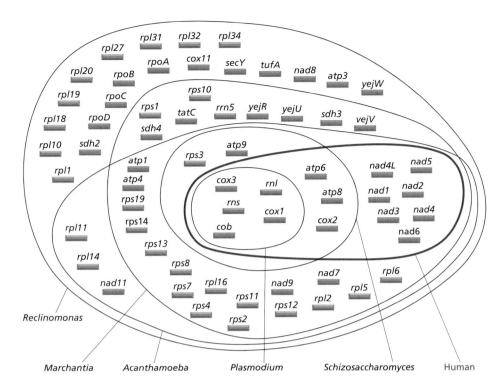

Figure 14–59 Comparison of mitochondrial genomes. Less complex mitochondrial genomes encode subsets of the proteins and ribosomal RNAs that are encoded by larger mitochondrial genomes. There are only four genes present in all known mitochondrial genomes; these encode ribosomal RNAs *(rns* and *rnl),* cytochrome *b (cob),* and a cytochrome oxidase subunit *(cox1).* (Adapted from M.W. Gray et al., *Science* 283:1476–1481, 1999. With permission from AAAS.)

Mitochondria Have a Relaxed Codon Usage and Can Have a Variant Genetic Code

The relatively small size of the human mitochondrial genome made it a particularly attractive target for early DNA-sequencing projects, and in 1981, researchers published the complete sequence of its 16,569 nucleotides. By comparing this sequence with known mitochondrial tRNA sequences and with the partial amino acid sequences available for proteins encoded by the mitochondrial DNA, all of the human mitochondrial genes were mapped on the circular DNA molecule (**Figure 14–60**).

Compared with nuclear, chloroplast, and bacterial genomes, the human mitochondrial genome has several surprising features:

1. *Dense gene packing.* Unlike other genomes, nearly every nucleotide seems to be part of a coding sequence, either for a protein or for one of the rRNAs or tRNAs. Since these coding sequences run directly into each other, there is very little room left for regulatory DNA sequences.

2. *Relaxed codon usage.* Whereas 30 or more tRNAs specify amino acids in the cytosol and in chloroplasts, only 22 tRNAs are required for mitochondrial protein synthesis. The normal codon–anticodon pairing rules are relaxed in mitochondria, so that many tRNA molecules recognize any one of the

Figure 14–60 The organization of the human mitochondrial genome. The genome contains 2 rRNA genes, 22 tRNA genes, and 13 protein-coding sequences. The DNAs of many other animal mitochondrial genomes have also been completely sequenced. Most of these animal mitochondrial DNAs encode precisely the same genes as humans, with the gene order being identical for animals that range from mammals to fish.

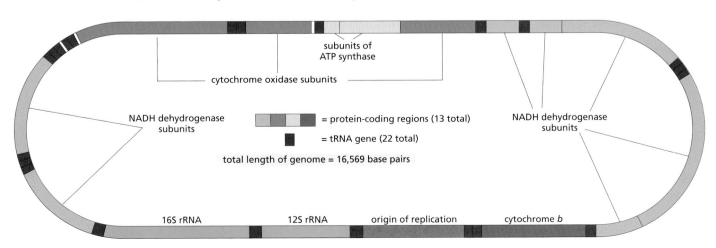

Table 14–3 Some Differences Between the "Universal" Code and Mitochondrial Genetic Codes*

		MITOCHONDRIAL CODES			
CODON	"UNIVERSAL" CODE	MAMMALS	INVERTEBRATES	YEASTS	PLANTS
UGA	STOP	*Trp*	*Trp*	*Trp*	STOP
AUA	Ile	*Met*	*Met*	*Met*	Ile
CUA	Leu	Leu	Leu	*Thr*	Leu
AGA	} Arg	*STOP*	*Ser*	Arg	Arg
AGG					

**Red italics indicate that the code differs from the "Universal" code.

four nucleotides in the third (wobble) position. Such "2 out of 3" pairing allows one tRNA to pair with any one of four codons and permits protein synthesis with fewer tRNA molecules.

3. *Variant genetic code.* Perhaps most surprising, comparisons of mitochondrial gene sequences and the amino acid sequences of the corresponding proteins indicate that the genetic code is different: 4 of the 64 codons have different "meanings" from those of the same codons in other genomes (**Table 14–3**).

The close similarity of the genetic code in all organisms provides strong evidence that all cells have evolved from a common ancestor. How, then, do we explain the few differences in the genetic code in many mitochondria? A hint comes from the finding that the mitochondrial genetic code is different in different organisms. In the mitochondrion with the largest number of genes in Figure 14–59, that of the protozoan *Reclinomonas*, the genetic code is unchanged from the standard genetic code of the cell nucleus. Yet UGA, which is a stop codon elsewhere, is read as tryptophan in mitochondria of mammals, fungi, and invertebrates. Similarly, the codon AGG normally codes for arginine, but it codes for *stop* in the mitochondria of mammals and codes for serine in the mitochondria of *Drosophila* (see Table 14–3). Such variation suggests that a random drift can occur in the genetic code in mitochondria. Presumably, the unusually small number of proteins encoded by the mitochondrial genome makes an occasional change in the meaning of a rare codon tolerable, whereas such a change in a large genome would alter the function of many proteins and thereby destroy the cell.

Animal Mitochondria Contain the Simplest Genetic Systems Known

Comparisons of DNA sequences in different organisms reveal that, in vertebrate animals (including ourselves), the rate of nucleotide substitution during evolution has been 10 times greater in the mitochondrial genome than in the nuclear genome. This difference is likely to be due to a reduced fidelity of mitochondrial DNA replication, inefficient DNA repair, or both. Because only about 16,500 DNA nucleotides need to be replicated and expressed as RNAs and proteins in animal cell mitochondria, the error rate per nucleotide copied by DNA replication, maintained by DNA repair, transcribed by RNA polymerase, or translated into protein by mitochondrial ribosomes can be relatively high without damaging one of the relatively few gene products. This could explain why the mechanisms that perform these processes are relatively simple compared with those used for the same purpose elsewhere in cells. The presence of only 22 tRNAs and the unusually small size of the rRNAs (less than two-thirds the size of the *E. coli* rRNAs), for example, would be expected to reduce the fidelity of protein synthesis in mitochondria, although this has not yet been tested adequately.

The relatively high rate of evolution of animal mitochondrial genes makes a comparison of mitochondrial DNA sequences especially useful for estimating the dates of relatively recent evolutionary events, such as the steps in primate evolution.

Some Organelle Genes Contain Introns

The processing of precursor RNAs has an important role in the two mitochondrial systems studied in most detail—human and yeast. In human cells, both strands of the mitochondrial DNA are transcribed at the same rate from a single promoter region on each strand, producing two different giant RNA molecules, each containing a full-length copy of one DNA strand. Transcription is therefore completely symmetric. The transcripts made on one strand are extensively processed by nuclease cleavage to yield the two rRNAs, most of the tRNAs, and about 10 poly-A-containing RNAs. In contrast, the transcript of the other strand is processed to produce only 8 tRNAs and 1 small poly-A-containing RNA; the remaining 90% of this transcript apparently contains no useful information (being complementary to coding sequences synthesized on the other strand) and is degraded. The poly-A-containing RNAs are the mitochondrial mRNAs: although they lack a cap structure at their 5′ end, they carry a poly-A tail at their 3′ end that is added posttranscriptionally by a mitochondrial poly-A polymerase.

Unlike human mitochondrial genes, some plant and fungal (including yeast) mitochondrial genes contain *introns*, which must be removed by RNA splicing. Introns also occur in some in plant chloroplast genes. Many of the introns in organelle genes consist of a family of related nucleotide sequences that are capable of splicing themselves out of the RNA transcripts by RNA-mediated catalysis (discussed in Chapter 6), although proteins generally aid these self-splicing reactions. The presence of introns in organelle genes is surprising, as introns are not common in the genes of the bacteria whose ancestors are thought to have given rise to mitochondria and plant chloroplasts.

In yeasts, the same mitochondrial gene may have an intron in one strain but not in another. Such "optional introns" seem to be able to move in and out of genomes like transposable elements. In contrast, introns in other yeast mitochondrial genes have also been found in a corresponding position in the mitochondria of *Aspergillus* and *Neurospora,* implying that they were inherited from a common ancestor of these three fungi. It is possible that these intron sequences are of ancient origin—tracing back to a bacterial ancestor—and that, although they have been lost from many bacteria, they have been preferentially retained in some organelle genomes where RNA splicing is regulated to help control gene expression.

The Chloroplast Genome of Higher Plants Contains About 120 Genes

More than 20 chloroplast genomes have now been sequenced. The genomes of even distantly related plants (such as tobacco and liverwort) are highly similar, and even those of green algae are closely related (**Figure 14–61**). Chloroplast genes are involved in four main types of processes: transcription, translation, photosynthesis, and the biosynthesis of small molecules such as amino acids, fatty acids, and pigments. Plant chloroplast genes also encode at least 40 proteins whose functions are as yet unknown. Paradoxically, all of the known proteins encoded in the chloroplast are part of larger protein complexes that also contain one or more subunits encoded in the nucleus. We discuss possible reasons for this paradox later.

The genomes of chloroplasts and bacteria have striking similarities. The basic regulatory sequences, such as transcription promoters and terminators, are virtually identical in the two cases. The amino acid sequences of the proteins encoded in chloroplasts are clearly recognizable as bacterial, and several clusters of genes with related functions (such as those encoding ribosomal proteins) are organized in the same way in the genomes of chloroplasts, *E. coli,* and cyanobacteria.

Further comparisons of large numbers of homologous nucleotide sequences should help clarify the exact evolutionary pathway from bacteria to chloroplasts, but some conclusions can already be drawn:

1. Chloroplasts in higher plants arose from photosynthetic bacteria.

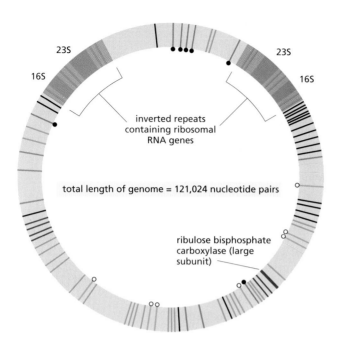

total length of genome = 121,024 nucleotide pairs

inverted repeats containing ribosomal RNA genes

ribulose bisphosphate carboxylase (large subunit)

23S

16S

23S

16S

Figure 14–61 The organization of the liverwort chloroplast genome. The chloroplast genome organization is similar in all higher plants, although the size varies from species to species—depending on how much of the DNA surrounding the genes encoding the chloroplast's 16S and 23S ribosomal RNAs is present in two copies.

KEY:

——————   tRNA genes
——————   ribosomal protein genes
——————   photosystem I genes
○—————   photosystem II genes
——————   ATP synthase genes
——————   genes for b_6-f complex
——————   RNA polymerase genes
●—————   genes for NADH dehydrogenase complex

2. Many of the genes of the original bacterium are now present in the nuclear genome, where they have been integrated and are stably maintained. In higher plants, for example, two-thirds of the 60 or so chloroplast ribosomal proteins are encoded in the cell nucleus; these genes have a clear bacterial ancestry, and the chloroplast ribosomes retain their original bacterial properties.

Mitochondrial Genes Are Inherited by a Non-Mendelian Mechanism

Many experiments on the mechanisms of mitochondrial biogenesis have been performed with *Saccharomyces cerevisiae* (baker's yeast). There are several reasons for this preference. First, when grown on glucose, this yeast has an ability to live by glycolysis alone and can therefore survive with defective mitochondria that cannot perform oxidative phosphorylation. This makes it possible to grow cells with mutations in mitochondrial or nuclear DNA that interfere with mitochondrial function; such mutations are lethal in many other eucaryotes. Second, yeasts are simple unicellular eucaryotes that are easy to grow and characterize biochemically. Finally, these yeast cells normally reproduce asexually by budding, but they can also reproduce sexually. During sexual reproduction two haploid cells mate and fuse to form a diploid zygote, which can either grow mitotically or divide by meiosis to produce new haploid cells.

The ability to control the alternation between asexual and sexual reproduction in the laboratory greatly facilitates genetic analyses. Mutations in mitochondrial genes are not inherited in accordance with the Mendelian rules that govern the inheritance of nuclear genes. Therefore, long before the mitochondrial genome could be sequenced, genetic studies revealed which of the genes involved in yeast mitochondrial function are located in the nucleus and which in the mitochondria. An example of non-Mendelian (cytoplasmic) inheritance of mitochondrial genes in a haploid yeast cell is shown in **Figure 14–62**. In this example, we trace the inheritance of a mutant gene that makes mitochondrial protein synthesis resistant to chloramphenicol.

When a chloramphenicol-resistant haploid cell mates with a chloramphenicol-sensitive wild-type haploid cell, the resulting diploid zygote contains a mixture of mutant and wild-type genomes. The two mitochondrial networks fuse in the zygote, creating one continuous reticulum that contains genomes of both parental cells. When the zygote undergoes mitosis, copies of both mutant and wild-type mitochondrial DNA are segregated to the diploid daughter cell. In the case of nuclear DNA, each daughter cell receives exactly two copies of each chromosome, one from each parent. By contrast, in the case of mitochondrial DNA, the daughter cell may inherit either more copies of the mutant DNA or more copies of the wild-type DNA. Successive mitotic divisions can further enrich for either DNA, so that subsequently many cells will arise that contain mitochondrial DNA of only one genotype. This stochastic process is called *mitotic segregation*.

When diploid cells that have segregated their mitochondrial genomes in this way undergo meiosis to form four haploid daughter cells, each of the four daughters receives the same mitochondrial genes. This type of inheritance is called non-Mendelian, or *cytoplasmic inheritance*, to contrast it with the Mendelian inheritance of nuclear genes (see Figure 14–62). When non-Mendelian inheritance occurs, it demonstrates that the gene in question is located outside the nuclear chromosomes.

Although clusters of mitochondrial DNA molecules (nucleoids) are relatively immobile in the mitochondrial reticulum because of their anchorage to the inner membrane, individual nucleoids occasionally come together. This is most likely to occur at sites where the two parental mitochondrial networks fuse during zygote formation. When different DNAs are present in the same nucleoid, genetic recombination can occur. This recombination can result in mitochondrial genomes that contain DNA from both parent cells, which are stably inherited after their mitotic segregation.

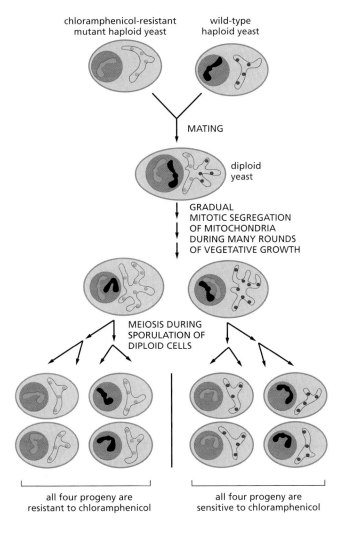

Figure 14–62 The difference in the patterns of inheritance between mitochondrial and nuclear genes of yeast cells. For nuclear genes (Mendelian inheritance), two of the four cells that result from meiosis inherit the gene from one of the original haploid parent cells (*green* chromosomes), and the remaining two cells inherit the gene from the other (*black* chromosomes). By contrast, for mitochondrial genes (non-Mendelian inheritance), it is possible for all four of the cells that result from meiosis to inherit their mitochondrial genes from only one of the two original haploid cells. In this example, the mitochondrial gene is one that, in its mutant form (mitochondrial DNA denoted by *blue dots*), makes protein synthesis in the mitochondrion resistant to chloramphenicol—a protein synthesis inhibitor that acts specifically on the procaryotic-like ribosomes in mitochondria and chloroplasts. We can detect yeast cells that contain the mutant gene by their ability to grow in the presence of chloramphenicol on a substrate, such as glycerol, that cannot be used for glycolysis. With glycolysis blocked, ATP must be provided by functional mitochondria, and therefore the cells that carry the normal (wild-type) mitochondrial DNA (*red dots*) cannot grow.

Organelle Genes Are Maternally Inherited in Many Organisms

The consequences of cytoplasmic inheritance are more profound for some organisms, including ourselves, than they are for yeasts. In yeasts, when two haploid cells mate, they are equal in size and contribute equal amounts of mitochondrial DNA to the zygote (see Figure 14–62). Mitochondrial inheritance in yeasts is therefore *biparental:* both parents contribute equally to the mitochondrial gene pool of the progeny (although, as we have just seen, after several generations of vegetative growth, the *individual* progeny often contain mitochondria from only one parent). In higher animals, by contrast, the egg cell always contributes much more cytoplasm to the zygote than does the sperm. One would therefore expect mitochondrial inheritance in higher animals to be nearly *uniparental*—or, more precisely, maternal. Such *maternal inheritance* has been demonstrated in laboratory animals. A cross between animals carrying type A mitochondrial DNA and animals carrying type B results in progeny that contain only the maternal type of mitochondrial DNA. Similarly, by following the distribution of variant mitochondrial DNA sequences in large families, we find that human mitochondrial DNA is maternally inherited.

In about two-thirds of higher plants, the chloroplasts from the male parent (contained in pollen grains) do not enter the zygote, so that chloroplast as well as mitochondrial DNA is maternally inherited. In other plants, the pollen chloroplasts enter the zygote, making chloroplast inheritance biparental. In such plants, defective chloroplasts are a cause of *variegation:* a mixture of normal and defective chloroplasts in a zygote may sort out by mitotic segregation during plant growth and development, thereby producing alternating green and white patches in leaves. The green patches contain normal chloroplasts, while the white patches contain defective chloroplasts (**Figure 14–63**).

A fertilized human egg carries perhaps 2000 copies of the human mitochondrial genome, all but one or two inherited from the mother. A human in which all of these genomes carried a deleterious mutation would generally not survive. But some mothers carry a mixed population of both mutant and normal mitochondrial genomes. Their daughters and sons inherit this mixture of normal and mutant mitochondrial DNAs and are healthy unless the process of mitotic segregation by chance results in a majority of defective mitochondria in a particular tissue. Muscle and nervous tissues are most at risk, because of their need for particularly large amounts of ATP.

We can identify an inherited disease in humans caused by a mutation in mitochondrial DNA by its passage from affected mothers to both their daughters and their sons, with the daughters but not the sons producing grandchildren with the disease. As expected from the random nature of mitotic segregation, the symptoms of these diseases vary greatly between different family members—including not only the severity and age of onset, but also which tissue is affected.

Consider, for example, the inherited disease *myoclonic epilepsy and ragged red fiber disease (MERRF)*, which can be caused by a mutation in one of the mitochondrial transfer RNA genes. This disease appears when, by chance, a particular tissue inherits a threshold amount of defective mitochondrial DNA genomes. Above this threshold, the pool of defective tRNA decreases the synthesis of the mitochondrial proteins required for electron transport and production of ATP. The result may be muscle weakness or heart problems (from effects on heart muscle), forms of epilepsy or dementia (from effects on nerve cells), or other symptoms. Not surprisingly, a similar variability in phenotypes is found for many other mitochondrial diseases.

Petite Mutants in Yeasts Demonstrate the Overwhelming Importance of the Cell Nucleus for Mitochondrial Biogenesis

Genetic studies of yeasts have had a crucial role in the analysis of mitochondrial biogenesis. A striking example is provided by studies of yeast mutants that contain large deletions in their mitochondrial DNA, so that all mitochondrial protein synthesis is abolished. Not surprisingly, these mutants cannot make

Figure 14–63 A variegated leaf. In the white patches, the plant cells have inherited a defective chloroplast. (Courtesy of John Innes Foundation.)

respiring mitochondria. Some of these mutants lack mitochondrial DNA altogether. Because they form unusually small colonies when grown in medium containing a low concentration of glucose, all mutants with such defective mitochondria are called *cytoplasmic petite mutants.*

Although petite mutants cannot synthesize proteins in their mitochondria and therefore cannot make mitochondria that produce ATP, they nevertheless contain mitochondria. These mitochondria have a normal outer membrane and an inner membrane with poorly developed cristae (**Figure 14–64**). They contain virtually all the mitochondrial proteins that are specified by nuclear genes and imported from the cytosol—including DNA and RNA polymerases, all of the citric acid cycle enzymes, and most inner membrane proteins—demonstrating the overwhelming importance of the nucleus in mitochondrial biogenesis. Petite mutants also show that an organelle that divides by fission can replicate indefinitely in the cytoplasm of proliferating eukaryotic cells, even in the complete absence of its own genome. It is possible that peroxisomes normally replicate in this way (see Figure 12–33).

For chloroplasts, the nearest equivalent to yeast mitochondrial petite mutants are mutants of unicellular algae such as *Euglena.* Mutant algae in which no chloroplast protein synthesis occurs still contain chloroplasts and are perfectly viable if they are provided with oxidizable substrates. If the development of mature chloroplasts is blocked in higher plants, however, either by raising the plants in the dark or because chloroplast DNA is defective or absent, the plants die.

Mitochondria and Plastids Contain Tissue-Specific Proteins That Are Encoded in the Cell Nucleus

Mitochondria can have specialized functions in particular types of cells. The *urea cycle,* for example, is the central metabolic pathway in mammals for disposing of cellular breakdown products that contain nitrogen. These products are excreted in the urine as urea. Nucleus-encoded enzymes in the mitochondrial matrix perform several steps in the cycle. Urea synthesis occurs in only a few tissues, such as the liver, and the required enzymes are synthesized and imported into mitochondria only in these tissues.

The respiratory enzyme complexes in the mitochondrial inner membrane of mammals also contain several tissue-specific, nucleus-encoded subunits that are thought to act as regulators of electron transport. Thus, some humans with a genetic muscle disease have a defective subunit of cytochrome oxidase; since the subunit is specific to skeletal muscle cells, their other cells, including their heart muscle cells, function normally, allowing the individuals to survive. As would be expected, tissue-specific differences are also found among the nucleus-encoded proteins in chloroplasts.

Mitochondria Import Most of Their Lipids; Chloroplasts Make Most of Theirs

The biosynthesis of new mitochondria and chloroplasts requires lipids in addition to nucleic acids and proteins. Chloroplasts tend to make the lipids they require. In spinach leaves, for example, all cellular fatty acid synthesis takes place in the chloroplast, although desaturation of the fatty acids occurs elsewhere. The major glycolipids of the chloroplast are also synthesized locally.

Mitochondria, in contrast, import most of their lipids. Animal cells synthesize the phospholipids phosphatidylcholine and phosphatidylserine in the endoplasmic reticulum and then transfer them to the outer membrane of mitochondria. In addition to decarboxylating imported phosphatidylserine to phosphatidylethanolamine, the main reaction of lipid biosynthesis catalyzed by the mitochondria themselves is the conversion of imported lipids to cardiolipin (bisphosphatidylglycerol). Cardiolipin is a "double" phospholipid that contains four fatty acid tails (**Figure 14–65**). It is found mainly in the mitochondrial inner membrane, where it constitutes about 20% of the total lipid.

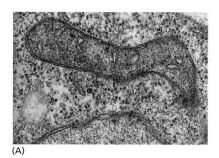

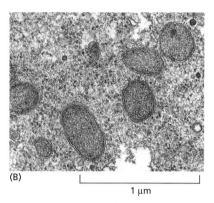

(A)

(B)

1 μm

Figure 14–64 Electron micrographs of yeast cells. (A) The structure of normal mitochondria. (B) Mitochondria in a petite mutant. In petite mutants, all the mitochondrion-encoded gene products are missing, and the organelle is constructed entirely from nucleus-encoded proteins. (Courtesy of Barbara Stevens.)

We discuss the important question of how specific cytosolic proteins are imported into mitochondria and chloroplasts in Chapter 12.

Mitochondria May Contribute to the Aging of Cells and Organisms

Oxygen gas, O_2, which has a strong affinity for electrons, becomes especially reactive as soon as it acquires a single electron and becomes partially reduced to superoxide (O_2^-). This explains why, at the very end of the respiratory chain, the two oxygen atoms of O_2^- remain tightly held by cytochrome oxidase until they have taken up four electrons and can be safely released as two water molecules (see Figure 14–27). As we learn more about the details of biological electron transport processes, we find that evolution has selected for many clever mechanisms that prevent high-energy electrons from escaping from electron transport chains. But accidents nevertheless occur frequently, with a molecule of O_2^- estimated to form in approximately 1 in every 2000 electron transfers from NADH to oxygen.

Cells therefore have backup systems that minimize the damage from superoxide and other reactive oxygen species once they form. Chief among these are *superoxide dismutase (SOD)* enzymes, which catalyze the reaction $O_2^- + O_2^- + 2H^+ \rightarrow H_2O_2 + O_2$. The still toxic hydrogen peroxide product is then further metabolized to water, either by a *catalase* enzyme (which carries out the reaction $H_2O_2 + H_2O_2 \rightarrow 2H_2O + O_2$) or by a *glutathione peroxidase* (which carries out the reaction $H_2O_2 + 2\,GSH \rightarrow 2\,H_2O + GSSG$, where GSH represents the tripeptide glutathione with its cysteine sulfhydryl group).

Approximately 90% of the O_2^- generated in cells is formed inside mitochondria. It is therefore not surprising that mitochondria contain their own superoxide dismutase and glutathione peroxidase enzymes, or that mice deficient in mitochondrial superoxide dismutase (MnSOD) die early in life, unlike mice that lack either the cytosolic or the extracellular versions of this enzyme. Even in normal animals, the mitochondrial DNA is observed to contain a ten-fold higher level of oxidized, abnormal nucleotides than the nuclear DNA. These types of observations, coupled with observations of poorly functioning mitochondria in older human beings, have led to a "vicious cycle" hypothesis that attempts to explain why organisms age. In this view, oxidation damage causes mutations to accumulate in the mitochondrial DNA of somatic tissues, which in turn causes an accelerating rate of oxidation errors in these tissues. This harmful feedback loop continues until the mitochondria of old mice or old humans are producing so many oxidation products that the entire organism decays—giving rise to a debilitating old age.

Is there any truth to this idea? It is clear that aging is a very complex process. Moreover, for lower multicellular organisms such as worms and flies, lifespans can be dramatically elongated through any one of a number of genetic changes. This data suggests that these organisms have been programmed to age and die when they do, rather than wearing out from any accumulation of mutations or damage. As a direct test of the vicious cycle hypothesis, mice engineered to produce half of the normal amount of MnSOD live as long as normal mice and do not appear to age prematurely, even though they accumulate an unusual amount of mitochondrial oxidative damage, show a premature decline in mitochondrial function, and have an increased formation of tumors.

Although the jury is still out with regard to humans, it seems unlikely at present that we shall be able to prolong our lives by simply popping pills of antioxidants.

Why Do Mitochondria and Chloroplasts Have Their Own Genetic Systems?

Why do mitochondria and chloroplasts require their own separate genetic systems, when other organelles that share the same cytoplasm, such as peroxisomes and lysosomes, do not? The question is not trivial, because maintaining a

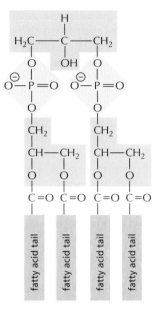

Figure 14–65 The structure of cardiolipin. Cardiolipin is an unusual lipid in the inner mitochondrial membrane.

separate genetic system is costly: more than 90 proteins—including many ribosomal proteins, aminoacyl-tRNA synthases, DNA and RNA polymerases, and RNA-processing and RNA-modifying enzymes—must be encoded by nuclear genes specifically for this purpose (**Figure 14–66**). The amino acid sequences of most of these proteins in mitochondria and chloroplasts differ from those of their counterparts in the nucleus and cytosol, and it appears that these organelles have relatively few proteins in common with the rest of the cell. This means that the nucleus must provide at least 90 genes just to maintain each organelle's genetic system.

The reason for such a costly arrangement is not clear. We cannot think of compelling reasons why the proteins made in mitochondria and chloroplasts should be made there rather than in the cytosol.

The individual protein subunits in the various mitochondrial enzyme complexes are highly conserved in evolution, but their site of synthesis is not (see Figure 14–59). Nevertheless, two protein-coding genes, *Cox1* and *Cob*, are present in all mitochondrial genomes. The proteins encoded by these genes are large and hydrophobic, with many membrane-spanning segments, and perhaps they need to be co-translationally inserted into the inner membrane by mitochondrial ribosomes.

Alternatively, the organelle genetic systems may simply be an evolutionary dead-end. In terms of the endosymbiont hypothesis, this would mean that the process whereby the endosymbionts transferred most of their genes to the nucleus stopped before it was complete. Further transfers may have been ruled out, for mitochondria, by alterations in the mitochondrial genetic code that made the remaining mitochondrial genes nonfunctional if they were transferred to the nucleus.

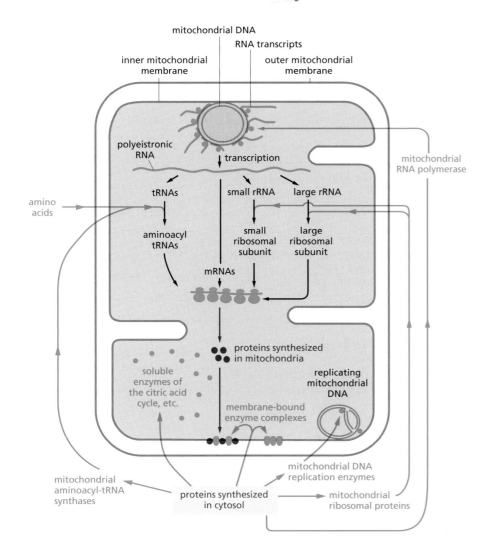

Figure 14–66 **The origins of mitochondrial RNAs and proteins.** The proteins encoded in the nucleus and imported from the cytosol have a major role in creating the genetic system of the mitochondrion, in addition to contributing most of the organelle's other proteins. Not indicated in this diagram are the additional nucleus-encoded proteins that regulate the expression of individual mitochondrial genes at posttranscriptional levels. The mitochondrion itself contributes only mRNAs, rRNAs, and tRNAs to its genetic system—and in the mitochondria of some organisms, even the tRNAs are imported.

Summary

Mitochondria are additions to cells that allow most eucaryotes to carry out oxidative phosphorylation, while chloroplasts are additions to cell that allow selected eucaryotes (plants and certain algae) to carry out photosynthesis. Presumably as a result of their procaryotic origins, each organelle grows in a coordinated process that requires the contribution of two separate genetic systems—one in the organelle and one in the cell nucleus. Most of the proteins in these organelles are encoded by nuclear DNA, synthesized in the cytosol, and then imported individually into the organelle. Some organelle proteins and RNAs are encoded by the organelle DNA and are synthesized in the organelle itself. The human mitochondrial genome contains about 16,500 nucleotides and encodes 2 ribosomal RNAs, 22 transfer RNAs, and 13 different polypeptide chains. Chloroplast genomes are about 10 times larger and contain about 120 genes. But partly functional organelles form in normal numbers even in mutants that lack a functional organelle genome, demonstrating the overwhelming importance of the nucleus for the biogenesis of both organelles.

The ribosomes of chloroplasts closely resemble bacterial ribosomes, while mitochondrial ribosomes show both similarities and differences that make their origin more difficult to trace. Protein similarities, however, suggest that both organelles originated when a primitive eucaryotic cell entered into a stable endosymbiotic relationship with a bacterium. A purple bacterium is thought to have given rise to the mitochondrion, and (later) a cyanobacterium is thought to have given rise to chloroplasts. Although many of the genes of these ancient bacteria still function to make organelle proteins, most of them have become integrated into the nuclear genome, where they encode bacterialike enzymes that are synthesized on cytosolic ribosomes and then imported into the organelle.

The damage that accumulates in the genomes of mitochondia over time may make a unique contribution to the aging of cells and organisms, although much remains to be done to prove this "vicious cycle" hypothesis.

THE EVOLUTION OF ELECTRON-TRANSPORT CHAINS

Much of the structure, function, and evolution of cells and organisms can be related to their need for energy. We have seen that the fundamental mechanisms for harnessing energy from such disparate sources as light and the oxidation of glucose are the same. Apparently, an effective method for synthesizing ATP arose early in evolution and has since been conserved with only small variations. How did the crucial individual components—ATP synthase, redox-driven H^+ pumps, and photosystems—first arise? Hypotheses about events occurring on an evolutionary time scale are difficult to test. But clues abound, both in the many different primitive electron-transport chains that survive in some present-day bacteria, and in geological evidence about the environment of the Earth billions of years ago.

The Earliest Cells Probably Used Fermentation to Produce ATP

As explained in Chapter 1, the first living cells on Earth are thought to have arisen more than 3×10^9 years ago, when the Earth was not more than about 10^9 years old. The environment lacked oxygen but was presumably rich in geochemically produced organic molecules, and some of the earliest metabolic pathways for producing ATP may have resembled present-day forms of fermentation.

In the process of fermentation, ATP is made by a phosphorylation event that harnesses the energy released when a hydrogen-rich organic molecule, such as glucose, is partly oxidized (see Figure 2–71). The electrons lost from the oxidized organic molecules are transferred (via NADH or NADPH) to a different organic molecule (or to a different part of the same molecule), which thereby becomes more reduced. At the end of the fermentation process, one or more of the organic molecules produced are excreted into the medium as metabolic waste products; others, such as pyruvate, are retained by the cell for biosynthesis.

The excreted end products are different in different organisms, but they tend to be organic acids (carbon compounds that carry a COOH group). Among the most important of such products in bacterial cells are lactic acid (which also accumulates in the anaerobic glycolysis that occurs in muscle) and formic, acetic, propionic, butyric, and succinic acids.

Electron-Transport Chains Enabled Anaerobic Bacteria to Use Nonfermentable Molecules as Their Major Source of Energy

The early fermentation processes would have provided not only the ATP but also the reducing power (as NADH or NADPH) required for essential biosyntheses. Thus, many of the major metabolic pathways could have evolved while fermentation was the only mode of energy production. With time, however, the metabolic activities of these procaryotic organisms must have changed the local environment, forcing organisms to evolve new biochemical pathways. The accumulation of waste products of fermentation, for example, might have resulted in the following series of changes:

Stage 1. The continuous excretion of organic acids lowered the pH of the environment, favoring the evolution of proteins that function as transmembrane H⁺ pumps that can pump H⁺ out of the cell to protect it from the dangerous effects of intracellular acidification. One of these pumps may have used the energy available from ATP hydrolysis and could have been the ancestor of the present-day ATP synthase.

Stage 2. At the same time as nonfermentable organic acids were accumulating in the environment and favoring the evolution of an ATP-consuming H⁺ pump, the supply of geochemically generated fermentable nutrients, which provided the energy for the pumps and for all other cellular processes, was dwindling. This favored bacteria that could excrete H⁺ without hydrolyzing ATP, allowing the ATP to be conserved for other cellular activities. Selective pressures of this kind might have led to the first membrane-bound proteins that could use electron transport between molecules of different redox potentials as the energy source for transporting H⁺ across the plasma membrane. Some of these proteins would have found their electron donors and electron acceptors among the nonfermentable organic acids that had accumulated. Present-day bacteria possess many such electron-transport proteins; some bacteria that grow on formic acid, for example, pump H⁺ by using the small amount of redox energy derived from the transfer of electrons from formic acid to fumarate (**Figure 14–67**). Others have similar electron-transport components devoted solely to the oxidation and reduction of inorganic substrates (see Figure 14–69, for example).

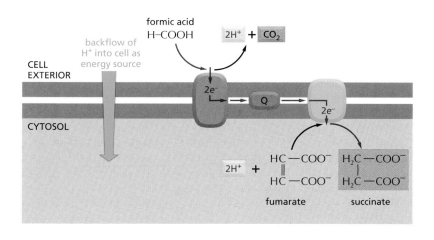

Figure 14–67 The oxidation of formic acid in some present-day bacteria. In such anaerobic bacteria, including *E. coli*, the oxidation is mediated by an energy-conserving electron-transport chain in the plasma membrane. As indicated, the starting materials are formic acid and fumarate, and the products are succinate and CO_2. Note that H⁺ is consumed inside the cell and generated outside the cell, which is equivalent to pumping H⁺ to the cell exterior. Thus, this membrane-bound electron-transport system can generate an electrochemical proton gradient across the plasma membrane. The redox potential of the formic acid–CO_2 pair is –420 mV, while that of the fumarate–succinate pair is +30 mV. (Compare to Figure 14–29.)

Figure 14–68 The evolution of oxidative phosphorylation mechanisms. One possible sequence is shown; the stages are described in the text.

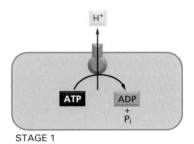

STAGE 1

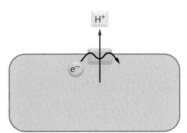

STAGE 2

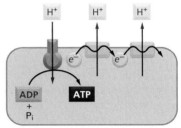

STAGE 3

Stage 3. Eventually some bacteria developed H⁺-pumping electron-transport systems that were efficient enough to harness more redox energy than they needed just to maintain their internal pH. Now, bacteria that carried both types of H⁺ pumps were at an advantage. In these cells, a large electrochemical proton gradient generated by excessive H⁺ pumping allowed protons to leak back into the cell through the ATP-driven H⁺ pumps, thereby running them in reverse, so that they functioned as ATP synthases to make ATP. Because such bacteria required much less of the increasingly scarce supply of fermentable nutrients, they proliferated at the expense of their neighbors.

Figure 14–68 shows these three hypothetical stages in the evolution of oxidative phosphorylation mechanisms.

By Providing an Inexhaustible Source of Reducing Power, Photosynthetic Bacteria Overcame a Major Evolutionary Obstacle

The evolutionary steps just described would have solved the problem of maintaining both a neutral intracellular pH and an abundant store of energy, but these steps would not have solved another equally serious problem. The depletion of organic nutrients from the environment meant that organisms had to find some alternative source of carbon to make the sugars that served as the precursors for so many other cellular molecules. Although the CO_2 in the atmosphere provided an abundant potential carbon source, to convert it into an organic molecule such as a carbohydrate requires reducing the fixed CO_2 with a strong electron donor, such as NADH or NADPH, which can provide the high-energy electrons needed to generate each (CH_2O) unit from CO_2 (see Figure 14–40). Early in cellular evolution, strong reducing agents (electron donors) would have been plentiful as products of fermentation. But as the supply of fermentable nutrients dwindled and a membrane-bound ATP synthase began to produce most of the ATP, the plentiful supply of NADH and other reducing agents would have disappeared. It thus became imperative for cells to evolve a new way of generating strong reducing agents.

Presumably, the main reducing agents still available were the organic acids produced by the anaerobic metabolism of carbohydrates, inorganic molecules such as hydrogen sulfide (H_2S) generated geochemically, and water. But the reducing power of these molecules is far too weak to be useful for CO_2 fixation. An early supply of strong electron donors could have been generated by using the electrochemical proton gradient across the plasma membrane to drive a *reverse electron flow*. This would have required the evolution of membrane-bound enzyme complexes resembling an NADH dehydrogenase, and mechanisms of this kind survive in the anaerobic metabolism of some present-day bacteria (**Figure 14–69**).

The major evolutionary breakthrough in energy metabolism, however, was almost certainly the development of photochemical reaction centers that could use the energy of sunlight to produce molecules such as NADH. It is thought that this occurred early in the process of cellular evolution—more than 3×10^9 years ago, in the ancestors of the green sulfur bacteria. Present-day green sulfur bacteria use light energy to transfer hydrogen atoms (as an electron plus a proton) from H_2S to NADPH, thereby creating the strong reducing power required for carbon fixation (**Figure 14–70**). Because the electrons removed from H_2S are at a much more negative redox potential than those of H_2O (– 230 mV for H_2S compared + 820 mV for H_2O), one quantum of light absorbed by the single photosystem in these bacteria is sufficient to achieve a high enough redox potential to generate NADPH via a relatively simple photosynthetic electron-transport chain.

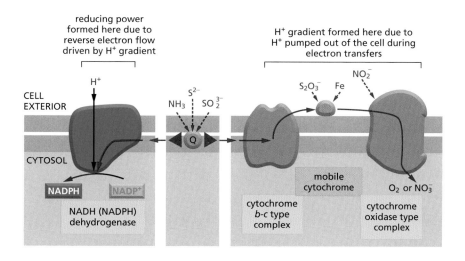

Figure 14–69 Some of the electron-transport pathways in present-day bacteria. Some species can grow anaerobically by substituting nitrate for oxygen as the terminal electron acceptor. Most use the carbon-fixation cycle and synthesize their organic molecules entirely from carbon dioxide. The indicated pathways generate all the ATP and reducing power required from the oxidation of inorganic molecules, such as iron, ammonia, nitrite, and sulfur compounds.

Note that both forward and reverse electron flows occur from the quinone (Q). As in the respiratory chain, the forward electron flows cause H^+ to be pumped out of the cell, and the resulting H^+ gradient drives the production of ATP by an ATP synthase (not shown). The NADPH required for carbon fixation is produced by an energy-requiring reverse electron flow; this energy is also derived from the H^+ gradient, as indicated.

The Photosynthetic Electron-Transport Chains of Cyanobacteria Produced Atmospheric Oxygen and Permitted New Life-Forms

The next step, which is thought to have occurred with the development of the cyanobacteria perhaps 3×10^9 years ago, was the evolution of organisms capable of using water as the electron source for CO_2 reduction. This entailed the evolution of a water-splitting enzyme and also required the addition of a second photosystem, acting in series with the first, to bridge the enormous gap in redox potential between H_2O and NADPH. Present-day structural homologies between photosystems suggest that this change involved the cooperation of a photosystem derived from green bacteria (photosystem I) with a photosystem derived from purple bacteria (photosystem II). The biological consequences of this evolutionary step were far-reaching. For the first time, there were organisms that made only very minimal chemical demands on their environment. These cells could spread and evolve in ways denied to the earlier photosynthetic bacteria, which needed H_2S or organic acids as a source of electrons. Consequently, large amounts of biologically synthesized, reduced organic materials accumulated. Moreover, oxygen entered the atmosphere for the first time.

Oxygen is highly toxic because the oxidation reactions it brings about can randomly alter biological molecules. Many present-day anaerobic bacteria, for example, are rapidly killed when exposed to air. Thus, organisms on the primitive Earth would have had to evolve protective mechanisms against the rising O_2 levels in the environment. Late evolutionary arrivals, such as ourselves, have numerous detoxifying mechanisms that protect our cells from the ill effects of

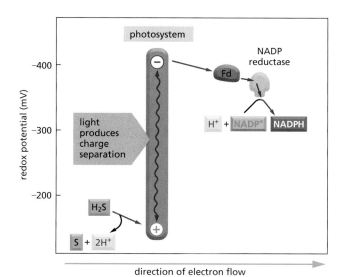

Figure 14–70 The general flow of electrons in a relatively primitive form of photosynthesis observed in present-day green sulfur bacteria. The photosystem in green sulfur bacteria resembles photosystem I in plants and cyanobacteria. Both photosystems use a series of iron–sulfur centers as the electron acceptors that eventually donate their high-energy electrons to ferredoxin (Fd). An example of a bacterium of this type is *Chlorobium tepidum,* which can thrive at high temperatures and low light intensities in hot springs.

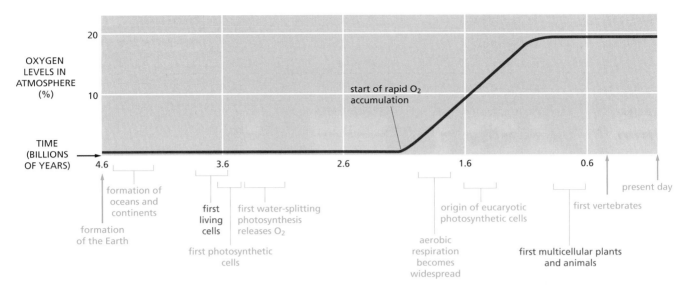

Figure 14–71 Some major events that are believed to have occurred during the evolution of living organisms on Earth. With the evolution of the membrane-based process of photosynthesis, organisms could make their own organic molecules from CO_2 gas. As explained in the text, the delay of more than 10^9 years between the appearance of bacteria that split water and released O_2 during photosynthesis and the accumulation of high levels of O_2 in the atmosphere is thought to be due to the initial reaction of the oxygen with the abundant ferrous iron (Fe^{2+}) that was dissolved in the early oceans. Only when the iron was used up would oxygen have started to accumulate. In response to the rising levels of oxygen in the atmosphere, nonphotosynthetic oxygen-using organisms appeared, and the concentration of oxygen in the atmosphere leveled out.

oxygen. Even so, an accumulation of oxidative damage to our macromolecules has been postulated to be a major cause of human aging, as previously discussed.

The increase in atmospheric O_2 was very slow at first and would have allowed a gradual evolution of protective devices. For example, the early seas contained large amounts of iron in its ferrous oxidation state (Fe^{2+}), and nearly all the O_2 produced by early photosynthetic bacteria could have been utilized in converting Fe^{2+} to Fe^{3+}. This conversion caused the precipitation of huge amounts of ferric oxides, and the extensive banded iron formations in sedimentary rocks beginning about 2.7×10^9 years ago help to date the spread of the cyanobacteria. By about 2×10^9 years ago, the supply of ferrous iron was exhausted, and the deposition of further iron precipitates ceased. Geological evidence suggests that O_2 levels in the atmosphere then began to rise steeply, reaching current levels between 0.5 and 1.5×10^9 years ago (**Figure 14–71**).

The availability of O_2 made possible the development of bacteria that relied on aerobic metabolism to make their ATP. As explained previously, these organisms could harness the large amount of energy released by breaking down carbohydrates and other reduced organic molecules all the way to CO_2 and H_2O. Components of preexisting electron-transport complexes were modified to produce a cytochrome oxidase, so that the electrons obtained from organic or inorganic substrates could be transported to O_2 as the terminal electron acceptor. Depending on the availability of light and O_2, many present-day purple photosynthetic bacteria can switch between photosynthesis and respiration, requiring only relatively minor reorganizations of their electron-transport chains.

As organic materials accumulated on Earth as a result of photosynthesis, some photosynthetic bacteria (including the precursors of *E. coli*) lost their ability to survive on light energy alone and came to rely entirely on respiration. As described previously (see Figure 14–58), it has been suggested that mitochondria first arose more than 1.5×10^9 years ago, perhaps when a primitive eucaryotic cell endocytosed such a respiration-dependent bacterium. Plants are believed to have evolved later, when a descendant of this early aerobic eucaryotic cell endocytosed a photosynthetic bacterium that became the precursor of chloroplasts.

In **Figure 14–72**, we relate these postulated pathways to the various types of bacteria discussed in this chapter. Evolution is always conservative, taking parts of the old and building on them to create something new. Thus, parts of

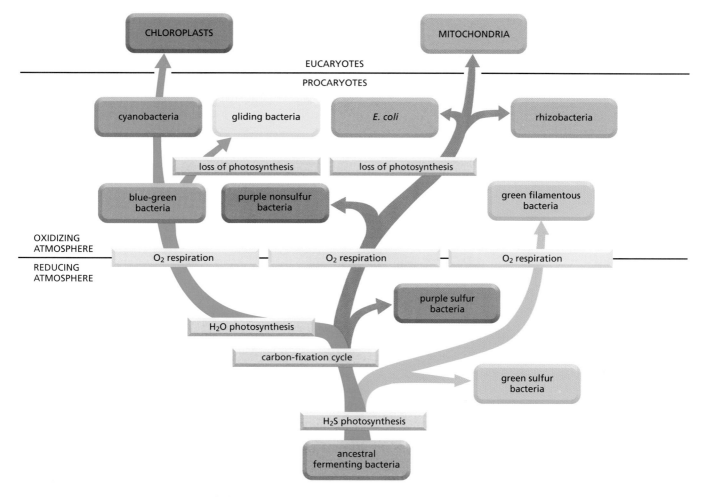

Figure 14–72 A phylogenetic tree of the proposed evolution of mitochondria and chloroplasts and their bacterial ancestors. Oxygen respiration is thought to have begun developing about 2×10^9 years ago. As indicated, it seems to have evolved independently in the green, purple, and blue-green (cyanobacterial) lines of photosynthetic bacteria. It is thought that an aerobic purple bacterium that had lost its ability to photosynthesize gave rise to the mitochondrion, while a cyanobacterium gave rise to chloroplasts. Nucleotide sequence analyses suggest that mitochondria arose from purple bacteria that resembled the rhizobacteria, agrobacteria, and rickettsias—three closely related species known to form intimate associations with present-day eucaryotic cells. Archaea are not known to contain the types of photosystems described in this chapter, and they are not included here.

the electron-transport chains that were derived to service anaerobic bacteria $3–4 \times 10^9$ years ago probably survive, in altered form, in the mitochondria and chloroplasts of today's higher eucaryotes. Consider, for example, the homology in structure and function between the enzyme complex that pumps H^+ in the central segment of the mitochondrial respiratory chain (the cytochrome b-c_1 complex) and its analogs in the electron-transport chains of both bacteria and chloroplasts (**Figure 14–73**).

Summary

Early cells are believed to have been bacteriumlike organisms living in an environment rich in highly reduced organic molecules that had been formed by geochemical processes over the course of hundreds of millions of years. They may have derived most of their ATP by converting these reduced organic molecules to a variety of organic acids, which were then released as waste products. By acidifying the environment, these fermentations may have led to the evolution of the first membrane-bound H^+ pumps, which could maintain a neutral pH in the cell interior by pumping out H^+. The properties of present-day bacteria suggest that an electron-transport-driven H^+ pump and an ATP-driven H^+ pump first arose in this anaerobic environment. Reversal of the ATP-driven pump would have allowed it to function as an ATP synthase. As

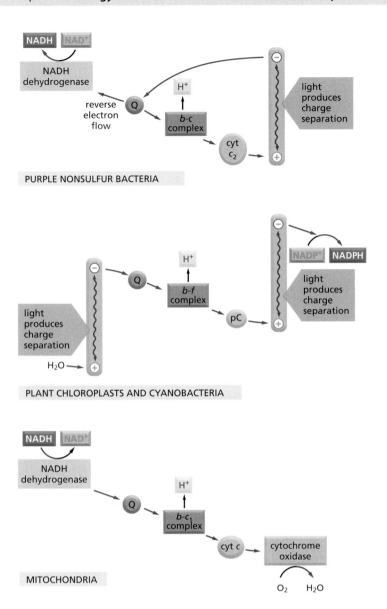

Figure 14–73 A comparison of three electron-transport chains discussed in this chapter. Bacteria, chloroplasts, and mitochondria all contain a membrane-bound enzyme complex that resembles the cytochrome b-c_1 complex of mitochondria. These complexes all accept electrons from a quinone carrier (Q) and pump H^+ across their respective membranes. Moreover, in reconstituted *in vitro* systems, the different complexes can substitute for one another, and the structures of their protein components reveal that they are evolutionarily related.

more effective electron-transport chains developed, the energy released by redox reactions between inorganic molecules and/or accumulated nonfermentable compounds produced a large electrochemical proton gradient, which could be harnessed by the ATP-driven pump for ATP production.

Because preformed organic molecules were replenished only very slowly by geochemical processes, the proliferation of bacteria that used them as the source of both carbon and reducing power could not go on forever. The depletion of fermentable organic nutrients presumably led to the evolution of bacteria that could use CO_2 to make carbohydrates. By combining parts of the electron-transport chains that had developed earlier, light energy was harvested by a single photosystem in photosynthetic bacteria to generate the NADPH required for carbon fixation. The subsequent appearance of the more complex photosynthetic electron-transport chains of the cyanobacteria allowed H_2O to be used as the electron donor for NADPH formation, rather than the much less abundant electron donors required by other photosynthetic bacteria. Life could then proliferate over large areas of the Earth, so that reduced organic molecules accumulated again.

About 2×10^9 years ago, the O_2 released by photosynthesis in cyanobacteria began to accumulate in the atmosphere. Once both organic molecules and O_2 had become abundant, electron-transport chains became adapted for the transport of electrons from NADH to O_2, and efficient aerobic metabolism developed in many bacteria. Exactly the same aerobic mechanisms operate today in the mitochondria of eucaryotes, and there is increasing evidence that both mitochondria and chloroplasts evolved from aerobic bacteria.

PROBLEMS

Which statements are true? Explain why or why not.

14–1 The three respiratory enzyme complexes in the mitochondrial inner membrane exist in structurally ordered arrays that facilitate the correct transfer of electrons between appropriate complexes.

14–2 Lipophilic weak acids short-circuit the normal flow of protons across the inner membrane, thereby eliminating the proton-motive force, stopping ATP synthesis, and blocking the flow of electrons.

14–3 Mutations that are inherited according to Mendelian rules affect nuclear genes; mutations whose inheritance violates Mendelian rules are likely to affect organelle genes.

Discuss the following problems.

14–4 In the 1860s Louis Pasteur noticed that when he added O_2 to a culture of yeast growing anaerobically on glucose, the rate of glucose consumption declined dramatically. Explain the basis for this result, which is known as the Pasteur effect.

14–5 In actively respiring liver mitochondria, the pH inside the matrix is about half a pH unit higher than that in the cytosol. Assuming that the cytosol is at pH 7 and the matrix is a sphere with a diameter of 1 μm [$V = (4/3)\pi r^3$], calculate the total number of protons in the matrix of a respiring liver mitochondrion. If the matrix began at pH 7 (equal to that in the cytosol), how many protons would have to be pumped out to establish a matrix pH of 7.5 (a difference of 0.5 pH units)?

14–6 Heart muscle gets most of the ATP needed to power its continual contractions through oxidative phosphorylation. When oxidizing glucose to CO_2, heart muscle consumes O_2 at a rate of 10 μmol/min per g of tissue, in order to replace the ATP used in contraction and give a steady-state ATP concentration of 5 μmol/g of tissue. At this rate, how many seconds would it take the heart to consume an amount of ATP equal to its steady-state levels? (Complete oxidation of one molecule of glucose to CO_2 yields 30 ATP, 26 of which are derived by oxidative phosphorylation using the 12 pairs of electrons captured in the electron carriers NADH and $FADH_2$.)

14–7 If isolated mitochondria are incubated with a source of electrons such as succinate, but without oxygen, electrons enter the respiratory chain, reducing each of the electron carriers almost completely. When oxygen is then introduced, the carriers become oxidized at different rates (**Figure Q14–1**). How does this result allow you to order the electron carriers in the respiratory chain? What is their order?

14–8 The uncoupler dinitrophenol was once prescribed as a diet drug to aid in weight loss. How would an uncoupler of oxidative phosphorylation promote weight loss? Why do you suppose that it is no longer prescribed?

14–9 How much energy is available in visible light? How much energy does sunlight deliver to the earth? How efficient are plants at converting light energy into chemical energy? The answers to these questions provide an important backdrop to the subject of photosynthesis.

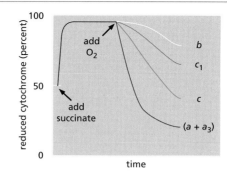

Figure Q14–1 Rapid spectrophotometric analysis of the rates of oxidation of electron carriers in the respiratory chain (Problem 14–7). Cytochromes a and a_3 cannot be distinguished and thus are listed as cytochrome $(a + a_3)$.

Each quantum or photon of light has energy hv, where h is Planck's constant (1.58×10^{-37} kcal sec/photon) and v is the frequency in sec^{-1}. The frequency of light is equal to c/λ, where c is the speed of light (3.0×10^{17} nm/sec) and λ is the wavelength in nm. Thus, the energy (E) of a photon is

$$E = hv = hc/\lambda$$

A. Calculate the energy of a mole of photons (6×10^{23} photons/mole) at 400 nm (violet light), at 680 nm (red light), and at 800 nm (near infrared light).

B. Bright sunlight strikes Earth at the rate of about 0.3 kcal/sec per square meter. Assuming for the sake of calculation that sunlight consists of monochromatic light of wavelength 680 nm, how many seconds would it take for a mole of photons to strike a square meter?

C. Assuming that it takes eight photons to fix one molecule of CO_2 as carbohydrate under optimal conditions (8–10 photons is the currently accepted value), calculate how long it would take a tomato plant with a leaf area of 1 square meter to make a mole of glucose from CO_2. Assume that photons strike the leaf at the rate calculated above and, furthermore, that all the photons are absorbed and used to fix CO_2.

D. If it takes 112 kcal/mole to fix a mole of CO_2 into carbohydrate, what is the efficiency of conversion of light energy into chemical energy after photon capture? Assume again that eight photons of red light (680 nm) are required to fix one molecule of CO_2.

14–10 Recalling Joseph Priestley's famous experiment in which a sprig of mint saved the life of a mouse in a sealed chamber, you decide to do an analogous experiment to see how a C_3 and a C_4 plant do when confined together in a sealed environment. You place a corn plant (C_4) and a geranium (C_3) in a sealed plastic chamber with normal air (300 parts per million CO_2) on a windowsill in your laboratory. What will happen to the two plants? Will they compete or collaborate? If they compete, which one will win and why?

14–11 Examine the variegated leaf shown in **Figure Q14–2**. Yellow patches surrounded by green are common, but there are no green patches surrounded by yellow. Propose an explanation for this phenomenon.

Figure Q14–2 A variegated leaf of *Aucuba japonica* with green and yellow patches (Problem 14–11).

REFERENCES

General

Cramer WA & Knaff DB (1990) Energy Transduction in Biological Membranes: A Textbook of Bioenergetics. New York: Springer-Verlag.

Mathews CK, van Holde KE & Ahern K-G (2000) Biochemistry, 3rd ed. San Francisco: Benjamin Cummings.

Nicholls DG & Ferguson SJ (2002) Bioenergetics, 3rd ed. London: Academic Press.

The Mitochondrion

Abrahams JP, Leslie AG, Lutter R & Walker JE (1994) Structure at 2.8Å resolution of F1-ATPase from bovine heart mitochondria. *Nature* 370:621–628.

Berg HC (2003) The rotary motor of bacterial flagella. *Annu Rev Biochem* 72:19-54.

Boyer PD (1997) The ATP synthase—a splendid molecular machine. *Annu Rev Biochem* 66:717–749.

Ernster L & Schatz G (1981) Mitochondria: a historical review. *J Cell Biol* 91:227s–255s.

Frey TG, Renken CW & Perkins GA (2002) Insight into mitochondrial structure and function from electron tomography. *Biochim Biophys Acta* 1555:196–203.

Pebay-Peyroula E & Brandolin G (2004) Nucleotide exchange in mitochondria: insight at a molecular level. *Curr Opin Struct Biol* 14:420–425.

Meier T, Polzer P, Diederichs K et al (2005) Structure of the rotor ring of F-type Na$^+$-ATPase from *Ilyobacter tartaricus. Science* 308:659–662.

Mitchell P (1961) Coupling of phosphorylation to electron and hydrogen transfer by a chemi-osmotic type of mechanism. *Nature* 191:144–148.

Nicholls DG (2006) The physiological regulation of uncoupling proteins. *Biochim Biophys Acta* 1757:459–466.

Racker E & Stoeckenius W (1974) Reconstitution of purple membrane vesicles catalyzing light-driven proton uptake and adenosine triphosphate formation. *J Biol Chem* 249:662–663.

Saraste M (1999) Oxidative phosphorylation at the fin de siecle. *Science* 283:1488–1493.

Scheffler IE (1999) Mitochondria. New York/Chichester: Wiley-Liss.

Stock D, Gibbons C, Arechaga I et al. (2000) The rotary mechanism of ATP synthase. *Curr Opin Struct Biol* 10:672–679.

Weber J (2007) ATP synthase—the structure of the stator stalk. *Trends Biochem Sci* 32:53–56.

Electron-Transport Chains and Their Proton Pumps

Beinert H, Holm RH & Munck E (1997) Iron-sulfur clusters: nature's modular, multipurpose structures. *Science* 277:653–659.

Berry EA, Guergova-Kuras M, Huang LS & Crofts AR (2000) Structure and function of cytochrome *bc* complexes. *Annu Rev Biochem* 69:1005–1075.

Brand MD (2005) The efficiency and plasticity of mitochondrial energy transduction. *Biochem Soc Trans* 33:897–904.

Brandt U (2006) Energy converting NADH:quinone oxidoreductase (complex I). *Annu Rev Biochem* 75:69–92.

Chance B & Williams GR (1955) A method for the localization of sites for oxidative phosphorylation. *Nature* 176:250–254.

Gottschalk G (1997) Bacterial Metabolism, 2nd ed. New York: Springer.

Gray HB & Winkler JR (1996) Electron transfer in proteins. *Annu Rev Biochem* 65:537–556.

Keilin D (1966) The History of Cell Respiration and Cytochromes. Cambridge: Cambridge University Press.

Sazanov LA (2007) Respiratory complex I: mechanistic and structural insights provided by the crystal structure of the hydrophilic domain. *Biochemistry* 46:2275–2288.

Subramaniam S & Henderson R (2000) Molecular mechanism of vectorial proton translocation by bacteriorhodopsin. *Nature* 406:653–657.

Tsukihara T, Aoyama H, Yamashita E et al (1996) The whole structure of the 13-subunit oxidized cytochrome *c* oxidase at 2.8Å. *Science* 272:1136–1144.

Wikstrom M & Verkhovsky MI (2006) Towards the mechanism of proton pumping by the haem-copper oxidases. *Biochim Biophys Acta* 1757:1047–1051.

Chloroplasts and Photosynthesis

Blankenship RE (2002) Molecular Mechanisms of Photosynthesis. Oxford, UK:Blackwell Scientific.

Bassham JA (1962) The path of carbon in photosynthesis. *Sci Am* 206:88–100.

Deisenhofer J & Michel H (1989) Nobel lecture. The photosynthetic reaction centre from the purple bacterium *Rhodopseudomonas viridis. EMBO J* 8:2149–2170.

Edwards GE, Furbank RT, Hatch MD & Osmond CB (2001) What does it take to be c(4)?—lessons from the evolution of c(4) photosynthesis. *Plant Physiol* 125:46–49.

Ferreira KN, Iverson TM, Maghlaoui K et al (2004) Architecture of the photosynthetic oxygen-evolving center. *Science* 303:1831–1838.

Iwata S & Barber J (2004) Structure of photosystem II and molecular architecture of the oxygen-evolving centre. *Curr Opin Struct Biol* 14:447–453.

Jordan P, Fromme P, Witt HT et al (2001) Three-dimensional structure of cyanobacterial photosystem I at 2.5Å resolution. *Nature* 411:909–917.

Merchant S & Sawaya MR (2005) The light reactions: A guide to recent acquisitions for the picture gallery. *The Plant Cell* 17:648–663.

Nelson N & Ben-Shem A (2004) The complex architecture of oxygenic photosynthesis. *Nature Rev Mol Cell Biol* 5:971–982.

The Genetic Systems of Mitochondria and Plastids

Anderson S, Bankier AT, Barrell BG et al (1981) Sequence and organization of the human mitochondrial genome. *Nature* 290:457–465.

Balaban RS, Nemoto S & Finkel T (2005) Mitochondria, oxidants, and aging. *Cell* 120:483–495.

Bendich AJ (2004) Circular chloroplast genomes:The grand illusion. *The Plant Cell* 16:1661–1666.

Birky CW, Jr (1995) Uniparental inheritance of mitochondrial and chloroplast genes: mechanisms and evolution. *Proc Natl Acad Sci USA* 92:11331–11338.

Bullerwell CE & Gray MW (2004) Evolution of the mitochondrial genome: protist connections to animals, fungi and plants. *Curr Opin Microbiol* 7:528–534.

Cavalier-Smith T (2002) Chloroplast evolution: secondary symbiogenesis and multiple losses. *Curr Biol* 12:R62–R64.

Chen XJ & Butow RA (2005) The organization and inheritance of the mitochondrial genome. *Nature Rev Genet* 6:815–825.

Clayton DA (2000) Vertebrate mitochondrial DNA—a circle of surprises. *Exp Cell Res* 255:4–9.

Daly DO & Whelan J (2005) Why genes persist in organelle genomes. *Genome Biol* 6:110.

de Duve C (2007) The origin of eukaryotes: A reappraisal. *Nature Rev Genet* 8:395–403.

Dyall SD, Brown MT & Johnson PJ (2004) Ancient invasions: From endosymbionts to organelles. *Science* 304:253–257.

Hoppins S, Lackner L & Nunnari J (2007) The Machines that Divide and Fuse Mitochondria. *Annu Rev Biochem* 76:751–80.

Wallace DC (1999) Mitochondrial diseases in man and mouse. *Science* 283:1482–1488.

The Evolution of Electron-Transport Chains

Blankenship RE & Bauer CE (eds) (1995) Anoxygenic Photosynthetic Bacteria. Dordrecht: Kluwer.

de Las Rivas J, Balsera M & Barber J (2004) Evolution of oxygenic photosynthesis: genome-wide analysis of the OEC extrinsic proteins. *Trends Plant Sci* 9:18–25.

Orgel LE (1998) The origin of life—a review of facts and speculations. *Trends Biochem Sci.* 23:491–495.

Schafer G, Purschke W & Schmidt CL (1996) On the origin of respiration: electron transport proteins from archaea to man. *FEMS Microbiol Rev* 18:173–188.

Skulachev VP (1994) Bioenergetics: the evolution of molecular mechanisms and the development of bioenergetic concepts. *Antonie Van Leeuwenhoek* 65:271–284.

Chapter 15

Mechanisms of Cell Communication

To make a multicellular organism, cells must communicate, just as humans must communicate if they are to organize themselves into a complex society. And just as human communication involves more than the passage of noises from mouth to ear, so cell–cell communication involves more than the transmission of chemical signals across the space between one cell and another. Complex intracellular mechanisms are needed to control which signals are emitted at what time and to enable the signal-receiving cell to interpret those signals and use them to guide its behavior. According to the fossil record, sophisticated multicellular organisms did not appear on Earth until unicellular organisms resembling present-day procaryotes had already been in existence for about 2.5 billion years. The long delay may reflect the difficulty of evolving the language systems of animal, plant, and fungal cells—the machinery that would enable cells sharing the same genome to collaborate and coordinate their behavior, specializing in different ways and subordinating their individual chances of survival to the interests of the multicellular organism as a whole. These highly evolved mechanisms of cell–cell communication are the topic of this chapter.

Communication between cells is mediated mainly by **extracellular signal molecules**. Some of these operate over long distances, signaling to cells far away; others signal only to immediate neighbors. Most cells in multicellular organisms both emit and receive signals. Reception of the signals depends on *receptor proteins*, usually (but not always) at the cell surface, which bind the signal molecule. The binding activates the receptor, which in turn activates one or more *intracellular signaling pathways*. These relay chains of molecules—mainly *intracellular signaling proteins*—process the signal inside the receiving cell and distribute it to the appropriate intracellular targets. These targets are generally *effector proteins*, which are altered when the signaling pathway is activated and implement the appropriate change of cell behavior. Depending on the signal and the nature and state of the receiving cell, these effectors can be gene regulatory proteins, ion channels, components of a metabolic pathway, or parts of the cytoskeleton—among other things (**Figure 15–1**).

We begin this chapter by discussing the general principles of cell communication. We then consider, in turn, the main families of cell-surface receptor proteins and the principal intracellular signaling pathways they activate. The main focus of the chapter is on animal cells, but we end by considering the special features of cell communication in plants.

GENERAL PRINCIPLES OF CELL COMMUNICATION

Long before multicellular organisms appeared on Earth, unicellular organisms had developed mechanisms for responding to physical and chemical changes in their environment. These almost certainly included mechanisms for response to the presence of other cells. Evidence comes from studies of present-day unicellular organisms such as bacteria and yeasts. Although these cells largely lead independent lives, they can communicate and influence one another's behavior. Many bacteria, for example, respond to chemical signals that are secreted by their neighbors and increase in concentration with increasing population density. This

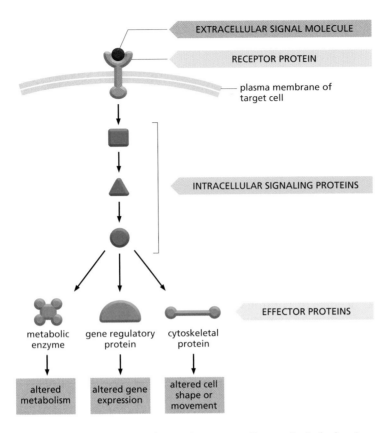

EXTRACELLULAR SIGNAL MOLECULE

RECEPTOR PROTEIN

plasma membrane of target cell

INTRACELLULAR SIGNALING PROTEINS

EFFECTOR PROTEINS

metabolic enzyme

gene regulatory protein

cytoskeletal protein

altered metabolism

altered gene expression

altered cell shape or movement

Figure 15–1 A simple intracellular signaling pathway activated by an extracellular signal molecule. The signal molecule usually binds to a receptor protein that is embedded in the plasma membrane of the target cell and activates one or more intracellular signaling pathways mediated by a series of signaling proteins. Finally, one or more of the intracellular signaling proteins alters the activity of effector proteins and thereby the behavior of the cell.

process, called *quorum sensing*, allows bacteria to coordinate their behavior, including their motility, antibiotic production, spore formation, and sexual conjugation.

Similarly, yeast cells communicate with one another in preparation for mating. The budding yeast *Saccharomyces cerevisiae* provides a well-studied example: when a haploid individual is ready to mate, it secretes a peptide *mating factor* that signals cells of the opposite mating type to stop proliferating and prepare to mate (**Figure 15–2**). The subsequent fusion of two haploid cells of opposite mating type produces a diploid cell, which can then undergo meiosis and sporulate, generating haploid cells with new assortments of genes (see Figure 21–3B). The reshuffling of genes through sexual reproduction helps a species survive in an unpredictably variable environment (as discussed in Chapter 21).

Studies of yeast mutants that are unable to mate have identified many proteins that are required in the signaling process. These proteins form a signaling network that includes cell-surface receptor proteins, GTP-binding proteins, and protein kinases, and each of these categories has close relatives among the receptors and intracellular signaling proteins in animal cells. Through gene duplication and divergence, however, the signaling systems in animals have become much more elaborate than those in yeasts; the human genome, for example, contains more than 1500 genes that encode receptor proteins, and the number of different receptor proteins is further increased by alternative RNA splicing and post-translational modifications.

The large numbers of signal proteins, receptors, and intracellular signaling proteins used by animals can be grouped into a much smaller number of protein families, most of which have been highly conserved in evolution. Flies, worms, and mammals all use essentially similar machinery for cell communication, and many of the key components and signaling pathways were first discovered through analysis of mutations in *Drosophila* and *C. elegans*.

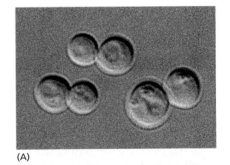

(A)

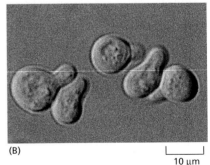

(B)

10 µm

Figure 15–2 Budding yeast cells responding to a mating factor. (A) The cells are normally spherical. (B) In response to the mating factor secreted by neighboring yeast cells, they put out a protrusion toward the source of the factor in preparation for mating. (Courtesy of Michael Snyder.)

Extracellular Signal Molecules Bind to Specific Receptors

Cells in multicellular animals communicate by means of hundreds of kinds of signal molecules. These include proteins, small peptides, amino acids, nucleotides, steroids, retinoids, fatty acid derivatives, and even dissolved gases such as nitric oxide and carbon monoxide. Most of these signal molecules are released into the extracellular space by exocytosis from the signaling cell, as discussed in Chapter 13. Some, however, are emitted by diffusion through the signaling cell's plasma membrane, whereas others are displayed on the external surface of the cell and remain attached to it, providing a signal to other cells only when they make contact. Transmembrane proteins may be used for signaling in this way; or their extracellular domains may be released from the signaling cell's surface by proteolytic cleavage and then act at a distance.

Regardless of the nature of the signal, the *target cell* responds by means of a **receptor**, which specifically binds the signal molecule and then initiates a response in the target cell. The extracellular signal molecules often act at very low concentrations (typically $\leq 10^{-8}$ M), and the receptors that recognize them usually bind them with high affinity (affinity constant $K_a \geq 10^8$ liters/mole; see Figure 3–43).

In most cases, the receptors are transmembrane proteins on the target cell surface. When these proteins bind an extracellular signal molecule (*a ligand*), they become activated and generate various intracellular signals that alter the behavior of the cell. In other cases, the receptor proteins are inside the target cell, and the signal molecule has to enter the cell to bind to them: this requires that the signal molecule be sufficiently small and hydrophobic to diffuse across the target cell's plasma membrane (**Figure 15–3**).

Extracellular Signal Molecules Can Act Over Either Short or Long Distances

Many signal molecules remain bound to the surface of the signaling cell and influence only cells that contact it (**Figure 15–4A**). Such **contact-dependent signaling** is especially important during development and in immune responses. Contact-dependent signaling during development can sometimes operate over relatively large distances, where the communicating cells extend long processes to make contact with one another.

In most cases, however, signaling cells secrete signal molecules into the extracellular fluid. The secreted molecules may be carried far afield to act on distant target cells, or they may act as **local mediators**, affecting only cells in the local environment of the signaling cell. The latter process is called **paracrine signaling** (Figure 15–4B). Usually, the signaling and target cells in paracrine signaling are of different cell types, but cells may also produce signals that they themselves respond to: this is referred to as *autocrine signaling*. Cancer cells, for example, often use this strategy to stimulate their own survival and proliferation.

For paracrine signals to act only locally, the secreted molecules must not be allowed to diffuse too far; for this reason they are often rapidly taken up by neighboring target cells, destroyed by extracellular enzymes, or immobilized by the extracellular matrix. *Heparan sulfate proteoglycans* (discussed in Chapter 19), either in the extracellular matrix or attached to cell surfaces, often play a part in localizing the action of secreted signal proteins. They contain long polysaccharide side chains that bind the signal proteins and immobilize them. They may also control the stability of these proteins, their transport through the extracellular space, or their interaction with cell-surface receptors. Secreted protein *antagonists* also affect the distance over which a paracrine signal protein acts. These antagonists bind to either the signal molecule itself or its cell-surface receptor and block its activity, and they play an important part in restricting the effective range of secreted signal proteins that influence the developmental decisions that embryonic cells make (discussed in Chapter 22).

Large, complex, multicellular organisms need long-range signaling mechanisms to coordinate the behavior of cells in remote parts of the body. Thus, they have evolved cell types specialized for intercellular communication over large distances. The most sophisticated of these are nerve cells, or neurons, which typically extend long branching processes (axons) that enable them to contact target cells far away, where the processes terminate at the specialized sites of

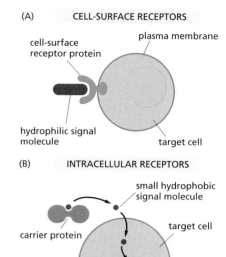

Figure 15–3 The binding of extracellular signal molecules to either cell-surface or intracellular receptors. (A) Most signal molecules are hydrophilic and are therefore unable to cross the target cell's plasma membrane directly; instead, they bind to cell-surface receptors, which in turn generate signals inside the target cell (see Figure 15–1). (B) Some small signal molecules, by contrast, diffuse across the plasma membrane and bind to receptor proteins inside the target cell— either in the cytosol or in the nucleus (as shown here). Many of these small signal molecules are hydrophobic and nearly insoluble in aqueous solutions; they are therefore transported in the bloodstream and other extracellular fluids bound to carrier proteins, from which they dissociate before entering the target cell.

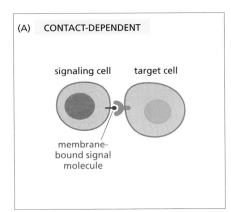

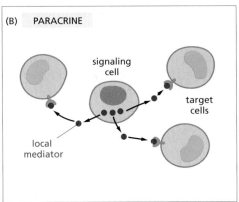

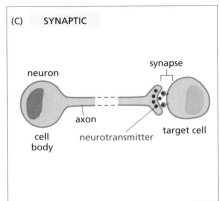

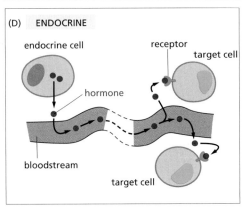

Figure 15–4 Four forms of intercellular signaling. (A) Contact-dependent signaling requires cells to be in direct membrane–membrane contact. (B) Paracrine signaling depends on signals that are released into the extracellular space and act locally on neighboring cells. (C) Synaptic signaling is performed by neurons that transmit signals electrically along their axons and release neurotransmitters at synapses, which are often located far away from the neuronal cell body. (D) Endocrine signaling depends on endocrine cells, which secrete hormones into the bloodstream for distribution throughout the body. Many of the same types of signaling molecules are used in paracrine, synaptic, and endocrine signaling; the crucial differences lie in the speed and selectivity with which the signals are delivered to their targets.

signal transmission known as *chemical synapses*. When activated by stimuli from the environment or from other nerve cells, the neuron sends electrical impulses (action potentials) rapidly along its axon; when such an impulse reaches the synapse at the end of the axon, it triggers secretion of a chemical signal that acts as a **neurotransmitter**. The tightly organized structure of the synapse ensures that the neurotransmitter is delivered specifically to receptors on the postsynaptic target cell (Figure 15–4C). The details of this **synaptic signaling** process are discussed in Chapter 11.

A quite different strategy for signaling over long distances makes use of **endocrine cells**. These secrete their signal molecules, called **hormones**, into the bloodstream, which carries the molecules far and wide, allowing them to act on target cells that may lie anywhere in the body (Figure 15–4D).

Figure 15–5 compares the mechanisms that allow endocrine cells and nerve cells to coordinate cell behavior over long distances in animals. Because endocrine signaling relies on diffusion and blood flow, it is relatively slow. Synaptic signaling, by contrast, is much faster, as well as more precise. Nerve cells can transmit information over long distances by electrical impulses that travel at speeds of up to 100 meters per second; once released from a nerve terminal, a neurotransmitter has to diffuse less than 100 nm to the target cell, a process that takes less than a millisecond. Another difference between endocrine and synaptic signaling is that, whereas hormones are greatly diluted in the bloodstream and interstitial fluid and therefore must be able to act at very low concentrations (typically $< 10^{-8}$ M), neurotransmitters are diluted much less and can achieve high local concentrations. The concentration of *acetylcholine* in the synaptic cleft of an active neuromuscular junction, for example, is about 5×10^{-4} M. Correspondingly, neurotransmitter receptors have a relatively low affinity for their ligand, which means that the neurotransmitter can dissociate rapidly from the receptor to help terminate a response. Moreover, after its release from a nerve terminal, a neurotransmitter is quickly removed from the synaptic cleft, either by specific hydrolytic enzymes that destroy it or by specific membrane transport proteins

Figure 15–5 **The contrast between endocrine and neuronal strategies for long-range signaling.** In complex animals, endocrine cells and nerve cells work together to coordinate the activities of cells in widely separated parts of the body. Whereas different endocrine cells must use different hormones to communicate specifically with their target cells, different nerve cells can use the same neurotransmitter and still communicate in a highly specific manner. (A) Endocrine cells secrete hormones into the blood, and these act only on those target cells that carry the appropriate receptors: the receptors bind the specific hormone, which the target cells thereby "pull" from the extracellular fluid. (B) In synaptic signaling, by contrast, specificity arises from the synaptic contacts between a nerve cell and the specific target cells it signals. <CTGA> Usually, only a target cell that is in synaptic communication with a nerve cell is exposed to the neurotransmitter released from the nerve terminal (although some neurotransmitters act in a paracrine mode, serving as local mediators that influence multiple target cells in the area).

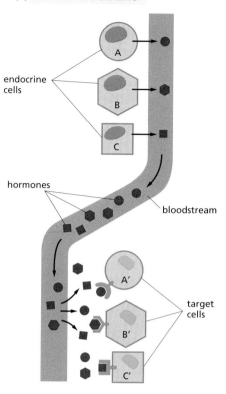

(A) ENDOCRINE SIGNALING

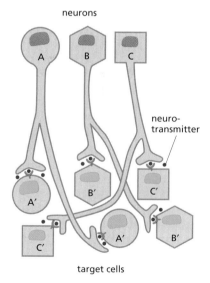

(B) SYNAPTIC SIGNALING

that pump it back into either the nerve terminal or neighboring glial cells. Thus, synaptic signaling is much more precise than endocrine signaling, both in time and in space.

The speed of a response to an extracellular signal depends not only on the mechanism of signal delivery but also on the nature of the target cell's response. When the response requires only changes in proteins already present in the cell, it can occur very rapidly: an allosteric change in a neurotransmitter-gated ion channel (discussed in Chapter 11), for example, can alter the plasma membrane electrical potential in milliseconds, and responses that depend solely on protein phosphorylation can occur within seconds. When the response involves changes in gene expression and the synthesis of new proteins, however, it usually requires many minutes or hours, regardless of the mode of signal delivery (**Figure 15–6**).

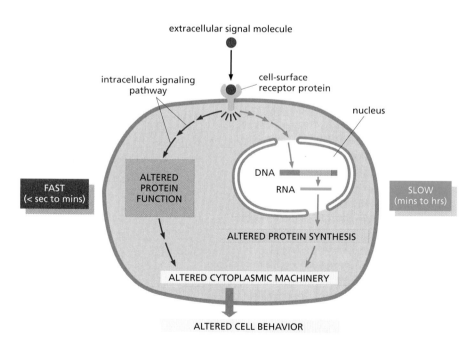

Figure 15–6 **Extracellular signals can act slowly or rapidly to change the behavior of a target cell.** Certain types of signaled responses, such as increased cell growth and division, involve changes in gene expression and the synthesis of new proteins; they therefore occur slowly, often starting after an hour or more. Other responses—such as changes in cell movement, secretion, or metabolism—need not involve changes in gene transcription and therefore occur much more quickly, often starting in seconds or minutes; they may involve the rapid phosphorylation of effector proteins in the cytoplasm, for example. Synaptic responses mediated by changes in membrane potential can occur in milliseconds (not shown).

Gap Junctions Allow Neighboring Cells to Share Signaling Information

Gap junctions are narrow water-filled channels that directly connect the cytoplasms of adjacent epithelial cells, as well as of some other cell types (see Figure 19–34). The channels allow the exchange of inorganic ions and other small water-soluble molecules, but not of macromolecules such as proteins or nucleic acids. Thus, cells connected by gap junctions can communicate with each other directly, without having to surmount the barrier presented by the intervening plasma membranes (**Figure 15–7**). In this way, gap junctions provide for the most intimate of all forms of cell communication, short of cytoplasmic bridges (see Figure 21–31) or cell fusion.

In contrast with other modes of cell signaling, gap junctions generally allow communications to pass in both directions symmetrically, and their typical effect is to homogenize conditions in the communicating cells. They can also be important in spreading the effect of extracellular signals that act through small intracellular mediators such as Ca^{2+} and cyclic AMP (discussed later), which pass readily through the gap-junctional channels. In the liver, for example, a fall in blood glucose levels releases *noradrenaline* (norepinephrine) from sympathetic nerve endings. The noradrenaline stimulates hepatocytes in the liver to increase glycogen breakdown and to release glucose into the blood, a response that depends on an increase in intracellular cyclic AMP. Not all of the hepatocytes are innervated by sympathetic nerves, however. By means of the gap junctions that connect hepatocytes, the innervated hepatocytes transmit the signal to the noninnervated ones, at least in part by the movement of cyclic AMP through gap junctions. As expected, mice with a mutation in the major gap junction gene expressed in the liver fail to mobilize glucose normally when blood glucose levels fall. Gap junctions are discussed in detail in Chapter 19.

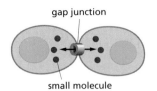

gap junction

small molecule

Figure 15–7 Signaling via gap junctions. Cells connected by gap junctions share small molecules, including small intracellular signaling molecules such as cyclic AMP and Ca^{2+}, and can therefore respond to extracellular signals in a coordinated way.

Each Cell Is Programmed to Respond to Specific Combinations of Extracellular Signal Molecules

A typical cell in a multicellular organism may be exposed to hundreds of different signal molecules in its environment. The molecules can be soluble, bound to the extracellular matrix, or bound to the surface of a neighboring cell; they can be stimulatory or inhibitory; they can act in innumerable different combinations; and they can influence almost any aspect of cell behavior. The cell must respond to this babel of signals selectively, according to its own specific character. This character is acquired through progressive cell specialization in the course of development. A cell may respond to one combination of signals by differentiating, to another combination by growing and dividing, and to yet another by performing some specialized function such as contraction or secretion. One of the great challenges in cell biology is to understand how a cell integrates all of this signaling information in order to make its crucial decisions—to divide, to move, to differentiate, and so on. For most of the cells in animal tissues, even the decision to continue living depends on correct interpretation of a specific combination of signals required for survival. When deprived of these signals (in a culture dish, for example), the cell activates a suicide program and kills itself—usually by *apoptosis*, a form of *programmed cell death* (**Figure 15–8**), as discussed in Chapter 18. Because different types of cells require different combinations of survival signals, each cell type is restricted to a specific set of environments in the body. Many epithelial cells, for example, require survival signals from the basal lamina on which they sit (discussed in Chapter 19); they die by apoptosis if they lose contact with this sheet of matrix.

In principle, the hundreds of signal molecules that an animal makes can be used in an almost unlimited number of signaling combinations so as to control the diverse behaviors of its cells in highly specific ways. Relatively small numbers of types of signal molecules and receptors are sufficient. The complexity lies in the ways in which cells respond to the combinations of signals that they receive.

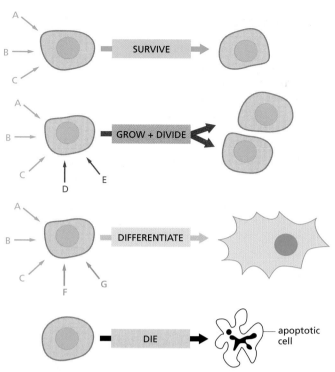

Figure 15–8 An animal cell's dependence on multiple extracellular signal molecules. Each cell type displays a set of receptors that enables it to respond to a corresponding set of signal molecules produced by other cells. These signal molecules work in combinations to regulate the behavior of the cell. As shown here, an individual cell often requires multiple signals to survive *(blue arrows)* and additional signals to grow and divide *(red arrow)* or differentiate *(green arrows)*. If deprived of appropriate survival signals, a cell will undergo a form of cell suicide known as apoptosis. The actual situation is even more complex. Although not shown, some extracellular signal molecules act to inhibit these and other cell behaviors, or even to induce apoptosis.

Different Types of Cells Usually Respond Differently to the Same Extracellular Signal Molecule

A cell's response to extracellular signals depends not only on the receptor proteins it possesses but also on the intracellular machinery by which it integrates and interprets the signals it receives. Thus, a single signal molecule usually has different effects on different types of target cells. The neurotransmitter acetylcholine (**Figure 15–9**A), for example, decreases the rate and force of contraction in heart muscle cells (Figure 15–9B), but it stimulates skeletal muscle cells to contract (see Figure 15–9C). In this case, the acetylcholine receptor proteins on skeletal muscle cells differ from those on heart muscle cells. But receptor

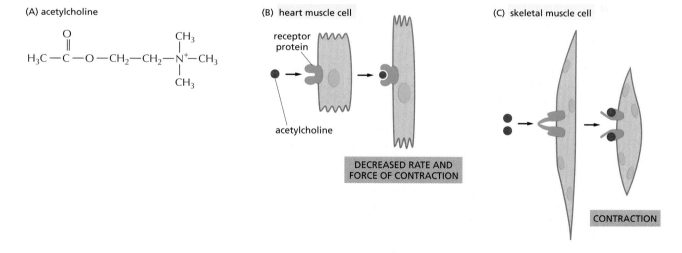

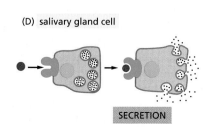

Figure 15–9 Various responses induced by the neurotransmitter acetylcholine. (A) The chemical structure of acetylcholine. (B–D) Different cell types are specialized to respond to acetylcholine in different ways. In some cases (B and C), the receptors for acetylcholine differ. In other cases (B and D), acetylcholine binds to similar receptor proteins, but the intracellular signals produced are interpreted differently in cells specialized for different functions.

differences are usually not the explanation for the different effects. The same signal molecule binding to identical receptor proteins usually produces very different responses in different types of target cells, as in the case of acetylcholine binding to heart muscle and to salivary gland cells (compare Figures 15–9B and D). In some cases, this reflects differences in the intracellular signaling proteins activated, whereas in others it reflects differences in the effector proteins or genes activated. Thus, an extracellular signal itself has little information content; it simply induces the cell to respond according to its predetermined state, which depends on the cell's developmental history and the specific genes it expresses.

The challenge of understanding how a cell integrates, processes, and reacts to the various inputs it receives is analogous in many ways to the challenge of understanding how the brain integrates and processes information to control behavior. In both cases, we need more than just a list of the components and connections in the system to understand how the process works. As a first step, we need to consider some basic principles concerning the way a cell responds to a simple signal of a given type.

The Fate of Some Developing Cells Depends on Their Position in Morphogen Gradients

The same signal acting on the same cell type can have qualitatively different effects depending on the signal's concentration. As we discuss in Chapter 22, such different responses of a cell to different concentrations of the same signal molecule are crucial in animal development, when cells are becoming different from one another.

The extracellular signal molecule in these cases during development is called a **morphogen**, and, in the simplest cases, it diffuses out from a localized cellular source (a *signaling center*), generating a signal concentration gradient. The responding cells adopt different cell fates in accordance with their position in the gradient: those cells closest to the signaling center that encounter the highest concentration of the morphogen have the highest number of receptors activated and follow one pathway of development, whereas those slightly further away follow another, and so on (**Figure 15–10**). As we discuss later (and in Chapter 22), the different levels of receptor activation lead to differences in the concentration or activity of one or more gene regulatory proteins in the nucleus of each cell, which in turn results in different patterns of gene expression. Further, more local signaling interactions between the cells in the gradient often help determine and stabilize the different fate choices.

A Cell Can Alter the Concentration of an Intracellular Molecule Quickly Only If the Lifetime of the Molecule Is Short

It is natural to think of signaling systems in terms of the changes produced when an extracellular signal is delivered. But it is just as important to consider what happens when the signal is withdrawn. During development, transient extracellular signals often produce lasting effects: they can trigger a change in the cell's development that persists indefinitely through cell memory mechanisms, as we discuss later (and in Chapters 7 and 22). In most cases in adult tissues, however, the response fades when a signal ceases. Often the effect is transient because the signal exerts its effects by altering the concentrations of intracellular molecules that are short-lived (unstable), undergoing continual turnover. Thus, once the extracellular signal is gone, the replacement of the old molecules by new ones wipes out all traces of the signal's action. It follows that the speed with which a cell responds to signal removal depends on the rate of destruction, or turnover, of the intracellular molecules that the signal affects.

It is also true, although much less obvious, that this turnover rate can determine the promptness of the response when an extracellular signal arrives. Consider, for example, two intracellular signaling molecules X and Y, both of which are normally maintained at a concentration of 1000 molecules per cell. The cell

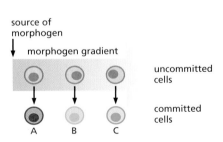

Figure 15–10 Cells adopt different fates depending on their position in a morphogen gradient. The different concentrations of morphogen induce the expression of different sets of genes, resulting in different cell fates (indicated by different letters and colors).

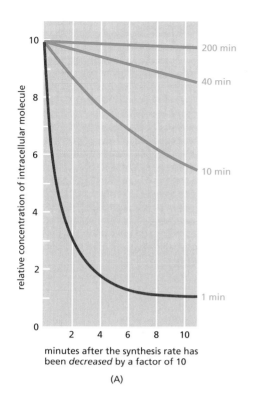

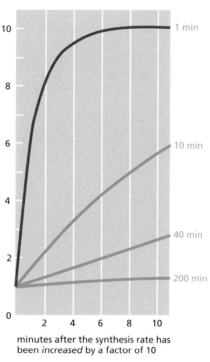

Figure 15–11 **The importance of rapid turnover.** The graphs show the predicted relative rates of change in the intracellular concentrations of molecules with differing turnover times when their synthesis rates are either (A) decreased or (B) increased suddenly by a factor of 10. In both cases, the concentrations of those molecules that are normally degraded rapidly in the cell *(red lines)* change quickly, whereas the concentrations of those that are normally degraded slowly *(green lines)* change proportionally more slowly. The numbers (in *blue*) on the right are the half-lives assumed for each of the different molecules.

synthesizes and degrades molecule Y at a rate of 100 molecules per second, with each molecule having an average lifetime of 10 seconds. Molecule X has a turnover rate that is 10 times slower than that of Y: it is both synthesized and degraded at a rate of 10 molecules per second, so that each molecule has an average lifetime in the cell of 100 seconds. If a signal acting on the cell causes a tenfold increase in the synthesis rates of both X and Y with no change in the molecular lifetimes, at the end of 1 second the concentration of Y will have increased by nearly 900 molecules per cell ($10 \times 100 - 100$), while the concentration of X will have increased by only 90 molecules per cell. In fact, after a molecule's synthesis rate has been either increased or decreased abruptly, the time required for the molecule to shift halfway from its old to its new equilibrium concentration is equal to its half-life—that is, equal to the time that would be required for its concentration to fall by half if all synthesis were stopped (**Figure 15–11**).

The same principles apply to proteins and small molecules, whether the molecules are in the extracellular space or inside cells. Many intracellular proteins have short half-lives, some surviving for less than 10 minutes. In most cases, these are key regulatory proteins whose concentrations are rapidly controlled in the cell by changes in their rates of synthesis.

We see later that many cell responses to extracellular signals depend on the conversion of intracellular signaling proteins from an inactive to an active form, rather than on their synthesis or degradation. Phosphorylation or the binding of GTP, for example, commonly activates signaling proteins. Even in these cases, however, the activation must be rapidly and continuously reversed (by dephosphorylation or GTP hydrolysis to GDP, respectively, in these examples) to make rapid signaling possible. These inactivation processes play a crucial part in determining the magnitude, rapidity, and duration of the response.

Nitric Oxide Gas Signals by Directly Regulating the Activity of Specific Proteins Inside the Target Cell

Most of this chapter is concerned with signaling pathways activated by cell-surface receptors. Before discussing these receptors and pathways, however, we briefly consider some important signal molecules that activate *intracellular receptors*. These molecules include nitric oxide and steroid hormones, which we discuss in turn. Although most extracellular signal molecules are hydrophilic

(A)

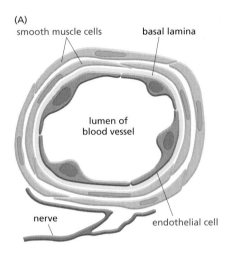

smooth muscle cells   basal lamina

lumen of
blood vessel

nerve   endothelial cell

Figure 15–12 **The role of nitric oxide
(NO) in smooth muscle relaxation in a
blood vessel wall.** (A) Simplified drawing
of an autonomic nerve contacting a
blood vessel. (B) Acetylcholine released
by nerve terminals in the blood vessel
wall activates NO synthase in endothelial
cells lining the blood vessel, causing the
endothelial cells to produce NO from
arginine. The NO diffuses out of the
endothelial cells and into the
neighboring smooth muscle cells, where
it binds to and activates guanylyl cyclase
to produce cyclic GMP. The cyclic GMP
triggers a response that causes the
smooth muscle cells to relax, enhancing
blood flow through the blood vessel.

(B)

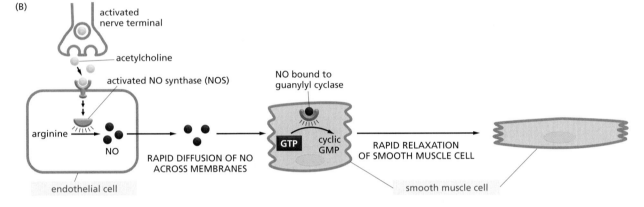

and bind to receptors on the surface of the target cell, some are hydrophobic
enough, small enough, or both, to pass readily across the target cell's plasma
membrane. Once inside, they directly regulate the activity of specific intracellular proteins. An important and remarkable example is the gas **nitric oxide (NO)**,
which acts as a signal molecule in both animals and plants. Even some bacteria
can detect a very low concentration of NO and move away from it.

In mammals, one of NO's many functions is to relax smooth muscle. It has
this role in the walls of blood vessels, for example (**Figure 15–12**A). Autonomic
nerves in the vessel wall release acetylcholine; the acetylcholine acts on the
nearby endothelial cells that line the interior of the vessel; and the endothelial
cells respond by releasing NO, which relaxes the smooth muscle cells in the wall,
allowing the vessel to dilate. This effect of NO on blood vessels provides an
explanation for the mechanism of action of nitroglycerine, which has been used
for about 100 years to treat patients with angina (pain resulting from inadequate
blood flow to the heart muscle). The nitroglycerine is converted to NO, which
relaxes blood vessels. This reduces the workload on the heart and, as a consequence, reduces the oxygen requirement of the heart muscle.

Many types of nerve cells use NO more directly to signal to their neighbors.
NO released by autonomic nerves in the penis, for example, causes the local
blood vessel dilation that is responsible for penile erection. NO is also produced
by activated macrophages and neutrophils to help them to kill invading
microorganisms. In plants, NO is involved in the defensive responses to injury or
infection.

NO is made by the deamination of the amino acid arginine, catalyzed by
enzymes called **NO synthases (NOS)** (Figure 15–12B). The NOS in endothelial
cells is called *eNOS*, while that in nerve and muscle cells is called *nNOS*. Nerve
and muscle cells constitutively make nNOS, which is activated to produce NO by
an influx of Ca^{2+} when the cells are stimulated. Macrophages, by contrast, make
yet another NOS, called inducible NOS (*iNOS*) because they make it only when
they are activated, usually in response to an infection.

Because dissolved NO passes readily across membranes, it rapidly diffuses out of the cell where it is produced and into neighboring cells (see Figure 15–12B). It acts only locally because it has a short half-life—about 5–10 seconds—in the extracellular space before oxygen and water convert it to nitrates and nitrites.

In some target cells, including smooth muscle cells, NO reversibly binds to iron in the active site of the enzyme *guanylyl cyclase*, stimulating this enzyme to produce the small intracellular signaling molecule *cyclic GMP*, which we discuss later. Thus, guanylyl cyclase acts both as an intracellular receptor for NO and as an intracellular signaling protein (see Figure 15–12B). NO can increase cyclic GMP in the cytosol within seconds, because the normal rate of turnover of cyclic GMP is high: a rapid degradation to GMP by a *phosphodiesterase* constantly balances the production of cyclic GMP from GTP by the guanylyl cyclase. The drug Viagra and its newer relatives inhibit the cyclic GMP phosphodiesterase in the penis, thereby increasing the amount of time that cyclic GMP levels remain elevated in the smooth muscle cells of penile blood vessels after NO production is induced by local nerve terminals. The cyclic GMP, in turn, keeps the blood vessels relaxed and thereby the penis erect.

NO can also signal cells independently of cyclic GMP. It can, for example, alter the activity of an intracellular protein by covalently nitrosylating thiol (–SH) groups on specific cysteines in the protein.

Carbon monoxide (CO) is another gas that is used as an extracellular signal molecule and like NO can act by stimulating guanylyl cyclase. But, these gases are not the only signal molecules that can pass directly across the target-cell plasma membrane, as we discuss next.

Nuclear Receptors Are Ligand-Modulated Gene Regulatory Proteins

Various small hydrophobic signal molecules diffuse directly across the plasma membrane of target cells and bind to intracellular receptors that are gene regulatory proteins. These signal molecules include steroid hormones, thyroid hormones, retinoids, and vitamin D. Although they differ greatly from one another in both chemical structure (**Figure 15–13**) and function, they all act by a similar mechanism. They bind to their respective intracellular receptor proteins and alter the ability of these proteins to control the transcription of specific genes. Thus, these proteins serve both as intracellular receptors and as intracellular effectors for the signal.

The receptors are all structurally related, being part of the very large **nuclear receptor superfamily**. Many family members have been identified by DNA sequencing only, and their ligand is not yet known; these proteins are therefore referred to as *orphan nuclear receptors*. Currently, more than half of the 48 nuclear receptors encoded in the human genome are orphans, as are 17 out of

Figure 15–13 **Some nongaseous signal molecules that bind to intracellular receptors.** Note that all of them are small and hydrophobic. The active, hydroxylated form of vitamin D$_3$ is shown. Estradiol and testosterone are steroid sex hormones.

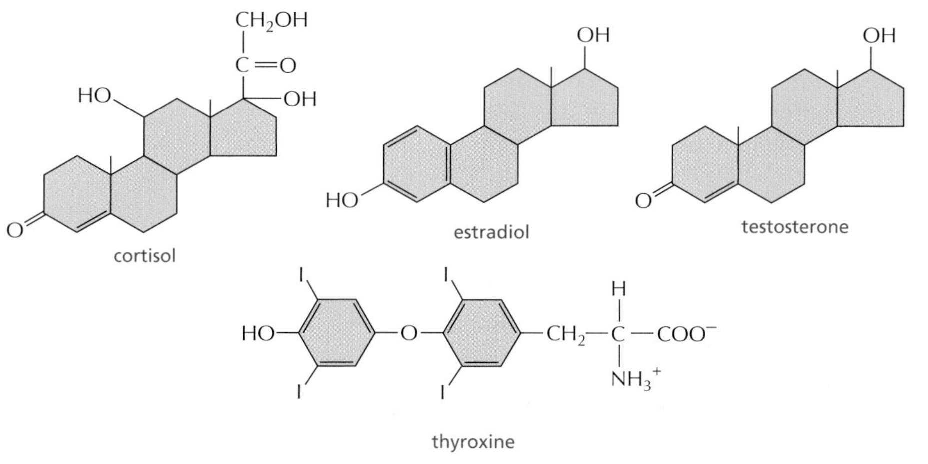

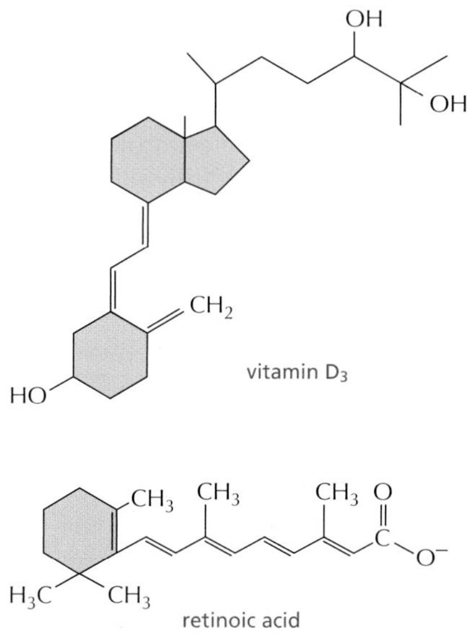

18 in *Drosophila* and all 278 in the nematode *C. elegans* (see Figure 4–85). Some mammalian nuclear receptors are regulated by intracellular metabolites rather than by secreted signal molecules; the *peroxisome proliferation-activated receptors (PPARs)*, for example, bind intracellular lipid metabolites and regulate the transcription of genes involved in lipid metabolism and fat cell differentiation (discussed in Chapter 23). It seems likely that the nuclear receptors for hormones evolved from such receptors for intracellular metabolites, which would help explain their intracellular location.

Steroid hormones—which include cortisol, the steroid sex hormones, vitamin D (in vertebrates), and the molting hormone *ecdysone* (in insects)—are all made from cholesterol. *Cortisol* is produced in the cortex of the adrenal glands and influences the metabolism of many types of cells. The *steroid sex hormones* are made in the testes and ovaries, and they are responsible for the secondary sex characteristics that distinguish males from females. *Vitamin D* is synthesized in the skin in response to sunlight; after it has been converted to its active form in the liver or kidneys, it regulates Ca^{2+} metabolism, promoting Ca^{2+} uptake in the gut and reducing its excretion in the kidneys. The *thyroid hormones,* which are made from the amino acid tyrosine, act to increase the metabolic rate of many cell types, while the *retinoids,* such as retinoic acid, are made from vitamin A and have important roles as local mediators in vertebrate development. Although all of these signal molecules are relatively insoluble in water, they are made soluble for transport in the bloodstream and other extracellular fluids by binding to specific carrier proteins, from which they dissociate before entering a target cell (see Figure 15–3B).

The nuclear receptors bind to specific DNA sequences adjacent to the genes that the ligand regulates. Some of the receptors, such as those for cortisol, are located primarily in the cytosol and enter the nucleus only after ligand binding; others, such as the thyroid and retinoid receptors, are bound to DNA in the nucleus even in the absence of ligand. In either case, the inactive receptors are usually bound to inhibitory protein complexes. Ligand binding alters the conformation of the receptor protein, causing the inhibitory complex to dissociate, while also causing the receptor to bind coactivator proteins that stimulate gene transcription (**Figure 15–14**). In other cases, however, ligand binding to a nuclear receptor inhibits transcription: some thyroid hormone receptors, for example, act as transcriptional activators in the absence of their hormone and become transcriptional repressors when hormone binds.

The transcriptional response usually takes place in multiple steps. In the cases in which ligand binding activates transcription, for example, the direct stimulation of a small number of specific genes occurs within about 30 minutes and constitutes the *primary response;* the protein products of these genes in turn activate other genes to produce a delayed, *secondary response;* and so on. In addition, some of the proteins produced in the primary response may act back to inhibit the transcription of primary response genes, thereby limiting the response—an example of negative feedback, which we discuss later. In this way, a simple hormonal trigger can cause a very complex change in the pattern of gene expression (**Figure 15–15**).

The responses to steroid and thyroid hormones, vitamin D, and retinoids are determined as much by the nature of the target cell as by the nature of the signal molecule. Many types of cells have the identical intracellular receptor, but the set of genes that the receptor regulates differs in each cell type. This is because more than one type of gene regulatory protein generally must bind to a eucaryotic gene to regulate its transcription. An intracellular receptor can therefore regulate a gene only if there is the right combination of other gene regulatory proteins, and many of these are cell-type specific.

In summary, each of these hydrophobic signal molecules induces a characteristic set of responses in an animal for two reasons. First, only certain types of cells have receptors for it. Second, each of these cell types contains a different combination of other cell-type-specific gene regulatory proteins that collaborate with the activated receptor to influence the transcription of specific sets of genes. This principle applies to all signaled responses that depend on gene regulatory proteins, including the many other examples we discuss in this chapter.

Figure 15–14 The nuclear receptor superfamily. All nuclear receptors bind to DNA as either homodimers or heterodimers, but for simplicity we show them as monomers here. (A) The receptors all have a related structure. Here, the short DNA-binding domain in each receptor is indicated in *light green*. (B) An inactive receptor protein is bound to inhibitory proteins. Domain-swap experiments suggest that many of the ligand-binding, transcription-activating, and DNA-binding domains in these receptors can function as interchangeable modules. (C) Receptor activation. Typically, the binding of ligand to the receptor causes the ligand-binding domain of the receptor to clamp shut around the ligand, the inhibitory proteins to dissociate, and coactivator proteins to bind to the receptor's transcription-activating domain, thereby increasing gene transcription. In other cases, ligand binding has the opposite effect, causing co-repressor proteins to bind to the receptor, thereby decreasing transcription (not shown). Though not shown here, activity can also be controlled through a change in the localization of the receptor: in its inactive form, it can be retained in the cytoplasm; ligand binding can then lead to the uncovering of nuclear localization signals that cause it to be imported into the nucleus to act on DNA. (D) The three-dimensional structure of a ligand-binding domain with *(right)* and without *(left)* ligand bound. Note that the *blue* α helix acts as a lid that snaps shut when the ligand (shown in *red*) binds, trapping the ligand in place.

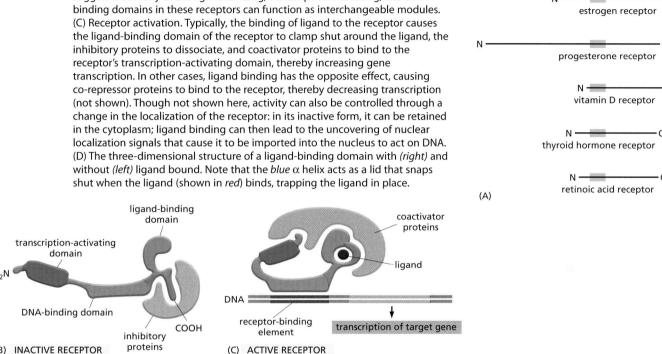

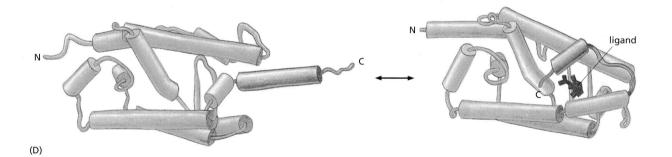

The molecular details of how nuclear receptors and other gene regulatory proteins control specific gene transcription are discussed in Chapter 7.

Nuclear receptor proteins are sometimes also present on the cell surface, where they function by mechanisms entirely different from that just described. In the remainder of the chapter, we consider various ways in which cell-surface receptors convert extracellular signals into intracellular ones, a process called *signal transduction*.

The Three Largest Classes of Cell-Surface Receptor Proteins Are Ion-Channel-Coupled, G-Protein-Coupled, and Enzyme-Coupled Receptors

In contrast to the small hydrophobic signal molecules just discussed that bind to intracellular receptors, most extracellular signal molecules bind to specific receptor proteins on the surface of the target cells they influence and do not enter the cytosol or nucleus. These cell-surface receptors act as *signal transducers* by converting an extracellular ligand-binding event into intracellular signals that alter the behavior of the target cell.

(A) PRIMARY (EARLY) RESPONSE TO STEROID HORMONE

(B) SECONDARY (DELAYED) RESPONSE TO STEROID HORMONE

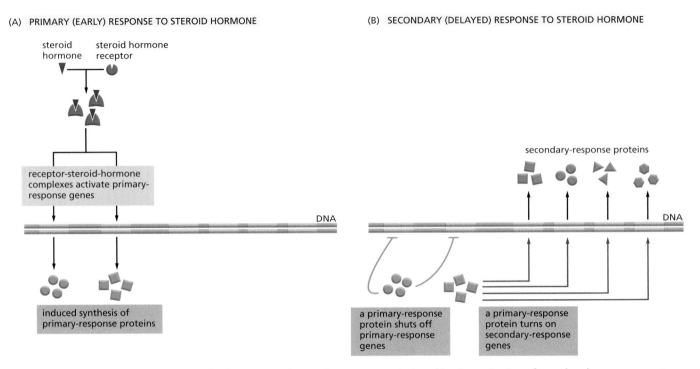

Figure 15–15 An example of primary and secondary responses induced by the activation of a nuclear hormone receptor. (A) Early primary response and (B) delayed secondary response. Some of the primary-response proteins turn on secondary-response genes, while others turn off the primary-response genes. The actual number of primary-response and secondary-response genes is greater than shown. Drugs that inhibit protein synthesis suppress the transcription of secondary-response genes but not primary-response genes, allowing these two classes of gene transcription responses to be readily distinguished. The figure shows the responses to a steroid hormone, but the same principles apply for many other ligands that activate nuclear receptor proteins.

Most cell-surface receptor proteins belong to one of three classes, defined by their transduction mechanism. **Ion-channel-coupled receptors**, also known as *transmitter-gated ion channels* or *ionotropic receptors*, are involved in rapid synaptic signaling between nerve cells and other electrically excitable target cells such as nerve and muscle cells (**Figure 15–16**A). This type of signaling is mediated by a small number of neurotransmitters that transiently open or close an ion channel formed by the protein to which they bind, briefly changing the ion permeability of the plasma membrane and thereby the excitability of the postsynaptic target cell. Most ion-channel-coupled receptors belong to a large family of homologous, multipass transmembrane proteins. Because they are discussed in detail in Chapter 11, we shall not consider them further here.

G-protein-coupled receptors act by indirectly regulating the activity of a separate plasma-membrane-bound target protein, which is generally either an enzyme or an ion channel. A *trimeric GTP-binding protein (G protein)* mediates the interaction between the activated receptor and this target protein (Figure 15–16B). The activation of the target protein can change the concentration of one or more small intracellular mediators (if the target protein is an enzyme), or it can change the ion permeability of the plasma membrane (if the target protein is an ion channel). The small intracellular mediators act in turn to alter the behavior of yet other signaling proteins in the cell. All of the G-protein-coupled receptors belong to a large family of homologous, multipass transmembrane proteins.

Enzyme-coupled receptors either function directly as enzymes or associate directly with enzymes that they activate (Figure 15–16C). They are usually single-pass transmembrane proteins that have their ligand-binding site outside the cell and their catalytic or enzyme-binding site inside. Enzyme-coupled receptors are heterogeneous in structure compared with the other two classes. The great majority, however, are either protein kinases or associate with protein kinases, which phosphorylate specific sets of proteins in the target cell when activated.

There are also some types of cell-surface receptors that do not fit easily into any of these classes but have important functions in controlling the specialization of different cell types during development and in tissue renewal and repair. We discuss these in a later section, after we explain how G-protein-coupled receptors and enzyme-coupled receptors operate. First, however, we consider some general principles of signaling via cell-surface receptors, in order to prepare for the detailed discussion of the major classes of cell-surface receptors that follows.

Most Activated Cell-Surface Receptors Relay Signals Via Small Molecules and a Network of Intracellular Signaling Proteins

A combination of small and large *intracellular signaling molecules* relays signals received at the cell surface by either G-protein-coupled or enzyme-coupled receptors into the cell interior. The resulting chain of intracellular signaling events ultimately alters effector proteins that are responsible for modifying the behavior of the cell (see Figure 15–1).

The small intracellular signaling molecules are called **small intracellular mediators**, or **second messengers** (the "first messengers" being the extracellular signals). They are generated in large numbers in response to receptor activation and often diffuse away from their source, spreading the signal to other parts of the cell. Some, such as *cyclic AMP* and Ca^{2+}, are water-soluble and diffuse in the cytosol, while others, such as *diacylglycerol*, are lipid-soluble and diffuse in the plane of the plasma membrane. In either case, they pass the signal on by binding to and altering the conformation and behavior of selected signaling proteins or effector proteins.

(A) ION-CHANNEL-COUPLED RECEPTORS

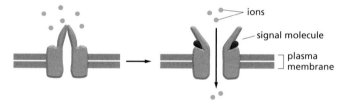

ions

signal molecule

plasma membrane

Figure 15–16 Three classes of cell-surface receptors. (A) Ion-channel-coupled receptors (also called transmitter-gated ion channels), (B) G-protein-coupled receptors, and (C) enzyme-coupled receptors. Although many enzyme-coupled receptors have intrinsic enzymatic activity, as shown on the left in (C), many others rely on associated enzymes, as shown on the right in (C).

(B) G-PROTEIN-COUPLED RECEPTORS

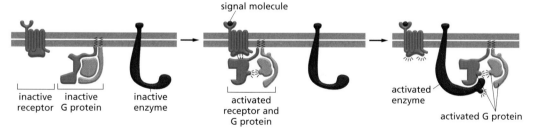

signal molecule

inactive receptor | inactive G protein | inactive enzyme

activated receptor and G protein

activated enzyme

activated G protein

(C) ENZYME-COUPLED RECEPTORS

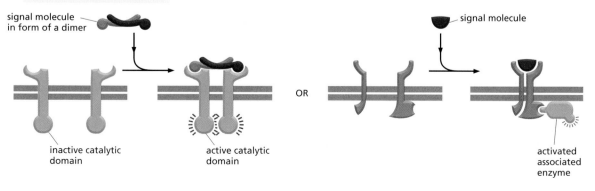

signal molecule in form of a dimer

signal molecule

OR

inactive catalytic domain

active catalytic domain

activated associated enzyme

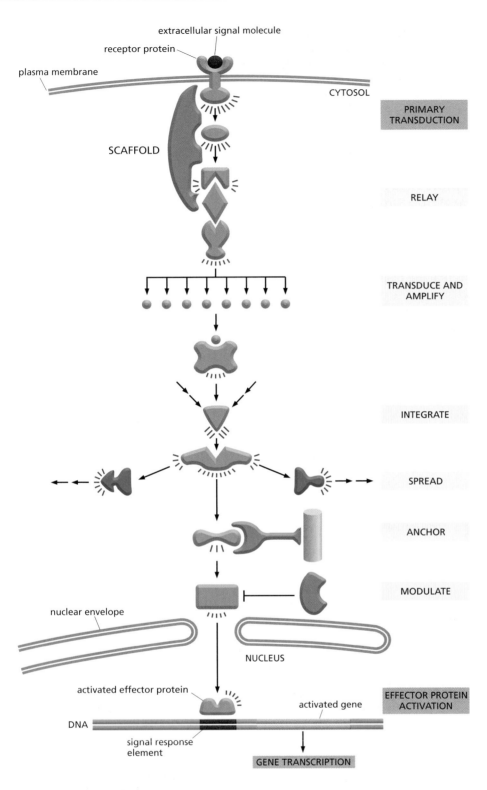

Figure 15–17 A hypothetical intracellular signaling pathway from a cell-surface receptor to the nucleus. In this example, a series of signaling proteins and small intracellular mediators relay the extracellular signal into the nucleus, causing a change in gene expression. The signal is altered (transduced), amplified, distributed, and modulated *en route*. Because many of the steps can be affected by other extracellular and intracellular signals, the final effect of one extracellular signal depends on multiple factors affecting the cell (see Figure 15–8). Ultimately, the signaling pathway activates (or inactivates) effector proteins that alter the cell's behavior. In this example, the effector is a gene regulatory protein that activates gene transcription. Although the figure shows individual signaling proteins performing a single function, in actuality they often have more than one function; scaffold proteins, for example, often also serve to anchor several signaling proteins to a particular intracellular structure.

Most signaling pathways to the nucleus are more direct than this one, which is not based on a known pathway.

The large intracellular signaling molecules are **intracellular signaling proteins**, which help relay the signal into the cell by either generating small intracellular mediators or activating the next signaling or effector protein in the pathway. These proteins form a functional network, in which each protein helps to process the signal in one or more of the following ways as it spreads the signal's influence through the cell (**Figure 15–17**):

1. The protein may simply *relay* the signal to the next signaling component in the chain.
2. It may act as a *scaffold* to bring two or more signaling proteins together so that they can interact more quickly and efficiently.

3. It may transform, or *transduce,* the signal into a different form, which is suitable for either passing the signal along or stimulating a cell response.

4. It may *amplify* the signal it receives, either by producing large amounts of a small intracellular mediator or by activating many copies of a downstream signaling protein. In this way, a small number of extracellular signal molecules can evoke a large intracellular response. When there are multiple amplification steps in a relay chain, the chain is often referred to as a **signaling cascade**.

5. It may receive signals from two or more signaling pathways and *integrate* them before relaying a signal onward. A protein that requires input from two or more signaling pathways to become activated, is often referred to as a coincidence detector.

6. It may *spread* the signal from one signaling pathway to another, creating branches in the signaling stream, thereby increasing the complexity of the response.

7. It may *anchor* one or more signaling proteins in a pathway to a particular structure in the cell where the signaling proteins are needed.

8. It may *modulate* the activity of other signaling proteins and thereby regulate the strength of signaling along a pathway.

We now consider in more detail some of the strategies that intracellular signaling proteins use in processing the signal as it is relayed along signaling pathways. We encounter these general strategies again in later sections of the chapter, when we discuss specific receptor classes and the signaling pathways they activate.

Many Intracellular Signaling Proteins Function as Molecular Switches That Are Activated by Phosphorylation or GTP Binding

Many intracellular signaling proteins behave like *molecular switches.* When they receive a signal, they switch from an inactive to an active conformation, until another process switches them off, returning them to their inactive conformation. As we mentioned earlier, the switching off is just as important as the switching on. If a signaling pathway is to recover after transmitting a signal so that it can be ready to transmit another, every activated molecule in the pathway must return to its original, unactivated state.

Two important classes of molecular switches that operate in intracellular signaling pathways depend on the gain or loss of phosphate groups for their activation or inactivation, although the way in which the phosphate is gained and lost is very different in the two classes. The largest class consists of proteins that are activated or inactivated by **phosphorylation** (discussed in Chapter 3). For these proteins, the switch is thrown in one direction by a **protein kinase**, which covalently adds one or more phosphate groups to the signaling protein, and in the other direction by a **protein phosphatase**, which removes the phosphate groups (**Figure 15–18A**). The activity of any protein regulated by

Figure 15–18 Two types of intracellular signaling proteins that act as molecular switches. Although one type is activated by phosphorylation and the other by GTP binding, in both cases the addition of a phosphate group switches the activation state of the protein and the removal of the phosphate switches it back again. (A) A protein kinase covalently adds a phosphate from ATP to the signaling protein, and a protein phosphatase removes the phosphate. Although not shown, some signaling proteins are activated by dephosphorylation rather than by phosphorylation. (B) A GTP-binding protein is induced to exchange its bound GDP for GTP, which activates the protein; the protein then inactivates itself by hydrolyzing its bound GTP to GDP.

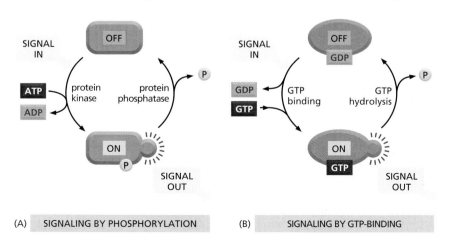

(A) SIGNALING BY PHOSPHORYLATION   (B)   SIGNALING BY GTP-BINDING

phosphorylation depends on the balance at any instant between the activities of the kinases that phosphorylate it and of the phosphatases that dephosphorylate it. About 30% of human proteins contain covalently attached phosphate, and the human genome encodes about 520 protein kinases and about 150 protein phosphatases. It is thought that a typical mammalian cell makes use of hundreds of distinct types of protein kinases at any one time.

Many signaling proteins controlled by phosphorylation are themselves protein kinases, and these are often organized into **phosphorylation cascades**. In such a cascade, one protein kinase, activated by phosphorylation, phosphorylates the next protein kinase in the sequence, and so on, relaying the signal onward and, in the process, amplifying it and sometimes spreading it to other signaling pathways. Two main types of protein kinases operate as intracellular signaling proteins. The great majority are **serine/threonine kinases**, which phosphorylate proteins on serines and (less often) threonines. Others are **tyrosine kinases**, which phosphorylate proteins on tyrosines. An occasional kinase can do both.

The other important class of molecular switches that function by gaining and losing phosphate groups consists of **GTP-binding proteins** (discussed in Chapter 3). These proteins switch between an "on" (actively signaling) state when GTP is bound and an "off" state when GDP is bound. In the "on" state, they have intrinsic GTPase activity and shut themselves off by hydrolyzing their bound GTP to GDP (Figure 15–18B). There are two major types of GTP-binding proteins. Large *trimeric GTP-binding proteins* (also called *G proteins*) help relay signals from G-protein-coupled receptors that activate them (see Figure 15–16B). Small **monomeric GTPases** (also called *monomeric GTP-binding proteins*) help relay signals from many classes of cell-surface receptors.

Specific regulatory proteins control both types of GTP-binding proteins. **GTPase-activating proteins** (**GAPs**) drive the proteins into an "off" state by increasing the rate of hydrolysis of bound GTP; the GAPs that function in this way are also called *regulators of G-protein signaling (RGS)* proteins. Conversely, G-protein-coupled receptors activate trimeric G proteins, and **guanine nucleotide exchange factors** (**GEFs**) activate monomeric GTPases, by promoting the release of bound GDP in exchange for binding of GTP. **Figure 15–19** illustrates the regulation of monomeric GTPases.

Both trimeric G proteins and monomeric GTPases also participate in many other processes in eucaryotic cells, including the regulation of vesicular traffic and aspects of cell division.

As discussed earlier, specific combinations of extracellular signals, rather than one signal molecule acting alone, are generally required to stimulate complex cell behaviors, such as cell survival and cell growth and proliferation (see Figure 15–8). The cell therefore has to integrate information coming from multiple signals if it is to make an appropriate response; many mammalian cells, for example, require both soluble signals and signals from the extracellular matrix

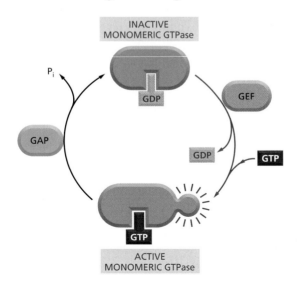

Figure 15–19 **The regulation of a monomeric GTPase.** GTPase-activating proteins (GAPs) inactivate the protein by stimulating it to hydrolyze its bound GTP to GDP, which remains tightly bound to the inactivated GTPase. Guanine nucleotide exchange factors (GEFs) activate the inactive protein by stimulating it to release its GDP; because the concentration of GTP in the cytosol is 10 times greater than the concentration of GDP, the protein rapidly binds GTP once it ejects GDP and is thereby activated.

(discussed in Chapter 19) to grow and proliferate. The integration depends in part on intracellular *coincidence detectors,* which are equivalent to *AND gates* in the microprocessor of a computer, in that they are only activated if they receive multiple converging signals (see Figure 15–17). **Figure 15–20** illustrates a simple hypothetical example of such a protein.

Not all molecular switches in signaling pathways depend on phosphorylation or GTP binding, however. We see later that some signaling proteins are switched on or off by the binding of another signaling protein or a small intracellular mediator such as cyclic AMP or Ca^{2+}, or by covalent modifications other than phosphorylation or dephosphorylation, such as ubiquitylation. Moreover, not all intracellular signaling proteins act as switches when they are phosphorylated or otherwise reversibly modified. As we discuss later, in many cases the covalently added group simply marks the protein so that it can interact with other signaling proteins that recognize the modification.

Intracellular Signaling Complexes Enhance the Speed, Efficiency, and Specificity of the Response

Even a single type of extracellular signal acting through a single type of cell-surface receptor often activates multiple parallel signaling pathways and can thereby influence multiple aspects of cell behavior—such as shape, movement, metabolism, and gene expression. Given the complexity of signal-response systems, which often involve multiple interacting relay chains of signaling proteins, how does an individual cell manage to make specific responses to so many different combinations of extracellular signals? The question is especially puzzling because many of the signals are closely related to one another and bind to closely related types of receptors. The same type of intracellular relay protein may couple one receptor subtype to one set of effectors and another receptor subtype to another set of effectors. In such cases, how is it possible to achieve specificity and avoid cross-talk? One strategy makes use of **scaffold proteins** (see Figure 15–17), which bind together groups of interacting signaling proteins into *signaling complexes,* often before a signal has been received (**Figure 15–21**A). Because the scaffold holds the signaling proteins in close proximity, the components can interact at high local concentrations and be sequentially activated speedily, efficiently, and selectively in response to an appropriate extracellular signal, avoiding unwanted cross-talk with other signaling pathways.

In other cases, signaling complexes form only transiently in response to an extracellular signal and rapidly disassemble when the signal is gone. Such transient complexes often assemble around a receptor after an extracellular signal molecule has activated it. In many of these cases, the cytoplasmic tail of the activated receptor is phosphorylated during the activation process, and the phosphorylated amino acids then serve as docking sites for the assembly of other signaling proteins (Figure 15–21B). In yet other cases, receptor activation leads to the production of modified phospholipid molecules (called phosphoinositides) in the adjacent plasma membrane, which then recruit specific intracellular signaling proteins to this region of membrane, where they are activated (Figure 15–21C). We discuss the roles of phosphoinositides in membrane trafficking events in Chapter 13.

Modular Interaction Domains Mediate Interactions Between Intracellular Signaling Proteins

Simply bringing intracellular signaling proteins together into close proximity is sometimes sufficient to activate them. Thus, *induced proximity,* where a signal triggers assembly of a signaling complex, is commonly used to relay signals from protein to protein along a signaling pathway. The assembly of such signaling complexes depends on various highly conserved, small **interaction domains,** which are found in many intracellular signaling proteins. Each of these compact

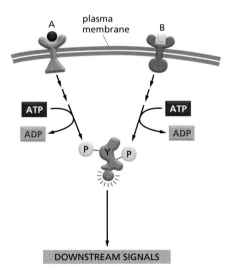

Figure 15–20 Signal integration. Extracellular signals A and B activate different intracellular signaling pathways, each of which leads to the phosphorylation of protein Y but at different sites on the protein. Protein Y is activated only when both of these sites are phosphorylated, and therefore it becomes active only when signals A and B are simultaneously present. Such proteins are often called coincidence detectors.

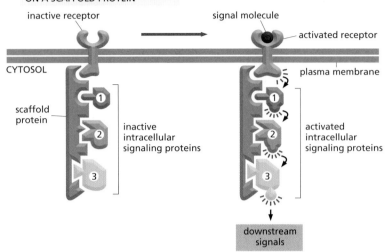

(A) PREFORMED SIGNALING COMPLEX ON A SCAFFOLD PROTEIN

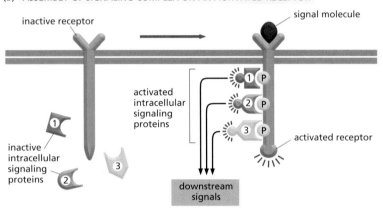

(B) ASSEMBLY OF SIGNALING COMPLEX ON AN ACTIVATED RECEPTOR

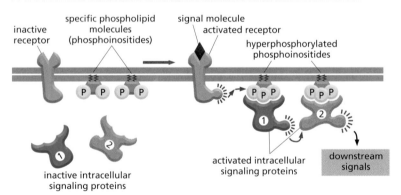

(C) ASSEMBLY OF SIGNALING COMPLEX ON PHOSPHOINOSITIDE DOCKING SITES

Figure 15–21 Three types of intracellular signaling complexes. (A) A receptor and some of the intracellular signaling proteins it activates in sequence are preassembled into a signaling complex on the inactive receptor by a large scaffold protein. In other cases, a preassembled complex binds to the receptor only after the receptor is activated (not shown). (B) A signaling complex assembles on a receptor only after the binding of an extracellular signal molecule has activated the receptor; here the activated receptor phosphorylates itself at multiple sites, which then act as docking sites for intracellular signaling proteins. (C) Activation of a receptor leads to the increased phosphorylation of specific phospholipids (phosphoinositides) in the adjacent plasma membrane, which then serve as docking sites for specific intracellular signaling proteins, which can now interact with each other.

protein modules binds to a particular structural motif in another protein (or lipid) molecule with which the signaling protein interacts. The recognized motif in the interacting protein can be a short peptide sequence, a covalent modification (such as phosphorylated or ubiquitylated amino acids), or another protein domain. The use of modular interaction domains presumably facilitated the evolution of new signaling pathways; because it can be inserted almost anywhere in a protein without disturbing the protein's folding or function, a new interaction domain added to an existing signaling protein could connect the protein to additional signaling pathways.

There are many types of interaction domains in signaling proteins. *Src homology 2 (SH2) domains* and *phosphotyrosine-binding (PTB) domains,* for example, bind to phosphorylated tyrosines in a particular peptide sequence on

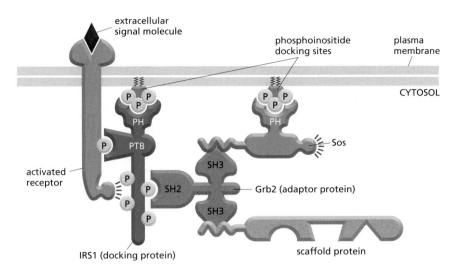

Figure 15–22 A specific signaling complex formed using modular interaction domains. This example is based on the insulin receptor, which is an enzyme-coupled receptor (a receptor tyrosine kinase, discussed later). First, the activated receptor phosphorylates itself on tyrosines, and one of the phosphotyrosines then recruits a docking protein called insulin receptor substrate-1 (IRS1) via a PTB domain of IRS1; the PH domain of IRS1 also binds to specific phosphoinositides on the inner surface of the plasma membrane. Then, the activated receptor phosphorylates IRS1 on tyrosines, and one of these phosphotyrosines recruits the adaptor protein (Grb2) via an SH2 domain of Grb2. Next, Grb2 uses one of its two SH3 domains to bind to a proline-rich region of the monomeric GTPase-activating protein called Sos (a Ras-GEF, discussed later), which also binds to phosphoinositides in the plasma membrane via its PH domain. Grb2 uses its other SH3 domain to bind to a proline-rich sequence in a scaffold protein. The scaffold protein binds several other signaling proteins, and the other phosphorylated tyrosines on IRS1 recruit additional signaling proteins that have SH2 domains (not shown).

activated receptors or intracellular signaling proteins. *Src homology 3 (SH3)* domains bind to short proline-rich amino acid sequences. Some *pleckstrin homology (PH)* domains bind to the charged head groups of specific phosphoinositides that are produced in the plasma membrane in response to an extracellular signal; they enable the protein they are part of to dock on the membrane and interact with other similarly recruited signaling proteins (see Figure 15–21C). Some signaling proteins consist solely of two or more interaction domains and function only as **adaptors** to link two other proteins together in a signaling pathway.

Interaction domains enable signaling proteins to bind to one another in multiple specific combinations. Like Lego bricks, the proteins can form linear or branching chains or three-dimensional networks, which determine the route followed by the signaling pathway. As an example, **Figure 15–22** illustrates how some interaction domains function in the case of the receptor for the hormone *insulin*.

Some cell-surface receptors and intracellular signaling proteins may cluster together transiently in specific microdomains in the lipid bilayer of the plasma membrane that are enriched in cholesterol and glycolipids (see Figure 10–13). These *lipid rafts* may promote efficient signaling by serving as sites where signaling molecules assemble and interact, but their importance in signaling remains controversial.

Another way of bringing receptors and intracellular signaling proteins together is to concentrate them in a specific region of the cell. An important example is the **primary cilium** that projects like an antenna from the surface of most vertebrate cells (discussed in Chapter 16). It is usually short and nonmotile and has microtubules in its core, and a number of surface receptors and signaling proteins are concentrated there. We shall see later that light and smell receptors are also highly concentrated in specialized cilia.

Cells Can Use Multiple Mechanisms to Respond Abruptly to a Gradually Increasing Concentration of an Extracellular Signal

Some cell responses to extracellular signal molecules are smoothly graded according to the concentration of the signal molecule. In other cases, the relationship between signal and response can be discontinuous or all-or-none, with an abrupt switch from one type of outcome to another as the signal concentration increases beyond a certain value (**Figure 15–23**). Such effects are often the result of positive

Figure 15–23 Smoothly graded versus switchlike signaling responses. Some cell responses increase gradually as the concentration of an extracellular signal molecule increases (*blue line*). In other cases, the responding cell is driven into an abruptly different state as the signal strength increases beyond a certain critical value (*red line*).

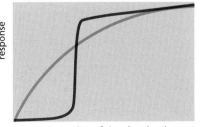

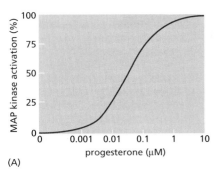

(A)

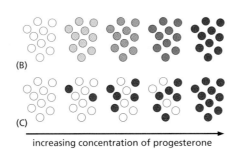

(B)

(C)

increasing concentration of progesterone

feedback in the response system, as we discuss below. Both types of responses are common, and it is not always easy to distinguish between them. When one measures the effect of a signal on a whole population of cells, it may appear smoothly graded, even though the individual cells may be responding in an all-or-none manner but with variation from cell to cell in the signal concentration at which the switch occurs; **Figure 15–24** shows an example.

Moreover, smoothly graded responses can sometimes be very steeply dependent on the strength of the signal, giving the appearance of almost switchlike behavior. Cells use a variety of molecular mechanisms to acheive such effects. In one mechanism, more than one intracellular signaling molecule must bind to its downstream target protein to induce a response. As we discuss later, four molecules of the small intracellular mediator cyclic AMP, for example, must be bound simultaneously to each molecule of *cyclic-AMP-dependent protein kinase (PKA)* to activate the kinase. A similar sharpening of response is seen when the activation of an intracellular signaling protein requires phosphorylation at more than one site. Such *cooperative* responses become sharper as the number of required cooperating molecules or phosphate groups increases, and if the number is large enough, responses become almost all-or-none (**Figure 15–25**).

Responses are also sharpened when an intracellular signaling molecule activates one enzyme and, at the same time, inhibits another enzyme that catalyzes the opposite reaction. A well-studied example of this common type of regulation is the stimulation of glycogen breakdown in skeletal muscle cells induced by the hormone *adrenaline* (epinephrine). Adrenaline's binding to a G-protein-coupled cell-surface receptor increases the intracellular concentration of cyclic AMP, which both activates an enzyme that promotes glycogen breakdown and inhibits an enzyme that promotes glycogen synthesis.

Figure 15–24 The importance of examining individual cells to detect all-or-none responses to increasing concentrations of an extracellular signal. In these experiments, immature frog eggs (oocytes) were stimulated with increasing concentrations of the hormone progesterone. The response was assessed by analyzing the activation of *MAP kinase* (discussed later), which is one of the protein kinases activated by phosphorylation in the response. The amount of phosphorylated (activated) MAP kinase in extracts of the oocytes was assessed biochemically. In (A), extracts of populations of stimulated oocytes were analyzed, and the activation of MAP kinase is seen to increase progressively with increasing progesterone concentration. There are two possible ways of explaining this result: (B) MAP kinase could have increased gradually in each individual cell with increasing progesterone concentration; (C) alternatively, individual cells could have responded in an all-or-none way, with the gradual increase in total MAP kinase activation reflecting the increasing number of cells responding with increasing progesterone concentration. When extracts of individual oocytes were analyzed, it was found that cells had either very low amounts or very high amounts, but not intermediate amounts, of the activated kinase, indicating that the response was essentially all-or-none at the level of individual cells, as diagrammed in (C). (Adapted from J.E. Ferrell and E.M. Machleder, *Science* 280:895–898, 1998. With permission from AAAS.)

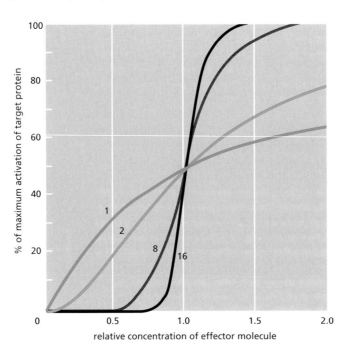

Figure 15–25 Activation curves for an allosteric protein as a function of effector molecule concentration. The curves show how the sharpness of the activation response increases with an increase in the number of allosteric effector molecules that must be bound simultaneously to activate the target protein. The curves shown are those expected, under certain conditions, if the activation requires the simultaneous binding of 1, 2, 8, or 16 effector molecules.

All of these mechanisms can produce very steep responses, but the responses are still smoothly graded according to the concentration of the extracellular signal molecule. To produce true all-or-none responses, one needs another mechanism, *positive feedback*, as we now discuss.

Intracellular Signaling Networks Usually Make Use of Feedback Loops

Like intracellular metabolic pathways (discussed in Chapter 2), most intracellular signaling networks incorporate feedback loops, in which the output of a process acts back to regulate that same process. In *positive feedback*, the output stimulates its own production; in *negative feedback*, the output inhibits its own production (**Figure 15–26**). Feedback loops are of great general importance in biology, and they regulate many chemical and physical processes in cells. They operate over an enormous range of time scales in cells, from milliseconds (in the case of an action potential, for example—see Figure 11–29) to many hours (in the case of circadian oscillations, for example—see Figure 7–73). Those that regulate cell signaling can either operate exclusively within the target cell or involve the secretion of extracellular signals. Here, we focus on those feedback loops that operate entirely within the target cell; even the simplest of these loops can produce complex and interesting effects.

A **positive feedback loop** in a signaling pathway can transform the behavior of the responding cell. If the positive feedback is of only moderate strength, its effect will be simply to steepen and increase the response to the signal. But if the feedback is strong enough, it can produce a qualitatively different result: a runaway increase in the quantity of product when the signal increases above a critical value, leading to a new steady level of production that is sharply different from that obtained when the signal was only slightly weaker (**Figure 15–27**).

This type of all-or-none response goes hand in hand with a further property: once the responding system has switched to the high level of activation, this condition is self-sustaining and can persist even after the signal strength drops back below its critical value. In such a case, the system is said to be *bistable:* it can exist in either a "switched-off" or a "switched-on" state, and a transient stimulus can flip it from the one state to the other (**Figure 15–28**A and B). Cells use positive feedback loops of this sort to make stable all-or-none decisions, especially during development, when cells in different positions must choose between alternative developmental pathways in response to smoothly graded positional signals (morphogens), as discussed earlier. Through positive feedback, a transient extracellular signal can often induce long-term changes in cells and their progeny that can persist for the lifetime of the organism. The signals that trigger muscle cell specification, for example, turn on the transcription of a series of genes that encode muscle-specific gene regulatory proteins, which stimulate the transcription of their own genes, as well as genes encoding various other muscle cell proteins; in this way, the decision to become a muscle cell is made permanent (see Figure 7–75). This type of cell memory dependent on positive feedback is one of the basic ways in which a cell can undergo a lasting change of character without any alteration in its DNA sequence; and this altered state can be passed on to daughter cells. Such mechanisms of inheritance are called *epigenetic,* in contrast to genetic mechanisms involving mutations of the DNA, and they are discussed more fully in Chapter 7 (see Figure 7–86).

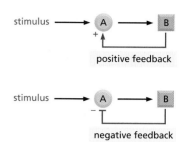

Figure 15–26 Positive and negative feedback. In these simple examples, a stimulus activates protein A, which, in turn, activates protein B. Protein B then acts back to either increase or decrease the activity of A.

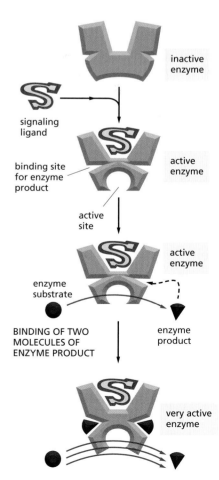

Figure 15–27 A positive feedback mechanism giving switch-like behavior. In this example, an intracellular signaling molecule (ligand) activates an enzyme located downstream in a signaling pathway. Two molecules of the product of the enzymatic reaction bind back to the same enzyme to activate it further. The consequence is a very low rate of synthesis of the product in the absence of the ligand. The rate increases slowly with the concentration of ligand until, at some threshold level of ligand, enough of the product has been synthesized to activate the enzyme in a self-accelerating, runaway fashion. The concentration of the product then rapidly increases to a much higher level.

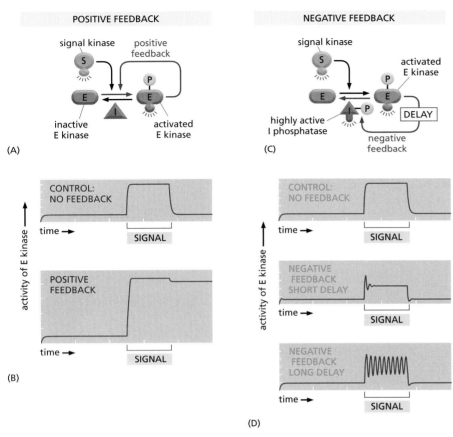

(A)

(B)

(C)

(D)

Figure 15–28 Some effects of simple feedback. The graphs show the computed effects of simple positive and negative feedback loops. In each case, the input signal is an activated protein kinase (S) that phosphorylates and thereby activates another protein kinase (E); a protein phosphatase (I) dephosphorylates and inactivates the activated E kinase. In the graphs, the *red* line indicates the activity of the E kinase over time; the underlying *blue* bar indicates the time for which the input signal (activated S kinase) is present. (A) Diagram of the positive feedback loop, in which the activated E kinase acts back to promote its own phosphorylation and activation; the basal activity of the I phosphatase dephosphorylates activated E at a steady low rate. (B) The top graph shows that, without feedback, the activity of the E kinase is simply proportional (with a short lag) to the level of stimulation by the S kinase. The bottom graph shows that, with the positive feedback loop, the system is bistable (i.e., it is capable of existing in either of two steady states): the transient stimulation by S kinase switches the system from an "off" state to an "on" state, which then persists after the stimulus has been removed. (C) Diagram of the negative feedback loop, in which the activated E kinase phosphorylates and activates the I phosphatase, thereby increasing the rate at which the phosphatase dephosphorylates and inactivates the phosphorylated E kinase. The top graph shows, again, the response in E kinase activity without feedback. The other graphs show the effects on E kinase activity of negative feedback operating after a short or long delay. With a short delay, the system shows a strong brief response when the signal is abruptly changed, and the feedback then drives the response back down to a lower level. With a long delay, the feedback produces sustained oscillations for as long as the stimulus is present.

Small changes in the details of the feedback can radically alter the way the system responds, even for this very simple example; the figure shows only a small sample of possible behaviors.

By contrast with positive feedback, a **negative feedback loop** counteracts the effect of a stimulus and thereby abbreviates and limits the level of the response, making the system less sensitive to perturbations. As with positive feedback, however, qualitatively different phenomena can be obtained when the feedback operates more powerfully. A delayed negative feedback with a long enough delay can produce responses that oscillate. The oscillations may persist for as long as the stimulus is present (Figure 15–28C) or they may even be generated spontaneously, without need of an external signal to drive them (see Figure 22–82). Later in this chapter, we shall encounter a number of examples of such oscillatory behavior in the intracellular responses to extracellular signals; all of them depend on negative feedback.

If negative feedback operates with a short delay, the system behaves like a change detector. It gives a strong response to a stimulus, but the response rapidly decays even while the stimulus persists; if the stimulus is suddenly increased, however, the system responds strongly again, but, again, the response rapidly decays. This is the phenomenon of *adaptation*, which we now discuss.

Cells Can Adjust Their Sensitivity to a Signal

In responding to many types of stimuli, cells and organisms are able to detect the same percentage of change in a signal over a very wide range of stimulus strengths. The target cells accomplish this through a reversible process of **adaptation**, or **desensitization**, whereby a prolonged exposure to a stimulus decreases the cells' response to that level of stimulus. In chemical signaling, adaptation enables cells to respond to *changes* in the concentration of an extracellular signal molecule (rather than to the absolute concentration of the signal) over a very wide range of signal concentrations. The underlying mechanism is negative feedback that operates with a short delay: a strong response modifies the signaling machinery involved, such that the machinery resets itself to become less responsive to the same level of signal (see Figure 15–28D, middle graph). Owing to the delay, however, a sudden increase in the signal is able to stimulate the cell again for a short period before the negative feedback has time to kick in.

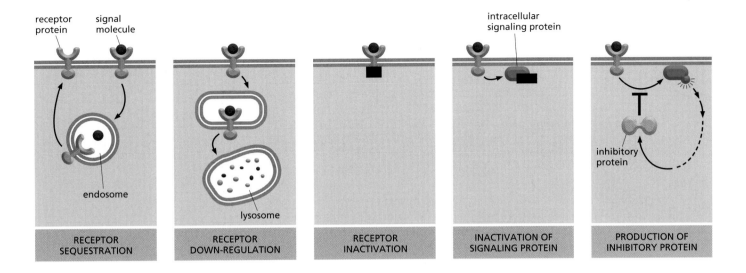

Figure 15–29 Some ways in which target cells can become desensitized (adapted) to an extracellular signal molecule. The mechanisms shown here that operate at the level of the receptor often involve phosphorylation or ubiquitylation of the receptor proteins. In bacterial chemotaxis, however, which we discuss later, adaptation depends on methylation of the receptor proteins.

Desensitization to a signal molecule can occur in various ways. It can result from an inactivation of the receptors themselves. The binding of signal molecules to cell-surface receptors, for example, may induce the endocytosis and temporary sequestration of the receptors in endosomes. In some cases, such signal-induced receptor endocytosis leads to the destruction of the receptors in lysosomes, a process referred to as *receptor down-regulation* (in other cases, however, activated receptors continue to signal after they have been endocytosed, as we discuss later). Receptors can also become desensitized (inactivated) on the cell surface—for example, by becoming phosphorylated or methylated—with a short delay following their activation. Desensitization can also occur at sites downstream of the receptors, either by a change in intracellular signaling proteins involved in transducing the extracellular signal or by the production of an inhibitor protein that blocks the signal transduction process. These various desensitization mechanisms are compared in **Figure 15–29**.

Having discussed some of the general principles of cell signaling, we next turn to the G-protein-coupled receptors, which are by far the largest class of cell-surface receptors, mediating responses to the most diverse types of extracellular signals.

Summary

Each cell in a multicellular animal has been programmed during development to respond to a specific set of extracellular signal molecules produced by other cells. The signal molecules act in various combinations to regulate the behavior of the cell. Most signal molecules act as local mediators, which are secreted and then rapidly taken up, destroyed, or immobilized, so that they act only on neighboring cells. Other signal molecules remain bound to the surface of the signaling cell and mediate contact-dependent signaling. There are also two distinct types of long-distance signaling. In endocrine signaling, hormones secreted by endocrine cells are carried in the blood to target cells throughout the body. In synaptic signaling, neurotransmitters secreted by nerve cell axons act locally on the postsynaptic cells that the axons contact.

Cell signaling requires not only extracellular signal molecules but also a complementary set of receptor proteins expressed by the target cells that specifically bind the signal molecules. Some small hydrophobic signal molecules, including steroid and thyroid hormones, diffuse across the plasma membrane of the target cell and activate intracellular receptor proteins that directly regulate the transcription of specific genes. The dissolved gases nitric oxide and carbon monoxide act as local mediators by diffusing across the plasma membrane of the target cell and activating an intracellular protein such as the enzyme guanylyl cyclase, which produces cyclic GMP in the target cell. But most extracellular signal molecules are hydrophilic and cannot pass through the plasma membrane. These activate cell-surface receptor proteins, which act as signal transducers, converting the extracellular signal into intracellular ones that alter the behavior of the target cell.

The three largest families of cell-surface receptors transduce extracellular signals in different ways. Ion-channel-coupled receptors are transmitter-gated ion channels that open or close briefly in response to the binding of a neurotransmitter. G-protein-coupled receptors indirectly activate or inactivate plasma-membrane-bound enzymes or ion channels via trimeric GTP-binding proteins (G proteins). Enzyme-coupled receptors either act directly as enzymes or are associated with enzymes; these enzymes are usually protein kinases that phosphorylate the receptors and specific signaling proteins in the target cell.

Once activated, G-protein-coupled receptors and enzyme-coupled receptors relay the signal into the cell interior by activating chains of intracellular signaling proteins. Some of these signaling proteins transduce, amplify, or spread the signal as they relay it, while others integrate signals from different signaling pathways. Some function as switches that are transiently activated by phosphorylation or GTP binding. Large signaling complexes form by means of modular interaction domains in the signaling proteins, which allow the proteins to form functional signaling networks.

Target cells use various intracellular mechanisms, including feedback loops, to adjust the ways in which they respond to extracellular signals. Positive feedback loops can help cells to respond in an all-or-none fashion to a gradually increasing concentration of an extracellular signal or to convert a short-lasting signal into a long-lasting, or even irreversible, response. Delayed negative feedback allows cells to desensitize to a signal molecule, which enables them to respond to small changes in the concentration of the signal molecule over a large concentration range.

SIGNALING THROUGH G-PROTEIN-COUPLED CELL-SURFACE RECEPTORS (GPCRs) AND SMALL INTRACELLULAR MEDIATORS

All eucaryotes use **G-protein-coupled receptors (GPCRs)**. These form the largest family of cell-surface receptors, and they mediate most responses to signals from the external world, as well as signals from other cells, including hormones, neurotransmitters, and local mediators. Our senses of sight, smell, and taste (with the possible exception of sour taste) depend on them. There are more than 700 GPCRs in humans, and in mice there are about 1000 concerned with the sense of smell alone. The signal molecules that act on GPCRs are as varied in structure as they are in function and include proteins and small peptides, as well as derivatives of amino acids and fatty acids, not to mention photons of light and all the molecules that we can smell or taste. The same signal molecule can activate many different GPCR family members; for example, adrenaline activates at least 9 distinct GPCRs, acetylcholine another 5, and the neurotransmitter serotonin at least 14. The different receptors for the same signal are usually expressed in different cell types and elicit different responses.

Despite the chemical and functional diversity of the signal molecules that activate them, all GPCRs have a similar structure. They consist of a single polypeptide chain that threads back and forth across the lipid bilayer seven times (**Figure 15–30**). In addition to their characteristic orientation in the plasma membrane, they all use G proteins to relay the signal into the cell interior.

The GPCR superfamily includes *rhodopsin*, the light-activated protein in the vertebrate eye, as well as the large number of olfactory receptors in the vertebrate nose. Other family members are found in unicellular organisms: the receptors in yeasts that recognize secreted mating factors are an example. It is likely that the GPCRs that mediate cell–cell signaling in multicellular organisms evolved from the sensory receptors in their unicellular eucaryotic ancestors.

It is remarkable that about half of all known drugs work through GPCRs or the signaling pathways GPCRs activate. Of the many hundreds of genes in the human genome that encode GPCRs, about 150 encode orphan receptors, for which the ligand is unknown. Many of them are likely targets for new drugs that remain to be discovered.

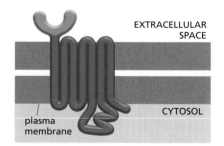

Figure 15–30 A G-protein-coupled receptor (GPCR). GPCRs that bind protein ligands have a large extracellular domain formed by the part of the polypeptide chain shown in *light green*. This domain, together with some of the transmembrane segments, binds the protein ligand. Receptors for small ligands such as adrenaline have small extracellular domains, and the ligand usually binds deep within the plane of the membrane to a site that is formed by amino acids from several transmembrane segments.

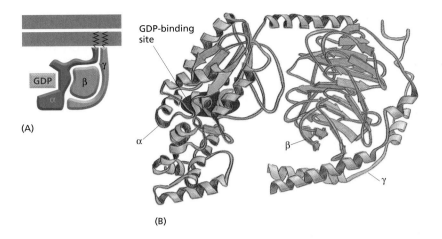

Figure 15–31 The structure of an inactive G protein. (A) Note that both the α and the γ subunits have covalently attached lipid molecules *(red)* that help bind them to the plasma membrane, and the α subunit has GDP bound. (B) The three-dimensional structure of an inactive G protein, based on transducin, the G protein that operates in visual transduction (discussed later). The α subunit contains the GTPase domain and binds to one side of the β subunit, which locks the GTPase domain in an inactive conformation that binds GDP. The γ subunit binds to the opposite side of the β subunit, and the β and γ subunits together form a single functional unit. (B, based on D.G. Lombright et al., *Nature* 379:311–319, 1996. With permission from Macmillan Publishers Ltd.)

Trimeric G Proteins Relay Signals from GPCRs

When an extracellular signal molecule binds to a GPCR, the receptor undergoes a conformational change that enables it to activate a **trimeric GTP-binding protein (G protein)**. The G protein is attached to the cytoplasmic face of the plasma membrane, where it functionally couples the receptor to either enzymes or ion channels in this membrane. In some cases, the G protein is physically associated with the receptor before the receptor is activated, whereas in others it binds only after receptor activation. There are various types of G proteins, each specific for a particular set of GPCRs and for a particular set of target proteins in the plasma membrane. They all have a similar structure, however, and operate similarly.

G proteins are composed of three protein subunits—α, β, and γ. In the unstimulated state, the α subunit has GDP bound and the G protein is inactive (**Figure 15–31**). When a GPCR is activated, it acts like a guanine nucleotide exchange factor (GEF) and induces the α subunit to release its bound GDP, allowing GTP to bind in its place. This exchange causes a large conformational change in the G protein, which activates it. It was originally thought that the activation always causes the trimer to dissociate into two activated components—an *α subunit* and a *βγ complex*. However, there is now evidence that, in some cases at least, the conformational change exposes previously buried surfaces between the α subunit and the βγ complex, so that the α subunit and βγ complex can each now interact with their targets without requiring the subunits to dissociate (**Figure 15–32**). These targets are either enzymes or ion channels in the plasma membrane that relay the signal onward.

The α subunit is a GTPase, and once it hydrolyzes its bound GTP to GDP it becomes inactive. The time for which the G protein remains active depends on how quickly the α subunit hydrolyzes its bound GTP. This time is usually short because the GTPase activity is greatly enhanced by the binding of the α subunit to a second protein, which can be either the target protein or a specific **regulator of G protein signaling** (**RGS**). RGS proteins act as α-subunit-specific GTPase-activating proteins (GAPs) (see Figure 15–19), and they help shut off G-protein-mediated responses in all eucaryotes. There are about 25 RGS proteins encoded in the human genome, each of which interacts with a particular set of G proteins.

GPCRs activate various intracellular signaling pathways, including some that are also activated by enzyme-coupled receptors. In this section, however, we focus on those GPCR-activated pathways that use small intracellular mediators.

Some G Proteins Regulate the Production of Cyclic AMP

Cyclic AMP (cAMP) acts as a small intracellular mediator in all procaryotic and animal cells that have been studied. Its normal concentration in the cytosol is about 10^{-7} M, but an extracellular signal can increase this concentration more than twentyfold in seconds (**Figure 15–33**). As explained earlier (see Figure

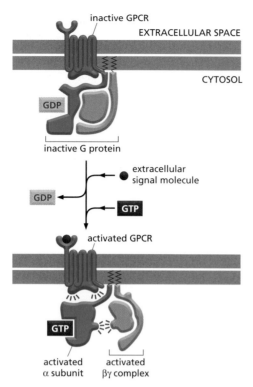

inactive GPCR

EXTRACELLULAR SPACE

CYTOSOL

GDP

inactive G protein

extracellular signal molecule

GDP

GTP

activated GPCR

GTP

activated α subunit   activated βγ complex

Figure 15–32 Activation of a G protein by an activated GPCR. Binding of an extracellular signal to a GPCR changes the conformation of the receptor, which in turn alters the conformation of the G protein. The alteration of the α subunit of the G protein allows it to exchange its GDP for GTP, activating both the α subunit and the βγ complex, both of which can regulate the activity of target proteins in the plasma membrane. The receptor stays active while the external signal molecule is bound to it, and it can therefore catalyze the activation of many molecules of G protein, which dissociate from the receptor once activated (not shown). In some cases, the α subunit and the βγ complex dissociate from each other when the G protein is activated.

15–11), such a rapid response requires balancing a rapid synthesis of the molecule with its rapid breakdown or removal. Cyclic AMP is synthesized from ATP by a plasma-membrane-bound enzyme **adenylyl cyclase**, and it is rapidly and continuously destroyed by **cyclic AMP phosphodiesterases** that hydrolyze cyclic AMP to adenosine 5′-monophosphate (5′-AMP) (**Figure 15–34**).

Many extracellular signal molecules work by increasing cyclic AMP concentration, and they do so by increasing the activity of adenylyl cyclase against a steady background of phosphodiesterase activity. Adenylyl cyclase is a large multipass transmembrane protein with its catalytic domain on the cytosolic side of the plasma membrane. There are at least eight isoforms in mammals, most of which are regulated by both G proteins and Ca^{2+}. GPCRs that act by increasing cyclic AMP are coupled to a **stimulatory G protein (G_s)**, which activates adenylyl cyclase and thereby increases cyclic AMP concentration. Another G protein, called **inhibitory G protein (G_i)**, inhibits adenylyl cyclase, but it acts mainly by directly regulating ion channels (as we discuss later).

Both G_s and G_i are targets for some medically important bacterial toxins. *Cholera toxin*, which is produced by the bacterium that causes cholera, is an enzyme that catalyzes the transfer of ADP ribose from intracellular NAD^+ to the α subunit of G_s. This ADP ribosylation alters the α subunit so that it can no longer hydrolyze its bound GTP, causing it to remain in an active state that stimulates adenylyl cyclase indefinitely. The resulting prolonged elevation in cyclic AMP concentration within intestinal epithelial cells causes a large efflux of Cl^- and water into the gut, thereby causing the severe diarrhea that characterizes cholera. *Pertussis toxin*, which is made by the bacterium that causes pertussis

Figure 15–33 An increase in cyclic AMP in response to an extracellular signal. This nerve cell in culture is responding to the neurotransmitter serotonin, which acts through a GPCR to cause a rapid rise in the intracellular concentration of cyclic AMP. To monitor the cyclic AMP level, the cell has been loaded with a fluorescent protein that changes its fluorescence when it binds cyclic AMP. *Blue* indicates a low level of cyclic AMP, *yellow* an intermediate level, and *red* a high level. (A) In the resting cell, the cyclic AMP level is about 5×10^{-8} M. (B) Twenty seconds after the addition of serotonin to the culture medium, the intracellular level of cyclic AMP has increased to more than 10^{-6} M in the relevant parts of the cell, an increase of more than twentyfold. (From Brian J. Bacskai et al., *Science* 260:222–226, 1993. With permission from AAAS.)

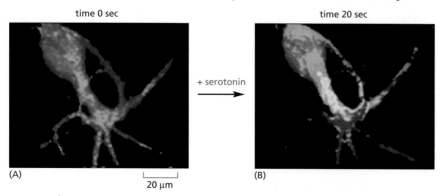

time 0 sec

time 20 sec

+ serotonin

(A)   20 µm   (B)

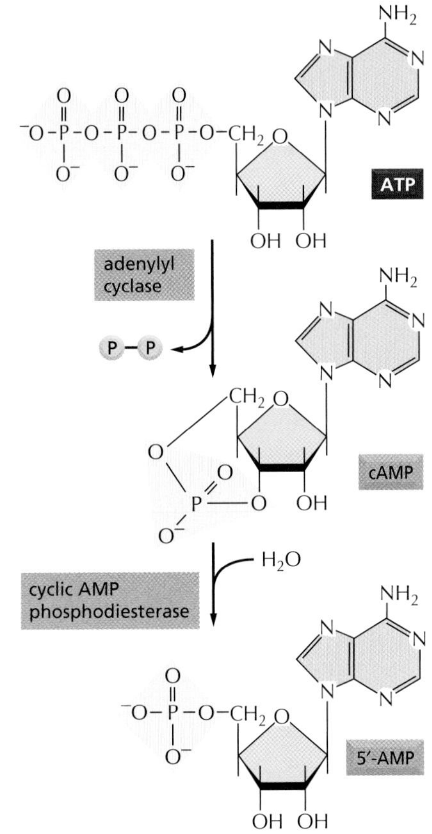

Figure 15–34 **The synthesis and degradation of cyclic AMP.** In a reaction catalyzed by the enzyme adenylyl cyclase, cyclic AMP (cAMP) is synthesized from ATP through a cyclization reaction that removes two phosphate groups as pyrophosphate ((P)–(P)); a pyrophosphatase drives this synthesis by hydrolyzing the released pyrophosphate to phosphate (not shown). Cyclic AMP is short-lived (unstable) in the cell because it is hydrolyzed by specific phosphodiesterases to form 5′-AMP, as indicated.

(whooping cough), catalyzes the ADP ribosylation of the α subunit of G_i, preventing the protein from interacting with receptors; as a result, the G protein retains its bound GDP and is unable to regulate its target proteins. These two toxins are widely used in experiments to determine whether a cell's GPCR-dependent response to a signal is mediated by G_s or by G_i.

Some of the responses mediated by a G_s-stimulated increase in cyclic AMP concentration are listed in **Table 15–1**. As the table shows, different cell types respond differently to an increase in cyclic AMP concentration, and one cell type often responds in the same way to such an increase, regardless of the extracellular signal that causes it. At least four hormones activate adenylyl cyclase in fat cells, for example, and all of them stimulate the breakdown of triglyceride (the storage form of fat) to fatty acids (see Table 15–1).

Individuals who are genetically deficient in a particular G_s α subunit show decreased responses to certain hormones. As a consequence, these people display metabolic abnormalities, have abnormal bone development, and are mentally retarded.

Table 15–1 **Some Hormone-induced Cell Responses Mediated by Cyclic AMP**

TARGET TISSUE	HORMONE	MAJOR RESPONSE
Thyroid gland	thyroid-stimulating hormone (TSH)	thyroid hormone synthesis and secretion
Adrenal cortex	adrenocorticotrophic hormone (ACTH)	cortisol secretion
Ovary	luteinizing hormone (LH)	progesterone secretion
Muscle	adrenaline	glycogen breakdown
Bone	parathormone	bone resorption
Heart	adrenaline	increase in heart rate and force of contraction
Liver	glucagon	glycogen breakdown
Kidney	vasopressin	water resorption
Fat	adrenaline, ACTH, glucagon, TSH	triglyceride breakdown

Cyclic-AMP-Dependent Protein Kinase (PKA) Mediates Most of the Effects of Cyclic AMP

In most animal cells, cyclic AMP exerts its effects mainly by activating **cyclic-AMP-dependent protein kinase (PKA)**. This kinase phosphorylates specific serines or threonines on selected target proteins, including intracellular signaling proteins and effector proteins, thereby regulating their activity. The target proteins differ from one cell type to another, which explains why the effects of cyclic AMP vary so markedly depending on the cell type (see Table 15–1).

In the inactive state, PKA consists of a complex of two catalytic subunits and two regulatory subunits. The binding of cyclic AMP to the regulatory subunits alters their conformation, causing them to dissociate from the complex. The released catalytic subunits are thereby activated to phosphorylate specific target proteins (**Figure 15–35**). The regulatory subunits of PKA (also called A-kinase) are important for localizing the kinase inside the cell: special *A-kinase anchoring proteins (AKAPs)* bind both to the regulatory subunits and to a component of the cytoskeleton or a membrane of an organelle, thereby tethering the enzyme complex to a particular subcellular compartment. Some AKAPs also bind other signaling proteins, forming a complex that functions as a signaling module. An AKAP located around the nucleus of heart muscle cells, for example, binds both PKA and a phosphodiesterase that hydrolyzes cyclic AMP. In unstimulated cells, the phosphodiesterase keeps the local cyclic AMP concentration low, so that the bound PKA is inactive; in stimulated cells, cyclic AMP concentration rapidly rises, overwhelming the phosphodiesterase and activating the PKA. Among the target proteins that PKA phosphorylates and activates in these cells is the adjacent phosphodiesterase, which rapidly lowers the cyclic AMP concentration again. This arrangement converts what might otherwise be a weak and prolonged PKA response into a strong, brief, local pulse of PKA activity.

Whereas some responses mediated by cyclic AMP occur within seconds and do not depend on changes in gene transcription (see Figure 15–33), others do depend on changes in the transcription of specific genes and take hours to develop fully. In cells that secrete the peptide hormone *somatostatin,* for example, cyclic AMP activates the gene that encodes this hormone. The regulatory region of the somatostatin gene contains a short DNA sequence, called the *cyclic AMP response element (CRE),* which is also found in the regulatory region of many other genes activated by cyclic AMP. A specific gene regulatory protein called **CRE-binding (CREB) protein** recognizes this sequence. When PKA is activated by cAMP, it phosphorylates CREB on a single serine; phosphorylated CREB then recruits a transcriptional coactivator called *CREB-binding protein (CBP),* which stimulates the transcription of the target genes (**Figure 15–36**). Thus, CREB can transform a short cyclic AMP signal into a long-term change in a cell, a process that, in the brain, is thought to play an important part in some forms of learning and memory.

PKA does not mediate all the effects of cyclic AMP in cells. As we discuss later, in olfactory neurons (responsible for the sense of smell), cyclic AMP also directly activates special ion channels in the plasma membrane. Moreover, in some other

Figure 15–35 The activation of cyclic-AMP-dependent protein kinase (PKA). The binding of cyclic AMP to the regulatory subunits of the PKA tetramer induces a conformational change, causing these subunits to dissociate from the catalytic subunits, thereby activating the kinase activity of the catalytic subunits. The release of the catalytic subunits requires the binding of more than two cyclic AMP molecules to the regulatory subunits in the tetramer. This requirement greatly sharpens the response of the kinase to changes in cyclic AMP concentration, as discussed earlier (see Figure 15–25). Mammalian cells have at least two types of PKAs: type I is mainly in the cytosol, whereas type II is bound via its regulatory subunits and special anchoring proteins to the plasma membrane, nuclear membrane, mitochondrial outer membrane, and microtubules. In both types, once the catalytic subunits are freed and active, they can migrate into the nucleus (where they can phosphorylate gene regulatory proteins), while the regulatory subunits remain in the cytoplasm. The three-dimensional structure of the protein kinase domain of the PKA catalytic subunit is shown in Figure 3–65.

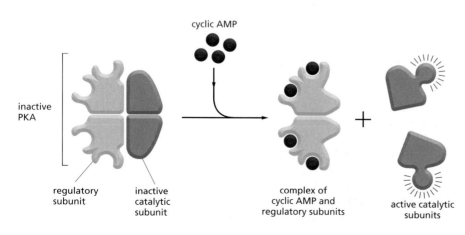

cyclic AMP

inactive PKA

regulatory subunit

inactive catalytic subunit

complex of cyclic AMP and regulatory subunits

active catalytic subunits

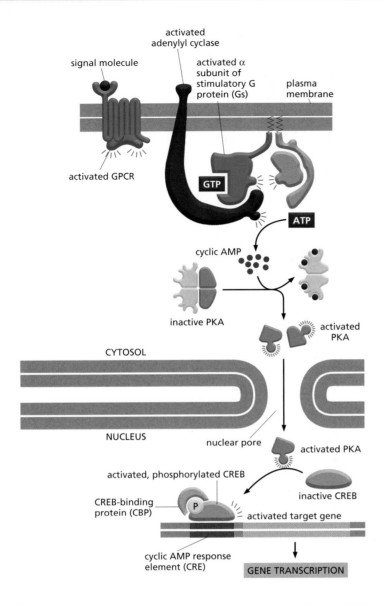

Figure 15–36 How a rise in intracellular cyclic AMP concentration can alter gene transcription. <AGAT> The binding of an extracellular signal molecule to its GPCR activates adenylyl cyclase via G_s and thereby increases cyclic AMP concentration in the cytosol. The rise in cyclic AMP concentration activates PKA, and the released catalytic subunits of PKA can then enter the nucleus, where they phosphorylate the gene regulatory protein CREB. Once phosphorylated, CREB recruits the coactivator CBP, which stimulates gene transcription. In some cases, at least, the inactive CREB protein is bound to the cyclic AMP response element (CRE) in DNA before it is phosphorylated (not shown).

This signaling pathway controls many processes in cells, ranging from hormone synthesis in endocrine cells to the production of proteins required for the induction of long-term memory in the brain. We will see later that CREB can also be activated by some other signaling pathways, independent of cAMP.

cells, it directly activates a guanine nucleotide exchange factor (GEF) that, in turn, activates a monomeric GTPase called *Rap1,* which often leads to increased cell adhesion through the activation of cell-surface *integrins* (discussed in Chapter 19).

Having discussed how trimeric G proteins link activated GPCRs to adenylyl cyclase, we now consider how they couple activated GPCRs to another crucial enzyme, *phospholipase C.* The activation of this enzyme increases the concentration of several small intracellular mediators, including Ca^{2+}, which help relay the signal onward. Ca^{2+} is even more widely used than cyclic AMP as a small intracellular mediator.

Some G Proteins Activate An Inositol Phospholipid Signaling Pathway by Activating Phospholipase C-β

Many GPCRs exert their effects mainly via G proteins that activate the plasma-membrane-bound enzyme **phospholipase C-β (PLCβ).** **Table 15–2** lists several examples of responses activated in this way. The phospholipase acts on a phosphorylated inositol phospholipid (a *phosphoinositide*) called **phosphatidylinositol 4,5-bisphosphate [PI(4,5)P_2, or PIP$_2$]**, which is present in small amounts in the inner half of the plasma membrane lipid bilayer (**Figure 15–37**). Receptors that activate this **inositol phospholipid signaling pathway** mainly do so via a G protein called **G_q**, which activates phospholipase C-β in much the same way that

Table 15–2 Some Cell Responses in Which GPCRs Activate PLCβ

TARGET TISSUE	SIGNAL MOLECULE	MAJOR RESPONSE
Liver	vasopressin	glycogen breakdown
Pancreas	acetylcholine	amylase secretion
Smooth muscle	acetylcholine	muscle contraction
Blood platelets	thrombin	platelet aggregation

G_s activates adenylyl cyclase. The activated phospholipase then cleaves the PIP_2 to generate two products: *inositol 1,4,5-trisphosphate (IP_3)* and *diacylglycerol* (**Figure 15–38**). At this step, the signaling pathway splits into two branches.

Inositol 1,4,5-trisphosphate (IP_3) is a water-soluble molecule that acts as a small intracellular mediator. It leaves the plasma membrane and diffuses rapidly through the cytosol. When it reaches the endoplasmic reticulum (ER), it binds to and opens **IP_3-gated Ca^{2+}-release channels** (also called **IP_3 receptors**) in the ER membrane. Ca^{2+} stored in the ER is released through the open channels, quickly raising the concentration of Ca^{2+} in the cytosol (**Figure 15–39**). Once the ER Ca^{2+} stores have been depleted, they are refilled by the activation of *store-operated Ca^{2+} channels* in the plasma membrane and a Ca^{2+} sensor protein in the ER membrane, in regions where the two membranes are closely apposed.

We discuss later how the increase in cytosolic Ca^{2+} propagates the signal by influencing the activity of Ca^{2+}-sensitive intracellular proteins. Several mechanisms operate to terminate the initial Ca^{2+} response: (1) IP_3 is rapidly dephosphorylated by specific lipid phosphatases to form IP_2; (2) IP_3 is phosphorylated by specific lipid kinases to form IP_4 (which may function as another small intracellular mediator); and (3) Ca^{2+} that enters the cytosol is rapidly pumped out, mainly to the exterior of the cell (see Figure 15–41).

At the same time that the IP_3 produced by the hydrolysis of PIP_2 is increasing the concentration of Ca^{2+} in the cytosol, the other cleavage product of the PIP_2, **diacylglycerol**, is exerting different effects. It also acts as a small intracellular mediator, but it remains embedded in the plasma membrane, where it has several potential signaling roles. It can be further cleaved to release arachidonic acid, which can either act as a signal in its own right or be used in the synthesis of other small lipid signal molecules called *eicosanoids*. Most vertebrate cell

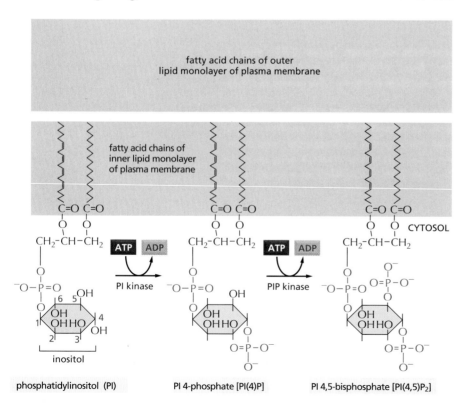

Figure 15–37 The synthesis of PI(4,5)P₂. The phosphoinositides PI(4)P and PI(4,5)P₂ are produced by the phosphorylation of phosphatidylinositol (PI) and PI(4)P, respectively. Although all three inositol phospholipids may be broken down in a signaling response, it is the breakdown of PI(4,5)P₂ that predominates and is most critical because it generates two intracellular mediators, as shown in Figures 15–38 and 15–39. Nevertheless, PI(4,5)P₂ is the least abundant, constituting less than 10% of the total inositol phospholipids and about 1% of the lipids in the plasma membrane. The conventional numbering of the carbon atoms in the inositol ring is shown in *red numbers* on the PI molecule.

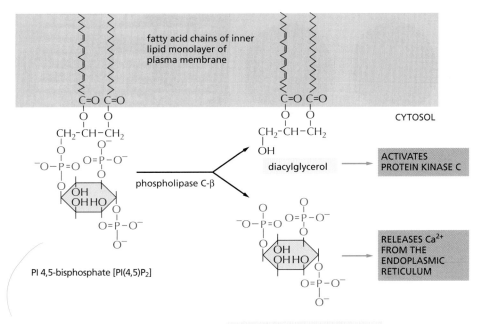

Figure 15–38 The hydrolysis of PI(4,5)P$_2$ by phospholipase C-β. Two small intracellular mediators are produced directly from the hydrolysis of the PIP$_2$: inositol 1,4,5-trisphosphate (IP$_3$), which diffuses through the cytosol and releases Ca^{2+} from the endoplasmic reticulum, and diacylglycerol, which remains in the membrane and helps to activate the enzyme protein kinase C (PKC; see Figure 15–39). There are several classes of phospholipase C; these include the β class, which is activated by GPCRs; as we see later, the γ class is activated by a class of enzyme-coupled receptors called receptor tyrosine kinases (RTKs).

types make eicosanoids, including *prostaglandins*, which have many biological activities. They participate in pain and inflammatory responses, for example, and most anti-inflammatory drugs (such as aspirin, ibuprofen, and cortisone) act—in part, at least—by inhibiting their synthesis.

A second function of diacylglycerol is to activate a crucial serine/threonine protein kinase called **protein kinase C (PKC)**, so named because it is Ca^{2+}-dependent. The initial rise in cytosolic Ca^{2+} induced by IP$_3$ alters the PKC so that it translocates from the cytosol to the cytoplasmic face of the plasma membrane. There it is activated by the combination of Ca^{2+}, diacylglycerol, and the negatively charged membrane phospholipid phosphatidylserine (see Figure 15–39). Once activated, PKC phosphorylates target proteins that vary depending on the cell type. The principles are the same as discussed earlier for PKA, although most of the target proteins are different.

There are various classes of PKCs, only some of which (called *conventional PKCs*) are activated by Ca^{2+} and diacylglycerol; the others are called *atypical PKCs*. Different PKCs phosphorylate different substrates mainly because different anchoring or scaffold proteins tether them to different compartments in the cell.

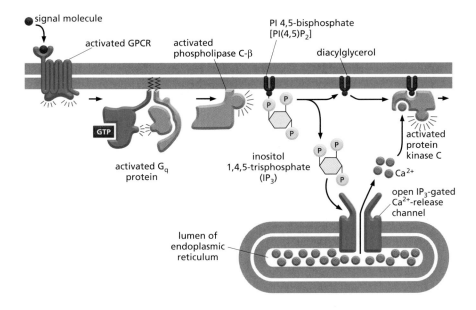

Figure 15–39 How GPCRs increase cytosolic Ca^{2+} and activate PKC. The activated GPCR stimulates the plasma-membrane-bound phospholipase PLCβ via a G protein. Depending on the isoform of the PLCβ, it may be activated by the α subunit of G$_q$ as shown, by the βγ subunits of another G protein, or by both. Two small intracellular messenger molecules are produced when PI(4,5)P$_2$ is hydrolyzed by activated PLCβ. Inositol 1,4,5-trisphosphate (IP$_3$) diffuses through the cytosol and releases Ca^{2+} from the ER by binding to and opening IP$_3$-gated Ca^{2+}-release channels (IP$_3$ receptors) in the ER membrane. The large electrochemical gradient for Ca^{2+} across this membrane causes Ca^{2+} to escape into the cytosol when the release channels are open. Diacylglycerol remains in the plasma membrane and, together with phosphatidylserine (not shown) and Ca^{2+}, helps to activate protein kinase C (PKC), which is recruited from the cytosol to the cytosolic face of the plasma membrane. Of the 10 or more distinct isoforms of PKC in humans, at least 4 are activated by diacylglycerol.

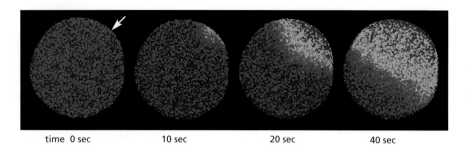

time 0 sec   10 sec   20 sec   40 sec

Figure 15–40 The fertilization of an egg by a sperm triggers an increase in cytosolic Ca²⁺. <AGGA> This starfish egg was injected with a Ca²⁺-sensitive fluorescent dye before it was fertilized. A wave of cytosolic Ca²⁺ *(red)*, released from the ER, sweeps across the egg from the site of sperm entry *(arrow)*. This Ca²⁺ wave changes the egg cell surface, preventing the entry of other sperm, and it also initiates embryonic development (discussed in Chapter 21). The initial increase in Ca²⁺ is thought to be caused by a sperm-specific form of PLC (PLCζ) that the sperm brings into the egg cytoplasm when it fuses with the egg; the PLCζ cleaves PI(4,5)P₂ to produce IP₃, which releases Ca²⁺ from the egg ER. (Courtesy of Stephen A. Stricker.)

Ca²⁺ Functions as a Ubiquitous Intracellular Mediator

Many extracellular signals trigger an increase in cytosolic Ca²⁺ concentration, not just those that work via G proteins. In egg cells, for example, a sudden rise in cytosolic Ca²⁺ concentration upon fertilization by a sperm triggers a Ca²⁺ wave that initiates embryonic development (**Figure 15–40**). In muscle cells, Ca²⁺ triggers contraction, and in many secretory cells, including nerve cells, it triggers secretion. Ca²⁺ can act as a signal in this way because its concentration in the cytosol is normally very low ($\sim 10^{-7}$ M), whereas its concentration in the extracellular fluid ($\sim 10^{-3}$ M) and in the lumen of the ER [and sarcoplasmic reticulum (SR) in muscle] is high. Thus, there is a large gradient tending to drive Ca²⁺ into the cytosol across both the plasma membrane and the ER or SR membrane. When a signal transiently opens Ca²⁺ channels in these membranes, Ca²⁺ rushes into the cytosol, increasing the local Ca²⁺ concentration 10–20-fold and activating Ca²⁺-responsive proteins in the cell. <CGTC>

The Ca²⁺ from outside the cell enters the cytosol through various Ca²⁺ channels in the plasma membrane, which open in response to ligand binding, stretch, or membrane depolarization. The Ca²⁺ from the ER enters the cytosol through either IP₃ receptors (see Figure 15–39) or **ryanodine receptors** (so called because they are sensitive to the plant alkaloid ryanodine). Ryanodine receptors are normally activated by Ca²⁺ binding and thereby amplify the Ca²⁺ signal. Ca²⁺ also activates IP₃ receptors but only in the presence of IP₃; and very high concentrations of Ca²⁺ inactivate them. We discuss how Ca²⁺ is released from the sarcoplasmic reticulum to cause the contraction of muscle cells in Chapter 16.

Several mechanisms keep the concentration of Ca²⁺ in the cytosol low in resting cells (**Figure 15–41**). All eucaryotic cells have a Ca²⁺-pump in their plasma membrane that uses the energy of ATP hydrolysis to pump Ca²⁺ out of the cytosol. Cells such as muscle and nerve cells, which make extensive use of Ca²⁺ signaling, have an additional Ca²⁺ transport protein (a Na⁺-driven Ca²⁺ exchanger) in their plasma membrane that couples the efflux of Ca²⁺ to the influx of Na⁺. A Ca²⁺ pump in the ER membrane also has an important role in keeping the cytosolic Ca²⁺ concentration low: this Ca²⁺-pump enables the ER to take up large amounts of Ca²⁺ from the cytosol against a steep concentration gradient, even when Ca²⁺ levels in the cytosol are low. In addition, a low-affinity, high-capacity Ca²⁺ pump in the inner mitochondrial membrane has an important role in limiting the Ca²⁺ signal and in terminating it; it uses the electrochemical gradient generated across this membrane during the electron-transfer steps of oxidative phosphorylation to take up Ca²⁺ from the cytosol. The resulting increase in Ca²⁺ concentration in the mitochondrion can activate some enzymes of the citric acid cycle, thereby increasing the synthesis of ATP and linking cell activation to energy production; an excessive increase in mitochondrial Ca²⁺, however, leads to cell death.

The Frequency of Ca²⁺ Oscillations Influences a Cell's Response

Researchers often use Ca²⁺-sensitive fluorescent indicators, such as *aequorin* or *fura-2* (discussed in Chapter 9), to monitor cytosolic Ca²⁺ in individual cells after activation of the inositol phospholipid signaling pathway. When viewed in this

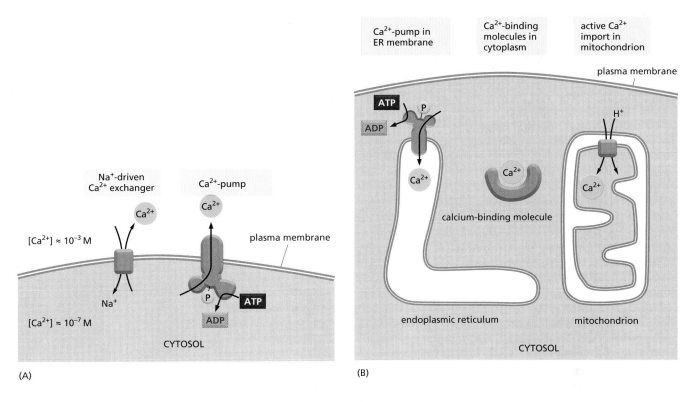

(A)

(B)

way, the initial Ca^{2+} signal appears small and localized to one or more discrete regions of the cell. These Ca^{2+} puffs or sparks reflect the local opening of individual or small groups of Ca^{2+}-release channels in the ER. Because various Ca^{2+}-binding proteins act as Ca^{2+} buffers and restrict the diffusion of Ca^{2+}, the signal often remains localized to the site where the Ca^{2+} entered the cytosol. If the extracellular signal is sufficiently strong and persistent, however, this localized Ca^{2+} signal can propagate as a regenerative Ca^{2+} wave through the cytosol (see Figure 15–40), much like an action potential in an axon. Such a Ca^{2+} "spike" is often followed by a series of further spikes, each usually lasting seconds (**Figure 15–42**). Such *Ca^{2+} oscillations* can persist for as long as receptors are activated at the cell surface. Both the waves and the oscillations are thought to depend, at least in part, on a combination of positive and negative feedback by Ca^{2+} on

Figure 15–41 The main ways in which eucaryotic cells maintain a very low concentration of free Ca^{2+} in their cytosol. (A) Ca^{2+} is actively pumped out of the cytosol to the cell exterior. (B) Ca^{2+} is pumped out of the cytosol into the ER and mitochondria, and various molecules in the cytosol bind free Ca^{2+} tightly.

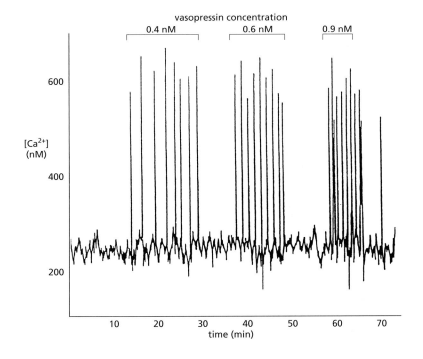

Figure 15–42 Vasopressin-induced Ca^{2+} oscillations in a liver cell. The cell was loaded with the Ca^{2+}-sensitive protein aequorin and then exposed to increasing concentrations of the peptide signal molecule *vasopressin*, which activates a GPCR and thereby PLCβ (see Tabe 15–2). Note that the frequency of the Ca^{2+} spikes increases with an increasing concentration of vasopressin but that the amplitude of the spikes is not affected. Each spike lasts about 7 seconds. (Adapted from N.M. Woods, K.S.R. Cuthbertson and P.H. Cobbold, *Nature* 319:600–602, 1986. With permission from Macmillan Publishers Ltd.)

both the IP_3 receptors and ryanodine receptors: the released Ca^{2+} initially stimulates more Ca^{2+} release from both receptors, a process known as *Ca²⁺-induced Ca²⁺ release;* but then, as its concentration gets high enough, the Ca^{2+} inhibits further release; and this delayed negative feedback gives rise to oscillations (see Figure 15–28D).

One can follow the effect of the Ca^{2+} oscillations on specific Ca^{2+}-sensitive proteins by using real-time imaging of individual cells expressing fluorescent reporter proteins. One can show, for example, that each Ca^{2+} spike in some signal-induced responses will recruit PKC transiently to the plasma membrane, where it will transiently phosphorylate a reporter protein.

The frequency of the Ca^{2+} oscillations reflects the strength of the extracellular stimulus (see Figure 15–42), and this frequency can be translated into a frequency-dependent cell response. In some cases, the frequency-dependent response itself is also oscillatory: in hormone-secreting pituitary cells, for example, stimulation by an extracellular signal induces repeated Ca^{2+} spikes, each of which is associated with a burst of hormone secretion. In other cases, the frequency-dependent response is non-oscillatory: in some types of cells, for instance, one frequency of Ca^{2+} spikes activates the transcription of one set of genes, while a higher frequency activates the transcription of a different set. How do cells sense the frequency of Ca^{2+} spikes and change their response accordingly? The mechanism presumably depends on Ca^{2+}-sensitive proteins that change their activity as a function of Ca^{2+} spike frequency. A protein kinase that acts as a molecular memory device seems to have this remarkable property, as we discuss next.

Ca²⁺/Calmodulin-Dependent Protein Kinases (CaM-Kinases) Mediate Many of the Responses to Ca²⁺ Signals in Animal Cells

Various Ca^{2+}-binding proteins help to relay the cytosolic Ca^{2+} signal. The most important is **calmodulin**, which is found in all eucaryotic cells and can constitute as much as 1% of the total protein mass. Calmodulin functions as a multipurpose intracellular Ca^{2+} receptor, governing many Ca^{2+}-regulated processes. It consists of a highly conserved, single polypeptide chain with four high-affinity Ca^{2+}-binding sites (**Figure 15–43**A). When activated by Ca^{2+} binding, it undergoes a conformational change. Because two or more Ca^{2+} ions must bind before calmodulin adopts its active conformation, the protein responds in an almost switchlike manner to increasing concentrations of Ca^{2+} (see Figure 15–25): a tenfold increase in Ca^{2+} concentration typically causes a fiftyfold increase in calmodulin activation.

The allosteric activation of calmodulin by Ca^{2+} is analogous to the allosteric activation of PKA by cyclic AMP, except that Ca^{2+}/calmodulin has no enzymatic

Figure 15–43 The structure of Ca²⁺/calmodulin based on x-ray diffraction and NMR studies. <CTTC> (A) The molecule has a dumbbell shape, with two globular ends, which can bind to many target proteins. The globular ends are connected by a long, exposed α helix, which allows the protein to adopt a number of very different conformations, depending on the target protein it interacts with. Each globular head has two Ca^{2+}-binding domains. (B) The major structural change in Ca^{2+}/calmodulin that occurs when it binds to a target protein (in this example, a peptide that consists of the Ca^{2+}/calmodulin-binding domain of a Ca^{2+}/calmodulin-dependent protein kinase). Note that the Ca^{2+}/calmodulin has "jack-knifed" to surround the peptide. When it binds to other targets, it can adopt different conformations. (A, based on x-ray crystallographic data from Y.S. Babu et al., *Nature* 315:37–40, 1985. With permission from Macmillan Publishers Ltd; B, based on x-ray crystallographic data from W.E. Meador, A.R. Means and F.A. Quiocho, *Science* 257:1251–1255, 1992, and on NMR data from M. Ikura et al., *Science* 256:632–638, 1992. With permission from AAAS.)

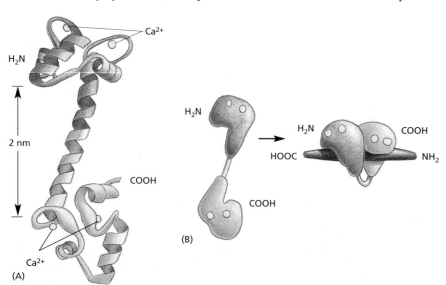

activity itself but instead acts by binding to and activating other proteins. In some cases, calmodulin serves as a permanent regulatory subunit of an enzyme complex, but usually the binding of Ca^{2+} instead enables calmodulin to bind to various target proteins in the cell to alter their activity.

When an activated molecule of Ca^{2+}/calmodulin binds to its target protein, the calmodulin further changes its conformation, the nature of which depends on the specific target protein (Figure 15–43B). Among the many targets calmodulin regulates are enzymes and membrane transport proteins. As one example, Ca^{2+}/calmodulin binds to and activates the plasma membrane Ca^{2+}-pump that uses ATP hydrolysis to pump Ca^{2+} out of cells (see Figure 15–41). Thus, whenever the concentration of Ca^{2+} in the cytosol rises, the pump is activated, which helps to return the cytosolic Ca^{2+} level to resting levels.

Many effects of Ca^{2+}, however, are more indirect and are mediated by protein phosphorylations catalyzed by a family of serine/threonine protein kinases called **Ca^{2+}/calmodulin-dependent kinases (CaM-kinases)**. Some CaM-kinases phosphorylate gene regulatory proteins, such as the CREB protein (see Figure 15–36), and in this way activate or inhibit the transcription of specific genes.

One of the best-studied CaM-kinases is **CaM-kinase II**, which is found in most animal cells but is especially enriched in the nervous system. It constitutes up to 2% of the total protein mass in some regions of the brain, and it is highly concentrated in synapses. CaM-kinase II has two remarkable properties that are related. First, it can function as a molecular memory device, switching to an active state when exposed to Ca^{2+}/calmodulin and then remaining active even after the Ca^{2+} signal has decayed. This is because the kinase phosphorylates itself (a process called *autophosphorylation*), as well as other cell proteins, when Ca^{2+}/calmodulin activates it. In its autophosphorylated state, the enzyme remains active even in the absence of Ca^{2+}, thereby prolonging the duration of the kinase activity beyond that of the initial activating Ca^{2+} signal. The enzyme maintains this activity until serine/threonine protein phosphatases overwhelm the autophosphorylation and shut the kinase off (**Figure 15–44**). CaM-kinase II activation can thereby serve as a memory trace of a prior Ca^{2+} pulse, and it seems to have a role in some types of memory and learning in the vertebrate nervous system. Mutant mice that lack a brain-specific form of the enzyme have specific defects in their ability to remember where things are.

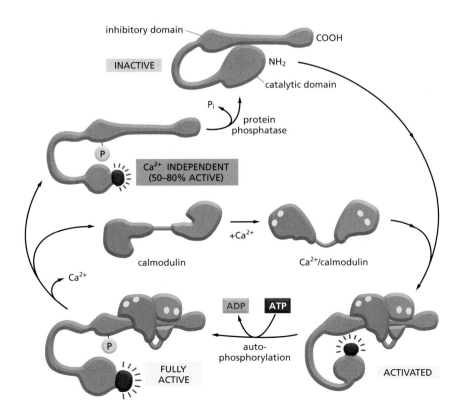

Figure 15–44 The stepwise activation of CaM-kinase II. The enzyme is a large protein complex of 12 subunits, although, for simplicity, only one subunit is shown (in *gray*). In the absence of Ca^{2+}/calmodulin, the enzyme is inactive as the result of an interaction between an inhibitory domain and the catalytic domain. Ca^{2+}/calmodulin binding alters the conformation of the protein, partially activating it. The catalytic domains in the complex phosphorylate the inhibitory domains of neighboring subunits, as well as other proteins in the cell (not shown). The autophosphorylation of the enzyme complex (by mutual phosphorylation of its subunits) fully activates the enzyme. It also prolongs the activity of the enzyme in two ways. First, it traps the bound Ca^{2+}/calmodulin so that it does not dissociate from the enzyme complex until cytosolic Ca^{2+} levels return to basal values for at least 10 seconds (not shown). Second, it converts the enzyme to a Ca^{2+}-independent form so that the kinase remains active even after the Ca^{2+}/calmodulin dissociates from it. This activity continues until the action of a protein phosphatase overrides the autophosphorylation activity of CaM-kinase II.

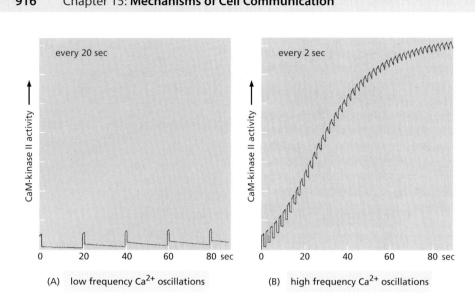

Figure 15–45 CaM-kinase II as a frequency decoder of Ca²⁺ oscillations. (A) At low frequencies of Ca²⁺ spikes, the enzyme becomes inactive after each spike, as the autophosphorylation induced by Ca²⁺/calmodulin binding does not maintain the enzyme's activity long enough for the enzyme to remain active until the next Ca²⁺ spike arrives. (B) At higher spike frequencies, however, the enzyme fails to inactivate completely between Ca²⁺ spikes, so its activity ratchets up with each spike. If the spike frequency is high enough, this progressive increase in enzyme activity will continue until the enzyme is autophosphorylated on all subunits and is therefore maximally activated. Although not shown, once enough of its subunits are autophosphorylated, the enzyme can be maintained in a highly active state even with a relatively low frequency of Ca²⁺ spikes (a form of cell memory).

The binding of Ca²⁺/calmodulin to the enzyme is enhanced by the CaM-kinase II autophosphorylation (an additional form of positive feedback), helping the response of the enzyme to repeated Ca²⁺ spikes to exhibit a steep threshold in its frequency response, as discussed earlier. (From P.I. Hanson, T. Meyer, L. Stryer and H. Schulman, *Neuron* 12:943–956, 1994. With permission from Elsevier.)

The second remarkable property of CaM-kinase II follows from the first, in that the enzyme can use its intrinsic memory mechanism to act as a frequency decoder of Ca²⁺ oscillations. This property is thought to be especially important at a nerve cell synapse, where changes in intracellular Ca²⁺ levels in a postsynaptic cell as a result of neural activity can lead to long-term changes in the subsequent effectiveness of that synapse (discussed in Chapter 11). When CaM-kinase II is immobilized on a solid surface and exposed to both a protein phosphatase and repetitive pulses of Ca²⁺/calmodulin at different frequencies that mimic those observed in stimulated cells, the enzyme's activity increases steeply as a function of pulse frequency (**Figure 15–45**). Moreover, the frequency response of this multisubunit enzyme depends on its exact subunit composition, so that a cell can tailor its response to Ca²⁺ oscillations to particular needs by adjusting the composition of the CaM-kinase II enzyme that it makes.

Some G Proteins Directly Regulate Ion Channels

G proteins do not act exclusively by regulating the activity of membrane-bound enzymes that alter the concentration of cyclic AMP or Ca²⁺ in the cytosol. The α subunit of one type of G protein (called G_{12}), for example, activates a guanine nucleotide exchange factor (GEF) that activates a monomeric GTPase of the *Rho family* (discussed later and in Chapter 16), which regulates the actin cytoskeleton.

In some other cases, G proteins directly activate or inactivate ion channels in the plasma membrane of the target cell, thereby altering the ion permeability—and hence the electrical excitability—of the membrane. As an example, acetylcholine released by the vagus nerve reduces both the rate and strength of heart muscle cell contraction (see Figure 15–9B). This effect is mediated by a special class of acetylcholine receptors that activate the G_i protein discussed earlier. Once activated, the α subunit of G_i inhibits adenylyl cyclase (as described previously), while the βγ subunits bind to K⁺ channels in the heart muscle cell plasma membrane and open them. The opening of these K⁺ channels makes it harder to depolarize the cell and thereby contributes to the inhibitory effect of acetylcholine on the heart. (These acetylcholine receptors, which can be activated by the fungal alkaloid muscarine, are called *muscarinic acetylcholine receptors* to distinguish them from the very different *nicotinic acetylcholine receptors*, which are ion-channel-coupled receptors on skeletal muscle and nerve cells that can be activated by the binding of nicotine, as well as by acetylcholine.)

Other G proteins regulate the activity of ion channels less directly, either by stimulating channel phosphorylation (by PKA, PKC, or CaM-kinase, for example) or by causing the production or destruction of cyclic nucleotides that directly activate or inactivate ion channels. These *cyclic-nucleotide-gated ion channels* have a crucial role in both smell (olfaction) and vision, as we now discuss.

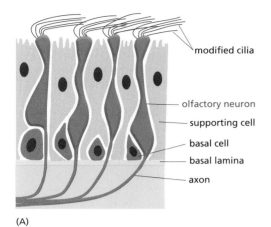

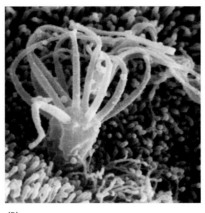

(A)   (B)

Figure 15–46 Olfactory receptor neurons.
(A) A section of olfactory epithelium in the nose. Olfactory receptor neurons possess modified cilia, which project from the surface of the epithelium and contain the olfactory receptors, as well as the signal transduction machinery. The axon, which extends from the opposite end of the receptor neuron, conveys electrical signals to the brain when an odorant activates the cell to produce an action potential. In rodents, at least, the basal cells act as stem cells, producing new receptor neurons throughout life, to replace the neurons that die. (B) A scanning electron micrograph of the cilia on the surface of an olfactory neuron. (B, from E.E. Morrison and R.M. Costanzo, *J. Comp. Neurol.* 297:1–13, 1990. With permission from Wiley-Liss.)

Smell and Vision Depend on GPCRs That Regulate Cyclic-Nucleotide-Gated Ion Channels

Humans can distinguish more than 10,000 distinct smells, which they detect using specialized olfactory receptor neurons in the lining of the nose. These cells use specific GPCRs called **olfactory receptors** to recognize odors; the receptors are displayed on the surface of the modified cilia that extend from each cell (**Figure 15–46**). The receptors act through cyclic AMP. When stimulated by odorant binding, they activate an olfactory-specific G protein (known as G_{olf}), which in turn activates adenylyl cyclase. The resulting increase in cyclic AMP opens *cyclic-AMP-gated cation channels*, thereby allowing an influx of Na^+, which depolarizes the olfactory receptor neuron and initiates a nerve impulse that travels along its axon to the brain.

There are about 1000 different olfactory receptors in a mouse and about 350 in a human, each encoded by a different gene and each recognizing a different set of odorants. Each olfactory receptor neuron produces only one of these receptors (see p. 453); the neuron responds to a specific set of odorants by means of the specific receptor it displays, and each odorant activates its own characteristic set of olfactory receptor neurons. The same receptor also helps direct the elongating axon of each developing olfactory neuron to the specific target neurons that it will connect to in the brain. A different set of GPCRs acts in a similar way to mediate responses to *pheromones,* chemical signals detected in a different part of the nose that are used in communication between members of the same species. Humans, however, lack functional pheromone receptors.

Vertebrate vision employs a similarly elaborate, highly sensitive, signal detection process. Cyclic-nucleotide-gated ion channels are also involved, but the crucial cyclic nucleotide is **cyclic GMP** (**Figure 15–47**) rather than cyclic AMP. As with cyclic AMP, a continuous rapid synthesis (by *guanylyl cyclase*) and rapid degradation (by *cyclic GMP phosphodiesterase*) controls the concentration of cyclic GMP in the cytosol.

In visual transduction responses, which are the fastest G-protein-mediated responses known in vertebrates, the receptor activation stimulated by light causes a fall rather than a rise in the level of the cyclic nucleotide. The pathway has been especially well studied in **rod photoreceptors** (**rods**) in the vertebrate retina. Rods are responsible for noncolor vision in dim light, whereas *cone photoreceptors (cones)* are responsible for color vision in bright light. A rod photoreceptor is a highly specialized cell with outer and inner segments, a cell body, and a synaptic region where the rod passes a chemical signal to a retinal nerve cell (**Figure 15–48**). This nerve cell relays the signal to another nerve cell in the retina, which in turn relays it to the brain (see Figure 23–16).

The phototransduction apparatus is in the outer segment of the rod, which contains a stack of *discs*, each formed by a closed sac of membrane that has many embedded photosensitive **rhodopsin** molecules. The plasma membrane surrounding the outer segment contains *cyclic-GMP-gated cation channels.* Cyclic GMP bound to these channels keeps them open in the dark. Paradoxically,

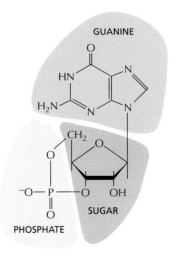

Figure 15–47 Cyclic GMP.

Figure 15–48 A rod photoreceptor cell. There are about 1000 discs in the outer segment. The disc membranes are not connected to the plasma membrane. The inner and outer segments are specialized parts of a *primary cilium* (discussed in Chapter 16); as mentioned earlier and discussed later, a primary cilium extends from the surface of most vertebrate cells, where it serves as a signaling organelle.

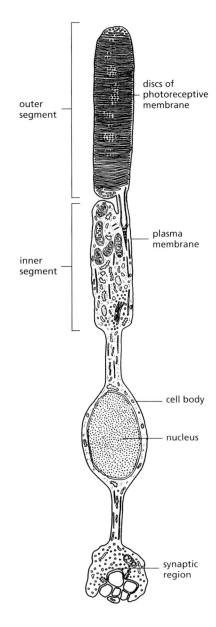

light causes a hyperpolarization (which inhibits synaptic signaling) rather than a depolarization of the plasma membrane (which would stimulate synaptic signaling). Hyperpolarization (that is, the membrane potential moves to a more negative value—discussed in Chapter 11) results because the light-induced activation of rhodopsin molecules in the disc membrane decreases the cyclic GMP concentration and closes the cation channels in the surrounding plasma membrane (**Figure 15–49**).

Rhodopsin is a member of the GPCR family, but the activating extracellular signal is not a molecule but a photon of light. Each rhodopsin molecule contains a covalently attached chromophore, 11-*cis* retinal, which isomerizes almost instantaneously to all-*trans* retinal when it absorbs a single photon. The isomerization alters the shape of the retinal, forcing a conformational change in the protein (opsin). The activated rhodopsin molecule then alters the conformation of the G-protein *transducin* (G_t), causing the transducin α subunit to activate **cyclic GMP phosphodiesterase**. The phosphodiesterase then hydrolyzes cyclic GMP, so that cyclic GMP levels in the cytosol fall. This drop in cyclic GMP concentration decreases the amount of cyclic GMP bound to the plasma membrane cation channels, allowing more of these cyclic-GMP-sensitive channels to close. In this way, the signal quickly passes from the disc membrane to the plasma membrane, and a light signal is converted into an electrical one, through a hyperpolarization of the rod cell plasma membrane.

Rods use several negative feedback loops to allow the cells to revert quickly to a resting, dark state in the aftermath of a flash of light—a requirement for perceiving the shortness of the flash. A rhodopsin-specific kinase called *rhodopsin kinase (RK)* phosphorylates the cytosolic tail of activated rhodopsin on multiple serines, partially inhibiting the ability of the rhodopsin to activate transducin. An

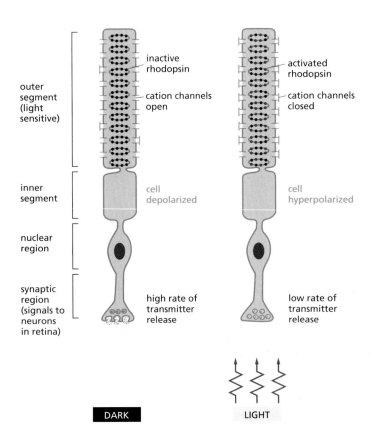

Figure 15–49 The response of a rod photoreceptor cell to light. Rhodopsin molecules in the outer-segment discs absorb photons. Photon absorption closes cation channels in the plasma membrane, which hyperpolarizes the membrane and reduces the rate of neurotransmitter release from the synaptic region. Because the neurotransmitter inhibits many of the postsynaptic retinal neurons, illumination serves to free the neurons from inhibition and thus, in effect, excites them.

Table 15–3 Four Major Families of Trimeric G Proteins*

FAMILY	SOME FAMILY MEMBERS	SUBUNITS THAT MEDIATE ACTION	SOME FUNCTIONS
I	G_s	α	activates adenylyl cyclase; activates Ca^{2+} channels
	G_{olf}	α	activates adenylyl cyclase in olfactory sensory neurons
II	G_i	α	inhibits adenylyl cyclase
		$\beta\gamma$	activates K^+ channels
	G_o	$\beta\gamma$	activates K^+ channels; inactivates Ca^{2+} channels
		α and $\beta\gamma$	activates phospholipase C-β
	G_t (transducin)	α	activates cyclic GMP phosphodiesterase in vertebrate rod photoreceptors
III	G_q	α	activates phospholipase C-β
IV	$G_{12/13}$	α	activates Rho family monomeric GTPases (via Rho-GEF) to regulate the actin cytoskeleton

*Families are determined by amino acid sequence relatedness of the α subunits. Only selected examples are included. About 20 α subunits and at least 6 β subunits and 11 γ subunits have been described in humans.

inhibitory protein called *arrestin* then binds to the phosphorylated rhodopsin, further inhibiting rhodopsin's activity. Mice or humans with a mutation that inactivates the gene encoding RK have a prolonged light response, and their rods eventually die.

At the same time as arrestin shuts off rhodopsin, an RGS protein (see p. 896) binds to activated transducin, stimulating the transducin to hydrolyze its bound GTP to GDP, which returns transducin to its inactive state. In addition, the cation channels that close in response to light are permeable to Ca^{2+}, as well as to Na^+, so that when they close, the normal influx of Ca^{2+} is inhibited, causing the Ca^{2+} concentration in the cytosol to fall. The decrease in Ca^{2+} concentration stimulates guanylyl cyclase to replenish the cyclic GMP, rapidly returning its level to where it was before the light was switched on. A specific Ca^{2+}-sensitive protein mediates the activation of guanylyl cyclase in response to a fall in Ca^{2+} levels. In contrast to calmodulin, this protein is inactive when Ca^{2+} is bound to it and active when it is Ca^{2+}-free. It therefore stimulates the cyclase when Ca^{2+} levels fall following a light response.

Negative feedback mechanisms do more than just return the rod to its resting state after a transient light flash; they also help the rod to *adapt*, stepping down the response when the rod is exposed to light continuously. Adaptation, as we discussed earlier, allows the receptor cell to function as a sensitive detector of *changes* in stimulus intensity over an enormously wide range of baseline levels of stimulation. It is why we can see a camera flash in broad daylight.

The various trimeric G proteins we have discussed in this chapter fall into four major families, as summarized in **Table 15–3**.

Intracellular Mediators and Enzymatic Cascades Amplify Extracellular Signals

Despite the differences in molecular details, the different intracellular signaling pathways that GPCRs trigger share certain features and obey similar general principles. They depend on relay chains of intracellular signaling proteins and small intracellular mediators. In contrast to the more direct signaling pathways used by nuclear receptors discussed earlier, these relay chains provide numerous opportunities for amplifying the responses to extracellular signals. In the visual transduction cascade, for example, a single activated rhodopsin molecule catalyzes the activation of hundreds of molecules of transducin at a rate of about 1000 transducin molecules per second. Each activated transducin molecule activates a molecule of cyclic GMP phosphodiesterase, each of which hydrolyzes about 4000 molecules of cyclic GMP per second. This catalytic cascade lasts for about 1 second and results in the hydrolysis of more than 10^5 cyclic GMP molecules for a single quantum of light absorbed, and the resulting drop in the concentration of cyclic GMP in turn transiently closes hundreds of cation channels in the plasma membrane (**Figure 15–50**). As a result, a rod cell can respond to even a single photon of light in a way that is highly reproducible in its timing and magnitude.

Likewise, when an extracellular signal molecule binds to a receptor that indirectly activates adenylyl cyclase via G_s, each receptor protein may activate many molecules of G_s protein, each of which can activate a cyclase molecule. Each cyclase molecule, in turn, can catalyze the conversion of a large number of ATP molecules to cyclic AMP molecules. A similar amplification operates in the inositol phospholipid signaling pathway. In these ways, a nanomolar (10^{-9} M) change in the concentration of an extracellular signal can induce micromolar (10^{-6} M) changes in the concentration of a small intracellular mediator such as cyclic AMP or Ca^{2+}. Because these mediators function as allosteric effectors to activate specific enzymes or ion channels, a single extracellular signal molecule can alter many thousands of protein molecules within the target cell.

Any such amplifying cascade of stimulatory signals requires counterbalancing mechanisms at every step of the cascade to restore the system to its resting state when stimulation ceases. As emphasized earlier, the response to stimulation can be rapid only if the inactivating mechanisms are also rapid. Cells therefore have efficient mechanisms for rapidly degrading (and resynthesizing) cyclic nucleotides and for buffering and removing cytosolic Ca^{2+}, as well as for inactivating the responding enzymes and ion channels once they have been activated. This is not only essential for turning a response off, it is also important for defining the resting state from which a response begins.

Each protein in the signaling relay chain can be a separate target for regulation, including the receptor itself, as we discuss next.

GPCR Desensitization Depends on Receptor Phosphorylation

When target cells are exposed to a high concentration of a stimulating ligand for a prolonged period, they can become *desensitized*, or *adapted*, in several different ways. An important class of desensitization mechanisms depend on alteration of the quantity or condition of the receptor molecules themselves.

For GPCRs, there are three general modes of desensitization (see Figure 15–29): (1) In *receptor inactivation*, they become altered so that they can no longer interact with G proteins. (2) In *receptor sequestration*, they are temporarily moved to the interior of the cell (internalized) so that they no longer have access to their ligand. (3) In *receptor down-regulation*, they are destroyed in lysosomes after internalization.

In each case, the desensitization of the GPCRs depends on their phosphorylation by PKA, PKC, or a member of the family of **GPCR kinases** (**GRKs**), which includes the rhodopsin-specific kinase RK involved in rod photoreceptor desensitization discussed earlier. The GRKs phosphorylate multiple serines and threonines on a GPCR, but they do so only after ligand binding has activated the receptor, because it is the activated receptor that allosterically activates the GRK. As with rhodopsin, once a receptor has been phosphorylated by a GRK, it binds with high affinity to a member of the **arrestin** family of proteins (**Figure 15–51**).

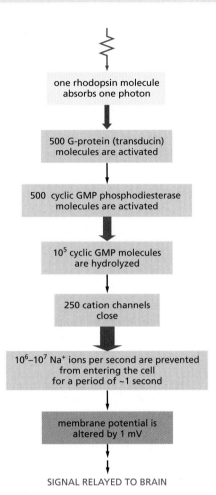

Figure 15–50 Amplification in the light-induced catalytic cascade in vertebrate rods. The *red arrows* indicate the steps where amplification occurs, with the thickness of the arrow roughly indicating the magnitude of the amplification.

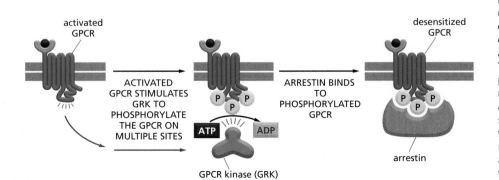

GPCR kinase (GRK)

Figure 15–51 The roles of GPCR kinases (GRKs) and arrestins in GPCR desensitization. A GRK phosphorylates only activated receptors because it is the activated GPCR that activates the GRK. The binding of an arrestin to the phosphorylated receptor prevents the receptor from binding to its G protein and also directs its endocytosis (not shown). Mice that are deficient in one form of arrestin fail to desensitize in response to morphine, for example, attesting to the importance of arrestins for desensitization.

The bound arrestin can contribute to the desensitization process in at least two ways. First, it prevents the activated receptor from interacting with G proteins. Second, it serves as an adaptor protein to help couple the receptor to the clathrin-dependent endocytosis machinery (discussed in Chapter 13), inducing receptor-mediated endocytosis. The fate of the internalized GPCR–arrestin complex depends on other proteins in the complex. In some cases, the receptor is dephosphorylated and recycled back to the plasma membrane for reuse. In others, it is ubiquitylated and degraded in lysosomes (discussed later).

Receptor endocytosis does not necessarily stop the receptor from signaling. In some cases, the bound arrestin recruits other signaling proteins to relay the signal onward along new pathways from the internalized GPCRs.

Summary

GPCRs can indirectly activate or inactivate either plasma-membrane-bound enzymes or ion channels via G proteins. When an activated receptor stimulates a G protein, the G protein undergoes a conformational change that activates both its α subunit and its βγ subunits, which can then directly regulate the activity of target proteins in the plasma membrane. Some GPCRs either activate or inactivate adenylyl cyclase, thereby altering the intracellular concentration of the small intracellular mediator cyclic AMP. Others activate a phosphoinositide-specific phospholipase C (PLCβ), which hydrolyzes phosphatidylinositol 4,5-bisphosphate [PI(4,5)P₂] to generate two small intracellular mediators. One is inositol 1,4,5-trisphosphate (IP₃), which releases Ca²⁺ from the ER and thereby increases the concentration of Ca²⁺ in the cytosol. The other is diacylglycerol, which remains in the plasma membrane and helps activate protein kinase C (PKC). An increase in cytosolic cyclic AMP or Ca²⁺ levels affects cells mainly by stimulating protein kinase A (PKA) and Ca²⁺/calmodulin-dependent protein kinases (CaM-kinases), respectively.

PKC, PKA, and CaM-kinases phosphorylate specific target proteins on serines or threonines and thereby alter the activity of the proteins. Each type of cell has its own characteristic set of target proteins that is regulated in these ways, enabling the cell to make its own distinctive responses to the small intracellular mediators. The intracellular signaling cascades activated by GPCRs greatly amplify the responses, so that many thousands of target protein molecules are changed for each molecule of extracellular signaling ligand bound to its receptor.

The responses mediated by GPCRs are rapidly turned off when the extracellular signal is removed. Thus, the G-protein α subunit is stimulated by its target protein or an RGS to inactivate itself by hydrolyzing its bound GTP to GDP, IP₃ is rapidly dephosphorylated by a lipid phosphatase or phosphorylated by a lipid kinase, cyclic nucleotides are hydrolyzed by phosphodiesterases, Ca²⁺ is rapidly pumped out of the cytosol, and phosphorylated proteins are dephosphorylated by protein phosphatases. Activated GPCRs themselves are phosphorylated by GRKs, thereby trigging arrestin binding. The bound arrestin uncouples the receptors from G proteins and promotes their endocytosis, resulting in either desensitization or continued signaling via arrestin-recruited signaling proteins.

SIGNALING THROUGH ENZYME-COUPLED CELL-SURFACE RECEPTORS

Like GPCRs, **enzyme-coupled receptors** are transmembrane proteins with their ligand-binding domain on the outer surface of the plasma membrane. Instead of having a cytosolic domain that associates with a trimeric G protein, however, their cytosolic domain either has intrinsic enzyme activity or associates directly with an enzyme. Whereas a GPCR has seven transmembrane segments, each subunit of an enzyme-coupled receptor usually has only one. GPCRs and enzyme-coupled receptors often activate some of the same signaling pathways, and there is usually no obvious reason why a particular extracellular signal utilizes one class of receptors rather than the other.

There are six principal classes of enzyme-coupled receptors:

1. *Receptor tyrosine kinases* directly phosphorylate specific tyrosines on themselves and on a small set of intracellular signaling proteins.
2. *Tyrosine-kinase-associated receptors* have no intrinsic enzyme activity but directly recruit cytoplasmic tyrosine kinases to relay the signal.
3. *Receptor serine/threonine kinases* directly phosphorylate specific serines or threonines on themselves and on latent gene regulatory proteins with which they are associated.
4. *Histidine-kinase-associated receptors* activate a two-component signaling pathway in which the kinase phosphorylates itself on histidine and then immediately transfers the phosphoryl group to a second intracellular signaling protein.
5. *Receptor guanylyl cyclases* directly catalyze the production of cyclic GMP in the cytosol, which acts as a small intracellular mediator in much the same way as cyclic AMP.
6. *Receptorlike tyrosine phosphatases* remove phosphate groups from tyrosines of specific intracellular signaling proteins. (They are called "receptorlike" because their presumptive ligands have not yet been identified, and so their receptor function is unproven.)

We focus our discussion on the first four classes, beginning with the receptor tyrosine kinases, the most numerous of the enzyme-coupled receptors.

Activated Receptor Tyrosine Kinases (RTKs) Phosphorylate Themselves

Many extracellular signal proteins act through **receptor tyrosine kinases** (**RTKs**). Notable examples discussed elsewhere in this book include *epidermal growth factor (EGF), platelet-derived growth factor (PDGF), fibroblast growth factors (FGFs), hepatocyte growth factor (HGF), insulin, insulinlike growth factor-1 (IGF1), vascular endothelial growth factor (VEGF), macrophage-colony-stimulating factor (MCSF)*, and the *neurotrophins*, including *nerve growth factor (NGF)*.

Many cell-surface-bound extracellular signal proteins also act through RTKs. **Ephrins** are the largest class of such membrane-bound ligands, with eight identified in humans. Among their many functions, they help guide the migration of cells and axons along specific pathways during animal development by stimulating responses that result in attraction or repulsion (discussed later and in Chapter 22). The receptors for ephrins, called **Eph receptors**, are also among the most numerous RTKs, with thirteen genes encoding them in humans. The ephrins and Eph receptors are unusual in that they can simultaneously act as both ligand and receptor: on binding to an Eph receptor, some ephrins not only activate the Eph receptor but also become activated themselves to transmit signals into the interior of the ephrin-expressing cell; in this way, the behavior of both the signaling cell and the target cell is altered. Such *bidirectional signaling* between ephrins and Eph receptors is required, for example, to keep some neighboring groups of cells from mixing with each other during development.

There are about 60 genes encoding human RTKs. These receptors can be classified into more than 16 structural subfamilies, each dedicated to its complementary family of protein ligands. **Figure 15–52** shows a number of the families that operate in mammals, and **Table 15–4** lists some of their ligands and functions. In all cases, the binding of the signal protein to the ligand-binding domain on the outside of the cell enables the intracellular tyrosine kinase domain to phosphorylate selected tyrosine side chains, both on the receptor proteins themselves and on intracellular signaling proteins that subsequently bind to the phosphorylated tyrosines on the receptors.

How does the binding of an extracellular ligand activate the kinase domain on the other side of the plasma membrane? For a GPCR, ligand binding is thought to change the relative orientation of several of the transmembrane α helices, thereby shifting the position of the cytoplasmic loops relative to one another. It is difficult to imagine, however, how a conformational change could propagate across the lipid bilayer through a single transmembrane α helix.

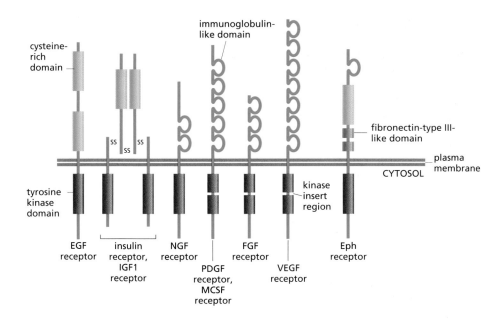

Figure 15–52 Some subfamilies of RTKs. Only one or two members of each subfamily are indicated. Note that the tyrosine kinase domain is interrupted by a "kinase insert region" in some of the subfamilies. The functional roles of most of the cysteine-rich, immunoglobulin-like, and fibronectin-type III-like domains are not known. Some of the ligands for the receptors shown are listed in Table 15–4, along with some representative responses that they mediate.

Instead, for many RTKs, ligand binding causes the receptor chains to dimerize, bringing the kinase domains of two receptor chains together (an example of *induced proximity*, discussed earlier) so that they can become activated and cross-phosphorylate each other on multiple tyrosines, a process referred to as *transautophosphorylation* (**Figure 15–53A**).

Because of the requirement for receptor dimerization, it is relatively easy to inactivate a specific RTK to determine its importance for a cell response. For this purpose, cells are transfected with DNA encoding a mutant form of the receptor that dimerizes normally but has an inactive kinase domain. When coexpressed at a high level with normal receptors, the mutant receptor acts in a *dominant-negative* way, disabling the normal receptors by forming inactive dimers with them (Figure 15–53B).

Phosphorylated Tyrosines on RTKs Serve as Docking Sites for Intracellular Signaling Proteins

The cross-phosphorylation of adjacent cytosolic tails of RTKs contributes to the receptor activation process in two ways. First, phosphorylation of tyrosines within the kinase domain increases the kinase activity of the enzyme. Second,

Table 15–4 Some Signal Proteins That Act Via RTKs

SIGNAL PROTEIN	RECEPTORS	SOME REPRESENTATIVE RESPONSES
Epidermal growth factor (EGF)	EGF receptors	stimulates cell survival, growth, proliferation, or differentiation of various cell types; acts as inductive signal in development
Insulin	insulin receptor	stimulates carbohydrate utilization and protein synthesis
Insulin-like growth factors (IGF1 and IGF2)	IGF receptor-1	stimulate cell growth and survival in many cell types
Nerve growth factor (NGF)	Trk A	stimulates survival and growth of some neurons
Platelet-derived growth factors (PDGF AA, BB, AB)	PDGF receptors (α and β)	stimulate survival, growth, proliferation, and migration of various cell types
Macrophage-colony-stimulating factor (MCSF)	MCSF receptor	stimulates monocyte/macrophage proliferation and differentiation
Fibroblast growth factors (FGF1 to FGF24)	FGF receptors (FGFR1–FGFR4, plus multiple isoforms of each)	stimulate proliferation of various cell types; inhibit differentiation of some precursor cells; act as inductive signals in development
Vascular endothelial growth factor (VEGF)	VEGF receptors	stimulates angiogenesis
Ephrins (A and B types)	Eph receptors (A and B types)	stimulate angiogenesis; guide cell and axon migration

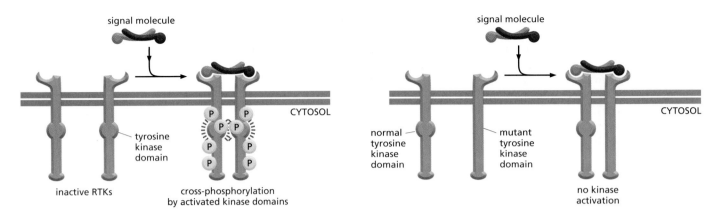

(A)   NORMAL RTK ACTIVATION

(B)   DOMINANT-NEGATIVE INHIBITION BY MUTANT RTK

phosphorylation of tyrosines outside the kinase domain creates high-affinity docking sites for the binding of specific intracellular signaling proteins. Each of the signaling proteins binds to specific phosphorylated sites on the activated receptors because it contains a specific phosphotyrosine-binding domain that recognizes surrounding features of the polypeptide chain in addition to the phosphotyrosine.

Once bound to the activated RTK, a signaling protein may itself become phosphorylated on tyrosines and thereby be activated. In many cases, however, the binding alone may be sufficient to activate the docked signaling protein, by either inducing a conformational change in the protein or simply bringing it near the protein that is next in the signaling pathway. Thus, transautophosphorylation serves as a switch to trigger the transient assembly of an intracellular signaling complex, which can then relay the signal onward, often along multiple routes, to various destinations in the cell (**Figure 15–54**). Because different RTKs bind different combinations of these signaling proteins, they activate different responses.

The receptors for insulin and IGF1 act in a slightly different way. They are tetramers (see Figure 15–52), and ligand binding is thought to rearrange their transmembrane receptor chains, moving the two kinase domains close together. Moreover, most of the phosphotyrosine docking sites generated by ligand binding are not on the receptor itself but on a specialized docking protein called *insulin receptor substrate-1 (IRS1)*. The activated receptor first transautophosphorylates its kinase domains, which then phosphorylate IRS1 on multiple tyrosines, thereby creating many more docking sites than could be accommodated on the receptor alone (see Figure 15–22). Some other RTKs use docking proteins in a similar way to enlarge the size of the signaling complex.

Figure 15–53 Activation and inactivation of RTKs by dimerization. (A) The normal receptors dimerize in response to ligand binding. The two kinase domains cross-phosphorylate each other, increasing the activity of the kinase domains, which now phosphorylate other sites on the receptors. (B) The mutant receptor with an inactivated kinase domain can dimerize normally, but it cannot cross-phosphorylate a normal receptor in a dimer. For this reason, the mutant receptors, if present in excess, will block signaling by the normal receptors—a process called *dominant-negative regulation*. Cell biologists frequently use this strategy to inhibit a specific type of RTK in a cell to determine its normal function. A similar approach can be used to inhibit the function of various other types of receptors and intracellular signaling proteins that function as dimers or larger oligomers.

Proteins with SH2 Domains Bind to Phosphorylated Tyrosines

A whole menagerie of intracellular signaling proteins can bind to the phosphotyrosines on activated RTKs (or on docking proteins such as IRS1). They help to

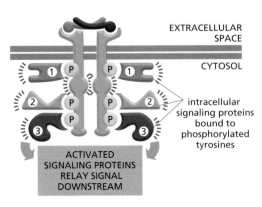

Figure 15–54 The docking of intracellular signaling proteins on phosphotyrosines on an activated RTK. The activated receptor and its bound signaling proteins form a signaling complex that can then broadcast signals along multiple signaling pathways.

relay the signal onward, mainly through chains of protein–protein interactions mediated by modular *interaction domains*, as discussed earlier. Some of the docked proteins are enzymes, such as **phospholipase C-γ (PLCγ)**, which functions in the same way as phospholipase C-β—activating the inositol phospholipid signaling pathway discussed earlier in connection with GPCRs (see Figures 15–38 and 15–39). Through this pathway, RTKs can increase cytosolic Ca^{2+} levels and activate PKC. Another enzyme that docks on these receptors is the cytoplasmic tyrosine kinase *Src*, which phosphorylates other signaling proteins on tyrosines. Yet another is *phosphoinositide 3-kinase (PI 3-kinase)*, which phosphorylates mainly lipids rather than proteins; as we discuss later, the phosphorylated lipids then serve as docking sites to attract various signaling proteins to the plasma membrane.

The intracellular signaling proteins that bind to phosphotyrosines, either on activated RTKs or on proteins docked on them, have varied structures and functions. However, they usually share highly conserved phosphotyrosine-binding domains. These can be either **SH2 domains** (for *Src homology region*) or, less commonly, *PTB domains* (for *phosphotyrosine-binding*). By recognizing specific phosphorylated tyrosines, these small interaction domains enable the proteins that contain them to bind to activated RTKs, as well as to many other intracellular signaling proteins that have been transiently phosphorylated on tyrosines (**Figure 15–55**). As previously discussed, many signaling proteins also contain other interaction domains that allow them to interact specifically with other proteins as part of the signaling process. These domains include the *SH3 domain*, which binds to proline-rich motifs in intracellular proteins (see Figure 15–22). There are about 115 SH2 domains and about 295 SH3 domains encoded in the human genome.

Not all proteins that bind to activated RTKs via SH2 domains help to relay the signal onward. Some act to decrease the signaling process, providing negative

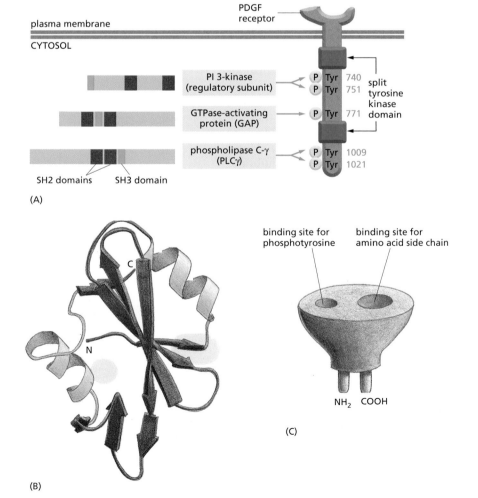

Figure 15–55 The binding of SH2-containing intracellular signaling proteins to an activated PDGF receptor. (A) This drawing of a PDGF receptor shows five phosphotyrosine docking sites, three in the kinase insert region and two on the C-terminal tail, to which the three signaling proteins shown bind as indicated. The numbers on the right indicate the positions of the tyrosines in the polypeptide chain. These binding sites have been identified by using recombinant DNA technology to mutate specific tyrosines in the receptor. Mutation of tyrosines 1009 and 1021, for example, prevents the binding and activation of PLCγ, so that receptor activation no longer stimulates the inositol phospholipid signaling pathway. The locations of the SH2 *(red)* and SH3 *(blue)* domains in the three signaling proteins are indicated. (Additional phosphotyrosine docking sites on this receptor are not shown, including those that bind the cytoplasmic tyrosine kinase Src and two adaptor proteins.) It is unclear how many signaling proteins can bind simultaneously to a single RTK. (B) The three-dimensional structure of an SH2 domain, as determined by x-ray crystallography. The binding pocket for phosphotyrosine is shown in *yellow* on the right, and a pocket for binding a specific amino acid side chain (isoleucine, in this case) is shown in *yellow* on the left (see also Figure 3–39). (C) The SH2 domain is a compact, "plug-in" module, which can be inserted almost anywhere in a protein without disturbing the protein's folding or function (discussed in Chapter 3). Because each domain has distinct sites for recognizing phosphotyrosine and for recognizing a particular amino acid side chain, different SH2 domains recognize phosphotyrosine in the context of different flanking amino acid sequences. (B, based on data from G. Waksman et al., *Cell* 72:779–790, 1993. With permission from Elsevier.)

feedback. One example is the *c-Cbl protein,* which can dock on some activated receptors and catalyze their ubiquitylation, covalently adding a single ubiquitin molecule to one or more sites on the receptor (called *monoubiquityl-ation* to distinguish it from *polyubiquitylation,* in which one or more long ubiquitin chains are added to a protein). Monoubiquitylation promotes the endocytosis and degradation of the receptors in lysosomes—an example of receptor down-regulation (see Figure 15–29). Endocytic proteins that contain *ubiquitin-interaction motifs (UIMs)* recognize the monoubiquitylated RTKs and direct them into clathrin-coated vesicles and, ultimately, into lysosomes (discussed in Chapter 13). Mutations that inactivate c-Cbl-dependent RTK down-regulation cause prolonged RTK signaling and thereby promote the development of cancer.

As is the case for GPCRs, ligand-induced endocytosis of RTKs does not always decrease signaling. In some cases, RTKs are endocytosed with their bound signaling proteins and continue to signal from endosomes or other intracellular compartments. This mechanism, for example, allows *nerve growth factor (NGF)* to bind to its specific RTK (called TrkA) at the end of a long nerve cell axon and signal to the cell body of the same cell a long distance away. Here, signaling endocytic vesicles containing TrkA, with NGF bound on the inside and signaling proteins docked on the cytosolic side, are transported along the axon to the cell body, where they signal the cell to survive.

Some signaling proteins are composed almost entirely of SH2 and SH3 domains and function as *adaptors* to couple tyrosine-phosphorylated proteins to other proteins that do not have their own SH2 domains (see Figure 15–22). Adaptor proteins of this type help to couple activated RTKs to the important signaling protein *Ras,* a monomeric GTPase that, in turn, can activate various downstream signaling pathways, as we now discuss.

Ras Belongs to a Large Superfamily of Monomeric GTPases

The **Ras superfamily** consists of various families of monomeric GTPases, but only the Ras and Rho families relay signals from cell-surface receptors (**Table 15–5**). By interacting with different intracellular signaling proteins, a single Ras or Rho family member can coordinately spread the signal along several distinct downstream signaling pathways, thereby acting as a *signaling hub.*

There are three major, closely related Ras proteins in humans (H-, K-, and N-Ras—see Table 15–5). Although they have subtly different functions, they are thought to work in the same way, and we will refer to them simply as **Ras.** Like many monomeric GTPases, Ras contains one or more covalently attached lipid groups that help anchor the protein to the cytoplasmic face of the membrane where the protein functions, which is mainly the plasma membrane, from where it relays signals to other parts of the cell. Ras is often required, for example, when RTKs signal to the nucleus to stimulate cell proliferation or differentiation, both of which require changes in gene expression. If one inhibits Ras function by the microinjection of either neutralizing anti-Ras antibodies or a dominant-negative mutant form of Ras, the cell proliferation or differentiation

Table 15–5 The Ras Superfamily of Monomeric GTPases

FAMILY	SOME FAMILY MEMBERS	SOME FUNCTIONS
Ras	H-Ras, K-Ras, N-Ras	relay signals from RTKs
	Rheb	activates mTOR to stimulate cell growth
	Rep1	activated by a cyclic-AMP-dependent GEF; influences cell adhesion by activating integrins
Rho*	Rho, Rac, Cdc42	relay signals from surface receptors to the cytoskeleton and elsewhere
ARF*	ARF1–ARF6	regulate assembly of protein coats on intracellular vesicles
Rab*	Rab1–60	regulate intracellular vesicle traffic
Ran*	Ran	regulates mitotic spindle assembly and nuclear transport of RNAs and proteins

*The Rho family is discussed in Chapter 16, the ARF and Rab proteins in Chapter 13, and Ran in Chapters 12 and 17. The three-dimensional structure of Ras is shown in Figure 3–72.

responses normally induced by the activated RTKs do not occur. Conversely, 30% of human tumors have hyperactive mutant forms of Ras, which contribute to the uncontrolled proliferation of the cancer cells.

Like other GTP-binding proteins, Ras functions as a molecular switch, cycling between two distinct conformational states—active when GTP is bound and inactive when GDP is bound (see Figure 15–18B). <GAAC> As discussed earlier for monomeric GTPases in general, two classes of signaling proteins regulate Ras activity by influencing its transition between active and inactive states (see Figure 15–19). *Ras guanine nucleotide exchange factors* (**Ras-GEFs**) stimulate the dissociation of GDP and the subsequent uptake of GTP from the cytosol, thereby activating Ras. *Ras GTPase-activating proteins* (**Ras-GAPs**) increase the rate of hydrolysis of bound GTP by Ras, thereby inactivating Ras. Hyperactive mutant forms of Ras are resistant to GAP-mediated GTPase stimulation and are locked permanently in the GTP-bound active state, which is why they promote the development of cancer.

But how do RTKs normally activate Ras? In principle, they could either activate a Ras-GEF or inhibit a Ras-GAP. Even though some GAPs bind directly (via their SH2 domains) to activated RTKs (see Figure 15–55A), whereas GEFs usually bind only indirectly, it is the indirect coupling of the receptor to a Ras-GEF that drives Ras into its active state. In fact, the loss of function of a Ras-GEF has a similar effect to the loss of function of Ras itself. Activation of the other Ras superfamily proteins, including those of the Rho family, also occurs through the activation of GEFs. The particular GEF determines in which membrane the GTPase is activated and, by acting as a scaffold, which downstream proteins the GTPase activates.

Ras proteins and the proteins that regulate them have been highly conserved in evolution, and genetic analyses in *Drosophila* and *C. elegans* provided the first clues to how RTKs activate Ras. Genetic studies of photoreceptor cell development in the *Drosophila* eye were particularly informative.

RTKs Activate Ras Via Adaptors and GEFs: Evidence from the Developing *Drosophila* Eye

The *Drosophila* compound eye consists of about 800 identical units called *ommatidia*, each composed of 8 photoreceptor cells (R1–R8) and 12 accessory cells (**Figure 15–56**). The eye develops from a simple epithelial sheet, and the cells that make up each ommatidium are recruited from the sheet in a fixed sequence, by a series of cell–cell interactions. Beginning with the development of the R8 photoreceptor, each differentiating cell induces its uncommitted immediate neighbors to adopt a specific fate and assemble into the developing ommatidium (**Figure 15–57**).

The development of the R7 photoreceptor, which is required for the detection of ultraviolet light, has been studied most intensively, beginning with the description of a mutant fly called *Sevenless (Sev)*, in which a deficiency of R7 is the only observed defect. Such mutants are easy to select on the basis of their blindness to ultraviolet light. The normal *Sev* gene was shown to encode an RTK that is expressed on R7 precursor cells. Further genetic analysis of mutants in which R7 development is blocked but the Sev protein itself is not affected led to the identification of the gene *Bride-of-sevenless (Boss)*, which encodes the ligand for the Sev RTK. Boss is a seven-pass transmembrane protein that is expressed exclusively on the surface of the adjacent R8 cell, and when it binds to and activates Sev, it induces the R7 precursor cell to differentiate into an R7 photoreceptor. The Sev protein is also expressed on several other precursor cells in the developing ommatidium, but none of these cells contact R8; therefore, the Sev protein is not activated, and these cells do not differentiate into R7 photoreceptors.

200 µm

Figure 15–56 Scanning electron micrograph of a compound eye of *Drosophila*. The eye is composed of about 800 identical units (ommatidia), each having a separate lens that focuses light onto eight photoreceptor cells at its base. (Courtesy of Kevin Moses.)

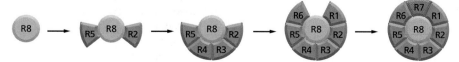

Figure 15–57 The assembly of photoreceptor cells in a developing *Drosophila* ommatidium. Cells are recruited sequentially to become photoreceptors, beginning with R8 and ending with R7, which is the last photoreceptor cell to develop.

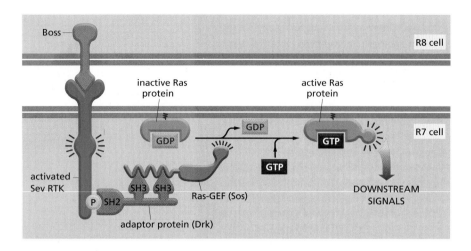

Figure 15–58 How the Sev RTK activates Ras in the fly eye. The activation of Sev on the surface of the R7 precursor cell by the Boss protein on the surface of R8 activates the Ras-GEF Sos via the adaptor protein Drk. Drk recognizes a specific phosphorylated tyrosine on the Sev protein by means of an SH2 domain and interacts with Sos by means of two SH3 domains. Sos stimulates the inactive Ras protein to replace its bound GDP by GTP, which activates Ras to relay the signal downstream, inducing the R7 precursor cell to differentiate into a UV-sensing photoreceptor cell.

The components of the intracellular signaling pathway activated by Sev in the R7 precursor cell proved more difficult to identify than the receptor or its ligand because mutations that inactivate them are lethal. The problem was solved by performing a genetic screen in flies with a partially inactive Sev protein. One of the genes identified encodes a Ras protein. Flies in which both copies of the *Ras* gene are inactivated by mutation die, whereas flies with only one inactivated copy survive. But when a fly carries a partially inactive Sev protein in its developing eyes, an inactivating mutation in one copy of the *Ras* gene results in the loss of R7. Moreover, if one of the *Ras* genes is rendered overactive by mutation, R7 develops even in mutants in which both *Sev* and *Boss* are inactive. These findings indicate that Ras acts downstream of Sev and that its activation in R7 precursor cells is necessary and sufficient to induce R7 differentiation.

A second gene identified in these genetic screens is called *Son-of-sevenless (Sos)*. It encodes a Ras-GEF, which is required for the Sev RTK to activate Ras. A third gene (called *Drk*) encodes an adaptor protein, which couples the Sev receptor to the Sos protein; the SH2 domain of the Drk adaptor binds to activated Sev, while its two SH3 domains bind to Sos (**Figure 15–58**). This type of genetic screen, in which organisms with a partially crippled component of a genetic pathway are used to identify genes encoding other proteins in the pathway, has now been widely used elsewhere with great success.

Biochemical and cell biological studies have shown that the coupling of RTKs to Ras occurs by a similar mechanism in mammalian cells, where the adaptor protein is called **Grb2** and the Ras-GEF is also called Sos (see Figure 15–22). Interestingly, when mammalian Sos activates Ras, Ras acts back to stimulate Sos further, forming a simple positive feedback loop.

RTKs are not the only means of activating Ras. Ca^{2+} and diacylglycerol, for example, activate a Ras-GEF that is found mainly in the brain; this Ras-GEF can couple GPCRs to Ras activation independently of Sos.

Once activated, Ras activates various other signaling proteins to relay the signal downstream, as we discuss next.

Ras Activates a MAP Kinase Signaling Module

Both the tyrosine phosphorylations and the activation of Ras triggered by activated RTKs are usually short-lived (**Figure 15–59**). *Tyrosine-specific protein phosphatases* quickly reverse the phosphorylations, and GAPs induce activated Ras to inactivate itself by hydrolyzing its bound GTP to GDP. To stimulate cells to proliferate or differentiate, these short-lived signaling events must be converted into longer-lasting ones that can sustain the signal and relay it downstream to the nucleus to alter the pattern of gene expression. One of the key mechanisms used for this purpose is a system of proteins called the *mitogen-activated protein kinase module* (**MAP kinase module**) (**Figure 15–60**). The three components of this system together form a functional signaling

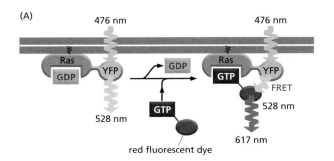

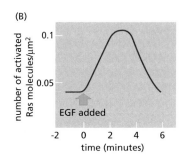

Figure 15–59 Transient activation of Ras revealed by single-molecule fluorescence resonance energy transfer (FRET). (A) Schematic drawing of the experimental strategy. Cells of a human cancer cell line are genetically engineered to express a Ras protein that is covalently linked to yellow fluorescent protein (YFP). GTP that is labeled with a red fluorescent dye is microinjected into some of the cells. The cells are then stimulated with the extracellular signal protein epidermal growth factor (EGF), and single fluorescent molecules of Ras-YFP at the inner surface of the plasma membrane are followed by video fluorescence microscopy in individual cells. When a fluorescent Ras-YFP molecule becomes activated, it exchanges unlabeled GDP for fluorescently labeled GTP; the yellow-green light emitted from the YFP now activates the fluorescent GTP to emit red light. Thus, the activation of single Ras molecules can be followed by the emission of red fluorescence from a previously yellow-green fluorescent spot at the plasma membrane. As shown in (B), activated Ras molecules can be detected after about 30 seconds of EGF stimulation. The red signal peaks at 3–4 minutes and then decreases to baseline by 6 minutes. As Ras-GAP is found to be recruited to the same spots at the plasma membrane as Ras, it presumably plays a major part in rapidly shutting off the Ras signal. (Modified from H. Murakoshi et al., *Proc. Natl Acad. Sci. U.S.A.* 101:7317–7322, 2004. With permission from National Academy of Sciences.)

module that has been highly conserved, from yeasts to humans, and is used, with variations, in many different signaling contexts.

The three components are all protein kinases. The final kinase in the series is called simply MAP kinase (MAPK). The next one upstream from this is MAP kinase kinase (MAPKK): it phosphorylates and thereby activates MAP kinase. And next above that, receiving an activating signal directly from Ras, is MAP kinase kinase kinase (MAPKKK): it phosphorylates and thereby activates MAPKK. In the mammalian **Ras-MAP-kinase signaling pathway**, these three kinases are known by shorter names: Raf (=MAPKKK), Mek (= MAPKK) and Erk (=MAPK).

Once activated, the MAP kinase relays the signal downstream by phosphorylating various proteins in the cell, including gene regulatory proteins and other protein kinases (see Figure 15–60). Erk MAP kinase, for example, enters the nucleus and phosphorylates one or more components of a gene regulatory complex. This activates the transcription of a set of *immediate early genes*, so named because they turn on within minutes after an RTK receives an extracellular signal,

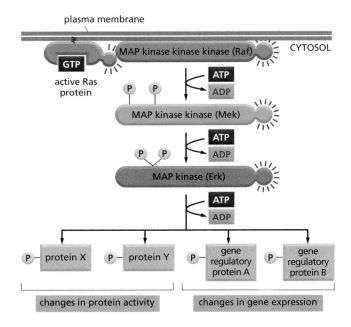

Figure 15–60 **The MAP kinase serine/threonine phosphorylation module activated by Ras.** The three-component module begins with a MAP kinase kinase kinase called *Raf*. Ras recruits Raf to the plasma membrane and helps activate it. Raf then activates the MAP kinase kinase *Mek*, which then activates the MAP kinase *Erk*. Erk in turn phosphorylates a variety of downstream proteins, including other protein kinases, as well as gene regulatory proteins in the nucleus. The resulting changes in gene expression and protein activity cause complex changes in cell behavior.

even if protein synthesis is experimentally blocked with drugs. Some of these genes encode other gene regulatory proteins that turn on other genes, a process that requires both protein synthesis and more time. (This relationship between the immediate early and later genes is similar to the relationship between primary and secondary response genes activated by the nuclear receptors discussed earlier—see Figure 15–15.)

In this way, the Ras–MAP-kinase signaling pathway conveys signals from the cell surface to the nucleus and alters the pattern of gene expression. Among the genes activated by this pathway are some that stimulate cell proliferation, such as the genes encoding G_1 *cyclins* (discussed in Chapter 17).

Extracellular signals usually activate MAP kinases only transiently, and the period during which the kinase remains active influences the response. When EGF activates its receptors in a neural precursor cell line, for example, Erk MAP kinase activity peaks at 5 minutes and rapidly declines, and the cells later go on to divide. By contrast, when NGF activates its receptors on the same cells, Erk activity remains high for many hours, and the cells stop proliferating and differentiate into neurons. Many factors can influence the duration of the signaling response, including positive and negative feedback loops (see Figure 15–28).

MAP kinases participate in both positive and negative feedback loops, which can combine to give responses that are either graded or switchlike and either brief or long lasting. In an example illustrated earlier, in Figure 15–24, MAP kinase activates a complex positive feedback loop to produce an all-or-none, irreversible response when frog oocytes are stimulated to mature by a brief exposure to the extracellular signal molecule progesterone. In many cells, MAP kinases activate a negative feedback loop by increasing the concentration of a *dual-specificity protein phosphatase* that removes the phosphate from both a tyrosine and a threonine to inactivate the MAP kinase (which is phosphorylated on the tyrosine and threonine by MAPKK). The increase in the phosphatase results from both an increase in the transcription of the phosphatase gene and the stabilization of the enzyme against degradation. In the Ras–MAP-kinase pathway shown in Figure 15–61, Erk also phosphorylates and inactivates Raf, providing another negative feedback loop that helps shut off the MAP kinase module.

Scaffold Proteins Help Prevent Cross-Talk Between Parallel MAP Kinase Modules

Three-component MAP kinase signaling modules operate in all eucaryotic cells, with different modules mediating different responses in the same cell. In budding yeast, for example, one such module mediates the response to mating pheromone, another the response to starvation, and yet another the response to osmotic shock. Some of these MAP kinase modules use one or more of the same kinases and yet manage to activate different effector proteins and hence different responses. As discussed earlier, one way in which cells avoid cross-talk between the different parallel signaling pathways and ensure that each response is specific is to use scaffold proteins (see Figure 15–21A). In yeast cells, such scaffolds bind all or some of the kinases in each MAP kinase module to form a complex and thereby help to ensure response specificity (**Figure 15–61**).

Mammalian cells also use this scaffold strategy to prevent cross-talk between different MAP kinase modules. At least 5 parallel MAP kinase modules can operate in a mammalian cell. These modules make use of at least 12 MAP kinases, 7 MAP kinase kinases, and 7 MAP kinase kinase kinases. Two of these modules (terminating in MAP kinases called JNK and p38) are activated by different kinds of cell stresses, such as UV irradiation, heat shock, and osmotic stress, as well as by inflammatory cytokines; others mainly mediate responses to signals from other cells.

Although the scaffold strategy provides precision and avoids cross-talk, it reduces the opportunities for amplification and spreading of the signal to different parts of the cell, which require at least some of the components to be diffusible (see Figure 15–17). It is unclear to what extent the individual components

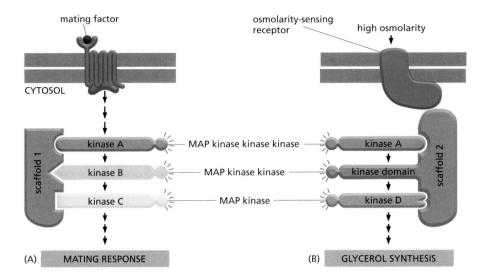

(A) MATING RESPONSE

(B) GLYCEROL SYNTHESIS

Figure 15–61 **The organization of two MAP kinase modules by scaffold proteins in budding yeast.** Budding yeast have at least six three-component MAP kinase modules involved in a variety of biological processes, including the two responses illustrated here—a mating response and the response to high osmolarity. (A) The mating response is triggered when a mating factor secreted by a yeast of opposite mating type binds to a GPCR. This activates a G protein, the βγ complex of which indirectly activates the MAP kinase kinase kinase (kinase A), which then relays the response onward. Once activated, the MAP kinase (kinase C) phosphorylates and thereby activates several proteins that mediate the mating response, in which the yeast cell stops dividing and prepares for fusion. The three kinases in this module are bound to scaffold protein 1. (B) In a second response, a yeast cell exposed to a high-osmolarity environment is induced to synthesize glycerol to increase its internal osmolarity. This response is mediated by a transmembrane, osmolarity-sensing, receptor protein and a different MAP kinase module bound to a second scaffold protein. (Note that the kinase domain of scaffold 2 provides the MAP kinase kinase activity of this module.) Although both pathways use the same MAP kinase kinase kinase (kinase A, *green*), there is no cross-talk between them, because the kinases in each module are bound to different scaffold proteins, and the osmosensor is bound to the same scaffold protein it activates.

Remarkably, scaffold 1 is able to function even if it is stripped (by genetic engineering) of any ability to align or allosterically regulate its three bound kinases. Its role is apparently just to bring the kinases into close proximity, thereby increasing the rate at which they encounter each other, as expected for reaction rate accelerations caused by protein tethering (see Figure 3–80C).

of MAP kinase modules can dissociate from the scaffold during the activation process to permit amplification, which is why we refer to them as modules rather than cascades.

Rho Family GTPases Functionally Couple Cell-Surface Receptors to the Cytoskeleton

Besides the Ras proteins, the other class of Ras superfamily GTPases that relays signals from cell-surface receptors is the large **Rho family** (see Table 15–5). Rho family monomeric GTPases regulate both the actin and microtubule cytoskeletons, controlling cell shape, polarity, motility, and adhesion (discussed in Chapter 16); they also regulate cell-cycle progression, gene transcription, and membrane transport. They play a key part in the guidance of cell migration and nerve axon outgrowth, mediating cytoskeletal responses to the activation of a special class of guidance receptors. We focus on this aspect of Rho family function here.

The three best-characterized members are **Rho** itself, **Rac**, and **Cdc42**, each of which affects multiple downstream target proteins. In the same way as for Ras, GEFs activate and GAPs inactivate the Rho family GTPases; there are more than 60 Rho-GEFs and more than 70 Rho-GAPs in humans. Some of the GEFs and GAPs are specific for one particular family member, whereas others are less specific. Unlike Ras, which is membrane-associated even when inactive (with GDP bound), inactive Rho family GTPases are often bound to *guanine nucleotide dissociation inhibitors* (*GDIs*) in the cytosol, which prevent the GTPases from interacting with their Rho-GEFs at the plasma membrane.

Although cell-surface receptors activate the Rho GTPases by activating Rho-GEFs, it is not known in most cases how a receptor activates its GEF. One of the exceptions is an *Eph* RTK (see Figure 15–52) on the surface of motor neurons that helps guide the migrating tip of the axon (called a *growth cone*) to its muscle target. The binding of a cell-surface *ephrin* protein activates the Eph receptor, causing the growth cones to collapse, thereby repelling them from inappropriate regions and keeping them on track. The response depends on a Rho-GEF called *ephexin*, which is stably associated with the cytosolic tail of the Eph receptor. When ephrin binding activates the Eph receptor, the receptor activates a cytoplasmic tyrosine kinase that phosphorylates ephexin on a tyrosine, enhancing the ability of ephexin to activate the Rho protein RhoA. The activated RhoA (RhoA-GTP) then regulates various downstream target proteins, including some effector proteins that control the actin cytoskeleton, causing the growth cone to collapse (**Figure 15–62**).

Ephrins are among the best-characterized extracellular *guidance proteins*, with many functions, both in the nervous system and outside it. One of their roles, for example, is to direct the way in which nerve axons grow from the eye to the optic tectum so as to create a neural 'map' of the visual field in the brain

(A)

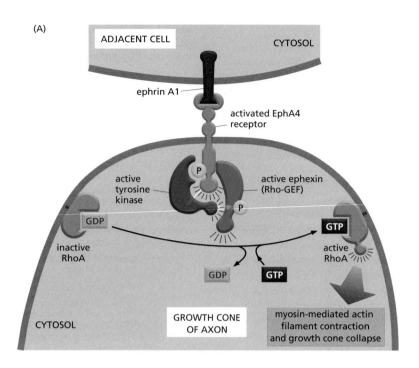

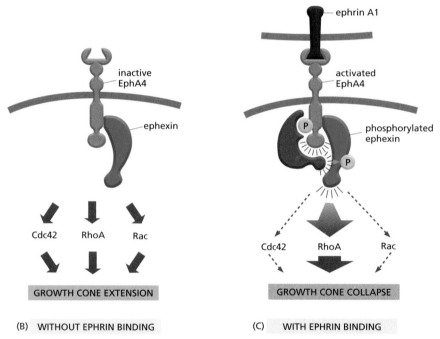

(B) **WITHOUT EPHRIN BINDING**

(C) **WITH EPHRIN BINDING**

Figure 15–62 Growth cone collapse mediated by Rho family GTPases.
(A) The binding of transmembrane ephrin A1 proteins on an adjacent cell activates EphA4 RTKs on the growth cone of an axon, by the mechanism illustrated in Figure 15–53A. Phosphotyrosines on the activated Eph receptors recruit and activate a cytoplasmic tyrosine kinase to phosphorylate the receptor-associated Rho-GEF ephexin on a tyrosine. This enhances the ability of the ephexin to activate the Rho family GTPase RhoA. RhoA then induces the growth cone to collapse by stimulating the myosin-dependent contraction of the actin cytoskeleton. (B) When ephrin A1 is not bound to the EphA4 receptor, ephexin activates three different Rho family members (Cdc42, Rac, and RhoA) about equally, promoting the forward advance of the growth cone. (C) EphrinA1 binding to EphA4 alters the activity of ephexin so that it now mainly activates RhoA, causing the growth cone to collapse.

(discussed in Chapter 22). But the guidance of axon outgrowth is a complex matter, and other types of guidance receptors are also involved, as discussed in Chapter 22. All of these receptors, however, are thought to guide cell movement by influencing the cytoskeleton, and to do so via members of the Rho family.

Having considered how RTKs use GEFs and monomeric GTPases to relay signals into the cell, we now consider a second major strategy that RTKs use that depends on a quite different intracellular relay mechanism.

PI 3-Kinase Produces Lipid Docking Sites in the Plasma Membrane

As mentioned earlier, one of the proteins that binds to the intracellular tail of RTK molecules is the plasma-membrane-bound enzyme **phosphoinositide 3-kinase (PI 3-kinase)**. This kinase principally phosphorylates inositol phospholipids rather than proteins, and both RTKs and GPCRs can activate it. It plays a central part in promoting cell survival and growth.

Phosphatidylinositol (PI) is unique among membrane lipids because it can undergo reversible phosphorylation at multiple sites on its inositol head group to generate a variety of phosphorylated PI lipids called **phosphoinositides** (see Figure 15–37). When activated, PI 3-kinase catalyzes phosphorylation at the 3 position of the inositol ring to generate several phosphoinositides (**Figure 15–63**). The production of $PI(3,4,5)P_3$ matters most because it can serve as a docking site for various intracellular signaling proteins, which assemble into signaling complexes that relay the signal into the cell from the cytosolic face of the plasma membrane (see Figure 15–21C).

Notice the difference between this use of phosphoinositides and their use described earlier, in which $PI(4,5)P_2$ is cleaved by PLCβ (in the case of GPCRs) or PLCγ (in the case of RTKs) to generate soluble IP_3 and membrane-bound diacylglycerol (see Figures 15–38 and 15–39). By contrast, $PI(3,4,5)P_3$ is not cleaved by PLC. It is made from $PI(4,5)P_2$ and then remains in the plasma membrane until specific *phosphoinositide phosphatases* dephosphorylate it. Prominent among these is the *PTEN* phosphatase, which dephosphorylates the 3 position of the inositol ring. Mutations in PTEN are found in many cancers: by prolonging signaling by PI 3-kinase, they promote uncontrolled cell growth.

There are various types of PI 3-kinases. Those activated by RTKs and GPCRs belong to class I. These are heterodimers composed of a common catalytic subunit and different regulatory subunits. RTKs activate *class Ia PI 3-kinases*, in which the regulatory subunit is an adaptor protein that binds to two phosphotyrosines on activated RTKs through its two SH2 domains (see Figure 15–55A). GPCRs activate *class Ib PI 3-kinases*, which have a regulatory subunit that binds to the βγ complex of an activated trimeric G protein when GPCRs are activated by their extracellular ligand. The direct binding of activated Ras can also activate the common class I catalytic subunit.

Intracellular signaling proteins bind to $PI(3,4,5)P_3$ produced by activated PI 3-kinase via a specific interaction domain, such as a **pleckstrin homology (PH) domain** mentioned earlier, first identified in the platelet protein pleckstrin. PH domains function mainly as protein–protein interaction domains, and it is only a small subset of them that bind to PIP_3; at least some of these also recognize a specific membrane-bound protein as well as the PIP_3, which greatly increases the specificity of the binding and helps to explain why the signaling proteins with PIP_3-binding PH domains do not all dock at all $PI(3,4,5)P_3$ sites. PH domains occur in about 200 human proteins, including the Ras-GEF Sos discussed earlier (see Figure 15–22).

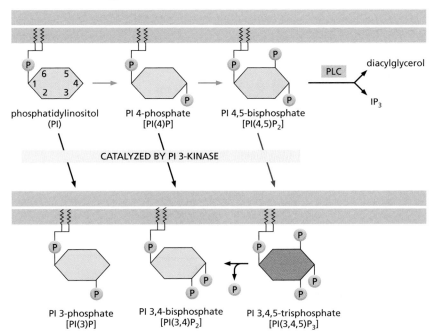

Figure 15–63 The generation of phosphoinositide docking sites by PI 3-kinase. PI 3-kinase phosphorylates the inositol ring on carbon atom 3 to generate the phosphoinositides shown at the bottom of the figure (diverting them away from the pathway leading to IP_3 and diacylglycerol). The most important phosphorylation (indicated in *red*) is of $PI(4,5)P_2$ to $PI(3,4,5)P_3$, which can serve as a docking site for signaling proteins with PIP_3-binding PH domains. Other inositol phospholipid kinases catalyze the phosphorylations indicated by the *green arrows*.

One especially important PH-domain-containing protein is the serine/threonine protein kinase *Akt*. The *PI-3-kinase–Akt signaling pathway* is the major pathway activated by the hormone *insulin*. It also plays a key part in promoting the survival and growth of many cell types in both invertebrates and vertebrates, as we now discuss.

The PI-3-Kinase–Akt Signaling Pathway Stimulates Animal Cells to Survive and Grow

For a multicellular organism to grow, its cells must grow (accumulate mass and enlarge). If the cells simply divided without growing, they would get progressively smaller and the organism as a whole would stay the same size. As discussed earlier, extracellular signals are usually required for animal cells to grow and divide, as well as to survive (see Figure 15–8). Members of the *insulin-like growth factor (IGF)* family of signal proteins, for example, stimulate many types of animal cells to survive and grow. They bind to specific RTKs (see Figure 15–52), which activate PI 3-kinase to produce PI(3,4,5)P₃. The PIP₃ recruits two protein kinases to the plasma membrane via their PH domains—**Akt** (also called *protein kinase B*, or *PKB)* and *phosphoinositide-dependent protein kinase 1 (PDK1)*, and this leads to the activation of Akt (**Figure 15–64**). Once activated, Akt phosphorylates various target proteins at the plasma membrane, as well as in the cytosol and nucleus. The effect on most of the known targets is to inactivate them; but the targets are such that these actions of Akt all conspire to enhance cell survival and growth.

One effect of Akt, for example, is to phosphorylate a cytosolic protein called *Bad*, which, in its nonphosphorylated state, promotes cell death by apoptosis (discussed in Chapter 18). The phosphorylation of Bad by Akt creates phosphoserine-binding sites for a scaffold protein called *14-3-3*, which sequesters phosphorylated Bad and keeps it out of action, thereby promoting cell survival (see Figure 15–64).

The **PI-3-kinase–Akt pathway** signals cells to grow through a more complex mechanism that depends on a large, serine/threonine protein kinase called **TOR** (named as the target of *rapamycin*, a bacterial toxin that inactivates the kinase and is used clinically as both an immunosuppressant and anti-cancer drug). TOR was originally identified in yeasts in genetic screens for rapamycin resistance; in mammalian cells, it is called **mTOR**. TOR exists in cells in two functionally distinct multiprotein complexes. In mammalian cells, *mTOR complex 1* contains the protein *raptor*; this complex is sensitive to rapamycin, and it stimulates cell

Figure 15–64 One way in which signaling through PI 3-kinase promotes cell survival. An extracellular survival signal activates an RTK, which recruits and activates PI 3-kinase. The PI 3-kinase produces PI(3,4,5)P₃, which serves as a docking site for two serine/threonine kinases with PH domains—Akt and the phosphoinositol-dependent kinase PDK1—which are brought into proximity at the plasma membrane. The Akt is phosphorylated on a serine by a third kinase (usually mTOR), which alters the conformation of the Akt so that it can be phosphorylated on a threonine by PDK1, which activates the Akt. The activated Akt now dissociates from the plasma membrane and phosphorylates various target proteins, including the Bad protein.

When unphosphorylated, Bad holds one or more apoptosis-inhibitory proteins (of the Bcl2 family—discussed in Chapter 18) in an inactive state. Once phosphorylated, Bad releases the inhibitory proteins, which now can block apoptosis and thereby promote cell survival. As shown, once phosphorylated, Bad binds to a ubiquitous cytosolic protein called *14-3-3*, which keeps Bad out of action.

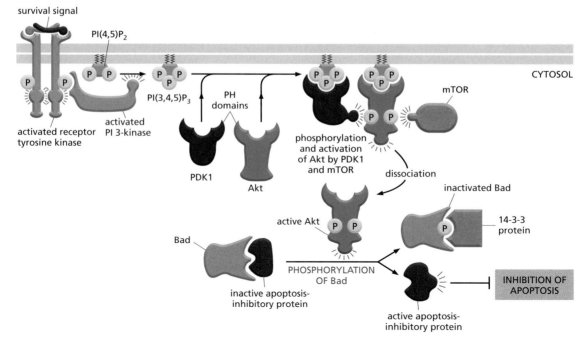

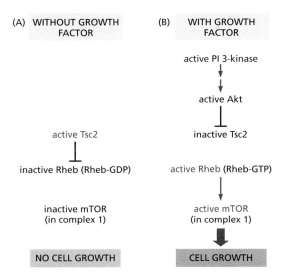

(A) WITHOUT GROWTH FACTOR

active Tsc2

inactive Rheb (Rheb-GDP)

inactive mTOR (in complex 1)

NO CELL GROWTH

(B) WITH GROWTH FACTOR

active PI 3-kinase

active Akt

inactive Tsc2

active Rheb (Rheb-GTP)

active mTOR (in complex 1)

CELL GROWTH

Figure 15–65 Activation of mTOR by the PI-3-kinase–Akt signaling pathway. (A) In the absence of extracellular growth factors, Tsc2 (a Rheb-GAP) keeps Rheb inactive; mTOR in complex I is inactive, and there is no cell growth. (B) In the presence of growth factors, activated Akt phosphorylates and inhibits Tsc2, thereby promoting the activation of Rheb. Activated Rheb (Rheb-GTP) helps activate mTOR in complex 1, which in turn stimulates cell growth. Figure 15–64 shows how growth factors (or survival signals) activate Akt.

The Erk MAP kinase (see Figure 15–60) can also phosphorylate and inhibit Tsc2 and thereby activate mTOR. Thus, both the PI-3-kinase–Akt and Ras–MAP-kinase signaling pathways converge on mTOR in complex 1 to stimulate cell growth.

Tsc2 is short for *tuberous sclerosis protein 2,* and it is one component of a heterodimer composed of Tsc1 and Tsc2 (not shown); these proteins are so called because mutations in either gene encoding them cause the genetic disease *tuberous sclerosis,* which is associated with benign tumors in the brain and elsewhere that contain abnormally large cells.

growth—both by promoting ribosome production and protein synthesis and by inhibiting protein degradation. Complex 1 also promotes both cell growth and cell survival by stimulating nutrient uptake and metabolism. *mTOR complex 2* contains the protein *rictor* and is insensitive to rapamycin; it helps to activate Akt (see Figure 15–64), and it regulates the actin cytoskeleton via Rho family GTPases.

The mTOR in complex 1 integrates inputs from various sources, including extracellular signal proteins referred to as *growth factors* and nutrients such as amino acids, both of which help activate mTOR and promote cell growth. The growth factors activate mTOR mainly via the PI-3-kinase–Akt pathway. Akt activates mTOR in complex 1 indirectly by phosphorylating, and thereby inhibiting, a GAP called Tsc2. Tsc2 acts on a monomeric Ras-related GTPase called **Rheb** (see Table 15–5, p. 926). Rheb in its active form (Rheb-GTP) activates mTOR. The net result is that Akt activates mTOR and thereby promotes cell growth (**Figure 15–65**). We discuss how mTOR stimulates ribosome production and protein synthesis in Chapter 17 (see Figure 17–65).

The Downstream Signaling Pathways Activated By RTKs and GPCRs Overlap

As mentioned earlier, RTKs and GPCRs activate some of the same intracellular signaling pathways. Both, for example, can activate the inositol phospholipid pathway triggered by phospholipase C. Moreover, even when they activate different pathways, the different pathways can converge on the same target proteins. **Figure 15–66** illustrates both of these types of signaling overlaps: it summarizes five parallel intracellular signaling pathways that we have discussed so far—one triggered by GPCRs, two triggered by RTKs, and two triggered by both kinds of receptors.

Tyrosine-Kinase-Associated Receptors Depend on Cytoplasmic Tyrosine Kinases

Many cell-surface receptors depend on tyrosine phosphorylation for their activity and yet lack a tyrosine kinase domain. These receptors act through **cytoplasmic tyrosine kinases**, which are associated with the receptors and phosphorylate various target proteins, often including the receptors themselves, when the receptors bind their ligand. These **tyrosine-kinase-associated receptors** thus function in much the same way as RTKs, except that their kinase domain is encoded by a separate gene and is noncovalently associated with the receptor polypeptide chain. A variety of receptor classes belong in this category, including the receptors for antigen and interleukins on lymphocytes (discussed in Chapter

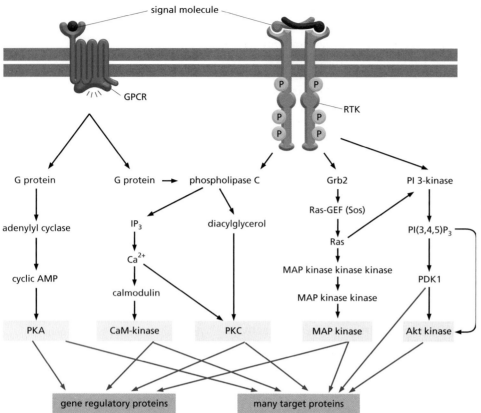

Figure 15–66 Five parallel intracellular signaling pathways activated by GPCRs, RTKs, or both. In this hypothetical example, the five kinases (shaded *yellow*) at the end of each signaling pathway phosphorylate target proteins (shaded *red*), many of which are phosphorylated by more than one of the kinases. The phospholipase C activated by the two types of receptors is different: GPCRs activate PLCβ, whereas RTKs activate PLCγ (not shown). Although not shown, some GPCRs can also activate Ras, but they do so independently of Grb2, via a Ras-GEF that is activated by Ca^{2+} and diacylglycerol.

25), integrins (discussed in Chapter 19), and receptors for various cytokines and some hormones. As with RTKs, many of these receptors are either preformed dimers (**Figure 15–67**) or are cross-linked into dimers by ligand binding.

Some of these receptors depend on members of the largest family of mammalian cytoplasmic tyrosine kinases, the **Src family** (see Figures 3–10 and 3–69), which includes *Src, Yes, Fgr, Fyn, Lck, Lyn, Hck,* and *Blk*. These protein kinases all contain SH2 and SH3 domains and are located on the cytoplasmic side of the plasma membrane, held there partly by their interaction with transmembrane receptor proteins and partly by covalently attached lipid chains. Different family members are associated with different receptors and phosphorylate overlapping but distinct sets of target proteins. Lyn, Fyn, and Lck, for example, are each associated with different sets of receptors in lymphocytes (discussed in Chapter 25). In each case, the kinase is activated when an extracellular ligand binds to the appropriate receptor protein. Src itself, as well as several other family members, can also bind to activated RTKs; in these cases, the receptor and cytoplasmic kinases mutually stimulate each other's catalytic activity, thereby strengthening and prolonging the signal (see Figure 15–62). There are even some G proteins (G_s and G_i) that can activate Src, which is one way that the activation of GPCRs can lead to tyrosine phosphorylation of intracellular signaling proteins and effector proteins.

Another type of cytoplasmic tyrosine kinase associates with *integrins*, the main receptors that cells use to bind to the extracellular matrix (discussed in Chapter 19). The binding of matrix components to integrins activates intracellu-

Figure 15–67 The three-dimensional structure of human growth hormone bound to its receptor. <CGCT> The receptor is a homodimer, and the hormone *(red)* binds to each of the two identical subunits (one shown in *green* and the other in *blue*), which recognize different parts of the monomeric hormone. Hormone binding is thought to cause a rearrangement of the subunits, which activates cytoplasmic tyrosine kinases that are tightly bound to the cytosolic tails of the receptors (not shown). The structure shown was determined by x-ray crystallographic studies of complexes formed between the hormone and extracellular receptor domains produced by recombinant DNA technology. (From A.M. deVos, M. Ultsch and A.A. Kossiakoff, *Science* 255:306–312, 1992. With permission from AAAS.)

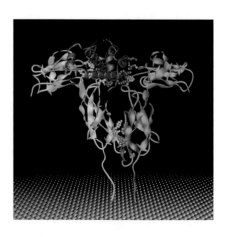

lar signaling pathways that influence the behavior of the cell. When integrins cluster at sites of matrix contact, they help trigger the assembly of cell–matrix junctions called *focal adhesions*. Among the many proteins recruited into these junctions is the cytoplasmic tyrosine kinase called **focal adhesion kinase (FAK)**, which binds to the cytosolic tail of one of the integrin subunits with the assistance of other proteins. The clustered FAK molecules cross-phosphorylate each other, creating phosphotyrosine-docking sites where the Src kinase can bind. Src and FAK now phosphorylate each other and other proteins that assemble in the junction, including many of the signaling proteins used by RTKs. In this way, the two tyrosine kinases signal to the cell that it has adhered to a suitable substratum, where the cell can now survive, grow, divide, migrate, and so on.

The largest and most diverse class of receptors that rely on cytoplasmic tyrosine kinases to relay signals into the cell is the class of *cytokine receptors*, which we consider next.

Cytokine Receptors Activate the JAK–STAT Signaling Pathway, Providing a Fast Track to the Nucleus

The large family of **cytokine receptors** includes receptors for many kinds of local mediators (collectively called *cytokines*), as well as receptors for some hormones, such as *growth hormone* (see Figure 15–67) and *prolactin*. These receptors are stably associated with cytoplasmic tyrosine kinases called **Janus kinases (JAKs)** (after the two-faced Roman god), which phosphorylate and activate gene regulatory proteins called **STATs** (**s**ignal **t**ransducers and **a**ctivators of **t**ranscription). STAT proteins are located in the cytosol and are referred to as *latent gene regulatory proteins* because they only migrate into the nucleus and regulate gene transcription after they are activated.

Although many intracellular signaling pathways lead from cell-surface receptors to the nucleus, where they alter gene transcription (see Figure 15–66), the **JAK–STAT signaling pathway** provides one of the more direct routes. Cytokine receptors are dimers or trimers and are stably associated with one or two of the four known JAKs (JAK1, JAK2, JAK3, and Tyk2). Cytokine binding alters the arrangement so as to bring two JAKs into close proximity so that they trans-phosphorylate each other, thereby increasing the activity of their tyrosine kinase domains. The JAKs then phosphorylate tyrosines on the cytokine receptors, creating phosphotyrosine docking sites for STATs (**Figure 15–68**). Some adaptor proteins can also bind to some of these sites and couple cytokine receptors to the Ras–MAP-kinase signaling pathway discussed earlier, but these will not be discussed here.

There are at least six STATs in mammals. Each has an SH2 domain that performs two functions. First, it mediates the binding of the STAT protein to a phosphotyrosine docking site on an activated cytokine receptor. Once bound, the JAKs phosphorylate the STAT on tyrosines, causing the STAT to dissociate from the receptor. Second, the SH2 domain on the released STAT now mediates its binding to a phosphotyrosine on another STAT molecule, forming either a STAT homodimer or a heterodimer. The STAT dimer then translocates to the nucleus, where, in combination with other gene regulatory proteins, it binds to a specific DNA response element in various genes and stimulates their transcription (see Figure 15–68). In response to the hormone prolactin, for example, which stimulates breast cells to produce milk, activated STAT5 stimulates the transcription of genes that encode milk proteins. **Table 15–6** lists some of the more than 30 cytokines and hormones that activate the JAK–STAT pathway by binding to cytokine receptors; it also shows the specific JAKs and STATs involved.

Some STATs also have an SH2 domain that enables them to dock onto specific phosphotyrosines on some activated RTKs. These receptors can directly activate the bound STAT, independently of JAKs. The nematode *C. elegans* uses STATs for signaling but does not make any JAKs or cytokine receptors, suggesting that STATs evolved before JAKs and cytokine receptors.

Negative feedback regulates the responses mediated by the JAK–STAT pathway. In addition to activating genes that encode proteins mediating the

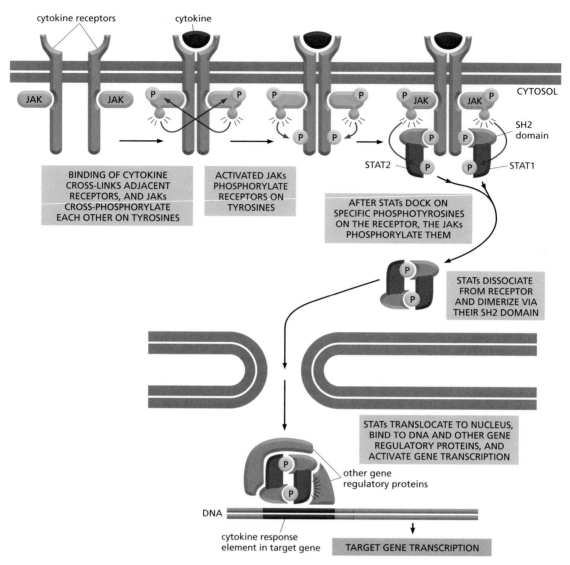

cytokine-induced response, the STAT dimers can also activate genes that encode inhibitory proteins that help shut off the response. Some of these proteins bind to and inactivate phosphorylated JAKs and their associated phosphorylated receptors; others bind to phosphorylated STAT dimers and prevent them from binding to their DNA targets. Such negative feedback mechanisms, however, are not enough on their own to turn off the response. Inactivation of the activated JAKs and STATs requires dephosphorylation of their phosphotyrosines.

Figure 15–68 The JAK–STAT signaling pathway activated by cytokines. The binding of the cytokine either causes two separate receptor polypeptide chains to dimerize (as shown) or reorients the receptor chains in a preformed dimer. In either case, the associated JAKs are brought together so that they can cross-phosphorylate each other on tyrosines, starting the signaling process shown. In some cases, the active receptor is a trimer rather than a dimer.

Protein Tyrosine Phosphatases Reverse Tyrosine Phosphorylations

In all signaling pathways that use tyrosine phosphorylation, the tyrosine phosphorylations are reversed by dephosphorylation performed by **protein tyrosine phosphatases**. These phosphatases are as important in the signaling process as the protein tyrosine kinases that add the phosphates. Whereas only a few types of *serine/threonine protein phosphatase* catalytic subunits are responsible for removing phosphate groups from phosphorylated serines and threonines on proteins, there are about 100 protein tyrosine phosphatases encoded in the human genome, including some *dual-specificity phosphatases* that also dephosphorylate serines and threonines.

Like tyrosine kinases, the tyrosine phosphatases occur in both cytoplasmic and transmembrane forms, none of which are structurally related to serine/threonine protein phosphatases. Some of the transmembrane forms are thought to act as cell-surface receptors, but as this has not been established, they are generally referred to as receptorlike tyrosine phosphatases.

Table 15–6 Some Extracellular Signal Proteins That Act Through Cytokine Receptors and the JAK–STAT Signaling Pathway

SIGNAL PROTEIN	RECEPTOR-ASSOCIATED JAKs	STATS ACTIVATED	SOME RESPONSES
γ-interferon	JAK1 and JAK2	STAT1	activates macrophages
α-interferon	Tyk2 and JAK2	STAT1 and STAT2	increases cell resistance to viral infection
Erythropoietin	JAK2	STAT5	stimulates production of erythrocytes
Prolactin	JAK1 and JAK2	STAT5	stimulates milk production
Growth hormone	JAK2	STAT1 and STAT5	stimulates growth by inducing IGF1 production
GMCSF	JAK2	STAT5	stimulates production of granulocytes and macrophages

Unlike serine/threonine protein phosphatases, which generally have broad specificity, most tyrosine phosphatases display exquisite specificity for their substrates, removing phosphate groups from only selected phosphotyrosines on a subset of proteins. Together, these phosphatases ensure that tyrosine phosphorylations are short-lived and that the level of tyrosine phosphorylation in resting cells is very low. They do not, however, simply continuously reverse the effects of protein tyrosine kinases; they are often regulated to act only at the appropriate time and place in a signaling response or in the cell-division cycle (discussed in Chapter 17).

Having discussed the crucial role of tyrosine phosphorylation and dephosphorylation in the intracellular signaling pathways activated by many enzyme-coupled receptors, we now turn to a class of enzyme-coupled receptors that rely entirely on serine/threonine phosphorylation. These *receptor serine/ threonine kinases* activate an even more direct signaling pathway to the nucleus than does the JAK–STAT pathway. They directly phosphorylate latent gene regulatory proteins called *Smads,* which then translocate into the nucleus to activate gene transcription.

Signal Proteins of the TGFβ Superfamily Act Through Receptor Serine/Threonine Kinases and Smads

The **transforming growth factor-β (TGFβ) superfamily** consists of a large number (30–40 in humans) of structurally related, secreted, dimeric proteins. They act either as hormones or, more commonly, as local mediators to regulate a wide range of biological functions in all animals. During development, they regulate pattern formation and influence various cell behaviors, including proliferation, specification and differentiation, extracellular matrix production, and cell death. In adults, they are involved in tissue repair and in immune regulation, as well as in many other processes. The superfamily consists of the TGFβ/*activin* family and the larger *bone morphogenetic protein (BMP)* family.

All of these proteins act through enzyme-coupled receptors that are single-pass transmembrane proteins with a serine/threonine kinase domain on the cytosolic side of the plasma membrane. There are two classes of these **receptor serine/threonine kinases**—*type I* and *type II*—which are structurally similar homodimers. Each member of the TGFβ superfamily binds to a characteristic combination of type-I and type-II receptor dimers, bringing the kinase domains together so that the type-II receptor can phosphorylate and activate the type-I receptor, forming an active tetrameric receptor complex.

Once activated, the receptor complex uses a strategy for rapidly relaying the signal to the nucleus that is very similar to the JAK–STAT strategy used by cytokine receptors. The activated type-I receptor directly binds and phosphorylates a latent gene regulatory protein of the **Smad family** (named after the first two identified, Sma in *C. elegans* and Mad in *Drosophila*). Activated TGFβ/activin receptors phosphorylate Smad2 or Smad3, while activated BMP receptors phosphorylate Smad1, Smad5, or Smad8. Once one of these *receptor-activated Smads (R-Smads)* has been phosphorylated, it dissociates from the receptor and binds to Smad4 (called a *co-Smad),* which can form a complex with any of the five R-Smads. The Smad complex then translocates into the nucleus, where it associates with other gene regulatory proteins and regulates the transcription of specific target genes (**Figure 15–69**). Because the partner proteins in the nucleus vary depending on the cell type and state of the cell, the genes affected vary.

Activated TGFβ receptors and their bound ligand are endocytosed by two distinct routes, one leading to further activation, and the other leading to inactivation. The activation route depends on clathrin-coated vesicles and leads to early endosomes (discussed in Chapter 13), where most of the Smad activation occurs. An anchoring protein called *SARA* (for *Smad anchor for receptor activation*) has an important role in this pathway; it is concentrated in early endosomes and binds to both activated TGFβ receptors and Smads, increasing the efficiency of receptor-mediated Smad phosphorylation. The inactivation route depends on *caveolae* (discussed in Chapter 13) and leads to receptor ubiquitylation and degradation in proteasomes.

Some TGFβ family members serve as graded morphogens during development, inducing different responses in a developing cell depending on the concentration of the morphogen (see Figure 15–10, and discussed in Chapter 22). Their effective extracellular concentrations are often regulated by secreted inhibitory proteins that bind directly to the signal molecules and prevent them from activating their receptors on target cells. *Noggin* and *chordin,* for example, inhibit BMPs, and *follistatin* inhibits activins. Some of these inhibitors, as well as most TGFβ family members themselves, are secreted as inactive precursors, which are activated by extracellular proteolytic cleavage.

During the signaling response, the Smads continuously shuttle between the cytoplasm and the nucleus: they are dephosphorylated in the nucleus and exported to the cytoplasm, where they can be rephosphorylated by activated receptors. In this way, the effect exerted on the target genes reflects both the concentration of the extracellular signal and the time for which it continues to act on the cell-surface receptors (often several hours). Cells exposed to a morphogen at

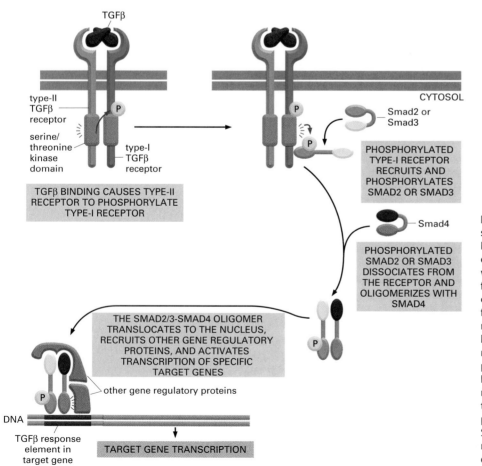

Figure 15–69 **The Smad-dependent signaling pathway activated by TGFβ.** Note that TGFβ is a dimer and that Smads open up to expose a dimerization surface when they are phosphorylated. Several features of the pathway have been omitted for simplicity, including the following. (1) The type-I and type-II receptor proteins are both thought to be homodimers. (2) The type-I receptors are normally associated with an inhibitory protein, which dissociates when the type-I receptor is phosphorylated by a type-II receptor. (3) The individual Smads are thought to be trimers. (4) An anchoring protein called SARA helps to recruit Smad2 or Smad3 to the activated type I receptor, mainly in endosomes, as discussed in the text.

high concentration, or for a long time, or both, will switch on one set of genes, while cells receiving a lower or more transient exposure will switch on another set.

As with the JAK–STAT pathway, negative feedback regulates the Smad pathway. Among the target genes activated by Smad complexes are those that encode *inhibitory Smads,* either Smad6 or Smad7. Smad7 (and possibly Smad6) binds to the activated receptor and inhibits its signaling ability in at least three ways: (1) it competes with R-Smads for binding sites on the receptor, decreasing R-Smad phosphorylation; (2) it recruits a ubiquitin ligase called *Smurf,* which ubiquitylates the receptor, leading to receptor internalization and degradation (it is because Smurfs also ubiquitylate and promote the degradation of Smads that they are called *Smad ubiquitylation regulatory factors,* or Smurfs); and (3) it recruits a protein phosphatase that dephosphorylates and inactivates the receptor. In addition, the inhibitory Smads bind to the co-Smad, Smad4, and inhibit it, either by preventing its binding to R-Smads or by promoting its ubiquitylation and degradation.

Although receptor serine/threonine kinases operate mainly through the Smad pathway just described, they can also affect other intracellular signaling pathways. Conversely, signaling proteins in other pathways can phosphorylate Smads and thereby influence signaling along the Smad pathway.

Serine/Threonine and Tyrosine Protein Kinases Are Structurally Related

All the signaling pathways activated by GPCRs and enzyme-coupled receptors that we have discussed so far depend on serine/threonine-specific protein kinases, tyrosine-specific protein kinases, or both. These kinases are all structurally related, as summarized in **Figure 15–70**.

The bewildering complexity of multiple cross-regulatory signaling pathways and feedback loops that we have discussed is not just a haphazard tangle, but a highly evolved system for processing and interpreting the mass of signals that impinge upon animal cells. The whole molecular control network, leading from the receptors at the cell surface to the genes in the nucleus, can be viewed as a computing device; and, like that other computing device, the brain, it presents one of the hardest problems in biology. We can identify the components and discover how they work individually. We can understand how small subsets of components work together as regulatory modules, as we have seen. But, it is a much more difficult task to understand how the system works as a whole. This is not only because the system is complex; it is also because the way it behaves is strongly dependent on the quantitative details of the molecular interactions, and for most animal cells we have only rough qualitative information.

In bacterial cells, the signaling pathways are simpler, and precise quantitative information is much easier to obtain. It is therefore possible to give a detailed account of how a complete signaling system works, at least for one particular bacterial cell behavior and the signals that control it. We discuss here one such example, in which bacteria respond to environmental signals delivered via enzyme-coupled receptors that are again kinases, but of a type unrelated to those we have discussed so far.

Bacterial Chemotaxis Depends on a Two-Component Signaling Pathway Activated by Histidine-Kinase-Associated Receptors

As pointed out earlier, many of the mechanisms involved in chemical signaling between cells in multicellular animals are thought to have evolved from mechanisms used by unicellular organisms to respond to chemical changes in their environment. In fact, both types of organisms use some of the same intracellular mediators, such as cyclic nucleotides and Ca^{2+}. Among the best-studied reactions of unicellular organisms to extracellular signals are their *chemotactic responses,* in which cell movement is oriented toward or away

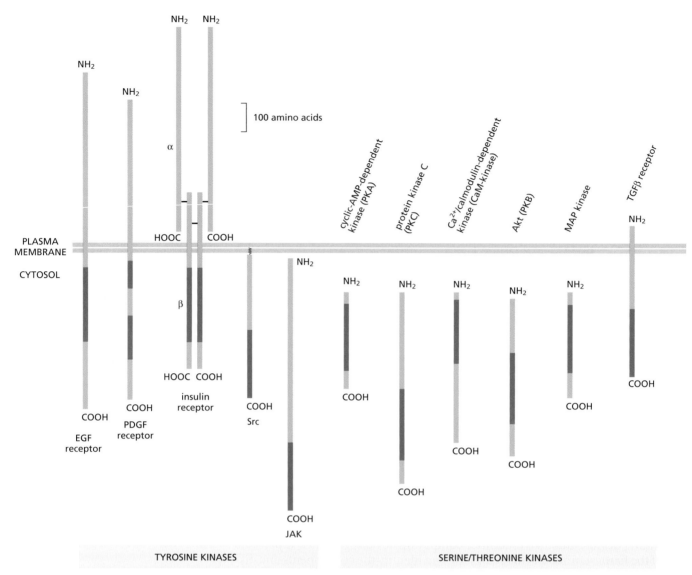

TYROSINE KINASES   SERINE/THREONINE KINASES

from a source of some chemical in the environment. We conclude this section on enzyme-coupled receptors with a brief account of **bacterial chemotaxis**, which provides a particularly well-understood illustration of the crucial role of adaptation in the response to chemical signals. This chemotaxis response is mediated by **histidine-kinase-associated receptors** that activate a *two-component signaling pathway*, which is also used by yeasts and plants, although apparently not by animals.

Motile bacteria such as *E. coli* will swim toward higher concentrations of nutrients *(attractants)*, including sugars, amino acids, and small peptides, and away from higher concentrations of various noxious chemicals *(repellents)*. They swim by means of four to six flagella, each of which is attached by a short, flexible hook at its base to a small protein disc embedded in the bacterial membrane. This disc is part of a tiny motor that uses the energy stored in the transmembrane H$^+$ gradient to rotate rapidly and turn the helical flagellum (**Figure 15–71**). Because the flagella on the bacterial surface have an intrinsic "handedness," different directions of rotation have different effects on movement. The flagella spend most of the time rotating counterclockwise, which draws all the flagella together into a coherent bundle, so that the bacterium swims uniformly in one direction. In the absence of any environmental stimulus, every second or so, one or more of the motors transiently reverses direction so that the attached flagellum breaks out of the bundle, causing the bacterium to tumble chaotically without moving forward (**Figure 15–72**). This sequence produces a characteristic pattern of movement in which smooth swimming in a straight line is interrupted by abrupt, random changes in direction caused by tumbling.

Figure 15–70 **Some of the protein kinases discussed in this chapter.** The size and location of their catalytic domains *(dark green)* are shown. In each case the catalytic domain is about 250 amino acids long. These domains are all similar in amino acid sequence, suggesting that they have all evolved from a common primordial kinase (see also Figure 3–66). Note that all of the tyrosine kinases shown are bound to the plasma membrane (JAKs are bound by their association with cytokine receptors—see Figure 15–68), whereas most of the serine/threonine kinases are in the cytosol.

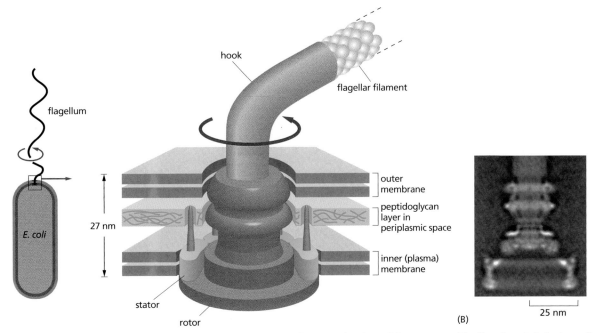

Figure 15–71 **The bacterial flagellar motor.** (A) A schematic drawing of the structure. The flagellum is linked to a flexible hook. The hook is attached to a series of protein rings (shown in *orange*), which are embedded in the outer and inner (plasma) membranes. The rings form a rotor, which rotates with the flagellum at more than 100 revolutions per second. The rotation is driven by a flow of protons through an outer ring of proteins, the stator *(blue)*, which is embedded in the inner membrane and tethered to the peptidoglycan layer. The stator also contains the proteins responsible for switching the direction of rotation. (B) A flagellar motor reconstructed from multiple electron microscopic images. (A, based on data from T. Kubori et al., *J. Mol. Biol.* 226:433–446,1992, with permission from Academic Press, and N.R. Francis et al., *Proc. Natl Acad. Sci. U.S.A.* 89:6304–6308, 1992, with permission from National Academy of Sciences; B, from D. Thomas, D.G. Morgan and D.J. DeRosier, *J. Bacteriol.* 183:6404–6412, 2001. With permission from American Society for Microbiology.)

The normal swimming behavior of bacteria is modified by chemotactic attractants or repellents, which bind to specific receptor proteins and affect the frequency of tumbling by increasing or decreasing the time that elapses between successive changes in direction of flagellar rotation. When bacteria are swimming in a favorable direction (toward a higher concentration of an attractant or away from a higher concentration of a repellent), they tumble less frequently than when they are swimming in an unfavorable direction (or when no gradient is present). Since the periods of smooth swimming are longer when a bacterium is traveling in a favorable direction, it will gradually progress in that direction—toward an attractant or away from a repellent.

Histidine-kinase-associated **chemotaxis receptors** mediate these responses. The receptors typically are dimeric transmembrane proteins that bind specific attractants and repellents on the outside of the plasma membrane. The cytoplasmic tail of the receptor is stably associated with a histidine kinase *CheA* via an adaptor protein *CheW* (**Figure 15–73**). Repellent binding activates the receptors, whereas attractant binding inactivates them; a single receptor can bind either type of molecule, with opposite consequences. The binding of a repellent to the receptor activates *CheA*, which phosphorylates itself on a histidine and almost immediately transfers the phosphoryl group to an aspartic acid on a response regulator protein *CheY*. The phosphorylated *CheY* dissociates from the receptor, diffuses through the cytosol, binds to a flagellar motor, and causes the motor to rotate clockwise, so that the bacterium tumbles. CheY has intrinsic phosphatase activity and dephosphorylates itself in a process that is greatly accelerated by the *CheZ* protein (see Figure 15–73).

Receptor Methylation Is Responsible for Adaptation in Bacterial Chemotaxis

The change in tumbling frequency in response to an increase in the concentration of an attractant or repellent occurs within less than a second, but it is only

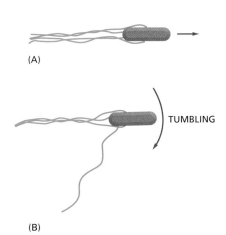

Figure 15–72 **Positions of the flagella on *E. coli* during swimming.** (A) When the flagella rotate counterclockwise (as seen looking into the cell from the flagellum), they are drawn together into a single bundle, which acts as a propeller to produce smooth swimming. (B) When one or more motors reverses direction, the attached flagellum breaks out of the bundle, causing the bacterium to tumble.

transient. Even if the higher level of ligand is maintained, within minutes, the bacteria *adapt* (desensitize), to the increased stimulus. The adaptation is a crucial part of the response, as it enables a bacterium to compare its present environment with that in its recent past and thereby respond to *changes* in the concentration of ligand rather than to steady-state levels.

The adaptation is mediated by the covalent methylation (catalyzed by a *methyl transferase*) or demethylation (catalyzed by a *methylase*) of the chemotaxis receptors, which change their responsiveness to ligand binding as a consequence of the covalent modification. When an attractant binds to a chemotaxis receptor, for example, it has two effects: (1) It decreases the ability of the receptor to activate CheA, resulting in a decreased rate of tumbling. (2) It slowly (over minutes) alters the receptor so that it can be methylated by the methyl transferase, which returns the receptor's ability to activate CheA to its original level. Thus, the unmethylated receptor without a bound ligand has the same activity as the methylated receptor with a bound ligand, and the tumbling frequency of the bacterium is therefore the same in both cases.

Each receptor dimer has eight methylation sites, and the number of sites methylated increases with increasing concentration of the attractant (as each receptor spends a longer time with a ligand bound at higher concentrations). When the attractant is removed, the methylase demethylates the receptor. Although the level of receptor methylation changes during a chemotactic response, it remains constant once a bacterium is adapted because an exact balance is reached between the rates of methylation and demethylation. A simple model of how ligand binding and methylation may operate in bacterial chemotaxis proposes that both receptor methylation and repellent binding tighten the structure of the multisubunit receptor and its associated signaling proteins, thereby increasing signaling and tumbling; by contrast, receptor demethylation and attractant binding loosen the structure of the complex, thereby decreasing signaling and tumbling. It is thought that the sensitivity of this response is greatly increased by cooperative effects that result from the clustering of the cytoplasmic tails of adjacent receptors in the membrane (**Figure 15–74**).

All of the genes and proteins involved in bacterial chemotaxis have now been identified, and their interactions have been measured in some detail. It therefore seems likely that it will be the first signaling system to be completely understood in molecular terms. Even in this relatively simple signaling network, however, computer-based simulations are required to comprehend how the system works as an integrated network. Cell signaling pathways will provide an especially rich area of investigation for a new generation of computational biologists, as the network properties of these pathways are not understandable without powerful computational tools.

As mentioned earlier, there are some cell-surface receptor proteins that do not fit into the three major classes we have discussed thus far—ion-channel-coupled, G-protein-coupled, and enzyme-coupled. In the next section, we consider cell-surface receptors that activate signaling pathways that depend on proteolysis to regulate the activity of latent gene regulatory proteins. These pathways have especially important roles in animal development and in tissue renewal and repair.

Summary

There are various classes of enzyme-coupled receptors, including receptor tyrosine kinases (RTKs), tyrosine-kinase-associated receptors, receptor serine/threonine kinases, and histidine-kinase-associated receptors. The first two classes are by far the most numerous in mammals.

Ligand binding to RTKs induces the receptors to cross-phosphorylate their cytoplasmic domains on multiple tyrosines. This transautophosphorylation both stimulates the kinases and produces a set of phosphotyrosines that serve as docking sites for a set of intracellular signaling proteins, which bind via their SH2 (or PTB) domains. One such signaling protein serves as an adaptor to couple some activated receptors to a Ras-GEF (Sos), which activates the monomeric GTPase Ras; Ras, in turn, activates a

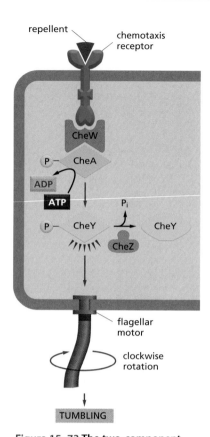

Figure 15–73 The two-component signaling pathway that enables chemotaxis receptors to control the flagellar motors during bacterial chemotaxis. The histidine kinase CheA is stably bound to the receptor via the adaptor protein CheW. The receptors and their associated proteins are all clustered at one end of the cell (see Figure 15–74). The binding of a repellent increases the activity of the receptor, which stimulates CheA to phosphorylate itself on histidine. CheA quickly transfers its covalently bound, high-energy phosphoryl group directly to the response regulator CheY to generate CheY-phosphate, which then diffuses away, binds to a flagellar motor, and causes the motor to rotate clockwise, which results in tumbling. The binding of an attractant has the opposite effect: it decreases the activity of the receptor and therefore decreases the phosphorylation of CheA and CheY, which results in counterclockwise flagellar rotation and smooth swimming. CheZ accelerates the autodephosphorylation of CheY-phosphate, thereby inactivating CheY. Each of the phosphorylated intermediates decays with a half life of less than a second, enabling the bacterium to respond very quickly to changes in its environment (see Figure 15–11).

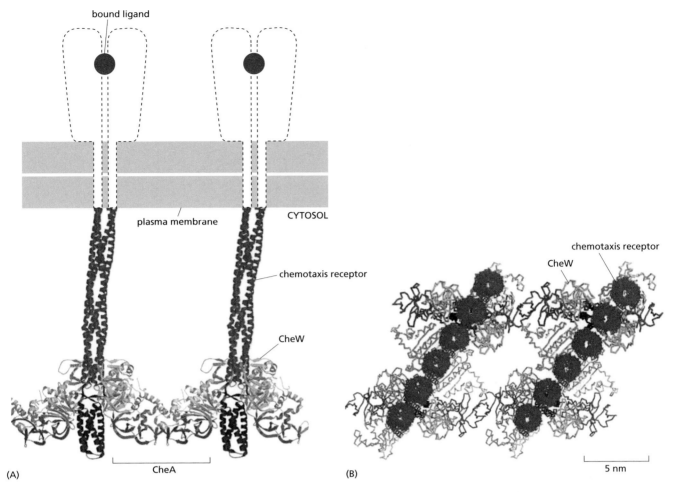

three-component MAP kinase signaling module, which relays the signal to the nucleus by phosphorylating gene regulatory proteins there. Another important signaling protein that can dock on activated RTKs is PI 3-kinase, which phosphorylates specific phosphoinositides to produce lipid docking sites in the plasma membrane for signaling proteins with phosphoinositide-binding PH domains, including the serine/threonine protein kinase Akt (PKB), which plays a key part in the control of cell survival and growth. Many receptor classes, including some RTKs, activate Rho family monomeric GTPases, which functionally couple the receptors to the cytoskeleton.

Tyrosine-kinase-associated receptors depend on various cytoplasmic tyrosine kinases for their action. These kinases include members of the Src family, which associate with many kinds of receptors, and the focal adhesion kinase (FAK), which associates with integrins at focal adhesions. The cytoplasmic tyrosine kinases then phosphorylate a variety of signaling proteins to relay the signal onward. The largest family of receptors in this class is the cytokine receptor family. When stimulated by ligand binding, these receptors activate JAK cytoplasmic tyrosine kinases, which phosphorylate STATs. The STATs then dimerize, translocate to the nucleus, and activate the transcription of specific genes. Receptor serine/threonine kinases, which are activated by signal proteins of the TGFβ superfamily, act similarly: they directly phosphorylate and activate Smads, which then oligomerize with another Smad, translocate to the nucleus, and activate gene transcription.

Bacterial chemotaxis is regulated by a signaling pathway that is understood exceptionally well. It is governed by histidine-kinase-associated chemotaxis receptors, which activate a two-component signaling pathway. When activated by a repellent, a chemotaxis receptor stimulates its associated histidine kinase to phosphorylate itself on histidine and then transfer the phosphoryl group to a response regulator protein, which relays the signal to a flagellar motor to alter the bacterium's swimming behavior. Attractants have the opposite effect on this kinase and therefore on swimming. These responses depend crucially on receptor adaptation, which is mediated by reversible receptor methylation.

Figure 15–74 A structural model for the clustering of chemotaxis receptors in the bacterial plasma membrane. Two clustered receptors are shown. Each is a homodimer, and the long α-helical cytoplasmic tail of each of the two subunits folds back on itself. Thus, each subunit contributes two helices to the four-helix bundle. Distance measurements between protein domains, determined by a technique known as pulsed electron-spin resonance, have been integrated with the known three-dimensional protein structure to produce the model shown. The clustering of receptor-CheA-CheW complexes into arrays of this type allows cooperative interactions between adjacent complexes, greatly increasing the sensitivity of the signaling process. (A) View of the structure just below the bacterial plasma membrane. The extracellular domain of the receptor is drawn schematically. (B) View of the same structure looking inward from the plasma membrane. (Adapted from S.Y. Park et al., *Nat. Struct. Biol.* 13:400–407, 2006. With permission from Macmillan Publishers Ltd.)

SIGNALING PATHWAYS DEPENDENT ON REGULATED PROTEOLYSIS OF LATENT GENE REGULATORY PROTEINS

The need for intercellular signaling is never greater than during animal development. Each cell in the embryo has to be guided along one developmental pathway or another according to its history, its position, and the character of its neighbors. At each step in the pathway, it must exchange signals with its neighbors to coordinate its behavior with theirs, ensuring the correct number and pattern of different cell types in each tissue and organ. Most of the signaling pathways already discussed are widely used for these developmental purposes, controlling cell survival, growth, proliferation, adhesion, specification, differentiation, and migration.

There are other signaling pathways, however, that are at least as important in controlling developmental processes but relay signals in other ways from cell-surface receptors to the interior of the cell. Several of these pathways depend on *regulated proteolysis* to control the activity and location of *latent gene regulatory proteins*, which enter the nucleus and activate the transcription of specific target genes only after they have been signaled to do so. Although the STAT and Smad proteins discussed earlier are also latent gene regulatory proteins, they are activated by phosphorylation in response to extracellular signals rather than by highly selective protein degradation. Signaling pathways that use latent gene regulatory proteins provide, as their primary function, a relatively direct linear pathway by which extracellular signals can control gene expression, which is presumably why they are so commonly used during development, especially to control cell fate decisions.

Although most of the pathways we discuss in this section were discovered through genetic studies in *Drosophila*, they have been highly conserved in evolution and are used over and over again during the development of different tissues and different animals. As we discuss in Chapter 23, they also have a crucial role in the many developmental processes that continue in adult tissues and organs, where new cells are constantly being produced.

We discuss four of these signaling pathways in this section: that mediated by the receptor protein *Notch*, that activated by secreted *Wnt* proteins, that activated by secreted *Hedgehog* proteins, and the pathway that activates the latent gene regulatory protein *NFκB*. All of these pathways have crucial roles in animal development, and we discuss the central roles of Notch, Wnt, and Hedgehog signaling in embryonic development in detail in Chapter 22.

The Receptor Protein Notch Is a Latent Gene Regulatory Protein

Signaling through the **Notch** receptor protein may be the most widely used signaling pathway in animal development. As discussed in Chapter 22, it has a general role in controlling cell fate choices and regulating pattern formation during the development of most tissues, as well as in the continual renewal of tissues such as the lining of the gut. It is best known, however, for its role in the production of nerve cells in *Drosophila*, which usually arise as isolated single cells within an epithelial sheet of precursor cells. During this process, when a precursor cell commits to becoming a nerve cell, it signals to its immediate neighbors not to do the same; the inhibited cells develop into epidermal cells instead. This process, called *lateral inhibition*, depends on a contact-dependent signaling mechanism that is activated by a single-pass transmembrane signal protein called **Delta**, displayed on the surface of the future neural cell. By binding to the Notch receptor protein on a neighboring cell, Delta signals to the neighbor not to become neural (**Figure 15–75**). When this signaling process is defective, the neighbors of neural cells also develop as neural cells, producing a huge excess of neurons at the expense of epidermal cells, which is lethal.

Signaling between adjacent cells via Notch and Delta (or Deltalike ligands) regulates cell fate choices in many tissues and animals. Often, it mediates lateral

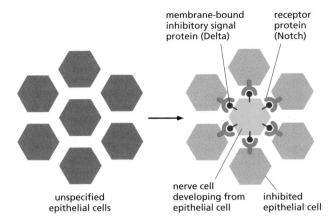

membrane-bound inhibitory signal protein (Delta)

receptor protein (Notch)

unspecified epithelial cells

nerve cell developing from epithelial cell

inhibited epithelial cell

Figure 15–75 **Lateral inhibition mediated by Notch and Delta during nerve cell development in *Drosophila*.** When individual cells in the epithelium begin to develop as neural cells, they signal to their neighbors not to do the same. This inhibitory, contact-dependent signaling is mediated by the ligand Delta, which appears on the surface of the future nerve cell and binds to Notch receptor proteins on the neighboring cells. In many tissues, all the cells in a cluster initially express both Delta and Notch, and a competition occurs, with one cell emerging as winner, expressing Delta strongly and inhibiting its neighbors from doing likewise (see Figure 22–60). In other cases, additional factors interact with Delta or Notch to make some cells susceptible to the lateral inhibition signal and others unresponsive to it.

inhibition to control the formation of mixtures of differentiated cell types within a tissue, as in the fly nervous system. In some other cases, however, it works in the opposite way, promoting rather than inhibiting a particular cell fate and driving neighboring cells to behave similarly. There is hardly any developmental cell behavior that is not regulated by Notch signaling in one tissue or another.

Notch is a single-pass transmembrane protein that requires proteolytic processing to function. It acts as a latent gene regulatory protein and provides the simplest and most direct signaling pathway known from a cell-surface receptor to the nucleus. When activated by the binding of Delta on another cell, a plasma-membrane-bound protease cleaves off the cytoplasmic tail of Notch, and the released tail translocates into the nucleus to activate the transcription of a set of Notch-response genes. The Notch tail fragment acts by binding to a DNA-binding protein, converting it from a transcriptional repressor into a transcriptional activator. We shall see that the Wnt and Hedgehog signaling pathways use this same strategy of switching a transcriptional repressor into a transcriptional activator to regulate cell fate. The set of genes regulated by Notch signaling varies depending on the tissue and circumstances, although the primary targets in most cells are members of a gene family, known (in mammals) as *Hes* genes, which encode inhibitory gene regulatory proteins. In the nervous system, for example, the products of the *Hes* genes block the expression of genes required for neural differentiation.

The Notch receptor undergoes three successive proteolytic cleavage steps, but only the last two depend on Delta binding. As part of its normal biosynthesis, it is cleaved in the Golgi apparatus to form a heterodimer, which is then transported to the cell surface as the mature receptor. The binding of Delta to Notch induces a second cleavage in the extracellular domain, mediated by an extracellular protease. A final cleavage quickly follows, cutting free the cytoplasmic tail of the activated receptor (**Figure 15–76**). Note that, unlike most receptors, the activation of Notch is irreversible; once activated by ligand binding, the protein cannot be used again.

This final cleavage of the Notch tail occurs just within the transmembrane segment, and it is mediated by a protease complex called *γ-secretase*, which is also responsible for the intramembrane cleavage of various other proteins. One of its essential subunits is *Presenilin*, so-called because mutations in the gene encoding it are a frequent cause of early-onset, familial Alzheimer's disease, a form of presenile dementia. The protease complex is thought to contribute to this and other forms of Alzheimer's disease by generating extracellular peptide fragments from a transmembrane neuronal protein; the fragments accumulate in excessive amounts and form aggregates of misfolded protein called amyloid plaques, which may injure nerve cells and contribute to their degeneration and loss.

Both Notch and Delta are glycoproteins, and their interaction is regulated by the glycosylation of Notch. The *Fringe family* of glycosyltransferases, in particular, adds extra sugars to the *O*-linked oligosaccharide (discussed in Chapter 13) on Notch, which alters the specificity of Notch for its ligands. This has provided the first example of the modulation of ligand-receptor signaling by differential receptor glycosylation.

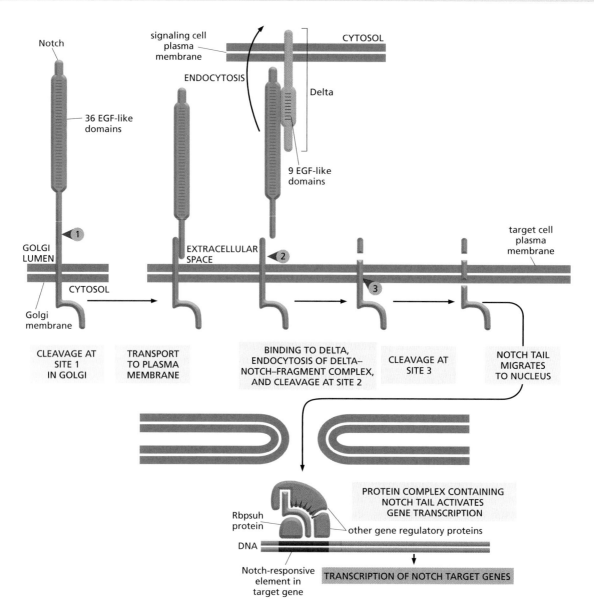

Figure 15–76 **The processing and activation of Notch by proteolytic cleavage.** The *numbered red arrowheads* indicate the sites of proteolytic cleavage. The first proteolytic processing step occurs within the *trans* Golgi network to generate the mature heterodimeric Notch receptor that is then displayed on the cell surface. The binding of Delta, which is displayed on a neighboring cell, triggers the next two proteolytic steps: the complex of Delta and the Notch subunit to which it is bound is endocytosed by the Delta-expressing cell, exposing the extracellular cleavage site in the transmembrane Notch subunit. Note that Notch and Delta interact through their repeated EGF-like domains.

The released Notch tail migrates into the nucleus, where it binds to the Rbpsuh protein and converts it from a transcriptional repressor to a transcriptional activator.

Wnt Proteins Bind to Frizzled Receptors and Inhibit the Degradation of β-Catenin

Wnt proteins are secreted signal molecules that act as local mediators and morphogens to control many aspects of development in all animals that have been studied. They were discovered independently in flies and in mice: in *Drosophila*, the *Wingless (Wg)* gene originally came to light because of its role as a morphogen in wing development (discussed in Chapter 22), while in mice, the *Int1* gene was found because it promoted the development of breast tumors when activated by the integration of a virus next to it. Both of these genes encode Wnt proteins. Wnts are unusual as secreted proteins in that they have a fatty acid chain covalently attached to their N-terminus, which increases their binding to cell surfaces. There are 19 Wnts in humans, each having distinct, but often overlapping, functions.

Wnts can activate at least three types of intracellular signaling pathways: (1) The *Wnt/β-catenin pathway* (also known as the *canonical Wnt pathway*) is centered on the latent gene regulatory protein *β-catenin*. (2) The *planar polarity pathway* coordinates the polarization of cells in the plane of a developing epithelium (discussed in Chapters 19 and 22) and depends on Rho family GTPases. (3) The Wnt/Ca^{2+} pathway stimulates an increase of intracellular Ca^{2+}, with consequences of the sort we described earlier for other pathways. All three pathways begin with the binding of Wnts to **Frizzled** family cell-surface receptors, which are seven-pass transmembrane proteins that resemble GPCRs in structure. There are seven such receptors in humans. When activated by Wnt binding, Frizzled proteins recruit the scaffold protein **Dishevelled**, which is required for relaying the signal down all three signaling pathways. We focus here on the first pathway.

The **Wnt/β-catenin pathway** acts by regulating the proteolysis of the multifunctional protein **β-catenin** (or *Armadillo* in flies), which functions both in cell–cell adhesion (discussed in Chapter 19) and in gene regulation. In this pathway (but not in the other Wnt pathways), Wnts act by binding to both a Frizzled protein and a co-receptor protein that is related to the low-density lipoprotein (LDL) receptor protein (discussed in Chapter 13) and is therefore called an **LDL-receptor-related protein** (**LRP**). In epithelial cells, most of a cell's β-catenin is located at cell–cell adherens junctions, where it is associated with *cadherins,* which are transmembrane cell–cell adhesion proteins. As discussed in Chapter 19, the β-catenin in these junctions helps link the cadherins to the actin cytoskeleton. In both epithelial and nonepithelial cells, the β-catenin that is not associated with cadherins is rapidly degraded in the cytoplasm.

The degradation of cytoplasmic β-catenin depends on a large protein *degradation complex*, which binds β-catenin and keeps it out of the nucleus while promoting its degradation. The complex contains at least four other proteins: a serine/threonine kinase called *casein kinase 1 (CK1)* phosphorylates the β-catenin on a serine, priming it for further phosphorylation by another serine/threonine kinase called *glycogen synthase kinase 3 (GSK3);* this final phosphorylation marks the protein for ubiquitylation and rapid degradation in proteasomes. Two scaffold proteins called *axin* and *Adenomatous polyposis coli (APC)* hold the protein complex together (**Figure 15–77**A). APC gets its name from the finding that the gene encoding it is often mutated in a type of benign tumor (adenoma) of the colon; the tumor projects into the lumen as a polyp and can eventually become malignant. (This APC is not to be confused with the Anaphase Promoting Complex, or APC, that plays a central part in selective protein degradation during the cell cycle—see Figure 17–20A.)

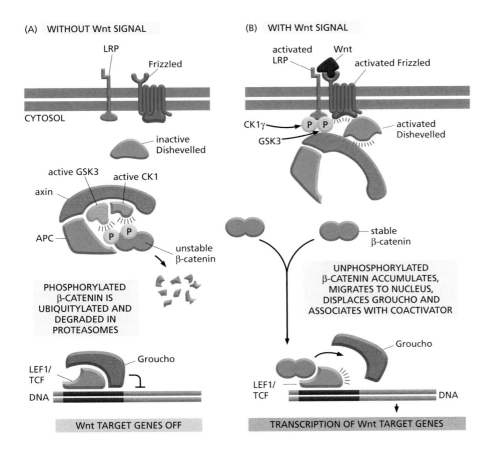

(A) WITHOUT Wnt SIGNAL

(B) WITH Wnt SIGNAL

Figure 15–77 The Wnt/β-catenin signaling pathway. (A) In the absence of a Wnt signal, β-catenin that is not bound to the cytosolic tail of cadherin proteins (not shown) becomes bound by a degradation complex containing APC, axin, GSK3, and CK1. In this complex, β-catenin is phosphorylated by CK1 and then by GSK3, triggering its ubiquitylation and degradation in proteasomes. Wnt-responsive genes are kept inactive by the Groucho co-repressor protein bound to the gene regulatory protein LEF1/TCF. (B) Wnt binding to Frizzled and LRP clusters the two types of receptors together, resulting in the recruitment of the degradation complex to the plasma membrane and the phosphorylation of the cytosolic tail of LRP by GSK3 and then by CK1γ. Axin binds to the phosphorylated LRP and is inactivated and/or degraded. The loss of axin from the degradation complex inactivates the complex and thereby blocks β-catenin phosphorylation and ubiquitylation, which allows unphosphorylated β-catenin to accumulate and translocate to the nucleus. Dishevelled and probably a G protein are required for the signaling pathway to operate; both bind to Frizzled and Dishevelled becomes phosphorylated (not shown), but their functional roles are unknown.

Once in In the nucleus, β-catenin binds to LEF1/TCF, displaces the co-repressor Groucho, and acts as a coactivator to stimulate the transcription of Wnt target genes.

The binding of a Wnt protein to both a Frizzled and LRP receptor brings the two receptors together to form a complex. In a poorly understood process, the two protein kinases, GSK3 and then CK1γ, phosphorylate the cytosolic tail of the LRP receptor, allowing the LRP tail to recruit and inactivate axin, thereby disrupting the degradation complex in the cytoplasm. In this way, the phosphorylation and degradation of β-catenin are inhibited, enabling unphosphorylated β-catenin gradually to accumulate and translocate to the nucleus, where it alters the pattern of gene transcription (Figure 15–77B).

In the absence of Wnt signaling, Wnt-responsive genes are kept silent by an inhibitory complex of gene regulatory proteins. The complex includes proteins of the *LEF1/TCF* family bound to a co-repressor protein of the *Groucho* family (see Figure 15–77A). In response to a Wnt signal, β-catenin enters the nucleus and binds to the LEF1/TCF proteins, displacing Groucho. The β-catenin now functions as a coactivator, inducing the transcription of the Wnt target genes (see Figure 15–77B). Thus, as in the case of Notch signaling, Wnt/β-catenin signaling triggers a switch from transcriptional repression to transcriptional activation.

Among the genes activated by β-catenin is *c-Myc*, which encodes a protein (c-Myc) that is a powerful stimulator of cell growth and proliferation (discussed in Chapter 17). Mutations of the *Apc* gene occur in 80% of human colon cancers (discussed in Chapter 20). These mutations inhibit the protein's ability to bind β-catenin, so that β-catenin accumulates in the nucleus and stimulates the transcription of *c-Myc* and other Wnt target genes, even in the absence of Wnt signaling. The resulting uncontrolled cell growth and proliferation promote the development of cancer.

Various secreted inhibitory proteins regulate Wnt signaling in development. Some bind to the LRP receptors and promote their down-regulation, whereas others compete with Frizzled receptors for secreted Wnts. In *Drosophila* at least, Wnts activate negative feedback loops, in which Wnt target genes encode proteins that help shut the response off; some of these proteins inhibit Dishevelled, and others are secreted inhibitors.

Hedgehog Proteins Bind to Patched, Relieving Its Inhibition of Smoothened

Hedgehog proteins and Wnt proteins act in similar ways. Both are secreted signal molecules, which act as local mediators and morphogens in many developing invertebrate and vertebrate tissues. Both proteins are modified by covalently attached lipids, depend on secreted or cell-surface-bound heparan sulfate proteoglycans (discussed in Chapter 19) for their action, and activate latent gene regulatory proteins by inhibiting their proteolysis. They both trigger a switch from transcriptional repression to transcriptional activation, and excessive signaling along either pathway in adult cells can lead to cancer. They even use some of the same intracellular signaling proteins and sometimes collaborate to mediate a response.

The **Hedgehog proteins** were discovered in *Drosophila,* where this protein family has only one member. Mutation of the *Hedgehog* gene produces a larva covered with spiky processes (denticles), like a hedgehog. At least three genes encode Hedgehog proteins in vertebrates—*Sonic, Desert,* and *Indian hedgehog.* The active forms of all Hedgehog proteins are covalently coupled to cholesterol, as well as to a fatty acid chain. The cholesterol is added during an unusual processing step, in which a precursor protein cleaves itself to produce a smaller, cholesterol-containing signal protein. Most of what we know about the downstream signaling pathway activated by Smoothened, however, came initially from genetic studies in flies, and it is the fly pathway that we summarize here.

Three transmembrane proteins—Patched, Smoothened, and iHog—mediate the responses to Hedgehog proteins. **Patched** is predicted to cross the plasma membrane 12 times, and, although much of it is in intracellular vesicles, some is on the cell surface where it binds the Hedgehog protein. **iHog** proteins have four or five immunoglobulin-like domains and two or three fibronectin-type-III-like domains; they are on the cell surface and are also thought to serve as receptors for Hedgehog proteins, probably acting as co-receptors with Patched. **Smoothened** is a seven-pass transmembrane protein with a structure very similar to a Frizzled protein. In the absence of a Hedgehog signal, Patched through some unknown mechanism, keeps Smoothened sequestered and inactive in intracellular vesicles. The binding of Hedgehog to iHog and Patched inhibits the activity of Patched and induces its endocytosis and degradation. The result is that Smoothened becomes phosphorylated, translocates to the cell surface, and relays the signal downstream.

The downstream effects are mediated by a latent gene regulatory protein called **Cubitus interruptus** (**Ci**). In the absence of a Hedgehog signal, Ci is ubiquitylated and proteolytically cleaved in proteasomes. Instead of being completely degraded, however, Ci is processed to form a smaller protein, which accumulates in the nucleus, where it acts as a transcriptional repressor, helping to keep Hedgehog-responsive genes silent. The proteolytic processing of the Ci protein depends on its phosphorylation by three serine/threonine protein kinases—PKA and two kinases also used in the Wnt pathway, namely GSK3 and CK1. As in the Wnt pathway, the proteolytic processing occurs in a multiprotein complex. The complex includes the serine/threonine kinase *Fused* and a scaffold protein *Costal2*, which stably associates with Ci, recruits the three other kinases, and binds the complex to microtubules, thereby keeping unprocessed Ci out of the nucleus (**Figure 15–78**A).

When the Hedgehog pathway is activated and Smoothened is thereby let loose in the plasma membrane, it recruits the protein complexes containing Ci, Fused, and Costal2. Costal2 is no longer able to bind the other three kinases, and so Ci is no longer cleaved. Unprocessed Ci protein can now enter the nucleus and activate the transcription of Hedgehog target genes (Figure 15–78B). Among the genes activated by Ci is *Patched* itself; the resulting increase in Patched protein on the cell surface inhibits further Hedgehog signaling—providing yet another example of a negative feedback loop.

Many gaps in the understanding of the Hedgehog signaling pathway remain to be filled in. It is not known, for example, how Patched keeps Smoothened inactive and intracellular. As the structure of Patched resembles a transmembrane transporter protein, it has been proposed that it may transport a small molecule into the cell that keeps Smoothened sequestered in vesicles.

Even less is known about the more complex Hedgehog pathway in vertebrate cells. In addition to there being at least three types of vertebrate Hedgehog proteins, there are three Ci-like gene regulatory proteins (*Gli1*, *Gli2*, and *Gli3*) downstream of Smoothened. Only Gli3 has been shown to undergo proteolytic processing like Ci and to act as either a transcriptional repressor or a transcriptional activator. Both Gli1 and Gli2 are thought to act exclusively as transcriptional activators. Moreover, in vertebrates, Smoothened, upon activation, becomes localized to a very specific site in the plasma membrane—the surface of the primary cilium, which projects from the surface of most types of vertebrate cells (as discussed in Chapter 16). The primary cilium thus acts as a Hedgehog signaling center, and the Gli proteins are also concentrated here. This arrangement presumably increases the speed and efficiency of the signaling process.

Hedgehog signaling can promote cell proliferation, and excessive Hedgehog signaling can lead to cancer. Inactivating mutations in one of the two human *Patched* genes, for example, which leads to excessive Hedgehog signaling, occurs frequently in *basal cell carcinoma* of the skin, the most common form of cancer in Caucasians. A small molecule called *cyclopamine*, made by a meadow lily, is being used to treat cancers associated with excessive Hedgehog signaling. It blocks Hedgehog signaling by binding tightly to Smoothened and inhibiting its activity. It was originally identified because it causes severe developmental defects in the progeny of sheep grazing on such lilies; these include the presence of a single central eye (a condition called *cyclopia)*, which is also seen in mice that are deficient in Hedgehog signaling.

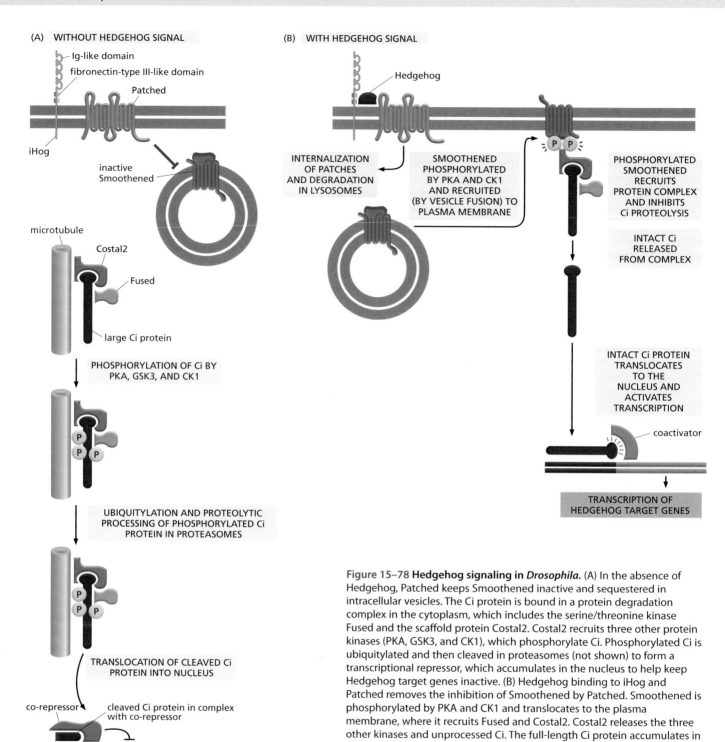

Figure 15–78 **Hedgehog signaling in *Drosophila*.** (A) In the absence of Hedgehog, Patched keeps Smoothened inactive and sequestered in intracellular vesicles. The Ci protein is bound in a protein degradation complex in the cytoplasm, which includes the serine/threonine kinase Fused and the scaffold protein Costal2. Costal2 recruits three other protein kinases (PKA, GSK3, and CK1), which phosphorylate Ci. Phosphorylated Ci is ubiquitylated and then cleaved in proteasomes (not shown) to form a transcriptional repressor, which accumulates in the nucleus to help keep Hedgehog target genes inactive. (B) Hedgehog binding to iHog and Patched removes the inhibition of Smoothened by Patched. Smoothened is phosphorylated by PKA and CK1 and translocates to the plasma membrane, where it recruits Fused and Costal2. Costal2 releases the three other kinases and unprocessed Ci. The full-length Ci protein accumulates in the nucleus and activates the transcription of Hedgehog target genes. The iHog proteins in mammals are called BOC and CDO. Many details in the pathway are poorly understood, including the role of Fused.

Many Stressful and Inflammatory Stimuli Act Through an NFκB-Dependent Signaling Pathway

The **NFκB proteins** are latent gene regulatory proteins that are present in most animal cells and are central to many stressful, inflammatory, and innate immune responses. These responses occur as a reaction to infection or injury and help protect stressed multicellular organisms and their cells (discussed in Chapter 24). An excessive or inappropriate inflammatory response in animals can also damage tissue and cause severe pain, and chronic inflammation can lead to cancer; as in the case of Wnt and Hedgehog signaling, excessive NFκB signaling is found in a number of human cancers. NFκB proteins also have important roles during normal animal development. The *Drosophila* NFκB family member *Dorsal*, for example, has a crucial role in specifying the dorsal–ventral axis of the developing fly embryo (discussed in Chapter 22).

Various cell-surface receptors activate the NFκB signaling pathway in animal cells. *Toll receptors* in *Drosophila* and *Toll-like receptors* in vertebrates, for example, recognize pathogens and activate this pathway in triggering innate immune responses (discussed in Chapter 24). The receptors for *tumor necrosis factor α (TNFα)* and *interleukin-1 (IL1)*, which are vertebrate cytokines especially important in inducing inflammatory responses, also activate this pathway. The Toll, Toll-like, and IL1 receptors belong to the same family of proteins, whereas TNF receptors belong to a different family; all of them, however, act in similar ways to activate NFκB. When activated, they trigger a multiprotein ubiquitylation and phosphorylation cascade that releases NFκB from an inhibitory protein complex, so that it can translocate to the nucleus and turn on the transcription of hundreds of genes that participate in inflammatory and innate immune responses.

There are five NFκB proteins in mammals (*RelA, RelB, c-Rel, NFκB1,* and *NFκB2*), and they form a variety of homodimers and heterodimers, each of which activates its own characteristic set of genes. Inhibitory proteins called **IκB** bind tightly to the dimers and hold them in an inactive state within the cytoplasm of unstimulated cells. There are three major IκB proteins in mammals (IκB α, β, and ε), and the signals that release NFκB dimers do so by triggering a signaling pathway that leads to the phosphorylation, ubiquitylation, and consequent degradation of the IκB proteins. The phosphorylation of IκB is mediated by *IκB kinase (IKK)*, which is a multiprotein complex containing two serine/threonine protein kinases (IKKα and IKKβ) and a regulatory protein called *NEMO* (for NFκB essential modifier), or IKKγ (**Figure 15–79**).

Among the genes activated by the released NFκB is the gene that encodes IκBα, one of three IκB isoforms that holds NFκB inactive in the cytosol of resting cells. This activation leads to the resynthesis of IκBα protein, which binds to NFκB and inactivates it, creating a negative feedback loop (**Figure 15–80A**). Experiments on TNFα-induced responses, as well as computer modeling studies of the responses, indicate that the negative feedback produces two types of

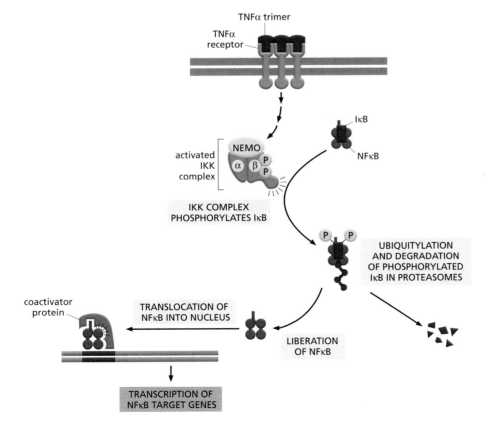

Figure 15–79 The activation of the NFκB pathway by TNFα. Both TNFα and its receptors are trimers. The binding of TNFα causes a rearrangement of the clustered cytosolic tails of the receptors, which now recruit various signaling proteins, resulting in the activation of a serine/threonine protein kinase that phosphorylates and activates IκB kinase kinase (IKK). IKK is a heterotrimer composed of two kinase subunits (IKKα and IKKβ) and a regulatory subunit called NEMO. IKKβ then phosphorylates IκB on two serines, which marks the protein for ubiquitylation and degradation in proteasomes. The released NFκB translocates into the nucleus, where, in collaboration with coactivator proteins, it stimulates the transcription of its target genes.

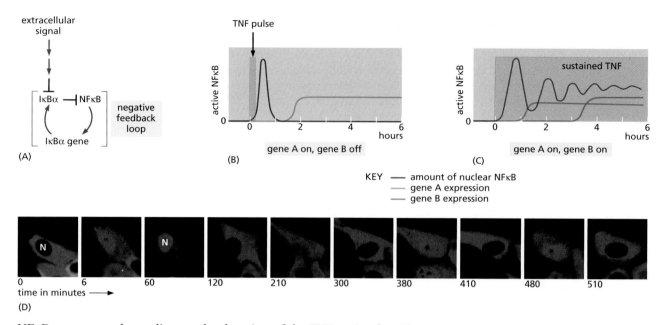

(A)

(B) gene A on, gene B off

(C) gene A on, gene B on

KEY —— amount of nuclear NFκB
—— gene A expression
—— gene B expression

(D)

NFκB responses, depending on the duration of the TNFα stimulus. Short exposure (less than an hour) to TNFα produces a short period of NFκB activation that is independent of the duration of the TNFα stimulus; the negative feedback through IκBα shuts off the response after about an hour. Prolonged exposure, by contrast, produces slow oscillations of NFκB activation, in which activation is followed by IκBα-mediated inactivation, which is followed by IκBα destruction and NFκB reactivation, and so on; the oscillations can persist for several hours before fading, even when the stimulus is maintained. Importantly, the two types of responses induce different patterns of gene expression, as some NFκB target genes turn on only in response to the prolonged oscillatory activation of NFκB (Figure 15–80B, C, and D). The negative feedback through IκBα is required for both types of responses: in cells deficient in IκBα, even a short exposure to TNFα induces a sustained activation of NFκB, without oscillations, and all of the NFκB-responsive genes are activated.

Thus far, we have discussed cell signaling mainly in animals, with a few diversions into yeasts and bacteria. But intercellular signaling is just as important for plants as it is for animals, although the mechanisms and molecules used are mainly different, as we discuss next.

Summary

Some signaling pathways that are especially important in animal development depend on proteolysis to control the activity and location of latent gene regulatory proteins. Notch receptors are themselves such latent gene regulatory proteins, which are activated by cleavage when Delta (or a related ligand) on another cell binds to them; the cleaved cytosolic tail of Notch migrates into the nucleus, where it stimulates the transcription of Notch-responsive genes. In the Wnt/β-catenin signaling pathway, by contrast, the proteolysis of the latent gene regulatory protein β-catenin is inhibited when a secreted Wnt protein binds to both a Frizzled and LRP receptor protein; as a result, β-catenin accumulates in the nucleus and activates the transcription of Wnt target genes.

Hedgehog signaling in flies works much like Wnt signaling. In the absence of a signal, a bifunctional, cytoplasmic gene regulatory protein, Ci, is proteolytically cleaved to form a transcriptional repressor that keeps Hedgehog target genes silenced. The binding of Hedgehog to its receptors (Patched and iHog) inhibits the proteolytic processing of Ci; as a result, the larger form of Ci accumulates in the nucleus and activates the transcription of Hedgehog-responsive genes. In Notch, Wnt, and Hedgehog signaling, the extracellular signal triggers a switch from transcriptional repression to transcriptional activation.

Signaling through the latent gene regulatory protein NFκB also depends on proteolysis. NFκB proteins are normally held in an inactive state by inhibitory IκB proteins

Figure 15–80 Negative feedback in the NFκB signaling pathway induces oscillations in NFκB activation.
(A) Drawing showing how activated NFκB stimulates the transcription of IκBα, the protein product of which acts back in the cytoplasm to sequester NFκB there; if the stimulus is persistent, the newly made IκBα protein will then be ubiquitylated and degraded, liberating active NFκB again so that it can return to the nucleus and activate transcription (see Figure 15–79). (B) A short exposure to TNFα produces a single, short pulse of NFκB activation, beginning within minutes and ending by 1 hour. This response turns on the transcription of gene A but not gene B. (C) A sustained exposure to TNFα for the entire 6 hours of the experiment produces oscillations in NFκB activation that damp down over time. This response turns on the transcription of both genes; gene B turns on only after several hours, indicating that gene B transcription requires prolonged activation of NFκB, for reasons that are not understood.
(D) These time-lapse confocal fluorescence micrographs from a different study of TNFα stimulation show the oscillations of NFκB in a cultured cell, as indicated by the periodic movement into the nucleus (N) of a fusion protein composed of NFκB fused to a red fluorescent protein. In the cell at the upper part of the micrographs, NFκB is active and in the nucleus at 6, 60, 210, 380, and 480 minutes, but it is exclusively in the cytoplasm at 0, 120, 300, 410, and 510 minutes. (A–C, based on data from A. Hoffmann et al., *Science* 298:1241–1245, 2002, and adapted from A.Y. Ting and D. Endy, *Science* 298:1189–1190, 2002; D, from D.E. Nelson et al., *Science* 306:704–708, 2004. All with permission from AAAS.)

in the cytoplasm. A variety of extracellular stimuli, including proinflammatory cytokines, trigger the phosphorylation and ubiquitylation of IκB, marking it for degradation; this enables the NFκB to translocate to the nucleus and activate the transcription of its target genes. NFκB also activates the transcription of the gene that encodes IκBα, creating a negative feedback loop, which can produce prolonged oscillations in NFκB activity with sustained extracellular signaling.

SIGNALING IN PLANTS

In plants, as in animals, cells are in constant communication with one another. Plant cells communicate to coordinate their activities in response to the changing conditions of light, dark, and temperature, which guide the plant's cycle of growth, flowering, and fruiting. Plant cells also communicate to coordinate activities in their roots, stems, and leaves. In this final section, we consider how plant cells signal to one another and how they respond to light. Less is known about the receptors and intracellular signaling mechanisms involved in cell communication in plants than is known in animals, and we will concentrate mainly on how the receptors and mechanisms differ from those used by animals. We discuss some of the details of plant development in Chapter 22.

Multicellularity and Cell Communication Evolved Independently in Plants and Animals

Although plants and animals are both eucaryotes, they have evolved separately for more than a billion years. Their last common ancestor is thought to have been a unicellular eucaryote that had mitochondria but no chloroplasts; the plant lineage acquired chloroplasts after plants and animals diverged. The earliest fossils of multicellular animals and plants date from almost 600 million years ago. Thus, it seems that plants and animals evolved multicellularity independently, each starting from a different unicellular eucaryote, sometime between 1.6 and 0.6 billion years ago (**Figure 15–81**).

If multicellularity evolved independently in plants and animals, the molecules and mechanisms used for cell communication will have evolved separately and would be expected to be different. There should be some degree of resemblance, however, because the genes in both plants and animals diverged from those contained by their last common unicellular ancestor. Thus, whereas both plants and animals use nitric oxide, cyclic GMP, Ca^{2+}, and Rho family GTPases for signaling, there are no homologs of the nuclear receptor family, Ras, JAK, STAT, TGFβ, Notch, Wnt, or Hedgehog encoded by the completely sequenced genome of *Arabidopsis thaliana*, the small flowering plant. Similarly, plants do not seem to use cyclic AMP for intracellular signaling.

Figure 15–81 The proposed divergence of plant and animal lineages from a common unicellular eucaryotic ancestor. The plant lineage acquired chloroplasts after the two lineages diverged. Both lineages independently gave rise to multicellular organisms—plants and animals. (Paintings courtesy of John Innes Foundation.)

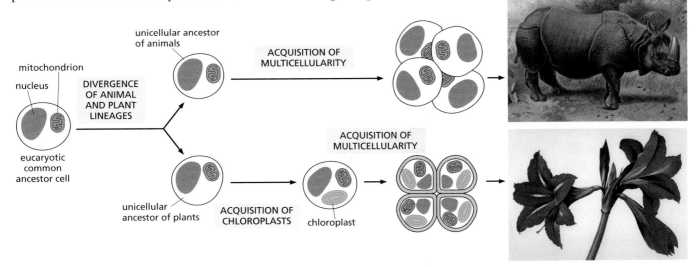

Much of what is known about the molecular mechanisms involved in signaling in plants has come from genetic studies on *Arabidopsis*. Although the specific molecules used in cell communication in plants often differ from those used in animals, the general strategies are frequently very similar. Both, for example, use enzyme-coupled cell-surface receptors, as we now discuss.

Receptor Serine/Threonine Kinases Are the Largest Class of Cell-Surface Receptors in Plants

Whereas most cell-surface receptors in animals are G-protein-coupled (GPCRs), most found so far in plants are enzyme-coupled. Moreover, whereas the largest class of enzyme-coupled receptors in animals is the receptor tyrosine kinase (RTK) class, this type of receptor is extremely rare in plants. Plants do, however, have many cytoplasmic tyrosine kinases, and tyrosine phosphorylation and dephosphorylation have important roles in plant cell signaling. Instead of RTKs, plants rely largely on a great diversity of transmembrane *receptor serine/threonine kinases*. Although very different from the corresponding animal receptors in most respects, they resemble them in having a typical serine/threonine kinase cytoplasmic domain and an extracellular ligand-binding domain. The most abundant types of these receptors have a tandem array of extracellular leucine-rich repeats (**Figure 15–82**) and are therefore called **leucine-rich repeat (LRR) receptor kinases**.

There are about 175 LRR receptor kinases encoded by the *Arabidopsis* genome. One of the best characterized is the *Clavata1/Clavata2 (Clv1/Clv2) receptor* complex. Mutations that inactivate either of the two receptor subunits cause the production of flowers with extra floral organs and a progressive enlargement of both the shoot and floral *meristems*, the groups of self-renewing stem cells that produce the cells that give rise to stems, leaves, and flowers (discussed in Chapter 22). The extracellular signal molecule that binds to the receptor is thought to be a small protein called *Clv3*, which is secreted by neighboring cells. The binding of Clv3 to the Clv1/Clv2 receptor suppresses meristem growth, either by inhibiting cell division there or, more probably, by stimulating cell differentiation (**Figure 15–83A**).

The intracellular signaling pathway from the Clv1/Clv2 receptor to the cell response is largely unknown, but it includes a serine/threonine protein phosphatase that inhibits the signaling. Other signaling proteins in the pathway include a Rho family GTPase and a nuclear gene regulatory protein that is distantly related to animal homeodomain proteins. Mutations that inactivate this gene regulatory protein have the opposite effect from those that inactivate the Clv1/Clv2 receptor: cell division is greatly decreased in the shoot meristem, and the plant produces flowers with too few organs. Thus, the intracellular signaling pathway activated by the Clv1/Clv2 receptor seems to stimulate cell differentiation by inhibiting the gene regulatory protein that normally inhibits cell differentiation (Figure 15–83B).

A different LRR receptor kinase in *Arabidopsis*, called *Bri1*, is part of a cell-surface steroid hormone receptor. Plants synthesize a class of steroids that are

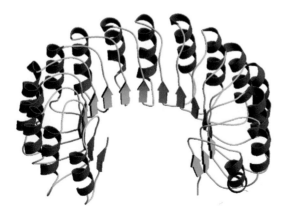

Figure 15–82 The three-dimensional structure of leucine-rich repeats, similar to those found in the LRR serine/threonine receptor kinases. Multiple copies of such repeats are present in the extracellular domain of LRR receptor kinases, where they participate in binding the signal molecule. (Courtesy of David Lawson.)

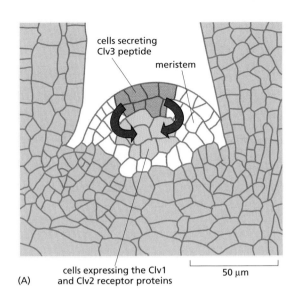

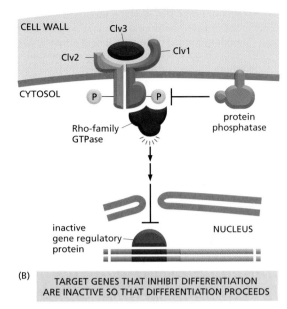

(B) **TARGET GENES THAT INHIBIT DIFFERENTIATION ARE INACTIVE SO THAT DIFFERENTIATION PROCEEDS**

called **brassinosteroids** because they were originally identified in the mustard family Brassicaceae, which includes *Arabidopsis*. These plant signal molecules regulate the growth and differentiation of plants throughout their life cycle. Binding of a brassinosteroid to a Bri1 cell-surface receptor kinase initiates a signaling cascade that uses a GSK3 protein kinase and a protein phosphatase to regulate the phosphorylation and degradation of specific gene regulatory proteins in the nucleus, and thereby specific gene transcription. Mutant plants that are deficient in the Bri1 receptor kinase are insensitive to brassinosteroids and are therefore dwarfs.

The LRR receptor kinases are only one of many classes of transmembrane receptor serine/threonine kinases in plants. There are at least six additional families, each with its own characteristic set of extracellular domains. The *lectin receptor kinases*, for example, have extracellular domains that bind carbohydrate signal molecules. The *Arabidopsis* genome encodes over 300 receptor serine/threonine kinases, which makes them the largest family of receptors known in plants. Many are involved in defense responses against pathogens.

Ethylene Blocks the Degradation of Specific Gene Regulatory Proteins in the Nucleus

Various **growth regulators** (also called **plant hormones**) help to coordinate plant development. They include *ethylene, auxin, cytokinins, gibberellins,* and *abscisic acid,* as well as brassinosteroids. Growth regulators are all small molecules made by most plant cells. They diffuse readily through cell walls and can either act locally or be transported to influence cells further away. Each growth regulator can have multiple effects. The specific effect depends on environmental conditions, the nutritional state of the plant, the responsiveness of the target cells, and which other growth regulators are acting.

Ethylene is an important example. This small gas molecule (**Figure 15–84**A) can influence plant development in various ways; it can, for example, promote fruit ripening, leaf abscission, and plant senescence. It also functions as a stress signal in response to wounding, infection, flooding, and so on. When the shoot of a germinating seedling, for instance, encounters an obstacle, such as a piece of gravel underground in the soil, the seedling responds to the encounter in three ways. First, it thickens its stem, which can then exert more force on the obstacle. Second, it shields the tip of the shoot by increasing the curvature of a specialized hook structure. Third, it reduces the shoot's tendency to grow away from the direction of gravity, so as to avoid the obstacle. This *triple response* is controlled by ethylene (Figure 15–84B and C).

Figure 15–83 A hypothetical model for how Clv3 and the Clv1/Clv2 receptor regulate cell proliferation and/or differentiation in the shoot meristem. (A) Cells in the outer layer of the meristem *(light red)* secrete Clv3 peptide, which binds to the Clv1/Clv2 receptor protein on target cells in an adjacent, more central region of the meristem *(green)*, stimulating the differentiation of the target cells. (B) Some parts of the intracellular signaling pathway activated by Clv3 binding. When activated by Clv3 binding, Clv1 phosphorylates the receptor proteins on serines and threonines, thereby activating the receptor complex and leading to the activation of a Rho family GTPase. The signaling pathway after this point is unclear, but it leads to the inhibition of a gene regulatory protein in the nucleus called Wuschel. Because Wuschel normally blocks the transcription of genes required for differentiation, its inhibition of Clv3 signaling allows differentiation to proceed. The protein phosphatase dephosphorylates the receptor proteins and thereby negatively regulates the signaling pathway.

Figure 15–84 The ethylene-mediated triple response that occurs when the growing shoot of a germinating seedling encounters an obstacle underground. (A) The structure of ethylene. (B) Before the encounter, the shoot is growing upward and is long and thin. (C) After the encounter, the shoot thickens, and the protective hook (at *top)* increases its curvature to protect the tip of the shoot. The shoot also alters its direction of growth to grow around the obstacle (not shown). (Courtesy of Melanie Webb.)

(A)

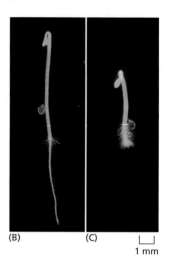

(B)   (C)

1 mm

Plants have various ethylene receptors, which are located in the endoplasmic reticulum and are all structurally related. They are dimeric, multipass transmembrane proteins, with a copper-containing ethylene-binding domain and a domain that interacts with a protein called *CTR1*, which is closely related in sequence to the Raf MAP kinase kinase kinase discussed earlier (see Figure 15–60). The function of CTR1 in ethylene signaling depends on its serine/threonine kinase activity and the association of its N-terminal domain with the ethylene receptors. Surprisingly, it is the empty receptors that are active and keep CTR1 active. By an unknown signaling mechanism, active CTR1 stimulates the ubiquitylation and degradation in proteasomes of a nuclear gene regulatory protein called *EIN3*, which is required for the transcription of ethylene-responsive genes. In this way, the empty but active receptors and active CTR1 keep ethylene-response genes off. The EIN protein gets its name from the finding that plants with inactivating mutations in the gene that encodes it are ethylene insensitive.

Ethylene binding inactivates the receptors, altering their conformation so that they no longer bind to CTR1. As a result, CTR1 is inactivated, and the downstream signaling pathway emanating from it is blocked; the EIN3 protein is no longer ubiquitylated and degraded and can now activate the transcription of the large number of ethylene-responsive genes (**Figure 15–85**).

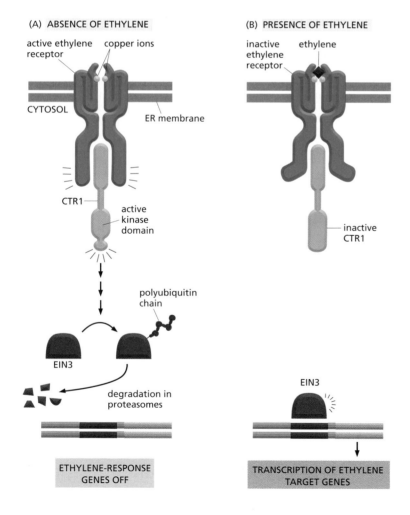

Figure 15–85 **A current view of the ethylene signaling pathway.** (A) In the absence of ethylene, both the receptors and CTR1 are active, causing the ubiquitylation and destruction of the EIN3 protein, the gene regulatory protein in the nucleus that is responsible for the transcription of ethylene-responsive genes. (B) The binding of ethylene inactivates the receptors and disrupts the interaction between the receptors and CTR1. The EIN3 protein is not degraded and can therefore activate the transcription of ethylene-responsive genes.

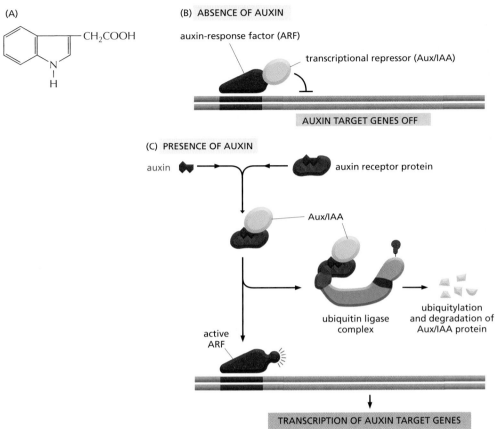

(A)

(B) ABSENCE OF AUXIN

auxin-response factor (ARF)

transcriptional repressor (Aux/IAA)

AUXIN TARGET GENES OFF

(C) PRESENCE OF AUXIN

auxin

auxin receptor protein

Aux/IAA

ubiquitin ligase complex

ubiquitylation and degradation of Aux/IAA protein

active ARF

TRANSCRIPTION OF AUXIN TARGET GENES

Figure 15–86 The auxin signaling pathway. (A) The structure of the auxin indole-3-acetic acid. (B) In the absence of auxin, a transcriptional repressor protein (called Aux/IAA) binds and suppresses a gene regulatory protein (called auxin-response factor, ARF), which is required for the transcription of auxin-responsive genes. (C) The auxin receptor proteins are mainly located in the nucleus and form part of ubiquitin ligase complexes (not shown). When activated by auxin binding, the receptor–auxin complexes recruit the ubiquitin ligase complexes, which ubiquitylate the Aux/IAA proteins, marking them for degradation in proteasomes. ARF is now free to activate the transcription of auxin-responsive genes. There are many ARFs, Aux/IAA proteins, and auxin receptors that work as illustrated.

A somewhat different strategy is involved in the regulation of *auxin*-responsive genes. Moreover, the way that auxin controls the direction and pattern of plant growth is unlike any mechanism observed in animals, as we now discuss.

Regulated Positioning of Auxin Transporters Patterns Plant Growth

The plant hormone **auxin**, which is generally indole-3-acetic acid (**Figure 15–86**A), binds to receptor proteins in the nucleus. It helps plants grow toward light, grow upward rather than branch out, and grow their roots down. It also regulates organ initiation and positioning and helps plants flower and bear fruit. Like ethylene, it influences gene expression by controlling the degradation of gene regulatory proteins in the nucleus, but instead of blocking the ubiquitylation and degradation of gene regulatory proteins required for the expression of auxin-responsive genes, it stimulates the ubiquitylation and degradation of repressor proteins that block the transcription of these genes in unstimulated cells (Figure 15–86B and C).

Auxin is unique in the way that it is transported. Unlike animal hormones, which are usually secreted by a specific endocrine organ and transported to target cells via the circulatory system, auxin has its own transport system. Specific plasma-membrane-bound *influx transporter proteins* and *efflux transporter proteins* move auxin into and out of plant cells, respectively. Distinct gene families encode the influx transporters and efflux transporters, and the two families of proteins are regulated independently. The efflux transporters are composed of *Pin proteins,* and cells can distribute them asymmetrically in the plasma membrane to make the efflux of auxin directional. A row of cells with their auxin efflux transporters confined to the basal plasma membrane, for example, will transport auxin from the top of the plant to the bottom.

In some regions of the plant, the localization of the auxin transporters, and therefore the direction of auxin flow, is highly dynamic and regulated. A cell can

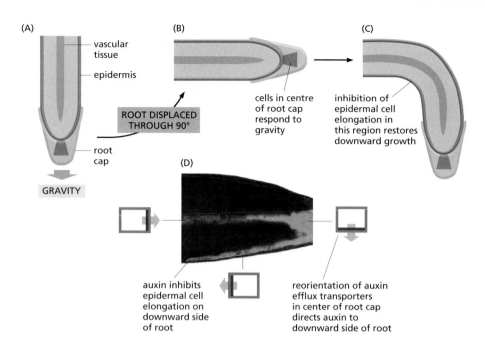

Figure 15–87 Auxin transport and root gravitropism. (A–C) Roots respond to a 90° change in the gravity vector and adjust their direction of growth so that they grow downward again. The cells that respond to gravity are in the center of the root cap, while it is the epidermal cells further back (on the lower side) that decrease their rate of elongation to restore downward growth. (D) The gravity-responsive cells in the root cap redistribute their auxin efflux transporters in response to the displacement of the root. This redirects the auxin flux mainly to the lower part of the displaced root, where it inhibits the elongation of the epidermal cells. The resulting asymmetrical distribution of auxin in the *Arabidopsis* root tip shown here is assessed indirectly, using an auxin-responsive reporter gene that encodes a protein fused to green fluorescent protein (GFP); the epidermal cells on the downward side of the root are green, whereas those on the upper side are not, reflecting the asymmetrical distribution of auxin. The distribution of auxin efflux transporters in the plasma membrane of cells in different regions of the root (shown as *gray rectangles*) is indicated in *red,* and the direction of auxin efflux is indicated by a *green arrow.* (The fluorescence photograph in D is from T. Paciorek et al., *Nature* 435:1251–1256, 2005. With permission from Macmillan Publishers Ltd.)

rapidly redistribute transporters by controlling the traffic of vesicles containing them. The auxin efflux transporters, for example, normally recycle between intracellular vesicles and the plasma membrane. A cell can redistribute these transporters on its surface by inhibiting their endocytosis in one domain of the plasma membrane, causing the transporters to accumulate there. One example occurs in the root, where gravity influences the direction of growth. The auxin efflux transporters are normally distributed symmetrically in the cap cells of the root. Within minutes of a change in the direction of the gravity vector, however, the efflux transporters redistribute to one side of the cells, so that auxin is pumped out toward the side of the root pointing downward. Because auxin inhibits root cell elongation, this redirection of auxin transport causes the root tip to reorient, so that it grows downward again (**Figure 15–87**).

In shoot apical meristems, the distribution of an auxin efflux transporter is also dynamic and regulated. Here, the directional transport of auxin helps determine the regular arrangement of leaves and flowers (see Figure 22–122).

Phytochromes Detect Red Light, and Cryptochromes Detect Blue Light

Plant development is greatly influenced by environmental conditions. Unlike animals, plants cannot move on when conditions become unfavorable; they have to adapt or they die. The most important environmental influence on plants is light, which is their energy source and has a major role throughout their entire life cycle—from germination, through seedling development, to flowering and senescence. Plants have thus evolved a large set of light-sensitive proteins to monitor the quantity, quality, direction, and duration of light. These are usually referred to as *photoreceptors*. However, because the term photoreceptor is also used for light-sensitive cells in the animal retina (see Figure 15–48), we shall use the term *photoprotein* instead.

All photoproteins sense light by means of a covalently attached light-absorbing chromophore, which changes its shape in response to light and then induces a change in the protein's conformation.

The best-known plant photoproteins are the **phytochromes**, which are present in all plants and in some algae but are absent in animals. These are dimeric, cytoplasmic serine/threonine kinases, which respond differentially and reversibly to red and far-red light: whereas red light usually activates the kinase

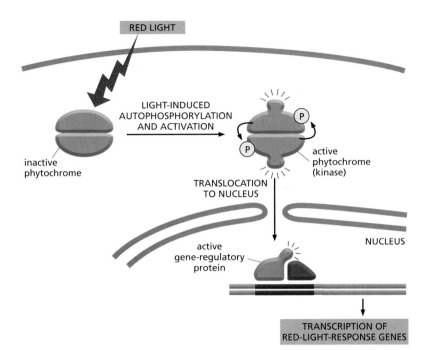

Figure 15–88 A current view of one way in which phytochromes mediate a light response in plant cells. When activated by red light, the phytochrome, which is a dimeric protein kinase, phosphorylates itself and then moves into the nucleus, where it activates gene regulatory proteins to stimulate the transcription of red-light-responsive genes.

activity of the phytochrome, far-red light inactivates it. When activated by red light, the phytochrome is thought to phosphorylate itself and then to phosphorylate one or more other proteins in the cell. In some light responses, the activated phytochrome translocates into the nucleus, where it activates gene regulatory proteins to alter gene transcription (**Figure 15–88**). In other cases, the activated phytochrome activates a latent gene regulatory protein in the cytoplasm, which then translocates into the nucleus to regulate gene transcription. In still other cases, the photoprotein triggers signaling pathways in the cytosol that alter the cell's behavior without involving the nucleus.

Although the phytochromes possess serine/threonine kinase activity, parts of their structure resemble the histidine kinases involved in bacterial chemotaxis discussed earlier. This finding suggests that the plant phytochromes originally evolved from bacterial histidine kinases and only later in evolution altered their substrate specificity from histidine to serine and threonine.

Plants sense blue light using photoproteins of two other sorts, phototropin and cryptochromes. **Phototropin** is associated with the plasma membrane and is partly responsible for *phototropism*, the tendency of plants to grow toward light. Phototropism occurs by directional cell elongation, which is stimulated by auxin, but the links between phototropin and auxin are unknown.

Cryptochromes are flavoproteins that are sensitive to blue light. They are structurally related to blue-light-sensitive enzymes called *photolyases*, which are involved in the repair of ultraviolet-induced DNA damage in all organisms, except most mammals. Unlike phytochromes, cryptochromes are also found in animals, where they have an important role in circadian clocks, which operate in most cells and cycle with a 24-hour rhythm (discussed in Chapter 7). Although cryptochromes are thought to have evolved from the photolyases, they do not have a role in DNA repair.

In this chapter, we have discussed how extracellular signals influence cell behavior. One crucial intracellular target of these signals is the cytoskeleton, which determines cell shape and is responsible for cell movements, as we discuss in the next chapter.

Summary

Plants and animals are thought to have evolved multicellularity and cell communication mechanisms independently, each starting from a different unicellular eucaryote, which in turn evolved from a common unicellular eucaryotic ancestor. Not surprisingly, therefore, the mechanisms used to signal between cells in animals and in plants have both similarities and differences. Whereas animals rely mainly on GPCRs, for example, plants rely mainly on enzyme-coupled receptors of the receptor serine/threonine kinase type, especially ones with extracellular leucine-rich repeats. Various plant hormones, or growth regulators, including ethylene and auxin, help coordinate plant development. Ethylene acts through intracellular receptors to stop the degradation of specific nuclear gene regulatory proteins, which can then activate the transcription of ethylene-responsive genes. The receptors for some other plant hormones, including auxin, also regulate the degradation of specific gene regulatory proteins, although the details vary. Auxin signaling is unusual in that it has its own highly regulated transport system, in which the dynamic positioning of plasma-membrane-bound auxin transporters controls the direction of auxin flow and thereby the direction of plant growth. Light has an important role in regulating plant development. These light responses are mediated by a variety of light-sensitive photoproteins, including phytochromes, which are responsive to red light, and cryptochromes and phototropin, which are sensitive to blue light.

PROBLEMS

Which statements are true? Explain why or why not.

15–1   The receptors involved in paracrine, synaptic, and endocrine signaling all have very high affinity for their respective signaling molecules.

15–2   All small intracellular mediators (second messengers) are water soluble and diffuse freely through the cytosol.

15–3   In the regulation of molecular switches, protein kinases and guanine nucleotide exchange factors (GEFs) always turn proteins on, whereas protein phosphatases and GTPase-activating proteins (GAPs) always turn proteins off.

15–4   In contrast to the more direct signaling pathways used by nuclear receptors, catalytic cascades of intracellular mediators provide numerous opportunities for amplifying the responses to extracellular signals.

15–5   Binding of extracellular ligands to receptor tyrosine kinases activates the intracellular catalytic domain by propagating a conformational change across the lipid bilayer through a single transmembrane α helix.

15–6   Protein tyrosine phosphatases display exquisite specificity for their substrates, unlike serine/threonine protein phosphatases, which have rather broad specificity.

15–7   Even though plants and animals independently evolved multicellularity, they use virtually all the same signaling proteins and second messengers for cell–cell communication.

Discuss the following problems.

15–8   Suppose that the circulating concentration of hormone is 10^{-10} M and the K_d for binding to its receptor is 10^{-8} M. What fraction of the receptors will have hormone bound? If a meaningful physiological response occurs when 50% of the receptors have bound a hormone molecule, how much will the concentration of hormone have to rise to elicit a response? The fraction of receptors (R) bound to hormone (H) to form a receptor–hormone complex (R–H) is $[R–H]/([R] + [R–H]) = [R–H]/[R]_{TOT} = [H]/([H] + K_d)$.

15–9   Cells communicate in ways that resemble human communication. Decide which of the following forms of human communication are analogous to autocrine, paracrine, endocrine, and synaptic signaling by cells.

A.   A telephone conversation
B.   Talking to people at a cocktail party
C.   A radio announcement
D.   Talking to yourself

15–10 Why do signaling responses that involve changes in proteins already present in the cell occur in milliseconds to seconds, whereas responses that require changes in gene expression require minutes to hours?

15–11 How is it that different cells can respond in different ways to exactly the same signaling molecule even when they have identical receptors?

15–12 Why do you suppose that phosphorylation/dephosphorylation, as opposed to allosteric binding of small molecules, for example, has evolved to play such a prominent role in switching proteins on and off in signaling pathways?

15–13 Consider a signaling pathway that proceeds through three protein kinases that are sequentially activated by phosphorylation. In one case, the kinases are held in a signaling complex by a scaffolding protein; in the other, the kinases are freely diffusible (**Figure Q15–1**). Discuss the properties of these two types of organization in terms of signal amplification, speed, and potential for cross-talk between signaling pathways.

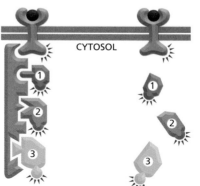

Figure Q15–1 A protein kinase cascade organized by a scaffolding protein or composed of freely diffusing components (Problem 15–13).

CYTOSOL

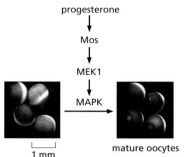

Figure Q15–2 Progesterone-induced MAP kinase activation, leading to oocyte maturation (Problem 15–15). (Courtesy of Helfrid Hochegger.)

15–14 Describe three ways in which a gradual increase in an extracellular signal can be sharpened to produce an abrupt or nearly all-or-none cellular response.

15–15 Activation ('maturation') of frog oocytes is signaled through a MAP kinase signaling module. An increase in the hormone progesterone triggers the module by stimulating the translation of Mos mRNA, which is the frog's MAP kinase kinase kinase (**Figure Q15–2**). Maturation is easy to score visually by the presence of a white spot in the middle of the brown surface of the oocyte (see Figure Q15–2). To determine the dose–response curve for progesterone-induced activation of MAP kinase, you place 16 oocytes in each of six plastic dishes and add various concentrations of progesterone. After an overnight incubation, you crush the oocytes, prepare an extract, and determine the state of MAP kinase phosphorylation (hence, activation) by SDS polyacrylamide-gel electrophoresis (**Figure Q15–3**A). This analysis shows a graded response of MAP kinase to increasing concentrations of progesterone.

Before you crushed the oocytes you noticed that not all oocytes in individual dishes had white spots. Had some oocytes undergone partial activation and not yet reached the white-spot stage? To answer this question, you repeat the experiment, but this time you analyze MAP kinase activation in individual oocytes. You are surprised to find that each oocyte has either a fully activated or a completely inactive MAP kinase (Figure Q15–3B). How can an all-or-none response in individual oocytes give rise to a graded response in the population?

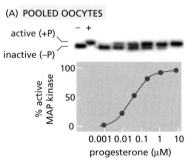

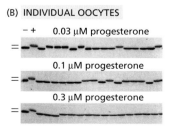

Figure Q15–3 Activation of frog oocytes (Problem 15–15). (A) Phosphorylation of MAP kinase in pooled oocytes. (B) Phosphorylation of MAP kinase in individual oocytes. MAP kinase was detected by immunoblotting using a MAP-kinase-specific antibody. The first two lanes in each gel contain nonphosphorylated, inactive MAP kinase (–) and phosphorylated, active MAP kinase (+). (From J.E. Ferrell, Jr., and E.M. Machleder, *Science* 280:895–898, 1998. With permission from AAAS.)

15–16 Propose specific types of mutations in the gene for the regulatory subunit of PKA that could lead to either a permanently active PKA or a permanently inactive PKA.

15–17 Phosphorylase kinase integrates signals from the cyclic-AMP-dependent and Ca^{2+}-dependent signaling pathways that control glycogen breakdown in liver and muscle cells (**Figure Q15–4**). Phosphorylase kinase is composed of four subunits. One is the protein kinase that catalyzes the addition of phosphate to glycogen phosphorylase to activate it for glycogen breakdown. The other three subunits are regulatory proteins that control the activity of the catalytic subunit. Two contain sites for phosphorylation by PKA, which is activated by cyclic AMP. The remaining subunit is calmodulin, which binds Ca^{2+} when the cytosolic concentration rises. The regulatory subunits control the equilibrium between the active and inactive conformations of the catalytic subunit. How does this arrangement allow phosphorylase kinase to serve its role as an integrator protein for the multiple pathways that stimulate glycogen breakdown?

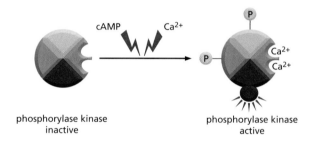

Figure Q15–4 Integration of cyclic-AMP-dependent and Ca^{2+}-dependent signaling pathways by phosphorylase kinase in liver and muscle cells (Problem 15–17).

15–18 In principle, the activated, GTP-bound form of Ras could be increased by activating a guanine-nucleotide exchange factor (GEF) or by inactivating a GTPase-activating protein (GAP). Why do you suppose that Ras-mediated signaling pathways always increase Ras–GTP by activating a GEF rather than inactivating a GAP?

15–19 The Wnt planar polarity signaling pathway normally ensures that each wing cell in *Drosophila* has a single hair. Overexpression of the *Frizzled* gene from a heat-shock promoter (hs-*Fz*) causes multiple hairs to grow from many cells (**Figure Q15–5**A). This phenotype is suppressed if hs-*Fz* is combined with a heterozygous deletion (*Dsh*$^\Delta$) of the *Dishevelled* gene (Figure Q15–5B). Do these results allow you to order the action of Frizzled and Dishevelled in the signaling pathway? If so, what is the order? Explain your reasoning.

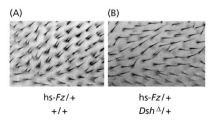

(A)　　　(B)

hs-*Fz*/+　　　hs-*Fz*/+
+/+　　　　*Dsh*$^\Delta$/+

Figure Q15–5 Pattern of hair growth on wing cells in genetically different *Drosophila* (Problem 15–19). (From C.G. Winter et al., *Cell* 105:81–91, 2001. With permission from Elsevier.)

REFERENCES

General

Bradshaw RA & Dennis EA (eds) (2003) Handbook of Cell Signaling. Elsevier: St Louis.

Science's Signal Transduction Knowledge Environment (Stke): www.stke.org

General Principles of Cell Communication

Ben-Shlomo I, Yu Hsu S, Rauch R et al (2003) Signaling receptome: a genomic and evolutionary perspective of plasma membrane receptors involved in signal transduction. *Sci STKE* 187:RE9.

Bourne HR (1995) GTPases: a family of molecular switches and clocks. *Philos Trans R Soc Lond B Biol Sci* 349:283–289.

Ferrell JE Jr (2002) Self-perpetuating states in signal transduction: positive feedback, double-negative feedback and bistability. *Curr Opin Cell Biol* 14:140–148.

Mangelsdorf DJ, Thummel C, Beato M et al (1995) The nuclear receptor superfamily: the second decade. *Cell* 83:835–839.

Murad F (2006) Shattuck Lecture. Nitric oxide and cyclic GMP in cell signaling and drug development. *N Engl J Med* 355:2003–2011.

Papin JA, Hunter T, Palsson BO & Subramaniam S (2005) Reconstruction of cellular signalling networks and analysis of their properties. *Nature Rev Mol Cell Biol* 6:99–111.

Pawson T & Scott JD (2005) Protein phosphorylation in signaling—50 years and counting. *Trends Biochem Sci* 30:286–290.

Pires-daSilva A & Sommer RJ (2003) The evolution of signalling pathways in animal development. *Nature Rev Genet* 4:39–49.

Robinson-Rechavi M, Escriva Garcia H & Laudet V (2003) The nuclear receptor superfamily. *J Cell Sci* 116:585–586.

Seet BT, Dikic I, Zhou MM & Pawson T (2006) Reading protein modifications with interaction domains. *Nature Rev Mol Cell Biol* 7:473–483.

Singla V & Reiter JF (2006) The primary cilium as the cell's antenna: signaling at a sensory organelle. *Science* 313:629–633.

Signaling Through G-protein-coupled Cell-surface Receptors and Small Intracellular Mediators

Berridge MJ (2005) Unlocking the secrets of cell signaling. *Annu Rev Physiol* 67:1–21.

Berridge MJ, Bootman MD & Roderick HL (2003) Calcium signalling: dynamics, homeostasis and remodelling. *Nature Rev Mol Cell Biol* 4:517–529.

Breer H (2003) Sense of smell: recognition and transduction of olfactory signals. *Biochem Soc Trans* 31:113–116.

Burns ME & Baylor DA (2001) Activation, deactivation, and adaptation in vertebrate photoreceptor cells. *Annu Rev Neurosci* 24:779–805.

Cooper DM (2005) Compartmentalization of adenylate cyclase and cAMP signalling. *Biochem Soc Trans* 33:1319–1322.

Hoeflich KP & Ikura M (2002) Calmodulin in action: diversity in target recognition and activation mechanisms. *Cell* 108:739–742.

Hudmon A & Schulman H (2002) Structure-function of the multifunctional Ca^{2+}/calmodulin-dependent protein kinase II. *Biochem J* 364:593–611.

Kamenetsky M, Middelhaufe S, Bank EM et al (2006) Molecular details of cAMP generation in mammalian cells: a tale of two systems. *J Mol Biol* 362:623–639.

Luttrell LM (2006) Transmembrane signaling by G protein-coupled receptors. *Methods Mol Biol* 332:3–49.

McConnachie G, Langeberg LK & Scott JD (2006) AKAP signaling complexes: getting to the heart of the matter. *Trends Mol Med* 12:317–323.

Parker PJ (2004) The ubiquitous phosphoinositides. *Biochem Soc Trans* 32:893–898.

Pierce KL, Premont RT & Lefkowitz RJ (2002) Seven-transmembrane receptors. *Nature Rev Mol Cell Biol* 3:639–650.

Reiter E & Lefkowitz RJ (2006) GRKs and beta-arrestins: roles in receptor silencing, trafficking and signaling. *Trends Endocrinol Metab* 17:159–165.

Rhee SG (2001) Regulation of phosphoinositide-specific phospholipase C. *Annu Rev Biochem* 70:281–312.

Robishaw JD & Berlot CH (2004) Translating G protein subunit diversity into functional specificity. *Curr Opin Cell Biol* 16:206–209.

Shaywitz AJ & Greenberg ME (1999) CREB: a stimulus-induced transcription factor activated by a diverse array of extracellular signals. *Annu Rev Biochem* 68:821–861.

Signaling Through Enzyme-coupled Cell-surface Receptors

Baker MD, Wolanin PM & Stock JB (2006) Signal transduction in bacterial chemotaxis. *BioEssays* 28:9–22.

Dard N & Peter M (2006) Scaffold proteins in MAP kinase signaling: more than simple passive activating platforms. *BioEssays* 28:146–156.

Darnell JE Jr, Kerr IM & Stark GR (1994) Jak-STAT pathways and transcriptional activation in response to IFNs and other extracellular signaling proteins. *Science* 264:1415–1421.

Downward J (2004) PI 3-kinase, Akt and cell survival. *Semin Cell Dev Biol* 15:177–182.

Jaffe AB & Hall A (2005) Rho GTPases: biochemistry and biology. *Annu Rev Cell Dev Biol* 21:247–269.

Massague J & Gomis RR (2006) The logic of TGFβ signaling. *FEBS Lett* 580:2811–2820.

Mitin N, Rossman KL & Der CJ (2005) Signaling interplay in Ras superfamily function. *Curr Biol* 15:R563–574.

Murai KK & Pasquale EB (2003) Eph'ective signaling: forward, reverse and crosstalk. *J Cell Sci* 116:2823–2832.

Pawson T (2004) Specificity in signal transduction: from phosphotyrosine-SH2 domain interactions to complex cellular systems. *Cell* 116:191–203.

Qi M & Elion EA (2005) MAP kinase pathways. *J Cell Sci* 118:3569–3572.

Rawlings JS, Rosler KM & Harrison DA (2004) The JAK/STAT signaling pathway. *J Cell Sci* 117:1281–1283.

Roskoski R Jr (2004) Src protein-tyrosine kinase structure and regulation. *Biochem Biophys Res Commun* 324:1155–1164.

Schlessinger J (2000) Cell signaling by receptor tyrosine kinases. *Cell* 103:211–225.

Sahin M, Greer PL, Lin MZ et al (2005) Eph-dependent tyrosine phosphorylation of ephexin1 modulates growth cone collapse. *Neuron* 46:191–204.

Schwartz MA & Madhani HD (2004) Principles of MAP kinase signaling specificity in *Saccharomyces cerevisiae*. *Annu Rev Genet* 38:725–748.

Shaw RJ & Cantley LC (2006) Ras, PI(3)K and mTOR signalling controls tumour cell growth. *Nature* 44:424–430.

Wassarman DA, Therrien M & Rubin GM (1995) The Ras signaling pathway in *Drosophila*. *Curr Opin Genet Dev* 5:44–50.

Wullschleger S, Loewith R & Hall MN (2006) TOR signaling in growth and metabolism. *Cell* 124:471–484.

Signaling Pathways Dependent on Regulated Proteolysis of Latent Gene Regulatory Proteins

Bray SJ (2006) Notch signalling: a simple pathway becomes complex. *Nature Rev Mol Cell Biol* 7:678–689.

Clevers H (2006) Wntβ-catenin signaling in development and disease. *Cell* 127:469–480.

Hoffmann A & Baltimore D (2006) Circuitry of nuclear factor κB signaling. *Immunol Rev* 210:171–186.

Huangfu D & Anderson KV (2006) Signaling from Smo to Ci/Gli: conservation and divergence of Hedgehog pathways from *Drosophila* to vertebrates. *Development* 133:3–14.

Signaling in Plants

Benavente LM & Alonso JM (2006) Molecular mechanisms of ethylene signaling in *Arabidopsis*. *Mol Biosyst* 2:165–173.

Chen M, Chory J & Fankhauser C (2004) Light signal transduction in higher plants. *Annu Rev Genet* 38:87–117.

Dievart A & Clark SE (2004) LRR-containing receptors regulating plant development and defense. *Development* 131:251–261.

Teale WD, Paponov IA & Palme K (2006) Auxin in action: signalling, transport and the control of plant growth and development. *Nature Rev Mol Cell Biol* 7:847–859.

Chapter 16

The Cytoskeleton

For cells to function properly, they must organize themselves in space and interact mechanically with their environment. They have to be correctly shaped, physically robust, and properly structured internally. Many have to change their shape and move from place to place. All cells have to be able to rearrange their internal components as they grow, divide, and adapt to changing circumstances. Eucaryotic cells have developed all these spatial and mechanical functions to a very high degree, and they depend on a remarkable system of filaments called the **cytoskeleton** (**Figure 16–1**).

The cytoskeleton pulls the chromosomes apart at mitosis and then splits the dividing cell into two. It drives and guides the intracellular traffic of organelles, ferrying materials from one part of the cell to another. It supports the fragile plasma membrane and provides the mechanical linkages that let the cell bear stresses and strains without being ripped apart as the environment shifts and changes. It enables cells such as sperm to swim and others, such as fibroblasts and white blood cells, to crawl across surfaces. It provides the machinery in the muscle cell for contraction and in the nerve cell to extend an axon and dendrites. It guides the growth of the plant cell wall and controls the amazing diversity of eucaryotic cell shapes.

The cytoskeleton's varied functions depend on the behavior of three families of protein molecules, which assemble to form three main types of filaments. Each type of filament has distinct mechanical properties, dynamics, and biological roles, but all three share certain fundamental principles. These principles provide the basis for a general understanding of how the cytoskeleton works and how the different elements cooperate. Just as we require our ligaments, bones, and muscles to work together, so all three cytoskeletal filament systems must normally function collectively to give a cell its strength, its shape, and its ability to move.

In this chapter, we begin by describing the three main types of filaments, the basic principles underlying their assembly and disassembly, and their individual peculiarities. We then describe how other proteins interact with the three main filament systems, enabling the cell to establish and maintain internal order, to shape and remodel its surface, to move organelles in a directed manner from one place to another, and—when appropriate—to move itself to new locations.

THE SELF-ASSEMBLY AND DYNAMIC STRUCTURE OF CYTOSKELETAL FILAMENTS

Most animal cells have three types of cytoskeletal filaments that are responsible for the cells' spatial organization and mechanical properties. *Intermediate filaments* provide mechanical strength. *Microtubules* determine the positions of membrane-enclosed organelles and direct intracellular transport. *Actin filaments* determine the shape of the cell's surface and are necessary for whole-cell locomotion. But these cytoskeletal filaments would be ineffective without the hundreds of accessory proteins that link the filaments to other cell components, as well as to each other. This large set of *accessory proteins* is essential for the

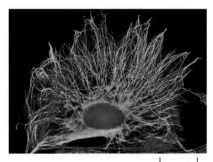

10 µm

Figure 16–1 **The cytoskeleton.** A cell in culture has been fixed and labeled to show two of its major cytoskeletal systems, the microtubules (*green*) and the actin filaments (*red*). The DNA in the nucleus is labeled in *blue*. (Courtesy of Albert Tousson.)

controlled assembly of the cytoskeletal filaments in particular locations, and it includes the *motor proteins*, remarkable molecular machines that convert the energy of ATP hydrolysis into mechanical force that can either move organelles along the filaments or move the filaments themselves.

In this section, we discuss the properties of the proteins that make up the filaments of the cytoskeleton. We focus on their ability to form intrinsically polarized and self-organized structures. We shall see that, because of the remarkable mechanisms that cause cytoskeletal filaments to be dynamic, the cell is able to respond rapidly to any eventuality that it may face.

Cytoskeletal Filaments Are Dynamic and Adaptable

Cytoskeletal systems are dynamic and adaptable, organized more like ant trails than interstate highways. A single trail of ants may persist for many hours, extending from the ant nest to a delectable picnic site, but the individual ants within the trail are anything but static. If the ant scouts find a new and better source of food, or if the picnickers clean up and leave, the dynamic structure rearranges itself with astonishing rapidity to deal with the new situation. In a similar way, large-scale cytoskeletal structures can change or persist, according to need, lasting for lengths of time ranging from less than a minute up to the cell's lifetime. But the individual macromolecular components that make up these structures are in a constant state of flux. Thus, like the alteration of an ant trail, a structural rearrangement in a cell requires little extra energy when conditions change.

Regulation of the dynamic behavior and assembly of the cytoskeletal filaments allows eucaryotic cells to build an enormous range of structures from the three basic filament systems. The micrographs in **Panel 16–1** reveal some of these structures. Microtubules, which are frequently found in a star-like cytoplasmic array emanating from the center of an interphase cell, can quickly rearrange themselves to form a bipolar *mitotic spindle* during cell division. They can also form motile whips called *cilia* and *flagella* on the surface of the cell, or tightly aligned bundles that serve as tracks for the transport of materials down long neuronal axons. In plant cells, organized arrays of microtubules help to direct the pattern of cell wall synthesis.

Actin filaments underlie the plasma membrane of animal cells, providing strength and shape to its thin lipid bilayer. They also form many types of cell-surface projections. Some of these are dynamic structures, such as the *lamellipodia* and *filopodia* that cells use to explore territory and pull themselves around. The actin-based *contractile ring* assembles transiently to divide cells in two; more stable arrays allow cells to brace themselves against an underlying substratum and enable muscle to contract. The regular bundles of *stereocilia* on the surface of hair cells in the inner ear contain stable bundles of actin filaments that tilt as rigid rods in response to sound, and similarly organized *microvilli* on the surface of intestinal epithelial cells vastly increase the apical cell surface area to enhance nutrient absorption.

Intermediate filaments line the inner face of the nuclear envelope, forming a protective cage for the cell's DNA; in the cytosol, they are twisted into strong cables that can hold epithelial cell sheets together or help nerve cells to extend long and robust axons, and they allow us to form tough appendages such as hair and fingernails.

An important and dramatic example of rapid reorganization of the cytoskeleton occurs during cell division, as shown in **Figure 16–2** for a fibroblast growing in a tissue culture dish. After the chromosomes have replicated, the interphase microtubule array that spans throughout the cytoplasm is reconfigured into the bipolar *mitotic spindle*, which serves the critical function of accurately segregating the two copies of each replicated chromosome into two separate daughter nuclei. At the same time, the specialized actin structures that enable the fibroblast to crawl across the surface of the dish disassemble so that the cell stops moving, rounds up, and assumes a more spherical shape. Actin and its associated motor protein myosin then form a belt around the middle of the cell, the *contractile ring*, which constricts like a tiny muscle to pinch the cell

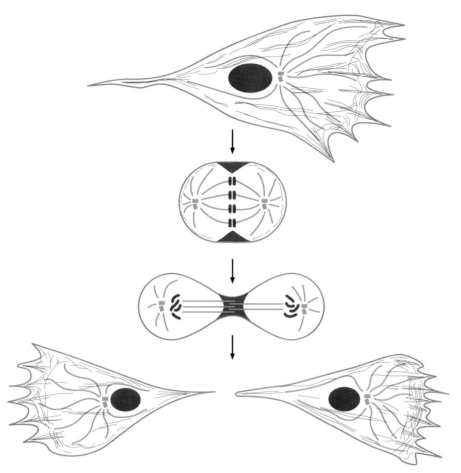

Figure 16–2 Rapid changes in cytoskeletal organization associated with cell division. The crawling fibroblast drawn here has a polarized, dynamic actin cytoskeleton (shown in *red*) that assembles to push its leading edge toward the right. The polarization of the actin cytoskeleton is assisted by the microtubule cytoskeleton (shown in *green*), consisting of long microtubules that emanate from a single microtubule organizing center located in front of the nucleus. When the cell divides, the polarized microtubule array rearranges to form a bipolar mitotic spindle, which is responsible for aligning and then separating the duplicated chromosomes (*brown*). The actin filaments form a contractile ring at the center of the cell that pinches the cell in two after the chromosomes have separated. After cell division is complete, the two daughter cells reorganize both the microtubule and actin cytoskeletons into smaller versions of those that were present in the mother cell, enabling them to crawl their separate ways.

in two. When division is complete, the cytoskeletons of the two daughter fibroblasts reassemble into their interphase structures to convert the two rounded up daughter cells into smaller versions of the flattened, crawling mother cell. In a fibroblast, this sequence of events takes about an hour; in some cases, such as the early nuclear divisions in a *Drosophila* embryo, the actin and microtubule cytoskeletons can completely rearrange themselves within less than five minutes (**Figure 16–3**).

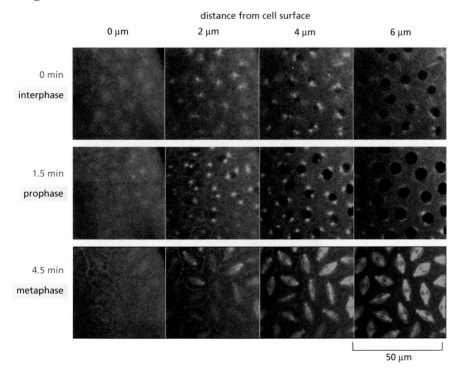

Figure 16–3 Rapid changes in cytoskeletal organization observed during the development of a *Drosophila* early embryo. <TTCT> In this giant multinuclear cell, the early nuclear divisions occur every 10 minutes or so in a common cytoplasm. The rapid rearrangements of the actin filaments *(red)* and microtubules *(green)*, seen here in a living embryo, are required to separate the chromosomes at mitosis, while keeping each nucleus from colliding with its neighbors. (Courtesy of William Sullivan.)

ACTIN FILAMENTS

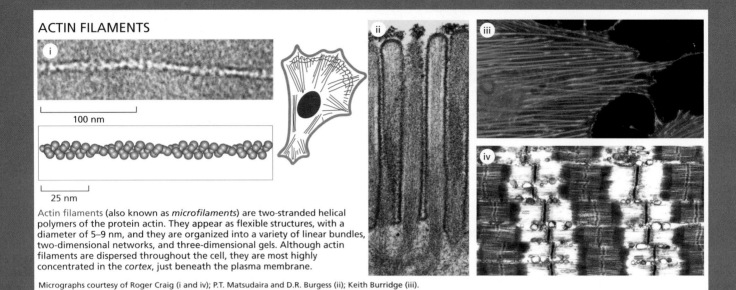

Actin filaments (also known as *microfilaments*) are two-stranded helical polymers of the protein actin. They appear as flexible structures, with a diameter of 5–9 nm, and they are organized into a variety of linear bundles, two-dimensional networks, and three-dimensional gels. Although actin filaments are dispersed throughout the cell, they are most highly concentrated in the *cortex*, just beneath the plasma membrane.

Micrographs courtesy of Roger Craig (i and iv); P.T. Matsudaira and D.R. Burgess (ii); Keith Burridge (iii).

MICROTUBULES

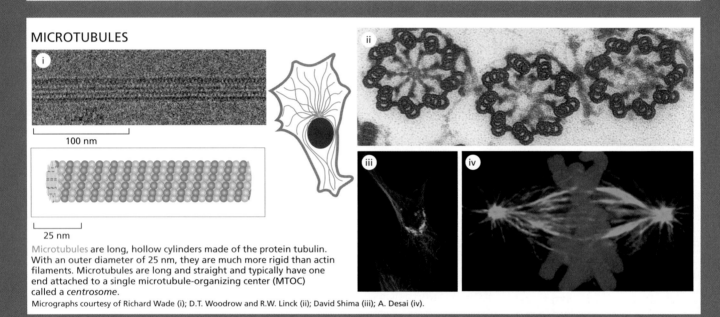

Microtubules are long, hollow cylinders made of the protein tubulin. With an outer diameter of 25 nm, they are much more rigid than actin filaments. Microtubules are long and straight and typically have one end attached to a single microtubule-organizing center (MTOC) called a *centrosome*.

Micrographs courtesy of Richard Wade (i); D.T. Woodrow and R.W. Linck (ii); David Shima (iii); A. Desai (iv).

INTERMEDIATE FILAMENTS

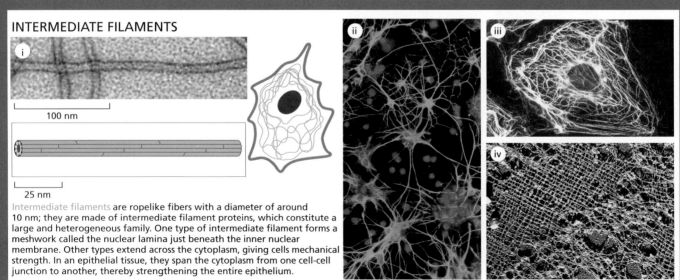

Intermediate filaments are ropelike fibers with a diameter of around 10 nm; they are made of intermediate filament proteins, which constitute a large and heterogeneous family. One type of intermediate filament forms a meshwork called the nuclear lamina just beneath the inner nuclear membrane. Other types extend across the cytoplasm, giving cells mechanical strength. In an epithelial tissue, they span the cytoplasm from one cell-cell junction to another, thereby strengthening the entire epithelium.

Micrographs courtesy of Roy Quinlan (i); Nancy L. Kedersha (ii); Mary Osborn (iii); Ueli Aebi (iv).

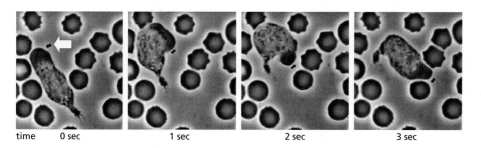

time   0 sec   1 sec   2 sec   3 sec

Figure 16–4 A neutrophil in pursuit of bacteria. <TGTA> In this preparation of human blood, a clump of bacteria (*white arrow*) is about to be captured by a neutrophil. As the bacteria move, the neutrophil quickly reassembles the dense actin network at the leading edge (*red*) to push toward the location of the bacteria. Rapid disassembly and reassembly of the actin cytoskeleton in this cell enables it to change its orientation and direction of movement within a few seconds. (From a video recorded by David Rogers.)

Many cells require rapid cytoskeletal rearrangements for their normal functioning during interphase as well. For example, the *neutrophil*, a type of white blood cell, chases and engulfs bacterial and fungal cells that accidentally gain access to the normally sterile parts of the body, as through cuts in the skin. Like most crawling cells, neutrophils advance by extending a protrusive structure at the leading edge filled with newly polymerized actin filaments. When the elusive bacterial prey moves in a different direction, the neutrophil is poised to reorganize its polarized protrusive structures within seconds (**Figure 16–4**). Both of these kinds of rapid cytoskeletal rearrangements will be discussed in more detail in the final section of this chapter.

The Cytoskeleton Can Also Form Stable Structures

In cells that have achieved a stable, differentiated morphology such as mature neurons or epithelial cells, the dynamic elements of the cytoskeleton must also provide stable, large-scale structures for cellular organization. On specialized epithelial cells that line tissues such as the intestine and the lung, cytoskeletal-based cell surface protrusions including microvilli and cilia are able to maintain a constant location, length, and diameter over the entire lifetime of the cell. For the actin bundles at the cores of microvilli on intestinal epithelial cells this is only a few days. But the actin bundles at the cores of stereocilia on the hair cells of the inner ear must maintain their stable organization for the entire lifetime of the animal, since these cells do not turn over. Nonetheless, the individual actin filaments remain strikingly dynamic and are continuously remodeled and replaced on average every 48 hours, even within these stable cell surface structures that persist for decades.

Besides forming stable specialized cell surface protrusions, the cytoskeleton is also responsible for large-scale cellular polarity, enabling cells to tell the difference between top and bottom, or front and back. The large-scale polarity information encoded by the organization of the cytoskeleton must also often be maintained over the lifetime of the cell. Polarized epithelial cells such as those found in the lining of the intestine, for example, use organized arrays of microtubules, actin filaments, and intermediate filaments to maintain the critical functional differences between the *apical surface* that absorbs nutrients from the lumen of the intestine where food passes by to the *basolateral surface* where the cells transfer nutrients through the plasma membrane to the bloodstream. They also must maintain strong adhesive contacts with one another to enable this single layer of cells to serve as an effective physical barrier (**Figure 16–5**).

Even small, morphologically simple cells such as the budding yeast *Saccharomyces cerevisiae* need stable large-scale polarity. The most notable feature of the structure of these cells is the marked asymmetry, evident in the way they divide by budding to create a small daughter cell and a large mother cell. This asymmetry derives from the polar orientation of the cell's actin cytoskeleton. There are two types of actin filament assemblies in these cells: actin cables (long bundles of actin filaments) and actin patches (small assemblies of filaments associated with the cell cortex, marking sites of actin-driven endocytosis). Proliferating budding yeast cells must be highly polarized to allow the cell to grow a bud from a single site on the cell surface, as opposed to simply growing uniformly larger. In this process, the actin patches become highly concentrated at the growing tip of the bud, with the actin cables aligned and pointing toward

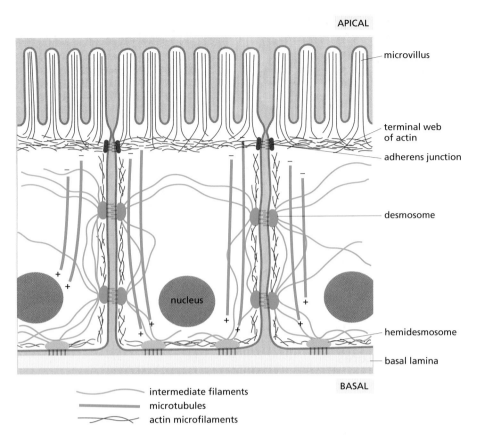

APICAL

microvillus

terminal web of actin

adherens junction

desmosome

nucleus

hemidesmosome

basal lamina

BASAL

intermediate filaments

microtubules

actin microfilaments

Figure 16–5 **Organization of the cytoskeleton in polarized epithelial cells.** All the components of the cytoskeleton cooperate to produce the characteristic shapes of specialized cells including the epithelial cells that line the small intestine. At the apical (upper) surface, facing the intestinal lumen, bundled actin filaments (*red*) form microvilli that increase the cell surface area available for absorbing nutrients from food. Just below the microvilli, a circumferential band of actin filaments contributes to forming cell-cell junctions that prevent the contents of the intestinal lumen from leaking into the body. Intermediate filaments (*blue*) are anchored to other kinds of adhesive structures including desmosomes and hemidesmosomes that connect the epithelial cells into a sturdy sheet and attach them to the underlying extracellular matrix on the basal side of the cell; these important adhesive structures will be discussed in Chapter 19. Microtubules (*green*) run vertically from the top of the cell to the bottom, and provide a global coordinate system that enables the cell to direct newly synthesized components to their proper locations.

them. This actin organization directs the secretion of new cell wall and other materials to the site of budding (**Figure 16–6**). The polarized organization of the actin structures in turn influences the orientation of the mitotic spindle, so that a complete set of replicated chromosomes can be delivered into the daughter cell at the end of the cell division process.

Each Type of Cytoskeletal Filament Is Constructed from Smaller Protein Subunits

Cytoskeletal structures frequently reach all the way from one end of the cell to the other, spanning tens or even hundreds of micrometers. Yet the individual protein molecules of the cytoskeleton are generally only a few nanometers in size. The cell builds the large structures by the repetitive assembly of large numbers of the small subunits, like building a skyscraper out of bricks. Because these subunits are small, they can diffuse rapidly within the cytoplasm, whereas the assembled filaments cannot. In this way, cells can undergo rapid

Figure 16–6 **Polarity of actin patches and cables throughout the yeast cell cycle.** Filamentous actin structures in the yeast cell, labeled here with fluorescent phalloidin, include actin patches (round bright spots) and actin cables (extended lines). (A) In a mother cell, before the formation of the bud, most of the patches become clustered at one end. The cables are lined up and point toward the cluster of patches, which is the site where the bud will emerge. (B) As the small bud grows, most patches remain within it. Cables in the mother cell continue to point toward this site of new cell wall growth. (C) Patches are almost uniformly distributed over the surface of a full-sized bud. Cables in the mother cell remain polarized. (D) Immediately after cell division, mother and daughter cells form new patches, which are concentrated near the division site, although both cells have randomly oriented cables. (From T.S. Karpova et al., *J. Cell Biol.* 142:1501–1517, 1998. With permission from The Rockefeller University Press.)

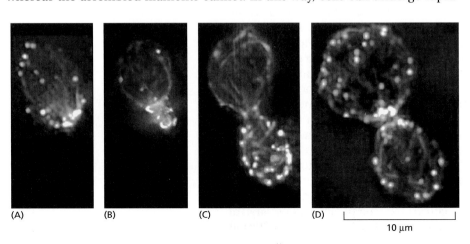

(A)   (B)   (C)   (D) |———————————|
                                          10 μm

Figure 16–7 The cytoskeleton during changes in cell shape. The formation of protein filaments from much smaller protein subunits allows regulated filament assembly and disassembly to reshape the cytoskeleton. (A) Filament formation from a small protein. (B) Rapid reorganization of the cytoskeleton in a cell in response to an external signal.

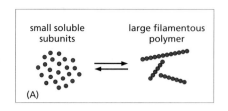

small soluble subunits → large filamentous polymer

(A)

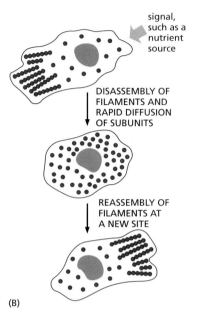

signal, such as a nutrient source

DISASSEMBLY OF FILAMENTS AND RAPID DIFFUSION OF SUBUNITS

REASSEMBLY OF FILAMENTS AT A NEW SITE

(B)

structural reorganizations, disassembling filaments at one site and reassembling them at another site far away (**Figure 16–7**).

Intermediate filaments are made up of smaller subunits that are themselves elongated and fibrous, whereas actin filaments and microtubules are made of subunits that are compact and globular—*actin subunits* for actin filaments, *tubulin subunits* for microtubules. All three types of cytoskeletal filaments form as helical assemblies of subunits (see Figure 3–26) that self-associate, using a combination of end-to-end and side-to-side protein contacts. Differences in the structures of the subunits and the strengths of the attractive forces between them produce critical differences in the stability and mechanical properties of each type of filament.

Covalent linkages between their subunits hold together many biological polymers—including DNA, RNA, and proteins. In contrast, weak noncovalent interactions hold together the three types of cytoskeletal "polymers". Consequently, their assembly and disassembly can occur rapidly, without covalent bonds being formed or broken.

Within the cell, hundreds of different cytoskeleton-associated accessory proteins regulate the spatial distribution and the dynamic behavior of the filaments, converting information received through signaling pathways into cytoskeletal action. These accessory proteins bind to the filaments or their subunits to determine the sites of assembly of new filaments, to regulate the partitioning of polymer proteins between filament and subunit forms, to change the kinetics of filament assembly and disassembly, to harness energy to generate force, and to link filaments to one another or to other cell structures such as organelles and the plasma membrane. In these processes, the accessory proteins bring cytoskeletal structure under the control of extracellular and intracellular signals, including those that trigger the dramatic transformations of the cytoskeleton that occur during each cell cycle. Acting together, the accessory proteins enable a eucaryotic cell to maintain a highly organized but flexible internal structure and, in many cases, to move.

Filaments Formed from Multiple Protofilaments Have Advantageous Properties

In general, we can view the linking of protein subunits together to form a filament as a simple association reaction. A free subunit binds to the end of a filament that contains n subunits to generate a filament of length $n + 1$. The addition of each subunit to the end of the polymer creates a new end to which yet another subunit can bind. However, the robust cytoskeletal filaments in living cells are not built by simply stringing subunits together in a single straight file. A thousand tubulin monomers, for example, lined up end to end, would span the diameter of a small eucaryotic cell, but a filament formed in this way would lack the strength to avoid breakage by ambient thermal energy, unless each subunit in the filament was bound extremely tightly to its two neighbors. Such tight binding would limit the rate at which the filaments could disassemble, making the cytoskeleton a static and less useful structure.

Cytoskeletal polymers combine strength with adaptability because they are built out of multiple **protofilaments**—long linear strings of subunits joined end to end—that associate with one another laterally. Typically, the protofilaments twist around one another in a helical lattice. The addition or loss of a subunit at the end of one protofilament makes or breaks one set of longitudinal bonds and either one or two sets of lateral bonds. In contrast, breakage of the composite filament in the middle requires breaking sets of longitudinal bonds in several

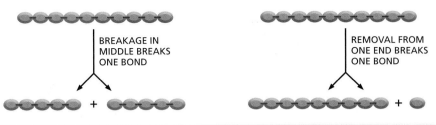

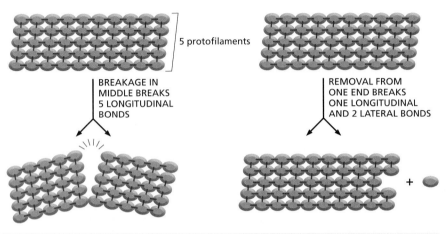

SINGLE PROTOFILAMENT: THERMALLY UNSTABLE

MULTIPLE PROTOFILAMENTS: THERMALLY STABLE

Figure 16–8 The thermal stability of cytoskeletal filaments with dynamic ends. Formation of a cytoskeletal filament from more than one protofilament allows the ends to be dynamic, while the filaments themselves are resistant to thermal breakage. In this hypothetical example, the stable filament is formed from five protofilaments. The bonds holding the subunits together in the filaments are shown in *red*.

protofilaments all at the same time (**Figure 16–8**). The large energy difference between these two processes allows most cytoskeletal filaments to resist thermal breakage, while leaving the filament ends as dynamic structures at which addition and loss of subunits can occur rapidly.

As with other specific protein–protein interactions, many hydrophobic interactions and weak noncovalent bonds hold the subunits in a cytoskeletal filament together (see Figure 3–4). The locations and types of subunit–subunit contacts differ for the different cytoskeletal filaments. Intermediate filaments, for example, assemble by forming strong lateral contacts between α-helical coiled coils, which extend over most of the length of each elongated fibrous subunit. Because the individual subunits are staggered in the filament, intermediate filaments tolerate stretching and bending, forming strong rope-like structures (**Figure 16–9**). Microtubules, by contrast, are built from globular subunits held together primarily by longitudinal bonds, and the lateral bonds holding the 13 protofilaments together are comparatively weak. For this reason, microtubules break much more easily when they are bent than do intermediate filaments.

staggered long subunits: lateral contacts dominate

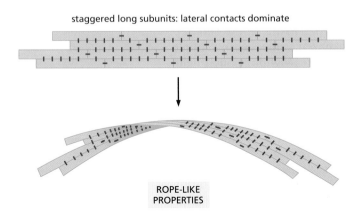

ROPE-LIKE
PROPERTIES

Figure 16–9 A strong filament formed from elongated fibrous subunits with strong lateral contacts. Intermediate filaments are formed in this way and are consequently especially resistant to stretching forces, although they are easily bent.

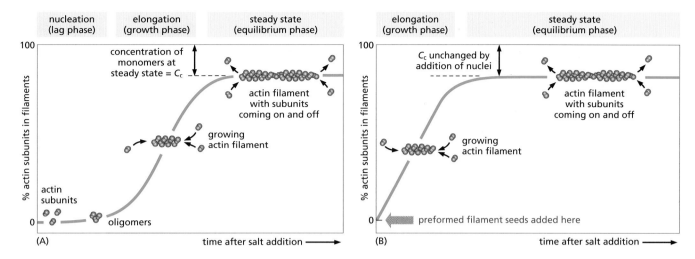

Figure 16–10 **The time course of actin polymerization in a test tube.**
(A) Polymerization is begun by raising the salt concentration in a solution of pure actin subunits. (B) Polymerization is begun in the same way, but with preformed fragments of actin filaments present to act as nuclei for filament growth. As indicated, the % free subunits reflects the critical concentration (C_C), the point at which there is no net change in polymer.

Nucleation Is the Rate-Limiting Step in the Formation of a Cytoskeletal Polymer

There is an important additional consequence of the multiple-protofilament organization of cytoskeletal polymers. Short oligomers composed of a few subunits can assemble spontaneously, but they are unstable and disassemble readily because each monomer is bonded only to a few other monomers. For a new large filament to form, subunits must assemble into an initial aggregate, or nucleus, that is stabilized by many subunit–subunit contacts and can then elongate rapidly by addition of more subunits. The initial process of nucleus assembly is called filament *nucleation*, and it can take quite a long time, depending on how many subunits must come together to form the nucleus.

The instability of smaller aggregates creates a kinetic barrier to nucleation, which is easily observed in a solution of pure actin or tubulin—the subunits of actin filaments and microtubules, respectively. When polymerization is initiated in a test tube containing a solution of pure individual subunits (by raising the temperature or raising the salt concentration), there is an initial lag phase, during which no filaments are observed. During this lag phase, however, a few of the small unstable aggregates succeed in making the transition to the more stable filament form, so that the lag phase is followed by a phase of rapid filament elongation, during which subunits add quickly onto the ends of the nucleated filaments (**Figure 16–10**A). Finally, the system approaches a steady state at which the rate of addition of new subunits to the filament ends exactly balances the rate of subunit dissociation from the ends. The concentration of free subunits left in solution at this point is called the *critical concentration, C_c*. As explained in **Panel 16–2** (pp. 978–979), the value of the critical concentration is equal to the rate constant for subunit loss divided by the rate constant for subunit addition— that is, $C_c = k_{off} / k_{on}$.

The lag phase in filament growth is eliminated if preexisting seeds (such as filament fragments that have been chemically cross-linked) are added to the solution at the beginning of the polymerization reaction (Figure 16–10B). The cell takes great advantage of this nucleation requirement: it uses special proteins to catalyze filament nucleation at specific sites, thereby determining the location at which new cytoskeletal filaments are assembled. Indeed, the regulation of filament nucleation is a primary way for cells to control their shape and their movement.

The Tubulin and Actin Subunits Assemble Head-to-Tail to Create Polar Filaments

Microtubules are formed from protein subunits of **tubulin**. The tubulin subunit is itself a heterodimer formed from two closely related globular proteins called *α-tubulin* and *β-tubulin*, tightly bound together by noncovalent bonds (**Figure**

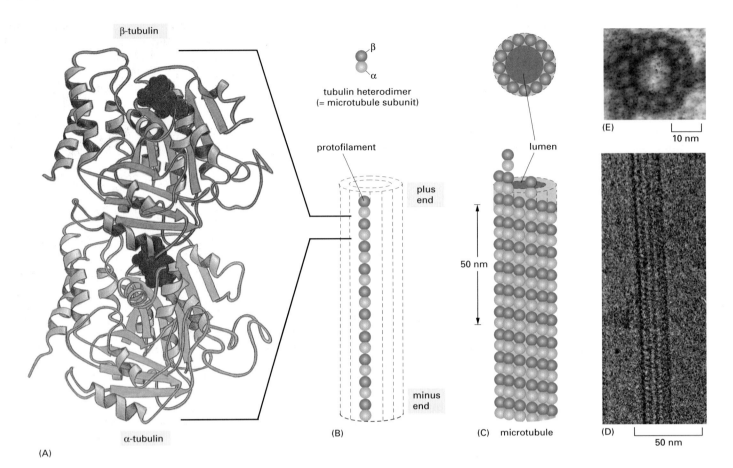

β-tubulin

β
α

tubulin heterodimer
(= microtubule subunit)

protofilament

plus end

50 nm

minus end

(A)

(B)

(C) microtubule

lumen

(E)

10 nm

(D)

50 nm

α-tubulin

16–11). These two tubulin proteins are found only in this heterodimer. Each α or β monomer has a binding site for one molecule of GTP. The GTP that is bound to the α-tubulin monomer is physically trapped at the dimer interface and is never hydrolyzed or exchanged; it can therefore be considered to be an integral part of the tubulin heterodimer structure. The nucleotide on the β-tubulin, in contrast, may be in either the GTP or the GDP form, and it is exchangeable. As we shall see, the hydrolysis of GTP at this site to produce GDP has an important effect on microtubule dynamics.

A microtubule is a hollow cylindrical structure built from 13 parallel protofilaments, each composed of alternating α-tubulin and β-tubulin molecules. When the tubulin heterodimers assemble to form the hollow cylindrical microtubule, they generate two new types of protein–protein contacts. Along the longitudinal axis of the microtubule, the "top" of one β-tubulin molecule forms an interface with the "bottom" of the α-tubulin molecule in the adjacent heterodimer. This interface is very similar to the interface holding the α and β monomers together in the dimer subunit, and the binding energy is strong. Perpendicular to these interactions, neighboring protofilaments form lateral contacts. In this dimension, the main lateral contacts are between monomers of the same type (α–α and β–β). Together, the longitudinal and lateral contacts are repeated in the regular helical lattice of the microtubule. Because multiple contacts within the lattice hold most of the subunits in a microtubule in place, the addition and loss of subunits occurs almost exclusively at the microtubule ends (see Figure 16–8). These multiple contacts among subunits make microtubules stiff and difficult to bend. The stiffness of a filament can be characterized by its *persistence length*, a property of the filament describing how long it must be before random thermal fluctuations are likely to cause it to bend. The persistence length of a microtubule is several millimeters, making microtubules the stiffest and straightest structural elements found in most animal cells.

The subunits in each protofilament in a microtubule all point in the same direction, and the protofilaments themselves are aligned in parallel (in Figure 16–11, for example, the α-tubulin is down and the β-tubulin up in each hetero-

Figure 16–11 The structure of a microtubule and its subunit. (A) The subunit of each protofilament is a tubulin heterodimer, formed from a very tightly linked pair of α- and β-tubulin monomers. The GTP molecule in the α-tubulin monomer is so tightly bound that it can be considered an integral part of the protein. The GTP molecule in the β-tubulin monomer, however, is less tightly bound and has an important role in filament dynamics. Both nucleotides are shown in *red*. **(B)** One tubulin subunit (α–β heterodimer) and one protofilament are shown schematically. Each protofilament consists of many adjacent subunits with the same orientation. **(C)** The microtubule is a stiff hollow tube formed from 13 protofilaments aligned in parallel. **(D)** A short segment of a microtubule viewed in an electron microscope. **(E)** Electron micrograph of a cross section of a microtubule showing a ring of 13 distinct protofilaments. (D, courtesy of Richard Wade; E, courtesy of Richard Linck.)

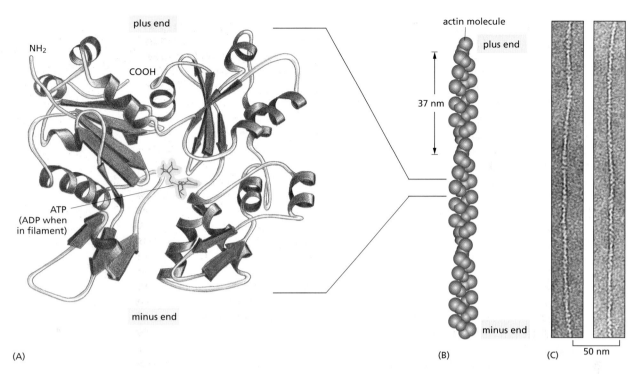

(A)   (B)   (C)   50 nm

Figure 16–12 The structures of an actin monomer and actin filament. (A) The actin monomer has a nucleotide (either ATP or ADP) bound in a deep cleft in the center of the molecule. (B) Arrangement of monomers in a filament. Although the filament is often described as a single helix of monomers, it can also be thought of as consisting of two protofilaments, held together by lateral contacts, which wind around each other as two parallel strands of a helix, with a twist repeating every 37 nm. All the subunits within the filament have the same orientation. (C) Electron micrographs of negatively stained actin filaments. (C, courtesy of Roger Craig.)

dimer). Therefore, the microtubule itself has a distinct structural polarity, with α-tubulins exposed at one end and β-tubulins exposed at the other end.

The actin subunit is a single globular polypeptide chain and is thus a monomer rather than a dimer. Like tubulin, each actin subunit has a binding site for a nucleotide, but for actin the nucleotide is ATP (or ADP) rather than GTP (or GDP) (**Figure 16–12**). As for tubulin, the actin subunits assemble head-to-tail to generate filaments with a distinct structural polarity. The actin filament can be considered to consist of two parallel protofilaments that twist around each other in a right-handed helix. Actin filaments are relatively flexible and easily bent compared with the hollow cylindrical microtubules, with a persistence length of only a few tens of micrometers. But in a living cell, accessory proteins (see below) crosslink and bundle them together, making these large-scale actin structures much stronger than an individual actin filament.

Microtubules and Actin Filaments Have Two Distinct Ends That Grow at Different Rates

The regular, parallel orientation of their subunits gives actin filaments and microtubules structural polarity. This orientation makes the two ends of each polymer different in ways that have a profound effect on filament growth rates. Addition of a subunit to either end of a filament of n subunits results in a filament of $n + 1$ subunits. In the absence of ATP or GTP hydrolysis, the free energy difference, and therefore the equilibrium constant (and the critical concentration), must be the same for addition of subunits at either end of the polymer. In this case, the ratio of the forward and backward rate constants, k_{on}/k_{off}, must be identical at the two ends, even though the absolute values of these rate constants may be very different at each end.

In a structurally polar filament, the kinetic rate constants for association and dissociation—k_{on} and k_{off}, respectively—are often much greater at one end than at the other. Thus, if an excess of purified subunits is allowed to assemble onto marked fragments of preformed filaments, one end of each fragment elongates much faster than the other (**Figure 16–13**). If filaments are rapidly diluted so that the free subunit concentration drops below the critical concentration, the fast-growing end also depolymerizes fastest. The more dynamic of the two ends of a filament, where both growth and shrinkage are fast, is called the **plus end**, and the other end is called the **minus end**.

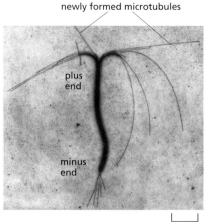

Figure 16–13 The preferential growth of microtubules at the plus end. Both microtubules and actin filaments grow faster at one end than at the other. In this case, a stable bundle of microtubules obtained from the core of a cilium (discussed later) was incubated for a short time with tubulin subunits under polymerizing conditions. Microtubules grow fastest from the plus end of the microtubule bundle, the end at the *top* in this micrograph. (Courtesy of Gary Borisy.)

On microtubules, α subunits are exposed at the minus end, and β subunits are exposed at the plus end. On actin filaments, the ATP-binding cleft on the monomer points toward the minus end. (For historical reasons, the plus ends of actin filaments are usually referred to as "barbed" ends, and minus ends as "pointed" ends, because of the arrowhead appearance of myosin heads when bound along the filament.)

Filament elongation proceeds spontaneously when the free energy change (ΔG) for addition of the soluble subunit is less than zero. This is the case when the concentration of subunits in solution exceeds the critical concentration. Likewise, filament depolymerization proceeds spontaneously when this free energy change is greater than zero. A cell can couple an energetically unfavorable process to these spontaneous processes; thus, the cell can use free energy released during spontaneous filament polymerization or depolymerization to do mechanical work—in particular, to push or pull an attached load. For example, elongating microtubules can help push out membranes, and shrinking microtubules can help pull mitotic chromosomes away from their sisters during anaphase. Similarly, elongating actin filaments help protrude the leading edge of motile cells, as we discuss later.

Filament Treadmilling and Dynamic Instability Are Consequences of Nucleotide Hydrolysis by Tubulin and Actin

Thus far, our discussion of filament dynamics has ignored a critical fact that applies to both actin filaments and microtubules. In addition to their ability to form noncovalent polymers, the actin and tubulin subunits are both enzymes that can catalyze the hydrolysis of a nucleoside triphosphate, ATP or GTP, respectively. For the free subunits, this hydrolysis proceeds very slowly; however, it is accelerated when the subunits are incorporated into filaments. Shortly after incorporation of an actin or tubulin subunit into a filament, nucleotide hydrolysis occurs; the free phosphate group is released from each subunit, but the nucleoside diphosphate remains trapped in the filament structure. (On tubulin, the nucleotide-binding site lies at the interface between two neighboring subunits—see Figure 16–11, whereas in actin, the nucleotide is deep in a cleft near the center of the subunit—see Figure 16–12.) Thus, two different types of filament structures can exist, one with the "T form" of the nucleotide bound (ATP for actin, GTP for tubulin), and one with the "D form" bound (ADP for actin, GDP for tubulin).

When the nucleotide is hydrolyzed, much of the free energy released by cleavage of the high-energy phosphate–phosphate bond is stored in the polymer lattice. This makes the free energy change for dissociation of a subunit from the D-form polymer more negative than the free energy change for dissociation of a subunit from the T-form polymer. Consequently, the ratio of k_{off} / k_{on} for the D-form polymer, which is numerically equal to its critical concentration [$C_c(D)$], is larger than the corresponding ratio for the T-form polymer. Thus, $C_c(D)$ is greater than $C_c(T)$. For certain concentrations of free subunits, D-form polymers will therefore shrink while T-form polymers grow.

In living cells, most of the free subunits are in the T form, as the free concentration of both ATP and GTP is about ten-fold higher than that of ADP and GDP. The longer the time that subunits have been in the polymer lattice, the more likely they are to have hydrolyzed their bound nucleotide. Whether the subunit at the very end of a filament is in the T or the D form depends on the rate of this hydrolysis compared with the rate of subunit addition. If the rate of subunit addition is high, that is if the filament is growing rapidly, then it is likely that

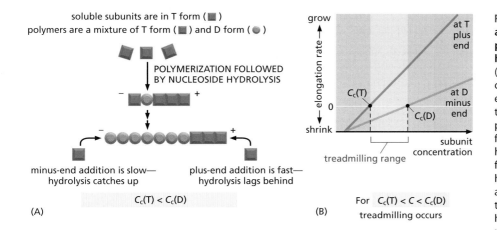

soluble subunits are in T form (■)
polymers are a mixture of T form (■) and D form (●)

POLYMERIZATION FOLLOWED
BY NUCLEOSIDE HYDROLYSIS

minus-end addition is slow—
hydrolysis catches up

plus-end addition is fast—
hydrolysis lags behind

$C_c(T) < C_c(D)$

(A)

grow

$C_c(T)$

at T
plus
end

0

at D
minus
end

$C_c(D)$

shrink

subunit
concentration

treadmilling range

(B)

For $C_c(T) < C < C_c(D)$
treadmilling occurs

Figure 16–14 The treadmilling of an actin filament or microtubule, made possible by the nucleoside triphosphate hydrolysis that follows subunit addition. (A) Explanation for the different critical concentrations (C_c) at the plus and minus ends. Subunits with bound nucleoside triphosphate (T-form subunits) polymerize at both ends of a growing filament, and then undergo nucleotide hydrolysis in the filament lattice. As the filament grows, elongation is faster than hydrolysis at the plus end in this example, and the terminal subunits at this end are therefore always in the T form. However, hydrolysis is faster than elongation at the minus end, and so terminal subunits at this end are in the D form.
(B) Treadmilling occurs at intermediate concentrations of free subunits. The critical concentration for polymerization on a filament end in the T form is lower than for a filament end in the D form. If the actual subunit concentration is somewhere between these two values, the plus end grows while the minus end shrinks, resulting in treadmilling.

a new subunit will add on to the polymer before the nucleotide in the previously added subunit has been hydrolyzed, so that the tip of the polymer remains in the T form, forming an *ATP cap* or *GTP cap*. However, if the rate of subunit addition is low, hydrolysis may occur before the next subunit is added, and the tip of the filament will then be in the D form.

The rate of subunit addition at the end of a filament is the product of the free subunit concentration and the rate constant k_{on}. The k_{on} is much faster for the plus end of a filament than for the minus end because of a structural difference between the two ends (see Panel 16–2). At an intermediate concentration of free subunits, it is therefore possible for the rate of subunit addition to be faster than nucleotide hydrolysis at the plus end, but slower than nucleotide hydrolysis at the minus end. In this case, the plus end of the filament remains in the T conformation, while the minus end adopts the D conformation. As just explained, the D form has a higher critical concentration than the T form. (In other words, the D form leans more readily toward disassembly, while the T form leans more readily toward assembly). If the concentration of free subunits in solution is in an intermediate range—higher than the critical concentration of the T form (that is, the plus end), but lower than the critical concentration of the D form (that is, the minus end)—the filament adds subunits at the plus end, and simultaneously loses subunits from the minus end. This leads to the remarkable property of filament **treadmilling** (**Figure 16–14** and Panel 16–2).

During treadmilling, subunits are recruited at the plus end of the polymer in the T form and shed from the minus end in the D form. The ATP or GTP hydrolysis that occurs along the way gives rise to the difference in the free energy of the association/dissociation reactions at the plus and minus ends of the actin filament or microtubule and thereby makes treadmilling possible. At a particular intermediate subunit concentration, the filament growth at the plus end exactly balances the filament shrinkage at the minus end. Now, the subunits cycle rapidly between the free and filamentous states, while the total length of the filament remains unchanged. This "steady-state treadmilling" requires a constant consumption of energy in the form of nucleoside triphosphate hydrolysis. While the extent of treadmilling inside the cell is uncertain, the treadmilling of single filaments has been observed *in vitro* for actin, and a phenomenon that looks like treadmilling can be observed in live cells for individual microtubules (**Figure 16–15**).

Figure 16–15 Treadmilling behavior of a microtubule, as observed in a living cell. A cell was injected with tubulin that had been covalently linked to the fluorescent dye rhodamine, so that approximately 1 tubulin subunit in 20 was fluorescent. The fluorescence of individual microtubules was then observed with a sensitive electronic camera. The microtubule shown appears to be sliding from *left* to *right*, but, in fact, the microtubule lattice remains stationary (as shown by a landmark within the microtubule lattice indicated by the *red* arrowhead), while the plus end (on the *right*) grows and the minus end (on the *left*) shrinks. The plus end also displays dynamic instability. (From C.M. Waterman-Storer and E.D. Salmon, *J. Cell Biol.* 139:417–434, 1997. With permission from The Rockefeller University Press.)

time 0 sec

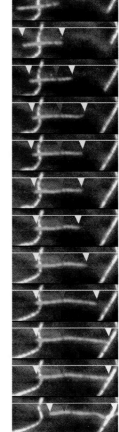

time 98 sec

10 μm

ON RATES AND OFF RATES

A linear polymer of protein molecules, such as an actin filament or a microtubule, assembles (polymerizes) and disassembles (depolymerizes) by the addition and removal of subunits at the ends of the polymer. The rate of addition of these subunits (called monomers) is given by the rate constant k_{on}, which has units of $M^{-1} sec^{-1}$. The rate of loss is given by k_{off} (units of sec^{-1}).

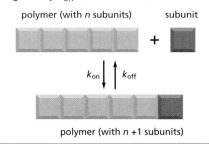

polymer (with n subunits)   subunit

k_{on} | k_{off}

polymer (with n +1 subunits)

NUCLEATION

A helical polymer is stabilized by multiple contacts between adjacent subunits. In the case of actin, two actin molecules bind relatively weakly to each other, but addition of a third actin monomer to form a trimer makes the entire group more stable.

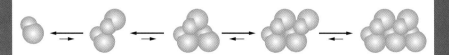

Further monomer addition can take place onto this trimer, which therefore acts as a nucleus for polymerization. For tubulin, the nucleus is larger and has a more complicated structure (possibly a ring of 13 or more tubulin molecules)—but the principle is the same.

The assembly of a nucleus is relatively slow, which explains the lag phase seen during polymerization. The lag phase can be reduced or abolished entirely by adding premade nuclei, such as fragments of already polymerized microtubules or actin filaments.

THE CRITICAL CONCENTRATION

The number of monomers that add to the polymer (actin filament or microtubule) per second will be proportional to the concentration of the free subunit ($k_{on}C$), but the subunits will leave the polymer end at a constant rate (k_{off}) that does not depend on C. As the polymer grows, subunits are used up, and C is observed to drop until it reaches a constant value, called the critical concentration (C_c). At this concentration the rate of subunit addition equals the rate of subunit loss.

At this equilibrium,

$$k_{on} C = k_{off}$$

so that

$$C_c = \frac{k_{off}}{k_{on}} = \frac{1}{K}$$

(where K is the equilibrium constant for subunit addition; see Figure 3–43).

TIME COURSE OF POLYMERIZATION

The assembly of a protein into a long helical polymer such as a cytoskeletal filament or a bacterial flagellum typically shows the following time course:

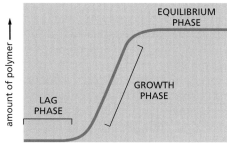

The lag phase corresponds to time taken for nucleation.

The growth phase occurs as monomers add to the exposed ends of the growing filament, causing filament elongation.

The equilibrium phase, or steady state, is reached when the growth of the polymer due to monomer addition precisely balances the shrinkage of the polymer due to disassembly back to monomers.

PLUS AND MINUS ENDS

The two ends of an actin filament or microtubule polymerize at different rates. The fast-growing end is called the plus end, whereas the slow-growing end is called the minus end. The difference in the rates of growth at the two ends is made possible by changes in the conformation of each subunit as it enters the polymer.

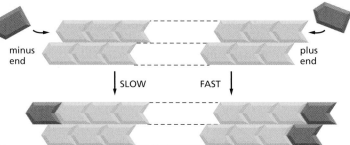

minus end   plus end

SLOW   FAST

free subunit → subunit in polymer

This conformational change affects the rates at which subunits add to the two ends.

Even though k_{on} and k_{off} will have different values for the plus and minus ends of the polymer, their ratio k_{off}/k_{on}—and hence C_c—must be the same at both ends for a simple polymerization reaction (no ATP or GTP hydrolysis). This is because exactly the same subunit interactions are broken when a subunit is lost at either end, and the final state of the subunit after dissociation is identical. Therefore, the ΔG for subunit

loss, which determines the equilibrium constant for its association with the end, is identical at both ends: if the plus end grows four times faster than the minus end, it must also shrink four times faster. Thus, for $C > C_c$, both ends grow; for $C < C_c$, both ends shrink.

The nucleoside triphosphate hydrolysis that accompanies actin and tubulin polymerization removes this constraint.

NUCLEOTIDE HYDROLYSIS

Each actin molecule carries a tightly bound ATP molecule that is hydrolyzed to a tightly bound ADP molecule soon after its assembly into the polymer. Similarly, each tubulin molecule carries a tightly bound GTP that is converted to a tightly bound GDP molecule soon after the molecule assembles into the polymer.

(T = monomer carrying ATP or GTP)

(D = monomer carrying ADP or GDP)

free monomer   subunit in polymer

Hydrolysis of the bound nucleotide reduces the binding affinity of the subunit for neighboring subunits and makes it more likely to dissociate from each end of the filament (see Figure 16–16 for a possible mechanism). It is usually the T form that adds to the filament and the D form that leaves.

Considering events at the plus end only:

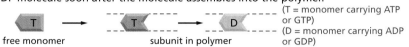

As before, the polymer will grow until $C = C_c$. For illustrative purposes, we can ignore k^D_{on} and k^T_{off} since they are usually very small, so that polymer growth ceases when

$$k^T_{on} C = k^D_{off} \qquad \text{or} \qquad C_c = \frac{k^D_{off}}{k^T_{on}}$$

This is a steady state and not a true equilibrium, because the ATP or GTP that is hydrolyzed must be replenished by a nucleotide exchange reaction of the free subunit (D → T).

ATP CAPS AND GTP CAPS

The rate of addition of subunits to a growing actin filament or microtubule can be faster than the rate at which their bound nucleotide is hydrolyzed. Under such conditions, the end has a "cap" of subunits containing the nucleoside triphosphate—an ATP cap on an actin filament or a GTP cap on a microtubule.

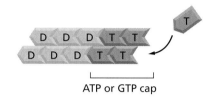

ATP or GTP cap

DYNAMIC INSTABILITY and TREADMILLING are two behaviors observed in cytoskeletal polymers. Both are associated with nucleoside triphosphate hydrolysis. Dynamic instability is believed to predominate in microtubules, whereas treadmilling may predominate in actin filaments.

TREADMILLING

One consequence of the nucleotide hydrolysis that accompanies polymer formation is to change the critical concentration at the two ends of the polymer. Since k^D_{off} and k^T_{on} refer to different reactions, their ratio k^D_{off}/k^T_{on} need not be the same at both ends of the polymer, so that:

$$C_c \text{ (minus end)} > C_c \text{ (plus end)}$$

Thus, if both ends of a polymer are exposed, polymerization proceeds until the concentration of free monomer reaches a value that is above C_c for the plus end but below C_c for the minus end. At this steady state, subunits undergo a net assembly at the plus end and a net disassembly at the minus end at an identical rate. The polymer maintains a constant length, even though there is a net flux of subunits through the polymer, known as treadmilling.

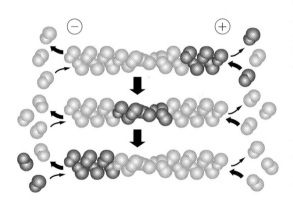

DYNAMIC INSTABILITY

Microtubules depolymerize about 100 times faster from an end containing GDP tubulin than from one containing GTP tubulin. A GTP cap favors growth, but if it is lost, then depolymerization ensues.

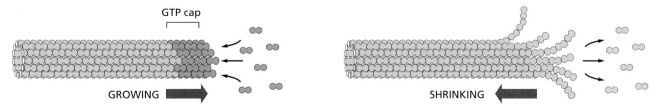

GTP cap

GROWING   SHRINKING

Individual microtubules can therefore alternate between a period of slow growth and a period of rapid disassembly, a phenomenon called dynamic instability.

The kinetic differences between the behavior of the T form and the D form have another important consequence for the behaviors of filaments. If the rate of subunit addition at one end is similar in magnitude to the rate of hydrolysis, there is a finite probability that this end will start out in a T form, but that hydrolysis will eventually "catch up" with the addition and transform the end to a D form. This transformation is sudden and random, with a certain probability per unit time.

Suppose that the concentration of free subunits is intermediate between the critical concentration for a T-form end and the critical concentration for a D-form end (that is, in the same range of concentrations where treadmilling is observed). Now, any end that happens to be in the T form will grow, whereas any end that happens to be in the D form will shrink. On a single filament, an end might grow for a certain length of time in a T form, but then suddenly change to the D form and begin to shrink rapidly, even while the free subunit concentration is held constant. At some later time, it might then regain a T-form end and begin to grow again. This rapid interconversion between a growing and shrinking state, at a uniform free subunit concentration, is called **dynamic instability** (**Figure 16–16**A). The change from growth to rapid shrinkage is called a *catastrophe*, while the change to growth is called a *rescue*. <CCCA>

In a population of microtubules, at any instant some of the ends are in the T form and some are in the D form, with the ratio depending on the hydrolysis rate and the free subunit concentration. The structural difference between a T-form end and a D-form end is dramatic. Tubulin subunits with GTP bound to the β-monomer produce straight protofilaments that make strong and regular lateral contacts with one another. But the hydrolysis of GTP to GDP is associated with a subtle conformational change in the protein, which makes the protofilaments curved (Figure 16–16B). On a rapidly growing microtubule, the GTP cap is thought to constrain the curvature of the protofilaments, and the ends appear straight. But when the terminal subunits have hydrolyzed their nucleotides, this constraint is removed, and the curved protofilaments spring apart. This cooperative release of the energy of hydrolysis stored in the microtubule lattice causes the curled protofilaments to peel off rapidly, and rings and curved oligomers of GDP-containing tubulin are seen near the ends of depolymerizing microtubules (Figure 16–16C).

Actin filaments also undergo length fluctuations but on a much smaller scale, so that at steady state the length fluctuates only a micrometer or so over several minutes, as compared to tens of micrometers for microtubules undergoing dynamic instability. In most eucaryotic cells, dynamic instability is thought to predominate in microtubules, whereas treadmilling may predominate in actin filaments.

Treadmilling and Dynamic Instability Aid Rapid Cytoskeletal Rearrangement <AAAT>

Both dynamic instability and treadmilling allow a cell to maintain the same overall filament content, while individual subunits constantly recycle between the filaments and the cytosol. How dynamic are the microtubules and actin filaments inside a living cell? Typically, a microtubule, with major structural differences between its growing and shrinking ends, switches between growth and shrinkage every few minutes. The ends of individual microtubules can therefore be seen in real time to exhibit dynamic instability (**Figure 16–17**). Because of their smaller size and denser packing, it is much more difficult to resolve the ends of individual actin filaments within living cells. With appropriate techniques based on fluorescence microscopy, however, one can show that actin filament turnover is typically rapid, with individual filaments persisting for only a few tens of seconds or minutes.

At first glance, the dynamic behavior of filaments seems like a waste of energy. To maintain a constant concentration of actin filaments and microtubules, most of which are undergoing a process of either treadmilling or dynamic instability, the cell must hydrolyze large amounts of nucleoside triphosphate. As we explained with our ant-trail analogy at the beginning of the chapter, the advantage to the cell seems to be the spatial and temporal flexibility that is inherent in a structural system with constant turnover. Individual subunits are small and can diffuse very rapidly; an actin or tubulin subunit can diffuse across the diameter of a typical eucaryotic cell in several seconds. As noted

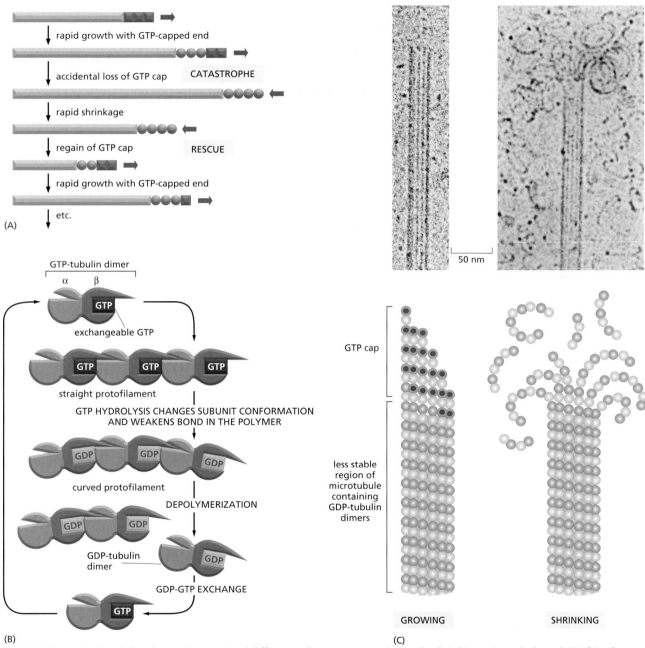

Figure 16–16 Dynamic instability due to the structural differences between a growing and a shrinking microtubule end. (A) If the free tubulin concentration in solution is between the critical values indicated in Figure 16–14B, a single microtubule end may undergo transitions between a growing state and a shrinking state. A growing microtubule has GTP-containing subunits at its end, forming a GTP cap. If nucleotide hydrolysis proceeds more rapidly than subunit addition, the cap is lost and the microtubule begins to shrink, an event called a "catastrophe." But GTP-containing subunits may still add to the shrinking end, and if enough add to form a new cap, then microtubule growth resumes, an event called "rescue." (B) Model for the structural consequences of GTP hydrolysis in the microtubule lattice. The addition of GTP-containing tubulin subunits to the end of a protofilament causes the end to grow in a linear conformation that can readily pack into the cylindrical wall of the microtubule. Hydrolysis of GTP after assembly changes the conformation of the subunits and tends to force the protofilament into a curved shape that is less able to pack into the microtubule wall. (C) In an intact microtubule, protofilaments made from GDP-containing subunits are forced into a linear conformation by the many lateral bonds within the microtubule wall, given a stable cap of GTP-containing subunits. Loss of the GTP cap, however, allows the GDP-containing protofilaments to relax into their more curved conformation. This leads to a progressive disruption of the microtubule. Above the drawings of a growing and a shrinking microtubule, electron micrographs show actual microtubules in each of these two states, as observed in preparations in vitreous ice. Note particularly the curling, disintegrating GDP-containing protofilaments at the end of the shrinking microtubule. (C, courtesy of E.M. Mandelkow, E. Mandelkow and R.A. Milligan, *J. Cell Biol.* 114:977–991, 1991. With permission from The Rockefeller University Press.)

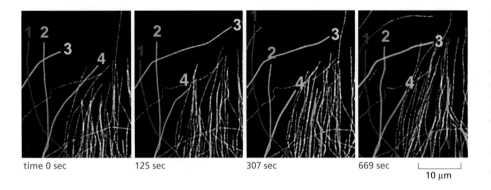

Figure 16–17 Direct observation of the dynamic instability of microtubules in a living cell. <AAAT> Microtubules in a newt lung epithelial cell were observed after the cell was injected with a small amount of rhodamine labeled tubulin, as in Figure 16–15. Notice the dynamic instability of microtubules at the edge of the cell. Four individual microtubules are highlighted for clarity; each of these shows alternating shrinkage and growth. (Courtesy of Wendy C. Salmon and Clare Waterman-Storer.)

time 0 sec   125 sec   307 sec   669 sec   10 µm

previously, the rate-limiting step in the formation of a new filament is nucleation, so these rapidly diffusing subunits tend to assemble either on the ends of preexisting filaments or at particular sites where special proteins catalyze the nucleation step. The new filaments in either case are highly dynamic, and unless specifically stabilized, they have only a fleeting existence. By controlling where filaments are nucleated and selectively stabilized, a cell can control the location of its filament systems, and hence its structure. It seems that the cell is continually testing a wide variety of internal structures and only preserving those that are useful. When external conditions change, or when internal signals arise (as during the transitions in the cell cycle), the cell is poised to change its structure rapidly (see Figures 16–2 to 16–4).

Actin and tubulin have independently evolved their nucleoside triphosphate hydrolysis to enable their filaments to depolymerize readily after they have polymerized. These two proteins are completely unrelated in amino acid sequence: actin is distantly related in structure to the glycolytic enzyme hexokinase, whereas tubulin is distantly related to a large family of GTPases that includes the heterotrimeric G proteins and monomeric GTPases such as Ras (discussed in Chapter 3). In both protein families, the coupling between nucleotide hydrolysis and a protein conformational change that alters protein function appears to be evolutionarily very ancient; however, the purposes of that structural coupling have diverged over time to include signal transmission, catalysis, and regulation of the polymerization/depolymerization cycle.

In certain specialized structures, parts of the cytoskeleton become less dynamic. In a terminally differentiated cell such as a neuron, for example, it is desirable to maintain a consistent structure over time, and many of the actin filaments and microtubules are stabilized by association with other proteins. However, when new connections are made in the brain, as when the information you are reading now is transferred into long-term memory, even a cell as stable as a neuron can grow new elongated processes to make new synapses. To do this, a neuron requires the dynamic, exploratory activities of its cytoskeletal filaments.

Tubulin and Actin Have Been Highly Conserved During Eucaryotic Evolution

Tubulin is found in all eucaryotic cells, and it exists in multiple isoforms. Yeast and human tubulins are 75% identical in amino acid sequence. In mammals, there are at least six forms of α-tubulin and a similar number of forms of β-tubulin, each encoded by a different gene. The different forms of tubulin are very similar, and they generally will copolymerize into mixed microtubules in the test tube. However, they can have distinct locations in a cell and perform subtly different functions. As a striking example, a specific form of β-tubulin forms the microtubules in six specialized touch-sensitive neurons in the nematode *Caenorhabditis elegans*. Mutations that eliminate this protein result in the loss of touch-sensitivity, with no apparent defect in other functions.

Like tubulin, actin is found in all eucaryotic cells. Most organisms have multiple genes encoding actin; humans have six. Actin is extraordinarily well conserved among eucaryotes. The amino acid sequences of actins from different species are usually about 90% identical. But, again like tubulin, small variations in actin amino acid sequence can cause significant functional differences. In vertebrates, there are three subtly different isoforms of actin, termed α, β, and γ, that differ slightly in their amino acid sequences. The α-actin is expressed only in muscle cells, while β and γ are found together in almost all nonmuscle cells. Yeast actin and *Drosophila* muscle actin are 89% identical, yet the expression of yeast actin in *Drosophila* results in a fly that looks normal but is unable to fly.

Why are the amino acid sequences of actin and tubulin so strictly conserved in eucaryotic evolution, whereas the sequences of most other cytoskeletal proteins, including intermediate filament proteins and the large families of accessory proteins that bind to actin or tubulin, are not? The likely explanation is that the need for large numbers of other proteins to interact with the entire surface of an actin filament or microtubule limits the variability of their structures. Genetic and biochemical studies in the yeast *Saccharomyces cerevisiae* have demonstrated that actin interacts directly with dozens of other proteins, and indirectly with even more (**Figure 16–18**). Thus, any mutation in actin that could result in a desirable change in its interaction with one protein might cause undesirable changes in its interactions with other proteins that bind at or near the same site. Over time, evolving organisms have found it more profitable to leave actin and tubulin alone, and to alter their binding partners instead.

Intermediate Filament Structure Depends on the Lateral Bundling and Twisting of Coiled Coils

All eucaryotic cells contain actin and tubulin. But the third major type of cytoskeletal protein, the *intermediate filament*, forms a cytplasmic filament in only some metazoans—including vertebrates, nematodes, and mollusks. Even in these organisms, intermediate filaments are not required in the cytoplasm of every cell type. The specialized glial cells (called oligodendrocytes) that make

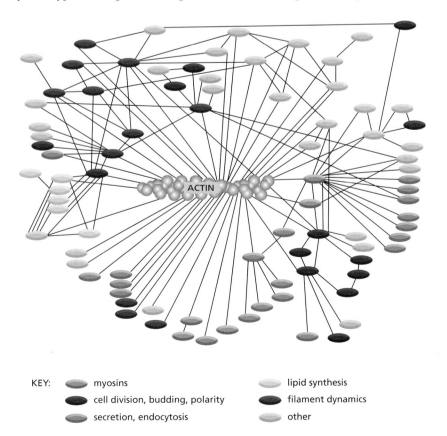

ACTIN

KEY:
myosins

cell division, budding, polarity

secretion, endocytosis

lipid synthesis

filament dynamics

other

Figure 16–18 Actin at the crossroads. Actin binds to a very large variety of accessory proteins in all eucaryotic cells. This diagram shows most of the interactions that have been demonstrated, using either genetic or biochemical techniques, in the yeast *Saccharomyces cerevisiae*. Accessory proteins that operate in the same intracellular process are shown in the same color, as indicated in the key. (Adapted from D. Botstein et al., in The Molecular and Cellular Biology of the Yeast *Saccharomyces* [J.R. Broach, J.R. Pringle, E.W. Jones, eds.], Cold Spring Harbor, NY: Cold Spring Harbor Laboratory Press, 1991.)

myelin in the vertebrate central nervous system, for example, do not contain such intermediate filaments. Intermediate filaments are particularly prominent in the cytoplasm of cells that are subject to mechanical stress, and are generally not found in animals that have rigid exoskeletons such as arthropods and echinoderms. It seems that intermediate filaments play an important role in imparting mechanical strength to tissues for the squishier animals.

Cytoplasmic intermediate filaments are closely related to their ancestors, the much more prevalent *nuclear lamins*. The nuclear lamins are intermediate filament proteins that form a meshwork lining the inner membrane of the eucaryotic nuclear envelope, where they provide anchorage sites for chromosomes and nuclear pores (their dynamic behavior during cell division is discussed in Chapter 12). Several times during metazoan evolution, lamin genes have apparently duplicated, and the duplicates have evolved to produce rope-like, cytoplasmic intermediate filaments.

The individual polypeptides of **intermediate filaments** are elongated molecules with an extended central α-helical domain that forms a parallel coiled coil with another monomer. A pair of parallel dimers then associates in an antiparallel fashion to form a staggered tetramer. This tetramer represents the soluble subunit that is analogous to the αβ-tubulin dimer, or to the actin monomer (**Figure 16–19**). Unlike the actin or tubulin, the intermediate filament subunits do not contain a binding site for a nucleoside triphosphate.

Figure 16–19 A model of intermediate filament construction. <GCCA> The monomer shown in (A) pairs with an identical monomer to form a dimer (B), in which the conserved central rod domains are aligned in parallel and wound together into a coiled coil. (C) Two dimers then line up side by side to form an antiparallel tetramer of four polypeptide chains. The tetramer is the soluble subunit of intermediate filaments. (D) Within each tetramer, the two dimers are offset with respect to one another, thereby allowing it to associate with another tetramer. (E) In the final 10-nm rope-like filament, tetramers are packed together in a helical array, which has 16 dimers (32 coiled coils) in cross-section. Half of these dimers are pointing in each direction. An electron micrograph of intermediate filaments is shown on the upper left. (Electron micrograph courtesy of Roy Quinlan.)

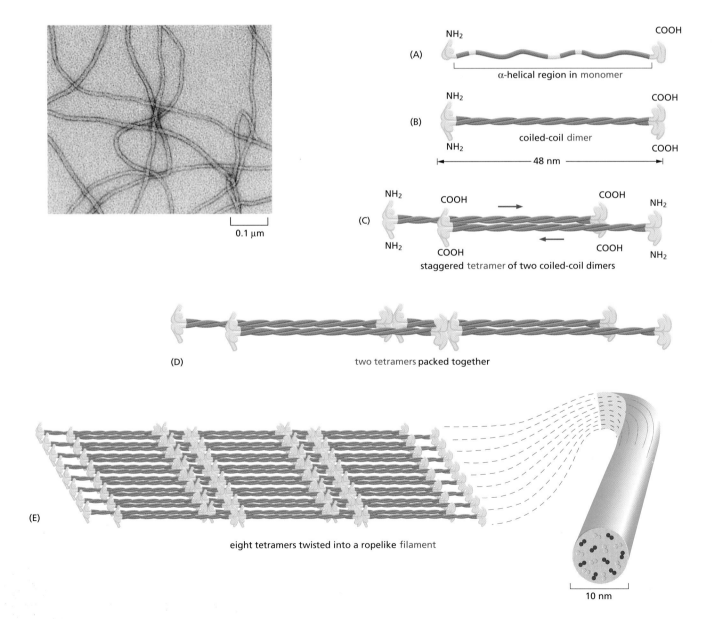

(A) α-helical region in monomer

(B) coiled-coil dimer — 48 nm

(C) staggered tetramer of two coiled-coil dimers

(D) two tetramers packed together

(E) eight tetramers twisted into a ropelike filament

0.1 µm

10 nm

Table 16–1 Major Types of Intermediate Filament Proteins in Vertebrate Cells

TYPES OF IF	COMPONENT POLYPEPTIDES	LOCATION
Nuclear	lamins A, B, and C	nuclear lamina (inner lining of nuclear envelope)
Vimentin-like	vimentin	many cells of mesenchymal origin
	desmin	muscle
	glial fibrillary acidic protein	glial cells (astrocytes and some Schwann cells)
	peripherin	some neurons
Epithelial	type I keratins (acidic) type II keratins (basic)	epithelial cells and their derivatives (e.g., hair and nails)
Axonal	neurofilament proteins (NF-L, NF-M, and NF-H)	neurons

Since the tetrameric subunit is made up of two dimers pointing in opposite directions, its two ends are the same. The assembled intermediate filament therefore lacks the overall structural polarity that is critical for actin filaments and microtubules. The tetramers pack together laterally to form the filament, which includes eight parallel protofilaments made up of tetramers. Each individual intermediate filament therefore has a cross section of 32 individual α-helical coils. This large number of polypeptides all lined up together, with the strong lateral hydrophobic interactions typical of coiled-coil proteins, gives intermediate filaments a rope-like character. They can be easily bent, with a persistence length of less than one micrometer (compared to several millimeters for microtubules and about ten micrometers for actin), but they are extremely difficult to break.

Less is understood about the mechanism of assembly and disassembly of intermediate filaments than of actin filaments and microtubules, but some types of intermediate filaments including *vimentin* form highly dynamic structures in cells such as fibroblasts. Under normal conditions, protein phosphorylation probably regulates their disassembly, in much the same way that phosphorylation regulates the disassembly of nuclear lamins in mitosis (see Figure 12–20). As evidence for rapid turnover, labeled subunits microinjected into tissue culture cells rapidly add themselves onto the existing intermediate filaments within a few minutes, while an injection of peptides derived from a conserved helical region of the subunit induces the rapid disassembly of the intermediate filament network. Interestingly, the latter injection can also induce the disassembly of the microtubule and actin filament networks in some cases, demonstrating that there is a fundamental mechanical integration of the three cytoskeletal systems in these cells.

Intermediate Filaments Impart Mechanical Stability to Animal Cells

Intermediate filaments come in a wide variety of types, with substantially more sequence variation in the subunit isoforms than occurs in the isoforms of actin or tubulin. A central α-helical domain has 40 or so heptad repeat motifs that form an extended coiled-coil structure (see Figure 3–9). This domain is similar in the different isoforms, but the N- and C-terminal globular domains can vary greatly.

Different families of intermediate filaments are expressed in different cell types (**Table 16–1**). **Keratins** are the most diverse intermediate filament family: there are about 20 found in different types of human epithelial cells, and about 10 more that are specific to hair and nails; analysis of the human genome sequence has revealed that there may be about 50 distinct keratins. Every keratin filament is made up of an equal mixture of type I (acidic) and type II (neutral/basic) keratin chains; these form heterodimers, two of which then join

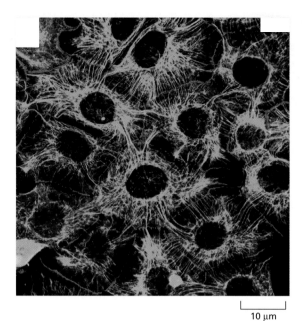

Figure 16–20 Keratin filaments in epithelial cells. Immunofluorescence micrograph of the network of keratin filaments *(green)* in a sheet of epithelial cells in culture. The filaments in each cell are indirectly connected to those of its neighbors by desmosomes (discussed in Chapter 19). A second protein *(blue)* has been stained to reveal the location of the cell boundaries. (Courtesy of Kathleen Green and Evangeline Amargo.)

10 µm

to form the fundamental tetrameric subunit (see Figure 16–19). Cross-linked keratin networks held together by disulfide bonds may survive even the death of their cells, forming tough coverings for animals, as in the outer layer of skin and in hair, nails, claws, and scales. The diversity in keratins is clinically useful in the diagnosis of epithelial cancers (carcinomas), as the particular set of keratins expressed gives an indication of the epithelial tissue in which the cancer originated and thus can help to guide the choice of treatment.

A single epithelial cell may produce multiple types of keratins, and these copolymerize into a single network (**Figure 16–20**). Keratin filaments impart mechanical strength to epithelial tissues in part by anchoring the intermediate filaments at sites of cell–cell contact, called *desmosomes*, or cell–matrix contact, called *hemidesmosomes* (see Figure 16–5). We discuss these important adhesive structures in more detail in Chapter 19.

Mutations in keratin genes cause several human genetic diseases. For example, when defective keratins are expressed in the basal cell layer of the epidermis, they produce a disorder called *epidermolysis bullosa simplex,* in which the skin blisters in response to even very slight mechanical stress, which ruptures the basal cells (**Figure 16–21**). Other types of blistering diseases, including disorders of the mouth, esophageal lining, and the cornea of the eye, are caused by mutations in the different keratins whose expression is specific to those tissues. All of these maladies are typified by cell rupture as a consequence of mechanical trauma and a disorganization or clumping of the keratin filament cytoskeleton. Many of the specific mutations that cause these diseases alter the ends of

Figure 16–21 Blistering of the skin caused by a mutant keratin gene. A mutant gene encoding a truncated keratin protein (lacking both the N- and C-terminal domains) was expressed in a transgenic mouse. The defective protein assembles with the normal keratins and thereby disrupts the keratin filament network in the basal cells of the skin. Light micrographs of cross sections of normal (A) and mutant (B) skin show that the blistering results from the rupturing of cells in the basal layer of the mutant epidermis (short *red* arrows). (C) A sketch of three cells in the basal layer of the mutant epidermis, as observed by electron microscopy. As indicated by the *red* arrow, the cells rupture between the nucleus and the hemidesmosomes (discussed in Chapter 19), which connect the keratin filaments to the underlying basal lamina. (From P.A. Coulombe et al., *J. Cell Biol.* 115:1661–1674, 1991. With permission from The Rockefeller University Press.)

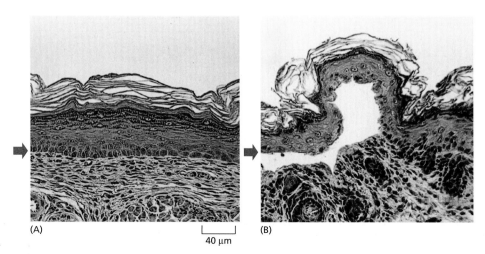

(A)   (B)

40 µm

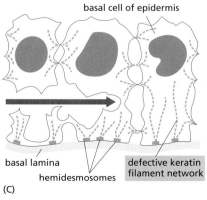

basal cell of epidermis

basal lamina   hemidesmosomes   defective keratin filament network

(C)

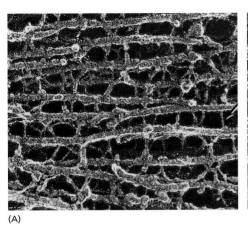

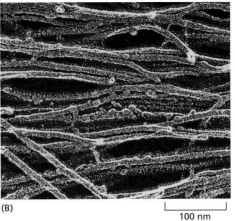

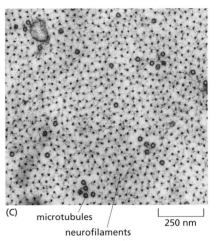

(A)

(B) 100 nm

(C) microtubules / neurofilaments 250 nm

the central rod domain, demonstrating the importance of this particular part of the protein for correct filament assembly.

A second family of intermediate filaments, called **neurofilaments**, is found in high concentrations along the axons of vertebrate neurons (**Figure 16–22**). Three types of neurofilament proteins (NF-L, NF-M, NF-H) coassemble *in vivo*, forming heteropolymers that contain NF-L plus one of the others. The NF-H and NF-M proteins have lengthy C-terminal tail domains that bind to neighboring filaments, generating aligned arrays with a uniform interfilament spacing. During axonal growth, new neurofilament subunits are incorporated all along the axon in a dynamic process that involves the addition of subunits along the filament length, as well as the addition of subunits at the filament ends. After an axon has grown and connected with its target cell, the diameter of the axon may increase as much as fivefold. The level of neurofilament gene expression seems to directly control axonal diameter, which in turn influences how fast electrical signals travel down the axon.

The neurodegenerative disease amyotrophic lateral sclerosis (ALS, or Lou Gehrig's Disease) is associated with an accumulation and abnormal assembly of neurofilaments in motor neuron cell bodies and in the axon, which may interfere with normal axonal transport. The degeneration of the axons leads to muscle weakness and atrophy, which is usually fatal. The over-expression of human NF-L or NF-H in mice results in mice that have an ALS-like disease.

The vimentin-like filaments are a third family of intermediate filaments. Desmin, a member of this family, is expressed in skeletal, cardiac, and smooth muscle. Mice lacking desmin show normal initial muscle development, but adults have various muscle cell abnormalities, including misaligned muscle fibers.

Drugs Can Alter Filament Polymerization

Because the survival of eucaryotic cells depends on a balanced assembly and disassembly of the highly conserved cytoskeletal filaments formed from actin and tubulin, the two types of filaments are frequent targets for natural toxins. These toxins are produced in self-defense by plants, fungi, or sponges that do not wish to be eaten but cannot run away from predators, and they generally disrupt the filament polymerization reaction. The toxin binds tightly to either the filament form or the free subunit form of a polymer, driving the assembly reaction in the direction that favors the form to which the toxin binds. For example, the drug *latrunculin*, extracted from the sea sponge *Latrunculia magnifica*, binds to actin monomers and prevents their assembly into filaments; it thereby causes a net depolymerization of actin filaments. In contrast, *phalloidin*, from the fungus *Amanita phalloides* (death cap), binds to and stabilizes actin filaments, causing a net increase in actin polymerization. (This attractive but inedible mushroom also expresses a second deadly toxin, the RNA polymerase II inhibitor α-amanitin.) Either change in actin filaments is very toxic for cells. Similarly, *colchicine*, from the autumn crocus (or meadow saffron), binds to and

Figure 16–22 Two types of intermediate filaments in cells of the nervous system. (A) Freeze-etch electron microscopic image of neurofilaments in a nerve cell axon, showing the extensive cross-linking through protein cross-bridges—an arrangement believed to give this long cell process great tensile strength. The cross-bridges are formed by the long, nonhelical extensions at the C-terminus of the largest neurofilament protein (NF-H). (B) Freeze-etch image of glial filaments in glial cells, showing that these intermediate filaments are smooth and have few cross-bridges. (C) Conventional electron micrograph of a cross section of an axon showing the regular side-to-side spacing of the neurofilaments, which greatly outnumber the microtubules. (A and B, courtesy of Nobutaka Hirokawa; C, courtesy of John Hopkins.)

Table 16–2 Drugs That Affect Actin Filaments and Microtubules

ACTIN-SPECIFIC DRUGS	
Phalloidin	binds and stabilizes filaments
Cytochalasin	caps filament plus ends
Swinholide	severs filaments
Latrunculin	binds subunits and prevents their polymerization

MICROTUBULE-SPECIFIC DRUGS	
Taxol	binds and stabilizes microtubules
Colchicine, colcemid	binds subunits and prevents their polymerization
Vinblastine, vincristine	binds subunits and prevents their polymerization
Nocodazole	binds subunits and prevents their polymerization

stabilizes free tubulin, causing microtubule depolymerization. In contrast, *taxol*, extracted from the bark of a rare species of yew tree, binds to and stabilizes microtubules, causing a net increase in tubulin polymerization. These and some other natural products that are commonly used by cell biologists to manipulate the cytoskeleton are listed in **Table 16–2**.

Drugs like these have a rapid and profound effect on the organization of the cytoskeleton in living cells (**Figure 16–23**). They provided early evidence that the cytoskeleton is a dynamic structure, maintained by a rapid and continual exchange of subunits between the soluble and filamentous forms, and they revealed that this subunit flux is necessary for normal cytoskeletal function.

The drugs listed in Table 16–2 have been useful to cell biologists trying to probe the roles of actin and microtubules in various cell processes. Some of them are also used to treat cancer. Both microtubule-depolymerizing drugs (such as vinblastine) and microtubule-polymerizing drugs (such as taxol) preferentially kill dividing cells, since both microtubule assembly and disassembly are crucial for correct function of the mitotic spindle (discussed later in this chapter). These drugs efficiently kill certain types of tumor cells in a human patient, although not without toxicity to rapidly dividing normal cells, including those in the bone marrow, intestine, and hair follicles. Taxol in particular has been widely used to treat cancers of the breast and lung, and it is frequently successful in treatment of tumors that are resistant to other chemotherapeutic agents.

Figure 16–23 Effect of the drug taxol on microtubule organization. (A) Molecular structure of taxol. Recently, organic chemists have succeeded in synthesizing this complex molecule, which is widely used for cancer treatment. (B) Immunofluorescence micrograph showing the microtubule organization in a liver epithelial cell before the addition of taxol. (C) Microtubule organization in the same type of cell after taxol treatment. Note the thick circumferential bundles of microtubules around the periphery of the cell. (D) A Pacific yew tree, the natural source of taxol. (B, C from N.A. Gloushankova et al., *Proc. Natl Acad. Sci. U.S.A.* 91:8597–8601, 1994. With permission from National Academy of Sciences; D, courtesy of A.K. Mitchell 2001. © Her Majesty the Queen in Right of Canada, Canadian Forest Service.)

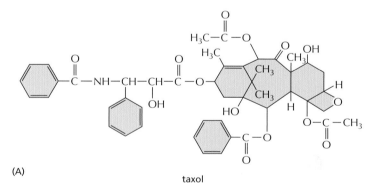

(A)   taxol

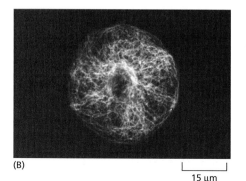

(B)

15 μm

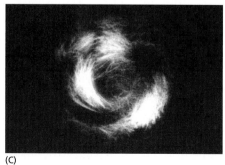

(C)

(D)

Figure 16–24 The bacterial FtsZ protein, a tubulin homolog in procaryotes. (A) A band of FtsZ protein forms a ring in a dividing bacterial cell. This ring has been labeled by fusing the FtsZ protein to the *green fluorescent protein* (GFP), which allows it to be observed in living *E. coli* cells with a fluorescence microscope. *Top*, side view shows the ring as a bar in the middle of the dividing cell. *Bottom*, rotated view showing the ring structure. (B) FtsZ filaments and rings, formed *in vitro*, as visualized using electron microscopy. Compare this image with that of the microtubule shown on the *right* in Figure 16–16C. (A, from X. Ma, D.W. Ehrhardt and W. Margolin, *Proc. Natl Acad. Sci. U.S.A.* 93:12998–13003, 1996; B, from H.A. Erickson et al., *Proc. Natl Acad. Sci. U.S.A.* 93:519–523, 1996. All with permission from National Academy of Sciences.)

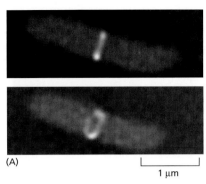

(A)
1 µm

Bacterial Cell Organization and Cell Division Depend on Homologs of the Eucaryotic Cytoskeleton

While eucaryotic cells are typically large and morphologically complex, bacterial cells are usually only a few micrometers long and assume simple, modest shapes such as spheres or rods. Bacteria also lack the elaborate networks of intracellular membrane-enclosed organelles such as the endoplasmic reticulum and Golgi apparatus. For many years, biologists assumed that the lack of a bacterial cytoskeleton was one reason for these striking differences between cell organization in the eucaryotic and bacterial kingdoms. This assumption was challenged with the discovery in the early 1990s that nearly all bacteria and many archaea contain a homolog of tubulin, FtsZ, that can polymerize into filaments and assemble into a ring (called the Z-ring) at the site where the septum forms during cell division (**Figure 16–24**).

The three-dimensional folded protein structure of FtsZ is remarkably similar to the structure of α or β tubulin and, like tubulin, hydrolysis of GTP is triggered by polymerization and causes a conformational change in the filament structure. Although the Z-ring itself persists for many minutes, the individual filaments within it are highly dynamic, with an average half-life of about thirty seconds. As the bacterium divides, the Z-ring becomes smaller until it has completely disassembled, and it is thought that the shrinkage of the Z-ring may contribute to the membrane invagination necessary for the completion of cell division. The Z-ring may also serve as a site for localization of specialized cell wall synthesis enzymes required for building the septum between the two daughter cells. The disassembled FtsZ subunits later reassemble at the new sites of septum formation in the daughter cells (**Figure 16–25**).

More recently, it has been found that many bacteria also contain homologs of actin. Two of these, MreB and Mbl, are found primarily in rod-shaped or spiral-shaped cells, and mutations disrupting their expression cause extreme abnormalities in cell shape and defects in chromosome segregation (**Figure 16–26**). MreB and Mbl filaments assemble *in vivo* to form large-scale spirals that

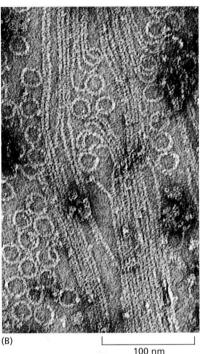

(B)
100 nm

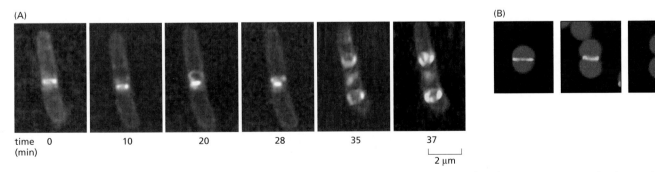

(A)

time 0   10   20   28   35   37
(min)
2 µm

(B)

1 µm

Figure 16–25 Rapid rearrangements of FtsZ through the bacterial cell cycle. (A) After chromosome segregation is complete, the ring formed by FtsZ at the middle of the cell becomes smaller as the cell pinches in two, much like the contractile ring formed by actin and myosin filaments in eucaryotic cells. The FtsZ filaments that have disassembled as the cells have separated then reassemble to form two new rings at the middle of the two daughter cells. (B) Dividing chloroplasts (*red*) from a red alga also make use of a protein ring made from FtsZ (*yellow*) for cleavage. (A, from Q. Sun and W. Margolin, *J. Bacteriol.* 180:2050–2056, 1998. With permission from American Society for Microbiology; B, from S. Miyagishima et al., *Plant Cell* 13:2257–2268, 2001. With permission from American Society of Plant Biologists.)

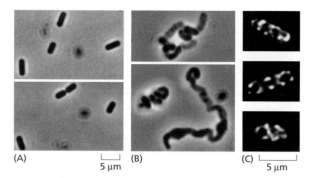

(A) |⎯⎯⎯| (B)
      5 μm

(C) |⎯⎯⎯|
      5 μm

Figure 16–26 Actin homologs in bacteria determine cell shape. (A) The common soil bacterium *Bacillus subtilis* normally forms cells with a regular rod-like shape. (B) *B. subtilis* cells lacking the actin homolog Mbl grow into irregular twisted tubes and eventually die. (C) The Mbl protein forms long helices made of up many short filaments that run the length of the bacterial cell and help to direct the sites of cell wall synthesis. (From L.J. Jones, R. Carbadillo-Lopez and J. Errington, *Cell* 104: 913-922, 2001. With permission from Elsevier.)

span the length of the cell and apparently contribute to cell shape determination by serving as a scaffold to direct the synthesis of the peptidoglycan cell wall, in much the same way that microtubules help organize the synthesis of the cellulose cell wall in higher plant cells (see Figure 19–82). As with FtsZ, the filaments within the MreB and Mbl spirals are highly dynamic, with half-lives of a few minutes; as for actin, ATP hydrolysis accompanies the polymerization process.

Diverse relatives of MreB and Mbl have more specialized roles. A particularly intriguing bacterial actin homolog is ParM, which is encoded on certain bacterial plasmids that also carry genes responsible for antibiotic resistance and frequently cause the spread of multi-drug resistance in epidemics. Bacterial plasmids typically encode all the gene products that are necessary for their own segregation, presumably as a strategy to ensure their faithful inheritance and propagation in their bacterial hosts. *In vivo*, ParM assembles into a filamentous structure that associates at each end with a copy of the plasmid that encodes it, and growth of the ParM filament appears to push the replicated plasmid copies apart, rather like a mitotic spindle operating in reverse (**Figure 16–27**). Although ParM is a structural homolog of actin, its dynamic behavior differs significantly. ParM filaments undergo dramatic dynamic instability *in vitro*, more closely resembling microtubules than actin filaments in the way that they grow and shrink. The spindle-like structure is apparently built by the selective stabilization of spontaneously nucleated filaments that bind to specialized proteins recruited to the origins of replication on the plasmids.

The various bacterial actin homologs share similar molecular structures but their amino acid sequence similarity to each other is quite low (~10–15% identical residues). They assemble into filaments with distinct helical packing patterns, which may also have very different dynamic behaviors. Rather than using the same well-conserved actin for many different purposes, as eucaryotic cells do, bacteria have apparently opted to proliferate and specialize their actin homologs for distinct purposes.

It is now clear that the general principle of organizing cell structure by the self-association of nucleotide-binding proteins into dynamic helical filaments is used in all cells, and that the two major families of actin and tubulin are very

Figure 16–27 Role of the actin homolog ParM in plasmid segregation. (A) Some bacterial drug-resistance plasmids (*yellow*) encode an actin homolog, ParM, that will spontaneously nucleate to form small, dynamic filaments (*green*) throughout the bacterial cytoplasm. A second plasmid-encoded protein (*blue*) binds to specific DNA sequences in the plasmid, and also stabilizes the dynamic ends of the ParM filaments. When the plasmid has duplicated, so that the ParM filaments can be stabilized at both ends, the filaments grow and push the duplicated plasmids to opposite ends of the cell. (B) In these bacterial cells harboring a drug-resistance plasmid, the plasmids are labeled in *red* and the ParM protein in *green*. Left, a short ParM bundle connects the two daughter plasmids shortly after their duplication. Right, the fully assembled ParM filament has pushed the duplicated plasmids to the cell poles. (A, adapted from E.C. Garner, C.S. Campbell and R.D. Mullins, *Science* 306:1021–1025, 2004. With permission from AAAS; B, from J. Moller-Jensen et al., *Mol. Cell* 12:1477–1487, 2003. With permission from Elsevier.)

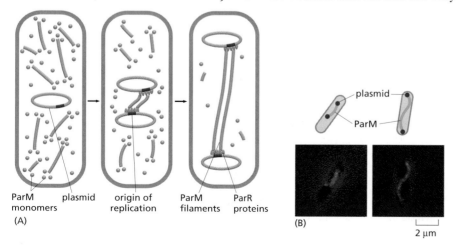

ParM   plasmid   origin of   ParM   ParR
monomers   replication   filaments   proteins

(A)

plasmid
ParM

(B) |⎯⎯⎯|
          2 μm

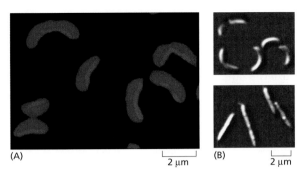

(A) 2 μm   (B) 2 μm

Figure 16–28 Caulobacter and crescentin. The sickle-shaped bacterium *Caulobacter crescentus* expresses a protein, crescentin, with a series of coiled-coil domains similar in size and organization to the domains of eucaryotic intermediate filaments. In cells, the crescentin protein forms a fiber that runs down the inner side of the curving bacterial cell wall. When the gene is disrupted, the bacteria are viable but grow in a straight rod-shaped form. (From N. Ausmees, J.R. Kuhn and C. Jacobs-Wagner, *Cell* 115:705–713, 2003. With permission from Elsevier.)

ancient, probably predating the split between the eucaryotic and bacterial kingdoms. However, the uses to which bacteria put their cytoskeletons appear somewhat different from their eucaryotic homologs. For example, in bacteria it is the tubulin (FtsZ) that is involved in *cytokinesis* (the pinching apart of a dividing cell into two daughters), while actin drives this process in eucaryotic cells. Conversely, eucaryotic microtubules are responsible for chromosome segregation, while bacterial actins (ParM and possibly MreB) help to segregate replicated DNA in bacteria.

At least one bacterial species with an unusual crescent shape, *Caulobacter crescentus*, even appears to harbor a protein with significant structural similarity to the third major class of cytoskeletal filaments found in animal cells, the intermediate filaments. A protein called crescentin forms a filamentous structure that seems to influence the cell shape; when the gene encoding crescentin is deleted, the *Caulobacter* cells grow as straight rods (**Figure 16–28**).

Since we now know that bacteria do in fact have sophisticated dynamic cytoskeletons, why then do they remain so small and morphologically simple? As yet there have been no *motor proteins* identified that walk along the bacterial filaments; perhaps the evolution of motor proteins was a critical step allowing morphological elaboration in the eucaryotes.

Summary

The cytoplasm of eucaryotic cells is spatially organized by a network of protein filaments known as the cytoskeleton. This network contains three principal types of filaments: microtubules, actin filaments, and intermediate filaments. All three types of filaments form as helical assemblies of subunits that self-associate using a combination of end-to-end and side-to-side protein contacts. Differences in the structure of the subunits and the manner of their self-assembly give the filaments different mechanical properties. Intermediate filaments are rope-like and easy to bend but hard to break. Microtubules are strong, rigid hollow tubes. Actin filaments are the thinnest of the three and are easy to break.

In living cells, the assembly and disassembly of their subunits constantly remodels all three types of cytoskeletal filaments. Microtubules and actin filaments add and lose subunits only at their ends, with one end (the plus end) growing faster than the other. Tubulin and actin (the subunits of microtubules and actin filaments, respectively) bind and hydrolyze nucleoside triphosphates (tubulin binds GTP and actin binds ATP). Nucleotide hydrolysis underlies the characteristic dynamic behavior of these two filaments. Actin filaments in cells seem to predominantly undergo treadmilling, where a filament assembles at one end while simultaneously disassembling at the other end. Microtubules in cells predominantly display dynamic instability, where a microtubule end undergoes alternating bouts of growth and shrinkage.

Whereas tubulin and actin have been strongly conserved in evolution, the family of intermediate filaments is very diverse. There are many tissue-specific forms found in the cytoplasm of animal cells, including keratin filaments in epithelial cells, neurofilaments in nerve cells, and desmin filaments in muscle cells. In all these cells, the primary job of intermediate filaments is to provide mechanical strength.

Bacterial cells also contain homologs of tubulin, actin and intermediate filaments that form dynamic filamentous structures involved in determining cell shape and in cell division.

HOW CELLS REGULATE THEIR CYTOSKELETAL FILAMENTS

Microtubules, actin filaments, and intermediate filaments are much more dynamic in cells than they are in the test tube. The cell regulates the length and stability of its cytoskeletal filaments, as well as their number and geometry. It does so largely by regulating their attachments to one another and to other components of the cell, so that the filaments can form a wide variety of higher-order structures. Direct covalent modification of the filament subunits regulates some filament properties, but most of the regulation is performed by a large array of accessory proteins that bind to either the filaments or their free subunits. Some of the most important accessory proteins associated with microtubules and actin filaments are outlined in **Panel 16–3** (pp. 994–995). This section describes how these accessory proteins modify the dynamics and structure of cytoskeletal filaments. We begin with a discussion of the way that microtubules and actin filaments are nucleated in cells, because this plays a major part in determining the overall organization of the cell's interior.

A Protein Complex Containing γ-Tubulin Nucleates Microtubules

While α- and β-tubulins are the regular building blocks of microtubules, another type of tubulin, called *γ-tubulin*, has a more specialized role. Present in much smaller amounts than α- and β-tubulin, this protein is involved in the nucleation of microtubule growth in organisms ranging from yeasts to humans. Microtubules are generally nucleated from a specific intracellular location known as a **microtubule-organizing center** (**MTOC**). Antibodies against γ-tubulin stain the MTOC in virtually all species and cell types thus far examined.

Microtubules are nucleated at their minus end, with the plus end growing outward from each MTOC to create various types of microtubule arrays. A **γ-tubulin ring complex** (**γ-TuRC**) that is capable of nucleating microtubule growth in a test tube has been isolated from both insect and vertebrate cells. Two proteins, conserved from yeasts to humans, bind directly to the γ-tubulin, along with several other proteins that help create a ring of γ-tubulin molecules. This ring can be seen at the minus ends of the microtubules nucleated by γ-TuRC, and it is therefore thought to serve as a template that creates a microtubule with 13 protofilaments (**Figure 16–29**).

Microtubules Emanate from the Centrosome in Animal Cells

Most animal cells have a single, well-defined MTOC called the **centrosome**, located near the nucleus. From this focal point, the cytoplasmic microtubules emanate in a star-like, "astral" conformation. Microtubules are nucleated at the centrosome at their minus ends, so the plus ends point outward and grow toward the cell periphery. Microtubules nucleated at the centrosome continuously grow and shrink by dynamic instability, probing the entire three-dimensional volume of the cell. A centrosome is composed of a fibrous *centrosome*

Figure 16–29 Polymerization of tubulin nucleated by γ-tubulin ring complexes. (A) Structure of the γ-tubulin ring complex, reconstructed from averaging electron micrographs of individual purified complexes. (B) Model for the nucleation of microtubule growth by the γ-TuRC. The red outline indicates a pair of proteins bound to two molecules of γ-tubulin; this group can be isolated as a separate subcomplex of the larger ring. Note the longitudinal discontinuity between two protofilaments. Microtubules generally have one such "seam" breaking the otherwise uniform helical packing of the protofilaments. (C) Electron micrograph of a single microtubule nucleated from the purified γ-tubulin ring complex. (A and C, from M. Moritz et al., *Nat. Cell Biol.* 2:365–370, 2000. With permission from Macmillan Publishers Ltd.)

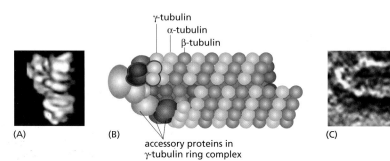

γ-tubulin
α-tubulin
β-tubulin

(A)   (B)   (C)

100 nm

accessory proteins in γ-tubulin ring complex

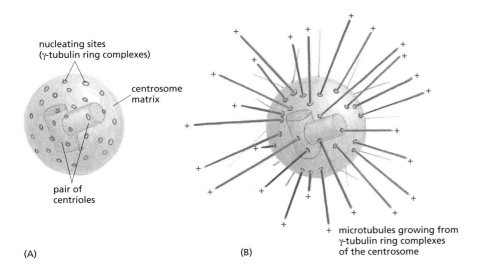

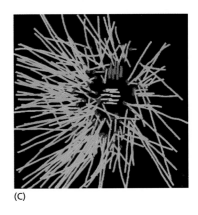

(A)

(B)

(C)

+ microtubules growing from γ-tubulin ring complexes of the centrosome

Figure 16–30 The centrosome. (A) The centrosome is the major MTOC of animal cells. Located in the cytoplasm next to the nucleus, it consists of an amorphous matrix of fibrous proteins to which the γ-tubulin ring complexes that nucleate microtubule growth are attached. This matrix is organized by a pair of centrioles, as described in the text. (B) A centrosome with attached microtubules. The minus end of each microtubule is embedded in the centrosome, having grown from a γ-tubulin ring complex, whereas the plus end of each microtubule is free in the cytoplasm. (C) In a reconstructed image of the MTOC from a *C. elegans* cell, a dense thicket of microtubules can be seen emanating from the centrosome. (C, from E.T. O'Toole et al., *J. Cell Biol.* 163:451–456, 2003. With permission from The Rockefeller University Press.)

matrix that contains more than fifty copies of γ-TuRC. Most of the proteins that form this matrix remain to be discovered, and it is not yet known how they recruit and activate the γ-TuRC.

Embedded in the centrosome is a pair of somewhat mysterious cylindrical structures arranged at right angles to each other in an L-shaped configuration (**Figure 16–30**). These are the **centrioles**, which become the basal bodies of cilia and flagella in motile cells (described later). The centrioles organize the centrosome matrix (also called the pericentriolar material), ensuring its duplication during each cell cycle as the centrioles themselves duplicate (**Figure 16–31**). As described in Chapter 17, the centrosome duplicates and splits into two equal parts during interphase, each half containing a duplicated centriole pair. These two daughter centrosomes move to opposite sides of the nucleus when mitosis begins, and they form the two poles of the mitotic spindle (see Panel 17–1, pp. 1072–1073). A centriole consists of a short cylinder of modified microtubules, plus a large number of accessory proteins. The molecular basis for its duplication is not well-understood.

In fungi and diatoms, microtubules are nucleated at an MTOC that is embedded in the nuclear envelope as a small plaque called the *spindle pole body*. Higher-plant cells seem to nucleate microtubules at sites distributed all around the nuclear envelope. Neither fungi nor most plant cells contain centrioles. Despite these differences, all these cells contain γ-tubulin and seem to use it to nucleate their microtubules.

In animal cells, the astral configuration of microtubules is very robust, with dynamic plus ends pointing outward toward the cell periphery and stable minus ends collected near the nucleus. The system of microtubules radiating from the centrosome acts as a device to survey the outlying regions of the cell and to position the centrosome at its center, and it does this even in artificial enclosures

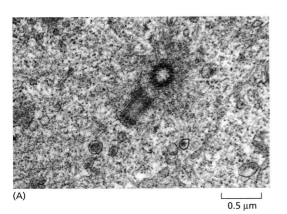

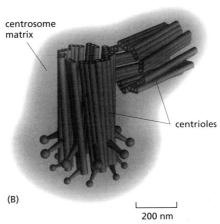

(A)

0.5 μm

(B)

200 nm

Figure 16–31 A centriole in the centrosome. (A) An electron micrograph of a thin section of a centrosome showing an end-on view of the mother centriole and a longitudinal section of the daughter centriole. Numerous microtubules are seen nearby. (B) Structure of the centriole pair. (A, from G.J. Mack, Y. Ou and J.B. Rattner, *Microsc. Res. Tech.* 49:409–419, 2000. With permission from John Wiley & Sons. B, adapted from D. Chrétien et al., *J. Struct. Biol.* 120:117–133, 1997. With permission from Elsevier.)

ACTIN FILAMENTS

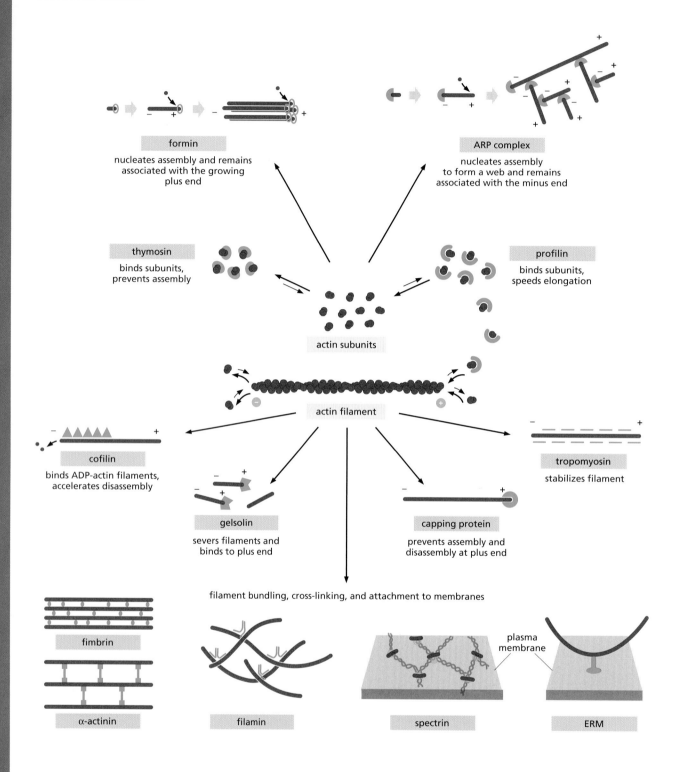

formin

nucleates assembly and remains associated with the growing plus end

ARP complex

nucleates assembly to form a web and remains associated with the minus end

thymosin

binds subunits, prevents assembly

profilin

binds subunits, speeds elongation

actin subunits

actin filament

cofilin

binds ADP-actin filaments, accelerates disassembly

gelsolin

severs filaments and binds to plus end

capping protein

prevents assembly and disassembly at plus end

tropomyosin

stabilizes filament

filament bundling, cross-linking, and attachment to membranes

fimbrin

α-actinin

filamin

plasma membrane

spectrin

ERM

Some of the major accessory proteins of the actin cytoskeleton. Except for the myosin motor proteins, to be discussed in a later section, an example of each major type is shown. Each of these is discussed in the text. However, most cells contain more than a hundred different actin-binding proteins, and it is likely that there are important types of actin-associated proteins that are not yet recognized.

MICROTUBULES

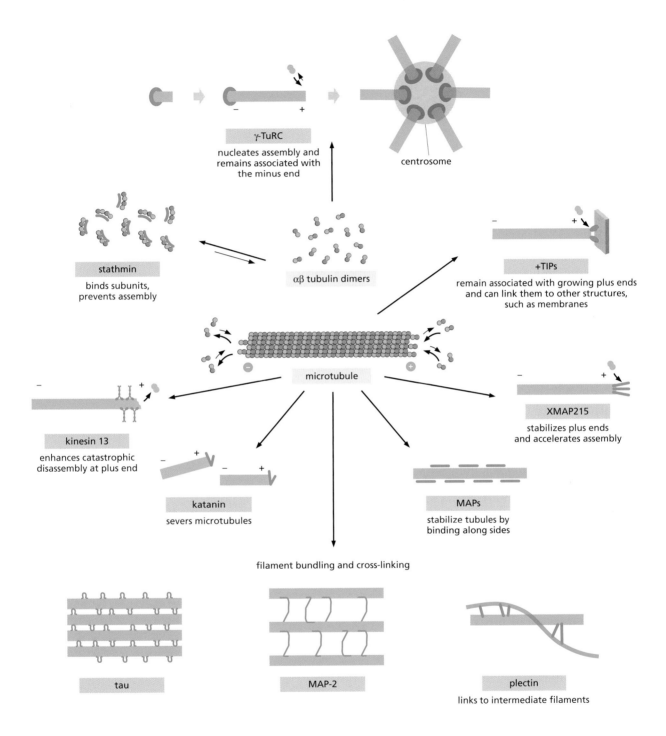

γ-TuRC
nucleates assembly and
remains associated with
the minus end

centrosome

stathmin
binds subunits,
prevents assembly

αβ tubulin dimers

+TIPs
remain associated with growing plus ends
and can link them to other structures,
such as membranes

microtubule

kinesin 13
enhances catastrophic
disassembly at plus end

XMAP215
stabilizes plus ends
and accelerates assembly

katanin
severs microtubules

MAPs
stabilize tubules by
binding along sides

filament bundling and cross-linking

tau

MAP-2

plectin
links to intermediate filaments

Some of the major accessory proteins of the microtubule cytoskeleton. Except for two classes of motor proteins, to be discussed in a later section, an example of each major type is shown. Each of these is discussed in the text. However, most cells contain more than a hundred different microtubule-binding proteins, and—as for the actin-associated proteins—it is likely that there are important types of microtubule-associated proteins that are not yet recognized.

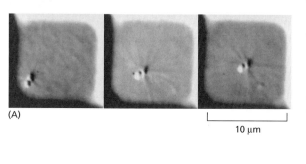

(A)

|← 10 µm →|

(B)

Figure 16–32 The center-seeking behavior of a centrosome. (A) Small square wells were micromachined into a plastic substrate. A single centrosome was placed into one of these wells, along with tubulin subunits in solution. As the microtubules polymerize, nucleated by the centrosome, they push against the walls of the well. The requirement for equal pushing in all directions to stabilize the position forces the centrosome to the center of the well. These pictures were taken at three-minute intervals. (B) A self-centered centrosome, fixed and stained to show the distribution of the microtubules pushing on all four walls of the enclosure. (From T.E. Holy et al., *Proc. Natl Acad. Sci. U.S.A.* 94:6228–6231, 1997. With permission from National Academy of Sciences.)

(**Figure 16–32**). Even in an isolated cell fragment lacking the centrosome, dynamic microtubules interacting with membraneous organelles arrange themselves into a star-shaped array with the microtubule minus ends clustered at the center, although this process may involve more components than just the simple pushing mechanism used by the isolated centrosome (**Figure 16–33**). This ability of the microtubule cytoskeleton to find the center of the cell establishes a general coordinate system, which is then used to position many organelles within the cell. Highly differentiated cells with complex morphologies such as neurons, muscles, and epithelial cells must use additional measuring mechanisms to establish their more elaborate internal coordinate systems. Thus, for example, when an epithelial cell forms cell–cell junctions and becomes highly polarized, the microtubule minus ends move to a region near the apical plasma membrane. From this asymmetric location, an array of nearly parallel microtubules forms along the long axis of the cell, with plus ends extending as far as the basal surface (see Figure 16–5).

Actin Filaments Are Often Nucleated at the Plasma Membrane

In contrast to microtubule nucleation, which occurs primarily deep within the cytoplasm near the nucleus, actin filament nucleation most frequently occurs at or near the plasma membrane. Consequently, the highest density of actin filaments in most cells is at the cell periphery. The layer just beneath the plasma membrane is called the **cell cortex**, and the actin filaments in it determine the shape and movement of the cell surface. For example, depending on their attachments to one another and to the plasma membrane, actin structures can form many strikingly different types of cell surface projections. These include spiky bundles such as *microvilli* or *filopodia*, flat protrusive veils called *lamellipodia* that help move cells over solid substrates, and the phagocytic cups in macrophages.

External signals frequently regulate the nucleation of actin filaments at the plasma membrane, allowing the cell to change its shape and stiffness rapidly in response to changes in its external environment. This nucleation can be catalyzed

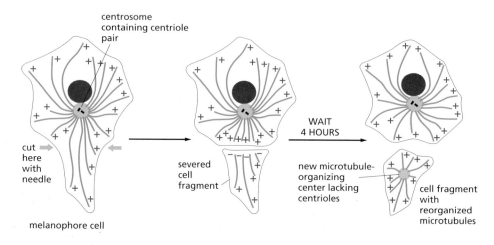

Figure 16–33 A microtubule array can find the center of a cell. After the arm of a fish pigment cell is cut off with a needle, the microtubules in the detached cell fragment reorganize so that their minus ends end up near the center of the fragment, buried in a new microtubule-organizing center.

by two different types of regulated factors, the ARP complex and the formins (discussed below). The first of these is a complex of proteins that includes two *actin-related proteins*, or *ARPs*, each of which is about 45% identical to actin. Analogous to the function of the γ-TuRC, the **ARP complex** (also known as the *Arp 2/3 complex*) nucleates actin filament growth from the minus end, allowing rapid elongation at the plus end (**Figure 16–34**A and B). The complex can also

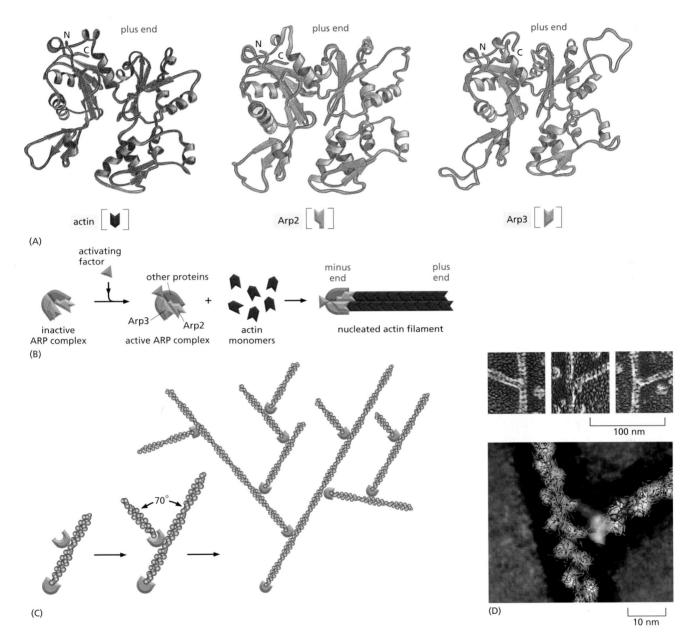

Figure 16–34 Nucleation and actin web formation by the ARP complex. (A)The structures of Arp2 and Arp3, compared to the structure of actin. Although the face of the molecule equivalent to the plus end *(top)* in both Arp2 and Arp3 is very similar to the plus end of actin itself, differences on the sides and minus end *(bottom)* prevent these actin-related proteins from forming filaments on their own or coassembling into filaments with actin. (B) A model for actin filament nucleation by the ARP complex. In the absence of an activating factor, Arp2 and Arp3 are held by their accessory proteins in an orientation that prevents them from nucleating a new actin filament. When an activating factor indicated by the *blue triangle* binds the complex, Arp2 and Arp3 are brought together into a new configuration that resembles the plus end of an actin filament. Actin subunits can then assemble onto this structure, bypassing the rate-limiting step of filament nucleation (see Figure 16–10). (C) The ARP complex nucleates filaments most efficiently when it is bound to the side of a preexisting actin filament. The result is a filament branch that grows at a 70° angle relative to the original filament. Repeated rounds of branching nucleation result in a treelike web of actin filaments. (D) Top, electron micrographs of branched actin filaments formed by mixing purified actin subunits with purified ARP complexes. Bottom, reconstructed image of a branch where the crystal structures of actin and the ARP complex have been fitted to the electron density. The mother filament runs from top to bottom, and the daughter filament branches off to the right where the ARP complex binds to three actin subunits in the mother filament (D, from R.D. Mullins et al., *Proc. Natl Acad. Sci. U.S.A.* 95:6181–6186, 1998. With permission from National Academy of Sciences, and from N. Volkmann et al., *Science* 293:2456–2459, 2001. With permission from Macmillan Publishers Ltd.)

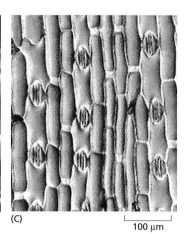

(A)   (B)   (C)

20 μm   100 μm

Figure 16–35 Function of the ARP complex in plant cells. (A) Cells in the maize leaf epidermis form small, actin-rich lobes that lock neighboring cells together like pieces of a jigsaw puzzle. (B) The regular pattern of interlocking cells covers the leaf surface. (C) Epidermal cells in a mutant plant lacking the ARP complex do not form the interlocking lobes. The brick-shaped cells are normal in size and spacing, but form leaves that appear too shiny to the naked eye. (From M.J. Frank, H.N. Cartwright and L.G. Smith, *Development* 130:753–762, 2003. With permission from the Company of Biologists.)

attach to the side of another actin filament while remaining bound to the minus end of the filament that it has nucleated, thereby building individual filaments into a treelike web (Figure 16–34C and D).

In animals, the ARP complex is associated with structures at the leading edge of migrating cells. The complex is localized in regions of rapid actin filament growth such as lamellipodia, and intracellular signaling molecules and components at the cytosolic face of the plasma membrane regulate its nucleating activity. This conserved complex is also involved in actin filament nucleation near the plasma membrane in yeast, where it is required to form cortical actin patches (see Figure 16–6), and in plant cells, where it directs the formation of actin bundles at the surface that are required for the growth of complex cell shapes in a variety of different tissues (**Figure 16–35**).

Both γ-tubulin and ARPs are evolutionarily ancient, and they are conserved among a wide variety of eucaryotic species. Their genes seem to have arisen by early duplication of the gene for the microtubule or actin filament subunit, respectively, followed by divergence and specialization of the gene copies so that they encode proteins with a special nucleating function. Thus, a similar strategy has evolved for two separate cytoskeletal systems. This underlines the central importance of regulated filament nucleation as a general organizing principle in cells.

The Mechanism of Nucleation Influences Large-Scale Filament Organization

Because the ARP complex nucleates the growth of a new actin filament most efficiently when it is bound to the side of an old actin filament, regulated activation of the ARP complex in animal cells tends to lead to the assembly of gel-like branched actin networks. However, many of the large-scale actin structures seen in cells are made up of parallel bundles of unbranched actin filaments, including the cleavage furrow found at the equator of dividing cells (see Figure 16–2) and the actin cables that point toward the site of bud growth in yeast (see Figure 16–6). The formation of many of these actin bundles is induced by a different set of nucleating proteins, the *formins*, which are able to nucleate the growth of straight, unbranched filaments that can be cross-linked by other proteins to form parallel bundles.

Formins are a large family of dimeric proteins (about 15 distinct formins are encoded in the mouse genome). Each formin subunit has a binding site for monomeric actin, and the formin dimer appears to nucleate actin filament polymerization by capturing two monomers. As the newly nucleated filament grows, the formin dimer remains associated with the rapidly growing plus end, while still allowing the binding of new subunits at that end to elongate the filament (**Figure 16–36**). This is very different from the behavior of the ARP complex or the γ-TuRC, which remain stably bound to the minus end of the actin filament or microtubule and prevent both subunit addition and subunit loss at this end.

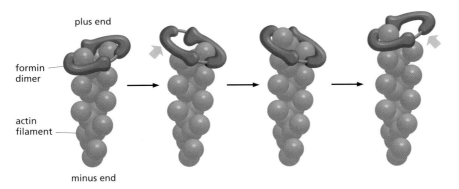

plus end

formin dimer

actin filament

minus end

Proteins That Bind to the Free Subunits Modify Filament Elongation

Once nucleated, cytoskeletal filaments generally elongate by the addition of soluble subunits. In most nonmuscle vertebrate cells, approximately 50% of the actin is in filaments and 50% is soluble, although this ratio can change rapidly in response to external signals. The soluble monomer concentration is typically 50–200 μM (2–8 mg/ml); this is surprisingly high, given the critical concentration of less than 1 μM observed for pure actin in a test tube. Why does so much of the actin remain soluble, rather than polymerizing into filaments? The reason is that the subunit pool contains special proteins that bind to the actin monomers, thereby making polymerization much less favorable (an action similar to that of the drug latrunculin). A small protein called *thymosin* is the most abundant of these proteins. Actin monomers bound to thymosin are in a locked state, where they cannot associate with either the plus or minus ends of actin filament and can neither hydrolyze nor exchange their bound nucleotide.

How do cells recruit actin monomers from this buffered storage pool and use them for polymerization? It might seem as if signal transduction pathways such as those discussed in Chapter 15 could regulate thymosin, but this has not been found to be the case. Instead, recruitment depends on another monomer-binding protein, *profilin*. Profilin binds to the face of the actin monomer opposite the ATP-binding cleft, blocking the side of the monomer that would normally associate with the filament minus end, while leaving exposed the site on the monomer that binds to the plus end (**Figure 16–37**). The profilin–actin complex can readily add onto a free plus end. This addition induces a conformational change in the actin that reduces its affinity for profilin, so the profilin falls off, leaving the actin filament one subunit longer. Because profilin competes with thymosin in binding to individual actin monomers, the net result of a local activation of profilin molecules is a movement of actin subunits from the sequestered thymosin-bound pool onto filament plus ends. Actin filament growth depends even more strongly on profilin activation for those filaments whose plus end is associated with certain formins (the family of actin-nucleating proteins discussed above); in these cases, actin filament elongation can require that the monomeric actin be bound to profilin (**Figure 16–38**).

Several intracellular mechanisms regulate profilin activity, including profilin phosphorylation and profilin binding to inositol phospholipids. These mechanisms can define the sites where profilin acts. For example, profilin's ability to move sequestered actin subunits onto the growing ends of filaments is critical for filament assembly at the plasma membrane. Profilin is localized at the cytosolic face of the plasma membrane because it binds to acidic membrane phospholipids there. At this location, extracellular signals can activate profilin to produce explosive local actin polymerization and the extension of actin-rich

Figure 16–36 Actin elongation mediated by formins. Formin proteins (*green*) form a dimeric complex that can nucleate the formation of a new actin filament (*red*) and remain associated with the rapidly-growing plus end as it elongates. The formin protein maintains its binding to one of the two actin subunits exposed at the plus end as it allows each new subunit to assemble. Only part of the large formin molecule is shown here. Other regions regulate its activity and link it to particular structures in the cell. Many formins are indirectly connected to the cell plasma membrane, and aid the insertional polymerization of the actin filament directly beneath the membrane surface.

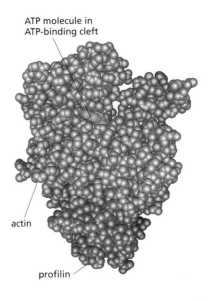

ATP molecule in ATP-binding cleft

actin

profilin

Figure 16–37 Profilin bound to an actin monomer. The profilin protein molecule is shown in *blue*, and the actin in *red*. ATP is shown in *green*. Profilin binds to the face of actin opposite the ATP-binding cleft. This profilin–actin heterodimer can therefore bind to and elongate the plus end of an actin filament, but it is sterically prevented from binding to the minus end. (Courtesy of Michael Rozycki and Clarence E. Schutt.)

Figure 16–38 Profilin and formins. Some members of the formin protein family have unstructured domains or "whiskers" that contain several binding sites for profilin or the profilin-actin complex. These flexible domains serve as a staging area for addition of actin to the growing plus end of the actin filament when formin is bound. Under some conditions, this can enhance the rate of actin filament elongation so that filament growth is faster than that expected for a diffusion-controlled reaction, and faster in the presence of formin and profilin than the rate for pure actin alone (see also Figure 3–80C).

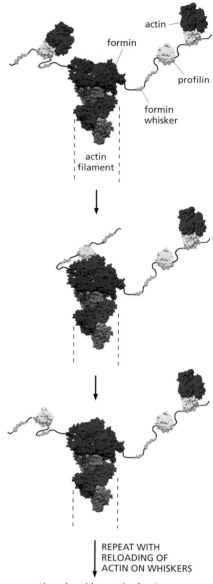

REPEAT WITH
RELOADING OF
ACTIN ON WHISKERS

continued rapid growth of actin
filament at plus end

motile structures such as filopodia and lamellipodia (see below). Besides binding to actin and phospholipids, profilin also binds to various other intracellular proteins that have domains rich in proline; these proteins can also help to localize profilin to sites that require rapid actin assembly.

As it does with actin monomers, the cell sequesters unpolymerized tubulin subunits to maintain the subunit pool at a level substantially higher than the critical concentration. One molecule of the small protein *stathmin* binds to two tubulin heterodimers and prevents their addition onto the ends of microtubules. Stathmin thus decreases the effective concentration of tubulin subunits that are available for polymerization (an action analogous to that of the drug colchicine). Furthermore, stathmin enhances the likelihood that a growing microtubule will undergo the catastrophic transition to the shrinking state. Phosphorylation of stathmin inhibits its binding to tubulin, and signals that cause stathmin phosphorylation can increase the rate of microtubule elongation and suppress dynamic instability. Cancer cells frequently overexpress stathmin, and the increased rate of microtubule turnover that results is thought to contribute to the characteristic change in cell shape associated with malignant transformation.

Severing Proteins Regulate the Length and Kinetic Behavior of Actin Filaments and Microtubules

In some situations, a cell may break an existing long filament into many smaller filaments. This generates a large number of new filament ends: one long filament with just one plus end and one minus end might be broken into dozens of short filaments, each with its own minus end and plus end. Under some intracellular conditions, these newly formed ends nucleate filament elongation, and in this case severing accelerates the assembly of new filament structures. Under other conditions, severing promotes the depolymerization of old filaments, speeding up the depolymerization rate by tenfold or more. In addition, severing filaments changes the physical and mechanical properties of the cytoplasm: stiff, large bundles and gels become more fluid when the filaments are severed.

To sever a microtubule, thirteen longitudinal bonds must be broken, one for each protofilament. The protein *katanin*, named after the Japanese word for "sword," accomplishes this demanding task (**Figure 16–39**). Katanin is made up of two subunits, a smaller subunit that hydrolyzes ATP and performs the actual severing, and a larger one that directs katanin to the centrosome. Katanin releases microtubules from their attachment to a microtubule organizing center, and it is thought to have an important role in the rapid microtubule depolymerization observed at the poles of spindles during meiosis and mitosis. It may also be involved in microtubule release and depolymerization in proliferating cells in interphase and in postmitotic cells such as neurons.

In contrast to microtubule severing by katanin, which requires ATP, the severing of actin filaments does not require an extra energy input. Most actin-severing proteins are members of the *gelsolin superfamily*, whose severing activity is activated by high levels of cytosolic Ca^{2+}. Gelsolin has subdomains that bind to two different sites on the actin subunit, one exposed on the surface of the filament and one that is normally hidden in the longitudinal bond to the next subunit in the protofilament. According to one model for gelsolin severing, gelsolin binds on the side of an actin filament and waits until a thermal fluctuation happens to create a small gap between neighboring subunits in the protofilament; gelsolin then insinuates its subdomain into the gap, breaking the filament.

Proteins That Bind Along the Sides of Filaments Can Either Stabilize or Destabilize Them

Once a cytoskeletal filament is formed by nucleation and elongated from the subunit pool, a set of proteins that bind along the sides of the polymer may alter the filament's stability and mechanical properties. Different filament-associated proteins use their binding energy to either lower or raise the free energy of the polymer state, and they thereby either stabilize or destabilize the polymer, respectively.

Proteins that bind along the sides of microtubules are collectively called **microtubule-associated proteins**, or **MAPs**. Like the drug taxol, MAPs can stabilize microtubules against disassembly. A subset of MAPs can also mediate the interaction of microtubules with other cell components. This subset is prominent in neurons, where stabilized microtubule bundles form the core of the axons and dendrites that extend from the cell body (**Figure 16–40**). These MAPs have at least one domain that binds to the microtubule surface and another that projects outward. The length of the projecting domain can determine how closely MAP-coated microtubules pack together, as demonstrated in cells engineered to overproduce different MAPs. Cells overexpressing *MAP2*, which has a long projecting domain, form bundles of stable microtubules that are kept widely spaced, while cells overexpressing *tau*, a MAP with a much shorter projecting domain, form bundles of more closely packed microtubules (**Figure 16–41**). Tau binding to filaments can also regulate the transport of membrane-enclosed organelles driven by molecular motors, which we will discuss later.

MAPs are the targets of several protein kinases, and the resulting phosphorylation of a MAP can have a primary role in controlling both its activity and localization inside cells. Among the important protein kinases that can regulate MAPs are those that are turned on and off as cells progress through the cell cycle (discussed in Chapter 17). In particular, MAP activities regulate the changes in microtubule dynamics that occur as the cell rearranges its microtubule cytoskeleton to form the mitotic spindle in preparation for chromosome segregation (see Figure 16–2).

In addition to binding along the sides of microtubules, tau protein forms its own helical filaments when present at sufficiently high concentrations. The nerve cell cytoplasm in the brains of people with Alzheimer's disease contains large aggregates of tau filaments, called neurofibrillary tangles. It is not yet clear whether these tangles of tau are a cause or a consequence of the neurodegeneration associated with this disease.

The binding of accessory proteins along their sides also affects actin filaments. Selected actin filaments in most cells are stabilized by the binding of *tropomyosin*, an elongated protein that binds simultaneously to seven adjacent actin subunits in one protofilament. The binding of tropomyosin along an actin filament can prevent the filament from interacting with other proteins; for this reason, the regulation of tropomyosin binding is an important step in muscle contraction, as we discuss later (see Figure 16–78).

Another important actin-filament binding protein present in all eucaryotic cells is *cofilin*, which destabilizes actin filaments. Also called *actin depolymerizing factor*, cofilin is unusual in that it binds to actin in both the filament and free subunit forms. Cofilin binds along the length of the actin filament, forcing the filament to twist a little more tightly (**Figure 16–42**). This mechanical stress

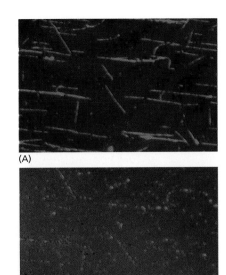

(A)

(B)

⊢————⊣
20 μm

Figure 16–39 Microtubule severing by katanin. Taxol-stabilized, rhodamine-labeled microtubules were adsorbed on the surface of a glass slide, and purified katanin was added along with ATP. (A) There are a few breaks in the microtubules 30 seconds after the addition of katanin. (B) The same field 3 minutes after the addition of katanin. The filaments have been severed in many places, leaving a series of small fragments at the previous locations of the long microtubules. (From J.J. Hartman et al., *Cell* 93:277–287, 1998. With permission from Elsevier.)

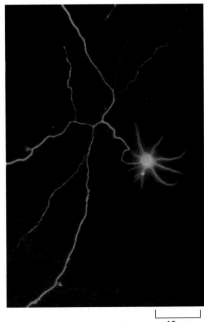

⊢————⊣
10 μm

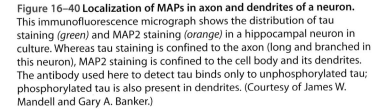

Figure 16–40 Localization of MAPs in axon and dendrites of a neuron. This immunofluorescence micrograph shows the distribution of tau staining *(green)* and MAP2 staining *(orange)* in a hippocampal neuron in culture. Whereas tau staining is confined to the axon (long and branched in this neuron), MAP2 staining is confined to the cell body and its dendrites. The antibody used here to detect tau binds only to unphosphorylated tau; phosphorylated tau is also present in dendrites. (Courtesy of James W. Mandell and Gary A. Banker.)

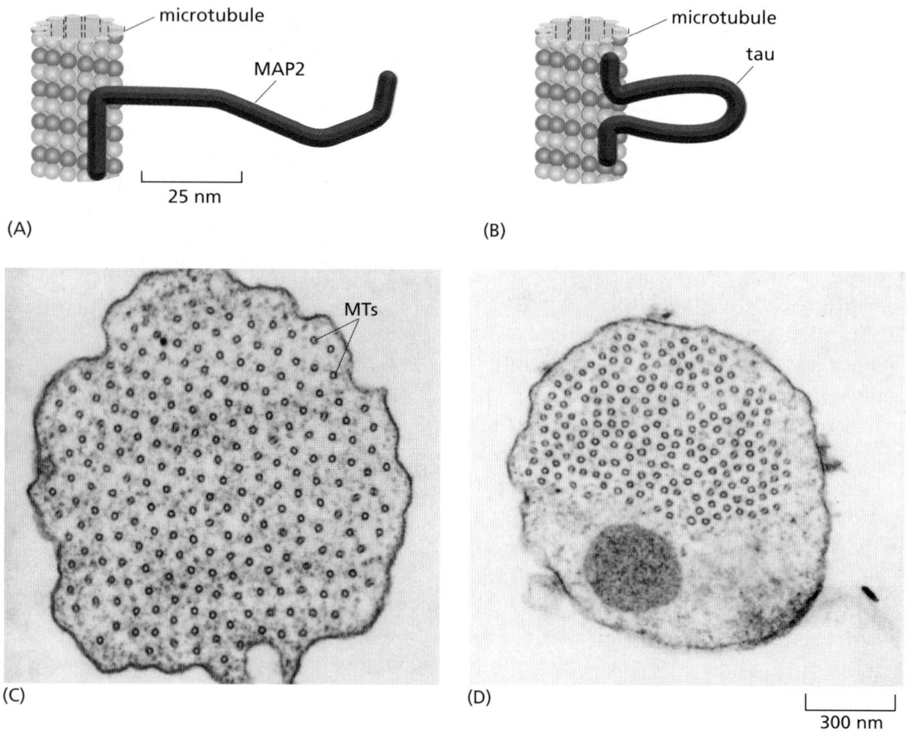

microtubule

MAP2

25 nm

(A)

microtubule

tau

(B)

MTs

(C)

(D)

300 nm

Figure 16–41 Organization of microtubule bundles by MAPs.
(A) MAP2 binds along the microtubule lattice at one of its ends and extends a long projecting arm with a second microtubule-binding domain at the other end. (B) Tau binds to the microtubule lattice at both its N- and C-termini, with a short projecting loop. (C) Electron micrograph showing a cross section through a microtubule bundle in a cell overexpressing MAP2. The regular spacing of the microtubules (MTs) in this bundle result from the constant length of the projecting arms of the MAP2.
(D) Similar cross section through a microtubule bundle in a cell overexpressing tau. Here the microtubules are spaced more closely together than they are in (C) because of tau's relatively short projecting arm.
(C and D, courtesy of V. Chen et al., *Nature* 360:674–647, 1992. With permission from Macmillan Publishers Ltd.)

weakens the contacts between actin subunits in the filament, making the filament brittle and more easily severed by thermal motions. In addition, it makes it much easier for an ADP-actin subunit to dissociate from the minus end of the filament. These activities greatly accelerate actin filament disassembly. As a result, most of the actin filaments inside cells are much shorter-lived than are filaments formed from pure actin in a test tube. Actin filaments can be protected from cofilin by tropomyosin binding.

Cofilin binds preferentially to ADP-containing actin filaments rather than to ATP-containing filaments. Since ATP hydrolysis is usually slower than filament assembly, the newest actin filaments in the cell still contain mostly ATP and are resistant to depolymerization by cofilin. Cofilin therefore efficiently dismantles the older filaments in the cell, ensuring that all actin filaments turn over rapidly. As we will discuss later, the cofilin-mediated disassembly of old but not new actin filaments is critical for the polarized, directed growth of the actin network responsible for unidirectional cell crawling.

Proteins That Interact with Filament Ends Can Dramatically Change Filament Dynamics

As we have just seen, proteins that bind along the side of a filament can change the filament's dynamic behavior. For maximum effect, however, these proteins often need to coat the filament completely, and this means they have to be present at fairly high stoichiometries (for example, about one tropomyosin for every

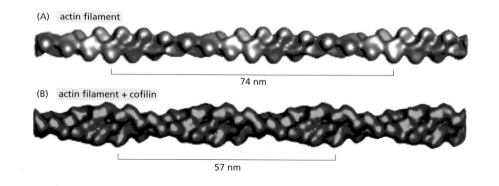

(A) actin filament

74 nm

(B) actin filament + cofilin

57 nm

Figure 16–42 Twisting of an actin filament induced by cofilin.
(A) Three-dimensional reconstruction from cryo-electron micrographs of filaments made of pure actin. The bracket shows the span of two twists of the actin helix. (B) Reconstruction of an actin filament coated with cofilin, which binds in a 1:1 stoichiometry to actin subunits all along the filament. Cofilin is a small protein (14 kilodaltons) compared to actin (43 kilodaltons), and so the filament appears only slightly thicker. The energy of cofilin binding serves to deform the actin filament lattice, twisting it more tightly and reducing the distance spanned by each twist of the helix. (From A. McGough et al., *J. Cell Biol.* 138:771–781, 1997. With permission from The Rockefeller University Press.)

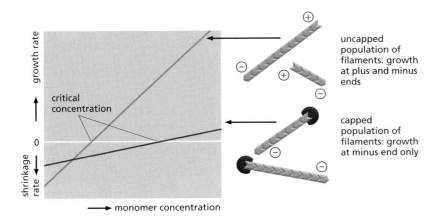

uncapped
population of
filaments: growth
at plus and minus
ends

capped
population of
filaments: growth
at minus end only

Figure 16–43 Filament capping and its effects on filament dynamics.
A population of uncapped filaments adds and loses subunits at both the plus and minus ends, resulting in rapid growth or shrinkage, depending on the concentration of available free monomers (*green* line). In the presence of a protein that caps the plus end (*red* line), only the minus end is able to add or lose subunits; consequently, filament growth will be slower at all monomer concentrations above the critical concentration, and filament shrinkage will be slower at all monomer concentrations below the critical concentration. In addition, the critical concentration for the population shifts to that of the filament minus end.

seven actin subunits, one tau for every four tubulin subunits, or one cofilin for every actin subunit). In contrast, proteins that bind preferentially to the ends of filaments can have dramatic effects on filament dynamics even when they are present at very low levels. Since subunit addition and loss occur primarily at filament ends, one molecule of such a protein per actin filament (typically one per about 200–500 actin monomers) can be enough to transform the architecture of an actin filament network.

As previously discussed, an actin filament that ceases elongation and is not specifically stabilized by the cell can depolymerize rapidly: it can lose subunits from either its plus or its minus end, once the actin molecules at that end have hydrolyzed their ATP to convert to the D form. The most rapid changes, however, occur at the plus end. The binding of a plus end *capping protein* stabilizes an actin filament at its plus end, which greatly slows the rates of both filament growth and filament depolymerization by making the plus end inactive (**Figure 16–43**). Indeed, most of the actin filaments in a cell are capped at their plus end by proteins such as *CapZ* (named for its location in the muscle Z band, see below; it is also called Capping Protein). At the minus end, an actin filament may be capped by remaining bound to the ARP complex that was responsible for its nucleation, although it is possible that many of the actin filament minus ends in typical cells are released from the ARP complex and are uncapped.

In muscle cells, where actin filaments are exceptionally long-lived, the filaments are specially capped at both ends—by CapZ at the plus end and by *tropomodulin* at the minus end. Tropomodulin binds only to the minus end of actin filaments that have been coated by tropomyosin and have thereby already been somewhat stabilized.

Different Kinds of Proteins Alter the Properties of Rapidly Growing Microtubule Ends

The end of a microtubule, with thirteen protofilaments in a hollow ring (see Figure 16–11), is a much larger and more complex structure than the end of an actin filament, with many more possibilities for accessory protein action. We have already discussed an important microtubule capper: the γ-tubulin ring complex (γ-TuRC), which both nucleates the growth of microtubules at an organizing center and caps their minus ends. Another true capping protein for microtubules is the special protein complex found at the ends of the microtubules in cilia (discussed later), where microtubules are both stable and uniform in length.

Some proteins that act at the ends of microtubules have crucial roles beyond those expected for a simple capping protein. In particular, they can have dramatic effects on the dynamic instability of microtubules (see Figure 16–16). They can influence the rate at which a microtubule switches from a growing to a shrinking state (the frequency of catastrophes) or from a shrinking to a growing state (the frequency of rescues). For example, a family of kinesin-related proteins known as *catastrophe factors* significantly increases the catastrophe rate

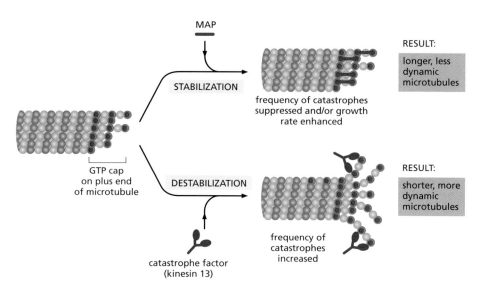

MAP

STABILIZATION

RESULT:

longer, less dynamic microtubules

frequency of catastrophes suppressed and/or growth rate enhanced

GTP cap on plus end of microtubule

DESTABILIZATION

RESULT:

shorter, more dynamic microtubules

frequency of catastrophes increased

catastrophe factor (kinesin 13)

Figure 16–44 The effects of proteins that bind to microtubule ends. The transition between microtubule growth and microtubule shrinking is controlled in cells by special proteins. A MAP such as XMAP215 stabilizes the end of a growing microtubule by its preferential binding there. Opposing its action are catastrophe factors such as kinesin-13, a member of the kinesin motor protein superfamily (discussed later).

(these proteins are members of the kinesin-13 family; see Figure 16–58). They bind specifically to microtubule ends and seem to pry protofilaments apart, lowering the normal activation energy barrier that prevents a microtubule from springing apart into the curved protofilament characteristic of the shrinking state (see Figure 16–16C). Opposing their actions are MAPs such as the ubiquitous *XMAP215* that has close homologs in organisms that range from yeast to humans (XMAP stands for *Xenopus* microtubule-associated protein, and the number refers to its molecular mass in kilodaltons). This protein has a special ability to stabilize free microtubule ends and inhibit their switch from a growing to a shrinking state. The phosphorylation of XMAP215 during mitosis inhibits its activity and shifts the balance of its competition with catastrophe factors (**Figure 16–44**). The shift results in a tenfold increase in the dynamic instability of microtubules observed during mitosis, a transition that is critical for the efficient construction of the mitotic spindle (see Figure 17–33).

In many cells, the minus ends of microtubules are stabilized by association with the centrosome, or else serve as microtubule depolymerization sites. The plus ends, in contrast, efficiently explore and probe the entire cell space. Microtubule-associated proteins called *plus-end tracking proteins (+TIPs)* accumulate at these active ends, and appear to rocket around the cell as passengers at the ends of rapidly growing microtubules, dissociating from the ends when the microtubules begin to shrink (**Figure 16–45**).

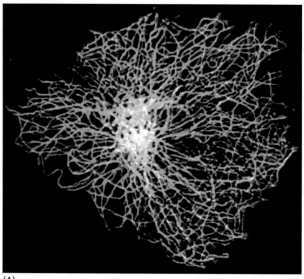

(A)

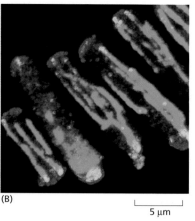

(B)

5 μm

Figure 16–45 +TIP proteins found at the growing plus ends of microtubules. <TAAT> (A) In an epithelial cell grown in tissue culture, each microtubule (*green*) has a growing plus end which is associated with the +TIP protein EB1 (*red*). (B) In the rod-shaped fission yeast *Schizosaccharomyces pombe*, the plus ends of the microtubules (*green*) are associated with the homolog of EB1 (*red*) at the two poles of the rod-shaped cells. (A, from A. Akhmanova and C.C. Hoogenraad, *Curr. Opin. Cell Biol.* 17:47–54, 2005. With permission from Elsevier; B, courtesy of Ken Sawin.)

Some of the +TIPs, such as the kinesin-related catastrophe factors and XMAP215 mentioned above, modulate the growth and shrinkage of the microtubule end to which they are attached. Others control microtubule positioning by helping to capture and stabilize the growing microtubule end at the location of specific target proteins in the cell cortex. EB1, a +TIP present in both yeasts and humans, for example, is essential for yeast mitotic spindle positioning, directing the growing plus ends of yeast spindle microtubules to a specific docking region in the yeast bud and then helping to anchor them there.

Filaments Are Organized into Higher-Order Structures in Cells

So far, we have described how cells use accessory proteins to regulate the location and dynamic behavior of cytoskeletal filaments. These proteins can nucleate filament assembly, bind to the ends or sides of the filaments, or bind to the free subunits of filaments. But in order for the cytoskeletal filaments to form a useful intracellular scaffold that gives the cell mechanical integrity and determines its shape, the individual filaments must be organized and attached to one another in larger-scale structures. The centrosome is one example of such a cytoskeletal organizer; in addition to nucleating the growth of microtubules, it holds them together in a defined geometry, with all of the minus ends buried in the centrosome and the plus ends pointing outward. In this way, the centrosome creates the astral array of microtubules that is able to find the center of each cell (see Figure 16–32).

Another mechanism that cells use to organize filaments into large structures is filament cross-linking. As described earlier, some MAPs can bundle microtubules together: they have two domains—one that binds along the microtubule side (and thereby stabilizes the filament) and another that projects outward to contact other MAP-coated microtubules. In the actin cytoskeleton, the stabilizing and cross-linking functions are separated. Tropomyosin binds along the sides of actin filaments, but it does not have an outward projecting domain. As we shall see shortly, filament cross-linking is instead mediated by a second group of actin-binding proteins that have only this function. Intermediate filaments are different yet again; they are organized both by a lateral self-association of the filaments themselves and by the cross-linking activity of accessory proteins, as we describe next.

Intermediate Filaments Are Cross-Linked and Bundled Into Strong Arrays

Each individual intermediate filament forms as a long bundle of tetrameric subunits (see Figure 16–19). Many intermediate filaments further bundle themselves by self-association; for example, the neurofilament proteins NF-M and NF-H (see Table 16–1, p. 985) contain a C-terminal domain that extends outward from the surface of the assembled intermediate filament and binds to a neighboring filament. Thus groups of neurofilaments form robust parallel arrays that are held together by multiple lateral contacts, giving strength and stability to the long cell processes of neurons (see Figure 16–22).

Other types of intermediate filament bundles are held together by accessory proteins, such as *filaggrin*, which bundles keratin filaments in differentiating cells of the epidermis to give the outermost layers of the skin their special toughness. *Plectin* is a particularly interesting cross-linking protein. Besides bundling intermediate filaments, it also links the intermediate filaments to microtubules, actin filament bundles, and filaments of the motor protein myosin II (discussed below), as well as helping to attach intermediate filament bundles to adhesive structures at the plasma membrane (**Figure 16–46**).

Mutations in the gene for plectin cause a devastating human disease that combines epidermolysis bullosa (caused by disruption of skin keratin filaments), muscular dystrophy (caused by disruption of desmin filaments), and neurodegeneration (caused by disruption of neurofilaments). Mice lacking a

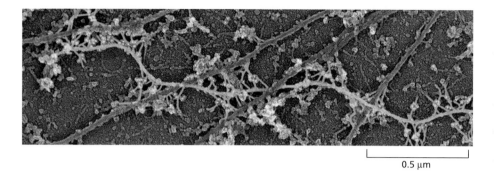

Figure 16–46 **Plectin cross-linking of diverse cytoskeletal elements.** Plectin *(green)* is seen here making cross-links from intermediate filaments *(blue)* to microtubules *(red)*. In this electron micrograph, the dots *(yellow)* are gold particles linked to anti-plectin antibodies. The entire actin filament network was removed to reveal these proteins. (From T.M. Svitkina and G.G.Borisy, *J. Cell Biol.* 135:991–1007, 1996. With permission from The Rockefeller University Press.)

0.5 μm

functional plectin gene die within a few days of birth, with blistered skin and abnormal skeletal and heart muscles. Thus, although plectin may not be necessary for the initial formation and assembly of intermediate filaments, its cross-linking action is required to provide cells with the strength they need to withstand the mechanical stresses inherent to vertebrate life.

Cross-linking Proteins with Distinct Properties Organize Different Assemblies of Actin Filaments

Actin filaments in animal cells are organized into two types of arrays: bundles and weblike (gel-like) networks (**Figure 16–47**). As described earlier, these different structures are initiated by the action of distinct nucleating proteins: the long straight filaments produced by formins make bundles and the ARP complex makes webs. The actin filament cross-linking proteins that help to stabilize and maintain these distinct structures are divided into two classes: *bundling proteins* and *gel-forming proteins*. Bundling proteins cross-link actin filaments into a parallel array, while gel-forming proteins hold two actin filaments together at a large angle to each other, thereby creating a looser meshwork. Both types of cross-linking protein generally have two similar actin-filament-binding sites, which can either be part of a single polypeptide chain or contributed by each of two polypeptide chains held together in a dimer (**Figure 16–48**). The spacing and arrangement of these two filament-binding domains determines the type of actin structure that a given cross-linking protein forms.

Each type of bundling protein also determines which other molecules can interact with an actin filament. Myosin II (discussed later) is the motor protein in stress fibers and other contractile arrays that enables them to contract. The

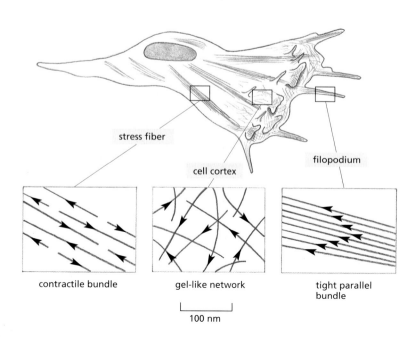

stress fiber

cell cortex

filopodium

contractile bundle

gel-like network

tight parallel bundle

100 nm

Figure 16–47 **Actin arrays in a cell.** A fibroblast crawling in a tissue culture dish is shown with three areas enlarged to show the arrangement of actin filaments. The actin filaments are shown in *red,* with arrowheads pointing toward the minus end. Stress fibers are contractile and exert tension. Filopodia are spike-like projections of the plasma membrane that allow a cell to explore its environment. The cortex underlies the plasma membrane.

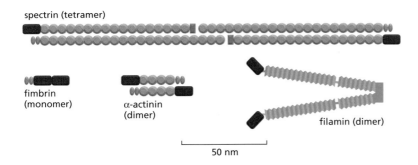

spectrin (tetramer)

fimbrin
(monomer)

α-actinin
(dimer)

filamin (dimer)

50 nm

Figure 16–48 The modular structures of four actin-cross-linking proteins. Each of the proteins shown has two actin-binding sites *(red)* that are related in sequence. Fimbrin has two directly adjacent actin-binding sites, so that it holds its two actin filaments very close together (14 nm apart), aligned with the same polarity (see Figure 16–49A). The two actin-binding sites in α-actinin are separated by a spacer around 30 nm long, so that it forms more loosely packed actin bundles (see Figure 16–49A). Filamin has two actin-binding sites with a V-shaped linkage between them, so that it cross-links actin filaments into a network with the filaments oriented almost at right angles to one another (see Figure 16–51). Spectrin is a tetramer of two α and two β subunits, and the tetramer has two actin-binding sites spaced about 200-nm apart (see Figure 10–41).

very close packing of actin filaments caused by the small monomeric bundling protein *fimbrin* apparently excludes myosin, and thus the parallel actin filaments held together by fimbrin are not contractile; on the other hand, the looser packing caused by the larger dimeric bundling protein *α-actinin* allows myosin molecules to enter, making stress fibers contractile (**Figure 16–49**). Because of the very different spacing between the actin filaments, bundling by fimbrin automatically discourages bundling by α-actinin, and vice-versa, so that the two types of bundling protein are themselves mutually exclusive.

Villin is another bundling protein that, like fimbrin, has two actin-filament-binding sites very close together in a single polypeptide chain. Villin (together with fimbrin) helps cross-link the 20 to 30 tightly bundled actin filaments found in microvilli, the finger-like extensions of the plasma membrane on the surface of many epithelial cells (**Figure 16–50**). A single absorptive epithelial cell in the human small intestine, for example, has several thousand microvilli on its apical surface. Each is about 0.08 μm wide and 1 μm long, making the cell's absorptive surface area about 20 times greater than it would be without microvilli. When villin is introduced into cultured fibroblasts, which do not normally contain villin and have only a few small microvilli, the existing microvilli become greatly elongated and stabilized, and new ones are induced. The actin filament core of the microvillus is attached to the plasma membrane along its sides by lateral sidearms made of *myosin I* (discussed later), which has a binding site for filamentous actin on one end and a domain that binds lipids on the other end. These two types of cross-linkers, one binding actin filaments to each other and the other binding these filaments to the membrane, seem to be sufficient to form microvilli on cells. Interestingly, when the gene for villin is disrupted in a mouse, the intestinal microvilli form with apparently normal morphology, indicating that other bundling proteins provide sufficient redundant function for this purpose. However, the remodeling of intestinal microvilli in response to certain kinds of stress or starvation is impaired.

Figure 16–49 The formation of two types of actin filament bundles. (A) α-actinin, which is a homodimer, cross-links actin filaments into loose bundles, which allow the motor protein myosin II (not shown) to participate in the assembly. Fimbrin cross-links actin filaments into tight bundles, which exclude myosin. Fimbrin and α-actinin tend to exclude one another because of the very different spacing of the actin filament bundles that they form. (B) Electron micrograph of purified α-actinin molecules. (B, courtesy of John Heuser.)

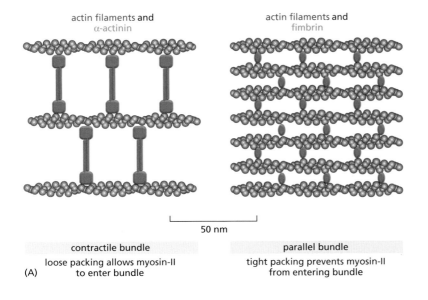

actin filaments and
α-actinin

actin filaments and
fimbrin

50 nm

contractile bundle
loose packing allows myosin-II
to enter bundle

(A)

parallel bundle
tight packing prevents myosin-II
from entering bundle

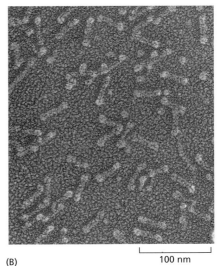

(B)

100 nm

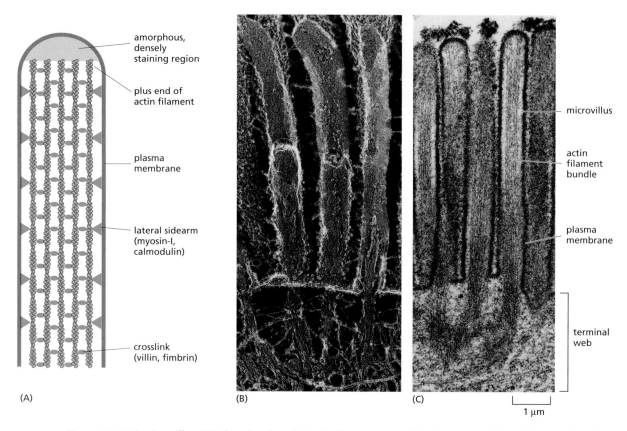

Figure 16–50 A microvillus. (A) A bundle of parallel actin filaments cross-linked by the actin-bundling proteins villin and fimbrin forms the core of a microvillus. Lateral sidearms (composed of myosin I and the Ca^{2+}-binding protein calmodulin) connect the sides of the actin filament bundle to the overlying plasma membrane. All the plus ends of the actin filaments are at the tip of the microvillus, where they are embedded in an amorphous, densely staining substance of unknown composition. (B) Freeze-fracture electron micrograph of the apical surface of an intestinal epithelial cell, showing microvilli. Actin bundles from the microvilli extend down into the cell and are rooted in the terminal web, where they are linked together by a complex set of proteins that includes spectrin and myosin II. Below the terminal web is a layer of intermediate filaments. (C) Thin section electron micrograph of microvilli. (B, courtesy of John Heuser; C, from P.T. Matsudaira and D.R. Burgess, *Cold Spring Harb. Symp. Quant. Biol.* 46:845–854, 1985. With permission from Cold Spring Harbor Laboratory Press.)

Filamin and Spectrin Form Actin Filament Webs

The various bundling proteins that we have discussed so far have straight, stiff connections between their two actin-filament-binding domains, and they tend to align filaments in parallel bundles. In contrast, those actin cross-linking proteins that have either a flexible or a stiff, bent connection between their two binding domains form actin filament webs or gels, rather than actin bundles.

Any cross-linking protein that has its two actin-binding domains joined by a long bent linkage can form three-dimensional actin gels. *Filamin* (see Figure 16–48) promotes the formation of a loose and highly viscous gel by clamping together two actin filaments roughly at right angles (**Figure 16–51**). Cells require the actin gels formed by filamin in order to extend the thin sheet-like membrane projections called *lamellipodia* that help them to crawl across solid surfaces. Filamin is lacking in some types of cancer cells, especially some malignant melanomas (pigment-cell cancers). These cells cannot crawl properly, and instead they protrude disorganized membrane blebs (**Figure 16–52**). Losing filamin is bad for the melanoma cells but good for the melanoma patient; because of the cells' inability to crawl, melanoma cells that have lost filamin expression are less invasive than similar melanoma cells that still express filamin, and, as a result, the cancer is much less likely to metastasize.

A very different well-studied web-forming protein is *spectrin*, which was first identified in red blood cells. Spectrin is a long, flexible protein made out of four elongated polypeptide chains (two α subunits and two β subunits), arranged so that the two actin-filament-binding sites are about 200 nm apart (compared

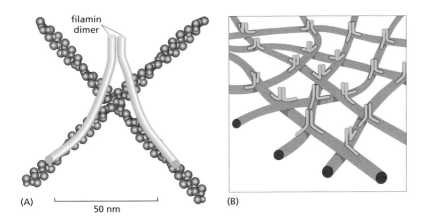

Figure 16–51 Filamin cross-links actin filaments into a three-dimensional network with the physical properties of a gel. (A) Each filamin homodimer is about 160 nm long when fully extended and forms a flexible, high-angle link between two adjacent actin filaments. (B) A set of actin filaments cross-linked by filamin forms a mechanically strong web or gel.

filamin dimer

(A)   50 nm

(B)

with 14 nm for fimbrin and about 30 nm for α-actinin, see Figure 16–48). In the red blood cell, spectrin is concentrated just beneath the plasma membrane, where it forms a two-dimensional web held together by short actin filaments; spectrin links this web to the plasma membrane because it has separate binding sites for peripheral membrane proteins, which are themselves positioned near the lipid bilayer by integral membrane proteins (see Figure 10–41). The resulting network creates a stiff cell cortex that provides mechanical support for the overlying plasma membrane, allowing the red blood cell to spring back to its original shape after squeezing through a capillary. Close relatives of spectrin are found in the cortex of most other vertebrate cell types, where they also help to shape and stiffen the surface membrane.

Cytoskeletal Elements Make Many Attachments to Membranes

Actin cytoskeletal structures both stiffen and change the shape of the plasma membrane. We have already discussed two examples: the spectrin-actin web that underlies the plasma membranes and the villin-actin bundles in microvilli that enlarge the absorptive surface area of epithelial cells. The effectiveness of these structures requires specific attachments between the actin filament network and proteins or lipids of the plasma membrane.

The connections of the cortical actin cytoskeleton to the plasma membrane are only partially understood. A widespread family of closely related intracellular proteins, the *ERM* family (named for its first three members, ezrin, radixin, and moesin), contains members that are required for the maintenance of cell polarity and involved in exocytosis and endocytosis. The C-terminal domain of an ERM protein binds directly to the sides of actin filaments. The N-terminal domain binds to the cytoplasmic face of one or more transmembrane glycoproteins, such as CD44, the receptor for the extracellular matrix component hyaluronan.

The attachments between actin and the plasma membrane mediated by ERM proteins are regulated by both intracellular and extracellular signals. ERM proteins can exist in two conformations, an active extended conformation that oligomerizes and binds to both actin and a transmembrane protein, and an inactive folded conformation, in which the N- and C-termini are held together by an intramolecular interaction. Switching to the active conformation can be

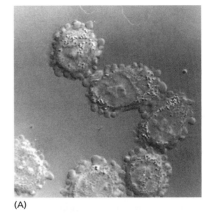

(A)

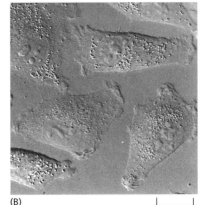

(B)   10 µm

Figure 16–52 Loss of filamin causes abnormal cell motility. (A) A group of melanoma cells that have an abnormally low level of filamin. These cells are not able to make normal lamellipodia and instead are covered with membrane "blebs." As a result, they crawl poorly and tend not to metastasize. (B) The same melanoma cells in which filamin expression has been artificially restored. The cells now make normal lamellipodia and are highly metastatic. This example is one of many demonstrating the profound effect that the presence or absence of a single structural protein can have on cell morphology and motility. (From C. Cunningham et al., *J. Cell Biol.* 136:845–857, 1997. With permission The Rockefeller University Press.)

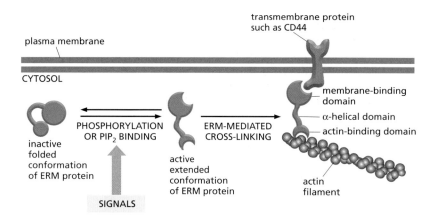

Figure 16–53 The role of ERM-family proteins in attaching actin filaments to the plasma membrane. Regulated unfolding of an ERM-family protein, caused by phosphorylation or by binding to PIP$_2$, exposes two binding sites, one for an actin filament and one for a transmembrane protein. Activation of ERM-family proteins can thereby generate and stabilize cell-surface protrusions that form in response to extracellular signals.

triggered by protein phosphorylation or by binding to PIP$_2$, either of which can occur, for example, in response to extracellular signals (the ERM proteins are direct targets of several receptor tyrosine kinases). In this way, the properties of the cell cortex become sensitive to a variety of signals received by the cell (**Figure 16–53**).

The loss of one of the members of the ERM family, called merlin, results in a form of the human genetic disease called *neurofibromatosis,* in which multiple benign tumors develop in the auditory nerves and certain other parts of the nervous system. This is one of many indications of a feedback system that connects cell structural elements to the control of cell growth (see Chapter 17).

The proteins, discussed in this section, that control the assembly and position of actin filaments and microtubules are reviewed in Panel 16–3 (pp. 994–995). Some of these proteins have the additional function of helping to connect the internal structure of a cell to other cells or to an extracellular basement membrane. Both actin filaments and intermediate filaments are critical for these connections, which require the specialized cell-cell junctions and cell-matrix junctions that we will discuss in Chapter 19.

Summary

The varied forms and functions of cytoskeletal filament structures in eucaryotic cells depend on a versatile repertoire of accessory proteins. Each of the three major filament classes (microtubules, intermediate filaments, and actin filaments) has a large dedicated subset of such accessory proteins.

A primary determinant of the sites of cytoskeletal structures is the regulation of the processes that initiate the nucleation of new filaments. In most animal cells, microtubules are nucleated at the centrosome, a complex assembly located near the center of the cell. In contrast, most actin filaments are nucleated near the plasma membrane.

The kinetics of filament assembly and disassembly can be either slowed or accelerated by accessory proteins that bind to either the free subunits or the filaments themselves. Some of these proteins alter filament dynamics by binding to the ends of filaments or by severing the filaments into smaller fragments. Another class of accessory proteins assembles the filaments into larger ordered structures by cross-linking them to one another in geometrically defined ways. Yet other accessory proteins determine the shape and adhesive properties of cells by attaching filaments to the plasma membrane.

MOLECULAR MOTORS

Among the most fascinating proteins that associate with the cytoskeleton are the molecular motors called **motor proteins**. These remarkable proteins bind to a polarized cytoskeletal filament and use the energy derived from repeated cycles of ATP hydrolysis to move steadily along it. Dozens of different motor proteins coexist in every eucaryotic cell. They differ in the type of filament they bind to (either actin or microtubules), the direction in which they move along the filament, and the "cargo" they carry. Many motor proteins carry membrane-enclosed organelles—such as mitochondria, Golgi stacks, or secretory vesicles—to their appropriate locations in the cell. Other motor proteins cause cytoskeletal filaments to exert tension or to slide against each other, generating the force that drives such phenomena as muscle contraction, ciliary beating, and cell division.

Cytoskeletal motor proteins that move unidirectionally along an oriented polymer track are reminiscent of some other proteins and protein complexes discussed elsewhere in this book, such as DNA and RNA polymerases, helicases, and ribosomes. All of these have the ability to use chemical energy to propel themselves along a linear track, with the direction of sliding dependent on the structural polarity of the track. All of them generate motion by coupling nucleoside triphosphate hydrolysis to a large-scale conformational change in a protein, as explained in Chapter 3 (see Figure 3–77).

The cytoskeletal motor proteins associate with their filament tracks through a "head" region, or *motor domain,* that binds and hydrolyzes ATP. Driven by cycles of nucleotide hydrolysis that produce conformational changes, the proteins cycle between states in which they are bound strongly to their filament tracks and states in which they are unbound. Through a mechanochemical cycle of filament binding, conformational change, filament release, conformational relaxation, and filament rebinding, the motor protein and its associated cargo move one step at a time along the filament (typically a distance of a few nanometers). The motor domain (head) determines the identity of the track and the direction of movement along it, whereas the tail of the motor protein determines the identity of the cargo (and therefore the biological function of the individual motor protein).

In this section, we begin by describing the three groups of cytoskeletal motor proteins: myosins, kinesins, and dyneins. We then describe how they can work to transport membrane-enclosed organelles and mRNAs or to change the shape of structures built from cytoskeletal filaments. In the final section of this chapter, we will examine how a collaboration between motor proteins and the dynamic cytoskeletal filaments described previously generates complex cell behaviors.

Actin-Based Motor Proteins Are Members of the Myosin Superfamily

The first motor protein identified was skeletal muscle **myosin**, which generates the force for muscle contraction. This myosin, called *myosin II* (see below) is an elongated protein that is formed from two heavy chains and two copies of each of two light chains. Each heavy chain has a globular head domain at its N-terminus that contains the force-generating machinery, followed by a very long amino acid sequence that forms an extended coiled-coil that mediates heavy chain dimerization (**Figure 16–54**). The two light chains bind close to the N-terminal

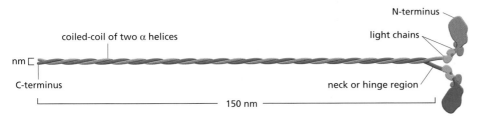

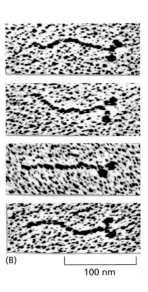

Figure 16–54 **Myosin II.** (A) A myosin II molecule is composed of two heavy chains (each about 2000 amino acids long *(green)* and four light chains *(blue)*. The light chains are of two distinct types, and one copy of each type is present on each myosin head. Dimerization occurs when the two α helices of the heavy chains wrap around each other to form a coiled-coil, driven by the association of regularly spaced hydrophobic amino acids (see Figure 3–9). The coiled-coil arrangement makes an extended rod in solution, and this part of the molecule is called the tail. (B) The two globular heads and the tail can be clearly seen in electron micrographs of myosin molecules shadowed with platinum. (B, courtesy of David Shotton.)

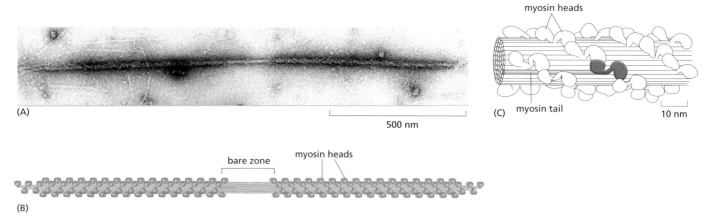

(A)

500 nm

(C)

myosin heads

myosin tail

10 nm

(B)

bare zone

myosin heads

Figure 16–55 The myosin II bipolar thick filament in muscle. (A) Electron micrograph of a myosin II thick filament isolated from frog muscle. Note the central bare zone, which is free of head domains. (B) Schematic diagram, not drawn to scale. The myosin II molecules aggregate by means of their tail regions, with their heads projecting to the outside of the filament. The bare zone in the center of the filament consists entirely of myosin II tails. (C) A small section of a myosin II filament as reconstructed from electron micrographs. An individual myosin molecule is highlighted in *green*. The cytoplasmic myosin II filaments in non-muscle cells are much smaller, although similarly organized (see Figure 16–72). (A, courtesy of Murray Stewart; C, based on R.A. Crowther, R. Padron and R. Craig, *J. Mol. Biol.* 184:429–439, 1985. With permission from Academic Press.)

head domain, while the long coiled-coil tail bundles itself with the tails of other myosin molecules. These tail–tail interactions form large bipolar "thick filaments" that have several hundred myosin heads, oriented in opposite directions at the two ends of the thick filament (**Figure 16–55**).

Each myosin head binds and hydrolyzes ATP, using the energy of ATP hydrolysis to walk toward the plus end of an actin filament. The opposing orientation of the heads in the thick filament makes the filament efficient at sliding pairs of oppositely oriented actin filaments past each other. In skeletal muscle, in which carefully arranged actin filaments are aligned in "thin filament" arrays surrounding the myosin thick filaments, the ATP-driven sliding of actin filaments results in muscle contraction (discussed later). Cardiac and smooth muscle contain myosin II molecules that are similarly arranged, although different genes encode them.

When a muscle myosin is digested by chymotrypsin and papain, the head domain is released as an intact fragment (called S1). The S1 fragment alone can generate filament sliding *in vitro*, proving that the motor activity is contained completely within the head (**Figure 16–56**).

It was initially thought that myosin was present only in muscle, but in the 1970s, researchers found that a similar two-headed myosin protein was also

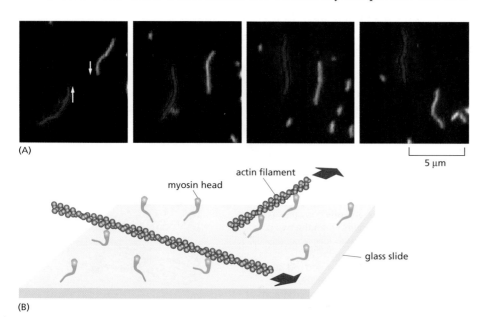

(A)

5 μm

actin filament

myosin head

glass slide

(B)

Figure 16–56 Direct evidence for the motor activity of the myosin head. <TTAT> In this experiment, purified S1 myosin heads were attached to a glass slide, and then actin filaments labeled with fluorescent phalloidin were added and allowed to bind to the myosin heads. (A) When ATP was added, the actin filaments began to glide along the surface, owing to the many individual steps taken by each of the dozens of myosin heads bound to each filament. The video frames shown in this sequence were recorded about 0.6 second apart; the two actin filaments shown (one *red* and one *green*) were moving in opposite directions at a rate of about 4 μm/sec. (B) Diagram of the experiment. The large *red* arrows indicate the direction of actin filament movement. (A, courtesy of James Spudich.)

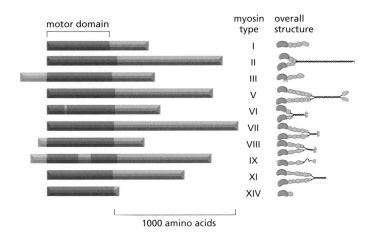

motor domain

myosin type

overall structure

I
II
III
V
VI
VII
VIII
IX
XI
XIV

1000 amino acids

Figure 16–57 **Myosin superfamily members.** Comparison of the domain structure of the heavy chains of some myosin types. All myosins share similar motor domains (shown in *dark green*), but their C-terminal tails *(light green)* and N-terminal extensions *(light blue)* are very diverse. On the right are depictions of the molecular structure for these family members. Many myosins form dimers, with two motor domains per molecule, but a few (such as I, IX, and XIV) seem to function as monomers, with just one motor domain. Myosin VI, despite its overall structural similarity to other family members, is unique in moving toward the minus end (instead of the plus end) of an actin filament. The small insertion within its motor head domain, not found in other myosins, is probably responsible for this change in direction.

present in nonmuscle cells, including protozoan cells. At about the same time, other researchers found a myosin in the freshwater amoeba *Acanthamoeba castellanii* that was unconventional in having a motor domain similar to the head of muscle myosin but a completely different tail. This molecule seemed to function as a monomer and was named *myosin I* (for one-headed); the conventional myosin was renamed *myosin II* (for two-headed).

Subsequently, many other myosin types were discovered. The heavy chains generally start with a recognizable myosin motor domain at the N-terminus, and then diverge widely with a variety of C-terminal tail domains (**Figure 16–57**). The newly identified types of myosins include a number of one-headed and two-headed varieties that are approximately equally related to myosin I and myosin II, and the nomenclature now reflects their approximate order of discovery (myosin III through at least myosin XVIII). Sequence comparisons among diverse eucaryotes indicate that there are at least 37 distinct myosin families in the superfamily. The myosin tails (and the tails of motor proteins generally) have apparently diversified during evolution to permit the proteins to bind other subunits and to interact with different cargoes.

Some myosins (such as VIII and XI) have been found only in plants, and some have been found only in vertebrates (IX). Most, however, are found in all eucaryotes, suggesting that myosins arose early in eucaryotic evolution. The yeast *Saccharomyces cerevisiae* contains five myosins: two myosin Is, one myosin II, and two myosin Vs. It is tempting to speculate that these three types of myosins are necessary for a eucaryotic cell to survive and that other myosins perform more specialized functions, particularly in multicellular organisms. The nematode *C. elegans*, for example, has at least 15 myosin genes, representing at least seven structural classes; the human genome includes about 40 myosin genes. Nine of the human myosins are expressed primarily or exclusively in the hair cells of the inner ear, and mutations in five of them are known to cause hereditary deafness. These extremely specialized myosins are important for the construction and function of the complex and beautiful bundles of actin-rich stereocilia that are found on the apical surface of these cells (see Figure 9–50); these tilt in response to sound and convert sound waves into electrical signals (discussed in Chapter 23).

All of the myosins except one move toward the plus end of an actin filament, although they do so at different speeds. The exception is myosin VI, which moves toward the minus end.

The exact functions for most of the myosins remain to be determined. Myosin V is involved in vesicle and organelle transport. Myosin II is associated with contractile activity in both muscle and nonmuscle cells. It is generally required for cytokinesis (the pinching apart of a dividing cell into two daughters), as well as for the forward translocation of the body of a cell during cell migration. The myosin I proteins often contain a second actin-binding site or a membrane-binding site in their tails, and they are generally involved in intracellular organization—including the protrusion of actin-rich structures at the cell surface, as discussed earlier for the construction of microvilli (see Figure 16–50).

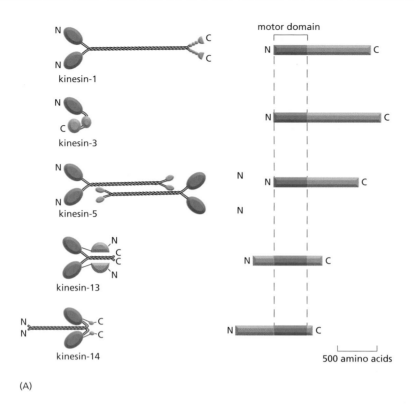

(A)

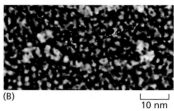

(B)

Figure 16–58 **Kinesin and kinesin-related proteins.** (A) Structures of five kinesin superfamily members. As in the myosin superfamily, only the motor domains are conserved. Kinesin-1 has the motor domain at the N-terminus of the heavy chain. The middle domain forms a long coiled-coil, mediating dimerization. The C-terminal domain forms a tail that attaches to cargo, such as a membrane-enclosed organelle. Kinesin-3 represents an unusual class of kinesins that seem to function as monomers and move membrane-enclosed organelles along microtubules. Kinesin-5 forms tetramers where two dimers associate by their tails. The bipolar kinesin-5 tetramer is able to slide two microtubules past each other, analogous to the activity of the bipolar thick filaments formed by myosin II. Kinesin-13 has its motor domain located in the middle of the heavy chain. It is a member of a family of kinesins that have lost typical motor activity and instead bind to microtubule ends to increase dynamic instability of microtubules; they are therefore called catastrophe factors (see p. 1003). Kinesin-14 is a C-terminal kinesin that includes the *Drosophila* protein Ncd and the yeast protein Kar3. These kinesins generally travel in the opposite direction from the majority of kinesins, toward the minus end instead of the plus end of a microtubule. (B) Freeze-etch electron micrograph of a kinesin molecule with the head domains on the left. (B, courtesy of John Heuser.)

There Are Two Types of Microtubule Motor Proteins: Kinesins and Dyneins

Kinesin is a motor protein that moves along microtubules. It was first identified in the giant axon of the squid, where it carries membrane-enclosed organelles away from the neuronal cell body toward the axon terminal by walking toward the plus end of microtubules. Kinesin is similar structurally to myosin II in having two heavy chains and two light chains per active motor; these form two globular head motor domains and an elongated coiled-coil tail responsible for heavy chain dimerization. Like myosin, kinesin is a member of a large protein superfamily, for which the motor domain is the only common element (**Figure 16–58**). The yeast *Saccharomyces cerevisiae* has six distinct kinesins. The nematode *C. elegans* has 16 kinesins, and humans have about 45.

There are at least fourteen distinct families in the kinesin superfamily. Most of them have the motor domain at the N-terminus of the heavy chain and walk toward the plus end of the microtubule. A particularly interesting family has the motor domain at the C-terminus and walks in the opposite direction, toward the minus end of the microtubule. Some kinesin heavy chains lack a coiled-coil sequence and seem to function as monomers, analogous to myosin I. Some others are homodimers, and yet others are heterodimers. Members of the kinesin-5 family can self-associate through the tail domain, forming a bipolar motor that slides oppositely oriented microtubules past one another, much as a myosin II thick filament does for actin filaments. Most kinesins carry a binding site in the tail for either a membrane-enclosed organelle or another microtubule. Many of the kinesin superfamily members have specific roles in mitotic and meiotic spindle formation, and in chromosome separation during cell division.

The **dyneins** are a family of minus-end-directed microtubule motors unrelated to the kinesin superfamily. They are composed of two or three heavy chains (that include the motor domain) and a large and variable number of associated intermediate chains and light chains. The dynein family has two major branches (**Figure 16–59**). The most ancient branch contains the *cytoplasmic dyneins*, which are typically heavy-chain homodimers, with two large motor domains as heads. Cytoplasmic dyneins are probably found in all eucaryotic cells, and they

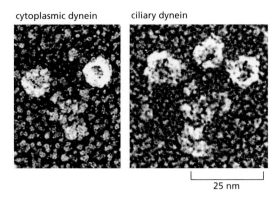

cytoplasmic dynein   ciliary dynein

25 nm

Figure 16–59 **Dyneins.** Freeze-etch electron micrographs of a molecule of cytoplasmic dynein and a molecule of ciliary (axonemal) dynein. Like myosin II and kinesin, cytoplasmic dynein is a two-headed molecule. The ciliary dynein shown is from a protozoan and has three heads; ciliary dynein from animals has two heads. Note that the dynein head is very large compared with the head of either myosin or kinesin. (Courtesy of John Heuser.)

are important for vesicle trafficking, as well as for localization of the Golgi apparatus near the center of the cell. *Axonemal dyneins,* the other large branch, include heterodimers and heterotrimers, with two or three motor-domain heads, respectively. They are highly specialized for the rapid and efficient sliding movements of microtubules that drive the beating of cilia and flagella (discussed later). A third, minor, branch shares greater sequence similarity with cytoplasmic than with axonemal dyneins but seems to be involved in the beating of cilia.

Dyneins are the largest of the known molecular motors, and they are also among the fastest: axonemal dyneins can move microtubules in a test tube at the remarkable rate of 14 μm/sec. In comparison, the fastest kinesins can move their microtubules at about 2–3 μm/sec. We shall discuss how they work below.

The Structural Similarity of Myosin and Kinesin Indicates a Common Evolutionary Origin

The motor domain of myosins is substantially larger than that of kinesins, about 850 amino acids compared with about 350. The two classes of motor proteins track along different filaments and have different kinetic properties, and they have no identifiable amino acid sequence similarities. However, determination of the three-dimensional structure of the motor domains of myosin and kinesin has revealed that these two motor domains are built around nearly identical cores (**Figure 16–60**). The central force-generating element that the two types of motor proteins have in common includes the site of ATP binding and the machinery necessary to translate ATP hydrolysis into an allosteric conformational change. Large loops extending outward from the central core cause the

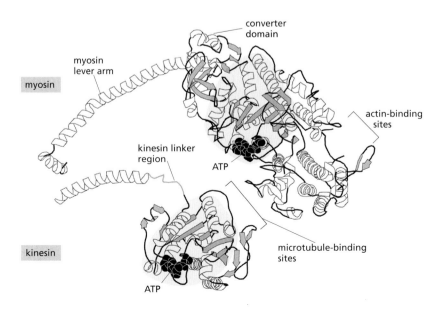

myosin
myosin lever arm
converter domain
kinesin linker region
ATP
actin-binding sites
kinesin
microtubule-binding sites
ATP

Figure 16–60 **X-ray crystal structures of myosin and kinesin heads.** The central nucleotide-binding domains of myosin and kinesin (shaded in *yellow*) are structurally very similar. The very different sizes and functions of the two motors are due to major differences in the polymer-binding and force-transduction portions of the motor domain. (Adapted from L.A. Amos and R.A. Cross, *Curr. Opin. Struct. Biol.* 7:239–246, 1997. With permission from Elsevier.)

difference in domain size and are responsible for the choice of track. These loops include the actin-binding and microtubule-binding sites in the myosin and kinesin, respectively. It is thought that both the kinesins and the myosins are descended from a common ancestral motor protein precursor, and that their various specialized functions arose via gene duplication and modification through evolution of the loops coming out from the central core.

An important clue to how the central core is involved in force generation has come from the observation that the motor core also bears some structural resemblance to the nucleotide binding site of the small GTPases of the Ras superfamily. As discussed in Chapter 3 (see Figure 3–72), these proteins exhibit distinct conformations in their GTP-bound (active) and GDP-bound (inactive) forms: mobile "switch" loops in the nucleotide-binding site are in close contact with the γ-phosphate in the GTP-bound state, but these loops swing out when the hydrolyzed γ-phosphate (the terminal phosphate) is released. Although the details of the movement differ for the two motor proteins, and ATP rather than GTP is hydrolyzed, a relatively small structural change in the active site—the presence or absence of a terminal phosphate—is similarly amplified to cause a rotation of a different part of the protein.

In kinesin and myosin, a switch loop interacts extensively with those regions of the protein involved in microtubule and actin binding, respectively, allowing the structural transitions caused by the ATP hydrolysis cycle to be relayed to the polymer-binding interface. The relay of structural changes between the polymer-binding site and the nucleotide hydrolysis site seems to work in both directions, since the ATPase activity of motor proteins is strongly activated by binding to their filament tracks.

Motor Proteins Generate Force by Coupling ATP Hydrolysis to Conformational Changes

Although the cytoskeletal motor proteins and GTP-binding proteins both use structural changes in their nucleoside-triphosphate-binding sites to produce cyclic interactions with a partner protein, the motor proteins have a further requirement: each cycle of binding and release must propel them forward in a single direction along a filament to a new binding site on the filament. For such unidirectional motion, a motor protein must use the energy derived from ATP binding and hydrolysis to force a large movement in part of the protein molecule. For myosin, each step of the movement along actin is generated by the swinging of an 8.5-nm-long α helix, or *lever arm*, which is structurally stabilized by the binding of light chains. At the base of this lever arm next to the head, there is a piston-like helix that connects movements at the ATP-binding cleft in the head to small rotations of the so-called converter domain (see Figure 16–60). A small change at this point can swing the helix like a long lever, causing the far end of the helix to move by about 5.0 nm.

These changes in the conformation of the myosin are coupled to changes in its binding affinity for actin, allowing the myosin head to release its grip on the actin filament at one point and snatch hold of it again at another. The full mechanochemical cycle of nucleotide binding, nucleotide hydrolysis, and phosphate release (which causes the "power stroke") produces a single step of movement (**Figure 16–61**).

In kinesin, instead of the rocking of a lever arm, the small movements of switch loops at the nucleotide-binding site regulate the docking and undocking of the motor head domain to a long linker region that connects this motor head at one end to the coiled-coil dimerization domain at the other end (see Figure 16–61). When the front (leading) kinesin head is bound to a microtubule before the power stroke, its linker region is relatively unstructured. On the binding of ATP to this bound head, its linker region docks along the side of the head; this throws the second head forward to a position where it will be able to bind a new attachment site on the protofilament, 8 nm closer to the microtubule plus end than the binding site for the first head. The nucleotide hydrolysis cycles in the two heads are closely coordinated, so that this cycle of linker docking and undocking allows

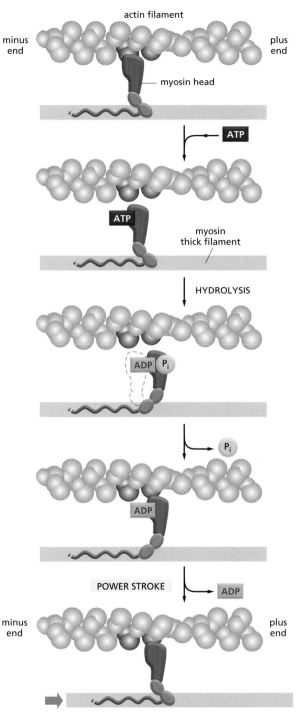

actin filament

minus end

plus end

myosin head

ATTACHED At the start of the cycle shown in this figure, a myosin head lacking a bound nucleotide is locked tightly onto an actin filament in a *rigor* configuration (so named because it is responsible for *rigor mortis*, the rigidity of death). In an actively contracting muscle, this state is very short-lived, being rapidly terminated by the binding of a molecule of ATP.

ATP

myosin thick filament

RELEASED A molecule of ATP binds to the large cleft on the "back" of the head (that is, on the side furthest from the actin filament) and immediately causes a slight change in the conformation of the domains that make up the actin-binding site. This reduces the affinity of the head for actin and allows it to move along the filament. (The space drawn here between the head and actin emphasizes this change, although in reality the head probably remains very close to the actin.)

HYDROLYSIS

ADP Pi

COCKED The cleft closes like a clam shell around the ATP molecule, triggering a large shape change that causes the head to be displaced along the filament by a distance of about 5 nm. Hydrolysis of ATP occurs, but the ADP and inorganic phosphate (P_i) produced remain tightly bound to the protein.

Pi

ADP

FORCE-GENERATING A weak binding of the myosin head to a new site on the actin filament causes release of the inorganic phosphate produced by ATP hydrolysis, concomitantly with the tight binding of the head to actin. This release triggers the power stroke—the force-generating change in shape during which the head regains its original conformation. In the course of the power stroke, the head loses its bound ADP, thereby returning to the start of a new cycle.

POWER STROKE

ADP

minus end

plus end

ATTACHED At the end of the cycle, the myosin head is again locked tightly to the actin filament in a rigor configuration. Note that the head has moved to a new position on the actin filament.

Figure 16–61 The cycle of structural changes used by myosin II to walk along an actin filament. <ATAT> In the myosin II cycle, the head remains bound to the actin filament for only about 5% of the entire cycle time, allowing many myosins to work together to move a single actin filament. (Based on I. Rayment et al., *Science* 261:50–58, 1993. With permission from AAAS.)

the two-headed motor to move in a hand-over-hand (or head-over-head) stepwise manner (**Figure 16–62**), each time taking a discrete 8-nm step.

The coiled-coil domain seems both to coordinate the mechanochemical cycles of the two heads (motor domains) of the kinesin dimer and to determine its directionality of movement. Recall that whereas most members of the kinesin superfamily, with their motor domains at the N-terminus, move toward the plus end of the microtubule, a few superfamily members have their motor domains at the C-terminus and move toward the minus end. Since the motor domains of these two types of kinesins are essentially identical, how can they move in opposite directions? The answer seems to lie in the way in which the heads are connected. In high-resolution images of forward-walking and backward-walking members of the kinesin superfamily bound to microtubules, the heads that are attached to the microtubule are essentially indistinguishable, but the second, unattached heads are oriented differently. This difference in tilt apparently

Figure 16–62 The mechanochemical cycle of kinesin. <GAAT> Kinesin-1 is a dimer of two nucleotide-binding motor domains (heads) that are connected through a long coiled-coil tail (see Figure 16–58). The two kinesin motor domains work in a coordinated manner; during a kinesin "step," the rear head detaches from its tubulin binding site, passes the partner motor domain, and then rebinds to the next available tubulin binding site. Using this "hand-over-hand" motion, the kinesin dimer can move for long distances on the microtubule without completely letting go of its track.

At the start of each step, one of the two kinesin heads, the rear or lagging head (*dark green*), is tightly bound to the microtubule and to ATP, while the front or leading head is loosely bound to the microtubule with ADP in its binding site. The forward displacement of the rear motor domain is driven by an exchange of ATP for ADP in the front motor domain (between panels 2 and 3 in this drawing). The binding of ATP to this motor domain causes a small peptide called the "neck linker" to shift from a rearward-pointing to a forward-pointing conformation (the neck linker is drawn here as a connecting line between the motor domain and the intertwined coiled coil). This shift pulls the rear motor domain forward, once it has detached from the microtubule with ADP bound (detachment requires ATP hydrolysis and phosphate (P$_i$) release). The kinesin molecule is now poised for the next step, which proceeds by an exact repeat of the process shown.

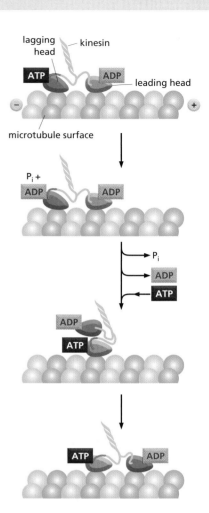

biases the next binding site for the second head, and thereby determines the directionality of motor movement (**Figure 16–63**).

The dynein motor is structurally unrelated to myosins and kinesins, but still follows the general rule of coupling the nucleotide hydrolysis to microtubule binding and unbinding as well as to a force-generating conformational change. A giant heavy chain of more than 500,000 daltons forms the basic structure that creates the movement. Its N-terminal portion forms a tail that binds a set of light chains and connects to the other heavy chains in the dynein molecule, while the major portion of the heavy chain is used to form an elaborate, ring-shaped head. The head consists of a planar ring formed from seven domains: six AAA domains plus the heavy-chain C-terminal domain; it is therefore a more complex relative of the hexameric ATPase discussed in Chapter 6 (see Figure 6–91). A hook-

Figure 16–63 Orientation of forward- and backward-walking kinesin superfamily proteins bound to microtubules. These images were generated by fitting the structures of the free motor-protein dimers (determined by x-ray crystallography) onto a lower resolution image of the dimers attached to microtubules (determined by cryoelectron microscopy). (A) Kinesin-1 (conventional kinesin) has its motor domain at the protein's N-terminus and moves toward the plus end of the microtubule. When one head of the dimer is bound to the microtubule in a post-stroke state (with ATP in the nucleotide binding site), the second, unbound head is pointing toward the microtubule plus end, poised to take the next step. (B) Kinesin-13 (called Ncd in *Drosophila*), a minus-end-directed motor with the motor domain at the C-terminus, forms dimers with the opposite orientation. (From E. Sablin et al., *Nature* 395:813–816, 1998. With permission from Macmillan Publishers Ltd.)

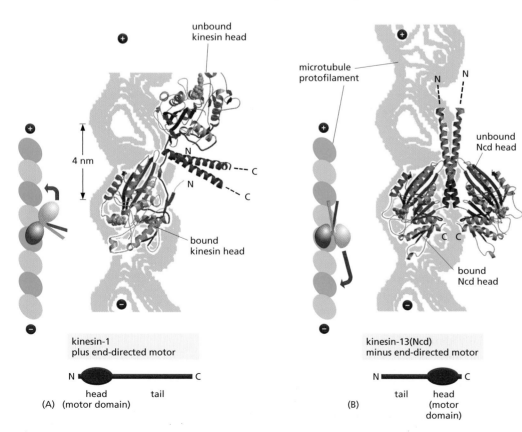

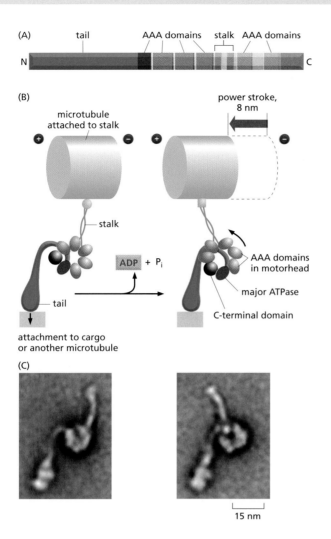

Figure 16–64 The power stroke of dynein.
(A) The organization of the domains in each dynein heavy chain. This is a huge molecule, containing nearly 5000 amino acids. The number of heavy chains in a dynein is equal to its number of motor heads. (B) Dynein c is a monomeric flagella dynein found in unicellular green alga *Chlamydomonas reinhardtii*. The large dynein motor head is a planar ring containing a C-terminal domain (*gray*) and six AAA domains, four of which retain ATP-binding sequences, but only one of which (*dark red*) has the major ATPase activity. Extending from the head are a long, coiled-coil stalk with the microtubule binding site at the tip, and a tail with a cargo-attachment site. In the ATP-bound state, the stalk is detached from the microtubule, but ATP hydrolysis causes stalk-microtubule attachment. Subsequent release of ADP and Pi then leads to a large conformational "power stroke" involving rotation of the head and stalk relative to the tail. Each cycle generates a step of about 8 nm along the microtubule towards its minus end. (C) Electron micrographs of purified dyneins in two different conformations representing different steps in the mechanochemical cycle. (B, from S.A. Burgess et al., *Nature* 421:715–718, 2003. With permission from Macmillan Publishers Ltd.)

shaped linker region connects the heavy-chain tail to the AAA domain that is most active as an ATPase. Between the fourth and fifth AAA domains is a heavy chain domain that forms a long anti-parallel coiled-coil stalk. This stalk extends from the top of the ring, with an ATP hydrolysis-regulated microtubule-binding site at its tip. Dynein's "power stroke" is driven by the release of ADP and inorganic phosphate, and it causes the ring to rotate relative to the tail (**Figure 16–64**).

Although kinesin, myosin, and dynein all undergo analogous mechanochemical cycles, the exact nature of the coupling between the mechanical and chemical cycles differs in the three cases. For example, myosin without any nucleotide is tightly bound to its actin track, in a so-called "rigor" state, and it is released from this track by the association of ATP. In contrast, kinesin forms a rigor-like tight association with a microtubule when ATP is bound to the kinesin, and it is hydrolysis of ATP that promotes release of the motor from its track. The mechanochemical cycle of dynein is more similar to myosin than to kinesin, in that nucleotide-free dynein is tightly bound to the microtubule and it is released by binding ATP. However, for dynein the inorganic phosphate and ADP appear to be released at the same time, causing the conformational change driving the power stroke, while for myosin the phosphate is released first and the power stroke does not occur until the ADP subsequently dissociates from the motor head.

Thus, cytoskeletal motor proteins work in a manner highly analogous to GTP-binding proteins, except that in motor proteins the small protein conformational changes (a few tenths of a nanometer) associated with nucleotide hydrolysis are amplified by special protein domains—the lever arm in the case of myosin, the linker in the case of kinesin, and the ring and stalk in the case of dynein—to generate large-scale (several nanometers) conformational changes that move the motor proteins stepwise along their filament tracks. The analogy between the GTPases and the cytoskeletal motor proteins has recently been extended by the observation that one of the GTP-binding proteins—the bacterial elongation factor G—translates the chemical energy of GTP hydrolysis into directional movement of the mRNA molecule on the ribosome.

Motor Protein Kinetics Are Adapted to Cell Functions

The motor proteins in the myosin and kinesin superfamilies exhibit a remarkable diversity of motile properties, well beyond their choice of different polymer tracks. Most strikingly, a single dimer of kinesin-1 moves in a highly *processive* fashion, traveling for hundreds of ATPase cycles along a microtubule without dissociating. Skeletal muscle myosin II, in contrast, cannot move processively and makes just one or a few steps along an actin filament before letting go. These differences are critical for the motors' various biological roles. A small number of kinesin-1 molecules must be able to transport an organelle all the way down a nerve cell axon, and therefore require a high level of processivity. Skeletal muscle myosin, in contrast, never operates as a single molecule but rather as part of a huge array of myosin II molecules in a thick filament. Here processivity would actually inhibit biological function, since efficient muscle contraction requires that each myosin head perform its power stroke and then quickly get out of the way—in order to avoid interfering with the actions of the other heads attached to the same actin filament.

There are two reasons for the high degree of processivity of kinesin-1 movement. The first is that the mechanochemical cycles of the two motor heads in a kinesin-1 dimer are coordinated with each other, so that one kinesin head does not let go until the other is poised to bind. This coordination allows the motor protein to operate in a hand-over-hand fashion, never allowing the organelle cargo to diffuse away from the microtubule track. In contrast, there is no apparent coordination between the myosin heads in a myosin II dimer. The second reason for the high processivity of kinesin-1 movement is that kinesin-1 spends a relatively large fraction of its ATPase cycle tightly bound to the microtubule. For both kinesin-1 and myosin II, the conformational change that produces the force-generating working stroke must occur while the motor protein is tightly bound to its polymer, and the recovery stroke in preparation for the next step must occur while the motor is unbound. But myosin II spends only about 5% of its ATPase cycle in the tightly bound state, and it is unbound the rest of the time.

What myosin loses in processivity it gains in speed; in an array in which many motor heads are interacting with the same actin filament, a set of linked myosins can move its filament a total distance equivalent to 20 steps during a single cycle time, while kinesins can move only two. Thus, myosin II can typically drive filament sliding much more rapidly than kinesin-1, even though the two different motor proteins hydrolyze ATP at comparable rates and take molecular steps of comparable length. This property is particularly important in the rapid contraction of skeletal muscle, as we will discuss later.

Within each motor protein class, movement speeds vary widely, from about 0.2 to 60 µm/sec for myosins, and from about 0.02 to 2 µm/sec for kinesins. These differences arise from a fine-tuning of the mechanochemical cycle. The number of steps that an individual motor molecule can take in a given time, and thereby the velocity, can be decreased by either decreasing the motor protein's intrinsic ATPase rate or by increasing the proportion of cycle time spent bound to the filament track. For example, myosin V (which acts as a processive vesicle motor) spends up to 90% of its nucleotide cycle tightly bound to the actin filament, in contrast to 5% for myosin II. Moreover, a motor protein can evolve to change the size of each step by either changing the length of the lever arm (for example, the lever arm of myosin V is about three times longer than the lever arm of myosin II) or the angle through which the helix swings (**Figure 16–65**). Each of these parameters varies slightly among different members of the myosin and kinesin families, corresponding to slightly different protein sequences and structures.

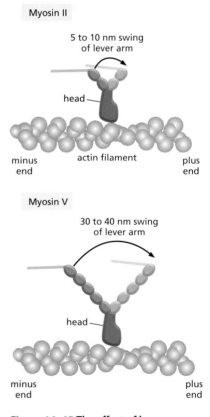

Figure 16–65 The effect of lever arm length on the step size for a motor protein. The lever arm of myosin II is much shorter than the lever arm of myosin V. The power stroke in the head swings their lever arms through the same angle, so myosin V is able to take a bigger step than myosin II.

It is assumed that evolution has fine-tuned the behavior of each motor protein, whose function is determined by the identity of the cargo attached through its tail domain, for speed and processivity according to the specific needs of the cell for the function of that particular family member. Whereas there are many different myosin and kinesin family members found in a typical eucaryotic cell, there is usually only one form of cytoplasmic dynein. It is not yet clear how or whether the mechanical properties of cytoplasmic dynein can be modified in response to differing needs of the cell.

Motor Proteins Mediate the Intracellular Transport of Membrane-Enclosed Organelles <CAAT> <AAAT>

A major function of cytoskeletal motors in interphase cells is the transport and positioning of membrane-enclosed organelles. Kinesin was originally identified as the protein responsible for fast axonal transport, the rapid movement of mitochondria, secretory vesicle precursors, and various synapse components down the microtubule highways of the axon to the distant nerve terminals. Although organelles in most cells need not cover such long distances, their polarized transport is equally necessary. A typical microtubule array in an interphase cell is oriented with the minus ends near the center of the cell at the centrosome, and the plus ends extending to the cell periphery. Thus, centripetal movements of organelles or vesicles toward the cell center require the action of minus-end-directed motor proteins such as cytoplasmic dynein, whereas centrifugal movements toward the periphery require plus-end-directed motors such as kinesins.

The clearest example of the effect of microtubules and microtubule motors on the behavior of intracellular membranes is their role in organizing the endoplasmic reticulum (ER) and the Golgi apparatus. The network of ER membrane tubules aligns with microtubules and extends almost to the edge of the cell, whereas the Golgi apparatus is located near the centrosome. When cells are treated with a drug that depolymerizes microtubules, such as colchicine or nocodazole, the ER collapses to the center of the cell, while the Golgi apparatus fragments and disperses throughout the cytoplasm (**Figure 16–66**). *In vitro*, kinesins can tether ER-derived membranes to preformed microtubule tracks, and walk toward the microtubule plus ends, dragging the ER membranes out into tubular protrusions and forming a membranous web very much like the ER in cells. Likewise, in living cells the outward movement of ER tubules toward the cell periphery is associated with microtubule growth. Conversely, dyneins are required for positioning the Golgi apparatus near the cell center, moving Golgi vesicles along microtubule tracks toward minus ends at the centrosome.

The different tails and their associated light chains on specific motor proteins allow the motors to attach to their appropriate organelle cargo. Membrane-associated motor receptors that are sorted to specific membrane-enclosed compartments interact directly or indirectly with the tails of the appropriate kinesin family members. One of these receptors seems to be the amyloid precursor protein, APP, which binds directly to a light chain on the tail of kinesin-1 and is proposed to be a transmembrane motor receptor in nerve-cell axons. The abnormal processing of this protein gives rise to Alzheimer's disease by producing large, stable

Figure 16–66 Effect of depolymerizing microtubules on the Golgi apparatus. (A) In this endothelial cell, the microtubules are labeled in *red,* and the Golgi apparatus is labeled in *green* (using an antibody against a Golgi protein). As long as the system of microtubules remains intact, the Golgi is localized near the centrosome, close to the nucleus at the center of the cell. The cell on the *right* is in interphase, with a single centrosome. The cell on the *left* is in prophase, and the duplicated centrosomes have moved to opposite sides of the nucleus. (B) After exposure to nocodazole, which causes microtubules to depolymerize (see Table 16–2), the Golgi apparatus fragments and is dispersed throughout the cell cytoplasm. (Courtesy of David Shima.)

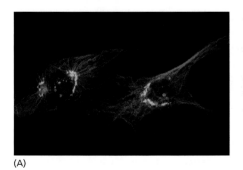
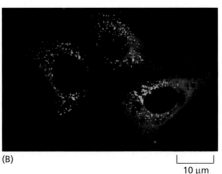

(A)   (B)

10 μm

protein aggregates in nerve cells of the brain (see Figure 6–95). Other receptors for specific kinesins have been identified on the endoplasmic reticulum, as well as on various other membrane-bound organelles that rely on microtubule-based transport for their localization. The JIPs (JNK-interacting proteins), are scaffold proteins associated with cell signaling. These kinesin receptors may provide a link between transport and cell signaling.

For dynein, a large macromolecular assembly often mediates attachment to membranes. Cytoplasmic dynein, itself a huge protein complex, requires association with a second large protein complex called *dynactin* to translocate organelles effectively. The dynactin complex includes a short actin-like filament that forms from the actin-related protein Arp1 (distinct from Arp2 and Arp3, the components of the ARP complex involved in the nucleation of conventional actin filaments). Membranes of the Golgi apparatus are coated with the proteins ankyrin and spectrin, which have been proposed to associate with the Arp1 filament in the dynactin complex to form a planar cytoskeletal array reminiscent of the erythrocyte membrane cytoskeleton (see Figure 10–41). The spectrin array probably gives structural stability to the Golgi membrane, and—via the Arp1 filament—it may mediate the regulated attachment of dynein to the organelle (**Figure 16–67**). In other cases, cytoplasmic dynein motors may interact directly with their cargo. The cytoplasmic tail of rhodopsin, the light-detecting protein found in the rod cells of the eye, binds directly to one of the dynein light chains, and this interaction is required for normal trafficking of rhodopsin in the rod cell.

Motor proteins also have a significant role in organelle transport along actin filaments. The first myosin shown to mediate organelle motility was myosin V, a two-headed myosin with a large step size (see Figure 16–65). In mice and humans, membrane-enclosed pigment granules, called *melanosomes*, are synthesized in cells called *melanocytes* beneath the skin surface. These melanosomes move out to the ends of dendritic processes in the melanocytes, where they are delivered to the overlying keratinocytes that form the skin (and the fur in mice). Myosin V is associated with the surface of melanosomes (**Figure 16–68**) and is able to mediate their actin-based movement in a test tube. In mice, mutations in the myosin V gene result in a "dilute" phenotype, in which fur color looks faded because the melanosomes are not delivered to the keratinocytes efficiently. Other myosins, including myosin I, are associated with endosomes and a variety of other organelles.

The Cytoskeleton Localizes Specific RNA Molecules

In order to concentrate proteins at their site of function, cells often restrict the synthesis of a particular protein by localizing its mRNA molecules, a process that establishes cellular asymmetries. This is particularly important when a parent cell divides to generate two daughters with distinct fates. As another example, several mRNAs encoding proteins involved in synapse function are specifically localized close to synapses in many neurons, and there is evidence that mRNA localization and translation regulation at the synaptic sites play important roles in regulating long-term memory and synaptic plasticity. Not surprisingly, the cytoskeleton and cytoskeletal motor proteins transport and position mRNA molecules in these types of situations.

The giant *Drosophila* oocyte localizes a large number of maternally encoded mRNAs to specific sites within the cell in anticipation of the rapid cell specification events in early embryogenesis (discussed in Chapter 22). A group of mRNAs that encode proteins necessary for proper development of the posterior region of the embryo, including development of the germ cells, is localized posteriorly in the oocyte, and a distinct group of mRNAs encoding proteins necessary for specification of anterior structures in the embryo is localized in the anterior region of the oocyte.

The oocyte takes advantage of its polarized microtubule cytoskeleton, where most microtubule minus ends are clustered in the anterior part of the cell and plus ends near the posterior, to establish these specialized mRNA distributions. For example, the mRNA encoding Bicoid, a transcription factor critical for

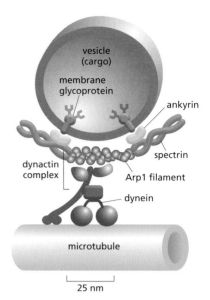

Figure 16–67 A model for the attachment of dynein to a membrane-enclosed organelle. Dynein requires the presence of a large number of accessory proteins to associate with membrane-enclosed organelles. Dynactin is a large complex *(red)* that includes components that bind weakly to microtubules, components that bind to dynein itself, and components that form a small actin-like filament made of the actin-related protein Arp1. It is thought that the Arp1 filament may mediate attachment of this large complex to membrane-enclosed organelles through a network of spectrin and ankyrin, similar to the membrane-associated cytoskeleton of the red blood cell (see Figure 10–41).

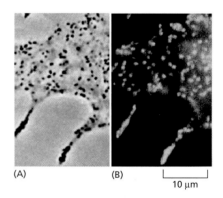

Figure 16–68 Myosin V on melanosomes. (A) Phase-contrast image of a portion of a melanocyte isolated from a mouse. The black spots are melanosomes, which are membrane-enclosed organelles filled with the skin pigment melanin. (B) The same cell labeled with a fluorescent antibody against myosin V. Every melanosome is associated with a large number of copies of this motor protein. (From X. Wu et al., *J. Cell Sci.* 110:847–859, 1997. With permission from The Company of Biologists.)

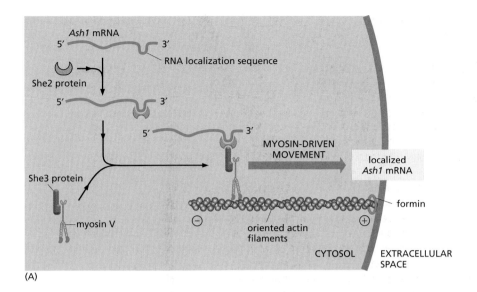

(A)

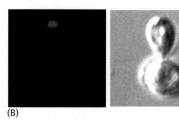

(B)

Figure 16–69 Polarized mRNA localization in the yeast bud tip. (A) The molecular mechanism of *Ash1* mRNA localization, as determined by genetics and biochemistry. (B) Fluorescent *in situ* hybridization (FISH) was used to localize the *Ash1* mRNA (*red*) in this dividing yeast cell. The mRNA is confined to the far tip of the daughter cell (here, still a large bud). Ash1 protein, transcribed from this localized mRNA, is also confined to the daughter cell. (B, courtesy of Peter Takizawa and Ron Vale.)

anterior development, has a structure within the 3′ UTR that binds a protein called Swallow, which in turn binds to a cytoplasmic dynein light chain, presumably enabling its transport to the microtubule minus ends at the cell anterior. Conversely, the transport of mRNA encoding Oskar, a protein necessary for germ cell development in the posterior of the embryo, requires kinesin-1 for its transport to the microtubule plus ends. The anchoring of the mRNAs to their appropriate locations after delivery via microtubules appears to involve the cortical actin cytoskeleton. The mRNA encoding Oskar, for example, binds directly to an actin-binding protein called moesin, a member of the ERM family.

In some cells, mRNA transport as well as anchoring is actin-dependent. The yeast mother and daughter cells retain distinct identities, as revealed by major differences in their subsequent ability to undergo mating-type switching (discussed in Chapter 7) and in the choice of their next bud site. Many of these differences are caused by a gene regulatory protein called *Ash1*. Both *Ash1* mRNA and protein are localized exclusively to the growing bud and therefore end up only in the daughter cell. One of the two type V myosins found in yeast, Myo4p, is required for this asymmetric distribution of *Ash1* mRNA. A genetic screen for other mutations that disrupt the mother/daughter difference has revealed that at least six other gene products that are associated with the cytoskeleton are required for normal polarity; these include one of the formins, tropomyosin, profilin, and actin itself, as well as a complex of two proteins that form a direct link between a specific sequence in the *Ash1* mRNA and the myosin V protein (**Figure 16–69**).

Cells Regulate Motor Protein Function

The cell can regulate the activity of motor proteins and thereby cause either a change in the positioning of its membrane-enclosed organelles or whole-cell movements. Fish melanocytes provide one of the most dramatic examples. These giant cells, which are responsible for rapid changes in skin coloration in several species of fish, contain large pigment granules that can alter their location in response to neuronal or hormonal stimulation (**Figure 16–70**). The pigment granules aggregate or disperse by moving along an extensive network of microtubules. The centrosome nucleates these microtubules, localizing their minus ends in the center of the cell, while the plus ends are distributed around the cell periphery.

The tracking of individual pigment granules (**Figure 16–71**) reveals that the inward movement is rapid and smooth, while the outward movement is jerky, with frequent backward steps. Both the microtubule motors dynein and kinesin are associated with the pigment granules, as well as the actin motor myosin V. The jerky outward movements apparently result from a tug-of-war between the two microtubule motor proteins, with the stronger kinesin winning out overall.

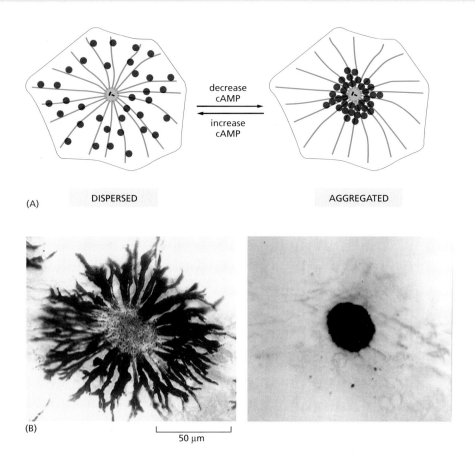

(A) DISPERSED   AGGREGATED

(B)

⊢———— 50 μm ————⊣

Figure 16–70 Regulated melanosome movements in fish pigment cells. These giant cells, which are responsible for changes in skin coloration in several species of fish, contain large pigment granules, or melanosomes *(brown)*. The melanosomes can change their location in the cell in response to a hormonal or neuronal stimulus. (A) Schematic view of a pigment cell, showing the dispersal and aggregation of melanosomes in response to an increase or decrease in intracellular cyclic AMP (cAMP), respectively. Both redistributions of melanosomes occur along microtubules. (B) Bright-field images of a single cell in a scale of an African cichlid fish, showing its melanosomes either dispersed throughout the cytoplasm *(left)* or aggregated in the center of the cell *(right)*. (B, courtesy of Leah Haimo.)

When the kinesin light chains become phosphorylated after a hormonal stimulation that signals skin color change, kinesin is inactivated, leaving dynein free to drag the pigment granules rapidly toward the cell center, changing the fish's color. In a similar way, the movement of other membrane organelles coated with particular motor proteins is controlled by a complex balance of competing signals that regulate both motor protein attachment and activity.

The cell can also use phosphorylation to regulate myosin activity. In non-muscle cells, myosin II can be phosphorylated on a variety of sites on both heavy and light chains, affecting both motor activity and thick filament assembly. The myosin II can exist in two different conformational states in such cells, an extended state that can form bipolar filaments, and a bent state in which the tail domain apparently interacts with the motor head. Phosphorylation of the regulatory light chain by the calcium-dependent *myosin light-chain kinase (MLCK)* causes the myosin II to preferentially assume the extended state, which promotes its assembly into a bipolar filament and leads to cell contraction (**Figure 16–72**). MLCK is also activated during mitosis, causing myosin II to assemble into the actin-based contractile ring that pinches the mitotic cell into two. As we will discuss below, myosin phosphorylation is also an important component of the control of contraction in smooth muscle cells. Regulation of other members

Figure 16–71 Bidirectional movement of a melanosome on a microtubule. An isolated melanosome *(yellow)* moves along a microtubule on a glass slide, from the plus end toward the minus end. Halfway through the video sequence, it abruptly switches direction and moves from the minus end toward the plus end. (From S.L. Rogers et al., *Proc. Natl Acad. Sci. U.S.A.* 94:3720–3725, 1997. With permission from National Academy of Sciences.)

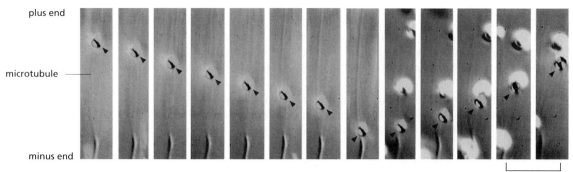

plus end

microtubule

minus end

⊢——— 10 μm ———⊣

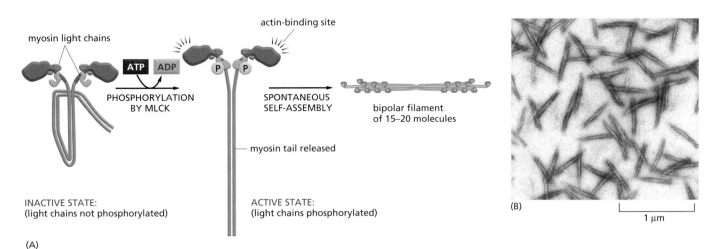

Figure 16–72 Light-chain phosphorylation and the regulation of the assembly of myosin II into thick filaments. (A) The controlled phosphorylation by the enzyme myosin light-chain kinase (MLCK) of one of the two light chains (the so-called regulatory light chain, shown in *light blue*) on nonmuscle myosin II in a test tube has at least two effects: it causes a change in the conformation of the myosin head, exposing its actin-binding site, and it releases the myosin tail from a "sticky patch" on the myosin head, thereby allowing the myosin molecules to assemble into short, bipolar, thick filaments. (B) Electron micrograph of negatively stained short filaments of myosin II that have been induced to assemble in a test tube by phosphorylation of their light chains. These myosin II filaments are much smaller than those found in skeletal muscle cells (see Figure 16–55). (B, courtesy of John Kendrick-Jones.)

of the myosin superfamily is not as well understood, but the control of these myosins is likewise thought to involve site-specific phosphorylations.

Summary

Motor proteins use the energy of ATP hydrolysis to move along microtubules or actin filaments. They mediate the sliding of filaments relative to one another and the transport of cargo along filament tracks. All known motor proteins that move on actin filaments are members of the myosin superfamily. The motor proteins that move on microtubules are either members of the kinesin superfamily or the dynein family. The myosin and kinesin superfamilies are diverse, with about 40 genes encoding each type of protein in humans. The only structural element shared among all members of each superfamily is the motor "head" domain. These heads are fused to a wide variety of different "tails," which attach to different types of cargo and enable the various family members to perform different functions in the cell. These functions include the transportation and localization of specific proteins, membrane-enclosed organelles, and mRNAs.

Although myosin and kinesin walk along different tracks and use different mechanisms to produce force and movement by ATP hydrolysis, they share a common structural core, suggesting that they are derived from a common ancestor. The dynein motor protein has independently evolved, and it has a distinct structure and mechanism of action.

THE CYTOSKELETON AND CELL BEHAVIOR

A central challenge in all areas of cell biology is to understand how the functions of many individual molecular components combine to produce complex cell behaviors. The cell behaviors that we describe in this final section all rely on a coordinated deployment of the components and processes that we have explored in the first three sections of the chapter: the dynamic assembly and disassembly of cytoskeletal polymers, the regulation and modification of their structure by polymer-associated proteins, and the actions of motor proteins moving along the polymers. How does the cell coordinate all these activities to define its shape, to enable it to crawl, or to divide it neatly into two at mitosis? These problems of cytoskeletal coordination will challenge scientists for many years to come.

To provide a sense of our present understanding, we first discuss examples where specialized cells build stable arrays of filaments and use highly ordered arrays of motor proteins sliding them relative to each other to generate the large-scale movements of muscle, cilia, and eucaryotic flagella. Next, we consider two important instances where filament dynamics collude with motor protein activity to generate complex, self-organized dynamic structures: the microtubule-based mitotic spindle and the actin arrays involved in cell crawling. Finally, we consider the extraordinary organization and behavior of the neuronal cytoskeleton.

Sliding of Myosin II and Actin Filaments Causes Muscles to Contract

Muscle contraction is the most familiar and the best understood form of movement in animals. In vertebrates, running, walking, swimming, and flying all depend on the rapid contraction of skeletal muscle on its scaffolding of bone, while involuntary movements such as heart pumping and gut peristalsis depend on the contraction of cardiac muscle and smooth muscle, respectively. All these forms of muscle contraction depend on the ATP-driven sliding of highly organized arrays of actin filaments against arrays of myosin II filaments.

Skeletal muscle was a relatively late evolutionary development, and muscle cells are highly specialized for rapid and efficient contraction. The long thin muscle fibers of skeletal muscle are actually huge single cells that form during development by the fusion of many separate cells, as discussed in Chapter 22. The large muscle cell retains the many nuclei of the contributing cells. These nuclei lie just beneath the plasma membrane (**Figure 16–73**). The bulk of the cytoplasm inside is made up of myofibrils, which is the name given to the basic contractile elements of the muscle cell. A **myofibril** is a cylindrical structure 1–2 μm in diameter that is often as long as the giant muscle cell itself. It consists of a long repeated chain of tiny contractile units—called *sarcomeres*, each about 2.2 μm long, which give the vertebrate myofibril its striated appearance (**Figure 16–74**).

Each sarcomere is formed from a miniature, precisely ordered array of parallel and partly overlapping thin and thick filaments. The *thin filaments* are composed of actin and associated proteins, and they are attached at their plus ends to a *Z disc* at each end of the sarcomere. The capped minus ends of the actin filaments extend in toward the middle of the sarcomere, where they overlap with *thick filaments*, the bipolar assemblies formed from specific muscle isoforms of myosin II (see Figure 16–55). When this region of overlap is examined in cross section by electron microscopy, the myosin filaments are seen to be arranged in a regular hexagonal lattice, with the actin filaments evenly spaced between them (**Figure 16–75**). Cardiac muscle and smooth muscle also contain sarcomeres, although the organization is not as regular as that in skeletal muscle.

Sarcomere shortening is caused by the myosin filaments sliding past the actin thin filaments, with no change in the length of either type of filament (Figure 16–74 C and D). Bipolar thick filaments walk toward the plus ends of two sets of thin filaments of opposite orientations, driven by dozens of independent myosin heads that are positioned to interact with each thin filament. Because there is no coordination among the movements of the myosin heads, it is critical

Figure 16–73 Skeletal muscle cells (also called muscle fibers). (A) These huge multinucleated cells form by the fusion of many muscle cell precursors, called myoblasts. In an adult human, a muscle cell is typically 50 μm in diameter and can be up to several centimeters long. (B) Fluorescence micrograph of rat muscle, showing the peripherally located nuclei *(blue)* in these giant cells. Myofibrils are stained *red*; see also Figure 23–46B. (B, courtesy of Nancy L. Kedersha.)

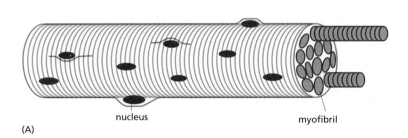

nucleus   myofibril

(A)

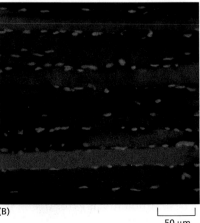

(B)   50 μm

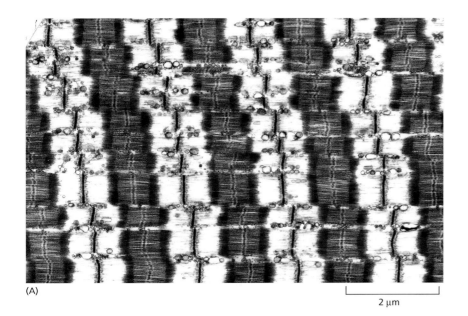

(A)

2 μm

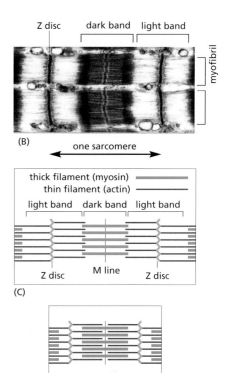

(B) one sarcomere

(C)

thick filament (myosin)
thin filament (actin)

light band   dark band   light band

Z disc   M line   Z disc

(D)

Figure 16–74 Skeletal muscle myofibrils. (A) Low-magnification electron micrograph of a longitudinal section through a skeletal muscle cell of a rabbit, showing the regular pattern of cross-striations. The cell contains many myofibrils aligned in parallel (see Figure 16–73). (B) Detail of the skeletal muscle shown in (A), showing portions of two adjacent myofibrils and the definition of a sarcomere (*black* arrow). (C) Schematic diagram of a single sarcomere, showing the origin of the dark and light bands seen in the electron micrographs. The Z discs, at each end of the sarcomere, are attachment sites for the plus ends of actin filaments (thin filaments); the M line, or midline, is the location of proteins that link adjacent myosin II filaments (thick filaments) to one another. The dark bands, which mark the location of the thick filaments, are sometimes called A bands because they appear anisotropic in polarized light (that is, their refractive index changes with the plane of polarization). The light bands, which contain only thin filaments and therefore have a lower density of protein, are relatively isotropic in polarized light and are sometimes called I bands. (D) When the sarcomere contracts, the actin and myosin filaments slide past one another without shortening. (A and B, courtesy of Roger Craig.)

that they operate with a low processivity, remaining tightly bound to the actin filament for only a small fraction of each ATPase cycle so that they do not hold one another back. Each myosin thick filament has about 300 heads (294 in frog muscle), and each head cycles about five times per second in the course of a

1 μm

Figure 16–75 Electron micrographs of an insect flight muscle viewed in cross section. The myosin and actin filaments are packed together with almost crystalline regularity. Unlike their vertebrate counterparts, these myosin filaments have a hollow center, as seen in the enlargement on the right. The geometry of the hexagonal lattice is slightly different in vertebrate muscle. (From J. Auber, *J. de Microsc.* 8:197–232, 1969. With permission from Societé française de microscopie électronique.)

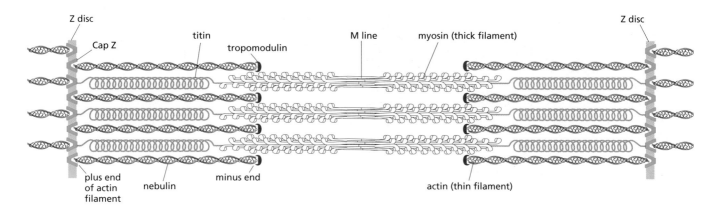

Z disc   Cap Z   titin   tropomodulin   M line   myosin (thick filament)   Z disc

plus end of actin filament   nebulin   minus end   actin (thin filament)

rapid contraction—sliding the myosin and actin filaments past one another at rates of up to 15 μm/sec and enabling the sarcomere to shorten by 10% of its length in less than 1/50th of a second. The rapid synchronized shortening of the thousands of sarcomeres lying end-to-end in each myofibril enables skeletal muscle to contract rapidly enough for running and flying, or for playing the piano.

Accessory proteins produce the remarkable uniformity in filament organization, length, and spacing in the sarcomere (**Figure 16–76**). The actin filament plus ends are anchored in the Z disc, which is built from CapZ and α-actinin; the Z disc caps the filaments (preventing depolymerization), while holding them together in a regularly spaced bundle. The precise length of each thin filament is determined by a template protein of enormous size, called *nebulin*, which consists almost entirely of a repeating 35-amino-acid actin-binding motif. Nebulin stretches from the Z disc to the minus end of each thin filament and acts as a "molecular ruler" to dictate the length of the filament. The minus ends of the thin filaments are capped and stabilized by tropomodulin. Although there is some slow exchange of actin subunits at both ends of the muscle thin filament, such that the components of the thin filament turn over with a half-life of several days, the actin filaments in sarcomeres are remarkably stable compared to the dynamic actin filaments characteristic of most other cell types that turn over with half-lives of a few minutes or less.

Opposing pairs of an even longer template protein, called *titin*, position the thick filaments midway between the Z discs. Titin acts as a molecular spring, with a long series of immunoglobulin-like domains that can unfold one by one as stress is applied to the protein. A springlike unfolding and refolding of these domains keeps the thick filaments poised in the middle of the sarcomere and allows the muscle fiber to recover after being overstretched. In *C. elegans*, whose sarcomeres are longer than those in vertebrates, titin is also longer, suggesting that it too serves as a molecular ruler, determining in this case the overall length of each sarcomere (see Figure 3–33).

A Sudden Rise in Cytosolic Ca^{2+} Concentration Initiates Muscle Contraction <CTGC>

The force-generating molecular interaction between myosin thick filaments and actin thin filaments takes place only when a signal passes to the skeletal muscle from its motor nerve. Immediately upon arrival of the signal, the muscle cell needs to be able to contract very rapidly, with all the sarcomeres shortening simultaneously. Two major features of the muscle cell are required for extremely rapid contraction. First, as previously discussed, the individual myosin motor heads in each thick filament spend only a small fraction of the ATP cycle time bound to the filament and actively generating force, so many myosin heads can act in rapid succession on the same thin filament without interfering with one another. Second, a specialized membrane system relays the incoming signal rapidly throughout the entire cell. The signal from the nerve triggers an action potential in the muscle cell plasma membrane (discussed in Chapter 11), and

Figure 16–76 Organization of accessory proteins in a sarcomere. <CTGC> Each giant titin molecule extends from the Z disc to the M line—a distance of over 1 μm. Part of each titin molecule is closely associated with a myosin thick filament (which switches polarity at the M line); the rest of the titin molecule is elastic and changes length as the sarcomere contracts and relaxes. Each nebulin molecule is exactly the length of a thin filament. The actin filaments are also coated with tropomyosin and troponin (not shown; see Figure 16–78) and are capped at both ends. Tropomodulin caps the minus end of the actin filaments, and CapZ anchors the plus end at the Z disc, which also contains α-actinin.

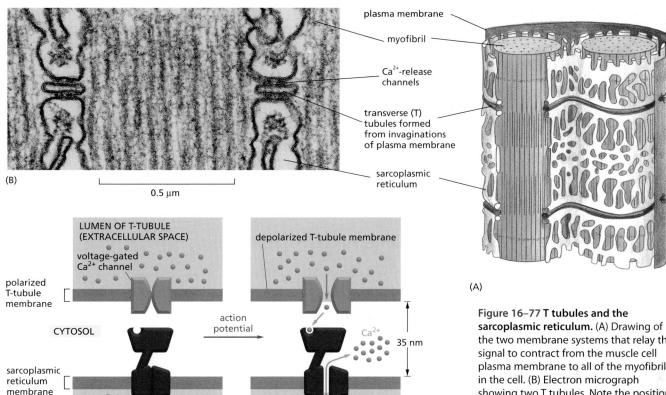

Figure 16–77 **T tubules and the sarcoplasmic reticulum.** (A) Drawing of the two membrane systems that relay the signal to contract from the muscle cell plasma membrane to all of the myofibrils in the cell. (B) Electron micrograph showing two T tubules. Note the position of the large Ca^{2+}-release channels in the sarcoplasmic reticulum membrane; they look like square-shaped "feet" that connect to the adjacent T-tubule membrane. (C) Schematic diagram showing how a Ca^{2+}-release channel in the sarcoplasmic reticulum membrane is thought to be opened by the activation of a voltage-gated Ca^{2+} channel. (B, courtesy of Clara Franzini-Armstrong.)

this electrical excitation spreads rapidly into a series of membraneous folds, the transverse tubules, or *T tubules*, that extend inward from the plasma membrane around each myofibril. The signal is then relayed across a small gap to the *sarcoplasmic reticulum*, an adjacent web-like sheath of modified endoplasmic reticulum that surrounds each myofibril like a net stocking (**Figure 16–77**A and B).

When the incoming action potential activates a Ca^{2+} channel in the T-tubule membrane, a Ca^{2+} influx triggers the opening of Ca^{2+}-release channels in the sarcoplasmic reticulum (Figure 16–77C). Ca^{2+} flooding into the cytosol then initiates the contraction of each myofibril. Because the signal from the muscle-cell plasma membrane is passed within milliseconds (via the T tubules and sarcoplasmic reticulum) to every sarcomere in the cell, all of the myofibrils in the cell contract at once. The increase in Ca^{2+} concentration is transient because the Ca^{2+} is rapidly pumped back into the sarcoplasmic reticulum by an abundant, ATP-dependent Ca^{2+}-pump (also called a Ca^{2+}-ATPase) in its membrane (see Figure 11–13). Typically, the cytoplasmic Ca^{2+} concentration is restored to resting levels within 30 msec, allowing the myofibrils to relax. Thus, muscle contraction depends on two processes that consume enormous amounts of ATP: filament sliding, driven by the ATPase of the myosin motor domain, and Ca^{2+} pumping, driven by the Ca^{2+}-pump.

The Ca^{2+} dependence of vertebrate skeletal muscle contraction, and hence its dependence on motor commands transmitted via nerves, is due entirely to a set of specialized accessory proteins that are closely associated with the actin thin filaments. One of these accessory proteins is a muscle form of *tropomyosin*, an elongated molecule that binds along the groove of the actin helix. The other is *troponin*, a complex of three polypeptides, troponins T, I, and C (named for their tropomyosin-binding, inhibitory, and Ca^{2+}-binding activities, respectively). Troponin I binds to actin as well as to troponin T. In a resting muscle, the troponin I–T complex pulls the tropomyosin out of its normal binding groove into a position along the actin filament that interferes with the binding of

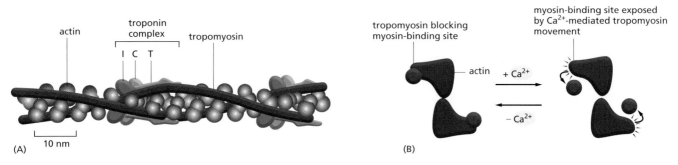

Figure 16–78 The control of skeletal muscle contraction by troponin. (A) A skeletal muscle cell thin filament, showing the positions of tropomyosin and troponin along the actin filament. Each tropomyosin molecule has seven evenly spaced regions with similar amino acid sequences, each of which is thought to bind to an actin subunit in the filament. (B) A thin filament shown end-on, illustrating how Ca^{2+} (binding to troponin) is thought to relieve the tropomyosin blockage of the interaction between actin and the myosin head. (A, adapted from G.N. Phillips, J.P. Fillers and C. Cohen, *J. Mol. Biol.* 192:111–131, 1986. With permission from Academic Press.)

myosin heads, thereby preventing any force-generating interaction. When the level of Ca^{2+} is raised, troponin C—which binds up to four molecules of Ca^{2+}—causes troponin I to release its hold on actin. This allows the tropomyosin molecules to slip back into their normal position so that the myosin heads can walk along the actin filaments (**Figure 16–78**). Troponin C is closely related to the ubiquitous Ca^{2+}-binding protein calmodulin (see Figure 15–44); it can be thought of as a specialized form of calmodulin that has acquired binding sites for troponin I and troponin T, thereby ensuring that the myofibril responds extremely rapidly to an increase in Ca^{2+} concentration.

In smooth muscle cells, so-called because they lack the regular striations of skeletal muscle, contraction is also triggered by an influx of calcium ions, but the regulatory mechanism is different. Smooth muscle forms the contractile portion of the stomach, intestine, and uterus, the walls of arteries, and many other structures requiring slow and sustained contractions. Smooth muscle is composed of sheets of highly elongated spindle-shaped cells, each with a single nucleus. Smooth muscle cells do not express the troponins. Instead, Ca^{2+} influx into the cell regulates contraction by two mechanisms that depend on the ubiquitous calcium binding protein calmodulin.

First, Ca^{2+}-bound calmodulin binds to an actin-binding protein, caldesmon, which blocks the actin sites where the myosin motor heads would normally bind. This causes the caldesmon to fall off of the actin filaments, preparing the filaments for contraction. Second, smooth muscle myosin is phosphorylated on one of its two light chains by myosin light chain kinase (MLCK), as described previously for regulation of nonmuscle myosin II (see Figure 16–72). When the light chain is phosphorylated, the myosin head can interact with actin filaments and cause contraction; when it is dephosphorylated, the myosin head tends to dissociate from actin and becomes inactive (in contrast to nonmuscle myosin II, light chain dephosphorylation does not cause thick filament disassembly in smooth muscle cells). MLCK requires bound Ca^{2+}/calmodulin to be fully active.

External signaling molecules such as adrenaline (epinephrine) can also regulate the contractile activity of smooth muscle. Adrenaline binding to its G-protein-coupled cell surface receptor causes an increase in the intracellular level of cyclic AMP, which in turn activates cyclic-AMP-dependent protein kinase (PKA) (see Figure 15–35). PKA phosphorylates and inactivates MLCK, thereby causing the smooth muscle cell to relax.

The phosphorylation events that regulate contraction in smooth muscle cells occur realtively slowly, so that maximum contraction often requires nearly a second (compared with the few milliseconds required for contraction of a skeletal muscle cell). But rapid activation of contraction is not important in smooth muscle: its myosin II hydrolyzes ATP about 10 times more slowly than skeletal muscle myosin, producing a slow cycle of myosin conformational changes that results in slow contraction.

Heart Muscle Is a Precisely Engineered Machine <AGGT>

The heart is the most heavily worked muscle in the body, contracting about 3 billion (3×10^9) times during the course of a human lifetime. This number is about the same as the average number of revolutions in the lifetime of an automobile's internal combustion engine. Heart cells express several specific isoforms of cardiac muscle myosin and cardiac muscle actin. Even subtle changes in these contractile proteins expressed in the heart—changes that would not cause any noticeable consequences in other tissues—can cause serious heart disease (**Figure 16–79**).

The normal cardiac contractile apparatus is such a highly tuned machine that a tiny abnormality anywhere in the works can be enough to gradually wear it down over years of repetitive motion. *Familial hypertrophic cardiomyopathy* is a frequent cause of sudden death in young athletes. It is a genetically dominant inherited condition that affects about two out of every thousand people, and it is associated with heart enlargement, abnormally small coronary vessels, and disturbances in heart rhythm (cardiac arrhythmias). The cause of this condition is either any one of over 40 subtle point mutations in the genes encoding cardiac β myosin heavy chain (almost all causing changes in or near the motor domain), or one of about a dozen mutations in other genes encoding contractile proteins—including myosin light chains, cardiac troponin, and tropomyosin. Minor missense mutations in the cardiac actin gene cause another type of heart condition, called *dilated cardiomyopathy*, that also frequently results in early heart failure.

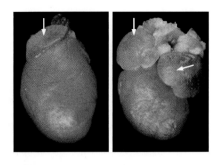

Figure 16–79 Effect on the heart of a subtle mutation in cardiac myosin. *Left*, normal heart from a 6-day old mouse pup. *Right*, heart from a pup with a point mutation in both copies of its cardiac myosin gene, changing Arg 403 to Gln. The arrows indicate the atria. In the heart from the pup with the cardiac myosin mutation, both atria are greatly enlarged (hypertrophic), and the mice die within a few weeks of birth. (From D. Fatkin et al., *J. Clin. Invest.* 103:147, 1999. With permission from The Rockefeller University Press.)

Cilia and Flagella Are Motile Structures Built from Microtubules and Dyneins

Just as myofibrils are highly specialized and efficient motility machines built from actin and myosin filaments, cilia and flagella are highly specialized and efficient motility structures built from microtubules and dynein. Both cilia and flagella are hair-like cell appendages that have a bundle of microtubules at their core. **Flagella** are found on sperm and many protozoa. By their undulating motion, they enable the cells to which they are attached to swim through liquid media (**Figure 16–80**A). **Cilia** tend to be shorter than flagella and are organized in a similar fashion, but they beat with a whip-like motion that resembles the breast stroke in swimming (Figure 16–80B). The cycles of adjacent cilia are almost but not quite in synchrony, creating the wave-like patterns that can be seen in fields of beating cilia under the microscope. Ciliary beating can either propel single cells through a fluid (as in the swimming of the protozoan *Paramecium*) or can move fluid over the surface of a group of cells in a tissue. In the human body, huge numbers of cilia (10^9/cm^2 or more) line our respiratory tract, sweeping layers of mucus, trapped particles of dust, and bacteria up to the mouth where they are swallowed and ultimately eliminated. Likewise, cilia along the oviduct help to sweep eggs toward the uterus.

The movement of a cilium or a flagellum is produced by the bending of its core, which is called the **axoneme**. The axoneme is composed of microtubules and their associated proteins, arranged in a distinctive and regular pattern. Nine special doublet microtubules (comprising one complete and one partial microtubule fused together so that they share a common tubule wall) are arranged in

Figure 16–80 The contrasting motions of flagella and cilia. (A) The wave-like motion of the flagellum of a sperm cell from a tunicate. The cell was photographed with stroboscopic illumination at 400 flashes per second. Note that waves of constant amplitude move continuously from the base to the tip of the flagellum. (B) The beat of a cilium, which resembles the breast stroke in swimming. A fast power stroke (*red* arrows), in which fluid is driven over the surface of the cell, is followed by a slow recovery stroke. Each cycle typically requires 0.1–0.2 sec and generates a force perpendicular to the axis of the axoneme (the ciliary core). (A, courtesy of C.J. Brokaw.)

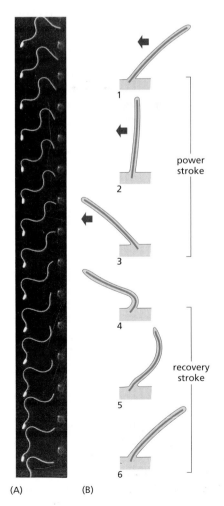

power stroke

recovery stroke

(A)   (B)

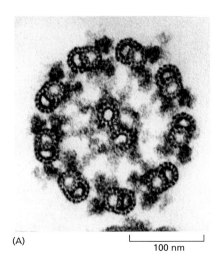

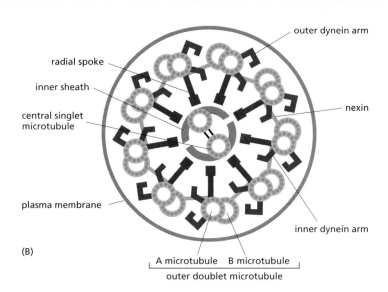

(A)

100 nm

(B)

outer dynein arm

radial spoke

inner sheath

central singlet
microtubule

nexin

plasma membrane

inner dynein arm

A microtubule   B microtubule

outer doublet microtubule

a ring around a pair of single microtubules (**Figure 16–81**). Almost all forms of eucaryotic flagella and cilia (from protozoans to humans) have this characteristic arrangement. The microtubules extend continuously for the length of the axoneme, which can be 10–200 μm. At regular positions along the length of the microtubules, accessory proteins cross-link the microtubules together.

Molecules of *ciliary dynein* form bridges between the neighboring doublet microtubules around the circumference of the axoneme (**Figure 16–82**). When the motor domain of this dynein is activated, the dynein molecules attached to one microtubule doublet (see Figure 16–64) attempt to walk along the adjacent microtubule doublet, tending to force the adjacent doublets to slide relative to one another, much as actin thin filaments slide during muscle contraction. However, the presence of other links between the microtubule doublets prevents this sliding, and the dynein force is instead converted into a bending motion (**Figure 16–83**).

The length of flagella is carefully regulated. If one of the two flagella on a *Chlamydomonas* cell is amputated, the remaining one will transiently shrink as the stump regrows until they reach the same length, and then the two shortened flagella will continue to elongate until both are as long as they were on the unperturbed cell. New flagellar components including tubulin and dynein are incorporated into the growing flagella at the distal tips. Thus, even in these

Figure 16–81 The arrangement of microtubules in a flagellum or cilium. (A) Electron micrograph of the flagellum of a green-alga cell *(Chlamydomonas)* shown in cross section, illustrating the distinctive "9 + 2" arrangement of microtubules. (B) Diagram of the parts of a flagellum or cilium. The various projections from the microtubules link the microtubules together and occur at regular intervals along the length of the axoneme. (A, courtesy of Lewis Tilney.)

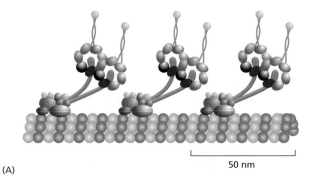

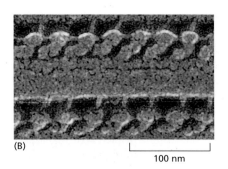

(A)

50 nm

(B)

100 nm

Figure 16–82 Ciliary dynein. Ciliary (axonemal) dynein is a large protein assembly (nearly 2 million daltons) composed of 9–12 polypeptide chains, the largest of which is the heavy chain of more than 500,000 daltons. (A) The heavy chains form the major portion of the globular head and stem domains, and many of the smaller chains are clustered around the base of the stem. There are two heads in the outer dynein in metazoans, but three heads in protozoa, each formed from their own heavy chain (see Figure 16–59B for a view of an isolated molecule). The tail of the molecule binds tightly to an A microtubule in an ATP-independent manner, while the large globular heads have an ATP-dependent binding site for a B microtubule (see Figure 16–81). When the heads hydrolyze their bound ATP, they move toward the minus end of the B microtubule, thereby producing a sliding force between the adjacent microtubule doublets in a cilium or flagellum. For details, see Figure 16–64. (B) Freeze-etch electron micrograph of a cilium showing the dynein arms projecting at regular intervals from the doublet microtubules. (B, courtesy of John Heuser.)

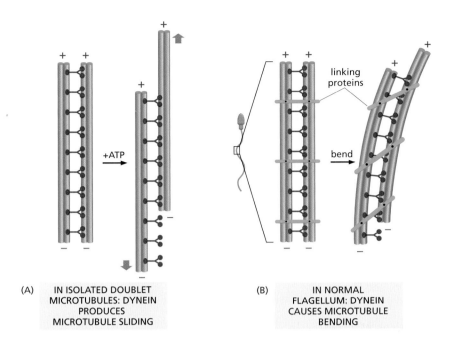

(A) IN ISOLATED DOUBLET MICROTUBULES: DYNEIN PRODUCES MICROTUBULE SLIDING

(B) IN NORMAL FLAGELLUM: DYNEIN CAUSES MICROTUBULE BENDING

Figure 16–83 The bending of an axoneme. (A) When axonemes are exposed to the proteolytic enzyme trypsin, the linkages holding neighboring doublet microtubules together are broken. In this case, the addition of ATP allows the motor action of the dynein heads to slide one pair of doublet microtubules against the other pair. (B) In an intact axoneme (such as in a sperm), flexible protein links prevent the sliding of the doublet. The motor action therefore causes a bending motion, creating waves or beating motions, as seen in Figure 16–80.

highly ordered and stable filament-motor arrays, cells use the intrinsic flexibility and adaptability of the cytoskeleton to respond rapidly and dynamically to the changes they experience.

In humans, hereditary defects in ciliary dynein cause Kartagener's syndrome. The syndrome is characterized by male sterility due to immotile sperm, a high susceptibility to lung infections owing to the paralyzed cilia in the respiratory tract that fail to clear debris and bacteria, and defects in determination of the left–right axis of the body during early embryonic development (discussed in Chapter 22).

Bacteria also swim using cell surface structures called flagella, but these do not contain microtubules or dynein and do not wave or beat. Instead, *bacterial flagella* are long, rigid helical filaments, made up of repeating subunits of the protein flagellin. The flagella rotate like propellers, driven by a special rotary motor embedded in the bacterial cell wall (see Figure 15–71). The use of the same name to denote these two very different types of swimming apparatus is an unfortunate historical accident.

Structures called *basal bodies* firmly root eucaryotic cilia and flagella at the cell surface. The basal bodies have the same form as the centrioles that are found embedded at the center of animal centrosomes, with nine groups of fused triplet microtubules arranged in a cartwheel (**Figure 16–84**). Indeed, in some organisms, basal bodies and centrioles are functionally interconvertible: during each mitosis in the unicellular alga *Chlamydomonas*, for example, the flagella are resorbed, and the basal bodies move into the cell interior and become part of the spindle poles. New centrioles and basal bodies arise by a curious replication process, in which a smaller daughter is formed perpendicular to the original structure by a still mysterious mechanism (see Figure 17–31).

Figure 16–84 Basal bodies. (A) Electron micrograph of a cross section through three basal bodies in the cortex of a protozoan. (B) Diagram of a basal body viewed from the side. Each basal body forms the lower portion of a ciliary axoneme and is composed of nine sets of triplet microtubules, each triplet containing one complete microtubule (the A microtubule) fused to two incomplete microtubules (the B and C microtubules). Other proteins (shown in red in B) form links that hold the cylindrical array of microtubules together. The arrangement of microtubules in a centriole is essentially the same (see Figure 16–31). (A, courtesy of D.T. Woodrow and R.W. Linck.)

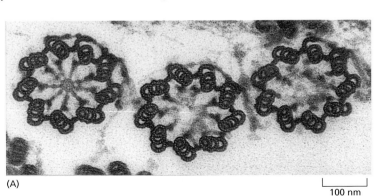

(A)

100 nm

A B C

(B)

Even in animal cells that lack fully developed beating cilia or flagella, centrioles frequently nucleate the growth of a non-motile, microtubule-rich surface projection called a *primary cilium*. Primary cilia are usually only a few micrometers in length and lack dynein. They are found on the surface of many different cell types including fibroblasts, epithelial cells, neurons, bone cells, and chondrocytes (cartilage cells). Many signaling proteins are concentrated in the primary cilium, including proteins involved in the Hedgehog signaling pathway (see p. 950), and receptors for neurotransmitters on neurons in the central nervous system. On kidney epithelial cells, primary cilia act as flow sensors that detect the movement of fluid through the kidney tubules. Mechanosensitive calcium channels are opened when the fluid flow bends the primary cilia, regulating kidney cell growth and proliferation. Loss of the calcium channel or other structural components of the primary cilium in the kidney cells causes polycystic kidney disease, a common genetic disorder that causes overproliferation of the kidney epithelial cells—resulting in the formation of large fluid-filled cysts throughout the organs and eventually in kidney failure. Another specialized kind of primary cilium that is unusual in being able to beat is required for establishing left-right asymmetry in the developing embryo (see Figure 22–87).

Construction of the Mitotic Spindle Requires Microtubule Dynamics and the Interactions of Many Motor Proteins

Myofibrils and cilia are relatively permanent structures specialized to produce repetitive movement. But most cell movements depend on labile structures that appear at specific stages of the cell cycle or in response to external signals and then disappear once they complete their jobs. The most familiar of these are the mitotic spindle and the contractile ring that form during cell division. In Chapter 17, we will describe in detail both the process of mitosis and the cell cycle control system that determines the timing of the events of cell division. Here, we briefly discuss a few of the cytoskeletal mechanisms that contribute to the construction and mechanical function of the mitotic spindle.

The construction of the mitotic spindle is a particularly important and fascinating example of the power of self-organization by teams of motor proteins interacting with dynamic cytoskeletal filaments. It also features the active participation of the chromosomes. In a rapid sequence of events that typically takes less than an hour in animal cells, the interphase array of microtubules is completely disassembled and reorganized to form the bipolar spindle structure that is responsible for segregating the replicated chromosomes with perfect fidelity to the two daughter cells. Because of the central importance of reliability in transmission of the genetic material, the construction and functioning of the mitotic spindle feature a tremendous degree of redundancy, so that if one set of mechanisms fails for any reason, there are backup mechanisms in place to ensure reliable chromosome partitioning.

In early mitosis, there are dramatic changes in the dynamic behavior and average length of the microtubules. In the interphase array the microtubules are typically long and undergo rare catastrophes, but during mitosis the microtubules are shorter and much more dynamic. Microtubule nucleation and assembly are enhanced in the regions around the condensed chromosomes. As microtubules assemble on condensed chromatin pointing in random directions, the coordinated actions of several motor proteins build a coherent bipolar spindle from the disorganized microtubule mass. First, the bipolar kinesin-5 (see Figure 16–58) bundles the microtubules into a parallel array and slides microtubules that are oriented in opposite directions away from each other. Next, another kinesin that is bound to chromosome arms, kinesin-4, walks toward the plus ends of chromosome-associated microtubules and pushes their minus ends away from the chromosome mass. Finally, the minus-end directed motors cytoplasmic dynein and kinesin-14 form oligomeric complexes with scaffold proteins that gather the microtubule minus ends together to form the spindle poles. In most animal cells, these processes are guided by a pair of centrosomes that help to nucleate and organise these microtubule minus ends. The final result is the elegantly balanced bipolar mitotic spindle (**Figure 16–85**).

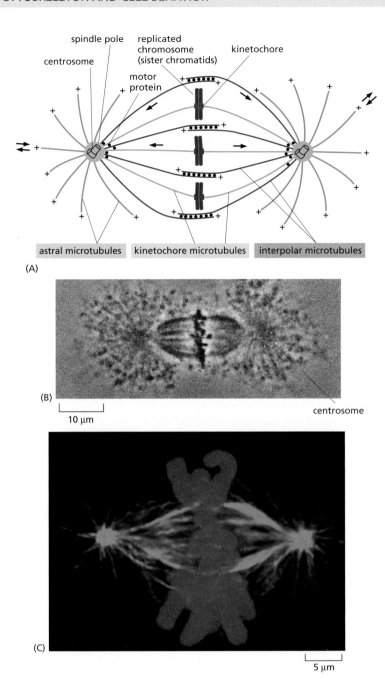

(A)

(B)

10 μm

centrosome

(C)

5 μm

Figure 16–85 The mitotic spindle in animal cells. <GTCT> (A) There are three classes of dynamic microtubules in the mitotic spindle at metaphase: kinetochore microtubules (*blue*) that attach each chromosome to the spindle pole, interpolar microtubules (*red*) that hold the two halves of the spindle together, and astral microtubules (*green*) that can interact with the cell cortex. All of the microtubules are oriented with their minus ends at the spindle poles where the centrosomes reside, and their plus ends projecting away. As indicated by the arrows, the astral microtubules undergo dynamic instability, growing and shrinking at their plus ends, while the kinetochore microtubules and interpolar microtubules both undergo continuous flux toward the spindle poles. (B) A phase-contrast micrograph of an isolated mitotic spindle at metaphase, with the chromosomes aligned at the spindle equator. (C) This fluorescence micrograph shows the microtubules of the spindle in *green* and the chromosomes in *blue*. The *red* spots mark the kinetochores, specialized structures that connect the microtubules to the chromosomes. (B, from E.D. Salmon and R.R. Segall, *J. Cell Biol.* 86:355–365, 1980. With permission from The Rockefeller University Press; C, from A. Desai, *Curr. Biol.* 10:R508, 2000. With permission from Elsevier.)

After the bipolar mitotic spindle has assembled, it can appear stable and quiescent for long periods of time. In many animals, the unfertilized egg arrests its cell cycle in meiotic metaphase, and the spindle waits for days or months until fertilization triggers the progression of the cell cycle (see Chapter 21). This steady appearance is deceptive, because the spindle is actually an extremely dynamic structure, tensed for action that will begin when the chromosomes suddenly begin to separate in anaphase. For example, many of the spindle microtubules exhibit a behavior called *poleward flux*, with a net addition of tubulin subunits at their plus ends, balancing a net loss at their minus ends near the spindle pole. Poleward flux is driven by the action of minus end-directed motor proteins at the spindle pole that are constantly reeling in the micro-tubules, and the bipolar plus end-directed kinesin-5 motors on the interpolar microtubules that are constantly pushing them apart (see Figure 16–85). As will be discussed in Chapter 17, the delicate balance between these two types of motor protein activities in the spindle also determines its length. Overall the mitotic spindle represents a collaborative effort combining the dynamic proper-ties of microtubules with the individual actions of dozens of molecular motors and other organizing components.

Many Cells Can Crawl Across a Solid Substratum

The process of cell crawling provides another instance where we can appreciate the dynamic integration of cytoskeletal filaments, filament regulators, and motor proteins. Many cells move by crawling over surfaces rather than by using cilia or flagella to swim. Predatory amoebae crawl continuously in search of food, and they can easily be observed to attack and devour smaller ciliates and flagellates in a drop of pond water. In animals, almost all cell locomotion occurs by crawling, with the notable exception of swimming sperm. During embryogenesis, the structure of an animal is created by the migrations of individual cells to specific target locations and by the coordinated movements of whole epithelial sheets (discussed in Chapter 23). In vertebrates, *neural crest cells* are remarkable for their long-distance migrations from their site of origin in the neural tube to a variety of sites throughout the embryo. These cells have diverse fates, becoming skin pigment cells, sensory and sympathetic neurons and glia, and various structures of the face. Long-distance crawling is fundamental to the construction of the entire nervous system: it is in this way that the actin-rich growth cones at the advancing tips of developing axons travel to their eventual synaptic targets, guided by combinations of soluble signals and signals bound to cell surfaces and extracellular matrix along the way.

The adult animal also seethes with crawling cells. Macrophages and neutrophils crawl to sites of infection and engulf foreign invaders as a critical part of the innate immune response. Osteoclasts tunnel into bone, forming channels that are filled in by the osteoblasts that follow after them, in a continuous process of bone remodeling and renewal. Similarly, fibroblasts can migrate through connective tissues, remodeling them where necessary and helping to rebuild damaged structures at sites of injury. In an ordered procession, the cells in the epithelial lining of the intestine travel up the sides of the intestinal villi, replacing absorptive cells lost at the tip of the villus. Unfortunately, cell crawling also has a role in many cancers, when cells in a primary tumor invade neighboring tissues and crawl into blood vessels or lymph vessels and then emerge at other sites in the body to form metastases.

Cell crawling is a highly complex integrated process, dependent on the actin-rich cortex beneath the plasma membrane. Three distinct activities are involved: *protrusion*, in which actin-rich structures are pushed out at the front of the cell; *attachment*, in which the actin cytoskeleton connects across the plasma membrane to the substratum; and *traction*, in which the bulk of the trailing cytoplasm is drawn forward (**Figure 16–86**). In some crawling cells, such as

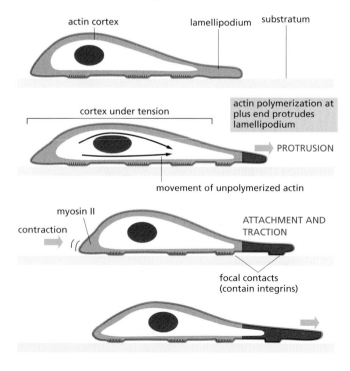

Figure 16–86 A model of how forces generated in the actin-rich cortex move a cell forward. The actin-polymerization-dependent protrusion and firm attachment of a lamellipodium at the leading edge of the cell moves the edge forward (*green* arrows at front) and stretches the actin cortex. Contraction at the rear of the cell propels the body of the cell forward (*green* arrow at back) to relax some of the tension (traction). New focal contacts are made at the front, and old ones are disassembled at the back as the cell crawls forward. The same cycle can be repeated, moving the cell forward in a stepwise fashion. Alternatively, all steps can be tightly coordinated, moving the cell forward smoothly. The newly polymerized cortical actin is shown in *red*.

keratocytes from the fish epidermis, these activities are closely coordinated, and the cells seem to glide forward smoothly without changing shape. In other cells, such as fibroblasts, these activities are more independent, and the locomotion is jerky and irregular.

Actin Polymerization Drives Plasma Membrane Protrusion

The first step in locomotion, protrusion of a leading edge, seems to rely primarily on forces generated by actin polymerization pushing the plasma membrane outward. Different cell types generate different types of protrusive structures, including filopodia (also known as microspikes), lamellipodia, and pseudopodia. All are filled with a dense core of filamentous actin, which excludes membrane-enclosed organelles. The three structures differ primarily in the way in which the actin is organized—in one, two, or three dimensions, respectively—and we have already discussed how this results from the presence of different actin-associated proteins.

Filopodia, formed by migrating growth cones and some types of fibroblasts, are essentially one-dimensional. They contain a core of long, bundled actin filaments, which are reminiscent of those in microvilli but longer and thinner, as well as more dynamic. **Lamellipodia**, formed by epithelial cells and fibroblasts, as well as by some neurons, are two-dimensional, sheet-like structures. They contain an orthogonally cross-linked mesh of actin filaments, most of which lie in a plane parallel to the solid substratum. **Pseudopodia**, formed by amoebae and neutrophils, are stubby three-dimensional projections filled with an actin-filament gel. <ATGG> Perhaps because their two-dimensional geometry is most convenient for examination with the light microscope, we have more information about the dynamic organization and protrusion mechanism of lamellipodia than we have for either filopodia or pseudopodia.

Lamellipodia contain all of the machinery that is required for cell motility. They have been especially well studied in the epithelial cells of the epidermis of fish and frogs, which are known as *keratocytes* because of their abundant keratin filaments. These cells normally cover the animal by forming an epithelial sheet, and they are specialized to close wounds very rapidly, moving at rates up to 30 μm/min. When cultured as individual cells, keratocytes assume a distinctive shape with a very large lamellipodium and a small, trailing cell body that is not attached to the substratum (**Figure 16–87**). Fragments of this lamellipodium can be sliced off with a micropipette. Although the fragments generally lack microtubules and membrane-enclosed organelles, they continue to crawl normally, looking like tiny keratocytes.

The dynamic behavior of actin filaments can be studied in keratocyte lamellipodia by marking a small patch of actin and examining its fate. This reveals

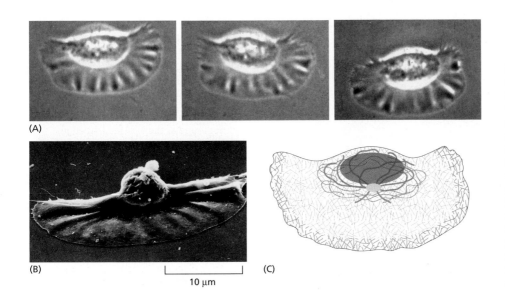

(A)

(B)

(C)

10 μm

Figure 16–87 Migratory keratocytes from a fish epidermis. <GTTA> (A) Light micrographs of a keratocyte in culture, taken about 15 sec apart. This cell is moving at about 15 μm/sec. (B) Keratocyte seen by scanning electron microscopy, showing its broad, flat lamellipodium and small cell body, including the nucleus, carried up above the substratum at the rear. (C) Distribution of cytoskeletal filaments in this cell. Actin filaments *(red)* fill the large lamellipodium and are responsible for the cell's rapid movement. Microtubules *(green)* and intermediate filaments *(blue)* are restricted to the regions close to the nucleus. (A and B, courtesy of Juliet Lee.)

Figure 16–88 Actin filament nucleation and web formation by the ARP complex in lamellipodia. (A) A keratocyte with actin filaments labeled in *red* by fluorescent phalloidin, and the ARP complex labeled in *green* with an antibody raised against one of its component proteins. The regions where the two overlap appear *yellow*. The ARP complex is highly concentrated near the front of the lamellipodium, where actin nucleation is most active. (B) Electron micrograph of a platinum-shadowed replica of the leading edge of a keratocyte, showing the dense actin filament meshwork. The labels denote areas enlarged in C. (C) Close-up views of the marked regions of the actin web at the leading edge shown in B. Numerous branched filaments can be seen, with the characteristic 70° angle formed when the ARP complex nucleates a new actin filament off the side of a preexisting filament (see Figure 16–34). (From T. Svitkina and G. Borisy, *J. Cell Biol.* 145:1009–1026, 1999. With permission from The Rockefeller University Press.)

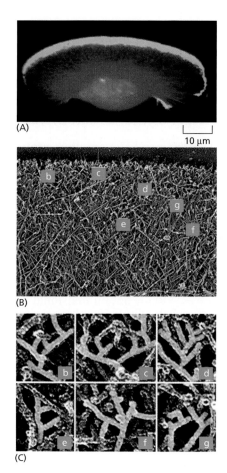

(A)

10 μm

(B)

(C)

that, while the lamellipodia crawl forward, the actin filaments remain stationary with respect to the substrate. The actin filaments in the meshwork are mostly oriented with their plus ends facing forward. The minus ends are frequently attached to the sides of other actin filaments by ARP complexes (see Figure 16–34), helping to form the two-dimensional web (**Figure 16–88**). The web as a whole seems to be undergoing treadmilling, assembling at the front and disassembling at the back, reminiscent of the treadmilling that occurs in individual actin filaments and microtubules discussed previously (see Figure 16–14). Treadmilling of a dendritic web built by the ARP complex is only one of several ways that cells can use dynamic actin filaments to drive the protrusion of the leading edge. Some slowly-moving cells including fibroblasts appear to use a mechanism that does not depend on the ARP complex, but still requires coordinated actin filament assembly and disassembly, possibly coordinated by formins.

Maintenance of unidirectional motion by lamellipodia is thought to require the cooperation and mechanical integration of several factors. Filament nucleation is localized at the leading edge, with new actin filament growth occurring primarily in that location to push the plasma membrane forward. Most filament depolymerization occurs at sites located well behind the leading edge. Because *cofilin* (see Figure 16–42) binds cooperatively and preferentially to actin filaments containing ADP-actin (the D form), the new T-form filaments generated at the leading edge should be resistant to depolymerization by cofilin (**Figure 16–89**). As the filaments age and ATP hydrolysis proceeds, cofilin can efficiently disassemble the older filaments. Thus, the delayed ATP hydrolysis by filamentous actin is thought to provide the basis for a mechanism that maintains an efficient, unidirectional treadmilling process in the lamellipodium (**Figure 16–90**). Finally, bipolar myosin II filaments seem to associate with the actin filaments in the web and pull them into a new orientation—from nearly perpendicular to the leading edge to an orientation almost parallel to the leading edge. This contraction prevents protrusion and it pinches in the sides of the locomoting lamellipodium, helping to gather in the sides of the cell as it moves forward (**Figure 16–91**).

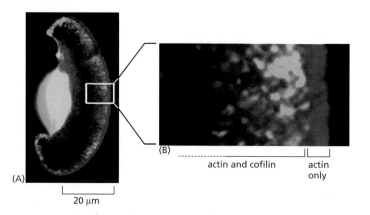

(A)

20 μm

(B)

actin and cofilin   actin only

Figure 16–89 Cofilin in lamellipodia. (A) A keratocyte with actin filaments labeled in *red* by fluorescent phalloidin and cofilin labeled in *green* with a fluorescent antibody. The regions where the two overlap appear *yellow*. Although the dense actin meshwork reaches all the way through the lamellipodium, cofilin is not found at the very leading edge. (B) Close-up view of the region marked with the *white* rectangle in A. The actin filaments closest to the leading edge, which are also the ones that have formed most recently and that are most likely to contain ATP actin (rather than ADP actin) in the filament lattice are generally not associated with cofilin. (From T. Svitkina and G. Borisy, *J. Cell Biol.* 145:1009–1026, 1999. With permission from The Rockefeller University Press.)

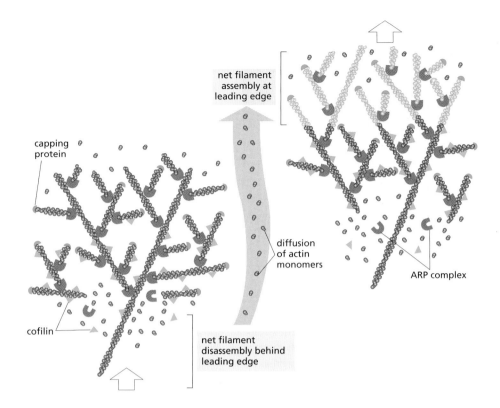

net filament
assembly at
leading edge

capping
protein

diffusion
of actin
monomers

ARP complex

cofilin

net filament
disassembly behind
leading edge

Figure 16–90 A model for protrusion of the actin meshwork at the leading edge. Two time points during advance of the lamellipodium are illustrated, with newly assembled structures at the later time point shown in a lighter color. Nucleation is mediated by the ARP complex at the front. Newly nucleated actin filaments are attached to the sides of preexisting filaments, primarily at a 70° angle. Filaments elongate, pushing the plasma membrane forward because of some sort of anchorage of the array behind. At a steady rate, actin filament plus ends become capped. After newly polymerized actin subunits hydrolyze their bound ATP in the filament lattice, the filaments become susceptible to depolymerization by cofilin. This cycle causes a spatial separation between net filament assembly at the front and net filament disassembly at the rear, so that the actin filament network as a whole can move forward, even though the individual filaments within it remain stationary with respect to the substratum.

The pushing force created by the polymerization of a branched web of actin filaments plays an important role in many cell processes. The polymerization at the plus end can push the plasma membrane outward, as in the example just discussed (see Figure 16–90), or it can propel vesicles or particles through the cell cytoplasm, as in the example of the bacterium *Listeria monocytogenes* discussed in Chapter 24 (see Figure 24–37 <GTAT>). Moreover, when anchored in a more complex way to the membrane, the same type of force drives plasma membrane invaginations, as it does during the endocytotic and phagocytotic processes discussed in Chapter 13.

It is interesting to compare the organization of the actin-rich lamellipodium to the organization of the microtubule-rich mitotic spindle. In both cases, the cell harnesses and amplifies the intrinsic dynamic behavior of the cytoskeletal filament systems to generate large-scale structures that determine the behavior of the whole cell. Both structures feature rapid turnover of their constituent cytoskeletal filaments, even though the structures themselves may remain intact at steady state for long periods of time. The leading edge plasma membrane in the lamellipodium fulfills an organizational role analogous to the condensed chromosomes in organizing and stimulating the dynamics of the mitotic spindle. In both cases, molecular motor proteins help to enhance cytoskeletal filament flux and turnover in the large-scale arrays.

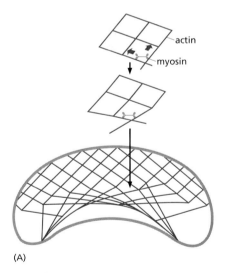

actin

myosin

(A)

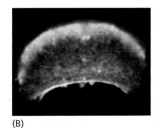

(B)

Figure 16–91 Contribution of myosin II to polarized cell motility.
(A) Myosin II bipolar filaments bind to actin filaments in the dendritic lamellipodial meshwork and cause network contraction. The myosin-driven reorientation of the actin filaments in the dendritic meshwork forms an actin bundle that recruits more myosin II and contributes to generating the contractile forces required for retraction of the trailing edge of the moving cell. (B) A fragment of the large lamellipodium of a keratocyte can be separated from the main cell body either by surgery with a micropipette or by treating the cell with certain drugs. Many of these fragments continue to move rapidly, with the same overall cytoskeletal organization as the intact keratocytes. Actin *(blue)* forms a protrusive meshwork at the front of the fragment. Myosin II *(pink)* is gathered into a band at the rear. (From A. Verkovsky et al., *Curr. Biol.* 9:11–20, 1999. With permission from Elsevier.)

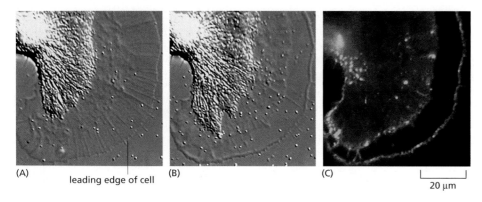

(A) leading edge of cell (B) (C) 20 μm

(D) cytochalasin B

Cell Adhesion and Traction Allow Cells to Pull Themselves Forward

Lamellipodia of all cells seem to share a basic, simple type of dynamic organization where actin filament assembly occurs preferentially at the leading edge and actin filament disassembly occurs preferentially at the rear. However, the interactions between the cell and its normal physical environment usually make the situation considerably more complex than for fish keratocytes crawling on a culture dish. Particularly important in locomotion is the intimate crosstalk between the cytoskeleton and the cell adhesion apparatus. Although some degree of adhesion to the substratum is necessary for any form of cell crawling, adhesion and locomotion rate seem generally to be inversely related, with highly adhesive cells moving more slowly than weakly adhesive ones. Keratocytes are so weakly adhesive to the substratum that the force of actin polymerization can push the leading edge forward very rapidly. In contrast, neurons from the sea slug *Aplysia* cultured on a sticky substratum form large lamellipodia that become stuck too tightly to move forward. In these lamellipodia, the same cycle of localized nucleation of new actin filaments, depolymerization of old filaments, and myosin-dependent contraction continues to operate. But because the leading edge is prevented physically from moving forward, the entire actin mesh moves backward toward the cell body instead, pulled by myosins (**Figure 16–92**). The adhesion of most cells lies somewhere between these two extremes, and most lamellipodia exhibit some combination of forward actin filament protrusion (like keratocytes) and rearward actin flux (like the *Aplysia* neurons).

As a lamellipodium, filopodium, or pseudopodium extends forward over a substratum, it can form new attachment sites at the cell front that remain stationary as the cell moves forward over them, persisting until the rear of the cell catches up with them. When an individual lamellipodium fails to adhere to the

Figure 16–92 Rearward movement of the actin network in a growth-cone lamellipodium. (A) A growth cone from a neuron of the sea slug *Aplysia* is cultured on a highly adhesive substratum and viewed by differential-interference-contrast microscopy. Microtubules and membrane-enclosed organelles are confined to the bright, rear area of the growth cone (to the *left*), while a meshwork of actin filaments fills the lamellipodium (on the *right*). (B) After brief treatment with the drug cytochalasin, which caps the plus ends of actin filaments (see Table 16–2, p. 988), the actin meshwork has detached from the front edge of the lamellipodium and has been pulled backward. (C) At the time point shown in B, the cell was fixed and labeled with fluorescent phalloidin to show the distribution of the actin filaments. Some actin filaments persist at the leading edge, but the region behind the leading edge is devoid of filaments. Note the sharp boundary of the rearward-moving actin meshwork. (D) The complex cyclic structure of cytochalasin B. (A–C, courtesy of Paul Forscher.)

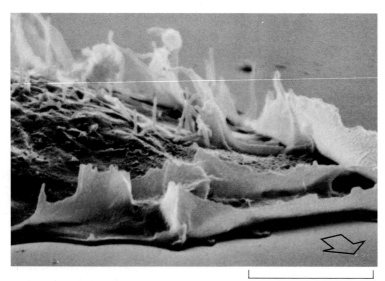

5 μm

Figure 16–93 Lamellipodia and ruffles at the leading edge of a human fibroblast migrating in culture. The *arrow* in this scanning electron micrograph shows the direction of cell movement. As the cell moves forward, lamellipodia that fail to attach to the substratum are swept backward over the dorsal surface of the cell, a movement known as ruffling. (Courtesy of Julian Heath.)

substratum, it is usually lifted up onto the dorsal surface of the cell and rapidly carried backward as a "ruffle" (**Figure 16–93**).

The attachment sites established at the leading edge serve as anchorage points, which allow the cell to generate traction on the substratum and pull its body forward. Myosin motor proteins, especially myosin II, seem to generate traction forces. In many locomoting cells, myosin II is highly concentrated at the posterior of the cell where it may help to push the cell body forward like toothpaste being squeezed out of a tube from the rear (**Figure 16–94**; see also Figure 16–91). *Dictyostelium* amoebae that are deficient in myosin II are able to protrude pseudopodia at normal speeds, but the translocation of their cell body is much slower than that of wild-type amoebae, indicating the importance of myosin II contraction in this part of the cell locomotion cycle. In addition to helping to push the cell body forward, contraction of the actin-rich cortex at the rear of the cell may selectively weaken the older adhesive interactions that tend to hold the cell back. Myosin II may also transport cell body components forward over a polarized array of actin filaments.

The traction forces generated by locomoting cells exert a significant pull on the substratum (**Figure 16–95**). In a living animal, most crawling cells move across a semiflexible substratum made of extracellular matrix, which can be deformed and rearranged by these cell forces. In culture, movement of fibroblasts through a gel of collagen fibrils aligns the collagen, generating an organized extracellular matrix that in turn affects the shape and direction of locomotion of the fibroblasts within it (**Figure 16–96**). Conversely, mechanical tension or stretching applied externally to a cell will cause it to assemble stress fibers and focal adhesions, and become more contractile. Although poorly understood, this two-way mechanical interaction between cells and their physical environment is thought to be a primary way that vertebrate tissues organize themselves.

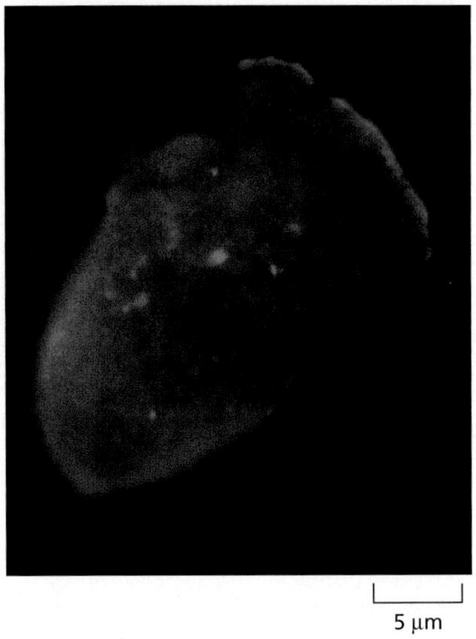

Members of the Rho Protein Family Cause Major Rearrangements of the Actin Cytoskeleton

Cell migration is one example of a process that requires long-distance communication and coordination between one end of a cell and the other. During directed migration, it is important that the front end of the cell remain structurally and functionally distinct from the back end. In addition to driving local mechanical processes such as protrusion at the front and retraction at the rear, the cytoskeleton is responsible for coordinating cell shape, organization, and mechanical properties from one end of the cell to the other, a distance which is typically several tens of micrometers for animal cells. In many cases, including but not limited to cell migration, large-scale cytoskeletal coordination takes the form of the establishment of cell polarity, where a cell builds different structures with distinct molecular components at the front vs. the back, or at the top vs. the

5 µm

Figure 16–94 The localization of myosin I and myosin II in a normal crawling *Dictyostelium* amoeba. This cell was crawling toward the upper right at the time that it was fixed and labeled with antibodies specific for two myosin isoforms. Myosin I *(green)* is mainly restricted to the leading edge of pseudopodia at the front of the cell. Myosin II *(red)* is highest in the posterior, actin-rich cortex. Contraction of the cortex at the posterior of the cell by myosin II may help to push the cell body forward. (Courtesy of Yoshio Fukui.)

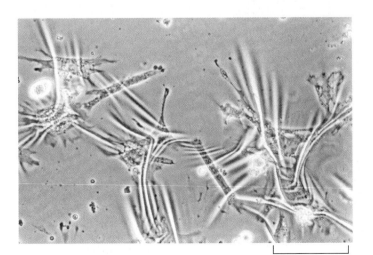

100 µm

Figure 16–95 Adhesive cells exert traction forces on the substratum. These fibroblasts have been cultured on a very thin sheet of silicon rubber. Attachment of the cells, followed by contraction of their cytoskeleton, has caused the rubber substratum to wrinkle. (From A.K. Harris, P. Wild and D. Stopak, *Science* 208:177–179, 1980. With permission from AAAS.)

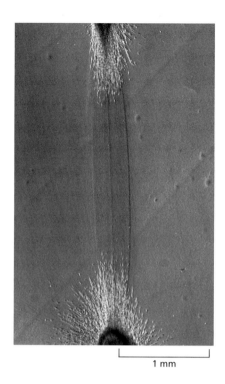

Figure 16–96 Shaping of the extracellular matrix by cell pulling. This micrograph shows a region between two pieces of embryonic chick heart (tissue explants rich in fibroblasts and heart muscle cells) that were grown in culture on a collagen gel for 4 days. A dense tract of aligned collagen fibers has formed between the two explants, apparently as a result of fibroblasts tugging on the collagen. (From D. Stopak and A.K. Harris, *Dev. Biol.* 90:383–398, 1982. With permission from Academic Press.)

1 mm

bottom. Cell locomotion requires an initial polarization of the cell to set it off in a particular direction. Carefully controlled cell polarization processes are also required for oriented cell divisions in tissues and for formation of a coherent, organized multicellular structure. Genetic studies in yeast, flies, and worms have provided most of our current understanding of the molecular basis of cell polarity. The mechanisms that generate cell polarity in vertebrates are only beginning to be explored. In all known cases, however, the cytoskeleton has a central role, and many of the molecular components have been evolutionarily conserved.

For the actin cytoskeleton, diverse cell-surface receptors trigger global structural rearrangements in response to external signals. But all of these signals seem to converge inside the cell on a group of closely related monomeric GTPases that are members of the **Rho protein family**—*Cdc42*, *Rac*, and *Rho*. The same Rho family proteins are also involved in the establishment of many kinds of cell polarity.

Like other members of the Ras superfamily, these Rho proteins act as molecular switches to control cell processes by cycling between an active, GTP-bound state and an inactive, GDP-bound state (see Figure 3–71). Activation of Cdc42 on the plasma membrane triggers actin polymerization and bundling to form either filopodia or shorter cell protrusions called microspikes. Activation of Rac promotes actin polymerization at the cell periphery leading to the formation of sheet-like lamellipodial extensions and membrane ruffles, which are actin-rich protrusions on the cell's dorsal surface (see Figure 16–93). Activation of Rho promotes both the bundling of actin filaments with myosin II filaments into stress fibers and the clustering of integrins and associated proteins to form focal contacts (**Figure 16–97**). These dramatic and complex structural changes occur because each of these three molecular switches has numerous downstream target proteins that affect actin organization and dynamics.

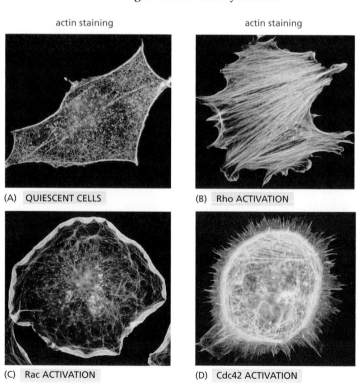

actin staining   actin staining

(A) QUIESCENT CELLS   (B) Rho ACTIVATION

(C) Rac ACTIVATION   (D) Cdc42 ACTIVATION

20 µm

Figure 16–97 The dramatic effects of Rac, Rho, and Cdc42 on actin organization in fibroblasts. In each case, the actin filaments have been labeled with fluorescent phalloidin. (A) Serum-starved fibroblasts have actin filaments primarily in the cortex, and relatively few stress fibers. (B) Microinjection of a constitutively activated form of Rho causes the rapid assembly of many prominent stress fibers. (C) Microinjection of a constitutively activated form of Rac, a closely related monomeric GTPase, causes the formation of an enormous lamellipodium that extends from the entire circumference of the cell. (D) Microinjection of a constitutively activated form of Cdc42, another Rho family member, causes the protrusion of many long filopodia at the cell periphery. The distinct global effects of these three GTPases on the organization of the actin cytoskeleton are mediated by the actions of dozens of other protein molecules that are regulated by the GTPases. These target proteins include some of the various actin-associated proteins that we have discussed in this chapter. (From A. Hall, *Science* 279:509–514, 1998. With permission from AAAS.)

Some key targets of activated Cdc42 are members of the WASp protein family. Human patients deficient in WASp suffer from Wiskott-Aldrich Syndrome, a severe form of immunodeficiency where immune system cells have abnormal actin-based motility and platelets do not form normally. Although WASp itself is expressed only in blood cells and immune system cells, other family members are expressed ubiquitously that enable activated Cdc42 to enhance actin polymerization. **WASp proteins** can exist in an inactive folded conformation and an activated open conformation. Association with Cdc42-GTP stabilizes the open form of WASp, enabling it to bind to the ARP complex and strongly enhancing this complex's actin-nucleating activity (see Figure 16–34). In this way, activation of Cdc42 increases actin nucleation.

Rac-GTP also activates WASp family members, as well as activating the crosslinking activity of the gel-forming protein filamin, and inhibiting the contractile activity of the motor protein myosin II, stabilizing the lamellipodia and inhibiting the formation of contractile stress fibers (**Figure 16–98**A).

Rho-GTP has a very different set of targets. Instead of activating the ARP complex to build actin networks, Rho-GTP turns on formin proteins to construct parallel actin bundles. At the same time, Rho-GTP activates a protein kinase that indirectly inhibits the activity of cofilin, leading to actin filament stabilization. The same protein kinase inhibits a phosphatase acting on myosin light chains (see Figure 16–72). The consequent increase in the net amount of myosin light chain phosphorylation increases the amount of contractile myosin motor protein activity in the cell, enhancing the formation of tension-dependent structures such as stress fibers (Figure 16–98B).

In some cell types, Rac-GTP activates Rho, usually with kinetics that are slow compared to Rac's activation of the ARP complex. This enables cells to use the Rac pathway to build a new actin structure while subsequently activating the Rho pathway to induce a contractility that builds up tension in this structure. This occurs, for example, during the formation and maturation of cell-cell contacts. As we will explore in more detail below, the communication between the Rac and Rho pathways also facilitates maintenance of the large-scale differences between the cell front and the cell rear during migration.

Extracellular Signals Can Activate the Three Rho Protein Family Members

The activation of the monomeric GTPases Rho, Rac, and Cdc42 occurs through an exchange of GTP for a tightly bound GDP molecule, catalyzed by guanine nucleotide exchange factors (GEFs). Of the 85 GEFs that have been identified in

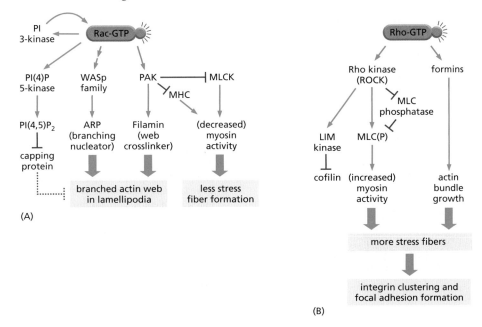

Figure 16–98 The contrasting effects of Rac and Rho activation on actin organization. (A) Activation of the small GTPase Rac leads to actin nucleation by the ARP complex and other alterations in actin accessory proteins that tend to favor the formation of actin networks, as in lamellipodia. Several different pathways contribute independently. Rac-GTP activates members of the WASp protein family, which in turn activate actin nucleation and branched web formation by the ARP complex. In a parallel pathway, Rac-GTP activates a protein kinase, PAK, which has several targets including the web-forming crosslinker filamin, which is activated by phosphorylation, and the myosin light chain kinase (MLCK), which is inhibited by phosphorylation. The resulting decrease in phosphorylation of the myosin regulatory light chain leads to myosin II filament disassembly and a decrease in contractile activity. In some cells, PAK also directly inhibits myosin II activity by phosphorylation of the myosin heavy chain (MHC). Another set of pathways downstream of Rac activation is mediated by phosphoinositide lipid signals. Local creation of PIP₂ [PI(4,5)P₂] may help to reduce the activity of capping protein, to further aid actin polymerization. Activation of PI 3-kinase, which generates PIP₃ from PIP₂, leads to further activation of Rac itself via a positive feedback loop.

(B) Activation of the related GTPase Rho leads to nucleation of actin filaments by formins and increases contraction by myosin II, promoting the formation of contractile actin bundles such as stress fibers. Activation of myosin II by Rho requires a Rho-dependent protein kinase called Rock. This kinase inhibits the phosphatase that removes the activating phosphate groups from myosin II light chains (MLC); it may also directly phosphorylate the myosin light chains in some cell types. Rock also activates other protein kinases, such as LIM kinase, which in turn contributes to the formation of stable contractile actin filament bundles by inhibiting the actin depolymerizing factor cofilin. A similar signaling pathway is important for forming the contractile ring necessary for cytokinesis (see Figure 17–52).

the human genome, some are specific for an individual Rho family GTPase, whereas others seem to act on all three family members. The number of GEFs exceeds the number of Rho GTPases that they regulate because different GEFs are restricted to specific tissues and even specific subcellular locations, and they are sensitive to distinct kinds of regulatory inputs. Various cell-surface receptors activate GEFs. An example is the Eph receptor tyrosine kinase involved in neurite growth cone guidance, which is discussed in detail in Chapter 15. Interestingly, several of the Rho family GEFs associate with the growing ends of microtubules by binding to one of the +TIPs. This provides a connection between the dynamics of the microtubule cytoskeleton and the large-scale organization of the actin cytoskeleton, which is important for the overall integration of cell shape and movement.

The Rho family GTPases are also primary determinants of cell polarity in budding yeast, where extensive genetic analyses have increased our understanding of the general mechanisms involved. On starvation, yeasts, like many other unicellular organisms, sporulate. But sporulation can occur only in diploid budding yeast cells, whereas budding yeasts mainly proliferate as haploid cells. A starving haploid individual must therefore locate a partner of the opposite mating type, woo it, and mate with it before sporulating. Yeast cells are unable to swim and, instead, reach their mates by polarized growth. The haploid form of budding yeast comes in two mating types, **a** and α, which secrete mating factors known as **a**-factor and α-factor, respectively. These secreted signal molecules act by binding to cell-surface receptors that belong to the G-protein-coupled receptor superfamily (discussed in Chapter 15). One consequence of the binding of α-factor to its receptor is to cause the recipient cell to become polarized, adopting a shape known as a "shmoo" (**Figure 16–99**). In the presence of an α-factor gradient, the **a**-cell shmoo tip is directed toward the highest concentration of the signal molecule, which under normal circumstances would direct it toward an amorous α cell located nearby.

This polarized cell growth requires alignment of the actin cytoskeleton in response to the mating factor signal. When the signal binds to its receptor, the receptor activates Cdc42, which in turn induces assembly of actin filaments at the location closest to the source of the signal. Local activation of Cdc42 is further enhanced by a positive feedback loop, requiring actin-dependent transport of Cdc42 itself as well as its GEF and other signaling components along the newly assembled actin structures toward the site of the signal. Subsequently, actin cables are assembled pointing toward the site of Cdc42 accumulation due to the activation of another Rho family GTPase that in turn stimulates a yeast formin. The actin cables serve as tracks for directed transport and exocytosis of new cell wall material, resulting in the polarized growth of the shmoo tip (**Figure 16–100**).

Haploid budding yeast cells use this same polarization machinery during vegetative growth. To form the bud that will grow out to become a daughter cell, the yeast must direct new plasma membrane and cell wall material primarily to a single site. As with shmoo formation, this requires an initial cytoskeletal polarity, with most actin patches in the growing bud and actin cables oriented along the bud axis. In haploid cells, a new bud site is always constructed immediately adjacent to the previous bud site. In this case, the spatial cues that set up cytoskeletal polarity are intrinsic to the cell, left behind from previous rounds of cell division. Cdc42 is once again involved in transducing the signal from the destined bud site to the cytoskeleton, and most of the proteins involved in the upstream and downstream pathways have been identified through genetic

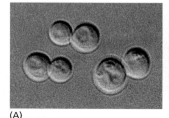
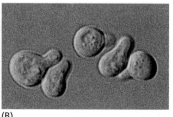

(A)   (B)   (C)

Figure 16–99 Morphological polarization of yeast cells in response to mating factor. (A) Cells of *Saccharomyces cerevisiae* are usually spherical. (B) They become polarized when treated with mating factor from cells of the opposite mating type. The polarized cells are called "shmoos." (C) Al Capp's famous cartoon character, the original Shmoo. (A and B, courtesy of Michael Snyder; C, © 1948 Capp Enterprises, Inc. Used by permission.)

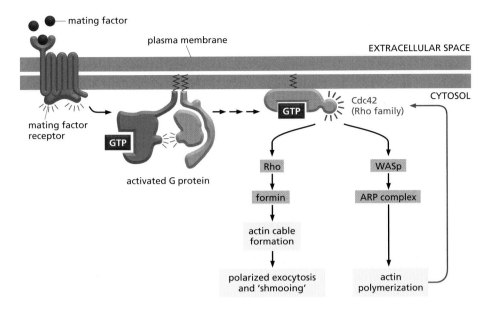

Figure 16–100 The signaling pathway in the yeast mating factor response. The extracellular mating factor binds to a G-protein-coupled receptor in the plasma membrane. Activation of the receptor triggers dissociation of the GTP-bound Gα subunit from a heterotrimeric G-protein (discussed in Chapter 15). This in turn activates the Rho family GTP-binding protein, Cdc42. As in mammalian cells, Cdc42 activates a WASp family protein that activates the ARP complex, leading to local actin nucleation at the site of mating factor binding. The local actin nucleation and filament growth create a positive feedback loop whereby Cdc42 activity is further enhanced. This leads to extensive Rho and formin activation, and finally to actin cable formation, polarized growth, and acquisition of a shmoo shape. In addition, receptor activation triggers other responses through a MAP kinase cascade (discussed in Chapter 15), preparing the haploid cell for mating (not shown).

experiments. Subsequent to their identification in yeast, many of these proteins have been found to have homologs in other organisms, where they are often likewise involved in the establishment of cell polarity.

External Signals Can Dictate the Direction of Cell Migration

Chemotaxis is defined as cell movement in a direction controlled by a gradient of a diffusible chemical. This is a particularly interesting case where external signals trigger the Rho family proteins to set up large-scale cell polarity by influencing the organization of the apparatus required for cell motility, described above. One well-studied example is the chemotactic movement of a class of white blood cells, called *neutrophils,* toward a source of bacterial infection. Receptor proteins on the surface of neutrophils enable them to detect the very low concentrations of the *N*-formylated peptides that are derived from bacterial proteins (only procaryotes begin protein synthesis with *N*-formylmethionine). Using these receptors, neutrophils are guided to bacterial targets by their ability to detect a difference of only 1% in the concentration of these diffusible peptides on one side of the cell versus the other (**Figure 16–101**).

Both in this case and in the similar chemotaxis of *Dictyostelium* amoebae toward a source of cyclic AMP, a local polymerization of actin near the receptors is stimulated when the receptors bind their ligands. This actin polymerization response depends on the monomeric Rho-family GTPases discussed earlier. As in the shmooing yeast (see Figure 16–99), the responding cell extends a protrusion toward the signal. For chemotactic cells, binding of the chemoattractant ligand to its G-protein coupled receptor activates phosphoinositide 3' kinases (PI3Ks), which generates a lipid-based signaling molecule (PI(3,4,5)P₃) that in turn activates the Rac GTPase. Rac then activates the ARP complex and lamellipodial protrusion results (see Figure 16–98). Through an unknown mechanism,

Figure 16–101 Neutrophil polarization and chemotaxis. <GTCG> <TGTA> The pipette tip at the right is leaking a small amount of the peptide formyl-Met-Leu-Phe. Only bacterial proteins have formylated methionine residues, so the human neutrophil recognizes this peptide as the product of a foreign invader (discussed in Chapter 24). The neutrophil quickly extends a new lamellipodium toward the source of the chemoattractant peptide *(top).* It then extends this lamellipodium and polarizes its cytoskeleton so that contractile myosin II is located primarily at the rear, opposite the position of the lamellipodium *(middle).* Finally, the cell crawls toward the source of this peptide *(bottom).* If a real bacterium were the source of the peptide, rather than an investigator's pipette, the neutrophil would engulf the bacterium and destroy it (see also Figure 16–4). (From O.D. Weiner et al., *Nat. Cell Biol.* 1:75–81, 1999. With permission from Macmillan Publishers Ltd.)

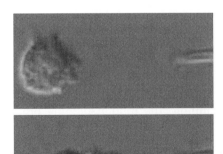

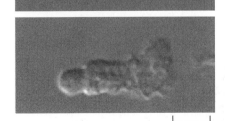

5 μm

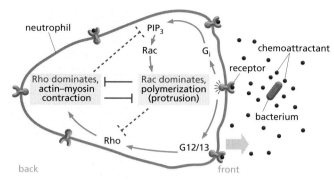

accumulation of the polarized actin web at the leading edge causes further local enhancement of PI3K activity in a positive feedback loop, strengthening the induction of protrusion. The PI(3,4,5)P_3 that activates Rac cannot diffuse far from its site of synthesis, since it is rapidly converted back into PIP$_2$ by a constitutively active lipid phosphatase. At the same time, binding of the chemoattractant ligand to its receptor activates another signaling pathway that turns on Rho and enhances myosin-based contractility. The two processes directly inhibit each other, and as a result, Rac activation dominates in the front of the cell and Rho activation dominates in the rear (**Figure 16–102**). This enables the cell to maintain its functional polarity with protrusion at the leading edge and contraction at the back.

Nondiffusible chemical cues attached to the extracellular matrix or to the surface of cells can also influence the direction of cell migration. When these signals activate receptors, they can cause increased cell adhesion and directed actin polymerization. Most long-distance cell migrations in animals, including neural-crest-cell migration and the travels of neuronal growth cones, depend on a combination of diffusible and non-diffusible signals to steer the locomoting cells or growth cones to their proper destinations (see Figure 15–62).

Communication Between the Microtubule and Actin Cytoskeletons Coordinates Whole-Cell Polarization and Locomotion

To help organize persistent movement in a particular direction cells use their microtubules along with their actin filaments. In many locomoting cells, the position of the centrosome is influenced by the location of protrusive actin polymerization, being found on the forward side of the nucleus. The mechanism of centrosome reorientation is not clear, although there is evidence that the Rho family protein Cdc42 may be involved. It is thought that the activation of receptors on one edge of a cell might not only stimulate actin polymerization there (and therefore local protrusion) but also locally activate dynein-like motor proteins that move the centrosome by pulling on its microtubules. Several effector proteins downstream of Rac and Rho modulate microtubule dynamics directly. For example, a protein kinase activated by Rac can phosphorylate (and therefore inhibit) the tubulin binding protein stathmin (see Panel 16–3, pp. 994–995), destabilizing microtubules, and Rho activation appears to stabilize microtubules.

In turn, microtubule dynamics influence actin rearrangements. The centrosome nucleates a large number of dynamic microtubules, and its repositioning means that many of these microtubules have their plus ends extending from the centrosome into the protrusive region of the cell. The dynamic microtubule plus ends may indirectly modulate local adhesion and also activate the Rac GTPase to further increase actin polymerization in the protrusive region by delivering Rac-GEFs that bind to the +TIPs traveling on growing microtubule ends. The increased concentration of microtubules would thereby encourage further protrusion, creating a positive feedback loop that enables protrusive motility to persist in the same direction for a prolonged period. Regardless of the exact mechanism, the orientation of the centrosome seems to reinforce the polarity information that the actin cytoskeleton receives from the outside world, allowing a sensitive response to weak signals.

Figure 16–102 Signaling during neutrophil polarization. Bacteria that have invaded the human body secrete molecules that are recognized as foreign by the cells of the immune system, including neutrophils. Binding of the bacterial molecules to G-protein-coupled receptors on the neutrophils stimulate directed motility. These receptors are found all over the surface of the cell, but are more likely to be bound to the bacterial ligand at the front. Two distinct signaling pathways contribute to the cell's polarization. At the front of the cell, close to the source of the bacterial signal, stimulation of the Rac pathway leads, via the trimeric G protein G_i, to growth of protrusive actin networks. Second messengers within this pathway are short-lived, so protrusion is limited to the region of the cell closest to the stimulant. The same receptor also stimulates a second signaling pathway, via the trimeric G proteins G12 and G13 (denoted G12/13), that triggers the activation of Rho. The two pathways are mutually antagonistic. Since Rac-based protrusion is active at the front of the cell, Rho is activated only at the rear of the cell, stimulating contraction of the cell rear and assisting directed movement. For a real-life example of the effectiveness of this signaling system, see Figure 16–4.

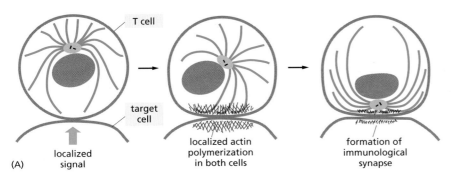

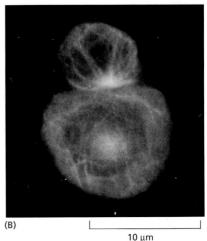

(A)   localized
        signal

        localized actin
        polymerization
        in both cells

        formation of
        immunological
        synapse

(B)

10 μm

Figure 16–103 The polarization of a cytotoxic T cell after target-cell recognition. (A) Changes in the cytoskeleton of a cytotoxic T cell after it has made contact with a target cell. The initial recognition event results in signals that cause actin polymerization in both cells at the site of contact. In the T cell, interactions between the actin-rich contact zone and microtubules emanating from the centrosome result in reorientation of the centrosome, so that the associated Golgi apparatus is directly apposed to the target cell. (B) Immuno-fluorescence micrograph in which both the T cell *(top)* and its target cell *(bottom)* have been stained with an antibody against microtubules. The centrosome and the microtubules radiating from it in the T cell are oriented toward the point of cell–cell contact. In contrast, the microtubule array in the target cell is not polarized. (B, from B. Geiger, D. Rosen and G. Berke, *J. Cell Biol.* 95:137–143, 1982. With permission from The Rockefeller University Press.)

A similar cooperative feedback loop seems to operate in many other instances of cell polarization. A particularly interesting example is the killing of specific target cells by T lymphocytes. These cells are a critical component of the vertebrate's adaptive immune response to infection by viruses. T cells, like neutrophils, use actin-based motility to crawl through the body's tissue and find infected target cells. When a T cell comes into contact with a virus-infected cell and its receptors recognize foreign viral antigens on the surface of the target cell, the same polarization machinery is engaged in a very different way to facilitate killing of the target cell. Rac is activated at the point of cell–cell contact and causes actin polymerization at this site, creating a specialized region of the cortex. This specialized site causes the centrosome to reorient, moving with its microtubules to the zone of T-cell-target contact (**Figure 16–103**). The microtubules, in turn, position the Golgi apparatus right under the contact zone, focusing the killing machinery onto the target cell. The mechanism of killing is discussed in Chapter 25 (see Figure 25–47).

The Complex Morphological Specialization of Neurons Depends on the Cytoskeleton

For our final case study of the ways that the intrinsic properties of the eucaryotic cytoskeleton enable specific and enormously complicated large-scale cell behaviors, we examine the neuron. Neurons begin life in the embryo as unremarkable cells, which use actin-based motility to migrate to specific locations. Once there, however, they send out a series of long specialized processes that will either receive electrical signals *(dendrites)* or transmit electrical signals *(axons)* to their target cells. <TACC> The beautiful and elaborate branching morphology of axons and dendrites enables neurons to form tremendously complex signaling networks, interacting with many other cells simultaneously and making possible the complicated and often unpredictable behavior of the higher animals. Both axons and dendrites (collectively called *neurites)* are filled with bundles of microtubules that are critical to both their structure and their function.

In axons, all the microtubules are oriented in the same direction, with their minus end pointing back toward the cell body and their plus end pointing forward toward the axon terminals (**Figure 16–104**). The microtubules do not reach

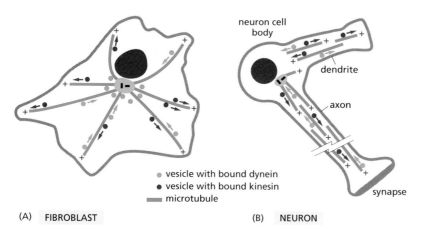

(A)   FIBROBLAST   (B)   NEURON

- vesicle with bound dynein
- vesicle with bound kinesin
- microtubule

Figure 16–104 Microtubule organization in fibroblasts and neurons. (A) In a fibroblast, microtubules emanate outward from the centrosome in the middle of the cell. Vesicles with plus-end-directed kinesin attached move outward, and vesicles with minus-end-directed dynein attached move inward. (B) In a neuron, microtubule organization is more complex. In the axon, all microtubules share the same polarity, with the plus ends pointing outward toward the axon terminus. No one microtubule stretches the entire length of the axon; instead, short overlapping segments of parallel microtubules make the tracks for fast axonal transport. In dendrites, the microtubules are of mixed polarity, with some plus ends pointing outward and some pointing inward.

from the cell body all the way to the axon terminals; each is typically only a few micrometers in length, but large numbers are staggered in an overlapping array. This set of aligned microtubule tracks acts as a highway to transport many specific proteins, protein-containing vesicles, and mRNAs to the axon terminals, where synapses must be constructed and maintained. The longest axon in the human body reaches from the base of the spinal cord to the foot, being up to a meter in length.

Mitochondria, large numbers of specific proteins in transport vesicles, and synaptic vesicle precursors make the long journey in the forward (anterograde) direction. They are carried there by plus-end-directed kinesin-family motor proteins that can move them a meter in as little as two or three days, which is a great improvement over diffusion, which would take approximately several decades to move a mitochondrion this distance. Many members of the kinesin superfamily contribute to this *anterograde axonal transport,* most carrying specific subsets of membrane-enclosed organelles along the microtubules. The great diversity of the kinesin family motor proteins used in axonal transport suggests that they are involved in targeting their cargo to specific structures near the terminus or along the way, as well as in cargo movement. Old components from the axon terminals are carried back to the cell body for degradation and recycling by a *retrograde axonal transport.* This transport occurs along the same set of oriented microtubules, but it relies on cytoplasmic dynein, which is a minus-end-directed motor protein. Retrograde transport is also critical for communicating the presence of growth and survival signals received by the nerve terminus back to the nucleus, in order to influence gene expression.

One form of human peripheral neuropathy, Charcot-Marie-Tooth disease, is caused by a point mutation in a particular kinesin family member that transports synaptic vesicle precursors down the axon. Other kinds of neurodegenerative diseases such as Alzheimer's disease may also be caused in part by disruptions in neuronal trafficking; as pointed out previously, the amyloid precursor protein APP is part of a protein complex that serves as a receptor for kinesin-1 binding to other axonal transport vesicles.

Axonal structure depends on the axonal microtubules, as well as on the contributions of the other two major cytoskeletal systems—actin filaments and intermediate filaments. Actin filaments line the cortex of the axon, just beneath the plasma membrane, and actin-based motor proteins such as myosin V are also abundant in the axon, presumably to help move materials. Neurofilaments, the specialized intermediate filaments of nerve cells, provide the most important structural support in the axon. A disruption in neurofilament structure, or in the cross-linking proteins that attach the neurofilaments to the microtubules and actin filaments distributed along the axon, can result in axonal disorganization and eventually axonal degeneration.

The construction of the elaborate branching architecture of the neuron during embryonic development requires actin-based motility. As mentioned earlier, the tips of growing axons and dendrites extend by means of a *growth cone,* a specialized motile structure rich in actin (**Figure 16–105**). Most neuronal growth cones produce filopodia, and some make lamellipodia as well. The protrusion

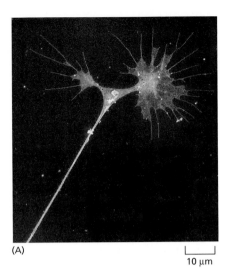 

(A)

10 μm

(B)

10 μm

Figure 16–105 Neuronal growth cones. <AAGA> (A) Scanning electron micrograph of two growth cones at the end of a neurite, put out by a chick sympathetic neuron in culture. Here, a previously single growth cone has recently split into two. Note the many filopodia and the large lamellipodia. The taut appearance of the neurite is due to tension generated by the forward movement of the growth cones, which are often the only firm points of attachment of the axon to the substratum. (B) Scanning electron micrograph of the growth cone of a sensory neuron crawling over the inner surface of the epidermis of a *Xenopus* tadpole. (A, from D. Bray, in Cell Behaviour [R. Bellairs, A. Curtis and G. Dunn, eds.]. Cambridge, UK: Cambridge University Press, 1982; B, from A. Roberts, *Brain Res*. 118:526–530, 1976. With permission from Elsevier.)

and stabilization of growth-cone filopodia are exquisitely sensitive to environmental cues. Some cells secrete soluble proteins such as netrin to attract or repel growth cones. These modulate the structure and motility of the growth cone cytoskeleton by altering the balance between Rac activity and Rho activity at the leading edge (see Figure 15–62). In addition, there are fixed guidance markers along the way, attached to the extracellular matrix or to the surfaces of cells. When a filopodium encounters such a "guidepost" in its exploration, it quickly forms adhesive contacts. It is thought that a myosin-dependent collapse of the actin meshwork in the unstabilized part of the growth cone then causes the developing axon to turn toward the guidepost.

Thus, a complex combination of positive and negative signals, both soluble and insoluble, accurately guide the growth cone to its final destination. Microtubules then reinforce the directional decisions made by the actin-rich protrusive structures at the leading edge of the growth cone. Microtubules from the axonal parallel array just behind the growth cone are constantly growing into the growth cone and shrinking back by dynamic instability. Adhesive guidance signals are somehow relayed to the dynamic microtubule ends, so that microtubules growing in the correct direction are stabilized against disassembly. In this way, a microtubule-rich axon is left behind, marking the path that the growth cone has traveled.

Dendrites are generally much shorter projections than axons, and they receive synaptic inputs rather than being specialized for sending signals like axons. The microtubules in dendrites all lie parallel to one another but their polarities are mixed, with some pointing their plus ends toward the dendrite tip, while others point back toward the cell body. Nevertheless, dendrites also form as the result of growth-cone activity. Therefore, it is the growth cones at the tips of axons and dendrites that create the intricate, highly individual morphology of each mature neuronal cell (**Figure 16–106**).

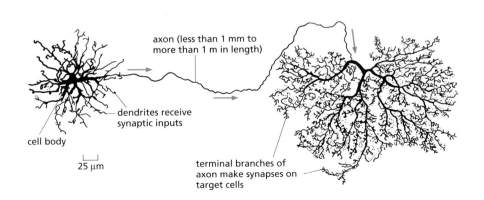

axon (less than 1 mm to more than 1 m in length)

dendrites receive synaptic inputs

cell body

25 μm

terminal branches of axon make synapses on target cells

Figure 16–106 The complex architecture of a vertebrate neuron. The neuron shown is from the retina of a monkey. The arrows indicate the direction of travel of the electrical signal along the axon. The longest and largest neurons in the human body extend for a distance of about 1 m (1 million μm), from the base of the spinal cord to the tip of the big toe, and have an axon diameter of 15 μm. (Adapted from B.B. Boycott, in Essays on the Nervous System [R. Bellairs and E.G. Gray, eds.]. Oxford, UK: Clarendon Press, 1974.)

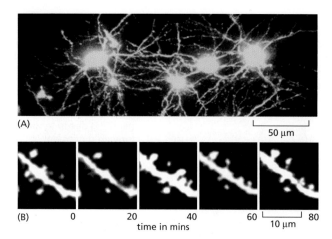

(A)

50 μm

(B)   0   20   40   60   80
            time in mins   10 μm

Figure 16–107 Rapid changes in dendrite structure within a living mouse brain. (A) Image of cortical neurons in a transgenic mouse that has been engineered to express green fluorescent protein in a small fraction of its brain cells. Changes in these brain neurons and their projections can be followed for months using highly sensitive fluorescence microscopy. To make this possible, the mouse is subjected to an operation that introduces a small transparent window through its skull, and it is anesthetized each time that an image is recorded. (B) A single dendrite, imaged over the period of 80 minutes, demonstrates that dendrites are constantly sending out and retracting tiny actin-dependent protrusions to create the dendritic spines that receive the vast majority of excitatory synapses from axons in the brain. Those spines that become stabilized and persist for months are thought to be important for brain function, and may be involved in long-term memory. (Courtesy of Karel Svoboda.)

Although the neurons of the central nervous system are long-lived cells, they are by no means static. Synapses are constantly being created, strengthened, weakened, and eliminated as the brain learns, evaluates, and forgets. High-resolution imaging of the structure of neurons in the brains of adult mice has revealed that neuronal morphology is undergoing constant rearrangement as synapses are forged and broken (**Figure 16–107**). These actin-dependent rearrangements are thought to be critical in learning and long-term memory. In this way, the cytoskeleton provides the engine for construction of the entire nervous system, as well as producing the supporting structures that strengthen, stabilize, and maintain its parts.

Summary

Two distinct types of specialized structures in eucaryotic cells are formed from highly ordered arrays of motor proteins that move on stabilized filament tracks. The myosin–actin system of the sarcomere powers the contraction of various types of muscle, including skeletal, smooth, and cardiac muscle. The dynein–microtubule system of the axoneme powers the beating of cilia and the undulations of flagella.

Whole-cell movements and the large-scale shaping and structuring of cells require the coordinated activities of all three basic filament systems along with a large variety of cytoskeletal accessory proteins, including motor proteins. During cell division, the functions of the microtubule-based mitotic spindle require spatial and temporal cooperation between dynamic cytoskeletal filaments, active molecular motor proteins, and a wide variety of accessory factors. Cell crawling—a widespread behavior important in embryonic development and also in wound healing, tissue maintenance, and immune system function in the adult animal—is another prime example of such complex, coordinated cytoskeletal action. For a cell to crawl, it must generate and maintain an overall structural polarity, which is influenced by external cues. In addition, the cell must coordinate protrusion at the leading edge (by assembly of new actin filaments), adhesion of the newly protruded part of the cell to the substratum, forces generated by molecular motors to bring the cell body forward.

Complex cells, such as neurons, require the coordinated assembly of microtubules, neurofilaments (neuronal intermediate filaments), and actin filaments, as well as the actions of dozens of highly specialized molecular motors that transport subcellular components to their appropriate destinations.

PROBLEMS

Which statements are true? Explain why or why not.

16–1   The role of ATP hydrolysis in actin polymerization is similar to the role of GTP hydrolysis in tubulin polymerization: both serve to weaken the bonds in the polymer and thereby promote depolymerization.

16–2   In most animal cells, minus end-directed microtubule motors deliver their cargo to the periphery of the cell, whereas plus end-directed microtubule motors deliver their cargo to the interior of the cell.

16–3   Motor neurons trigger action potentials in muscle cell membranes that open voltage-sensitive Ca^{2+} channels in T-tubules, allowing extracellular Ca^{2+} to enter the cytosol, bind to troponin C, and initiate rapid muscle contraction.

Discuss the following problems.

16–4 At 1.4 mg/mL pure tubulin, microtubules grow at a rate of about 2 μm/min. At this growth rate how many αβ-tubulin dimers (8 nm in length) are added to the ends of a microtubule each second?

16–5 A solution of pure αβ-tubulin dimers is thought to nucleate microtubules by forming a linear protofilament about seven dimers in length. At that point, the probabilities that the next αβ-dimer will bind laterally or to the end of the protofilament are about equal. The critical event for microtubule formation is thought to be the first lateral association (**Figure Q16–1**). How does lateral association promote the subsequent rapid formation of a microtubule?

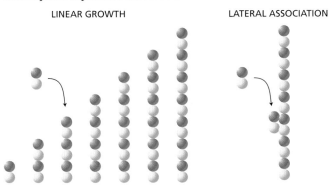

LINEAR GROWTH LATERAL ASSOCIATION

Figure Q16–1 Model for microtubule nucleation by pure αβ-tubulin dimers (Problem 16–5).

16–6 How does a centrosome "know" when it has found the center of the cell?

16–7 The concentration of actin in cells is 50–100 times greater than the critical concentration observed for pure actin in a test tube. How is this possible? What prevents the actin subunits in cells from polymerizing into filaments? Why is it advantageous to the cell to maintain such a large pool of actin subunits?

16–8 The movements of single motor-protein molecules can be analyzed directly. Using polarized laser light, it is possible to create interference patterns that exert a centrally directed force, ranging from zero at the center to a few piconewtons at the periphery (about 200 nm from the center). Individual molecules that enter the interference pattern are rapidly pushed to the center, allowing them to be captured and moved at the experimenter's discretion.

Using such "optical tweezers," single kinesin molecules can be positioned on a microtubule that is fixed to a coverslip. Although a single kinesin molecule cannot be seen optically, it can be tagged with a silica bead and tracked indirectly by following the bead (**Figure Q16–2A**). In the absence of ATP, the kinesin molecule remains at the center of the interference pattern, but with ATP it moves toward the plus end of the microtubule. As kinesin moves along the microtubule, it encounters the force of the interference pattern, which simulates the load kinesin carries during its actual function in the cell. Moreover, the pressure against the silica bead counters the effects of Brownian (thermal) motion, so that the position of the bead more accurately reflects the position of the kinesin molecule on the microtubule.

Traces of the movements of a kinesin molecule along a microtubule are shown in Figure Q16–2B.

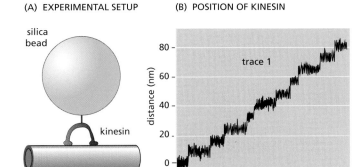

(A) EXPERIMENTAL SETUP (B) POSITION OF KINESIN

Figure Q16–2 Movement of kinesin along a microtubule (Problem 16–8). (A) Experimental setup with kinesin linked to a silica bead, moving along a microtubule. (B) Position of kinesin (as visualized by position of silica bead) relative to center of interference pattern, as a function of time of movement along the microtubule. The jagged nature of the trace results from Brownian motion of the bead.

A. As shown in Figure Q16–2B, all movement of kinesin is in one direction (toward the plus end of the microtubule). What supplies the free energy needed to ensure a unidirectional movement along the microtubule?
B. What is the average rate of movement of kinesin along the microtubule?
C. What is the length of each step that a kinesin takes as it moves along a microtubule?
D. From other studies it is known that kinesin has two globular domains that each can bind to β-tubulin, and that kinesin moves along a single protofilament in a microtubule. In each protofilament the β-tubulin subunit repeats at 8-nm intervals. Given the step length and the interval between β-tubulin subunits, how do you suppose a kinesin molecule moves along a microtubule?
E. Is there anything in the data in Figure Q16–2B that tells you how many ATP molecules are hydrolyzed per step?

16–9 How is the unidirectional motion of a lamellipodium maintained?

16–10 Detailed measurements of sarcomere length and tension during isometric contraction in striated muscle provided crucial early support for the sliding filament model of muscle contraction. Based on your understanding of the sliding filament model and the structure of a sarcomere, propose a molecular explanation for the relationship of tension to sarcomere length in the portions of **Figure Q16–3** marked I, II, III, and IV. (In this muscle, the length of the myosin filament is 1.6 μm and the lengths of the actin thin filaments that project from the Z discs are 1.0 μm.)

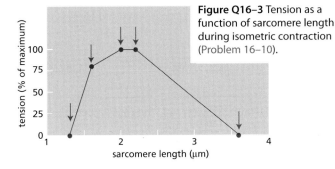

Figure Q16–3 Tension as a function of sarcomere length during isometric contraction (Problem 16–10).

REFERENCES

General

Bray D (2001) Cell Movements: From Molecules to Motility, 2nd ed. New York: Garland Science.

Howard J (2001) Mechanics of Motor Proteins and the Cytoskeleton. Sunderland, MA: Sinauer.

The Self-Assembly and Dynamic Structure of Cytoskeletal Filaments

Dogterom M & Yurke B (1997) Measurement of the force-velocity relation for growing microtubules. *Science* 278:856–860.

Garner EC, Campbell CS & Mullins RD (2004) Dynamic instability in a DNA-segregating prokaryotic actin homolog. *Science* 306:1021–1025.

Helfand BT, Chang L & Goldman RD (2003) The dynamic and motile properties of intermediate filaments. *Annu Rev Cell Dev Biol* 19:445–467.

Hill TL & Kirschner MW (1982) Bioenergetics and kinetics of microtubule and actin filament assembly-disassembly. *Int Rev Cytol* 78:1–125.

Hotani H & Horio T (1988) Dynamics of microtubules visualized by darkfield microscopy: treadmilling and dynamic instability. *Cell Motil Cytoskeleton* 10:229–236.

Jones LJ, Carballido-Lopez R & Errington J (2001) Control of cell shape in bacteria: helical, actin-like filaments in *Bacillus subtilis*. *Cell* 104:913–922.

Kerssemakers JW, Munteanu EL, Laan L et al (2006) Assembly dynamics of microtubules at molecular resolution. *Nature* 442:709–712.

Luby-Phelps K (2000) Cytoarchitecture and physical properties of cytoplasm: volume, viscosity, diffusion, intracellular surface area. *Int Rev Cytol* 192:189–221.

Mitchison T & Kirschner M (1984) Dynamic instability of microtubule growth. *Nature* 312:237–242.

Mitchison TJ (1995) Evolution of a dynamic cytoskeleton. *Philos Trans R Soc Lond B Biol Sci* 349:299–304.

Mukherjee A & Lutkenhaus J (1994) Guanine nucleotide-dependent assembly of FtsZ into filaments. *J Bacteriol* 176:2754–2758.

Oosawa F & Asakura S (1975) Thermodynamics of the polymerization of protein. New York: Academic Press, pp 41–55, pp 90–108.

Pauling L (1953) Aggregation of Globular Proteins. *Discuss Faraday Soc* 13:170–176

Rodionov VI & Borisy GG (1997) Microtubule treadmilling *in vivo*. *Science*. 275:215–218.

Shih YL & Rothfield L (2006) The bacterial cytoskeleton. *Microbiol Mol Biol Rev* 70:729–754.

Theriot JA (2000) The polymerization motor. *Traffic* 1:19–28.

How Cells Regulate Their Cytoskeletal Filaments

Aldaz H, Rice LM, Stearns T & Agard DA (2005) Insights into microtubule nucleation from the crystal structure of human gamma-tubulin. *Nature* 435:523–527

Bretscher A, Chambers D, Nguyen R & Reczek D (2000) ERM-Merlin and EBP50 protein families in plasma membrane organization and function. *Annu Rev Cell Dev Biol* 16:113–143.

Doxsey S, McCollum D & Theurkauf W (2005) Centrosomes in cellular regulation. *Annu Rev Cell Dev Biol* 21:411–434.

Garcia ML & Cleveland DW (2001) Going new places using an old MAP: tau, microtubules and human neurodegenerative disease. *Curr Opin Cell Biol* 13:41–48.

Holy TE, Dogterom M, Yurke B & Leibler S (1997) Assembly and positioning of microtubule asters in microfabricated chambers. *Proc Natl Acad Sci USA* 94:6228–6231.

Mullins RD, Heuser JA & Pollard TD (1998) The interaction of Arp2/3 complex with actin: nucleation, high affinity pointed end capping, and formation of branching networks of filaments. *Proc Natl Acad Sci USA* 95:6181–6186.

Robinson RC, Turbedsky K, Kaiser DA et al (2001) Crystal structure of Arp2/3 complex. *Science* 294:1679–1684.

Quarmby L (2000) Cellular Samurai: katanin and the severing of microtubules. *J Cell Sci* 113:2821–2827.

Stearns T & Kirschner M (1994) *In vitro* reconstitution of centrosome assembly and function: the central role of gamma-tubulin. *Cell* 76:623–637.

Wiese C & Zheng Y (2006) Microtubule nucleation: gamma-tubulin and beyond. *J Cell Sci* 119:4143–4153.

Zheng Y, Wong ML, Alberts B & Mitchison T (1995) Nucleation of microtubule assembly by a gamma-tubulin-containing ring complex. *Nature* 378:578–583.

Zigmond SH (2004) Formin-induced nucleation of actin filaments. *Curr Opin Cell Biol* 16:99–105.

Molecular Motors

Burgess SA, Walker ML, Sakakibara H et al (2003) Dynein structure and power stroke. *Nature* 421:715–718.

Hirokawa N (1998) Kinesin and dynein superfamily proteins and the mechanism of organelle transport. *Science* 279:519–526.

Howard J, Hudspeth AJ & Vale RD (1989) Movement of microtubules by single kinesin molecules. *Nature* 342:154–158.

Howard J (1997) Molecular motors: structural adaptations to cellular functions. *Nature* 389:561–567.

Rayment I, Rypniewski WR, Schmidt-Base K et al (1993) Three-dimensional structure of myosin subfragment-1: a molecular motor. *Science* 261:50–58.

Reck-Peterson SL, Yildiz A, Carter AP et al (2006) Single-molecule analysis of dynein processivity and stepping behavior. *Cell* 126:335–348.

Rice S, Lin AW, Safer D et al (1999) A structural change in the kinesin motor protein that drives motility. *Nature* 402:778–784.

Richards TA & Cavalier-Smith T (2005) Myosin domain evolution and the primary divergence of eukaryotes. *Nature* 436:1113–1118.

Svoboda K, Schmidt CF, Schnapp BJ & Block SM (1993) Direct observation of kinesin stepping by optical trapping interferometry. *Nature* 365:721–727.

Vikstrom KL & Leinwand LA (1996) Contractile protein mutations and heart disease. *Curr Opin Cell Biol* 8:97–105.

Wells AL, Lin AW, Chen LQ et al (1999) Myosin VI is an actin-based motor that moves backwards. *Nature* 401:505–508.

Yildiz A, Forkey JN, McKinney SA et al (2003) Myosin V walks hand-over-hand: single fluorophore imaging with 1.5-nm localization. *Science* 300:2061–2065.

Yildiz A & Selvin PR (2005) Kinesin: walking, crawling or sliding along? *Trends Cell Biol* 15:112–120.

The Cytoskeleton and Cell Behavior

Abercrombie M (1980) The crawling movement of metazoan cells. *Proc Roy Soc B* 207:129–147.

Cooke R (2004) The sliding filament model: 1972–2004. *J Gen Physiol* 123:643–656.

Dent EW & Gertler FB (2003) Cytoskeletal dynamics and transport in growth cone motility and axon guidance. *Neuron* 40:209–227.

Lauffenburger DA & Horwitz AF (1996) Cell migration: a physically integrated molecular process. *Cell* 84:359–369.

Lo CM, Wang HB, Dembo M & Wang YL (2000) Cell movement is guided by the rigidity of the substrate. *Biophys J* 79:144–152.

Madden K & Snyder M (1998) Cell polarity and morphogenesis in budding yeast. *Annu Rev Microbiol* 52:687–744.

Ridley AJ, Schwartz MA, Burridge K et al (2003) Cell migration: integrating signals from front to back. *Science* 302:1704–1709.

Rafelski SM & Theriot JA (2004) Crawling toward a unified model of cell motility: spatial and temporal regulation of actin dynamics. *Annu Rev Biochem* 73:209–239.

Parent CA & Devreotes PN (1999) A cell's sense of direction. *Science* 284:765–770.

Pollard TD & Borisy GG (2003) Cellular motility driven by assembly and disassembly of actin filaments. *Cell* 112:453–465.

Purcell EM (1977) Life at low Reynolds' number. *Am J Phys* 45:3–11.

Wittmann T, Hyman A & Desai A (2001) The spindle: a dynamic assembly of microtubules and motors. *Nature Cell Biol* 3:E28–E34.

Chapter 17

The Cell Cycle

17

The only way to make a new cell is to duplicate a cell that already exists. This simple fact, first established in the middle of the nineteenth century, carries with it a profound message for the continuity of life. All living organisms, from the unicellular bacterium to the multicellular mammal, are products of repeated rounds of cell growth and division extending back in time to the beginnings of life on Earth over three billion years ago.

A cell reproduces by performing an orderly sequence of events in which it duplicates its contents and then divides in two. This cycle of duplication and division, known as the **cell cycle**, is the essential mechanism by which all living things reproduce. In unicellular species, such as bacteria and yeasts, each cell division produces a complete new organism. In multicellular species, long and complex sequences of cell divisions are required to produce a functioning organism. Even in the adult body, cell division is usually needed to replace cells that die. In fact, each of us must manufacture many millions of cells every second simply to survive: if all cell division were stopped—by exposure to a very large dose of x-rays, for example—we would die within a few days.

The details of the cell cycle vary from organism to organism and at different times in an organism's life. Certain characteristics, however, are universal. The minimum set of processes that a cell has to perform are those that allow it to accomplish its most fundamental task: the passing on of its genetic information to the next generation of cells. To produce two genetically identical daughter cells, the DNA in each chromosome must first be faithfully replicated to produce two complete copies, and the replicated chromosomes must then be accurately distributed *(segregated)* to the two daughter cells, so that each receives a copy of the entire genome (**Figure 17–1**).

Eucaryotic cells have evolved a complex network of regulatory proteins, known as the *cell-cycle control system*, that governs progression through the cell cycle. The core of this system is an ordered series of biochemical switches that initiate the main events of the cycle, including chromosome duplication and segregation. In most cells, additional layers of regulation enhance the fidelity of cell division and allow the control system to respond to various signals from both inside and outside the cell. Inside the cell, the control system monitors progression through the cell cycle and delays later events until earlier events have been completed. It does not permit preparations for the segregation of duplicated chromosomes, for example, until DNA replication is complete. The control system also monitors conditions outside the cell. In a multicellular animal, the system is highly responsive to signals from other cells, stimulating cell division when more cells are needed and blocking it when they are not. The cell-cycle control system therefore has a central role in regulating cell numbers in the tissues of the body. When the system malfunctions, excessive cell divisions can result in cancer.

In addition to duplicating their genome, most cells also duplicate their other organelles and macromolecules; otherwise, daughter cells would get smaller with each division. To maintain their size, dividing cells must coordinate their growth (that is, their increase in cell mass) with their division.

This chapter describes the various events of the cell cycle and how they are controlled and coordinated. We begin with a brief overview of the cell cycle. We then describe the cell-cycle control system and explain how it triggers the

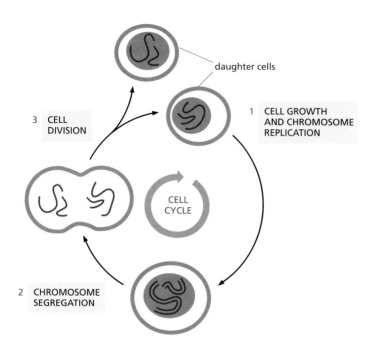

daughter cells

3 CELL DIVISION

1 CELL GROWTH AND CHROMOSOME REPLICATION

CELL CYCLE

2 CHROMOSOME SEGREGATION

Figure 17–1 The cell cycle. The division of a hypothetical eucaryotic cell with two chromosomes is shown to illustrate how two genetically identical daughter cells are produced in each cycle. Each of the daughter cells will often continue to divide by going through additional cell cycles.

different events of the cycle. We next consider in detail the major stages of the cell cycle, in which the chromosomes are duplicated and then segregated into the two daughter cells. Finally, we consider how extracellular signals govern the rates of cell growth and division and how these two processes are coordinated.

OVERVIEW OF THE CELL CYCLE

We begin this section with a brief description of the four *phases* of the eucaryotic cell cycle. We then consider some of the methods and model cell systems used to study the cell cycle.

The Eucaryotic Cell Cycle Is Divided into Four Phases

The most basic function of the cell cycle is to duplicate accurately the vast amount of DNA in the chromosomes and then segregate the copies precisely into two genetically identical daughter cells. These processes define the two major phases of the cell cycle. Chromosome duplication occurs during *S phase* (S for DNA *s*ynthesis), which requires 10–12 hours and occupies about half of the cell-cycle time in a typical mammalian cell. After S phase, chromosome segregation and cell division occur in *M phase* (M for *m*itosis), which requires much less time (less than an hour in a mammalian cell). M phase comprises two major events: nuclear division, or *mitosis*, during which the copied chromosomes are distributed into a pair of daughter nuclei; and cytoplasmic division, or *cytokinesis*, when the cell itself divides in two (**Figure 17–2**).

At the end of S phase, the DNA molecules in each pair of duplicated chromosomes are intertwined and held tightly together by specialized protein linkages. Early in mitosis at a stage called *prophase*, the two DNA molecules are gradually disentangled and condensed into pairs of rigid and compact rods called **sister chromatids**, which remain linked together by *sister-chromatid cohesion*. When the nuclear envelope disassembles later in mitosis, the sister chromatid pairs become attached to the *mitotic spindle*, a giant bipolar array of microtubules (discussed in Chapter 16). Sister chromatids are attached to opposite poles of the spindle, and, eventually, all sisters align at the spindle equator in a stage called *metaphase*. The destruction of sister-chromatid cohesion at the start of *anaphase* separates the sister chromatids, which are pulled to opposite poles of the spindle. The spindle is then disassembled, and the segregated

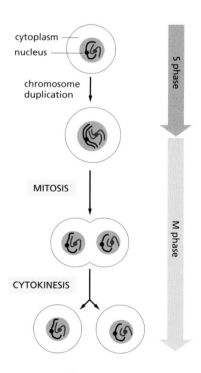

cytoplasm

nucleus

chromosome duplication

S phase

MITOSIS

M phase

CYTOKINESIS

Figure 17–2 The major events of the cell cycle. The major chromosomal events of the cell cycle occur in S phase, when the chromosomes are duplicated, and M phase, when the duplicated chromosomes are segregated into a pair of daughter nuclei (in mitosis), after which the cell itself divides into two (cytokinesis).

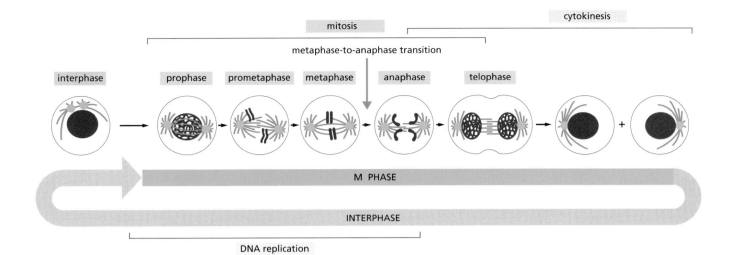

Figure 17–3 **The events of eucaryotic cell division as seen under a microscope.** The easily visible processes of nuclear division (mitosis) and cell division (cytokinesis), collectively called M phase, typically occupy only a small fraction of the cell cycle. The other, much longer, part of the cycle is known as interphase, which includes S phase and the gap phases (discussed in text). The five stages of mitosis are shown: an abrupt change in the biochemical state of the cell occurs at the transition from metaphase to anaphase. A cell can pause in metaphase before this transition point, but once it passes this point, the cell carries on to the end of mitosis and through cytokinesis into interphase.

chromosomes are packaged into separate nuclei at *telophase.* Cytokinesis then cleaves the cell in two, so that each daughter cell inherits one of the two nuclei (**Figure 17–3**). <TACT> <TCAA>

Most cells require much more time to grow and double their mass of proteins and organelles than they require to duplicate their chromosomes and divide. Partly to allow more time for growth, most cell cycles have extra *gap phases*—a **G$_1$ phase** between M phase and S phase and a **G$_2$ phase** between S phase and mitosis. Thus, the eucaryotic cell cycle is traditionally divided into four sequential phases: G$_1$, S, G$_2$, and M. G$_1$, S, and G$_2$ together are called **interphase** (**Figure 17–4**, and see Figure 17–3). In a typical human cell proliferating in culture, interphase might occupy 23 hours of a 24-hour cycle, with 1 hour for M phase. Cell growth occurs throughout the cell cycle, except during mitosis.

The two gap phases are more than simple time delays to allow cell growth. They also provide time for the cell to monitor the internal and external environment to ensure that conditions are suitable and preparations are complete before the cell commits itself to the major upheavals of S phase and mitosis. The G$_1$ phase is especially important in this respect. Its length can vary greatly depending on external conditions and extracellular signals from other cells. If extracellular conditions are unfavorable, for example, cells delay progress through G$_1$ and may even enter a specialized resting state known as G$_0$ (G zero), in which they can remain for days, weeks, or even years before resuming proliferation. Indeed, many cells remain permanently in G$_0$ until they or the organism dies. If extracellular conditions are favorable and signals to grow and divide are present, cells in early G$_1$ or G$_0$ progress through a commitment point near the end of G$_1$ known as **Start** (in yeasts) or the **restriction point** (in mammalian cells). We will use the term Start for both yeast and animal cells. After passing this point, cells are committed to DNA replication, even if the extracellular signals that stimulate cell growth and division are removed.

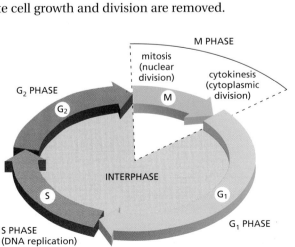

Figure 17–4 **The four phases of the cell cycle.** In most cells, gap phases separate the major events of S phase and M phase. G$_1$ is the gap between M phase and S phase, while G$_2$ is the gap between S phase and M phase.

Cell-Cycle Control Is Similar in All Eucaryotes

Some features of the cell cycle, including the time required to complete certain events, vary greatly from one cell type to another, even in the same organism. The basic organization of the cycle, however, is essentially the same in all eucaryotic cells, and all eucaryotes appear to use similar machinery and control mechanisms to drive and regulate cell-cycle events. The proteins of the cell-cycle control system, for example, first appeared over a billion years ago. Remarkably, they have been so well conserved over the course of evolution that many of them function perfectly when transferred from a human cell to a yeast cell. We can therefore study the cell cycle and its regulation in a variety of organisms and use the findings from all of them to assemble a unified picture of how eucaryotic cells divide. In the rest of this section, we briefly review the three eucaryotic systems most commonly used to study cell-cycle organization and control: yeasts, animal embryos, and cultured mammalian cells.

Cell-Cycle Control Can Be Dissected Genetically by Analysis of Yeast Mutants

Yeasts are tiny, single-celled fungi, with a cell-cycle control system remarkably similar to our own. Two species are generally used in studies of the cell cycle. The **fission yeast** *Schizosaccharomyces pombe* is named after the African beer it is used to produce. It is a rod-shaped cell that grows by elongation at its ends. Division occurs when a septum, or cell plate, forms midway along the rod (**Figure 17–5A**). The **budding yeast** *Saccharomyces cerevisiae* is used by both brewers and bakers. It is an oval cell that divides by forming a bud, which first appears during G_1 and grows steadily until it separates from the mother cell after mitosis (Figure 17–5B).

Despite their outward differences, the two yeast species share many features that are extremely useful for genetic studies. They reproduce almost as rapidly as bacteria and have a genome size less than 1% that of a mammal. They are amenable to rapid molecular genetic manipulation, in which genes can be deleted, replaced, or altered. Most importantly, they have the ability to proliferate in a *haploid* state, with only a single copy of each gene present in the cell. When cells are haploid, it is easy to isolate and study mutations that inactivate a gene, because we avoid the complication of having a second copy of the gene in the cell.

Many important discoveries about cell-cycle control have come from systematic searches for mutations in yeasts that inactivate genes encoding essential components of the cell-cycle control system. The genes affected by some of

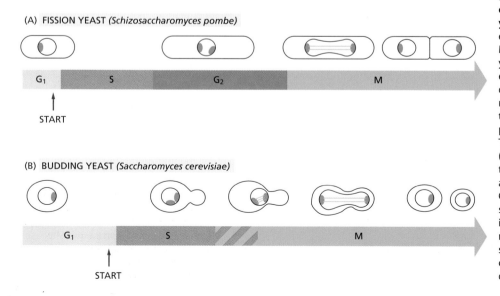

(A) FISSION YEAST *(Schizosaccharomyces pombe)*

G_1 S G_2 M

START

(B) BUDDING YEAST *(Saccharomyces cerevisiae)*

G_1 S   M

START

Figure 17–5 A comparison of the cell cycles of fission yeasts and budding yeasts. (A) The fission yeast has a typical eucaryotic cell cycle with G_1, S, G_2, and M phases. The nuclear envelope of the yeast cell, unlike that of a higher eucaryotic cell, does not break down during M phase. The microtubules of the mitotic spindle *(light green)* form inside the nucleus and are attached to spindle pole bodies *(dark green)* at its periphery. The cell divides by forming a partition (known as the cell plate) and splitting in two. (B) The budding yeast has normal G_1 and S phases but does not have a normal G_2 phase. Instead, a microtubule-based spindle begins to form late in S phase; as in fission yeasts, the nuclear envelope remains intact during mitosis, and the spindle forms within the nucleus. In contrast with a fission yeast cell, the cell divides by budding.

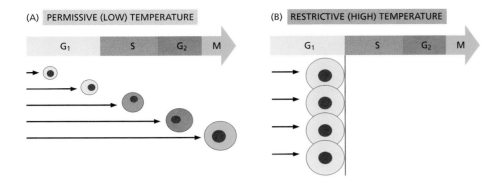

(A) PERMISSIVE (LOW) TEMPERATURE

(B) RESTRICTIVE (HIGH) TEMPERATURE

Figure 17–6 The behavior of a temperature-sensitive Cdc mutant. (A) At the permissive (low) temperature, the cells divide normally and are found in all phases of the cycle (the phase of the cell is indicated by its color). (B) On warming to the restrictive (high) temperature, at which the mutant gene product functions abnormally, the mutant cells continue to progress through the cycle until they come to the specific step that they are unable to complete (initiation of S phase, in this example). Because the Cdc mutants still continue to grow, they become abnormally large. By contrast, non-Cdc mutants, if deficient in a process that is necessary throughout the cycle for biosynthesis and growth (such as ATP production), halt haphazardly at any stage of the cycle—depending on when their biochemical reserves run out (not shown).

these mutations are known as **cell-division-cycle genes**, or **Cdc genes**. Many of these mutations cause cells to arrest at a specific point in the cell cycle, suggesting that the normal gene product is required to get the cell past this point.

A mutant that cannot complete the cell cycle, however, cannot be propagated. Thus, Cdc mutants can be selected and maintained only if their phenotype is *conditional*—that is, if the mutant gene product fails to function only in certain specific conditions. Most conditional cell-cycle mutations are *temperature-sensitive mutations*, in which the mutant protein fails to function at high temperatures but functions well enough to allow cell division at low temperatures. A temperature-sensitive Cdc mutant can be propagated at a low temperature (the *permissive condition*) and then raised to a higher temperature (the *restrictive condition*) to switch off the function of the mutant gene. At the higher temperature, cells continue through the cell cycle until they reach the point where the function of the mutant gene is required for further progress, and at this point they halt (**Figure 17–6**). In budding yeasts, we can detect a uniform cell-cycle arrest of this type by just looking at the cells: the presence or absence of a bud, and bud size, indicate the point in the cycle at which the mutant is arrested (**Figure 17–7**).

Cell-Cycle Control Can Be Analyzed Biochemically in Animal Embryos

The biochemical features of the cell cycle are readily analyzed in the giant fertilized eggs of many animals, which carry large stockpiles of the proteins needed for cell division. The egg of the frog *Xenopus*, for example, is over 1 mm in diameter and contains 100,000 times more cytoplasm than an average cell in the human body (**Figure 17–8**). Fertilization of the *Xenopus* egg triggers an astonishingly rapid sequence of cell divisions, called *cleavage divisions*, in which the single giant cell divides, without growing, to generate an embryo containing

0.5 mm

Figure 17–8 A mature Xenopus egg, ready for fertilization. The pale spot near the top shows the site of the nucleus, which has displaced the *brown* pigment in the surface layer of the egg cytoplasm. Although this cannot be seen in the picture, the nuclear envelope has broken down during the process of egg maturation. (Courtesy of Tony Mills.)

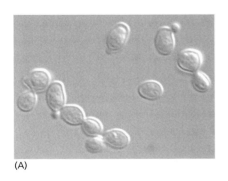

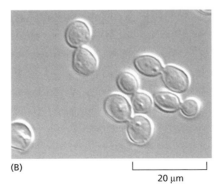

(A)

(B)

20 μm

Figure 17–7 The morphology of budding yeast cells arrested by a Cdc mutation. (A) In a normal population of proliferating yeast cells, buds vary in size according to the cell-cycle stage. (B) In a Cdc15 mutant grown at the restrictive temperature, cells complete anaphase but cannot complete the exit from mitosis and cytokinesis. As a result, they arrest uniformly with large buds, which are characteristic of late M phase. (Courtesy of Jeff Ubersax.)

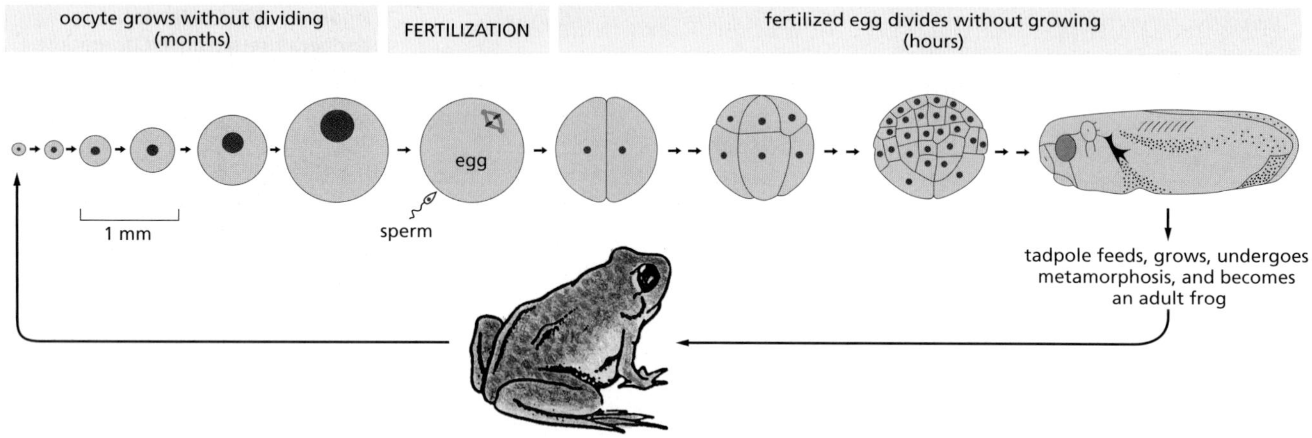

oocyte grows without dividing
(months)

FERTILIZATION

fertilized egg divides without growing
(hours)

1 mm

egg

sperm

tadpole feeds, grows, undergoes
metamorphosis, and becomes
an adult frog

Figure 17–9 Oocyte growth and egg cleavage in *Xenopus*. The oocyte grows without dividing for many months in the ovary of the mother frog and finally matures into an egg (discussed in Chapter 21). Upon fertilization, the egg cleaves very rapidly—initially at a rate of one division cycle every 30 minutes—forming a multicellular tadpole within a day or two. The cells get progressively smaller with each division, and the embryo remains the same size. Growth starts only when the tadpole begins feeding. The drawings in the top row are all on the same scale (but the frog below is not).

thousands of smaller cells (**Figure 17–9**). <ATTT> After a first division that takes about 90 minutes, the next 11 divisions occur, more or less synchronously, at 30-minute intervals, producing about 4096 (2^{12}) cells within 7 hours. Each cycle is divided into S and M phases of about 15 minutes each, without detectable G_1 or G_2 phases.

The early embryonic cells of *Xenopus*, as well as those of the clam *Spisula* and the fruit fly *Drosophila*, are thus capable of exceedingly rapid division in the absence of either growth or many of the control mechanisms (discussed later) that operate in more complex cell cycles. These *early embryonic cell cycles* therefore reveal the workings of the cell-cycle control system stripped down and simplified to the minimum needed to achieve the most fundamental requirements— the duplication of the genome and its segregation into two daughter cells. Another advantage of these early embryos for cell-cycle analysis is their large size. It is relatively easy to inject test substances into an egg to determine their effect on cell-cycle progression. It is also possible to prepare almost pure cytoplasm from *Xenopus* eggs and reconstitute many events of the cell cycle in a test tube (**Figure 17–10**). In such cell extracts, we can observe and manipulate cell-cycle events under highly simplified and controllable conditions.

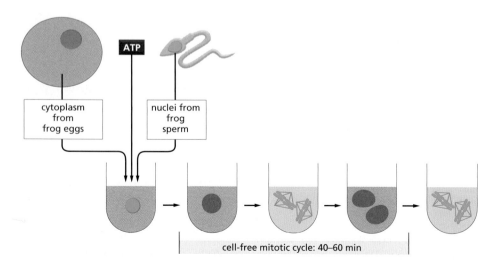

ATP

cytoplasm
from
frog eggs

nuclei from
frog
sperm

cell-free mitotic cycle: 40–60 min

Figure 17–10 Studying the cell cycle in a cell-free system. Gentle centrifugation is used to break open a large batch of frog eggs and separate the cytoplasm from other cell components. The undiluted cytoplasm is collected, and sperm nuclei are added to it, together with ATP. The sperm nuclei decondense and then go through repeated cycles of DNA replication and mitosis, indicating that the cell-cycle control system is operating in this cell-free cytoplasmic extract.

Cell-Cycle Control Can Be Studied in Cultured Mammalian Cells

It is not easy to observe individual cells in an intact mammal. Most studies on mammalian cell-cycle control therefore use cells that have been isolated from normal tissues or tumors and grown in plastic culture dishes in the presence of essential nutrients and other factors (**Figure 17–11**). There is a complication, however. When cells from normal mammalian tissues are cultured in standard conditions, they often stop dividing after a limited number of division cycles. Human fibroblasts, for example, permanently cease dividing after 25–40 divisions, a process called *replicative cell senescence*, as we discuss later.

Mammalian cells occasionally undergo mutations that help them proliferate indefinitely in culture as "immortalized" *cell lines*. Although they are not normal, such cell lines are widely used for cell-cycle studies—and for cell biology generally—because they provide an unlimited source of genetically homogeneous cells. In addition, these cells are sufficiently large to allow detailed cytological observations of cell-cycle events, and they are amenable to biochemical analysis of the proteins involved in cell-cycle control.

Studies of cultured mammalian cells have been especially useful for examining the molecular mechanisms governing the control of cell proliferation in multicellular organisms. Such studies are important not only for understanding the normal controls of cell numbers in tissues but also for understanding the loss of these controls in cancer (discussed in Chapter 20).

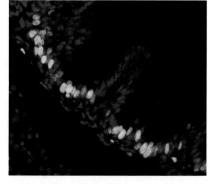

Figure 17–11 Mammalian cells proliferating in culture. The cells in this scanning electron micrograph are rat fibroblasts. (Courtesy of Guenter Albrecht-Buehler.)

Cell-Cycle Progression Can Be Studied in Various Ways

How can we tell what stage an animal cell has reached in the cell cycle? One way is simply to look at living cells with a microscope. A glance at a population of mammalian cells proliferating in culture reveals that a fraction of the cells have rounded up and are in mitosis (see Figure 17–11). Others can be observed in the process of cytokinesis. We can gain additional clues about cell-cycle position by staining cells with DNA-binding fluorescent dyes (which reveal the condensation of chromosomes in mitosis) or with antibodies that recognize specific cellular components such as the microtubules (revealing the mitotic spindle). Similarly, S-phase cells can be identified in the microscope by supplying them with visualizable molecules that are incorporated into newly synthesized DNA, such as the artificial thymidine analog bromo-deoxyuridine (BrdU). Cell nuclei that have incorporated BrdU are then visualized by staining with anti-BrdU antibodies (**Figure 17–12**).

Typically, in a population of cells that are all proliferating rapidly but asynchronously, about 30–40% will be in S phase at any instant and become labeled by a brief pulse of BrdU. From the proportion of cells in such a population that are labeled (the *labeling index*), we can estimate the duration of S phase as a fraction of the whole cell-cycle duration. Similarly, from the proportion of cells in mitosis (the *mitotic index*), we can estimate the duration of M phase. In addition, by giving a pulse of BrdU and allowing the cells to continue around the cycle for measured lengths of time, we can determine how long it takes for an S-phase cell to progress through G_2 into M phase, through M phase into G_1, and finally through G_1 back into S phase.

Another way to assess the stage that a cell has reached in the cell cycle is by measuring its DNA content, which doubles during S phase. This approach is greatly facilitated by the use of fluorescent DNA-binding dyes and a *flow cytometer*, which allows the rapid and automatic analysis of large numbers of cells (**Figure 17–13**). We can also use flow cytometry to determine the lengths of G_1, S,

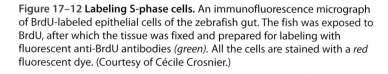

Figure 17–12 Labeling S-phase cells. An immunofluorescence micrograph of BrdU-labeled epithelial cells of the zebrafish gut. The fish was exposed to BrdU, after which the tissue was fixed and prepared for labeling with fluorescent anti-BrdU antibodies *(green)*. All the cells are stained with a *red* fluorescent dye. (Courtesy of Cécile Crosnier.)

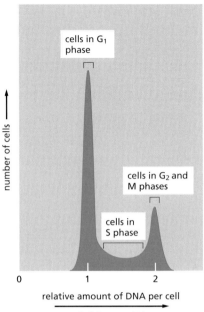

Figure 17–13 Analysis of DNA content with a flow cytometer. This graph shows typical results obtained for a proliferating cell population when the DNA content of its individual cells is determined in a flow cytometer. (A flow cytometer, also called a fluorescence-activated cell sorter, or FACS, can also be used to sort cells according to their fluorescence—see Figure 8–2). The cells analyzed here were stained with a dye that becomes fluorescent when it binds to DNA, so that the amount of fluorescence is directly proportional to the amount of DNA in each cell. The cells fall into three categories: those that have an unreplicated complement of DNA and are therefore in G_1, those that have a fully replicated complement of DNA (twice the G_1 DNA content) and are in G_2 or M phase, and those that have an intermediate amount of DNA and are in S phase. The distribution of cells in the case illustrated indicates that there are greater numbers of cells in G_1 than in G_2 + M phase, showing that G_1 is longer than G_2 + M in this population.

and G_2 + M phases, by following over time a population of DNA-labeled cells that have been preselected to be in one particular phase of the cell cycle: DNA content measurements on such a synchronized population of cells reveal how the cells progress through the cycle.

Summary

Cell division usually begins with duplication of the cell's contents, followed by distribution of those contents into two daughter cells. Chromosome duplication occurs during S phase of the cell cycle, whereas most other cell components are duplicated continuously throughout the cycle. During M phase, the replicated chromosomes are segregated into individual nuclei (mitosis), and the cell then splits in two (cytokinesis). S phase and M phase are usually separated by gap phases called G_1 and G_2, when various intracellular and extracellular signals regulate cell-cycle progression. Cell-cycle organization and control have been highly conserved during evolution, and studies in a wide range of systems—including yeasts, animal embryos, and mammalian cells in culture—have led to a unified view of eucaryotic cell-cycle control.

THE CELL-CYCLE CONTROL SYSTEM

For many years cell biologists watched the puppet show of DNA synthesis, mitosis, and cytokinesis but had no idea of what lay behind the curtain controlling these events. The cell-cycle control system was simply a black box inside the cell. It was not even clear whether there was a separate control system, or whether the processes of DNA synthesis, mitosis, and cytokinesis somehow controlled themselves. A major breakthrough came in the late 1980s with the identification of the key proteins of the control system, along with the realization that they are distinct from the proteins that perform the processes of DNA replication, chromosome segregation, and so on.

In this section, we first consider the basic principles upon which the cell-cycle control system operates. We then discuss the protein components of the control system and how they work together to time and coordinate the events of the cell cycle.

The Cell-Cycle Control System Triggers the Major Events of the Cell Cycle

The **cell-cycle control system** operates much like a timer or oscillator that triggers the events of the cell cycle in a set sequence (**Figure 17–14**). In its simplest form—as seen in the stripped-down embryonic cell cycles discussed earlier—the control system is like a rigidly programmed timer that provides a fixed amount of time for the completion of each cell-cycle event. The control system

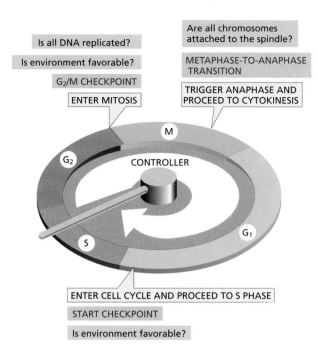

Is all DNA replicated?

Is environment favorable?

G₂/M CHECKPOINT

ENTER MITOSIS

Are all chromosomes attached to the spindle?

METAPHASE-TO-ANAPHASE TRANSITION

TRIGGER ANAPHASE AND PROCEED TO CYTOKINESIS

M

G₂

CONTROLLER

G₁

S

ENTER CELL CYCLE AND PROCEED TO S PHASE

START CHECKPOINT

Is environment favorable?

Figure 17–14 The control of the cell cycle. A cell-cycle control system triggers the essential processes of the cell cycle—such as DNA replication, mitosis, and cytokinesis. The control system is represented here as a central arm—the controller—that rotates clockwise, triggering essential processes when it reaches specific checkpoints on the outer dial. Information about the completion of cell-cycle events, as well as signals from the environment, can cause the control system to arrest the cycle at these checkpoints. The most prominent checkpoints occur at locations marked with *yellow boxes*.

in these cells is independent of the events it controls, so that its timing mechanisms continue to operate even if those events fail. In most cells, however, the control system does respond to information received back from the processes it controls. Sensors, for example, detect the completion of DNA synthesis, and if some malfunction prevents the successful completion of this process, signals are sent to the control system to delay progression to M phase. Such delays provide time for the machinery to be repaired and also prevent the disaster that might result if the cycle progressed prematurely to the next stage—and segregated incompletely replicated chromosomes, for example.

The cell-cycle control system is based on a connected series of biochemical switches, each of which initiates a specific cell-cycle event. This system of switches possesses many important engineering features that increase the accuracy and reliability of cell-cycle progression. First, the switches are generally *binary* (on/off) and launch events in a complete, irreversible fashion. It would clearly be disastrous, for example, if events like chromosome condensation or nuclear envelope breakdown were only partially initiated or started but not completed. Second, the cell-cycle control system is remarkably robust and reliable, partly because backup mechanisms and other features allow the system to operate effectively under a variety of conditions and even if some components fail. Finally, the control system is highly adaptable and can be modified to suit specific cell types or to respond to specific intracellular or extracellular signals.

In most eukaryotic cells, the cell-cycle control system triggers cell-cycle progression at three major regulatory transitions, or **checkpoints** (see Figure 17–14). The first checkpoint is Start (or the restriction point) in late G₁, where the cell commits to cell-cycle entry and chromosome duplication, as mentioned earlier. The second is the **G₂/M checkpoint**, where the control system triggers the early mitotic events that lead to chromosome alignment on the spindle in metaphase. The third is the **metaphase-to-anaphase transition**, where the control system stimulates sister-chromatid separation, leading to the completion of mitosis and cytokinesis. The control system blocks progression through each of these checkpoints if it detects problems inside or outside the cell. If the control system senses problems in the completion of DNA replication, for example, it will hold the cell at the G₂/M checkpoint until those problems are solved. Similarly, if extracellular conditions are not appropriate for cell proliferation, the control system blocks progression through Start, thereby preventing cell division until conditions become favorable.

The Cell-Cycle Control System Depends on Cyclically Activated Cyclin-Dependent Protein Kinases (Cdks)

Central components of the cell-cycle control system are members of a family of protein kinases known as **cyclin-dependent kinases** (**Cdks**). The activities of these kinases rise and fall as the cell progresses through the cycle, leading to cyclical changes in the phosphorylation of intracellular proteins that initiate or regulate the major events of the cell cycle. An increase in Cdk activity at the G_2/M checkpoint, for example, increases the phosphorylation of proteins that control chromosome condensation, nuclear envelope breakdown, spindle assembly, and other events that occur at the onset of mitosis.

Cyclical changes in Cdk activity are controlled by a complex array of enzymes and other proteins that regulate these kinases. The most important of these Cdk regulators are proteins known as **cyclins**. Cdks, as their name implies, are dependent on cyclins for their activity: unless they are tightly bound to a cyclin, they have no protein kinase activity (**Figure 17–15**). Cyclins were originally named because they undergo a cycle of synthesis and degradation in each cell cycle. The levels of the Cdk proteins, by contrast, are constant, at least in the simplest cell cycles. Cyclical changes in cyclin protein levels result in the cyclic assembly and activation of the **cyclin–Cdk complexes**; this activation in turn triggers cell-cycle events.

There are four classes of cyclins, each defined by the stage of the cell cycle at which they bind Cdks and function. All eucaryotic cells require three of these classes (**Figure 17–16**):

1. **G_1/S-cyclins** activate Cdks in late G_1 and thereby help trigger progression through Start, resulting in a commitment to cell-cycle entry. Their levels fall in S phase.

2. **S-cyclins** bind Cdks soon after progression through Start and help stimulate chromosome duplication. S-cyclin levels remain elevated until mitosis, and these cyclins also contribute to the control of some early mitotic events.

3. **M-cyclins** activate Cdks that stimulate entry into mitosis at the G_2/M checkpoint. Mechanisms that we discuss later destroy M-cyclins in midmitosis.

In most cells, a fourth class of cyclins, the **G_1-cyclins**, helps govern the activities of the G_1/S cyclins, which control progression through Start in late G_1.

In yeast cells, a single Cdk protein binds all classes of cyclins and triggers different cell-cycle events by changing cyclin partners at different stages of the cycle. In vertebrate cells, by contrast, there are four Cdks. Two interact with G_1-cyclins, one with G_1/S- and S-cyclins, and one with M-cyclins. In this chapter, we simply refer to the different cyclin–Cdk complexes as **G_1-Cdk**, **G_1/S-Cdk**, **S-Cdk**, and **M-Cdk**. **Table 17–1** lists the names of the individual Cdks and cyclins.

How do different cyclin–Cdk complexes trigger different cell-cycle events? The answer, at least in part, seems to be that the cyclin protein does not simply activate its Cdk partner but also directs it to specific target proteins. As a result,

Figure 17–15 Two key components of the cell-cycle control system. When cyclin forms a complex with Cdk, the protein kinase is activated to trigger specific cell-cycle events. Without cyclin, Cdk is inactive.

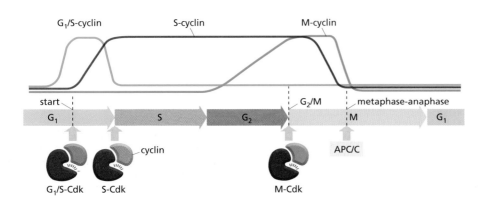

Figure 17–16 Cyclin–Cdk complexes of the cell-cycle control system. The concentrations of the three major cyclin types oscillate during the cell cycle, while the concentrations of Cdks (not shown) do not change and exceed the amounts of cyclins. In late G_1, rising G_1/S-cyclin levels lead to the formation of G_1/S-Cdk complexes that trigger progression through the Start checkpoint. S-Cdk complexes form at the start of S phase and trigger DNA replication, as well as some early mitotic events. M-Cdk complexes form during G_2 but are held in an inactive state by mechanisms we describe later. These complexes are activated at the end of G_2 and trigger the early events of mitosis. A separate regulatory protein, the APC/C, which we discuss later, initiates the metaphase-to-anaphase transition.

Table 17–1 The Major Cyclins and Cdks of Vertebrates and Budding Yeast

CYCLIN–CDK COMPLEX	VERTEBRATES		BUDDING YEAST	
	CYCLIN	CDK PARTNER	CYCLIN	CDK PARTNER
G_1-Cdk	cyclin D*	Cdk4, Cdk6	Cln3	Cdk1**
G_1/S-Cdk	cyclin E	Cdk2	Cln1, 2	Cdk1
S-Cdk	cyclin A	Cdk2, Cdk1**	Clb5, 6	Cdk1
M-Cdk	cyclin B	Cdk1	Clb1, 2, 3, 4	Cdk1

* There are three D cyclins in mammals (cyclins D1, D2, and D3).
** The original name of Cdk1 was Cdc2 in both vertebrates and fission yeast, and Cdc28 in budding yeast.

each cyclin–Cdk complex phosphorylates a different set of substrate proteins. The same cyclin–Cdk complex can also induce different effects at different times in the cycle, probably because the accessibility of some Cdk substrates changes during the cell cycle. Certain proteins that function in mitosis, for example, may become available for phosphorylation only in G_2.

Studies of the three-dimensional structures of Cdk and cyclin proteins have revealed that, in the absence of cyclin, the active site in the Cdk protein is partly obscured by a slab of protein, like a stone blocking the entrance to a cave (**Figure 17–17**A). Cyclin binding causes the slab to move away from the active site, resulting in partial activation of the Cdk enzyme (Figure 17–17B). Full activation of the cyclin–Cdk complex then occurs when a separate kinase, the **Cdk-activating kinase (CAK)**, phosphorylates an amino acid near the entrance of the Cdk active site. This causes a small conformational change that further increases the activity of the Cdk, allowing the kinase to phosphorylate its target proteins effectively and thereby induce specific cell-cycle events (Figure 17–17C). <TAGA>

Inhibitory Phosphorylation and Cdk Inhibitory Proteins (CKIs) Can Suppress Cdk Activity

The rise and fall of cyclin levels is the primary determinant of Cdk activity during the cell cycle. Several additional mechanisms, however, fine-tune Cdk activity at specific stages of the cycle.

Phosphorylation at a pair of amino acids in the roof of the kinase active site inhibits the activity of a cyclin–Cdk complex. Phosphorylation of these sites by a protein kinase known as **Wee1** inhibits Cdk activity, while dephosphorylation of these sites by a phosphatase known as **Cdc25** increases Cdk activity (**Figure 17–18**). We will see later that this regulatory mechanism is particularly important in the control of M-Cdk activity at the onset of mitosis.

Binding of **Cdk inhibitor proteins (CKIs)** also regulates cyclin–Cdk complexes. The three-dimensional structure of a cyclin–Cdk–CKI complex reveals

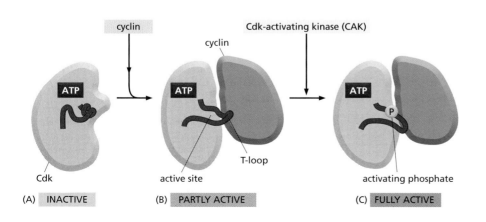

(A) INACTIVE   (B) PARTLY ACTIVE   (C) FULLY ACTIVE

Figure 17–17 The structural basis of Cdk activation. These drawings are based on three-dimensional structures of human Cdk2, as determined by x-ray crystallography. The location of the bound ATP is indicated. The enzyme is shown in three states. (A) In the inactive state, without cyclin bound, the active site is blocked by a region of the protein called the T-loop *(red)*. (B) The binding of cyclin causes the T-loop to move out of the active site, resulting in partial activation of the Cdk2. (C) Phosphorylation of Cdk2 (by CAK) at a threonine residue in the T-loop further activates the enzyme by changing the shape of the T-loop, improving the ability of the enzyme to bind its protein substrates.

that CKI binding stimulates a large rearrangement in the structure of the Cdk active site, rendering it inactive (**Figure 17–19**). Cells use CKIs primarily to help govern the activities of G$_1$/S- and S-Cdks early in the cell cycle.

The Cell-Cycle Control System Depends on Cyclical Proteolysis

Whereas activation of specific cyclin-Cdk complexes drives progression through the Start and G$_2$/M checkpoints (see Figure 17–16), progression through the metaphase-to-anaphase transition is triggered not by protein phosphorylation but by protein destruction, leading to the final stages of cell division.

The key regulator of the metaphase-to-anaphase transition is the **anaphase-promoting complex**, or **cyclosome** (**APC/C**), a member of the ubiquitin ligase family of enzymes. As discussed in Chapter 3, many of these enzymes are used in numerous cell processes to stimulate the proteolytic destruction of specific regulatory proteins. They transfer multiple copies of the small protein ubiquitin to specific target proteins, resulting in their proteolytic destruction by the proteasomes. Other ubiquitin ligases mark proteins for purposes other than destruction.

The APC/C catalyzes the ubiquitylation and destruction of two major proteins. The first is *securin,* which normally protects the protein linkages that hold sister chromatid pairs together in early mitosis. Destruction of securin at the metaphase-to-anaphase transition activates a protease that separates the sisters and unleashes anaphase. The S- and M-cyclins are the second major targets of the APC/C. Destroying these cyclins inactivates most Cdks in the cell (see Figure 17–16). As a result, the many proteins phosphorylated by Cdks from S phase to early mitosis are dephosphorylated by various phosphatases that are present in the anaphase cell. This dephosphorylation of Cdk targets is required for the completion of M phase, including the final steps in mitosis and the process of cytokinesis. Following its activation in mid-mitosis, the APC/C remains active in G$_1$, thereby providing a stable period of Cdk inactivity. When G$_1$/S-Cdks are activated in late G$_1$, the APC/C is turned off, thereby allowing cyclin accumulation to start the next cell cycle.

The cell-cycle control system also uses another ubiquitin ligase called **SCF** (after the names of its three subunits). It ubiquitylates certain CKI proteins in late G$_1$ and thereby helps control the activation of S-Cdks and DNA replication.

The APC/C and SCF are both large, multisubunit complexes with some related components, but they are regulated differently. APC/C activity changes during the cell cycle, primarily as a result of changes in its association with an activating subunit—either **Cdc20** during anaphase or **Cdh1** from late mitosis through early G$_1$. These subunits help the APC/C recognize its target proteins (**Figure 17–20**A). SCF activity also depends on subunits called F-box proteins, which help the complex recognize its target proteins. Unlike APC/C activity, however, SCF activity is constant during the cell cycle. Ubiquitylation by SCF is controlled instead by changes in the phosphorylation state of its target proteins, as F-box subunits recognize only specifically phosphorylated proteins (Figure 17–20B).

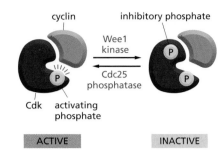

Figure 17–18 The regulation of Cdk activity by inhibitory phosphorylation. The active cyclin–Cdk complex is turned off when the kinase Wee1 phosphorylates two closely spaced sites above the active site. Removal of these phosphates by the phosphatase Cdc25 activates the cyclin–Cdk complex. For simplicity, only one inhibitory phosphate is shown. CAK adds the activating phosphate, as shown in Figure 17–17.

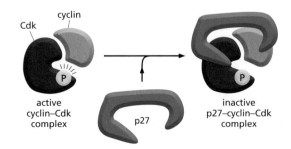

Figure 17–19 The inhibition of a cyclin–Cdk complex by a CKI. This drawing is based on the three-dimensional structure of the human cyclin A–Cdk2 complex bound to the CKI p27, as determined by x-ray crystallography. The p27 binds to both the cyclin and Cdk in the complex, distorting the active site of the Cdk. It also inserts into the ATP-binding site, further inhibiting the enzyme activity.

(A) control of proteolysis by APC/C

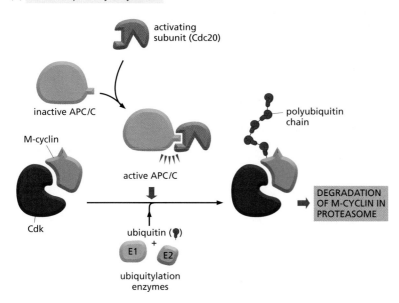

(B) control of proteolysis by SCF

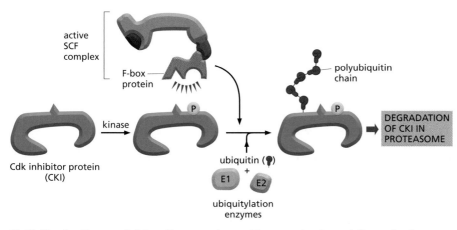

Figure 17–20 The control of proteolysis by APC/C and SCF during the cell cycle. (A) The APC/C is activated in mitosis by association with the activating subunit Cdc20, which recognizes specific amino acid sequences on M-cyclin and other target proteins. With the help of two additional proteins called E1 and E2, the APC/C transfers multiple ubiquitin molecules onto the target protein. The polyubiquitylated target is then recognized and degraded in a proteasome. (B) The activity of the ubiquitin ligase SCF depends on substrate-binding subunits called F-box proteins, of which there are many different types. The phosphorylation of a target protein, such as the CKI shown, allows the target to be recognized by a specific F-box subunit.

Cell-Cycle Control Also Depends on Transcriptional Regulation

In the frog embryonic cell cycle discussed earlier, gene transcription does not occur. Cell-cycle control depends exclusively on post-transcriptional mechanisms that involve the regulation of Cdks and ubiquitin ligases and their target proteins. In the more complex cell cycles of most cell types, however, transcriptional control provides an additional level of regulation. Changes in cyclin gene transcription, for example, help control cyclin levels in most cells.

We can use DNA microarrays (discussed in Chapter 8) to analyze changes in the expression of all of the genes in the genome as the cell progresses through the cell cycle. The results of these studies are surprising. In budding yeast, for example, about 10% of the genes encode mRNAs whose levels oscillate during the cell cycle. Some of these genes encode proteins with known cell-cycle functions, but the functions of many others are unknown.

The Cell-Cycle Control System Functions as a Network of Biochemical Switches

Table 17–2 summarizes some of the major components of the cell-cycle control system. These proteins are functionally linked together to form a robust network, which operates essentially autonomously to activate a series of biochemical switches, each of which triggers a specific cell-cycle event.

Table 17–2 Summary of the Major Cell-Cycle Regulatory Proteins

GENERAL NAME	FUNCTIONS AND COMMENTS
Protein kinases and protein phosphatases that modify Cdks	
Cdk-activating kinase (CAK)	phosphorylates an activating site in Cdks
Wee1 kinase	phosphorylates inhibitory sites in Cdks; primarily involved in suppressing Cdk1 activity before mitosis
Cdc25 phosphatase	removes inhibitory phosphates from Cdks; three family members (Cdc25A, B, C) in mammals; primarily involved in controlling Cdk1 activation at the onset of mitosis
Cdk inhibitor proteins (CKIs)	
Sic1 (budding yeast)	suppresses Cdk1 activity in G_1; phosphorylation by Cdk1 at the end of G_1 triggers its destruction
p27 (mammals)	suppresses G_1/S-Cdk and S-Cdk activities in G_1; helps cells withdraw from cell cycle when they terminally differentiate; phosphorylation by Cdk2 triggers its ubiquitylation by SCF
p21 (mammals)	suppresses G_1/S-Cdk and S-Cdk activities following DNA damage
p16 (mammals)	suppresses G_1-Cdk activity in G_1; frequently inactivated in cancer
Ubiquitin ligases and their activators	
APC/C	catalyzes ubiquitylation of regulatory proteins involved primarily in exit from mitosis, including securin and S- and M-cyclins; regulated by association with activating subunits
Cdc20	APC/C-activating subunit in all cells; triggers initial activation of APC/C at metaphase-to-anaphase transition; stimulated by M-Cdk activity
Cdh1	APC/C-activating subunit that maintains APC/C activity after anaphase and throughout G_1; inhibited by Cdk activity
SCF	catalyzes ubiquitylation of regulatory proteins involved in G_1 control, including some CKIs (Sic1 in budding yeast, p27 in mammals); phosphorylation of target protein usually required for this activity

When conditions for cell proliferation are right, various external and internal signals stimulate the activation of G_1-Cdk, which in turn stimulates the expression of genes encoding G_1/S- and S-cyclins. The resulting activation of G_1/S-Cdk then drives progression through the Start checkpoint. By mechanisms we discuss later, G_1/S-Cdks unleash a wave of S-Cdk activity, which initiates chromosome duplication in S phase and also contributes to some early events of mitosis. M-Cdk activation then triggers progression through the G_2/M checkpoint and the events of early mitosis, leading to the alignment of sister chromatids at the equator of the mitotic spindle. Finally, the APC/C, together with its activator Cdc20, triggers the destruction of securin and cyclins at the metaphase-to-anaphase transition, thereby unleashing sister-chromatid segregation and the completion of mitosis (**Figure 17–21**). When mitosis is complete, multiple mechanisms collaborate to suppress Cdk activity after mitosis, resulting in a stable G_1 period, as we discuss later. We are now ready to discuss these cell-cycle stages in more detail, starting with S phase.

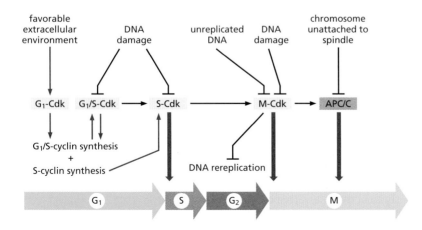

Figure 17–21 An overview of the cell-cycle control system. The core of the cell-cycle control system consists of a series of cyclin–Cdk complexes (*yellow*). As discussed in more detail later, the activity of each complex is also influenced by various inhibitory mechanisms, which provide information about the extracellular environment, cell damage, and incomplete cell-cycle events (*top*). These mechanisms are not present in all cell types; many are missing in early embryonic cell cycles, for example.

Summary

The cell-cycle control system triggers the events of the cell cycle and ensures that these events are properly timed and occur in the correct order. The control system responds to various intracellular and extracellular signals and arrests the cycle when the cell either fails to complete an essential cell-cycle process or encounters unfavorable environmental or intracellular conditions.

Central components of the cell-cycle control system are cyclin-dependent protein kinases (Cdks), which depend on cyclin subunits for their activity. Oscillations in the activities of various cyclin–Cdk complexes control various cell-cycle events. Thus, activation of S-phase cyclin–Cdk complexes (S-Cdk) initiates S phase, while activation of M-phase cyclin–Cdk complexes (M-Cdk) triggers mitosis. The mechanisms that control the activities of cyclin–Cdk complexes include phosphorylation of the Cdk subunit, binding of Cdk inhibitor proteins (CKIs), proteolysis of cyclins, and changes in the transcription of genes encoding Cdk regulators. The cell-cycle control system also depends crucially on two additional enzyme complexes, the APC/C and SCF ubiquitin ligases, which catalyze the ubiquitylation and consequent destruction of specific regulatory proteins that control critical events in the cycle.

S PHASE

The linear chromosomes of eucaryotic cells are vast and dynamic assemblies of DNA and protein, and their duplication is a complex process that takes up a major fraction of the cell cycle. Not only must the long DNA molecule of each chromosome be duplicated accurately—a remarkable feat in itself—but the protein packaging surrounding each region of that DNA must also be reproduced, ensuring that the daughter cells inherit all features of chromosome structure.

The central event of chromosome duplication is replication of the DNA. A cell must solve two problems when initiating and completing DNA replication. First, replication must occur with extreme accuracy to minimize the risk of mutations in the next cell generation. Second, every nucleotide in the genome must be copied once, and only once, to prevent the damaging effects of gene amplification. In Chapter 5, we discuss the sophisticated protein machinery that performs DNA replication with astonishing speed and accuracy. In this section, we consider the elegant mechanisms by which the cell-cycle control system initiates the replication process and, at the same time, prevents it from happening more than once per cycle.

S-Cdk Initiates DNA Replication Once Per Cycle

DNA replication begins at *origins of replication*, which are scattered at numerous locations in every chromosome. During S phase, the *initiation* of DNA replication occurs at these origins when specialized protein machines (sometimes called *initiator proteins*) unwind the double helix at the origin and load DNA replication enzymes onto the two single-stranded templates. This leads to the *elongation* phase of replication, when the replication machinery moves outward from the origin at two *replication forks* (discussed in Chapter 5).

To ensure that chromosome duplication occurs only once per cell cycle, the initiation phase of DNA replication is divided into two distinct steps that occur at different times in the cell cycle. The first step occurs in late mitosis and early G$_1$, when a large complex of initiator proteins, called the **prereplicative complex**, or **pre-RC**, assembles at origins of replication. This step is sometimes called *licensing* of replication origins because initiation of DNA synthesis is permitted only at origins containing a pre-RC. The second step occurs at the onset of S phase, when components of the pre-RC nucleate the formation of a larger protein complex called the **preinitiation complex**. This complex then unwinds the DNA helix and loads DNA polymerases and other replication enzymes onto the DNA strands, thereby initiating DNA synthesis, as described in Chapter 5. Once the replication origin has been activated in this way, the

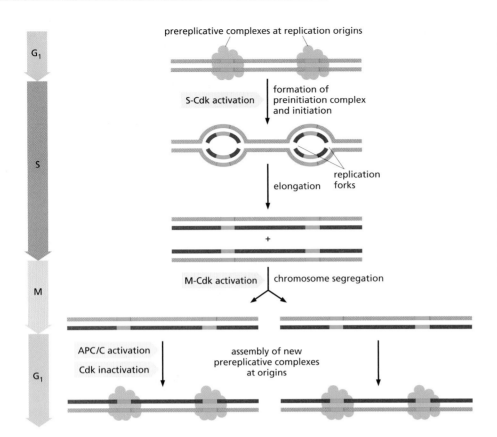

G$_1$

S

M

G$_1$

prereplicative complexes at replication origins

S-Cdk activation | formation of preinitiation complex and initiation

replication forks

elongation

+

M-Cdk activation | chromosome segregation

APC/C activation

Cdk inactivation

assembly of new prereplicative complexes at origins

Figure 17–22 Control of chromosome duplication. Preparations for DNA replication begin in G$_1$ with the assembly of prereplicative complexes (pre-RCs) at replication origins. S-Cdk activation leads to the formation of multiprotein preinitiation complexes that unwind the DNA at origins and begin the process of DNA replication. Two replication forks move out from each origin until the entire chromosome is duplicated. Duplicated chromosomes are then segregated in M phase. The activation of replication origins in S phase also causes disassembly of the prereplicative complex, which does not reform at the origin until the following G$_1$—thereby ensuring that each origin is activated only once in each cell cycle.

pre-RC is dismantled and cannot be reassembled at that origin until the following G$_1$. As a result, origins can be activated only once per cell cycle.

The cell-cycle control system governs both assembly of the pre-RC and assembly of the pre-initiation complex (**Figure 17–22**). Assembly of the pre-RC is inhibited by Cdk activity, and, in most cells, is stimulated by the APC/C. Pre-RC assembly therefore occurs only in late mitosis and early G$_1$, when Cdk activity is low and APC/C activity is high. At the onset of S phase, activation of S-Cdk then triggers the formation of a preinitiation complex, which initiates DNA synthesis. In addition, the pre-RC is partly dismantled. Because S-Cdk and M-Cdk activities remain high (and APC/C activity remains low) until late mitosis, new pre-RCs cannot be assembled at fired origins until the cell cycle is complete.

Figure 17–23 illustrates some of the proteins involved in the initiation of DNA replication. A key player is a large, multiprotein complex called the **origin recognition complex** (**ORC**), which binds to replication origins throughout the cell cycle. In late mitosis and early G$_1$, the proteins **Cdc6** and **Cdt1** bind to the ORC at origins and help load a group of six related proteins called the **Mcm proteins**. The resulting large complex is the pre-RC, and the origin is now licensed for replication.

The six Mcm proteins of the pre-RC form a ring around the DNA that is thought to serve as the major DNA helicase that unwinds the origin DNA when DNA synthesis begins and as the replication forks move out from the origin. Thus, the central purpose of the pre-RC is to load the helicase that will play a central part in the subsequent DNA replication process.

Once the pre-RC has assembled in G$_1$, the replication origin is ready to fire. The activation of S-Cdk in late G$_1$ triggers the assembly of several additional protein complexes at the origin, leading to the formation of a giant preinitiation complex that unwinds the helix and begins DNA synthesis.

At the same time as it initiates DNA replication, S-Cdk triggers the disassembly of some pre-RC components at the origin. Cdks phosphorylate both the ORC and Cdc6, resulting in their inhibition by various mechanisms. Furthermore, inactivation of the APC/C in late G$_1$ also helps turn off pre-RC assembly. In late mitosis and early G$_1$, the APC/C triggers the destruction of a protein,

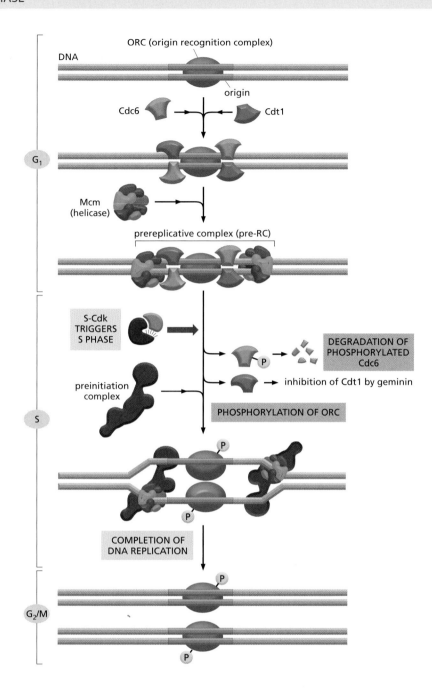

Figure 17–23 Control of the initiation of DNA replication. The ORC remains associated with a replication origin throughout the cell cycle. In early G$_1$, Cdc6 and Cdt1 associate with the ORC. The resulting protein complex then assembles Mcm ring complexes on the adjacent DNA, resulting in the formation of the prereplicative complex (pre-RC). S-Cdk (with assistance from another protein kinase, not shown) then stimulates the assembly of several additional proteins at the origin to form the preinitiation complex. DNA polymerase and other replication proteins are recruited to the origin, the Mcm protein rings are activated as DNA helicases, and DNA unwinding allows DNA replication to begin. S-Cdk also blocks rereplication by triggering the destruction of Cdc6 and the inactivation of the ORC. Cdt1 is inactivated by the protein geminin. Geminin is an APC/C target and its levels therefore increase in S and M phases, when APC/C is inactive. Thus, the components of the pre-RC (Cdc6, Cdt1, Mcm) cannot form a new pre-RC at the origins until M-Cdk is inactivated and the APC/C is activated at the end of mitosis (see text).

geminin, that binds and inhibits the pre-RC component Cdt1. Thus, when the APC/C is turned off in late G$_1$, geminin accumulates and inhibits Cdt1. In these various ways, S- and M-Cdk activities, combined with low APC/C activity, block pre-RC formation during S phase and thereafter. How, then, is the cell-cycle control system reset to allow replication to occur in the next cell cycle? The answer is simple. At the end of mitosis, APC/C activation leads to the inactivation of Cdks and the destruction of geminin. Pre-RC components are dephosphorylated and Cdt1 is activated, allowing pre-RC assembly to prepare the cell for the next S phase.

Chromosome Duplication Requires Duplication of Chromatin Structure

The DNA of the chromosomes is extensively packaged in a variety of protein components, including histones and various regulatory proteins involved in the control of gene expression (discussed in Chapter 4). Thus, duplication of a chromosome is not simply a matter of duplicating the DNA at its core but also requires the duplication of these chromatin proteins and their proper assembly on the DNA.

The production of chromatin proteins increases during S phase to provide the raw materials needed to package the newly synthesized DNA. Most importantly, S-Cdks stimulate a large increase in the synthesis of the four histone subunits that form the histone octamers at the core of each nucleosome. These subunits are assembled into nucleosomes on the DNA by nucleosome assembly factors, which typically associate with the replication fork and distribute nucleosomes on both strands of the DNA as they emerge from the DNA synthesis machinery.

Chromatin packaging helps to control gene expression. In some parts of the chromosome, the chromatin is highly condensed and is called *heterochromatin*, whereas in other regions it has a more open structure and is called *euchromatin*. These differences in chromatin structure depend on a variety of mechanisms, including modification of histone tails and the presence of non-histone proteins (discussed in Chapter 4). Because these differences are important in gene regulation, it is crucial that chromatin structure, like the DNA within, is reproduced accurately during S phase. How chromatin structure is duplicated is not well understood, however. During DNA synthesis, histone-modifying enzymes and various non-histone proteins are probably deposited onto the two new DNA strands as they emerge from the replication fork, and these proteins are thought to help reproduce the local chromatin structure of the parent chromosome.

Cohesins Help Hold Sister Chromatids Together

At the end of S phase, each replicated chromosome consists of a pair of identical sister chromatids glued together along their length. This sister-chromatid cohesion sets the stage for a successful mitosis because it greatly facilitates the attachment of the two sister chromatids in a pair to opposite poles of the mitotic spindle. Imagine how difficult it would be to achieve this bipolar attachment if sister chromatids were allowed to drift apart after S phase. Indeed, defects in sister-chromatid cohesion—in yeast mutants, for example—lead inevitably to major errors in chromosome segregation.

Sister-chromatid cohesion depends on a large protein complex called **cohesin**, which is deposited at many locations along the length of each sister chromatid as the DNA is replicated in S phase. Two of the subunits of cohesin are members of a large family of proteins called *SMC proteins* (for Structural Maintenance of Chromosomes). Cohesin forms giant ring-like structures, and it has been proposed that these might form rings that surround the two sister chromatids (**Figure 17–24**).

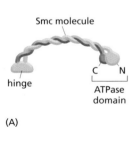

(A)

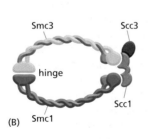

(B)

(C)

Figure 17–24 **Cohesin.** Cohesin is a protein complex with four subunits. Two subunits, Smc1 and Smc3, are coiled-coil proteins with an ATPase domain at one end; together, they form a large V-shaped structure as shown. Two additional subunits, Scc1 and Scc3, connect the ATPase head domains forming a ring structure that may encircle the sister chromatids as shown.

Sister-chromatid cohesion also results, at least in part, from *DNA catenation*, the intertwining of sister DNA molecules that occurs when two replication forks meet during DNA synthesis. The enzyme topoisomerase II gradually disentangles the catenated sister DNAs between S phase and early mitosis by cutting one DNA molecule, passing the other through the break, and then resealing the cut DNA (see Figure 5–23). Once the catenation has been removed, sister-chromatid cohesion depends primarily on cohesin complexes. The loss of sister cohesion at the metaphase-to-anaphase transition therefore depends primarily on disruption of these complexes, as we describe later.

Summary

Duplication of the chromosomes in S phase involves the accurate copying of the entire DNA molecule in each chromosome, as well as the duplication of the chromatin proteins that associate with the DNA and govern various aspects of chromosome function. Chromosome duplication is triggered by the activation of S-Cdk, which activates proteins that unwind the DNA and initiate its replication at sites in the DNA called replication origins. Once a replication origin is activated during S phase, S-Cdk also inhibits proteins that are required to allow that origin to initiate DNA replication again. Thus, each origin is fired once and only once in each S phase and cannot be reused until the next cell cycle.

MITOSIS

Following the completion of S phase and transition through G_2, the cell undergoes the dramatic upheaval of M phase. This begins with mitosis, during which the sister chromatids are separated and distributed (*segregated*) to a pair of identical daughter nuclei, each with its own copy of the genome. Mitosis is traditionally divided into five stages—*prophase, prometaphase, metaphase, anaphase,* and *telophase*—defined primarily on the basis of chromosome behavior as seen in a microscope. <TACT> <TCAA> As mitosis is completed, the second major event of M phase—cytokinesis—divides the cell into two halves, each with an identical nucleus. **Panel 17–1** summarizes the major events of M phase.

From a regulatory point of view, mitosis can be divided into two major parts, each governed by distinct components of the cell-cycle control system. First, an abrupt increase in M-Cdk activity at the G_2/M checkpoint triggers the events of early mitosis (prophase, prometaphase, and metaphase). During this period, M-Cdk and several other mitotic protein kinases phosphorylate a variety of proteins, leading to the assembly of the mitotic spindle and its attachment to the sister chromatid pairs. The second major part of mitosis begins at the metaphase-to-anaphase transition, when the APC/C triggers the destruction of securin, liberating a protease that cleaves cohesin and thereby initiates separation of the sister chromatids. The APC/C also triggers the destruction of cyclins, which leads to Cdk inactivation and the dephosphorylation of Cdk targets, which is required for all events of late M phase, including the completion of anaphase, the disassembly of the mitotic spindle, and the division of the cell by cytokinesis.

In this section, we describe the key mechanical events of mitosis and how M-Cdk and the APC/C orchestrate them.

M-Cdk Drives Entry Into Mitosis

One of the most remarkable features of cell-cycle control is that a single protein kinase, M-Cdk, brings about all of the diverse and complex cell rearrangements that occur in the early stages of mitosis. At a minimum, M-Cdk must induce the assembly of the mitotic spindle and ensure that each sister chromatid in a pair is attached to the opposite pole of the spindle. It also triggers *chromosome condensation*, the large-scale reorganization of the intertwined sister chromatids into compact, rod-like structures. In animal cells, M-Cdk also promotes the breakdown of the nuclear envelope and rearrangements of

1 PROPHASE

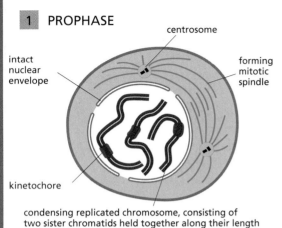

centrosome

intact nuclear envelope

forming mitotic spindle

kinetochore

condensing replicated chromosome, consisting of two sister chromatids held together along their length

At prophase, the replicated chromosomes, each consisting of two closely associated sister chromatids, condense. Outside the nucleus, the mitotic spindle assembles between the two centrosomes, which have replicated and moved apart. For simplicity, only three chromosomes are shown. In diploid cells, there would be two copies of each chromosome present. In the photomicrograph, chromosomes are stained *orange* and microtubules are *green*.

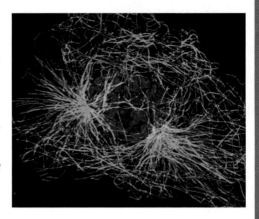

2 PROMETAPHASE

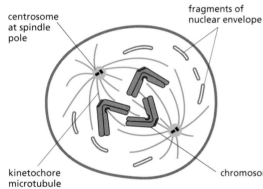

centrosome at spindle pole

fragments of nuclear envelope

kinetochore microtubule

chromosome in active motion

Prometaphase starts abruptly with the breakdown of the nuclear envelope. Chromosomes can now attach to spindle microtubules via their kinetochores and undergo active movement.

3 METAPHASE

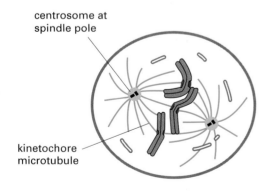

centrosome at spindle pole

kinetochore microtubule

At metaphase, the chromosomes are aligned at the equator of the spindle, midway between the spindle poles. The kinetochore microtubules attach sister chromatids to opposite poles of the spindle.

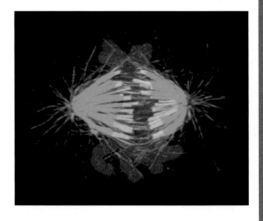

4 ANAPHASE

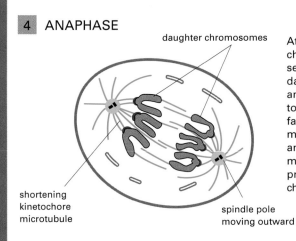

daughter chromosomes

shortening kinetochore microtubule

spindle pole moving outward

At anaphase, the sister chromatids synchronously separate to form two daughter chromosomes, and each is pulled slowly toward the spindle pole it faces. The kinetochore microtubules get shorter, and the spindle poles also move apart; both processes contribute to chromosome segregation.

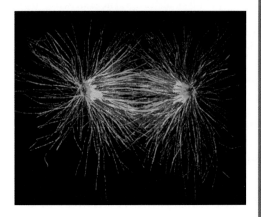

5 TELOPHASE

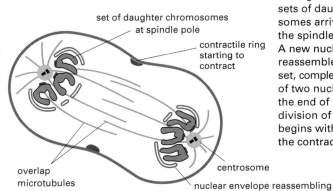

set of daughter chromosomes at spindle pole

contractile ring starting to contract

overlap microtubules

centrosome

nuclear envelope reassembling around individual chromosomes

During telophase, the two sets of daughter chromosomes arrive at the poles of the spindle and decondense. A new nuclear envelope reassembles around each set, completing the formation of two nuclei and marking the end of mitosis. The division of the cytoplasm begins with contraction of the contractile ring.

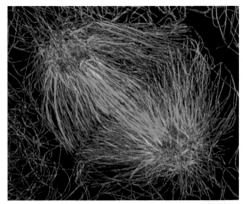

6 CYTOKINESIS

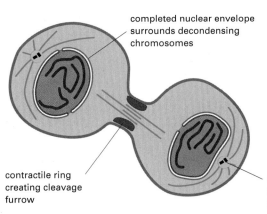

completed nuclear envelope surrounds decondensing chromosomes

contractile ring creating cleavage furrow

re-formation of interphase array of microtubules nucleated by the centrosome

During cytokinesis, the cytoplasm is divided in two by a contractile ring of actin and myosin filaments, which pinches the cell in two to create two daughters, each with one nucleus.

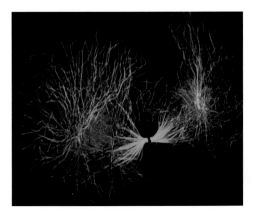

(Micrographs courtesy of Julie Canman and Ted Salmon.)

the actin cytoskeleton and the Golgi apparatus. Each of these processes is thought to be triggered when M-Cdk phosphorylates specific proteins involved in the process, although most of these proteins have not yet been identified.

M-Cdk does not act alone to phosphorylate key proteins involved in early mitosis. Two additional families of protein kinases, the *Polo-like kinases* and the *Aurora kinases*, also make important contributions to the control of early mitotic events. The Polo-like kinase Plk, for example, is required for the normal assembly of a bipolar mitotic spindle, in part because it phosphorylates proteins involved in separation of the spindle poles early in mitosis. The Aurora kinase Aurora-A also helps control proteins that govern the assembly and stability of the spindle, whereas Aurora-B controls attachment of sister chromatids to the spindle, as we discuss later. Activation of Polo-like kinases and Aurora kinases depends on M-Cdk activity, but the precise activation mechanisms are not clear.

Dephosphorylation Activates M-Cdk at the Onset of Mitosis

M-Cdk activation begins with the accumulation of M-cyclin (cyclin B in vertebrate cells; see Table 17–1). In embryonic cell cycles, the synthesis of M-cyclin is constant throughout the cell cycle, and M-cyclin accumulation results from the high stability of the protein in interphase. In most cell types, however, M-cyclin synthesis increases during G_2 and M, owing primarily to an increase in *M-cyclin* gene transcription. The increase in M-cyclin protein leads to a corresponding accumulation of M-Cdk (the complex of Cdk1 and M-cyclin) as the cell approaches mitosis. Although the Cdk in these complexes is phosphorylated at an activating site by the Cdk-activating kinase (CAK), as discussed earlier, the protein kinase Wee1 holds it in an inactive state by inhibitory phosphorylation at two neighboring sites (see Figure 17–18). Thus, by the time the cell reaches the end of G_2, it contains an abundant stockpile of M-Cdk that is primed and ready to act but is suppressed by phosphates that block the active site of the kinase.

What, then, triggers the activation of the M-Cdk stockpile? The crucial event is the activation of the protein phosphatase Cdc25, which removes the inhibitory phosphates that restrain M-Cdk (**Figure 17–25**). At the same time, the inhibitory activity of the kinase Wee1 is suppressed, further ensuring that M-Cdk activity increases. The mechanisms that unleash Cdc25 activity (and suppress Wee1) in early mitosis are not well understood. One possibility is that the S-Cdks that are active in G_2 and early prophase stimulate Cdc25.

Interestingly, Cdc25 can also be activated, at least in part, by its target, M-Cdk. M-Cdk may also inhibit the inhibitory kinase Wee1. The ability of M-Cdk to activate its own activator (Cdc25) and inhibit its own inhibitor (Wee1) suggests that M-Cdk activation in mitosis involves positive feedback loops (see Figure 17–25). According to this attractive model, the partial activation of Cdc25 (perhaps by S-Cdk) leads to the partial activation of a subpopulation of M-Cdk

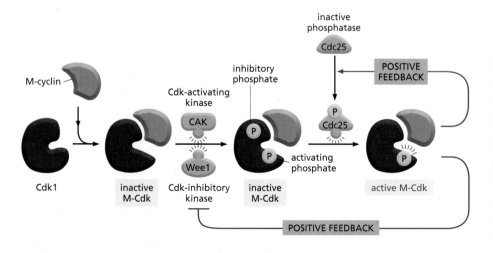

Figure 17–25 The activation of M-Cdk. Cdk1 associates with M-cyclin as the levels of M-cyclin gradually rise. The resulting M-Cdk complex is phosphorylated on an activating site by the Cdk-activating kinase (CAK) and on a pair of inhibitory sites by the Wee1 kinase. The resulting inactive M-Cdk complex is then activated at the end of G_2 by the phosphatase Cdc25. Cdc25 is further stimulated by active M-Cdk, resulting in positive feedback. This feedback is enhanced by the ability of M-Cdk to inhibit Wee1.

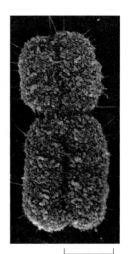

Figure 17–26 **The mitotic chromosome.** Scanning electron micrograph of a human mitotic chromosome, consisting of two sister chromatids joined along their length. The constricted regions are the centromeres. (Courtesy of Terry D. Allen.)

1 µm

complexes, which then phosphorylate more Cdc25 and Wee1 molecules. This leads to more M-Cdk activation, and so on. Such a mechanism would quickly promote the complete activation of all the M-Cdk complexes in the cell. As mentioned earlier, similar molecular switches operate at various points in the cell cycle to promote the abrupt and complete transition from one cell-cycle state to the next.

Condensin Helps Configure Duplicated Chromosomes for Separation

At the end of S phase, the immensely long DNA molecules of the sister chromatids are tangled in a mass of partially catenated DNA and proteins. Any attempt to pull the sisters apart in this state would undoubtedly lead to breaks in the chromosomes. To avoid this disaster, the cell devotes a great deal of energy in early mitosis to gradually reorganizing the sister chromatids into relatively short, distinct structures that can be pulled apart more easily in anaphase. These chromosomal changes involve two processes: *chromosome condensation*, in which the chromatids are dramatically compacted; and *sister-chromatid resolution*, whereby the two sisters are resolved into distinct, separable units (**Figure 17–26**). Resolution results from the decatenation of the sister DNAs, accompanied by the partial removal of cohesin molecules along the chromosome arms. As a result, when the cell reaches metaphase, the sister chromatids appear in the microscope as compact, rod-like structures that are joined tightly at their centromeric regions and only loosely along their arms.

The condensation and resolution of sister chromatids depends, at least in part, on a five-subunit protein complex called **condensin**. Condensin structure is related to that of the cohesin complex that holds sister chromatids together (see Figure 17–24). It contains two SMC subunits like those of cohesin, plus three non-SMC subunits (**Figure 17–27**). Condensin may form a ring-like structure that somehow uses the energy provided by ATP hydrolysis to promote the compaction and resolution of sister chromatids. Condensin is able to change the coiling of DNA molecules in a test tube, and this coiling activity is thought to be important for chromosome condensation during mitosis. Interestingly, phosphorylation of condensin subunits by M-Cdk stimulates this coiling activity, providing one mechanism by which M-Cdk may promote chromosome restructuring in early mitosis.

The Mitotic Spindle Is a Microtubule-Based Machine

The central event of mitosis—chromosome segregation—depends in all eucaryotes on a complex and beautiful machine called the **mitotic spindle**. The spindle is a bipolar array of microtubules, which pulls sister chromatids apart in anaphase, thereby segregating the two sets of chromosomes to opposite ends of the cell, where they are packaged into daughter nuclei. M-Cdk triggers the assembly of the spindle early in mitosis, in parallel with the chromosome restructuring just described. Before we consider how the spindle assembles and how its microtubules attach to sister chromatids, we briefly review the basic features of spindle structure.

As discussed in Chapter 16, the core of the mitotic spindle is a bipolar array of microtubules, the minus ends of which are focused at the two spindle poles, and the plus ends of which radiate outward from the poles (**Figure 17–28**). <GTCT> The plus ends of some microtubules—called the **interpolar microtubules**—interact with the plus ends of microtubules from the other pole, resulting in an antiparallel array in the spindle midzone. The plus ends of other

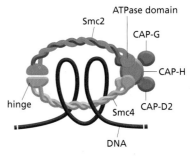

Figure 17–27 **Condensin.** Condensin is a five-subunit protein complex that resembles cohesin (see Figure 17–24). The head domains of its two major subunits, Smc2 and Smc4, are held together by three additional subunits. It is not clear how condensin catalyzes the restructuring and compaction of chromosome DNA, but it may form a ring structure that encircles loops of DNA as shown; it can hydrolyze ATP and coil DNA molecules in a test tube.

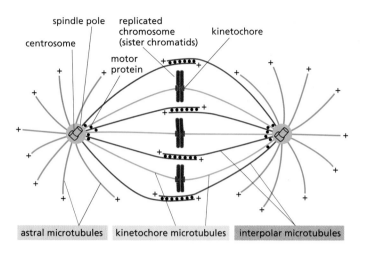

Figure 17–28 The three classes of microtubules of the mitotic spindle in an animal cell. The plus ends of the microtubules project away from the centrosomes, while the minus ends are anchored at the spindle poles, which in this example are organized by centrosomes. Kinetochore microtubules connect the spindle poles with the kinetochores of sister chromatids, while interpolar microtubules from the two poles interdigitate at the spindle equator. Astral microtubules radiate out from the poles into the cytoplasm and usually interact with the cell cortex, helping to position the spindle in the cell.

microtubules—the **kinetochore microtubules**—are attached to sister chromatid pairs at large protein structures called *kinetochores*, which are located at the *centromere* of each sister chromatid. Finally, many spindles also contain **astral microtubules** that radiate outward from the poles and contact the cell cortex, helping to position the spindle in the cell.

In most somatic animal cells, each spindle pole is focused at a protein organelle called the **centrosome** (discussed in Chapter 16). Each centrosome consists of a cloud of amorphous material (called the *pericentriolar matrix)* that surrounds a pair of *centrioles* (**Figure 17–29**). The pericentriolar matrix nucleates a radial array of microtubules, with their fast-growing plus ends projecting outward and their minus ends associated with the centrosome. The matrix contains a variety of proteins, including microtubule-dependent motor proteins,

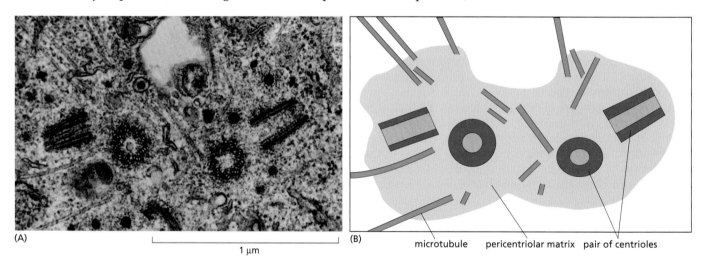

(A)   1 μm

(B)   microtubule   pericentriolar matrix   pair of centrioles

Figure 17–29 The centrosome. (A) Electron micrograph of an S-phase mammalian cell in culture, showing a duplicated centrosome. Each centrosome contains a pair of centrioles; although the centrioles have duplicated, they remain together in a single complex, as shown in the drawing of the micrograph in (B). One centriole of each centriole pair has been cut in cross section, while the other is cut in longitudinal section, indicating that the two members of each pair are aligned at right angles to each other. The two halves of the replicated centrosome, each consisting of a centriole pair surrounded by pericentriolar matrix, will split and migrate apart to initiate the formation of the two poles of the mitotic spindle when the cell enters M phase. (C) Electron micrograph of a centriole pair that has been isolated from a cell. The two centrioles have partly separated during the isolation procedure but remain tethered together by fine fibers, which keep the centriole pair together until it is time for them to separate. Both centrioles are cut longitudinally, and it can now be seen that the two have different structures: the mother centriole is larger and more complex than the daughter centriole, and only the mother centriole is associated with pericentriolar matrix that nucleates microtubules. Each daughter centriole will mature during the next cell cycle, when it will replicate to give rise to its own daughter centriole. (A, from M. McGill, D.P. Highfield, T.M. Monahan and B.R. Brinkley, *J. Ultrastruct. Res.* 57:43–53, 1976. With permission from Academic Press; C, from M. Paintrand et al., *J. Struct. Biol.* 108:107–128, 1992. With permission from Elsevier.)

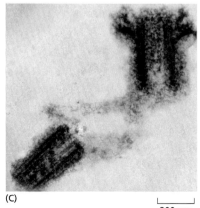

(C)   200 nm

coiled-coil proteins that link the motors to the centrosome, structural proteins, and components of the cell-cycle control system. Most important, it contains the *γ-tubulin ring complex*, which is the component mainly responsible for nucleating microtubules (discussed in Chapter 16).

Some cells—notably the cells of higher plants and the oocytes of many vertebrates—do not have centrosomes, and microtubule-dependent motor proteins and other proteins associated with microtubule minus ends organize and focus the spindle poles.

Microtubule-Dependent Motor Proteins Govern Spindle Assembly and Function

The assembly and function of the mitotic spindle depend on numerous microtubule-dependent motor proteins. As discussed in Chapter 16, these proteins belong to two families—the kinesin-related proteins, which usually move toward the plus end of microtubules, and dyneins, which move toward the minus end. In the mitotic spindle, these motor proteins generally operate at or near the ends of the microtubules. Four major types of motor proteins—*kinesin-5, kinesin-14, kinesins-4 and 10*, and *dynein*—are particularly important in spindle assembly and function (**Figure 17–30**).

Kinesin-5 proteins contain two motor domains that interact with the plus ends of antiparallel microtubules in the spindle midzone. Because the two motor domains move toward the plus ends of the microtubules, they slide the two antiparallel microtubules past each other toward the spindle poles, forcing the poles apart. Kinesin-14 proteins, by contrast, are minus-end directed motors with a single motor domain and other domains that can interact with a different microtubule. They can cross-link antiparallel interpolar microtubules at the spindle midzone and tend to pull the poles together. Kinesin-4 and kinesin-10 proteins, also called *chromokinesins*, are plus-end directed motors that associate with chromosome arms and push the attached chromosome away from the pole (or the pole away from the chromosome). Finally, dyneins are minus-end directed motors that, together with associated proteins, organize microtubules at various cellular locations. They link the plus ends of astral microtubules to components of the actin cytoskeleton at the cell cortex, for example; by moving toward the minus end of the microtubules, the dynein motors pull the spindle poles toward the cell cortex and away from each other.

Two Mechanisms Collaborate in the Assembly of a Bipolar Mitotic Spindle

The mitotic spindle must have two poles if it is to pull the two sets of sister chromatids to opposite ends of the cell in anaphase. In animal cells, the primary focus of this chapter, two mechanisms collaborate to ensure the bipolarity of the spindle. One depends on the ability of mitotic chromosomes to nucleate and stabilize microtubules and on the ability of the various motor proteins just described to organize microtubules into a bipolar array, with minus ends

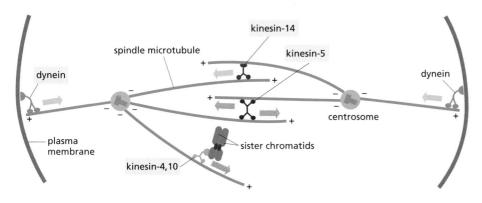

Figure 17–30 Major motor proteins of the spindle. Four major classes of microtubule-dependent motor proteins (*yellow* boxes) contribute to spindle assembly and function (see text). The colored arrows indicate the direction of motor movement along a microtubule— *blue* toward the minus end, and *red* toward the plus end.

focused at two spindle poles and plus ends interacting with each other in the spindle midzone. The other depends on the ability of centrosomes to help form the spindle poles. A typical animal cell enters mitosis with a pair of centrosomes, each of which nucleates a radial array of microtubules. The two centrosomes facilitate bipolar spindle assembly by providing a pair of prefabricated spindle poles. Centrosomes are not essential for the assembly of a bipolar spindle, however, since a functional spindle forms in cells that normally lack centrosomes and in cultured cells in which a laser beam has destroyed the centrosome.

We now describe the steps of spindle assembly, beginning with centrosome-dependent assembly in early mitosis. We then consider the self-organization mechanism that does not require centrosomes and becomes particularly important after nuclear envelope breakdown.

Centrosome Duplication Occurs Early in the Cell Cycle

Most animal cells contain a single centrosome that nucleates most of the cell's cytoplasmic microtubules. The centrosome duplicates when the cell enters the cell cycle, so that by the time the cell reaches mitosis there are two centrosomes. Centrosome duplication begins at about the same time as the cell enters S phase. The G_1/S-Cdk (a complex of cyclin E and Cdk2 in animal cells; see Table 17–1) that triggers cell cycle entry also initiates centrosome duplication. The two centrioles in the centrosome separate, and each nucleates the formation of a single new centriole, resulting in two centriole pairs within an enlarged pericentriolar matrix (**Figure 17–31**). This centrosome pair remains together on one side of the nucleus until the cell enters mitosis.

There are interesting parallels between centrosome duplication and chromosome duplication. Both use a semi-conservative mechanism of duplication, in which the two halves separate and serve as templates for construction of a new half. Centrosomes, like chromosomes, must replicate once and only once per cell cycle, to ensure that the cell enters mitosis with only two copies: an incorrect number of centrosomes could lead to defects in spindle assembly and thus errors in chromosome segregation.

The mechanisms that limit centrosome duplication to once per cell cycle are uncertain. In many cell types, experimental inhibition of DNA synthesis blocks centrosome duplication, providing one mechanism by which centrosome number is kept in check. Other cell types, however, including those in the early embryos of flies, sea urchins, and frogs, do not contain such a mechanism and centrosome duplication continues if chromosome duplication is blocked. It is not known how such cells limit centrosome duplication to once per cell cycle.

M-Cdk Initiates Spindle Assembly in Prophase

At the beginning of mitosis, the sudden rise in M-Cdk activity initiates spindle assembly. In animal cells, the two centrosomes move apart along the nuclear

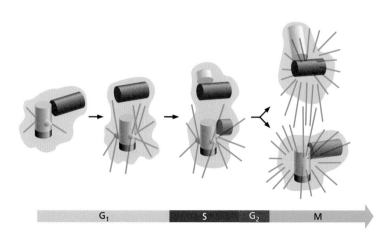

Figure 17–31 Centriole replication. The centrosome consists of a centriole pair and associated pericentriolar matrix *(green)*. At a certain point in G_1, the two centrioles of the pair separate by a few micrometers. During S phase, a daughter centriole begins to grow near the base of each mother centriole and at a right angle to it. The elongation of the daughter centriole is usually completed by G_2. The two centriole pairs remain close together in a single centrosomal complex until the beginning of M phase, when the complex splits in two and the two halves begin to separate. Each centrosome now nucleates its own radial array of microtubules called an aster.

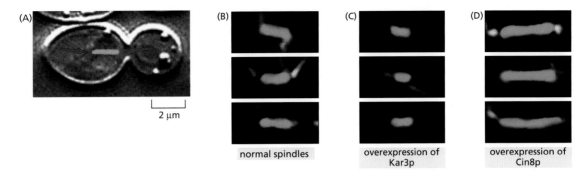

Figure 17–32 The influence of opposing motor proteins on spindle length in budding yeast. (A) A differential-interference-contrast micrograph of a mitotic yeast cell. The spindle is highlighted in *green,* and the position of the spindle poles is indicated by *red arrows.* The nuclear envelope does not break down during mitosis in yeasts, and the spindle forms inside the nucleus. In (B–D), the mitotic spindles have been stained with fluorescent anti-tubulin antibodies. (B) Normal yeast cells. (C) Overexpression of the minus-end-directed motor protein Kar3 (a kinesin-14 protein) leads to abnormally short spindles. (D) Overexpression of the plus-end-directed motor protein Cin8 (a kinesin-5 protein) leads to abnormally long spindles. Thus, it seems that a balance between opposing motor proteins determines spindle length in these cells. (A, courtesy of Kerry Bloom; B–D, from W. Saunders, V. Lengyel and M.A. Hoyt, *Mol. Biol. Cell* 8:1025–1033, 1997. With permission from American Society for Cell Biology.)

envelope, and the plus ends of the microtubules between them interdigitate to form the interpolar microtubules of the developing spindle. At the same time, the amount of γ-tubulin ring complexes in each centrosome increases greatly, increasing the ability of the centrosomes to nucleate new microtubules, a process called *centrosome maturation.*

Multiple motor proteins drive the separation of centrosomes in early mitosis. In prophase, minus-end directed dynein motor proteins at the plus ends of astral microtubules provide the major force. These motors are anchored at the cell cortex or on the nuclear envelope, and their movement toward the microtubule minus end pulls the centrosomes apart (see Figure 17–30). Following nuclear envelope breakdown at the end of prophase, interactions between the centrosomal microtubules and the cell cortex allow actin–myosin bundles in the cortex to pull the centrosomes further apart. Finally, kinesin-5 motors cross-link the overlapping, antiparallel ends of interpolar microtubules and push the poles apart (see Figure 17–30).

The balance of opposing forces generated by different types of motor proteins determines the final length of the spindle. Dynein and kinesin-5 motors generally promote centrosome separation and increase spindle length. Kinesin-14 proteins do the opposite: they are minus-end directed motors and interact with a microtubule from one pole while traveling toward the minus end of an antiparallel microtubule from the other pole; as a result, they tend to pull the poles together. It is not clear how the cell regulates the balance of opposing forces to generate the appropriate spindle length (**Figure 17–32**).

M-Cdk and other mitotic protein kinases are required for centrosome separation and maturation. M-Cdk and aurora-A phosphorylate kinesin-5 motors and stimulate them to drive centrosome separation. Aurora-A and Plk also phosphorylate components of the centrosome and thereby promote its maturation.

The Completion of Spindle Assembly in Animal Cells Requires Nuclear Envelope Breakdown

The centrosomes and microtubules of animal cells are located in the cytoplasm, separated from the chromosomes by the double membrane barrier of the nuclear envelope (discussed in Chapter 12). Clearly, the attachment of sister chromatids to the spindle requires the removal of this barrier. In addition, many of the motor proteins and microtubule regulators that promote spindle assembly are associated with the chromosomes inside the nucleus. Nuclear envelope breakdown allows these proteins to carry out their important functions in spindle assembly.

Nuclear envelope breakdown is a complex, multi-step process that is thought to begin when M-Cdk phosphorylates several subunits of the giant nuclear pore complexes in the nuclear envelope. This initiates the disassembly of nuclear pore complexes and their dissociation from the envelope. M-Cdk also phosphorylates components of the nuclear lamina, the structural framework that lies beneath the envelope. The phosphorylation of these lamina components and of several inner nuclear envelope proteins leads to disassembly of the nuclear lamina and the breakdown of the envelope membranes into small vesicles.

Microtubule Instability Increases Greatly in Mitosis

Most animal cells in interphase contain a cytoplasmic array of microtubules radiating out from the single centrosome. As discussed in Chapter 16, the microtubules of this interphase array are in a state of *dynamic instability*, in which individual microtubules are either growing or shrinking and stochastically switch between the two states. The switch from growth to shrinkage is called a *catastrophe*, and the switch from shrinkage to growth is called a *rescue* (see Figure 16–16). New microtubules are continually being created to balance the loss of those that disappear completely by depolymerization.

Entry into mitosis signals an abrupt change in the cell's microtubules. The interphase array of few, long microtubules is converted to a larger number of shorter and more dynamic microtubules surrounding each centrosome. During prophase, and particularly in prometaphase and metaphase (see Panel 17–1), the half-life of microtubules decreases dramatically. This increase in microtubule instability, coupled with the increased ability of centrosomes to nucleate microtubules as mentioned earlier, results in remarkably dense and dynamic arrays of spindle microtubules that are ideally suited for capturing sister chromatids.

M-Cdk initiates these changes in microtubule behavior, at least in part, by phosphorylating two classes of proteins that control microtubule dynamics (discussed in Chapter 16). These include microtubule-dependent motor proteins and **microtubule-associated proteins (MAPs)**. Experiments using cell-free *Xenopus* egg extracts, which reproduce many of the changes that occur in intact cells during the cell cycle, have revealed the roles of these regulators in controlling microtubule dynamics. If centrosomes and fluorescent tubulin are added to these extracts, fluorescent microtubules nucleate from the centrosomes, and we can observe the behavior of individual microtubules by time-lapse fluorescence video microscopy. The microtubules in mitotic extracts differ from those in interphase extracts primarily by the increased rate of catastrophes, in which the microtubules switch abruptly from slow growth to rapid shortening.

Two classes of proteins govern microtubule dynamics in mitosis. Proteins called **catastrophe factors** destabilize microtubule arrays by increasing the frequency of catastrophes (see Figure 16–16). One of these proteins is a kinesin-related protein that does not function as a motor. MAPs, by contrast, have the opposite effect, stabilizing microtubules in various ways: they can increase the frequency of rescues, in which microtubules switch from shrinkage to growth, or they can either increase the growth rate or decrease the shrinkage rate of microtubules. Thus, in principle, changes in catastrophe factors and MAPs can make microtubules much more dynamic in M phase by increasing total microtubule depolymerization rates, decreasing total microtubule polymerization rates, or both.

In *Xenopus* egg extracts, the balance between a single type of catastrophe factor and a single type of MAP determines the catastrophe rate and the steady-state length of microtubules (**Figure 17–33**). This balance, in turn, governs the assembly of the mitotic spindle, as microtubules that are either too long or too short cannot form a normal spindle. One way in which M-Cdk may control microtubule length is by phosphorylating this MAP and reducing its ability to stabilize microtubules. Even if the activity of the catastrophe factor remained constant throughout the cell cycle, the balance between the two opposing activities of the MAP and catastrophe factor would shift, increasing the dynamic instability of the microtubules.

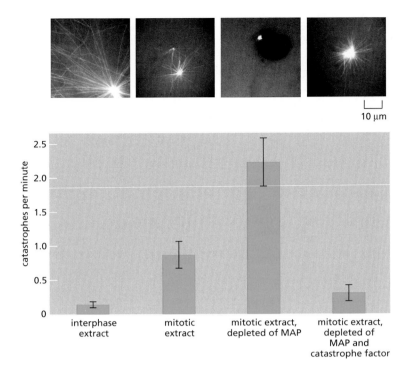

10 μm

Figure 17–33 Experimental evidence that the balance between catastrophe factors and MAPs influences the frequency of microtubule catastrophes and microtubule length. Interphase or mitotic *Xenopus* egg extracts were incubated with centrosomes and fluorescent tubulin, and the behavior of individual microtubules nucleated from the centrosomes was followed by fluorescence video microscopy. As expected, the catastrophe rate is higher in mitotic than in interphase extracts. The depletion of a specific MAP (called Xmap215) from the mitotic extracts increases the catastrophe rate, indicating that this MAP inhibits catastrophes in mitotic extracts. Inhibition of a specific catastrophe factor (the kinesin-related protein Mcak) greatly reduces the catastrophe rate in the MAP-depleted mitotic extracts, indicating that this factor is responsible for stimulating catastrophes in mitotic extracts. Thus, the catastrophe rate depends on the balance between the MAP and the catastrophe factor. Fluorescence micrographs of the asters formed in the different experimental conditions are shown in the top panels; note that the higher the catastrophe rates, the shorter the microtubules. (From R. Tournebize et al., *Nat. Cell Biol.* 2:13–19, 2000. With permission from Macmillan Publishers Ltd.)

Mitotic Chromosomes Promote Bipolar Spindle Assembly

Chromosomes are not just passive passengers in the process of spindle assembly. By creating a local environment that favors both microtubule nucleation and microtubule stabilization, they play an active part in spindle formation. The influence of the chromosomes can be demonstrated by using a fine glass needle to reposition them after the spindle has formed. For some cells in metaphase, if a single chromosome is tugged out of alignment, a mass of new spindle microtubules rapidly appears around the newly positioned chromosome, while the spindle microtubules at the chromosome's former position depolymerize. This property of the chromosomes seems to depend, at least in part, on a guanine-nucleotide exchange factor (GEF) that is bound to chromatin; the GEF stimulates a small GTPase in the cytosol called *Ran* to bind GTP in place of GDP. The activated Ran–GTP, which is also involved in nuclear transport (discussed in Chapter 12), releases microtubule-stabilizing proteins from protein complexes in the cytosol, thereby stimulating the local nucleation and stabilization of microtubules around chromosomes.

It is this ability of chromosomes to stabilize and organize microtubules that enables cells to form bipolar spindles in the absence of centrosomes, as discussed earlier. Acentrosomal spindle assembly is thought to begin with the nucleation and stabilization of microtubules around the chromosomes. Motor proteins, particularly members of the kinesin-5 family (see Figure 17–30), then cross-link microtubules in an antiparallel orientation and push their minus ends apart. Kinesins-4 and 10 on the chromosome arms also help push the minus ends away from the chromosomes. Dynein, kinesin-14, and various minus-end binding proteins then cross-link and focus the minus ends of the microtubules to form the two spindle poles (**Figure 17–34**).

Cells that normally lack centrosomes, such as those of higher plants and many animal oocytes, use this chromosome-based self-organization process to form spindles. It is also the process used to assemble spindles in certain insect embryos that have been induced to develop from eggs without fertilization (that is, *parthenogenetically*); as the sperm normally provides the centrosome when it fertilizes an egg (discussed in Chapter 21), the mitotic spindles in these parthenogenic embryos develop without centrosomes (**Figure 17–35**). Even in cells that normally contain centrosomes, the chromosomes help organize the

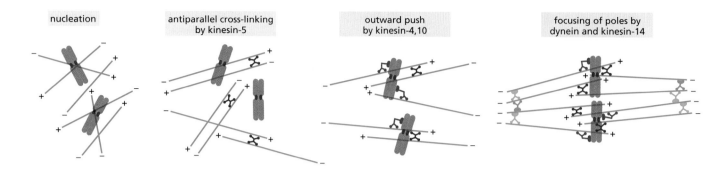

| nucleation | antiparallel cross-linking by kinesin-5 | outward push by kinesin-4,10 | focusing of poles by dynein and kinesin-14 |

Figure 17–34 Spindle self-organization by motor proteins. Mitotic chromosomes stimulate the local production of RanGTP (not shown), which activates proteins that nucleate and promote the formation of microtubules in the vicinity of the chromosomes. Kinesin-5 motor proteins (see Figure 17–30) organize these microtubules into antiparallel bundles, while plus-end directed kinesins-4 and 10 link the microtubules to chromosome arms and push minus ends away from the chromosomes. Dynein and kinesin-14 motors, together with numerous other proteins, focus these minus ends into a pair of spindle poles.

spindle microtubules and, with the help of various motor proteins, can promote the assembly of a bipolar mitotic spindle if the centrosomes are removed. Although the resulting acentrosomal spindle can segregate chromosomes normally, it lacks astral microtubules, which are responsible for positioning the spindle in animal cells; as a result, the spindle is often mispositioned, resulting in abnormalities in cytokinesis.

Kinetochores Attach Sister Chromatids to the Spindle

Following the assembly of a bipolar microtubule array, the second major step in spindle formation is the attachment of the array to the chromosomes. Spindle microtubules are attached to each sister chromatid at the *kinetochore*, a giant, multilayered protein structure that is built on the heterochromatin that forms at the centromeric region of the chromosome (**Figure 17–36**). The plus ends of kinetochore microtubules are embedded head-on in specialized microtubule-attachment sites within the kinetochore. Animal cell kinetochores contain 10–40 of these attachment sites, whereas yeast kinetochores contain just one. Each attachment site contains a protein collar that surrounds the microtubule near its end, thereby holding the microtubule tightly to the kinetochore while still allowing the addition or removal of tubulin subunits at this end (**Figure 17–37**). Regulation of plus end polymerization and depolymerization at the kinetochore is critical for the control of chromosome movement on the spindle, as we discuss later.

Cells containing centrosomes employ a "search and capture" mechanism to attach their mitotic chromosomes to the spindle. The dynamic plus ends of microtubules radiate outward from the centrosomes and eventually capture the kinetochore of one sister chromatid. In newt lung cells, where we can observe the initial capture event in a microscope, the kinetochore is seen first to bind to the side of the microtubule and then to slide rapidly along it toward the centrosome. The lateral attachment to the chromosome is rapidly converted to an end-on attachment. At the same time, microtubules growing from

Figure 17–35 Bipolar spindle assembly without centrosomes in parthenogenetic embryos of the insect *Sciara* (or fungus gnat). The microtubules are stained *green*, the chromosomes *red*. The *top* fluorescence micrograph shows a normal spindle formed with centrosomes in a normally fertilized *Sciara* embryo. The *bottom* micrograph shows a spindle formed without centrosomes in an embryo that initiated development without fertilization. Note that the spindle with centrosomes has an aster at each pole of the spindle, whereas the spindle formed without centrosomes does not. Both types of spindles are able to segregate the replicated chromosomes. (From B. de Saint Phalle and W. Sullivan, *J. Cell Biol.* 141:1383–1391, 1998. With permission from The Rockefeller University Press.)

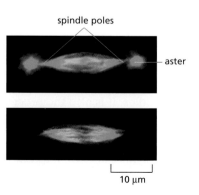

spindle poles

aster

10 μm

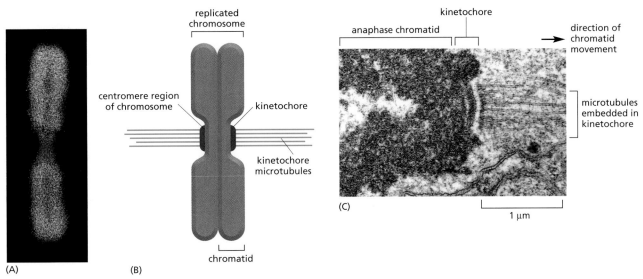

Figure 17–36 The kinetochore. (A) A fluorescence micrograph of a metaphase chromosome stained with a DNA-binding fluorescent dye and with human autoantibodies that react with specific kinetochore proteins. The two kinetochores, one associated with each chromatid, are stained *red*. (B) A drawing of a metaphase chromosome showing its two sister chromatids attached to the plus ends of kinetochore microtubules. Each kinetochore forms a plaque on the surface of the centromere. The number of microtubules bound to a metaphase kinetochore varies from 1 in budding yeast to 40 in some mammalian cells. (C) Electron micrograph of an anaphase chromatid with microtubules attached to its kinetochore. While most kinetochores have a trilaminar structure, the one shown here (from a green alga) has an unusually complex structure with additional layers. (A, courtesy of B.R. Brinkley; C, from J.D. Pickett-Heaps and L.C. Fowke, *Aust. J. Biol. Sci.* 23:71–92, 1970. With permission from CSIRO.)

the opposite spindle pole attach to the kinetochore on the opposite side of the chromosome, forming a bipolar attachment (**Figure 17–38**).

How does chromosome attachment occur in the absence of centrosomes? One possibility is that short microtubules in the vicinity of the chromosomes interact with kinetochores and become embedded in the plus-end-binding collars of the kinetochore. Polymerization at these plus ends would then result in growth of the microtubules away from the kinetochore. The minus ends of these kinetochore microtubules, like other minus ends in centrosome-free spindles, would eventually become cross-linked to other minus ends and focused by motor proteins at the spindle pole (see Figure 17–34).

Bi-Orientation Is Achieved by Trial and Error

The success of mitosis demands that sister chromatids in a pair attach to opposite poles of the mitotic spindle, so that they move to opposite ends of the cell when they separate in anaphase. How is this mode of attachment, called **bi-orientation**, achieved? What prevents the attachment of both kinetochores to the same spindle pole or the attachment of one kinetochore to both spindle poles? Part of the answer is that sister kinetochores are constructed in a back-to-back orientation that reduces the likelihood that both kinetochores can face the same spindle pole. Nevertheless, incorrect attachments do occur, and elegant regulatory mechanisms have evolved to correct them.

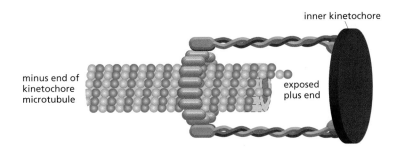

Figure 17–37 A microtubule attachment site in a kinetochore. Each site is thought to contain a collar structure (*yellow*) that surrounds the microtubule plus end, allowing polymerization and depolymerization to occur at the exposed plus end while the microtubule remains attached to the kinetochore.

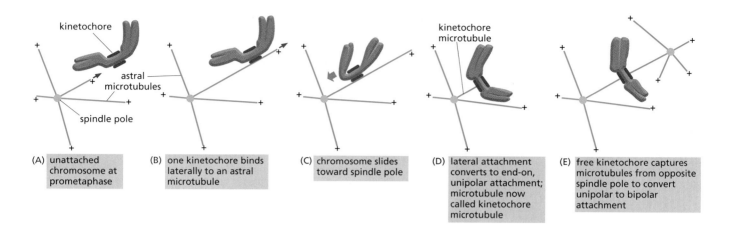

(A) unattached chromosome at prometaphase

(B) one kinetochore binds laterally to an astral microtubule

(C) chromosome slides toward spindle pole

(D) lateral attachment converts to end-on, unipolar attachment; microtubule now called kinetochore microtubule

(E) free kinetochore captures microtubules from opposite spindle pole to convert unipolar to bipolar attachment

Incorrect attachments are corrected by a system of trial and error that is based on a simple principle: incorrect attachments are highly unstable and do not last, while correct attachments are locked in place. But, how does the kinetochore sense a correct attachment? The answer appears to be tension (**Figure 17–39**). When a sister chromatid pair is properly bi-oriented on the spindle, the two kinetochores are pulled in opposite directions by strong poleward forces. Sister-chromatid cohesion resists these poleward forces, creating high levels of tension within the kinetochores. When chromosomes are incorrectly attached—when both sister chromatids are attached to the same spindle pole, for example—tension is low and the kinetochore sends an inhibitory signal that loosens the grip of its microtubule attachment site, allowing detachment to occur. When bi-orientation occurs, the high tension at the kinetochore shuts off the inhibitory signal, strengthening microtubule attachment. In animal cells, tension not only increases the affinity of the attachment site but also leads to the attachment of additional microtubules to the kinetochore. This results in the formation of a thick *kinetochore fiber* composed of multiple microtubules.

The tension-sensing mechanism depends on the protein kinase aurora-B, which is associated with the kinetochore. Aurora-B is thought to generate the inhibitory signal that reduces the strength of microtubule attachment in the absence of tension. It phosphorylates several components of the microtubule attachment site, decreasing the site's affinity for a microtubule plus end. Aurora-B is inactivated when bi-orientation occurs, thereby reducing kinetochore phosphorylation and increasing the affinity of the attachment site.

Figure 17–38 The capture of centrosome microtubules by kinetochores. The *red arrow* in (A) indicates the direction of microtubule growth, while the *gray arrow* in (C) indicates the direction of chromosome sliding.

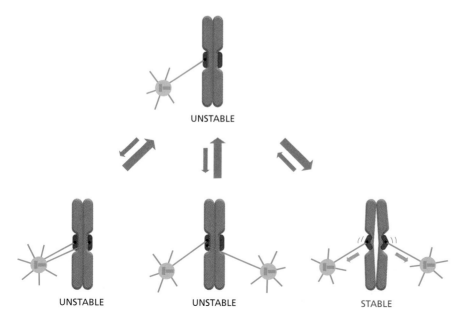

UNSTABLE

UNSTABLE

UNSTABLE

STABLE

Figure 17–39 Alternative forms of chromosome attachment. Initially, a single microtubule from a spindle pole binds to one kinetochore in a sister chromatid pair. Additional microtubules can then bind to the chromosome in various ways. A microtubule from the same spindle pole can attach to the other sister kinetochore, or microtubules from both spindle poles can attach to one kinetochore. These incorrect attachments are unstable, however, so that one of the two microtubules tends to dissociate. When a second microtubule from the opposite pole binds to the second kinetochore, the sister kinetochores are thought to sense tension across their microtubule-binding sites, which triggers an increase in microtubule binding affinity. This correct attachment is thereby locked in place.

Following their attachment to the two spindle poles, the chromosomes are tugged back and forth, eventually assuming a position equidistant between the two spindle poles, a position called the **metaphase plate**. In vertebrate cells, the chromosomes then oscillate gently at the metaphase plate, awaiting the signal for the sister chromatids to separate. The signal is produced, with a predictable lag time, after the bipolar attachment of the last of the chromosomes, as we discuss later.

Multiple Forces Move Chromosomes on the Spindle

Motor proteins and other mechanisms generate the forces that move chromosomes on the microtubules of the mitotic spindle. Three major forces are thought to be particularly important.

The first major force pulls the kinetochore and its associated chromosome along the kinetochore microtubule toward the spindle pole. It is produced by proteins at the kinetochore itself. By an uncertain mechanism, depolymerization at the plus end of the microtubule somehow generates a force that pulls the kinetochore poleward (**Figure 17–40**). This force pulls on chromosomes during prometaphase and metaphase and is particularly important for moving sister chromatids toward the poles after they separate in anaphase, as we discuss later. Interestingly, this kinetochore-generated poleward force does not require ATP. This might seem implausible at first, but it has been shown that purified kinetochores in a test tube, with no ATP present, can remain attached to depolymerizing microtubules and thereby move. The energy that drives the movement is stored in the microtubule and is released when the microtubule depolymerizes; it ultimately comes from the hydrolysis of GTP that occurs after a tubulin subunit adds to the end of a microtubule (discussed in Chapter 16).

A second poleward force is provided in some cell types by **microtubule flux**, whereby the microtubules themselves are moved toward the spindle poles and dismantled at their minus ends. Until the onset of anaphase, the addition of new tubulin at the plus end of a microtubule compensates for the loss of tubulin at the minus end, so that microtubule length remains constant despite the movement of microtubules toward the spindle pole. Microtubule flux in metaphase spindles can be seen by an ingenious method in which very small amounts of fluorescent tubulin are injected into living cells (**Figure 17–41**). This results in the appearance of tiny fluorescent speckles that travel toward the poles on both kinetochore and interpolar microtubules. Any kinetochore that is attached to a microtubule undergoing such flux experiences a poleward force, which contributes to the generation of tension at the kinetochore and the poleward movement of sister chromatids after they separate in anaphase.

A third force acting on chromosomes is the *polar ejection force*. Plus-end-directed kinesin-4 and 10 motors on chromosome arms interact with interpolar microtubules and transport the chromosomes away from the spindle poles. This force is particularly important in prometaphase and metaphase, when it helps align the bi-oriented sister chromatid pairs at the metaphase plate (**Figure 17–42**).

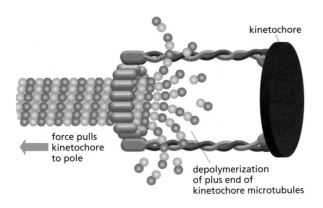

force pulls kinetochore to pole

kinetochore

depolymerization of plus end of kinetochore microtubules

Figure 17–40 How depolymerization may pull the kinetochore toward the spindle pole. When depolymerization occurs, the protofilaments of the microtubule curl outward (see Figure 16–16) and push against the collar structure that surrounds the microtubule plus end. In principle, this will move the kinetochore toward the microtubule minus end at the spindle pole.

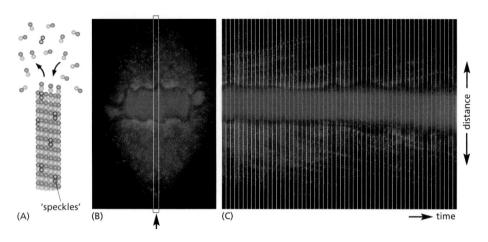

(A) 'speckles' (B) (C) → time

distance

spindle pole

TUBULIN REMOVAL

TUBULIN ADDITION

TUBULIN ADDITION

speckles moving poleward

TUBULIN REMOVAL

(D)

Figure 17–41 Microtubule flux in the metaphase spindle. (A) To observe microtubule flux, a very small amount of fluorescent tubulin is injected into living cells so that individual microtubules form with a very small proportion of fluorescent tubulin. Such microtubules have a speckled appearance when viewed by fluorescence microscopy. (B) Fluorescence micrographs of a mitotic spindle in a living newt lung epithelial cell. The chromosomes are colored *brown,* and the tubulin speckles are *red.* (C) The movement of individual speckles can be followed by time-lapse video microscopy. Images of the long, thin, rectangular, boxed region *(arrow)* in (B) were taken at sequential times and pasted side by side to make a montage of the region over time. Individual speckles can be seen to move toward the poles at a rate of about 0.75 μm/min, indicating that the microtubules are moving poleward. (D) Microtubule length in the metaphase spindle does not change significantly because new tubulin subunits are added at the microtubule plus end at the same rate as tubulin subunits are removed from the minus end. (B and C, from T.J. Mitchison and E.D. Salmon, *Nat. Cell Biol.* 3:E17–21, 2001. With permission from Macmillan Publishers Ltd.)

One of the most striking aspects of mitosis in vertebrate cells is the continuous oscillatory movement of the chromosomes in prometaphase and metaphase. When studied by video microscopy in newt lung cells, the movements are seen to switch between two states—a poleward (P) state, when the chromosomes are pulled toward the pole, and an away-from-the-pole (AP), or neutral, state, when the poleward forces are turned off and the polar ejection force pushes the chromosomes away from the pole. The switch between the two states may depend on the degree of tension in the kinetochore. It has been proposed, for example, that as chromosomes move toward the spindle pole, an increasing polar ejection force generates tension in the kinetochore nearest the

Figure 17–42 How opposing forces may drive chromosomes to the metaphase plate. (A) Evidence for a polar ejection force that pushes chromosomes away from the spindle poles toward the spindle equator. In this experiment, a laser beam severs a prometaphase chromosome that is attached to a single pole by kinetochore microtubules. The part of the severed chromosome without a kinetochore is pushed rapidly away from the pole, whereas the part with the kinetochore moves toward the pole, reflecting a decreased repulsion. (B) A model of how two opposing forces may cooperate to move chromosomes to the metaphase plate. Plus-end-directed motor proteins (kinesin-4 and kinesin-10) on the chromosome arms are thought to interact with microtubules to generate the polar ejection force, which pushes chromosomes toward the spindle equator (see Figure 17–30). Poleward forces generated by depolymerization at the kinetochore, together with microtubule flux, are thought to pull chromosomes toward the pole.

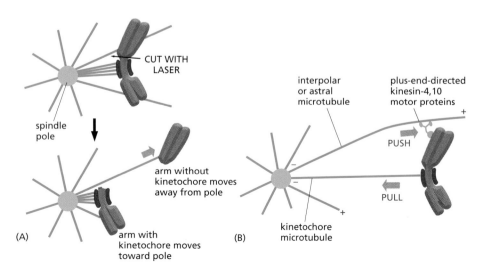

CUT WITH LASER

spindle pole

arm without kinetochore moves away from pole

arm with kinetochore moves toward pole

(A)

interpolar or astral microtubule

plus-end-directed kinesin-4,10 motor proteins

PUSH

PULL

kinetochore microtubule

(B)

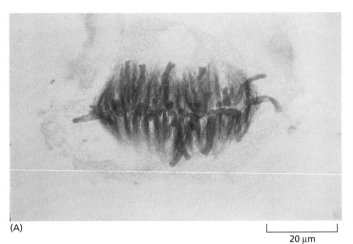
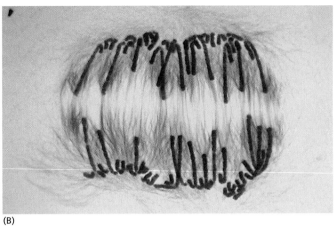

(A)   (B)

|_____|
20 μm

Figure 17–43 Sister-chromatid separation at anaphase. In the transition from metaphase (A) to anaphase (B), sister chromatids suddenly separate and move toward opposite poles of the mitotic spindle—as shown in these light micrographs of *Haemanthus* (lily) endosperm cells that were stained with gold-labeled antibodies against tubulin. (Courtesy of Andrew Bajer.)

pole, triggering a switch to the away-from-the-pole state and gradually resulting in the accumulation of chromosomes at the equator of the spindle.

The APC/C Triggers Sister-Chromatid Separation and the Completion of Mitosis

After M-Cdk has triggered the complex rearrangements that occur in early mitosis, the cell cycle reaches its climax with the separation of the sister chromatids at the metaphase-to-anaphase transition (**Figure 17–43**). Although M-Cdk activity sets the stage for this event, the anaphase-promoting complex (APC/C) discussed earlier throws the switch that initiates sister-chromatid separation by ubiquitylating several mitotic regulatory proteins and thereby triggering their destruction (see Figure 17–20A).

During metaphase, cohesins holding the sister chromatids together resist the poleward forces that pull the sister chromatids apart. Anaphase begins with a sudden disruption of sister-chromatid cohesion, which allows the sisters to separate and move to opposite poles of the spindle. The APC/C initiates the process by targeting the inhibitory protein **securin** for destruction. Before anaphase, securin binds to and inhibits the activity of a protease called **separase**. The destruction of securin at the end of metaphase releases separase, which is then free to cleave one of the subunits of cohesin. The cohesins fall away, and the sister chromatids abruptly and synchronously separate (**Figure 17–44**).

In addition to securin, the APC/C also targets the S- and M-cyclins for destruction, leading to the loss of most Cdk activity in anaphase. Cdk inactivation allows phosphatases to dephosphorylate the many Cdk target substrates in the cell, as required for the completion of mitosis and cytokinesis (discussed later).

If the APC/C triggers anaphase, what activates the APC/C? The answer is only partly known. As mentioned earlier, APC/C activation requires the protein Cdc20, which binds to and activates the APC/C in mitosis (see Figure 17–20A). At least two processes regulate Cdc20 and its association with the APC/C. First, Cdc20 synthesis increases as the cell approaches mitosis, owing to an increase in the transcription of its gene. Second, phosphorylation of the APC/C helps Cdc20 bind to the APC/C, thereby helping to create an active complex. Among the kinases that phosphorylate and thus activate the APC/C is M-Cdk. Thus, M-Cdk not only triggers the early mitotic events leading up to metaphase, but it also sets the stage for progression into anaphase. The ability of M-Cdk to promote Cdc20–APC/C activity creates a negative feedback loop: M-Cdk sets in motion a regulatory process that leads to cyclin destruction and thus its own inactivation.

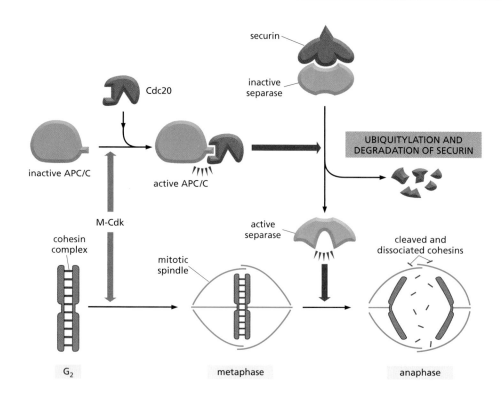

Figure 17–44 The initiation of sister-chromatid separation by the APC/C. The activation of APC/C by Cdc20 leads to the ubiquitylation and destruction of securin, which normally holds separase in an inactive state. The destruction of securin allows separase to cleave Scc1, a subunit of the cohesin complex holding the sister chromatids together (see Figure 17–24). The pulling forces of the mitotic spindle then pull the sister chromatids apart. In animal cells, phosphorylation by Cdks also inhibits separase (not shown). Thus, Cdk inactivation in anaphase (resulting from cyclin destruction) also promotes separase activation by allowing its dephosphorylation.

Unattached Chromosomes Block Sister-Chromatid Separation: The Spindle Assembly Checkpoint

Cells usually spend about half of mitosis in metaphase, with the chromosomes aligned on the metaphase plate, jostling about, awaiting the APC/C signal that induces sister chromatids to separate. Drugs that destabilize microtubules, such as colchicine or vinblastine (discussed in Chapter 16), arrest cells in mitosis for hours or even days. This observation led to the identification of a **spindle assembly checkpoint** mechanism that is activated by the drug treatment and blocks progression through the metaphase-to-anaphase transition. The checkpoint mechanism ensures that cells do not enter anaphase until all chromosomes are correctly bi-oriented on the mitotic spindle.

The spindle assembly checkpoint depends on a sensor mechanism that monitors the strength of microtubule attachment, and possibly tension, at the kinetochore. Any kinetochore that is not properly attached to the spindle sends out a negative signal that blocks Cdc20–APC/C activation and thus blocks the metaphase-to-anaphase transition. Only when the last kinetochore is properly attached is this block removed, allowing sister-chromatid separation to occur.

It is thought that inappropriately attached kinetochores somehow generate a diffusible signal that inhibits Cdc20–APC/C activity throughout the cell. The molecular basis of this signal is not clear, although several proteins, including *Mad2*, are recruited to unattached kinetochores and are required for the spindle assembly checkpoint to function (**Figure 17–45**). One appealing possibility, based primarily on detailed structural analyses of Mad2, is that the unattached kinetochore acts like an enzyme that catalyzes a change in the conformation of Mad2, so that Mad2 can bind and inhibit Cdc20–APC/C.

In mammalian somatic cells, the spindle assembly checkpoint determines the normal timing of anaphase. The destruction of securin in these cells begins moments after the last sister chromatid pair becomes bi-oriented on the spindle, and anaphase begins about 20 minutes later. Experimental inhibition of the checkpoint mechanism causes premature sister-chromatid separation and anaphase. Surprisingly, the normal timing of anaphase does not depend on the spindle assembly checkpoint in some cells, such as yeasts and the cells of early frog and fly embryos. Some other mechanism, as yet unknown, must determine the timing of anaphase in these cells.

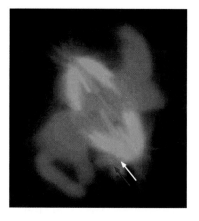

Figure 17–45 Mad2 protein on unattached kinetochores. This fluorescence micrograph shows a mammalian cell in prometaphase, with the mitotic spindle in *green* and the sister chromatids in *blue*. One sister chromatid pair is attached to only one pole of the spindle. Staining with anti-Mad2 antibodies indicates that Mad2 is bound to the kinetochore of the unattached sister chromatid (*red dot*, indicated by *red arrow*). A small amount of Mad2 is associated with the kinetochore of the sister chromatid that is attached to the spindle pole (*pale dot*, indicated by *white arrow*). (From J.C. Waters et al., *J. Cell Biol.* 141:1181–1191, 1998. With permission from The Rockefeller University Press.)

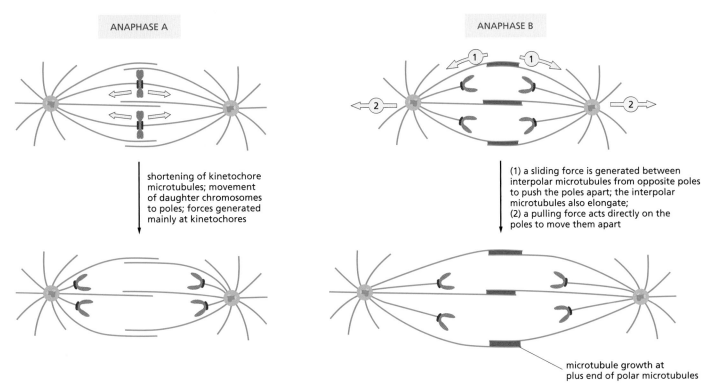

ANAPHASE A

shortening of kinetochore
microtubules; movement
of daughter chromosomes
to poles; forces generated
mainly at kinetochores

ANAPHASE B

(1) a sliding force is generated between
interpolar microtubules from opposite poles
to push the poles apart; the interpolar
microtubules also elongate;
(2) a pulling force acts directly on the
poles to move them apart

microtubule growth at
plus end of polar microtubules

Chromosomes Segregate in Anaphase A and B

The sudden loss of sister-chromatid cohesion at the onset of anaphase leads to sister-chromatid separation, which allows the forces of the mitotic spindle to pull the sisters to opposite poles of the cell—called *chromosome segregation*. The chromosomes move by two independent and overlapping processes. The first, referred to as **anaphase A**, is the initial poleward movement of the chromosomes, which is accompanied by shortening of the kinetochore microtubules. The second, referred to as **anaphase B**, is the separation of the spindle poles themselves, which begins after the sister chromatids have separated and the daughter chromosomes have moved some distance apart (**Figure 17–46**).

Chromosome movement in anaphase A depends on a combination of the two major poleward forces described earlier. The first is the force generated by microtubule depolymerization at the kinetochore, which results in the loss of tubulin subunits at the plus end as the kinetochore moves toward the pole. The second is provided by microtubule flux, which is the poleward movement of the microtubules toward the spindle pole, where minus end depolymerization occurs. The relative importance of these two forces during anaphase varies in different cell types: in embryonic cells, chromosome movement depends mainly on microtubule flux, for example, whereas movement in yeast and vertebrate somatic cells results primarily from forces generated at the kinetochore.

Spindle pole separation during anaphase B depends on motor-driven mechanisms similar to those that separate the two centrosomes in early mitosis (see Figure 17–30). Plus-end directed kinesin-5 motor proteins, which cross-link the overlapping plus ends of the interpolar microtubules, push the poles apart. In addition, dynein motors that anchor astral microtubule plus ends to the cell cortex pull the poles apart.

Although sister-chromatid separation initiates the chromosome movements of anaphase A, other mechanisms also ensure correct chromosome movements in anaphase A and spindle elongation in anaphase B. Most importantly, the completion of a normal anaphase depends on the dephosphorylation of Cdk substrates, which in most cells results from the APC/C-dependent destruction of cyclins. If M-cyclin destruction is prevented—by the production of a mutant form that is not recognized by the APC/C, for example—sister-chromatid separation generally occurs but the chromosome movements and microtubule behavior of anaphase are abnormal.

Figure 17–46 **The major forces that separate sister chromatids at anaphase in mammalian cells.** Chromosome movement toward the poles in anaphase A depends on the depolymerization of kinetochore microtubules and poleward microtubule flux. In anaphase B, the two spindle poles move apart. Two separate forces are thought to be responsible for anaphase B: the elongation and sliding of the interpolar microtubules past one another in the central spindle push the two poles apart, and motor proteins attached to the plasma membrane near each spindle pole act on astral microtubules to pull the poles away from each other, toward the cell surface.

The relative contributions of anaphase A and anaphase B to chromosome segregation vary greatly, depending on the cell type. In mammalian cells, anaphase B begins shortly after anaphase A and stops when the spindle is about twice its metaphase length; in contrast, the spindles of yeasts and certain protozoa primarily use anaphase B to separate the chromosomes at anaphase, and their spindles elongate to up to 15 times the metaphase length in the process.

Segregated Chromosomes Are Packaged in Daughter Nuclei at Telophase

By the end of anaphase, the daughter chromosomes have segregated into two equal groups at opposite ends of the cell. In **telophase**, the final stage of mitosis, the two sets of chromosomes are packaged into a pair of daughter nuclei. The first major event of telophase is the disassembly of the mitotic spindle, followed by the re-formation of the nuclear envelope. Initially, nuclear membrane fragments associate with the surface of individual chromosomes. These membrane fragments fuse to partly enclose clusters of chromosomes and then coalesce to re-form the complete nuclear envelope. Nuclear pore complexes are incorporated into the envelope, the nuclear lamina re-forms, and the envelope once again becomes continuous with the endoplasmic reticulum. Once the nuclear envelope has re-formed, the pore complexes pump in nuclear proteins, the nucleus expands, and the condensed mitotic chromosomes are reorganized into their interphase state, allowing gene transcription to resume. A new nucleus has been created, and mitosis is complete. All that remains is for the cell to complete its division into two.

We saw earlier that phosphorylation of various proteins by M-Cdk promotes spindle assembly, chromosome condensation, and nuclear envelope breakdown in early mitosis. It is thus not surprising that the dephosphorylation of these same proteins is required for spindle disassembly and the re-formation of daughter nuclei in telophase. In principle, these dephosphorylations and the completion of mitosis could be triggered by the inactivation of Cdks, the activation of phosphatases, or both. Although Cdk inactivation—resulting primarily from cyclin destruction—is mainly responsible in most cells, some cells also rely on activation of phosphatases. In budding yeast, for example, the completion of mitosis depends on the activation of a phosphatase called *Cdc14*, which dephosphorylates a subset of Cdk substrates involved in anaphase and telophase.

Meiosis Is a Special Form of Nuclear Division Involved in Sexual Reproduction

Most eucaryotic organisms reproduce sexually: the genomes of two parents mix to generate offspring that are genetically distinct from either parent (discussed in Chapter 21). The cells of these organisms are generally *diploid*: that is, they contain two slightly different copies, or *homologs*, of each chromosome, one from each parent. Sexual reproduction depends on a specialized nuclear division process called *meiosis*, which produces *haploid* cells carrying only a single copy of each chromosome. In many organisms, the haploid cells differentiate into specialized reproductive cells called *gametes*—eggs and sperm in most species. In these species, the reproductive cycle ends when a sperm and egg fuse to form a diploid zygote with the potential to form a new individual. Here, we consider the basic mechanisms and regulation of meiosis, with an emphasis on how they compare with those of mitosis. Meiosis is discussed in more detail in Chapter 21.

Meiosis begins with a round of chromosome duplication, called *meiotic S phase*, followed by two rounds of chromosome segregation, called *meiosis I* and *II*. Meiosis I segregates the homologs (each composed of a tightly linked pair of sister chromatids). Meiosis II, like conventional mitosis, segregates the sister chromatids of each homolog (**Figure 17–47**).

The first meiotic division solves the central problem of meiosis: how to segregate the homologous chromosomes. Like sister-chromatid segregation in

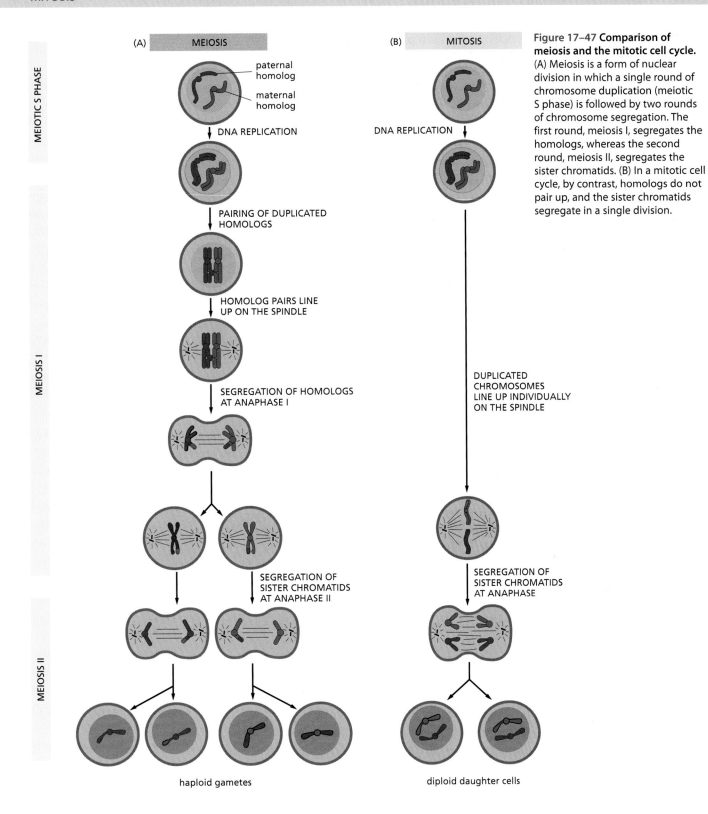

(A) MEIOSIS

MEIOTIC S PHASE

paternal homolog

maternal homolog

DNA REPLICATION

PAIRING OF DUPLICATED HOMOLOGS

HOMOLOG PAIRS LINE UP ON THE SPINDLE

MEIOSIS I

SEGREGATION OF HOMOLOGS AT ANAPHASE I

SEGREGATION OF SISTER CHROMATIDS AT ANAPHASE II

MEIOSIS II

haploid gametes

(B) MITOSIS

DNA REPLICATION

DUPLICATED CHROMOSOMES LINE UP INDIVIDUALLY ON THE SPINDLE

SEGREGATION OF SISTER CHROMATIDS AT ANAPHASE

diploid daughter cells

Figure 17–47 Comparison of meiosis and the mitotic cell cycle. (A) Meiosis is a form of nuclear division in which a single round of chromosome duplication (meiotic S phase) is followed by two rounds of chromosome segregation. The first round, meiosis I, segregates the homologs, whereas the second round, meiosis II, segregates the sister chromatids. (B) In a mitotic cell cycle, by contrast, homologs do not pair up, and the sister chromatids segregate in a single division.

mitosis, homolog segregation in meiosis I depends on the formation of linkages between homologs. These linkages allow the homolog pairs to be bi-oriented on the first meiotic spindle, with the homologs in a pair attached to opposite poles. Homolog linkage is removed at the onset of anaphase I, allowing the spindle to pull the homologs to opposite ends of the cell.

Linkages form between homologs by a remarkably complex and lengthy process that occurs after meiotic S phase, in a period called *meiotic prophase* or *prophase I*. This process begins with homolog *pairing*, whereby the homologs gradually move close to each other in the nucleus, primarily as a result of inter-actions between complementary DNA sequences in the two homologs. Homolog linkages are then locked in place by homologous recombination between nonsister chromatids in each homolog pair: DNA double-strand breaks are formed at several locations in each sister chromatid, resulting in large num-bers of DNA recombination events between the homologs. Some of these events lead to reciprocal DNA exchanges called *crossovers*, where the DNA of a chro-matid crosses over to become continuous with the DNA of a homologous chro-matid (**Figure 17–48**). At least one of these crossovers occurs in each homolog pair, ensuring that the homologs in every pair are physically connected when the cell enters the first meiotic division.

Another uniquely meiotic problem must be solved when the homolog pairs are attached to the first meiotic spindle. Each homolog contains two tightly linked sister chromatids, and thus the attachment of a homolog to a spindle pole requires that both sister kinetochores in a homolog attach to the same pole. This type of attachment is normally avoided during mitosis (see Figure 17–39). In meiosis I, however, the two sister kinetochores are somehow fused into a single microtubule-binding unit that attaches to just one pole (see Figure 21–12A). These mechanisms are reversed after meiosis I, so that in meiosis II the sister chromatid pairs can be bi-oriented on the spindle as they are in mitosis.

Crossovers hold homolog pairs together only because the arms of the sister chromatids are connected by sister-chromatid cohesion (see Figure 17–48). The loss of cohesion from sister-chromatid arms therefore triggers homolog separa-tion at the onset of anaphase I. In most species, the loss of arm cohesion in meiosis I depends on APC/C activation, which leads to securin destruction, sep-arase activation, and cohesin cleavage along the arms (see Figure 17–44). In con-trast to mitosis, however, cohesin complexes near the centromeres remain uncleaved in meiosis I because cohesin in that region is protected from separase (discussed in Chapter 21). Sister-chromatid pairs therefore remain linked at their centromeres throughout meiosis I, allowing their correct bi-orientation on the spindle in meiosis II. The mechanisms that block cohesin cleavage at the centromere in meiosis I are removed in meiosis II. At the onset of anaphase II, APC/C activation therefore triggers centromeric cohesin cleavage and sister-chromatid separation—much as it does in mitosis.

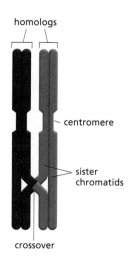

Figure 17–48 A crossover between homologs. As in mitosis, the sister chromatids in each homolog are tightly connected along their entire lengths. A single crossover has occurred between two nonsister chromatids in this example, but either of the two chromatids of a homolog can form a crossover with either chromatid of the other homolog, and it is usual for multiple crossovers to be formed.

Summary

M-Cdk triggers the events of early mitosis, including chromosome condensation, assembly of the mitotic spindle, and bipolar attachment of the sister chromatid pairs to microtubules of the spindle. Spindle formation in animal cells depends largely on the ability of mitotic chromosomes to stimulate local microtubule nucleation and sta-bility, as well as on the ability of motor proteins to organize microtubules into a bipo-lar array. Many cells also use centrosomes to facilitate spindle assembly. Anaphase is triggered by the APC/C, which stimulates the destruction of the proteins that hold the sister chromatids together. APC/C also promotes cyclin destruction and thus the inac-tivation of M-Cdk. The resulting dephosphorylation of Cdk targets is required for the events that complete mitosis, including the disassembly of the spindle and the re-for-mation of the nuclear envelope. Meiosis is a specialized form of nuclear division in which a single round of chromosome duplication is followed by two rounds of chro-mosome segregation, resulting in the formation of haploid nuclei.

CYTOKINESIS

The final step in the cell cycle is **cytokinesis**, the division of the cytoplasm. In a typical cell, cytokinesis accompanies every mitosis, although some cells, such as early *Drosophila* embryos (discussed later) and some mammalian hepatocytes and heart muscle cells, undergo mitosis without cytokinesis and thereby acquire multiple nuclei. In most animal cells, cytokinesis begins in anaphase and ends shortly after the completion of mitosis in telophase.

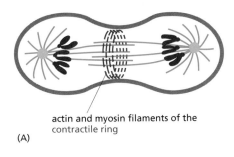

actin and myosin filaments of the contractile ring

(A)

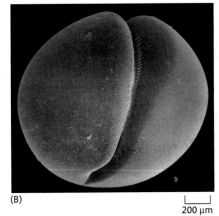

(B) ⊢—⊣ 200 μm

(C) ⊢—⊣ 25 μm

The first visible change of cytokinesis in an animal cell is the sudden appearance of a pucker, or *cleavage furrow*, on the cell surface. The furrow rapidly deepens and spreads around the cell until it completely divides the cell in two. In animal cells and many unicellular eucaryotes, the structure underlying this process is the *contractile ring*—a dynamic assembly composed of actin filaments, myosin II filaments, and many structural and regulatory proteins. During anaphase, the ring assembles just beneath the plasma membrane (**Figure 17–49**; see also Panel 17–1). The ring gradually contracts, and, at the same time, fusion of intracellular vesicles with the plasma membrane inserts new membrane adjacent to the ring. This addition of membrane compensates for the increase in surface area that accompanies cytoplasmic division. When ring contraction is completed, membrane insertion and fusion seal the gap between the daughter cells. Thus, cytokinesis can be considered to occur in four stages—initiation, contraction, membrane insertion, and completion.

Figure 17–49 Cytokinesis. (A) The actin-myosin bundles of the contractile ring are oriented as shown, so that their contraction pulls the membrane inward. (B) In this low-magnification scanning electron micrograph of a cleaving frog egg, the cleavage furrow is especially obvious and well defined, as the cell is unusually large. The furrowing of the cell membrane is caused by the activity of the contractile ring underneath it. (C) The surface of a furrow at higher magnification. (B and C, from H.W. Beams and R.G. Kessel, *Am. Sci.* 64:279–290, 1976. With permission from Sigma Xi.)

Actin and Myosin II in the Contractile Ring Generate the Force for Cytokinesis

In interphase cells, actin and myosin filaments form a cortical network underlying the plasma membrane. In some cells, they also form large cytoplasmic bundles called *stress fibers* (discussed in Chapter 16). As cells enter mitosis, these arrays of actin and myosin disassemble; much of the actin reorganizes, and myosin II filaments are released. As the sister chromatids separate in anaphase, actin and myosin II begin to accumulate in the rapidly assembling **contractile ring** (**Figure 17–50**), which also contains numerous other proteins that provide

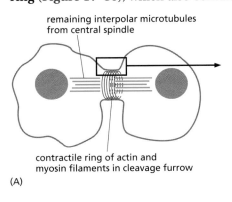

remaining interpolar microtubules from central spindle

contractile ring of actin and myosin filaments in cleavage furrow

(A)

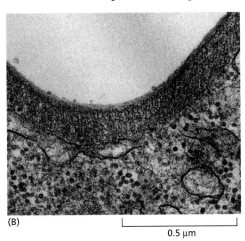

(B) ⊢—⊣ 0.5 μm

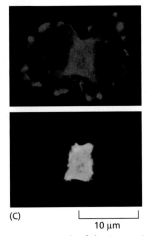

(C) ⊢—⊣ 10 μm

Figure 17–50 The contractile ring. (A) A drawing of the cleavage furrow in a dividing cell. (B) An electron micrograph of the ingrowing edge of a cleavage furrow of a dividing animal cell. (C) Fluorescence micrographs of a dividing slime mold amoeba stained for actin *(red)* and myosin II *(green)*. Whereas all of the visible myosin II has redistributed to the contractile ring, only some of the actin has done so; the rest remains in the cortex of the nascent daughter cells. (B, from H.W. Beams and R.G. Kessel, *Am. Sci.* 64:279–290, 1976. With permission from Sigma Xi; C, courtesy of Yoshio Fukui.)

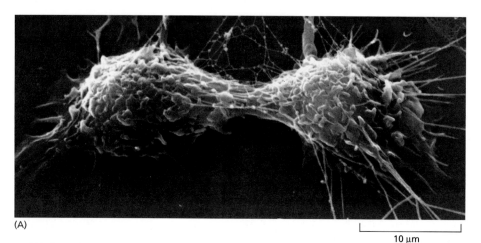

(A)

|_____10 µm_____|

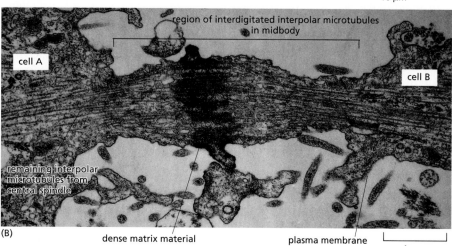

region of interdigitated interpolar microtubules in midbody

cell A

cell B

remaining interpolar microtubules from central spindle

(B)

dense matrix material

plasma membrane

|_____1 µm_____|

Figure 17–51 The midbody. (A) A scanning electron micrograph of a cultured animal cell in the process of dividing; the midbody still joins the two daughter cells. (B) A conventional electron micrograph of the midbody of a dividing animal cell. Cleavage is almost complete, but the daughter cells remain attached by this thin strand of cytoplasm containing the remains of the central spindle. (A, courtesy of Guenter Albrecht-Buehler; B, courtesy of J.M. Mullins.)

structural support or assist in ring assembly. Assembly of the contractile ring results in part from the local formation of new actin filaments, which depends on *formin* proteins that nucleate the assembly of parallel arrays of linear, unbranched actin filaments (discussed in Chapter 16). After anaphase, the overlapping arrays of actin and myosin II filaments contract to generate the force that divides the cytoplasm in two. Once contraction begins, the ring exerts a force large enough to bend a fine glass needle that is inserted in its path. As the ring constricts, it maintains the same thickness, suggesting that its total volume and the number of filaments it contains decrease steadily. Moreover, unlike actin in muscle, the actin filaments in the ring are highly dynamic, and their arrangement changes continually during cytokinesis.

The contractile ring is finally dispensed with altogether when cleavage ends, as the plasma membrane of the cleavage furrow narrows to form the **midbody**. The midbody persists as a tether between the two daughter cells and contains the remains of the central spindle, a large protein structure derived from the antiparallel interpolar microtubules of the spindle midzone, packed tightly together within a dense matrix material (**Figure 17–51**). After the daughter cells separate completely, some of the components of the residual midbody often remain on the inside of the plasma membrane of each cell, where they may serve as a mark on the cortex that helps to orient the spindle in the subsequent cell division.

Local Activation of RhoA Triggers Assembly and Contraction of the Contractile Ring

RhoA, a small GTPase of the Ras superfamily (see Table 15–5), controls the assembly and function of the contractile ring at the site of cleavage. RhoA is activated at the cell cortex at the future division site, where it promotes actin filament formation, myosin II assembly, and ring contraction. It promotes actin filament formation by activating formins, and it promotes myosin II assembly and contractions by activating multiple protein kinases, including the Rho-activated kinase Rock (**Figure 17–52**). These kinases phosphorylate the regulatory myosin light chain (RMLC), which is one of the subunits of myosin II. Phosphorylation of the RMLC stimulates bipolar myosin II filament formation and motor activity, thereby promoting the assembly and contraction of the actin–myosin ring.

Like other GTPases, RhoA is inactive when bound to GDP and active when bound to GTP (discussed in Chapter 15). The local activation of RhoA at the cleavage furrow is thought to depend on a Rho guanine nucleotide exchange factor (RhoGEF), which is found at the cell cortex at the future division site and stimulates the release of GDP and binding of GTP to RhoA. We know little about how the RhoGEF is localized or activated at the division site, although the microtubules of the anaphase spindle seem to be involved, as we discuss next.

The Microtubules of the Mitotic Spindle Determine the Plane of Animal Cell Division

The central problem in cytokinesis is how to ensure that division occurs at the right time and in the right place. Cytokinesis must occur only after the two sets of chromosomes are fully segregated from each other, and the site of division must be placed between the two sets of daughter chromosomes, thereby ensuring that each daughter cell receives a complete set. The correct timing and positioning of cytokinesis in animal cells are achieved by elegant mechanisms that depend on the mitotic spindle. During anaphase, the spindle generates signals that initiate furrow formation at a position midway between the spindle poles, thereby ensuring that division occurs between the two sets of separated chromosomes. Because these signals originate in the anaphase spindle, this mechanism also contributes to the correct timing of cytokinesis in late mitosis. Cytokinesis also occurs at the correct time because dephosphorylation of Cdk

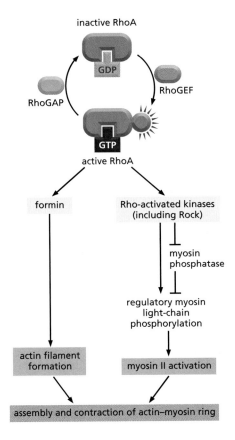

Figure 17–52 **Regulation of the contractile ring by the GTPase RhoA.** Like other Rho family GTPases, RhoA is activated by a RhoGEF protein and inactivated by a RhoGAP protein. The active GTP-bound form of RhoA is focused at the future cleavage site. By binding formins, activated RhoA promotes the assembly of actin filaments in the contractile ring. By activating Rho-activated protein kinases, such as Rock, it stimulates myosin II filament formation and activity, thereby promoting contraction of the ring.

 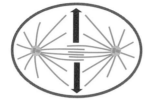 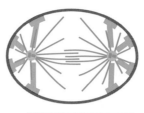

(A) astral stimulation model   (B) central spindle stimulation model   (C) astral relaxation model

Figure 17–53 Three current models of how the microtubules of the anaphase spindle generate signals that influence the positioning of the contractile ring. No single model explains all the observations, and it is likely that furrow positioning is determined by a combination of these mechanisms, with the importance of the different mechanisms varying in different organisms.

substrates, which depends on cyclin destruction in metaphase and anaphase, initiates cytokinesis. We now describe these regulatory mechanisms in more detail, with an emphasis on cytokinesis in animal cells.

Studies of the fertilized eggs of marine invertebrates first revealed the importance of spindle microtubules in determining the placement of the contractile ring. After fertilization, these embryos cleave rapidly without intervening periods of growth. In this way, the original egg is progressively divided up into smaller and smaller cells. Because the cytoplasm is clear, the spindle can be observed in real time with a microscope. If the spindle is tugged into a new position with a fine glass needle in early anaphase, the incipient cleavage furrow disappears, and a new one develops in accord with the new spindle site—supporting the idea that signals generated by the spindle induce local furrow formation.

How does the mitotic spindle specify the site of division? Three general mechanisms have been proposed, and most cells appear to employ a combination of these (**Figure 17–53**). The first is termed the *astral stimulation model*, which postulates that the astral microtubules carry furrow-inducing signals to the cell cortex, where they are somehow focused into a ring halfway between the spindle poles. Evidence for this model comes from ingenious experiments in large embryonic cells, which demonstrate that a cleavage furrow forms midway between two asters, even when the two centrosomes nucleating the asters are not connected to each other by a mitotic spindle (**Figure 17–54**).

A second possibility, called the *central spindle stimulation model*, is that the spindle midzone, or central spindle, generates a furrow-inducing signal that specifies the site of furrow formation at the cell cortex. The overlapping interpolar microtubules of the central spindle associate with numerous signaling proteins, including proteins that may stimulate RhoA (**Figure 17–55**). Defects in the function of these proteins (in *Drosophila* mutants, for example) result in failure of cytokinesis.

A third model proposes that, in some cell types, the astral microtubules promote the local relaxation of actin–myosin bundles at the cell cortex. According to this *astral relaxation model*, the cortical relaxation is minimal at the spindle equator, thus promoting cortical contraction at that site. In the early embryos of

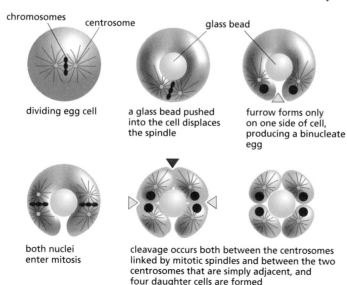

chromosomes   centrosome   glass bead

dividing egg cell

a glass bead pushed into the cell displaces the spindle

furrow forms only on one side of cell, producing a binucleate egg

both nuclei enter mitosis

cleavage occurs both between the centrosomes linked by mitotic spindles and between the two centrosomes that are simply adjacent, and four daughter cells are formed

Figure 17–54 An experiment demonstrating the influence of the position of microtubule asters on the subsequent plane of cleavage in a large egg cell. If the mitotic spindle is mechanically pushed to one side of the cell with a glass bead, the membrane furrowing is incomplete, failing to occur on the opposite side of the cell. Subsequent cleavages occur not only at the midzone of each of the two subsequent mitotic spindles *(yellow arrowheads)*, but also between the two adjacent asters that are not linked by a mitotic spindle—but in this abnormal cell share the same cytoplasm *(red arrowhead)*. Apparently, the contractile ring that produces the cleavage furrow in these cells always forms in the region midway between two asters, suggesting that the asters somehow alter the adjacent region of cell cortex to induce furrow formation between them.

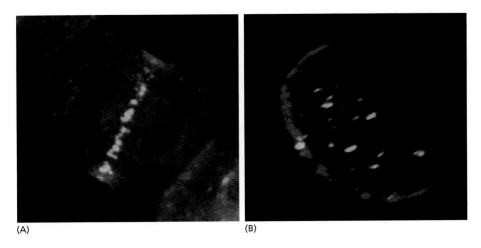

(A)   (B)

Figure 17–55 Localization of cytokinesis regulators at the central spindle of the human cell. (A) Fluorescence micrograph of a cultured human cell at the beginning of cytokinesis reveals the locations of the GTPase RhoA *(red)* and a protein called Cyk4 *(green)*, which is one of several regulatory proteins that form complexes at the overlapping plus ends of interpolar microtubules. These proteins are thought to generate signals that help control RhoA activity at the cell cortex (see Figures 17–52 and 17–53B). (B) When the cell is cross-sectioned in the plane of the contractile ring as shown here, RhoA *(red)* forms a ring beneath the cell surface, while the central spindle protein Cyk4 *(green)* is associated with microtubule bundles scattered throughout the equatorial plane of the cell. (Courtesy of Alisa Piekny and Michael Glotzer.)

C. elegans, for example, treatments that result in the loss of astral microtubules lead to increased contractile activity throughout the cell cortex, consistent with this model.

In some cell types, the site of ring assembly is chosen before mitosis. In budding yeasts, for example, a ring of proteins called *septins* assembles in late G_1 at the future division site. The septins are thought to form a scaffold onto which other components of the contractile ring, including myosin II, assemble. In plant cells, an organized band of microtubules and actin filaments, called the **preprophase band**, assembles just before mitosis and marks the site where the cell wall will assemble and divide the cell in two, as we now discuss.

The Phragmoplast Guides Cytokinesis in Higher Plants

In most animal cells, the inward movement of the cleavage furrow depends on an increase in the surface area of the plasma membrane. New membrane is added at the inner edge of the cleavage furrow and is generally provided by small membrane vesicles that are transported on microtubules from the Golgi apparatus to the furrow.

Membrane deposition is particularly important for cytokinesis in higher-plant cells. These cells are enclosed by a semirigid *cell wall*. Rather than a contractile ring dividing the cytoplasm from the outside in, the cytoplasm of the plant cell is partitioned from the inside out by the construction of a new cell wall, called the **cell plate**, between the two daughter nuclei (**Figure 17–56**). The assembly of the cell plate begins in late anaphase and is guided by a structure called the **phragmoplast**, which contains microtubules derived from the mitotic

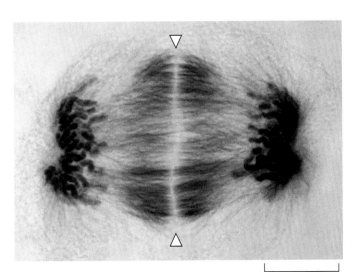

50 μm

Figure 17–56 Cytokinesis in a plant cell in telophase. In this light micrograph, the early cell plate (between the two *arrowheads*) has formed in a plane perpendicular to the plane of the page. The microtubules of the spindle are stained with gold-labeled antibodies against tubulin, and the DNA in the two sets of daughter chromosomes is stained with a fluorescent dye. Note that there are no astral microtubules, because there are no centrosomes in higher-plant cells. (Courtesy of Andrew Bajer.)

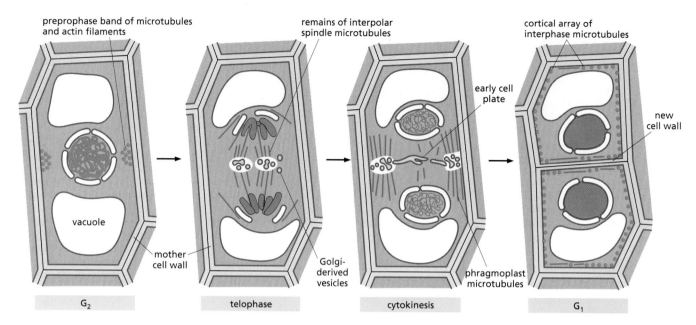

Figure 17–57 The special features of cytokinesis in a higher plant cell. The division plane is established before M phase by a band of microtubules and actin filaments (the preprophase band) at the cell cortex. At the beginning of telophase, after the chromosomes have segregated, a new cell wall starts to assemble inside the cell at the equator of the old spindle. The interpolar microtubules of the mitotic spindle remaining at telophase form the phragmoplast. The plus ends of these microtubules no longer overlap but end at the cell equator. Golgi-derived vesicles, filled with cell-wall material, are transported along these microtubules and fuse to form the new cell wall, which grows outward to reach the plasma membrane and original cell wall. The plasma membrane and the membrane surrounding the new cell wall fuse, completely separating the two daughter cells.

spindle. Motor proteins transport small vesicles along these microtubules from the Golgi apparatus to the cell center. These vesicles, filled with polysaccharide and glycoproteins required for the synthesis of the new cell wall, fuse to form a disclike, membrane-enclosed structure called the *early cell plate*. The plate expands outward by further vesicle fusion until it reaches the plasma membrane and the original cell wall and divides the cell in two. Later, cellulose microfibrils are laid down within the matrix of the cell plate to complete the construction of the new cell wall (**Figure 17–57**).

Membrane-Enclosed Organelles Must Be Distributed to Daughter Cells During Cytokinesis

The process of mitosis ensures that each daughter cell receives a full complement of chromosomes. When a eucaryotic cell divides, however, each daughter cell must also inherit all of the other essential cell components, including the membrane-enclosed organelles. As discussed in Chapter 12, organelles such as mitochondria and chloroplasts cannot be assembled de novo from their individual components; they can arise only by the growth and division of the preexisting organelles. Similarly, cells cannot make a new endoplasmic reticulum (ER) unless some part of it is already present.

How, then, do the various membrane-enclosed organelles segregate when a cell divides? Organelles such as mitochondria and chloroplasts are usually present in large enough numbers to be safely inherited if, on average, their numbers roughly double once each cycle. The ER in interphase cells is continuous with the nuclear membrane and is organized by the microtubule cytoskeleton. Upon entry into M phase, the reorganization of the microtubules and breakdown of the nuclear envelope releases the ER. In most cells, the ER remains largely intact and is cut in two during cytokinesis. The Golgi apparatus is reorganized and fragmented during mitosis. Golgi fragments associate with the spindle poles and are thereby distributed to opposite ends of the spindle, ensuring that each daughter cell inherits the materials needed to reconstruct the Golgi in telophase.

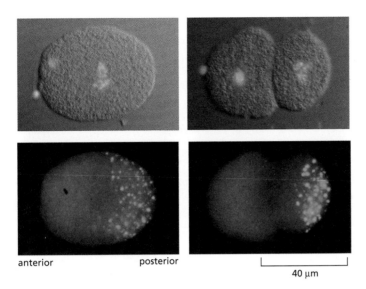

anterior   posterior

40 μm

Figure 17–58 An asymmetric cell division segregating cytoplasmic components to only one daughter cell. These light micrographs illustrate the controlled asymmetric segregation of specific cytoplasmic components to one daughter cell during the first division of a fertilized egg of the nematode *C. elegans*. The cells *above* have been stained with a *blue*, DNA-binding, fluorescent dye to show the nucleus (and polar bodies); they are viewed by both differential-interference-contrast and fluorescence microscopy. The cells *below* are the same cells stained with an antibody against P-granules and viewed by fluorescence microscopy. These small granules are made of RNA and proteins and determine which cells become the germ cells. They are distributed randomly throughout the cytoplasm of the unfertilized egg (not shown) but become segregated to the posterior pole of the fertilized egg, as shown on the *left*. The cleavage plane is oriented to ensure that only the posterior daughter cell receives the P-granules when the egg divides, as shown on the *right*. The same segregation process is repeated in several subsequent cell divisions, so that the P-granules end up only in the cells that give rise to eggs and sperm. (Courtesy of Susan Strome.)

Some Cells Reposition Their Spindle to Divide Asymmetrically

Most animal cells divide symmetrically: the contractile ring forms around the equator of the parent cell, producing two daughter cells of equal size and with the same components. This symmetry results from the placement of the mitotic spindle, which in most cases tends to center itself in the cytoplasm. Astral microtubules and motor proteins that either push or pull on these microtubules contribute to the centering process.

There are many instances in development, however, when cells divide asymmetrically to produce two cells that differ in size, in the cytoplasmic contents they inherit, or in both. Usually, the two different daughter cells are destined to develop along different pathways. To create daughter cells with different fates in this way, the mother cell must first segregate certain components (called *cell fate determinants*) to one side of the cell and then position the plane of division so that the appropriate daughter cell inherits these components (**Figure 17–58**). To position the plane of division asymmetrically, the spindle has to be moved in a controlled manner within the dividing cell. It seems likely that changes in local regions of the cell cortex direct such spindle movements and that motor proteins localized there pull one of the spindle poles, via its astral microtubules, to the appropriate region. Genetic analyses in *C. elegans* and *Drosophila* have identified some of the proteins required for such asymmetric divisions (discussed in Chapter 22), and some of these proteins seem to have a similar role in vertebrates.

Mitosis Can Occur Without Cytokinesis

Although nuclear division is usually followed by cytoplasmic division, there are exceptions. Some cells undergo multiple rounds of nuclear division without intervening cytoplasmic division. In the early *Drosophila* embryo, for example, the first 13 rounds of nuclear division occur without cytoplasmic division, resulting in the formation of a single large cell containing several thousand nuclei, arranged in a monolayer near the surface. A cell in which multiple nuclei share the same cytoplasm is called a **syncytium**. This arrangement greatly speeds up early development, as the cells do not have to take the time to go through all the steps of cytokinesis for each division. After these rapid nuclear divisions, membranes are created around each nucleus in one round of coordinated cytokinesis called *cellularization*. The plasma membrane extends inward and, with the help of an actin–myosin ring, pinches off to enclose each nucleus (**Figure 17–59**). <TTCT>

Nuclear division without cytokinesis also occurs in some types of mammalian cells. Megakaryocytes, which produce blood platelets, and some hepatocytes and heart muscle cells, for example, become multinucleated in this way.

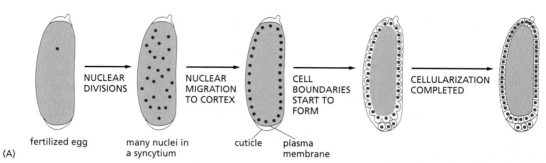

fertilized egg   many nuclei in   cuticle   plasma
(A)   a syncytium   membrane

Figure 17–59 Mitosis without cytokinesis in the early *Drosophila* embryo. (A) The first 13 nuclear divisions occur synchronously and without cytoplasmic division to create a large syncytium. Most of the nuclei then migrate to the cortex, and the plasma membrane extends inward and pinches off to surround each nucleus to form individual cells in a process called cellularization. (B) Fluorescence micrograph of multiple mitotic spindles in a *Drosophila* embryo before cellularization. The microtubules are stained *green* and the centrosomes *red*. Note that all the nuclei go through the cycle synchronously; here, they are all in metaphase. (B, courtesy of Kristina Yu and William Sullivan.)

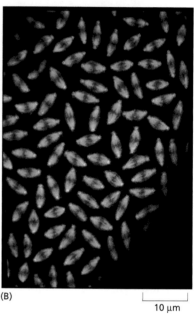

(B)

⊢————⊣
10 μm

After cytokinesis, most cells enter G$_1$, in which Cdks are mostly inactive. We end this section by discussing how this state is achieved at the end of M phase.

The G$_1$ Phase Is a Stable State of Cdk Inactivity

A key regulatory event in late M phase is the inactivation of Cdks, which is driven primarily by APC/C-dependent cyclin destruction. As described earlier in this chapter, the inactivation of Cdks in late M phase has many functions: it triggers the events of late mitosis, promotes cytokinesis, and enables the synthesis of prereplicative complexes at DNA replication origins. It also provides a mechanism for resetting the cell-cycle control system to a state of Cdk inactivity as the cell prepares to enter a new cell cycle. In most cells, this state of Cdk inactivity generates a G$_1$ gap phase, during which the cell grows and monitors its environment before committing to a new division.

In early animal embryos, the inactivation of M-Cdk in late mitosis is due almost entirely to the action of Cdc20–APC/C, discussed earlier. Recall, however, that M-Cdk stimulates Cdc20–APC/C activity. Thus, the destruction of M-cyclin in late mitosis soon leads to the inactivation of all APC/C activity in an embryonic cell. This APC/C inactivation immediately after mitosis is especially useful in rapid embryonic cell cycles, as it allows the cell to quickly begin accumulating new M-cyclin for the next cycle (**Figure 17–60**A).

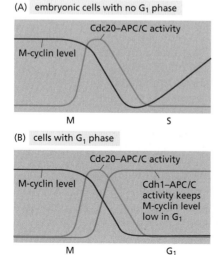

(A) embryonic cells with no G$_1$ phase

Cdc20–APC/C activity
M-cyclin level

M   S

(B) cells with G$_1$ phase

Cdc20–APC/C activity
M-cyclin level
Cdh1–APC/C activity keeps M-cyclin level low in G$_1$

M   G$_1$

Figure 17–60 The creation of a G$_1$ phase by stable Cdk inhibition after mitosis. (A) In early embryonic cell cycles, Cdc20–APC/C activity rises at the end of metaphase, triggering M-cyclin destruction. Because M-Cdk activity stimulates Cdc20–APC/C activity, the loss of M-cyclin leads to APC/C inactivation after mitosis, which allows M-cyclins to begin accumulating again. (B) In cells containing a G$_1$ phase, the drop in M-Cdk activity in late mitosis leads to the activation of Cdh1–APC/C (as well as to the accumulation of Cdk inhibitor proteins; not shown). This ensures a continued suppression of Cdk activity after mitosis, as required for a G$_1$ phase.

Rapid cyclin accumulation immediately after mitosis is not useful, however, for cells with cell cycles containing a G_1 phase. These cells employ several mechanisms to prevent Cdk reactivation after mitosis. One mechanism uses another APC/C-activating protein called Cdh1, a close relative of Cdc20. Although both Cdh1 and Cdc20 bind to and activate the APC/C, they differ in one important respect. Whereas M-Cdk activates the Cdc20–APC/C complex, it inhibits the Cdh1–APC/C complex by directly phosphorylating Cdh1. As a result of this relationship, Cdh1–APC/C activity increases in late mitosis after the Cdc20–APC/C complex has initiated the destruction of M-cyclin. M-cyclin destruction therefore continues after mitosis: although Cdc20–APC/C activity has declined, Cdh1–APC/C activity is high (Figure 17–60B).

A second mechanism that suppresses Cdk activity in G_1 depends on the increased production of CKIs, the Cdk inhibitory proteins discussed earlier. Budding yeast cells, in which this mechanism is best understood, contain a CKI protein called *Sic1*, which binds to and inactivates M-Cdk in late mitosis and G_1. Like Cdh1, Sic1 is inhibited by M-Cdk, which phosphorylates Sic1 during mitosis and thereby promotes its ubiquitylation by SCF. Thus, Sic1 and M-Cdk, like Cdh1 and M-Cdk, inhibit each other. As a result, the decline in M-Cdk activity that occurs in late mitosis causes the Sic1 protein to accumulate, and this CKI helps keep M-Cdk activity low after mitosis. A CKI protein called *p27* (see Figure 17–19/17–19) may serve similar functions in animal cells.

In most cells, decreased transcription of *M-cyclin* genes also inactivates M-Cdks in late mitosis. In budding yeast, for example, M-Cdk promotes the expression of these genes, resulting in a positive feedback loop. This loop is turned off as cells exit from mitosis: the inactivation of M-Cdk by Cdh1 and Sic1 leads to decreased M-cyclin gene transcription and thus decreased M-cyclin synthesis. Gene regulatory proteins that promote the expression of G_1/S- and S-cyclins are also inhibited during G_1.

Thus, Cdh1-APC/C activation, CKI accumulation, and decreased cyclin gene expression act together to ensure that the early G_1 phase is a time when essentially all Cdk activity is suppressed. As in many other aspects of cell-cycle control, the use of multiple regulatory mechanisms makes the suppression system robust, so that it still operates with reasonable efficiency even if one mechanism fails. So how does the cell escape from this stable G_1 state to initiate a new cell cycle? The answer is that G_1/S-Cdk activity, which rises in late G_1, releases all the braking mechanisms that suppress Cdk activity, as we describe in the next section.

Summary

After mitosis completes the formation of a pair of daughter nuclei, cytokinesis finishes the cell cycle by dividing the cell itself. Cytokinesis depends on a ring of actin and myosin that contracts in late mitosis at a site midway between the segregated chromosomes. In animal cells, the positioning of the contractile ring is determined by signals emanating from the microtubules of the anaphase spindle. Dephosphorylation of Cdk targets, which results from Cdk inactivation in anaphase, triggers cytokinesis at the correct time after anaphase. After cytokinesis, the cell enters a stable G_1 state of low Cdk activity, where it awaits signals to enter a new cell cycle.

CONTROL OF CELL DIVISION AND CELL GROWTH

A fertilized mouse egg and a fertilized human egg are similar in size, yet they produce animals of very different sizes. What factors in the control of cell behavior in humans and mice are responsible for these size differences? The same fundamental question can be asked for each organ and tissue in an animal's body. What factors in the control of cell behavior explain the length of an elephant's trunk or the size of its brain or its liver? These questions are largely unanswered, at least in part because they have received relatively little attention compared with other questions in cell and developmental biology. It is nevertheless possible to say what the ingredients of an answer must be.

The size of an organ or organism depends mainly on its total cell mass, which depends on both the total number of cells and the size of the cells. Cell number, in turn, depends on the amounts of cell division and cell death. Organ and body size are therefore determined by three fundamental processes: cell growth, cell division, and cell death. Each is tightly regulated—both by intracellular programs and by extracellular signal molecules that control these programs.

The extracellular signal molecules that regulate cell size and cell number are generally soluble secreted proteins, proteins bound to the surface of cells, or components of the extracellular matrix. They can be divided operationally into three major classes:

1. *Mitogens*, which stimulate cell division, primarily by triggering a wave of G_1/S-Cdk activity that relieves intracellular negative controls that otherwise block progress through the cell cycle.
2. *Growth factors*, which stimulate cell growth (an increase in cell mass) by promoting the synthesis of proteins and other macromolecules and by inhibiting their degradation.
3. *Survival factors*, which promote cell survival by suppressing the form of programmed cell death known as *apoptosis*.

Many extracellular signal molecules promote all of these processes, while others promote one or two of them. Indeed, the term *growth factor* is often used inappropriately to describe a factor that has any of these activities. Even worse, the term *cell growth* is often used to mean an increase in cell number, or *cell proliferation*.

In addition to these three classes of stimulating signals, there are extracellular signal molecules that suppress cell proliferation, cell growth, or both; in general, less is known about them. There are also extracellular signal molecules that activate apoptosis.

In this section, we focus primarily on how mitogens and other factors, such as DNA damage, control the rate of cell division. We then turn to the important but poorly understood problem of how a proliferating cell coordinates its growth with cell division so as to maintain its appropriate size. We discuss the control of cell survival and cell death by apoptosis in Chapter 18.

Mitogens Stimulate Cell Division

Unicellular organisms tend to grow and divide as fast as they can, and their rate of proliferation depends largely on the availability of nutrients in the environment. The cells of a multicellular organism, however, divide only when the organism needs more cells. Thus, for an animal cell to proliferate, it must receive stimulatory extracellular signals, in the form of **mitogens**, from other cells, usually its neighbors. Mitogens overcome intracellular braking mechanisms that block progress through the cell cycle.

One of the first mitogens to be identified was *platelet-derived growth factor (PDGF)*, and it is typical of many others discovered since. The path to its isolation began with the observation that fibroblasts in a culture dish proliferate when provided with *serum* but not when provided with *plasma*. Plasma is prepared by removing the cells from blood without allowing clotting to occur; serum is prepared by allowing blood to clot and taking the cell-free liquid that remains. When blood clots, platelets incorporated in the clot are stimulated to release the contents of their secretory vesicles (**Figure 17–61**). The superior ability of serum to support cell proliferation suggested that platelets contain one or more mitogens. This hypothesis was confirmed by showing that extracts of platelets could serve instead of serum to stimulate fibroblast proliferation. The crucial factor in the extracts was shown to be a protein, which was subsequently purified and named PDGF. In the body, PDGF liberated from blood clots helps stimulate cell division during wound healing.

PDGF is only one of over 50 proteins that are known to act as mitogens. Most of these proteins have a broad specificity. PDGF, for example, can stimulate many types of cells to divide, including fibroblasts, smooth muscle cells, and

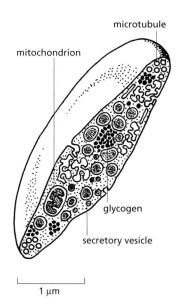

Figure 17–61 A platelet. Platelets are miniature cells without a nucleus. They circulate in the blood and help stimulate blood clotting at sites of tissue damage, thereby preventing excessive bleeding. They also release various factors that stimulate healing. The platelet shown here has been cut in half to show its secretory vesicles, some of which contain platelet-derived growth factor (PDGF).

neuroglial cells. Similarly, *epidermal growth factor* (*EGF*) acts not only on epidermal cells but also on many other cell types, including both epithelial and nonepithelial cells. Some mitogens, however, have a narrow specificity; *erythropoietin*, for example, only induces the proliferation of red blood cell precursors. Many mitogens, including PDGF, also have other actions beside the stimulation of cell division: they can stimulate cell growth, survival, differentiation, or migration, depending on the circumstances and the cell type.

In some tissues, inhibitory extracellular signal proteins oppose the positive regulators and thereby inhibit organ growth. The best-understood inhibitory signal proteins are TGFβ and its relatives. TGFβ inhibits the proliferation of several cell types, either by blocking cell-cycle progression in G_1 or by stimulating apoptosis.

Cells Can Delay Division by Entering a Specialized Nondividing State

In the absence of a mitogenic signal to proliferate, Cdk inhibition in G_1 is maintained by the multiple mechanisms discussed earlier, and progression into a new cell cycle is blocked. In some cases, cells partly disassemble their cell-cycle control system and exit from the cycle to a specialized, nondividing state called G_0.

Most cells in our body are in G_0, but the molecular basis and reversibility of this state vary in different cell types. Most of our neurons and skeletal muscle cells, for example, are in a *terminally differentiated* G_0 state, in which their cell-cycle control system is completely dismantled: the expression of the genes encoding various Cdks and cyclins are permanently turned off, and cell division rarely occurs. Other cell types withdraw from the cell cycle only transiently and retain the ability to reassemble the cell-cycle control system quickly and reenter the cycle. Most liver cells, for example, are in G_0, but they can be stimulated to divide if the liver is damaged. Still other types of cells, including fibroblasts and some lymphocytes, withdraw from and re-enter the cell cycle repeatedly throughout their lifetime.

Almost all the variation in cell-cycle length in the adult body occurs during the time the cell spends in G_1 or G_0. By contrast, the time a cell takes to progress from the beginning of S phase through mitosis is usually brief (typically 12–24 hours in mammals) and relatively constant, regardless of the interval from one division to the next.

Mitogens Stimulate G_1-Cdk and G_1/S-Cdk Activities

For the vast majority of animal cells, mitogens control the rate of cell division by acting in the G_1 phase of the cell cycle. As discussed earlier, multiple mechanisms act during G_1 to suppress Cdk activity and thereby block entry into S phase. Mitogens release these brakes on Cdk activity, thereby allowing S phase to begin.

As we discuss in Chapter 15, mitogens interact with cell-surface receptors to trigger multiple intracellular signaling pathways. One major pathway acts through the small GTPase **Ras**, which leads to the activation of a *MAP kinase cascade*. This leads to an increase in the production of gene regulatory proteins, including **Myc**. Myc is thought to promote cell-cycle entry by several mechanisms, one of which is to increase the expression of genes encoding G_1 cyclins (D cyclins), thereby increasing G_1-Cdk (cyclin D–Cdk4) activity. As we discuss later, Myc also has a major role in stimulating the transcription of genes that increase cell growth.

The key function of G_1-Cdk complexes in animal cells is to activate a group of gene regulatory factors called the **E2F proteins**, which bind to specific DNA sequences in the promoters of a wide variety of genes that encode proteins required for S-phase entry, including G_1/S-cyclins, S-cyclins, and proteins involved in DNA synthesis and chromosome duplication. In the absence of mitogenic stimulation, E2F-dependent gene expression is inhibited by an interaction between E2F and members of the **retinoblastoma protein** (**Rb**) family.

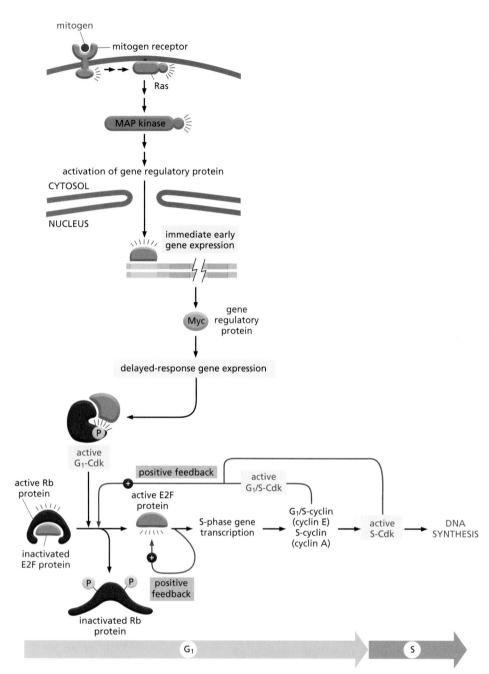

Figure 17–62 Mechanisms controlling cell-cycle entry and S-phase initiation in animal cells. As discussed in Chapter 15, mitogens bind to cell-surface receptors to initiate intracellular signaling pathways. One of the major pathways involves activation of the small GTPase Ras, which activates a MAP kinase cascade, leading to increased expression of numerous *immediate early* genes, including the gene encoding the gene regulatory protein Myc. Myc increases the expression of many *delayed-response* genes, including some that lead to increased G_1-Cdk activity (cyclin D–Cdk4), which triggers the phosphorylation of members of the Rb family of proteins. This inactivates the Rb proteins, freeing the gene regulatory protein E2F to activate the transcription of G_1/S genes, including the genes for a G_1/S-cyclin (cyclin E) and S-cyclin (cyclin A). The resulting G_1/S-Cdk and S-Cdk activities further enhance Rb protein phosphorylation, forming a positive feedback loop. E2F proteins also stimulate the transcription of their own genes, forming another positive feedback loop.

When cells are stimulated to divide by mitogens, active G_1-Cdk accumulates and phosphorylates Rb family members, reducing their binding to E2F. The liberated E2F proteins then activate expression of their target genes (**Figure 17–62**).

This transcriptional control system, like so many other control systems that regulate the cell cycle, includes feedback loops that sharpen the G_1/S transition. The liberated E2F proteins, for example, increase the transcription of their own genes. In addition, E2F-dependent transcription of G_1/S-cyclin (cyclin E) and S-cyclin (cyclin A) genes leads to increased G_1/S-Cdk and S-Cdk activities, which in turn increase Rb protein phosphorylation and promote further E2F release (see Figure 17–62).

The central member of the Rb family, the Rb protein itself, was identified originally through studies of an inherited form of eye cancer in children, known as *retinoblastoma* (discussed in Chapter 20). The loss of both copies of the Rb gene leads to excessive cell proliferation in the developing retina, suggesting that the Rb protein is particularly important for restraining cell division in this tissue. The complete loss of Rb does not immediately cause increased proliferation of retinal or other cell types, in part because Cdh1 and CKIs also help inhibit progression through G_1 and in part because other cell types contain Rb-related proteins that provide backup support in the absence of Rb. It is also likely that other proteins, unrelated to Rb, help to regulate the activity of E2F.

Additional layers of control promote an overwhelming increase in S-Cdk activity at the beginning of S phase. We mentioned earlier that the APC/C activator Cdh1 suppresses cyclin levels after mitosis. In animal cells, however, G_1- and G_1/S-cyclins are resistant to Cdh1 and can therefore act unopposed by the APC/C to promote Rb protein phosphorylation and E2F-dependent gene expression. S-cyclin, by contrast, is not resistant to Cdh1, and its level is initially restrained by Cdh1-APC/C activity. However, G_1/S-Cdk also phosphorylates and inactivates Cdh1-APC/C, thereby allowing the accumulation of S-cyclin, further promoting S-Cdk activation. G_1/S-Cdk also inactivates CKI proteins that suppress S-Cdk activity. The overall effect of all these interactions is the rapid and complete activation of the S-Cdk complexes required for S-phase initiation.

DNA Damage Blocks Cell Division: The DNA Damage Response

Progression through the cell cycle, and thus the rate of cell proliferation, is controlled not only by extracellular mitogens but also by other extracellular and intracellular mechanisms. One of the most important influences is DNA damage, which can occur as a result of spontaneous chemical reactions in DNA, errors in DNA replication, or exposure to radiation or certain chemicals. It is essential that the cell repair damaged chromosomes before attempting to duplicate or segregate them. The cell-cycle control system can readily detect DNA damage and arrest the cycle at either of two checkpoints—one at Start in late G_1, which prevents entry into the cell cycle and into S phase, and one at the G_2/M checkpoint, which prevents entry into mitosis (see Figure 17–21).

DNA damage initiates a signaling pathway by activating one of a pair of related protein kinases called **ATM** and **ATR**, which associate with the site of damage and phosphorylate various target proteins, including two other protein kinases called *Chk1* and *Chk2*. Together these various kinases phosphorylate other target proteins that lead to cell-cycle arrest. A major target is the gene regulatory protein **p53**, which stimulates transcription of the gene encoding a CKI protein called *p21;* this protein binds to G_1/S-Cdk and S-Cdk complexes and inhibits their activities, thereby helping to block entry into the cell cycle (**Figure 17–63**). <TGAA>

DNA damage activates p53 by an indirect mechanism. In undamaged cells, p53 is highly unstable and is present at very low concentrations. This is largely because it interacts with another protein, *Mdm2*, which acts as a ubiquitin ligase that targets p53 for destruction by proteasomes. Phosphorylation of p53 after DNA damage reduces its binding to Mdm2. This decreases p53 degradation, which results in a marked increase in p53 concentration in the cell. In addition, the decreased binding to Mdm2 enhances the ability of p53 to stimulate gene transcription (see Figure 17–63).

The protein kinases Chk1 and Chk2 also block cell cycle progression by phosphorylating members of the Cdc25 family of protein phosphatases, thereby inhibiting their function. As described earlier, these kinases are particularly important in the activation of M-Cdk at the beginning of mitosis (see Figure 17–25). Thus, the inhibition of Cdc25 activity by DNA damage helps block entry into mitosis (see Figure 17–21).

The DNA-damage response also detects problems that arise when a replication fork fails during DNA replication. When nucleotides are depleted, for example, replication forks stall during the elongation phase of DNA synthesis. To prevent the cell from attempting to segregate partially replicated chromosomes, the same mechanisms that respond to DNA damage detect the stalled replication forks and block entry into mitosis until the problems at the replication fork are resolved.

The DNA damage response is not essential for normal cell division if environmental conditions are ideal. Conditions are rarely ideal, however: a low level of DNA damage occurs in the normal life of any cell, and this damage accumulates in the cell's progeny if the damage response is not functioning. Over the long term, the accumulation of genetic damage in cells lacking the DNA damage response leads to an increased frequency of cancer-promoting mutations. Indeed, mutations in the *p53* gene occur in at least half of all human cancers (discussed in Chapter 20). This loss of p53 function allows the cancer cell to accumulate mutations more readily. Similarly, a rare genetic disease known as *ataxia telangiectasia* is caused by a defect in ATM, one of the protein kinases that is activated in response to x-ray-induced DNA damage; patients with this disease are very sensitive to x-rays and suffer from increased rates of cancer.

What happens if DNA damage is so severe that repair is not possible? The answer differs in different organisms. Unicellular organisms such as budding yeast transiently arrest their cell cycle to try to repair the damage, but the cycle resumes even if the repair cannot be completed. For a single-celled organism, life with mutations is apparently better than no life at all. In multicellular organisms, however, the health of the organism takes precedence over the life of an individual cell. Cells that divide with severe DNA damage threaten the life of the organism, since genetic damage can often lead to cancer and other diseases. Thus, animal cells with severe DNA damage do not attempt to continue division, but instead commit suicide by undergoing apoptosis. Thus, unless the DNA damage is repaired, the DNA damage response can lead to either cell-cycle

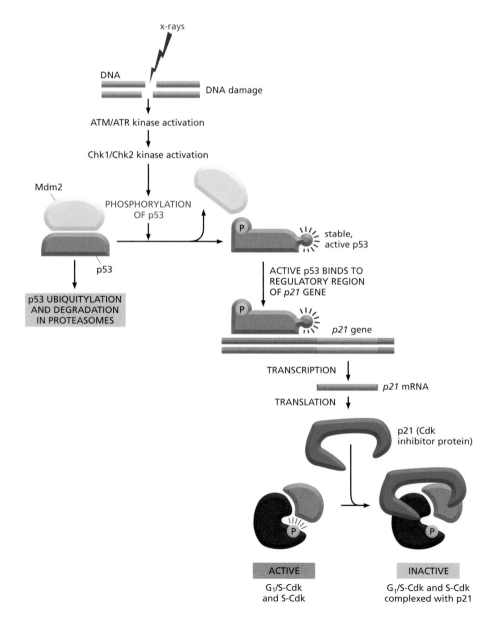

Figure 17–63 How DNA damage arrests the cell cycle in G_1. When DNA is damaged, various protein kinases are recruited to the site of damage and initiate a signaling pathway that causes cell-cycle arrest. The first kinase at the damage site is either ATM or ATR, depending on the type of damage. Additional protein kinases, called Chk1 and Chk2, are then recruited and activated, resulting in the phosphorylation of the gene regulatory protein p53. Mdm2 normally binds to p53 and promotes its ubiquitylation and destruction in proteasomes. Phosphorylation of p53 blocks its binding to Mdm2; as a result, p53 accumulates to high levels and stimulates transcription of the gene that encodes the CKI protein p21. The p21 binds and inactivates G_1/S-Cdk and S-Cdk complexes, arresting the cell in G_1. In some cases, DNA damage also induces either the phosphorylation of Mdm2 or a decrease in Mdm2 production, which causes a further increase in p53 (not shown).

arrest or cell death. As we discuss in the next chapter, DNA damage-induced apoptosis often depends on the activation of p53. Indeed, it is this apoptosis-promoting function of p53 that is apparently most important in protecting us against cancer.

Many Human Cells Have a Built-In Limitation on the Number of Times They Can Divide

Many human cells divide a limited number of times before they stop and undergo a permanent cell-cycle arrest. Fibroblasts taken from normal human tissue, for example, go through only about 25–50 population doublings when cultured in a standard mitogenic medium. Toward the end of this time, proliferation slows down and finally halts, and the cells enter a nondividing state from which they never recover. This phenomenon is called **replicative cell senescence**, although it is unlikely to be responsible for the senescence (aging) of the organism. Organism senescence is thought to depend, in part, on progressive oxidative damage to long-lived macromolecules, as strategies that reduce metabolism (such as reduced food intake), and thereby reduce the production of reactive oxygen species, can extend the lifespan of experimental animals.

Replicative cell senescence in human fibroblasts seems to be caused by changes in the structure of the **telomeres**, the repetitive DNA sequences and associated proteins at the ends of chromosomes. As discussed in Chapter 5, when a cell divides, telomeric DNA sequences are not replicated in the same manner as the rest of the genome but instead are synthesized by the enzyme **telomerase**. Telomerase also promotes the formation of protein cap structures that protect the chromosome ends. Because human fibroblasts, and many other human somatic cells, are deficient in telomerase, their telomeres become shorter with every cell division, and their protective protein caps progressively deteriorate. Eventually, the exposed chromosome ends are sensed as DNA damage, which activates a p53-dependent cell-cycle arrest that resembles the arrest caused by other types of DNA damage (see Figure 17–63). Rodent cells, by contrast, maintain telomerase activity when they proliferate in culture and therefore do not have such a telomere-dependent mechanism for limiting proliferation. The forced expression of telomerase in normal human fibroblasts, using genetic engineering techniques, blocks this form of senescence. Unfortunately, most cancer cells have regained the ability to produce telomerase and therefore maintain telomere function as they proliferate; as a result, they do not undergo replicative cell senescence.

Abnormal Proliferation Signals Cause Cell-Cycle Arrest or Apoptosis, Except in Cancer Cells

Many of the components of mitogenic signaling pathways are encoded by genes that were originally identified as cancer-promoting genes, or *oncogenes*, because mutations in them contribute to the development of cancer. The mutation of a single amino acid in the small GTPase Ras, for example, causes the protein to become permanently overactive, leading to constant stimulation of Ras-dependent signaling pathways, even in the absence of mitogenic stimulation. Similarly, mutations that cause an overexpression of Myc stimulate excessive cell growth and proliferation and thereby promote the development of cancer.

Surprisingly, however, when a hyperactivated form of Ras or Myc is experimentally overproduced in most normal cells, the result is not excessive proliferation but the opposite: the cells undergo either cell-cycle arrest or apoptosis. The normal cell seems able to detect abnormal mitogenic stimulation, and it responds by preventing further division. Such responses help prevent the survival and proliferation of cells with various cancer-promoting mutations.

Although it is not known how a cell detects excessive mitogenic stimulation, such stimulation often leads to the production of a cell-cycle inhibitor protein called *Arf*, which binds and inhibits Mdm2. As discussed earlier, Mdm2 normally

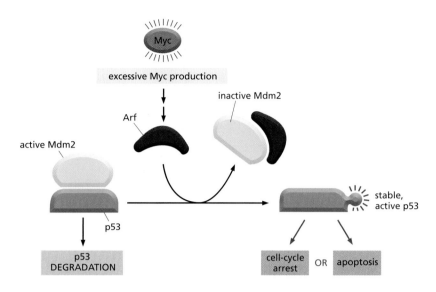

Figure 17–64 Cell-cycle arrest or apoptosis induced by excessive stimulation of mitogenic pathways. Abnormally high levels of Myc cause the activation of Arf, which binds and inhibits Mdm2 and thereby increases p53 levels (see Figure 17–60). Depending on the cell type and extracellular conditions, p53 then causes either cell-cycle arrest or apoptosis.

promotes p53 degradation. Activation of Arf therefore causes p53 levels to increase, inducing either cell-cycle arrest or apoptosis (**Figure 17–64**).

How do cancer cells ever arise if these mechanisms block the division or survival of mutant cells with overactive proliferation signals? The answer is that the protective system is often inactivated in cancer cells by mutations in the genes that encode essential components of the checkpoint responses, such as Arf or p53 or the proteins that help activate them.

Organism and Organ Growth Depend on Cell Growth

For an organism or organ to grow, cell division is not enough. If cells proliferated without growing, they would get progressively smaller and there would be no net increase in total cell mass. In most proliferating cell populations, therefore, cell growth accompanies cell division. In single-celled organisms such as yeasts, both cell growth and cell division require only nutrients. In animals, by contrast, both cell growth and cell proliferation depend on extracellular signal molecules, produced by other cells, which we call **growth factors** and mitogens, respectively.

Like mitogens, the extracellular growth factors that stimulate animal cell growth bind to receptors on the cell surface and activate intracellular signaling pathways. These pathways stimulate the accumulation of proteins and other macromolecules, and they do so by both increasing their rate of synthesis and decreasing their rate of degradation. They also trigger increased uptake of nutrients and production of the ATP required to fuel increased protein synthesis. One of the most important intracellular signaling pathways activated by growth factor receptors involves the enzyme *PI 3-kinase*, which adds a phosphate from ATP to the 3 position of inositol phospholipids in the plasma membrane. As discussed in Chapter 15, the activation of PI 3-kinase leads to the activation of a kinase called *TOR*, which lies at the heart of growth regulatory pathways in all eucaryotes. TOR activates many targets in the cell that stimulate metabolic processes and increase protein synthesis. One target is a protein kinase called *S6 kinase (S6K)*, which phosphorylates ribosomal protein S6, increasing the ability of ribosomes to translate a subset of mRNAs that mostly encode ribosomal components. TOR also indirectly activates a translation initiation factor called *eIF4E* and directly activates gene regulatory proteins that promote the increased expression of genes encoding ribosomal subunits (**Figure 17–65**).

Proliferating Cells Usually Coordinate Their Growth and Division

For proliferating cells to maintain a constant size, they must coordinate their growth with cell division to ensure that cell size doubles with each division: if cells grow too slowly, they will get smaller with each division, and if they grow

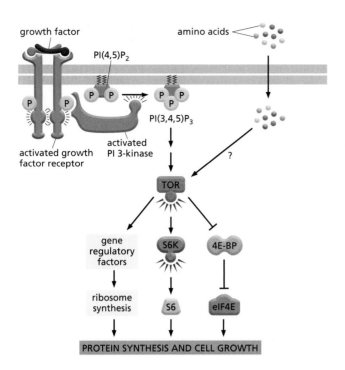

Figure 17–65 Stimulation of cell growth by extracellular growth factors and nutrients. As discussed in Chapter 15, the occupation of cell-surface receptors by growth factors leads to the activation of PI 3-kinase, which promotes protein synthesis through a complex signaling pathway that leads to the activation of the protein kinase TOR; extracellular nutrients such as amino acids also help activate TOR by an unknown pathway. TOR employs multiple mechanisms to stimulate protein synthesis, as shown; it also inhibits protein degradation (not shown). Growth factors also stimulate increased production of the gene regulatory protein Myc (not shown), which activates the transcription of various genes that promote cell metabolism and growth. 4E-BP is an inhibitor of the translation initiation factor eIF4E.

too fast, they will get larger with each division. It is not clear how cells achieve this coordination, but it is likely to involve multiple mechanisms that vary in different organisms and even in different cell types of the same organism (**Figure 17–66**).

Animal cell growth and division are not always coordinated, however. In many cases, they are completely uncoupled to allow growth without division or division without growth. Muscle cells and nerve cells, for example, can grow dramatically after they have permanently withdrawn from the cell cycle. Similarly, the eggs of many animals grow to an extremely large size without dividing; after fertilization, however, this relationship is reversed, and many rounds of division occur without growth (see Figure 17–9).

Compared to cell division, there has been surprisingly little study of how cell size is controlled in animals. As a result, it remains a mystery how cell size is determined and why different cell types in the same animal grow to be so different in size (**Figure 17–67**). One of the best-understood cases in mammals is the adult *sympathetic neuron,* which has permanently withdrawn from the cell cycle. Its size depends on the amount of *nerve growth factor* (NGF) secreted by the target cells it innervates; the greater the amount of NGF the neuron has access to, the larger it becomes. It seems likely that the genes a cell expresses set limits on the size it can be, while extracellular signal molecules and nutrients

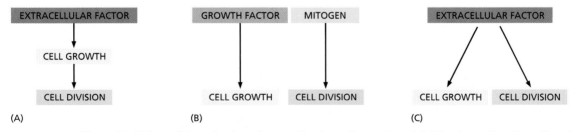

(A) (B) (C)

Figure 17–66 Potential mechanisms for coordinating cell growth and division. In proliferating cells, cell size is maintained by mechanisms that coordinate rates of cell division and cell growth. Numerous alternative coupling mechanisms are thought to exist, and different cell types appear to employ different combinations of these mechanisms. (A) In many cell types—particularly yeast—the rate of cell division is governed by the rate of cell growth, so that division occurs only when growth rate achieves some minimal threshold; in yeasts, it is mainly the levels of extracellular nutrients that regulate the rate of cell growth and thereby the rate of cell division. (B) In some animal cell types, growth and division can each be controlled by separate extracellular factors (growth factors and mitogens, respectively), and cell size depends on the relative levels of the two types of factors. (C) Some extracellular factors can stimulate both cell growth and cell division by simultaneously activating signaling pathways that promote growth and other pathways that promote cell-cycle progression.

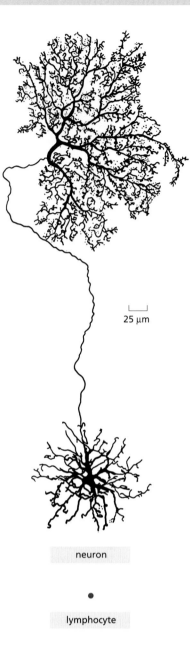

Figure 17–67 The size difference between a neuron (from the retina) and a lymphocyte in a mammal. Both cells contain the same amount of DNA. A neuron grows progressively larger after it has permanently withdrawn from the cell cycle. During this time, the ratio of cytoplasm to DNA increases enormously (by a factor of more than 10^5 for some neurons). (Neuron from B.B. Boycott, in Essays on the Nervous System [R. Bellairs and E.G. Gray, eds]. Oxford, UK: Clarendon Press, 1974.)

25 μm

neuron

lymphocyte

regulate the size within these limits. The challenge is to identify the relevant genes and signal molecules for each cell type.

Neighboring Cells Compete for Extracellular Signal Proteins

When most types of mammalian cells are cultured in a dish in the presence of serum, they adhere to the bottom of the dish, spread out, and divide until they form a confluent monolayer. Each cell is attached to the dish and contacts its neighbors on all sides. At this point, normal cells, unlike cancer cells, stop proliferating—a phenomenon known as *density-dependent inhibition of cell division.* This phenomenon was originally described in terms of "contact inhibition" of cell division, but it is unlikely that cell–cell contact interactions are solely responsible. The cell population density at which cell proliferation ceases in the confluent monolayer increases as the concentration of serum in the medium increases. Moreover, if a stream of fresh culture medium is passed over a confluent layer of fibroblasts to increase the supply of mitogens, the cells under the stream are induced to divide (**Figure 17–68**). Thus, density-dependent inhibition of cell proliferation seems to reflect, in part at least, the ability of a cell to deplete the medium around itself of extracellular mitogens, thereby depriving its neighbors.

This type of competition could be important for cells in tissues as well as in culture, because it prevents them from proliferating beyond a certain population density, determined by the available amounts of mitogens, growth factors, and survival factors. The amounts of these factors in tissues are usually limiting, in that increasing their amounts results in an increase in cell number, cell size, or both. Thus, the amounts of these factors in tissues have important roles in determining cell size and number, and possibly the final size of the organ or tissue.

The overall size of a tissue may also be governed in some cases by extracellular inhibitory factors. *Myostatin,* for example, is a TGFβ family member that normally inhibits the proliferation of myoblasts that fuse to form skeletal muscle cells. When the gene that encodes myostatin is deleted in mice, muscles grow to be several times larger than normal. Remarkably, two breeds of cattle that were bred for large muscles have both turned out to have mutations in the gene encoding myostatin (**Figure 17–69**).

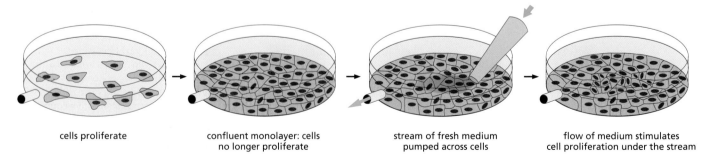

cells proliferate

confluent monolayer: cells no longer proliferate

stream of fresh medium pumped across cells

flow of medium stimulates cell proliferation under the stream

Figure 17–68 The effect of fresh medium on a confluent cell monolayer. Cells in a confluent monolayer do not divide *(gray).* The cells resume dividing *(green)* when exposed directly to fresh culture medium. Apparently, in the undisturbed confluent monolayer, proliferation has halted because the medium close to the cells is depleted of mitogens, for which the cells compete.

Figure 17–69 **The effects of a myostatin mutation on muscle size.** The mutation leads to a great increase in the mass of muscle tissue, as illustrated in this Belgian Blue bull. The Belgian Blue was produced by cattle breeders and was only recently found to have a mutation in the *Myostatin* gene. (From H.L. Sweeney, *Sci. Am.* 291:62, 2004. With permission from Scientific American.)

Animals Control Total Cell Mass by Unknown Mechanisms

The size of an animal or one of its organs depends largely on the number and size of the cells it contains—that is, on total cell mass. Remarkably, animals can somehow assess the total cell mass in a tissue or organ and regulate it: in many circumstances, for example, if cell size is experimentally increased or decreased in an organ, cell numbers adjust to maintain a normal organ size. This has been most dramatically illustrated by experiments in salamanders, in which cell size was manipulated by altering cell ploidy (in all organisms, the size of a cell is proportional to its ploidy, or genome content). Salamanders of different ploidies are the same size but have different numbers of cells. Individual cells in a pentaploid salamander are about five times the volume of those in a haploid salamander, and in each organ the pentaploids have only one-fifth as many cells as their haploid cousins, so that the organs are about the same size in the two animals (**Figure 17–70** and **Figure 17–71**). Evidently, in this case (and in many others) the size of organs and organisms depends on mechanisms that can somehow measure total cell mass. How animals measure and adjust total mass remains a mystery, however.

The development of limbs and organs of specific size and shape depends on complex positional controls, as well as on local concentrations of extracellular signal proteins that stimulate or inhibit cell growth, division, and survival. As we discuss in Chapter 22, we now know many of the genes that help pattern these processes in the embryo. A great deal remains to be learned, however, about how these genes regulate cell growth, division, survival, and differentiation to generate a complex organism.

The controls that govern these processes in an adult body are also poorly understood. When a skin wound heals in a vertebrate, for example, about a dozen cell types, ranging from fibroblasts to Schwann cells, must be regenerated in appropriate numbers, sizes, and positions to reconstruct the lost tissue. The

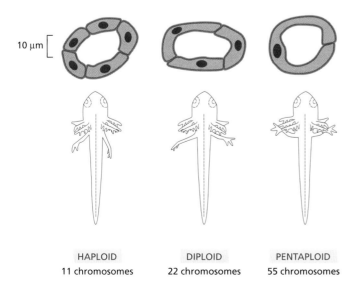

| HAPLOID | DIPLOID | PENTAPLOID |
| 11 chromosomes | 22 chromosomes | 55 chromosomes |

Figure 17–70 **Sections of kidney tubules from salamander larvae of different ploidies.** In all organisms, from bacteria to humans, cell size is proportional to ploidy. Pentaploid salamanders, for example, have cells that are much larger than those of haploid salamanders. The animals and their individual organs, however, are the same size because each tissue in the pentaploid animal contains fewer cells. This indicates that the size of an organism or organ is not controlled simply by counting cell divisions or cell numbers; total cell mass must somehow be regulated. (Adapted from G. Fankhauser, in Analysis of Development [B.H. Willier, P.A. Weiss, and V. Hamburger, eds.], pp. 126–150. Philadelphia: Saunders, 1955.)

Figure 17–71 **The hindbrain in a haploid and in a tetraploid salamander.** (A) This light micrograph shows a cross section of the hindbrain of a haploid salamander. (B) A corresponding cross section of the hindbrain of a tetraploid salamander, revealing how reduced cell numbers compensate for increased cell size, so that the overall size of the hindbrain is the same in the two animals. (From G. Fankhauser, *Int. Rev. Cytol.* 1:165–193, 1952. With permission from Elsevier.)

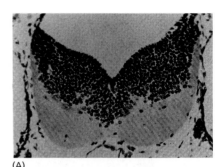

(A)

mechanisms that control cell growth and proliferation in tissues are likewise central to understanding cancer, a disease in which the controls go wrong, as discussed in Chapter 20.

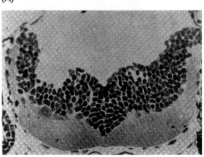

(B)

100 μm

Summary

In multicellular animals, cell size, cell division, and cell death are carefully controlled to ensure that the organism and its organs achieve and maintain an appropriate size. Mitogens stimulate the rate of cell division by removing intracellular molecular brakes that restrain cell-cycle progression in G_1. Growth factors promote cell growth (an increase in cell mass) by stimulating the synthesis and inhibiting the degradation of macromolecules. For proliferating cells to maintain a constant cell size, they employ multiple mechanisms to ensure that cell growth is coordinated with cell division. Animals maintain the normal size of their tissues and organs by adjusting cell size to compensate for changes in cell number, or vice versa. The mechanisms that make this possible are not known.

PROBLEMS

Which statements are true? Explain why or why not.

17–1   Since there are about 10^{13} cells in an adult human, and about 10^{10} cells die and are replaced each day, we become new people every three years.

17–2   The regulation of cyclin–Cdk complexes depends entirely on phosphorylation and dephosphorylation.

17–3   In order for proliferating cells to maintain a relatively constant size, the length of the cell cycle must match the time it takes for the cell to double in size.

17–4   While other proteins come and go during the cell cycle, the proteins of the origin recognition complex remain bound to the DNA throughout.

17–5   Chromosomes are positioned on the metaphase plate by equal and opposite forces that pull them toward the two poles of the spindle.

17–6   If we could turn on telomerase activity in all our cells, we could prevent aging.

Discuss the following problems.

17–7   Many cell-cycle genes from human cells function perfectly well when expressed in yeast cells. Why do you suppose that is considered remarkable? After all, many human genes encoding enzymes for metabolic reactions also function in yeast, and no one thinks that is remarkable.

17–8   You have isolated a new *Cdc* mutant of budding yeast that forms colonies at 25°C but not at 37°C. You would now

like to isolate the wild-type gene that corresponds to the defective gene in your *Cdc* mutant. How might you isolate the wild-type gene using a plasmid-based DNA library prepared from wild-type yeast cells?

17–9   You have isolated a temperature-sensitive mutant of budding yeast. It proliferates well at 25°C, but at 35°C all the cells develop a large bud and then halt their progression through the cell cycle. The characteristic morphology of the cells at the time they stop cycling is known as the landmark morphology.

It is very difficult to obtain synchronous cultures of this yeast, but you would like to know exactly where in the cell cycle the temperature-sensitive gene product must function—its execution point, in the terminology of the field—in order for the cell to complete the cycle. A clever friend, who has a good microscope with a heated stage and a video camera, suggests that you take movies of a field of cells as they experience the temperature increase, and follow the morphology of the cells as they stop cycling. Since the cells do not move much, it is relatively simple to study individual cells. To make sense of what you see, you arrange a circle of pictures of cells at the start of the experiment in order of the size of their daughter buds. You then find the corresponding pictures of those same cells 6 hours later, when growth and division has completely stopped. The results with your mutant are shown in **Figure Q17–1**.

A.   Indicate on the diagram in Figure Q17–1 where the execution point for your mutant lies.

B.   Does the execution point correspond to the time at which the cell cycle is arrested in your mutant? How can you tell?

17–10   The yeast cohesin subunit Scc1, which is essential for sister-chromatid pairing, can be artificially regulated for

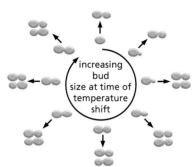

Figure Q17–1 Time-lapse photography of a temperature-sensitive mutant of yeast (Problem 17–9). Cells on the *inner ring* are arranged in order of their bud size, which corresponds to their position in the cell cycle. After 6 hours at 37°C, they have given rise to the cells shown on the *outer ring*. No further growth or division occurs.

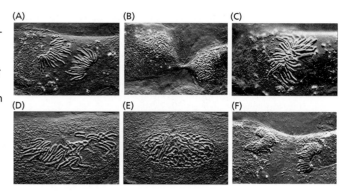

Figure Q17–3 Light micrographs of a single cell at different stages of M phase (Problem 17–13). (Courtesy of Conly L. Rieder.)

expression at any point in the cell cycle. If expression is turned on at the beginning of S phase, all the cells divide satisfactorily and survive. By contrast, if Scc1 expression is turned on only after S phase is completed, the cells fail to divide and they die, even though Scc1 accumulates in the nucleus and interacts efficiently with chromosomes. Why do you suppose that cohesin must be present during S phase for cells to divide normally?

17–11 If cohesins join sister chromatids all along their length, how is it possible for condensins to generate mitotic chromosomes such as that shown in **Figure Q17–2**, which clearly shows the two sister chromatids as separate domains?

Figure Q17–2 A scanning electron micrograph of a fully condensed mitotic chromosome from vertebrate cells (Problem 17–11). (Courtesy of Terry D. Allen.)

1 µm

17–12 High doses of caffeine interfere with the DNA replication checkpoint mechanism in mammalian cells. Why then do you suppose the Surgeon General has not yet issued an appropriate warning to heavy coffee and cola drinkers? A typical cup of coffee (150 mL) contains 100 mg of caffeine (196 g/mole). How many cups of coffee would you have to drink to reach the dose (10 mM) required to interfere with the DNA replication checkpoint mechanism? (A typical adult contains about 40 liters of water.)

17–13 A living cell from the lung epithelium of a newt is shown at different stages in M phase in **Figure Q17–3**. Order these light micrographs into the correct sequence and identify the stage in M phase that each represents.

17–14 How many kinetochores are there in a human cell at mitosis?

17–15 A classic paper clearly distinguished the properties of astral microtubules from those of kinetochore microtubules.

Centrosomes were used to initiate microtubule growth, and then chromosomes were added. The chromosomes bound to the free ends of the microtubules, as illustrated in **Figure Q17–4**. The complexes were then diluted to very low tubulin concentration (well below the critical concentration for microtubule assembly) and examined again (Figure Q17–4). As is evident, only the kinetochore microtubules were stable to dilution.

A. Why do you think kinetochore microtubules are stable?
B. How would you explain the disappearance of the astral microtubules after dilution? Do they detach from the centrosome, depolymerize from an end, or disintegrate along their length at random?
C. How would a time course after dilution help to distinguish among these possible mechanisms for disappearance of the astral microtubules?

17–16 What are the two distinct cytoskeletal machines that are assembled to carry out the mechanical processes of mitosis and cytokinesis in animal cells?

17–17 How do mitogens, growth factors, and survival factors differ from one another?

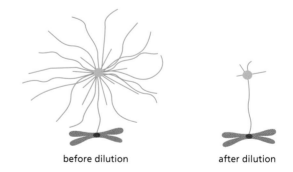

before dilution   after dilution

Figure Q17–4 Arrangements of centrosomes, chromosomes, and microtubules before and after dilution to low tubulin concentration (Problem 17–15).

REFERENCES

General

Morgan DO (2007) The Cell Cycle: Principles of Control. London: New Science Press.
Murray AW & Hunt T (1993) The Cell Cycle: An Introduction. New York: WH Freeman and Co.

Overview of the Cell Cycle

Forsburg SL & Nurse P (1991) Cell cycle regulation in the yeasts *Saccharomyces cerevisiae* and *Schizosaccharomyces pombe*. *Annu Rev Cell Biol* 7:227–256.
Hartwell LH, Culotti J, Pringle JR et al (1974) Genetic control of the cell division cycle in yeast. *Science* 183:46–51.
Kirschner M, Newport J & Gerhart J (1985) The timing of early developmental events in *Xenopus*. *Trends Genet* 1:41–47.

Nurse P, Thuriaux P & Nasmyth K (1976) Genetic control of the cell division cycle in the fission yeast *Schizosaccharomyces pombe. Mol Gen Genet* 146:167–178.

The Cell-Cycle Control System

Evans T, Rosenthal ET, Youngblom J et al (1983) Cyclin: a protein specified by maternal mRNA in sea urchin eggs that is destroyed at each cleavage division. *Cell* 33:389–396.

Lohka MJ, Hayes MK & Maller JL (1988) Purification of maturation-promoting factor, an intracellular regulator of early mitotic events. *Proc Natl Acad Sci USA* 85:3009–3013.

Masui Y and Markert CL (1971) Cytoplasmic control of nuclear behavior during meiotic maturation of frog oocytes. *J Exp Zool* 177:129–146.

Morgan DO (1997) Cyclin-dependent kinases: engines, clocks, and microprocessors. *Annu Rev Cell Dev Biol* 13:261–291.

Murray AW & Kirschner MW (1989) Cyclin synthesis drives the early embryonic cell cycle. *Nature* 339:275–280.

Pavletich NP (1999) Mechanisms of cyclin-dependent kinase regulation: structures of Cdks, their cyclin activators, and CIP and Ink4 inhibitors. *J Mol Biol* 287:821–828.

Peters JM (2006) The anaphase promoting complex/cyclosome: a machine designed to destroy. *Nature Rev Mol Cell Biol* 7:644–656.

Petroski MD & Deshaies RJ (2005) Function and regulation of cullin-RING ubiquitin ligases. *Nature Rev Mol Cell Biol* 6:9–20.

Wittenberg C & Reed SI (2005) Cell cycle-dependent transcription in yeast: promoters, transcription factors, and transcriptomes. *Oncogene* 24:2746–2755.

S Phase

Arias EE & Walter JC (2007) Strength in numbers: preventing rereplication via multiple mechanisms in eukaryotic cells. *Genes Dev* 21:497–518.

Bell SP & Dutta A (2002) DNA replication in eukaryotic cells. *Annu Rev Biochem* 71:333–374.

Bell SP & Stillman B (1992) ATP-dependent recognition of eukaryotic origins of DNA replication by a multiprotein complex. *Nature* 357:128–134.

Diffley JF (2004) Regulation of early events in chromosome replication. *Curr Biol* 14:R778–R786.

Groth A, Rocha W, Verreault A et al (2007) Chromatin challenges during DNA replication and repair. *Cell* 128:721–733.

Tanaka S, Umemori T, Hirai K et al (2007) CDK-dependent phosphorylation of Sld2 and Sld3 initiates DNA replication in budding yeast. *Nature* 445:328–332.

Zegerman P & Diffley JF (2007) Phosphorylation of Sld2 and Sld3 by cyclin-dependent kinases promotes DNA replication in budding yeast. *Nature* 445:281–285.

Mitosis

Cheeseman IM, Chappie JS, Wilson-Kubalek EM et al (2006) The conserved KMN network constitutes the core microtubule-binding site of the kinetochore. *Cell* 127:983–997.

Dong Y, Vanden Beldt KJ, Meng X et al (2007) The outer plate in vertebrate kinetochores is a flexible network with multiple microtubule interactions. *Nature Cell Biol* 9:516–522.

Heald R, Tournebize R, Blank T et al (1996) Self-organization of microtubules into bipolar spindles around artificial chromosomes in *Xenopus* egg extracts. *Nature* 382:420–425.

Hirano T (2005) Condensins: organizing and segregating the genome. *Curr Biol* 15:R265–R275.

Kapoor TM, Lampson MA, Hergert P et al (2006) Chromosomes can congress to the metaphase plate before biorientation. *Science* 311:388–391.

Mitchison T & Kirschner M (1984) Dynamic instability of microtubule growth. *Nature* 312:237–242.

Mitchison TJ (1989) Polewards microtubule flux in the mitotic spindle: evidence from photoactivation of fluorescence. *J Cell Biol* 109:637–652.

Mitchison TJ & Salmon ED (2001) Mitosis: a history of division. *Nature Cell Biol* 3:E17–E21.

Musacchio A & Salmon ED (2007) The spindle-assembly checkpoint in space and time. *Nature Rev Mol Cell Biol* 8:379–393.

Nasmyth K (2002) Segregating sister genomes: the molecular biology of chromosome separation. *Science* 297:559–565.

Nigg EA (2007) Centrosome duplication: of rules and licenses. *Trends Cell Biol* 17:215–221.

Nurse P (1990) Universal control mechanism regulating onset of M-phase. *Nature* 344:503–508.

Page SL & Hawley RS (2003) Chromosome choreography: the meiotic ballet. *Science* 301:785–789.

Petronczki M, Siomos MF & Nasmyth K (2003) Un ménage à quatre: the molecular biology of chromosome segregation in meiosis. *Cell* 112:423–440.

Salmon ED (2005) Microtubules: a ring for the depolymerization motor. *Curr Biol* 15:R299–R302.

Tanaka TU, Stark MJ & Tanaka K (2005) Kinetochore capture and bi-orientation on the mitotic spindle. *Nature Rev Mol Cell Biol* 6:929–942.

Uhlmann F, Lottspeich F & Nasmyth K (1999) Sister-chromatid separation at anaphase onset is promoted by cleavage of the cohesin subunit Scc1. *Nature* 400:37–42.

Wadsworth P & Khodjakov A (2004) E pluribus unum: towards a universal mechanism for spindle assembly. *Trends Cell Biol* 14:413–419.

Cytokinesis

Albertson R, Riggs B & Sullivan W (2005) Membrane traffic: a driving force in cytokinesis. *Trends Cell Biol* 15:92–101.

Burgess DR & Chang F (2005) Site selection for the cleavage furrow at cytokinesis. *Trends Cell Biol.* 15:156–162.

Dechant R & Glotzer M (2003) Centrosome separation and central spindle assembly act in redundant pathways that regulate microtubule density and trigger cleavage furrow formation. *Dev Cell* 4:333–344.

Eggert US, Mitchison TJ & Field CM (2006) Animal cytokinesis: from parts list to mechanisms. *Annu Rev Biochem* 75:543–566.

Glotzer M (2005) The molecular requirements for cytokinesis. *Science* 307:1735–1739.

Grill SW, Gonczy P, Stelzer EH et al (2001) Polarity controls forces governing asymmetric spindle positioning in the *Caenorhabditis elegans* embryo. *Nature* 409:630–633.

Jurgens G (2005) Plant cytokinesis: fission by fusion. *Trends Cell Biol* 15:277–283.

Rappaport R (1986) Establishment of the mechanism of cytokinesis in animal cells. *Int Rev Cytol* 105:245–281.

Control of Cell Division and Cell Growth

Adhikary S & Eilers M (2005) Transcriptional regulation and transformation by Myc proteins. *Nature Rev Mol Cell Biol* 6:635–645.

Campisi J (2005) Senescent cells, tumor suppression, and organismal aging: good citizens, bad neighbors. *Cell* 120:513–522.

Conlon I & Raff M (1999) Size control in animal development. *Cell* 96:235–244.

Frolov MV, Huen DS, Stevaux O et al (2001) Functional antagonism between E2F family members. *Genes Dev* 15:2146–2160.

Harrison JC & Haber JE (2006) Surviving the breakup: the DNA damage checkpoint. *Annu Rev Genet* 40:209–235.

Jorgensen P & Tyers M (2004) How cells coordinate growth and division. *Curr Biol* 14:R1014–R1027.

Levine AJ (1997) p53, the cellular gatekeeper for growth and division. *Cell* 88:323–331.

Raff MC (1992) Social controls on cell survival and cell death. *Nature* 356:397–400.

Sherr CJ & DePinho RA (2000) Cellular senescence: mitotic clock or culture shock? *Cell* 102:407–410.

Sherr CJ & Roberts JM (1999) CDK inhibitors: positive and negative regulators of G1-phase progression. *Genes Dev* 13:1501–1512.

Trimarchi JM & Lees JA (2002) Sibling rivalry in the E2F family. *Nature Rev Mol Cell Biol* 3:11–20.

Vousden KH & Lu X (2002) Live or let die: the cell's response to p53. *Nature Rev Cancer* 2:594–604.

Zetterberg A & Larsson O (1985) Kinetic analysis of regulatory events in G1 leading to proliferation or quiescence of Swiss 3T3 cells. *Proc Natl Acad Sci USA* 82:5365–5369.

Chapter 18

Apoptosis

Cell death plays a crucially important part in animal and plant development, and it usually continues into adulthood. In a healthy adult human, billions of cells die in the bone marrow and intestine every hour. Our tissues do not shrink because, by unknown regulatory mechanisms, cell division exactly balances the cell death. We now know that these "normal" cell deaths are suicides, in which the cells activate an intracellular death program and kill themselves in a controlled way—a process known as **programmed cell death**. The idea that animal cells have a built-in death program was proposed in the 1970s, but its general acceptance took another 20 years and depended on genetic studies in the nematode *C. elegans* that identified the first genes dedicated to programmed cell death and its control.

Programmed cell death in animals usually, but not exclusively, occurs by **apoptosis** (from the Greek word meaning "falling off," as leaves from a tree). Although apoptosis is only one form of programmed cell death, it is by far the most common and best understood, and, confusingly, biologists often use the terms programmed cell death and apoptosis interchangeably. Cells dying by apoptosis undergo characteristic morphological changes. <GCCC> They shrink and condense, the cytoskeleton collapses, the nuclear envelope disassembles, and the nuclear chromatin condenses and breaks up into fragments (**Figure 18–1**A). The cell surface often blebs and, if the cell is large, often breaks up into membrane-enclosed fragments called *apoptotic bodies*. Most importantly, the surface of the cell or apoptotic bodies becomes chemically altered, so that a neighboring cell or a macrophage (a specialized phagocytic cell, discussed in Chapter 23) rapidly engulfs them, before they can spill their contents (Figure 18–1B). In this way, the cell dies neatly and is rapidly cleared away, without causing a damaging inflammatory response. Because the cells are eaten and digested so quickly, there are usually few dead cells to be seen, even when large numbers of cells have died by apoptosis. This is probably why biologists overlooked apoptosis for many years and still probably underestimate its extent.

By contrast to apoptosis and other less well characterized forms of programmed cell death (which implies the operation of an intracellular death program), animal cells that die accidentally in response to an acute insult, such as trauma or a lack of blood supply, usually do so by a process called *cell necrosis*. Necrotic cells swell and burst, spilling their contents over their neighbors and eliciting an inflammatory response (Figure 18–1C).

Programmed cell death is not confined to animals. In plants, it occurs during development and in the senescence of flowers and leaves, as well as in the response to injury and infection. Programmed cell death even occurs in unicellular organisms, including yeasts and bacteria. The molecular mechanisms involved in these cases are distinct from those that mediate apoptosis in animal cells, and we shall not consider them. In this chapter, we discuss the functions of programmed cell death in animals, the molecular mechanism of apoptosis and its regulation, and how excessive or insufficient apoptosis can contribute to human disease.

Programmed Cell Death Eliminates Unwanted Cells

The amount of programmed cell death that occurs in developing and adult animal tissues can be astonishing. In the developing vertebrate nervous system, for

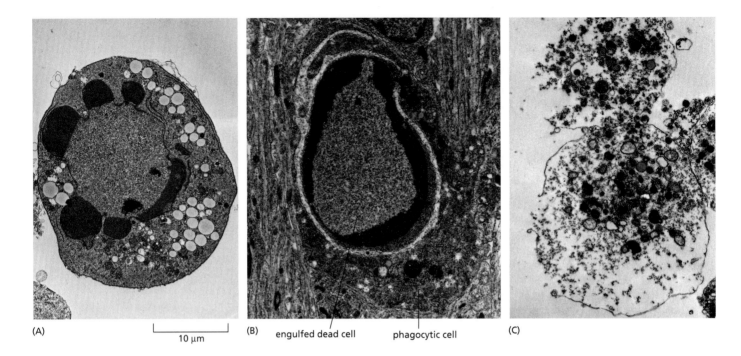

(A)

⌞_____⌟
10 μm

(B) engulfed dead cell   phagocytic cell

(C)

Figure 18–1 Two distinct forms of cell death. These electron micrographs show cells that have died by apoptosis (A and B) or by a type of accidental cell death called necrosis (C). The cells in (A) and (C) died in a culture dish, whereas the cell in (B) died in a developing tissue and has been engulfed by a phagocytic cell. Note that the cells in (A) and (B) have condensed but seem relatively intact, whereas the cell in (C) seems to have exploded. The large vacuoles visible in the cytoplasm of the cell in (A) are a variable feature of apoptosis. (Courtesy of Julia Burne.)

example, more than half of many types of nerve cells normally die soon after they are formed. It seems remarkably wasteful for so many cells to die, especially as the vast majority are perfectly healthy at the time they kill themselves. What purposes does this massive cell death serve?

In some cases, the answer is clear. In animal development, programmed cell death eliminates unwanted cells, usually by apoptosis. Cell death, for example, helps sculpt hands and feet during embryonic development: they start out as spade-like structures, and the individual digits separate only as the cells between them die, as illustrated for a mouse paw in **Figure 18–2**. In other cases, cells die when the structure they form is no longer needed. When a tadpole changes into a frog at metamorphosis, the cells in the tail die, and the tail, which is not needed in the frog, disappears (**Figure 18–3**). In many other cases, cell death helps regulate cell numbers. In the developing nervous system, for example, cell death adjusts the number of nerve cells to match the number of target cells that the nerve cells connect to, as we discuss later.

Programmed cell death also functions as a quality-control process in development, eliminating cells that are abnormal, misplaced, nonfunctional, or potentially dangerous to the animal. Striking examples occur in the vertebrate adaptive immune system, where apoptosis eliminates developing T and B lymphocytes that either fail to produce potentially useful antigen-specific receptors or produce self-reactive receptors that make the cells potentially dangerous; it also eliminates most of the lymphocytes activated by an infection, after they have helped destroy the responsible microbes (discussed in Chapter 25).

In adult tissues that are neither growing nor shrinking, cell death and cell division must be tightly regulated to ensure that they are exactly in balance. If

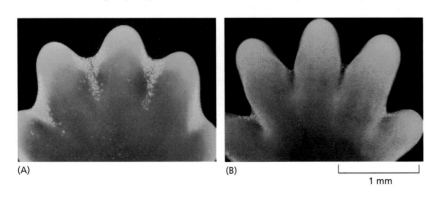

(A)   (B)

⌞_____⌟
1 mm

Figure 18–2 Sculpting the digits in the developing mouse paw by apoptosis. (A) The paw in this mouse fetus has been stained with a dye that specifically labels cells that have undergone apoptosis. The apoptotic cells appear as *bright green* dots between the developing digits. (B) The interdigital cell death has eliminated the tissue between the developing digits, as seen one day later, when there are very few apoptotic cells. (From W. Wood et al., *Development* 127:5245–5252, 2000. With permission from The Company of Biologists.)

Figure 18–3 **Apoptosis during the metamorphosis of a tadpole into a frog.** As a tadpole changes into a frog, the cells in the tadpole tail are induced to undergo apoptosis; as a consequence, the tail is lost. An increase in thyroid hormone in the blood stimulates all the changes that occur during metamorphosis, including apoptosis in the tail.

part of the liver is removed in an adult rat, for example, liver cell proliferation increases to make up the loss. Conversely, if a rat is treated with the drug phenobarbital—which stimulates liver cell division (and thereby liver enlargement)—and then the phenobarbital treatment is stopped, apoptosis in the liver greatly increases until the liver has returned to its original size, usually within a week or so. Thus, the liver is kept at a constant size through the regulation of both the cell death rate and the cell birth rate, although the control mechanisms responsible for such regulation are largely unknown.

Apoptosis occurs at a staggeringly high rate in the adult human bone marrow, where most blood cells are produced. Here, for example, *neutrophils* (a type of white blood cell discussed in Chapter 23) are produced continuously in very large numbers, but the vast majority die by apoptosis in the bone marrow within a few days without ever functioning. This apparently futile cycle of production and destruction serves to maintain a ready supply of short-lived neutrophils that can be rapidly mobilized to fight infection wherever it occurs in the body. Compared with the life of the organism, cells are evidently cheap.

Animal cells can recognize damage in their various organelles and, if the damage is great enough, they can kill themselves by undergoing apoptosis. An important example is DNA damage, which can produce cancer-promoting mutations if not repaired. Cells have various ways of detecting DNA damage, and, if they cannot repair it, they often kill themselves by undergoing apoptosis.

Apoptotic Cells Are Biochemically Recognizable

Cells undergoing apoptosis not only have a characteristic morphology but also display characteristic biochemical changes, which can be used to identify apoptotic cells. During apoptosis, for example, an endonuclease cleaves the chromosomal DNA into fragments of distinctive sizes; because the cleavages occur in the linker regions between nucleosomes, the fragments separate into a characteristic ladder pattern when analyzed by gel electrophoresis (**Figure 18–4**A). Moreover, the cleavage of DNA generates many new DNA ends, which can be marked in apoptotic nuclei by using a labeled nucleotide in the so-called TUNEL technique (Figure 18–4B).

An especially important change occurs in the plasma membrane of apoptotic cells. The negatively charged phospholipid *phosphatidylserine* is normally exclusively located in the inner leaflet of the lipid bilayer of the plasma membrane (see Figures 10–3 and 10–16), but it flips to the outer leaflet in apoptotic cells, where it can serve as a marker of these cells. The phosphatidylserine on the surface of apoptotic cells can be visualized with a labeled form of the *Annexin V* protein, which specifically binds to this phospholipid. The cell-surface phosphatidylserine is more than a convenient marker of apoptosis for biologists; it helps signal to neighboring cells and macrophages to phagocytose the dying cell. In addition to serving as an "eat me" signal, it also blocks the inflammation often associated with phagocytosis: the phosphatidylserine-dependent engulfment of apoptotic cells inhibits the production of inflammation-inducing signal proteins (cytokines) by the phagocytic cell.

Macrophages will phagocytose most types of small particles, including oil droplets and glass beads, but they do not phagocytose any healthy cells in the animal, presumably because healthy cells express "don't eat me" signal molecules on their surface. Thus, in addition to expressing cell-surface "eat me" signals such as phosphatidylserine that stimulate phagocytosis, apoptotic cells must lose or inactivate their "don't eat me" signals in order for macrophages to ingest them.

time (hr)

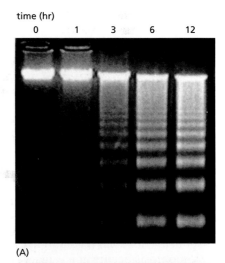

Figure 18–4 **Markers of apoptosis.** (A) Cleavage of nuclear DNA into a characteristic ladder pattern of fragments. Mouse thymus lymphocytes were treated with an antibody against the cell-surface death receptor Fas (discussed later), inducing the cells to undergo apoptosis. After various times (indicated in hours at the top of the figure), DNA was extracted, and the fragments were separated by size by electrophoresis in an agarose gel and stained with ethidium bromide. (B) The TUNEL technique was used to label the cut ends of DNA fragments in the nuclei of apoptotic cells in a tissue section of a developing chick leg bud; this cross section through the skin and underlying tissue is from a region between two developing digits, as indicated in the underlying drawing. The procedure is called the TUNEL (<u>T</u>dT-mediated d<u>U</u>TP <u>n</u>ick <u>e</u>nd <u>l</u>abeling) technique because the enzyme terminal deoxynucleotidyl transferase (TdT) adds chains of labeled deoxynucleotide (dUTP) to the 3′-OH ends of DNA fragments. (A, from D. McIlroy et al., *Genes Dev.* 14:549–558, 2000. With permisison from Cold Spring Harbor Laboratory Press; B, from V. Zuzarte-Luís and J.M. Hurlé, *Int. J. Dev. Biol.* 46:871–876, 2002. With permission from UBC Press.)

(A)

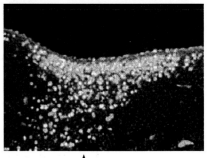

Cells undergoing apoptosis often lose the electrical potential that normally exists across the inner membrane of their mitochondria (discussed in Chapter 14). This membrane potential can be measured by the use of positively charged fluorescent dyes that accumulate in mitochondria, driven by the negative charge on the inside of the inner membrane. A decrease in the labeling of mitochondria with these dyes helps to identify cells that are undergoing apoptosis. As we discuss later, proteins such as *cytochrome c* are usually released from the space between the inner and outer membrane (the *intermembrane space*) of mitochondria during apoptosis, and the relocation of cytochrome *c* from mitochondria to the cytosol can be used as another marker of apoptosis (see Figure 18–7).

Apoptosis Depends on an Intracellular Proteolytic Cascade That Is Mediated by Caspases

The intracellular machinery responsible for apoptosis is similar in all animal cells. It depends on a family of proteases that have a cysteine at their active site and cleave their target proteins at specific aspartic acids. They are therefore called **caspases** (c for cysteine and asp for aspartic acid). Caspases are synthesized in the cell as inactive precursors, or **procaspases**, which are typically activated by proteolytic cleavage. Procaspase cleavage occurs at one or two specific aspartic acids and is catalyzed by other (already active) caspases; the procaspase is split into a large and a small subunit that form a heterodimer, and two such dimers assemble to form the active tetramer (**Figure 18–5**A). Once activated, caspases cleave, and thereby activate, other procaspases, resulting in an amplifying proteolytic cascade (Figure 18–5B).

(B)

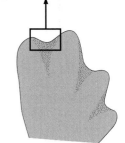

Not all caspases mediate apoptosis. Indeed, the first caspase identified was a human protein called *interleukin-1-converting enzyme (ICE)*, which is concerned with inflammatory responses rather than with cell death; ICE cuts out the inflammation-inducing cytokine *interleukin-1 (IL1)* from a larger precursor protein. Subsequent to the discovery of ICE, a gene required for apoptosis in *C. elegans* was shown to encode a protein that is structurally and functionally similar to ICE, providing the first evidence that proteolysis and caspases are involved in apoptosis. It is now clear that, whereas several human caspases are involved in inflammatory and immune responses, most are involved in apoptosis (**Table 18–1**).

As shown in Figure 18–5B and Table 18–1, some of the procaspases that operate in apoptosis act at the start of the proteolytic cascade and are called **initiator procaspases**; when activated, they cleave and activate downstream **executioner procaspases**, which, then cleave and activate other executioner procaspases, as well as specific *target proteins* in the cell. Among the many target proteins cleaved by executioner caspases are the nuclear lamins (see Figure 18–5B), the cleavage of which causes the irreversible breakdown of the nuclear lamina (discussed in Chapter 16). Another target is a protein that normally holds

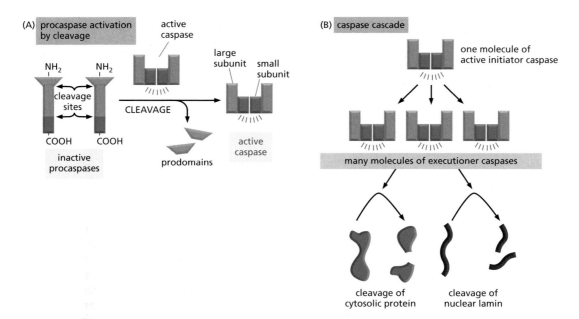

Figure 18–5 **Procaspase activation during apoptosis.** (A) Each caspase is initially made as an inactive proenzyme (procaspase). Some procaspases are activated by proteolytic cleavage by an activated caspase: two cleaved fragments from each of two procaspase molecules associate to form an active caspase, which is a tetramer of two small and two large subunits; the prodomains are usually discarded, as indicated. (B) The first procaspases activated are called *initiator procaspases,* which then cleave and activate many *executioner procaspase* molecules, producing an amplifying chain reaction (a proteolytic caspase cascade). The executioner caspases then cleave a variety of key proteins in the cell, including specific cytosolic proteins and nuclear lamins, as shown here, leading to the controlled death of the cell. Although not shown, the initiator procaspases are activated by adaptor proteins that bring the procaspases together in close proximity within an activation complex; although the initiator procaspases cleave each other within the complex, the cleavage serves only to stabilize the active protease.

the DNA-degrading enzyme mentioned earlier (an endonuclease) in an inactive form; its cleavage frees the endonuclease to cut up the DNA in the cell nucleus. Other target proteins include components of the cytoskeleton and cell–cell adhesion proteins that attach cells to their neighbors; the cleavage of these proteins helps the apoptotic cell to round up and detach from its neighbors, making it easier for a healthy neighboring cell to engulf it, or, in the case of an epithelial cell, for the neighbors to extrude the apoptotic cell from the cell sheet. The caspase cascade is not only destructive and self-amplifying but also irreversible, so that once a cell reaches a critical point along the path to destruction, it cannot turn back.

The caspases required for apoptosis vary depending on the cell type and stimulus. Inactivation of the mouse gene encoding caspase-3, an executioner caspase, for example, reduces normal apoptosis in the developing brain. As a result, the mouse often dies around birth with a deformed brain that contains too many cells. Apoptosis occurs normally, however, in many other organs of such mice.

From the earliest stages of an animal's development, healthy cells continuously make the procaspases and other proteins required for apoptosis. Thus, the apoptosis machinery is always in place; all that is needed is a trigger to activate it. How, then, is a caspase cascade initiated? In particular, how is the first procaspase in the cascade activated? Initiator procaspases have a long *prodomain,* which contains a *caspase recruitment domain (CARD)* that enables them to assemble with adaptor proteins into *activation complexes* when the cell receives a signal to undergo apoptosis. Once incorporated into such a complex, the initiator procaspases are brought into close proximity, which is sufficient to activate them; they then cleave each other to make the process irreversible. The activated initiator caspases then cleave and activate executioner procaspases, thereby initiating a proteolytic caspase cascade, which amplifies the death signal and spreads it throughout the cell.

The two best understood signaling pathways that can activate a caspase cascade leading to apoptosis in mammalian cells are called the *extrinsic pathway* and the *intrinsic pathway.* Each uses its own initiator procaspases and activation complex, as we now discuss.

Table 18–1 Some Human Caspases

Caspases involved in inflammation	caspases 1 (ICE), 4, 5
Caspases involved in apoptosis	
Initiator caspases	caspases 2, 8, 9, 10
Executioner caspases	caspases 3, 6, 7

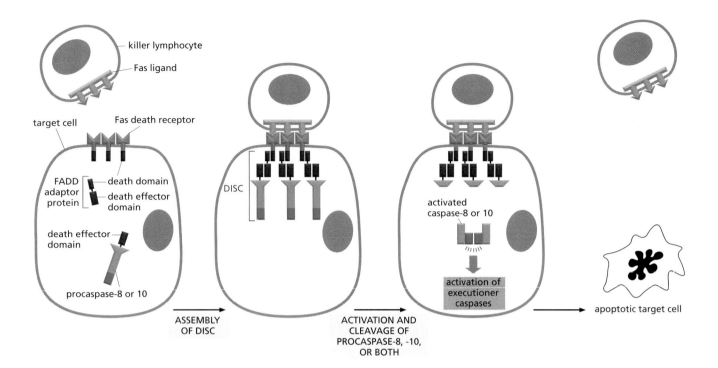

Cell-Surface Death Receptors Activate the Extrinsic Pathway of Apoptosis

Extracellular signal proteins binding to cell-surface **death receptors** trigger the **extrinsic pathway** of apoptosis. Death receptors are transmembrane proteins containing an extracellular ligand-binding domain, a single transmembrane domain, and an intracellular *death domain*, which is required for the receptors to activate the apoptotic program. The receptors are homotrimers and belong to the *tumor necrosis factor (TNF) receptor* family, which includes a receptor for TNF itself (discussed in Chapter 15) and the *Fas* death receptor. The ligands that activate the death receptors are also homotrimers; they are structurally related to one another and belong to the *TNF family* of signal proteins.

A well-understood example of how death receptors trigger the extrinsic pathway of apoptosis is the activation of **Fas** on the surface of a target cell by **Fas ligand** on the surface of a killer (cytotoxic) lymphocyte (discussed in Chapter 25). When activated by the binding of Fas ligand, the death domains on the cytosolic tails of the Fas death receptors recruit intracellular adaptor proteins, which in turn recruit initiator procaspases (*procaspase-8, procaspase-10*, or both), forming a **death-inducing signaling complex** (**DISC**). Once activated in the DISC, the initiator caspases activate downstream executioner procaspases to induce apoptosis (**Figure 18–6**). As we discuss later, in some cells the extrinsic pathway must recruit the intrinsic apoptotic pathway to amplify the caspase cascade in order to kill the cell.

Many cells produce inhibitory proteins that act either extracellularly or intracellularly to restrain the extrinsic pathway. For example, some produce cell-surface *decoy receptors,* which have a ligand-binding domain but not a death domain; because they can bind a death ligand but cannot activate apoptosis, the decoys competitively inhibit the death receptors. Cells can also produce intra-cellular blocking proteins such as *FLIP,* which resembles an initiator procaspase but lacks the proteolytic domain; it competes with procaspase-8 and procaspase-10 for binding sites in the DISC and thereby inhibits the activation of these initiator procaspases. Such inhibitory mechanisms help prevent the inappropriate activation of the extrinsic pathway of apoptosis.

In some circumstances, death receptors activate other intracellular signaling pathways that do not lead to apoptosis. TNF receptors, for example, can also activate the NFκB pathway (discussed in Chapter 15), which can promote cell

Figure 18–6 **The extrinsic pathway of apoptosis activated through Fas death receptors.** Fas ligand on the surface of a killer lymphocyte activates Fas death receptors on the surface of the target cell. Both the ligand and receptor are homotrimers. The cytosolic tail of Fas then recruits the adaptor protein FADD via the death domain on each protein (FADD stands for Fas-associated death domain). Each FADD protein then recruits an initiator procaspase (procaspase-8, procaspase-10, or both) via a death effector domain on both FADD and the procaspase, forming a death-inducing signaling complex (DISC). Within the DISC, the initiator procaspase molecules are brought into close proximity, which activates them; the activated procaspases then cleave one another to stabilize the activated protease, which is now a caspase. Activated caspase-8 and caspase-10 then cleave and activate executioner procaspases, producing a caspase cascade, which leads to apoptosis.

survival and activate genes involved in inflammatory responses. Which responses dominate depends on the type of cell and the other signals acting on it.

The Intrinsic Pathway of Apoptosis Depends on Mitochondria

Cells can also activate their apoptosis program from inside the cell, usually in response to injury or other stresses, such as DNA damage or lack of oxygen, nutrients, or extracellular survival signals (discussed later). In vertebrate cells, such intracellular activation of the apoptotic death program occurs via the **intrinsic pathway** of apoptosis, which depends on the release into the cytosol of mitochondrial proteins that normally reside in the intermembrane space of these organelles (see Figure 12–21A). Some of the released proteins activate a caspase proteolytic cascade in the cytoplasm, leading to apoptosis.

A crucial protein released from mitochondria in the intrinsic pathway is **cytochrome _c_**, a water-soluble component of the mitochondrial electron-transport chain. When released into the cytosol (**Figure 18–7**), it has an entirely different function: it binds to a procaspase-activating adaptor protein called **Apaf1** _(apoptotic protease activating factor-1)_, causing the Apaf1 to oligomerize into a wheel-like heptamer called an **apoptosome**. The Apaf1 proteins in the apoptosome then recruit initiator procaspase proteins _(procaspase-9)_, which are activated by proximity in the apoptosome, just as procaspase-8 and -10 proteins are activated in the DISC. The activated caspase-9 molecules then activate downstream executioner procaspases to induce apoptosis (**Figure 18–8**).

As mentioned earlier, in some cells, the extrinsic pathway must recruit the intrinsic pathway to amplify the apoptotic signal to kill the cell. It does so by activating a member of the _Bcl2_ family of proteins, which we now discuss.

Bcl2 Proteins Regulate the Intrinsic Pathway of Apoptosis

The intrinsic pathway of apoptosis is tightly regulated to ensure that cells kill themselves only when it is appropriate. A major class of intracellular regulators of apoptosis is the **Bcl2** family of proteins, which, like the caspase family, has been conserved in evolution from worms to humans; a human Bcl2 protein, for example, can suppress apoptosis when expressed in _C. elegans_.

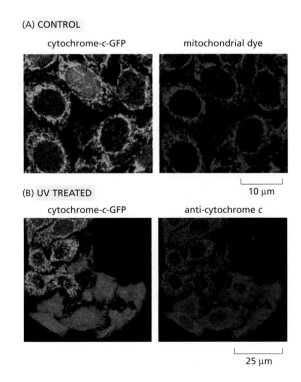

(A) CONTROL

cytochrome-_c_-GFP   mitochondrial dye

(B) UV TREATED

cytochrome-_c_-GFP   anti-cytochrome _c_

10 μm

25 μm

Figure 18–7 Release of cytochrome _c_ from mitochondria during apoptosis. Fluorescence micrographs of human cancer cells in culture. (A) The control cells were transfected with a gene encoding a fusion protein consisting of cytochrome _c_ linked to green fluorescent protein (cytochrome-_c_-GFP); they were also treated with a positively charged red dye that accumulates in mitochondria. The overlapping distribution of the _green_ and _red_ indicate that the cytochrome-_c_-GFP is located in mitochondria. (B) Cells expressing cytochrome-_c_-GFP were irradiated with ultraviolet light to induce apoptosis, and after 5 hours they were stained with antibodies (in _red_) against cytochrome _c_; the cytochrome-_c_-GFP is also shown (in _green_). The six cells in the bottom half of the micrographs in B have released their cytochrome _c_ from mitochondria into the cytosol, whereas the cells in the upper half of the micrographs have not yet done so. (From J.C. Goldstein et al., _Nat. Cell Biol._ 2:156–162, 2000. With permission from Macmillan Publishers Ltd.)

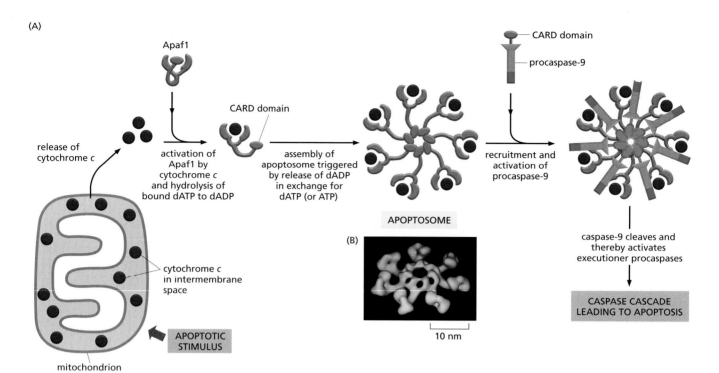

(A)

(B)

10 nm

APOPTOSOME

CASPASE CASCADE LEADING TO APOPTOSIS

Mammalian Bcl2 proteins regulate the intrinsic pathway of apoptosis mainly by controlling the release of cytochrome *c* and other intermembrane mitochondrial proteins into the cytosol. Some Bcl2 proteins are *pro-apoptotic* and promote apoptosis by enhancing the release, whereas others are *anti-apoptotic* and inhibit apoptosis by blocking the release. The pro-apoptotic and anti-apoptotic Bcl2 proteins can bind to each other in various combinations to form heterodimers, in which the two proteins inhibit each other's function. The balance between the activities of these two functional classes of Bcl2 proteins largely determines whether a mammalian cell lives or dies by the intrinsic pathway of apoptosis.

As illustrated in **Figure 18–9**, the anti-apoptotic Bcl2 proteins, including *Bcl2* itself (the founding member of the Bcl2 family) and *Bcl-X$_L$*, share four distinctive *Bcl2 homology (BH) domains* (BH1–4). The pro-apoptotic Bcl2 proteins consist of two subfamilies—the *BH123* proteins and the *BH3-only* proteins. The main BH123 proteins are *Bax* and *Bak*, which are structurally similar to Bcl2 but lack the BH4 domain. The BH3-only proteins share sequence homology with Bcl2 in only the BH3 domain (see Figure 18–9).

When an apoptotic stimulus triggers the intrinsic pathway, the pro-apoptotic **BH123 proteins** become activated and aggregate to form oligomers in the mitochondrial outer membrane, inducing the release of cytochrome *c* and other intermembrane proteins by an unknown mechanism (**Figure 18–10**). In mammalian cells, **Bax** and **Bak** are the main BH123 proteins, and at least one of them is required for the intrinsic pathway of apoptosis to operate: mutant mouse cells that lack both proteins are resistant to all pro-apoptotic signals that normally activate this pathway. Whereas Bak is tightly bound to the mitochondrial outer membrane even in the absence of an apoptotic signal, Bax is mainly located in the cytosol and translocates to the mitochondria only after an apoptotic signal activates it. As we discuss below, the activation of Bax and Bak usually depends on activated pro-apoptotic BH3-only proteins. Both Bax and Bak also operate on the surface of the endoplasmic reticulum (ER) and nuclear membranes; when activated in response to ER stress, they are thought to release Ca^{2+} into the cytosol, which helps activate the mitochondrial-dependent intrinsic pathway of apoptosis by a poorly understood mechanism.

The **anti-apoptotic Bcl2 proteins** such as **Bcl2** itself and **Bcl-X$_L$** are also mainly located on the cytosolic surface of the outer mitochondrial membrane, the ER, and the nuclear envelope, where they help preserve the integrity of the

Figure 18–8 The intrinsic pathway of apoptosis. (A) A schematic drawing of how cytochrome *c* released from mitochondria activates Apaf1. The binding of cytochrome *c* causes the Apaf1 to hydrolyze its bound dATP to dADP (not shown). The replacement of the dADP with dATP or ATP (not shown) then induces the complex of Apaf1 and cytochrome *c* to aggregate to form a large, heptameric apoptosome, which then recruits procaspase-9 through a caspase recruitment domain (CARD) in each protein. The procaspase-9 molecules are activated within the apoptosome and are now able to cleave and activate downstream executioner procaspases, which leads to the cleavage and activation of these molecules in a caspase cascade. Other proteins released from the mitochondrial intermembrane space are not shown. (B) A model of the three-dimensional structure of an apoptosome. Note that some scientists use the term "apoptosome" to refer to the complex containing procaspase-9. (B, from D. Aceham et al., *Mol. Cell* 9:423–432, 2002. With permission from Elsevier.)

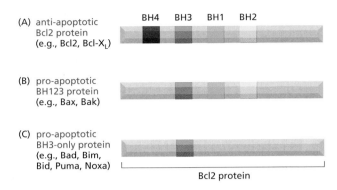

(A) anti-apoptotic
Bcl2 protein
(e.g., Bcl2, Bcl-X$_L$)

(B) pro-apoptotic
BH123 protein
(e.g., Bax, Bak)

(C) pro-apoptotic
BH3-only protein
(e.g., Bad, Bim,
Bid, Puma, Noxa)

BH4 BH3 BH1 BH2

Bcl2 protein

Figure 18–9 The three classes of Bcl2 proteins. Note that the BH3 domain is the only BH domain shared by all Bcl2 family members; it mediates the direct interactions between pro-apoptotic and anti-apoptotic family members.

membrane—preventing, for example, inappropriate release of intermembrane proteins from mitochondria and of Ca^{2+} from the ER. These proteins inhibit apoptosis mainly by binding to and inhibiting pro-apoptotic Bcl2 proteins—either on these membranes or in the cytosol. On the outer mitochondrial membrane, for example, they bind to Bak and prevent it from oligomerizing, thereby inhibiting the release of cytochrome c and other intermembrane proteins. There are at least five mammalian anti-apoptotic Bcl2 proteins, and every mammalian cell requires at least one to survive. Moreover, a number of these proteins must be inhibited for the intrinsic pathway to induce apoptosis; the BH3-only proteins mediate the inhibition.

The **BH3-only proteins** are the largest subclass of Bcl2 family proteins. The cell either produces or activates them in response to an apoptotic stimulus, and they are thought to promote apoptosis mainly by inhibiting anti-apoptotic Bcl2 proteins. Their BH3 domain binds to a long hydrophobic groove on anti-apoptotic Bcl2 proteins, neutralizing their activity. By a poorly understood mechanism, this binding and inhibition enables the aggregation of Bax and Bak on the surface of mitochondria, which triggers the release of the intermembrane mitochondrial proteins that induce apoptosis (**Figure 18–11**). Some BH3-only proteins may bind directly to Bax and Bak to help trigger the activation and aggregation of these BH123 pro-apoptotic proteins on mitochondria and thereby help release the intermembrane proteins.

BH3-only proteins provide the crucial link between apoptotic stimuli and the intrinsic pathway of apoptosis, with different stimuli activating different BH3-only proteins. When some cells are deprived of extracellular survival signals, for example, an intracellular signaling pathway that depends on the MAP kinase *JNK* activates the transcription of the gene encoding the BH3-only protein *Bim*, which then triggers the intrinsic pathway. Similarly, in response to DNA damage that cannot be repaired, the tumor suppressor protein **p53** accumulates (discussed in Chapters 17 and 20) and activates the transcription of genes that encode the BH3-only proteins *Puma* and *Noxa*; these BH3-only proteins then trigger the intrinsic pathway, thereby eliminating a potentially dangerous cell that could otherwise become cancerous.

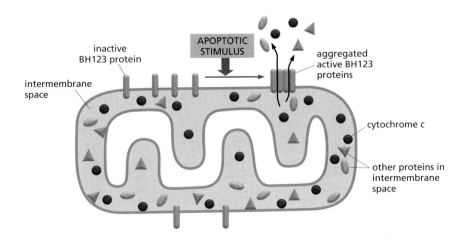

inactive
BH123 protein

APOPTOTIC
STIMULUS

aggregated
active BH123
proteins

intermembrane
space

cytochrome c

other proteins in
intermembrane
space

Figure 18–10 The role of BH123 pro-apoptotic Bcl2 proteins (mainly Bax and Bak) in the release of mitochondrial intermembrane proteins in the intrinsic pathway of apoptosis. When activated by an apoptotic stimulus, the BH123 proteins aggregate on the outer mitochondrial membrane and release cytochrome c and other proteins from the intermembrane space into the cytosol by an unknown mechanism.

(A) INACTIVE INTRINSIC PATHWAY

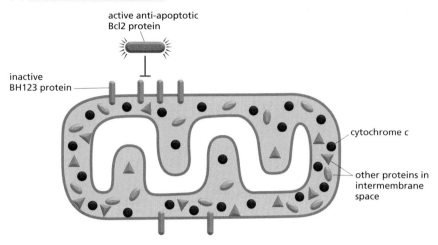

(B) ACTIVATION OF INTRINSIC PATHWAY

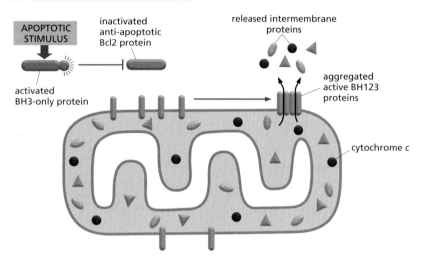

Figure 18–11 How pro-apoptotic BH3-only and anti-apoptotic Bcl2 proteins regulate the intrinsic pathway of apoptosis. (A) In the absence of an apoptotic stimulus, anti-apoptotic Bcl2 proteins bind to and inhibit the BH123 proteins on the mitochondrial outer membrane (and in the cytosol—not shown). (B) In the presence of an apoptotic stimulus, BH3-only proteins are activated and bind to the anti-apoptotic Bcl2 proteins so that they can no longer inhibit the BH123 proteins, which now become activated and aggregate in the outer mitochondrial membrane and promote the release of intermembrane mitochondrial proteins into the cytosol. Some activated BH3-only proteins may stimulate mitochondrial protein release more directly by binding to and activating the BH123 proteins. Although not shown, the anti-apoptotic Bcl2 proteins are bound to the mitochondrial surface.

As mentioned earlier, in some cells the extrinsic apoptotic pathway recruits the intrinsic pathway to amplify the caspase cascade to kill the cell. The BH3-only protein *Bid* is the link between the two pathways. When death receptors activate the extrinsic pathway in these cells, the initiator caspase, caspase-8, cleaves Bid, producing a truncated form of Bid called *tBid*. tBid translocates to mitochondria, where it inhibits anti-apoptotic Bcl2 proteins and triggers the aggregation of pro-apoptotic BH123 proteins to release cytochrome *c* and other intermembrane proteins, thereby amplifying the death signal.

The BH3-only proteins Bid, Bim, and Puma (see Figure 18–9) can inhibit all of the anti-apoptotic Bcl2 proteins, whereas the other BH3-only proteins can inhibit only a small subset of the anti-apoptotic proteins. Thus, Bid, Bim, and Puma are the most potent activators of apoptosis in the BH3-only subfamily of Bcl2 proteins.

Bcl2 proteins are not the only intracellular regulators of apoptosis. The IAP (inhibitor of apoptosis) proteins also play an important part in suppressing apoptosis, especially in *Drosophila*.

IAPs Inhibit Caspases

Inhibitors of apoptosis (IAPs) were first identified in certain insect viruses (baculoviruses), which encode IAP proteins to prevent a host cell that is infected by the virus from killing itself by apoptosis. (Virus-infected animal cells frequently kill themselves to prevent the virus from replicating and infecting other cells.) It is now known that most animal cells also make IAP proteins.

All IAPs have one or more BIR (baculovirus IAP repeat) domains, which enable them to bind to and inhibit activated caspases. Some IAPs also polyubiquitylate caspases, marking the caspases for destruction by proteasomes. In this way, the IAPs set an inhibitory threshold that activated caspases must overcome to trigger apoptosis.

In *Drosophila* at least, this inhibitory barrier provided by IAPs can be neutralized by **anti-IAP** proteins, which are produced in response to various apoptotic stimuli. There are five anti-IAPs in flies, including *Reaper, Grim,* and *Hid,* and their only structural similarity is their short, N-terminal, IAP-binding motif, which binds to the BIR domain of IAPs, preventing the domain from binding to a caspase. Deletion of the three genes encoding Reaper, Grim, and Hid blocks apoptosis in flies. Conversely, inactivation of one of the two genes that encode IAPs in *Drosophila* causes all of the cells in the developing fly embryo to undergo apoptosis. Clearly, the balance between IAPs and anti-IAPs is tightly regulated and is crucial for controlling apoptosis in the fly.

The role of mammalian anti-IAP proteins in apoptosis is more controversial. As illustrated in **Figure 18–12**, anti-IAPs are released from the mitochondrial intermembrane space when the intrinsic pathway of apoptosis is activated, blocking IAPs in the cytosol and thereby promoting apoptosis. When, however, the genes encoding two known mammalian anti-IAPs called *Smac* (also called *DIABLO*) and *Omi* are inactivated in mouse cells, apoptosis is apparently unaffected—hence the controversy about their normal roles in regulating apoptosis.

In summary, the combined activities of the Bcl2 proteins, IAPs, and anti-IAPs determine the sensitivity of an animal cell to apoptosis-inducing stimuli, with IAPs and anti-IAPs dominant in flies and Bcl2 proteins dominant in mammals.

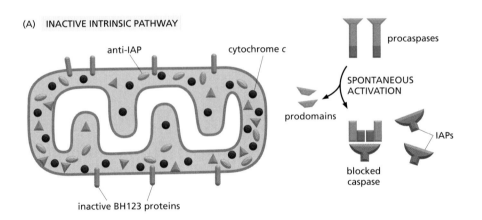

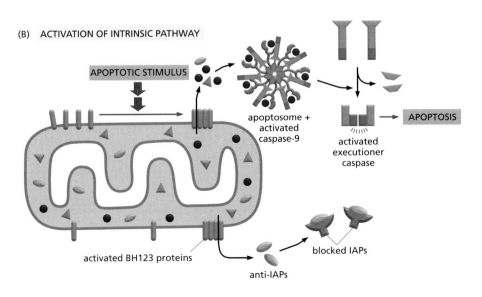

Figure 18–12 **A proposed model for the roles of IAPs and anti-IAPs in the control of apoptosis in mammalian cells.** (A) In the absence of an apoptotic stimulus, IAPs prevent accidental apoptosis caused by the spontaneous activation of procaspases. The IAPs are located in the cytosol and bind to and inhibit any spontaneously activated caspases. Some IAPs are also ubiquitin ligases that ubiquitylate the caspases they bind to, marking them for degradation in proteasomes (not shown). (B) When an apoptotic stimulus activates the intrinsic pathway, among the proteins released from the mitochondrial intermembrane space are anti-IAP proteins, which bind to and block the inhibitory activity of the IAPs. At the same time, the released cytochrome *c* triggers the assembly of apoptosomes, which can now activate a caspase cascade, leading to apoptosis.

Extracellular Survival Factors Inhibit Apoptosis in Various Ways

As discussed in Chapter 15, intercellular signals regulate most activities of animal cells, including apoptosis. These extracellular signals are part of the normal "social" controls that ensure that individual cells behave for the good of the organism as a whole—in this case, by surviving when they are needed and killing themselves when they are not. Some extracellular signal molecules stimulate apoptosis, whereas others inhibit it. We have discussed signal proteins such as Fas ligand that activate death receptors and thereby trigger the extrinsic pathway of apoptosis. Other extracellular signal molecules that stimulate apoptosis are especially important during animal development: a surge of thyroid hormone in the bloodstream, for example, signals cells in the tadpole tail to undergo apoptosis at metamorphosis (see Figure 18–3), while locally produced bone morphogenic proteins (BMPs, discussed in Chapters 15 and 22) stimulate cells between developing fingers and toes to kill themselves (see Figure 18–2). Here, however, we focus on extracellular signal molecules that inhibit apoptosis, which are collectively called **survival factors**.

Most animal cells require continuous signaling from other cells to avoid apoptosis. This surprising arrangement apparently helps ensure that cells survive only when and where they are needed. Nerve cells, for example, are produced in excess in the developing nervous system and then compete for limited amounts of survival factors that are secreted by the target cells that they normally connect to. Nerve cells that receive enough of the appropriate type of survival signal live, while the others die. In this way, the number of surviving neurons is automatically adjusted so that it is appropriate for the number of target cells they connect with (**Figure 18–13**). A similar competition for limited amounts of survival factors produced by neighboring cells is thought to control cell numbers in other tissues, both during development and in adulthood.

Survival factors usually bind to cell-surface receptors, which activate intracellular signaling pathways that suppress the apoptotic program, often by regulating members of the Bcl2 family of proteins. Some survival factors, for example, stimulate an increased production of anti-apoptotic Bcl2 proteins such as Bcl2 itself or Bcl-X$_L$ (**Figure 18–14**A). Others act by inhibiting the function of BH3-only pro-apoptotic Bcl2 proteins such as *Bad* (Figure 18–14B). In *Drosophila*, some survival factors act by phosphorylating and inactivating anti-IAP proteins, thereby enabling IAP proteins to suppress apoptosis (Figure 18–14C).

When mammalian cells are deprived of survival factors, they kill themselves by producing and activating pro-apoptotic BH3-only proteins, which activate the intrinsic pathway of apoptosis by overriding the anti-apoptotic Bcl2 proteins that are required to keep the cells alive. Mouse cells that lack both Bax and Bak are unable to activate the intrinsic pathway and can therefore live for weeks in culture in the absence of survival factors; without survival signals, however, the cells cannot efficiently import nutrients. Such cells fuel their metabolic needs through *autophagy*, in which the cell sequesters organelles and bits of its cytoplasm within autophagosomes, which then fuse with lysosomes (discussed in Chapter 13). The cells eventually die from starvation, but not by apoptosis.

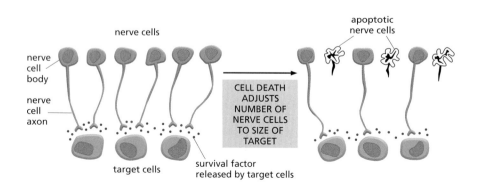

Figure 18–13 The role of survival factors and cell death in adjusting the number of developing nerve cells to the amount of target tissue. More nerve cells are produced than can be supported by the limited amount of survival factors released by the target cells. Therefore, some cells receive an insufficient amount of survival factors to avoid apoptosis. This strategy of overproduction followed by culling ensures that all target cells are contacted by nerve cells and that the extra nerve cells are automatically eliminated.

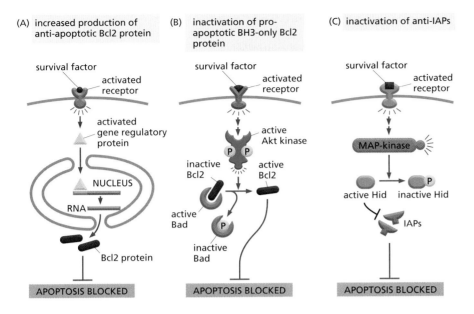

(A) increased production of anti-apoptotic Bcl2 protein

(B) inactivation of pro-apoptotic BH3-only Bcl2 protein

(C) inactivation of anti-IAPs

Figure 18–14 Three ways that extracellular survival factors can inhibit apoptosis. (A) Some survival factors suppress apoptosis by stimulating the transcription of genes that encode anti-apoptotic Bcl2 proteins such as Bcl2 itself or Bcl-X$_L$. (B) Many others activate the serine/threonine protein kinase Akt, which, among many other targets, phosphorylates and inactivates the BH3-only pro-apoptotic Bcl2 protein Bad (see Figure 15–64). When not phosphorylated, Bad promotes apoptosis by binding to and inhibiting Bcl2; once phosphorylated, Bad dissociates, freeing Bcl2 to suppress apoptosis. Akt also suppresses apoptosis by phosphorylating and inactivating gene regulatory proteins of the Forkhead family that stimulate the transcription of genes encoding proteins that promote apoptosis (not shown). (C) In *Drosophila*, some survival factors inhibit apoptosis by stimulating the phosphorylation of the anti-IAP protein Hid. When not phosphorylated, Hid promotes cell death by inhibiting IAPs. Once phosphorylated, Hid no longer inhibits IAPs, which become active and block apoptosis.

Either Excessive or Insufficient Apoptosis Can Contribute to Disease

There are many human disorders in which excessive numbers of cells undergo apoptosis and thereby contribute to tissue damage. Among the most dramatic examples are heart attacks and strokes. In these acute conditions, many cells die by necrosis as a result of ischemia (inadequate blood supply), but some of the less affected cells die by apoptosis. It is hoped that, in the future, drugs such as caspase inhibitors that block apoptosis will prove useful in saving cells in these conditions.

There are other conditions where too few cells die by apoptosis. Mutations in mice and humans, for example, that inactivate the genes that encode the Fas death receptor or the Fas ligand prevent the normal death of some lymphocytes, causing these cells to accumulate in excessive numbers in the spleen and lymph glands. In many cases, this leads to autoimmune disease, in which the lymphocytes react against the individual's own tissues.

Decreased apoptosis also makes an important contribution to many tumors, as cancer cells often regulate the apoptotic program abnormally. <TGAA> The *Bcl2* gene, for example, was first identified in a common form of lymphocyte cancer in humans, where a chromosome translocation causes excessive production of the Bcl2 protein; indeed, Bcl2 gets its name from this *B cell lymphoma*. The high level of Bcl2 protein in the lymphocytes that carry the translocation promotes the development of cancer by inhibiting apoptosis, thereby prolonging cell survival and increasing cell numbers; it also decreases the cells' sensitivity to anticancer drugs, which commonly work by causing cancer cells to undergo apoptosis.

Similarly, the gene encoding the tumor suppressor protein p53 is mutated in 50% of human cancers so that it no longer promotes apoptosis or cell-cycle arrest in response to DNA damage. The lack of p53 function therefore enables the cancer cells to survive and proliferate even when their DNA is damaged; in this way, the cells accumulate more mutations, some of which make the cancer more malignant (discussed in Chapter 20). As many anticancer drugs induce apoptosis (and cell-cycle arrest) by a p53-dependent mechanism (discussed in Chapters 17 and 20), the loss of p53 function also makes cancer cells less sensitive to these drugs.

Most human cancers arise in epithelial tissues such as those in the lung, intestinal tract, breast, and prostate. Such cancer cells display many abnormalities in their behavior, including a decreased ability to adhere to the extracellular matrix and to adhere to one another at specialized cell–cell junctions. In the next chapter, we discuss the remarkable structures and functions of the extracellular matrix and cell junctions.

Summary

Cells can activate an intracellular death program and kill themselves in a controlled way—a process called programmed cell death. In this way, animal cells that are irreversibly damaged, no longer needed, or are a threat to the organism can be eliminated quickly and neatly. In most cases, these deaths occur by apoptosis: the cells shrink, condense, and frequently fragment, and neighboring cells or macrophages rapidly phagocytose the cells or fragments before there is any leakage of cytoplasmic contents. Apoptosis depends on proteolytic enzymes called caspases, which cleave specific intracellular proteins to help kill the cell. Caspases are present in all nucleated animal cells as inactive precursors called procaspases. Initiator procaspases are activated when brought into proximity in activation complexes: once activated, they cleave and activate downstream executioner procaspases, which activate other executioner procaspases (and various other target proteins in the cell), producing an amplifying, irreversible proteolytic cascade.

Cells use at least two distinct pathways to activate initiator procaspases and trigger a caspase cascade leading to apoptosis: the extrinsic pathway is activated by extracellular ligands binding to cell-surface death receptors; the intrinsic pathway is activated by intracellular signals generated when cells are stressed. Each pathway uses its own initiator procaspases, which are activated in distinct activation complexes, called the DISC and the apoptosome, respectively. In the extrinsic pathway, the death receptors recruit procaspases-8 and 10 via adaptor proteins to form the DISC; in the intrinsic pathway, cytochrome c released from the intermembrane space of mitochondria activates Apaf1, which assembles into an apoptosome and recruits and activates procaspase-9.

Both extracellular signal proteins and intracellular Bcl2 proteins and IAP proteins tightly regulate the apoptotic program to ensure that cells normally kill themselves only when it benefits the animal. Both anti-apoptotic and pro-apoptotic Bcl2 proteins regulate the intrinsic pathway by controlling the release of mitochondrial intermembrane proteins, while IAP proteins inhibit activated caspases and promote their degradation.

PROBLEMS

Which statements are true? Explain why or why not.

18–1 In normal adult tissues, cell death usually balances cell division.

18–2 Mammalian cells that do not have cytochrome c should be resistant to apoptosis induced by UV light.

Discuss the following problems.

18–3 One important role of Fas and Fas ligand is to mediate elimination of tumor cells by killer lymphocytes. In a study of 35 primary lung and colon tumors, half the tumors were found to have amplified and overexpressed a gene for a secreted protein that binds to Fas ligand. How do you suppose that overexpression of this protein might contribute to the survival of these tumor cells? Explain your reasoning.

18–4 Development of the nematode *Caenorhabditis elegans* generates exactly 959 somatic cells; it also produces an additional 131 cells that are later eliminated by programmed cell death. Classical genetic experiments in *C. elegans* isolated mutants that identified the first genes involved in apoptosis. Of the many mutant genes affecting apoptosis in the nematode, none have ever been found in the gene for cytochrome c. Why do you suppose that such a central effector molecule in apoptosis was not found in the many genetic screens for "death" genes that have been carried out in *C. elegans*?

18–5 Imagine that you could microinject cytochrome c into the cytosol of wild-type cells and of cells that were doubly defective for Bax and Bak. Would you expect one, both, or neither type of cell to undergo apoptosis? Explain your reasoning.

18–6 In contrast to their similar brain abnormalities, newborn mice deficient in Apaf1 or caspase-9 have distinctive abnormalities in their paws. Apaf1-deficient mice fail to eliminate the webs between their developing digits, whereas caspase-9-deficient mice have normally formed digits (**Figure Q18–1**). If Apaf1 and caspase-9 function in the same apoptotic pathway, how is it possible for these deficient mice to differ in web-cell apoptosis?

Figure Q18–1 Appearance of paws in *Apaf1*$^{-/-}$ and *Casp9*$^{-/-}$ newborn mice relative to normal newborn mice (Problem 18–6). (From H. Yoshida et al., *Cell* 94:739–750, 1998. With permission from Elsevier.)

18–7 When human cancer (HeLa) cells are exposed to UV light at 90 mJ/cm^2, most of the cells undergo apoptosis within 24 hours. Release of cytochrome *c* from mitochondria can be detected as early as 6 hours after exposure of a population of cells to UV light, and it continues to increase for more than 10 hours thereafter. Does this mean that individual cells slowly release their cytochrome *c* over this time period? Or, alternatively, do individual cells release their cytochrome *c* rapidly but with different cells being triggered over the longer time period?

To answer this fundamental question, you have fused the gene for green fluorescent protein (GFP) to the gene for cytochrome *c*, so that you can observe the behavior of individual cells by confocal fluorescence microscopy. In cells that are expressing the cytochrome *c*–GFP fusion, fluorescence shows the punctate pattern typical of mitochondrial proteins. You then irradiate these cells with UV light and observe individual cells for changes in the punctate pattern. Two such cells (outlined in white) are shown in **Figure Q18–2**A and B. Release of cytochrome *c*–GFP is detected as a change from a punctate to a diffuse pattern of fluorescence. Times after UV exposure are indicated as hours:minutes below the individual panels.

Which model for cytochrome c release do these observations support? Explain your reasoning.

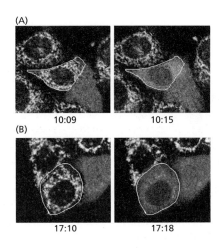

Figure Q18–2 Time-lapse video, fluorescence microscopic analysis of cytochrome *c*–GFP release from mitochondria of individual cells (Problem 18–7). (A) Cells observed for 8 minutes, 10 hours after UV irradiation. (B) Cells observed for 6 minutes, 17 hours after UV irradiation. One cell in (A) and one in (B), each *outlined in white*, have released their cytochrome *c*–GFP during the time frame of the observation, which is shown as hours:minutes below each panel. (From J.C. Goldstein et al., *Nat. Cell Biol.* 2:156–162, 2000. With permission from Macmillan Publishers Ltd.)

REFERENCES

Adams JM, Huang DC, Strasser A et al (2005) Subversion of the Bcl-2 life/death switch in cancer development and therapy. *Cold Spring Harb Symp Quant Biol* 70:469–77.

Boatright KM & Salvesen GS (2003) Mechanisms of caspase activation. *Curr Opin Cell Biol* 15:725–731.

Danial NN & Korsmeyer SJ (2004) Cell death: critical control points. *Cell* 116:205–219.

Ellis RE, Yuan JY & Horvitz RA (1991) Mechanisms and functions of cell death. *Annu Rev Cell Biol* 7:663–698.

Fadok VA & Henson PM (2003) Apoptosis: giving phosphatidylserine recognition an assist—with a twist. *Curr Biol* 13:R655–R657.

Galonek HL & Hardwick JM (2006) Upgrading the BCL-2 network. *Nature Cell Biol* 8:1317–1319.

Green DR (2005) Apoptotic pathways: ten minutes to dead. *Cell* 121:671–674.

Horvitz HR (2003) Worms, life, and death (Nobel lecture). *Chembiochem* 4:697–711.

Hyun-Eui K, Fenghe D, Fang M & Wang X (2005) Formation of apoptosome is initiated by cytochrome c-induced dATP hydrolysis and subsequent nucleotide exchange on Apaf-1. *Proc Natl Acad Sci USA* 102:17545–17550.

Jacobson MD, Weil M & Raff MC (1997) Programmed cell death in animal development. *Cell* 88:347–354.

Jiang X & Wang X (2004) Cytochrome C-mediated apoptosis. *Annu Rev Biochem* 73:87–106.

Kerr JF, Wyllie AH & Currie AR (1972) Apoptosis: a basic biological phenomenon with wide-ranging implications in tissue kinetics. *B J Cancer* 26:239–257.

Kumar S (2007) Caspase function in programmed cell death. *Cell Death Differ* 14:32–43.

Lavrik I, Golks A & Krammer PH (2005) Death receptor signaling. *J Cell Sci* 118:265–267.

Lowe SW, Cepero E & Evan G (2004) Intrinsic tumour suppression. *Nature* 432:307–315.

Lum JJ, Bauer DE, Kong M et al (2005) Growth factor regulation of autophagy and cell survival in the absence of apoptosis. *Cell* 120:237–48.

McCall K & Steller H (1997) Facing death in the fly: genetic analysis of apoptosis in *Drosophila*. *Trends Genet* 13:222–226.

Nagata S (1999) Fas ligand-induced apoptosis. *Annu Rev Genet* 33:29–55.

Nagata S (2005) DNA degradation in development and programmed cell death. *Annu Rev Immunol* 23:853–875.

Pop C, Timmer J, Sperandio S & Salvesen GS (2006) The apoptosome activates caspase-9 by dimerization. *Mol Cell* 22:269–275.

Raff MC (1999) Cell suicide for beginners. *Nature* 396:119–122.

Rathmell JC & Thompson CB (2002) Pathways of apoptosis in lymphocyte development, homeostasis, and disease. *Cell* 109:S97–107.

Tittel JN & Steller H (2000) A comparison of programmed cell death between species. *Genome Biol* 1.

Verhagen AM, Coulson EJ & Vaux DL (2001) Inhibitor of apoptosis proteins and their relatives: IAPs and other BIRPs. *Genome Biol* 2:3009.1–3009.10.

Vousden KH (2005) Apoptosis. p53 and PUMA: a deadly duo. *Science* 309:1685–1686.

Willis SN & Adams JM (2005) Life in the balance: how BH3-only proteins induce apoptosis. *Curr Opin Cell Biol* 17:617–625.

V CELLS IN THEIR SOCIAL CONTEXT

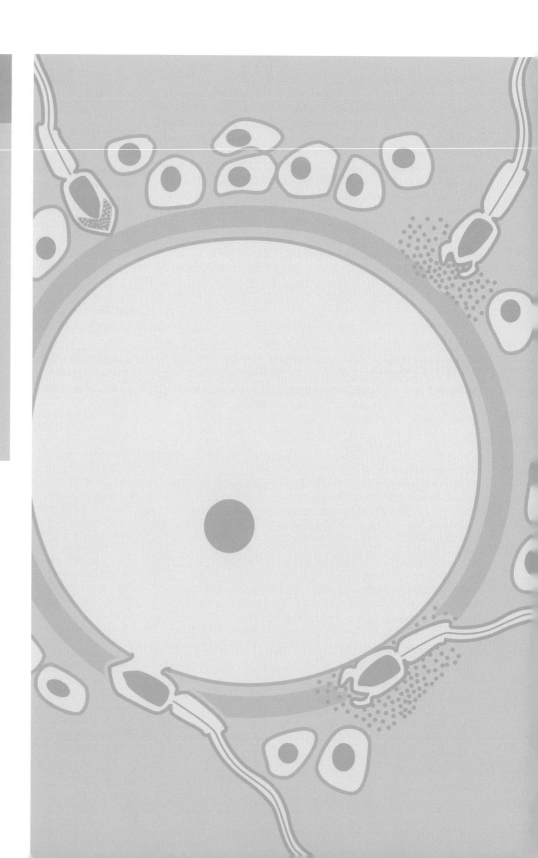

Chapter 19

Cell Junctions, Cell Adhesion, and the Extracellular Matrix

19

Of all the social interactions between cells in a multicellular organism, the most fundamental are those that hold the cells together. Cells may cling to one another through direct cell–cell junctions, or they may be bound together by extracellular materials that they secrete; but by one means or another, they must cohere if they are to form an organized multicellular structure.

The mechanisms of cohesion govern the architecture of the body—its shape, its strength, and the arrangement of its different cell types. The junctions between cells create pathways for communication, allowing the cells to exchange the signals that coordinate their behavior and regulate their patterns of gene expression. Attachments to other cells and to extracellular matrix control the orientation of each cell's internal structure. The making and breaking of the attachments and the modeling of the matrix govern the way cells move within the organism, guiding them as the body grows, develops, and repairs itself. Thus, the apparatus of cell junctions, cell adhesion mechanisms, and extracellular matrix is critical for every aspect of the organization, function, and dynamics of multicellular structures. Defects in this apparatus underlie an enormous variety of diseases.

As examples of structural engineering, large multicellular organisms represent a most surprising feat. Cells are small, squishy, and often motile objects, filled with an aqueous medium and enclosed in a flimsy plasma membrane; yet they can combine in their millions to form a structure as massive, as strong, and as stable as a horse or a tree. How is this possible?

The answer lies in two basic building strategies, corresponding to two ways in which stresses can be transmitted across a multicellular structure. One strategy depends on the strength of the *extracellular matrix*, a complex network of proteins and polysaccharide chains that the cells secrete. The other strategy depends on the strength of the cytoskeleton inside the cells and on *cell–cell adhesions* that tie the cytoskeletons of neighboring cells together. In plants, the extracellular matrix is all-important: plant tissues owe their strength to the cell walls that surround each cell. In animals, both architectural strategies are used, but to different extents in different tissues.

Animal tissues are extraordinarily varied, as we shall see in Chapter 23, but most fall into one or other of two broad categories, representing two architectural extremes (**Figure 19–1**). In **connective tissues**, such as bone or tendon, the extracellular matrix is plentiful, and cells are sparsely distributed within it. The matrix is rich in fibrous polymers, especially *collagen*, and it is the matrix—rather than the cells—that bears most of the mechanical stress to which the tissue is subjected. Direct attachments between one cell and another are relatively rare, but the cells have important attachments to the matrix, allowing them to pull on it and to be pulled by it.

By contrast, in **epithelial tissues**, such as the lining of the gut or the epidermal covering of the skin, cells are closely bound together into sheets called **epithelia**. The extracellular matrix is scanty, consisting mainly of a thin mat called the *basal lamina* (or *basement membrane*), underlying one face of the sheet. Within the epithelium, the cells are attached to each other directly by cell–cell adhesions, where cytoskeletal filaments are anchored, transmitting stresses across the interior of each cell, from adhesion site to adhesion site.

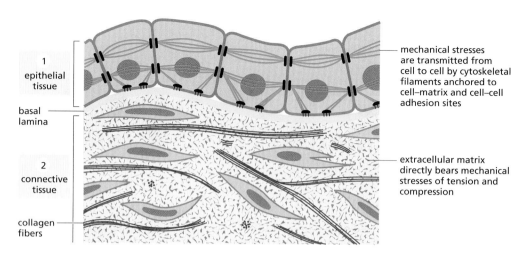

1 epithelial tissue

basal lamina

2 connective tissue

collagen fibers

mechanical stresses are transmitted from cell to cell by cytoskeletal filaments anchored to cell–matrix and cell–cell adhesion sites

extracellular matrix directly bears mechanical stresses of tension and compression

Figure 19–1 Two main ways in which animal cells are bound together. In connective tissue, the main stress-bearing component is the extracellular matrix. In epithelial tissue, it is the cytoskeletons of the cells themselves, linked from cell to cell by anchoring junctions. Cell–matrix attachments bond epithelial tissue to the connective tissue beneath it.

Physical attachment is critical, both in epithelia and in nonepithelial tissues, but junctions between cell and cell or between cells and matrix are diverse in structure and do more than just transmit physical forces. Four main functions can be distinguished, each with a different molecular basis (**Figure 19–2** and **Table 19–1**):

1. **Anchoring junctions**, including both *cell–cell adhesions* and *cell–matrix adhesions*, transmit stresses and are tethered to cytoskeletal filaments inside the cell.

2. **Occluding junctions** seal the gaps between cells in epithelia so as to make the cell sheet into an impermeable (or selectively permeable) barrier.

3. **Channel-forming junctions** create passageways linking the cytoplasms of adjacent cells.

4. **Signal-relaying junctions** allow signals to be relayed from cell to cell across their plasma membranes at sites of cell-to-cell contact.

Chemical synapses in the nervous system (discussed in Chapter 11) and immunological synapses, where T lymphocytes interact with antigen-presenting cells (discussed in Chapter 25), are the most obvious examples of signal-relaying junctions, but they are not the only ones. Sites of cell–cell communication via transmembrane ligand–receptor pairs such as Delta and Notch, or ephrins and Eph receptors, as discussed in Chapter 15, fall under this heading: the cell membranes must be held in contact with one another for the ligands to activate the receptors. Moreover, we shall see that anchoring junctions, occluding junctions, and channel-forming junctions, in different ways, all can have important roles in signal transmission.

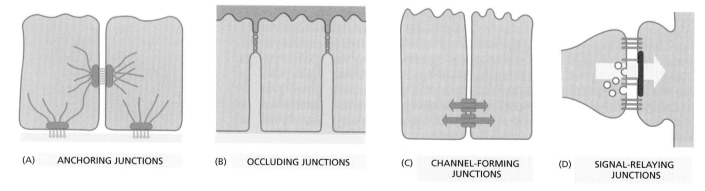

(A) ANCHORING JUNCTIONS (B) OCCLUDING JUNCTIONS (C) CHANNEL-FORMING JUNCTIONS (D) SIGNAL-RELAYING JUNCTIONS

Figure 19–2 Four functional classes of cell junctions in animal tissues. (A) Anchoring junctions link cell to cell (typically via transmembrane *cadherin* proteins) or cell to matrix (typically via transmembrane *integrin* proteins). (B) Occluding junctions (involving *claudin* proteins) seal gaps between epithelial cells. (C) Channel-forming junctions (composed of *connexin* or *innexin* proteins) form passageways for small molecules and ions to pass from cell to cell. (D) Signal-relaying junctions are complex structures, typically involving anchorage proteins alongside proteins mediating signal transduction.

Table 19–1 A Functional Classification of Cell Junctions

ANCHORING JUNCTIONS
Actin filament attachment sites
1.   cell–cell junctions (adherens junctions)
2.   cell–matrix junctions (actin-linked cell–matrix adhesions)
Intermediate filament attachment sites
1.   cell–cell junctions (desmosomes)
2.   cell–matrix junctions (hemidesmosomes)

OCCLUDING JUNCTIONS
1.   tight junctions (in vertebrates)
2.   septate junctions (in invertebrates)

CHANNEL-FORMING JUNCTIONS
1.   gap junctions (in animals)
2.   plasmodesmata (in plants)

SIGNAL-RELAYING JUNCTIONS
1.   chemical synapses (in the nervous system)
2.   immunological synapses (in the immune system)
3.   transmembrane ligand–receptor cell–cell signaling contacts (Delta-Notch, ephrin-Eph, etc.). Anchoring, occluding, and channel-forming junctions can all have signaling functions in addition to their structural roles

The first part of this chapter will focus on animal cells and tissues, beginning with the cell–cell adhesions, occluding junctions, and channel-forming junctions that link cell to cell directly. As examples of signal-relaying junctions, we shall briefly examine neuronal synapses from the point of view of their adhesion mechanisms and assembly. We shall see how the different kinds of junctions together organize cells into polarized epithelial sheets. We shall then discuss the extracellular matrix in animals and the ways in which the cells interact with it through cell–matrix adhesions. Last, we shall turn to plants and the central role of the plant cell wall in their construction.

CADHERINS AND CELL–CELL ADHESION

The structures of **cell–cell adhesions** are most clearly seen in mature epithelia and in some other tissues, such as heart muscle, that are held together by strong direct anchorage of cell to cell. Study of these tissues by electron microscopy provided the first general classification of cell junctions. Biochemistry and molecular biology have since shown that the different structures seen in the electron microscope relate to distinct systems of molecules, important not only in adult epithelia but also in other tissues where the junctional specializations are not always so plainly visible.

Figure 19–3 illustrates schematically the types of junctions that the electron microscope reveals in a section of mature epithelium and shows how the cell–cell adhesions (anchoring junctions) that will concern us in this section are distributed in relation to other types of junctions to be discussed later. The diagram shows the typical arrangement in a *simple columnar* epithelium such as the lining of the small intestine of a vertebrate. Here, a single layer of tall cells all stand on a basal lamina, with their uppermost surface, or *apex*, free and exposed to the extracellular medium. On their sides, or *lateral* surfaces, the cells make junctions with one another. Closest to the apex lie occluding junctions (known as *tight junctions* in vertebrates), preventing molecules from leaking across the epithelium through gaps between the cells. Below these are two types of cell–cell

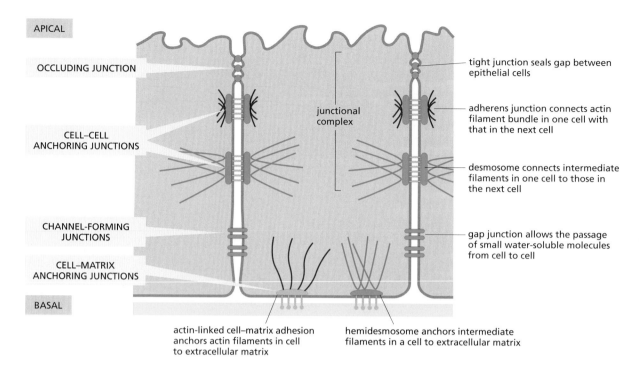

APICAL

OCCLUDING JUNCTION

CELL–CELL
ANCHORING JUNCTIONS

CHANNEL-FORMING
JUNCTIONS

CELL–MATRIX
ANCHORING JUNCTIONS

BASAL

junctional
complex

tight junction seals gap between
epithelial cells

adherens junction connects actin
filament bundle in one cell with
that in the next cell

desmosome connects intermediate
filaments in one cell to those in
the next cell

gap junction allows the passage
of small water-soluble molecules
from cell to cell

actin-linked cell–matrix adhesion
anchors actin filaments in cell
to extracellular matrix

hemidesmosome anchors intermediate
filaments in a cell to extracellular matrix

adhesions. **Adherens junctions** are anchorage sites for actin filaments; **desmosome junctions** are anchorage sites for intermediate filaments. Still lower, often mingled with additional desmosome junctions, lie channel-forming junctions, called *gap junctions.*

Additional sets of adhesions attach the epithelial cells to the basal lamina and will be discussed in a later section. We classify these cell–matrix adhesions, like the cell–cell adhesions, according to their cytoskeletal connections: *actin-linked cell–matrix adhesions* (indistinct in the small intestine, but prominent elsewhere) anchor actin filaments to the matrix, while *hemidesmosomes* anchor intermediate filaments to it.

At each of the four types of anchoring junctions, the central role is played by **transmembrane adhesion proteins** that span the membrane, with one end linking to the cytoskeleton inside the cell and the other end linking to other structures outside it (**Figure 19–4**). These cytoskeleton-linked transmembrane molecules fall neatly into two superfamilies, corresponding to the two basic kinds of external attachment (**Table 19–2**). Proteins of the **cadherin** superfamily chiefly mediate attachment of cell to cell. Proteins of the **integrin** superfamily chiefly mediate attachment of cells to matrix. Within each family, there is specialization: some cadherins link to actin and form adherens junctions, while others link to intermediate filaments and form desmosome junctions; likewise,

Figure 19–3 A summary of the various cell junctions found in a vertebrate epithelial cell, classified according to their primary functions. In the most apical portion of the cell, the relative positions of the junctions are the same in nearly all vertebrate epithelia. The tight junction occupies the most apical position, followed by the adherens junction (adhesion belt) and then by a special parallel row of desmosomes; together these form a structure called a junctional complex. Gap junctions and additional desmosomes are less regularly organized. The drawing is based on epithelial cells of the small intestine. Specialized signal-relaying junctions are discussed later in the chapter.

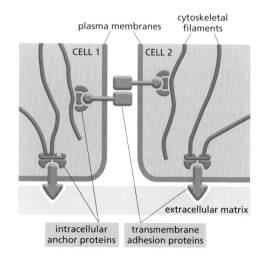

plasma membranes

cytoskeletal
filaments

CELL 1   CELL 2

extracellular matrix

intracellular
anchor proteins

transmembrane
adhesion proteins

Figure 19–4 Transmembrane adhesion proteins link the cytoskeleton to extracellular structures. The external linkage may be either to parts of other cells (cell–cell anchorage, mediated typically by cadherins) or to extracellular matrix (cell–matrix anchorage, mediated typically by integrins). The internal linkage to the cytoskeleton is generally indirect, via intracellular anchor proteins, to be discussed later.

Table 19–2 Anchoring Junctions

JUNCTION	TRANSMEMBRANE ADHESION PROTEIN	EXTRACELLULAR LIGAND	INTRACELLULAR CYTOSKELETAL ATTACHMENT	INTRACELLULAR ANCHOR PROTEINS
Cell–Cell				
adherens junction	cadherin (classical cadherin)	cadherin in neighboring cell	actin filaments	α-catenin, β-catenin, plakoglobin (γ-catenin), p120-catenin, vinculin, α-actinin
desmosome	cadherin (desmoglein, desmocollin)	desmoglein and desmocollin in neighboring cell	intermediate filaments	plakoglobin (γ-catenin), plakophilin, desmoplakin
Cell–Matrix				
actin-linked cell–matrix adhesion	integrin	extracellular matrix proteins	actin filaments	talin, vinculin, α-actinin, filamin, paxillin, focal adhesion kinase (FAK)
hemidesmosome	integrin α6β4, type XVII collagen (BP180)	extracellular matrix proteins	intermediate filaments	plectin, dystonin (BP230)

some integrins link to actin and form actin-linked cell–matrix adhesions, while others link to intermediate filaments and form hemidesmosomes.

There are some exceptions to these rules. Some integrins, for example, mediate cell–cell rather than cell–matrix attachment. Moreover, there are other types of cell adhesion molecules that can provide attachments more flimsy than anchoring junctions, but sufficient to stick cells together in special circumstances. Cell–cell adhesions based on cadherins, however, seem to be the most fundamentally important class, and we begin our account of cell–cell adhesion with them. <CGAA>

Cadherins Mediate Ca^{2+}-Dependent Cell–Cell Adhesion in All Animals

Cadherins are present in all multicellular animals whose genomes have been analyzed, and in one other known group, the choanoflagellates. These creatures can exist either as free-living unicellular organisms or as multicellular colonies and are thought to be representatives of the group of protists from which all animals evolved. Other eucaryotes, including fungi and plants, lack cadherins, and they are absent from bacteria and archaea also. Cadherins therefore seem to be part of the essence of what it is to be an animal.

The cadherins take their name from their dependence on Ca^{2+} ions: removing Ca^{2+} from the extracellular medium causes adhesions mediated by cadherins to come adrift. Sometimes, especially for embryonic tissues, this is enough to let the cells be easily separated. In other cases, a more severe treatment is required, combining Ca^{2+} removal with exposure to a protease such as trypsin. The protease loosens additional connections mediated by extracellular matrix and by other cell–cell adhesion molecules that do not depend on Ca^{2+}. In either case, when the dissociated cells are put back into a normal culture medium, they will generally stick together again by reconstructing their adhesions.

This type of cell–cell association provided one of the first assays that allowed cell–cell adhesion molecules to be identified. In these experiments, monoclonal antibodies were raised against the cells of interest, and each antibody was tested for its ability to prevent the cells from sticking together again after they had been dissociated. Rare antibodies that bound to the cell–cell adhesion molecules showed this blocking effect. These antibodies then were used to isolate the adhesion molecule that they recognized.

Virtually all cells in vertebrates, and probably in other animals too, seem to express one or more proteins of the cadherin family, according to the cell type.

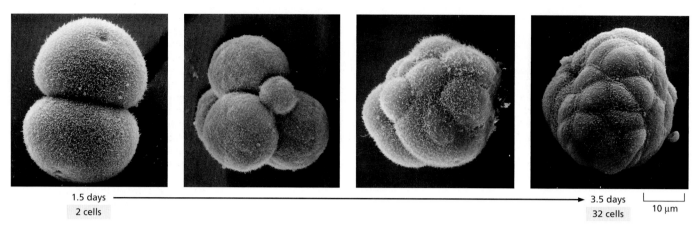

1.5 days — ————————————————————————————————————→ 3.5 days

2 cells   32 cells   10 μm

Several lines of evidence indicate that they are the main adhesion molecules holding cells together in early embryonic tissues. For example, embryonic tissues in culture disintegrate when treated with anti-cadherin antibodies, and if cadherin-mediated adhesion is left intact, antibodies against other adhesion molecules have little effect. Studies of the early mouse embryo illustrate the role of cadherins in development. Up to the eight-cell stage, the mouse embryo cells are only very loosely held together, remaining individually more or less spherical; then, rather suddenly, in a process called compaction, they become tightly packed together and joined by cell–cell junctions, so that the outer surface of the embryo becomes smoother (**Figure 19–5**). Antibodies against a specific cadherin, called *E-cadherin*, block compaction, whereas antibodies that react with various other cell-surface molecules on these cells do not. Mutations that inactivate E-cadherin cause the embryos to fall apart and die early in development.

Figure 19–5 Compaction of an early mouse embryo. The cells of the early embryo at first stick together only weakly. At about the eight-cell stage, they begin to express E-cadherin and as a result become strongly and closely adherent to one another. (Scanning electron micrographs courtesy of Patricia Calarco; 16–32-cell stage is from P. Calarco and C.J. Epstein, *Dev. Biol.* 32:208–213, 1973. With permission from Academic Press.)

The Cadherin Superfamily in Vertebrates Includes Hundreds of Different Proteins, Including Many with Signaling Functions

The first three cadherins that were discovered were named according to the main tissues in which they were found: *E-cadherin* is present on many types of epithelial cells; *N-cadherin* on nerve, muscle, and lens cells; and *P-cadherin* on cells in the placenta and epidermis. All are also found in various other tissues; N-cadherin, for example, is expressed in fibroblasts, and E-cadherin is expressed in parts of the brain (**Figure 19–6**). These and other **classical cadherins** are closely related in sequence throughout their extracellular and intracellular domains. While all of them have well-defined adhesive functions, they are also important in signaling. Through their intracellular domains, as we shall see later, they relay information into the cell interior, enabling the cell to adapt its behavior according to whether it is attached or detached from other cells.

There are also a large number of **nonclassical cadherins** more distantly related in sequence, with more than 50 expressed in the brain alone. The nonclassical cadherins include proteins with known adhesive function, such as the diverse *protocadherins* found in the brain, and the *desmocollins* and *desmogleins* that form desmosome junctions. They also include proteins that appear to be primarily involved in signaling, such as *T-cadherin*, which lacks a transmembrane domain and is attached to the plasma membrane of nerve and muscle

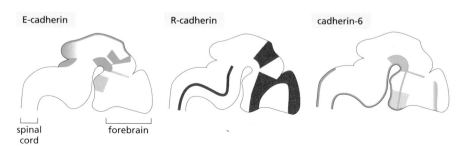

E-cadherin   R-cadherin   cadherin-6

spinal   forebrain
cord

Figure 19–6 Cadherin diversity in the central nervous system. The diagram shows the expression patterns of three different classical cadherins in the embryonic mouse brain. More than 70 other cadherins, both classical and nonclassical, are also expressed in the brain, in complex patterns that are thought to reflect their roles in guiding and maintaining the organization of this intricate organ.

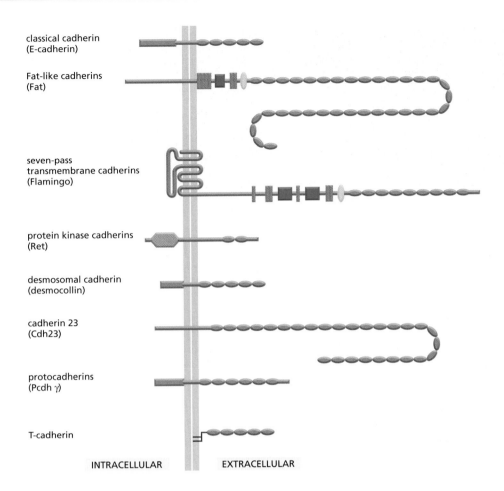

Figure 19–7 The cadherin superfamily. The diagram shows some of the diversity among cadherin superfamily members. These proteins all have extracellular portions containing multiple copies of the cadherin domain motif (*green ovals*), but their intracellular portions are more varied, reflecting interactions with a wide variety of intracellular ligands, including signaling molecules as well as components that anchor the cadherin to the cytoskeleton. The differently colored motifs in Fat, Flamingo, and Ret represent conserved domains that are also found in other protein families.

cells by a glycosylphosphatidylinositol (GPI) anchor, and the *Fat* and *Flamingo* proteins, which were first identified as the products of genes in *Drosophila* that regulate, respectively, epithelial growth and cell polarity. Together, the classical and nonclassical cadherin proteins constitute the **cadherin superfamily** (**Figure 19–7** and **Table 19–3**), with more than 180 members in humans. How do the structures of these proteins relate to their functions, and why are there so many of them?

Cadherins Mediate Homophilic Adhesion

Anchoring junctions between cells are usually symmetrical: if the linkage is to actin, for example, in the cell on one side of the junction, it will be to actin in the cell on the other side also. In fact, the binding between cadherins is generally **homophilic** (like-to-like, **Figure 19–8**): cadherin molecules of a specific subtype on one cell bind to cadherin molecules of the same or closely related subtype on adjacent cells. According to a current model, the binding occurs at the N-terminal tips of the cadherin molecules—the ends that lie furthest from the membrane. The protein chain here forms a terminal knob and a nearby pocket, and the cadherin molecules protruding from opposite cell membranes bind by insertion of the knob of each one in the pocket of the other (**Figure 19–9A**).

The spacing between the cell membranes at an anchoring junction is precisely defined and depends on the structure of the participating cadherin molecules. All the members of the superfamily, by definition, have an extracellular portion consisting of several copies of a motif called the *cadherin domain*. In the classical cadherins of vertebrates there are 5 of these repeats, and in desmogleins and desmocollins there are 4 or 5, but some nonclassical cadherins have more than 30. Each cadherin domain forms a more or less rigid unit, joined to the next cadherin domain by a hinge (Figure 19–9B). Ca^{2+} ions bind to sites

Table 19–3 Some Members of the Cadherin Superfamily

NAME	MAIN LOCATION	JUNCTION ASSOCIATION	PHENOTYPE WHEN INACTIVATED IN MICE
Classical cadherins			
E-cadherin	many epithelia	adherens junctions	death at blastocyst stage; embryos fail to undergo compaction
N-cadherin	neurons, heart, skeletal muscle, lens, and fibroblasts	adherens junctions and chemical synapses	embryos die from heart defects
P-cadherin	placenta, epidermis, breast epithelium	adherens junctions	abnormal mammary gland development
VE-cadherin	endothelial cells	adherens junctions	abnormal vascular development (apoptosis of endothelial cells)
Nonclassical cadherins			
Desmocollin	skin	desmosomes	blistering of skin
Desmoglein	skin	desmosomes	blistering skin disease due to loss of keratinocyte cell–cell adhesion
T-cadherin	neurons, muscle, heart	none	unknown
Cadherin 23	inner ear, other epithelia	links between stereocilia in sensory hair cells	deafness
Fat (in *Drosophila*)	epithelia and central nervous system	signal-relaying junction (planar cell polarity)	enlarged imaginal discs and tumors; disrupted planar cell polarity
Fat1 (in mammals)	various epithelia and central nervous system	slit diaphragm in kidney glomerulus and other cell junctions	loss of slit diaphragm; malformation of forebrain and eye
α, β, and γ-Protocadherins	neurons	chemical synapses and nonsynaptic membranes	neuronal degeneration
Flamingo	sensory and some other epithelia	cell–cell junctions	disrupted planar cell polarity; neural tube defects

near each hinge and prevent it from flexing, so that the whole string of cadherin domains behaves as a rigid, slightly curved, rod. When Ca^{2+} is removed, the hinges can flex, and the structure becomes floppy. At the same time, the conformation at the N terminus is thought to change slightly, weakening the binding affinity for the matching cadherin molecule on the opposite cell. Cadherin molecules destabilized in this way by loss of Ca^{2+} are rapidly degraded by proteolytic enzymes.

Unlike receptors for soluble signal molecules, which bind their specific ligand with high affinity, cadherins (and most other cell–cell adhesion proteins) typically bind to their partners with relatively low affinity. Strong attachments result from the formation of many such weak bonds in parallel. When binding to oppositely oriented partners on another cell, cadherin molecules are often clustered side-to-side with many other cadherin molecules on the same cell. Many cadherin molecules packed side by side in this way collaborate to form an anchoring junction (Figure 19–9C). The strength of this junction is far greater than that of any individual intermolecular bond, and yet it can be easily disassembled by separating the molecules sequentially, just as two pieces of fabric can be strongly joined by Velcro and yet easily peeled apart. A similar "Velcro principle" also operates at cell–cell and cell–matrix adhesions formed by other

HOMOPHILIC BINDING

HETEROPHILIC BINDING

Figure 19–8 Homophilic versus heterophilic binding. Cadherins in general bind homophilically; some other cell adhesion molecules, discussed later, bind heterophilically.

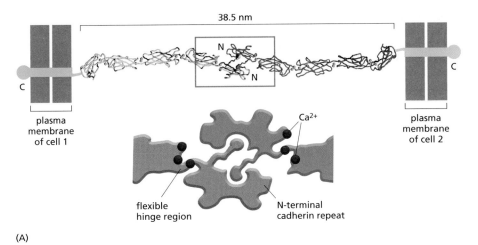

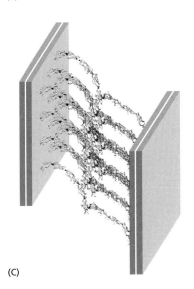

(A)

(B)

Figure 19–9 Cadherin structure and function. (A) The extracellular domain of a classical cadherin (C-cadherin) is shown here, illustrating how two such molecules on opposite cells are thought to bind homophilically, end-to-end. The structure was determined by x-ray diffraction of the crystallized C-cadherin extracellular domain. (B) The extracellular part of each polypeptide consists of a series of compact domains called cadherin repeats, joined by flexible hinge regions. Ca^{2+} binds in the neighborhood of each hinge, preventing it from flexing. In the absence of Ca^{2+}, the molecule becomes floppy and adhesion fails. (C) At a typical junction, many cadherin molecules are arrayed in parallel, functioning like Velcro to hold cells together. Cadherins on the same cell are thought to be coupled by side-to-side interactions between their N-terminal head regions, and via the attachments of their intracellular tails to a mat of other proteins (not shown here). (Based on T.J. Boggon et al., *Science* 296:1308–1313, 2002. With permission from AAAS.)

(C)

types of transmembrane adhesion proteins. The making and breaking of anchoring junctions plays a vital part in development and in the constant turnover of tissues in many parts of the mature body. <CGAA>

Selective Cell–Cell Adhesion Enables Dissociated Vertebrate Cells to Reassemble into Organized Tissues

Cadherins form specific homophilic attachments, and this explains why there are so many different family members. Cadherins are not like glue, making cell surfaces generally sticky. Rather, they mediate highly selective recognition, enabling cells of a similar type to stick together and to stay segregated from other types of cells.

This selectivity in the way that animal cells consort with one another was demonstrated more than 50 years ago, long before the discovery of cadherins, in experiments in which amphibian embryos were dissociated into single cells. These cells were then mixed up and allowed to reassociate. Remarkably, the dissociated cells often reassembled *in vitro* into structures resembling those of the original embryo (**Figure 19–10**). The same phenomenon occurs when dissociated cells from two embryonic vertebrate organs, such as the liver and the retina, are mixed together and artificially formed into a pellet: the mixed aggregates

Figure 19–10 Sorting out. Cells from different parts of an early amphibian embryo will sort out according to their origins. In the classical experiment shown here, mesoderm cells (*green*), neural plate cells (*blue*), and epidermal cells (*red*) have been disaggregated and then reaggregated in a random mixture. They sort out into an arrangement reminiscent of a normal embryo, with a "neural tube" internally, epidermis externally, and mesoderm in between. (Modified from P.L. Townes and J. Holtfreter, *J. Exp. Zool.* 128:53–120, 1955. With permission from Wiley-Liss.)

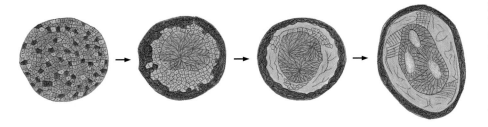

Figure 19–11 Selective dispersal and reassembly of cells to form tissues in a vertebrate embryo. Some cells that are initially part of the epithelial neural tube alter their adhesive properties and disengage from the epithelium to form the neural crest on the upper surface of the neural tube. The cells then migrate away and form a variety of cell types and tissues throughout the embryo. Here they are shown assembling and differentiating to form two clusters of nerve cells, called ganglia, in the peripheral nervous system. While some of the neural crest cells differentiate in the ganglion to become the neurons, others become satellite cells (specialized glial supporting cells) wrapped around the neurons. Changing patterns of expression of cell adhesion molecules underlie all these architectural rearrangements.

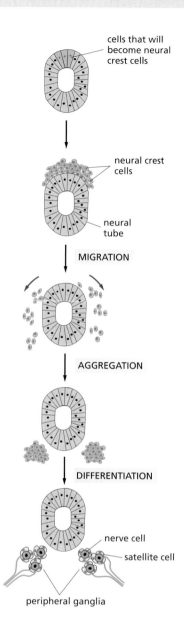

gradually sort out according to their organ of origin. More generally, disaggregated cells are found to adhere more readily to aggregates of their own organ than to aggregates of other organs. Evidently there are cell–cell recognition systems that make cells of the same differentiated tissue preferentially adhere to one another.

Such findings suggest that tissue architecture in animals is not just a product of history but is actively organized and maintained by the system of affinities that cells have for one another and for the extracellular matrix. In the developing embryo, we can indeed watch the cells as they differentiate, and see how they move and regroup to form new structures, guided by selective adhesion. Some of these movements are subtle, others more far-reaching, involving long-range migrations, as we shall describe in Chapter 22. In vertebrate embryos, for example, cells from the *neural crest* break away from the epithelial neural tube, of which they are initially a part, and migrate along specific paths to many other regions. There they reaggregate with other cells and with one another to form a variety of tissues, including those of the peripheral nervous system (**Figure 19–11**). To find their way, the cells depend on guidance from the embryonic tissues along the path. This may involve *chemotaxis* or *chemorepulsion*, that is, movement under the influence of soluble chemicals that attract or repel migrating cells. It may also involve *contact guidance*, in which the migrant cell touches other cells or extracellular matrix components, making transient adhesions that govern the track taken. Then, once the migrating cell has reached its destination, it must recognize and join other cells of the appropriate type to assemble into a tissue. In all these processes of sorting out, contact guidance, and tissue assembly, cadherins play a crucial part.

Cadherins Control the Selective Assortment of Cells

The appearance and disappearance of specific cadherins correlate with steps in embryonic development where cells regroup and change their contacts to create new tissue structures. As the neural tube forms and pinches off from the overlying ectoderm, for example, neural tube cells lose E-cadherin and acquire other cadherins, including N-cadherin, while the cells in the overlying ectoderm continue to express E-cadherin (**Figure 19–12**A, B). Then, when the neural crest cells migrate away from the neural tube, these cadherins become scarcely detectable, and another cadherin (cadherin-7) appears that helps hold the migrating cells together as loosely associated cell groups (Figure 19–12C). Finally, when the cells aggregate to form a ganglion, they switch on expression of N-cadherin again (see Figure 19–11). If N-cadherin is artificially overexpressed in the emerging neural crest cells, the cells fail to escape from the neural tube.

Studies with cultured cells support the suggestion that the homophilic binding of cadherins controls these processes of tissue segregation. In a line of cultured fibroblasts called *L cells*, for example, cadherins are not expressed and the cells do not adhere to one another. When these cells are transfected with DNA encoding E-cadherin, however, they become adherent to one another, and the adhesion is inhibited by anti-E-cadherin antibodies. Since the transfected cells do not stick to untransfected L cells, we can conclude that the attachment depends on E-cadherin on one cell binding to E-cadherin on another. If L cells expressing different cadherins are mixed together, they sort out and aggregate separately, indicating that different cadherins preferentially bind to their own type (**Figure 19–13**A), mimicking what happens when cells derived from tissues that express different cadherins are mixed together. A similar segregation of cells occurs if L cells expressing different amounts of the same cadherin are mixed together (Figure 19–13B). It therefore seems likely that both qualitative and quantitative differences in the expression of cadherins have a role in organizing tissues.

Twist Regulates Epithelial–Mesenchymal Transitions

The assembly of cells into an epithelium is a reversible process. By switching on expression of adhesion molecules, dispersed unattached cells—often called *mesenchymal cells*—can come together to form an epithelium. Conversely, epithelial cells can change their character, disassemble, and migrate away from their parent epithelium as separate individuals. Such *epithelial–mesenchymal transitions* play an important part in normal embryonic development; the origin of the neural crest is one example (see Figure 19–11). A control system involving a set of gene regulatory components called Slug, Snail, and Twist, with E-cadherin as a downstream component, seems to be critical for such transitions: in several tissues, both in flies and vertebrates, switching on expression of Twist, for example, converts epithelial cells to a mesenchymal character, and switching it off does the opposite.

Epithelial–mesenchymal transitions also occur as pathological events during adult life, in cancer. Most cancers originate in epithelia, but become dangerously prone to spread—that is, *malignant*—only when the cancer cells escape from the epithelium of origin and invade other tissues. Experiments with malignant breast cancer cells in culture show that blocking expression of Twist can convert them back toward a nonmalignant character. Conversely, by forcing Twist expression, one can make normal epithelial cells undergo an epithelial–mesenchymal transition and behave like malignant cells. Twist exerts its effects, in part at least, by inhibiting expression of the cadherins that hold epithelial cells together. E-cadherin, in particular, is a target. Mutations that disrupt the production or function of E-cadherin are in fact often found in cancer cells and are thought to help make them malignant, as we shall discuss in Chapter 20.

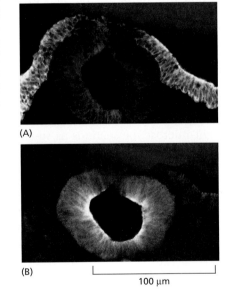

(A)

(B)

100 μm

Figure 19–12 Changing patterns of cadherin expression during construction of the nervous system. The figure shows cross-sections of the early chick embryo, as the neural tube detaches from the ectoderm and then as neural crest cells detach from the neural tube.
(A,B) Immunofluorescence micrographs showing the developing neural tube labeled with antibodies against (A) E-cadherin and (B) N-cadherin.
(C) As the patterns of gene expression change, the different groups of cells segregate from one another according to the cadherins they express.
(Micrographs courtesy of Kohei Hatta and Masatoshi Takeichi.)

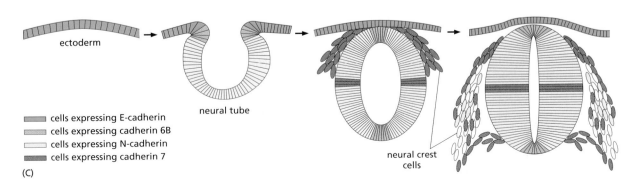

ectoderm

neural tube

cells expressing E-cadherin
cells expressing cadherin 6B
cells expressing N-cadherin
cells expressing cadherin 7

neural crest cells

(C)

Catenins Link Classical Cadherins to the Actin Cytoskeleton

The extracellular domains of cadherins mediate homophilic binding. The intracellular domains of typical cadherins, including all classical and some nonclassical ones, provide anchorage for filaments of the cytoskeleton: anchorage to actin at adherens junctions, and to intermediate filaments at desmosome junctions, as mentioned earlier (see Figure 19–3). The linkage to the cytoskeleton is indirect and depends on a cluster of accessory *intracellular anchor proteins* that assemble on the tail of the cadherin. This linkage, connecting the cadherin family member to actin or intermediate filaments, includes several different components (**Figure 19–14**). These components vary somewhat according to the type of anchorage—but in general a central part is played by *β-catenin* and/or its close relative *γ-catenin (plakoglobin)*.

At adherens junctions, a remote relative of this pair of proteins, *p120-catenin*, is also present and helps to regulate assembly of the whole complex. When p120-catenin is artificially depleted, cadherin proteins are rapidly degraded, and cell–cell adhesion is lost. An artificial increase in the level of p120-catenin has an opposite effect. It is possible that cells use changes in the level of p120-catenin or in its phosphorylation state as one way to regulate their strength of adhesion. In any case, it seems that the linkage to actin is essential for efficient cell–cell adhesion, as classical cadherins that lack their cytoplasmic domain cannot hold cells strongly together.

Adherens Junctions Coordinate the Actin-Based Motility of Adjacent Cells

Adherens junctions are an essential part of the machinery for modeling the shapes of multicellular structures in the animal body. By indirectly linking the actin filaments in one cell to those in its neighbors, they enable the cells in the tissue to use their actin cytoskeletons in a coordinated way.

Adherens junctions occur in various forms. In many nonepithelial tissues, they appear as small punctate or streaklike attachments that indirectly connect the cortical actin filaments beneath the plasma membranes of two interacting cells. In heart muscle (discussed in Chapter 23), they anchor the actin bundles of the contractile apparatus and act in parallel with desmosome junctions to link the contractile cells end-to-end. (The cell–cell interfaces in the muscle where these adhesions occur are so substantial that they show up clearly in stained light-microscope sections as so-called *intercalated discs*.) But the prototypical examples of adherens junctions occur in epithelia, where they often form a continuous **adhesion belt** (or *zonula adherens*) close beneath the apical face of the epithelium, encircling each of the interacting cells in the sheet (**Figure 19–15**). Within each cell, a contractile bundle of actin filaments lies adjacent to the adhesion belt, oriented parallel to the plasma membrane and tethered to it by the cadherins and their associated intracellular anchor proteins. The actin bundles are thus linked, via the cadherins and anchor proteins, into an extensive transcellular network. This network can contract with the help of myosin motor proteins (discussed in Chapter 16), providing the motile force for a fundamental process in animal morphogenesis—the folding of epithelial cell sheets into tubes, vesicles, and other related structures (**Figure 19–16**).

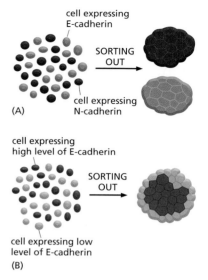

Figure 19–13 Cadherin-dependent cell sorting. Cells in culture can sort themselves out according to the type and level of cadherins they express. This can be visualized by labeling different populations of cells with dyes of different colors. (A) Cells expressing N-cadherin sort out from cells expressing E-cadherin. (B) Cells expressing high levels of E-cadherin sort out from cells expressing low levels of E-cadherin.

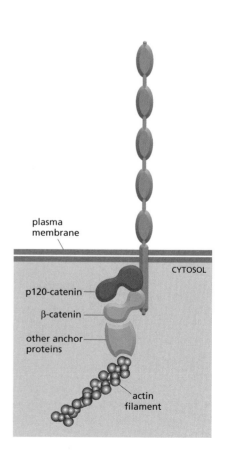

Figure 19–14 The linkage of classical cadherins to actin filaments. The cadherins are coupled indirectly to actin filaments via β-catenin and other anchor proteins. α-Catenin, vinculin, and plakoglobin (a relative of β-catenin, also called γ-catenin) are probably also present in the linkage or involved in control of its assembly, but the details of the anchorage are not well understood. Another intracellular protein, called p120-catenin, also binds to the cadherin cytoplasmic tail and regulates cadherin function. β-Catenin has a second, and very important, function in intracellular signaling, as we discuss in Chapter 15 (see Figure 15–77).

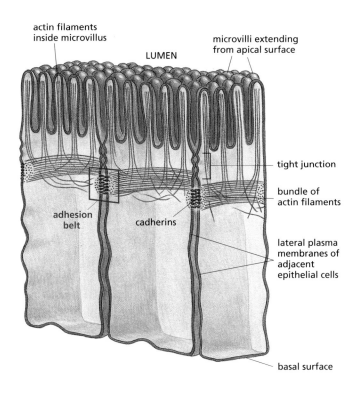

Figure 19–15 Adherens junctions between epithelial cells in the small intestine. These cells are specialized for absorption of nutrients; at their apex, facing the lumen of the gut, they have many microvilli (protrusions that serve to increase the absorptive surface area). The adherens junction takes the form of an *adhesion belt*, encircling each of the interacting cells. Its most obvious feature is a contractile bundle of actin filaments running along the cytoplasmic surface of the junctional plasma membrane. The actin filament bundles are tethered by intracellular anchor proteins to cadherins. The cadherins span the plasma membrane, and their extracellular domains bind homophilically to those of the cadherins on the adjacent cell. In this way, the actin filament bundles in adjacent cells are tied together.

Desmosome Junctions Give Epithelia Mechanical Strength

Desmosome junctions are structurally similar to adherens junctions but link to intermediate filaments instead of actin. Their main function is to provide mechanical strength. Desmosome junctions are important in vertebrates but are not found, for example, in *Drosophila*. They are present in most mature vertebrate epithelia, and are extremely plentiful in the epidermis, the epithelium that forms the outer layer of the skin; a favorite source for biochemical studies is the epidermis of the snout of cows, which has to withstand constant battering as the animal grazes.

Figure 19–17A shows the general structure of a desmosome, and Figure 19–17B shows some of the proteins that form it. Desmosomes typically appear as buttonlike spots of intercellular adhesion, riveting the cells together (Figure 19–17C). Inside the cell, the bundles of ropelike intermediate filaments that are

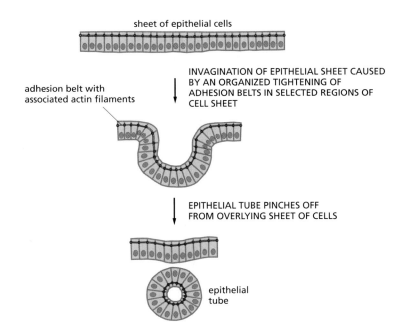

Figure 19–16 The folding of an epithelial sheet to form an epithelial tube. The oriented contraction of the bundles of actin filaments running along adhesion belts causes the epithelial cells to narrow at their apex and helps the epithelial sheet to roll up into a tube. An example is the formation of the neural tube in early vertebrate development (see Figure 19–12 and Chapter 22). Although not shown here, rearrangements of the cells within the epithelial sheet are also thought to have an important role in the process.

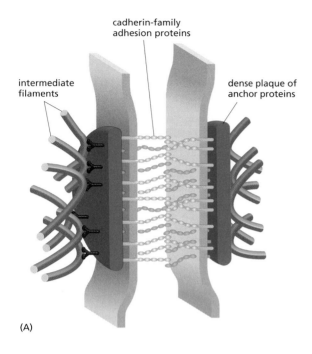

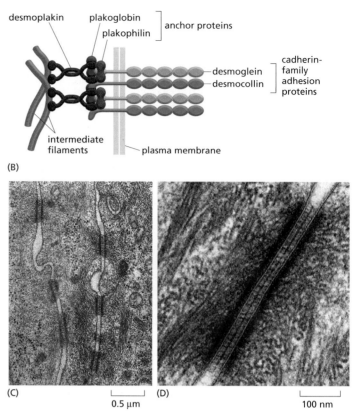

(B)

(C)   (D)

0.5 μm   100 nm

Figure 19–17 Desmosomes. (A) The structural components of a desmosome. On the cytoplasmic surface of each interacting plasma membrane is a dense plaque composed of a mixture of intracellular anchor proteins. A bundle of keratin intermediate filaments is attached to the surface of each plaque. Transmembrane adhesion proteins of the cadherin family bind to the plaques and interact through their extracellular domains to hold the adjacent membranes together by a Ca^{2+}-dependent mechanism. (B) Some of the molecular components of a desmosome. Desmoglein and desmocollin are members of the cadherin family of adhesion proteins. Their cytoplasmic tails bind *plakoglobin* (γ-catenin) and *plakophilin* (a distant relative of p120-catenin), which in turn bind to *desmoplakin*. Desmoplakin binds to the sides of intermediate filaments, thereby tying the desmosome to these filaments. (C) An electron micrograph of desmosome junctions between epidermal cells in the skin of a baby mouse. (D) Part of the same tissue at higher magnification, showing a single desmosome, with intermediate filaments attached to it. (C and D, from W. He, P. Cowin and D.L. Stokes, *Science* 302:109–113, 2003. With permission from AAAS.)

anchored to the desmosomes form a structural framework of great tensile strength (Figure 19–17D), with linkage to similar bundles in adjacent cells, creating a network that extends throughout the tissue (**Figure 19–18**). The particular type of intermediate filaments attached to the desmosomes depends on the cell type: they are *keratin filaments* in most epithelial cells, for example, and *desmin filaments* in heart muscle cells.

The importance of desmosome junctions is demonstrated by some forms of the potentially fatal skin disease *pemphigus*. Affected individuals make antibodies against one of their own desmosomal cadherin proteins. These antibodies bind to and disrupt the desmosomes that hold their epidermal cells (keratinocytes) together. This results in a severe blistering of the skin, with leakage of body fluids into the loosened epithelium.

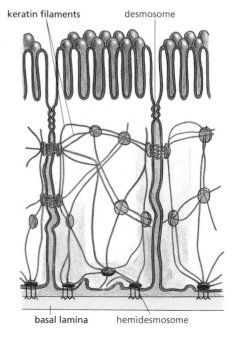

Figure 19–18 Desmosomes, hemidesmosomes, and the intermediate filament network. The keratin intermediate filament networks of adjacent cells—in this example, epithelial cells of the small intestine—are indirectly connected to one another through desmosomes, and to the basal lamina through hemidesmosomes.

Cell–Cell Junctions Send Signals into the Cell Interior

The making and breaking of attachments are important events in the lives of cells and provoke large changes in their internal affairs. Conversely, changes in the internal state of a cell must be able to trigger the making or breaking of attachments. Thus there is a complex cross-talk between the adhesion machinery and chemical signaling pathways. We have described, for example, how changes in p120-catenin may regulate the formation of adherens junctions, and several intracellular signaling pathways can control junction formation by phosphorylating this and other junctional proteins. Later, we shall discuss how the making and breaking of adhesions can send signals into the cell interior through mechanisms involving *scaffold proteins* on the intracellular side of the junction.

Another of the central players in the two-way interaction between adhesion and signaling is thought to be β-catenin. In this chapter, we have mentioned it as an essential intracellular anchor protein at adherens junctions, linking cadherins to actin filaments. In Chapter 15, we encountered it in another guise, as a component of the Wnt cell–cell signaling pathway, moving from the cytoplasm to the nucleus to activate the transcription of target genes. Separate parts of the molecule are responsible for the adhesive and gene-regulatory functions, but an individual molecule cannot do both things at once. Disintegration of an adherens junction can set β-catenin molecules free to move from the cell surface into the nucleus as signaling molecules, and, conversely, the activities of components of the Wnt signaling pathway (which regulate the phosphorylation and degradation of β-catenin) may control the availability of β-catenin to form adherens junctions.

Some nonclassical cadherins transmit signals into the cell interior in yet other ways. Members of the Flamingo subfamily, for example, have a seven-pass transmembrane domain suggesting that they might function as G-protein-coupled receptors. Vascular endothelial cadherin (VE-cadherin) provides another example. This protein not only mediates adhesion between endothelial cells but also is required for endothelial cell survival. Although endothelial cells that do not express VE-cadherin still adhere to one another via N-cadherin, they fail to survive, because they are unable to respond to an extracellular protein called *vascular endothelial growth factor (VEGF)* that acts as a survival signal. VEGF binds to a receptor tyrosine kinase (discussed in Chapter 15) that requires VE-cadherin as a co-receptor.

Selectins Mediate Transient Cell–Cell Adhesions in the Bloodstream

The cadherin superfamily is central to cell–cell adhesion in animals, but at least three other superfamilies of cell–cell adhesion proteins are also important: the *integrins*, the *selectins*, and the adhesive *immunoglobulin (Ig)-superfamily* members. We shall discuss integrins in more detail later: their main function is in cell–matrix adhesion, but a few of them mediate cell–cell adhesion in specialized circumstances. Ca^{2+} dependence provides one simple way to distinguish among these classes of proteins experimentally. Selectins, like cadherins and integrins, require Ca^{2+} for their adhesive function; Ig-superfamily members do not.

Selectins are cell-surface carbohydrate-binding proteins *(lectins)* that mediate a variety of transient, cell–cell adhesion interactions in the bloodstream. Their main role, in vertebrates at least, is in inflammatory responses and in governing the traffic of white blood cells. White blood cells lead a nomadic life, roving between the bloodstream and the tissues, and this necessitates special adhesive behavior. The selectins control the binding of white blood cells to the endothelial cells lining blood vessels, thereby enabling the blood cells to migrate out of the bloodstream into a tissue.

Each selectin is a transmembrane protein with a conserved lectin domain that binds to a specific oligosaccharide on another cell (**Figure 19–19**A). There are at least three types: *L-selectin* on white blood cells, *P-selectin* on blood platelets and on endothelial cells that have been locally activated by an inflammatory

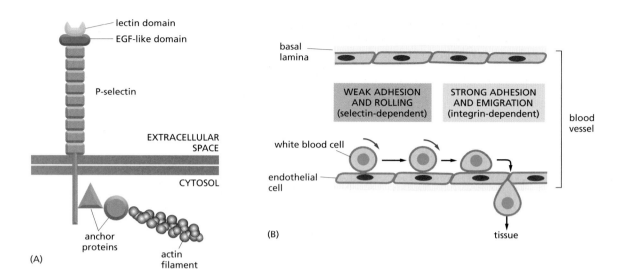

response, and *E-selectin* on activated endothelial cells. In a lymphoid organ, such as a lymph node or a tonsil, the endothelial cells express oligosaccharides that are recognized by L-selectin on lymphocytes, causing the lymphocytes to loiter and become trapped. At sites of inflammation, the roles are reversed: the endothelial cells switch on expression of selectins that recognize the oligosaccharides on white blood cells and platelets, flagging the cells down to help deal with the local emergency. Selectins do not act alone, however; they collaborate with integrins, which strengthen the binding of the blood cells to the endothelium. The cell–cell adhesions mediated by both selectins and integrins are *heterophilic*—that is, the binding is to a molecule of a different type: selectins bind to specific oligosaccharides on glycoproteins and glycolipids, while integrins bind to other specific proteins.

Selectins and integrins act in sequence to let white blood cells leave the bloodstream and enter tissues (Figure 19–19B). The selectins mediate a weak adhesion because the binding of the lectin domain of the selectin to its carbohydrate ligand is of low affinity. This allows the white blood cell to adhere weakly and reversibly to the endothelium, rolling along the surface of the blood vessel, propelled by the flow of blood. The rolling continues until the blood cell activates its integrins. As we discuss later, these transmembrane molecules can be switched into an adhesive conformation that enables them to latch onto other molecules external to the cell—in the present case, proteins on the surfaces of the endothelial cells. Once it has attached in this way, the white blood cell escapes from the blood stream into the tissue by crawling out of the blood vessel between adjacent endothelial cells.

Figure 19–19 The structure and function of selectins. (A) The structure of P-selectin. The selectin attaches to the actin cytoskeleton through anchor proteins that are still poorly characterized. (B) How selectins and integrins mediate the cell–cell adhesions required for a white blood cell to migrate out of the bloodstream into a tissue. First, selectins on endothelial cells bind to oligosaccharides on the white blood cell, so that it becomes loosely attached to the vessel wall. Then the white blood cell activates an integrin (usually one called LFA1) in its plasma membrane, enabling this integrin to bind to a protein called ICAM1, belonging to the immunoglobulin superfamily, in the membrane of the endothelial cell. This creates a stronger attachment that allows the white blood cell to crawl out of the vessel.

Members of the Immunoglobulin Superfamily of Proteins Mediate Ca²⁺-Independent Cell–Cell Adhesion

The chief endothelial cell proteins that are recognized by the white blood cell integrins are called *ICAMs (intercellular cell adhesion molecules)* or *VCAMs (vascular cell adhesion molecules)*. They are members of another large and ancient family of cell surface molecules—the **immunoglobulin (Ig) superfamily**. These contain one or more of the extracellular Ig-like domains that are characteristic of antibody molecules (discussed in Chapter 25). They have many functions outside the immune system that are unrelated to immune defenses.

While ICAMs and VCAMs on endothelial cells both mediate heterophilic binding to integrins, many other Ig superfamily members appear to mediate homophilic binding. An example is the *neural cell adhesion molecule (NCAM)*, which is expressed by various cell types, including most nerve cells, and can take different forms, generated by alternative splicing of an RNA transcript produced from a single gene (**Figure 19–20**). Some forms of NCAM carry an unusually large quantity of sialic acid (with chains containing hundreds of repeating sialic

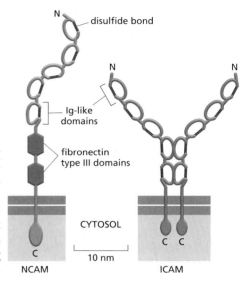

Figure 19–20 **Two members of the Ig superfamily of cell–cell adhesion molecules.** NCAM is expressed on neurons and many other cell types, and mediates homophilic binding. Only the protein backbone of NCAM is shown here; this often has side chains of sialic acid (a polysaccharide) covalently attached to it, hindering adhesion. ICAM is expressed on endothelial cells and some other cell types and binds heterophilically to an integrin on white blood cells.

acid units). By virtue of their negative charge, the long polysialic acid chains can interfere with cell adhesion (because like charges repel one another); NCAM heavily loaded with sialic acid may even serve to inhibit adhesion, rather than cause it.

A cell of a given type generally uses an assortment of different adhesion proteins to interact with other cells, just as each cell uses an assortment of different receptors to respond to the many soluble extracellular signal molecules, such as hormones and growth factors, in its environment. Although cadherins and Ig family members are frequently expressed on the same cells, the adhesions mediated by cadherins are much stronger, and they are largely responsible for holding cells together, segregating cell collectives into discrete tissues, and maintaining tissue integrity. Molecules such as NCAM seem to contribute more to the fine-tuning of these adhesive interactions during development and regeneration, playing a part in various specialized adhesive phenomena, such as that discussed for blood and endothelial cells. Thus, while mutant mice that lack N-cadherin die early in development, those that lack NCAM develop relatively normally but show some mild abnormalities in the development of certain specific tissues, including parts of the nervous system.

Many Types of Cell Adhesion Molecules Act in Parallel to Create a Synapse

Cells of the nervous system, especially, rely on complex systems of adhesion molecules, as well as chemotaxis and soluble signal factors, to guide axon outgrowth along precise pathways and to direct the formation of specific nerve connections (discussed in Chapter 22). Adhesion proteins of the Ig superfamily, along with many other classes of adhesion and signaling molecules, have important roles in these processes. Thus, for example, in flies with a mutation of *Fasciclin2*, related to NCAM, some axons follow aberrant pathways and fail to reach their proper targets.

Another member of the Ig superfamily, *Fasciclin3*, enables the neuronal growth cones to recognize their proper targets when they meet them. This protein is expressed transiently on some motor neurons in *Drosophila*, as well as on the muscle cells they normally innervate. If Fasciclin3 is genetically removed from these motor neurons, they fail to recognize their muscle targets and do not make synapses with them. Conversely, if motor neurons that normally do not express Fasciclin3 are made to express this protein, they will synapse with Fasciclin3-expressing muscle cells to which they normally do not connect. It seems that Fasciclin3 mediates these synaptic connections by a homophilic "matchmaking" mechanism. Ig superfamily proteins have similar roles in vertebrates. Proteins of the *Sidekicks* subfamily, for example, mediate homophilic adhesion, and different Sidekicks proteins are expressed in different layers of the retina, with synapses forming between sets of retinal neurons that share expression of the same family member. When the pattern of expression of the proteins is artificially altered, the pattern of synaptic connections changes accordingly.

These Ig superfamily members are by no means the only adhesion molecules involved in initiating synapse formation. Misexpression of certain other synaptic adhesion proteins, unrelated to any of the types we have mentioned so far, can even trick growth cones into synapsing on non-neuronal cells that would never normally be innervated. Thus, if non-neuronal cells are forced to express *neuroligin*, a transmembrane protein evolutionarily related to the enzyme acetylcholinesterase, neurons will synapse on them, as a consequence of binding of neuroligin to a protein called *neurexin* in the membrane of the presynaptic neuron.

Scaffold Proteins Organize Junctional Complexes

To make a synapse, the pre- and postsynaptic cells have to do more than recognize one another and adhere: they have to assemble a complex system of signal receptors, ion channels, synaptic vesicles, docking proteins, and other components, as described in Chapter 11. This apparatus for synaptic signaling could not exist without cell adhesion molecules to join the pre- and postsynaptic membranes firmly together and to help hold all the components of the signaling machinery in their proper positions. Thus, cadherins are generally present, concentrated at spots around the periphery of the synapse and within it, as well as Ig superfamily members and various other types of adhesion molecules. In fact, about 20 different classical cadherins are expressed in the vertebrate nervous system, in different combinations in different subsets of neurons, and it is likely that selective binding of these molecules also plays a part in ensuring that neurons synapse with their correct partners.

But how does the array of adhesion molecules recruit the other components of the synapse and hold them in place? **Scaffold proteins** are thought to have a central role here. These intracellular molecules consist of strings of protein-binding domains, typically including several **PDZ domains**—segments about 70 amino acids long that can recognize and bind the C-terminal intracellular tails of specific transmembrane molecules (**Figure 19–21**). One domain of a scaffold protein may attach to a cell–cell adhesion protein, for example, while another latches onto a ligand-gated ion channel, and yet another binds a protein that regulates exocytosis or endocytosis or provides attachment to the cytoskeleton. Moreover, one molecule of scaffold protein can bind to another. In this way, the cell can assemble a mat of proteins, with all the components that are needed at the synapse woven into its fabric (**Figure 19–22**). Several hundred different types of proteins participate in this complex structure. Mutations in synaptic scaffold proteins alter the size and structure of synapses and can have severe consequences for the function of the nervous system. Among other things, such mutations can damage the molecular machinery underlying learning and memory, which depend on the ability of electrical activity to leave a long-lasting trace in the form of alterations of synaptic architecture.

The scaffold proteins, with their many potential binding partners, are involved in organizing other structures and functions beside synapses and synaptic signaling. The *Discs large (Dlg)* protein of *Drosophila* is an example (see Figure 19–21). Dlg is needed for the construction of normal synapses; but we shall see that it, along with a set of other related scaffold proteins, also plays an

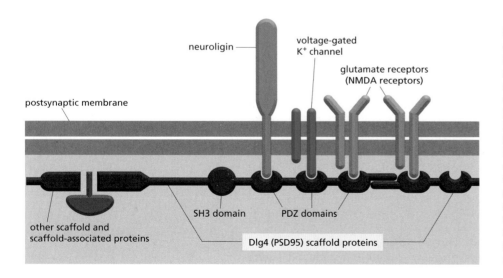

Figure 19–21 **A scaffold protein.** The diagram shows the domain structure of Dlg4, a mammalian homolog of the *Drosophila* protein Discs-large, along with some of its binding partners. Dlg4 is concentrated beneath the postsynaptic membrane at synapses, and is also known as postsynaptic density protein 95, or PSD95. With its multiple protein-binding domains, it can link together different components of the synapse. One molecule of Dlg4 can also bind to another or to scaffolding molecules of other types, thereby creating an extensive framework that holds together all the components of the synapse. Scaffold proteins also have important roles at other types of cell junctions.

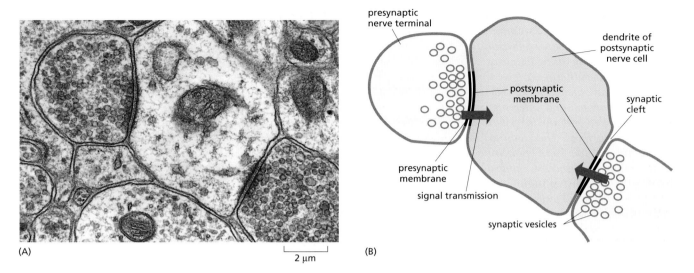

(A)   2 μm   (B)

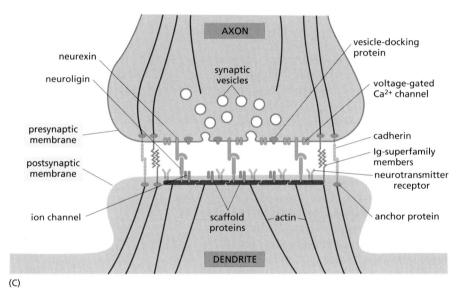

(C)

Figure 19–22 Organization of a synapse. (A) Electron micrograph and (B) line drawing of a cross section of two nerve terminals synapsing on a dendrite in the mammalian brain. Note the synaptic vesicles in the two nerve terminals and the dark-staining material associated with the pre- and postsynaptic membranes. (C) Schematic diagram showing some of the synaptic components that are assembled at these sites. Cell–cell adhesion molecules, including cadherins and neuroligins and neurexins, hold the pre- and postsynaptic membranes together. Scaffolding proteins help to form a mat (corresponding to the dark-staining material seen in (A)) that links the adhesion molecules by their intracellular tails to the components of the synaptic signal-transmission machinery, such as ion channels and neurotransmitter receptors. The structure of this large, complex multiprotein assembly is not yet known in detail. It includes anchorage sites for hundreds of additional components, not shown here, including cytoskeletal molecules and various regulatory kinases and phosphatases. (A, courtesy of Cedric Raine.)

essential part in almost every aspect of the organization of epithelia, including the formation of occluding junctions between the cells, the control of cell polarity, and even the control of cell proliferation. All these processes have a shared dependence on the same machinery, not only in flies, but also in vertebrates.

Summary

In epithelia, as well as in some other types of tissue, cells are directly attached to one another through strong cell–cell adhesions, mediated by transmembrane proteins that are anchored intracellularly to the cytoskeleton. At adherens junctions, the anchorage is to actin filaments; at desmosome junctions, it is to intermediate filaments. In both these structures, and in many less conspicuous cell–cell junctions, the adhesive transmembrane proteins are members of the cadherin superfamily. Cadherins generally bind to one another homophilically: the head of one cadherin molecule binds to the head of a similar cadherin on an opposite cell. This selectivity enables mixed populations of cells of different types to sort out from one another according to the specific cadherins they express, and it helps to control cell rearrangements during development, where many different cadherins are expressed in complex, changing patterns. Changes in cadherin expression can cause cells to undergo transitions between a cohesive epithelial state and a detached mesenchymal state—a phenomenon important in cancer as well as in embryonic development.

The "classical" cadherins are linked to the actin cytoskeleton by intracellular proteins called catenins. These form an anchoring complex on the intracellular tail of the cadherin molecule, and are involved not only in physical anchorage but also in the genesis of intracellular signals. Conversely, intracellular signals can regulate the formation of cadherin-mediated adhesions. β-Catenin, for example, is also a key component of the Wnt cell signaling pathway.

In addition to cadherins, at least three other classes of transmembrane molecules are also important mediators of cell–cell adhesion: selectins, immunoglobulin (Ig) superfamily members, and integrins. Selectins are expressed on white blood cells, blood platelets, and endothelial cells, and they bind heterophilically to carbohydrate groups on cell surfaces. They help to trap circulating white blood cells at sites of inflammation. Ig-superfamily proteins also play a part in this trapping, as well as in many other adhesive processes; some of them bind homophilically, some heterophilically. Integrins, though they mainly serve to attach cells to the extracellular matrix, can also mediate cell–cell adhesion by binding to the Ig-superfamily members.

Many different Ig-superfamily members, cadherins, and other cell–cell adhesion molecules guide the formation of nerve connections and hold neuronal membranes together at synapses. In these complicated structures, as well as at other types of cell–cell junctions, intracellular scaffold proteins containing multiple PDZ protein-binding domains have an important role in holding the many different adhesive and signaling molecules in their proper arrangements.

TIGHT JUNCTIONS AND THE ORGANIZATION OF EPITHELIA

An epithelial sheet, with its cells joined side by side and standing on a basal lamina, may seem a specialized type of structure, but it is central to the construction of multicellular animals. In fact, more than 60% of the cell types in the vertebrate body are epithelial. Just as cell membranes enclose and partition the interior of the eucaryotic cell, so epithelia enclose and partition the animal body, lining all its surfaces and cavities, and creating internal compartments where specialized processes occur. The epithelial sheet seems to be one of the inventions that lie at the origin of animal evolution, diversifying in a huge variety of ways (as we see in Chapter 23), but retaining an organization based on a set of conserved molecular mechanisms that practically all epithelia have in common.

Essentially all epithelia are anchored to other tissue on one side—the **basal** side—and free of such attachment on their opposite side—the **apical** side. A basal lamina lies at the interface with the underlying tissue, mediating the attachment, while the apical surface of the epithelium is generally bathed by extracellular fluid (but sometimes covered by material that the cells have secreted at their apices). Thus all epithelia are structurally **polarized**, and so are their individual cells: the basal end of a cell, adherent to the basal lamina below, differs from the apical end, exposed to the medium above.

Correspondingly, all epithelia have at least one function in common: they serve as selective permeability barriers, separating the fluid that permeates the tissue on their basal side from fluid with a different chemical composition on their apical side. This barrier function requires that the adjacent cells be sealed together by **occluding junctions**, so that molecules cannot leak freely across the cell sheet. In this section we consider how the occluding junctions are formed, and how the polarized architecture of the epithelium is maintained. These two fundamental aspects of epithelia are closely linked: the junctions play a key part in organizing and maintaining the polarity of the cells in the sheet.

Tight Junctions Form a Seal Between Cells and a Fence Between Membrane Domains

The occluding junctions found in vertebrate epithelia are called **tight junctions**. The epithelium of the small intestine provides a good illustration of their structure and function (see Figure 19–3). This epithelium has a *simple columnar* structure; that is, it consists of a single layer of tall (columnar) cells. These are of several differentiated types, but the majority are absorptive cells, specialized for uptake of nutrients from the internal cavity, or *lumen*, of the gut.

The absorptive cells have to transport selected nutrients across the epithelium from the lumen into the extracellular fluid that permeates the connective tissue on the other side. From there, these nutrients diffuse into small blood vessels to provide nourishment to the organism. This *transcellular transport* depends on two sets of transport proteins in the plasma membrane of the absorptive cell. One set is confined to the apical surface of the cell (facing the lumen) and actively transports selected molecules into the cell from the gut. The other set is confined to the *basolateral* (basal and lateral) surfaces of the cell, and it allows the same molecules to leave the cell by facilitated diffusion into the extracellular fluid on the other side of the epithelium. For this transport activity to be effective, the spaces between the epithelial cells must be tightly sealed, so that the transported molecules cannot leak back into the gut lumen through these spaces (**Figure 19–23**). Moreover, the proteins that form the pumps and channels must be correctly distributed in the cell membranes: the apical set of active transport proteins must be delivered to the cell apex (as discussed in Chapter 13) and must not be allowed to drift to the basolateral surface, and the basolateral set of channel proteins must be delivered to the basolateral surface and must not be allowed to drift to the apical surface. The tight junctions between epithelial cells, besides sealing the gaps between the cells, may also function as "fences" helping to separate domains within the plasma membrane of each cell, so as to hinder apical proteins (and lipids) from diffusing into the basal region, and vice versa (see Figure 19–23).

The sealing function of tight junctions is easy to demonstrate experimentally: a low-molecular-weight tracer added to one side of an epithelium will generally not pass beyond the tight junction (**Figure 19–24**). This seal is not absolute, however. Although all tight junctions are impermeable to macromolecules, their permeability to small molecules varies. Tight junctions in the epithelium lining the small intestine, for example, are 10,000 times more permeable to inorganic ions,

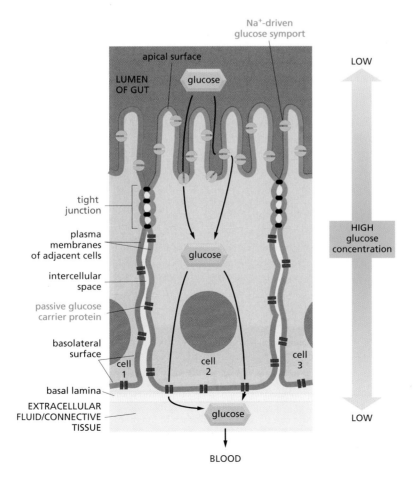

Figure 19–23 The role of tight junctions in transcellular transport. Transport proteins are confined to different regions of the plasma membrane in epithelial cells of the small intestine. This segregation permits a vectorial transfer of nutrients across the epithelium from the gut lumen to the blood. In the example shown, glucose is actively transported into the cell by Na⁺-driven glucose symports at its apical surface, and it diffuses out of the cell by facilitated diffusion mediated by glucose carriers in its basolateral membrane. Tight junctions are thought to confine the transport proteins to their appropriate membrane domains by acting as diffusion barriers or "fences" within the lipid bilayer of the plasma membrane; these junctions also block the backflow of glucose from the basal side of the epithelium into the gut lumen.

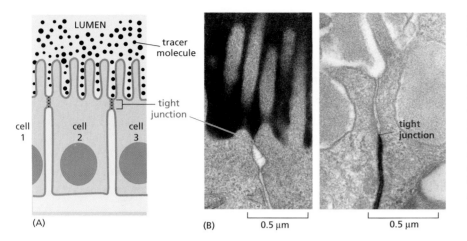

(A)   (B)   0.5 µm   0.5 µm

Figure 19–24 The role of tight junctions in allowing epithelia to serve as barriers to solute diffusion. (A) The drawing shows how a small extracellular tracer molecule added on one side of an epithelium is prevented from crossing the epithelium by the tight junctions that seal adjacent cells together. (B) Electron micrographs of cells in an epithelium in which a small, extracellular, electron-dense tracer molecule has been added to either the apical side (on the *left*) or the basolateral side (on the *right*). In both cases, the tight junction blocks passage of the tracer. (B, courtesy of Daniel Friend.)

such as Na⁺, than the tight junctions in the epithelium lining the urinary bladder. These differences reflect differences in the proteins that form the junctions.

Epithelial cells can also alter their tight junctions transiently to permit an increased flow of solutes and water through breaches in the junctional barriers. Such *paracellular transport* is especially important in the absorption of amino acids and monosaccharides from the lumen of the intestine, where the concentration of these nutrients can increase enough after a meal to drive passive transport in the proper direction.

When tight junctions are visualized by freeze-fracture electron microscopy, they seem to consist of a branching network of *sealing strands* that completely encircles the apical end of each cell in the epithelial sheet (**Figure 19–25A and B**). In conventional electron micrographs, the outer leaflets of the two interacting

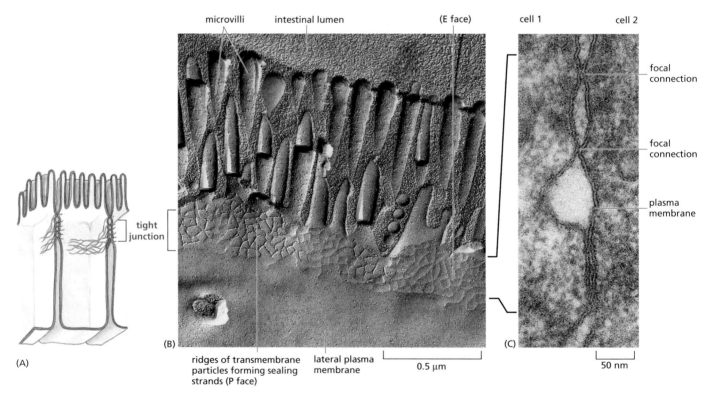

Figure 19–25 The structure of a tight junction between epithelial cells of the small intestine. The junctions are shown (A) schematically, (B) in a freeze-fracture electron micrograph, and (C) in a conventional electron micrograph. In (B), the plane of the micrograph is parallel to the plane of the membrane, and the tight junction appears as a band of branching sealing strands that encircle each cell in the epithelium. The sealing strands are seen as ridges of intramembrane particles on the cytoplasmic fracture face of the membrane (the P face) or as complementary grooves on the external face of the membrane (the E face) (see Figure 19–26A). In (C), the junction is seen in cross section as a series of focal connections between the outer leaflets of the two interacting plasma membranes, each connection corresponding to a sealing strand in cross section. (B and C, from N.B. Gilula, in Cell Communication [R.P. Cox, ed.], pp. 1–29. New York: Wiley, 1974.)

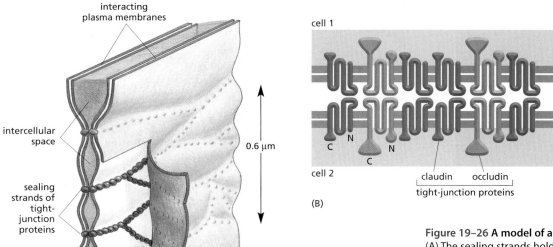

interacting
plasma membranes

intercellular
space

0.6 µm

sealing
strands of
tight-
junction
proteins

cytoplasmic
half of
lipid bilayer

cell
1

cell
2

(A)

Figure 19–26 A model of a tight junction.
(A) The sealing strands hold adjacent plasma membranes together. The strands are composed of transmembrane proteins that make contact across the intercellular space and create a seal. (B) The molecular composition of a sealing strand. The claudins are the main functional components; the role of the occludins is uncertain.

plasma membranes are seen to be tightly apposed where sealing strands are present (Figure 19–25C). Each tight junction sealing strand is composed of a long row of transmembrane adhesion proteins embedded in each of the two interacting plasma membranes. The extracellular domains of these proteins adhere directly to one another to occlude the intercellular space (**Figure 19–26**).

The main transmembrane proteins forming these strands are the *claudins,* which are essential for tight junction formation and function. Mice that lack the *claudin-1* gene, for example, fail to make tight junctions between the cells in the epidermal layer of the skin; as a result, the baby mice lose water rapidly by evaporation through the skin and die within a day after birth. Conversely, if nonepithelial cells such as fibroblasts are artificially caused to express claudin genes, they will form tight-junctional connections with one another. Normal tight junctions also contain a second major transmembrane protein called *occludin,* but the function of this protein is uncertain, and it does not seem to be as essential as the claudins. A third transmembrane protein, *tricellulin* (related to occludin), is required to seal cell membranes together and prevent transepithelial leakage at the points where three cells meet.

The claudin protein family has many members (24 in humans), and these are expressed in different combinations in different epithelia to confer particular permeability properties on the epithelial sheet. They are thought to form *paracellular pores*—selective channels allowing specific ions to cross the tight-junctional barrier, from one extracellular space to another. A specific claudin found in kidney epithelial cells, for example, is needed to let Mg^{2+} pass between the cells of the sheet so that this ion can be resorbed from the urine into the blood. A mutation in the gene encoding this claudin results in excessive loss of Mg^{2+} in the urine.

Scaffold Proteins in Junctional Complexes Play a Key Part in the Control of Cell Proliferation

The claudins and occludins have to be held in the right position in the cell, so as to form the tight-junctional network of sealing strands. This network usually lies just apical to the adherens and desmosome junctions that bond the cells together mechanically, and the whole assembly is called a *junctional complex* (**Figure 19–27**). The parts of this junctional complex depend on each other for

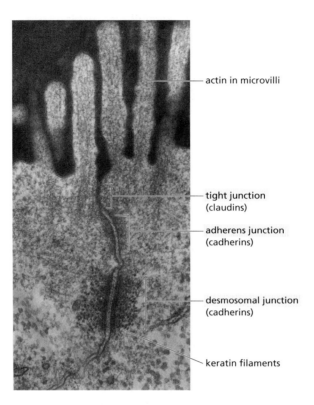

— actin in microvilli

— tight junction (claudins)

— adherens junction (cadherins)

— desmosomal junction (cadherins)

— keratin filaments

Figure 19–27 A junctional complex between two epithelial cells in the lining of the gut. Most apically, there is a tight junction; beneath this, an adherens junction; and beneath the adherens junction, a desmosomal junction. This example is from a vertebrate; in insects, the arrangement is different. (Courtesy of Daniel S. Friend.)

their formation. For example, anti-cadherin antibodies that block the formation of adherens junctions also block the formation of tight junctions. The positioning and organization of tight junctions in relation to these other structures is thought to depend on association with intracellular scaffold proteins of the *Tjp (Tight junction protein)* family, also called *ZO proteins* (a tight junction is also known as a *zonula occludens*). The vertebrate Tjp proteins belong to the same family as the Discs-large proteins that we mentioned earlier for their role at synapses, and they anchor the tight-junctional strands to other components including the actin cytoskeleton.

In invertebrates such as insects and mollusks, occluding junctions have a different appearance and are called **septate junctions**. Like tight junctions, these form a continuous band around each epithelial cell, but the structure is more regular, and the interacting plasma membranes are joined by proteins that are arranged in parallel rows with a regular periodicity (**Figure 19–28**). Septate junctions are nevertheless based on proteins homologous to the vertebrate claudins, and they depend on scaffold proteins in a similar way, including in particular the same Discs-large protein that is present at synapses. Mutant flies that are deficient in Discs-large have defective septate junctions.

Strikingly, these mutants also develop epithelial tumors, in the form of large overgrowths of the imaginal discs—the structures in the fly larva from which most of the adult body derives (as described in Chapter 22). The gene takes its name from this remarkable effect, which depends on the presence of binding sites for growth regulators on the Discs-large protein. But why should the apparatus of cell–cell adhesion be linked in this way with the control of cell proliferation? The relationship seems to be fundamental: in vertebrates also, genes

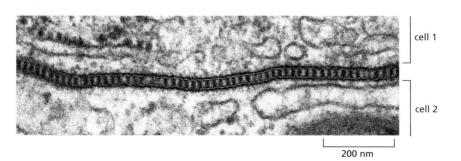

cell 1

cell 2

200 nm

Figure 19–28 A septate junction. A conventional electron micrograph of a septate junction between two epithelial cells in a mollusk. The interacting plasma membranes, seen in cross section, are connected by parallel rows of junctional proteins. The rows, which have a regular periodicity, are seen as dense bars, or septa. (From N.B. Gilula, in Cell Communication [R.P. Cox, ed.], pp. 1–29. New York: Wiley, 1974.)

homologous to *Discs large* have this dual involvement. One possibility is that it reflects a basic mechanism for repair and maintenance of epithelia. If an epithelial cell is deprived of adhesive contacts with neighbors, its program of growth and proliferation is activated, thereby creating new cells to reconstruct a continuous multicellular sheet. In fact, a large body of evidence indicates that junctional complexes are important sites of cell–cell signaling not only via Discs-large but also through other components of these structures, including cadherins as we have seen.

Cell–Cell Junctions and the Basal Lamina Govern Apico-Basal Polarity in Epithelia

Most cells in animal tissues are strongly polarized: they have a front that differs from the back, or a top that differs from the bottom. Examples include virtually all epithelial cells, as we have discussed, as well as neurons with their dendrite–axon polarity, migrating fibroblasts and white blood cells, with their locomotor leading edge and trailing rear end, and many other cells in embryos as they prepare to divide asymmetrically to create daughter cells that are different. A core set of components is critical for cell polarity in all these cases, throughout the animal kingdom, from worms and flies to mammals.

In the case of epithelial cells, these fundamental generators of cell polarity have to establish the difference between the apical and basal poles, and they have to do so in a properly oriented way, in accordance with the cell's surroundings. The basic phenomenon is nicely illustrated by experiments with a cultured line of epithelial cells, called MDCK cells (**Figure 19–29**A). These can be separated from one another and cultured in suspension in a collagen gel. A single isolated cell in these circumstances does not show any obvious polarity, but if it is allowed to divide to form a small colony of cells, these cells will organize themselves into a hollow epithelial vesicle where the polarity of each cell is clearly apparent. The vesicle becomes surrounded by a basal lamina, and all the cells orient themselves in the same way, with apex-specific marker molecules facing the lumen. Evidently, the MDCK cells have a spontaneous tendency to become polarized, but the mechanism is cooperative and depends on contacts with neighbors.

To discover how the underlying molecular mechanism works, the first step is to identify its components. Studies in the worm *C. elegans* and in *Drosophila* have been most informative here. In the worm, a screen for mutations upsetting the organization of the early embryo has revealed a set of genes essential for normal cell polarity and asymmetry of cell division (as discussed in Chapter 22). There are at least six of these genes, called *Par (partitioning defective)* genes. In all animal species studied, they and their homologs (along with other genes discovered through studies in *Drosophila* and vertebrates) have a fundamental role not just in asymmetric cell division in the early embryo, but in many other processes of cell polarization, including the polarization of epithelial cells. The *Par4* gene of *C. elegans*, for example, is homologous to a gene called *Lkb1* in mammals and *Drosophila*, coding for a serine/threonine kinase. In the fly, mutations of

Figure 19–29 Cooperative polarization of a cluster of epithelial cells in culture and its dependence on Rac and laminin. Cells of the MDCK line, derived from dog kidney epithelium, were dissociated, embedded in a collagen matrix, and allowed to proliferate, creating small isolated colonies, shown here schematically in cross section. (A) The cells in such a colony will normally organize themselves spontaneously into an epithelium surrounding a central cavity. Staining for actin (which marks apical microvilli), ZO1 protein (a tight-junction protein), Golgi apparatus, and laminin (a basal lamina component) shows that the cells have all cooperatively become polarized, with apical components facing the lumen of the cavity and basal components facing the surrounding collagen gel. (B) When Rac function is blocked by expression of a dominant-negative form of the protein, the cells show inverted polarity, fail to form a cyst with a central cavity, and cease to deposit laminin in the normal manner around the periphery of the cell cluster. (C) When the cyst is embedded in a matrix rich in exogenous laminin, near-normal polarity is restored even though Rac function is still blocked. (Based on L.E. O'Brien et al., *Nat. Cell Biol.* 3:831–838, 2001. With permission from Macmillan Publishers Ltd.)

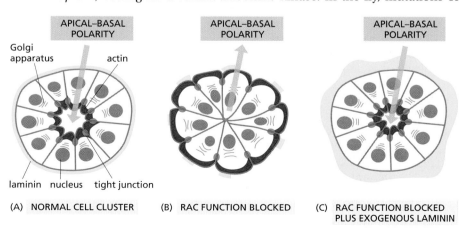

(A) NORMAL CELL CLUSTER   (B) RAC FUNCTION BLOCKED   (C) RAC FUNCTION BLOCKED PLUS EXOGENOUS LAMININ

(A) NON-POLARIZED CELL   (B)   POLARIZED CELL

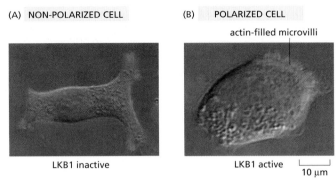

actin-filled microvilli

LKB1 inactive   LKB1 active   ⎿_____⎤
                                               10 µm

Figure 19–30 Development of polarity in single isolated epithelial cells. Cells of a line derived from intestinal epithelium were transfected with DNA constructs coding for regulatory components through which the activity of the LKB1 protein could be switched on or off by a change in the composition of the culture medium. When LKB1 activity is low, the cells appear unpolarized; when it is high, they become individually polarized. Their polarity is manifest in the distribution of tight-junction proteins (ZO1) and adherens-junction proteins (p120-catenin), which accumulate on one side of the cell, around a cap of actin-filled microvilli, even though the cells are isolated from one another and make no cell–cell junctions. This cell-autonomous polarization occurs even when the cells are cultured in suspension, without contact with any substratum that could tell them which way was up. (From A.F. Baas et al., *Cell* 116:457–466, 2004. With permission from Elsevier.)

this gene disrupt the polarity of the egg cell and of cells in epithelia. In humans, such mutations give rise to *Peutz–Jeghers syndrome*, involving disorderly abnormal growths of the lining of the gut and a predisposition to certain rare types of cancer. When cultured human colon epithelial cells are prevented from expressing LKB1, they fail to polarize normally. Moreover, when such cells in culture are artificially driven to express abnormally high levels of LKB1 activity, they can become individually polarized, even when isolated from other cells, and surrounded on all sides by a uniform medium (**Figure 19–30**). This suggests that normal epithelial polarity depends on two interlocking mechanisms: one that endows individual cells with a tendency to become polarized cell-autonomously, and another that orients their polarity axis in relation to their neighbors and the basal lamina. The latter mechanism would be peculiar to epithelia; the former could be much more general, operating also in other polarized cell types.

The molecules known to be needed for epithelial polarity can be classified in relation to these two mechanisms. Central to the polarity of individual animal cells in general is a set of three membrane-associated proteins: **Par3**, **Par6**, and **atypical protein kinase C (aPKC)**. Par3 and Par6 are both scaffold proteins containing PDZ domains, and they bind to one another and to aPKC. The complex of these three components also has binding sites for various other molecules, including the small GTPases Rac and Cdc42. These latter molecules play a crucial part. Thus, for example, when Rac function is blocked in a cluster of MDCK cells, the cells develop with inverted polarity (see Figure 19–29B). Rac and Cdc42 are key regulators of actin assembly, as explained in Chapter 16; through them, it seems, assembly of a Par3-Par6-aPKC complex in a specific region of the cell cortex is associated with polarization of the cytoskeleton towards that region. The assembly process is evidently cooperative and involves some positive feedback and spatial signaling, so that a small initial cluster of these components is able to recruit more of them and to inhibit the development of clusters of the same type elsewhere in the cell. One source of positive feedback may lie in the behavior of Cdc42 and Rac: a high activity of these molecules at a particular site, by organizing the cytoskeleton, may direct intracellular transport so as to bring still more Cdc42 or Rac, or more of their activators, to the same site. This is suspected to be an essential part of the polarization mechanism in budding yeast cells, and it may be the way in which cells such as migrating fibroblasts establish the difference between their leading edge and the rest of their periphery. It could be the core of the eucaryotic cell polarization machinery, at least in evolutionary terms.

The Par3-Par6-aPKC complex, combined with Cdc42 or Rac, seems to control the organization of other protein complexes associated with the internal face of the cell membrane. In particular, in epithelial cells, it causes the *Crumbs complex*, held together by the PDZ-domain scaffold proteins Discs-lost and Stardust, to become localized toward the apex of the cell, while a third such complex, called the *Scribble complex*, held together by the scaffold proteins Scribble and Discs-large (the same protein that we encountered previously) is localized more basally (**Figure 19–31**). These various protein assemblies interact with one another and with other cell components in ways that are only beginning to be understood.

But how is this whole elaborate system oriented correctly in relation to neighboring cells? In an epithelium, the Par3-Par6-aPKC complex assembles at

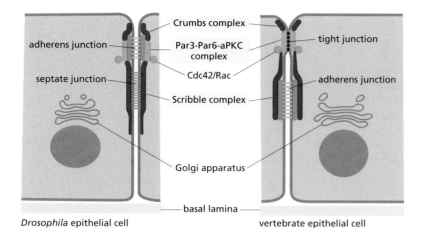

Figure 19–31 The coordinated arrangement of three membrane-associated protein complexes thought to be critical for epithelial polarity. A *Drosophila* epithelial cell is shown schematically on the left, and a vertebrate epithelial cell on the right. All three complexes—the Par3-Par6-aPkc complex, the Crumbs complex, and the Scribble complex—are organized around scaffold proteins containing PDZ domains. The detailed distribution of the complexes varies somewhat according to cell type.

cell–cell junctions—tight junctions in vertebrates, adherens junctions in *Drosophila*—because the scaffold proteins in the complex bind to the tails of certain of the junctional transmembrane adhesion proteins. Meanwhile, the cytoskeleton, under the influence of Rac or its relatives, directs the delivery of basal lamina components to the opposite end of the cell. These extracellular matrix molecules then act back on the cell to give that region a basal character (see Figure 19–29C). In this way, the polarity of the cell is coupled to its orientation in the epithelial sheet and its relation to the basal lamina.

A Separate Signaling System Controls Planar Cell Polarity

Apico-basal polarity is a universal feature of epithelia, but the cells of some epithelia show an additional polarity at right angles to this axis: it is as if they had an arrow written on them, pointing in a specific direction in the plane of the epithelium. This type of polarity is called **planar cell polarity** (**Figure 19–32**A and B). In the wing of a fly, for example, each epithelial cell has a tiny asymmetrical projection, called a wing-hair, on its surface, and the hairs all point toward the tip of the wing. Similarly, in the inner ear of a vertebrate, each mechanosensory hair cell has an asymmetric bundle of stereocilia (actin-filled rod-like protrusions) sticking up from its apical surface: tilting the bundle in one direction causes ion channels to open, stimulating the cell electrically; tilting in the opposite direction has the contrary effect. For the ear to function correctly, the hair cells must be correctly oriented. Planar cell polarity is important also in the respiratory tract, for example, where every ciliated cell must orient its beating so as to sweep mucus up out of the lungs, and not down into them (see Chapter 23).

Screens for mutants with disorderly wing hairs in *Drosophila* have identified a set of genes that are critical for planar cell polarity in the fly. Some of these,

Figure 19–32 Planar cell polarity. (A) Wing hairs on the wing of a fly. Each cell in the wing epithelium forms one of these little spiky protrusions or "hairs" at its apex, and all the hairs point the same way, toward the tip of the wing. This reflects a planar polarity in the structure of each cell. (B) Sensory hair cells in the inner ear of a mouse similarly have a well-defined planar polarity, manifest in the oriented pattern of stereocilia (actin-filled protrusions) on their surface. The detection of sound depends on the correct, coordinated orientation of the hair cells. (C) A mutation in the gene *Flamingo* in the fly, coding for a non-classical cadherin, disrupts the pattern of planar cell polarity in the wing. (D) A mutation in a homologous *Flamingo* gene in the mouse randomizes the orientation of the planar cell polarity vector of the hair cells in the ear. The mutant mice are deaf. (A and C, from J. Chae et al., *Development* 126:5421–5429, 1999. With permission from The Company of Biologists; B and D, from J.A. Curtin et al., *Curr. Biol.* 13:1129–1133, 2003. With permission from Elsevier.)

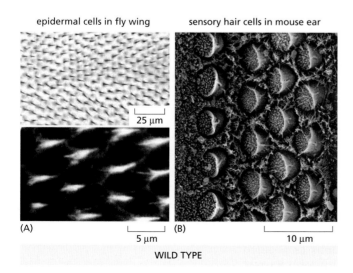

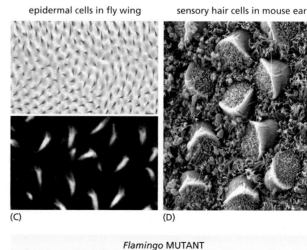

epidermal cells in fly wing   sensory hair cells in mouse ear   epidermal cells in fly wing   sensory hair cells in mouse ear

25 µm

(A)   5 µm   (B)   10 µm

(C)   (D)

WILD TYPE   *Flamingo* MUTANT

such as *Frizzled*, for example, and *Dishevelled*, code for proteins that have since been shown to be components of the Wnt signaling pathway (discussed in Chapter 15). Two others, *Flamingo* (see Figure 19–32C) and *Dachsous*, code for members of the cadherin superfamily. Still others are less easily classified functionally, but it is clear that planar cell polarity is organized by machinery formed from these components and assembled at cell–cell junctions in such a way that a polarizing influence can propagate from cell to cell. Essentially the same system of proteins controls planar cell polarity in vertebrates. Mice with mutations in a *Flamingo* homolog, for example, have incorrectly oriented hair cells in their ears (among other defects) and thus are deaf (see Figure 19–32D).

Summary

Occluding junctions—tight junctions in vertebrates, septate junctions in insects and molluscs—seal the gaps between cells in epithelia, creating a barrier to the diffusion of molecules across the cell sheet. They also form a bar to the diffusion of proteins in the plane of the membrane, and so help to maintain a difference between the populations of proteins in the apical and basolateral membrane domains of the epithelial cell. The major transmembrane proteins forming occluding junctions are called claudins; different members of the family are expressed in different tissues, conferring different permeability properties on the various epithelial sheets.

Intracellular scaffold proteins bind to the transmembrane components at occluding junctions and coordinate these junctions with cadherin-based anchoring junctions, so as to create junctional complexes. The junctional scaffold proteins have at least two other crucial functions. They play a part in the control of epithelial cell proliferation; and, in conjunction with other regulatory molecules such as Rac and Cdc42, they govern cell polarity. Epithelial cells have an intrinsic tendency to develop a polarized apico-basal axis. The orientation of this axis in relation to the cell's neighbors in an epithelial sheet depends on protein complexes involving scaffold proteins that assemble at cell–cell junctions, as well as on cytoskeletal polarization controlled by Rac/Cdc42 and on influences from the basal lamina.

The cells of some epithelia have an additional polarity in the plane of the epithelium, at right angles to the apico-basal axis. A separate set of conserved proteins, operating in a similar way in vertebrates and in insects, governs this planar cell polarity through poorly understood signaling processes that are likewise based on cell–cell junctions.

PASSAGEWAYS FROM CELL TO CELL: GAP JUNCTIONS AND PLASMODESMATA

Tight junctions block the passageways through the gaps between cells, preventing extracellular molecules from leaking from one side of an epithelium to the other. Another type of junctional structure has a radically different function: it bridges gaps between adjacent cells so as to create direct passageways from the cytoplasm of one into that of the other. These passageways take quite different forms in animal tissues, where they are called *gap junctions*, and in plants, where they are called *plasmodesmata* (singular *plasmodesma*). In both cases, however, the function is similar: the connections allow neighboring cells to exchange small molecules but not macromolecules (with some exceptions for plasmodesmata). Many of the implications of this cell coupling are only beginning to be understood.

Gap Junctions Couple Cells Both Electrically and Metabolically

Gap junctions are present in most animal tissues, including connective tissues as well as epithelia, allowing the cells to communicate with their neighbors. Each gap junction appears in conventional electron micrographs as a patch where the membranes of two adjacent cells are separated by a uniform narrow gap of about 2–4 nm. The gap is spanned by channel-forming proteins, of which there are two distinct families, called the *connexins* and the *innexins*. These are unrelated in sequence but similar in shape and function: in vertebrates, both families are present, but connexins predominate, with 21 members in humans. In *Drosophila* and *C. elegans*, only innexins are present, with 15 family members in the fly and 25 in the worm.

The channels formed by the gap-junction proteins allow inorganic ions and other small water-soluble molecules to pass directly from the cytoplasm of one cell to the cytoplasm of the other, thereby coupling the cells both electrically and metabolically. Thus, when a suitable dye is injected into one cell, it diffuses readily into the other, without escaping into the extracellular space. Similarly, an electric current injected into one cell through a microelectrode causes an almost instantaneous electrical disturbance in the neighboring cell, due to the flow of ions carrying electric charge through gap junctions. With microelectrodes inserted into both cells, one can easily monitor this effect and measure properties of the gap junctions, such as their electrical resistance and the ways in which the coupling changes as conditions change. In fact, some of the earliest evidence of gap-junctional communication came from electrophysiological studies that demonstrated this type of rapid, direct electrical coupling between some types of neurons. Similar methods were used to identify connexins as the proteins that mediate the gap-junctional communication: when connexin mRNA is injected into either frog oocytes or gap-junction-deficient cultured cells, channels with the properties expected of gap-junction channels can be demonstrated electrophysiologically where pairs of injected cells make contact.

From experiments with injected dye molecules of different sizes, it seems that the largest functional pore size for gap-junctional channels is about 1.5 nm. Thus, the coupled cells share their small molecules (such as inorganic ions, sugars, amino acids, nucleotides, vitamins, and the intracellular mediators cyclic AMP and inositol trisphosphate) but not their macromolecules (proteins, nucleic acids, and polysaccharides) (**Figure 19–33**).

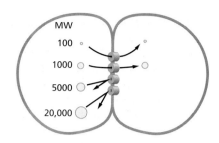

Figure 19–33 Determining the size of a gap-junction channel. When fluorescent molecules of various sizes are injected into one of two cells coupled by gap junctions, molecules with a mass of less than about 1000 daltons can pass into the other cell, but larger molecules cannot.

A Gap-Junction Connexon Is Made Up of Six Transmembrane Connexin Subunits

Connexins are four-pass transmembrane proteins, six of which assemble to form a *hemichannel*, or **connexon**. When the connexons in the plasma membranes of two cells in contact are aligned, they form a continuous aqueous channel that connects the two cell interiors (**Figure 19–34**A and **Figure 19–35**). A gap junction consists of many such connexon pairs in parallel, forming a sort of molecular sieve. The connexons hold the interacting plasma membranes a fixed distance apart—hence the gap.

Gap junctions in different tissues can have different properties because they are formed from different combinations of connexins, creating channels that differ in permeability. Most cell types express more than one type of connexin, and two different connexin proteins can assemble into a heteromeric connexon, with its own distinct properties. Moreover, adjacent cells expressing different

Figure 19–34 Gap junctions. (A) A three-dimensional drawing showing the interacting plasma membranes of two adjacent cells connected by gap junctions. Each lipid bilayer is shown as a pair of *red* sheets. Protein assemblies called connexons (*green*), each of which is formed by six connexin subunits, penetrate the apposed lipid bilayers (*red*). Two connexons join across the intercellular gap to form a continuous aqueous channel connecting the two cells. (B) The organization of connexins into connexons and connexons into intercellular channels. The connexons can be homomeric or heteromeric, and the intercellular channels can be homotypic or heterotypic.

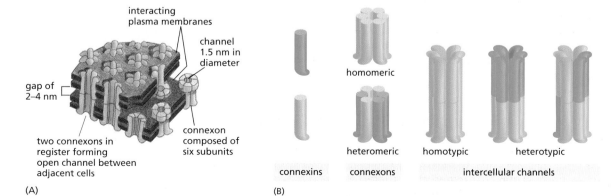

(A) interacting plasma membranes

gap of 2–4 nm

two connexons in register forming open channel between adjacent cells

channel 1.5 nm in diameter

connexon composed of six subunits

(B) homomeric · heteromeric · connexins · connexons · homotypic · heterotypic · intercellular channels

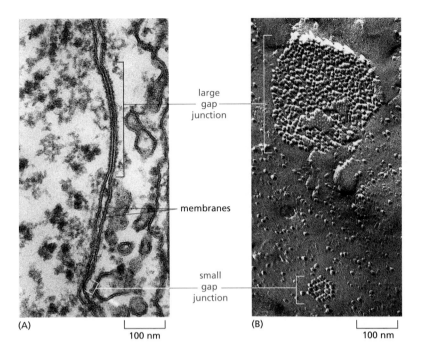

(A)   (B)

L__100 nm__J   L__100 nm__J

Figure 19–35 Gap junctions as seen in the electron microscope. (A) Thin-section and (B) freeze-fracture electron micrographs of a large and a small gap junction between fibroblasts in culture. In (B), each gap junction is seen as a cluster of homogeneous intramembrane particles. Each intramembrane particle corresponds to a connexon. (From N.B. Gilula, in Cell Communication [R.P. Cox, ed.], pp. 1–29. New York: Wiley, 1974.)

connexins can form intercellular channels in which the two aligned half-channels are different (Figure 19–34B).

Each gap-junctional plaque is a dynamic structure that can readily assemble, disassemble, or be remodelled, and it can contain a cluster of a few to many thousands of connexons (see Figure 19–35B). Studies with fluorescently labeled connexins in living cells show that new connexons are continually added around the periphery of an existing junctional plaque, while old connexons are removed from the middle of it and destroyed (**Figure 19–36**). This turnover is rapid: the connexin molecules have a half-life of a few hours.

The mechanism of removal of old connexons from the middle of the plaque is not known, but the route of delivery of new connexons to its periphery seems clear: they are inserted into the plasma membrane by exocytosis, like other integral membrane proteins, and then diffuse in the plane of the membrane until they bump into the periphery of a plaque and become trapped. This has a corollary: the plasma membrane away from the gap junction should contain connexons—hemichannels—that have not yet paired with their counterparts on another cell. It is thought that these unpaired hemichannels are normally held

Figure 19–36 Connexin turnover at a gap junction. Cells were transfected with a slightly modified connexin gene, coding for a connexin with a short amino-acid tag containing four cysteines in the sequence ...Cys-Cys-X-X-Cys-Cys (where X denotes an arbitrary amino acid). This *tetracysteine tag* can bind strongly, and in effect irreversibly, to certain small fluorescent dye molecules that can be added to the culture medium and will readily enter cells by diffusing across the plasma membrane. In the experiment shown, a green dye was added first, and the cells were then washed and incubated for 4 or 8 hours. At the end of this time, a red dye was added to the medium and the cells were washed again and fixed. Connexin molecules already present at the beginning of the experiment are labeled green (and take up no red dye because their tetracysteine tags are already saturated with green dye), while connexins synthesized subsequently, during the 4- or 8-hour incubation, are labeled red. The fluorescence images show optical sections of gap junctions between pairs of cells prepared in this way. The central part of the gap-junction plaque is *green*, indicating that it consists of old connexin molecules, while the periphery is *red*, indicating that it consists of connexins synthesized during the past 4 or 8 hours. The longer the time of incubation, the smaller the green central patch of old molecules, and the larger the peripheral ring of new molecules that have been recruited to replace them. (From G. Gaietta et al., *Science* 296:503–507, 2002. With permission from AAAS.)

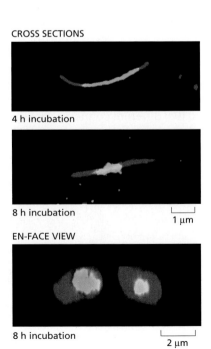

CROSS SECTIONS

4 h incubation

8 h incubation   L__1 μm__J

EN-FACE VIEW

8 h incubation   L__2 μm__J

in a closed conformation, preventing the cell from losing its small molecules by leakage through them. But there is also evidence that in some physiological circumstances they can open and serve as channels for the release of small molecules, such as the neurotransmitter glutamate, to the exterior, or for the entry of small molecules into the cell.

Gap Junctions Have Diverse Functions

In tissues containing electrically excitable cells, cell–cell coupling via gap junctions serves an obvious purpose. Some nerve cells, for example, are electrically coupled, allowing action potentials to spread rapidly from cell to cell, without the delay that occurs at chemical synapses. This is advantageous when speed and reliability are crucial, as in certain escape responses in fish and insects, or where a set of neurons need to act in synchrony. Similarly, in vertebrates, electrical coupling through gap junctions synchronizes the contractions of heart muscle cells as well as those of the smooth muscle cells responsible for the peristaltic movements of the intestine.

Gap junctions also occur in many tissues whose cells are not electrically excitable. In principle, the sharing of small metabolites and ions provides a mechanism for coordinating the activities of individual cells in such tissues and for smoothing out random fluctuations in small-molecule concentrations in different cells. Gap junctions are required in the liver, for example, to coordinate the response of the liver cells to signals from nerve terminals that contact only a part of the cell population (see Figure 15–7). The normal development of ovarian follicles also depends on gap-junction-mediated communication—in this case, between the oocyte and the surrounding granulosa cells. A mutation in the gene that encodes the connexin that normally couples these two cell types causes infertility.

Mutations in connexins, especially connexin-26, are the commonest of all genetic causes of congenital deafness: they result in the death of cells in the organ of Corti, probably because they disrupt functionally important pathways for the flow of ions from cell to cell in this electrically active sensory epithelium. Connexin mutations are responsible for many other disorders besides deafness, ranging from cataracts in the lens of the eye to a form of demyelinating disease in peripheral nerves.

Cell coupling via gap junctions also seems to play a part in embryogenesis. In early vertebrate embryos (beginning with the late eight-cell stage in mouse embryos), most cells are electrically coupled to one another. As specific groups of cells in the embryo develop their distinct identities and begin to differentiate, they commonly uncouple from surrounding tissue. As the neural plate folds up and pinches off to form the neural tube, for instance (see Figure 19–16), its cells uncouple from the overlying ectoderm. Meanwhile, the cells within each group remain coupled with one another and therefore tend to behave as a cooperative assembly, all following a similar developmental pathway in a coordinated fashion.

Cells Can Regulate the Permeability of Their Gap Junctions

Like conventional ion channels (discussed in Chapter 11), individual gap-junction channels do not remain continuously open; instead, they flip between open and closed states. Moreover, the permeability of gap junctions is rapidly (within seconds) and reversibly reduced by experimental manipulations that decrease the cytosolic pH or increase the cytosolic concentration of free Ca^{2+} to very high levels.

The purpose of the pH regulation of gap-junction permeability is unknown. In one case, however, the purpose of Ca^{2+} control seems clear. When a cell is damaged, its plasma membrane can become leaky. Ions present at high concentration in the extracellular fluid, such as Ca^{2+} and Na^+, then move into the cell, and valuable metabolites leak out. If the cell were to remain coupled to its

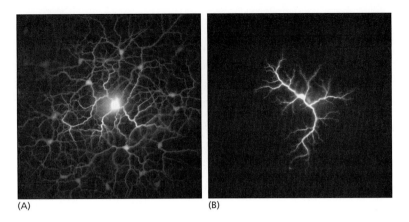

(A)   (B)

Figure 19–37 The regulation of gap-junction coupling by a neurotransmitter. (A) A neuron in a rabbit retina was injected with the dye Lucifer yellow, which passes readily through gap junctions and labels other neurons of the same type that are connected to the injected cell by gap junctions. (B) The retina was first treated with the neurotransmitter dopamine, before the neuron was injected with dye. As can be seen, the dopamine treatment greatly decreased the permeability of the gap junctions. Dopamine acts by increasing intracellular cyclic AMP levels. (Courtesy of David Vaney.)

healthy neighbors, these too would suffer a dangerous disturbance of their internal chemistry. But the large influx of Ca^{2+} into the damaged cell causes its gap-junction channels to close immediately, effectively isolating the cell and preventing the damage from spreading to other cells.

Gap-junction communication can also be regulated by extracellular signals. The neurotransmitter *dopamine*, for example, reduces gap-junction communication between a class of neurons in the retina in response to an increase in light intensity (**Figure 19–37**). This reduction in gap-junction permeability helps the retina switch from using rod photoreceptors, which are good detectors of low light, to cone photoreceptors, which detect color and fine detail in bright light.

In Plants, Plasmodesmata Perform Many of the Same Functions as Gap Junctions

The tissues of a plant are organized on different principles from those of an animal. This is because plant cells are imprisoned within tough *cell walls* composed of an extracellular matrix rich in cellulose and other polysacharides, as we discuss later. The cell walls of adjacent cells are firmly cemented to those of their neighbors, which eliminates the need for anchoring junctions to hold the cells in place. But a need for direct cell–cell communication remains. Thus, plant cells have only one class of intercellular junctions, **plasmodesmata**. Like gap junctions, they directly connect the cytoplasms of adjacent cells.

In plants, the cell wall between a typical pair of adjacent cells is at least 0.1 μm thick, and so a structure very different from a gap junction is required to mediate communication across it. Plasmodesmata solve the problem. With a few specialized exceptions, every living cell in a higher plant is connected to its living neighbors by these structures, which form fine cytoplasmic channels through the intervening cell walls. As shown in **Figure 19–38**A, the plasma membrane of one cell is continuous with that of its neighbor at each plasmodesma, which connects the cytoplasms of the two cells by a roughly cylindrical channel with a diameter of 20–40 nm.

Running through the center of the channel in most plasmodesmata is a narrower cylindrical structure, the *desmotubule*, which is continuous with elements of the smooth endoplasmic reticulum in each of the connected cells (Figure 19–38B–D). Between the outside of the desmotubule and the inner face of the cylindrical channel formed by plasma membrane is an annulus of cytosol through which small molecules can pass from cell to cell. As each new cell wall is assembled during the cytokinesis phase of cell division, plasmodesmata are created within it. They form around elements of smooth ER that become trapped across the developing cell plate (discussed in Chapter 17). They can also be inserted *de novo* through preexisting cell walls, where they are commonly found in dense clusters called *pit fields*. When no longer required, plasmodesmata can be readily removed.

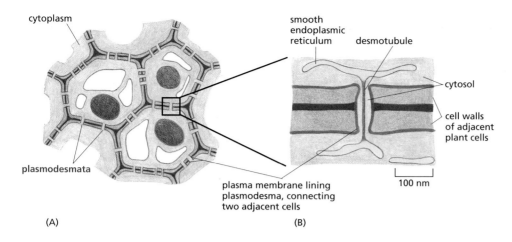

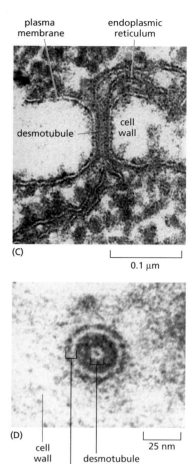

Figure 19–38 Plasmodesmata. (A) The cytoplasmic channels of plasmodesmata pierce the plant cell wall and connect all cells in a plant together. (B) Each plasmodesma is lined with plasma membrane that is common to two connected cells. It usually also contains a fine tubular structure, the desmotubule, derived from smooth endoplasmic reticulum. (C) Electron micrograph of a longitudinal section of a plasmodesma from a water fern. The plasma membrane lines the pore and is continuous from one cell to the next. Endoplasmic reticulum and its association with the central desmotubule can also be seen. (D) A similar plasmodesma seen in cross section. (C and D, from R. Overall, J. Wolfe and B.E.S. Gunning, in Protoplasma 9, pp. 137 and 140. Heidelberg: Springer-Verlag, 1982.)

In spite of the radical difference in structure between plasmodesmata and gap junctions, they seem to function in remarkably similar ways. Evidence obtained by injecting tracer molecules of different sizes suggests that plasmodesmata allow the passage of molecules with a molecular weight of less than about 800, which is similar to the molecular-weight cutoff for gap junctions. As with gap junctions, transport through plasmodesmata is regulated. Dye-injection experiments, for example, show that there can be barriers to the movement of even low-molecular-weight molecules between certain cells, or groups of cells, that are connected by apparently normal plasmodesmata; the mechanisms that restrict communication in these cases are not understood.

During plant development, groups of cells within the shoot and root meristems signal to one another in the process of defining their future fates (discussed in Chapter 22). Some gene regulatory proteins involved in this process of cell fate determination pass from cell to cell through plasmodesmata. They bind to components of the plasmodesmata and override the size exclusion mechanism that would otherwise prevent their passage. In some cases, the mRNA that encodes the protein can also pass through. Some plant viruses also exploit this route: infectious viral RNA, or even intact virus particles, can pass from cell to cell in this way. These viruses produce proteins that bind to components of the plasmodesmata in ways that increase the effective pore size of the channel dramatically. The functional components of plasmodesmata that mediate this effect are unknown, and it is still a mystery how endogenous or viral macromolecules regulate the transport properties of the channel so as to pass through it.

Summary

The cells of many animal tissues, both epithelial and nonepithelial, are coupled by channel-forming junctions called gap junctions. These take the form of plaques of clustered connexons that usually allow molecules smaller than about 1000 daltons to pass directly from the inside of one cell to the inside of the next. The cells can regulate the permeability of these junctions. Gap junctions are dynamic structures: new connexons are continually recruited to the periphery of the plaque, while old connexons are continually removed from its center.

Cells connected by gap junctions share many of their inorganic ions and other small molecules and are therefore chemically and electrically coupled. Gap junctions are important in coordinating the activities of electrically active cells, and they have a coordinating role in other groups of cells as well. In plants, cells are linked by communicating junctions called plasmodesmata. Although their structure is entirely different from that of gap junctions, and they can sometimes transport informational macromolecules, they function remarkably like gap junctions in permitting small molecules to pass from cell to cell while blocking the passage of most large molecules.

THE BASAL LAMINA

Tissues are not made up solely of cells. A part of their volume—sometimes a very large part—is *extracellular space*, which is occupied by an intricate network of macromolecules constituting the *extracellular matrix*. This matrix is composed of various proteins and polysaccharides that are secreted locally and assembled into an organized meshwork in close association with the surfaces of the cells that produced them.

In our own bodies, the most plentiful forms of extracellular matrix are found in bulky connective tissues such as bone, tendon, and the dermal layer of the skin. For animals in general, however, from an evolutionary point of view, pride of place goes to the extracellular matrix that forms a much less obvious structure—the **basal lamina** (also referred to as the **basement membrane**). This exceedingly thin, tough, flexible sheet of matrix molecules is an essential underpinning of all epithelia. Small as it is in volume, it has a critical role in the architecture of the body. Like the cadherins, it seems to be one of the defining features common to all multicellular animals. Other forms of extracellular matrix are more variable from one animal phylum to another, in both composition and quantity.

In this section, we discuss the basal lamina itself. In the next section we shall consider how epithelial cells and basal lamina interact with one another, through *integrin proteins* in the epithelial cell membranes, and we shall see that integrins are also present in other cell types, mediating their interactions with the varied types of extracellular matrix found in connective tissues. These other forms of extracellular matrix will be discussed in detail later.

Basal Laminae Underlie All Epithelia and Surround Some Nonepithelial Cell Types

Basal laminae are typically 40–120 nm thick. A sheet of basal lamina not only underlies all epithelia but also surrounds individual muscle cells, fat cells, and Schwann cells (which wrap around peripheral nerve cell axons to form myelin). The basal lamina thus separates these cells and epithelia from the underlying or surrounding connective tissue and forms the mechanical connection between them. In other locations, such as the kidney glomerulus, a basal lamina lies between two cell sheets and functions as a selective filter (**Figure 19–39**). Basal laminae have more than simple structural and filtering roles, however. They are able to determine cell polarity, influence cell metabolism, organize the proteins

Figure 19–39 Three ways in which basal laminae are organized. Basal laminae *(yellow)* surround certain cells (such as skeletal muscle cells), underlie epithelia, and are interposed between two cell sheets (as in the kidney glomerulus). Note that, in the kidney glomerulus, both cell sheets have gaps in them, and the basal lamina has a filtering as well as a supportive function, helping to determine which molecules will pass into the urine from the blood. The filtration also depends on other protein-based structures, called *slit diaphragms,* that span the intercellular gaps in the epithelial sheet.

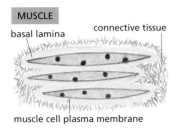

MUSCLE
basal lamina   connective tissue

muscle cell plasma membrane

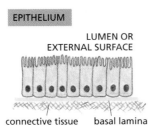

EPITHELIUM

LUMEN OR
EXTERNAL SURFACE

connective tissue   basal lamina

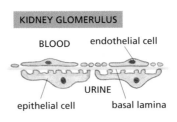

KIDNEY GLOMERULUS

BLOOD   endothelial cell

URINE

epithelial cell   basal lamina

in adjacent plasma membranes, promote cell survival, proliferation, or differentiation, and serve as highways for cell migration.

The mechanical role is nevertheless essential. In the skin, for example, the epithelial outer layer—the epidermis—depends on the strength of the basal lamina to keep it attached to the underlying connective tissue—the dermis. In people with genetic defects in certain basal lamina proteins or in a special type of collagen that anchors the basal lamina to the underlying connective tissue, the epidermis becomes detached from the dermis. This causes a blistering disease called *junctional epidermolysis bullosa,* a severe and sometimes lethal condition.

Laminin Is a Primary Component of the Basal Lamina

The basal lamina is synthesized by the cells on each side of it: the epithelial cells contribute one set of basal lamina components, while cells of the underlying bed of connective tissue (called the *stroma,* Greek for "bedding") contribute another set (**Figure 19–40**). Like other extracellular matrices in animal tissues, the basal lamina consists of two main classes of extracellular macromolecules: (1) fibrous proteins (usually glycoproteins, which have short oligosaccharide side chains attached) and (2) polysaccharide chains of the type called *glycosaminoglycans (GAGs),* which are usually found covalently linked to specific *core proteins* to form *proteoglycans* (**Figure 19–41**). In a later section, we shall discuss these two large and varied classes of matrix molecules in greater detail. We introduce them here through the special subset that are found in basal laminae.

Although the precise composition of the mature basal lamina varies from tissue to tissue and even from region to region in the same lamina, it typically contains the glycoproteins *laminin, type IV collagen,* and *nidogen,* along with the proteoglycan *perlecan.* Together with these key components, present in the basal laminae of virtually all animals from jellyfish to mammals, it holds in its meshes, or is closely associated with, various other molecules. These include *collagen XVIII* (an atypical member of the collagen family, forming the core protein of a proteoglycan) and *fibronectin,* a fibrous protein important in the adhesion of connective-tissue cells to matrix.

The laminin is thought to be the primary organizer of the sheet structure, and early in development, basal laminae consist mainly of laminin molecules. **Laminin-1** (classical laminin) is a large, flexible protein composed of three very

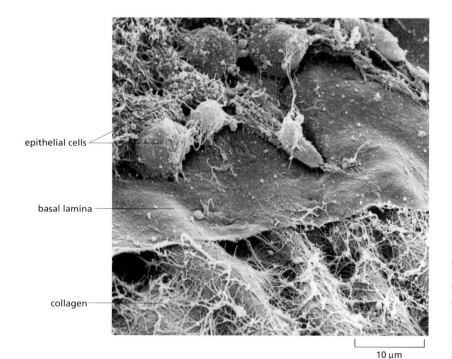

epithelial cells

basal lamina

collagen

10 μm

Figure 19–40 **The basal lamina in the cornea of a chick embryo.** In this scanning electron micrograph, some of the epithelial cells have been removed to expose the upper surface of the matlike basal lamina. A network of collagen fibrils in the underlying connective tissue interacts with the lower face of the lamina. (Courtesy of Robert Trelstad.)

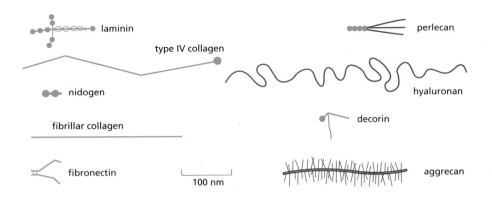

Figure 19–41 The comparative shapes and sizes of some of the major extracellular matrix macromolecules. Protein is shown in *green*, and glycosaminoglycan in *red*.

long polypeptide chains (α, β, and γ) held together by disulfide bonds and arranged in the shape of an asymmetric bouquet, like a bunch of three flowers whose stems are twisted together at the foot but whose heads remain separate (**Figure 19–42**). These heterotrimers can self-assemble *in vitro* into a network, largely through interactions between their heads, although interaction with cells is needed to organize the the network into an orderly sheet. Since there are several isoforms of each type of chain, and these can associate in different combinations, many different laminins can be produced, creating basal laminae with distinctive properties. The laminin γ-1 chain is, however, a component of most laminin heterotrimers; mice lacking it die during embryogenesis because they are unable to make basal lamina.

Type IV Collagen Gives the Basal Lamina Tensile Strength

Type IV collagen is a second essential component of mature basal laminae, and it, too, exists in several isoforms. Like the *fibrillar collagens* that constitute the bulk of the protein in connective tissues such as bone or tendon (discussed later), type IV collagen molecules consist of three separately synthesized long protein chains that twist together to form a ropelike superhelix; but they differ from the fibrillar collagens in that the triple-stranded helical structure is interrupted in more than 20 regions, allowing multiple bends. The type IV collagen molecules interact via their terminal domains to assemble extracellularly into a flexible, felt-like network. In this way, type IV collagen gives the basal lamina tensile strength.

But how do the networks of laminin and type IV collagen bond to one another and to the surfaces of the cells that sit on the basal lamina? Why do they form a two-dimensional sheet, rather than a three-dimensional gel? The molecules of laminin have several functional domains, including one that binds to the perlecan proteoglycan, one that binds to the nidogen protein, and two or more that bind to laminin receptor proteins on the surface of cells. Type IV collagen also has domains that bind nidogen and perlecan. It is thought, therefore,

Figure 19–42 The structure of laminin. (A) The subunits of a laminin-1 molecule, and some of their binding sites for other molecules *(yellow boxes)*. Laminin is a multidomain glycoprotein composed of three polypeptides (α, β, and γ) that are disulfide-bonded into an asymmetric crosslike structure. Each of the polypeptide chains is more than 1500 amino acids long. Five types of α chains, three types of β chains, and three types of γ chains are known; in principle, they can assemble to form 45 (5 × 3 × 3) laminin isoforms. Several such isoforms have been found, each with a characteristic tissue distribution. Through their binding sites for other proteins, laminin molecules play a central part in organizing the assembly of basal laminas and anchoring them to cells. (B) Electron micrographs of laminin molecules shadowed with platinum. (B, from J. Engel et al., *J. Mol. Biol.* 150:97–120, 1981. With permission from Academic Press.)

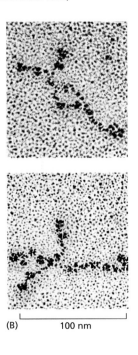

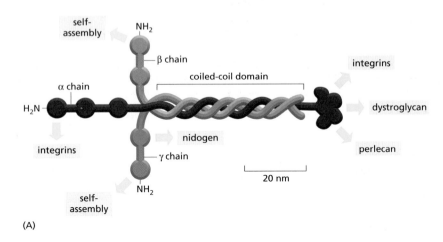

(A)

(B)

(A)

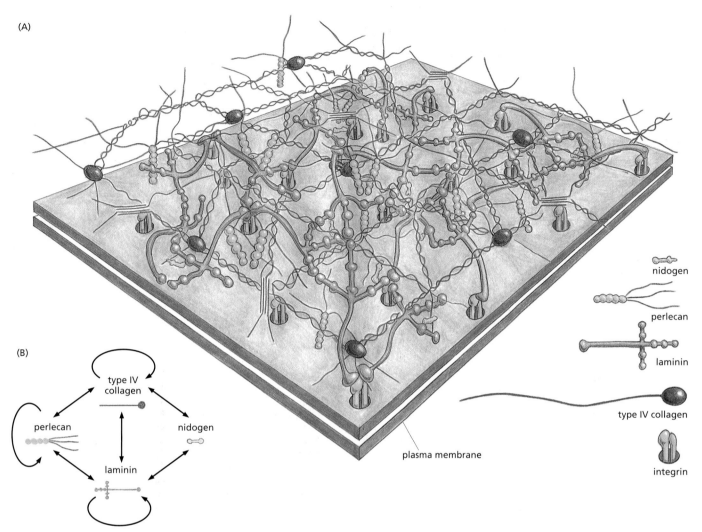

(B)

type IV collagen

perlecan

nidogen

laminin

plasma membrane

nidogen

perlecan

laminin

type IV collagen

integrin

Figure 19–43 A model of the molecular structure of a basal lamina. (A) The basal lamina is formed by specific interactions (B) between the proteins laminin, type IV collagen, and nidogen, and the proteoglycan perlecan. *Arrows* in (B) connect molecules that can bind directly to each other. There are various isoforms of type IV collagen and laminin, each with a distinctive tissue distribution. Transmembrane laminin receptors (integrins and dystroglycan) in the plasma membrane are thought to organize the assembly of the basal lamina; only the integrins are shown. (Based on H. Colognato and P.D. Yurchenco, *Dev. Dyn.* 218:213–234, 2000. With permission from Wiley-Liss.)

that nidogen and perlecan serve as linkers to connect the laminin and type IV collagen networks once the laminin is in place (**Figure 19–43**).

The laminin molecules that generate the initial sheet structure first join to each other while bound to receptors on the surface of the cells that produce them. The cell-surface receptors are of several sorts. Many of them are members of the integrin family; another important type of laminin receptor is *dystroglycan*, a proteoglycan with a core protein that spans the cell membrane, dangling its glycosaminoglycan polysaccharide chains in the extracellular space. Together, these receptors organize basal lamina assembly: they hold the laminin molecules by their feet, leaving the laminin heads positioned to interact so as to form a two-dimensional network. This laminin network then presumably coordinates the assembly of the other basal lamina components.

Basal Laminae Have Diverse Functions

As we have mentioned, in the kidney glomerulus, an unusually thick basal lamina acts as one of the layers of a molecular filter, helping to prevent the passage of macromolecules from the blood into the urine as urine is formed (see Figure 19–39). The proteoglycan in the basal lamina seems to be important for this function: when its GAG chains are removed by specific enzymes, the filtering properties of the lamina are destroyed. Type IV collagen also has a role: in a human hereditary kidney disorder *(Alport syndrome)*, mutations in type IV collagen genes result in an irregularly thickened and dysfunctional glomerular filter. Laminin mutations, too, can disrupt the function of the kidney filter, but in a different way—by interfering with the differentiation of the cells that contact it and support it.

The basal lamina can act as a selective barrier to the movement of cells, as well as a filter for molecules. The lamina beneath an epithelium, for example, usually prevents fibroblasts in the underlying connective tissue from making contact with the epithelial cells. It does not, however, stop macrophages, lymphocytes, or nerve processes from passing through it, using specialized protease enzymes to cut a hole for their transit. The basal lamina is also important in tissue regeneration after injury. When cells in tissues such as muscles, nerves, and epithelia are damaged or killed, the basal lamina often survives and provides a scaffold along which regenerating cells can migrate. In this way, the original tissue architecture is readily reconstructed.

A particularly striking example of the role of the basal lamina in regeneration comes from studies on the *neuromuscular junction*, the site where the nerve terminals of a motor neuron form a chemical synapse with a skeletal muscle cell (discussed in Chapter 11). In vertebrates, the basal lamina that surrounds the muscle cell separates the nerve and muscle cell plasma membranes at the synapse, and the synaptic region of the lamina has a distinctive chemical character, with special isoforms of type IV collagen and laminin and a proteoglycan called *agrin*.

This basal lamina at the synapse has a central role in reconstructing the synapse after nerve or muscle injury. If a frog muscle and its motor nerve are destroyed, the basal lamina around each muscle cell remains intact and the sites of the old neuromuscular junctions are still recognizable. If the motor nerve, but not the muscle, is allowed to regenerate, the nerve axons seek out the original synaptic sites on the empty basal lamina and differentiate there to form normal-looking nerve terminals. Thus, the junctional basal lamina by itself can guide the regeneration of motor nerve terminals.

Similar experiments show that the basal lamina also controls the localization of the acetylcholine receptors that cluster in the muscle cell plasma membrane at a neuromuscular junction. If the muscle and nerve are both destroyed, but now the muscle is allowed to regenerate while the nerve is prevented from doing so, the acetylcholine receptors synthesized by the regenerated muscle localize predominantly in the region of the old junctions, even though the nerve is absent (**Figure 19–44**). Thus, the junctional basal lamina apparently coordinates the local spatial organization of the components in each of the two cells that form a neuromuscular junction. Some of the molecules responsible for these effects have been identified. Motor neuron axons, for example, deposit agrin in the junctional basal lamina, where it regulates the assembly of acetyl-

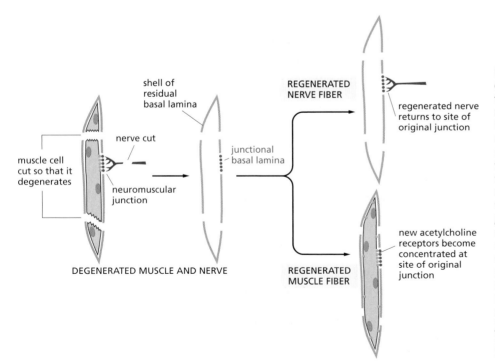

Figure 19–44 Regeneration experiments demonstrating the special character of the junctional basal lamina at a neuromuscular junction. When the nerve, but not the muscle, is allowed to regenerate after both the nerve and muscle have been damaged (*upper part* of figure), the junctional basal lamina directs the regenerating nerve to the original synaptic site. When the muscle, but not the nerve, is allowed to regenerate (*lower part* of figure), the junctional basal lamina causes newly made acetylcholine receptors (*blue*) to accumulate at the original synaptic site. The muscle regenerates from satellite cells (discussed in Chapter 23) located between the basal lamina and the original muscle cell (not shown). These experiments show that the junctional basal lamina controls the localization of synaptic components on both sides of the lamina.

choline receptors and other proteins in the junctional plasma membrane of the muscle cell. Reciprocally, muscle cells deposit a particular isoform of laminin in the junctional basal lamina, and some evidence suggests that this binds directly to the extracellular domain of voltage-gated Ca^{2+} channels in the presynaptic membrane of the nerve cell, helping to hold them at the synapse where they are needed. Both agrin and the synaptic isoform of laminin are essential for the formation of normal neuromuscular junctions. Defects in components of the basal lamina or in proteins that tether muscle cell components to it at the synapse are responsible for many of the forms of muscular dystrophy, in which muscles at first develop normally but then degenerate in later years of life.

Summary

The basal lamina is a thin tough sheet of extracellular matrix that closely underlies epithelia in all multicellular animals. It also wraps around certain other cell types, such as muscle cells. All basal laminae are organized on a framework of laminin molecules, linked together by their side-arms and held close beneath the basal ends of the epithelial cells by attachment to integrins and other receptors in the basal plasma membrane. Type IV collagen molecules are recruited into this structure, assembling into a sheetlike mesh that is an essential component of all mature basal laminae. The collagen and laminin networks in mature basal laminae are bridged by the protein nidogen and the large heparan sulfate proteoglycan perlecan.

Basal laminae provide mechanical support for epithelia; they form the interface and the attachment between epithelia and connective tissue; they serve as filters in the kidney; they act as barriers to keep cells in their proper compartments; they influence cell polarity and cell differentiation; they guide cell migrations; and molecules embedded in them help to organize elaborate structures such as neuromuscular synapses. When cells are damaged or killed, basal laminae often survive and can help guide tissue regeneration.

INTEGRINS AND CELL–MATRIX ADHESION

Cells make extracellular matrix, organize it, and degrade it. The matrix in its turn exerts powerful influences on the cells. The influences are exerted chiefly through transmembrane cell adhesion proteins that act as *matrix receptors*. These tie the matrix outside the cell to the cytoskeleton inside it, but their role goes far beyond simple passive mechanical attachment. Through them, components of the matrix can affect almost any aspect of a cell's behavior. The matrix receptors have a crucial role in epithelial cells, mediating their interactions with the basal lamina beneath them; and they are no less important in connective-tissue cells, for their interactions with the matrix that surrounds them.

Several types of molecules can function as matrix receptors or co-receptors, including the transmembrane proteoglycans. But the principal receptors on animal cells for binding most extracellular matrix proteins are the **integrins**. Like the cadherins and the key components of the basal lamina, integrins are part of the fundamental architectural toolkit that is characteristic of multicellular animals. The members of this large family of homologous transmembrane adhesion molecules have a remarkable ability to transmit signals in both directions across the cell membrane. The binding of a matrix component to an integrin can send a message into the interior of the cell, and conditions in the cell interior can send a signal outward to control binding of the integrin to matrix (or, in some cases, to a cell-surface molecule on another cell, as we saw in the case of white blood cells binding to endothelial cells). Tension applied to an integrin can cause it to tighten its grip on intracellular and extracellular structures, and loss of tension can loosen its hold, so that molecular signaling complexes fall apart on either side of the membrane. In this way, integrins can also serve not only to transmit mechanical and molecular signals, but also to convert the one type of signal into the other. Studies of the structure of integrin molecules have begun to reveal how they perform these tasks.

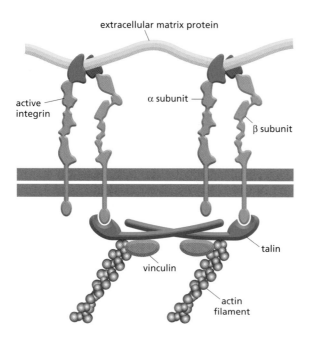

Figure 19–45 The subunit structure of an active integrin molecule, linking extracellular matrix to the actin cytoskeleton. The head of the integrin molecule attaches directly to an extracellular protein such as fibronectin; the intracellular tail of the integrin binds to talin, which in turn binds to filamentous actin. A set of other intracellular anchor proteins, including α-actinin, filamin, and vinculin, help to reinforce the linkage.

Integrins Are Transmembrane Heterodimers That Link to the Cytoskeleton

There are many varieties of integrins—at least 24 in humans—but they all conform to a common plan. An integrin molecule is composed of two noncovalently associated glycoprotein subunits called α and β. Both subunits span the cell membrane, with short intracellular C-terminal tails and large N-terminal extracellular domains. The extracellular portion of the integrin dimer binds to specific amino acid sequences in extracellular matrix proteins such as laminin or fibronectin or, in some cases, to ligands on the surfaces of other cells. The intracellular portion binds to a complex of proteins that form a linkage to the cytoskeleton.

For all but one of the 24 varieties of human integrins, this intracellular linkage is to actin filaments, via *talin* and a set of other intracellular anchorage proteins (**Figure 19–45**); talin, as we shall see later, seems to be the key component of the linkage. Like the actin-linked cell–cell junctions formed by cadherins, the actin-linked cell–matrix junctions formed by integrins may be small, inconspicuous and transient, or large, prominent, and durable. Examples of the latter are the *focal adhesions* that form when fibroblasts have sufficient time to form strong attachments to the rigid surface of a culture dish, and the *myotendinous junctions* that attach muscle cells to their tendons.

In epithelia, the most prominent cell–matrix attachment sites are the hemidesmosomes, where a specific type of integrin (α6β4) anchors the cells to laminin in the basal lamina. Here, uniquely, the intracellular attachment is to keratin filaments, via the intracellular anchor proteins plectin and dystonin (**Figure 19–46**).

Integrins Can Switch Between an Active and an Inactive Conformation

A cell crawling through a tissue—a fibroblast or a macrophage, for example, or an epithelial cell migrating along a basal lamina—has to be able both to make and to break attachments to the matrix, and to do so rapidly if it is to travel quickly. <TGAT> Similarly, a circulating white blood cell has to be able to switch on or off its tendency to bind to endothelial cells in order to crawl out of a blood vessel at a site of inflammation under the appropriate circumstances. Furthermore, if

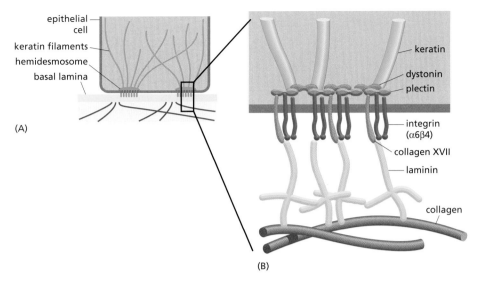

(A)

(B)

keratin
dystonin
plectin
integrin (α6β4)
collagen XVII
laminin
collagen

epithelial cell
keratin filaments
hemidesmosome
basal lamina

Figure 19–46 Hemidesmosomes. (A) Hemidesmosomes spot-weld epithelial cells to the basal lamina, linking laminin outside the cell to keratin filaments inside it. (B) Molecular components of a hemisdesmosome. A specialized integrin (α6β4 integrin) spans the membrane, attaching to keratin filaments intracellularly via anchor proteins called plectin and dystonin, and to laminin extracellularly. The adhesive complex also contains, in parallel with the integrin, an unusual collagen family member, collagen type XVII; this has a membrane-spanning domain attached to its extracellular collagenous portion. Defects in any of these components can give rise to a blistering disease of the skin.

force is to be applied where it is needed, the making and breaking of the extracellular attachments in all these cases has to be coupled to the prompt assembly and disassembly of cytoskeletal attachments inside the cell. The integrin molecules that span the membrane and mediate the attachments cannot simply be passive, rigid objects with sticky patches at their two ends. They must be able to switch between an active state, where they readily form attachments, and an inactive state, where they do not; and the binding of their ligands on one side of the membrane must alter their propensity to bind to a different set of ligands on the opposite side.

The basis for these dynamic phenomena is allosteric regulation: as an integrin binds to or detaches from its ligands, it undergoes conformational changes that affect both the intracellular and the extracellular ends of the molecule. Structural change at one end is coupled to structural change at the other, so that influences can be transmitted in either direction across the cell membrane. The timber tongs that lumberjacks use to grab hold of logs of wood provide a simple mechanical analogy (**Figure 19–47**).

The structural changes in integrins can be demonstrated by taking a purified preparation of integrin molecules and examining them at high resolution by electron microscopy. If the integrins are kept in a calcium-rich medium similar to normal extracellular fluid, but without any extracellular ligand, and then rapidly prepared for microscopy, they appear as tightly folded V-shaped objects. But if a small synthetic peptide containing a sequence that mimics the integrin-binding domain of a natural extracellular matrix protein is added to the medium, the integrins bind this molecule and extend into a different shape, with two legs that are no longer tightly bent, but are now straightened and separated from each other, supporting a head region high above them (**Figure 19–48A**). This pair of structures can be compared with more detailed data from x-ray crystallography, which reveals that the two legs correspond to the integrin α and β chains. The head region, where they meet, contains the binding site for the extracellular ligand. Binding of the ligand distorts this region so as to favor adoption of the extended, "active" conformation; conversely, adoption of the extended conformation creates a more favorable binding site, with a higher affinity for ligand (Figure 19–48B).

But how do these changes in the extracellular region of the integrin relate to events at the intracellular end of the integrin molecule? In its folded, inactive

Figure 19–47 Timber tongs. Holding the handles together causes the claws to grip the log; and closing the claws on the log causes the handles to come together. Moreover, the greater the pull on the tongs, the tighter the grip at both ends. In an integrin molecule, the details of the linkage are different, but the mechanical principles are similar: conformational changes at opposite ends of the molecule are coupled, and pulling tightens the grip.

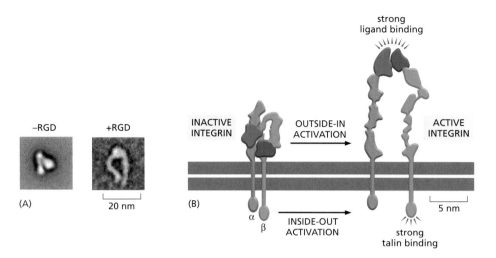

Figure 19–48 Change in conformation of an integrin molecule when it binds its ligand. (A) Images were produced by averaging many similarly aligned electron micrographs of individual integrin molecules. In the absence of extracellular ligand, the integrin molecules appear small and tightly folded. When incubated with an RGD peptide, the integrins unfold into an extended structure with two distinct legs. (B) Active (extended) and inactive (folded) structures of an integrin molecule, based on data from x-ray crystallography. Although it is difficult to crystallize the intact molecule in its natural conformations, with and without ligand bound, the complete structure can be inferred with reasonable confidence from x-ray crystallography of defined molecular fragments. (A, From J. Takagi et al., *Cell* 110:599–611, 2002. With permission from Elsevier; B, based on T. Xiao at al., *Nature* 432:59–67, 2004. With permission from Macmillan Publishers Ltd.)

state, the intracellular portions of its α and β chains lie close together and adhere to one another. When the extracellular domain unfolds, this contact is broken and the intracellular (and transmembrane) portions of these chains move apart. As a result, a binding site for talin on the tail of the β chain is exposed. The binding of talin then leads to assembly of actin filaments anchored to the intracellular end of the integrin molecule (see Figure 19–45). In this way, when an integrin catches hold of its ligand outside the cell, the cell reacts by tying its cytoskeleton to the integrin molecule, so that force can be applied at the point of attachment. This is referred to as "outside-in activation".

The chain of cause and effect can also operate in reverse, from inside to outside instead of outside to inside. Talin competes with the integrin α chain for its binding site on the tail of the β chain. Thus when talin binds to the β chain it undoes the intracellular α–β linkage, allowing the two legs of the integrin molecule to spring apart. This drives the extracellular portion of the integrin into its extended, active conformation.

This "inside-out activation" is triggered by intracellular regulatory molecules. These include the phosphoinositide PIP_2 (discussed in Chapter 15), which is thought to be capable of activating talin so that it binds to the integrin β chain strongly. In this way, a signal generated inside the cell can trigger its integrin molecules to reach out and grab hold of their extracellular ligands.

Intracellular signal molecules such as PIP_2 are themselves produced in response to signals received from outside the cell via other types of cell-surface receptors, such as G-protein-coupled receptors and receptor tyrosine kinases, which can thus control integrin activation (**Figure 19–49**). Conversely, the activation of integrins by attachment to matrix can influence the reception of signals by other pathways. The cross-talk between all these communication pathways, transmitting signals in both directions across the cell membrane, allows for some complex interactions between the cell and its physical and chemical environment.

Integrin Defects Are Responsible for Many Different Genetic Diseases

The 24 types of integrins found in a human are formed from the products of 8 different β-chain genes and 18 different α-chain genes, dimerized in different combinations. Each integrin has distinctive properties and functions. Moreover, because the same integrin molecule in different cell types can have different ligand-binding specificities, it seems that additional cell-type-specific factors can interact with integrins to modulate their binding activity. The binding of integrins to their ligands is also affected by the concentration of Ca^{2+} and Mg^{2+} in the extracellular medium, reflecting the presence of divalent-cation-binding domains in the α and β subunits. The divalent cations can influence both the affinity and the specificity of the binding of an integrin to its ligands.

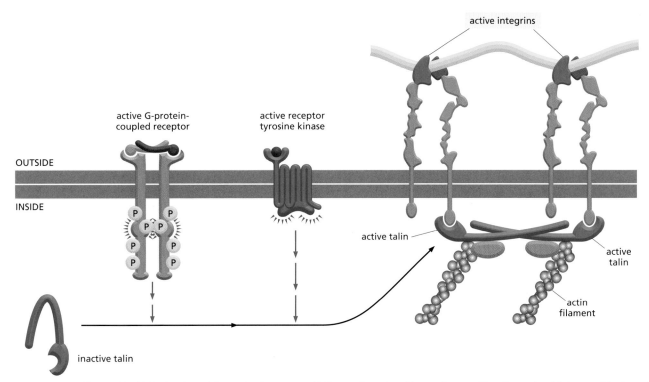

Figure 19–49 Activation of integrins by cross-talk from other signaling pathways. Signals received from outside the cell via other types of cell surface receptors, such as G-protein-coupled receptors and receptor tyrosine kinases, can alter the conformation of talin and thereby activate the cell's integrins.

Although there is some overlap in the activities of the different integrins—at least five bind laminin, for example—it is the diversity of integrin functions that is more remarkable. **Table 19–4** lists some of varieties of integrins and the problems that result when individual integrin α or β chains are defective.

The β1 subunits form dimers with at least 12 distinct α subunits and are found on almost all vertebrate cells: α5β1 is a fibronectin receptor and α6β1 a laminin receptor on many types of cells. Mutant mice that cannot make any β1 integrins die at implantation (very early in embryonic development). Mice that are only unable to make the α7 subunit (the partner for β1 in muscle) survive but develop muscular dystrophy (as do mice that cannot make the laminin ligand for the α7β1 integrin).

Table 19–4 Some Types of Integrins

INTEGRIN	LIGAND*	DISTRIBUTION	PHENOTYPE WHEN α SUBINUT IS MUTATED	PHENOTYPE WHEN β SUBUNIT IS MUTATED
α5β1	fibronectin	ubiquitous	death of embryo; defects in blood vessels, somites, neural crest	early death of embryo (at implantation)
α6β1	laminin	ubiquitous	severe skin blistering; defects in other epithelia also	early death of embryo (at implantation)
α7β1	laminin	muscle	muscular dystrophy; defective myotendinous junctions	early death of embryo (at implantation)
αLβ2 (LFA1)	Ig superfamily counterreceptors (ICAM)	white blood cells	impaired recruitment of leucocytes	leucocyte adhesion deficiency (LAD) impaired inflammatory responses; recurrent life-threatening infections
αIIbβ3	fibrinogen	platelets	bleeding; no platelet aggregation (Glanzmann's disease)	bleeding; no platelet aggregation (Glanzmann's disease); mild osteopetrosis
α6β4	laminin	hemidesmosomes in epithelia	severe skin blistering; defects in other epithelia also	severe skin blistering; defects in other epithelia also

*Not all ligands are listed.

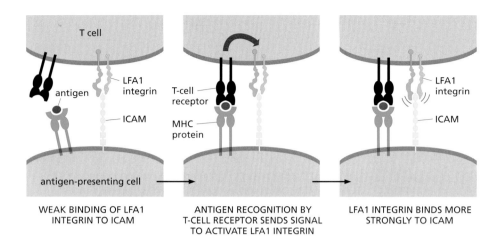

WEAK BINDING OF LFA1
INTEGRIN TO ICAM

ANTIGEN RECOGNITION BY
T-CELL RECEPTOR SENDS SIGNAL
TO ACTIVATE LFA1 INTEGRIN

LFA1 INTEGRIN BINDS MORE
STRONGLY TO ICAM

Figure 19–50 Integrin activation in the encounter of a T lymphocyte with an antigen-presenting cell. The two cells at first adhere weakly through binding of the LFA1 integrin in the T cell to the Ig-superfamily molecule ICAM in the membrane of the antigen-presenting cell. If the T-cell receptor at the same time recognizes its specific antigen, presented to it by the MHC molecule on the antigen-presenting cell, an intracellular signal is generated from the T cell receptor to activate the LFA1 integrin. As a result, LFA1 binds more strongly and persistently to ICAM. This gives the antigen-presenting cell time to activate the T cell and thereby elicit a specific immune response. (Adapted from K. Murphy et al., Janeway's Immunobiology, 7th ed. New York: Garland Science, 2008.)

The β2 subunits form dimers with at least four types of α subunit and are expressed exclusively on the surface of white blood cells, where they have an essential role in enabling these cells to fight infection. The β2 integrins mainly mediate cell–cell rather than cell–matrix interactions, binding to specific ligands on another cell, such as an endothelial cell. The ligands, sometimes referred to as *counterreceptors,* are members of the Ig superfamily of cell–cell adhesion molecules. We have already described an example earlier in the chapter: an integrin of this class (αLβ2, also known as LFA1) on white blood cells enables them to attach firmly to the Ig-family protein ICAM on endothelial cells at sites of infection and, through this attachment, to migrate out of the bloodstream into the infected tissue (see Figure 19–19B). People with the genetic disease called *leucocyte adhesion deficiency* fail to synthesize functional β2 subunits. As a consequence, their white blood cells lack the entire family of β2 receptors, and they suffer repeated bacterial infections.

The β3 integrins are found on blood platelets (as well as various other cells), and they bind several matrix proteins, including the blood clotting factor *fibrinogen.* Platelets have to interact with fibrinogen to mediate normal blood clotting, and humans with *Glanzmann's disease,* who are genetically deficient in β3 integrins, suffer from defective clotting and bleed excessively.

In both white blood cells and platelets, the ability to regulate integrin activity via inside-out signaling is particularly important. Regulated adhesion allows the cells to circulate unimpeded until they are activated by an appropriate stimulus. Because the integrins do not need to be synthesized *de novo,* the signaled adhesion response can be rapid. Platelets, for example, respond to contact with the wall of a damaged blood vessel and to various soluble signaling molecules, triggering activation of the β3 integrin in the platelet membrane. The resulting interaction of platelets with fibrinogen leads to formation of a platelet plug, which helps to stop the bleeding at just the site where it is needed. Similarly, the binding of a T lymphocyte to its specific antigen on the surface of an antigen-presenting cell (discussed in Chapter 25) switches on intracellular signaling pathways in the T cell that activate its β2 integrins (**Figure 19–50**). The activated integrins then enable the T cell to adhere strongly to the antigen-presenting cell so that it remains in contact long enough to become stimulated fully. The integrins may then return to an inactive state, allowing the T cell to disengage.

Integrins Cluster to Form Strong Adhesions

Integrins, like other cell adhesion molecules, differ from cell-surface receptors for hormones and for other extracellular soluble signal molecules in that they usually bind their ligand with lower affinity and are usually present at a 10- to 100-fold higher concentration on the cell surface. The Velcro principle, mentioned earlier, operates here too. Strong adhesion depends on clustering of integrins, creating a plaque in which many cytoskeletal filaments are anchored, as at a hemidesmosome in the epidermis or at a focal adhesion made by a fibroblast on a culture dish. At focal adhesions, and probably also in the less prominent actin-linked cell–matrix adhesions that cells mainly make in normal tissues, activation of the small GTPase Rho plays a part in the maturation of the adhesive complex, by promoting recruitment of actin filaments and integrins to the contact site. Artificially mutated integrins that lack an intracellular tail fail to connect with cytoskeletal filaments, fail to cluster, and are unable to form strong adhesions.

Extracellular Matrix Attachments Act Through Integrins to Control Cell Proliferation and Survival

Like other transmembrane cell adhesion proteins, integrins do more than just create attachments. They also activate intracellular signaling pathways and thereby allow control of almost any aspect of the cell's behavior according to the nature of the surrounding matrix and the state of the cell's attachments to it.

Studies in culture show that many cells will not grow or proliferate unless they are attached to extracellular matrix; nutrients and soluble growth factors in the culture medium are not enough. For some cell types, including epithelial, endothelial, and muscle cells, even cell survival depends on such attachments. When these cells lose contact with the extracellular matrix, they undergo programmed cell death, or apoptosis. This dependence of cell growth, proliferation, and survival on attachment to a substratum is known as **anchorage dependence**, and it is mediated mainly by integrins and the intracellular signals they generate. Anchorage dependence is thought to help ensure that each type of cell survives and proliferates only when it is in an appropriate situation. Mutations that disrupt or override this form of control, allowing cells to escape from anchorage dependence, occur in cancer cells and play a major part in their invasive behavior.

The physical spreading of a cell on the matrix also has a strong influence on intracellular events. Cells that are forced to spread over a large surface area by the formation of multiple adhesions at widely separate sites survive better and proliferate faster than cells that are not so spread out (**Figure 19–51**). The stimulatory effect of cell spreading presumably helps tissues to regenerate after injury. If cells are lost from an epithelium, for example, the spreading of the remaining cells into the vacated space will help stimulate these survivors to proliferate until they fill the gap. It is uncertain how a cell senses its extent of spreading so as to adjust its behavior accordingly, but the ability to spread depends on integrins, and signals generated by integrins at the sites of adhesion must play a part in providing the spread cells with stimulation.

Our understanding of anchorage dependence and of the effects of cell spreading has come mainly from studies of cells living on the surface of matrix-coated culture dishes. For connective-tissue cells that are normally surrounded

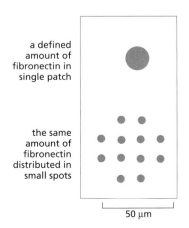

a defined amount of fibronectin in single patch

the same amount of fibronectin distributed in small spots

50 µm

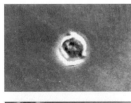

CELL DIES BY APOPTOSIS

CELL SPREADS, SURVIVES, AND GROWS

Figure 19–51 The importance of cell spreading. In this experiment, cell growth and survival are shown to depend on the extent of cell spreading on a substratum, rather than the mere fact of attachment or the number of matrix molecules the cell contacts. (Based on C.S. Chen et al., *Science* 276:1425–1428, 1997. With permission from AAAS.)

by matrix on all sides, this is a far cry from the natural environment. Walking over a plain is very different from clambering through a jungle. The types of contacts that cells make with a rigid substratum are not the same as those, much less well studied, that they make with the deformable web of fibers of the extracellular matrix, and there are substantial differences of cell behavior between the two contexts. Nevertheless, it is likely that the same basic principles apply. Both *in vitro* and *in vivo*, intracellular signals generated at cell–matrix adhesion sites, by molecular complexes organized around integrins, are crucial for cell proliferation and survival.

Integrins Recruit Intracellular Signaling Proteins at Sites of Cell–Substratum Adhesion

The mechanisms by which integrins signal into the cell interior are complex, involving several different pathways, and integrins and conventional signaling receptors often influence one another and work together to regulate cell behavior, as we have already emphasized. The Ras/MAP kinase pathway (see Figure 15–61), for example, can be activated both by conventional signaling receptors and by integrins, but cells often need both kinds of stimulation of this pathway at the same time to give sufficient activation to induce cell proliferation. Integrins and conventional signaling receptors also cooperate in activating similar pathways to promote cell survival (discussed in Chapters 15 and 17).

One of the best-studied modes of integrin signaling depends on a cytoplasmic protein tyrosine kinase called **focal adhesion kinase** (**FAK**). In studies of cells cultured in the normal way on rigid substrata, focal adhesions are often prominent sites of tyrosine phosphorylation (**Figure 19–52**A), and FAK is one of the major tyrosine-phosphorylated proteins found at these sites. When integrins cluster at cell–matrix contacts, FAK is recruited by intracellular anchor proteins such as talin (binding to the integrin β subunit) or *paxillin* (which binds to one type of integrin α subunit). The clustered FAK molecules cross-phosphorylate each other on a specific tyrosine, creating a phosphotyrosine docking site for members of the Src family of cytoplasmic tyrosine kinases. In addition to phosphorylating other proteins at the adhesion sites, these kinases then phosphorylate FAK on additional tyrosines, creating docking sites for a variety of additional intracellular signaling proteins. In this way, outside-in signaling from integrins, via FAK and Src-family kinases, is relayed into the cell (as discussed in Chapter 15).

One way to analyze the function of FAK is to examine focal adhesions in cells from mutant mice that lack the protein. FAK-deficient fibroblasts still adhere to

Figure 19–52 Focal adhesions and the role of focal adhesion kinase (FAK). (A) A fibroblast cultured on a fibronectin-coated substratum and stained with fluorescent antibodies: actin filaments are stained *green* and activated proteins that contain phosphotyrosine are *red*, giving *orange* where the two components overlap. The actin filaments terminate at focal adhesions, where the cell attaches to the substratum by means of integrins. Proteins containing phosphotyrosine are also concentrated at these sites, reflecting the local activation of FAK and other protein kinases. Signals generated at such adhesion sites help regulate cell division, growth, and survival. (B, C) The influence of FAK on formation of focal adhesions is shown by a comparison of normal and FAK-deficient fibroblasts, stained with an antibody against vinculin to reveal the focal adhesions. (B) The normal fibroblasts have fewer focal adhesions and have spread after 2 hours in culture. (C) At the same time point, the FAK-deficient fibroblasts have more focal adhesions and have not spread. (A, courtesy of Keith Burridge; B, C, from D. Ilic et al., *Nature* 377:539–544, 1995. With permission from Macmillan Publishers Ltd.)

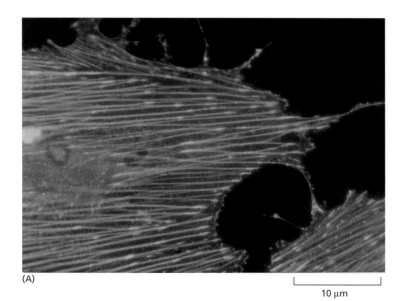

(A)

10 μm

normal fibroblasts

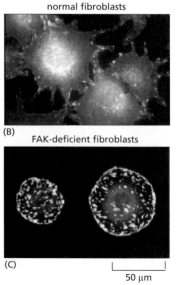

(B)

FAK-deficient fibroblasts

(C)

50 μm

fibronectin and form focal adhesions. In fact, they form too many focal adhesions; as a result, cell spreading and migration are slowed (Figure 19–52B and C). This unexpected finding suggests that FAK normally helps disassemble focal adhesions and that this loss of adhesions is required for normal cell migration. Many cancer cells have elevated levels of FAK, which may help explain why they are often more motile than their normal counterparts.

Integrins Can Produce Localized Intracellular Effects

Through FAK and other pathways, activated integrins, like other signaling receptors, can induce global cell responses, often including changes in gene expression. But the integrins are especially adept at stimulating localized changes in the cytoplasm close to the cell–matrix contact. We have already mentioned an important example in our discussion of epithelial cell polarity: it is through integrins that the basal lamina plays its part in directing the internal apico-basal organization of epithelial cells.

Localized intracellular effects may be a common feature of signaling by transmembrane cell adhesion proteins in general. In the developing nervous system, for example, the growing tip of an axon is guided mainly by its responses to local adhesive (and repellent) cues in the environment that are recognized by transmembrane cell adhesion proteins, as discussed in Chapter 22. The primary effects of the adhesion proteins are thought to result from the activation of intracellular signaling pathways that act locally in the axon tip, rather than through cell–cell adhesion itself or signals conveyed to the cell body. Through localized activation of the Rho family of small GTPases, for example (as discussed in Chapters 15 and 16), the transmembrane adhesion proteins can control motility and guide forward movement. In this and other ways, practically all the classes of cell-cell and cell-matrix adhesion molecules that we have mentioned, including integrins, are deployed to help guide axon outgrowth in the developing nervous system.

Table 19–5 summarizes the categories of cell adhesion molecules that we have considered in this chapter. In the next section, we turn from the adhesion molecules in cell membranes to look in detail at the extracellular matrix that surrounds cells in connective tissues.

Table 19–5 Cell Adhesion Molecule Families

	SOME FAMILY MEMBERS	Ca^{2+} OR Mg^{2+} DEPENDENCE	HOMOPHILIC OR HETEROPHILIC	CYTOSKELETON ASSOCIATIONS	CELL JUNCTION ASSOCIATIONS
Cell–Cell Adhesion					
Classical cadherins	E, N, P, VE	yes	homophilic	actin filaments (via catenins)	adherens junctions, synapses
Desmosomal cadherins	desmoglein, desmocollin	yes	homophilic	intermediate filaments (via desmoplakin, plakoglobin, and plakophilin)	desmosomes
Ig family members	N-CAM, ICAM	no	both	unknown	neuronal and immunological synapses
Selectins (blood cells and endothelial cells only)	L-, E-, and P-selectins	yes	heterophilic	actin filaments	(no prominent junctional structure)
Integrins on blood cells	αLβ2 (LFA1)	yes	heterophilic	actin filaments	immunological synapses
Cell–Matrix Adhesion					
Integrins	many types	yes	heterophilic	actin filaments (via talin, paxillin, filamin, α-actinin, and vinculin)	focal adhesions
	α6β4	yes	heterophilic	intermediate filaments (via plectin and dystonin)	hemidesmosomes
Transmembrane proteoglycans	syndecans	no	heterophilic	actin filaments	(no prominent junctional structure)

Summary

Integrins are the principal receptors used by animal cells to bind to the extracellular matrix: they function as transmembrane linkers between the extracellular matrix and the cytoskeleton connecting usually to actin, but to intermediate filaments for the specialized integrins at hemidesmosomes. Integrin molecules are heterodimers, and the binding of ligands is associated with dramatic changes of conformation. This creates an allosteric coupling between binding to matrix outside the cell and binding to the cytoskeleton inside it, allowing the integrin to convey signals in both directions across the plasma membrane—from inside to out and from outside to in. Binding of the intracellular anchor protein talin to the tail of an integrin molecule tends to drive the integrin into an extended conformation with increased affinity for its extracellular ligand. Conversely, binding to an extracellular ligand, by promoting the same conformational change, leads to binding of talin and formation of a linkage to the actin cytoskeleton. Complex assemblies of proteins become organized around the intracellular tails of integrins, producing intracellular signals that can influence almost any aspect of cell behavior, from proliferation and survival, as in the phenomenon of anchorage dependence, to cell polarity and guidance of migration.

THE EXTRACELLULAR MATRIX OF ANIMAL CONNECTIVE TISSUES

We have already discussed the basal lamina as an archetypal example of extracellular matrix, common to practically all multicellular animals and an essential feature of epithelial tissues. We now turn to the much more varied and bulky forms of extracellular matrix found in connective tissues (**Figure 19–53**). Here, the extracellular matrix is generally more plentiful than the cells it surrounds, and it determines the tissue's physical properties.

The classes of macromolecules constituting the extracellular matrix in animal tissues are broadly similar, whether we consider the basal lamina or the other forms that matrix can take, but variations in the relative amounts of these different classes of molecules and in the ways in which they are organized give rise to an amazing diversity of materials. The matrix can become calcified to form the rock-hard structures of bone or teeth, or it can form the transparent substance of the cornea, or it can adopt the ropelike organization that gives tendons their enormous tensile strength. It forms the jelly in a jellyfish. Covering the body of a beetle or a lobster, it forms a rigid carapace. Moreover, the extracellular matrix is more than a passive scaffold to provide physical support. It has an active and complex role in regulating the behavior of the cells that touch it, inhabit it, or crawl through its meshes, influencing their survival, development, migration, proliferation, shape, and function.

In this section, we focus our discussion on the extracellular matrix of connective tissues in vertebrates, but bulky forms of extracellular matrix play an

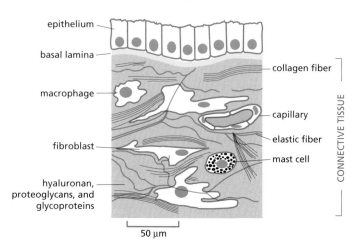

Figure 19–53 **The connective tissue underlying an epithelium.** This tissue contains a variety of cells and extracellular matrix components. The predominant cell type is the fibroblast, which secretes abundant extracellular matrix.

important part in virtually all multicellular organisms; examples include the cuticles of worms and insects, the shells of mollusks, the cell walls of fungi, and, as we discuss later, the cell walls of plants.

The Extracellular Matrix Is Made and Oriented by the Cells Within It

The macromolecules that constitute the extracellular matrix are mainly produced locally by cells in the matrix. As we discuss later, these cells also help to organize the matrix: the orientation of the cytoskeleton inside the cell can control the orientation of the matrix produced outside. In most connective tissues, the matrix macromolecules are secreted largely by cells called **fibroblasts (Figure 19–54)**. In certain specialized types of connective tissues, such as cartilage and bone, however, they are secreted by cells of the fibroblast family that have more specific names: *chondroblasts*, for example, form cartilage, and *osteoblasts* form bone.

The matrix in connective tissue is constructed from the same two main classes of macromolecules as in basal laminae: (1) glycosaminoglycan polysaccharide chains, usually covalently linked to protein in the form of proteoglycans, and (2) fibrous proteins such as collagen. We shall see that the members of both classes come in a great variety of shapes and sizes.

The proteoglycan molecules in connective tissue typically form a highly hydrated, gel-like "ground substance" in which the fibrous proteins are embedded. The polysaccharide gel resists compressive forces on the matrix while permitting the rapid diffusion of nutrients, metabolites, and hormones between the blood and the tissue cells. The collagen fibers strengthen and help organize the matrix, while other fibrous proteins, such as the rubberlike *elastin*, give it resilience. Finally, many matrix proteins help cells migrate, settle, and differentiate in the appropriate locations.

Glycosaminoglycan (GAG) Chains Occupy Large Amounts of Space and Form Hydrated Gels

Glycosaminoglycans (GAGs) are unbranched polysaccharide chains composed of repeating disaccharide units. They are called GAGs because one of the two sugars in the repeating disaccharide is always an amino sugar (*N*-acetylglucosamine or *N*-acetylgalactosamine), which in most cases is sulfated. The second sugar is usually a uronic acid (glucuronic or iduronic). Because there are sulfate or carboxyl groups on most of their sugars, GAGs are highly negatively charged (**Figure 19–55**). Indeed, they are the most anionic molecules produced by animal cells. Four main groups of GAGs are distinguished by their sugars, the type of linkage between the sugars, and the number and location of sulfate groups: (1) *hyaluronan*, (2) *chondroitin sulfate* and *dermatan sulfate*, (3) *heparan sulfate*, and (4) *keratan sulfate*.

Polysaccharide chains are too stiff to fold up into the compact globular structures that polypeptide chains typically form. Moreover, they are strongly hydrophilic. Thus, GAGs tend to adopt highly extended conformations that occupy a huge volume relative to their mass (**Figure 19–56**), and they form gels

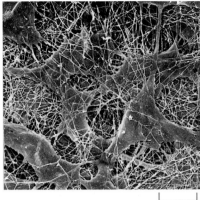

Figure 19–54 Fibroblasts in connective tissue. This scanning electron micrograph shows tissue from the cornea of a rat. The extracellular matrix surrounding the fibroblasts is here composed largely of collagen fibrils. The glycoproteins, hyaluronan, and proteoglycans, which normally form a hydrated gel filling the interstices of the fibrous network, have been removed by enzyme and acid treatment. (From T. Nishida et al., *Invest. Ophthalmol. Vis. Sci.* 29:1887–1890, 1988. With permission from Association for Research in Vision and Opthalmology.)

Figure 19–55 The repeating disaccharide sequence of a heparan sulfate glycosaminoglycan (GAG) chain. These chains can consist of as many as 200 disaccharide units, but are typically less than half that size. There is a high density of negative charges along the chain due to the presence of both carboxyl and sulfate groups. The proteoglycans of the basal lamina—perlecan, dystroglycan, and collagen XVIII—all carry heparan sulfate GAGs. The molecule is shown here with its maximal number of sulfate groups. *In vivo*, the proportion of sulfated and nonsulfated groups is variable. Heparin typically has >70% sulfation, while heparan sulfate has <50%.

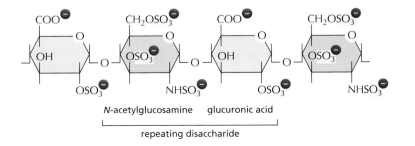

N-acetylglucosamine glucuronic acid

repeating disaccharide

even at very low concentrations. The weight of GAGs in connective tissue is usually less than 10% of the weight of the fibrous proteins. But, because they form porous hydrated gels, the GAG chains fill most of the extracellular space. Their high density of negative charges attracts a cloud of cations, especially Na^+, that are osmotically active, causing large amounts of water to be sucked into the matrix. This creates a swelling pressure, or turgor, that enables the matrix to withstand compressive forces (in contrast to collagen fibrils, which resist stretching forces). The cartilage matrix that lines the knee joint, for example, can support pressures of hundreds of atmospheres in this way.

Defects in the production of GAGs can affect many different body systems. In one rare human genetic disease, for example, there is a severe deficiency in the synthesis of dermatan sulfate disaccharide. The affected individuals have a short stature, prematurely aged appearance, and generalized defects in their skin, joints, muscles, and bones.

In invertebrates, plants, and fungi, other types of polysaccharides, rather than GAGs, often dominate the extracellular matrix. Thus, in higher plants, as we discuss later, cellulose (polyglucose) chains are packed tightly together in ribbonlike arrays to form the main component of the cell wall. In insects, crustaceans, and other arthropods, chitin (poly-N-acetylglucosamine) similarly forms the main component of the exoskeleton. Fungi, too, make their cell walls mainly out of chitin. Together, cellulose and chitin are the most abundant biopolymers on Earth.

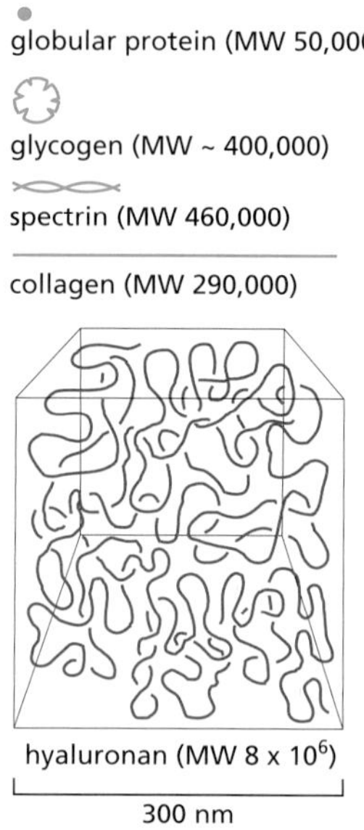

Hyaluronan Acts as a Space Filler and a Facilitator of Cell Migration During Tissue Morphogenesis and Repair

Hyaluronan (also called *hyaluronic acid* or *hyaluronate*) is the simplest of the GAGs (**Figure 19–57**). It consists of a regular repeating sequence of up to 25,000 disaccharide units, is found in variable amounts in all tissues and fluids in adult animals, and is especially abundant in early embryos. Hyaluronan is not typical of the majority of GAGs. In contrast with all of the others, it contains no sulfated sugars, all its disaccharide units are identical, its chain length is enormous (thousands of sugar monomers), and it is not generally linked covalently to any core protein. Moreover, whereas other GAGs are synthesized inside the cell and released by exocytosis, hyaluronan is spun out directly from the cell surface by an enzyme complex embedded in the plasma membrane.

Hyaluronan is thought to have a role in resisting compressive forces in tissues and joints. It is also important as a space filler during embryonic development, where it can be used to force a change in the shape of a structure, as a small quantity expands with water to occupy a large volume (see Figure 19–56). Hyaluronan synthesized locally from the basal side of an epithelium can deform the epithelium by creating a cell-free space beneath it, into which cells subsequently migrate. In the developing heart, for example, hyaluronan synthesis helps in this way to drive formation of the valves and septa that separate the heart's chambers. Similar processes occur in several other organs. When cell

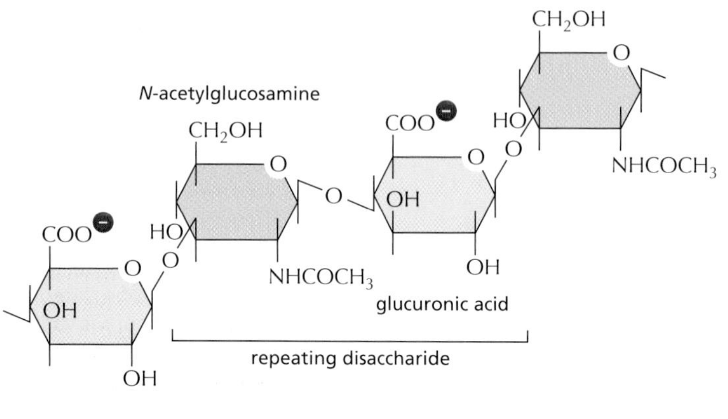

globular protein (MW 50,000)

glycogen (MW ~ 400,000)

spectrin (MW 460,000)

collagen (MW 290,000)

hyaluronan (MW 8×10^6)

300 nm

Figure 19–56 The relative dimensions and volumes occupied by various macromolecules. Several proteins, a glycogen granule, and a single hydrated molecule of hyaluronan are shown.

Figure 19–57 The repeating disaccharide sequence in hyaluronan, a relatively simple GAG. This ubiquitous molecule in vertebrates consists of a single long chain of up to 25,000 sugar monomers. Note the absence of sulfate groups.

migration ends, the excess hyaluronan is generally degraded by the enzyme *hyaluronidase*. Hyaluronan is also produced in large quantities during wound healing, and it is an important constituent of joint fluid, in which it serves as a lubricant.

Many of the functions of hyaluronan depend on specific interactions with other molecules, including both proteins and proteoglycans. Some of these molecules that bind to hyaluronan are constituents of the extracellular matrix, while others are integral components of cell surfaces.

Proteoglycans Are Composed of GAG Chains Covalently Linked to a Core Protein

Except for hyaluronan, all GAGs are covalently attached to protein as **proteoglycans**, which are produced by most animal cells. Membrane-bound ribosomes make the polypeptide chain, or *core protein*, of a proteoglycan, which is then threaded into the lumen of the endoplasmic reticulum. The polysaccharide chains are mainly assembled on this core protein in the Golgi apparatus before delivery to the exterior of the cell by exocytosis. First, a special *linkage tetrasaccharide* is attached to a serine side chain on the core protein to serve as a primer for polysaccharide growth; then, one sugar at a time is added by specific glycosyl transferases (**Figure 19–58**). While still in the Golgi apparatus, many of the polymerized sugars are covalently modified by a sequential and coordinated series of reactions. Epimerizations alter the configuration of the substituents around individual carbon atoms in the sugar molecule; sulfations increase the negative charge.

Proteoglycans are usually clearly distinguished from other glycoproteins by the nature, quantity, and arrangement of their sugar side chains. By definition, at least one of the sugar side chains of a proteoglycan must be a GAG. Whereas glycoproteins contain 1–60% carbohydrate by weight, and usually only a few percent, in the form of numerous relatively short, branched oligosaccharide chains, proteoglycans can contain as much as 95% carbohydrate by weight, mostly in the form of long, unbranched GAG chains, each typically about 80 sugars long.

In principle, proteoglycans have the potential for almost limitless heterogeneity. Even a single type of core protein can carry highly variable numbers and types of attached GAG chains. Moreover, the underlying repeating sequence of disaccharides in each GAG can be modified by a complex pattern of sulfate groups. The core proteins, too, are diverse, though many of them share some characteristic domains such as the LINK domain, involved in binding to GAGs.

Proteoglycans can be huge. The proteoglycan *aggrecan*, for example, which is a major component of cartilage, has a mass of about 3×10^6 daltons with over 100 GAG chains. Other proteoglycans are much smaller and have only 1–10 GAG chains; an example is *decorin*, which is secreted by fibroblasts and has a single

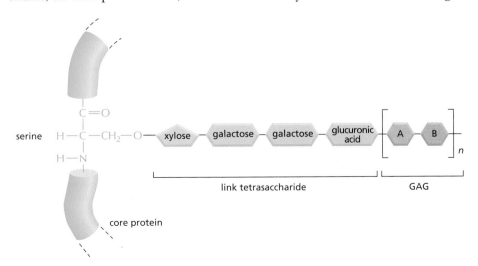

Figure 19–58 The linkage between a GAG chain and its core protein in a proteoglycan molecule. A specific link tetrasaccharide is first assembled on a serine side chain. In most cases, it is unclear how the particular serine is selected, but it seems that a specific local conformation of the polypeptide chain, rather than a specific linear sequence of amino acids, is recognized. The rest of the GAG chain, consisting mainly of a repeating disaccharide unit, is then synthesized, with one sugar being added at a time. In chondroitin sulfate, the disaccharide is composed of D-glucuronic acid and N-acetyl-D-galactosamine; in heparan sulfate, it is D-glucosamine (or L-iduronic acid) and N-acetyl-D-glucosamine; in keratan sulfate, it is D-galactose and N-acetyl-D-glucosamine.

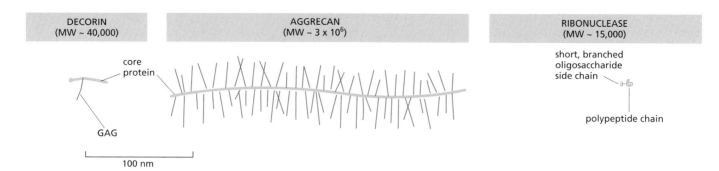

| DECORIN (MW ~ 40,000) | AGGRECAN (MW ~ 3 x 10⁶) | RIBONUCLEASE (MW ~ 15,000) |

GAG chain (**Figure 19–59**). Decorin binds to collagen fibrils and regulates fibril assembly and fibril diameter; mice that cannot make decorin have fragile skin that has reduced tensile strength. The GAGs and proteoglycans of these various types can associate to form even larger polymeric complexes in the extracellular matrix. Molecules of aggrecan, for example, assemble with hyaluronan in cartilage matrix to form aggregates that are as big as a bacterium (**Figure 19–60**). Moreover, besides associating with one another, GAGs and proteoglycans associate with fibrous matrix proteins such as collagen and with protein meshworks such as the basal lamina, creating extremely complex composites (**Figure 19–61**).

Proteoglycans Can Regulate the Activities of Secreted Proteins

Proeteoglycans are as diverse in function as they are in chemistry and structure. Their GAG chains, for example, can form gels of varying pore size and charge density; one possible function, therefore, is to serve as selective sieves

Figure 19–59 Examples of a small (decorin) and a large (aggrecan) proteoglycan found in the extracellular matrix. The figure compares these two proteoglycans with a typical secreted glycoprotein molecule, pancreatic ribonuclease B. All three are drawn to scale. The core proteins of both aggrecan and decorin contain oligosaccharide chains as well as the GAG chains, but these are not shown. Aggrecan typically consists of about 100 chondroitin sulfate chains and about 30 keratan sulfate chains linked to a serine-rich core protein of almost 3000 amino acids. Decorin "decorates" the surface of collagen fibrils, hence its name.

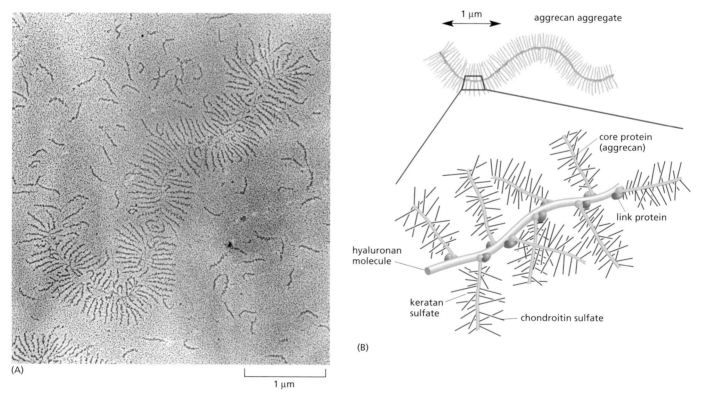

(A)

(B)

Figure 19–60 An aggrecan aggregate from fetal bovine cartilage. (A) An electron micrograph of an aggrecan aggregate shadowed with platinum. Many free aggrecan molecules are also visible. (B) A drawing of the giant aggrecan aggregate shown in (A). It consists of about 100 aggrecan monomers (each like the one shown in Figure 19–59) noncovalently bound through the N-terminal domain of the core protein to a single hyaluronan chain. A link protein binds both to the core protein of the proteoglycan and to the hyaluronan chain, thereby stabilizing the aggregate. The link proteins are members of a family of hyaluronan-binding proteins, some of which are cell-surface proteins. The molecular weight of such a complex can be 10⁸ or more, and it occupies a volume equivalent to that of a bacterium, which is about 2×10^{-12} cm³. (A, courtesy of Lawrence Rosenberg.)

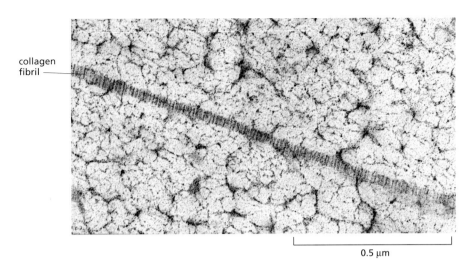

collagen fibril

0.5 μm

Figure 19–61 Proteoglycans in the extracellular matrix of rat cartilage. The tissue was rapidly frozen at –196°C, and fixed and stained while still frozen (a process called freeze substitution) to prevent the GAG chains from collapsing. In this electron micrograph, the proteoglycan molecules are seen to form a fine filamentous network in which a single striated collagen fibril is embedded. The more darkly stained parts of the proteoglycan molecules are the core proteins; the faintly stained threads are the GAG chains. (Reproduced from E.B. Hunziker and R.K. Schenk, *J. Cell Biol.* 98:277–282, 1984. With permission from The Rockefeller University Press.)

to regulate the traffic of molecules and cells according to their size and charge, as in the thick basal lamina of the kidney glomerulus (see p. 1167).

Proteoglycans have an important role in chemical signaling between cells. They bind various secreted signal molecules, such as certain protein growth factors—controlling their diffusion through the matrix, their range of action, and their lifetime, as well as enhancing or inhibiting their signaling activity. For example, the heparan sulfate chains of proteoglycans bind to *fibroblast growth factors (FGFs)*, which (among other effects) stimulate a variety of cell types to proliferate; this interaction oligomerizes the growth factor molecules, enabling them to cross-link and activate their cell-surface receptors, which are transmembrane tyrosine kinases (see Figure 15–54A). In inflammatory responses, heparan sulfate proteoglycans immobilize secreted chemotactic attractants called *chemokines* (discussed in Chapter 25) on the endothelial surface of a blood vessel at an inflammatory site. This allows the chemokines to remain there for a prolonged period, stimulating white blood cells to leave the bloodstream and migrate into the inflamed tissue. Whereas in most cases the signal molecules bind to the GAG chains of the proteoglycan, this is not always so. Some members of the *transforming growth factor β (TGFβ)* family bind to the core proteins of several matrix proteoglycans, including decorin; binding to decorin inhibits the growth factor activity. Signal molecules (including TGFβ) can also bind to fibrous proteins in the matrix. *Vascular endothelial growth factor (VEGF)*, for example, binds to fibronectin.

Proteoglycans also bind, and regulate the activities of, other types of secreted proteins, including proteolytic enzymes (proteases) and protease inhibitors. Thus they play a part in controlling both the assembly and the degradation of other components of the extracellular matrix, including collagen.

Cell-Surface Proteoglycans Act as Co-Receptors

Not all proteoglycans are secreted components of the extracellular matrix. Some are integral components of plasma membranes and have their core protein either inserted across the lipid bilayer or attached to the lipid bilayer by a glycosylphosphatidylinositol (GPI) anchor. Some of these plasma membrane proteoglycans act as *co-receptors* that collaborate with conventional cell-surface receptor proteins. In addition, some conventional receptors have one or more GAG chains and are therefore proteoglycans themselves.

Among the best-characterized plasma membrane proteoglycans are the *syndecans*, which have a membrane-spanning core protein whose intracellular domain is thought to interact with the actin cytoskeleton and with signaling molecules in the cell cortex. Syndecans are located on the surface of many types of cells, including fibroblasts and epithelial cells. In fibroblasts, syndecans can be found in cell-matrix adhesions, where they modulate integrin function by

Table 19–6 **Some Common Proteoglycans**

PROTEOGLYCAN	APPROXIMATE MOLECULAR WEIGHT OF CORE PROTEIN	TYPE OF GAG CHAINS	NUMBER OF GAG CHAINS	LOCATION	FUNCTIONS
Aggrecan	210,000	chondroitin sulfate + keratan sulfate (in separate chains)	~130	cartilage	mechanical support; forms large aggregates with hyaluronan
Betaglycan	36,000	chondroitin sulfate/ dermatan sulfate	1	cell surface and matrix	binds TGFβ
Decorin	40,000	chondroitin sulfate/ dermatan sulfate	1	widespread in connective tissues	binds to type I collagen fibrils and TGFβ
Perlecan	600,000	heparan sulfate	2–15	basal laminae	structural and filtering function in basal lamina
Syndecan-1	32,000	chondroitin sulfate + heparan sulfate (in separate chains)	1–3	cell surface	cell adhesion; binds FGF and other growth factors
Dally (in *Drosophila*)	60,000	heparan sulfate	1–3	cell surface	co-receptor for Wingless and Decapentaplegic signaling proteins

interacting with fibronectin on the cell surface and with cytoskeletal and signaling proteins inside the cell. Syndecans also bind FGFs and present them to FGF receptor proteins on the same cell. Similarly, another plasma membrane proteoglycan, called *betaglycan*, binds TGFβ and presents it to TGFβ receptors.

The importance of proteoglycans as co-receptors and as regulators of the distribution and activity of signal molecules is illustrated by the severe developmental defects that can occur when specific proteoglycans are inactivated by mutation. In *Drosophila*, for example, the products of the *Dally* and *Dally-like* genes, coding for members of the *glypican* family, with heparan sulfate side chains, are needed for signaling by no less than four of the main signal proteins that govern the patterning of the embryo (Wingless, Hedgehog, FGF, and Decapentaplegic (Dpp), as discussed in Chapter 22). For some of these signal pathways, mutations in *Dally* or *Dally-like* mimic the effects of mutations in the genes that code for the signal proteins themselves.

The structure, function, and location of some of the proteoglycans discussed in this chapter are summarized in **Table 19–6**.

Collagens Are the Major Proteins of the Extracellular Matrix

The fibrous proteins are no less important than the proteoglycans as components of the extracellular matrix. Foremost among them are the **collagens,** a family of fibrous proteins found in all multicellular animals. Collagens are secreted in large quantities by connective tissue cells, and in smaller quantities by many other cell types. As a major component of skin and bone, they are the most abundant proteins in mammals, where they constitute 25% of the total protein mass.

The primary feature of a typical collagen molecule is its long, stiff, triple-stranded helical structure, in which three collagen polypeptide chains, called α *chains,* are wound around one another in a ropelike superhelix (**Figure 19–62**). Collagens are extremely rich in proline and glycine, both of which are important in the formation of the triple-stranded helix. Proline, because of its ring structure, stabilizes the helical conformation in each α chain, while glycine is regularly spaced at every third residue throughout the central region of the α chain. Being the smallest amino acid (having only a hydrogen atom as a side chain), glycine allows the three helical α chains to pack tightly together to form the final collagen superhelix.

The human genome contains 42 distinct genes coding for different collagen α chains. Different combinations of these genes are expressed in different tissues. Although in principle thousands of types of triple-stranded collagen

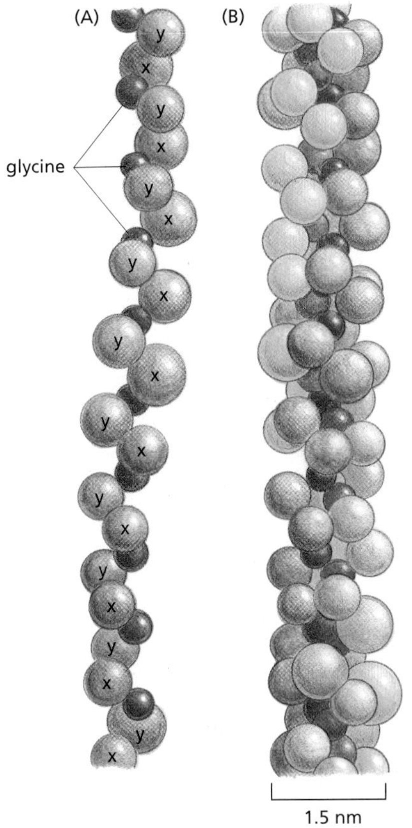

Figure 19–62 **The structure of a typical collagen molecule.** (A) A model of part of a single collagen α chain in which each amino acid is represented by a sphere. The chain is about 1000 amino acids long. It is arranged as a left-handed helix, with three amino acids per turn and with glycine as every third amino acid. Therefore, an α chain is composed of a series of triplet Gly-X-Y sequences, in which X and Y can be any amino acid (although X is commonly proline and Y is commonly hydroxyproline). (B) A model of part of a collagen molecule in which three α chains, each shown in a different color, are wrapped around one another to form a triple-stranded helical rod. Glycine is the only amino acid small enough to occupy the crowded interior of the triple helix. Only a short length of the molecule is shown; the entire molecule is 300 nm long. (From a model by B.L. Trus.)

molecules could be assembled from various combinations of the 42 α chains, only a limited number of triple-helical combinations are possible, and less than 40 types of collagen molecules have been found. Type I is by far the most common, being the principal collagen of skin and bone. It belongs to the class of **fibrillar collagens**, or fibril-forming collagens, which have long ropelike structures with few or no interruptions. After being secreted into the extracellular space, these collagen molecules assemble into higher-order polymers called *collagen fibrils,* which are thin structures (10–300 nm in diameter) many hundreds of micrometers long in mature tissues and clearly visible in electron micrographs (**Figure 19–63**; see also Figure 19–61). Collagen fibrils often aggregate into larger, cablelike bundles, several micrometers in diameter, that are visible in the light microscope as *collagen fibers.*

Collagen types IX and XII are called *fibril-associated collagens* because they decorate the surface of collagen fibrils. They are thought to link these fibrils to one another and to other components in the extracellular matrix. Type IV, as we have already seen, is a *network-forming collagen,* forming a major part of basal laminae, while type VII molecules form dimers that assemble into specialized structures called *anchoring fibrils.* Anchoring fibrils help attach the basal lamina of multilayered epithelia to the underlying connective tissue and therefore are especially abundant in the skin.

There are also a number of "collagen-like" proteins, including type XVII, which has a transmembrane domain and is found in hemidesmosomes, and type XVIII, which we mentioned earlier as the core protein of a proteoglycan in basal laminae.

Many proteins appear to have evolved by repeated duplications of an original DNA sequence, giving rise to a repetitive pattern of amino acids. The genes that encode the α chains of most of the fibrillar collagens provide a good example: they are very large (up to 44 kilobases in length) and contain about 50 exons. Most of the exons are 54, or multiples of 54, nucleotides long, suggesting that these collagens originated through multiple duplications of a primordial gene

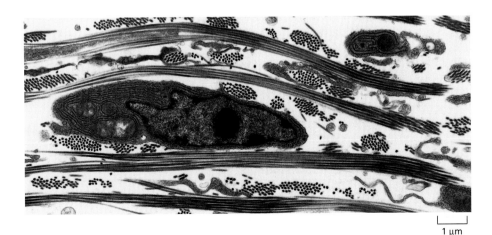

Figure 19–63 **A fibroblast surrounded by collagen fibrils in the connective tissue of embryonic chick skin.** In this electron micrograph, the fibrils are organized into bundles that run approximately at right angles to one another. Therefore, some bundles are oriented longitudinally, whereas others are seen in cross section. The collagen fibrils are produced by the fibroblasts, which contain abundant endoplasmic reticulum, where secreted proteins such as collagen are synthesized. (From C. Ploetz, E.I. Zycband and D.E. Birk, *J. Struct. Biol.* 106:73–81, 1991. With permission from Elsevier.)

Table 19–7 Some Types of Collagen and Their Properties

	TYPE	POLYMERIZED FORM	TISSUE DISTRIBUTION	MUTANT PHENOTYPE
Fibril-forming (fibrillar)	I	fibril	bone, skin, tendons, ligaments, cornea, internal organs (accounts for 90% of body collagen)	severe bone defects, fractures
	II	fibril	cartilage, invertebral disc, notochord, vitreous humor of the eye	cartilage deficiency, dwarfism
	III	fibril	skin, blood vessels, internal organs	fragile skin, loose joints, blood vessels prone to rupture
	V	fibril (with type I)	as for type I	fragile skin, loose joints, blood vessels prone to rupture
	XI	fibril (with type II)	as for type II	myopia, blindness
Fibril-associated	IX	lateral association with type II fibrils	cartilage	osteoarthritis
Network-forming	IV	sheetlike network	basal lamina	kidney disease (glomerulonephritis), deafness
	VII	anchoring fibrils	beneath stratified squamous epithelia	skin blistering
Transmembrane	XVII	non-fibrillar	hemidesmosomes	skin blistering
Proteoglycan core protein	XVIII	non-fibrillar	basal lamina	myopia, detached retina, hydrocephalus

Note that types I, IV, V, IX, and XI are each composed of two or three types of α chains (distinct, nonoverlapping sets in each case), whereas types II, III, VII, XII, XVII, and XVIII are composed of only one type of α chain each. Only 10 types of collagen are shown, but about 27 types of collagen and 42 types of α chains have been identified in humans.

containing 54 nucleotides and encoding exactly 6 Gly-X-Y repeats (see Figure 19–62).

Table 19–7 provides additional details for some of the collagen types discussed in this chapter.

Collagen Chains Undergo a Series of Post-Translational Modifications

Individual collagen polypeptide chains are synthesized on membrane-bound ribosomes and injected into the lumen of the endoplasmic reticulum (ER) as larger precursors, called *pro-α chains*. These precursors not only have the short amino-terminal signal peptide required to direct the nascent polypeptide to the ER, but also have, at both their N- and C-terminal ends, additional amino acids, called *propeptides*, that are clipped off at a later step of collagen assembly. Moreover, in the lumen of the ER, selected prolines and lysines are hydroxylated to form hydroxyproline and hydroxylysine, respectively, and some of the hydroxy-lysines are glycosylated. Each pro-α chain then combines with two others to form a hydrogen-bonded, triple-stranded, helical molecule known as *procollagen*.

Hydroxylysines and *hydroxyprolines* (**Figure 19–64**) are infrequently found in other animal proteins, although hydroxyproline is abundant in some proteins in the plant cell wall. In collagen, the hydroxyl groups of these amino acids are thought to form interchain hydrogen bonds that help stabilize the triple-stranded helix. Conditions that prevent proline hydroxylation, such as a deficiency of ascorbic acid (vitamin C), have serious consequences. In *scurvy*, the often fatal disease caused by a dietary deficiency of vitamin C that was common in sailors until the nineteenth century, the defective pro-α chains that are synthesized fail to form a stable triple helix and are immediately degraded within the cell, and synthesis of new collagen is inhibited. In healthy tissues, collagen is continually degraded and replaced (with a turnover time of months or years, depending on the tissue). In scurvy, replacement fails, and within a few months, with the gradual loss of the preexisting normal collagen in the matrix, blood vessels become fragile, teeth become loose in their sockets, and wounds cease to heal.

Figure 19–64 **Hydroxylysine and hydroxyproline.** These modified amino acids are common in collagen. They are formed by enzymes that act after the lysine and proline have been incorporated into procollagen molecules.

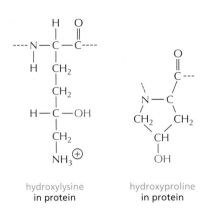

hydroxylysine
in protein

hydroxyproline
in protein

Propeptides Are Clipped Off Procollagen After its Secretion, to Allow Assembly of Fibrils

After secretion, the propeptides of the fibrillar procollagen molecules are removed by specific proteolytic enzymes outside the cell. This converts the procollagen molecules to collagen molecules, which assemble in the extracellular space to form much larger **collagen fibrils**. The propeptides have at least two functions. First, they guide the intracellular formation of the triple-stranded collagen molecules. Second, because they are retained until after secretion, they prevent the intracellular formation of large collagen fibrils, which could be catastrophic for the cell.

The process of fibril formation is driven, in part, by the tendency of the collagen molecules, which are more than a thousandfold less soluble than procollagen molecules, to self-assemble. The fibrils begin to form close to the cell surface, often in deep infoldings of the plasma membrane formed by the fusion of secretory vesicles with the cell surface. The underlying cortical cytoskeleton can influence the sites, rates, and orientation of fibril assembly.

When viewed in an electron microscope, collagen fibrils have characteristic cross-striations every 67 nm, reflecting the regularly staggered packing of the individual collagen molecules in the fibril. After the fibrils have formed in the extracellular space, they are greatly strengthened by the formation of covalent cross-links between lysine residues of the constituent collagen molecules (**Figure 19–65**). The types of covalent bonds involved are found only in collagen and elastin. If cross-linking is inhibited, the tensile strength of the fibrils is drastically reduced: collagenous tissues become fragile, and structures such as skin, tendons, and blood vessels tend to tear. The extent and type of cross-linking vary from tissue to tissue. Collagen is especially highly cross-linked in the Achilles tendon, for example, where tensile strength is crucial.

Figure 19–66 summarizes the steps in the synthesis and assembly of collagen fibrils. Given the large number of enzymatic processes involved, it is not surprising that there are many human genetic diseases that affect fibril formation. Mutations affecting type I collagen cause *osteogenesis imperfecta*, characterized by weak bones that fracture easily. Mutations affecting type II collagen cause *chondrodysplasias*, characterized by abnormal cartilage, which leads to bone and joint deformities. And mutations affecting type III collagen cause *Ehlers–Danlos syndrome*, characterized by fragile skin and blood vessels and hypermobile joints.

Secreted Fibril-Associated Collagens Help Organize the Fibrils

In contrast to GAGs, which resist compressive forces, collagen fibrils form structures that resist tensile forces. The fibrils have various diameters and are organized in different ways in different tissues. In mammalian skin, for example, they are woven in a wickerwork pattern so that they resist tensile stress in multiple directions; leather consists of this material, suitably preserved. In tendons, collagen fibrils are organized in parallel bundles aligned along the major axis of

Figure 19–65 **Cross-links formed between modified lysine side chains within a collagen fibril.** Covalent intramolecular and intermolecular cross-links are formed in several steps. First, the extracellular enzyme lysyl oxidase deaminates certain lysines and hydroxylysines, to yield highly reactive aldehyde groups. The aldehydes then react spontaneously to form covalent bonds with each other or with other lysines or hydroxylysines. Most of the cross-links form between the short nonhelical segments at each end of the collagen molecules.

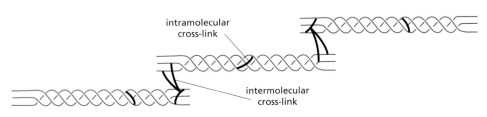

intramolecular
cross-link

intermolecular
cross-link

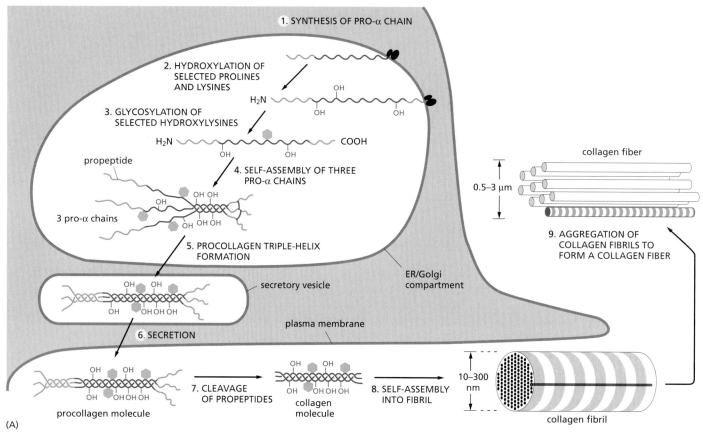

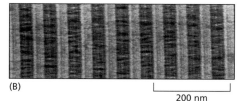

Figure 19–66 The intracellular and extracellular events in the formation of a collagen fibril. (A) Note that procollagen assembles into collagen fibrils in the extracellular space, often within large infoldings in the plasma membrane (not shown). As one example of how collagen fibrils can form ordered arrays in the extracellular space, they are shown further assembling into large collagen fibers, which are visible in the light microscope. The covalent cross-links that stabilize the extracellular assemblies are omitted in this picture. (B) Electron micrograph of a negatively stained collagen fibril reveals its typical striated appearance. (B, courtesy of Robert Horne.)

tension. In mature bone and in the cornea, they are arranged in orderly ply-woodlike layers, with the fibrils in each layer lying parallel to one another but nearly at right angles to the fibrils in the layers on either side. The same arrangement occurs in tadpole skin (**Figure 19–67**).

The connective tissue cells themselves determine the size and arrangement of the collagen fibrils. The cells can express one or more genes for the different types of fibrillar procollagen molecules. But even fibrils composed of the same mixture of fibrillar collagen molecules have different arrangements in different tissues. How is this achieved? Part of the answer is that cells can regulate the disposition of the collagen molecules after secretion by guiding collagen fibril formation in close association with the plasma membrane (see Figure 19–66). In addition, cells can influence this organization by secreting, along with their fibrillar collagens, different kinds and amounts of other matrix macromolecules. In particular, they secrete the fibrous protein fibronectin, as we shall discuss later, and this precedes the formation of collagen fibrils and helps guide their organization.

Fibril-associated collagens, such as types IX and XII collagens, are thought to be especially important in this regard. They differ from fibrillar collagens in several ways.

1. Their triple-stranded helical structure is interrupted by one or two short nonhelical domains, which makes the molecules more flexible than fibrillar collagen molecules.

2. They are not cleaved after secretion and therefore retain their propeptides.
3. They do not aggregate with one another to form fibrils in the extracellular space. Instead, they bind in a periodic manner to the surface of fibrils formed by the fibrillar collagens. Type IX molecules bind to type-II-collagen-containing fibrils in cartilage, the cornea, and the vitreous of the eye (**Figure 19–68**), whereas type XII molecules bind to type-I-collagen-containing fibrils in tendons and various other tissues.

Fibril-associated collagens are thought to mediate the interactions of collagen fibrils with one another and with other matrix macromolecules to help determine the organization of the fibrils in the matrix.

Cells Help Organize the Collagen Fibrils They Secrete by Exerting Tension on the Matrix

Cells interact with the extracellular matrix mechanically as well as chemically, and studies in culture suggest that this mechanical interaction can have dramatic effects on the architecture of connective tissue. Thus, when fibroblasts are mixed with a meshwork of randomly oriented collagen fibrils that form a gel in a culture dish, the fibroblasts tug on the meshwork, drawing in collagen from their surroundings and thereby causing the gel to contract to a small fraction of its initial volume. By similar activities, a cluster of fibroblasts surrounds itself with a capsule of densely packed and circumferentially oriented collagen fibers.

If two small pieces of embryonic tissue containing fibroblasts are placed far apart on a collagen gel, the intervening collagen becomes organized into a compact band of aligned fibers that connect the two explants (**Figure 19–69**). The fibroblasts subsequently migrate out from the explants along the aligned collagen fibers. Thus, the fibroblasts influence the alignment of the collagen fibers, and the collagen fibers in turn affect the distribution of the fibroblasts.

Fibroblasts may have a similar role in organizing the extracellular matrix inside the body, first of all synthesizing the collagen fibrils and depositing them in the correct orientation, then working on the matrix they have secreted, crawling over it and tugging on it so as to create tendons and ligaments and the tough, dense layers of connective tissue that ensheathe and bind together most organs.

Elastin Gives Tissues Their Elasticity

Many vertebrate tissues, such as skin, blood vessels, and lungs, need to be both strong and elastic in order to function. A network of **elastic fibers** in the extracellular matrix of these tissues gives them the required resilience so that they

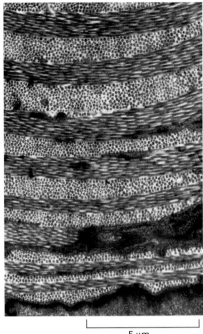

5 μm

Figure 19–67 Collagen fibrils in the tadpole skin. This electron micrograph shows the plywoodlike arrangement of the fibrils: successive layers of fibrils are laid down nearly at right angles to each other. This organization is also found in mature bone and in the cornea. (Courtesy of Jerome Gross.)

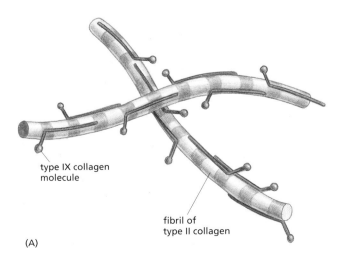

(A)

type IX collagen molecule

fibril of type II collagen

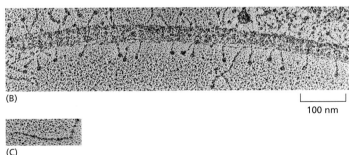

(B)

100 nm

(C)

Figure 19–68 Type IX collagen. (A) Type IX collagen molecules binding in a periodic pattern to the surface of a fibril containing type II collagen. (B) Electron micrograph of a rotary-shadowed type-II-collagen-containing fibril in cartilage, sheathed in type IX collagen molecules. (C) An individual type IX collagen molecule. (B and C, from L. Vaughan et al., *J. Cell Biol.* 106:991–997, 1988. With permission from The Rockefeller University Press.)

can recoil after transient stretch (**Figure 19–70**). Elastic fibers are at least five times more extensible than a rubber band of the same cross-sectional area. Long, inelastic collagen fibrils are interwoven with the elastic fibers to limit the extent of stretching and prevent the tissue from tearing.

The main component of elastic fibers is **elastin**, a highly hydrophobic protein (about 750 amino acids long), which, like collagen, is unusually rich in proline and glycine but, unlike collagen, is not glycosylated and contains some hydroxyproline but no hydroxylysine. Soluble *tropoelastin* (the biosynthetic precursor of elastin) is secreted into the extracellular space and assembled into elastic fibers close to the plasma membrane, generally in cell-surface infoldings. After secretion, the tropoelastin molecules become highly cross-linked to one another, generating an extensive network of elastin fibers and sheets. A mechanism similar to the one that operates in cross-linking collagen molecules forms cross-links between the lysines.

The elastin protein is composed largely of two types of short segments that alternate along the polypeptide chain: hydrophobic segments, which are responsible for the elastic properties of the molecule; and alanine- and lysine-rich α-helical segments, which form cross-links between adjacent molecules. Each segment is encoded by a separate exon. There is still uncertainty concerning the conformation of elastin molecules in elastic fibers and how the structure of these fibers accounts for their rubberlike properties. However, it seems that parts of the elastin polypeptide chain, like the polymer chains in ordinary rubber, adopt a loose "random coil" conformation, and it is the random coil nature of the component molecules cross-linked into the elastic fiber network that allows the network to stretch and recoil like a rubber band (**Figure 19–71**).

Elastin is the dominant extracellular matrix protein in arteries, comprising 50% of the dry weight of the largest artery—the aorta. Mutations in the elastin gene causing a deficiency of the protein in mice or humans result in narrowing of the aorta or other arteries and excessive proliferation of smooth muscle cells in the arterial wall. Apparently, the normal elasticity of an artery is required to restrain the proliferation of these cells.

Elastic fibers do not consist solely of elastin. The elastin core is covered with a sheath of *microfibrils,* each of which has a diameter of about 10 nm. The microfibrils appear before elastin in developing tissues and seem to provide scaffolding to guide elastin deposition. Arrays of microfibrils are elastic in their own right, and in some places they persist in the absence of elastin: they help to hold the lens in its place in the eye, for example. Microfibrils are composed of a number of distinct glycoproteins, including the large glycoprotein *fibrillin,* which binds to elastin and is essential for the integrity of elastic fibers. Mutations in the fibrillin gene result in *Marfan's syndrome,* a relatively common human genetic disorder. In the most severely affected individuals the aorta is

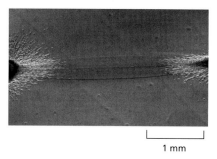

1 mm

Figure 19–69 The shaping of the extracellular matrix by cells. This micrograph shows a region between two pieces of embryonic chick heart (rich in fibroblasts as well as heart muscle cells) that were cultured on a collagen gel for 4 days. A dense tract of aligned collagen fibers has formed between the explants, presumably as a result of the fibroblasts in the explants tugging on the collagen. (From D. Stopak and A.K. Harris, *Dev. Biol.* 90:383–398, 1982. With permission from Academic Press.)

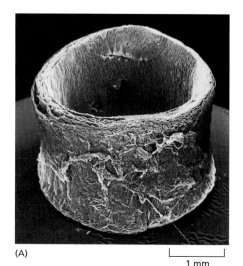

(A)

1 mm

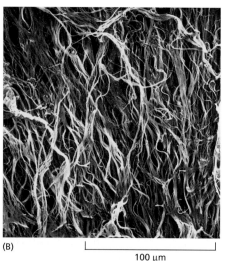

(B)

100 μm

Figure 19–70 Elastic fibers. These scanning electron micrographs show (A) a low-power view of a segment of a dog's aorta and (B) a high-power view of the dense network of longitudinally oriented elastic fibers in the outer layer of the same blood vessel. All the other components have been digested away with enzymes and formic acid. (From K.S. Haas et al., *Anat. Rec.* 230:86–96, 1991. With permission from Wiley-Liss.)

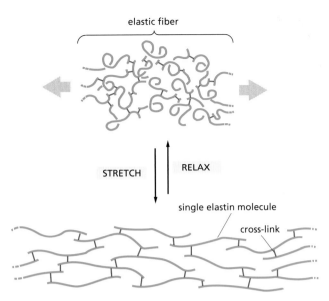

elastic fiber

STRETCH | RELAX

single elastin molecule

cross-link

Figure 19–71 **Stretching a network of elastin molecules.** The molecules are joined together by covalent bonds *(red)* to generate a cross-linked network. In this model, each elastin molecule in the network can extend and contract in a manner resembling a random coil, so that the entire assembly can stretch and recoil like a rubber band.

prone to rupture; other common effects include displacement of the lens and abnormalities of the skeleton and joints. Affected individuals are often unusually tall and lanky: Abraham Lincoln is suspected to have had the condition.

Fibronectin Is an Extracellular Protein That Helps Cells Attach to the Matrix

The extracellular matrix contains a number of noncollagen proteins that typically have multiple domains, each with specific binding sites for other matrix macromolecules and for receptors on the surface of cells. These proteins therefore contribute to both organizing the matrix and helping cells attach to it. Like the proteoglycans, they also guide cell movements in developing tissues, by serving as tracks along which cells can migrate or as repellents that keep cells out of forbidden areas.

The first of this class of matrix proteins to be well characterized was **fibronectin**, a large glycoprotein found in all vertebrates and important for many cell–matrix interactions. Thus, for example, mutant mice that are unable to make fibronectin die early in embryogenesis because their endothelial cells fail to form proper blood vessels. The defect is thought to result from abnormalities in the interactions of these cells with the surrounding extracellular matrix, which normally contains fibronectin.

Fibronectin is a dimer composed of two very large subunits joined by disulfide bonds at one end. Each subunit is folded into a series of functionally distinct domains separated by regions of flexible polypeptide chain (**Figure 19–72**). The domains in turn consist of smaller modules, each of which is serially repeated and usually encoded by a separate exon, suggesting that the fibronectin gene, like the collagen genes, evolved by multiple exon duplications. In the human genome, there is only one fibronectin gene, containing about 50 exons of similar size, but the transcripts can be spliced in different ways to produce many different fibronectin isoforms. Sites in the main type of module, called the **type III fibronectin repeat**, bind to integrins and thereby to cell surfaces. This module is about 90 amino acids long and occurs at least 15 times in each subunit. The type III fibronectin repeat is among the most common of all protein domains in vertebrates.

Tension Exerted by Cells Regulates the Assembly of Fibronectin Fibrils

Fibronectin can exist both in a soluble form, circulating in the blood and other body fluids, and as insoluble *fibronectin fibrils*, in which fibronectin dimers are

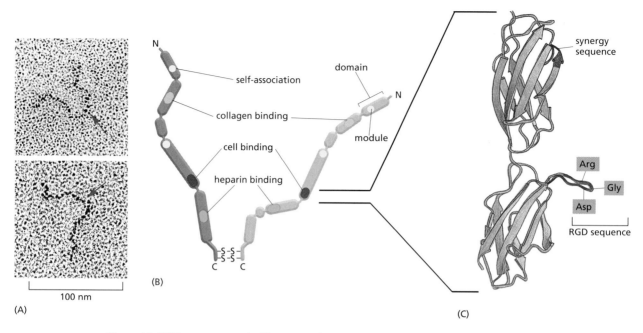

(A)   (B)   (C)

Figure 19–72 The structure of a fibronectin dimer. (A) Electron micrographs of individual fibronectin dimer molecules shadowed with platinum; *red arrows* mark the C-termini. (B) The two polypeptide chains are similar but generally not identical (being made from the same gene but from differently spliced mRNAs). They are joined by two disulfide bonds near the C-termini. Each chain is almost 2500 amino acids long and is folded into five or six domains connected by flexible polypeptide segments. Individual domains contain modules specialized for binding to a particular molecule or to a cell, as indicated for five of the domains. For simplicity, not all of the known binding sites are shown (there are other cell-binding sites, for example). (C) The three-dimensional structure of two type III fibronectin repeat modules as determined by x-ray crystallography. The type III repeat is the main repeating module in fibronectin. Both the Arg-Gly-Asp (RGD) and the "synergy" sequences shown in *red* form part of the major cell-binding site, and are important for binding. (A, from J. Engel et al., *J. Mol. Biol.* 150:97–120, 1981. With permission from Academic Press; C, from Daniel J. Leahy, *Annu. Rev. Cell Dev. Biol.* 13:363–393, 1997. With permission from Annual Reviews.)

cross-linked to one another by additional disulfide bonds and form part of the extracellular matrix. Unlike fibrillar collagen molecules, however, which can self-assemble into fibrils in a test tube, fibronectin molecules assemble into fibrils only on the surface of cells, and only where those cells possess appropriate fibronectin-binding proteins—in particular, integrins. The integrins provide a linkage from the fibronectin outside the cell to the actin cytoskeleton inside it. The linkage transmits tension to the fibronectin molecules—provided that they also have an attachment to some other structure—and stretches them, exposing a cryptic (hidden) binding site in the fibronectin molecules (**Figure 19–73**). This allows them to bind directly to one another and to recruit additional fibronectin molecules to form a fibril (**Figure 19–74**). This dependence on tension and interaction with cell surfaces ensures that fibronectin fibrils assemble where there is a mechanical need for them and not in inappropriate locations such as the bloodstream.

Figure 19–73 Unfolding of a type III fibronectin domain in response to tension. The stretching of fibronectin in this way exposes cryptic binding sites that cause the stretched molecules to assemble into filaments, as shown in Figure 19–74. (From V. Vogel and M. Sheetz, *Nat. Rev. Mol. Cell Biol.* 7:265–275, 2006. With permission from Macmillan Publishers Ltd.)

Many other extracellular matrix proteins have multiple repeats similar to the type III fibronectin repeat, and it is possible that tension exerted on these proteins also uncovers cryptic binding sites and thereby influences their polymerization.

Fibronectin Binds to Integrins Through an RGD Motif

One way to analyze a complex multifunctional protein molecule such as fibronectin is to chop it into pieces and determine the function of its individual domains. When fibronectin is treated with a low concentration of a proteolytic enzyme, the polypeptide chain is cut in the connecting regions between the domains, leaving the domains themselves intact. One can then show that one of its domains binds to collagen, another to heparin, another to specific receptors on the surface of various types of cells, and so on (see Figure 19–72B). Synthetic peptides corresponding to different segments of the cell-binding domain have been used to identify a specific tripeptide sequence (*Arg-Gly-Asp*, or *RGD*), which is found in one of the type III repeats (see Figure 19–72C), as a central feature of the cell-binding site. Even very short peptides containing this **RGD sequence** can compete with fibronectin for the binding site on cells, thereby inhibiting the attachment of the cells to a fibronectin matrix. If these peptides are coupled to a solid surface, they cause cells to adhere to it.

Several extracellular proteins besides fibronectin also have an RGD sequence that mediates cell-surface binding. Some of these are involved in blood clotting, and peptides containing the RGD sequence have been useful in the development of anti-clotting drugs. Some snakes use a similar strategy to cause their victims to bleed: they secrete RGD-containing anti-clotting proteins called *disintegrins* into their venom.

The cell-surface receptors that bind RGD-containing proteins are members of the integrin family. Each integrin, however, specifically recognizes its own small set of matrix molecules, indicating that tight binding requires more than just the RGD sequence. Moreover, RGD sequences are not the only sequence motifs used for binding to integrins: many integrins recognize and bind to other motifs instead.

Cells Have to Be Able to Degrade Matrix, as Well as Make it

The ability of cells to degrade and destroy extracellular matrix is as important as their ability to make it and bind to it. Rapid matrix degradation is required in processes such as tissue repair, and even in the seemingly static extracellular matrix of adult animals there is a slow, continuous turnover, with matrix macromolecules being degraded and resynthesized. This allows bone, for example, to be remodelled so as to adapt to the stresses on it, as discussed in Chapter 23.

From the point of view of individual cells, the ability to cut through matrix is crucial in two ways: it enables them to divide while embedded in matrix, and it enables them to travel through it. As we have already mentioned, cells in connective tissues generally need to be able to stretch out in order to divide. If a cell lacks the enzymes needed to cut through the surrounding matrix or is embedded in a matrix that resists their action, it remains rounded, unable to extend processes because the matrix is impenetrable; as a result, the cell is strongly inhibited from dividing, as well as being hindered from migrating.

Localized degradation of matrix components is also required wherever cells have to escape from confinement by a basal lamina. It is needed during normal branching growth of epithelial structures such as glands, for example, to allow the population of epithelial cells to expand, and needed also when white blood cells migrate across the basal lamina of a blood vessel into tissues in response to infection or injury. Less benignly, matrix degradation is important both for the spread of cancer cells through the body and for the ability of the cancer cells to proliferate in the tissues that they invade (discussed in Chapter 20).

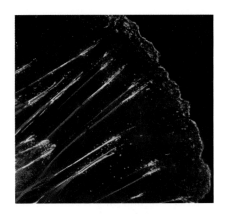

Figure 19–74 Organization of fibronectin into fibrils at the cell surface. The picture shows the front end of a migrating mouse fibroblast. The extracellular adhesion protein fibronectin is stained *green*, and the intracellular filaments of actin, *red*. The fibronectin is initially present as small dot-like aggregates near the leading edge of the cell. It accumulates at focal adhesions (sites of anchorage of actin filaments) and becomes organized into fibrils parallel to the actin filaments. Integrin molecules spanning the cell membrane link the fibronectin outside the cell to the actin filaments inside it. Tension exerted on the fibronectin molecules through this linkage is thought to stretch them, exposing binding sites that promote fibril formation. (Courtesy of Roumen Pankov and Kenneth Yamada.)

Matrix Degradation Is Localized to the Vicinity of Cells

In general, matrix components are degraded by extracellular proteolytic enzymes (proteases) that act close to the cells that produce them. Thus, antibodies that recognize the products of proteolytic cleavage stain matrix only around cells. Many of these proteases belong to one of two general classes. Most are **matrix metalloproteases**, which depend on bound Ca^{2+} or Zn^{2+} for activity; the others are **serine proteases**, which have a highly reactive serine in their active site. Together, metalloproteases and serine proteases cooperate to degrade matrix proteins such as collagen, laminin, and fibronectin. Some metalloproteases, such as the *collagenases*, are highly specific, cleaving particular proteins at a small number of sites. In this way, the structural integrity of the matrix is largely retained, while the limited amount of proteolysis that occurs is sufficient for cell migration. Other metalloproteases may be less specific, but, because they are anchored to the plasma membrane, they can act just where they are needed; it is this type of matrix metalloprotease that is crucial for a cell's ability to divide when embedded in matrix.

Clearly, the activity of the proteases that degrade the matrix components has to be tightly controlled, if the fabric of the body is not to collapse in a heap. Three basic control mechanisms operate:

Local activation: Many proteases are secreted as inactive precursors that can be activated locally when needed. An example is *plasminogen*, an inactive protease precursor that is abundant in the blood. Proteases called *plasminogen activators* cleave plasminogen locally to yield the active serine protease *plasmin*, which helps break up blood clots. *Tissue-type plasminogen activator (tPA)* is often given to patients who have just had a heart attack or thrombotic stroke; it helps dissolve the arterial clot that caused the attack, thereby restoring blood-flow to the tissue.

Confinement by cell-surface receptors: Many cells have receptors on their surface that bind proteases, thereby confining the enzyme to the sites where it is needed. *Urokinase-type plasminogen activator (uPA)* is an example. It is found bound to receptors on the growing tips of axons and at the leading edge of some migrating cells, where it may serve to clear a pathway for their migration. Receptor-bound uPA may also help some cancer cells metastasize (**Figure 19–75**).

Secretion of inhibitors: The action of proteases is confined to specific areas by various secreted protease inhibitors, including the *tissue inhibitors of metalloproteases (TIMPs)* and the serine protease inhibitors known as *serpins*. The inhibitors are protease-specific and bind tightly to the activated enzyme, blocking its activity. An attractive idea is that the inhibitors are secreted by cells at the margins of areas of active protein degradation to protect uninvolved matrix; they may also protect cell-surface proteins required for cell adhesion and migration. The overexpression of TIMPs inhibits the migration of some cell types, emphasizing the importance of metalloproteases for the migration.

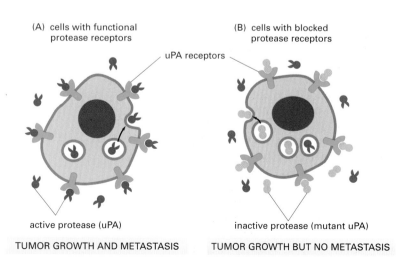

(A) cells with functional protease receptors

(B) cells with blocked protease receptors

uPA receptors

active protease (uPA)

inactive protease (mutant uPA)

TUMOR GROWTH AND METASTASIS   **TUMOR GROWTH BUT NO METASTASIS**

Figure 19–75 The importance of proteases bound to cell-surface receptors. (A) Human prostate cancer cells make and secrete the serine protease uPA, which binds to cell-surface uPA receptor proteins. (B) The same cells have been transfected with DNA that encodes an excess of an inactive form of uPA, which binds to the uPA receptors but has no protease activity. By occupying most of the uPA receptors, the inactive uPA prevents the active protease from binding to the cell surface. Both types of cells secrete active uPA, grow rapidly, and produce tumors when injected into experimental animals. But the cells in (A) metastasize widely, whereas the cells in (B) do not. To metastasize via the blood, tumor cells have to crawl through basal laminae and other extracellular matrices on the way into and out of the bloodstream. This experiment suggests that proteases must be bound to the cell surface to facilitate migration through the matrix.

Summary

Cells in connective tissues are embedded in an intricate extracellular matrix that not only binds the cells together but also influences their survival, development, shape, polarity, and migratory behavior. The matrix contains various protein fibers interwoven in a hydrated gel composed of a network of glycosaminoglycan (GAG) chains.

GAGs are a heterogeneous group of negatively charged polysaccharide chains that (except for hyaluronan) are covalently linked to protein to form proteoglycan molecules. The negative charges attract counterions, which have a powerful osmotic effect, drawing water into the matrix and keeping it swollen so as to occupy a large volume of extracellular space. Proteoglycans are also found on the surface of cells, where they often function as co-receptors to help cells respond to secreted signal proteins.

Fiber-forming proteins give the matrix strength and resilience. They also form structures to which cells can be anchored, often via large multidomain glycoproteins such as laminin and fibronectin that have multiple binding sites for integrins on the cell surface. Elasticity is provided by elastin molecules, which form an extensive crosslinked network of fibers and sheets that can stretch and recoil. The fibrillar collagens (types I, II, III, V, and XI) provide tensile strength. They are ropelike, triple-stranded helical molecules that aggregate into long fibrils in the extracellular space. The fibrils in turn can assemble into various highly ordered arrays. Fibril-associated collagen molecules, such as types IX and XII, decorate the surface of collagen fibrils and influence the interactions of the fibrils with one another and with other matrix components.

Matrix components are degraded by extracellular proteolytic enzymes. Most of these are matrix metalloproteases, which depend on bound Ca^{2+} or Zn^{2+} for activity, while others are serine proteases, which have a reactive serine in their active site. The degradation of matrix components is subject to complex controls, and cells can, for example, cause a localized degradation of matrix components to clear a path through the matrix.

THE PLANT CELL WALL

Each cell in a plant deposits, and is in turn completely enclosed by, an elaborate extracellular matrix called the *plant cell wall*. It was the thick cell walls of cork, visible in a primitive microscope, that in 1663 enabled Robert Hooke to distinguish and name cells for the first time. The walls of neighboring plant cells, cemented together to form the intact plant (**Figure 19–76**), are generally thicker, stronger, and, most important of all, more rigid than the extracellular matrix produced by animal cells. In evolving relatively rigid walls, which can be up to many micrometers thick, early plant cells forfeited the ability to crawl about and adopted a sedentary lifestyle that has persisted in all present-day plants.

The Composition of the Cell Wall Depends on the Cell Type

All cell walls in plants have their origin in dividing cells, as the cell plate forms during cytokinesis to create a new partition wall between the daughter cells (discussed in Chapter 17). The new cells are usually produced in special regions called *meristems* (discussed in Chapter 22), and they are generally small in comparison with their final size. To accommodate subsequent cell growth, the walls of the newborn cells, called **primary cell walls**, are thin and extensible, although tough. Once growth stops, the wall no longer needs to be extensible: sometimes the primary wall is retained without major modification, but, more commonly, a rigid, **secondary cell wall** is produced by depositing new layers of matrix inside the old ones. These new layers generally have a composition that is significantly different from that of the primary wall. The most common additional polymer in secondary walls is **lignin**, a complex network of covalently linked phenolic compounds found in the walls of the xylem vessels and fiber cells of woody tissues.

Although the cell walls of higher plants vary in both composition and organization, they are all constructed, like animal extracellular matrices, using a

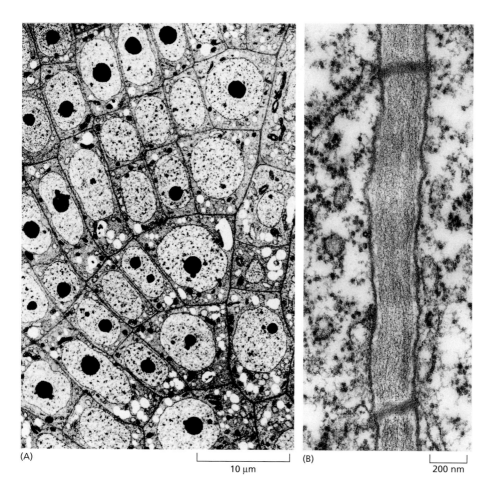

(A)

10 µm

(B)

200 nm

Figure 19–76 Plant cell walls. (A) Electron micrograph of the root tip of a rush, showing the organized pattern of cells that results from an ordered sequence of cell divisions in cells with relatively rigid cell walls. In this growing tissue, the cell walls are still relatively thin, appearing as fine black lines between the cells in the micrograph. (B) Section of a typical cell wall separating two adjacent plant cells. The two dark transverse bands correspond to plasmodesmata that span the wall (see Figure 19–38). (A, courtesy of C. Busby and B. Gunning, *Eur. J. Cell Biol.* 21:214–233, 1980. With permission from Elsevier; B, courtesy of Jeremy Burgess.)

structural principle common to all fiber-composites, including fiberglass and reinforced concrete. One component provides tensile strength, while another, in which the first is embedded, provides resistance to compression. While the principle is the same in plants and animals, the chemistry is different. Unlike the animal extracellular matrix, which is rich in protein and other nitrogen-containing polymers, the plant cell wall is made almost entirely of polymers that contain no nitrogen, including *cellulose* and lignin. For a sedentary organism that depends on CO_2, H_2O, and sunlight, these two abundant biopolymers represent "cheap," carbon-based, structural materials, helping to conserve the scarce fixed nitrogen available in the soil that generally limits plant growth. Thus trees, for example, make a huge investment in the cellulose and lignin that comprise the bulk of their biomass.

In the cell walls of higher plants, the tensile fibers are made from the polysaccharide cellulose, the most abundant organic macromolecule on Earth, tightly linked into a network by *cross-linking glycans*. In primary cell walls, the matrix in which the cross-linked cellulose network is embedded is composed of *pectin*, a highly hydrated network of polysaccharides rich in galacturonic acid. Secondary cell walls contain additional molecules to make them rigid and permanent; lignin, in particular, forms a hard, waterproof filler in the interstices between the other components. All of these molecules are held together by a combination of covalent and noncovalent bonds to form a highly complex structure, whose composition, thickness and architecture depend on the cell type.

The plant cell wall thus has a "skeletal" role in supporting the structure of the plant as a whole, a protective role as an enclosure for each cell individually, and a transport role, helping to form channels for the movement of fluid in the plant. When plant cells become specialized, they generally adopt a specific shape and produce specially adapted types of walls, according to which the different types of cells in a plant can be recognized and classified (**Figure 19–77**;

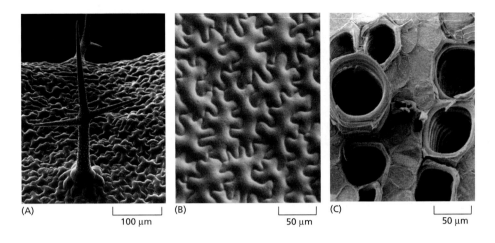

(A) |—————| 100 μm (B) |—————| 50 μm (C) |—————| 50 μm

Figure 19–77 **Some specialized plant cell types with appropriately modified cell walls.** (A) A trichome, or hair, on the upper surface of an *Arabidopsis* leaf. This spiky, protective single cell is shaped by the local deposition of a tough, cellulose-rich wall. (B) Surface view of tomato leaf epidermal cells. The cells fit together snugly like the pieces of a jigsaw puzzle, providing a strong outer covering for the leaf. The outer cell wall is reinforced with a cuticle and with waxes that waterproof the leaf and help defend it against pathogens. (C) This view into young xylem elements shows the thick, lignified, hoop-reinforced secondary cell wall that creates robust tubes for the transport of water throughout the plant. (A, courtesy of Paul Linstead; B and C, courtesy of Kim Findlay.)

see also Panel 22–2, pp. 1404–1405).We focus here, however, on the primary cell wall and the molecular architecture that underlies its remarkable combination of strength, resilience, and plasticity, as seen in the growing parts of a plant.

The Tensile Strength of the Cell Wall Allows Plant Cells to Develop Turgor Pressure

The aqueous extracellular environment of a plant cell consists of the fluid contained in the walls that surround the cell. Although the fluid in the plant cell wall contains more solutes than does the water in the plant's external milieu (for example, soil), it is still hypotonic in comparison with the cell interior. This osmotic imbalance causes the cell to develop a large internal hydrostatic pressure, or **turgor pressure**, which pushes outward on the cell wall, just as an inner tube pushes outward on a tire. The turgor pressure increases just to the point where the cell is in osmotic equilibrium, with no net influx of water despite the salt imbalance (see Panel 11–1, p. 664). The turgor pressure generated in this way may reach 10 or more atmospheres, about five times that in the average car tire. This pressure is vital to plants because it is the main driving force for cell expansion during growth, and it provides much of the mechanical rigidity of living plant tissues. Compare the wilted leaf of a dehydrated plant, for example, with the turgid leaf of a well-watered one. It is the mechanical strength of the cell wall that allows plant cells to sustain this internal pressure.

The Primary Cell Wall Is Built from Cellulose Microfibrils Interwoven with a Network of Pectic Polysaccharides

Cellulose gives the primary cell wall tensile strength. Each cellulose molecule consists of a linear chain of at least 500 glucose residues that are covalently linked to one another to form a ribbonlike structure, which is stabilized by hydrogen bonds within the chain (**Figure 19–78**). In addition, hydrogen bonds between adjacent cellulose molecules cause them to stick together in overlapping parallel arrays, forming bundles of about 40 cellulose chains, all of which have the same polarity. These highly ordered crystalline aggregates, many micrometers long, are called **cellulose microfibrils**, and they have a tensile strength comparable to steel (see Figure 19–78). Sets of microfibrils are arranged in layers, or lamellae, with each microfibril about 20–40 nm from its neighbors and connected to them by long cross-linking glycan molecules that are attached by hydrogen bonds to the surface of the microfibrils. The primary cell wall consists of several such lamellae arranged in a plywoodlike network (**Figure 19–79**).

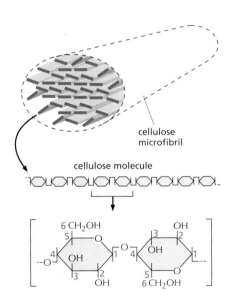

cellulose microfibril

cellulose molecule

Figure 19–78 **Cellulose.** Cellulose molecules are long, unbranched chains of β1,4-linked glucose units. Each glucose residue is inverted with respect to its neighbors, and the resulting disaccharide repeat occurs hundreds of times in a single cellulose molecule. About 40 individual cellulose molecules assemble to form a strong, hydrogen-bonded cellulose microfibril.

The **cross-linking glycans** are a heterogeneous group of branched polysaccharides that bind tightly to the surface of each cellulose microfibril and thereby help to cross-link the microfibrils into a complex network. Their function is analogous to that of the fibril-associated collagens discussed earlier (see Figure 19–68). There are many classes of cross-linking glycans, but they all have a long linear backbone composed of one type of sugar (glucose, xylose, or mannose) from which short side chains of other sugars protrude. It is the backbone sugar molecules that form hydrogen bonds with the surface of cellulose microfibrils, cross-linking them in the process. Both the backbone and the side-chain sugars vary according to the plant species and its stage of development.

Coextensive with this network of cellulose microfibrils and cross-linking glycans is another cross-linked polysaccharide network based on **pectins** (see Figure19–79). Pectins are a heterogeneous group of branched polysaccharides that contain many negatively charged galacturonic acid units. Because of their negative charge, pectins are highly hydrated and associated with a cloud of cations, resembling the glycosaminoglycans of animal cells in the large amount of space they occupy (see Figure 19–56). When Ca^{2+} is added to a solution of pectin molecules, it cross-links them to produce a semirigid gel (it is pectin that is added to fruit juice to make jam set). Certain pectins are particularly abundant in the *middle lamella*, the specialized region that cements together the walls of adjacent cells (see Figure 19–79); here, Ca^{2+} cross-links are thought to help hold cell-wall components together. Although covalent bonds also play a part in linking the components, very little is known about their nature. Regulated separation of cells at the middle lamella underlies such processes as the ripening of tomatoes and the abscission (detachment) of leaves in the fall.

In addition to the two polysaccharide-based networks that form the bulk of all plant primary cell walls, proteins are present, contributing up to about 5% of the wall's dry mass. Many of these proteins are enzymes, responsible for wall turnover and remodeling, particularly during growth. Another class of wall proteins, like collagen, contain high levels of hydroxyproline. These proteins are thought to strengthen the wall, and they are produced in greatly increased amounts as a local response to attack by pathogens. From the genome sequence of *Arabidopsis*, it has been estimated that more than 700 genes are required to synthesize, assemble, and remodel the plant cell wall. Some of the main polymers found in the primary and secondary cell wall are listed in **Table 19–8**.

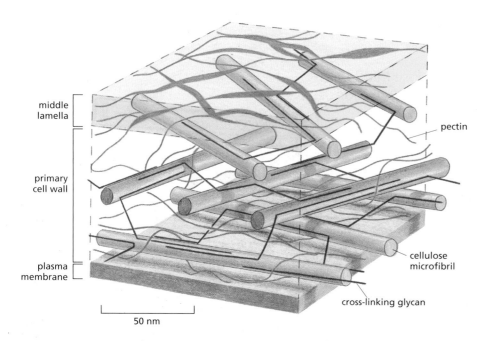

middle lamella

primary cell wall

plasma membrane

pectin

cellulose microfibril

cross-linking glycan

50 nm

Figure 19–79 Scale model of a portion of a primary plant cell wall showing the two major polysaccharide networks. The orthogonally arranged layers of cellulose microfibrils *(green)* are tied into a network by the cross-linking glycans *(red)* that form hydrogen bonds with the microfibrils. This network is coextensive with a network or matrix of pectic polysaccharides *(blue)*. The network of cellulose and cross-linking glycans provides tensile strength, while the pectin network resists compression. Cellulose, cross-linking glycans, and pectin are typically present in roughly equal amounts in a primary cell wall. The middle lamella is especially rich in pectin, and it cements adjacent cells together.

Table 19–8 The Polymers of the Plant Cell Wall

POLYMER	COMPOSITION	FUNCTIONS
Cellulose	linear polymer of glucose	fibrils confer tensile strength on all walls
Cross-linking glycans	xyloglucan, glucuronoarabinoxylan, and mannan	cross-link cellulose fibrils into robust network
Pectin	homogalacturonans and rhamnogalacturonans	forms negatively charged, hydrophilic network that gives compressive strength to primary walls; cell–cell adhesion
Lignin	cross-linked coumaryl, coniferyl, and sinapyl alcohols	forms strong waterproof polymer that reinforces secondary cell walls
Proteins and glycoproteins	enzymes, hydroxyproline-rich proteins	responsible for wall turnover and remodeling; help defend against pathogens

Oriented Cell-Wall Deposition Controls Plant Cell Growth

Once a plant cell has left the meristem where it is generated, it can grow dramatically, commonly by more than a thousand times in volume. The manner of this expansion determines the final shape of each cell, and hence the final form of the plant as a whole. Turgor pressure inside the cell drives the expansion, but it is the behavior of the cell wall that governs its direction and extent. Complex wall-remodeling activities are required, as well as the deposition of new wall materials. Because of their crystalline structure, the individual cellulose microfibrils in the wall are unable to stretch, and this gives them a crucial role in the process. For the cell wall to stretch or deform, the microfibrils must either slide past one another or become more widely separated, or both. The orientation of the microfibrils in the innermost layers of the wall governs the direction in which the cell expands. Cells in plants therefore anticipate their future morphology by controlling the orientation of the cellulose microfibrils that they deposit in the wall (**Figure 19–80**).

Unlike most other matrix macromolecules, which are made in the endoplasmic reticulum and Golgi apparatus and are secreted, cellulose, like hyaluronan in animals, is spun out from the surface of the cell by a plasma-membrane-bound enzyme complex (cellulose synthase), which uses as its substrate the sugar nucleotide UDP-glucose supplied from the cytosol. Each enzyme complex, or *rosette*, has a six-fold symmetry and contains the protein products of three separate cellulose synthase *(CESA)* genes. Each CESA protein

Figure 19–80 Cellulose microfibrils influence the direction of cell elongation. (A) The orientation of cellulose microfibrils in the primary cell wall of an elongating carrot cell is shown in this electron micrograph of a shadowed replica from a rapidly frozen and deep-etched cell wall. The cellulose microfibrils are aligned parallel to one another and perpendicular to the axis of cell elongation. The microfibrils are cross-linked by, and interwoven with, a complex web of matrix molecules (compare with Figure 19–79). (B,C) The cells in (A) and (B) start off with identical shapes (shown here as cubes) but with different net orientations of cellulose microfibrils in their walls. Although turgor pressure is uniform in all directions, cell wall loosening allows each cell to elongate only in a direction perpendicular to the orientation of the innermost layer of microfibrils, which have great tensile strength. Cell expansion occurs in concert with the insertion of new wall material. The final shape of an organ, such as a shoot, is determined in part by the direction in which its component cells can expand. (A, courtesy of Brian Wells and Keith Roberts.)

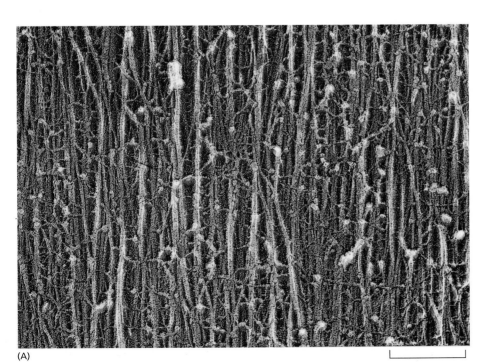

(A)

200 nm

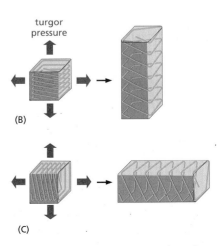

turgor pressure

(B)

(C)

is essential for the production of a cellulose microfibril. Three CESA genes are required for primary cell wall synthesis and a different three for secondary cell wall synthesis.

As they are being synthesized, the nascent cellulose chains assemble into microfibrils. These are spun out on the extracellular surface of the plasma membrane, forming a layer, or lamella, in which all the microfibrils have more or less the same alignment (see Figure 19–79). Each new lamella is deposited internally to the previous one, so that the wall consists of concentrically arranged lamellae, with the oldest on the outside. The most recently deposited microfibrils in elongating cells commonly lie perpendicular to the axis of cell elongation, although the orientation of the microfibrils in the outer lamellae that were laid down earlier may be different (see Figure 19–80B and C).

Microtubules Orient Cell-Wall Deposition

An important clue to the mechanism that dictates this orientation came from observations of the microtubules in plant cells. These are frequently arranged in the cortical cytoplasm with the same orientation as the cellulose microfibrils that are currently being deposited in the cell wall in that region. These cortical microtubules form a *cortical array* close to the cytosolic face of the plasma membrane, held there by poorly characterized proteins (**Figure 19–81**). The

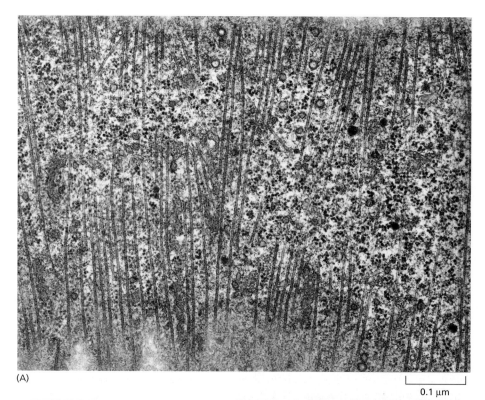

(A)

|_____|
0.1 μm

(B)

|_____|
10 μm

Figure 19–81 **The cortical array of microtubules in a plant cell.** (A) A grazing section of a root-tip cell from timothy grass, showing a cortical array of microtubules lying just below the plasma membrane. These microtubules are oriented perpendicularly to the long axis of the cell. (B) Epidermal cells in a root tip of *Arabidopsis* that has been stained, by immunofluorescence, to reveal the entire cortical microtubule arrays in these rapidly elongating cells. The clear transverse arrays are at right angles to the axis of root elongation. (A, courtesy of Brian Gunning; B, courtesy of Keiko Sugimoto-Shirasu.)

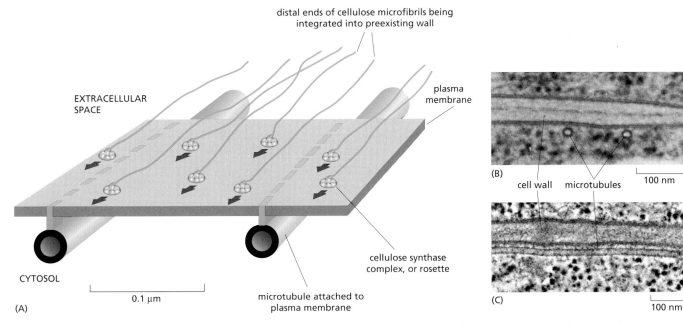

EXTRACELLULAR
SPACE

plasma
membrane

CYTOSOL

0.1 μm

(A)

cellulose synthase
complex, or rosette

microtubule attached to
plasma membrane

(B)

cell wall   microtubules

100 nm

(C)

100 nm

congruent orientation of the cortical array of microtubules (lying just inside the plasma membrane) and cellulose microfibrils (lying just outside) is seen in many types and shapes of plant cells and is present during both primary and secondary cell-wall deposition, suggesting a causal relationship.

This relationship can be tested by treating a plant tissue with a microtubule-depolymerizing drug so as to disassemble the entire system of cortical micro-tubules. The consequences for subsequent cellulose deposition, however, are not as straightforward as might be expected. The drug treatment does not disrupt the production of new cellulose microfibrils, and in some cases cells can continue to deposit new microfibrils in the preexisting orientation. Any developmental switch in the orientation of the microfibril pattern that would normally occur between successive lamellae, however, is invariably blocked. It seems that a pre-existing orientation of microfibrils can be propagated even in the absence of microtubules, but any change in the deposition of cellulose microfibrils requires that intact microtubules be present to determine the new orientation.

These observations are consistent with the following model. The cellulose-synthesizing rosettes embedded in the plasma membrane spin out long cellulose molecules. As the synthesis of cellulose molecules and their self-assembly into microfibrils proceeds, the distal end of each microfibril presumably forms indirect cross-links to the previous layer of wall material, orienting the new microfibril in parallel with the old ones as it becomes integrated into the texture of the wall. Since the microfibril is stiff, the rosette at its growing, proximal end has to move as it deposits the new material, traveling in the plane of the membrane in the direction defined by the way in which the far end of the microfibril is anchored in the existing wall. In this way, each layer of microfibrils would tend to be spun out from the membrane in the same orientation as the layer laid down previously, with the rosettes following the direction of the preexisting oriented microfibrils outside the cell. Oriented microtubules inside the cell, however, can force a change in the direction in which the rosettes move: they can create boundaries in the plasma membrane that act like the banks of a canal to constrain rosette movement (**Figure 19–82**). In this view, cellulose synthesis can occur independently of microtubules; but it is constrained spatially when corti-cal microtubules are present to define membrane microdomains within which the enzyme complex can move.

In this way, plant cells can change their direction of expansion by a sudden change in the orientation of their cortical array of microtubules. Because plant cells cannot move (being constrained by their walls), the entire morphology of a multicellular plant presumably depends on a coordinated, highly patterned deployment of cortical microtubule orientations during plant development. It is not known how these orientations are controlled, although it has been shown

Figure 19–82 One model of how the orientation of newly deposited cellulose microfibrils might be determined by the orientation of cortical microtubules. (A) The large cellulose synthase complexes are integral membrane proteins that continuously synthesize cellulose microfibrils on the outer face of the plasma membrane. The distal ends of the stiff microfibrils become integrated into the texture of the wall, and their elongation at the proximal end pushes the synthase complex along in the plane of the membrane. Because the cortical array of microtubules is attached to the plasma membrane in a way that confines this complex to defined membrane channels, the orientation of these microtubules—when they are present—determines the axis along which the new microfibrils are laid down. (B,C) Two electron micrographs show the tight association of the cortical microtubules with the plasma membrane. One shows the microtubules in cross section while the other shows a longitudinal section. Both emphasize the constant gap of about 20 nm between membrane and microtubule; the connecting molecules responsible remain obscure. (B and C, courtesy of Andrew Staehelin.)

that the microtubules can reorient rapidly in response to extracellular stimuli, including low-molecular-weight plant growth regulators such as ethylene and gibberellic acid (see Figure 22–119).

Microtubules are not, however, the only cytoskeletal elements that influence wall deposition. Local foci of cortical actin filaments can also direct the deposition of new wall material at specific sites on the cell surface, contributing to the elaborate final shaping of many differentiated plant cells, such as the leaf trichomes and epidermal cells shown in Figure 19–77. In cells that cannot organize their cortical actin network, as in *Arp2/3* mutants, the cells lose their characteristic shapes, as illustrated in Figure 16–35.

Summary

Plant cells are surrounded by a tough extracellular matrix, or cell wall, which is responsible for many of the unique features of a plant's life style. The wall is composed of a network of cellulose microfibrils and cross-linking glycans, embedded in a highly cross-linked matrix of pectic polysaccharides. In secondary cell walls, lignin may be deposited making them waterproof, hard, and woody. A cortical array of microtubules can control the orientation of newly deposited cellulose microfibrils, which in turn determine the direction of cell expansion and therefore the final shape of the cell and, ultimately, of the plant as a whole.

PROBLEMS

Which statements are true? Explain why or why not.

19–1   Given the numerous processes inside cells that are regulated by changes in Ca^{2+} concentration, it seems likely that Ca^{2+}-dependent cell–cell adhesions are also regulated by changes in Ca^{2+} concentration.

19–2   Tight junctions perform two distinct functions: they seal the space between cells to restrict paracellular flow and they fence off membrane domains to prevent mixing of apical and basolateral proteins.

19–3   Integrins can convert mechanical signals into molecular signals.

19–4   The elasticity of elastin derives from its high content of α helices, which act as molecular springs.

Discuss the following problems.

19–5   Comment on the following (1922) quote from Warren Lewis, who was one of the pioneers of cell biology. "Were the various types of cells to lose their stickiness for one another and for the supporting extracellular matrix, our bodies would at once disintegrate and flow off into the ground in a mixed stream of cells."

19–6   Cell adhesion molecules were originally identified using antibodies raised against cell-surface components to block cell aggregation. In the adhesion-blocking assays, the researchers found it necessary to use antibody fragments, each with a single binding site (so-called Fab fragments), rather than intact IgG antibodies, which are Y-shaped molecules with two identical binding sites. The Fab fragments were generated by digesting the IgG antibodies with papain, a protease, to separate the two binding sites (**Figure Q19–1**). Why do you suppose it was necessary to use Fab fragments to block cell aggregation?

19–7   The food poisoning bacterium *Clostridium perfringens* makes a toxin that binds to members of the claudin family of proteins, which are the main constituents of tight junctions. When bound to a claudin, the C-terminus of the toxin allows the N-terminus to insert into the adjacent cell membrane, forming holes that kill the cell. The portion of the toxin that binds to the claudins has proven to be a valuable reagent for investigating the properties of tight junctions. MDCK cells are a common choice for studies of tight junctions because they can form an intact epithelial sheet with high transepithelial resistance. MDCK cells express two claudins: claudin-1, which is not bound by the toxin, and claudin-4, which is.

When an intact MDCK epithelial sheet is incubated with the C-terminal toxin fragment, claudin-4 disappears, becoming undetectable within 24 hours. In the absence of claudin-4, the cells remain healthy and the epithelial sheet appears intact. The mean number of strands in the tight junctions that link the cells also decreases over 24 hours from about four to about two, and they are less highly branched. A functional assay for the integrity of the tight junctions shows that transepithelial resistance decreases dramatically in the presence of the toxin, but the resistance can be restored by washing it out (**Figure Q19–2A**). Curiously, the toxin produces these effects only when it is added to the basolateral side of the sheet; it has no effect when added to the apical surface (Figure Q19–2B).

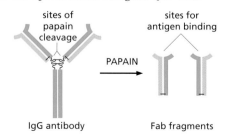

Figure Q19–1 Production of Fab fragments from IgG antibodies by digestion with papain (Problem 19–6).

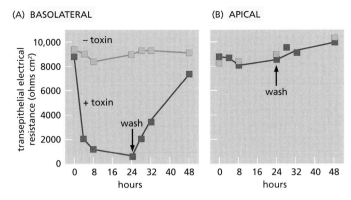

Figure Q19–2 Effects of *Clostridium* toxin on the barrier function of MDCK cells (Problem 19–7). (A) Addition of toxin from the basolateral side of the epithelial sheet. (B) Addition of toxin from the apical side of the epithelial sheet. The higher the resistance (ohms cm²), the less the paracellular current for a given voltage.

A. How can it be that two tight junction strands remain, even though all of the claudin-4 has disappeared?

B. How do you suppose the toxin fragment causes the tight junction strands to disintegrate?

C. Why do you suppose the toxin works when it is added to the basolateral side of the epithelial sheet, but not when added to the apical side?

19–8 It is not an easy matter to assign particular functions to specific components of the basal lamina, since the overall structure is a complicated composite material with both mechanical and signaling properties. Nidogen, for example, cross-links two central components of the basal lamina by binding to the laminin-γ1 chain and to type IV collagen. Given such a key role, it was surprising that mice with a homozygous knockout of the gene for nidogen-1 were entirely healthy, with no abnormal phenotype. Similarly, mice homozygous for a knockout of the gene for nidogen-2 also appeared completely normal. By contrast, mice that were homozygous for a defined mutation in the gene for laminin-γ1, which eliminated just the binding site for nidogen, died at birth with severe defects in lung and kidney formation. The mutant portion of the laminin-γ1 chain is thought to have no other function than to bind nidogen, and does not affect laminin structure or its ability to assemble into basal lamina. How would you explain these genetic observations, which are summarized in **Table Q19–1**? What would you predict would be the phenotype of a mouse that was homozygous for knockouts of both nidogen genes?

19–9 Discuss the following statement: "The basal lamina of muscle fibers serves as a molecular bulletin board, in which adjoining cells can post messages that direct the differentiation and function of the underlying cells."

Table Q19–1 Phenotypes of mice with genetic defects in components of the basal lamina (Problem 19–8).

PROTEIN	GENETIC DEFECT	PHENOTYPE
nidogen-1	gene knockout (–/–)	none
nidogen-2	gene knockout (–/–)	none
laminin-γ1	nidogen binding-site deletion (+/–)	none
laminin-γ1	nidogen binding-site deletion (–/–)	dead at birth

+/– stands for heterozygous, –/– stands for homozygous.

19–10 The affinity of integrins for matrix components can be modulated by changes to their cytoplasmic domains: a process known as inside-out signaling. You have identified a key region in the cytoplasmic domains of αIIbβ3 integrin that seems to be required for inside-out signaling (**Figure Q19–3**). Substitution of alanine for either D723 in the β chain or R995 in the α chain leads to a high level of spontaneous activation, under conditions where the wild-type chains are inactive. Your advisor suggests that you convert the aspartate in the β chain to an arginine (D723R) and the arginine in the α chain to an aspartate (R995D). You compare all three α chains (R995, R995A, and R995D) against all three β chains (D723, D723A, and D723R). You find that all pairs have a high level of spontaneous activation, except D723 vs R995 (the wild type) and D723R vs R995D, which have low levels. Based on these results, how do you think the αIIbβ3 integrin is held in its inactive state?

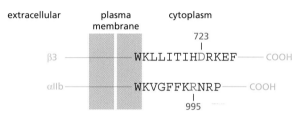

Figure Q19–3 Schematic representation of αIIbβ3 integrin (Problem 19–10). The D723 and R995 residues are indicated. (From P.E. Hughes et al. *J. Biol. Chem.* 271:6571–6574, 1996. With permission from the American Society for Biochemistry and Molecular Biology.)

19–11 Carboxymethyl Sephadex is a negatively charged crosslinked dextran that is commonly used for purifying proteins. It comes in the form of dry beads that swell tremendously when added to water. You have packed a chromatography column with the swollen gel. When you start equilibrating the column with a buffer that contains 50 mM NaCl at neutral pH, you are alarmed to see a massive shrinkage in gel volume. Why does the dry Sephadex swell so dramatically when it is placed in water? Why does the swollen gel shrink so much when a salt solution is added?

19–12 At body temperature, L-aspartate in proteins racemizes to D-aspartate at an appreciable rate. Most proteins in the body have a very low level of D-aspartate, if it can be detected at all. Elastin, however, has a fairly high level of D-aspartate. Moreover, the amount of D-aspartate increases in direct proportion to the age of the person from whom the sample was taken. Why do you suppose that most proteins have little if any D-aspartate, while elastin has high, age-dependent levels?

19–13 Your boss is coming to dinner! All you have for a salad is some wilted, day-old lettuce. You vaguely recall that there is a trick to rejuvenating wilted lettuce, but you can't remember what it is. Should you soak the lettuce in salt water, soak it in tap water, or soak it in sugar water, or maybe just shine a bright light on it and hope that photosynthesis will perk it up?

19–14 The hydraulic conductivity of a single water channel is 4.4×10^{-22} m³ per second per MPa (megapascal) of pressure. What does this correspond to in terms of water molecules per second at atmospheric pressure? [Atmospheric pressure is 0.1 MPa (1 bar) and the concentration of water is 55.5 M.]

REFERENCES

General

Beckerle M (ed) (2002) Cell Adhesion. Oxford: Oxford University Press.

Kreis T & Vale R (ed) (1999) Guidebook to the Extracellular Matrix, Anchor, and Adhesion Proteins, 2nd ed. Oxford: Oxford University Press.

Miner JH (ed) (2005) Extracellular Matrix in Development and Disease (Advances in Developmental Biology, Vol 15). Elsevier Science.

Cadherins and Cell–Cell Adhesion

Garrod DR, Merritt AJ & Nie Z (2002) Desmosomal cadherins. *Curr Opin Cell Biol* 14:537–545.

Gumbiner BM (2005) Regulation of cadherin-mediated adhesion in morphogenesis. *Nature Rev Mol Cell Biol* 6:622–634.

King N, Hittinger CT & Carroll SB (2003) Evolution of key cell signaling and adhesion protein families predates animal origins. *Science* 301:361–363.

Kowalczyk AP & Reynolds AB (2004) Protecting your tail: regulation of cadherin degradation by p120-catenin. *Curr Opin Cell Biol* 16:522–527.

Nelson WJ & Nusse R (2004) Convergence of Wnt, beta-catenin, and cadherin pathways. *Science* 303:1483–1487.

Pokutta S & Weis WI (2007) Structure and mechanism of cadherins and catenins in cell-cell contacts. *Annu Rev Cell Dev Biol.* in press.

Rosen SD (2004) Ligands for L-selectin: homing, inflammation, and beyond. *Annu Rev Immunol* 22:129–156.

Scheiffele P, Fan J, Choih J, Fetter R & Serafini T (2000) Neuroligin expressed in nonneuronal cells triggers presynaptic development in contacting axons. *Cell* 101:657–669.

Shapiro L, Love J & Colman DR (2007) Adhesion molecules in the nervous system: structural insights into function and diversity. *Annu Rev Neurosci* 30:451–474.

Sheng M & Hoogenraad CC (2007) The postsynaptic architecture of excitatory synapses: a more quantitative view. *Annu Rev Biochem* 76:823–847.

Takeichi M (2007) The cadherin superfamily in neuronal connections and interactions. *Nature Rev Neurosci* 8:11–20.

Vestweber D (2002) Regulation of endothelial cell contacts during leukocyte extravasation. *Curr Opin Cell Biol* 14:587–593.

Yamagata M, Sanes JR & Weiner JA (2003) Synaptic adhesion molecules. *Curr Opin Cell Biol* 15:621–632.

Yang J, Mani SA, Donaher JL et al (2004) Twist, a master regulator of morphogenesis, plays an essential role in tumor metastasis. *Cell* 117:927–939.

Tight Junctions and the Organization of Epithelia

Furuse M, Hata M, Furuse K et al (2002) Claudin-based tight junctions are crucial for the mammalian epidermal barrier: a lesson from claudin-1-deficient mice. *J Cell Biol* 156:1099–1111.

Ikenouchi J, Furuse M, Furuse K et al (2005) Tricellulin constitutes a novel barrier at tricellular contacts of epithelial cells. *J Cell Biol* 171:939–945.

Henrique D & Schweisguth F (2003) Cell polarity: the ups and downs of the Par6/aPKC complex. *Curr Opin Genet Dev* 13:341–350.

Nelson WJ (2003) Adaptation of core mechanisms to generate cell polarity. *Nature* 422:766–774.

O'Brien LE, Jou TS, Pollack AL et al (2001) Rac1 orientates epithelial apical polarity through effects on basolateral laminin assembly. *Nature Cell Biol* 3:831–838.

Seifert JR & Mlodzik M (2007) Frizzled/PCP signalling: a conserved mechanism regulating cell polarity and directed motility. *Nature Rev Genet* 8:126–138.

Shin K, Fogg VC & Margolis B (2006) Tight junctions and cell polarity. *Annu Rev Cell Dev Biol* 22:207–236.

Passageways from Cell to Cell: Gap Junctions and Plasmodesmata

Cilia ML & Jackson D (2004) Plasmodesmata form and function. *Curr Opin Cell Biol* 16:500–506.

Gaietta G, Deerinck TJ, Adams SR et al (2002) Multicolor and electron microscopic imaging of connexin trafficking. *Science* 296:503–507.

Goodenough DA & Paul DL (2003) Beyond the gap: functions of unpaired connexon channels. *Nature Rev Mol Cell Biol* 4:285–294.

Segretain D & Falk MM (2004) Regulation of connexin biosynthesis, assembly, gap junction formation, and removal. *Biochim Biophys Acta* 1662:3–21.

Wei CJ, Xu X & Lo CW (2004) Connexins and cell signaling in development and disease. *Annu Rev Cell Dev Biol* 20:811–838.

The Basal Lamina

Miner JH & Yurchenco PD (2004) Laminin functions in tissue morphogenesis. *Annu Rev Cell Dev Biol* 20:255–284.

Sanes JR. (2003) The basement membrane/basal lamina of skeletal muscle. *J Biol Chem* 278:12601–12604.

Sasaki T, Fassler R & Hohenester E (2004) Laminin: the crux of basement membrane assembly. *J Cell Biol* 164:959–963.

Yurchenco PD, Amenta PS & Patton BL (2004) Basement membrane assembly, stability and activities observed through a developmental lens. *Matrix Biol* 22:521–538.

Integrins and Cell–Matrix Adhesion

Galbraith CG, Yamada KM & Sheetz MP (2002) The relationship between force and focal complex development. *J Cell Biol* 159:695–705.

Hynes RO (2002) Integrins: bidirectional, allosteric signaling machines. *Cell* 110:673–687.

Luo BH & Springer TA (2006) Integrin structures and conformational signaling. *Curr Opin Cell Biol* 18:579–586.

Nayal A, Webb DJ & Horwitz AF (2004) Talin: an emerging focal point of adhesion dynamics. *Curr Opin Cell Biol* 16:94–98.

Vogel V & Sheetz M (2006) Local force and geometry sensing regulate cell functions. *Nature Rev Mol Cell Biol* 7:265–275.

The Extracellular Matrix of Animal Connective Tissues

Aszodi A, Legate KR, Nakchbandi I & Fassler R (2006) What mouse mutants teach us about extracellular matrix function. *Annu Rev Cell Dev Biol* 22:591–621.

Bulow HE & Hobert O (2006) The molecular diversity of glycosaminoglycans shapes animal development. *Annu Rev Cell Dev Biol* 22:375–407.

Couchman JR (2003) Syndecans: proteoglycan regulators of cell-surface microdomains? *Nature Rev Mol Cell Biol* 4:926–937.

Hotary KB, Allen ED, Brooks PC et al (2003) Membrane type I matrix metalloproteinase usurps tumor growth control imposed by the three-dimensional extracellular matrix. *Cell* 114:33–45.

Kielty CM, Sherratt MJ & Shuttleworth CA (2002) Elastic fibres. *J Cell Sci* 115:2817–2828.

Larsen M, Artym VV, Green JA & Yamada KM (2006) The matrix reorganized: extracellular matrix remodeling and integrin signaling. *Curr Opin Cell Biol* 18:463–471.

Lin X (2004) Functions of heparan sulfate proteoglycans in cell signaling during development. *Development* 131:6009–6021.

Mott JD & Werb Z (2004) Regulation of matrix biology by matrix metalloproteases. *Curr Opin Cell Biol* 16:558–564.

Prockop DJ & Kivirikko KI (1995) Collagens: molecular biology, diseases, and potentials for therapy. *Annu Rev Biochem* 64:403–434.

Toole BP (2001) Hyaluronan in morphogenesis. *Semin Cell Dev Biol* 12:79–87.

The Plant Cell Wall

Carpita NC & McCann M (2000) The Cell Wall. In Biochemistry and Molecular Biology of Plants (Buchanan BB, Gruissem W & Jones RL eds), pp 52–108. Rockville, MD: ASPB.

Carpita NC, Campbell M & Tierney M (eds) (2001) Plant cell walls. *Plant Mol Biol* 47:1–340. (special issue)

Dhugga KS (2001) Building the wall: genes and enzyme complexes for polysaccharide synthases. *Curr Opin Plant Biol* 4:488–493.

McCann MC & Roberts K (1991) Architecture of the primary cell wall. In The Cytoskeletal Basis of Plant Growth and Form (Lloyd CW ed), pp 109–129. London: Academic Press.

Smith LG & Oppenheimer DG (2005) Spatial control of cell expansion by the plant cytoskeleton. *Annu Rev Cell Dev Biol* 21:271–295.

Somerville C (2006) Cellulose synthesis in higher plants. *Annu Rev Cell Dev Biol* 22:53–78.

Chapter 20

Cancer

About one in five of us will die of cancer, but that is not why we devote a chapter to this disease. Cancer cells break the most basic rules of cell behavior by which multicellular organisms are built and maintained, and they exploit every kind of opportunity to do so. In studying these transgressions, we discover what the normal rules are and how they are enforced. Thus, in the context of cell biology, cancer has a unique importance, and the emphasis given to cancer research has profoundly benefited a much wider area of biomedical science than that of cancer alone.

The effort to combat cancer has led to many fundamental discoveries in cell biology. Many proteins have been discovered because abnormalities in their function can lead to uncontrolled growth, increased division, decreased death, or other aberrant characteristics of cancer cells. Among them are proteins involved in DNA repair (Chapter 5), cell signaling (Chapter 15), the cell cycle and cell growth (Chapter 17), programmed cell death (apoptosis, Chapter 18), and tissue architecture (Chapter 19).

In this chapter, we first examine the nature of cancer by describing the natural history of the disease from a cellular standpoint. In a second section, we discuss the molecular changes that make a cell cancerous. Finally, we end the chapter by discussing how our enhanced understanding of the molecular basis of cancer is leading to improved methods for its prevention and treatment.

CANCER AS A MICROEVOLUTIONARY PROCESS

The body of an animal operates as a society or ecosystem. The individual members are cells that reproduce by cell division and organize into collaborative assemblies called tissues. This society is very peculiar, however, because self-sacrifice—as opposed to survival of the fittest—is the rule. Ultimately, all of the somatic cell lineages in animals are committed to die: they leave no progeny and instead dedicate their existence to support of the germ cells, which alone have a chance of continued survival (discussed in Chapter 21). There is no particular mystery in this, for the body is a clone derived from a fertilized egg, and the genome of the somatic cells is the same as that of the germ cell lineage that gives rise to sperm or eggs. By their self-sacrifice for the sake of the germ cells, the somatic cells help to propagate copies of their own genes.

Thus, unlike free-living cells such as bacteria, which compete to survive, the cells of a multicellular organism are committed to collaboration. To coordinate their behavior, the cells send, receive, and interpret an elaborate set of extracellular signals that serve as *social controls,* directing each of them how to act (discussed in Chapter 15). As a result, each cell behaves in a socially responsible manner—resting, growing, dividing, differentiating, or dying—as needed for the good of the organism. Molecular disturbances that upset this harmony mean trouble for a multicellular society. In a human body with more than 10^{14} cells, billions of cells experience mutations every day, potentially disrupting the social controls. Most dangerously, a mutation may give one cell a selective advantage, allowing it to grow and divide more vigorously and survive more readily than its neighbors and to become a founder of a growing mutant clone. A mutation that promotes such selfish behavior by individual members of the cooperative can jeopardize the future of the whole enterprise. Over time, repeated rounds of mutation, competition, and natural selection operating within the population of somatic cells can cause matters to go from bad to worse. These are the basic ingredients of cancer: it is a disease in which an individual mutant clone of cells begins by prospering at the expense of its neighbors, but in the end the descendants of this clone can destroy the whole cellular society.

In this section, we discuss the development of cancer as a microevolutionary process. In a human, this process occurs on a time scale of years or decades, in a subpopulation of cells in the body. But the final result depends on the same principles of mutation and natural selection that have driven the evolution of living organisms for billions of years.

Cancer Cells Reproduce Without Restraint and Colonize Other Tissues

Cancer cells are defined by two heritable properties: (1) they reproduce in defiance of the normal restraints on cell growth and division, and (2) they invade and colonize territories normally reserved for other cells. It is the combination of these properties that makes cancers particularly dangerous. An abnormal cell that grows (increases in mass) and proliferates (divides) out of control will give rise to a tumor, or *neoplasm*—literally, a new growth. As long as the neoplastic cells do not become invasive, however, the tumor is said to be **benign**, and removing or destroying the mass locally usually achieves a complete cure. A tumor is considered a cancer only if it is **malignant**, that is, only if its cells have acquired the ability to invade surrounding tissue. Invasiveness is an essential characteristic of cancer cells. It allows them to break loose, enter blood or lymphatic vessels, and form secondary tumors, called **metastases**, at other sites in the body (**Figure 20–1**). The more widely a cancer spreads, the harder it becomes to eradicate, and it is generally metastases that kill the cancer patient.

Cancers are classified according to the tissue and cell type from which they arise. **Carcinomas** are cancers arising from epithelial cells, and they are by far the most common cancers in humans. **Sarcomas** arise from connective tissue or muscle cells. Cancers that do not fit in either of these two broad categories include the various **leukemias** and **lymphomas**, derived from white blood cells and their precursors (hemopoietic cells), as well as cancers derived from cells of the nervous system. **Figure 20–2** shows the types of cancers that are common in the United States, together with their incidence and death rate. Each broad category has many subdivisions according to the specific cell type, location in the body, and the microscopic appearance of the tumor.

In parallel with the set of names for malignant tumors, there is a related set of names for benign tumors: an *adenoma*, for example, is a benign epithelial tumor with a glandular organization; the corresponding type of malignant tumor is an *adenocarcinoma* (**Figure 20–3**). Similarly a *chondroma* and a *chondrosarcoma* are, respectively, benign and malignant tumors of cartilage.

Most cancers have characteristics that reflect their origin. Thus, for example, the cells of a *basal-cell carcinoma*, derived from a keratinocyte stem cell in the skin, generally continue to synthesize cytokeratin intermediate filaments, whereas the cells of a *melanoma*, derived from a pigment cell in the skin, will

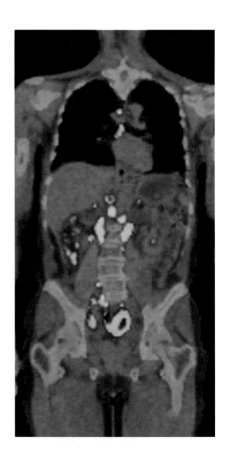

Figure 20–1 Metastasis. Malignant tumors typically give rise to metastases, making the cancer hard to eradicate. Shown in this fusion image is a whole-body scan of a patient with metastatic non-Hodgkins lymphoma (NHL). The background image of the body's tissues was obtained by CT (computed X-ray tomography) scanning. Overlaid on this image, a PET (positron emission tomography) scan that detects the uptake of radioactively labeled fluorodeoxyglucose (FDG) in various tissues reveals the tumor tissue (*yellow*). High FDG uptake indicates cells with unusually active glucose uptake and metabolism, a characteristic of tumors. The yellow spots in the abdominal region reveal multiple NHL metastases. (Courtesy of S. Gambhir.)

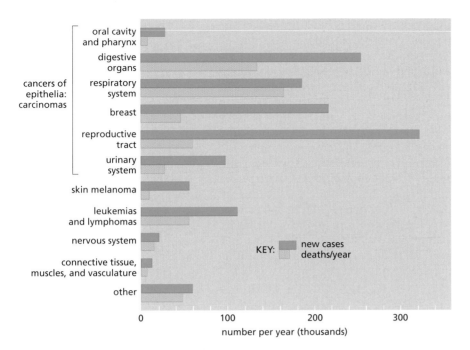

Figure 20–2 **Cancer incidence and mortality in the United States.** The total number of new cases diagnosed in 2004 in the United States was 1,368,030, and total cancer deaths were 563,700. Note that deaths reflect cases diagnosed at many different times and that somewhat less than half of the people who develop cancer die of it. In the world as a whole, the five most common cancers are those of the lung, stomach, breast, colon/rectum, and uterine cervix (included in the figure under the heading of reproductive tract), and the total number of new cancer cases recorded per year is just over 6 million. Skin cancers other than melanomas are not included in these figures, since almost all are cured easily and many are unrecorded. (Data from American Cancer Society, Cancer Facts and Figures, 2004.)

often (but not always) continue to make pigment granules. Cancers originating from different cell types are, in general, very different diseases. Basal-cell carcinomas of the skin, for example, are only locally invasive and rarely metastasize, whereas melanomas can become much more malignant and often form metastases (reflecting the migratory behavior of normal pigment-cell precursors during development, discussed in Chapter 22). Basal-cell carcinomas are readily cured by surgery or local irradiation, whereas malignant melanomas, once they have metastasized widely, are usually fatal.

About 80% of human cancers are carcinomas, perhaps because most of the cell proliferation in adults occurs in epithelia or because epithelial tissues are most frequently exposed to the various forms of physical and chemical damage that favor the development of cancer.

Most Cancers Derive from a Single Abnormal Cell

Even when a cancer has metastasized, we can usually trace its origins to a single **primary tumor**, arising in a specific organ. The primary tumor is thought to derive by cell division from a single cell that initially experienced some heritable change. Subsequently, additional changes accumulate in some of the descendants of this cell, allowing them to outgrow, out-divide, and often outlive their neighbors. By the time it is first detected, a typical human cancer will have been developing for many years and will already contain about a billion cancer cells or more (**Figure 20–4**). Tumors will usually also contain a variety of other cell types—fibroblasts, for example, in the supporting connective tissue associated

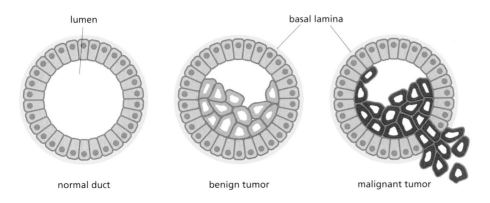

Figure 20–3 **Benign versus malignant tumors.** A benign glandular tumor (an adenoma) remains inside the basal lamina that marks the boundary of the normal structure (a duct, in this example), whereas a malignant glandular tumor (an adenocarcinoma) destroys duct integrity as shown. There are many forms that such tumors may take.

with a carcinoma, as well as inflammatory cells. How can we be sure that the cancer cells are a clone descended from a single abnormal cell?

Molecular analyses of chromosomes in tumor cells clearly demonstrate the clonal origin of those cells. In almost all patients with *chronic myelogenous leukemia (CML)*, for example, we can distinguish leukemic white blood cells from normal cells by a specific chromosomal abnormality: the so-called *Philadelphia chromosome*, created by a translocation between the long arms of chromosomes 9 and 22 (**Figure 20–5**). When the DNA at the site of translocation is cloned and sequenced, it is found that the site of breakage and rejoining of the translocated fragments is identical in all the leukemic cells in any given patient, but that this site differs slightly (by a few hundred or thousand base pairs) from one patient to another. This is the expected result if, and only if, the cancer in each patient arises from a unique accident occurring in a single cell. We will see later how this translocation promotes the development of CML by creating a novel hybrid gene encoding a protein that promotes cell proliferation.

Many other lines of evidence, from a variety of cancers, point to the same conclusion: most cancers originate from a single aberrant cell (**Figure 20–6**).

Cancer Cells Contain Somatic Mutations

If a single abnormal cell is to give rise to a tumor, it must pass on its abnormality to its progeny: the aberration has to be heritable. One problem in understanding a cancer is to discover whether a particular heritable aberration is due to a genetic change—that is, an alteration in the cell's DNA sequence—or to an *epigenetic* change—that is, a persistent change in the pattern of gene expression without a change in the DNA sequence. Epigenetic changes occur during normal development, and they are responsible for the stability of the differentiated state and for such phenomena as X-chromosome inactivation and genomic imprinting (discussed in Chapter 7). As we shall discuss later, epigenetic changes are now known to play an important part in the development of cancers.

But it is also clear that the development of a clone of cancer cells also involves genetic changes. The cells of many cancers can be shown to have one or more shared detectable abnormalities in their DNA sequence that distinguish them from the normal cells surrounding the tumor, as in the example of CML just described, and advances in DNA analysis have identified such genetic faults in an increasing proportion of cancer types. This agrees with the finding that many of the agents that provoke the development of cancer also cause genetic changes. Thus, **carcinogenesis** (the generation of cancer) appears to be linked to *mutagenesis* (the production of a change in the DNA sequence). This correlation is particularly clear for two classes of agents: (1) **chemical carcinogens** (which typically cause simple local changes in the nucleotide sequence), and (2) *radiation* such as x-rays (which typically cause chromosome breaks and translocations) or UV light (which causes specific DNA base alterations). We shall discuss these carcinogenic agents in detail later.

The conclusion that the development of cancer depends on somatic mutations is further supported by the finding that people who have inherited a genetic defect in one of several DNA repair mechanisms, causing their cells to accumulate mutations at an elevated rate, show a strong predisposition to develop cancer. People with the disease *xeroderma pigmentosum*, for example, have defects in the system that repairs DNA damage induced by UV light, and they have a greatly increased incidence of skin cancers. Likewise, mice lacking specific DNA repair genes are abnormally prone to cancer.

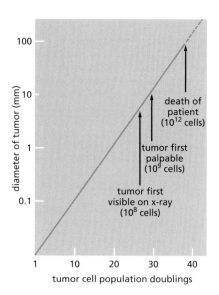

Figure 20–4 The growth of a typical human tumor, such as a tumor of the breast. The diameter of the tumor is plotted on a logarithmic scale. Years may elapse before the tumor becomes noticeable. The doubling time of a typical breast tumor, for example, is about 100 days.

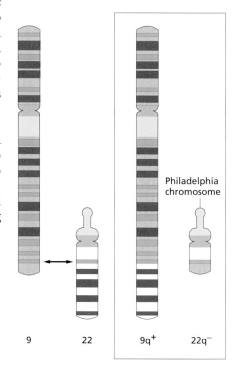

Figure 20–5 The translocation between chromosomes 9 and 22 responsible for chronic myelogenous leukemia. The normal structures of chromosomes 9 and 22 are shown at the left. When a translocation occurs between them at the indicated site, the result is the abnormal pair at the right. The smaller of the two resulting abnormal chromosomes ($22q^-$) is called the Philadelphia chromosome, after the city where the abnormality was first recorded.

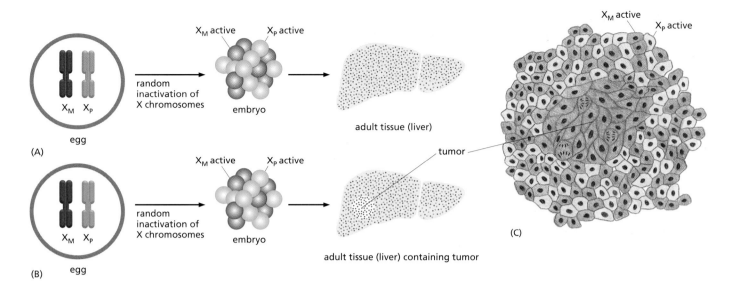

(A)

(B)

(C)

A Single Mutation Is Not Enough to Cause Cancer

An estimated 10^{16} cell divisions occur in a normal human body in the course of a typical lifetime; in a mouse, with its smaller number of cells and its shorter life span, the number is about 10^{12}. Even in an environment that is free of mutagens, mutations would occur spontaneously at an estimated rate of about 10^{-6} mutations per gene per cell division—a value set by fundamental limitations on the accuracy of DNA replication and repair (see p. 263). Thus, in a typical lifetime, every single gene is likely to have undergone mutation on about 10^{10} separate occasions in a human, or on about 10^6 occasions in a mouse. Among the resulting mutant cells, we might expect a large number that have sustained deleterious mutations in genes that regulate cell growth and division, causing the cells to disobey the normal restrictions on cell proliferation. From this point of view, the problem of cancer seems to be not why it occurs, but why it occurs so infrequently.

Clearly, if a single mutation were enough to convert a typical healthy cell into a cancer cell that proliferates without restraint, we would not be viable organisms. Many lines of evidence indicate that the development of a cancer typically requires that a substantial number of independent, rare genetic accidents occur in the lineage of one cell. One comes from epidemiological studies of the incidence of cancer as a function of age (**Figure 20–7**). If a single mutation were responsible, occurring with a fixed probability per year, the chance of developing cancer in any given year of life should be independent of age. In fact, for most types of cancer, the incidence rises steeply with age—as would be expected if cancer is caused by a progressive accumulation of random mutations in a single lineage of cells.

Now that many of the specific mutations responsible for the development of cancer have been identified, we can test directly for their presence in the genome of a particular type of cancer cell. Such tests indicate that a human malignant cell generally harbors many mutations. Animal models confirm that a single genetic alteration is insufficient to cause cancer: in genetically engineered mice carrying one cancer-causing mutation, the disease occurs only

Figure 20–6 Evidence from X-inactivation mosaics that demonstrates the monoclonal origin of cancers. (A) As a result of a random process that occurs in the early embryo, practically every normal tissue in a woman's body is a mixture of cells with different X chromosomes heritably inactivated (indicated here by the mixture of *red* cells and *blue* cells in the normal tissue). (B) When the cells of a cancer are tested for their expression of an X-linked marker gene (a specific form of the enzyme G6PD), however, they are usually all found to have the same X chromosome inactivated, as shown in (C). This implies that they are all derived from a single cancerous founder cell.

Figure 20–7 Cancer incidence as a function of age. The number of newly diagnosed cases of colon cancer in women in England and Wales in 1 year is plotted as a function of age at diagnosis and expressed relative to the total number of individuals in each age group. The incidence of cancer rises steeply as a function of age. If only a single mutation were required to trigger the cancer and this mutation had an equal chance of occurring at any time, the incidence would be independent of age. (Data from C. Muir et al., Cancer Incidence in Five Continents, Vol. V. Lyon: International Agency for Research on Cancer, 1987.)

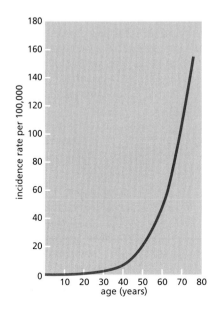

after a lag period of months and arises only in a small scattered subset of cells in the body, implying that other genetic events are necessary for cancer to develop. Furthermore, engineered mice with mutations in more than one cancer-critical gene develop cancer faster than mice with single mutations, demonstrating that multiple genetic events in a single cell can cooperate to promote cancerous cell growth and the development of a tumor.

Cancers Develop Gradually from Increasingly Aberrant Cells

For those cancers known to have a specific external cause, the disease does not usually become apparent until long after exposure to the causal agent. The incidence of lung cancer, for example, does not begin to rise steeply until after 20 years of heavy smoking. Similarly, the incidence of leukemias in Hiroshima and Nagasaki did not show a marked rise until about 5 years after the explosion of the atomic bombs, and industrial workers exposed for a limited period to chemical carcinogens do not usually develop the cancers characteristic of their occupation until 10, 20, or even more years after the exposure (**Figure 20–8**). During this long incubation period, the prospective cancer cells undergo a succession of genetic and epigenetic changes. The same seems to apply to cancers where the initial genetic lesion has no such obvious external cause.

The concept that the development of a cancer requires a gradual accumulation of mutations in a number of different genes—different for different cancers, but usually at least five—helps to explain the well-known phenomenon of **tumor progression**, whereby an initial mild disorder of cell behavior evolves gradually into a full-blown cancer. Chronic myelogenous leukemia provides a clear example. It begins as a disorder characterized by a nonlethal overproduction of white blood cells and continues in this form for several years before changing into a much more rapidly progressing illness that usually ends in death within a few months. In the early chronic phase, the leukemic cells are distinguished mainly by the chromosomal translocation (the Philadelphia chromosome) mentioned previously, although there may well be other less visible genetic or epigenetic changes. In the subsequent acute phase, cells that show not only the translocation but also several other chromosomal abnormalities overrun the hemopoietic (blood-forming) system. It appears as though cells of the initial mutant clone have undergone further mutations that make them proliferate even more extensively, so that they come to outnumber both the normal blood cells and the ancestor cells that have only the primary chromosomal translocation.

Carcinomas and other solid tumors are thought to evolve in a similar way. Although many such cancers in humans are not diagnosed until a relatively late stage, in some cases it is possible to observe the earlier steps. Cancers of the *uterine cervix* (the neck of the womb) are one such example, because of the routine screening for this cancer by cervical scraping.

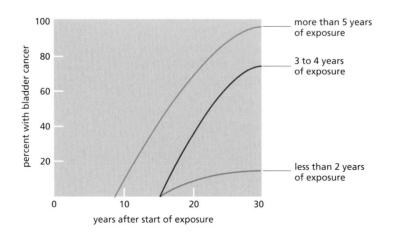

Figure 20–8 Delayed onset of cancer following exposure to a carcinogen. The graph shows the length of the delay before onset of bladder cancer in a set of 78 workers in the chemical industry who had been exposed to the carcinogen 2-naphthylamine, grouped according to the duration of their exposure. (Modified from J. Cairns, Cancer: Science and Society. San Francisco: W.H. Freeman, 1978. After M.H.C. Williams, in Cancer, Vol. III [R.W. Raven, ed.]. London: Butterworth Heinemann, 1958.)

NORMAL EPITHELIUM   LOW-GRADE INTRAEPITHELIAL NEOPLASIA   HIGH-GRADE INTRAEPITHELIAL NEOPLASIA   INVASIVE CARCINOMA

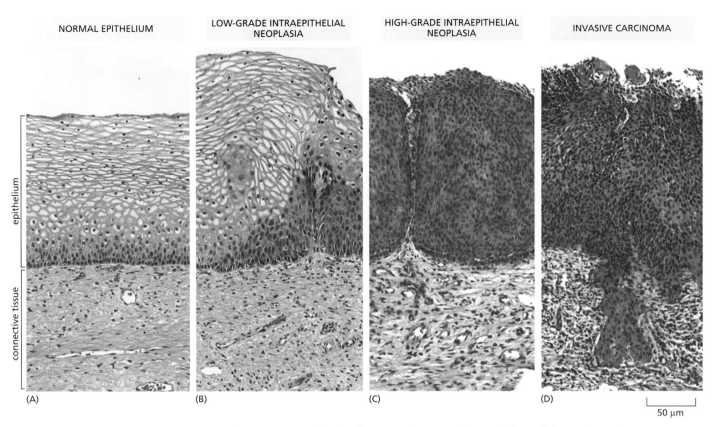

(A)   (B)   (C)   (D)

50 µm

Figure 20–9 The stages of progression in the development of cancer of the epithelium of the uterine cervix. Pathologists use standardized terminology to classify the types of disorders they see, so as to guide the choice of treatment. (A) In the normal stratified squamous epithelium, dividing cells are confined to the basal layer. (B) In low-grade intraepithelial neoplasia, dividing cells can be found throughout the lower third of the epithelium; the superficial cells are still flattened and show signs of differentiation, but this is incomplete. (C) In high-grade intraepithelial neoplasia, cells in all the epithelial layers are proliferating and exhibit defective differentiation. (D) True malignancy begins when the cells move through or destroy the basal lamina that underlies the basal layer of epithelium and invade the underlying connective tissue. (Photographs courtesy of Andrew J. Connolly.)

Cervical Cancers Are Prevented by Early Detection

The epithelium covering the cervix is initially organized as a stratified (multilayered) squamous epithelium (**Figure 20–9**A), which is similar in structure to the epidermis of the skin (see Figure 23–3). In such stratified epithelia, cell proliferation normally occurs only in the basal layer, generating cells that stop dividing and then move out toward the surface, differentiating as they move to form flattened, keratin-rich cells that are eventually sloughed off when they reach the surface. When specimens of cervical epithelium from different women are examined, however, it is not unusual to find patches in which this organization is disturbed in a way that suggests the beginnings of a cancerous transformation. Pathologists describe these changes as *intraepithelial neoplasia,* and classify them as low-grade (mild) or high-grade (moderate to severe).

In the low-grade lesions, the undifferentiated dividing cells are no longer confined to the basal layer but occupy other layers in the lower third of the epithelium; although differentiation proceeds in the upper epithelial layers, it is somewhat disordered (Figure 20–9B). Left alone, most of these mild lesions will spontaneously regress, but about 10% progress to become high-grade lesions, in which most or all of the epithelial layers are occupied by undifferentiated dividing cells, which are usually highly variable in size and shape. Abnormal mitotic structures are frequently seen, and the *karyotype* (the display of the full set of chromosomes) is usually abnormal. But the aberrant cells are still confined to the epithelial side of the basal lamina (Figure 20–9C). At this stage, it is still easy to cure the condition by destroying or surgically removing the abnormal tissue. Fortunately, the presence of such lesions can be detected by

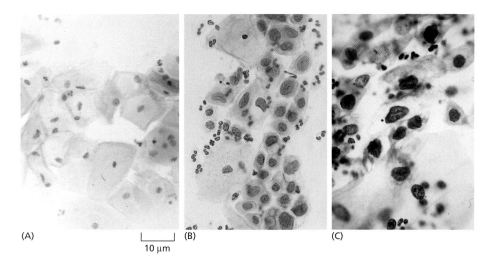

(A)

(B)

(C)

10 µm

Figure 20–10 **Photographs of cells collected by scraping the surface of the uterine cervix (the Papanicolaou or "Pap smear" technique).** (A) Normal: the cells are large and well differentiated, with highly condensed nuclei. (B) Precancerous lesion: differentiation and proliferation are abnormal but the lesion is not yet invasive; the cells are in various stages of differentiation, some quite immature. (C) Invasive carcinoma: the cells all appear undifferentiated, with scanty cytoplasm and a relatively large nucleus. For all three panels, debris in the background includes some white blood cells. (Courtesy of Winifred Gray.)

scraping off a sample of cells from the surface of the cervix and viewing it under the microscope (the "Pap smear" technique—**Figure 20–10**), thereby saving many lives.

Without treatment, the abnormal patch of tissue may simply persist and progress no further or may even regress spontaneously. In at least 30–40% of cases, however, progression will occur, giving rise, over a period of several years, to a frank invasive carcinoma (see Figure 20–9D): the cancer cells cross or destroy the basal lamina, invade the underlying tissue, and may metastasize to local lymph nodes via lymphatic vessels. Surgical cure becomes progressively more difficult as the invasive growth spreads.

Tumor Progression Involves Successive Rounds of Random Inherited Change Followed by Natural Selection

From all the evidence, it seems that cancers arise by a process in which an initial population of slightly abnormal cells, descendants of a single abnormal ancestor, evolve from bad to worse through successive cycles of random inherited change followed by natural selection. At each stage, one cell acquires an additional mutation or epigenetic change that gives it a selective advantage over its neighbors, making it better able to thrive in its environment—an environment that, inside a tumor, may be harsh, with low levels of oxygen, scarce nutrients, and the natural barriers to growth presented by the surrounding normal tissues. The offspring of the best-adapted cell continue to divide, eventually taking over the tumor and becoming the dominant clone in the developing lesion (**Figure 20–11**). Tumors grow in fits and starts, as additional advantageous inherited changes arise and the cells bearing them flourish. Tumor progression involves a large element of chance and usually takes many years, which is why the majority of us will die of causes other than cancer.

Why are so many changes needed? Clearly, large animals have had to evolve a very complex set of regulatory mechanisms to keep their cells in check. Without multiple controls, inevitable errors in the maintenance of DNA sequences would

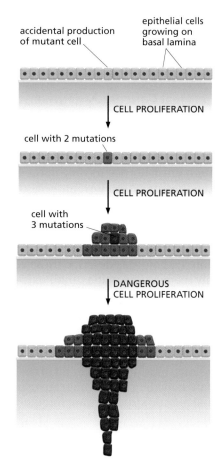

accidental production of mutant cell

epithelial cells growing on basal lamina

CELL PROLIFERATION

cell with 2 mutations

CELL PROLIFERATION

cell with 3 mutations

DANGEROUS CELL PROLIFERATION

Figure 20–11 **Clonal evolution.** In this schematic diagram, a tumor develops through repeated rounds of mutation and proliferation, giving rise eventually to a clone of fully malignant cancer cells. At each step, a single cell undergoes a mutation that either enhances cell proliferation or decreases cell death, so that its progeny become the dominant clone in the tumor. Proliferation of each clone hastens the occurrence of the next step of tumor progression by increasing the size of the cell population at risk of undergoing an additional mutation. The final step depicted here is invasion through the basement membrane, an initial step in metastasis. In reality, there are more than the three steps shown here, and a combination of genetic and epigenetic changes are involved.

produce numerous tumors early in life and quickly destroy any large multicellular organism. Thus, we should not be surprised that cells employ multiple regulatory mechanisms to help them maintain tight and precise control over their behavior, and that many different regulatory systems have to be disrupted before a cell can throw off its normal restraints and behave as an asocial cancer cell. The restraints are of many types, and tumor cells meet new barriers to further expansion at each stage of their evolution. Oxygen and nutrients, for example, do not become limiting until a tumor is one or two millimeters in diameter, at which point the cells in the tumor interior may not have adequate access to these necessary resources. Cells must acquire additional mutations, epigenetic changes (or both) to overcome each new barrier, whether physical or physiological.

In general, the rate of evolution in any population of organisms on Earth would be expected to depend mainly on four parameters: (1) the *mutation rate*, that is, the probability per gene per unit time that any given member of the population will undergo genetic change; (2) the *number of reproducing individuals* in the population; (3) the *rate of reproduction*, that is, the average number of generations of progeny produced per unit time; and (4) the *selective advantage* enjoyed by successful mutant individuals, that is, the ratio of the number of surviving fertile progeny they produce per unit time to the number of surviving fertile progeny produced by nonmutant individuals. The same types of factors are also crucial for the evolution of cancer cells in a multicellular organism, except that both genetic and epigenetic changes help drive the evolutionary process.

The Epigenetic Changes That Accumulate in Cancer Cells Involve Inherited Chromatin Structures and DNA Methylation

As just stated, the progression toward cancer differs from normal biological evolution in one important respect: epigenetic changes also occur that give the cells a selective advantage.

For many years, pathologists have used an abnormal appearance of the cell nucleus to identify and classify cancer cells in tumor biopsies. For example, cancer cells sometimes contain an unusually large amount of heterochromatin—a condensed form of interphase chromatin that silences genes (see p. 238). We now understand some of the molecular mechanisms involved in forming this chromatin, and it is possible to associate new heterochromatin formation with the epigenetic silencing of specific genes that would otherwise block tumor progression.

Heterochromatin formation and maintenance involve specific covalent modifications of histones; these in turn attract complexes of chromatin-binding proteins that are stably maintained following DNA replication (discussed in Chapter 4). In this way, genes can be turned off in a cell-to-cell inherited manner without any change in the DNA sequence. This epigenetic form of gene regulation plays a major role in driving the orderly pattern of cell specializations that occur during embryonic development (discussed in Chapter 22). Errors that occur in this process are potentially dangerous because they can be transmitted to the progeny of the cell in which the original change occurred.

We now know that many of the mutations that make a cell cancerous alter the proteins that determine chromatin structures. These include not only enzymes that modify the histones in nucleosomes, but also proteins in the "code reader–writer" complexes that interpret the histone code (see Figure 4–43). These findings provide strong evidence for the importance of chromatin-based epigenetic changes in tumor progression.

During normal developmental processes, an inherited pattern of DNA methylation reinforces many of the gene silencing events caused by the packaging of genes into heterochromatin (see p. 467). Analyses reveal that a large amount of DNA methylation also occurs on selected genes during tumor progression. In summary, although recognized relatively recently, it now appears that the abnormal epigenetic silencing of genes is no less important than mutations in DNA sequences for the development of most cancers (**Figure 20–12**).

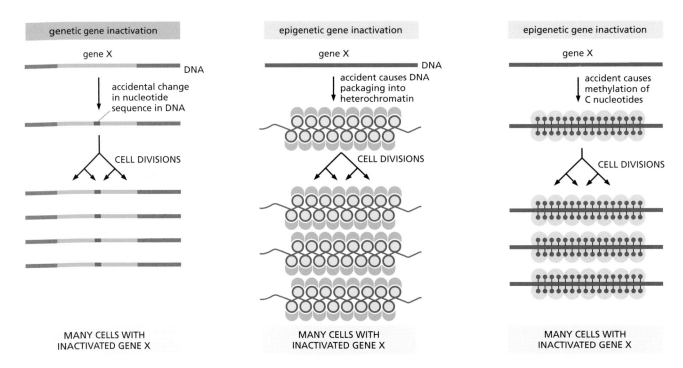

Figure 20–12 **Comparison of the genetic and epigenetic changes observed in tumors.** A mutation results from an irreversible change in DNA sequence. In contrast to such genetic changes, epigenetic changes are based on alterations that, while heritable from cell to cell, can be reversed either by site-specific changes in histone modification (heterochromatin pathway) or by site-specific DNA demethylation (methylation pathway). Because the epigenetic marks on genes are generally reversed during the formation of eggs and sperm, they are not inherited between generations and therefore have not been extensively studied by geneticists.
   As discussed in Chapter 7, it is thought that DNA methylation (an inherited pattern of methylation of C nucleotides in CpG sequences) is a mechanism that is used to more permanently silence genes that have already been turned off, and DNA methylation seems to be a silencing mechanism that normally follows the formation of heterochromatin. But the misregulation that arises during tumor progression may be able to produce gene silencing independently of heterochromatinization, as schematically illustrated here.

Human Cancer Cells Are Genetically Unstable

Most human cancer cells accumulate genetic changes at an abnormally rapid rate and are said to be **genetically unstable**. This instability can take various forms. Some cancer cells are unable to repair certain types of DNA damage or to correct replication errors of various kinds. These cells tend to accumulate more point mutations or other DNA sequence changes than do normal cells. Other cancer cells fail to maintain either the number or the integrity of their chromosomes, and they consequently accumulate gross abnormalities in their karyotype that are visible at mitosis (**Figure 20–13**). The genetic instability is further amplified when some of the DNA changes alter epigenetic control mechanisms in ways that produce extra heterochromatin and DNA methylation. And epigenetic changes by themselves, arising accidentally and independently of prior genetic changes, may also lead—in principle at least—to destabilization of normal patterns of cellular inheritance by facilitating other changes either at the genetic or the epigenetic level. From an evolutionary perspective, none of this should be a surprise: anything that increases the probability of random changes in gene function that are heritable from one cell generation to the next is likely to speed the evolution of a clone of cells toward malignancy.

Although the epigenetic changes in cancer cells have not been so thoroughly analyzed, a great deal is now known about the genetic changes. Different tumors—even those from the same tissues—can show different kinds of genetic instability, caused by heritable alterations in any one of a large number of so-called *DNA maintenance genes* involved in propagating DNA or chromosomes. It is rare for a person to inherit a mutation in one of these genes, but those people that do so have a raised incidence of cancer, confirming that a loss of genetic stability can cause cancer. Generally, such destabilizing mutations

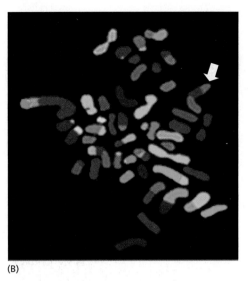

(A)   (B)

Figure 20–13 **Chromosomes from a breast tumor displaying abnormalities in structure and number.** Chromosomes were prepared from a breast tumor cell in metaphase, spread on a glass slide, and stained with (A) a general DNA stain or (B) a combination of fluorescent stains that color each normal human chromosome differently. The staining (displayed in false color) shows multiple translocations, including a doubly translocated chromosome *(white arrow)* made up of two pieces of chromosome 8 *(brown)* and a piece of chromosome 17 *(purple)*. The karyotype also contains 48 chromosomes, instead of the normal 46. (Courtesy of Joanne Davidson and Paul Edwards.)

are not inherited but arise *de novo* in the clone of cells in which a tumor develops, helping the cancer cell to accumulate mutations much more rapidly than its neighbors. Recent analyses show that the cells in a variety of human cancers experience single nucleotide substitutions at a rate that is 10–20 times higher than the rate observed in normal cells. As a result, the sequencing of more than 10,000 genes in a set of breast and colorectal cancers reveals that cancer cells have accumulated enough mutations to cause an amino acid change in the proteins of roughly 100 genes. Most are random changes that affect different genes in different individual tumors. But a subset of genes are found to be repeatedly mutated in a particular type of cancer, suggesting that alterations in as many as 20 genes are needed to drive tumor progression.

Genetic instability does not in itself give a cell a selective advantage. It seems that there is some optimum level of genetic instability for the development of cancer, making a cell mutable enough to evolve rapidly, but not so mutable that it accumulates too many harmful changes and dies. To be favored by selection, a genetically unstable cell must acquire properties that confer some competitive advantage.

Cancerous Growth Often Depends on Defective Control of Cell Death, Cell Differentiation, or Both

Just as an increased mutation rate per cell can raise the probability of cancer, so can any circumstance that increases the number of proliferating cells available for mutating. People who are clinically obese, for example, have a strongly increased risk of many types of cancer, compared with people of normal weight; and this is presumably due, in part at least, to an increase in both the number of cells in the body and the rate at which these cells divide when overnourished or overstimulated by growth factors. The same principle applies to both cancer initiation and cancer progression: the bigger the clone of altered cells resulting from an early inherited change, the greater the chance that at least one of these cells will undergo an additional mutation or epigenetic change that will allow the cancer to progress. Thus, at every stage in the development of cancer, any condition that helps the altered cells to increase in number favors the progression of the tumor.

An early mutation or epigenetic change can have this effect by increasing the rate at which a clone of cells proliferates, as we discuss in detail later. Such changes, however, are not the only—or necessarily the most important—mechanism for increasing cell number. In normal adult tissues, especially those at risk of cancer, cells may proliferate continually; but their numbers remain steady because cell production is balanced by cell loss, as part of the body's homeostatic control mechanisms. Programmed cell death by *apoptosis* usually plays an essential part in this balance, as we discuss in Chapters 18 and 23. If too many

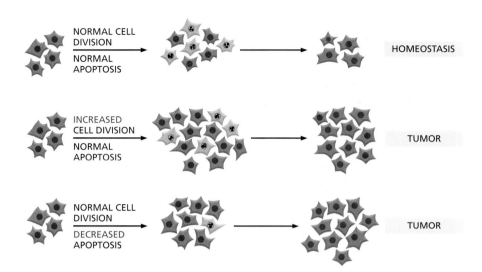

Figure 20–14 **Both increased cell division and decreased apoptosis can contribute to tumorigenesis.** In normal tissues, apoptosis balances cell division to maintain homeostasis. <GCCC> During the development of cancer, either an increase in cell division or an inhibition of apoptosis can lead to the increased cell numbers important for tumorigenesis. Cells fated to undergo apoptosis are shown in *gray*.

cells are generated, the rate of apoptosis increases to dispose of the surplus. One of the most important properties of many types of cancer cells is that they fail to undergo apoptosis when a normal cell would do so. This greatly contributes to the growth of a tumor (**Figure 20–14**).

Inherited changes can also increase the size of a clone of mutant cells by altering their ability to differentiate, as illustrated by the situation in the uterine cervix, discussed earlier. When a stem cell in the basal layer divides, each daughter cell has a choice—it can either remain a stem cell or commit to a pathway leading to differentiation; the committed cells initially proliferate and then stop dividing and terminally differentiate (the committed, proliferating cells are called *transit amplifying cells*). If the differentiation program is blocked at some stage, proliferating cells accumulate, contributing to the progression from low-grade intraepithelial neoplasia of the cervix to high-grade intraepithelial neoplasia and malignant cancer (see Figure 20–9). Similar considerations apply to the development of cancer in other tissues that rely on stem cells, such as the skin, the lining of the gut, and the hemopoietic system. Several forms of leukemia, for example, seem to arise from a disruption of the normal program of differentiation, such that a committed progenitor of a particular type of blood cell eventually becomes able to divide indefinitely, instead of undergoing terminal differentiation in the normal way and dying after a limited number of cell divisions (as discussed in Chapter 23).

Thus, the accumulation of mutations and epigenetic changes that lead to defects in the normal controls of cell division, apoptosis, and differentiation can all contribute to the development and progression of cancers.

Cancer Cells Are Usually Altered in Their Responses to DNA Damage and Other Forms of Stress

As just discussed, many normal cells permanently stop dividing when they differentiate into specialized cells. Differentiation, however, is not the only reason that proliferating cells stop dividing; they can also do so in response to stress or to damage to their DNA. As described in Chapter 17, normal cells contain a set of *checkpoint control* mechanisms. These arrest the cell cycle temporarily when something goes awry, providing time to correct the problem. Chromosome breakage and other types of damage to DNA generate intracellular signals that activate these checkpoint mechanisms, causing a normal cell to halt its cycle, and thereby giving it a chance to repair the damage before it attempts to progress through the cycle and divide. And if the damage is irreparable, a normal cell, rather than forge ahead and generate daughters with a damaged genome, will either permanently withdraw from the cell cycle or commit suicide by apoptosis.

Cancer cells usually acquire mutations and epigenetic changes that inactivate such responses to DNA damage. From the selfish standpoint of the cancer cells, the loss of these restraints confers an advantage, allowing them to continue to multiply even with damaged DNA. At the same time, this accelerates the rate at which mutations accumulate in the tumor cell population and can thereby hasten the progression to greater malignancy. The cancer cells are also often defective in other checkpoint control mechanisms that help regulate the cell cycle. Thus, they will often stagger on through the cycle when things go wrong, and in this way will do themselves still further genetic damage, contributing to the chromosome abnormalities that they often display. Even though such genetically damaged cells usually take longer to complete each cell cycle than normal cycling cells, they accumulate over time because they display a greater propensity to proliferate. It is perhaps remarkable that these cells can manage to outcompete their normal neighbors when they are so messed up.

Human Cancer Cells Escape a Built-in Limit to Cell Proliferation

Many normal human cells have a built-in limit to the number of times they can divide when stimulated to proliferate in culture: they permanently stop dividing after a certain number of population doublings (25–50 for human fibroblasts, for example), a phenomenon termed **replicative cell senescence** (discussed in Chapter 17). This cell-division-counting mechanism generally depends on the progressive shortening of the telomeres at the ends of chromosomes, which eventually changes their structure. As discussed in Chapter 5, the replication of telomere DNA during S phase depends on the enzyme **telomerase**, which maintains the special telomeric DNA sequence and promotes the formation of protein cap structures that protect the chromosome ends. Because many proliferating human cells (but not stem cells) are deficient in telomerase, their telomeres shorten with every division, and their protective caps deteriorate. Eventually, the altered chromosome ends trigger a permanent cell-cycle arrest.

The cell-division-counting mechanism just described sets a physiological limit to cell proliferation for most of the types of cells in the body. Human cancer cells must somehow avoid or overcome this barrier to form large tumors. Some rodent cells, by contrast, maintain telomerase activity and normal telomeres as they proliferate and therefore lack this barrier. It has been proposed that humans need replicative cell senescence to help prevent cancer because our comparatively long lifespan would otherwise provide enormous opportunities for tumor progression.

Human cancer cells avoid replicative cell senescence in two ways. First, they acquire genetic and epigenetic changes that disable the checkpoint control so that the cells continue to cycle even when telomeres become uncapped. Mutations that inactivate the p53 pathway have this effect and are very common in cancer cells, as we shall discuss below. As another strategy to escape replicative senescence, cancer cells often maintain telomerase activity as they proliferate, so that their telomeres do not shorten or become uncapped. In some cases, the cancer cells may maintain telomerase activity because the cancer originated in stem cells that have this activity. In other cases, although the cancer originated in cells without appreciable telomerase activity, the cancer cells have acquired the activity as a result of genetic or epigenetic changes that were selected for as their telomeres shortened. Yet other cancer cells evolve alternate mechanisms for elongating chromosome ends. Regardless of the strategy used, the result is that the cancer cells continue to proliferate under conditions when normal cells would stop.

A Small Population of Cancer Stem Cells Maintains Many Tumors

As we discuss in Chapter 23, in a normal adult cell lineage with substantial cell turnover, such as the hemopoietic cell lineage, the cells are typically organized as a hierarchy: rare, slowly dividing stem cells produce cells (the transit amplifying cells) that proliferate extensively and eventually differentiate into the specialized cells characteristic of the lineage. The stem cells maintain the lineage by producing some daughter cells that can remain stem cells (a process called *self-renewal*), while producing others that commit to a pathway leading to differentiation. There is increasing evidence that many cancers are organized in the same hierarchical way, with rare **cancer stem cells** at the top of the hierarchy responsible for maintaining the population of cells in a tumor. These stem cells are capable of indefinite self-renewal, but they also give rise to rapidly dividing transit amplifying cells that have a limited capacity for self-renewal. It has only recently become widely recognized that these cells with a limited growth potential make up the vast majority of the cells in many cancers.

It has been known for over 40 years that there is only a small chance—usually much less than 1%—that a randomly chosen individual cell from a cancer will be able to generate a new tumor when tested for its ability to do so—for example, by implanting it into an immunodeficient mouse. New technologies for sorting cells have shown that, in many cases, this is because the cancer cells are heterogeneous, and only a small subset of them have the special property needed for tumor propagation. One way to enrich the small population of presumptive cancer stem cells in some cancers uses a cell-permeable fluorescent dye and a fluorescence-activated cell sorter (see Figure 8–2). When a cell suspension prepared from a normal tissue is exposed to the dye and then sorted in this way, the stem cells of the tissue are often found to be highly enriched in a small *side population* of weakly fluorescent cells. The stem cells are only weakly fluorescent because they have an ABC transporter (discussed in Chapter 11) in their plasma membrane that actively pumps the dye out of the cell.

When some cancers are analyzed in the same way, a small side population with an ABC transporter is found to contain most, if not all, of the cells capable of forming new cancers after transplantation into immunodeficient mice. Remarkably, this is also the case for some cancer cell lines that have been propagated for more than thirty years in culture: whereas the side population of each of these lines readily forms tumors after transplantation, the remaining highly fluorescent population, which includes the vast majority of the cells in the line, does not (**Figure 20–15**). Moreover, when the side population isolated from such a cancer cell line is returned to culture, a small side population rapidly reappears and is maintained indefinitely, whereas a side population does not develop in cultures of the remaining cells.

In other studies, cell fractionations using monoclonal antibodies that recognize normal stem cells in the tissue of origin of the cancer have given similar results.

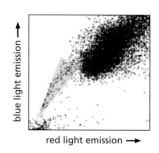

Figure 20–15 Cell sorting reveals cancer stem cells in a small side population of cells. In this example, the cells in an established cell line obtained from a rat glioma tumor were treated with a dye that can pass through the plasma membrane (Hoechst 33342) and then analyzed by cell sorting. About 0.4% of the cells pump out the dye and are therefore only lightly stained, as indicated in *yellow*. Experiments comparing these side population (SP) cells with the remaining cells showed that only the side population cells could proliferate extensively and form tumors. Moreover, whereas the SP cells produced both SP and non-SP cells, the non-SP cells could not produce SP cells. (Adapted from T. Kondo, T. Setoguchi and T. Taga, *Proc. Natl Acad. Sci. U.S.A.* 101:781–786, 2004. With permission from National Academy of Sciences.)

How Do Cancer Stem Cells Arise?

The finding that many cancers are maintained by a small population of cancer stem cells has extremely important implications, both for understanding cancer and for treating it, and it raises a number of intriguing questions. How do these stem cells arise? What is the nature of the cancer stem cell, and what distinguishes it in molecular terms from the majority of cells in the cancer? From the results just described with cultured cancer cell lines, many of these differences must result from epigenetic rather than from genetic changes.

It is clear for some tumors that the cancer originates in a normal tissue stem cell, which gradually accumulates the mutations and epigenetic changes responsible for the asocial behavior of the cancer. For example, this sequence of events is almost certainly the way chronic myelogenous leukemias arise. Thus, when the multipotential hemopoietic stem cells (see p. 1456) are purified from the bone marrow of these patients, many contain the Philadelphia chromosome, with the translocation characteristic of this cancer. Likewise, many human cancers probably arise from normal stem cells, because they develop in epithelia that undergo substantial cell turnover. In these tissues, only stem cells remain in the body and proliferate for long enough to accumulate the number of mutations required to develop into a cancer.

The second way that a cancer stem cell can arise is through a change in a proliferating cell that is more differentiated, such as a transit amplifying cell. The

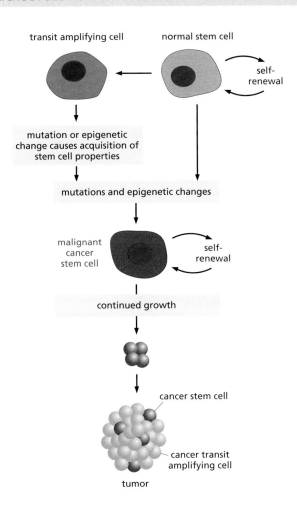

Figure 20–16 Cancers may arise from cancer stem cells. Cancer stem cells are defined as those cancer cells that can self-renew to produce additional malignant stem cells, and at the same time generate non-tumorigenic cells such as transit amplifying cells. Cancer stem cells can arise from normal stem cells that have sustained mutations to make them cancerous; or from more differentiated cells that have undergone mutations or epigenetic changes that give them stem cell properties.

changes must confer the two crucial stem-cell properties: the ability to be retained in the body, and the capacity for prolonged self-renewal. Then, and only then, the altered cell will have the time to accumulate the other mutations and epigenetic changes needed to become a full-blown cancer cell. Some B lymphocyte leukemias, for example, seem to arise in this way. All of the cells in such a leukemia contain the same genetic rearrangements in their antibody genes (discussed in Chapter 25), suggesting that the cancer originated in a committed B lymphocyte precursor cell, rather than in an uncommitted stem cell. Thus, it seems likely that, while most human cancers arise from a stem cell, others arise from a cell that has acquired the capacity for sustained self-renewal through mutation, epigenetic change, or both (**Figure 20–16**).

Most current cancer therapies such as radiation and cytotoxic drugs are likely to preferentially kill the most rapidly proliferating cancer cells in a tumor. Because the cancer stem cells usually divide much more slowly, they may be less sensitive to these treatments. Moreover, in the 1970s, it was proposed that evolution would have selected for mechanisms that protect stem cells from accumulating mutations. In particular, since many mutations arise during DNA replication, could each division of a stem cell utilize a special mechanism that allows the old DNA template strand in each chromosome to be segregated into the one daughter cell that will remain a stem cell? Recent experiments suggest that stem cells, and only stem cells, can indeed coordinate the segregation of their chromosomes in this way (see Figure 23–16). If this is also the case for cancer stem cells, it could make them unusually resistant to many treatments that kill other cells.

If the stem cells are not eradicated, they will very likely regenerate the cancer. This may be a major reason why cancers commonly recur after a dramatic initial response to therapy. To cure cancers, it is clearly crucial to find better ways to target the cancer stem cells and to kill them.

To Metastasize, Malignant Cancer Cells Must Survive and Proliferate in a Foreign Environment

Metastasis is the most deadly—and least understood—aspect of cancer, being responsible for 90% of cancer-associated deaths. By spreading throughout the body, a cancer becomes almost impossible to eradicate by either surgery or localized irradiation. Metastasis is itself a multistep process: the cancer cells have to invade local tissues and vessels, move through the circulation, leave the vessels, and then establish new cellular colonies at distant sites. Each of these events is, in itself, complex, and most of the molecular mechanisms involved are not yet clear.

For a cancer cell to metastasize, it must break free of constraints that keep normal cells in their proper places and prevent them from invading neighboring tissues. Invasiveness is thus one of the defining properties of malignant tumors, which show a disorganized pattern of growth and ragged borders, with extensions into the surrounding tissue (see, for example, Figure 20–9). Although the underlying molecular changes are not well understood, invasiveness almost certainly requires a disruption of the adhesive mechanisms that normally keep cells tethered to their proper neighbors and to the extracellular matrix. As we discuss later, for carcinomas, this change resembles the *epithelial-to-mesenchymal transition (EMT)*, which occurs in some epithelial tissues during normal development (see Figure 19–12).

The next step in metastasis—the establishment of colonies in distant organs—is a complex, slow, and inefficient operation; few cells achieve it. Before it can metastasize successfully, a cell must penetrate a blood vessel or a lymphatic vessel by crossing the basal lamina and the endothelial lining of the vessel so as to enter the blood or lymph, exit from a vessel elsewhere in the body, and grow in the new site, forming at first a small clump of cells known as a micrometastasis. To complete the metastatic process, some of these micrometastases must produce cells that survive and proliferate extensively in the new environment, a difficult process known as colonization (**Figure 20–17**).

The vasculature that grows into tumors is leaky, and it seems that many of the cells in a tumor manage to escape into lymphatic or blood vessels. However, experiments show that only a tiny proportion of these, fewer than one in thousands, perhaps one in millions, survive to form anything more than micrometastases. Many cancers are discovered before they have managed to found metastatic colonies and can be cured if the primary tumor is removed.

While it is possible in principle that each cancer cell entering the bloodstream might have the same tiny chance of surviving, settling, and growing in an alien site, this seems unlikely. More likely, the low rate of colonization reflects the rarity of cells that have the special properties needed. That is, it may be that only a very small population of the cells in a typical cancer have both the stem-cell character that will enable them to divide without limit, and the ability to settle and survive in an alien environment. The rarity of cancer stem cells would explain why so few cells achieve colonization, even after forming a micrometastasis.

The need for special survival capabilities is likely to be an additional restraint. Normal cells depend on extracellular survival signals that are relatively abundant in their normal environment; when deprived of these signals, they activate their cell death machinery and undergo apoptosis (discussed in Chapter 18). Cancer cells capable of metastasis are often resistant to apoptosis compared with normal cells and therefore can more readily survive after escaping from their proper home. They are also less dependent on signals from other cells to grow and divide.

Tumors Induce Angiogenesis

In addition to all the requirements just described, to grow large, a tumor must recruit an adequate blood supply to ensure that it gets sufficient oxygen and nutrients. Thus, *angiogenesis,* the formation of new blood vessels, is required for tumor growth beyond a certain size. Like normal tissues, tumors attract a blood

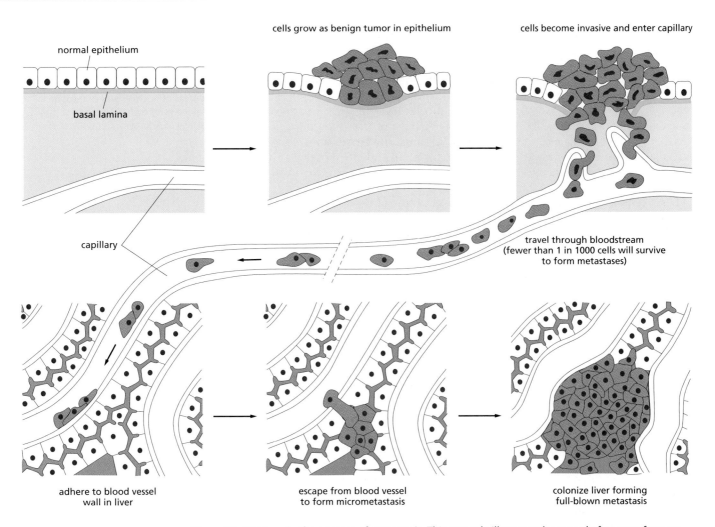

cells grow as benign tumor in epithelium

cells become invasive and enter capillary

normal epithelium

basal lamina

capillary

travel through bloodstream
(fewer than 1 in 1000 cells will survive
to form metastases)

adhere to blood vessel
wall in liver

escape from blood vessel
to form micrometastasis

colonize liver forming
full-blown metastasis

Figure 20–17 **Steps in the process of metastasis.** This example illustrates the spread of a tumor from an organ such as the bladder to the liver. Tumor cells may enter the bloodstream directly by crossing the wall of a blood vessel, as diagrammed here, or, more commonly perhaps, by crossing the wall of a lymphatic vessel that ultimately discharges its contents (lymph) into the bloodstream. Tumor cells that have entered a lymphatic vessel often become trapped in lymph nodes along the way, giving rise to lymph-node metastases. Studies in animals show that typically far fewer than one in every thousand malignant tumor cells that enter the bloodstream will colonize a new tissue so as to produce a detectable tumor at a new site.

supply by secreting angiogenic signals. These signals are produced in response to hypoxia, which begins to affect the cells as the tumor enlarges beyond a millimeter or two in diameter. The hypoxia activates an *angiogenic switch* to increase the blood supply by provoking an increase in the level of *Hypoxia Inducible Factor-1α (HIF-1α)*, a gene regulatory protein described in Chapter 23; this protein, in turn, activates the transcription of genes that encode pro-angiogenic factors, such as *vascular endothelial growth factor (VEGF)*. These are secreted proteins that attract endothelial cells and stimulate the growth of new blood vessels (see Figure 23–34). The vessels not only help supply the tumor with nutrients and oxygen, but also provide an escape route for its cells to metastasize.

However, the induced vessels are tortuous, heterogeneous in diameter, and leaky, and they have many branches with dead ends. These abnormalities, which probably result from an abnormal balance of signaling molecules, lead to irregular blood flow within the tumor, helping to create additional regions of hypoxia (**Figure 20–18**). Hypoxia, in turn, selects for mutant cancer cells that are better able to survive in these harsh and stressful conditions, which usually means that they are more malignant. Ultimately, however, tumor growth relies on an adequate blood supply, and the abnormal vessels that are attracted into tumors are one obvious target for drug therapy, as we discuss later.

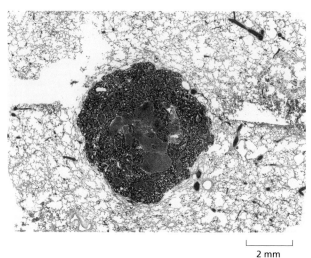

Figure 20–18 **Colon adenocarcinoma metastasis in the lung.** This tissue slice shows well-differentiated colorectal cancer cells, still forming cohesive glands in the lung. This metastasis has central pink areas of necrosis where the cancer cells have outgrown their blood supply. (Courtesy of Andrew J. Connolly.)

2 mm

The Tumor Microenvironment Influences Cancer Development

As mentioned earlier, carcinomas are complex entities, containing many cell types in addition to the cancer cells. While the cancer cells are grossly abnormal, being the bearers of mutations in these tumors, the other cells in the tumor—especially those of the supporting connective tissue, or **stroma**—are far from innocent bystanders. This is because the development of a tumor relies on cross-talk between the tumor cells and the tumor stroma, just as the normal development of epithelial organs relies on cross-talk between epithelial cells and mesenchymal cells (discussed in Chapter 22).

The stroma provides the framework for the tumor and is composed of normal connective tissue containing fibroblasts, myofibroblasts, inflammatory white blood cells, the endothelial cells of blood and lymphatic vessels, and their attendant pericytes and smooth muscle cells (**Figure 20–19**). As a carcinoma progresses, the cancer cells induce changes in the stroma by secreting both signal proteins that alter the behavior of the stromal cells and proteolytic enzymes that modify the extracellular matrix. The stromal cells in turn act back on the tumor cells in various ways. They secrete signal proteins that stimulate cancer cell growth and division, and they secrete proteases that further remodel the extracellular matrix. In these ways, the tumor and its stroma evolve together, and the tumor becomes dependent on its particular stromal cells. Experiments in mice indicate that the growth of some transplanted carcinomas depends on the tumor-associated fibroblasts, and normal fibroblasts will not do. As a second example, the growth of skin tumors is significantly retarded in mice lacking a specific matrix metalloproteinase that is normally secreted by stromal mast cells; the protease is responsible for releasing angiogenic factors from the extracellular matrix that stimulate the angiogenesis required for the skin tumor to grow. These and other results suggest that cancer treatments could be designed to inhibit activated stromal cells, in addition to targeting the tumor cells themselves.

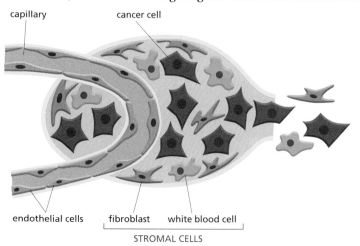

capillary   cancer cell

endothelial cells   fibroblast   white blood cell

STROMAL CELLS

Figure 20–19 **The tumor microenvironment plays a role in tumorigenesis.** Tumors consist of many cell types including cancer cells, vascular epithelial cells, fibroblasts, and inflammatory white blood cells. Cross-talk between these and other cell types plays an important part in tumor development.

Many Properties Typically Contribute to Cancerous Growth

Clearly, to produce a cancer, a cell must acquire a range of aberrant properties—a collection of subversive new skills—as it evolves. Different cancers require different combinations of these properties. Nevertheless, we can draw up a short list of the key behaviors of cancer cells in general:

1. They are more self-sufficient than normal cells for their growth and proliferation. For example, unlike most normal cells, they can often survive and proliferate in cell culture even when not adherent to a substratum and floating free in suspension.
2. They are relatively insensitive to anti-proliferative extracellular signals.
3. They are less prone than normal cells to undergo apoptosis.
4. They are defective in the intracellular control mechanisms that normally stop cell division permanently in response to stress (such as hypoxia) or DNA damage.
5. They induce help from the normal stromal cells in their local environment.
6. They induce angiogenesis.
7. They escape from their home tissues (that is, they are invasive) and survive and proliferate in foreign sites (that is, they metastasize).
8. They are genetically unstable.
9. They either produce telomerase, or acquire another way of stabilizing their telomeres.

In the later sections of the chapter, we examine the mutations and molecular mechanisms that underlie some of these properties, as well as the environmental factors that promote the development of cancer.

Summary

Cancer cells, by definition, grow and proliferate in defiance of normal controls (that is, they are neoplastic) and are able to invade surrounding tissues and colonize distant organs (that is, they are malignant). By giving rise to secondary tumors, or metastases, they become difficult to eradicate by surgery or local irradiation. Cancers are thought to originate from a single cell that has experienced an initial mutation, but the progeny of this cell must undergo many further changes, requiring numerous additional mutations and epigenetic events, to become cancerous. The cell of origin of a cancer may be either a tissue stem cell, which is already endowed with the ability to self-renew indefinitely, or a more differentiated cell that acquires the capacity for indefinite self-renewal. Tumor progression usually takes many years and reflects the operation of a Darwinian-like process of evolution, in which somatic cells undergo mutation and epigenetic changes accompanied by natural selection.

Cancer cells acquire a variety of special properties as they evolve, multiply, and spread. These include alterations in cell signaling pathways, enabling the cells to ignore the signals from their environment that normally keep cell proliferation under tight control. As part of the evolutionary process of tumor progression, cancer cells acquire defects in differentiation and in the control mechanisms that permanently stop cell division or induce apoptosis in response to cell stress or DNA damage. All of these changes increase the ability of cancer cells to survive, grow, and divide in their original tissue and then to metastasize—which requires survival and proliferation in foreign environments. However, the evolution of a tumor does not simply rely on changes in the cancer cells themselves; it also depends on other cells present in the tumor microenvironment, collectively called stromal cells, including new blood vessels that allow the tumor to grow large and metastasize via the bloodstream.

Since many mutations and epigenetic changes are needed to confer this collection of asocial behaviors, it is not surprising that nearly all cancer cells are genetically unstable. The genetic instability can arise from defects in the ability to repair DNA damage or to correct replication errors of various kinds, which lead to changes in DNA sequence. Also common are defects in chromosome segregation during mitosis, which lead to chromosome instability and changes in karyotype. This genetic instability is selected for in the clones of aberrant cells that are able to produce tumors, because it greatly accelerates the accumulation of the further genetic and epigenetic changes that are required for tumor progression.

THE PREVENTABLE CAUSES OF CANCER

The development of a cancer generally requires many steps, each governed by multiple factors—some dependent on the genetic constitution of the individual, others dependent on his or her environment and way of life. A certain irreducible background incidence of cancer is to be expected regardless of circumstances: mutations can never be absolutely avoided because they are an inescapable consequence of fundamental limitations on the accuracy of DNA replication and repair, as discussed in Chapter 5. If a human could live long enough, it is inevitable that at least one of his or her cells would eventually accumulate a set of mutations sufficient for cancer to develop.

Nevertheless, there is evidence that environmental factors, including the food we eat, accelerate the onset of most cancers. This is demonstrated most clearly by a comparison of cancer incidence in different countries: for almost every cancer that is common in one country, there is another country where the incidence is much lower. Because migrant populations tend to adopt the pattern of cancer incidence typical of the host country, the differences appear to be due mostly to environmental, not genetic, factors. From such findings, it is estimated that 80–90% of cancers should be avoidable, or at least postponable (**Figure 20–20**).

Unfortunately, different cancers have different environmental risk factors, and a population that escapes one such danger is usually exposed to another. This is not, however, inevitable. There are some human subgroups whose way of life substantially reduces the total cancer death rate among individuals of a given age. Under the current conditions in the United States and Europe,

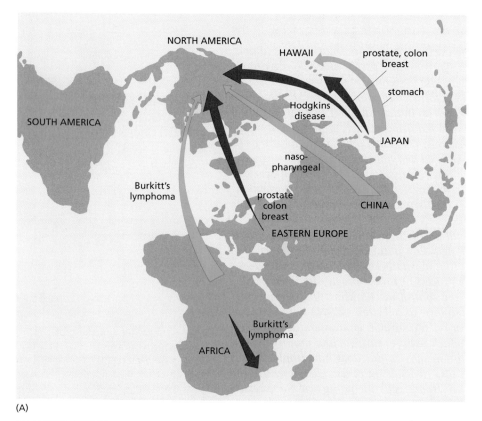

(A)

environmental and lifestyle factors	cancer	% total cases
• occupational exposure	various types	1–2
• tobacco related	lung, kidney, bladder	24
• diet: low in vegetables, high salt, high nitrate	stomach, esophagus	5
• diet: high fat, low fiber, fried and broiled foods	bowel, pancreas, prostate, breast	37
• tobacco and alcohol	mouth, throat	2

(B)

Figure 20–20 Cancer incidence is related to environmental influences. (A) This map of the world shows the rates of cancer increasing (*red arrows*) or decreasing (*blue arrows*) when specific populations move from one location to another. These observations suggest the importance of environmental factors, including diet, in dictating cancer risk. (B) Some estimated effects of environment and lifestyle on cancer in the United States. The table shows the percentage of total cancer cases attributable to each specified factor. (B, adapted from Cancer Facts and Figures, American Cancer Society, 1990. With permission from American Cancer Society.)

approximately one in five people will die of cancer. But the incidence of cancer among strict Mormons in Utah (but, importantly, not among non-practicing members of the same family), who avoid alcohol, coffee, cigarettes, drugs, and casual sex, is only about half the incidence among Americans in general. Cancer incidence is also low in certain relatively affluent populations in Africa.

Although such observations on human populations indicate that cancer can often be avoided, it has been difficult in most cases—with the exception of tobacco—to identify the specific environmental risk factors or to establish how they act. We will first look at what has been learned about the external agents that are known to cause cancer. We will then consider some of the triumphs, and difficulties, in finding ways to prevent human cancer. We will discuss the treatment of cancer in the last section, after we have examined the molecular biology of the disease.

Many, But Not All, Cancer-Causing Agents Damage DNA

The agents that can cause cancer, known as **carcinogens**, are many and varied, but the easiest to understand are those that damage DNA and so generate mutations. These cancer-causing mutagens include chemicals and various forms of radiation—UV light from sunshine, as well as ionizing radiation, such as γ rays and α particles from radioactive decay.

Many quite disparate chemicals are carcinogenic when they are fed to experimental animals or painted repeatedly on their skin. Examples include a range of aromatic hydrocarbons and derivatives of them such as aromatic amines, nitrosamines, and alkylating agents such as mustard gas. Although these chemical carcinogens have diverse structures, they have at least one common property—they cause mutations. In one common test for mutagenicity, the carcinogen is mixed with an activating extract prepared from rat liver cells (to mimic the biochemical processing that occurs in an intact animal—discussed below) and is added to a culture of specially designed test bacteria; the resulting mutation rate of the bacteria is then measured (**Figure 20–21**). Most of the compounds scored as mutagenic by this rapid and convenient assay in bacteria also cause mutations or chromosome aberrations when tested on mammalian cells. When mutagenicity data from various sources are analyzed, one finds that many carcinogens are mutagens.

A few of these carcinogens act directly on DNA, but generally the more potent ones are relatively inert chemically; they become damaging only after they have been changed to a more reactive form by metabolic processes in the liver, catalyzed by a set of intracellular enzymes known as the *cytochrome P-450 oxidases*. These enzymes normally help to convert ingested toxins into harmless and easily excreted compounds. Unfortunately, their activity on certain chemicals generates products that are highly mutagenic. Examples of carcinogens activated in this way include the fungal toxin *aflatoxin B1* and *benzo[a]pyrene*, a cancer-causing chemical present in coal tar and tobacco smoke (**Figure 20–22**).

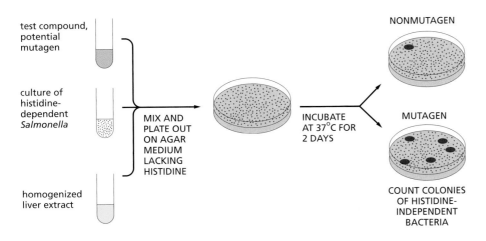

Figure 20–21 **The Ames test for mutagenicity.** The test uses a strain of *Salmonella* bacteria that require histidine in the medium because of a defect in a gene necessary for histidine synthesis. Mutagens can cause a further change in this gene that reverses the defect, creating revertant bacteria that do not require histidine. To increase the sensitivity of the test, the bacteria also have a defect in their DNA repair machinery that makes them especially susceptible to agents that damage DNA. A majority of compounds that are mutagenic in tests such as this are also carcinogenic.

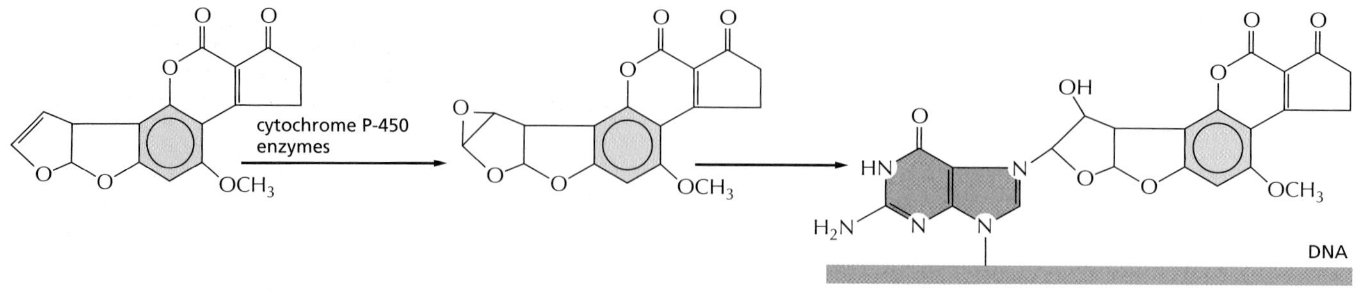

(A)   AFLATOXIN   AFLATOXIN-2,3-EPOXIDE   CARCINOGEN BOUND TO GUANINE IN DNA

Figure 20–22 **Some known carcinogens.** (A) Carcinogen activation. A metabolic transformation must activate many chemical carcinogens before they will cause mutations by reacting with DNA. The compound illustrated here is *aflatoxin B1,* a toxin from a mold *(Aspergillus flavus oryzae)* that grows on grain and peanuts when they are stored under humid tropical conditions. It is thought to be a contributory cause of liver cancer in the tropics and is associated with characteristic mutations of the *p53* tumor suppressor gene. (B) Different carcinogens cause different types of cancer. (B, data from Cancer and the Environment: Gene Environment Interactions, National Academies Press, 2002. With permission from National Academies Press.)

- VINYL CHLORIDE:
  liver angiosarcoma

- BENZENE:
  acute leukemias

- ARSENIC:
  skin carcinomas, bladder cancer

- ASBESTOS:
  mesothelioma

- RADIUM:
  osteosarcoma

(B)

Tumor Initiators Damage DNA; Tumor Promoters Do Not

Not all substances that favor the development of cancer are mutagens. Some of the clearest evidence comes from studies done long ago on the effects of cancer-causing chemicals on mouse skin, where it is easy to observe the stages of tumor progression. Skin cancers can be elicited in mice by repeatedly painting the skin with a DNA-damaging carcinogen such as benzo[a]pyrene or the related compound dimethylbenz[a]anthracene (DMBA). A single application of these mutagens, however, does not by itself give rise to a tumor or any other obvious lasting abnormality. Yet, it does cause latent genetic damage—mutations that set the stage for a greatly increased incidence of cancer when the cells are exposed either to further treatments with the carcinogen or to certain other, quite different, insults. A carcinogen that sows the seeds of cancer in this way acts as a **tumor initiator**.

Simply wounding skin that has been exposed once to such an initiator can, by stimulating cell proliferation, cause cancers to develop from some of the cells at the edge of the wound. Inflammatory white blood cells are attracted to a wound, and it is thought that inflammatory cells play a part in the cancer-promoting effect, as we mentioned earlier. Alternatively, repeated exposure over a period of months to substances known as **tumor promoters**, chemicals that are not themselves mutagenic, can cause cancer selectively in skin previously exposed to a tumor initiator. The most widely studied tumor promoters are *phorbol esters,* such as tetradecanoylphorbol acetate (TPA), which behave as artificial activators of *protein kinase C (PKC)* and hence activate part of the phosphatidylinositol intracellular signaling pathway (discussed in Chapter 15). These substances cause cancers at high frequency only if they are applied after a treatment with a mutagenic initiator (**Figure 20–23**).

The immediate effect of a tumor promoter is apparently to induce an inflammatory response, causing the secretion of both growth factors and proteases in the local environment. These act both directly and indirectly on cells to stimulate cell division (or to cause cells that would normally differentiate and stop dividing to continue dividing instead). In a region of skin that has previously been exposed to an initiator, this proliferation results in the development of many small, benign, wartlike tumors called *papillomas.* The greater the prior dose of initiator, the larger the number of papillomas induced; it is thought that each papilloma consists of a single clone of cells descended from a mutant cell that the initiator has produced. As one might expect for genetic damage, the hidden changes

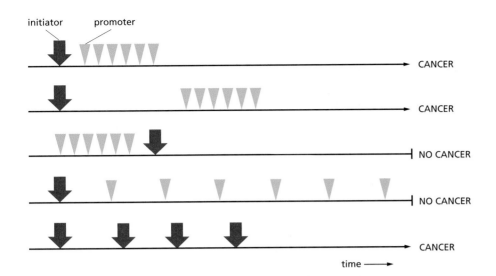

Figure 20–23 Some possible schedules of exposure to a tumor initiator (mutagenic) and a tumor promoter (nonmutagenic) and their outcomes. Cancer ensues only if the exposure to the promoter follows exposure to the initiator and only if the intensity of exposure to the promoter exceeds a certain threshold. Cancer can also occur as a result of repeated exposure to the initiator alone.

caused by a tumor initiator are irreversible; for this reason, they can be uncovered by treatment with a tumor promoter even after a long delay.

It is still not certain how tumor promoters work, and different promoters are likely to work in different ways. One possibility is that they simply provoke the expression of proliferation-inducing genes that had been mutated before the promoter was applied but were not expressed: a mutation that makes a gene product hyperactive will not show its effects until the gene is expressed. Alternatively, the tumor promoter may temporarily alter the way the cell reacts to the product of the mutated gene, either by releasing the cell from a counteracting inhibitory influence or by triggering production of a co-factor necessary for proliferation of the mutated gene product. Whichever the mechanism, the result is that the mutant cell is enabled to grow and divide and produce a large cluster of cells (**Figure 20–24**).

A typical papilloma might contain about 10^5 cells. If exposure to the tumor promoter is stopped, almost all the papillomas regress, and the skin regains a largely normal appearance. In a few of the papillomas, however, further changes occur that enable cell growth and division to continue in an uncontrolled way, even after the promoter has been withdrawn. These changes seem to originate in an occasional single papilloma cell, at about the frequency expected for spontaneous mutations. In this way, a small proportion of the papillomas progress to become cancers. Thus, the tumor promoter apparently favors the development of cancer by expanding the population of cells that carry an initial mutation: the more such cells there are and the more times they divide, the greater is the chance that at least one of them will sustain another mutation, or an epigenetic change, that carries it one step further toward malignancy.

Although naturally occurring cancers do not necessarily arise through the specific sequence of distinct initiation and promotion steps just described, their evolution must be governed by similar principles. They too will evolve at a rate that depends not only on the frequency of genetic or epigenetic changes, but also on the local influences affecting the survival, growth, proliferation, and spread of these altered cells.

Viruses and Other Infections Contribute to a Significant Proportion of Human Cancers

A small but significant proportion of human cancers, perhaps 15% in the world as a whole, are thought to arise by mechanisms that involve viruses, bacteria, or parasites. The main culprits, as shown in **Table 20–1**, are the DNA viruses. Evidence for their involvement comes partly from the detection of viruses in cancer patients and partly from epidemiology. Liver cancer, for example, is common in parts of the world (Africa and Southeast Asia) where hepatitis-B viral infections

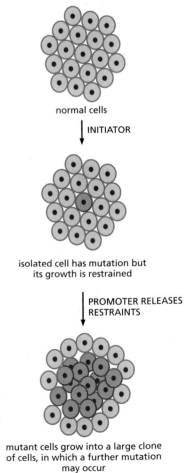

normal cells

INITIATOR

isolated cell has mutation but its growth is restrained

PROMOTER RELEASES RESTRAINTS

mutant cells grow into a large clone of cells, in which a further mutation may occur

Figure 20–24 The effect of a tumor promoter. The tumor promoter creates a local environment that expands the population of mutant cells, thereby increasing the probability of tumor progression by further genetic change.

Table 20–1 Viruses Associated with Human Cancers

VIRUS	ASSOCIATED CANCER	AREAS OF HIGH INCIDENCE
DNA viruses		
Papovavirus family		
Papillomavirus (many distinct strains)	warts (benign) carcinoma of the uterine cervix	worldwide worldwide
Hepadnavirus family		
Hepatitis-B virus	liver cancer (hepatocellular carcinoma)	Southeast Asia, tropical Africa
Hepatitis-C virus	liver cancer (hepatocellular carcinoma)	worldwide
Herpesvirus family		
Epstein–Barr virus	Burkitt's lymphoma (cancer of B lymphocytes) nasopharyngeal carcinoma	West Africa, Papua New Guinea Southern China, Greenland
RNA viruses		
Retrovirus family		
Human T-cell leukemia virus type I (HTLV-1)	adult T-cell leukemia/ lymphoma	Japan, West Indies
Human immuno-deficiency virus (HIV, the AIDS virus)	Kaposi's sarcoma	Central and Southern Africa

For all these viruses, the number of people infected is much larger than the numbers who develop cancer: the viruses must act in conjunction with other factors. Moreover, some of the viruses contribute to cancer only indirectly; HIV, for example, destroys helper T lymphocytes, which allows a herpes virus to transform endothelial cells. Similarly, hepatitis-C virus causes chronic hepatitis, which promotes the development of liver cancer.

are common, and in those regions the cancer occurs almost exclusively in people who show signs of chronic hepatitis-B infection. Chronic infection with hepatitis-C virus, which has infected 170 million people worldwide, is also clearly associated with the development of liver cancer.

The precise role of a cancer-associated virus is often hard to decipher because there is a delay of many years from the initial viral infection to the development of the cancer. Moreover, the virus is responsible for only one of a series of steps in the progression to cancer, and other environmental factors and genetic accidents are also involved. As we explain later, DNA viruses frequently carry genes that can subvert the control of cell division in the host cell, causing uncontrolled cell proliferation. DNA viruses that operate in this manner include the human papillomaviruses. Some of these cause harmless warts, but others infect the uterine cervix and are implicated in the development of carcinomas of the cervix.

In some cancers, viruses seem to have additional, indirect tumor-promoting actions. The hepatitis-B and C viruses may, for example, favor the development of liver cancer by causing chronic inflammation (hepatitis), which stimulates cell division in the liver, and enhances the viruses' more direct effects on cell proliferation. In AIDS, the human immunodeficiency virus (HIV) promotes development of an otherwise rare cancer called Kaposi's sarcoma by destroying the immune system, thereby permitting a secondary infection with a human herpes virus (HHV-8) that has a direct carcinogenic action. Chronic infection with parasites and bacteria can also promote the development of some cancers. For example, chronic infection of the stomach with the bacterium *Helicobacter pylori*, which causes ulcers, appears to be a major cause of stomach cancer. And bladder cancer in some parts of the world is associated with chronic infection by the blood fluke, *Schistosoma haematobium*, a parasitic flatworm.

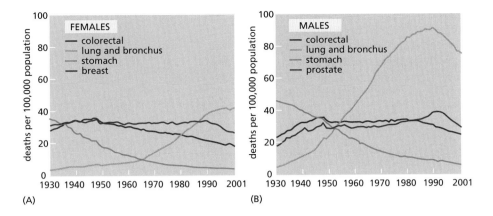

(A)   (B)

Figure 20–25 Age-adjusted cancer death rates, United States, 1930–2001. Selected death rates adjusted to the age distribution of the US population in 1970, are plotted for (A) females and (B) males. Note the dramatic rise in lung cancer for both sexes, following the pattern of tobacco smoking, and the fall in deaths from stomach cancer, possibly related to changes in diet or in patterns of infection with *Helicobacter*. Recent reductions in other cancer death rates may correspond to improvements in detection and treatment. Age-adjusted data like these are needed to compensate for the inevitable increase in cancer rates as people live longer, on average. (Adapted from Cancer Facts and Figures, 2005. With permission from American Cancer Society.)

Identification of Carcinogens Reveals Ways to Avoid Cancer

Tobacco smoke is by far the most important environmental cause of cancer in the world today. It contains both carcinogens and tumor promoters. Other comparably important chemical causes of cancer in humans remain to be identified. It is sometimes thought that the main environmental causes of cancer are the products of a highly industrialized way of life—the rise in pollution, the enhanced use of food additives, and so on—but there is little evidence to support this view. The idea may have come in part from the identification of some highly carcinogenic materials used in industry, such as 2-naphthylamine and asbestos. Except for the increase in cancers caused by smoking, however, and a remarkable decrease in stomach cancer that may reflect decreasing rates of infection with *Helicobacter pylori,* the incidence of the most common cancers for individuals of a given age has not changed very much in the past hundred years (**Figure 20–25**).

Most of the carcinogenic factors that are known to be significant are by no means specific to the modern world. The most potent known carcinogen, by certain assays at least, is aflatoxin B1 (see Figure 20–22). It is produced by fungi that naturally contaminate foods such as tropical peanuts and is an important cause of liver cancer in Africa and Asia. Some factors that accelerate cancer are intrinsic to our bodies. Thus, in women, the risk of cancer is powerfully influenced by the reproductive hormones produced at different stages of life, and there is a striking correlation between reproductive history and the occurrence of breast cancer (**Figure 20–26**). The reproductive hormones presumably affect breast cancer incidence through their influence on cell proliferation in the breast. Clearly, when attempting to identify the environmental causes of cancer, we need to keep an open mind.

Epidemiology—the analysis of disease frequency in populations—remains the principal tool for finding environmental causes of human cancer. The approach has enjoyed some notable successes, and it promises more to come. By revealing the role of smoking, for example, epidemiology has shown a way to reduce the total cancer death rate in North America and Europe by as much as 25%. The approach works best when applied to a fairly homogeneous population in which it is easy to distinguish between individuals who were exposed to the agent and those who were not, and when the agent under investigation is responsible for most of the cases of a certain kind of cancer. In the early part of the twentieth century, for example, in one British factory all of the men who had been employed in distilling 2-naphthylamine (and were thereby subjected to prolonged exposure) eventually developed bladder cancer (see Figure 20–8); the connection was relatively easy to establish because both the chemical and the form of cancer were uncommon in the general population.

In contrast, it is very hard to identify, by epidemiology alone, everyday environmental factors that favor the development of common cancers: most of these factors are probably agents to which we are all exposed to some extent, and

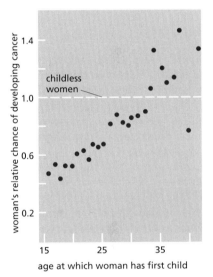

childless women

age at which woman has first child

Figure 20–26 Effects of childbearing on the risk of breast cancer. The relative probability of breast cancer developing at some time in a woman's life is plotted as a function of the age at which she gives birth to her first child. The graph shows the probability relative to that for a childless woman. The longer the period of exposure to reproductive hormones up to birth of the first child, the greater the risk. It is thought that the first full-term pregnancy may result in a permanent change in the state of differentiation of the cells of the breast, altering their subsequent responses to hormones. Several other lines of epidemiological evidence also support the view that exposure to certain combinations of reproductive hormones, especially estrogen, can promote development of breast cancer. (From J. Cairns, Cancer: Science and Society. San Francisco: W.H. Freeman, 1978. After B. MacMahon, P. Cole and J. Brown, *J. Natl Cancer Inst.* 50:21–42, 1973. With permission from Oxford University Press.)

many of them probably contribute together to a given cancer's incidence. If, say, eating oranges either doubled or halved the risk of colorectal cancer, it is unlikely that we would find it out, unless we had some prior reason to suspect a connection. The same goes for the countless other substances that we eat, drink, breathe, and put on our bodies. The epidemiological evidence is, however, fairly clear on one dietary point: as mentioned earlier, too much food, leading to obesity, significantly increases the risk of cancer.

Even when evidence is obtained that a substance may be carcinogenic, either from epidemiology or from laboratory tests, it may be difficult to decide what level of human exposure is acceptable. Estimating how many cases of human cancer a certain amount of a substance is likely to cause is difficult; balancing this risk against the utility of the substance is more difficult still. Certain agricultural fungicides, for example, appear to be mildly carcinogenic at high doses in animal tests. But it has been calculated that if they were not used in agriculture, the contamination of food by fungal metabolites such as aflatoxin B1 would cause far more cases of cancer than the fungicide residues in food ever would.

Nevertheless, efforts to identify potential carcinogens still have a central place in our struggle to prevent cancer. Prevention is not only better than a cure, but for many types of cancer, it may be more readily attainable.

Summary

Environmental agents accelerate the rate of tumor development and progression. These can act either as tumor initiators or tumor promoters. Tumor promoters induce an inflammatory response, and they create a local environment that alters gene expression, stimulates cell proliferation, and increases the population of mutant cells created by the initiator. The majority of the known environmental agents that increase cancer development are mutagens, including chemical carcinogens and various forms of radiation, including UV light and ionizing radiation. Because many environmental factors contribute to the development of a given cancer and some of these are under our control, a large proportion of cancers are in principle preventable.

The finding that immigrant populations adopt the cancer patterns of their new host country strongly suggests that the majority of cancers could be avoided by changing the food we eat and other environmental exposures. Epidemiology can be a powerful tool for identifying these environmental effects on human cancer. The epidemiological approach does not require knowing how the environmental agents work, and it can uncover factors that are not chemical entities, such as lifestyle and certain patterns of child-bearing. Individuals can avoid many of the environmental risk factors that are identified in this way. These include smoking tobacco and falling prey to infection with cancer-causing viruses such as papillomaviruses or hepatitis B or C. However, because many factors that affect cancer risk are very difficult to identify, we remain largely ignorant of the principal environmental factors that affect cancer incidence.

FINDING THE CANCER-CRITICAL GENES

As we have seen, cancer depends on the accumulation of inherited changes in somatic cells. To understand it at a molecular level we need to identify the mutations and epigenetic changes involved and to discover how they give rise to cancerous cell behavior. Finding the relevant cells is often easy; they are favored by natural selection and call attention to themselves by giving rise to tumors. But how do we identify the small number of genes with the cancer-promoting changes among all the other genes in the cancerous cells? A similar needle-in-haystack problem arises in any search for a gene underlying a given mutant phenotype, but for cancer the task is particularly complex. A typical cancer depends on a whole set of mutations and epigenetic changes—usually a somewhat different set in each individual patient—and introduction of any single one of these into a normal cell is not enough to make the cell cancerous. In addition, the cooperation between different altered genes makes it harder to test the significance of an inherited change in any individual gene. To make matters worse, a given cancer cell will also contain a large number of somatic mutations that are accidental by-products of its genetic instability, and it can be difficult to distinguish these meaningless changes from those changes that have a causative role in the disease.

Despite these difficulties, many genes that are repeatedly altered in human cancers have been identified—several hundred of them—although it is clear that many more remain to be discovered. We will call such genes, for want of a better term, **cancer-critical genes**, meaning all genes whose alteration frequently contributes to the causation of cancer. Our knowledge of these genes has accumulated piecemeal through many different and sometimes circuitous approaches, ranging from early studies of cancer-causing infections in chickens to investigations of embryonic development. Analyses of rare but highly revealing inherited forms of cancer have also added to our understanding. More recently, sequencing of DNA from multiple cases of specific types of cancers has begun to give a more systematic picture of the genetic changes that are a regular feature of those diseases.

In this section, we discuss both the methods used for identifying cancer-critical genes and the varied kinds of inherited changes that occur in them during the development of cancer.

The Identification of Gain-of-Function and Loss-of-Function Mutations Requires Different Methods

Cancer-critical genes are grouped into two broad classes, according to whether the cancer risk arises from too much activity of the gene product, or too little. Genes of the first class, in which a gain-of-function mutation can drive a cell toward cancer, are called **proto-oncogenes**; their mutant, overactive or overexpressed forms are called **oncogenes**. Genes of the second class, in which a loss-of-function mutation can contribute to cancer, are called **tumor suppressor genes**. A third class, whose effects are more indirect, are those genes whose mutation results in genomic instability, a class we describe as *DNA maintenance genes*.

As we shall see, mutations in both oncogenes and tumor suppressor genes can have similar effects in enhancing cell proliferation and cell survival, as well as in promoting tumor development. Thus, from the point of view of a cancer cell, oncogenes and tumor suppressor genes—and the mutations that affect them—are flip sides of the same coin. The techniques needed to find these genes, however, differ—depending on whether the genes are overactive or underactive in cancer.

Mutation of a single copy of a proto-oncogene that converts it to an oncogene has a dominant, growth-promoting effect on a cell (**Figure 20–27A**). Thus, we can identify the oncogene by its effect when it is *added*—by DNA transfection, for example, or through infection with a viral vector—to the genome of a suitable type of tester cell. In the case of the tumor suppressor gene, on the other hand, the cancer-causing alleles produced by the change are generally recessive: often (but not always) both copies of the normal gene must be removed or inactivated in the diploid somatic cell before an effect is seen (Figure 20–27B). This calls for a different approach, aimed at detecting what is *missing* in the cancer cell.

In some cases, a specific gross chromosomal abnormality, visible under the microscope, is repeatedly associated with a particular type of cancer. This can give a clue to the location of an oncogene that is activated as a result of the chromosomal rearrangement (as in the example of the chromosomal translocation responsible for chronic myelogenous leukemia, discussed earlier). Alternatively, a visible deletion of a chromosomal segment may reveal the site of a deleted tumor suppressor gene.

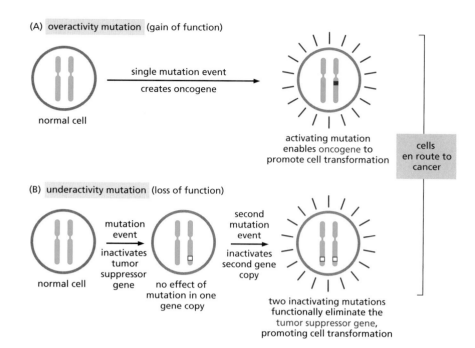

(A) overactivity mutation (gain of function)

normal cell

single mutation event
creates oncogene

activating mutation
enables oncogene to
promote cell transformation

cells
en route to
cancer

(B) underactivity mutation (loss of function)

normal cell

mutation
event

inactivates
tumor
suppressor
gene

no effect of
mutation in one
gene copy

second
mutation
event

inactivates
second gene
copy

two inactivating mutations
functionally eliminate the
tumor suppressor gene,
promoting cell transformation

Figure 20–27 Cancer-critical mutations fall into two readily distinguishable categories, dominant and recessive. In this diagram, activating mutations are represented by *solid red boxes*, inactivating mutations by *hollow red boxes*. (A) Oncogenes act in a dominant manner: a gain-of-function mutation in a single copy of the cancer-critical gene can drive a cell toward cancer. (B) Mutations in tumor suppressor genes, on the other hand, generally act in a recessive manner: the function of both alleles of the cancer-critical gene must be lost to drive a cell toward cancer. Although in this diagram the second allele of the tumor suppressor gene is inactivated by mutation, it is often inactivated instead by loss of the second chromosome. Not shown is the fact that mutation of some tumor suppressor genes can have an effect even when only one of the two gene copies is damaged.

Retroviruses Can Act as Vectors for Oncogenes That Alter Cell Behavior

Tumor viruses have played a remarkable part in the search for the genetic causes of human cancer. Although viruses have no role in the majority of common human cancers, they are more prominent as causes of cancer in some animal species. Analysis of these animal tumor viruses provided a critical key to understanding the mechanisms of cancer in general and to the discovery of oncogenes in particular.

One of the first animal viruses to be implicated in cancer was discovered nearly 100 years ago in chickens, which are subject to infections that cause connective-tissue tumors, or sarcomas. The infectious agent was characterized as a virus—the *Rous sarcoma virus,* which we now know to be an RNA virus. Like all the other *RNA tumor viruses* discovered since, it is a **retrovirus**. When it infects a cell, its RNA is copied into DNA by reverse transcription, and the DNA is inserted into the host genome, where it can persist and be inherited by subsequent generations of cells. The Rous sarcoma virus carries an oncogene that causes cancer in chickens. This oncogene is not necessary for the virus's own survival or reproduction, as demonstrated by the discovery of mutant forms of the virus that multiply normally but no longer make their host cells cancerous—a process called cell *transformation.* Some of these mutants were found to have deleted all or part of a gene that codes for a protein called **Src** (pronounced "Sarc"). Other mutations in this gene made the transforming effect of the virus temperature-sensitive: infected cells show a transformed phenotype at 34°C, but return to the normal phenotype within a few hours when the temperature is raised to 39°C.

What was the origin of this gene? In 1975, when a radioactive DNA copy of the viral *Src* gene sequences was used as a probe to search for closely related sequences by DNA–DNA hybridization (see Figure 8–36), the genomes of vertebrate cells were found to contain sequences that are closely similar, but not identical, to this viral gene. The viral oncogene had evidently been picked up accidentally by the retrovirus from the genome of a previous infected host cell, but the host cell proto-oncogene (*c-Src*) had undergone mutation in the process to become an oncogene (*v-Src*) (**Figure 20–28**). This finding that cancer could originate from mutation of a specific gene that was present in the genome of a normal animal profoundly altered the field of cancer research.

A large number of other oncogenes have since been identified in other retroviruses and analyzed in similar ways. Each has led to the discovery of a corresponding proto-oncogene that is present in normal animal cells.

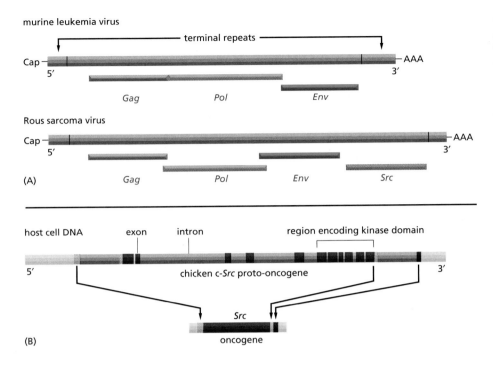

murine leukemia virus

Cap — 5'

terminal repeats

AAA
3'

Gag   Pol   Env

Rous sarcoma virus

Cap — 5'

AAA
3'

(A)   Gag   Pol   Env   Src

host cell DNA   exon   intron   region encoding kinase domain

5'   chicken c-Src proto-oncogene   3'

Src

(B)   oncogene

Figure 20–28 **The structure of the Rous sarcoma virus.** (A) The organization of the viral genome as compared with that of a more typical retrovirus (murine leukemia virus). Rous sarcoma virus is unusual among the retroviruses that carry oncogenes in that it has retained the three viral genes required for the ordinary viral life cycle: *Gag* (which produces a polyprotein that is cleaved to generate the capsid proteins), *Pol* (which produces reverse transcriptase and an enzyme involved in integrating the viral chromosome into the host genome), and *Env* (which produces the envelope glycoprotein). In other oncogenic retroviruses, one or more of these viral genes are wholly or partly lost when the retrovirus acquires the transforming oncogene. Therefore the transforming virus can generate infectious particles only in a cell that is simultaneously infected with a nondefective, nontransforming *helper virus,* which supplies the missing functions. (B) The relationship between the *v-Src* oncogene and the cellular *Src* proto-oncogene from which it has been derived. The introns present in cellular *Src* have been spliced out of *v-Src;* in addition, *v-Src* contains mutations that alter the amino acid sequence of the protein (not shown), making it hyperactive and unregulated as a tyrosine-specific protein kinase (see Figure 3–69). The v-Src protein is also greatly overproduced because the virus makes large amounts of RNA.

Different Searches for Oncogenes Have Converged on the Same Gene—Ras

While some researchers searched for oncogenes in retroviruses, others took a more direct approach and searched for DNA sequences in human cancer cells that could provoke uncontrolled proliferation when introduced into noncancerous cell lines. As tester cells for the assay, they usually used cell lines derived from mouse fibroblasts. These cells, previously selected to proliferate indefinitely in culture, are thought to already contain genetic alterations that take them part of the way toward malignancy. For this reason, the addition of a single oncogene can be enough to produce a dramatic effect.

To detect an oncogene in this way, DNA is extracted from the tumor cells, broken into fragments, and introduced into the cultured cells. If any of the fragments contains an oncogene, small colonies of abnormally proliferating cells begin to appear; because the cells forming such colonies display cancerous properties, they are said to be *transformed* or to have undergone **transformation**. <CCGA> Each colony is a clone of cells that originated from a single cell, carrying an added gene that released the cell from some of the social controls on cell proliferation; the transformed cells forming the colony outgrow the untransformed cells in the culture and pile up layer upon layer as they proliferate (**Figure 20–29**).

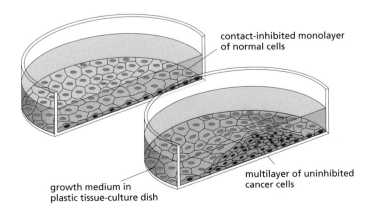

contact-inhibited monolayer of normal cells

growth medium in plastic tissue-culture dish

multilayer of uninhibited cancer cells

Figure 20–29 **Loss of contact inhibition in cell culture.** Most normal cells stop proliferating once they have carpeted the dish with a single layer of cells: proliferation seems to depend on contact with the dish, and to be inhibited by contacts with other cells—a phenomenon known as "contact inhibition." Cancer cells, in contrast, usually disregard these restraints and continue to grow, so that they pile up on top of one another.

This assay led to the isolation and sequencing of the first human oncogene, a mutant version of the proto-oncogene **Ras**. The *Ras* gene is now known to be mutated in about one in five human cancers. The discovery of the human *Ras* oncogene was all the more dramatic because, shortly before, a mutated *Ras* gene had been found to be the tumor-causing gene in a retrovirus that causes sarcomas in rodents. This discovery in the early 1980s of the same oncogene in human tumor cells and in an animal tumor virus was electrifying. The implication that cancers are caused by mutations in a limited number of cancer-critical genes changed our understanding of the molecular biology of cancer.

As discussed in Chapter 15, normal Ras proteins are monomeric GTPases that help transmit signals from cell-surface receptors to the cell interior. <GAAC> The *Ras* oncogenes isolated from human tumors contain point mutations that create a hyperactive Ras protein that cannot shut itself off by hydrolyzing its bound GTP to GDP. Because this makes the protein hyperactive, its effect is dominant—that is, only one of the cell's two gene copies needs to change to have an effect. Because the *Ras* genes are mutated in a wide range of human cancers, they remain one of the most important examples of cancer-critical genes.

Through methods such as those we have just described, and various other routes, we now know of hundreds of proto-oncogenes—that is, genes that can be converted by activating mutations into oncogenes that help to cause cancer.

Studies of Rare Hereditary Cancer Syndromes First Identified Tumor Suppressor Genes

Identifying a gene that has been inactivated in the genome of a cancer cell requires a different strategy from finding a gene that has become hyperactive: one cannot, for example, use a cell transformation assay to identify something that simply is not there. The key insight that led to the discovery of the first tumor suppressor gene came from studies of a rare type of human cancer, **retinoblastoma**, which arises from cells in the retina of the eye that are converted to a cancerous state by an unusually small number of mutations. As often happens in biology, the discovery arose from examination of a special case, but it turned out to reveal a gene of widespread relevance. In fact, genes identified in rare familial cancer syndromes are commonly found to be relevant in more common sporadic cancers, where they often serve as tumor suppressors.

Retinoblastoma occurs in childhood, and tumors develop from neural precursor cells in the immature retina. About one child in 20,000 is afflicted. One form of the disease is hereditary, and the other is not. In the hereditary form, multiple tumors usually arise independently, affecting both eyes; in the nonhereditary form, only one eye is affected, and by only one tumor. A few individuals with retinoblastoma have a visibly abnormal karyotype, with a deletion of a specific band on chromosome 13 that, if inherited, predisposes an individual to the disease. Deletions of this same region are also encountered in tumor cells from some patients with the nonhereditary disease, which suggested that the cancer was caused by loss of a critical gene in that location.

Using the location of this chromosomal deletion, it was possible to clone and sequence the **Rb gene**. It was then discovered that those who suffer from the hereditary form of the disease have a deletion or loss-of-function mutation present in one copy of the *Rb* gene in every somatic cell. These cells are predisposed to becoming cancerous, but do not do so if they retain one good copy of the gene. The retinal cells that are cancerous are defective in both copies of *Rb* because of a somatic event that has eliminated the function of the previously good copy.

In patients with the nonhereditary form of the disease, by contrast, the non-cancerous cells show no defect in either copy of *Rb*, while the cancerous cells have become defective in both copies. These nonhereditary retinoblastomas are very rare because they require two independent events that inactivate the same gene on two chromosomes in a single retinal cell lineage (**Figure 20–30**). The *Rb* gene is also missing in several common types of sporadic cancer, including carcinomas of

NORMAL, HEALTHY INDIVIDUAL

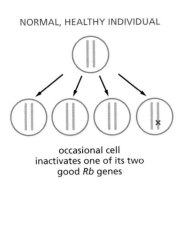

occasional cell
inactivates one of its two
good *Rb* genes

RESULT: NO TUMOR

HEREDITARY RETINOBLASTOMA

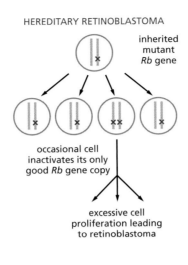

inherited
mutant
Rb gene

occasional cell
inactivates its only
good *Rb* gene copy

excessive cell
proliferation leading
to retinoblastoma

RESULT: MOST PEOPLE WITH INHERITED
MUTATION DEVELOP MULTIPLE TUMORS
IN BOTH EYES

NONHEREDITARY RETINOBLASTOMA

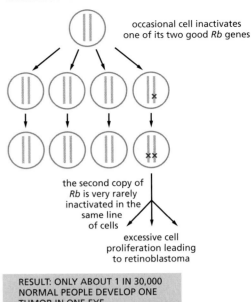

occasional cell inactivates
one of its two good *Rb* genes

the second copy of
Rb is very rarely
inactivated in the
same line
of cells

excessive cell
proliferation leading
to retinoblastoma

RESULT: ONLY ABOUT 1 IN 30,000
NORMAL PEOPLE DEVELOP ONE
TUMOR IN ONE EYE

lung, breast, and bladder. These more common cancers arise by a more complex series of genetic changes than does retinoblastoma, and they make their appearance much later in life. But in all of them, it seems, loss of *Rb* function is frequently a major step in the progression toward malignancy.

The *Rb* gene encodes the **Rb protein**, which is a universal regulator of the cell cycle present in almost all cells of the body (see Figure 17–62). It acts as one of the main brakes on progress through the cell-division cycle, and its loss can allow cells to enter the cell cycle inappropriately, as we discuss later.

Figure 20–30 The genetic mechanisms that cause retinoblastoma. In the hereditary form, all cells in the body lack one of the normal two functional copies of the *Rb* tumor suppressor gene, and tumors occur where the remaining copy is lost or inactivated by a somatic event (either mutation or epigenetic silencing). In the nonhereditary form, all cells initially contain two functional copies of the gene, and the tumor arises because both copies are lost or inactivated through the coincidence of two somatic events in one line of cells.

Tumor Suppressor Genes Can Also Be Identified from Studies of Tumors

The Rb story illustrates how rare hereditary cancer syndromes can be used to uncover tumor suppressor genes that are relevant to common sporadic cancers. However, only a few of the tumor suppressor genes now known to be important have been discovered in this way. A more direct approach to the identification of such genes involves comparing tumor cells with non-cancerous cells from the same patient so as to discover what exactly, out of the 3 billion nucleotides of the human genome, is missing, functionally defective, or abnormally silenced. Because of the genetic instability of cancer cells, there is usually a great deal changed. Most of the alterations will be random, accidental by-products of the genetic instability. Therefore, a relevant tumor suppressor gene can only be identified by the criterion that it is repeatedly missing, defective, or otherwise silenced in many independent cases of a particular type of cancer. Tracking down tumor suppressor genes in this way is a hard task, but feasible with modern techniques for large-scale DNA analysis (as we explain below). Dozens of tumor suppressor genes have already been well characterized and many more are known.

Both Genetic and Epigenetic Mechanisms Can Inactivate Tumor Suppressor Genes

It is the inactivation of tumor suppressor genes that is dangerous. This inactivation can occur in many ways, with different combinations of mishaps converging to eliminate or cripple both gene copies. The first copy may, for example, be lost by a small chromosomal deletion or inactivated by a point mutation. The second copy is commonly eliminated by a less specific and more probable

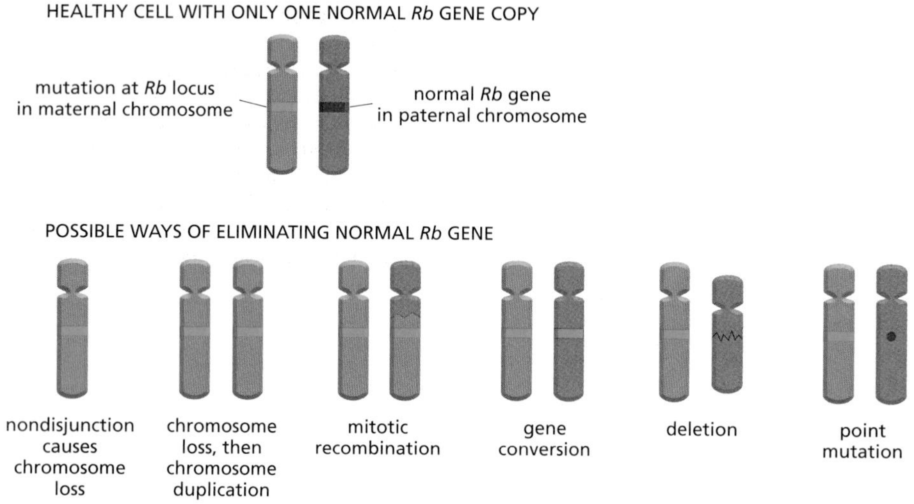

HEALTHY CELL WITH ONLY ONE NORMAL *Rb* GENE COPY

mutation at *Rb* locus
in maternal chromosome

normal *Rb* gene
in paternal chromosome

POSSIBLE WAYS OF ELIMINATING NORMAL *Rb* GENE

nondisjunction
causes
chromosome
loss

chromosome
loss, then
chromosome
duplication

mitotic
recombination

gene
conversion

deletion

point
mutation

Figure 20–31 Six ways of losing the remaining good copy of a tumor suppressor gene through changes in DNA sequences. A cell that is defective in only one of its two copies of a tumor suppressor gene—for example, the *Rb* gene—usually behaves as a normal, healthy cell; the diagrams below show how this cell may lose the function of the other gene copy as well and thereby progress toward cancer. A seventh possibility, frequently encountered with some tumor suppressors, is that the gene may be silenced by an epigenetic change, without alteration of the DNA sequence, as illustrated in Figure 20–32. (After W.K. Cavenee et al., *Nature* 305:779–784, 1983. With permission from Macmillan Publishers Ltd.)

mechanism: the chromosome carrying the remaining normal copy may be lost from the cell through errors in chromosome segregation, or the normal gene may be replaced by a mutant version through either mitotic recombination or a gene conversion event.

Figure 20–31 summarizes the range of possible ways to lose the remaining good copy of a tumor suppressor gene through DNA sequence changes, using the *Rb* gene as an example. It is important to note that, except for the point mutation mechanism illustrated at the far right, these pathways all produce cells that carry only a single type of DNA sequence in the chromosomal region containing their *Rb* genes—a sequence that is identical to the sequence in the original mutant chromosome.

As discussed in Chapter 4, normal human genetic variation makes each of our maternal and paternal chromosome sets distinguishably different. On average, human DNA sequences differ—that is, we are heterozygous—at roughly one in every thousand nucleotides. Where a large segment of one chromosome has been either lost or converted to the DNA sequence in its homologous chromosome, as in Figure 20–31, there is a *loss of heterozygosity (LOH)*: only one version of each variable DNA sequence in that neighborhood remains. Millions of common sites of heterozygosity in the human genome have been mapped as part of the Human Genome Project: each of these sites is characterized by a specific DNA sequence that is known to be polymorphic—that is, to occur commonly in two or more slightly different versions in the human population. Given a sample of tumor DNA, one can check which of the versions of these polymorphic sequences are present. The same can be done with a sample of DNA from noncancerous tissue from the same patient, for comparison. A loss of heterozygosity throughout a region of the genome containing one or more polymorphic sites, or loss of a genetic marker sequence that is seen in the non-cancerous control DNA, can point the way toward a chromosomal region that contains a relevant tumor suppressor gene. However, because of their genetic instability, cancer cells often exhibit a loss of heterogeneity for many different chromosomal regions. Therefore the detection of tumor suppressor genes by this approach generally requires the subtraction of a large amount of random noise.

Epigenetic changes provide another important way to permanently inactivate a tumor suppressor gene. Most commonly, the gene may become packaged into heterochromatin and the C nucleotides in CpG sequences in its promoter may become methylated in a heritable manner (see Figure 20–12). These mechanisms can irreversibly silence the gene in a cell and in all of its progeny. Given a catalog of possible tumor suppressor genes, it is relatively easy to test their promoters for abnormal amounts of DNA methylation. Studies of this type suggest that epigenetic gene silencing is a frequent event in tumor progression, and epigenetic mechanisms are now thought to help inactivate several different tumor suppressor genes in most human cancers (**Figure 20–32**).

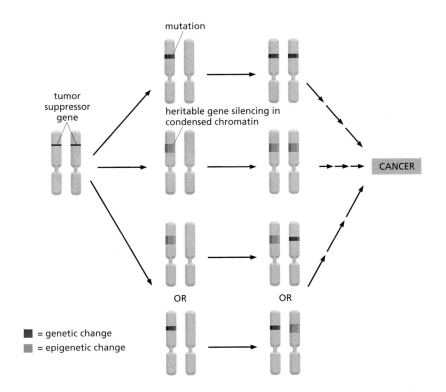

mutation

tumor
suppressor
gene

heritable gene silencing in
condensed chromatin

OR   OR

CANCER

■ = genetic change
■ = epigenetic change

Figure 20–32 The pathways leading to loss of tumor suppressor gene function in cancer involve both genetic and epigenetic changes. As discussed in Chapter 4, the packaging of a gene into condensed chromatin can prevent its expression in a way that is inherited when a cell divides (see Figure 4–52). As indicated, the changes that silence tumor suppressor genes can occur in any order.

Genes Mutated in Cancer Can Be Made Overactive in Many Ways

In the case of a proto-oncogenes it is gene activation that leads toward cancer. **Figure 20–33** summarizes the types of accidents that can convert a proto-oncogene into an oncogene. (1) A small change in DNA sequence such as a point mutation or deletion may produce a hyperactive protein when it occurs within a protein-coding sequence, or lead to protein overproduction when it occurs within a regulatory region for that gene. (2) Gene amplification events, such as those that can be caused by errors in DNA replication, may produce extra gene copies; this can lead to overproduction of the protein (**Figure 20–34**). (3) A chromosomal rearrangement, involving the breakage and rejoining of the DNA helix, may either change the protein-coding region, resulting in a hyperactive fusion protein, or alter the control regions for a gene so that a normal protein is overproduced.

Specific types of abnormality are characteristic of particular genes and of the responses to particular carcinogens. For example, 90% of the skin tumors

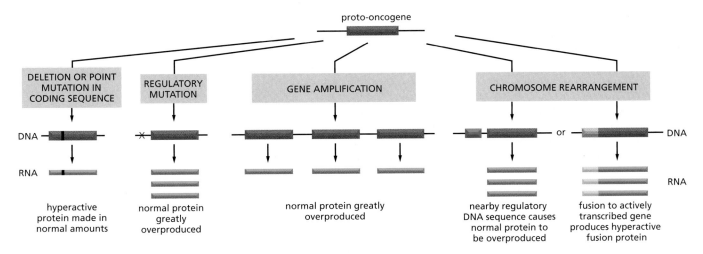

proto-oncogene

| DELETION OR POINT MUTATION IN CODING SEQUENCE | REGULATORY MUTATION | GENE AMPLIFICATION | CHROMOSOME REARRANGEMENT |

DNA

or   DNA

RNA

RNA

hyperactive
protein made in
normal amounts

normal protein
greatly
overproduced

normal protein greatly
overproduced

nearby regulatory
DNA sequence causes
normal protein to
be overproduced

fusion to actively
transcribed gene
produces hyperactive
fusion protein

Figure 20–33 The types of accidents that can make a proto-oncogene overactive and convert it into an oncogene.

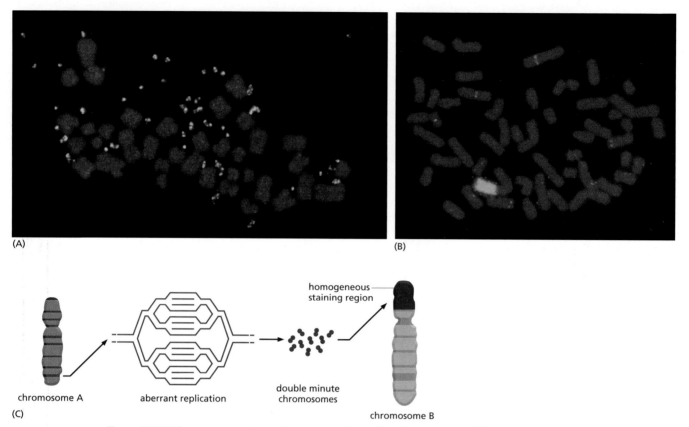

(A)

(B)

chromosome A   aberrant replication

homogeneous staining region

double minute chromosomes

chromosome B

(C)

Figure 20–34 **Chromosomal changes in cancer cells resulting from gene amplification.** In these examples, the numbers of copies of a *Myc* proto-oncogene have been amplified. Amplification of oncogenes is common in carcinomas and is often visible as a curious change in the karyotype: the cell is seen to contain additional pairs of miniature chromosomes—so-called *double minute chromosomes*—or to have a *homogeneously staining region* interpolated in the normal banding pattern of one of its regular chromosomes. Both these aberrations consist of massively amplified numbers of copies of a small segment of the genome. The chromosomes are stained with a *red* fluorescent dye, while the multiple copies of the *Myc* gene are detected by *in situ* hybridization with a *yellow* fluorescent probe. (A) Karyotype of a cell in which the *Myc* gene copies are present as double minute chromosomes (paired *yellow specks*). (B) Karyotype of a cell in which the multiple *Myc* gene copies appear as a homogeneously staining region (*yellow*) interpolated in one of the regular chromosomes. (Ordinary single-copy *Myc* genes can be detected as tiny *yellow dots* elsewhere in the genome.) (C) Schematic of how gene amplification occurs. It is thought that a rare, abnormal DNA replication event produces a chromosome with extra copies of one chromosomal region, as shown. The repair of this structure releases DNA circles that can replicate to form long tandemly repeated sequences, producing double minute chromosomes. As the result of a second rare event, the DNA from one of these chromosomes can become integrated into a new site on a normal chromosome to produce a homogeneously staining region. Other pathways can also amplify genes, such as that described later in Figure 20–41. (A and B, courtesy of Denise Sheer.)

evoked in mice by painting the skin with the tumor initiator dimethylbenz[a]anthracene (DMBA) have an A-to-T alteration at exactly the same site in a mutant *Ras* gene; presumably, of the many mutations caused by DMBA, only those at this site efficiently stimulate skin cells to form a tumor.

The receptor for the extracellular signal protein *epidermal growth factor (EGF)*, by contrast, can be activated by a deletion that removes part of its extracellular domain. These mutant receptors are able to form active dimers even in the absence of EGF, and thus they produce an inappropriate stimulatory signal, analogous to a faulty doorbell that rings even when nobody is pressing the button. Mutations of this type are found in the most common type of human brain tumor, called glioblastoma.

The Myc protein, on the other hand, generally contributes to cancer by being overproduced in its normal form. The Myc protein acts in the nucleus to stimulate cell growth and division, as discussed in Chapter 17; thus, excessive quantities of Myc cause cells to proliferate in circumstances where a normal cell would halt. Overproduction of Myc can occur in various ways. In some cases, the gene is amplified—that is, errors of DNA replication lead to the creation of large numbers of gene copies in a single cell (see Figure 20–34). More commonly, the overproduction appears to be due to a change in a regulatory element that acts on the gene. For example, a chromosomal translocation can inappropriately bring powerful gene regulatory sequences next to the Myc protein-coding sequence, so as to produce unusually large amounts of *Myc* mRNA. Thus, in Burkitt's lymphoma a translocation brings the *Myc* gene under the control of sequences that normally drive the expression of antibody genes in B lymphocytes. As a result, the mutant B cells tend to proliferate excessively and form a tumor. Similar specific chromosome translocations are common in other lymphomas and leukemias.

The Hunt for Cancer-critical Genes Continues

The sequencing of the human genome has opened up new avenues for the systematic discovery of cancer-critical genes. It is now possible, in principle, to examine every one of the approximately 25,000 human genes in a given cancer cell line, or in samples of tissue from a set of cases of cancer of a given type, looking for potentially significant abnormalities, using automated analysis of either their genomic DNA or the mRNAs the cells produce. By analyzing substantial numbers of different cancers, it should be possible eventually to identify all the genes that are commonly altered in human cancer. Although enormously costly, large-scale DNA sequencing efforts have already begun to identify new human oncogenes. One example is the finding that a hyperactive form of the Raf protein kinase (discussed in Chapter 15) is present in a high percentage of melanomas and, at lower frequency, in other cancers.

Straightforward DNA sequencing is not the only way to tackle the problem, however, and there are other powerful new methods for identifying new oncogenes and tumor suppressor genes that seem more efficient. Three different types of approaches appear particularly promising.

1. *Comparative genomic hybridization (CGH)* uses fluorescent labeling of fragments of DNA extracted from normal cells and from cancer cells to identify regions of the genome that are amplified or deleted in a given type of cancer (**Figure 20–35**). The labeled fragments are allowed to hybridize to DNA microarrays in which each spot corresponds to a known location in the normal genome. Different spots light up in different fluorescent colors,

Figure 20–35 Comparative genomic hybridization for detection of the DNA changes in tumor cells. (A) DNA fragments from tumor cells and normal cells are labeled with two different fluorescent molecules (*red* for the tumor, *green* for the normal control) and hybridized to a DNA microarray in which each spot corresponds to a defined position in the normal genome. (B) The ratios of red to green fluorescence for each spot is plotted as shown below to define those regions amplified or deleted in the tumor cells. A *red* signal indicates an amplification, while a *green* signal signifies a deletion.

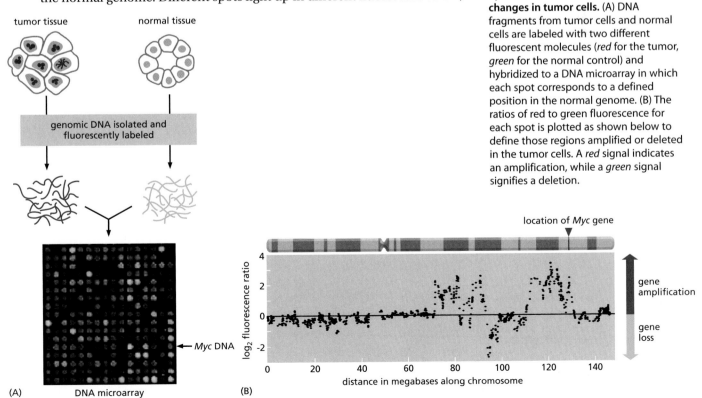

(A) DNA microarray

(B)

according to the ratio of normal to cancerous labeled fragments that bind to them. In this way, one can locate regions of the cancer cell genome that have been amplified or deleted. One then searches more narrowly for candidate genes within these regions that may contribute to cancer development.

2. DNA microarrays can also be used to reveal specific changes in gene expression associated with cancer. In this case, a cell's population of mRNAs is used to prepare the probe for hybridization, rather than its chromosomal DNA (see Figure 8–76).

3. Finally, large genetic screens based on RNA interference (RNAi) technology (discussed in Chapter 8) provide a powerful new functional approach for identifying tumor suppressor genes. The expression of small interfering RNAs (siRNA) in a cell can inactivate both copies of a gene by either destroying the corresponding mRNA or inhibiting its translation (see p. 571). This provides an efficient way, in principle, to identify any tumor suppressor gene whose loss promotes cancerous transformation in the particular animal or cell line being tested.

Whatever the approach used to identify a new candidate cancer-critical gene, it remains a challenge to determine whether the gene really contributes to cancer causation. Testing for the effects of an overexpression of candidate oncogenes or an inhibition of candidate tumor suppressor genes in cell culture assays can help. But more convincing tests use transgenic mice that overexpress candidate oncogenes and knockout mice that lack candidate tumor suppressor genes. In both types of mice, cancer development should be accelerated if the genes are real culprits.

To search for novel cancer-critical genes in the ways just described, there is in principle no need to know at the outset what the normal functions of the gene are or how altering them promotes cancer development. As our understanding of cell biology improves, however, it becomes easier to guess which genes are likely suspects and to test their culpability by directly determining if they are frequently altered in specific cancers. The task of finding cancer-critical genes is therefore closely entangled with the problem of discovering what they normally do and how, when altered, they contribute to the development of cancer. This is the topic of the next section.

Summary

Cancer-critical genes can be classified into two groups, according to whether their gain of function or their loss of function contributes to cancer development. Gain-of-function mutations that convert proto-oncogenes to oncogenes stimulate cells to increase their numbers when they should not; loss-of-function mutations of tumor suppressor genes abolish the inhibitory controls that normally help to hold cell numbers in check. Oncogenes have a dominant genetic effect, and many of them were first discovered because they cause cancer in animals when introduced by a retrovirus that originally picked up a normal form of the gene (a proto-oncogene) from a previous host cell. Oncogenes can also be identified by characteristic chromosomal aberrations that can activate a proto-oncogene.

Mutations in tumor suppressor genes are generally recessive, in that cells tend to behave normally until both gene copies are deleted, inactivated, or silenced epigenetically. Several of these genes were initially identified in rare, hereditary cancer syndromes, but their loss or inactivation is a common feature of many sporadic cancers. Individuals who inherit one defective and one functional copy of a tumor suppressor gene have an increased predisposition toward developing cancer because a single change, anywhere in the body, that eliminates or inactivates the remaining good gene copy is enough to produce a cell that totally lacks the tumor-suppressor function.

The quest for cancer-critical genes continues, with increasingly powerful tools available for systematically searching the DNA or mRNAs of cancer cells for either significant mutations or altered expression. Once a candidate gene has been identified, its importance for cancer development can be assessed in mice—by overexpressing it in the case of a candidate oncogene, or by inactivating it in the case of a candidate tumor suppressor gene.

THE MOLECULAR BASIS OF CANCER-CELL BEHAVIOR

The cancer-gene hunter views oncogenes and tumor suppressor genes—and the mutations and epigenetic changes that affect them—differently. But, from a cancer cell's point of view, they are two sides of the same coin. The same kinds of effects on cell behavior can result from mutations in either class of genes, because most cellular control mechanisms have both stimulatory (proto-oncogene) and inhibitory (tumor suppressor) components. If the aim is to understand how cancer cells function, and to do this in terms of molecular genetics, the important distinction is not between tumor-suppressor genes and proto-oncogenes, but between cancer-critical genes that act in different biochemical and regulatory pathways.

Some of the pathways important in cancer convey signals from a cell's environment (discussed in Chapter 15); others are responsible for the cell's internal programs, such as those that control the cell cycle (discussed in Chapter 17) or cell death (discussed in Chapter 18); still others govern the cell's movements (discussed in Chapter 16) or mechanical interactions with its neighbors (discussed in Chapter 19). The various pathways are linked and interdependent in complex ways. Much of what we know about them has been learned as a byproduct of cancer research; conversely, research on basic aspects of cell biology has transformed our understanding of cancer.

In the first section of this chapter, we summarized the general properties that make a cell cancerous and listed the kinds of misbehaviors that a cancer cell displays. In this section, we consider how these characteristic misbehaviors arise from mutations in cancer-critical genes and how the functions of these genes in the context of cancer can be determined. We end the section by discussing colon cancer as an extended example, showing how a succession of changes in cancer-critical genes enables a tumor to evolve from one pattern of bad behavior to another that is worse.

Studies of Both Developing Embryos and Genetically Engineered Mice Have Helped to Uncover the Function of Cancer-critical Genes

Given a gene that is mutated in a cancer, we need to understand both how the gene functions in normal cells and how mutations in the gene contribute to the aberrant behaviors characteristic of cancer cells. When the *Rb* gene was originally cloned, for example, all that was known was that it was mutated in retinoblastomas. In the case of *Ras*, the mutant gene was known to direct cells to proliferate excessively and inappropriately in culture, but how the Ras protein functioned in either normal or cancer cells was a mystery. For both Rb and Ras, cancer research was the starting point for studies that revealed the key role of these proteins in normal cells—Rb as a cell-cycle inhibitor and Ras as a central component of intracellular signaling pathways.

Today, we know much more about cells, so that when a new gene is identified as critical for cancer, it often turns out to be familiar from studies in another context. Many oncogenes and tumor suppressor genes, for example, are homologs of genes already known for their role in embryonic development. Examples include components of practically all the major signaling pathways through which cells communicate during development (discussed in Chapters 15 and 22): the Wnt, Hedgehog, TGFβ, Notch, and receptor tyrosine kinase signaling pathways all include components encoded by cancer-critical genes—with Ras being part of the last of these pathways.

With hindsight, this is no surprise. As we discuss in Chapter 23, the same signaling mechanisms that control embryonic development operate in the normal adult body to control cell turnover and maintain homeostasis. Both the development of a multicellular animal and its maintenance as an adult depend on cell–cell communication and on regulated cell growth, cell division, cell differentiation, cell death, cell movement, and cell adhesion—in other words, on all the aspects of cell behavior that are deranged in cancer. Developmental biology, often using model animals such as *Drosophila* and *C. elegans,* thus provides a key to the normal functions of many cancer-critical genes.

Ultimately, however, we want to know what alterations in these genes do to cells in the tissues that give rise to the cancer. Some information can be obtained by studying cells in culture or by examining samples from human cancer patients. But to investigate how mutations in cancer-critical genes affect tissues in a whole organism, transgenic and knockout mice have proven particularly useful.

Transgenic mice that carry an oncogene in all their cells can be generated by methods described in Chapter 8. Oncogenes introduced in this way may be expressed in many tissues or in only a select few, depending on the regulatory DNA associated with the transgene. Studies of such mice confirm that a single oncogene is generally not sufficient to turn a normal cell into a cancer cell. Typically, in mice engineered to express a *Myc* or *Ras* oncogene, some of the tissues that express the oncogene may show enhanced cell proliferation, and, over time, occasional cells will undergo further changes to give rise to cancers. Most cells expressing the transgene, however, do not give rise to cancers. Nevertheless, from the point of view of the whole animal, the inherited oncogene is a serious menace because it creates a high risk that a cancer will arise somewhere in the body. Mice that express both *Myc* and *Ras* oncogenes (bred by mating a transgenic mouse carrying a *Myc* oncogene with one carrying a *Ras* oncogene) develop cancers earlier and at a much higher rate than either parental strain (**Figure 20–36**); but, again, the cancers originate as scattered isolated tumors among noncancerous cells. Thus, even cells expressing these two oncogenes must undergo further, randomly generated changes to become cancerous, strongly suggesting that multiple mutations are required for tumorigenesis, as supported by other evidence and discussed earlier.

Numerous tumor suppressor genes have been knocked out in mice, including *Rb.* As anticipated, many of the mutant strains that are missing one copy of a tumor suppressor gene are cancer prone, and the tumors that develop typically show a loss of heterozygosity—that is, they have lost or inactivated the second copy of the gene. Deletion of both copies of a tumor suppressor gene often leads to death at an embryonic stage, reflecting the essential roles of these genes in normal development. To overcome this problem and test the effect of homozygous mutations in an adult tissue, one can engineer mice to carry conditional mutations (see Figure 5–79), such that only one organ or tissue—say, the liver—displays the genetic defect. Alternatively, mice can be engineered so that the defect is created in response to an experimental signal (a drug) delivered at a chosen time. Through the use of these techniques, mice not only provide insights into the mechanisms of tumor formation, but also models in which to test and develop new cancer therapies.

Many Cancer-Critical Genes Regulate Cell Proliferation

Many cancer-critical genes encode components of the pathways that regulate the social behaviors of cells in an animal—in particular, components of the signaling pathways through which influences from neighboring cells control whether a cell grows, divides, differentiates, or dies (**Figure 20–37**). In fact, many of the components of these signaling pathways were first identified through searches for cancer-causing genes, and a full list of the proteins encoded by proto-oncogenes and tumor suppressor genes includes examples of practically every type of protein involved in cell signaling: secreted signal proteins, transmembrane receptors, intracellular GTP-binding proteins, protein kinases, gene regulatory proteins, and many other classes of proteins besides. Cancer mutations alter these components in a way that creates proliferative signals even when more cells are not needed, switching on cell growth, DNA replication, and cell division inappropriately. Examples include the gain-of-function mutations that inappropriately activate the EGF receptor and the Ras protein, discussed earlier.

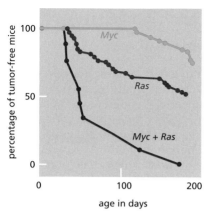

Figure 20–36 Oncogene collaboration in transgenic mice. The graphs show the incidence of tumors in three types of transgenic mouse strains, one carrying a *Myc* oncogene, one carrying a *Ras* oncogene, and one carrying both oncogenes. For these experiments two lines of transgenic mice were first generated. One carries an inserted copy of an oncogene created by fusing the proto-oncogene *Myc* with the mouse mammary tumor virus regulatory DNA (which then drives *Myc* overexpression in the mammary gland). The other line carries an inserted copy of the *Ras* oncogene under control of the same regulatory element. Both strains of mice develop tumors much more frequently than normal, most often in the mammary or salivary glands. Mice that carry both oncogenes together are obtained by crossing the two strains. These hybrids develop tumors at a far higher rate still, much greater than the sum of the rates for the two oncogenes separately. Nevertheless, the tumors arise only after a delay and only from a small proportion of the cells in the tissues where the two genes are expressed. Further accidental changes, in addition to the two oncogenes, are apparently required for the development of cancer. (After E. Sinn et al., *Cell* 49:465–475, 1987. With permission from Elsevier.)

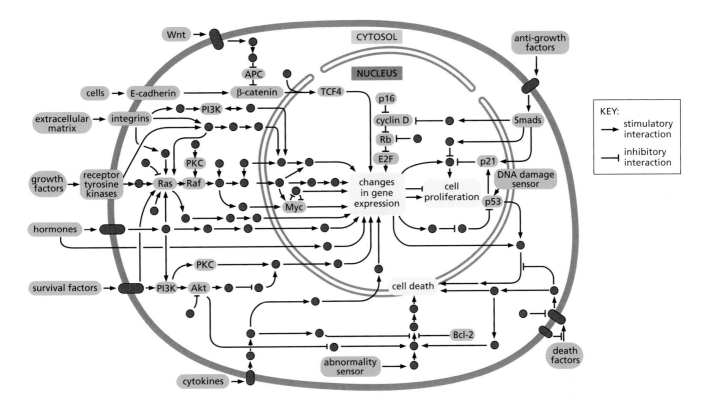

Figure 20–37 Chart of the major signaling pathways relevant to cancer in human cells, indicating the cellular locations of some of the proteins modified by mutation in cancers. Products of both oncogenes and tumor suppressor genes often act within the same pathways. Individual signaling proteins are indicated by *red circles*, with the cancer-critical components and control mechanisms discussed in this chapter in *green*. Stimulatory and inhibitory interactions between proteins are designated as shown in the key. (Adapted from D. Hanahan and R.A. Weinberg, *Cell* 100:57–70, 2000. With permission from Elsevier.)

The components in signaling pathways that normally function to inhibit cell proliferation often appear as tumor suppressors. A well-studied example is the TGFβ signaling pathway (discussed in Chapter 15). Loss-of-function mutations in this pathway contribute to the development of several types of human cancers. The receptor TGFβ-RII, for example, is mutated in some cancers of the colon, as is Smad4—a key intracellular signaling protein in the pathway—which is also often inactivated in cancers of the pancreas and some other organs. These findings reflect the normal function of the TGFβ pathway in restricting cell proliferation (see Figure 23–26).

Not surprisingly, mutations that directly affect the central cell-cycle control system feature prominently in many cancers. The tumor suppressor protein Rb, discussed earlier, controls a key point at which cells decide to enter the cell cycle and replicate their DNA. Rb serves as a brake that restricts entry into S phase by binding to and inhibiting gene regulatory proteins of the E2F family, which are needed to transcribe genes that encode proteins required for entrance into S phase. Normally, this inhibition by Rb is relieved at the appropriate time by the phosphorylation of Rb by various *cyclin-dependent kinases (Cdks)*, which cause Rb to release its inhibitory grip on the E2F proteins (discussed in Chapter 17).

Many cancer cells proliferate inappropriately by eliminating Rb entirely, as we have already seen. Others achieve the same effect by acquiring mutations that alter other components of the Rb regulatory pathway (**Figure 20–38**). In normal cells, a complex of cyclin D and the cyclin-dependent kinase Cdk4 (G₁-Cdk) is responsible for phosphorylating Rb so as to allow progress through the cycle (see Figure 17–62 and Table 17–1, p. 1063). The *p16 (INK4)* protein—which is produced when cells are stressed—inhibits cell-cycle progression by preventing the formation of an active cyclin D–Cdk4 complex; it is an important component of the normal cell-cycle-arrest response to stress. Some glioblastomas and breast cancers have amplified the genes encoding Cdk4 or cyclin D, thus favoring cell proliferation. And deletion or inactivation of the *p16* gene is common in many forms of human cancer. In cancers in which mutations do not inactivate the *p16* gene, DNA methylation of its regulatory region often silences the gene; this is an example of an epigenetic change contributing to the development of cancer. Mutation in any one component of a given pathway is sufficient to inactivate the pathway and promote cancer. As expected, therefore, a

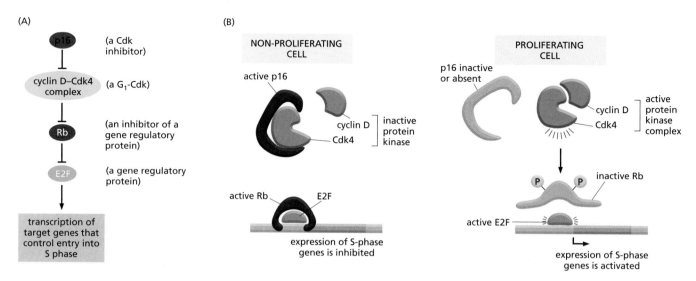

Figure 20–38 The pathway by which Rb controls cell cycle entry contains both proto-oncogenes and tumor suppressor genes. All the components of this pathway have been found to be altered by mutation in human cancers (products of proto-oncogenes, *green;* products of tumor suppressor genes, *red;* E2F is shown in *blue* because it has both inhibitory and stimulatory actions, depending on the other proteins that are bound to it). In most cases, only one of the components is altered in any individual tumor. (A) A simplified view of the dependency relationships in this pathway; see Figure 17–62 for further details. (B) The Rb protein inhibits entry into the cell-division cycle when it is unphosphorylated. The complex of Cdk4 and cyclin D phosphorylates Rb, thereby encouraging cell proliferation. When a cell is stressed, p16 inhibits the formation of an active Cdk4–cyclin D complex, preventing proliferation. Inactivation of Rb or p16 by mutation encourages cell division (thus each can be regarded as a tumor suppressor), while overactivity of Cdk4 or cyclin D encourages cell division (thus each can be regarded as a proto-oncogene).

cancer rarely inactivates more than one component in a pathway: this would bring no additional benefit for the cancer's evolution.

Distinct Pathways May Mediate the Disregulation of Cell-Cycle Progression and the Disregulation of Cell Growth in Cancer Cells

As described in Chapter 17, a special cell-cycle control system ensures that eucaryotic cells make a copy of each of their chromosomes and segregate exactly one copy into each of the two daughter cells created by cell division. The initiation of this process requires cell-cycle progression signals that are carefully regulated, as the default state for each cell in a multicellular organism is to remain in a quiescent G_0 state (see p. 1103). But cell proliferation requires more than progression through the cell cycle; it also requires cell growth, which involves complex anabolic processes through which the cell synthesizes all the necessary macromolecules from small-molecule precursors. If a cell divides inappropriately without growing first, it will get smaller at each division and will ultimately halt or die. The continued growth of a cancer therefore requires heritable changes that not only disregulate cell-cycle progression but also provoke cell growth (**Figure 20–39**).

The phosphoinositide 3-kinase (PI 3-kinase)/Akt intracellular signaling pathway is critical for cell growth control. As described in Chapter 15, various extracellular signal proteins, including insulin and insulin-like growth factors, normally stimulate this pathway. In cancer cells, however, the pathway is activated by mutation so that the cell can grow in the absence of such signals. The resulting abnormal activation of the protein kinase Akt is central to this disregulated growth process, since this activation not only stimulates protein synthesis (see Figure 17–65), but also greatly increases both glucose uptake (via mTOR activation) and the production of the acetyl CoA in the cytosol required for cell lipid synthesis (via ACL activation), as outlined in Figure 20–39. Thus, a common tumor-suppressor gene mutation, in many different cancers, is loss of the PTEN phosphatase, whose normal function is to limit Akt activation by dephosphorylating the molecules that PI 3-kinase phosphorylates.

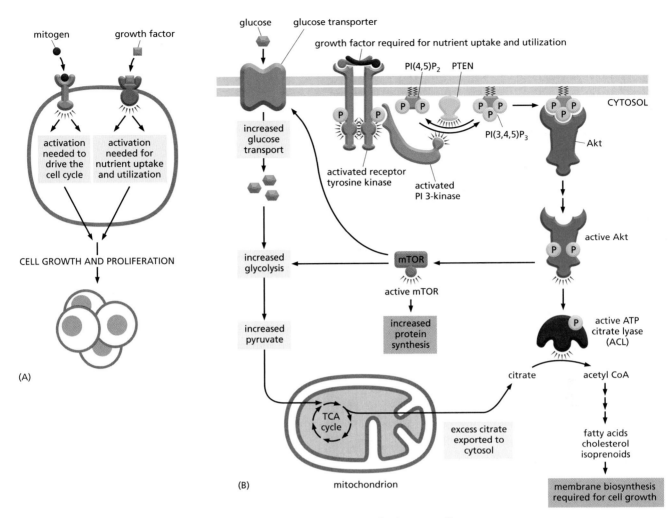

Figure 20–39 Cells may require two types of signals to proliferate. (A) In order to multiply successfully, most normal cells are suspected to require both extracellular signals that drive cell cycle progression (here *red* mitogen) and extracellular signals that drive cell growth (here *blue* growth factor). (B) Diagram of the signaling system containing Akt that drives cell growth through greatly stimulating glucose uptake and utilization, including a conversion of the excess citric acid produced from sugar intermediates in mitochondria into the acetyl CoA that is needed in the cytosol for lipid synthesis and new membrane production. As indicated, protein synthesis is also increased. This system becomes abnormally activated early in tumor progression.

The abnormal activation of the PI 3-kinase/Akt pathway, which normally occurs early in the process of tumor progression, explains the excessive rate of glycolysis that is observed in tumor cells, known as the Warburg effect. Accompanied by excretion of excess pyruvate as lactate, the excessive glucose uptake by cancer cells is used to locate tumors by modern whole-body imaging techniques (see Figure 20–1).

Mutations in Genes That Regulate Apoptosis Allow Cancer Cells to Survive When They Should Not

Control of cell numbers depends on maintaining a balance between cell proliferation and cell death. In the germinal centers of lymph nodes, for example, B cells proliferate rapidly, but most of their progeny are eliminated by apoptosis. Correct regulation of apoptosis is thus essential in maintaining the normal balance of cell birth and death in tissues that undergo cell turnover. It also has a vital role in eliminating damaged or stressed cells. As described in Chapter 18, animal cells commit suicide by undergoing apoptosis when they sense that something has gone drastically wrong—when their DNA is severely damaged, for example, or when they are deprived of extracellular survival signals that tell them they are in their proper place. As previously discussed (see Figure 20–14), cancer cells are relatively resistant to apoptosis, which allows them to increase in number and survive where they should not.

Mutations in apoptosis-control genes usually account for this resistance. One protein that normally inhibits apoptosis, called Bcl2, was discovered and named because its expression is activated by a chromosomal translocation in a B-cell lymphoma. The translocation places the *Bcl2* gene under the control of a regulatory DNA sequence that drives Bcl2 overexpression. This allows the survival of B lymphocytes that would normally have died, greatly increasing the number of B cells and contributing to the development of the B-cell cancer.

One of the genes involved in the control of apoptosis is mutated in an exceptionally wide range of cancers. This tumor suppressor gene encodes a protein that stands at a crucial intersection in the network of intracellular control pathways governing a cell's responses to DNA damage and various other cell stresses, including low oxygen (hypoxia) and growth factor deprivation. The protein is tumor protein p53, and it is produced by the *Tp53* gene, more commonly known as *p53*. As we now explain, when *p53* is defective, genetically damaged dividing cells do not merely fail to die; they persist in proliferating, accumulating yet more genetic damage that can increase malignancy.

Mutations in the *p53* Gene Allow Many Cancer Cells to Survive and Proliferate Despite DNA Damage

The *p53* gene—named for the molecular mass of its protein product—may be the most important gene preventing human cancer. Either this tumor suppressor gene or other components in the p53 pathway are mutated in nearly all human cancers. Why is *p53* so critical? The answer lies in its multifaceted function in cell-cycle control, in apoptosis, and in maintenance of genetic stability—all aspects of the fundamental role of the p53 protein in protecting the organism against the consequences of cell damage and the risk of cancer.

In contrast to Rb, most body cells have very little p53 protein under normal conditions: although the protein is synthesized, it is rapidly degraded. Moreover, p53 is not essential for normal development. Mice in which both copies of the gene have been deleted or inactivated typically appear normal in all respects except one—they universally develop cancer before 10 months of age. These observations suggest that p53 may have a function that is required only in special circumstances. Indeed, when normal cells are deprived of oxygen or exposed to treatments that damage DNA, such as ultraviolet light or gamma rays, they raise their concentration of p53 protein by blocking the degradation of the molecule. The accumulation of p53 protein is seen also in cells where oncogenes such as *Ras* and *Myc* are unusually active, generating an abnormal stimulus for cell division.

In all these cases, the high level of p53 protein acts to limit the harm done. Depending on circumstances and the severity of the damage, p53 may either drive the damaged or abnormally proliferating cell to commit suicide by apoptosis—a relatively harmless event for the multicellular organism—or may trigger a mechanism that bars the cell from dividing so long as the damage remains unrepaired. Similarly, when telomeres get too short, p53 is activated and inhibits further cell division, producing the phenomenon of replicative cell senescence (**Figure 20–40**). The protection that p53 provides is part of the reason why mutations that activate oncogenes such as *Ras* and *Myc* are not sufficient to create a tumor.

The p53 protein performs its job mainly by acting as a gene regulatory protein. <TGAA> Indeed, the most common mutations observed in p53 in human tumors are in its DNA-binding domain, where they cripple the ability of p53 to bind to its DNA target sequences. As discussed in Chapter 17, the p53 protein

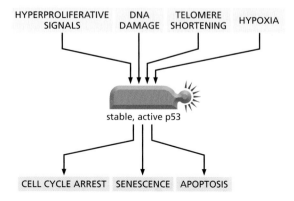

Figure 20–40 **Modes of action of the p53 tumor suppressor.** The p53 protein is a cellular stress sensor. In response to hyperproliferative signals, DNA damage, hypoxia, and/or telomere shortening, the p53 levels in the cell rise and cause cells to undergo cell cycle arrest, apoptosis, or replicative cell senescence. (As discussed in Chapter 17, a senescent cell progressively loses the ability to divide.) All of these outcomes keep tumor growth in check.

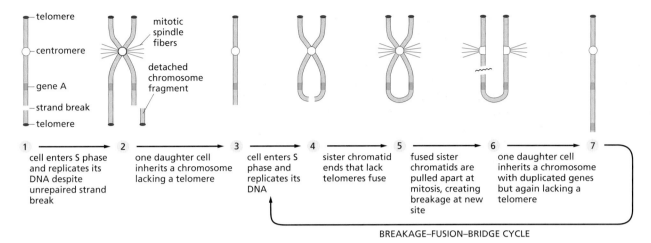

1 — cell enters S phase and replicates its DNA despite unrepaired strand break

2 — one daughter cell inherits a chromosome lacking a telomere

3 — cell enters S phase and replicates its DNA

4 — sister chromatid ends that lack telomeres fuse

5 — fused sister chromatids are pulled apart at mitosis, creating breakage at new site

6 — one daughter cell inherits a chromosome with duplicated genes but again lacking a telomere

7

BREAKAGE–FUSION–BRIDGE CYCLE

exerts its inhibitory effects on the cell cycle, in part at least, by binding to DNA to induce the transcription of *p21*, which encodes a protein that binds to and inhibits the Cdk complexes required for progression through the cell cycle. By blocking the kinase activity of these Cdk complexes, the p21 protein prevents the cell from entering S phase and replicating its DNA.

The mechanisms by which p53 induces apoptosis are more complex; it can stimulate the expression of many pro-apoptotic genes, but it may also bind to and inactivate anti-apoptotic Bcl2 proteins on the surface of mitochondria, thereby promoting apoptosis (discussed in Chapter 18).

Cells defective in *p53* fail to show these p53-dependent stress responses. They tend to escape apoptosis, and, if their DNA is damaged, they may carry on dividing, plunging into DNA replication without pausing to repair the damage. As a result, they may either die or, far worse, survive and proliferate with a corrupted genome (**Figure 20–41**). Such genetic mayhem can lead both to loss of tumor suppressor genes and to activation of oncogenes. The loss of p53 also makes some cancer cells much less sensitive to irradiation and to many anticancer drugs, which would otherwise cause them to die or to halt their proliferation.

In summary, p53 helps a multicellular organism cope safely with DNA damage and other cell stresses, acting as a check on cell proliferation in circumstances where it would be dangerous. The loss of *p53* activity is unusually dangerous because it can promote cancer for four different reasons. First, it allows cells with DNA damage to continue through the cell cycle. Second, it allows them to escape apoptosis. Third, by permitting division of cells with damaged chromosomes, it leads to the genetic instability characteristic of cancer cells, allowing further cancer-promoting mutations to accumulate as they divide. Fourth, it makes cells relatively resistant to anticancer drugs and irradiation in some tumor types. Many other mutations can contribute to one or another of these types of misbehavior, but p53 mutations contribute to them all.

DNA Tumor Viruses Block the Action of Key Tumor Suppressor Proteins

DNA tumor viruses cause cancer mainly by interfering with the cell's controls on the cell cycle and apoptosis, including those that depend on p53. To understand this type of viral carcinogenesis, it is important to understand the life history of viruses. Many DNA viruses use the host cell's DNA replication machinery to replicate their own genomes. To produce numerous infectious virus particles within a single host cell, the DNA virus has to commandeer this machinery and drive it hard, breaking through the normal constraints on DNA replication and usually killing the host cell in the process. Many DNA viruses reproduce only in this way. But some of them have a second option: they can propagate their genome as a quiet, well-behaved passenger in the host cell, replicating in parallel with the host cell's DNA in the course of ordinary cell division cycles. These viruses can switch between two modes of existence according to circumstances,

Figure 20–41 How the replication of damaged DNA can lead to chromosome abnormalities, such as gene amplification and gene loss. The diagram shows one of several possible mechanisms. The process begins with accidental DNA damage in a cell that lacks functional p53 protein. Instead of halting at the p53-dependent checkpoint in the division cycle, where a normal cell with damaged DNA would pause until it repaired the damage, the p53-defective cell enters S phase, with the consequences shown. Once a chromosome carrying a duplication and lacking a telomere has been generated, repeated rounds of replication, chromatid fusion, and unequal breakage (the so-called breakage–fusion–bridge cycle) can increase the number of copies of the duplicated region still further. Selection in favor of cells with increased numbers of copies of a gene in the affected chromosomal region will thus lead to mutants in which the gene is amplified to a high copy number. The multiple copies may—either through a recombination event or through unrepaired DNA strand breakage—become excised from their original locus and so appear as independent double minute chromosomes (see Figure 20–34). The chromosomal disorder can also lead to a loss of genes, with selection in favor of cells that have lost tumor suppressors.

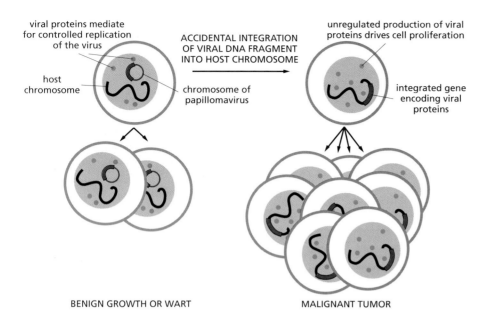

viral proteins mediate for controlled replication of the virus

host chromosome

chromosome of papillomavirus

ACCIDENTAL INTEGRATION OF VIRAL DNA FRAGMENT INTO HOST CHROMOSOME

unregulated production of viral proteins drives cell proliferation

integrated gene encoding viral proteins

BENIGN GROWTH OR WART

MALIGNANT TUMOR

Figure 20–42 How certain papillomaviruses are thought to give rise to cancer of the uterine cervix. Papillomaviruses have double-stranded circular DNA chromosomes of about 8000 nucleotide pairs. These chromosomes are normally stably maintained in the basal cells of the epithelium as plasmids whose replication is regulated so as to keep step with the chromosomes of the host *(left)*. Rare accidents can cause the integration of a fragment of such a plasmid into a chromosome of the host, altering the environment of the viral genes in the basal cells. This (or possibly some other cause) disrupts the normal control of viral gene expression. The unregulated production of viral proteins interferes with the control of cell division in the basal cells, thereby helping to generate a cancer *(right)*.

remaining latent and harmless for a long time, but then proliferating in occasional cells in a process that kills the host cell and generates large numbers of infectious particles.

Those DNA viruses that have the ability to replicate their DNA as quiescent passengers in a latent phase of viral replication either integrate their viral genome into one or more host chromosomes, or they have the ability to form an extrachromosomal plasmid that replicates in step with the host chromosomes during the latent phase.

No matter which way of life a DNA virus is following, it is not in its interest to kill the host organism. But for viruses with a latent phase, accidents can occur that prematurely activate some of the viral proteins that the virus would normally use in its replicative phase to allow the viral DNA to replicate independently of the cell cycle. This type of accident can switch on the persistent proliferation of the host cell, leading to cancer, and the DNA viruses susceptible to such accidents have been designated as **DNA tumor viruses**.

The **papillomaviruses**, for example, are the cause of human warts and are especially important as a key causative factor in carcinomas of the uterine cervix (about 6% of all human cancers). Papillomaviruses infect the cervical epithelium and maintain themselves in a latent phase in the basal layer of cells as extrachromosomal plasmids, which replicate in step with the chromosomes. Infectious virus particles are generated through a switch to a replicative phase in the outer epithelial layers, as progeny of these cells begin to differentiate before being sloughed from the surface. Here, cell division should normally stop, but the virus interferes with this cell-cycle arrest so as to allow replication of its own genome. Usually, the effect is restricted to the outer layers of cells and is relatively harmless, as in a wart. Occasionally, however, a genetic accident causes the viral genes that encode the proteins that prevent cell-cycle arrest to integrate into the host chromosome and become active in the basal layer, where the stem cells of the epithelium reside. This can lead to cancer, with the viral genes acting as oncogenes (**Figure 20–42**).

In papillomaviruses, the viral genes that are mainly to blame are called *E6* and *E7*. The protein products of these viral oncogenes interact with many host cell proteins, but, in particular, they bind to two key tumor suppressor proteins of the host cell, putting them out of action and so permitting the cell to replicate its DNA and divide in an uncontrolled way. One of these host proteins is Rb: by binding to Rb, the viral E7 protein prevents Rb from binding to and inhibiting the E2F proteins mentioned earlier, which allows uncontrolled entry into S phase. The other is p53; by binding to p53, the viral E6 protein triggers p53 destruction (**Figure 20–43**), allowing the abnormal cell to survive, divide, and accumulate yet more abnormalities. Other DNA tumor viruses use similar mechanisms to inhibit Rb and p53, underlining the importance of inactivating

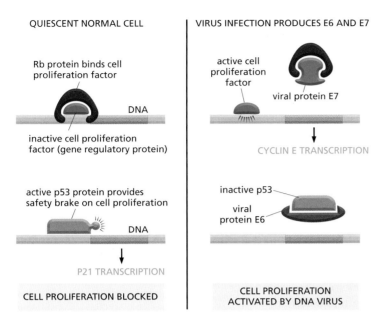

QUIESCENT NORMAL CELL

Rb protein binds cell proliferation factor

DNA

inactive cell proliferation factor (gene regulatory protein)

active p53 protein provides safety brake on cell proliferation

DNA

P21 TRANSCRIPTION

CELL PROLIFERATION BLOCKED

VIRUS INFECTION PRODUCES E6 AND E7

active cell proliferation factor

viral protein E7

CYCLIN E TRANSCRIPTION

inactive p53

viral protein E6

CELL PROLIFERATION ACTIVATED BY DNA VIRUS

Figure 20–43 Activation of cell proliferation by a DNA tumor virus. Papillomavirus produces two viral proteins, E6 and E7, which sequester the host cell's p53 and Rb proteins, respectively. E6 protein binding leads to ubiquitylation of its p53 partner, inducing p53 proteolysis (not shown). The SV40 virus (a related virus that infects monkeys) uses a single dual-purpose protein, called large T antigen, to sequester these same two proteins.

both of these tumor suppressor pathways for releasing the normal constraints on cell proliferation. It is striking that virtually all human cancers have also inactivated both pathways, usually affecting only one component in each.

The Changes in Tumor Cells That Lead to Metastasis Are Still Largely a Mystery

Perhaps the most significant gap in our understanding of cancer concerns invasiveness and metastasis. We have yet to clearly identify mutations that specifically permit cells to invade surrounding tissues, spread through the body, and form metastases. Indeed, it is not even clear exactly what properties a cancer cell must acquire to become metastatic. One extreme view is that the ability of cancer cells to metastasize requires no further genetic changes beyond those needed to weaken the normal controls on cell growth, cell division, and cell death. Some experiments using DNA microarrays to compare the mRNAs expressed in a metastasis and in the primary tumor support this view. The alternative view, more widely held, is that metastasis is a difficult and complex task that requires additional mutations, but that these may be so varied according to circumstance that they are hard to discover individually.

One molecular change in metastatic cancer cells from a variety of tumors—including melanomas and carcinomas of breast, stomach, and liver—is an overexpression of the Rho-family GTPase *RhoC*, which helps to mediate actin-based cell motility. In some cases, inhibition of RhoC activity in these cells is sufficient to block their metastatic ability. It is not yet clear, however, exactly how RhoC contributes to the ability to metastasize.

Metastasis presents different problems for different types of cells. For a cancerous white blood cell already roaming the body via the circulation, metastasis should be easier than for a carcinoma cell that has to escape from an epithelium. As we discussed earlier, it is helpful to distinguish two phases of tumor progression required for a carcinoma to metastasize. First, the cells must escape the normal confines of their parent epithelium and begin to invade the tissue immediately beneath. Second, they must travel to distant sites via the blood or lymph and establish *metastases*.

The first step, local invasiveness, requires a loss of the mechanisms that normally hold epithelial cells together. As mentioned earlier, this step resembles the normal developmental process known as the *epithelial-to-mesenchymal transition (EMT)*, in which epithelial cells undergo a shift in character, becoming less adhesive and more migratory (discussed in Chapter 19). At the heart of this transition are changes in expression of the *E-cadherin* gene. The primary function of the transmembrane E-cadherin protein is in cell–cell adhesion, and it is especially

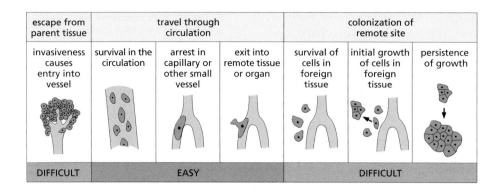

escape from parent tissue	travel through circulation			colonization of remote site		
invasiveness causes entry into vessel	survival in the circulation	arrest in capillary or other small vessel	exit into remote tissue or organ	survival of cells in foreign tissue	initial growth of cells in foreign tissue	persistence of growth
DIFFICULT	EASY			DIFFICULT		

Figure 20–44 The barriers to metastasis. Studies of labeled tumor cells leaving a tumor site, entering the circulation, and establishing metastases show which steps in the metastatic process, outlined in Figure 20–17, are difficult or 'inefficient,' in the sense that they are steps in which large numbers of cells fail and are lost. It is in these difficult steps that highly metastatic cells are observed to have much greater success than nonmetastatic cells. It seems that the ability to escape from the parent tissue, and an ability to survive and grow in the foreign tissue, are key properties that cells must acquire to become metastatic. (Adapted from A.F. Chambers et al., *Breast Cancer Res.* 2:400–407, 2000. With permission from BioMed Central Ltd.)

important in binding epithelial cells together through adherens junctions (see Figure 19–15). In some carcinomas of the stomach and of the breast, the *E-cadherin* gene has been identified as a tumor suppressor gene. The abnormal behavior of tumor cells lacking the protein can be partially rescued, at least in culture: if a functional *E-cadherin* gene is put back into these cancer cells, they lose some of their invasive characteristics and begin to cohere more like normal epithelial cells. Loss of *E-cadherin*, therefore, may promote cancer development by facilitating local invasiveness.

The initial entry of tumor cells into the circulation is helped by the presence of a dense supply of blood vessels and sometimes lymphatic vessels, which tumors attract to themselves by secreting angiogenic factors such as VEGF, as discussed earlier; the abnormal fragility and leakiness of these new vessels may help the cells that have become invasive to enter and then move through the circulation with relative ease. The remaining steps in metastasis, involving exit from a blood or lymphatic vessel and the effective colonization of remote sites, are much harder to study. To discover which of the later steps in metastasis present cancer cells with the greatest difficulties, one can label the cells with a fluorescent dye or green fluorescent protein (GFP), inject them into the bloodstream of a mouse, and then monitor their fate. In such experiments one observes that many cells survive in the circulation, lodge in small vessels, and exit into the surrounding tissue, regardless of whether they come from a tumor that metastasizes or one that does not. Some cells die immediately after they enter foreign tissue; others survive entry into the foreign tissue but fail to proliferate; still others divide a few times and then stop, forming micrometastases containing ten to several thousand cells. At this point, the metastasis-competent cells outperform their nonmetastatic relatives, as only they form growing tumors, raising the possibility that the ability to continue to proliferate in a foreign tissue (the process called *colonization*) is a key property that cancer cells must acquire to become metastatic (**Figure 20–44**).

The challenge is to use assays such as this to discover at a molecular level what, if anything, distinguishes the cells that colonize from those that do not. One possibility, as discussed earlier, is that the cells capable of colonization are just those rare members of the parent tumor population that have the special character of cancer stem cells. It seems that this must be at least part of the story, as we have already explained. But although we know of markers that help identify cancer stem cells, at least for some cancers, we do not yet understand what it is that fundamentally gives them their special stem-cell properties. This is part of the wider problem of understanding stem cells in general, discussed at length in Chapter 23. In any case, it remains very hard to pinpoint what additional genetic or epigenetic changes, if any, are needed to enable cancer stem cells to colonize foreign tissue.

Colorectal Cancers Evolve Slowly Via a Succession of Visible Changes

At the beginning of this chapter, we saw that most cancers develop gradually from a single aberrant cell, progressing from benign to malignant tumors by the accumulation of a number of independent genetic accidents and epigenetic changes. We have discussed what some of these accidents and changes are in molecular terms and seen how they contribute to cancerous behavior. We now examine one of the common human cancers more closely, using it to illustrate and enlarge upon some of the general principles and molecular mechanisms we have introduced, attempting to make sense of the natural history of the disease in these terms. We take **colorectal cancer** as our example, in which the steps of tumor progression have been followed *in vivo* with a colonoscope (a fiber-optic device for viewing the interior of the colon and rectum) and carefully studied at the molecular level. This model, based on studies of both sporadic and inherited forms of the human disease, serves as a paradigm for illustrating the stepwise nature of tumor development, and it allows us to associate specific morphological stages with inherited changes in the function of particular genes.

Colorectal cancers arise from the epithelium lining the colon (the large intestine) and rectum (the terminal segment of the gut). The organization of this tissue is broadly similar to that of the small intestine, discussed in detail in Chapter 23 (pp. 1436–1442). Both regions of gut epithelium are renewed at an extraordinarily rapid rate, taking about a week to completely replace most of the epithelial sheet. In both regions the renewal depends on stem cells that lie in deep pockets of the epithelium, called intestinal crypts. The signals that maintain the stem cells and control the normal organization and renewal of the epithelium are beginning to be quite well understood, as explained in Chapter 23. This is in large part thanks to insights that have come from cancer research. Mutations that disrupt the normal organizing signals begin the process of tumor progression for most colorectal cancers.

Colorectal cancers are common, currently causing nearly 60,000 deaths a year in the United States, or about 10% of total deaths from cancer. Like most cancers, they are not usually diagnosed until late in life (90% occur after the age of 55). However, routine examination of normal adults with a colonoscope often reveals a small benign tumor, or adenoma, of the gut epithelium in the form of a protruding mass of tissue called a *polyp* (**Figure 20–45**A; see also Figure 23–23). These adenomatous polyps are believed to be the precursors of a large proportion of colorectal cancers. Because the progression of the disease is usually very slow, there is typically a period of about 10 years in which the slowly growing tumor is detectable but has not yet turned malignant. Thus, when people are screened by colonoscopy in their fifties and the polyps are removed through the colonoscope—a quick and easy surgical procedure—the subsequent incidence of colorectal cancer is very low, according to some studies, less than a quarter of what it would be otherwise.

Colon cancer provides a clear example of the phenomenon of tumor progression discussed previously. In microscopic sections of polyps smaller than 1 cm in diameter, the cells and their arrangement in the epithelium usually appear almost normal. The larger the polyp, the more likely it is to contain cells that look abnormally undifferentiated and form abnormally organized structures. Sometimes, two or more distinct areas can be distinguished within a single polyp, with the cells in one area appearing relatively normal and those in the other appearing clearly cancerous, as though they have arisen as a mutant subclone within the original clone of adenomatous cells. At later stages in the disease, the tumor cells become invasive, first breaking through the epithelial basal lamina, then spreading through the layer of muscle that surrounds the gut (Figure 20–45B), and finally metastasizing to lymph nodes via lymphatic vessels and to liver, lung, and other organs via blood vessels.

A Few Key Genetic Lesions Are Common to a Large Fraction of Colorectal Cancers

What are the mutations that accumulate with time to produce this chain of events? Of those genes so far discovered to be involved in colorectal cancer, three stand out as most frequently mutated—the proto-oncogene *K-Ras* (a member of the *Ras* gene family) and the tumor suppressor genes *p53* and *Apc*

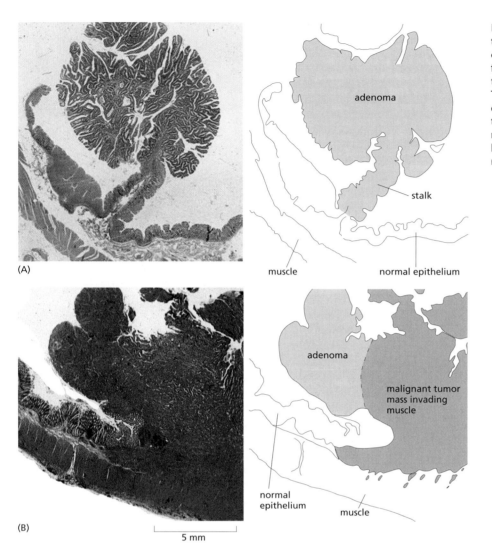

Figure 20–45 Cross sections showing the stages in development of a typical colon cancer. (A) An adenomatous polyp from the colon. The polyp protrudes into the lumen—the space inside the colon. The rest of the wall of the colon is covered with normal colonic epithelium; the epithelium on the polyp appears mildly abnormal. (B) A carcinoma that is beginning to invade the underlying muscle layer. (Courtesy of Paul Edwards.)

(discussed below). Others are involved in smaller numbers of colon cancers (**Table 20–2**). Still others remain to be identified.

One approach to discovery of the mutations responsible for colorectal cancer is to screen the cells for abnormalities in genes already known or suspected to be involved in cancers elsewhere. This type of genetic screening has revealed that about 40% of colorectal cancers have a specific point mutation in *K-Ras*, activating it to become an oncogene, and about 60% have inactivating mutations or deletions of *p53*.

As discussed earlier, another approach is to track down the genetic defects in those rare families that show a hereditary predisposition to colorectal cancer.

Table 20–2 **Some Genetic Abnormalities Detected in Colorectal Cancer Cells**

GENE	CLASS	PATHWAY AFFECTED	HUMAN COLON CANCERS (%)
K-Ras	oncogene	receptor tyrosine-kinase signaling	40
β-catenin [1]	oncogene	Wnt signaling	5–10
Apc [1]	tumor suppressor	Wnt signaling	> 80
p53	tumor suppressor	response to stress and DNA damage	60
TGFβ receptor II [2]	tumor suppressor	TGFβ signaling	10
Smad4 [2]	tumor suppressor	TGFβ signaling	30
MLH1 and other DNA mismatch repair genes	tumor suppressor (genetic stability)	DNA mismatch repair	15 (often silenced by methylation)

The genes with the same superscript act in the same pathway, and therefore only one of the components is mutated in an individual cancer.

Figure 20–46 Colon of familial adenomatous polyposis coli patient compared with normal colon. (A) The normal colon wall is a gently undulating but smooth surface. (B) The polyposis colon is completely covered by hundreds of projecting polyps (shown in section in Figure 20–45), each resembling a tiny cauliflower when viewed with the naked eye. (Courtesy of Andrew Wyllie and Mark Arends.)

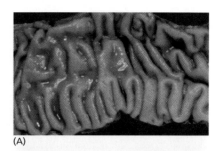
(A)

(B)

The first hereditary colorectal cancer syndrome to be analyzed was *familial adenomatous polyposis coli (FAP)*, in which hundreds or thousands of polyps develop along the length of the colon (**Figure 20–46**). These polyps start to appear in early adult life, and if they are not removed, one or more will almost always progress to become malignant; the average time from the first detection of polyps to the diagnosis of cancer is 12 years. The disease can be traced to a deletion or inactivation of the tumor suppressor gene *Apc*, named after the syndrome. Individuals with FAP have inactivating mutations or deletions of one copy of the *Apc* gene in all their cells and show loss of heterozygosity in tumors, even in the benign polyps. Most patients with colorectal cancer do not have the hereditary condition. But more than 80% of their cancers (but not their normal cells) have inactivated both copies of the *Apc* gene through mutations acquired during their lifetime. Thus, by a route similar to that we discussed for retinoblastoma, mutation of the *Apc* gene has been identified as one of the central ingredients of colorectal cancer.

As explained earlier, even where there is no known hereditary syndrome, tumor suppressor genes can be tracked down by searching for genetic changes acquired during the evolution of the tumor cells. A systematic scan of a large number of colorectal cancers reveals that selected regions of certain chromosomes frequently have precisely the same DNA sequences on both the maternally and paternally inherited chromosomes. This loss of heterozygosity suggests that those regions may harbor tumor suppressor genes. One of them is the region that includes *Apc*. Another includes the *Smad4* gene, which is mutated in about 30% of colon cancers; its removal blocks the growth-inhibitory effects of the TGFβ pathway. Specific parts of other chromosomes also show frequent losses or gains in colorectal cancers and are likely to be fruitful targets for finding additional cancer-critical genes.

As our knowledge of genes and their functions has expanded, another useful approach has been to look at genes that encode proteins that interact with the protein encoded by a known cancer-critical gene, in the expectation that these genes, too, may be targets for mutation. The Apc protein is now known to be an inhibitory component of the *Wnt signaling pathway* (discussed in Chapter 15). It binds to the *β-catenin* protein, another component of the Wnt pathway, and helps to induce its degradation. By inhibiting β-catenin in this way, APC prevents the β-catenin from migrating to the nucleus, where it acts as a transcriptional regulator to maintain the stem-cell state, as discussed in Chapter 23. Loss of Apc results in an excess of free β-catenin and thus to an uncontrolled expansion of the stem-cell population and a massive increase in the number and size of the intestinal crypts (see Figure 23–23).

When the *β-catenin* gene was sequenced in a collection of colorectal tumors, among the few tumors that did not have *Apc* mutations, a high proportion had activating mutations in *β-catenin* instead. Thus, it is excessive activity in the Wnt signaling pathway that is critical for the initiation of this cancer, rather than any single oncogene or tumor suppressor gene that the pathway contains. This is in accord with the general principle that an individual cancer rarely acquires mutations in more than one component in a pathway that is critical to tumorigenesis, as there is no competitive advantage for it to do so.

The being so, why is the *Apc* gene in particular so much the most common culprit in colorectal cancer? The Apc protein is large and it interacts not only with the β-catenin but also with various other cell components, including microtubules. Loss of Apc appears to increase the frequency of mitotic spindle defects, leading to chromosome abnormalities when cells divide. This additional, independent cancer-promoting effect could explain why *Apc* mutations feature so prominently in the causation of colorectal cancer.

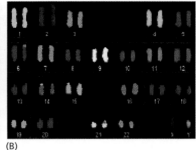

(A)   (B)

Figure 20–47 **Chromosome complements (karyotypes) of colon cancers showing different kinds of genetic instability.** (A) The karyotype of a typical cancer shows many gross abnormalities in chromosome number and structure. Considerable variation can also exist from cell to cell (not shown). (B) The karyotype of a tumor that has a stable chromosome complement with few chromosomal anomalies. Its defects are mostly invisible, having been created by defects in DNA mismatch repair. All of the chromosomes in this figure were stained as in Figure 20–13, the DNA of each human chromosome being stained with a different combination of fluorescent dyes. (Courtesy Wael Abdel-Rahman and Paul Edwards.)

Some Colorectal Cancers Have Defects in DNA Mismatch Repair

In addition to the hereditary disease (FAP) associated with *Apc* mutations, there is a second, more common, kind of hereditary predisposition to colon carcinoma in which the course of events differs from the one we have described for FAP. In this condition, called *hereditary nonpolyposis colorectal cancer (HNPCC),* the probability of colon cancer is increased without any increase in the number of colorectal polyps (adenomas). Moreover, the cancer cells are unusual, in that they have a normal (or almost normal) karyotype; the majority of colorectal tumors in non-HNPCC patients, in contrast, have gross chromosomal abnormalities, with multiple translocations, deletions, and other aberrations, as well as having many more chromosomes than normal (**Figure 20–47**).

The mutations that predispose HNPCC individuals to colorectal cancer occur in one of several genes that code for central components of the DNA *mismatch repair system,* which are homologous in structure and function to the *MutL* and *MutS* genes in bacteria and yeast (see Figure 5–20). Only one of the two copies of the involved gene is defective, so the repair system is still able to remove the inevitable DNA replication errors that occur in the patient's cells. However, as discussed previously, these individuals are at risk, because the accidental loss or inactivation of the remaining good gene copy will immediately elevate the spontaneous mutation rate by a hundredfold or more (discussed in Chapter 5). These genetically unstable cells now can presumably speed through the standard processes of mutation and natural selection that allow clones of cells to progress to malignancy.

This particular type of genetic instability produces invisible changes in the chromosomes—most notably changes in individual nucleotides and short expansions and contractions of mono- and dinucleotide repeats such as AAAA… or CACACA…. Once the defect in HNPCC patients was recognized, mutations in mismatch repair genes were found in about 15% of the colorectal cancers occurring in people with no inherited predisposing mutation.

Thus, the genetic instability found in many colorectal cancers can be acquired in at least two ways. The majority of the cancers display a form of chromosomal instability that leads to visibly altered chromosomes, whereas in the others the instability occurs on a much smaller scale and reflects a defect in DNA mismatch repair. Indeed, many carcinomas show either chromosomal instability or defective mismatch repair—but rarely both. These findings clearly demonstrate that genetic instability is not an accidental byproduct of malignant behavior but a contributory cause—and that cancer cells can acquire this instability in multiple ways.

The Steps of Tumor Progression Can Often Be Correlated with Specific Mutations

In what order do *K-Ras, p53, Apc,* and the other identified colorectal cancer-critical genes mutate, and what contribution does each of them make to the asocial behavior of the cancer cell? There is no single answer, because colorectal cancer can arise by more than one route: thus, we know that in some cases, the first mutation can be in a DNA mismatch repair gene; in others, it can be in a gene regulating cell proliferation. Moreover, as previously discussed, a general feature

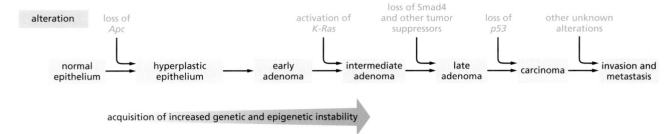

such as genetic instability or a tendency to proliferate abnormally can arise in a variety of ways, through mutations in different genes.

Nevertheless, certain sets of mutations are particularly common in colorectal cancer, and they occur in a characteristic order. Thus, in most cases, mutations inactivating the *Apc* gene appear to be the first, or at least a very early step, as they are detected at the same high frequency in small benign polyps as in large malignant tumors. Inherited changes that lead to genetic and epigenetic instability are likely also to arise early in tumor progression, since they are needed to drive the later steps.

Activating mutations in the *K-Ras* gene occur later, as they are rare in small polyps but common in larger ones that show disturbances in cell differentiation and histological pattern. Cultured colorectal carcinoma cells at this stage show the typical features of transformed cells, such as the ability to proliferate without anchorage to a substratum. Loss of the Smad4 tumor suppressor gene and inactivating mutations in *p53* come later still, as they are rare in polyps but common in carcinomas (**Figure 20–48**). As discussed earlier, loss of *p53* function allows cancer cells to accumulate additional mutations and to avoid apoptosis and cell-cycle arrest.

In summary, although the exact set of mutations will vary from one colorectal cancer to the next, colorectal carcinogenesis requires certain genetic alterations that disable particular control mechanisms, and the order in which these alterations occur is not random (**Figure 20–49**).

Figure 20–48 Suggested typical sequence of genetic changes underlying the development of a colorectal carcinoma. This over-simplified diagram provides a general idea of the way mutation and tumor development can fit together. But there are certainly other mutant genes of which we are not yet aware, and different colon cancers can progress through different sequences of mutations.

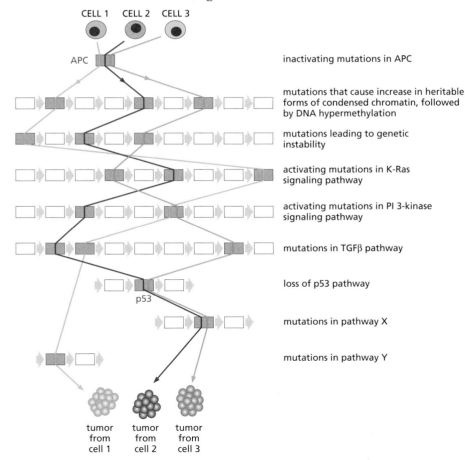

Figure 20–49 The general nature of tumor progression at the molecular level. Each row in this diagram represents a biochemical pathway, so that changing any one protein *(rectangular box)* by altering its gene is often enough to gain the full effect. The three paths *(colored vertical lines)* have been drawn to indicate the mutations that one evolving clone of tumor cells accumulates as it progresses toward cancer. This diagram is modeled after some of the known events in the development of colon cancer, but it should not be taken literally. The major points are that, while many different genes will be altered in individual colon tumors as they develop to metastasis, so that tumors are not expected to be exactly alike, there are some common patterns. In particular, both the signaling pathways that are altered and the general order in which these changes arise are often similar for different cancers.

Each Case of Cancer Is Characterized by Its Own Array of Genetic Lesions

As illustrated by colorectal cancer, the traditional classification of cancers is simplistic. Even a single type of cancer is a heterogeneous collection of disorders; despite some common features, each disorder will be characterized by its own array of genetic lesions. In the form of lung cancer known as *small-cell carcinoma,* for example, one finds mutations in *p53, Rb, Myc,* and at least five other known proto-oncogenes and tumor suppressor genes. Different patients have different combinations of mutations, and these genetic differences often correspond to different responses to treatment.

In principle, molecular biology provides the tools to find out precisely which genes are amplified, deleted, mutated, or misregulated by epigenetic mechanisms in the tumor cells of any given patient. As we discuss next, such information may soon prove to be as important for the diagnosis and treatment of cancer as was the identification of microorganisms for patients with infectious diseases.

Summary

Studies of developing embryos and genetically engineered mice have helped to reveal the functions of a large number of cancer-critical genes. Many of the oncogenes and tumor suppressor genes mutated in cancer code for components of the social control pathways that regulate when cells grow, divide, differentiate, or die. Other genes, loosely categorized as "DNA maintenance genes," help maintain the integrity of the genome. The molecular changes that allow cancers to metastasize are still largely unknown.

DNA tumor viruses such as papillomaviruses can promote the development of cancer by encoding proteins that inhibit the products of some tumor suppressor genes. The viral proteins bind and inhibit the Rb protein, which normally acts as a brake on cell division, and the p53 protein, which normally induces cell-cycle arrest or apoptosis in response to DNA damage and other cell stresses. Loss or inactivation of the p53 pathway, which occurs in nearly all spontaneous human cancers, is especially dangerous, as it allows genetically damaged cells to escape apoptosis and continue to proliferate. Inactivation of the Rb pathway also occurs in most human cancers, illustrating how fundamental each of these pathways is in protecting us against cancer.

We can often correlate the steps of tumor progression with mutations that activate specific oncogenes and inactivate specific tumor suppressor genes, with colon cancer providing the best-understood example. But different combinations of mutations and epigenetic changes are found in different types of cancer and even in different patients with the same type of cancer, reflecting the random way in which these inherited changes occur. Nevertheless, many of the same types of changes are encountered repeatedly, suggesting that there is only a limited number of ways in which our defenses against cancer can be breached.

CANCER TREATMENT: PRESENT AND FUTURE

We can apply our growing understanding of the biology of cancer to sharpen our attack on the disease at three levels: prevention, diagnosis, and treatment. Prevention is always better than cure, and, as discussed earlier, many cancers can be prevented, especially by avoiding smoking. Moreover, cancers can often be nipped in the bud by screening for small primary tumors, which can be removed before they have metastasized, as we saw for cervical cancers. Highly sensitive molecular assays promise to create new opportunities for better prevention and treatment through earlier and more precise diagnosis. Advances in screening methods along with changes in lifestyle probably offer the most immediate prospects for reducing the cancer death rate, but they can never be perfectly effective. Thus, for many years to come, full-blown malignant disease will continue to be common and cancer treatments will continue to be needed.

The Search for Cancer Cures Is Difficult but Not Hopeless

The difficulty of curing a cancer is similar to the difficulty of getting rid of weeds. Cancer cells can be removed surgically or destroyed with toxic chemicals or radiation, but it is hard to eradicate every single one of them. Surgery can rarely ferret out every metastasis, and treatments that kill cancer cells are generally toxic to normal cells as well. In addition, as described earlier, whereas the great majority of the cancer cells can often be killed by irradiation or chemotherapy, the small population of slowly dividing cancer stem cells may be harder to eliminate in these ways; if even a few cancer stem cells remain, they can regenerate the tumor. Moreover, unlike normal cells, cancer cells mutate rapidly and will often evolve resistance to the poisons and irradiation used against them.

In spite of these difficulties, effective cures using anticancer drugs (alone or in combination with other treatments) have already been found for some formerly highly lethal cancers, including Hodgkin's lymphoma, testicular cancer, choriocarcinoma, and some leukemias and other cancers of childhood. Even for types of cancer where a cure at present seems beyond our reach, there are treatments that will prolong life or at least relieve distress. But what prospect is there of doing better and finding cures for the most common forms of cancer, which still cause great suffering and so many tragic deaths?

Traditional Therapies Exploit the Genetic Instability and Loss of Cell-Cycle Checkpoint Responses in Cancer Cells

Anticancer therapies need to take advantage of some molecular abnormality of cancer cells that distinguishes them from normal cells. One such property is genetic instability, caused by an abnormality in chromosome maintenance, cell cycle checkpoints, or DNA repair. Remarkably, most current cancer therapies work by exploiting these abnormalities, although this was not known by the scientists who developed the treatments. Most anticancer drugs and ionizing radiation damage DNA. They preferentially kill certain kinds of cancer cells because these mutant cells have a diminished ability to survive the damage. Normal cells treated with radiation, for example, arrest their cell cycle until they have repaired the damage to their DNA. This cell-cycle arrest is an example of a *cell-cycle checkpoint response* discussed in Chapter 17. Cancer cells generally have defects in many of these checkpoint responses and often continue to divide after irradiation; this causes many to die after a few days because of the severe genetic damage they experience.

Unfortunately, while some of the molecular defects present in cancer cells enhance their sensitivity to cytotoxic agents, others increase their resistance. Some of the cell death that DNA damage induces, for example, occurs by apoptosis, and cancer cells often acquire defects in the control systems that activate apoptosis in response to such damage. For example, as we discussed earlier, DNA damage induced by anticancer drugs or irradiation normally activates p53, which can then trigger apoptosis. Thus the inactivation of the *p53* pathway that occurs in most cancers makes certain types of cancer cells less sensitive to these agents. Cancer cells vary widely in their response to various treatments, probably reflecting the particular kinds of defects they have in DNA repair, cell-cycle checkpoints, and the control of apoptosis.

New Drugs Can Exploit the Specific Cause of a Tumor's Genetic Instability

As we become increasingly able to pinpoint the specific alterations in cancer cells that make them different from their normal neighbors, we can use this knowledge to design therapies that kill the cells in the cancer without killing normal cells. One of the characteristics of cancer cells is their genetic instability; as we have explained, this is one of the features that helps them to evolve and proliferate dangerously. But it is at the same time a defect—a vulnerability that we can exploit to kill them.

We have seen that cancer cells are forced to walk a difficult line as they evolve toward metastasis: they need to have a defect in their DNA maintenance processes that is severe enough to allow them to accumulate new mutations at a significantly increased rate, but not so severe that they kill themselves by frequent loss or genes necessary for cell survival. Since there are hundreds of different genes required to maintain DNA sequences and chromosome structures with high fidelity (discussed in Chapters 4 and 5), we would expect there to be at least dozens of different ways for a particular tumor cell to acquire its genetic instability. Moreover, these ways should be mutually exclusive: once a cell has become genetically unstable to a modest extent, it is likely to increase its risk of death if it inactivates additional DNA maintenance genes. Those cells that do so will die and be lost from the tumor population.

Detailed studies of the mechanisms for DNA maintenance discussed in Chapter 5 reveal a surprising amount of apparent redundancy, with multiple pathways for repairing each type of DNA damage. Thus, knocking out a particular pathway for DNA repair is generally less disastrous than one might expect, because alternate repair pathways exist. We have seen, for example, how stalled DNA replication forks can arise when the fork encounters a single-strand break in a template strand, and how cells avoid the disaster that would otherwise result. First, they have machinery to escape the problem by directly repairing single-strand breaks; and then, if that fails, they can repair stalled forks by homologous recombination (see Figure 5–53). Suppose that the cells in a particular cancer have become genetically unstable by acquiring a mutation that reduces their ability to repair stalled forks by homologous recombination. Might it be possible to eradicate that cancer by treating it with a drug that inhibits the repair of single-strand breaks, thereby greatly increasing the number of forks that stall? The consequences would be expected to be harmless for normal cells, which can repair stalled forks, but lethal for the cancer cells, which cannot.

While such a possibility might seem too good to be true, precisely this strategy appears to work to kill the cells in cancers that have inactivated both of their *Brca1* or *Brca2* tumor suppressor genes. As described in Chapter 5, Brca2 is an accessory protein that interacts with the Rad51 protein (the RecA analog in humans) in the initiation of general recombination events. Brca2 is another protein that is also required for this repair process. Like *Rb*, *Brca1* and *Brca2* were discovered as mutations that predispose humans to cancer—in this case, cancers of the breast and ovaries (though unlike *Rb*, they seem to be involved in only a small proportion of such cancers). Individuals who inherit one mutant copy of *Brca1* or *Brca2* develop tumors that have inactivated the second copy of the same gene, presumably because this change makes the cells genetically unstable and speeds tumor progression.

Drugs that inhibit an enzyme called PARP (poly ADP-ribose polymerase) have a dramatic effect on cells of such tumors, killing them with enormous selectivity. This is attributed to the fact that PARP is required for the repair of single-strand breaks in DNA. Perhaps surprisingly, PARP inhibition has very little effect on normal cells; in fact, mice that have been engineered to lack PARP1—the major PARP family member involved in DNA repair—remain healthy under laboratory conditions. This result suggests that, while the repair pathway requiring PARP provides a first line of defense against breaks in a DNA strand, these breaks can easily be repaired by a genetic recombination pathway in normal cells (**Figure 20–50**). In contrast, tumor cells that have acquired their genetic instability by the loss of Brca1 or Brca2 have lost this second line of defense, and they are therefore uniquely sensitive to PARP inhibitors (that is, they are missing repair pathway 2 in Figure 20–50).

PARP inhibition is still only under trial as a cancer treatment in humans, and is likely to be applicable to only a small proportion of cancer cases. But it exemplifies the type of rational, highly selective approach to cancer therapy that is beginning to be possible, and thus it offers hope for many other cancers. To extend this approach widely, we will need new tools for determining the precise

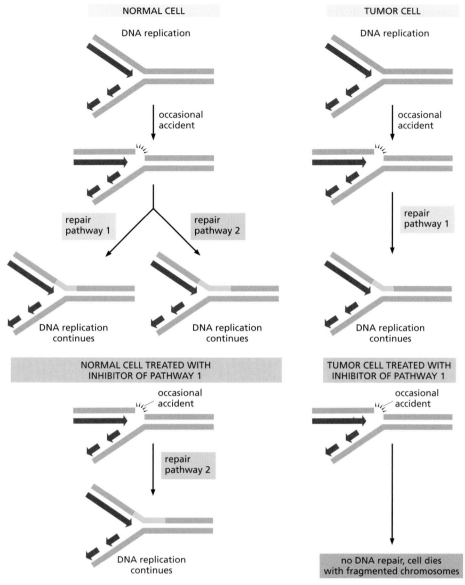

NORMAL CELL

DNA replication

occasional accident

repair pathway 1   repair pathway 2

DNA replication continues   DNA replication continues

NORMAL CELL TREATED WITH INHIBITOR OF PATHWAY 1

occasional accident

repair pathway 2

DNA replication continues

TUMOR CELL

DNA replication

occasional accident

repair pathway 1

DNA replication continues

TUMOR CELL TREATED WITH INHIBITOR OF PATHWAY 1

occasional accident

no DNA repair, cell dies with fragmented chromosomes

Figure 20–50 How a tumor's genetic instability can be exploited for cancer therapy. As explained in Chapter 5, the maintenance of DNA sequences is so critical for life that cells have evolved multiple pathways for repairing DNA damage and avoiding DNA replication errors. As illustrated, a replication fork will stall whenever it encounters a break in a DNA template strand. In this example, normal cells have two different repair pathways that can fix the problem and thereby prevent a mutation from arising in the newly synthesized DNA sequences. They are therefore not harmed by treatment with a drug that blocks repair pathway 1. In contrast, the inactivation of repair pathway 2 was selected for in the evolution of the tumor cells (because it made them genetically unstable). Consequently, only the tumor cells are killed by a drug treatment that blocks repair pathway 1. If a treated cell does not die by apoptosis, its daughters will die because they inherit fragmented, incomplete chromosome sets.

In the actual case that underlies this more general schematic example, the function of repair pathway 1 (requiring the PARP protein discussed in the text) is to remove accidental breaks that occur in a DNA single strand *before* they are encountered by a moving replication fork, and pathway 2 is the recombination-dependent process for repairing stalled replication forks illustrated in Figure 5–53 (requiring the Brca2 and Brca1 proteins; for details, see H.E. Bryant et al., *Nature* 434:913–916, 2005 and H. Farmer et al., *Nature* 434:917–921, 2005.)

cause of the genetic instability in individual tumors, as well as the development of many more drugs that target alternate DNA repair pathways.

Genetic Instability Helps Cancers Become Progressively More Resistant to Therapies

Genetic instability itself can be both good and bad for anticancer therapy. Although it seems to provide an Achilles' heel that therapies can exploit, it can also make eradicating cancer more difficult. An abnormally high mutation rate tends to make cancer cell populations heterogeneous, which may make it difficult to kill off the entire population with a single type of treatment. Moreover, it allows many cancers to evolve resistance to therapeutic drugs at an alarming rate.

To make matters worse, cells that are exposed to one anticancer drug often develop a resistance not only to that drug but also to other drugs to which they have never been exposed. This phenomenon of **multidrug resistance** frequently correlates with amplification of a part of the genome that contains a gene called *Mdr1*. This gene encodes a plasma-membrane-bound transport ATPase of the ABC transporter superfamily (discussed in Chapter 11), which pumps lipophilic drugs out of the cell. The overproduction of this protein (or some of its other family members) by a cancer cell can prevent the intracellular accumulation of many cytotoxic drugs, making the cell insensitive to them. The amplification of other types of genes can likewise give the cancer cell a selective advantage. The gene that encodes the enzyme dihydrofolate reductase (DHFR), for example, can become amplified in cancer cells treated with the anticancer drug *methotrexate*. Methotrexate binds to and inhibits the ability of DHFR to bind folic acid, and the amplification greatly increases the amount of enzyme, reducing the cells' sensitivity to the drug.

New Therapies Are Emerging from Our Knowledge of Cancer Biology

Our growing understanding of cancer cell biology and tumor progression is gradually leading to better methods for treating the disease, and not only by targeting defects in cell-cycle checkpoints and DNA repair processes. As an example, estrogen receptor antagonists (such as *tamoxifen*) and drugs that block estrogen synthesis are widely used to prevent or delay recurrence of breast cancers that have been screened and found to express estrogen receptors. These antiestrogen treatments are also being tested for their ability to prevent the development of new breast cancers. These drugs do not directly kill the tumor cells, but instead prevent estrogen from promoting their proliferation.

The greatest hopes lie, however, in finding more powerful and selective ways to exterminate cancer cells directly. A variety of adventurous new ways to attack tumor cells have been shown to work in model systems—typically reducing or preventing the growth of human tumors transplanted into immunodeficient mice. Many of these strategies will turn out to be of no medical use, because they do not work in humans, have bad side effects, or are simply too difficult to implement. But some have turned out to be highly successful in the clinic. One strategy depends on the reliance of some cancer cells on a particular hyperactive protein they produce, a phenomenon known as *oncogene addiction*. Blocking the activity of the protein may be an effective means of treating the cancer if it does not unduly damage normal tissues. About 25% of breast cancers, for example, express unusually high levels of the Her2 protein, a receptor tyrosine kinase related to the EGF receptor that plays a part in the normal development of mammary epithelium. Monoclonal antibodies that inhibit Her2 function slow the growth of breast tumors in humans that overexpress Her2, and they are now an approved therapy for these cancers.

A similar approach uses antibodies to deliver toxic molecules directly to the cancer cells. Antibodies against proteins like Her2 that are abundant on the cancer cell surface can be armed with a toxin or made to carry an enzyme that cleaves a harmless 'prodrug' and converts this into a toxic molecule. In the latter case, one molecule of enzyme can then generate a large number of toxic molecules on the surface of a tumor cell; these molecules can also diffuse to neighboring tumor cells, increasing the odds that they too will be killed, even if the antibody did not bind to them directly.

Antibodies are hard to make in large amounts, are very expensive to produce and buy, and must be given by injection. The ultimate goal in cancer therapy is to develop small molecules that kill cancer cells specifically. The PARP inhibitors discussed above are one example; but to tackle the majority of cancers with simple drug treatments, we shall need a large collection of different small molecules, tailored to the many different types of cancer.

Small Molecules Can Be Designed to Inhibit Specific Oncogenic Proteins

As our knowledge of specific molecules involved in the genesis of particular cancers builds, there is an increasing effort to devise targeted therapies directed against the oncogenic proteins that are essential for a cancer cell to survive and proliferate—thereby exploiting the phenomenon of oncogene addiction mentioned previously. A particular dramatic success of this type has raised high hopes for the general utility of such targeted therapies in the future.

As we saw earlier, chronic myelogenous leukemia (CML) is usually associated with a particular chromosomal translocation, visible as the Philadelphia chromosome (see Figure 20–5). This is the consequence of chromosome breakage and rejoining at the sites of two specific genes, called *Abl* and *Bcr.* The fusion of these genes creates a hybrid gene that codes for a chimeric protein called *Bcr-Abl*, consisting of the N-terminal fragment of Bcr fused to the C-terminal portion of Abl (**Figure 20–51**). Abl is a tyrosine kinase involved in cell signaling. The substitution of the Bcr fragment for the normal N-terminus of Abl makes it hyperactive, so that it stimulates inappropriate proliferation of the hemopoietic precursor cells that contain it and prevents these cells from dying by apoptosis—which many of them would normally do. As a result, excessive numbers of white blood cells accumulate in the bloodstream, producing CML.

The chimeric Bcr-Abl protein is an obvious target for therapeutic attack. Searches for synthetic drug molecules that can inhibit the activity of tyrosine kinases discovered one, called **Gleevec**, that blocks Bcr-Abl (**Figure 20–52**). When the drug was first given to patients with CML, nearly all of them showed a dramatic response, with an apparent disappearance of the cells carrying the Philadelphia chromosome in over 80% of patients. The response appears relatively durable: after years of continuous treatment, many patients have not progressed to later stages of the disease—although Gleevec-resistant cancers emerge with a probability of at least 5% per year.

Results are not so good for those patients who have already progressed to the acute phase of myeloid leukemia, known as blast crisis, where genetic instability has set in and the march of the disease is far more rapid. These patients show a response at first and then relapse because the cancer cells develop a resistance to Gleevec. This resistance is usually associated with secondary mutations in the part of the *Bcr-Abl* gene that encodes the kinase domain, disrupting the ability of Gleevec to bind to Bcr-Abl kinase. Second-generation inhibitors that function effectively against these Gleevec-resistant mutants have now been developed. Ultimately a cocktail of multiple agents that cooperatively block Bcr-Abl action may provide the key to successful treatment, by preventing the selection for drug-resistant cancer cells at all stages of the disease.

Despite the complications with resistance, the extraordinary success of Gleevec for the patients in the chronic (early) stage of the disease is enough to prove the principle: once we understand precisely what genetic lesions have occurred in a cancer, we can begin to design effective rational methods to treat it. This success story has fueled efforts to identify small-molecule inhibitors for other oncogenic protein kinases that could be effective targets for new anticancer drugs. A second example of such a therapy is provided by a small-molecule inhibitor of the EGF receptor, currently approved for the treatment of some lung cancers.

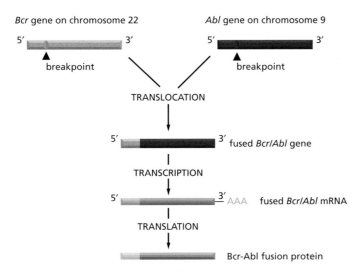

Figure 20–51 **The conversion of the *Abl* proto-oncogene into an oncogene in patients with chronic myelogenous leukemia.** The chromosome translocation responsible joins the *Bcr* gene on chromosome 22 to the *Abl* gene from chromosome 9, thereby generating a Philadelphia chromosome (see Figure 20–5). The resulting fusion protein has the N-terminus of the Bcr protein joined to the C-terminus of the Abl tyrosine protein kinase; in consequence, the Abl kinase domain becomes inappropriately active, driving excessive proliferation of a clone of hemopoietic cells in the bone marrow.

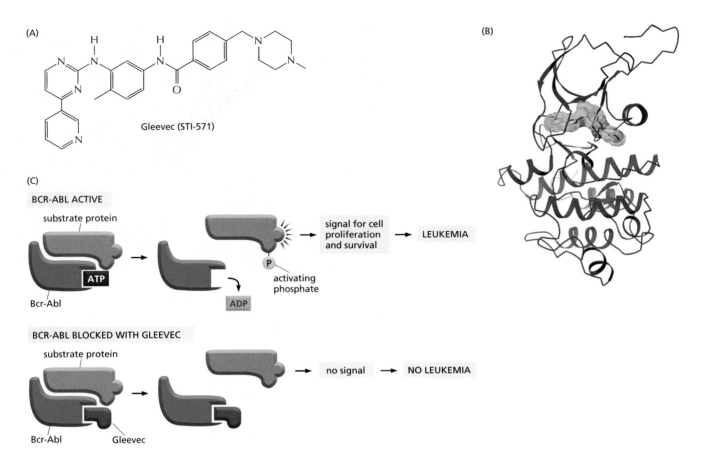

Figure 20–52 How Gleevec blocks the activity of Bcr-Abl protein and halts chronic myeloid leukemia.
(A) The chemical structure of Gleevec (imatinib). The drug can be given by mouth; it has side effects but they are usually quite tolerable. (B) The structure of the complex of Gleevec (solid *green* object) with the tyrosine kinase domain of the Abl protein (ribbon diagram), as determined by X-ray crystallography.
(C) Gleevec sits in the ATP-binding pocket of the tyrosine kinase domain of Bcr-Abl and thereby prevents Bcr-Abl from transferring a phosphate group from ATP onto a tyrosine residue in a substrate protein. This blocks onward transmission of a signal for cell proliferation and survival. (B, from T. Schindler et al., *Science* 289:1938–1942, 2000. With permission from AAAS.)

Tumor Blood Vessels Are Logical Targets for Cancer Therapy

Another approach to destroying tumors does not directly target cancer cells at all but instead targets the blood vessels on which large tumors depend. As discussed earlier, the growth of these vessels requires angiogenic signals such as VEGF, which, in animal models at least, can be blocked to prevent further tumor growth. Moreover, endothelial cells in the process of forming new vessels express distinctive cell-surface markers, which might provide an opportunity to attack developing tumor vessels specifically, without harming existing blood vessels in normal tissues. Clinical trials with various inhibitors of angiogenic signals are now taking place, and several drugs that inhibit the VEGF receptor have recently been approved for treating kidney cancer. Similarly, a monoclonal antibody against VEGF has been approved for treatment of colon cancer in combination with chemotherapy, even though the added benefit is only modest.

Many Cancers May Be Treatable by Enhancing the Immune Response Against a Specific Tumor

In recent years, we have realized that the body's normal immune responses help to protect us against cancer. Mice lacking important parts of their immune system have elevated levels of several types of cancer. Similarly, humans who are immunocompromised have two- to three-fold higher rates of certain solid tumors. The self-antigens that are recognized in these cases may attract the attention of the immune system because they are expressed in unusually large amounts in the particular tumor cells. Alternatively, it is possible that these proteins provoke an autoimmune response because they are normally expressed only in embryos or in another immunologically privileged area such as the brain, so that the immune system has never had the opportunity to develop tolerance towards them.

Variant tumor cells with a reduced exposure to immune attack can defeat tumor immunosurveillance, and the process of tumor progression is thought to include heritable changes that reduce the antigenicity of the tumor. However, through our increased understanding of the complex mechanisms of both adaptive and innate immunity (see Chapters 24 and 25), oncologists are learning how to manipulate the immune response to intensify the attack on specific tumors. The hope is that this will allow these cancer cells to be eliminated as if they were foreign tissue.

Treating Patients with Several Drugs Simultaneously Has Potential Advantages for Cancer Therapy

As already described, some cancers can be dramatically reduced in size by an initial drug treatment, with all of the detectable tumor cells seeming to disappear. But months or years later the cancer will reappear in an altered form that is resistant to the drug that was at first so successful. Given the hypermutable nature of tumor cells, this is not surprising. The initial drug treatment has presumably failed to affect a tiny fraction of mutant cells in the original tumor cell population (some of the cancer stem cells, according to that view of tumors), and these cells eventually recreate the tumor by continuing to proliferate.

In some cases, this problem can be prevented by treating patients with two or more drugs simultaneously. The logic is the same as that behind the current treatment of HIV-AIDS with a cocktail of three different protease inhibitors: whereas there may always be some cells in the initial populations resistant to any one drug treatment, there should be no cell resistant to two drugs that are different. In contrast, sequential drug treatments can allow the few cells resistant to the first drug to multiply to large numbers, thereby making it likely that cells with double resistance will arise (**Figure 20–53**).

As our armament of possible drugs against cancer cells increases, it becomes increasingly likely that effective treatments involving simultaneous treatment with multiple drugs can be designed for many cancers.

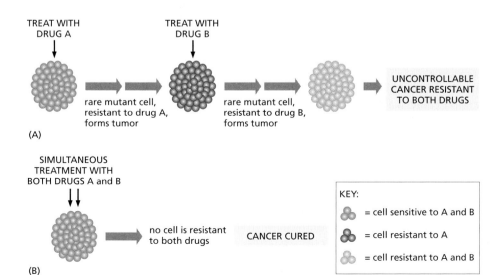

Figure 20–53 **Why multidrug treatments may be more effective than sequential treatments for cancer therapy.** Because tumor cells are hypermutable, single drug treatments that are given sequentially often allow for the selection of mutant cell clones that are resistant to both drugs.

Gene Expression Profiling Can Help Classify Cancers into Clinically Meaningful Subgroups

It is now possible to characterize each individual cancer at a molecular level in unprecedented detail. For example, DNA microarray technology (discussed in Chapter 8) can be used to determine the pattern of expression of thousands of genes simultaneously in a single sample of a cancer and compare it with the pattern of expression of the same genes in normal control tissue. Each case of a given type of cancer, such as breast cancer, will have its own **gene expression profile**. However, comparing the profiles of many patients shows that they can be grouped into a small number of distinct classes, whose members share common profile features.

Why might such molecular profiles be useful? All medical progress depends on accurate diagnosis. If we cannot identify a disease correctly, we cannot discover its causes, predict its outcome, select the appropriate treatment for a given patient, or conduct effective trials on a population of patients to judge whether a treatment is effective. Cancers, as we have seen, are an extraordinarily heterogeneous collection of diseases. Gene expression profiling represents a tool to make diagnosis more precise through better tumor classification. Standard classification schemes mainly rely on histological analysis and we now know that they can lump together cancers that behave differently. Thus, gene expression profiling can not only help to identify a tumor in uncertain cases, but also help to classify tumors more precisely, with important consequences for the patient. Analysis of a collection of diffuse large B-cell lymphomas, for example, demonstrated that there are two separate classes based on gene expression profiles, even though the classes could not be distinguished histologically. One class was associated with a poor prognosis and the other with a better prognosis, establishing the usefulness of the classification and providing an explanation for the previous observation that some patients with this disease respond effectively to therapy and survive, whereas others do not. Gene expression profiling is still being developed for most cancers, but it is likely to find widespread use in the future for guiding treatments.

There Is Still Much More to Do

The molecular analysis of cancer promises to transform cancer treatment by enabling us to tailor therapy much more accurately to the individual patient. The discovery of a large number of cancer-critical genes brought an end to an era of groping in the dark for clues to the molecular basis of cancer. It has been encouraging to find that there are, after all, some general principles and that many forms of the disease share some key genetic abnormalities. Because we know the identities of many cancer-critical genes and their normal functions, it is beginning to be possible to devise precisely targeted, rational treatments. But we are still far from fully understanding the most common human cancers, and we still need a much better understanding of nearly all of the processes described in this chapter before we can conquer this lethal cell-biological disease. Among other things, we need better ways to define cancer stem cells and isolate them, so that we can determine their special characteristics and the molecular basis of their "stemness;" most crucially, we need to find ways to kill them—otherwise, our therapies will ultimately fail.

Cancer survivors and their families have long been important advocates for public investment in basic biomedical research. These groups occasionally express understandable frustration with the slow pace at which our greatly increased knowledge of cancer has been translated into effective cancer therapies. But the accumulation of knowledge brings an accelerating improvement in understanding, and there is a real sense in the cancer research community today that major progress in the treatment of this group of diseases is imminent. To take one example, in recent years we have obtained a much deeper understanding of the detailed molecular processes that underlie the all-or-none decision that cells make to commit suicide through apoptosis. How long will it be before we can analyze the abnormal signaling network in an individual tumor well enough to allow us to select a tailor-made cocktail of drugs and growth factors that specifically causes all cells in that cancer to die?

Looking back on the history of cell biology and contemplating the speed of recent progress, we can be hopeful. The desire to understand, which drives basic research, will surely reveal new ways to use our knowledge of the cell for humanitarian goals, not only in relation to cancer, but also with regard to infectious disease, mental illness, agriculture, and other areas that we can scarcely foresee.

Summary

Our growing understanding of the cell biology of cancers has already begun to lead to better ways of preventing, diagnosing and treating these diseases. Anticancer therapies can be designed to destroy cancer cells preferentially by exploiting the properties that distinguish them from normal cells, including their dependence on oncogenic proteins and the defects they harbor in their DNA repair mechanisms, cell-cycle checkpoint mechanisms, and apoptosis control pathways. It may also be possible to control the growth of tumors by attacking their blood supply and depriving them of the help they require from stromal cells. We now have proof that, by understanding normal cell control mechanisms and exactly how they are subverted in specific cancers, we can devise drugs to kill cancers precisely by attacking specific molecules critical for the growth and survival of the cancer cells. And, as we become better able to determine which genes are altered in the cells of any given tumor, we can begin to tailor treatments more accurately to each individual patient.

PROBLEMS

Which statements are true? Explain why or why not.

20–1 Genetic instability in the form of point mutations, chromosome rearrangements, and epigenetic changes needs to be maximal to allow the development of cancer.

20–2 Cancer therapies directed solely at killing the rapidly dividing cells that make up the bulk of a tumor are unlikely to eliminate the cancer from many patients.

20–3 The main environmental causes of cancer are the products of our highly industrialized way of life such as pollution and food additives.

20–4 Dimethylbenz[a]anthracene (DMBA) must be an extraordinarily specific mutagen since 90% of the skin tumors it causes have an A-to-T alteration at exactly the same site in the mutant *Ras* gene.

20–5 In the cellular regulatory pathways that control cell growth and proliferation, the products of oncogenes are stimulatory components and the products of tumor suppressor genes are inhibitory components.

Discuss the following problems.

20–6 In contrast to colon cancer, whose incidence increases dramatically with age, osteosarcoma—a tumor that occurs most commonly in the long bones— peaks during adolescence. Osteo-

sarcomas are relatively rare in young children (up to age 9) and in adults (over 20). Why do you suppose that the incidence of osteosarcoma does not show the same sort of age-dependence as colon cancer?

20–7 As shown in **Figure Q20–1**, plots of deaths due to breast cancer and cervical cancer in women differ dramatically from the same plot for colon cancer. At around age 50 the age-dependent increase in death rates for breast and cervical cancer slows markedly, whereas death rates due to colon cancer (and most other cancers) continue to increase. Why do you suppose that the age-dependent increase in death rates for breast and cervical cancer slows after age 50?

20–8 By analogy with automobiles, defects in cancer-critical genes have been likened to broken brakes and stuck accelerators, which are caused in some cases through faulty service by bad mechanics. Using this analogy, decide how oncogenes, tumor suppressor genes, and DNA maintenance genes relate to broken brakes, stuck accelerators, and bad mechanics. Explain the basis for each of your choices.

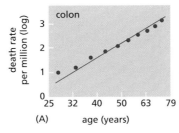
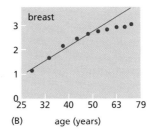
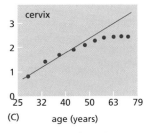

Figure Q20–1 Cancer death rates as a function of age (Problem 20–7). (A) Death rates for colon cancer in females. (B) Death rates for breast cancer in females. (C) Death rates for cervical cancer. The data in all cases are plotted as log of the death rate versus the patient age (on a log scale) at death. The straight lines in B and C are fitted to the data for the earlier age groups, whereas the line in A is fitted to all the data points. (Data from P. Armitage and R. Doll, *Br. J. Cancer* 91:1983–1989, 2004. With permission from Macmillan Publishers Ltd.)

Figure Q20–2 Cumulative risk of lung cancer mortality for nonsmokers, smokers, and former smokers (Problem 20–9). Cumulative risk is the running total of deaths, as a percentage, for each group. Thus, for continuing smokers, 1% died of lung cancer between ages 45 and 55; an additional 4% died between 55 and 65 (giving a cumulative risk of 5%); and 11% more died between 65 and 75 (for a cumulative risk of 16%).

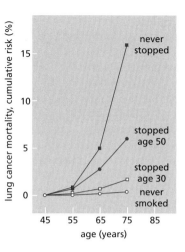

20–9 Mortality due to lung cancer was followed in groups of males in the United Kingdom for 50 years. **Figure Q20–2** shows the cumulative risk of dying from lung cancer as a function of age and smoking habits for four groups of males: those who never smoked, those who stopped at age 30, those who stopped at age 50, and those who continued to smoke. These data show clearly that an individual can substantially reduce his cumulative risk of dying from lung cancer by stopping smoking. What do you suppose is the biological basis for this observation?

20–10 The Tasmanian devil, a carnivorous Australian marsupial, is threatened with extinction by the spread of a fatal disease in which a malignant oral–facial tumor interferes with the animal's ability to feed. You have been called in to analyze the source of this unusual cancer. It seems clear to you that the cancer is somehow spread from devil to devil, very likely by their frequent fighting, which is accompanied by biting around the face and mouth. To uncover the source of the cancer, you isolate tumors from 11 devils captured in widely separated regions and examine them. As might be expected, the karyotypes of the tumor cells are highly rearranged relative to that of the wild-type devil (**Figure Q20–3**). Surprisingly, you find that the karyotypes from all 11 tumor samples are very similar. Moreover, one of the Tasmanian devils has an inversion on chromosome 5 that is not present in its facial tumor. How do you suppose this cancer is transmitted from devil to devil? Is it likely to arise as a consequence of an infection by a virus or microorganism? Explain your reasoning.

20–11 Now that DNA sequencing is so inexpensive, reliable, and fast, your mentor has set up a consortium of investigators to pursue the ambitious goal of tracking down *all* the mutations in a set of human tumors. He has decided to focus on breast cancer and colorectal cancer because they cause 14% of all cancer deaths. For each of 11 breast cancers and 11 colorectal cancers, you design primers to amplify 120,839 exons in 14,661 transcripts from 13,023 genes. As controls, you amplify the same regions from DNA samples taken from two normal individuals. You sequence the PCR products and use analytical software to compare the 456 Mb of tumor sequence with the published human genome sequence. You are astounded to find 816,986 putative mutations. This represents more than 37,000 mutations per tumor! Surely that can't be right.

Once you think about it for a while, you realize the computer sometimes makes mistakes in calling bases. To test for that source of error, you visually inspect every sequencing read and

find that you can exclude 353,738 changes, leaving you with 463,248, or about 21,000 mutations per tumor. Still a lot!

A. Can you suggest at least three other sources of apparent mutations that do not actually contribute to the tumor?

B. After applying a number of criteria to filter out irrelevant sequence changes, you find a total of 1307 mutations in the 22 breast and colorectal cancers, or about 59 mutations per tumor. How might you go about deciding which of these sequence changes are likely to be cancer-related mutations and which are probably "passenger" mutations that occurred in genes with nothing to do with cancer (but were found in the tumors because they happened to occur in the same cells with true cancer mutations)?

C. Will your comprehensive sequencing strategy detect all possible genetic changes that affect the targeted genes in the cancer cells?

20–12 Virtually all cancer treatments are designed to kill cancer cells, usually by inducing apoptosis. However, one particular cancer—acute promyelocytic leukemia (APL)—has been successfully treated with all-*trans*-retinoic acid, which causes the promyelocytes to differentiate into neutrophils. How might a change in the state of differentiation of APL cancer cells help the patient?

20–13 One major goal of modern cancer therapy is to identify small molecules—anticancer drugs—that can be used to inhibit the products of specific cancer-critical genes. If you were searching for such molecules, would you design inhibitors for the products of oncogenes, tumor suppressor genes, or DNA maintenance genes? Explain why you would (or would not) select each type of gene.

(A)   Tasmanian devil (*Sarcophilus harrisii*)

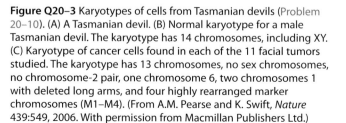

Figure Q20–3 Karyotypes of cells from Tasmanian devils (Problem 20–10). (A) A Tasmanian devil. (B) Normal karyotype for a male Tasmanian devil. The karyotype has 14 chromosomes, including XY. (C) Karyotype of cancer cells found in each of the 11 facial tumors studied. The karyotype has 13 chromosomes, no sex chromosomes, no chromosome-2 pair, one chromosome 6, two chromosomes 1 with deleted long arms, and four highly rearranged marker chromosomes (M1–M4). (From A.M. Pearse and K. Swift, *Nature* 439:549, 2006. With permission from Macmillan Publishers Ltd.)

REFERENCES

General

Bishop JM (2004) How to Win the Nobel Prize: An Unexpected Life in Science. Harvard University Press: Cambridge MA.

Stillman B & Stewart D (eds) (2005) Molecular Approaches to Controlling Cancer. Cold Spring Harbor Laboratory Press: Cold Spring Harbor.

Weinberg RA (2007) The Biology of Cancer. Garland Science: New York.

Cancer as a Microevolutionary Process

Al-Hajj M, Becker MW, Wicha M et al (2004) Therapeutic implications of cancer stem cells. Curr Opin Genet Dev 14:43–47.

Bielas JH, Loeb KR, Rubin BP et al (2006) Human cancers express a mutator phenotype. Proc Natl Acad Sci USA 103:18238–18242.

Cairns J (1975) Mutation, selection and the natural history of cancer. Nature 255:197–200.

Campesi J (2005) Senescent cells, tumor supression, and organismal aging: good citizens, bad neighbors. Cell 120:513–522.

Evan GI & Vousden KH (2001) Proliferation, cell cycle and apoptosis in cancer. Nature 411:342–348.

Fialkow PJ (1976) Clonal origin of human tumors. Biochim Biophys Acta 458:283–321.

Fidler IJ (2003) The pathogenesis of cancer metastasis: the 'seed and soil' hypothesis revisited. Nature Rev Cancer 3:453–458.

Jones PA & Baylin SB (2002) The fundamental role of epigenetic events in cancer. Nature Rev Genet 3:415–428.

Loeb LA, Loeb KR & Anderson JP (2003) Multiple mutations and cancer. Proc Natl Acad Sci USA 100:776–781.

Lowe SW, Cepero E & Evan G (2004) Intrinsic tumour suppression. Nature 432:307–315.

Nowell PC (1976) The clonal evolution of tumor cell populations. Science 194:23–28.

Pardal R, Clarke MF & Morrison SJ (2003) Applying the principles of stem-cell biology to cancer. Nature Rev Cancer 3:895–902.

Sjoblom T, Jones S, Wood LD et al (2006) The consensus coding sequences of human breast and colorectal cancers. Science 314:268–274.

Thiery JP (2002) Epithelial-mesenchymal transitions in tumour progression. Nature Rev Cancer 2:442–454.

Tlsty TD & Coussens LM (2006) Tumor Stroma and Regulation of Cancer Development. Annu Rev Pathol Mech Dis 1:119–150.

Ward RJ & Dirks PB (2007) Cancer stem cells: at the headwaters of tumor development. Annu Rev Pathol Mech Dis 2:175–189.

The Preventable Causes of Cancer

Ames B, Durston WE, Yamasaki E & Lee FD (1973) Carcinogens are mutagens: a simple test system combining liver homogenates for activation and bacteria for detection. Proc Natl Acad Sci USA 70:2281–2285.

Berenblum I (1954) A speculative review: the probable nature of promoting action and its significance in the understanding of the mechanism of carcinogenesis. Cancer Res 14:471–477.

Cairns J (1985) The treatment of diseases and the war against cancer. Sci Am 253:51–59.

Doll R (1977) Strategy for detection of cancer hazards to man. Nature 265:589–597.

Doll R & Peto R (1981) The causes of cancer: quantitative estimates of avoidable risks of cancer in the United States today. J Natl Cancer Inst 66:1191–1308.

Hsu IC, Metcalf RA, Sun T et al (1991) Mutational hotspot in the p53 gene in human hepatocellular carcinomas. Nature 350:427–428.

Newton R, Beral V & Weiss RA (eds) (1999) Infections and Human Cancer. Cold Spring Harbor Laboratory Press: Cold Spring Harbor.

Peto J (2001) Cancer epidemiology in the last century and the next decade. Nature 411:390–395.

Finding the Cancer-Critical Genes

Davies H, Bignell GR, Cox C et al (2002) Mutations of the BRAF gene in human cancer. Nature 417:949–954.

Easton DF, Pooley KA, Dunning AM et al (2007) Genome-wide association study identifies novel breast cancer susceptibility loci. Nature 447:1087–1095.

Feinberg AP (2007) Phenotypic plasticity and the epigenetics of human disease. Nature 447:433–440.

Herman JG & Baylin SB (2003) Gene silencing in cancer in association with promoter hypermethylation. N Engl J Med 349:2042–2054.

Lowe SW, Cepero E & Evan G (2004) Intrinsic tumour suppression. Nature 432:307–315.

Mitelman F, Johansson B & Mertens F (2007) The impact of translocations and gene fusions on cancer causation. Nature Rev Cancer 7:233–245.

Pinkel D & Albertson DG (2005). Comparative genomic hybridization. Annu Rev Genomics Hum Genet 6:331–354.

Rowley JD (2001) Chromosome translocations: dangerous liasisons revisited. Nature Rev Cancer 1:245–250.

The Wellcome Trust Case Control Consortium (2007) Genome-wide association study of 14,000 cases of seven common diseases and 3,000 shared controls. Nature 447:661–683.

Westbrook TF, Stegmeier F & Elledge SJ (2005) Dissecting cancer pathways and vulnerabilities with RNAi. Cold Spring Harb Symp Quant Biol 70:435–444.

The Molecular Basis of Cancer Cell Behavior

Artandi SE & DePinho RA (2000) Mice without telomerase: what can they teach us about human cancer? Nature Med 6:852–855.

Brown JM & Attardi LD (2005) The role of apoptosis in cancer development and treatment response. Nature Rev Cancer 5:231–237.

Chambers AF, Naumov GN, Vantyghem S & Tuck AB (2000) Molecular biology of breast cancer metastasis: clinical implications of experimental studies on metastatic inefficiency. Breast Cancer Res 2:400–407.

Edwards PAW (1999) The impact of developmental biology on cancer research: an overview. Cancer Metastasis Rev 18:175–180.

Futreal PA, Coin L, Marshall M et al (2004) A census of human cancer genes. Nature Rev Cancer 4:177–183.

Hanahan D & Weinberg RA (2000) The hallmarks of cancer. Cell 100:57–70.

Hoffman RM (2005) The multiple uses of fluorescent proteins to visualize cancer in vivo. Nature Rev Cancer 5:796–806.

Kinzler KW & Vogelstein B (1996) Lessons from hereditary colorectal cancer. Cell 87:159–170.

Levine AJ, Hu W & Feng Z (2006) The P53 pathway: what questions remain to be explored? Cell Death Differ 13:1027–1036.

Macleod KF & Jacks T (1999) Insights into cancer from transgenic mouse models. J Pathol 187:43–60.

Mani SA, Yang J, Brooks M et al (2007) Mesenchyme Forkhead 1 (FOXC2) plays a key role in metastasis and is associated with aggressive basal-like breast cancers. Proc Natl Acad Sci USA 104:10069–10074.

Mueller MM & Fusenig NE (2004) Friends or foes—bipolar effects of the tumour stroma in cancer. Nature Rev Cancer 4:839–849.

Plas DR & Thompson CB (2005) Akt-dependent transformation: there is more to growth than just surviving. Oncogene 24:7435–7442.

Ridley A (2000) Molecular switches in metastasis. Nature 406:466–467.

Shaw RJ & Cantley LC (2006) Ras, PI(3)K and mTOR signalling controls tumour cell growth. Nature 441:424–430.

Sherr CJ (2006) Divorcing ARF and p53: an unsettled case. Nature Rev Cancer 6:663–673.

Vogelstein B & Kinzler KW (2004) Cancer genes and the pathways they control. Nature Med 10:789–799.

Weinberg RA (1995) The retinoblastoma protein and cell-cycle control. Cell 81:323–330.

Cancer Treatment: Present and Future

Alizadeh AA, Eisen MB, Davis RE et al (2000). Distinct types of diffuse large B-cell lymphoma identified by gene expression profiling. Nature 403:503–511.

Baselga J & Albanell J (2001) Mechanism of action of anti-HER2 monoclonal antibodies. Annu Oncol 12:S35–S41.

Druker BJ & Lydon NB (2000) Lessons learned from the development of an abl tyrosine kinase inhibitor for chronic myelogenous leukemia. *J Clin Invest* 105:3–7.

Folkman J (1996) Fighting cancer by attacking its blood supply. *Sci Am* 275:150–154.

Golub TR, Slonim DK, Tamayo P et al (1999) Molecular classification of cancer: class discovery and class prediction by gene expression monitoring. *Science* 286:531–537.

Huang P & Oliff A (2001) Signaling pathways in apoptosis as potential targets for cancer therapy. *Trends Cell Biol* 11:343–348.

Garraway LA & Sellers WR (2006) Lineage dependency and lineage-survival oncogenes in human cancer. *Nature Rev Cancer* 6:593–602.

Jonkers J & Berns A (2004) Oncogene addiction: sometimes a temporary slavery. *Cancer Cell* 6:535–538.

Jain RK (2005) Normalization of tumor vasculature: an emerging concept in antiangiogenic therapy. *Science* 307:58–62.

Kim R, Emi M & Tanabe K (2007) Cancer immunoediting from immune surveillance to immune escape. *Immunology* 121:1–14.

Madhusudan S & Middleton MR (2005) The emerging role of DNA repair proteins as predictive, prognostic and therapeutic targets in cancer. *Cancer Treat Rev* 31:603–617.

Roninson IB, Abelson HT, Housman DE et al (1984) Amplification of specific DNA sequences correlates with multi-drug resistance in Chinese hamster cells. *Nature* 309:626–628.

Sarkadi B, Homolya L, Szakács G & Váradi A (2006) Human multidrug resistance ABCB and ABCG transporters: participation in a chemoimmunity defense system. *Physiol Rev* 86:1179–236.

Sawyers C (2004) Targeted cancer therapy. *Nature* 432:294–297.

van 't Veer LJ, Dai H, van de Vijver MJ et al (2002) Gene expression profiling predicts clinical outcome of breast cancer. *Nature* 415:530–536.

Varmus H, Pao W, Politi K et al (2005) Oncogenes come of age. *Cold Spring Harb Symp Quant Biol* 70:1–9.

Ventura A, Kirsch DG, McLaughlin ME et al (2007) Restoration of p53 function leads to tumour regression *in vivo*. *Nature* 445:661–665.

Glossary

ABC transporter protein
A large family of membrane transport proteins that use the energy of ATP hydrolysis to transfer peptides or small molecules across membranes. (Figure 11–7)

acetyl (–COCH₃)
Chemical group derived from acetic acid (CH_3COOH). Acetyl groups are important in metabolism and are also added covalently to some proteins as a post-translational modification.

acetyl CoA
Small water-soluble activated carrier molecule. Consists of an acetyl group linked to coenzyme A (CoA) by an easily hydrolyzable thioester bond. (Figure 2–62)

acetylcholine (ACh)
Neurotransmitter that functions at cholinergic synapses. Found in the brain and peripheral nervous system. The neurotransmitter at vertebrate neuromuscular junctions. (Figure 15–9)

acetylcholine receptor (AChR)
Membrane protein that responds to binding of acetylcholine (ACh). The nicotinic AChR is a transmitter-gated ion channel that opens in response to ACh. The muscarinic AChR is not an ion channel, but a G-protein-coupled cell-surface receptor.

acid
A proton donor. Substance that releases protons (H^+) when dissolved in water, forming hydronium ions (H_3O^+) and lowering the pH. (Panel 2–2, pp. 108–109)

acrosomal vesicle
Region at the head end of a sperm cell that contains a sac of hydrolytic enzymes used to digest the protective coating of the egg. When a sperm starts to enter an egg, the contents of the vesicle are released (the acrosome reaction), helping sperm penetrate the zona pellucida. (Figures 21–33 and 21–27)

actin
Abundant protein that forms actin filaments in all eucaryotic cells. The monomeric form is sometimes called globular or G-actin; the polymeric form is filamentous or F-actin. (Panel 16–1, p. 968, and Figure 16–12)

actin-binding protein
Protein that associates with either actin monomers or actin filaments in cells and modifies their properties. Examples include myosin, α-actinin, and profilin. (Panel 16–3, pp. 994–995)

actin filament (microfilament)
Helical protein filament formed by polymerization of globular actin molecules. A major constituent of the cytoskeleton of all eucaryotic cells and part of the contractile apparatus of skeletal muscle. (Panel 16–1, p. 968)

action potential
Rapid, transient, self-propagating electrical excitation in the plasma membrane of a cell such as a neuron or muscle cell. Action potentials, or nerve impulses, make possible long-distance signaling in the nervous system. (Figure 11–30)

activated carrier
Small diffusible molecule that stores easily exchangeable energy in the form of one or more energy-rich covalent bonds. Examples are ATP, acetyl CoA, $FADH_2$, NADH, and NADPH. (Figure 2–55)

activation energy
Extra energy that must be acquired by atoms or molecules in addition to their ground-state energy in order to reach a transition state and undergo a particular chemical reaction. (Figure 2–44)

activator (gene activator protein, transcriptional activator)
Gene regulatory protein that when bound to its regulatory sequence in DNA activates transcription.

active site
Region of an enzyme surface to which a substrate molecule binds in order to undergo a catalyzed reaction. (Figure 1–7)

active transport
Movement of a molecule across a membrane or other barrier driven by energy other than that stored in the electrochemical or concentration gradient of the transported molecule.

acyl group (–CO–R)
Functional group derived from a carboxylic acid ($R-C{\overset{O}{\underset{OH}{\lessgtr}}}$).

adaptation
(1) adaptation (desensitization): Adjustment of sensitivity following repeated stimulation. The mechanism that allows a cell to react to small changes in stimuli even against a high background level of stimulation. (2) evolutionary adaptation: an evolved trait.

adaptive immune response
Response of the vertebrate adaptive immune system to a specific antigen that typically generates immunological memory. (Figures 25–1 and 25–2)

adaptor protein
General term for a protein that functions solely to link two or more different proteins together in an intracellular signaling pathway or protein complex. (Figure 15–22)

adenomatous polyposis coli (APC) protein
Tumor suppressor protein that forms part of a protein complex in the canonical Wnt signaling pathway that recruits free cytoplasmic β-catenin and degrades it.

adenosine triphosphate—see ATP

adenylyl cyclase (adenylate cyclase)
Membrane-bound enzyme that catalyzes the formation of cyclic AMP from ATP. An important component of some intracellular signaling pathways.

adherens junction
Cell junction in which the cytoplasmic face of the plasma

membrane is attached to actin filaments. Examples include adhesion belts linking adjacent epithelial cells and focal contacts on the lower surface of cultured fibroblasts.

ADP (adenosine 5′-diphosphate)
Nucleotide produced by hydrolysis of the terminal phosphate of ATP. Regenerates ATP when phosphorylated by an energy-generating process such as oxidative phosphorylation. (Figure 2–57)

adrenaline (epinephrine)
Hormone released by adrenal gland chromaffin cells, especially in response to stress, that binds to specific GPCRs. It can initiate and coordinate a "fight or flight" response, which includes an increase in heart rate and blood sugar levels. It is also a catecholamine neurotransmitter.

aerobic
Occurring in, or requiring, the presence of molecular oxygen (O_2).

affinity
The strength of binding of a molecule to its ligand at a single binding site.

affinity chromatography
Type of chromatography in which the protein mixture to be purified is passed over a matrix to which specific ligands for the required protein are attached, so that the protein is retained on the matrix. (Figure 8-13)

affinity constant (association constant) (K_a)
Measure of the strength of binding of the components in a complex. For components A and B and a binding equilibrium A + B ⇌ AB, the association constant is given by [AB]/[A][B], and is larger the tighter the binding between A and B. (Figure 3–43)

affinity maturation
Progressive increase in the affinity of antibodies for the immunizing antigen with the passage of time after immunization.

Akt (protein kinase B, PKB)
Serine/threonine protein kinase that acts in the PI 3-kinase/Akt intracellular signaling pathway involved especially in signaling cells to grow and survive. Also called protein kinase B (PKB). (Figure 15–64)

aldehyde
Organic compound that contains the $[-C{\stackrel{O}{\scriptstyle\diagdown}}_H]$ group. Example: glyceraldehyde. Can be oxidized to an acid or reduced to an alcohol. (Panel 2–1, p. 107)

alga (plural algae)
Informal term used to describe a wide range of simple unicellular and multicellular eucaryotic photosynthetic organisms. Examples include *Nitella, Volvox,* and *Fucus.*

alkaline—*see* **basic**

alkyl group (C_nH_{2n+1})
General term for a group of covalently linked carbon and hydrogen atoms such as methyl (–CH_3) or ethyl (–CH_2CH_3) groups. Usually exist as part of larger organic molecules. On their own they form extremely reactive free radicals.

allele
One of several alternative forms of a gene. In a diploid cell each gene will typically have two alleles, occupying the corresponding position (locus) on homologous chromosomes.

allelic exclusion
The expression of a protein from only one of the two alleles of the gene encoding the protein in the cell, as occurs, for example, in the expression of an immunoglobulin or T cell receptor chain or an olfactory receptor.

allostery (adjective allosteric)
Change in a protein's conformation brought about by the binding of a regulatory ligand (at a site other than the protein's catalytic site), or by covalent modification. The change

in conformation alters the activity of the protein and can form the basis of directed movement. (Figures 3–58 and 16–61)

alpha helix (α helix)
Common folding pattern in proteins, in which a linear sequence of amino acids folds into a right-handed helix stabilized by internal hydrogen bonding between backbone atoms. (Figure 3–7)

alternative RNA splicing
Production of different RNAs from the same gene by splicing the transcript in different ways. (Figure 7–94)

alveoli (singular alveolus)
Small dilated outpocketings of an epithelium, especially the epithelium of the lung, where they form millions of air-filled sacs. Similar structures are found in the milk-secreting glandular epithelium of the breast.

amide
Molecule containing a carbonyl group linked to an amine. (Panel 2–1, p. 107)

amine
Chemical group containing nitrogen and hydrogen. Becomes positively charged in water. (Panel 2–1, p. 107)

amino acid
Organic molecule containing both an amino group and a carboxyl group. Those that serve as building blocks of proteins are alpha amino acids, having both the amino and carboxyl groups linked to the same carbon atom. ($NH_2CHRCOOH$, Panel 3–1, pp. 128–129)

aminoacyl-tRNA synthetase
Enzyme that attaches the correct amino acid to a tRNA molecule to form an aminoacyl-tRNA. (Figure 6–57)

amino group (–NH_2)
Weakly basic functional group derived from ammonia (NH_3) in which one or more hydrogen atoms are replaced by another atom. In aqueous solution it can accept a proton and carry a positive charge (–NH_3^+).

amino terminus—*see* **N terminus**

amoeba (plural amoebae)
Carnivorous unicellular protozoan that crawls using pseudopodia.

AMP (adenosine 5′-monophosphate)
One of the four nucleotides in an RNA molecule. Two phosphates are added to AMP to form ATP. (Panel 2–6, pp. 116–117)

amphipathic
Having both hydrophobic and hydrophilic regions, as in a phospholipid or a detergent molecule.

anabolism (biosynthesis)
Formation of complex molecules from simple substances by living cells. (Figure 2–36)

anaerobic
Requiring, or occurring in, the *absence* of molecular oxygen (O_2).

anaphase
(1) Stage of mitosis during which sister chromatids separate and move away from each other. Composed of anaphase A (chromosomes move toward the two spindle poles) and anaphase B (spindle poles move apart). (2) Anaphase I and II: stages of meiosis during which chromosome homolog pairs separate (I), and then sister chromatids separate (II). (Panel 17–1, pp. 1072–1073)

anaphase-promoting complex (APC/C) (cyclosome)
Ubiquitin ligase that catalyzes the ubiquitylation and destruction of securin and M- and S-cyclins, initiating the separation of sister chromatids in the metaphase-to-anaphase transition during mitosis.

anchorage dependence
Dependence of cell growth, proliferation, and survival on attachment to a substratum.

anchoring junction
Cell junction that attaches cells to neighboring cells or to the extracellular matrix. (Figure 19–2, and Table 19–1, p. 1133)

angiogenesis
Growth of new blood vessels by sprouting from existing ones.

Ångstrom (Å)
Unit of length used to measure atoms and molecules. Equal to 10^{-10} meter or 0.1 nanometer (nm).

animal pole
In yolky eggs, the end opposite the yolk. Cells derived from the animal region will envelop those derived from the yolky (vegetal) region. (Figure 22–68)

anion
Negatively charged ion.

antenna complex
Part of a photosystem that captures light energy and channels it into the photochemical reaction center. It consists of protein complexes that bind large numbers of chlorophyll molecules and other pigments.

antibiotic
Substance such as penicillin or streptomycin that is toxic to microorganisms. Often a product of a particular microorganism or plant.

antibody (immunoglobulin, Ig)
Protein produced by B cells in response to a foreign molecule or invading microorganism. Binds tightly to the foreign molecule or cell, inactivating it or marking it for destruction by phagocytosis or complement-induced lysis.

anticodon
Sequence of three nucleotides in a transfer RNA (tRNA) molecule that is complementary to a three-nucleotide codon in a messenger RNA (mRNA) molecule.

antigen
A molecule that can induce an adaptive immune response or that can bind to an antibody or T cell receptor.

antigenic determinant (epitope)
Specific region of an antigen that binds to an antibody or a T cell receptor.

antigenic variation
Ability to change the antigens displayed on the cell surface; a property of some pathogenic microorganisms that enables them to evade attack by the adaptive immune system.

antigen-presenting cell
Cell that displays foreign antigen complexed with an MHC protein on the cell surface for presentation to T lymphocytes.

antiparallel
Describes the relative orientation of the two strands in a DNA double helix or two paired regions of a polypeptide chain; the polarity of one strand is oriented in the opposite direction to that of the other.

antiporter
Carrier protein that transports two different ions or small molecules across a membrane in opposite directions, either simultaneously or in sequence. (Figure 11–8)

antisense RNA
RNA complementary to an RNA transcript of a gene. Can hybridize to the specific RNA and block its function.

APC—*see* **adenomatous polyposis coli**

APC/C—*see* **anaphase-promoting complex**

apical
Referring to the tip of a cell, a structure, or an organ. The apical surface of an epithelial cell is the exposed free surface, opposite to the basal surface. The basal surface rests on the basal lamina that separates the epithelium from other tissue.

apical meristem
The growing tip of a plant shoot or root, composed of dividing undifferentiated cells. (Panel 22–1, p. 1401)

apoptosis
Form of programmed cell death, in which a "suicide" program is activated within an animal cell, leading to rapid cell death mediated by intracellular proteolytic enzymes called caspases.

aqueous
Pertaining to water, as in an aqueous solution.

***Arabidopsis thaliana* (common Thale cress)**
Small flowering weed related to mustard. Model organism for flowering plants and the primary model for studies of plant molecular genetics.

archaeon (plural arch[a]ea) (archaebacterium).
Single-celled organism without a nucleus, superficially similar to bacteria. At a molecular level, more closely related to bacteria in metabolic machinery, but more similar to eucaryotes in genetic machinery. Archaea and Bacteria together make up the Procaryotes. (Figure 1–21)

ARF (ADP-ribosylation factor, ARF protein)
Monomeric GTPase in the Ras superfamily responsible for regulating both COPI coat assembly and clathrin coat assembly at Golgi membranes. (Table 15–5, p. 926)

aromatic
Molecule that contains carbon atoms in a ring drawn as having alternating single and double bonds. Often a molecule related to benzene.

ARP (actin-related protein) complex (ARP2/3 complex)
Complex of proteins that nucleates actin filament growth from the minus end.

ARS—*see* **autonomously replicating sequence**

asexual reproduction
Any type of reproduction (such as budding in *Hydra*, binary fission in bacteria, or mitotic division in eucaryotic microorganisms) that does not involve the mixing of two different genomes. Produces individuals that are genetically identical to the parent.

association constant—*see* **affinity constant**

aster
Star-shaped system of microtubules emanating from a centrosome or from a pole of a mitotic spindle.

astral microtubule
In the mitotic spindle, any of the microtubules radiating from the aster which are not attached to a kinetochore of a chromosome.

ATM (ataxia telangiectasia mutated protein)
Protein kinase activated by double-strand DNA breaks. If breaks are not repaired, ATM initiates a signal cascade that culminates in cell cycle arrest. Related to ATR.

ATP (adenosine 5′-triphosphate)
Nucleoside triphosphate composed of adenine, ribose, and three phosphate groups. The principal carrier of chemical energy in cells. The terminal phosphate groups are highly reactive in the sense that their hydrolysis, or transfer to another molecule, takes place with the release of a large amount of free energy. (Figure 2–26)

ATPase
Enzyme that catalyzes the hydrolysis of ATP. Many proteins have ATPase activity.

ATP synthase (F_0F_1 ATPase)
Transmembrane enzyme complex in the inner membrane of

mitochondria and the thylakoid membrane of chloroplasts. Catalyzes the formation of ATP from ADP and inorganic phosphate during oxidative phosphorylation and photosynthesis, respectively. Also present in the plasma membrane of bacteria.

ATR (ataxia telangiectasia and Rad3 related protein)
Protein kinase activated by DNA damage. If damage remains unrepaired, ATR helps initiate a signal cascade that culminates in cell cycle arrest. Related to ATM.

atypical protein kinase (aPKC)
An atypical form of protein kinase C (PKC) that does not require both Ca^{2a} and phosphatidylserine for activation. One such aPKC is involved in the specification of polarity in some individual animal cells.

auditory hair cell (sensory hair cell)
Sensory cells in the inner ear, responsible for detecting sound by converting a mechanical stimulus (the vibrations caused by sound waves) into a release of neurotransmitter. (Figures 23–13 to 23–15)

autocrine signaling
Where a cell secretes signal molecules that act back on itself.

autoimmune disease, autoimmune response
Pathological state in which the body mounts a disabling adaptive immune response against one or more of its own molecules.

autonomously replicating sequence (ARS)
Origin of replication in yeast DNA.

autophagy
Digestion of worn-out organelles by the cell's own lysosomes.

autoradiography
Technique in which a radioactive object produces an image of itself on a photographic film or emulsion.

autosome
Any chromosome other than a sex chromosome.

auxin
Plant hormone, commonly indole-3-acetic acid, with numerous roles in plant growth and development.

avidity
Total binding strength of a polyvalent antibody with a polyvalent antigen.

axon
Long nerve cell projection that can rapidly conduct nerve impulses over long distances so as to deliver signals to other cells.

axonal transport
Directed intracellular transport of organelles and molecules along a nerve cell axon. Can be anterograde (outward from the cell body) or retrograde (back toward the cell body).

axoneme
Bundle of microtubules and associated proteins that forms the core of a cilium or a flagellum in eucaryotic cells and is responsible for their movements.

BAC—*see* **bacterial artificial chromosome**

bacterium (plural bacteria) (eubacterium)
Member of the domain Bacteria, one of the three main branches of the tree of life (Archaea, Bacteria, and Eucaryotes). Bacteria and Archaea both lack a distinct nuclear compartment, and together comprise the Procaryotes. (Figure 1–21)

bacterial artificial chromosome (BAC)
Cloning vector that can accommodate large pieces of DNA up to 1 million base pairs.

bacteriophage (phage)
Any virus that infects bacteria. Phages were the first organisms used to analyse the molecular basis of genetics, and are now widely used as cloning vectors. *See also* bacteriophage lambda.

bacteriophage lambda (bacteriophage λ, lambda)
Virus that infects *E. coli*. Widely used as a DNA cloning vector.

bacteriorhodopsin
Pigmented protein found in the plasma membrane of a salt-loving archaean, *Halobacterium salinarium (Halobacterium halobium)*. Pumps protons out of the cell in response to light. (Figure 10–33)

basal
Situated near the base. Opposite the apical surface.

basal body
Short cylindrical array of microtubules and their associated proteins found at the base of a eucaryotic cell cilium or flagellum. Serves as a nucleation site for growth of the axoneme. Closely similar in structure to a centriole.

basal lamina (plural basal laminae)
Thin mat of extracellular matrix that separates epithelial sheets, and many other types of cells such as muscle or fat cells, from connective tissue. Sometimes called basement membrane. (Figure 19–40)

base
A substance that can reduce the number of protons in solution, either by accepting H^+ ions directly, or by releasing OH^- ions, which then combine with H^+ to form H_2O. The purines and pyrimidines in DNA and RNA are organic nitrogenous bases and are often referred to simply as bases. (Panel 2–2, pp. 108–109)

base excision repair
DNA repair pathway in which single faulty bases are removed from the DNA helix and replaced. *Compare* nucleotide excision repair. (Figure 5–48)

basement membrane—*see* **basal lamina**

base pair
Two nucleotides in an RNA or DNA molecule that are held together by hydrogen bonds—for example, G paired with C, and A paired with T or U.

basic (alkaline)
Having the properties of a base.

B cell (B lymphocyte)
Type of lymphocyte that makes antibodies.

Bcl2 family
Family of intracellular proteins that either promote or inhibit apoptosis by regulating the release of cytochrome *c* and other mitochondrial proteins from the intermembrane space into the cytosol.

benign
Of tumors: self-limiting in growth, and noninvasive.

beta-catenin (β-catenin)
Multifunctional cytoplasmic protein involved in cadherin-mediated cell–cell adhesion, linking cadherins to the actin cytoskeleton. Can also act independently as a gene regulatory protein. Has an important role in animal development as part of a Wnt signaling pathway.

beta sheet (β sheet)
Common structural motif in proteins in which different sections of the polypeptide chain run alongside each other, joined together by hydrogen bonding between atoms of the polypeptide backbone. Also known as a β-pleated sheet. (Figure 3–7)

binding site
Region on the surface of one molecule (usually a protein or nucleic acid) that can interact with another molecule through noncovalent bonding.

bi-orientation
The attachment of sister chromatids to opposite poles of the mitotic spindle, so that they move to opposite ends of the cell when they separate in anaphase.

biosphere
All of the living organisms on Earth.

biosynthesis—see anabolism

biotin
Low-molecular-weight compound used as a coenzyme. Also useful technically as a covalent label for proteins, allowing them to be detected by the egg protein avidin, which binds extremely tightly to biotin. (Figure 2–63)

bivalent
A four-chromatid structure formed during meiosis, consisting of a duplicated chromosome tightly paired with its homologous duplicated chromosome.

blastomere
One of the many cells formed by the cleavage of a fertilized egg. (Figure 22–69)

blastula
Early stage of an animal embryo, usually consisting of a hollow ball of epithelial cells surrounding a fluid-filled cavity, before gastrulation begins.

blotting
Biochemical technique in which macromolecules separated on a gel are transferred to a nylon membrane or sheet of paper, thereby immobilizing them for further analysis. [See Northern, Southern, and Western (immuno-) blotting.] (Figure 8–38)

B lymphocyte—see B cell

bond energy
Strength of the chemical linkage between two atoms, measured by the energy in kilocalories or kilojoules needed to break it.

bright-field microscope
Normal light microscope in which the image is obtained by simple transmission of light through the object being viewed.

brush border
Dense covering of microvilli on the apical surface of epithelial cells in the intestine and kidney.

budding yeast
Common name given to the baker's yeast *Saccharomyces cerevisiae*, a model experimental organism, which divides by budding off a smaller cell.

buffer
Solution of weak acid or weak base that resists pH change when small quantities of acid or base are added, or when solution is diluted.

Ca^{2+}/calmodulin-dependent protein kinase—see (CaM-kinase)

cadherin
Member of the large cadherin superfamily of transmembrane adhesion proteins. Mediates homophilic Ca^{2+}-dependent cell–cell adhesion in animal tissues. (Figure 19–4, and Table 19–2, p. 1135)

Caenorhabditis elegans
A small (~1mm) nematode worm used extensively in molecular and developmental biology as a model organism.

caged molecule
Organic molecule designed to change into an active form when irradiated with light of a specific wavelength. Example: caged ATP.

CAK—see Cdk-activating kinase

calcium pump—see Ca^{2+} pump

calmodulin
Ubiquitous intracellular Ca^{2+}-binding protein that undergoes a large conformation change when it binds Ca^{2+}, allowing it to regulate the activity of many target proteins. In its activated (Ca^{2+}-bound) form, it is called Ca^{2+}/calmodulin. (Figure 15–43)

calorie
Unit of heat energy, equal to 4.2 joules. One calorie (small "c") is the amount of heat needed to raise the temperature of 1 gram of water by 1°C. A kilocalorie (1000 calories) is the unit used to describe the energy content of foods.

Calvin cycle —see carbon-fixation cycle

CAM (cell adhesion molecule)
Protein on the surface of an animal cell that mediates cell–cell binding or cell–matrix binding.

CaM-kinase
Serine/threonine protein kinase that is activated by Ca^{2+}/calmodulin. Indirectly mediates the effects of an increase in cytosolic Ca^{2+} by phosphorylating specific target proteins. (Figure 15–43)

CaM-kinase II
Multifunctional Ca^{2+}/calmodulin-dependent protein kinase that phosphorylates itself and various target proteins when activated. Found in most animal cells but is especially abundant at synapses in the brain, and is involved in some forms of synaptic plasticity in vertebrates. (Figure 15–44)

cAMP—see cyclic AMP

cAMP-dependent protein kinase—see protein kinase A

cancer
Disease featuring abnormal and improperly controlled cell division resulting in invasive growths, or tumors, that may spread throughout the body. (Figure 20–37)

capsid
Protein coat of a virus, formed by the self-assembly of one or more types of protein subunit into a geometrically regular structure. (Figure 3–30)

Ca^{2+} pump (calcium pump, Ca^{2+} ATPase)
Transport protein in the membrane of sarcoplasmic reticulum of muscle cells (and elsewhere). Pumps Ca^{2+} out of the cytoplasm into the sarcoplasmic reticulum using the energy of ATP hydrolysis.

carbohydrate
General term for sugars and related compounds containing carbon, hydrogen, and oxygen, usually with the empirical formula $(CH_2O)_n$.

carbon fixation reaction
Process by which inorganic carbon (as atmospheric CO_2) is incorporated into organic molecules. The second stage of photosynthesis. (Figure 14–39)

carbon-fixation cycle (Calvin cycle)
Major metabolic pathway in photosynthetic organisms by which CO_2 and H_2O are converted into carbohydrates. Requires both ATP and NADPH. (Figure 14–40)

carbonyl group (C=O)
Carbon atom linked to an oxygen atom by a double bond. (Panel 2–1, p. 107)

carboxyl group (–COOH)
Carbon atom linked both to an oxygen atom by a double bond and to a hydroxyl group. Molecules containing a carboxyl group are weak acids—carboxylic acids. (Panel 2–1, p. 107)

carboxyl terminus—see C terminus

carcinogen
Any agent, such as a chemical or a form of radiation, that causes cancer.

carcinoma
Cancer of epithelial cells. The most common form of human cancer.

cardiac muscle
Specialized form of striated muscle found in the heart, consisting of individual heart muscle cells linked together by cell junctions.

carrier protein—*see* transporter

cartilage
Form of connective tissue composed of cells (chondrocytes) embedded in a matrix rich in type II collagen and chondroitin sulfate proteoglycan.

cascade—*see* signaling cascade

caspase
Intracellular protease that is involved in mediating the intracellular events of apoptosis.

catabolism
General term for the enzyme-catalyzed reactions in a cell by which complex molecules are degraded to simpler ones with release of energy. (Figure 2–36)

catalyst
Substance that can lower the activation energy of a reaction (thus increasing its rate), without itself being consumed by the reaction.

catastrophe factor
Protein that destabilizes microtubule arrays by increasing the frequency of rapid disassembly of tubulin subunits from one end (catastrophe). (Figure 16–16)

β-catenin—*see* beta catenin

cation
Positively-charged ion.

caveola (plural **caveolae**)
Invaginations at the cell surface that bud off internally to form pinocytic vesicles. Thought to form from lipid rafts, regions of membrane rich in certain lipids.

CD4
Co-receptor protein found on helper T cells, regulatory T cells, and macrophages. It binds to class II MHC proteins (on antigen presenting cells) outside the peptide-binding groove.

CD8
Co-receptor protein found on cytotoxic T cells. It binds to class I MHC proteins (on antigen-presenting cell) outside the peptide-binding groove.

CD28
Co-receptor protein on T cells that binds a co-stimulatory B7 protein on dendritic cells, providing an additional signal required for the activation of a naïve T cell by antigen.

Cdc6
Protein essential in the preparation of DNA for replication. With Cdt1 it binds to an origin recognition complex on chromosomal DNA and helps load the Mcm proteins onto the complex to form the prereplicative complex.

Cdc20
Activating subunit of the anaphase-promoting complex (APC/C).

Cdc25
Protein phosphatase that dephosphorylates Cdks and increases their activity.

***Cdc* gene (cell-division-cycle gene)**
Gene whose product (a Cdc protein) controls a specific step or set of steps in the eucaryotic cell cycle. Originally identified in yeasts.

Cdk—*see* cyclin-dependent kinase

Cdk-activating kinase (CAK)
Protein kinase that phosphorylates Cdks in cyclin-Cdk complexes, activating the Cdk.

Cdk inhibitor protein (CKI)
Protein that binds to and inhibits cyclin-Cdk complexes, primarily involved in the control of G1 and S phases.

cDNA
DNA molecule made as a copy of mRNA and therefore lacking the introns that are present in genomic DNA.

Cdt1
Protein essential in the preparation of DNA for replication. With Cdc6 it binds to origin recognition complexes on chromosomes and helps load the Mcm proteins on to the complex, forming the prereplicative complex.

cell adhesion molecule—*see* CAM

cell cortex
Specialized layer of cytoplasm on the inner face of the plasma membrane. In animal cells it is an actin-rich layer responsible for movements of the cell surface.

cell cycle (cell-division cycle)
Reproductive cycle of a cell: the orderly sequence of events by which a cell duplicates its chromosomes and, usually, the other cell contents, and divides into two. (Figure 17–4)

cell-cycle control system
Network of regulatory proteins that governs progression of a eucaryotic cell through the cell cycle.

cell division
Separation of a cell into two daughter cells. In eucaryotic cells it entails division of the nucleus (mitosis) closely followed by division of the cytoplasm (cytokinesis).

cell-division-cycle gene—*see* *Cdc* gene

cell fate
In developmental biology, describes what a particular cell at a given stage of development will normally give rise to.

cell-free system
Fractionated cell homogenate that retains a particular biological function of the intact cell, and in which biochemical reactions and cell processes can be more easily studied.

cell line
Population of cells of plant or animal origin capable of dividing indefinitely in culture.

cell memory
Retention by cells and their descendants of persistently altered patterns of gene expression, without any change in DNA sequence. *See also* epigenetic inheritance.

cell plate
Flattened membrane-bounded structure that forms by fusing vesicles in the cytoplasm of a dividing plant cell and is the precursor of the new cell wall.

cell senescence—*see* replicative cell senescence

cell signaling
The processes in which cells are stimulated or inhibited by extracellular signals, usually chemical signals produced by other cells.

cell transformation—*see* transformation

cellularization
The formation of cells around each nucleus in a multinucleate cytoplasm, transforming it into a multicellular structure.

cellulose
Structural polysaccharide consisting of long chains of covalently linked glucose units. Provides tensile strength in plant cell walls. (Figures 19–78 and 19–79)

cell wall
Mechanically strong extracellular matrix deposited by a cell outside its plasma membrane. Prominent in most plants, bacteria, archaea, algae, and fungi. Not present in most animal cells.

central lymphoid organ (primary lymphoid organ)
Organ in which lymphocytes are produced from precursor cells. In adult mammals, these are the thymus and bone marrow.

centriole
Short cylindrical array of microtubules, closely similar in structure to a basal body. A pair of centrioles is usually found at the center of a centrosome in animal cells. (Figure 16–31)

centromere
Constricted region of a mitotic chromosome that holds sister chromatids together. Also the site on the DNA where the kinetochore forms that captures microtubules from the mitotic spindle. (Figure 4–50)

centrosome
Centrally located organelle of animal cells that is the primary microtubule-organizing center (MTOC) and acts as the spindle pole during mitosis. In most animal cells it contains a pair of centrioles. (Figures 16–30 and 17–29)

CG island
Region of DNA with a greater than average density of CG sequences; these regions generally remain unmethylated.

CGN—*see cis* **Golgi network**

channel (membrane channel)
Transmembrane protein complex that allows inorganic ions or other small molecules to diffuse passively across the lipid bilayer. (Figures 11–3 and 11–4)

channel-forming junction
Cell-cell junction that links the cytoplasm of adjacent cells and provides a passageway for small molecules and ions to pass from cell to cell. In animal tissues, composed of connexin or innexin proteins. In plants, a similar function is performed by plasmodesmata. (Figure 19–2, and Table 19–1, p. 1133)

channel protein
Membrane transport protein that forms an aqueous pore in the membrane through which a specific solute, usually an ion, can pass. *Compare* transporter.

chaperone (molecular chaperone)
Protein that helps guide the proper folding of other proteins, or helps them avoid misfolding. Includes heat shock proteins (Hsp).

checkpoint
Point in the eucaryotic cell-division cycle where progress through the cycle can be halted until conditions are suitable for the cell to proceed.

chelate
To combine reversibly, usually with high affinity, with a metal ion such as iron, calcium, or magnesium.

chemical biology
Name given to a strategy that uses large-scale screening of hundreds of thousands of small molecules in biological assays to identify chemicals that affect a particular biological process and that can then be used to study it.

chemical bond
Chemical affinity between two atoms that holds them together. Types found in living cells include covalent bonds and noncovalent bonds. (*See also* ionic bond, hydrogen bond.)

chemiosmotic coupling (chemiosmosis)
Mechanism in which a gradient of hydrogen ions (a pH gradient, or proton gradient) across a membrane is used to drive an energy-requiring process, such as ATP production or the rotation of bacterial flagella.

chemokine
Chemotactic cytokine. Small secreted protein that attracts cells, such as white blood cells, to move towards its source. Important in the functioning of the immune system.

chemotaxis
Directed movement of a cell or organism towards or away from a diffusible chemical.

chiasma (plural chiasmata)
X-shaped connection visible between paired homologous chromosomes during meiosis. Represents a site of chromosomal crossing-over, a form of genetic recombination.

chimera
Whole organism formed from an aggregate of two or more genetically different populations of cells (two or more genotypes), originating from different zygotes. *Compare* mosaic.

ChIP—*see* **chromatin immunoprecipitation**

chitin
Abundant organic polymeric polysaccharide of *N*-acetylglucosamine. A major component of insect exoskeletons and the cell walls of fungi.

chlorophyll
Light-absorbing green pigment that plays a central part in photosynthesis in bacteria, plants, and algae.

chloroplast
Organelle in green algae and plants that contains chlorophyll and carries out photosynthesis. A specialized form of plastid.

cholesterol
An abundant lipid molecule with a characteristic four-ring steroid structure. An important component of the plasma membranes of animal cells. (Figure 10–4)

chondrocyte (cartilage cell)
Connective-tissue cell that secretes the matrix of cartilage.

chromatid
One of the two copies of a duplicated chromosome formed by DNA replication during S phase. The two chromatids, called sister chromatids, are joined at the centromere.

chromatin
Complex of DNA, histones, and nonhistone proteins found in the nucleus of a eucaryotic cell. The material of which chromosomes are made.

chromatin immunoprecipitation (ChIP)
Technique by which chromosomal DNA bound by a particular protein can be isolated and identified by precipitating it by means of an antibody against the bound protein. (Figure 7–32)

chromatin remodeling complex
Enzyme complex that alters histone-DNA configurations in eucaryotic chromosomes, changing the accessibility of the DNA to other proteins, notably those involved in transcription.

chromatography
Broad class of biochemical techniques in which a mixture of substances is separated by charge, size, hydrophobicity, noncovalent binding affinities, or some other property by allowing the mixture to partition between a moving phase and a stationary phase. Used to separate mixtures of proteins or nucleic acids. *See also* affinity-, DNA affinity-, and high-performance liquid chromatography. (Figures 8–13 and 8–14)

chromosome
Structure composed of a very long DNA molecule and associated proteins that carries part (or all) of the hereditary information of an organism. Especially evident in plant and animal cells undergoing mitosis or meiosis, during which each chromosome becomes condensed into a compact rod-like structure visible in the light microscope.

ciliate
Single-celled eucaryotic organism (protozoan) characterized by numerous cilia on its surface.

cilium (plural cilia)
Hairlike extension of a eucaryotic cell containing a core bundle of microtubules. Many cells contain a single non-motile cilium, while others contain large numbers that perform repeated beating movements. *Compare* flagellum.

circadian clock (circadian rhythm)
Internal cyclical process that produces a particular change in a cell or organism with a period of around 24 hours, for example the sleep-wakefulness cycle in humans.

cis
On the same or near side.

cisterna (plural cisternae)
Flattened membrane-bounded compartment, as found in the endoplasmic reticulum or Golgi apparatus. (Figures 13–3 and 13–25)

citric acid cycle (tricarboxylic acid (TCA) cycle, Krebs cycle)
Central metabolic pathway found in aerobic organisms. Oxidizes acetyl groups derived from food molecules, generating the activated carriers NADH and $FADH_2$, some GTP, and waste CO_2. In eucaryotic cells it occurs in the mitochondria. (Panel 2–9, pp. 122–123)

CKI—*see* Cdk inhibitor protein

class I MHC protein
One of two classes of MHC protein. Found on the surface of almost all vertebrate cell types, where it can present peptides derived from an infecting intracellular microbe (such as a virus) to cytotoxic T cells. (Figure 25–50)

class II MHC protein
One of two classes of MHC protein. Found on the surface of various antigen-presenting cells, where it presents foreign peptides to helper T cells. (Figure 25–50)

class switch
Change from making one class of immunoglobulin (for example, IgM) to making another class (for example, IgG) that many B cells undergo during the course of an adaptive immune response. Involves DNA rearrangement called class-switch recombination. (Figure 25–41)

clathrin
Protein that assembles into a polyhedral cage on the cytosolic side of a membrane so as to form a clathrin-coated pit, which buds off by endocytosis to form an intracellular clathrin-coated vesicle. (Figure 13–6)

cleavage
(1) Physical splitting of a cell into two. (2) Specialized type of cell division seen in many early embryos whereby a large cell becomes subdivided into many smaller cells without growth.

clonal selection theory
Theory that explains how the adaptive immune system can respond to millions of different antigens in a highly specific way. From a population of lymphocytes with a vast repertoire of randomly generated antigen-specific receptors, a given foreign antigen activates (selects) only those lymphocyte clones that display a receptor that fits the antigen. (Figure 25–8)

clone
Population of identical individuals (cells or organisms) formed by repeated (asexual) division from a common ancestor. Also used as a verb: "to clone a gene" means to create multiple copies of a gene by growing a clone of carrier cells (such as *E. coli*) into which the gene has been introduced, and from which it can be recovered, by recombinant DNA techniques.

cloning vector
Small DNA molecule, usually derived from a bacteriophage or plasmid, which is used to carry the fragment of DNA to be cloned into the recipient cell, and which enables the DNA fragment to be replicated. (Figure 8–39)

coactivator
Protein that does not itself bind DNA but assembles on other DNA-bound gene regulatory proteins to activate transcription of a gene. (Figure 7–51)

coated vesicle
Small membrane-bounded organelle with a cage of proteins (the coat) on its cytosolic surface. Formed by the pinching off of a coated region of membrane (coated pit). Some coats are made of clathrin, others are made from other proteins.

codon
Sequence of three nucleotides in a DNA or mRNA molecule that represents the instruction for incorporation of a specific amino acid into a growing polypeptide chain.

coenzyme
Small molecule tightly associated with an enzyme that participates in the reaction that the enzyme catalyzes, often by forming a covalent bond to the substrate. Examples include biotin, NAD^+, and coenzyme A.

coenzyme A (CoA)
Small molecule used in the enzymatic transfer of acyl groups. *See also* acetyl CoA, and Figure 2–62.

cofactor
Inorganic ion or coenzyme required for an enzyme's activity.

cohesin, cohesin complex
Complex of proteins that holds sister chomatids together along their length before their separation. (Figure 17–24)

coiled-coil
Especially stable rodlike protein structure formed by two or more alpha helices coiled around each other. (Figure 3–9)

co-immunoprecipitation (co-IP)
Method of isolating proteins that form a complex with each other by using an antibody specific for one of the partners.

collagen
Fibrous protein rich in glycine and proline that is a major component of the extracellular matrix in animals, conferring tensile strength. Exists in many forms: type I, the most common, is found in skin, tendon, and bone; type II is found in cartilage; type IV is present in basal laminae. (Figures 3–23 and 19–66)

colony-stimulating factor (CSF)
General name for numerous signal molecules that control differentiation of blood cells.

complement system
System of serum proteins activated by antibody–antigen complexes or by microorganisms. Helps eliminate pathogenic microorganisms by directly causing their lysis or by promoting their phagocytosis.

complementary
(1) Of nucleic acid sequences: capable of forming a perfect base-paired duplex with each other. (FIgure 4–4) (2) Of other interacting molecules, such as an enzyme and its substrate: having biochemical or structural features that marry up, so that noncovalent bonding is facilitated. (FIgure 2–16)

complementary DNA—*see* cDNA

complementation (genetic complementation)
Phenomenon in which the mating of two individuals, each showing an abnormal phenotype, results in offspring in which the normal (wild-type) phenotype has been restored. Basis of a test of whether two mutations are in the same or different genes. (Panel 8–1, pp. 554–555)

complex trait
Heritable characteristic whose transmission to progeny does not obey simple Mendelian laws. Such traits are due to the interaction of multiple genes and/or gene–environment interactions.

condensation reaction (dehydration reaction)
Chemical reaction in which two molecules are covalently linked through –OH groups with the removal of a molecule of water.

condensin (condensin complex)
Complex of proteins involved in chromosome condensation prior to mitosis. Target for M-Cdk. (Figure 17–27)

conditional mutation
Mutation that changes a protein or RNA molecule so that its function is altered only under some conditions, such as at an unusually high or unusually low temperature. (Panel 8–1, pp. 554–555)

confocal microscope
Type of light microscope that produces a clear image of a given plane within a solid object. It uses a laser beam as a pinpoint source of illumination and scans across the plane to produce a two-dimensional "optical section."

connective tissue
Any supporting tissue that lies between other tissues and consists of cells embedded in a relatively large amount of extracellular matrix. Includes bone, cartilage, and loose connective tissue.

connexin
Protein component of gap junctions, a four-pass transmembrane protein. Six connexins assemble in the plasma membrane to form a connexon, or 'hemichannel.' (Figure 19–34)

connexon
Water-filled pore in the plasma membrane formed by a ring of six connexin protein subunits. Half of a gap junction: connexons from two adjoining cells join to form a continuous channel through which ions and small molecules can pass. (Figure 19–34)

consensus sequence
Average or most typical form of a sequence that is reproduced with minor variations in a group of related DNA, RNA, or protein sequences. Indicates the nucleotide or amino acid most often found at each position. Preservation of a sequence implies that it is functionally important. (Figure 6–12)

constitutive
Occurring steadily, regardless of circumstances; opposite of regulated.

constitutive secretory pathway (default pathway)
Pathway present in all cells by which molecules such as plasma membrane proteins are continually delivered to the plasma membrane from the Golgi apparatus in vesicles that fuse with the plasma membrane. The default route to the plasma membrane if no other sorting signals are present. (Figure 13–63)

contractile ring
Ring containing actin and myosin that forms under the surface of animal cells undergoing cell division. Contracts to pinch the two daughter cells apart. (Figure 17–50)

convergent extension
Rearrangement of cells within a tissue that causes it to extend in one dimension and shrink in another. (Figure 22–76)

cooperativity
Phenomenon in which the binding of one ligand molecule to a target molecule promotes the binding of successive ligand molecules. Seen in the assembly of large complexes, as well as in enzymes and receptors composed of multiple allosteric subunits, where it sharpens the response to a ligand. (Figure 15–25)

co-receptor
In immunology: receptor on B cells or T cells that does not bind antigen but binds to other molecules and helps the antigen-binding receptors activate the lymphocyte. (Figure 25–57) More generally: a receptor that collaborates with another conventional cell-surface receptor, helping the cell respond to secreted signal proteins. Examples are LRP (in the Wnt/β-catenin signaling pathway) and cell-surface proteoglycans. (Figure 15–77)

co-repressor
Protein that does not itself bind DNA but assembles on other DNA-bound gene regulatory proteins to inhibit the expression of a gene. (Figure 7–51)

cortical granule
Specialized secretory vesicle present under the plasma membrane of unfertilized eggs, including those of mammals. The contents of the cortical granules, released by exocytosis after fertilization, alter the egg coat so as to prevent the entry of further sperm.

co-translational
Occurring as translation proceeds. Examples include the import of a protein into the endoplasmic reticulum before the polypeptide chain is completely synthesized (co-translational translocation, Figure 12–35), and the folding of a nascent protein into its secondary and tertiary structure as it emerges from a ribosome. (Figure 6–84)

co-transport (coupled transport)
Membrane transport process in which the transfer of one molecule depends on the simultaneous or sequential transfer of a second molecule. (Figure 11–8)

coupled reaction
Linked pair of chemical reactions in which the free energy released by one serves to drive the other. (Figure 2–51)

covalent bond
Stable chemical link between two atoms produced by sharing one or more pairs of electrons. (Figure 2–5 and Panel 2–1, pp. 106–107)

Cre/lox
Site-specific recombination system used to produce conditional mutants in which the target gene can be excised in a specific tissue or at a specific time. A site-specific recombinase (Cre) is introduced under the control of a promoter that will activate it as required. The gene to be disrupted is flanked by introduced lox sequences, at which the activated Cre operates to excise the gene. (Figure 5–79)

crista (plural cristae)
Invagination of the inner mitochondrial membrane.

critical concentration
Concentration of a protein monomer, such as actin or tubulin, that is in equilibrium with the assembled form of the protein (i.e., assembled into actin filaments or microtubules, respectively). (Panel 16–2, pp. 978–979)

crossover (chiasma)
In meiotic recombination, a site on the paired chromosomes where a segment of a maternal chromatid is exchanged for a corresponding segment of a homologous paternal chromatid (Figures 21–6 and 21–10)

cross-strand exchange—*see* **Holliday junction**

cryptochrome
Flavoprotein responsive to blue light, found in both plants and animals. In animals, it is involved in circadian rhythms.

C terminus (carboxyl terminus)
The end of a polypeptide chain that carries a free carboxyl (–COOH) group. (Figure 3–1)

cut-and-paste transposition
Type of movement of a transposable element in which the element is cut out of the DNA and inserted into a new site by a special transposase enzyme. (Figure 5–69)

cyclic AMP (cAMP)
Nucleotide that is generated from ATP by adenylyl cyclase in response to various extracellular signals. It acts as small

intracellular signaling molecule, mainly by activating cAMP-dependent protein kinase (PKA). It is hydrolyzed to AMP by a phosphodiesterase. (Figure 15–34)

cyclic AMP-dependent protein kinase (protein kinase A, PKA)
Enzyme that phosphorylates target proteins in response to a rise in intracellular cyclic AMP. (Figure 15–35)

cyclic GMP (cGMP)
Nucleotide that is generated from GTP by guanylyl cyclase in response to various extracellular signals.

cyclin
Protein that periodically rises and falls in concentration in step with the eucaryotic cell cycle. Cyclins activate crucial protein kinases (called cyclin-dependent protein kinases, or Cdks) and thereby help control progression from one stage of the cell cycle to the next.

cyclin-Cdk complex
Protein complex formed periodically during the eucaryotic cell cycle as the level of a particular cyclin increases. A cyclin-dependent kinase (Cdk) then becomes partially activated. (Figures 17–15 and 17–16, and Table 17–1, p. 1063)

cyclin-dependent kinase (Cdk)
Protein kinase that has to be complexed with a cyclin protein in order to act. Different Cdk-cyclin complexes trigger different steps in the cell-division cycle by phosphorylating specific target proteins. (Figure 17–15)

cyclosome—*see* **anaphase-promoting complex**

cytochrome
Colored heme-containing protein that transfers electrons during respiration and photosynthesis.

cytochrome b-c_1 complex
Second of the three electron-driven proton pumps in the respiratory chain. Accepts electrons from ubiquinone. (Figure 14–26)

cytochrome c
Soluble component of the mitochondrial electron-transport chain. Its release into the cytosol from the mitochondrial intermembrane space also initiates apoptosis. (Figure 14–26)

cytochrome oxidase complex
Third of the three electron-driven proton pumps in the respiratory chain. It accepts electrons from cytochrome c and generates water using molecular oxygen as an electron acceptor. (Figure 14–26)

cytokine
Extracellular signal protein or peptide that acts as a local mediator in cell–cell communication.

cytokine receptor
Cell-surface receptor that binds a specific cytokine or hormone and acts through the Jak–STAT signaling pathway. (Figure 15–68)

cytokinesis
Division of the cytoplasm of a plant or animal cell into two, as distinct from the associated division of its nucleus (which is mitosis). Part of M phase. (Panel 17–1, pp. 1072–1073)

cytoplasm
Contents of a cell that are contained within its plasma membrane but, in the case of eucaryotic cells, outside the nucleus.

cytoplasmic tyrosine kinase
Enzymes activated by certain cell-surface receptors (tyrosine-kinase-associated receptors) that transmit the receptor signal onwards by phosphorylating target cytoplasmic proteins on tyrosine side chains.

cytoskeleton
System of protein filaments in the cytoplasm of a eucaryotic cell that gives the cell shape and the capacity for directed movement. Its most abundant components are actin filaments, microtubules, and intermediate filaments.

cytosol
Contents of the main compartment of the cytoplasm, excluding membrane-bounded organelles such as endoplasmic reticulum and mitochondria.

cytotoxic T cell (killer T cell)
Type of T cell responsible for killing host cells infected with a virus or another type of intracellular pathogen.

DAG—*see* **diacylglycerol**

dalton
Unit of molecular mass. Approximately equal to the mass of a hydrogen atom (1.66×10^{-24} g).

death receptor
Transmembrane receptor protein that can signal the cell to undergo apoptosis when it binds its extracellular ligand. (Figure 18–6)

default secretory pathway—*see* **constitutive secretory pathway**

degenerate
Not a moral judgment but an adjective describing multiple states that amount to the same thing: different triplet combinations of nucleotide bases (codons) that code for the same amino acid, for example.

dehydration reaction—*see* **condensation reaction**

dehydrogenase
Enzyme that removes a hydride ion (H⁻), equivalent to a proton plus two electrons from a substrate molecule.

delta G—*see* **free-energy change**

delta G^o—*see* **standard free-energy change**

denaturation
Dramatic change in conformation of a protein or nucleic acid caused by heating or by exposure to chemicals. Usually results in the loss of biological function.

dendrite
Extension of a nerve cell, often elaborately branched, that receives stimuli from other nerve cells.

dendritic cell
The most potent type of antigen-presenting cell, which takes up antigen and processes it for presentation to T cells.

deoxyribonucleic acid—*see* **DNA**

deoxyribose
The five-carbon monosaccharide component of DNA. Differs from ribose in having H at the 2-carbon position rather than OH. $C_5H_{10}O_4$. *Compare* ribose.

depolarization
Shift in a cell's membrane potential to a less negative value inside.

dermis
Thick underlying layer of connective tissue in the skin, beneath the epidermis. Rich in collagen.

desensitization—*see* **adaptation**

desmosome
Anchoring cell–cell junction, usually formed between two epithelial cells. Characterized by dense plaques of protein into which intermediate filaments in the two adjoining cells insert. (Figure 19–3)

determination
In developmental biology, an embryonic cell is said to be determined if it has become committed to a particular specialized path of development. Determination reflects a change in the internal character of the cell, and it precedes the much more readily detected process of cell differentiation.

diacylglycerol (DAG)
Lipid produced by the cleavage of inositol phospholipids in response to extracellular signals. Composed of two fatty acid

chains linked to glycerol, it serves as a small signaling molecule to help activate protein kinase C (PKC). (Figure 15–38)

dideoxy method
The standard enzymatic method of DNA sequencing. (Figure 8–50)

differential-interference-contrast microscope
Type of light microscope that exploits the interference effects that occur when light passes through parts of a cell of different refractive indexes. Used to view unstained living cells.

differentiation
Process by which a cell undergoes a change to an overtly specialized cell type.

diffusion
Net drift of molecules in the direction of lower concentration due to random thermal movement.

diploid
Containing a double genome (two sets of homologous chromosomes and hence two copies of each gene or genetic locus). *Compare* haploid.

dissociation constant (K_d)
Measure of the tendency of a complex to dissociate. For components A and B and the binding equilibrium A + B $\rightleftharpoons$ AB, the dissociation constant is given by [A][B]/[AB], and it is smaller the tighter the binding between A and B. The dissociation constant (K_d) is the reciprocal of K_a. *See also* affinity constant, equilibrium constant. (Figure 3–43)

disulfide bond (–S–S–)
Covalent linkage formed between two sulfhydryl groups on cysteines. For extracellular proteins, a common way of joining two proteins together or linking different parts of the same protein. Formed in the endoplasmic reticulum of eucaryotic cells. (Panel 2–1, p. 107, and Figure 3–28)

DNA (deoxyribonucleic acid)
Polynucleotide formed from covalently linked deoxyribonucleotide units. The store of hereditary information within a cell and the carrier of this information from generation to generation. (Figure 4–3 and Panel 2–6, pp. 116–117)

DNA affinity chromatography
Technique for purifying sequence-specific DNA-binding proteins by their binding to a matrix to which the appropriate DNA fragments are attached. (Figure 7–28)

DNA footprinting
Technique for determining the DNA sequence to which a DNA-binding protein binds. (Figure 7–29)

DNA helicase
Enzyme that is involved in opening the DNA helix into its single strands for DNA replication.

DNA library
Collection of cloned DNA molecules, representing either an entire genome (genomic library) or complementary DNA copies of the mRNA produced by a cell (cDNA library).

DNA ligase
Enzyme that joins the ends of two strands of DNA together with a covalent bond to make a continuous DNA strand.

DNA methylation
Addition of methyl groups to DNA. Extensive methylation of the cytosine base in CG sequences is used in vertebrates to keep genes in an inactive state.

DNA microarray
A large array of short DNA molecules (each of known sequence) bound to a glass microscope slide or other suitable support. Used to monitor expression of thousands of genes simultaneously: mRNA isolated from test cells is converted to cDNA, which in turn is hybridized to the microarray. (Figure 8–73)

DNA-only transposon
Transposable element that exists as DNA throughout its life cycle. Many move by cut-and-paste transposition. *See also* transposon.

DNA polymerase
Enzyme that synthesizes DNA by joining nucleotides together using a DNA template as a guide.

DNA primase
Enzyme that synthesizes a short strand of RNA on a DNA template, producing a primer for DNA synthesis. (Figure 5–11)

DNA replication
Process by which a copy of a DNA molecule is made.

DNA tumor virus
General term for a variety of different DNA viruses that can cause tumors. (Figure 20–43)

domain (protein domain)
Portion of a protein that has a tertiary structure of its own. Larger proteins are generally composed of several domains, each connected to the next by short flexible regions of polypeptide chain. Homologous domains are recognized in many different proteins.

dominant
In genetics, the member of a pair of alleles that is expressed in the phenotype of an organism while the other allele is not, even though both alleles are present. Opposite of recessive. (Panel 8–1, pp. 554–555)

dominant negative mutation
Mutation that dominantly affects the phenotype, blocking gene activity and causing a loss-of-function phenotype even in the presence of a normal copy of the gene. (Panel 8–1, pp. 554–555)

double helix
The three-dimensional structure of DNA, in which two antiparallel DNA chains, held together by hydrogen bonding between the bases, are wound into a helix. (Figure 4–5)

Drosophila melanogaster
Species of small fly, commonly called a fruit fly. A model organism in molecular genetics.

duplex DNA
Double-stranded DNA.

dynamic instability
Sudden conversion from growth to shrinkage, and vice versa, in a protein filament such as a microtubule or actin filament. (Panel 16–2, pp. 978–979)

dynamin
Cytosolic GTPase that binds to the neck of a clathrin-coated vesicle in the process of budding from the membrane, and which is involved in completing vesicle formation.

dynein
Large motor protein that undergoes ATP-dependent movement along microtubules.

E. coli—see Escherichia coli

ectoderm
Embryonic epithelial tissue that is the precursor of the epidermis and nervous system.

effector cell
Cell that carries out the final response or function in a particular process. The main effector cells of the immune system, for example, are activated lymphocytes and phagocytes that help eliminate pathogens.

E2F protein
Gene regulatory protein that switches on many genes that encode proteins required for entry into the S-phase of the cell cycle.

eIF—*see* **eucaryotic initiation factor**

elastic fiber
Extensible fiber formed by the protein elastin in many animal connective tissues, such as in skin, blood vessels, and lungs, which gives them their stretchability and resilience.

elastin
Extracellular protein that forms extensible fibers (elastic fibers) in connective tissues.

electrochemical gradient
Combined influence of a difference in the concentration of an ion on two sides of a membrane and the electrical charge difference across the membrane (membrane potential). Ions or charged molecules can move passively only down their electrochemical gradient.

electron
Negatively charged subatomic particle that orbits the nucleus in an atom. (Figure 2–1)

electron acceptor
Atom or molecule that takes up electrons readily, thereby gaining an electron and becoming reduced.

electron carrier
Molecule such as cytochrome *c*, NADH, NADPH, and $FADH_2$ which carries electrons and transfers them from donor molecules to acceptor molecules. *See also* electron transport chain, and Figure 2–60.

electron donor
Molecule that easily gives up an electron, becoming oxidized in the process.

electron microscope
Microscope that uses a beam of electrons to create the image.

electron-microscope tomography (EM tomography)
Technique for viewing three-dimensional specimens in the electron microscope in which multiple views are taken from different directions by tilting the specimen holder. The views are combined computationally to give a three-dimensional image.

electron-transport chain
Series of reactions in which electron carrier molecules pass electrons 'down the chain' from a higher to successively lower energy levels, to a final acceptor molecule. The energy released during electron movement can be used to power various processes. Electron-transport chains present in the inner mitochondrial membrane (called the respiratory chain) and in the thylakoid membrane of chloroplasts generate a proton gradient across the membrane that is used to drive ATP synthesis. *See especially* Figures 14–3 and 14–10.

electrophoresis
Technique for separating molecules (typically proteins or nucleic acids) on the basis of their speed of migration through a porous medium when subjected to a strong electric field.

electroporation
Method for introducing DNA into cells, especially bacteria, in which a brief electric shock makes the cell membrane temporarily permeable to the foreign DNA.

elongation factor (EF)
Nomenclature used in both transcription and translation. In transcription, elongation factors associate with RNA polymerase and allow it to transcribe long stretches of DNA without dissociating. In translation, elongation factors bind to the ribosome and, by hybridizing GTP, drive the addition of amino acids to the growing polypeptide chain.

EM tomography—*see* **electron-microscope tomography**

embryonic stem cell (ES cell)
Cell derived from the inner cell mass of the early mammalian embryo. Capable of giving rise to all the cells in the body. Can be grown in culture, genetically modified, and inserted into a blastocyst to develop a transgenic animal.

endocrine
Relating to hormones or the glands that secrete them.

endocrine cell
Specialized animal cell that secretes a hormone into the blood. Usually part of a gland, such as the thyroid or pituitary gland.

endocrine signaling
Signaling via hormones released by endocrine glands into the bloodstream and carried to distant target cells that have receptors that bind the specific hormone. (Figures 15–4 and 15–5)

endocytosis
Uptake of material into a cell by an invagination of the plasma membrane and its internalization in a membrane-bounded vesicle. *See also* pinocytosis and phagocytosis.

endoderm
Embryonic tissue that is the precursor of the gut and associated organs.

endonuclease
Enzyme that cleaves nucleic acids *within* the polynucleotide chain. *Compare* exonuclease.

endoplasmic reticulum (ER)
Labyrinthine membrane-bounded compartment in the cytoplasm of eucaryotic cells, where lipids are synthesized and membrane-bound proteins and secretory proteins are made. (Figure 12–36)

endosome
Membrane-bounded organelle in animal cells that carries materials newly ingested by endocytosis and passes many of them on to lysosomes for degradation.

endothelial cell
Flattened cell type that forms a sheet (the endothelium) lining all blood and lymphatic vessels.

enhancer
Regulatory DNA sequence to which gene regulatory proteins bind, increasing the rate of transcription of a structural gene that can be many thousands of base pairs away.

entropy (*S*)
Thermodynamic quantity that measures the degree of disorder or randomness in a system; the higher the entropy, the greater the disorder. (Panel 2–7, pp. 118–119)

enveloped virus
Virus with a capsid surrounded by a lipid bilayer membrane (the envelope), which is derived from the host cell plasma membrane when the virus buds from the cell. (Figure 24–15)

enzyme
Protein that catalyzes a specific chemical reaction.

enzyme-coupled receptor
A major type of cell-surface receptor that has a cytoplasmic domain that either has enzymatic activity or is associated with an intracellular enzyme. In either case, the enzymatic activity is stimulated by an extracellular ligand binding to the receptor. (Figure 15–16)

Eph receptor
The most numerous type of receptor tyrosine kinase (RTK) that recognizes Ephrins. (Figure 15–52)

Ephrin
One of a family of membrane-bound protein ligands for the Eph receptor tyrosine kinases (RTKs) that, among many other functions, stimulate repulsion or attraction responses that guide the migration of cells and nerve cell axons during animal development.

epidermis
Epithelial layer covering the outer surface of the body. Has different structures in different animal groups. The outer layer of plant tissue is also called the epidermis.

epigenetic inheritance
Inheritance of phenotypic changes in a cell or organism that do not result from changes in the nucleotide sequence of DNA. Can be due to positive feedback loops of gene regulatory proteins or to heritable modifications in chromatin such as DNA methylation or histone modifications causing heterochromatin formation. (Figures 4–35 and 7–86)

epinephrine—*see* **adrenaline**

epistatic
Describes a mutation in one gene that masks the effect of a mutation in another gene when both mutations are present in the same organism or cell.

epithelium (plural epithelia)
Sheet of cells covering the outer surface of a structure or lining a cavity.

epitope—*see* **antigenic determinant**

equilibrium constant (K)
Ratio of forward and reverse rate constants for a reaction. Equal to the association or affinity constant (K_a) for a simple binding reaction (A + B $\rightleftharpoons$ AB). *See* also affinity constant, dissociation constant. (Figure 3–43)

ER—*see* **endoplasmic reticulum**

ER lumen
Space enclosed by the membrane of the endoplasmic reticulum (ER).

ER resident protein
Protein that remains in the endoplasmic reticulum (ER) or its membranes and carries out its function there, as opposed to proteins that are present in the ER only in transit.

ER retention signal
Short amino acid sequence on a protein that prevents it from moving out of the endoplasmic reticulum (ER). Found on proteins that are resident in the ER and function there.

ER signal sequence
N-terminal signal sequence that directs proteins to enter the endoplasmic reticulum (ER). Cleaved off by signal peptidase after entry.

erythrocyte (red blood cell)
Small hemoglobin-containing blood cell of vertebrates that transports oxygen to, and carbon dioxide from, tissues. (Figure 11–40)

erythropoietin
A hormone produced by the kidney that stimulates the production of red blood cells in bone marrow.

ES cell—*see* **embryonic stem cell**

Escherichia coli (E. coli)
Rodlike bacterium normally found in the colon of humans and other mammals and widely used in biomedical research.

ester
Molecule formed by the condensation reaction of an alcohol group with an acidic group. Phosphate groups usually form esters when linked to a second molecule. (Panel 2–1, pp. 106–107)

eubacteria
True bacteria, in contradistinction to archaea (archaebacteria). (Figure 1–21)

eucaryote (eukaryote)
Organism composed of one or more cells that have a distinct nucleus. Member of one of the three main divisions of the living world, the other two being Bacteria and Archaea. (Figure 1–21)

eucaryotic initiation factor (eIF)
Protein that helps load initiator tRNA on to the ribosome, thus initiating translation.

euchromatin
Region of an interphase chromosome that stains diffusely; "normal" chromatin, as opposed to the more condensed heterochromatin.

excision repair—*see* **base excision repair**

exocytosis
Process by which most molecules are secreted from a eucaryotic cell. These molecules are packaged into membrane-bounded vesicles that fuse with the plasma membrane and release their contents to the outside.

exon
Segment of a eucaryotic gene that consists of a sequence of nucleotides that will be represented in mRNA or in the final transfer, ribosomal, or other mature RNA molecule. In protein-coding genes, exons encode the amino acids in the protein. An exon is usually adjacent to a noncoding DNA segment called an intron. (Figure 4–15)

exonuclease
Enzyme that cleaves nucleotides one at a time from the ends of polynucleotides. *Compare* endonuclease.

expression vector
A virus or plasmid that carries a DNA sequence into a suitable host cell and there directs the synthesis of the protein encoded by the sequence. (Figure 8–48)

extracellular matrix
Complex network of polysaccharides (such as glycosaminoglycans or cellulose) and proteins (such as collagen) secreted by cells, and in which the cells are embedded.

extracellular signal molecule
Any secreted or cell-surface chemical signal that binds to receptors and regulates the activity of the cell expressing the receptor.

facilitated diffusion—*see* **passive transport**

FAD/FADH$_2$ (flavin adenine dinucleotide/reduced flavin adenine dinucleotide)
Electron carrier system that functions in the citric acid cycle. One molecule of FAD gains two electrons plus two protons in becoming the activated carrier FADH$_2$. (Figure 2–83)

FAK—*see* **focal adhesion kinase**

Fas (Fas protein, Fas death receptor)
Transmembrane death receptor that initiates apoptosis when it binds its extracellular ligand (Fas ligand). (Figure 18–6)

fat
Energy-storage lipid in cells. Composed of triglycerides—fatty acids esterified with glycerol.

fatty acid
Carboxylic acid with a long hydrocarbon tail. A major source of energy during metabolism and a starting point for the synthesis of phospholipids. (Panel 2–5, pp. 114–115)

Fc receptor
One of a family of cell-surface receptors that bind the tail region (Fc region) of an antibody (immunoglobulin) molecule. Different Fc receptors are specific for different classes of antibodies such as IgG, IgA, or IgE.

feedback inhibition
The process in which a product of a reaction feeds back to inhibit a previous reaction in the same pathway. (Figures 3–56 and 3–57)

fermentation
Anaerobic energy-yielding metabolic pathway. In anaerobic glycolysis, for instance, pyruvate is converted into lactate or ethanol, with the conversion of NADH to NAD$^+$. (Figure 2–71)

fertilization
Fusion of a male and a female gamete (both haploid) to form a diploid zygote, which develops into a new individual.

F$_0$F$_1$ ATPase—*see* **ATP synthase**

fibroblast
Common cell type found in connective tissue. Secretes an extracellular matrix rich in collagen and other extracellular matrix macromolecules. Migrates and proliferates readily in wounded tissue and in tissue culture.

fibronectin
Extracellular matrix protein involved in adhesion of cells to the matrix and guidance of migrating cells during embryogenesis. Integrins on the cell surface are receptors for fibronectin.

filopodium (plural **filopodia**) (**microspike**)
Thin, spike-like protrusion with an actin filament core, generated on the leading edge of a crawling animal cell. (Figure 16–47)

FISH—*see* **fluorescence** *in situ* **hybridization**

fission yeast
Common name for the yeast model organism *Schizosaccharomyces pombe*. It divides to give two equal-sized cells.

fixative
Chemical reagent such as formaldehyde, glutaraldehyde, or osmium tetroxide used to preserve cells for microscropy. Samples treated with these reagents are said to be 'fixed,' and the process is called fixation.

flagellum (plural **flagella**)
Long, whiplike protrusion whose undulations drive a cell through a fluid medium. Eucaryotic flagella are longer versions of cilia. Bacterial flagella are smaller and completely different in construction and mechanism of action. *Compare* **cilium**.

flavin adenine dinucleotide—*see* **FAD/FADH$_2$**

fluorescence
Light emission exhibited by some substances (fluorochromes) as their electrons, having been excited by absorption of light, return to their normal lower energy state. The emitted radiation is always at a lower energy (longer wavelength) than the absorbed radiation.

fluorescence-activated cell sorter (FACS)
Machine that sorts cells according to their fluorescence. (Figure 8–2)

fluorescence *in situ* **hybridization (FISH)**
Technique in which fluorescently labeled nucleic acid probes hybridize to specific DNA or RNA sequences in situ.

fluorescence microscope
Microscope designed to view material stained with fluorescent dyes. Similar to a light microscope but the illuminating light is passed through one set of filters before the specimen, to select those wavelengths that excite the dye, and through another set of filters before it reaches the eye, to select only those wavelengths emitted when the dye fluoresces.

fluorescence recovery after photobleaching (FRAP)
Technique for monitoring the kinetic parameters of a protein by analyzing how fluorescent protein molecules move into an area of the cell bleached by a beam of laser light.

fluorescence resonance energy transfer (FRET)
Technique for monitoring the closeness of two fluorescently labeled molecules (and thus their interaction) in cells. (Figure 8–26)

focal adhesion, (focal contact, adhesion plaque)
Anchoring cell junction, forming a small region on the surface of a fibroblast or other cell that is anchored to the extracellular matrix. Attachment is mediated by transmembrane proteins such as integrins, which are linked, through other proteins, to actin filaments in the cytoplasm.

focal adhesion kinase (FAK)
Cytoplasmic tyrosine kinase present at cell-matrix junctions (focal adhesions) in association with the cytoplasmic tails of integrins.

follicle cell
One of the cell types that surround a developing oocyte or egg. (Figure 21–24)

footprinting—*see* **DNA footprinting**

FRAP—*see* **fluorescence recovery after photobleaching**

free energy (*G*) (Gibbs free energy)
The energy that can be extracted from a system to drive reactions. Takes into account changes in both energy and entropy. (Panel 2–7, pp. 118–119)

free-energy change (Δ*G*)
Change in the free energy during a reaction: the free energy of the product molecules minus the free energy of the starting molecules. A large negative value of Δ*G* indicates that the reaction has a strong tendency to occur. (Panel 2–7, pp. 118–119)

free radical
Atom or molecule which is extremely reactive by virtue of its having at least one unpaired electron. Responsible for much intracellular DNA damage.

free ribosome
Ribosome that is free in the cytosol, unattached to any membrane.

freeze-fracture electron microscopy
Technique for studying membrane structure, in which the membrane of a frozen cell is fractured along the interior of the bilayer, separating it into the two monolayers with the interior faces exposed.

FRET—*see* **fluorescence resonance energy transfer**

fungus (plural **fungi**)
Kingdom of eucaryotic organisms that includes the yeasts, molds, and mushrooms. Many plant diseases and a relatively small number of animal diseases are caused by fungi.

fusion protein
Engineered protein that combines two or more normally separate polypeptides. Produced from a recombinant gene.

G—*see* **free energy**

Δ*G*—*see* **free energy change**

Δ*G*°—*see* **standard free energy change**

G$_i$—*see* **inhibitory G protein**

G$_q$
Class of G protein that couples GPCRs to phospholipase C-β to activate the inositol phospholipid signaling pathway.

G$_s$—*see* **stimulatory G protein**

GAG—*see* **glycosaminoglycan**

gain-of-function mutation
Mutation that increases the activity of a gene, or makes it active in inappropriate circumstances. Usually dominant. (Panel 8–1, pp. 554–555)

gamete
Specialized haploid cell (a sperm or egg) formed from primordial germ cells by meiosis and specialized for sexual reproduction. *See also* **germ cell**.

gamma-tubulin ring complex (γTuRC)
Protein complex containing γ-tubulin and other proteins that is an efficient nucleator of microtubules and caps their minus ends.

ganglion (plural **ganglia**)
Cluster consisting mainly of neuronal cell bodies and associated glial cells, located outside the central nervous system.

ganglioside
Any glycolipid having one or more sialic acid residues in its structure. Found in the plasma membrane of eucaryotic cells and especially abundant in nerve cells. (Figure 10–18)

GAP—*see* **GTPase-activating protein**

gap gene
In *Drosophila* development, a gene that is expressed in specific broad regions along the anteroposterior axis of the early embryo, and which helps designate the main divisions of the insect body. (Figure 22–37)

gap junction
Communicating channel-forming cell–cell junction present in most animal tissues that allows ions and small molecules to pass from the cytoplasm of one cell to the cytoplasm of the next.

gastrulation
Stage in animal embryogenesis during which the embryo is transformed from a ball of cells to a structure with a gut (a gastrula). (Figure 22–3)

G_1-Cdk
Cyclin-Cdk complex formed in vertebrate cells by a G_1-cyclin and the corresponding cyclin-dependent kinase (Cdk). (Table 17–1, p. 1063)

G_1-cyclin
Cyclin present in the G_1 phase of the eucaryotic cell cycle. Forms complexes with Cdks that help govern the activity of the G_1/S-cyclins, which control progression to S-phase.

GEF—*see* **guanine nucleotide exchange factor**

gel-mobility shift assay
Technique for detecting proteins bound to a specific DNA sequence by the fact that the bound protein slows down the migration of the DNA fragment through a gel during gel electrophoresis. (Figure 7–27)

gel-transfer hybridization—*see* **blotting**

geminin
Protein that prevents the formation of new prereplicative complexes during S phase and mitosis, thus ensuring that the chromosomes are replicated only once in each cell cycle.

gene
Region of DNA that is transcribed as a single unit and carries information for a discrete hereditary characteristic, usually corresponding to (1) a single protein (or set of related proteins generated by variant post-transcriptional processing), or (2) a single RNA (or set of closely related RNAs).

gene activator protein—*see* **activator**

gene control region
The set of linked DNA sequences regulating expression of a particular gene. Includes promoter and regulatory sequences required to initiate transcription of the gene and control the rate of initiation. [Figures 7–37 (procaryotes) and 7–44 (eucaryotes)]

gene conversion
Process by which DNA sequence information can be transferred from one DNA helix (which remains unchanged) to another DNA helix whose sequence is altered. It often accompanies general recombination events. (Figure 5–66)

gene expression
Production of an observable molecular product (RNA or protein) by a gene.

general recombination, general genetic recombination—*see* **homologous recombination**

general transcription factor
Any of the proteins whose assembly at a promoter is required for the binding and activation of RNA polymerase and the initiation of transcription. (Table 6–3, p. 341)

gene regulatory protein
General name for any protein that binds to a specific DNA sequence to influence the transcription of a gene.

gene repressor protein—*see* **repressor**

genetic code
Set of rules specifying the correspondence between nucleotide triplets (codons) in DNA or RNA and amino acids in proteins. (Figure 6–50)

genetic engineering (recombinant DNA technology)
Collection of techniques by which DNA segments from different sources are combined to make a new DNA, often called a recombinant DNA. Recombinant DNAs are widely used in the cloning of genes, in the genetic modification of organisms, and in molecular biology generally.

genetic instability
Abnormally increased spontaneous mutation rate, such as occurs in cancer cells.

genetic mosaic—*see* **mosaic**

genetic recombination—*see* **recombination**

genetic redundancy
The presence of two or more similar genes with overlapping functions.

genetics
The study of the genes of an organism on the basis of heredity and variation.

genetic screen
Procedure for discovery of genes affecting specific aspects of the phenotype by surveying large numbers of mutagenized individuals.

genome
The totality of genetic information belonging to a cell or an organism; in particular, the DNA that carries this information.

genomic DNA
DNA constituting the genome of a cell or an organism. Often used in contrast to cDNA (DNA prepared by reverse transcription from mRNA). Genomic DNA clones represent DNA cloned directly from chromosomal DNA, and a collection of such clones from a given genome is a genomic DNA library.

genomic imprinting
Phenomenon in which a gene is either expressed or not expressed in the offspring depending on which parent it is inherited from. (Figure 7–82)

genomics
Study of the DNA sequences and properties of entire genomes.

genotype
Genetic constitution of an individual cell or organism. The particular combination of alleles found in a specific individual. (Panel 8–1, pp. 554–555)

germ cell
A cell in the germ line of an organism, which includes the haploid gametes and their specified diploid precursor cells. Germ cells contribute to the formation of a new generation of organisms and are distinct from somatic cells, which form the body and leave no descendants.

germ layer
One of the three primary tissue layers (endoderm, mesoderm, and ectoderm) of an animal embryo. (Figure 22–70)

germ line
The cell lineage that consists of the haploid gametes and their specified diploid precursor cells.

GFP—*see* **green fluorescent protein**

Gibbs free energy—*see* **free energy**

giga-
Prefix denoting one billion (10^9).

glial cell
Supporting nonneural cell of the nervous system. Includes oligodendrocytes and astrocytes in the vertebrate central nervous system and Schwann cells in the peripheral nervous system.

glucagon
Hormone produced in the pancreas that stimulates liver cells to convert stored glycogen to glucose, leading to a rise in blood sugar levels. Effect is opposite to that of insulin.

glucose
Six-carbon sugar that plays a major role in the metabolism of living cells. Stored in polymeric form as glycogen in animal cells and as starch in plant cells. (Panel 2–4, pp. 112–113)

glycerol
Small organic molecule that is the parent compound of many small molecules in the cell, including phospholipids. (Panel 2–5, pp. 114–115)

glycogen
Polysaccharide composed exclusively of glucose units. Used to store energy in animal cells. Large granules of glycogen are especially abundant in liver and muscle cells. (Figure 2–75 and Panel 2–4, pp. 112–113)

glycolipid
Lipid molecule with a sugar residue or oligosaccharide attached. (Panel 2–5, pp. 114–115)

glycolysis
Ubiquitous metabolic pathway in the cytosol in which sugars are incompletely degraded with production of ATP. Literally, "sugar splitting." (Figure 2–70 and Panel 2–8 , pp. 120–121)

glycoprotein
Any protein with one or more oligosaccharide chains covalently linked to amino-acid side chains. Most secreted proteins and most proteins exposed on the outer surface of the plasma membrane are glycoproteins.

glycosaminoglycan (GAG)
Long, linear, highly charged polysaccharide composed of a repeating pair of sugars, one of which is always an amino sugar. Mainly found covalently linked to a protein core in extracellular matrix proteoglycans. Examples include chondroitin sulfate, hyaluronan, and heparin. (Figure 19–55)

glycosidic bond
Chemical bond formed by a condensation reaction between a hydroxyl group on the sugar carbon atom that carries the aldehyde or the ketone and another molecule, often another sugar. Glycosidic bonds are found in numerous structures including polysaccharides, nucleic acids, and glycoproteins. See *O*-linked and *N*-linked oligosaccharides. (Panel 2–4, pp. 112–113)

glycosylase—*see* **DNA glycosylase**

glycosylation
Addition of one or more sugars to a protein or lipid molecule.

glycosylphosphatidylinositol anchor—*see* **GPI anchor**

G₂/M checkpoint
Point in the eucaryotic cell cycle at which the cell checks for completion of DNA replication before triggering the early mitotic events that lead to chromosome alignment on the spindle. (Figure 17–14)

Golgi apparatus (Golgi complex)
Complex organelle in eucaryotic cells, centered on a stack of flattened membrane-bounded spaces, in which proteins and lipids transferred from the endoplasmic reticulum are modified and sorted. It is the site of synthesis of many cell wall polysaccharides in plants and extracellular matrix glycosaminoglycans in animal cells. (Figure 13–25)

G₀ phase
State of withdrawal from the eucaroytic cell-division cycle by entry into a quiescent digression from the G₁ phase. A common, sometimes permanent, state for differentiated cells.

G₁ phase
Gap 1 phase of the eucaryotic cell-division cycle, between the end of mitosis and the start of DNA synthesis. (Figure 17–4)

G₂ phase
Gap 2 phase of the eucaryotic cell-division cycle, between the end of DNA synthesis and the beginning of mitosis. (Figure 17–4)

GPI anchor (glycosylphosphatidylinositol anchor)
Lipid linkage by which some membrane proteins are bound to the membrane. The protein is joined, via an oligosaccharide linker, to a phosphatidylinositol anchor during its travel through the endoplasmic reticulum. (Figure 10–19(6))

G protein (trimeric GTP-binding protein)
A trimeric GTP-binding protein with intrinsic GTPase activity that couples GPCRs to enzymes or ion channels in the plasma membrane. (Table 15–3, p. 919)

G-protein-coupled receptor (GPCR)
A seven-pass cell-surface receptor that, when activated by its extracellular ligand, activates a G protein, which in turn, activates either an enzyme or ion channel in the plasma membrane. (Figures 15–16 and 15–30)

grana (singular granum)
Stacked membrane discs (thylakoids) in chloroplasts that contain chlorophyll and are the site of the light-trapping reactions of photosynthesis. (Figures 14–35 and 14–36)

granulocyte
Category of white blood cell distinguished by conspicuous cytoplasmic granules. Includes neutrophils, basophils, and eosinophils. Arises from a granulocyte/macrophage (GM) progenitor cell. (Figure 23–37)

green fluorescent protein (GFP)
Fluorescent protein isolated from a jellyfish. Widely used as a marker in cell biology.

growth cone
Migrating motile tip of a growing nerve cell axon or dendrite. (Figure 16–105)

growth factor
Extracellular signal protein that can stimulate a cell to grow. They often have other functions as well, including stimulating cells to survive or proliferate. Examples include epidermal growth factor (EGF) and platelet-derived growth factor (PDGF).

G₁/S-Cdk
Cyclin-Cdk complex formed in vertebrate cells by a G₁/S-cyclin and the corresponding cyclin-dependent kinase (Cdk). (Figure 17–16 and Table 17–1, p. 1063)

G₁/S-cyclin
Cyclin that activates Cdks in late G₁ of the eucaryotic cell cycle and thereby helps trigger progression through Start, resulting in a commitment to cell-cycle entry. Its level falls at the start of S phase. (Figure 17–16)

GTP (guanosine 5′-triphosphate)
Nucleoside triphosphate produced by the phosphorylation of GDP (guanosine diphosphate). Like ATP, it releases a large amount of free energy on hydrolysis of its terminal phosphate group. Has a special role in microtubule assembly, protein synthesis, and cell signaling. (Figure 2–83)

GTPase
An enzyme that converts GTP to GDP. GTPases fall into two large families. Large **trimeric G proteins** are composed of three different subunits and mainly couple GPCRs to enzymes or ion channels in the plasma membrane. Small

monomeric GTP-binding proteins (also called **monomeric GTPases**) consist of a single subunit and help relay signals from many types of cell-surface receptors and have roles in intracellular signaling pathways, regulating intracellular vesicle trafficking, and signaling to the cytoskeleton. Both trimeric G proteins and monomeric GTPases cycle between an active GTP-bound form and an inactive GDP-bound form and frequently act as molecular switches in intracellular signaling pathways. (Figure 15–19)

GTPase-activating protein (GAP)
Protein that binds to a GTPase and inhibits it by stimulating its GTPase activity, causing the enzyme to hydrolyzes its bound GTP to GDP. (Figure 3–71)

GTP-binding protein—*see* **GTPase**

guanine nucleotide exchange factor (GEF)
Protein that binds to a GTPase and activates it by stimulating it to release its tightly bound GDP, thereby allowing it to bind GTP in its place. (Figure 3–73)

guanosine triphosphate—*see* **GTP**

H+—*see* **proton**

H−—*see* **hydride ion**

hair cell—*see* **auditory hair cell**

haploid
Having only a single copy of the genome (one set of chromosomes), as in a sperm cell, unfertilized egg, or bacterium. *Compare* diploid.

haplotype
Haploid genotype.

haplotype block
Combination of alleles and DNA markers that has been inherited in a large, linked block on one chromosome of a homologous pair—undisturbed by genetic recombination—across many generations.

haplotype map (hapmap)
Human genome map based on haplotype blocks. Intended to help identify and catalog human genetic variation.

H chain—*see* **heavy chain**

heart muscle cell—*see* **muscle cell**

heat shock protein (Hsp, stress-response protein)
One of a large family of highly conserved molecular chaperone proteins, so named because they are synthesized in increased amounts in response to an elevated temperature or other stressful treatment. Hsps have important roles in aiding correct protein folding or refolding. Prominent examples are Hsp60 and Hsp70.

heavy chain (H chain)
The larger of the two types of polypeptide chain in an immunoglobulin molecule.

Hedgehog protein
Secreted extracellular signal molecule that has many different roles controlling cell differentiation and gene expression in animal embryos and adult tissues. Excessive Hedgehog signaling can lead to cancer.

HeLa cells
'Immortal' line of human epithelial cells that grows vigorously in culture. Derived in 1951 from a human cervical carcinoma.

helicase—*see* **DNA helicase**

α helix—*see* **alpha helix**

helix-loop-helix (HLH)
DNA-binding structural motif present in many gene regulatory proteins, consisting of a short alpha helix connected by a flexible loop to a second, longer alpha helix. Its structure enables two HLH-containing proteins to dimerize, forming a complex that binds to DNA. Distinct from the helix-turn-helix motif. (Figure 7–23)

helix-turn-helix
DNA-binding structural motif present in many gene regulatory proteins, consisting of two alpha helices held at a fixed angle and connected by a short chain of amino acids, constituting the turn. Proteins containing this motif frequently form symmetric dimers and bind to DNA sequences that are themselves similar and arranged symmetrically. Distinct from the helix-loop-helix motif. (Figures 7–10, 7–11, 7–12)

helper T cell
Type of T cell that helps stimulate B cells to make antibodies and macrophages to kill ingested microorganisms. Also helps activate dendritic cells and cytotoxic T cells.

heme
Cyclic organic molecule containing an iron atom that carries oxygen in hemoglobin and carries an electron in cytochromes. (Figure 14–22)

hemidesmosome
Specialized anchoring cell junction between an epithelial cell and the underlying basal lamina.

hemopoiesis (hematopoiesis)
Generation of blood cells, mainly in the bone marrow. (Figure 23–42)

hemopoietic stem cell
Self-renewing bone marrow cell that gives rise to all the various types of blood cells, as well as some other cell types.

hepatocyte
The major cell type in the liver.

heterocaryon
Cell with two or more genetically different nuclei; produced by the fusion of two or more different cells.

heterochromatin
Region of a chromosome that remains in the form of unusually condensed chromatin; generally transcriptionally inactive. *Compare* euchromatin.

heterodimer
Protein complex composed of two different polypeptide chains.

heterophilic binding
Binding between molecules of different kinds, especially those involved in cell–cell adhesion. (Figure 19–8)

heterozygote
Diploid cell or individual having two different alleles of one or more specified genes.

high-energy bond
Covalent bond whose hydrolysis releases an unusually large amount of free energy under the conditions existing in a cell. A group linked to a molecule by such a bond is readily transferred from one molecule to another. Examples include the phosphodiester bonds in ATP and the thioester linkage in acetyl CoA.

high-performance liquid chromatography (HPLC)
Type of chromatography that uses columns packed with tiny beads of matrix; the solution to be separated is pushed through under high pressure.

histidine-kinase-associated receptor
Transmembrane receptor found in the plasma membrane of bacteria, yeast, and plant cells, and involved, for example, in sensing stimuli that cause bacterial chemotaxis. Associated with a histidine protein kinase on its cytoplasmic side.

histone
One of a group of small abundant proteins, rich in arginine

and lysine. Histones form the nucleosome cores around which DNA is wrapped in eucaryotic chromosomes. (Figure 4–25)

histone chaperone (chromatin assembly factor)
Protein that binds free histones, releasing them once they have been incorporated into newly replicated chromatin. (Figure 4–30)

histone code
Combinations of chemical modifications to histones (e.g., actylation, methylation) that are thought to determine how and when the DNA packaged in nucleosomes can be accessed (e.g., for replication or transcription). (Figure 4–44)

histone H1
'Linker' (as opposed to 'core') histone protein that binds to DNA where it exits from a nucleosome and helps package nucleosomes into the 30 nm chromatin fiber. (Figure 4–34)

HIV
Human immunodeficiency virus, the retrovirus that is the cause of AIDS (acquired immune deficiency syndrome). (Figures 13–19 and 24–16)

HLH—*see* **helix-loop-helix**

hnRNP protein (heterogeneous nuclear ribonuclear protein)
Any of a group of proteins that assemble on newly synthesized RNA, organizing it into a more compact form. (Figure 6–33)

Holliday junction (cross-strand exchange)
X-shaped structure observed in DNA undergoing recombination, in which the two DNA molecules are held together at the site of crossing-over, also called a cross-strand exchange. (Figure 5–61)

homeobox
Short (180 base pairs long) conserved DNA sequence that encodes a DNA-binding protein motif (homeodomain) famous for its presence in genes that are involved in orchestrating development in a wide range of organisms.

homeodomain
DNA-binding domain that defines a class of gene regulatory proteins important in animal development. (Figures 3–13 and 7–13)

homeotic mutation
Mutation that causes cells in one region of the body to behave as though they were located in another, causing a bizarre disturbance of the body plan. (Figures 22–42 and 22–127)

homeotic selector gene
In *Drosophila* development, a gene that defines and preserves the differences between body segments.

homolog
One of two or more genes that are similar in sequence as a result of derivation from the same ancestral gene. The term covers both orthologs and paralogs. (Figure 1–25) *See* homologous chromosomes.

homologous
Genes, proteins, or body structures that are similar as a result of a shared evolutionary origin.

homologous chromosomes (homologs)
The maternal and paternal copies of a particular chromosome in a diploid cell.

homologous recombination (general recombination)
Genetic exchange between a pair of identical or very similar DNA sequences, typically those located on two copies of the same chromosome. (Figures 5–51, 5–53, 5–59, and 5–64)

homophilic binding
Binding between molecules of the same kind, especially those involved in cell–cell adhesion. (Figure 19–8)

homozygote
Diploid cell or organism having two identical alleles of a specified gene or set of genes.

hormone
Signal molecule secreted by an endocrine cell into the bloodstream, which can then carry it to distant target cells.

housekeeping gene
Gene serving a function required in all the cell types of an organism, regardless of their specialized role.

***Hox* gene complex**
Cluster of genes coding for gene regulatory factors, each gene containing a homeodomain, and specifying body-region differences. *Hox* mutations typically cause homeotic transformations.

HPLC—*see* **high-performance liquid chromatography**

hyaluronan (hyaluronic acid)
Type of nonsulfated glycosoaminoglycan with a regular repeating sequence of up to 25,000 identical disaccharide units, not linked to a core protein. Found in the fluid lubricating joints and in many other tissues. (Figures 19–56 and 19–57)

hybridization
In molecular biology, the process whereby two complementary nucleic acid strands form a base-paired duplex DNA-DNA, DNA-RNA, or RNA-RNA molecule. Forms the basis of a powerful technique for detecting specific nucleotide sequences. (Figures 5–54 and 8–36)

hybridoma
Cell line used in the production of monoclonal antibodies. Obtained by fusing antibody-secreting B cells with cells of a B lymphocyte tumor. (Figure 8–8)

hydride ion (H^-)
A proton (H^+) plus two electrons ($2e^-$). Equivalent to a hydrogen atom with one extra electron.

hydrocarbon
Compound that has only carbon and hydrogen atoms. (Panel 2–1, pp. 106–107)

hydrogen bond
Noncovalent bond in which an electropositive hydrogen atom is partially shared by two electronegative atoms. (Panel 2–3, pp. 110–111)

hydrogen ion
A proton (H^+) in an aqueous solution. The basis of acidity. Since the proton readily combines with a water molecule to form H_3O^+, it is more accurate to call it a hydronium ion.

hydrolase
General term for enzyme that catalyses a hydrolytic cleavage reaction. Includes nucleases and proteases.

hydrolysis (adjective hydrolytic)
Cleavage of a covalent bond with accompanying addition of water. General formula: $AB + H_2O \rightarrow AOH + BH$.

hydronium ion (H_3O^+)
Water molecule associated with an additional proton. The form generally taken by protons in aqueous solution.

hydrophilic
Dissolving readily in water. Literally, "water loving."

hydrophobic (lipophilic)
Not dissolving readily in water. Literally, "water hating."

hydrophobic force
Force exerted by the hydrogen-bonded network of water molecules that brings two nonpolar surfaces together by excluding water between them. (Panel 2–3 , pp. 110–111)

hydroxyl (–OH)
Chemical group consisting of a hydrogen atom linked to an oxygen, as in an alcohol. (Panel 2–1, pp. 106–107)

hypertonic
Having a sufficiently high concentration of solutes to cause water to move out of a cell by osmosis.

hypervariable region
Any of three small regions within the variable region of an immunoglobulin or T cell receptor chain that shows the highest variability from molecule to molecule. These regions determine the specificity of the antigen-binding site. (Figure 25–31)

hypotonic
Having a sufficiently low concentration of solutes to cause water to move into a cell by osmosis.

IAP (inhibitor of apoptosis) family
Intracellular protein inhibitors of apoptosis.

IF—see **initiation factor**

Ig—see **antibody**

Ig superfamily
Large and diverse family of proteins that contain immunoglobulin domains or immunoglobulin-like domains. Most are involved in cell–cell interactions or antigen recognition. (Figure 25–74)

imaginal disc
Group of cells that are set aside, apparently undifferentiated, in the *Drosophila* embryo and which will develop into an adult structure, e.g., eye, leg, wing. Overt differentiation occurrs at metamorphosis. (Figure 22–51)

immortalization
Production of a cell line capable of an apparently unlimited number of cell divisions. Can be the result of mutations or viral transformation or of fusion of the original cells with cells of a tumor line.

immune response
Response made by the immune system when a foreign substance or microorganism enters the body; usually refers to an adaptive immune response. *See also* innate, adaptive, primary, and secondary immune responses.

immune system
System of lymphocytes and other cells in the body that provides defense against infection. There are two types of immune systems in vertebrates—innate and adaptive.

immunoblotting—see **blotting**

immunoglobulin (Ig)
An antibody molecule. Higher vertebrates have five classes of immunoglobulin—IgA, IgD, IgE, IgG, and IgM—each with a different role in adaptive immune responses.

immunoglobulin domain (Ig domain)
Characteristic protein domain of about 100 amino acids that is found in immunoglobulin light and heavy chains. Similar domains, known as immunoglobulin-like (Ig-like) domains, are present in many other proteins involved in cell–cell interactions and antigen recognition and define the Ig superfamily. (Figure 25–32)

immunological memory
Long-lived property of the adaptive immune system that follows a primary immune response to many antigens, in which subsequent encounter with the same antigen will provoke a more rapid and stronger secondary immune response. (Figure 25–10)

immunoprecipitation (IP)
Use of a specific antibody to draw the corresponding protein antigen out of solution. The technique can identify complexes of interacting proteins in cell extracts by using an antibody specific for one of the proteins to precipitate the complex. *See also* chromatin immunoprecipitation and co-immunoprecipitation.

imprinting—see **genomic imprinting**

inducible promoter
A regulatory DNA sequence that allows expression of an associated gene to be switched on by a particular molecular or physical stimulus (e.g., heat shock). (Figure 22–49)

induction (inductive interaction)
In developmental biology, a change in the developmental fate of one tissue caused by an interaction with another tissue. Such an effect is called an inductive interaction. (Figures 22–10 and 22–16)

inflammatory response
Local response of a tissue to injury or infection—characterized by tissue redness, swelling, heat, and pain. Caused by invasion of white blood cells, which release various local mediators such as histamine.

inhibitor of apoptosis family—see **IAP family**

inhibitory G protein (G_i)
Trimeric G protein that can regulate ion channels and inhibit the enzyme adenylyl cyclase in the plasma membrane. *See also* G protein. (Table 15–3, p. 919)

inhibitory neurotransmitter
Neurotransmitter that opens transmitter-gated Cl^- or K^+ channels in the post-synaptic membrane of a nerve or muscle cell and thus tends to inhibit the generation of an action potential.

initiation factor (IF and eIF)
Protein that promotes the proper association of ribosomes with mRNA and is required for the initiation of protein synthesis. Abbreviated eIF in eucaryotes, IF in procaryotes. eIFs help load Met-tRNAi on to the ribosome, thus initiating translation. (Figure 6–72)

initiator tRNA
Special tRNA that intiates translation. It always carries the amino acid methionine, forming the complex Met-tRNAi. (Figure 6–72)

innate immune response
Immune response (of both animals and plants) to a pathogen that involves the pre-existing defenses of the body (the innate immune system), which include antimicrobial molecules and phagocytic cells. Such a response is not specific for the pathogen, in contrast to an adaptive immune response.

inner cell mass
Cluster of undifferentiated cells in the early mammalian embryo from which the whole of the adult body is derived. (Figure 22–88)

inositol
Ring-shaped sugar molecule forming part of inositol phospholipids.

inositol phospholipid (phosphoinositide)
A lipid containing a phosphorylated inositol derivative. Minor component of the plasma membrane, but important in demarking different membranes and for intracellular signal transduction in eucaryotic cells. (Figure 15–37)

inositol phospholipid signaling pathway
Intracellular signaling pathway that starts with the activation of phospholipase C and the generation of IP_3 and diacylglycerol (DAG) from inositol phospholipids in the plasma membrane. The DAG helps to activate protein kinase C. (Figures 15–38 and 15–39)

inositol 1,4,5-trisphosphate (IP_3)
Small intracellular signaling molecule produced during activation of the inositol phospholipid signaling pathway. Acts to release Ca^{2+} from the endoplasmic reticulum. (Figures 15–38 and 15–39)

in situ hybridization
Technique in which a single-stranded RNA or DNA probe is used to locate a gene or mRNA molecule in a cell or tissue by hybridization.

insulator element

DNA sequence that prevents a gene regulatory protein bound to DNA in the control region of one gene from influencing the transcription of adjacent genes. (Figure 7–62)

insulin

Polypeptide hormone that is secreted by β cells in the pancreas to help regulate glucose metabolism in animals. (Figure 3–35)

integral membrane protein

Protein that is retained in a membrane by virtue of one or more domains that span or are embedded in the lipid bilayer. (Figure 10–19)

integrin

Transmembrane adhesion protein that is involved in the attachment of cells to the extracellular matrix and to each other. (Figure 19–4, and Table 19–2, p. 1135)

intercalary regeneration

Type of regeneration that fills in the missing tissues when the cells in two mismatched parts of a structure are grafted together. (Figure 22–56)

interferon (IFN)

Member of a class of cytokines secreted by virus-infected cells and certain types of activated T cells. Interferons induce antiviral responses, inhibiting viral replication and stimulating macrophages and natural killer cells to kill virus-infected cells. (Figure 25–60)

interleukin (IL)

Secreted cytokine that mainly mediates local interactions between white blood cells (leucocytes) during inflammation and immune responses. (Table 25–4, p. 1598)

intermediate filament

Fibrous protein filament (~10 nm diameter) that forms rope-like networks in animal cells. One of the three most prominent types of cytoskeletal filaments. (Panel 16–1, pp. 968–969)

internal ribosome entry site (IRES)

Specific site in a eucaryotic mRNA, other than at the 5′ end, at which translation can be initiated. (Figure 7–108)

interphase

Long period of the cell cycle between one mitosis and the next. Includes G_1 phase, S phase, and G_2 phase. (Figure 17–4)

interpolar microtubule

In the mitotic or meiotic spindle, a microtubule interdigitating at the equator with the microtubules emanating from the other pole. (Figure 16–85)

intracellular signaling protein

Protein involved in a signaling pathway inside the cell. It usually activates the next protein in the pathway or generates a small intracellular mediator. (Figure 15–1)

intron

Noncoding region of a eucaryotic gene that is transcribed into an RNA molecule but is then excised by RNA splicing during production of the mRNA or other functional RNA. (Figure 4–15)

inversion

Type of mutation in which a segment of chromosome is inverted. (Panel 8–1, pp. 554–555)

in vitro

Taking place in an isolated cell-free extract, as opposed to in a living cell; also sometimes used to distinguish studies in cell cultures from studies in intact organisms. (Latin for "in glass.")

***in vitro* fertilization (IVF)**

An infertility treatment in which eggs are fertilized with sperm outside the mother's body; successfully fertilized eggs are cultured for a few days, and the early embryos are then transferred into the mother's uterus.

in vivo

In an intact cell or organism. Latin for "in life."

ion

An atom that has either gained or lost electrons to acquire a charge; for example Na^+ and Cl^-.

ion channel

Transmembrane protein complex that forms a water-filled channel across the lipid bilayer through which specific inorganic ions can diffuse down their electrochemical gradients. (Figure 11–21)

ion-channel-coupled receptor—*see* transmitter-gated ion channel

ionic bond (ionic interaction)

Cohesion due to electrostatic attraction between two atoms, one with a positive charge, the other with a negative charge. One type of noncovalent bond. (Figure 2–5, and Panel 2–3, pp. 110–111)

ionophore

Small hydrophobic molecule that dissolves in lipid bilayers and increases their permeability to specific inorganic ions.

IP_3—*see* inositol 1,4,5-trisphosphate

IP_3 receptor (IP_3-gated Ca^{2+}–release channel)

Gated Ca^{2+} channel in the ER membrane that opens on binding cytosolic IP_3, releasing stored Ca^{2+} into the cytosol. (Figure 15–39)

IRES—*see* internal ribosome entry site

iron-sulfur center

Electron-transporting group consisting of either two or four iron atoms bound to an equal number of sulfur atoms, found in a class of electron-transport proteins. (Figure 14–23)

isoelectric point (pI)

The pH at which a molecule in solution has no net electric charge and therefore does not move in an electric field. (Figure 8–22)

isoform

One of a set of variant forms of a protein, derived either by alternative splicing of a common transcript or as products of different members of a set of closely homologous genes.

isomer

Molecule formed from the same atoms and linkages as another but having a different three-dimensional conformation. (Panel 2–4, pp. 112–113)

isomerase

Enzyme that catalyzes the rearrangement of bonds within a single molecule. *See also* topoisomerase.

isoprenoid (polyisoprenoid)

Lipid molecule with a carbon skeleton based on multiple five-carbon isoprene units. Examples include retinoic acid and dolichol. (Panel 2–5 , pp. 114 –115)

isotope

One of a number of forms of an atom that differ in atomic weight but have the same number of protons and electrons, and therefore the same chemistry. May be either stable or radioactive.

JAK–STAT signaling pathway

Signaling pathway activated by cytokines and some hormones, providing a rapid route from the plasma membrane to the nucleus to alter gene transcription. Involves cytoplasmic Janus kinases (JAKs), and signal transducers and activators of transcription (STATs).

joule

Standard unit of energy in the meter-kilogram-second system. One joule is the energy delivered in one second by a 1-watt power source. Approximately equal to 0.24 calories.

K—*see* **equilibrium constant**

*K*ₐ—*see* **affinity constant**

*K*d—*see* **dissociation constant**

K_M
The Michaelis-Menten constant. Equal to the concentration of substrate at which an enzyme works at half its maximum rate. Large values of K_M usually indicate that the enzyme binds to its substrate with relatively low affinity. (Panel 3–3)

karyotype
Display of the full set of chromosomes of a cell, arranged with respect to size, shape, and number.

keratin
Protein that forms keratin intermediate filaments, mainly in epithelial cells. Specialized keratins are found in hair, nails, and feathers.

ketone
Organic molecule containing a carbonyl group linked to two alkyl groups. (Panel 2–1, p 107)

killer cell
Any eucaryotic cell capable of directly killing another eucaryotic cell.

kinase
Enzyme that catalyzes the addition of phosphate groups to molecules. *See also* protein kinase.

kinesin
Member of one of the two main classes of motor proteins that use the energy of ATP hydrolysis to move along microtubules. (Figure 16–58)

kinetochore
Complex structure formed from proteins on a mitotic chromosome to which microtubules attach. Plays an active part in the movement of chromosomes to the poles. Forms on the chromosome centromere. (Figure 17–36)

kinetochore microtubule
In the mitotic or meiotic spindle, a microtubule that connects the spindle pole to the kinetochore of a chromosome.

K⁺ leak channel
K⁺-transporting ion channel in the plasma membrane of animal cells that remains open even in a "resting" cell. (Panel 11–3, p. 679)

knockout
An engineered deletion or inactivating mutation of a gene.

Krebs cycle—*see* **citric acid cycle**

lagging strand
One of the two newly synthesized strands of DNA found at a replication fork. The lagging strand is made in discontinuous lengths that are later joined covalently. (Figure 5–7)

lambda—*see* **bacteriophage lambda**

lamellipodium (plural **lamellipodia**)
Flattened, sheetlike protrusion supported by a meshwork of actin filaments, which is extended at the leading edge of a crawling animal cell. (Figures 16–86 and 16–87)

lamin—*see* **nuclear lamin**

laminin
Extracellular matrix fibrous protein found in basal laminae, where it forms a sheetlike network. (Figures 19–42 and 19–43)

lampbrush chromosome
Huge paired chromosome in meiosis in immature amphibian eggs, in which the chromatin forms large stiff loops extending out from the linear axis of the chromosome. (Figure 4–54)

L chain—*see* **light chain**

leading strand
One of the two newly synthesized strands of DNA found at a replication fork. The leading strand is made by continuous synthesis in the 5′-to-3′ direction. (Figure 5–7)

lecithin—*see* **phosphatidylcholine**

lectin
Protein that binds tightly to a specific sugar. Abundant lectins from plant seeds are used as affinity reagents to purify glycoproteins or to detect them on the surface of cells.

leptin
Peptide hormone secreted by fat cells. Helps regulate the desire to eat by suppressing appetite.

lethal mutation
Mutation that causes the death of the cell or the organism that contains it. (Panel 8–1, pp. 554–555)

leucine-rich repeat protein (LRR protein)
Common type of receptor serine/threonine kinase in plants that contains a tandem array of leucine-rich repeat sequences in its extracellular portion.

leucine zipper
Structural motif seen in many DNA-binding proteins in which two alpha helices from separate proteins are joined together in a coiled-coil (rather like a zipper), forming a protein dimer. (Figure 7–19)

leucocyte—*see* **white blood cell**

leukemia
Cancer of white blood cells.

ligand
Any molecule that binds to a specific site on a protein or other molecule. From Latin *ligare*, to bind.

ligase
Enzyme that joins together (ligates) two molecules in an energy-dependent process. DNA ligase, for example, joins two DNA molecules together end-to-end through phosphodiester bonds.

light chain (L chain)
One of the smaller polypeptides of a multisubunit protein such as myosin (Figure 16–72) or immunoglobulin. (Figure 25–21)

lignin
Network of cross-linked phenolic compounds that forms a supporting network throughout the cell walls of xylem and woody tissue in plants.

limit of resolution
In microscopy, the smallest distance apart at which two point objects can be resolved as separate. Just under 0.2 mm for conventional light microscopy, a limit determined by the wavelength of light.

lipase
Enzyme that catalyzes the cleavage of fatty acids from the glycerol moiety of a triglyceride.

lipid
Organic molecule that is insoluble in water but tends to dissolve in nonpolar organic solvents. A special class, the phospholipids, forms the structural basis of biological membranes. (Panel 2–5, pp. 114–115)

lipid bilayer (phospholipid bilayer)
Thin double sheet of phospholipid molecules that forms the core structure of all cell membranes. The two layers of lipid molecules are packed with their hydrophobic tails pointing inward and their hydrophilic heads outward, exposed to water. (Figure 2–22 and Panel 2–5, pp.114–115)

lipid raft
Small region of the plasma membrane enriched in sphingolipids and cholesterol. (Figure 10–14)

lipophilic—*see* **hydrophobic**

liposome
Artificial phospholipid bilayer vesicle formed from an aqueous suspension of phospholipid molecules. (Figure 10–9)

local mediator
Secreted signal molecule that acts locally on nearby cells. (Figure 15–4)

locus
In genetics, the position on a chromosome. For example, in a diploid cell different alleles of the same gene occupy the same locus.

long-term potentiation
Long-lasting increase (days to weeks) in the sensitivity of certain synapses in the brain, induced by a short burst of repetitive firing in the presynaptic neurons. (Figure 11–42)

loss-of-function mutation
Mutation which reduces or abolishes the activity of a gene. Usually recessive. (Panel 8–1, pp. 554–555)

low-density lipoprotein (LDL)
Large complex composed of a single protein molecule and many esterified cholesterol molecules, together with other lipids. The form in which cholesterol is transported in the blood and taken up into cells. (Figure 13–50)

LTP—*see* **long-term potentiation**

lymph
Colorless fluid containing lymphocytes found in lymphatic vessels. (Figure 25–3)

lymphocyte
White blood cell responsible for the specificity of adaptive immune responses. Two main types: B cells, which produce antibody, and T cells, which interact directly with other effector cells of the immune system and with infected cells. T cells develop in the thymus and are responsible for cell-mediated immunity. B cells develop in the bone marrow in mammals and are responsible for the production of circulating antibodies. (Figure 25–7)

lymphoid organ
Organ involved in the production or function of lymphocytes. Lymphocytes are produced in primary lymphoid organs and respond to antigen in peripheral lymphoid organs. (Figure 25–3)

lymphoma
Cancer of lymphocytes, in which the cancer cells are mainly found in lymphoid organs (rather than in the blood, as in leukemias).

lysis
Rupture of a cell's plasma membrane, leading to the release of cytoplasm and the death of the cell.

lysosome
Membrane-bounded organelle in eucaryotic cells containing digestive enzymes, which are typically most active at the acid pH found in the lumen of lysosomes. (Figures 12–2 and 13–37)

lysozyme
Enzyme that catalyzes the cutting of polysaccharide chains in the cell walls of bacteria.

M—*see* **M phase**

macrophage
Phagocytic cell derived from blood monocytes, resident in most tissues but able to roam. It has both scavenger and antigen-presenting functions in immune responses. (Figure 24–53)

major histocompatibility complex—*see* **MHC complex**

malignant
Of tumors and tumor cells: invasive and/or able to undergo metastasis. A malignant tumor is a cancer. (Figure 20–3)

mannose 6-phosphate (M6P)
Unique marker attached to the oligosaccharides on some glycoproteins destined for lysosomes. (Figure 13–43)

MAP—*see* **microtubule-associated protein**

MAP kinase (mitogen-activated protein kinase)
Protein kinase at the end of a three-component signaling module involved in relaying signals from the plasma membrane to the nucleus.

MAP-kinase module (mitogen-activated protein kinase module)
An intracellular signaling module composed of three protein kinases, acting in sequence, with MAP kinase as the third. Typically activated by a Ras protein in response to extracellular signals. (Figure 15–78)

mass spectrometry (MS)
Technique for identifying compounds on the basis of their precise mass-to-charge ratio. Powerful tool for identifying proteins and sequencing polypeptides. (Figure 8–21)

maternal-effect gene
Gene that acts in the mother to specify maternal mRNAs and proteins in the egg. Maternal-effect mutations affect the development of the embryo even if the embryo itself has not inherited the mutant gene.

mating-type locus (Mat locus)
In budding yeast, the locus that determines the mating type (α or a) of the haploid yeast cell. (Figure 7–65)

matrix
Space or supporting medium within which something is formed. (1) Large internal compartment of the mitochondrion. (2) The corresponding compartment in a chloroplast, more commonly known as the stroma. (3) Extracellular matrix. The extracellular composite of secreted proteins and polysaccharides in which cells are embedded. (Figure 14–37)

matrix metalloprotease
Ca^{2+}- or Zn^{2+}-dependent proteolytic enzyme present in the extracellular matrix that degrades matrix proteins. Includes the collagenases.

M-Cdk (M-phase Cdk)
Cyclin-Cdk complex formed in vertebrate cells by an M-cyclin and the corresponding cyclin-dependent kinase (Cdk). (Figure 17–16 and Table 17–1, p. 1063)

Mcm proteins
Proteins in the eucaryotic cell needed for the initiation of DNA replication; thought to form the helicase present at moving DNA replication forks.

M-cyclin
A cyclin found in all eucaryotic cells that promotes the events of mitosis. (Figure 17–16)

MDR protein—*see* **multidrug resistance protein**

megakaryocyte
Large myeloid cell with a multilobed nucleus that remains in the bone marrow when mature. Buds off platelets from long cytoplasmic processes. (Figures 23–40 and 23–42)

meiosis
Special type of cell division that occurs in sexual reproduction. It involves two successive nuclear divisions with only one round of DNA replication, thereby producing haploid cells from a diploid cell. (Figure 21–5)

melanocyte
Cell that produces the dark pigment melanin. Responsible for the pigmentation of skin and hair. (Figure 23–1)

membrane
The lipid bilayer plus associated proteins that encloses all cells and, in eucaryotic cells, many organelles as well. (Figure 10–1)

membrane-bound ribosome
Ribosome attached to the cytosolic face of the endoplasmic reticulum. The site of synthesis of proteins that enter the endoplasmic reticulum. (Figure 12–41)

membrane channel—*see* **channel**

membrane potential
Voltage difference across a membrane due to a slight excess of positive ions on one side and of negative ions on the other. A typical membrane potential for an animal cell plasma membrane is –60 mV (inside negative relative to the surrounding fluid). (Figure 11–22)

membrane transport
Movement of molecules across a membrane, mediated by a membrane transport protein. (Figures 11–3 and 11–4)

membrane transport protein
Membrane protein that mediates the passage of ions or molecules across a membrane. The two main classes are transporters (also called carriers or permeases) and channels. (Figures 11–3 and 11–4)

meristem
Organized group of dividing cells whose derivatives give rise to the tissues and organs of a flowering plant. Examples are the apical meristems at the tips of shoots and roots. (Panel 22–2, pp. 1404–1405, and Figure 22–118)

mesenchyme
Immature, unspecialized form of connective tissue in animals, consisting of cells embedded in a thin extracellular matrix.

mesoderm
Embryonic tissue that is the precursor to muscle, connective tissue, skeleton, and many of the internal organs. (Figures 22–3 and 22–70)

messenger RNA (mRNA)
RNA molecule that specifies the amino acid sequence of a protein. Produced in eucaryotes by processing of an RNA molecule made by RNA polymerase as a complementary copy of DNA. It is translated into protein in a process catalyzed by ribosomes. (Figures 6–21 and 6–22)

metabolism
The sum total of the chemical processes that take place in living cells. All of catabolism plus anabolism. (Figure 2–36)

metaphase
Stage of mitosis at which chromosomes are firmly attached to the mitotic spindle at its equator but have not yet segregated toward opposite poles. (Panel 17–1, pp. 1072–1073)

metaphase plate
Imaginary plane at right angles to the mitotic spindle and midway between the spindle poles; the plane in which chromosomes are positioned at metaphase. (Panel 17–1, pp. 1072–1073)

metaphase-to-anaphase transition
Checkpoint in the eucaryotic cell cycle preceding sister-chromatid separation at anaphase. If the cell is not ready to proceed to anaphase, the cell cycle is halted at this point. (Figure 17–14, and Panel 17–1, pp. 1072–1073)

metastasis
Spread of cancer cells from their site of origin to other sites in the body. (Figures 20–1 and 20–17)

methyl group (–CH$_3$)
Hydrophobic chemical group derived from methane (CH$_4$).

MHC complex (major histocompatibility complex)
Complex of genes in vertebrates coding for a large family of cell-surface glycoproteins (MHC proteins). (Figure 25–51)

MHC protein
One of a large family of vertebrate cell-surface glycoproteins, which are members of the Ig superfamily. MHC proteins bind peptide fragments of foreign antigens and present them to T cells to induce an adaptive immune response. *See also* class I MHC protein, class II MHC protein. (Figures 25–49 and 25–50)

Michaelis-Menten constant—*see* **K_M**

microarray—*see* **DNA microarray**

microfilament—*see* **actin filament**

micron (μm or micrometer)
Unit of measurement equal to 10^{-6} meter or 10^{-3} millimeter.

micro RNA (miRNA)
Short (21–26 nucleotide) eucaryotic RNAs, produced by the processing of specialized RNA transcripts coded in the genome, that regulate gene expression through complementary base-pairing with mRNA. Depending on the extent of base pairing, mRNAs can lead either to destruction of the mRNA or to a block in its translation. (Figures 7–112)

microsome
Small vesicle derived from endoplasmic reticulum that is produced by fragmentation when cells are homogenized. (Figure 12–37)

microtubule
Long hollow cylindrical structure composed of the protein tubulin. It is one of the three major classes of filaments of the cytoskeleton. (Panel 16–1, p. 968)

microtubule-associated protein (MAP)
Any protein that binds to microtubules and modifies their properties. Many different kinds have been found, including structural proteins, such as MAP2, and motor proteins, such as dynein. [Not to be confused with the 'MAP' (mitogen-activated protein kinase) of 'MAP kinase.']

microtubule flux
Movement of individual tubulin molecules in the microtubules of the spindle towards the poles by loss of tubulin at their minus ends. Helps to generate the poleward movement of sister chromatids after they separate in anaphase. (Figure 17–41)

microtubule-organizing center (MTOC)
Region in a cell, such as a centrosome or a basal body, from which microtubules grow.

microvillus (plural microvilli)
Thin cylindrical membrane-covered projection on the surface of an animal cell containing a core bundle of actin filaments. Present in especially large numbers on the absorptive surface of intestinal epithelial cells. (Figure 16–50)

midbody
Structure formed at the end of cleavage that can persist for some time as a tether between the two daughter cells in animals. (Figure 17–51)

minus end
End of a microtubule or actin filament at which the addition of monomers occurs least readily; the "slow-growing" end of the microtubule or actin filament. The minus end of an actin filament is also known as the pointed end. (Panel 16–2, pp. 978–979)

miRNA—*see* **micro RNA**

mismatch repair
DNA repair process that corrects mismatched nucleotides inserted during DNA replication. A short stretch of newly synthesized DNA including the mismatched nucleotide is removed and replaced with the correct sequence with reference to the template strand. (Figure 5–20)

mitochondrion (plural mitochondria)
Membrane-bounded organelle, about the size of a bacterium, that carries out oxidative phosphorylation and produces most of the ATP in eucaryotic cells. (Figure 1–33)

mitogen

Extracellular signal molecule that stimulates cells to proliferate.

mitogen-activated protein kinase—see **MAP kinase**

mitosis

Division of the nucleus of a eucaryotic cell, involving condensation of the DNA into visible chromosomes, and separation of the duplicated chromosomes to form two identical sets. From Greek *mitos,* a thread, referring to the threadlike appearance of the condensed chromosomes. (Panel 17–1, pp. 1072–1073)

mitotic chromosome

Highly condensed duplicated chromosome with the two new chromosomes still held together at the centromere as sister chromatids.

mitotic spindle

Array of microtubules and associated molecules that forms between the opposite poles of a eucaryotic cell during mitosis and serves to move the duplicated chromosomes apart. (Figure 17–28 and Panel 17–1, pp. 1072–1073)

mobile genetic element

Genetic element (DNA segment) that can move or copy itself to another position in the genome.

model organism

A species that has been studied intensively over a long period and thus serves as a "model" for deriving fundamental biological principles.

molarity (M)

Number of moles of a solute per liter of solution. A 'one molar' (1M) solution contains 1 mole of solute dissolved in 1 liter of solution.

mole

X grams of a substance, where X is its relative molecular mass (molecular weight). A mole consists of 6.02×10^{23} molecules of the substance.

molecular chaperone—see **chaperone**

molecular weight (relative molecular mass)

Mass of a molecule relative to the mass of a hydrogen atom (strictly, relative to 1/12 of the mass of an atom of ^{12}C, i.e., one dalton).

monoclonal antibody

Antibody secreted by a hybridoma cell line. Because the hybridoma is generated by the fusion of a single B cell with a single tumor cell, each hybridoma produces antibodies that are all identical. (Figure 8–8)

monocyte

Type of white blood cell that leaves the bloodstream and matures into a macrophage in tissues. (Figure 23–37)

monomer

Small molecular building block that can serve as a subunit, being linked to others of the same type to form a larger molecule (a polymer).

monomeric GTPase—see **GTPase**

monosaccharide

Simple sugar with the general formula $(CH_2O)n$, where $n = 3$ to 8. (Panel 2–4, pp. 112–113)

morphogen

Signal molecule that can impose a pattern on a field of cells by causing cells in different places to adopt different fates. (Figure 22–13)

mosaic (genetic mosaic)

In developmental biology, an individual organism made of a mixture of cells with different genotypes, but developed from a single zygote. Mosaics can arise naturally, as a result of a mutation in cells that give rise to new tissues, or can be made deliberately to aid genetic analysis. *Compare* chimera.

motif

Element of structure or pattern that recurs in many contexts. Specifically, a small structural domain that can be recognized in a variety of proteins.

motor protein

Protein that uses energy derived from nucleoside triphosphate hydrolysis to propel itself along a linear track (protein filament or other polymeric molecule).

M6P—see **mannose 6-phosphate**

M phase

Period of the eucaryotic cell cycle during which chromosomes are condensed and the nucleus and cytoplasm divide. (Figure 17–4, and Panel 17–1, pp. 1072–1073)

M-phase Cdk—see **M-Cdk**

mRNA—see **messenger RNA**

MTOC—see **microtubule-organizing center**

mTOR (mammalian target of rapamycin)

Mammalian serine/threonine protein kinase that is a downstream component of the PI 3-kinase/Akt signaling pathway. mTOR (the yeast homolog is called TOR) is part of two distinct protein complexes, one of which is sensitive to the drug rapamycin and stimulates cell growth, while the other helps activate the protein kinase Akt. (Figures 15–64 and 15–65)

multidrug resistance protein (MDR protein)

Type of ABC transporter protein that can pump hydrophobic drugs (such as some anti-cancer drugs) out of the cytoplasm of eucaryotic cells.

multimeric

Of proteins: containing more than one subunit.

multipass transmembrane protein

Membrane protein in which the polypeptide chain crosses the lipid bilayer more than once. (Figure 10–19)

muscle cell

Cell type specialized for contraction. The three main classes are skeletal, heart, and smooth muscle cells. (Figure 23–47)

mutation

Heritable change in the nucleotide sequence of a chromosome. (Panel 8–1, pp. 554–555)

Myc

Gene regulatory protein that is activated when a cell is stimulated to grow and divide by extracellular signals. It activates the transcription of many genes, including those that stimulate cell growth. (Figure 17–62)

myelin sheath

Insulating layer of specialized cell membrane wrapped around vertebrate axons. Produced by oligodendrocytes in the central nervous system and by Schwann cells in the peripheral nervous system. (Figure 11–32)

myeloid cell

Any white blood cell other than a lymphocyte. (Figure 23–42)

myoblast

Mononucleated, undifferentiated muscle precursor cell. A skeletal muscle cell is formed by the fusion of multiple myoblasts. (Figure 23–48)

myoepithelial cell

Type of unstriated muscle cell found in epithelia, e.g. in the iris of the eye and in glandular tissue. (Figure 23–47)

myofibril

Long, highly organized bundle of actin, myosin, and other proteins in the cytoplasm of muscle cells that contracts by a sliding filament mechanism.

myosin

Any of a large class of motor proteins that move along actin filaments. (Figure 16–57)

NAD⁺/NADH (nicotinamide adenine dinucleotide/reduced nicotinamide adenine dinucleotide)

Electron carrier system that participates in oxidation-reduction reactions, such as the oxidation of food molecules. NAD^+ accepts the equivalent of a hydride ion (H^-, a proton plus two electrons) to become the activated carrier NADH. The NADH formed donates its high-energy electrons to the ATP-generating process of oxidative phosphorylation. (Figure 2–86)

NADH dehydrogenase complex

First of the three electron-driven proton pumps in the mitochondrial respiratory chain. It accepts electrons from NADH. (Figure 14–26)

NADP⁺/NADPH (nicotinamide adenine dinucleotide phosphate/reduced nicotinamide adenine dinucleotide phosphate)

Electron carrier system closely related to NAD⁺/NADH, but used almost exclusively in reductive biosynthetic, rather than catabolic, pathways. (Figure 2–60)

Na⁺-K⁺ pump (Na⁺-K⁺ ATPase)

Transmembrane carrier protein found in the plasma membrane of most animal cells that pumps Na^+ out of and K^+ into the cell, using energy derived from ATP hydrolysis. (Figure 11–14)

nanometer (nm)

Unit of length commonly used to measure molecules and cell organelles. $1 \text{ nm} = 10^{-3}$ micrometer (μm) $= 10^{-9}$ meter.

natural killer cell (NK cell)

Cytotoxic cell of the innate immune system that can kill virus-infected cells and some cancer cells. (Figure 24–57)

negative feedback

Control mechanism whereby the output of a reaction or pathway inhibits an earlier step in the same pathway.

Nernst equation

Quantitative expression that relates the equilibrium ratio of concentrations of an ion on either side of a permeable membrane to the voltage difference across the membrane. (Panel 11–2, p. 670)

nerve cell—*see* **neuron**

nerve impulse—*see* **action potential**

neural cell adhesion molecule (NCAM)

Cell adhesion molecule of the immunoglobulin superfamily, expressed by many cell types including most nerve cells. A mediator of Ca^{2+}-independent cell-cell attachment in vertebrates. (Figure 19–20)

neural crest

Collection of cells located along the line where the neural tube pinches off from the surrounding epidermis in the vertebrate embryo. Neural crest cells migrate to give rise to a variety of tissues, including neurons and glia of the peripheral nervous system, pigment cells of the skin, and the bones of the face and jaws. (Figures 19–11, 22–84, and 22–97)

neural tube

Tube of ectoderm that will form the brain and spinal cord in a vertebrate embryo. (Figure 22–78)

neurite

Long process growing from a nerve cell in culture. A generic term that does not specify whether the process is an axon or a dendrite. (Figure 16–105)

neuroblast

Embryonic nerve cell precursor. (Figure 22–66)

neurofilament

Type of intermediate filament found in nerve cells. (Figure 16–22)

neuromuscular junction

Specialized chemical synapse between an axon terminal of a motor neuron and a skeletal muscle cell. (Figures 11–36 and 11–39)

neuron (nerve cell)

Impulse-conducting cell of the nervous system, with extensive processes specialized to receive, conduct, and transmit signals. (Figures 22–93 and 22–94)

neuropeptide

Peptide secreted by neurons as a signal molecule either at synapses or elsewhere.

neurotransmitter

Small signal molecule secreted by the presynaptic nerve cell at a chemical synapse to relay the signal to the postsynaptic cell. Examples include acetylcholine, glutamate, GABA, glycine, and many neuropeptides.

neutron

Uncharged heavy subatomic particle that forms part of an atomic nucleus. (Figure 2–1)

neutrophil

White blood cell that is specialized for the uptake of particulate material by phagocytosis. Enters tissues that become infected or inflamed. (Figures 24–55 and 25–24)

NFκB protein

Latent gene regulatory protein that is activated by various intracellular signaling pathways when cells are stimulated during immune, inflammatory, or stress responses. Also has important roles in animal development. (Figure 15–79)

nicotinamide adenine dinucleotide—*see* **NAD⁺/NADH**

nicotinamide adenine dinucleotide phosphate—*see* **NADP⁺/NADPH**

nitric oxide (NO)

Gaseous signal molecule that is widely used in cell–cell communication in both animals and plants. (Figure 15–12)

nitrogen fixation

Biochemical process carried out by certain bacteria that reduces atmospheric nitrogen (N_2) to ammonia, leading eventually to various nitrogen-containing metabolites.

NK cell—*see* **natural killer cell**

N-linked oligosaccharide

Chain of sugars attached to a protein through the NH_2 group of the side chain of an asparagine residue. *Compare* O-linked oligosaccharide. (Figures 12–50 and 12–51)

nm—*see* **nanometer**

NMR (nuclear magnetic resonance, NMR spectroscopy)

NMR is the resonant absorption of electromagnetic radiation at a specific frequency by atomic nuclei in a magnetic field, due to flipping of the orientation of their magnetic dipole moments. The NMR spectrum provides information about the chemical environment of the nuclei. NMR is used widely to determine the three-dimensional structure of small proteins and other small molecules. The principles of NMR are also used for medical diagnostic purposes in magnetic resonance imaging (MRI). (Figure 8–29)

NO—*see* **nitric oxide**

noncovalent bond (noncovalent attraction, noncovalent interaction)

Chemical bond in which, in contrast to a covalent bond, no electrons are shared. Noncovalent bonds are relatively weak, but they can sum together to produce strong, highly specific interactions between molecules. (Panel 2–3, pp. 110–111)

noncyclic photophosphorylation

Photosynthetic process that produces both ATP and NADPH in plants and cyanobacteria. (Figure 14–49)

nondisjunction

Event occurring occasionally during meiosis in which a pair

of homologous chromosomes fails to separate so that the resulting germ cell has either too many or too few chromosomes.

nonenveloped virus
Virus consisting of a nucleic acid core and a protein capsid only. (Figure 24–24C and D)

nonpolar (apolar)
Lacking any asymmetric accumulation of positive and negative charge. Nonpolar molecules are generally insoluble in water. (Panels 2–2 and 2–3, pp. 108–111)

nonretroviral retrotransposon—*see* retrotransposon

nonsense-mediated mRNA decay
Mechanism for degrading aberrant mRNAs containing in-frame internal stop codons before they can be translated into protein. (Figure 6–80)

Northern blotting
Technique in which RNA fragments separated by electrophoresis are immobilized on a paper sheet, and a specific RNA is then detected by hybridization with a labeled nucleic acid probe.

NO synthase (NOS)
Enzyme that synthesizes nitric oxide (NO) by the deamination of arginine. (Figure 15–12B)

Notch
Transmembrane receptor protein (and latent gene regulatory protein) involved in many cell-fate choices in animal development, for example in the specification of nerve cells from ectodermal epithelium. Its ligands are cell-surface proteins such as Delta and Serrate. (Figure 15–76)

notochord
Stiff rod of cells defining the central axis of all chordate embryos. In vertebrates becomes incorporated into the vertebral column. (Figure 22–97)

NPC—*see* nuclear pore complex

NSF (N-ethylmaleimide sensitive factor)
Protein with ATPase activity that disassembles a complex of a v-SNARE and a t-SNARE. (Figures 13–18 and 13–22)

N terminus (amino terminus)
The end of a polypeptide chain that carries a free α-amino group. (Figure 3–1)

nuclear envelope (nuclear membrane)
Double membrane (two bilayers) surrounding the nucleus. Consists of an outer and inner membrane and is perforated by nuclear pores. The outer membrane is continuous with the endoplasmic reticulum. (Figures 4–9 and 12–8)

nuclear export signal
Sorting signal contained in the structure of molecules and complexes, such as RNAs and new ribosomal subunits, that are transported from the nucleus to the cytosol through nuclear pore complexes. (Figure 12–15)

nuclear lamin
Protein subunit of the intermediate filaments that form the nuclear lamina.

nuclear lamina
Fibrous meshwork of proteins on the inner surface of the inner nuclear membrane. It is made up of a network of intermediate filaments formed from nuclear lamins.

nuclear localization signal (NLS)
Signal sequence or signal patch found in proteins destined for the nucleus that enables their selective transport into the nucleus from the cytosol through the nuclear pore complexes. (Figures 12–11 and 12–15)

nuclear magnetic resonance—*see* NMR

nuclear pore complex (NPC)
Large multiprotein structure forming an aqueous channel (the nuclear pore) through the nuclear envelope that allows selected molecules to move between nucleus and cytoplasm. (Figure 12–9)

nuclear receptor superfamily
Intracellular receptors for hydrophobic signal molecules such as steroid and thyroid hormones and retinoic acid. The receptor-ligand complex acts as a transcription factor in the nucleus. (Figure 15–14)

nuclear transplantation
Transfer of a nucleus from one cell to another by microinjection. (Figure 8–6)

nuclear transport receptor (karyopherin)
Protein that escorts macromolecules either into or out of the nucleus: nuclear import receptor or nuclear export receptor. (Figure 12–15)

nuclease
Enzyme that splits nucleic acids by hydrolyzing bonds between nucleotides. *See also* endonuclease and exonuclease.

nucleation
Critical stage in the assembly of a polymeric structure, such as a microtubule, in which a small cluster of monomers aggregates in the correct arrangement to initiate rapid polymerization. (Panel 16–2, pp. 978–979) More generally, the rate-limiting step in an assembly process.

nucleic acid
RNA or DNA, a macromolecule consisting of a chain of nucleotides joined together by phosphodiester bonds.

nucleic acid hybridization—*see* hybridization

nucleocapsid
Viral nucleic acid plus its surrounding protein capsid. If a viral envelope is present, the envelope surrounds the nucleocapsid. (Figure 13–19)

nucleolar organizer
Region of a chromosome containing a cluster of rRNA genes that gives rise to part of a nucleolus. (Figure 6–47)

nucleolus
Structure in the nucleus where rRNA is transcribed and ribosomal subunits are assembled. (Figure 4–9)

nucleoporin
Any of a number of different proteins that make up nuclear pore complexes.

nucleoside
Purine or pyrimidine base covalently linked to a ribose or deoxyribose sugar. (Panel 2–6, pp. 116–117)

nucleosome
Beadlike structure in eukaryotic chromatin, composed of a short length of DNA wrapped around an octameric core of histone proteins. The fundamental structural unit of chromatin. (Figures 4–23 and 4–24)

nucleotide
Nucleoside with one or more phosphate groups joined in ester linkages to the sugar moiety. DNA and RNA are polymers of nucleotides. (Panel 2–6, pp. 116–117)

nucleotide excision repair
Type of DNA repair that corrects damage of the DNA double helix, such as those caused by chemical or UV damage, by cutting out the damaged region on one strand and resynthesizing it using the undamaged strand as template. *Compare* base excision repair. (Figure 5–48)

nucleus
Prominent membrane-bounded organelle in eukaryotic cells, containing DNA organized into chromosomes.

null mutation
Loss-of-function mutation that completely abolishes the activity of a gene. (Panel 8–1, pp. 554–555)

nurse cell
Cell in the invertebrate ovary that is connected by cytoplasmic bridges to a developing oocyte and thereby supplies the oocyte with ribosomes, mRNAs, and proteins needed for the development of the early embryo. (Figure 21–24)

occluding junction
Type of cell junction that seals cells together in an epithelium, forming a barrier through which even small molecules cannot pass—making the cell sheet an impermeable (or selectively permeable) barrier. (Figure 19–2, and Table 19–1, p. 1133)

Okazaki fragments
Short lengths of DNA produced on the lagging strand during DNA replication. Rapidly joined by DNA ligase to form a continuous DNA strand. (Figure 5–7)

olfactory sensory neuron
The sensory cell in the nasal olfactory epithelium responsible for detecting odors.

oligodendrocyte
Glial cell in the vertebrate central nervous system that forms a myelin sheath around axons. *Compare* Schwann cell.

oligomer
Short polymer.

oligosaccharide
Short linear or branched chain of covalently linked sugars. (Panel 2–4, pp. 112–113)

O-linked oligosaccharide
Chain of sugars attached to a protein through the OH group of serine or threonine residues. *Compare* N-linked oligosaccharide. (Figure 13–32)

oncogene
An altered gene whose product can act in a dominant fashion to help make a cell cancerous. Typically, an oncogene is a mutant form of a normal gene (proto-oncogene) involved in the control of cell growth or division. (Figure 20–27)

oocyte
Developing egg, before it has completed meiosis. (Figures 21–25 and 21–26)

oogenesis
Formation and maturation of oocytes in the ovary. (Figure 21–23)

open reading frame (ORF)
A continuous nucleotide sequence free from stop codons in at least one of the three reading frames (and thus with the potential to code for protein).

operator
Short region of DNA in a bacterial chromosome that controls the transcription of an adjacent gene. (Figure 7–34)

operon
In a bacterial chromosome, a group of contiguous genes that are transcribed into a single mRNA molecule. (Figure 7–34)

ORC—see **origin recognition complex**

ORF—see **open reading frame**

organelle
Subcellular compartment or large macromolecular complex, often membrane-enclosed, that has a distinct structure, composition, and function. Examples are nucleus, nucleolus, mitochondrion, Golgi apparatus, and centrosomes. (Figure 1–30)

Organizer (Spemann's Organizer)
Specialized tissue at the dorsal lip of the blastopore in an amphibian embryo; a source of signals that help to orchestrate formation of the embryonic body axis. (Named after H. Spemann and H. Mangold, co-discoverers) (Figure 22–74)

origin of replication
Site in a chromosome where DNA replication starts.

origin recognition complex (ORC)
Large protein complex that is bound to the DNA at origins of replication in eucaryotic chromosomes throughout the cell cycle. (Figure 5–36)

orthologs
Genes or proteins from different species that are similar in sequence because they are descendants of the same gene in the last common ancestor of those species. *Compare* paralogs. (Figure 1–25)

osmosis
Net movement of water molecules across a semipermeable membrane driven by a difference in concentration of solute on either side. The membrane must be permeable to water but not to the solute molecules. (Panel 11–1, p. 664)

osteoblast
Cell that secretes matrix of bone. (Figure 23–55)

osteoclast
Macrophage-like cell that erodes bone, enabling it to be remodeled during growth and in response to stresses throughout life. (Figure 23–59)

osteocyte
Nondividing cell in bone that develops from an osteoblast and is embedded in bone matrix. (Figure 23–55)

ovulation
Release of an egg from the ovary. (Figure 21–26)

ovum
Mature egg.

oxidase
Enzyme that catalyzes an oxidation reaction, especially one in which molecular oxygen is the electron acceptor.

oxidation (verb oxidize)
Loss of electrons from an atom, as occurs during the addition of oxygen to a molecule or when a hydrogen is removed. Opposite of reduction. (Figure 2–43)

oxidative phosphorylation
Process in bacteria and mitochondria in which ATP formation is driven by the transfer of electrons through the electron transport chain to molecular oxygen. Involves the intermediate generation of a proton gradient (pH gradient) across a membrane and a chemiosmotic coupling of that gradient to the ATP synthase. (Figures 14–10 and 14–14)

p53
Tumor suppressor gene found mutated in about half of human cancers. Encodes a gene regulatory protein that is activated by damage to DNA and is involved in blocking further progression through the cell cycle. (Figures 20–37 and 20–40)

pairing (homolog pairing)
In meiosis, the lining up of the two homologous chromosomes along their length. (Figure 21–6)

pair-rule gene
In *Drosophila* development, a gene expressed in a series of regular transverse stripes along the body of the embryo and which helps to determine its different segments. (Figure 22–37)

palindromic sequence
Nucleotide sequence that is identical to its complementary strand when each is read in the same chemical direction—e.g., GATC. (Figure 8–31)

Par3, Par6
Scaffold proteins involved in the specification of polarity in individual animal cells; Par3 and Par6 form a complex with atypical protein kinase C (aPKC). (Figure 19–31)

paracrine signaling
Short-range cell–cell communication via secreted signal molecules that act on neighboring cells. (Figure 15–4)

paralogs
Genes or proteins that are similar in sequence because they are the result of a gene duplication event occurring in an ancestral organism. *Compare* orthologs. (Figure 1–25)

parthenogenesis
Production of a new individual from an egg cell in the absence of fertilization by a sperm.

passive transport (facilitated diffusion)
Transport of a solute across a membrane down its concentration gradient or its electrochemical gradient, using only the energy stored in the gradient. (Figure 11–4)

patch-clamp recording
Electrophysiological technique in which a tiny electrode tip is sealed onto a patch of cell membrane, thereby making it possible to record the flow of current through individual ion channels in the patch. (Figure 11–33)

pathogen (adjective **pathogenic**)
An organism, cell, virus, or prion that causes disease.

pattern recognition receptor
Receptor present on or in cells of the innate immune system that recognizes and responds to pathogen-associated molecular patterns (PAMPs)—such as surface carbohydrates on bacteria and viruses and unmethylated GC sequences in bacterial DNA.

PCR (polymerase chain reaction)
Technique for amplifying specific regions of DNA by the use of sequence-specific primers and multiple cycles of DNA synthesis, each cycle being followed by a brief heat treatment to separate complementary strands. (Figure 8–45)

PDZ domain
Protein-binding domain present in many scaffold proteins, and often used as a docking site for intracellular tails of transmembrane proteins. (Figure 19–21)

pectin
Mixture of polysaccharides rich in galacturonic acid which forms a highly hydrated matrix in which cellulose is embedded in plant cell walls. (Figure 19–79)

pentose
Five-carbon sugar.

peptide
Short polymer of amino acids.

peptide bond
Chemical bond between the carbonyl group of one amino acid and the amino group of a second amino acid—a special form of amide linkage. Peptide bonds link amino acids together in proteins. (Panel 3–1, pp. 128–129)

peripheral lymphoid organ (secondary lymphoid organ)
Lymphoid organ in which T cells and B cells interact with foreign antigens. Examples are spleen, lymph nodes, and mucosal-associated lymphoid tissue. (Figure 25–3)

peripheral membrane protein
Protein that is attached to one face of a membrane only by noncovalent interactions with other membrane proteins, and which can be removed by relatively gentle treatments that leave the lipid bilayer intact. (Figure 10–19)

permease—*see* **transporter**

permissive (nonrestrictive) conditions
Circumstances (such as temperature or nutrient availability) in which the phenotypic effect of a conditional mutation will be absent: that is, the phenotype will be normal. (Figure 8–55 and Panel 8–1, pp. 554–555)

peroxisome
Small membrane-bounded organelle that uses molecular oxygen to oxidize organic molecules. Contains some enzymes that produce and others that degrade hydrogen peroxide (H_2O_2). (Figure 12–30)

pH
Common measure of the acidity of a solution: "p" refers to power of 10, "H" to hydrogen. Defined as the negative logarithm of the hydrogen ion concentration in moles per liter (M). $pH = -\log [H^+]$. Thus a solution of pH 3 will contain 10^{-3} M hydrogen ions. pH less than 7 is acidic and pH greater than 7 is alkaline.

phage—*see* **bacteriophage**

phagocyte
General term for a professional phagocytic cell—that is, a cell such as a macrophage or neutrophil that is specialized to take up particles and microorganisms by phagocytosis. (Figures 13–46 and 13–47)

phagocytosis
Process by which unwanted cells, debris, and other bulky particulate material is endocytosed ("eaten") by a cell. Prominent in carnivorous cells, such as *Amoeba proteus*, and in vertebrate macrophages and neutrophils. From Greek *phagein*, to eat. (Figure 24–53)

phagosome
Large intracellular membrane-bounded vesicle that is formed as a result of phagocytosis. Contains ingested extracellular material. (Figure 24–30)

phase-contrast microscope
Type of light microscope that exploits the interference effects that occur when light passes through material of different refractive indexes. Used to view living cells. (Figures 9–7 and 22–101)

PH domain—*see* **pleckstrin homology domain**

phenotype
The observable character (including both physical appearance and behavior) of a cell or organism. (Panel 8–1, pp. 554–555)

phosphatase
Enzyme that catalyzes the hydrolytic removal of phosphate groups from a molecule.

phosphatidylcholine (lecithin)
Common phospholipid present in abundance in most biological mebranes. (Figure 10–3)

phosphatidylinositol
An inositol phospholipid. (Figure 15–37)

phosphatidylinositol 4,5-bisphosphate ($PI(4,5)P_2$, PIP_2)
Membrane inositol phospholipid (a phosphoinositide) that is cleaved by phospholipase C into IP_3 and diacylglycerol at the beginning of the inositol phospholipid signaling pathway. It can also be phosphorylated by PI 3-kinase to produce PIP_3 docking sites for signaling poteins in the PI 3-kinase/Akt signaling pathway. (Figures 15–38 and 15–64)

phosphoanhydride bond
High-energy bond linking phosphate groups in, for instance, ATP and GTP. (Panel 2–6, pp. 116–117)

phosphodiester bond
A covalent chemical bond formed when two hydroxyl groups form ester linkages to the same phosphate group, such as between adjacent nucleotides in RNA or DNA. (Figure 2–28)

phosphoglyceride
Phospholipid derived from glycerol, abundant in biomembranes. (Figures 10–2 and 10–3)

phosphoinositide—*see* **inositol phospholipid**

phosphoinositide 3-kinase (PI 3-kinase)
Membrane-bound enzyme that is a component of the PI 3-kinase/Akt intracellular signaling pathway. It phosphorylates phosphatidylinositol 4,5-bisphosphate at the 3 position on the inositol ring to produce PIP_3 docking sites in the membrane for other intracellular signaling proteins. (Figure 15–64)

phospholipase C (PLC)
Membrane-bound enzyme that cleaves inositol phospholipids to produce IP_3 and diacylglycerol in the inositol phospholipid signaling pathway. PLCβ is activated by GPCRs via specific G proteins, while PLCγ is activated by RTKs. (Figures 25–39 and 15–55)

phospholipid
The main category of lipids used to construct biomembranes. Generally composed of two fatty acids linked through glycerol (or sphingosine) phosphate to one of a variety of polar groups. (FIgure 10–3, and Panel 2–5, pp. 114–115)

phosphorylation
Reaction in which a phosphate group is covalently coupled to another molecule.

phosphorylation cascade
Series of sequential protein phosphorylations mediated by a series of protein kinases, each of which phosphorylates and activates the next kinase in the chain. Such cascades are common in intracellular signaling pathways. (Figure 15–60)

photoactivation
Technique for studying intracellular processes in which an inactive form of a molecule of interest is introduced into the cell, and is then activated by a focused beam of light at a precise spot in the cell. (Figure 9–30)

photochemical reaction center
The part of a photosystem that converts light energy into chemical energy in photosynthesis. (Figure 14–43)

photoreceptor
Cell or molecule that is sensitive to light.

photorespiration
Wasteful metabolic process that occurs in plants in conditions of low CO_2 in which O_2 is used up and CO_2 liberated without the production of carbohydrate for storage.

photosynthesis
Process by which plants, algae and some bacteria use the energy of sunlight to drive the synthesis of organic molecules from carbon dioxide and water. (Figures 2–40 and 14–38)

photosynthetic electron transfer
Light-driven reactions in photosynthesis in which electrons move along an electron-transport chain in a membrane, generating ATP and NADPH. (Figure 14–38)

photosystem
Multiprotein complex involved in photosynthesis that captures the energy of sunlight and converts it to useful forms of energy. (Figure 14–43)

phragmoplast
Structure made of microtubules and actin filaments that forms in the prospective plane of division of a plant cell and guides formation of the cell plate. (Figure 17–57)

phylogeny
Evolutionary history of an organism or group of organisms, often presented in chart form as a phylogenetic tree. (Figures 4–75 and 14–72)

pI—*see* **isoelectric point**

PI 3-kinase—*see* **phosphoinositide 3-kinase**

pinocytosis
Literally, "cell drinking." Type of endocytosis in which soluble materials are continually taken up from the environment in small vesicles and moved into endosomes along with the membrane-bound molecules. *Compare* phagocytosis. (Figure 13–48)

PKA—*see* **cyclic-AMP-dependent protein kinase**

PKB—*see* **Akt**

PKC—*see* **protein kinase C**

planar cell polarity
Type of cellular asymmetry seen in some epithelia, such that each cell has a polarity vector oriented in the plane of the epithelium. (Figure 19–32)

plant growth regulator (plant hormone)
Signal molecule that helps coordinate growth and development. Examples are ethylene, auxins, gibberellins, cytokinins, abscisic acid, and the brassinosteroids.

plasma membrane
The membrane that surrounds a living cell. (Figure 10–1)

plasmid
Small circular extrachromosomal DNA molecule that replicates independently of the genome. Modified plasmids are used extensively as vectors for DNA cloning.

plasmodesma (plural plasmodesmata)
Plant equivalent of a gap junction. Communicating cell–cell junction in plants in which a channel of cytoplasm lined by plasma membrane connects two adjacent cells through a small pore in their cell walls.

plastid
Cytoplasmic organelle in plants, bounded by a double membrane, that carries its own DNA and is often pigmented. Chloroplasts are plastids.

platelet
Cell fragment, lacking a nucleus, that breaks off from a megakaryocyte in the bone marrow and is found in large numbers in the bloodstream. Helps initiate blood clotting when blood vessels are injured.

PLC—*see* **phospholipase C**

pleckstrin homology domain (PH domain)
Protein domain found in some intracellular signaling proteins. Some PH domains in intracellular signaling proteins bind to phosphatidylinositol 3,4,5-trisphosphate produced by PI 3-kinase, bringing the signaling protein to the plasma membrane when PI 3-kinase is activated.

ploidy
The number of complete homologous sets of chromosomes in a genome. Diploid organisms have two sets in their somatic cells; haploid organisms have a single set; polyploid organisms have more than two.

plus end
End of a microtubule or actin filament at which addition of monomers occurs most readily; the "fast-growing" end of a microtubule or actin filament. The plus end of an actin filament is also known as the barbed end. (Panel 16–2, pp. 978–979)

point mutation
Change of a single nucleotide pair, or a very small part of a single gene, in DNA. (Panel 8–1, pp. 554–555)

polar
In the electrical sense, describes a structure (for example, a chemical bond, chemical group, or molecule) with positive charge concentrated toward one end and negative charge toward the other as a result of an uneven distribution of electrons. Polar molecules are likely to be soluble in water. (Figure 2–10 and Panel 2–2, pp. 108–109)

polar body
Smaller of two daughter cells produced during meiosis by asymmetric division of a primary or secondary oocyte, the other (large) daughter being the oocyte or ovum itself. The polar bodies eventually degenerate. (Figure 21–23)

polyadenylation
Addition of a long sequence of A nucleotides (the poly-A tail) to the 3′ end of a nascent mRNA molecule. (Figures 6–21 and 6–38)

poly-A tail—*see* polyadenylation

polycistronic mRNA
Individual mRNA that encodes several different proteins—commonly found in bacteria but not in eucaryotes. (Figure 6–73)

polygenic trait
Heritable characteristic that is influenced by multiple genes, each of which makes a small contribution to the phenotype.

polymer
Large molecule made by covalently linking multiple identical or similar units (monomers) together.

polymerase
Enzyme that catalyzes polymerization reactions such as the synthesis of DNA and RNA. *See also* DNA polymerase, RNA polymerase.

polymerase chain reaction—*see* PCR

polymorphic
Describes a gene with two or more alleles that coexist at high frequency in a population.

polypeptide
Linear polymer of amino acids. Proteins are large polypeptides, and the two terms can be used interchangeably. (Panel 3–1, pp. 128–129)

polyploid
Cell or organism that contains more than two sets of homologous chromosomes.

polyribosome (polysome)
Messenger RNA molecule to which are attached a number of ribosomes engaged in protein synthesis. (Figure 6–76)

polysaccharide
Linear or branched polymer of monosaccharides. Examples include glycogen, starch, hyaluronic acid, and cellulose. (Panel 2–4, p. 113)

polytene chromosome
Giant chromosome in which the DNA has undergone repeated replication and the many copies have stayed together. (Figures 4–58 and 4–59)

porin
Channel-forming proteins of the outer membranes of bacteria, mitochondria, and chloroplasts.

position effect
Difference in gene expression that depends on the position of the gene on the chromosome and probably reflects differences in the state of the chromatin along the chromosome. When an active gene is placed next to heterochromatin, the inactivating influence of the heterochromatin can spread to affect the gene to a variable degree, giving rise to *position effect variegation*. (Figure 4–36)

positional information
Information supplied to or possessed by cells according to their position in a multicellular organism. A cell's internal record of its positional information is called its positional value.

positive feedback
Control mechanism whereby the end product of a reaction or pathway stimulates its own production or activation.

post-transcriptional control
Any control on gene expression that is exerted at a stage after transcription has begun. (Figure 7–92)

post-translational modification
An enzyme-catalyzed change to a protein made after it is

synthesized. Examples are acetylation, cleavage, glycosylation, methylation, phosphorylation, and prenylation.

pre-B cell
Immediate precursor of a B cell. (Figure 25–22)

preinitiation complex
A multiprotein complex that is assembled on the origin of replication at the onset of the S phase of the eucaryotic cell cycle. Initiates DNA synthesis by unwinding the DNA helix and loading DNA polymerases and other replication enzymes onto the DNA strands. (Figure 17–23)

pre-mRNA
Precursor to messenger RNA. In eucaryotes, includes all intermediate stages of RNA processing.

prenylation
Covalent attachment of an isoprenoid lipid group to a protein. (Figure 10–20)

preprophase band
Circumferential band of microtubules and actin filaments that forms around a plant cell under the plasma membrane prior to mitosis and cell division. (Figure 17–57)

prereplicative complex (pre-RC)
Multiprotein complex that is assembled at origins of replication during late mitosis and early G_1 phases of the cell cycle; a prerequisite to license the assembly of a preinitiation complex, and the subsequent initiation of DNA replication. (Figures 17–22 and 17–23)

primary cell wall
The first cell wall produced by a developing plant cell; it is thin and flexible, allowing room for cell growth. (Figure 19–79)

primary cilium
Short, single, nonmotile, cilium lacking dynein that arises from a centriole and projects from the surface of many animal cell types. Some signaling proteins are concentrated in the primary cilium. (Figure 15–48)

primary immune response
Adaptive immune response to an antigen that is made on first encounter with that antigen. (Figure 25–10)

primary structure
Linear sequence of monomer units in a polymer, such as the amino acid sequence of a protein.

primary transcript (primary RNA transcript)
Freshly synthesized transcript, before it has undergone splicing or other modifications. (Figure 6–21)

primary tumor
Tumor at the original site at which a cancer first arose. Secondary tumors develop elsewhere by metastasis.

primase—*see* DNA primase

primer
Oligonucleotide that pairs with a template DNA or RNA strand and promotes the synthesis of a new complementary strand by a polymerase.

primordial germ cell
Cell in an embryo that is a precursor to cells that give rise to gametes. (Figures 21–17, 21–23, and 21–30)

prion
An infectious, abnormally folded form of a protein that is replicated in the host by inducing normal proteins of the same type to adopt the aberrant structure. (Figures 6–95 and 6–96)

prion disease
Transmissible spongiform encephalopathy—such as Kuru and Creutzfeld-Jakob disease (CJD) in humans, scrapie in sheep, and bovine spongiform encephalopathy (BSE, or 'mad cow disease') in cows—that is caused and transmitted by an infectious, abnormally folded protein (prion). (Figure 24–18)

probe
Defined fragment of RNA or DNA, radioactively or chemically labeled, used to locate specific nucleic acid sequences by hybridization.

procaryote (prokaryote)
Single-celled microorganism whose cells lack a well-defined, membrane-enclosed nucleus. Either a bacterium or an archaeon. (Figure 1–21)

procaspase
Inactive precursor of a caspase, a proteolytic enzyme usually involved in apoptosis. (Figure 18–5)

processive
Of an enzyme: able to proceed along a polymer chain catalyzing the same reaction repeatedly without detaching from the chain.

programmed cell death
A form of cell death in which a cell kills itself by activating an intracellular death program.

prometaphase
Phase of mitosis preceding metaphase in which the nuclear envelope breaks down and chromosomes first attach to the spindle. (Panel 17–1, pp. 1072–1073)

promoter
Nucleotide sequence in DNA to which RNA polymerase binds to begin transcription. *See also* inducible promoter. (Figure 7–44)

proneural gene
Gene whose expression defines cells with the potential to develop as neural tissue.

proofreading
Process by which potential errors in DNA replication, transcription, and translation are detected and corrected.

prophase
First stage of mitosis, during which the chromosomes are condensed but not yet attached to a mitotic spindle. (Panel 17–1, pp. 1072–1073)

protease (proteinase, proteolytic enzyme)
Enzyme that degrades proteins by hydrolyzing some of the peptide bonds between amino acids.

proteasome
Large protein complex in the cytosol with proteolytic activity that is responsible for degrading proteins that have been marked for destruction by ubiquitylation or by some other means. (Figures 6–89 and 6–90)

protein
The major macromolecular constituent of cells. A linear polymer of amino acids linked together by peptide bonds in a specific sequence. (Figure 3–1)

protein activity control
The selective activation, inactivation, degradation, or compartmentalization of specific proteins after they have been made. One of the means by which a cell controls which proteins are active at a given time or location in the cell.

protein domain—*see* **domain**

protein kinase
Enzyme that transfers the terminal phosphate group of ATP to one or more specific amino acids (serine, threonine, or tyrosine) of a target protein.

protein kinase A (PKA)—*see* **cyclic-AMP-dependent protein kinase**

protein kinase B—*see* **Akt**

protein kinase C (PKC)
Ca^{2+}-dependent protein kinase that, when activated by diacylglycerol and an increase in the concentration of cytosolic Ca^{2+}, phosphorylates target proteins on specific serine and threonine residues. (Figure 15–39)

protein subunit
An individual protein chain in a protein composed of more than one chain.

protein translocator
Membrane-bound protein that mediates the transport of another protein across a membrane. (Figure 12–23)

protein tyrosine phosphatase
Enzyme that removes phosphate groups from phosphorylated tyrosine residues on proteins. (Figure 25–71)

proteoglycan
Molecule consisting of one or more glycosaminoglycan chains attached to a core protein. (Figure 19–58)

proteolysis
Degradation of a protein by hydrolysis at one or more of its peptide bonds.

proteolytic enzyme—*see* **protease**

proteomics
Study of all the proteins, including all the covalently modified forms of each, produced by a cell, tissue, or organism. Proteomics often investigates changes in this larger set of proteins in 'the proteome'—caused by changes in the environment or by extracellular signals.

protist
Single-celled eucaryote. Includes protozoa, algae, yeasts. (Figure 1–41)

proton
Positively charged subatomic particle that forms part of an atomic nucleus. Hydrogen has a nucleus composed of a single proton (H^+). (Figure 2–1)

proton-motive force
The force exerted by the electrochemical proton gradient that moves protons across a membrane. (Figure 14–13)

proto-oncogene
Normal gene, usually concerned with the regulation of cell proliferation, that can be converted into a cancer-promoting oncogene by mutation. (Figure 20–34)

protozoa
Free-living or parasitic, nonphotosynthetic, single-celled, motile eucaryotic organisms, such as *Paramecium* and *Amoeba*. Free-living protozoa feed on bacteria or other microorganisms. (Figure 1–41)

pseudogene
Nucleotide sequence of DNA that has accumulated multiple mutations that have rendered an ancestral gene inactive and nonfunctional.

pseudopodium (plural pseudopodia)
Large, thick cell-surface protrusion formed by amoeboid cells as they crawl. More generally, any similarly shaped dynamic actin-rich extension of the surface of an animal cell. *Compare* filopodium, lamellipodium. (Figure 16–94)

pump
Transmembrane protein that drives the active transport of ions or small molecules across the lipid bilayer.

purifying selection
Natural selection operating to retard divergence in gene sequences within a population in the course of evolution by eliminating individuals carrying deleterious mutations.

purine
Nitrogen-containing ring compound found in DNA and RNA: adenine or guanine. (Panel 2–6, pp. 116–117)

pyrimidine
Nitrogen-containing ring compound found in DNA and

RNA: cytosine, thymine, or uracil. (Panel 2–6, pp. 116–117)

pyruvate (CH₃COCOO⁻)
End-product of the glycolytic pathway. Enters mitochondria and feeds into the citric acid cycle and other biosynthetic pathways.

quaternary structure
Three-dimensional relationship of the different polypeptide chains in a multisubunit protein or protein complex.

quinone (Q)
Small, lipid-soluble mobile electron carrier molecule found in the respiratory and photosynthetic electron-transport chains. (Figure 14–24)

Rab (Rab protein)
Monomeric GTPase in the Ras superfamily present in the plasma membrane and organelle membranes. Involved in conferring specificity on vesicle docking. (Table 15–5, p. 926)

Ran (Ran protein)
Monomeric GTPase in the Ras superfamily present in both cytosol and nucleus. Required for the active transport of macromolecules into and out of the nucleus through nuclear pore complexes. (Table 15–5, p. 926)

Ras (Ras protein)
Monomeric GTPase of the Ras superfamily that helps to relay signals from cell-surface RTK receptors to the nucleus, frequently in response to signals that stimulate cell division. Named for the *ras* gene, first identified in viruses that cause rat sarcomas. (Figure 3–72)

Ras superfamily
Large superfamily of monomeric GTPases (also called small GTP-binding proteins) of which Ras is the prototypical member. (Table 15–5, p. 926)

Rb—*see* **retinoblastoma protein**

reading frame
Phase in which nucleotides are read in sets of three to encode a protein. A mRNA molecule can be read in any one of three reading frames, only one of which will give the required protein. (Figure 6–51)

RecA (RecA protein)
Prototype for a class of DNA-binding proteins that catalyze synapsis of DNA strands during genetic recombination. (Figure 5–56)

receptor
Any protein that binds a specific signal molecule (ligand) and initiates a response in the cell. Some are on the cell surface, while others are inside the cell. (Figure 15–3)

receptor-mediated endocytosis
Internalization of receptor-ligand complexes from the plasma membrane by endocytosis. (Figure 13–53)

receptor serine/threonine kinase
Cell-surface receptor with an extracellular ligand-binding domain and an intracellular kinase domain that phosphorylates signaling proteins on serine or threonine residues in response to ligand binding. The TGFβ receptor is an example. (Figure 15–69)

receptor tyrosine kinase (RTK)
Cell-surface receptor with an extracellular ligand-binding domain and an intracellular kinase domain that phosphorylates signaling proteins on tyrosine residues in response to ligand binding. (Figure 15–52 and Table 15–4, p. 923)

recessive
In genetics, the member of a pair of alleles that fails to be expressed in the phenotype of the organism when the dominant allele is present. Also refers to the phenotype of an individual that has only the recessive allele. (Panel 8–1, pp. 554–555)

recombinant DNA
Any DNA molecule formed by joining DNA segments from different sources.

recombinant DNA technology—*see* **genetic engineering**

recombination (genetic recombination)
Process in which DNA molecules are broken and the fragments are rejoined in new combinations. Can occur naturally in the living cell—for example, through crossing-over during meiosis—or *in vitro* using purified DNA and enzymes that break and ligate DNA strands. Three broad classes are homologous (general), conservative site-specific, and transpositional recombination.

recombination complex
In meiosis, a protein complex that assembles at a DNA double-strand break and helps mediate homologous recombination.

recycling endosome
Large intracellular membrane-bounded vesicle formed from a fragment of an endosome; an intermediate stage on the passage of recycled receptors back to the cell membrane. (Figure 13–60)

red blood cell—*see* **erythrocyte**

redox pair
Pair of molecules in which one acts as an electron donor and one as an electron acceptor in an oxidation–reduction reaction: for example, NADH (electron donor) and NAD⁺ (electron acceptor). (Panel 14–1, p. 830)

redox potential
The affinity of a redox pair for electrons, generally measured as the voltage difference between an equimolar mixture of the pair and a standard reference. NADH/NAD⁺ has a low redox potential and O_2/H_2 has a high redox potential (high affinity for electrons). (Panel 14–1, p. 830)

redox reaction
Reaction in which one component becomes oxidized and the other reduced; an oxidation–reduction reaction. (Panel 14–1, p. 830)

reduction (verb reduce)
Addition of electrons to an atom, as occurs during the addition of hydrogen to a biological molecule or the removal of oxygen from it. Opposite of oxidation. (Figure 2–43)

regulative
Of embryos or parts of embryos: self-adjusting, so that a normal structure emerges even if the starting conditions are perturbed.

regulator of G protein signaling (RGS)
A GAP protein that binds to a trimeric G protein and enhances its GTPase activity, thus helping to limit G-protein-mediated signaling. (Figure 15–19)

regulatory sequence
DNA sequence to which a gene regulatory protein binds to control the rate of assembly of the transcriptional complex at the promoter. (Figure 7–44)

release factor
Protein that enables release of a newly synthesized protein from the ribosome by binding to the ribosome in the place of tRNA (whose structure it mimics). (Figure 16–74)

replication—*see* **DNA replication**

replication fork
Y-shaped region of a replicating DNA molecule at which the two strands of the DNA are being separated and the daughter strands are being formed. (Figures 5–7 and 5–19)

replication origin
Location on a DNA molecule at which duplication of the DNA begins. (Figures 4–21 and 5–25)

replicative cell senescence—*see also* **senescence**

Phenomenon observed in primary cell cultures in which cell proliferation slows down and finally irreversibly halts.

reporter gene

Genetic construct, usually artificial, in which a copy of the regulatory DNA of a gene of interest is linked to a sequence coding for an easily-detectable product. The presence or absence of this product (the 'reporter protein') in a cell containing the construct indicates whether the gene of interest is active or inactive. (Figure 8–70)

repressor (gene repressor protein, transcriptional repressor)

Protein that binds to a specific region of DNA to prevent transcription of an adjacent gene.

respiration

General term for an energy-generating process in cells that involves the oxidative breakdown of sugars or other organic molecules and requires the uptake of O_2 while producing CO_2 and H_2O as waste products. (Figure 2–41)

respiratory chain—*see* **electron transport chain**

respiratory enzyme complex

Any of the major protein complexes of the mitochondrial respiratory chain that act as electron-driven proton pumps to generate the proton gradient across the inner membrane. (Figures 14–14 and 14–26)

resting membrane potential

Membrane potential in equilibrium conditions in which there is no net flow of ions across the plasma membrane. *See also* membrane potential.

restriction fragment

Fragment of DNA generated by the action of restriction enzyme(s).

restriction map

Diagrammatic representation of a DNA molecule indicating the sites of cleavage by various restriction enzymes.

restriction nuclease (restriction enzyme)

One of a large number of nucleases that can cleave a DNA molecule at any site where a specific short sequence of nucleotides occurs. Extensively used in recombinant DNA technology. (Figures 8–31 and 8–32)

restriction point—*see* **Start**

restrictive (nonpermissive) conditions

Circumstances (such as temperature or nutrient availability) in which the phenotypic effect of a conditional mutation will be evident. (Figure 8–55, and Panel 8–1, pp. 554–555)

retinoblastoma protein (Rb)

Tumor suppressor protein involved in the regulation of cell division. Mutated in the cancer retinoblastoma, as well as in many other tumors. Its normal activity is to regulate the eucaryotic cell cycle by binding to and inhibiting the E2F proteins, thus blocking progression to DNA replication and cell division. (Figure 17–62)

retrotransposon

Type of transposable element that moves by being first transcribed into an RNA copy that is then reconverted to DNA by reverse transcriptase and inserted ('retro-transposed') elsewhere in the genome. There are two types: retroviral-like retrotransposons and nonretroviral retrotransposons. (Table 5–3, p. 318)

retrovirus

RNA-containing virus that replicates in a cell by first making an RNA-DNA intermediate and then a double-stranded DNA molecule that becomes integrated into the cell's DNA. (Figure 5–71)

reverse genetics

Approach to discovering gene function that starts from the DNA (gene) and its protein product and then creates mutants to analyze the gene's function.

reverse transcriptase

Enzyme first discovered in retroviruses that makes a double-stranded DNA copy from a single-stranded RNA template molecule.

reverse transcription

Transcription from RNA to DNA. This is in the opposite direction to that prescribed by central dogma, which holds that DNA is transcribed into RNA and RNA is translated into protein.

RGD sequence

Tripeptide sequence of arginine-glycine-aspartic acid that forms a binding site for integrins; present in fibronectin and some other extracellular proteins. (Figure 19–72C)

RGS—*see* **regulator of G protein signaling**

Rho (Rho protein family)

Family of monomeric GTPases within the Ras superfamily involved in signaling the rearrangement of the cytoskeleton. Includes Rho, Rac, and Cdc42. (Table 15–5, p. 926)

rhodopsin

Seven-span membrane protein of the GPCR family that acts as a light sensor in rod photoreceptor cells in the vertebrate retina. Contains the light-sensitive prosthetic group retinol. (Figure 15–49)

ribonuclease

Enzyme that cuts an RNA molecule by hydrolyzing one or more of its phosphodiester bonds.

ribonucleic acid—*see* **RNA**

ribose

The five-carbon monosaccharide component of RNA. $C_5H_{10}O_5$. *Compare* deoxyribose.

ribosomal RNA (rRNA)

Any one of a number of specific RNA molecules that form part of the structure of a ribosome and participate in the synthesis of proteins. Often distinguished by their sedimentation coefficient (e.g., 28S rRNA, 5S rRNA).

ribosome

Particle composed of rRNAs and ribosomal proteins that catalyzes the synthesis of protein using information provided by mRNA. (Figure 1–10)

ribozyme

RNA with catalytic activity.

RNA (ribonucleic acid)

Polymer formed from covalently linked ribonucleotide monomers. *See also* messenger RNA, ribosomal RNA, transfer RNA. (Table 6–1, p. 336, and Panel 2–6, pp. 116–117)

RNA editing

Type of RNA processing that alters the nucleotide sequence of a pre-mRNA transcript after it is synthesized by inserting, deleting, or altering individual nucleotides.

RNA interference (RNAi)

As originally described, mechanism by which an experimentally introduced double-stranded RNA induces sequence-specific destruction of complementary mRNAs. The mechanism, which is highly conserved in eucaryotes, proceeds through short double-stranded small interfering RNAs (siRNAs) produced by endonucleolytic cleavage. The term RNAi is often used broadly to also include the inhibition of gene expression by microRNAs (miRNAs), which are encoded in the cells own genome. RNA interference is widely used experimentally to study the effects of inactivating specific genes. (Figure 7–115)

RNA polymerase

Enzyme that catalyzes the synthesis of an RNA molecule on a DNA template from ribonucleoside triphosphate precursors. (Figure 6–8)

RNA primer

Short stretch of RNA synthesized on a DNA template. It is

required by DNA polymerases to start their DNA synthesis.

RNA processing
Broad term for the various modifications an RNA transcript undergoes as it reaches its mature form. May include 5′ capping, 3′ polyadenylation, 3′ cleavage, splicing, and editing.

RNA splicing
Process in which intron sequences are excised from RNA transcripts in the nucleus during formation of messenger and other RNAs.

rod photoreceptor (rod)
Photoreceptor cell in the vertebrate retina that is responsible for noncolor vision in dim light. (Figure 23–17)

rough endoplasmic reticulum (rough ER)
Endoplasmic reticulum with ribosomes on its cytosolic surface. Involved in the synthesis of secreted and membrane-bound proteins.

rRNA—*see* **ribosomal RNA**

rRNA gene
Gene that specifies a ribosomal RNA (rRNA).

RTK—*see* **receptor tyrosine kinase**

RT-PCR (reverse transcription–polymerase chain reaction)
Technique in which a population of mRNAs is converted into cDNAs via reverse transcription, and the cDNAs are then amplified by PCR.

S—*see* **S phase**

saccharide
Sugar.

Saccharomyces
Genus of yeasts that reproduce asexually by budding or sexually by conjugation. Economically important in brewing and baking. *Saccharomyces cerevisiae* is widely used as a simple model organism in the study of eucaryotic cell biology. *See also Schizosaccharomyces.*

sarcoma
Cancer of connective tissue.

sarcomere
Repeating unit of a myofibril in a muscle cell, composed of an array of overlapping thick (myosin) and thin (actin) filaments between two adjacent Z discs. (Figure 16–74)

sarcoplasmic reticulum
Specialized type of endoplasmic reticulum in the cytoplasm of muscle cells that contains high concentrations of sequestered Ca^{2+} that is released into the cytosol during muscle excitation. (Figure 16–77)

satellite DNA
Region of highly repetitive DNA from a eucaryotic chromosome, identifiable by its unusual nucleotide composition. Typically present at centromeres (as well as other sites) in higher eucaryotes, and thought to play a part in centromere function. (Figure 4–49)

scaffold protein
Protein that binds groups of intracelluar signaling proteins into a signaling complex, often anchoring the complex at a specific location in the cell. (Figure 15–17)

scanning electron microscope
Type of electron microscope that produces an image of the surface of an object.

S-Cdk
Cyclin-Cdk complex formed in vertebrate cells by an S-cyclin and the corresponding cyclin-dependent kinase (Cdk). (Figure 17–16 and Table 17–1, p. 1063)

S. cerevisiae—*see* **Saccharomyces**

Schizosaccharomyces
Genus of rod-shaped yeasts that reproduce by binary fission. *S. pombe*, along with the budding yeast *Saccharomyces cerevisiae*, is a model organism used in many different studies.

SCF (SCF protein)
Family of ubiquitin ligases formed as a complex of several different proteins. One is involved in regulating the eucaryotic cell cycle, directing the destruction of inhibitors of S-Cdks in late G_1 and thus promoting the activation of S-Cdks and DNA replication. (Figures 3–79 and 17–20)

Schwann cell
Glial cell responsible for forming myelin sheaths in the peripheral nervous system. *Compare* oligodendrocyte. (Figure 11–32)

S-cyclin
Member of a class of cyclins that accumulate during late G_1 phase and bind Cdks soon after progression through Start; they help stimulate DNA replication and chromosome duplication. Levels remain high until late mitosis, after which these cyclins are destroyed. (Figure 17–16)

SDS-PAGE (sodium dodecyl sulfate–polyacrylamide gel electrophoresis)
Type of electrophoresis used to separate proteins by size. The protein mixture to be separated is first treated with a powerful negatively charged detergent (SDS) and with a reducing agent (β mercaptoethanol), before being run through a polyacrylamide gel. The detergent and reducing agent unfold the proteins, free them from association with other molecules, and separate the polypeptide subunits. *See also* electrophoresis.

secondary cell wall
Permanent rigid cell wall that is laid down underneath the thin primary cell wall in certain plant cells that have completed their growth. (Figure 19–77C)

secondary immune response
Adaptive immune response to an antigen that is made on a second or subsequent encounter with a given antigen. More rapid in onset and stronger than the primary immune response. (Figure 25–10)

secondary structure
Regular local folding pattern of a polymeric molecule; in proteins, α-helices and β-sheets.

second messenger (small intracellular mediator)
Small intracellular signaling molecule that is formed or released for action in response to an extracellular signal and helps to relay the signal within the cell. Examples include cyclic AMP, cyclic GMP, IP_3, Ca^{2+}, and diacylglycerol. (Figure 15–17)

secretory vesicle
Membrane-bounded organelle in which molecules destined for secretion are stored prior to release. Sometimes called secretory granule because darkly staining contents make the organelle visible as a small solid object. (Figures 13–63 and 13–66)

securin
Protein that binds to the protease separase and thereby prevents its cleavage of the protein linkages that hold sister chromatids together in early mitosis. Securin is destroyed at the metaphase-to-anaphase transition. (Figure 17–44)

seed
In plants, the structure containing the dormant embryo, along with a food store, enclosed in a hard protective coat. (Panel 22–1, p. 1401)

segment-polarity gene
In *Drosophila* development, a gene involved in specifying the anteroposterior organization of each body segment. (Figures 22–37 and 22–41)

selectable marker gene
Gene included in a DNA construct to signal presence of that construct in a cell, and making it possible to select cells according to whether they contain the construct.

selectin
Member of a family of cell-surface carbohydrate-binding proteins that mediate transient, Ca^{2+}-dependent cell–cell adhesion in the bloodstream—for example between white blood cells and the endothelium of the blood vessel wall. (Figure 19–19)

selectivity filter
The part of an ion channel structure that determines which ions it can transport. (Figures 11–23 and 11–24)

senescence
(1) aging of an organism. (2) replicative cell senescence: phenomenon observed in primary cell cultures in which cell proliferation slows down and finally halts irreversibly.

sensory hair cell—*see* **auditory hair cell**

separase
Protease that cleaves the cohesin protein linkages that hold sister chromatids together. Acts at anaphase, enabling chromatid separation and segregation. (Figure 17–44)

septate junction
Main type of occluding cell junction in invertebrates; its structure is distinct from that of vertebrate tight junctions. (Figure 19–28)

sequencing
Determination of the order of nucleotides or amino acids in a nucleic acid or protein molecule. (Figure 8–50)

serine protease
Type of protease that has a reactive serine in the active site. (Figures 3–12 and 3–38)

serine/threonine kinase
Enzyme that phosphorylates specific proteins on serine or threonines. (Figure 15–70)

sex chromosome
Chromosome that may be present or absent, or present in a variable number of copies, determining the sex of the individual; in mammals, the X and Y chromosomes.

SH2 domain
Src homology region 2, a protein domain present in many signaling proteins. Binds a short amino acid sequence containing a phosphotyrosine. (Panel 3–2, pp. 132–133)

β-sheet—*see* **beta sheet**

side chain
The part of an amino acid that differs between amino acid types. The side chains give each type of amino acid its unique physical and chemical properties. (Panel 3–1, pp. 128–129)

signaling cascade
Sequence of linked intracellular reactions, typically involving multiple amplification steps in a relay chain, triggered by an activated cell-surface receptor.

signal molecule
Extracellular chemical produced by a cell that signals to other cells in the organism to alter the cells' behavior. (Figure 15–1)

signal patch
Protein-sorting signal that consists of a specific three-dimensional arrangement of atoms on the folded protein's surface. (Figure 13–45)

signal peptidase
Enzyme that removes a terminal signal sequence from a protein once the sorting process is complete. (Figure 12–25)

signal-recognition particle (SRP)
Ribonucleoprotein particle that binds an ER signal sequence on a partially synthesized polypeptide chain and directs the polypeptide and its attached ribosome to the endoplasmic reticulum. (Figure 12–39)

signal-relaying junction
Complex type of cell–cell junction that alows signals to be relayed from one cell to another across their plasma membranes at sites of cell-to-cell contact. Typically includes anchorage proteins as well as proteins mediating signal transduction. (Figure 19–2, and Table 19–1, p. 1133)

signal sequence
Short continuous sequence of amino acids that determines the eventual location of a protein in the cell. An example is the N-terminal sequence of 20 or so amino acids that directs nascent secretory and transmembrane proteins to the endoplasmic reticulum. (Table 12–3, p. 702)

signal transduction
Conversion of a signal from one physical or chemical form to another (e.g., conversion of light to a chemical signal or of extracellular signals to intracellular ones).

single-nucleotide polymorphism (SNP)
Variation between individuals in a population at a specific nucleotide in their DNA sequence.

single-pass transmembrane protein
Membrane protein in which the polypeptide chain crosses the lipid bilayer only once. (Figure 10–19)

single-strand DNA-binding protein
Protein that binds to the single strands of the opened-up DNA double helix, preventing helical structures from reforming while the DNA is being replicated. (Figure 5–16)

siRNA—*see* **small interfering RNA**

sister chromatids
Tightly linked pair of chromosomes that arise from chromosome duplication during S phase. They separate during M phase and segregate into different daughter cells. (Figure 17–26)

site-directed mutagenesis
Technique by which a mutation can be made at a particular site in DNA. (Figure 8–63)

site-specific recombination
Type of recombination that occurs at specific DNA sequences and is carried out by specific proteins that recognize these sequences. Can occur between two different DNA molecules or within a single DNA molecule.

skeletal muscle cell—*see* **muscle cell**

sliding clamp
Protein complex that holds the DNA polymerase on DNA during DNA replication. (Figure 5–18)

Smad protein
Latent gene regulatory protein that is phosphorylated and activated by receptor serine/threonine kinases and carries the signal from the cell surface to the nucleus. (Figure 15–69)

small interfering RNA (siRNA)
Short (21–26 nucleotides) double-stranded RNAs that inhibit gene expression by directing destruction of complementary mRNAs. Production of siRNAs is triggered by exogenously introduced double-stranded RNA. (Figure 7–115)

small intracellular mediator—*see* **second messenger**

small nuclear ribonucleoprotein (snRNP)
Complex of an snRNA with proteins that forms part of a spliceosome. (Figure 6–29)

small nuclear RNA (snRNA)
Small RNA molecules that are complexed with proteins to

form the ribonucleoprotein particles (snRNPs) involved in RNA splicing. (Figures 6–29 and 6–30)

small nucleolar RNA (snoRNA)
Small RNAs found in the nucleolus, with various functions, including guiding the modifications of precursor rRNA. (Table 6–1, p. 336, and Figure 6–43)

smooth endoplasmic reticulum (smooth ER)
Region of the endoplasmic reticulum not associated with ribosomes. Involved in lipid synthesis. (Figure 12–36)

smooth muscle cell—*see* **muscle cell**

SNARE
Member of a large family of transmembrane proteins present in organelle membranes and the vesicles derived from them. SNAREs catalyze the many membrane fusion events in cells. They exist in pairs—a v-SNARE in the vesicle membrane that binds specifically to a complementary t-SNARE in the target membrane.

SNP—*see* **single-nucleotide polymorphism**

snRNA—*see* **small nuclear RNA**

solute
Any molecule that is dissolved in a liquid. The liquid is called a solvent.

somatic cell
Any cell of a plant or animal other than cells of the germ line. From Greek *soma*, body.

somatic hypermutation
Accumulation of point mutations in the assembled variable-region-coding sequences of immunoglobulin genes that occurs when B cells are activated to form memory cells. Results in the production of antibodies with altered antigen-binding sites.

somite
One of a series of paired blocks of mesoderm that form during early development and lie on either side of the notochord in a vertebrate embryo. They give rise to the segments of the body axis, including the vertebrae, muscles, and associated connective tissue. (Figure 22–81)

sorting signal
Amino acid sequence that directs the delivery of a protein to a specific location, such as a particular intracellular compartment.

Southern blotting
Technique in which DNA fragments separated by electrophoresis are immobilized on a paper sheet. Specific fragments are then detected with a labeled nucleic acid probe. (Named after E.M. Southern, inventor of the technique.)

spectrin
Abundant protein associated with the cytosolic side of the plasma membrane in red blood cells, forming a network that supports the membrane. Also present in other cells. (Figure 10–41)

Spemann's Organizer—*see* **Organizer**

sperm (spermatozoon, plural spermatozoa)
Mature male gamete in animals. Motile and usually small compared with the egg. (Figure 21–27)

spermatogenesis
Development of sperm in the testes. (Figure 21–30)

S phase
Period of a eucaryotic cell cycle in which DNA is synthesized. (Figure 17–4)

sphingolipid
Phospholipid derived from sphingosine. (Figure 10–3)

spindle assembly checkpoint (metaphase-to-anaphase transition checkpoint)
Checkpoint that operates during mitosis to ensure that all chromosomes are properly attached to the spindle before sister-chromatid separation starts. (Figure 17–14, and Panel 17–1, pp. 1072–1073)

spliceosome
Large assembly of RNA and protein molecules that performs pre-mRNA splicing in eucaryotic cells. (Figures 6–29 and 6–30)

splicing
Removal of introns from a pre-mRNA transcript by splicing together the exons that lie on either side of each intron. *See also* alternative RNA splicing, *and* trans-splicing.

S. pombe—*see* **Schizosaccharomyces**

Src (Src protein family)
Family of cytoplasmic tyrosine kinases (pronounced "sark") that associate with the cytoplasmic domains of some enzyme-linked cell-surface receptors (for example, the T cell antigen receptor) that lack intrinsic tyrosine kinase activity. They transmit a signal onwards by phosphorylating the receptor itself and specific intracellular signaling proteins on tyrosines. (Figures 3–10 and 15–70)

SRP—*see* **signal-recognition particle**

standard free-energy change (ΔG°)
Free-energy change of two reacting molecules at standard temperature and pressure when all components are present at a concentration of 1 mole per liter. (Table 2–4, p. 77, and Figure 14–18)

starch
Polysaccharide composed exclusively of glucose units, used as an energy storage material in plant cells. (Figure 2–75)

Start (Start checkpoint, restriction point)
Important checkpoint at the end of G_1 in the eucaryotic cell cycle. Passage through Start commits the cell to enter S phase. The term was originally used for this checkpoint in the yeast cell cycle only; the equivalent point in the mammalian cell cycle was called the restriction point. In this book we use Start for both. (Figure 17–14)

start-transfer signal
Short amino acid sequence that enables a polypeptide chain to start being translocated across the endoplasmic reticulum membrane through a protein translocator. Multipass membrane proteins have both N-terminal (signal sequence) and internal start-transfer signals. (Figures 12–45–12–48)

STAT (signal transducer and activator of transcription)
Latent gene regulatory protein that is activated by phosphorylation by JAK kinases and enters the nucleus in response to signaling from receptors of the cytokine receptor family. (Figure 15–68)

stem cell
Undifferentiated cell that can continue dividing indefinitely, throwing off daughter cells that can either commit to differentiation or remain a stem cell (in the process of self-renewal). (Figure 23–5)

stem-cell niche
The specialized microenvironment in a tissue in which self-renewing stem cells can be maintained. (Figure 23–27)

stereocilium
A large, rigid microvillus found in "organ pipe" arrays on the apical surface of hair cells in the ear. A stereocilium contains a bundle of actin filaments, rather than microtubules, and is thus not a true cilium. (Figures 23–13 and 23–15)

steroid
Hydrophobic lipid molecule with a characteristic four-ringed structure; derived from cholesterol. Many important

hormones, including cortisol, estrogen, and testosterone, are steroids that activate intracellular nuclear receptors. (Panel 2–5, pp. 114–115)

stimulatory G protein (G_s)
G protein that, when activated, activates the enzyme adenylyl cyclase and thus stimulates the production of cyclic AMP. (*See also* G protein, and Table 15–3, p. 919.)

stochastic
Random. Involving chance, probability, or random variables.

stop-transfer signal
Hydrophobic amino acid sequence that halts translocation of a polypeptide chain through the endoplasmic reticulum membrane, thus anchoring the protein chain in the membrane. (Figure 12–48 and 12–49)

striated muscle
Muscle composed of transversely striped (striated) myofibrils. Skeletal and heart muscle of vertebrates are examples. (Figure 16–74)

stroma
(1) 'Bedding': the connective tissue in which a glandular or other epithelium is embedded. Stromal cells provide the environment necessary for the development of other cells within the tissue. (Figure 20–19) (2) The large interior space of a chloroplast, containing enzymes that incorporate CO_2 into sugars. (Figure 12–21)

structural gene
Region of DNA that codes for a protein or for an RNA molecule that forms part of a structure or has an enzymatic function. Distinguished from regions of DNA that regulate gene expression.

substrate
Molecule on which an enzyme acts.

substratum
Solid surface to which a cell adheres.

sucrose
Disaccharide composed of one glucose unit and one fructose unit. The major form in which glucose is transported between plant cells. (Panel 2–4, pp.112–113)

sugar
Small carbohydrate with a monomer unit of general formula $(CH_2O)n$. Examples are the monosaccharides glucose, fructose and mannose, and the disaccharide sucrose (glucose and fructose linked). (Panel 2–4, pp.112–113)

sulfhydryl (thiol, –SH)
Chemical group containing sulfur and hydrogen; found in the amino acid cysteine and other molecules. Two sulfhydryls can join to produce a disulfide bond. (Panel 2–1, pp. 106–107, and Figure 3–28)

suppressor mutation
Mutation that suppresses the phenotypic effect of another mutation, so that the double mutant individual seems normal. (Panel 8–1, pp. 554–555)

surface plasmon resonance (SPR)
Technique used to characterize molecular interactions, such as antibody-antigen binding, ligand-receptor coupling, and the binding of proteins to DNA. Binding interactions are detected by monitoring the reflection of a beam of light off the interface between an aqueous solution of potential binding molecules and a biosensor surface carrying the immobilized bait protein.

survival factor
Extracellular signal that promotes cell survival by inhibiting apoptosis. (Figure 18–14)

symbiosis
Intimate association between two organisms of different species from which both derive a long-term selective advantage. (Figure 1–16)

symporter
Carrier protein that transports two types of solute across the membrane in the same direction. (Figure 11–8)

synapse
Communicating cell–cell junction that allows signals to pass from a nerve cell to another cell. In a chemical synapse, the signal is carried by a diffusible neurotransmitter. (Figure 19–22) In an electrical synapse, a direct connection is made between the cytoplasms of the two cells via gap junctions. (Figure 19–34)

synapsis
(1) In genetic recombination, the initial formation of base pairs between complementary DNA strands in different DNA molecules that occurs at sites of crossing-over between chromosomes. (Figure 5–56) (2) In meiosis, the formation of a synaptonemal complex between two tightly aligned homologous chromosomes. (Figure 21–9)

synaptic vesicle
Small neurotransmitter-filled secretory vesicle found at the axon terminals of nerve cells. Its contents are released into the synaptic cleft by exocytosis when an action potential reaches the axon terminal.

synaptonemal complex
Structure that holds paired homologous chromosomes tightly together in pachytene of prophase I in meiosis and promotes the final steps of crossing over. (Figures 21–8 and 21–9)

syncytium
Mass of cytoplasm containing many nuclei enclosed by a single plasma membrane. Typically the result either of cell fusion or of a series of incomplete division cycles in which the nuclei divide but the cell does not.

synteny
The presence, in different species, of regions of chromosomes with the same genes in the same order.

synthetic lethality
An interaction between two mutant genes in which the two mutant genes together result in cell death, whereas either single mutation alone does not.

tap-tagging (tandem affinity purification tagging)
Highly efficient method for protein purification, based on construction of a fusion protein in which the protein of interest is linked to two protein domains in tandem that act as tags for affinity chromatography purification. Two rounds of affinity purification, first using one tag and then the other, result in a very pure preparation.

TATA box
Sequence in the promoter region of many eucaryotic genes that binds a general transcription factor and hence specifies the position at which transcription is initiated. (Figures 6–16 and 6–17)

TCA (tricarboxylic acid) cycle—*see* citric acid cycle

T cell (T lymphocyte)
Type of lymphocyte responsible for T-cell-mediated adaptive immune responses; the class includes cytotoxic T cells, helper T cells, and regulatory T cells.

T-cell-mediated immune response
Any adaptive immune response mediated by antigen-specific T cells.

telomerase
Enzyme that elongates telomere sequences in DNA, which occur at the ends of eucaryotic chromosomes.

telomere
End of a chromosome, associated with a characteristic DNA sequence that is replicated in a special way. Counteracts the tendency of the chromosome otherwise to shorten with each round of replication. From Greek *telos*, end.

telophase

Final stage of mitosis in which the two sets of separated chromosomes decondense and become enclosed by nuclear envelopes. (Panel 17–1, pp. 1072–1073)

temperature-sensitive (ts) mutant

Organism or cell carrying a mutation that shows its phenotypic effect in one temperature range (usually high temperature) but not at other (usually low) temperatures. (Panel 8–1, pp. 554–555, and Figure 8–55)

template

Single strand of DNA or RNA whose nucleotide sequence acts as a guide for the synthesis of a complementary strand. (Figure 1–3)

terminator

Signal in bacterial DNA that halts transcription.

tertiary structure

Complex three-dimensional form of a folded polymer chain, especially a protein or RNA molecule.

TGFβ superfamily (transforming growth factor-β superfamily)

Large family of structurally related secreted proteins that act as hormones and local mediators to control a wide range of functions in animals, including during development. It includes the TGFβ/activin and bone morphogenetic protein (BMP) subfamilies. (Figure 15–69)

TGN—see trans Golgi network

thioester bond

High-energy bond formed by a condensation reaction between an acid (acyl) group and a thiol group (–SH). Seen, for example, in acetyl CoA and in many enzyme-substrate complexes. (Figure 2–62)

thiol—see sulfhydryl

thylakoid

Flattened sac of membrane in a chloroplast that contains chlorophyll and other pigments and carries out the light-trapping reactions of photosynthesis. Stacks of thylakoids form the grana of chloroplasts. (Figures 14–35 and 14–36)

tight junction

Cell–cell junction that seals adjacent epithelial cells together, preventing the passage of most dissolved molecules from one side of the epithelial sheet to the other. (Figures 19–3 and 19–26)

TIM complexes

Protein translocators in the mitochondrial inner membrane. The TIM23 complex mediates the transport of proteins into the matrix and the insertion of some proteins into the inner membrane; the TIM22 complex mediates the insertion of a subgroup of proteins into the inner membrane. (Figure 12–23)

T lymphocyte—see T cell

Toll-like receptor family (TLR)

Important family of mammalian pattern recognition receptors abundant on or in cells of the innate immune system. They recognize pathogen-associated immunostimulants such as lipopolysacharide and peptidoglycan. (Figure 24–51)

TOM complex

Multisubunit protein complex that transports proteins across the mitochondrial outer membrane. (Figure 12–23)

topoisomerase (DNA topoisomerase)

Enzyme that binds to DNA and reversibly breaks a phosphodiester bond in one or both strands. Topoisomerase I creates transient single-strand breaks, allowing the double helix to swivel and relieving superhelical tension. Topoisomerase II creates transient double-strand breaks, allowing one double helix to pass through another and thus resolving tangles. (Figures 5–22 and 5–23)

totipotent

Describes a cell that is able to give rise to all the different cell types in an organism.

trans

On the other (far) side.

transcellular transport

Transport of solutes, such as nutrients, across an epithelium, by means of membrane transport proteins in the apical and basal faces of the epithelial cells. (Figure 11–11)

transcript

RNA product of DNA transcription. (Figure 6–21)

transcription (DNA transcription)

Copying of one strand of DNA into a complementary RNA sequence by the enzyme RNA polymerase. (Figure 6–21)

transcriptional activator—see activator

transcriptional repressor—see repressor

transcription attenuation

Inhibition of gene expression by the premature termination of transcription.

transcription factor

Term loosely applied to any protein required to initiate or regulate transcription in eucaryotes. Includes gene regulatory proteins, the general transcription factors, coactivators, co-repressors, histone-modifying enzymes, and chromatin remodeling complexes. (Figures 6–19 and 7–44)

transcytosis

Uptake of material at one face of a cell by endocytosis, its transfer across a cell in vesicles, and discharge from another face by exocytosis. (Figure 13–60)

transfection

Introduction of a foreign DNA molecule into a cell. Usually followed by expression of one or more genes in the newly introduced DNA.

transfer RNA (tRNA)

Set of small RNA molecules used in protein synthesis as an interface (adaptor) between mRNA and amino acids. Each type of tRNA molecule is covalently linked to a particular amino acid. (Figures 1–9 and 6–52)

transformation

(1) Insertion of new DNA (e.g., a plasmid) into a cell or organism, such as into competent *E.coli*. (2) Conversion of a normal cell into one that behaves in many ways like a cancer cell (i.e., unregulated proliferation, anchorage-independent growth in culture).

transforming growth factor-β superfamily—see TGFβ superfamily

transgenic organism

Plant or animal that has stably incorporated one or more genes from another cell or organism (through insertion, deletion, and/or replacement) and can pass them on to successive generations. The gene that has been added is called a transgene. (Figures 8–64 and 8–65)

transit amplifying cell

Cell derived from a stem cell that divides a limited number of cycles before terminally differentiating. (Figure 23–7)

transition state

Structure that forms transiently in the course of a chemical reaction and has the highest free energy of any reaction intermediate. Its formation is a rate-limiting step in the reaction. (Figure 3–46)

translation (RNA translation)

Process by which the sequence of nucleotides in a mRNA molecule directs the incorporation of amino acids into protein. Occurs on a ribosome. (Figures 6–66 and 6–67)

translocation
(1) Type of mutation in which a portion of one chromosome is broken off and attached to another. (Panel 8–1 and Figure 20–5) (2) The process of transferring a protein across a membrane.

transmembrane protein
Membrane protein that extends through the lipid bilayer, with part of its mass on either side of the membrane. (Figure 10–19)

transmitter-gated ion channel (ion-channel-coupled receptor, ionotropic receptor)
Ion channel found at chemical synapses in the postsynaptic plasma membranes of nerve and muscle cells. Opens only in response to the binding of a specific extracellular neurotransmitter. The resulting inflow of ions leads to the generation of a local electrical signal in the postsynaptic cell. (Figures 15–16 and 11–35)

transporter (carrier protein, permease)
Membrane transport protein that binds to a solute and transports it across the membrane by undergoing a series of conformational changes. Transporters can transport ions or molecules passively down an electrochemical gradient or can link the conformational changes to a source of metabolic energy such as ATP hydrolysis to drive active transport. *Compare* channel protein. *See also* membrane transport protein. (Figure 11–3)

transposable element (transposon)
Segment of DNA that can move from one genome position to another by transposition. (Table 5–3, p. 318)

transposase
Enzyme that cuts a transposon sequence in a chromosome and causes the DNA sequence to be inserted into a new target site. The transposase is usually encoded by the transposon that it acts upon. (FIgure 5–69)

transposition (transpositional recombination)
Movement of a DNA sequence from one genome site to another. (Table 5–3, p. 318)

transposon—*see* **transposable element**

trans-splicing
Type of RNA splicing present in a few eucaryotic organisms in which exons from two separate RNA transcripts are joined together to form an mRNA. (Figure 6–34)

treadmilling
Process by which a polymeric protein filament is maintained at constant length by addition of protein subunits at one end and loss of subunits at the other. (Panel 16–2, pp. 978–979)

triacylglycerol (triglyceride)
Molecule composed of three fatty acids esterified to glycerol. The main constituent of fat droplets in animal tissues (where the fatty acids are saturated) and of vegetable oils (where the fatty acids are mainly unsaturated). (Panel 2–5, pp. 114–115)

tricarboxylic acid (TCA) cycle—*see* **citric acid cycle**

trimeric G protein (trimeric GTP-binding protein)—*see* **G protein**

triglyceride—*see* **triacylglycerol**

tRNA—*see* **transfer RNA**

ts mutant—*see* **temperature-sensitive mutant**

t-SNARE—*see* **SNARE**

tubulin
The protein subunit of microtubules. (Panel 16–1, p. 968, and Figure 16–11)

γ-tubulin ring complex (γTuRC)—*see* **gamma tubulin ring complex**

tumor
Abnormal mass of cells resulting from a defect in cell-prolif-eration control. A tumor can be noninvasive (benign) or invasive (cancerous, malignant). (Figure 20–3)

tumor necrosis factor (TNFα)
Cytokine that is especially important in inducing inflammatory responses. (Figure 15–79)

tumor progression
Process by which an initial mildly disordered cell behavior gradually evolves into a full-blown cancer. (Figures 20–9 and 20–11)

tumor suppressor gene
Gene that appears to help prevent formation of a cancer. Loss-of-function mutations in such genes favor the development of cancer. (Figure 20–27)

tumor virus
Virus that can make the cell it infects cancerous.

turgor pressure
Large hydrostatic pressure developed inside a plant cell as the result of the intake of water by osmosis; it is the force driving cell expansion in plant growth and it maintains the rigidity of plant stems and leaves. (Panel 11–1, p. 664)

two-hybrid system—*see* **yeast two-hybrid system**

type III secretion system
Bacterial system for delivering toxic proteins into the cells of their host. (Figure 24–8)

tyrosine kinase
Enzyme that phosphorylates specific proteins on tyrosines. *See also* cytoplasmic tyrosine kinase. (Figure 15–70)

ubiquitin
Small, highly conserved protein present in all eucaryotic cells that becomes covalently attached to lysines of other proteins. Attachment of a short chain of ubiquitins to such a lysine can tag a protein for intracellular proteolytic destruction by a proteasome. (Figure 6–92)

ubiquitin ligase
Any one of a large number of enzymes that attach ubiquitin to a protein, often marking it for destruction in a proteasome. The process catalyzed by a ubiquitin ligase is called ubiquitylation. (Figure 3–79)

unfolded protein response
Cellular response triggered by an accumulation of misfolded proteins in the endoplasmic reticulum. Involves increased transcription of ER chaperones and degradative enzymes. (Figure 12–55)

uniporter
Carrier protein that transports a single solute from one side of the membrane to the other. (Figure 11–8)

UTR (untranslated region)
Noncoding region of an mRNA molecule. The 3′ UTR extends from the stop codon that terminates protein synthesis to the start of the poly-A tail. (Figure 7–105) The 5′ UTR extends from the 5′ cap to the start codon that initiates protein synthesis.

V_{MAX}
Maximum rate of an enzymatic reaction. (Figure 3–45 and Panel 3–3, pp. 162–163)

vacuole
Very large fluid-filled vesicle found in most plant and fungal cells, typically occupying more than a third of the cell volume. (Figure 13–39)

valence
Number of electrons that an atom must gain or lose (either by sharing or by transfer) to fill its outer shell. The valence of an atom is equal to the number of single bonds that the atom can form.

van der Waals attraction
Type of (individually weak) noncovalent bond that is formed

at close range between nonpolar atoms. (Table 2–1, p. 53 and Panel 2–3, pp. 110–111)

variable region
Region of an immunoglobulin light or heavy chain that differs from molecule to molecule and forms the antigen-binding site. (Figures 25–30 and 25–31)

vascular endothelial growth factor (VEGF)
Secreted protein that stimulates the growth of blood vessels. (Table 15–4, p. 923, and Figure 23–35)

V(D)J recombination
Somatic recombination process by which gene segments are brought together to form a functional gene for a polypeptide chain of an immunoglobulin or T cell receptor. (Figures 25–36, 25–37, and 25–38)

vector
In cell biology, the DNA of an agent (virus or plasmid) used to transmit genetic material to a cell or organism. *See also* cloning vector, expression vector. (Figure 8–39)

vegetal pole
The end at which most of the yolk is located in an animal egg. The end opposite the animal pole. (Figure 22–68)

VEGF—*see* **vascular endothelial growth factor**

vesicle
Small, membrane-bounded, organelle in the cytoplasm of a eucaryotic cell; often spherical. (Figure 11–35)

vesicular transport
Transport of proteins from one cell compartment to another by means of membrane-bounded intermediaries such as vesicles or organelle fragments.

V gene segment
Gene segment encoding most of the variable region of an immunoglobulin or T cell receptor polypeptide chain.

viral envelope
Phospholipid bilayer, derived from host-cell plasma membrane, covering an enveloped virus. Encloses the nucleocapsid. (Figure 24–15)

virion
A single virus particle. (Figure 24–13)

virulence gene
Gene that contributes to an organism's ability to cause disease.

virus
Particle consisting of nucleic acid (RNA or DNA) enclosed in a protein coat and capable of replicating within a host cell and spreading from cell to cell. (Figure 24–13)

voltage-gated cation channel
Type of ion channel found in the membranes of electrically excitable cells (such as nerve, endocrine, egg, and muscle cells). Opens in response to a shift in membrane potential past a threshold value.

v-SNARE—*see* **SNARE**

Wee1
Protein kinase that inhibits Cdk activity by phosphorylating amino acids in the Cdk active site. Important in regulating entry into M phase of the cell cycle.

Western blotting (immunoblotting)
Technique by which proteins are separated by electrophoresis and immobilized on a paper sheet and then analyzed, usually by means of a labeled antibody.

white blood cell (leucocyte)
General name for all the nucleated blood cells lacking hemoglobin. Includes lymphocytes, granulocytes, and monocytes.

wild-type
Normal, nonmutant form of an organism; the form found in nature (in the wild).

Wnt (Wnt protein)
Member of a family of secreted signal proteins that have many different roles in controlling cell differentiation, proliferation, and gene expression in animal embryos and adult tissues.

Wnt signaling pathway
Signaling pathway activated by binding of a Wnt protein to its cell-surface receptors. The pathway has several branches. In the major (canonical) branch, activation causes increased amounts of β-catenin to enter the nucleus, where it regulates the transcription of genes controlling cell differentiation and proliferation. Overactivation of the Wnt/β-catenin pathway can lead to cancer. (Figure 15–77)

X chromosome
One of the two sex chromosomes of mammals. The cells of women contain two X chromosomes, while those of men contain only one.

***Xenopus laevis* (South African clawed toad)**
Species of frog (not toad) frequently used in studies of early vertebrate development.

XIC—*see* **X-inactivation center**

X-inactivation
Inactivation of one copy of the X chromosome in the somatic cells of female mammals.

X-inactivation center (XIC)
Site in an X chromosome at which inactivation is initiated and spreads outwards.

X-ray crystallography (X-ray diffraction)
Technique for determining the three-dimensional arrangement of atoms in a molecule based on the diffraction pattern of X-rays passing through a crystal of the molecule. (Figure 8–28)

Y chromosome
One of the two sex chromosomes of mammals. The cells of men contain one Y and one X chromosome.

yeast
Common name for several families of unicellular fungi. Includes species used for brewing and bread-making, as well as pathogenic species. Among the simplest of eucaryotes.

yeast two-hybrid system
Molecular genetic technique for identifying protein-protein and protein-DNA interactions. (Figure 8–24)

yolk
Nutritional reserves rich in lipids, proteins and polysaccharides, present in the eggs of many animals.

Z disc (Z line)
Platelike region of a muscle sarcomere to which the plus ends of actin filaments are attached. Seen as a dark transverse line in micrographs.

zinc finger
DNA-binding structural motif present in many gene regulatory proteins. All zinc finger motifs incorporate one or more zinc atoms that help hold the protein conformation together.

zona pellucida
Glycoprotein layer on the surface of the unfertilized egg. It is often a barrier to fertilization across species. (Figure 21–22)

zonula adherens—*see* **adhesion belt**

zygote
Diploid cell produced by fusion of a male and female gamete. A fertilized egg.

Index

Page numbers in **boldface** refer to a major text discussion of the entry; page numbers with an F refer to a figure, with an FF to figures that follow consecutively; page numbers with a T refer to a table; *vs* means compare/comparison.

Page numbers in **boldface** refer to a major text discussion of the entry; page numbers with an F refer to a figure, with an FF to figures that follow consecutively; page numbers with a T refer to a table; *vs* means compare/comparison.

Page numbers in **boldface** refer to a major text discussion of the entry; page numbers with an F refer to a figure, with an FF to figures that follow consecutively; page numbers with a T refer to a table; vs means compare/comparison.

Page numbers in **boldface** refer to a major text discussion of the entry; page numbers with an F refer to a figure, with an FF to figures that follow consecutively; page numbers with a T refer to a table; vs means compare/comparison.

Page numbers in **boldface** refer to a major text discussion of the entry; page numbers with an F refer to a figure, with an FF to figures that follow consecutively; page numbers with a T refer to a table; vs means compare/comparison.

Page numbers in **boldface** refer to a major text discussion of the entry; page numbers with an F refer to a figure, with an FF to figures that follow consecutively; page numbers with a T refer to a table; *vs* means compare/comparison.

Page numbers in **boldface** refer to a major text discussion of the entry; page numbers with an F refer to a figure, with an FF to figures that follow consecutively; page numbers with a T refer to a table; *vs* means compare/comparison.

Page numbers in **boldface** refer to a major text discussion of the entry; page numbers with an F refer to a figure, with an FF to figures that follow consecutively; page numbers with a T refer to a table; vs means compare/comparison.

Page numbers in **boldface** refer to a major text discussion of the entry; page numbers with an F refer to a figure, with an FF to figures that follow consecutively; page numbers with a T refer to a table; *vs* means compare/comparison.

mass spectrometry, **519–521**
see also Protein analysis; Protein purification
Protein interaction maps, 188–190, 189F, 190F
Protein kinase(s), 176–178, **895**
  Cdk inactivation, 1063–1064, 1064F, 1066T
  evolution, 176, 177F
  regulation, "microchips," 177–178, 177F, 179F
  sequence homology, 176, 177F
  structure, 141, 176F
  tyrosine kinases *see* Tyrosine kinase(s)
  see also individual enzymes
Protein kinase B (Akt; PKB), 934, 1244
Protein kinase C (PKC), 185, 626, 911
Protein–ligand interactions, 153
  amino acid side-chains, 154–155, 154F, 155F
  antigen–antibody *see* Antigen–antibody binding
  binding sites, 153, 153F, 154F, 155F
  binding strength, 157–158
    see also Equilibrium constant *(K)*
  cooperativity, 172–173, 173F
    see also Allosteric regulation
  enzyme–substrate *see* Enzyme–substrate interactions
  linkage, 171–172, 172FF
  noncovalent interactions, 153, 153F, 154, 154F, 155F
  regulation, 169–170
    allostery *see* Allosteric regulation
  specificity, 153
  water exclusion, 154
  see also Protein–DNA interactions; Protein–protein interactions
Protein machines, **184–186**
  activation, **185–186**
  gene minimization, 184–185
  interchangeable parts, 184–185
  position in cell, **185–186**, 186F
  scaffold proteins, 184–185, **185–186**
  structure, 184
  see also Protein complexes (assemblies)
Protein microarrays, 188
Protein modules, 140–141, 175–176
Protein motifs, DNA-binding motifs *see* DNA-binding motifs (proteins)
'Protein-only inheritance,' yeast, 398, 398F
Protein phosphatase(s), 176–177, 176F, **895**
  Cdk activation, 1063, 1063F, 1066T, **1074–1075**
  5' mRNA capping, 347F, 3446
  protein tyrosine phosphatases (PTPs), **938–939**
Protein phosphorylation, **175–178**
  ATP as phosphate donor, 84F
  energetics, 177
  enzymes involved, 176, 176F
    see also Protein kinase(s); Protein phosphatase(s)
  functional effects, 175–176, 186–187, 187F
  gene regulatory proteins, 450, 451F
  GTP-GDP exchange as alternative to phosphorylation, 178–179
  GTP-mediated *see* GTP-binding proteins (GTPases)
  phosphorylation cycles, 177
  phosphorylation site as recognition signal for protein binding, 175–176
  Rb protein regulation, 1104
Protein–protein interactions, **187–190**
  analysis methods, **523–527**
  dimerization, 142
  interface types, 156, 156F
  map construction, 188–190, 524
  optical methods, **524–526**
  quaternary structure, 136
  SH2 domains *see* SH2 domain (Src homology 2 domain)
  single molecules, 526–527
  subunits, 140F, 141F, 142–143, 148F
Protein purification
  chromatography, **512–514**, 515F, 535F
  electrophoresis, 517, 518FF, 521–522, 522FF
  Western blotting, 519F
  see also Protein analysis; Recombinant proteins
Protein sorting
  apical domain, **806–807**
  basolateral domain, 807
  endosomes, 807F
  glycosphingolipids, **807**
  lipid rafts, 807
  signals, 699, **701–702**, 701F, 702T, 703F
  trans Golgi network, 807F
Protein stains, Coomassie blue, 517, 518F

Protein structure, 59–60, **125–152**
  amino acid sequence, **125–137**
    see also Amino acid(s)
  analysis, 139
  components, 126F
  computational analysis, 139, 139F
  conformation, **130–131**, 154–155, 154F, 155F
    see also Allosteric regulation; Conformational changes
  core structure, 131, 135, 135F
  C-terminus, 59, 126F, **482–483**
  depicting, 132–133FF, 144F
  determination, 139, **517–532**
    SDS-PAGE, 517
    X-ray diffraction, 139, **527–529**, 528F
  domains, 131, 136, 348
    3-D models, 131, 132–133FF, 137F
    evolutionary tracing, 155–156, 155F, 257F
    fusion, 560F
    large kinase domain, 136F
    modules, 140–141, 140F
    paired, 141
    SH2 domain, 131, 132–133FF, 136F, 155, 156
    SH3 domain, 136F
  helical filaments, 143, 144F, 145, 145F
    actin, 145F
      see also Actin/actin filaments
    assembly, 145F
    electron microscopy, 610F
    handedness, 143
  limited fold numbers, 136–137
  membrane spanning regions, 135
  noncovalent forces, 126–127
  N-terminus, 59, 126F
  polypeptide backbone, 125
  primary structure, **136**
    coding, 199–200, 199F, 200F
      see also Genetic code
    prediction, 550–551
  quaternary structure, 136
  secondary structure, 136
    α helix, **131**, 134F, **135**, 135F
    β sheets, **131**, 134F, **135**, 135F, **422**, 423F
  stability, 137, 147–148, 147F
  structure–function relationship, 154–155
  subunits, 140F, 141F, 142–143, 143F
  tertiary structure, 136
  unstructured polypeptide chains, 146–147
  see also Peptide bonds; Protein folding
Protein structure–function relationship, 154–155
Protein synthesis, 85, 85F, **366–400**
  elongation cycle, 373, 373F, 376F, 377F
    elongation factors, 179–180, **377–378**, 377F
    peptidyl transferase, 376, 379, 379F
    steps, 375–376, 376F
  ER and *see under* Endoplasmic reticulum (ER)
  eucaryotic, 345F, 380, 380F, 399F
  evolution, 407–408
  experimental analysis, cell-free systems, 512
  inhibitors, 384, 385T
    see also Antibiotics
  initiation, **379–380**, 380F
  levels of regulation, 399
  location *see* Ribosome(s)
  peptide bond formation, 59, 60F, 125, 373F
  post-translational alterations *see* Post-translational modification
  procaryotic, 345F, 380
  reading frames, 368, 368F
  relation to DNA *see* Gene expression; Genetic code; Messenger RNA
  termination, **381**, 381F
  see also Protein folding; Translation
Protein tags
  affinity chromatography, 515F
  cleavable, 515
  GST, 514–515, 516F
  histidine tag, 548
  see also Affinity chromatography; Epitope tagging
Protein translocation, 703F
  across chloroplast membranes, **719–720**
    see also Chloroplast(s),protein import
  across mitochondrial membranes, **713–718**
    see also Mitochondrial protein import
  in endoplasmic reticulum *see* Endoplasmic reticulum (ER)
Protein translocators, **714–715**
Protein transport
  across membranes, **695–748**

chloroplasts *see* Chloroplast(s), protein import
endoplasmic reticulum *see* Endoplasmic reticulum (ER), protein transport
gated, 700, 701F
general flow, 701F
mitochondria *see* Mitochondrial protein import
nucleus *see* Nuclear–cytoplasmic transport
between organelles, **699–701**
peroxisomes *see* Peroxisome(s), protein import
signal hypothesis, 726–727, 727F, 728F
studies, 703F
through secretory pathway, **749–787**
see also Protein sorting; Transmembrane protein(s); Vesicular transport
Protein tyrosine kinases *see* Tyrosine kinase(s)
Protein tyrosine phosphatases (PTPs), **938–939**
Proteoglycan(s), 1184T
  aggregates, 1182, 1182F, 1183F
  assembly/synthesis, **775–776**, 1181, 1181F
  basal lamina, 1165
  cell surface co-receptors, **1183–1184**
  connective tissues, 1179, **1181–1182**
  membrane proteins, 636
  as molecular sieves, 1182–1183
  protein regulation by, **1182–1183**
  sizes, 1181–1182, 1181F
  see also specific types
Proteolysis, 391
  apoptosis (programmed cell death), caspases *see* Caspase(s)
  cell cycle control, **1064**
  degradation signals, 396F
  insulin, 151, 152F
  protein aggregate resistance, 397
  regulated destruction, **395–396**, 396F
  regulatory function, **395–396**
  see also Proteasome(s); Ubiquitin pathway
Proteomics, **188**
Protists, **32–34**, 33F
Protocadherins, 1136, 1138T
Protofilament, 971–972, 972F
Proton *see* Hydrogen ion (H⁺, proton)
Proton gradients *see* Electrochemical proton gradients
Proton-motive force, 821, 821F, 853–854
Proton pumps, 827–828, 829F
  allostery, 836
  atomic detail, 835
  bacteriorhodopsin, **640–642**
  evolution, 871
  general mechanism, 837F
  lysosomes, 780, 780F
  photosynthesis, 851, 854
  redox potentials, 835, 835F
  see also Respiratory enzyme complexes
Proto-oncogenes, 1231, 1237, 1237F
Protozoa, 28F, 32–33
  as parasites, **1494–1495**, **1508–1511**
Protrusion, in cell movement, of actin meshwork at leading edge, 1039F
Prox1 protein, 1448
P-selectin, 1145–1146
Pseudogenes, 248, 255, 257
Pseudoknot, RNA structure, 403F
Pseudomonas aeruginosa, 1491
Pseudopodia, 1037
Pseudouridine, transfer RNA modification, 368F
P-site (ribosome binding), 375F, 376
Psychoactive drugs, 686, 690
PTB domain (phosphotyrosine-binding domain), 898, 899F, 925
PTEN phosphatase (tumor suppressor), 933
P-type pumps (ATPases), 659, 660F
  see also Calcium pump; Sodium–potassium pump (ATPase)
Puberty, 1288–1289, 1292, 1293
Puffer fish *(Fugu rubripes)*, 30, 31F, 251F
Pulse-chase experiments, 602, 602F
Pulsed-field gel electrophoresis, 534
Puma, apoptosis, 1123
Pumps *see* Carrier protein(s)
Pupa, *Drosophila*, 1329
Purifying selection, 247
Purine base(s), 61, 116F
  depurination, 296, 297F
  see also Adenine; Guanine
Purkinje cell, intracellular calcium, 597F
Puromycin, 384, 385T
Purple bacteria
  mitochondrial evolution, role, 860

Page numbers in **boldface** refer to a major text discussion of the entry; page numbers with an F refer to a figure, with an FF to figures that follow consecutively; page numbers with a T refer to a table; *vs* means compare/comparison.

Page numbers in **boldface** refer to a major text discussion of the entry; page numbers with an F refer to a figure, with an FF to figures that follow consecutively; page numbers with a T refer to a table; vs means compare/comparison.

The Genetic Code

1st position (5' end)	2nd Position				3rd Position (3' end)
↓	U	C	A	G	↓
U	Phe	Ser	Tyr	Cys	U
	Phe	Ser	Tyr	Cys	C
	Leu	Ser	**STOP**	**STOP**	A
	Leu	Ser	**STOP**	Trp	G
C	Leu	Pro	His	Arg	U
	Leu	Pro	His	Arg	C
	Leu	Pro	Gln	Arg	A
	Leu	Pro	Gln	Arg	G
A	Ile	Thr	Asn	Ser	U
	Ile	Thr	Asn	Ser	C
	Ile	Thr	Lys	Arg	A
	Met	Thr	Lys	Arg	G
G	Val	Ala	Asp	Gly	U
	Val	Ala	Asp	Gly	C
	Val	Ala	Glu	Gly	A
	Val	Ala	Glu	Gly	G

AMINO ACIDS AND THEIR SYMBOLS			CODONS
A	Ala	Alanine	GCA GCC GCG GCU
C	Cys	Cysteine	UGC UGU
D	Asp	Aspartic acid	GAC GAU
E	Glu	Glutamic acid	GAA GAG
F	Phe	Phenylalanine	UUC UUU
G	Gly	Glycine	GGA GGC GGG GGU
H	His	Histidine	CAC CAU
I	Ile	Isoleucine	AUA AUC AUU
K	Lys	Lysine	AAA AAG
L	Leu	Leucine	UUA UUG CUA CUC CUG CUU
M	Met	Methionine	AUG
N	Asn	Asparagine	AAC AAU
P	Pro	Proline	CCA CCC CCG CCU
Q	Gln	Glutamine	CAA CAG
R	Arg	Arginine	AGA AGG CGA CGC CGG CGU
S	Ser	Serine	AGC AGU UCA UCC UCG UCU
T	Thr	Threonine	ACA ACC ACG ACU
V	Val	Valine	GUA GUC GUG GUU
W	Trp	Tryptophan	UGG
Y	Tyr	Tyrosine	UAC UAU